FORMULAS FROM GEOMETRY

Area A; circumference C; volume V; curved surface area S; altitude h; radius r

RIGHT TRIANGLE

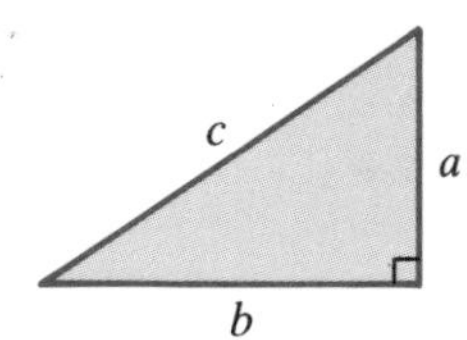

Pythagorean Theorem: $c^2 = a^2 + b^2$

TRIANGLE

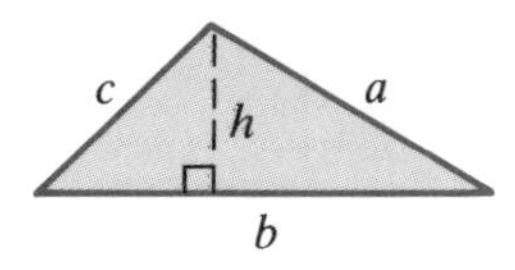

$A = \frac{1}{2}bh \qquad C = a + b + c$

EQUILATERAL TRIANGLE

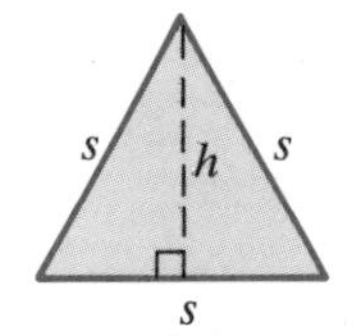

RECTANGLE

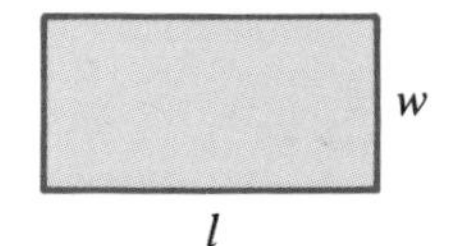

$A = lw \qquad C = 2l + 2w$

PARALLELOGRAM

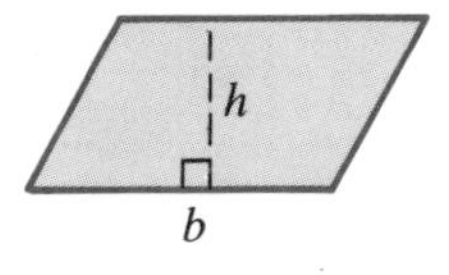

$A = bh$

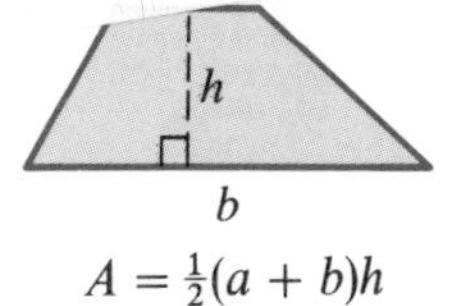

$A = \frac{1}{2}(a + b)h$

CIRCLE

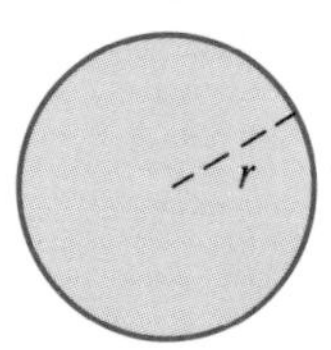

$A = \pi r^2 \qquad C = 2\pi r$

CIRCULAR SECTOR

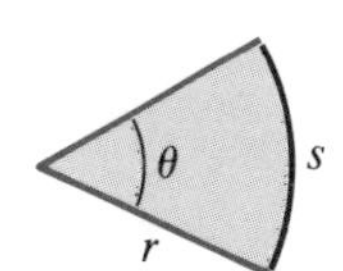

$A = \frac{1}{2}r^2\theta \qquad s = r\theta$

CIRCULAR RING

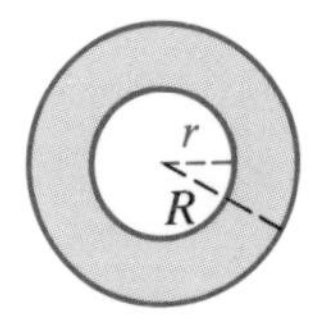

$A = \pi(R^2 - r^2)$

RECTANGULAR BOX

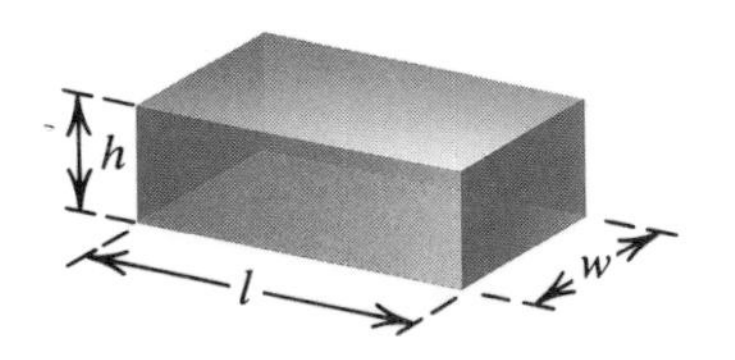

$V = lwh \qquad S = 2(hl + lw + hw)$

SPHERE

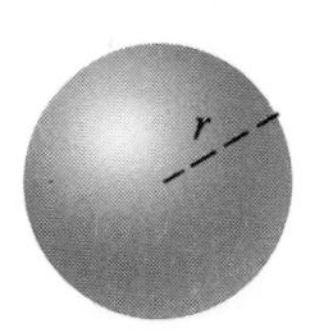

$V = \frac{4}{3}\pi r^3 \qquad S = 4\pi r^2$

RIGHT CIRCULAR CYLINDER

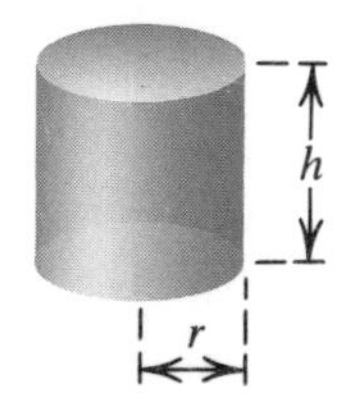

$V = \pi r^2 h \qquad S = 2\pi rh$

RIGHT CIRCULAR CONE

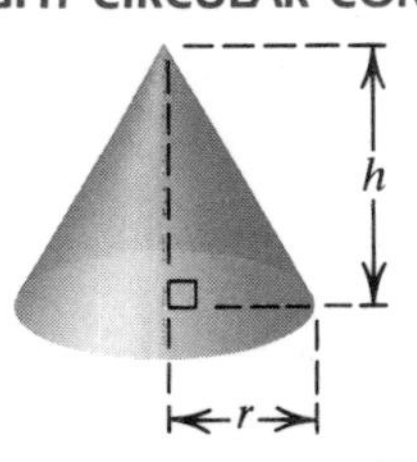

$V = \frac{1}{3}\pi r^2 h \qquad S = \pi r\sqrt{r^2 + h^2}$

FRUSTUM OF A CONE

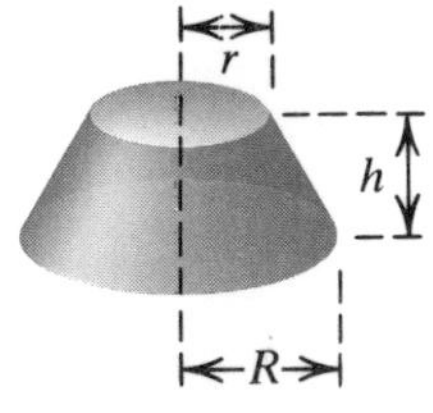

$V = \frac{1}{3}\pi h(r^2 + rR + R^2)$

PRISM

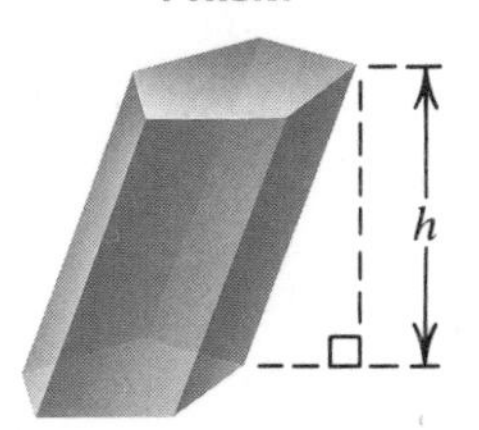

$V = Bh$ with B the area of the base

CALCULUS

EARL W. SWOKOWSKI
Marquette University

Australia • Brazil • Japan • Korea • Mexico • Singapore • Spain • United Kingdom • United States

Calculus, Fifth Edition
Earl W. Swokowski

Sponsoring Editor: **David Geggis**
Production Supervisor: **Elise Kaiser**
Manufacturing Coordinator: **Marcia A. Locke**
Production: **Lifand et al., Bookmakers**
Composition: **Syntax International Pte. Ltd.**
Technical Artwork: **Scientific Illustrators**
Interior Design: **Elise Kaiser**
Cover Design: **Vernon Boes**

ISBN-13: 978-0-534-43538-7

ISBN-10: 0-534-43538-6

Classic Edition
ISBN-13: 978-0-534-38212-4
ISBN-10: 0-534-38212-6

International Student Version
ISBN-13: 978-0-534-98392-5
ISBN-10: 0-534-98392-8

Brooks/Cole
10 Davis Drive
Belmont Drive, CA 94002-3098
USA

Cengage Learning is a leading provider of customized learning solutions with office locations around the globe, including Singapore, the United Kingdom, Australia, Mexico, Brazil, and Japan. Locate your local office at:
international.cengage.com/region

Cengage Learning products are represented in Canada by
Nelson Education, Ltd.

For your course and learning solutions, visit **academic.cengage.com**

Purchase any of our products at your local college store or at our preferred online store **www.ichapters.com**

Printed in the United States of America
10 11 12 13 14 13 12

Dedicated to the memory of
my mother and father,
Sophia and John Swokowski

CONTENTS

CHAPTER 6 APPLICATIONS OF THE DEFINITE INTEGRAL 303

CHAPTER 7 LOGARITHMIC AND EXPONENTIAL FUNCTIONS 373

CHAPTER 8 INVERSE TRIGONOMETRIC AND HYPERBOLIC FUNCTIONS 425

CHAPTER 12 TOPICS FROM ANALYTIC GEOMETRY 601

CHAPTER 13 PLANE CURVES AND POLAR COORDINATES 641

CHAPTER 14 VECTORS AND SURFACES 683

CHAPTER 15 VECTOR-VALUED FUNCTIONS 747

CHAPTER 16 PARTIAL DIFFERENTIATION 793

CHAPTER 17 MULTIPLE INTEGRALS 885

ADDITIONAL CALCULUS TEXTBOOKS BY EARL W. SWOKOWSKI

CALCULUS, FIFTH EDITION, LATE TRIGONOMETRY VERSION This text provides a review of trigonometry and complete coverage of the trigonometric functions in Chapter 8. Consequently, there is slightly different treatment of limits, derivatives, and integrals.

CALCULUS OF A SINGLE VARIABLE Designed for a two-semester course, this volume consists of the first thirteen chapters of *Calculus, Fifth Edition*.

PREFACE

This revision of what was previously called *Calculus with Analytic Geometry, Alternate Edition* was written with three objectives in mind. The first was to make the book more student-oriented by expanding discussions and providing more examples and figures to help clarify concepts. To further aid students, guidelines for solving problems were added in many sections of the text. The second objective was to stress the usefulness of calculus by means of modern applications of derivatives and integrals. The third objective, to make the text as error-free as possible, was accomplished by a careful examination of the exposition, combined with a thorough checking of each example and exercise.

CHANGES FOR THE FIFTH EDITION

Suggestions for improvements from instructors and reviewers resulted in a great deal of rewriting and reorganization. The principal changes are highlighted in the following list.

CHAPTER 1 The number of review sections has been reduced from six to three, with proofs of results from precalculus replaced by examples on inequalities, equations, and graphs.

CHAPTER 2 There is more emphasis on the graphical significance of limits. A physical application and tolerance statements are used to motivate the ϵ-δ definition. Limits involving ∞ (formerly in Chapter 4) are discussed in Section 2.4.

CHAPTER 3 The interpretations of the derivative as the slope of a tangent line and as the rate of change of a function are considered simultaneously instead of in separate sections. The power rule for rational numbers and the concept of higher derivatives are introduced in Section 3.2. Greater emphasis is given to the use of differentials for linear approximations of function values.

CHAPTER 4 The definition of concavity has been changed to make it easier to relate the sign of a second derivative to the shape of a graph. A new section titled *Summary of Graphical Methods* includes a list of guidelines for sketching the graph of a function.

CHAPTER 5 Antiderivatives and indefinite integrals are discussed in the first two sections instead of in different chapters. There are fifteen new examples pertaining to definite integrals.

CHAPTER 6 Almost every example on applications of definite integrals has been rewritten so as to replace formal limits of sums with a more intuitive method using differentials. Guidelines are stated that contain strategies for finding areas and volumes.

CHAPTER 7 The formula for the derivative of an inverse function is proved in the first section instead of the last. Integrals of the tangent, cotangent, secant, and cosecant functions (formerly in Chapter 8) are discussed in Section 7.4.

CHAPTER 8 The topics considered are restricted to inverse trigonometric and hyperbolic functions.

CHAPTER 9 Explanations and guidelines for methods of integration have been improved.

CHAPTER 10 For easy reference, definitions and notations for indeterminate forms are stated in tables. The discussion of Taylor's formula has been moved to Chapter 11.

CHAPTER 11 More emphasis is given to the differences between sequences, partial sums, sums of infinite series, and the fact that tests for convergence do not determine the sum of a series. The material on power series representations of functions and Taylor series has been completely reorganized.

CHAPTER 12 Additional calculus applications involving conic sections are included so that the discussion is not simply a review of precalculus topics.

CHAPTER 13 The concept of *orientation* of a curve has been added and incorporated into examples and exercises.

CHAPTER 14 To help readers visualize and sketch surfaces, many examples contain charts that show the trace on each coordinate plane. The discussion of cylindrical and spherical coordinates has been moved to Chapter 17.

CHAPTER 15 The introduction to vector-valued functions has been rewritten and integrated with the notion of space curves. The significance of using an arc length parameter has been given more prominence.

CHAPTER 16 Sixteen new figures provide a greater emphasis on graphs and geometric interpretations of functions of several variables. Newton's method for a system of two nonlinear equations is explained in Section 16.4. The discussion of Lagrange multipliers has been expanded.

CHAPTER 17 The definition of double integral and methods of evaluation appear in one section instead of two. The discussions of triple integrals in cylindrical coordinates and in spherical coordinates are presented in two separate sections.

CHAPTER 18 There is a more detailed explanation of conservative vector fields and of the technique of determining a potential function from its gradient. Two new examples apply Stokes' theorem and the concept of circulation to the analysis of winds inside a tornado.

CHAPTER 19 Series solutions of differential equations are discussed in the last section.

FEATURES OF THE TEXT

APPLICATIONS The previous edition contained applied problems from fields such as engineering, physics, chemistry, biology, economics, physiology, sociology, psychology, ecology, oceanography, meteorology, radiotherapy, astronautics, and transportation. This already extensive list was augmented with examples and exercises that include modern applications of calculus to the design of computers, the analysis of temperature grids, and the measurement of the thickness of the ozone layer, the greenhouse effect, vertical wind shear, the circulation of winds inside tornados and hurricanes, the energy released by earthquakes, the density of the atmosphere, the movement of robot arms, and the health hazards of radon gas.

EXAMPLES Well-structured and graded examples provide detailed solutions of problems similar to those that appear in exercise sets. Many examples contain graphs, charts, or tables to help readers understand procedures and solutions. There are also labeled *illustrations*, which are brief demonstrations of the use of definitions, laws, or theorems. Whenever feasible, applications are included that indicate the usefulness of a topic.

EXERCISES Exercise sets begin with routine drill problems and progress gradually to more difficult types. Many exercises containing graphs were added to this edition. Applied problems generally appear near the end of a set, to allow students to gain confidence in manipulations and new ideas before attempting questions that require analyses of practical situations.

A new feature in this edition is over 300 exercises, designated by the symbol [c], specifically designed to be solved with the aid of a scientific calculator or computer. Graphics capabilities are required for some of these exercises (see the remarks that follow under *Calculators*).

Review exercises at the end of each chapter except the first may be used to prepare for examinations.

ANSWERS The answer section at the end of the text provides answers for most of the odd-numbered exercises. Considerable thought and effort were devoted to making this section a learning device instead of merely a place to check answers. To illustrate, if an answer is the area of a region or the volume of a solid, a suitable (set-up) definite integral is often given along with its value. Numerical answers for many exercises are stated in both an exact and an approximate form. Graphs, proofs, and hints are included whenever appropriate.

CALCULATORS Since students may have access to a variety of calculators or computers, no attempt was made to categorize the exercises that are designated by [c]. The statement of a problem should provide sufficient information for determining whether a particular calculator or available computer software can be used to obtain a numerical solution. For example, if an exercise states that the trapezoidal rule with $n = 4$ should be applied, then almost any calculator is suitable, provided the function is not too complicated. However, for $n = 20$, a programmable calculator or computer is recommended. If the solution of an exercise involves a graph, a graphics calculator may be adequate; however, complicated functions or surfaces may require sophisticated computer software. Since numerical accuracy is also dependent on the type of computational device used, some answers are rounded to two decimal places, but in other cases eight-decimal-place accuracy may be given.

TEXT DESIGN AND FIGURES The text has been completely redesigned to make discussions easier to follow and to highlight important concepts. All graphics have been redone. Graphs of functions of one or two variables were computer-generated and plotted to a high degree of accuracy using the latest technology. Colors are used to distinguish between different parts of figures. For example, the graph of a function may be shown in blue and a tangent line to the graph in red. Labels are in the same color as the parts of the figure they identify. The use of several colors should help readers visualize surfaces and solids more readily than in previous editions.

FLEXIBILITY Syllabi from schools that used previous editions attest to the flexibility of the text. Sections and chapters can be rearranged in different ways, depending on the objectives and the length of the course.

Earl W. Swokowski

ACKNOWLEDGMENTS

I am indebted to Jeffery Cole, of Anoka-Ramsey Community College, for his excellent work on the exercise sets, the answer section, and artwork preparation for the figures. Jeff also carefully read the manuscript and offered many suggestions that helped make discussions and examples more student-oriented. He often reminded me of the motto I adopted when I started on this revision: "*Teach the Class.*"

I wish to thank Gary Rockswold, of Mankato State University, for supplying many of the new applied problems and calculator exercises.

I am also grateful to the following individuals, who reviewed the manuscript:

Stephen Andrilli, *LaSalle University*
Robert Beezer, *University of Puget Sound*
Clifton Clarridge, *Santa Monica College*
Andrew Demetropoulos, *Montclair State College*
Richard Grassl, *Muhlenberg College*
Harvey B. Keynes, *University of Minnesota*
Karl R. Klose, *Susquehanna University*
James E. McKenna, *SUNY College at Fredonia*
James R. McKinney, *California State Polytechnic University, Pomona*
Richard A. Quint, *Ventura College*
Leon F. Sagan, *Anne Arundel Community College*
David J. Sprows, *Villanova University*
Ronald Stoltenberg, *Sam Houston State University*
Jan Vandever, *South Dakota State University*
Kenneth L. Wiggins, *Walla Walla College*

The following mathematics educators met with me and representatives from PWS–KENT for several days in the summer of 1989 and later reviewed the manuscript:

Richard D. Armstrong, *St. Louis Community College at Florissant Valley*
Paul W. Britt, *Louisiana State University*
Jeffery A. Cole, *Anoka-Ramsey Community College*
William James Lewis, *University of Nebraska–Lincoln*
Michael A. Penna, *Indiana University–Purdue University at Indianapolis*

Their comments on pedagogy and their specific recommendations about the content of calculus courses helped me to improve the book.

I am thankful for the excellent cooperation of the staff of PWS–KENT. Two people in the company deserve special mention. Managing Editor Dave Geggis supervised the project, contacted many reviewers and users of the text, and was a constant source of information and advice. Dave is also responsible for finding the authors who developed the excellent ancillaries that accompany this book. Elise Kaiser redesigned the book, coordinated the production of the text, and kept the schedule running smoothly and on time. Sally Lifland, of Lifland et al., Bookmakers, took exceptional care in seeing that no editorial inconsistencies occurred.

Finally, I am deeply grateful to my wife Maureen—for her understanding, patience, support, and love.

In addition to all the persons named here, I express my sincere appreciation to the many students and teachers who have helped shape my views on how calculus should be presented in the classroom. I am grateful to many people who have reviewed the various editions of this book. My thanks go to these reviewers:

Alfred D. Andrew
Georgia Institute of Technology
Jan Frederick Andrus
University of New Orleans

Jacqueline E. Barab
University of Nevada, Las Vegas
Phillip W. Bean
Mercer University
Daniel D. Benice
Montgomery College
Delmar L. Boyar
University of Texas–El Paso
Christian C. Braunschweiger
Marquette University
Robert M. Brooks
University of Utah
Ronald E. Bruck
University of Southern California
David C. Buchthal
University of Akron
Dawson Carr
Sandhills Community College
James L. Cornette
Iowa State University
John E. Derwent
University of Notre Dame
Daniel Drucker
Wayne State University
Joseph M. Egar
Cleveland State University
Ronald D. Ferguson
San Antonio State College
William R. Fuller
Purdue University
August J. Garver
University of Missouri
Stuart Goldenberg
California State Polytechnic University
Joe A. Guthrie
University of Texas–El Paso
Gary Haggard
Bucknell University
Mark P. Hale, Jr.
University of Florida
Clemens B. Hanneken
Marquette University
Alan Heckenbach
Iowa State University
Simon Hellerstein
University of Wisconsin
John C. Higgins
Brigham Young University
Arthur M. Hobbs
Texas A & M University
David Hoff
Indiana University
Dale T. Hoffman
Bellevue Community College
Adam J. Hulin
University of New Orleans
Michael Iannone
Trenton State College
George Johnson
University of South Carolina
Elgin H. Johnston
Iowa State University
Herbert M. Kamowitz
University of Massachusetts–Boston
Andrew Karantinos
University of South Dakota
Eleanor Killam
University of Massachusetts
Margaret Lial
American River College
James T. Loats
Metropolitan State College
Phil Locke
University of Maine–Orono
Robert H. Lohman
Kent State University
Stanley M. Lukawecki
Clemson University
Francis E. Masat
Glassboro State College
Wayne McDaniel
University of Missouri–St. Louis
Judith R. McKinney
California State Polytechnic University, Pomona
Burnett Meyer
University of Colorado at Boulder
Joseph Miles
University of Illinois
David Minda
University of Cincinnati
Chester L. Miracle
University of Minnesota
John A. Nohel
University of Wisconsin

Norman K. Nystrom
American River College
James Osterburg
University of Cincinnati
Richard R. Patterson
Indiana University–Purdue University at Indianapolis
Charles V. Peele
Marshall University
Ada Peluso
CUNY–Hunter College
David A. Petrie
Cypress College
Neal C. Raber
University of Akron
William H. Robinson
Ventura College
Jean E. Rubin
Purdue University
John T. Scheick
Ohio State University
Eugene P. Schlereth
University of Tennessee
Jon W. Scott
Montgomery College
Richard D. Semmler
Northern Virginia Community College–Annandale
Leonard Shapiro
University of Minnesota
Donald Sherbert
University of Illinois
Eugene R. Speer
Rutgers University
Monty J. Strauss
Texas Tech University
John Tung
Miami University
Charles Van Gordon
Millersville State College
Richard G. Vinson
University of South Alabama
Roman W. Voronka
New Jersey Institute of Technology
Dale E. Walston
University of Texas at Austin
Frederick R. Ward
Boise State University
Alan Wiederhold
San Jacinto College
Loyd V. Wilcox
Golden West College
T. J. Worosz
Metropolitan State College
Dennis Wortman
University of Massachusetts–Boston

EWS

SUPPLEMENTS

STUDENT STUDY GUIDE Richard M. Grassl's *Student Study Guide* leads students through all major topics in the text and reinforces their understanding through review sections, drills, and self-tests.

STUDENT SOLUTIONS MANUAL, VOLUMES I AND II Jeffery A. Cole and Gary K. Rockswold's *Student Solutions Manual* contains solutions, worked out in detail, to a subset of the odd-numbered exercises from the text.

INSTRUCTOR'S SOLUTIONS MANUAL, VOLUMES I AND II Jeffery A. Cole and Gary K. Rockswold's *Instructor's Solutions Manual* gives solutions or answers for all exercises in the text.

EVEN-NUMBERED ANSWER BOOK Jeffery A. Cole and Gary K. Rockswold's *Even-Numbered Answer Book* is for both instructor and student use.

PRINTED TEST BANK Available to instructors, the *Printed Test Bank* contains sample tests for each chapter.

TRANSPARENCIES Also available to instructors are four-color acetate transparencies of selected figures from the text.

SOFTWARE

MATHEMATICS PLOTTING PACKAGE Available for IBM-PCs and compatibles, the Mathematics Plotting Package combines superb plotting and graphing software. This public domain program developed at the United States Naval Academy is accompanied by a text-specific **Instructor's Resource Guide to MPP**, by Howard Penn and Craig Bailey. The guide provides instructors with notes on how to teach key concepts by using MPP to illustrate examples and problems from the text. Also included are **Example Files for MPP**—disks containing examples and problems from the Resource Guide.

PC:SOLVE Available for IBM-PCs and compatibles, PC:SOLVE is an interactive mathematical language with scratchpad that supports calculations and graphics, combined with a calculus curriculum library of key concepts to promote critical thinking and problem solving. A course license for PC:SOLVE is available upon adoption of this text.

GRAPHER Steve Scarborough's Grapher, for the Macintosh, is a flexible program that can be used to generate several types of graphs. In addition to plotting rectangular and polar curves and interpolating polynomials, it handles parametric equations, systems of two first-order differential equations, series, and direction fields.

TrueBASIC CALCULUS Available for IBM-PCs and compatibles, TrueBASIC Calculus is a disk and manual package that is equally useful for self-study, free exploration of topics, and solutions of problems. The record and playback feature make the package ideal for classroom demonstrations.

COMPUTERIZED TESTING

EXPTest This testing program for IBM-PCs and compatibles allows users to view and edit all tests, adding to, deleting from, and modifying existing questions. Any number of student tests can be created—multiple forms for larger sections or single tests for individual use. Users can create a question bank using mathematical symbols and notation. A graphics importation feature permits display and printing of graphs, diagrams, and maps provided with the test banks. The package includes easy-to-follow documentation, with a quick-start guide. A demonstration disk is available.

LXRTest This Macintosh testing program allows users to create, view, and edit tests. Questions can be stored by objectives, and the user can create questions using multiple-choice, true/false, fill-in-the-blank, essay, and matching formats. The order of alternatives and the order of questions can be scrambled. A demonstration disk is available (the user must have HyperCard to run the demo disk).

GRAPHIC CALCULATOR SUPPLEMENTS

CALCULUS ACTIVITIES FOR GRAPHIC CALCULATORS

Dennis Pence's *Calculus Activities for Graphic Calculators*, for students and instructors, offers exercises and examples utilizing the graphing calculator (Sharp, Casio, and HP28S). Classroom tested, these varied activities will enhance understanding of calculus topics for those using graphing calculators.

CALCULUS ACTIVITIES FOR THE TI-81 GRAPHIC CALCULATOR

For students and instructors using the TI-81 graphic calculator, Dennis Pence provides calculus exercises and examples in *Calculus Activities for the TI-81 Graphic Calculator*.

FOR THE STUDENT

Calculus was invented in the seventeenth century as a tool for investigating problems that involve motion. Algebra and trigonometry may be used to study objects that move at constant speeds along linear or circular paths, but calculus is needed if the speed varies or if the path is irregular. An accurate description of motion requires precise definitions of *velocity* (the rate at which distance changes per unit time) and *acceleration* (the rate at which velocity changes). These definitions may be obtained by using one of the fundamental concepts of calculus—the *derivative*.

Although calculus was developed to solve problems in physics, its power and versatility have led to uses in many diverse fields of study. Modern-day applications of the derivative include investigating the rate of growth of bacteria in a culture, predicting the outcome of a chemical reaction, measuring instantaneous changes in electrical current, describing the behavior of atomic particles, estimating tumor shrinkage in radiation therapy, forecasting economic profits and losses, and analyzing vibrations in mechanical systems.

The derivative is also useful in solving problems that involve maximum or minimum values, such as manufacturing the least expensive rectangular box that has a given volume, calculating the greatest distance a rocket will travel, obtaining the maximum safe flow of traffic across a long bridge, determining the number of wells to drill in an oil field for the most efficient production, finding the point between two light sources at which illumination will be greatest, and maximizing corporate revenue for a particular product. Mathematicians often employ derivatives to find tangent lines to curves and to help analyze graphs of complicated functions.

Another fundamental concept of calculus—the *definite integral*—is motivated by the problem of finding areas of regions that have curved boundaries. Definite integrals are employed as extensively as derivatives and in as many different fields. Some applications are finding the center of mass or moment of inertia of a solid, determining the work required to send a space probe to another planet, calculating the blood flow through an arteriole, estimating depreciation of equipment in a manufacturing plant, and interpreting the amount of dye dilution in physiological tests that involve tracer methods. We can also use definite integrals to investigate mathematical concepts such as area of a curved surface, volume of a geometric solid, or length of a curve.

The concepts of derivative and definite integral are defined by limiting processes. The notion of *limit* is the initial idea that separates calculus from elementary mathematics. Sir Isaac Newton (1642–1727) and Gottfried Wilhelm Leibniz (1646–1716) independently discovered the connection between derivatives and integrals and are both credited with the invention of calculus. Many other mathematicians have added greatly to its development in the last 300 years.

The applications of calculus mentioned here represent just a few of the many considered in this book. We can't possibly discuss all the uses of calculus, and more are being developed with every advance in technology. Whatever your field of interest, calculus is probably used in some pure or applied investigations. Perhaps *you* will discover a new application for this branch of science.

CHAPTER

1

PRECALCULUS REVIEW

INTRODUCTION

In this chapter we review topics from precalculus mathematics that are essential for the study of calculus. After a brief discussion of inequalities, equations, absolute values, and graphs, we turn our attention to functions. To say that the concept of function is important in mathematics is an understatement. It is literally the foundation of calculus and the backbone of the entire subject. You will find the word *function* and the symbol f or $f(x)$ used frequently on almost every page of this text.

In precalculus courses we study properties of functions by using algebra and graphical methods that include plotting points, determining symmetry, and making horizontal or vertical shifts. These techniques are adequate for obtaining a rough sketch of a graph; however, calculus is required to find precisely where graphs of functions rise or fall, exact coordinates of high or low points, slopes of tangent lines, and many other useful facts. Applied problems that cannot be solved by means of algebra, geometry, or trigonometry can often be attacked by representing physical quantities in terms of functions and then applying the tools developed in calculus.

With the preceding remarks in mind, read Section 1.2 carefully. A good understanding of this material is required before beginning the next chapter.

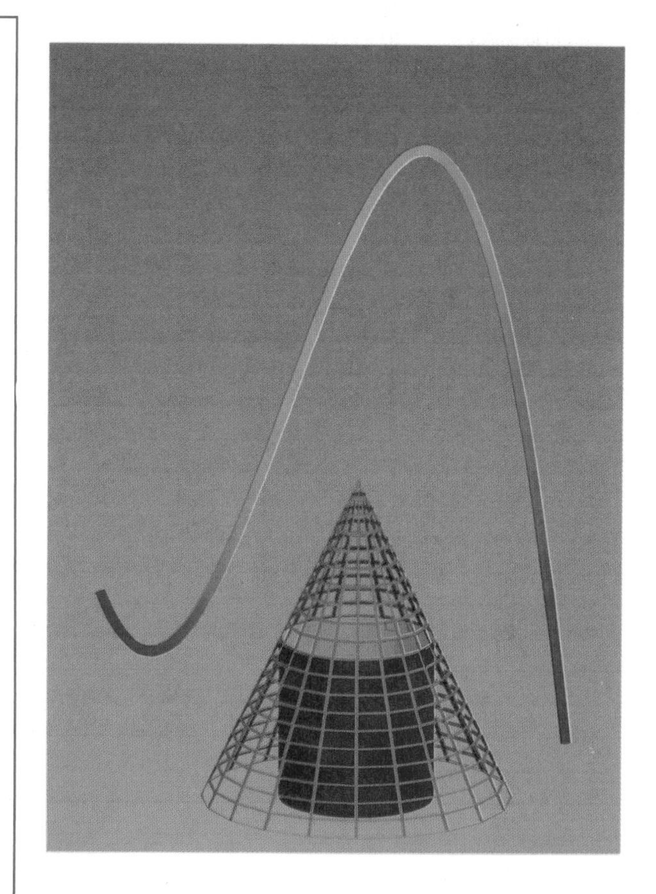

1.1 ALGEBRA

This section contains a review of topics from algebra that are prerequisites for calculus. We shall state important facts and work examples without supplying detailed reasons to justify our work. A more extensive coverage of this material can be found in texts on precalculus mathematics.

All concepts in calculus are based on properties of the set $\mathbb{R}$ of real numbers. There is a one-to-one correspondence between $\mathbb{R}$ and points on a *coordinate line* (or *real line*) l as illustrated in Figure 1.1, where O is the *origin*. The number 0 (*zero*) is neither positive nor negative.

FIGURE 1.1

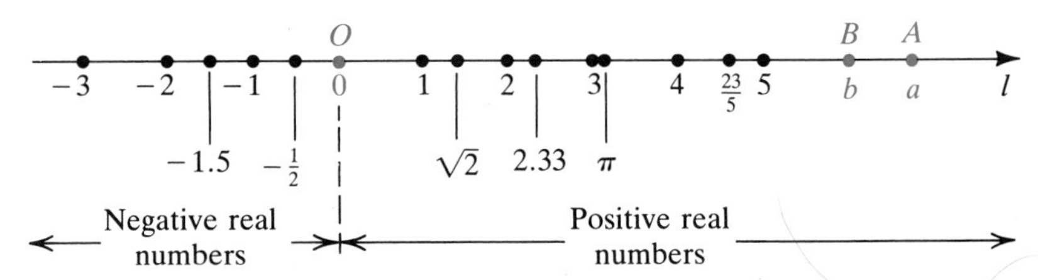

If a and b are real numbers, then $a > b$ (**a is greater than b**) if $a - b$ is positive. An equivalent statement is $b < a$ (**b is less than a**). Referring to the coordinate line in Figure 1.1, we see that $a > b$ if and only if the point A corresponding to a lies to the right of the point B corresponding to b. Other types of inequality symbols include $a \leq b$, which means $a < b$ or $a = b$, and $a < b \leq c$, which means $a < b$ and $b \leq c$.

ILLUSTRATION

- $5 > 3$
- $-7 < -2$
- $(-3)^2 > 0$
- $a^2 \geq 0$ for every real number a

The following properties can be proved.

Properties of inequalities **(1.1)**

(i) If $a > b$ and $b > c$, then $a > c$.
(ii) If $a > b$, then $a + c > b + c$.
(iii) If $a > b$, then $a - c > b - c$.
(iv) If $a > b$ and c is positive, then $ac > bc$.
(v) If $a > b$ and c is negative, then $ac < bc$.

Analogous properties are true if the inequality signs are reversed. Thus, if $a < b$ and $b < c$, then $a < c$; if $a < b$, then $a + c < b + c$; and so on.

The **absolute value** $|a|$ of a real number a is defined as follows:

$$|a| = \begin{cases} a & \text{if } a \geq 0 \\ -a & \text{if } a < 0 \end{cases}$$

If a is the coordinate of the point A on the coordinate line in Figure 1.1, then $|a|$ is the number of units (that is, the distance) between A and the origin O.

ILLUSTRATION

- $|3| = 3$ ■ $|-3| = -(-3) = 3$
- $|0| = 0$ ■ $|3 - \pi| = -(3 - \pi) = \pi - 3$

The following can be proved.

Properties of absolute values $(b > 0)$ **(1.2)**

(i) $|a| < b$ if and only if $-b < a < b$

(ii) $|a| > b$ if and only if either $a > b$ or $a < -b$

(iii) $|a| = b$ if and only if $a = b$ or $a = -b$

An **equation** (in x) is a statement such as

$$x^2 = 3x - 4 \quad \text{or} \quad 5x^3 + 2\sin x - \sqrt{x} = 0.$$

A **solution** (or **root**) is a number a that produces a true statement when a is substituted for x. To *solve an equation* means to find all the solutions.

EXAMPLE 1 Solve

(a) $x^3 + 3x^2 - 10x = 0$ **(b)** $2x^2 + 5x - 6 = 0$

SOLUTION

(a) Factoring the left-hand side yields

$$x(x^2 + 3x - 10) = 0, \quad \text{or} \quad x(x - 2)(x + 5) = 0.$$

Setting each factor equal to zero gives us the solutions 0, 2, and -5.

(b) Using the *quadratic formula*,

$$x = \frac{-b \pm \sqrt{b^2 - 4ac}}{2a},$$

with $a = 2$, $b = 5$, and $c = -6$, we obtain

$$x = \frac{-5 \pm \sqrt{25 - 4 \cdot 2 \cdot (-6)}}{2 \cdot 2} = \frac{-5 \pm \sqrt{73}}{4}.$$

Hence the solutions are $-\frac{5}{4} + \frac{1}{4}\sqrt{73}$ and $-\frac{5}{4} - \frac{1}{4}\sqrt{73}$.

An **inequality** (in x) is a statement that contains at least one of the symbols $<$, $>$, $\leq$, or $\geq$, such as

$$5x - 4 > x^2 \quad \text{or} \quad -3 < 4x + 2 \leq 5.$$

The notions of **solution** of an inequality and *solving* an inequality are similar to the analogous concepts for equations.

We shall often refer to *intervals*. In the definitions that follow we employ the set notation $\{x: \quad \}$, where the space after the colon is used to specify restrictions on the variable x. In (1.3) we call (a, b) an **open interval,** $[a, b]$ a **closed interval,** $[a, b)$ and $(a, b]$ **half-open intervals,** and intervals defined in terms of ∞ (*infinity*) or $-\infty$ (*minus infinity*) **infinite intervals**.

Intervals **(1.3)**

NOTATION	DEFINITION	GRAPH
(a, b)	$\{x: a < x < b\}$	
$[a, b]$	$\{x: a \leq x \leq b\}$	
$[a, b)$	$\{x: a \leq x < b\}$	
$(a, b]$	$\{x: a < x \leq b\}$	
(a, ∞)	$\{x: x > a\}$	
$[a, \infty)$	$\{x: x \geq a\}$	
$(-\infty, b)$	$\{x: x < b\}$	
$(-\infty, b]$	$\{x: x \leq b\}$	
$(-\infty, \infty)$	$\mathbb{R}$	

EXAMPLE 2 Solve each inequality and sketch the graph of its solution.

(a) $-5 \leq \dfrac{4-3x}{2} < 1$ **(b)** $x^2 - 10 > 3x$

SOLUTION

(a)

$$-5 \leq \frac{4-3x}{2} < 1 \quad \text{(given)}$$

$$-10 \leq 4 - 3x < 2 \quad \text{(multiply by 2)}$$

$$-14 \leq -3x < -2 \quad \text{(subtract 4)}$$

$$\tfrac{14}{3} \geq x > \tfrac{2}{3} \quad \text{(divide by } -3\text{)}$$

$$\tfrac{2}{3} < x \leq \tfrac{14}{3} \quad \text{(equivalent inequality)}$$

FIGURE 1.2

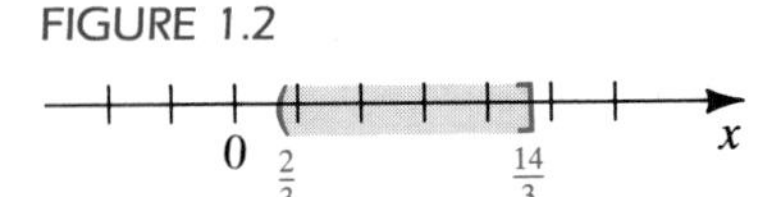

Hence the solutions are the numbers in the half-open interval $(\frac{2}{3}, \frac{14}{3}]$. The graph is sketched in Figure 1.2.

(b)

$$x^2 - 10 > 3x \quad \text{(given)}$$

$$x^2 - 3x - 10 > 0 \quad \text{(subtract } 3x\text{)}$$

$$(x-5)(x+2) > 0 \quad \text{(factor)}$$

FIGURE 1.3

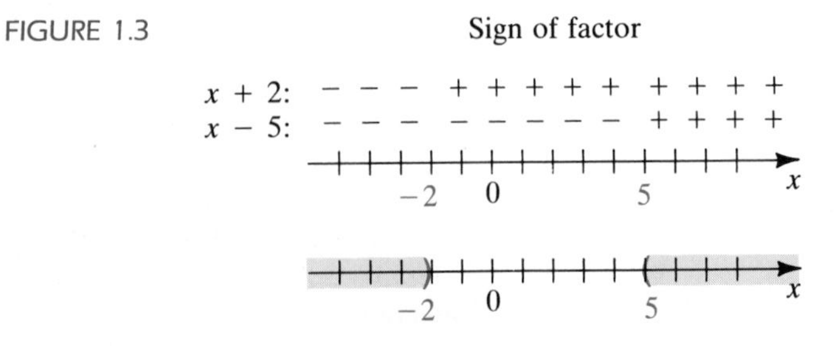

We next examine the signs of the factors $x - 5$ and $x + 2$, as shown in Figure 1.3. Since $(x - 5)(x + 2) > 0$ if both factors have the same sign, the solutions are the real numbers in the union $(-\infty, -2) \cup (5, \infty)$, as illustrated in Figure 1.3.

Inequalities involving absolute value occur frequently in calculus.

EXAMPLE 3 Solve each inequality and sketch the graph of its solution.

(a) $|x - 3| < 0.5$ **(b)** $|2x - 7| > 3$

SOLUTION

(a)

$$|x - 3| < 0.5 \quad \text{(given)}$$

$$-0.5 < x - 3 < 0.5 \quad \text{(property of absolute value)}$$

$$2.5 < x < 3.5 \quad \text{(add 3)}$$

FIGURE 1.4

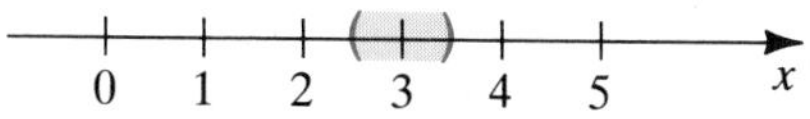

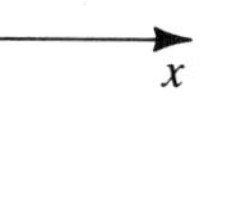

The solutions are the real numbers in the open interval (2.5, 3.5), as shown in Figure 1.4.

(b) $|2x - 7| > 3$ (given)

$$2x - 7 < -3 \quad \text{or} \quad 2x - 7 > 3 \quad \text{(property of absolute value)}$$

$$2x < 4 \quad \text{or} \quad 2x > 10 \quad \text{(add 7)}$$

$$x < 2 \quad \text{or} \quad x > 5 \quad \text{(divide by 2)}$$

FIGURE 1.5

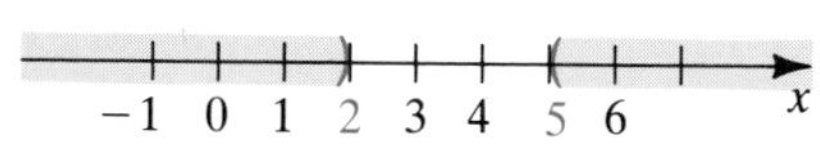

The solutions are given by $(-\infty, 2) \cup (5, \infty)$. The graph is sketched in Figure 1.5.

A **rectangular coordinate system** is an assignment of *ordered pairs* (a, b) to points in a plane, as illustrated in Figure 1.6. The plane is called a **coordinate plane**, or an ***xy*-plane**. Note that in this context (a, b) is not an open interval. It should always be clear from our discussion whether (a, b) represents a point or an interval.

FIGURE 1.6

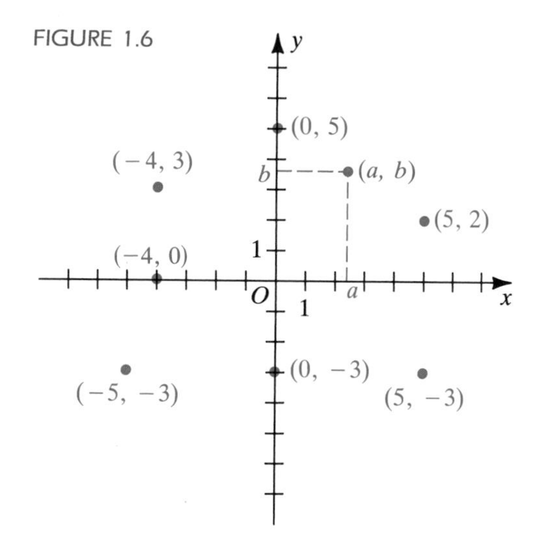

The following formulas can be proved.

Distance formula (1.4)

The distance between P_1 and P_2 is

$$d(P_1, P_2) = \sqrt{(x_2 - x_1)^2 + (y_2 - y_1)^2}.$$

Midpoint formula (1.5)

The midpoint M of segment P_1P_2 is

$$M\left(\frac{x_1 + x_2}{2}, \frac{y_1 + y_2}{2}\right).$$

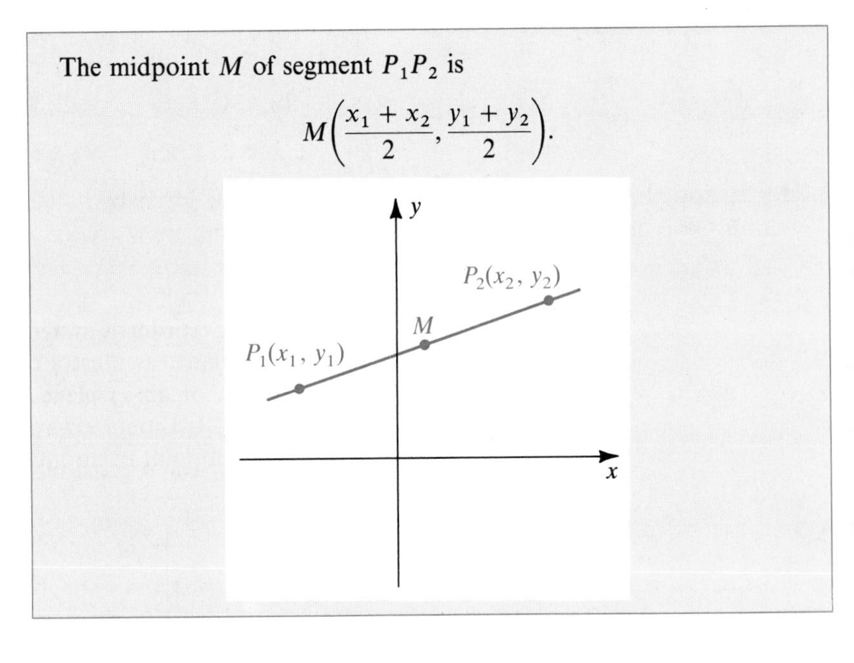

FIGURE 1.7

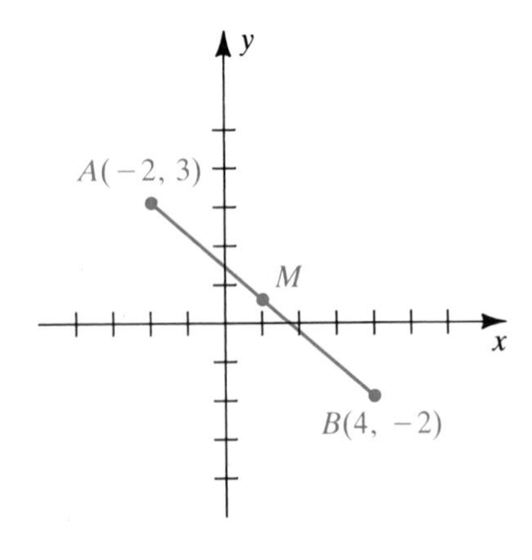

EXAMPLE 4 Given $A(-2, 3)$ and $B(4, -2)$, find

(a) $d(A, B)$ **(b)** the midpoint M of segment AB

SOLUTION The points are plotted in Figure 1.7. Using the formulas in (1.4) and (1.5), we obtain the following:

(a) $d(A, B) = \sqrt{(4 + 2)^2 + (-2 - 3)^2} = \sqrt{36 + 25} = \sqrt{61}$

(b) $M\left(\frac{-2 + 4}{2}, \frac{3 + (-2)}{2}\right) = M(1, \frac{1}{2})$

An **equation in x and y** is an equality such as

$$2x + 3y = 5, \quad y = x^2 - 5x + 2, \quad \text{or} \quad y^2 + \sin x = 8.$$

A **solution** is an ordered pair (a, b) that produces a true statement when $x = a$ and $y = b$. The **graph** of the equation consists of all points (a, b) in a plane that correspond to the solutions. We shall assume that you have experience in sketching graphs of basic equations in x and y. Certain graphs have symmetries as indicated in (1.6), which states tests that can be applied to an equation in x and y to determine a symmetry.

Symmetries of graphs **(1.6)**

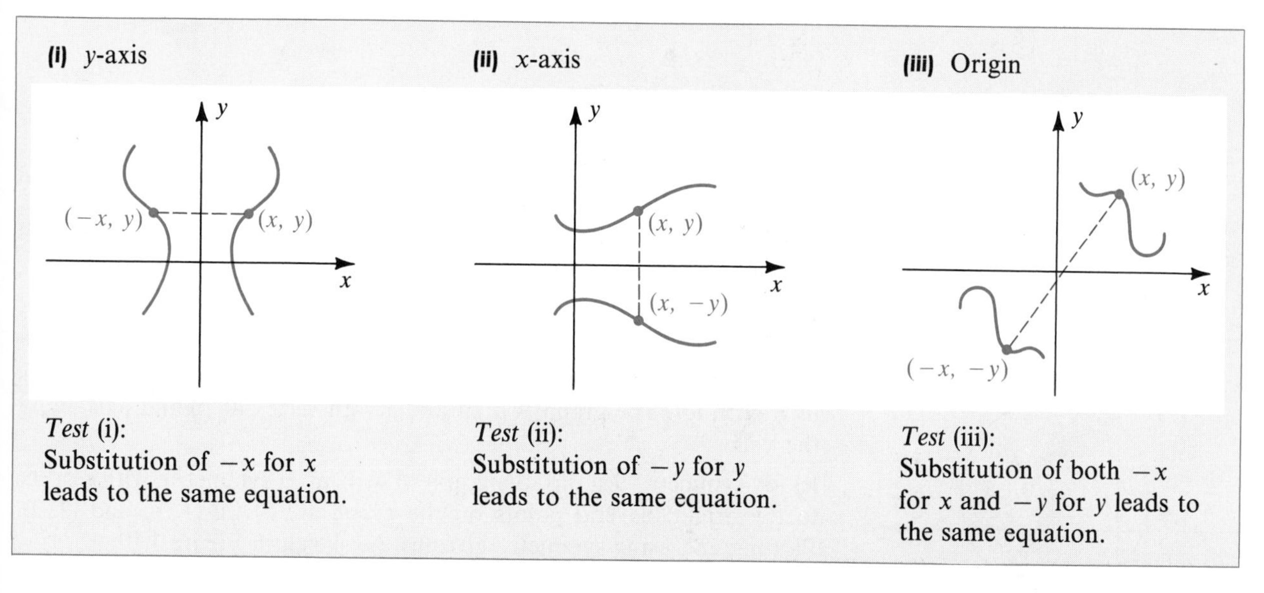

Test (i): Substitution of $-x$ for x leads to the same equation.

Test (ii): Substitution of $-y$ for y leads to the same equation.

Test (iii): Substitution of both $-x$ for x and $-y$ for y leads to the same equation.

Tests for symmetry are useful in the next example, because they enable us to sketch only half of a graph and then reflect that half through an axis or the origin as shown in (1.6). We shall plot several points on each graph to illustrate solutions of the equation; however, *a principal objective in graphing is to obtain an accurate sketch without plotting many (or any) points.*

EXAMPLE 5 Sketch the graph of

(a) $y = \frac{1}{2}x^2$ **(b)** $y^2 = x$ **(c)** $4y = x^3$

SOLUTION

(a) By symmetry test (i), the graph of $y = \frac{1}{2}x^2$ is symmetric with respect to the y-axis. Some points (x, y) on the graph are listed in the following table.

x	0	1	2	3	4
y	0	$\frac{1}{2}$	2	$\frac{9}{2}$	8

FIGURE 1.8

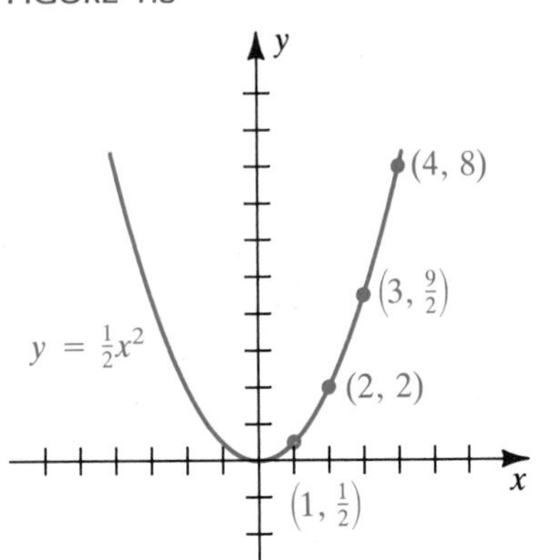

FIGURE 1.9

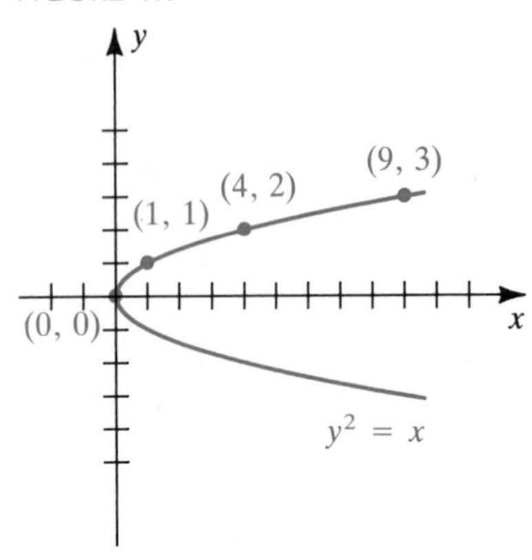

FIGURE 1.10

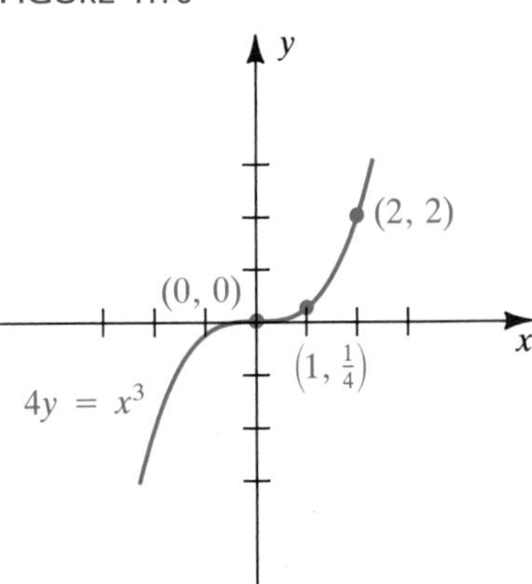

Plotting, drawing a smooth curve through the points, and then using symmetry gives us the sketch in Figure 1.8. The graph is a *parabola*, with *vertex* (0, 0) and *axis* along the y-axis. Parabolas are discussed in detail in Chapter 12.

(b) By symmetry test (ii), the graph of $y^2 = x$ is symmetric with respect to the x-axis. Points above the x-axis are given by $y = \sqrt{x}$. Several such points are (0, 0), (1, 1), (4, 2), and (9, 3). Plotting and using symmetry gives us Figure 1.9. The graph is a parabola with vertex (0, 0) and axis along the x-axis.

(c) By symmetry test (iii), the graph of $4y = x^3$ is symmetric with respect to the origin. Several points on the graph are (0, 0), $(1, \frac{1}{4})$, and (2, 2). Plotting and using symmetry gives us the sketch in Figure 1.10.

FIGURE 1.11

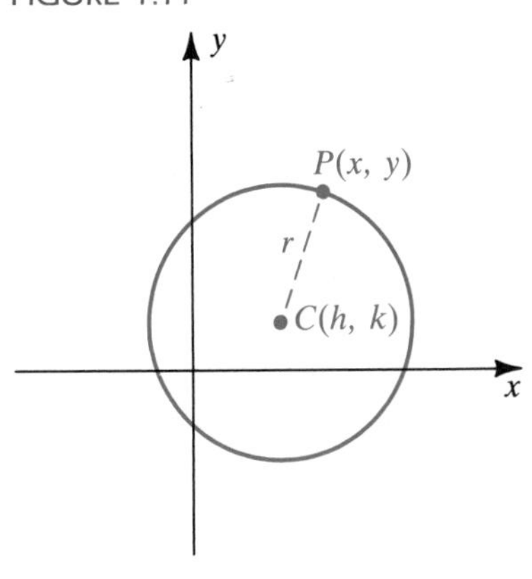

A circle with center $C(h, k)$ and radius r is illustrated in Figure 1.11. If $P(x, y)$ is any point on the circle, then by the distance formula (1.4), $d(P, C) = r$, or $[d(P, C)]^2 = r^2$. This leads us to the following equation.

Equation of a circle **(1.7)**

$$(x - h)^2 + (y - k)^2 = r^2$$

If the radius r is 1, then the circle is called a **unit circle**. A unit circle U with center at the origin has the equation $x^2 + y^2 = 1$.

FIGURE 1.12

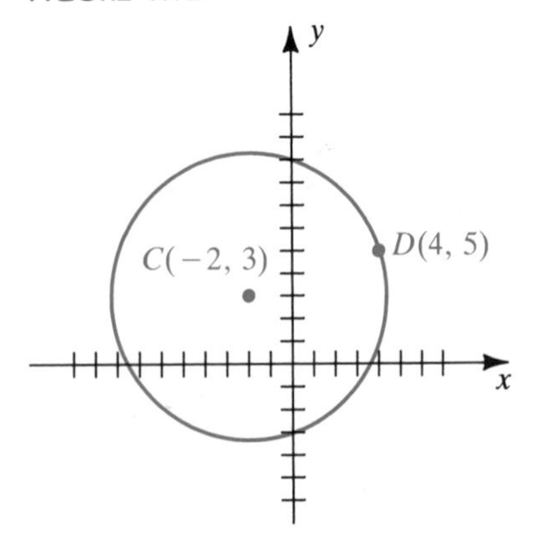

EXAMPLE 6 Find an equation of the circle that has center $C(-2, 3)$ and contains the point $D(4, 5)$.

SOLUTION The circle is illustrated in Figure 1.12. Since D is on the circle, the radius r is $d(C, D)$. By the distance formula,

$$r = \sqrt{(-2 - 4)^2 + (3 - 5)^2} = \sqrt{36 + 4} = \sqrt{40}.$$

Using the equation of a circle with $h = -2$, $k = 3$, and $r = \sqrt{40}$ gives us

$$(x + 2)^2 + (y - 3)^2 = 40,$$

or

$$x^2 + y^2 + 4x - 6y - 27 = 0.$$

In calculus we often consider lines in a coordinate plane. The following formulas are used for finding their equations.

Lines **(1.8)**

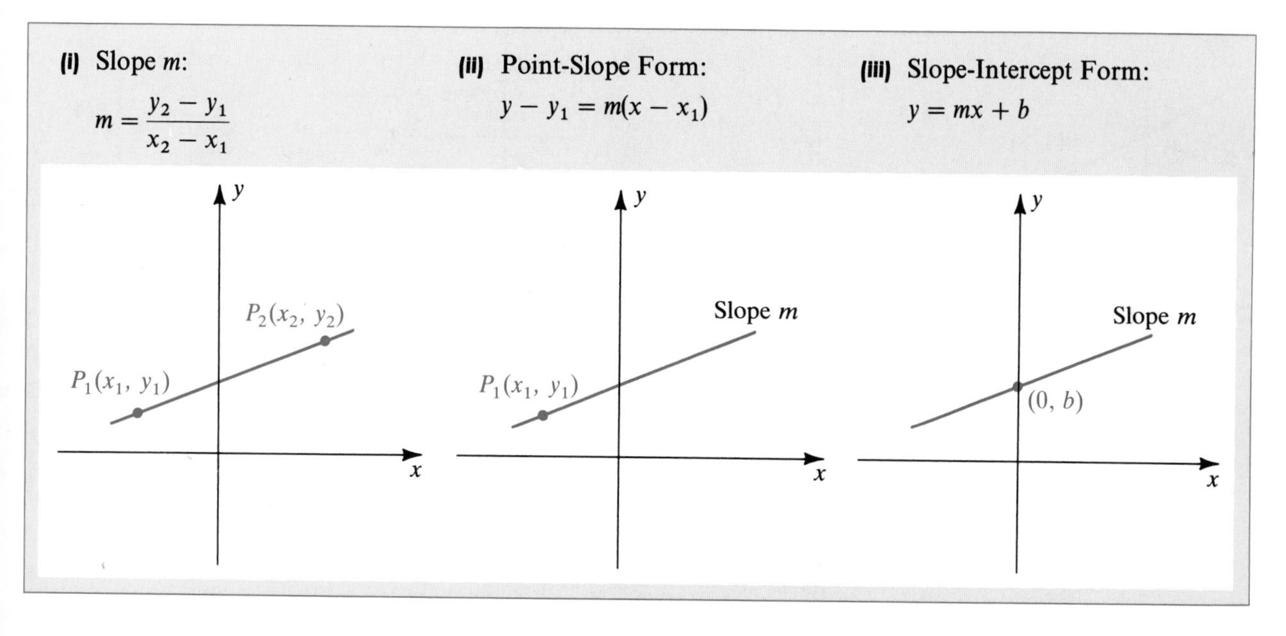

Some special types of lines and properties of their slopes are given in (1.9).

Special lines **(1.9)**

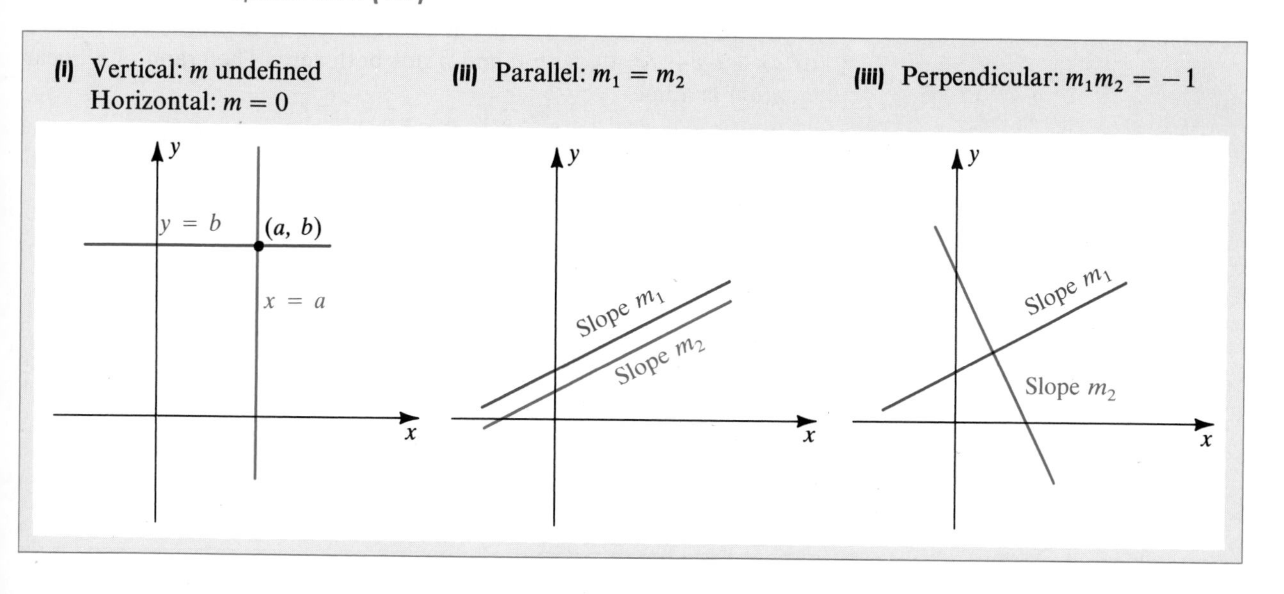

EXAMPLE 7 Sketch the line through each pair of points and find its slope.

(a) $A(-1, 4)$ and $B(3, 2)$ **(b)** $A(2, 5)$ and $B(-2, -1)$
(c) $A(4, 3)$ and $B(-2, 3)$ **(d)** $A(4, -1)$ and $B(4, 4)$

SOLUTION The lines are sketched in Figure 1.13.

FIGURE 1.13

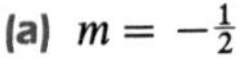
(a) $m = -\frac{1}{2}$

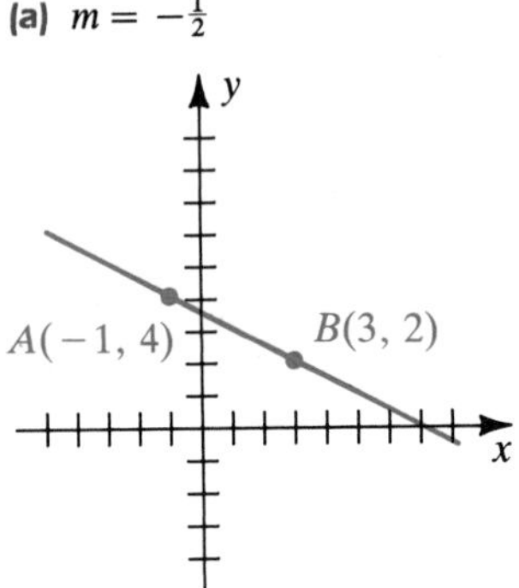

(b) $m = \frac{3}{2}$

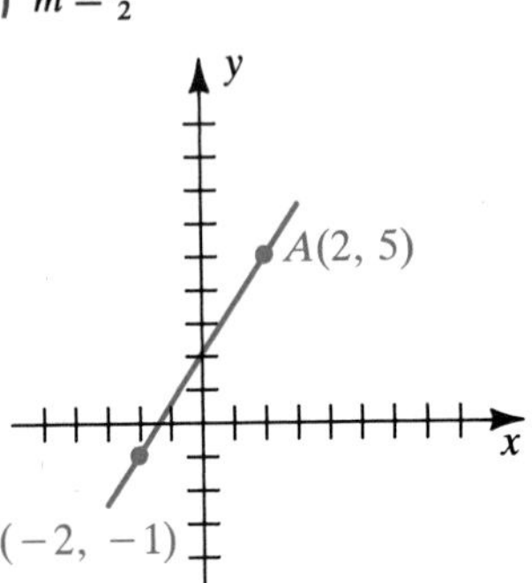

(c) $m = 0$

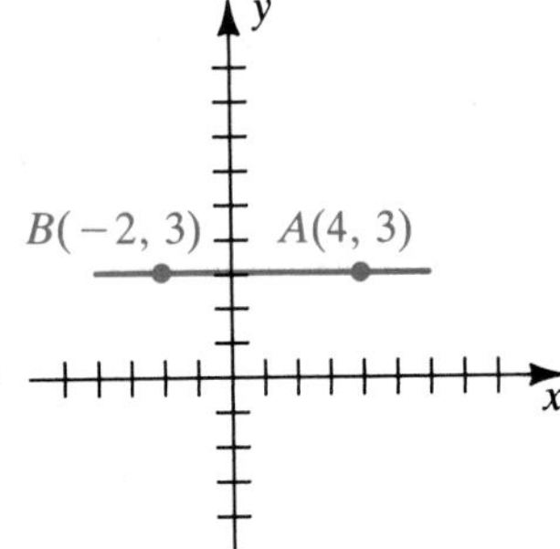

(d) m undefined

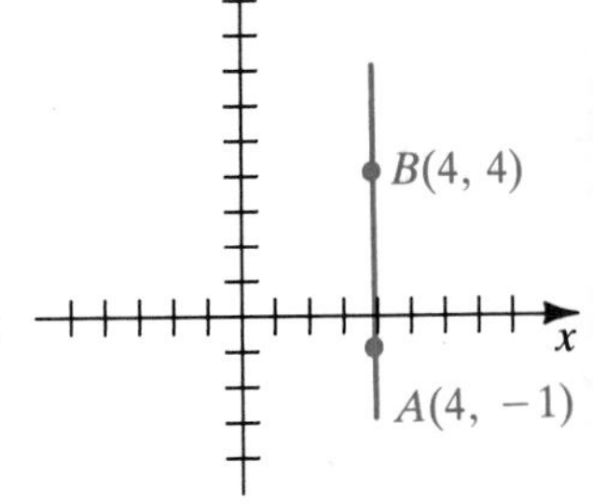

From the slope formula (1.8)(i):

(a) $m = \dfrac{2-4}{3-(-1)} = \dfrac{-2}{4} = -\dfrac{1}{2}$

(b) $m = \dfrac{5-(-1)}{2-(-2)} = \dfrac{6}{4} = \dfrac{3}{2}$

(c) $m = \dfrac{3-3}{4-(-2)} = \dfrac{0}{6} = 0$

(d) $m = \dfrac{4-(-1)}{4-4} = \dfrac{5}{0}$, which is undefined. Note that the line is vertical.

A **linear equation** in x and y is an equation of the form $ax + by = c$ (or $ax + by + d = 0$), with a and b not both zero. The graph of a linear equation is a line.

EXAMPLE 8 Find a linear equation for the line through $A(1, 7)$ and $B(-3, 2)$.

SOLUTION The slope m of the line is

$$m = \frac{7-2}{1-(-3)} = \frac{5}{4}.$$

We may use the coordinates of either A or B for (x_1, y_1) in the point-slope form (1.8)(ii). Using $A(1, 7)$ gives us

$$y - 7 = \tfrac{5}{4}(x - 1),$$

which is equivalent to

$$4y - 28 = 5x - 5, \quad \text{or} \quad 5x - 4y = -23.$$

EXAMPLE 9

(a) Find the slope of the line l with equation $2x - 5y = 9$.

(b) Find linear equations for the lines through $P(3, -4)$ that are parallel to l and perpendicular to l.

SOLUTION

(a) If we rewrite the equation as $5y = 2x - 9$ and divide both sides by 5, we obtain

$$y = \tfrac{2}{5}x - \tfrac{9}{5}.$$

Comparing this equation with the slope-intercept form $y = mx + b$, we see that the slope is $m = \frac{2}{5}$.

(b) By (ii) and (iii) of (1.9), the line through $P(3, -4)$ parallel to l has slope $\frac{2}{5}$ and the line perpendicular to l has slope $-\frac{5}{2}$. The corresponding equations are

$$y + 4 = \tfrac{2}{5}(x - 3), \quad \text{or} \quad 2x - 5y = 26$$

and

$$y + 4 = -\tfrac{5}{2}(x - 3), \quad \text{or} \quad 5x + 2y = 7.$$

EXAMPLE 10 Sketch the graphs of $4x + 3y = 5$ and $3x - 2y = 8$, and find their point of intersection.

SOLUTION Both equations are linear, and hence the graphs are lines. To sketch the graphs (see Figure 1.14), we can use the x-intercepts and the y-intercepts, found by letting $y = 0$ and $x = 0$, respectively.

The coordinates of the point P of intersection of the lines are obtained from the solution of the following system of equations:

$$\begin{cases} 4x + 3y = 5 \\ 3x - 2y = 8 \end{cases}$$

FIGURE 1.14

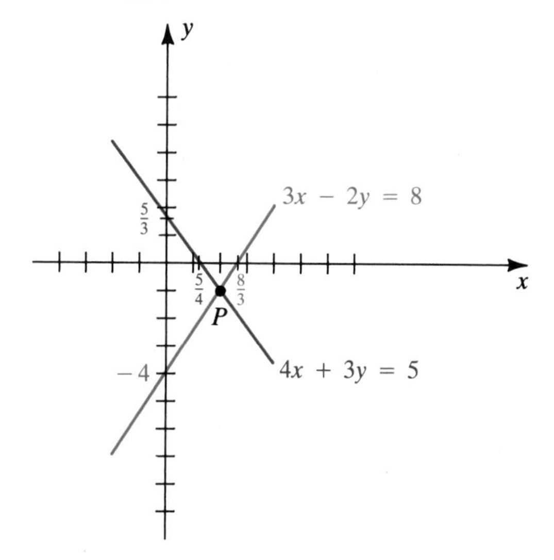

To eliminate y from the system, we begin by multiplying the first equation by 2 and the second by 3:

$$\begin{cases} 8x + 6y = 10 \\ 9x - 6y = 24 \end{cases}$$

Next we add both sides of the equations, to obtain

$$17x = 34, \quad \text{or} \quad x = 2.$$

This is the x-coordinate of the point of intersection. To find the y-coordinate of P, we let $x = 2$ in $4x + 3y = 5$, obtaining

$$4(2) + 3y = 5, \quad \text{or} \quad y = -1.$$

Hence P has coordinates $(2, -1)$.

EXERCISES 1.1

Exer. 1–8: Rewrite without using the absolute value symbol.

1 (a) $(-5)|3-6|$ (b) $|-6|/(-2)$ (c) $|-7| + |4|$

2 (a) $(4)|6-7|$ (b) $5/|-2|$ (c) $|-1| + |-9|$

3 (a) $|4-\pi|$ (b) $|\pi - 4|$ (c) $|\sqrt{2} - 1.5|$

4 (a) $|\sqrt{3} - 1.7|$ (b) $|1.7 - \sqrt{3}|$ (c) $|\frac{1}{5} - \frac{1}{3}|$

5 $|3 + x|$ if $x < -3$

6 $|5 - x|$ if $x > 5$

7 $|2 - x|$ if $x < 2$

8 $|7 + x|$ if $x \geq -7$

Exer. 9–12: Solve the equation by factoring.

9 $15x^2 - 12 = -8x$

10 $15x^2 - 14 = 29x$

11 $2x(4x + 15) = 27$

12 $x(3x + 10) = 77$

Exer. 13–16: Solve the equation by using the quadratic formula.

13 $x^2 + 4x + 2 = 0$

14 $x^2 - 6x - 3 = 0$

15 $2x^2 - 3x - 4 = 0$

16 $3x^2 + 5x + 1 = 0$

Exer. 17–38: Solve the inequality and express the solution in terms of intervals whenever possible.

17 $2x + 5 < 3x - 7$

18 $x - 8 > 5x + 3$

19 $3 \leq \dfrac{2x - 3}{5} < 7$

20 $-2 < \dfrac{4x + 1}{3} \leq 0$

21 $x^2 - x - 6 < 0$

22 $x^2 + 4x + 3 \geq 0$

23 $x^2 - 2x - 5 > 3$

24 $x^2 - 4x - 17 \leq 4$

25 $x(2x + 3) \geq 5$

26 $x(3x - 1) \leq 4$

27 $\dfrac{x + 1}{2x - 3} > 2$

28 $\dfrac{x - 2}{3x + 5} \leq 4$

29 $\dfrac{1}{x - 2} \geq \dfrac{3}{x + 1}$

30 $\dfrac{2}{2x + 3} \leq \dfrac{2}{x - 5}$

31 $|x + 3| < 0.01$

32 $|x - 4| \leq 0.03$

33 $|x + 2| \geq 0.001$

34 $|x - 3| > 0.002$

35 $|2x + 5| < 4$

36 $|3x - 7| \geq 5$

37 $|6 - 5x| \leq 3$

38 $|-11 - 7x| > 6$

Exer. 39–40: Describe the set of all points $P(x, y)$ in a coordinate plane that satisfy the given condition.

39 (a) $x = -2$ (b) $y = 3$ (c) $x \geq 0$ (d) $xy > 0$
(e) $y < 0$ (f) $|x| \leq 2$ and $|y| \leq 1$

40 (a) $y = -2$ (b) $x = -4$ (c) $x/y < 0$ (d) $xy = 0$
(e) $y > 1$ (f) $|x| \geq 2$ and $|y| \geq 3$

Exer. 41–42: Find (a) $d(A, B)$ and (b) the midpoint of AB.

41 $A(4, -3)$, $B(6, 2)$

42 $A(-2, -5)$, $B(4, 6)$

43 Show that the triangle with vertices $A(8, 5)$, $B(1, -2)$, and $C(-3, 2)$ is a right triangle, and find its area.

44 Show that the points $A(-4, 2)$, $B(1, 4)$, $C(3, -1)$, and $D(-2, -3)$ are vertices of a square.

Exer. 45–56: Sketch the graph of the equation.

45 $y = 2x^2 - 1$

46 $y = -x^2 + 2$

47 $x = \frac{1}{4}y^2$

48 $x = -2y^2$

49 $y = x^3 - 8$

50 $y = -x^3 + 1$

51 $y = \sqrt{x} - 4$

52 $y = \sqrt{x - 4}$

53 $(x + 3)^2 + (y - 2)^2 = 9$

54 $x^2 + (y - 2)^2 = 25$

55 $y = -\sqrt{16 - x^2}$

56 $y = \sqrt{4 - x^2}$

Exer. 57–60: Find an equation of the circle that satisfies the given conditions.

57 Center $C(2, -3)$; radius 5

58 Center $C(-4, 6)$; passing through $P(1, 2)$

59 Tangent to both axes; center in the second quadrant; radius 4

60 Endpoints of a diameter $A(4, -3)$ and $B(-2, 7)$

Exer. 61–66: Find an equation of the line that satisfies the given conditions.

61 Through $A(5, -3)$; slope -4

62 Through $A(-1, 4)$; slope $\frac{2}{3}$

63 x-intercept 4; y-intercept -3

64 Through $A(5, 2)$ and $B(-1, 4)$

65 Through $A(2, -4)$; parallel to the line $5x - 2y = 4$

66 Through $A(7, -3)$; perpendicular to the line $2x - 5y = 8$

Exer. 67–68: Find an equation for the perpendicular bisector of AB.

67 $A(3, -1)$, $B(-2, 6)$

68 $A(4, 2)$, $B(-2, 10)$

Exer. 69–72: Sketch the graphs of the lines and find their point of intersection.

69 $2x + 3y = 2$; $x - 2y = 8$

70 $4x + 5y = 13$; $3x + y = -4$

71 $2x + 5y = 16$; $3x - 7y = 24$

72 $7x - 8y = 9$; $4x + 3y = -10$

73 Approximate the coordinates of the point of intersection of the lines

$$(\sqrt{1.25} - 0.1)x + (0.11)^{2/3}y = 1/\sqrt{5}$$
$$(2.51)^{2/3}x + (6.27 - \sqrt{3})y = \sqrt{2}.$$

74 Approximate the smallest root of the following equation: $x^2 - (6.7 \times 10^6)x + 1.08 = 0$. To avoid calculating a zero value for this root, rewrite the quadratic formula as

$$x = \frac{2c}{-b \pm \sqrt{b^2 - 4ac}}.$$

75 The rate at which a tablet of vitamin C begins to dissolve depends on the surface area of the tablet. One brand of tablet is 2 centimeters long and is in the shape of a cylinder with hemispheres of diameter 0.5 centimeter attached to both ends (see figure). A second brand of tablet is to be manufactured in the shape of a right circular cylinder of altitude 0.5 centimeter.

EXERCISE 75

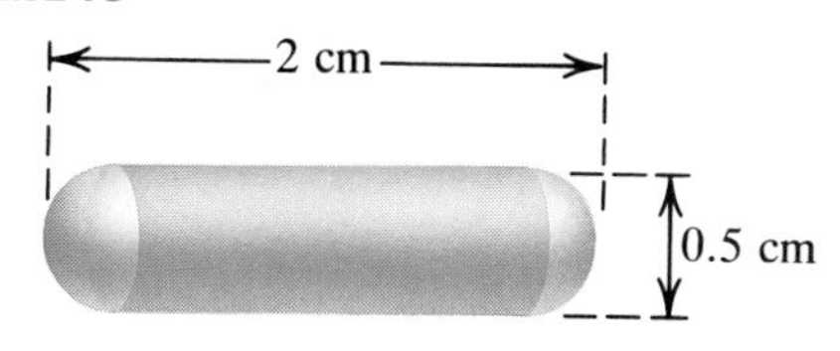

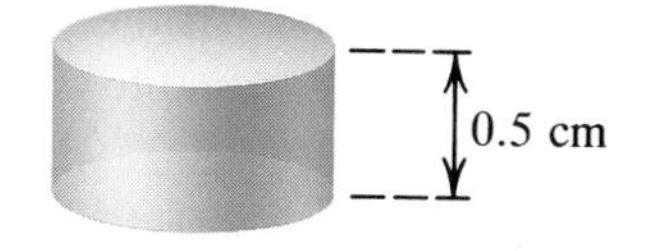

(a) Find the diameter of the second tablet so that its surface area is equal to that of the first tablet.

(b) Find the volume of each tablet.

76 A manufacturer of tin cans wishes to construct a right circular cylindrical can of height 20 centimeters and of capacity 3000 cm^3 (see figure). Find the inner radius r of the can.

EXERCISE 76

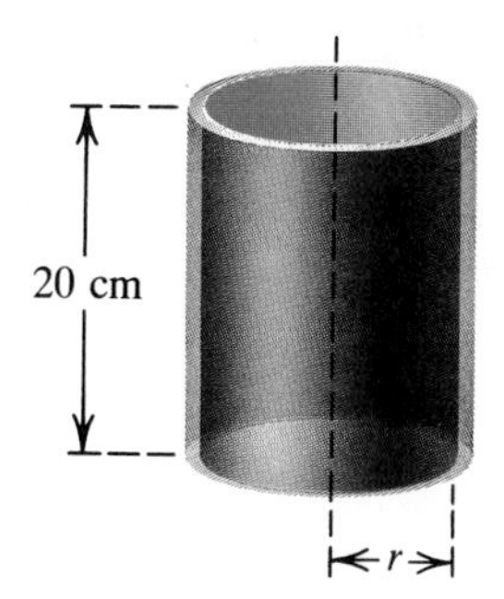

77 Shown in the figure is a simple magnifier consisting of a convex lens. The object to be magnified is positioned so that its distance p from the lens is less than the focal length f. The linear magnification M is the ratio of the image size to the object size. It is shown in physics that $M = f/(f - p)$. If $f = 6$ cm, how far should the object be placed from the lens so that its image appears at least three times as large?

EXERCISE 77

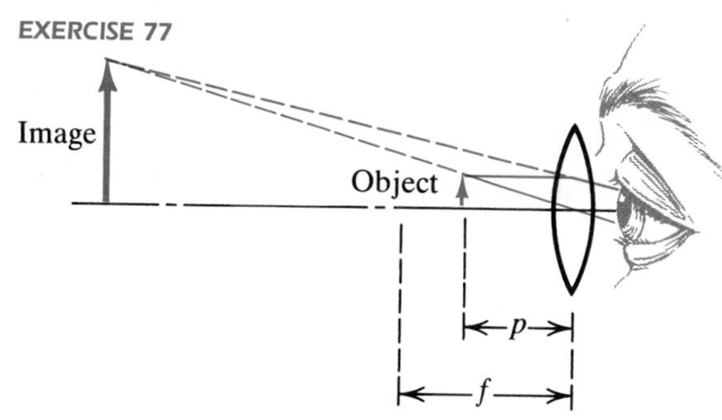

78 As the altitude of a space shuttle increases, an astronaut's weight decreases until a state of weightlessness is achieved. The weight of a 125-pound astronaut at an altitude of x kilometers above sea level is given by

$$W = 125\left(\frac{6400}{6400 + x}\right)^2.$$

At what altitudes is the astronaut's weight less than 5 pounds?

79 The braking distance d (in feet) of a car traveling v mi/hr is approximated by $d = v + (v^2/20)$. Determine velocities that result in braking distances of less than 75 feet.

80 For a drug to have a beneficial effect, its concentration in the bloodstream must exceed a certain value, the

minimum therapeutic level. Suppose that the concentration c of a drug t hours after it is taken orally is given by $c = 20t/(t^2 + 4)$ mg/L. If the minimum therapeutic level is 4 mg/L, determine when this level is exceeded.

81 The electrical resistance R (in ohms) for a pure metal wire is related to its temperature T (in °C) by the formula $R = R_0(1 + aT)$ for positive constants a and R_0.

(a) For what temperature is $R = R_0$?

(b) Assuming that the resistance is 0 if $T = -273\,°C$ (absolute zero), find a.

(c) Silver wire has a resistance of 1.25 ohms at 0 °C. At what temperature is the resistance 2 ohms?

82 Pharmacological products must specify recommended dosages for adults and children. Two formulas for modification of adult dosage levels for young children are

$$\text{Cowling's rule:} \quad y = \tfrac{1}{24}(t + 1)a$$

$$\text{Friend's rule:} \quad y = \tfrac{2}{25}ta$$

where a denotes the adult dose (in milligrams) and t denotes the age of the child (in years).

(a) If $a = 100$, graph the two linear equations on the same axes for $0 \le t \le 12$.

(b) For what age do the two formulas specify the same dosage?

1.2 FUNCTIONS

The notion of *function* is basic for much of our work in calculus. We may define a function as follows.

Definition **(1.10)**

> A **function** f from a set D to a set E is a correspondence that assigns to each element x of D exactly one element y of E.

The element y of E is the **value** of f at x and is denoted by $f(x)$, read *f of x*. The set D is the **domain** of the function. The **range** of f is the subset of E consisting of all possible function values $f(x)$ for x in D.

FIGURE 1.15

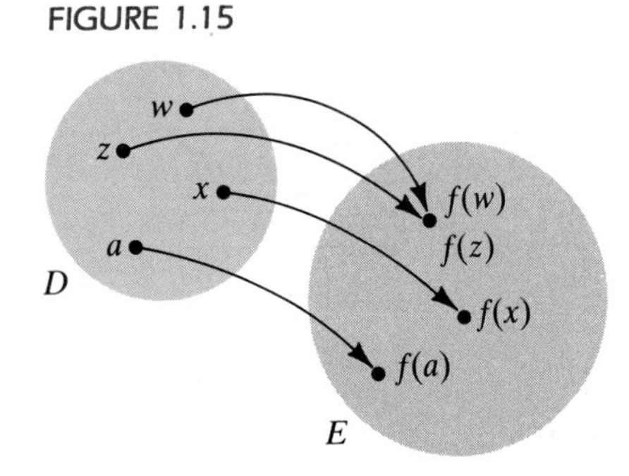

We sometimes depict functions as in Figure 1.15, where the sets D and E are represented by points within regions in a plane. The curved arrows indicate that the elements $f(x)$, $f(w)$, $f(z)$, and $f(a)$ of E correspond to the elements x, w, z, and a of D. It is important to remember that *to each x in D there is assigned exactly one function value $f(x)$ in E*; however, different elements of D, such as w and z in Figure 1.15, may yield the same function value in E. Throughout Chapters 1–14, the phrase *f is a function* will mean that the domain and range of f are sets of real numbers.

We usually define a function f by stating a formula or rule for finding $f(x)$, such as $f(x) = \sqrt{x-2}$. The domain is then assumed to be the set of all real numbers such that $f(x)$ is real. Thus, for $f(x) = \sqrt{x-2}$, the domain is the infinite interval $[2, \infty)$. If x is in the domain, we say that **f is defined at** x, or that **$f(x)$ exists**. If S is a subset of the domain, then **f is defined on** S. The terminology **f is undefined at** x means that x is not in the domain of f.

EXAMPLE 1 Let $f(x) = \dfrac{\sqrt{4+x}}{1-x}$.

(a) Find the domain of f. **(b)** Find $f(5)$, $f(-2)$, $f(-a)$, and $-f(a)$.

SOLUTION

(a) Note that $f(x)$ is a real number if and only if the radicand $4 + x$ is nonnegative and the denominator $1 - x$ is not equal to 0. Thus, $f(x)$ exists if and only if

$$4 + x \ge 0 \quad \text{and} \quad 1 - x \ne 0$$

or, equivalently, $\qquad x \geq -4 \quad \text{and} \quad x \neq 1.$

Hence the domain is $[-4, 1) \cup (1, \infty)$.

(b) To find values of f, we substitute for x:

$$f(5) = \frac{\sqrt{4+5}}{1-5} = \frac{\sqrt{9}}{-4} = -\frac{3}{4}$$

$$f(-2) = \frac{\sqrt{4+(-2)}}{1-(-2)} = \frac{\sqrt{2}}{3}$$

$$f(-a) = \frac{\sqrt{4+(-a)}}{1-(-a)} = \frac{\sqrt{4-a}}{1+a}$$

$$-f(a) = -\frac{\sqrt{4+a}}{1-a} = \frac{\sqrt{4+a}}{a-1}$$

Many formulas that occur in mathematics and the sciences determine functions. For instance, the formula $A = \pi r^2$ for the area A of a circle of radius r assigns to each positive real number r exactly one value of A. The letter r, which represents an arbitrary number from the domain, is an **independent variable**. The letter A, which represents a number from the range, is a **dependent variable**, since its value *depends* on the number assigned to r. If two variables r and A are related in this manner, we say that *A is a function of r*. As another example, if an automobile travels at a uniform rate of 50 mi/hr, then the distance d (in miles) traveled in time t (in hours) is given by $d = 50t$, and hence the distance *d is a function of time t*.

EXAMPLE 2 A steel storage tank for propane gas is to be constructed in the shape of a right circular cylinder of altitude 10 feet with a hemisphere attached to each end. The radius r is yet to be determined. Express the volume V of the tank as a function of r.

FIGURE 1.16

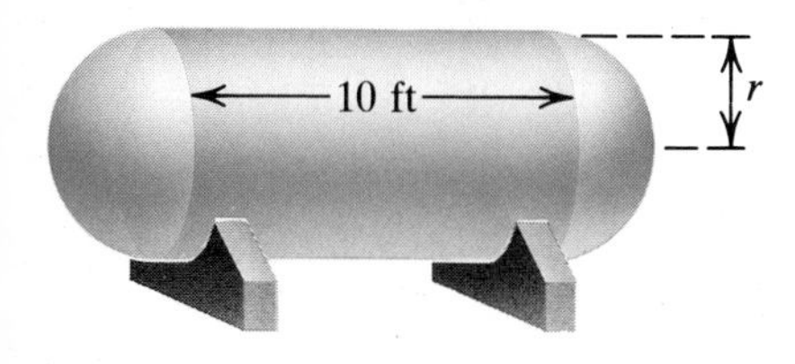

SOLUTION The tank is sketched in Figure 1.16. We may find the volume of the cylindrical part of the tank by multiplying the altitude 10 by the area πr^2 of the base of the cylinder:

$$\text{volume of cylinder} = 10(\pi r^2) = 10\pi r^2$$

The two hemispherical ends, taken together, form a sphere of radius r. Using the formula for the volume of a sphere, we obtain

$$\text{volume of the two ends} = \tfrac{4}{3}\pi r^3.$$

Thus, the volume V of the tank is

$$V = \tfrac{4}{3}\pi r^3 + 10\pi r^2 = \tfrac{2}{3}\pi r^2(2r + 15).$$

This formula expresses V as a function of r.

If f is a function, we may use a graph to illustrate the change in the function value $f(x)$ as x varies through the domain of f. By definition,

FIGURE 1.17

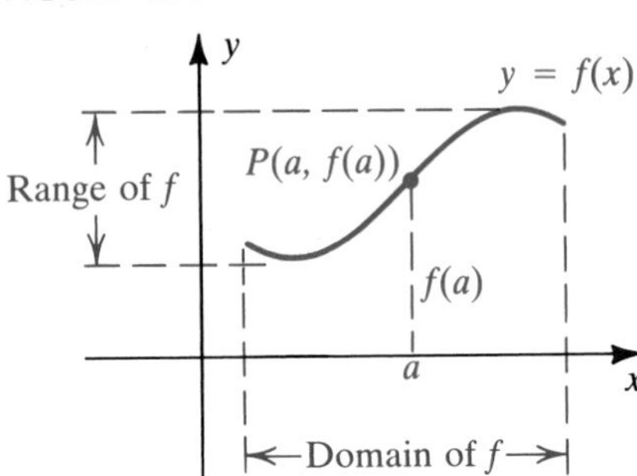

the **graph of a function** f is the graph of the equation $y = f(x)$ for x in the domain of f. As shown in Figure 1.17, we often attach the label $y = f(x)$ to a sketch of the graph. Note that if $P(a, b)$ is on the graph, then the y-coordinate b is the function value $f(a)$. The figure exhibits the domain of f (the set of possible values of x) and the range of f (the corresponding values of y). Although we have pictured the domain and range as closed intervals, they may be infinite intervals or other sets of real numbers.

It is important to note that since there is exactly one value $f(a)$ for each a in the domain, only *one* point on the graph has x-coordinate a. Thus, *every vertical line intersects the graph of a function in at most one point*. Consequently, the graph of a function cannot be a figure such as a circle, which can be intersected by a vertical line in more than one point.

The x-intercepts of the graph of a function f are the solutions of the equation $f(x) = 0$. These numbers are the **zeros** of the function. The y-intercept of the graph is $f(0)$, if it exists.

If f is an **even function**—that is, if $f(-x) = f(x)$ for every x in the domain of f—then the graph of f is symmetric with respect to the y-axis, by symmetry test (i) of (1.6). If f is an **odd function**—that is, if $f(-x) = -f(x)$ for every x in the domain of f—then the graph of f is symmetric with respect to the origin, by symmetry test (iii). Most functions in calculus are neither even nor odd.

The next illustration contains sketches of graphs of some common functions. You should check each for the indicated symmetry, domain, and range.

ILLUSTRATION

FUNCTION f	GRAPH	SYMMETRY	DOMAIN D, RANGE R
$f(x) = \sqrt{x}$		none	$D = [0, \infty)$ $R = [0, \infty)$
$f(x) = x^2$		y-axis (even function)	$D = (-\infty, \infty)$ $R = [0, \infty)$
$f(x) = x^3$		origin (odd function)	$D = (-\infty, \infty)$ $R = (-\infty, \infty)$

(continued)

FUNCTION f	GRAPH	SYMMETRY	DOMAIN D, RANGE R
$f(x) = x^{2/3}$	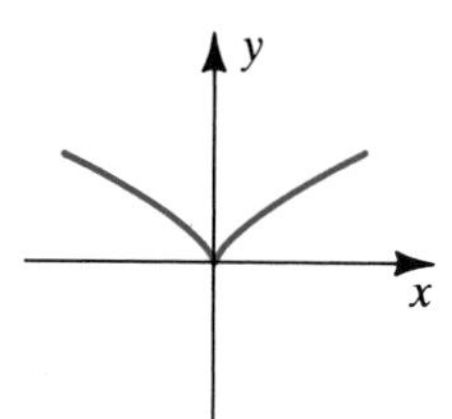	y-axis (even function)	$D = (-\infty, \infty)$ $R = [0, \infty)$
$f(x) = x^{1/3}$	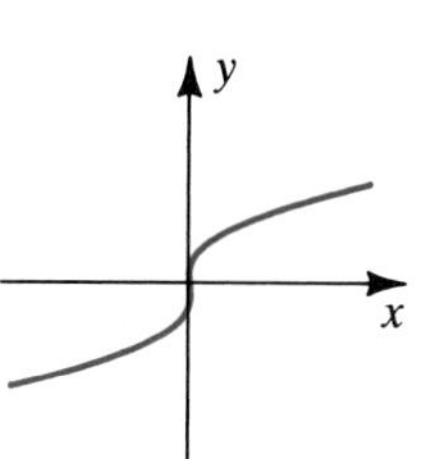	origin (odd function)	$D = (-\infty, \infty)$ $R = (-\infty, \infty)$
$f(x) = \lvert x \rvert$	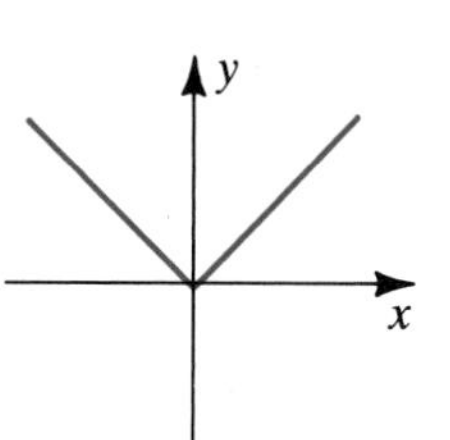	y-axis (even function)	$D = (-\infty, \infty)$ $R = [0, \infty)$
$f(x) = \dfrac{1}{x}$	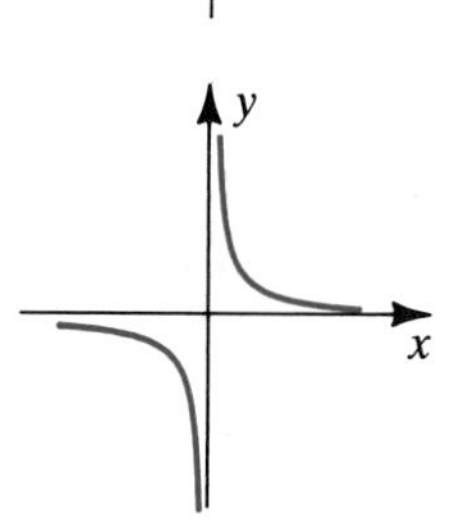	origin (odd function)	$D = (-\infty, 0) \cup (0, \infty)$ $R = (-\infty, 0) \cup (0, \infty)$

Functions that are described by more than one expression, as in the next example, are called **piecewise-defined functions**.

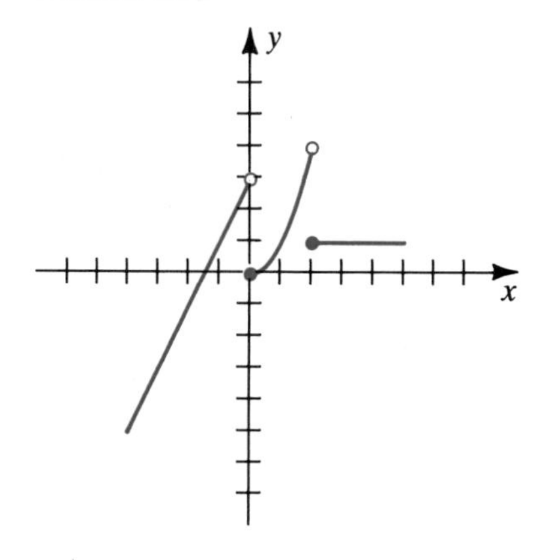

FIGURE 1.18

EXAMPLE 3 Sketch the graph of the function f defined as follows:

$$f(x) = \begin{cases} 2x + 3 & \text{if } x < 0 \\ x^2 & \text{if } 0 \leq x < 2 \\ 1 & \text{if } x \geq 2 \end{cases}$$

SOLUTION If $x < 0$, then $f(x) = 2x + 3$, and the graph of f is part of the line $y = 2x + 3$, as indicated in Figure 1.18. The small circle indicates that $(0, 3)$ is not on the graph.

If $0 \leq x < 2$, then $f(x) = x^2$, and the graph of f is part of the parabola $y = x^2$. Note that $(2, 4)$ is not on the graph.

If $x \geq 2$, the function values are always 1, and the graph is a horizontal half-line with endpoint $(2, 1)$.

If x is a real number, we define the symbol $[\![x]\!]$ as follows:

$$[\![x]\!] = n, \quad \text{where } n \text{ is the greatest integer such that } n \le x$$

If we identify $\mathbb{R}$ with points on a coordinate line, then n is the first integer to the *left* of (or *equal* to) x.

ILLUSTRATION

- $[\![0.5]\!] = 0$
- $[\![1.8]\!] = 1$
- $[\![\sqrt{5}]\!] = 2$
- $[\![3]\!] = 3$
- $[\![-3]\!] = -3$
- $[\![-2.7]\!] = -3$
- $[\![-\sqrt{3}]\!] = -2$
- $[\![-0.5]\!] = -1$

The **greatest integer function** f is defined by $f(x) = [\![x]\!]$.

EXAMPLE 4 Sketch the graph of the greatest integer function.

SOLUTION The x- and y-coordinates of some points on the graph may be listed as follows:

FIGURE 1.19

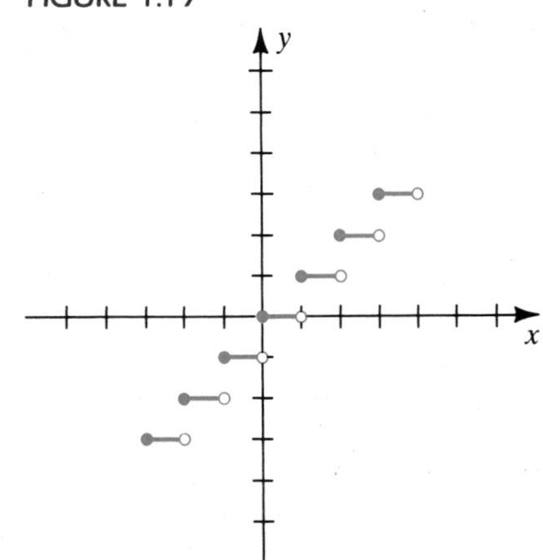

Values of x	$f(x) = [\![x]\!]$
$\vdots$	$\vdots$
$-2 \le x < -1$	-2
$-1 \le x < 0$	-1
$0 \le x < 1$	0
$1 \le x < 2$	1
$2 \le x < 3$	2
$\vdots$	$\vdots$

Whenever x is between successive integers, the corresponding part of the graph is a segment of a horizontal line. Part of the graph is sketched in Figure 1.19. The graph continues indefinitely to the right and to the left.

The graphs in Figure 1.20 illustrate **vertical shifts** of the graph of $y = f(x)$ resulting from adding a positive constant c to each function value $f(x)$ or subtracting c from $f(x)$. **Horizontal shifts** are illustrated in Figure 1.21.

We can sometimes obtain the graph of a function by applying a shift to a known graph, as shown in Figures 1.22 and 1.23 for $f(x) = x^2$.

The graph of $y = cf(x)$ can be obtained by multiplying the y-coordinate of each point on the graph of $y = f(x)$ by c. If $c < 0$, the graphs of $y = cf(x)$ and $y = |c| f(x)$ are called **reflections** of each other about the x-axis. Some special cases with $f(x) = x^2$ are sketched in Figures 1.24 and 1.25.

FIGURE 1.20
Vertical shifts, $c > 0$

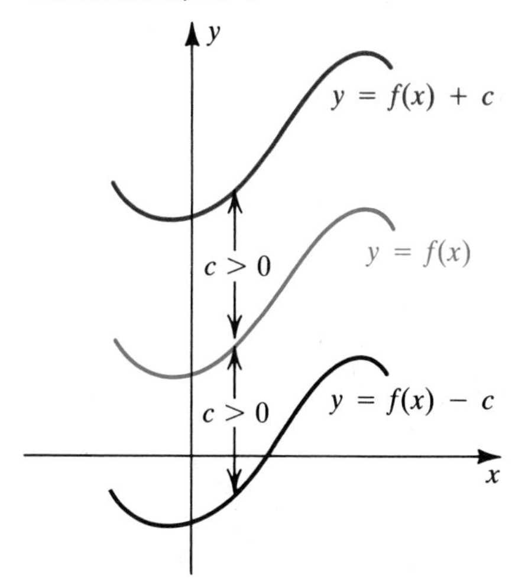

FIGURE 1.21
Horizontal shifts, $c > 0$

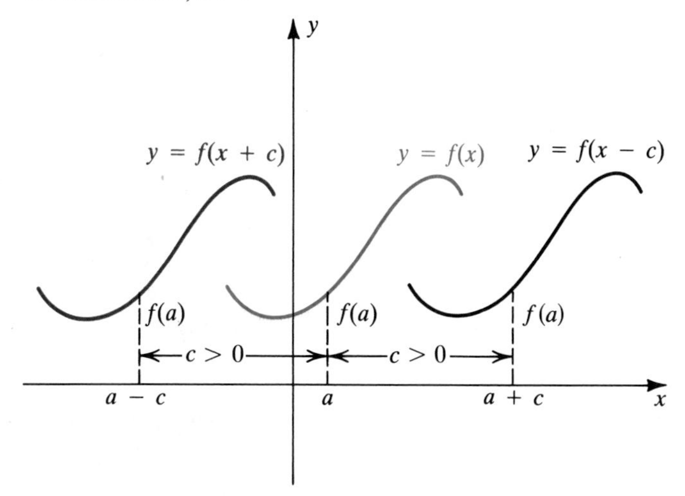

FIGURE 1.22

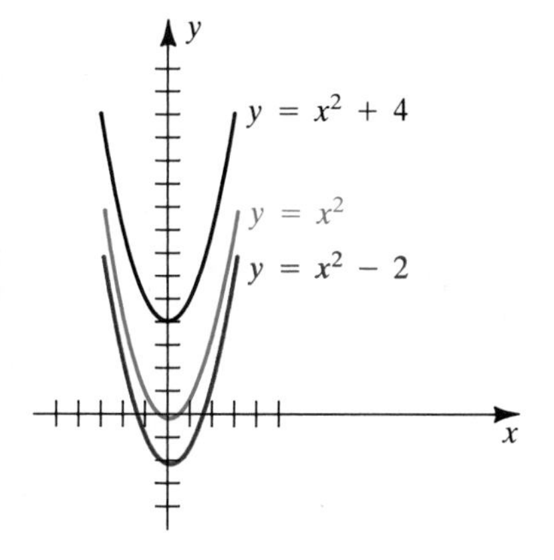

FIGURE 1.23

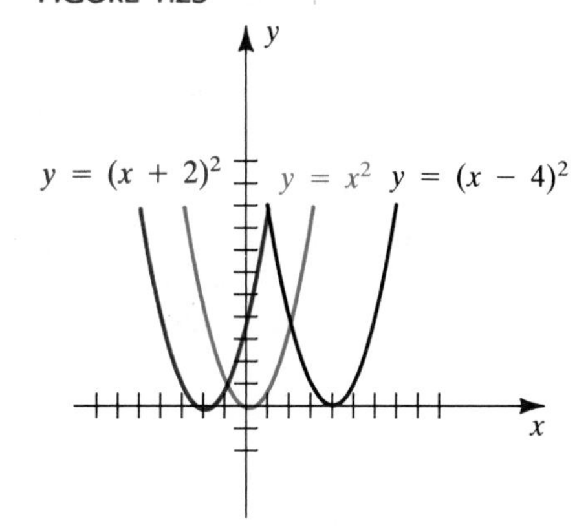

FIGURE 1.24

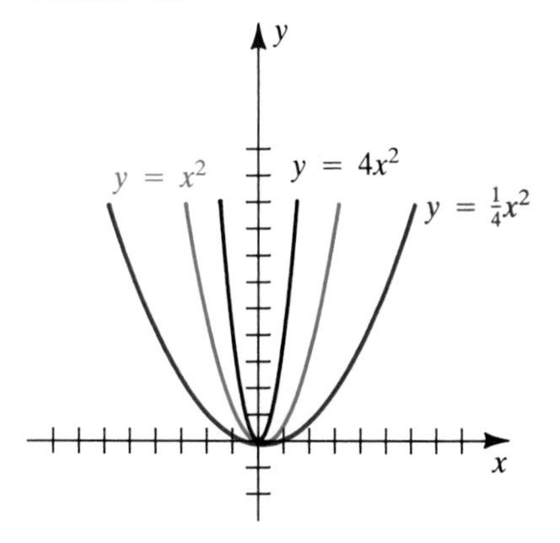

FIGURE 1.25

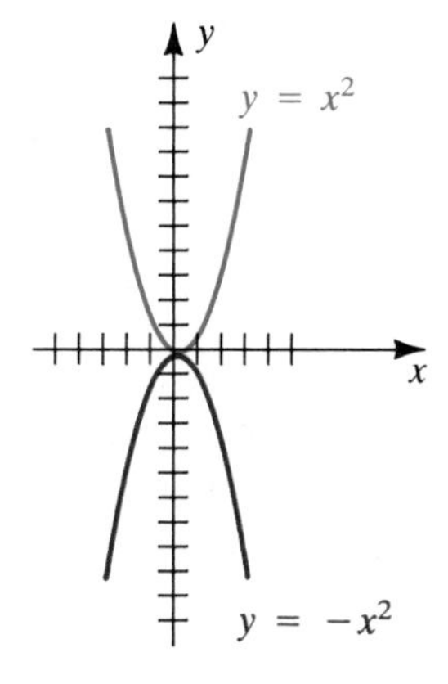

If f and g are functions, we define the **sum** $f + g$, **difference** $f - g$, **product** fg, and **quotient** f/g as follows:

$$(f + g)(x) = f(x) + g(x)$$
$$(f - g)(x) = f(x) - g(x)$$
$$(fg)(x) = f(x)g(x)$$
$$\left(\frac{f}{g}\right)(x) = \frac{f(x)}{g(x)}$$

The domain of $f + g$, $f - g$, and fg is the *intersection* of the domains of f and g—that is, the numbers that are *common* to both domains. The domain of f/g consists of all numbers x in the intersection such that $g(x) \neq 0$.

EXAMPLE 5 Let $f(x) = \sqrt{4 - x^2}$ and $g(x) = 3x + 1$. Find the sum, difference, product, and quotient of f and g, and specify the domain of each.

SOLUTION The domain of f is the closed interval $[-2, 2]$, and the domain of g is $\mathbb{R}$. Consequently, the intersection of their domains is $[-2, 2]$, and we obtain the following:

$$(f + g)(x) = \sqrt{4 - x^2} + (3x + 1), \qquad -2 \le x \le 2$$
$$(f - g)(x) = \sqrt{4 - x^2} - (3x + 1), \qquad -2 \le x \le 2$$
$$(fg)(x) = \sqrt{4 - x^2}(3x + 1), \qquad -2 \le x \le 2$$
$$\left(\frac{f}{g}\right)(x) = \frac{\sqrt{4 - x^2}}{3x + 1}, \qquad -2 \le x \le 2 \text{ and } x \neq -\tfrac{1}{3}$$

A function f is a **polynomial function** if $f(x)$ is a polynomial—that is, if

$$f(x) = a_n x^n + a_{n-1}x^{n-1} + \cdots + a_1 x + a_0,$$

where the coefficients $a_0, a_1, \ldots, a_n$ are real numbers and the exponents are nonnegative integers. If $a_n \neq 0$, then f has **degree** n. The following are special cases, where $a \neq 0$:

degree 0:	$f(x) = a$	(constant function)
degree 1:	$f(x) = ax + b$	(linear function)
degree 2:	$f(x) = ax^2 + bx + c$	(quadratic function)

A **rational function** is a quotient of two polynomial functions. Later in the text we shall use methods of calculus to investigate graphs of polynomial and rational functions.

An **algebraic function** is a function that can be expressed in terms of sums, differences, products, quotients, or rational powers of polynomials. For example, if

$$f(x) = 5x^4 - 2\sqrt[3]{x} + \frac{x(x^2 + 5)}{\sqrt{x^3 + \sqrt{x}}},$$

then f is an algebraic function. Functions that are not algebraic are termed **transcendental**. The trigonometric, exponential, and logarithmic functions considered later are examples of transcendental functions.

In the remainder of this section we shall discuss how two functions f and g may be used to obtain the *composite functions* $f \circ g$ and $g \circ f$ (read f *circle* g and g *circle* f, respectively). The function $f \circ g$ is defined as follows.

Definition **(1.11)**

> The **composite function** $f \circ g$ of f and g is defined by
> $$(f \circ g)(x) = f(g(x)).$$
> The domain of $f \circ g$ is the set of all x in the domain of g such that $g(x)$ is in the domain of f.

FIGURE 1.26

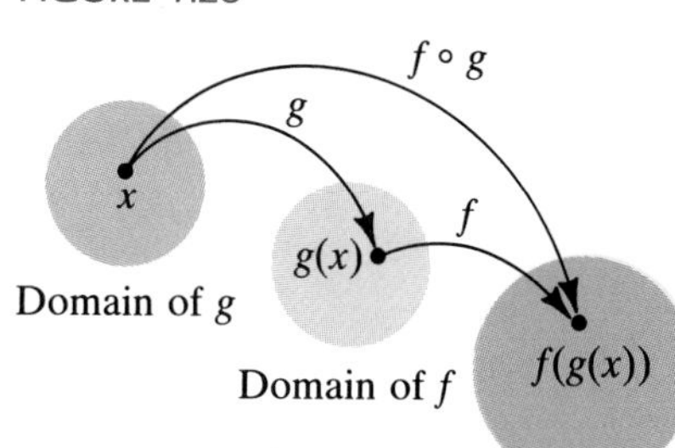

Figure 1.26 illustrates relationships between f, g, and $f \circ g$. Note that for x in the domain of g, we *first find* $g(x)$ (which must be in the domain of f) and then, *second*, *find* $f(g(x))$.

For the composite function $g \circ f$, we reverse this order, first finding $f(x)$ and then finding $g(f(x))$. The domain of $g \circ f$ is the set of all x in the domain of f such that $f(x)$ is in the domain of g.

EXAMPLE 6 If $f(x) = x^2 - 1$ and $g(x) = 3x + 5$, find

(a) $(f \circ g)(x)$ and the domain of $f \circ g$

(b) $(g \circ f)(x)$ and the domain of $g \circ f$

SOLUTION

(a)
$$\begin{aligned}(f \circ g)(x) &= f(g(x)) && \text{(definition of } f \circ g)\\ &= f(3x + 5) && \text{(definition of } g)\\ &= (3x + 5)^2 - 1 && \text{(definition of } f)\\ &= 9x^2 + 30x + 24 && \text{(simplifying)}\end{aligned}$$

The domain of both f and g is $\mathbb{R}$. Since for each x in $\mathbb{R}$ (the domain of g) the function value $g(x)$ is in $\mathbb{R}$ (the domain of f), the domain of $f \circ g$ is also $\mathbb{R}$.

(b)
$$\begin{aligned}(g \circ f)(x) &= g(f(x)) && \text{(definition of } g \circ f)\\ &= g(x^2 - 1) && \text{(definition of } f)\\ &= 3(x^2 - 1) + 5 && \text{(definition of } g)\\ &= 3x^2 + 2 && \text{(simplifying)}\end{aligned}$$

Since for each x in $\mathbb{R}$ (the domain of f) the function value $f(x)$ is in $\mathbb{R}$ (the domain of g), the domain of $g \circ f$ is $\mathbb{R}$.

Note that in Example 6, $f(g(x))$ and $g(f(x))$ are not always the same; that is, $f \circ g \neq g \circ f$.

If two functions f and g both have domain $\mathbb{R}$, then the domain of $f \circ g$ and $g \circ f$ is also $\mathbb{R}$. This was illustrated in Example 6. The next example shows that the domain of a composite function may differ from those of the two given functions.

EXAMPLE 7 If $f(x) = x^2 - 16$ and $g(x) = \sqrt{x}$, find

(a) $(f \circ g)(x)$ and the domain of $f \circ g$

(b) $(g \circ f)(x)$ and the domain of $g \circ f$

SOLUTION We first note that the domain of f is $\mathbb{R}$ and the domain of g is the set of all nonnegative real numbers—that is, the interval $[0, \infty)$. We may proceed as follows.

(a)
$$\begin{aligned}(f \circ g)(x) &= f(g(x)) && \text{(definition of } f \circ g\text{)}\\ &= f(\sqrt{x}) && \text{(definition of } g\text{)}\\ &= (\sqrt{x})^2 - 16 && \text{(definition of } f\text{)}\\ &= x - 16 && \text{(simplifying)}\end{aligned}$$

If we considered only the final expression $x - 16$, we might be led to believe that the domain of $f \circ g$ was $\mathbb{R}$, since $x - 16$ is defined for every real number x. However, this is not the case. By definition, the domain of $f \circ g$ is the set of all x in $[0, \infty)$ (the domain of g) such that $g(x)$ is in $\mathbb{R}$ (the domain of f). Since $g(x) = \sqrt{x}$ is in $\mathbb{R}$ for every x in $[0, \infty)$, it follows that the domain of $f \circ g$ is $[0, \infty)$.

(b)
$$\begin{aligned}(g \circ f)(x) &= g(f(x)) && \text{(definition of } g \circ f\text{)}\\ &= g(x^2 - 16) && \text{(definition of } f\text{)}\\ &= \sqrt{x^2 - 16} && \text{(definition of } g\text{)}\end{aligned}$$

By definition, the domain of $g \circ f$ is the set of all x in $\mathbb{R}$ (the domain of f) such that $f(x) = x^2 - 16$ is in $[0, \infty)$ (the domain of g). The statement $x^2 - 16$ is in $[0, \infty)$ is equivalent to each of the inequalities

$$x^2 - 16 \geq 0, \quad x^2 \geq 16, \quad \text{and} \quad |x| \geq 4.$$

Thus, the domain of $g \circ f$ is the union $(-\infty, -4] \cup [4, \infty)$. Note that this is different from the domains of both f and g.

If f and g are functions such that

$$y = f(u) \quad \text{and} \quad u = g(x),$$

then substituting for u in $y = f(u)$ yields

$$y = f(g(x)).$$

For certain problems in calculus we *reverse* this procedure; that is, given $y = h(x)$ for some function h, we find a *composite function form* $y = f(u)$ and $u = g(x)$ such that $h(x) = f(g(x))$.

EXAMPLE 8 Express $y = (2x + 5)^8$ in a composite function form.

SOLUTION Suppose, for a real number x, we wanted to evaluate $(2x + 5)^8$ by using a calculator. We would first calculate $2x + 5$ and then raise the result to the eighth power. This suggests that we let

$$u = 2x + 5 \quad \text{and} \quad y = u^8,$$

which is a composite function form for $y = (2x + 5)^8$.

The method used in the preceding example can be extended to other functions. In general, suppose we are given $y = h(x)$. To choose the *inside* expression $u = g(x)$ in a composite function form, ask the following question: If a calculator were being used, which part of the expression $h(x)$ would be evaluated first? This often leads to a suitable choice for $u = g(x)$. After choosing u, refer to $h(x)$ to determine $y = f(u)$. The following illustration contains typical problems.

ILLUSTRATION

FUNCTION VALUE	CHOICE FOR $u = g(x)$	CHOICE FOR $y = f(u)$
$y = (x^3 - 5x + 1)^4$	$u = x^3 - 5x + 1$	$y = u^4$
$y = \sqrt{x^2 - 4}$	$u = x^2 - 4$	$y = \sqrt{u}$
$y = \dfrac{2}{3x + 7}$	$u = 3x + 7$	$y = \dfrac{2}{u}$

The composite function form is never unique. For example, consider the first expression in the preceding illustration:

$$y = (x^3 - 5x + 1)^4$$

If n is any nonzero integer, we could choose

$$u = (x^3 - 5x + 1)^n \quad \text{and} \quad y = u^{4/n}.$$

Thus, there are an *unlimited* number of composite function forms. Generally, our goal is to choose a form such that the expression for y is simple, as we did in the illustration.

EXERCISES 1.2

1 If $f(x) = \sqrt{x - 4} - 3x$, find $f(4)$, $f(8)$, and $f(13)$.

2 If $f(x) = \dfrac{x}{x - 3}$, find $f(-2)$, $f(0)$, and $f(3.01)$.

Exer. 3–6: If a and h are real numbers, find and simplify (a) $f(a)$, (b) $f(-a)$, (c) $-f(a)$, (d) $f(a + h)$, (e) $f(a) + f(h)$, and (f) $\dfrac{f(a + h) - f(a)}{h}$, provided $h \neq 0$.

3 $f(x) = 5x - 2$

4 $f(x) = 3 - 4x$

5 $f(x) = x^2 - x + 3$

6 $f(x) = 2x^2 + 3x - 7$

Exer. 7–10: Find the domain of f.

7 $f(x) = \dfrac{x + 1}{x^3 - 4x}$

8 $f(x) = \dfrac{4x}{6x^2 + 13x - 5}$

9 $f(x) = \dfrac{\sqrt{2x - 3}}{x^2 - 5x + 4}$

10 $f(x) = \dfrac{\sqrt{4x - 3}}{x^2 - 4}$

Exer. 11–12: Determine whether f is even, odd, or neither even nor odd.

11 (a) $f(x) = 5x^3 + 2x$

(b) $f(x) = |x| - 3$

(c) $f(x) = (8x^3 - 3x^2)^3$

12 (a) $f(x) = \sqrt{3x^4 + 2x^2 - 5}$

(b) $f(x) = 6x^5 - 4x^3 + 2x$

(c) $f(x) = x(x - 5)$

Exer. 13–22: Sketch, on the same coordinate plane, the graphs of f for the given values of c. (Make use of symmetry, vertical shifts, horizontal shifts, stretching, or reflecting.)

13 $f(x) = |x| + c$; $c = 0, 1, -3$

14 $f(x) = |x - c|$; $c = 0, 2, -3$

15 $f(x) = 2\sqrt{x} + c$; $c = 0, 3, -2$

16 $f(x) = \sqrt{9 - x^2} + c$; $c = 0, 1, -3$

17 $f(x) = 2\sqrt{x - c}$; $c = 0, 1, -2$

18 $f(x) = -2(x - c)^2$; $c = 0, 1, -2$

19 $f(x) = c\sqrt{4 - x^2}$; $c = 1, 3, -2$

20 $f(x) = (x + c)^3$; $c = 0, 1, -2$

21 $f(x) = (x - c)^{2/3} + 2$; $c = 0, 4, -3$

22 $f(x) = (x - 1)^{1/3} - c$; $c = 0, 2, -1$

Exer. 23–24: The graph of a function f with domain $0 \leq x \leq 4$ is shown in the figure. Sketch the graph of the given equation.

23

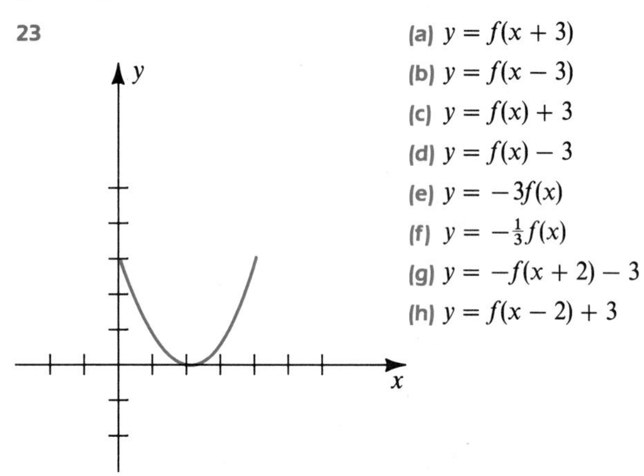

(a) $y = f(x + 3)$
(b) $y = f(x - 3)$
(c) $y = f(x) + 3$
(d) $y = f(x) - 3$
(e) $y = -3f(x)$
(f) $y = -\frac{1}{3}f(x)$
(g) $y = -f(x + 2) - 3$
(h) $y = f(x - 2) + 3$

24

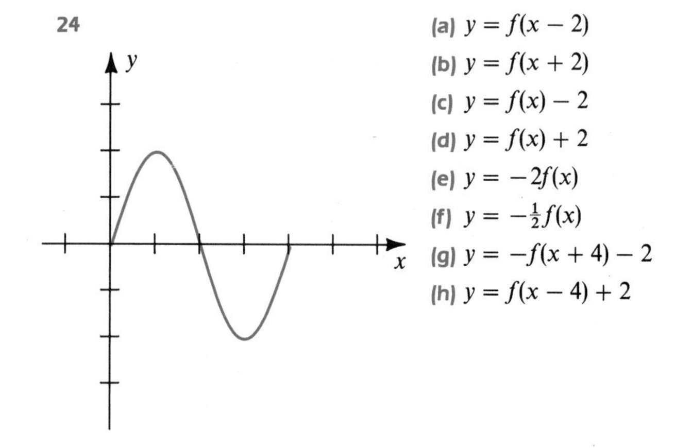

(a) $y = f(x - 2)$
(b) $y = f(x + 2)$
(c) $y = f(x) - 2$
(d) $y = f(x) + 2$
(e) $y = -2f(x)$
(f) $y = -\frac{1}{2}f(x)$
(g) $y = -f(x + 4) - 2$
(h) $y = f(x - 4) + 2$

Exer. 25–30: Sketch the graph of f.

25 $f(x) = \begin{cases} x + 2 & \text{if } x \leq -1 \\ x^3 & \text{if } |x| < 1 \\ -x + 3 & \text{if } x \geq 1 \end{cases}$

26 $f(x) = \begin{cases} x - 3 & \text{if } x \leq -2 \\ -x^2 & \text{if } -2 < x < 1 \\ -x + 4 & \text{if } x \geq 1 \end{cases}$

27 $f(x) = \begin{cases} \dfrac{x^2 - 1}{x + 1} & \text{if } x \neq -1 \\ 2 & \text{if } x = -1 \end{cases}$

28 $f(x) = \begin{cases} \dfrac{x^2 - 4}{2 - x} & \text{if } x \neq 2 \\ 1 & \text{if } x = 2 \end{cases}$

29 (a) $f(x) = [\![x - 3]\!]$ (b) $f(x) = [\![x]\!] - 3$
(c) $f(x) = 2[\![x]\!]$ (d) $f(x) = [\![2x]\!]$

30 (a) $f(x) = [\![x + 2]\!]$ (b) $f(x) = [\![x]\!] + 2$
(c) $f(x) = \frac{1}{2}[\![x]\!]$ (d) $f(x) = [\![\frac{1}{2}x]\!]$

Exer. 31–34: (a) Find $(f + g)(x)$, $(f - g)(x)$, $(fg)(x)$, and $(f/g)(x)$. (b) Find the domain of $f + g$, $f - g$, and fg; and find the domain of f/g.

31 $f(x) = \sqrt{x + 5}$; $g(x) = \sqrt{x + 5}$

32 $f(x) = \sqrt{3 - 2x}$; $g(x) = \sqrt{x + 4}$

33 $f(x) = \dfrac{2x}{x - 4}$; $g(x) = \dfrac{x}{x + 5}$

34 $f(x) = \dfrac{x}{x - 2}$; $g(x) = \dfrac{3x}{x + 4}$

Exer. 35–42: (a) Find $(f \circ g)(x)$ and the domain of $f \circ g$. (b) Find $(g \circ f)(x)$ and the domain of $g \circ f$.

35 $f(x) = x^2 - 3x$; $g(x) = \sqrt{x + 2}$

36 $f(x) = \sqrt{x - 15}$; $g(x) = x^2 + 2x$

37 $f(x) = \sqrt{x - 2}$; $g(x) = \sqrt{x + 5}$

38 $f(x) = \sqrt{3 - x}$; $g(x) = \sqrt{x + 2}$

39 $f(x) = \sqrt{25 - x^2}$; $g(x) = \sqrt{x - 3}$

40 $f(x) = \sqrt{3 - x}$; $g(x) = \sqrt{x^2 - 16}$

41 $f(x) = \dfrac{x}{3x + 2}$; $g(x) = \dfrac{2}{x}$

42 $f(x) = \dfrac{x}{x - 2}$; $g(x) = \dfrac{3}{x}$

Exer. 43–50: Find a composite function form for y.

43 $y = (x^2 + 3x)^{1/3}$

44 $y = \sqrt[4]{x^4 - 16}$

45 $y = \dfrac{1}{(x - 3)^4}$

46 $y = 4 + \sqrt{x^2 + 1}$

47 $y = (x^4 - 2x^2 + 5)^5$

48 $y = \dfrac{1}{(x^2 + 3x - 5)^3}$

49 $y = \dfrac{\sqrt{x + 4} - 2}{\sqrt{x + 4} + 2}$

50 $y = \dfrac{\sqrt[3]{x}}{1 + \sqrt[3]{x}}$

c 51 If $f(x) = \sqrt{x^2 - 1.7}$ and $g(x) = \dfrac{x^3 - x + 1}{\sqrt{x}}$, approximate $(f \circ g)(2.4)$ and $(g \circ f)(2.4)$.

c 52 If $f(x) = \sqrt{x^3 + 1} - 1$, approximate $f(0.0001)$. In order to avoid calculating a zero value for $f(0.0001)$, rewrite the formula for f as

$$f(x) = \frac{x^3}{\sqrt{x^3 + 1} + 1}.$$

53 An open box is to be made from a rectangular piece of cardboard 20 inches × 30 inches by cutting out identical squares of area x^2 from each corner and turning up the sides (see figure). Express the volume V of the box as a function of x.

EXERCISE 53

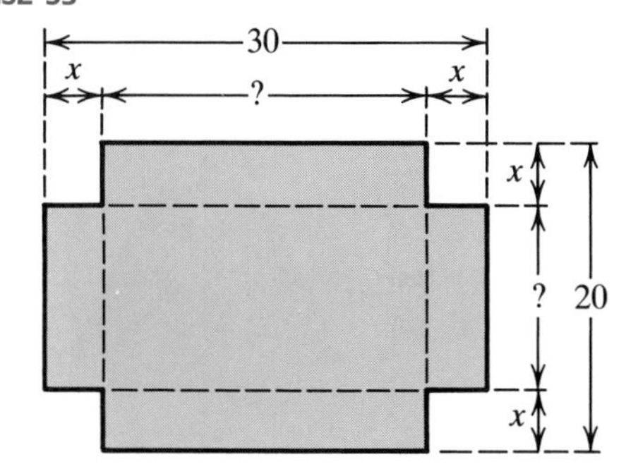

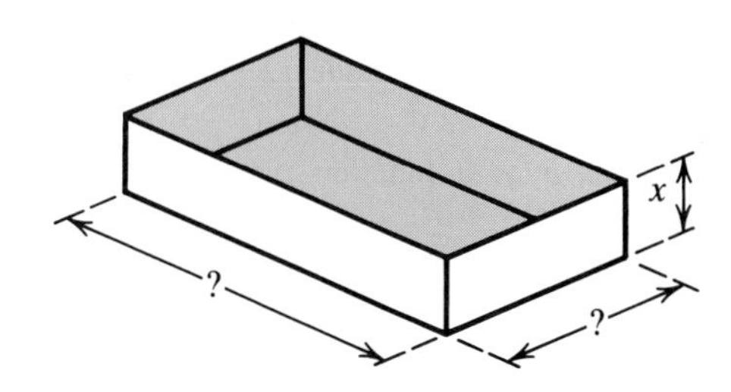

54 An open-top aquarium of height 1.5 feet is to have a volume of 6 ft^3. Let x denote the length of the base, and let y denote the width (see figure).

(a) Express y as a function of x.

(b) Express the total number of square feet S of glass needed as a function of x.

EXERCISE 54

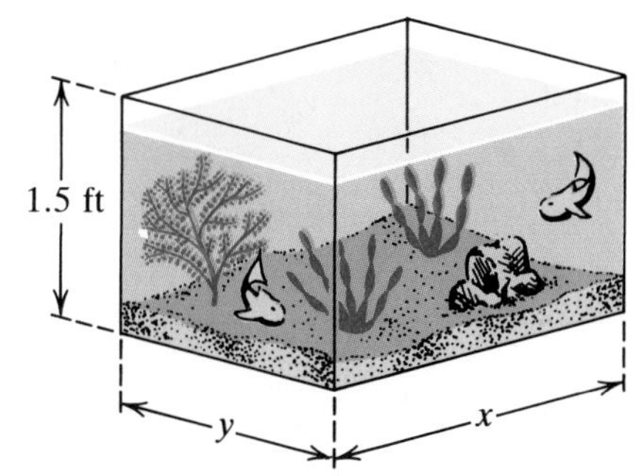

55 A hot-air balloon is released at 1:00 P.M. and rises vertically at a rate of 2 m/sec. An observation point is situated 100 meters from a point on the ground directly below the balloon (see figure). If t denotes the time (in seconds) after 1:00 P.M., express the distance d between the balloon and the observation point as a function of t.

EXERCISE 55

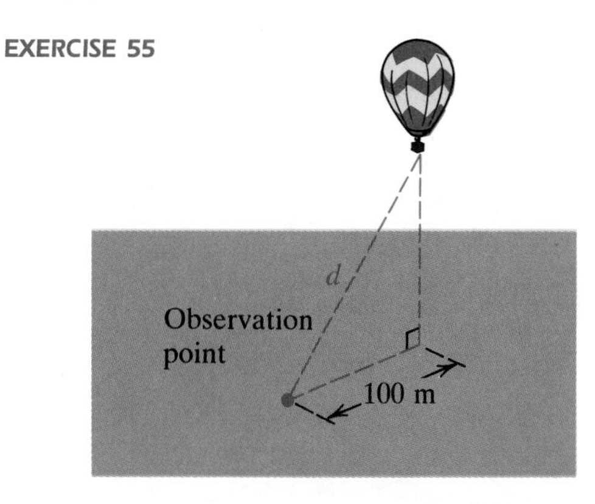

56 Refer to Example 2. A steel storage tank for propane gas is to be constructed in the shape of a right circular cylinder of altitude 10 feet with a hemisphere attached to each end. The radius r is yet to be determined. Express the surface area S of the tank as a function of r.

57 From an exterior point P that is h units from a circle of radius r, a tangent line is drawn to the circle (see figure). Let y denote the distance from the point P to the point of tangency T.

(a) Express y as a function of h. (*Hint:* If C is the center of the circle, then PT is perpendicular to CT.)

(b) If r is the radius of the earth and h is the altitude of a space shuttle, then we can derive a formula for the maximum distance (to the earth) that an astronaut can see from the shuttle. In particular, if $h = 200$ mi and $r \approx 4000$ mi, approximate y.

EXERCISE 57

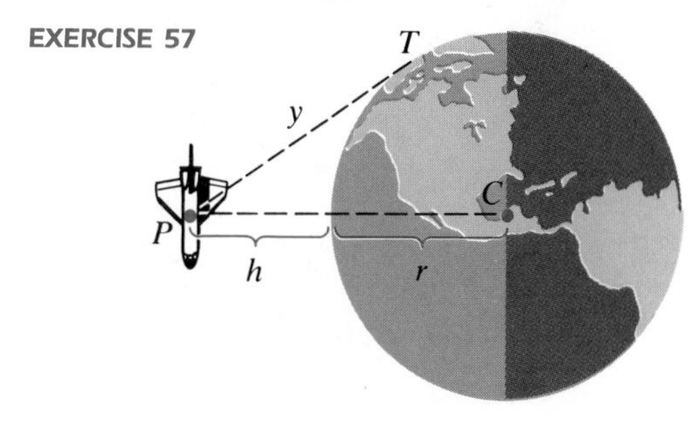

58 Triangle ABC is inscribed in a semicircle of diameter 15 (see the figure on the following page).

(a) If x denotes the length of side AC, express the length y of side BC as a function of x, and state its domain. (*Hint:* Angle ACB is a right angle.)

(b) Express the area of triangle ABC as a function of x.

EXERCISE 58

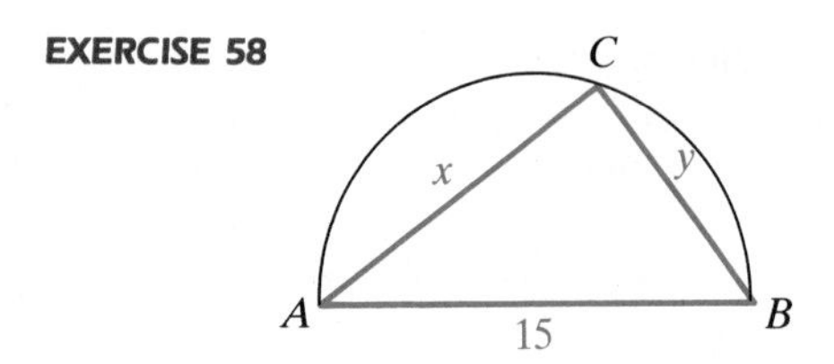

59 The relative positions of an airport runway and a 20-foot-tall control tower are shown in the figure. The beginning of the runway is at a perpendicular distance of 300 feet from the base of the tower. If x denotes the distance an airplane has moved down the runway, express the distance d between the airplane and the control booth as a function of x.

EXERCISE 59

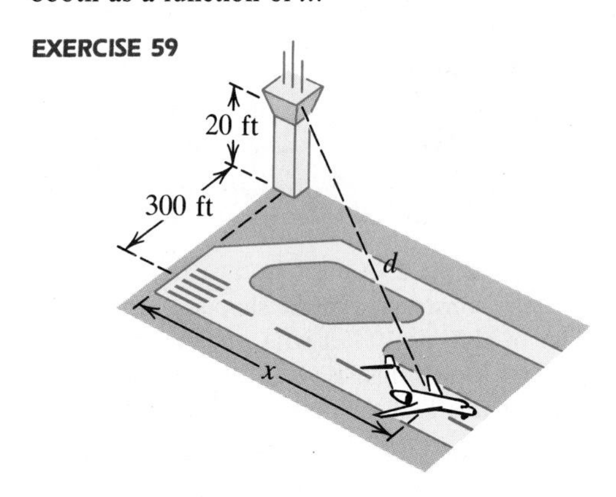

60 An open rectangular storage shelter consisting of two vertical sides, 4 feet wide, and a flat roof is to be attached to an existing structure as illustrated in the figure. The flat roof is made of tin that costs \$5 per square foot, and the other two sides are made of plywood costing \$2 per square foot.

(a) If \$400 is available for construction, express the length y as a function of the height x.

EXERCISE 60

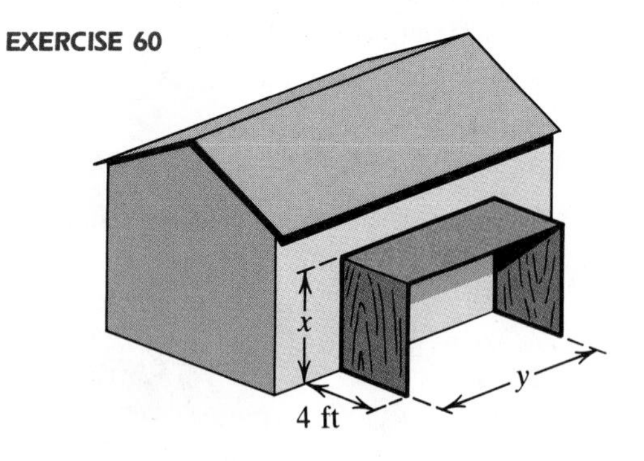

(b) Express the volume V inside the shelter as a function of x.

61 The shape of the first spacecraft in the Apollo program was a frustum of a right circular cone, a solid formed by truncating a cone by a plane parallel to its base. For the frustum shown in the figure, the radii a and b have already been determined.

(a) Use similar triangles to express y as a function of h.

(b) Express the volume of the frustum as a function of h.

(c) If $a = 6$ ft and $b = 3$ ft, for what value of h is the volume of the frustum 600 ft^3?

EXERCISE 61

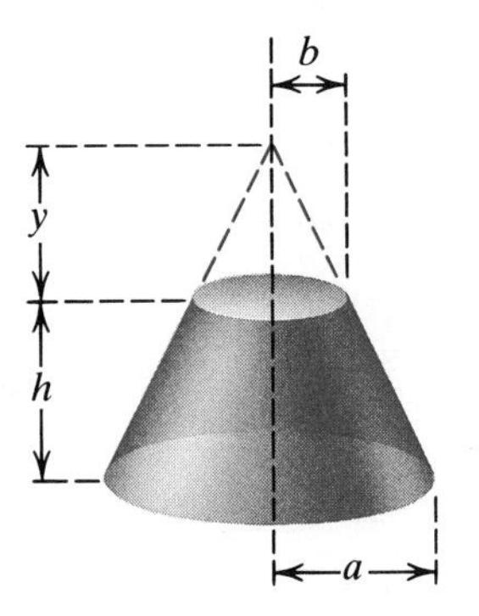

62 A right circular cylinder of radius r and height h is inscribed in a cone of altitude 12 and base radius 4, as illustrated in the figure.

(a) Express h as a function of r. (*Hint:* Use similar triangles.)

(b) Express the volume V of the cylinder as a function of r.

EXERCISE 62

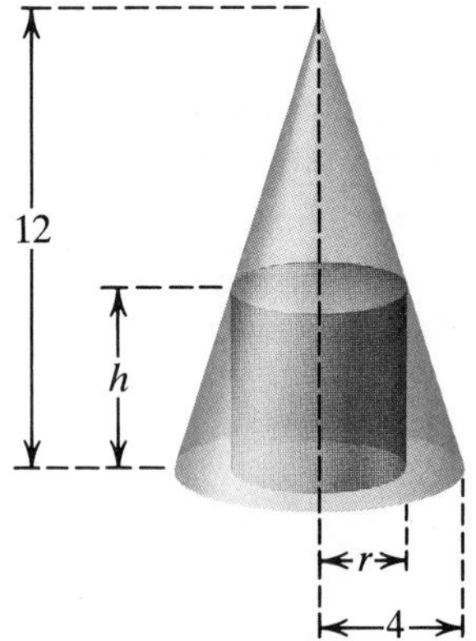

1.3 TRIGONOMETRY

In geometry an angle is determined by two rays, or line segments, having the same initial point O (the **vertex** of the angle). If A and B are points on the rays l_1 and l_2 in Figure 1.27, we refer to **angle** $\boldsymbol{AOB}$, or $\angle AOB$. We

FIGURE 1.27

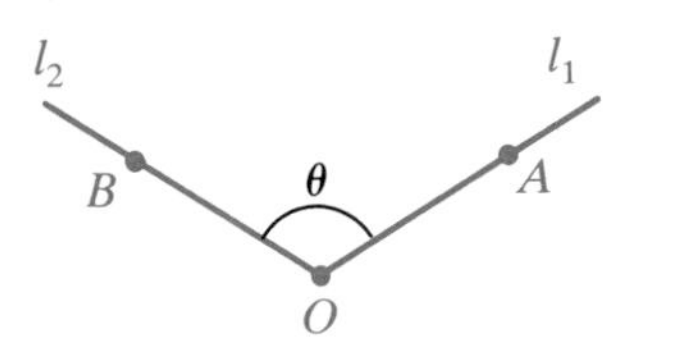

often use a lower-case Greek letter, such as θ, α, or β, to denote an angle. In trigonometry we may also interpret $\angle AOB$ as a rotation of ray l_1 in the plane determined by A, O, and B: We rotate l_1 (the **initial side** of the angle) about O, to a position specified by l_2 (the **terminal side**). The amount or direction of rotation is not restricted; we can let l_1 make several revolutions in either direction about O before stopping at l_2. Thus, many angles may have the same initial and terminal sides.

FIGURE 1.28

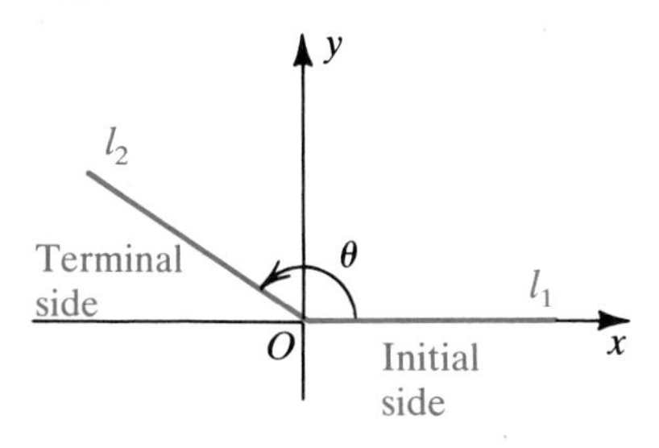

If we introduce a rectangular coordinate system, then the **standard position** of an angle θ is obtained by taking the vertex at the origin and the initial side along the positive x-axis (see Figure 1.28). The angle θ is **positive** for a counterclockwise rotation and **negative** for a clockwise rotation.

The magnitude of an angle may be expressed in either degrees or radians. An angle of **degree measure** 1° corresponds to $\frac{1}{360}$ of a complete revolution in the counterclockwise direction. One **minute** (1′) is $\frac{1}{60}$ of a degree, and one **second** (1″) is $\frac{1}{60}$ of a minute. In calculus the most important unit of angular measure is the *radian*. To define radian measure, consider the *unit circle U* with center at the origin of a rectangular coordinate system, and let θ be an angle in standard position (see Figure 1.29). If we rotate the x-axis to the terminal side of θ, its point of intersection with U travels a certain distance t before arriving at its final position $P(x, y)$. If t is considered positive for a counterclockwise rotation and negative for a clockwise rotation, then θ is an **angle of *t* radians**, and we write either $\theta = t$ or $\theta = t$ *radians*. In Figure 1.29, t is the length of the arc $\widehat{AP}$. If $\theta = 1$ (that is, if θ is an angle of 1 radian), then the length of arc $\widehat{AP}$ on U is 1 (see Figure 1.30).

FIGURE 1.29
$\theta = t$ radians

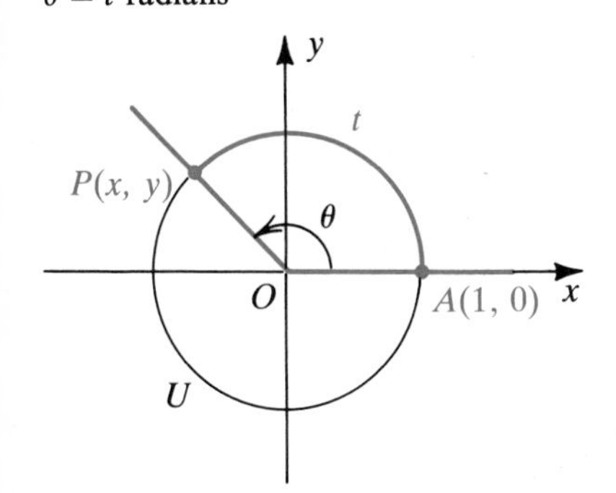

FIGURE 1.30

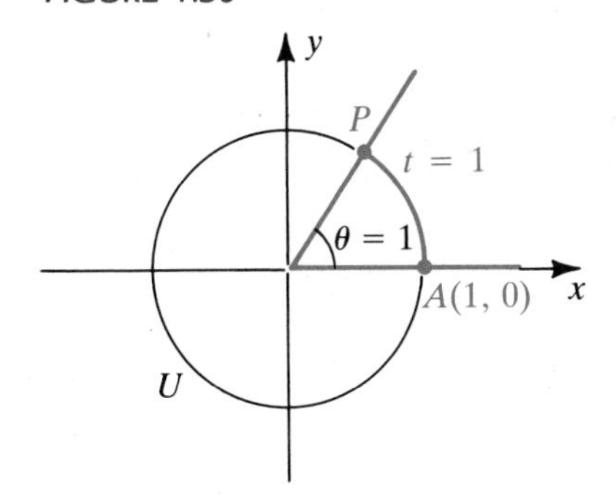

Since the circumference of the unit circle is 2π, it follows that

$$2\pi \text{ radians} = 360°.$$

This fact gives us the following relationships.

Relationships between degrees and radians **(1.12)**

$$180° = \pi \text{ radians} \qquad 1° = \frac{\pi}{180} \text{ radian} \qquad 1 \text{ radian} = \left(\frac{180}{\pi}\right)^{\circ}$$

If we approximate $\pi/180$ and $180/\pi$, we obtain

$$1° \approx 0.01745 \text{ radian} \quad \text{and} \quad 1 \text{ radian} \approx 57.29578°.$$

The next theorem is a consequence of the formulas in (1.12).

Theorem **(1.13)**

(i) To change radian measure to degrees, multiply by $180/\pi$.
(ii) To change degree measure to radians, multiply by $\pi/180$.

When radian measure of an angle is used, no units will be indicated. Thus, if an angle has radian measure 5, we write $\theta = 5$ instead of $\theta = 5$ *radians*. There should be no confusion as to whether radian or degree measure is being used, since if θ has degree measure 5°, we write $\theta = 5°$, *not* $\theta = 5$.

Radians	0	$\frac{\pi}{6}$	$\frac{\pi}{4}$	$\frac{\pi}{3}$	$\frac{\pi}{2}$	$\frac{2\pi}{3}$	$\frac{3\pi}{4}$	$\frac{5\pi}{6}$	π	$\frac{7\pi}{6}$	$\frac{5\pi}{4}$	$\frac{4\pi}{3}$	$\frac{3\pi}{2}$	$\frac{5\pi}{3}$	$\frac{7\pi}{4}$	$\frac{11\pi}{6}$	2π
Degrees	0°	30°	45°	60°	90°	120°	135°	150°	180°	210°	225°	240°	270°	300°	315°	330°	360°

FIGURE 1.31

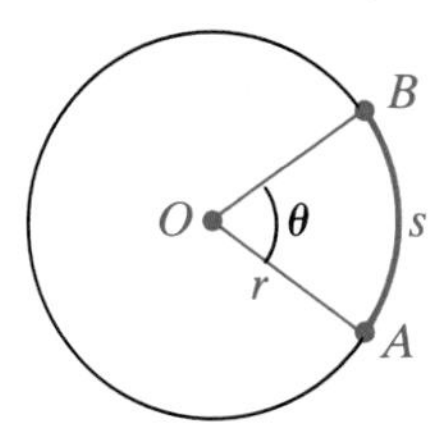

The above table displays the relationship between the radian and degree measures of several common angles. The entries may be checked by using Theorem (1.13).

A **central angle** of a circle is an angle θ whose vertex is at the center of the circle, as illustrated in Figure 1.31. We say that arc $\overset{\frown}{AB}$ *subtends* θ, or that θ is *subtended by* $\overset{\frown}{AB}$. The relationship between the length s of $\overset{\frown}{AB}$, the radian measure of θ, and the radius r of the circle follows.

Length of a circular arc **(1.14)**

If an arc of length s on a circle of radius r subtends a central angle of radian measure θ, then

$$s = r\theta.$$

PROOF If s_1 is the length of any other arc on the circle and if θ_1 is the radian measure of the corresponding central angle, then, from plane geometry, the ratio of the arcs is the same as the ratio of the angular measures; that is, $s/s_1 = \theta/\theta_1$, and hence $s = s_1\theta/\theta_1$. If we consider the special case in which $\theta_1 = 2\pi$, then $s_1 = 2\pi r$, and we obtain $s = 2\pi r\theta/(2\pi) = r\theta$. ■

We will make use of the next result later in the text.

Area of a circular sector **(1.15)**

If θ is the radian measure of a central angle of a circle of radius r and if A is the area of the circular sector determined by θ, then

$$A = \tfrac{1}{2}r^2\theta.$$

PROOF A typical angle and circular sector are shown in Figure 1.31. If θ_1 is any other central angle and A_1 the area of the corresponding sector, then, from plane geometry, $A/A_1 = \theta/\theta_1$, or $A = A_1\theta/\theta_1$. If we consider the special case $\theta_1 = 2\pi$, then $A_1 = \pi r^2$ and $A = \pi r^2\theta/(2\pi) = \frac{1}{2}r^2\theta$. ■

The six trigonometric functions are the **sine**, **cosine**, **tangent**, **cosecant**, **secant**, and **cotangent**, denoted by **sin**, **cos**, **tan**, **csc**, **sec**, and **cot**, respectively. We may define the trigonometric functions in terms of either an angle θ or a real number x. There are two standard methods that employ angles:

1. If θ is *acute* $(0 < \theta < \pi/2)$, we may use a right triangle.
2. If θ is *any* angle (in standard position), we may use the point $P(a, b)$ at which the terminal side of θ intersects the circle $x^2 + y^2 = r^2$.

In the following definitions, the abbreviations *adj*, *opp*, and *hyp* are used for the lengths of the adjacent side, opposite side, and hypotenuse of a right triangle having θ as an angle.

The trigonometric functions **(1.16)**

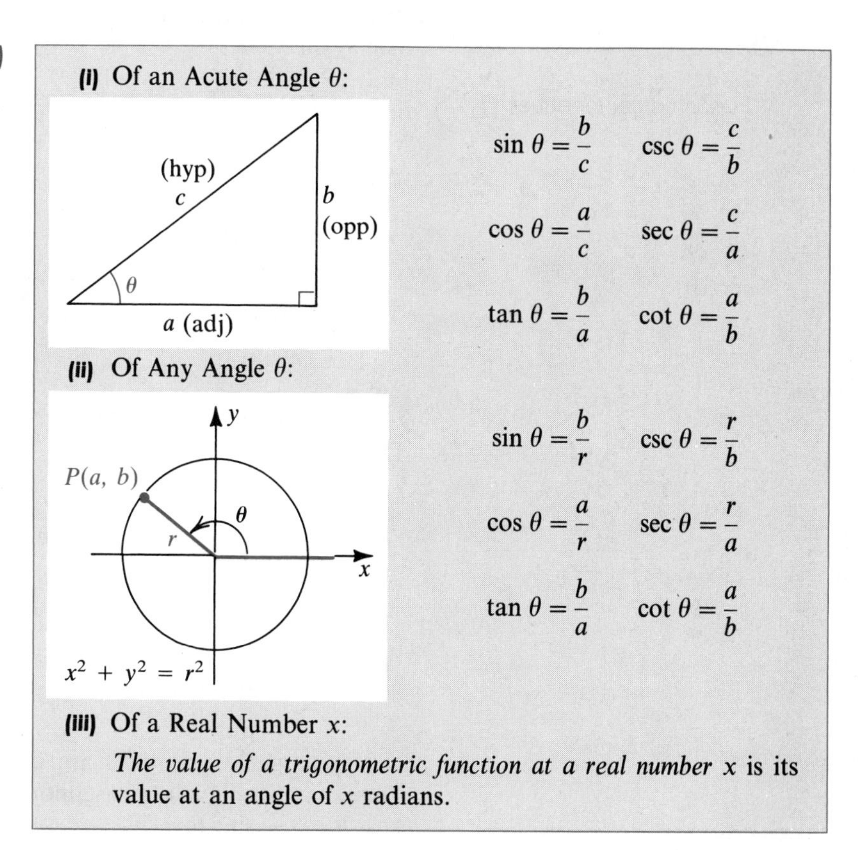

FIGURE 1.32

II	I
$\sin\theta > 0$, $\csc\theta > 0$	All positive
$\tan\theta > 0$, $\cot\theta > 0$	$\cos\theta > 0$, $\sec\theta > 0$
III	IV

Note that, by (iii), there is no difference between trigonometric functions of angles measured in radians and trigonometric functions of real numbers. For example, we can interpret sin 2 as *either* the sine of an angle of 2 radians or the sine of the real number 2.

The values of the trigonometric functions of acute angles in (i) are ratios of sides of a right triangle, and hence they are positive real numbers. For the general case (ii), the sign of the function value depends on the quadrant containing the terminal side of θ. For example, if θ is in quadrant II, then $a < 0$, $b > 0$, and hence $\sin\theta = b/r > 0$ and $\csc\theta = r/b > 0$. The other four functions are negative. These facts are indicated schematically in Figure 1.32. You should check the signs in the remaining quadrants.

We see from (ii) of Definition (1.16) that the domain of sin and cos consists of all angles θ. Since $\tan\theta$ and $\sec\theta$ are undefined if $a = 0$ (that is, if the terminal side of θ is on the y-axis), the domain of tan and sec consists of all angles *except* those of radian measure $(\pi/2) + \pi n$, where n is an integer. The domain of cot and csc consists of all angles except those of radian measure πn, since $\cot\theta$ and $\csc\theta$ are undefined if $b = 0$.

Note from (ii) that

$$|\sin\theta| \le 1, \quad |\cos\theta| \le 1, \quad |\csc\theta| \ge 1, \quad \text{and} \quad |\sec\theta| \ge 1$$

for every θ in the domains of these functions.

We shall next list some important relationships that exist among the trigonometric functions. Recall that an expression such as $\sin^2\theta$ means $(\sin\theta)(\sin\theta)$.

Fundamental identities **(1.17)**

$$\csc\theta = \frac{1}{\sin\theta} \qquad \tan\theta = \frac{\sin\theta}{\cos\theta} \qquad \sin^2\theta + \cos^2\theta = 1$$

$$\sec\theta = \frac{1}{\cos\theta} \qquad \cot\theta = \frac{\cos\theta}{\sin\theta} \qquad 1 + \tan^2\theta = \sec^2\theta$$

$$\cot\theta = \frac{1}{\tan\theta} \qquad 1 + \cot^2\theta = \csc^2\theta$$

Each fundamental identity can be proved by referring to (ii) of Definition (1.16). For example:

$$\csc\theta = \frac{r}{b} = \frac{1}{(b/r)} = \frac{1}{\sin\theta}$$

$$\tan\theta = \frac{b}{a} = \frac{(b/r)}{(a/r)} = \frac{\sin\theta}{\cos\theta}$$

$$\sin^2\theta + \cos^2\theta = \frac{b^2}{r^2} + \frac{a^2}{r^2} = \frac{a^2 + b^2}{r^2} = \frac{r^2}{r^2} = 1$$

Fundamental identities are useful for changing the form of an expression that involves trigonometric functions. To illustrate, since $\cos^2\theta = 1 - \sin^2\theta$,

$$\tan\theta = \frac{\sin\theta}{\cos\theta} = \frac{\sin\theta}{\pm\sqrt{1 - \sin^2\theta}}.$$

In Chapter 9 we will use *trigonometric substitutions* of the type illustrated in the next example.

EXAMPLE 1 If $a > 0$, express $\sqrt{a^2 - x^2}$ in terms of a trigonometric function of θ without radicals by making the trigonometric substitution

$$x = a\sin\theta \quad \text{for } -\frac{\pi}{2} \le \theta \le \frac{\pi}{2}.$$

SOLUTION We let $x = a \sin \theta$:

$$\begin{aligned}\sqrt{a^2 - x^2} &= \sqrt{a^2 - (a \sin \theta)^2}\\ &= \sqrt{a^2 - a^2 \sin^2 \theta}\\ &= \sqrt{a^2(1 - \sin^2 \theta)}\\ &= \sqrt{a^2 \cos^2 \theta}\\ &= a \cos \theta\end{aligned}$$

The last equality is true because, first, $\sqrt{a^2} = a$ if $a > 0$, and, second, if $-\pi/2 \le \theta \le \pi/2$, then $\cos \theta \ge 0$ and hence $\sqrt{\cos^2 \theta} = \cos \theta$.

There are a variety of methods for finding values of trigonometric functions. For certain special cases we can refer to the right triangles in Figure 1.33. Applying (i) of Definition (1.16) gives us the following.

Special values of the trigonometric functions **(1.18)**

θ (RADIANS)	θ (DEGREES)	$\sin \theta$	$\cos \theta$	$\tan \theta$	$\cot \theta$	$\sec \theta$	$\csc \theta$
$\frac{\pi}{6}$	30°	$\frac{1}{2}$	$\frac{\sqrt{3}}{2}$	$\frac{\sqrt{3}}{3}$	$\sqrt{3}$	$\frac{2\sqrt{3}}{3}$	2
$\frac{\pi}{4}$	45°	$\frac{\sqrt{2}}{2}$	$\frac{\sqrt{2}}{2}$	1	1	$\sqrt{2}$	$\sqrt{2}$
$\frac{\pi}{3}$	60°	$\frac{\sqrt{3}}{2}$	$\frac{1}{2}$	$\sqrt{3}$	$\frac{\sqrt{3}}{3}$	2	$\frac{2\sqrt{3}}{3}$

FIGURE 1.33

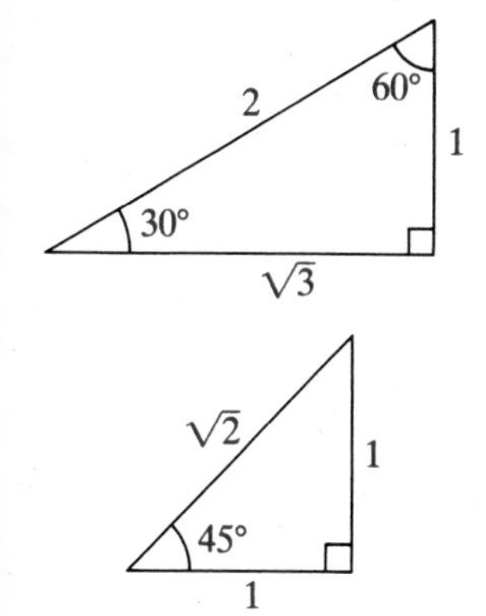

Two reasons for stressing these special values are that (1) they are *exact* and (2) they occur frequently in work involving trigonometry. Because of their importance, it is a good idea either to memorize the table or to be able to find the values quickly by using the triangles in Figure 1.33.

It is possible to approximate, to any degree of accuracy, the values of the trigonometric functions for any angle. Table A in Appendix III gives four-decimal-place approximations to some values.

Scientific calculators have keys labeled [SIN], [COS], and [TAN] that can be used to approximate values of these functions. The values of csc, sec, and cot may then be found by means of the reciprocal key [1/x]. *Before using a calculator to find function values that correspond to the radian measure of an angle or a real number, be sure that the calculator is in radian mode. For values corresponding to degree measure, select degree mode.*

As an illustration, to find $\sin 30^\circ$ on a typical calculator, place the calculator in degree mode, enter the number 30, and press the [SIN] key. You will obtain $\sin 30^\circ = 0.5$, which is the exact value. Using the same procedure for 60°, we obtain a decimal approximation to $\sqrt{3}/2$, such as

$$\sin 60^\circ \approx 0.8660254.$$

Similarly, to find a value such as $\cos 1.3$, where 1.3 is a real number or the radian measure of an angle, we place the calculator in radian mode, enter 1.3, and press [COS], obtaining

$$\cos 1.3 \approx 0.2674988.$$

To find exact values of trigonometric functions for an angle θ in (ii) of Definition (1.16), we sometimes use the **reference angle** of θ—that is, the acute angle θ_R that the terminal side of θ makes with the x-axis. Figure 1.34 illustrates the reference angle θ_R for an angle in each of the four quadrants.

FIGURE 1.34
Reference angles

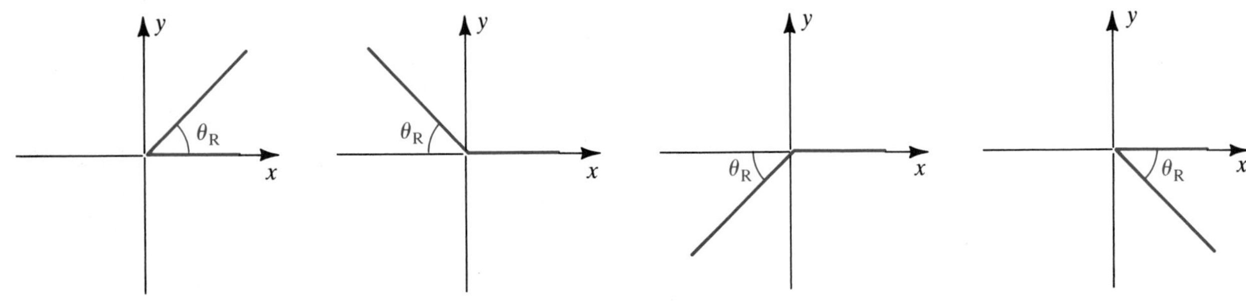

It can be shown that to find the value of a trigonometric function at θ, we may determine its value for the reference angle θ_R of θ and then prefix the proper sign, by referring to the quadrant containing θ (see Figure 1.32).

EXAMPLE 2 Find $\sin\theta$, $\cos\theta$, and $\tan\theta$ if

(a) $\theta = \dfrac{5\pi}{6}$ **(b)** $\theta = 315°$

FIGURE 1.35

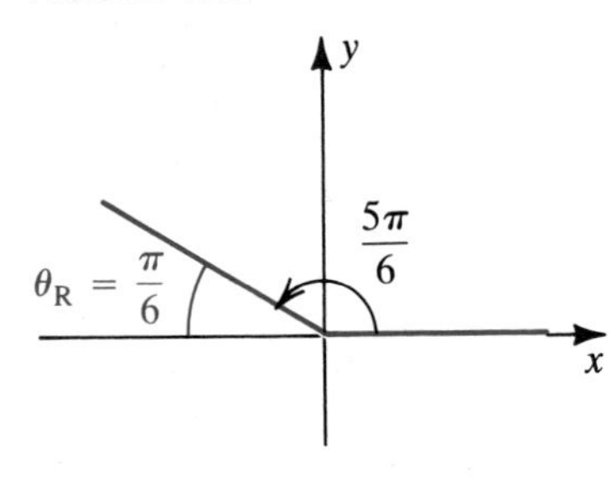

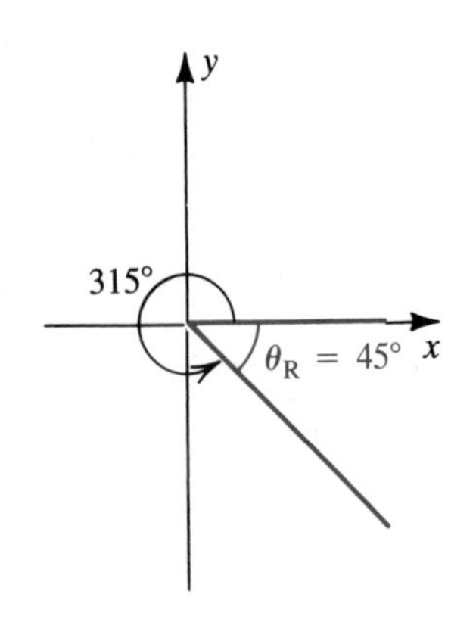

SOLUTION The angles and their reference angles are sketched in Figure 1.35. Using function values of special angles (1.18), we obtain the following:

(a) $\sin\dfrac{5\pi}{6} = \sin\dfrac{\pi}{6} = \dfrac{1}{2}$

$\cos\dfrac{5\pi}{6} = -\cos\dfrac{\pi}{6} = -\dfrac{\sqrt{3}}{2}$

$\tan\dfrac{5\pi}{6} = -\tan\dfrac{\pi}{6} = -\dfrac{\sqrt{3}}{3}$

(b) $\sin 315° = -\sin 45° = -\dfrac{\sqrt{2}}{2}$

$\cos 315° = \cos 45° = \dfrac{\sqrt{2}}{2}$

$\tan 315° = -\tan 45° = -1$

If we use a calculator to approximate function values, reference angles are unnecessary. As an illustration, to find $\sin 210°$, we place the calculator in degree mode, enter the number 210, and press the SIN key, obtaining $\sin 210° = -0.5$, which is the exact value. Using the same procedure for

240°, we obtain a decimal approximation:

$$\sin 240° \approx -0.8660254$$

If we want to find the *exact* value of sin 240°, a calculator should not be used. In this case we find the reference angle 60° of 240° and use the theorem on reference angles, together with known results about special angles, to obtain

$$\sin 240° = -\sin 60° = -\frac{\sqrt{3}}{2}.$$

To obtain the graphs of the sine and cosine functions we may study the variation of $\sin\theta$ and $\cos\theta$ as θ changes by using a unit circle U in (ii) of Definition (1.16). If we let $r = 1$, then the formulas $\cos\theta = a/r$ and $\sin\theta = b/r$ take on the simple forms $\cos\theta = a$ and $\sin\theta = b$. Hence the point $P(a, b)$ on U may be denoted by $P(\cos\theta, \sin\theta)$, as illustrated in Figure 1.36. If we let θ increase from 0 to 2π, the point $P(\cos\theta, \sin\theta)$ travels around the unit circle once in the counterclockwise direction. By observing the y-coordinate, $\sin\theta$, of P, we obtain the following facts, where arrows are used to indicate the variations of θ and $\sin\theta$. (For example, $0 \to \pi/2$ means that θ increases from 0 to $\pi/2$, and $0 \to 1$ means that $\sin\theta$ increases from 0 to 1.)

FIGURE 1.36

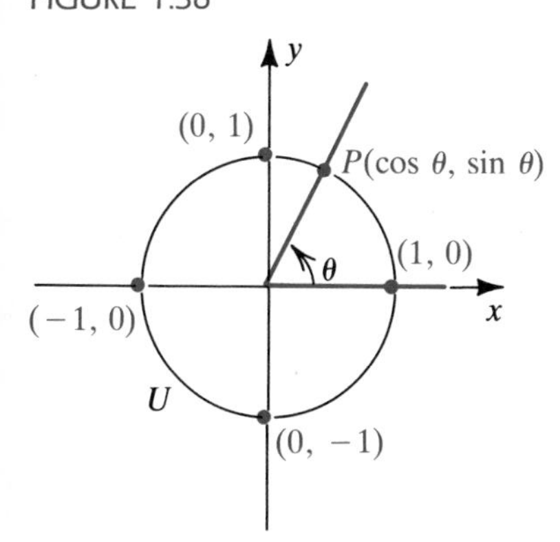

$$\theta\colon\quad 0 \to \frac{\pi}{2} \to \pi \to \frac{3\pi}{2} \to 2\pi$$

$$\sin\theta\colon\quad 0 \to 1 \to 0 \to -1 \to 0$$

If we let P continue to travel around U, the same pattern is repeated in the θ-intervals $[2\pi, 4\pi]$ and $[4\pi, 6\pi]$. In general, the values of $\sin\theta$ repeat in all successive intervals of length 2π. A function f with domain D is **periodic** if there exists a positive real number k such that $x + k$ is in D and $f(x + k) = f(x)$ for every x in D. This implies that the graph of f repeats itself over successive intervals of length k. If a least such positive real number k exists, it is called the **period** of f. It follows that the sine function is periodic with period 2π. Using these facts and plotting several points, employing special values of θ such as $\pi/6$, $\pi/4$, and $2\pi/3$, gives us the graph in Figure 1.37(i), where we have used $\theta = x$ for the independent variable (measured in radians or real numbers).

The graph of $y = \cos\theta$ may be obtained in similar fashion by studying the variation of the x-coordinate, $\cos\theta$, of P in Figure 1.36 as θ increases. You should check the remaining graphs in Figure 1.37. Note that the period of the tangent and cotangent functions is π.

A **trigonometric equation** is an equation that contains trigonometric expressions. Each fundamental identity is an example of a trigonometric equation where every number (or angle) in the domain of the variable is a solution of the equation. If a trigonometric equation is not an identity, we often find solutions by using techniques similar to those used for algebraic equations. The main difference is that we first solve the trigonometric equation for $\sin x$, $\cos\theta$, and so on, and then find values of x or θ that satisfy the equation. *If degree measure is not specified, then solutions of a trigonometric equation should be expressed in radian measure (or as real numbers).*

FIGURE 1.37

(i) $y = \sin x$

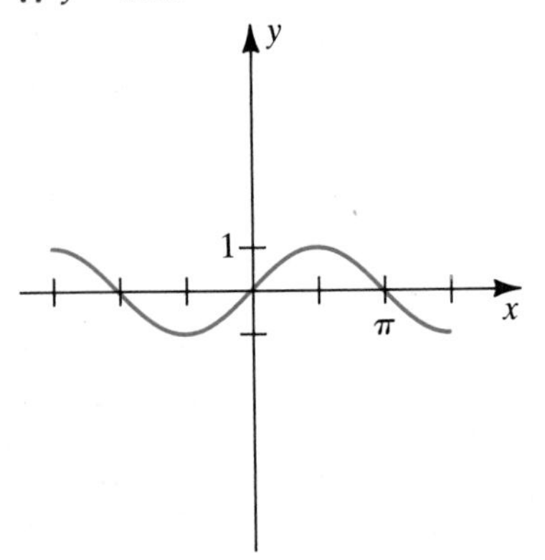

(ii) $y = \cos x$

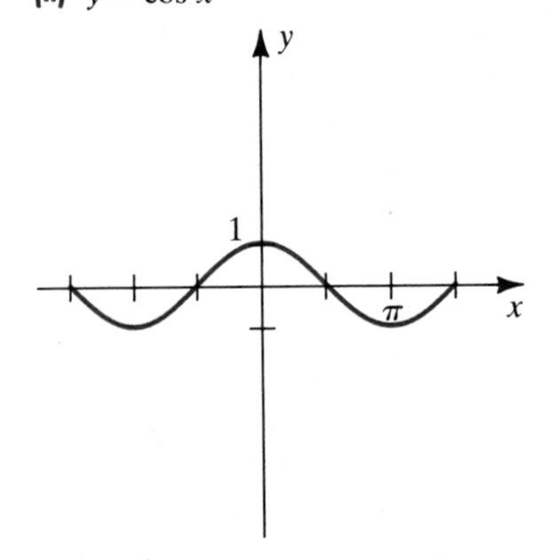

(iii) $y = \tan x$

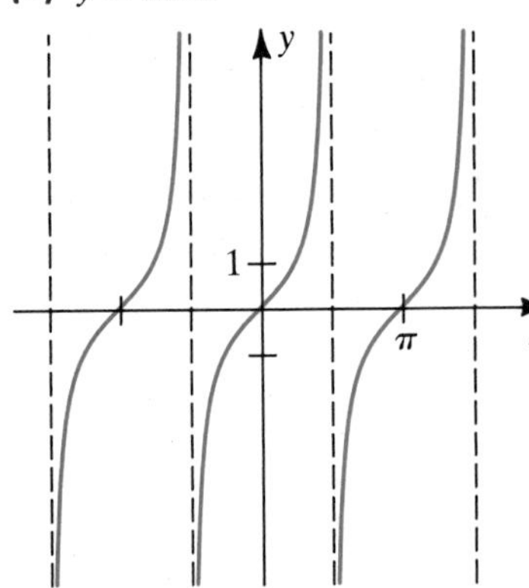

(iv) $y = \csc x$

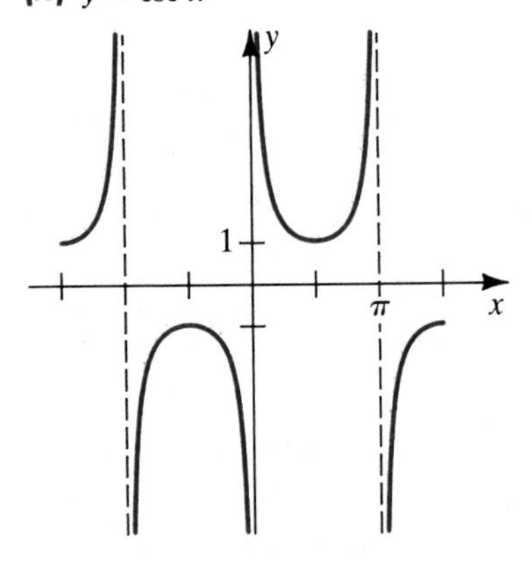

(v) $y = \sec x$

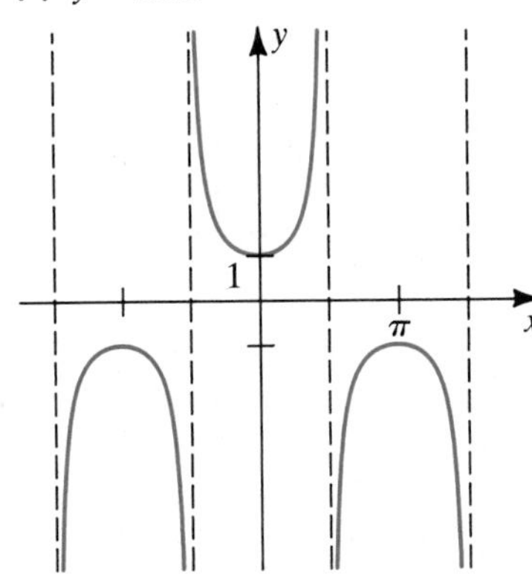

(vi) $y = \cot x$

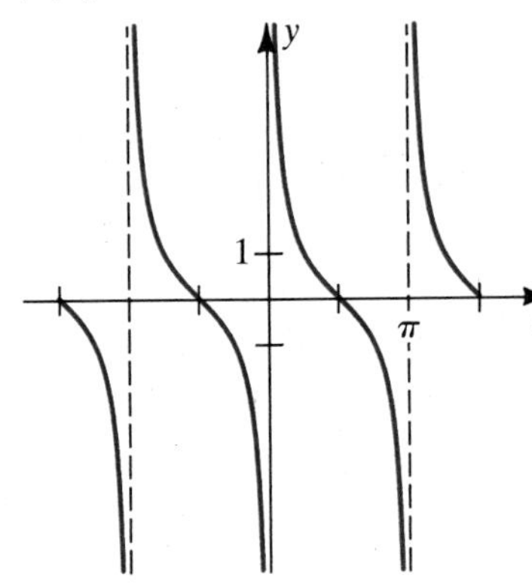

EXAMPLE 3 Find the solutions of the equation $\sin\theta = \frac{1}{2}$ if

(a) θ is in the interval $[0, 2\pi)$ **(b)** θ is any real number

FIGURE 1.38

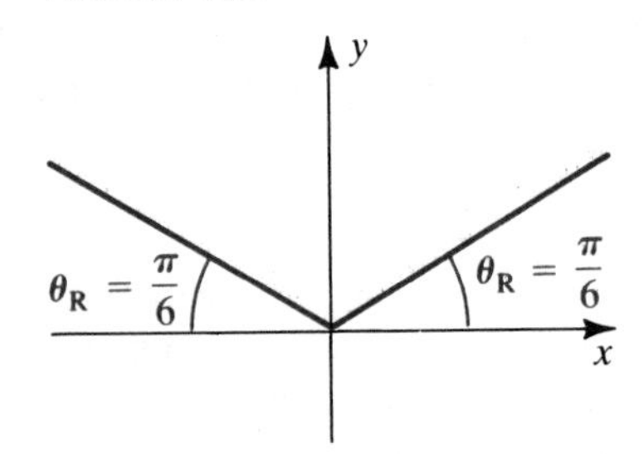

SOLUTION

(a) If $\sin\theta = \frac{1}{2}$, then the reference angle for θ is $\theta_R = \pi/6$. If we regard θ as an angle in standard position, then, since $\sin\theta > 0$, the terminal side is in either quadrant I or quadrant II, as illustrated in Figure 1.38. Thus, there are two solutions for $0 \le \theta < 2\pi$:

$$\theta = \frac{\pi}{6} \quad \text{and} \quad \theta = \pi - \frac{\pi}{6} = \frac{5\pi}{6}$$

(b) Since the sine function has period 2π, we may obtain all solutions by adding multiples of 2π to $\pi/6$ and $5\pi/6$. This gives us

$$\theta = \frac{\pi}{6} + 2\pi n \quad \text{and} \quad \theta = \frac{5\pi}{6} + 2\pi n \quad \text{for every integer } n.$$

An alternative (graphical) solution involves determining where the graph of $y = \sin\theta$ intersects the horizontal line $y = \frac{1}{2}$, as illustrated in Figure 1.39, on the next page.

FIGURE 1.39

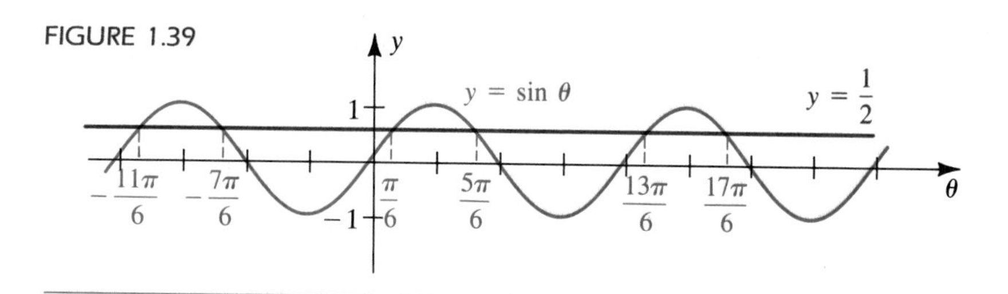

Given a trigonometric equation such as $\sin \theta = 0.6635$, we can approximate θ by using either a calculator or a table. Certain calculators have a key labeled [SIN⁻¹] or [ASIN] for this purpose. With other calculators it is necessary to press [INV] and then [SIN]. These notations are based on the *inverse trigonometric functions*, which are discussed in Section 8.2. As we shall see, there is a function, denoted by $\sin^{-1}$, such that

$$\sin^{-1}(\sin \theta) = \theta \quad \text{if} \quad -\frac{\pi}{2} \leq \theta \leq \frac{\pi}{2} \quad (\text{or} \quad -90^\circ \leq \theta \leq 90^\circ).$$

Note that this formula states that by applying $\sin^{-1}$ to $\sin \theta$, we obtain θ, provided θ satisfies the indicated restrictions.

The next example illustrates the use of a calculator in solving a trigonometric equation.

EXAMPLE 4 If $\sin \theta = 0.5$ and θ is an acute angle, use a calculator to approximate the measure of θ in

(a) degrees **(b)** radians

SOLUTION

(a) Place the calculator in degree mode:

Enter 0.5: 0.5 (the value of $\sin \theta$)

Press [INV] [SIN]: 30 (the degree measure of θ)

(b) Place the calculator in radian mode:

Enter 0.5: 0.5 (the value of $\sin \theta$)

Press [INV] [SIN]: 0.5235988 (the radian measure of θ)

The last number is a decimal approximation for an angle of radian measure $\pi/6$.

It is important to observe that there are many values of θ such that $\sin \theta = 0.5$; however, a calculator gives only the one value between 0 and $\pi/2$ (or between 0° and 90°). Similarly, if $\sin \theta = -0.5$, a calculator will give an approximation to the value $\theta = -\pi/6$ (or $\theta = -30^\circ$) between $-\pi/2$ and 0 (or between -90° and 0°).

In Section 8.2 we will also define functions denoted by $\cos^{-1}$ and $\tan^{-1}$, with the following properties:

$$\cos^{-1}(\cos \theta) = \theta \quad \text{if} \quad 0 \leq \theta \leq \pi \quad (\text{or} \quad 0^\circ \leq \theta \leq 180^\circ)$$

$$\tan^{-1}(\tan \theta) = \theta \quad \text{if} \quad -\frac{\pi}{2} < \theta < \frac{\pi}{2} \quad (\text{or} \quad -90^\circ < \theta < 90^\circ)$$

These functions may be employed in the same manner as [SIN⁻¹] (that is, [INV] [SIN]) was used in Example 4. When using a calculator to find θ, be aware of the restrictions on θ. For example, there are many values of θ such that $\tan\theta = -1$; however, a calculator gives only the value that is between $-\pi/2$ and 0 (or between $-90°$ and $0°$). If other values are desired, then we may proceed as in the next example.

EXAMPLE 5 If $\tan\theta = -0.4623$ and $0° \le \theta < 360°$, find θ to the nearest $0.1°$.

SOLUTION If we use a calculator (in degree mode) to find θ when $\tan\theta$ is negative, then the degree measure is in the interval $(-90°, 0°]$. In particular, we have the following:

Enter -0.4623:	-0.4623	(the value of $\tan\theta$)
Press [INV] [TAN]:	-24.811101	(a value of θ)

FIGURE 1.40

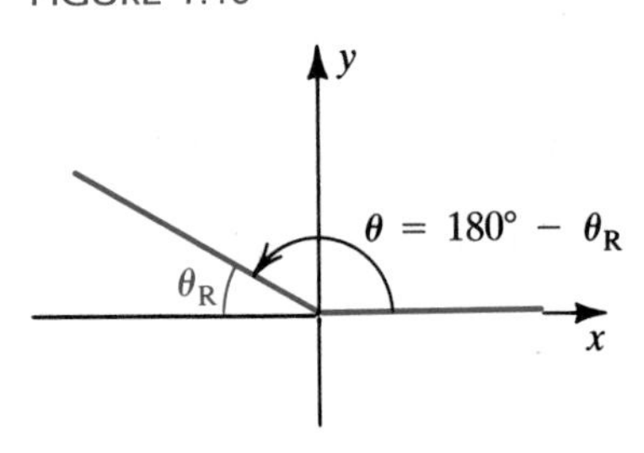

Thus, a degree approximation is $\theta \approx -24.8°$.

Since we wish to find values of θ between $0°$ and $360°$, we use the (approximate) reference angle $\theta_R \approx 24.8°$. There are two possible values of θ such that $\tan\theta$ is negative—one in quadrant II, the other in quadrant IV. If θ is in quadrant II and $0° \le \theta < 360°$, we have the situation shown in Figure 1.40, and

$$\theta = 180° - \theta_R \approx 180° - 24.8°, \quad \text{or} \quad \theta \approx 155.2°.$$

FIGURE 1.41

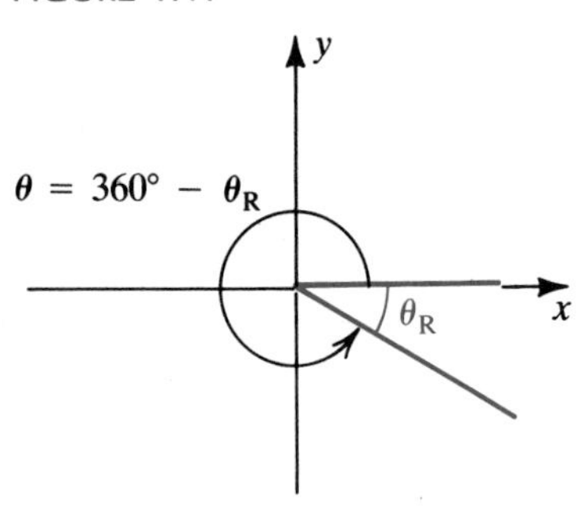

If θ is in quadrant IV and $0° \le \theta < 360°$, then, as in Figure 1.41,

$$\theta = 360° - \theta_R \approx 360° - 24.8°, \quad \text{or} \quad \theta \approx 335.2°.$$

A method of solution that does not involve reference angles is to use the fact that the tangent function has period π, or $180°$. Thus, after obtaining $\theta \approx -24.8°$, we may find appropriate angles between $0°$ and $360°$ by adding $180°$ and $360°$ as follows:

$$-24.8° + 180° = 155.2°$$

$$-24.8° + 360° = 335.2°$$

Many important relationships exist among the trigonometric functions. The *formulas for negatives* are

$$\sin(-u) = -\sin u \qquad \cos(-u) = \cos u \qquad \tan(-u) = -\tan u$$

$$\csc(-u) = -\csc u \qquad \sec(-u) = \sec u \qquad \cot(-u) = -\cot u.$$

These formulas show that the sine, tangent, cosecant, and cotangent functions are odd and the cosine and secant functions are even, as also indicated by the symmetries of their graphs in Figure 1.37.

The *addition and subtraction formulas* for the sine and cosine are

$$\sin(u \pm v) = \sin u \cos v \pm \cos u \sin v$$

$$\cos(u \pm v) = \cos u \cos v \mp \sin u \sin v.$$

The *double-angle formulas* for the sine and cosine are

$$\sin 2u = 2 \sin u \cos u$$

$$\cos 2u = \cos^2 u - \sin^2 u$$
$$= 1 - 2 \sin^2 u = 2 \cos^2 u - 1.$$

The *half-angle formulas* are

$$\sin^2 u = \frac{1 - \cos 2u}{2}$$

$$\cos^2 u = \frac{1 + \cos 2u}{2}.$$

These and other formulas useful in calculus are listed on the back inside cover of this text.

EXERCISES 1.3

Exer. 1–2: Find the exact radian measure of the angle.

1 (a) $150°$ (b) $120°$ (c) $450°$ (d) $-60°$

2 (a) $225°$ (b) $210°$ (c) $630°$ (d) $-135°$

Exer. 3–4: Find the exact degree measure of the angle.

3 (a) $\frac{2\pi}{3}$ (b) $\frac{5\pi}{6}$ (c) $\frac{3\pi}{4}$ (d) $-\frac{7\pi}{2}$

4 (a) $\frac{11\pi}{6}$ (b) $\frac{4\pi}{3}$ (c) $\frac{11\pi}{4}$ (d) $-\frac{5\pi}{2}$

Exer. 5–6: Find the length of arc that subtends a central angle θ on a circle of diameter d.

5 $\theta = 50°$; $d = 16$

6 $\theta = 2.2$; $d = 120$

Exer. 7–8: Find the values of x and y in the figure.

7

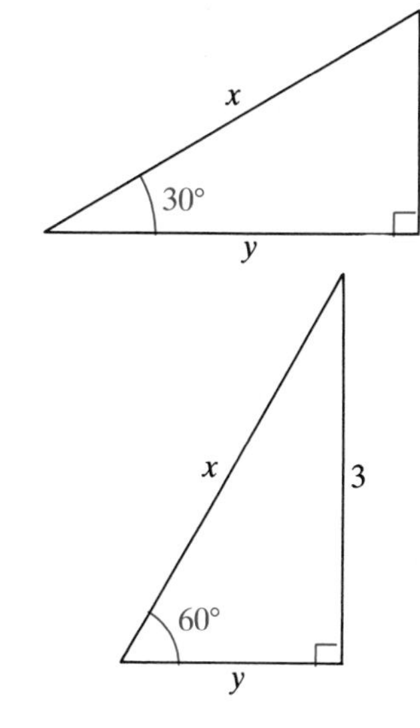

8

Exer. 9–12: Find the values of the trigonometric functions if θ is an acute angle.

9 $\sin \theta = \frac{3}{5}$

10 $\cos \theta = \frac{8}{17}$

11 $\tan \theta = \frac{5}{12}$

12 $\cot \theta = 1$

Exer. 13–14: If θ is in standard position and Q is on the terminal side of θ, find the values of the trigonometric functions of θ.

13 $Q(4, -3)$

14 $Q(-8, -15)$

Exer. 15–16: Let θ be in standard position, with terminal side in the specified quadrant and satisfying the given condition. Find the values of the trigonometric functions of θ.

15 III; parallel to the line $2y - 7x + 2 = 0$

16 IV; perpendicular to the line through $A(5, 12)$ and $B(-3, -3)$

Exer. 17–20: If θ is an acute angle, use fundamental identities to write the first expression in terms of the second.

17 (a) $\cot \theta$, $\sin \theta$ (b) $\sec \theta$, $\sin \theta$

18 (a) $\tan \theta$, $\cos \theta$ (b) $\csc \theta$, $\cos \theta$

19 (a) $\tan \theta$, $\sec \theta$ (b) $\sin \theta$, $\sec \theta$

20 (a) $\cot \theta$, $\csc \theta$ (b) $\cos \theta$, $\cot \theta$

Exer. 21–26: Refer to Example 1. Make the indicated trigonometric substitution and use fundamental identities to obtain a simplified trigonometric expression that contains no radicals.

21 $\sqrt{16 - x^2}$; $x = 4 \sin \theta$ for $-\frac{\pi}{2} \le \theta \le \frac{\pi}{2}$

22 $\dfrac{x^2}{\sqrt{9-x^2}}$; $x = 3 \sin \theta$ for $-\dfrac{\pi}{2} < \theta < \dfrac{\pi}{2}$

23 $\dfrac{x}{\sqrt{25+x^2}}$; $x = 5 \tan \theta$ for $-\dfrac{\pi}{2} < \theta < \dfrac{\pi}{2}$

24 $\dfrac{\sqrt{x^2+4}}{x^2}$; $x = 2 \tan \theta$ for $-\dfrac{\pi}{2} < \theta < \dfrac{\pi}{2}$

25 $\dfrac{\sqrt{x^2-9}}{x}$; $x = 3 \sec \theta$ for $0 < \theta < \dfrac{\pi}{2}$

26 $x^3\sqrt{x^2-25}$; $x = 5 \sec \theta$ for $0 < \theta < \dfrac{\pi}{2}$

Exer. 27–32: Find the exact value.

27 (a) $\sin(2\pi/3)$ (b) $\sin(-5\pi/4)$

28 (a) $\cos 150°$ (b) $\cos(-60°)$

29 (a) $\tan(5\pi/6)$ (b) $\tan(-\pi/3)$

30 (a) $\cot 120°$ (b) $\cot(-150°)$

31 (a) $\sec(2\pi/3)$ (b) $\sec(-\pi/6)$

32 (a) $\csc 240°$ (b) $\csc(-330°)$

Exer. 33–38: Sketch the graph of f, making use of stretching, reflecting, or shifting.

33 (a) $f(x) = \frac{1}{4} \sin x$ (b) $f(x) = -4 \sin x$

34 (a) $f(x) = \sin(x - \pi/2)$ (b) $f(x) = \sin x - \pi/2$

35 (a) $f(x) = 2 \cos(x + \pi)$ (b) $f(x) = 2 \cos x + \pi$

36 (a) $f(x) = \frac{1}{3} \cos x$ (b) $f(x) = -3 \cos x$

37 (a) $f(x) = 4 \tan x$ (b) $f(x) = \tan(x - \pi/4)$

38 (a) $f(x) = \frac{1}{4} \tan x$ (b) $f(x) = \tan(x + 3\pi/4)$

Exer. 39–42: Find a composite function form for y.

39 $y = \sqrt{\tan^2 x + 4}$ 40 $y = \cot^3(2x)$

41 $y = \sec(x + \pi/4)$ 42 $y = \csc\sqrt{x - \pi}$

43 If $f(x) = \cos x$, show that

$$\frac{f(x+h) - f(x)}{h} = \cos x\left(\frac{\cos h - 1}{h}\right) - \sin x\left(\frac{\sin h}{h}\right).$$

44 If $f(x) = \sin x$, show that

$$\frac{f(x+h) - f(x)}{h} = \sin x\left(\frac{\cos h - 1}{h}\right) + \cos x\left(\frac{\sin h}{h}\right).$$

Exer. 45–54: Verify the identity.

45 $(1 - \sin^2 t)(1 + \tan^2 t) = 1$

46 $\sec\beta - \cos\beta = \tan\beta \sin\beta$

47 $\dfrac{\csc^2\theta}{1 + \tan^2\theta} = \cot^2\theta$

48 $\cot t + \tan t = \csc t \sec t$

49 $\dfrac{1 + \csc\beta}{\sec\beta} - \cot\beta = \cos\beta$

50 $\dfrac{1}{\csc z - \cot z} = \csc z + \cot z$

51 $\sin 3u = \sin u(3 - 4\sin^2 u)$

52 $2\sin^2 2t + \cos 4t = 1$

53 $\cos^4(\theta/2) = \frac{3}{8} + \frac{1}{2}\cos\theta + \frac{1}{8}\cos 2\theta$

54 $\sin^4 2x = \frac{3}{8} - \frac{1}{2}\cos 4x + \frac{1}{8}\cos 8x$

Exer. 55–56: Find all solutions of the equation.

55 $2\cos 2\theta - \sqrt{3} = 0$ 56 $2\sin 3\theta + \sqrt{2} = 0$

Exer. 57–64: Find the solutions of the equation in $[0, 2\pi)$.

57 $2\sin^2 u = 1 - \sin u$ 58 $\cos\theta - \sin\theta = 1$

59 $2\tan t - \sec^2 t = 0$

60 $\sin x + \cos x \cot x = \csc x$

61 $\sin 2t + \sin t = 0$ 62 $\cos u + \cos 2u = 0$

63 $\tan 2x = \tan x$ 64 $\sin\frac{1}{2}u + \cos u = 1$

c **Exer. 65–70: Approximate, to the nearest 10′, the solutions of the equation that are in [0°, 360°).**

65 $\sin\theta = -0.5640$ 66 $\cos\theta = 0.7490$

67 $\tan\theta = 2.798$ 68 $\cot\theta = -0.9601$

69 $\sec\theta = -1.116$ 70 $\csc\theta = 1.485$

c 71 Graph $y = (\sin x - \cos \pi x)/\cos x$ for $-1 \le x \le 1$ and estimate the x-intercepts.

c 72 Approximate the solution of the equation $x = \frac{1}{2}\cos x$ by using the following procedure.

(1) Graph $y = x$ and $y = \frac{1}{2}\cos x$ on the same coordinate axes.

(2) Use the graphs in (1) to find a first approximation x_1 to the solution.

(3) Find successive approximations $x_2, x_3, \ldots$ by using the formulas $x_2 = \frac{1}{2}\cos x_1$, $x_3 = \frac{1}{2}\cos x_2, \ldots$ until 6-decimal-place accuracy is obtained.

CHAPTER 2

LIMITS OF FUNCTIONS

INTRODUCTION

The concept of *limit of a function f* is one of the fundamental ideas that distinguishes calculus from algebra and trigonometry. In the development of calculus in the 18th century, the limit concept was treated intuitively, as is done in Section 2.1, where we regard the function value $f(x)$ as getting close to some number L as x gets close to a number a. The flaw in using this definition is the word *close*. A scientist may consider a measurement as being close to an exact value L if it is within 10^{-6} cm of L. A marathon runner is close to the finish line when there are 100 yards left in the race. An astronomer sometimes measures closeness in terms of light-years. Thus, to avoid ambiguities, it is necessary to formulate a definition of limit that does not contain the word *close*. We do so in Section 2.2, by stating what is traditionally called the *ϵ-δ definition of limit of a function*. The definition is precise and applicable to every situation we wish to consider. Later in the chapter we discuss properties that enable us to find many limits easily, without applying the ϵ-δ definition.

In the last section we use limits to define *continuous function*, a concept that is employed extensively throughout calculus.

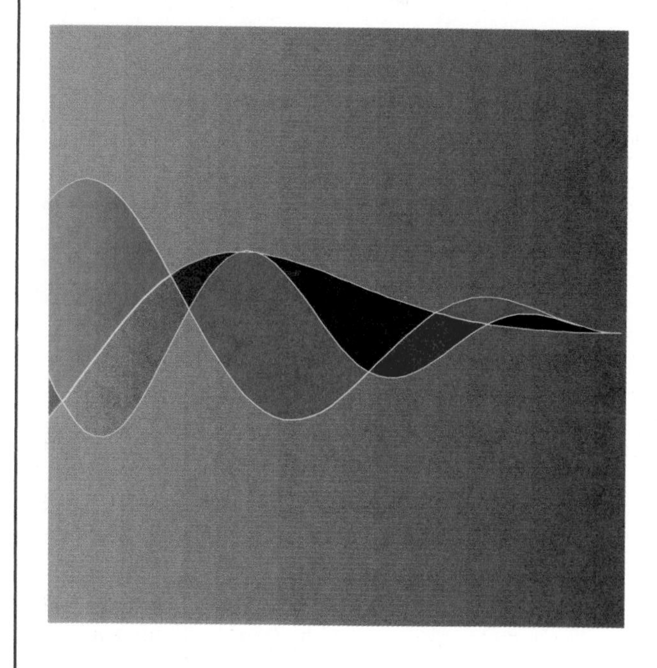

2.1 INTRODUCTION TO LIMITS

In calculus and its applications we are often interested in function values $f(x)$ of a function f when x is *close* to a number a, *but not necessarily equal to a*. In fact, there are many instances where a is not in the domain of f; that is, $f(a)$ is undefined. As an illustration, consider

$$f(x) = \frac{x^3 - 2x^2}{3x - 6}$$

with $a = 2$. Note that 2 is not in the domain of f, since substituting $x = 2$ gives us the undefined expression 0/0. The following tables, obtained with a calculator, list some function values (to eight-decimal-place accuracy) for x close to 2.

x	$f(x)$
1.9	1.20333333
1.99	1.32003333
1.999	1.33200033
1.9999	1.33320000
1.99999	1.33332000
1.999999	1.33333200

x	$f(x)$
2.1	1.47000000
2.01	1.34670000
2.001	1.33466700
2.0001	1.33346667
2.00001	1.33334667
2.000001	1.33333467

It *appears* that the closer x is to 2, the closer $f(x)$ is to $\frac{4}{3}$; however, we cannot be certain of this because we have merely calculated several function values for x near 2. To give a more convincing argument, let us factor the numerator and denominator of $f(x)$ as follows:

$$f(x) = \frac{x^2(x - 2)}{3(x - 2)}$$

If $x \neq 2$, then we may cancel the common factor $x - 2$ and observe that $f(x)$ is given by $\frac{1}{3}x^2$. Thus, the graph of f is the parabola $y = \frac{1}{3}x^2$ with the point $(2, \frac{4}{3})$ deleted, as shown in Figure 2.1. It is geometrically evident that as x gets closer to 2, $f(x)$ gets closer to $\frac{4}{3}$, as indicated in the preceding tables.

FIGURE 2.1

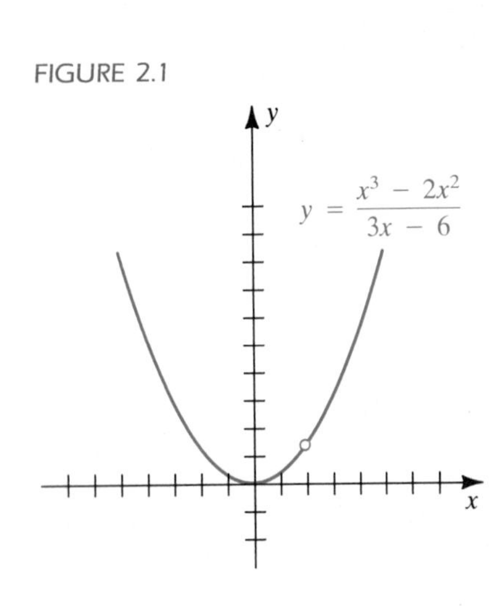

In general, if a function f is defined throughout an open interval containing a real number a, except possibly at a itself, we may ask the following questions:

1. As x gets closer to a (but $x \neq a$), does the function value $f(x)$ get closer to some real number L?

2. Can we make the function value $f(x)$ *as close to L as desired* by choosing x sufficiently close to a (but $x \neq a$)?

If the answers to these questions are *yes*, we write

$$\lim_{x \to a} f(x) = L$$

and say that *the limit of $f(x)$, as x approaches a, is L*, or that *$f(x)$ approaches L as x approaches a*. We may also use the notation

$$f(x) \to L \quad \text{as} \quad x \to a.$$

This means that the point $(x, f(x))$ on the graph of f approaches the point (a, L) as x approaches a. We shall use the phrases *close to* and *approaches* in an intuitive manner. The next section contains a formal definition of limit that avoids this terminology. Using this *limit notation*, we may denote the result of our illustration as follows:

$$\lim_{x \to 2} \frac{x^3 - 2x^2}{3x - 6} = \frac{4}{3}$$

The following chart summarizes the preceding discussion and provides a graphical illustration.

Limit of a function **(2.1)**

NOTATION	INTUITIVE MEANING	GRAPHICAL INTERPRETATION
$\lim_{x \to a} f(x) = L$	We can make $f(x)$ as close to L as desired by choosing x sufficiently close to a, and $x \neq a$.	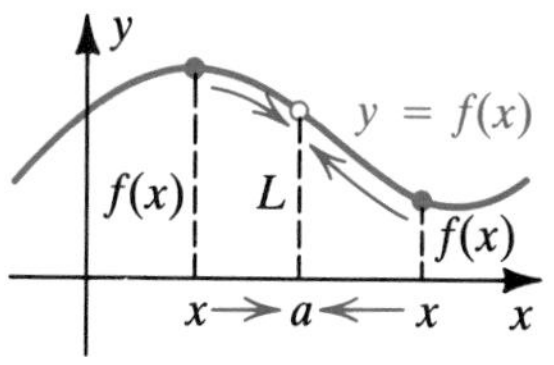 y; $y = f(x)$; $f(x)$; L; $f(x)$; $x \to a \leftarrow x$; x

If $f(x)$ approaches some number as x approaches a, but we do not know what that number is, we use the phrase *$\lim_{x \to a} f(x)$ exists.*

The graph of f shown in the preceding chart illustrates only one way in which $f(x)$ might approach L as x approaches a. We have not shown a point with x-coordinate a because, when using the limit concept (2.1), we always assume that $x \neq a$; that is, *the function value $f(a)$ is completely irrelevant*. As we shall see, $f(a)$ may be different from L, may equal L, or may not exist, depending on the nature of the function f.

In our discussion of $f(x) = (x^3 - 2x^2)/(3x - 6)$ it was possible to simplify $f(x)$ by factoring the numerator and denominator. In many cases such algebraic simplifications are impossible. In particular, when we consider derivatives of trigonometric functions later in the text, it will be necessary to answer the following question.

$$\text{Question: Does } \lim_{x \to 0} \frac{\sin x}{x} \text{ exist?}$$

Note that substituting 0 for x gives us the undefined expression $0/0$. The table on the following page, obtained with a calculator, lists some approximations of $f(x) = (\sin x)/x$ for x near 0, *where x is a real number or the radian measure of an angle*. The graph of f is sketched in Figure 2.2, beside the table.

x	$f(x) = \dfrac{\sin x}{x}$
± 2.0	0.454648713
± 1.0	0.841470985
± 0.5	0.958851077
± 0.4	0.973545856
± 0.3	0.985067356
± 0.2	0.993346654
± 0.1	0.998334166
± 0.01	0.999983333
± 0.001	0.999999833
± 0.0001	0.999999995

FIGURE 2.2

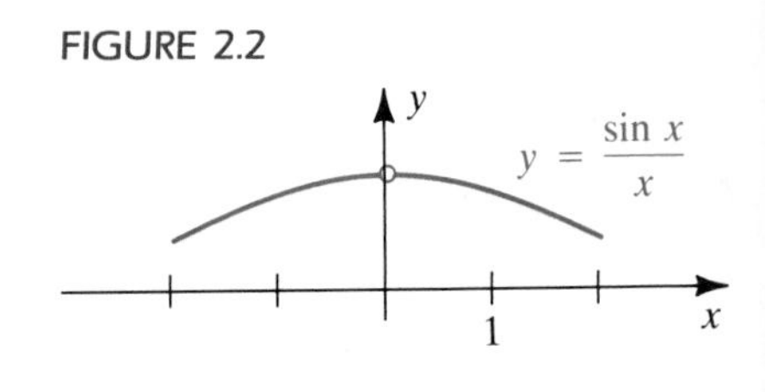

Referring to the table or the graph, we arrive at the following conjecture.

$$\text{Educated guess:} \quad \lim_{x \to 0} \frac{\sin x}{x} = 1$$

As indicated, *we have merely guessed at the answer*. The table indicates that $(\sin x)/x$ gets closer to 1 as x gets closer to 0; however, we cannot be absolutely sure of this fact. Conceivably the function values could deviate from 1 if x were closer to 0 than are those x-values listed in the table. Although a calculator may help us *guess* if a limit exists, it cannot be used in proofs. We will return to this limit in Section 3.4, where we will *prove* that our guess is correct.

It is easy to find $\lim_{x \to a} f(x)$ if $f(x)$ is a simple algebraic expression. For example, if $f(x) = 2x - 3$ and $a = 4$, it is evident that the closer x is to 4, the closer $f(x)$ is to $2(4) - 3$, or 5. This gives us the first limit in the following illustration. The remaining two limits may be obtained in the same intuitive manner.

ILLUSTRATION

- $\lim_{x \to 4} (2x - 3) = 2(4) - 3 = 8 - 3 = 5$
- $\lim_{x \to -3} (x^2 + 1) = (-3)^2 + 1 = 9 + 1 = 10$
- $\lim_{x \to 7} \sqrt{x + 2} = \sqrt{7 + 2} = \sqrt{9} = 3$

In the preceding illustration the limits as $x \to a$ can be found by merely substituting the number a for x. This is a property that holds for special functions called *continuous functions*, to be discussed in Section 2.5. The next illustration shows that we cannot use this technique for *every* algebraic function f. In the illustration it is important to note that

$$\frac{x^2 + x - 2}{x - 1} = \frac{(x - 1)(x + 2)}{x - 1} = x + 2, \quad \textit{provided } x \neq 1.$$

(If $x \neq 1$, then $x - 1 \neq 0$, and it is permissible to cancel the common factor $x - 1$ in the numerator and denominator.) It follows that the graphs of the equations $y = (x^2 + x - 2)/(x - 1)$ and $y = x + 2$ are the same *except* for $x = 1$. Specifically, the point $(1, 3)$ is on the graph of $y = x + 2$, but is not on the graph of $y = (x^2 + x - 2)/(x - 1)$, as indicated in the illustration.

ILLUSTRATION

FUNCTION VALUE	GRAPH	LIMIT AS $x \to 1$
$f(x) = x + 2$	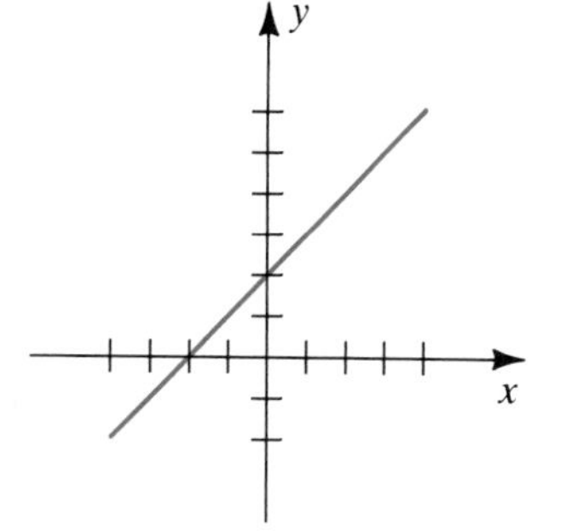	$\lim_{x \to 1} f(x) = 3$
$g(x) = \dfrac{x^2 + x - 2}{x - 1}$	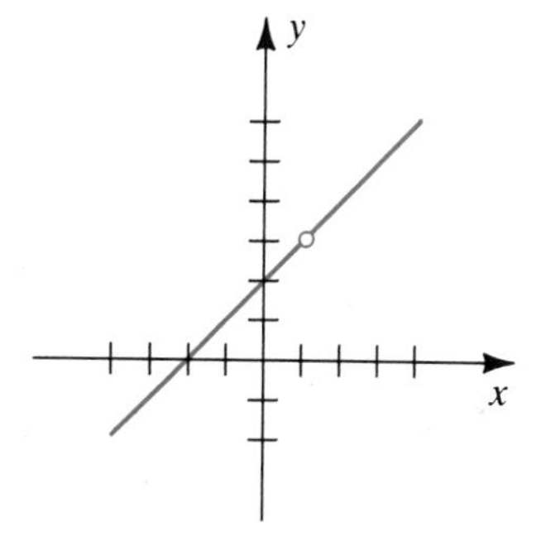	$\lim_{x \to 1} g(x) = 3$
$h(x) = \begin{cases} \dfrac{x^2 + x - 2}{x - 1} & \text{if } x \neq 1 \\ 2 & \text{if } x = 1 \end{cases}$	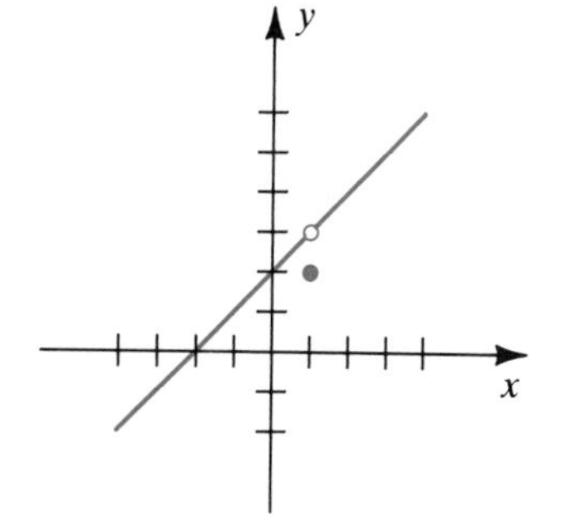	$\lim_{x \to 1} h(x) = 3$

In the preceding illustration the limit of each function as x approaches 1 is 3, but in the first case $f(1) = 3$, in the second $g(1)$ does not exist, and in the third $h(1) = 2 \neq 3$.

The following two examples illustrate how algebraic manipulations may be used to help find certain limits.

EXAMPLE 1 If $f(x) = \dfrac{2x^2 - 5x + 2}{5x^2 - 7x - 6}$, find $\lim_{x \to 2} f(x)$.

SOLUTION The number 2 is not in the domain of f since the meaningless expression 0/0 is obtained if 2 is substituted for x. Factoring the numerator and denominator gives us

$$f(x) = \frac{(x-2)(2x-1)}{(x-2)(5x+3)}.$$

We cannot cancel the factor $x - 2$ at this stage; however, if we take the *limit* of $f(x)$ as $x \to 2$ this cancellation *is* allowed, because by (2.1) $x \neq 2$ and hence $x - 2 \neq 0$. Thus,

$$\begin{aligned} \lim_{x \to 2} f(x) &= \lim_{x \to 2} \frac{2x^2 - 5x + 2}{5x^2 - 7x - 6} \\ &= \lim_{x \to 2} \frac{(x-2)(2x-1)}{(x-2)(5x+3)} \\ &= \lim_{x \to 2} \frac{2x-1}{5x+3} = \frac{3}{13}. \end{aligned}$$

EXAMPLE 2 Let $f(x) = \dfrac{x - 9}{\sqrt{x} - 3}$.

(a) Find $\lim_{x \to 9} f(x)$.

(b) Sketch the graph of f and illustrate the limit in part (a) graphically.

SOLUTION

(a) Note that the number 9 is not in the domain of f. To find the limit we shall change the form of $f(x)$ by rationalizing the denominator as follows:

$$\begin{aligned} \lim_{x \to 9} f(x) &= \lim_{x \to 9} \frac{x-9}{\sqrt{x}-3} \\ &= \lim_{x \to 9} \left(\frac{x-9}{\sqrt{x}-3} \cdot \frac{\sqrt{x}+3}{\sqrt{x}+3} \right) \\ &= \lim_{x \to 9} \frac{(x-9)(\sqrt{x}+3)}{x-9} \end{aligned}$$

By (2.1), when investigating the limit as $x \to 9$, we assume that $x \neq 9$. Hence $x - 9 \neq 0$, and we can divide numerator and denominator by $x - 9$; that is, we can *cancel* the expression $x - 9$. This gives us

$$\lim_{x \to 9} f(x) = \lim_{x \to 9} (\sqrt{x} + 3) = \sqrt{9} + 3 = 6.$$

(b) If we rationalize the denominator of $f(x)$ as in part (a), we see that the graph of f is the same as the graph of the equation $y = \sqrt{x} + 3$, *except for the point* (9, 6), as illustrated in Figure 2.3. As x gets closer to 9, the point $(x, f(x))$ on the graph of f gets closer to the point (9, 6). Note that $f(x)$ never actually attains the value 6; however, $f(x)$ can be made as close to 6 as desired by choosing x sufficiently close to 9.

FIGURE 2.3

$f(x) = \dfrac{x - 9}{\sqrt{x} - 3}$

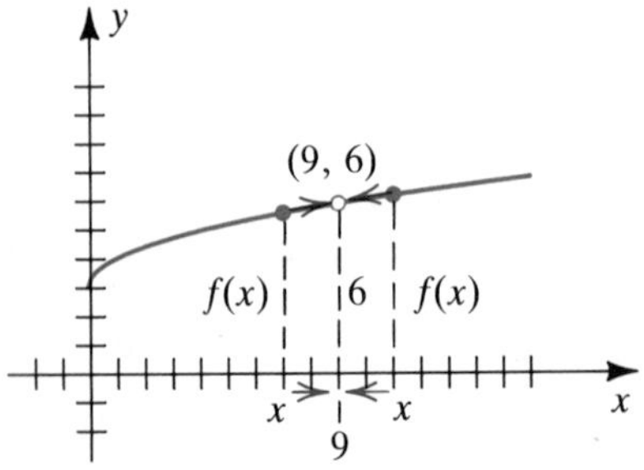

y
(9, 6)
f(x)
6
f(x)
x
9
x
x

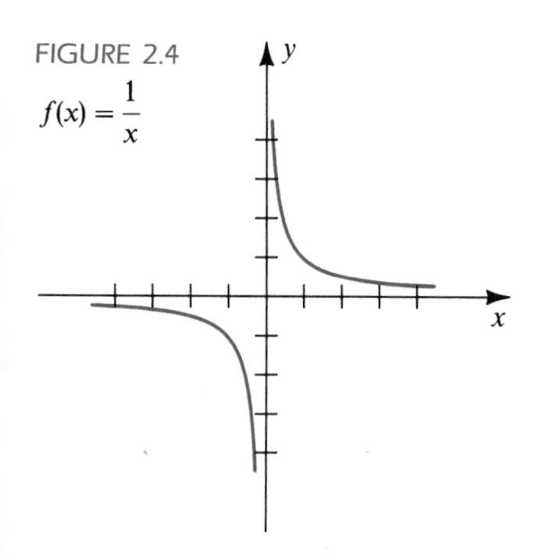

FIGURE 2.4
$f(x) = \frac{1}{x}$

The next two examples involve functions that have no limit as x approaches 0. The solutions are intuitive in nature. Rigorous proofs require the formal definition of limit discussed in the next section.

EXAMPLE 3 Show that $\lim_{x \to 0} \frac{1}{x}$ does not exist.

SOLUTION The graph of $f(x) = 1/x$ is sketched in Figure 2.4. Note that we can make $|f(x)|$ as large as desired by choosing x sufficiently close to 0 (but $x \neq 0$). For example, if we want $f(x) = -1{,}000{,}000$, we choose $x = -0.000001$. For $f(x) = 10^9$, we choose $x = 10^{-9}$. Since $f(x)$ does not approach a specific number L as x approaches 0, the limit does not exist.

EXAMPLE 4 Show that $\lim_{x \to 0} \sin \frac{1}{x}$ does not exist.

SOLUTION Let us first determine some of the characteristics of the graph of $y = \sin(1/x)$. To find the x-intercepts we note that the following statement is true for every integer n:

$$\sin \frac{1}{x} = 0 \quad \text{if and only if} \quad \frac{1}{x} = \pi n, \quad \text{or} \quad x = \frac{1}{\pi n}$$

Some special x-intercepts are

$$\pm\frac{1}{\pi}, \pm\frac{1}{2\pi}, \pm\frac{1}{3\pi}, \ldots, \pm\frac{1}{100\pi},$$

and so on. If we let x approach 0, then the distance between successive x-intercepts decreases and, in fact, approaches 0. Similarly,

$$\sin \frac{1}{x} = 1 \quad \text{if and only if} \quad x = \frac{1}{(\pi/2) + 2\pi n}$$

and

$$\sin \frac{1}{x} = -1 \quad \text{if and only if} \quad x = \frac{1}{(3\pi/2) + 2\pi n},$$

where n is any integer. Thus, as x approaches 0, the function values $\sin(1/x)$ oscillate between -1 and 1, and the corresponding waves on the graph become very compressed, as illustrated in Figure 2.5. Hence

FIGURE 2.5
$f(x) = \sin \frac{1}{x}$

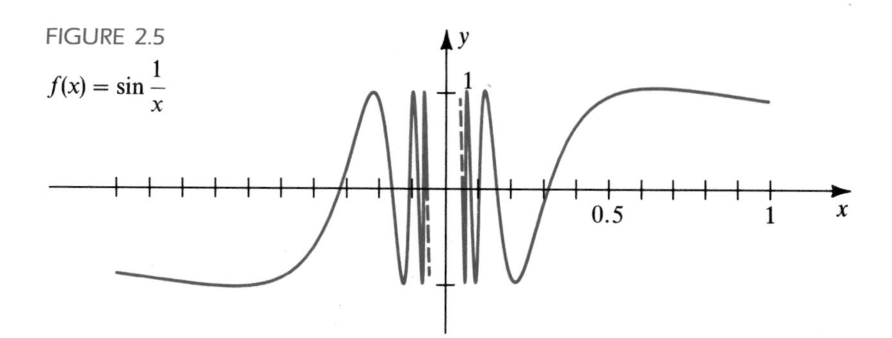

$\lim_{x \to 0} \sin(1/x)$ does not exist, because the function values do not approach a specific number L as x approaches 0.

We sometimes employ *one-sided* limits of the types listed in the next chart.

One-sided limits **(2.2)**

NOTATION	INTUITIVE MEANING	GRAPHICAL INTERPRETATION
$\lim_{x \to a^-} f(x) = L$ (left-hand limit)	We can make $f(x)$ as close to L as desired by choosing x sufficiently close to a, and $x < a$.	y; $y = f(x)$; $f(x)$; L; $x \to a$; x
$\lim_{x \to a^+} f(x) = L$ (right-hand limit)	We can make $f(x)$ as close to L as desired by choosing x sufficiently close to a, and $x > a$.	y; $y = f(x)$; $f(x)$; L; $a \leftarrow x$; x

For a left-hand limit, the function f must be defined in (at least) an open interval of the form (c, a) for some real number c. For a right-hand limit, f must be defined in (a, c) for some c. The notation $x \to a^-$ is read *x approaches a from the left*, and $x \to a^+$ is read *x approaches a from the right*.

EXAMPLE 5 If $f(x) = \sqrt{x - 2}$, sketch the graph of f and find, if possible,

(a) $\lim_{x \to 2^+} f(x)$ **(b)** $\lim_{x \to 2^-} f(x)$ **(c)** $\lim_{x \to 2} f(x)$

FIGURE 2.6

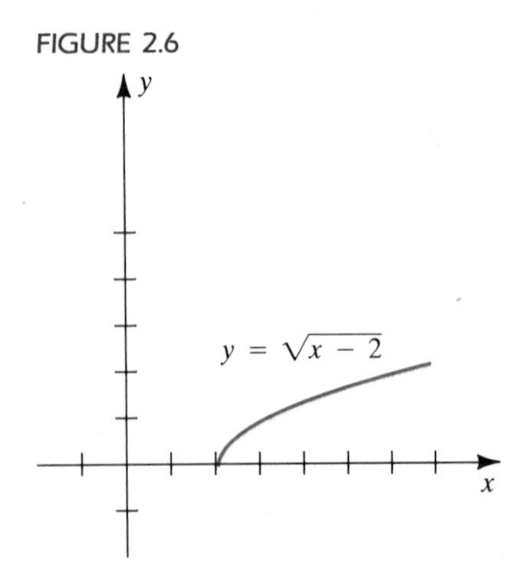

SOLUTION The graph of f is sketched in Figure 2.6.

(a) If $x > 2$, then $x - 2 > 0$ and hence $f(x) = \sqrt{x - 2}$ is a real number; that is, $f(x)$ is defined. Thus,

$$\lim_{x \to 2^+} \sqrt{x - 2} = \sqrt{2 - 2} = 0.$$

(b) The left-hand limit does not exist because $f(x) = \sqrt{x - 2}$ is not a real number if $x < 2$.

(c) The limit does not exist because $f(x) = \sqrt{x - 2}$ is not defined throughout an open interval containing 2.

The relationship between one-sided limits and limits is described in the next theorem.

Theorem **(2.3)**

$$\lim_{x \to a} f(x) = L \quad \text{if and only if} \quad \lim_{x \to a^-} f(x) = L = \lim_{x \to a^+} f(x)$$

The preceding theorem (which can be proved using definitions in Section 2.2) tells us that *the limit of $f(x)$ as x approaches a exists if and only if both the right-hand and left-hand limits exist and are equal.*

FIGURE 2.7
$f(x) = \dfrac{|x|}{x}$

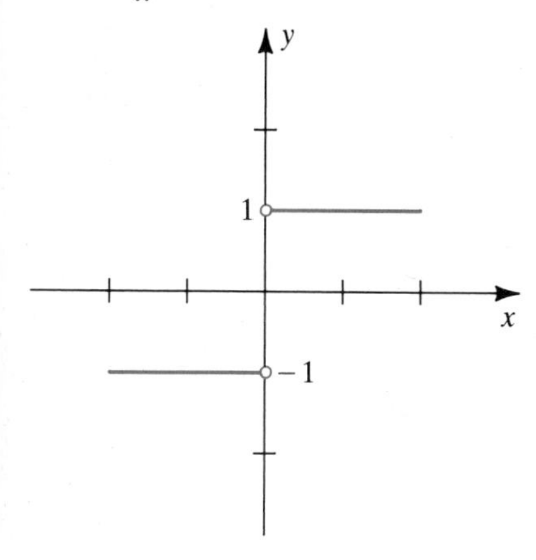

EXAMPLE 6 If $f(x) = \dfrac{|x|}{x}$, sketch the graph of f and find, if possible,

(a) $\lim_{x \to 0^-} f(x)$ **(b)** $\lim_{x \to 0^+} f(x)$ **(c)** $\lim_{x \to 0} f(x)$

SOLUTION The function f is undefined at $x = 0$. If $x > 0$, then $|x| = x$ and $f(x) = x/x = 1$. Hence for $x > 0$ the graph coincides with the horizontal line $y = 1$. If $x < 0$, then $|x| = -x$ and $f(x) = -x/x = -1$. This gives us the sketch in Figure 2.7. Referring to the graph, we see that

(a) $\lim_{x \to 0^-} f(x) = -1$

(b) $\lim_{x \to 0^+} f(x) = 1$

(c) Since the left-hand and right-hand limits are not equal, it follows from Theorem (2.3) that $\lim_{x \to 0} f(x)$ does not exist.

In the next example we consider a piecewise-defined function.

EXAMPLE 7 Sketch the graph of the function f defined as follows:

$$f(x) = \begin{cases} 3 - x & \text{for } x < 1 \\ 4 & \text{for } x = 1 \\ x^2 + 1 & \text{for } x > 1 \end{cases}$$

Find $\lim_{x \to 1^-} f(x)$, $\lim_{x \to 1^+} f(x)$, and $\lim_{x \to 1} f(x)$.

FIGURE 2.8

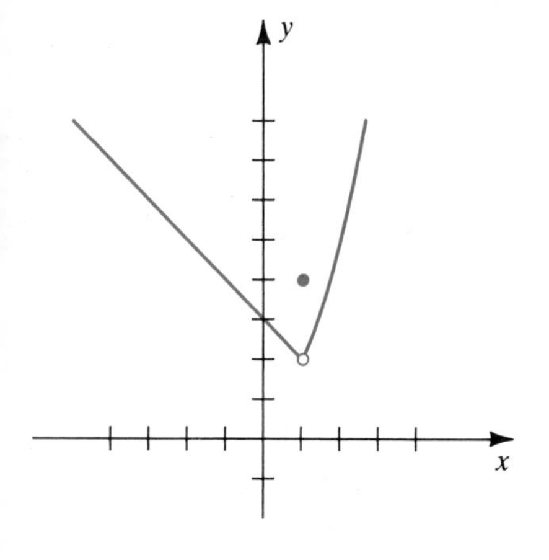

SOLUTION The graph is sketched in Figure 2.8. The one-sided limits are

$$\lim_{x \to 1^-} f(x) = \lim_{x \to 1^-} (3 - x) = 2$$

and

$$\lim_{x \to 1^+} f(x) = \lim_{x \to 1^+} (x^2 + 1) = 2.$$

Since the right-hand and left-hand limits are equal, it follows from Theorem (2.3) that

$$\lim_{x \to 1} f(x) = 2.$$

Note that the function value $f(1) = 4$ is irrelevant in finding the limit.

The following application involves one-sided limits.

FIGURE 2.9

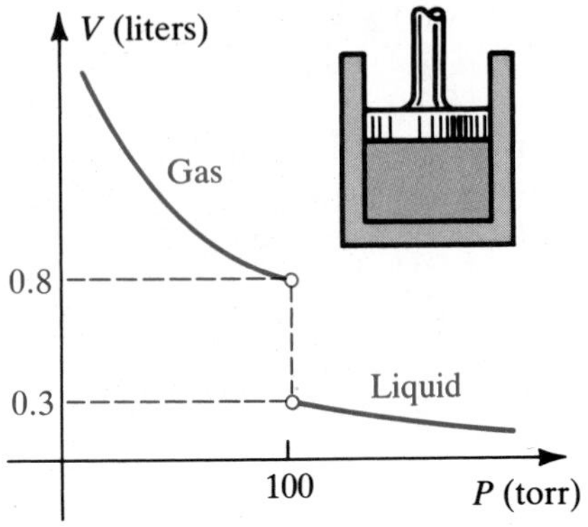

EXAMPLE 8 A gas (such as water vapor or oxygen) is held at a constant temperature in the piston shown in Figure 2.9. As the gas is compressed, the volume V decreases until a certain critical pressure is reached. Beyond this pressure, the gas will assume liquid form. Use the graph in Figure 2.9 to find and interpret

(a) $\lim\limits_{P \to 100^-} V$ **(b)** $\lim\limits_{P \to 100^+} V$ **(c)** $\lim\limits_{P \to 100} V$

SOLUTION

(a) We see from Figure 2.9 that when the pressure P (in torrs) is low, the substance is a gas and the volume V (in liters) is large. (The definition of the unit of pressure, the *torr*, may be found in textbooks on physics.) If P approaches 100 through values less than 100, V decreases and approaches 0.8; that is,

$$\lim_{P \to 100^-} V = 0.8.$$

The limit 0.8 represents the volume at which the substance begins to change from a gas to a liquid.

(b) If $P > 100$, the substance is a liquid. If P approaches 100 through values greater than 100, the volume V increases very slowly (since liquids are nearly incompressible), and

$$\lim_{P \to 100^+} V = 0.3.$$

The limit 0.3 represents the volume at which the substance begins to change from a liquid to a gas.

(c) $\lim_{P \to 100}$ does not exist since the left-hand and right-hand limits in (a) and (b) are not equal. (At $P = 100$ the gas and liquid forms exist together in equilibrium, and the substance cannot be classified as either a gas or a liquid.)

EXERCISES 2.1

Exer. 1–10: Find the limit.

1 $\lim\limits_{x \to -2} (3x - 1)$

2 $\lim\limits_{x \to 3} (x^2 + 2)$

3 $\lim\limits_{x \to 4} x$

4 $\lim\limits_{x \to -3} (-x)$

5 $\lim\limits_{x \to 100} 7$

6 $\lim\limits_{x \to 7} 100$

7 $\lim\limits_{x \to -1} \pi$

8 $\lim\limits_{x \to \pi} (-1)$

9 $\lim\limits_{x \to -1} \dfrac{x + 4}{2x + 1}$

10 $\lim\limits_{x \to 5} \dfrac{x + 2}{x - 4}$

Exer. 11–24: Use an algebraic simplification to help find the limit, if it exists.

11 $\lim\limits_{x \to -3} \dfrac{(x + 3)(x - 4)}{(x + 3)(x + 1)}$

12 $\lim\limits_{x \to -1} \dfrac{(x + 1)(x^2 + 3)}{x + 1}$

13 $\lim\limits_{x \to 2} \dfrac{x^2 - 4}{x - 2}$

14 $\lim\limits_{x \to 3} \dfrac{2x^3 - 6x^2 + x - 3}{x - 3}$

15 $\lim\limits_{r \to 1} \dfrac{r^2 - r}{2r^2 + 5r - 7}$

16 $\lim\limits_{r \to -3} \dfrac{r^2 + 2r - 3}{r^2 + 7r + 12}$

17 $\lim\limits_{k \to 4} \dfrac{k^2 - 16}{\sqrt{k} - 2}$

18 $\lim\limits_{x \to 25} \dfrac{\sqrt{x} - 5}{x - 25}$

The *double-angle formulas* for the sine and cosine are

$$\sin 2u = 2 \sin u \cos u$$

$$\cos 2u = \cos^2 u - \sin^2 u = 1 - 2\sin^2 u = 2\cos^2 u - 1.$$

The *half-angle formulas* are

$$\sin^2 u = \frac{1 - \cos 2u}{2}$$

$$\cos^2 u = \frac{1 + \cos 2u}{2}.$$

These and other formulas useful in calculus are listed on the back inside cover of this text.

EXERCISES 1.3

Exer. 1–2: Find the exact radian measure of the angle.

1 (a) $150°$ (b) $120°$ (c) $450°$ (d) $-60°$

2 (a) $225°$ (b) $210°$ (c) $630°$ (d) $-135°$

Exer. 3–4: Find the exact degree measure of the angle.

3 (a) $\frac{2\pi}{3}$ (b) $\frac{5\pi}{6}$ (c) $\frac{3\pi}{4}$ (d) $-\frac{7\pi}{2}$

4 (a) $\frac{11\pi}{6}$ (b) $\frac{4\pi}{3}$ (c) $\frac{11\pi}{4}$ (d) $-\frac{5\pi}{2}$

Exer. 5–6: Find the length of arc that subtends a central angle θ on a circle of diameter d.

5 $\theta = 50°$; $d = 16$

6 $\theta = 2.2$; $d = 120$

Exer. 7–8: Find the values of x and y in the figure.

7

8

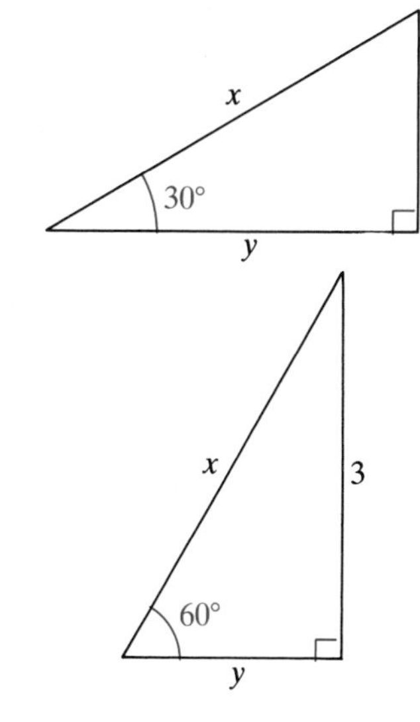

Exer. 9–12: Find the values of the trigonometric functions if θ is an acute angle.

9 $\sin \theta = \frac{3}{5}$

10 $\cos \theta = \frac{8}{17}$

11 $\tan \theta = \frac{5}{12}$

12 $\cot \theta = 1$

Exer. 13–14: If θ is in standard position and Q is on the terminal side of θ, find the values of the trigonometric functions of θ.

13 $Q(4, -3)$

14 $Q(-8, -15)$

Exer. 15–16: Let θ be in standard position, with terminal side in the specified quadrant and satisfying the given condition. Find the values of the trigonometric functions of θ.

15 III; parallel to the line $2y - 7x + 2 = 0$

16 IV; perpendicular to the line through $A(5, 12)$ and $B(-3, -3)$

Exer. 17–20: If θ is an acute angle, use fundamental identities to write the first expression in terms of the second.

17 (a) $\cot \theta$, $\sin \theta$ (b) $\sec \theta$, $\sin \theta$

18 (a) $\tan \theta$, $\cos \theta$ (b) $\csc \theta$, $\cos \theta$

19 (a) $\tan \theta$, $\sec \theta$ (b) $\sin \theta$, $\sec \theta$

20 (a) $\cot \theta$, $\csc \theta$ (b) $\cos \theta$, $\cot \theta$

Exer. 21–26: Refer to Example 1. Make the indicated trigonometric substitution and use fundamental identities to obtain a simplified trigonometric expression that contains no radicals.

21 $\sqrt{16 - x^2}$; $x = 4 \sin \theta$ for $-\frac{\pi}{2} \leq \theta \leq \frac{\pi}{2}$

22 $\dfrac{x^2}{\sqrt{9-x^2}}$; $x = 3 \sin \theta$ for $-\dfrac{\pi}{2} < \theta < \dfrac{\pi}{2}$

23 $\dfrac{x}{\sqrt{25+x^2}}$; $x = 5 \tan \theta$ for $-\dfrac{\pi}{2} < \theta < \dfrac{\pi}{2}$

24 $\dfrac{\sqrt{x^2+4}}{x^2}$; $x = 2 \tan \theta$ for $-\dfrac{\pi}{2} < \theta < \dfrac{\pi}{2}$

25 $\dfrac{\sqrt{x^2-9}}{x}$; $x = 3 \sec \theta$ for $0 < \theta < \dfrac{\pi}{2}$

26 $x^3\sqrt{x^2-25}$; $x = 5 \sec \theta$ for $0 < \theta < \dfrac{\pi}{2}$

Exer. 27–32: Find the exact value.

27 (a) $\sin (2\pi/3)$ (b) $\sin (-5\pi/4)$

28 (a) $\cos 150°$ (b) $\cos (-60°)$

29 (a) $\tan (5\pi/6)$ (b) $\tan (-\pi/3)$

30 (a) $\cot 120°$ (b) $\cot (-150°)$

31 (a) $\sec (2\pi/3)$ (b) $\sec (-\pi/6)$

32 (a) $\csc 240°$ (b) $\csc (-330°)$

Exer. 33–38: Sketch the graph of f, making use of stretching, reflecting, or shifting.

33 (a) $f(x) = \frac{1}{4} \sin x$ (b) $f(x) = -4 \sin x$

34 (a) $f(x) = \sin (x - \pi/2)$ (b) $f(x) = \sin x - \pi/2$

35 (a) $f(x) = 2 \cos (x + \pi)$ (b) $f(x) = 2 \cos x + \pi$

36 (a) $f(x) = \frac{1}{3} \cos x$ (b) $f(x) = -3 \cos x$

37 (a) $f(x) = 4 \tan x$ (b) $f(x) = \tan (x - \pi/4)$

38 (a) $f(x) = \frac{1}{4} \tan x$ (b) $f(x) = \tan (x + 3\pi/4)$

Exer. 39–42: Find a composite function form for y.

39 $y = \sqrt{\tan^2 x + 4}$ 40 $y = \cot^3 (2x)$

41 $y = \sec (x + \pi/4)$ 42 $y = \csc \sqrt{x - \pi}$

43 If $f(x) = \cos x$, show that

$$\frac{f(x+h) - f(x)}{h} = \cos x\left(\frac{\cos h - 1}{h}\right) - \sin x\left(\frac{\sin h}{h}\right).$$

44 If $f(x) = \sin x$, show that

$$\frac{f(x+h) - f(x)}{h} = \sin x\left(\frac{\cos h - 1}{h}\right) + \cos x\left(\frac{\sin h}{h}\right).$$

Exer. 45–54: Verify the identity.

45 $(1 - \sin^2 t)(1 + \tan^2 t) = 1$

46 $\sec \beta - \cos \beta = \tan \beta \sin \beta$

47 $\dfrac{\csc^2 \theta}{1 + \tan^2 \theta} = \cot^2 \theta$

48 $\cot t + \tan t = \csc t \sec t$

49 $\dfrac{1 + \csc \beta}{\sec \beta} - \cot \beta = \cos \beta$

50 $\dfrac{1}{\csc z - \cot z} = \csc z + \cot z$

51 $\sin 3u = \sin u(3 - 4 \sin^2 u)$

52 $2 \sin^2 2t + \cos 4t = 1$

53 $\cos^4 (\theta/2) = \frac{3}{8} + \frac{1}{2} \cos \theta + \frac{1}{8} \cos 2\theta$

54 $\sin^4 2x = \frac{3}{8} - \frac{1}{2} \cos 4x + \frac{1}{8} \cos 8x$

Exer. 55–56: Find all solutions of the equation.

55 $2 \cos 2\theta - \sqrt{3} = 0$ 56 $2 \sin 3\theta + \sqrt{2} = 0$

Exer. 57–64: Find the solutions of the equation in $[0, 2\pi)$.

57 $2 \sin^2 u = 1 - \sin u$ 58 $\cos \theta - \sin \theta = 1$

59 $2 \tan t - \sec^2 t = 0$

60 $\sin x + \cos x \cot x = \csc x$

61 $\sin 2t + \sin t = 0$ 62 $\cos u + \cos 2u = 0$

63 $\tan 2x = \tan x$ 64 $\sin \frac{1}{2}u + \cos u = 1$

c **Exer. 65–70: Approximate, to the nearest 10′, the solutions of the equation that are in [0°, 360°).**

65 $\sin \theta = -0.5640$ 66 $\cos \theta = 0.7490$

67 $\tan \theta = 2.798$ 68 $\cot \theta = -0.9601$

69 $\sec \theta = -1.116$ 70 $\csc \theta = 1.485$

c 71 Graph $y = (\sin x - \cos \pi x)/\cos x$ for $-1 \le x \le 1$ and estimate the x-intercepts.

c 72 Approximate the solution of the equation $x = \frac{1}{2} \cos x$ by using the following procedure.

(1) Graph $y = x$ and $y = \frac{1}{2} \cos x$ on the same coordinate axes.

(2) Use the graphs in (1) to find a first approximation x_1 to the solution.

(3) Find successive approximations $x_2, x_3, \ldots$ by using the formulas $x_2 = \frac{1}{2} \cos x_1$, $x_3 = \frac{1}{2} \cos x_2, \ldots$ until 6-decimal-place accuracy is obtained.

19 $\lim\limits_{h \to 0} \dfrac{(x+h)^2 - x^2}{h}$

20 $\lim\limits_{h \to 0} \dfrac{(x+h)^3 - x^3}{h}$

21 $\lim\limits_{h \to -2} \dfrac{h^3 + 8}{h + 2}$

22 $\lim\limits_{h \to 2} \dfrac{h^3 - 8}{h^2 - 4}$

23 $\lim\limits_{z \to -2} \dfrac{z - 4}{z^2 - 2z - 8}$

24 $\lim\limits_{z \to 5} \dfrac{z - 5}{z^2 - 10z + 25}$

Exer. 25–30: Find each limit, if it exists:

(a) $\lim\limits_{x \to a^-} f(x)$ (b) $\lim\limits_{x \to a^+} f(x)$ (c) $\lim\limits_{x \to a} f(x)$

25 $f(x) = \dfrac{|x - 4|}{x - 4}$; $a = 4$

26 $f(x) = \dfrac{x + 5}{|x + 5|}$; $a = -5$

27 $f(x) = \sqrt{x + 6} + x$; $a = -6$

28 $f(x) = \sqrt{5 - 2x} - x^2$; $a = \frac{5}{2}$

29 $f(x) = \dfrac{1}{x^3}$; $a = 0$

30 $f(x) = \dfrac{1}{x - 8}$; $a = 8$

Exer. 31–40: Refer to the graph to find each limit, if it exists:

(a) $\lim\limits_{x \to 2^-} f(x)$ (b) $\lim\limits_{x \to 2^+} f(x)$ (c) $\lim\limits_{x \to 2} f(x)$

(d) $\lim\limits_{x \to 0^-} f(x)$ (e) $\lim\limits_{x \to 0^+} f(x)$ (f) $\lim\limits_{x \to 0} f(x)$

31

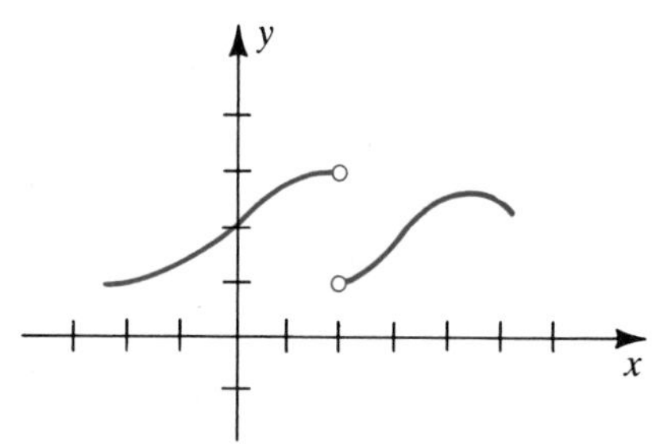

32

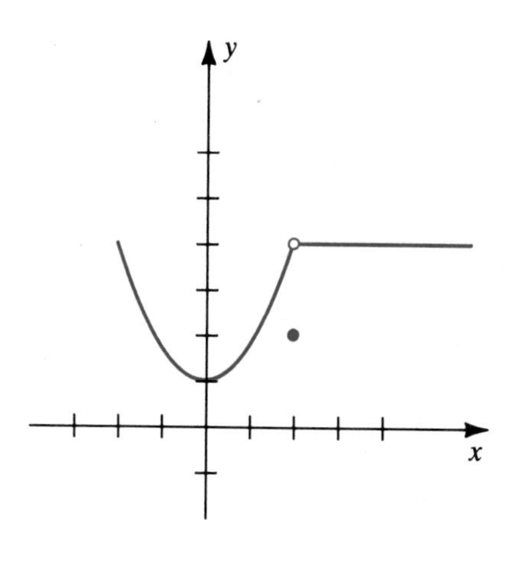

33

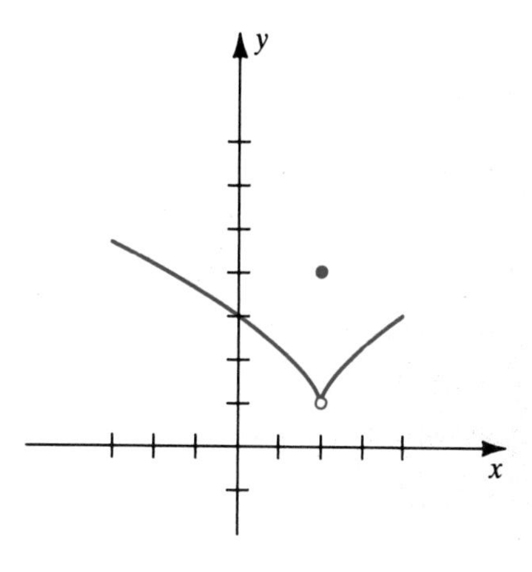

34

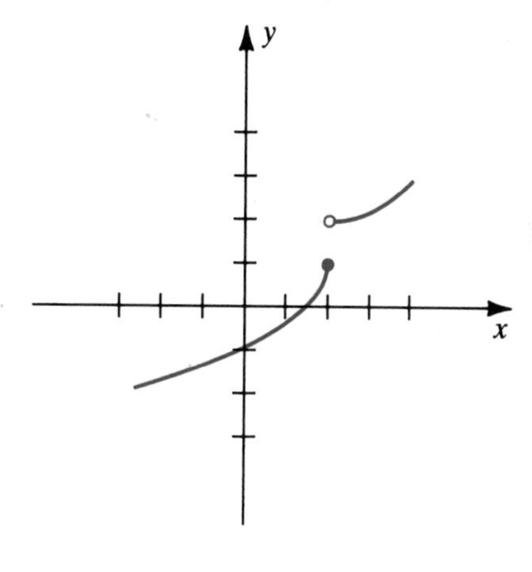

35

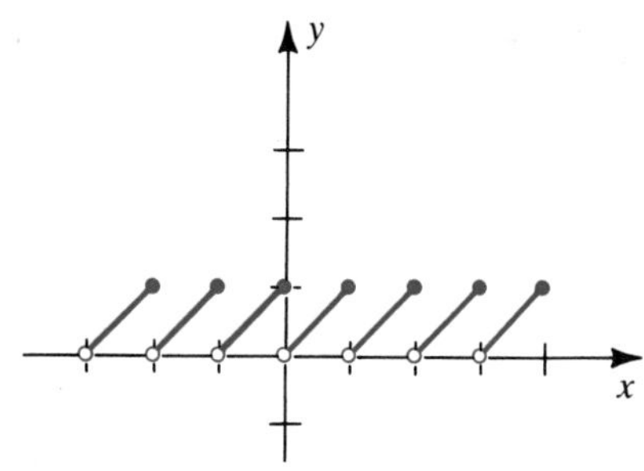

36

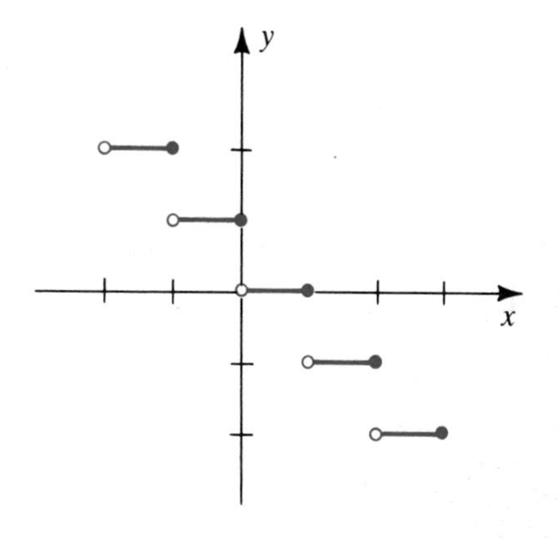

37

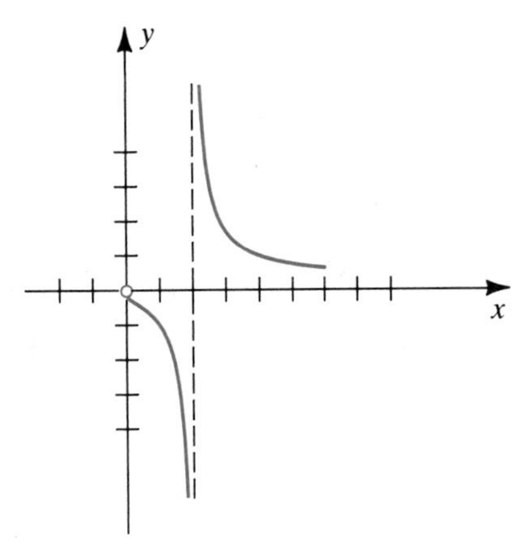

38

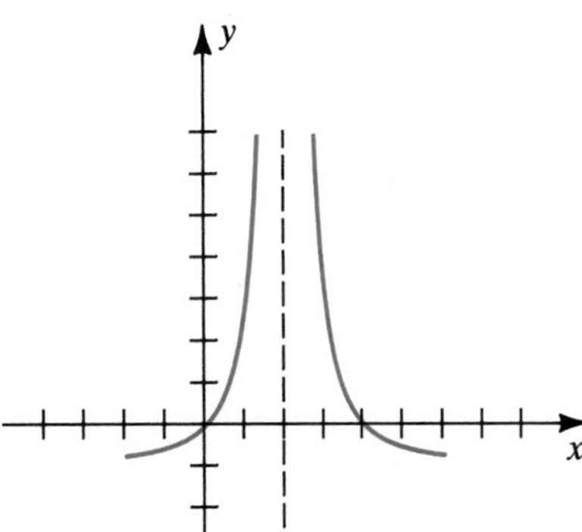

39

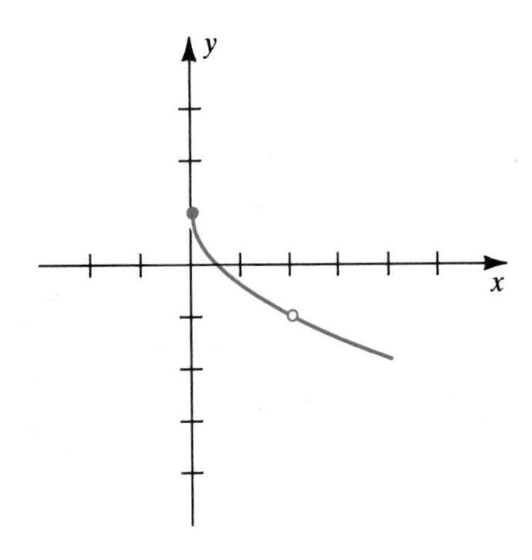

40

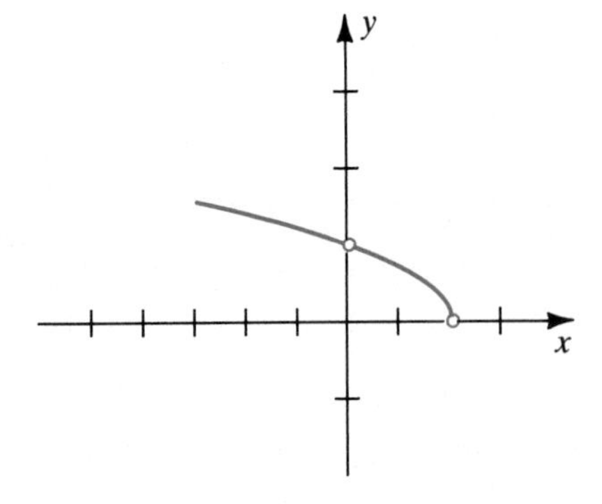

Exer. 41–46: Sketch the graph of f and find each limit, if it exists:

(a) $\lim_{x\to 1^-} f(x)$ (b) $\lim_{x\to 1^+} f(x)$ (c) $\lim_{x\to 1} f(x)$

41 $f(x) = \begin{cases} x^2 - 1 & \text{if } x < 1 \\ 4 - x & \text{if } x \geq 1 \end{cases}$

42 $f(x) = \begin{cases} x^3 & \text{if } x \leq 1 \\ 3 - x & \text{if } x > 1 \end{cases}$

43 $f(x) = \begin{cases} 3x - 1 & \text{if } x \leq 1 \\ 3 - x & \text{if } x > 1 \end{cases}$

44 $f(x) = \begin{cases} |x - 1| & \text{if } x \neq 1 \\ 1 & \text{if } x = 1 \end{cases}$

45 $f(x) = \begin{cases} x^2 + 1 & \text{if } x < 1 \\ 1 & \text{if } x = 1 \\ x + 1 & \text{if } x > 1 \end{cases}$

46 $f(x) = \begin{cases} -x^2 & \text{if } x < 1 \\ 2 & \text{if } x = 1 \\ x - 2 & \text{if } x > 1 \end{cases}$

47 A country taxes the first \$20,000 of an individual's income at a rate of 15%, and all income over \$20,000 is taxed at 20%.

(a) Find a piecewise-defined function T for the total tax on an income of x dollars.

(b) Find

$$\lim_{x\to 20{,}000^-} T(x) \quad \text{and} \quad \lim_{x\to 20{,}000^+} T(x).$$

48 A telephone company charges 25 cents for the first minute of a long-distance call and 15 cents for each additional minute.

(a) Find a piecewise-defined function C for the total cost of a long-distance call of x minutes.

(b) If n is an integer greater than 1, find

$$\lim_{x\to n^-} C(x) \quad \text{and} \quad \lim_{x\to n^+} C(x).$$

49 Shown in the figure on the next page is a graph of the g-forces experienced by astronauts during the takeoff of a spacecraft with two rocket boosters. (A force of $2g$'s is twice that of gravity, $3g$'s is three times that of gravity, etc.) If $F(t)$ denotes the g-force t minutes into the flight, find and interpret

(a) $\lim_{t\to 0^+} F(t)$

(b) $\lim_{t\to 3.5^-} F(t)$ and $\lim_{t\to 3.5^+} F(t)$

(c) $\lim_{t\to 5^-} F(t)$ and $\lim_{t\to 5^+} F(t)$

EXERCISE 49

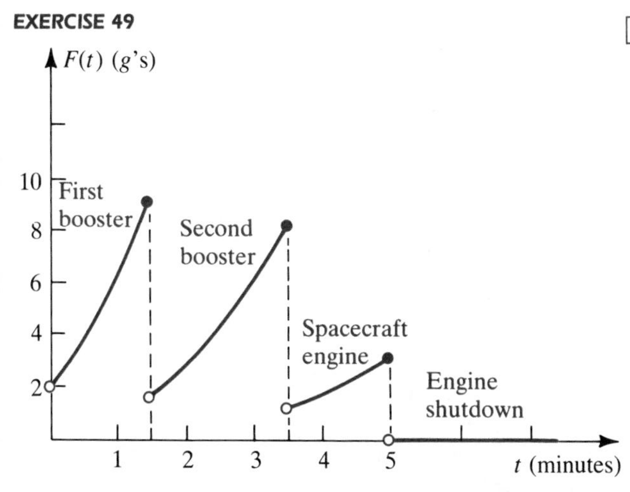

50 A hospital patient receives an initial 200-milligram dose of a drug. Additional doses of 100 milligrams each are then administered every 4 hours. The amount $f(t)$ of the drug present in the bloodstream after t hours is shown in the figure. Find and interpret $\lim_{t \to 8^-} f(t)$ and $\lim_{t \to 8^+} f(t)$.

EXERCISE 50

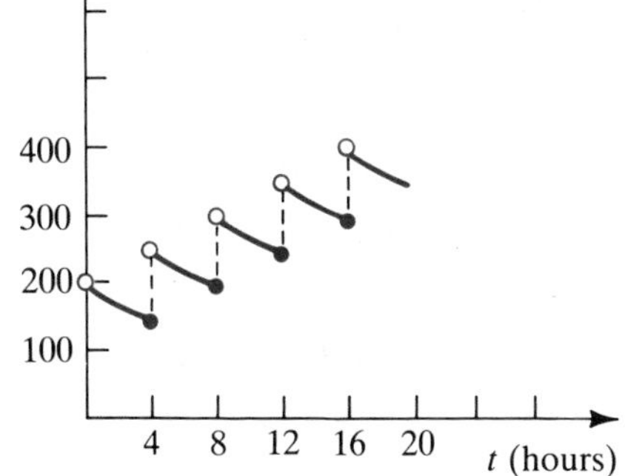

c **Exer. 51–56: The stated limit may be verified by methods developed later in the text. Lend support to the result by substituting appropriate real numbers for x. Why does this fail to *prove* that the limit exists?**

51 $\lim_{x \to 0} (1 + x)^{1/x} \approx 2.72$

52 $\lim_{x \to 0} (1 + 2x)^{3/x} \approx 403.4$

53 $\lim_{x \to 2} \dfrac{3^x - 9}{x - 2} \approx 9.89$

54 $\lim_{x \to 1} \dfrac{2^x - 2}{x - 1} \approx 1.39$

55 $\lim_{x \to 0} \left(\dfrac{4^{|x|} + 9^{|x|}}{2} \right)^{1/|x|} = 6$

56 $\lim_{x \to 0} |x|^x = 1$

c **57** **(a)** If $f(x) = \cos(1/x) - \sin(1/x)$, investigate $\lim_{x \to 0} f(x)$ by first letting $x = 3.1830989 \times 10^{-n}$ for $n = 2, 3, 4$ and then letting $x = 3 \times 10^{-n}$ for $n = 2, 3, 4$.

(b) What is the limit in (a)?

c **58** **(a)** If $f(x) = x^{1/100} - 0.933$, investigate $\lim_{x \to 0^+} f(x)$ by letting $x = 10^{-n}$ for $n = 1, 2, 3$.

(b) What *appears* to be the limit in (a), and what is the *actual* limit?

2.2 DEFINITION OF LIMIT

The precise meaning of limit of a function is stated in Definition (2.4) of this section. Because the definition is rather abstract, let us begin with a physical illustration that may make it easier to understand. Scientists often investigate the manner in which quantities vary and whether they approach specific values under certain conditions. The apparatus illustrated in Figure 2.10(i) is used to study the flow of liquids or gases. It consists of a *Venturi tube* (a cylindrical pipe with a narrow constriction) and two meters that measure the pressure of a flowing liquid or gas in the nonconstricted and constricted parts of the tube. (Other types of meters may measure the speed at which the liquid or gas flows through the tube.) Let

FIGURE 2.10

(i)

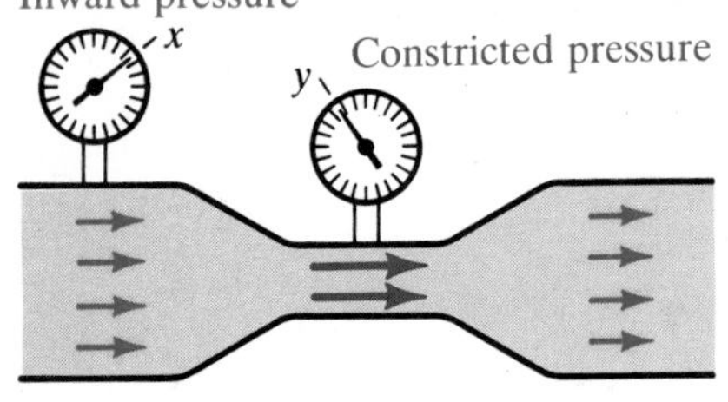

(ii)

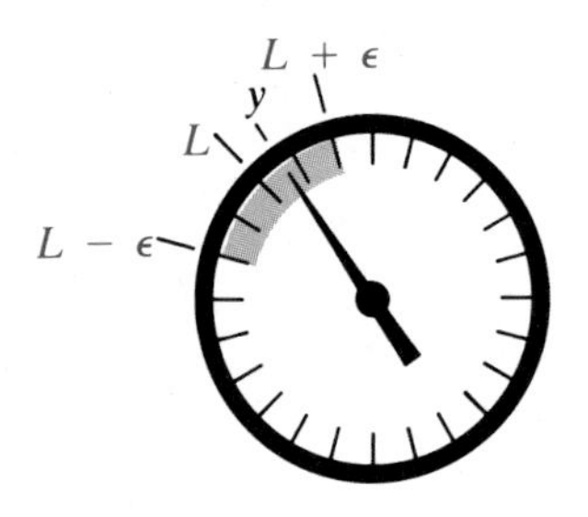

us suppose that a liquid enters the tube from the left, with a certain velocity. (The notion of velocity is defined precisely in Chapter 3.) The inward pressure meter displays a measurement x of the pressure in the nonconstricted part of the tube. As the liquid passes through the constriction it speeds up, and the pressure decreases to a value y, as indicated by the constricted pressure meter. Let us fix our attention on the two meters, as illustrated in Figure 2.10(ii). We will use these meters to give a precise meaning to the statement *y approaches L as x approaches a*, or, symbolically,

$$\lim_{x \to a} y = L.$$

When using this laboratory equipment, we would not expect the pressure y to remain *exactly* at L over a long period of time. Instead, our goal might be to force y to remain *very close* to L by restricting x to values *near a*. In particular, if ϵ (epsilon) denotes a small positive real number, let us suppose it is sufficient that

$$L - \epsilon < y < L + \epsilon,$$

as indicated on the constricted pressure meter in Figure 2.10(ii). An equivalent statement using absolute values is

$$|y - L| < \epsilon.$$

If these inequalities are true, we say that **y has ϵ-tolerance at L**. For example, the statement *y has 0.01-tolerance at L* means $|y - L| < 0.01$; that is, y is within 0.01 units of L. This tolerance may be sufficiently accurate for experimental purposes.

Similarly, let us consider a small positive number δ (delta) and define *δ-tolerance* at a on the inward pressure meter in Figure 2.10(ii). In our later work with functions it will be important that $x \neq a$. Anticipating this restriction, we say that **x has δ-tolerance at a** if

$$0 < |x - a| < \delta,$$

or, equivalently, if

$$a - \delta < x < a + \delta \quad \text{and} \quad x \neq a.$$

Let us now consider the following question.

Question: Given any $\epsilon > 0$, is there a $\delta > 0$ such that if x has δ-tolerance at a, then y has ϵ-tolerance at L?

If the answer to this question is *yes*, we write

$$\lim_{x \to a} y = L.$$

It is important to note that if $\lim_{x \to a} y = L$, then *no matter how small the number ϵ, we can always find a $\delta > 0$* such that if x is restricted to the interval $(a - \delta, a + \delta)$ on the inward pressure meter (and $x \neq a$), then y will lie in the interval $(L - \epsilon, L + \epsilon)$ on the constricted pressure meter.

This is a more precise interpretation of the limit concept than that used in Section 2.1, where we used words such as *closer* and *approaches*. If we rephrase the last question and its answer in terms of inequalities, we obtain the following statement:

$$\lim_{x \to a} y = L$$

means that for every $\epsilon > 0$, there is a $\delta > 0$ such that

$$\text{if} \quad 0 < |x - a| < \delta, \quad \text{then} \quad |y - L| < \epsilon$$

It is now a small step to formulate the definition of a limit of a function f. All that is required is to let $y = f(x)$ in the preceding discussion. This gives us the following definition, which also states the conditions required for the function f.

Definition of limit of a function **(2.4)**

> Let a function f be defined on an open interval containing a, except possibly at a itself, and let L be a real number. The statement
>
> $$\lim_{x \to a} f(x) = L$$
>
> means that for every $\epsilon > 0$, there is a $\delta > 0$ such that
>
> $$\text{if} \quad 0 < |x - a| < \delta, \quad \text{then} \quad |f(x) - L| < \epsilon.$$

FIGURE 2.11

(i) $0 < |x - a| < \delta$

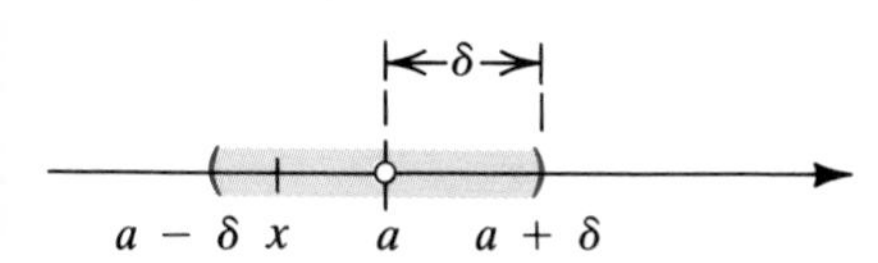

(ii) $|f(x) - L| < \epsilon$

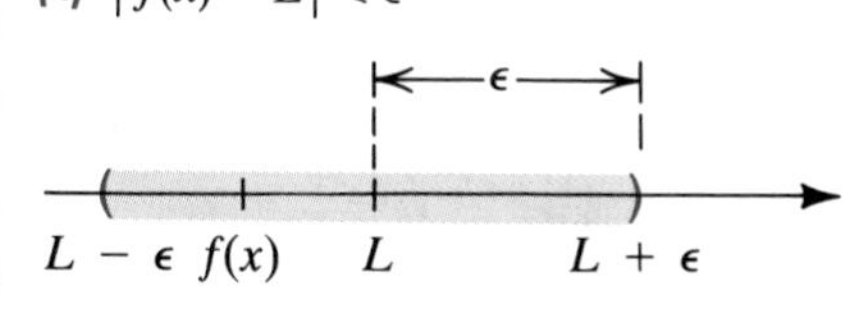

We sometimes call the inequality $0 < |x - a| < \delta$ a *δ-tolerance statement* and the inequality $|f(x) - L| < \epsilon$ an *ϵ-tolerance statement*.

If we wish to use a form of Definition (2.4) that does not contain absolute value symbols, we note that

(i) $0 < |x - a| < \delta$ is equivalent to $a - \delta < x < a + \delta$ and $x \neq a$

(ii) $|f(x) - L| < \epsilon$ is equivalent to $L - \epsilon < f(x) < L + \epsilon$

Inequalities (i) and (ii) are represented graphically on real lines in Figure 2.11. We may restate Definition (2.4) as follows.

Alternative definition of limit **(2.5)**

> $$\lim_{x \to a} f(x) = L$$
>
> means that for every $\epsilon > 0$, there is a $\delta > 0$ such that if x is in the open interval $(a - \delta, a + \delta)$ and $x \neq a$, then $f(x)$ is in the open interval $(L - \epsilon, L + \epsilon)$.

If $f(x)$ *has a limit as x approaches a, then that limit is unique*. A proof of this fact is given at the beginning of Appendix II.

In using either Definition (2.4) or (2.5) to show that $\lim_{x \to a} f(x) = L$, *it is very important to remember the order in which we consider the numbers* ϵ *and* δ. Always use the following two steps:

Step 1 Consider *any* $\epsilon > 0$.

Step 2 Show that there is a $\delta > 0$ such that if x has δ-tolerance at a, then $f(x)$ has ϵ-tolerance at L.

The number δ in the limit definitions is not unique, for if a specific δ can be found, then any *smaller* positive number δ_1 will also satisfy the requirements.

Before considering examples, let us rephrase the preceding discussion in terms of the graph of the function f. In particular, for $\epsilon > 0$ and $\delta > 0$ we have the following graphical interpretations of tolerances, where $P(x, f(x))$ denotes a point on the graph of f.

Tolerance statement	Graphical interpretation
$f(x)$ has ϵ-tolerance at L.	$P(x, f(x))$ lies between the horizontal lines $y = L \pm \epsilon$.
x has δ-tolerance at a.	x is in the interval $(a - \delta, a + \delta)$ on the x-axis, and $x \neq a$.

The two steps in showing that $\lim_{x \to a} f(x) = L$ may now be interpreted graphically as follows:

Step 1 For any $\epsilon > 0$, consider the horizontal lines $y = L \pm \epsilon$ (shown in Figure 2.12).

Step 2 Show that there is a $\delta > 0$ such that if x is in the open interval $(a - \delta, a + \delta)$ and $x \neq a$, then $P(x, f(x))$ lies between the horizontal lines $y = L \pm \epsilon$ (that is, inside the shaded rectangular region shown in Figure 2.13).

FIGURE 2.12

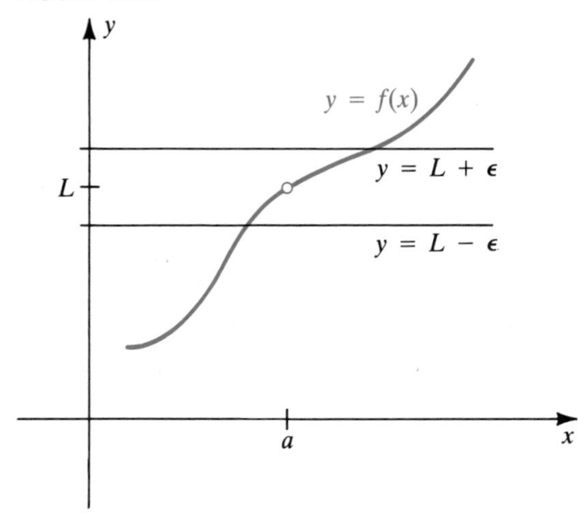

FIGURE 2.13

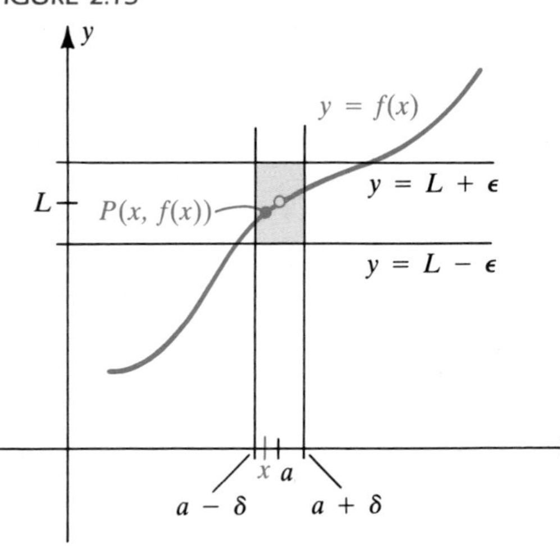

EXAMPLE 1 Use Definition (2.4) to prove that $\lim_{x \to 4} (3x - 5) = 7$.

SOLUTION If, in Definition (2.4), we let $f(x) = 3x - 5$, $a = 4$, and $L = 7$, then we must show that given any $\epsilon > 0$, we can find a $\delta > 0$ such that

$$(*) \qquad \text{if} \quad 0 < |x - 4| < \delta, \quad \text{then} \quad |(3x - 5) - 7| < \epsilon.$$

To solve inequality problems of this type we can often obtain a clue to a proper choice for δ by examining the ϵ-tolerance statement. Doing so leads to the following list of equivalent inequalities:

$$|(3x - 5) - 7| < \epsilon \quad (\epsilon\text{-tolerance statement})$$
$$|3x - 12| < \epsilon \quad (\text{simplifying})$$
$$|3(x - 4)| < \epsilon \quad (\text{common factor } 3)$$
$$3|x - 4| < \epsilon \quad (\text{properties of absolute value})$$
$$|x - 4| < \tfrac{1}{3}\epsilon \quad (\text{multiply by } \tfrac{1}{3})$$

The final inequality in the list gives us the needed clue. Specifically, we choose δ such that $\delta \le \frac{1}{3}\epsilon$ and obtain the following equivalent inequalities:

$$0 < |x - 4| < \delta \quad (\delta\text{-tolerance statement})$$
$$0 < |x - 4| < \tfrac{1}{3}\epsilon \quad (\text{choice of } \delta \le \tfrac{1}{3}\epsilon)$$
$$0 < 3|x - 4| < \epsilon \quad (\text{multiply by } 3)$$
$$0 < |3x - 12| < \epsilon \quad (\text{properties of absolute value})$$
$$0 < |(3x - 5) - 7| < \epsilon \quad (\text{equivalent form})$$

This verifies (*) and hence completes the proof.

FIGURE 2.14

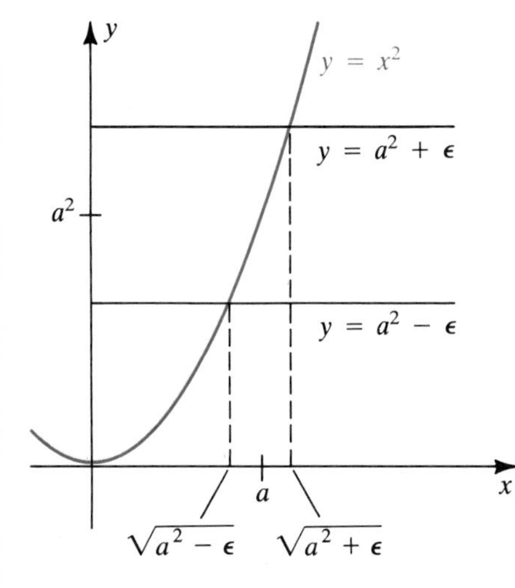

The next example illustrates how the geometric process shown in Figures 2.12 and 2.13 may be applied to a specific function.

EXAMPLE 2 Prove that $\lim_{x \to a} x^2 = a^2$.

SOLUTION Let us consider the case $a > 0$. We shall apply the alternative definition (2.5) with $f(x) = x^2$ and $L = a^2$. Thus, given any $\epsilon > 0$, we must find a $\delta > 0$ such that

(*) if x is in $(a - \delta, a + \delta)$ and $x \ne a$, then x^2 is in $(a^2 - \epsilon, a^2 + \epsilon)$.

FIGURE 2.15

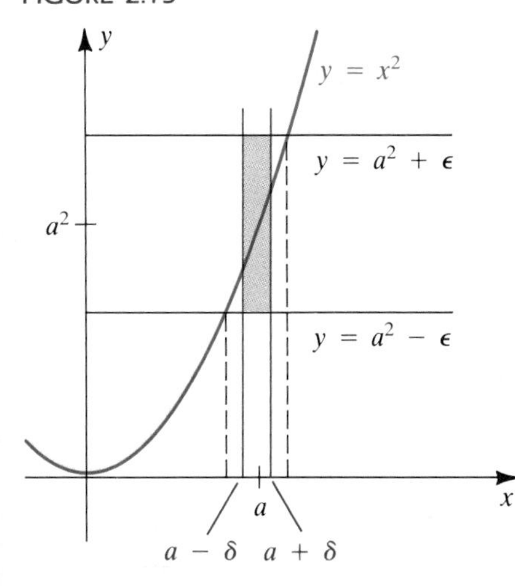

We can obtain a clue to a proper choice for δ by examining graphical interpretations of tolerance statements. Thus, as in step 1 on page 54, consider the horizontal lines $y = a^2 \pm \epsilon$. As shown in Figure 2.14, these lines intersect the graph of $y = x^2$ at points with x-coordinates $\sqrt{a^2 - \epsilon}$ and $\sqrt{a^2 + \epsilon}$. Note that if x is in the open interval $(\sqrt{a^2 - \epsilon}, \sqrt{a^2 + \epsilon})$, then the point (x, x^2) on the graph of f lies between the horizontal lines. If we choose a positive number δ *smaller* than both $\sqrt{a^2 + \epsilon} - a$ and $a - \sqrt{a^2 - \epsilon}$, with $a^2 - \epsilon > 0$, as illustrated in Figure 2.15, then when x has δ-tolerance at a, the point (x, x^2) lies between the horizontal lines $y = a^2 \pm \epsilon$ (that is, x^2 has ϵ-tolerance at a^2). This proves (*). Although we have considered only $a > 0$, a similar argument applies if $a \le 0$.

The next two examples, which were also discussed in Section 2.1, indicate how the geometric process illustrated in Figure 2.13 may be used to show that certain limits do *not* exist.

EXAMPLE 3 Show that $\lim_{x\to 0} \dfrac{1}{x}$ does not exist.

SOLUTION Let us proceed in an indirect manner. Thus, *suppose* that

$$\lim_{x\to 0} \frac{1}{x} = L$$

for some number L. Consider any pair of horizontal lines $y = L \pm \epsilon$, as illustrated in Figure 2.16. Since we are assuming that the limit exists, it should be possible to find an open interval $(0 - \delta, 0 + \delta)$, or, equivalently, $(-\delta, \delta)$, such that if $-\delta < x < \delta$ and $x \neq 0$, then the point $(x, 1/x)$ on the graph lies between the horizontal lines. However, since $|1/x|$ can be made as large as desired by choosing x close to 0, some points on the graph will lie either above or below the lines. Hence our supposition is false; that is, $\lim_{x\to 0} (1/x) \neq L$ for any real number L. Thus, the limit does not exist.

FIGURE 2.16
$f(x) = \dfrac{1}{x}$

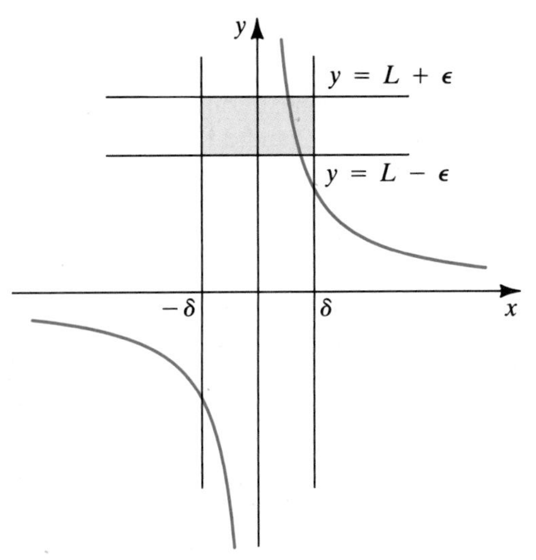

FIGURE 2.17
$f(x) = \dfrac{|x|}{x}$

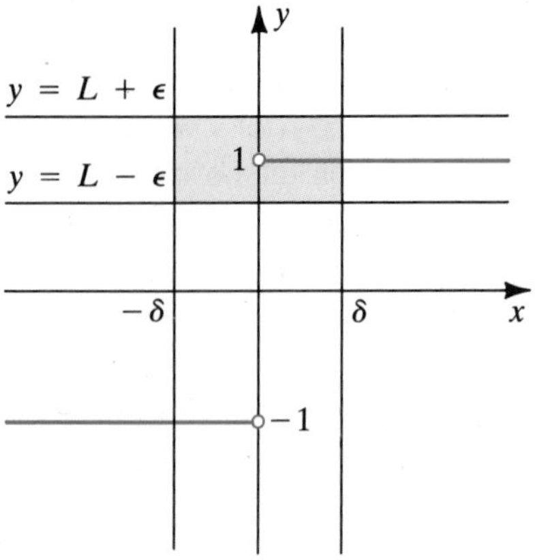

EXAMPLE 4 If $f(x) = \dfrac{|x|}{x}$, show that $\lim_{x\to 0} f(x)$ does not exist.

SOLUTION The graph of f is sketched in Figure 2.17. If we consider any pair of horizontal lines $y = L \pm \epsilon$, with $0 < \epsilon < 1$, then there are always some points on the graph that do not lie between these lines. In the figure we have illustrated a case for $L > 0$; however, our proof is valid for every L. Since we cannot find a $\delta > 0$ such that step 2 on page 54 is true, the limit does not exist.

The following theorem states that *if a function f has a positive limit as x approaches a, then $f(x)$ is positive throughout some open interval containing a, with the possible exception of a.*

Theorem (2.6)

> If $\lim_{x \to a} f(x) = L$ and $L > 0$, then there is an open interval $(a - \delta, a + \delta)$ containing a such that $f(x) > 0$ for every x in $(a - \delta, a + \delta)$, except possibly $x = a$.

PROOF If $L > 0$ and we let $\epsilon = \frac{1}{2}L$, then the horizontal lines $y = L \pm \epsilon$ are above the x-axis, as illustrated in Figure 2.18. By Definition (2.5) there is a $\delta > 0$ such that if $a - \delta < x < a + \delta$ and $x \neq a$, then $L - \epsilon < f(x) < L + \epsilon$. Since $f(x) > L - \epsilon$ and $L - \epsilon > 0$, it follows that $f(x) > 0$ for these values of x. ■

FIGURE 2.18

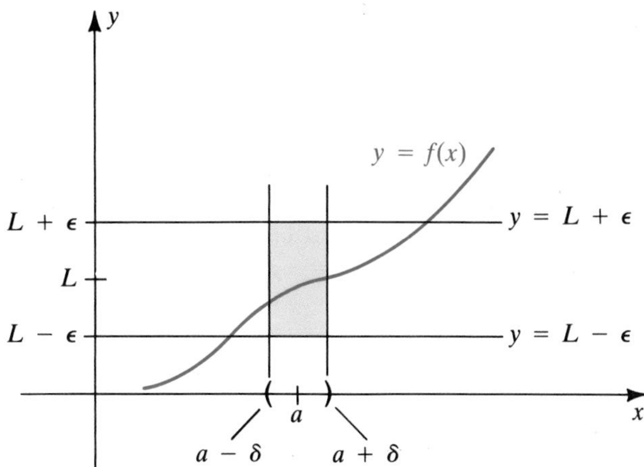

We can also prove that *if f has a negative limit as x approaches a, then there is an open interval containing a such that $f(x) < 0$ for every x in the interval, with the possible exception of $x = a$.*

Formal definitions can be given for one-sided limits. For the right-hand limit $x \to a^+$, we replace the condition $0 < |x - a| < \delta$ in Definition (2.4) by $a < x < a + \delta$. In terms of the alternative definition (2.5), we restrict x to the *right* half $(a, a + \delta)$ of the interval $(a - \delta, a + \delta)$. Similarly, for the left-hand limit $x \to a^-$, we replace $0 < |x - a| < \delta$ in (2.4) by $a - \delta < x < a$. This is equivalent to restricting x to the *left* half $(a - \delta, a)$ of the interval $(a - \delta, a + \delta)$ in (2.5).

EXERCISES 2.2

Exer. 1–2: Express the limit statement in the form of (a) Definition (2.4) and (b) Alternative Definition (2.5).

1 $\lim_{t \to c} v(t) = K$

2 $\lim_{t \to b} f(t) = M$

Exer. 3–6: Express the one-sided limit statement in a form similar to (a) Definition (2.4) and (b) Alternative Definition (2.5).

3 $\lim_{x \to p^-} g(x) = C$

4 $\lim_{z \to a^-} h(z) = L$

5 $\lim_{z\to t^+} f(z) = N$

6 $\lim_{x\to c^+} s(x) = D$

Exer. 7–12: For the given $\lim_{x\to a} f(x) = L$ and ϵ, use the graph of f to find the largest δ such that if $0 < |x - a| < \delta$, then $|f(x) - L| < \epsilon$.

7 $\lim_{x\to 3/2} \frac{4x^2 - 9}{2x - 3} = 6$; $\epsilon = 0.01$

8 $\lim_{x\to -2/3} \frac{9x^2 - 4}{3x + 2} = -4$; $\epsilon = 0.1$

9 $\lim_{x\to 4} x^2 = 16$; $\epsilon = 0.1$

10 $\lim_{x\to 3} x^3 = 27$; $\epsilon = 0.01$

11 $\lim_{x\to 16} \sqrt{x} = 4$; $\epsilon = 0.1$

12 $\lim_{x\to 27} \sqrt[3]{x} = 3$; $\epsilon = 0.1$

Exer. 13–24: Use Definition (2.4) to prove that the limit exists.

13 $\lim_{x\to 3} 5x = 15$

14 $\lim_{x\to 5} (-4x) = -20$

15 $\lim_{x\to -3} (2x + 1) = -5$

16 $\lim_{x\to 2} (5x - 3) = 7$

17 $\lim_{x\to -6} (10 - 9x) = 64$

18 $\lim_{x\to 4} (15 - 8x) = -17$

19 $\lim_{x\to 6} (3 - \frac{1}{2}x) = 0$

20 $\lim_{x\to 6} (9 - \frac{1}{6}x) = 8$

21 $\lim_{x\to 3} 5 = 5$

22 $\lim_{x\to 5} 3 = 3$

23 $\lim_{x\to a} c = c$ for all real numbers a and c

24 $\lim_{x\to a} (mx + b) = ma + b$ for all real numbers m, b, and a

Exer. 25–30: Use the graphical method illustrated in Example 2 to verify the limit for $a > 0$.

25 $\lim_{x\to -a} x^2 = a^2$

26 $\lim_{x\to a} (x^2 + 1) = a^2 + 1$

27 $\lim_{x\to a} x^3 = a^3$

28 $\lim_{x\to a} x^4 = a^4$

29 $\lim_{x\to a} \sqrt{x} = \sqrt{a}$

30 $\lim_{x\to a} \sqrt[3]{x} = \sqrt[3]{a}$

Exer. 31–38: Use the method illustrated in Examples 3 and 4 to show that the limit does not exist.

31 $\lim_{x\to 3} \frac{|x - 3|}{x - 3}$

32 $\lim_{x\to -2} \frac{x + 2}{|x + 2|}$

33 $\lim_{x\to -1} \frac{3x + 3}{|x + 1|}$

34 $\lim_{x\to 5} \frac{2x - 10}{|x - 5|}$

35 $\lim_{x\to 0} \frac{1}{x^2}$

36 $\lim_{x\to 4} \frac{7}{x - 4}$

37 $\lim_{x\to -5} \frac{1}{x + 5}$

38 $\lim_{x\to 1} \frac{1}{(x - 1)^2}$

39 Give an example of a function f that is defined at a such that $\lim_{x\to a} f(x)$ exists and $\lim_{x\to a} f(x) \neq f(a)$.

40 If f is the greatest integer function (see Example 4 of Section 1.2) and a is any integer, show that $\lim_{x\to a} f(x)$ does not exist.

41 Let f be defined as follows:

$$f(x) = \begin{cases} 0 & \text{if } x \text{ is rational} \\ 1 & \text{if } x \text{ is irrational} \end{cases}$$

Prove that for every real number a, $\lim_{x\to a} f(x)$ does not exist.

42 Why is it impossible to investigate $\lim_{x\to 0} \sqrt{x}$ by means of Definition (2.4)?

2.3 TECHNIQUES FOR FINDING LIMITS

It would be an excruciating task to verify every limit by means of Definition (2.4) or (2.5). The purpose of this section is to introduce theorems that may be used to simplify problems involving limits. Before stating the first theorem, let us consider the limits of two very simple functions:

(i) The constant function f given by $f(x) = c$

(ii) The linear function g given by $g(x) = x$

The graph of f is the horizontal line $y = c$ sketched in Figure 2.19 for the case $c > 0$. Since

$$|f(x) - c| = |c - c| = 0 \quad \text{for every } x$$

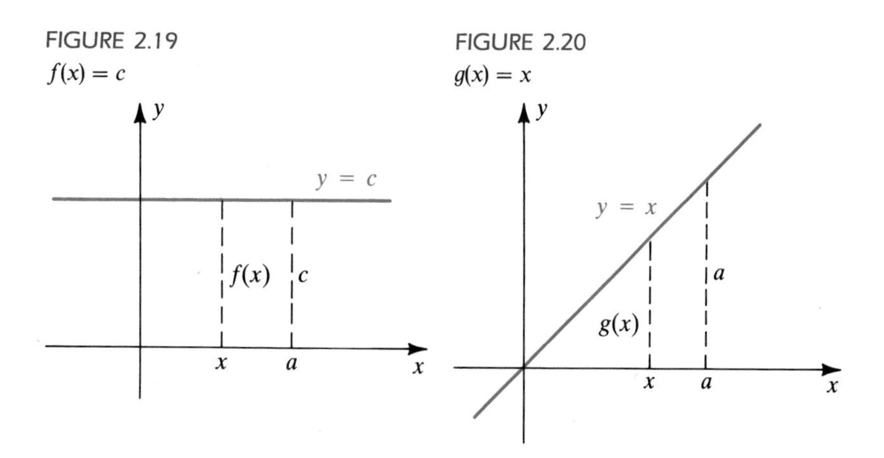

FIGURE 2.19
$f(x) = c$

FIGURE 2.20
$g(x) = x$

and since 0 is less than any $\epsilon > 0$, it follows from Definition (2.4) that $f(x)$ has the limit c as x approaches a. Thus,

$$\lim_{x \to a} f(x) = \lim_{x \to a} c = c.$$

This limit is often described by the phrase *the limit of a constant is the constant.*

The graph of the linear function g given in (ii) is sketched in Figure 2.20. As x approaches a, $g(x)$ approaches a; that is,

$$\lim_{x \to a} g(x) = \lim_{x \to a} x = a.$$

The preceding limit can also be established by means of Definition (2.4). We state these facts for reference in the next theorem.

Theorem **(2.7)**

(i) $\lim_{x \to a} c = c$

(ii) $\lim_{x \to a} x = a$

As we shall see, the limits in Theorem (2.7) can be used as building blocks for finding limits of very complicated expressions.

ILLUSTRATION

- $\lim_{x \to 3} 8 = 8$
- $\lim_{x \to 8} 3 = 3$
- $\lim_{x \to \sqrt{2}} x = \sqrt{2}$
- $\lim_{x \to -4} x = -4$

Many functions can be expressed as sums, differences, products, and quotients of other functions. Suppose f and g are functions and L and M are real numbers. If

$$f(x) \to L \quad \text{and} \quad g(x) \to M \quad \text{as} \quad x \to a,$$

we would expect that

$$f(x) + g(x) \to L + M \quad \text{as} \quad x \to a.$$

The next theorem states that this expectation is true and gives analogous results for products and quotients.

Theorem **(2.8)**

If $\lim_{x \to a} f(x)$ and $\lim_{x \to a} g(x)$ both exist, then

(i) $\lim_{x \to a} [f(x) + g(x)] = \lim_{x \to a} f(x) + \lim_{x \to a} g(x)$

(ii) $\lim_{x \to a} [f(x) \cdot g(x)] = \lim_{x \to a} f(x) \cdot \lim_{x \to a} g(x)$

(iii) $\lim_{x \to a} \left[\frac{f(x)}{g(x)}\right] = \frac{\lim_{x \to a} f(x)}{\lim_{x \to a} g(x)}$, provided $\lim_{x \to a} g(x) \neq 0$

(iv) $\lim_{x \to a} [cf(x)] = c\left[\lim_{x \to a} f(x)\right]$

(v) $\lim_{x \to a} [f(x) - g(x)] = \lim_{x \to a} f(x) - \lim_{x \to a} g(x)$

We may state the properties in Theorem (2.8) as follows:

(i) The limit of a sum is the sum of the limits.

(ii) The limit of a product is the product of the limits.

(iii) The limit of a quotient is the quotient of the limits, provided the denominator has a nonzero limit.

(iv) The limit of a constant times a function is the constant times the limit of the function.

(v) The limit of a difference is the difference of the limits.

Proofs for (i)–(iii), based on Definition (2.4), are given in Appendix II. Part (iv) of the theorem follows readily from part (ii) and from Theorem (2.7)(i):

$$\lim_{x \to a} [cf(x)] = \left[\lim_{x \to a} c\right]\left[\lim_{x \to a} f(x)\right] = c\left[\lim_{x \to a} f(x)\right]$$

To prove (v) we may write

$$f(x) - g(x) = f(x) + (-1)g(x)$$

and then use parts (i) and (iv) (with $c = -1$).

We shall use the preceding theorems to establish the following.

Theorem **(2.9)**

If m, b, and a are real numbers, then

$$\lim_{x \to a} (mx + b) = ma + b.$$

PROOF By Theorem (2.7),

$$\lim_{x \to a} x = a \quad \text{and} \quad \lim_{x \to a} b = b.$$

We next use (i) and (iv) of Theorem (2.8):

$$\begin{aligned}\lim_{x \to a} (mx + b) &= \lim_{x \to a} (mx) + \lim_{x \to a} b \\ &= m\left(\lim_{x \to a} x\right) + b \\ &= ma + b \quad \blacksquare\end{aligned}$$

This result can also be proved directly from Definition (2.4).

ILLUSTRATION

- $\lim_{x \to -2} (5x + 2) = 5(-2) + 2 = -8$
- $\lim_{x \to 6} (4x - 11) = 4(6) - 11 = 13$

It is easy to find the limit in the next example by means of Theorems (2.8) and (2.9). To obtain a better appreciation of the power of these theorems, you could try to verify the limit by using only Definition (2.4).

EXAMPLE 1 Find $\lim_{x \to 2} \dfrac{3x + 4}{5x + 7}$.

SOLUTION From Theorem (2.9) we know that the limits of the numerator and denominator exist. Morever, the limit of the denominator is not 0. Hence, by Theorem (2.8)(iii) and Theorem (2.9),

$$\lim_{x \to 2} \frac{3x + 4}{5x + 7} = \frac{\lim_{x \to 2} (3x + 4)}{\lim_{x \to 2} (5x + 7)} = \frac{3(2) + 4}{5(2) + 7} = \frac{10}{17}.$$

Theorem (2.8) can be extended to limits of sums, differences, products, and quotients that involve any number of functions. In the next example we use part (ii) for a product of three (equal) functions.

EXAMPLE 2 Prove that $\lim_{x \to a} x^3 = a^3$.

SOLUTION Since $\lim_{x \to a} x = a$,

$$\begin{aligned}\lim_{x \to a} x^3 &= \lim_{x \to a} (x \cdot x \cdot x) \\ &= \left(\lim_{x \to a} x\right) \cdot \left(\lim_{x \to a} x\right) \cdot \left(\lim_{x \to a} x\right) \\ &= a \cdot a \cdot a = a^3.\end{aligned}$$

The method used in Example 2 can be extended to x^n for any positive integer n. We merely write x^n as a product $x \cdot x \cdot \cdots \cdot x$ of n factors and then take the limit of each factor. This gives us (i) of the next theorem. Part (ii) may be proved in similar fashion by using Theorem (2.8)(ii). Another method of proof is to use mathematical induction (see Appendix I).

Theorem (2.10)

If n is a positive integer, then

(i) $\lim_{x \to a} x^n = a^n$

(ii) $\lim_{x \to a} \left[f(x) \right]^n = \left[\lim_{x \to a} f(x) \right]^n$, provided $\lim_{x \to a} f(x)$ exists

EXAMPLE 3 Find $\lim_{x \to 2} (3x + 4)^5$.

SOLUTION Applying Theorems (2.10)(ii) and (2.9), we have

$$\begin{aligned} \lim_{x \to 2} (3x + 4)^5 &= \left[\lim_{x \to 2} (3x + 4) \right]^5 \\ &= [3(2) + 4]^5 \\ &= 10^5 = 100{,}000. \end{aligned}$$

EXAMPLE 4 Find $\lim_{x \to -2} (5x^3 + 3x^2 - 6)$.

SOLUTION We may proceed as follows (supply reasons):

$$\begin{aligned} \lim_{x \to -2} (5x^3 + 3x^2 - 6) &= \lim_{x \to -2} (5x^3) + \lim_{x \to -2} (3x^2) - \lim_{x \to -2} (6) \\ &= 5 \lim_{x \to -2} (x^3) + 3 \lim_{x \to -2} (x^2) - 6 \\ &= 5(-2)^3 + 3(-2)^2 - 6 \\ &= 5(-8) + 3(4) - 6 = -34. \end{aligned}$$

The limit in Example 4 is the number obtained by substituting -2 for x in $5x^3 + 3x^2 - 6$. The next theorem states that the same is true for the limit of *every* polynomial.

Theorem (2.11)

If f is a polynomial function and a is a real number, then

$$\lim_{x \to a} f(x) = f(a).$$

PROOF Since f is a polynomial function,

$$f(x) = b_n x^n + b_{n-1} x^{n-1} + \cdots + b_0$$

for real numbers $b_n, b_{n-1}, \ldots, b_0$. As in Example 4,

$$\begin{aligned}\lim_{x\to a} f(x) &= \lim_{x\to a} (b_n x^n) + \lim_{x\to a} (b_{n-1}x^{n-1}) + \cdots + \lim_{x\to a} b_0 \\ &= b_n \lim_{x\to a} (x^n) + b_{n-1} \lim_{x\to a} (x^{n-1}) + \cdots + \lim_{x\to a} b_0 \\ &= b_n a^n + b_{n-1}a^{n-1} + \cdots + b_0 = f(a).\end{aligned}$$

Corollary (2.12)

If q is a rational function and a is in the domain of q, then

$$\lim_{x\to a} q(x) = q(a).$$

PROOF Since q is a rational function, $q(x) = f(x)/h(x)$, where f and h are polynomial functions. If a is in the domain of q, then $h(a) \neq 0$. Using Theorems (2.8)(iii) and (2.11) gives us

$$\lim_{x\to a} q(x) = \frac{\lim_{x\to a} f(x)}{\lim_{x\to a} h(x)} = \frac{f(a)}{h(a)} = q(a).$$

EXAMPLE 5 Find $\displaystyle\lim_{x\to 3} \frac{5x^2 - 2x + 1}{4x^3 - 7}$.

SOLUTION Applying Corollary (2.12) yields

$$\begin{aligned}\lim_{x\to 3} \frac{5x^2 - 2x + 1}{4x^3 - 7} &= \frac{5(3)^2 - 2(3) + 1}{4(3)^3 - 7} \\ &= \frac{45 - 6 + 1}{108 - 7} = \frac{40}{101}.\end{aligned}$$

The next theorem states that for positive integral roots of x, we may determine a limit by substitution. A proof, using Definition (2.4), may be found in Appendix II.

Theorem (2.13)

If $a > 0$ and n is a positive integer, or if $a \le 0$ and n is an odd positive integer, then

$$\lim_{x\to a} \sqrt[n]{x} = \sqrt[n]{a}.$$

If m and n are positive integers and $a > 0$, then using Theorems (2.10)(ii) and (2.13) gives us

$$\lim_{x\to a} \left(\sqrt[n]{x}\right)^m = \left(\lim_{x\to a} \sqrt[n]{x}\right)^m = \left(\sqrt[n]{a}\right)^m.$$

In terms of rational exponents,

$$\lim_{x\to a} x^{m/n} = a^{m/n}.$$

This limit formula may be extended to negative exponents by writing $x^{-r} = 1/x^r$ and then using Theorem (2.8)(iii).

EXAMPLE 6 Find $\lim_{x \to 8} \dfrac{x^{2/3} + 3\sqrt{x}}{4 - (16/x)}$.

SOLUTION We may proceed as follows (supply reasons):

$$\begin{aligned}\lim_{x \to 8} \frac{x^{2/3} + 3\sqrt{x}}{4 - (16/x)} &= \frac{\lim_{x \to 8} (x^{2/3} + 3\sqrt{x})}{\lim_{x \to 8} [4 - (16/x)]} \\ &= \frac{\lim_{x \to 8} x^{2/3} + \lim_{x \to 8} 3\sqrt{x}}{\lim_{x \to 8} 4 - \lim_{x \to 8} (16/x)} \\ &= \frac{8^{2/3} + 3\sqrt{8}}{4 - (16/8)} \\ &= \frac{4 + 6\sqrt{2}}{4 - 2} = 2 + 3\sqrt{2}\end{aligned}$$

Theorem **(2.14)**

If a function f has a limit as x approaches a, then

$$\lim_{x \to a} \sqrt[n]{f(x)} = \sqrt[n]{\lim_{x \to a} f(x)},$$

provided either n is an odd positive integer or n is an even positive integer and $\lim_{x \to a} f(x) > 0$.

The preceding theorem will be proved in Section 2.5. In the meantime we shall use it whenever applicable to gain experience in finding limits that involve roots of algebraic expressions.

EXAMPLE 7 Find $\lim_{x \to 5} \sqrt[3]{3x^2 - 4x + 9}$.

SOLUTION Using Theorems (2.14) and (2.11), we obtain

$$\begin{aligned}\lim_{x \to 5} \sqrt[3]{3x^2 - 4x + 9} &= \sqrt[3]{\lim_{x \to 5} (3x^2 - 4x + 9)} \\ &= \sqrt[3]{75 - 20 + 9} = \sqrt[3]{64} = 4.\end{aligned}$$

The next theorem concerns three functions f, h, and g such that $h(x)$ is "sandwiched" between $f(x)$ and $g(x)$. If f and g have a common limit L as x approaches a, then, as stated in the theorem, h must have the same limit.

Sandwich theorem **(2.15)**

Suppose $f(x) \le h(x) \le g(x)$ for every x in an open interval containing a, except possibly at a.

If $\lim_{x \to a} f(x) = L = \lim_{x \to a} g(x)$, then $\lim_{x \to a} h(x) = L$.

If $f(x) \le h(x) \le g(x)$ for every x in an open interval containing x, then the graph of h lies between the graphs of f and g in that interval, as illus-

FIGURE 2.21

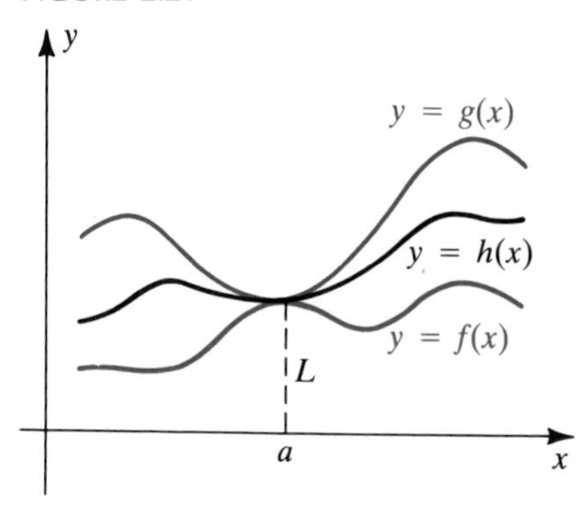

trated in Figure 2.21. If f and g have the same limit L as x approaches a, then it appears from the graphs that h also has the limit L. A proof of the sandwich theorem based on the definition of limit may be found in Appendix II.

FIGURE 2.22

EXAMPLE 8 Use the sandwich theorem to prove that

$$\lim_{x\to 0} x^2 \sin\frac{1}{x^2} = 0.$$

SOLUTION Since $-1 \le \sin t \le 1$ for every real number t,

$$-1 \le \sin\frac{1}{x^2} \le 1$$

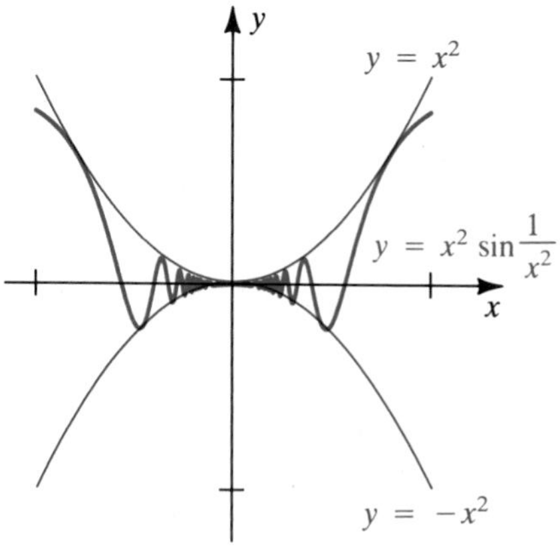

for every $x \ne 0$. Multiplying by x^2 (which is positive if $x \ne 0$), we obtain

$$-x^2 \le x^2 \sin\frac{1}{x^2} \le x^2.$$

This inequality implies that the graph of $y = x^2 \sin(1/x^2)$ lies between the parabolas $y = -x^2$ and $y = x^2$ (see Figure 2.22). Since

$$\lim_{x\to 0} (-x^2) = 0 \quad \text{and} \quad \lim_{x\to 0} x^2 = 0,$$

it follows from the sandwich theorem (2.15), with $f(x) = -x^2$ and $g(x) = x^2$, that

$$\lim_{x\to 0} x^2 \sin\frac{1}{x^2} = 0.$$

Theorems similar to the limit theorems in this section can be proved for one-sided limits. For example,

$$\lim_{x\to a^+} [f(x) + g(x)] = \lim_{x\to a^+} f(x) + \lim_{x\to a^+} g(x)$$

and

$$\lim_{x\to a^+} \sqrt[n]{f(x)} = \sqrt[n]{\lim_{x\to a^+} f(x)}$$

with the usual restrictions on the existence of limits and nth roots. Analogous results are true for left-hand limits.

FIGURE 2.23

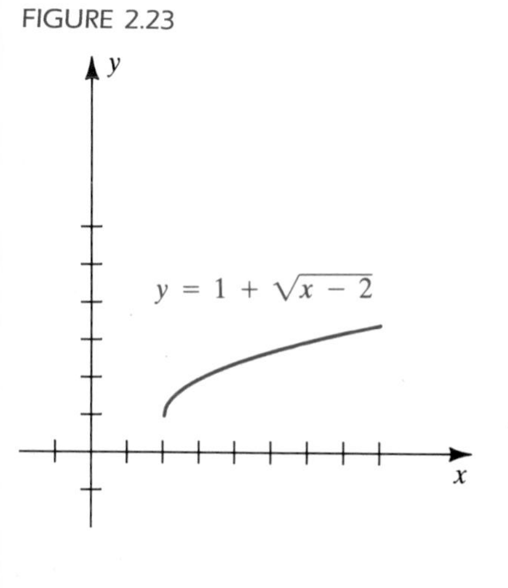

EXAMPLE 9 Find $\lim_{x\to 2^+} (1 + \sqrt{x-2})$.

SOLUTION The graph of $f(x) = 1 + \sqrt{x-2}$ is sketched in Figure 2.23. Using (one-sided) limit theorems, we obtain

$$\begin{aligned}\lim_{x\to 2^+} (1 + \sqrt{x-2}) &= \lim_{x\to 2^+} 1 + \lim_{x\to 2^+} \sqrt{x-2}\\ &= 1 + \sqrt{\lim_{x\to 2^+} (x-2)}\\ &= 1 + 0 = 1.\end{aligned}$$

Note that there is no left-hand limit, or limit as x approaches 2, since $\sqrt{x-2}$ is not a real number if $x < 2$.

FIGURE 2.24

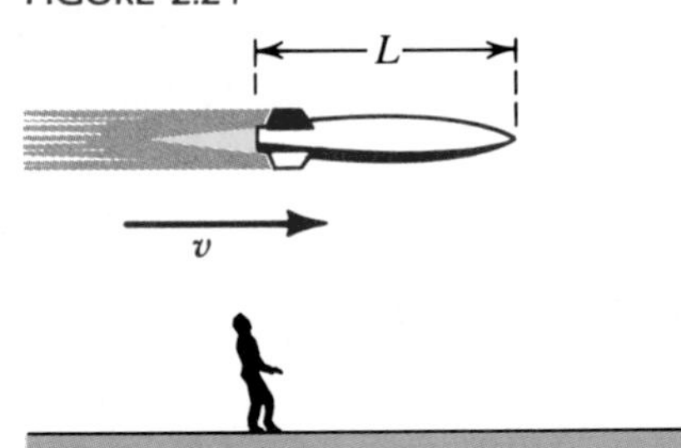

EXAMPLE 10 Let c denote the speed of light (approximately 3.0×10^8 m/sec, or 186,000 mi/sec). In Einstein's theory of relativity, the *Lorentz contraction formula*

$$L = L_0\sqrt{1 - \frac{v^2}{c^2}}$$

specifies the relationship between (1) the length L of an object that is moving at a velocity v with respect to an observer and (2) its length L_0 at rest (see Figure 2.24). The formula implies that the length of the object measured by the observer is shorter when it is moving than when it is at rest. Find and interpret $\lim_{v \to c^-} L$, and explain why a left-hand limit is necessary.

SOLUTION Using (one-sided) limit theorems yields

$$\begin{aligned}\lim_{v \to c^-} L &= \lim_{v \to c^-} L_0\sqrt{1 - \frac{v^2}{c^2}}\\ &= L_0 \lim_{v \to c^-} \sqrt{1 - \frac{v^2}{c^2}}\\ &= L_0\sqrt{\lim_{v \to c^-}\left(1 - \frac{v^2}{c^2}\right)}\\ &= L_0\sqrt{0} = 0.\end{aligned}$$

Thus, if the velocity of an object could approach the speed of light, then its length, as measured by an observer at rest, would approach zero. This result is sometimes used to help justify the theory that the speed of light is the ultimate speed in the universe; that is, no object can have a velocity that is greater than or equal to c.

A left-hand limit is necessary because if $v > c$, then $\sqrt{1 - (v^2/c^2)}$ is not a real number.

EXERCISES 2.3

Exer. 1–48: Use theorems on limits to find the limit, if it exists.

1 $\lim_{x \to \sqrt{2}} 15$

2 $\lim_{x \to 15} \sqrt{2}$

3 $\lim_{x \to -2} x$

4 $\lim_{x \to 3} x$

5 $\lim_{x \to 4} (3x - 4)$

6 $\lim_{x \to -2} (-3x + 1)$

7 $\lim_{x \to -2} \dfrac{x - 5}{4x + 3}$

8 $\lim_{x \to 4} \dfrac{2x - 1}{3x + 1}$

9 $\lim_{x \to 1} (-2x + 5)^4$

10 $\lim_{x \to -2} (3x - 1)^5$

11 $\lim_{x \to 3} (3x - 9)^{100}$

12 $\lim_{x \to 1/2} (4x - 1)^{50}$

13 $\lim_{x \to -2} (3x^3 - 2x + 7)$

14 $\lim_{x \to 4} (5x^2 - 9x - 8)$

15 $\lim_{x \to \sqrt{2}} (x^2 + 3)(x - 4)$

16 $\lim_{t \to -3} (3t + 4)(7t - 9)$

17 $\lim_{x \to \pi} (x - 3.1416)$

18 $\lim_{x \to \pi} (\frac{1}{2}x - \frac{11}{7})$

19 $\lim_{s \to 4} \dfrac{6s - 1}{2s - 9}$

20 $\lim_{x \to 1/2} \dfrac{4x^2 - 6x + 3}{16x^3 + 8x - 7}$

21 $\lim_{x \to 1/2} \dfrac{2x^2 + 5x - 3}{6x^2 - 7x + 2}$

22 $\lim_{x \to 2} \dfrac{x - 2}{x^3 - 8}$

23 $\lim_{x \to 2} \dfrac{x^2 - x - 2}{(x - 2)^2}$

24 $\lim_{x \to -2} \dfrac{x^2 + 2x - 3}{x^2 + 5x + 6}$

25 $\lim_{x \to -2} \dfrac{x^3 + 8}{x^4 - 16}$

26 $\lim_{x \to 16} \dfrac{x - 16}{\sqrt{x} - 4}$

27 $\lim_{x \to 2} \dfrac{(1/x) - (1/2)}{x - 2}$

28 $\lim_{x \to -3} \dfrac{x + 3}{(1/x) + (1/3)}$

29 $\lim_{x\to 1}\left(\frac{x^2}{x-1}-\frac{1}{x-1}\right)$

30 $\lim_{x\to 1}\left(\sqrt{x}+\frac{1}{\sqrt{x}}\right)^6$

31 $\lim_{x\to 16}\frac{2\sqrt{x}+x^{3/2}}{\sqrt[4]{x}+5}$

32 $\lim_{x\to -8}\frac{16x^{2/3}}{4-x^{4/3}}$

33 $\lim_{x\to 4}\sqrt[3]{x^2-5x-4}$

34 $\lim_{x\to -2}\sqrt{x^4-4x+1}$

35 $\lim_{x\to 3}\sqrt[3]{\frac{2+5x-3x^3}{x^2-1}}$

36 $\lim_{x\to \pi}\sqrt[5]{\frac{x-\pi}{x+\pi}}$

37 $\lim_{h\to 0}\frac{4-\sqrt{16+h}}{h}$

38 $\lim_{h\to 0}\left(\frac{1}{h}\right)\left(\frac{1}{\sqrt{1+h}}-1\right)$

39 $\lim_{x\to 1}\frac{x^2+x-2}{x^5-1}$

40 $\lim_{x\to 2}\frac{x^2-7x+10}{x^6-64}$

41 $\lim_{v\to 3}v^2(3v-4)(9-v^3)$

42 $\lim_{k\to 2}\sqrt{3k^2+4}\sqrt[3]{3k+2}$

43 $\lim_{x\to 5^+}(\sqrt{x^2-25}+3)$

44 $\lim_{x\to 3^-}x\sqrt{9-x^2}$

45 $\lim_{x\to 3^+}\frac{\sqrt{(x-3)^2}}{x-3}$

46 $\lim_{x\to -10^-}\frac{x+10}{\sqrt{(x+10)^2}}$

47 $\lim_{x\to 5^+}\frac{1+\sqrt{2x-10}}{x+3}$

48 $\lim_{x\to 4^+}\frac{\sqrt[4]{x^2-16}}{x+4}$

Exer. 49–52: Find each limit, if it exists:

(a) $\lim_{x\to a^-}f(x)$ **(b)** $\lim_{x\to a^+}f(x)$ **(c)** $\lim_{x\to a}f(x)$

49 $f(x)=\sqrt{5-x};\quad a=5$

50 $f(x)=\sqrt{8-x^3};\quad a=2$

51 $f(x)=\sqrt[3]{x^3-1};\quad a=1$

52 $f(x)=x^{2/3};\quad a=-8$

Exer. 53–56: Let n denote an arbitary integer. Sketch the graph of f and find $\lim_{x\to n^-}f(x)$ and $\lim_{x\to n^+}f(x)$.

53 $f(x)=(-1)^n$ if $n\le x<n+1$

54 $f(x)=n$ if $n\le x<n+1$

55 $f(x)=\begin{cases}x & \text{if } x=n\\ 0 & \text{if } x\ne n\end{cases}$

56 $f(x)=\begin{cases}0 & \text{if } x=n\\ 1 & \text{if } x\ne n\end{cases}$

Exer. 57–60: Let $[\![\ ]\!]$ denote the greatest integer function and n an arbitrary integer. Find

(a) $\lim_{x\to n^-}f(x)$ **(b)** $\lim_{x\to n^+}f(x)$

57 $f(x)=[\![x]\!]$

58 $f(x)=x-[\![x]\!]$

59 $f(x)=-[\![-x]\!]$

60 $f(x)=[\![x]\!]-x^2$

Exer. 61–64: Use the sandwich theorem to verify the limit.

61 $\lim_{x\to 0}(x^2+1)=1$ (*Hint:* Use $\lim_{x\to 0}(|x|+1)=1$.)

62 $\lim_{x\to 0}\frac{|x|}{\sqrt{x^4+4x^2+7}}=0$
(*Hint:* Use $f(x)=0$ and $g(x)=|x|$.)

63 $\lim_{x\to 0}x\sin(1/x)=0$
(*Hint:* Use $f(x)=-|x|$ and $g(x)=|x|$.)

64 $\lim_{x\to 0}x^4\sin(1/\sqrt[3]{x})=0$ (*Hint:* See Example 8.)

65 If $0\le f(x)\le c$ for some real number c, prove that $\lim_{x\to 0}x^2f(x)=0$.

66 If $\lim_{x\to a}f(x)=L\ne 0$ and $\lim_{x\to a}g(x)=0$, prove that $\lim_{x\to a}[f(x)/g(x)]$ does not exist. (*Hint:* Assume there is a number M such that $\lim_{x\to a}[f(x)/g(x)]=M$ and consider $\lim_{x\to a}f(x)=\lim_{x\to a}[g(x)\cdot f(x)/g(x)]$.)

67 Explain why $\lim_{x\to 0}\left(x\sin\frac{1}{x}\right)\ne\left(\lim_{x\to 0}x\right)\left(\lim_{x\to 0}\sin\frac{1}{x}\right)$.

68 Explain why $\lim_{x\to 0}\left(\frac{1}{x}+x\right)\ne\lim_{x\to 0}\frac{1}{x}+\lim_{x\to 0}x$.

69 Charles' law for gases states that if the pressure remains constant, then the relationship between the volume V that a gas occupies and its temperature T (in °C) is given by $V=V_0(1+\frac{1}{273}T)$. The temperature $T=-273$°C is *absolute zero.*

(a) Find $\lim_{T\to -273^+}V$.

(b) Why is a right-hand limit necessary?

70 According to the theory of relativity, the length of an object depends on its velocity v (see Example 10). Einstein also proved that the mass m of an object is related to v by the formula

$$m=\frac{m_0}{\sqrt{1-(v^2/c^2)}},$$

where m_0 is the mass of the object at rest.

(a) Investigate $\lim_{v\to c^-}m$.

(b) Why is a left-hand limit necessary?

71 A convex lens has focal length f centimeters. If an object is placed a distance p centimeters from the lens, then the distance q centimeters of the image from the lens is related to p and f by the *lens equation*

$$\frac{1}{p}+\frac{1}{q}=\frac{1}{f}.$$

As shown in the figure on the following page, p must be greater than f for the rays to converge.

(a) Investigate $\lim_{p\to f^+}q$.

(b) What is happening to the image as $p\to f^+$?

EXERCISE 71

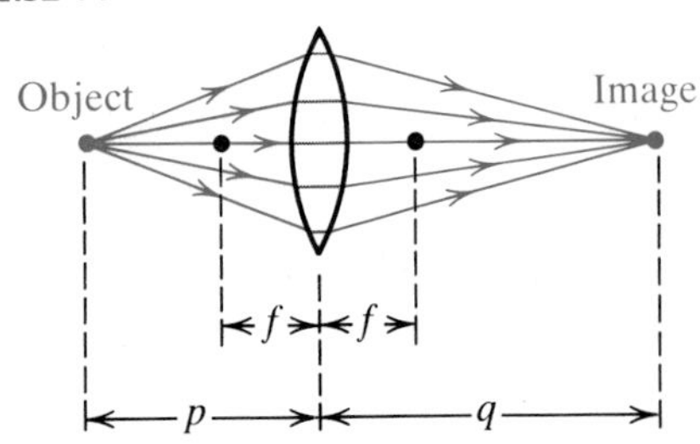

72 Shown in the figure is a simple magnifier consisting of a convex lens. The object to be magnified is positioned so that its distance p from the lens is less than the focal length f. The linear magnification M is the ratio of the image size to the object size. Using similar triangles, we obtain $M = q/p$, where q is the distance between the image and the lens.

(a) Find $\lim_{p \to 0^+} M$ and explain why a right-hand limit is necessary.

(b) Investigate $\lim_{p \to f^-} M$ and explain what is happening to the image size as $p \to f^-$.

EXERCISE 72

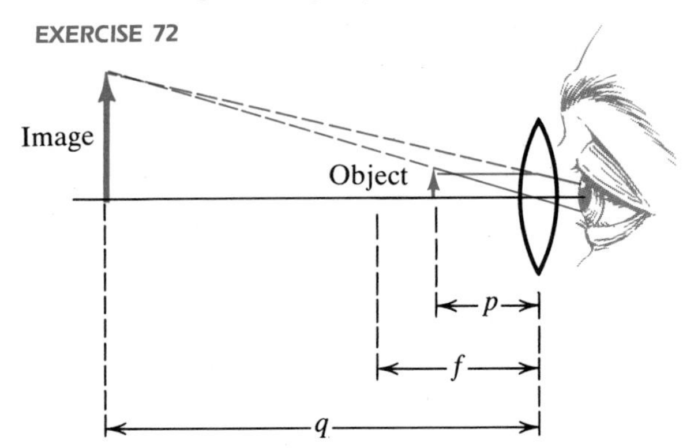

2.4 LIMITS INVOLVING INFINITY

When investigating $\lim_{x \to a^-} f(x)$ or $\lim_{x \to a^+} f(x)$, we may find that as x approaches a, the function value $f(x)$ either increases without bound or decreases without bound. To illustrate, let us consider

$$f(x) = \frac{1}{x - 2}.$$

FIGURE 2.25
$f(x) = \dfrac{1}{x - 2}$

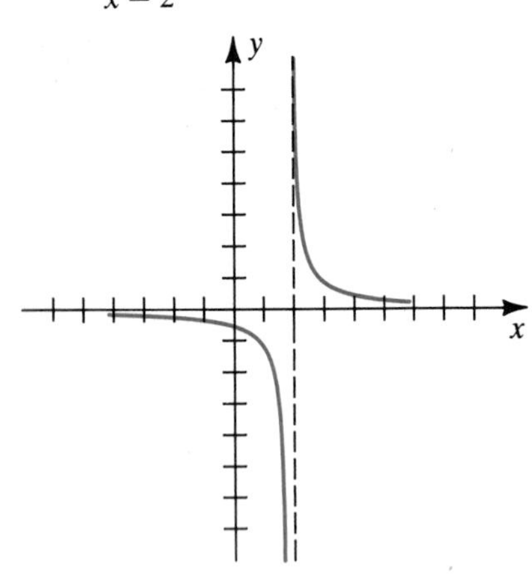

The graph of f is sketched in Figure 2.25. We can show, as in Example 3 of Sections 2.1 and 2.2, that

$$\lim_{x \to 2} \frac{1}{x - 2} \quad \text{does not exist.}$$

Some function values for x near 2, with $x > 2$, are listed in the following table.

x	2.1	2.01	2.001	2.0001	2.00001	2.000001
$f(x)$	10	100	1000	10,000	100,000	1,000,000

As x approaches 2 from the right, $f(x)$ *increases without bound* in the sense that *we can make* $f(x)$ *as large as desired by choosing* x *sufficiently close to* 2 *and* $x > 2$. We denote this by writing

$$\lim_{x \to 2^+} \frac{1}{x - 2} = \infty, \quad \text{or} \quad \frac{1}{x - 2} \to \infty \quad \text{as} \quad x \to 2^+.$$

The symbol ∞ (**infinity**) *does not represent a real number*. It is a notation we use to denote how certain functions behave. Thus, although we may state that as x *approaches* 2 *from the right*, $1/(x - 2)$ *approaches* ∞ (*or tends to* ∞), or that *the limit of* $1/(x - 2)$ *equals* ∞, we do *not* mean that $1/(x - 2)$ gets closer to some specific real number or that $\lim_{x \to 2^+} [1/(x - 2)]$ *exists*.

The symbol $-\infty$ (**minus infinity**) is used in similar fashion to denote that $f(x)$ *decreases without bound* (takes on very large *negative* values) as x approaches a real number. Thus, for $f(x) = 1/(x - 2)$ (see Figure 2.25), we write

$$\lim_{x\to 2^-}\frac{1}{x-2}=-\infty, \quad \text{or} \quad \frac{1}{x-2}\to -\infty \quad \text{as} \quad x\to 2^-.$$

Figure 2.26 contains typical (partial) graphs of arbitrary functions that approach ∞ or $-\infty$ in various ways. We have pictured a as positive; however, we can also have $a \leq 0$.

FIGURE 2.26

(i) $\lim_{x\to a^-} f(x)=\infty$, or $f(x)\to\infty$ as $x\to a^-$

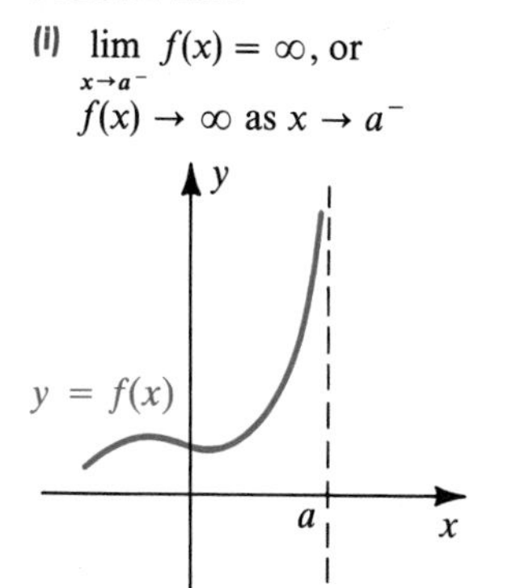

(ii) $\lim_{x\to a^+} f(x)=\infty$, or $f(x)\to\infty$ as $x\to a^+$

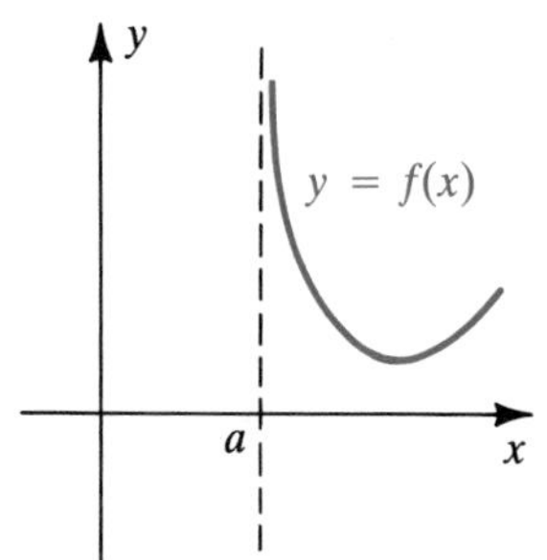

(iii) $\lim_{x\to a^-} f(x)=-\infty$, or $f(x)\to-\infty$ as $x\to a^-$

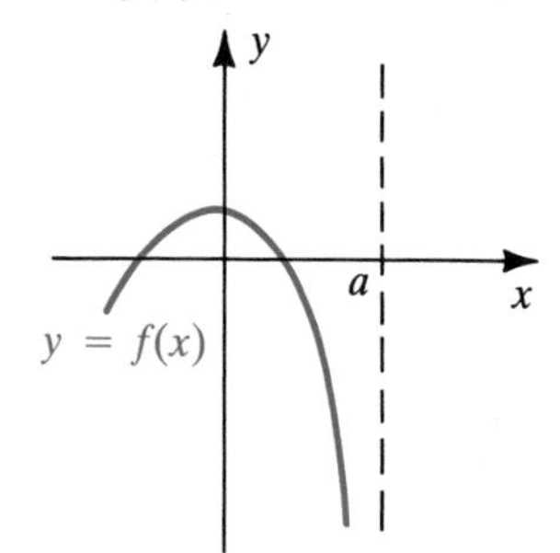

(iv) $\lim_{x\to a^+} f(x)=-\infty$, or $f(x)\to-\infty$ as $x\to a^+$

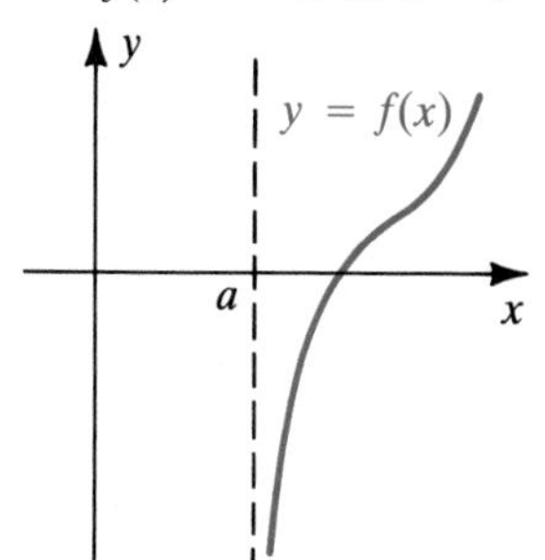

We shall also consider the two-sided limits illustrated in Figure 2.27. The line $x = a$ in Figures 2.26 and 2.27 is called a **vertical asymptote** for the graph of f.

FIGURE 2.27

(i) $\lim_{x\to a} f(x)=\infty$, or $f(x)\to\infty$ as $x\to a$

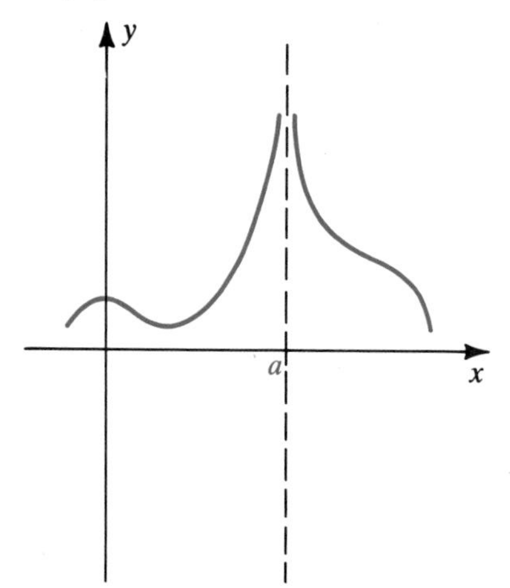

(ii) $\lim_{x\to a} f(x)=-\infty$, or $f(x)\to-\infty$ as $x\to a$

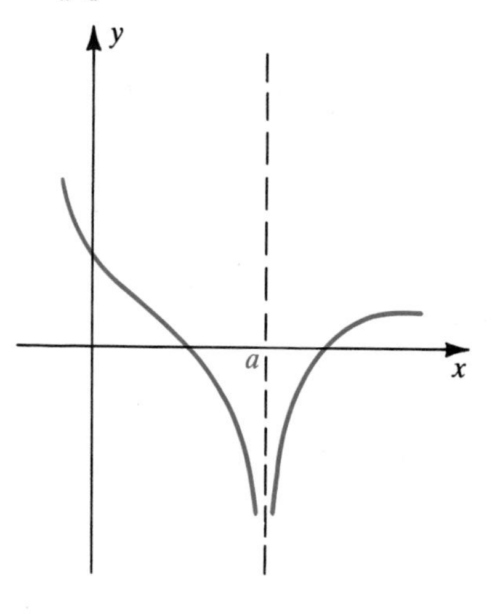

Note that for $f(x)$ to approach ∞ as x approaches a, *both* the right-hand limit and the left-hand limit must be ∞. For $f(x)$ to approach $-\infty$, *both* one-sided limits must be $-\infty$. If the limit of $f(x)$ from one side of a is ∞ and from the other side of a is $-\infty$, as in Figure 2.25, we say that $\lim_{x\to a} f(x)$ *does not exist.*

It is possible to investigate many algebraic functions that approach ∞ or $-\infty$ by reasoning intuitively, as in the following examples. A formal definition that can be used for rigorous proofs is stated at the end of this section.

FIGURE 2.28
$f(x) = \dfrac{1}{(x-2)^2}$

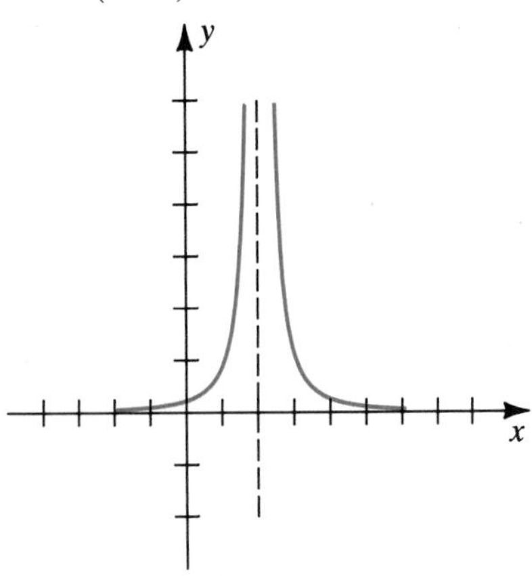

EXAMPLE 1 Find $\lim_{x \to 2} \dfrac{1}{(x-2)^2}$, if it exists.

SOLUTION If x is close to 2 and $x \neq 2$, then $(x-2)^2$ is positive and close to zero. Hence $1/(x-2)^2$ is large and positive. Since we can make $1/(x-2)^2$ as large as desired by choosing x sufficiently close to 2, we see that the limit equals ∞. Thus, the limit does not exist.

The graph of $y = 1/(x-2)^2$ is sketched in Figure 2.28. The line $x = 2$ is a vertical asymptote for the graph.

FIGURE 2.29
$f(x) = \dfrac{1}{(x-4)^3}$

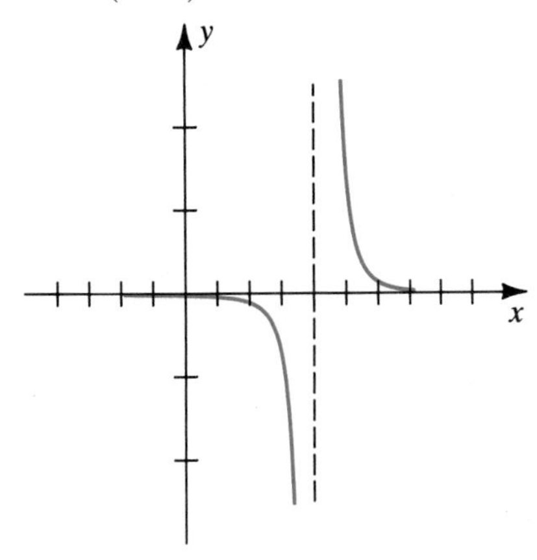

EXAMPLE 2 Find each limit, if it exists.

(a) $\lim_{x \to 4^-} \dfrac{1}{(x-4)^3}$ **(b)** $\lim_{x \to 4^+} \dfrac{1}{(x-4)^3}$ **(c)** $\lim_{x \to 4} \dfrac{1}{(x-4)^3}$

SOLUTION

(a) If x is close to 4 and $x < 4$, then $x - 4$ is close to 0 and *negative*, and

$$\lim_{x \to 4^-} \frac{1}{(x-4)^3} = -\infty.$$

(b) If x is close to 4 and $x > 4$, then $x - 4$ is a small positive number and hence $1/(x-4)^3$ is a large positive number. Thus,

$$\lim_{x \to 4^+} \frac{1}{(x-4)^3} = \infty.$$

(c) Since the one-sided limits are not both ∞ or both $-\infty$,

$$\lim_{x \to 4} \frac{1}{(x-4)^3} \quad \text{does not exist.}$$

The graph of $y = 1/(x-4)^3$ is sketched in Figure 2.29. The line $x = 4$ is a vertical asymptote.

Formulas that represent physical quantities may lead to limits involving infinity. Obviously, a physical quantity cannot approach infinity, but an analysis of a hypothetical situation in which that *could* occur may suggest uses for other related quantities. For example, consider Ohm's law in electrical theory, which states that $I = V/R$, where R is the resistance (in ohms) of a conductor, V is the potential difference (in volts) across the conductor, and I is the current (in amperes) that flows through the conductor (see Figure 2.30). The resistance of certain alloys approaches zero as the temperature approaches absolute zero (approximately $-273\,^\circ\text{C}$), and the alloy becomes a *superconductor* of electricity. If the voltage V is fixed, then, for such a superconductor,

FIGURE 2.30

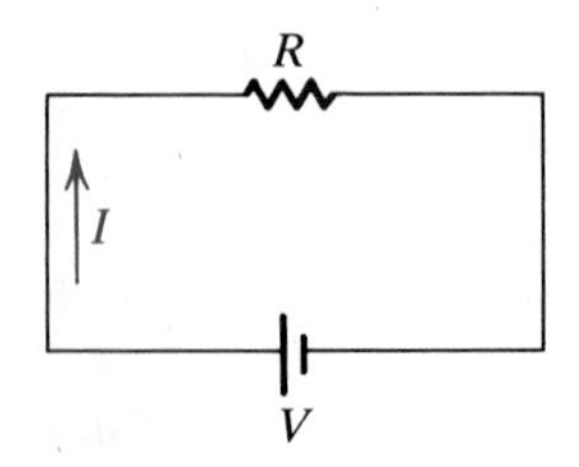

$$\lim_{R \to 0^+} I = \lim_{R \to 0^+} \frac{V}{R} = \infty;$$

that is, the current increases without bound. Superconductors allow very large currents to be used in generating plants or motors. They also have applications in experimental high-speed ground transportation, where the strong magnetic fields produced by superconducting magnets enable trains to levitate so that there is essentially no friction between the wheels and the track. Perhaps the most important use for superconductors is in circuits for computers, because such circuits produce very little heat.

Let us next discuss functions whose values approach some number L as $|x|$ becomes very large. Consider

$$f(x) = 2 + \frac{1}{x}.$$

FIGURE 2.31
$f(x) = 2 + \dfrac{1}{x}$

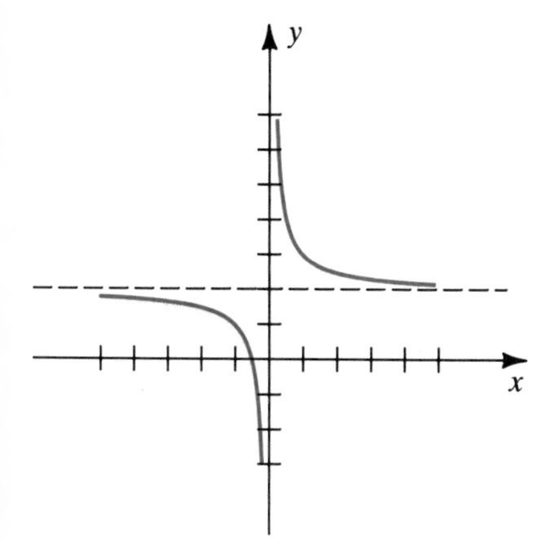

The graph of f is sketched in Figure 2.31. Some values of $f(x)$ if x is large are listed in the following table.

x	100	1000	10,000	100,000	1,000,000
$f(x)$	2.01	2.001	2.0001	2.00001	2.000001

We can make $f(x)$ as close to 2 as desired by choosing x sufficiently large. This is denoted by

$$\lim_{x \to \infty} \left(2 + \frac{1}{x}\right) = 2$$

which may be read *the limit of* $2 + (1/x)$ *as* x *approaches* ∞ *is* 2.

Once again, remember that ∞ is not a real number, and hence ∞ *should never be substituted for the variable* x. The terminology x *approaches* ∞ does not mean that x gets close to some real number. Intuitively, we think of x as increasing without bound, or being assigned arbitrarily large values.

If we let x *decrease* without bound—that is, if we let x take on very large *negative* values—then, as indicated by the second-quadrant portion of the graph in Figure 2.31, $2 + (1/x)$ again approaches 2, and we write

$$\lim_{x \to -\infty} \left(2 + \frac{1}{x}\right) = 2.$$

Before considering additional examples, let us state definitions for such limits involving infinity, using ϵ-tolerances for $f(x)$ at L. When we considered $\lim_{x \to a} f(x) = L$ in Section 2.2, we wanted $|f(x) - L| < \epsilon$ whenever x was close to a and $x \neq a$. In the present situation, we want $|f(x) - L| < \epsilon$ whenever x is *sufficiently large*, say larger than any given positive number M. This results in the next definition, in which we assume that f is defined on an infinite interval (c, ∞) for a real number c.

Definition **(2.16)**

$$\lim_{x \to \infty} f(x) = L$$

means that for every $\epsilon > 0$, there is a number $M > 0$ such that

if $x > M$, then $|f(x) - L| < \epsilon$.

If $\lim_{x\to\infty} f(x) = L$, we say that *the limit of $f(x)$ as x approaches ∞ is L*, or that *$f(x)$ approaches L as x approaches ∞*. We sometimes write

$$f(x) \to L \quad \text{as} \quad x \to \infty.$$

We may give a graphical interpretation of $\lim_{x\to\infty} f(x) = L$ as follows. Consider any horizontal lines $y = L \pm \epsilon$, as in Figure 2.32. According to Definition (2.16), if x is larger than some positive number M, the point $P(x, f(x))$ on the graph lies between these horizontal lines. Intuitively we know that the graph of f gets closer to the line $y = L$ as x gets larger. We call the line $y = L$ a **horizontal asymptote** for the graph of f. As illustrated in Figure 2.32, *a graph may cross a horizontal asymptote*. The line $y = 2$ in Figure 2.31 is a horizontal asymptote for the graph of $f(x) = 2 + (1/x)$.

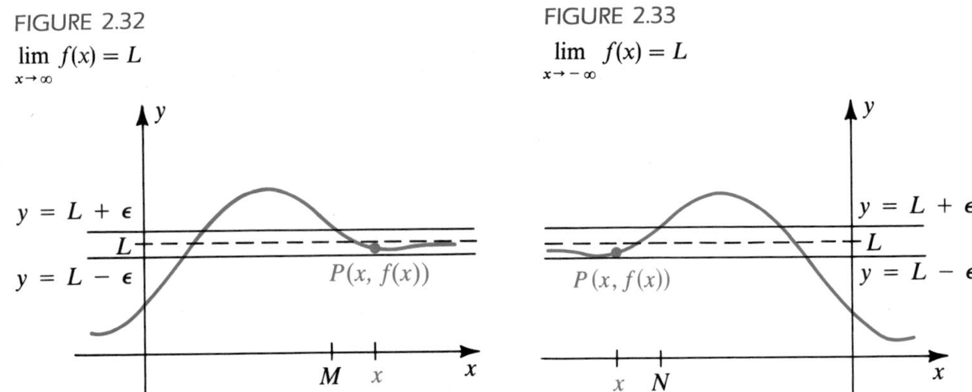

FIGURE 2.32
$\lim_{x\to\infty} f(x) = L$

FIGURE 2.33
$\lim_{x\to-\infty} f(x) = L$

In Figure 2.32 the graph of f approaches the asymptote $y = L$ from below—that is, with $f(x) < L$. A graph can also approach $y = L$ from above—that is, with $f(x) > L$—or other ways, such as with $f(x)$ alternately greater than and less than L as $x \to \infty$.

The next definition covers the case in which x is *large negative*. We shall assume that f is defined on an infinite interval $(-\infty, c)$ for some real number c.

Definition **(2.17)**

$$\lim_{x\to-\infty} f(x) = L$$

means that for every $\epsilon > 0$, there is a number $N < 0$ such that

$$\text{if} \quad x < N, \quad \text{then} \quad |f(x) - L| < \epsilon.$$

If $\lim_{x\to-\infty} f(x) = L$, we say that *the limit of $f(x)$ as x approaches $-\infty$ is L*, or that *$f(x)$ approaches L as x approaches $-\infty$*.

Definition (2.17) is illustrated in Figure 2.33. If we consider any horizontal lines $y = L \pm \epsilon$, then every point $P(x, f(x))$ on the graph lies between these lines if x is *less* than some negative number N. The line $y = L$ is a horizontal asymptote for the graph of f.

Limit theorems that are analogous to those in Section 2.3 may be established for limits involving infinity. In particular, Theorem (2.8) concerning limits of sums, products, and quotients is true for $x \to \infty$ or $x \to -\infty$. Similarly, Theorem (2.14) on the limit of $\sqrt[n]{f(x)}$ holds if $x \to \infty$

or $x \to -\infty$. We can also show that

$$\lim_{x \to \infty} c = c \quad \text{and} \quad \lim_{x \to -\infty} c = c.$$

A proof of the next theorem, using Definition (2.16), is given in Appendix II.

Theorem **(2.18)**

> If k is a positive rational number and c is any real number, then
>
> $$\lim_{x \to \infty} \frac{c}{x^k} = 0 \quad \text{and} \quad \lim_{x \to -\infty} \frac{c}{x^k} = 0,$$
>
> provided x^k is always defined.

Theorem (2.18) is useful for investigating limits of rational functions. Specifically, *to find* $\lim_{x \to \infty} f(x)$ *or* $\lim_{x \to -\infty} f(x)$ *for a rational function* f, *first divide the numerator and denominator of* $f(x)$ *by* x^n, *where* n *is the highest power of* x *that appears in the denominator, and then use limit theorems*. This technique is illustrated in the next three examples.

EXAMPLE 3 Find $\displaystyle\lim_{x \to -\infty} \frac{2x^2 - 5}{3x^2 + x + 2}$.

SOLUTION The highest power of x in the denominator is 2. Hence, by the rule stated in the preceding paragraph, we divide numerator and denominator by x^2 and then use limit theorems. Thus,

$$\lim_{x \to -\infty} \frac{2x^2 - 5}{3x^2 + x + 2} = \lim_{x \to -\infty} \frac{\dfrac{2x^2 - 5}{x^2}}{\dfrac{3x^2 + x + 2}{x^2}}$$

$$= \lim_{x \to -\infty} \frac{2 - \dfrac{5}{x^2}}{3 + \dfrac{1}{x} + \dfrac{2}{x^2}}$$

$$= \frac{\displaystyle\lim_{x \to -\infty} \left(2 - \frac{5}{x^2}\right)}{\displaystyle\lim_{x \to -\infty} \left(3 + \frac{1}{x} + \frac{2}{x^2}\right)}$$

$$= \frac{\displaystyle\lim_{x \to -\infty} 2 - \lim_{x \to -\infty} \frac{5}{x^2}}{\displaystyle\lim_{x \to -\infty} 3 + \lim_{x \to -\infty} \frac{1}{x} + \lim_{x \to -\infty} \frac{2}{x^2}}$$

$$= \frac{2 - 0}{3 + 0 + 0} = \frac{2}{3}.$$

It follows that the line $y = \frac{2}{3}$ is a horizontal asymptote for the graph of f.

EXAMPLE 4 Find $\lim_{x\to\infty} \dfrac{2x^2 - 5}{3x^4 + x + 2}$.

SOLUTION Since the highest power of x in the denominator is 4, we first divide numerator and denominator by x^4, obtaining

$$\lim_{x\to\infty} \frac{2x^2 - 5}{3x^4 + x + 2} = \lim_{x\to\infty} \frac{\dfrac{2}{x^2} - \dfrac{5}{x^4}}{3 + \dfrac{1}{x^3} + \dfrac{2}{x^4}}$$

$$= \frac{0 - 0}{3 + 0 + 0} = \frac{0}{3} = 0.$$

(We omitted several steps in the use of limit theorems.)

EXAMPLE 5 Find $\lim_{x\to\infty} \dfrac{2x^3 - 5}{3x^2 + x + 2}$.

SOLUTION Since the highest power of x in the denominator is 2, we first divide numerator and denominator by x^2, obtaining

$$\lim_{x\to\infty} \frac{2x^3 - 5}{3x^2 + x + 2} = \lim_{x\to\infty} \frac{2x - \dfrac{5}{x^2}}{3 + \dfrac{1}{x} + \dfrac{2}{x^2}}.$$

Since each term of the form c/x^k approaches 0 as $x \to \infty$, we see that

$$\lim_{x\to\infty} \left(2x - \frac{5}{x^2}\right) = \infty \quad \text{and} \quad \lim_{x\to\infty} \left(3 + \frac{1}{x} + \frac{2}{x^2}\right) = 3.$$

It follows that

$$\lim_{x\to\infty} \frac{2x^3 - 5}{3x^2 + x + 2} = \infty.$$

EXAMPLE 6 If $f(x) = \dfrac{\sqrt{9x^2 + 2}}{4x + 3}$, find

(a) $\lim_{x\to\infty} f(x)$ **(b)** $\lim_{x\to-\infty} f(x)$

SOLUTION

(a) If x is large and positive, then

$$\sqrt{9x^2 + 2} \approx \sqrt{9x^2} = 3x \quad \text{and} \quad 4x + 3 \approx 4x$$

and hence

$$f(x) = \frac{\sqrt{9x^2 + 2}}{4x + 3} \approx \frac{3x}{4x} = \frac{3}{4}.$$

This *suggests* that $\lim_{x\to\infty} f(x) = \frac{3}{4}$. To give a rigorous proof we may write

$$\lim_{x\to\infty} \frac{\sqrt{9x^2+2}}{4x+3} = \lim_{x\to\infty} \frac{\sqrt{x^2\left(9+\frac{2}{x^2}\right)}}{4x+3}$$

$$= \lim_{x\to\infty} \frac{\sqrt{x^2}\sqrt{9+\frac{2}{x^2}}}{4x+3}.$$

If x is positive, then $\sqrt{x^2} = x$, and dividing numerator and denominator of the last fraction by x gives us

$$\lim_{x\to\infty} \frac{\sqrt{9x^2+2}}{4x+3} = \lim_{x\to\infty} \frac{\sqrt{9+\frac{2}{x^2}}}{4+\frac{3}{x}}$$

$$= \frac{\sqrt{9+0}}{4+0} = \frac{3}{4}.$$

(b) If x is large *negative*, then $\sqrt{x^2} = -x$. If we use the same steps as in part (a), we obtain

$$\lim_{x\to-\infty} \frac{\sqrt{9x^2+2}}{4x+3} = \lim_{x\to-\infty} \frac{\sqrt{x^2}\sqrt{9+\frac{2}{x^2}}}{4x+3}$$

$$= \lim_{x\to-\infty} \frac{(-x)\sqrt{9+\frac{2}{x^2}}}{4x+3}$$

$$= \lim_{x\to-\infty} \frac{-\sqrt{9+\frac{2}{x^2}}}{4+\frac{3}{x}}$$

$$= \frac{-\sqrt{9+0}}{4+0} = -\frac{3}{4}.$$

We may also consider cases where *both* x and $f(x)$ approach ∞ or $-\infty$. For example, the limit statement

$$\lim_{x\to-\infty} f(x) = \infty$$

means that $f(x)$ increases without bound as x decreases without bound, as would be the case for $f(x) = x^2$.

The preceding types of limits involving ∞ occur in applications. To illustrate, Newton's law of universal gravitation may be stated: *Every particle in the universe attracts every other particle with a force that is proportional to the product of their masses and inversely proportional to the square of the distance between them.* In symbols,

$$F = G\frac{m_1 m_2}{r^2},$$

where F is the force on each particle, m_1 and m_2 are their masses, r is the distance between them, and G is a gravitational constant. Assuming that m_1 and m_2 are constant, we obtain

$$\lim_{r\to\infty} F = \lim_{r\to\infty} G\frac{m_1 m_2}{r^2} = 0.$$

This tells us that as the distance between the particles increases without bound, the force of attraction approaches 0. Theoretically, there is always *some* attraction; however, if r is very large, the attraction cannot be measured with conventional laboratory equipment.

We shall conclude this section by stating a formal definition of $\lim_{x\to a} f(x) = \infty$. The main difference from our work in Section 2.2 is that instead of showing that $|f(x) - L| < \epsilon$ whenever x is near a, we consider any (large) positive number M and show that $f(x) > M$ whenever x is near a.

Definition (2.19)

$$\lim_{x\to a} f(x) = \infty$$

means that for every $M > 0$, there is a $\delta > 0$ such that

$$\text{if} \quad 0 < |x - a| < \delta, \quad \text{then} \quad f(x) > M.$$

FIGURE 2.34
$\lim_{x\to a} f(x) = \infty$

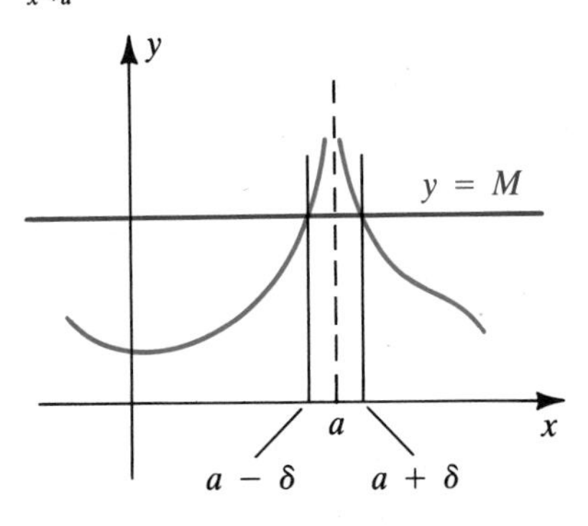

For a graphical interpretation of Definition (2.19), consider any horizontal line $y = M$, as in Figure 2.34. If $\lim_{x\to a} f(x) = \infty$, then whenever x is in a suitable interval $(a - \delta, a + \delta)$ and $x \neq a$, the points on the graph of f lie *above* the horizontal line.

To define $\lim_{x\to a} f(x) = -\infty$ we may alter Definition (2.19), replacing $M > 0$ by $N < 0$ and $f(x) > M$ by $f(x) < N$. In this case, if we consider any horizontal line $y = N$ (with N negative), the graph of f lies *below* this line whenever x is in a suitable interval $(a - \delta, a + \delta)$ and $x \neq a$.

EXERCISES 2.4

Exer. 1–10: For the given $f(x)$, express each of the following limits as ∞, $-\infty$, or DNE (Does Not Exist):

(a) $\lim_{x\to a^-} f(x)$ (b) $\lim_{x\to a^+} f(x)$ (c) $\lim_{x\to a} f(x)$

1 $f(x) = \dfrac{5}{x-4}$; $a = 4$

2 $f(x) = \dfrac{5}{4-x}$; $a = 4$

3 $f(x) = \dfrac{8}{(2x+5)^3}$; $a = -\dfrac{5}{2}$

4 $f(x) = \dfrac{-4}{7x+3}$; $a = -\dfrac{3}{7}$

5 $f(x) = \dfrac{3x}{(x+8)^2}$; $a = -8$

6 $f(x) = \dfrac{3x^2}{(2x-9)^2}$; $a = \dfrac{9}{2}$

7 $f(x) = \dfrac{2x^2}{x^2-x-2}$; $a = -1$

8 $f(x) = \dfrac{4x}{x^2-4x+3}$; $a = 1$

9 $f(x) = \dfrac{1}{x(x-3)^2}$; $a = 3$

10 $f(x) = \dfrac{-1}{(x+1)^2}$; $a = -1$

Exer. 11–24: Find the limit, if it exists.

11 $\lim_{x\to\infty} \dfrac{5x^2 - 3x + 1}{2x^2 + 4x - 7}$

12 $\lim_{x\to\infty} \dfrac{3x^3 - x + 1}{6x^3 + 2x^2 - 7}$

13 $\lim_{x\to-\infty} \dfrac{4 - 7x}{2 + 3x}$

14 $\lim_{x\to-\infty} \dfrac{(3x + 4)(x - 1)}{(2x + 7)(x + 2)}$

15 $\lim_{x\to-\infty} \dfrac{2x^2 - 3}{4x^3 + 5x}$

16 $\lim_{x\to\infty} \dfrac{2x^2 - x + 3}{x^3 + 1}$

17 $\lim_{x\to\infty} \dfrac{-x^3 + 2x}{2x^2 - 3}$

18 $\lim_{x\to-\infty} \dfrac{x^2 + 2}{x - 1}$

19 $\lim_{x\to-\infty} \dfrac{2 - x^2}{x + 3}$

20 $\lim_{x\to\infty} \dfrac{3x^4 + x + 1}{x^2 - 5}$

21 $\lim_{x\to\infty} \sqrt[3]{\dfrac{8 + x^2}{x(x + 1)}}$

22 $\lim_{x\to-\infty} \dfrac{4x - 3}{\sqrt{x^2 + 1}}$

23 $\lim_{x\to\infty} \sin x$

24 $\lim_{x\to\infty} \cos x$

c **Exer. 25–26: Investigate the limit by letting $x = 10^n$ for $n = 1, 2, 3$, and 4.**

25 $\lim_{x\to\infty} \dfrac{1}{x} \tan\left(\dfrac{\pi}{2} - \dfrac{1}{x}\right)$

26 $\lim_{x\to\infty} \sqrt{x} \sin \dfrac{1}{x}$

Exer. 27–36: Find the vertical and horizontal asymptotes for the graph of f.

27 $f(x) = \dfrac{1}{x^2 - 4}$

28 $f(x) = \dfrac{5x}{4 - x^2}$

29 $f(x) = \dfrac{2x^2}{x^2 + 1}$

30 $f(x) = \dfrac{3x}{x^2 + 1}$

31 $f(x) = \dfrac{1}{x^3 + x^2 - 6x}$

32 $f(x) = \dfrac{x^2 - x}{16 - x^2}$

33 $f(x) = \dfrac{x^2 + 3x + 2}{x^2 + 2x - 3}$

34 $f(x) = \dfrac{x^2 - 5x}{x^2 - 25}$

35 $f(x) = \dfrac{x + 4}{x^2 - 16}$

36 $f(x) = \dfrac{\sqrt[3]{16 - x^2}}{4 - x}$

Exer. 37–40: A function f satisfies the given conditions. Sketch a possible graph for f, assuming that it does not cross a horizontal asymptote.

37 $\lim_{x\to-\infty} f(x) = 1$; $\lim_{x\to\infty} f(x) = 1$;
$\lim_{x\to3^-} f(x) = -\infty$; $\lim_{x\to3^+} f(x) = \infty$

38 $\lim_{x\to-\infty} f(x) = -1$; $\lim_{x\to\infty} f(x) = -1$;
$\lim_{x\to2^-} f(x) = \infty$; $\lim_{x\to2^+} f(x) = -\infty$

39 $\lim_{x\to-\infty} f(x) = -2$; $\lim_{x\to\infty} f(x) = -2$;
$\lim_{x\to3^-} f(x) = \infty$; $\lim_{x\to3^+} f(x) = -\infty$;
$\lim_{x\to-1^-} f(x) = -\infty$; $\lim_{x\to-1^+} f(x) = \infty$

40 $\lim_{x\to-\infty} f(x) = 3$; $\lim_{x\to\infty} f(x) = 3$;
$\lim_{x\to1^-} f(x) = \infty$; $\lim_{x\to1^+} f(x) = -\infty$;
$\lim_{x\to-2^-} f(x) = -\infty$; $\lim_{x\to-2^+} f(x) = \infty$

41 Salt water of concentration 0.1 pound of salt per gallon flows into a large tank that initially contains 50 gallons of pure water.

(a) If the flow rate of salt water into the tank is 5 gallons per minute, find the volume $V(t)$ of water and the amount $A(t)$ of salt in the tank after t minutes.

(b) Find a formula for the salt concentration $c(t)$ (in lb/gal) after t minutes.

(c) What happens to $c(t)$ over a long period of time?

42 An important problem in fishery science is predicting next year's adult breeding population R (the recruits) from the number S that are presently spawning. For some species (such as North Sea herring), the relationship between R and S is given by $R = aS/(S + b)$, where a and b are positive constants. What happens as the number of spawners increases?

2.5 CONTINUOUS FUNCTIONS

In everyday usage we say that time is continuous, since it proceeds in an uninterrupted manner. On any given day, time does not jump from 1:00 P.M. to 1:01 P.M., leaving a gap of one minute. If an object is dropped from a hot air balloon, we regard its subsequent motion as continuous. If the initial altitude is 500 feet above ground, the object passes through every altitude between 500 and 0 feet before it hits the ground.

In mathematics we use the phrase *continuous function* in a similar sense. Intuitively, we regard a continuous function as a function whose graph has no breaks, holes, or vertical asymptotes. To illustrate, the graph of each function in Figure 2.35 is *not continuous* at the number c.

FIGURE 2.35

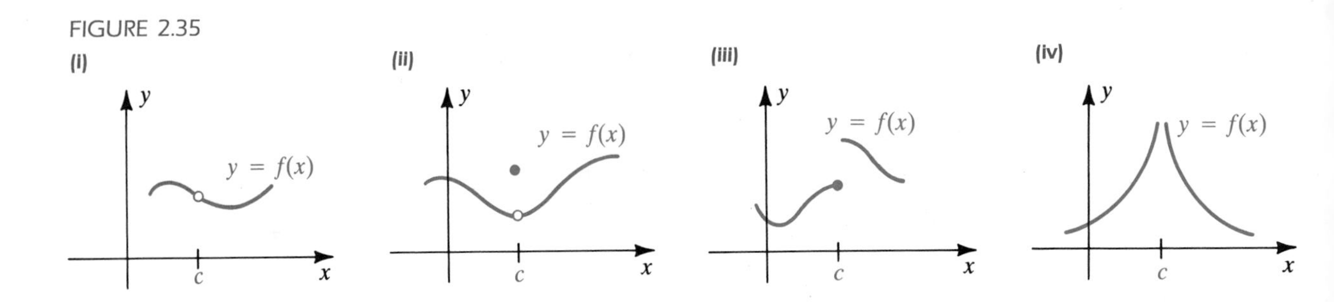

Note that in (i) of the figure, $f(c)$ is not defined. In (ii), $f(c)$ is defined; however, $\lim_{x\to c} f(x) \neq f(c)$. In (iii), $\lim_{x\to c} f(x)$ does not exist. In (iv), $f(c)$ is undefined and, in addition, $\lim_{x\to c} f(x) = \infty$. The graph of a function f is *not* one of these types if f satisfies the three conditions listed in the next definition.

Definition **(2.20)**

A function f is **continuous** at a number c if the following conditions are satisfied:

(i) $f(c)$ is defined

(ii) $\lim_{x\to c} f(x)$ exists

(iii) $\lim_{x\to c} f(x) = f(c)$

Whenever this definition is used to show that a function f is continuous at c, it is sufficient to verify only the third condition, because if $\lim_{x\to c} f(x) = f(c)$, then $f(c)$ must be defined and also $\lim_{x\to c} f(x)$ must exist; that is, the first two conditions are satisfied automatically.

Intuitively we know that condition (iii) implies that as x gets closer to c, the function value $f(x)$ gets closer to $f(c)$. More precisely, we can make *$f(x)$ as close to $f(c)$ as desired* by choosing x sufficiently close to c.

If one (or more) of the three conditions in Definition (2.20) is not satisfied, we say that f is **discontinuous** at c, or that f has a **discontinuity** at c. Certain types of discontinuities are given special names. The discontinuities in (i) and (ii) of Figure 2.35 are **removable discontinuities**, because we could remove each discontinuity by defining the function value $f(c)$ appropriately. The discontinuity in (iii) of the figure is a **jump discontinuity**, so named because of the appearance of the graph. If $f(x)$ approaches ∞ or $-\infty$ as x approaches c from either side, as, for example, in (iv) of the figure, we say that f has an **infinite discontinuity** at c.

In the following illustration, we reconsider some specific functions that were discussed in Sections 2.1 and 2.2.

ILLUSTRATION

FUNCTION VALUE | **GRAPH**  | **DISCONTINUITIES**

■ $f(x) = x + 2$

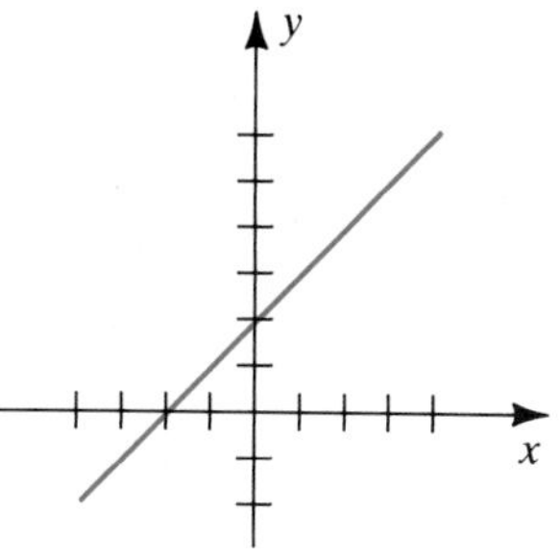

None, since for every c, $\lim_{x \to c} f(x) = c + 2 = f(c)$.

■ 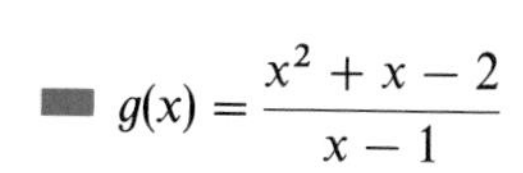 $g(x) = \dfrac{x^2 + x - 2}{x - 1}$

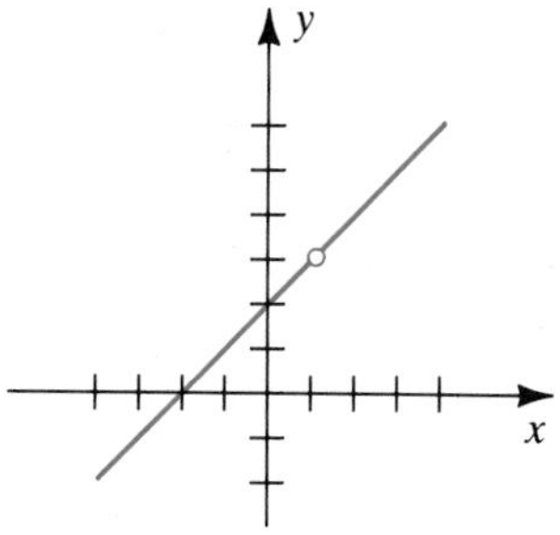

$c = 1$ since $g(1)$ is undefined (removable discontinuity).

■ 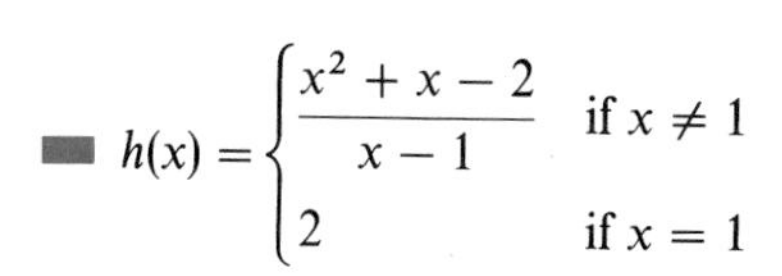 $h(x) = \begin{cases} \dfrac{x^2 + x - 2}{x - 1} & \text{if } x \neq 1 \\ 2 & \text{if } x = 1 \end{cases}$

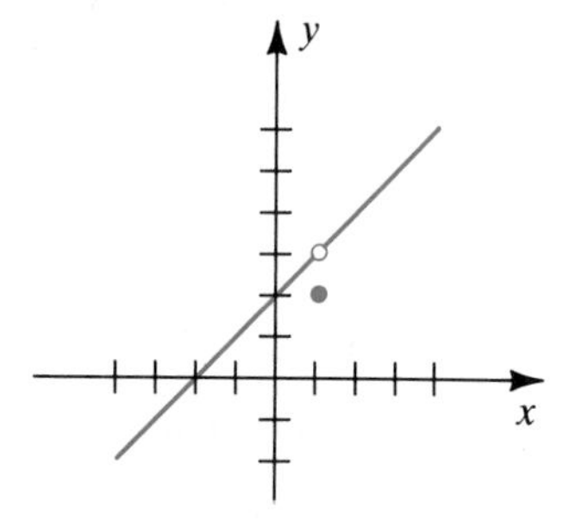

$c = 1$ since $\lim_{x \to 1} h(x) = 3 \neq h(1)$ (removable discontinuity).

■ 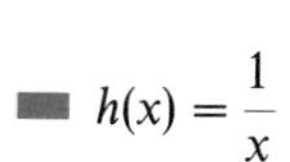 $h(x) = \dfrac{1}{x}$

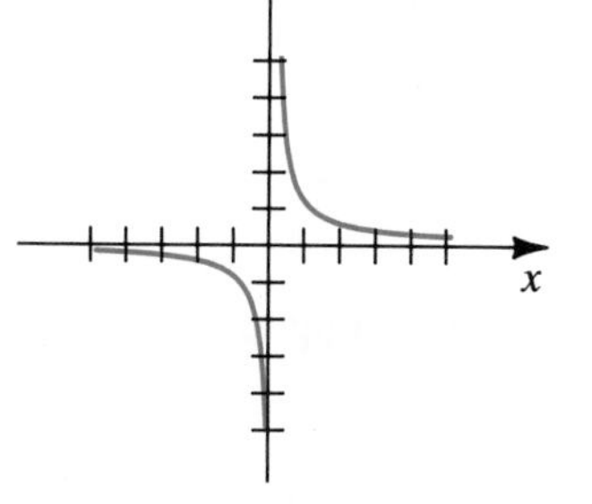

$c = 0$ since $h(0)$ does not exist and also $\lim_{x \to 0} h(x)$ does not exist (infinite discontinuity).

■ 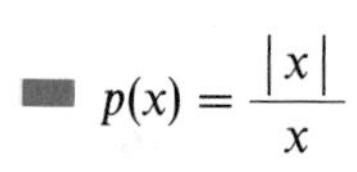 $p(x) = \dfrac{|x|}{x}$

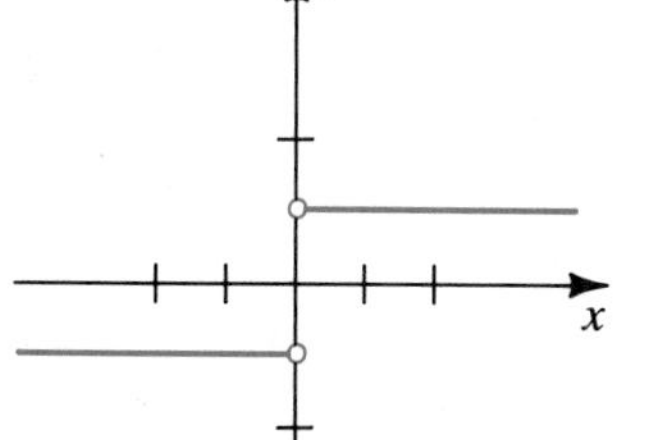

$c = 0$ since $p(0)$ is undefined and also $\lim_{x \to 0} p(x)$ does not exist (jump discontinuity).

The next theorem states that polynomial functions and rational functions (quotients of polynomial functions) are continuous at every number in their domains.

Theorem (2.21)

(i) A polynomial function f is continuous at every real number c.

(ii) A rational function $q = f/g$ is continuous at every number except the numbers c such that $g(c) = 0$.

PROOF (i) If f is a polynomial function and c is a real number, then, by Theorem (2.11), $\lim_{x \to c} f(x) = f(c)$. Hence f is continuous at every real number.

(ii) If $g(c) \neq 0$, then c is in the domain of $q = f/g$ and, by Theorem (2.12), $\lim_{x \to c} q(x) = q(c)$; that is, q is continuous at c. ■

EXAMPLE 1 If $f(x) = |x|$, show that f is continuous at every real number c.

SOLUTION The graph of f is sketched in Figure 2.36. If $x > 0$, then $f(x) = x$. If $x < 0$, then $f(x) = -x$. Since x and $-x$ are polynomials, it follows from Theorem (2.21)(i) that f is continuous at every nonzero real number. It remains to be shown that f is continuous at 0. The one-sided limits of $f(x)$ at 0 are

FIGURE 2.36
$f(x) = |x|$

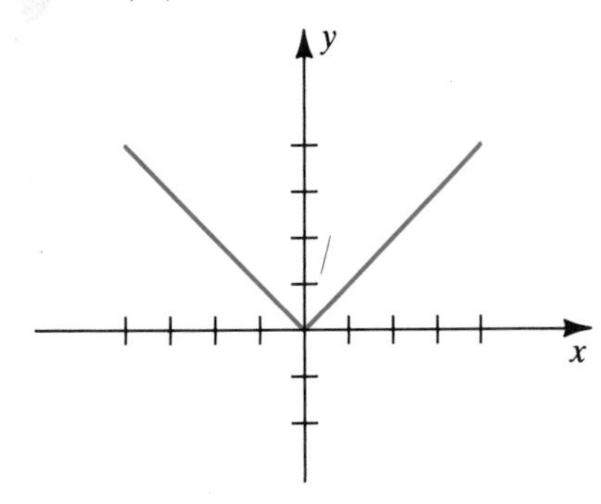

$$\lim_{x \to 0^+} |x| = \lim_{x \to 0^+} x = 0$$

and

$$\lim_{x \to 0^-} |x| = \lim_{x \to 0^-} (-x) = 0.$$

Since the right-hand and left-hand limits are equal, it follows from Theorem (2.3) that

$$\lim_{x \to 0} |x| = 0 = |0| = f(0).$$

Hence f is continuous at 0.

EXAMPLE 2 If $f(x) = \dfrac{x^2 - 1}{x^3 + x^2 - 2x}$, find the discontinuities of f.

SOLUTION Since f is a rational function, it follows from Theorem (2.21) that the only discontinuities occur at the zeros of the denominator $x^3 + x^2 - 2x$. By factoring we obtain

$$x^3 + x^2 - 2x = x(x^2 + x - 2) = x(x + 2)(x - 1).$$

Setting each factor equal to zero, we see that the discontinuities of f are at 0, -2, and 1.

If a function f is continuous at every number in an open interval (a, b), we say that **f is continuous on the interval (a, b)**. Similarly, a function is continuous on an infinite interval of the form (a, ∞) or $(-\infty, b)$ if it is continuous at every number in the interval. The next definition covers the case of a closed interval.

Definition **(2.22)**

Let a function f be defined on a closed interval $[a, b]$. The **function f is continuous on $[a, b]$** if it is continuous on (a, b) and if, in addition,

$$\lim_{x \to a^+} f(x) = f(a) \quad \text{and} \quad \lim_{x \to b^-} f(x) = f(b).$$

If a function f has either a right-hand or a left-hand limit of the type indicated in Definition (2.22), we say that **f is continuous from the right at a** or that **f is continuous from the left at b**, respectively.

EXAMPLE 3 If $f(x) = \sqrt{9 - x^2}$, sketch the graph of f and prove that f is continuous on the closed interval $[-3, 3]$.

FIGURE 2.37

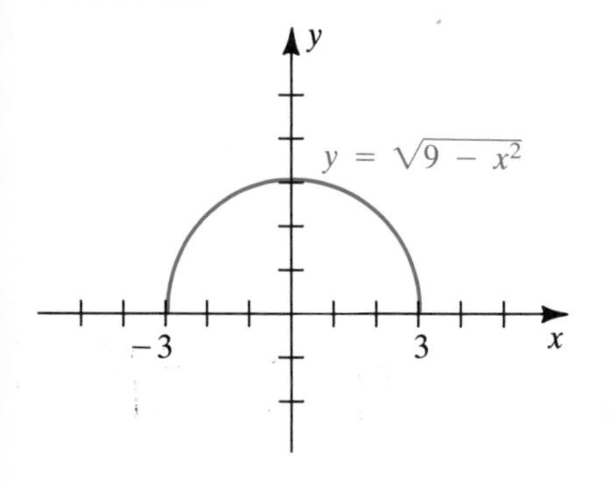

SOLUTION The graph of $x^2 + y^2 = 9$ is a circle with center at the origin and radius 3. Solving for y gives us $y = \pm\sqrt{9 - x^2}$, and hence the graph of $y = \sqrt{9 - x^2}$ is the upper half of that circle (see Figure 2.37).

If $-3 < c < 3$, then, using Theorem (2.14), we obtain

$$\lim_{x \to c} f(x) = \lim_{x \to c} \sqrt{9 - x^2} = \sqrt{9 - c^2} = f(c).$$

Hence f is continuous at c by Definition (2.20). All that remains is to check the endpoints of the interval $[-3, 3]$ using one-sided limits as follows:

$$\lim_{x \to -3^+} f(x) = \lim_{x \to -3^+} \sqrt{9 - x^2} = \sqrt{9 - 9} = 0 = f(-3)$$

$$\lim_{x \to 3^-} f(x) = \lim_{x \to 3^-} \sqrt{9 - x^2} = \sqrt{9 - 9} = 0 = f(3)$$

Thus, f is continuous from the right at -3 and from the left at 3. By Definition (2.22), f is continuous on $[-3, 3]$.

Strictly speaking, the function f in Example 3 is discontinuous at every number c *outside* of the interval $[-3, 3]$, because $f(c)$ is not a real number if $x < -3$ or $x > 3$. However, it is *not* customary to use the phrase *discontinuous at c* if c is in an open interval throughout which f is undefined.

We may also define continuity on other types of intervals. For example, a function f is continuous on $[a, b)$ or $[a, \infty)$ if it is continuous at every number greater than a in the interval and if, in addition, f is continuous from the right at a. For intervals of the form $(a, b]$ or $(-\infty, b]$, we require continuity at every number less than b in the interval and also continuity from the left at b.

Using facts stated in Theorem (2.8), we can prove the following.

Theorem (2.23)

If f and g are continuous at c, then the following are also continuous at c:

(i) the sum $f + g$

(ii) the difference $f - g$

(iii) the product fg

(iv) the quotient f/g, provided $g(c) \neq 0$

PROOF If f and g are continuous at c, then

$$\lim_{x \to c} f(x) = f(c) \quad \text{and} \quad \lim_{x \to c} g(x) = g(c).$$

By the definition of the sum of two functions,

$$(f + g)(x) = f(x) + g(x).$$

Consequently,

$$\begin{aligned} \lim_{x \to c} (f + g)(x) &= \lim_{x \to c} [f(x) + g(x)] \\ &= \lim_{x \to c} f(x) + \lim_{x \to c} g(x) \\ &= f(c) + g(c) \\ &= (f + g)(c). \end{aligned}$$

This proves that $f + g$ is continuous at c. Parts (ii)–(iv) are proved in similar fashion. ■

If f and g are continuous on an interval, then $f + g$, $f - g$, and fg are continuous on the interval. If, in addition, $g(c) \neq 0$ for every c in the interval, then f/g is continuous on the interval. These results may be extended to more than two functions; that is, sums, differences, products, or quotients involving any number of continuous functions are continuous (provided zero denominators do not occur).

EXAMPLE 4 If $k(x) = \dfrac{\sqrt{9 - x^2}}{3x^4 + 5x^2 + 1}$, prove that k is continuous on the closed interval $[-3, 3]$.

SOLUTION Let $f(x) = \sqrt{9 - x^2}$ and $g(x) = 3x^4 + 5x^2 + 1$. From Example 3, f is continuous on $[-3, 3]$, and from Theorem (2.21), g is continuous at every real number. Moreover $g(c) \neq 0$ for every c in $[-3, 3]$. Hence, by Theorem (2.23)(iv), the quotient $k = f/g$ is continuous on $[-3, 3]$.

A proof of the next result on the limit of a composite function $f \circ g$ is given in Appendix II.

Theorem **(2.24)**

If $\lim_{x \to c} g(x) = b$ and if f is continuous at b, then

$$\lim_{x \to c} f(g(x)) = f(b) = f\left(\lim_{x \to c} g(x)\right).$$

The principal use of Theorem (2.24) is to prove other theorems. To illustrate, let us use Theorem (2.24) to prove Theorem (2.14) from Section 2.3, in which we assumed that $\lim_{x \to c} g(x)$ and the indicated nth roots exist.

Conclusion of theorem **(2.14)**

$$\lim_{x \to c} \sqrt[n]{g(x)} = \sqrt[n]{\lim_{x \to c} g(x)}$$

PROOF Let $f(x) = \sqrt[n]{x}$. Applying Theorem (2.24), which states that

$$\lim_{x \to c} f(g(x)) = f\left(\lim_{x \to c} g(x)\right),$$

we obtain

$$\lim_{x \to c} \sqrt[n]{g(x)} = \sqrt[n]{\lim_{x \to c} g(x)}. \blacksquare$$

Part (i) of the next theorem follows from Theorem (2.24) and the definition of a continuous function. Part (ii) is a restatement of (i) using the composite function $f \circ g$.

Theorem **(2.25)**

If g is continuous at c and if f is continuous at $b = g(c)$, then

(i) $\lim_{x \to c} f(g(x)) = f\left(\lim_{x \to c} g(x)\right) = f(g(c))$

(ii) the composite function $f \circ g$ is continuous at c

EXAMPLE 5 If $k(x) = |3x^2 - 7x - 12|$, show that k is continuous at every real number.

SOLUTION If we let

$$f(x) = |x| \quad \text{and} \quad g(x) = 3x^2 - 7x - 12,$$

then $k(x) = f(g(x)) = (f \circ g)(x)$. Since both f and g are continuous functions (see Example 1 and (i) of Theorem (2.21)), it follows from (ii) of Theorem (2.25) that the composite function $k = f \circ g$ is continuous at c.

A proof of the following property of continuous functions may be found in more advanced texts on calculus.

Intermediate value theorem **(2.26)**

> If f is continuous on a closed interval $[a, b]$ and if w is any number between $f(a)$ and $f(b)$, then there is at least one number c in $[a, b]$ such that $f(c) = w$.

The intermediate value theorem states that *as x varies from a to b, the continuous function f takes on every value between $f(a)$ and $f(b)$*. If the graph of the continuous function f is regarded as extending in an unbroken manner from the point $(a, f(a))$ to the point $(b, f(b))$, as illustrated in Figure 2.38, then for any number w between $f(a)$ and $f(b)$, the horizontal line with y-intercept w intersects the graph in at least one point P. The x-coordinate c of P is a number such that $f(c) = w$.

FIGURE 2.38

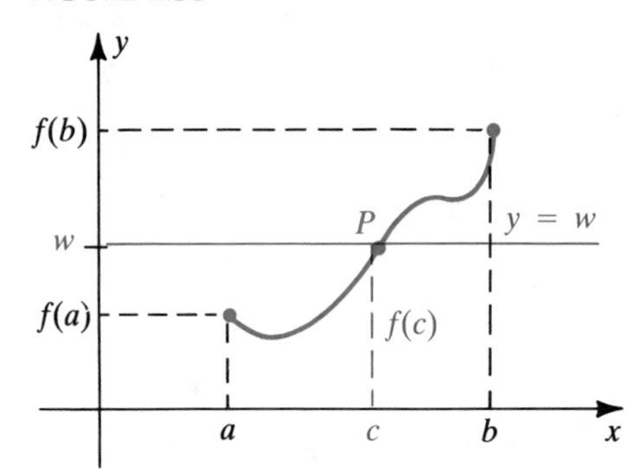

A consequence of the intermediate value theorem is that *if $f(a)$ and $f(b)$ have opposite signs, then there is a number c between a and b such that $f(c) = 0$; that is, f has a zero at c.* Thus, if the point $(a, f(a))$ on the graph of a continuous function lies below the x-axis and the point $(b, f(b))$ lies above the x-axis, or vice versa, then the graph crosses the x-axis at some point $(c, 0)$ for $a < c < b$.

EXAMPLE 6 Show that $f(x) = x^5 + 2x^4 - 6x^3 + 2x - 3$ has a zero between 1 and 2.

SOLUTION Substituting 1 and 2 for x gives us the function values:

$$f(1) = 1 + 2 - 6 + 2 - 3 = -4$$

$$f(2) = 32 + 32 - 48 + 4 - 3 = 17$$

Since $f(1)$ and $f(2)$ have opposite signs, it follows from the intermediate value theorem that $f(c) = 0$ for at least one real number c between 1 and 2.

Example 6 illustrates a scheme for locating real zeros of polynomials. By using a method of *successive approximation*, we can approximate each zero to any degree of accuracy by locating it in smaller and smaller intervals.

Another useful consequence of the intermediate value theorem is the following. The interval referred to may be either closed, open, half-open, or infinite.

Theorem **(2.27)**

> If a function f is continuous and has no zeros on an interval, then either $f(x) > 0$ or $f(x) < 0$ for every x in the interval.

PROOF The conclusion of the theorem states that under the given hypothesis, $f(x)$ has the same sign throughout the interval. If this conclusion were *false*, then there would exist numbers x_1 and x_2 in the interval such that $f(x_1) > 0$ and $f(x_2) < 0$. By our preceding remarks, this, in turn, would imply that $f(c) = 0$ for some number c between x_1 and x_2, contrary to hypothesis. Thus, the conclusion must be true. ■

In Chapter 4 we shall apply Theorem (2.27) to the derivative of a function f to help determine the manner in which $f(x)$ varies on various intervals.

EXERCISES 2.5

Exer. 1–10: The graph of a function f is given. Classify the discontinuities of f as removable, jump, or infinite.

1

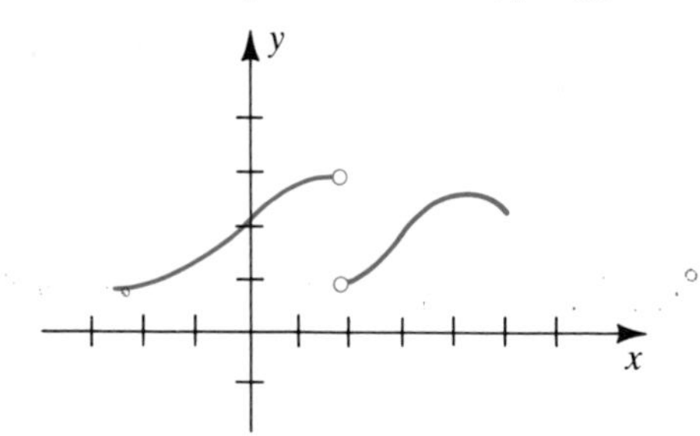

2

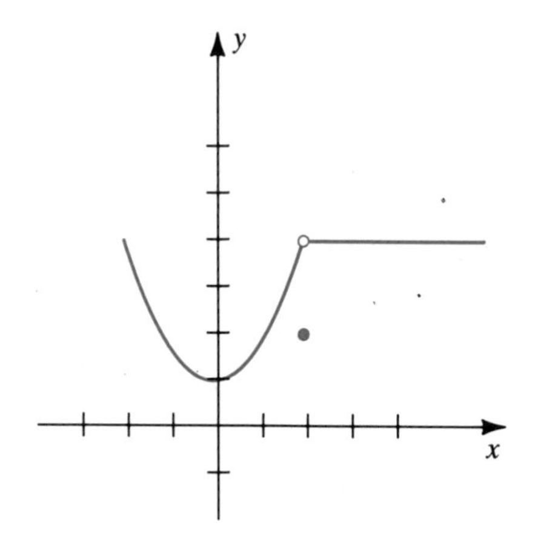

3

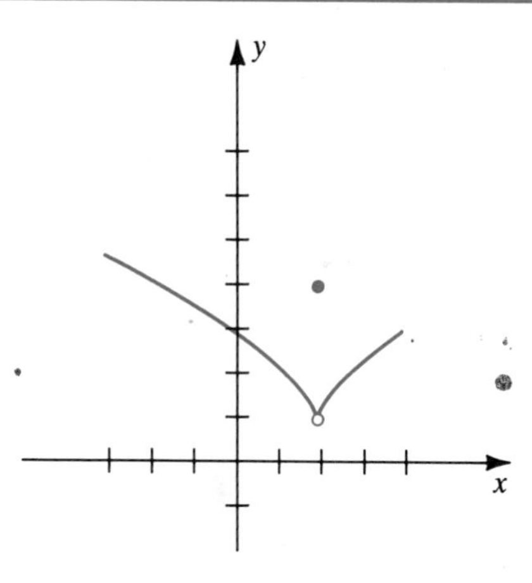

4

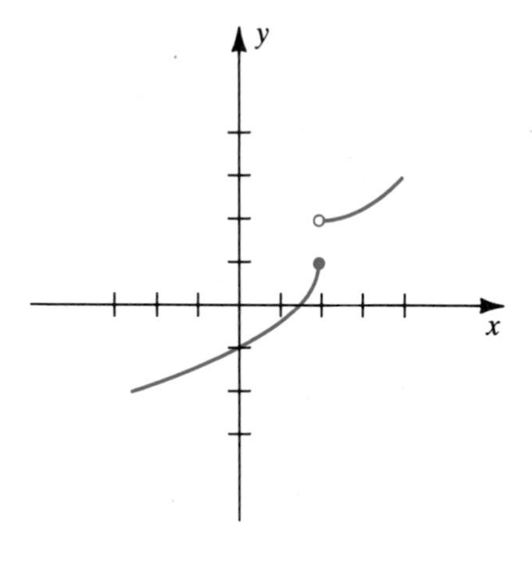

5

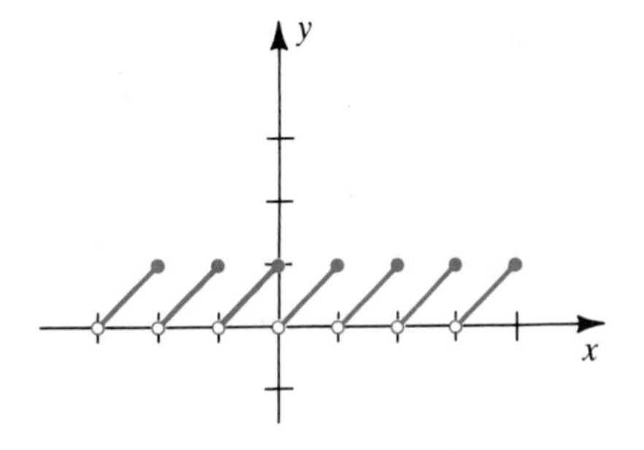

6

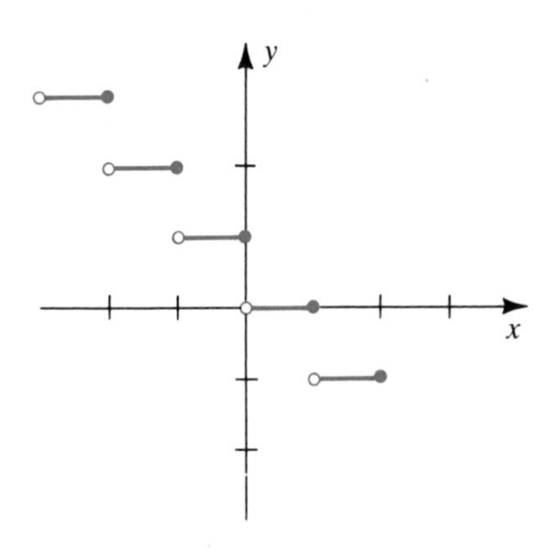

7

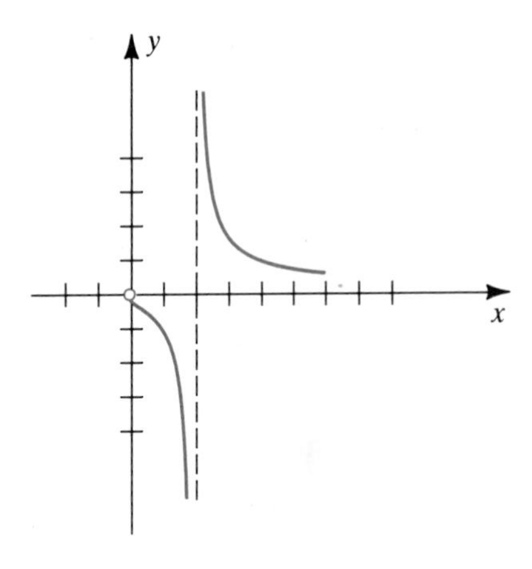

8

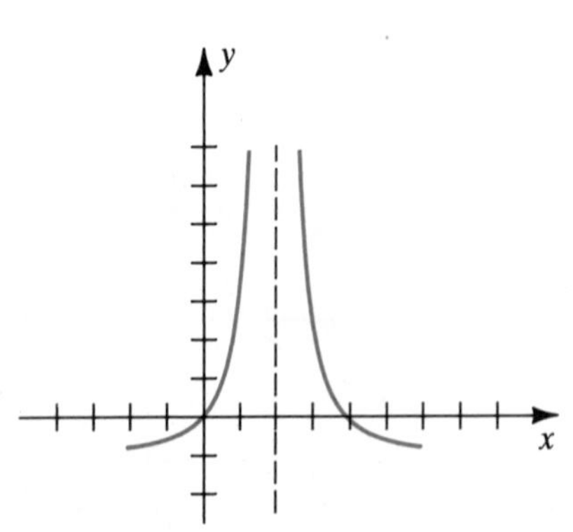

9

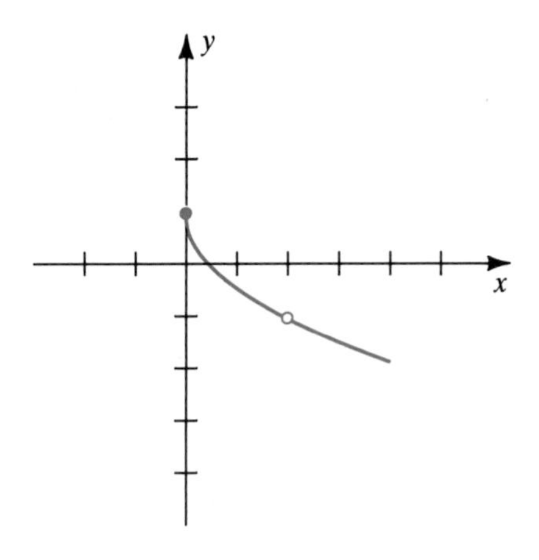

10

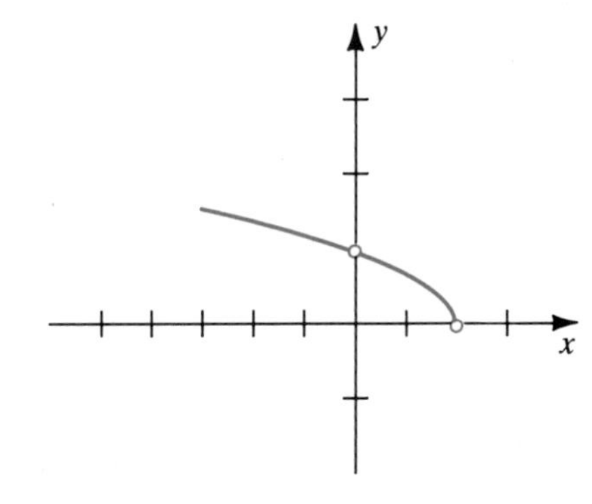

Exer. 11–18: Classify the discontinuities of f as removable, jump, or infinite.

11 $f(x) = \begin{cases} x^2 - 1 & \text{if } x < 1 \\ 4 - x & \text{if } x \geq 1 \end{cases}$

12 $f(x) = \begin{cases} x^3 & \text{if } x \leq 1 \\ 3 - x & \text{if } x > 1 \end{cases}$

13 $f(x) = \begin{cases} |x + 3| & \text{if } x \neq -2 \\ 2 & \text{if } x = -2 \end{cases}$

14 $f(x) = \begin{cases} |x - 1| & \text{if } x \neq 1 \\ 1 & \text{if } x = 1 \end{cases}$

15 $f(x) = \begin{cases} x^2 + 1 & \text{if } x < 1 \\ 1 & \text{if } x = 1 \\ x + 1 & \text{if } x > 1 \end{cases}$

16 $f(x) = \begin{cases} -x^2 & \text{if } x < 1 \\ 2 & \text{if } x = 1 \\ x - 2 & \text{if } x > 1 \end{cases}$

c 17 $f(x) = x^{-1/3} \sin \left[\cos \left(\frac{\pi}{2} - x^2 \right) \right]$

c 18 $f(x) = \dfrac{\sin (x^2 - 1)}{(x - 1)^2}$

Exer. 19–22: Show that f is continuous at a.

19 $f(x) = \sqrt{2x - 5} + 3x$; $a = 4$

20 $f(x) = \sqrt[3]{x^2 + 2}$; $a = -5$

21 $f(x) = 3x^2 + 7 - \dfrac{1}{\sqrt{-x}}$; $a = -2$

22 $f(x) = \dfrac{\sqrt[3]{x}}{2x + 1}$; $a = 8$

Exer. 23–30: Explain why f is not continuous at a.

23 $f(x) = \dfrac{3}{x + 2}$; $a = -2$

24 $f(x) = \dfrac{1}{x - 1}$; $a = 1$

25 $f(x) = \begin{cases} \dfrac{x^2 - 9}{x - 3} & \text{if } x \neq 3 \\ 4 & \text{if } x = 3 \end{cases}$ $a = 3$

26 $f(x) = \begin{cases} \dfrac{x^2 - 9}{x + 3} & \text{if } x \neq -3 \\ 2 & \text{if } x = -3 \end{cases}$ $a = -3$

27 $f(x) = \begin{cases} 1 & \text{if } x \neq 3 \\ 0 & \text{if } x = 3 \end{cases}$ $a = 3$

28 $f(x) = \begin{cases} \dfrac{|x - 3|}{x - 3} & \text{if } x \neq 3 \\ 1 & \text{if } x = 3 \end{cases}$ $a = 3$

c 29 $f(x) = \begin{cases} \dfrac{\sin x}{x} & \text{if } x \neq 0 \\ 0 & \text{if } x = 0 \end{cases}$ $a = 0$

c 30 $f(x) = \begin{cases} \dfrac{1 - \cos x}{x} & \text{if } x \neq 0 \\ 1 & \text{if } x = 0 \end{cases}$ $a = 0$

Exer. 31–34: Find all numbers at which f is discontinuous.

31 $f(x) = \dfrac{3}{x^2 + x - 6}$

32 $f(x) = \dfrac{5}{x^2 - 4x - 12}$

33 $f(x) = \dfrac{x - 1}{x^2 + x - 2}$

34 $f(x) = \dfrac{x - 4}{x^2 - x - 12}$

Exer. 35–38: Show that f is continuous on the given interval.

35 $f(x) = \sqrt{x - 4}$; $[4, 8]$

36 $f(x) = \sqrt{16 - x}$; $(-\infty, 16]$

37 $f(x) = \dfrac{1}{x^2}$; $(0, \infty)$

38 $f(x) = \dfrac{1}{x - 1}$; $(1, 3)$

Exer. 39–54: Find all numbers at which f is continuous.

39 $f(x) = \dfrac{3x - 5}{2x^2 - x - 3}$

40 $f(x) = \dfrac{x^2 - 9}{x - 3}$

41 $f(x) = \sqrt{2x - 3} + x^2$

42 $f(x) = \dfrac{x}{\sqrt[3]{x - 4}}$

43 $f(x) = \dfrac{x - 1}{\sqrt{x^2 - 1}}$

44 $f(x) = \dfrac{x}{\sqrt{1 - x^2}}$

45 $f(x) = \dfrac{|x + 9|}{x + 9}$

46 $f(x) = \dfrac{x}{x^2 + 1}$

47 $f(x) = \dfrac{5}{x^3 - x^2}$

48 $f(x) = \dfrac{4x - 7}{(x + 3)(x^2 + 2x - 8)}$

49 $f(x) = \dfrac{\sqrt{x^2 - 9}\,\sqrt{25 - x^2}}{x - 4}$

50 $f(x) = \dfrac{\sqrt{9 - x}}{\sqrt{x - 6}}$

51 $f(x) = \tan 2x$

52 $f(x) = \cot \frac{1}{3}x$

53 $f(x) = \csc \frac{1}{2}x$

54 $f(x) = \sec 3x$

Exer. 55–58: Verify the intermediate value theorem (2.26) for f on the stated interval $[a, b]$ by showing that if $f(a) \leq w \leq f(b)$, then $f(c) = w$ for some c in $[a, b]$.

55 $f(x) = x^3 + 1$; $[-1, 2]$

56 $f(x) = -x^3$; $[0, 2]$

57 $f(x) = x^2 - x$; $[1, 3]$

58 $f(x) = 2x - x^2$; $[-2, -1]$

59 If $f(x) = x^3 - 5x^2 + 7x - 9$, use the intermediate value theorem (2.26) to prove that there is a real number a such that $f(a) = 100$.

60 Prove that the equation $x^5 - 3x^4 - 2x^3 - x + 1 = 0$ has a solution between 0 and 1.

c 61 In models for free fall, it is usually assumed that the gravitational acceleration g is the constant 9.8 m/sec^2 (or 32 ft/sec^2). Actually, g varies with latitude. If θ is the latitude (in degrees), then a formula that approximates g is

$$g(\theta) = 9.78049(1 + 0.005264 \sin^2 \theta + 0.000024 \sin^4 \theta).$$

Use the intermediate value theorem to show that $g = 9.8$ somewhere between latitudes 35° and 40°.

c 62 The temperature T (in °C) at which water boils may be approximated by the formula

$$T(h) = 100.862 - 0.0415\sqrt{h + 431.03},$$

where h is the elevation (in meters above sea level). Use the intermediate value theorem to show that water boils at 98 °C at an elevation somewhere between 4000 and 4500 meters.

2.6 REVIEW EXERCISES

Exer. 1–26: Find the limit, if it exists.

1 $\lim_{x \to 3} \dfrac{5x + 11}{\sqrt{x + 1}}$

2 $\lim_{x \to -2} \dfrac{6 - 7x}{(3 + 2x)^4}$

3 $\lim_{x \to -2} (2x - \sqrt{4x^2 + x})$

4 $\lim_{x \to 4^-} (x - \sqrt{16 - x^2})$

5 $\lim_{x \to 3/2} \dfrac{2x^2 + x - 6}{4x^2 - 4x - 3}$

6 $\lim_{x \to 2} \dfrac{3x^2 - x - 10}{x^2 - x - 2}$

7 $\lim_{x \to 2} \dfrac{x^4 - 16}{x^2 - x - 2}$

8 $\lim_{x \to 3^+} \dfrac{1}{x - 3}$

9 $\lim_{x \to 0^+} \dfrac{1}{\sqrt{x}}$

10 $\lim_{x \to 5} \dfrac{(1/x) - (1/5)}{x - 5}$

11 $\lim_{x \to 1/2} \dfrac{8x^3 - 1}{2x - 1}$

12 $\lim_{x \to 2} 5$

13 $\lim_{x \to 3^+} \dfrac{3 - x}{|3 - x|}$

14 $\lim_{x \to 2} \dfrac{\sqrt{x} - \sqrt{2}}{x - 2}$

15 $\lim_{h \to 0} \dfrac{(a + h)^4 - a^4}{h}$

16 $\lim_{h \to 0} \dfrac{(2 + h)^{-3} - 2^{-3}}{h}$

17 $\lim_{x \to -3} \sqrt[3]{\dfrac{x + 3}{x^3 + 27}}$

18 $\lim_{x \to 5/2^-} (\sqrt{5 - 2x} - x^2)$

19 $\lim_{x \to -\infty} \dfrac{(2x - 5)(3x + 1)}{(x + 7)(4x - 9)}$

20 $\lim_{x \to \infty} \dfrac{2x + 11}{\sqrt{x + 1}}$

21 $\lim_{x \to -\infty} \dfrac{6 - 7x}{(3 + 2x)^4}$

22 $\lim_{x \to \infty} \dfrac{x - 100}{\sqrt{x^2 + 100}}$

23 $\lim_{x \to 2/3^+} \dfrac{x^2}{4 - 9x^2}$

24 $\lim_{x \to 3/5^-} \dfrac{1}{5x - 3}$

25 $\lim_{x \to 0^+} \left(\sqrt{x} - \dfrac{1}{\sqrt{x}}\right)$

26 $\lim_{x \to 1} \dfrac{x - 1}{\sqrt{(x - 1)^2}}$

Exer. 27–32: Sketch the graph of the piecewise-defined function f and, for the indicated value of a, find each limit, if it exists:

(a) $\lim_{x \to a^-} f(x)$ (b) $\lim_{x \to a^+} f(x)$ (c) $\lim_{x \to a} f(x)$

27 $f(x) = \begin{cases} 3x & \text{if } x \le 2 \\ x^2 & \text{if } x > 2 \end{cases}$ $a = 2$

28 $f(x) = \begin{cases} x^3 & \text{if } x \le 2 \\ 4 - 2x & \text{if } x > 2 \end{cases}$ $a = 2$

29 $f(x) = \begin{cases} \dfrac{1}{2 - 3x} & \text{if } x < -3 \\ \sqrt[3]{x + 2} & \text{if } x \ge -3 \end{cases}$ $a = -3$

30 $f(x) = \begin{cases} \dfrac{9}{x^2} & \text{if } x \le -3 \\ 4 + x & \text{if } x > -3 \end{cases}$ $a = -3$

31 $f(x) = \begin{cases} x^2 & \text{if } x < 1 \\ 2 & \text{if } x = 1 \\ 4 - x^2 & \text{if } x > 1 \end{cases}$ $a = 1$

32 $f(x) = \begin{cases} \dfrac{x^4 + x}{x} & \text{if } x \ne 0 \\ 2 & \text{if } x = 0 \end{cases}$ $a = 0$

33 Use Definition (2.4) to prove that $\lim_{x \to 6} (5x - 21) = 9$.

34 Let f be defined as follows:

$$f(x) = \begin{cases} 1 & \text{if } x \text{ is rational} \\ -1 & \text{if } x \text{ is irrational} \end{cases}$$

Prove that for every real number a, $\lim_{x \to a} f(x)$ does not exist.

Exer. 35–38: Find all numbers at which f is discontinuous.

35 $f(x) = \dfrac{|x^2 - 16|}{x^2 - 16}$

36 $f(x) = \dfrac{1}{x^2 - 16}$

37 $f(x) = \dfrac{x^2 - x - 2}{x^2 - 2x}$

38 $f(x) = \dfrac{x + 2}{x^3 - 8}$

Exer. 39–42: Find all numbers at which f is continuous.

39 $f(x) = 2x^4 - \sqrt[3]{x} + 1$

40 $f(x) = \sqrt{(2 + x)(3 - x)}$

41 $f(x) = \dfrac{\sqrt{9 - x^2}}{x^4 - 16}$

42 $f(x) = \dfrac{\sqrt{x}}{x^2 - 1}$

Exer. 43–44: Show that f is continuous at the number a.

43 $f(x) = \sqrt{5x + 9}$; $a = 8$

44 $f(x) = \sqrt[3]{x^2 - 4}$; $a = 27$

CHAPTER

3

THE DERIVATIVE

INTRODUCTION

We begin this chapter by considering two applied problems. The first is to find the slope of the tangent line at a point on the graph of a function, and the second is to define the velocity of an object moving along a line. Remarkably, these seemingly diverse applications lead us to the same concept: the *derivative*. Our discussion provides insight into the power and generality of mathematics. Specifically, in Section 3.2 we strip away the geometric and physical aspects of the two problems and define the derivative as the limit of an expression involving a function f. This allows us, in later sections, to apply the derivative concept to any quantity that can be represented by a function. Since quantities of this type occur in almost every field of knowledge, applications of derivatives are numerous and varied, but each concerns a *rate of change*. Thus, returning to the two problems that started it all, the slope of a tangent line may be used to describe the rate at which a graph rises (or falls), and velocity is the rate at which distance changes with respect to time.

Our main objective in this chapter is to introduce derivatives and develop rules that can be used to find them without employing limits. We shall consider some applications here and many more in subsequent chapters.

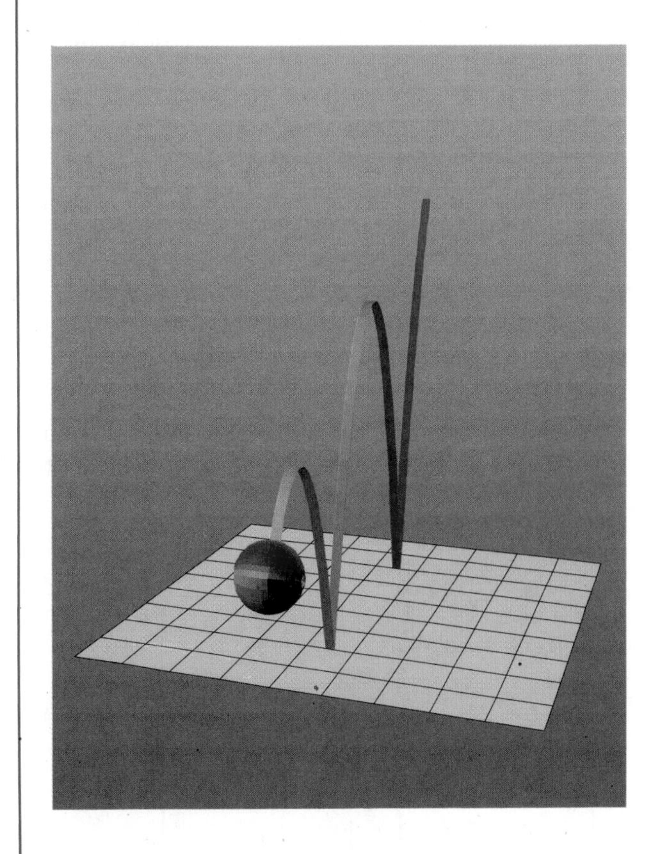

3.1 TANGENT LINES AND RATES OF CHANGE

FIGURE 3.1

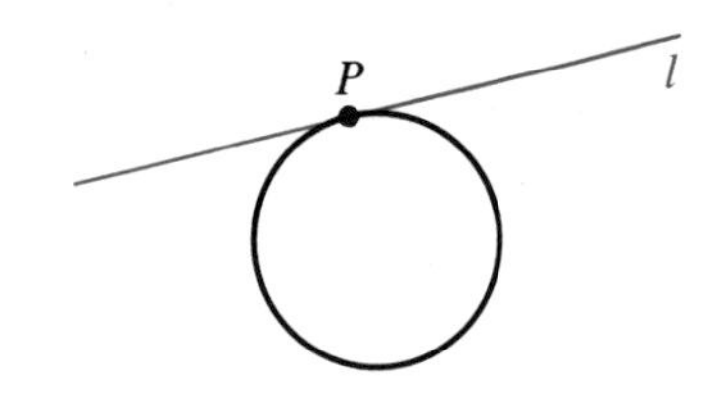

FIGURE 3.2

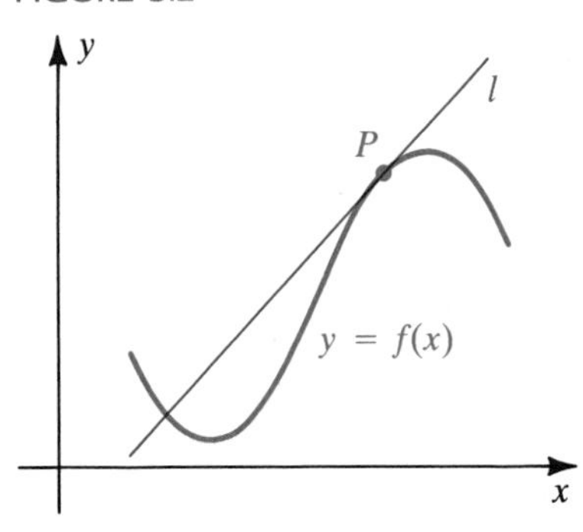

Tangent lines to graphs are useful in many applications of calculus. In geometry the tangent line l at a point P on a circle may be interpreted as the line that intersects the circle only at P, as illustrated in Figure 3.1. We cannot extend this interpretation to the graph of a function f, since a line may "touch" the graph of f at some isolated point P and then intersect it again at another point, as illustrated in Figure 3.2. Our plan is to define the *slope* of the tangent line at P, for if the slope is known, we can find an equation for l by using the point-slope form (1.8)(ii).

To define the slope of the tangent line l at $P(a, f(a))$ on the graph of f, we first choose another point $Q(x, f(x))$ (see Figure 3.3(i)) and consider the line through P and Q. This line is called a **secant line** for the graph.

We shall use the following notation:

l_{PQ}: the secant line through P and Q

m_{PQ}: the slope of l_{PQ}

m_a: the slope of the tangent line l at $P(a, f(a))$

If Q is close to P, it appears that m_{PQ} is an approximation to m_a. Moreover, we would expect this approximation to improve if we take Q closer to P. With this in mind, we let Q *approach* P—that is, we (intuitively) let Q get closer to P—but $Q \neq P$. If Q approaches P from the right, we have the situation illustrated in Figure 3.3(ii), where dashed lines indicate possible positions for l_{PQ}. In Figure 3.3(iii), Q approaches P from the left. We could also let Q approach P in other ways, such as by taking points on the graph that are alternately to the left and to the right of P. If m_{PQ} has a limiting value—that is, if m_{PQ} gets closer to some number as Q approaches P—then that number is the slope m_a of the tangent line l.

Let us rephrase this discussion in terms of the function f. Referring to Figure 3.3 and using the coordinates of $P(a, f(a))$ and $Q(x, f(x))$, we see that the slope of the secant line l_{PQ} is

$$m_{PQ} = \frac{f(x) - f(a)}{x - a}.$$

If f is continuous at a, we can make $Q(x, f(x))$ approach $P(a, f(a))$ by letting x approach a. This motivates the following definition for the slope

FIGURE 3.3

(i)

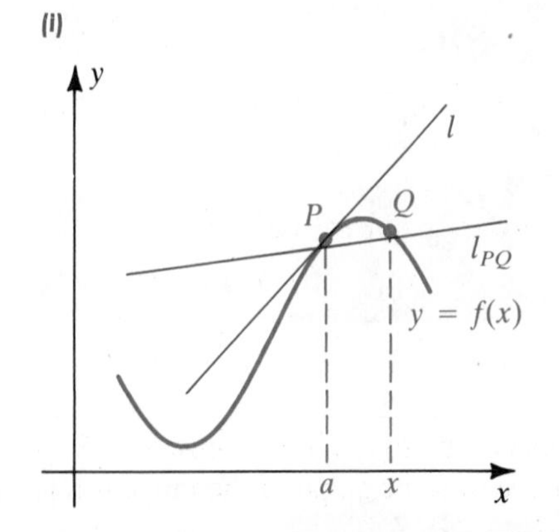

(ii)

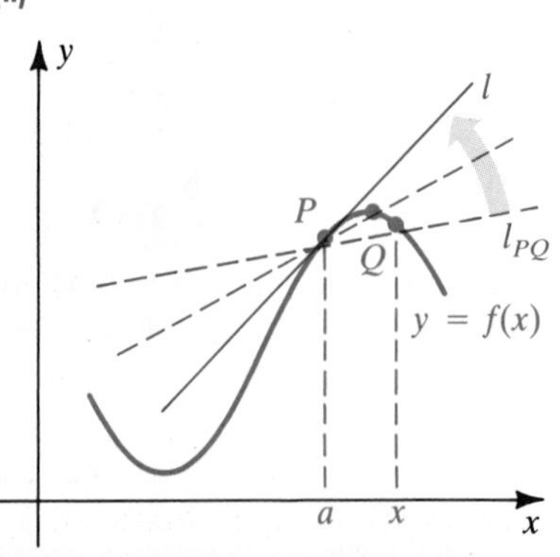

(iii)

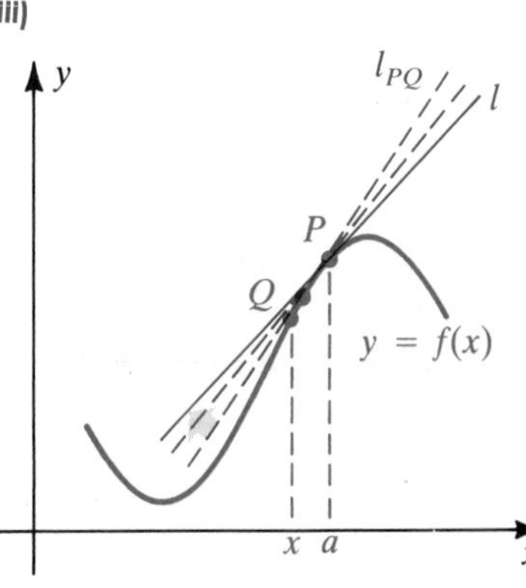

m_a of l at $P(a, f(a))$:

$$m_a = \lim_{x \to a} \frac{f(x) - f(a)}{x - a},$$

provided the limit exists.

It is often desirable to use an alternative form for m_a obtained by changing from the variable x to a variable h as follows.

$$\text{Let} \quad h = x - a, \quad \text{or, equivalently,} \quad x = a + h.$$

Referring to Figure 3.4 and using the coordinates $P(a, f(a))$ and $Q(a + h, f(a + h))$, we see that the slope m_{PQ} of the secant line is

$$m_{PQ} = \frac{f(a + h) - f(a)}{h}.$$

FIGURE 3.4

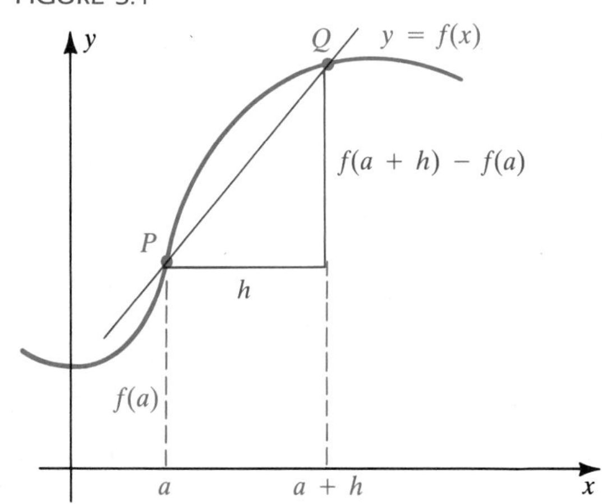

Since $x \to a$ is equivalent to $h \to 0$, our definition of the slope m_a of the tangent line l may be stated as follows.

Definition **(3.1)**

> The **slope m_a of the tangent line** to the graph of a function f at $P(a, f(a))$ is
>
> $$m_a = \lim_{h \to 0} \frac{f(a + h) - f(a)}{h},$$
>
> provided the limit exists.

FIGURE 3.5

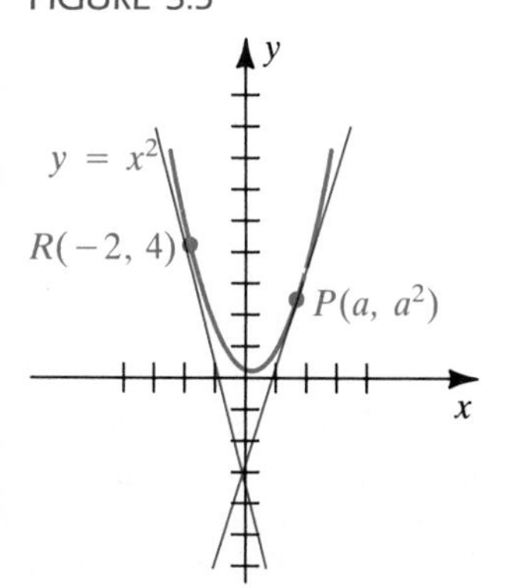

If the limit in Definition (3.1) does not exist, then the slope of the tangent line at $P(a, f(a))$ is undefined.

EXAMPLE 1 Let $f(x) = x^2$, and let a be any real number.

(a) Find the slope of the tangent line to the graph of f at $P(a, a^2)$.

(b) Find an equation of the tangent line at $R(-2, 4)$.

SOLUTION

(a) The graph of $y = x^2$ and a typical point $P(a, a^2)$ are shown in Figure 3.5. Applying Definition (3.1), we see that the slope of the tangent line at P is

$$\begin{aligned} m_a &= \lim_{h \to 0} \frac{f(a+h) - f(a)}{h} \\ &= \lim_{h \to 0} \frac{(a+h)^2 - a^2}{h} \\ &= \lim_{h \to 0} \frac{a^2 + 2ah + h^2 - a^2}{h} \\ &= \lim_{h \to 0} \frac{2ah + h^2}{h} \\ &= \lim_{h \to 0} (2a + h) = 2a. \end{aligned}$$

(b) The slope of the tangent line at the point $R(-2, 4)$ is the special case of the formula $m_a = 2a$ with $a = -2$; that is,

$$m_{-2} = 2(-2) = -4.$$

Using the point-slope form (1.8)(ii), we can express an equation for the tangent line as

$$y - 4 = -4(x + 2), \quad \text{or} \quad y = -4x - 4.$$

The limit in Definition (3.1) arises in a variety of applications. One of the most familiar is the determination of the speed, or velocity, of a moving object. Let us consider the case of **rectilinear motion**, in which the object travels along a line. To find the *average velocity* v_{av} during an interval of time, we use the formula $d = rt$, where r is the rate, t is the length of the time interval, and d is the net distance traveled. Solving for r gives us the following definition.

Definition **(3.2)**

> The **average velocity** v_{av} of an object that travels a distance d in time t is
>
> $$v_{av} = \frac{d}{t}.$$

FIGURE 3.6

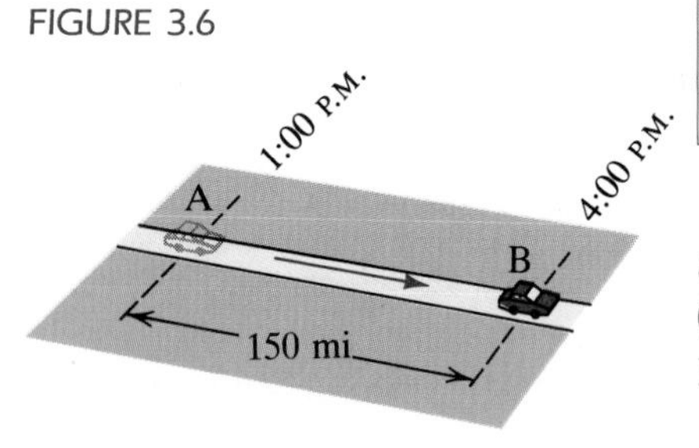

To illustrate, if an automobile leaves city A at 1:00 P.M. and travels along a straight highway, arriving at city B, 150 miles from A, at 4:00 P.M. (see Figure 3.6), then using Definition (3.2) with $d = 150$ and $t = 3$ (hours) yields the average velocity during the time interval $[1, 4]$:

$$v_{av} = \tfrac{150}{3} = 50 \text{ mi/hr}$$

FIGURE 3.7

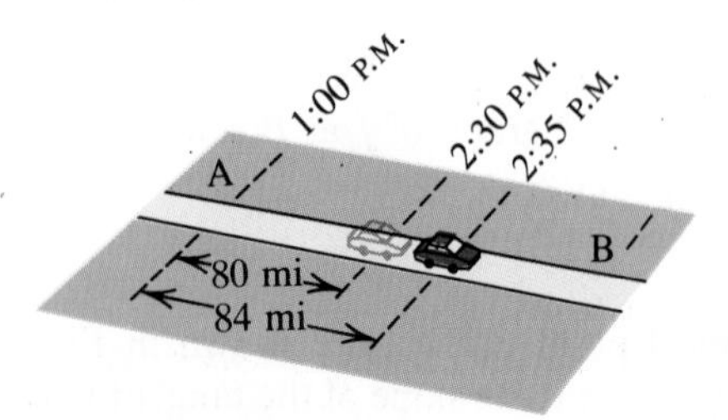

This is the velocity that, if maintained for 3 hours, would enable the automobile to travel the 150 miles from A to B.

The average velocity gives no information whatsoever about the velocity at any instant. For example, at 2:30 P.M. the automobile's speedometer may have registered 40 or 30, or the automobile may have been standing still. If we wish to determine the rate at which the automobile is traveling at 2:30 P.M., information is needed about its motion or position *near* this time. For example, suppose at 2:30 P.M. the automobile is 80 miles from A and at 2:35 P.M. it is 84 miles from A, as illustrated in Figure 3.7. The length of the time interval from 2:30 P.M. to 2:35 P.M. is 5 minutes, or $\frac{1}{12}$ hour, and the distance d is 4 miles. Substituting these numbers in Definition (3.2), we find that the average velocity during this time interval is

$$v_{\text{av}} = \frac{4}{\frac{1}{12}} = 48 \text{ mi/hr}.$$

This result is still not an accurate indication of the velocity at 2:30 P.M. since, for example, the automobile may have been traveling very slowly at 2:30 P.M. and then increased speed considerably to arrive at the point 84 miles from A at 2:35 P.M. Evidently, we obtain a better approximation by using the average velocity during a smaller time interval, say from 2:30 P.M. to 2:31 P.M. It appears that the best procedure would be to take smaller and smaller time intervals near 2:30 P.M. and study the average velocity in each time interval. This leads us into a limiting process similar to that discussed for tangent lines.

In order to make our discussion more precise, let us represent the position of an object moving rectilinearly by a point P on a coordinate line l. We sometimes refer to the motion of the *point* P on l, or the motion of an object whose position is specified by P. We shall assume that we know the position of P at every instant in a given interval of time. If $s(t)$ denotes the coordinate of P at time t, then the function s is called the **position function** for P. If we keep track of time by means of a clock, then, as illustrated in Figure 3.8, for each t the point P is $s(t)$ units from the origin.

FIGURE 3.8

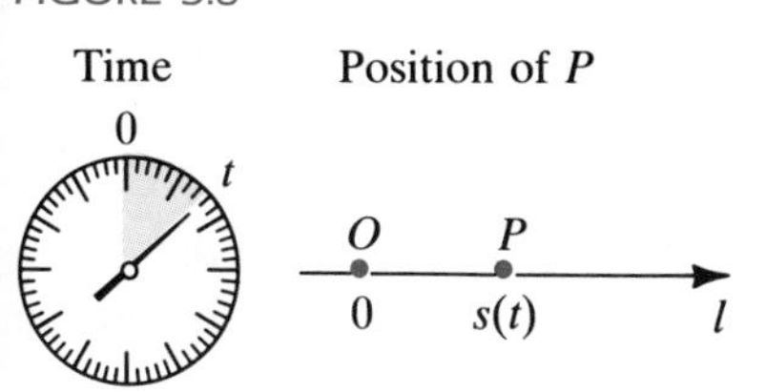

To define the velocity of P at time a, we first determine the average velocity in a (small) time interval near a. Thus, we consider times a and $a + h$, where h is a (small) nonzero real number. The corresponding positions of P are $s(a)$ and $s(a + h)$, as illustrated in Figure 3.9. The amount of change in the position of P is $s(a + h) - s(a)$. This number may be positive, negative, or zero. Note that $s(a + h) - s(a)$ is not necessarily the distance traveled by P between times a and $a + h$ since, for example, P may have moved beyond the point corresponding to $s(a + h)$ and then returned to that point at time a.

FIGURE 3.9

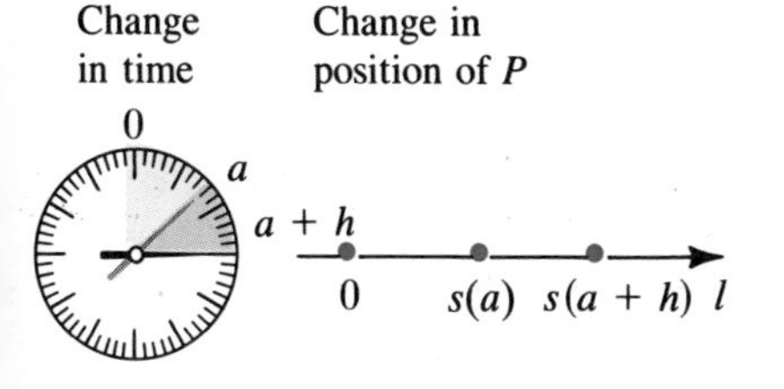

By Definition (3.2), the average velocity of P between times a and $a + h$ is

$$v_{\text{av}} = \frac{\text{change in distance}}{\text{change in time}} = \frac{s(a + h) - s(a)}{h}.$$

As in our previous discussion, we assume that the closer h is to 0, the closer v_{av} is to the velocity of P at time a. Thus, we *define* the velocity as the limit, as h approaches 0, of v_{av}, as in the following definition.

Definition **(3.3)**

Suppose a point P moves on a coordinate line l such that its coordinate at time t is $s(t)$. The **velocity** v_a of P at time a is

$$v_a = \lim_{h \to 0} \frac{s(a + h) - s(a)}{h},$$

provided the limit exists.

The limit in Definition (3.3) is also called the **instantaneous velocity** of P at time a.

If $s(t)$ is measured in centimeters and t in seconds, then the unit of velocity is centimeters per second (cm/sec). If $s(t)$ is in miles and t in hours, then velocity is in miles per hour. Other units of measurement may, of course, be used.

We shall return to the velocity concept in Chapter 4, where we will show that if the velocity is positive in a given time interval, then the point is moving in the positive direction on l. If the velocity is negative, the point is moving in the negative direction. Although these facts have not been proved, we shall use them in the following example.

EXAMPLE 2 A sandbag is dropped from a hot-air balloon that is hovering at a height of 512 feet above the ground. If air resistance is disregarded, then the distance $s(t)$ from the ground to the sandbag after t seconds is given by

$$s(t) = -16t^2 + 512.$$

Find the velocity of the sandbag at

(a) $t = a$ sec **(b)** $t = 2$ sec **(c)** the instant it strikes the ground

SOLUTION

FIGURE 3.10

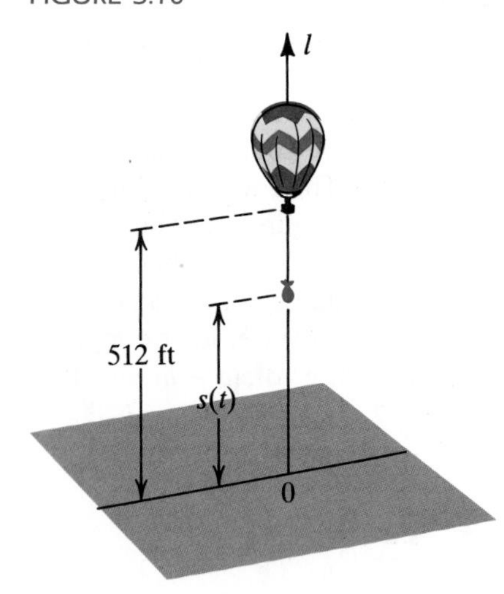

(a) As shown in Figure 3.10, we consider the sandbag to be moving along a vertical coordinate line l with origin at ground level. Note that at the instant it is dropped, $t = 0$ and

$$s(0) = -16(0) + 512 = 512 \text{ ft}.$$

To find the velocity of the sandbag at $t = a$, we use Definition (3.3), obtaining

$$\begin{aligned} v_a &= \lim_{h\to 0} \frac{s(a+h) - s(a)}{h} \\ &= \lim_{h\to 0} \frac{[-16(a+h)^2 + 512] - (-16a^2 + 512)}{h} \\ &= \lim_{h\to 0} \frac{-16(a^2 + 2ah + h^2) + 512 + 16a^2 - 512}{h} \\ &= \lim_{h\to 0} \frac{-32ah - 16h^2}{h} \\ &= \lim_{h\to 0} (-32a - 16h) = -32a \text{ ft/sec}. \end{aligned}$$

The negative sign indicates that the motion of the sandbag is in the negative direction (downward) on l.

(b) To find the velocity at $t = 2$, we substitute 2 for a in the formula $v_a = -32a$, obtaining

$$v_2 = -32(2) = -64 \text{ ft/sec}.$$

(c) The sandbag strikes the ground when the distance above the ground is zero—that is, when

$$s(t) = -16t^2 + 512 = 0, \quad \text{or} \quad t^2 = \tfrac{512}{16} = 32.$$

This gives us $t = \sqrt{32} = 4\sqrt{2} \approx 5.7$ sec. Using the formula $v_a = -32a$ from part (a) with $a = 4\sqrt{2}$, we obtain the following impact velocity:

$$-32(4\sqrt{2}) = -128\sqrt{2} \approx -181 \text{ ft/sec}$$

There are many other applications that require limits similar to those in (3.1) and (3.3). In some, the independent variable is time t, as in the definition of velocity. For example, over a period of time, a chemist may be interested in the rate at which a certain substance dissolves in water; an electrical engineer may wish to know the rate of change of current in part of an electrical circuit; a biologist may be concerned with the rate at which the bacteria in a culture increase or decrease. We can also consider rates of change with respect to quantities other than time. To illustrate, Boyle's law for a confined gas states that if the temperature remains constant, then the volume v and pressure p are related by the formula $v = c/p$ for some constant c. If the pressure is changing, a typical problem is to find the rate at which the volume is changing per unit change in pressure. This rate is known as *the instantaneous rate of change of v with respect to p*. To develop general methods that can be applied to different problems of this type, let us use x and y for variables and suppose that $y = f(x)$ for some function f. (In the preceding illustration $y = v$, $x = p$, and $f(x) = c/x$.) We define rates of change of a variable y with respect to a variable x as follows.

Definition **(3.4)**

Let $y = f(x)$, where f is defined on an open interval containing a.

(i) The **average rate of change** of $y = f(x)$ with respect to x on the interval $[a, a + h]$ is

$$y_{av} = \frac{f(a+h) - f(a)}{h}.$$

(ii) The **instantaneous rate of change** of y with respect to x at a is

$$y_a = \lim_{h \to 0} \frac{f(a+h) - f(a)}{h},$$

provided the limit exists.

We shall use the phrase *rate of change* interchangeably with *instantaneous rate of change*.

If, in Definition (3.4), we consider the special case $x = t$ (time) and $y = s(t)$ (position on a coordinate line), we obtain the following interpretations for rectilinear motion:

average velocity (v_{av}): the average rate of change of s with respect to t in some interval of time.

velocity (v_a): the instantaneous rate of change of s with respect to t at time a.

To interpret (ii) of Definition (3.4) geometrically, imagine a point P traveling from left to right along the graph of $y = f(x)$ in Figure 3.11. The

FIGURE 3.11

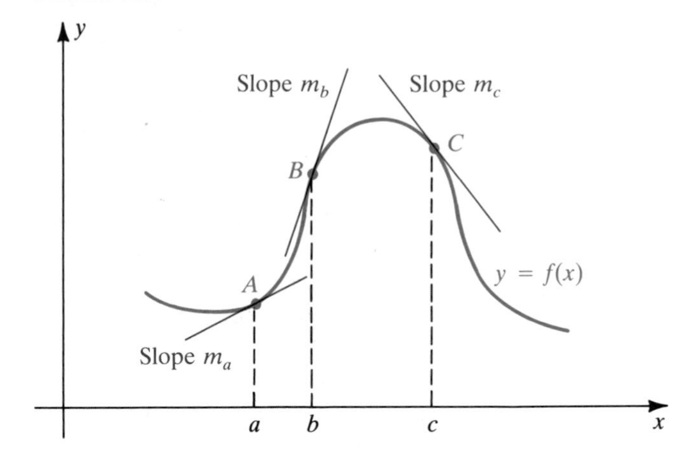

instantaneous rate of change of y with respect to x gives us information about the rate at which the graph rises or falls per unit change in x. In Figure 3.11, m_a (the slope of the tangent line at A) is less than m_b (the slope of the tangent line at B), and the rate y_a at which y changes with respect to x at a is less than the rate y_b at which y changes with respect to x at b. Also note that since $m_c < 0$, the slope of the tangent line at C is negative, and y *decreases* as x increases.

A physical application of Definition (3.4) is given in the following example.

EXAMPLE 3 The voltage in a certain electrical circuit is 100 volts. If the current (in amperes) is I and the resistance (in ohms) is R, then by Ohm's law, $I = 100/R$. If R is increasing, find the instantaneous rate of change of I with respect to R at

(a) any resistance R **(b)** a resistance of 20 ohms

SOLUTION

(a) Using Definition (3.4)(ii) with $y = I$, $x = R$, and $f(R) = 100/R$ yields the instantaneous rate of change of I with respect to R at a resistance of R ohms:

$$\begin{aligned} I_R &= \lim_{h \to 0} \frac{f(R+h) - f(R)}{h} \\ &= \lim_{h \to 0} \frac{\dfrac{100}{R+h} - \dfrac{100}{R}}{h} \\ &= \lim_{h \to 0} \frac{100R - 100(R+h)}{h(R+h)R} \\ &= \lim_{h \to 0} \frac{-100h}{h(R+h)R} \\ &= \lim_{h \to 0} \frac{-100}{(R+h)R} = -\frac{100}{R^2} \end{aligned}$$

The negative sign indicates that the current is decreasing.

(b) Using the formula $I_R = -100/R^2$ from part (a), we find the instantaneous rate of change of I with respect to R at $R = 20$ to be

$$I_{20} = -\frac{100}{20^2} = -\frac{1}{4}.$$

Thus, when $R = 20$, the current is *decreasing* at a rate of $\frac{1}{4}$ ampere per ohm.

EXERCISES 3.1

Exer. 1–6: (a) Use Definition (3.1) to find the slope of the tangent line to the graph of f at $P(a, f(a))$. (b) Find an equation of the tangent line at $P(2, f(2))$.

1 $f(x) = 5x^2 - 4x$

2 $f(x) = 3 - 2x^2$

3 $f(x) = x^3$

4 $f(x) = x^4$

5 $f(x) = 3x + 2$

6 $f(x) = 4 - 2x$

Exer. 7–10: (a) Use Definition (3.1) to find the slope of the tangent line to the graph of the equation at the point with x-coordinate a. (b) Find an equation of the tangent line at P. (c) Sketch the graph and the tangent line at P.

7 $y = \sqrt{x}$; $P(4, 2)$

8 $y = \sqrt[3]{x}$; $P(-8, -2)$

9 $y = 1/x$; $P(2, \frac{1}{2})$

10 $y = 1/x^2$; $P(2, \frac{1}{4})$

Exer. 11–12: (a) Sketch the graph of the equation and the tangent lines at the points with x-coordinates -2, -1, 1, and 2. (b) Find the point on the graph at which the slope of the tangent line is the given number m.

11 $y = x^2$; $m = 6$

12 $y = x^3$; $m = 9$

Exer. 13–14: The position function s of a point P moving on a coordinate line l is given, with t in seconds and $s(t)$ in centimeters. (a) Find the average velocity of P in the following time intervals: [1, 1.2], [1, 1.1], and [1, 1.01]. (b) Find the velocity of P at $t = 1$.

13 $s(t) = 4t^2 + 3t$

14 $s(t) = 2t - 3t^2$

15 A balloonist drops a sandbag from a balloon 160 feet above the ground. After t seconds, the sandbag is $160 - 16t^2$ feet above the ground.

(a) Find the velocity of the sandbag at $t = 1$.

(b) With what velocity does the sandbag strike the ground?

16 A projectile is fired directly upward from the ground with an initial velocity of 112 ft/sec. Its distance above the ground after t seconds is $112t - 16t^2$ feet.

(a) Find the velocity of the projectile at $t = 2$, $t = 3$, and $t = 4$.

(b) When does the projectile strike the ground?

(c) Find the velocity at the instant it strikes the ground.

17 In the video game shown in the figure, airplanes fly from left to right along the path $y = 1 + (1/x)$ and can shoot their bullets in the tangent direction at creatures placed along the x-axis at $x = 1$, 2, 3, 4, and 5. Determine whether a creature will be hit if the player shoots when the plane is at

(a) $P(1, 2)$ (b) $Q(\frac{3}{2}, \frac{5}{3})$

EXERCISE 17

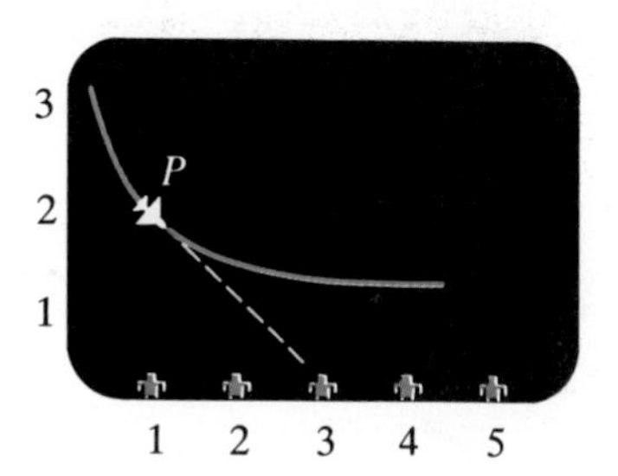

18 An athlete runs the hundred-meter dash in such a way that the distance $s(t)$ run after t seconds is given by $s(t) = \frac{1}{5}t^2 + 8t$ m (see figure). Find the athlete's velocity at

(a) the start of the dash (b) $t = 5$ sec

(c) the finish line

EXERCISE 18

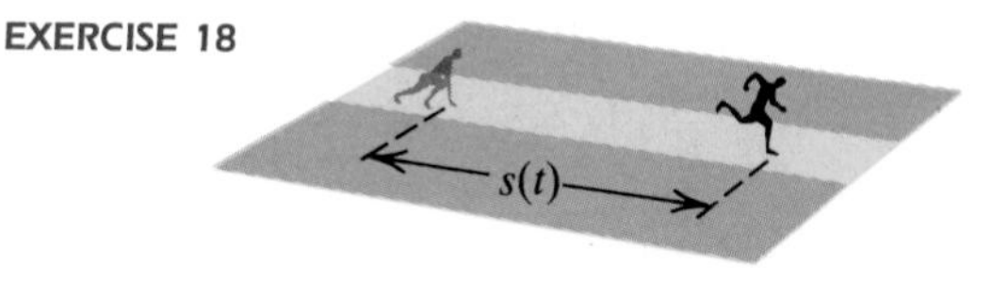

Exer. 19–20: (a) Find the average rate of change of y with respect to x on the given interval. (b) Find the instantaneous rate of change of y with respect to x at the left endpoint of the interval.

19 $y = x^2 + 2$; [3, 3.5]

20 $y = 3 - 2x^2$; [2, 2.4]

21 Boyle's law states that if the temperature remains constant, the pressure p and volume v of a confined gas are related by $p = c/v$ for some constant c. If, for a certain gas, $c = 200$ and v is increasing, find the instantaneous rate of change of p with respect to v at

(a) any volume v (b) a volume of 10

22 A spherical balloon is being inflated. Find the instantaneous rate of change of the surface area S of the balloon with respect to the radius r at

(a) any value of r (b) $r = 3$ ft

c 23 Graph $f(x) = \sin \pi x$ on the interval $[0, 2]$.

(a) Use the graph to estimate the slope of the tangent line at $P(1.4, f(1.4))$.

(b) Use Definition (3.1) with $h = \pm 0.0001$ to approximate the slope in (a).

c 24 Graph $f(x) = \dfrac{10 \cos x}{x^2 + 4}$ on the interval $[-2, 2]$.

(a) Use the graph to estimate the slope of the tangent line at $P(-0.5, f(-0.5))$.

(b) Use Definition (3.1) with $h = \pm 0.0001$ to approximate the slope in (a).

(c) Find an (approximate) equation of the tangent line to the graph at P.

c 25 An object's position on a coordinate line is given by

$$s(t) = \frac{\cos^2 t + t^2 \sin t}{t^2 + 1},$$

where $s(t)$ is in feet and t is in seconds. Approximate its velocity at $t = 2$ by using Definition (3.3) with $h = 0.01$, 0.001, and 0.0001.

c 26 The position function s of an object moving on a coordinate line is given by

$$s(t) = \frac{t - t^2 \sin t}{t^2 + 1},$$

where $s(t)$ is in meters and t is in minutes.

(a) Graph s for $0 \le t \le 10$.

(b) Approximate the time intervals in which its velocity is positive.

3.2 DEFINITION OF DERIVATIVE

In the preceding section we used the limit

$$\lim_{h \to 0} \frac{f(a + h) - f(a)}{h} \quad \text{or, equivalently,} \quad \lim_{x \to a} \frac{f(x) - f(a)}{x - a},$$

in several different applications. This limit is the basis for one of the fundamental concepts of calculus, *the derivative*, defined next.

Definition **(3.5)**

> The **derivative** of a function f is the function f' defined by
>
> $$f'(x) = \lim_{h \to 0} \frac{f(x + h) - f(x)}{h},$$
>
> provided the limit exists.

The symbol f' in Definition (3.5) is read f *prime*. It is important to note that in determining $f'(x)$ we regard x as an arbitrary real number and consider the limit as h approaches zero. Once we have obtained $f'(x)$, we can find $f'(a)$ for a specific real number a by substituting a for x.

The statement $f'(x)$ *exists* means that the limit in Definition (3.5) exists. If $f'(x)$ exists, we say that $\boldsymbol{f}$ **is differentiable at** $\boldsymbol{x}$, or that $\boldsymbol{f}$ **has a derivative at** $\boldsymbol{x}$. If the limit does not exist, then f is not differentiable at x. The terminology *differentiate* $f(x)$ or *find the derivative of* $f(x)$ means to find $f'(x)$.

Occasionally we will find it convenient to use the following alternative form of Definition (3.5) to find $f'(a)$.

Alternative definition of derivative **(3.6)**

$$f'(a) = \lim_{x \to a} \frac{f(x) - f(a)}{x - a}$$

This was the first formula used to define m_a on page 91.

The following applications are restatements of Definitions (3.1) and (3.4)(ii) using $f'(x)$. These interpretations of the derivative are very important and will be used in many examples and exercises throughout the text.

Applications of the derivative **(3.7)**

(i) **Tangent line:** The slope of the tangent line to the graph of $y = f(x)$ at the point $(a, f(a))$ is $f'(a)$.

(ii) **Rate of change:** If $y = f(x)$, the instantaneous rate of change of y with respect to x at a is $f'(a)$.

As a special case of (3.7)(ii), recall from Definition (3.3) that if $x = t$ denotes time and $y = s(t)$ is the position of a point P on a coordinate line, then $s'(t)$ is the velocity of P at time a.

A function f **is differentiable on an open interval** (a, b) if $f'(x)$ exists for every x in (a, b). We shall also consider functions that are differentiable on an infinite interval (a, ∞), $(-\infty, a)$, or $(-\infty, \infty)$. For closed intervals we use the following convention, which is analogous to the definition of continuity on a closed interval given in (2.22).

Definition **(3.8)**

A function f **is differentiable on a closed interval** $[a, b]$ if f is differentiable on the open interval (a, b) and if the following limits exist:

$$\lim_{h \to 0^+} \frac{f(a + h) - f(a)}{h} \quad \text{and} \quad \lim_{h \to 0^-} \frac{f(b + h) - f(b)}{h}$$

The one-sided limits in Definition (3.8) are sometimes referred to as the **right-hand derivative** and **left-hand derivative** of f at a and b, respectively. Note that for the right-hand derivative, $h \to 0^+$ and $a + h$ approaches a *from the right*. For the left-hand derivative, $h \to 0^-$ and $b + h$ approaches b *from the left*.

If f is defined on a closed interval $[a, b]$ and is undefined elsewhere, then the right-hand and left-hand derivatives define the slopes of the tangent lines at the points $P(a, f(a))$ and $R(b, f(b))$, respectively, as illustrated in Figure 3.12 on the following page. For the slope of the tangent line at P, we take the limiting value of the slope of the secant line through P and Q as Q approaches P from the right. For the tangent line at R, the point Q approaches R from the left.

Differentiability on an interval of the form $[a, b)$, $[a, \infty)$, $(a, b]$, or $(-\infty, b]$ is defined in similar fashion, using a one-sided limit at an endpoint.

The **domain of the derivative** f' consists of all numbers at which f is differentiable and also possible endpoints of the domain of f, whenever one-sided limits exist as indicated in Figure 3.12.

FIGURE 3.12

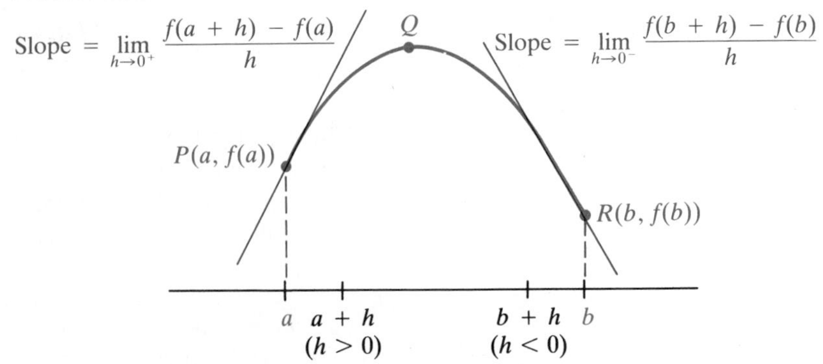

If f is defined on an open interval containing a, then $f'(a)$ *exists if and only if both right-hand and left-hand derivatives at a exist and are equal.* The functions whose graphs are sketched in Figure 3.13 have right-hand and left-hand derivatives at a that give the slopes of the lines l_1 and l_2, respectively. Since the slopes of l_1 and l_2 are unequal, $f'(a)$ does not exist. The graph of f has a **corner** at $P(a, f(a))$ if f is continuous at a and if the right-hand and left-hand derivatives at a exist and are unequal or if *one* of those derivatives exists at a and $|f'(x)| \to \infty$ as $x \to a^-$ or $x \to a^+$.

FIGURE 3.13

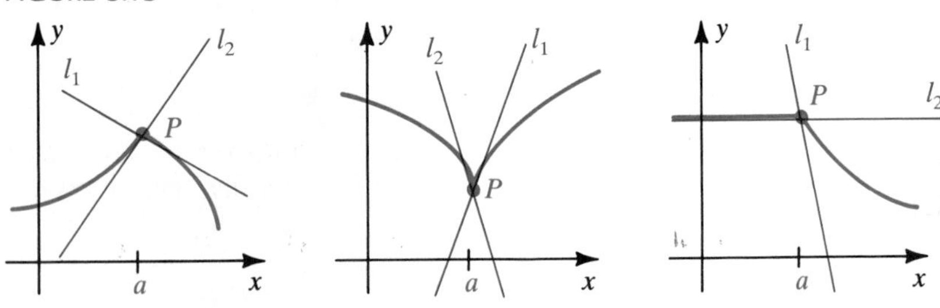

As indicated in the next definition, a *vertical tangent line* may occur at $P(a, f(a))$ if $f'(a)$ does not exist.

Definition **(3.9)**

> The graph of a function f has a **vertical tangent line** $x = a$ at the point $P(a, f(a))$ if f is continuous at a and if
>
> $$\lim_{x \to a} |f'(x)| = \infty.$$

If P is an endpoint of the domain of f, we can state a similar definition using a right-hand or left-hand derivative. Some typical vertical tangent lines are illustrated in Figure 3.14. As indicated in the next definition, the point P in Figure 3.14(iii) is called a *cusp*.

Definition **(3.10)**

> A point $P(a, f(a))$ on the graph of a function f is a **cusp** if f is continuous at a and if the following two conditions hold:
>
> **(i)** $f'(x) \to \infty$ as x approaches a from one side
>
> **(ii)** $f'(x) \to -\infty$ as x approaches a from the other side

FIGURE 3.14

(i)

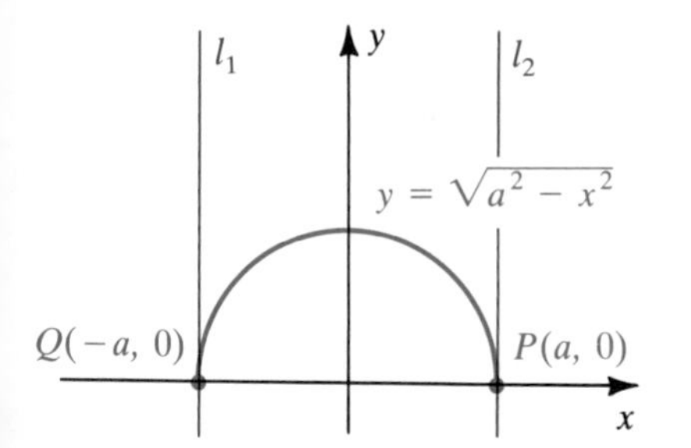

(ii)

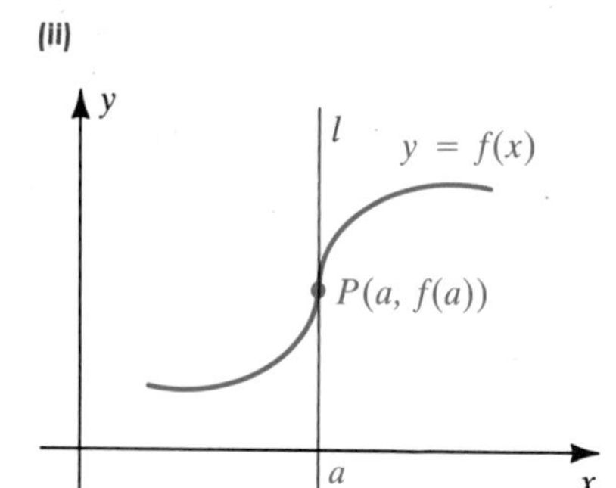

(iii)

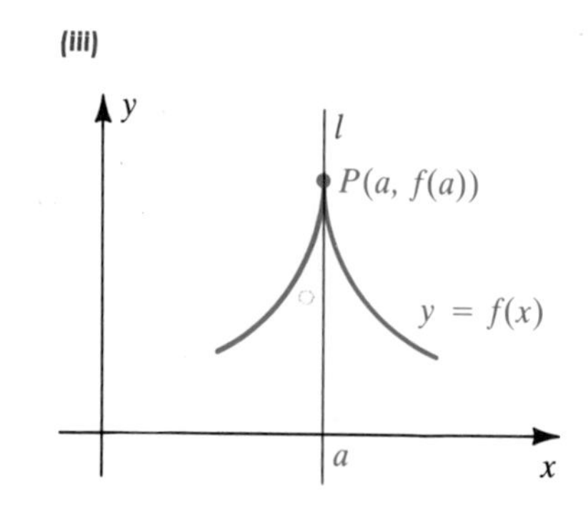

EXAMPLE 1 If $f(x) = 3x^2 - 12x + 8$, find

(a) $f'(x)$ **(b)** $f'(4)$, $f'(-2)$, and $f'(a)$ **(c)** the domain of f'

SOLUTION

(a) By Definition (3.5),

$$\begin{aligned}
f'(x) &= \lim_{h\to 0} \frac{f(x+h) - f(x)}{h} \\
&= \lim_{h\to 0} \frac{[3(x+h)^2 - 12(x+h) + 8] - (3x^2 - 12x + 8)}{h} \\
&= \lim_{h\to 0} \frac{(3x^2 + 6xh + 3h^2 - 12x - 12h + 8) - (3x^2 - 12x + 8)}{h} \\
&= \lim_{h\to 0} \frac{6xh + 3h^2 - 12h}{h} \\
&= \lim_{h\to 0} (6x + 3h - 12) \\
&= 6x - 12.
\end{aligned}$$

(b) Substituting for x in $f'(x) = 6x - 12$, we obtain

$$f'(4) = 6(4) - 12 = 12,$$

$$f'(-2) = 6(-2) - 12 = -24,$$

and

$$f'(a) = 6a - 12.$$

(c) Since $f'(x) = 6x - 12$, the derivative exists for every real number x. Hence the domain of f' is $\mathbb{R}$.

EXAMPLE 2 If $y = 3x^2 - 12x + 8$, use the result of Example 1 to find

(a) the slope of the tangent line to the graph of this equation at the point $P(3, -1)$

(b) the point on the graph at which the tangent line is horizontal

SOLUTION

(a) If we let $f(x) = 3x^2 - 12x + 8$, then, by (3.7)(i) and Example 1, the slope of the tangent line at $(x, f(x))$ is $f'(x) = 6x - 12$. In particular, the slope at $P(3, -1)$ is

$$f'(3) = 6(3) - 12 = 6.$$

FIGURE 3.15

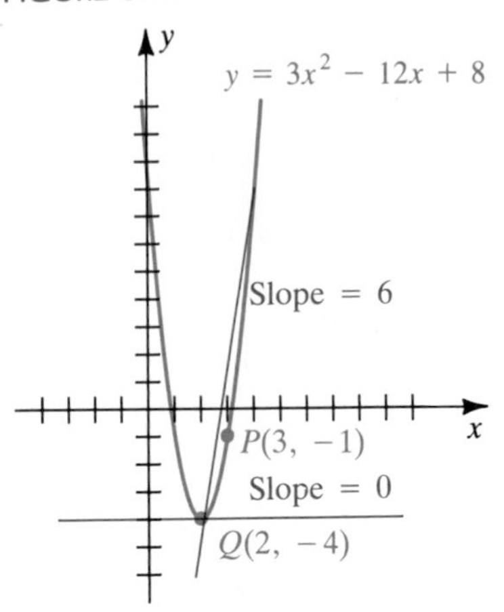

(b) Since the tangent line is horizontal if the slope $f'(x)$ is zero, we solve $6x - 12 = 0$, obtaining $x = 2$. The corresponding value of y is -4. Hence the tangent line is horizontal at the point $Q(2, -4)$.

The graph of f (a parabola) and the tangent lines at P and Q are sketched in Figure 3.15. Note that the vertex of the parabola is the point $Q(2, -4)$.

EXAMPLE 3 If $f(x) = \sqrt{x}$,

(a) sketch the graph of f **(b)** find $f'(x)$ and the domain of f'

SOLUTION

(a) The graph of f is sketched in Figure 3.16. Note that the domain of f consists of all nonnegative numbers.

FIGURE 3.16

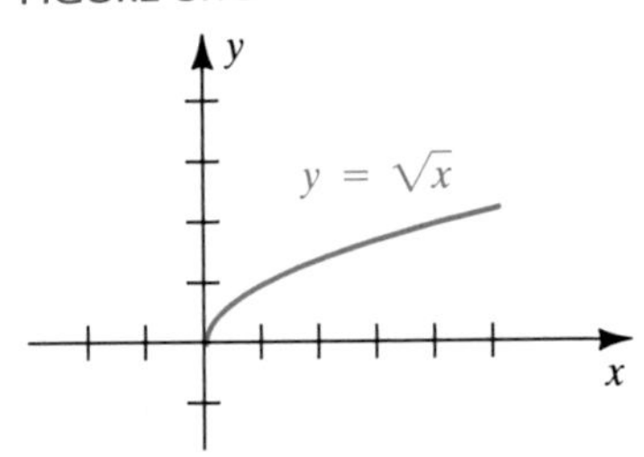

(b) Since $x = 0$ is an endpoint of the domain of f, we shall examine the cases $x > 0$ and $x = 0$ separately.

If $x > 0$, then, by Definition (3.5),

$$f'(x) = \lim_{h \to 0} \frac{\sqrt{x+h} - \sqrt{x}}{h}.$$

To find the limit, we first rationalize the numerator of the quotient and then simplify:

$$\begin{aligned} f'(x) &= \lim_{h \to 0} \frac{\sqrt{x+h} - \sqrt{x}}{h} \cdot \frac{\sqrt{x+h} + \sqrt{x}}{\sqrt{x+h} + \sqrt{x}} \\ &= \lim_{h \to 0} \frac{(x+h) - x}{h(\sqrt{x+h} + \sqrt{x})} \\ &= \lim_{h \to 0} \frac{1}{\sqrt{x+h} + \sqrt{x}} \\ &= \frac{1}{\sqrt{x} + \sqrt{x}} = \frac{1}{2\sqrt{x}} \end{aligned}$$

Since $x = 0$ is an endpoint of the domain of f, we must use a one-sided limit to determine if $f'(0)$ exists. Using Definition (3.8) with $x = 0$, we obtain

$$\begin{aligned} \lim_{h \to 0^+} \frac{f(0+h) - f(0)}{h} &= \lim_{h \to 0^+} \frac{\sqrt{0+h} - \sqrt{0}}{h} \\ &= \lim_{h \to 0^+} \frac{\sqrt{h}}{h} = \lim_{h \to 0^+} \frac{1}{\sqrt{h}} = \infty. \end{aligned}$$

Since the limit does not exist, the domain of f' is the set of positive real numbers. The last limit shows that the graph of f has a vertical tangent line (the y-axis) at the point $(0, 0)$.

FIGURE 3.17

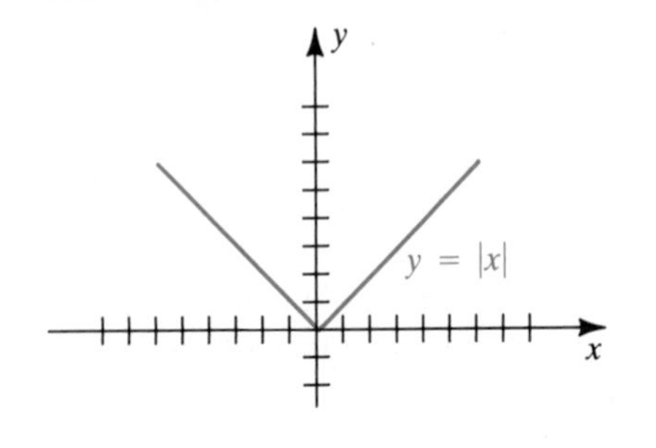

EXAMPLE 4 If $f(x) = |x|$, show that f is not differentiable at 0.

SOLUTION The graph of f is sketched in Figure 3.17. We can prove that $f'(0)$ does not exist by showing that the right-hand and left-hand derivatives of f at 0 are not equal. Using the limits in Definition (3.8) with $a = 0$ and $b = 0$ yields

$$\lim_{h\to 0^+} \frac{f(0+h)-f(0)}{h} = \lim_{h\to 0^+} \frac{|0+h|-|0|}{h} = \lim_{h\to 0^+} \frac{|h|}{h} = 1$$

$$\lim_{h\to 0^-} \frac{f(0+h)-f(0)}{h} = \lim_{h\to 0^-} \frac{|0+h|-|0|}{h} = \lim_{h\to 0^-} \frac{|h|}{h} = -1$$

Thus $f'(0)$ does not exist, and hence f is not differentiable at 0.

Note that the graph of $y = |x|$ in Figure 3.17 has a corner and therefore no tangent line at the point $P(0, 0)$.

The function f in Example 4 is continuous at $x = 0$ (see Example 1 of Section 2.5); however, $f'(0)$ does not exist. Thus, *not every continuous function is differentiable*. In contrast, the next theorem states that *every differentiable function is continuous.*

Theorem **(3.11)**

If a function f is differentiable at a, then f is continuous at a.

PROOF We shall use Alternative Definition (3.6):

$$f'(a) = \lim_{x\to a} \frac{f(x)-f(a)}{x-a}$$

We may write $f(x)$ in a form that contains $[f(x) - f(a)]/(x - a)$ as follows, provided $x \neq a$:

$$f(x) = \frac{f(x)-f(a)}{x-a}(x-a) + f(a)$$

Employing limit theorems, we find that

$$\lim_{x\to a} f(x) = \lim_{x\to a} \frac{f(x)-f(a)}{x-a} \cdot \lim_{x\to a}(x-a) + \lim_{x\to a} f(a)$$
$$= f'(a)\cdot 0 + f(a) = f(a).$$

Thus, by Definition (2.20), f is continuous at a. ■

By using one-sided limits, we can extend Theorem (3.11) to functions that are differentiable on a closed interval.

The process of finding a derivative by means of Definition (3.5) can be very tedious if $f(x)$ is a complicated expression. Fortunately, we can establish general formulas and rules that enable us to find $f'(x)$ without using limits.

FIGURE 3.18

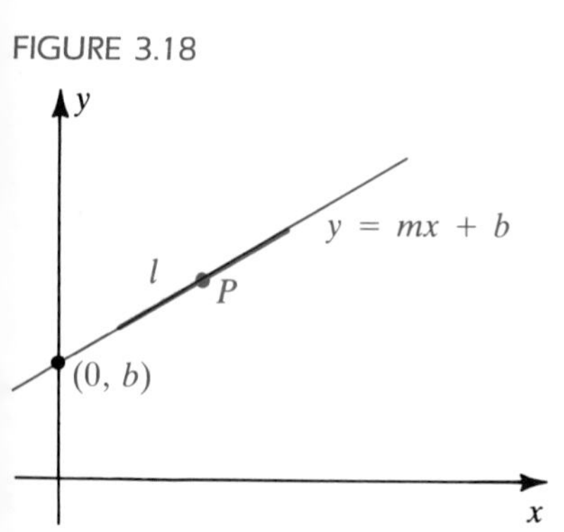

If f is a linear function, then $f(x) = mx + b$ for real numbers m and b. The graph of f is the line that has slope m and y-intercept b (see Figure 3.18). As indicated in the figure, the tangent line l at any point P coincides with the graph of f and hence has slope m. Thus, from (i) of (3.7), $f'(x) = m$ for every x. We can also prove this fact directly from Definition (3.5). This gives us the following rule.

Derivative of a linear function **(3.12)**

> If $f(x) = mx + b$, then $f'(x) = m$.

The following result is the special case of (3.12) with $m = 0$.

Derivative of a constant function **(3.13)**

> If $f(x) = b$, then $f'(x) = 0$.

The preceding result is also graphically evident, because the graph of a constant function is a horizontal line and hence has slope 0.

Some special cases of (3.12) and (3.13) are given in the following illustration.

ILLUSTRATION

$f(x)$	$3x - 7$	$-4x + 2$	$7x$	x	13	π^2	$\sqrt[3]{10}$
$f'(x)$	3	-4	7	1	0	0	0

Many algebraic expressions contain a variable x raised to some power n. The next result, appropriately called the *power rule*, provides a simple formula for finding the derivative if n is an integer.

Power rule **(3.14)**

> Let n be an integer.
>
> If $\quad f(x) = x^n, \quad$ then $\quad f'(x) = nx^{n-1}$,
>
> provided $x \neq 0$ when $n \leq 0$.

PROOF By Definition (3.5),

$$f'(x) = \lim_{h \to 0} \frac{f(x+h) - f(x)}{h} = \lim_{h \to 0} \frac{(x+h)^n - x^n}{h}.$$

If n is a positive integer, then we can expand $(x + h)^n$ by using the binomial theorem, obtaining

$$f'(x) = \lim_{h \to 0} \frac{\left[x^n + nx^{n-1}h + \dfrac{n(n-1)}{2!}x^{n-2}h^2 + \cdots + nxh^{n-1} + h^n\right] - x^n}{h}$$

$$= \lim_{h \to 0} \left[nx^{n-1} + \frac{n(n-2)}{2!}x^{n-2}h + \cdots + nxh^{n-2} + h^{n-1}\right].$$

Since each term within the brackets, except the first, contains a power of h, we see that $f'(x) = nx^{n-1}$.

If n is negative and $x \neq 0$, then we can write $n = -k$ with k positive. Thus,

$$f'(x) = \lim_{h \to 0} \frac{(x+h)^{-k} - x^{-k}}{h} = \lim_{h \to 0} \frac{x^k - (x+h)^k}{h(x+h)^k x^k}.$$

As before, if we use the binomial theorem to expand $(x + h)^k$, then simplify and take the limit, we obtain

$$f'(x) = -kx^{-k-1} = nx^{n-1}.$$

If $n = 0$ and $x \neq 0$, then the power rule is also true, for in this case $f(x) = x^0 = 1$ and, by (3.13), $f'(x) = 0 = 0 \cdot x^{0-1}$. ■

Some special cases of the power rule are listed in the next illustration.

ILLUSTRATION

$f(x)$	x^2	x^3	x^4	x^{100}	$x^{-1} = \frac{1}{x}$	$x^{-2} = \frac{1}{x^2}$	$x^{-10} = \frac{1}{x^{10}}$
$f'(x)$	$2x$	$3x^2$	$4x^3$	$100x^{99}$	$(-1)x^{-2} = -\frac{1}{x^2}$	$-2x^{-3} = -\frac{2}{x^3}$	$-10x^{-11} = -\frac{10}{x^{11}}$

We can extend the power rule to rational exponents. In particular, in Appendix II we show that for every positive integer n,

$$\text{if} \quad f(x) = x^{1/n}, \quad \text{then} \quad f'(x) = \frac{1}{n} x^{(1/n)-1},$$

provided these expressions are defined. By using rule (3.34) proved in Section 3.6, we can then show that for any rational number m/n,

$$\text{if} \quad f(x) = x^{m/n}, \quad \text{then} \quad f'(x) = \frac{m}{n} x^{(m/n)-1}.$$

In Chapter 7 we will prove that the power rule holds for every *real* number n. Some special cases of the power rule for rational exponents are given in the next illustration.

ILLUSTRATION

$f(x)$	$f'(x)$
$\sqrt{x} = x^{1/2}$	$\frac{1}{2}x^{-1/2} = \frac{1}{2\sqrt{x}}$
$\sqrt[3]{x^2} = x^{2/3}$	$\frac{2}{3}x^{-1/3} = \frac{2}{3\sqrt[3]{x}}$
$\sqrt[4]{x^5} = x^{5/4}$	$\frac{5}{4}x^{1/4} = \frac{5}{4}\sqrt[4]{x}$
$\frac{1}{x^{1/3}} = x^{-1/3}$	$-\frac{1}{3}x^{-4/3} = -\frac{1}{3x^{4/3}}$
$\frac{1}{x^{3/2}} = x^{-3/2}$	$-\frac{3}{2}x^{-5/2} = -\frac{3}{2x^{5/2}}$

By using the same type of proof that was used for the power rule (3.14), we can prove the following for any real number c.

Theorem **(3.15)**

If $f(x) = cx^n$, then $f'(x) = (cn)x^{n-1}$.

In words, *to differentiate* cx^n, *multiply the coefficient* c *by the exponent* n *and reduce the exponent by* 1.

We shall conclude this section by introducing additional notations for derivatives.

Notations for the derivative of $y = f(x)$ **(3.16)**

$$f'(x) = D_x\, f(x) = D_x\, y = y' = \frac{dy}{dx} = \frac{d}{dx}\, f(x)$$

All of these notations are used in mathematics and applications, and you should become familiar with the different forms. For example, we can now write

$$D_x\, f(x) = \frac{d}{dx}\, f(x) = \lim_{h \to 0} \frac{f(x + h) - f(x)}{h}.$$

The letter x in D_x and d/dx denotes the independent variable. If we use a different independent variable, say t, then we write

$$f'(t) = D_t\, f(t) = \frac{d}{dt}\, f(t).$$

Each of the symbols D_x and d/dx is called a **differential operator**. Standing alone, D_x or d/dx has no practical significance; however, when either symbol has an expression to its right, it denotes a derivative. We say that D_x or d/dx *operates* on the expression, and we call $D_x\, y$ or dy/dx **the derivative of** y **with respect to** x. We shall justify the notation dy/dx in Section 3.5, where the concept of a *differential* is defined.

The next illustration contains some examples of the use of (3.16) and (3.15).

ILLUSTRATION

- $D_x\,(3x^7) = (3 \cdot 7)x^6 = 21x^6$
- $D_t\,(\frac{1}{2}t^{12}) = (\frac{1}{2} \cdot 12)t^{11} = 6t^{11}$
- $\dfrac{d}{dx}(4x^{3/2}) = (4 \cdot \frac{3}{2})x^{1/2} = 6x^{1/2}$
- $\dfrac{d}{dr}(2r^{-4}) = 2(-4)r^{-5} = -\dfrac{8}{r^5}$

Note that in (3.16), $D_x\, y$, y', and dy/dx are used for the derivative of y with respect to x. If we wish to denote the *value* of the derivative $D_x\, y$, y', or dy/dx at some number $x = a$, we often use a single or double bracket and write

$$D_x\, y\big]_{x=a}, \quad \left.\frac{dy}{dx}\right]_{x=a}, \quad [D_x\, y]_{x=a}, \quad \text{or} \quad \left[\frac{dy}{dx}\right]_{x=a}.$$

ILLUSTRATION

- $D_x(x^3)]_{x=5} = 3x^2]_{x=5} = 3(5^2) = 75$
- $[D_x(9x^{4/3})]_{x=8} = [12x^{1/3}]_{x=8} = 12(8^{1/3}) = 24$

We may also consider *derivatives of derivatives*. Specifically, if we differentiate a function f, we obtain another function f'. If f' has a derivative, it is denoted by f'' (read *f double prime*) and is called the **second derivative** of f. Thus,

$$f''(x) = D_x[f'(x)] = D_x[D_x f(x))] = D_x^2 f(x).$$

As we have indicated, the operator symbol D_x^2 is used for second derivatives. The **third derivative** f''' of f is the derivative of the second derivative. Thus,

$$f'''(x) = D_x[f''(x)] = D_x[D_x^2 f(x)] = D_x^3 f(x).$$

In general, if n is a positive integer, then $f^{(n)}$ denotes the ***n*th derivative** of f and is found by starting with f and differentiating, successively, n times. In operator notation, $f^{(n)}(x) = D_x^n f(x)$. The integer n is called the **order** of the derivative $f^{(n)}(x)$. The following summarizes various notations that are used for these **higher derivatives**, with $y = f(x)$.

Notations for higher derivatives **(3.17)**

$$f'(x),\quad f''(x),\quad f'''(x),\quad f^{(4)}(x),\quad \ldots,\quad f^{(n)}(x)$$

$$D_x y,\quad D_x^2 y,\quad D_x^3 y,\quad D_x^4 y,\quad \ldots,\quad D_x^n y$$

$$y',\quad y'',\quad y''',\quad y^{(4)},\quad \ldots,\quad y^{(n)}$$

$$\frac{dy}{dx},\quad \frac{d^2y}{dx^2},\quad \frac{d^3y}{dx^3},\quad \frac{d^4y}{dx^4},\quad \ldots,\quad \frac{d^ny}{dx^n}$$

EXAMPLE 5 Find the first four derivatives of $f(x) = 4x^{3/2}$.

SOLUTION We use (3.15) four times:

$$f'(x) = (4 \cdot \tfrac{3}{2})x^{1/2} = 6x^{1/2}$$

$$f''(x) = (6 \cdot \tfrac{1}{2})x^{-1/2} = 3x^{-1/2}$$

$$f'''(x) = 3(-\tfrac{1}{2})x^{-3/2} = -\tfrac{3}{2}x^{-3/2}$$

$$f^{(4)}(x) = (-\tfrac{3}{2})(-\tfrac{3}{2})x^{-5/2} = \tfrac{9}{4}x^{-5/2}$$

EXERCISES 3.2

Exer. 1–4: (a) Use Definition (3.5) to find $f'(x)$. (b) Find the domain of f'. (c) Find an equation of the tangent line to the graph of f at P. (d) Find the points on the graph at which the tangent line is horizontal.

1 $f(x) = -5x^2 + 8x + 2;\quad P(-1, -11)$

2 $f(x) = 3x^2 - 2x - 4;\quad P(2, 4)$

3 $f(x) = x^3 + x;\quad P(1, 2)$

4 $f(x) = x^3 - 4x;\quad P(2, 0)$

Exer. 5–12: (a) Use (3.12)–(3.15) to find $f'(x)$. (b) Find the domain of f'. (c) Find an equation of the tangent line to the graph of f at P. (d) Find the points on the graph at which the tangent line is horizontal.

5 $f(x) = 9x - 2;\quad P(3, 25)$

6 $f(x) = -4x + 3;\quad P(-2, 11)$

7 $f(x) = 37;\quad P(0, 37)$

8 $f(x) = \pi^2;\quad P(5, \pi^2)$

9 $f(x) = 1/x^3$; $P(2, \frac{1}{8})$

10 $f(x) = 1/x^4$; $P(1, 1)$

11 $f(x) = 4x^{1/4}$; $P(81, 12)$

12 $f(x) = 12x^{1/3}$; $P(-27, -36)$

Exer. 13–16: Find the first three derivatives.

13 $f(x) = 3x^6$

14 $f(x) = 6x^4$

15 $f(x) = 9\sqrt[3]{x^2}$

16 $f(x) = 3x^{7/3}$

17 If $z = 25t^{9/5}$, find $D_t^2\, z$.

18 If $y = 3x + 5$, find $D_x^3\, y$.

19 If $y = -4x + 7$, find $\dfrac{d^3y}{dx^3}$.

20 If $z = 64\sqrt[4]{t^3}$, find $\dfrac{d^2z}{dt^2}$.

Exer. 21–22: Is f differentiable on the given interval? Explain.

21 $f(x) = 1/x$ (a) $[0, 2]$ (b) $[1, 3)$

22 $f(x) = \sqrt[3]{x}$ (a) $[-1, 1]$ (b) $[-2, -1)$

Exer. 23–24: Use the graph of f to determine if f is differentiable on the given interval.

23 $f(x) = \sqrt{4 - x}$ (a) $[0, 4]$ (b) $[-5, 0]$

24 $f(x) = \sqrt{4 - x^2}$ (a) $[-2, 2]$ (b) $[-1, 1]$

Exer. 25–30: Determine whether f has (a) a vertical tangent line at (0, 0) and (b) a cusp at (0, 0).

25 $f(x) = x^{1/3}$ 26 $f(x) = x^{5/3}$ 27 $f(x) = x^{2/5}$

28 $f(x) = x^{1/4}$ 29 $f(x) = 5x^{3/2}$ 30 $f(x) = 7x^{4/3}$

Exer. 31–32: Estimate $f'(-1)$, $f'(1)$, $f'(2)$, and $f'(3)$, whenever they exist.

31

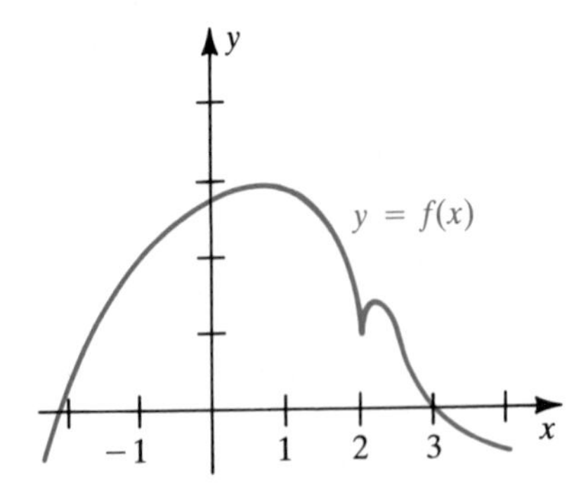

32

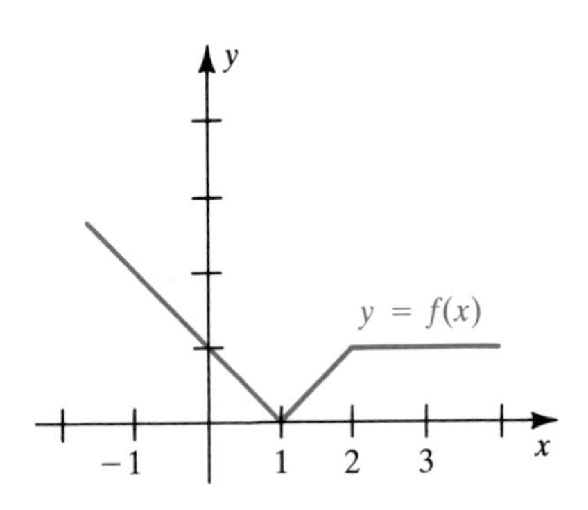

Exer. 33–36: Use right-hand and left-hand derivatives to prove that f is not differentiable at a.

33 $f(x) = |x - 5|$; $a = 5$

34 $f(x) = |x + 2|$; $a = -2$

35 $f(x) = [\![x - 2]\!]$; $a = 2$

36 $f(x) = [\![x]\!] - 2$; $a = 2$

Exer. 37–40: Use the graph of f to find the domain of f'.

37 $f(x) = \begin{cases} 2x & \text{if } x \le 0 \\ x^2 & \text{if } x > 0 \end{cases}$

38 $f(x) = \begin{cases} 2x - 1 & \text{if } x \le 1 \\ x^2 & \text{if } x > 1 \end{cases}$

39 $f(x) = \begin{cases} -x^2 & \text{if } x < -1 \\ 2x + 3 & \text{if } x \ge -1 \end{cases}$

40 $f(x) = \begin{cases} x^2 - 2 & \text{if } x < 0 \\ -3 & \text{if } x \ge 0 \end{cases}$

Exer. 41–42: Each figure is the graph of a function f. Sketch the graph of f' and determine where f is not differentiable.

41

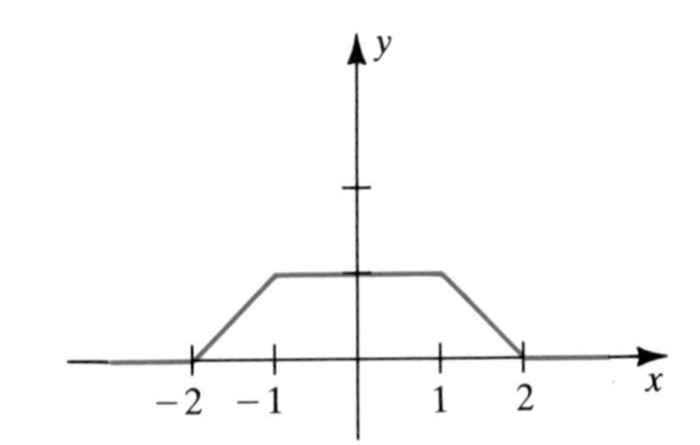

42

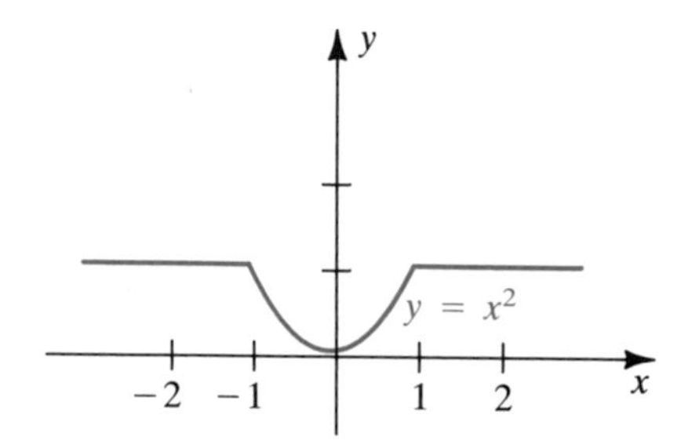

Exer. 43–44: Given the position function s of a point P moving on a coordinate line l, find the times at which the velocity is the given value k.

43 $s(t) = 3t^{2/3}$; $k = 4$

44 $s(t) = 4t^3$; $k = 300$

45 The relationship between the temperature F on the Fahrenheit scale and the temperature C on the Celsius scale is given by $C = \frac{5}{9}(F - 32)$. Find the rate of change of F with respect to C.

46 Charles' law for gases states that if the pressure remains constant, then the relationship between the volume V that a gas occupies and its temperature T (in °C) is given by $V = V_0(1 + \frac{1}{273}T)$. Find the rate of change of T with respect to V.

47 Show that the rate of change of the volume of a sphere with respect to its radius is numerically equal to the surface area of the sphere.

48 Show that the rate of change of the radius of a circle with respect to its circumference is independent of the size of the circle.

49 An oil spill is increasing such that the surface covered by the spill is always circular. Find the rate at which the area A of the surface is changing with respect to the radius r of the circle at

(a) any value of r (b) $r = 500$ ft

50 A spherical balloon is being inflated. Find the rate at which its volume V is changing with respect to the radius r of the balloon at

(a) any value of r (b) $r = 10$ ft

51 In some applications, function values $f(x)$ may be known only for several values of x near a. In these situations $f'(a)$ is frequently approximated by the formula

$$f'(a) \approx \frac{f(a+h) - f(a-h)}{2h}.$$

(a) Interpret this formula graphically.

(b) Show that $\lim_{h \to 0} \frac{f(a+h) - f(a-h)}{2h} = f'(a)$.

(c) If $f(x) = 1/x^2$, use the approximation formula to estimate $f'(1)$ with $h = 0.1$, 0.01, and 0.001.

(d) Find the exact value of $f'(1)$.

52 (a) Use the approximation formula in Exercise 51 to show that if $h \approx 0$, then

$$f''(a) \approx \frac{f(a+h) - 2f(a) + f(a-h)}{h^2}.$$

(b) If $f(x) = 1/x^2$, use part (a) to estimate $f''(1)$ with $h = 0.1$, 0.01, and 0.001.

(c) Find the exact value of $f''(1)$.

c **Exer. 53–54: Use the following table, which lists the approximate number of feet $s(t)$ that a car travels in t seconds to reach a velocity of 60 mi/hr in 6 seconds.**

t	0	1	2	3	4	5	6	7
$s(t)$	0	11.7	42.6	89.1	149.0	220.1	303.7	396.7

53 Use Exercise 51 to approximate the velocity of the car at

(a) $t = 3$ (b) $t = 6$

54 Use Exercise 52 to approximate the rate of change of the velocity of the car with respect to t at

(a) $t = 3$ (b) $t = 6$

c 55 Graph $f(x) = |x^5 - 2x^4 + 3x^3 - x + 1|$ on the interval $[-1, 1]$ and estimate where f is not differentiable.

c 56 Graph $f(x) = x^4 - 3x^3 + 2x - 1$ on the interval $[-1, 3]$ and estimate the x-coordinates of points at which the tangent line is horizontal.

3.3 TECHNIQUES OF DIFFERENTIATION

This section contains some general rules that simplify the task of finding derivatives. The rules are stated in terms of the differential operator D_x, where $D_x f(x) = f'(x)$. In the rules, f and g denote differentiable functions,

c, m, and b are real numbers, and n is a rational number. The first three parts of the following theorem were proved in Section 3.2 and are restated here for completeness.

Theorem (3.18)

(i) $D_x\, c = 0$

(ii) $D_x\,(mx + b) = m$

(iii) $D_x\,(x^n) = nx^{n-1}$

(iv) $D_x\,[cf(x)] = c\, D_x\, f(x)$

(v) $D_x\,[f(x) + g(x)] = D_x\, f(x) + D_x\, g(x)$

(vi) $D_x\,[f(x) - g(x)] = D_x\, f(x) - D_x\, g(x)$

PROOF (iv) Applying the definition of the derivative to $cf(x)$, we have

$$\begin{aligned} D_x\,[cf(x)] &= \lim_{h\to 0} \frac{cf(x+h) - cf(x)}{h} \\ &= \lim_{h\to 0} c\, \frac{f(x+h) - f(x)}{h} \\ &= c \lim_{h\to 0} \frac{f(x+h) - f(x)}{h} \\ &= c\, D_x\, f(x). \end{aligned}$$

(v) Applying the definition of derivative to $f(x) + g(x)$ yields

$$\begin{aligned} D_x\,[f(x) + g(x)] &= \lim_{h\to 0} \frac{[f(x+h) + g(x+h)] - [f(x) + g(x)]}{h} \\ &= \lim_{h\to 0} \left[\frac{f(x+h) - f(x)}{h} + \frac{g(x+h) - g(x)}{h}\right] \\ &= \lim_{h\to 0} \frac{f(x+h) - f(x)}{h} + \lim_{h\to 0} \frac{g(x+h) - g(x)}{h} \\ &= D_x\, f(x) + D_x\, g(x) \end{aligned}$$

We can prove (vi) in similar fashion, or we can write

$$f(x) - g(x) = f(x) + (-1)g(x)$$

and then use (v) and (iv). ■

If we use the differential operator d/dx in place of D_x, the rules in Theorem (3.18) take on the following forms:

$$\frac{d}{dx}\, c = 0 \qquad \frac{d}{dx}\,(mx + b) = m$$

$$\frac{d}{dx}\, x^n = nx^{n-1} \qquad \frac{d}{dx}\,[cf(x)] = c\,\frac{d}{dx}\, f(x)$$

$$\frac{d}{dx}\,[f(x) \pm g(x)] = \frac{d}{dx}\, f(x) \pm \frac{d}{dx}\, g(x)$$

Parts (v) and (vi) of Theorem (3.18) may be stated as follows:

(v) *The derivative of a sum is the sum of the derivatives.*

(vi) *The derivative of a difference is the difference of the derivatives.*

These results can be extended to sums or differences of any number of functions. Since a polynomial is a sum of terms of the form cx^n, where c is a real number and n is a nonnegative integer, we may use results on sums and differences to obtain the derivative, as illustrated in the next example.

EXAMPLE 1 If $f(x) = 2x^4 - 5x^3 + x^2 - 4x + 1$, find $f'(x)$.

SOLUTION

$$\begin{aligned} f'(x) &= D_x\,(2x^4 - 5x^3 + x^2 - 4x + 1) \\ &= D_x\,(2x^4) - D_x\,(5x^3) + D_x\,(x^2) - D_x\,(4x) + D_x\,(1) \\ &= 8x^3 - 15x^2 + 2x - 4. \end{aligned}$$

EXAMPLE 2 Find an equation of the tangent line to the graph of $y = 6\sqrt[3]{x^2} - \dfrac{4}{\sqrt{x}}$ at $P(1, 2)$.

SOLUTION We first express y in terms of rational exponents and then find dy/dx:

$$\begin{aligned} y &= 6x^{2/3} - 4x^{-1/2} \\ \frac{dy}{dx} &= \frac{d}{dx}\,(6x^{2/3}) - \frac{d}{dx}\,(4x^{-1/2}) \\ &= 4x^{-1/3} - (-2x^{-3/2}) \\ &= \frac{4}{x^{1/3}} + \frac{2}{x^{3/2}} \end{aligned}$$

To find the slope of the tangent line at $P(1, 2)$, we evaluate dy/dx at $x = 1$:

$$\left.\frac{dy}{dx}\right]_{x=1} = \frac{4}{1} + \frac{2}{1} = 6$$

Using the point-slope form, we can express an equation of the tangent line as

$$y - 2 = 6(x - 1), \quad \text{or} \quad 6x - y = 4.$$

Formulas for derivatives of products or quotients are more complicated than those for sums and differences. In particular, *the derivative of a product generally is not equal to the product of the derivatives.* We may illustrate this by using the product $x^2 \cdot x^5$ as follows:

$$D_x\,(x^2 \cdot x^5) = D_x\,(x^7) = 7x^6$$

$$D_x\,(x^2) \cdot D_x\,(x^5) = (2x) \cdot (5x^4) = 10x^5$$

Hence $$D_x\,(x^2 \cdot x^5) \neq D_x\,(x^2) \cdot D_x\,(x^5).$$

The derivative of any product $f(x)\,g(x)$ may be expressed in terms of derivatives of $f(x)$ and $g(x)$ as in the following rule.

Product rule **(3.19)**

$$D_x\, f(x)g(x) = f(x)\, D_x\, g(x) + g(x)\, D_x\, f(x)$$

PROOF Let $y = f(x)g(x)$. Using the definition of the derivative, we write

$$D_x\, y = \lim_{h \to 0} \frac{f(x+h)g(x+h) - f(x)g(x)}{h}.$$

To change the form of the quotient so that the limit may be evaluated, we subtract and add the expression $f(x+h)g(x)$ in the numerator. Thus,

$$\begin{aligned} D_x\, y &= \lim_{h \to 0} \frac{f(x+h)g(x+h) - f(x+h)g(x) + f(x+h)g(x) - f(x)g(x)}{h} \\ &= \lim_{h \to 0} \left[f(x+h) \cdot \frac{g(x+h) - g(x)}{h} + g(x) \cdot \frac{f(x+h) - f(x)}{h} \right] \\ &= \lim_{h \to 0} f(x+h) \cdot \lim_{h \to 0} \frac{g(x+h) - g(x)}{h} + \lim_{h \to 0} g(x) \cdot \lim_{h \to 0} \frac{f(x+h) - f(x)}{h}. \end{aligned}$$

Since f is differentiable at x, it is continuous at x (see Theorem (3.11)). Hence $\lim_{h \to 0} f(x+h) = f(x)$. Also, $\lim_{h \to 0} g(x) = g(x)$, since x is fixed in this limiting process. Finally, applying the definition of derivative to $f(x)$ and $g(x)$, we obtain

$$D_x\, y = f(x)g'(x) + g(x)f'(x). \quad \blacksquare$$

The product rule may be phrased as follows: *The derivative of a product equals the first factor times the derivative of the second factor, plus the second times the derivative of the first.*

EXAMPLE 3 If $y = (x^3 + 1)(2x^2 + 8x - 5)$, find $D_x\, y$.

SOLUTION Using the product rule (3.19), we have

$$\begin{aligned} D_x\, y &= (x^3 + 1)\, D_x\,(2x^2 + 8x - 5) + (2x^2 + 8x - 5)\, D_x\,(x^3 + 1) \\ &= (x^3 + 1)(4x + 8) + (2x^2 + 8x - 5)(3x^2) \\ &= (4x^4 + 8x^3 + 4x + 8) + (6x^4 + 24x^3 - 15x^2) \\ &= 10x^4 + 32x^3 - 15x^2 + 4x + 8. \end{aligned}$$

EXAMPLE 4 If $f(x) = x^{1/3}(x^2 - 3x + 2)$, find

(a) $f'(x)$ **(b)** the x-coordinate of the points on the graph of f at which the tangent line is either horizontal or vertical

SOLUTION

(a) By the product rule (3.19),

$$\begin{aligned} f'(x) &= x^{1/3}\, D_x\,(x^2 - 3x + 2) + (x^2 - 3x + 2)\, D_x\,(x^{1/3}) \\ &= x^{1/3}(2x - 3) + (x^2 - 3x + 2)(\tfrac{1}{3}x^{-2/3}) \\ &= \frac{3x(2x - 3) + (x^2 - 3x + 2)}{3x^{2/3}} \\ &= \frac{7x^2 - 12x + 2}{3x^{2/3}}. \end{aligned}$$

(b) The tangent line to the graph of f is horizontal if its slope is zero. Setting $f'(x) = 0$ and using the quadratic formula, we obtain

$$x = \frac{12 \pm \sqrt{144 - 56}}{2(7)} = \frac{12 \pm \sqrt{88}}{14} = \frac{6 \pm \sqrt{22}}{7}.$$

Referring to $f'(x)$, we see that the denominator $3x^{2/3}$ is zero at $x = 0$. Since f is continuous at 0 and $\lim_{x\to 0} |f'(x)| = \infty$, it follows from Definition (3.9) that the graph of f has a vertical tangent line at $x = 0$—that is, the point (0, 0) (the origin).

We shall next obtain a formula for the derivative of a quotient. Note that *the derivative of a quotient generally is not equal to the quotient of the derivatives.* We may illustrate this with the quotient x^5/x^2 as follows:

$$D_x\left(\frac{x^5}{x^2}\right) = D_x\,(x^3) = 3x^2$$

$$\frac{D_x\,(x^5)}{D_x\,(x^2)} = \frac{5x^4}{2x} = \frac{5}{2}x^3$$

Hence
$$D_x\left(\frac{x^5}{x^2}\right) \neq \frac{D_x\,(x^5)}{D_x\,(x^2)}.$$

The derivative of any quotient $f(x)/g(x)$ may be expressed in terms of the derivatives of $f(x)$ and $g(x)$ as in the following rule.

Quotient rule **(3.20)**

$$D_x\left[\frac{f(x)}{g(x)}\right] = \frac{g(x)\, D_x\, f(x) - f(x)\, D_x\, g(x)}{[g(x)]^2}$$

PROOF Let $y = f(x)/g(x)$. From the definition of derivative,

$$\begin{aligned} D_x\, y &= \lim_{h\to 0} \frac{\dfrac{f(x+h)}{g(x+h)} - \dfrac{f(x)}{g(x)}}{h} \\ &= \lim_{h\to 0} \frac{g(x)f(x+h) - f(x)g(x+h)}{hg(x+h)g(x)}. \end{aligned}$$

Subtracting and adding $g(x)f(x)$ in the numerator of the last quotient, we obtain

$$\begin{aligned} D_x\, y &= \lim_{h\to 0} \frac{g(x)f(x+h) - g(x)f(x) + g(x)f(x) - f(x)g(x+h)}{hg(x+h)g(x)} \\ &= \lim_{h\to 0} \frac{g(x)[f(x+h) - f(x)] - f(x)[g(x+h) - g(x)]}{hg(x+h)g(x)} \\ &= \lim_{h\to 0} \frac{g(x)\left[\dfrac{f(x+h) - f(x)}{h}\right] - f(x)\left[\dfrac{g(x+h) - g(x)}{h}\right]}{g(x+h)g(x)}. \end{aligned}$$

Taking the limit of the numerator and denominator gives us the quotient rule. ■

The quotient rule may be stated as follows: *The derivative of a quotient is equal to the denominator times the derivative of the numerator minus the numerator times the derivative of the denominator, divided by the square of the denominator.*

EXAMPLE 5 Find $\dfrac{dy}{dx}$ if $y = \dfrac{3x^2 - x + 2}{4x^2 + 5}$.

SOLUTION By the quotient rule (3.20),

$$\begin{aligned} \frac{dy}{dx} &= \frac{(4x^2 + 5)\, D_x\, (3x^2 - x + 2) - (3x^2 - x + 2)\, D_x\, (4x^2 + 5)}{(4x^2 + 5)^2} \\ &= \frac{(4x^2 + 5)(6x - 1) - (3x^2 - x + 2)(8x)}{(4x^2 + 5)^2} \\ &= \frac{(24x^3 - 4x^2 + 30x - 5) - (24x^3 - 8x^2 + 16x)}{(4x^2 + 5)^2} \\ &= \frac{4x^2 + 14x - 5}{(4x^2 + 5)^2}. \end{aligned}$$

If we let $f(x) = 1$ in the quotient rule (3.20), then, since $D_x\,(1) = 0$, we obtain the following.

Reciprocal rule **(3.21)**

$$D_x \left[\frac{1}{g(x)}\right] = -\frac{D_x\, g(x)}{[g(x)]^2}$$

ILLUSTRATION

- $D_x \dfrac{1}{x} = -\dfrac{D_x\,(x)}{(x)^2} = -\dfrac{1}{x^2}$
- $D_x \dfrac{1}{3x^2 - 5x + 4} = -\dfrac{D_x\,(3x^2 - 5x + 4)}{(3x^2 - 5x + 4)^2} = -\dfrac{6x - 5}{(3x^2 - 5x + 4)^2}$

The differentiation formulas in (3.18) are stated in terms of the function values $f(x)$ and $g(x)$. If we wish to state such rules without referring to the variable x, we may write

$$(cf)' = cf', \quad (f + g)' = f' + g', \quad \text{and} \quad (f - g)' = f' - g'.$$

Using this notation for the product and quotient rules and at the same time commuting some of the factors that appear in (3.19) and (3.20), we obtain

$$(fg)' = f'g + fg' \quad \text{and} \quad \left(\frac{f}{g}\right)' = \frac{f'g - fg'}{g^2}.$$

You may find it helpful to memorize these formulas. To obtain the quotient rule, change the + sign in the formula for $(fg)'$ to $-$ and divide by g^2.

We conclude this section with an application that makes use of the quotient rule.

FIGURE 3.19

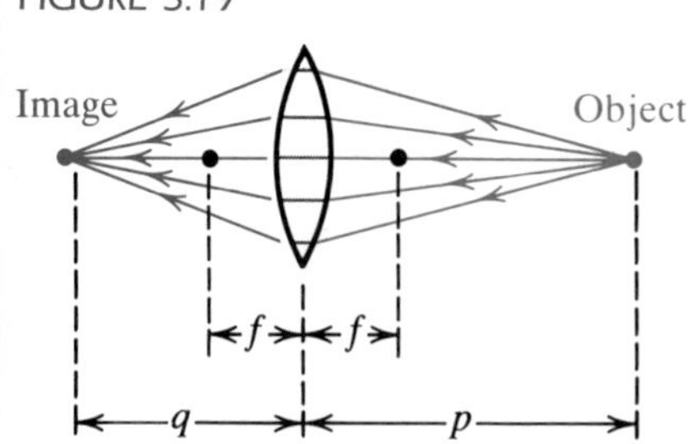

EXAMPLE 6 A convex lens of focal length f is shown in Figure 3.19. If an object is a distance p from the lens as shown, then the distance q from the lens to the image is related to p and f by the *lens equation*

$$\frac{1}{f} = \frac{1}{p} + \frac{1}{q}.$$

If, for a particular lens, $f = 2$ cm and p is increasing, find

(a) a general formula for the rate of change of q with respect to p

(b) the rate of change of q with respect to p if $p = 22$ cm

SOLUTION

(a) By (3.7)(ii), the rate of change of q with respect to p is given by the derivative $D_p\, q$. If $f = 2$, then the lens equation gives us

$$\frac{1}{2} = \frac{1}{p} + \frac{1}{q}, \quad \text{or} \quad \frac{1}{q} = \frac{1}{2} - \frac{1}{p} = \frac{p - 2}{2p}.$$

Hence

$$q = \frac{2p}{p - 2}.$$

Applying the quotient rule (with $x = p$) yields

$$\begin{aligned} D_p\, q &= \frac{(p - 2)\, D_p\, (2p) - (2p)\, D_p\, (p - 2)}{(p - 2)^2} \\ &= \frac{(p - 2)(2) - (2p)(1)}{(p - 2)^2} \\ &= \frac{-4}{(p - 2)^2}. \end{aligned}$$

(b) Substituting $p = 22$ in the formula obtained in part (a), we get

$$D_p\, q\big]_{p=22} = \frac{-4}{(22 - 2)^2} = \frac{-4}{400} = -\frac{1}{100}.$$

Thus, if $p = 22$ cm, the image distance q is decreasing at a rate of $\frac{1}{100}$ centimeter per centimeter change in p.

EXERCISES 3.3

Exer. 1–40: Find the derivative.

1 $g(t) = 6t^{5/3}$

2 $h(z) = 8z^{3/2}$

3 $f(s) = 15 - s + 4s^2 - 5s^4$

4 $f(t) = 12 - 3t^4 + 4t^6$

5 $f(x) = 3x^2 + \sqrt[3]{x^4}$

6 $g(x) = x^4 - \sqrt[4]{x^3}$

7 $g(x) = (x^3 - 7)(2x^2 + 3)$

8 $k(x) = (2x^2 - 4x + 1)(6x - 5)$

9 $f(x) = x^{1/2}(x^2 + x - 4)$

10 $h(x) = x^{2/3}(3x^2 - 2x + 5)$

11 $h(r) = r^2(3r^4 - 7r + 2)$

12 $k(v) = v^3(-2v^3 + v - 3)$

13 $g(x) = (8x^2 - 5x)(13x^2 + 4)$

14 $H(z) = (z^5 - 2z^3)(7z^2 + z - 8)$

15 $f(x) = \dfrac{4x - 5}{3x + 2}$

16 $h(x) = \dfrac{8x^2 - 6x + 11}{x - 1}$

17 $h(z) = \dfrac{8 - z + 3z^2}{2 - 9z}$

18 $f(w) = \dfrac{2w}{w^3 - 7}$

19 $G(v) = \dfrac{v^3 - 1}{v^3 + 1}$

20 $f(t) = \dfrac{8t + 15}{t^2 - 2t + 3}$

21 $g(t) = \dfrac{\sqrt[3]{t^2}}{3t - 5}$

22 $f(x) = \dfrac{\sqrt{x}}{2x^2 - 4x + 8}$

23 $f(x) = \dfrac{1}{1 + x + x^2 + x^3}$

24 $p(x) = 1 + \dfrac{1}{x} + \dfrac{1}{x^2} + \dfrac{1}{x^3}$

25 $h(x) = \dfrac{7}{x^2 + 5}$

26 $k(z) = \dfrac{6}{z^2 + z - 1}$

27 $F(t) = t^2 + \dfrac{1}{t^2}$

28 $s(x) = 2x + \dfrac{1}{2x}$

29 $K(s) = (3s)^{-4}$

30 $W(s) = (3s)^4$

31 $h(x) = (5x - 4)^2$

32 $S(w) = (2w + 1)^3$

33 $g(r) = (5r - 4)^{-2}$

34 $S(x) = (3x + 1)^{-2}$

35 $f(t) = \dfrac{3/(5t) - 1}{(2/t^2) + 7}$

36 $N(z) = \dfrac{4/z^2}{(3/z) + 2}$

37 $M(x) = \dfrac{2x^3 - 7x^2 + 4x + 3}{x^2}$

38 $T(z) = \dfrac{5z^4 + z^3 - 2z}{z^3}$

39 $f(x) = \dfrac{3x^2 - 5x + 8}{7}$

40 $h(t) = \dfrac{3t^5 + 2t}{5}$

Exer. 41–44: Solve the equation $D_x\, y = 0$.

41 $y = 2x^3 - 3x^2 - 36x + 4$

42 $y = 4x^3 + 21x^2 - 24x + 11$

43 $y = \dfrac{2x^2 + 3x - 6}{x - 2}$

44 $y = \dfrac{x^2 + 2x + 5}{x + 1}$

Exer. 45–46: Solve the equation $D_x^2\, y = 0$.

45 $y = 6x^4 + 24x^3 - 540x^2 + 7$

46 $y = 6x^5 - 5x^4 - 30x^3 + 11x$

Exer. 47–50: Find dy/dx by (a) using the quotient rule, (b) using the product rule, and (c) simplifying algebraically and using (3.18).

47 $y = \dfrac{3x - 1}{x^2}$

48 $y = \dfrac{x^2 + 1}{x^4}$

49 $y = \dfrac{x^2 - 3x}{\sqrt[3]{x^2}}$

50 $y = \dfrac{2x + 3}{\sqrt{x^3}}$

Exer. 51–52: Find d^2y/dx^2.

51 $y = \dfrac{3x + 4}{x + 1}$

52 $y = \dfrac{x + 3}{2x + 3}$

Exer. 53–54: Find an equation of the tangent line to the graph of f at P.

53 $f(x) = \dfrac{5}{1 + x^2}$; $P(-2, 1)$

54 $f(x) = 3x^2 - 2\sqrt{x}$; $P(4, 44)$

55 Find the x-coordinates of all points on the graph of $y = x^3 + 2x^2 - 4x + 5$ at which the tangent line is

(a) horizontal (b) parallel to the line $2y + 8x = 5$

56 Find the point P on the graph of $y = x^3$ such that the tangent line at P has x-intercept 4.

57 Find the points on the graph of $y = x^{3/2} - x^{1/2}$ at which the tangent line is parallel to the line $y - x = 3$.

58 Find the points on the graph of $y = x^{5/3} + x^{1/3}$ at which the tangent line is perpendicular to the line $2y + x = 7$.

Exer. 59–60: Sketch the graph of the equation and find the vertical tangent lines.

59 $y = \sqrt{x - 4}$

60 $y = x^{1/3} + 2$

61 A weather balloon is released and rises vertically such that its distance $s(t)$ above the ground during the first 10 seconds of flight is given by $s(t) = 6 + 2t + t^2$, where $s(t)$ is in feet and t is in seconds. Find the velocity of the

balloon at

(a) $t = 1$, $t = 4$, and $t = 8$

(b) the instant the balloon is 50 feet above the ground

62 A ball rolls down an inclined plane such that the distance (in centimeters) that it rolls in t seconds is given by $s(t) = 2t^3 + 3t^2 + 4$ for $0 \le t \le 3$ (see figure).

(a) Find the velocity of the ball at $t = 2$.

(b) At what time is the velocity 30 cm/sec?

EXERCISE 62

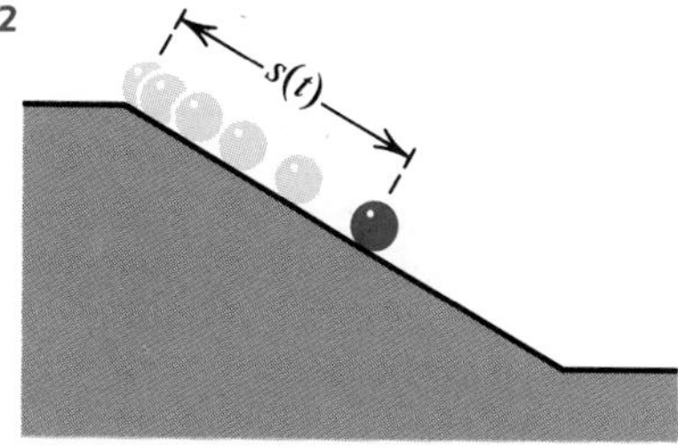

Exer. 63–64: An equation of a classical curve and its graph are given for positive constants a and b. (Consult books on analytic geometry for further information.) Find the slope of the tangent line at the point P.

63 *Witch of Agnesi:* $y = \dfrac{a^3}{a^2 + x^2}$; $P(a, a/2)$

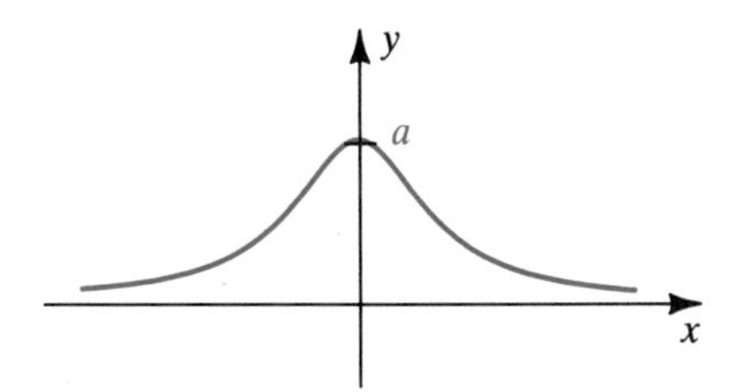

64 *Serpentine curve:* $y = \dfrac{abx}{a^2 + x^2}$; $P(a, b/2)$

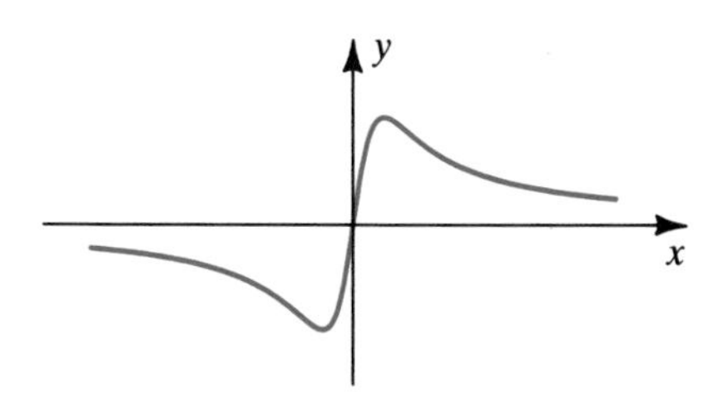

Exer. 65–66: Find equations of the lines through P that are tangent to the graph of the equation.

65 $P(5, 9)$; $y = x^2$

66 $P(3, 1)$; $xy = 4$

Exer. 67–70: If f and g are functions such that $f(2) = 3$, $f'(2) = -1$, $g(2) = -5$, and $g'(2) = 2$, evaluate the expression.

67 (a) $(f + g)'(2)$ (b) $(f - g)'(2)$ (c) $(4f)'(2)$
(d) $(fg)'(2)$ (e) $(f/g)'(2)$ (f) $(1/f)'(2)$

68 (a) $(g - f)'(2)$ (b) $(g/f)'(2)$
(c) $(4g)'(2)$ (d) $(ff)'(2)$

69 (a) $(2f - g)'(2)$ (b) $(5f + 3g)'(2)$
(c) $(gg)'(2)$ (d) $\left(\dfrac{1}{f + g}\right)'(2)$

70 (a) $(3f - 2g)'(2)$ (b) $(5/g)'(2)$
(c) $(6f)'(2)$ (d) $\left(\dfrac{f}{f + g}\right)'(2)$

71 If f, g, and h are differentiable, use the product rule to prove that

$$D_x [f(x)g(x)h(x)] = f(x)g(x)h'(x) + f(x)g'(x)h(x) + f'(x)g(x)h(x).$$

As a corollary, let $f = g = h$ to prove that

$$D_x [f(x)]^3 = 3[f(x)]^2 f'(x).$$

72 Extend Exercise 71 to the derivative of a product of four functions, and then find a formula for $D_x [f(x)]^4$

Exer. 73–76: Use Exercise 71 to find dy/dx.

73 $y = (8x - 1)(x^2 + 4x + 7)(x^3 - 5)$

74 $y = (3x^4 - 10x^2 + 8)(2x^2 - 10)(6x + 7)$

75 $y = x(2x^3 - 5x - 1)(6x^2 + 7)$

76 $y = 4x(x - 1)(2x - 3)$

77 As a spherical balloon is being inflated, its radius r (in centimeters) after t minutes is given by $r = 3\sqrt[3]{t}$ for $0 \le t \le 10$. Find the rate of change for each of the following with respect to t at $t = 8$:

(a) the radius r

(b) the volume V of the balloon

(c) the surface area S of the balloon

78 The volume V (in ft^3) of water in a small reservoir during spring runoff is given by $V = 5000(t + 1)^2$ for t in months and $0 \le t \le 3$. The rate of change of volume with respect to time is the instantaneous *flow rate* into the reservoir. Find the flow rate at times $t = 0$ and $t = 2$. What is the flow rate when the volume is 11,250 ft^3?

79 A stone is dropped into a pond, causing water waves that form concentric circles. If, after t seconds, the radius of one of the waves is $40t$ centimeters, find the rate of change, with respect to t, of the area of the circle caused by the wave at

(a) $t = 1$ (b) $t = 2$ (c) $t = 3$

80 Boyle's law for confined gases states that if the temperature remains constant, then $pv = c$, where p is the pressure, v is the volume, and c is a constant. Suppose that at time t (in minutes) the pressure is $20 + 2t$ centimeters of mercury for $0 \le t \le 10$. If the volume is 60 cm^3 at $t = 0$, find the rate at which the volume is changing with respect to t at $t = 5$.

c 81 (a) If $f(x) = x^3 - 2x + 2$, approximate $f'(1)$ using Exercise 51 of Section 3.2 with $h = 0.1$.

(b) Graph the following on the same coordinate axes: $y = f(x)$, the secant line l_1 through $(1, f(1))$ and $(1.1, f(1.1))$, and the secant line l_2 through $(0.9, f(0.9))$ and $(1.1, f(1.1))$.

(c) Find $f'(1)$ and explain why the slope of l_2 is a better approximation to $f'(1)$ than is the slope of l_1.

c 82 (a) If $f(x) = x^{2/3} + 1$, approximate $f'(0)$ using Exercise 51 of Section 3.2 with $h = 0.1$.

(b) Graph the following on the same coordinate axes: $y = f(x)$, the secant line through $(0, f(0))$ and $(0.1, f(0.1))$, and the secant line through $(-0.1, f(-0.1))$ and $(0.1, f(0.1))$.

(c) Why don't the slopes of the secant lines in (b) approximate $f'(0)$?

3.4 DERIVATIVES OF THE TRIGONOMETRIC FUNCTIONS

To obtain formulas for derivatives of trigonometric functions, it will be necessary to first prove several results about limits. Whenever we discuss limits of trigonometric expressions involving $\sin \theta$, $\cos t$, $\tan x$, and so on, *we shall assume that each variable represents the radian measure of an angle or a real number.*

Let θ denote an angle in standard position on a rectangular coordinate system, and consider the unit circle U in Figure 3.20. According to the definition of the sine and cosine functions, the coordinates of the indicated point P are $(\cos \theta, \sin \theta)$. It appears that if $\theta \to 0$, then $\sin \theta \to 0$ and $\cos \theta \to 1$. This suggests the following theorem.

FIGURE 3.20

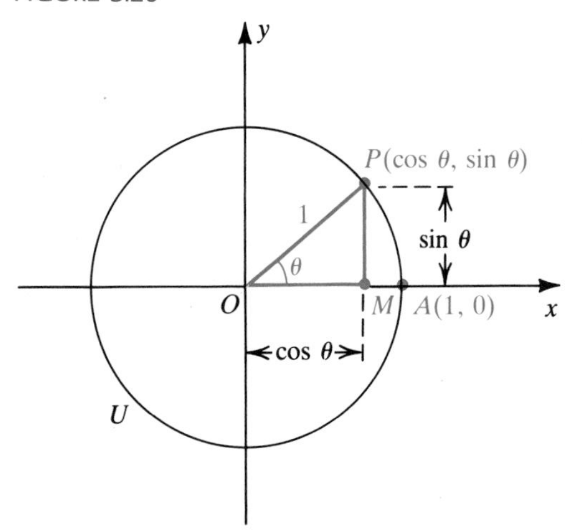

Theorem (3.22)

(i) $\lim_{\theta \to 0} \sin \theta = 0$ (ii) $\lim_{\theta \to 0} \cos \theta = 1$

PROOF (i) Let us first show that $\lim_{\theta \to 0^+} \sin \theta = 0$. If $0 < \theta < \pi/2$, then, referring to Figure 3.20, we see that

$$0 < MP < \widehat{AP},$$

where MP denotes the length of the line segment joining M to P and $\widehat{AP}$ denotes the length of the circular arc between A and P. By the definition of radian measure of an angle (see Section 1.3), $\widehat{AP} = \theta$, and therefore the preceding inequality can be written

$$0 < \sin\theta < \theta.$$

Since $\lim_{\theta\to 0^+} \theta = 0$ and $\lim_{\theta\to 0^+} 0 = 0$, it follows from the sandwich theorem (2.15) that $\lim_{\theta\to 0^+} \sin\theta = 0$.

To complete the proof of (i), it is sufficient to show that $\lim_{\theta\to 0^-} \sin\theta = 0$. If $-\pi/2 < \theta < 0$, then $0 < -\theta < \pi/2$ and hence, from the first part of the proof,

$$0 < \sin(-\theta) < -\theta.$$

Using the trigonometric identity $\sin(-\theta) = -\sin\theta$ and then multiplying by -1 gives us

$$\theta < \sin\theta < 0.$$

Since $\lim_{\theta\to 0^-} \theta = 0$ and $\lim_{\theta\to 0^-} 0 = 0$, it follows from the sandwich theorem that $\lim_{\theta\to 0^-} \sin\theta = 0$.

(ii) Using $\sin^2\theta + \cos^2\theta = 1$, we obtain $\cos\theta = \pm\sqrt{1 - \sin^2\theta}$. If $-\pi/2 < \theta < \pi/2$, then $\cos\theta$ is positive, and hence $\cos\theta = \sqrt{1 - \sin^2\theta}$. Consequently,

$$\lim_{\theta\to 0} \cos\theta = \lim_{\theta\to 0} \sqrt{1 - \sin^2\theta} = \sqrt{\lim_{\theta\to 0} (1 - \sin^2\theta)}$$
$$= \sqrt{1 - 0} = 1. \quad ■$$

In Section 2.1 we used a calculator and a graph to guess the limit stated in the next theorem (see page 42). We shall now give a rigorous proof.

Theorem **(3.23)**

$$\lim_{\theta\to 0} \frac{\sin\theta}{\theta} = 1$$

FIGURE 3.21

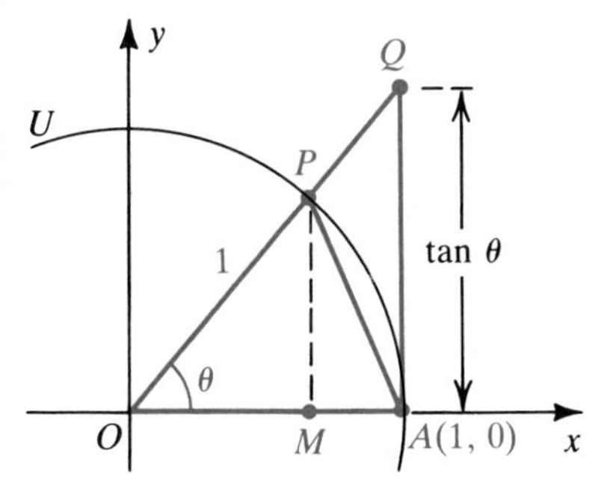

PROOF If $0 < \theta < \pi/2$, we have the situation illustrated in Figure 3.21, where U is a unit circle. Note that

$$MP = \sin\theta \quad \text{and} \quad AQ = \tan\theta.$$

From the figure we see that

$$\text{area of } \triangle AOP < \text{area of sector } AOP < \text{area of } \triangle AOQ.$$

From geometry and Theorem (1.15),

$$\text{area of } \triangle AOP = \tfrac{1}{2}bh = \tfrac{1}{2}(1)(MP) = \tfrac{1}{2}\sin\theta,$$
$$\text{area of sector } AOP = \tfrac{1}{2}r^2\theta = \tfrac{1}{2}(1)^2\theta = \tfrac{1}{2}\theta,$$
$$\text{area of } \triangle AOQ = \tfrac{1}{2}bh = \tfrac{1}{2}(1)(AQ) = \tfrac{1}{2}\tan\theta.$$

Hence the preceding inequality may be written

$$\tfrac{1}{2}\sin\theta < \tfrac{1}{2}\theta < \tfrac{1}{2}\tan\theta.$$

Using the identity $\tan\theta = (\sin\theta)/(\cos\theta)$ and then dividing by $\frac{1}{2}\sin\theta$ leads to the following equivalent inequalities:

$$1 < \frac{\theta}{\sin\theta} < \frac{1}{\cos\theta}$$

$$1 > \frac{\sin\theta}{\theta} > \cos\theta$$

$$\cos\theta < \frac{\sin\theta}{\theta} < 1$$

The last inequality is also true if $-\pi/2 < \theta < 0$, for in this case we have $0 < -\theta < \pi/2$ and hence

$$\cos(-\theta) < \frac{\sin(-\theta)}{-\theta} < 1.$$

Using the identities $\cos(-\theta) = \cos\theta$ and $\sin(-\theta) = -\sin\theta$, we again obtain

$$\cos\theta < \frac{\sin\theta}{\theta} < 1.$$

Since $\lim_{\theta\to 0}\cos\theta = 1$ and $\lim_{\theta\to 0} 1 = 1$, the statement of the theorem follows from the sandwich theorem. ■

We shall also make use of the following result.

Theorem **(3.24)**

$$\lim_{\theta\to 0}\frac{1-\cos\theta}{\theta} = 0$$

PROOF If we let $\theta = 0$ in the expression $(1-\cos\theta)/\theta$, we obtain $0/0$. Hence we must change the form of the quotient. Remembering from trigonometry that $1-\cos^2\theta = \sin^2\theta$, we multiply the numerator and denominator of the expression by $1+\cos\theta$ and then simplify as follows:

$$\begin{aligned}\frac{1-\cos\theta}{\theta} &= \frac{1-\cos\theta}{\theta}\cdot\frac{1+\cos\theta}{1+\cos\theta}\\ &= \frac{1-\cos^2\theta}{\theta(1+\cos\theta)}\\ &= \frac{\sin^2\theta}{\theta(1+\cos\theta)} = \frac{\sin\theta}{\theta}\cdot\frac{\sin\theta}{1+\cos\theta}\end{aligned}$$

Consequently,

$$\begin{aligned}\lim_{\theta\to 0}\frac{1-\cos\theta}{\theta} &= \lim_{\theta\to 0}\left(\frac{\sin\theta}{\theta}\cdot\frac{\sin\theta}{1+\cos\theta}\right)\\ &= \left(\lim_{\theta\to 0}\frac{\sin\theta}{\theta}\right)\left(\lim_{\theta\to 0}\frac{\sin\theta}{1+\cos\theta}\right)\\ &= 1\cdot\left(\frac{0}{1+1}\right) = 1\cdot 0 = 0.\end{aligned}$$ ■

We may now establish the formulas listed in the following theorem, where x *denotes a real number or the radian measure of an angle.*

Derivatives of the trigonometric functions **(3.25)**

$D_x \sin x = \cos x$	$D_x \cos x = -\sin x$
$D_x \tan x = \sec^2 x$	$D_x \cot x = -\csc^2 x$
$D_x \sec x = \sec x \tan x$	$D_x \csc x = -\csc x \cot x$

PROOF Applying Definition (3.5) with $f(x) = \sin x$ and then using the addition formula for the sine function, we obtain

$$\begin{aligned} D_x \sin x &= \lim_{h \to 0} \frac{\sin (x + h) - \sin x}{h} \\ &= \lim_{h \to 0} \frac{\sin x \cos h + \cos x \sin h - \sin x}{h} \\ &= \lim_{h \to 0} \frac{\sin x (\cos h - 1) + \cos x \sin h}{h} \\ &= \lim_{h \to 0} \left[\sin x \left(\frac{\cos h - 1}{h} \right) + \cos x \left(\frac{\sin h}{h} \right) \right]. \end{aligned}$$

By Theorems (3.24) and (3.23),

$$\lim_{h \to 0} \left(\frac{\cos h - 1}{h} \right) = 0 \quad \text{and} \quad \lim_{h \to 0} \frac{\sin h}{h} = 1,$$

and hence $\qquad D_x \sin x = (\sin x)(0) + (\cos x)(1) = \cos x.$

We have shown that the derivative of the sine function is the cosine function. We may obtain the derivative of the cosine function in similar fashion. Thus,

$$\begin{aligned} D_x \cos x &= \lim_{h \to 0} \frac{\cos (x + h) - \cos x}{h} \\ &= \lim_{h \to 0} \frac{\cos x \cos h - \sin x \sin h - \cos x}{h} \\ &= \lim_{h \to 0} \frac{\cos x (\cos h - 1) - \sin x \sin h}{h} \\ &= \lim_{h \to 0} \left[\cos x \left(\frac{\cos h - 1}{h} \right) - \sin x \left(\frac{\sin h}{h} \right) \right] \\ &= (\cos x)(0) - (\sin x)(1) = -\sin x. \end{aligned}$$

Thus, the derivative of the cosine function is the *negative* of the sine function.

To find the derivative of the tangent function, we begin with the fundamental identity $\tan x = \sin x/\cos x$ and then apply the quotient rule as follows:

$$\begin{aligned}
D_x \tan x &= D_x\left(\frac{\sin x}{\cos x}\right)\\
&= \frac{\cos x\,(D_x \sin x) - \sin x\,(D_x \cos x)}{\cos^2 x}\\
&= \frac{\cos x\,(\cos x) - \sin x\,(-\sin x)}{\cos^2 x}\\
&= \frac{\cos^2 x + \sin^2 x}{\cos^2 x} = \frac{1}{\cos^2 x} = \sec^2 x
\end{aligned}$$

For the secant function, we first write $\sec x = 1/\cos x$ and then use the reciprocal rule (3.21):

$$\begin{aligned}
D_x \sec x &= D_x\left(\frac{1}{\cos x}\right)\\
&= -\frac{D_x \cos x}{\cos^2 x}\\
&= -\frac{-\sin x}{\cos^2 x}\\
&= \frac{\sin x}{\cos^2 x} = \frac{1}{\cos x}\,\frac{\sin x}{\cos x}\\
&= \sec x \tan x
\end{aligned}$$

Proofs of the formulas for $D_x \cot x$ and $D_x \csc x$ are left as exercises. ■

We can use (3.25) to obtain information about the continuity of the trigonometric functions. For example, since the sine and cosine functions are differentiable at every real number, it follows from Theorem (3.11) that these functions are continuous throughout $\mathbb{R}$. Similarly, the tangent function is continuous on the open intervals $(-\pi/2, \pi/2)$, $(\pi/2, 3\pi/2)$, and so on, since it is differentiable at each number in these intervals.

EXAMPLE 1 Find y' if $y = \dfrac{\sin x}{1 + \cos x}$.

SOLUTION By the quotient rule and (3.25),

$$\begin{aligned}
y' &= \frac{(1 + \cos x)(D_x \sin x) - (\sin x)\,D_x(1 + \cos x)}{(1 + \cos x)^2}\\
&= \frac{(1 + \cos x)(\cos x) - (\sin x)(0 - \sin x)}{(1 + \cos x)^2}\\
&= \frac{\cos x + \cos^2 x + \sin^2 x}{(1 + \cos x)^2}\\
&= \frac{\cos x + 1}{(1 + \cos x)^2}\\
&= \frac{1}{1 + \cos x}.
\end{aligned}$$

In the solution to Example 1 we used the fundamental identity $\cos^2 x + \sin^2 x = 1$. This and other trigonometric identities are often useful in simplifying problems that involve derivatives of trigonometric functions.

EXAMPLE 2 Find $g'(x)$ if $g(x) = \sec x \tan x$.

SOLUTION By the product rule and (3.25),

$$\begin{aligned} g'(x) &= (\sec x)(D_x \tan x) + (\tan x)(D_x \sec x) \\ &= (\sec x)(\sec^2 x) + (\tan x)(\sec x \tan x) \\ &= \sec^3 x + \sec x \tan^2 x \\ &= \sec x\,(\sec^2 x + \tan^2 x). \end{aligned}$$

The formula for $g'(x)$ can be written in many other ways. For example, because

$$\sec^2 x = \tan^2 x + 1, \quad \text{or} \quad \tan^2 x = \sec^2 x - 1,$$

we can write

$$g'(x) = \sec x\,(2\tan^2 x + 1), \quad \text{or} \quad g'(x) = \sec x\,(2\sec^2 x - 1).$$

EXAMPLE 3 Find $dy/d\theta$ if $y = \sec\theta \cot\theta$.

SOLUTION We could use the product rule as in Example 2; however, it is simpler to first change the form of y by using fundamental identities as follows:

$$y = \sec\theta\cot\theta = \frac{1}{\cos\theta}\frac{\cos\theta}{\sin\theta} = \frac{1}{\sin\theta} = \csc\theta$$

Applying (3.25) yields

$$\frac{dy}{d\theta} = \frac{d}{d\theta}\csc\theta = -\csc\theta\cot\theta.$$

EXAMPLE 4

(a) Find the slopes of the tangent lines to the graph of $y = \sin x$ at the points with x-coordinates 0, $\pi/3$, $\pi/2$, $2\pi/3$, and π.

(b) Sketch the graph of $y = \sin x$ and the tangent lines of part (a).

(c) For what values of x is the tangent line horizontal?

SOLUTION

(a) The slope of the tangent line at the point (x, y) on the graph of the equation $y = \sin x$ is given by the derivative $y' = \cos x$. The slopes at the desired points are listed in the table on the following page.

x	0	$\frac{\pi}{3}$	$\frac{\pi}{2}$	$\frac{2\pi}{3}$	π
$y' = \cos x$	1	$\frac{1}{2}$	0	$-\frac{1}{2}$	-1

(b) A portion of the graph of $y = \sin x$ and the tangent lines of part (a) are sketched in Figure 3.22.

FIGURE 3.22

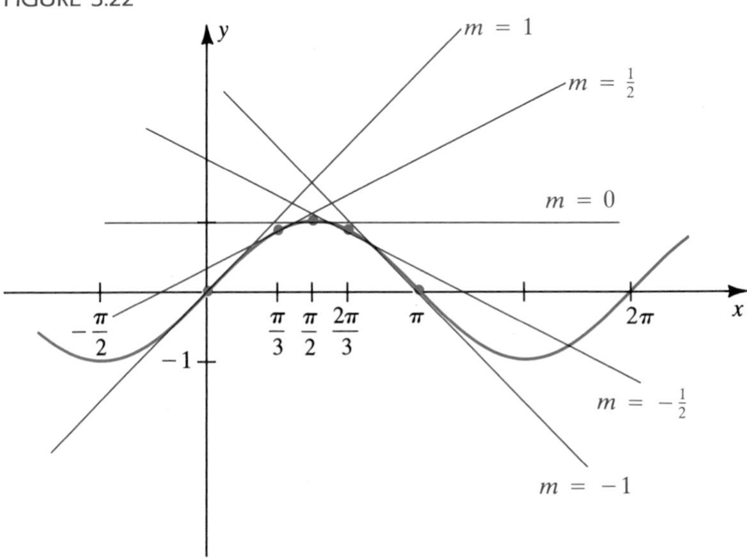

(c) A tangent line is horizontal if its slope is zero. Since the slope of the tangent line at the point (x, y) is y', we must solve the equation

$$y' = 0; \quad \text{that is,} \quad \cos x = 0.$$

Thus the tangent line is horizontal if $x = \pm\pi/2$, $x = \pm 3\pi/2$, and, in general, if $x = (\pi/2) + \pi n$ for any integer n.

FIGURE 3.23

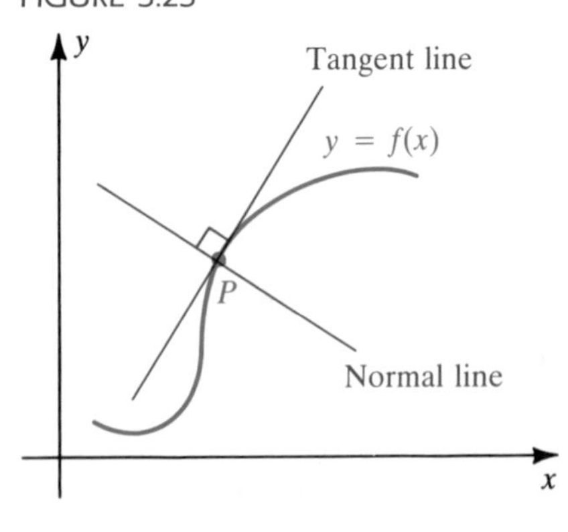

If f is a differentiable function, then the **normal line** at a point $P(a, f(a))$ on the graph of f is the line through P that is perpendicular to the tangent line, as illustrated in Figure 3.23. If $f'(a) \neq 0$, then, by (1.9)(iii), the slope of the normal line is $-1/f'(a)$. If $f'(a) = 0$, then the tangent line is horizontal, and in this case the normal line is vertical and has the equation $x = a$.

EXAMPLE 5 Find an equation of the normal line to the graph of $y = \tan x$ at the point $P(\pi/4, 1)$, and illustrate it graphically.

SOLUTION Since $y' = \sec^2 x$, the slope m of the tangent line at P is

$$m = \sec^2 (\pi/4) = (\sqrt{2})^2 = 2$$

and hence the slope of the normal line is $-1/m = -1/2$.

FIGURE 3.24

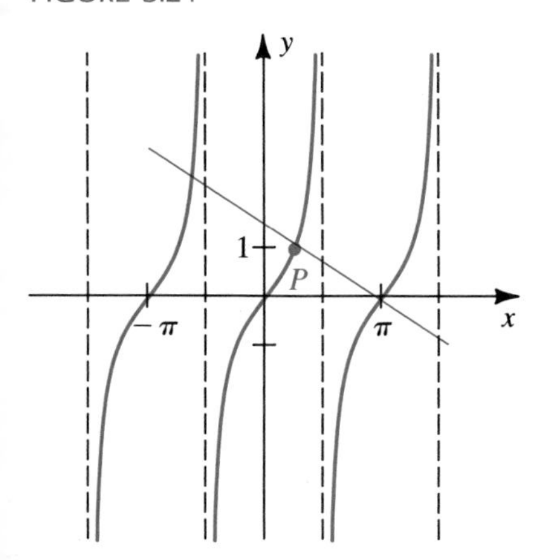

Using the point-slope form, we can express an equation for the normal line as

$$y - 1 = -\frac{1}{2}\left(x - \frac{\pi}{4}\right),$$

or

$$y = -\frac{1}{2}x + \frac{\pi}{8} + 1.$$

The graph of $y = \tan x$ for $-3\pi/2 < x < 3\pi/2$ and the normal line at P are sketched in Figure 3.24.

When we study *Taylor series* in Chapter 11, it will be necessary to find many higher derivatives of functions. The next example illustrates that these derivatives are easy to find for the sine function.

EXAMPLE 6 Find the first eight derivatives of $f(x) = \sin x$.

SOLUTION Applying (3.25) yields

$$f'(x) = D_x \sin x = \cos x$$
$$f''(x) = D_x \cos x = -\sin x$$
$$f'''(x) = D_x(-\sin x) = -D_x(\sin x) = -\cos x$$
$$f^{(4)}(x) = D_x(-\cos x) = -D_x(\cos x) = -(-\sin x) = \sin x.$$

Since $f^{(4)}(x) = \sin x$, it follows that if we continue differentiating, the same pattern repeats; that is,

$$f^{(5)}(x) = \cos x, \quad f^{(6)}(x) = -\sin x,$$
$$f^{(7)}(x) = -\cos x, \quad f^{(8)}(x) = \sin x.$$

EXERCISES 3.4

Exer. 1–28: Find the derivative.

1 $f(x) = 4 \cos x$

2 $H(z) = 7 \tan z$

3 $G(v) = 5v \csc v$

4 $f(x) = 3x \sin x$

5 $k(t) = t - t^2 \cos t$

6 $p(w) = w^2 + w \sin w$

7 $f(\theta) = \dfrac{\sin \theta}{\theta}$

8 $g(\alpha) = \dfrac{1 - \cos \alpha}{\alpha}$

9 $g(t) = t^3 \sin t$

10 $T(r) = r^2 \sec r$

11 $f(x) = 2x \cot x + x^2 \tan x$

12 $f(x) = 3x^2 \sec x - x^3 \tan x$

13 $h(z) = \dfrac{1 - \cos z}{1 + \cos z}$

14 $R(w) = \dfrac{\cos w}{1 - \sin w}$

15 $g(x) = \dfrac{1}{\sin x \tan x}$

16 $k(x) = \dfrac{1}{\cos x \cot x}$

17 $g(x) = (x + \csc x) \cot x$

18 $K(\theta) = (\sin \theta + \cos \theta)^2$

19 $p(x) = \sin x \cot x$

20 $g(t) = \csc t \sin t$

21 $f(x) = \dfrac{\tan x}{1 + x^2}$

22 $h(\theta) = \dfrac{1 + \sec \theta}{1 - \sec \theta}$

23 $k(v) = \dfrac{\csc v}{\sec v}$

24 $q(t) = \sin t \sec t$

25 $g(x) = \sin(-x) + \cos(-x)$

26 $s(z) = \tan(-z) + \sec(-z)$

27 $H(\phi) = (\cot \phi + \csc \phi)(\tan \phi - \sin \phi)$

28 $f(x) = \dfrac{1 + \sec x}{\tan x + \sin x}$

Exer. 29–30: Find equations of the tangent line and the normal line to the graph of f at the point $(\pi/4, f(\pi/4))$.

29 $f(x) = \sec x$

30 $f(x) = \csc x + \cot x$

Exer. 31–34: Shown is a graph of the function f with restricted domain. Find the points at which the tangent line is horizontal.

31 $f(x) = \cos x + \sin x; \quad 0 \le x \le 2\pi$

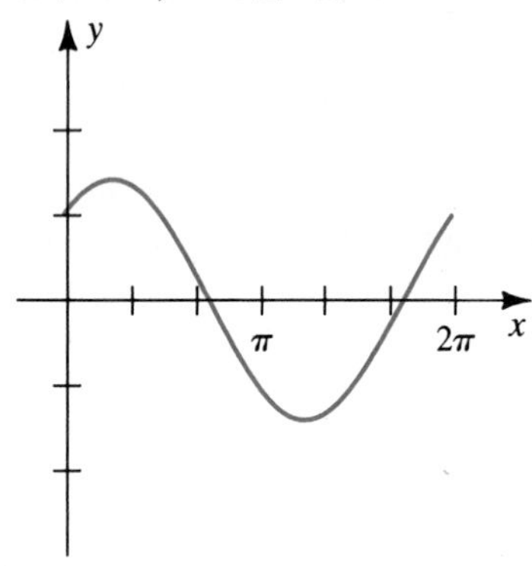

32 $f(x) = \cos x - \sin x; \quad 0 \le x \le 2\pi$

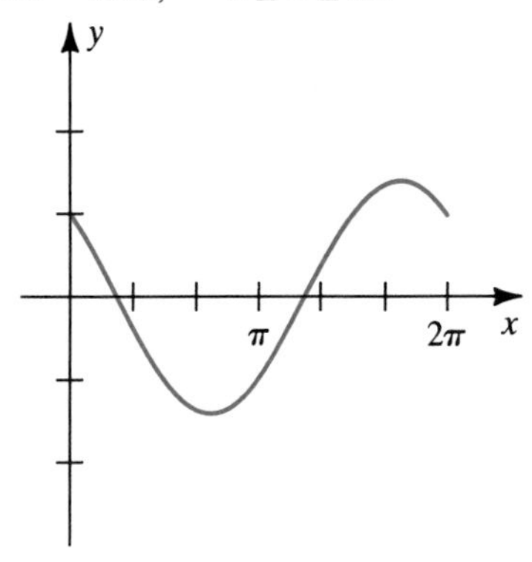

33 $f(x) = \csc x + \sec x; \quad 0 < x < \pi/2$

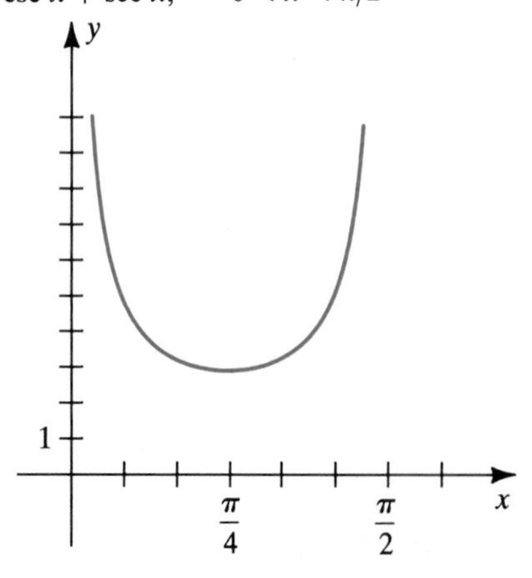

34 $f(x) = 2 \sec x - \tan x; \quad -\pi/2 < x < \pi/2$

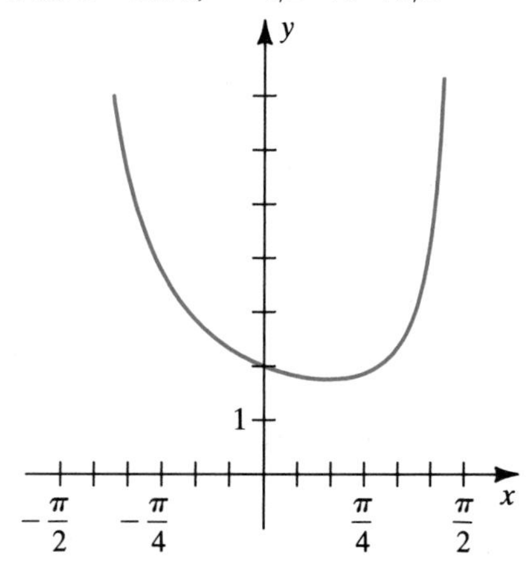

Exer. 35–36: (a) Find the x-coordinates of all points on the graph of f at which the tangent line is horizontal. (b) Find an equation of the tangent line to the graph of f at P.

35 $f(x) = x + 2 \cos x; \quad P(0, f(0))$

36 $f(x) = x + \sin x; \quad P(\pi/2, f(\pi/2))$

37 If $y = 3 + 2 \sin x$, find

(a) the x-coordinates of all points on the graph at which the tangent line is parallel to the line $y = \sqrt{2}x - 5$

(b) an equation of the tangent line to the graph at the point on the graph with x-coordinate $\pi/6$

38 If $y = 1 + 2 \cos x$, find

(a) the x-coordinates of all points on the graph at which the tangent line is perpendicular to the line

$$y = \frac{1}{\sqrt{3}}x + 4$$

(b) an equation of the tangent line to the graph at the point where the graph crosses the y-axis

c 39 Graph $f(x) = |\sin^2 x - \cos x \sin(\frac{1}{2}\pi x)|$ on the interval $[0, 5]$ and estimate where f is not differentiable.

c 40 Graph $f(x) = \dfrac{4}{16 \sin 2x - x}$ on the interval $[0, 4]$ and estimate the x-coordinates of points at which the tangent line is horizontal.

Exer. 41–42: A point P moving on a coordinate line l has the given position function s. When is its velocity 0?

41 $s(t) = t + 2 \cos t$

42 $s(t) = t - \sqrt{2} \sin t$

Exer. 43–44: A point $P(x, y)$ is moving from left to right along the graph of the equation. Where is the rate of change of y with respect to x equal to the given number a?

43 $y = x^{3/2} + 2x; \quad a = 8$

44 $y = x^{5/3} - 10x; \quad a = 5$

45 (a) Find the first four derivatives of $f(x) = \cos x$.
(b) Find $f^{(99)}(x)$.

46 Find $f'''(x)$ if $f(x) = \cot x$.

47 Find $D_x^3\, y$ if $y = \tan x$.

48 Find $\dfrac{d^3y}{dx^3}$ if $y = \sec x$.

Exer. 49–52: Prove each formula.

49 $D_x \cot x = -\csc^2 x$

50 $D_x \csc x = -\csc x \cot x$

51 $D_x \sin 2x = 2 \cos 2x$ (*Hint:* $\sin 2x = 2 \sin x \cos x$.)

52 $D_x \cos 2x = -2 \sin 2x$ (*Hint:* $\cos 2x = 1 - 2 \sin^2 x$.)

3.5 INCREMENTS AND DIFFERENTIALS

In this section we shall introduce additional notation and terminology that are used in problems involving differentiation. The new notation will allow us to regard dy/dx as a *quotient* instead of merely a symbol for the derivative of y with respect to x. We will also use it to estimate changes in quantities.

Let us consider the equation $y = f(x)$, where f is a function. If the variable x has an initial value x_0 and then is assigned a different value x_1, the difference $x_1 - x_0$ is called an **increment** of x. In calculus it is traditional to denote an increment of x by the symbol Δx (read *delta* x). Thus,

$$\Delta x = x_1 - x_0.$$

The corresponding increment of $y = f(x)$, denoted by Δy, is

$$\Delta y = f(x_1) - f(x_0).$$

Since $x_1 = x_0 + \Delta x$, we may also write

$$\Delta y = f(x_0 + \Delta x) - f(x_0).$$

The graph in Figure 3.25 illustrates a case in which both increments are positive; however, Δx may be either positive or negative, and Δy may be positive, negative, or zero. In applications Δx and Δy are usually numerically very small.

FIGURE 3.25

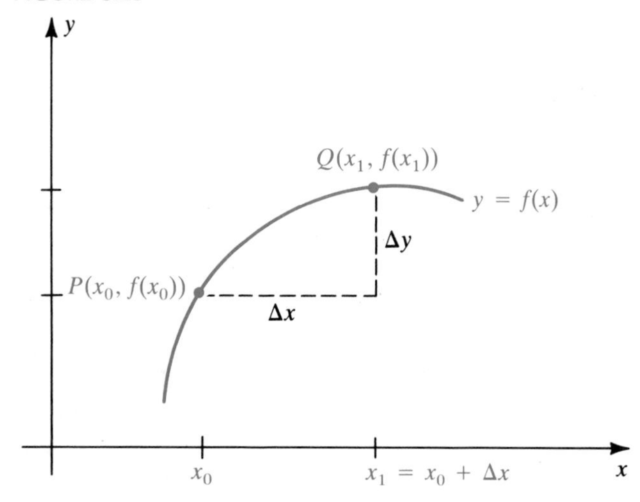

If we let x represent the initial value x_0 of the independent variable, then the definition of Δy has the following form.

Definition **(3.26)**

> Let $y = f(x)$ and let Δx be an increment of x. The **increment Δy of y** is
>
> $$\Delta y = f(x + \Delta x) - f(x).$$

The increment notation may be used in the definition of the derivative of a function. All that is necessary is to substitute Δx for h in (3.5) as follows.

Increment definition of derivative **(3.27)**

$$f'(x) = \lim_{\Delta x \to 0} \frac{f(x + \Delta x) - f(x)}{\Delta x} = \lim_{\Delta x \to 0} \frac{\Delta y}{\Delta x}$$

If f is differentiable, then, as illustrated in Figure 3.26, $\Delta y/\Delta x$ is the slope m_{PQ} of the secant line through P and Q, and as Δx approaches 0, $\Delta y/\Delta x$ approaches the slope $f'(x)$ of the tangent line at P; that is,

$$\frac{\Delta y}{\Delta x} \approx f'(x) \quad \text{if} \quad \Delta x \approx 0.$$

FIGURE 3.26

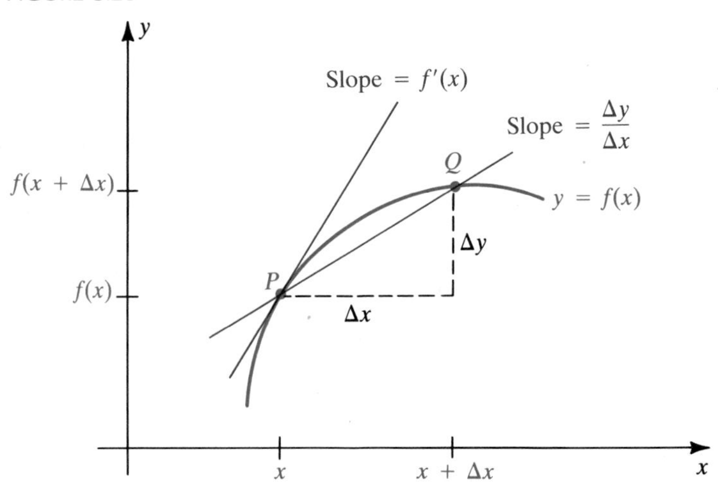

This gives us the following *approximation formula* for Δy:

$$\Delta y \approx f'(x)\,\Delta x \quad \text{if} \quad \Delta x \approx 0$$

We give $f'(x)\,\Delta x$ a special name in (ii) of the next definition.

Definition **(3.28)**

> Let $y = f(x)$, where f is a differentiable function, and let Δx be an increment of x.
>
> **(i)** The **differential dx** of the independent variable x is $dx = \Delta x$.
>
> **(ii)** The **differential dy** of the dependent variable y is
>
> $$dy = f'(x)\,\Delta x = f'(x)\,dx = (D_x y)\,dx.$$

In (3.28)(i) we see that for the *independent* variable x, *there is no difference between the increment* Δx *and the differential* dx. However, for the *dependent* variable y in (ii), *the value of* dy *depends on both* x *and* dx.

The discussion preceding Definition (3.28) gives us the following formula.

Approximation formula for Δy **(3.29)**

$$\text{If}\quad \Delta x \approx 0, \quad \text{then} \quad \Delta y \approx dy.$$

By Definition (3.28)(ii), $dy = f'(x)\,dx$. If both sides of this equation are divided by dx, we obtain the following, which justifies the quotient notation dy/dx introduced in (3.16) for the derivative of $y = f(x)$ with respect to x.

Derivative as a quotient of differentials **(3.30)**

$$\text{If}\quad y = f(x), \quad \text{then} \quad \frac{dy}{dx} = \frac{f'(x)\,dx}{dx} = f'(x).$$

It is important to recognize the *graphical* distinction between dy and Δy. If l is the tangent line at the point $P(x, y)$ on the graph of $y = f(x)$, then, from (3.30), the quotient dy/dx is the slope of l. Hence, as illustrated in Figure 3.27 (with $\Delta x > 0$), dy is the amount that the *tangent line* at P rises or falls when the independent variable changes from x to $x + \Delta x$. This is in contrast to the amount Δy that the *graph* rises (or falls).

FIGURE 3.27

(i)

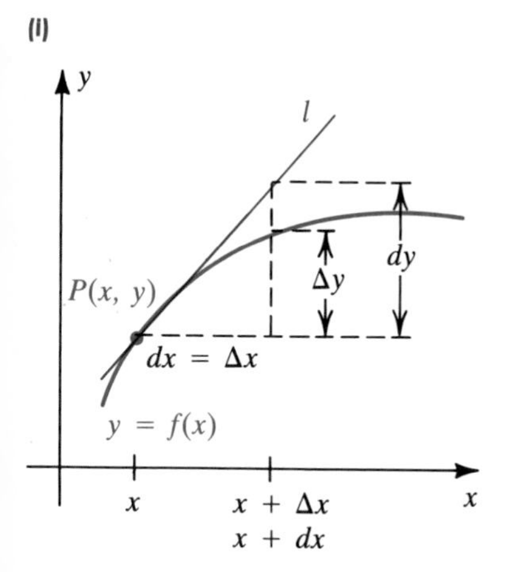

(ii)

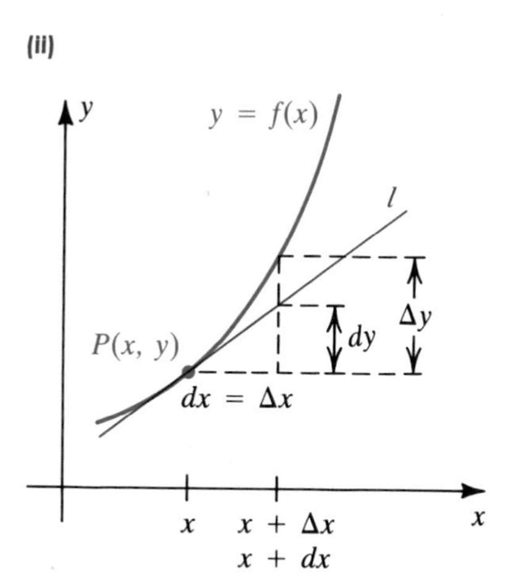

(iii)

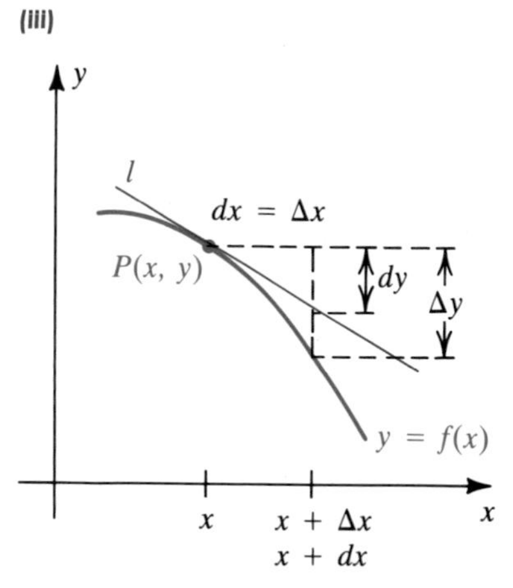

If we rewrite the formula in (3.26) as

$$f(x + \Delta x) = f(x) + \Delta y$$

and use $\Delta y \approx dy$, we obtain the following formula.

Linear approximation formula **(3.31)**

If $y = f(x)$, with f differentiable, and if Δx is an increment of x, then

$$f(x + \Delta x) \approx f(x) + dy.$$

The formula in (3.31) is called a **linear approximation** to $f(x + \Delta x)$ because, as illustrated in Figure 3.27, we can approximate the function value $f(x + \Delta x)$ by using the point $(x + \Delta x, y + dy)$ on the tangent line instead of using the point $(x + \Delta x, y + \Delta y)$ on the graph of f. Thus, *for points near* (x, y), *we can approximate the graph of* f *by means of the tangent line.*

EXAMPLE 1 Let $y = 3x^2 - 5$ and let Δx be an increment of x.

(a) Find general formulas for Δy and dy.

(b) If x changes from 2 to 2.1, find the values of Δy and dy.

SOLUTION

(a) If $y = f(x) = 3x^2 - 5$, then, by Definition (3.26),

$$\begin{aligned}\Delta y &= f(x + \Delta x) - f(x)\\ &= [3(x + \Delta x)^2 - 5] - (3x^2 - 5)\\ &= [3(x^2 + 2x(\Delta x) + (\Delta x)^2) - 5] - (3x^2 - 5)\\ &= 3x^2 + 6x(\Delta x) + 3(\Delta x)^2 - 5 - 3x^2 + 5\\ &= 6x(\Delta x) + 3(\Delta x)^2.\end{aligned}$$

To find dy, we use Definition (3.28)(ii):

$$dy = f'(x)\,dx = 6x\,dx.$$

(b) We wish to find Δy and dy if $x = 2$ and $\Delta x = 0.1$. Substituting in the formula for Δy obtained in (a) gives us

$$\Delta y = 6(2)(0.1) + 3(0.1)^2 = 1.23.$$

Thus, y changes by 1.23 if x changes from 2 to 2.1. We could also find Δy directly as follows:

$$\begin{aligned}\Delta y &= f(2.1) - f(2)\\ &= [3(2.1)^2 - 5] - [3(2)^2 - 5] = 1.23\end{aligned}$$

Similarly, using the formula $dy = 6x\,dx$, with $x = 2$ and $dx = \Delta x = 0.1$, yields

$$dy = (6)(2)(0.1) = 1.2.$$

Note that the approximation 1.2 is correct to the nearest tenth.

EXAMPLE 2 If $y = x^3$ and Δx is an increment of x, find

(a) Δy **(b)** dy

(c) $\Delta y - dy$ **(d)** the value of $\Delta y - dy$ if $x = 1$ and $\Delta x = 0.02$

SOLUTION

(a) Using (3.26) with $f(x) = x^3$ gives us

$$\begin{aligned}\Delta y &= f(x + \Delta x) - f(x)\\ &= (x + \Delta x)^3 - x^3\\ &= x^3 + 3x^2(\Delta x) + 3x(\Delta x)^2 + (\Delta x)^3 - x^3\\ &= 3x^2(\Delta x) + 3x(\Delta x)^2 + (\Delta x)^3.\end{aligned}$$

(b) By Definition (3.28)(ii),

$$dy = f'(x)\,dx = 3x^2\,dx = 3x^2(\Delta x).$$

(c) From parts (a) and (b),

$$\begin{aligned}\Delta y - dy &= [3x^2(\Delta x) + 3x(\Delta x)^2 + (\Delta x)^3] - 3x^2(\Delta x)\\ &= 3x(\Delta x)^2 + (\Delta x)^3.\end{aligned}$$

(d) Substituting $x = 1$ and $\Delta x = 0.02$ in part (c), we obtain

$$\Delta y - dy = 3(1)(0.02)^2 + (0.02)^3 \approx 0.001.$$

This shows that if dy is used to approximate Δy when x changes from 1 to 1.02, then the error involved is approximately 0.001.

If $y = f(x)$, then by (3.29), dy can be used as an approximation to the exact change Δy of the dependent variable corresponding to a small change Δx in the variable x. This observation is useful in applications where only a rough estimate of the change in y is desired.

EXAMPLE 3

(a) Use differentials to approximate the change in $\sin\theta$ if θ changes from 60° to 61°.

(b) Find a linear approximation to $\sin 61°$.

SOLUTION

(a) If $y = \sin\theta = f(\theta)$, then

$$dy = f'(\theta)\,d\theta = \cos\theta\,d\theta.$$

When we use derivatives or differentials of trigonometric functions of angles, we must employ radian measure. Thus, we let $\theta = 60° = \pi/3$ and $\Delta\theta = 1° = \pi/180$ in the formula for dy, obtaining

$$dy = \left(\cos\frac{\pi}{3}\right)\left(\frac{\pi}{180}\right) = \left(\frac{1}{2}\right)\left(\frac{\pi}{180}\right) = \frac{\pi}{360} \approx 0.0087.$$

(b) If we use the linear approximation formula (3.31) with $x = \theta$ and $y = \sin\theta$, we have

$$\sin(\theta + \Delta\theta) \approx \sin\theta + dy.$$

Letting $\theta = 60°$, $\Delta\theta = 1°$, and $dy \approx 0.0087$ (see part (a)), we obtain

$$\begin{aligned}\sin 61° &\approx \sin 60° + dy\\ &\approx \frac{\sqrt{3}}{2} + 0.0087\\ &\approx 0.8660 + 0.0087 = 0.8747.\end{aligned}$$

If we use a calculator, then to four decimal places, $\sin 61° \approx 0.8746$. Hence the error involved in using the linear approximation is roughly 0.0001.

You may be asking yourself: why use differentials in Example 3 when a calculator can do the job more efficiently and accurately? The answer is that *our objective was to demonstrate the use of differentials* in an elementary problem. Finding a numerical value of sin 61° was secondary and unimportant. We often do this sort of thing in mathematics to illustrate new concepts. Keep in mind that there may be other types of problems where differentials are more efficient than calculators.

The next example illustrates the use of differentials in estimating errors that may arise because of approximate measurements. As indicated in the solution, *it is important to first consider general formulas involving the variables* that are being considered. Specific values should *not* be substituted for variables until the final steps of the solutions.

EXAMPLE 4 The radius of a spherical balloon is measured as 12 inches, with a maximum error in measurement of ± 0.06 inch. Approximate the maximum error in the calculated volume of the sphere.

SOLUTION We begin by considering *general* formulas involving the radius and the volume Thus, we let

$$x = \text{measured value of the radius}$$

and

$$dx = \Delta x = \text{maximum error in } x.$$

FIGURE 3.28

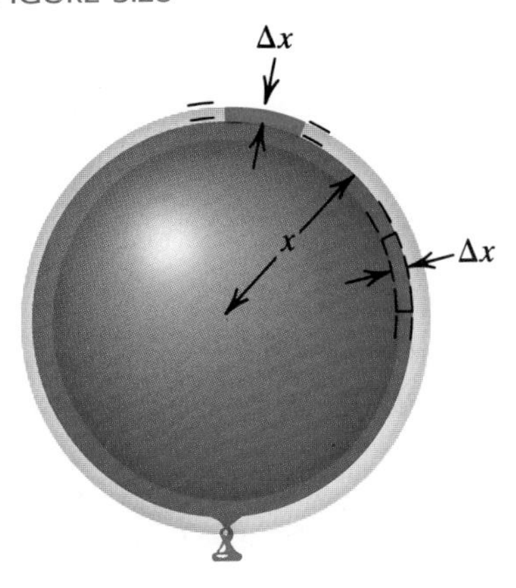

Assuming that Δx is positive, we have

$$x - \Delta x \leq \text{exact radius} \leq x + \Delta x.$$

If Δx is negative, we may use $|\Delta x|$ in place of Δx. A cross-sectional view of the balloon, indicating the possible error Δx, is shown in Figure 3.28. If the volume V of the balloon is calculated using the measured value x, then $V = \frac{4}{3}\pi x^3$.

Let ΔV be the change in V that corresponds to Δx. We may interpret ΔV as *the error in the calculated volume* caused by the error Δx. We approximate ΔV by means of dV as follows:

$$\Delta V \approx dV = (D_x V)\, dx = 4\pi x^2\, dx$$

Finally, we substitute specific values for x and dx. If $x = 12$ and if $\Delta x = dx = \pm 0.06$, then

$$dV = 4\pi(12^2)(\pm 0.06) = \pm(34.56)\pi \approx \pm 109.$$

Thus, the maximum error in the calculated volume due to the error in measurement of the radius is approximately ± 109 in.^{3}

The radius of the balloon in Example 4 was measured as 12 inches, with a maximum error of ± 0.06 inch. The ratio of ± 0.06 to 12 is called the *average error* in the measurement of the radius. Thus,

$$\text{average error} = \frac{\pm 0.06}{12} = \pm 0.005.$$

The significance of this number is that the error in measurement of the radius is, *on the average*, ± 0.005 inch per inch. The *percentage error* is defined as the average error multiplied by 100%. In this illustration,

$$\text{percentage error} = (\pm 0.005)(100\%) = \pm 0.5\%.$$

The general definition of these concepts follows.

Definition **(3.32)**

If w denotes a measurement with a maximum error Δw, then

(i) **average error** $= \dfrac{\Delta w}{w}$

(ii) **percentage error** $=$ (average error) $\times$ (100%)

In terms of differentials, if w represents a measurement with a possible error of dw, then the average error is $(dw)/w$. Of course, if dw is an *approximation* to the error in w, then $(dw)/w$ is an *approximation* to the average error. These remarks are illustrated in the next example.

EXAMPLE 5 The radius of a spherical balloon is measured as 12 inches, with a maximum error in measurement of ± 0.06 inch. Approximate the average error and the percentage error for the calculated value of the volume.

SOLUTION As in Figure 3.28, let x denote the measured radius of the balloon and Δx the maximum error in x. Let V denote the calculated volume and ΔV the error in V caused by Δx. Applying Definition (3.32)(i) to the volume $V = \frac{4}{3}\pi x^3$ yields

$$\text{average error} = \frac{\Delta V}{V} \approx \frac{dV}{V} = \frac{4\pi x^2\,dx}{\frac{4}{3}\pi x^3} = \frac{3\,dx}{x}.$$

For the special case $x = 12$ and $dx = \pm 0.06$, we obtain

$$\text{average error} \approx \frac{3(\pm 0.06)}{12} = \pm 0.015.$$

From Definition (3.32)(ii),

$$\text{percentage error} \approx (\pm 0.015) \times (100\%) = \pm 1.5\%.$$

Thus, *on the average*, there is an error of ± 0.015 in.^{3} per in.^{3} of calculated volume. Note that this leads to a percentage error of $\pm 1.5\%$ for the volume.

EXAMPLE 6 A sperm whale is spotted by a merchant ship, and crew members estimate its length L to be 32 feet, with a possible error of ± 2 feet. Whale research has shown that the weight W (in metric tons) is related to L by means of the formula $W = 0.000137L^{3.18}$. Use differentials to approximate

(a) the error in estimating the weight of the whale (to the nearest tenth of a metric ton)

(b) the average and percentage errors

SOLUTION Let ΔL denote the error in the estimation of L, and let ΔW be the corresponding error in the calculated value of W. These errors may be approximated by dL and dW.

(a) Applying Definition (3.28) yields

$$\Delta W \approx dW = (0.000137)(3.18)L^{2.18}\, dL.$$

Substituting $L = 32$ and $dL = \pm 2$, we obtain

$$\Delta W \approx (0.000137)(3.18)(32)^{2.18}(\pm 2) \approx \pm 1.7 \text{ metric tons.}$$

(b) By Definition (3.32)(i),

$$\text{average error} = \frac{\Delta W}{W} \approx \frac{dW}{W} = \frac{(0.000137)(3.18)L^{2.18}\, dL}{(0.000137)L^{3.18}} = \frac{3.18\, dL}{L}.$$

Substituting $dL = \pm 2$ and $L = 32$, we have

$$\text{average error} \approx \frac{3.18(\pm 2)}{32} \approx \pm 0.20.$$

By Definition (3.32)(ii),

$$\text{percentage error} \approx (\pm 0.20) \times (100\%) = \pm 20\%.$$

Estimates of vertical wind shear are of great importance to pilots during take-offs and landings. If we assume that the wind speed v at a height h above the ground is given by $v = f(h)$, where f is a differentiable function, then **vertical (scalar) wind shear** is defined as dv/dh (the instantaneous rate of change of v with respect to h). Since it is impossible to know the wind speed v at every height h, the wind shear must be estimated by using only a finite number of function values. Consider the situation illustrated in Figure 3.29, where we know only the wind speeds v_0 and v_1 at heights h_0 and h_1, respectively. An estimate of wind shear at height h_1 may be obtained by using the approximation formula

FIGURE 3.29

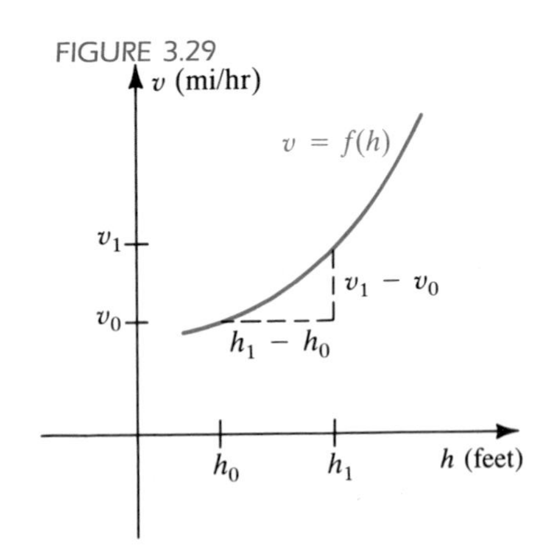

$$\left.\frac{dv}{dh}\right]_{h=h_1} \approx \frac{v_1 - v_0}{h_1 - h_0}.$$

The empirical relation

$$\left(\frac{v_0}{v_1}\right) = \left(\frac{h_0}{h_1}\right)^P$$

may also be employed, where the exponent P is determined by observation and depends on many factors. For strong winds, the value $P = \frac{1}{7}$ is sometimes used.

EXAMPLE 7 Suppose that at a height of 20 feet above the ground the wind speed is 28 mi/hr. Based on the preceding discussion (with $P = \frac{1}{7}$), estimate the vertical wind shear 200 feet above the ground.

SOLUTION Using the notation of the preceding discussion, we let

$$h_0 = 20, \quad v_0 = 28, \quad \text{and} \quad h_1 = 200.$$

Solving $(v_0/v_1) = (h_0/h_1)^P$ for v_1 and then substituting values, we obtain

$$v_1 = v_0\left(\frac{h_1}{h_0}\right)^P = 28\left(\frac{200}{20}\right)^{1/7} \approx 39 \text{ mi/hr}.$$

At $h_1 = 200$,

$$\left.\frac{dv}{dh}\right]_{h=h_1} \approx \frac{v_1 - v_0}{h_1 - h_0} \approx \frac{39 - 28}{200 - 20} \approx 0.06.$$

Thus, at a height of 200 feet, the vertical wind shear is approximately 0.06 (mi/hr)/ft. This is a common value. Wind shear values greater than 0.1 are considered high.

EXERCISES 3.5

Exer. 1–4: (a) Find general formulas for Δy and dy. (b) If, for the given values of a and Δx, x changes from a to $a + \Delta x$, find the values of Δy and dy.

1 $y = 2x^2 - 4x + 5$; $a = 2$, $\Delta x = -0.2$

2 $y = x^3 - 4$; $a = -1$, $\Delta x = 0.1$

3 $y = 1/x^2$; $a = 3$, $\Delta x = 0.3$

4 $y = \dfrac{1}{2 + x}$; $a = 0$, $\Delta x = -0.03$

Exer. 5–10: Find (a) Δy, (b) dy, and (c) $dy - \Delta y$.

5 $y = 4 - 9x$

6 $y = 7x + 12$

7 $y = 3x^2 + 5x - 2$

8 $y = 4 - 7x - 2x^2$

9 $y = 1/x$

10 $y = 1/x^2$

Exer. 11–18: Find a linear approximation for $f(b)$ if the independent variable changes from a to b.

11 $f(x) = 4x^5 - 6x^4 + 3x^2 - 5$; $a = 1$, $b = 1.03$

12 $f(x) = -3x^3 + 8x - 7$; $a = 4$, $b = 3.96$

13 $f(x) = x^4$; $a = 1$, $b = 0.98$

14 $f(x) = x^4 - 3x^3 + 4x^2 - 5$; $a = 2$, $b = 2.01$

15 $f(\theta) = 2 \sin \theta + \cos \theta$; $a = 30°$, $b = 27°$

16 $f(\phi) = \csc \phi + \cot \phi$; $a = 45°$, $b = 46°$

17 $f(\alpha) = \sec \alpha$; $a = 60°$, $b = 62°$

18 $f(\beta) = \tan \beta$; $a = 30°$, $b = 28°$

19 (a) If $f(x) = \sin(\tan x - 1)$, find an (approximate) equation of the tangent line to the graph of f at $(2.5, f(2.5))$ using Exercise 51 in Section 3.2.

(b) Use the equation found in (a) to approximate $f(2.6)$.

(c) Use (3.31) with $x = 2.5$ to approximate $f(2.6)$.

(d) Compare the two approximations in (b) and (c).

c 20 (a) If $f(x) = x^3 + 3x^2 - 2x + 5$, find an (approximate) equation of the tangent line to the graph of f at $(0.4, f(0.4))$.

(b) Use the equation found in (a) to approximate $f(0.43)$.

(c) Use (3.31) with $x = 0.4$ to approximate $f(0.43)$.

(d) Compare the two approximations in (b) and (c).

Exer. 21–24: Let x denote a measurement with a maximum error of Δx. Use differentials to approximate the average error and the percentage error for the calculated value of y.

21 $y = 3x^4$; $x = 2$, $\Delta x = \pm 0.01$

22 $y = x^3 + 5x$; $x = 1$, $\Delta x = \pm 0.1$

23 $y = 4\sqrt{x} + 3x$; $x = 4$, $\Delta x = \pm 0.2$

24 $y = 6\sqrt[3]{x}$; $x = 8$, $\Delta x = \pm 0.03$

25 If $A = 3x^2 - x$, find dA for $x = 2$ and $dx = 0.1$.

26 If $P = 6t^{2/3} + t^2$, find dP for $t = 8$ and $dt = 0.2$.

27 If $y = 4x^3$ and the maximum percentage error in x is $\pm 15\%$, approximate the maximum percentage error in y.

28 If $z = 40\sqrt[5]{w^2}$ and the maximum average error in w is ± 0.08, approximate the maximum average error in z.

29 If $A = 15\sqrt[3]{s^2}$ and the allowable maximum average error in A is to be ± 0.04, determine the allowable maximum average error in s.

30 If $S = 10\pi x^2$ and the allowable maximum percentage error in S is to be $\pm 10\%$, determine the allowable maximum percentage error in x.

31 The radius of a circular manhole cover is estimated to be 16 inches, with a maximum error in measurement of ± 0.06 inch. Use differentials to estimate the maximum error in the calculated area of one side of the cover. Approximate the average error and the percentage error.

32 The length of a side of a square floor tile is estimated as 1 foot, with a maximum error in measurement of $\pm\frac{1}{16}$ inch. Use differentials to estimate the maximum error in the calculated area. Approximate the average error and the percentage error.

33 Use differentials to approximate the increase in volume of a cube if the length of each edge changes from 10 inches to 10.1 inches. What is the exact change in volume?

34 A spherical balloon is being inflated with gas. Use differentials to approximate the increase in surface area of the balloon if the diameter changes from 2 feet to 2.02 feet.

35 One side of a house has the shape of a square surmounted by an equilateral triangle. If the length of the base is measured as 48 feet, with a maximum error in measurement of ± 1 inch, calculate the area of the side. Use differentials to estimate the maximum error in the calculation. Approximate the average error and the percentage error.

36 Small errors in measurements of dimensions of large containers can have a marked effect on calculated volumes. A silo has the shape of a right circular cylinder surmounted by a hemisphere (see figure). The altitude of the cylinder is exactly 50 feet. The circumference of the base is measured as 30 feet, with a maximum error in measurement of ± 6 inches. Calculate the volume of the silo from these measurements, and use differentials to estimate the maximum error in the calculation. Approximate the average error and the percentage error.

EXERCISE 36

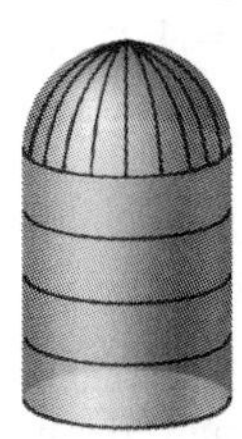

37 As sand leaks out of a container, it forms a conical pile whose altitude is always the same as the radius. If, at a certain instant, the radius is 10 centimeters, use differentials to approximate the change in radius that will increase the volume of the pile by 2 cm^3.

38 An isosceles triangle has equal sides of length 12 inches. If the angle θ between these sides is increased from 30° to 33°, use differentials to approximate the change in the area of the triangle.

39 Newton's law of gravitation states that the force F of attraction between two particles having masses m_1 and m_2 is given by $F = Gm_1m_2/s^2$, where G is a constant and s is the distance between the particles. If $s = 20$ cm, use differentials to approximate the change in s that will increase F by 10%.

40 The formula $T = 2\pi\sqrt{l/g}$ relates the length l of a pendulum to its period T, where g is a gravitational constant. What percentage change in the length corresponds to a 30% increase in the period?

41 Constriction of arterioles is a cause of high blood pressure. It has been verified experimentally that as blood flows through an arteriole of fixed length, the pressure difference between the two ends of the arteriole is inversely proportional to the fourth power of the radius. If the radius of an arteriole decreases by 10%, use differentials to find the percentage change in the pressure difference.

42 The electrical resistance R of a wire is directly proportional to its length and inversely proportional to the square of its diameter. If the length is fixed, how accurately must the diameter be measured (in terms of percentage error) to keep the percentage error in R between -3% and 3%?

43 If an object of weight W pounds is pulled along a horizontal plane by a force applied to a rope that is attached to the object and if the rope makes an angle θ with the horizontal, then the magnitude of the force is given by

$$F(\theta) = \frac{\mu W}{\mu \sin\theta + \cos\theta},$$

where μ is a constant called the *coefficient of friction*. Suppose that a 100-pound box is being pulled along a floor and that $\mu = 0.2$ (see figure). If θ is changed from 45° to 46°, use differentials to approximate the change in the force that must be applied.

EXERCISE 43

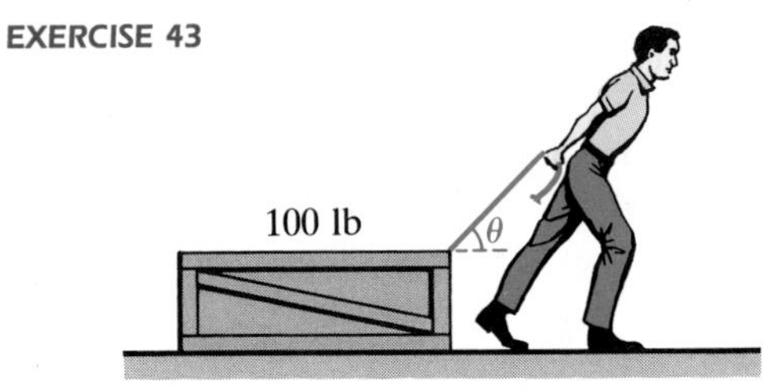

44 It will be shown in Chapter 15 that if a projectile is fired from a cannon with an initial velocity v_0 and at an angle α to the horizontal, then its maximum height h and range R are given by

$$h = \frac{v_0^2 \sin^2\alpha}{2g} \quad \text{and} \quad R = \frac{2v_0^2 \sin\alpha \cos\alpha}{g}.$$

Suppose $v_0 = 100$ ft/sec and $g = 32$ ft/sec^2. If α is increased from 30° to 30°30′, use differentials to estimate the changes in h and R.

45 At a point 20 feet from the base of a flagpole, the angle of elevation of the top of the pole is measured as 60°, with a possible error of $\pm 15'$. Use differentials to approximate the error in the calculated height of the pole.

46 A spacelab circles the earth at an altitude of 380 miles. When an astronaut views the horizon, the angle θ shown in the figure is $65.8°$, with a possible maximum error of $\pm 0.5°$. Use differentials to approximate the error in the astronaut's calculation of the radius of the earth.

EXERCISE 46

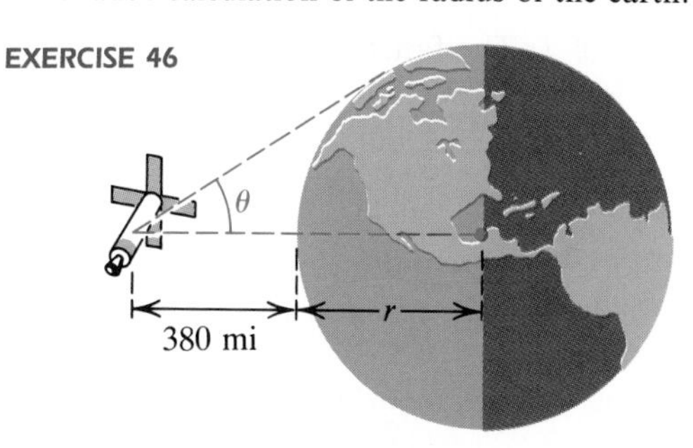

47 The Great Pyramid of Egypt has a square base of 230 meters (see figure). To estimate the height h of this massive structure, an observer stands at the midpoint of one of the sides and views the apex of the pyramid. The angle of elevation ϕ is found to be $52°$. How accurate must this measurement be to keep the error in h between -1 meter and 1 meter?

EXERCISE 47

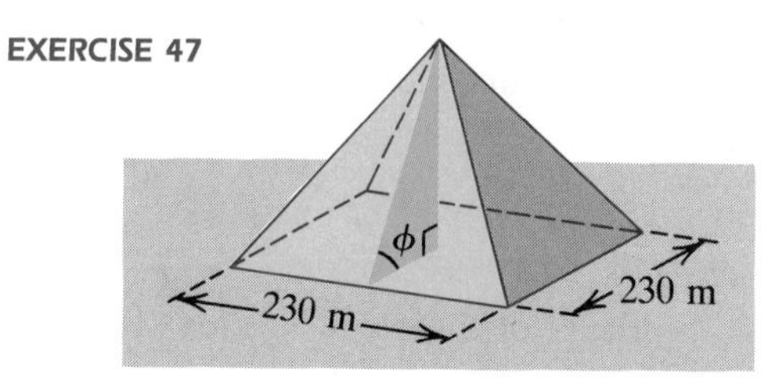

48 As a point light source moves on a semicircular track, as shown in the figure, the illuminance E on the surface is inversely proportional to the square of the distance s from the source and is directly proportional to the cosine of the angle θ between the direction of light flow and the normal to the surface. If θ is decreased from $21°$ to $20°$ and s is constant, use differentials to approximate the percentage increase in illuminance.

EXERCISE 48

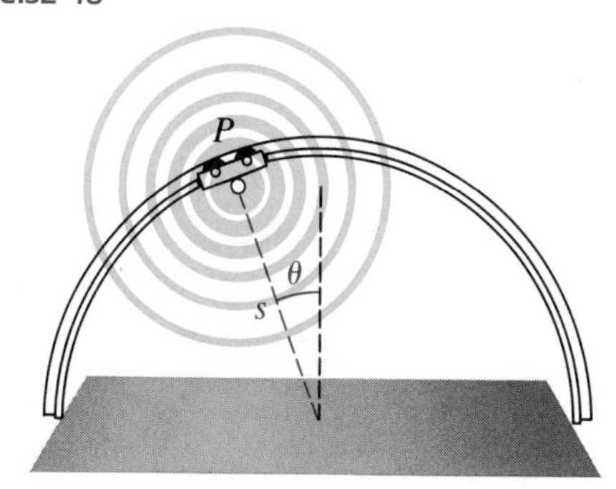

49 Boyle's law states that if the temperature is constant, the pressure p and volume v of a confined gas are related by the formula $pv = c$, where c is a constant or, equivalently, by $p = c/v$ with $v \neq 0$. Show that dp and dv are related by means of the formula $p\,dv + v\,dp = 0$.

50 In electrical theory Ohm's law states that $I = V/R$, where I is the current (in amperes), V is the electromotive force (in volts), and R is the resistance (in ohms). Show that dI and dR are related by the formula $R\,dI + I\,dR = 0$.

51 The area A of a square of side s is given by $A = s^2$. If s increases by an amount Δs, illustrate dA and $\Delta A - dA$ geometrically.

52 The volume V of a cube of edge s is given by $V = s^3$. If s increases by an amount Δs, illustrate dV and $\Delta V - dV$ geometrically.

3.6 THE CHAIN RULE

The rules for derivatives obtained in previous sections are limited in scope because they can be used only for sums, differences, products, and quotients that involve x^n, $\sin x$, $\cos x$, $\tan x$, and so on. There is no rule that can be applied *directly* to expressions such as $\sin 2x$ or $\sqrt{x^2 + 1}$. Note that

$$D_x \sin 2x \neq \cos 2x,$$

for if we use the identity $\sin 2x = 2 \sin x \cos x$ and apply the product rule,

$$\begin{aligned} D_x \sin 2x &= D_x\,(2 \sin x \cos x) \\ &= 2\,D_x\,(\sin x \cos x) \\ &= 2[\sin x\,(D_x \cos x) + \cos x\,(D_x \sin x)] \\ &= 2[\sin x\,(-\sin x) + \cos x\,(\cos x)] \\ &= 2(-\sin^2 x + \cos^2 x) \\ &= 2 \cos 2x. \end{aligned}$$

Since these manipulations are rather cumbersome, let us seek a more direct method of finding the derivative of $y = \sin 2x$. The key is to regard y as a *composite* function of x. Thus, for functions f and g,

$$\text{if} \quad y = f(u) \quad \text{and} \quad u = g(x), \quad \text{then} \quad y = f(g(x)),$$

provided $g(x)$ is in the domain of f. The function given by $y = f(g(x))$ is the composite function $f \circ g$ defined in Section 1.2. Note that $y = \sin 2x$ may be expressed in this way because

$$\text{if} \quad y = \sin u \quad \text{and} \quad u = 2x, \quad \text{then} \quad y = \sin 2x.$$

If we can find a general rule for differentiating $y = f(g(x))$, then, as a special case, we may apply it to $y = \sin 2x$ and, in fact, to $y = \sin g(x)$ for any differentiable function g.

To get an idea of the type of rule to expect, let us return to the equations

$$y = f(u), \quad u = g(x), \quad \text{and} \quad y = f(g(x)) = (f \circ g)(x).$$

We shall consider the following derivatives:

$$\frac{dy}{du} = f'(u), \quad \frac{du}{dx} = g'(x), \quad \text{and} \quad \frac{dy}{dx} = (f \circ g)'(x)$$

It is important to note that dy/du is the derivative with respect to u when y is regarded as a function of u and that dy/dx is the derivative with respect to x when y is regarded as a (composite) function of x. If we consider the product

$$\frac{dy}{du}\frac{du}{dx}$$

and treat the derivatives as quotients of differentials, then the product *suggests* the following rule:

$$\frac{dy}{dx} = \frac{dy}{du}\frac{du}{dx} = f'(u)g'(x)$$

Note that this rule *does* lead to the correct derivative of $y = \sin 2x$, for if we write

$$y = \sin u \quad \text{and} \quad u = 2x$$

and use the rule, we obtain

$$\frac{dy}{dx} = \frac{dy}{du}\frac{du}{dx} = (\cos u)(2) = 2 \cos u = 2 \cos 2x.$$

Although we have not *proved* that this rule is valid, it makes the next theorem plausible. We assume that variables are chosen such that the composite function $f \circ g$ is defined, and that if g has a derivative at x, then f has a derivative at $g(x)$.

Chain rule (3.33)

If $y = f(u)$, $u = g(x)$, and the derivatives dy/du and du/dx both exist, then the composite function defined by $y = f(g(x))$ has a derivative given by

$$\frac{dy}{dx} = \frac{dy}{du}\frac{du}{dx} = f'(u)g'(x) = f'(g(x))g'(x).$$

PARTIAL PROOF Let Δx be an increment such that both x and $x + \Delta x$ are in the domain of the composite function. Since $y = f(g(x))$, the corresponding increment of y is given by

$$\Delta y = f(g(x + \Delta x)) - f(g(x)).$$

If the composite function has a derivative at x, then, by (3.27),

$$\frac{dy}{dx} = \lim_{\Delta x \to 0} \frac{\Delta y}{\Delta x}.$$

Next consider $u = g(x)$ and let Δu be the increment of u that corresponds to Δx; that is,

$$\Delta u = g(x + \Delta x) - g(x).$$

Since

$$g(x + \Delta x) = g(x) + \Delta u = u + \Delta u,$$

we may express the formula $\Delta y = f(g(x + \Delta x)) - f(g(x))$ as

$$\Delta y = f(u + \Delta u) - f(u).$$

If $y = f(u)$ is differentiable at u, then, as in (3.27),

$$\frac{dy}{du} = f'(u) = \lim_{\Delta u \to 0} \frac{\Delta y}{\Delta u}.$$

Similarly, if $u = g(x)$ is differentiable at x, then

$$\frac{du}{dx} = g'(x) = \lim_{\Delta x \to 0} \frac{\Delta u}{\Delta x}.$$

Let us assume that there exists an open interval I containing x such that whenever $x + \Delta x$ is in I and $\Delta x \neq 0$, then $\Delta u \neq 0$. In this case we may write

$$\frac{dy}{dx} = \lim_{\Delta x \to 0} \frac{\Delta y}{\Delta x} = \lim_{\Delta x \to 0} \left(\frac{\Delta y}{\Delta u}\frac{\Delta u}{\Delta x}\right) = \left(\lim_{\Delta x \to 0} \frac{\Delta y}{\Delta u}\right)\left(\lim_{\Delta x \to 0} \frac{\Delta u}{\Delta x}\right),$$

provided the limits exist. Since g is differentiable at x, it is continuous at x. Hence if $\Delta x \to 0$, then $g(x + \Delta x)$ approaches $g(x)$ and, therefore, $\Delta u \to 0$. It follows that the last limit formula may be written

$$\begin{aligned}\frac{dy}{dx} &= \left(\lim_{\Delta u \to 0} \frac{\Delta y}{\Delta u}\right)\left(\lim_{\Delta x \to 0} \frac{\Delta u}{\Delta x}\right) \\ &= \left(\frac{dy}{du}\right)\left(\frac{du}{dx}\right) = f'(u)g'(x) = f'(g(x))g'(x),\end{aligned}$$

which is what we wished to prove.

In many applications of the chain rule, $u = g(x)$ has the property that if $\Delta x \neq 0$, then $\Delta u \neq 0$, which we assumed at the beginning of the preceding paragraph. If g does not satisfy this property, then every open interval containing x contains a number $x + \Delta x$, with $\Delta x \neq 0$, such that $\Delta u = 0$. In this case our proof is invalid, since Δu occurs in a denominator. To construct a proof that takes functions of this type into account, it is necessary to introduce additional techniques. *A complete proof of the chain rule is given in Appendix II.* ■

EXAMPLE 1 Find $\dfrac{dy}{dx}$ if $y = \sqrt{u}$ and $u = x^2 + 1$.

SOLUTION If we substitute $x^2 + 1$ for u in $y = \sqrt{u} = u^{1/2}$, we obtain

$$y = \sqrt{x^2 + 1} = (x^2 + 1)^{1/2}.$$

We cannot find dy/dx by using previous differentiation formulas; however, using the chain rule (3.33), we have

$$\frac{dy}{dx} = \frac{dy}{du}\frac{du}{dx} = \left(\frac{1}{2}u^{-1/2}\right)(2x) = \frac{x}{\sqrt{u}}$$

and hence $$\frac{dy}{dx} = \frac{x}{\sqrt{x^2 + 1}}.$$

In Example 1, the composite function was given by a power of $x^2 + 1$. Since powers of functions occur frequently in calculus, it will save us time to state a general differentiation rule that can be applied to such special cases. In the following we assume that n is any rational number, g is a differentiable function, and zero denominators do not occur. We shall see later that the rule can be used for any *real* number n.

Power rule for functions **(3.34)**

> If $y = u^n$ and $u = g(x)$, then
> $$D_x\,(u^n) = nu^{n-1}\,D_x\,u,$$
> or, equivalently,
> $$D_x\,[g(x)]^n = n[g(x)]^{n-1}\,D_x\,g(x).$$

PROOF By the chain rule,

$$\frac{dy}{dx} = \frac{dy}{du}\frac{du}{dx} = nu^{n-1}\,D_x\,u = n[g(x)]^{n-1}\,D_x\,g(x).$$ ■

Note that if $u = x$, then $D_x\,u = 1$ and (3.34) reduces to (3.14).

EXAMPLE 2 Find $f'(x)$ if $f(x) = (x^5 - 4x + 8)^7$.

SOLUTION Using the power rule (3.34) with $u = x^5 - 4x + 8$ and $n = 7$ yields

$$\begin{aligned} f'(x) &= D_x\,(x^5 - 4x + 8)^7 \\ &= 7(x^5 - 4x + 8)^6\,D_x\,(x^5 - 4x + 8) \\ &= 7(x^5 - 4x + 8)^6(5x^4 - 4). \end{aligned}$$

EXAMPLE 3 Find $\dfrac{dy}{dx}$ if $y = \dfrac{1}{(4x^2 + 6x - 7)^3}$.

SOLUTION Writing $y = (4x^2 + 6x - 7)^{-3}$ and using the power rule with $u = 4x^2 + 6x - 7$ and $n = -3$, we have

$$\begin{aligned}\frac{dy}{dx} &= \frac{d}{dx}(4x^2 + 6x - 7)^{-3} \\ &= -3(4x^2 + 6x - 7)^{-4}\frac{d}{dx}(4x^2 + 6x - 7) \\ &= -3(4x^2 + 6x - 7)^{-4}(8x + 6) \\ &= \frac{-6(4x + 3)}{(4x^2 + 6x - 7)^4}.\end{aligned}$$

EXAMPLE 4 Find $f'(x)$ if $f(x) = \sqrt[3]{5x^2 - x + 4}$.

SOLUTION Writing $f(x) = (5x^2 - x + 4)^{1/3}$ and using the power rule with $u = 5x^2 - x + 4$ and $n = \frac{1}{3}$, we obtain

$$\begin{aligned}f'(x) &= \tfrac{1}{3}(5x^2 - x + 4)^{-2/3}\, D_x\,(5x^2 - x + 4) \\ &= \left(\frac{1}{3}\right)\frac{1}{(5x^2 - x + 4)^{2/3}}(10x - 1) \\ &= \frac{10x - 1}{3\sqrt[3]{(5x^2 - x + 4)^2}}.\end{aligned}$$

EXAMPLE 5 Find $F'(z)$ if $F(z) = (2z + 5)^3(3z - 1)^4$.

SOLUTION Using first the product rule, second the power rule, and then factoring the result gives us

$$\begin{aligned}F'(z) &= (2z + 5)^3\, D_z\,(3z - 1)^4 + (3z - 1)^4\, D_z\,(2z + 5)^3 \\ &= (2z + 5)^3 \cdot 4(3z - 1)^3(3) + (3z - 1)^4 \cdot 3(2z + 5)^2(2) \\ &= 6(2z + 5)^2(3z - 1)^3[2(2z + 5) + (3z - 1)] \\ &= 6(2z + 5)^2(3z - 1)^3(7z + 9).\end{aligned}$$

EXAMPLE 6 Find y' if $y = (3x + 1)^6\sqrt{2x - 5}$.

SOLUTION Since $y = (3x + 1)^6(2x - 5)^{1/2}$, we have, by the product and power rules,

$$\begin{aligned}y' &= (3x + 1)^6\tfrac{1}{2}(2x - 5)^{-1/2}(2) + (2x - 5)^{1/2}6(3x + 1)^5(3) \\ &= \frac{(3x + 1)^6}{\sqrt{2x - 5}} + 18(3x + 1)^5\sqrt{2x - 5} \\ &= \frac{(3x + 1)^6 + 18(3x + 1)^5(2x - 5)}{\sqrt{2x - 5}} \\ &= \frac{(3x + 1)^5(39x - 89)}{\sqrt{2x - 5}}.\end{aligned}$$

The next example is of interest because it illustrates the fact that after the power rule is applied to $[g(x)]^r$, it may be necessary to apply it again in order to find $g'(x)$.

EXAMPLE 7 Find $f'(x)$ if $f(x) = (7x + \sqrt{x^2 + 6})^4$.

SOLUTION Applying the power rule yields

$$\begin{aligned} f'(x) &= 4(7x + \sqrt{x^2 + 6})^3\, D_x\, (7x + \sqrt{x^2 + 6}) \\ &= 4(7x + \sqrt{x^2 + 6})^3[D_x\, (7x) + D_x \sqrt{x^2 + 6}]. \end{aligned}$$

Again applying the power rule, we have

$$\begin{aligned} D_x \sqrt{x^2 + 6} &= D_x\, (x^2 + 6)^{1/2} = \tfrac{1}{2}(x^2 + 6)^{-1/2}\, D_x\, (x^2 + 6) \\ &= \frac{1}{2\sqrt{x^2 + 6}}(2x) = \frac{x}{\sqrt{x^2 + 6}}. \end{aligned}$$

Therefore, $$f'(x) = 4(7x + \sqrt{x^2 + 6})^3\left(7 + \frac{x}{\sqrt{x^2 + 6}}\right).$$

As another application of the chain rule, we can prove the following.

Theorem (3.35)

If $u = g(x)$ and g is differentiable, then

$$\begin{aligned} D_x \sin u &= (\cos u)\, D_x\, u & D_x \cos u &= (-\sin u)\, D_x\, u \\ D_x \tan u &= (\sec^2 u)\, D_x\, u & D_x \cot u &= (-\csc^2 u)\, D_x\, u \\ D_x \sec u &= (\sec u \tan u)\, D_x\, u & D_x \csc u &= (-\csc u \cot u)\, D_x\, u \end{aligned}$$

PROOF If we let $y = \sin u$, then, by (3.25),

$$\frac{dy}{du} = \cos u.$$

Applying the chain rule (3.33) yields

$$\frac{dy}{dx} = \frac{dy}{du}\frac{du}{dx} = \cos u\, D_x\, u.$$

The remaining formulas may be obtained in similar fashion. ■

Note that Theorem (3.25) is the special case of Theorem (3.35) in which $u = x$.

EXAMPLE 8 If $y = \cos\,(5x^3)$, find $D_x\, y$ and $D_x^2\, y$.

SOLUTION Using the formula for $D_x \cos u$ in Theorem (3.35) with $u = 5x^3$, we have

$$\begin{aligned} D_x\, y &= D_x \cos\,(5x^3) \\ &= [-\sin\,(5x^3)]\, D_x\, (5x^3) \\ &= [-\sin\,(5x^3)](15x^2) \\ &= -15x^2 \sin\,(5x^3). \end{aligned}$$

To find $D_x^2 y$, we differentiate $D_x y = -15x^2 \sin(5x^3)$. Using the product rule and Theorem (3.35) gives us

$$\begin{aligned} D_x^2 y &= -15x^2 D_x \sin(5x^3) + \sin(5x^3) D_x(-15x^2) \\ &= -15x^2 \cos(5x^3) D_x(5x^3) + [\sin(5x^3)](-30x) \\ &= -15x^2[\cos(5x^3)](15x^2) - 30x \sin(5x^3) \\ &= -225x^4 \cos(5x^3) - 30x \sin(5x^3). \end{aligned}$$

EXAMPLE 9 Find $f'(x)$ if $f(x) = \tan^3 4x$.

SOLUTION First note that $f(x) = \tan^3 4x = (\tan 4x)^3$. Applying the power rule with $u = \tan 4x$ and $n = 3$ yields

$$f'(x) = 3(\tan 4x)^2 D_x \tan 4x = (3 \tan^2 4x) D_x \tan 4x.$$

Next, by Theorem (3.35),

$$D_x \tan 4x = (\sec^2 4x) D_x(4x) = (\sec^2 4x)(4) = 4 \sec^2 4x.$$

Thus $$f'(x) = (3 \tan^2 4x)(4 \sec^2 4x) = 12 \tan^2 4x \sec^2 4x.$$

EXAMPLE 10 Find y' if $y = \sqrt{\sin 6x}$.

SOLUTION Writing $y = (\sin 6x)^{1/2}$ and using the power rule, we obtain

$$y' = \tfrac{1}{2}(\sin 6x)^{-1/2} D_x \sin 6x.$$

Next, by Theorem (3.35),

$$D_x \sin 6x = (\cos 6x) D_x(6x) = (\cos 6x)(6) = 6 \cos 6x.$$

Consequently,

$$\begin{aligned} y' &= \tfrac{1}{2}(\sin 6x)^{-1/2}(6 \cos 6x) \\ &= \frac{3 \cos 6x}{(\sin 6x)^{1/2}} = \frac{3 \cos 6x}{\sqrt{\sin 6x}}. \end{aligned}$$

FIGURE 3.30

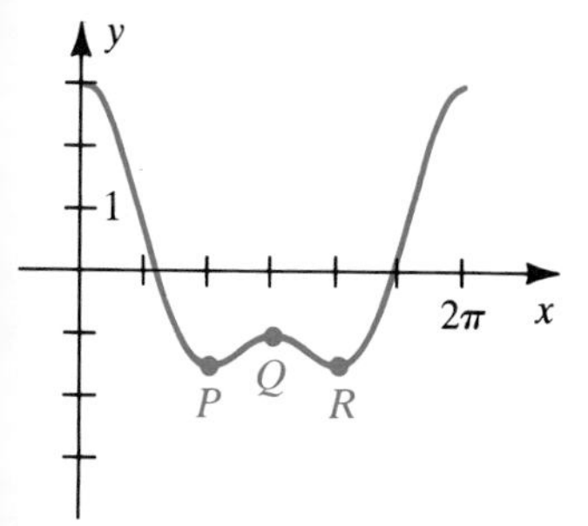

EXAMPLE 11 A graph of $y = \cos 2x + 2 \cos x$ for $0 \le x \le 2\pi$ is shown in Figure 3.30. Find the points at which the tangent line is horizontal.

SOLUTION Differentiating, we obtain

$$\begin{aligned} D_x y &= (-\sin 2x) D_x(2x) + 2(-\sin x) \\ &= -2 \sin 2x - 2 \sin x. \end{aligned}$$

The tangent line is horizontal if its slope $D_x y$ is 0—that is, if

$$-2 \sin 2x - 2 \sin x = 0, \quad \text{or} \quad \sin 2x + \sin x = 0.$$

Using the double-angle formula $\sin 2x = 2 \sin x \cos x$ gives us

$$2 \sin x \cos x + \sin x = 0,$$

or, equivalently, $$\sin x (2 \cos x + 1) = 0.$$

Thus, either

$$\sin x = 0 \quad \text{or} \quad 2 \cos x + 1 = 0;$$

that is,

$$\sin x = 0 \quad \text{or} \quad \cos x = -\tfrac{1}{2}.$$

The solutions of these equations for $0 \le x \le 2\pi$ are

$$0, \quad \pi, \quad 2\pi, \quad 2\pi/3, \quad 4\pi/3.$$

The two solutions $x = 0$ and $x = 2\pi$ tell us that there are horizontal tangent lines at the endpoints of the interval $[0, 2\pi]$. The remaining solutions $2\pi/3$, π, and $4\pi/3$ are the x-coordinates of the points P, Q, and R shown in Figure 3.30. Using $y = \cos 2x + 2 \cos x$, we see that horizontal tangent lines occur at the points

$$(0, 3), \quad (2\pi/3, -1.5), \quad (\pi, -1), \quad (4\pi/3, -1.5), \quad (2\pi, 3).$$

If only approximate solutions are desired, then, to the nearest tenth, we obtain

$$(0, 3), \quad (2.1, -1.5), \quad (3.1, -1), \quad (4.2, -1.5), \quad (6.3, 3).$$

EXERCISES 3.6

Exer. 1–6: Use the chain rule to find $\frac{dy}{dx}$, and express the answer in terms of x.

1 $y = u^2$; $u = x^3 - 4$

2 $y = \sqrt[3]{u}$; $u = x^2 + 5x$

3 $y = 1/u$; $u = \sqrt{3x - 2}$

4 $y = 3u^2 + 2u$; $u = 4x$

5 $y = \tan 3u$; $u = x^2$

6 $y = u \sin u$; $u = x^3$

Exer. 7–62: Find the derivative.

7 $f(x) = (x^2 - 3x + 8)^3$

8 $f(x) = (4x^3 + 2x^2 - x - 3)^2$

9 $g(x) = (8x - 7)^{-5}$

10 $k(x) = (5x^2 - 2x + 1)^{-3}$

11 $f(x) = \dfrac{x}{(x^2 - 1)^4}$

12 $g(x) = \dfrac{x^4 - 3x^2 + 1}{(2x + 3)^4}$

13 $f(x) = (8x^3 - 2x^2 + x - 7)^5$

14 $g(w) = (w^4 - 8w^2 + 15)^4$

15 $F(v) = (17v - 5)^{1000}$

16 $s(t) = (4t^5 - 3t^3 + 2t)^{-2}$

17 $N(x) = (6x - 7)^3(8x^2 + 9)^2$

18 $f(w) = (2w^2 - 3w + 1)(3w + 2)^4$

19 $g(z) = \left(z^2 - \dfrac{1}{z^2}\right)^6$

20 $S(t) = \left(\dfrac{3t + 4}{6t - 7}\right)^3$

21 $k(r) = \sqrt[3]{8r^3 + 27}$

22 $h(z) = (2z^2 - 9z + 8)^{-2/3}$

23 $F(v) = \dfrac{5}{\sqrt[5]{v^5 - 32}}$

24 $k(s) = \dfrac{1}{\sqrt{3s - 4}}$

25 $g(w) = \dfrac{w^2 - 4w + 3}{w^{3/2}}$

26 $K(x) = \sqrt{4x^2 + 2x + 3}$

27 $H(x) = \dfrac{2x + 3}{\sqrt{4x^2 + 9}}$

28 $f(x) = (7x + \sqrt{x^2 + 3})^6$

29 $k(x) = \sin (x^2 + 2)$

30 $f(t) = \cos (4 - 3t)$

31 $H(\theta) = \cos^5 3\theta$

32 $g(x) = \sin^4 (x^3)$

33 $g(z) = \sec (2z + 1)^2$

34 $k(z) = \csc (z^2 + 4)$

35 $H(s) = \cot (s^3 - 2s)$

36 $f(x) = \tan (2x^2 + 3)$

37 $f(x) = \cos (3x^2) + \cos^2 3x$

38 $g(w) = \tan^3 6w$

39 $F(\phi) = \csc^2 2\phi$

40 $M(x) = \sec (1/x^2)$

41 $K(z) = z^2 \cot 5z$

42 $G(s) = s \csc (s^2)$

43 $h(\theta) = \tan^2 \theta \sec^3 \theta$

44 $H(u) = u^2 \sec^3 4u$

45 $N(x) = (\sin 5x - \cos 5x)^5$

46 $p(v) = \sin 4v \csc 4v$

47 $T(w) = \cot^3 (3w + 1)$

48 $g(r) = \sin (2r + 3)^4$

49 $h(w) = \dfrac{\cos 4w}{1 - \sin 4w}$

50 $f(x) = \dfrac{\sec 2x}{1 + \tan 2x}$

51 $f(x) = \tan^3 2x - \sec^3 2x$

52 $h(\phi) = (\tan 2\phi - \sec 2\phi)^3$

53 $f(x) = \sin \sqrt{x} + \sqrt{\sin x}$

54 $f(x) = \tan \sqrt[3]{5 - 6x}$

55 $k(\theta) = \cos^2 \sqrt{3 - 8\theta}$

56 $r(t) = \sqrt{\sin 2t - \cos 2t}$

57 $g(x) = \sqrt{x^2 + 1} \tan \sqrt{x^2 + 1}$

58 $h(\phi) = \dfrac{\cot 4\phi}{\sqrt{\phi^2 + 4}}$

59 $M(x) = \sec \sqrt{4x + 1}$

60 $F(s) = \sqrt{\csc 2s}$

61 $h(x) = \sqrt{4 + \csc^2 3x}$

62 $f(t) = \sin^2 2t \sqrt{\cos 2t}$

Exer. 63–68: (a) Find equations of the tangent line and the normal line to the graph of the equation at P. (b) Find the x-coordinates on the graph at which the tangent line is horizontal.

63 $y = (4x^2 - 8x + 3)^4$; $P(2, 81)$

64 $y = (2x - 1)^{10}$; $P(1, 1)$

65 $y = \left(x + \dfrac{1}{x}\right)^5$; $P(1, 32)$

66 $y = \sqrt{2x^2 + 1}$; $P(-1, \sqrt{3})$

67 $y = 3x + \sin 3x$; $P(0, 0)$

68 $y = x + \cos 2x$; $P(0, 1)$

Exer. 69–74: Find the first and second derivatives.

69 $g(z) = \sqrt{3z + 1}$

70 $k(s) = (s^2 + 4)^{2/3}$

71 $k(r) = (4r + 7)^5$

72 $f(x) = \sqrt[5]{10x + 7}$

73 $f(x) = \sin^3 x$

74 $G(t) = \sec^2 4t$

Exer. 75–76: Use differentials to approximate the value.

75 $\sqrt[3]{65}$ (*Hint:* Let $y = \sqrt[3]{x}$.)

76 $\sqrt[5]{35}$

77 If an object of mass m has velocity v, then its *kinetic energy* K is given by $K = \frac{1}{2}mv^2$. If v is a function of time t, use the chain rule to find a formula for dK/dt.

78 As a spherical weather balloon is being inflated, its radius r is a function of time t. If V is the volume of the balloon, use the chain rule to find a formula for dV/dt.

79 When a space shuttle is launched into space, an astronaut's body weight decreases until a state of weightlessness is achieved. The weight W of a 150-pound astronaut at an altitude of x kilometers above sea level is given by

$$W = 150\left(\frac{6400}{6400 + x}\right)^2.$$

If the space shuttle is moving away from the earth's surface at the rate of 6 km/sec, at what rate is W decreasing when $x = 1000$ km?

80 The length-weight relationship for Pacific halibut is well described by the formula $W = 10.375L^3$, where L is the length in meters and W is the weight in kilograms. The rate of growth in length dL/dt is given by $0.18(2 - L)$, where t is time in years.

(a) Find a formula for the rate of growth in weight dW/dt in terms of L.

(b) Use the formula in part (a) to estimate the rate of growth in weight of a halibut weighing 20 kilograms.

81 If $k(x) = f(g(x))$ and if $f(2) = -4$, $g(2) = 2$, $f'(2) = 3$, and $g'(2) = 5$, find $k(2)$ and $k'(2)$.

82 Let p, q, and r be functions such that $p(z) = q(r(z))$. If $r(3) = 3$, $q(3) = -2$, $r'(3) = 4$, and $q'(3) = 6$, find $p(3)$ and $p'(3)$.

83 If $f(t) = g(h(t))$ and if $f(4) = 3$, $g(4) = 3$, $h(4) = 4$, $f'(4) = 2$, and $g'(4) = -5$, find $h'(4)$.

84 If $u(x) = v(w(x))$ and if $v(0) = -1$, $w(0) = 0$, $u(0) = -1$, $v'(0) = -3$, and $u'(0) = 2$, find $w'(0)$.

c 85 Let $h = f \circ g$ be a differentiable function. The following tables list some values of f and g. Use Exercise 51 of Section 3.2 to approximate $h'(1.12)$.

x	2.2210	2.2320	2.2430
$f(x)$	4.9328	4.9818	5.0310

x	1.1100	1.1200	1.1300
$g(x)$	2.2210	2.2320	2.2430

c 86 Let $h = f \circ g$ be a differentiable function. The following tables list some values of f and g. Use Exercise 51 of Section 3.2 to approximate $h'(-2)$.

x	−8.48092	−8.46000	−8.43908
$f(x)$	−2.03930	−2.03762	−2.03594

x	−2.00400	−2.00000	−1.99600
$g(x)$	−8.48092	−8.46000	−8.43908

87 Let f be differentiable. Use the chain rule to prove that

(a) if f is even, then f' is odd

(b) if f is odd, then f' is even

Use polynomial functions to give examples of (a) and (b).

88 Use the chain rule, the derivative formula for $D_x \sin u$, together with the identities

$$\cos x = \sin\left(\frac{\pi}{2} - x\right) \quad \text{and} \quad \sin x = \cos\left(\frac{\pi}{2} - x\right)$$

to obtain the formula for $D_x \cos x$.

89 Pinnipeds are a suborder of aquatic carnivorous mammals, such as seals and walruses, whose limbs are modified into flippers. The length-weight relationship during fetal growth is well described by the formula $W = (6 \times 10^{-5})L^{2.74}$, where L is the length in centimeters and W is the weight in kilograms.

(a) Use the chain rule to find a formula for the rate of growth in weight with respect to time t.

(b) If the weight of a seal is 0.5 kilogram and is changing at a rate of 0.4 kilogram per month, how fast is the length changing?

90 The formula for the adiabatic expansion of air is $pv^{1.4} = c$, where p is the pressure, v is the volume, and c is a constant. Find a formula for the rate of change of pressure with respect to volume.

91 The curved surface area S of a right circular cone having altitude h and base radius r is given by $S = \pi r\sqrt{r^2 + h^2}$. For a certain cone, $r = 6$ cm. The altitude is measured as 8 centimeters, with a maximum error in measurement of ± 0.1 centimeter.

(a) Calculate S from the measurements and use differentials to estimate the maximum error in the calculation.

(b) Approximate the percentage error.

92 The period T of a simple pendulum of length l may be calculated by means of the formula $T = 2\pi\sqrt{l/g}$, where g is a gravitational constant. Use differentials to approximate the change in l that will increase T by 1%.

3.7 IMPLICIT DIFFERENTIATION

Given the equation

$$y = 2x^2 - 3,$$

we sometimes say that y is an **explicit function** of x, since we can write

$$y = f(x) \quad \text{with} \quad f(x) = 2x^2 - 3.$$

The equation

$$4x^2 - 2y = 6$$

determines the same function f, since solving for y gives us

$$-2y = -4x^2 + 6, \quad \text{or} \quad y = 2x^2 - 3.$$

For the case $4x^2 - 2y = 6$ we say that y (or f) is an **implicit function** of x, or that f is determined *implicitly* by the equation. If we substitute $f(x)$ for y in $4x^2 - 2y = 6$, we obtain

$$4x^2 - 2f(x) = 6$$
$$4x^2 - 2(2x^2 - 3) = 6$$
$$4x^2 - 4x^2 + 6 = 6.$$

The last equation is an identity, since it is true for every x in the domain of f. This is a characteristic of every function f determined implicitly by an equation in x and y; that is, *f is implicit if and only if substitution of $f(x)$ for y leads to an identity*. Since $(x, f(x))$ is a point on the graph of f, the last statement implies that *the graph of the implicit function f coincides with a portion (or all) of the graph of the equation.*

In the next example we show that an equation in x and y may determine more than one implicit function.

EXAMPLE 1 How many different functions are determined implicitly by the equation $x^2 + y^2 = 1$?

SOLUTION The graph of $x^2 + y^2 = 1$ is the unit circle with center at the origin. Solving the equation for y in terms of x, we obtain

$$y = \pm\sqrt{1 - x^2}.$$

Two functions f and g determined implicitly by the equation are given by

$$f(x) = \sqrt{1 - x^2} \quad \text{and} \quad g(x) = -\sqrt{1 - x^2}.$$

The graphs of f and g are the upper and lower halves, respectively, of the unit circle (see Figure 3.31(i) and (ii)). To find other implicit functions, we may let a be any number between -1 and 1 and then define the function k by

$$k(x) = \begin{cases} \sqrt{1 - x^2} & \text{if } -1 \le x \le a \\ -\sqrt{1 - x^2} & \text{if } a < x \le 1 \end{cases}$$

FIGURE 3.31

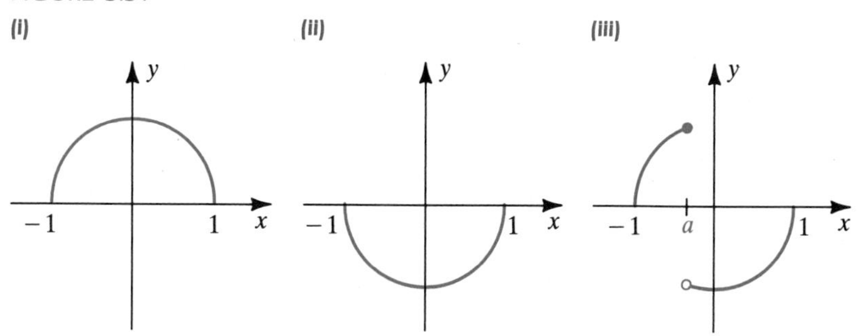

The graph of k is sketched in Figure 3.31(iii). Note that there is a jump discontinuity at $x = a$. The function k is determined implicitly by the equation $x^2 + y^2 = 1$, since

$$x^2 + [k(x)]^2 = 1$$

for every x in the domain of k. By letting a take on different values, we can obtain as many implicit functions as desired. Many other functions are determined implicitly by $x^2 + y^2 = 1$, and the graph of each is a portion of the graph of the equation.

If the equation

$$y^4 + 3y - 4x^3 = 5x + 1$$

determines an implicit function f, then

$$[f(x)]^4 + 3[f(x)] - 4x^3 = 5x + 1$$

for every x in the domain of f; however, there is no obvious way to solve for y in terms of x to obtain $f(x)$. It is possible to state conditions under which an implicit function exists and is differentiable at numbers in its domain; however, the proof requires advanced methods and hence is omitted. In the examples that follow we will assume that a given equation in x and y determines a differentiable function f such that if $f(x)$ is substituted for y, the equation is an identity for every x in the domain of f. The derivative of f may then be found by the method of **implicit differentiation**, in which we differentiate each term of the equation with respect to x. In using implicit differentiation it is often necessary to consider $D_x(y^n)$ for some unknown function y of x, say $y = f(x)$. By the power rule (3.34) with $y = u$, we can write $D_x(y^n)$ in any of the following forms:

$$D_x(y^n) = ny^{n-1} D_x y = ny^{n-1} y' = ny^{n-1} \frac{dy}{dx}$$

Since the dependent variable y represents the expression $f(x)$, it is *essential* to multiply ny^{n-1} by the derivative y' when we differentiate y with respect to x. Thus,

$$D_x(y^n) \neq ny^{n-1}, \quad \text{unless} \quad y = x.$$

EXAMPLE 2 Assuming that the equation $y^4 + 3y - 4x^3 = 5x + 1$ determines, implicitly, a differentiable function f such that $y = f(x)$, find its derivative.

SOLUTION We regard y as a symbol that denotes $f(x)$ and consider the equation as an identity for every x in the domain of f. Since derivatives of both sides are equal, we obtain the following:

$$D_x(y^4 + 3y - 4x^3) = D_x(5x + 1)$$

$$D_x(y^4) + D_x(3y) - D_x(4x^3) = D_x(5x) + D_x(1)$$

$$4y^3y' + 3y' - 12x^2 = 5 + 0$$

We now solve for y', obtaining

$$(4y^3 + 3)y' = 12x^2 + 5,$$

or

$$y' = \frac{12x^2 + 5}{4y^3 + 3},$$

provided $4y^3 + 3 \neq 0$. Thus, if $y = f(x)$, then

$$f'(x) = \frac{12x^2 + 5}{4(f(x))^3 + 3}.$$

The last two equations in the solution of Example 2 bring out a disadvantage of using the method of implicit differentiation: the formula for y' (or $f'(x)$) may contain the expression y (or $f(x)$). However, these formulas can still be very useful in analyzing f and its graph.

In the next example we use implicit differentiation to find the slope of the tangent line at a point $P(a, b)$ on the graph of an equation. In problems of this type we shall assume that the equation determines an implicit function f whose graph coincides with the graph of the equation for every x in some open interval containing a. Note that, since $P(a, b)$ is a point on the graph, the ordered pair (a, b) must be a solution of the equation.

EXAMPLE 3 Find the slope of the tangent line to the graph of

$$y^4 + 3y - 4x^3 = 5x + 1$$

at the point $P(1, -2)$.

SOLUTION The point $P(1, -2)$ is on the graph, since substituting $x = 1$ and $y = -2$ gives us

$$(-2)^4 + 3(-2) - 4(1)^3 = 5(1) + 1, \quad \text{or} \quad 6 = 6.$$

The slope of the tangent line at $P(1, -2)$ is the value of the derivative y' when $x = 1$ and $y = -2$. The given equation is the same as that in Example 2, where we found that $y' = (12x^2 + 5)/(4y^3 + 3)$. Substituting 1 for x and -2 for y gives us the following, where $y'\rbrack_{(1,-2)}$ denotes the value of y' when $x = 1$ and $y = -2$:

$$y'\rbrack_{(1,-2)} = \frac{12(1)^2 + 5}{4(-2)^3 + 3} = -\frac{17}{29}$$

EXAMPLE 4 If $y = f(x)$, where f is determined implicitly by the equation $x^2 + y^2 = 1$, find y'.

SOLUTION In Example 1 we showed that there are an unlimited number of implicit functions determined by $x^2 + y^2 = 1$. As in Example 2, we differentiate both sides of the equation with respect to x, obtaining

$$\begin{aligned} D_x(x^2) + D_x(y^2) &= D_x(1) \\ 2x + 2yy' &= 0 \\ yy' &= -x \\ y' &= -\frac{x}{y} \quad \text{if} \quad y \neq 0. \end{aligned}$$

The method of implicit differentiation provides the derivative of *any* differentiable function determined by an equation in two variables. For example, the equation $x^2 + y^2 = 1$ determines many implicit functions (see Example 1). From Example 4, the slope of the tangent line at the point (x, y) on any of the graphs in Figure 3.31 is given by $y' = -x/y$, provided the derivative exists.

EXAMPLE 5 Find y' if $4xy^3 - x^2y + x^3 - 5x + 6 = 0$.

SOLUTION Differentiating both sides of the equation with respect to x yields

$$D_x(4xy^3) - D_x(x^2y) + D_x(x^3) - D_x(5x) + D_x(6) = D_x(0).$$

Since y denotes $f(x)$ for some function f, the product rule must be applied to $D_x(4xy^3)$ and $D_x(x^2y)$. Thus,

$$\begin{aligned} D_x(4xy^3) &= 4x\, D_x(y^3) + y^3\, D_x(4x) \\ &= 4x(3y^2y') + y^3(4) \\ &= 12xy^2y' + 4y^3 \end{aligned}$$

and

$$\begin{aligned} D_x(x^2y) &= x^2\, D_x\, y + y\, D_x(x^2) \\ &= x^2y' + y(2x). \end{aligned}$$

Substituting these expressions in the first equation of the solution and differentiating the other terms leads to

$$(12xy^2y' + 4y^3) - (x^2y' + 2xy) + 3x^2 - 5 = 0.$$

Collecting the terms containing y' and transposing the remaining terms to the right side of the equation gives us

$$(12xy^2 - x^2)y' = 5 - 3x^2 + 2xy - 4y^3.$$

Consequently,
$$y' = \frac{5 - 3x^2 + 2xy - 4y^3}{12xy^2 - x^2},$$

provided $12xy^2 - x^2 \neq 0$.

EXAMPLE 6 Find y' if $y = x^2 \sin y$.

SOLUTION Differentiating both sides of the equation with respect to x and using the product rule, we obtain

$$D_x\, y = x^2\, D_x\,(\sin y) + (\sin y)\, D_x\,(x^2).$$

Since $y = f(x)$ for some (implicit) function f, we have, by Theorem (3.35),

$$D_x \sin y = \cos y\; D_x\, y.$$

Using this equation and the fact that $D_x\,(x^2) = 2x$, we may rewrite the first equation of our solution as

$$D_x\, y = (x^2 \cos y)\, D_x\, y + (\sin y)(2x),$$

or
$$y' = (x^2 \cos y)y' + 2x \sin y.$$

Finally, we solve for y' as follows:

$$y' - (x^2 \cos y)y' = 2x \sin y$$

$$(1 - x^2 \cos y)y' = 2x \sin y$$

$$y' = \frac{2x \sin y}{1 - x^2 \cos y},$$

provided $1 - x^2 \cos y \neq 0$.

In the next example we find the second derivative of an implicit function.

EXAMPLE 7 Find y'' if $y^4 + 3y - 4x^3 = 5x + 1$.

SOLUTION The equation was considered in Example 2, where we found that

$$y' = \frac{12x^2 + 5}{4y^3 + 3}.$$

Hence
$$y'' = D_x\,(y') = D_x\left(\frac{12x^2 + 5}{4y^3 + 3}\right).$$

We now use the quotient rule, differentiating implicitly as follows:

$$y'' = \frac{(4y^3 + 3)\, D_x\,(12x^2 + 5) - (12x^2 + 5)\, D_x\,(4y^3 + 3)}{(4y^3 + 3)^2}$$
$$= \frac{(4y^3 + 3)(24x) - (12x^2 + 5)(12y^2 y')}{(4y^3 + 3)^2}$$

Substituting for y' yields

$$y'' = \frac{(4y^3 + 3)(24x) - (12x^2 + 5) \cdot 12y^2\left(\dfrac{12x^2 + 5}{4y^3 + 3}\right)}{(4y^3 + 3)^2}$$
$$= \frac{(4y^3 + 3)^2(24x) - 12y^2(12x^2 + 5)^2}{(4y^3 + 3)^3}.$$

EXERCISES 3.7

Exer. 1–18: Assuming that the equation determines a differentiable function f such that $y = f(x)$, find y'.

1 $8x^2 + y^2 = 10$

2 $4x^3 - 2y^3 = x$

3 $2x^3 + x^2y + y^3 = 1$

4 $5x^2 + 2x^2y + y^2 = 8$

5 $5x^2 - xy - 4y^2 = 0$

6 $x^4 + 4x^2y^2 - 3xy^3 + 2x = 0$

7 $\sqrt{x} + \sqrt{y} = 100$

8 $x^{2/3} + y^{2/3} = 4$

9 $x^2 + \sqrt{xy} = 7$

10 $2x - \sqrt{xy} + y^3 = 16$

11 $\sin^2 3y = x + y - 1$

12 $x = \sin(xy)$

13 $y = \csc(xy)$

14 $y^2 + 1 = x^2 \sec y$

15 $y^2 = x \cos y$

16 $xy = \tan y$

17 $x^2 + \sqrt{\sin y} - y^2 = 1$

18 $\sin \sqrt{y} - 3x = 2$

Exer. 19–22: The equation of a classical curve and its graph are given for positive constants a and b. (Consult books on analytic geometry for further information.) Find the slope of the tangent line at the point P for the stated values of a and b.

19 *Ovals of Cassini:* $(x^2 + y^2 + a^2)^2 - 4a^2x^2 = b^4$; $a = 2$, $b = \sqrt{6}$, $P(2, \sqrt{2})$

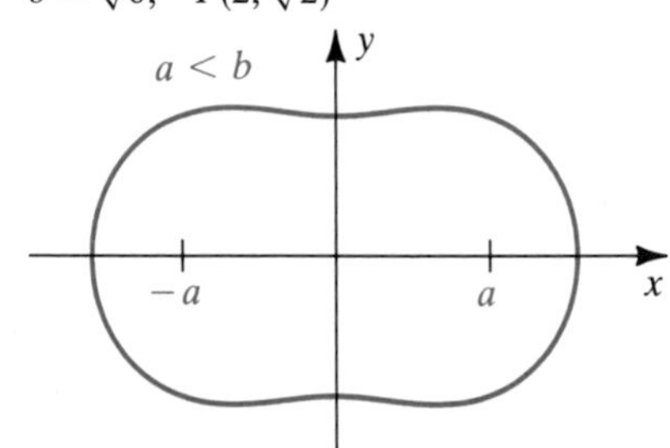

20 *Folium of Descartes:* $x^3 + y^3 - 3axy = 0$; $a = 4$, $P(6, 6)$

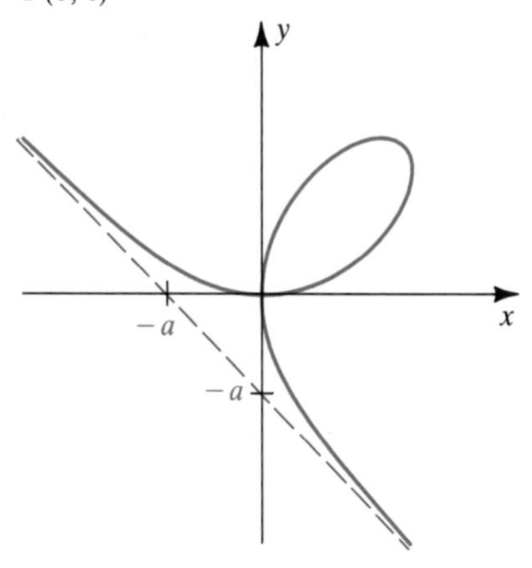

21 *Lemniscate of Bernoulli:* $(x^2 + y^2)^2 = 2a^2xy$; $a = \sqrt{2}$, $P(1, 1)$

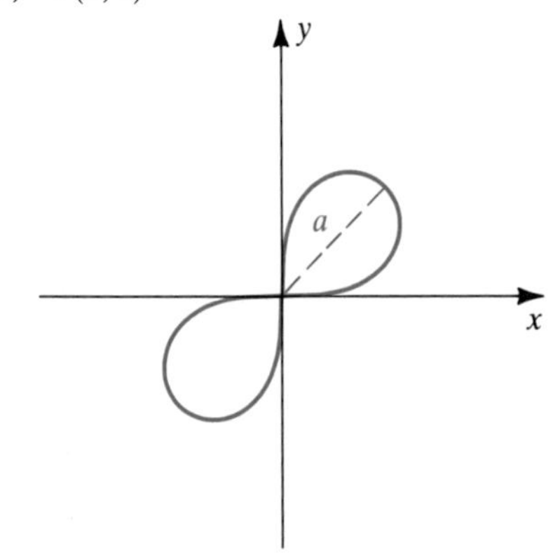

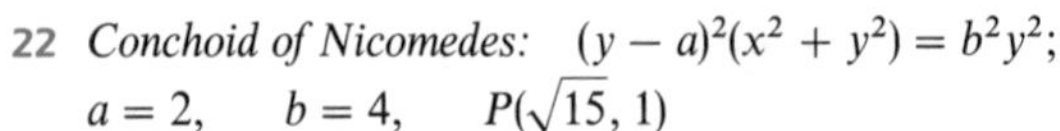

22 *Conchoid of Nicomedes:* $(y - a)^2(x^2 + y^2) = b^2y^2$;
$a = 2, \quad b = 4, \quad P(\sqrt{15}, 1)$

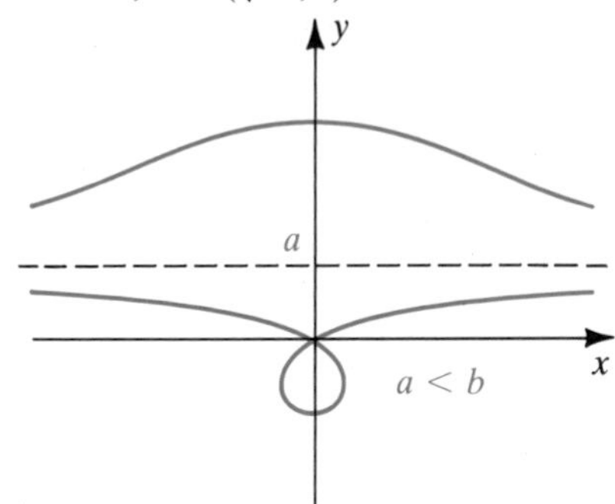

Exer. 23–28: Find the slope of the tangent line to the graph of the equation at P.

23 $xy + 16 = 0$; $\quad P(-2, 8)$

24 $y^2 - 4x^2 = 5$; $\quad P(-1, 3)$

25 $2x^3 - x^2y + y^3 - 1 = 0$; $\quad P(2, -3)$

26 $3y^4 + 4x - x^2 \sin y - 4 = 0$; $\quad P(1, 0)$

27 $x^2y + \sin y = 2\pi$; $\quad P(1, 2\pi)$

28 $xy^2 + 3y = 27$; $\quad P(2, 3)$

Exer. 29–34: Assuming that the equation determines a function f such that $y = f(x)$, find y'', if it exists.

29 $3x^2 + 4y^2 = 4$

30 $5x^2 - 2y^2 = 4$

31 $x^3 - y^3 = 1$

32 $x^2y^3 = 1$

33 $\sin y + y = x$

34 $\cos y = x$

Exer. 35–38: How many implicit functions are determined by the equation?

35 $x^4 + y^4 - 1 = 0$

36 $x^4 + y^4 = 0$

37 $x^2 + y^2 + 1 = 0$

38 $\cos x + \sin y = 3$

39 Show that the equation $y^2 = x$ determines an infinite number of implicit functions.

40 Use implicit differentiation to show that if P is any point on the circle $x^2 + y^2 = a^2$, then the tangent line at P is perpendicular to OP.

41 Suppose that $3x^2 - x^2y^3 + 4y = 12$ determines a differentiable function f such that $y = f(x)$. If $f(2) = 0$, use differentials to approximate the change in $f(x)$ if x changes from 2 to 1.97.

42 Suppose that $x^3 + xy + y^4 = 19$ determines a differentiable function f such that $y = f(x)$. If $P(1, 2)$ is a point on the graph of f, use differentials to approximate the y-coordinate b of the point $Q(1.1, b)$ on the graph.

c 43 Suppose that $x^2 + xy^3 = 4.0764$ determines a differentiable function f such that $y = f(x)$.

(a) If $P(1.2, 1.3)$ and $Q(1.23, b)$ are on the graph of f, use (3.31) to approximate b.

(b) Apply the method in (a), using $Q(1.23, b)$ to approximate the y-coordinate of $R(1.26, c)$. (This process, called *Euler's method*, can be repeated to approximate additional points on the graph.)

c 44 Suppose that $\sin x + y \cos y = -2.395$ determines a differentiable function f such that $y = f(x)$.

(a) If $P(2.1, 3.3)$ is an approximation to a point on the graph of f, use (3.31) to approximate the y-coordinate of $Q(2.12, b)$.

(b) Apply the method in (a), using $Q(2.12, b)$ to approximate the y-coordinate of $R(2.14, c)$.

3.8 RELATED RATES

Suppose that two variables x and y are functions of another variable t, say

$$x = f(t) \quad \text{and} \quad y = g(t).$$

By (3.7)(ii), we may interpret the derivatives dx/dt and dy/dt as the rates of change of x and y with respect to t. As a special case, if f and g are position functions for points moving on coordinate lines, then dx/dt and dy/dt are the velocities of these points (see (3.2)). In other situations these derivatives may represent rates of change of physical quantities.

In certain applications x and y may be related by means of an equation, such as

$$x^2 - y^3 - 2x + 7y^2 - 2 = 0.$$

If we differentiate this equation implicitly *with respect to t*, we obtain

$$\frac{d}{dt}(x^2) - \frac{d}{dt}(y^3) - \frac{d}{dt}(2x) + \frac{d}{dt}(7y^2) - \frac{d}{dt}(2) = \frac{d}{dt}(0).$$

Using the power rule (3.34) with t as the independent variable gives us

$$2x\frac{dx}{dt} - 3y^2\frac{dy}{dt} - 2\frac{dx}{dt} + 14y\frac{dy}{dt} = 0.$$

The derivatives dx/dt and dy/dt are called **related rates**, since they are related by means of an equation. This equation can be used to find one of the rates when the other is known. The following examples give several illustrations.

EXAMPLE 1 Two variables x and y are functions of a variable t and are related by the equation

$$x^3 - 2y^2 + 5x = 16.$$

If $dx/dt = 4$ when $x = 2$ and $y = -1$, find the corresponding value of dy/dt.

SOLUTION We differentiate the given equation implicitly with respect to t as follows:

$$\frac{d}{dt}(x^3) - \frac{d}{dt}(2y^2) + \frac{d}{dt}(5x) = \frac{d}{dt}(16)$$

$$3x^2\frac{dx}{dt} - 4y\frac{dy}{dt} + 5\frac{dx}{dt} = 0$$

$$(3x^2 + 5)\frac{dx}{dt} = 4y\frac{dy}{dt}$$

$$\frac{dy}{dt} = \frac{3x^2 + 5}{4y}\frac{dx}{dt}$$

The last equation is a *general* formula relating dy/dt and dx/dt. For the special case $dx/dt = 4$, $x = 2$, and $y = -1$, we obtain

$$\frac{dy}{dt} = \frac{3(2)^2 + 5}{4(-1)} \cdot 4 = -17.$$

EXAMPLE 2 A ladder 20 feet long leans against a vertical building. If the bottom of the ladder slides away from the building horizontally at a rate of 2 ft/sec, how fast is the ladder sliding down the building when the top of the ladder is 12 feet above the ground?

SOLUTION We begin by sketching a general position of the ladder as in Figure 3.32, where x denotes the distance from the base of the building to the bottom of the ladder and y denotes the distance from the ground to the top of the ladder.

FIGURE 3.32

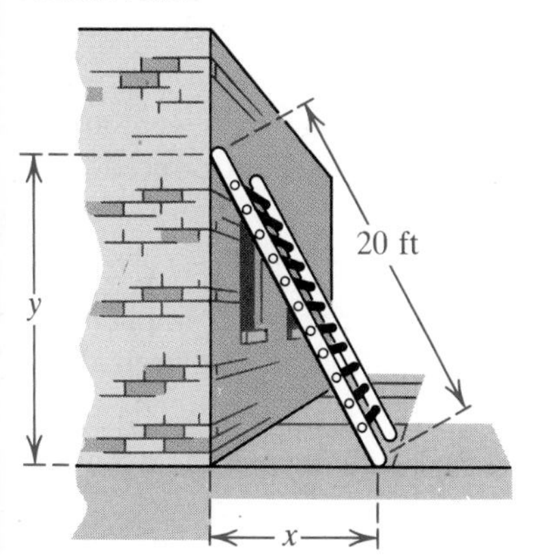

We next consider the following problem involving the rates of change of x and y with respect to t:

$$\textit{Given:} \quad \frac{dx}{dt} = 2 \text{ ft/sec}$$

$$\textit{Find:} \quad \frac{dy}{dt} \text{ when } y = 12 \text{ ft}$$

An equation that relates the variables x and y can be obtained by applying the Pythagorean theorem to the right triangle formed by the building, the ground, and the ladder (see Figure 3.32). This gives us

$$x^2 + y^2 = 400.$$

Differentiating both sides of this equation implicitly with respect to t, we obtain

$$\frac{d}{dt}(x^2) + \frac{d}{dt}(y^2) = \frac{d}{dt}(400)$$

$$2x\frac{dx}{dt} + 2y\frac{dy}{dt} = 0$$

$$\frac{dy}{dt} = -\frac{x}{y}\frac{dx}{dt},$$

provided $y \neq 0$.

The last equation is a general formula relating the two rates of change dx/dt and dy/dt. Let us now consider the special case $y = 12$. The corresponding value of x may be determined from

$$x^2 + 12^2 = 400, \quad \text{or} \quad x^2 = 256.$$

Thus, $x = \sqrt{256} = 16$ when $y = 12$. Substituting these values into the general formula for dy/dt, we obtain

$$\frac{dy}{dt} = -\frac{16}{12}(2) = -\frac{8}{3} \text{ ft/sec}.$$

The following guidelines may be helpful for solving related rate problems of the type illustrated in Example 2.

Guidelines for solving related rate problems **(3.36)**

1. Read the problem carefully several times, and think about the given facts and the unknown quantities that are to be found.
2. Sketch a picture or diagram and label it appropriately, introducing variables for unknown quantities.
3. Write down all the known facts, expressing the given and unknown rates as derivatives of the variables introduced in guideline 2.
4. Formulate a *general* equation that relates the variables.
5. Differentiate the equation formulated in guideline 4 implicitly with respect to t, obtaining a *general* relationship between the rates.
6. Substitute the *known* values and rates, and then find the unknown rate of change.

A common error is introducing specific values for the rates and variable quantities too early in the solution. Always remember to obtain a *general* formula that involves the rates of change at *any* time t. *Specific values should not be substituted for variables until the final steps of the solution.*

EXAMPLE 3 At 1:00 P.M., ship A is 25 miles due south of ship B. If ship A is sailing west at a rate of 16 mi/hr and ship B is sailing south at a rate of 20 mi/hr, find the rate at which the distance between the ships is changing at 1:30 P.M.

FIGURE 3.33

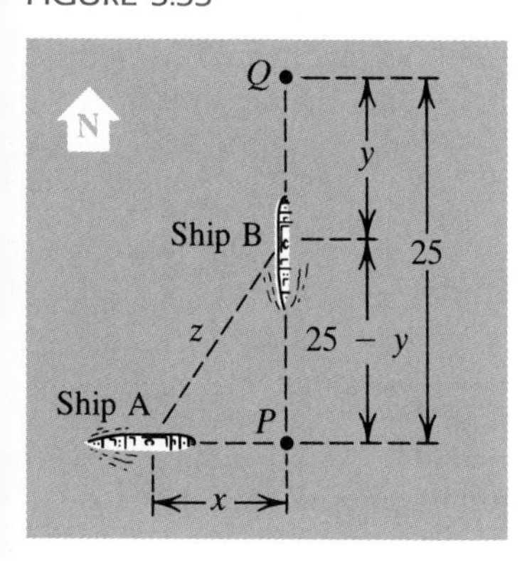

SOLUTION Let t denote the number of hours after 1:00 P.M. In Figure 3.33, P and Q are the positions of the ships at 1:00 P.M., x and y are the number of miles they have traveled in t hours, and z is the distance between the ships after t hours. Our problem may be stated as follows:

$$\textit{Given:}\quad \frac{dx}{dt} = 16 \text{ mi/hr and } \frac{dy}{dt} = 20 \text{ mi/hr}$$

$$\textit{Find:}\quad \frac{dz}{dt} \text{ when } t = \frac{1}{2} \text{ hr}$$

Applying the Pythagorean theorem to the triangle in Figure 3.33 gives us the following general equation relating the variables x, y, and z:

$$z^2 = x^2 + (25 - y)^2$$

Differentiating implicitly with respect to t and using the power rule and chain rule, we obtain

$$\frac{d}{dt}(z^2) = \frac{d}{dt}(x^2) + \frac{d}{dt}(25 - y)^2$$

$$2z\frac{dz}{dt} = 2x\frac{dx}{dt} + 2(25 - y)\left(0 - \frac{dy}{dt}\right)$$

$$z\frac{dz}{dt} = x\frac{dx}{dt} + (y - 25)\frac{dy}{dt}.$$

At 1:30 P.M. the ships have traveled for half an hour and

$$x = \tfrac{1}{2}(16) = 8, \quad y = \tfrac{1}{2}(20) = 10, \quad \text{and} \quad 25 - y = 15.$$

Consequently,

$$z^2 = 64 + 225 = 289, \quad \text{or} \quad z = \sqrt{289} = 17.$$

Substituting into the last equation involving dz/dt, we have

$$17\frac{dz}{dt} = 8(16) + (-15)(20),$$

or

$$\frac{dz}{dt} = -\frac{172}{17} \approx -10.12 \text{ mi/hr}.$$

The negative sign indicates that the distance between the ships is decreasing at 1:30 P.M.

Another method of solution is to write $x = 16t$, $y = 20t$, and

$$z = [x^2 + (25 - y)^2]^{1/2} = [256t^2 + (25 - 20t)^2]^{1/2}.$$

The derivative dz/dt may then be found, and substitution of $\frac{1}{2}$ for t produces the desired rate of change.

EXAMPLE 4 A water tank has the shape of an inverted right circular cone of altitude 12 feet and base radius 6 feet. If water is being pumped into the tank at a rate of 10 gal/min, approximate the rate at which the water level is rising when the water is 3 feet deep (1 gal $\approx$ 0.1337 ft^3).

FIGURE 3.34

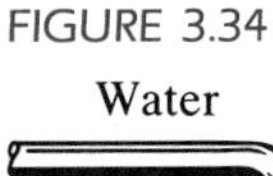

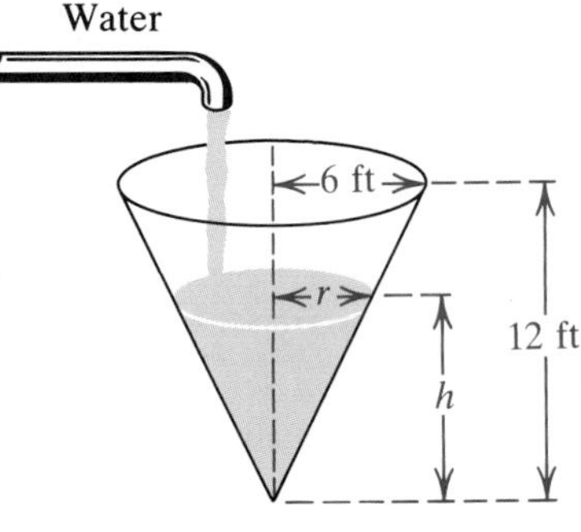

SOLUTION We begin by sketching the tank as in Figure 3.34, letting r denote the radius of the surface of the water when the depth is h. Note that r and h are functions of time t.

We next consider the following:

$$\textit{Given:} \quad \frac{dV}{dt} = 10 \text{ gal/min}$$

$$\textit{Find:} \quad \frac{dh}{dt} \text{ when } h = 3 \text{ ft}$$

The volume V of water in the tank corresponding to depth h is

$$V = \tfrac{1}{3}\pi r^2 h.$$

This formula for V relates V, r, and h. Before differentiating implicitly with respect to t, let us express V in terms of one variable. Referring to Figure 3.34 and using similar triangles, we obtain

$$\frac{r}{h} = \frac{6}{12}, \quad \text{or} \quad r = \frac{h}{2}.$$

Consequently, at depth h,

$$V = \frac{1}{3}\pi\left(\frac{h}{2}\right)^2 h = \frac{1}{12}\pi h^3.$$

Differentiating the last equation implicitly with respect to t gives us the following general relationship between the rates of change of V and h at any time t:

$$\frac{dV}{dt} = \frac{1}{4}\pi h^2 \frac{dh}{dt}$$

If $h \neq 0$, an equivalent formula is

$$\frac{dh}{dt} = \frac{4}{\pi h^2}\frac{dV}{dt}.$$

Finally, we let $h = 3$ and $dV/dt = 10$ gal/min ≈ 1.337 ft^3/min, obtaining

$$\frac{dh}{dt} \approx \frac{4}{\pi(9)}(1.337) \approx 0.189 \text{ ft/min}.$$

EXAMPLE 5 A revolving beacon in a lighthouse makes one revolution every 15 seconds. The beacon is 200 feet from the nearest point P on a straight shoreline. Find the rate at which a ray from the light moves along the shore at a point 400 feet from P.

SOLUTION The problem is diagrammed in Figure 3.35, where B denotes the position of the beacon and ϕ is the angle between BP and a light ray to a point S on the shore x units from P.

FIGURE 3.35

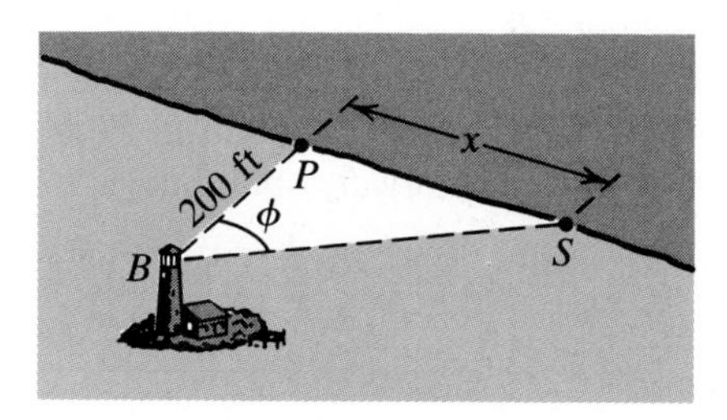

Since the light revolves four times per minute, the angle ϕ changes at a rate of $4 \cdot 2\pi$ radians per minute; that is, $d\phi/dt = 8\pi$. Using triangle PBS, we see that

$$\tan \phi = \frac{x}{200},$$

or

$$x = 200 \tan \phi.$$

The rate at which the ray of light moves along the shore is

$$\frac{dx}{dt} = 200 \sec^2 \phi \frac{d\phi}{dt} = (200 \sec^2 \phi)(8\pi) = 1600\pi \sec^2 \phi.$$

If $x = 400$, then $BS = \sqrt{200^2 + 400^2} = 200\sqrt{5}$, and

$$\sec \phi = \frac{200\sqrt{5}}{200} = \sqrt{5}.$$

Hence

$$\frac{dx}{dt} = 1600\pi(\sqrt{5})^2 = 8000\pi \approx 25{,}133 \text{ ft/min}.$$

FIGURE 3.36

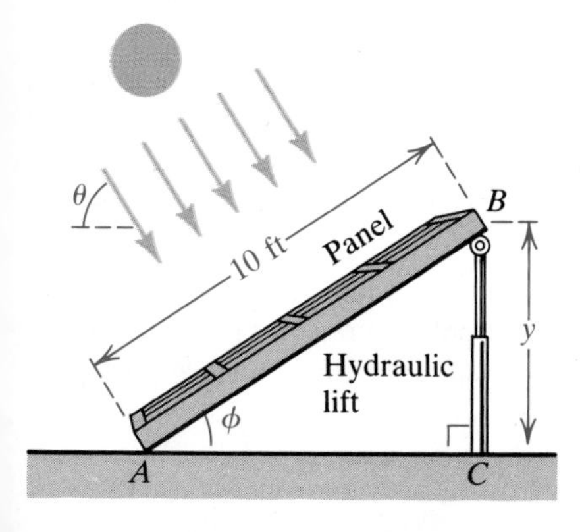

EXAMPLE 6 Figure 3.36 shows a solar panel that is 10 feet in width and is equipped with a hydraulic lift. As the sun rises, the panel is adjusted so that the sun's rays are perpendicular to the panel's surface.

(a) Find the relationship between the rate dy/dt at which the panel should be lowered and the rate $d\theta/dt$ at which the angle of inclination of the sun increases.

(b) If, when $\theta = 30°$, $d\theta/dt = 15°/\text{hr}$, find dy/dt.

SOLUTION

(a) If we let ϕ denote angle BAC in Figure 3.36, then, from plane geometry, $\phi = 90° - \theta = \frac{1}{2}\pi - \theta$. Since $d\phi/dt = -d\theta/dt$, ϕ decreases at the rate at which θ increases.

Referring to right triangle BAC, we see that

$$\sin \phi = \frac{y}{10},$$

or

$$y = 10 \sin \phi = 10 \sin\left(\tfrac{1}{2}\pi - \theta\right).$$

Differentiating implicitly with respect to t and using the cofunction identity $\cos\left(\tfrac{1}{2}\pi - \theta\right) = \sin \theta$ yields

$$\frac{dy}{dt} = 10 \cos\left(\tfrac{1}{2}\pi - \theta\right)\left(0 - \frac{d\theta}{dt}\right) = -10 \sin \theta \frac{d\theta}{dt}.$$

(b) We must use radian measure for $d\theta/dt$. Since $15° = 15(\pi/180) = \pi/12$ radians, we substitute $d\theta/dt = \pi/12$ rad/hr and $\theta = 30° = \pi/6$ in the formula for dy/dt, obtaining

$$\frac{dy}{dt} = -10\left(\frac{1}{2}\right)\left(\frac{\pi}{12}\right) = -\frac{5\pi}{12} \approx -1.3 \text{ ft/hr}.$$

EXERCISES 3.8

Exer. 1–8: Assume that all variables are functions of t.

1 If $A = x^2$ and $\dfrac{dx}{dt} = 3$ when $x = 10$, find $\dfrac{dA}{dt}$.

2 If $S = z^3$ and $\dfrac{dz}{dt} = -2$ when $z = 3$, find $\dfrac{dS}{dt}$.

3 If $V = -5p^{3/2}$ and $\dfrac{dV}{dt} = -4$ when $V = -40$, find $\dfrac{dp}{dt}$.

4 If $P = 3/w$ and $\dfrac{dP}{dt} = 5$ when $P = 9$, find $\dfrac{dw}{dt}$.

5 If $x^2 + 3y^2 + 2y = 10$ and $\dfrac{dx}{dt} = 2$ when $x = 3$ and $y = -1$, find $\dfrac{dy}{dt}$.

6 If $2y^3 - x^2 + 4x = -10$ and $\dfrac{dy}{dt} = -3$ when $x = -2$ and $y = 1$, find $\dfrac{dx}{dt}$.

7 If $3x^2y + 2x = -32$ and $\dfrac{dy}{dt} = -4$ when $x = 2$ and $y = -3$, find $\dfrac{dx}{dt}$.

8 If $-x^2y^2 - 4y = -44$ and $\dfrac{dx}{dt} = 5$ when $x = -3$ and $y = 2$, find $\dfrac{dy}{dt}$.

9 As a circular metal griddle is being heated, its diameter changes at a rate of 0.01 cm/min. Find the rate at which the area of one side is changing when the diameter is 30 centimeters.

10 A fire has started in a dry, open field and spreads in the form of a circle. The radius of the circle increases at the rate of 6 ft/min. Find the rate at which the fire area is increasing when the radius is 150 feet.

11 Gas is being pumped into a spherical balloon at a rate of 5 ft^3/min. Find the rate at which the radius is changing when the diameter is 18 inches.

12 Suppose a spherical snowball is melting and the radius is decreasing at a constant rate, changing from 12 inches to 8 inches in 45 minutes. How fast was the volume changing when the radius was 10 inches?

13 A ladder 20 feet long leans against a vertical building. If the bottom of the ladder slides away from the building horizontally at a rate of 3 ft/sec, how fast is the ladder sliding down the building when the top of the ladder is 8 feet from the ground?

14 A girl starts at a point A and runs east at a rate of 10 ft/sec. One minute later, another girl starts at A and runs north at a rate of 8 ft/sec. At what rate is the distance between them changing 1 minute after the second girl starts?

15 A light is at the top of a 16-foot pole. A boy 5 feet tall walks away from the pole at a rate of 4 ft/sec (see figure). At what rate is the tip of his shadow moving when he

is 18 feet from the pole? At what rate is the length of his shadow increasing?

EXERCISE 15

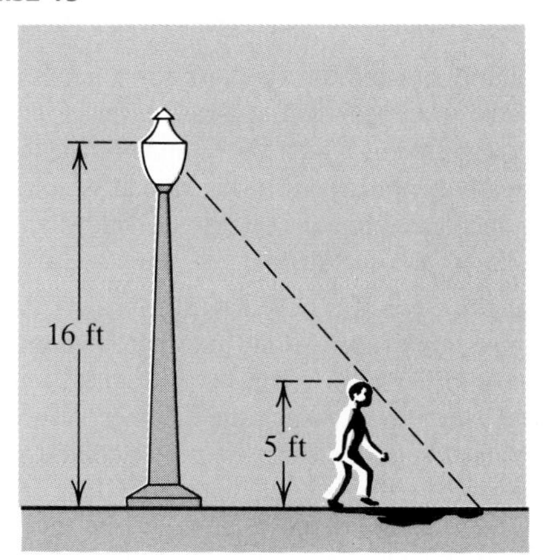

16 A man on a dock is pulling in a boat by means of a rope attached to the bow of the boat 1 foot above water level and passing through a simple pulley located on the dock 8 feet above water level (see figure). If he pulls in the rope at a rate of 2 ft/sec, how fast is the boat approaching the dock when the bow of the boat is 25 feet from a point that is directly below the pulley?

EXERCISE 16

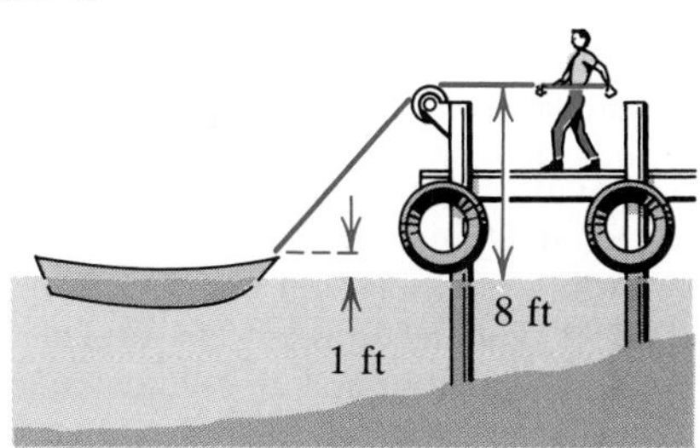

17 The top of a silo has the shape of a hemisphere of diameter 20 feet. If it is coated uniformly with a layer of ice and if the thickness is decreasing at a rate of $\frac{1}{4}$ in./hr, how fast is the volume of the ice changing when the ice is 2 inches thick?

18 As sand leaks out of a hole in a container, it forms a conical pile whose altitude is always the same as its radius. If the height of the pile is increasing at a rate of 6 in./min, find the rate at which the sand is leaking out when the altitude is 10 inches.

19 A person flying a kite holds the string 5 feet above ground level, and the string is payed out at a rate of 2 ft/sec as the kite moves horizontally at an altitude of 105 feet (see figure). Assuming there is no sag in the string, find the rate at which the kite is moving when 125 feet of string has been payed out.

EXERCISE 19

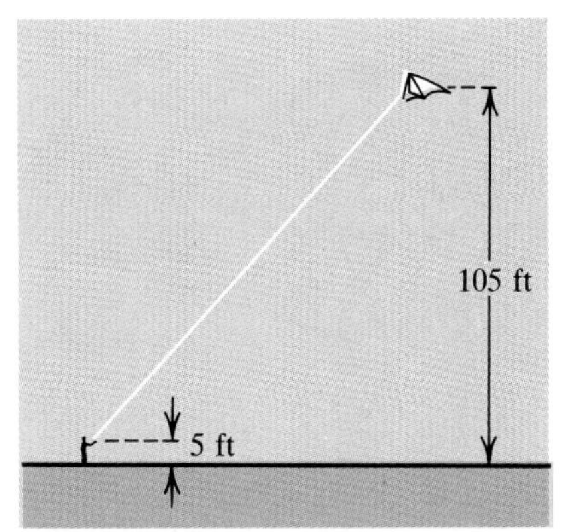

20 A hot-air balloon rises vertically as a rope attached to the base of the balloon is released at the rate of 5 ft/sec. The pulley that releases the rope is 20 feet from the platform where passengers board (see figure). At what rate is the balloon rising when 500 feet of rope has been payed out?

EXERCISE 20

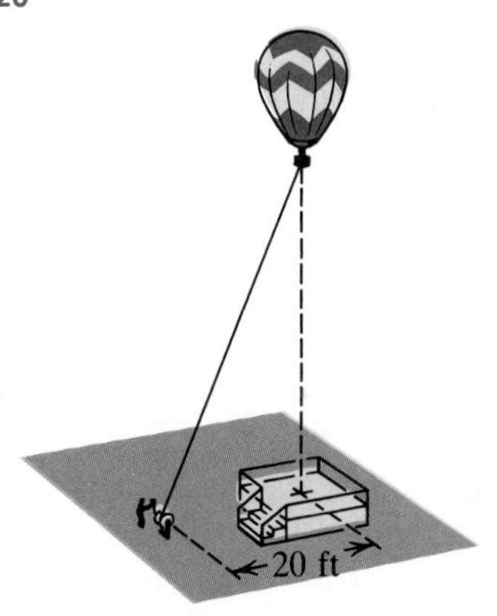

21 Boyle's law for confined gases states that if the temperature is constant, $pv = c$, where p is pressure, v is volume, and c is a constant. At a certain instant the volume is 75 in.^{3}, the pressure is 30 lb/in.^{2}, and the pressure is decreasing at a rate of 2 lb/in.^{2} every minute. At what rate is the volume changing at this instant?

22 A 100-foot-long cable of diameter 4 inches is submerged in seawater. Because of corrosion, the surface area of the cable decreases at a rate of 750 in.^{2}/year. Ignoring the corrosion at the ends of the cable, find the rate at which the diameter is decreasing.

23 The ends of a water trough 8 feet long are equilateral triangles whose sides are 2 feet long (see figure on following page). If water is being pumped into the trough at a rate of 5 ft^3/min, find the rate at which the water level is rising when the depth of the water is 8 inches.

EXERCISE 23

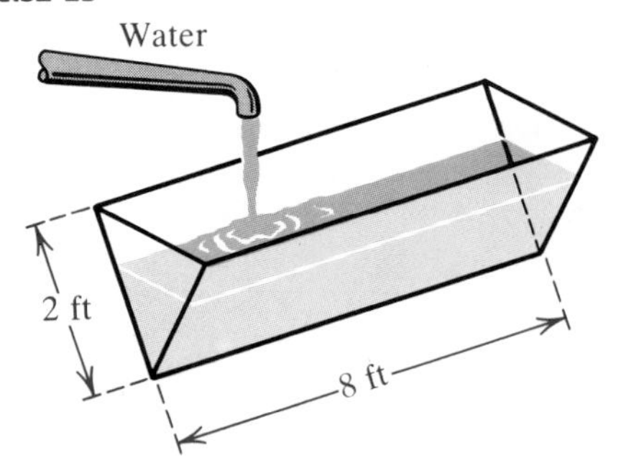

24 Work Exercise 23 if the ends of the trough have the shape of the graph of $y = 2|x|$ between the points $(-1, 2)$ and $(1, 2)$.

25 The area of an equilateral triangle is decreasing at a rate of 4 cm^2/min. Find the rate at which the length of a side is changing when the area of the triangle is 200 cm^2.

26 Gas is escaping from a spherical balloon at a rate of 10 ft^3/hr. At what rate is the radius changing when the volume is 400 ft^3?

27 A stone is dropped into a lake, causing circular waves whose radii increase at a constant rate of 0.5 m/sec. At what rate is the circumference of a wave changing when its radius is 4 meters?

28 A softball diamond has the shape of a square with sides 60 feet long. If a player is running from second base to third at a speed of 24 ft/sec, at what rate is her distance from home plate changing when she is 20 feet from third?

29 When two resistors R_1 and R_2 are connected in parallel (see figure), the total resistance R is given by the equation $1/R = (1/R_1) + (1/R_2)$. If R_1 and R_2 are increasing at rates of 0.01 ohm/sec and 0.02 ohm/sec, respectively, at what rate is R changing at the instant that $R_1 = 30$ ohms and $R_2 = 90$ ohms?

EXERCISE 29

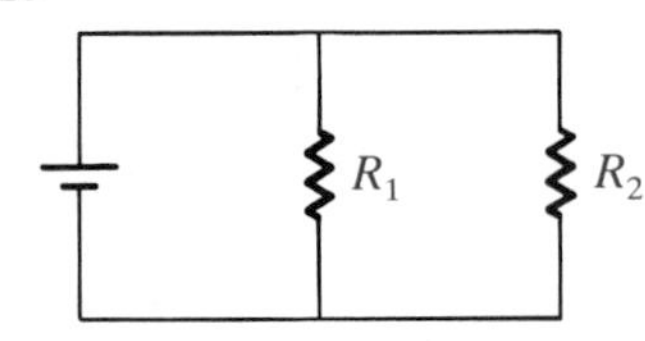

30 The formula for the adiabatic expansion of air is $pv^{1.4} = c$, where p is the pressure, v is the volume, and c is a constant. At a certain instant the pressure is 40 dyn/cm^2 and is increasing at a rate of 3 dyn/cm^2 per second. If, at that same instant, the volume is 60 cm^3, find the rate at which the volume is changing.

31 If a spherical tank of radius a contains water that has a maximum depth h, then the volume V of water in the tank is given by $V = \frac{1}{3}\pi h^2(3a - h)$. Suppose a spherical tank of radius 16 feet is being filled at a rate of 100 gal/min. Approximate the rate at which the water level is rising when $h = 4$ feet (1 gal $\approx 0.1337\ \text{ft}^3$).

32 A spherical water storage tank for a small community is coated uniformly with a 2-inch layer of ice. As the ice melts, the rate at which the volume of the ice decreases is directly proportional to the rate at which the surface area decreases. Show that the outside diameter is decreasing at a constant rate.

33 From the edge of a cliff that overlooks a lake 200 feet below, a boy drops a stone and then, two seconds later, drops another stone from exactly the same position. Discuss the rate at which the distance between the two stones is changing during the next second. (Assume that the distance an object falls in t seconds is $16t^2$ feet.)

34 A metal rod has the shape of a right circular cylinder. As it is being heated, its length is increasing at a rate of 0.005 cm/min and its diameter is increasing at 0.002 cm/min. At what rate is the volume changing when the rod has length 40 centimeters and diameter 3 centimeters?

35 An airplane is flying at a constant speed of 360 mi/hr and climbing at an angle of 45°. At the moment the plane's altitude is 10,560 feet, it passes directly over an air traffic control tower on the ground. Find the rate at which the airplane's distance from the tower is changing one minute later (neglect the height of the tower).

36 A North-South highway A and an East-West highway B intersect at a point P. At 10:00 A.M. an automobile crosses P traveling north on highway A at a speed of 50 mi/hr. At that same instant, an airplane flying east at a speed of 200 mi/hr and an altitude of 26,400 feet is directly above the point on highway B that is 100 miles west of P. If the airplane and the automobile maintain the same speed and direction, at what rate is the distance between them changing at 10:15 A.M.?

37 A paper cup containing water has the shape of a frustum of a right circular cone of altitude 6 inches and lower and upper base radii 1 inch and 2 inches, respectively. If water is leaking out of the cup at a rate of 3 in.^{3}/hr, at what rate is the water level decreasing when the depth of the water is 4 inches? (*Note:* The volume V of a frustum of a right circular cone of altitude h and base radii a and b is given by $V = \frac{1}{3}\pi h(a^2 + b^2 + ab)$.)

38 The top part of a swimming pool is a rectangle of length 60 feet and width 30 feet. The depth of the pool varies uniformly from 4 feet to 9 feet through a horizontal distance of 40 feet and then is level for the remaining 20 feet, as illustrated by the cross-sectional view in the figure. If the pool is being filled with water at a rate of 500 gal/min, approximate the rate at which the water

level is rising when the depth of the water at the deep end is 4 feet (1 gal $\approx$ 0.1337 ft^3).

EXERCISE 38

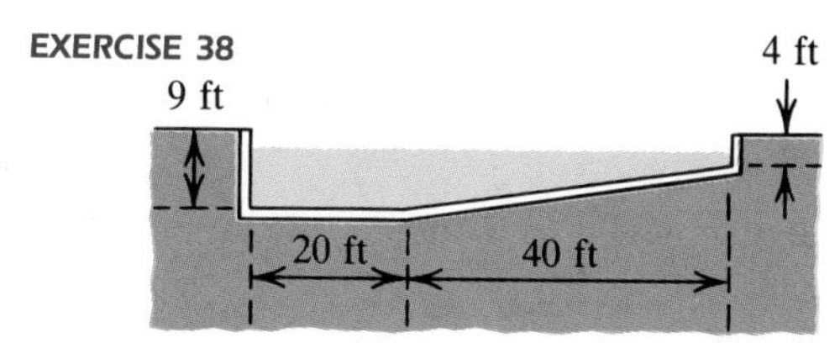

39 An airplane at an altitude of 10,000 feet is flying at a constant speed on a line that will take it directly over an observer on the ground. If, at a given instant, the observer notes that the angle of elevation of the airplane is 60° and is increasing at a rate of 1° per second, find the speed of the airplane.

40 In Exercise 16, let θ be the angle that the rope makes with the horizontal. Find the rate at which θ is changing at the instant that $\theta = 30°$.

41 An isosceles triangle has equal sides 6 inches long. If the angle θ between the equal sides is changing at a rate of 2° per minute, how fast is the area of the triangle changing when $\theta = 30°$?

42 A ladder 20 feet long leans against a vertical building. If the bottom of the ladder slides away from the building horizontally at a rate of 2 ft/sec, at what rate is the angle between the ladder and the ground changing when the top of the ladder is 12 feet above the ground?

43 The relative positions of an airport runway and a 20-foot-tall control tower are shown in the figure. The beginning of the runway is at a perpendicular distance of 300 feet from the base of the tower. If an airplane reaches a speed of 100 mi/hr after traveling 300 feet down the runway, at approximately what rate is the distance between the airplane and the control booth increasing at this time?

EXERCISE 43

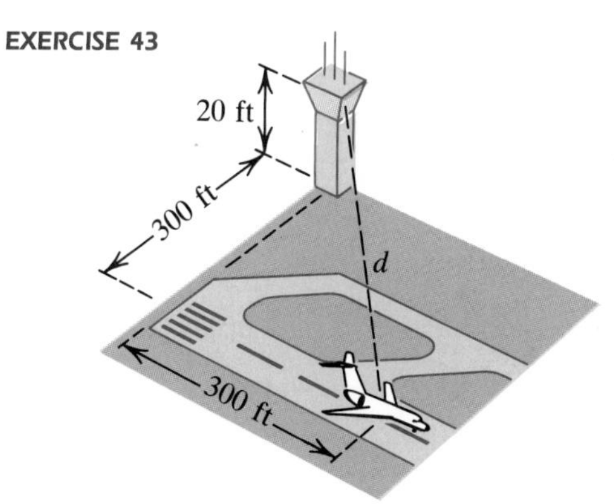

44 The speed of sound in air at 0 °C (or 273 °K) is 1087 ft/sec, but this speed increases as the temperature rises. If T is temperature in °K, the speed of sound v at this temperature is given by $v = 1087\sqrt{T/273}$. If the temperature increases at the rate of 3 °C per hour, approximate the rate at which the speed of sound is increasing when $T =$ 30 °C (or 303 °K).

45 An airplane is flying at a constant speed and altitude, on a line that will take it directly over a radar station located on the ground. At the instant that the airplane is 60,000 feet from the station, an observer in the station notes that its angle of elevation is 30° and is increasing at a rate of 0.5° per second. Find the speed of the airplane.

46 A missile is fired vertically from a point that is 5 miles from a tracking station and at the same elevation (see figure). For the first 20 seconds of flight, its angle of elevation θ changes at a constant rate of 2° per second. Find the velocity of the missile when the angle of elevation is 30°.

EXERCISE 46

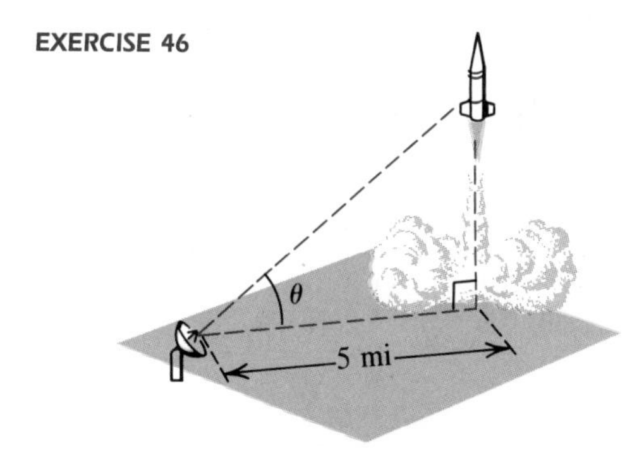

47 The sprocket assembly for a 28-inch bicycle is shown in the figure. Find the relationship between the angular velocity $d\theta/dt$ (in radians/sec) of the pedal assembly and the ground speed of the bicycle (in mi/hr).

EXERCISE 47

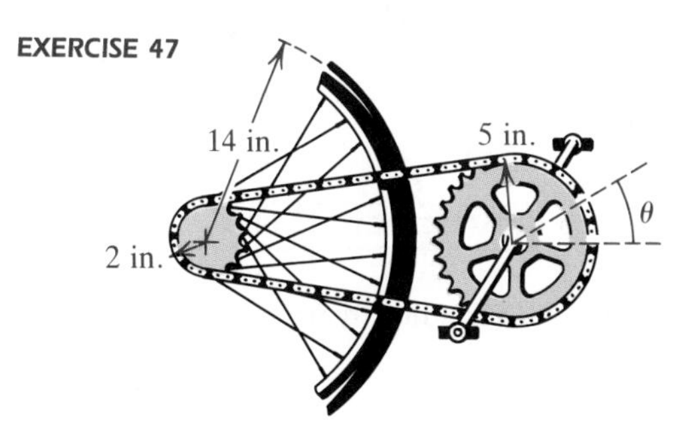

48 A 100-candlepower spotlight is located 20 feet above a stage (see figure on following page). The illuminance E (in footcandles) in the small lighted area of the stage is given by $E = (I \cos \theta)/s^2$, where I is the intensity of the light, s is the distance the light must travel, and θ is the indicated angle. As the spotlight is rotated through ϕ degrees, find the relationship between the rate of change in illumination dE/dt and the rate of rotation $d\phi/dt$.

EXERCISE 48

49 A *conical pendulum* consists of a mass m, attached to a string of fixed length l, that travels around a circle of radius r at a fixed velocity v (see figure). As the velocity of the mass is increased, both the radius and the angle θ increase. Given that $v^2 = rg \tan \theta$, where g is a gravitational constant, find a relationship between the related rates

(a) dv/dt and $d\theta/dt$ (b) dv/dt and dr/dt

EXERCISE 49

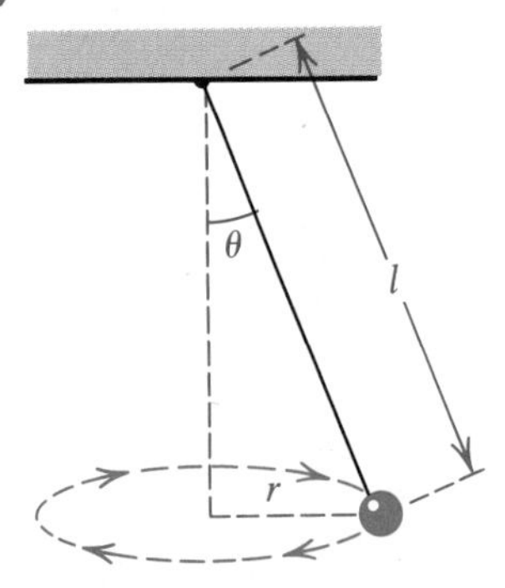

50 Water in a paper conical filter drips into a cup as shown in the figure. Let x denote the height of the water in the filter and y the height of the water in the cup. If 10 in.^{3} of water is poured into the filter, find the relationship between dy/dt and dx/dt.

EXERCISE 50

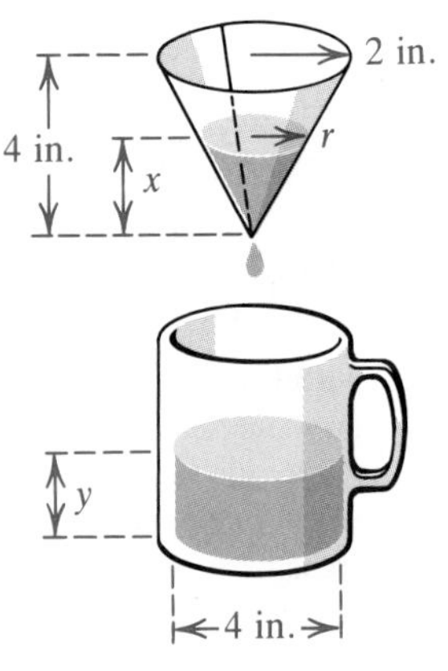

c 51 Ship A is sailing north, and ship B is sailing west. Using an xy-plane, radar records the coordinates (in miles) of each ship at intervals of 1.25 minutes as shown in the following tables. Approximate the rate (in mi/hr) at which the distance between the ships is changing at $t = 5$.

Ship A:

t (min)	1.25	2.50	3.75	5.00
x (mi)	1.77	1.77	1.77	1.77
y (mi)	2.71	3.03	3.35	3.67

Ship B:

t (min)	1.25	2.50	3.75	5.00
x (mi)	5.24	5.52	5.80	6.08
y (mi)	1.24	1.24	1.24	1.24

c 52 Two variables x and y are functions of a variable t and are related by the formula

$$1.31 \sin(2.56x) + \sqrt{y} = (x - 1)^2.$$

If $dy/dt \approx 3.68$ when $x \approx 1.71$ and $y \approx 3.03$, approximate the corresponding value of dx/dt.

3.9 REVIEW EXERCISES

Exer. 1–2: Find $f'(x)$ directly from the definition of the derivative.

1 $f(x) = \dfrac{4}{3x^2 + 2}$

2 $f(x) = \sqrt{5 - 7x}$

Exer. 3–42: Find the first derivative.

3 $f(x) = 2x^3 - 7x + 2$

4 $k(x) = \dfrac{1}{x^4 - x^2 + 1}$

5 $g(t) = \sqrt{6t + 5}$

6 $h(t) = \dfrac{1}{\sqrt{6t + 5}}$

7 $F(z) = \sqrt[3]{7z^2 - 4z + 3}$

8 $f(w) = \sqrt[5]{3w^2}$

9 $G(x) = \dfrac{6}{(3x^2 - 1)^4}$

10 $H(x) = \dfrac{(3x^2 - 1)^4}{6}$

11 $F(r) = (r^2 - r^{-2})^{-2}$

12 $h(z) = [(z^2 - 1)^5 - 1]^5$

13 $g(x) = \sqrt[5]{(3x + 2)^4}$

14 $P(x) = (x + x^{-1})^2$

15 $r(s) = \left(\dfrac{8s^2 - 4}{1 - 9s^3}\right)^4$

16 $g(w) = \dfrac{(w - 1)(w - 3)}{(w + 1)(w + 3)}$

17 $F(x) = (x^6 + 1)^5(3x + 2)^3$

18 $k(z) = [z^2 + (z^2 + 9)^{1/2}]^{1/2}$

19 $k(s) = (2s^2 - 3s + 1)(9s - 1)^4$

20 $p(x) = \dfrac{2x^4 + 3x^2 - 1}{x^2}$

21 $f(x) = 6x^2 - \dfrac{5}{x} + \dfrac{2}{\sqrt[3]{x^2}}$

22 $F(t) = \dfrac{5t^2 - 7}{t^2 + 2}$

23 $f(w) = \sqrt{\dfrac{2w + 5}{7w - 9}}$

24 $S(t) = \sqrt{t^2 + t + 1}\sqrt[3]{4t - 9}$

25 $g(r) = \sqrt{1 + \cos 2r}$

26 $g(z) = \csc\left(\dfrac{1}{z}\right) + \dfrac{1}{\sec z}$

27 $f(x) = \sin^2(4x^3)$

28 $H(t) = (1 + \sin 3t)^3$

29 $h(x) = (\sec x + \tan x)^5$

30 $K(r) = \sqrt[3]{r^3 + \csc 6r}$

31 $f(x) = x^2 \cot 2x$

32 $P(\theta) = \theta^2 \tan^2(\theta^2)$

33 $K(\theta) = \dfrac{\sin 2\theta}{1 + \cos 2\theta}$

34 $g(v) = \dfrac{1}{1 + \cos^2 2v}$

35 $g(x) = (\cos \sqrt[3]{x} - \sin \sqrt[3]{x})^3$

36 $f(x) = \dfrac{x}{2x + \sec^2 x}$

37 $G(u) = \dfrac{\csc u + 1}{\cot u + 1}$

38 $k(\phi) = \dfrac{\sin \phi}{\cos \phi - \sin \phi}$

39 $F(x) = \sec 5x \tan 5x \sin 5x$

40 $H(z) = \sqrt{\sin \sqrt{z}}$

41 $g(\theta) = \tan^4(\sqrt[4]{\theta})$

42 $f(x) = \csc^3 3x \cot^2 3x$

Exer. 43–48: Assuming that the equation determines a differentiable function f such that $y = f(x)$, find y'.

43 $5x^3 - 2x^2y^2 + 4y^3 - 7 = 0$

44 $3x^2 - xy^2 + y^{-1} = 1$

45 $\dfrac{\sqrt{x} + 1}{\sqrt{y} + 1} = y$

46 $y^2 - \sqrt{xy} + 3x = 2$

47 $xy^2 = \sin(x + 2y)$

48 $y = \cot(xy)$

Exer. 49–50: Find equations of the tangent line and the normal line to the graph of f at P.

49 $y = 2x - \dfrac{4}{\sqrt{x}}$; $P(4, 6)$

50 $x^2y - y^3 = 8$; $P(-3, 1)$

51 Find the x-coordinates of all points on the graph of $y = 3x - \cos 2x$ at which the tangent line is perpendicular to the line $2x + 4y = 5$.

52 If $f(x) = \sin 2x - \cos 2x$ for $0 \le x \le 2\pi$, find the x-coordinates of all points on the graph of f at which the tangent line is horizontal.

Exer. 53–54: Find y', y'', and y'''.

53 $y = 5x^3 + 4\sqrt{x}$

54 $y = 2x^2 - 3x - \cos 5x$

55 If $x^2 + 4xy - y^2 = 8$, find y'' by implicit differentiation.

56 If $f(x) = x^3 - x^2 - 5x + 2$, find

(a) the x-coordinates of all points on the graph of f at which the tangent line is parallel to the line through $A(-3, 2)$ and $B(1, 14)$

(b) the value of f'' at each zero of f'

57 If $y = 3x^2 - 7$, find

(a) Δy (b) dy (c) $dy - \Delta y$

58 If $y = 5x/(x^2 + 1)$, find dy and use it to approximate the change in y if x changes from 2 to 1.98. What is the exact change in y?

59 The side of an equilateral triangle is estimated to be 4 inches, with a maximum error of ± 0.03 inch. Use differentials to estimate the maximum error in the calculated area of the triangle. Approximate the percentage error.

60 If $s = 3r^2 - 2\sqrt{r + 1}$ and $r = t^3 + t^2 + 1$, use the chain rule to find the value of ds/dt at $t = 1$.

61 If $f(x) = 2x^3 + x^2 - x + 1$ and $g(x) = x^5 + 4x^3 + 2x$, use differentials to approximate the change in $g(f(x))$ if x changes from -1 to -1.01.

62 Use differentials to find a linear approximation of $\sqrt[3]{64.2}$.

63 Suppose f and g are functions such that $f(2) = -1$, $f'(2) = 4$, $f''(2) = -2$, $g(2) = -3$, $g'(2) = 2$, and $g''(2) = 1$. Find the value of each of the following at $x = 2$.

(a) $(2f - 3g)'$ (b) $(2f - 3g)''$ (c) $(fg)'$

(d) $(fg)''$ (e) $\left(\dfrac{f}{g}\right)'$ (f) $\left(\dfrac{f}{g}\right)''$

64 Refer to Exercise 87 in Section 3.6. Let f be an odd function and g an even function such that $f(3) = -3$, $f'(3) = 7$, $g(3) = -3$, and $g'(3) = -5$. Find $(f \circ g)'(3)$ and $(g \circ f)'(3)$.

65 Determine where the graph of f has a vertical tangent line or a cusp.

(a) $f(x) = 3(x + 1)^{1/3} - 4$ (b) $f(x) = 2(x - 8)^{2/3} - 1$

66 Let $f(x) = \begin{cases} (2x - 1)^3 & \text{if } x \ge 2 \\ 5x^2 + 34x - 61 & \text{if } x < 2 \end{cases}$

Determine if f is differentiable at 2.

67 The Stefan-Boltzmann law states that the radiant energy emitted from a unit area of a black surface is given by $R = kT^4$, where R is the rate of emission per unit area, T is the temperature (in °K), and k is a constant. If the error in the measurement of T is 0.5%, find the resulting percentage error in the calculated value of R.

68 Let V and S denote the volume and surface area, respectively, of a spherical balloon. If the diameter is 8 centimeters and the volume increases by 12 cm^3, use differentials to approximate the change in S.

69 A right circular cone has height $h = 8$ ft, and the base radius r is increasing. Find the rate of change of its surface area S with respect to r when $r = 6$ ft.

70 The intensity of illumination from a source of light is inversely proportional to the square of the distance from the source. If a student works at a desk that is a certain distance from a lamp, use differentials to find the percentage change in distance that will increase the intensity by 10%.

71 The ends of a horizontal water trough 10 feet long are isosceles trapezoids with lower base 3 feet, upper base 5 feet, and altitude 2 feet. If the water level is rising at a rate of $\frac{1}{4}$ in./min when the depth of the water is 1 foot, how fast is water entering the trough?

72 Two cars are approaching the same intersection along roads that run at right angles to each other. Car A is traveling at 20 mi/hr, and car B is traveling at 40 mi/hr. If, at a certain instant, A is $\frac{1}{4}$ mile from the intersection and B is $\frac{1}{2}$ mile from the intersection, find the rate at which they are approaching each other at that instant.

73 Boyle's law states that $pv = c$, where p is pressure, v is volume, and c is a constant. Find a formula for the rate of change of p with respect to v.

74 A railroad bridge is 20 feet above, and at right angles to, a river. A man in a train traveling 60 mi/hr passes over the center of the bridge at the same instant that a man in a motor boat traveling 20 mi/hr passes under the center of the bridge (see figure). How fast are the two men moving away from each other 10 seconds later?

EXERCISE 74

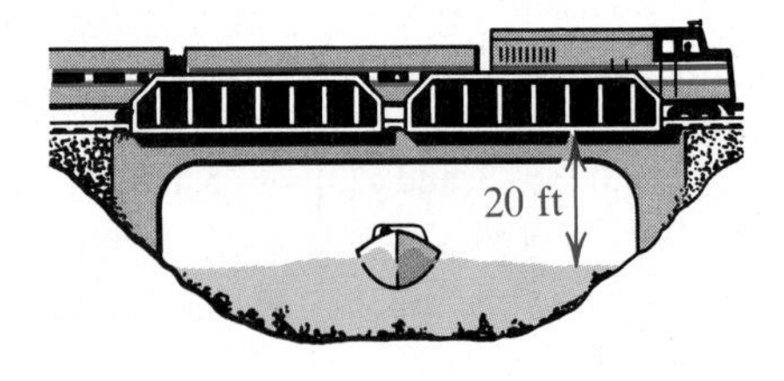

75 A large ferris wheel is 100 feet in diameter and rises 110 feet off the ground, as illustrated in the figure. Each revolution of the wheel takes 30 seconds.

(a) Express the distance h of a seat from the ground as a function of time t (in seconds) if $t = 0$ corresponds to a time when the seat is at the bottom.

(b) If a seat is rising, how fast is the distance changing when $h = 55$ feet?

EXERCISE 75

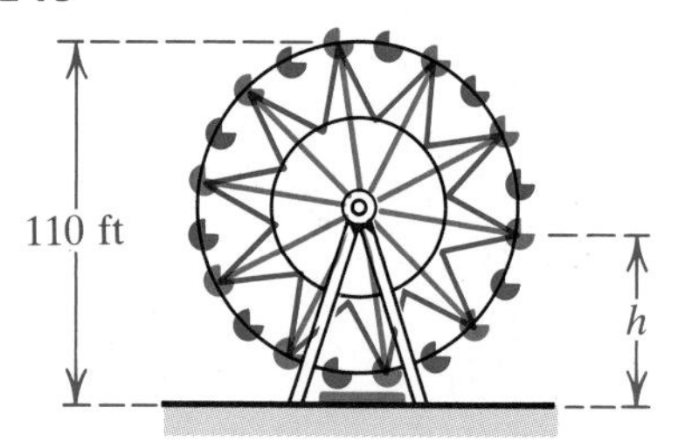

76 A piston is attached to a crankshaft as shown in the figure. The connecting rod AB has length 6 inches, and the radius of the crankshaft is 2 inches.

(a) If the crankshaft rotates counterclockwise 2 times per second, find formulas for the position of point A at t seconds after A has coordinates (2, 0).

(b) Find a formula for the position of point B at time t.

(c) How fast is B moving when A has coordinates (0, 2)?

EXERCISE 76

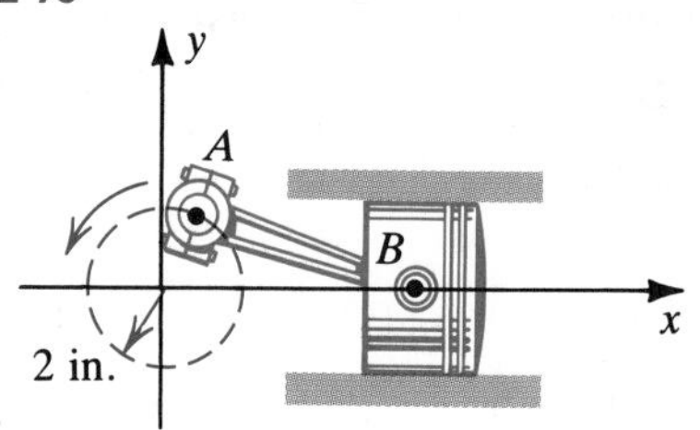

CHAPTER 4

APPLICATIONS OF THE DERIVATIVE

INTRODUCTION

In this chapter we use derivatives to obtain important details about function values $f(x)$ as x varies through an interval. These facts enable us to sketch the graph of a function accurately and to describe precisely where it rises or falls—something that cannot usually be accomplished with precalculus methods. Of major importance is finding the high and low points on the graph, since they provide the largest and smallest function values. The determination of these *extreme values* plays a significant role in applications that involve time, temperature, volume, pressure, gasoline consumption, air pollution, business profits, corporate expenses, and, in general, *any* quantity that can be represented by a function. Such *optimization problems* are considered in Section 4.6, after we have developed the theory that relates derivatives to extreme values. This relationship is obtained by employing one of the most important results in calculus: the *mean value theorem*, proved in Section 4.2. This theorem is also used in a crucial way in Chapter 5, where we introduce another concept of calculus—the *definite integral*.

The chapter closes with *Newton's method*, a tool for approximating solutions of equations. It is easily programmable for use with calculators or computers and is one of the basic techniques used in the field of mathematics called *numerical analysis*.

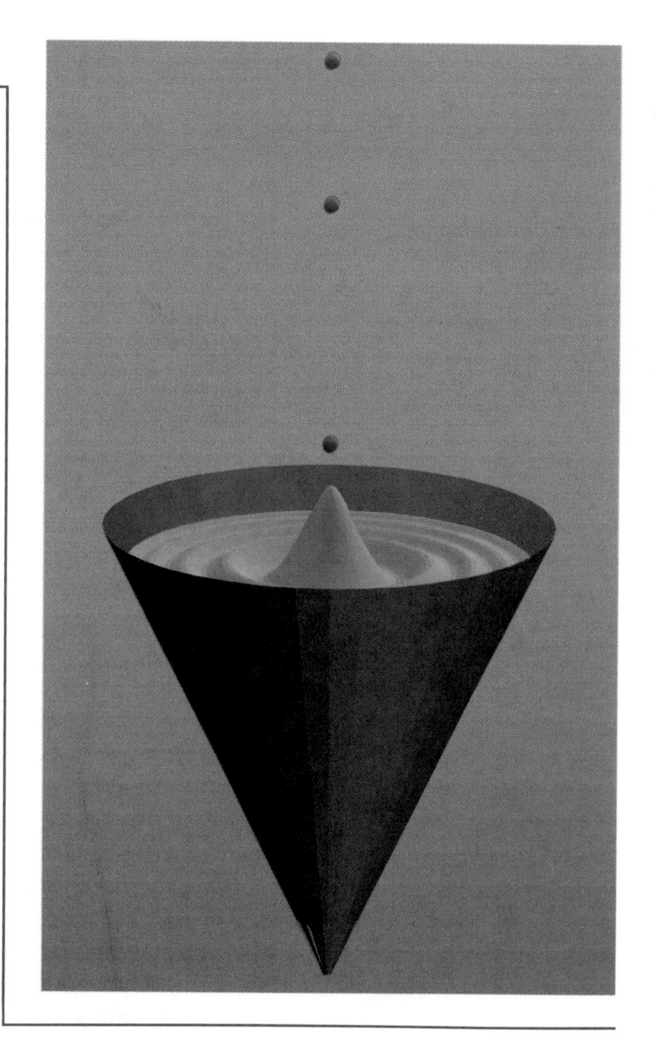

4.1 EXTREMA OF FUNCTIONS

Suppose that the graph in Figure 4.1 was made by a recording instrument that measures the variation of a physical quantity. The x-axis represents time and the y-values represent measurements such as temperature, resistance in an electrical circuit, blood pressure of an individual, the amount of chemical in a solution, or the bacteria count in a culture.

FIGURE 4.1

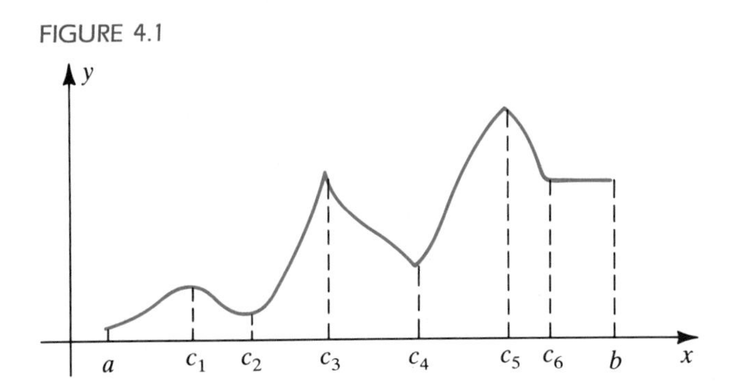

The graph indicates that the quantity increased in the time interval $[a, c_1]$, decreased in $[c_1, c_2]$, increased in $[c_2, c_3]$, decreased in $[c_3, c_4]$, and so on. If we restrict our attention to the interval $[c_1, c_4]$, the quantity had its largest (or maximum) value at c_3 and its smallest (or minimum) value at c_2. In other intervals there were different largest or smallest values. Over the *entire* interval $[a, b]$, the maximum value occurred at c_5 and the minimum value at a.

The terminology for describing the variation of physical quantities is also used for functions.

Definition **(4.1)**

Let a function f be defined on an interval I, and let x_1, x_2 denote numbers in I.

(i) f is **increasing** on I if $f(x_1) < f(x_2)$ whenever $x_1 < x_2$.

(ii) f is **decreasing** on I if $f(x_1) > f(x_2)$ whenever $x_1 < x_2$.

(iii) f is **constant** on I if $f(x_1) = f(x_2)$ for every x_1 and x_2.

Figure 4.2 contains graphical illustrations of Definition (4.1).

We shall use the phrases *f is increasing* and *$f(x)$ is increasing* interchangeably. This will also be done for *decreasing*. If a function is increasing, then its graph rises as x increases. If a function is decreasing, then its graph falls as x increases. If Figure 4.1 is the graph of a function f, then f is increasing on the intervals $[a, c_1]$, $[c_2, c_3]$, and $[c_4, c_5]$. It is decreasing on $[c_1, c_2]$, $[c_3, c_4]$, and $[c_5, c_6]$. The function is constant on the interval $[c_6, b]$.

The next definition introduces the terms we shall use for largest and smallest values of functions.

FIGURE 4.2

(i) Increasing function

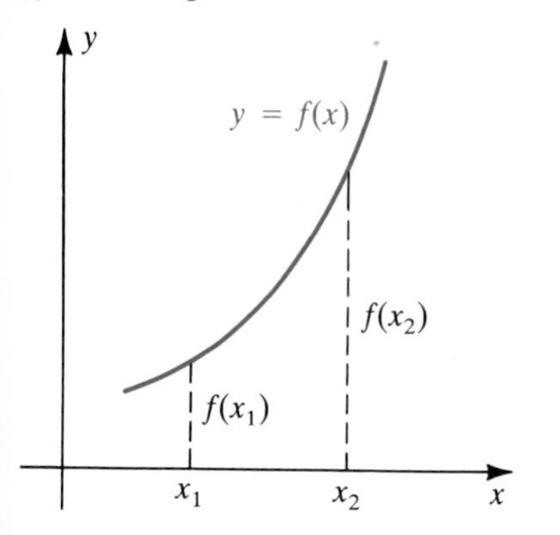

(ii) Decreasing function

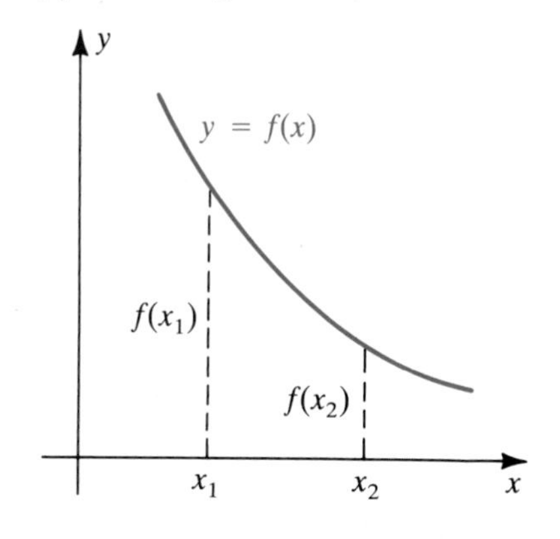

(iii) Constant function

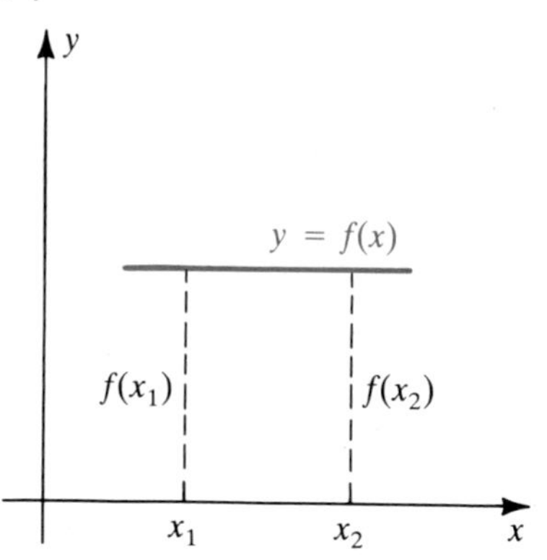

Definition **(4.2)**

Let a function f be defined on a set S of real numbers, and let c be a number in S.

(i) $f(c)$ is the **maximum value** of f on S if $f(x) \leq f(c)$ for every x in S.

(ii) $f(c)$ is the **minimum value** of f on S if $f(x) \geq f(c)$ for every x in S.

Maximum and minimum values are illustrated in Figures 4.3 and 4.4. In the figures we have pictured S as a closed interval; however, Definition (4.2) may be applied to other types of intervals or sets of real numbers. Each graph shown has a maximum or minimum value $f(c)$ with $a < c < b$, but such values may also occur at endpoints of intervals or domains of functions.

If $f(c)$ is the maximum value of f on S, we say that f *takes on* its maximum value at c. The point $(c, f(c))$ is a highest point on the graph of f. If $f(c)$ is the minimum value of f, we say that f *takes on* its minimum value at c, and $(c, f(c))$ is a lowest point on the graph of f. Maximum and minimum values are sometimes called **extreme values**, or **extrema**, of f. A function can take on a maximum or minimum value more than once. If f is a constant function, then $f(c)$ is both a maximum and a minimum value of f for *every* real number c.

FIGURE 4.3

Maximum value $f(c)$

(i)

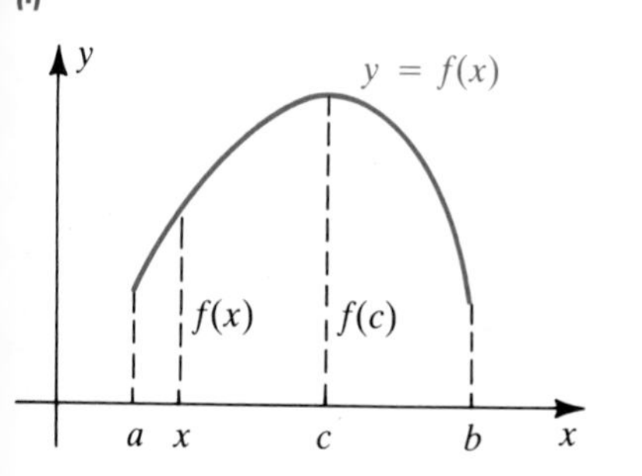

(ii)

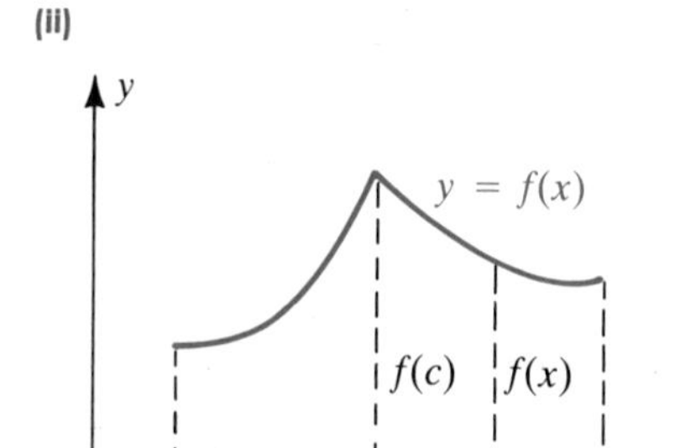

(iii)

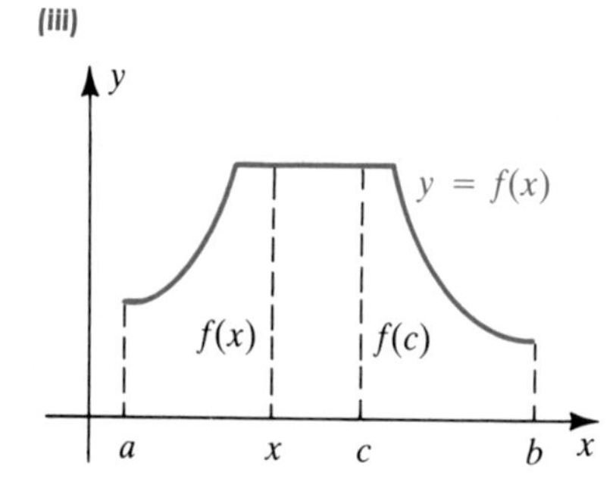

FIGURE 4.4
Minimum value $f(c)$

(i)

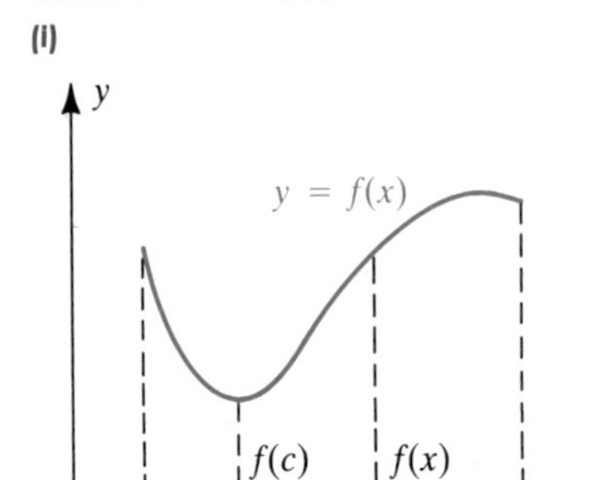

(ii)

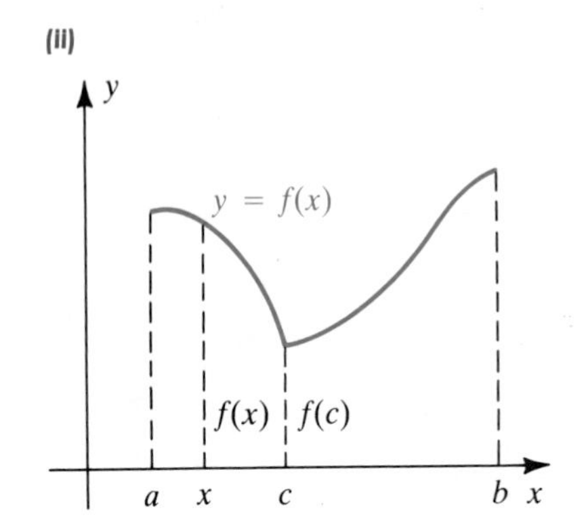

(iii)

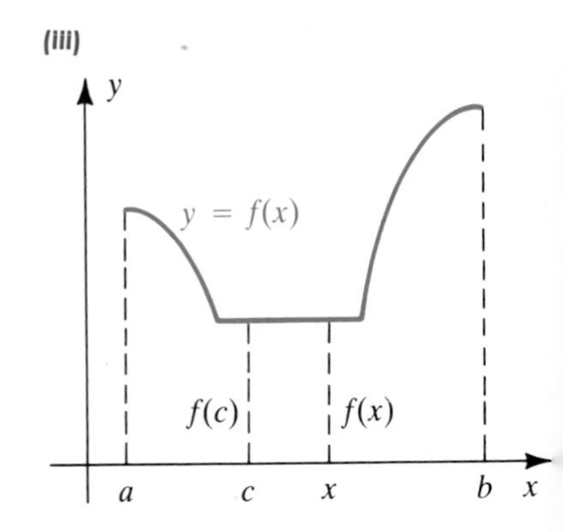

If D is the domain of f, then the maximum and minimum values of f on D, if they exist, are called the **absolute maximum** and **absolute minimum** of f.

EXAMPLE 1 Let $f(x) = 4 - x^2$. Find the extrema of f on the following intervals:

(a) $[-2, 1]$ **(b)** $(-2, 1)$ **(c)** $(1, 2]$ **(d)** $(1, 2)$

SOLUTION The graph of f (a parabola) is sketched with dashes in Figure 4.5, where the solid portions correspond to the intervals (a)–(d). The extrema in each interval (denoted by Max and Min) are listed under each graph.

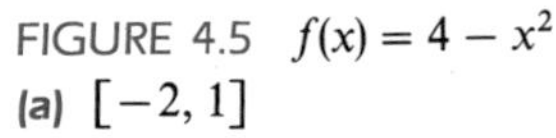

FIGURE 4.5 $f(x) = 4 - x^2$

(a) $[-2, 1]$

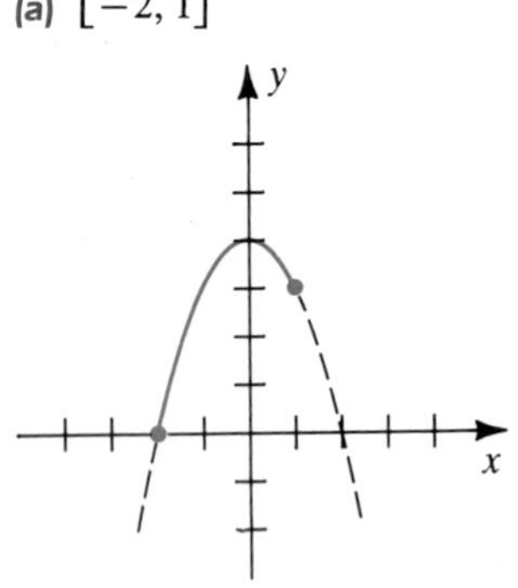

Max: $f(0) = 4$
Min: $f(-2) = 0$

(b) $(-2, 1)$

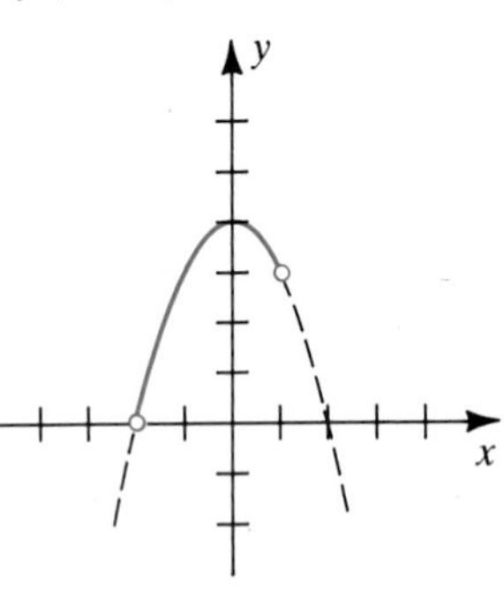

Max: $f(0) = 4$
Min: none

(c) $(1, 2]$

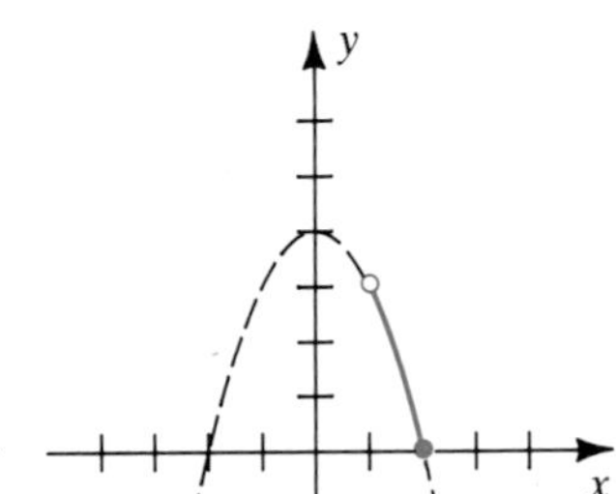

Max: none
Min: $f(2) = 0$

(d) $(1, 2)$

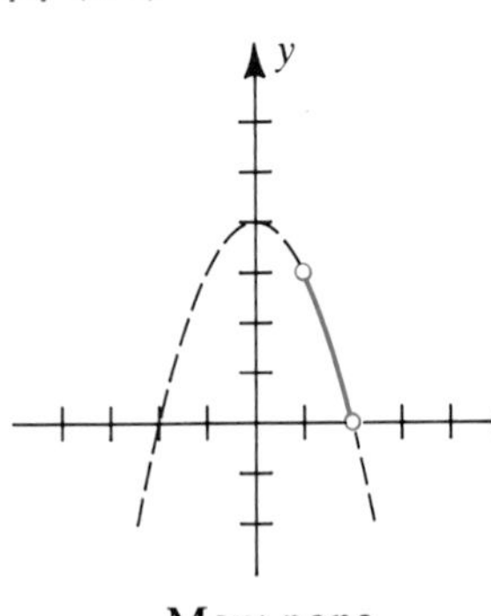

Max: none
Min: none

There is no minimum value of f in (b); if c is any number in $(-2, 1)$, we can find a number a in $(-2, 1)$ such that $f(a) < f(c)$. Similarly, there is no maximum value in (c) and neither a maximum nor a minimum value in (d).

EXAMPLE 2 Let $f(x) = \dfrac{1}{x^2}$. Find the extrema of f on

(a) $[-1, 2]$ **(b)** $[-1, 2)$

SOLUTION Portions of the graph of f and the extrema on the given intervals are shown in Figure 4.6. Note that f is not continuous at 0.

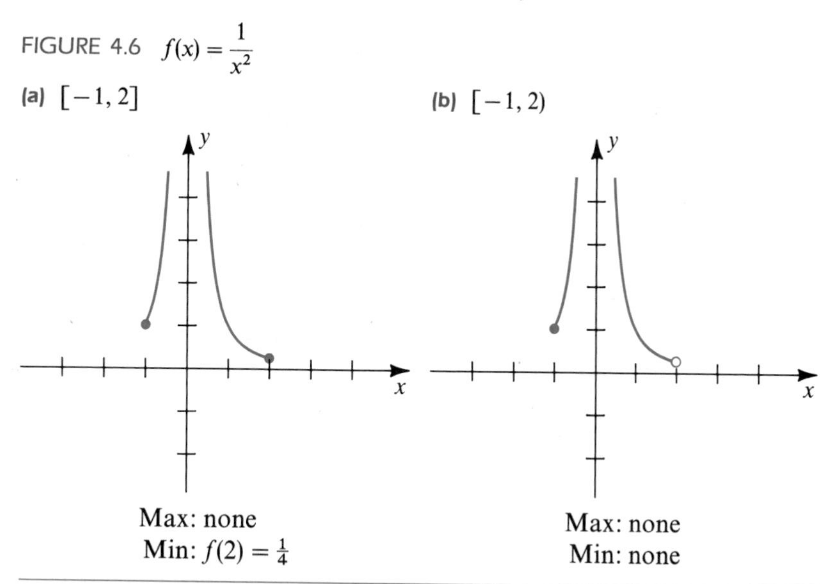

FIGURE 4.6 $f(x) = \dfrac{1}{x^2}$

The preceding examples indicate that the existence of maximum or minimum values may depend on the type of interval and on the continuity of the function. The next theorem states conditions under which a function takes on a maximum value and a minimum value on an interval. More advanced texts on calculus may be consulted for the proof.

Extreme value theorem **(4.3)**

> If a function f is continuous on a closed interval $[a, b]$, then f takes on a minimum value and a maximum value at least once in $[a, b]$.

The importance of this theorem is that it *guarantees* the existence of extrema if f is continuous on a closed interval. However, as indicated by our examples, extrema *may* occur on intervals that are not closed and for functions that are not continuous.

We shall also be interested in *local extrema* of f, defined as follows.

Definition **(4.4)**

> Let c be a number in the domain of a function f.
>
> **(i)** $f(c)$ is a **local maximum** of f if there exists an open interval (a, b) containing c such that $f(x) \le f(c)$ for every x in (a, b).
>
> **(ii)** $f(c)$ is a **local minimum** of f if there exists an open interval (a, b) containing c such that $f(x) \ge f(c)$ for every x in (a, b).

The term *local* is used because we *localize* our attention to a sufficiently small *open* interval containing c such that f takes on a largest (or smallest) value at c. Outside of that open interval, f may take on larger (or smaller) values. The word *relative* is sometimes used in place of *local*. Each local maximum or minimum is called a **local extremum** of f, and the totality of such numbers are the **local extrema** of f. Several examples of local extrema are illustrated in Figure 4.7. As indicated in the figure, it is possible for a local minimum to be *larger* than a local maximum.

FIGURE 4.7

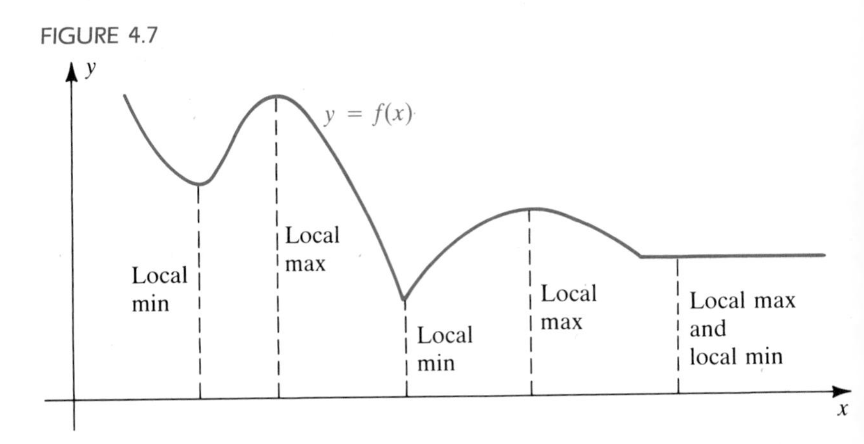

In this text, whenever we use the phrases *maximum value of f* and *minimum value of f* without including the adjective *local*, we mean the function values in Definition (4.2). When referring to the extrema in Definition (4.4), we shall always include the word *local*.

The local extrema on an interval may not include the maximum or minimum values of f. For example, in Figure 4.1, $f(a)$ is the minimum value of f on $[a, b]$; however, it is not a local minimum, since there is no *open* interval I contained in $[a, b]$ such that $f(a)$ is the least value of f on I.

At a point corresponding to a local extremum of the function graphed in Figure 4.7, the tangent line is horizontal or the graph has a corner. The x-coordinates of these points are numbers at which the derivative is zero or does not exist. The next theorem specifies that this is generally true. A proof is given at the end of this section.

Theorem **(4.5)**

If a function f has a local extremum at a number c in an open interval, then either $f'(c) = 0$ or $f'(c)$ does not exist.

The following is an immediate consequence of Theorem (4.5).

Corollary **(4.6)**

If $f'(c)$ exists and $f'(c) \neq 0$, then $f(c)$ is not a local extremum of the function f.

A result similar to Theorem (4.5) is true for the maximum and minimum values of a function that is continuous on a closed interval $[a, b]$, provided the extrema occur on the *open* interval (a, b). The theorem may be stated as follows.

Theorem (4.7)

> If a function f is continuous on a closed interval $[a, b]$ and has its maximum or minimum value at a number c in the open interval (a, b), then either $f'(c) = 0$ or $f'(c)$ does not exist.

The proof of Theorem (4.7) is exactly the same as that of (4.5) with the word *local* deleted.

It follows from Theorems (4.5) and (4.7) that the numbers at which the derivative either is zero or does not exist play a crucial role in the search for extrema of a function. Because of this, we give these numbers a special name.

Definition (4.8)

> A number c in the domain of a function f is a **critical number** of f if either $f'(c) = 0$ or $f'(c)$ does not exist.

If c is a critical number of f such that $f'(c) = 0$, then c must be in an open interval of the domain of f, because the limit in Definition (3.6) exists, with $a = c$. A critical number c such that $f'(c)$ does not exist may occur at an endpoint of the domain of f.

Referring to Theorem (4.7), we see that if f is continuous on a closed interval $[a, b]$, then the maximum and minimum values of f occur either at a critical number of f in (a, b) or at the endpoints a or b of the interval. It is also possible for a critical number c to be one of the endpoints of the interval $[a, b]$. If either $f(a)$ or $f(b)$ is an extremum of f on $[a, b]$, it is called an **endpoint extremum**. The sketches in Figure 4.8 illustrate this concept.

FIGURE 4.8
Endpoint extrema of f on $[a, b]$

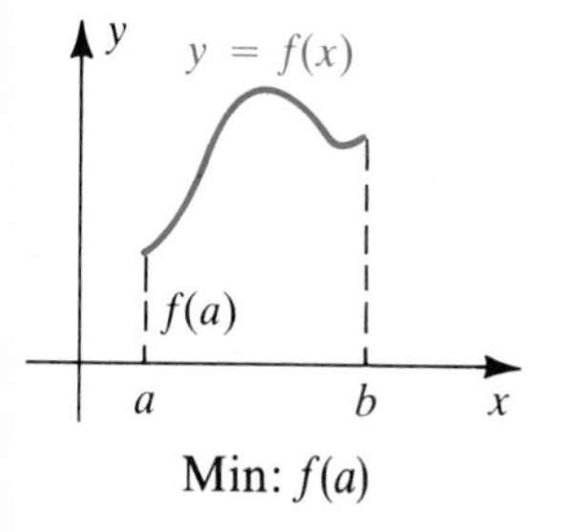

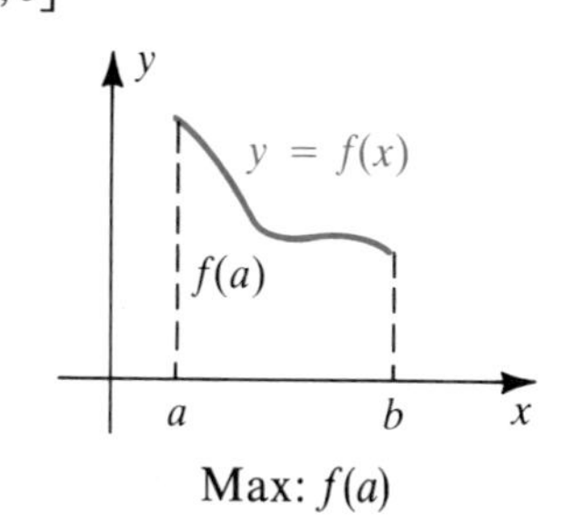

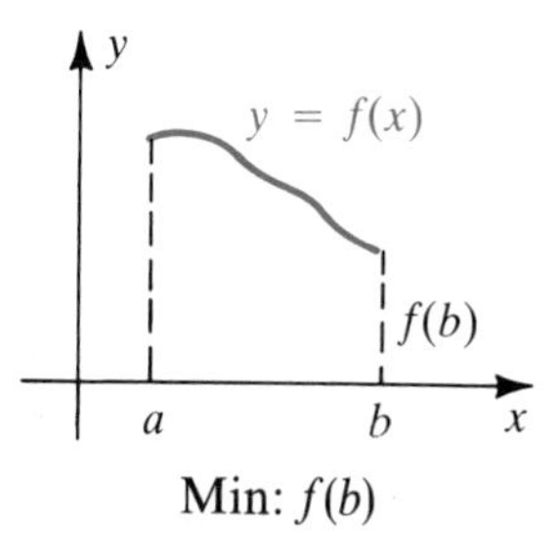

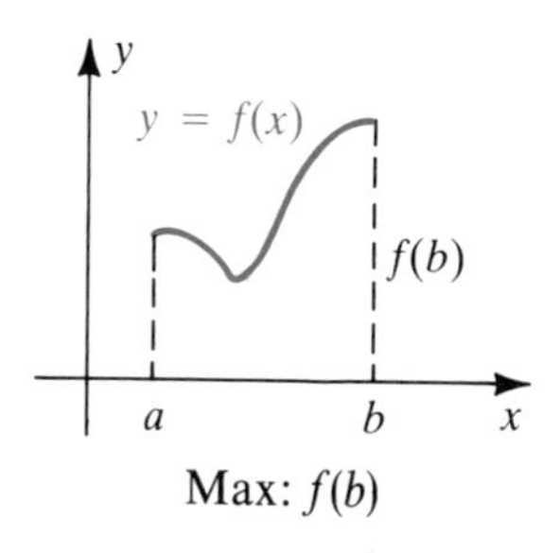

The preceding discussion gives us the following:

Guidelines for finding the extrema of a continuous function f on $[a, b]$ (4.9)

> 1 Find all the critical numbers of f in (a, b).
> 2 Calculate $f(c)$ for each critical number c found in guideline 1.
> 3 Calculate the endpoint values $f(a)$ and $f(b)$.
> 4 The maximum and minimum values of f on $[a, b]$ are the largest and smallest function values calculated in guidelines 2 and 3.

EXAMPLE 3 If $f(x) = x^3 - 12x$, find the maximum and minimum values of f on the closed interval $[-3, 5]$ and sketch the graph of f.

SOLUTION Using Guidelines (4.9), we begin by finding the critical numbers of f. Differentiating yields

$$f'(x) = 3x^2 - 12 = 3(x^2 - 4) = 3(x + 2)(x - 2).$$

Since the derivative exists for every x, the only critical numbers are those for which the derivative is zero—that is, -2 and 2. Since f is continuous on $[-3, 5]$, it follows from our discussion that the maximum and minimum values are among the numbers $f(-2)$, $f(2)$, $f(-3)$, and $f(5)$. Calculating these values (see guidelines 2 and 3), we obtain the following table.

Value of x	Classification of x	Function value $f(x)$
-2	critical number of f	$f(-2) = 16$
2	critical number of f	$f(2) = -16$
-3	endpoint of $[-3, 5]$	$f(-3) = 9$
5	endpoint of $[-3, 5]$	$f(5) = 65$

FIGURE 4.9

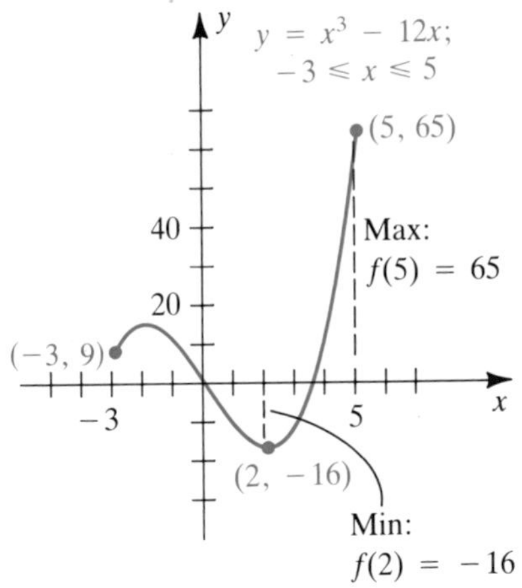

By guideline 4, the minimum value of f on $[-3, 5]$ is the smallest function value $f(2) = -16$, and the maximum value is the endpoint extremum $f(5) = 65$.

The graph of f is sketched in Figure 4.9, with different scales on the x- and y-axes. The tangent line is horizontal at the point corresponding to each of the critical numbers, -2 and 2. It will follow from our work in Section 4.4 that $f(-2) = 16$ is a *local* maximum for f, as indicated by the graph.

Maximum or minimum values are sometimes referred to incorrectly. Note that in Example 3 the minimum occurs *at* $x = 2$, and the minimum *is* $f(2) = -16$. The maximum occurs *at* $x = 5$, and the maximum *is* $f(5) = 65$. When asked to find a minimum (or maximum), do not merely find the value of x at which it occurs: *Be sure to complete the problem by calculating the function value.*

EXAMPLE 4 If $f(x) = (x - 1)^{2/3} + 2$, find the maximum and minimum values of f on $[0, 9]$, and sketch the graph of f.

SOLUTION We first differentiate $f(x)$:

$$f'(x) = \frac{2}{3}(x - 1)^{-1/3} = \frac{2}{3(x - 1)^{1/3}}$$

To find the critical numbers, we note that $f'(x) \neq 0$ for every x and that $f'(x)$ does not exist at $x = 1$. Hence 1 is the only critical number in $[0, 9]$.

Let us tabulate our work as in Example 3:

Value of x	Classification of x	Function value $f(x)$
1	critical number of f	$f(1) = 2$
0	endpoint of $[0, 9]$	$f(0) = 3$
9	endpoint of $[0, 9]$	$f(9) = 6$

FIGURE 4.10

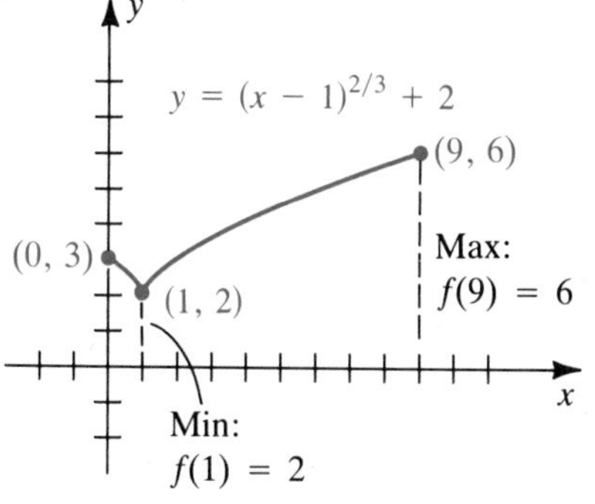

Thus, by Guidelines (4.9), f has a minimum value $f(1) = 2$ and a maximum value $f(9) = 6$ on the interval $[0, 9]$.

The graph of f is sketched in Figure 4.10. Note that

$$\lim_{x \to 1^-} f'(x) = -\infty \quad \text{and} \quad \lim_{x \to 1^+} f'(x) = \infty.$$

Since f is continuous at $x = 1$, the graph has a cusp at $(1, 2)$ by Definition (3.10).

FIGURE 4.11

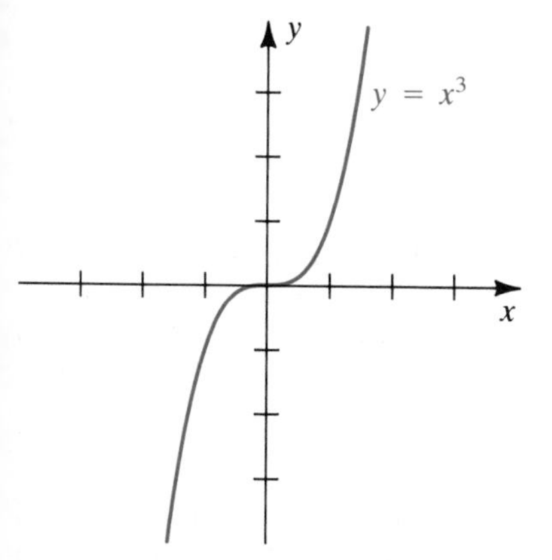

We see from Theorem (4.5) that if a function has a *local* extremum, then it *must* occur at a critical number; however, not every critical number leads to a local extremum, as illustrated by the following example.

EXAMPLE 5 If $f(x) = x^3$, prove that f has no local extremum.

SOLUTION The graph of f is sketched in Figure 4.11. The derivative is $f'(x) = 3x^2$, which exists for every x and is zero only if $x = 0$. Consequently, 0 is the only critical number. However, if $x < 0$, then $f(x)$ is negative, and if $x > 0$, then $f(x)$ is positive. Thus, $f(0)$ is neither a local maximum nor a local minimum. Since a local extremum *must* occur at a critical number (see Theorem (4.5)), it follows that f has no local extrema. Note that the tangent line is horizontal and crosses the graph at the point $(0, 0)$.

It sometimes requires considerable effort to find the critical numbers of a function. In the next two examples we shall find these numbers for functions that are more complicated than those considered in the preceding examples. We could again use Guidelines (4.9) to determine extrema on a closed interval, but to do so would merely be repetitive of our previous work. In Sections 4.3 and 4.4 we discuss methods for determining the local extrema of a function by using critical numbers and derivatives.

EXAMPLE 6 Find the critical numbers of f if $f(x) = (x + 5)^2 \sqrt[3]{x - 4}$.

SOLUTION Differentiating $f(x) = (x + 5)^2(x - 4)^{1/3}$, we obtain

$$f'(x) = (x + 5)^2 \tfrac{1}{3}(x - 4)^{-2/3} + 2(x + 5)(x - 4)^{1/3}.$$

To find the critical numbers, we simplify $f'(x)$ as follows:

$$\begin{aligned} f'(x) &= \frac{(x+5)^2}{3(x-4)^{2/3}} + 2(x+5)(x-4)^{1/3} \\ &= \frac{(x+5)^2 + 6(x+5)(x-4)}{3(x-4)^{2/3}} \\ &= \frac{(x+5)[(x+5)+6(x-4)]}{3(x-4)^{2/3}} \\ &= \frac{(x+5)(7x-19)}{3(x-4)^{2/3}} \end{aligned}$$

Hence $f'(x) = 0$ if $x = -5$ or $x = \frac{19}{7}$. The derivative $f'(x)$ does not exist at $x = 4$. Thus, f has the three critical numbers -5, $\frac{19}{7}$, and 4.

EXAMPLE 7 If $f(x) = 2 \sin x + \cos 2x$, find the critical numbers of f that are in the interval $[0, 2\pi]$.

SOLUTION Differentiating $f(x)$ gives us

$$f'(x) = 2 \cos x - 2 \sin 2x.$$

Since $\sin 2x = 2 \sin x \cos x$, this may be rewritten

$$\begin{aligned} f'(x) &= 2 \cos x - 4 \sin x \cos x \\ &= 2 \cos x\,(1 - 2 \sin x). \end{aligned}$$

The derivative exists for every x, and $f'(x) = 0$ if either $\sin x = \frac{1}{2}$ or $\cos x = 0$. Hence the critical numbers of f in the interval $[0, 2\pi]$ are $\pi/6$, $5\pi/6$, $\pi/2$, and $3\pi/2$. We shall return to this function and sketch the graph of f in Example 7 of Section 4.4.

PROOF OF THEOREM (4.5) Suppose f has a local extremum at c. If $f'(c)$ does not exist, there is nothing more to prove. If $f'(c)$ exists, then precisely one of the following occurs:

$$\text{(i) } f'(c) > 0 \quad \text{(ii) } f'(c) < 0 \quad \text{(iii) } f'(c) = 0$$

We shall arrive at (iii) by proving that neither (i) nor (ii) can occur. Thus, suppose $f'(c) > 0$. Employing Definition (3.6), yields

$$f'(c) = \lim_{x \to c} \frac{f(x) - f(c)}{x - c} > 0,$$

and hence, by Theorem (2.6), there exists an open interval (a, b) containing c such that

$$\frac{f(x) - f(c)}{x - c} > 0$$

for every x in (a, b) different from c. The last inequality implies that if $a < x < b$ and $x \neq c$, then $f(x) - f(c)$ and $x - c$ are either both positive or both negative:

$$f(x) - f(c) < 0 \quad \text{whenever } x - c < 0$$
$$f(x) - f(c) > 0 \quad \text{whenever } x - c > 0$$

Equivalently, if x is in (a, b) and $x \neq c$:

$$f(x) < f(c) \quad \text{whenever } x < c$$
$$f(x) > f(c) \quad \text{whenever } x > c$$

It follows that $f(c)$ is neither a local maximum nor a local minimum for f, contrary to hypothesis. Consequently, (i) cannot occur. Similarly, the assumption that $f'(c) < 0$ leads to a contradiction. Hence (iii) must hold and the theorem is proved. ■

EXERCISES 4.1

Exer. 1–2: Use the graph to estimate the absolute maximum, the absolute minimum, and the local extrema of f.

1

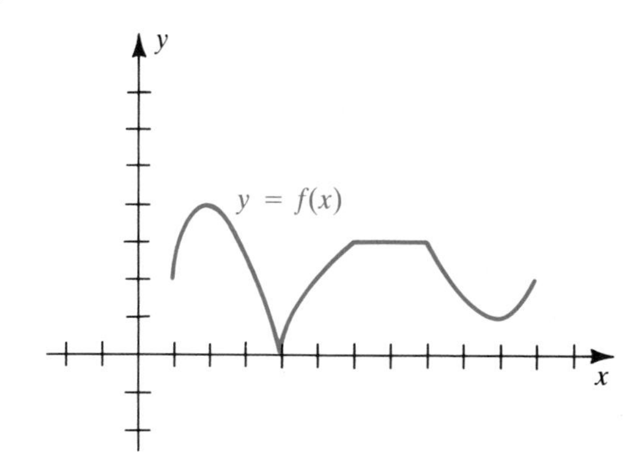

2

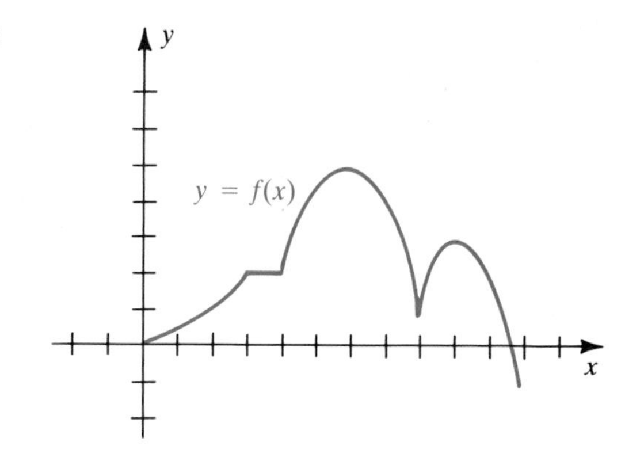

Exer. 3–4: Use the graph to estimate the extrema of f on each interval.

3 (a) $[-3, 3)$ (b) $(-3, \sqrt{3})$ (c) $[-\sqrt{3}, 1)$ (d) $[0, 3]$

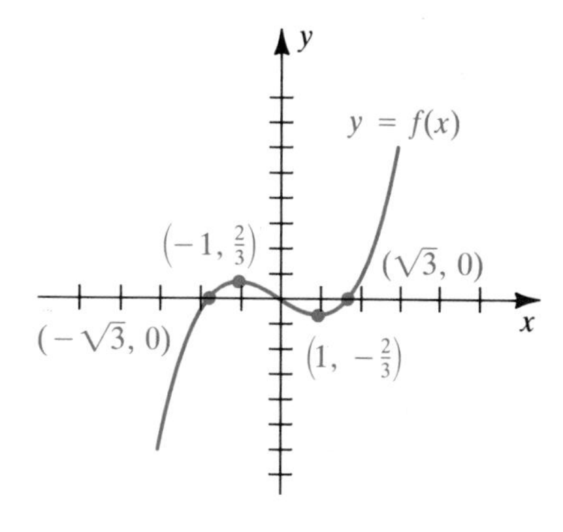

4 (a) $[-2, 2]$ (b) $(0, 2)$ (c) $(-1, 1]$ (d) $[-2, -1)$

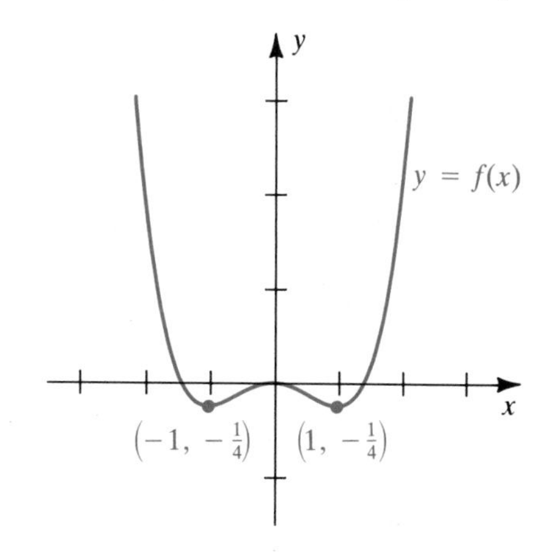

Exer. 5–6: Sketch the graph of f, and find the extrema on each interval.

5 $f(x) = \frac{1}{2}x^2 - 2x$

(a) $[0, 5)$ (b) $(0, 2)$ (c) $(0, 4)$ (d) $[2, 5]$

6 $f(x) = (x - 1)^{2/3} - 4$

(a) $[0, 9]$ (b) $(1, 2]$ (c) $(-7, 2)$ (d) $[0, 1)$

Exer. 7–10: Find the extrema of f on the given interval.

7 $f(x) = 5 - 6x^2 - 2x^3$; $[-3, 1]$

8 $f(x) = 3x^2 - 10x + 7$; $[-1, 3]$

9 $f(x) = 1 - x^{2/3}$; $[-1, 8]$

10 $f(x) = x^4 - 5x^2 + 4$; $[0, 2]$

Exer. 11–36: Find the critical numbers of the function.

11 $f(x) = 4x^2 - 3x + 2$

12 $g(x) = 2x + 5$

13 $s(t) = 2t^3 + t^2 - 20t + 4$

14 $K(z) = 4z^3 + 5z^2 - 42z + 7$

15 $F(w) = w^4 - 32w$

16 $k(r) = r^5 - 2r^3 + r - 12$

17 $f(z) = \sqrt{z^2 - 16}$

18 $M(x) = \sqrt[3]{x^2 - x - 2}$

19 $h(x) = (2x - 5)\sqrt{x^2 - 4}$

20 $T(v) = (4v + 1)\sqrt{v^2 - 16}$

21 $g(t) = t^2 \sqrt[3]{2t - 5}$

22 $g(x) = (x - 3)\sqrt{9 - x^2}$

23 $G(x) = \dfrac{2x - 3}{x^2 - 9}$

24 $f(s) = \dfrac{s^2}{5s + 4}$

25 $f(t) = \sin^2 t - \cos t$

26 $g(t) = 4 \sin^3 t + 3\sqrt{2} \cos^2 t$

27 $K(\theta) = \sin 2\theta + 2 \cos \theta$

28 $f(x) = 8 \cos^3 x - 3 \sin 2x - 6x$

29 $f(x) = \dfrac{1 + \sin x}{1 - \sin x}$

30 $g(\theta) = 2\sqrt{3}\theta + \sin 4\theta$

31 $k(u) = u - \tan u$

32 $p(z) = 3 \tan z - 4z$

33 $H(\phi) = \cot \phi + \csc \phi$

34 $g(x) = 2x + \cot x$

35 $f(x) = \sec (x^2 + 1)$

36 $s(t) = \dfrac{\sec t + 1}{\sec t - 1}$

37 (a) If $f(x) = x^{1/3}$, prove that 0 is the only critical number of f and that $f(0)$ is not a local extremum.

(b) If $f(x) = x^{2/3}$, prove that 0 is the only critical number of f and that $f(0)$ is a local minimum of f.

38 If $f(x) = |x|$, prove that 0 is the only critical number of f, that $f(0)$ is a local minimum of f, and that the graph of f has no tangent line at the point $(0, 0)$.

Exer. 39–40: (a) Prove that f has no local extrema. (b) Sketch the graph of f. (c) Prove that f is continuous on the interval $(0, 1)$, but f has neither a maximum nor a minimum value on $(0, 1)$. (d) Explain why (c) does not contradict Theorem (4.3).

39 $f(x) = x^3 + 1$

40 $f(x) = \dfrac{1}{x^2}$

41 (a) Prove that a polynomial function f of degree 1 has no extrema on the interval $(-\infty, \infty)$.

(b) Discuss the extrema of f on a closed interval $[a, b]$.

42 If f is a constant function and (a, b) is any open interval, prove that $f(c)$ is both a local and an absolute extremum of f for every number c in (a, b).

43 If f is the greatest integer function, prove that every number is a critical number of f.

44 Let f be defined by the following conditions: $f(x) = 0$ if x is rational and $f(x) = 1$ if x is irrational. Prove that every number is a critical number of f.

45 Prove that a quadratic function has exactly one critical number on $(-\infty, \infty)$.

46 Prove that a polynomial function of degree 3 has either two, one, or no critical numbers on $(-\infty, \infty)$, and sketch graphs that illustrate how each of these possibilities can occur.

47 Let $f(x) = x^n$ for a positive integer n. Prove that f has either one or no local extremum on $(-\infty, \infty)$ depending on whether n is even or odd, respectively, and sketch a typical graph illustrating each case.

48 Prove that a polynomial function of degree n can have at most $n - 1$ local extrema on $(-\infty, \infty)$.

c **Exer. 49–50: Graph f on the given interval and use the graph to estimate the extrema of f.**

49 $f(x) = x^6 - x^5 + 3x^3 - 2x + 1$; $[-1, 1]$

50 $f(x) = \dfrac{x^2 - \cos 3x}{\sin 2x - x + 2}$; $[-\pi, 1]$

c **Exer. 51–52: Graph f on the given interval and use the graph to estimate the critical numbers of f.**

51 $f(x) = |\frac{1}{3}x^3 + x^2 - x - 1.5|$; $[-3, 2]$

52 $f(x) = x \sin^2 x \cos^2 x$; $[-2, 2]$

4.2 THE MEAN VALUE THEOREM

Finding the critical numbers of a function can sometimes be extremely difficult. As a matter of fact, there is no guarantee that critical numbers even exist. The next theorem, credited to the French mathematician Michel Rolle (1652–1719), provides sufficient conditions for the existence of a critical number. The theorem is stated for a function f that is continuous on a closed interval $[a, b]$, is differentiable on the open interval (a, b) and satisfies the condition $f(a) = f(b)$. Some typical graphs of functions of this type are sketched in Figure 4.12.

FIGURE 4.12

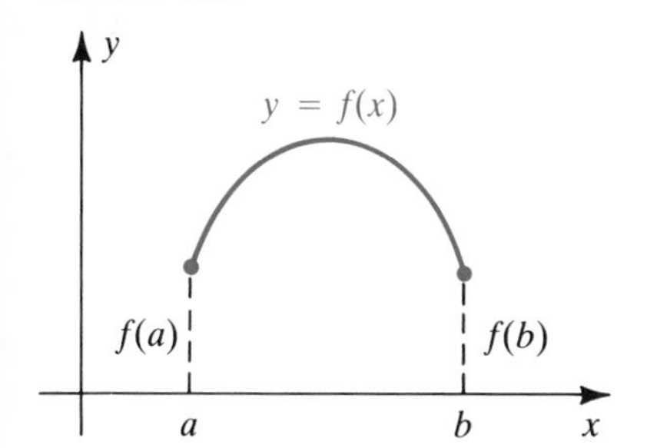

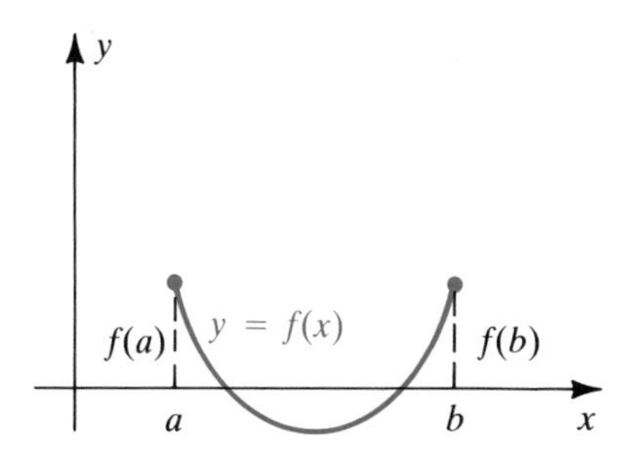

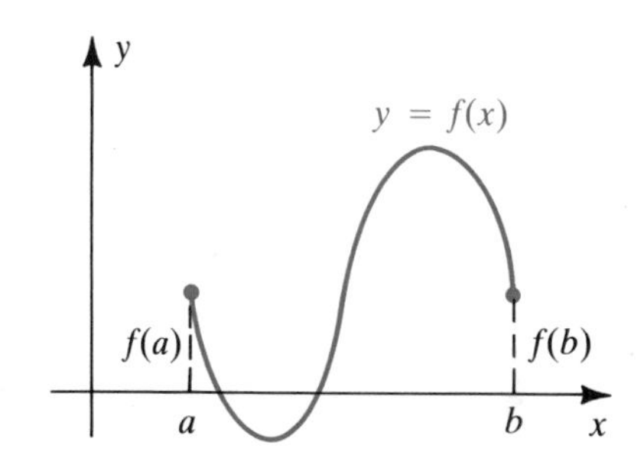

From the sketches in Figure 4.12 we expect that there is at least one number c between a and b such that the tangent line at the point $(c, f(c))$ is horizontal; that is, $f'(c) = 0$. This is precisely the conclusion of Rolle's theorem.

Rolle's theorem (4.10)

> If f is continuous on a closed interval $[a, b]$ and differentiable on the open interval (a, b) and if $f(a) = f(b)$, then $f'(c) = 0$ for at least one number c in (a, b).

PROOF The function f must fall into at least one of the following three categories:

(i) *$f(x) = f(a)$ for every x in (a, b).* In this case f is a constant function, and hence $f'(x) = 0$ for every x. Consequently, *every* number c in (a, b) is a critical number.

(ii) *$f(x) > f(a)$ for some x in (a, b).* In this case the maximum value of f in $[a, b]$ is greater than $f(a)$ or $f(b)$ and, therefore, must occur at some number c in the *open* interval (a, b). Since the derivative exists throughout (a, b), we conclude from Theorem (4.7) that $f'(c) = 0$.

(iii) *$f(x) < f(a)$ for some x in (a, b).* In this case the minimum value of f in $[a, b]$ is less than $f(a)$ or $f(b)$ and must occur at some number c in (a, b). As in (ii), $f'(c) = 0$. ■

Corollary (4.11)

> If f is continuous on a closed interval $[a, b]$ and if $f(a) = f(b)$, then f has at least one critical number in the open interval (a, b).

PROOF If f' does not exist at some number c in (a, b), then, by Definition (4.8), c is a critical number. Alternatively, if f' exists throughout (a, b), then, by Rolle's theorem, a critical number exists. ■

EXAMPLE 1 Let $f(x) = 4x^2 - 20x + 29$. Show that f satisfies the hypotheses of Rolle's theorem on the interval $[1, 4]$, and find all real numbers c in the open interval $(1, 4)$ such that $f'(c) = 0$. Illustrate the results graphically.

SOLUTION Since f is a polynomial function, it is continuous and differentiable for every x. In particular, it is continuous on $[1, 4]$ and differentiable on $(1, 4)$. Moreover,

$$f(1) = 4 - 20 + 29 = 13$$

$$f(4) = 64 - 80 + 29 = 13$$

and hence $f(1) = f(4)$. Thus, f satisfies the hypotheses of Rolle's theorem on $[1, 4]$.

Differentiating $f(x)$, we have

$$f'(x) = 8x - 20.$$

Setting $f'(x) = 0$ gives us $8x = 20$, or $x = \frac{5}{2}$. Hence

$$f'(\tfrac{5}{2}) = 0 \quad \text{and} \quad 1 < \tfrac{5}{2} < 4.$$

The graph of f (a parabola) is sketched in Figure 4.13. Since $f'(\frac{5}{2}) = 0$, the tangent line is horizontal at the vertex $(\frac{5}{2}, 4)$.

FIGURE 4.13

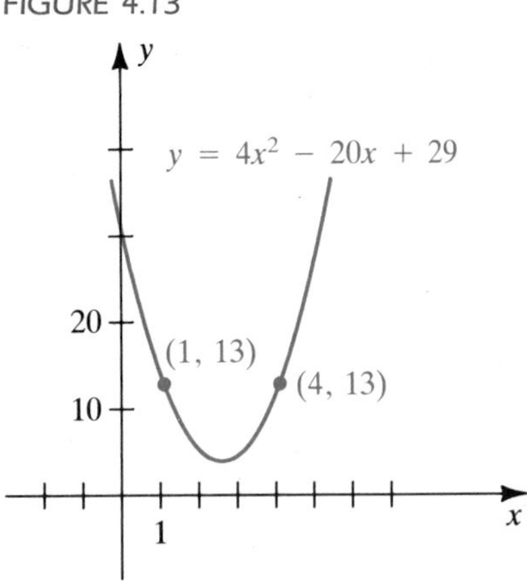

Although Rolle's theorem is of interest in itself, our principal use for it is in the proof of one of the most important tools in calculus: the **mean value theorem.** Later in the text we shall use this theorem to establish many fundamental results in calculus. In order to show the graphical significance of the mean value theorem, let f be any function that is continuous on $[a, b]$ and differentiable on (a, b) (possibly $f(a) \neq f(b)$). Consider the points $P(a, f(a))$ and $Q(b, f(b))$ on the graph of f, as illustrated by any of the sketches in Figure 4.14.

FIGURE 4.14

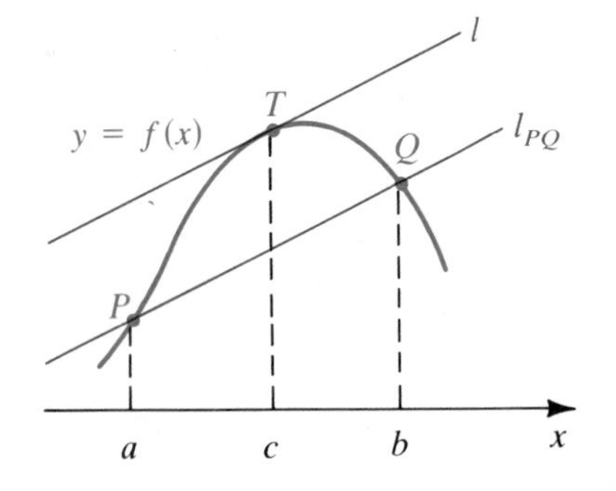

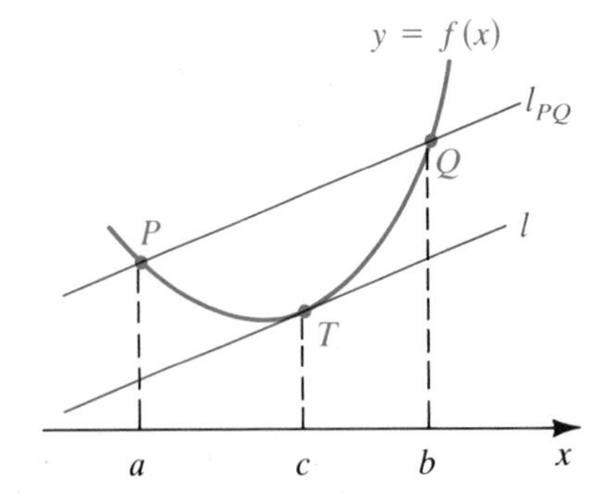

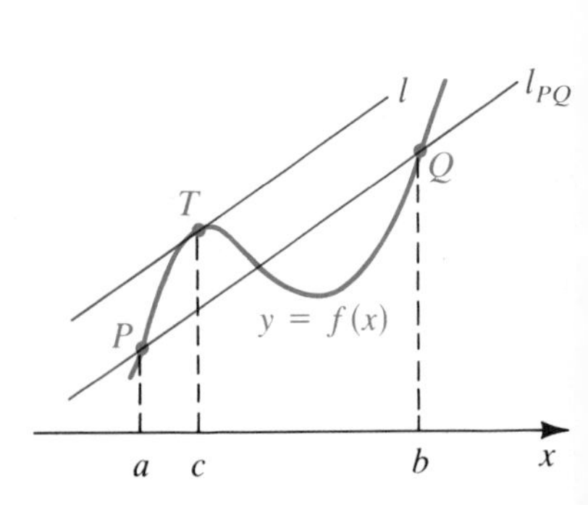

If f' exists throughout the open interval (a, b), then there appears to be at least one point $T(c, f(c))$ on the graph at which the tangent line l is parallel to the secant line l_{PQ} through P and Q. In terms of slopes,

$$\text{slope of } l = \text{slope of } l_{PQ},$$

or

$$f'(c) = \frac{f(b) - f(a)}{b - a}.$$

If we multiply both sides of the last equation by $b - a$, we obtain the second formula stated in the mean value theorem.

Mean value theorem **(4.12)**

If f is continuous on a closed interval $[a, b]$ and differentiable on the open interval (a, b), then there exists a number c in (a, b) such that

$$f'(c) = \frac{f(b) - f(a)}{b - a},$$

or, equivalently,

$$f(b) - f(a) = f'(c)(b - a).$$

FIGURE 4.15

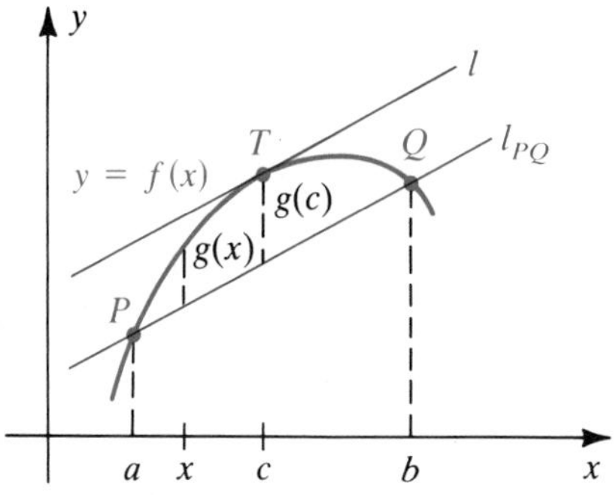

PROOF For any x in the interval $[a, b]$, let $g(x)$ be the vertical (signed) distance from the secant line l_{PQ} to the graph of f, as illustrated in Figure 4.15. (The term *signed distance* means that $g(x)$ is positive or negative if the graph of f lies above or below l_{PQ}, respectively.)

It appears that if $T(c, f(c))$ is a point at which the tangent line l is parallel to l_{PQ}, then the distance $g(c)$ is a local extremum of g. This suggests that we find the critical numbers of the function g.

We may obtain a formula for $g(x)$ as follows. First, by the point-slope form, an equation of the secant line l_{PQ} is

$$y - f(a) = \frac{f(b) - f(a)}{b - a}(x - a),$$

or, equivalently,

$$y = f(a) + \frac{f(b) - f(a)}{b - a}(x - a).$$

As illustrated in Figure 4.15, $g(x)$ is the difference of the distances from the x-axis to the graph of f and to the line l_{PQ}; that is,

$$g(x) = f(x) - \left[f(a) + \frac{f(b) - f(a)}{b - a}(x - a) \right].$$

We shall use Rolle's theorem to find a critical number of g. We first observe that since f is continuous on $[a, b]$ and differentiable on (a, b), the same is true for the function g. Differentiating, we obtain

$$g'(x) = f'(x) - \frac{f(b) - f(a)}{b - a}.$$

Moreover, by direct substitution we see that $g(a) = g(b) = 0$, and hence the function g satisfies the hypotheses of Rolle's theorem. Consequently, there exists a number c in the open interval (a, b) such that $g'(c) = 0$, or, equivalently,

$$f'(c) - \frac{f(b) - f(a)}{b - a} = 0.$$

The last equation may be written in the forms stated in the conclusion of the theorem. ■

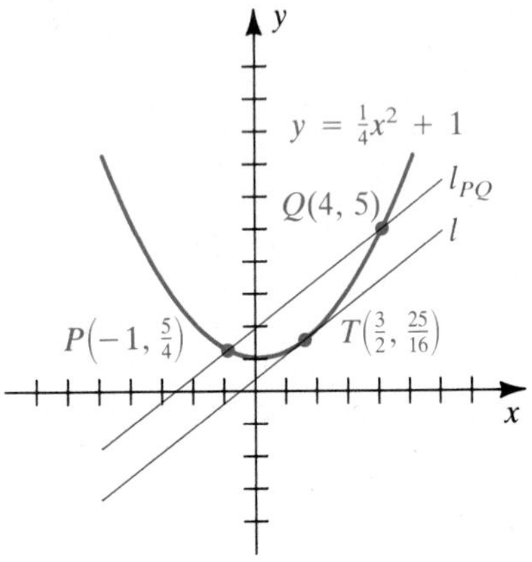

EXAMPLE 2 If $f(x) = \frac{1}{4}x^2 + 1$, show that f satisfies the hypotheses of the mean value theorem on the interval $[-1, 4]$, and find a number c in $(-1, 4)$ that satisfies the conclusion of the theorem. Illustrate the results graphically.

SOLUTION The quadratic function f is continuous on $[-1, 4]$ and differentiable on $(-1, 4)$; hence, by the mean value theorem, there is a number c in $(-1, 4)$ such that

FIGURE 4.16

$$\frac{f(4) - f(-1)}{4 - (-1)} = f'(c).$$

Using $f(4) = 5$, $f(-1) = \frac{5}{4}$, and $f'(x) = \frac{1}{2}x$ gives us

$$\frac{5 - \frac{5}{4}}{5} = \frac{1}{2}c, \quad \text{or} \quad \frac{3}{4} = \frac{1}{2}c.$$

Thus, $c = \frac{3}{2}$.

The graph of f (a parabola) is sketched in Figure 4.16. The points $P(-1, \frac{5}{4})$ and $Q(4, 5)$ correspond to the endpoints of the interval $[-1, 4]$. The point $T(\frac{3}{2}, \frac{25}{16})$ is obtained by using $c = \frac{3}{2}$ and is the point at which the tangent line l is parallel to the secant line l_{PQ}.

EXAMPLE 3 If $f(x) = x^3 - 8x - 5$, show that f satisfies the hypotheses of the mean value theorem on the interval $[1, 4]$, and find a number c in the open interval $(1, 4)$ that satisfies the conclusion of the theorem.

SOLUTION Since f is a polynomial function, it is continuous and differentiable for all real numbers. In particular, it is continuous on $[1, 4]$ and differentiable on the open interval $(1, 4)$. By the mean value theorem, there is a number c in $(1, 4)$ such that

$$\frac{f(4) - f(1)}{4 - 1} = f'(c).$$

Since $f(4) = 27$, $f(1) = -12$, and $f'(x) = 3x^2 - 8$, this is equivalent to

$$\frac{27 - (-12)}{3} = 3c^2 - 8, \quad \text{or} \quad 13 = 3c^2 - 8.$$

Thus, $c^2 = 7$, or $c = \pm\sqrt{7}$. Since $-\sqrt{7}$ is not in the interval $(1, 4)$, the desired number is $c = \sqrt{7}$.

The mean value theorem is also called the **theorem of the mean**. In the study of statistics the term *mean* is used for the *average* of a collection of numbers. The word has a similar connotation in *mean value theorem*. As an illustration, if a point P is moving on a coordinate line and if $s(t)$ denotes the coordinate of P at time t, then, by Definition (3.2), the average velocity of P during the time interval $[a, b]$ is

$$v_{\text{av}} = \frac{s(b) - s(a)}{b - a}.$$

According to the mean value theorem, this average (mean) velocity is equal to the velocity $s'(c)$ at some time c between a and b.

EXAMPLE 4 The speedometer of an automobile registers 50 mi/hr as it passes a mileage marker along a highway. Four minutes later, as the automobile passes a marker that is five miles from the first, the speedometer registers 55 mi/hr. Use the mean value theorem to prove that the velocity exceeded 70 mi/hr at some time while the automobile was traveling between the two markers.

SOLUTION We may assume that the automobile is moving along a straight highway with the two mileage markers at A and B, as illustrated in Figure 4.17. Let t denote the elapsed time (in hours) after the automobile passes A, and let $s(t)$ denote its distance from A (in miles) at time t. Since the automobile passes B at $t = \frac{4}{60} = \frac{1}{15}$ hr, the average velocity during the trip from A to B is

FIGURE 4.17

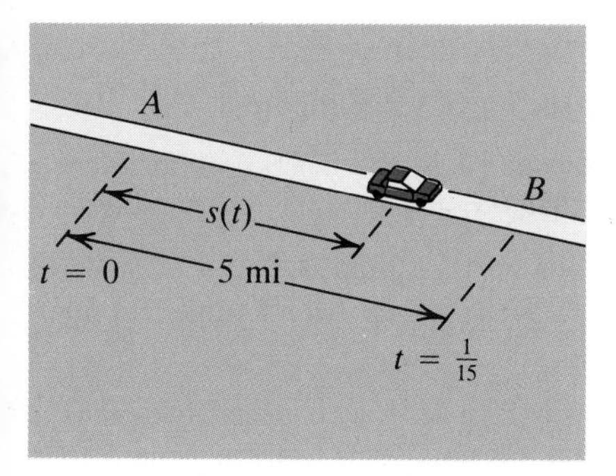

$$v_{\text{av}} = \frac{s(\frac{1}{15}) - s(0)}{\frac{1}{15} - 0} = \frac{5}{\frac{1}{15}} = 75 \text{ mi/hr}.$$

If we assume that s is a differentiable function, then applying the mean value theorem (4.12) to s, with $a = 0$ and $b = \frac{1}{15}$, proves that there is a time c in the time interval $[0, \frac{1}{15}]$ at which the velocity $s'(c)$ of the automobile is 75 mi/hr. Note that the velocity at A or B is irrelevant.

EXERCISES 4.2

Exer. 1–2: Use the graph of f to estimate the numbers in [1, 10] that satisfy the conclusion of the mean value theorem.

1

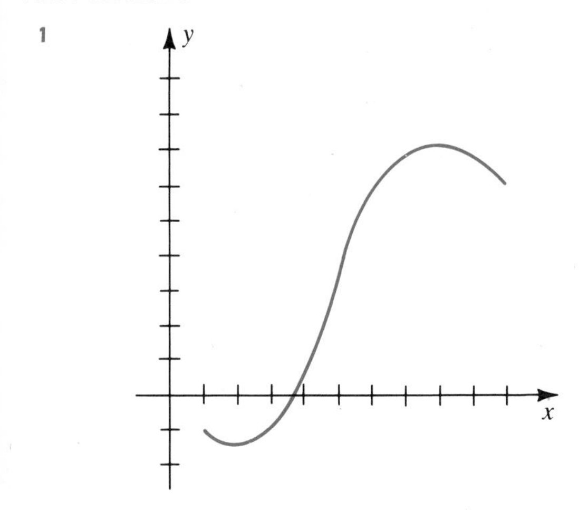

2

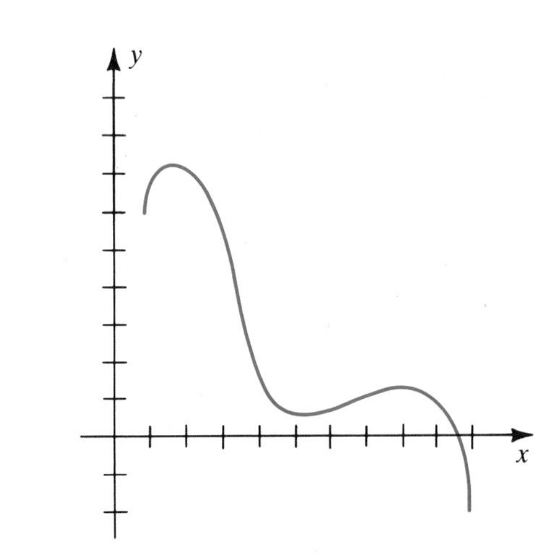

Exer. 3–8: Show that f satisfies the hypotheses of Rolle's theorem on $[a, b]$, and find all numbers c in (a, b) such that $f'(c) = 0$.

3 $f(x) = 3x^2 - 12x + 11$; $[0, 4]$

4 $f(x) = 5 - 12x - 2x^2$; $[-7, 1]$

5 $f(x) = x^4 + 4x^2 + 1$; $[-3, 3]$

6 $f(x) = x^3 - x$; $[-1, 1]$

7 $f(x) = \sin 2x$; $[0, \pi]$

8 $f(x) = \cos 2x + 2 \cos x$; $[0, 2\pi]$

Exer. 9–24: Determine whether f satisfies the hypotheses of the mean value theorem on $[a, b]$, and, if so, find all numbers c in (a, b) such that

$$f(b) - f(a) = f'(c)(b - a).$$

9 $f(x) = 5x^2 - 3x + 1$; $[1, 3]$

10 $f(x) = 3x^2 + x - 4$; $[1, 5]$

11 $f(x) = \dfrac{1}{(x-1)^2}$; $[0, 2]$

12 $f(x) = \dfrac{x+3}{x-2}$; $[-2, 3]$

13 $f(x) = x^{2/3}$; $[-8, 8]$

14 $f(x) = |x - 3|$; $[-1, 4]$

15 $f(x) = x + (4/x)$; $[1, 4]$

16 $f(x) = 3x^5 + 5x^3 + 15x$; $[-1, 1]$

17 $f(x) = x^3 - 2x^2 + x + 3$; $[-1, 1]$

18 $f(x) = 1 - 3x^{1/3}$; $[-8, -1]$

19 $f(x) = 4 + \sqrt{x - 1}$; $[1, 5]$

20 $f(x) = (x + 2)^{2/3}$; $[-1, 6]$

21 $f(x) = x^3 + 1$; $[-2, 4]$

22 $f(x) = x^3 + 4x$; $[-3, 6]$

23 $f(x) = \sin x$; $[0, \pi/2]$

24 $f(x) = \tan x$; $[0, \pi/4]$

25 If $f(x) = |x|$, show that $f(-1) = f(1)$ but $f'(c) \neq 0$ for every number c in the open interval $(-1, 1)$. Why doesn't this contradict Rolle's theorem?

26 If $f(x) = 5 + 3(x - 1)^{2/3}$, show that $f(0) = f(2)$ but $f'(c) \neq 0$ for every number c in the open interval $(0, 2)$. Why doesn't this contradict Rolle's theorem?

27 If $f(x) = 4/x$, prove that there is no real number c such that $f(4) - f(-1) = f'(c)[4 - (-1)]$. Why doesn't this contradict the mean value theorem applied to the interval $[-1, 4]$?

28 If f is the greatest integer function and if a and b are real numbers such that $b - a \geq 1$, prove that there is no real number c such that $f(b) - f(a) = f'(c)(b - a)$. Why doesn't this contradict the mean value theorem?

29 If f is a linear function, prove that f satisfies the hypotheses of the mean value theorem on every closed interval $[a, b]$ and that *every* number c satisfies the conclusion of the theorem.

30 If f is a quadratic function and $[a, b]$ is any closed interval, prove that precisely one number c in the interval (a, b) satisfies the conclusion of the mean value theorem.

31 If f is a polynomial function of degree 3 and $[a, b]$ is any closed interval, prove that at most two numbers in (a, b) satisfy the conclusion of the mean value theorem. Generalize to polynomial functions of degree n for any positive integer n.

32 If $f(x)$ is a polynomial of degree 3, use Rolle's theorem to prove that f has at most three real zeros. Extend this result to polynomials of degree n.

Exer. 33–40: Use the mean value theorem.

33 If f is continuous on $[a, b]$ and if $f'(x) = 0$ for every x in (a, b), prove that $f(x) = k$ for some real number k.

34 If f is continuous on $[a, b]$ and if $f'(x) = c$ for every x in (a, b), prove that $f(x) = cx + d$ for some real number d.

35 A straight highway 50 miles long connects two cities A and B. Prove that it is impossible to travel from A to B by automobile in exactly one hour without having the speedometer register 50 mi/hr at least once.

36 Let T denote the temperature (in °F) at time t (in hours). If the temperature is decreasing, then dT/dt is the *rate of cooling*. The greatest temperature variation during a twelve-hour period occurred in Montana in 1916, when the temperature dropped from 44 °F to a chilling −56 °F. Show that the rate of cooling exceeded −8 °F/hr at some time during the period of change.

37 If W denotes the weight (in pounds) of an individual and t denotes time (in months), then dW/dt is the rate of weight gain or loss (in lb/mo). The current speed record for weight loss is a drop in weight from 487 pounds to 130 pounds over an eight-month period. Show that the rate of weight loss exceeded 44 lb/mo at some time during the eight-month period.

38 The electrical charge Q on a capacitor increases from 2 millicoulombs to 10 millicoulombs in 15 milliseconds. Show that the current $I = dQ/dt$ exceeded $\frac{1}{2}$ ampere at some instant during this short time interval. (*Note:* 1 ampere = 1 coulomb/sec.)

39 Prove that if u and v are any real numbers, then

$$|\sin u - \sin v| \le |u - v|.$$

(*Hint:* Apply the mean value theorem with $f(x) = \sin x$.)

40 Prove that $\sqrt{1+h} < 1 + \frac{1}{2}h$ for $h > 0$. (*Hint:* Apply the mean value theorem with $f(x) = \sqrt{1+x}$.)

c **Exer. 41–42: Graph, on the same coordinate axes, $y = f(x)$ for $0 \le x \le 1$ and the line through $(0, f(0))$ and $(1, f(1))$. Use the graph to estimate the numbers in [0, 1] that satisfy the conclusion of the mean value theorem.**

41 $f(x) = 2.1x^4 - 1.4x^3 + 0.8x^2 - 1$

42 $f(x) = \dfrac{\sin 2x + \cos x}{2 + \cos \pi x}$

4.3 THE FIRST DERIVATIVE TEST

In this section we show how the sign of the derivative f' may be used to determine where a function f is increasing and where it is decreasing. This information will enable us to classify the local extrema of the function.

The graph of $y = f(x)$ in Figure 4.18 indicates that if the slope of the tangent line is positive in an open interval I (that is, if $f'(x) > 0$ for every x in I), then f is increasing on I. Similarly, it appears that if the slope is negative (that is, if $f'(x) < 0$), then f is decreasing.

FIGURE 4.18

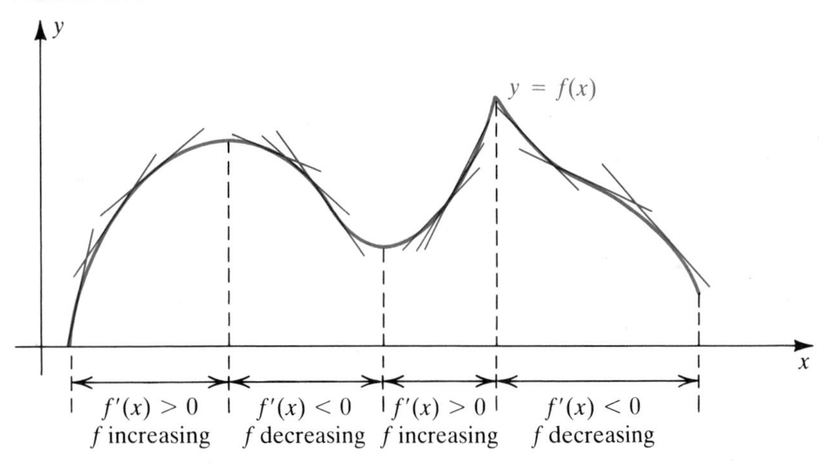

That these intuitive observations are actually true is a consequence of the following theorem.

Theorem **(4.13)**

> Let f be continuous on $[a, b]$ and differentiable on (a, b).
>
> **(i)** If $f'(x) > 0$ for every x in (a, b), then f is increasing on $[a, b]$.
>
> **(ii)** If $f'(x) < 0$ for every x in (a, b), then f is decreasing on $[a, b]$.

PROOF To prove (i), suppose that $f'(x) > 0$ for every x in (a, b), and consider any numbers x_1, x_2 in $[a, b]$ such that $x_1 < x_2$. We wish to show that $f(x_1) < f(x_2)$. Applying the mean value theorem (4.12) to the interval $[x_1, x_2]$ gives us

$$f(x_2) - f(x_1) = f'(c)(x_2 - x_1)$$

for some number c in the open interval (x_1, x_2). Since $x_2 - x_1 > 0$ and since, by hypothesis, $f'(c) > 0$, the right-hand side of the previous equation is positive; that is, $f(x_2) - f(x_1) > 0$. Hence $f(x_2) > f(x_1)$, which is what we wished to show. The proof of (ii) is similar. ■

We may also show that if $f'(x) > 0$ throughout an infinite interval $(-\infty, a)$ or (b, ∞), then f is increasing on $(-\infty, a]$ or $[b, \infty)$, respectively, provided f is continuous on those intervals. An analogous result holds for decreasing functions if $f'(x) < 0$.

To apply Theorem (4.13), we must determine intervals in which $f'(x)$ is either always positive or always negative. Theorem (2.7) is useful in this respect. Specifically, if the *derivative* f' is continuous and has no zeros on an interval, then either $f'(x) > 0$ or $f'(x) < 0$ for every x in the interval. Thus, if we choose *any* number k in the interval and if $f'(k) > 0$, then $f'(x) > 0$ for *every* x in the interval. Similarly, if $f'(k) < 0$, then $f'(x) < 0$ throughout the interval. We shall call $f'(k)$ a **test value** of $f'(x)$ for the interval.

EXAMPLE 1 If $f(x) = x^3 + x^2 - 5x - 5$,

(a) find the intervals on which f is increasing and the intervals on which f is decreasing

(b) sketch the graph of f

SOLUTION

(a) First we differentiate $f(x)$:

$$f'(x) = 3x^2 + 2x - 5 = (3x + 5)(x - 1)$$

By Theorem (4.13), it is sufficient to find the intervals in which $f'(x) > 0$ and those in which $f'(x) < 0$. The factored form of $f'(x)$ and the critical numbers $-\frac{5}{3}$ and 1 suggest the open intervals $(-\infty, -\frac{5}{3})$, $(-\frac{5}{3}, 1)$, and $(1, \infty)$. On each of these intervals f' is continuous and has no zeros, and therefore $f'(x)$ has the same sign throughout the interval. This sign can be determined by choosing a suitable test value for the interval.

The following table displays our work. The values of k were chosen for convenience. We chose $k = -2$ in the interval $(-\infty, -\frac{5}{3})$, but *any* number, such as -3 or -10, could have been used.

Interval	$(-\infty, -\frac{5}{3})$	$(-\frac{5}{3}, 1)$	$(1, \infty)$
k	-2	0	2
Test value $f'(k)$	$f'(-2) = 3 > 0$	$f'(0) = -5 < 0$	$f'(2) = 11 > 0$
Sign of $f'(x)$	$+$	$-$	$+$
Conclusion	f is increasing on $(-\infty, -\frac{5}{3}]$	f is decreasing on $[-\frac{5}{3}, 1]$	f is increasing on $[1, \infty)$

FIGURE 4.19

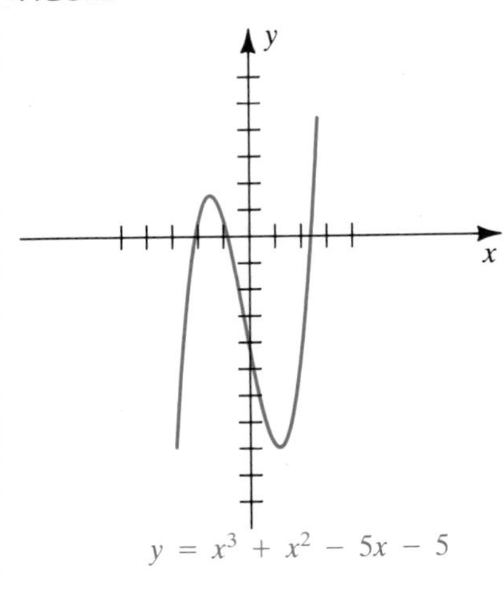

$y = x^3 + x^2 - 5x - 5$

(b) As an aid to sketching the graph of f, we shall find the x-intercepts by solving the equation $f(x) = 0$. Since

$$\begin{aligned} f(x) &= x^3 + x^2 - 5x - 5 \\ &= x^2(x + 1) - 5(x + 1) \\ &= (x^2 - 5)(x + 1), \end{aligned}$$

we see that the x-intercepts are $\sqrt{5}$, $-\sqrt{5}$, and -1. The y-intercept is $f(0) = -5$. The points corresponding to the critical numbers are $(-\frac{5}{3}, \frac{40}{27})$ and $(1, -8)$. Plotting these six points and using the information in the table gives us the sketch in Figure 4.19.

We saw in Section 4.1 that if a function has a local extremum, then it must occur at a critical number; however, not every critical number leads to a local extremum (see Example 5 of Section 4.1). To find the local extrema, we begin by locating all the critical numbers of the function. Next, we test each critical number to determine whether or not a local extremum occurs. There are several methods for conducting this test. The following theorem is based on the sign of the first derivative of f. In the statement of the theorem, the terminology *f' changes from positive to negative at c* means that there is an open interval (a, b) containing c such that $f'(x) > 0$ if $a < x < c$ and $f'(x) < 0$ if $c < x < b$. Analogous meanings are applied to the phrases *f' changes from negative to positive at c* and *f' does not change sign at c*.

First derivative test **(4.14)**

Let c be a critical number for f, and suppose f is continuous at c and differentiable on an open interval I containing c, except possibly at c itself.

(i) If f' changes from positive to negative at c, then $f(c)$ is a local maximum of f.

(ii) If f' changes from negative to positive at c, then $f(c)$ is a local minimum of f.

(iii) If $f'(x) > 0$ or if $f'(x) < 0$ for every x in I except $x = c$, then $f(c)$ is not a local extremum of f.

PROOF If f' changes from positive to negative at c, then there is an open interval (a, b) containing c such that $f'(x) > 0$ for $a < x < c$ and $f'(x) < 0$ for $c < x < b$. Furthermore, we may choose (a, b) such that f is continuous on $[a, b]$. Hence, by Theorem (4.13), f is increasing on $[a, c]$ and decreasing on $[c, b]$. Thus, $f(x) < f(c)$ for every x in (a, b) different from c; that is, $f(c)$ is a local maximum for f. This proves (i). Similar proofs may be given for (ii) and (iii). ■

To remember the first derivative test, think of the graphs in Figures 4.20 and 4.21, on the next page. For a local maximum, the slope $f'(x)$ of the tangent line at $P(x, f(x))$ is positive if $x < c$ and negative if $x > c$. For a local minimum, the opposite situation occurs. Figure 4.22 illustrates (iii) of (4.14).

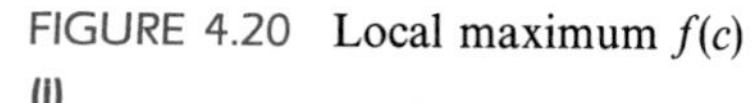

FIGURE 4.20 Local maximum $f(c)$

(i)

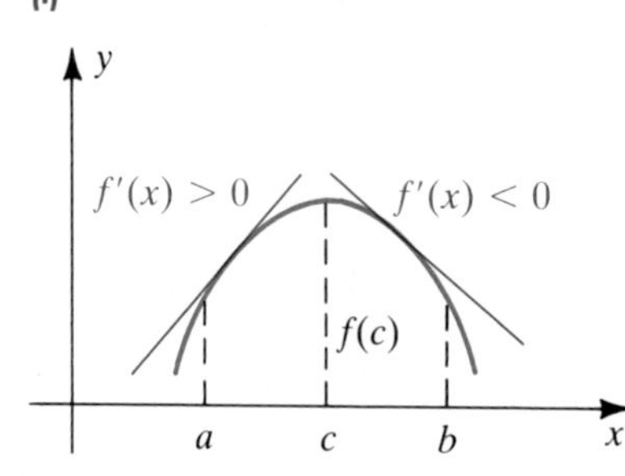

(ii)

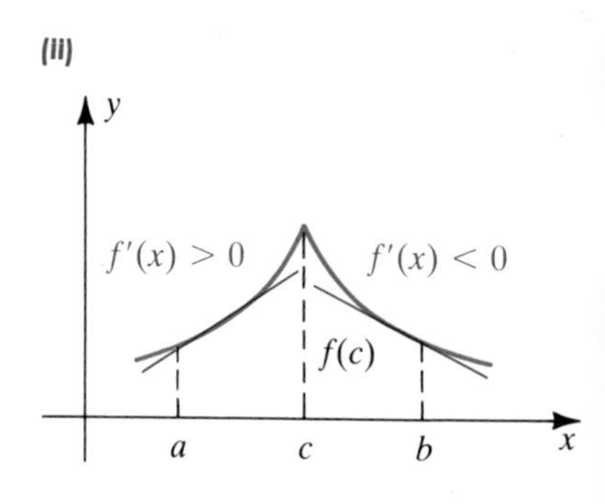

FIGURE 4.21 Local minimum $f(c)$

(i)

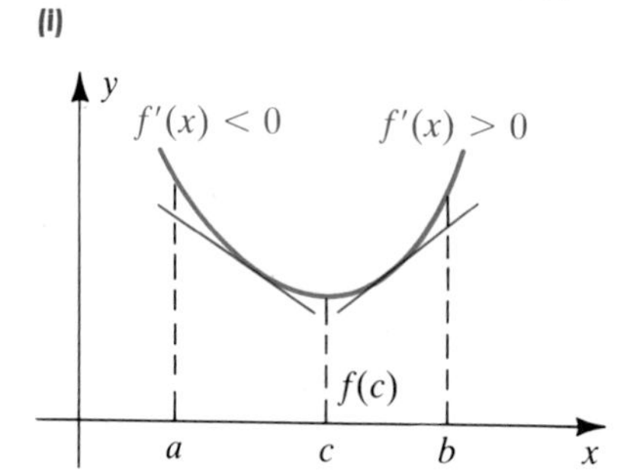

(ii)

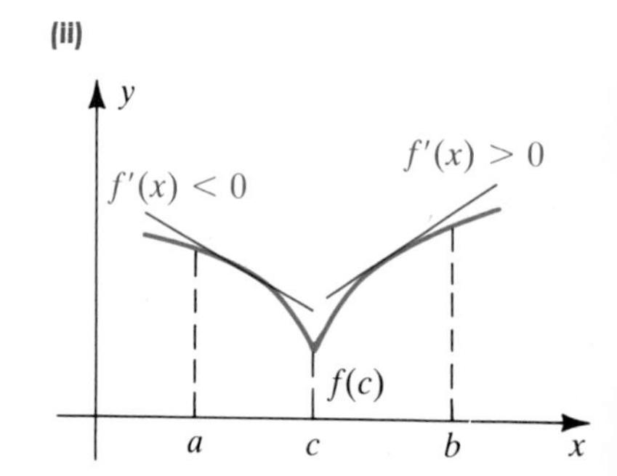

FIGURE 4.22 $f(c)$ is not a local extremum

(i)

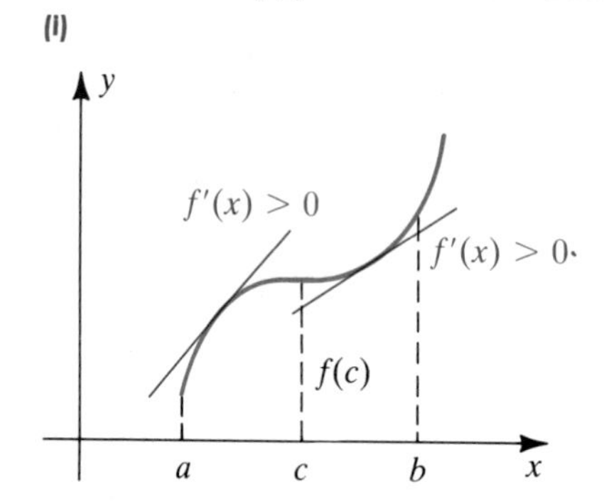

(ii)

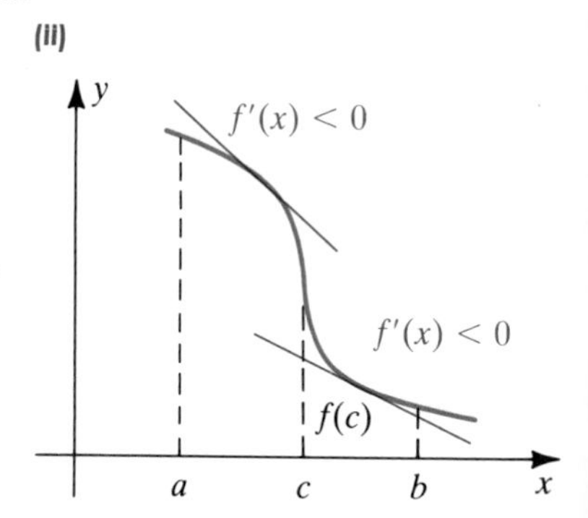

EXAMPLE 2 If $f(x) = x^3 + x^2 - 5x - 5$, find the local extrema of f.

SOLUTION This is the function considered in Example 1. The critical numbers are $-\frac{5}{3}$ and 1. We see from the table in Example 1 that the sign of $f'(x)$ changes from positive to negative as x increases through $-\frac{5}{3}$. Hence, by the first derivative test, f has a local maximum at $-\frac{5}{3}$. This maximum value is $f(-\frac{5}{3}) = \frac{40}{27}$ (see Figure 4.19).

A local minimum occurs at 1, since the sign of $f'(x)$ changes from negative to positive as x increases through 1. This minimum value is $f(1) = -8$.

EXAMPLE 3 If $f(x) = x^{1/3}(8 - x)$, find the local extrema of f, and sketch the graph of f.

SOLUTION By the product rule,

$$f'(x) = x^{1/3}(-1) + (8 - x)\tfrac{1}{3}x^{-2/3}$$
$$= \frac{-3x + (8 - x)}{3x^{2/3}} = \frac{4(2 - x)}{3x^{2/3}}.$$

Hence the critical numbers of f are 0 and 2. As in Example 1, this suggests that we consider the sign of $f'(x)$ in each of the intervals $(-\infty, 0)$, $(0, 2)$, and $(2, \infty)$. Since f' is continuous and has no zeros on each interval, we may determine the sign of $f'(x)$ by using a suitable test value $f'(k)$. It is unnecessary to actually evaluate $f'(k)$; all we need to know is its sign. Thus, if we choose $k = 3$ in $(2, \infty)$, then

$$f'(3) = \frac{4(2 - 3)}{3(3^{2/3})}$$

and we can tell, *without evaluating*, that the numerator is negative and the denominator is positive. Hence $f'(3) < 0$, as shown in the following table.

Interval	$(-\infty, 0)$	$(0, 2)$	$(2, \infty)$
$\boldsymbol{k}$	-1	1	3
Test value $f'(k)$	$f'(-1) > 0$	$f'(1) > 0$	$f'(3) < 0$
Sign of $f'(x)$	$+$	$+$	$-$
Conclusion	f is increasing on $(-\infty, 0]$	f is increasing on $[0, 2]$	f is decreasing on $[2, \infty)$

FIGURE 4.23

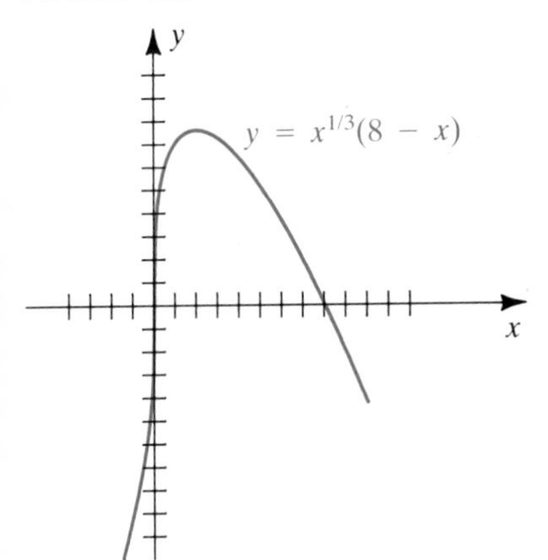

By the first derivative test, f has a local maximum at 2, since f' changes from positive to negative at 2. Thus, we have

$$\text{Local max:} \quad f(2) = 2^{1/3}(8 - 2) = 6\sqrt[3]{2} \approx 7.6.$$

The function does not have an extremum at 0, since f' does not change sign at 0.

To sketch the graph, we first plot points corresponding to the critical numbers. From $f(x) = x^{1/3}(8 - x)$ we see that the x-intercepts of the graph are 0 and 8. The graph is sketched in Figure 4.23. Note that

$$\lim_{x \to 0} f'(x) = \infty.$$

Since f is continuous at $x = 0$, the graph has a vertical tangent line at $(0, 0)$ by Definition (3.9).

EXAMPLE 4 If $f(x) = x^{2/3}(x^2 - 8)$, find the local extrema, and sketch the graph of f.

SOLUTION Applying the product rule, we obtain

$$f'(x) = x^{2/3}(2x) + (x^2 - 8)(\tfrac{2}{3}x^{-1/3}) = \frac{6x^2 + 2(x^2 - 8)}{3x^{1/3}} = \frac{8(x^2 - 2)}{3x^{1/3}}.$$

The critical numbers are the solutions of $x^2 - 2 = 0$ and $x^{1/3} = 0$—that is, $-\sqrt{2}$, 0, and $\sqrt{2}$. This suggests that we find the sign of $f'(x)$ in each of the intervals $(-\infty, -\sqrt{2})$, $(-\sqrt{2}, 0)$, $(0, \sqrt{2})$, and $(\sqrt{2}, \infty)$. Arranging our work in tabular form as in previous examples, we obtain the following.

Interval	$(-\infty, -\sqrt{2})$	$(-\sqrt{2}, 0)$	$(0, \sqrt{2})$	$(\sqrt{2}, \infty)$
k	-2	-1	1	2
Test value $f'(k)$	$f'(-2) < 0$	$f'(-1) > 0$	$f'(1) < 0$	$f'(2) > 0$
Sign of $f'(x)$	$-$	$+$	$-$	$+$
Conclusion	f is decreasing on $(-\infty, -\sqrt{2}]$	f is increasing on $[-\sqrt{2}, 0]$	f is decreasing on $[0, \sqrt{2}]$	f is increasing on $[\sqrt{2}, \infty)$

FIGURE 4.24

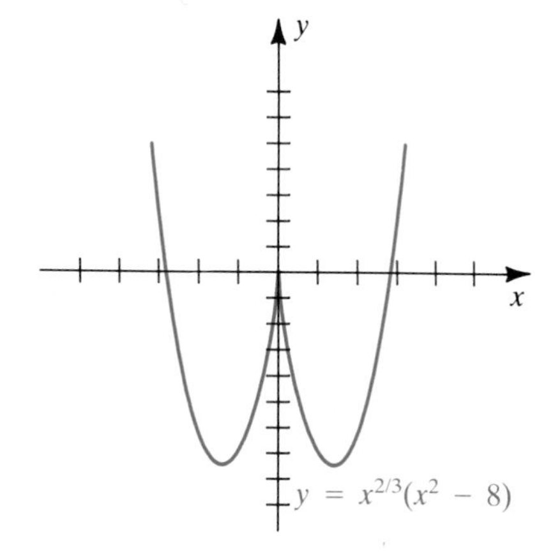

By the first derivative test, f has local minima at $-\sqrt{2}$ and $\sqrt{2}$ and a local maximum at 0. The corresponding function values give us the following results.

$$\text{Local max:} \quad f(0) = 0$$

$$\text{Local min:} \quad f(\sqrt{2}) = f(-\sqrt{2}) = -6\sqrt[3]{2} \approx -7.6$$

Note that $f'(0)$ does not exist. Since

$$\lim_{x \to 0^-} f'(x) = \infty \quad \text{and} \quad \lim_{x \to 0^+} f'(x) = -\infty$$

and since f is continuous at $x = 0$, the graph has a cusp at (0, 0), by Definition (3.10).

The graph of f is sketched in Figure 4.24.

Recall from (4.9) that if a function f is continuous on a closed interval $[a, b]$ and we wish to determine the maximum and minimum values, then in addition to finding all the local extrema, we should calculate the values $f(a)$ and $f(b)$ of f at the endpoints a and b of the interval $[a, b]$. The largest number among the local extrema and the values $f(a)$ and $f(b)$ is the maximum of f on $[a, b]$. The smallest of these numbers is the minimum of f on $[a, b]$. To illustrate these remarks, we shall consider the function discussed in the previous example and restrict our attention to certain intervals.

EXAMPLE 5 If $f(x) = x^{2/3}(x^2 - 8)$, find the maximum and minimum values of f on each of the following intervals:

(a) $[-1, \frac{1}{2}]$ **(b)** $[-1, 3]$ **(c)** $[-3, -2]$

SOLUTION The graph in Figure 4.24 indicates the local extrema and the intervals on which f is increasing or decreasing. Figure 4.25 illustrates the part of the graph of f that corresponds to each of the intervals (a), (b), and (c).

FIGURE 4.25

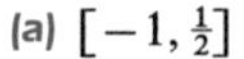

(a) $[-1, \frac{1}{2}]$

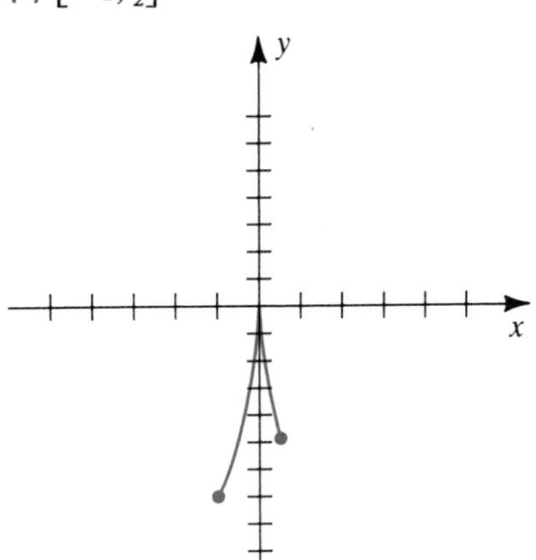

(b) $[-1, 3]$

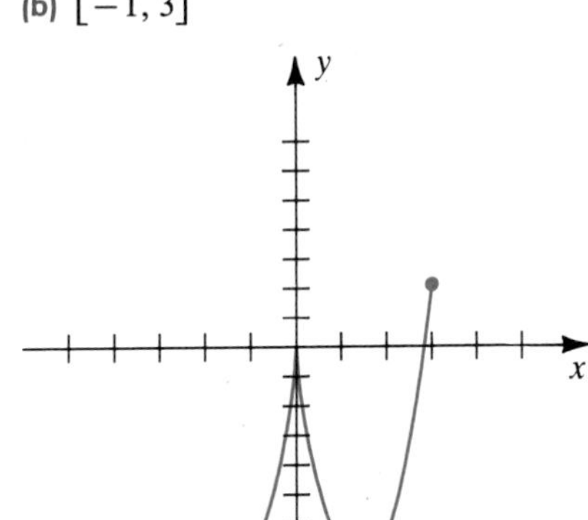

(c) $[-3, -2]$

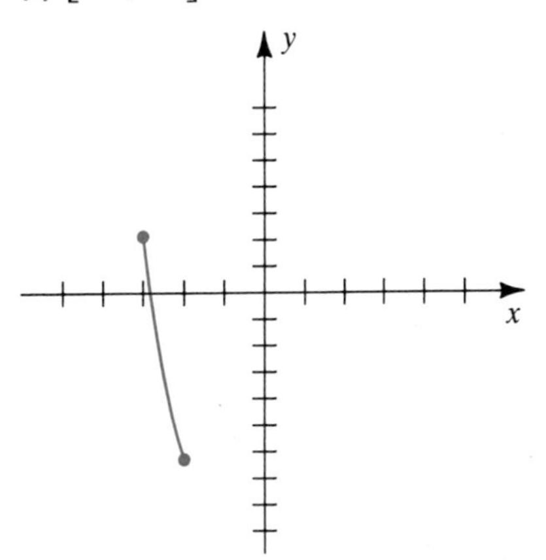

Referring to these sketches, we obtain the following table (check each entry).

Interval	Minimum value	Maximum value
$[-1, \frac{1}{2}]$	$f(-1) = -7$	$f(0) = 0$
$[-1, 3]$	$f(\sqrt{2}) = -6\sqrt[3]{2}$	$f(3) = \sqrt[3]{9}$
$[-3, -2]$	$f(-2) = -4\sqrt[3]{4}$	$f(-3) = \sqrt[3]{9}$

Note that on some intervals the maximum or minimum value of f is also a local extremum; on other intervals this is not the case.

FIGURE 4.26

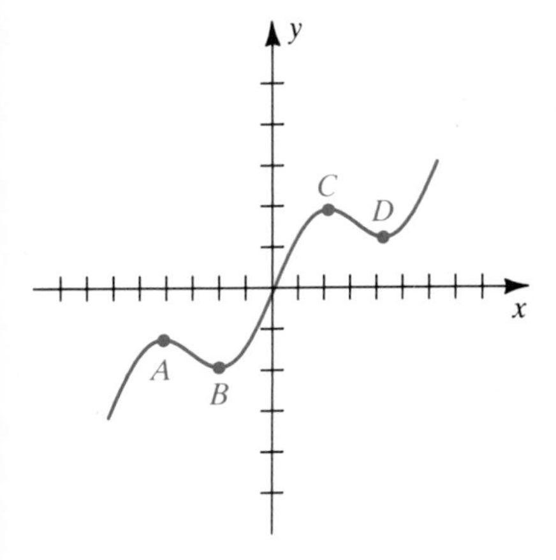

EXAMPLE 6 The graph of $y = \frac{1}{2}x + \sin x$ for $-2\pi \leq x \leq 2\pi$ is sketched in Figure 4.26. Determine the coordinates of the points A, B, C, and D, which correspond to local extrema.

SOLUTION Letting $f(x) = \frac{1}{2}x + \sin x$ and differentiating, we obtain

$$f'(x) = \tfrac{1}{2} + \cos x.$$

The local extrema occur if $f'(x) = 0$—that is, if

$$\tfrac{1}{2} + \cos x = 0, \quad \text{or} \quad \cos x = -\tfrac{1}{2}.$$

The solutions of the last equation in $[0, 2\pi]$ are $2\pi/3$ and $4\pi/3$. These are the x-coordinates of C and D. By symmetry (with respect to the origin), the x-coordinates of A and B are $-4\pi/3$ and $-2\pi/3$.

We could now apply the first derivative test; however, it is evident from the graph that points A and C correspond to local maxima and points B and D correspond to local minima. The following table lists the coordinates of these points.

Point	x-coordinate	y-coordinate
A	$-\frac{4\pi}{3} \approx -4.2$	$-\frac{2\pi}{3} + \frac{\sqrt{3}}{2} \approx -1.2$
B	$-\frac{2\pi}{3} \approx -2.1$	$-\frac{\pi}{3} + \left(-\frac{\sqrt{3}}{2}\right) \approx -1.9$
C	$\frac{2\pi}{3} \approx 2.1$	$\frac{\pi}{3} + \frac{\sqrt{3}}{2} \approx 1.9$
D	$\frac{4\pi}{3} \approx 4.2$	$\frac{2\pi}{3} + \left(-\frac{\sqrt{3}}{2}\right) \approx 1.2$

EXERCISES 4.3

Exer. 1–16: Find the local extrema of f and the intervals on which f is increasing or is decreasing, and sketch the graph of f.

1 $f(x) = 5 - 7x - 4x^2$

2 $f(x) = 6x^2 - 9x + 5$

3 $f(x) = 2x^3 + x^2 - 20x + 1$

4 $f(x) = x^3 - x^2 - 40x + 8$

5 $f(x) = x^4 - 8x^2 + 1$

6 $f(x) = 4x^3 - 3x^4$

7 $f(x) = 10x^3(x - 1)^2$

8 $f(x) = (x^2 - 10x)^4$

9 $f(x) = x^{4/3} + 4x^{1/3}$

10 $f(x) = x(x - 5)^{1/3}$

11 $f(x) = x^{2/3}(x - 7)^2 + 2$

12 $f(x) = x^{2/3}(8 - x)$

13 $f(x) = x^2\sqrt[3]{x^2 - 4}$

14 $f(x) = 8 - \sqrt[3]{x^2 - 2x + 1}$

15 $f(x) = x\sqrt{x^2 - 9}$

16 $f(x) = x\sqrt{4 - x^2}$

Exer. 17–22: Find the local extrema of f on $[0, 2\pi]$ and the subintervals on which f is increasing or is decreasing. Sketch the graph of f.

17 $f(x) = \cos x + \sin x$

18 $f(x) = \cos x - \sin x$

19 $f(x) = \frac{1}{2}x - \sin x$

20 $f(x) = x + 2\cos x$

21 $f(x) = 2\cos x + \sin 2x$

22 $f(x) = 2\cos x + \cos 2x$

Exer. 23–28: Find the local extrema of f.

23 $f(x) = \sqrt[3]{x^3 - 9x}$

24 $f(x) = \sqrt{x^2 + 4}$

25 $f(x) = (x - 2)^3(x + 1)^4$

26 $f(x) = x^2(x - 5)^4$

27 $f(x) = \dfrac{\sqrt{x - 3}}{x^2}$

28 $f(x) = \dfrac{x^2}{\sqrt{x + 7}}$

Exer. 29–32: Find the local extrema of f on the given interval.

29 $f(x) = \sec \frac{1}{2}x$; $[-\pi/2, \pi/2]$

30 $f(x) = \cot^2 x + 2\cot x$; $[\pi/6, 5\pi/6]$

31 $f(x) = 2\tan x - \tan^2 x$; $[-\pi/3, \pi/3]$

32 $f(x) = \tan x - 2\sec x$; $[-\pi/4, \pi/4]$

Exer. 33–34: Shown in the figure is a graph of the equation for $-2\pi \le x \le 2\pi$. Determine the x-coordinates of the points that correspond to local extrema.

33 $y = \frac{1}{2}x + \cos x$

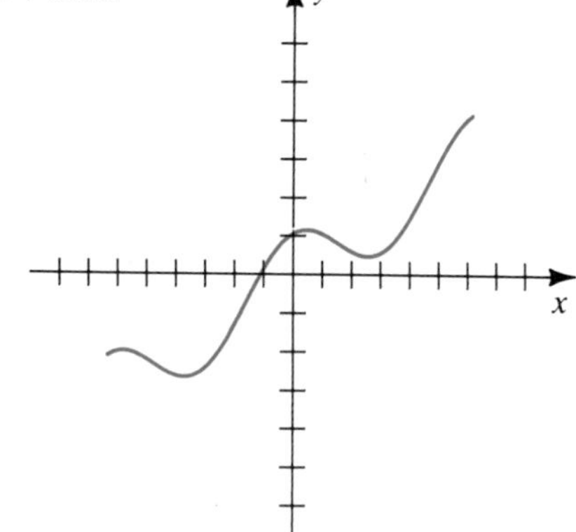

34 $y = \dfrac{\sqrt{3}}{2}x - \sin x$

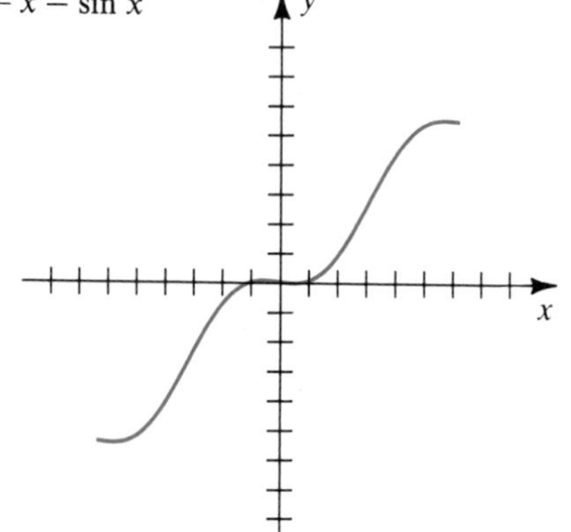

Exer. 35–40: Sketch the graph of a differentiable function f that satisfies the given conditions.

35 $f(0) = 3$; $f(-2) = f(2) = -4$; $f'(0)$ is undefined; $f'(-2) = f'(2) = 0$; $f'(x) > 0$ if $-2 < x < 0$ or $x > 2$; $f'(x) < 0$ if $x < -2$ or $0 < x < 2$

36 $f(3) = 5$; $f(5) = 0$; $f'(5)$ is undefined; $f'(3) = 0$; $f'(x) > 0$ if $x < 3$ or $x > 5$; $f'(x) < 0$ if $3 < x < 5$

37 $f(3) = 5$; $f(5) = 0$; $f'(3) = f'(5) = 0$; $f'(x) > 0$ if $x < 3$ or $x > 5$; $f'(x) < 0$ if $3 < x < 5$

38 $f(0) = 3$; $f(-2) = f(2) = -4$; $f'(-2) = f'(0) = f'(2) = 0$; $f'(x) > 0$ if $-2 < x < 0$ or $x > 2$; $f'(x) < 0$ if $x < -2$ or $0 < x < 2$

39 $f(-5) = 4$; $f(0) = 0$; $f(5) = -4$; $f'(-5) = f'(0) = f'(5) = 0$; $f'(x) > 0$ if $|x| > 5$; $f'(x) < 0$ if $0 < |x| < 5$

40 $f(a) = a$ and $f'(a) = 0$ for $a = 0, \pm 1, \pm 2, \pm 3$; $f'(x) > 0$ for all other values of x

c **Exer. 41–42: Graph f on $[-2, 2]$. (a) Use the graph to estimate the local extrema of f. (b) Estimate where f is increasing or is decreasing.**

41 $f(x) = \dfrac{x^2 - 1.5x + 2.1}{0.3x^4 + 2.3x + 2.7}$

42 $f(x) = \dfrac{10 \cos 2x}{x^2 + 4}$

c **Exer. 43–44: Graph f' on $[-1, 1]$. Estimate the x-coordinates of the local extrema of f and classify each local extremum.**

43 $f'(x) = x - \cos \pi x - \sin x$

44 $f'(x) = 6x^3 - 3x^2 - 1.3x + 0.5$

4.4 CONCAVITY AND THE SECOND DERIVATIVE TEST

In the preceding section we used the sign of the derivative f' to determine where a function f was increasing and where it was decreasing. Similarly, we can use the sign of the *second* derivative f'' to determine where the *derivative* f' is increasing and where it is decreasing. Specifically, if $f''(x) > 0$ on an open interval I, then $f'(x)$ is increasing on I; that is, the slope of the tangent line to the graph of f increases as x increases (see Figure 4.27). As the next definition states, the graph is *concave upward* on I.

FIGURE 4.27 Concave upward graph

FIGURE 4.28 Concave downward graph

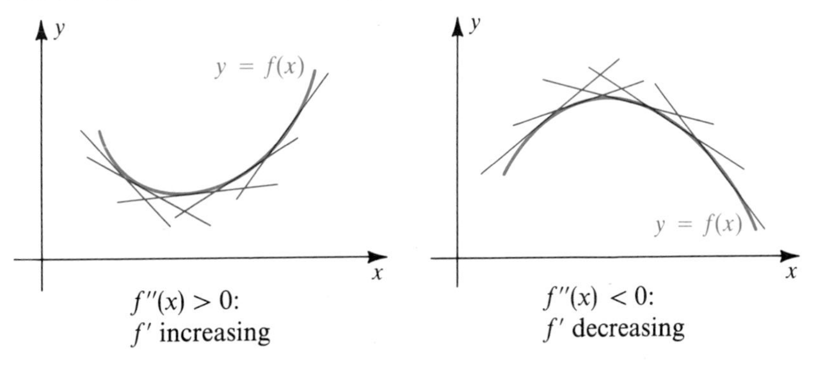

If $f''(x) < 0$ on an open interval I, then $f'(x)$ is decreasing on I. In this case, the slope of the tangent line to the graph of f *decreases* as x increases (see Figure 4.28), and we say that the graph is *concave downward* on I.

Definition (4.15)

Let f be differentiable on an open interval I. The graph of f is

(i) **concave upward** on I if f' is increasing on I

(ii) **concave downward** on I if f' is decreasing on I

It can be proved that for upward concavity on an interval, the graph lies *above* every tangent line to the graph (see Figure 4.27) and for downward concavity, the graph lies *below* every tangent line (see Figure 4.28).

The next theorem is a restatement, in terms of concavity, of our discussion about the sign of $f''(x)$.

Test for concavity **(4.16)**

> If the second derivative f'' of f exists on an open interval I, then the graph of f is
>
> **(i)** **concave upward** on I if $f''(x) > 0$ on I
>
> **(ii)** **concave downward** on I if $f''(x) < 0$ on I

EXAMPLE 1 If $f(x) = x^3 + x^2 - 5x - 5$, determine intervals on which the graph of f is concave upward or is concave downward, and illustrate the results graphically.

SOLUTION The function f was considered in Examples 1 and 2 of the preceding section and is resketched in Figure 4.29. Since $f'(x) = 3x^2 + 2x - 5$,

$$f''(x) = 6x + 2 = 2(3x + 1).$$

Hence $f''(x) < 0$ if $3x + 1 < 0$—that is, if $x < -\frac{1}{3}$. Similarly, $f''(x) > 0$ if $x > -\frac{1}{3}$.

Applying the test for concavity (4.16) gives us the following.

Interval	$(-\infty, -\frac{1}{3})$	$(-\frac{1}{3}, \infty)$
Sign of $f''(x)$	$-$	$+$
Concavity	downward	upward

These facts are illustrated in Figure 4.29, where P is the point with x-coordinate $-\frac{1}{3}$.

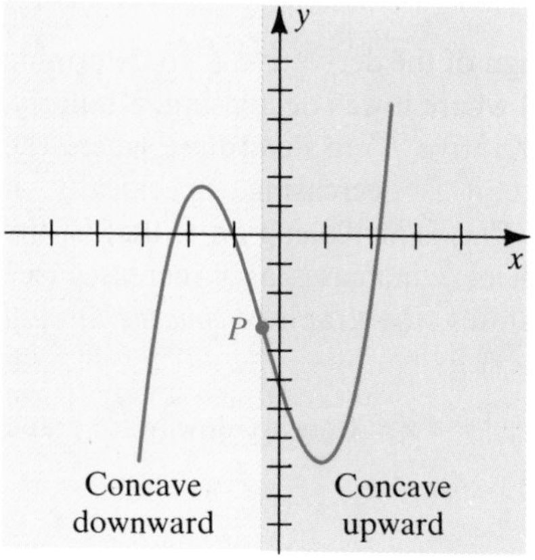

FIGURE 4.29

$y = x^3 + x^2 - 5x - 5$

EXAMPLE 2 If $f(x) = \sin x$, determine where the graph of f is concave upward and where it is concave downward, and illustrate the results graphically.

SOLUTION Differentiating $f(x)$ twice, we obtain

$$f'(x) = \cos x, \quad f''(x) = -\sin x.$$

Since $f''(x) = -f(x)$, we see that $f''(x) < 0$ whenever $f(x) > 0$; hence, by the test for concavity (4.16), the graph is concave downward whenever it lies above the x-axis. Similarly, $f''(x) > 0$ whenever $f(x) < 0$, so the graph is concave upward whenever it lies below the x-axis. These facts are partially illustrated in Figure 4.30, in which the abbreviations CD and CU are used for concave downward and concave upward, respectively.

FIGURE 4.30

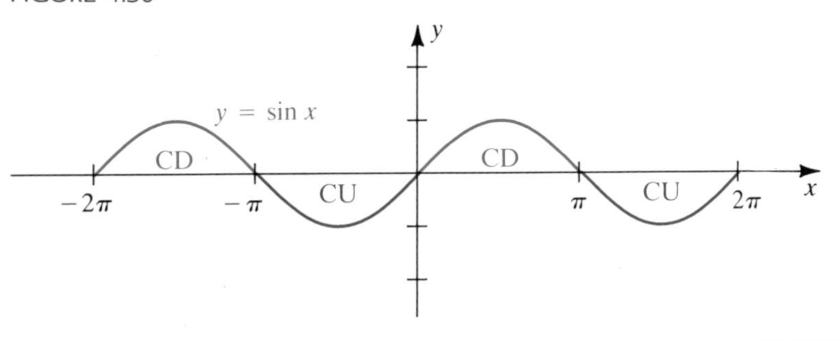

Each point on the graph of f at which the concavity changes from upward to downward, or vice versa, is called a *point of inflection (P.I.)*. The precise definition may be stated as follows.

Definition **(4.17)**

> A point $(c, f(c))$ on the graph of f is a **point of inflection** if the following two conditions are satisfied:
>
> **(i)** f is continuous at c.
>
> **(ii)** There is an open interval (a, b) containing c such that the graph is concave upward on (a, c) and concave downward on (c, b), or vice versa.

Statement (ii) of Definition (4.17) is sometimes abbreviated by stating that *the concavity changes* at $P(c, f(c))$.

The point P in Figure 4.29 is a point of inflection. In Figure 4.30, every point $P(\pi n, 0)$, where n is an integer, is a point of inflection. The sketch in Figure 4.31 displays typical points of inflection on the graph of a function. Observe that a corner or cusp may or may not be a point of inflection.

FIGURE 4.31

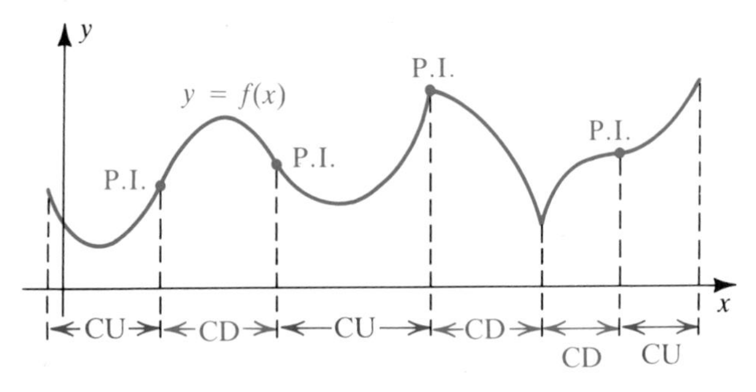

By the test for concavity (4.16), if the second derivative f'' changes sign at a number c, then the point $P(c, f(c))$ is a point of inflection. If, in addition, f'' is continuous at c, then we must have $f''(c) = 0$. However, it is also possible that $f''(c)$ may not exist at a point of inflection. Thus, *to find points of inflection, we begin by finding the zeros of f'' and the numbers at which f'' does not exist*. Each of these numbers is then tested to determine if it is an x-coordinate of a point of inflection.

The following test for local extrema is based on the second derivative.

Second derivative test (4.18)

> Suppose that f is differentiable on an open interval containing c and that $f'(c) = 0$.
> **(i)** If $f''(c) < 0$, then f has a local maximum at c.
> **(ii)** If $f''(c) > 0$, then f has a local minimum at c.

FIGURE 4.32
Local maximum $f(c)$

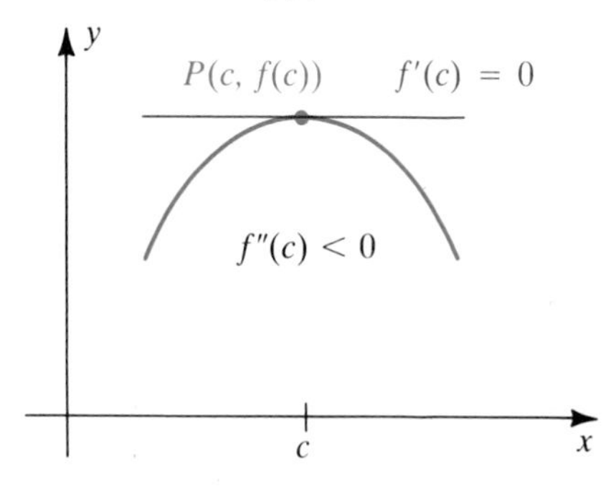

PROOF If $f'(c) = 0$, then the tangent line to the graph at $P(c, f(c))$ is horizontal. If, in addition, $f''(c) < 0$, then the graph is concave downward at c, and hence there is an interval (a, b) containing c such that the graph lies below the tangent lines. It follows that $f(c)$ is a local maximum for f, as illustrated in Figure 4.32. This proves (i). A similar proof may be given for (ii), which is illustrated in Figure 4.33. ■

If $f''(c) = 0$, the second derivative test is not applicable. In such cases the first derivative test should be employed.

FIGURE 4.33
Local minimum $f(c)$

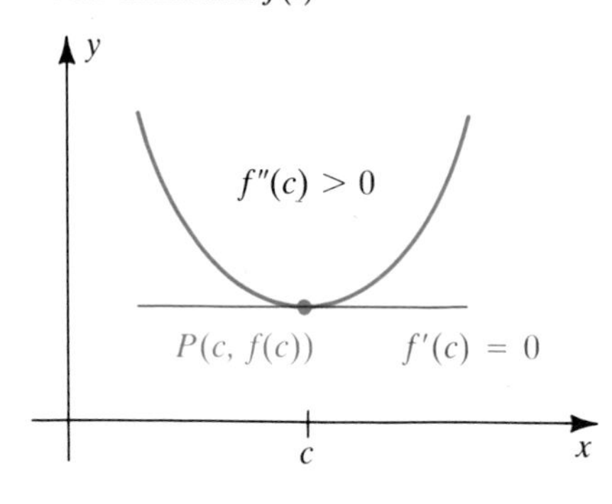

EXAMPLE 3 If $f(x) = 12 + 2x^2 - x^4$, use the second derivative test to find the local extrema of f. Discuss concavity, find the points of inflection, and sketch the graph of f.

SOLUTION Differentiating $f(x)$ twice yields

$$f'(x) = 4x - 4x^3 = 4x(1 - x^2)$$
$$f''(x) = 4 - 12x^2 = 4(1 - 3x^2).$$

The expression for $f'(x)$ is used to find the critical numbers 0, 1, and -1. The values of f'' at these numbers are

$$f''(0) = 4 > 0, \quad f''(1) = -8 < 0, \quad \text{and} \quad f''(-1) = -8 < 0.$$

Hence, by the second derivative test, the function has a local minimum at 0 and local maxima at 1 and -1. The corresponding function values are $f(0) = 12$ and $f(1) = 13 = f(-1)$. The following table summarizes our discussion.

Critical number c	$f''(c)$	Sign of $f''(c)$	Conclusion
-1	-8	$-$	Local max: $f(-1) = 13$
0	4	+	Local min: $f(0) = 12$
1	-8	$-$	Local max: $f(1) = 13$

To locate the possible points of inflection, we solve the equation $f''(x) = 0$ (that is, $4(1 - 3x^2) = 0$), obtaining the solutions $-\sqrt{3}/3$ and $\sqrt{3}/3$. We next examine the sign of $f''(x)$ in each of the intervals

$$(-\infty, -\sqrt{3}/3), \quad (-\sqrt{3}/3, \sqrt{3}/3), \quad \text{and} \quad (\sqrt{3}/3, \infty).$$

Since f'' is continuous and has no zeros on each interval, we may use test values to determine the sign of $f''(x)$. Let us arrange our work in tabular form as follows. The last row is a consequence of (4.16).

Interval	$(-\infty, -\sqrt{3}/3)$	$(-\sqrt{3}/3, \sqrt{3}/3)$	$(\sqrt{3}/3, \infty)$
k	-1	0	1
Test value $f''(k)$	$f''(-1) = -8$	$f''(0) = 4$	$f''(1) = -8$
Sign of $f''(x)$	$-$	$+$	$-$
Concavity	downward	upward	downward

FIGURE 4.34

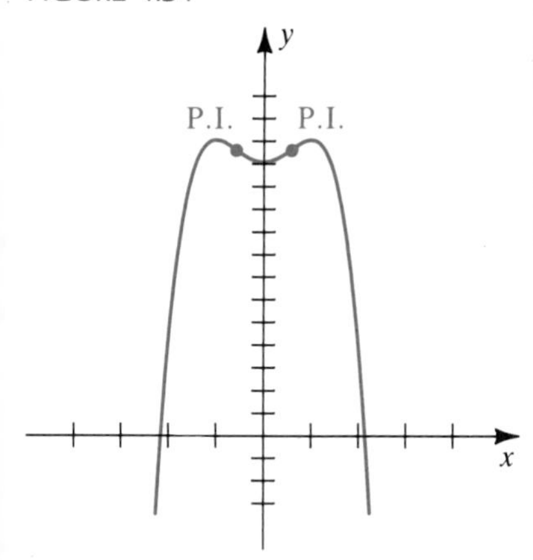

Since $f''(x)$ changes sign at $-\sqrt{3}/3$ and $\sqrt{3}/3$, the corresponding points $(\pm\sqrt{3}/3, 113/9)$ on the graph are points of inflection. These are the points at which the concavity changes. As shown in the table, the graph is concave upward on the open interval $(-\sqrt{3}/3, \sqrt{3}/3)$ and concave downward outside of $[-\sqrt{3}/3, \sqrt{3}/3]$. The graph is sketched in Figure 4.34.

EXAMPLE 4 If $f(x) = x^5 - 5x^3$, find the local extrema of f. Discuss concavity, find the points of inflection, and sketch the graph of f.

SOLUTION We begin by differentiating $f(x)$ twice:

$$f'(x) = 5x^4 - 15x^2 = 5x^2(x^2 - 3)$$

$$f''(x) = 20x^3 - 30x = 10x(2x^2 - 3)$$

Solving the equation $f'(x) = 0$ gives us the critical numbers 0, $-\sqrt{3}$, and $\sqrt{3}$. As in Example 3, we obtain the following table (check each entry).

Critical number c	$f''(c)$	**Sign of $f''(c)$**	**Conclusion**
$-\sqrt{3}$	$-30\sqrt{3}$	$-$	Local max: $f(-\sqrt{3}) = 6\sqrt{3}$
0	0	none	No conclusion
$\sqrt{3}$	$30\sqrt{3}$	$+$	Local min: $f(\sqrt{3}) = -6\sqrt{3}$

Since $f''(0) = 0$, the second derivative test is not applicable at 0, and so we apply the first derivative test. We can show, using test values, that if $-\sqrt{3} < x < 0$, then $f'(x) < 0$, and if $0 < x < \sqrt{3}$, then $f'(x) < 0$. Since $f'(x)$ does not change sign, there is no extremum at $x = 0$.

To find possible points of inflection, we consider the equation $f''(x) = 0$—that is, $10x(2x^2 - 3) = 0$. The solutions of this equation, in order of magnitude, are $-\sqrt{6}/2$, 0, and $\sqrt{6}/2$. As in Example 3, we construct a table.

Interval	$(-\infty, -\sqrt{6}/2)$	$(-\sqrt{6}/2, 0)$	$(0, \sqrt{6}/2)$	$(\sqrt{6}/2, \infty)$
k	-2	-1	1	2
Test value $f''(k)$	-100	10	-10	100
Sign of $f''(x)$	$-$	$+$	$-$	$+$
Concavity	downward	upward	downward	upward

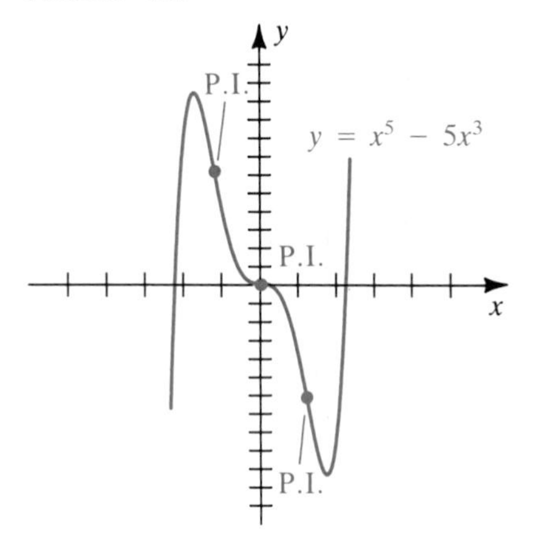

FIGURE 4.35

The sign of $f''(x)$ changes at $-\sqrt{6}/2$, 0, and $\sqrt{6}/2$, so it follows that the points $(0, 0)$, $(-\sqrt{6}/2, 21\sqrt{6}/8)$, and $(\sqrt{6}/2, -21\sqrt{6}/8)$ are points of inflection. The graph is sketched in Figure 4.35, with different scales on the x- and y-axes. The x-intercepts are 0 and $\pm\sqrt{5} \approx 2.2$.

In all of the preceding examples f'' is continuous. It is also possible for $(c, f(c))$ to be a point of inflection if either $f'(c)$ or $f''(c)$ does not exist, as illustrated in the next example.

EXAMPLE 5 If $f(x) = 1 - x^{1/3}$, find the local extrema, discuss concavity, find the points of inflection, and sketch the graph of f.

SOLUTION Differentiating $f(x)$ twice yields

$$f'(x) = -\frac{1}{3}x^{-2/3} = -\frac{1}{3x^{2/3}}$$

$$f''(x) = \frac{2}{9}x^{-5/3} = \frac{2}{9x^{5/3}}.$$

The first derivative does not exist at $x = 0$, and 0 is the only critical number for f. Since $f''(0)$ is undefined, the second derivative test is not applicable. However, if $x \neq 0$, then $x^{2/3} > 0$ and $f'(x) = -1/(3x^{2/3}) < 0$, which means that f is decreasing throughout its domain. Consequently, $f(0)$ is not a local extremum.

Using test values gives us the following table.

Interval	$(-\infty, 0)$	$(0, \infty)$
Sign of $f''(x)$	$-$	$+$
Concavity	downward	upward

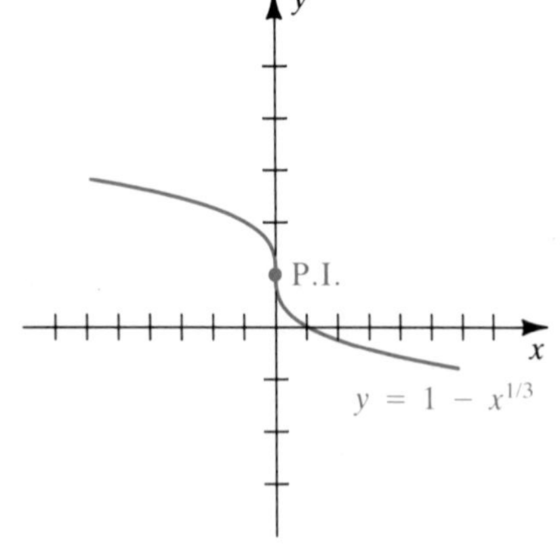

FIGURE 4.36

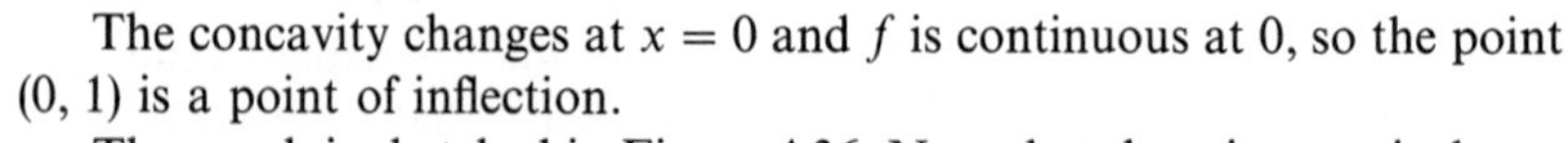
The concavity changes at $x = 0$ and f is continuous at 0, so the point $(0, 1)$ is a point of inflection.

The graph is sketched in Figure 4.36. Note that there is a vertical tangent line at $(0, 1)$, since f is continuous at $x = 0$ and $\lim_{x\to 0} |f'(x)| = \infty$.

EXAMPLE 6 If $f(x) = x^{2/3}(5 + x)$, find the local extrema, discuss concavity, find points of inflection, and sketch the graph of f.

SOLUTION Writing $f(x) = 5x^{2/3} + x^{5/3}$ and differentiating twice gives us the following:

$$f'(x) = \frac{10}{3}x^{-1/3} + \frac{5}{3}x^{2/3} = \frac{5}{3}\left(\frac{2 + x}{x^{1/3}}\right)$$

$$f''(x) = -\frac{10}{9}x^{-4/3} + \frac{10}{9}x^{-1/3} = \frac{10}{9}\left(\frac{x - 1}{x^{4/3}}\right)$$

Referring to $f'(x)$ we see that the critical numbers for f are -2 and 0. We apply the second derivative test, as indicated in the following table.

Critical number c	-2	0
Sign of $f''(c)$	$-$	none
Conclusion	Local max: $f(-2) = (-2)^{2/3}(3) \approx 4.8$	No conclusion

Since the second derivative test is not applicable at $c = 0$, let us apply the first derivative test. Using test values, we see that the sign of $f'(x)$ changes from $-$ to $+$ at $c = 0$. Hence f has a local minimum at (0, 0).

To determine concavity, first we note that $f''(x) = 0$ at $x = 1$ and $f''(x)$ does not exist at $x = 0$. Next we examine the sign of $f''(x)$ for the cases $x < 0$, $0 < x < 1$, and $x > 1$. Using test values for f'' leads to the following table.

FIGURE 4.37

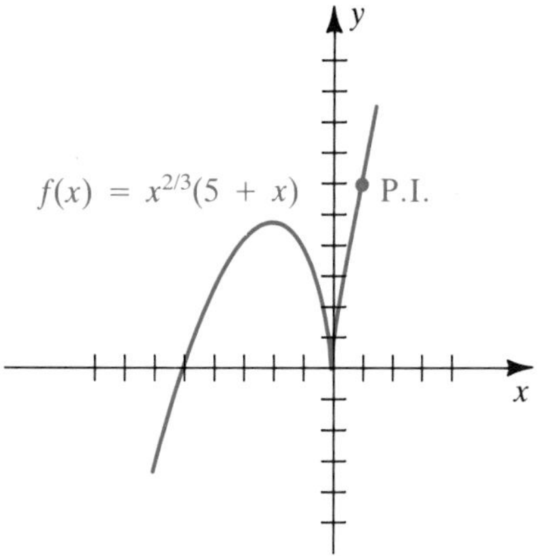

$f(x) = x^{2/3}(5 + x)$ P.I.

Interval	$(-\infty, 0)$	$(0, 1)$	$(1, \infty)$
Sign of $f''(x)$	$-$	$-$	$+$
Concavity	downward	downward	upward

We see from the table that the graph of f has a point of inflection at (1, 6), but not at (0, 0). The graph is sketched in Figure 4.37. Note that

$$\lim_{x \to 0^-} f'(x) = -\infty \quad \text{and} \quad \lim_{x \to 0^+} f'(x) = \infty.$$

Since f is continuous at $x = 0$, the graph has a cusp at (0, 0) by Definition (3.10).

EXAMPLE 7 If $f(x) = 2 \sin x + \cos 2x$, find the local extrema and sketch the graph of f on the interval $[0, 2\pi]$.

SOLUTION We differentiate $f(x)$ twice:

$$f'(x) = 2 \cos x - 2 \sin 2x$$

$$f''(x) = -2 \sin x - 4 \cos 2x$$

FIGURE 4.38

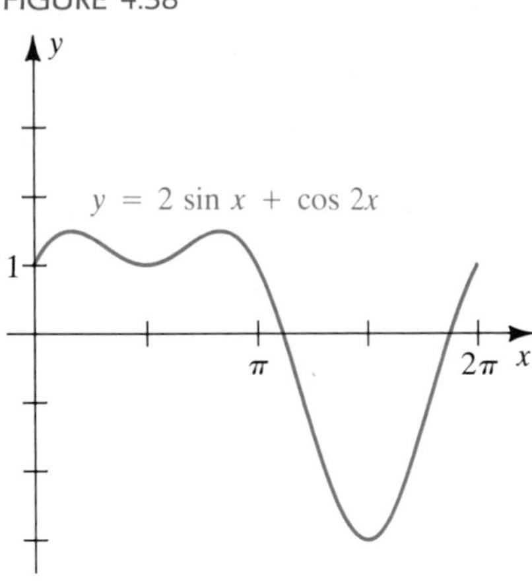

In Example 7 of Section 4.1 we found that the critical numbers of f in the interval $[0, 2\pi]$ are $\pi/6$, $5\pi/6$, $\pi/2$, and $3\pi/2$. Substituting these critical numbers for x in $f''(x)$, we obtain

$$f''(\pi/6) = -3, \quad f''(5\pi/6) = -3, \quad f''(\pi/2) = 2, \quad f''(3\pi/2) = 6.$$

Applying the second derivative test, we see that there are local maxima at $\pi/6$ and $5\pi/6$ and local mimima at $\pi/2$ and $3\pi/2$. Thus, we have

$$\text{Local max:} \quad f(\pi/6) = 3/2 \quad \text{and} \quad f(5\pi/6) = 3/2$$

$$\text{Local min:} \quad f(\pi/2) = 1 \quad \text{and} \quad f(3\pi/2) = -3$$

Using this information and plotting several more points gives us the sketch in Figure 4.38.

EXERCISES 4.4

Exer. 1–18: Find the local extrema of f, using the second derivative test whenever applicable. Find the intervals on which the graph of f is concave upward or is concave downward, and find the x-coordinates of the points of inflection. Sketch the graph of f.

1 $f(x) = x^3 - 2x^2 + x + 1$

2 $f(x) = x^3 + 10x^2 + 25x - 50$

3 $f(x) = 3x^4 - 4x^3 + 6$

4 $f(x) = 8x^2 - 2x^4$

5 $f(x) = 2x^6 - 6x^4$

6 $f(x) = 3x^5 - 5x^3$

7 $f(x) = (x^2 - 1)^2$

8 $f(x) = x^4 - 4x^3 + 10$

9 $f(x) = \sqrt[5]{x} - 1$

10 $f(x) = 2 - \sqrt[3]{x^2}$

11 $f(x) = \sqrt[3]{x^2}(3x + 10)$

12 $f(x) = x^{2/3}(1 - x)$

13 $f(x) = x^2(3x - 5)^{1/3}$

14 $f(x) = x\sqrt[3]{3x + 2}$

15 $f(x) = 8x^{1/3} + x^{4/3}$

16 $f(x) = 6x^{1/2} + x^{3/2}$

17 $f(x) = x^2\sqrt{9 - x^2}$

18 $f(x) = x\sqrt{4 - x^2}$

Exer. 19–24: Use the second derivative test to find the local extrema of f on the interval $[0, 2\pi]$. (These exercises are the same as Exercises 17–22 in Section 4.3, for which the method of solution involved the first derivative test.)

19 $f(x) = \cos x + \sin x$

20 $f(x) = \cos x - \sin x$

21 $f(x) = \frac{1}{2}x - \sin x$

22 $f(x) = x + 2\cos x$

23 $f(x) = 2\cos x + \sin 2x$

24 $f(x) = 2\cos x + \cos 2x$

Exer. 25–28: Use the second derivative test to find the local extrema of f on the given interval. (See Exercises 29–32 of Section 4.3.)

25 $f(x) = \sec \frac{1}{2}x$; $[-\pi/2, \pi/2]$

26 $f(x) = \cot^2 x + 2\cot x$; $[\pi/6, 5\pi/6]$

27 $f(x) = 2\tan x - \tan^2 x$; $[-\pi/3, \pi/3]$

28 $f(x) = \tan x - 2\sec x$; $[-\pi/4, \pi/4]$

Exer. 29–30: Shown in the figure is a graph of the equation for $-2\pi \le x \le 2\pi$. (See Exercises 33–34 of Section 4.3.) Use the second derivative test to find the local extrema of f.

29 $y = \frac{1}{2}x + \cos x$

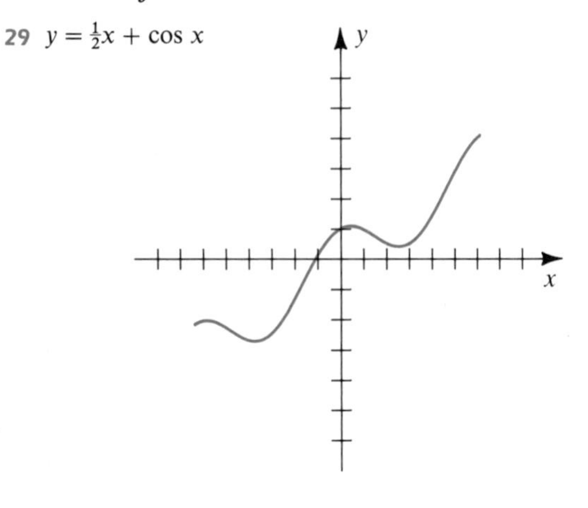

30 $y = \frac{\sqrt{3}}{2}x - \sin x$

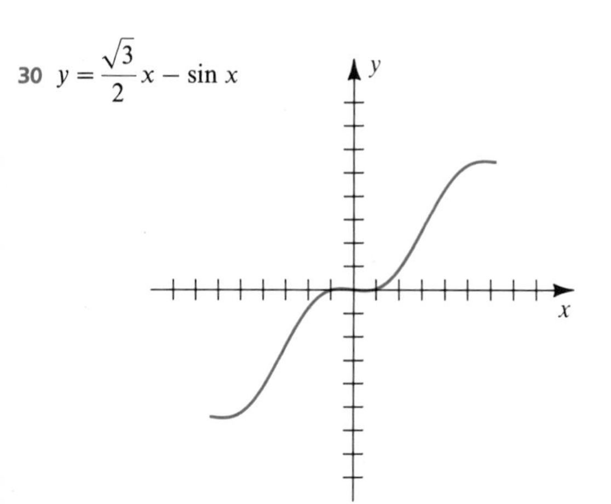

Exer. 31–38: Sketch the graph of a continuous function f that satisfies all of the stated conditions.

31 $f(0) = 1$; $f(2) = 3$; $f'(0) = f'(2) = 0$;
$f'(x) < 0$ if $|x - 1| > 1$; $f'(x) > 0$ if $|x - 1| < 1$;
$f''(x) > 0$ if $x < 1$; $f''(x) < 0$ if $x > 1$

32 $f(0) = 4$; $f(2) = 2$; $f(5) = 6$; $f'(0) = f'(2) = 0$;
$f'(x) > 0$ if $|x - 1| > 1$; $f'(x) < 0$ if $|x - 1| < 1$;
$f''(x) < 0$ if $x < 1$ or if $|x - 4| < 1$;
$f''(x) > 0$ if $|x - 2| < 1$ or if $x > 5$

33 $f(0) = 2$; $f(2) = f(-2) = 1$; $f'(0) = 0$;
$f'(x) > 0$ if $x < 0$; $f'(x) < 0$ if $x > 0$;
$f''(x) < 0$ if $|x| < 2$; $f''(x) > 0$ if $|x| > 2$

34 $f(1) = 4$; $f'(x) > 0$ if $x < 1$; $f'(x) < 0$ if $x > 1$;
$f''(x) > 0$ for every $x \neq 1$

35 $f(-2) = f(6) = -2$; $f(0) = f(4) = 0$; $f(2) = f(8) = 3$;
f' is undefined at 2 and 6; $f'(0) = 1$;
$f'(x) > 0$ throughout $(-\infty, 2)$ and $(6, \infty)$;
$f'(x) < 0$ if $|x - 4| < 2$;
$f''(x) < 0$ throughout $(-\infty, 0)$, $(4, 6)$, and $(6, \infty)$;
$f''(x) > 0$ throughout $(0, 2)$ and $(2, 4)$

36 $f(0) = 2$; $f(2) = 1$; $f(4) = f(10) = 0$; $f(6) = -4$;
$f'(2) = f'(6) = 0$;
$f'(x) < 0$ throughout $(-\infty, 2)$, $(2, 4)$, $(4, 6)$, and $(10, \infty)$;
$f'(x) > 0$ throughout $(6, 10)$;
$f'(4)$ and $f'(10)$ do not exist;
$f''(x) > 0$ throughout $(-\infty, 2)$, $(4, 10)$, and $(10, \infty)$;
$f''(x) < 0$ throughout $(2, 4)$

37 If n is an odd integer, then $f(n) = 1$ and $f'(n) = 0$; if n is an even integer, then $f(n) = 0$ and $f'(n)$ does not exist; if n is any integer, then

(a) $f'(x) > 0$ whenever $2n < x < 2n + 1$

(b) $f'(x) < 0$ whenever $2n - 1 < x < 2n$

(c) $f''(x) < 0$ whenever $2n < x < 2n + 2$

38 $f(x) = x$ if $x = -1, 2, 4$, or 8;
$f'(x) = 0$ if $x = -1, 4, 6$, or 8;
$f'(x) < 0$ throughout $(-\infty, -1)$, $(4, 6)$, and $(8, \infty)$;
$f'(x) > 0$ throughout $(-1, 4)$ and $(6, 8)$;
$f''(x) > 0$ throughout $(-\infty, 0)$, $(2, 3)$, and $(5, 7)$;
$f''(x) < 0$ throughout $(0, 2)$, $(3, 5)$, and $(7, \infty)$

39 Prove that the graph of a quadratic function has no point of inflection. State conditions for which the graph is always

(a) concave upward (b) concave downward

40 Prove that the graph of a polynomial function of degree 3 has exactly one point of inflection.

c **Exer. 41–42: Graph f on $[-1, 1]$. (a) Estimate where the graph of f is concave upward or is concave downward. (b) Estimate the x-coordinate of each point of inflection.**

41 $f(x) = 4x^5 + x^4 + 3x^2 - 2x + 1$

42 $f(x) = (x - 0.1)^2\sqrt{1.08 - 0.9x^2}$

c **Exer. 43–44: Graph f'' on $[0, 3]$. (a) Estimate where the graph of f is concave upward or is concave downward. (b) Estimate the x-coordinate of each point of inflection.**

43 $f''(x) = x^4 - 5x^3 + 7.57x^2 - 3.3x + 0.4356$

44 $f''(x) = 2.1 \sin \pi x + 1.4 \cos x - 0.6$

4.5 SUMMARY OF GRAPHICAL METHODS

Many applications of calculus are based on the behavior of the function values $f(x)$ of a function f as x varies throughout a set of real numbers. One of the best ways to exhibit this behavior is to sketch the graph of f, because a graph displays properties of f that may otherwise be hidden or unclear. Up to this point in the text we have discussed a variety of topics that are useful for sketching graphs. These discussions have been scattered throughout the first four chapters, so we shall summarize this work here and consider several additional examples.

Guidelines for sketching the graph of $y = f(x)$ **(4.19)**

1 **Domain of f** Find the domain of f—that is, all real numbers x such that $f(x)$ is defined.

2 **Continuity of f** Determine whether f is continuous on its domain, and, if not, find and classify the discontinuities.

3 **x- and y-intercepts** The x-intercepts are the solutions of the equation $f(x) = 0$; the y-intercept is the function value $f(0)$, if it exists.

4 **Symmetry** If f is an even function, the graph is symmetric with respect to the y-axis. If f is an odd function, the graph is symmetric with respect to the origin.

5 **Critical numbers and local extrema** Find $f'(x)$ and determine the critical numbers—that is, the values of x such that $f'(x) = 0$ or $f'(x)$ does not exist. Use the first derivative test to help find local extrema. Employ the sign of $f'(x)$ to find intervals on which f is increasing ($f'(x) > 0$) or is decreasing ($f'(x) < 0$). Determine whether there are corners or cusps on the graph.

6 **Concavity and points of inflection** Find $f''(x)$, and use the second derivative test whenever appropriate. If $f''(x) > 0$ on an open interval I, the graph is concave upward. If $f''(x) < 0$, the graph is concave downward. If f is continuous at c and if $f''(x)$ changes sign at c, then $P(c, f(c))$ is a point of inflection.

7 **Asymptotes**
Horizontal: If $\lim_{x\to\infty} f(x) = L$ or $\lim_{x\to-\infty} f(x) = L$, then the line $y = L$ is a horizontal asymptote.
Vertical: If $\lim_{x\to a^+} f(x)$ or $\lim_{x\to a^-} f(x)$ is either ∞ or $-\infty$, then the line $x = a$ is a vertical asymptote.

FIGURE 4.39

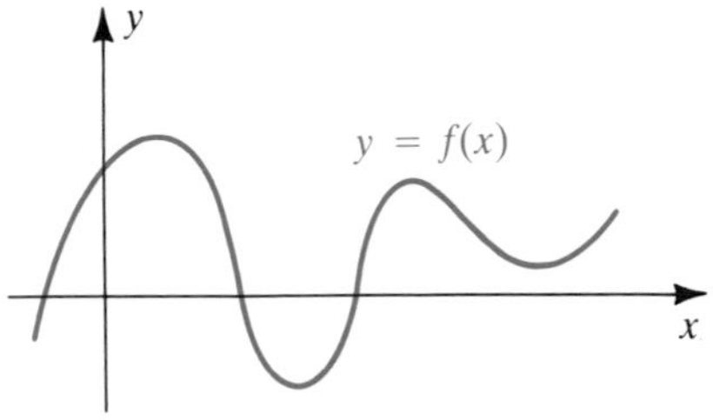

Among the simplest functions to graph is a polynomial function f; however, if the degree of $f(x)$ is large, it may take considerable effort to find the zeros of f, f', and f''. Every derivative of f is continuous; therefore, the graph has a smooth appearance and possibly many high and low points, as illustrated in Figure 4.39. Each critical number c determines a local extremum $f(c)$ or a point of inflection. We have considered graphs of polynomial functions previously in numerous examples and exercises, so we shall not give another example in this section.

Recall that a function f is a *rational function* if $f(x) = g(x)/h(x)$, where g and h are polynomial functions. If g and h have no common factors, then for every zero c of the denominator $h(x)$, the line $x = c$ is a vertical asymptote. In the next example we shall use Guidelines (4.19) to sketch the graph of a rational function.

EXAMPLE 1 If $f(x) = \dfrac{2x^2}{9 - x^2}$, discuss and sketch the graph of f.

SOLUTION We shall follow Guidelines (4.19).

Guideline 1 The domain of f consists of all real numbers except -3 and 3.

Guideline 2 The function f has infinite discontinuities at -3 and 3 and is continuous at all other real numbers.

Guideline 3 To find the x-intercepts, we solve the equation $f(x) = 0$, obtaining $x = 0$. The y-intercept is $f(0) = 0$. Therefore, the graph intersects both the x-axis and the y-axis at the origin.

Guideline 4 Since $f(-x) = f(x)$, f is an even function and the graph is symmetric with respect to the y-axis.

Guideline 5 We differentiate $f(x)$:

$$f'(x) = \frac{(9 - x^2)(4x) - 2x^2(-2x)}{(9 - x^2)^2} = \frac{36x}{(9 - x^2)^2}$$

Since $f'(x) = 0$ if $x = 0$, 0 is a critical number. The numbers -3 and 3 are not critical numbers because they are not in the domain of f.

Using test values gives us the following table.

Interval	$(-\infty, -3)$	$(-3, 0)$	$(0, 3)$	$(3, \infty)$
Sign of $f'(x)$	$-$	$-$	$+$	$+$
Conclusion	f is decreasing	f is decreasing	f is increasing	f is increasing

The sign of the derivative $f'(x)$ changes from negative to positive at $x = 0$, so, by the first derivative test, $f(0) = 0$ is a local minimum for f.

Guideline 6 We differentiate $f'(x)$:

$$f''(x) = \frac{(9 - x^2)^2(36) - (36x)(2)(9 - x^2)(-2x)}{(9 - x^2)^4} = \frac{108(x^2 + 3)}{(9 - x^2)^3}$$

The numerator is always positive, so the sign of $f''(x)$ is determined by $(9 - x^2)^3$. Using test values gives us the following.

Interval	$(-\infty, -3)$	$(-3, 3)$	$(3, \infty)$
Sign of $f''(x)$	$-$	$+$	$-$
Concavity	downward	upward	downward

FIGURE 4.40

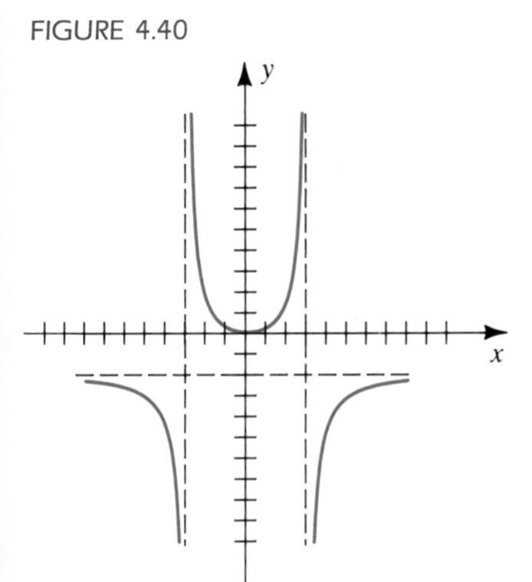

Since f is not continuous at -3 or 3, there are no points of inflection. As a check on local extrema (see guideline 5), we note that $f''(0) > 0$, and hence, by the second derivative test, $f(0) = 0$ is a local minimum.

Guideline 7 To find horizontal asymptotes, we use the methods in Section 2.4, obtaining

$$\lim_{x \to \infty} \frac{2x^2}{9 - x^2} = -2 \quad \text{and} \quad \lim_{x \to -\infty} \frac{2x^2}{9 - x^2} = -2.$$

Thus, the line $y = -2$ is a horizontal asymptote.

The vertical asymptotes correspond to the zeros of the denominator $9 - x^2$ and hence are $x = -3$ and $x = 3$.

Using the results of the guidelines and referring to the table developed from guideline 5 to obtain the behavior of f near the vertical asymptotes $(x = \pm 3)$ gives us the sketch in Figure 4.40.

The graph of $y = f(x)$ in the preceding example had no points of inflection. It is often difficult to find the points of inflection, whenever they exist, for the graph of a rational function. The reason is that the quotient rule is used to find $f'(x)$ and again to find $f''(x)$. Hence finding the solutions of $f''(x) = 0$ may require solving a polynomial equation of degree greater than 2. Solutions of such equations are often approximated by using Newton's method or a computer program. Because of this difficulty, in the next example and in most of the exercises we shall not find the points of inflection or discuss concavity for graphs of rational functions.

EXAMPLE 2 If $f(x) = \dfrac{x^2}{x^2 - x - 2}$, discuss and sketch the graph of f.

SOLUTION We again follow the guidelines.

Guideline 1 The denominator equals $(x - 2)(x + 1)$, so the domain of f consists of all numbers except -1 and 2.

Guideline 2 The function has infinite discontinuities at -1 and 2 and is continuous at all other numbers.

Guideline 3 To find the x-intercepts, we solve the equation $f(x) = 0$, obtaining $x = 0$. The y-intercept is $f(0) = 0$. Hence, as in Example 1, the graph intersects the x-axis and y-axis at the origin.

Guideline 4 Since f is neither even nor odd, the graph is not symmetric to the y-axis or to the origin.

Guideline 5 We differentiate $f(x)$:

$$f'(x) = \frac{(x^2 - x - 2)(2x) - x^2(2x - 1)}{(x^2 - x - 2)^2} = -\frac{x(x + 4)}{(x^2 - x - 2)^2}$$

Solving $f'(x) = 0$ gives us the critical numbers 0 and -4. The zeros of the denominator, -1 and 2, are not critical numbers since $f(-1)$ and $f(2)$ do not exist. Using test values we obtain the following table.

Interval	$(-\infty, -4)$	$(-4, -1)$	$(-1, 0)$	$(0, 2)$	$(2, \infty)$
Sign of $f'(x)$	$-$	$+$	$+$	$-$	$-$
Conclusion	f is decreasing	f is increasing	f is increasing	f is decreasing	f is decreasing

By the first derivative test, f has a local minimum $f(-4) = \frac{8}{9}$ and a local maximum $f(0) = 0$.

Guideline 6 We will not discuss concavity or find points of inflection. You may wish to verify that

$$f''(x) = \frac{2(x^3 + 6x^2 + 4)}{(x^2 - x - 2)^3}.$$

To solve the equation $f''(x) = 0$, we must solve the cubic equation

$$x^3 + 6x^2 + 4 = 0.$$

It can be shown that this equation has one real root r. To the nearest tenth, $r \approx -6.1$. The point of inflection is *approximately* $(-6.1, 0.9)$, slightly higher than the low (minimum) point $(-4, \frac{8}{9})$.

Guideline 7 To find horizontal asymptotes, we consider

$$\lim_{x\to\infty} \frac{x^2}{x^2 - x - 2} = 1 \quad \text{and} \quad \lim_{x\to-\infty} \frac{x^2}{x^2 - x - 2} = 1.$$

Thus, the graph has a horizontal asymptote $y = 1$.

The vertical asymptotes are obtained from the solutions of the equation $x^2 - x - 2 = 0$, so they are $x = -1$ and $x = 2$.

Using the preceding results and referring to the table developed from guideline 5 to obtain the behavior of f near the vertical asymptotes leads to the sketch in Figure 4.41.

FIGURE 4.41

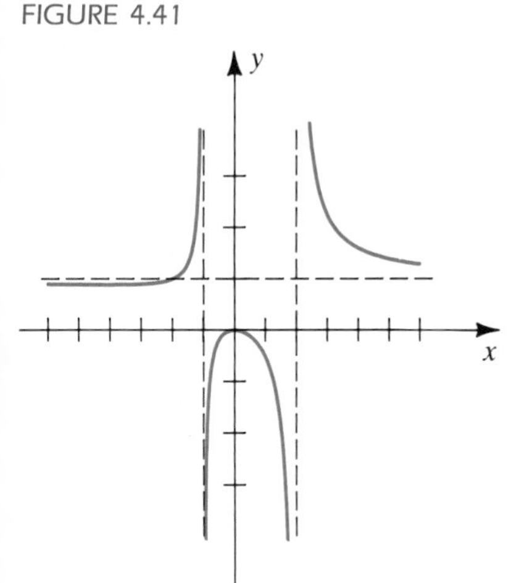

The graph intersects the horizontal asymptote $y = 1$. To find the x-coordinate of the point of intersection, we solve the equation $f(x) = 1$ as follows:

$$\frac{x^2}{x^2 - x - 2} = 1$$

$$x^2 = x^2 - x - 2$$

$$x = -2$$

Hence the point of intersection is $(-2, 1)$.

If $f(x) = g(x)/h(x)$ for polynomials $g(x)$ and $h(x)$ and *if the degree of $g(x)$ is 1 greater than the degree of $h(x)$*, then the graph of f has an **oblique asymptote** $y = ax + b$; that is, the vertical distance between the graph of f and this line approaches 0 as $x \to \infty$ or $x \to -\infty$. To establish this fact, we may use long division to express $f(x)$ in the form

$$f(x) = \frac{g(x)}{h(x)} = (ax + b) + \frac{r(x)}{h(x)},$$

where either $r(x) = 0$ or the degree of $r(x)$ is less than the degree of $h(x)$. It follows that $\lim_{x\to\infty} r(x)/h(x) = 0$ and $\lim_{x\to-\infty} r(x)/h(x) = 0$. Consequently, $f(x)$ gets closer to $ax + b$ as x approaches either ∞ or $-\infty$. If the graph of f has an oblique asymptote, then it has no horizontal asymptote.

EXAMPLE 3 If $f(x) = \dfrac{x^2 - 9}{2x - 4}$, discuss and sketch the graph of f.

SOLUTION

Guidelines 1 and 2 The domain of f consists of all real numbers except $x = 2$, where there is an infinite discontinuity.

Guideline 3 The x-intercepts are -3 and 3, and the y-intercept is $f(0) = \frac{9}{4}$.

Guideline 4 The graph is symmetric with respect to neither the y-axis nor the origin.

Guidelines 5 and 6 You should verify that

$$f'(x) = \frac{x^2 - 4x + 9}{2(x - 2)^2} \quad \text{and} \quad f''(x) = -\frac{5}{(x - 2)^3}.$$

Since $f'(x) \neq 0$ for every $x \neq 2$, there are no local extrema (see Corollary (4.6)).

Since $f''(x) > 0$ if $x < 2$ and $f''(x) < 0$ if $x > 2$, the graph of f is concave upward on $(-\infty, 2)$ and concave downward on $(2, \infty)$. There is no point of inflection.

FIGURE 4.42

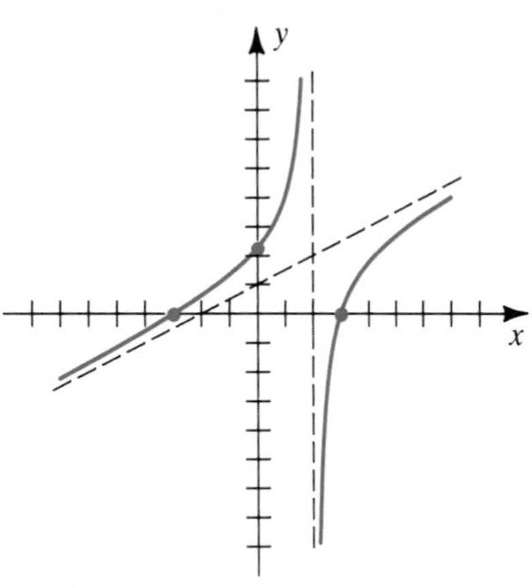

Guideline 7 The degree of the numerator $x^2 - 9$ is one greater than that of the denominator $2x - 4$, so there is an oblique asymptote and we use long division to express $f(x)$ as follows:

$$\frac{x^2 - 9}{2x - 4} = \left(\frac{1}{2}x + 1\right) - \frac{5}{2x - 4}$$

From the discussion preceding this example, the line $y = \frac{1}{2}x + 1$ is an oblique asymptote.

Since the graph has an oblique asymptote, there is no horizontal asymptote. There is a vertical asymptote $x = 2$ corresponding to the zero of the denominator $2x - 4$.

Representing the asymptotes by dashed lines, plotting the intercepts, and using the other information obtained by following the guidelines gives us the sketch in Figure 4.42.

In previous sections we considered expressions $f(x)$ that contain rational powers, such as $x^{2/3}$ and $(x^2 - 9)^{3/2}$, or the equivalent form that contains radicals. In some cases the graph of f has a vertical tangent line at some point P, with possibly a cusp at P (see Figures 4.23 and 4.24). The next example illustrates that horizontal tangent lines also may occur when radicals are involved.

EXAMPLE 4 If $f(x) = \dfrac{2x}{\sqrt{x^2 + 1}}$, discuss and sketch the graph of f.

SOLUTION

Guidelines 1 and 2 The domain of f is $\mathbb{R}$, and f is continuous at every real number.

Guideline 3 The x-intercept is 0, and the y-intercept is $f(0) = 0$.

Guideline 4 Since $f(-x) = -f(x)$, the function f is odd, and hence the graph is symmetric with respect to the origin.

Guidelines 5 and 6 Differentiating $f(x) = \dfrac{2x}{(x^2 + 1)^{1/2}}$ twice gives us

$$f'(x) = \frac{(x^2 + 1)^{1/2}(2) - (2x)(\frac{1}{2})(x^2 + 1)^{-1/2}(2x)}{x^2 + 1} = \frac{2}{(x^2 + 1)^{3/2}},$$

$$f''(x) = \frac{-6x}{(x^2 + 1)^{5/2}}.$$

Since $f'(x) > 0$ for every x, the function f is increasing on $(-\infty, \infty)$ and there are no critical numbers.

Since $f''(x) > 0$ for $x < 0$ and $f''(x) < 0$ for $x > 0$, the graph is concave upward on $(-\infty, 0)$ and concave downward on $(0, \infty)$, with a point of inflection $(0, 0)$.

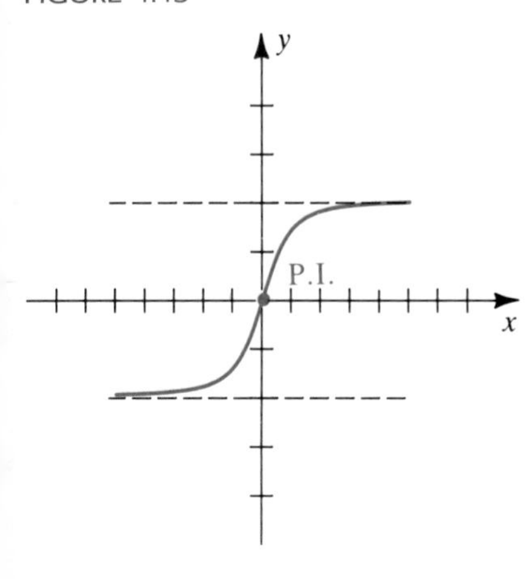

FIGURE 4.43

Guideline 7 You should check that

$$\lim_{x \to \infty} \frac{2x}{\sqrt{x^2+1}} = 2 \quad \text{and} \quad \lim_{x \to -\infty} \frac{2x}{\sqrt{x^2+1}} = -2.$$

There is a horizontal asymptote $y = 2$ (for the first-quadrant part of the graph) and a horizontal asymptote $y = -2$ (for the third-quadrant part). There are no vertical asymptotes.

Using the preceding results leads to the sketch in Figure 4.43.

In the examples we have considered so far in this section, the numerator and denominator of a quotient have had no common factor. When there *is* a common factor, say $p(x)$, the graph of f has a *hole* at each point that corresponds to a zero of $p(x)$. To illustrate, if

$$f(x) = \frac{2x(x-1)}{\sqrt{x^2+1}(x-1)},$$

then the graph of f would be the same as that in Figure 4.43, with one exception: there would be a hole at $(1, \sqrt{2})$. (See Exercises 25–28.)

The next example involves absolute values.

EXAMPLE 5 Sketch the graph of $y = |4 - x^2|$.

SOLUTION The difficult way to proceed is to use the fact that $|a| = \sqrt{a^2}$ to write $y = \sqrt{(4 - x^2)}$, and then use the derivatives to find local extrema and points of inflection. A simpler method is to sketch the graph of $y = 4 - x^2$, as shown in Figure 4.44(i). This graph is the same as that of $y = |4 - x^2|$ if $-2 \le x \le 2$. If $x > 2$ or $x < -2$, then $4 - x^2 < 0$, and the graph of $y = |4 - x^2|$ is the reflection of the graph of $y = 4 - x^2$ through the x-axis, as shown in Figure 4.44(ii). The extrema, concavity, and points of inflection should be self-evident.

FIGURE 4.44

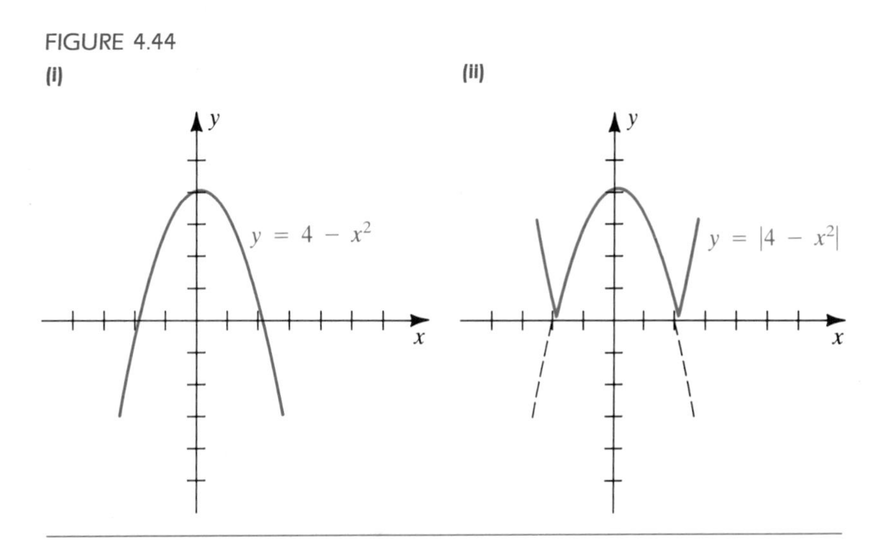

EXERCISES 4.5

Exer. 1–18: Find the extrema and sketch the graph of f.

1 $f(x) = \dfrac{2x - 5}{x + 3}$

2 $f(x) = \dfrac{3 - x}{x + 2}$

3 $f(x) = \dfrac{x^2 + x - 6}{x^2 - 1}$

4 $f(x) = \dfrac{4x}{x^2 - 4x + 3}$

5 $f(x) = \dfrac{3x^2 - 6x}{x^2 - x - 12}$

6 $f(x) = \dfrac{-2x^2 + 14x - 24}{x^2 + 2x}$

7 $f(x) = \dfrac{2x^2}{x^2 + 1}$

8 $f(x) = \dfrac{x - 1}{x^2 + 1}$

9 $f(x) = \dfrac{x + 4}{\sqrt{x}}$

10 $f(x) = \dfrac{x - 4}{\sqrt[3]{x^2}}$

11 $f(x) = \dfrac{-3x}{\sqrt{x^2 + 4}}$

12 $f(x) = \dfrac{2x}{\sqrt{x^2 + x + 2}}$

13 $f(x) = \dfrac{x^2 - x - 6}{x + 1}$

14 $f(x) = \dfrac{2x^2 - x - 3}{x - 2}$

15 $f(x) = \dfrac{x^2}{x + 1}$

16 $f(x) = \dfrac{-x^2 - 4x - 4}{x + 1}$

17 $f(x) = \dfrac{4 - x^2}{x + 3}$

18 $f(x) = \dfrac{8 - x^3}{2x^2}$

Exer. 19–24: Find the extrema and points of inflection, and sketch the graph of f.

19 $f(x) = \dfrac{3x}{(x + 8)^2}$

20 $f(x) = \dfrac{3x^2}{(2x - 9)^2}$

21 $f(x) = \dfrac{3x}{x^2 + 1}$

22 $f(x) = \dfrac{-4}{x^2 + 1}$

23 $f(x) = x^2 - \dfrac{27}{x^2}$

24 $f(x) = x^3 + \dfrac{3}{x}$

Exer. 25–28: Simplify $f(x)$ and sketch the graph of f.

25 $f(x) = \dfrac{2x^2 + x - 6}{x^2 + 3x + 2}$

26 $f(x) = \dfrac{x^2 - x - 6}{x^2 - 2x - 3}$

27 $f(x) = \dfrac{x - 1}{1 - x^2}$

28 $f(x) = \dfrac{x + 2}{x^2 - 4}$

Exer. 29–34: Sketch the graph of f.

29 $f(x) = |x^2 - 6x + 5|$

30 $f(x) = |8 + 2x - x^2|$

31 $f(x) = |x^3 + 1|$

32 $f(x) = |x^3 - x|$

33 $f(x) = -|\sin x|$

34 $f(x) = |\cos x| + 2$

Exer. 35–38: Use the information given to sketch the graph of f.

35 $f(x) = \dfrac{x - 3}{x^2 - 1} = \dfrac{x - 3}{(x + 1)(x - 1)}$;

$f'(x) = \dfrac{-x^2 + 6x - 1}{(x^2 - 1)^2}$;

$f'(x) = 0$ if $x = 3 \pm 2\sqrt{2}$ (approximately 5.83, 0.17);

$f''(x) = \dfrac{2x^3 - 18x^2 + 6x - 6}{(x^2 - 1)^3}$;

$f''(x) = 0$ if $x = 2\sqrt[3]{2} + 2\sqrt[3]{4} + 3 \approx 8.69$;
$f(5.83) \approx 0.09$; $f(0.17) \approx 2.91$; $f(8.69) \approx 0.08$

36 $f(x) = \dfrac{x + 4}{x^2 - 4} = \dfrac{x + 4}{(x + 2)(x - 2)}$;

$f'(x) = \dfrac{-x^2 - 8x - 4}{(x^2 - 4)^2}$;

$f'(x) = 0$ if $x = -4 \pm 2\sqrt{3}$ (approximately -0.54, -7.46);

$f''(x) = \dfrac{2x^3 + 24x^2 + 24x + 32}{(x^2 - 4)^3}$;

$f''(x) = 0$ if $x = -2\sqrt[3]{3} - 2\sqrt[3]{9} - 4 \approx -11.04$;
$f(-0.54) \approx -0.93$; $f(-7.46) \approx -0.07$;
$f(-11.04) \approx -0.06$

37 $f(x) = \dfrac{2x^2 - 2x - 4}{x^2 + x - 12} = \dfrac{2(x + 1)(x - 2)}{(x + 4)(x - 3)}$;

$f'(x) = \dfrac{4x^2 - 40x + 28}{(x^2 + x - 12)^2}$;

$f'(x) = 0$ if $x = 5 \pm 3\sqrt{2}$ (approximately 9.24, 0.76);

$f''(x) = \dfrac{-8x^3 + 120x^2 - 168x + 424}{(x^2 + x - 12)^3}$;

$f''(x) = 0$ if $x = \sqrt[3]{36} + 3\sqrt[3]{6} + 5 \approx 13.75$;
$f(9.24) \approx 1.79$; $f(0.76) \approx 0.41$; $f(13.75) \approx 1.82$

38 $f(x) = \dfrac{-3x^2 - 3x + 6}{x^2 - 9} = \dfrac{-3(x + 2)(x - 1)}{(x + 3)(x - 3)}$;

$f'(x) = \dfrac{3x^2 + 42x + 27}{(x^2 - 9)^2}$;

$f'(x) = 0$ if $x = -7 \pm 2\sqrt{10}$
(approximately -0.68, -13.32);

$f''(x) = \dfrac{-6x^3 - 126x^2 - 162x - 378}{(x^2 - 9)^3}$;

$f''(x) = 0$ if $x = -2\sqrt[3]{20} - 2\sqrt[3]{50} - 7 \approx -19.80$;
$f(-0.68) \approx -0.78$; $f(-13.32) \approx -2.89$;
$f(-19.80) \approx -2.90$

39 Coulomb's law in electricity asserts that the force of attraction F between two charged particles is directly proportional to the product of the charges and inversely proportional to the square of the distance between the

particles. Suppose a particle of charge $+1$ is placed on a coordinate line between two particles of charge -1 (see figure).

(a) Show that the net force acting on the particle of charge $+1$ is given by

$$F(x) = -\frac{k}{x^2} + \frac{k}{(x-2)^2},$$

where k is a positive constant.

(b) Let $k = 1$ and sketch the graph of F for $0 < x < 2$.

EXERCISE 39

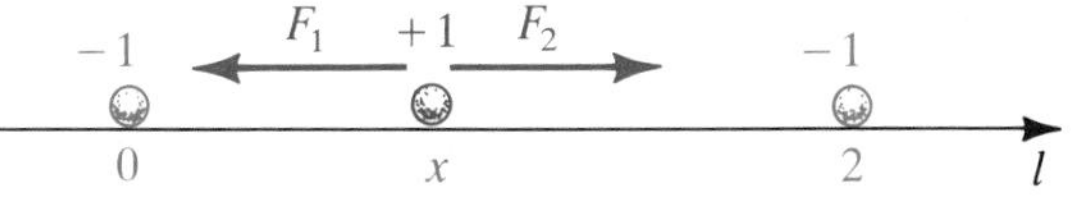

40 Biomathematicians have proposed many different functions for describing the effect of light on the rate at which photosynthesis can take place. If the function is to be realistic, then it must exhibit the *photoinhibition effect*; that is, the rate of production P of photosynthesis must decrease to 0 as the light intensity I reaches high levels (see figure). Which of the following formulas for P, where a and b are constants, may be used? Give reasons for your answer.

(a) $P = \dfrac{aI}{b + I}$ (b) $P = \dfrac{aI}{b + I^2}$

EXERCISE 40

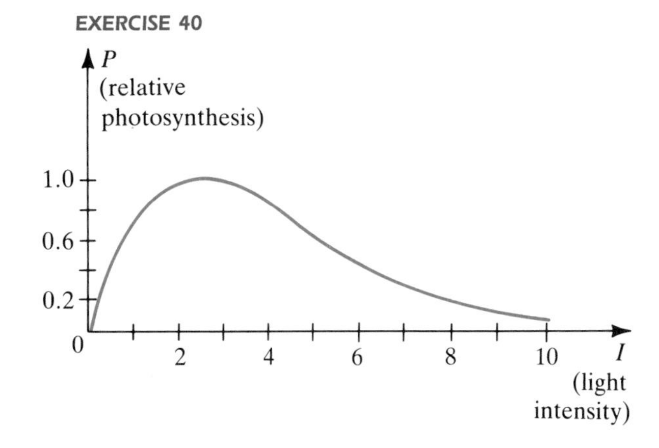

c **Exer. 41–42: Graph f on the given interval. (a) Estimate where the graph of f is concave upward or is concave downward. (b) Estimate the x-coordinate of each point of inflection.**

41 $f(x) = \dfrac{0.5x^3 + 3x + 7.1}{6 - x^2}$; $[-2, 2]$

42 $f(x) = \dfrac{x^2 + 2x + 1}{0.7x^2 - x + 1}$; $[-6, 6]$

4.6 OPTIMIZATION PROBLEMS

In applications, a physical or geometric quantity Q is often described by means of some formula $Q = f(x)$, where f is a function. Thus, Q might represent the temperature of a substance at time x, the current in an electrical circuit when the resistance is x, or the volume of gas in a spherical balloon of radius x. Of course, we often use other symbols for variables, such as T for temperature, t for time, I for current, R for resistance, V for volume, and r for radius. If $Q = f(x)$ and f is differentiable, then the derivative $D_x Q = f'(x)$ can be used to help find the maximum or minimum values of Q. In applications, these extreme values are sometimes called **optimal values**, because they are, in a sense, the best or most favorable values of the quantity Q. The task of finding these values is called an **optimization problem**.

If an optimization problem is stated in words, then it is often necessary to convert the statement into an appropriate formula such as $Q = f(x)$ in order to find critical numbers. In most cases there will be only one critical number c. If, in addition, f is continuous on a closed interval $[a, b]$ containing c, then, by Guidelines (4.9), the extrema of f are the largest and smallest of the values $f(a)$, $f(b)$, and $f(c)$. Hence it is often unnecessary to apply a derivative test. However, if it is easy to calculate $f''(x)$, we sometimes apply the second derivative test to verify an extremum, as illustrated in the next example.

EXAMPLE 1 A long rectangular sheet of metal, 12 inches wide, is to be made into a rain gutter by turning up two sides so that they are perpendicular to the sheet. How many inches should be turned up to give the gutter its greatest capacity?

FIGURE 4.45

SOLUTION The gutter is illustrated in Figure 4.45, where x denotes the number of inches turned up on each side. The width of the base of the gutter is $12 - 2x$ inches. The capacity of the gutter will be greatest when the area of the rectangle with sides of lengths x and $12 - 2x$ has its greatest value. Letting $f(x)$ denote this area, we obtain

$$f(x) = x(12 - 2x) = 12x - 2x^2.$$

Since $0 \le 2x \le 12$, the domain of f is $0 \le x \le 6$. If $x = 0$ or $x = 6$, no gutter is formed (the area of the rectangle would be $f(0) = 0 = f(6)$).

Differentiating yields

$$f'(x) = 12 - 4x = 4(3 - x),$$

and hence the only critical number of f is 3. Since $f''(x) = -4 < 0$, $f(3)$ is a local maximum for f. It follows that 3 inches should be turned up to achieve maximum capacity.

Because the types of optimization problems are unlimited, it is difficult to state specific rules for finding solutions. However, we can develop a general strategy for attacking such problems. The following guidelines are often helpful. When using the guidelines, don't become discouraged if you are unable to solve a given problem quickly. It takes a great deal of effort and practice to become proficient in solving optimization problems. Keep trying!

Guidelines for solving optimization problems **(4.20)**

1. Read the problem carefully several times, and think about the given facts as well as the unknown quantities that are to be found.
2. If possible, sketch a picture or diagram and label it appropriately, introducing variables for unknown quantities. Words such as *what*, *find*, *how much*, *how far*, or *when* should alert you to the unknown quantities.
3. Write down the known facts together with any relationships involving the variables.
4. Determine which variable is to be maximized or minimized, and express this variable as a function of *one* of the other variables.
5. Find the critical numbers of the function obtained in guideline 4.
6. Determine the extrema by using Guidelines (4.9) or the first or second derivative test. Check for endpoint extrema whenever appropriate.

The use of Guidelines (4.20) is illustrated in the next example.

EXAMPLE 2 An open box with a rectangular base is to be constructed from a rectangular piece of cardboard 16 inches wide and 21 inches long by cutting a square from each corner and then bending up the resulting sides. Find the size of the corner square that will produce a box having the largest possible volume. (Disregard the thickness of the cardboard.)

SOLUTION

Guideline 1 Read the problem at least one more time.

Guideline 2 Sketch the cardboard, as in Figure 4.46(i), introducing a variable x for the length of the side of the square to be cut from each corner.

FIGURE 4.46

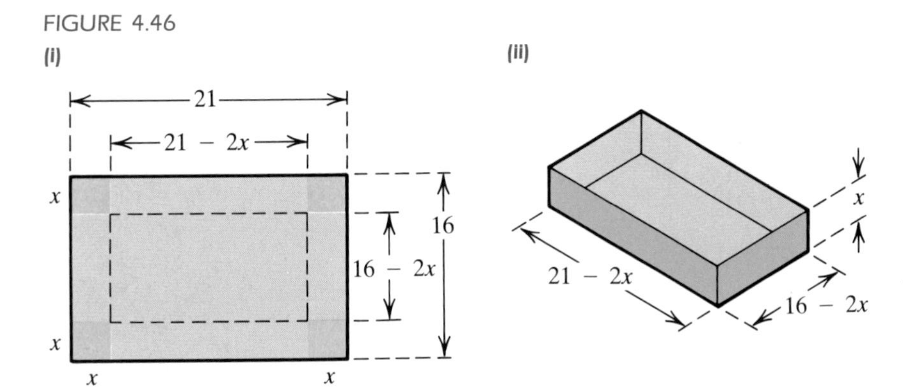

Guideline 3 If the cardboard is folded along the dashed lines in the sketch (see Figure 4.46(ii)), the base of the resulting box has dimensions $21 - 2x$ and $16 - 2x$.

Guideline 4 The quantity to be maximized is the volume V of the box. Referring to Figure 4.46(ii), we express V as a function of x:

$$V = x(16 - 2x)(21 - 2x) = 2(168x - 37x^2 + 2x^3)$$

Since $0 \le 2x \le 16$, the domain of V is $0 \le x \le 8$.

Guideline 5 To find the critical numbers for the function in guideline 4, differentiate V with respect to x:

$$\begin{aligned} D_x V &= 2(168 - 74x + 6x^2) \\ &= 4(3x^2 - 37x + 84) \\ &= 4(3x - 28)(x - 3) \end{aligned}$$

Thus, the possible critical numbers are $\frac{28}{3}$ and 3. Since $\frac{28}{3}$ is outside the domain of V, the only critical number is 3.

Guideline 6 Since V is continuous on $[0, 8]$, we shall use Guidelines (4.9) to determine the extrema. The endpoints $x = 0$ and $x = 8$ of the domain yield the minimum value $V = 0$. For the critical number $x = 3$, we obtain $V = 450$, which is a maximum value. Consequently, a 3-inch square should be cut from each corner of the cardboard in order to maximize the volume of the resulting box.

In the remaining examples we shall not always point out the guidelines employed. You should be able to determine specific guidelines by studying the solutions.

EXAMPLE 3 A circular cylindrical metal container, open at the top, is to have a capacity of 24π in.^{3} The cost of the material used for the bottom of the container is 15 cents per in.^{2}, and that of the material used for the curved part is 5 cents per in.^{2} If there is no waste of material, find the dimensions that will minimize the cost of the material.

FIGURE 4.47

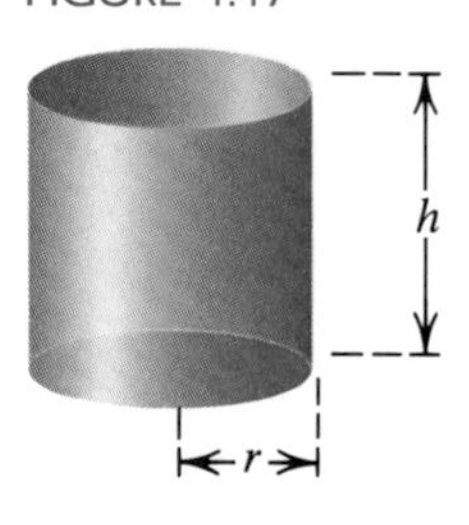

SOLUTION We begin by sketching a typical container, as in Figure 4.47, letting r denote the radius of the base and h the altitude (both in inches). *The quantity we wish to minimize is the cost C of the material.* Since the costs per square inch for the base and the curved part are 15 cents and 5 cents, respectively, we have, in terms of cents,

$$\text{cost of container} = 15(\text{area of base}) + 5(\text{area of curved part}).$$

Thus,

$$C = 15(\pi r^2) + 5(2\pi rh)$$

or

$$C = 5\pi(3r^2 + 2rh).$$

We can express C as a function of one variable r by expressing h in terms of r. Since the volume of the container is 24π in.^{3}, we see that

$$\pi r^2 h = 24\pi, \quad \text{or} \quad h = \frac{24}{r^2}.$$

Substituting $24/r^2$ for h in the last formula for C gives us

$$C = 5\pi\left(3r^2 + 2r \cdot \frac{24}{r^2}\right) = 5\pi\left(3r^2 + \frac{48}{r}\right).$$

The domain of C is $(0, \infty)$.

Next we find critical numbers by differentiating C with respect to r:

$$D_r C = 5\pi\left(6r - \frac{48}{r^2}\right) = 30\pi\left(\frac{r^3 - 8}{r^2}\right)$$

Since $D_r C = 0$ if $r = 2$, we see that 2 is the only critical number. Since $D_r C < 0$ if $r < 2$ and $D_r C > 0$ if $r > 2$, it follows from the first derivative test that C has its minimum value if the radius of the cylinder is 2 inches. The corresponding value for the altitude (obtained from $h = 24/r^2$) is $\frac{24}{4}$, or 6 inches.

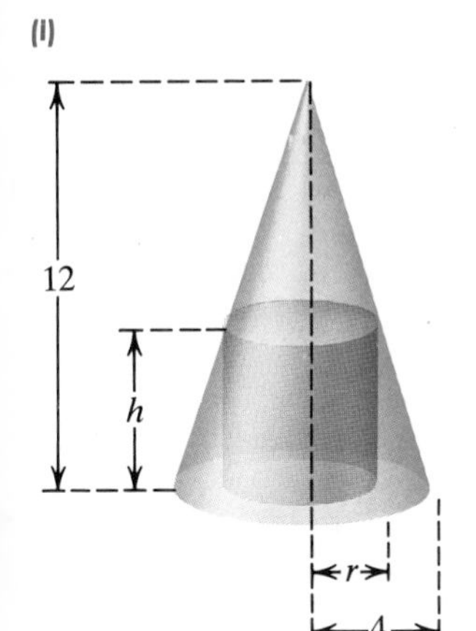

FIGURE 4.48 (i)

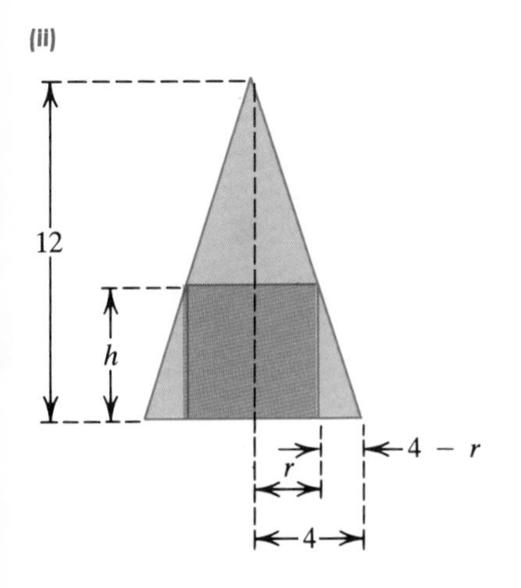

(ii)

EXAMPLE 4 Find the maximum volume of a right circular cylinder that can be inscribed in a cone of altitude 12 centimeters, and base radius 4 centimeters, if the axes of the cylinder and cone coincide.

SOLUTION The problem is sketched in Figure 4.48, where (ii) represents a cross section through the axes of the cone and cylinder. *The quantity we wish to maximize is the volume V of the cylinder.* From geometry,

$$V = \pi r^2 h.$$

Next we express V in terms of one variable by finding a relationship between r and h. Referring to Figure 4.48(ii) and using similar triangles, we see that

$$\frac{h}{4-r} = \frac{12}{4} = 3, \quad \text{or} \quad h = 3(4-r).$$

Consequently,

$$V = \pi r^2 h = \pi r^2 \cdot 3(4-r) = 3\pi r^2(4-r).$$

The domain of V is $0 \le r \le 4$.

If either $r = 0$ or $r = 4$, we see that $V = 0$, and hence the maximum volume is not an endpoint extremum. It is sufficient, therefore, to search for local maxima. Since $V = 3\pi(4r^2 - r^3)$,

$$D_r V = 3\pi(8r - 3r^2) = 3\pi r(8 - 3r).$$

Thus, the critical numbers for V are $r = 0$ and $r = \frac{8}{3}$. At $r = \frac{8}{3}$, we have

$$V = \pi\left(\frac{8}{3}\right)^2(4) = \frac{256\pi}{9} \approx 89.4 \text{ cm}^3,$$

which, by Guidelines (4.9), is a maximum value for the volume of the inscribed cylinder.

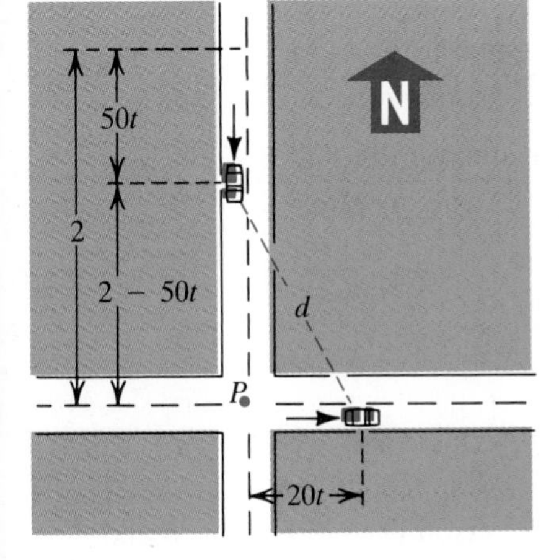

FIGURE 4.49

EXAMPLE 5 A North-South highway intersects an East-West highway at a point P. An automobile crosses P at 10:00 A.M., traveling east at a constant speed of 20 mi/hr. At that same instant another automobile is 2 miles north of P, traveling south at 50 mi/hr. Find the time at which they are closest to each other and approximate the minimum distance between the automobiles.

SOLUTION Typical positions of the automobiles are illustrated in Figure 4.49. If t denotes the number of hours after 10:00 A.M., then the slower automobile is $20t$ miles east of P. The faster automobile is $50t$ miles south of its position at 10:00 A.M., and hence its distance from P is $2 - 50t$. By the Pythagorean theorem, the distance d between the automobiles is

$$\begin{aligned} d &= \sqrt{(2-50t)^2 + (20t)^2} \\ &= \sqrt{4 - 200t + 2500t^2 + 400t^2} \\ &= \sqrt{4 - 200t + 2900t^2}. \end{aligned}$$

We wish to find the time t at which d has its smallest value. This will occur when the expression under the radical is minimal because d increases if and only if $4 - 200t + 2900t^2$ increases. Thus, we may simplify our work by letting

$$f(t) = 4 - 200t + 2900t^2$$

and finding the value of t for which f has a minimum. Since

$$f'(t) = -200 + 5800t,$$

the only critical number for f is

$$t = \frac{200}{5800} = \frac{1}{29}.$$

Moreover, $f''(t) = 5800$, so the second derivative is always positive. Therefore, f has a local minimum at $t = \frac{1}{29}$, and $f(\frac{1}{29}) = \frac{16}{29}$. Since the domain of t is $[0, \infty)$ and since $f(0) = 4$, there is no endpoint extremum. Consequently, the automobiles will be closest at $\frac{1}{29}$ hour (or approximately 2.07 minutes) after 10:00 A.M. The minimal distance is

$$\sqrt{f(\tfrac{1}{29})} = \sqrt{\tfrac{16}{29}} \approx 0.74 \text{ mi}.$$

EXAMPLE 6 A person in a rowboat 2 miles from the nearest point on a straight shoreline wishes to reach a house 6 miles farther down the shore. If the person can row at a rate of 3 mi/hr and walk at a rate of 5 mi/hr, find the least amount of time required to reach the house.

FIGURE 4.50

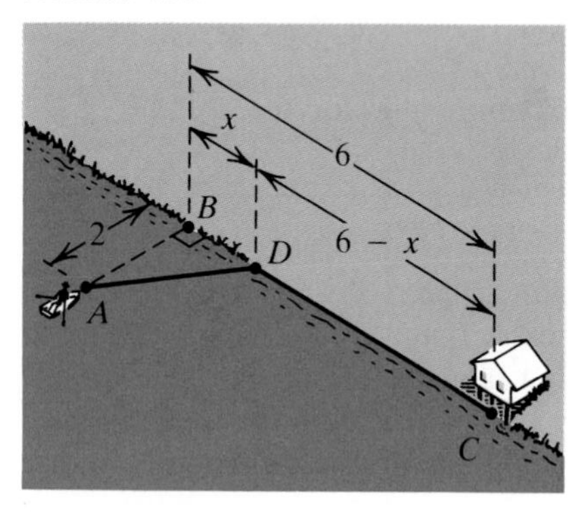

SOLUTION Figure 4.50 illustrates the problem: A denotes the position of the boat, B the nearest point on shore, C the house, D the point at which the boat reaches shore, and x the distance between B and D. By the Pythagorean theorem, the distance between A and D is $\sqrt{x^2 + 4}$, where $0 \le x \le 6$. Using the formula

$$\text{time} = \frac{\text{distance}}{\text{rate}},$$

we obtain

$$\text{time to row from } A \text{ to } D = \frac{\text{distance from } A \text{ to } D}{\text{rowing rate}} = \frac{\sqrt{x^2 + 4}}{3}$$

$$\text{time to walk from } D \text{ to } C = \frac{\text{distance from } D \text{ to } C}{\text{walking rate}} = \frac{6 - x}{5}.$$

Hence the total time T for the trip is

$$T = \frac{\sqrt{x^2 + 4}}{3} + \frac{6 - x}{5},$$

or, equivalently, $$T = \tfrac{1}{3}(x^2 + 4)^{1/2} + \tfrac{6}{5} - \tfrac{1}{5}x.$$

We wish to find the minimum value for T. Note that $x = 0$ corresponds to the extreme situation in which the person rows directly to B and then walks the entire distance from B to C. If $x = 6$, then the person rows directly from A to C. These numbers may be considered as endpoints of the domain of T. If $x = 0$, then, from the formula for T,

$$T = \frac{\sqrt{4}}{3} + \frac{6}{5} - 0 = \frac{28}{15},$$

which is 1 hour 52 minutes. If $x = 6$, then

$$T = \frac{\sqrt{40}}{3} + \frac{6}{5} - \frac{6}{5} = \frac{2\sqrt{10}}{3} \approx 2.11,$$

or approximately 2 hours 7 minutes.

Differentiating the general formula for T, we see that

$$D_x T = \tfrac{1}{3} \cdot \tfrac{1}{2}(x^2 + 4)^{-1/2}(2x) - \tfrac{1}{5},$$

or

$$D_x T = \frac{x}{3(x^2 + 4)^{1/2}} - \frac{1}{5}.$$

In order to find the critical numbers, we let $D_x T = 0$, obtaining the following equations:

$$\frac{x}{3(x^2 + 4)^{1/2}} = \frac{1}{5}$$

$$5x = 3(x^2 + 4)^{1/2}$$

$$25x^2 = 9(x^2 + 4)$$

$$x^2 = \tfrac{36}{16}$$

$$x = \tfrac{6}{4} = \tfrac{3}{2}$$

Thus, $\frac{3}{2}$ is the only critical number. The time T that corresponds to $x = \frac{3}{2}$ is

$$T = \tfrac{1}{3}(\tfrac{9}{4} + 4)^{1/2} + \tfrac{6}{5} - \tfrac{3}{10} = \tfrac{26}{15},$$

or, equivalently, 1 hour 44 minutes.

We have already examined the values of T at the endpoints of the domain, obtaining 1 hour 52 minutes and approximately 2 hours 7 minutes, respectively. Hence the minimum time of 1 hour 44 minutes occurs at $x = \frac{3}{2}$. Therefore, the boat should land at D, $1\frac{1}{2}$ miles from B, in order to minimize T. For a similar problem, but one in which the endpoints of the domain lead to minimum time, see Exercise 6.

EXAMPLE 7 A wire 60 inches long is to be cut into two pieces. One of the pieces will be bent into the shape of a circle and the other into the shape of an equilateral triangle. Where should the wire be cut so that the sum of the areas of the circle and triangle is minimized? maximized?

SOLUTION If x denotes the length of one of the cut pieces of wire, then the length of the other piece is $60 - x$. Let the piece of length x be bent to form a circle of radius r so that $2\pi r = x$, or $r = x/(2\pi)$ (see Figure 4.51).

FIGURE 4.51

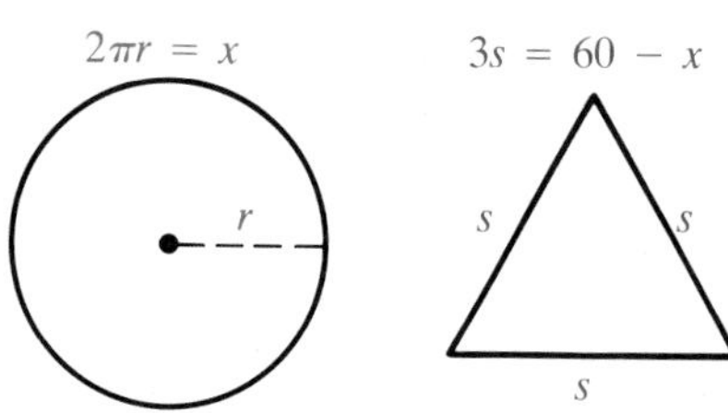

If the remaining piece is bent into an equilateral triangle of side s, then $3s = 60 - x$, or $s = (60 - x)/3$. *We wish to minimize and maximize the sum A of the areas of the circle and the triangle.* Referring to Figure 4.51, we see that

$$\text{area of circle} = \pi r^2 = \pi\left(\frac{x}{2\pi}\right)^2 = \frac{1}{4\pi}x^2$$

$$\text{area of triangle} = \frac{\sqrt{3}}{4}s^2 = \frac{\sqrt{3}}{4}\left(\frac{60-x}{3}\right)^2 = \frac{\sqrt{3}}{36}(60-x)^2.$$

Hence the sum A of the areas is

$$A = \frac{1}{4\pi}x^2 + \frac{\sqrt{3}}{36}(60-x)^2.$$

We now find the critical numbers. Differentiating, we obtain

$$D_x A = \frac{1}{2\pi}x - \frac{\sqrt{3}}{18}(60-x)$$
$$= \left(\frac{1}{2\pi} + \frac{\sqrt{3}}{18}\right)x - \frac{10\sqrt{3}}{3}.$$

Thus, $D_x A = 0$ if and only if

$$x = \frac{10\sqrt{3}/3}{1/(2\pi) + (\sqrt{3}/18)} \approx 22.61.$$

The second derivative of A with respect to x is

$$D_x^2 A = \frac{1}{2\pi} + \frac{\sqrt{3}}{18},$$

which is positive. Hence the critical number yields a minimum value for A, and the wire should be cut (approximately) 22.61 inches from one end. The minimum value of A is approximated by

$$A \approx \frac{1}{4\pi}(22.61)^2 + \frac{\sqrt{3}}{36}(60 - 22.61)^2 \approx 107.94 \text{ in.}^2$$

Since there are no other critical numbers, the maximum value of A must be an endpoint extremum. If $x = 0$, then all the wire is used to form a triangle and

$$A = \frac{\sqrt{3}}{36}(60)^2 \approx 173.21 \text{ in.}^2$$

If $x = 60$, then all the wire is used for the circle, and

$$A = \frac{1}{4\pi}(60)^2 \approx 286.48 \text{ in.}^2$$

Thus, the maximum value of A occurs if the wire is not cut and the entire length of wire is bent into the shape of a circle.

FIGURE 4.52

(i)

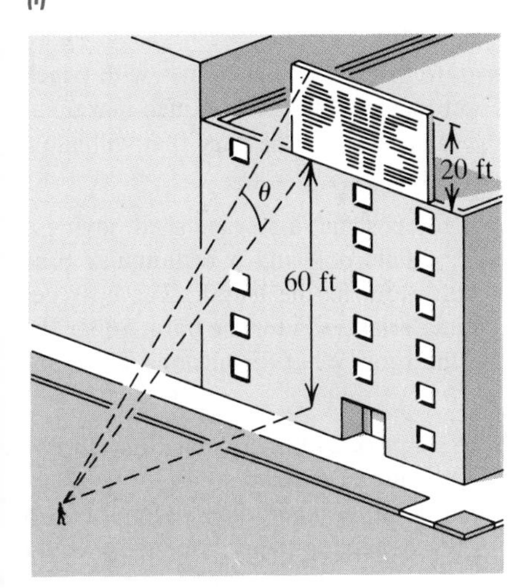

(ii)

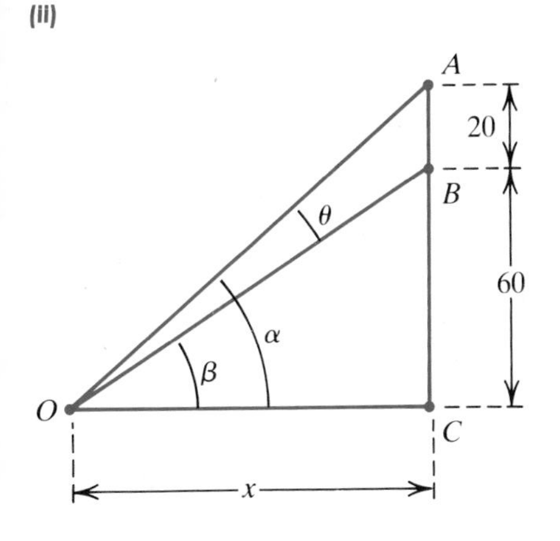

EXAMPLE 8 A billboard 20 feet tall is located on top of a building, with its lower edge 60 feet above the level of a viewer's eye, as shown in Figure 4.52(i). How far from a point directly below the sign should a viewer stand to maximize the angle θ between the lines of sight of the top and bottom of the billboard? (This angle should result in the best view of the billboard.)

SOLUTION The problem is sketched in Figure 4.52(ii), using right triangles AOC and BOC having common side OC of (variable) length x. We see that

$$\tan \alpha = \frac{80}{x} \quad \text{and} \quad \tan \beta = \frac{60}{x}.$$

The angle $\theta = \alpha - \beta$ is a function of x and

$$\tan \theta = \tan (\alpha - \beta) = \frac{\tan \alpha - \tan \beta}{1 + \tan \alpha \tan \beta}.$$

Substituting for $\tan \alpha$ and $\tan \beta$ and simplifying, we obtain

$$\tan \theta = \frac{\dfrac{80}{x} - \dfrac{60}{x}}{1 + \left(\dfrac{80}{x}\right)\left(\dfrac{60}{x}\right)} = \frac{20x}{x^2 + 4800}.$$

The extrema of θ occur if $D_x\, \theta = 0$. Differentiating implicitly with respect to x and using the quotient rule gives us

$$\sec^2 \theta\, D_x\, \theta = \frac{(x^2 + 4800)(20) - 20x(2x)}{(x^2 + 4800)^2} = \frac{96{,}000 - 20x^2}{(x^2 + 4800)^2}.$$

Since $\sec^2 \theta > 0$, it follows that $D_x\, \theta = 0$ if and only if

$$96{,}000 - 20x^2 = 0, \quad \text{or} \quad x^2 = 4800.$$

Thus, the only critical number of θ is

$$x = \sqrt{4800} = 40\sqrt{3}.$$

We may verify that the sign of $D_x\, \theta$ changes from positive to negative at $\sqrt{4800}$, and hence a maximum value of θ occurs at $x = 40\sqrt{3}$ ft ≈ 69.3 ft.

EXERCISES 4.6

1 If a box with a square base and open top is to have a volume of 4 ft^3, find the dimensions that require the least material. (Disregard the thickness of the material and waste in construction.)

2 Work Exercise 1 if the box has a closed top.

3 A metal cylindrical container with an open top is to hold one cubic foot. If there is no waste in construction, find the dimensions that require the least amount of material. (Compare with Example 3.)

4 If the circular base of the container in Exercise 3 is cut from a square sheet and the remaining metal is discarded, find the dimensions that require the least amount of material.

5 One thousand feet of chain link fence will be used to construct six cages for a zoo exhibit, as shown in the figure on the following page. Find the dimensions that maximize the enclosed area A. (*Hint:* First express y as a function of x, and then express A as a function of x.)

EXERCISE 5

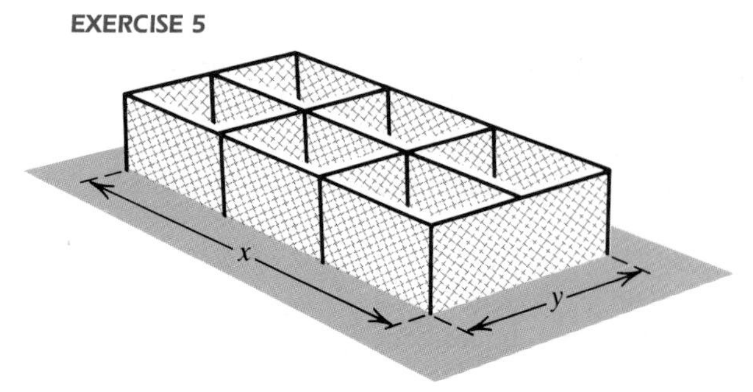

6 Refer to Example 6. If the person is in a motorboat that can travel at an average rate of 15 mi/hr, what route should be taken to arrive at the house in the least amount of time?

7 At 1:00 P.M. ship A is 30 miles due south of ship B and is sailing north at a rate of 15 mi/hr. If ship B is sailing west at a rate of 10 mi/hr, find the time at which the distance d between the ships is minimal (see figure).

EXERCISE 7

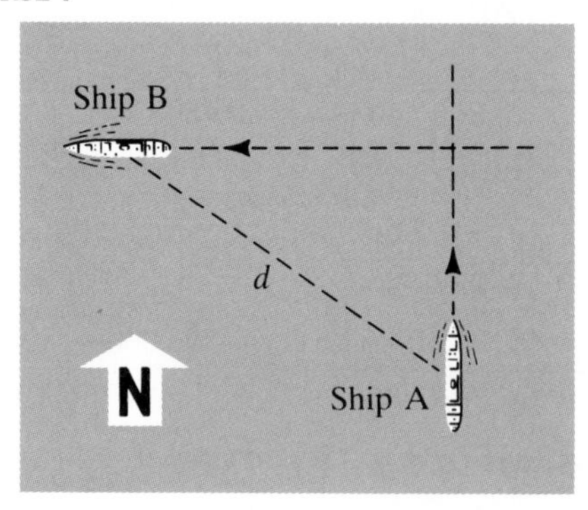

8 A window has the shape of a rectangle surmounted by a semicircle. If the perimeter of the window is 15 feet, find the dimensions that will allow the maximum amount of light to enter.

9 A fence 8 feet tall stands on level ground and runs parallel to a tall building (see figure). If the fence is 1 foot from the building, find the length of the shortest ladder that will extend from the ground over the fence to the wall of the building. (*Hint:* Use similar triangles.)

EXERCISE 9

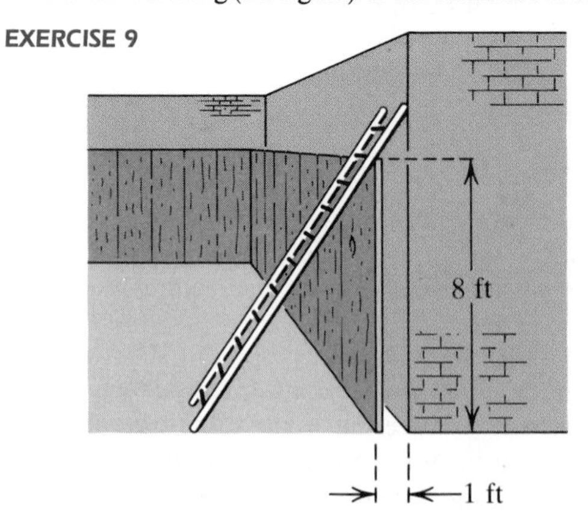

10 A page of a book is to have an area of 90 in.^{2}, with 1-inch margins at the bottom and sides and a $\frac{1}{2}$-inch margin at the top. Find the dimensions of the page that will allow the largest printed area.

11 A builder intends to construct a storage shed having a volume of 900 ft^3, a flat roof, and a rectangular base whose width is three-fourths the length. The cost per square foot of the materials is \$4 for the floor, \$6 for the sides, and \$3 for the roof. What dimensions will minimize the cost?

12 A water cup in the shape of a right circular cone is to be constructed by removing a circular sector from a circular sheet of paper of radius a and then joining the two straight edges of the remaining paper (see figure). Find the volume of the largest cup that can be constructed.

EXERCISE 12

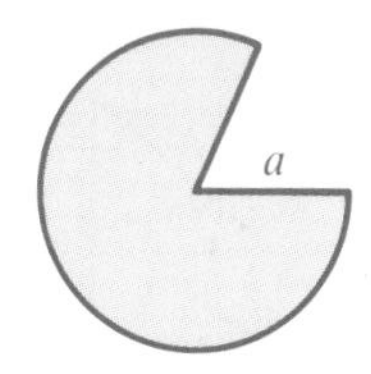

13 A farmer has 500 feet of fencing to enclose a rectangular field. A barn will be used as part of one side of the field (see figure). Prove that the area of the field is greatest when the rectangle is a square.

EXERCISE 13

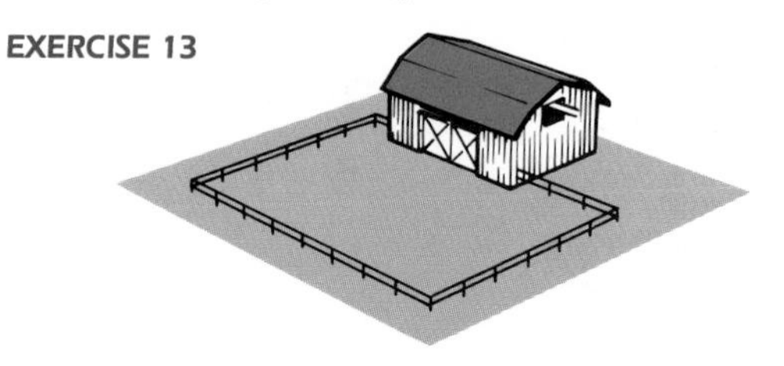

14 Refer to Exercise 13. Suppose the farmer wants the area of the rectangular field to be A ft^2. Prove that the least amount of fencing is required when the rectangle is a square.

15 A hotel that charges \$80 per day for a room gives special rates to organizations that reserve between 30 and 60 rooms. If more than 30 rooms are reserved, the charge per room is decreased by \$1 times the number of rooms over 30. Under these conditions, how many rooms must be rented if the hotel is to receive the maximum income per day?

16 Refer to Exercise 15. Suppose that for each room rented it costs the hotel \$6 per day for cleaning and mainte-

nance. In this case, how many rooms must be rented to obtain the greatest net income?

17 A steel storage tank for propane gas is to be constructed in the shape of a right circular cylinder with a hemisphere at each end (see figure). The construction cost per square foot for the end pieces is twice that for the cylindrical piece. If the desired capacity is 10π ft^3, what dimensions will minimize the cost of construction?

EXERCISE 17

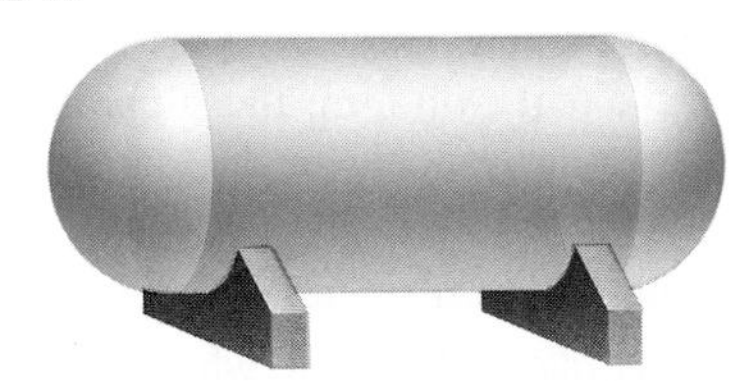

18 A pipeline for transporting oil will connect two points A and B that are 3 miles apart and on opposite banks of a straight river 1 mile wide (see figure). Part of the pipeline will run under water from A to a point C on the opposite bank, and then above ground from C to B. If the cost per mile of running the pipeline under water is four times the cost per mile of running it above ground, find the location of C that will minimize the cost (disregard the slope of the river bed).

EXERCISE 18

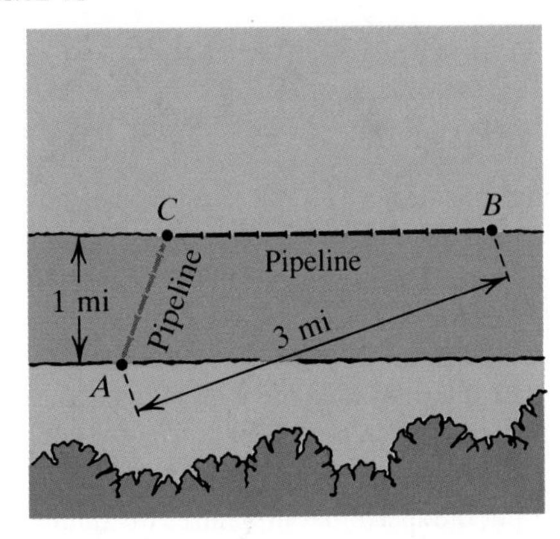

19 Find the dimensions of the rectangle of maximum area that can be inscribed in a semicircle of radius a, if two vertices lie on the diameter (see figure).

EXERCISE 19

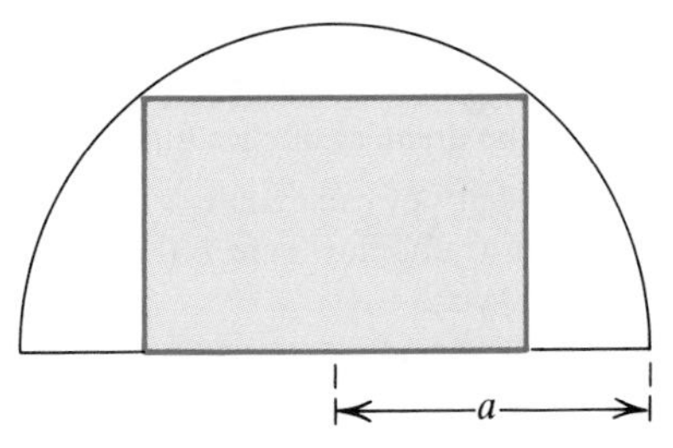

20 Find the dimensions of the rectangle of maximum area that can be inscribed in an equilateral triangle of side a, if two vertices of the rectangle lie on one of the sides of the triangle.

21 Of all possible right circular cones that can be inscribed in a sphere of radius a, find the volume of the one that has maximum volume.

22 Find the dimensions of the right circular cylinder of maximum volume that can be inscribed in a sphere of radius a.

23 Find the point on the graph of $y = x^2 + 1$ that is closest to the point $(3, 1)$.

24 Find the point on the graph of $y = x^3$ that is closest to the point $(4, 0)$.

25 The strength of a rectangular beam is directly proportional to the product of its width and the square of the depth of a cross section. Find the dimensions of the strongest beam that can be cut from a cylindrical log of radius a (see figure).

EXERCISE 25

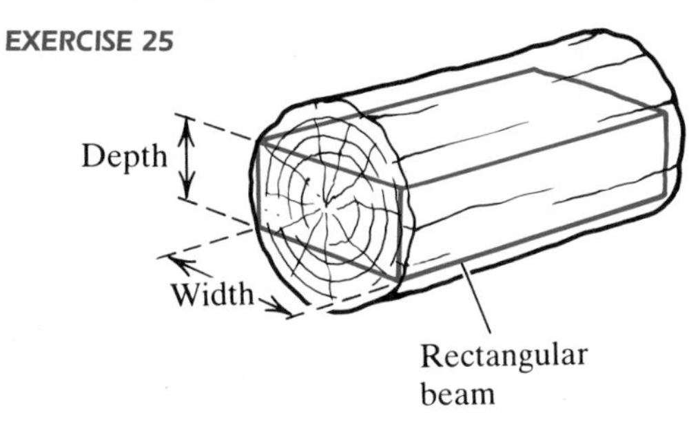

26 The illumination from a light source is directly proportional to the strength of the source and inversely proportional to the square of the distance from the source. If two light sources of strengths S_1 and S_2 are d units apart, at what point on the line segment joining the two sources is the illumination minimal?

27 A wholesaler sells running shoes at \$20 per pair if less than 50 pairs are ordered. If 50 or more pairs are ordered (up to 600), the price per pair is reduced by 2 cents times the number ordered. What size order will produce the maximum amount of money for the wholesaler?

28 A paper cup is to be constructed in the shape of a right circular cone. If the volume desired is 36π in.^{3}, find the dimensions that require the least amount of paper. (Disregard any waste that may occur in the construction.)

29 A wire 36 centimeters long is to be cut into two pieces. One of the pieces will be bent into the shape of an equilateral triangle and the other into the shape of a rectangle whose length is twice its width. Where should the wire be cut if the combined area of the triangle and rectangle is to be **(a)** minimized? **(b)** maximized?

30 An isosceles triangle has base b and equal sides of length a. Find the dimensions of the rectangle of maximum area that can be inscribed in the triangle if one side of the rectangle lies on the base of the triangle.

31 A window has the shape of a rectangle surmounted by an equilateral triangle. If the perimeter of the window is 12 feet, find the dimensions of the rectangle that will produce the largest area for the window.

32 Two vertical poles of lengths 6 feet and 8 feet stand on level ground, with their bases 10 feet apart. Approximate the minimal length of cable that can reach from the top of one pole to some point on the ground between the poles and then to the top of the other pole.

33 Prove that the rectangle of largest area having a given perimeter p is a square.

34 A right circular cylinder is generated by rotating a rectangle of perimeter p about one of its sides. What dimensions of the rectangle will generate the cylinder of maximum volume?

35 The owner of an apple orchard estimates that if 24 trees are planted per acre, then each mature tree will yield 600 apples per year. For each additional tree planted per acre, the number of apples produced by each tree decreases by 12 per year. How many trees should be planted per acre to obtain the most apples per year?

36 A real estate company owns 180 efficiency apartments, which are fully occupied when the rent is \$300 per month. The company estimates that for each \$10 increase in rent, 5 apartments will become unoccupied. What rent should be charged in order to obtain the largest gross income?

37 A package can be sent by parcel post only if the sum of its length and girth (the perimeter of the base) is not more than 108 inches. Find the dimensions of the box of maximum volume that can be sent, if the base of the box is a square.

38 A North-South highway A and an East-West highway B intersect at a point P. At 10:00 A.M. an automobile crosses P traveling north on highway A at a speed of 50 mi/hr. At that same instant, an airplane flying east at a speed of 200 mi/hr and an altitude of 26,400 feet is directly above the point on highway B that is 100 miles west of P. If the automobile and the airplane maintain the same speeds and directions, at what time will they be closest to each other?

39 Two factories A and B that are 4 miles apart emit particles in smoke that pollute the area between the factories. Suppose that the number of particles emitted from each factory is directly proportional to the amount of smoke and inversely proportional to the cube of the distance from the factory. If factory A emits twice as much smoke as factory B, at what point between A and B is the pollution minimal?

40 An oil field contains 8 wells, which produce a total of 1600 barrels of oil per day. For each additional well that is drilled, the average production per well decreases by 10 barrels per day. How many additional wells should be drilled to obtain the maximum amount of oil per day?

41 A canvas tent is to be constructed in the shape of a pyramid with a square base. A steel pole, placed in the center of the tent, will form the support (see figure). If S ft^2 of canvas is available for the four sides and x is the length of the base, show that

(a) the volume V of the tent is $V = \frac{1}{6}x\sqrt{S^2 - x^4}$

(b) V has its maximum value when x equals $\sqrt{2}$ times the length of the pole

EXERCISE 41

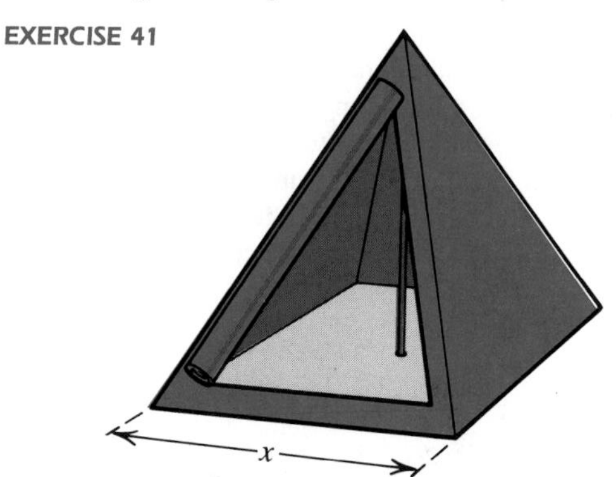

42 A boat must travel 100 miles upstream against a 10 mi/hr current. When the boat's velocity is v mi/hr, the number of gallons of gasoline consumed each hour is directly proportional to v^2.

(a) If a constant velocity of v mi/hr is maintained, show that the total number y of gallons of gasoline consumed is given by $y = 100kv^2/(v - 10)$ for $v > 10$ and for some positive constant k.

(b) Find the speed that minimizes the number of gallons of gasoline consumed during the trip.

43 Cars are crossing a bridge that is 1 mile long. Each car is 12 feet long and is required to stay a distance of at least d feet from the car in front of it (see figure).

(a) Show that the largest number of cars that can be on the bridge at one time is $[\![5280/(12 + d)]\!]$, where $[\![\ \ ]\!]$ denotes the greatest integer function.

(b) If the velocity of each car is v mi/hr, show that the maximum traffic flow rate F (in cars/hr) is given by $F = [\![5280v/(12 + d)]\!]$.

(c) The stopping distance (in feet) of a car traveling v mi/hr is approximately $0.05v^2$. If $d = 0.025v^2$, find the speed that maximizes the traffic flow across the bridge.

EXERCISE 43

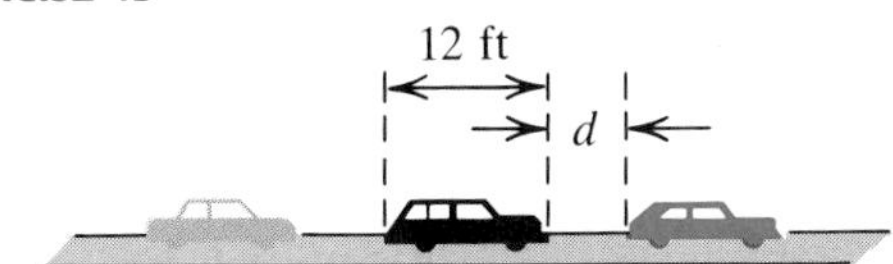

44 Prove that the shortest distance from a point (x_1, y_1) to the graph of a differentiable function f is measured along a normal line to the graph—that is, a line perpendicular to the tangent line.

45 A railroad route is to be constructed from town A to town C, branching out from a point B toward C at an angle of θ degrees (see figure). Because of the mountains between A and C, the branching point B must be at least 20 miles east of A. If the construction costs are \$50,000 per mile between A and B and \$100,000 per mile between B and C, find the branching angle θ that minimizes the total construction cost.

EXERCISE 45

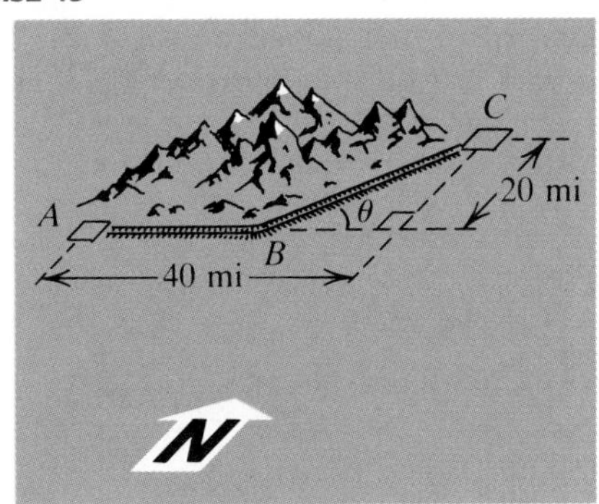

46 A long rectangular sheet of metal, 12 inches wide, is to be made into a rain gutter by turning up two sides at angles of 120° to the sheet. How many inches should be turned up to give the gutter its greatest capacity?

47 Refer to Exercise 12. Find the central angle of the sector that will maximize the volume of the cup.

48 A square picture having sides 2 feet long is hung on a wall such that the base is 6 feet above the floor. If a person whose eye level is 5 feet above the floor looks at the picture and if θ is the angle between the line of sight and the top and bottom of the picture, find the person's distance from the wall at which θ has its maximum value.

49 A rectangle made of elastic material will be made into a cylinder by joining edges AD and BC (see figure). To support the structure, a wire of fixed length L is placed along the diagonal of the rectangle. Find the angle θ that will result in the cylinder of maximum volume.

EXERCISE 49

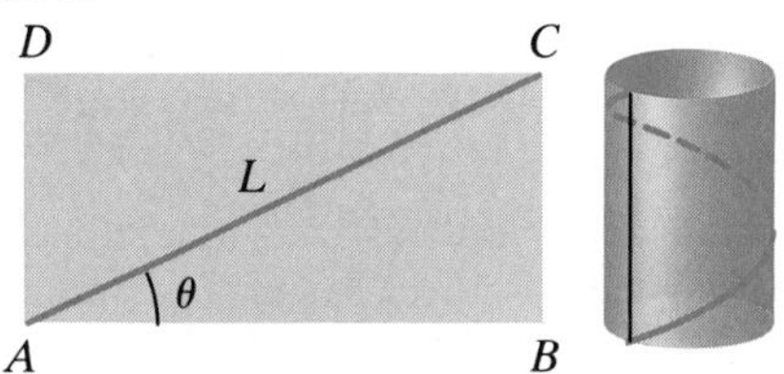

50 When a person is walking, the magnitude F of the vertical force of one foot on the ground (see figure) can be approximated by $F = A(\cos bt - a \cos 3bt)$ for time t (in seconds), with $A > 0$, $b > 0$, and $0 < a < 1$.

(a) Show that $F = 0$ when $t = -\pi/(2b)$ and $t = \pi/(2b)$. (The time $t = -\pi/(2b)$ corresponds to the instant when the foot first touches the ground and the weight of the body is being supported by the other foot.)

(b) Show that the maximum force occurs when $t = 0$ or when $\sin^2 bt = (9a - 1)/(12a)$.

(c) If $a = \frac{1}{3}$, express the maximum force in terms of A.

(d) If $0 < a \le \frac{1}{9}$, express the maximum force in terms of A.

EXERCISE 50

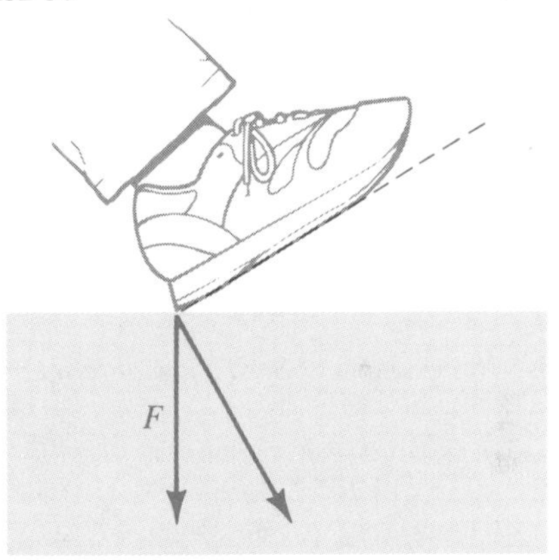

51 A battery having fixed voltage V and fixed internal resistance r is connected to a circuit that has variable resistance R (see figure). By Ohm's law, the current I in the circuit is $I = V/(R + r)$. If the power output P is given by $P = I^2R$, show that the maximum power occurs if $R = r$.

EXERCISE 51

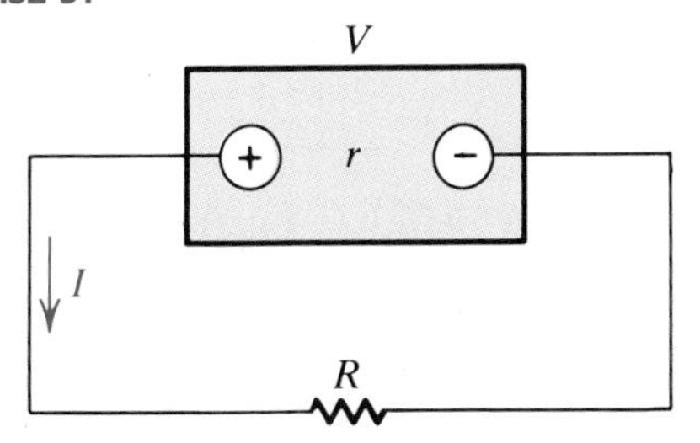

52 The power output P of an automobile battery is given by $P = VI - I^2r$ for voltage V, current I, and internal resistance r of the battery. What current corresponds to the maximum power?

53 Two corridors 3 feet and 4 feet wide, respectively, meet at a right angle. Find the length of the longest nonbendable rod that can be carried horizontally around the corner, as shown in the figure. (Disregard the thickness of the rod.)

EXERCISE 53

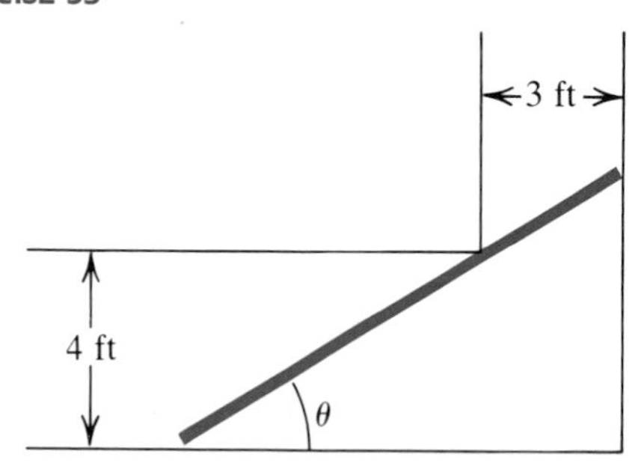

54 Light travels from one point to another along the path that requires the least amount of time. Suppose that light has velocity v_1 in air and v_2 in water, where $v_1 > v_2$. If light travels from a point P in air to a point Q in water (see figure), show that the path requires the least amount of time if

$$\frac{\sin \theta_1}{\sin \theta_2} = \frac{v_1}{v_2}.$$

(This is an example of *Snell's law of refraction.*)

EXERCISE 54

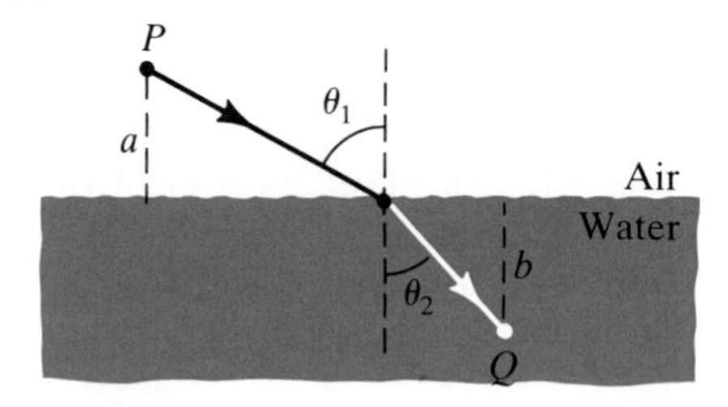

55 A circular cylinder of fixed radius R is surmounted by a cone (see figure). The ends of the cylinder are open, and the total volume is to be a specified constant V.

(a) Show that the total surface area S is given by

$$S = \frac{2V}{R} + \pi R^2\left(\csc \theta - \frac{2}{3}\cot \theta\right).$$

(b) Show that S is minimized when $\theta \approx 48.2^\circ$.

EXERCISE 55

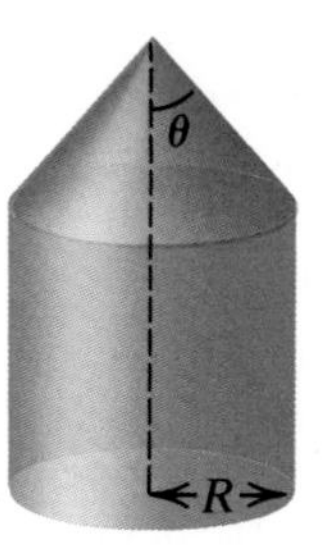

56 In the classic honeycomb-structure problem, a hexagonal prism of fixed radius (and side) R is surmounted by adding three identical rhombuses that meet in a common vertex (see figure). The bottom of the prism is open, and the total volume is to be a specified constant V. A more elaborate geometric argument than that in Exercise 55 establishes that the total surface area S is given by

$$S = \frac{4}{3}\sqrt{3}\,\frac{V}{R} - \frac{3}{2}R^2 \cot \theta + \frac{3\sqrt{3}}{2}R^2 \csc \theta.$$

Show that S is minimized when $\theta \approx 54.7^\circ$. (Remarkably, bees construct their honeycombs so that the amount of wax S is minimized.)

EXERCISE 56

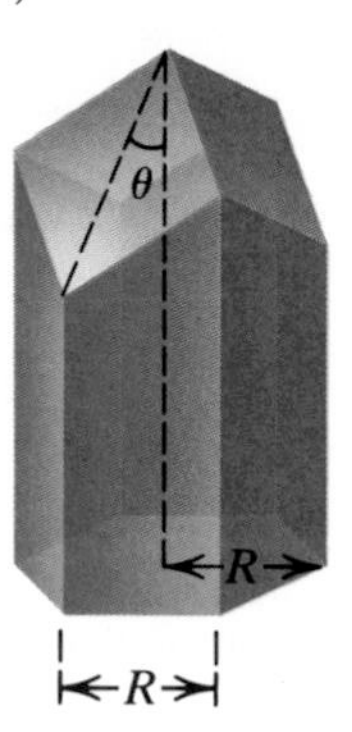

4.7 RECTILINEAR MOTION AND OTHER APPLICATIONS

In this section we use derivatives to describe several important types of motion that occur in physical situations. We also discuss how derivatives may be applied to problems in economics.

FIGURE 4.53

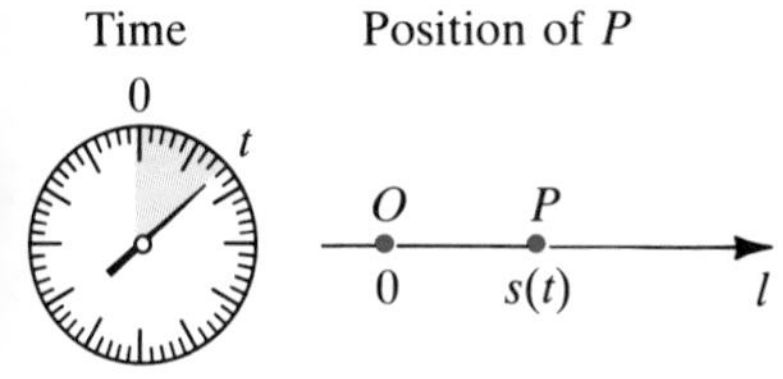

Time   Position of P

0   t O P 0 $s(t)$ l

Recall that if a point P is moving along a line l, its motion is *rectilinear* (see Section 3.1). Moreover, if l is a *coordinate line* and if the coordinate of P at time t is $s(t)$, as illustrated in Figure 4.53, then s is the *position function* of P. By Definition (3.3), the *velocity* of P at time t—that is, the rate of change of P with respect to t—is the derivative $s'(t)$.

The *acceleration* $a(t)$ of P at time t is defined as the rate of change of velocity with respect to time: $a(t) = v'(t)$. Thus, the acceleration is the second derivative $D_t\,[s'(t)] = s''(t)$. The next definition summarizes this discussion and also introduces the notion of the *speed* of P.

Definition **(4.21)**

Let $s(t)$ be the coordinate of a point P on a coordinate line l at time t.

(i) The **velocity** of P is $v(t) = s'(t)$.

(ii) The **speed** of P is $|v(t)|$.

(iii) The **acceleration** of P is $a(t) = v'(t) = s''(t)$.

We shall call v the **velocity function** of P and a the **acceleration function** of P. We sometimes use the notation

$$v = \frac{ds}{dt} \quad \text{and} \quad a = \frac{dv}{dt}.$$

If t is in seconds and $s(t)$ is in centimeters, then $v(t)$ is in cm/sec and $a(t)$ is in cm/sec^2 (centimeters per second per second). If t is in hours and $s(t)$ is in miles, then $v(t)$ is in mi/hr and $a(t)$ is in mi/hr^2 (miles per hour per hour).

If $v(t)$ is positive in a time interval, then $s'(t) > 0$, and, by Theorem (4.13), $s(t)$ is increasing; that is, the point P is moving in the positive direction on l. If $v(t)$ is negative, the motion is in the negative direction. The velocity is zero at a point where P changes direction. If the acceleration $a(t) = v'(t)$ is positive, the velocity is increasing. If $a(t)$ is negative, the velocity is decreasing.

EXAMPLE 1 The position function s of a point P on a coordinate line is given by

$$s(t) = t^3 - 12t^2 + 36t - 20,$$

with t in seconds and $s(t)$ in centimeters. Describe the motion of P during the time interval $[-1, 9]$.

SOLUTION Differentiating, we obtain

$$v(t) = s'(t) = 3t^2 - 24t + 36 = 3(t-2)(t-6)$$

$$a(t) = v'(t) = 6t - 24 = 6(t-4).$$

Let us determine when $v(t) > 0$ and when $v(t) < 0$, since this will tell us when P is moving to the right and to the left, respectively. Since $v(t) = 0$ at $t = 2$ and $t = 6$, we examine the following time subintervals of $[-1, 9]$:

$$(-1, 2), \quad (2, 6), \quad \text{and} \quad (6, 9)$$

We may determine the sign of $v(t)$ by using test values, as indicated in the table (check each entry):

Time interval	$(-1, 2)$	$(2, 6)$	$(6, 9)$
k	0	3	7
Test value $v(k)$	36	-9	15
Sign of $v(t)$	+	−	+
Direction of motion	right	left	right

The next table lists the values of the position, velocity, and acceleration functions at the endpoints of the time interval $[-1, 9]$ and the times at which the velocity or acceleration is zero.

t	-1	2	4	6	9
$s(t)$	-69	12	-4	-20	61
$v(t)$	63	0	-12	0	63
$a(t)$	-30	-12	0	12	30

It is convenient to represent the motion of P schematically, as in Figure 4.54. *The curve above the coordinate line is not the path of the point*, but rather a scheme for showing the manner in which P moves on the line l.

FIGURE 4.54

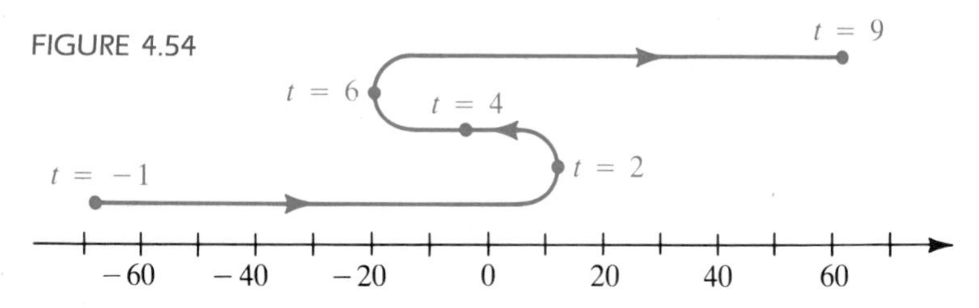

As indicated by the tables and Figure 4.54, at $t = -1$ the point is 69 centimeters to the left of the origin and is moving to the right with a velocity of 63 cm/sec. The negative acceleration -30 cm/sec^2 indicates that the velocity is decreasing at a rate of 30 cm/sec, each second. The point continues to move to the right, slowing down until it has zero velocity at $t = 2$, 12 centimeters to the right of the origin. The point P then reverses direction and moves until, at $t = 6$, it is 20 centimeters to the left of the origin. It then again reverses direction and moves to the right for the remainder of the time interval, with increasing velocity. The direction of motion is indicated by the arrows on the curve in Figure 4.54.

EXAMPLE 2 A projectile is fired straight upward with a velocity of 400 ft/sec. From physics, its distance above the ground after t seconds is $s(t) = -16t^2 + 400t$.

(a) Find the time and the velocity at which the projectile hits the ground.

(b) Find the maximum altitude achieved by the projectile.

(c) Find the acceleration at any time t.

FIGURE 4.55

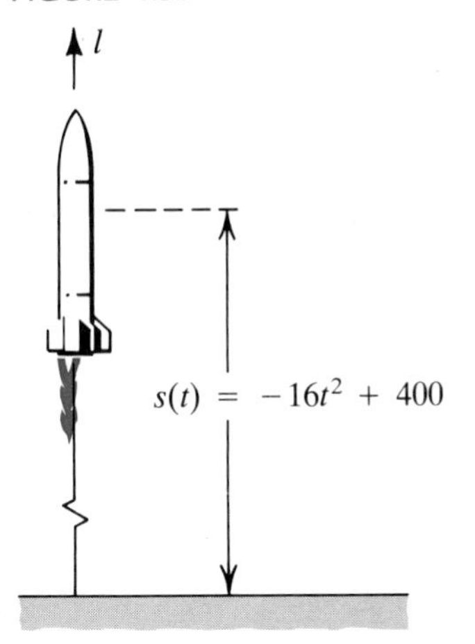

SOLUTION

(a) Let us represent the path of the projectile on a vertical coordinate line l with origin at ground level and positive direction upward, as illustrated in Figure 4.55. The projectile is on the ground when $-16t^2 + 400t = 0$—that is, when $-16t(t - 25) = 0$. This gives us $t = 0$ and $t = 25$. Hence the projectile hits the ground after 25 seconds. The velocity at time t is

$$v(t) = s'(t) = -32t + 400.$$

In particular, at $t = 25$, we obtain the *impact velocity*:

$$v(25) = -32(25) + 400 = -400 \text{ ft/sec}$$

The negative velocity indicates that the projectile is moving in the negative direction on l (downward) at the instant that it strikes the ground. Note that the *speed* at this time is

$$|v(25)| = |-400| = 400 \text{ ft/sec}.$$

(b) The maximum altitude occurs when the velocity is zero—that is, when $s'(t) = -32t + 400 = 0$. Solving for t gives us $t = \frac{400}{32} = \frac{25}{2}$, and hence the maximum altitude is

$$s(\tfrac{25}{2}) = -16(\tfrac{25}{2})^2 + 400(\tfrac{25}{2}) = 2500 \text{ ft}.$$

(c) The acceleration at any time t is

$$a(t) = v'(t) = -32 \text{ ft/sec}^2.$$

This constant acceleration is caused by the force of gravity.

The following type of motion involves trigonometric functions.

Definition **(4.22)**

> A point P moving on a coordinate line l is in **simple harmonic motion** if its distance $s(t)$ from the origin at time t is given by either
>
> $$s(t) = k \cos(\omega t + b) \quad \text{or} \quad s(t) = k \sin(\omega t + b),$$
>
> where k, ω, and b are constants, with $\omega > 0$.

Simple harmonic motion may also be defined by requiring that the acceleration $a(t)$ satisfy the condition

$$a(t) = -\omega^2 s(t)$$

for every t. It can be shown that this condition is equivalent to Definition (4.22).

In simple harmonic motion the point P oscillates between the points on l with coordinates $-k$ and k. The **amplitude** of the motion is the maximum displacement $|k|$ of the point from the origin. The **period** is the time $2\pi/\omega$ required for one complete oscillation. The **frequency** $\omega/2\pi$ is the number of oscillations per unit of time.

Simple harmonic motion takes place in many different types of waves, such as water waves, sound waves, radio waves, light waves, and distortional waves present in vibrating bodies. Functions of the type defined in (4.22) also occur in the analysis of electrical circuits that contain an alternating electromotive force and current.

FIGURE 4.56

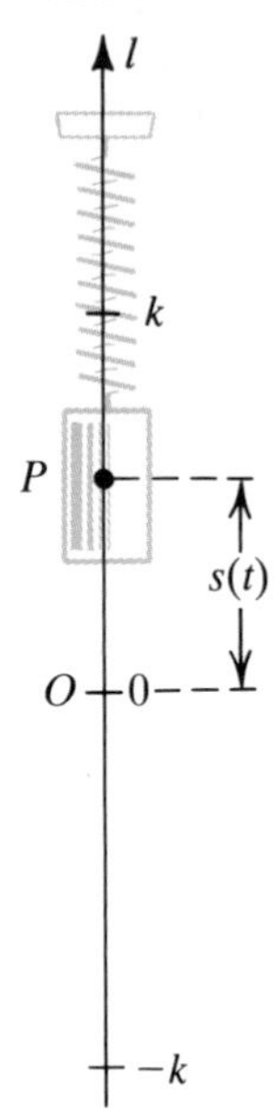

As another example of simple harmonic motion, consider a spring with an attached weight, which is oscillating vertically relative to a coordinate line, as illustrated in Figure 4.56. The number $s(t)$ represents the coordinate of a fixed point P in the weight, and we assume that the amplitude $|k|$ of the motion is constant. In this case, there is no frictional force retarding the motion. If friction is present, then the amplitude decreases with time, and the motion is *damped*.

EXAMPLE 3 Suppose the weight shown in Figure 4.56 is oscillating and

$$s(t) = 10 \cos \frac{\pi}{6} t,$$

where t is in seconds and $s(t)$ is in centimeters. Discuss the motion of the weight.

SOLUTION Comparing the given equation with $s(t) = k \cos(\omega t + b)$ in Definition (4.22), we obtain $k = 10$, $\omega = \pi/6$, and $b = 0$. This gives us the following:

$$\text{amplitude:} \quad k = 10 \text{ cm}$$

$$\text{period:} \quad \frac{2\pi}{\omega} = \frac{2\pi}{\pi/6} = 12 \text{ sec}$$

$$\text{frequency:} \quad \frac{\omega}{2\pi} = \frac{1}{12} \text{ oscillation/sec}$$

Let us examine the motion during the time interval $[0, 12]$. The velocity and acceleration functions are given by the following:

$$v(t) = s'(t) = 10\left(-\sin \frac{\pi}{6} t\right) \cdot \frac{\pi}{6} = -\frac{5\pi}{3} \sin \frac{\pi}{6} t$$

$$a(t) = v'(t) = -\frac{5\pi}{3}\left(\cos \frac{\pi}{6} t\right) \cdot \frac{\pi}{6} = -\frac{5\pi^2}{18} \cos \frac{\pi}{6} t$$

The velocity is 0 at $t = 0$, $t = 6$, and $t = 12$, since $\sin [(\pi/6)t] = 0$ for these values of t. The acceleration is 0 at $t = 3$ and $t = 9$, since in these cases $\cos [(\pi/6)t] = 0$. The times at which the velocity and acceleration are zero lead us to examine the time intervals (0, 3), (3, 6), (6, 9), and (9, 12). The following table displays the main characteristics of the motion. The signs of $v(t)$ and $a(t)$ in the intervals can be determined using test values (verify each entry).

Time interval	Sign of $v(t)$	Direction of motion	Sign of $a(t)$	Variation of $v(t)$	Speed $\lvert v(t) \rvert$
(0, 3)	−	downward	−	decreasing	increasing
(3, 6)	−	downward	+	increasing	decreasing
(6, 9)	+	upward	+	increasing	increasing
(9, 12)	+	upward	−	decreasing	decreasing

Note that if $0 < t < 3$, the velocity $v(t)$ is negative and decreasing; that is, $v(t)$ becomes *more* negative. Hence the absolute value $|v(t)|$, the speed, is *increasing*. If $3 < t < 6$, the velocity is negative and increasing ($v(t)$ becomes *less* negative); that is, the speed of P is *decreasing* in the time interval $(3, 6)$. Similar remarks can be made for the intervals $(6, 9)$ and $(9, 12)$.

We may summarize the motion of P as follows: At $t = 0$, $s(0) = 10$ and the point P is 10 centimeters above the origin O. It then moves downward, gaining speed until it reaches the origin O at $t = 3$. It then slows down until it reaches a point 10 centimeters below O at the end of 6 seconds. The direction of motion is then reversed, and the weight moves upward, gaining speed until it reaches O at $t = 9$, after which it slows down until it returns to its original position at the end of 12 seconds. The direction of motion is then reversed again, and the same pattern is repeated indefinitely.

Calculus has become an important tool for solving problems that occur in economics. If a function f is used to describe some economic entity, the adjective *marginal* is employed to specify the derivative f'.

If x is the number of units of a commodity, economists often use the functions C, c, R, and P, defined as follows:

Cost function: $C(x) = \text{cost of producing } x \text{ units}$

Average cost function: $c(x) = \dfrac{C(x)}{x}$

$= \text{average cost of producing one unit}$

Revenue function: $R(x) = \text{revenue received for selling } x \text{ units}$

Profit function: $P(x) = R(x) - C(x) = \text{profit in selling } x \text{ units}$

To use the techniques of calculus, we regard x as a real number, even though this variable may take on only integer values. We always assume that $x \geq 0$, since the production of a negative number of units has no practical significance.

EXAMPLE 4 A manufacturer of miniature tape decks has a monthly fixed cost of \$10,000, a production cost of \$12 per unit, and a selling price of \$20 per unit.

(a) Find $C(x)$, $c(x)$, $R(x)$, and $P(x)$.

(b) Find the function values in part (a) if $x = 1000$.

(c) How many units must be manufactured in order to break even?

SOLUTION

(a) The production costs of manufacturing x units is $12x$. Since there is also a fixed monthly cost of \$10,000, the total monthly cost of manufacturing x units is

$$C(x) = 12x + 10{,}000.$$

The remaining functions are given by

$$c(x) = \frac{C(x)}{x} = 12 + \frac{10{,}000}{x}$$

$$R(x) = 20x$$

$$P(x) = R(x) - C(x) = 8x - 10{,}000.$$

(b) Substituting $x = 1000$ in part (a) gives us the following values:

$C(1000) = 22{,}000$ (cost of manufacturing 1000 units)

$c(1000) = 22$ (average cost of manufacturing one unit)

$R(1000) = 20{,}000$ (total revenue received for 1000 units)

$P(1000) = -2000$ (profit in manufacturing 1000 units)

Note that the manufacturer incurs a loss of $2000 per month if only 1000 units are produced and sold.

(c) The break-even point corresponds to zero profit—that is, when we have $8x - 10{,}000 = 0$. This gives us

$$8x = 10{,}000, \quad \text{or} \quad x = 1250.$$

Thus, to break even it is necessary to produce and sell 1250 units per month.

The derivatives C', c', R', and P' are called the **marginal cost function**, the **marginal average cost function**, the **marginal revenue function**, and the **marginal profit function**, respectively. The number $C'(x)$ is referred to as the **marginal cost** associated with the production of x units. If we interpret the derivative as a rate of change, then $C'(x)$ is the rate at which the cost changes with respect to the number x of units produced. Similar statements can be made for $c'(x)$, $R'(x)$, and $P'(x)$.

If C is a cost function and n is a positive integer, then, by Definition (3.5),

$$C'(n) = \lim_{h \to 0} \frac{C(n + h) - C(n)}{h}.$$

Hence, if h is small, then

$$C'(n) \approx \frac{C(n + h) - C(n)}{h}.$$

If the number n of units produced is large, economists often let $h = 1$ in the last formula to approximate the marginal cost, obtaining

$$C'(n) \approx C(n + 1) - C(n).$$

In this context, *the marginal cost associated with the production of n units is (approximately) the cost of producing one more unit.*

Some companies find that the cost $C(x)$ of producing x units of a commodity is given by a formula such as

$$C(x) = a + bx + dx^2 + kx^3.$$

The constant a represents a fixed overhead charge for items like rent, heat, and light that are independent of the number of units produced. If the cost of producing one unit were b dollars and no other factors were involved, then the second term bx in the formula would represent the cost of producing x units. If x becomes very large, then the terms dx^2 and kx^3 may significantly affect production costs.

EXAMPLE 5 An electronics company estimates that the cost (in dollars) of producing x components used in electronic toys is given by

$$C(x) = 200 + 0.05x + 0.0001x^2.$$

(a) Find the cost, the average cost, and the marginal cost of producing 500 units, 1000 units, and 5000 units.

(b) Compare the marginal cost of producing 1000 units with the cost of producing the 1001st unit.

SOLUTION

(a) The average cost of producing x components is

$$c(x) = \frac{C(x)}{x} = \frac{200}{x} + 0.05 + 0.0001x.$$

The marginal cost is

$$C'(x) = 0.05 + 0.0002x.$$

You should verify the entries in the following table, where numbers in the last three columns represent dollars.

Units x	Cost $C(x)$	Average cost $c(x) = \frac{C(x)}{x}$	Marginal cost $C'(x)$
500	250.00	0.50	0.15
1000	350.00	0.35	0.25
5000	2950.00	0.59	1.05

(b) Using the cost function yields

$$C(1001) = 200 + 0.05(1001) + (0.0001)(1001)^2 \approx 350.25.$$

Hence the cost of producing the 1001st unit is

$$C(1001) - C(1000) \approx 350.25 - 350.00 = 0.25,$$

which is approximately the same as the marginal cost $C'(1000)$.

EXAMPLE 6 A furniture company estimates that the weekly cost (in dollars) of manufacturing x hand-finished reproductions of a colonial desk is given by $C(x) = x^3 - 3x^2 - 80x + 500$. Each desk produced is sold for \$2800. What weekly production rate will maximize the profit? What is the largest possible profit per week?

SOLUTION Since the income obtained from selling x desks is $2800x$, the revenue function R is given by $R(x) = 2800x$. The profit function P is the difference between the revenue function R and the cost function

C—that is,

$$P(x) = R(x) - C(x) = 2800x - (x^3 - 3x^2 - 80x + 500),$$

or

$$P(x) = -x^3 + 3x^2 + 2880x - 500.$$

To find the maximum profit, we differentiate, obtaining

$$P'(x) = -3x^2 + 6x + 2880 = -3(x^2 - 2x - 960).$$

The critical numbers of P are found by solving

$$x^2 - 2x - 960 = 0, \quad \text{or} \quad (x - 32)(x + 30) = 0,$$

which yields $x = 32$ or $x = -30$. Since the negative solution is extraneous, it suffices to check $x = 32$.

The second derivative of the profit function P is

$$P''(x) = -6x + 6.$$

Consequently,

$$P''(32) = -6(32) + 6 = -186 < 0.$$

Thus, by the second derivative test, a maximum profit occurs if 32 desks per week are manufactured and sold. The maximum weekly profit is

$$P(32) = -(32)^3 + 3(32)^2 + 2880(32) - 500 = \$61{,}964.$$

A company must consider many factors in order to determine a selling price for each product. In addition to the cost of production and the profit desired, the company should be aware of the manner in which consumer demand will vary if the price increases. For some products there is a constant demand, and changes in price have little effect on sales. For items that are not necessities of life, a price increase will probably lead to a decrease in the number of units sold. Suppose a company knows from past experience that it can sell x units when the price per unit is given by $p(x)$ for some function p. We sometimes say that $p(x)$ is the price per unit when there is a **demand** for x units, and we refer to p as the **demand function** for the commodity. The total income, or revenue, is the number of units sold times the price per unit—that is, $x \cdot p(x)$. Thus,

$$R(x) = xp(x).$$

The derivative p' is called the **marginal demand function.**

If $S = p(x)$, then S is the selling price per unit associated with a demand of x units. Since a decrease in S would ordinarily be associated with an increase in x, a demand function p is usually decreasing; that is, $p'(x) < 0$ for every x. Demand functions are sometimes defined implicitly by an equation involving S and x, as in the next example.

EXAMPLE 7 The demand for x units of a product is related to a selling price of S dollars per unit by the equation $2x + S^2 - 12{,}000 = 0$.

(a) Find the demand function, the marginal demand function, the revenue function, and the marginal revenue function.

(b) Find the number of units and the price per unit that yield the maximum revenue.

(c) Find the maximum revenue.

FIGURE 4.57

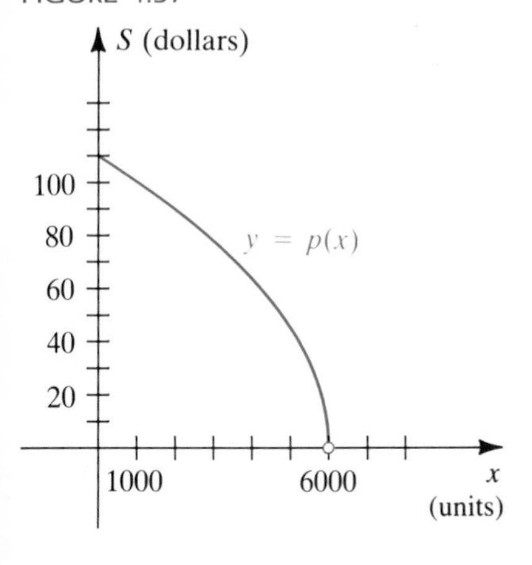

SOLUTION

(a) Since $S^2 = 12{,}000 - 2x$ and S is positive, we see that the demand function p is given by

$$S = p(x) = \sqrt{12{,}000 - 2x}.$$

The domain of p consists of every x such that $12{,}000 - 2x > 0$, or, equivalently, $2x < 12{,}000$. Thus, $0 \le x < 6000$. The graph of p is sketched in Figure 4.57. In theory, there are no sales if the selling price is $\sqrt{12{,}000}$, or approximately \$109.54, and when the selling price is close to \$0 the demand is close to 6000.

The marginal demand function p' is given by

$$p'(x) = \frac{-1}{\sqrt{12{,}000 - 2x}}.$$

The negative sign indicates that a decrease in price is associated with an increase in demand.

The revenue function R is given by

$$R(x) = xp(x) = x\sqrt{12{,}000 - 2x}.$$

Differentiating and simplifying gives us the marginal revenue function R':

$$R'(x) = \frac{12{,}000 - 3x}{\sqrt{12{,}000 - 2x}}$$

(b) A critical number for the revenue function R is $x = 12{,}000/3 = 4000$. Since $R'(x)$ is positive if $0 \le x < 4000$ and negative if $4000 < x < 6000$, the maximum revenue occurs when 4000 units are produced and sold. This corresponds to a selling price per unit of

$$p(4000) = \sqrt{12{,}000 - 2(4000)} \approx \$63.25.$$

(c) The maximum revenue, obtained from selling 4000 units at \$63.25 per unit, is

$$4000(63.25) = \$253{,}000.$$

EXERCISES 4.7

Exer. 1–8: A point moving on a coordinate line has position function s. Find the velocity and acceleration at time t, and describe the motion of the point during the indicated time interval. Illustrate the motion by means of a diagram of the type shown in Figure 4.54.

1 $s(t) = 3t^2 - 12t + 1$; $[0, 5]$

2 $s(t) = t^2 + 3t - 6$; $[-2, 2]$

3 $s(t) = t^3 - 9t + 1$; $[-3, 3]$

4 $s(t) = 24 + 6t - t^3$; $[-2, 3]$

5 $s(t) = -2t^3 + 15t^2 - 24t - 6$; $[0, 5]$

6 $s(t) = 2t^3 - 12t^2 + 6$; $[-1, 6]$

7 $s(t) = 2t^4 - 6t^2$; $[-2, 2]$

8 $s(t) = 2t^3 - 6t^5$; $[-1, 1]$

Exer. 9–10: An automobile rolls down an incline, traveling $s(t)$ feet in t seconds. (a) Find its velocity at $t = 3$. (b) After how many seconds will the velocity be k ft/sec?

9 $s(t) = 5t^2 + 2; \quad k = 28$

10 $s(t) = 3t^2 + 7; \quad k = 88$

Exer. 11–12: A projectile is fired directly upward with an initial velocity of v_0 ft/sec, and its height (in feet) above the ground after t seconds is given by $s(t)$. Find (a) the velocity and acceleration after t seconds, (b) the maximum height, and (c) the duration of the flight.

11 $v_0 = 144; \quad s(t) = 144t - 16t^2$

12 $v_0 = 192; \quad s(t) = 100 + 192t - 16t^2$

Exer. 13–16: A particle in simple harmonic motion has position function s, and t is the time in seconds. Find the amplitude, period, and frequency.

13 $s(t) = 5 \cos \frac{\pi}{4} t$

14 $s(t) = 4 \sin \pi t$

15 $s(t) = 6 \sin \frac{2\pi}{3} t$

16 $s(t) = 3 \cos 2t$

17 The electromotive force V and current I in an alternating-current circuit are given by

$$V = 220 \sin 360\pi t$$
$$I = 20 \sin \left(360\pi t - \frac{\pi}{4} \right).$$

Find the rates of change of V and I with respect to time at $t = 1$.

18 The annual variation in temperature T (in °C) in Vancouver, B.C., may be approximated by the formula

$$T = 14.8 \sin \left[\frac{\pi}{6} (t - 3) \right] + 10,$$

where t is in months, with $t = 0$ corresponding to January 1. Approximate the rate at which the temperature is changing at time $t = 3$ (April 1) and at time $t = 10$ (November 1). At what time of the year is the temperature changing most rapidly?

19 The graph in the figure shows the rise and fall of the water level in Boston Harbor during a particular 24-hour period.

(a) Approximate the water level y by means of an expression of the form

$$y = a \sin (bt + c) + d,$$

with $t = 0$ corresponding to midnight.

(b) Approximately how fast is the water level rising at 12 noon?

EXERCISE 19

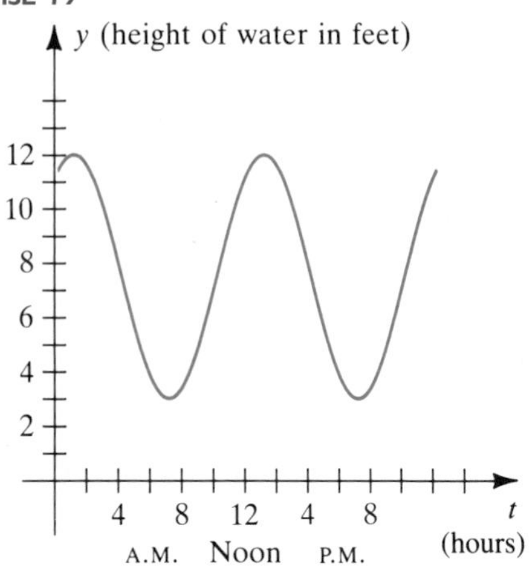

20 A *tsunami* is a tidal wave caused by an earthquake beneath the sea. These waves can be more than a hundred feet in height and can travel at great speeds. Engineers sometimes represent tsunamis by an equation of the form $y = a \cos bt$. Suppose that a wave has height $h = 25$ ft and period 30 minutes and is traveling at the rate of 180 ft/sec.

(a) Let (x, y) be a point on the wave represented in the figure. Express y as a function of t if $y = 25$ ft when $t = 0$.

(b) How fast is the wave rising (or falling) when $y = 10$ ft?

EXERCISE 20

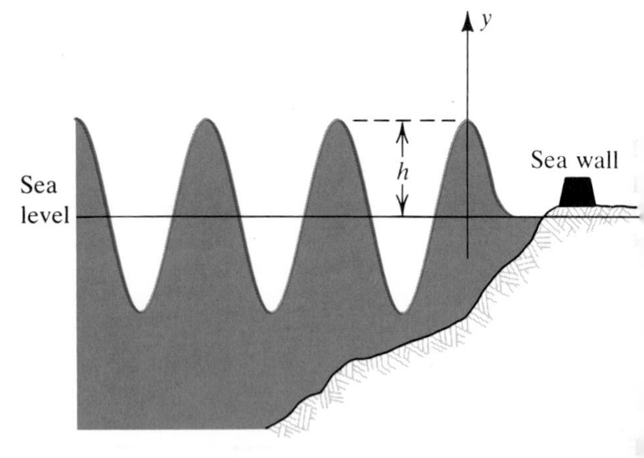

21 A cork bobs up and down in a lake. The distance from the bottom of the lake to the center of the cork at time $t \geq 0$ is given by $s(t) = \cos \pi t + 12$, where $s(t)$ is in inches and t is in seconds. (See figure on page 231.)

(a) Find the velocity of the cork at $t = 0, \frac{1}{2}, 1, \frac{3}{2}$, and 2.

(b) During what time intervals is the cork rising?

EXERCISE 21

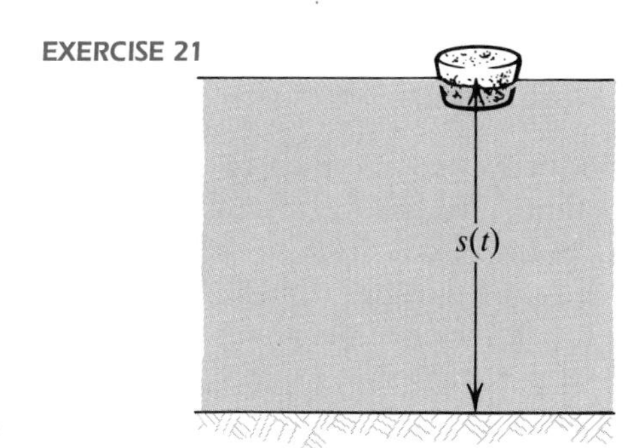

22 A particle in a vibrating spring is moving vertically such that its distance $s(t)$ from a fixed point on the line of vibration is given by $s(t) = 4 + \frac{1}{25} \sin 100\pi t$, where $s(t)$ is in centimeters and t is in seconds.

(a) How long does it take the particle to make one complete vibration?

(b) Find the velocity of the particle at $t = 1, 1.005, 1.01$, and 1.015.

Exer. 23–24: Show that $s''(t) = -\omega^2 s(t)$.

23 $s(t) = k \cos(\omega t + b)$

24 $s(t) = k \sin(\omega t + b)$

25 A point $P(x, y)$ is moving at a constant rate around the circle $x^2 + y^2 = a^2$. Prove that the projection $Q(x, 0)$ of P onto the x-axis is in simple harmonic motion.

26 If a point P moves on a coordinate line such that

$$s(t) = a \cos \omega t + b \sin \omega t,$$

show that P is in simple harmonic motion by

(a) using the remark following Definition (4.22)

(b) using only trigonometric methods (*Hint:* Show that $s(t) = A \cos(\omega t - c)$ for some constants A and c.)

Exer. 27–28: A point moving on a coordinate line has position function s. (a) Graph $y = s(t)$ for $0 \le t \le 5$. (b) Approximate the point's position, velocity, and acceleration at $t = 0, 1, 2, 3, 4$, and 5.

27 $s(t) = \dfrac{10 \sin t}{t^2 + 1}$

28 $s(t) = \dfrac{5 \tan(\frac{1}{4}t)}{2t + 1}$

Exer. 29–32: If C is the cost function for a particular product, find (a) the cost of producing 100 units and (b) the average and marginal cost functions and their values at $x = 100$.

29 $C(x) = 800 + 0.04x + 0.0002x^2$

30 $C(x) = 6400 + 6.5x + 0.003x^2$

31 $C(x) = 250 + 100x + 0.001x^3$

32 $C(x) = 200 + 0.01x + \dfrac{100}{x}$

33 A manufacturer of small motors estimates that the cost (in dollars) of producing x motors per day is given by $C(x) = 100 + 50x + (100/x)$. Compare the marginal cost of producing five motors with the cost of producing the sixth motor.

34 A company conducts a pilot test for production of a new industrial solvent and finds that the cost of producing x liters of each pilot run is given by the formula $C(x) = 3 + x + (10/x)$. Compare the marginal cost of producing 10 liters with the cost of producing the 11th liter.

Exer. 35–36: For the given demand and cost functions, find (a) the marginal demand function, (b) the revenue function, (c) the profit function, (d) the marginal profit function, (e) the maximum profit, and (f) the marginal cost when the demand is 10 units.

35 $p(x) = 50 - 0.1x$; $C(x) = 10 + 2x$

36 $p(x) = 80 - \sqrt{x - 1}$; $C(x) = 75x + 2\sqrt{x - 1}$

37 A travel agency estimates that, in order to sell x package-deal vacations, it must charge a price per vacation of $1800 - 2x$ dollars for $1 \le x \le 100$. If the cost to the agency for x vacations is $1000 + x + 0.01x^2$ dollars, find

(a) the revenue function

(b) the profit function

(c) the number of vacations that will maximize the profit

(d) the maximum profit

38 A manufacturer determines that x units of a product will be sold if the selling price is $400 - 0.05x$ dollars for each unit. If the production cost for x units is $500 + 10x$, find

(a) the revenue function

(b) the profit function

(c) the number of units that will maximize the profit

(d) the price per unit when the marginal revenue is 300

39 A kitchen specialty company determines that the cost of manufacturing and packaging x pepper mills per day is $500 + 0.02x + 0.001x^2$. If each mill is sold for \$8.00, find

(a) the rate of production that will maximize the profit

(b) the maximum daily profit

40 A company that conducts bus tours found that when the price was \$9.00 per person, the average number of customers was 1000 per week. When the company reduced the price to \$7.00 per person, the average number of customers increased to 1500 per week. Assuming that the demand function is linear, what price should be charged to obtain the greatest weekly revenue?

4.8 NEWTON'S METHOD

In this section we show how derivatives can be used to approximate a real zero of a differentiable function f. To use the method, we begin by making a first approximation x_1 to the zero r. Since $f(r) = 0$, the number r is an x-intercept of the graph of f, and the approximation x_1 can sometimes be made by referring to a rough sketch of the graph. If we consider the tangent line l to the graph of f at the point $(x_1, f(x_1))$ and if x_1 is sufficiently close to r, then, as illustrated in Figure 4.58, the x-intercept x_2 of l should be a better approximation to r.

FIGURE 4.58

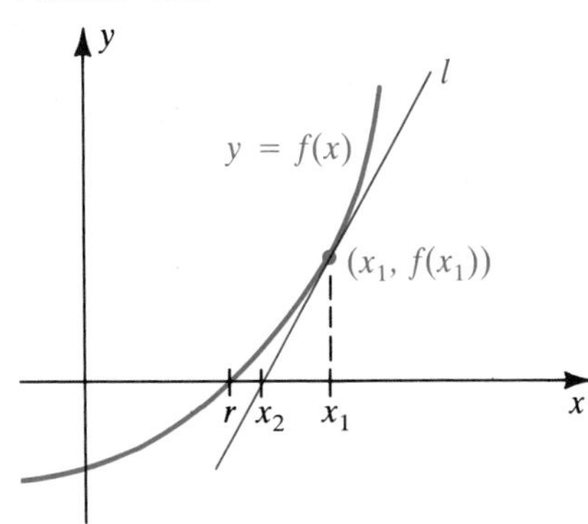

Since the slope of l is $f'(x_1)$, an equation of the tangent line is

$$y - f(x_1) = f'(x_1)(x - x_1).$$

The x-intercept x_2 of l corresponds to the point $(x_2, 0)$ on l, so

$$0 - f(x_1) = f'(x_1)(x_2 - x_1).$$

If $f'(x_1) \neq 0$, the preceding equation is equivalent to

$$x_2 = x_1 - \frac{f(x_1)}{f'(x_1)}.$$

If we take x_2 as a second approximation to r, then the process may be repeated by using the tangent line at $(x_2, f(x_2))$. If $f'(x_2) \neq 0$, a third approximation x_3 is given by

$$x_3 = x_2 - \frac{f(x_2)}{f'(x_2)}.$$

The process is continued until the desired degree of accuracy is obtained. This technique of successive approximations of real zeros is referred to as **Newton's method**, which we state as follows.

Newton's method **(4.23)**

Let f be a differentiable function, and suppose r is a real zero of f. If x_n is an approximation to r, then the next approximation x_{n+1} is given by

$$x_{n+1} = x_n - \frac{f(x_n)}{f'(x_n)},$$

provided $f'(x_n) \neq 0$.

FIGURE 4.59

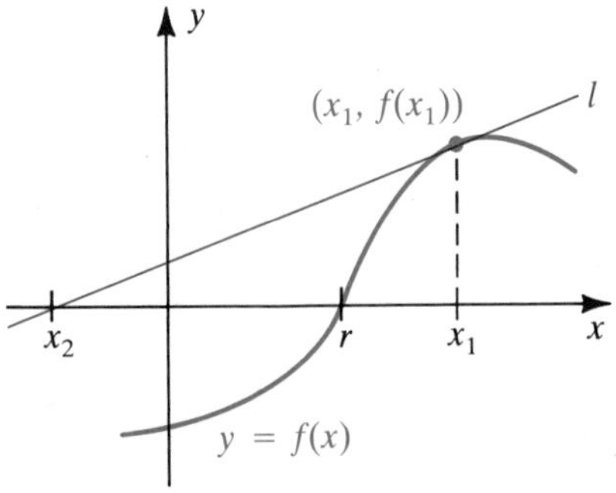

Newton's method does not guarantee that x_{n+1} is a better approximation to r than x_n for every n. In particular, we must be careful in choosing the first approximation x_1. If x_1 is not sufficiently close to r, it is possible for the second approximation x_2 to be worse than x_1, as illustrated in Figure 4.59. It is evident that we should not choose a number x_n such that $f'(x_n)$ is close to 0, for then the tangent line l is almost horizontal.

It can be shown that if $x_1 \approx r$ and if f'' is continuous near r and if $f'(r) \neq 0$, then the approximations $x_2, x_3, \ldots$ approach r rapidly, with the number of decimal places of accuracy nearly doubling with each successive approximation. If $f(x)$ has a factor $(x - r)^k$ with $k > 1$ and if $x_n \neq r$ for each n, then the approximations approach r more slowly, because $f'(r) = 0$.

We shall use the following rule when applying Newton's method: *If an approximation to k decimal places is required, we shall approximate each of the numbers $x_2, x_3, \ldots$ to k decimal places, continuing the process until two consecutive approximations are the same*. This process is illustrated in the following examples.

EXAMPLE 1 Use Newton's method to approximate $\sqrt{7}$ to five decimal places.

FIGURE 4.60

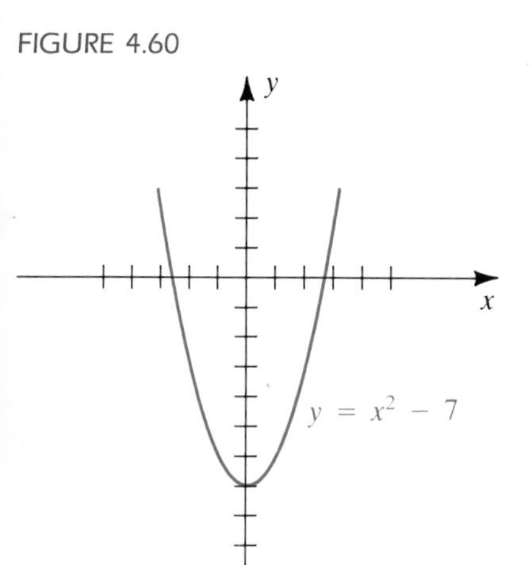

SOLUTION The stated problem is equivalent to that of approximating the positive real zero r of $f(x) = x^2 - 7$. The graph of f is sketched in Figure 4.60. Since $f(2) = -3$ and $f(3) = 2$, it follows from the continuity of f that $2 < r < 3$. Moreover, since f is increasing, there can be only one zero in the open interval (2, 3).

If x_n is any approximation to r, then, by (4.23), the next approximation x_{n+1} is given by

$$x_{n+1} = x_n - \frac{f(x_n)}{f'(x_n)} = x_n - \frac{x_n^2 - 7}{2x_n}.$$

Let us choose $x_1 = 2.5$ as a first approximation. Using the formula for x_{n+1} with $n = 1$ gives us

$$x_2 = 2.5 - \frac{(2.5)^2 - 7}{2(2.5)} = 2.65000.$$

Again using the formula (with $n = 2$), we obtain the next approximation,

$$x_3 = 2.65000 - \frac{(2.65000)^2 - 7}{2(2.65000)} \approx 2.64575.$$

Repeating the procedure (with $n = 3$) yields

$$x_4 = 2.64575 - \frac{(2.64575)^2 - 7}{2(2.64575)} \approx 2.64575.$$

Since two consecutive values of x_n are the same (to the desired degree of accuracy), we have $\sqrt{7} \approx 2.64575$. It can be shown that, to nine decimal places, $\sqrt{7} \approx 2.645751311$.

FIGURE 4.61

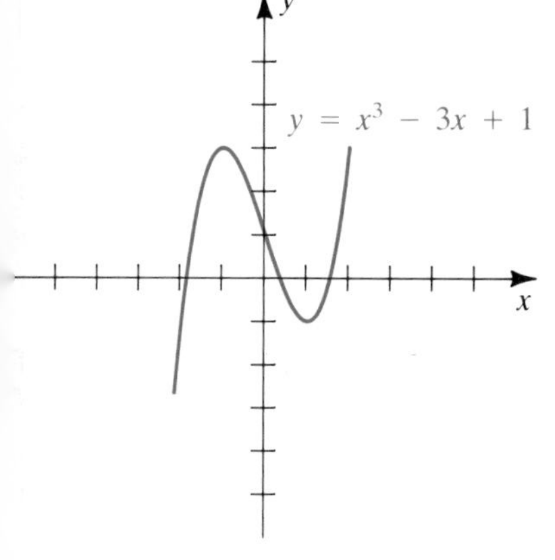

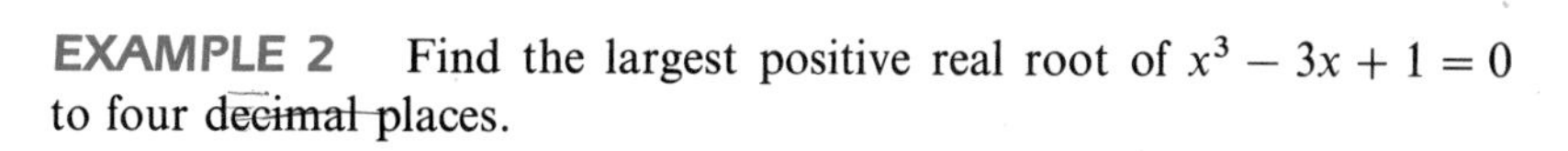

EXAMPLE 2 Find the largest positive real root of $x^3 - 3x + 1 = 0$ to four decimal places.

SOLUTION If we let $f(x) = x^3 - 3x + 1$, then the problem is equivalent to finding the largest positive real zero of f. The graph of f is sketched in Figure 4.61. Note that f has three real zeros. We wish to find the zero that lies between 1 and 2. Since $f'(x) = 3x^2 - 3$, the formula for x_{n+1} in Newton's method is

$$x_{n+1} = x_n - \frac{x_n^3 - 3x_n + 1}{3x_n^2 - 3}.$$

Referring to the graph, we take $x_1 = 1.5$ as a first approximation and proceed as follows:

$$x_2 = 1.5 - \frac{(1.5)^3 - 3(1.5) + 1}{3(1.5)^2 - 3} \approx 1.5333$$

$$x_3 = 1.5333 - \frac{(1.5333)^3 - 3(1.5333) + 1}{3(1.5333)^2 - 3} \approx 1.5321$$

$$x_4 = 1.5321 - \frac{(1.5321)^3 - 3(1.5321) + 1}{3(1.5321)^2 - 3} \approx 1.5321$$

Thus, the desired approximation is 1.5321. The remaining two real roots can be approximated in similar fashion (see Exercise 17).

EXAMPLE 3 Approximate the real root of $x - \cos x = 0$ to three decimal places.

FIGURE 4.62

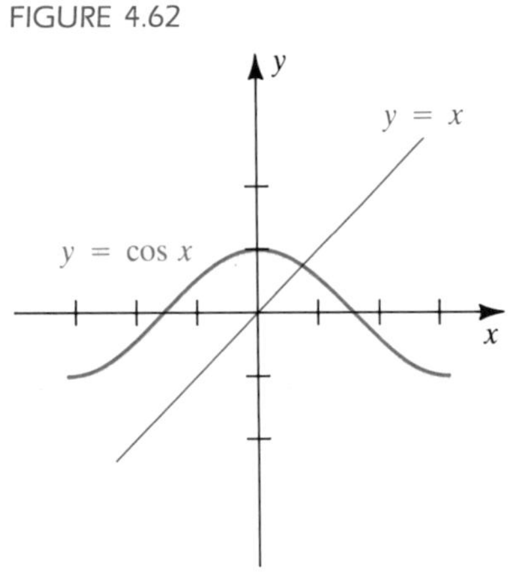

SOLUTION We wish to find a value of x such that $\cos x = x$. This coincides with the x-coordinate of the point of intersection of the graphs of $y = \cos x$ and $y = x$. It appears, from Figure 4.62, that $x_1 = 0.8$ is a reasonable first approximation. (Note that the figure also indicates that there is only one real root of the given equation.)

If we let $f(x) = x - \cos x$, then $f'(x) = 1 + \sin x$ and the formula in Newton's method is

$$x_{n+1} = x_n - \frac{x_n - \cos x_n}{1 + \sin x_n}.$$

Following the usual procedure, we obtain

$$x_2 = 0.8 - \frac{0.8 - \cos 0.8}{1 + \sin 0.8} \approx 0.740$$

$$x_3 = 0.740 - \frac{0.740 - \cos 0.740}{1 + \sin 0.740} \approx 0.739$$

$$x_4 = 0.739 - \frac{0.739 - \cos 0.739}{1 + \sin 0.739} \approx 0.739.$$

Hence 0.739 is the desired approximation.

EXERCISES 4.8

c Use Newton's method for Exercises 1–24.

Exer. 1–2: Approximate to four decimal places.

1 $\sqrt[3]{2}$

2 $\sqrt[5]{3}$

Exer. 3–6: Approximate, to four decimal places, the root of the equation that lies in the interval.

3 $x^4 + 2x^3 - 5x^2 + 1 = 0$; $[1, 2]$

4 $x^4 - 5x^2 + 2x - 5 = 0$; $[2, 3]$

5 $x^5 + x^2 - 9x - 3 = 0$; $[-2, -1]$

6 $\sin x + x \cos x = \cos x$; $[0, 1]$

Exer. 7–8: Approximate the largest zero of $f(x)$ to four decimal places.

7 $f(x) = x^4 - 11x^2 - 44x - 24$

8 $f(x) = x^3 - 36x - 84$

Exer. 9–12: Approximate the real root to two decimal places.

9 The root of $x^3 + 5x - 3 = 0$

10 The largest root of $2x^3 - 4x^2 - 3x + 1 = 0$

11 The positive root of $2x - 3 \sin x = 0$

12 The root of $\cos x + x = 2$

Exer. 13–20: Approximate all real roots of the equation to two decimal places.

13 $x^4 = 125$

14 $10x^2 - 1 = 0$

15 $x^4 - x - 2 = 0$

16 $x^5 - 2x^2 + 4 = 0$

17 $x^3 - 3x + 1 = 0$

18 $x^3 + 2x^2 - 8x - 3 = 0$

19 $2x - 5 - \sin x = 0$

20 $x^2 - \cos 2x = 0$

Exer. 21–24: Approximate, to two decimal places, the *x*-coordinates of the points of intersection of the graphs of the equations.

21 $y = x^2$; $\quad y = \sqrt{x + 3}$

22 $y = x^3$; $\quad y = 7 - x^2$

23 $y = \cos \frac{1}{2}x$; $\quad y = 9 - x^2$

24 $y = \sin 2x$; $\quad y = 6x - 6$

25 Approximations to π may be obtained by applying Newton's method to $f(x) = \sin x$ and letting $x_1 = 3$.

(a) Find the first five approximations to π.

(b) What happens to the approximations if $x_1 = 6$?

26 A dramatic example of the phenomenon of *resonance* occurs when a singer adjusts the pitch of her voice to shatter a wine glass. Functions given by $f(x) = ax \cos bx$ occur in the mathematical analysis of such vibrations. Shown in the figure is a graph of $f(x) = x \cos 2x$. Use Newton's method to approximate, to three decimal places, the critical number of f that lies between 1 and 2.

EXERCISE 26

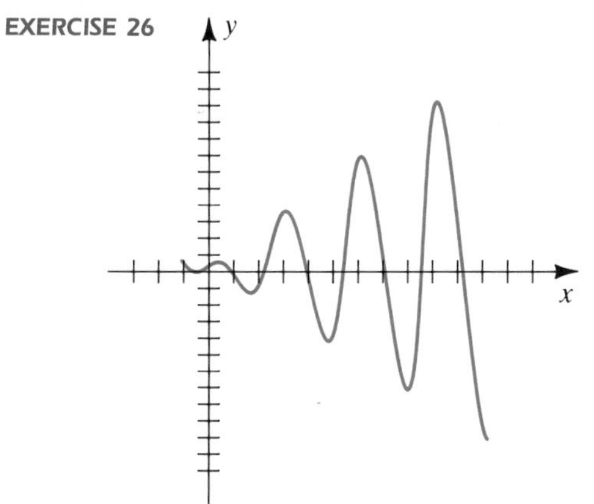

27 The graph of a function f is shown. Explain why Newton's method fails to approximate the zero of f if $x_1 = 0.5$.

EXERCISE 27

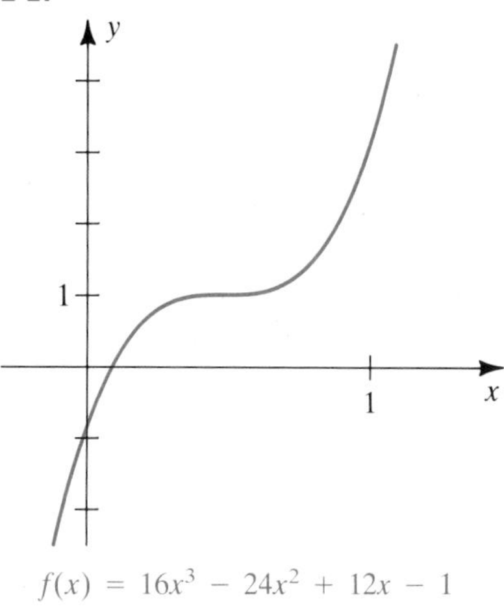

28 If $f(x) = x^{1/3}$, show that Newton's method fails for any first approximation $x_1 \neq 0$.

c **Exer. 29–30: The functions f and g have a zero at $x = 1$. (a) Let $x_1 = 1.1$ in (4.23) and find x_2, x_3, and x_4 for each function. (b) Why are the approximations for the zero of g more accurate than those for the zero of f?**

29 $f(x) = (x - 1)^3(x^2 - 3x + 7)$; $g(x) = (x - 1)(x^2 - 3x + 7)$

30 $f(x) = (x - 1)^2\sqrt{x + 7}$; $\quad g(x) = (x - 1)\sqrt{x + 7}$

c **Exer. 31–32: If it is difficult to calculate $f'(x)$, the formula in (4.23) may be replaced by**

$$x_{n+1} = x_n - \frac{f(x_n)}{m},$$

where

$$m = \frac{f(x_n) - f(x_{n-1})}{x_n - x_{n-1}} \approx f'(x_n).$$

Two initial values, x_1 and x_2, are required to use this method (called the *secant method*). Use the secant method to approximate, to three decimal places, the zero of f that is in [0, 1].

31 $f(x) = \tan^2 (\cos^2 x - x + 0.25) - 0.5x$; $\quad x_1 = 0.5, \quad x_2 = 0.55$

32 $f(x) = \dfrac{1}{\cos^2 x - x} - 5\sqrt{x}$; $\quad x_1 = 0.4, \quad x_2 = 0.5$

c **Exer. 33–34: Graph f and g on the same coordinate axes. (a) Estimate, to one decimal place, the *x*-coordinate x_1 of the point of intersection of the graphs. (b) Use Newton's method (4.23) to approximate the *x*-coordinate in (a) to two decimal places.**

33 $f(x) = \frac{1}{4}x^3 + x - 1$; $\quad g(x) = \sin^2 x$

34 $f(x) = x^3 - x^2 + x - 1$; $g(x) = -x^3 - 1.1x^2 - x - 1.9$

4.9 REVIEW EXERCISES

Exer. 1–2: Find the extrema of f on the given interval.

1 $f(x) = -x^2 + 6x - 8;\quad [1, 6]$

2 $f(x) = 3x^3 + x^2;\quad (-1, 0]$

Exer. 3–4: Find the critical numbers of f.

3 $f(x) = (x + 2)^3(3x - 1)^4$

4 $f(x) = \sqrt{x - 1}\,(x - 2)^3$

Exer. 5–8: Use the first derivative test to find the local extrema of f. Find the intervals on which f is increasing or is decreasing, and sketch the graph of f.

5 $f(x) = -4x^3 + 9x^2 + 12x$

6 $f(x) = \dfrac{1}{x^2 + 1}$

7 $f(x) = (4 - x)x^{1/3}$

8 $f(x) = \sqrt[3]{x^2 - 9}$

Exer. 9–12: Use the second derivative test (whenever applicable) to find the local extrema of f. Find the intervals on which the graph of f is concave upward or is concave downward, and find the x-coordinates of the points of inflection. Sketch the graph of f.

9 $f(x) = \sqrt[3]{8 - x^3}$

10 $f(x) = -x^3 + 4x^2 - 3x$

11 $f(x) = \dfrac{1}{x^2 + 1}$

12 $f(x) = 40x^3 - x^6$

13 If $f(x) = 2 \sin x - \cos 2x$, find the local extrema, and sketch the graph of f for $0 \le x \le 2\pi$.

14 If $f(x) = 2 \sin x - \cos 2x$, find equations of the tangent and normal lines to the graph of f at the point $(\pi/6, 1/2)$.

Exer. 15–16: Sketch the graph of a continuous function f that satisfies all of the stated conditions.

15 $f(0) = 2$; $f(-2) = f(2) = 0$; $f'(-2) = f'(0) = f'(2) = 0$; $f'(x) > 0$ if $-2 < x < 0$; $f'(x) < 0$ if $x < -2$ or $x > 0$; $f''(x) > 0$ if $x < -1$ or $1 < x < 2$; $f''(x) < 0$ if $-1 < x < 1$ or $x > 2$

16 $f(0) = 4$; $f(-3) = f(3) = 0$; $f'(-3) = 0$; $f'(0)$ is undefined; $f'(x) > 0$ if $-3 < x < 0$; $f'(x) < 0$ if $x < -3$ or $x > 0$; $f''(x) > 0$ if $x < 0$ or $0 < x < 2$; $f''(x) < 0$ if $x > 2$

Exer. 17–22: Find the extrema and sketch the graph of f.

17 $f(x) = \dfrac{3x^2}{9x^2 - 25}$

18 $f(x) = \dfrac{x^2}{(x - 1)^2}$

19 $f(x) = \dfrac{x^2 + 2x - 8}{x + 3}$

20 $f(x) = \dfrac{x^4 - 16}{x^3}$

21 $f(x) = \dfrac{x - 3}{x^2 + 2x - 8}$

22 $f(x) = \dfrac{x}{\sqrt{x + 4}}$

23 If $f(x) = x^3 + x^2 + x + 1$, find a number c that satisfies the conclusion of the mean value theorem on the interval $[0, 4]$.

24 The posted speed limit on a 125-mile toll highway is 65 mi/hr. When an automobile enters the toll road, the driver is issued a ticket on which is printed the exact time. If the driver completes the trip in 1 hour 40 minutes or less, a speeding citation is issued when the toll is paid. Use the mean value theorem to explain why this citation is justified.

25 A man wishes to put a fence around a rectangular field and then subdivide this field into three smaller rectangular plots by placing two fences parallel to one of the sides. If he can afford only 1000 yards of fencing, what dimensions will give him the maximum area?

26 An open rectangular storage shelter 4 feet deep, consisting of two vertical sides and a flat roof, is to be attached to an existing structure, as illustrated in the figure. The flat roof is made of tin and costs \$5 per ft^2. The two sides are made of plywood costing \$2 per ft^2. If \$400 is available for construction, what dimensions will maximize the volume of the shelter?

EXERCISE 26

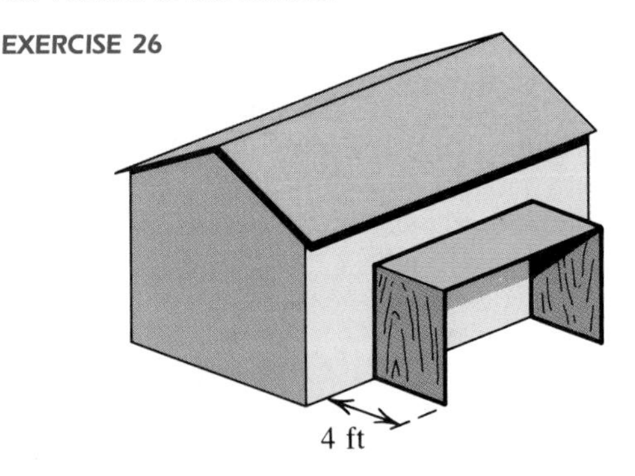

27 A V-shaped water gutter is to be constructed from two rectangular sheets of metal 10 inches wide. Find the angle between the sheets that will maximize the carrying capacity of the gutter.

28 Find the altitude of the right circular cylinder of maximum curved surface area that can be inscribed in a sphere of radius a.

29 The interior of a half-mile race track consists of a rectangle with semicircles at two opposite ends. Find the dimensions that will maximize the area of the rectangle.

30 A cable television firm presently serves 5000 households and charges $20 per month. A marketing survey indicates that each decrease of $1 in monthly charge will result in 500 new customers. Find the monthly charge that will result in maximum monthly revenue.

31 A wire 5 feet long is to be cut into two pieces. One piece is to be bent into the shape of a circle and the other into the shape of a square. Where should the wire be cut so that the sum of the areas of the circle and square is

(a) a maximum

(b) a minimum

32 In biochemistry, the general threshold-response curve is given by $R = kS^n/(S^n + a^n)$, where R is the chemical response that corresponds to a concentration S of a substance for positive constants k, n, and a. An example is the rate R at which the liver removes alcohol from the bloodstream when the concentration of alcohol is S. Show that R is an increasing function of S and that $R = k$ is a horizontal asymptote for the curve.

33 The position function of a point moving on a coordinate line is given by $s(t) = (t^2 + 3t + 1)/(t^2 + 1)$. Find the velocity and acceleration at time t, and describe the motion of the point during the time interval $[-2, 2]$.

34 The position of a moving point on a coordinate line is given by

$$s(t) = a \sin(kt + m) + b \cos(kt + m)$$

for constants a, b, k, and m. Prove that the magnitude of the acceleration is directly proportional to the distance from the origin.

35 A manufacturer of microwave ovens determines that the cost of producing x units is given by

$$C(x) = 4000 + 100x + 0.05x^2 + 0.0002x^3.$$

Compare the marginal cost of producing 100 ovens with the cost of producing the 101st oven.

36 The cost function for producing a microprocessor component is given by $C(x) = 1000 + 2x + 0.005x^2$. If 2000 units are produced, find the cost, the average cost, the marginal cost, and the marginal average cost.

37 An electronics company estimates that the cost of producing x calculators per day is $C(x) = 500 + 6x + 0.02x^2$. If each calculator is sold for $18, find

(a) the revenue function

(b) the profit function

(c) the daily production that will maximize the profit

(d) the maximum daily profit

38 A small office building is to contain 500 ft^2 of floor space. Simplified floor plans are shown in the figure. If the walls cost $100 per running foot and if the wall space above the doors is disregarded,

(a) show that the cost $C(x)$ of the walls is

$$C(x) = 100[3x - 6 + (1000/x)]$$

(b) find the vertical and oblique asymptotes, and sketch the graph of $C(x)$ for $x > 0$

(c) find the design that minimizes the cost

EXERCISE 38

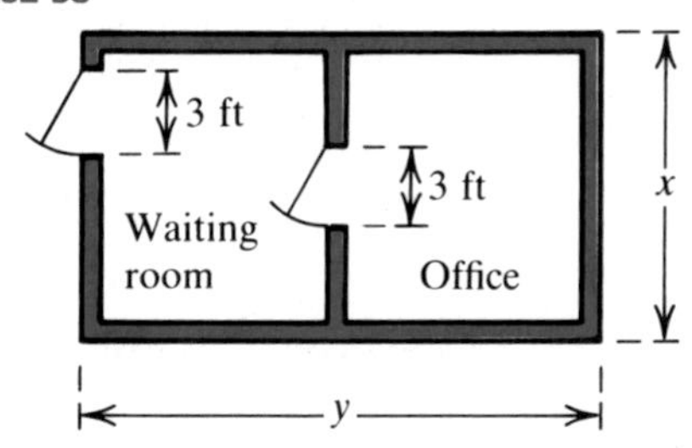

C 39 Use Newton's method to approximate the root of the equation $\sin x - x \cos x = 0$ between π and $3\pi/2$ to three decimal places.

C 40 Use Newton's method to approximate $\sqrt[4]{5}$ to three decimal places.

CHAPTER 5

INTEGRALS

INTRODUCTION

The two most important tools in calculus are the derivative, considered in previous chapters, and the *definite integral*, defined in Section 5.4. The derivative was motivated by the problems of finding the slope of a tangent line and defining velocity. The definite integral arises naturally when we consider the problem of finding the area of a region in the xy-plane. However, this is merely one application. As we shall see in later chapters, the uses for definite integrals are as abundant and varied as those for derivatives.

The principal result in this chapter is the *fundamental theorem of calculus*, proved in Section 5.6. This outstanding theorem enables us to find exact values of definite integrals by using an *antiderivative* or *indefinite integral*. Each of these concepts is defined in Section 5.1; the procedure may be regarded as a reverse procedure to finding the derivative of a function. Thus, in addition to providing an important evaluation process, the fundamental theorem shows that there is a relationship between derivatives and integrals—a key result in calculus.

The chapter closes with a discussion of methods of *numerical integration*, used for approximating definite integrals that cannot be evaluated by means of the fundamental theorem. These methods are readily programmable for use with calculators and computers and are employed in a wide variety of applied fields.

5.1 ANTIDERIVATIVES AND INDEFINITE INTEGRALS

In our previous work we solved problems of the following type: *Given a function f, find the derivative f'.* We shall now consider a related problem: *Given a function f, find a function F such that $F' = f$.* In the next definition we give F a special name.

Definition (5.1)

A function F is an **antiderivative** of f on an interval I if $F'(x) = f(x)$ for every x in I.

FIGURE 5.1

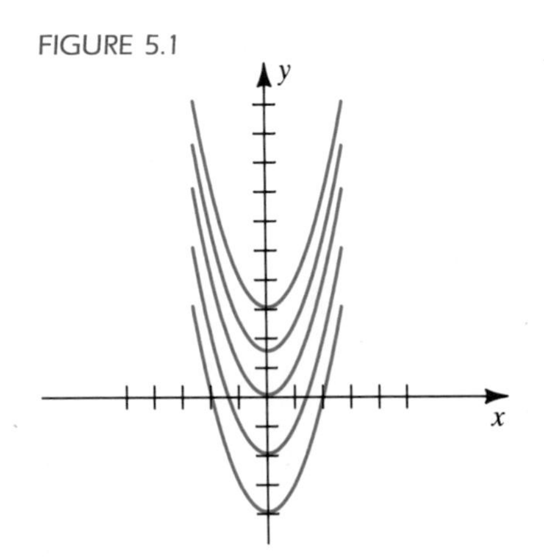

We shall also call $F(x)$ an antiderivative of $f(x)$. The process of finding F, or $F(x)$, is called **antidifferentiation**.

To illustrate, $F(x) = x^2$ is an antiderivative of $f(x) = 2x$, because

$$F'(x) = D_x(x^2) = 2x = f(x).$$

There are many other antiderivatives of $2x$, such as $x^2 + 2$, $x^2 - \frac{5}{3}$, and $x^2 + \sqrt{3}$. In general, if C is *any* constant, then $x^2 + C$ is an antiderivative of $2x$, because

$$D_x(x^2 + C) = 2x + 0 = 2x.$$

Thus, there is a *family of antiderivatives* of $2x$ of the form $F(x) = x^2 + C$, where C is any constant. Graphs of several members of this family are sketched in Figure 5.1.

The next illustration contains other examples of antiderivatives, where C is a constant.

ILLUSTRATION

$f(x)$	ANTIDERIVATIVES OF $f(x)$
x^2	$\frac{1}{3}x^3$, $\frac{1}{3}x^3 + 8$, $\frac{1}{3}x^3 + C$
$8x^3$	$2x^4$, $2x^4 - \sqrt[3]{7}$, $2x^4 + C$
$\cos x$	$\sin x$, $\sin x + \frac{4}{9}$, $\sin x + C$

As in the preceding illustration, if $F(x)$ is an antiderivative of $f(x)$, then so is $F(x) + C$ for any constant C. The next theorem states that *every* antiderivative is of this form.

Theorem (5.2)

Let F be an antiderivative of f on an interval I. If G is any antiderivative of f on I, then

$$G(x) = F(x) + C$$

for some constant C and every x in I.

PROOF If F and G are antiderivatives of f, let H be the function defined by

$$H(x) = G(x) - F(x)$$

for every x in I. We will show that H is a constant function on I; that is, $G(x) - F(x) = C$ for some C, or, equivalently, $G(x) = F(x) + C$.

Let a and b be any numbers in I such that $a < b$. To show that H is constant on I, it suffices to prove that $H(a) = H(b)$. Since F and G are antiderivatives of f,

$$H'(x) = G'(x) - F'(x) = f(x) - f(x) = 0$$

for every x in I. Since $H(x)$ is differentiable, H is continuous, by Theorem (3.11). Applying the mean value theorem (4.12) to H on the interval $[a, b]$, there exists a number c in (a, b) such that

$$H'(c) = \frac{H(b) - H(a)}{b - a}.$$

Since c is in I, $H'(c) = 0$, and thus

$$H(b) - H(a) = 0, \quad \text{or} \quad H(a) = H(b),$$

which is what we wished to prove. ■

We refer to the constant C in Theorem (5.2) as an **arbitrary constant**. If $F(x)$ is an antiderivative of $f(x)$, then *all* antiderivatives of $f(x)$ can be obtained from $F(x) + C$ by letting C range through the set of real numbers. We shall employ the following notation for a family of antiderivatives of this type.

Definition **(5.3)**

The notation

$$\int f(x)\, dx = F(x) + C,$$

where $F'(x) = f(x)$ and C is an arbitrary constant, denotes the family of all antiderivatives of $f(x)$ on an interval I.

The symbol $\int$ used in Definition (5.3) is an **integral sign**. We call $\int f(x)\, dx$ the **indefinite integral** of $f(x)$. The expression $f(x)$ is the **integrand**, and C is the **constant of integration**. The process of finding $F(x) + C$, when given $\int f(x)\, dx$, is referred to as **indefinite integration**, **evaluating the integral**, or **integrating $f(x)$**. The adjective *indefinite* is used because $\int f(x)\, dx$ represents a *family* of antiderivatives, not any *specific* function. Later in the chapter, when we discuss definite integrals, we shall give reasons for using the integral sign and the differential dx that appears to the right of the integrand $f(x)$. At present we shall not interpret $f(x)\, dx$ as the product of $f(x)$ and the differential dx. We shall regard dx merely as a symbol that specifies the independent variable x, which we refer to as the **variable of integration**. If we use a different variable of integration, such as t, we write

$$\int f(t)\, dt = F(t) + C,$$

where $F'(t) = f(t)$.

ILLUSTRATION

- $\int x^4\,dx = \frac{1}{5}x^5 + C$ because $D_x(\frac{1}{5}x^5) = x^4$.
- $\int t^{-3}\,dt = -\frac{1}{2}t^{-2} + C$ because $D_t(-\frac{1}{2}t^{-2}) = t^{-3}$.
- $\int \cos u\,du = \sin u + C$ because $D_u \sin u = \cos u$.

Note that, in general,

$$\int [D_x f(x)]\,dx = f(x) + C$$

because $f'(x) = D_x f(x)$. This allows us to use any derivative formula to obtain a corresponding formula for an indefinite integral, as illustrated in the next table. As shown in Formula (1), it is customary to abbreviate $\int 1\,dx$ by $\int dx$.

Brief table of indefinite integrals **(5.4)**

DERIVATIVE $D_x[f(x)]$	INDEFINITE INTEGRAL $\int D_x[f(x)]\,dx = f(x) + C$
$D_x(x) = 1$	**(1)** $\int 1\,dx = \int dx = x + C$
$D_x\left(\dfrac{x^{r+1}}{r+1}\right) = x^r\ (r \neq -1)$	**(2)** $\int x^r\,dx = \dfrac{x^{r+1}}{r+1} + C\ (r \neq -1)$
$D_x(\sin x) = \cos x$	**(3)** $\int \cos x\,dx = \sin x + C$
$D_x(-\cos x) = \sin x$	**(4)** $\int \sin x\,dx = -\cos x + C$
$D_x(\tan x) = \sec^2 x$	**(5)** $\int \sec^2 x\,dx = \tan x + C$
$D_x(-\cot x) = \csc^2 x$	**(6)** $\int \csc^2 x\,dx = -\cot x + C$
$D_x(\sec x) = \sec x \tan x$	**(7)** $\int \sec x \tan x\,dx = \sec x + C$
$D_x(-\csc x) = \csc x \cot x$	**(8)** $\int \csc x \cot x\,dx = -\csc x + C$

Formula (2) is called the *power rule for indefinite integration*. As in the following illustration, it is often necessary to rewrite an integrand before applying the power rule or one of the trigonometric formulas.

ILLUSTRATION

- $\int x^3 \cdot x^5\,dx = \int x^8\,dx = \dfrac{x^{8+1}}{8+1} = \dfrac{1}{9}x^9 + C$
- $\int \dfrac{1}{x^3}\,dx = \int x^{-3}\,dx = \dfrac{x^{-3+1}}{-3+1} = -\dfrac{1}{2x^2} + C$
- $\int \sqrt[3]{x^2}\,dx = \int x^{2/3}\,dx = \dfrac{x^{2/3+1}}{\frac{2}{3}+1} = \dfrac{3}{5}x^{5/3} + C$
- $\int \dfrac{\tan x}{\sec x}\,dx = \int \cos x\,\dfrac{\sin x}{\cos x}\,dx = \int \sin x\,dx = -\cos x + C$

It is a good idea to check indefinite integrations (such as those in the preceding illustration) by differentiating the final expression to see if either the integrand or an equivalent form of the integrand is obtained.

The next theorem indicates that differentiation and indefinite integration are inverse processes, because each, in a sense, undoes the other. In statement (i) it is assumed that f is differentiable, and in (ii) that f has an antiderivative on some interval.

Theorem **(5.5)**

(i) $\int [D_x f(x)]\, dx = f(x) + C$

(ii) $D_x\left[\int f(x)\, dx\right] = f(x)$

PROOF We have already proved (i). To prove (ii), let F be an antiderivative of f and write

$$D_x\left[\int f(x)\, dx\right] = D_x[F(x) + C] = F'(x) + 0 = f(x). \blacksquare$$

EXAMPLE 1 Verify Theorem (5.5) for the special case $f(x) = x^2$.

SOLUTION

(i) If we first differentiate x^2 and then integrate,

$$\int D_x(x^2)\, dx = \int 2x\, dx = x^2 + C.$$

(ii) If we first integrate x^2 and then differentiate,

$$D_x \int x^2\, dx = D_x\left(\frac{x^3}{3} + C\right) = x^2.$$

The next theorem is useful for evaluating many types of indefinite integrals. In the statements we assume that $f(x)$ and $g(x)$ have antiderivatives on an interval I.

Theorem **(5.6)**

(i) $\int cf(x)\, dx = c\int f(x)\, dx$ for any constant c

(ii) $\int [f(x) + g(x)]\, dx = \int f(x)\, dx + \int g(x)\, dx$

(iii) $\int [f(x) - g(x)]\, dx = \int f(x)\, dx - \int g(x)\, dx$

PROOF We shall prove (ii). The proofs of (i) and (iii) are similar. If F and G are antiderivatives of f and g, respectively,

$$D_x[F(x) + G(x)] = F'(x) + G'(x) = f(x) + g(x).$$

Hence, by Definition (5.3),

$$\int [f(x) + g(x)]\, dx = F(x) + G(x) + C,$$

where C is an arbitrary constant. Similarly,

$$\int f(x)\,dx + \int g(x)\,dx = F(x) + C_1 + G(x) + C_2$$

for arbitrary constants C_1 and C_2. These give us the same family of antiderivatives, since for any special case we can choose values of the constants such that $C = C_1 + C_2$. This proves (ii). ■

EXAMPLE 2 Evaluate $\int (5x^3 + 2\cos x)\,dx$.

SOLUTION We first use (ii) and (i) of Theorem (5.6) and then formulas from (5.4):

$$\begin{aligned}
\int (5x^3 + 2\cos x)\,dx &= \int 5x^3\,dx + \int 2\cos x\,dx \\
&= 5\int x^3\,dx + 2\int \cos x\,dx \\
&= 5\left(\frac{x^4}{4} + C_1\right) + 2(\sin x + C_2) \\
&= \tfrac{5}{4}x^4 + 5C_1 + 2\sin x + 2C_2 \\
&= \tfrac{5}{4}x^4 + 2\sin x + C
\end{aligned}$$

where $C = 5C_1 + 2C_2$.

In Example 2 we added the two constants $5C_1$ and $2C_2$ to obtain one arbitrary constant C. We can always manipulate arbitrary constants in this way, so it is not necessary to introduce a constant for each indefinite integration as we did in Example 2. Instead, if an integrand is a sum, *we integrate each term of the sum without introducing constants and then add one arbitrary constant C after the last integration.* We also often bypass the step $\int cf(x)\,dx = c\int f(x)\,dx$, as in the next example.

EXAMPLE 3 Evaluate $\int \left(8t^3 - 6\sqrt{t} + \frac{1}{t^3}\right) dt$.

SOLUTION First we find an antiderivative for each of the three terms in the integrand and then add an arbitrary constant C. We rewrite $\sqrt{t}$ as $t^{1/2}$ and $1/t^3$ as t^{-3} and then use the power rule for integration:

$$\begin{aligned}
\int \left(8t^3 - 6\sqrt{t} + \frac{1}{t^3}\right) dt &= \int (8t^3 - 6t^{1/2} + t^{-3})\,dt \\
&= 8\cdot\frac{t^4}{4} - 6\cdot\frac{t^{3/2}}{\frac{3}{2}} + \frac{t^{-2}}{-2} + C \\
&= 2t^4 - 4t^{3/2} - \frac{1}{t^2} + C
\end{aligned}$$

EXAMPLE 4 Evaluate $\int \frac{(x^2 - 1)^2}{x^2}\,dx$.

SOLUTION First we change the form of the integrand, because the degree of the numerator is greater than or equal to the degree of the denominator. We then find an antiderivative for each term, adding an arbitrary constant C after the last integration:

$$\begin{aligned}\int \frac{(x^2-1)^2}{x^2}\,dx &= \int \frac{x^4-2x^2+1}{x^2}\,dx\\ &= \int (x^2-2+x^{-2})\,dx\\ &= \frac{x^3}{3}-2x+\frac{x^{-1}}{-1}+C\\ &= \frac{1}{3}x^3-2x-\frac{1}{x}+C\end{aligned}$$

EXAMPLE 5 Evaluate $\displaystyle\int \frac{1}{\cos u \cot u}\,du$.

SOLUTION We use trigonometric identities to change the integrand and then apply formula (7) from Table (5.4):

$$\begin{aligned}\int \frac{1}{\cos u \cot u}\,du &= \int \sec u \tan u\,du\\ &= \sec u + C\end{aligned}$$

An applied problem may be stated in terms of a **differential equation**—that is, an equation that involves derivatives or differentials of an unknown function. A function f is a **solution** of a differential equation if it satisfies the equation—that is, if substitution of f for the unknown function produces a true statement. To **solve** a differential equation means to find all solutions. Sometimes, in addition to the differential equation, we may know certain values of f or f', called **initial conditions**.

Indefinite integrals are useful for solving certain differential equations, because if we are given a derivative $f'(x)$ we can integrate and use Theorem (5.5)(i) to obtain an equation involving the unknown function f:

$$\int f'(x)\,dx = f(x) + C$$

If we are also given an initial condition for f, it may be possible to find $f(x)$ explicitly, as in the next example.

EXAMPLE 6 Solve the differential equation

$$f'(x) = 6x^2 + x - 5$$

subject to the initial condition $f(0) = 2$.

SOLUTION We proceed as follows:

$$\begin{aligned}f'(x) &= 6x^2 + x - 5\\ \int f'(x)\,dx &= \int (6x^2 + x - 5)\,dx\\ f(x) &= 2x^3 + \tfrac{1}{2}x^2 - 5x + C\end{aligned}$$

for some number C. (It is unnecessary to add a constant of integration to *each* side of the equation.) Letting $x = 0$ and using the given initial condition $f(0) = 2$ gives us

$$f(0) = 0 + 0 - 0 + C, \quad \text{or} \quad 2 = C.$$

Hence the solution f of the differential equation with the initial condition $f(0) = 2$ is

$$f(x) = 2x^3 + \tfrac{1}{2}x^2 - 5x + 2.$$

The given equation can also be stated in terms of the differentials of $y = f(x)$ by writing

$$\frac{dy}{dx} = 6x^2 + x - 5, \quad \text{or} \quad dy = (6x^2 + x - 5)\,dx.$$

In this case we integrate as follows:

$$\int dy = \int (6x^2 + x - 5)\,dx$$
$$y = 2x^3 + \tfrac{1}{2}x^2 - 5x + C$$

The constant C may be found by letting $x = 0$ and using $y = f(0) = 2$.

If we are given a *second* derivative $f''(x)$, then we must employ two successive indefinite integrals to find $f(x)$. First we use Theorem (5.5)(i) as follows:

$$\int f''(x)\,dx = \int [D_x f'(x)]\,dx = f'(x) + C$$

After finding $f'(x)$, we proceed as in Example 6.

EXAMPLE 7 Solve the differential equation

$$f''(x) = 5\cos x + 2\sin x$$

subject to the initial conditions $f(0) = 3$ and $f'(0) = 4$.

SOLUTION We proceed as follows:

$$f''(x) = 5\cos x + 2\sin x$$
$$\int f''(x)\,dx = \int (5\cos x + 2\sin x)\,dx$$
$$f'(x) = 5\sin x - 2\cos x + C$$

Letting $x = 0$ and using the initial condition $f'(0) = 4$ gives us

$$f'(0) = 5\sin 0 - 2\cos 0 + C$$
$$4 = 0 - 2 \cdot 1 + C, \quad \text{or} \quad C = 6.$$

Hence

$$f'(x) = 5\sin x - 2\cos x + 6.$$

We integrate a second time:

$$\int f'(x)\,dx = \int (5 \sin x - 2 \cos x + 6)\,dx$$

$$f(x) = -5 \cos x - 2 \sin x + 6x + D$$

Letting $x = 0$ and using the initial condition $f(0) = 3$, we find that

$$f(0) = -5 \cos 0 - 2 \sin 0 + 6 \cdot 0 + D$$

$$3 = -5 - 0 + 0 + D, \quad \text{or} \quad D = 8.$$

Therefore, the solution of the differential equation with the given initial conditions is

$$f(x) = -5 \cos x - 2 \sin x + 6x + 8.$$

Suppose a point P is moving on a coordinate line with an acceleration $a(t)$ at time t, and the corresponding velocity is $v(t)$. By Definition (4.21), $a(t) = v'(t)$ and hence

$$\int a(t)\,dt = \int v'(t)\,dt = v(t) + C$$

for some constant C.

Similarly, if we know $v(t)$, then since $v(t) = s'(t)$, where s is the position function of P, we can find a formula that involves $s(t)$ by indefinite integration:

$$\int v(t)\,dt = \int s'(t)\,dt = s(t) + D$$

for some constant D. In the next example we shall use this technique to find the position function for an object that is moving under the influence of gravity. Understanding the problem requires knowledge of a fact from physics: An object on or near the surface of the earth is acted upon by a force—*gravity*—that produces a constant acceleration, denoted by g. The approximation to g that is employed for most problems is 32 ft/sec^2, or 980 cm/sec^2.

FIGURE 5.2

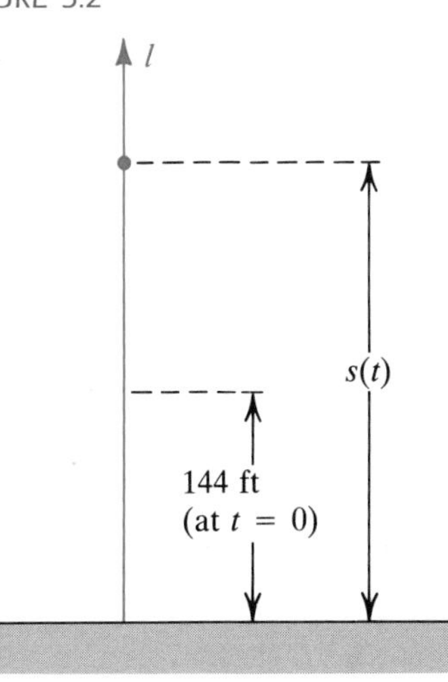

EXAMPLE 8 A stone is thrown vertically upward from a position 144 feet above the ground with an initial velocity of 96 ft/sec. Disregarding air resistance, find

(a) the stone's distance above the ground after t seconds

(b) the length of time that the stone rises

(c) when and with what velocity the stone strikes the ground

SOLUTION The motion of the stone may be represented by a point moving on a vertical coordinate line l with origin at ground level and positive direction upward (see Figure 5.2).

(a) The stone's distance above the ground at time t is $s(t)$, and the initial conditions are $s(0) = 144$ and $v(0) = 96$. Since the velocity is decreasing,

$v'(t) < 0$; that is, the acceleration is negative. Hence, by the remarks preceding this example;

$$a(t) = v'(t) = -32$$
$$\int v'(t)\, dt = \int -32\, dt$$
$$v(t) = -32t + C$$

for some number C. Substituting 0 for t and using the fact that $v(0) = 96$ gives us $96 = 0 + C = C$ and, consequently,

$$v(t) = -32t + 96.$$

Since $s'(t) = v(t)$, we obtain

$$s'(t) = -32t + 96$$
$$\int s'(t)\, dt = \int (-32t + 96)\, dt$$
$$s(t) = -16t^2 + 96t + D$$

for some number D. Letting $t = 0$ and using the fact that $s(0) = 144$ leads to $144 = 0 + 0 + D$, or $D = 144$. It follows that the distance from the ground to the stone at time t is given by

$$s(t) = -16t^2 + 96t + 144.$$

(b) The stone will rise until $v(t) = 0$—that is, until

$$-32t + 96 = 0, \quad \text{or} \quad t = 3.$$

(c) The stone will strike the ground when $s(t) = 0$—that is, when

$$-16t^2 + 96t + 144 = 0.$$

An equivalent equation is $t^2 - 6t - 9 = 0$. Applying the quadratic formula, we obtain $t = 3 \pm 3\sqrt{2}$. The solution $3 - 3\sqrt{2}$ is extraneous, since t is nonnegative. Hence the stone strikes the ground after $3 + 3\sqrt{2}$ sec. The velocity at that time is

$$\begin{aligned} v(3 + 3\sqrt{2}) &= -32(3 + 3\sqrt{2}) + 96 \\ &= -96\sqrt{2} \approx -135.8 \text{ ft/sec.} \end{aligned}$$

In economic applications, if a marginal function is known (see page 226), then we can use indefinite integration to find the function, as illustrated in the next example.

EXAMPLE 9 A manufacturer finds that the marginal cost (in dollars) associated with the production of x units of a photocopier component is given by $30 - 0.02x$. If the cost of producing one unit is \$35, find the cost function and the cost of producing 100 units.

SOLUTION If C is the cost function, then the marginal cost is the rate of change of C with respect to x—that is,

$$C'(x) = 30 - 0.02x.$$

Hence

$$\int C'(x)\,dx = \int (30 - 0.02x)\,dx$$

and $$C(x) = 30x - 0.01x^2 + K$$

for some K. Letting $x = 1$ and using $C(1) = 35$, we obtain

$$35 = 30 - 0.01 + K, \quad \text{or} \quad K = 5.01.$$

Consequently,

$$C(x) = 30x - 0.01x^2 + 5.01.$$

In particular, the cost of producing 100 units is

$$C(100) = 3000 - 100 + 5.01 = \$2905.01.$$

EXERCISES 5.1

Exer. 1–42: Evaluate.

1 $\int (4x + 3)\,dx$

2 $\int (4x^2 - 8x + 1)\,dx$

3 $\int (9t^2 - 4t + 3)\,dt$

4 $\int (2t^3 - t^2 + 3t - 7)\,dt$

5 $\int \left(\dfrac{1}{z^3} - \dfrac{3}{z^2}\right) dz$

6 $\int \left(\dfrac{4}{z^7} - \dfrac{7}{z^4} + z\right) dz$

7 $\int \left(3\sqrt{u} + \dfrac{1}{\sqrt{u}}\right) du$

8 $\int (\sqrt{u^3} - \frac{1}{2}u^{-2} + 5)\,du$

9 $\int (2v^{5/4} + 6v^{1/4} + 3v^{-4})\,dv$

10 $\int (3v^5 - v^{5/3})\,dv$

11 $\int (3x - 1)^2\,dx$

12 $\int \left(x - \dfrac{1}{x}\right)^2 dx$

13 $\int x(2x + 3)\,dx$

14 $\int (2x - 5)(3x + 1)\,dx$

15 $\int \dfrac{8x - 5}{\sqrt[3]{x}}\,dx$

16 $\int \dfrac{2x^2 - x + 3}{\sqrt{x}}\,dx$

17 $\int \dfrac{x^3 - 1}{x - 1}\,dx, \quad x \neq 1$

18 $\int \dfrac{x^3 + 3x^2 - 9x - 2}{x - 2}\,dx, \quad x \neq 2$

19 $\int \dfrac{(t^2 + 3)^2}{t^6}\,dt$

20 $\int \dfrac{(\sqrt{t} + 2)^2}{t^3}\,dt$

21 $\int \frac{3}{4}\cos u\,du$

22 $\int -\frac{1}{5}\sin u\,du$

23 $\int \dfrac{7}{\csc x}\,dx$

24 $\int \dfrac{1}{4\sec x}\,dx$

25 $\int (\sqrt{t} + \cos t)\,dt$

26 $\int (\sqrt[3]{t^2} - \sin t)\,dt$

27 $\int \dfrac{\sec t}{\cos t}\,dt$

28 $\int \dfrac{1}{\sin^2 t}\,dt$

29 $\int (\csc v \cot v \sec v)\,dv$

30 $\int (4 + 4\tan^2 v)\,dv$

31 $\int \dfrac{\sec w \sin w}{\cos w}\,dw$

32 $\int \dfrac{\csc w \cos w}{\sin w}\,dw$

33 $\int \dfrac{(1 + \cot^2 z)\cot z}{\csc z}\,dz$

34 $\int \dfrac{\tan z}{\cos z}\,dz$

35 $\int D_x \sqrt{x^2 + 4}\,dx$

36 $\int D_x \sqrt[3]{x^3 - 8}\,dx$

37 $\int \dfrac{d}{dx}(\sin \sqrt[3]{x})\,dx$

38 $\int \dfrac{d}{dx}(\sqrt{\tan x})\,dx$

39 $D_x \int (x^3\sqrt{x - 4})\,dx$

40 $D_x \int (x^4 \sqrt[3]{x^2 + 9})\,dx$

41 $\dfrac{d}{dx}\int \cot x^3\,dx$

42 $\dfrac{d}{dx}\int \cos\sqrt{x^2 + 1}\,dx$

Exer. 43–48: Evaluate the integral if a and b are constants.

43 $\int a^2\,dx$

44 $\int ab\,dx$

45 $\int (at + b)\,dt$

46 $\int \left(\dfrac{a}{b^2}t\right) dt$

47 $\int (a + b)\,du$

48 $\int (b - a^2)\,du$

Exer. 49–56: Solve the differential equation subject to the given conditions.

49 $f'(x) = 12x^2 - 6x + 1; \quad f(1) = 5$

50 $f'(x) = 9x^2 + x - 8; \quad f(-1) = 1$

51 $\dfrac{dy}{dx} = 4x^{1/2}; \quad y = 21 \text{ if } x = 4$

52 $\dfrac{dy}{dx} = 5x^{-1/3}$; $y = 70$ if $x = 27$

53 $f''(x) = 4x - 1$; $f'(2) = -2$; $f(1) = 3$

54 $f''(x) = 6x - 4$; $f'(2) = 5$; $f(2) = 4$

55 $\dfrac{d^2y}{dx^2} = 3 \sin x - 4 \cos x$; $y = 7$ and $y' = 2$ if $x = 0$

56 $\dfrac{d^2y}{dx^2} = 2 \cos x - 5 \sin x$; $y = 2 + 6\pi$ and $y' = 3$ if $x = \pi$

Exer. 57–58: If a point is moving on a coordinate line with the given acceleration $a(t)$ and initial conditions, find $s(t)$.

57 $a(t) = 2 - 6t$; $v(0) = -5$; $s(0) = 4$

58 $a(t) = 3t^2$; $v(0) = 20$; $s(0) = 5$

59 A projectile is fired vertically upward from ground level with a velocity of 1600 ft/sec. Disregarding air resistance, find

(a) its distance $s(t)$ above ground at time t

(b) its maximum height

60 An object is dropped from a height of 1000 feet. Disregarding air resistance, find

(a) the distance it falls in t seconds

(b) its velocity at the end of 3 seconds

(c) when it strikes the ground

61 A stone is thrown directly downward from a height of 96 feet with an initial velocity of 16 ft/sec. Find

(a) its distance above the ground after t seconds

(b) when it strikes the ground

(c) the velocity at which it strikes the ground

62 A gravitational constant for objects near the surface of the moon is 5.3 ft/sec^2.

(a) If an astronaut on the moon throws a stone directly upward with an initial velocity of 60 ft/sec, find its maximum altitude.

(b) If, after returning to Earth, the astronaut throws the same stone directly upward with the same initial velocity, find the maximum altitude.

63 If a projectile is fired vertically upward from a height of s_0 feet above the ground with a velocity of v_0 ft/sec, prove that if air resistance is disregarded, its distance $s(t)$ above the ground after t seconds is given by $s(t) = -\frac{1}{2}gt^2 + v_0t + s_0$, where g is a gravitational constant.

64 A ball rolls down an inclined plane with an acceleration of 2 ft/sec^2.

(a) If the ball is given no initial velocity, how far will it roll in t seconds?

(b) What initial velocity must be given for the ball to roll 100 feet in 5 seconds?

65 If an automobile starts from rest, what constant acceleration will enable it to travel 500 feet in 10 seconds?

66 If a car is traveling at a speed of 60 mi/hr, what constant (negative) acceleration will enable it to stop in 9 seconds?

67 A small country has natural gas reserves of 100 billion ft^3. If $A(t)$ denotes the total amount of natural gas consumed after t years, then dA/dt is the *rate of consumption.* If the rate of consumption is predicted to be $5 + 0.01t$ billion ft^3/year, in approximately how many years will the country's natural gas reserves be depleted?

68 Refer to Exercise 67. Based on U.S. Department of Energy statistics, the rate of consumption of gasoline in the United States (in billions of gallons per year) is approximated by $dA/dt = 2.74 - 0.11t - 0.01t^2$, with $t = 0$ corresponding to the year 1980. Estimate the number of gallons of gasoline consumed in the United States between 1980 and 1984.

69 A sportswear manufacturer determines that the marginal cost in dollars of producing x warm-up suits is given by $20 - 0.015x$. If the cost of producing one suit is \$25, find the cost function and the cost of producing 50 suits.

70 If the marginal cost function of a product is given by $2/x^{1/3}$ and if the cost of producing 8 units is \$20, find the cost function and the cost of producing 64 units.

5.2 CHANGE OF VARIABLES IN INDEFINITE INTEGRALS

The formulas for indefinite integrals in Table (5.4) are limited in scope, because we cannot use them directly to evaluate integrals such as

$$\int \sqrt{5x + 7}\, dx \quad \text{or} \quad \int \cos 4x\, dx.$$

In this section we shall develop a simple but powerful method for changing the variable of integration so that these integrals (and many others) can be evaluated by using the formulas in Table (5.4).

To justify this method, we shall apply formula (i) of Theorem (5.5) to a *composite* function. We intend to consider several functions f, g, and F, so it will simplify our work if we state the formula in terms of a function h as follows:

$$\int [D_x\, h(x)]\, dx = h(x) + C$$

Suppose that F is an antiderivative of a function f and that g is a differentiable function such that $g(x)$ is in the domain of F for every x in some interval. If we let h denote the composite function $F \circ g$, then $h(x) = F(g(x))$ and hence

$$\int [D_x\, F(g(x))]\, dx = F(g(x)) + C.$$

Applying the chain rule (3.33) to the integrand $D_x\, F(g(x))$ and using the fact that $F' = f$, we obtain

$$D_x\, F(g(x)) = F'(g(x))g'(x) = f(g(x))g'(x).$$

Substitution in the preceding indefinite integral gives us

$$(*) \qquad \int f(g(x))g'(x)\, dx = F(g(x)) + C.$$

We can employ the following device to help remember this formula:

$$\text{Let} \quad u = g(x) \quad \text{and} \quad du = g'(x)\, dx.$$

Note that once we have introduced the variable $u = g(x)$, the differential du of u is determined by using (ii) of Definition (3.28). If we *formally substitute* into the last integration formula, we obtain

$$\int f(u)\, du = F(u) + C.$$

This has the same *form* as the integral in Definition (5.3); however, u represents a *function*, not an independent variable x, as before. This indicates that $g'(x)\, dx$ in $(*)$ may be regarded as the product of $g'(x)$ and dx. Since the variable x has been replaced by a new variable u, finding indefinite integrals in this way is referred to as a **change of variable**, or as the **method of substitution**. We may summarize our discussion as follows, where we assume that f and g have the properties described previously.

Method of substitution **(5.7)**

If F is an antiderivative of f, then

$$\int f(g(x))g'(x)\, dx = F(g(x)) + C.$$

If $u = g(x)$ and $du = g'(x)\, dx$, then

$$\int f(u)\, du = F(u) + C.$$

After making the substitution $u = g(x)$ as indicated in (5.7), it may be necessary to insert a constant factor k into the integrand in order to arrive at the proper form $\int f(u)\, du$. We must then also multiply by $1/k$ to maintain equality, as illustrated in the following examples.

EXAMPLE 1 Evaluate $\int \sqrt{5x+7}\,dx$.

SOLUTION We let $u = 5x + 7$ and calculate du:

$$u = 5x + 7, \quad du = 5\,dx$$

Since du contains the factor 5, the integral is not in the proper form $\int f(u)\,du$ required by (5.7). However, we can *introduce* the factor 5 into the integrand, provided we also multiply by $\frac{1}{5}$. Doing this and using (i) of Theorem (5.6) gives us

$$\begin{aligned}\int \sqrt{5x+7}\,dx &= \int \sqrt{5x+7}\left(\tfrac{1}{5}\right)5\,dx \\ &= \tfrac{1}{5}\int \sqrt{5x+7}\;5\,dx.\end{aligned}$$

We now substitute and use the power rule for integration:

$$\begin{aligned}\int \sqrt{5x+7}\,dx &= \tfrac{1}{5}\int \sqrt{u}\,du \\ &= \tfrac{1}{5}\int u^{1/2}\,du \\ &= \tfrac{1}{5}\frac{u^{3/2}}{\frac{3}{2}} + C \\ &= \tfrac{2}{15}u^{3/2} + C \\ &= \tfrac{2}{15}(5x+7)^{3/2} + C\end{aligned}$$

In the future, after inserting a factor k into an integrand, as in Example 1, we shall simply multiply the integral by $1/k$, skipping the intermediate steps of first writing $(1/k)k$ and then bringing $1/k$ *outside*—that is, to the left of—the integral sign.

EXAMPLE 2 Evaluate $\int \cos 4x\,dx$.

SOLUTION We make the substitution

$$u = 4x, \quad du = 4\,dx.$$

Since du contains the factor 4, we adjust the integrand by multiplying by 4 and compensate by multiplying the integral by $\frac{1}{4}$ before substituting:

$$\begin{aligned}\int \cos 4x\,dx &= \tfrac{1}{4}\int (\cos 4x)4\,dx \\ &= \tfrac{1}{4}\int \cos u\,du \\ &= \tfrac{1}{4}\sin u + C \\ &= \tfrac{1}{4}\sin 4x + C\end{aligned}$$

It is not always easy to decide what substitution $u = g(x)$ is needed to transform an indefinite integral into a form that can be readily evaluated.

It may be necessary to try several different possibilities before finding a suitable substitution. In most cases *no* substitution will simplify the integrand properly. The following guidelines may be helpful.

Guidelines for changing variables in indefinite integrals **(5.8)**

1. Decide on a reasonable substitution $u = g(x)$.
2. Calculate $du = g'(x)\,dx$.
3. Using 1 and 2, try to transform the integral into a form that involves only the variable u. If necessary, introduce a *constant* factor k into the integrand and compensate by multiplying the integral by $1/k$. If any part of the resulting integrand contains the variable x, use a different substitution in 1.
4. Evaluate the integral obtained in 3, obtaining an antiderivative involving u.
5. Replace u in the antiderivative obtained in guideline 4 by $g(x)$. The final result should contain only the variable x.

The following examples illustrate the use of the guidelines.

EXAMPLE 3 Evaluate $\int (2x^3 + 1)^7 x^2\,dx$.

SOLUTION If an integrand involves an expression raised to a power, such as $(2x^3 + 1)^7$, we often substitute u for the expression. Thus, we let

$$u = 2x^3 + 1, \quad du = 6x^2\,dx.$$

Comparing $du = 6x^2\,dx$ with $x^2\,dx$ in the integral suggests that we introduce the factor 6 into the integrand. Doing this and compensating by multiplying the integral by $\frac{1}{6}$, we obtain the following:

$$\begin{aligned}\int (2x^3 + 1)^7 x^2\,dx &= \tfrac{1}{6}\int (2x^3 + 1)^7 6x^2\,dx \\ &= \tfrac{1}{6}\int u^7\,du \\ &= \frac{1}{6}\left(\frac{u^8}{8}\right) + C \\ &= \tfrac{1}{48}(2x^3 + 1)^8 + C\end{aligned}$$

A substitution in an indefinite integral can sometimes be made in several different ways. To illustrate, another method for evaluating the integral in Example 3 is to consider

$$u = 2x^3 + 1, \quad du = 6x^2\,dx, \quad \tfrac{1}{6}\,du = x^2\,dx.$$

We then substitute $\frac{1}{6}\,du$ for $x^2\,dx$,

$$\int (2x^3 + 1)^7 x^2\,dx = \int u^7 \tfrac{1}{6}\,du = \tfrac{1}{6}\int u^7\,du,$$

and integrate as before.

EXAMPLE 4 Evaluate $\int x\sqrt[3]{7-6x^2}\,dx$.

SOLUTION Note that the integrand contains the term $x\,dx$. If the factor x were missing or if x were raised to a higher power, the problem would be more complicated. For integrands that involve a radical, we often substitute for the expression under the radical sign. Thus, we let

$$u = 7 - 6x^2, \quad du = -12x\,dx.$$

Next we introduce the factor -12 into the integrand, compensate by multiplying the integral by $-\frac{1}{12}$, and proceed as follows:

$$\begin{aligned}\int x\sqrt[3]{7-6x^2}\,dx &= -\tfrac{1}{12}\int \sqrt[3]{7-6x^2}(-12)x\,dx\\ &= -\tfrac{1}{12}\int \sqrt[3]{u}\,du = -\tfrac{1}{12}\int u^{1/3}\,du\\ &= -\frac{1}{12}\left(\frac{u^{4/3}}{4/3}\right) + C = -\frac{1}{16}u^{4/3} + C\\ &= -\tfrac{1}{16}(7-6x^2)^{4/3} + C\end{aligned}$$

We could also have written

$$u = 7 - 6x^2 \quad du = -12x\,dx, \quad -\tfrac{1}{12}\,du = x\,dx$$

and substituted directly for $x\,dx$. Thus,

$$\int \sqrt[3]{7-6x^2}\,x\,dx = \int \sqrt[3]{u}\left(-\tfrac{1}{12}\right)du = -\tfrac{1}{12}\int \sqrt[3]{u}\,du.$$

The remainder of the solution would proceed exactly as before.

EXAMPLE 5 Evaluate $\int \dfrac{x^2-1}{(x^3-3x+1)^6}\,dx$.

SOLUTION Let

$$u = x^3 - 3x + 1, \quad du = (3x^2-3)\,dx = 3(x^2-1)\,dx$$

and proceed as follows:

$$\begin{aligned}\int \frac{x^2-1}{(x^3-3x+1)^6}\,dx &= \frac{1}{3}\int \frac{3(x^2-1)}{(x^3-3x+1)^6}\,dx\\ &= \frac{1}{3}\int \frac{1}{u^6}\,du = \frac{1}{3}\int u^{-6}\,du\\ &= \frac{1}{3}\left(\frac{u^{-5}}{-5}\right) + C = -\frac{1}{15}\left(\frac{1}{u^5}\right) + C\\ &= -\frac{1}{15}\frac{1}{(x^3-3x+1)^5} + C\end{aligned}$$

EXAMPLE 6 Evaluate $\int \dfrac{\cos\sqrt{x}}{\sqrt{x}}\,dx$.

SOLUTION We wish to use the formula $\int \cos u\, du = \sin u + C$, so let us make the substitution

$$u = \sqrt{x} = x^{1/2}, \quad du = \frac{1}{2}x^{-1/2}\, dx = \frac{1}{2\sqrt{x}}\, dx.$$

If we introduce the factor $\frac{1}{2}$ into the integrand and compensate by multiplying the integral by 2, we obtain

$$\begin{aligned}\int \frac{\cos \sqrt{x}}{\sqrt{x}}\, dx &= 2 \int \cos \sqrt{x}\left(\frac{1}{2} \cdot \frac{1}{\sqrt{x}}\right) dx \\ &= 2 \int \cos u\, du = 2 \sin u + C \\ &= 2 \sin \sqrt{x} + C.\end{aligned}$$

EXAMPLE 7 Evaluate $\int \cos^3 5x \sin 5x\, dx$.

SOLUTION The form of the integrand suggests that we use the power rule $\int u^3\, du = \frac{1}{4}u^4 + C$. Thus, we let

$$u = \cos 5x, \quad du = -5 \sin 5x\, dx.$$

The form of du indicates that we should introduce the factor -5 into the integrand, multiply the integral by $-\frac{1}{5}$, and then integrate as follows:

$$\begin{aligned}\int \cos^3 5x \sin 5x\, dx &= -\tfrac{1}{5} \int \cos^3 5x(-5 \sin 5x)\, dx \\ &= -\tfrac{1}{5} \int u^3\, du \\ &= -\frac{1}{5}\left(\frac{u^4}{4}\right) + C \\ &= -\tfrac{1}{20} \cos^4 5x + C\end{aligned}$$

EXERCISES 5.2

Exer. 1–8: Evaluate the integral using the given substitution, and express the answer in terms of x.

1 $\int x(2x^2 + 3)^{10}\, dx; \quad u = 2x^2 + 3$

2 $\int \frac{x}{(x^2 + 5)^3}\, dx; \quad u = x^2 + 5$

3 $\int x^2 \sqrt[3]{3x^3 + 7}\, dx; \quad u = 3x^3 + 7$

4 $\int \frac{5x}{\sqrt{x^2 - 3}}\, dx; \quad u = x^2 - 3$

5 $\int \frac{(1 + \sqrt{x})^3}{\sqrt{x}}\, dx; \quad u = 1 + \sqrt{x}$

6 $\int \frac{1}{(5x - 4)^{10}}\, dx; \quad u = 5x - 4$

7 $\int \sqrt{x} \cos \sqrt{x^3}\, dx; \quad u = x^{3/2}$

8 $\int \tan x \sec^2 x\, dx; \quad u = \tan x$

Exer. 9–48: Evaluate the integral.

9 $\int \sqrt{3x - 2}\, dx$

10 $\int \sqrt[4]{2x + 5}\, dx$

11 $\int \sqrt[3]{8t + 5}\, dt$

12 $\int \frac{1}{\sqrt{4 - 5t}}\, dt$

13 $\int (3z + 1)^4\, dz$

14 $\int (2z^2 - 3)^5 z\, dz$

15 $\int v^2 \sqrt{v^3 - 1}\, dv$

16 $\int v\sqrt{9 - v^2}\, dv$

17 $\int \frac{x}{\sqrt[3]{1 - 2x^2}}\, dx$

18 $\int (3 - x^4)^3 x^3\, dx$

19 $\int (s^2 + 1)^2 \, ds$

20 $\int (3 - s^3)^2 s \, ds$

21 $\int \frac{(\sqrt{x} + 3)^4}{\sqrt{x}} \, dx$

22 $\int \left(1 + \frac{1}{x}\right)^{-3} \left(\frac{1}{x^2}\right) dx$

23 $\int \frac{t - 2}{(t^2 - 4t + 3)^3} \, dt$

24 $\int \frac{t^2 + t}{(4 - 3t^2 - 2t^3)^4} \, dt$

25 $\int 3 \sin 4x \, dx$

26 $\int 4 \cos \frac{1}{2}x \, dx$

27 $\int \cos (4x - 3) \, dx$

28 $\int \sin (1 + 6x) \, dx$

29 $\int v \sin (v^2) \, dv$

30 $\int \frac{\cos \sqrt[3]{v}}{\sqrt[3]{v^2}} \, dv$

31 $\int \cos 3x \sqrt[3]{\sin 3x} \, dx$

32 $\int \frac{\sin 2x}{\sqrt{1 - \cos 2x}} \, dx$

33 $\int (\sin x + \cos x)^2 \, dx$ (*Hint:* $\sin 2\theta = 2 \sin \theta \cos \theta$)

34 $\int \frac{\sin 4x}{\cos 2x} \, dx$ (*Hint:* $\sin 2\theta = 2 \sin \theta \cos \theta$)

35 $\int \sin x(1 + \cos x)^2 \, dx$

36 $\int \sin^3 x \cos x \, dx$

37 $\int \frac{\sin x}{\cos^4 x} \, dx$

38 $\int \sin 2x \sec^5 2x \, dx$

39 $\int \frac{\cos t}{(1 - \sin t)^2} \, dt$

40 $\int (2 + 5 \cos t)^3 \sin t \, dt$

41 $\int \sec^2 (3x - 4) \, dx$

42 $\int \frac{\csc 2x}{\sin 2x} \, dx$

43 $\int \sec^2 3x \tan 3x \, dx$

44 $\int \frac{1}{\tan 4x \sin 4x} \, dx$

45 $\int \frac{1}{\sin^2 5x} \, dx$

46 $\int \frac{x}{\cos^2 (x^2)} \, dx$

47 $\int x \cot (x^2) \csc (x^2) \, dx$

48 $\int \sec \left(\frac{x}{3}\right) \tan \left(\frac{x}{3}\right) dx$

Exer. 49–52: Solve the differential equation subject to the given conditions.

49 $f'(x) = \sqrt[3]{3x + 2}$; $\quad f(2) = 9$

50 $\frac{dy}{dx} = x\sqrt{x^2 + 5}$; $\quad y = 12$ if $x = 2$

51 $f''(x) = 16 \cos 2x - 3 \sin x$; $\quad f(0) = -2$; $\quad f'(0) = 4$

52 $f''(x) = 4 \sin 2x + 16 \cos 4x$; $\quad f(0) = 6$; $\quad f'(0) = 1$

Exer. 53–56: Evaluate the integral by (a) the method of substitution and (b) expanding the integrand. In what way do the constants of integration differ?

53 $\int (x + 4)^2 \, dx$

54 $\int (x^2 + 4)^2 x \, dx$

55 $\int \frac{(\sqrt{x} + 3)^2}{\sqrt{x}} \, dx$

56 $\int \left(1 + \frac{1}{x}\right)^2 \frac{1}{x^2} \, dx$

57 A charged particle is moving on a coordinate line in a magnetic field such that its velocity (in cm/sec) at time t is given by $v(t) = \frac{1}{2} \sin (3t - \frac{1}{4}\pi)$. Show that the motion is simple harmonic (see page 223).

58 The acceleration of a particle that is moving on a coordinate line is given by $a(t) = k \cos (\omega t + \phi)$ for constants k, ω, and ϕ and time t (in seconds). Show that the motion is simple harmonic (see page 223).

59 A reservoir supplies water to a community. In summer the demand A for water (in ft^3/day) changes according to the formula $dA/dt = 4000 + 2000 \sin (\frac{1}{90}\pi t)$ for time t (in days), with $t = 0$ corresponding to the beginning of summer. Estimate the total water consumption during 90 days of summer.

60 The pumping action of the heart consists of the systolic phase, in which blood rushes from the left ventricle into the aorta, and the diastolic phase, during which the heart muscle relaxes. The graph shown in the figure is sometimes used to model one complete cycle of the process. For a particular individual, the systolic phase lasts $\frac{1}{4}$ second and has a maximum flow rate dV/dt of 8 L/min, where V is the volume of blood in the heart at time t.

(a) Show that $dV/dt = 8 \sin (240\pi t)$ L/min.

(b) Estimate the total amount of blood pumped into the aorta during a systolic phase.

EXERCISE 60

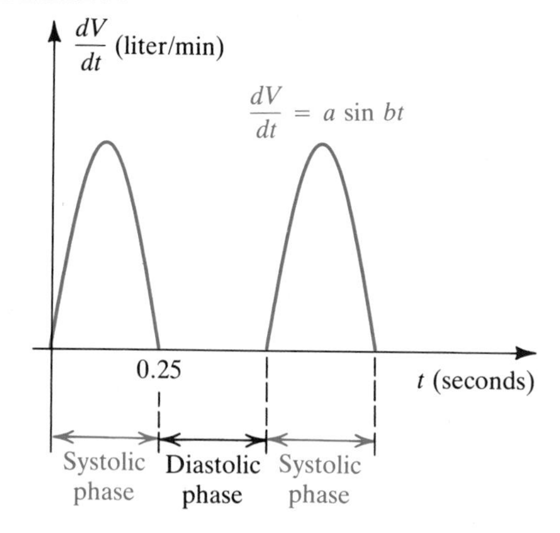

61 The rhythmic process of breathing consists of alternating periods of inhaling and exhaling. For an adult, one complete cycle normally takes place every 5 seconds. If V denotes the volume of air in the lungs at time t, then dV/dt is the flow rate.

(a) If the maximum flow rate is 0.6 L/sec, find a formula $dV/dt = a \sin bt$ that fits the given information.

(b) Use part (a) to estimate the amount of air inhaled during one cycle.

62 Many animal populations fluctuate over 10-year cycles. Suppose that the rate of growth of a rabbit population is given by $dN/dt = 1000 \cos(\frac{1}{5}\pi t)$ rabbits/yr, where N denotes the number in the population at time t (in years) and $t = 0$ corresponds to the beginning of a cycle. If the population after 5 years is estimated to be 3000 rabbits, find a formula for N at time t and estimate the maximum population.

63 Show, by evaluating in three different ways, that

$$\begin{aligned}\int \sin x \cos x \, dx &= \tfrac{1}{2}\sin^2 x + C \\ &= -\tfrac{1}{2}\cos^2 x + D \\ &= -\tfrac{1}{4}\cos 2x + E.\end{aligned}$$

How can all three answers be correct?

5.3 SUMMATION NOTATION AND AREA

In this section we shall lay the foundation for the definition of the *definite integral*. At the outset, it is virtually impossible to see any connection between definite integrals and indefinite integrals. In Section 5.6, however, we show that there is a very close relationship: *Indefinite integrals can be used to evaluate definite integrals.*

In our development of the definite integral we shall employ sums of many numbers. To express such sums compactly, it is convenient to use **summation notation**. Given a collection of numbers $\{a_1, a_2, \ldots, a_n\}$, the symbol $\sum_{k=1}^{n} a_k$ represents their sum as follows.

Summation notation **(5.9)**

$$\sum_{k=1}^{n} a_k = a_1 + a_2 + a_3 + \cdots + a_n$$

The Greek capital letter Σ (sigma) indicates a sum, and a_k represents the kth term of the sum. The letter k is the **index of summation**, or the **summation variable**, and assumes successive integer values. The integers 1 and n indicate the extreme values of the summation variable.

EXAMPLE 1 Evaluate $\sum_{k=1}^{4} k^2(k-3)$.

SOLUTION Comparing the sum with (5.9), we see that $a_k = k^2(k-3)$ and $n = 4$. To find the sum, we substitute 1, 2, 3, and 4 for k and add the resulting terms. Thus,

$$\begin{aligned}\sum_{k=1}^{4} k^2(k-3) &= 1^2(1-3) + 2^2(2-3) + 3^2(3-3) + 4^2(4-3) \\ &= (-2) + (-4) + 0 + 16 = 10.\end{aligned}$$

Letters other than k can be used for the summation variable. To illustrate,

$$\sum_{k=1}^{4} k^2(k-3) = \sum_{i=1}^{4} i^2(i-3) = \sum_{j=1}^{4} j^2(j-3) = 10.$$

If $a_k = c$ for every k, then

$$\sum_{k=1}^{2} a_k = a_1 + a_2 = c + c = 2c = \sum_{k=1}^{2} c$$

$$\sum_{k=1}^{3} a_k = a_1 + a_2 + a_3 = c + c + c = 3c = \sum_{k=1}^{3} c.$$

In general, the following result is true for every positive integer n.

Theorem **(5.10)**

$$\sum_{k=1}^{n} c = nc$$

The domain of the summation variable does not have to begin at 1. For example,

$$\sum_{k=4}^{8} a_k = a_4 + a_5 + a_6 + a_7 + a_8.$$

EXAMPLE 2 Evaluate $\sum_{k=0}^{3} \frac{2^k}{(k+1)}$.

SOLUTION

$$\sum_{k=0}^{3} \frac{2^k}{(k+1)} = \frac{2^0}{(0+1)} + \frac{2^1}{(1+1)} + \frac{2^2}{(2+1)} + \frac{2^3}{(3+1)}$$
$$= 1 + 1 + \tfrac{4}{3} + 2 = \tfrac{16}{3}$$

Theorem **(5.11)**

If n is any positive integer and $\{a_1, a_2, \ldots, a_n\}$ and $\{b_1, b_2, \ldots, b_n\}$ are sets of real numbers, then

(i) $\sum_{k=1}^{n} (a_k + b_k) = \sum_{k=1}^{n} a_k + \sum_{k=1}^{n} b_k$

(ii) $\sum_{k=1}^{n} ca_k = c\left(\sum_{k=1}^{n} a_k\right)$ for every real number c

(iii) $\sum_{k=1}^{n} (a_k - b_k) = \sum_{k=1}^{n} a_k - \sum_{k=1}^{n} b_k$

PROOF To prove (i), we begin with

$$\sum_{k=1}^{n} (a_k + b_k) = (a_1 + b_1) + (a_2 + b_2) + (a_3 + b_3) + \cdots + (a_n + b_n).$$

Rearranging terms on the right we obtain

$$\sum_{k=1}^{n} (a_k + b_k) = (a_1 + a_2 + a_3 + \cdots + a_n) + (b_1 + b_2 + b_3 + \cdots + b_n)$$
$$= \sum_{k=1}^{n} a_k + \sum_{k=1}^{n} b_k.$$

For (ii),

$$\sum_{k=1}^{n} (ca_k) = ca_1 + ca_2 + ca_3 + \cdots + ca_n$$

$$= c(a_1 + a_2 + a_3 + \cdots + a_n) = c\left(\sum_{k=1}^{n} a_k\right).$$

To prove (iii), write $a_k - b_k = a_k + (-1)b_k$ and use (i) and (ii). ■

The formulas in the following theorem will be useful later in this section. They may be proved by mathematical induction (see Appendix I).

Theorem **(5.12)**

(i) $\sum_{k=1}^{n} k = 1 + 2 + \cdots + n = \dfrac{n(n+1)}{2}$

(ii) $\sum_{k=1}^{n} k^2 = 1^2 + 2^2 + \cdots + n^2 = \dfrac{n(n+1)(2n+1)}{6}$

(iii) $\sum_{k=1}^{n} k^3 = 1^3 + 2^3 + \cdots + n^3 = \left[\dfrac{n(n+1)}{2}\right]^2$

EXAMPLE 3 Evaluate $\sum_{k=1}^{100} k$ and $\sum_{k=1}^{20} k^2$.

SOLUTION Using (i) and (ii) of Theorem (5.12), we obtain

$$\sum_{k=1}^{100} k = 1 + 2 + \cdots + 100 = \frac{100(101)}{2} = 5050$$

and

$$\sum_{k=1}^{20} k^2 = 1^2 + 2^2 + \cdots + 20^2 = \frac{20(21)(41)}{6} = 2870.$$

EXAMPLE 4 Express $\sum_{k=1}^{n} (k^2 - 4k + 3)$ in terms of n.

SOLUTION We use Theorems (5.11), (5.12), and (5.10):

$$\sum_{k=1}^{n} (k^2 - 4k + 3) = \sum_{k=1}^{n} k^2 - 4\sum_{k=1}^{n} k + \sum_{k=1}^{n} 3$$

$$= \frac{n(n+1)(2n+1)}{6} - 4\,\frac{n(n+1)}{2} + 3n$$

$$= \tfrac{1}{3}n^3 - \tfrac{3}{2}n^2 + \tfrac{7}{6}n$$

The definition of the definite integral (to be given in Section 5.4) is closely related to the areas of certain regions in a coordinate plane. We can easily calculate the area if the region is bounded by lines. For example, the area of a rectangle is the product of its length and width. The area of a triangle is one-half the product of an altitude and the corresponding base. The area of any polygon can be found by subdividing it into triangles.

In order to find areas of regions whose boundaries involve graphs of functions, however, we utilize a limiting process and then use methods of calculus. In particular, let us consider a region R in a coordinate plane, bounded by the vertical lines $x = a$ and $x = b$, by the x-axis, and by the graph of a function f that is continuous and nonnegative on the closed interval $[a, b]$. A region of this type is illustrated in Figure 5.3. Since $f(x) \geq 0$ for every x in $[a, b]$, no part of the graph lies below the x-axis. For convenience, we shall refer to R as **the region under the graph of f from a to b**. We wish to define the area A of R.

FIGURE 5.3 Region under the graph of f

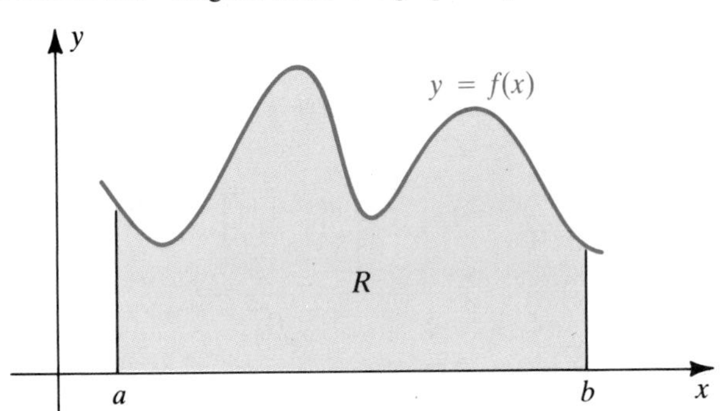

FIGURE 5.4 An inscribed rectangular polygon

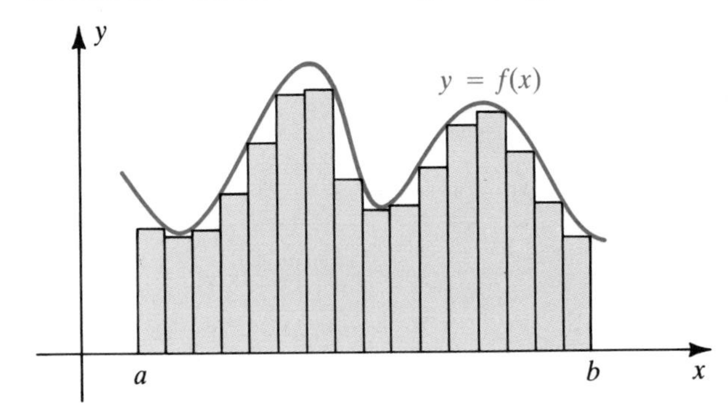

To arrive at a satisfactory definition of A, we shall consider many rectangles of equal width such that each rectangle lies completely under the graph of f and intersects the graph in at least one point, as illustrated in Figure 5.4. The boundary of the region formed by the totality of these rectangles is called an **inscribed rectangular polygon**. We shall use the following notation:

$$A_{\text{IP}} = \text{area of an inscribed rectangular polygon}$$

If the width of the rectangles in Figure 5.4 is small, then it appears that

$$A_{\text{IP}} \approx A.$$

This suggests that we let the width of the rectangles approach zero and define A as a limiting value of the areas A_{IP} of the corresponding inscribed rectangular polygons. The notation discussed next will allow us to carry out this procedure rigorously.

If n is any positive integer, divide the interval $[a, b]$ into n subintervals, all having the same length $\Delta x = (b - a)/n$. We can do this by choosing numbers $x_0, x_1, x_2, \ldots, x_n$, with $a = x_0$, $b = x_n$, and

$$x_k - x_{k-1} = \frac{b - a}{n} = \Delta x$$

for $k = 1, 2, \ldots, n$, as indicated in Figure 5.5. Note that

$$x_0 = a, \quad x_1 = a + \Delta x, \quad x_2 = a + 2\,\Delta x, \quad x_3 = a + 3\,\Delta x, \quad \ldots$$
$$x_k = a + k\,\Delta x, \quad \ldots, \quad x_n = a + n\,\Delta x = b.$$

FIGURE 5.5

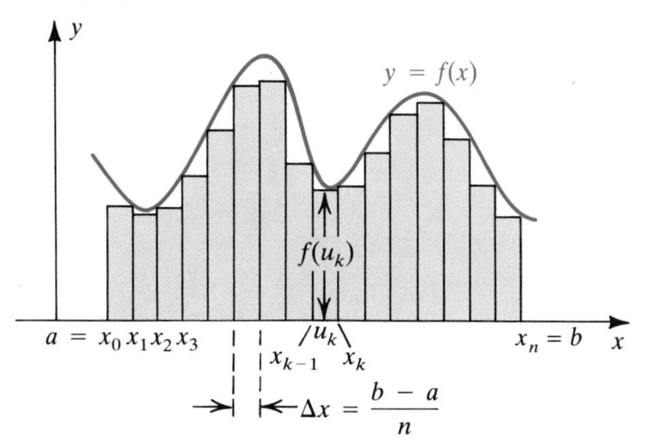

The function f is continuous on each subinterval $[x_{k-1}, x_k]$, and hence, by the extreme value theorem (4.3), f takes on a minimum value at some number u_k in $[x_{k-1}, x_k]$. For each k, let us construct a rectangle of width $\Delta x = x_k - x_{k-1}$ and height equal to the minimum distance $f(u_k)$ from the x-axis to the graph of f (see Figure 5.5). The area of the kth rectangle is $f(u_k)\,\Delta x$. The area A_{IP} of the resulting inscribed rectangular polygon is the sum of the areas of the n rectangles; that is,

$$A_{\text{IP}} = f(u_1)\,\Delta x + f(u_2)\,\Delta x + \cdots + f(u_n)\,\Delta x.$$

Using summation notation, we may write

$$A_{\text{IP}} = \sum_{k=1}^{n} f(u_k)\,\Delta x,$$

where $f(u_k)$ is the minimum value of f on $[x_{k-1}, x_k]$.

If n is very large, or, equivalently, if Δx is very small, then the sum A_{IP} of the rectangular areas should approximate the area of the region R. Intuitively we know that if there exists a number A such that $\sum_{k=1}^{n} f(u_k)\,\Delta x$ gets closer to A as Δx gets closer to 0 (but $\Delta x \neq 0$), we can call A the **area** of R and write

$$A = \lim_{\Delta x \to 0} A_{\text{IP}} = \lim_{\Delta x \to 0} \sum_{k=1}^{n} f(u_k)\,\Delta x.$$

The meaning of this *limit of sums* is not the same as that of the limit of a function, introduced in Chapter 2. To eliminate the word *closer* and arrive at a satisfactory definition of A, let us take a slightly different point of view. If A denotes the area of the region R, then the difference

$$A - \sum_{k=1}^{n} f(u_k)\,\Delta x$$

is the area of the portion in Figure 5.5 that lies *under* the graph of f and *over* the inscribed rectangular polygon. This number may be regarded as the error in using the area of the inscribed rectangular polygon to approximate A. We should be able to make this error as small as desired by choosing the width Δx of the rectangles sufficiently small. This is the motivation for the following definition of the area A of R. The notation is the same as that used in the preceding discussion.

Definition **(5.13)**

Let f be continuous and nonnegative on $[a, b]$. Let A be a real number, and let $f(u_k)$ be the minimum value of f on $[x_{k-1}, x_k]$. The notation

$$A = \lim_{\Delta x \to 0} \sum_{k=1}^{n} f(u_k)\,\Delta x$$

means that for every $\epsilon > 0$ there is a $\delta > 0$ such that if $0 < \Delta x < \delta$, then

$$A - \sum_{k=1}^{n} f(u_k)\,\Delta x < \epsilon.$$

If A is the indicated limit and we let $\epsilon = 10^{-9}$, then Definition (5.13) states that by using rectangles of sufficiently small width Δx, we can make the difference between A and the area of the inscribed polygon less than one-billionth of a square unit. Similarly, if $\epsilon = 10^{-12}$, we can make this difference less than one-trillionth of a square unit. In general, the difference can be made less than *any* preassigned ϵ.

If f is continuous on $[a, b]$, it is shown in more advanced texts that a number A satisfying Definition (5.13) actually exists. We shall call A **the area under the graph of f from a to b**.

The area A may also be obtained by means of **circumscribed rectangular polygons** of the type illustrated in Figure 5.6. In this case we select the number v_k in each interval $[x_{k-1}, x_k]$ such that $f(v_k)$ is the *maximum* value of f on $[x_{k-1}, x_k]$.

FIGURE 5.6 A circumscribed rectangular polygon

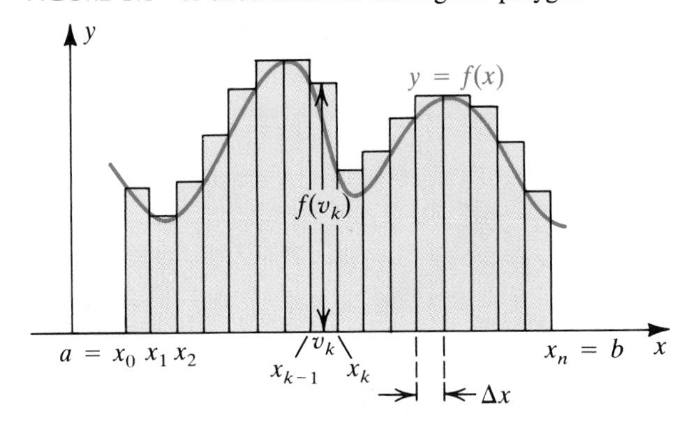

Let

$$A_{\text{CP}} = \text{area of a circumscribed rectangular polygon.}$$

Using summation notation, we have

$$A_{\text{CP}} = \sum_{k=1}^{n} f(v_k)\,\Delta x,$$

where $f(v_k)$ is the maximum value of f on $[x_{k-1}, x_k]$. Note that

$$\sum_{k=1}^{n} f(u_k)\,\Delta x \le A \le \sum_{k=1}^{n} f(v_k)\,\Delta x.$$

The limit of A_{CP} as $\Delta x \to 0$ is defined as in (5.13). The only change is that we use

$$\sum_{k=1}^{n} f(v_k)\,\Delta x - A < \epsilon,$$

since we want this difference to be nonnegative. It can be proved that the same number A is obtained using either inscribed or circumscribed rectangles.

EXAMPLE 5 Let $f(x) = 16 - x^2$, and let R be the region under the graph of f from 0 to 3. Approximate the area A of R using

(a) an inscribed rectangular polygon with $\Delta x = \frac{1}{2}$

(b) a circumscribed rectangular polygon with $\Delta x = \frac{1}{2}$

FIGURE 5.7

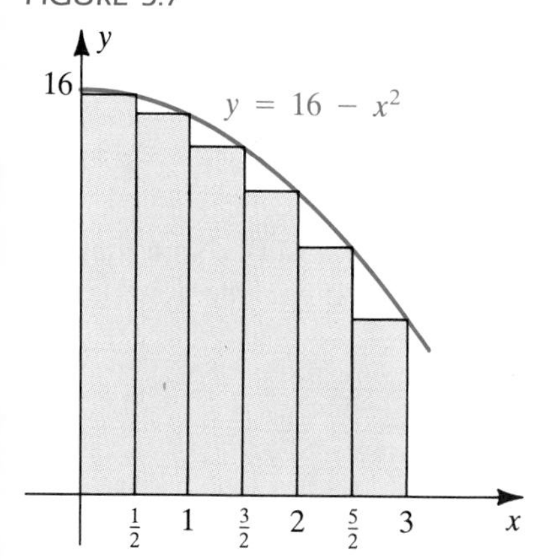

SOLUTION

(a) The graph of f and the inscribed rectangular polygon with $\Delta x = \frac{1}{2}$ are sketched in Figure 5.7 (with different scales on the x- and y-axes). Note that f is decreasing on $[0, 3]$, and hence the minimum value $f(u_k)$ on the kth subinterval occurs at the right-hand endpoint of the subinterval. Since there are six rectangles to consider, the formula for A_{IP} is

$$\begin{aligned}
A_{IP} &= \sum_{k=1}^{6} f(u_k)\,\Delta x \\
&= f(\tfrac{1}{2})\cdot\tfrac{1}{2} + f(1)\cdot\tfrac{1}{2} + f(\tfrac{3}{2})\cdot\tfrac{1}{2} + f(2)\cdot\tfrac{1}{2} + f(\tfrac{5}{2})\cdot\tfrac{1}{2} + f(3)\cdot\tfrac{1}{2} \\
&= \tfrac{63}{4}\cdot\tfrac{1}{2} + 15\cdot\tfrac{1}{2} + \tfrac{55}{4}\cdot\tfrac{1}{2} + 12\cdot\tfrac{1}{2} + \tfrac{39}{4}\cdot\tfrac{1}{2} + 7\cdot\tfrac{1}{2} \\
&= \tfrac{293}{8} = 36.625.
\end{aligned}$$

FIGURE 5.8

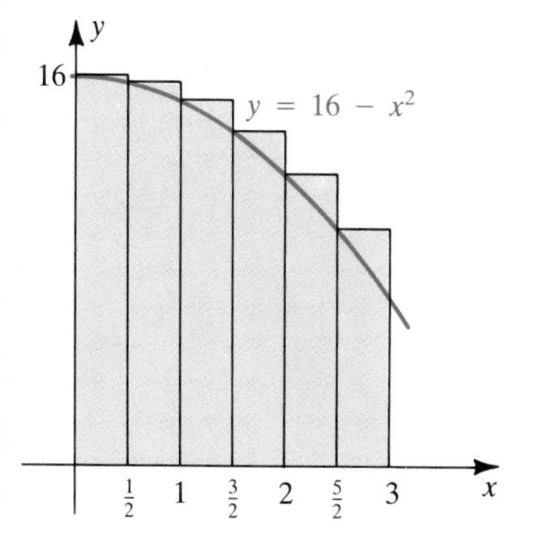

(b) The graph of f and the circumscribed rectangular polygon are sketched in Figure 5.8. Since f is decreasing on $[0, 3]$, the maximum value $f(v_k)$ occurs at the left-hand endpoint of the kth subinterval. Hence

$$\begin{aligned}
A_{CP} &= \sum_{k=1}^{6} f(v_k)\,\Delta x \\
&= f(0)\cdot\tfrac{1}{2} + f(\tfrac{1}{2})\cdot\tfrac{1}{2} + f(1)\cdot\tfrac{1}{2} + f(\tfrac{3}{2})\cdot\tfrac{1}{2} + f(2)\cdot\tfrac{1}{2} + f(\tfrac{5}{2})\cdot\tfrac{1}{2} \\
&= 16\cdot\tfrac{1}{2} + \tfrac{63}{4}\cdot\tfrac{1}{2} + 15\cdot\tfrac{1}{2} + \tfrac{55}{4}\cdot\tfrac{1}{2} + 12\cdot\tfrac{1}{2} + \tfrac{39}{4}\cdot\tfrac{1}{2} \\
&= \tfrac{329}{8} = 41.125.
\end{aligned}$$

It follows that $36.625 < A < 41.125$. In the next example we prove that $A = 39$.

EXAMPLE 6 If $f(x) = 16 - x^2$, find the area of the region under the graph of f from 0 to 3.

SOLUTION The region was considered in Example 5 and is resketched in Figure 5.9, on the following page. If the interval $[0, 3]$ is divided into n equal subintervals, then the length Δx of each subinterval is $3/n$. Employing the notation used in Figure 5.5, with $a = 0$ and $b = 3$, we have

$$x_0 = 0, \quad x_1 = \Delta x, \quad x_2 = 2\,\Delta x, \quad \ldots, \quad x_k = k\,\Delta x, \quad \ldots, \quad x_n = n\,\Delta x = 3.$$

FIGURE 5.9

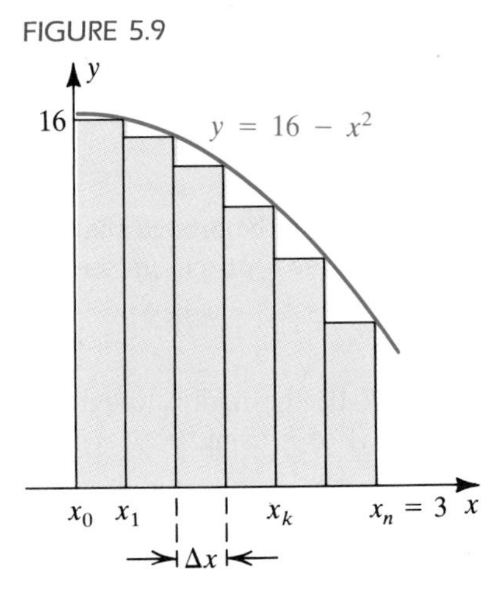

Since $\Delta x = 3/n$,

$$x_k = k\,\Delta x = k\frac{3}{n} = \frac{3k}{n}.$$

Since f is decreasing on $[0, 3]$, the number u_k in $[x_{k-1}, x_k]$ at which f takes on its minimum value is always the right-hand endpoint x_k of the subinterval; that is, $u_k = x_k = 3k/n$. Thus,

$$f(u_k) = f\left(\frac{3k}{n}\right) = 16 - \left(\frac{3k}{n}\right)^2 = 16 - \frac{9k^2}{n^2},$$

and the summation in Definition (5.13) is

$$\begin{aligned}\sum_{k=1}^{n} f(u_k)\,\Delta x &= \sum_{k=1}^{n}\left[\left(16 - \frac{9k^2}{n^2}\right)\cdot\frac{3}{n}\right]\\ &= \frac{3}{n}\sum_{k=1}^{n}\left(16 - \frac{9k^2}{n^2}\right),\end{aligned}$$

where the last equality follows from (ii) of Theorem (5.11). (Note that $3/n$ does not contain the summation variable k.) We next use Theorems (5.11), (5.10), and (5.12) to obtain

$$\begin{aligned}\sum_{k=1}^{n} f(u_k)\,\Delta x &= \frac{3}{n}\left(\sum_{k=1}^{n} 16 - \frac{9}{n^2}\sum_{k=1}^{n} k^2\right)\\ &= \frac{3}{n}\left[n\cdot 16 - \frac{9}{n^2}\,\frac{n(n+1)(2n+1)}{6}\right]\\ &= 48 - \frac{9}{2}\,\frac{(n+1)(2n+1)}{n^2}.\end{aligned}$$

To find the area of the region, we let Δx approach 0. Since $\Delta x = 3/n$, we can accomplish this by letting n increase without bound. Although our discussion of limits involving infinity in Section 2.4 was concerned with a real variable x, a similar discussion can be given if the variable is an integer n. Assuming that this is true and that we can replace $\Delta x \to 0$ by $n \to \infty$, we obtain

$$\begin{aligned}\lim_{\Delta x\to 0}\sum_{k=1}^{n} f(u_k)\,\Delta x &= \lim_{n\to\infty}\left[48 - \frac{9}{2}\,\frac{(n+1)(2n+1)}{n^2}\right]\\ &= 48 - \tfrac{9}{2}\cdot 2 = 39.\end{aligned}$$

Thus, the area of the region is 39 square units.

The area in the preceding example may also be found by using circumscribed rectangular polygons. In this case we select, in each subinterval $[x_{k-1}, x_k]$, the number $v_k = (k-1)(3/n)$ at which f takes on its maximum value.

The next example illustrates the use of circumscribed rectangles in finding an area.

EXAMPLE 7 If $f(x) = x^3$, find the area under the graph of f from 0 to b for any $b > 0$.

SOLUTION Subdividing the interval $[0, b]$ into n equal parts (see Figure 5.10), we obtain a circumscribed rectangular polygon such that $\Delta x = b/n$ and $x_k = k\,\Delta x$.

FIGURE 5.10

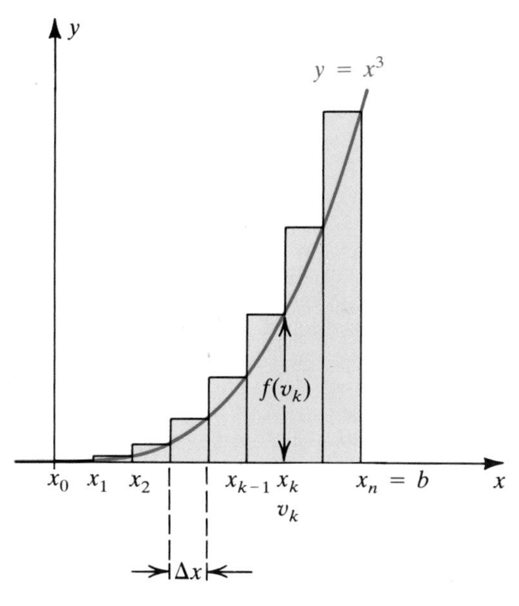

Since f is an increasing function, the maximum value $f(v_k)$ in the interval $[x_{k-1}, x_k]$ occurs at the right-hand endpoint; that is,

$$v_k = x_k = k\,\Delta x = k\frac{b}{n} = \frac{bk}{n}.$$

The sum of the areas of the circumscribed rectangles is

$$\begin{aligned}\sum_{k=1}^{n} f(v_k)\,\Delta x &= \sum_{k=1}^{n}\left[\left(\frac{bk}{n}\right)^3\cdot\frac{b}{n}\right] = \sum_{k=1}^{n}\frac{b^4}{n^4}k^3\\ &= \frac{b^4}{n^4}\sum_{k=1}^{n}k^3 = \frac{b^4}{n^4}\left[\frac{n(n+1)}{2}\right]^2\\ &= \frac{b^4}{4}\cdot\frac{n^2(n+1)^2}{n^4},\end{aligned}$$

where we have used Theorem (5.12)(iii). If we let Δx approach 0, then n increases without bound and the expression involving n approaches 1. It follows that the area under the graph is

$$\lim_{\Delta x\to 0}\sum_{k=1}^{n} f(v_k)\,\Delta x = \frac{b^4}{4}.$$

FIGURE 5.11

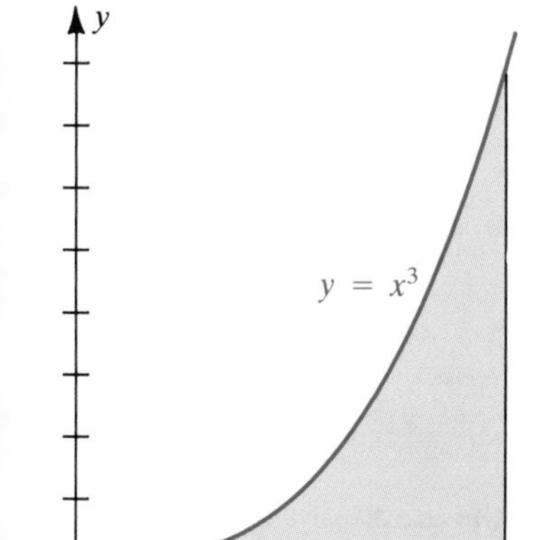

EXAMPLE 8 If $f(x) = x^3$, find the area A of the region under the graph of f from $\frac{1}{2}$ to 2.

SOLUTION The region is sketched in Figure 5.11.

If we let A_1 = area under the graph of f from 0 to $\frac{1}{2}$

and A_2 = area under the graph of f from 0 to 2,

the area A can be found by subtracting A_1 from A_2:

$$A = A_2 - A_1$$

In Example 7 we found that the area under the graph of $y = x^3$ from 0 to b is $\frac{1}{4}b^4$. Hence, using $b = \frac{1}{2}$ for A_1 and $b = 2$ for A_2 yields

$$A = \frac{2^4}{4} - \frac{(\frac{1}{2})^4}{4} = 4 - \frac{1}{64} \approx 3.98.$$

EXERCISES 5.3

Exer. 1–8: Evaluate the sum.

1 $\sum_{j=1}^{4} (j^2 + 1)$

2 $\sum_{j=1}^{4} (2^j + 1)$

3 $\sum_{k=0}^{5} k(k-1)$

4 $\sum_{k=0}^{4} (k-2)(k-3)$

5 $\sum_{n=1}^{10} [1 + (-1)^n]$

6 $\sum_{n=1}^{4} (-1)^n\left(\frac{1}{n}\right)$

7 $\sum_{i=1}^{50} 10$

8 $\sum_{k=1}^{1000} 2$

Exer. 9–12: Express the sum in terms of n (see Example 4).

9 $\sum_{k=1}^{n} (k^2 + 3k + 5)$

10 $\sum_{k=1}^{n} (3k^2 - 2k + 1)$

11 $\sum_{k=1}^{n} (k^3 + 2k^2 - k + 4)$

12 $\sum_{k=1}^{n} (3k^3 + k)$

Exer. 13–18: Express in summation notation.

13 $1 + 5 + 9 + 13 + 17$

14 $2 + 5 + 8 + 11 + 14$

15 $\frac{1}{2} + \frac{2}{5} + \frac{3}{8} + \frac{4}{11}$

16 $\frac{1}{4} + \frac{2}{9} + \frac{3}{14} + \frac{4}{19}$

17 $1 - \frac{x^2}{2} + \frac{x^4}{4} - \frac{x^6}{6} + \cdots + (-1)^n \frac{x^{2n}}{2n}$

18 $1 + x + \frac{x^2}{2} + \frac{x^3}{3} + \cdots + \frac{x^n}{n}$

Exer. 19–22: Let A be the area under the graph of the given function f from a to b. Approximate A by dividing $[a, b]$ into subintervals of equal length Δx and using (a) A_{IP} and (b) A_{CP}.

19 $f(x) = 3 - x$; $a = -2$, $b = 2$; $\Delta x = 1$

20 $f(x) = x + 2$; $a = -1$, $b = 4$; $\Delta x = 1$

21 $f(x) = x^2 + 1$; $a = 1$, $b = 3$; $\Delta x = \frac{1}{2}$

22 $f(x) = 4 - x^2$; $a = 0$, $b = 2$; $\Delta x = \frac{1}{2}$

c 23 $f(x) = \sqrt{\sin x}$; $a = 0$, $b = 1.5$; $\Delta x = 0.15$

c 24 $f(x) = \frac{1}{\sqrt{x^3 + 1}}$; $a = 0$, $b = 3$; $\Delta x = 0.3$

Exer. 25–30: Refer to Examples 6 and 7. Find the area under the graph of the given function f from 0 to b using (a) inscribed rectangles and (b) circumscribed rectangles.

25 $f(x) = 2x + 3$; $b = 4$

26 $f(x) = 8 - 3x$; $b = 2$

27 $f(x) = 9 - x^2$; $b = 3$

28 $f(x) = x^2$; $b = 5$

29 $f(x) = x^3 + 1$; $b = 2$

30 $f(x) = 4x + x^3$; $b = 2$

Exer. 31–32: Refer to Example 7. Find the area under the graph of f corresponding to the interval (a) $[1, 3]$ and (b) $[a, b]$.

31 $f(x) = x^3$

32 $f(x) = x^3 + 2$

5.4 THE DEFINITE INTEGRAL

In Section 5.3 we defined the area under the graph of a function f from a to b as a limit of the form

$$\lim_{\Delta x \to 0} \sum_{k=1}^{n} f(w_k)\,\Delta x.$$

In our discussion we restricted f and Δx as follows:

1. The function f is continuous on the closed interval $[a, b]$.
2. $f(x)$ is nonnegative for every x in $[a, b]$.
3. All the subintervals $[x_{k-1}, x_k]$ have the same length Δx.
4. The number w_k is chosen such that $f(w_k)$ is always the minimum (or maximum) value of f on $[x_{k-1}, x_k]$.

There are many applications involving this type of limit in which one or more of these conditions is not satisfied. Thus, it is desirable to allow the following changes in 1–4:

1′. The function f may be discontinuous at some numbers in $[a, b]$.
2′. $f(x)$ may be negative for some x in $[a, b]$.
3′. The lengths of the subintervals $[x_{k-1}, x_k]$ may be different.
4′. The number w_k may be *any* number in $[x_{k-1}, x_k]$.

Note that if 2′ is true, part of the graph lies under the x-axis, and therefore the limit is no longer the area under the graph of f.

Let us introduce some new terminology and notation. A **partition** P of a closed interval $[a, b]$ is any decomposition of $[a, b]$ into subintervals of the form

$$[x_0, x_1], \quad [x_1, x_2], \quad [x_2, x_3], \quad \ldots, \quad [x_{n-1}, x_n]$$

for a positive integer n and numbers x_k such that

$$a = x_0 < x_1 < x_2 < x_3 < \cdots < x_{n-1} < x_n = b.$$

The length of the kth subinterval $[x_{k-1}, x_k]$ will be denoted by Δx_k; that is,

$$\Delta x_k = x_k - x_{k-1}.$$

A typical partition of $[a, b]$ is illustrated in Figure 5.12. The largest of the numbers $\Delta x_1, \Delta x_2, \ldots, \Delta x_n$ is the **norm** of the partition P and is denoted by $\|P\|$.

FIGURE 5.12 A partition of $[a, b]$

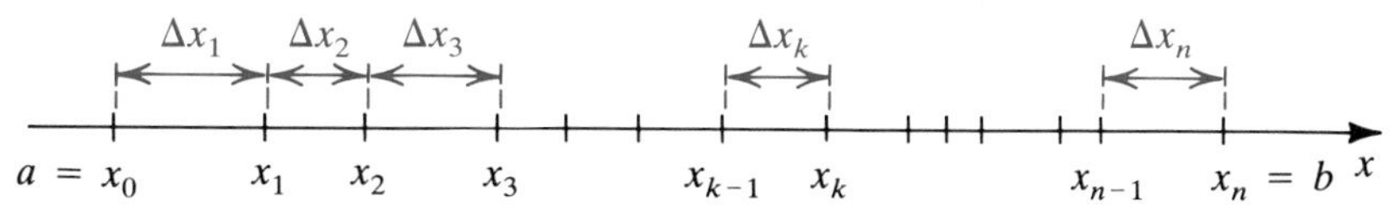

EXAMPLE 1 The numbers $\{1, 1.7, 2.2, 3.3, 4.1, 4.5, 5, 6\}$ determine a partition P of the interval $[1, 6]$. Find the lengths $\Delta x_1, \Delta x_2, \ldots, \Delta x_n$ of the subintervals in P and the norm of the partition.

SOLUTION The lengths Δx_k of the subintervals are found by subtracting successive numbers in P. Thus,

$$\Delta x_1 = 0.7, \quad \Delta x_2 = 0.5, \quad \Delta x_3 = 1.1, \quad \Delta x_4 = 0.8, \quad \Delta x_5 = 0.4,$$
$$\Delta x_6 = 0.5, \quad \Delta x_7 = 1.0.$$

The norm of P is the largest of these numbers. Hence

$$\|P\| = \Delta x_3 = 1.1.$$

The following concept, named after the mathematician G. F. B. Riemann (1826–1866), is fundamental to the definition of the definite integral.

Definition **(5.14)**

> Let f be defined on a closed interval $[a, b]$, and let P be a partition of $[a, b]$. A **Riemann sum** of f (or $f(x)$) for P is any expression R_P of the form
>
> $$R_P = \sum_{k=1}^{n} f(w_k)\,\Delta x_k,$$
>
> where w_k is in $[x_{k-1}, x_k]$ and $k = 1, 2, \ldots, n$.

FIGURE 5.13

In Definition (5.14), $f(w_k)$ is not necessarily a maximum or minimum value of f on $[x_{k-1}, x_k]$. If we construct a rectangle of length $|f(w_k)|$ and width Δx_k, as illustrated in Figure 5.13, the rectangle may be neither inscribed nor circumscribed. Moreover, since $f(x)$ may be negative, certain terms of the Riemann sum R_P may be negative. Consequently, R_P does not always represent a sum of areas of rectangles.

We may interpret the Riemann sum R_P in (5.14) geometrically, as follows. For each subinterval $[x_{k-1}, x_k]$, construct a horizontal line segment through the point $(w_k, f(w_k))$, thereby obtaining a collection of rectangles. If $f(w_k)$ is positive, the rectangle lies above the x-axis, as illustrated by the lighter rectangles in Figure 5.14, and the product $f(w_k)\,\Delta x_k$ is the area of this rectangle. If $f(w_k)$ is negative, then the rectangle lies below the x-axis, as illustrated by the darker rectangles in Figure 5.14. In this case the product $f(w_k)\,\Delta x_k$ is the *negative* of the area of a rectangle. It follows that R_P is the sum of the areas of the rectangles that lie above the x-axis and the *negatives* of the areas of the rectangles that lie below the x-axis.

FIGURE 5.14

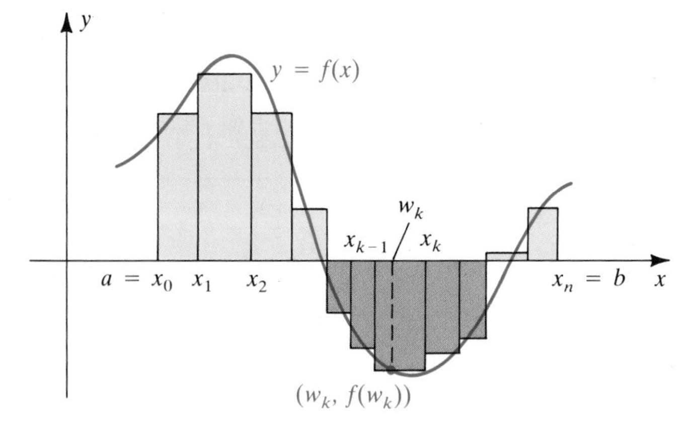

EXAMPLE 2 Let $f(x) = 8 - \frac{1}{2}x^2$, and let P be the partition of $[0, 6]$ into the five subintervals determined by

$$x_0 = 0, \quad x_1 = 1.5, \quad x_2 = 2.5, \quad x_3 = 4.5, \quad x_4 = 5, \quad x_5 = 6.$$

Find the norm of the partition and the Riemann sum R_P if

$$w_1 = 1, \quad w_2 = 2, \quad w_3 = 3.5, \quad w_4 = 5, \quad w_5 = 5.5.$$

FIGURE 5.15

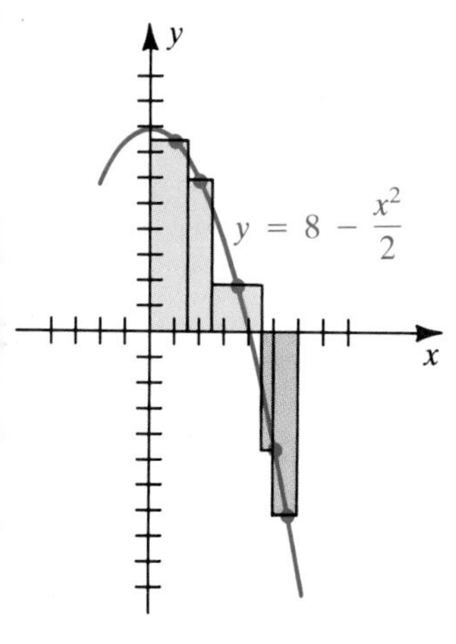

SOLUTION The graph of f is sketched in Figure 5.15, where we have also shown the points that correspond to w_k and the rectangles of lengths $|f(w_k)|$ for $k = 1, 2, 3, 4$, and 5. Thus,

$$\Delta x_1 = 1.5, \quad \Delta x_2 = 1, \quad \Delta x_3 = 2, \quad \Delta x_4 = 0.5, \quad \Delta x_5 = 1.$$

The norm $\|P\|$ of the partition is Δx_3, or 2.

Using Definition (5.14) with $n = 5$, we have

$$\begin{aligned} R_P &= \sum_{k=1}^{5} f(w_k)\,\Delta x_k \\ &= f(w_1)\,\Delta x_1 + f(w_2)\,\Delta x_2 + f(w_3)\,\Delta x_3 + f(w_4)\,\Delta x_4 + f(w_5)\,\Delta x_5 \\ &= f(1)(1.5) + f(2)(1) + f(3.5)(2) + f(5)(0.5) + f(5.5)(1) \\ &= (7.5)(1.5) + (6)(1) + (1.875)(2) + (-4.5)(0.5) + (-7.125)(1) \\ &= 11.625. \end{aligned}$$

We shall not always specify the number n of subintervals in a partition P of $[a, b]$. A Riemann sum (5.14) will then be written

$$R_P = \sum_k f(w_k)\,\Delta x_k,$$

and we will assume that terms of the form $f(w_k)\,\Delta x_k$ are to be summed over all subintervals $[x_{k-1}, x_k]$ of the partition P.

Using the same approach as in Definition (5.13), we next define

$$\lim_{\|P\| \to 0} \sum_k f(w_k)\,\Delta x_k = L$$

for a real number L.

Definition **(5.15)**

> Let f be defined on a closed interval $[a, b]$, and let L be a real number. The statement
>
> $$\lim_{\|P\| \to 0} \sum_k f(w_k)\,\Delta x_k = L$$
>
> means that for every $\epsilon > 0$ there is a $\delta > 0$ such that if P is a partition of $[a, b]$ with $\|P\| < \delta$, then
>
> $$\left| \sum_k f(w_k)\,\Delta x_k - L \right| < \epsilon$$
>
> for any choice of numbers w_k in the subintervals $[x_{k-1}, x_k]$ of P. The number L is a **limit of (Riemann) sums**.

For every $\delta > 0$ there are infinitely many partitions P of $[a, b]$ with $\|P\| < \delta$. Moreover, for each such partition P there are infinitely many ways of choosing the number w_k in $[x_{k-1}, x_k]$. Consequently, an infinite number of different Riemann sums may be associated with *each* partition P. However, if the limit L exists, then for any $\epsilon > 0$, every Riemann sum is within ϵ units of L, provided a small enough norm is chosen. Although Definition (5.15) differs from the definition of the limit of a function, we may use a proof similar to that given for the uniqueness theorem in Appendix II to show that if the limit L exists, then it is unique.

We next define the definite integral as a limit of a sum, where w_k and Δx_k have the same meanings as in Definition (5.15).

Definition **(5.16)**

Let f be defined on a closed interval $[a, b]$. The **definite integral of f from a to b**, denoted by $\int_a^b f(x)\,dx$, is

$$\int_a^b f(x)\,dx = \lim_{\|P\| \to 0} \sum_k f(w_k)\,\Delta x_k,$$

provided the limit exists.

If the limit in Definition (5.16) exists, then f is **integrable** on $[a, b]$, and we say that the definite integral $\int_a^b f(x)\,dx$ **exists**. The process of finding the limit is called **evaluating the integral**. Note that the value of a definite integral is a *real number*, not a family of antiderivatives, as was the case for indefinite integrals.

The integral sign in Definition (5.16), which may be thought of as an elongated letter S (the first letter of the word *sum*), is used to indicate the connection between definite integrals and Riemann sums. The numbers a and b are the **limits of integration**, a being the **lower limit** and b the **upper limit**. In this context *limit* refers to the smallest or largest number in the interval $[a, b]$ and is not related to definitions of limits given earlier in the text. The expression $f(x)$, which appears to the right of the integral sign, is the *integrand*, as it is with indefinite integrals. The differential symbol dx that follows $f(x)$ may be associated with the increment Δx_k of a Riemann sum of f. This association will be useful in later applications.

EXAMPLE 3 Express the following limit of sums as a definite integral on the interval $[3, 8]$:

$$\lim_{\|P\| \to 0} \sum_{k=1}^{n} (5w_k^3 + \sqrt{w_k} - 4 \sin w_k)\,\Delta x_k,$$

where w_k and Δx_k are as in Definition (5.15).

SOLUTION The given limit of sums has the form stated in Definition (5.16), with

$$f(x) = 5x^3 + \sqrt{x} - 4 \sin x.$$

Hence the limit can be expressed as the definite integral

$$\int_3^8 (5x^3 + \sqrt{x} - 4 \sin x)\,dx.$$

Letters other than x may be used in the notation for the definite integral. If f is integrable on $[a, b]$, then

$$\int_a^b f(x)\,dx = \int_a^b f(s)\,ds = \int_a^b f(t)\,dt$$

and so on. For this reason the letter x in Definition (5.16) is called a **dummy variable**.

Whenever an interval $[a, b]$ is employed, we assume that $a < b$. Consequently, Definition (5.16) does not take into account the cases in which the lower limit of integration is greater than or equal to the upper limit. The definition may be extended to include the case where the lower limit is greater than the upper limit, as follows.

Definition **(5.17)**

> If $c > d$, then $\int_c^d f(x)\,dx = -\int_d^c f(x)\,dx$.

Definition (5.17) may be phrased as follows: *Interchanging the limits of integration changes the sign of the integral.*

The case in which the lower and upper limits of integration are equal is covered by the next definition.

Definition **(5.18)**

> If $f(a)$ exists, then $\int_a^a f(x)\,dx = 0$.

If f is integrable, then the limit in Definition (5.16) exists for every choice of w_k in $[x_{k-1}, x_k]$. This allows us to specialize w_k if we wish to do so. For example, we could always choose w_k as the smallest number x_{k-1} in the subinterval or as the largest number x_k or as the midpoint of the subinterval or as the number that always produces the minimum or maximum value in $[x_{k-1}, x_k]$. Moreover, since the limit is independent of the partition P of $[a, b]$ (provided that $\|P\|$ is sufficiently small), we may specialize the partitions to the case in which all the subintervals $[x_{k-1}, x_k]$ have the same length Δx. A partition of this type is a **regular partition**.

If a regular partition of $[a, b]$ contains n subintervals, then $\Delta x = (b - a)/n$. In this case the symbol $\|P\| \to 0$ is equivalent to $\Delta x \to 0$ or $n \to \infty$, and Definition (5.16) takes on the form

$$\int_a^b f(x)\,dx = \lim_{n\to\infty} \sum_{k=1}^{n} f(w_k)\,\Delta x.$$

The following theorem is a first application of these special Riemann sums. Many other applications will be discussed in Chapter 6.

Theorem **(5.19)**

> If f is integrable and $f(x) \geq 0$ for every x in $[a, b]$, then the area A of the region under the graph of f from a to b is
>
> $$A = \int_a^b f(x)\,dx.$$

FIGURE 5.16

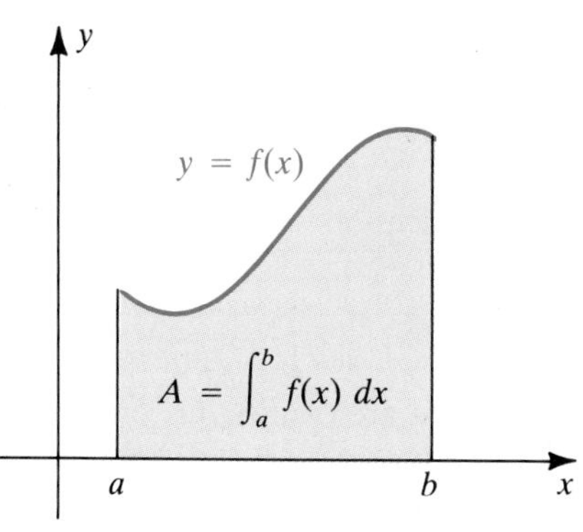

PROOF From the preceding section we know that the area A is a limit of sums $\sum_k f(u_k)\,\Delta x$, where $f(u_k)$ is the minimum value of f on $[x_{k-1}, x_k]$. Since these are Riemann sums, the conclusion follows from Definition (5.16). ■

Theorem (5.19) is illustrated in Figure 5.16. It is important to keep in mind that area is merely our first application of the definite integral. *There are many instances where $\int_a^b f(x)\,dx$ does not represent the area of a region.*

In fact, if $f(x) < 0$ for some x in $[a, b]$, then the definite integral may be negative or zero.

If f is continuous and $f(x) \geq 0$ on $[a, b]$, then Theorem (5.19) can be used to evaluate $\int_a^b f(x)\,dx$, provided we can find the area of the region under the graph of f from a to b. This will be true if the graph is a line or part of a circle, as in the following examples. (We consider more complicated definite integrals later in this chapter.) When evaluating a definite integral using this empirical technique, remember that the area of the region and the value of the integral are *numerically equal*; that is, if the area is A square units, the value of the integral is the real number A.

FIGURE 5.17

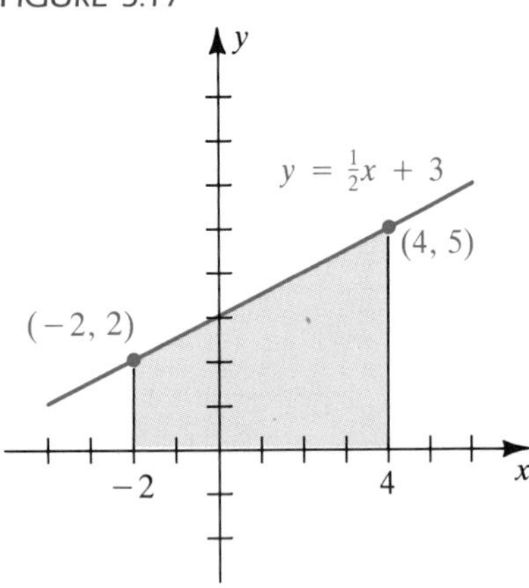

EXAMPLE 4 Evaluate $\int_{-2}^{4} (\frac{1}{2}x + 3)\,dx$.

SOLUTION If $f(x) = \frac{1}{2}x + 3$, then the graph of f is the line sketched in Figure 5.17. By Theorem (5.19), the value of the integral is numerically equal to the area of the region under this line from $x = -2$ to $x = 4$. The region is a trapezoid with bases parallel to the y-axis of length 2 and 5 and altitude on the x-axis of length 6. Using the formula for the area of a trapezoid, we obtain

$$\int_{-2}^{4} (\tfrac{1}{2}x + 3)\,dx = \tfrac{1}{2}(2 + 5)6 = 21.$$

FIGURE 5.18

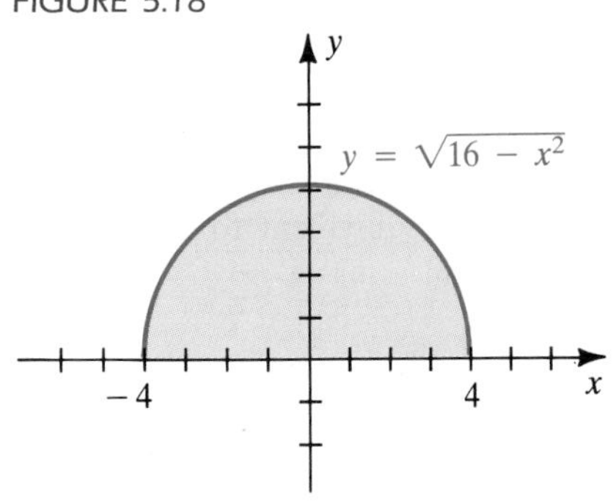

EXAMPLE 5 Evaluate $\int_{-4}^{4} \sqrt{16 - x^2}\,dx$.

SOLUTION If $f(x) = \sqrt{16 - x^2}$, then the graph of f is the semicircle shown in Figure 5.18. By Theorem (5.19), the value of the integral is numerically equal to the area of the region under this semicircle from $x = -4$ to $x = 4$. Hence

$$\int_{-4}^{4} \sqrt{16 - x^2}\,dx = \tfrac{1}{2} \cdot \pi(4)^2 = 8\pi.$$

EXAMPLE 6 Evaluate

(a) $\int_{4}^{-4} \sqrt{16 - x^2}\,dx$ **(b)** $\int_{4}^{4} \sqrt{16 - x^2}\,dx$

SOLUTION

(a) Using Definition (5.17) and Example 5, we have

$$\int_{4}^{-4} \sqrt{16 - x^2}\,dx = -\int_{-4}^{4} \sqrt{16 - x^2}\,dx = -8\pi.$$

(b) By Definition (5.18),

$$\int_{4}^{4} \sqrt{16 - x^2}\,dx = 0.$$

The next theorem states that functions that are continuous on closed intervals are integrable. This fact will play a crucial role in the proof of the fundamental theorem of calculus in Section 5.6.

Theorem (5.20)

If f is continuous on $[a, b]$, then f is integrable on $[a, b]$.

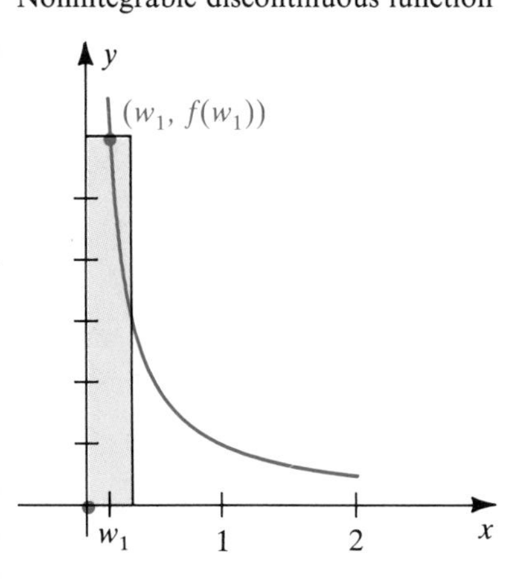

FIGURE 5.19
Nonintegrable discontinuous function

A proof of Theorem (5.20) may be found in texts on advanced calculus.

Definite integrals of discontinuous functions may or may not exist, depending on the types of discontinuities. In particular, *functions that have infinite discontinuities on a closed interval are not integrable on that interval.* To illustrate, consider the piecewise-defined function f with domain $[0, 2]$ such that

$$f(x) = \begin{cases} 0 & \text{if } x = 0 \\ \dfrac{1}{x} & \text{if } 0 < x \le 2 \end{cases}.$$

The graph of f is sketched in Figure 5.19. Note that $\lim_{x \to 0^+} f(x) = \infty$. If M is any (large) positive number, then in the first subinterval $[x_0, x_1]$ of any partition P of $[a, b]$, we can find a number w_1 such that $f(w_1) > M/\Delta x_1$, or, equivalently, $f(w_1)\,\Delta x_1 > M$. It follows that there are Riemann sums $\sum_k f(w_k)\,\Delta x_k$ that are arbitrarily large, and hence the limit in Definition (5.16) cannot exist. Thus, f is not integrable. A similar argument can be given for *any* function that has an infinite discontinuity in $[a, b]$. Consequently, *if a function f is integrable on $[a, b]$, then it is bounded on $[a, b]$; that is, there is a real number M such that $|f(x)| \le M$ for every x in $[a, b]$.*

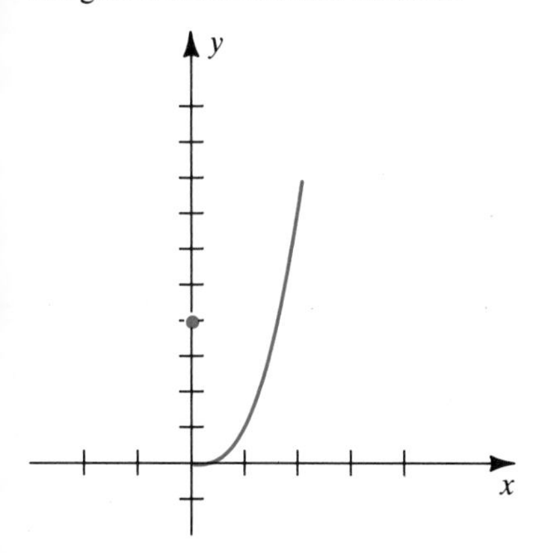

FIGURE 5.20
Integrable discontinuous function

As an illustration of a discontinuous function that *is* integrable, consider the piecewise-defined function f with domain $[0, 2]$ such that

$$f(x) = \begin{cases} 4 & \text{if } x = 0 \\ x^3 & \text{if } 0 < x \le 2 \end{cases}$$

The graph of f is sketched in Figure 5.20. Note that f has a jump discontinuity at $x = 0$. From Example 7 of Section 5.3, the area under the graph of $y = x^3$ from 0 to 2 is $2^4/4 = 4$. Thus, by Theorem (5.19), $\int_0^2 x^3\,dx = 4$. We can also show that $\int_0^2 f(x)\,dx = 4$. Hence f is integrable.

We have shown that a function that is discontinuous on a closed interval may or may not be integrable. However, by Theorem (5.20), functions that are *continuous* on a closed interval are *always* integrable.

EXERCISES 5.4

Exer. 1–4: The given numbers determine a partition P of an interval. (a) Find the length of each subinterval of P. (b) Find the norm $\|P\|$ of the partition.

1 $\{0, 1.1, 2.6, 3.7, 4.1, 5\}$

2 $\{2, 3, 3.7, 4, 5.2, 6\}$

3 $\{-3, -2.7, -1, 0.4, 0.9, 1\}$

4 $\{1, 1.6, 2, 3.5, 4\}$

Exer. 5–6: Let P be the partition of $[1, 5]$ determined by $\{1, 3, 4, 5\}$. Find a Riemann sum R_P for the given function f by choosing, in each subinterval of P, (a) the right-hand endpoint, (b) the left-hand endpoint, and (c) the midpoint.

5 $f(x) = 2x + 3$

6 $f(x) = 3 - 4x$

7 If $f(x) = 8 - \frac{1}{2}x^2$ and P is the regular partition of $[0, 6]$ into six subintervals, find a Riemann sum R_P of f by choosing the midpoint of each subinterval.

8 If $f(x) = 8 - \frac{1}{2}x^2$ and P is the partition of $[0, 6]$ determined by $\{0, 1.5, 3, 4.5, 6\}$, find a Riemann sum R_P of f by choosing the numbers 1, 2, 4, and 5 in the subintervals of P.

9 If $f(x) = x^3$ and P is the partition of $[-2, 4]$ determined by $\{-2, 0, 1, 3, 4\}$, find a Riemann sum R_P of f by choosing the numbers -1, 1, 2, and 4 in the subintervals of P.

10 If $f(x) = \sqrt{x}$ and P is the partition of $[1, 16]$ determined by $\{1, 3, 5, 7, 9, 16\}$, find a Riemann sum R_P of f by choosing the numbers 1, 4, 5, 9, and 9 in the subintervals of P.

c 11 If $f(x) = x^2\sqrt{\cos x}$ and P is the regular partition of $[0, 1]$ into ten subintervals, find a Riemann sum R_P of f by choosing the midpoint of each subinterval of P.

c 12 If $f(x) = \sin(\cos x)$ and P is the partition of $[-1, 1]$ determined by $\{-1, -0.65, -0.23, 0.51, 0.85, 1\}$, find a Riemann sum R_P of f by choosing the numbers -0.75, -0.5, 0, 0.6, 0.9 in the subintervals of P.

Exer. 13–16: Use Definition (5.16) to express each limit as a definite integral on the given interval $[a, b]$.

13 $\lim_{\|P\| \to 0} \sum_{k=1}^{n} (3w_k^2 - 2w_k + 5)\,\Delta x_k$; $[-1, 2]$

14 $\lim_{\|P\| \to 0} \sum_{k=1}^{n} \pi(w_k^2 - 4)\,\Delta x_k$; $[2, 3]$

15 $\lim_{\|P\| \to 0} \sum_{k=1}^{n} 2\pi w_k(1 + w_k^3)\,\Delta x_k$; $[0, 4]$

16 $\lim_{\|P\| \to 0} \sum_{k=1}^{n} (\sqrt[3]{w_k} + 4w_k)\,\Delta x_k$; $[-4, -3]$

Exer. 17–22: Given $\int_1^4 \sqrt{x}\,dx = \frac{14}{3}$, evaluate the integral.

17 $\int_4^1 \sqrt{x}\,dx$

18 $\int_1^4 \sqrt{s}\,ds$

19 $\int_1^4 \sqrt{t}\,dt$

20 $\int_1^4 \sqrt{x}\,dx + \int_4^1 \sqrt{x}\,dx$

21 $\int_4^4 \sqrt{x}\,dx + \int_4^1 \sqrt{x}\,dx$

22 $\int_4^4 \sqrt{x}\,dx$

Exer. 23–26: Express the area of the region in the figure as a definite integral.

23

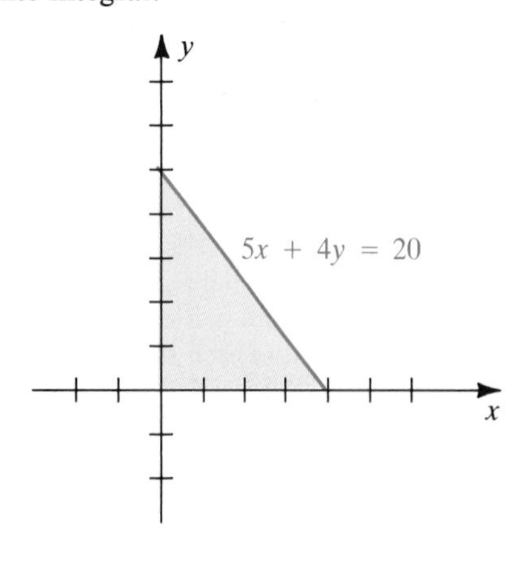

24

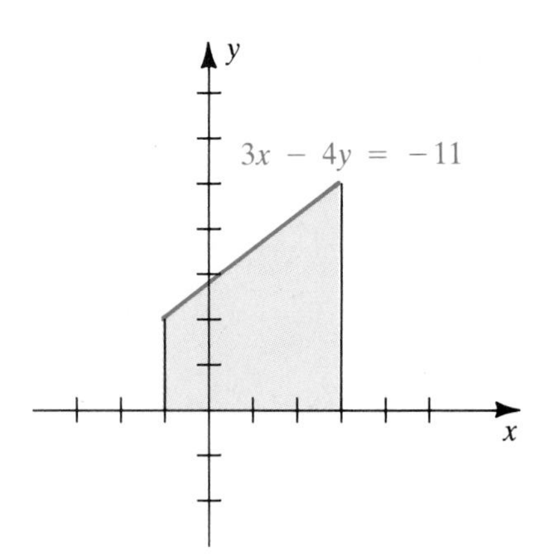

25

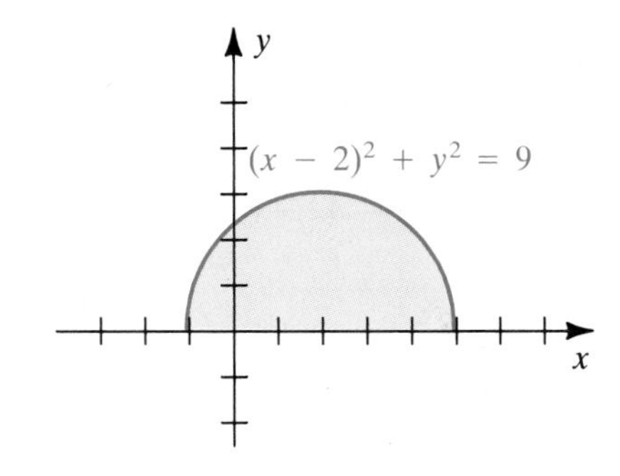

26

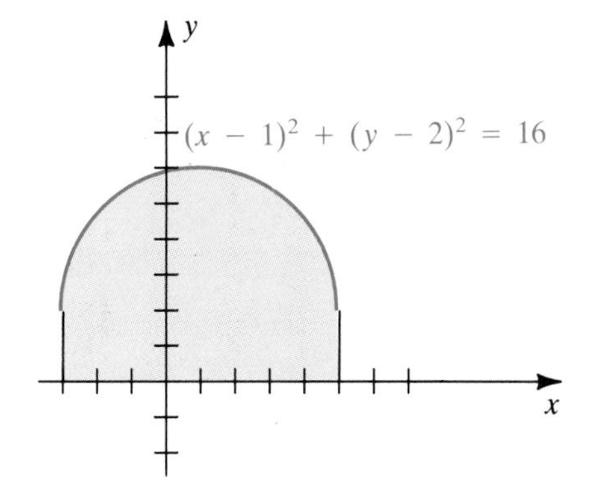

Exer. 27–36: Evaluate the definite integral by regarding it as the area under the graph of a function.

27 $\int_{-1}^{5} 6\,dx$

28 $\int_{-2}^{3} 4\,dx$

29 $\int_{-3}^{2} (2x + 6)\,dx$

30 $\int_{-1}^{2} (7 - 3x)\,dx$

31 $\int_0^3 |x - 1|\,dx$

32 $\int_{-1}^{4} |x|\,dx$

33 $\int_0^3 \sqrt{9 - x^2}\,dx$

34 $\int_0^a \sqrt{a^2 - x^2}\,dx, \quad a > 0$

35 $\int_{-2}^{2} (3 + \sqrt{4 - x^2})\,dx$

36 $\int_{-2}^{2} (3 - \sqrt{4 - x^2})\,dx$

5.5 PROPERTIES OF THE DEFINITE INTEGRAL

This section contains some fundamental properties of the definite integral. Most of the proofs are difficult and have been placed in Appendix II.

Theorem (5.21)

> If c is a real number, then
> $$\int_a^b c\,dx = c(b-a).$$

PROOF Let f be the constant function defined by $f(x) = c$ for every x in $[a, b]$. If P is a partition of $[a, b]$, then for every Riemann sum of f,

$$\sum_k f(w_k)\,\Delta x_k = \sum_k c\,\Delta x_k = c\sum_k \Delta x_k = c(b-a).$$

(The last equality is true because the sum $\sum_k \Delta x_k$ is the length of the interval $[a, b]$.) Consequently,

$$\left|\sum_k f(w_k)\,\Delta x_k - c(b-a)\right| = |c(b-a) - c(b-a)| = 0,$$

which is less than any positive number ϵ *regardless* of the size of $\|P\|$. Thus, by Definition (5.15), with $L = c(b-a)$,

$$\lim_{\|P\|\to 0} \sum_k f(w_k)\,\Delta x_k = \lim_{\|P\|\to 0} \sum_k c\,\Delta x_k = c(b-a).$$

By Definition (5.16), this means that

$$\int_a^b f(x)\,dx = \int_a^b c\,dx = c(b-a). \quad ■$$

FIGURE 5.21

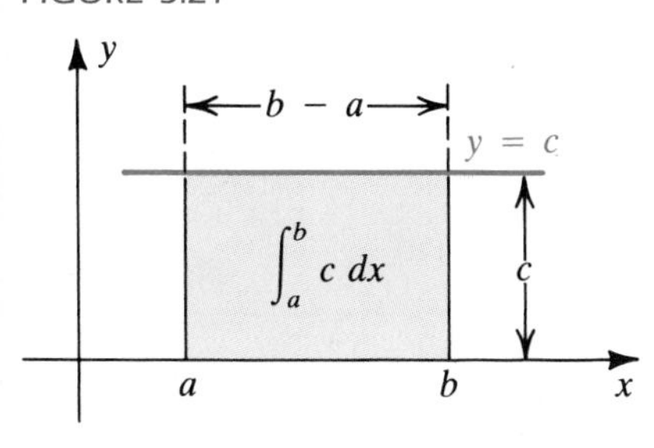

Note that if $c > 0$, then Theorem (5.21) agrees with Theorem (5.19): As illustrated in Figure 5.21, the graph of f is the horizontal line $y = c$, and the region under the graph from a to b is a rectangle with sides of lengths c and $b - a$. Hence the area $\int_a^b f(x)\,dx$ of the rectangle is $c(b-a)$.

EXAMPLE 1 Evaluate $\int_{-2}^{3} 7\,dx$.

SOLUTION Using Theorem (5.21) yields

$$\int_{-2}^{3} 7\,dx = 7[3-(-2)] = 7(5) = 35.$$

If $c = 1$ in Theorem (5.21), we shall abbreviate the integrand as follows:

$$\int_a^b dx = b - a$$

If a function f is integrable on $[a, b]$ and c is a real number, then, by Theorem (5.11)(ii), a Riemann sum of the function cf may be written

$$\sum_k cf(w_k)\,\Delta x_k = c\sum_k f(w_k)\,\Delta x_k.$$

We can prove that the limit of the sums on the left of the last equation is equal to c times the limit of the sums on the right. This gives us the next theorem. A proof may be found in Appendix II.

Theorem **(5.22)**

If f is integrable on $[a, b]$ and c is any real number, then cf is integrable on $[a, b]$ and

$$\int_a^b cf(x)\, dx = c\int_a^b f(x)\, dx.$$

Theorem (5.22) is sometimes stated as follows: *A constant factor in the integrand may be taken outside the integral sign.* It is *not* permissible to take expressions involving variables outside the integral sign in this manner.

If two functions f and g are defined on $[a, b]$, then, by Theorem (5.11)(i), a Riemann sum of $f + g$ may be written

$$\sum_k [f(w_k) + g(w_k)]\, \Delta x_k = \sum_k f(w_k)\, \Delta x_k + \sum_k g(w_k)\, \Delta x_k.$$

We can show that if f and g are integrable, then the limit of the sums on the left may be found by adding the limits of the two sums on the right. This fact is stated in integral form in (i) of the next theorem. A proof of (i) may be found in Appendix II. The analogous result for differences is stated in (ii) of the theorem.

Theorem **(5.23)**

If f and g are integrable on $[a, b]$, then $f + g$ and $f - g$ are integrable on $[a, b]$ and

(i) $\displaystyle\int_a^b [f(x) + g(x)]\, dx = \int_a^b f(x)\, dx + \int_a^b g(x)\, dx$

(ii) $\displaystyle\int_a^b [f(x) - g(x)]\, dx = \int_a^b f(x)\, dx - \int_a^b g(x)\, dx$

Theorem (5.23)(i) may be extended to any finite number of functions. Thus, if $f_1, f_2, \ldots, f_n$ are integrable on $[a, b]$, then so is their sum and

$$\int_a^b [f_1(x) + f_2(x) + \cdots + f_n(x)]\, dx$$
$$= \int_a^b f_1(x)\, dx + \int_a^b f_2(x)\, dx + \cdots + \int_a^b f_n(x)\, dx.$$

EXAMPLE 2 It will follow from the results in Section 5.6 that

$$\int_0^2 x^3\, dx = 4 \quad \text{and} \quad \int_0^2 x\, dx = 2.$$

Use these facts to evaluate $\int_0^2 (5x^3 - 3x + 6)\, dx$.

SOLUTION We may proceed as follows:

$$\begin{aligned}\int_0^2 (5x^3 - 3x + 6)\, dx &= \int_0^2 5x^3\, dx - \int_0^2 3x\, dx + \int_0^2 6\, dx \\ &= 5\int_0^2 x^3\, dx - 3\int_0^2 x\, dx + 6(2 - 0) \\ &= 5(4) - 3(2) + 12 = 26\end{aligned}$$

If f is continuous on $[a, b]$ and $f(x) \geq 0$ for every x in $[a, b]$, then, by Theorem (5.19), the integral $\int_a^b f(x)\,dx$ is the area under the graph of f from a to b. Similarly, if $a < c < b$, then the integrals $\int_a^c f(x)\,dx$ and $\int_c^b f(x)\,dx$ are the areas under the graph of f from a to c and from c to b, respectively, as illustrated in Figure 5.22. Since the area from a to b is the sum of the two smaller areas, we have

$$\int_a^b f(x)\,dx = \int_a^c f(x)\,dx + \int_c^b f(x)\,dx.$$

The next theorem shows that the preceding equality is true under a more general hypothesis. The proof is given in Appendix II.

FIGURE 5.22

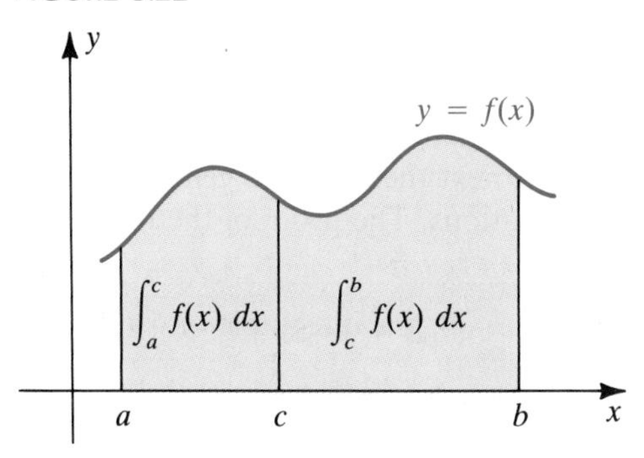

Theorem **(5.24)**

> If $a < c < b$ and if f is integrable on both $[a, c]$ and $[c, b]$, then f is integrable on $[a, b]$ and
>
> $$\int_a^b f(x)\,dx = \int_a^c f(x)\,dx + \int_c^b f(x)\,dx.$$

The following result is a generalization of Theorem (5.24) to the case where c is not necessarily between a and b.

Theorem **(5.25)**

> If f is integrable on a closed interval and if a, b, and c are any three numbers in the interval, then
>
> $$\int_a^b f(x)\,dx = \int_a^c f(x)\,dx + \int_c^b f(x)\,dx.$$

PROOF If a, b, and c are all different, then there are six possible ways of arranging these three numbers. The theorem should be verified for each of these cases and also for the cases in which two or all three of the numbers are equal. We shall verify one case. Suppose $c < a < b$. By Theorem (5.24),

$$\int_c^b f(x)\,dx = \int_c^a f(x)\,dx + \int_a^b f(x)\,dx,$$

which, in turn, may be written

$$\int_a^b f(x)\,dx = -\int_c^a f(x)\,dx + \int_c^b f(x)\,dx.$$

The conclusion of the theorem now follows from the fact that interchanging the limits of integration changes the sign of the integral (see Definition (5.17)). ■

EXAMPLE 3 Express as one integral:

$$\int_2^7 f(x)\,dx - \int_5^7 f(x)\,dx$$

SOLUTION First we interchange the limits of the second integral using Definition (5.17), and then we use Theorem (5.25) with $a = 2$, $b = 5$, and $c = 7$:

$$\begin{aligned}\int_2^7 f(x)\,dx - \int_5^7 f(x)\,dx &= \int_2^7 f(x)\,dx + \int_7^5 f(x)\,dx \\ &= \int_2^5 f(x)\,dx\end{aligned}$$

If f and g are continuous on $[a, b]$ and $f(x) \geq g(x) \geq 0$ for every x in $[a, b]$, then the area under the graph of f from a to b is greater than or equal to the area under the graph of g from a to b. The corollary to the next theorem is a generalization of this fact to arbitrary integrable functions. The proof of the theorem is given in Appendix II.

Theorem (5.26)

If f is integrable on $[a, b]$ and $f(x) \geq 0$ for every x in $[a, b]$, then

$$\int_a^b f(x)\,dx \geq 0.$$

Corollary (5.27)

If f and g are integrable on $[a, b]$ and $f(x) \geq g(x)$ for every x in $[a, b]$, then

$$\int_a^b f(x)\,dx \geq \int_a^b g(x)\,dx.$$

PROOF By Theorem (5.23), $f - g$ is integrable. Moreover, since $f(x) \geq g(x)$, $f(x) - g(x) \geq 0$ for every x in $[a, b]$. Hence, by Theorem (5.26),

$$\int_a^b [f(x) - g(x)]\,dx \geq 0.$$

Applying Theorem (5.23)(ii) leads to the desired conclusion. ■

FIGURE 5.23

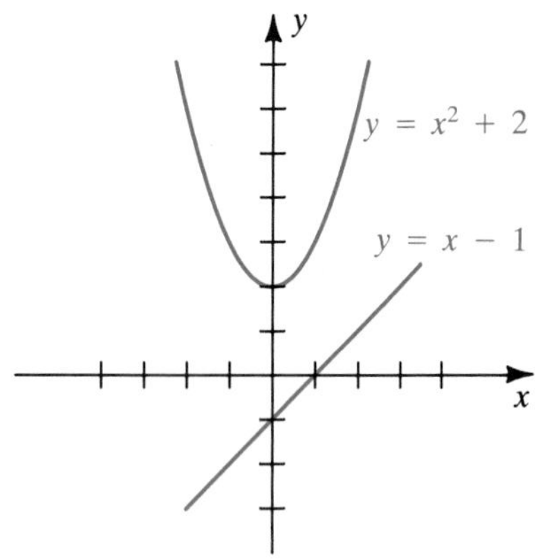

EXAMPLE 4 Show that $\int_{-1}^2 (x^2 + 2)\,dx \geq \int_{-1}^2 (x - 1)\,dx$.

SOLUTION The graphs of $y = x^2 + 2$ and $y = x - 1$ are sketched in Figure 5.23. Since

$$x^2 + 2 \geq x - 1$$

for every x in $[-1, 2]$, the conclusion follows from Corollary (5.27).

Suppose, in Theorem (5.26), that f is continuous and that, in addition to the condition $f(x) \geq 0$, we have $f(c) > 0$ for some c in $[a, b]$. In this case $\lim_{x\to c} f(x) > 0$, and, by Theorem (2.6), there is a subinterval $[a', b']$

of $[a, b]$ throughout which $f(x)$ is positive. If $f(u)$ is the minimum value of f on $[a', b']$ (see Figure 5.24), then the area under the graph of f from a to b is at least as large as the area $f(u)(b' - a')$ of the pictured rectangle. Consequently, $\int_a^b f(x)\,dx > 0$. It now follows, as in the proof of (5.27), that if f and g are continuous on $[a, b]$, if $f(x) \geq g(x)$ throughout $[a, b]$, and if $f(x) > g(x)$ for some x in $[a, b]$, then $\int_a^b f(x)\,dx > \int_a^b g(x)\,dx$. This fact will be used in the proof of the next theorem.

Mean value theorem for definite integrals **(5.28)**

> If f is continuous on a closed interval $[a, b]$, then there is a number z in the open interval (a, b) such that
>
> $$\int_a^b f(x)\,dx = f(z)(b - a).$$

FIGURE 5.24

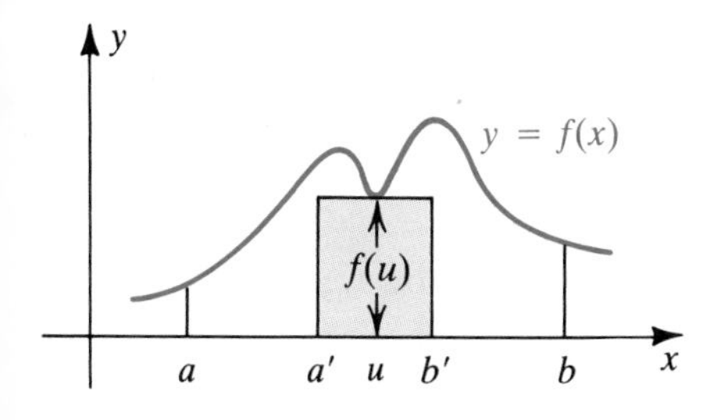

PROOF If f is a constant function, then $f(x) = c$ for some number c, and by Theorem (5.21),

$$\int_a^b f(x)\,dx = \int_a^b c\,dx = c(b - a) = f(z)(b - a)$$

for every number z in (a, b).

Next assume that f is not a constant function and suppose that m and M are the minimum and maximum values of f, respectively, on $[a, b]$. Let $f(u) = m$ and $f(v) = M$ for some u and v in $[a, b]$. This is illustrated in Figure 5.25 for the case in which $f(x)$ is positive throughout $[a, b]$. Since f is not a constant function, $m < f(x) < M$ for some x in $[a, b]$. Hence, by the remark immediately preceding this theorem,

FIGURE 5.25

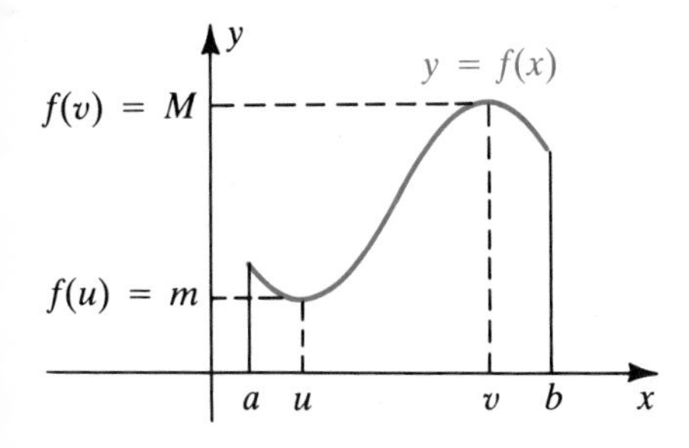

$$\int_a^b m\,dx < \int_a^b f(x)\,dx < \int_a^b M\,dx.$$

Applying Theorem (5.21) yields

$$m(b - a) < \int_a^b f(x)\,dx < M(b - a).$$

Dividing by $b - a$ and recalling that $m = f(u)$ and $M = f(v)$ gives us

$$f(u) < \frac{1}{b - a}\int_a^b f(x)\,dx < f(v).$$

Since $[1/(b - a)]\int_a^b f(x)\,dx$ is a number between $f(u)$ and $f(v)$, it follows from the intermediate value theorem (2.26) that there is a number z, with $u < z < v$, such that

$$f(z) = \frac{1}{b - a}\int_a^b f(x)\,dx.$$

Multiplying both sides by $b - a$ gives us the conclusion of the theorem. ■

FIGURE 5.26

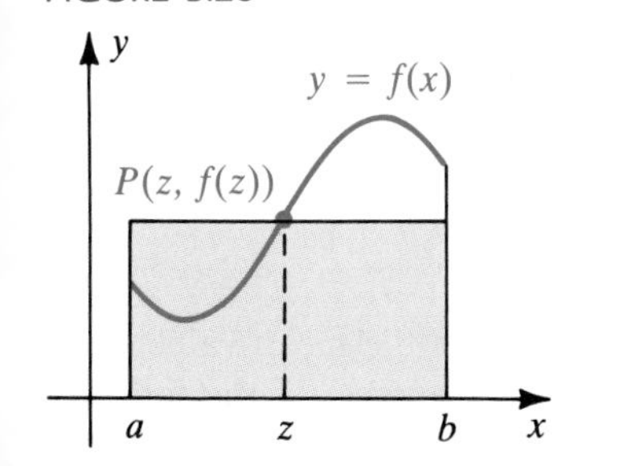

The number z of Theorem (5.28) is not necessarily unique; however, the theorem guarantees that *at least* one number z will produce the desired result.

The mean value theorem has an interesting geometric interpretation if $f(x) \geq 0$ on $[a, b]$. In this case $\int_a^b f(x)\,dx$ is the area under the graph of f from a to b. If, as in Figure 5.26, a horizontal line is drawn through the point $P(z, f(z))$, then the area of the rectangular region bounded by this line, the x-axis, and the lines $x = a$ and $x = b$ is $f(z)(b - a)$, which, according to Theorem (5.28), is the same as the area under the graph of f from a to b.

EXAMPLE 5 It will follow from the results of Section 5.6 that $\int_0^3 x^2\,dx = 9$. Find a number z that satisfies the conclusion of the mean value theorem (5.28) for this definite integral.

SOLUTION The graph of $f(x) = x^2$ for $0 \le x \le 3$ is sketched in Figure 5.27. By the mean value theorem, there is a number z between 0 and 3 such that

FIGURE 5.27

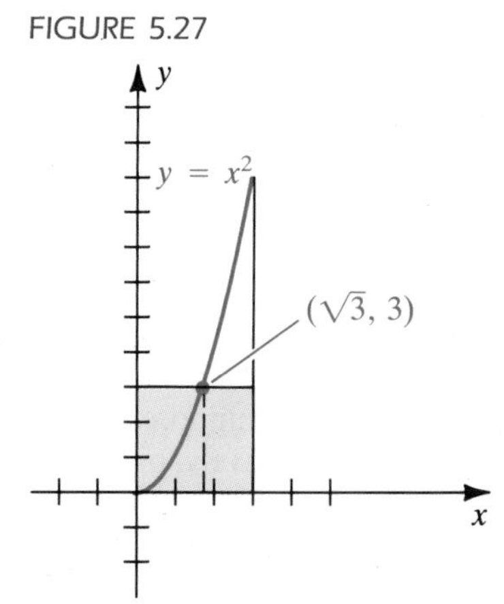

$$\int_0^3 x^2\,dx = f(z)(3 - 0) = z^2(3).$$

This implies that

$$9 = 3z^2, \quad \text{or} \quad z^2 = 3.$$

The solutions of the last equation are $z = \pm\sqrt{3}$; however, $-\sqrt{3}$ is not in $[0, 3]$. The number $z = \sqrt{3}$ satisfies the conclusion of the theorem.

If we consider the horizontal line through $P(\sqrt{3}, 3)$, then the area of the rectangle bounded by this line, the x-axis, and the lines $x = 0$ and $x = 3$ is equal to the area under the graph of f from $x = 0$ to $x = 3$ (see Figure 5.27).

In statistics the term **arithmetic mean** is used for the **average** of a set of numbers. Thus, the arithmetic mean of two numbers a and b is $(a + b)/2$, the arithmetic mean of three numbers a, b, and c is $(a + b + c)/3$, and so on. To see the relationship between arithmetic means and the word *mean* used in mean value theorem, let us rewrite the conclusion of (5.28) as

$$f(z) = \frac{1}{b - a}\int_a^b f(x)\,dx$$

and express the definite integral as a limit of sums. If we specialize Definition (5.16) by using a regular partition P with n subintervals, then

$$f(z) = \frac{1}{b - a}\lim_{n\to\infty}\sum_{k=1}^{n} f(w_k)\,\Delta x = \lim_{n\to\infty}\sum_{k=1}^{n}\left[f(w_k)\frac{\Delta x}{b - a}\right]$$

for any number w_k in the kth subinterval of P and $\Delta x = (b - a)/n$. Since $\Delta x/(b - a) = 1/n$, we obtain

$$f(z) = \lim_{n\to\infty}\sum_{k=1}^{n}\left[f(w_k)\frac{\Delta x}{b - a}\right] = \lim_{n\to\infty}\sum_{k=1}^{n}\left[f(w_k)\frac{1}{n}\right],$$

or

$$f(z) = \lim_{n\to\infty}\left[\frac{f(w_1) + f(w_2) + \cdots + f(w_n)}{n}\right].$$

This shows that we may regard the number $f(z)$ in the mean value theorem (5.28) as a limit of the arithmetic means (averages) of the function values $f(w_1), f(w_2), \ldots, f(w_n)$ as n increases without bound. This is the motivation for the next definition.

Definition (5.29)

> Let f be continuous on $[a, b]$ The **average value** f_{av} of f on $[a, b]$ is
> $$f_{av} = \frac{1}{b-a}\int_a^b f(x)\,dx.$$

Note that, by the mean value theorem for definite integrals, if f is continuous on $[a, b]$, then

$$f_{av} = f(z) \quad \text{for some } z \text{ in } [a, b].$$

EXAMPLE 6 Given $\int_0^3 x^2\,dx = 9$, find the average value of f on $[0, 3]$.

SOLUTION By Definition (5.29), with $a = 0$, $b = 3$, and $f(x) = x^2$,

$$f_{av} = \frac{1}{3-0}\int_0^3 x^2\,dx = \frac{1}{3}\cdot 9 = 3.$$

In the interval $[0, 3]$, the function values $f(x) = x^2$ range from $f(0) = 0$ to $f(3) = 9$. Note that the function f *takes on* its average value 3 at the number $z = \sqrt{3}$.

EXERCISES 5.5

Exer. 1–6: Evaluate the integral.

1 $\int_{-2}^{4} 5\,dx$ 2 $\int_{1}^{10} \sqrt{2}\,dx$ 3 $\int_{6}^{2} 3\,dx$

4 $\int_{4}^{-3} dx$ 5 $\int_{-1}^{1} dx$ 6 $\int_{2}^{2} 100\,dx$

Exer. 7–10: It will follow from the results in Section 5.6 that

$$\int_1^4 x^2\,dx = 21 \quad \text{and} \quad \int_1^4 x\,dx = \tfrac{15}{2}.$$

Use these facts to evaluate the integral.

7 $\int_1^4 (3x^2 + 5)\,dx$ 8 $\int_1^4 (6x - 1)\,dx$

9 $\int_1^4 (2 - 9x - 4x^2)\,dx$ 10 $\int_1^4 (3x + 2)^2\,dx$

Exer. 11–16: Verify the inequality without evaluating the integrals.

11 $\int_1^2 (3x^2 + 4)\,dx \geq \int_1^2 (2x^2 + 5)\,dx$

12 $\int_1^4 (2x + 2)\,dx \leq \int_1^4 (3x + 1)\,dx$

13 $\int_2^4 (x^2 - 6x + 8)\,dx \leq 0$

14 $\int_2^4 (5x^2 - x + 1)\,dx \geq 0$

15 $\int_0^{2\pi} (1 + \sin x)\,dx \geq 0$

16 $\int_{-\pi/3}^{\pi/3} (\sec x - 2)\,dx \leq 0$

Exer. 17–22: Express as one integral.

17 $\int_5^1 f(x)\,dx + \int_{-3}^5 f(x)\,dx$

18 $\int_4^1 f(x)\,dx + \int_6^4 f(x)\,dx$

19 $\int_c^d f(x)\,dx + \int_e^c f(x)\,dx$

20 $\int_{-2}^6 f(x)\,dx - \int_{-2}^2 f(x)\,dx$

21 $\int_c^{c+h} f(x)\,dx - \int_c^h f(x)\,dx$

22 $\int_c^m f(x)\,dx - \int_d^m f(x)\,dx$

Exer. 23–30: The given integral $\int_a^b f(x)\,dx$ may be verified using the results in Section 5.6. (a) Find a number z that satisfies the conclusion of the mean value theorem (5.28). (b) Find the average value of f on $[a, b]$.

23 $\int_0^3 3x^2\,dx = 27$ 24 $\int_{-4}^{-1} \frac{3}{x^2}\,dx = \frac{9}{4}$

25 $\int_{-2}^1 (x^2 + 1)\,dx = 6$

26 $\int_{-1}^3 (3x^2 - 2x + 3)\,dx = 32$

27 $\int_{-1}^8 3\sqrt{x+1}\,dx = 54$ 28 $\int_{-2}^{-1} \frac{8}{x^3}\,dx = -3$

29 $\int_1^2 (4x^3 - 1)\,dx = 14$ 30 $\int_1^4 (2 + 3\sqrt{x})\,dx = 20$

Exer. 31–32: The given integral may be verified using results in Section 5.6. Use Newton's method to approximate, to three decimal places, a number z that satisfies the conclusion of the mean value theorem (5.28).

31 $\int_{-2}^{3} (8x^3 + 3x - 1)\, dx = 132.5$

32 $\int_{\pi/6}^{\pi/4} (1 - \cos 4x)\, dx = \frac{\pi}{12} + \frac{\sqrt{3}}{8}$

33 Let f and g be integrable on $[a, b]$. If c and d are any real numbers, prove that

$$\int_a^b [cf(x) + dg(x)]\, dx = c \int_a^b f(x)\, dx + d \int_a^b g(x)\, dx.$$

34 If f is continuous on $[a, b]$, prove that

$$\left| \int_a^b f(x)\, dx \right| \le \int_a^b |f(x)|\, dx.$$

(*Hint:* $-|f(x)| \le f(x) \le |f(x)|$.)

5.6 THE FUNDAMENTAL THEOREM OF CALCULUS

This section contains one of the most important theorems in calculus. In addition to being useful in evaluating definite integrals, the theorem also exhibits the relationship between derivatives and definite integrals. This theorem, aptly called *the fundamental theorem of calculus*, was discovered independently by Sir Isaac Newton (1642–1727) in England and by Gottfried Wilhelm Leibniz (1646–1716) in Germany. It is primarily because of this discovery that both men are credited with the invention of calculus.

To avoid confusion in the following discussion, we shall use t as the independent variable and denote the definite integral of f from a to b by $\int_a^b f(t)\, dt$. If f is continuous on $[a, b]$ and $a \le x \le b$, then f is continuous on $[a, x]$; therefore, by Theorem (5.20), f is integrable on $[a, x]$. Consequently, the formula

$$G(x) = \int_a^x f(t)\, dt$$

determines a function G with domain $[a, b]$, since for each x in $[a, b]$ there corresponds a unique number $G(x)$.

FIGURE 5.28

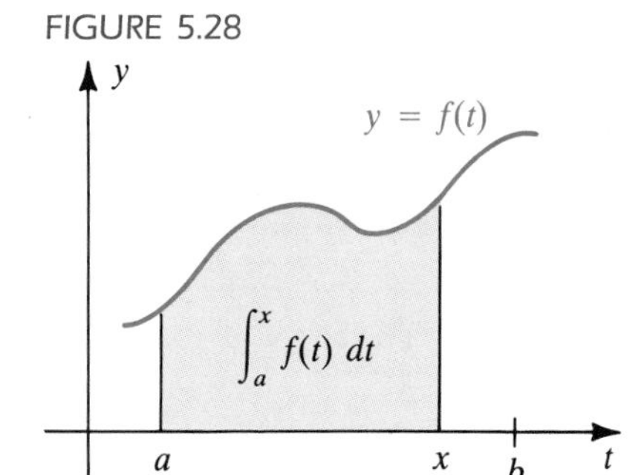

To obtain a geometric interpretation of $G(x)$, suppose that $f(t) \ge 0$ for every t in $[a, b]$. In this case we see from Theorem (5.19) that $G(x)$ is the area of the region under the graph of f from a to x (see Figure 5.28).

FIGURE 5.29

As a specific illustration, consider $f(t) = t^3$ with $a = 0$ and $b > 0$ (see Figure 5.29). In Example 7 of Section 5.3 we proved that the area under the graph of f from 0 to b is $\frac{1}{4}b^4$. Hence the area from 0 to x is

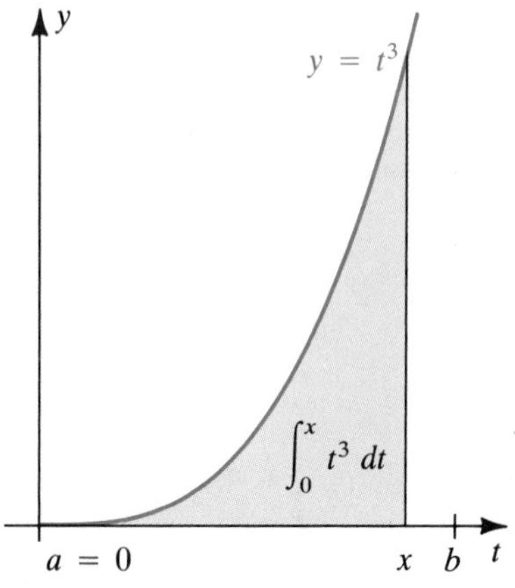

$$G(x) = \int_0^x t^3\, dt = \tfrac{1}{4}x^4.$$

This gives us an explicit form for the function G if $f(t) = t^3$. Note that in this illustration,

$$G'(x) = D_x\left(\tfrac{1}{4}x^4\right) = x^3 = f(x).$$

Thus, by Definition (5.1), G is an antiderivative of f. This result is not an accident. Part I of the next theorem brings out the remarkable fact that if f is *any* continuous function and $G(x) = \int_a^x f(t)\, dt$, then G is an antiderivative of f. Part II of the theorem shows how *any* antiderivative may be used to find the value of $\int_a^b f(x)\, dx$.

Fundamental theorem of calculus (5.30)

> Suppose f is continuous on a closed interval $[a, b]$.
>
> **Part I** If the function G is defined by
>
> $$G(x) = \int_a^x f(t)\,dt$$
>
> for every x in $[a, b]$, then G is an antiderivative of f on $[a, b]$.
>
> **Part II** If F is any antiderivative of f on $[a, b]$, then
>
> $$\int_a^b f(x)\,dx = F(b) - F(a).$$

PROOF To establish Part I, we must show that if x is in $[a, b]$, then $G'(x) = f(x)$; that is,

$$\lim_{h \to 0} \frac{G(x + h) - G(x)}{h} = f(x).$$

FIGURE 5.30

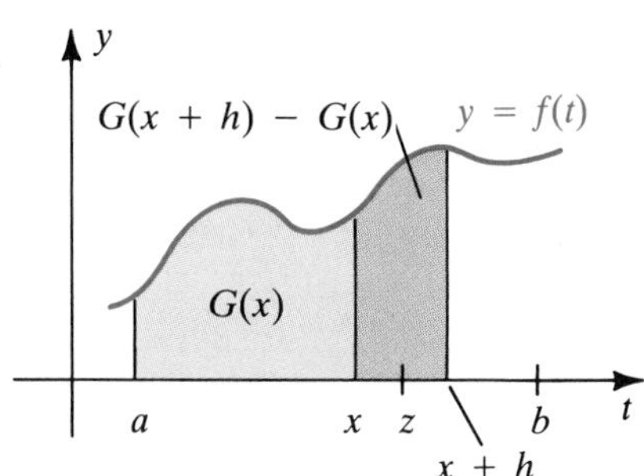

Before giving a formal proof, let us consider some geometric aspects of this limit. If $f(x) \geq 0$ throughout $[a, b]$, then $G(x)$ is the area under the graph of f from a to x, as illustrated in Figure 5.30. If $h > 0$, then the difference $G(x + h) - G(x)$ is the area under the graph of f from x to $x + h$, and the number h is the length of the interval $[x, x + h]$. We shall show that

$$\frac{G(x + h) - G(x)}{h} = f(z)$$

for some number z between x and $x + h$. Apparently, if $h \to 0$, then $z \to x$ and $f(z) \to f(x)$, which is what we wish to prove.

Let us now give a rigorous proof that $G'(x) = f(x)$. If x and $x + h$ are in $[a, b]$, then using the definition of G together with Definition (5.17) and Theorem (5.24) yields

$$\begin{aligned} G(x + h) - G(x) &= \int_a^{x+h} f(t)\,dt - \int_a^x f(t)\,dt \\ &= \int_a^{x+h} f(t)\,dt + \int_x^a f(t)\,dt \\ &= \int_x^{x+h} f(t)\,dt. \end{aligned}$$

Consequently, if $h \neq 0$,

$$\frac{G(x + h) - G(x)}{h} = \frac{1}{h}\int_x^{x+h} f(t)\,dt.$$

If $h > 0$, then, by the mean value theorem (5.28), there is a number z in the open interval $(x, x + h)$ such that

$$\int_x^{x+h} f(t)\,dt = f(z)h$$

and, therefore,

$$\frac{G(x + h) - G(x)}{h} = f(z).$$

Since $x < z < x + h$, it follows from the continuity of f that

$$\lim_{h \to 0^+} f(z) = \lim_{z \to x^+} f(z) = f(x)$$

and hence $$\lim_{h \to 0^+} \frac{G(x+h) - G(x)}{h} = \lim_{h \to 0^+} f(z) = f(x).$$

If $h < 0$, then we may prove in similar fashion that

$$\lim_{h \to 0^-} \frac{G(x+h) - G(x)}{h} = f(x).$$

The two preceding one-sided limits imply that

$$G'(x) = \lim_{h \to 0} \frac{G(x+h) - G(x)}{h} = f(x).$$

This completes the proof of Part I.

To prove Part II, let F be any antiderivative of f and let G be the special antiderivative defined in Part I. From Theorem (5.2) we know that there is a constant C such that

$$G(x) = F(x) + C$$

for every x in $[a, b]$. Hence, from the definition of G,

$$\int_a^x f(t)\, dt = F(x) + C$$

for every x in $[a, b]$. If we let $x = a$ and use the fact that $\int_a^a f(t)\, dt = 0$, we obtain $0 = F(a) + C$, or $C = -F(a)$. Consequently,

$$\int_a^x f(t)\, dt = F(x) - F(a).$$

This is an identity for every x in $[a, b]$, so we may substitute b for x, obtaining

$$\int_a^b f(t)\, dt = F(b) - F(a).$$

Replacing the variable t by x gives us the conclusion of Part II. ■

We often denote the difference $F(b) - F(a)$ either by $F(x)]_a^b$ or by $[F(x)]_a^b$. Part II of the fundamental theorem may then be expressed as follows.

Corollary **(5.31)**

If f is continuous on $[a, b]$ and F is any antiderivative of f, then

$$\int_a^b f(x)\, dx = F(x)\Big]_a^b = F(b) - F(a).$$

The formula in Corollary (5.31) is also valid if $a \geq b$. If $a > b$, then, by Definition (5.17),

$$\begin{aligned} \int_a^b f(x)\, dx &= -\int_b^a f(x)\, dx \\ &= -[F(a) - F(b)] \\ &= F(b) - F(a). \end{aligned}$$

If $a = b$, then, by Definition (5.18),

$$\int_a^a f(x)\, dx = 0 = F(a) - F(a).$$

Corollary (5.31) allows us to evaluate a definite integral very easily if we can find an antiderivative of the integrand. For example, since an antiderivative of x^3 is $\frac{1}{4}x^4$, we have

$$\int_0^b x^3\,dx = \tfrac{1}{4}x^4\Big]_0^b = \tfrac{1}{4}b^4 - \tfrac{1}{4}(0)^4 = \tfrac{1}{4}b^4.$$

Those who doubt the importance of the fundamental theorem should compare this simple computation with the limit of sums calculation discussed in Example 7 of Section 5.3.

EXAMPLE 1 Evaluate $\int_{-2}^{3} (6x^2 - 5)\,dx$.

SOLUTION An antiderivative of $6x^2 - 5$ is $F(x) = 2x^3 - 5x$. Applying Corollary (5.31), we get

$$\begin{aligned}\int_{-2}^{3} (6x^2 - 5)\,dx &= \Big[2x^3 - 5x\Big]_{-2}^{3}\\ &= [2(3)^3 - 5(3)] - [2(-2)^3 - 5(-2)]\\ &= [54 - 15] - [-16 + 10] = 45.\end{aligned}$$

Note that if $F(x) + C$ is used in place of $F(x)$ in Corollary (5.31), the same result is obtained, since

$$\begin{aligned}\Big[F(x) + C\Big]_a^b &= [F(b) + C] - [F(a) + C]\\ &= F(b) - F(a) = \Big[F(x)\Big]_a^b.\end{aligned}$$

In particular, since

$$\int f(x)\,dx = F(x) + C,$$

where $F'(x) = f(x)$, we obtain the following theorem.

Theorem **(5.32)**

$$\int_a^b f(x)\,dx = \left[\int f(x)\,dx\right]_a^b$$

Theorem (5.32) states that *a definite integral can be evaluated by evaluating the corresponding indefinite integral.* As with previous cases, when using Theorem (5.32) it is unnecessary to include the constant of integration C for the indefinite integral.

EXAMPLE 2 Find the area A of the region between the graph of $y = \sin x$ and the x-axis from $x = 0$ to $x = \pi$.

FIGURE 5.31

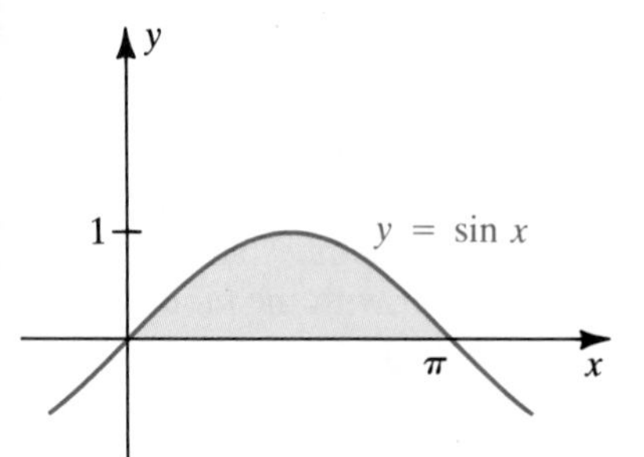

SOLUTION The region is sketched in Figure 5.31. Applying Theorems (5.19) and (5.32) gives us the following:

$$\begin{aligned}A = \int_0^{\pi} \sin x\,dx &= \left[\int \sin x\,dx\right]_0^{\pi}\\ &= \Big[-\cos x\Big]_0^{\pi}\\ &= -\cos \pi - (-\cos 0)\\ &= -(-1) + 1 = 2\end{aligned}$$

By Theorem (5.32), we can use any formula for indefinite integration to obtain a formula for definite integrals. To illustrate, using Table (5.4), we obtain

$$\int_a^b x^r\,dx = \left[\frac{x^{r+1}}{r+1}\right]_a^b \quad \text{if } r \neq -1$$

$$\int_a^b \sin x\,dx = \Big[-\cos x\Big]_a^b$$

$$\int_a^b \sec^2 x\,dx = \Big[\tan x\Big]_a^b.$$

EXAMPLE 3 Evaluate $\int_{-1}^{2} (x^3 + 1)^2\,dx$.

SOLUTION We first square the integrand and then apply the power rule to each term as follows:

$$\begin{aligned}
\int_{-1}^{2} (x^3+1)^2\,dx &= \int_{-1}^{2} (x^6 + 2x^3 + 1)\,dx \\
&= \left[\frac{x^7}{7} + 2\cdot\frac{x^4}{4} + x\right]_{-1}^{2} \\
&= \left[\frac{2^7}{7} + 2\cdot\frac{2^4}{4} + 2\right] - \left[\frac{(-1)^7}{7} + 2\frac{(-1)^4}{4} + (-1)\right] \\
&= \tfrac{405}{14}
\end{aligned}$$

EXAMPLE 4 Evaluate $\int_1^4 \left(5x - 2\sqrt{x} + \frac{32}{x^3}\right) dx$.

SOLUTION We begin by changing the form of the integrand so that the power rule may be applied to each term. Thus,

$$\begin{aligned}
\int_1^4 (5x - 2x^{1/2} + 32x^{-3})\,dx &= \left[5\left(\frac{x^2}{2}\right) - 2\left(\frac{x^{3/2}}{3/2}\right) + 32\left(\frac{x^{-2}}{-2}\right)\right]_1^4 \\
&= \left[\frac{5}{2}x^2 - \frac{4}{3}x^{3/2} - \frac{16}{x^2}\right]_1^4 \\
&= \left[\frac{5}{2}(4)^2 - \frac{4}{3}(4)^{3/2} - \frac{16}{4^2}\right] - \left[\frac{5}{2} - \frac{4}{3} - 16\right] \\
&= \tfrac{259}{6}.
\end{aligned}$$

The method of substitution developed for indefinite integrals may also be used to evaluate a definite integral. We could use (5.7) to find an indefinite integral (that is, an antiderivative) and then apply the fundamental theorem of calculus. Another method, which is sometimes shorter, is to change the limits of integration. Using (5.7) together with the fundamental theorem gives us the following formula, with $F' = f$:

$$\int_a^b f(g(x))g'(x)\,dx = F(g(x))\Big]_a^b$$

The number on the right may be written

$$F(g(b)) - F(g(a)) = F(u)\Big]_{g(a)}^{g(b)} = \int_{g(a)}^{g(b)} f(u)\,du.$$

This gives us the following result, provided f and g' are integrable.

Theorem **(5.33)**

$$\text{If } u = g(x), \text{ then } \int_a^b f(g(x))g'(x)\,dx = \int_{g(a)}^{g(b)} f(u)\,du.$$

Theorem (5.33) states that after making the substitution $u = g(x)$ and $du = g'(x)\,dx$, we may use the values of g that correspond to $x = a$ and $x = b$, respectively, as the limits of the integral involving u. It is then unnecessary to return to the variable x after integrating. This technique is illustrated in the next example.

EXAMPLE 5 Evaluate $\displaystyle\int_2^{10} \frac{3}{\sqrt{5x - 1}}\,dx.$

SOLUTION Let us begin by writing the integral as

$$3\int_2^{10} \frac{1}{\sqrt{5x - 1}}\,dx.$$

The expression $\sqrt{5x - 1}$ in the integrand suggests the following substitution:

$$u = 5x - 1, \quad du = 5\,dx$$

The form of du indicates that we should introduce the factor 5 into the integrand and then compensate by multiplying the integral by $\frac{1}{5}$, as follows:

$$3\int_2^{10} \frac{1}{\sqrt{5x - 1}}\,dx = \frac{3}{5}\int_2^{10} \frac{1}{\sqrt{5x - 1}}\,5\,dx$$

We next calculate the values of $u = 5x - 1$ that correspond to the limits of integration $x = 2$ and $x = 10$:

(i) If $x = 2$, then $u = 5(2) - 1 = 9$.

(ii) If $x = 10$, then $u = 5(10) - 1 = 49$.

Substituting in the integrand and changing the limits of integration as in Theorem (5.33) gives us

$$\begin{aligned} 3\int_2^{10} \frac{1}{\sqrt{5x - 1}}\,dx &= \frac{3}{5}\int_2^{10} \frac{1}{\sqrt{5x - 1}}\,5\,dx \\ &= \frac{3}{5}\int_9^{49} \frac{1}{\sqrt{u}}\,du = \frac{3}{5}\int_9^{49} u^{-1/2}\,du \\ &= \left[\left(\frac{3}{5}\right)\frac{u^{1/2}}{1/2}\right]_9^{49} = \frac{6}{5}[49^{1/2} - 9^{1/2}] = \frac{24}{5}. \end{aligned}$$

EXAMPLE 6 Evaluate $\int_0^{\pi/4} (1 + \sin 2x)^3 \cos 2x\, dx$.

SOLUTION The integrand suggests the power rule $\int_a^b u^3\, du = [\frac{1}{4}u^4]_a^b$. Thus, we let

$$u = 1 + \sin 2x, \quad du = 2 \cos 2x\, dx.$$

The form of du indicates that we should introduce the factor 2 into the integrand and multiply the integral by $\frac{1}{2}$, as follows:

$$\int_0^{\pi/4} (1 + \sin 2x)^3 \cos 2x\, dx = \frac{1}{2} \int_0^{\pi/4} (1 + \sin 2x)^3\, 2 \cos 2x\, dx$$

We next calculate the values of $u = 1 + \sin 2x$ that correspond to the limits of integration $x = 0$ and $x = \pi/4$:

(i) If $x = 0$, then $u = 1 + \sin 0 = 1 + 0 = 1$.

(ii) If $x = \dfrac{\pi}{4}$, then $u = 1 + \sin \dfrac{\pi}{2} = 1 + 1 = 2$.

Substituting in the integrand and changing the limits of integration gives us

$$\begin{aligned}\int_0^{\pi/4} (1 + \sin 2x)^3 \cos 2x\, dx &= \frac{1}{2} \int_1^2 u^3\, du \\ &= \frac{1}{2}\left[\frac{u^4}{4}\right]_1^2 \\ &= \frac{1}{8}[16 - 1] = \frac{15}{8}.\end{aligned}$$

The following theorem illustrates a useful technique for evaluating certain definite integrals.

Theorem **(5.34)**

Let f be continuous on $[-a, a]$.

(i) If f is an even function,

$$\int_{-a}^{a} f(x)\, dx = 2 \int_0^a f(x)\, dx.$$

(ii) If f is an odd function,

$$\int_{-a}^{a} f(x)\, dx = 0.$$

FIGURE 5.32

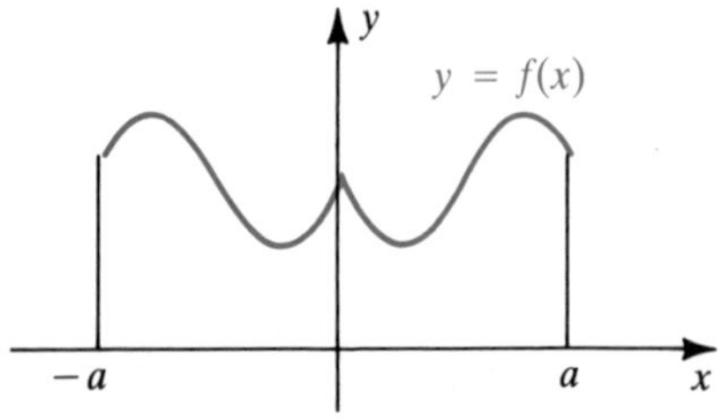

PROOF We shall prove (i). If f is an even function, then the graph of f is symmetric with respect to the y-axis. As a special case, if $f(x) \geq 0$ for every x in $[0, a]$, we have a situation similar to that in Figure 5.32, and hence the area under the graph of f from $x = -a$ to $x = a$ is twice that from $x = 0$ to $x = a$. This gives us the formula in (i).

To show that the formula is true if $f(x) < 0$ for some x, we may proceed as follows. Using, successively, Theorem (5.24), Definition (5.17), and Theorem (5.22), we have

$$\begin{aligned}\int_{-a}^{a} f(x)\,dx &= \int_{-a}^{0} f(x)\,dx + \int_{0}^{a} f(x)\,dx\\ &= -\int_{0}^{-a} f(x)\,dx + \int_{0}^{a} f(x)\,dx\\ &= \int_{0}^{-a} f(x)(-dx) + \int_{0}^{a} f(x)\,dx.\end{aligned}$$

Since f is even, $f(-x) = f(x)$, and the last equality may be written

$$\int_{-a}^{a} f(x)\,dx = \int_{0}^{-a} f(-x)(-dx) + \int_{0}^{a} f(x)\,dx.$$

If, in the first integral on the right, we substitute $u = -x$, $du = -dx$ and observe that $u = a$ when $x = -a$, we obtain

$$\int_{-a}^{a} f(x)\,dx = \int_{0}^{a} f(u)\,du + \int_{0}^{a} f(x)\,dx.$$

The last two integrals on the right are equal, since the variables are *dummy variables*, and, therefore,

$$\int_{-a}^{a} f(x)\,dx = 2\int_{0}^{a} f(x)\,dx. \quad ■$$

EXAMPLE 7 Evaluate

(a) $\int_{-1}^{1} (x^4 + 3x^2 + 1)\,dx$ **(b)** $\int_{-1}^{1} (x^5 + 3x^3 + x)\,dx$

(c) $\int_{-5}^{5} (2x^3 + 3x^2 + 7x)\,dx$

SOLUTION

(a) Since the integrand determines an even function, we may apply Theorem (5.34)(i):

$$\begin{aligned}\int_{-1}^{1} (x^4 + 3x^2 + 1)\,dx &= 2\int_{0}^{1} (x^4 + 3x^2 + 1)\,dx\\ &= 2\left[\frac{x^5}{5} + x^3 + x\right]_0^1 = \frac{22}{5}\end{aligned}$$

(b) The integrand is odd, so we apply Theorem (5.34)(ii):

$$\int_{-1}^{1} (x^5 + 3x^3 + x)\,dx = 0$$

(c) The function given by $2x^3 + 7x$ is odd but the function given by $3x^2$ is even, so we apply Theorem (5.34)(ii) *and* (i):

$$\begin{aligned}\int_{-5}^{5} (2x^3 + 3x^2 + 7x)\,dx &= \int_{-5}^{5} (2x^3 + 7x)\,dx + \int_{-5}^{5} 3x^2\,dx\\ &= 0 + 2\int_{0}^{5} 3x^2\,dx\\ &= 2\Big[x^3\Big]_0^5 = 250\end{aligned}$$

The technique of defining a function by means of a definite integral, as in Part I of the fundamental theorem of calculus (5.30), will have a very important application in Chapter 7, when we consider logarithmic functions. Recall, from (5.30), that if f is continuous on $[a, b]$ and $G(x) = \int_a^x f(t)\,dt$ for $a \le x \le b$, then G is an antiderivative of f; that is, $D_x\, G(x) = f(x)$. This may be stated in integral form as follows:

$$D_x \int_a^x f(t)\,dt = f(x)$$

The preceding formula is generalized in the next theorem.

Theorem (5.35)

Let f be continuous on $[a, b]$. If $a \le c \le b$, then for every x in $[a, b]$,

$$D_x \int_c^x f(t)\,dt = f(x).$$

PROOF If F is an antiderivative of f, then

$$\begin{aligned} D_x \int_c^x f(t)\,dt &= D_x\,[F(x) - F(c)] \\ &= D_x\, F(x) - D_x\, F(c) \\ &= f(x) - 0 = f(x). \end{aligned}$$

EXAMPLE 8 If $G(x) = \int_1^x \frac{1}{t}\,dt$ and $x > 0$, find $G'(x)$.

SOLUTION We shall apply Theorem (5.35) with $c = 1$ and $f(x) = 1/x$. If we choose a and b such that $0 < a \le 1 \le b$, then f is continuous on $[a, b]$. Hence, by Theorem (5.35), for every x in $[a, b]$,

$$G'(x) = D_x \int_1^x \frac{1}{t}\,dt = \frac{1}{x}.$$

In (5.29) we defined the *average value* f_{av} of a function f on $[a, b]$ as follows:

$$f_{av} = \frac{1}{b-a} \int_a^b f(x)\,dx$$

The next example indicates why this terminology is appropriate in applications.

EXAMPLE 9 Suppose a point P moving on a coordinate line has a continuous velocity function v. Show that the average value of v on $[a, b]$ equals the average velocity during the time interval $[a, b]$.

SOLUTION By Definition (5.29) with $f = v$,

$$v_{av} = \frac{1}{b-a} \int_a^b v(t)\,dt.$$

If s is the position function of P, then $s'(t) = v(t)$; that is, $s(t)$ is an antiderivative of $v(t)$. Hence, by the fundamental theorem of calculus,

$$\int_a^b v(t)\,dt = \int_a^b s'(t)\,dt = s(t)\Big]_a^b = s(b) - s(a).$$

Substituting in the formula for v_{av} gives us

$$v_{\text{av}} = \frac{s(b) - s(a)}{b - a},$$

which is the average velocity of P on $[a, b]$ (see (3.2)).

Results similar to that in Example 9 occur in discussions of average acceleration, average marginal cost, average marginal revenue, and many other applications of the derivative (see Exercises 45–50).

EXERCISES 5.6

Exer. 1–36: Evaluate the integral.

1 $\int_1^4 (x^2 - 4x - 3)\,dx$

2 $\int_{-2}^3 (5 + x - 6x^2)\,dx$

3 $\int_{-2}^3 (8z^3 + 3z - 1)\,dz$

4 $\int_0^2 (z^4 - 2z^3)\,dz$

5 $\int_7^{12} dx$

6 $\int_{-6}^{-1} 8\,dx$

7 $\int_1^2 \frac{5}{x^6}\,dx$

8 $\int_1^4 \sqrt{16x^5}\,dx$

9 $\int_4^9 \frac{t - 3}{\sqrt{t}}\,dt$

10 $\int_{-1}^{-2} \frac{2t - 7}{t^3}\,dt$

11 $\int_{-8}^8 (\sqrt[3]{s^2} + 2)\,ds$

12 $\int_1^0 s^2(\sqrt[3]{s} - \sqrt{s})\,ds$

13 $\int_{-1}^0 (2x + 3)^2\,dx$

14 $\int_1^2 (4x^{-5} - 5x^4)\,dx$

15 $\int_3^2 \frac{x^2 - 1}{x - 1}\,dx$

16 $\int_0^{-1} \frac{x^3 + 8}{x + 2}\,dx$

17 $\int_1^1 (4x^2 - 5)^{100}\,dx$

18 $\int_5^5 \sqrt[3]{x^2 + \sqrt{x^5 + 1}}\,dx$

19 $\int_1^3 \frac{2x^3 - 4x^2 + 5}{x^2}\,dx$

20 $\int_{-2}^{-1} \left(x - \frac{1}{x}\right)^2 dx$

21 $\int_{-3}^6 |x - 4|\,dx$

22 $\int_{-1}^5 |2x - 3|\,dx$

23 $\int_1^4 \sqrt{5 - x}\,dx$

24 $\int_1^5 \sqrt[3]{2x - 1}\,dx$

25 $\int_{-1}^1 (v^2 - 1)^3\,v\,dv$

26 $\int_{-2}^0 \frac{v^2}{(v^3 - 2)^2}\,dv$

27 $\int_0^1 \frac{1}{(3 - 2x)^2}\,dx$

28 $\int_0^4 \frac{x}{\sqrt{x^2 + 9}}\,dx$

29 $\int_1^4 \frac{1}{\sqrt{x}(\sqrt{x} + 1)^3}\,dx$

30 $\int_0^1 (3 - x^4)^3\,x^3\,dx$

31 $\int_{\pi/2}^{\pi} \cos\left(\tfrac{1}{3}x\right)dx$

32 $\int_0^{\pi/2} 3\sin\left(\tfrac{1}{2}x\right)dx$

33 $\int_{\pi/4}^{\pi/3} (4\sin 2\theta + 6\cos 3\theta)\,d\theta$

34 $\int_{\pi/6}^{\pi/4} (1 - \cos 4\theta)\,d\theta$

35 $\int_{-\pi/6}^{\pi/6} (x + \sin 5x)\,dx$

36 $\int_0^{\pi/3} \frac{\sin x}{\cos^2 x}\,dx$

Exer. 37–40: (a) Find a number z that satisfies the conclusion of the mean value theorem (5.28) for the given integral $\int_a^b f(x)\,dx$. (b) Find the average value of f on $[a, b]$.

37 $\int_0^4 \frac{x}{\sqrt{x^2 + 9}}\,dx$

38 $\int_{-2}^0 \sqrt[3]{x + 1}\,dx$

39 $\int_0^5 \sqrt{x + 4}\,dx$

40 $\int_{-3}^2 \sqrt{6 - x}\,dx$

Exer. 41–44: Find the derivative without integrating.

41 $D_x \int_0^3 \sqrt{x^2 + 16}\,dx$

42 $D_x \int_0^1 x\sqrt{x^2 + 4}\,dx$

43 $D_x \int_0^x \frac{1}{t + 1}\,dt$

44 $D_x \int_0^x \frac{1}{\sqrt{1 - t^2}}\,dt, \quad |x| < 1$

45 A point P is moving on a coordinate line with a continuous acceleration function a. If v is the velocity function, then the *average acceleration* on a time interval $[t_1, t_2]$ is

$$\frac{v(t_2) - v(t_1)}{t_2 - t_1}.$$

Show that the average acceleration is equal to the average value of a on $[t_1, t_2]$.

46 If a function f has a continuous derivative on $[a, b]$, show that the average rate of change of $f(x)$ with respect to x on $[a, b]$ (see Definition (3.4)) is equal to the average value of f' on $[a, b]$.

47 The vertical distribution of velocity of the water in a river may be approximated by $v = c(d - y)^{1/6}$, where v is the velocity (in m/sec) at a depth of y meters below the water surface, d is the depth of the river, and c is a positive constant.

(a) Find a formula for the average velocity v_{av} in terms of d and c.

(b) If v_0 is the velocity at the surface, show that $v_{av} = \frac{6}{7}v_0$.

48 In the electrical circuit shown in the figure, the alternating current I is given by $I = I_M \sin \omega t$, where t is the time and I_M is the maximum current. The rate P at which heat is being produced in the resistor of R ohms is given by $P = I^2R$. Compute the *average rate* of production of heat over one complete cycle (from $t = 0$ to $t = 2\pi/\omega$). (*Hint:* Use the half-angle formula for the sine.)

EXERCISE 48

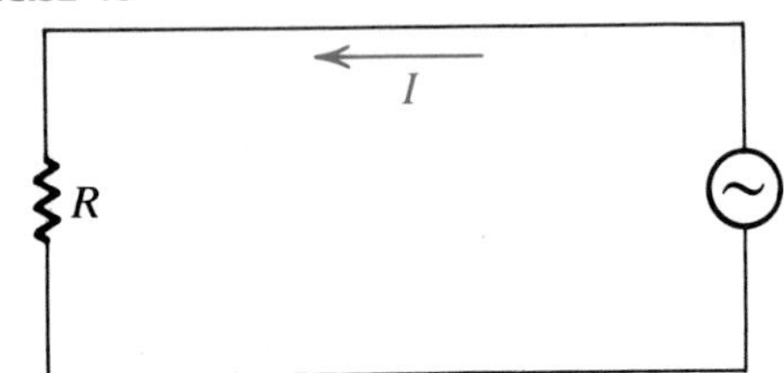

49 If a ball is dropped from a height of s_0 feet above the ground and air resistance is negligible, then the distance it falls in t seconds is $16t^2$ feet. Use Definition (5.29) to show that the average velocity for the ball's journey to the ground is $4\sqrt{s_0}$ ft/sec.

50 A meteorologist determines that the temperature T (in °F) on a cold winter day is given by

$$T = \tfrac{1}{20}t(t - 12)(t - 24),$$

where t is time (in hours) and $t = 0$ corresponds to midnight. Find the average temperature between 6 A.M. and 12 noon.

51 If g is differentiable and f is continuous for every x, prove that

$$D_x \int_a^{g(x)} f(t)\, dt = f(g(x))g'(x).$$

52 Extend the formula in Exercise 51 to

$$D_x \int_{k(x)}^{g(x)} f(t)\, dt = f(g(x))g'(x) - f(k(x))k'(x).$$

Exer. 53–56: Use Exercises 51 and 52 to find the derivative.

53 $D_x \int_2^{x^4} \dfrac{t}{\sqrt{t^3 + 2}}\, dt$

54 $D_x \int_0^{x^2} \sqrt[3]{t^4 + 1}\, dt$

55 $D_x \int_{3x}^{x^3} (t^3 + 1)^{10}\, dt$

56 $D_x \int_{1/x}^{\sqrt{x}} \sqrt{t^4 + t^2 + 4}\, dt$

5.7 NUMERICAL INTEGRATION

To evaluate a definite integral $\int_a^b f(x)\, dx$ by means of the fundamental theorem of calculus, we need an antiderivative of f. If an antiderivative cannot be found, then we may use numerical methods to approximate the integral to any desired degree of accuracy. For example, if the norm of a partition of $[a, b]$ is small, then, by Definition (5.16), the definite integral can be approximated by any Riemann sum of f. In particular, if we use a regular partition with $\Delta x = (b - a)/n$, then

$$\int_a^b f(x)\, dx \approx \sum_{k=1}^{n} f(w_k)\, \Delta x,$$

where w_k is any number in the kth subinterval $[x_{k-1}, x_k]$ of the partition. Of course, the accuracy of the approximation depends on the nature of f and the magnitude of Δx. It may be necessary to make Δx very small to obtain a desired degree of accuracy. This, in turn, means that n is large, and hence the preceding sum contains many terms. Figure 5.5 illustrates the case in which $f(w_k)$ is the minimum value of f on $[x_{k-1}, x_k]$. In this case the error involved in the approximation is numerically the same as the area of the region that lies under the graph of f and over the inscribed rectangles.

If we let $w_k = x_{k-1}$ (that is, if f is evaluated at the left-hand endpoint of each subinterval $[x_{k-1}, x_k]$), then

$$\int_a^b f(x)\,dx \approx \sum_{k=1}^{n} f(x_{k-1})\,\Delta x.$$

If we let $w_k = x_k$ (that is, if f is evaluated at the right-hand endpoint of $[x_{k-1}, x_k]$), then

$$\int_a^b f(x)\,dx \approx \sum_{k=1}^{n} f(x_k)\,\Delta x.$$

Another and usually more accurate approximation can be obtained by using the average of the last two approximations—that is,

$$\frac{1}{2}\left[\sum_{k=1}^{n} f(x_{k-1})\,\Delta x + \sum_{k=1}^{n} f(x_k)\,\Delta x\right].$$

With the exception of $f(x_0)$ and $f(x_n)$, each function value $f(x_k)$ appears twice, and hence we may write the last expression as

$$\frac{\Delta x}{2}\left[f(x_0) + \sum_{k=1}^{n-1} 2f(x_k) + f(x_n)\right].$$

Since $\Delta x = (b - a)/n$, this gives us the following rule.

Trapezoidal rule **(5.36)**

Let f be continuous on $[a, b]$. If a regular partition of $[a, b]$ is determined by $a = x_0, x_1, \ldots, x_n = b$, then

$$\int_a^b f(x)\,dx \approx \frac{b-a}{2n}\left[f(x_0) + 2f(x_1) + 2f(x_2) + \cdots + 2f(x_{n-1}) + f(x_n)\right].$$

The term *trapezoidal* comes from the case in which $f(x)$ is nonnegative on $[a, b]$. As illustrated in Figure 5.33, if P_k is the point with x-coordinate x_k on the graph of $y = f(x)$, then for each $k = 1, 2, \ldots, n$, the points on the x-axis with x-coordinates x_{k-1} and x_k, together with P_{k-1} and P_k, are vertices of a trapezoid having area

$$\frac{\Delta x}{2}\left[f(x_{k-1}) + f(x_k)\right].$$

FIGURE 5.33

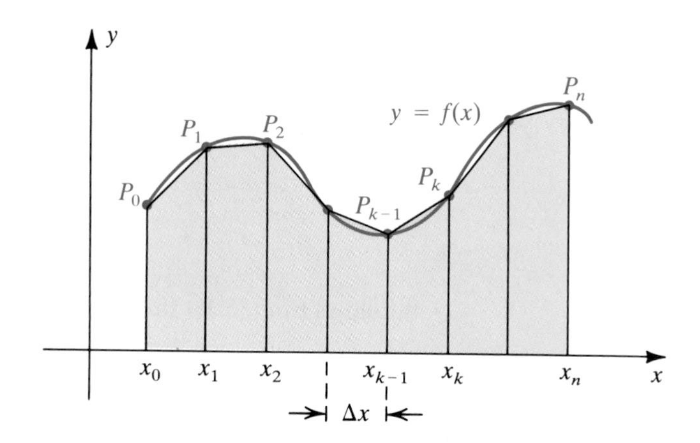

The sum of the areas of these trapezoids is the same as the sum in Rule (5.36). Hence, in geometric terms, the trapezoidal rule gives us an approximation to the area under the graph of f from a to b by means of trapezoids instead of the rectangles associated with Riemann sums.

The next result provides information about the maximum error that can occur if the trapezoidal rule is used to approximate a definite integral. The proof is omitted.

Error estimate for the trapezoidal rule **(5.37)**

If f'' is continuous and if M is a positive real number such that $|f''(x)| \le M$ for every x in $[a, b]$, then the error involved in using the trapezoidal rule (5.36) is not greater than

$$\frac{M(b-a)^3}{12n^2}.$$

EXAMPLE 1 Use the trapezoidal rule with $n = 10$ to approximate $\int_1^2 \frac{1}{x}\,dx$. Estimate the maximum error in the approximation.

SOLUTION It is convenient to arrange our work as follows. Each $f(x_k)$ was obtained with a calculator and is accurate to nine decimal places. The column labeled m contains the coefficient of $f(x_k)$ in the trapezoidal rule (5.36). Thus, $m = 1$ for $f(x_0)$ or $f(x_n)$, and $m = 2$ for the remaining $f(x_k)$.

k	x_k	$f(x_k)$	m	$mf(x_k)$
0	1.0	1.000000000	1	1.000000000
1	1.1	0.909090909	2	1.818181818
2	1.2	0.833333333	2	1.666666666
3	1.3	0.769230769	2	1.538461538
4	1.4	0.714285714	2	1.428571428
5	1.5	0.666666667	2	1.333333334
6	1.6	0.625000000	2	1.250000000
7	1.7	0.588235294	2	1.176470588
8	1.8	0.555555556	2	1.111111112
9	1.9	0.526315789	2	1.052631578
10	2.0	0.500000000	1	0.500000000

The sum of the numbers in the last column is 13.875428062.

Since

$$\frac{b-a}{2n} = \frac{2-1}{2(10)} = \frac{1}{20},$$

it follows from (5.36) that

$$\int_1^2 \frac{1}{x}\,dx \approx \frac{1}{20}(13.875428062) \approx 0.693771403.$$

The error in the approximation may be estimated by means of (5.37). Since $f(x) = 1/x$, we have $f'(x) = -1/x^2$ and $f''(x) = 2/x^3$. The maximum value of $f''(x)$ on the interval $[1, 2]$ occurs at $x = 1$ because f'' is a decreasing function. Hence

$$|f''(x)| \leq \frac{2}{(1)^3} = 2.$$

Applying (5.37) with $M = 2$, we see that the maximum error is not greater than

$$\frac{2(2-1)^3}{12(10)^2} = \frac{1}{600} < 0.002.$$

In Chapter 7 we shall see that the integral in Example 1 equals the natural logarithm of 2, denoted by ln 2, which to nine decimal places is 0.693147181. To obtain this approximation by means of the trapezoidal rule, it is necessary to use a very large value of n.

The following rule is often more accurate than the trapezoidal rule.

Simpson's rule **(5.38)**

> Let f be continuous on $[a, b]$, and let n be an even integer. If a regular partition is determined by $a = x_0, x_1, \ldots, x_n = b$, then
>
> $$\int_a^b f(x)\,dx \approx \frac{b-a}{3n}[f(x_0) + 4f(x_1) + 2f(x_2) + 4f(x_3) + \cdots + 2f(x_{n-2}) + 4f(x_{n-1}) + f(x_n)].$$

FIGURE 5.34

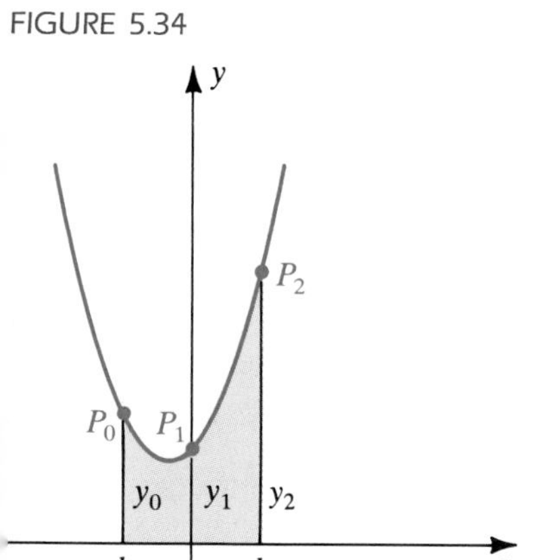

The idea behind the proof of Simpson's rule is that instead of using trapezoids to approximate the graph of f, we use portions of graphs of equations of the form $y = cx^2 + dx + e$ for constants c, d, and e; that is, we use portions of parabolas or lines. If $P_0(x_0, y_0)$, $P_1(x_1, y_1)$, and $P_2(x_2, y_2)$ are points on the parabola such that $x_0 < x_1 < x_2$, then substituting the coordinates of P_0, P_1, and P_2, respectively, into the equation $y = cx^2 + dx + e$ gives us three equations that may be solved for c, d, and e. As a special case, suppose h, y_0, y_1, and y_2 are positive, and consider the points $P_0(-h, y_0)$, $P_1(0, y_1)$, and $P_2(h, y_2)$, as illustrated in Figure 5.34.

The area A under the graph of the equation from $-h$ to h is

$$A = \int_{-h}^{h} (cx^2 + dx + e)\,dx = \left[\frac{cx^3}{3} + \frac{dx^2}{2} + ex\right]_{-h}^{h} = \frac{h}{3}(2ch^2 + 6e).$$

Since the coordinates of $P_0(-h, y_0)$, $P_1(0, y_1)$, and $P_2(h, y_2)$ are solutions of $y = cx^2 + dx + e$, we have, by substitution,

$$y_0 = ch^2 - dh + e$$

$$y_1 = e$$

$$y_2 = ch^2 + dh + e.$$

Thus,

$$y_0 + 4y_1 + y_2 = 2ch^2 + 6e$$

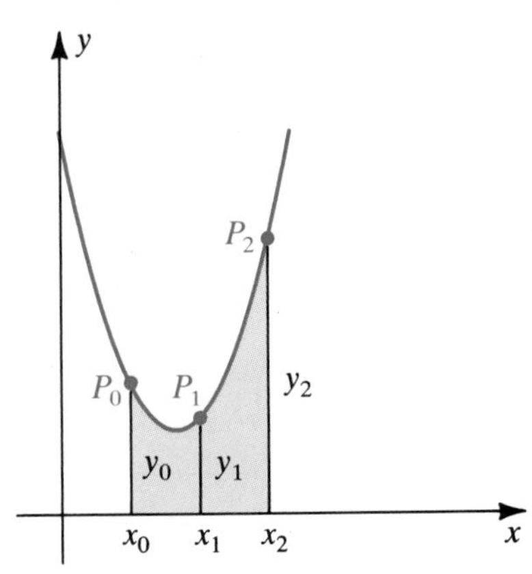

FIGURE 5.35

and
$$A = \frac{h}{3}(y_0 + 4y_1 + y_2).$$

If the points P_0, P_1, and P_2 are translated horizontally, as illustrated in Figure 5.35, then the area under the graph remains the same. Consequently, the preceding formula for A is true for *any* points P_0, P_1, and P_2, provided $x_1 - x_0 = x_2 - x_1$.

If $f(x) \geq 0$ on $[a, b]$, then Simpson's rule is obtained by regarding the definite integral as the area under the graph of f from a to b. Thus, suppose n is an even integer and $h = (b - a)/n$. We divide $[a, b]$ into n subintervals, each of length h, by choosing numbers $a = x_0, x_1, \ldots, x_n = b$. Let $P_k(x_k, y_k)$ be the point on the graph of f with x-coordinate x_k, as illustrated in Figure 5.36.

FIGURE 5.36

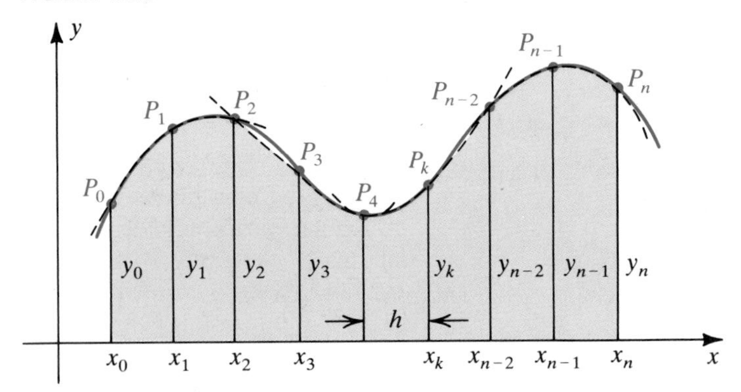

If the arc through P_0, P_1, and P_2 is approximated by the graph of an equation $y = cx^2 + dx + e$, then, as we have seen, the area under the graph of f from x_0 to x_2 is approximated by

$$\frac{h}{3}(y_0 + 4y_1 + y_2).$$

If we use the arc through P_2, P_3, and P_4, then the area under the graph of f from x_2 to x_4 is approximately

$$\frac{h}{3}(y_2 + 4y_3 + y_4).$$

We continue in this manner until we reach the last triple of points P_{n-2}, P_{n-1}, P_n and the corresponding approximation to the area under the graph:

$$\frac{h}{3}(y_{n-2} + 4y_{n-1} + y_n)$$

Summing these approximations give us

$$\int_a^b f(x)\, dx \approx \frac{h}{3}(y_0 + 4y_1 + 2y_2 + 4y_3 + \cdots + 2y_{n-2} + 4y_{n-1} + y_n),$$

which is the same as the sum in (5.38). If f is negative for some x in $[a, b]$, then negatives of areas may be used to establish Simpson's rule.

The following result is analogous to (5.37).

Error estimate for Simpson's rule **(5.39)**

> If $f^{(4)}$ is continuous and if M is a positive real number such that $|f^{(4)}(x)| \leq M$ for every x in $[a, b]$, then the error involved in using Simpson's rule (5.38) is not greater than
>
> $$\frac{M(b-a)^5}{180n^4}.$$

EXAMPLE 2 Use Simpson's rule with $n = 10$ to approximate $\int_1^2 \frac{1}{x}\,dx$. Estimate the maximum error in the approximation.

SOLUTION This is the same integral considered in Example 1. We arrange our work as follows. The column labeled m contains the coefficient of $f(x_k)$ in Simpson's rule (5.38).

k	x_k	$f(x_k)$	m	$mf(x_k)$
0	1.0	1.000000000	1	1.000000000
1	1.1	0.909090909	4	3.636363636
2	1.2	0.833333333	2	1.666666666
3	1.3	0.769230769	4	3.076923076
4	1.4	0.714285714	2	1.428571428
5	1.5	0.666666667	4	2.666666668
6	1.6	0.625000000	2	1.250000000
7	1.7	0.588235294	4	2.352941176
8	1.8	0.555555556	2	1.111111112
9	1.9	0.526315789	4	2.105263156
10	2.0	0.500000000	1	0.500000000

The sum of the numbers in the last column is 20.794506918. Since

$$\frac{b-a}{3n} = \frac{2-1}{30},$$

it follows from (5.38) that

$$\int_1^2 \frac{1}{x}\,dx \approx \left(\frac{1}{30}\right)(20.794506918) \approx 0.693150231.$$

We shall use (5.39) to estimate the error in the approximation. If $f(x) = 1/x$, we can verify that $f^{(4)}(x) = 24/x^5$. The maximum value of $f^{(4)}(x)$ on the interval $[1, 2]$ occurs at $x = 1$, and hence

$$|f^{(4)}(x)| \leq \frac{24}{(1)^5} = 24.$$

Applying (5.39) with $M = 24$, we see that the maximum error in the approximation is not greater than

$$\frac{24(2-1)^5}{180(10)^4} = \frac{2}{150{,}000} < 0.00002.$$

Note that this estimated error is much less than that obtained using the trapezoidal rule in Example 1.

An important aspect of numerical integration is that it can be used to approximate the definite integral of a function that is described by means of a table or graph. To illustrate, suppose it is found experimentally that two physical variables x and y are related as shown in the following table.

x	1.0	1.5	2.0	2.5	3.0	3.5	4.0
y	3.1	4.0	4.2	3.8	2.9	2.8	2.7

FIGURE 5.37

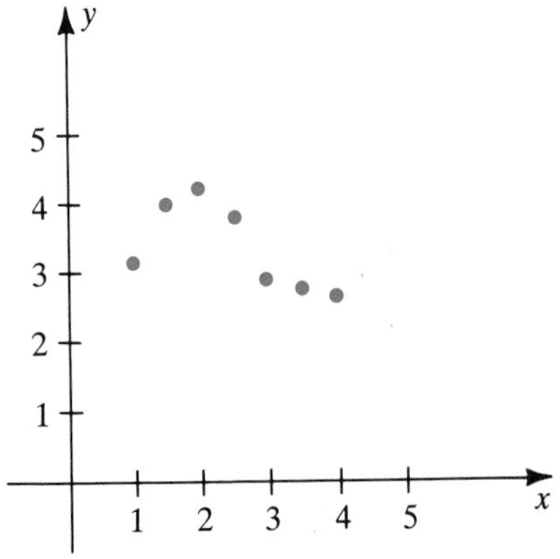

The points (x, y) are plotted in Figure 5.37. If we regard y as a function of x, say $y = f(x)$ with f continuous, then the definite integral $\int_1^4 f(x)\,dx$ may represent a physical quantity. In the present illustration, *the integral may be approximated without knowing an explicit form for* $f(x)$. In particular, if the trapezoidal rule (5.36) is used, with $n = 6$ and $(b-a)/(2n) = (4-1)/12 = 0.25$, then

$$\int_1^4 f(x)\,dx \approx 0.25[3.1 + 2(4.0) + 2(4.2) + 2(3.8) + 2(2.9) + 2(2.8) + 2.7],$$

or

$$\int_1^4 f(x)\,dx \approx 10.3.$$

The number of subdivisions is even, so we could also approximate the integral by means of Simpson's rule.

EXERCISES 5.7

Exer. 1–12: Approximate the definite integral for the stated value of n by using (a) the trapezoidal rule and (b) Simpson's rule. (Approximate each $f(x_k)$ to four decimal places, and round off answers to two decimal places, whenever appropriate.)

1. $\int_1^3 (x^2 + 1)\,dx$; $n = 4$
2. $\int_1^5 x^3\,dx$; $n = 4$
3. $\int_1^{1.6} (2x - 1)\,dx$; $n = 6$
4. $\int_2^{3.2} (\frac{1}{2}x + 1)\,dx$; $n = 6$
5. [c] $\int_1^4 \frac{1}{x}\,dx$; $n = 6$
6. [c] $\int_0^3 \frac{1}{1+x}\,dx$; $n = 8$
7. [c] $\int_0^1 \frac{1}{\sqrt{1+x^2}}\,dx$; $n = 4$
8. [c] $\int_2^3 \sqrt{1+x^3}\,dx$; $n = 4$
9. [c] $\int_0^2 \frac{1}{4+x^2}\,dx$; $n = 6$
10. [c] $\int_0^{0.6} \frac{1}{\sqrt{4-x^2}}\,dx$; $n = 6$
11. [c] $\int_0^{\pi} \sqrt{\sin x}\,dx$; $n = 6$
12. [c] $\int_0^{\pi} \sin \sqrt{x}\,dx$; $n = 4$

Exer. 13–16: Estimate the maximum error in approximating the definite integral for the stated value of n when using (a) the trapezoidal rule and (b) Simpson's rule.

13 $\int_{-2}^{3} (\frac{1}{360}x^6 + \frac{1}{60}x^5)\,dx;\quad n = 4$

14 $\int_{0}^{3} (-\frac{1}{12}x^4 + \frac{2}{3}x^3)\,dx;\quad n = 4$

15 $\int_{1}^{5} \frac{1}{x^2}\,dx;\quad n = 8$

16 $\int_{1}^{4} \frac{1}{35}x^{7/2}\,dx;\quad n = 6$

Exer. 17–20: Find the least integer n such that the error in approximating the definite integral is less than E when using (a) the trapezoidal rule and (b) Simpson's rule.

17 $\int_{1}^{8} 81x^{8/3}\,dx;\quad E = 0.001$

18 $\int_{1}^{2} \frac{1}{120x^2}\,dx;\quad E = 0.001$

19 $\int_{1/2}^{1} \frac{1}{x}\,dx;\quad E = 0.0001$

20 $\int_{0}^{3} \frac{1}{x+1}\,dx;\quad E = 0.001$

Exer. 21–24: Suppose the table of values for x and y was obtained empirically. Assuming that $y = f(x)$ and f is continuous, approximate $\int_2^4 f(x)\,dx$ by means of (a) the trapezoidal rule and (b) Simpson's rule.

21

x	2.0	2.5	3.0	3.5	4.0
y	3	2	4	3	5

22

x	2.0	3.0	4.0
y	5	4	3

23

x	y
2.00	4.12
2.25	3.76
2.50	3.21
2.75	3.58
3.00	3.94
3.25	4.15
3.50	4.69
3.75	5.44
4.00	7.52

c 24

x	y
2.0	12.1
2.2	11.4
2.4	9.7
2.6	8.4
2.8	6.3
3.0	6.2
3.2	5.8
3.4	5.4
3.6	5.1
3.8	5.9
4.0	5.6

c 25 Use the trapezoidal rule with $(b - a)/n = 0.1$ to show that

$$\int_1^{2.7} \frac{1}{x}\,dx < 1 < \int_1^{2.8} \frac{1}{x}\,dx.$$

c 26 It will follow from our work in Chapter 8 that

$$\int_0^1 \frac{1}{x^2+1}\,dx = \frac{\pi}{4}.$$

Use this fact and Simpson's rule with $n = 8$ to approximate π to four decimal places.

27 If $f(x)$ is a polynomial of degree less than 4, prove that Simpson's rule gives the exact value of $\int_a^b f(x)\,dx$.

28 Suppose that f is continuous and that both f and f'' are nonnegative throughout $[a, b]$. Prove that $\int_a^b f(x)\,dx$ is less than the approximation given by the trapezoidal rule.

29 The graph in the figure was recorded by an instrument used to measure a physical quantity. Estimate y-coordinates of points on the graph, and approximate the area of the shaded region by using (with $n = 6$) (a) the trapezoidal rule and (b) Simpson's rule.

EXERCISE 29

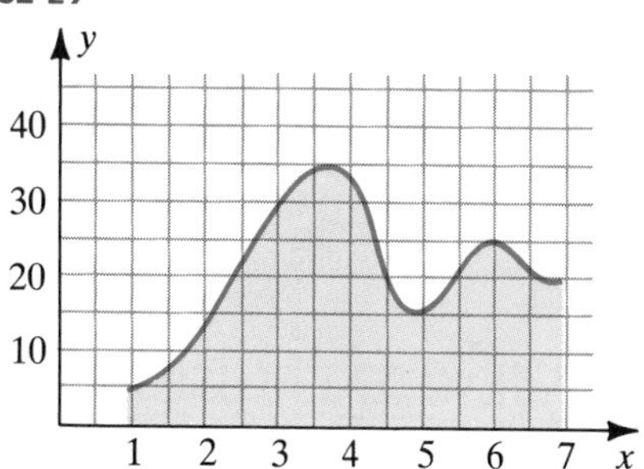

30 A man-made lake has the shape illustrated in the figure, with adjacent measurements 20 feet apart. Use the trapezoidal rule to estimate the surface area of the lake.

EXERCISE 30

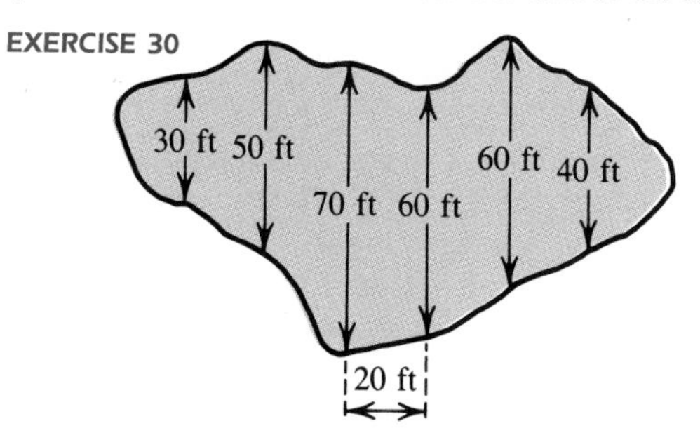

c 31 An important aspect of water management is the production of reliable data on *streamflow*, the number of cubic meters of water passing through a cross section of a stream or river per second. A first step in this computation is the determination of the average velocity $\bar{v}_x$ at a distance x meters from the river bank. If k is the depth of the stream at a point x meters from the bank and $v(y)$ is the velocity (in m/sec) at a depth of y meters (see figure), then

$$\bar{v}_x = \frac{1}{k}\int_0^k v(y)\,dy$$

(see Definition (5.29)). With the *six-point method*, velocity readings are taken at the surface; at depths $0.2k$, $0.4k$, $0.6k$, and $0.8k$; and near the river bottom. The trapezoidal rule is then used to estimate $\bar{v}_x$. Given the data in the following table, estimate $\bar{v}_x$.

y (m)	0	$0.2k$	$0.4k$	$0.6k$	$0.8k$	k
$v(y)$ (m/sec)	0.28	0.23	0.19	0.17	0.13	0.02

EXERCISE 31

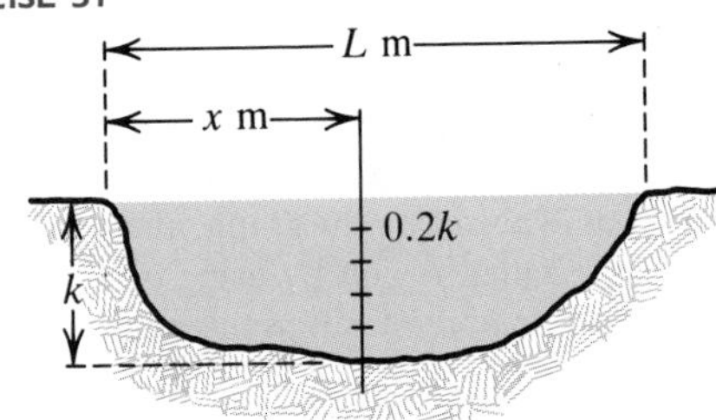

C 32 Refer to Exercise 31. The streamflow F (in m^3/sec) can be approximated using the formula

$$F = \int_0^L \bar{v}_x\, h(x)\,dx,$$

where $h(x)$ is the depth of the stream at a distance x meters from the bank and L is the length of the cross section. Given the data in the following table, use Simpson's rule to estimate F.

x (m)	0	3	6	9	12
$h(x)$ (m)	0	0.51	0.73	1.61	2.11
$\bar{v}_x$ (m/sec)	0	0.09	0.18	0.21	0.36

x (m)	15	18	21	24
$h(x)$ (m)	2.02	1.53	0.64	0
$\bar{v}_x$ (m/sec)	0.32	0.19	0.11	0

C **Exer. 33–34: Use Simpson's rule with $n = 8$ to approximate the average value of f on the given interval.**

33 $f(x) = \dfrac{1}{x^4 + 1}$; $[0, 4]$

34 $f(x) = \sqrt{\cos x}$; $[-1, 1]$

C **Exer. 35–36: If f is determined by the given differential equation and initial condition $f(0)$, approximate $f(1)$ using the trapezoidal rule with $n = 10$.**

35 $f'(x) = \dfrac{\sqrt{x}}{x^2 + 1}$; $f(0) = 1$

36 $f'(x) = \sqrt{\tan x}$; $f(0) = 2$

5.8 REVIEW EXERCISES

Exer. 1–42: Evaluate.

1 $\displaystyle\int \frac{8x^2 - 4x + 5}{x^4}\,dx$

2 $\displaystyle\int (3x^5 + 2x^3 - x)\,dx$

3 $\displaystyle\int 100\,dx$

4 $\displaystyle\int x^{3/5}(2x - \sqrt{x})\,dx$

5 $\displaystyle\int (2x + 1)^7\,dx$

6 $\displaystyle\int \sqrt[3]{5x + 1}\,dx$

$\displaystyle\int (1 - 2x^2)^3\,x\,dx$

8 $\displaystyle\int \frac{(1 + \sqrt{x})^2}{\sqrt[3]{x}}\,dx$

9 $\displaystyle\int \frac{1}{\sqrt{x}(1 + \sqrt{x})^2}\,dx$

10 $\displaystyle\int (x^2 + 4)^2\,dx$

11 $\displaystyle\int (3 - 2x - 5x^3)\,dx$

12 $\displaystyle\int (x + x^{-1})^2\,dx$

13 $\displaystyle\int (4x + 1)(4x^2 + 2x - 7)^2\,dx$

14 $\displaystyle\int \frac{\sqrt[4]{1 - (1/x)}}{x^2}\,dx$

15 $\displaystyle\int (2x^{-3} - 3x^2)\,dx$

16 $\displaystyle\int (x^{3/2} + x^{-3/2})\,dx$

17 $\displaystyle\int_0^1 \sqrt[3]{8x^7}\,dx$

18 $\displaystyle\int_1^2 \frac{x^2 - x - 6}{x + 2}\,dx$

19 $\displaystyle\int_0^1 \frac{x^2}{(1 + x^3)^2}\,dx$

20 $\displaystyle\int_1^9 \sqrt{2x + 7}\,dx$

21 $\displaystyle\int_1^2 \frac{x + 1}{\sqrt{x^2 + 2x}}\,dx$

22 $\displaystyle\int_1^2 \frac{x^2 + 2}{x^2}\,dx$

23 $\displaystyle\int_0^2 x^2\sqrt{x^3 + 1}\,dx$

24 $\displaystyle\int_1^1 3x^2\sqrt{x^3 + x}\,dx$

25 $\displaystyle\int_0^1 (2x - 3)(5x + 1)\,dx$

26 $\displaystyle\int_{-1}^1 (x^2 + 1)^2\,dx$

27 $\displaystyle\int_0^4 \sqrt{3x}(\sqrt{x} + \sqrt{3})\,dx$

28 $\displaystyle\int_{-1}^1 (x + 1)(x + 2)(x + 3)\,dx$

29 $\displaystyle\int \sin(3 - 5x)\,dx$

30 $\displaystyle\int x^2\cos(2x^3)\,dx$

31 $\displaystyle\int \cos 3x \sin^4 3x\,dx$

32 $\displaystyle\int \frac{\sin(1/x)}{x^2}\,dx$

33 $\displaystyle\int \frac{\cos 3x}{\sin^3 3x}\,dx$

34 $\int (3 \cos 2\pi t - 5 \sin 4\pi t)\, dt$

35 $\int_0^{\pi/2} \cos x\sqrt{3 + 5 \sin x}\, dx$

36 $\int_{-\pi/4}^{0} (\sin x + \cos x)^2\, dx$

37 $\int_0^{\pi/4} \sin 2x \cos^2 2x\, dx$

38 $\int_{\pi/6}^{\pi/4} (\sec x + \tan x)(1 - \sin x)\, dx$

39 $\int D_x \sqrt[5]{x^4 + 2x^2 + 1}\, dx$

40 $\int_0^{\pi/2} D_x (x \sin^3 x)\, dx$

41 $D_x \int_0^1 (x^3 + x^2 - 7)^5\, dx$

42 $D_x \int_0^x (t^2 + 1)^{10}\, dt$

Exer. 43–44: Solve the differential equation subject to the given conditions.

43 $\dfrac{d^2y}{dx^2} = 6x - 4; \quad y = 4 \text{ and } y' = 5 \text{ if } x = 2$

44 $f''(x) = x^{1/3} - 5; \quad f'(1) = 2; \quad f(1) = -8$

Exer. 45–46: Let $f(x) = 9 - x^2$ for $-2 \le x \le 3$, and let P be the regular partition of $[-2, 3]$ into five subintervals.

45 Find the Riemann sum R_P if f is evaluated at the midpoint of each subinterval of P.

46 Find (a) A_{IP} and (b) A_{CP}.

Exer. 47–48: Verify the inequality without evaluating the integrals.

47 $\int_0^1 x^2\, dx \ge \int_0^1 x^3\, dx$

48 $\int_1^2 x^2\, dx \le \int_1^2 x^3\, dx$

Exer. 49–50: Express as one integral.

49 $\int_c^e f(x)\, dx + \int_a^b f(x)\, dx - \int_c^b f(x)\, dx - \int_d^a f(x)\, dx$

50 $\int_a^d f(x)\, dx - \int_t^b f(x)\, dx - \int_g^g f(x)\, dx + \int_m^b f(x)\, dx + \int_t^a f(x)\, dx$

51 A stone is thrown directly downward from a height of 900 feet with an initial velocity of 30 ft/sec.

(a) Determine the stone's distance above ground after t seconds.

(b) Find its velocity after 5 seconds.

(c) Determine when it strikes the ground.

52 Given $\int_1^4 (x^2 + 2x - 5)\, dx$, find

(a) a number z that satisfies the conclusion of the mean value theorem for integrals (5.28)

(b) the average value of $x^2 + 2x - 5$ on $[1, 4]$

c 53 Evaluate $\int_0^{10} \sqrt{1 + x^4}\, dx$ by using

(a) the trapezoidal rule, with $n = 5$

(b) Simpson's rule, with $n = 8$

(Use approximations to four decimal places for $f(x_k)$, and round off answers to two decimal places.)

c 54 To monitor the thermal pollution of a river, a biologist takes hourly temperature readings (in °F) from 9 A.M. to 5 P.M. The results are shown in the following table.

Time of day	9	10	11	12	1
Temperature	75.3	77.0	83.1	84.8	86.5

Time of day	2	3	4	5
Temperature	86.4	81.1	78.6	75.1

Use Simpson's rule and Definition (5.29) to estimate the average water temperature between 9 A.M. and 5 P.M.

CHAPTER

6

APPLICATIONS OF THE DEFINITE INTEGRAL

INTRODUCTION

In this chapter we discuss some of the uses for the definite integral. We begin by reconsidering the application that motivated the definition of this mathematical concept—determining the area of a region in the xy-plane. Then, in turn, we use definite integrals to find volumes, lengths of graphs, surface areas of solids, work done by a variable force, and moments and the center of mass (the balance point) of a flat plate. The reason definite integrals are applicable is that each of these quantities is expressible as a limit of sums. Moreover, because of the multitude of other quantities that can be similarly expressed, there is no end to the types of applications. In the last section we further illustrate the versatility of the definite integral by considering a variety of uses that include the following: finding the force exerted by oil on one end of a storage tank, measuring cardiac output and blood flow in arteries, estimating the future wealth of a corporation, calculating the thickness of the ozone layer, determining the amount of radon gas in a home, and finding the number of calories burned during a workout on an exercise bicycle.

As you proceed through this chapter and whenever you encounter definite integrals in applied courses, keep the following nine words in mind: *limit of sums, limit of sums, limit of sums*.

6.1 AREA

If a function f is continuous and $f(x) \geq 0$ on $[a, b]$, then, by Theorem (5.19), the area of the region under the graph of f from a to b is given by the definite integral $\int_a^b f(x)\,dx$. In this section we shall consider the region that lies *between* the graphs of two functions.

FIGURE 6.1

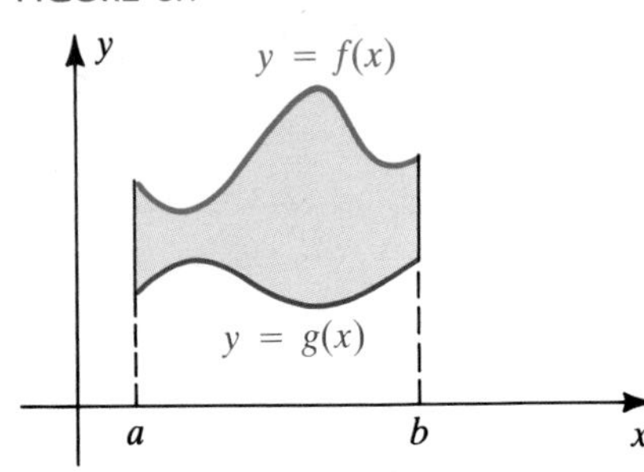

If f and g are continuous and $f(x) \geq g(x) \geq 0$ for every x in $[a, b]$, then the area A of the region R bounded by the graphs of f, g, $x = a$, and $x = b$ (see Figure 6.1) can be found by subtracting the area of the region under the graph of g (the **lower boundary** of R) from the area of the region under the graph of f (the **upper boundary** of R), as follows:

$$A = \int_a^b f(x)\,dx - \int_a^b g(x)\,dx = \int_a^b [f(x) - g(x)]\,dx$$

This formula for A is also true if f or g is negative for some x in $[a, b]$. To verify this fact, choose a *negative* number d that is less than the minimum value of g on $[a, b]$, as illustrated in Figure 6.2(i). Next, consider the functions f_1 and g_1, defined as follows:

$$f_1(x) = f(x) - d = f(x) + |d|$$
$$g_1(x) = g(x) - d = g(x) + |d|$$

FIGURE 6.2

(i)

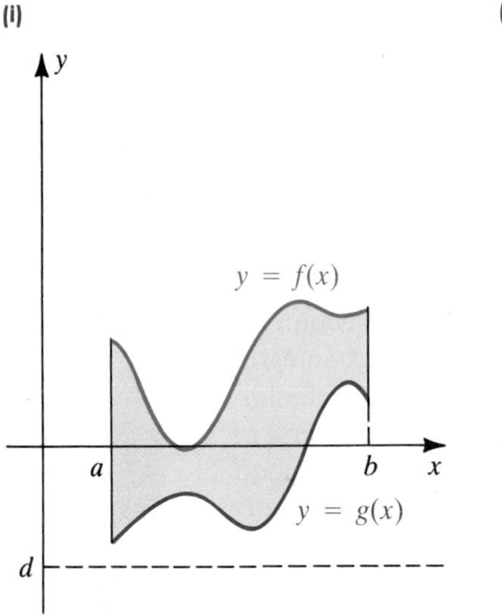

(ii)

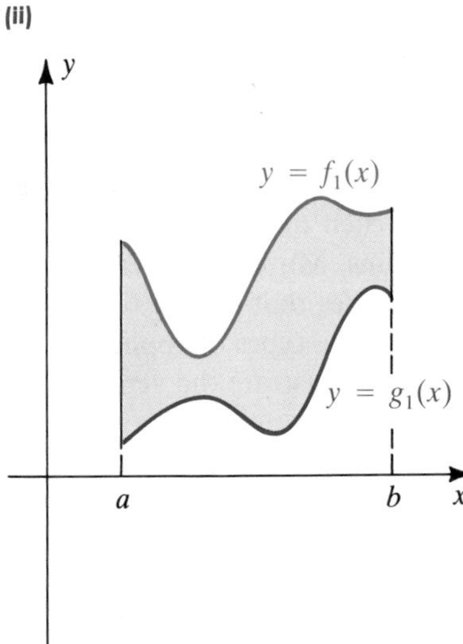

The graphs of f_1 and g_1 can be obtained by vertically shifting the graphs of f and g a distance $|d|$. If A is the area of the region in Figure 6.2(ii), then

$$\begin{aligned} A &= \int_a^b [f_1(x) - g_1(x)]\,dx \\ &= \int_a^b \{[f(x) - d] - [g(x) - d]\}\,dx \\ &= \int_a^b [f(x) - g(x)]\,dx. \end{aligned}$$

We may summarize our discussion as follows.

Theorem **(6.1)**

> If f and g are continuous and $f(x) \geq g(x)$ for every x in $[a, b]$, then the area A of the region bounded by the graphs of f, g, $x = a$, and $x = b$ is
>
> $$A = \int_a^b [f(x) - g(x)]\, dx.$$

FIGURE 6.3

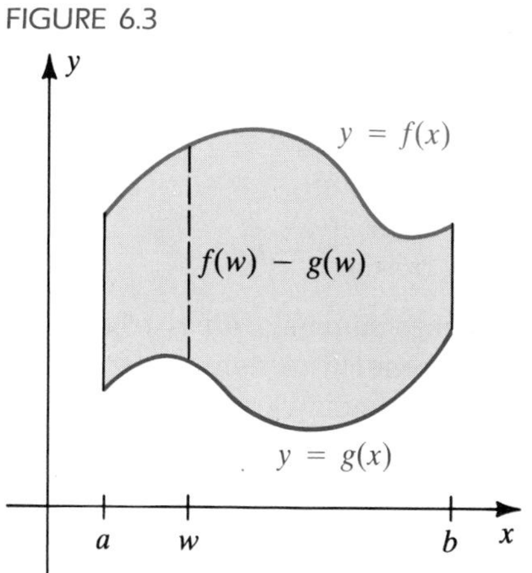

We may interpret the formula for A in Theorem (6.1) as a limit of sums. If we let $h(x) = f(x) - g(x)$ and if w is in $[a, b]$, then $h(w)$ is the vertical distance between the graphs of f and g for $x = w$ (see Figure 6.3). As in our discussion of Riemann sums in Chapter 5, let P denote a partition of $[a, b]$ determined by $a = x_0, x_1, \ldots, x_n = b$. For each k, let $\Delta x_k = x_k - x_{k-1}$, and let w_k be any number in the kth subinterval $[x_{k-1}, x_k]$ of P. By the definition of h,

$$h(w_k)\,\Delta x_k = [f(w_k) - g(w_k)]\,\Delta x_k,$$

which is the area of the rectangle of length $f(w_k) - g(w_k)$ and width Δx_k shown in Figure 6.4.

FIGURE 6.4

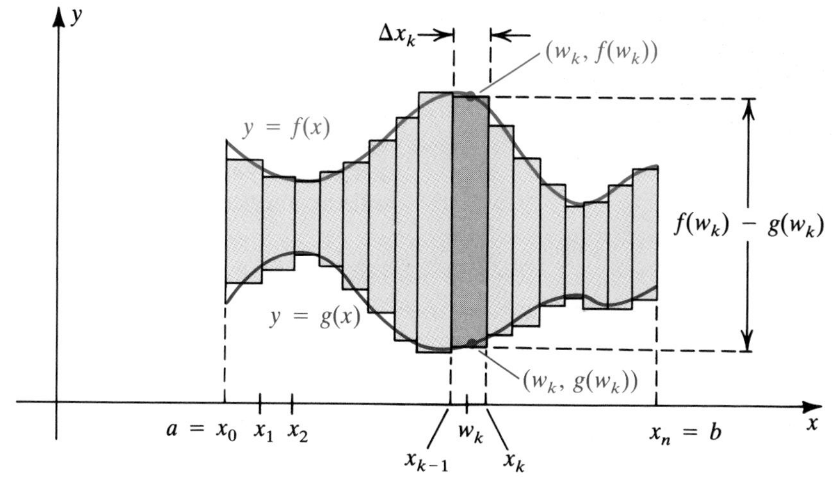

The Riemann sum

$$\sum_k h(w_k)\,\Delta x_k = \sum_k [f(w_k) - g(w_k)]\,\Delta x_k$$

is the sum of the areas of the rectangles in Figure 6.4 and is therefore an approximation to the area of the region between the graphs of f and g from a to b. By the definition of the definite integral,

$$\lim_{\|P\| \to 0} \sum_k h(w_k)\,\Delta x_k = \int_a^b h(x)\, dx.$$

Since $h(x) = f(x) - g(x)$, we obtain the following corollary of Theorem (6.1).

Corollary **(6.2)**

> $$A = \lim_{\|P\| \to 0} \sum_k [f(w_k) - g(w_k)]\,\Delta x_k = \int_a^b [f(x) - g(x)]\, dx$$

FIGURE 6.5

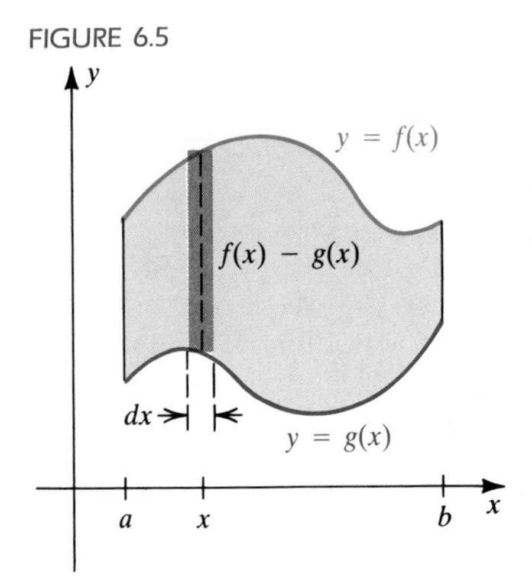

We may employ the following intuitive method for remembering this limit of sums formula (see Figure 6.5):

1. Use dx for the width Δx_k of a typical vertical rectangle.

2. Use $f(x) - g(x)$ for the length $f(w_k) - g(w_k)$ of the rectangle.

3. Regard the symbol $\int_a^b$ as an operator that takes a limit of sums of the rectangular areas $[f(x) - g(x)]\,dx$.

This method allows us to interpret the area formula in Theorem (6.1) as follows:

$$A = \int_a^b [f(x) - g(x)]\,dx$$

limit of sums — length of a rectangle — width of a rectangle

When using this technique, we visualize summing areas of vertical rectangles by moving through the region from left to right. Later in this section we consider different types of regions, finding areas by using *horizontal* rectangles and integrating with respect to y.

Let us call a region an ***R_x* region** (for integration with respect to x) if it lies between the graphs of two equations $y = f(x)$ and $y = g(x)$, with f and g continuous, and $f(x) \geq g(x)$ for every x in $[a, b]$, where a and b are the smallest and largest x-coordinates, respectively, of the points (x, y) in the region. The regions in Figures 6.1 through 6.5 are R_x regions. Several others are sketched in Figure 6.6. Note that the graphs of $y = f(x)$ and $y = g(x)$ may intersect one or more times; however, $f(x) \geq g(x)$ throughout the interval.

FIGURE 6.6 R_x regions

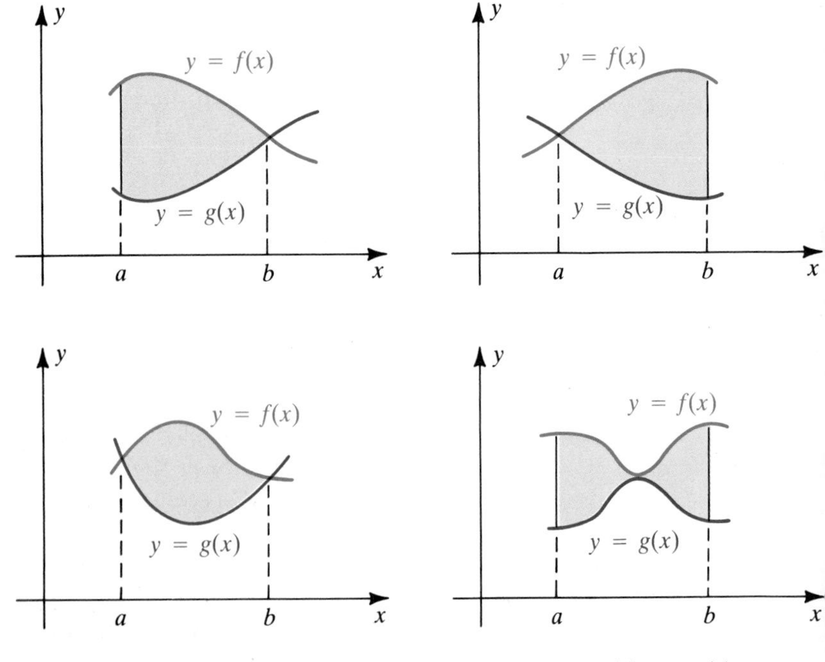

The following guidelines may be helpful when working problems.

Guidelines for finding the area of an R_x region **(6.3)**

1. Sketch the region, labeling the upper boundary $y = f(x)$ and the lower boundary $y = g(x)$. Find the smallest value $x = a$ and the largest value $x = b$ for points (x, y) in the region.
2. Sketch a typical vertical rectangle and label its width dx.
3. Express the area of the rectangle in guideline 2 as $[f(x) - g(x)]\,dx$.
4. Apply the limit of sums operator $\int_a^b$ to the expression in guideline 3 and evaluate the integral.

EXAMPLE 1 Find the area of the region bounded by the graphs of the equations $y = x^2$ and $y = \sqrt{x}$.

FIGURE 6.7

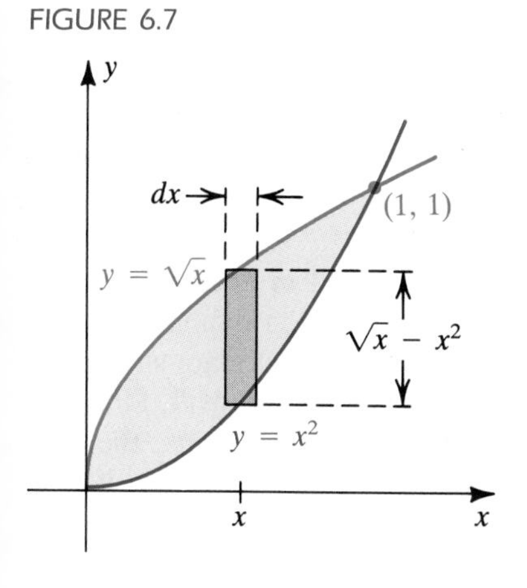

SOLUTION Following guidelines 1–3, we sketch and label the region and show a typical vertical rectangle (see Figure 6.7). The points (0, 0) and (1, 1) at which the graphs intersect can be found by solving the equations $y = x^2$ and $y = \sqrt{x}$ simultaneously. Referring to the figure, we obtain the following facts:

upper boundary: $y = \sqrt{x}$
lower boundary: $y = x^2$
width of rectangle: dx
length of rectangle: $\sqrt{x} - x^2$
area of rectangle: $(\sqrt{x} - x^2)\,dx$

Next, we follow guideline 4 with $a = 0$ and $b = 1$, remembering that applying $\int_0^1$ to the expression $(\sqrt{x} - x^2)\,dx$ represents taking a limit of sums of areas of vertical rectangles. This gives us

$$\begin{aligned} A &= \int_0^1 (\sqrt{x} - x^2)\,dx = \int_0^1 (x^{1/2} - x^2)\,dx \\ &= \left[\frac{x^{3/2}}{\frac{3}{2}} - \frac{x^3}{3}\right]_0^1 = \frac{2}{3} - \frac{1}{3} = \frac{1}{3}. \end{aligned}$$

EXAMPLE 2 Find the area of the region bounded by the graphs of $y + x^2 = 6$ and $y + 2x - 3 = 0$.

FIGURE 6.8

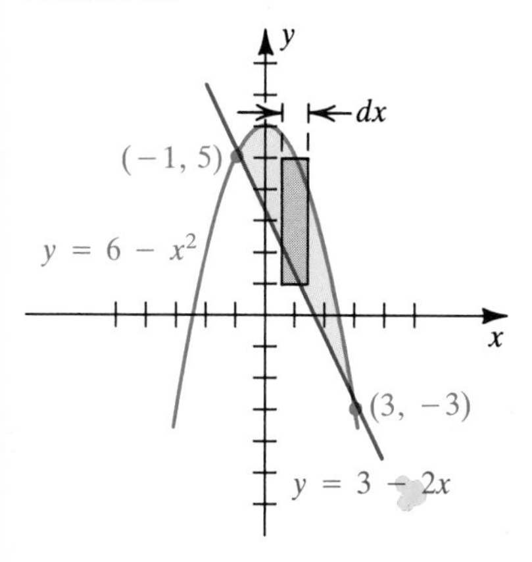

SOLUTION The region and a typical rectangle are sketched in Figure 6.8. The points of intersection $(-1, 5)$ and $(3, -3)$ of the two graphs may be found by solving the two given equations simultaneously. To apply guideline 1 we must label the upper and lower boundaries $y = f(x)$ and $y = g(x)$, respectively, and hence we solve each of the given equations for y in terms of x, as shown in Figure 6.8. This gives us the following:

upper boundary: $y = 6 - x^2$
lower boundary: $y = 3 - 2x$
width of rectangle: dx
length of rectangle: $(6 - x^2) - (3 - 2x)$
area of rectangle: $[(6 - x^2) - (3 - 2x)]\,dx$

We next use guideline 4, with $a = -1$ and $b = 3$, regarding $\int_{-1}^{3}$ as an operator that takes a limit of sums of areas of rectangles. Thus,

$$\begin{aligned} A &= \int_{-1}^{3} [(6 - x^2) - (3 - 2x)]\, dx \\ &= \int_{-1}^{3} (3 - x^2 + 2x)\, dx \\ &= \left[3x - \frac{x^3}{3} + x^2 \right]_{-1}^{3} \\ &= [9 - \tfrac{27}{3} + 9] - [-3 - (-\tfrac{1}{3}) + 1] = \tfrac{32}{3}. \end{aligned}$$

The following example illustrates that it is sometimes necessary to subdivide a region into several R_x regions and then use more than one definite integral to find the area.

EXAMPLE 3 Find the area of the region R bounded by the graphs of $y - x = 6$, $y - x^3 = 0$, and $2y + x = 0$.

FIGURE 6.9

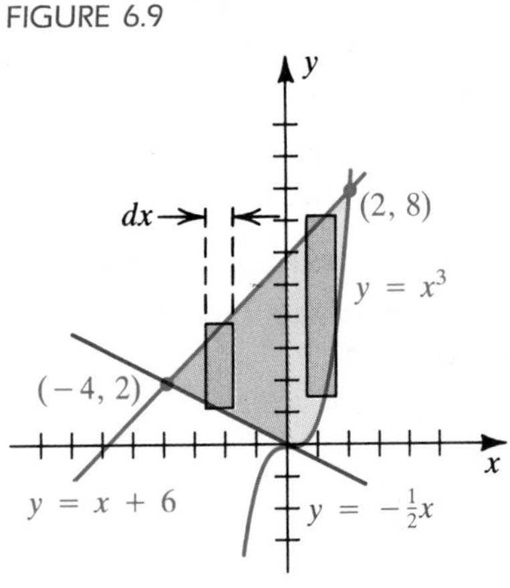

FIGURE 6.10

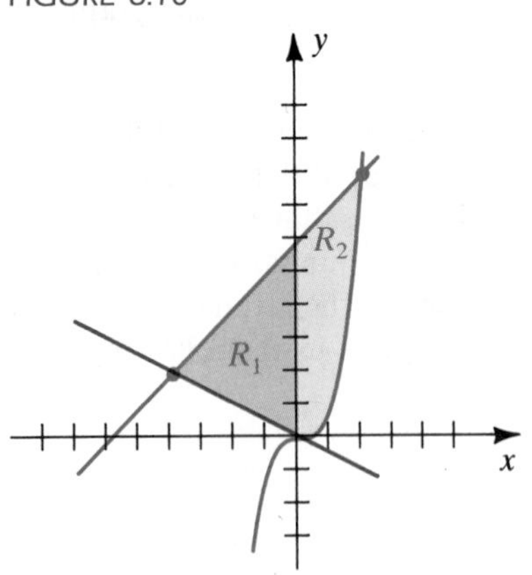

SOLUTION The graphs and region are sketched in Figure 6.9. Each equation has been solved for y in terms of x, and the boundaries have been labeled as in guideline 1. Typical vertical rectangles are shown extending from the lower boundary to the upper boundary of R. Since the lower boundary consists of portions of two different graphs, the area cannot be found by using only one definite integral. However, if R is divided into two R_x regions, R_1 and R_2, as shown in Figure 6.10, then we can determine the area of each and add them together. Let us arrange our work as follows.

	REGION R_1	REGION R_2
upper boundary:	$y = x + 6$	$y = x + 6$
lower boundary:	$y = -\frac{1}{2}x$	$y = x^3$
width of rectangle:	dx	dx
length of rectangle:	$(x + 6) - (-\frac{1}{2}x)$	$(x + 6) - x^3$
area of rectangle:	$[(x + 6) - (-\frac{1}{2}x)]\, dx$	$[(x + 6) - x^3]\, dx$

Applying guideline 4, we find the areas A_1 and A_2 of R_1 and R_2:

$$\begin{aligned} A_1 &= \int_{-4}^{0} [(x + 6) - (-\tfrac{1}{2}x)]\, dx \\ &= \int_{-4}^{0} \left(\frac{3}{2}x + 6 \right) dx = \left[\frac{3}{2}\left(\frac{x^2}{2} \right) + 6x \right]_{-4}^{0} \\ &= 0 - (12 - 24) = 12 \end{aligned}$$

$$\begin{aligned} A_2 &= \int_{0}^{2} [(x + 6) - x^3]\, dx \\ &= \left[\frac{x^2}{2} + 6x - \frac{x^4}{4} \right]_{0}^{2} \\ &= (2 + 12 - 4) - 0 = 10 \end{aligned}$$

The area A of the entire region R is

$$A = A_1 + A_2 = 12 + 10 = 22.$$

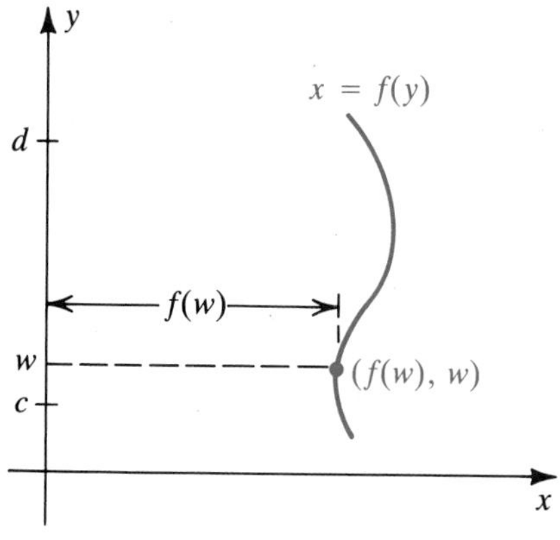

FIGURE 6.11

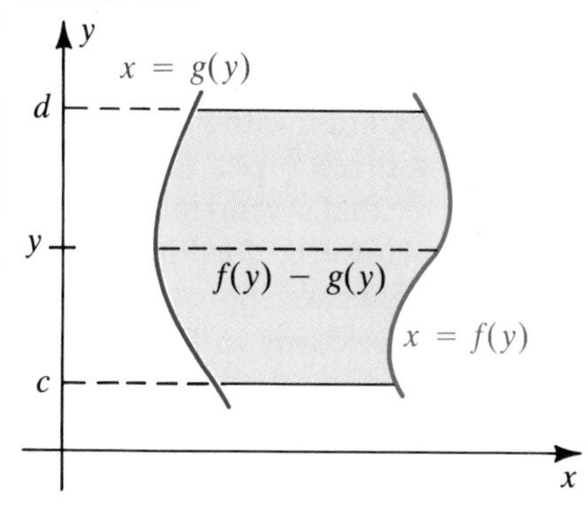

FIGURE 6.12

We have now evaluated many integrals similar to those in Example 3. For this reason we sometimes merely *set up* an integral; that is, we express it in the proper form but do not find its numerical value.

If we consider an equation of the form $x = f(y)$, where f is continuous for $c \le y \le d$, then we *reverse the roles of x and y in the previous discussion, treating y as the independent variable and x as the dependent variable.* A typical graph of $x = f(y)$ is sketched in Figure 6.11. Note that if a value w is assigned to y, then $f(w)$ *is an x-coordinate* of the corresponding point on the graph.

An $\boldsymbol{R_y}$ **region** is a region that lies between the graphs of two equations of the form $x = f(y)$ and $x = g(y)$, with f and g continuous, and $f(y) \ge g(y)$ for every y in $[c, d]$, where c and d are the smallest and largest y-coordinates, respectively, of points in the region. One such region is illustrated in Figure 6.12. We call the graph of f the **right boundary** of the region and the graph of g the **left boundary**. For any y, the number $f(y) - g(y)$ is the horizontal distance between these boundaries, as shown in Figure 6.12.

We can use limits of sums to find the area A of an R_y region. We begin by selecting points on the y-axis with y-coordinates $c = y_0, y_1, \ldots, y_n = d$, obtaining a partition of the interval $[c, d]$ into subintervals of width $\Delta y_k = y_k - y_{k-1}$. For each k we choose a number w_k in $[y_{k-1}, y_k]$ and consider horizontal rectangles that have areas $[f(w_k) - g(w_k)]\,\Delta y_k$, as illustrated in Figure 6.13. This leads to

$$A = \lim_{\|P\| \to 0} \sum_k [f(w_k) - g(w_k)]\,\Delta y_k = \int_c^d [f(y) - g(y)]\,dy.$$

The last equality follows from the definition of the definite integral.

FIGURE 6.13

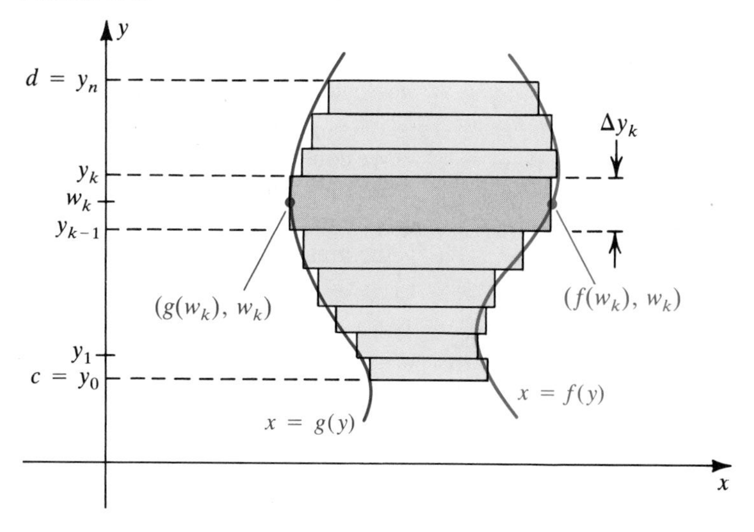

Using notation similar to that for R_x regions, we represent the width Δy_k of a horizontal rectangle by dy and the length $f(w_k) - g(w_k)$ of the rectangle by $f(y) - g(y)$ in the following guidelines.

Guidelines for finding the area of an R_y region **(6.4)**

1. Sketch the region, labeling the right boundary $x = f(y)$ and the left boundary $x = g(y)$. Find the smallest value $y = c$ and the largest value $y = d$ for points (x, y) in the region.
2. Sketch a typical horizontal rectangle and label its width dy.
3. Express the area of the rectangle in guideline 2 as $[f(y) - g(y)]\, dy$.
4. Apply the limit of sums operator $\int_c^d$ to the expression in guideline 3 and evaluate the integral.

In guideline 4, we visualize summing areas of horizontal rectangles by moving from the lowest point of the region to the highest point.

EXAMPLE 4 Find the area of the region bounded by the graphs of the equations $2y^2 = x + 4$ and $y^2 = x$.

FIGURE 6.14

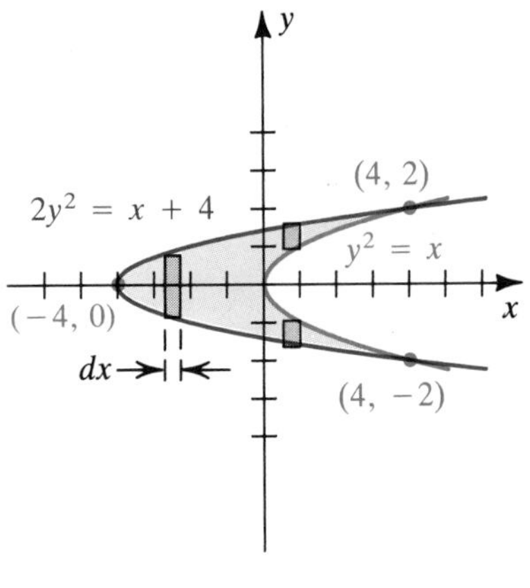

SOLUTION The region is sketched in Figures 6.14 and 6.15. Figure 6.14 illustrates the use of vertical rectangles (integration with respect to x), and Figure 6.15 illustrates the use of horizontal rectangles (integration with respect to y). Referring to Figure 6.14, we see that several integrations with respect to x are required to find the area. However, for Figure 6.15, we need only one integration with respect to y. Thus we apply Guidelines (6.4), solving each equation for x in terms of y. Referring to Figure 6.15, we obtain the following:

$$\begin{aligned} \text{right boundary:}&\quad x = y^2 \\ \text{left boundary:}&\quad x = 2y^2 - 4 \\ \text{width of rectangle:}&\quad dy \\ \text{length of rectangle:}&\quad y^2 - (2y^2 - 4) \\ \text{area of rectangle:}&\quad [y^2 - (2y^2 - 4)]\, dy \end{aligned}$$

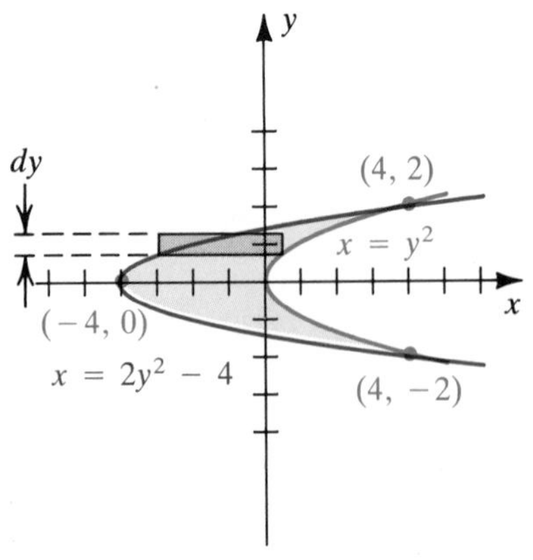

FIGURE 6.15

We could now use guideline 4 with $c = -2$ and $d = 2$, finding A by applying the operator $\int_{-2}^{2}$ to $[y^2 - (2y^2 - 4)]\, dy$. Another method is to use the symmetry of the region with respect to the x-axis and find A by doubling the area of the part that lies above the x-axis. Thus,

$$\begin{aligned} A &= \int_{-2}^{2} [y^2 - (2y^2 - 4)]\, dy \\ &= 2\int_0^2 (4 - y^2)\, dy \\ &= 2\left[4y - \frac{y^3}{3}\right]_0^2 \\ &= 2(8 - \tfrac{8}{3}) = \tfrac{32}{3} \end{aligned}$$

FIGURE 6.16

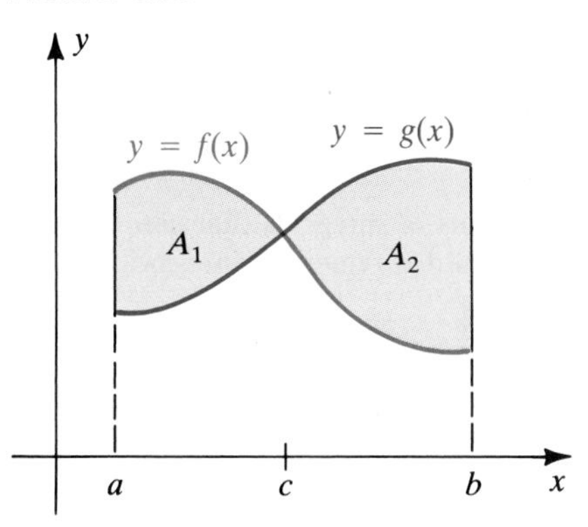

Throughout this section we have assumed that the graphs of the functions (or equations) do not cross one another in the interval under discussion. If the graphs of f and g cross at one point $P(c, d)$, with $a < c < b$, and we wish to find the area bounded by the graphs from $x = a$ to $x = b$, then the methods developed in this section may still be used; however, *two* integrations are required, one corresponding to the interval $[a, c]$ and the other to $[c, b]$. This is illustrated in Figure 6.16, with $f(x) \geq g(x)$ on $[a, c]$ and $g(x) \geq f(x)$ on $[c, b]$. The area A is given by

$$A = A_1 + A_2 = \int_a^c [f(x) - g(x)]\, dx + \int_c^b [g(x) - f(x)]\, dx.$$

If the graphs cross *several* times, then several integrals may be necessary. Problems in which graphs cross one or more times appear in Exercises 31–36.

In scientific investigations, a physical quantity is often interpreted as an area. One illustration of this occurs in the *theory of elasticity*. To test the strength of a material, an investigator records values of strain that correspond to different loads (stresses). The sketch in Figure 6.17 is a typical stress-strain diagram for a sample of an elastic material, such as vulcanized rubber. (Note that stress values are assigned in the vertical direction.) Referring to the figure, we see that as the load applied to the material (the stress) increases, the strain (indicated by the arrows on the red graph) increases until the material is stretched to six times its original length. As the load decreases, the elastic material returns to its original length; however, the same graph is not retraced. Instead, the graph shown in blue is obtained. This phenomenon is called *elastic hysteresis*. (A similar occurrence takes place in the study of magnetic materials, where it is called *magnetic hysteresis*.) The two curves in the figure make up a *hysteresis loop* for the material. The area of the region enclosed by this loop is numerically equal to the energy dissipated within the elastic (or magnetic) material during the test. In the case of vulcanized rubber, the larger the area, the better the material is for absorbing vibrations.

FIGURE 6.17
Stress-strain diagram for an elastic material

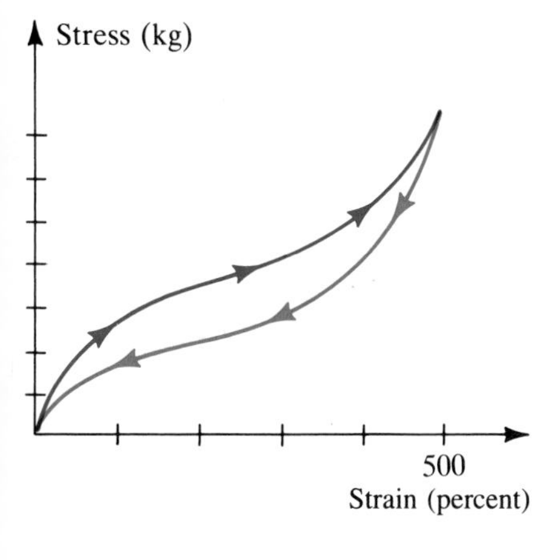

EXERCISES 6.1

Exer. 1–4: Set up an integral that can be used to find the area of the shaded region.

1

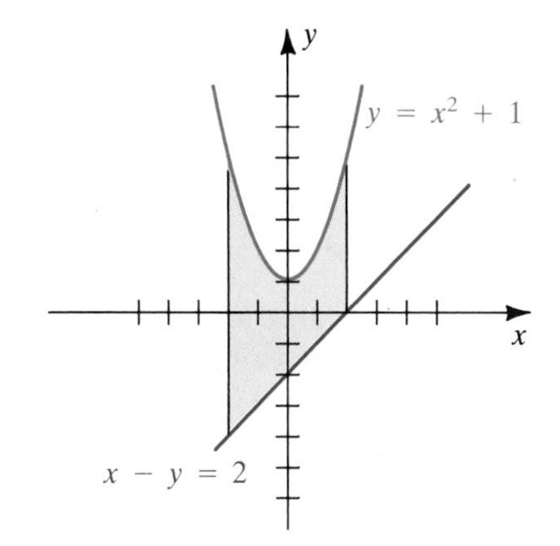

2

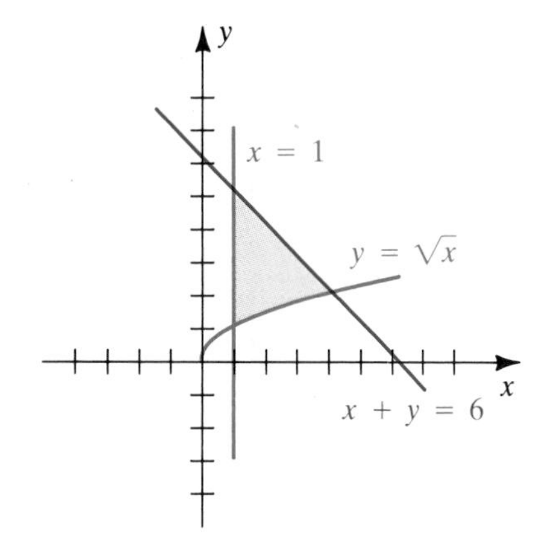

3

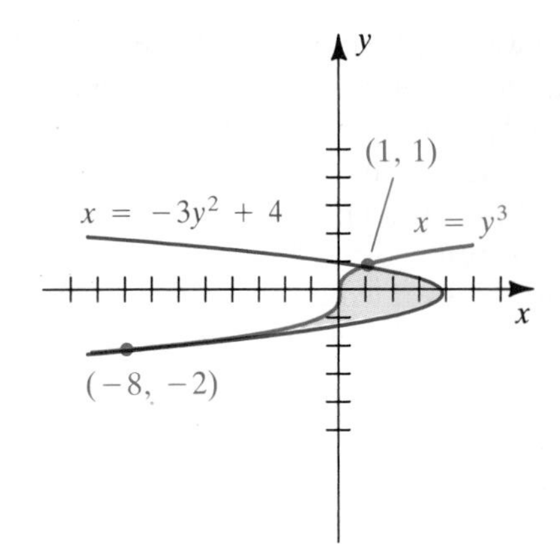

4

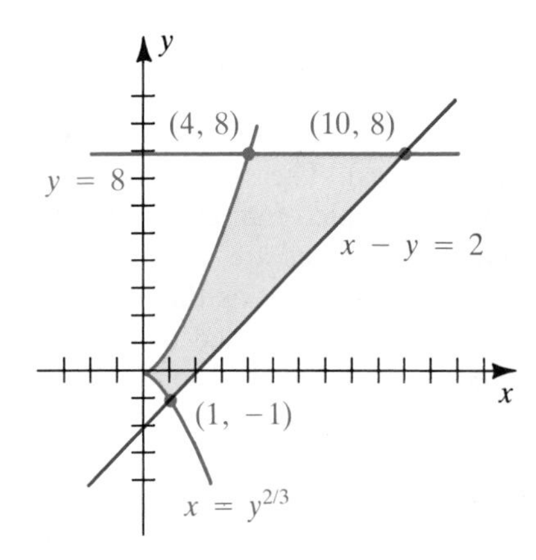

Exer. 5–22: Sketch the region bounded by the graphs of the equations and find its area.

5 $y = x^2$; $y = 4x$

6 $x + y = 3$; $y + x^2 = 3$

7 $y = x^2 + 1$; $y = 5$

8 $y = 4 - x^2$; $y = -4$

9 $y = 1/x^2$; $y = -x^2$; $x = 1$; $x = 2$

10 $y = x^3$; $y = x^2$

11 $y^2 = -x$; $x - y = 4$; $y = -1$; $y = 2$

12 $x = y^2$; $y - x = 2$; $y = -2$; $y = 3$

13 $y^2 = 4 + x$; $y^2 + x = 2$

14 $x = y^2$; $x - y = 2$

15 $x = 4y - y^3$; $x = 0$

16 $x = y^{2/3}$; $x = y^2$

17 $y = x^3 - x$; $y = 0$

18 $y = x^3 - x^2 - 6x$; $y = 0$

19 $x = y^3 + 2y^2 - 3y$; $x = 0$

20 $x = 9y - y^3$; $x = 0$

21 $y = x\sqrt{4 - x^2}$; $y = 0$

22 $y = x\sqrt{x^2 - 9}$; $y = 0$; $x = 5$

Exer. 23–24: Find the area of the region between the graphs of the two equations from $x = 0$ to $x = \pi$.

23 $y = \sin 4x$; $y = 1 + \cos \frac{1}{3}x$

24 $y = 4 + \cos 2x$; $y = 3 \sin \frac{1}{2}x$

Exer. 25–26: Set up sums of integrals that can be used to find the area of the shaded region by integrating with respect to (a) x and (b) y.

25

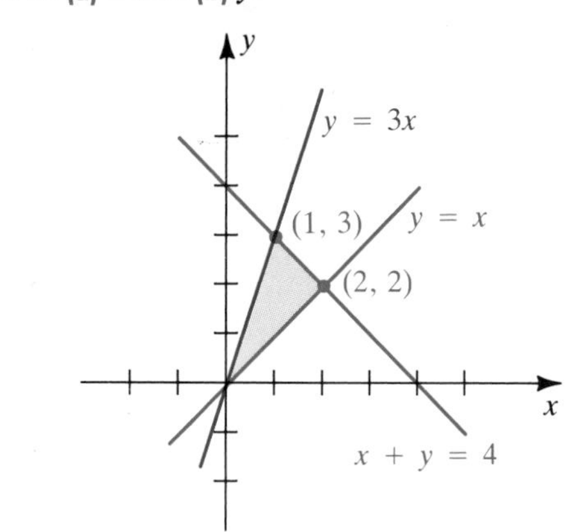

26

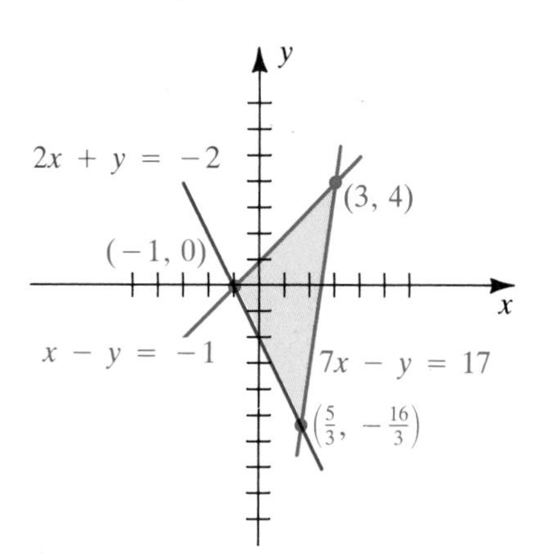

Exer. 27–30: Set up sums of integrals that can be used to find the area of the region bounded by the graphs of the equations by integrating with respect to (a) x and (b) y.

27 $y = \sqrt{x}$; $y = -x$; $x = 1$; $x = 4$

28 $y = 1 - x^2$; $y = x - 1$

29 $y = x + 3$; $x = -y^2 + 3$

30 $x = y^2$; $x = 2y^2 - 4$

Exer. 31–36: Find the area of the region between the graphs of f and g if x is restricted to the given interval.

31 $f(x) = 6 - 3x^2$; $g(x) = 3x$; $[0, 2]$

32 $f(x) = x^2 - 4$; $g(x) = x + 2$; $[1, 4]$

33 $f(x) = x^3 - 4x + 2$; $g(x) = 2$; $[-1, 3]$

34 $f(x) = x^2$; $g(x) = x^3$; $[-1, 2]$

35 $f(x) = \sin x;\quad g(x) = \cos x;\quad [0, 2\pi]$

36 $f(x) = \sin x;\quad g(x) = \frac{1}{2};\quad [0, \pi/2]$

Exer. 37–38: Let R be the region bounded by the graph of f and the x-axis, from $x = a$ to $x = b$. Set up a sum of integrals, not containing the absolute value symbol, that can be used to find the area of R.

37 $f(x) = |x^2 - 6x + 5|;\quad a = 0,\quad b = 7$

38 $f(x) = |-x^2 + 2x + 3|;\quad a = -3,\quad b = 4$

39 The shape of a particular stress-strain diagram is shown in the figure (see the last paragraph of this section).

EXERCISE 39

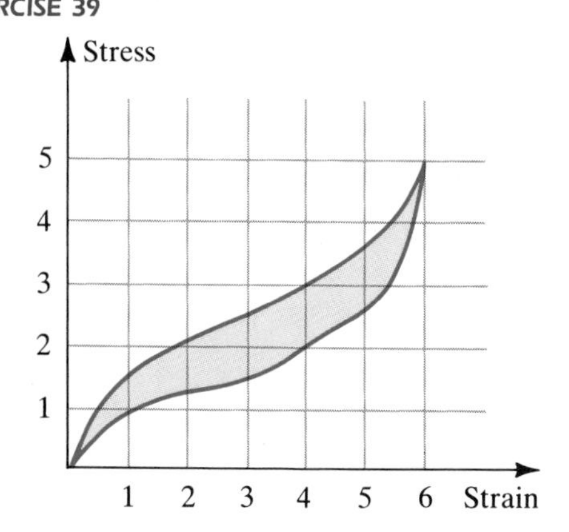

Estimate y-coordinates and approximate the area of the region enclosed by the hysteresis loop using, with $n = 6$,

(a) the trapezoidal rule

(b) Simpson's rule

40 Suppose the function values of f and g in the table below were obtained empirically. Assuming that f and g are continuous, approximate the area between their graphs from $x = 1$ to $x = 5$ using, with $n = 8$,

(a) the trapezoidal rule

(b) Simpson's rule

x	1	1.5	2	2.5	3	3.5	4	4.5	5
$f(x)$	3.5	2.5	3	4	3.5	2.5	2	2	3
$g(x)$	1.5	2	2	1.5	1	0.5	1	1.5	1

c 41 Graph $f(x) = |x^3 - 0.7x^2 - 0.8x + 1.3|$ on $[-1.5, 1.5]$. Set up a sum of integrals not containing the absolute value symbol that can be used to approximate the area of the region bounded by the graph of f, the x-axis, and the lines $x = -1.5$ and $x = 1.5$.

c 42 Graph, on the same coordinate axes, $f(x) = \sin x$ and $g(x) = x^3 - x + 0.2$ for $-2 \le x \le 2$. Set up a sum of integrals that can be used to approximate the area of the region bounded by the graphs.

6.2 SOLIDS OF REVOLUTION

The volume of an object plays an important role in many problems in the physical sciences, such as finding centers of mass and moments of inertia. (These concepts will be discussed later in the text.) Since it is difficult to determine the volume of an irregularly shaped object, we shall begin with objects that have simple shapes. Included in this category are the solids of revolution discussed here and in the next section.

If a region in a plane is revolved about a line in the plane, the resulting solid is a **solid of revolution**, and we say that the solid is **generated** by the region. The line is an **axis of revolution**. In particular, if the R_x region shown in Figure 6.18(i), on the following page, is revolved about the x-axis, we obtain the solid illustrated in (ii) of the figure. As a special case, if f is a constant function, say $f(x) = k$, then the region is rectangular and the solid generated is a right circular cylinder. If the graph of f is a semicircle with endpoints of a diameter at the points $(a, 0)$ and $(b, 0)$, then the solid of revolution is a sphere. If the region is a right triangle with base on the x-axis and two vertices at the points $(a, 0)$ and $(b, 0)$ with the right angle at one of these points, then a right circular cone is generated.

If a plane perpendicular to the x-axis intersects the solid shown in Figure 6.18(ii), a circular cross section is obtained. If, as indicated in the figure, the plane passes through the point on the axis with x-coordinate w, then the radius of the circle is $f(w)$, and hence its area is $\pi[f(w)]^2$. We shall arrive at a definition for the volume of such a solid of revolution by using Riemann sums.

FIGURE 6.18

(i)

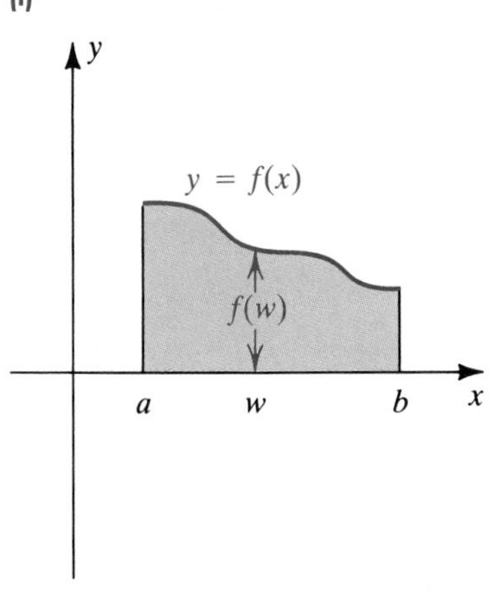

(ii)

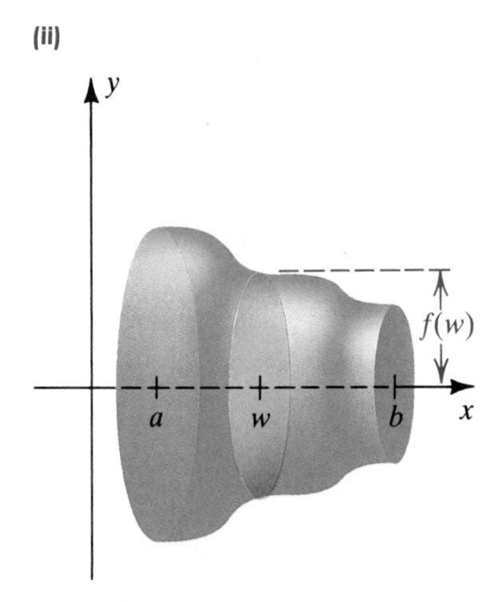

Let us partition the interval $[a, b]$, as we did for areas in the preceding section, and consider the rectangles in Figure 6.19(i). The solid of revolution generated by these rectangles has the shape indicated in (ii) of the figure. Beginning with Figure 6.23, we shall remove, or cut out, parts of solids of revolution to help us visualize portions generated by typical rectangles. When referring to such figures, remember that the entire solid is obtained by one *complete* revolution about an axis, not a partial one.

Observe that the kth rectangle generates a **circular disk** (a flat right circular cylinder) of base radius $f(w_k)$ and altitude (thickness) $\Delta x_k = x_k - x_{k-1}$. The volume of this disk is the area of the base times the altitude—that is, $\pi[f(w_k)]^2\,\Delta x_k$. The volume of the solid shown in Figure 6.19(ii) is the sum of the volumes of all such disks:

$$\sum_k \pi[f(w_k)]^2\,\Delta x_k$$

This sum may be regarded as a Riemann sum for $\pi[f(x)]^2$. If the norm $\|P\|$ of the partition is close to zero, then the sum should be close to the volume of the solid. Hence we define the volume of the solid of revolution as a limit of these sums.

FIGURE 6.19

(i)

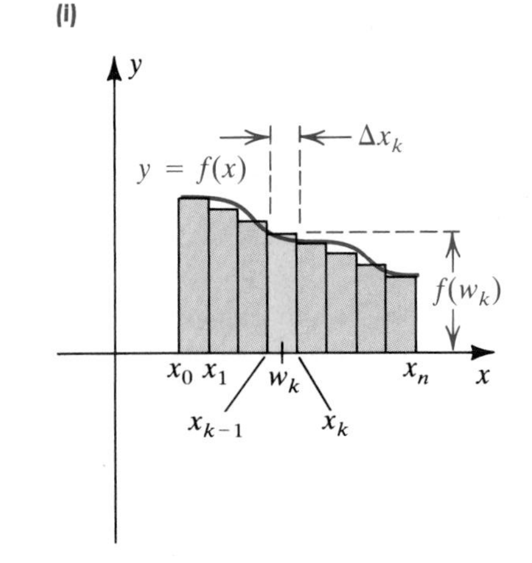

(ii)

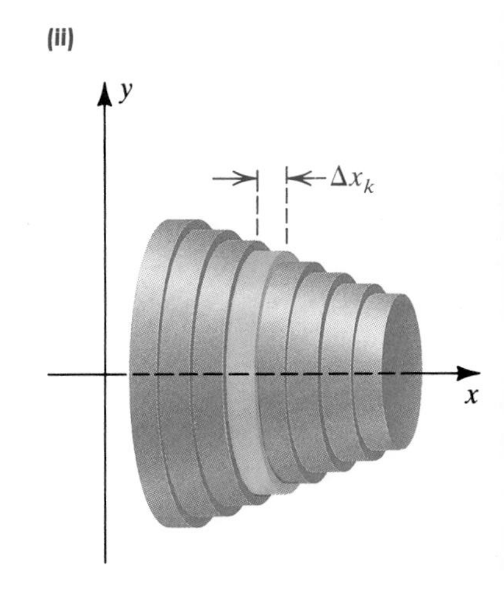

Definition **(6.5)**

> Let f be continuous on $[a, b]$, and let R be the region bounded by the graph of f, the x-axis, and the vertical lines $x = a$ and $x = b$. The **volume** V of the solid of revolution generated by revolving R about the x-axis is
>
> $$V = \lim_{\|P\|\to 0} \sum_k \pi[f(w_k)]^2\,\Delta x_k = \int_a^b \pi[f(x)]^2\,dx.$$

The fact that the limit of sums in this definition equals $\int_a^b \pi[f(x)]^2\,dx$ follows from the definition of the definite integral. We shall not ordinarily specify the units of measure for volume. If the linear measurement is inches, the volume is in cubic inches (in.^{3}). If x is measured in centimeters, then V is in cubic centimeters (cm^3), and so on.

FIGURE 6.20

(i)

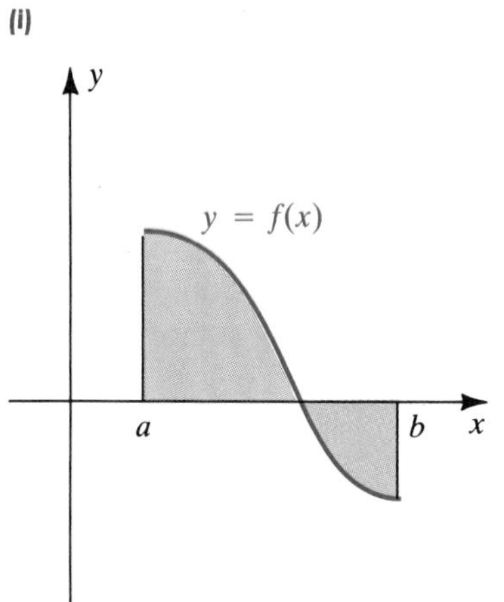

(ii)

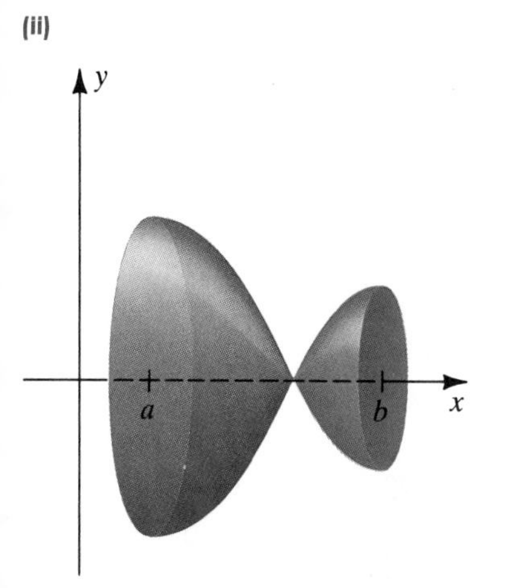

The requirement that $f(x) \geq 0$ was omitted intentionally in Definition (6.5). If f is negative for some x, as in Figure 6.20(i), and if the region bounded by the graphs of f, $x = a$, $x = b$, and the x-axis is revolved about the x-axis, we obtain the solid shown in (ii) of the figure. This solid is the same as that generated by revolving the region under the graph of $y = |f(x)|$ from a to b about the x-axis. Since $|f(x)|^2 = [f(x)]^2$, the limit in Definition (6.5) gives us the volume.

Let us interchange the roles of x and y and revolve the R_y region in Figure 6.21(i) about the y-axis, obtaining the solid illustrated in (ii) of the figure. If we partition the y-interval $[c, d]$ and use *horizontal* rectangles of width Δy_k and length $g(w_k)$, the same type of reasoning that gave us (6.5) leads to the following definition.

FIGURE 6.21

(i)

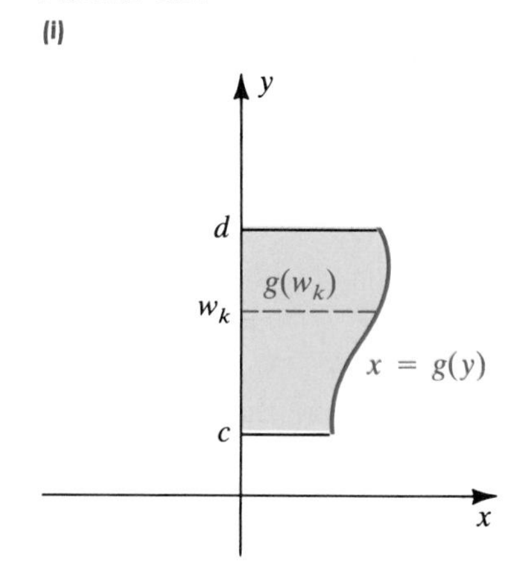

(ii)

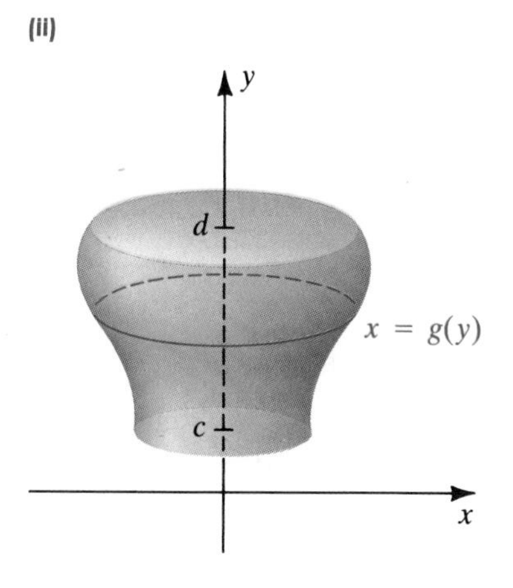

Definition **(6.6)**

$$V = \lim_{\|P\| \to 0} \sum_k \pi[g(w_k)]^2 \, \Delta y_k = \int_c^d \pi[g(y)]^2 \, dy$$

Since we may revolve a region about the x-axis, the y-axis, or some other line, *it is not advisable to merely memorize the formulas in* (6.5) *and* (6.6). It is better to remember the following general rule for finding the volume of a circular disk (see Figure 6.22).

FIGURE 6.22

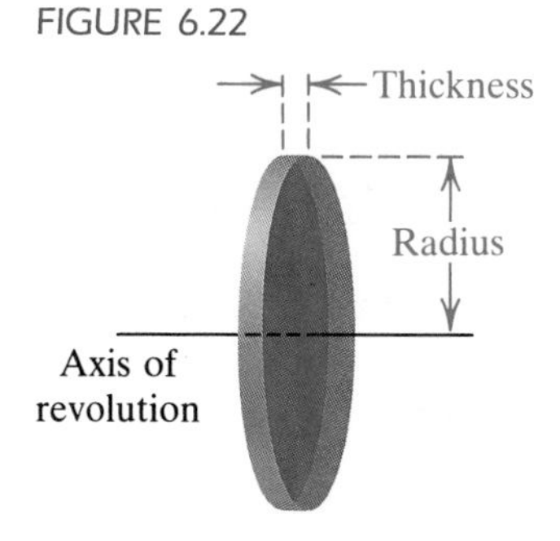

Volume V of a circular disk **(6.7)**

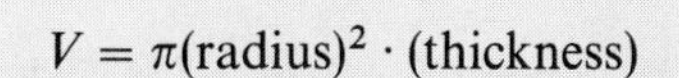

$$V = \pi(\text{radius})^2 \cdot (\text{thickness})$$

When working problems we shall use the intuitive method developed in Section 6.1, replacing Δx_k or Δy_k by dx or dy, and so on. The following guidelines may be helpful.

Guidelines for finding the volume of a solid of revolution using disks **(6.8)**

1. Sketch the region R to be revolved, and label the boundaries. Show a typical vertical rectangle of width dx or a horizontal rectangle of width dy.
2. Sketch the solid generated by R and the disk generated by the rectangle in guideline 1.
3. Express the radius of the disk in terms of x or y, depending on whether its thickness is dx or dy.
4. Use (6.7) to find a formula for the volume of the disk.
5. Apply the limit of sums operator $\int_a^b$ or $\int_c^d$ to the expression in guideline 4 and evaluate the integral.

FIGURE 6.23

(i)

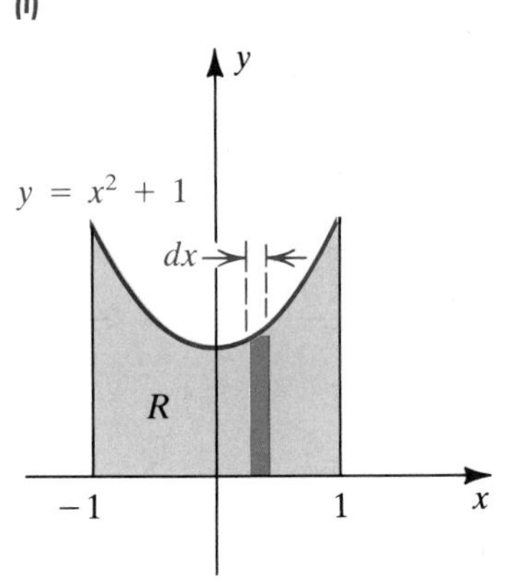

(ii)

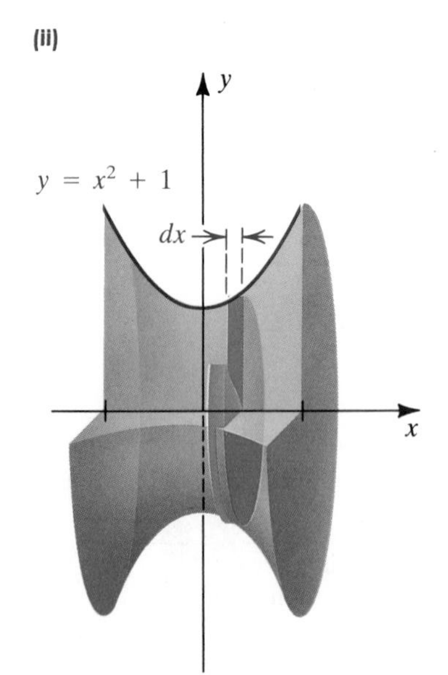

EXAMPLE 1 The region bounded by the x-axis, the graph of the equation $y = x^2 + 1$, and the lines $x = -1$ and $x = 1$ is revolved about the x-axis. Find the volume of the resulting solid.

SOLUTION As specified in guideline 1, we sketch the region and show a vertical rectangle of width dx (see Figure 6.23(i)). Following guideline 2, we sketch the solid generated by R and the disk generated by the rectangle (see Figure 6.23(ii)). As specified in guidelines 3 and 4, we note the following:

$$\text{thickness of disk: } dx$$
$$\text{radius of disk: } x^2 + 1$$
$$\text{volume of disk: } \pi(x^2 + 1)^2\,dx$$

We could next apply guideline 5 with $a = -1$ and $b = 1$, finding the volume V by regarding $\int_{-1}^{1}$ as an operator that takes a limit of sums of volumes of disks. Another method is to use the symmetry of the region with respect to the y-axis and find V by applying $\int_0^1$ to $\pi(x^2 + 1)^2\,dx$ and doubling the result. Thus,

$$\begin{aligned} V &= \int_{-1}^{1} \pi(x^2 + 1)^2\,dx \\ &= 2\int_0^1 \pi(x^4 + 2x^2 + 1)\,dx \\ &= 2\pi\left[\frac{x^5}{5} + 2\left(\frac{x^3}{3}\right) + x\right]_0^1 \\ &= 2\pi\left(\tfrac{1}{5} + \tfrac{2}{3} + 1\right) = \tfrac{56}{15}\pi \approx 11.7. \end{aligned}$$

EXAMPLE 2 The region bounded by the y-axis and the graphs of $y = x^3$, $y = 1$, and $y = 8$ is revolved about the y-axis. Find the volume of the resulting solid.

FIGURE 6.24

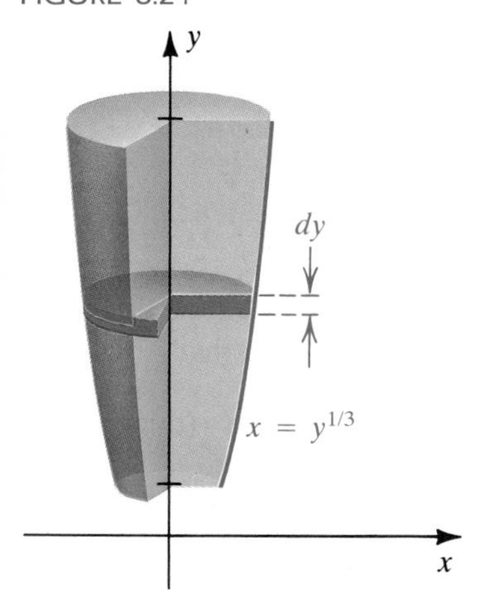

SOLUTION The region and the solid are sketched in Figure 6.24, together with a disk generated by a typical horizontal rectangle. Since we plan to integrate with respect to y, we solve the equation $y = x^3$ for x in terms of y, obtaining $x = y^{1/3}$. We note the following facts (see guidelines 3 and 4):

thickness of disk: dy

radius of disk: $y^{1/3}$

volume of disk: $\pi(y^{1/3})^2\,dy$

Finally, we apply guideline 5, with $c = 1$ and $d = 8$, regarding $\int_1^8$ as an operator that takes a limit of sums of disks:

$$V = \int_1^8 \pi(y^{1/3})^2\,dy = \pi \int_1^8 y^{2/3}\,dy = \pi\left[\frac{y^{5/3}}{\frac{5}{3}}\right]_1^8$$

$$= \tfrac{3}{5}\pi\left[y^{5/3}\right]_1^8 = \tfrac{3}{5}\pi[32 - 1] = \tfrac{93}{5}\pi \approx 58.4$$

Let us next consider an R_x region of the type illustrated in Figure 6.25(i). If this region is revolved about the x-axis, we obtain the solid illustrated in (ii) of the figure. Note that if $g(x) > 0$ for every x in $[a, b]$, there is a hole through the solid.

The volume V of the solid may be found by subtracting the volume of the solid generated by the smaller region from the volume of the solid generated by the larger region. Using Definition (6.5) gives us

$$V = \int_a^b \pi[f(x)]^2\,dx - \int_a^b \pi[g(x)]^2\,dx = \int_a^b \pi\{[f(x)]^2 - [g(x)]^2\}\,dx.$$

The last integral has an interesting interpretation as a limit of sums. As illustrated in Figure 6.25(iii), a vertical rectangle extending from the graph

FIGURE 6.25

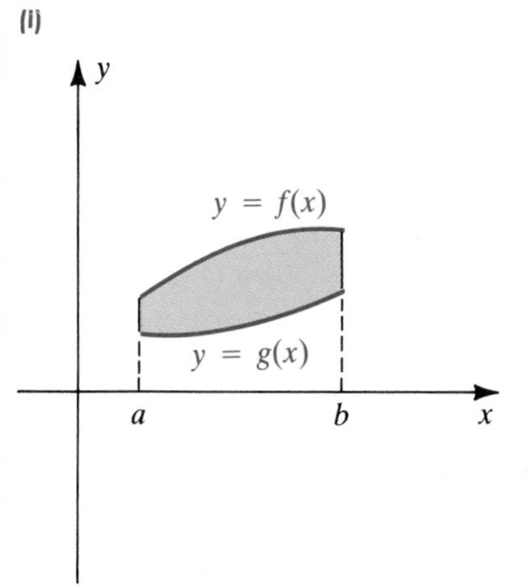

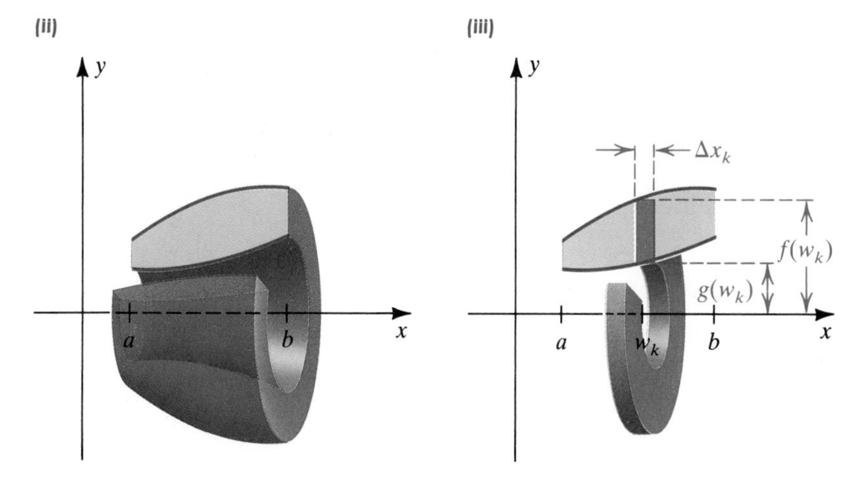

of g to the graph of f, through the points with x-coordinate w_k, generates a washer-shaped solid whose volume is

$$\pi[f(w_k)]^2\,\Delta x_k - \pi[g(w_k)]^2\,\Delta x_k = \pi\{[f(w_k)]^2 - [g(w_k)]^2\}\,\Delta x_k.$$

Summing the volumes of all such washers and taking the limit gives us the desired definite integral. When working problems of this type it is convenient to use the following general rule.

Volume V of a washer **(6.9)**

$$V = \pi[(\text{outer radius})^2 - (\text{inner radius})^2]\cdot(\text{thickness})$$

In applying (6.9), a common error is to use the square of the difference of the radii instead of the difference of the squares. Note that

$$\text{volume of a washer} \neq \pi[(\text{outer radius}) - (\text{inner radius})]^2\cdot(\text{thickness}).$$

Guidelines similar to (6.8) can be stated for problems involving washers. The principal differences are that in guideline 3 we find expressions for the outer radius and inner radius of a typical washer, and in guideline 4 we use (6.9) to find a formula for the volume of the washer.

FIGURE 6.26

(i)

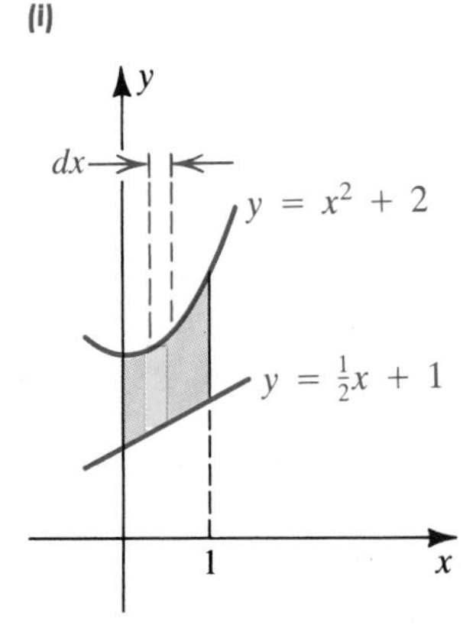

(ii)

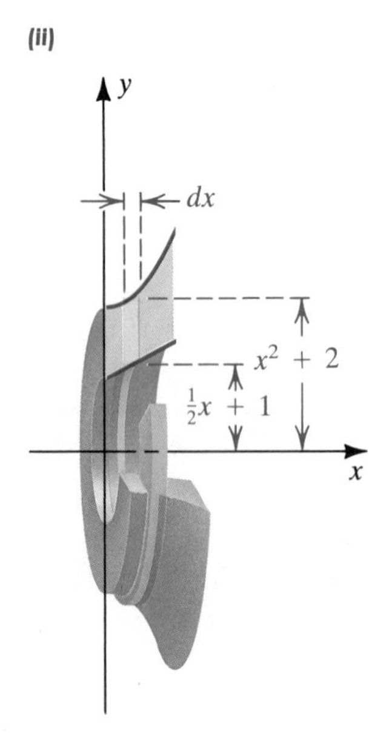

EXAMPLE 3 The region bounded by the graphs of the equations $x^2 = y - 2$ and $2y - x - 2 = 0$ and by the vertical lines $x = 0$ and $x = 1$ is revolved about the x-axis. Find the volume of the resulting solid.

SOLUTION The region and a typical vertical rectangle are sketched in Figure 6.26(i). Since we wish to integrate with respect to x, we solve the first two equations for y in terms of x, obtaining $y = x^2 + 2$ and $y = \frac{1}{2}x + 1$. The solid and a washer generated by the rectangle are illustrated in (ii) of the figure. Using (6.9), we obtain the following:

thickness of washer: dx

outer radius: $x^2 + 2$

inner radius: $\frac{1}{2}x + 1$

volume: $\pi[(x^2+2)^2 - (\frac{1}{2}x+1)^2]\,dx$

We take a limit of sums of volumes of washers by applying $\int_0^1$:

$$\begin{aligned} V &= \int_0^1 \pi[(x^2+2)^2 - (\tfrac{1}{2}x+1)^2]\,dx \\ &= \pi\int_0^1 (x^4 + \tfrac{15}{4}x^2 - x + 3)\,dx \\ &= \pi\left[\frac{x^5}{5} + \frac{15}{4}\left(\frac{x^3}{3}\right) - \frac{x^2}{2} + 3x\right]_0^1 = \frac{79\pi}{20} \approx 12.4 \end{aligned}$$

EXAMPLE 4 Find the volume of the solid generated by revolving the region described in Example 3 about the line $y = 3$.

SOLUTION The region and a typical vertical rectangle are resketched in Figure 6.27(i), together with the axis of revolution $y = 3$. The solid and a washer generated by the rectangle are illustrated in (ii) of the figure. We note the following:

$$\begin{aligned} \text{thickness of washer:}&\quad dx \\ \text{outer radius:}&\quad 3 - (\tfrac{1}{2}x + 1) = 2 - \tfrac{1}{2}x \\ \text{inner radius:}&\quad 3 - (x^2 + 2) = 1 - x^2 \\ \text{volume:}&\quad \pi[(2 - \tfrac{1}{2}x)^2 - (1 - x^2)^2]\,dx \end{aligned}$$

FIGURE 6.27

(i) (ii)

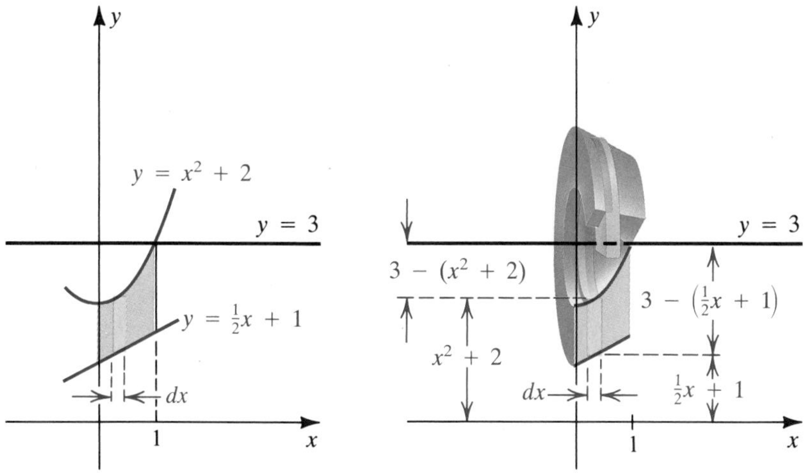

Applying the limit of sums operator $\int_0^1$ gives us the volume:

$$\begin{aligned} V &= \int_0^1 \pi[(2 - \tfrac{1}{2}x)^2 - (1 - x^2)^2]\,dx \\ &= \pi \int_0^1 (3 - 2x + \tfrac{9}{4}x^2 - x^4)\,dx \\ &= \pi\left[3x - x^2 + \frac{9}{4}\left(\frac{x^3}{3}\right) - \frac{x^5}{5}\right]_0^1 = \frac{51\pi}{20} \approx 8.01 \end{aligned}$$

FIGURE 6.28

(i)

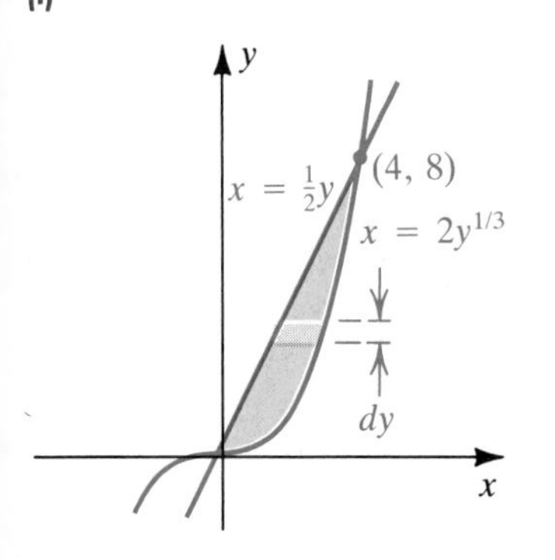

EXAMPLE 5 The region in the first quadrant bounded by the graphs of $y = \frac{1}{8}x^3$ and $y = 2x$ is revolved about the y-axis. Find the volume of the resulting solid.

SOLUTION The region and a typical horizontal rectangle are shown in Figure 6.28(i). We wish to integrate with respect to y, so we solve the given equations for x in terms of y, obtaining

$$x = \tfrac{1}{2}y \quad \text{and} \quad x = 2y^{1/3}.$$

FIGURE 6.28
(ii)

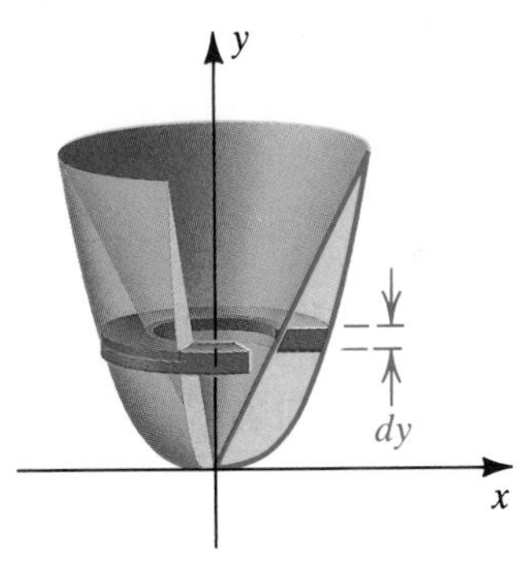

Figure 6.28(ii) illustrates the volume generated by the region and the washer generated by the rectangle. We note the following:

thickness of washer: dy

outer radius: $2y^{1/3}$

inner radius: $\frac{1}{2}y$

volume: $\pi[(2y^{1/3})^2 - (\frac{1}{2}y)^2]\,dy = \pi(4y^{2/3} - \frac{1}{4}y^2)\,dy$

Applying the limit of sums operator $\int_0^8$ gives us the volume:

$$V = \int_0^8 \pi(4y^{2/3} - \tfrac{1}{4}y^2)\,dy$$
$$= \pi\left[\tfrac{12}{5}y^{5/3} - \tfrac{1}{12}y^3\right]_0^8 = \tfrac{512}{15}\pi \approx 107.2$$

EXERCISES 6.2

Exer. 1–4: Set up an integral that can be used to find the volume of the solid obtained by revolving the shaded region about the indicated axis.

1

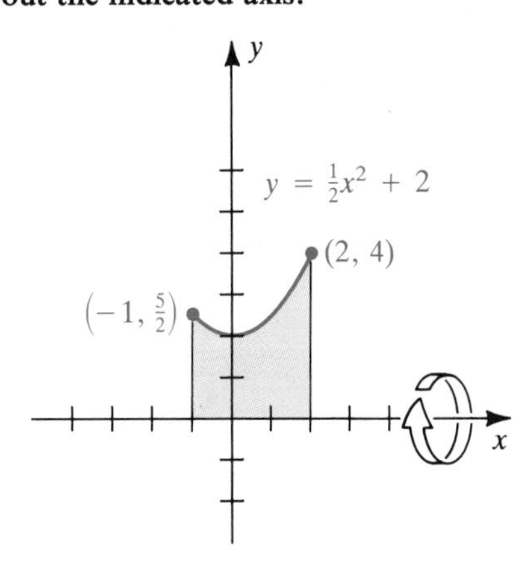

2

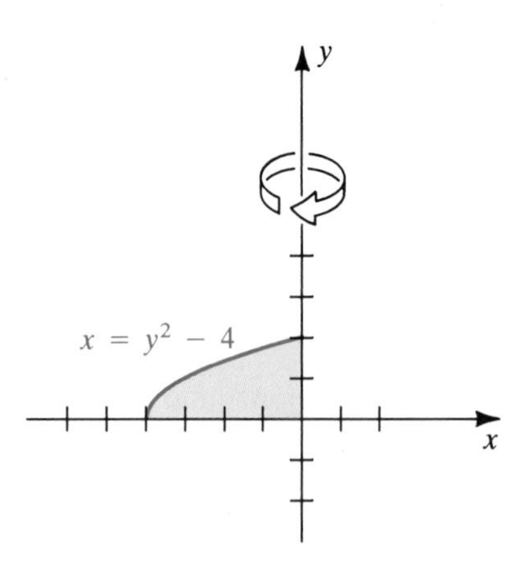

3

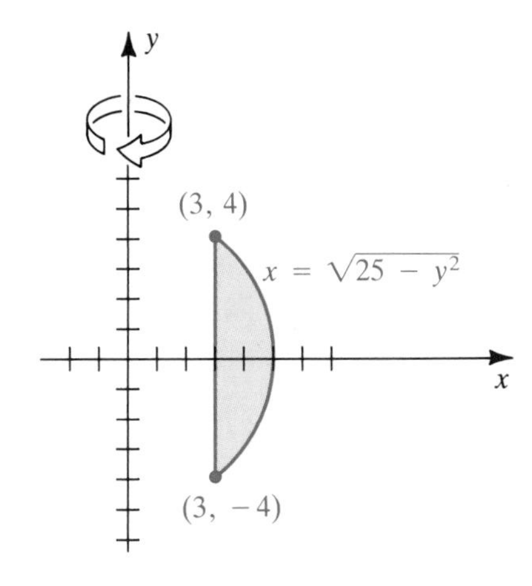

4

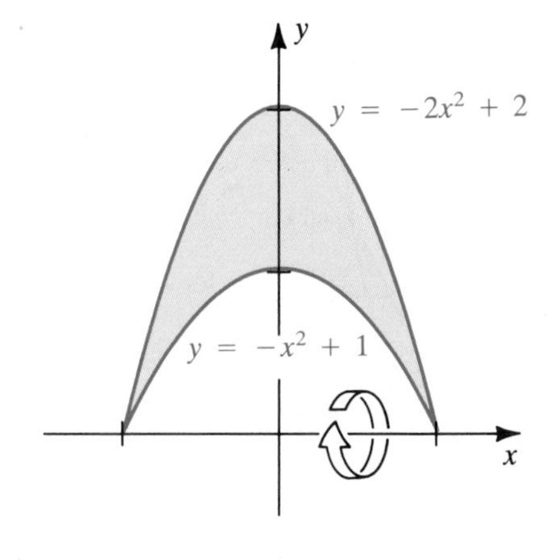

Exer. 5–24: Sketch the region R bounded by the graphs of the equations, and find the volume of the solid generated if R is revolved about the indicated axis.

5 $y = 1/x$, $x = 1$, $x = 3$, $y = 0$; x-axis

6 $y = \sqrt{x}$, $x = 4$, $y = 0$; x-axis

7 $y = x^2 - 4x$, $y = 0$; x-axis

8 $y = x^3$, $x = -2$, $y = 0$; x-axis

9 $y = x^2$, $y = 2$; y-axis

10 $y = 1/x$, $y = 1$, $y = 3$, $x = 0$; y-axis

11 $x = 4y - y^2$, $x = 0$; y-axis

12 $y = x$, $y = 3$, $x = 0$; y-axis

13 $y = x^2$, $y = 4 - x^2$; x-axis

14 $x = y^3$, $x^2 + y = 0$; x-axis

15 $y = x$, $x + y = 4$, $x = 0$; x-axis

16 $y = (x - 1)^2 + 1$, $y = -(x - 1)^2 + 3$; x-axis

17 $y^2 = x$, $2y = x$; y-axis

18 $y = 2x$, $y = 4x^2$; y-axis

19 $x = y^2$, $x - y = 2$; y-axis

20 $x + y = 1$, $x - y = -1$, $x = 2$; y-axis

21 $y = \sin 2x$, $x = 0$, $x = \pi$, $y = 0$; x-axis
(*Hint:* Use a half-angle formula.)

22 $y = 1 + \cos 3x$, $x = 0$, $x = 2\pi$, $y = 0$; x-axis
(*Hint:* Use a half-angle formula.)

23 $y = \sin x$, $y = \cos x$, $x = 0$, $x = \pi/4$; x-axis
(*Hint:* Use a double angle formula.)

24 $y = \sec x$, $y = \sin x$, $x = 0$, $x = \pi/4$; x-axis

Exer. 25–26: Sketch the region R bounded by the graphs of the equations, and find the volume of the solid generated if R is revolved about the given line.

25 $y = x^2$, $y = 4$
(a) $y = 4$ (b) $y = 5$
(c) $x = 2$ (d) $x = 3$

26 $y = \sqrt{x}$, $y = 0$, $x = 4$
(a) $x = 4$ (b) $x = 6$
(c) $y = 2$ (d) $y = 4$

Exer. 27–28: Set up an integral that can be used to find the volume of the solid generated by revolving the shaded region about the line (a) $y = -2$, (b) $y = 5$, (c) $x = 7$, and (d) $x = -4$.

27

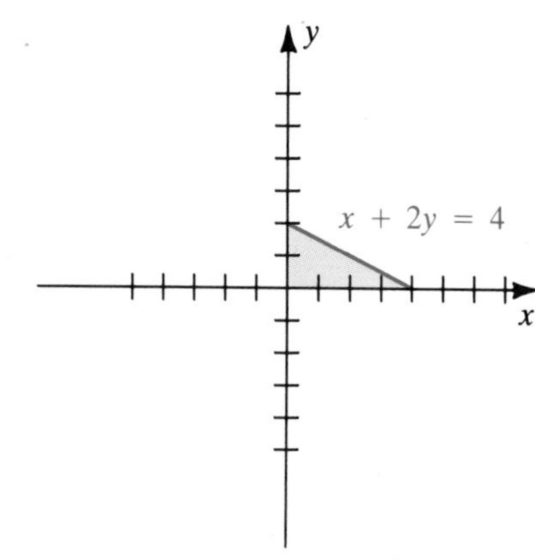

28

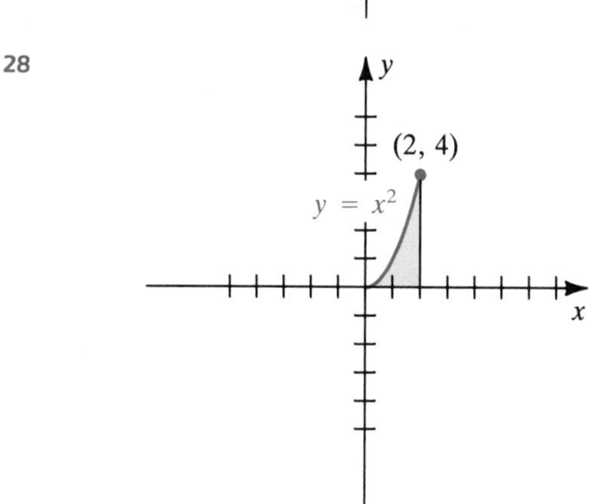

Exer. 29–34: Sketch the region R bounded by the graphs of the equations, and set up integrals that can be used to find the volume of the solid generated if R is revolved about the given line.

29 $y = x^3$, $y = 4x$; $y = 8$

30 $y = x^3$, $y = 4x$; $x = 4$

31 $x + y = 3$, $y + x^2 = 3$; $x = 2$

32 $y = 1 - x^2$, $x - y = 1$; $y = 3$

33 $x^2 + y^2 = 1$; $x = 5$

34 $y = x^{2/3}$, $y = x^2$; $y = -1$

Exer. 35–40: Use a definite integral to derive a formula for the volume of the indicated solid.

35 A right circular cylinder of altitude h and radius r

36 A cylindrical shell of altitude h, outer radius R, and inner radius r

37 A right circular cone of altitude h and base radius r

38 A sphere of radius r

39 A frustum of a right circular cone of altitude h, lower base radius R, and upper base radius r

40 A spherical segment of altitude h in a sphere of radius r

41 If the region shown in the figure is revolved about the x-axis, use the trapezoidal rule with $n = 6$ to approximate the volume of the resulting solid.

EXERCISE 41

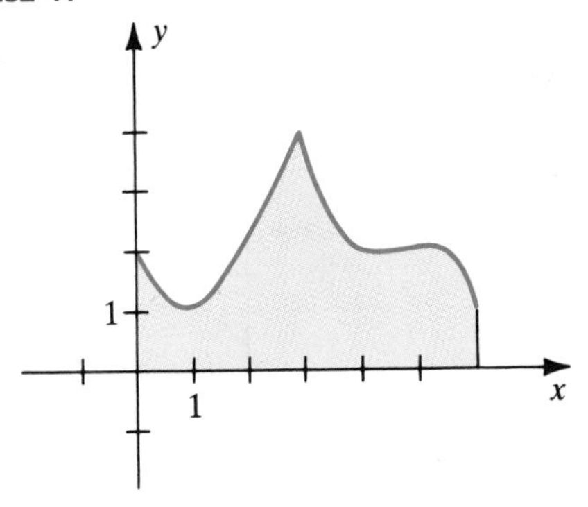

42 If the region shown in the figure is revolved about the x-axis, use Simpson's rule with $n = 8$ to approximate the volume of the resulting solid.

EXERCISE 42

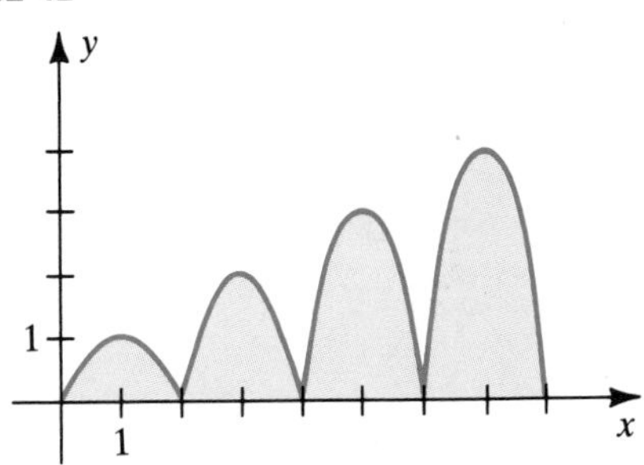

c **Exer. 43–44: Graph f and g on the same coordinate axes for $0 \leq x \leq \pi$. (a) Estimate the x-coordinates of the points of intersection of the graphs. (b) If the region bounded by the graphs of f and g is revolved about the x-axis, use Simpson's rule with $n = 4$ to approximate the volume of the resulting solid.**

43 $f(x) = \dfrac{\sin x}{1 + x}$; $\quad g(x) = 0.3$

44 $f(x) = \sqrt[4]{|\sin x|}$; $\quad g(x) = 0.2x + 0.7$

6.3 VOLUMES BY CYLINDRICAL SHELLS

In the preceding section we found volumes of solids of revolution by using circular disks or washers. For certain types of solids it is convenient to use hollow circular cylinders—that is, thin **cylindrical shells** of the type illustrated in Figure 6.29, where r_1 is the *outer radius*, r_2 is the *inner radius*, h is the *altitude*, and $\Delta r = r_1 - r_2$ is the *thickness* of the shell. The **average radius** of the shell is $r = \frac{1}{2}(r_1 + r_2)$. We can find the volume of the shell by subtracting the volume $\pi r_2^2 h$ of the inner cylinder from the volume $\pi r_1^2 h$ of the outer cylinder. If we do this and change the form of the resulting expression, we obtain

FIGURE 6.29

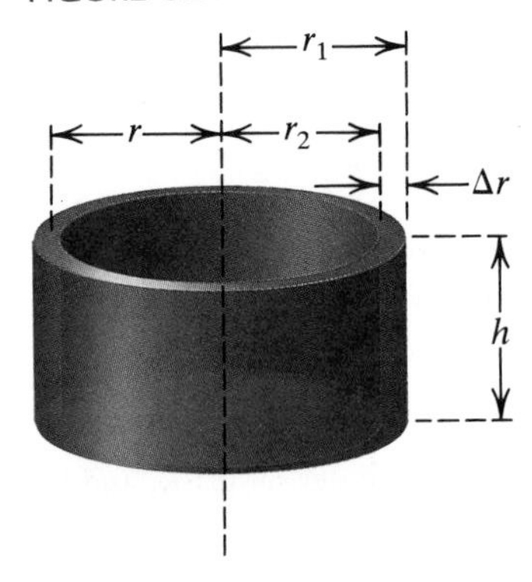

$$\begin{aligned}
\pi r_1^2 h - \pi r_2^2 h &= \pi(r_1^2 - r_2^2)h \\
&= \pi(r_1 + r_2)(r_1 - r_2)h \\
&= 2\pi \cdot \tfrac{1}{2}(r_1 + r_2)h(r_1 - r_2) \\
&= 2\pi r h \,\Delta r,
\end{aligned}$$

which gives us the following general rule.

Volume V of a cylindrical shell **(6.10)**

$$V = 2\pi(\text{average radius})(\text{altitude})(\text{thickness})$$

If the R_x region in Figure 6.30(i) is revolved about the y-axis, we obtain the solid illustrated in (ii) of the figure.

FIGURE 6.30

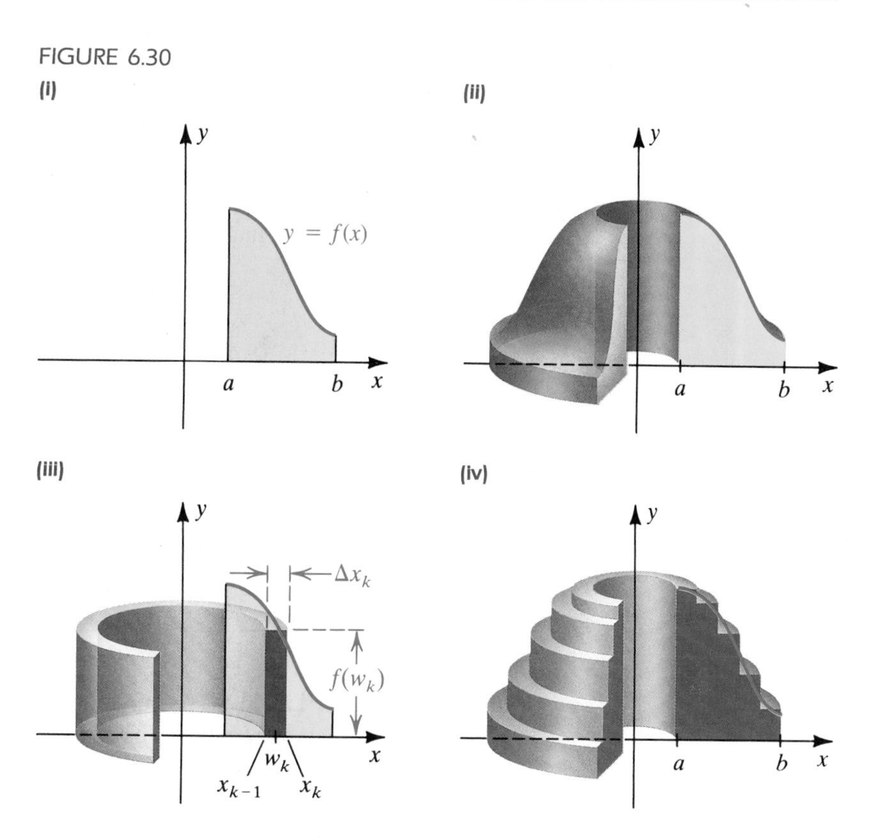

Let P be a partition of $[a, b]$, and consider the typical vertical rectangle in Figure 6.30(iii), where w_k is the midpoint of $[x_{k-1}, x_k]$. If we revolve this rectangle about the y-axis, we obtain a cylindrical shell of average radius w_k, altitude $f(w_k)$, and thickness Δx_k. Hence, by (6.10), the volume of the shell is

$$2\pi w_k f(w_k)\,\Delta x_k.$$

Revolving the rectangular polygon formed by *all* the rectangles determined by P gives us the solid illustrated in Figure 6.30(iv). The volume of this solid is a Riemann sum:

$$\sum_k 2\pi w_k f(w_k)\,\Delta x_k$$

The smaller the norm $\|P\|$ of the partition, the better the sum approximates the volume V of the solid shown in (ii) of the figure. This is the motivation for the following definition.

Definition **(6.11)**

Let f be continuous and suppose $f(x) \geq 0$ on $[a, b]$, where $0 \leq a \leq b$. Let R be the region under the graph of f from a to b. The volume V of the solid of revolution generated by revolving R about the y-axis is

$$V = \lim_{\|P\| \to 0} \sum_k 2\pi w_k f(w_k)\,\Delta x_k = \int_a^b 2\pi x f(x)\,dx.$$

It can be proved that if the methods of Section 6.2 are also applicable, then both methods lead to the same answer.

We may also consider solids obtained by revolving a region about the y-axis or some other line. The following guidelines may be useful.

Guidelines for finding the volume of a solid of revolution using cylindrical shells **(6.12)**

1. Sketch the region R to be revolved, and label the boundaries. Show a typical vertical rectangle of width dx or a horizontal rectangle of width dy.
2. Sketch the cylindrical shell generated by the rectangle in guideline 1.
3. Express the average radius of the shell in terms of x or y, depending on whether its thickness is dx or dy. *Remember that x represents a distance from the y-axis to a vertical rectangle, and y represents a distance from the x-axis to a horizontal rectangle.*
4. Express the altitude of the shell in terms of x or y, depending on whether its thickness is dx or dy.
5. Use (6.10) to find a formula for the volume of the shell.
6. Apply the limit of sums operator $\int_a^b$ or $\int_c^d$ to the expression in guideline 5 and evaluate the integral.

EXAMPLE 1 The region bounded by the graph of $y = 2x - x^2$ and the x-axis is revolved about the y-axis. Find the volume of the resulting solid.

SOLUTION The region to be revolved is sketched in Figure 6.31(i), together with a typical vertical rectangle of width dx. Figure 6.31(ii) shows the cylindrical shell generated by revolving the rectangle about the y-axis. Note that x represents the distance from the y-axis to the midpoint of the rectangle (the average radius of the shell). Referring to the figure and using (6.10) gives us the following:

$$\begin{aligned} \text{thickness of shell:}\quad & dx \\ \text{average radius:}\quad & x \\ \text{altitude:}\quad & 2x - x^2 \\ \text{volume:}\quad & 2\pi x(2x - x^2)\,dx \end{aligned}$$

FIGURE 6.31

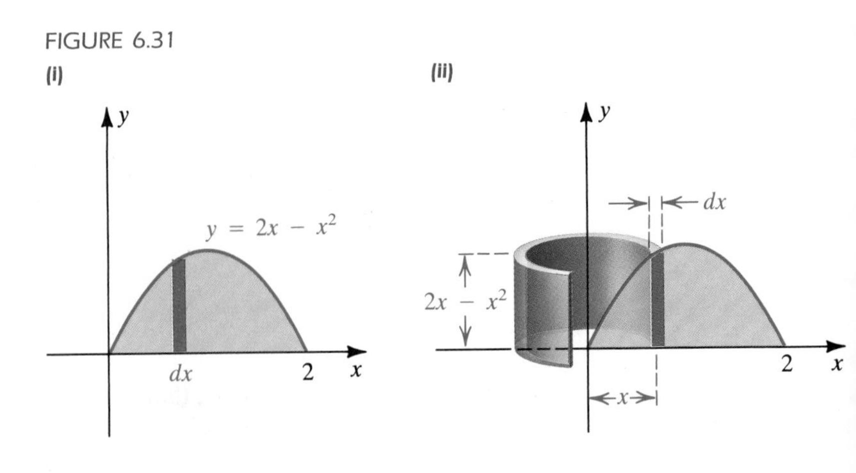

To sum all such shells, we move from left to right through the region from $a = 0$ to $b = 2$ (do *not* sum from -2 to 2). Hence the limit of sums is

$$V = \int_0^2 2\pi x(2x - x^2)\,dx = 2\pi \int_0^2 (2x^2 - x^3)\,dx$$
$$= 2\pi\left[2\left(\frac{x^3}{3}\right) - \frac{x^4}{4}\right]_0^2 = \frac{8\pi}{3} \approx 8.4.$$

The volume V can also be found using washers; however, the calculations would be much more involved, since the equation $y = 2x - x^2$ would have to be solved for x in terms of y.

FIGURE 6.32

(i)

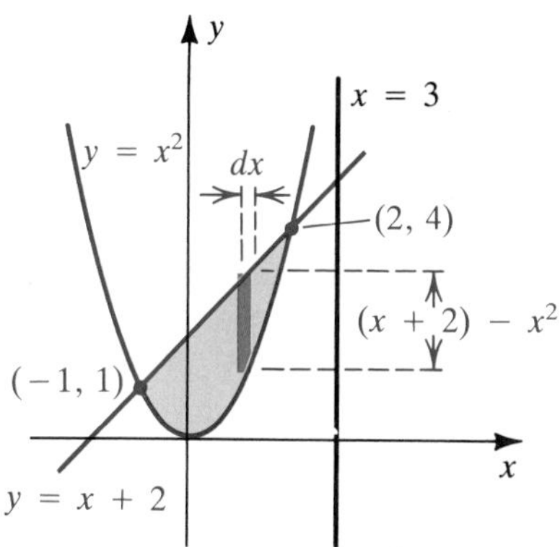

(ii)

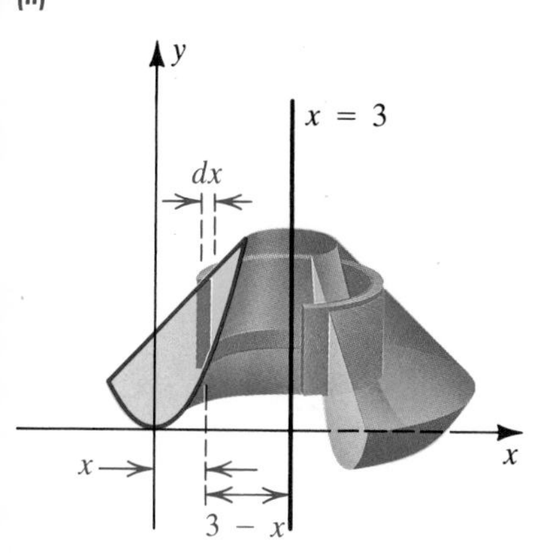

EXAMPLE 2 The region bounded by the graphs of $y = x^2$ and $y = x + 2$ is revolved about the line $x = 3$. Set up the integral for the volume of the resulting solid.

SOLUTION The region is sketched in Figure 6.32(i), together with a typical vertical rectangle extending from the lower boundary $y = x^2$ to the upper boundary $y = x + 2$. Also shown is the axis of revolution $x = 3$. In (ii) of the figure we have illustrated both the cylindrical shell and the solid that are generated by revolving the rectangle and the region about the line $x = 3$. It is important to note that since x is the distance from the y-axis to the rectangle, the radius of the shell is $3 - x$. Referring to Figure 6.32 and using (6.10) gives us the following:

thickness of shell: dx

average radius: $3 - x$

altitude: $(x + 2) - x^2$

volume: $2\pi(3 - x)(x + 2 - x^2)\,dx$

To sum all such shells, we move from left to right through the region from $a = -1$ to $b = 2$. Hence the limit of sums is

$$V = \int_{-1}^{2} 2\pi(3 - x)(x + 2 - x^2)\,dx.$$

EXAMPLE 3 The region in the first quadrant bounded by the graph of the equation $x = 2y^3 - y^4$ and the y-axis is revolved about the x-axis. Set up the integral for the volume of the resulting solid.

SOLUTION The region is sketched in Figure 6.33(i) on the following page, together with a typical horizontal rectangle. Part (ii) of the figure shows the cylindrical shell and the solid that are generated by the revolution about the x-axis. Referring to the figure and using (6.10) gives the following:

thickness of shell: dy

average radius: y

altitude: $2y^3 - y^4$

volume: $2\pi y(2y^3 - y^4)\,dy$

FIGURE 6.33

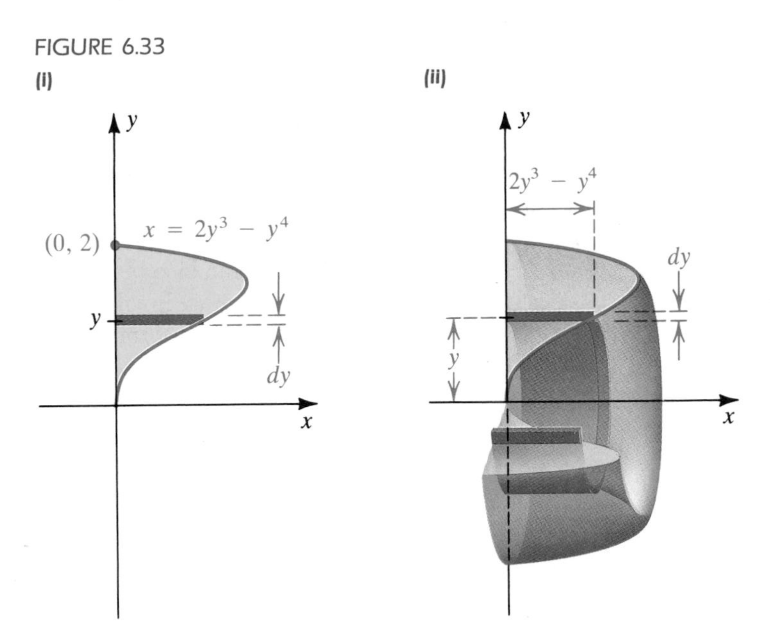

To sum all such shells, we move upward through the region from $c = 0$ to $d = 2$. Hence the limit of sums is

$$V = \int_0^2 2\pi y(2y^3 - y^4)\, dy.$$

It is worth noting that in the preceding example we were forced to use shells and to integrate with respect to y, since the use of washers and integration with respect to x would require that we solve the equation $x = 2y^3 - y^4$ for y in terms of x, a rather formidable task.

EXERCISES 6.3

Use cylindrical shells for each exercise.

Exer. 1–4: Set up an integral that can be used to find the volume of the solid obtained by revolving the shaded region about the indicated axis.

1

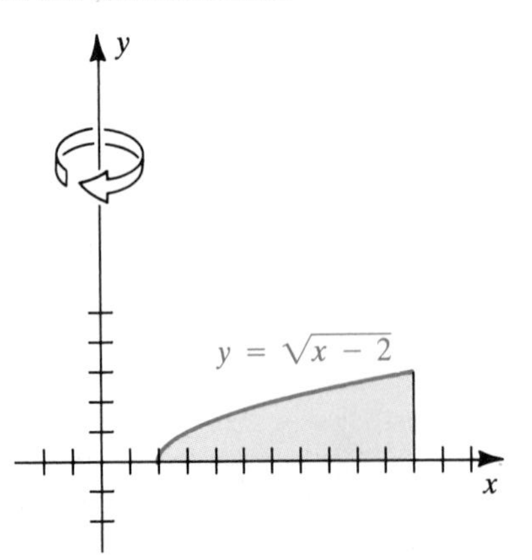

2

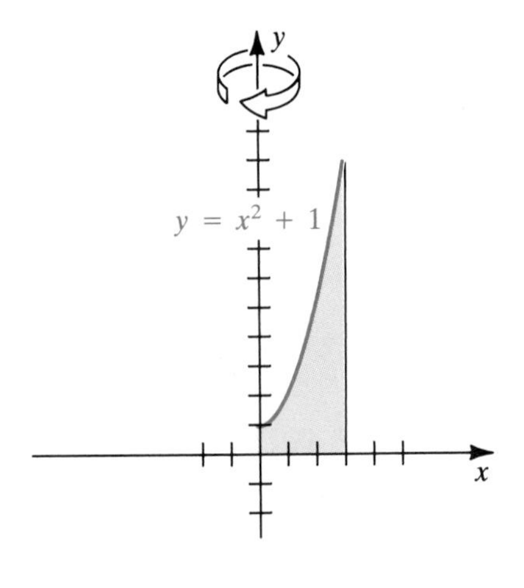

3

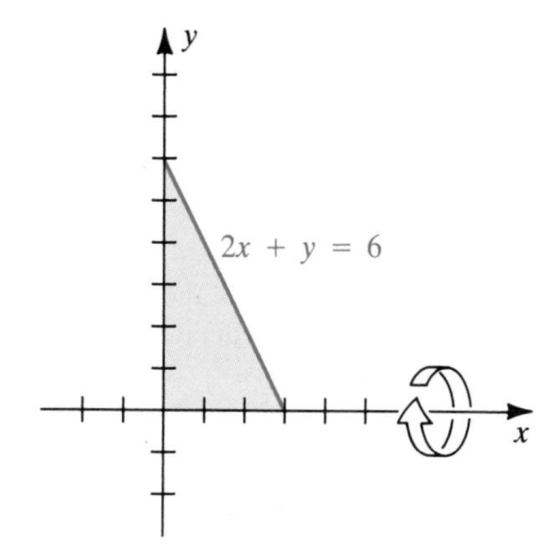

4

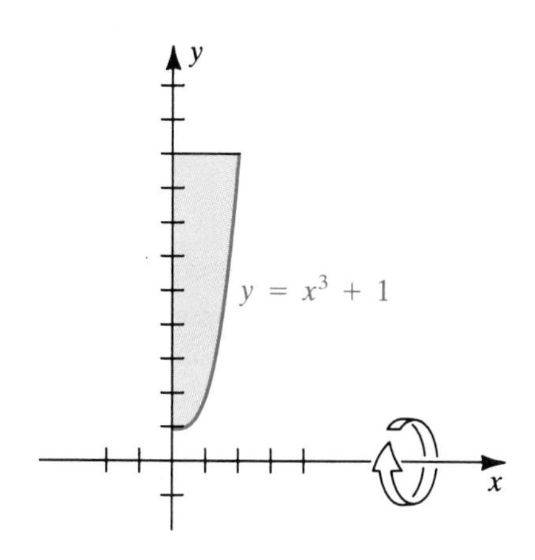

Exer. 5–18: Sketch the region R bounded by the graphs of the equations, and find the volume of the solid generated if R is revolved about the indicated axis.

5 $y = \sqrt{x}$, $x = 4$, $y = 0$; y-axis

6 $y = 1/x$, $x = 1$, $x = 2$, $y = 0$; y-axis

7 $y = x^2$, $y^2 = 8x$; y-axis

8 $16y = x^2$, $y^2 = 2x$; y-axis

9 $2x - y = 12$, $x - 2y = 3$, $x = 4$; y-axis

10 $y = x^3 + 1$, $x + 2y = 2$, $x = 1$; y-axis

11 $2x - y = 4$, $x = 0$, $y = 0$; y-axis

12 $y = x^2 - 5x$, $y = 0$; y-axis

13 $x^2 = 4y$, $y = 4$; x-axis

14 $y^3 = x$, $y = 3$, $x = 0$; x-axis

15 $y = 2x$, $y = 6$, $x = 0$; x-axis

16 $2y = x$, $y = 4$, $x = 1$; x-axis

17 $y = \sqrt{x + 4}$, $y = 0$, $x = 0$; x-axis

18 $y = -x$, $x - y = -4$, $y = 0$; x-axis

Exer. 19–26: Let R be the region bounded by the graphs of the equations. Set up an integral that can be used to find the volume of the solid generated if R is revolved about the given line.

19 $y = x^2 + 1$, $x = 0$, $x = 2$, $y = 0$

(a) $x = 3$ (b) $x = -1$

20 $y = 4 - x^2$, $y = 0$

(a) $x = 2$ (b) $x = -3$

21 $y = x^2$, $y = 4$

(a) $y = 4$ (b) $y = 5$ (c) $x = 2$ (d) $x = -3$

22 $y = \sqrt{x}$, $y = 0$, $x = 4$

(a) $x = 4$ (b) $x = 6$ (c) $y = 2$ (d) $y = -4$

23 $x + y = 3$, $y + x^2 = 3$; $x = 2$

24 $y = 1 - x^2$, $x - y = 1$; $y = 3$

25 $x^2 + y^2 = 1$; $x = 5$

26 $y = x^{2/3}$, $y = x^2$; $y = -1$

Exer. 27–30: Let R be the region bounded by the graphs of the equations. Set up integrals that can be used to find the volume of the solid generated if R is revolved about the given axis using (a) cylindrical shells and (b) disks or washers.

27 $y = 1/\sqrt{x}$, $x = 1$, $x = 4$, $y = 0$; x-axis

28 $y = 9 - x^2$, $x = 0$, $x = 2$, $y = 0$; x-axis

29 $y = x^2 + 2$, $x = 0$, $x = 1$, $y = 0$; y-axis

30 $y = x + 1$, $x = 0$, $x = 1$, $y = 0$; y-axis

31 If the region shown in the figure is revolved about the y-axis, use the trapezoidal rule, with $n = 6$, to approximate the volume of the resulting solid.

EXERCISE 31

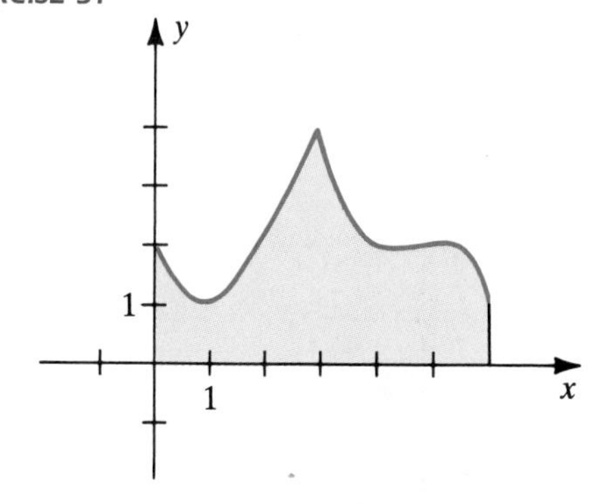

32 If the region shown in the figure on the following page is revolved about the y-axis, use Simpson's rule, with $n = 8$, to approximate the volume of the resulting solid.

EXERCISE 32

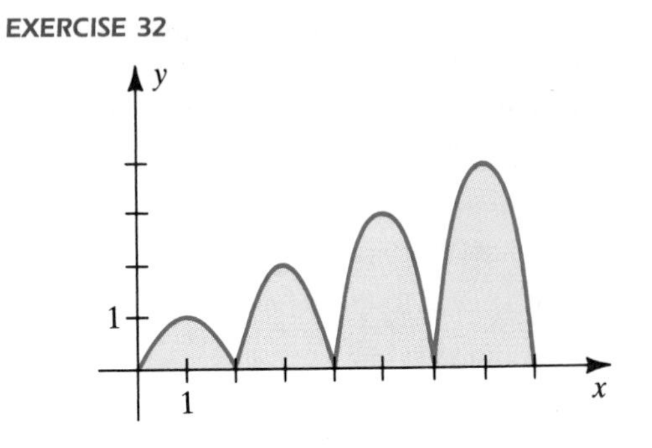

[c] 33 Graph $f(x) = -x^4 + 2.21x^3 - 3.21x^2 + 4.42x - 2$.

(a) Estimate the x-intercepts of the graph.

(b) If the region bounded by the graph of f and the x-axis is revolved about the y-axis, set up an integral that can be used to approximate the volume of the resulting solid.

[c] 34 Graph, on the same coordinate axes, $f(x) = \csc x$ and $g(x) = x + 1$ for $0 < x < \pi$.

(a) Use Newton's method to approximate, to two decimal places, the x-coordinates of the points of intersection of the graphs.

(b) If the region bounded by the graphs is revolved about the y-axis, use the trapezoidal rule with $n = 6$ to approximate the volume of the resulting solid.

6.4 VOLUMES BY CROSS SECTIONS

If a plane intersects a solid, then the region common to the plane and the solid is a **cross section** of the solid. In Section 6.2 we used circular and washer-shaped cross sections to find volumes of solids of revolution. Let us now consider a solid that has the following property (see Figure 6.34): For every x in $[a, b]$, the plane perpendicular to the x-axis at x intersects the solid in a cross section whose area is $A(x)$, where A is a continuous function on $[a, b]$.

FIGURE 6.34

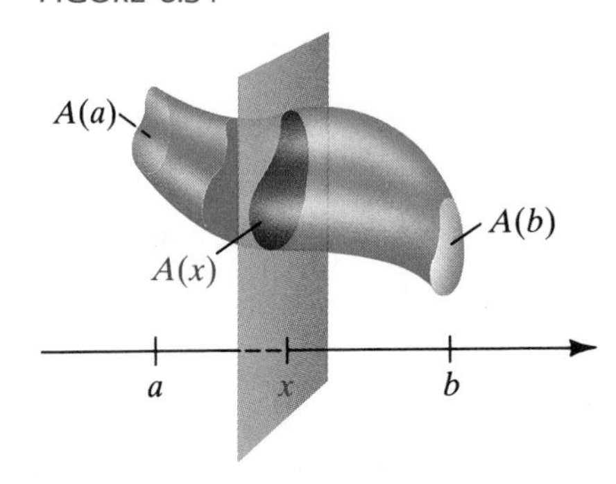

FIGURE 6.35

The solid is called a **cylinder** if, as illustrated in Figure 6.35, a line parallel to the x-axis that traces the boundary of the cross section corresponding to a also traces the boundary of the cross section corresponding to every x in $[a, b]$. The cross sections determined by the planes through $x = a$ and $x = b$ are the **bases** of the cylinder. The distance between the bases is the **altitude** of the cylinder. By definition, the volume of the cylinder is the area of a base multiplied by the altitude. Thus, the volume of the solid in Figure 6.35 is $A(a) \cdot (b - a)$.

To find the volume of a noncylindrical solid of the type illustrated in Figure 6.36, we begin with a partition P of $[a, b]$. Planes perpendicular

FIGURE 6.36

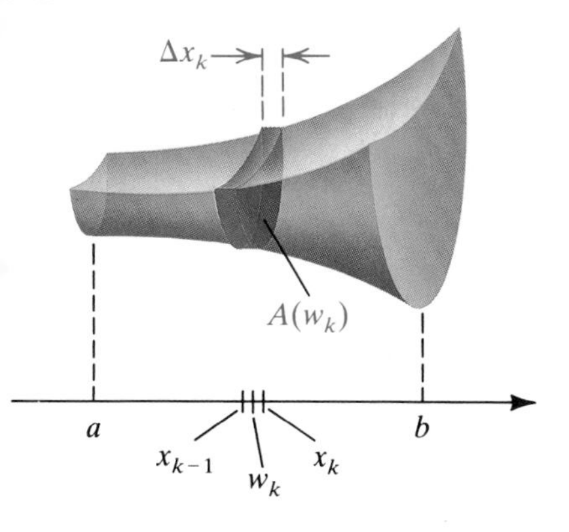

to the x-axis at each x_k in the partition slice the solid into smaller pieces. If we choose any number w_k in $[x_{k-1}, x_k]$, the volume of a typical slice can be approximated by the volume $A(w_k)\,\Delta x_k$ of the red cylinder shown in Figure 6.36. If V is the volume of the solid and if the norm $\|P\|$ is small, then

$$V \approx \sum_k A(w_k)\,\Delta x_k.$$

Since this approximation improves as $\|P\|$ gets smaller, we define the volume of the solid by

$$V = \lim_{\|P\|\to 0} \sum_k A(w_k)\,\Delta x_k = \int_a^b A(x)\,dx,$$

where the last equality follows from the definition of the definite integral. We may summarize our discussion as follows.

Volumes by cross sections (6.13)

Let S be a solid bounded by planes that are perpendicular to the x-axis at a and b. If, for every x in $[a, b]$, the cross-sectional area of S is given by $A(x)$, where A is continuous on $[a, b]$, then the volume of S is

$$V = \int_a^b A(x)\,dx.$$

An analogous result can be stated for a y-interval $[c, d]$ and a cross-sectional area $A(y)$.

EXAMPLE 1 Find the volume of a right pyramid with a square base of side a and altitude h.

SOLUTION As in Figure 6.37(i), let us take the vertex of the pyramid at the origin, with the x-axis passing through the center of the square base, a distance h from O. Cross sections by planes perpendicular to the x-axis are squares. Figure 6.37(ii) is a side view of the pyramid. Since $2y$ is the length of the side of the square cross section corresponding to x, the cross-sectional area $A(x)$ is

$$A(x) = (2y)^2 = 4y^2.$$

FIGURE 6.37

(i)

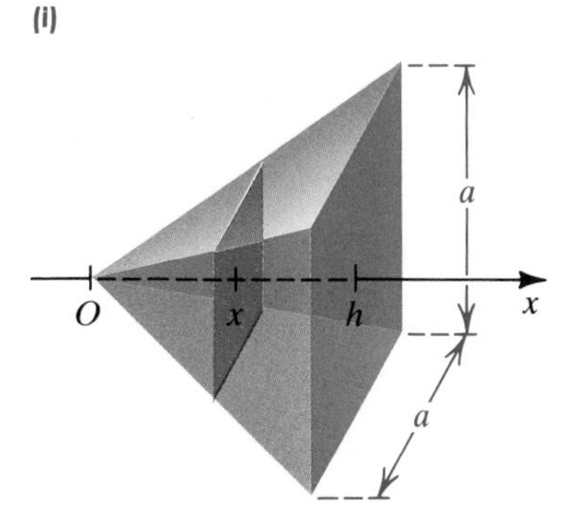

(ii)

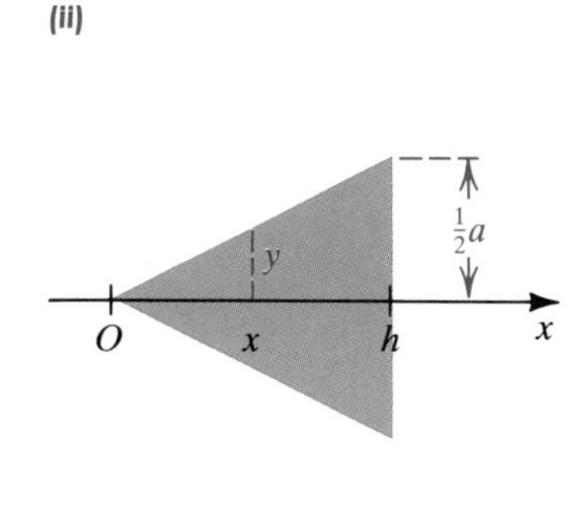

Using similar triangles in Figure 6.37(ii), we have

$$\frac{y}{x} = \frac{\frac{1}{2}a}{h}, \quad \text{or} \quad y = \frac{ax}{2h}.$$

Hence
$$A(x) = 4y^2 = \frac{4a^2x^2}{4h^2} = \frac{a^2}{h^2}x^2.$$

Applying (6.13) yields

$$V = \int_0^h A(x)\,dx = \int_0^h \left(\frac{a^2}{h^2}\right)x^2\,dx$$
$$= \left(\frac{a^2}{h^2}\right)\left[\frac{x^3}{3}\right]_0^h = \frac{a^2}{h^2}\frac{h^3}{3} = \frac{1}{3}a^2h.$$

EXAMPLE 2 A solid has, as its base, the circular region in the xy-plane bounded by the graph of $x^2 + y^2 = a^2$ with $a > 0$. Find the volume of the solid if every cross section by a plane perpendicular to the x-axis is an equilateral triangle with one side in the base.

SOLUTION A triangular cross section by a plane x units from the origin is illustrated in Figure 6.38(i). If the point $P(x, y)$ is on the circle and $y > 0$, then the lengths of the sides of this equilateral triangle are $2y$. Referring to (ii) of the figure, we see, by the Pythagorean theorem, that the altitude of the triangle is

$$\sqrt{(2y)^2 - y^2} = \sqrt{3y^2} = \sqrt{3}\,y.$$

FIGURE 6.38

(i)

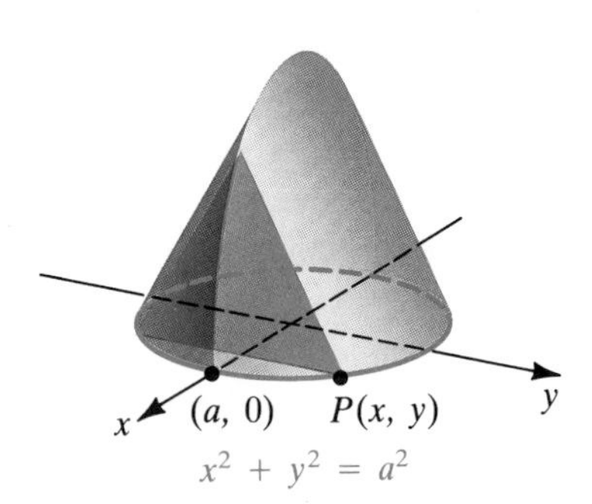

(ii)

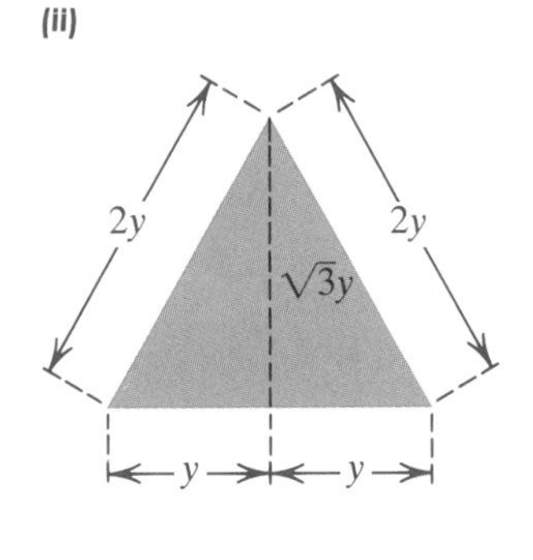

Hence the area $A(x)$ of the cross section is

$$A(x) = \tfrac{1}{2}(2y)(\sqrt{3}\,y) = \sqrt{3}\,y^2 = \sqrt{3}(a^2 - x^2).$$

Applying (6.13) gives us

$$V = \int_{-a}^{a} A(x)\,dx = \int_{-a}^{a} \sqrt{3}(a^2 - x^2)\,dx$$
$$= \sqrt{3}\left[a^2x - \frac{x^3}{3}\right]_{-a}^{a} = \frac{4\sqrt{3}}{3}a^3.$$

EXERCISES 6.4

Exer. 1–8: Let R be the region bounded by the graphs of $x = y^2$ and $x = 9$. Find the volume of the solid that has R as its base if every cross section by a plane perpendicular to the x-axis has the given shape.

1 A square

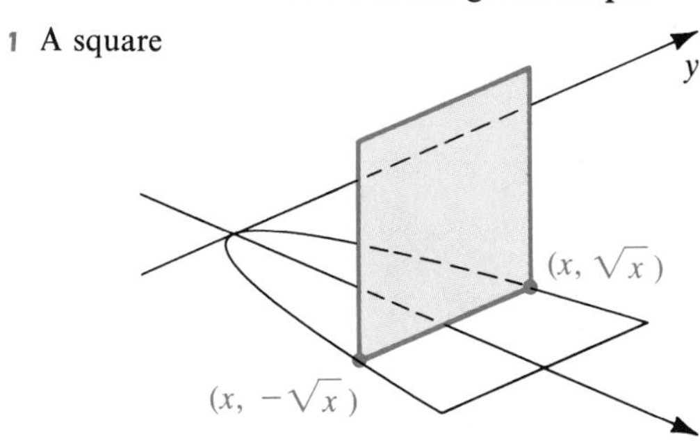

2 A rectangle of height 2

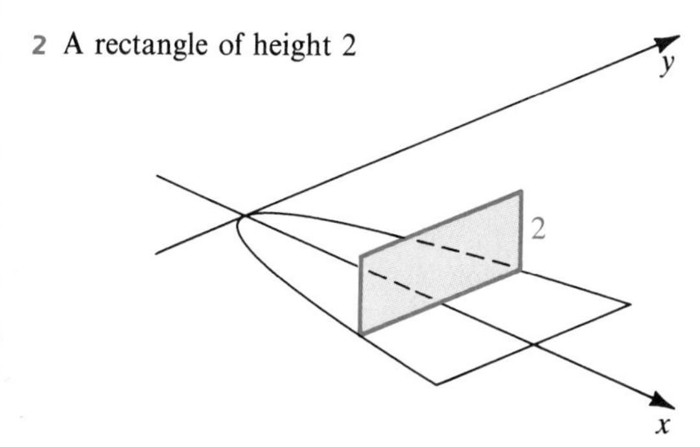

3 A semicircle

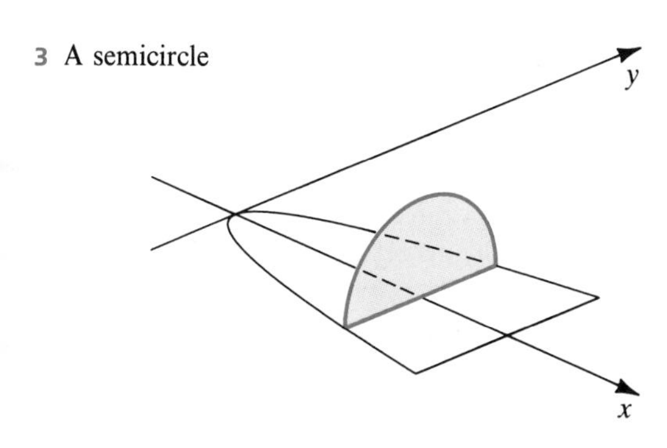

4 A quartercircle

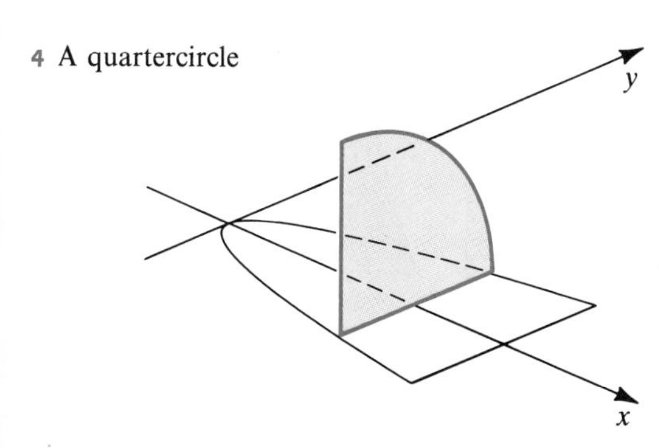

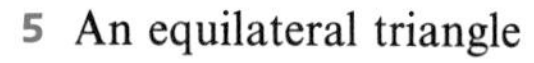

5 An equilateral triangle

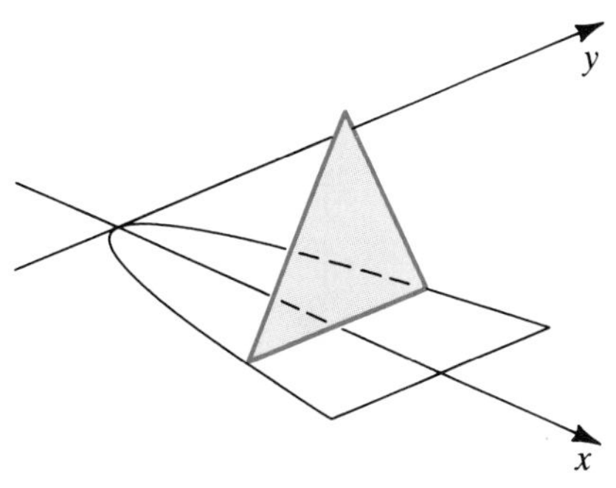

6 A triangle with height equal to $\frac{1}{4}$ the length of the base

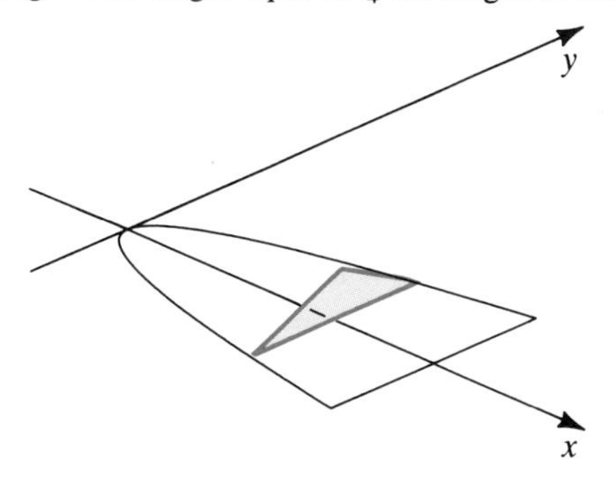

7 A trapezoid with lower base in the xy-plane, upper base equal to $\frac{1}{2}$ the length of the lower base, and height equal to $\frac{1}{4}$ the length of the lower base

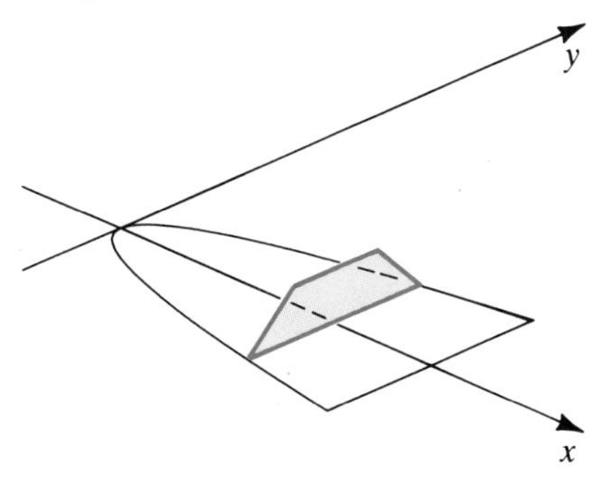

8 A parallelogram with base in the xy-plane and height equal to twice the length of the base

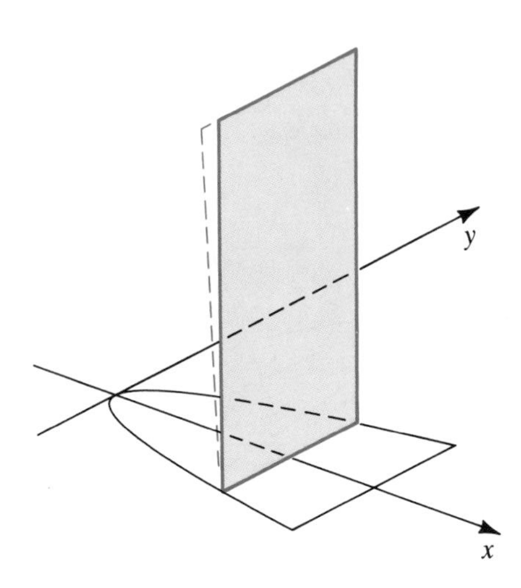

9 A solid has as its base the circular region in the xy-plane bounded by the graph of $x^2 + y^2 = a^2$ with $a > 0$. Find the volume of the solid if every cross section by a plane perpendicular to the x-axis is a square.

10 Work Exercise 9 if every cross section is an isosceles triangle with base on the xy-plane and altitude equal to the length of the base.

11 A solid has as its base the region in the xy-plane bounded by the graphs of $y = 4$ and $y = x^2$. Find the volume of the solid if every cross section by a plane perpendicular to the x-axis is an isosceles right triangle with hypotenuse on the xy-plane.

12 Work Exercise 11 if every cross section is a square.

13 Find the volume of a pyramid of the type illustrated in Figure 6.37 if the altitude is h and the base is a rectangle of dimensions a and $2a$.

14 A solid has as its base the region in the xy-plane bounded by the graphs of $y = x$ and $y^2 = x$. Find the volume of the solid if every cross section by a plane perpendicular to the x-axis is a semicircle with diameter in the xy-plane.

15 A solid has as its base the region in the xy-plane bounded by the graphs of $y^2 = 4x$ and $x = 4$. If every cross section by a plane perpendicular to the y-axis is a semicircle, find the volume of the solid.

16 A solid has as its base the region in the xy-plane bounded by the graphs of $x^2 = 16y$ and $y = 2$. Every cross section by a plane perpendicular to the y-axis is a rectangle whose height is twice that of the side in the xy-plane. Find the volume of the solid.

17 A log having the shape of a right circular cylinder of radius a is lying on its side. A wedge is removed from the log by making a vertical cut and another cut at an angle of 45°, both cuts intersecting at the center of the log (see figure). Find the volume of the wedge.

EXERCISE 17

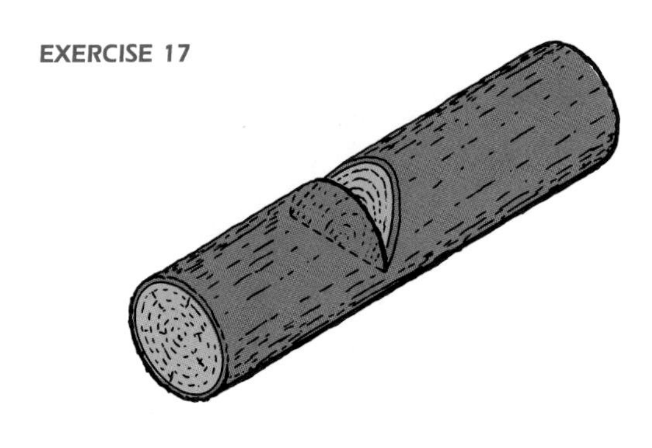

18 The axes of two right circular cylinders of radius a intersect at right angles. Find the volume of the solid bounded by the cylinders.

19 The base of a solid is the circular region in the xy-plane bounded by the graph of $x^2 + y^2 = a^2$ with $a > 0$. Find the volume of the solid if every cross section by a plane perpendicular to the x-axis is an isosceles triangle of constant altitude h. (*Hint:* Interpret $\int_{-a}^{a} \sqrt{a^2 - x^2}\, dx$ as an area.)

20 Cross sections of a horn-shaped solid by planes perpendicular to its axis are circles. If a cross section that is s inches from the smaller end of the solid has diameter $6 + \frac{1}{36}s^2$ inches and if the length of the solid is 2 feet, find its volume.

21 A tetrahedron has three mutually perpendicular faces and three mutually perpendicular edges of lengths 2, 3, and 4 centimeters, respectively. Find its volume.

22 *Cavalieri's theorem* states that if two solids have equal altitudes and if all cross sections by planes parallel to their bases and at the same distances from their bases have equal areas, then the solids have the same volume (see figure). Prove Cavalieri's theorem.

EXERCISE 22

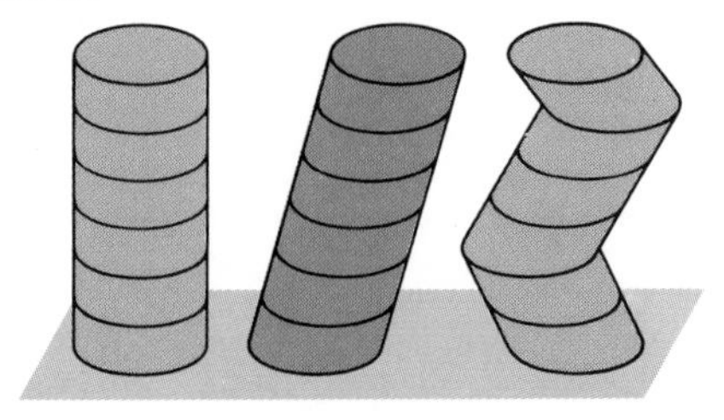

23 The base of a solid is an isosceles right triangle whose equal sides have length a. Find the volume if cross sections that are perpendicular to the base and to one of the equal sides are semicircular.

24 Work Exercise 23 if the cross sections are regular hexagons with one side in the base.

25 Show that the disk and washer methods discussed in Section 6.2 are special cases of (6.13).

26 A circular swimming pool has diameter 28 feet. The depth of the water changes slowly from 3 feet at a point A on one side of the pool to 9 feet at a point B diametrically opposite A (see figure). Depth readings $h(x)$ (in feet) taken along the diameter AB are given in the following table, where x is the distance (in feet) from A.

x	0	4	8	12	16	20	24	28
$h(x)$	3	3.5	4	5	6.5	8	8.5	9

Use the trapezoidal rule, with $n = 7$, to estimate the volume of water in the pool. Approximate the number of gallons of water contained in the pool (1 gal ≈ 0.134 ft^3).

EXERCISE 26

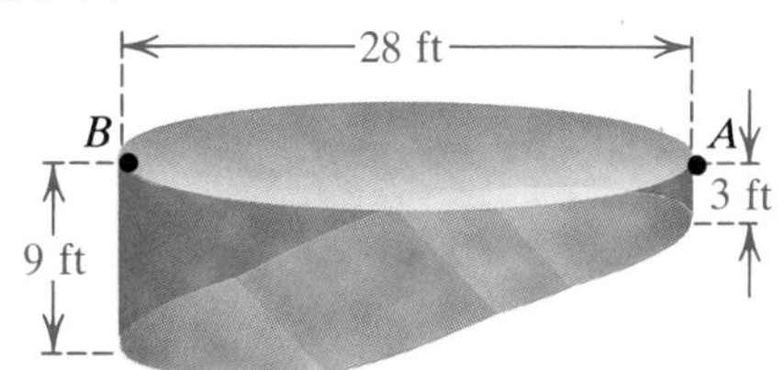

6.5 ARC LENGTH AND SURFACES OF REVOLUTION

For some applications we must determine the *length* of the graph of a function. To obtain a suitable formula, we shall employ a process similar to one that could be used to approximate the length of a twisted wire. Let us imagine dividing the wire into many small pieces by placing dots at $Q_0, Q_1, Q_2, \ldots, Q_n$, as illustrated in Figure 6.39. We may then approximate the length of the wire between Q_{k-1} and Q_k (for each k) by measuring the distance $d(Q_{k-1}, Q_k)$ with a ruler. The sum of all these distances is an approximation for the total length of the wire. Evidently, the closer together we place the dots, the better the approximation. The process we shall use for the graph of a function is similar; however, we shall find the *exact* length by taking a *limit of sums* of lengths of line segments. This process leads to a definite integral. To guarantee that the integral exists, we must place restrictions on the function, as indicated in the following discussion.

FIGURE 6.39 Twisted wire

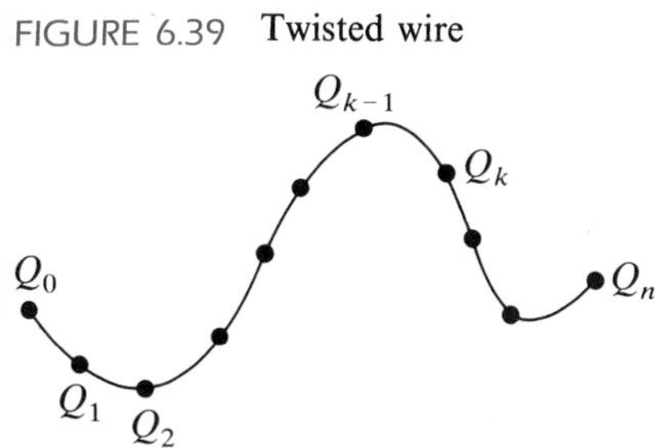

A function f is **smooth** on an interval if it has a derivative f' that is continuous throughout the interval. Intuitively, this means that a small change in x produces a small change in the slope $f'(x)$ of the tangent line to the graph of f. Thus, the graph has no corners or cusps. We shall define the **length of arc** between two points A and B on the graph of a smooth function.

If f is smooth on a closed interval $[a, b]$, the points $A(a, f(a))$ and $B(b, f(b))$ are called the **endpoints** of the graph of f. Let P be the partition of $[a, b]$ determined by $a = x_0, x_1, x_2, \ldots, x_n = b$, and let Q_k denote the point with coordinates $(x_k, f(x_k))$ on the graph of f, as illustrated in

FIGURE 6.40

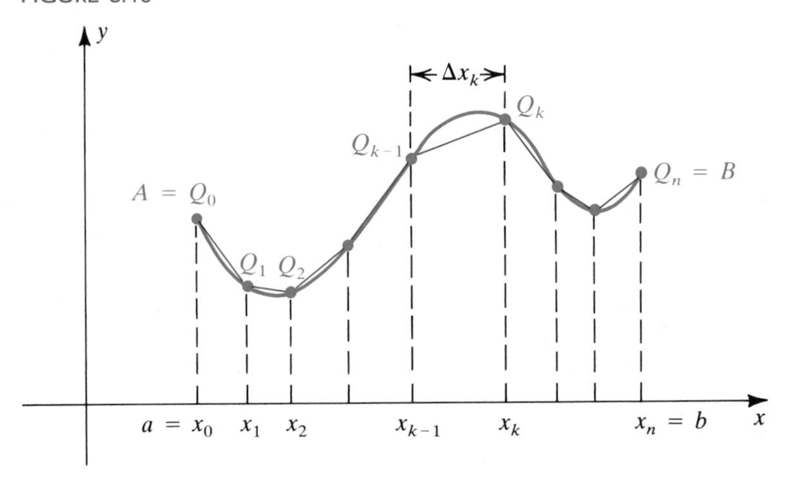

Figure 6.40. If we connect each Q_{k-1} to Q_k by a line segment of length $d(Q_{k-1}, Q_k)$, the length L_P of the resulting broken line is

$$L_P = \sum_{k=1}^{n} d(Q_{k-1}, Q_k).$$

Using the distance formula, we get

$$d(Q_{k-1}, Q_k) = \sqrt{(x_k - x_{k-1})^2 + [f(x_k) - f(x_{k-1})]^2}.$$

By the mean value theorem (4.12),

$$f(x_k) - f(x_{k-1}) = f'(w_k)(x_k - x_{k-1})$$

for some number w_k in the open interval (x_{k-1}, x_k). Substituting for $f(x_k) - f(x_{k-1})$ in the preceding formula and letting $\Delta x_k = x_k - x_{k-1}$, we obtain

$$\begin{aligned} d(Q_{k-1}, Q_k) &= \sqrt{(\Delta x_k)^2 + [f'(w_k)\,\Delta x_k]^2} \\ &= \sqrt{1 + [f'(w_k)]^2}\,\Delta x_k. \end{aligned}$$

Consequently,

$$L_P = \sum_{k=1}^{n} \sqrt{1 + [f'(w_k)]^2}\,\Delta x_k.$$

Observe that L_P is a Riemann sum for $g(x) = \sqrt{1 + [f'(x)]^2}$. Moreover, g is continuous on $[a, b]$, since f' is continuous. If the norm $\|P\|$ is small, then the length L_P of the broken line approximates the length of the graph of f from A to B. This approximation should improve as $\|P\|$ decreases, so we define the *length* (also called the *arc length*) of the graph of f from A to B as the limit of sums L_P. Since $g = \sqrt{1 + (f')^2}$ is a continuous function, the limit exists and equals the definite integral $\int_a^b \sqrt{1 + [f'(x)]^2}\,dx$. This arc length will be denoted by L_a^b.

Definition **(6.14)**

Let f be smooth on $[a, b]$. The **arc length of the graph** of f from $A(a, f(a))$ to $B(b, f(b))$ is

$$L_a^b = \int_a^b \sqrt{1 + [f'(x)]^2}\,dx.$$

Definition (6.14) will be extended to more general graphs in Chapter 13. If a function f is defined implicitly by an equation in x and y, then we shall also refer to the *arc length of the graph of the equation.*

EXAMPLE 1 If $f(x) = 3x^{2/3} - 10$, find the arc length of the graph of f from the point $A(8, 2)$ to $B(27, 17)$.

FIGURE 6.41

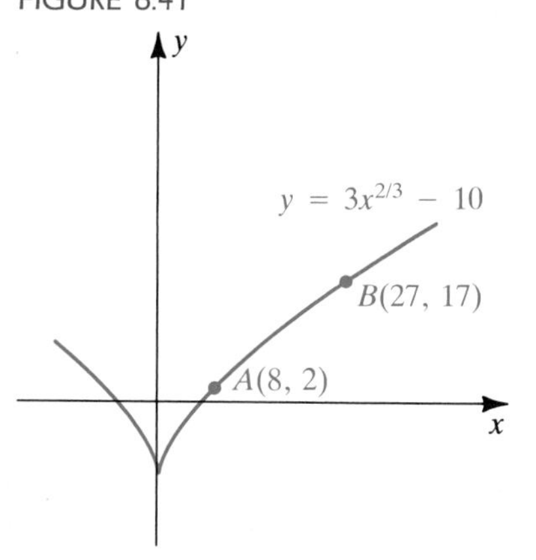

SOLUTION The graph of f is sketched in Figure 6.41. Since

$$f'(x) = 2x^{-1/3} = \frac{2}{x^{1/3}},$$

we have, by Definition (6.14),

$$\begin{aligned} L_8^{27} &= \int_8^{27} \sqrt{1 + \left(\frac{2}{x^{1/3}}\right)^2}\, dx = \int_8^{27} \sqrt{1 + \frac{4}{x^{2/3}}}\, dx \\ &= \int_8^{27} \sqrt{\frac{x^{2/3} + 4}{x^{2/3}}}\, dx \\ &= \int_8^{27} \sqrt{x^{2/3} + 4}\, \frac{1}{x^{1/3}}\, dx. \end{aligned}$$

To evaluate this integral, we make the substitution

$$u = x^{2/3} + 4, \qquad du = \frac{2}{3}x^{-1/3}\, dx = \frac{2}{3}\frac{1}{x^{1/3}}\, dx.$$

The integral can be expressed in a suitable form for integration by introducing the factor $\frac{2}{3}$ in the integrand and compensating by multiplying the integral by $\frac{3}{2}$:

$$L_8^{27} = \frac{3}{2}\int_8^{27} \sqrt{x^{2/3} + 4}\left(\frac{2}{3}\frac{1}{x^{1/3}}\right) dx$$

We next calculate the values of $u = x^{2/3} + 4$ that correspond to the limits of integration $x = 8$ and $x = 27$:

(i) If $x = 8$, then $u = 8^{2/3} + 4 = 8$.

(ii) If $x = 27$, then $u = 27^{2/3} + 4 = 13$.

Substituting in the integrand and changing the limits of integration gives us the arc length:

$$L_8^{27} = \tfrac{3}{2}\int_8^{13} \sqrt{u}\, du = u^{3/2}\Big]_8^{13} = 13^{3/2} - 8^{3/2} \approx 24.2$$

FIGURE 6.42

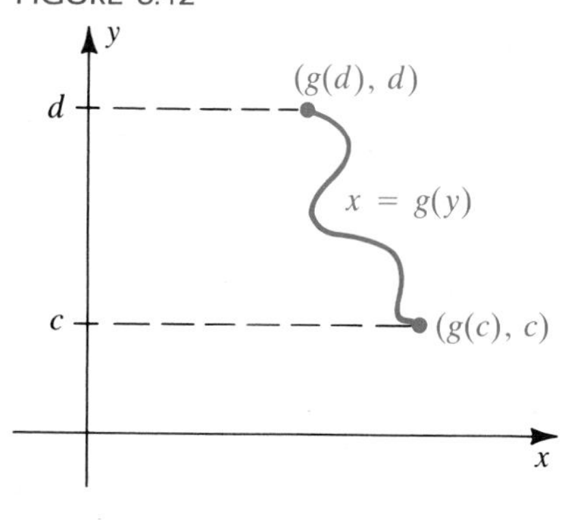

Interchanging the roles of x and y in Definition (6.14) gives us the following formula for integration with respect to y.

Definition **(6.15)**

> Let $x = g(y)$ with g smooth on the interval $[c, d]$. The arc length of the graph of g from $(g(c), c)$ to $(g(d), d)$ (see Figure 6.42) is
>
> $$L_c^d = \int_c^d \sqrt{1 + [g'(y)]^2}\, dy.$$

The integrands $\sqrt{1 + [f'(x)]^2}$ and $\sqrt{1 + [g'(y)]^2}$ in formulas (6.14) and (6.15) often result in expressions that have no obvious antiderivatives. In such cases, numerical integration may be used to approximate arc length, as illustrated in the next example.

EXAMPLE 2

(a) Set up an integral for finding the arc length of the graph of the equation $y^3 - y - x = 0$ from $A(0, -1)$ to $B(6, 2)$.

(b) Approximate the integral in (a) by using Simpson's rule (5.38), with $n = 6$, and round the answer to one decimal place.

SOLUTION

FIGURE 6.43

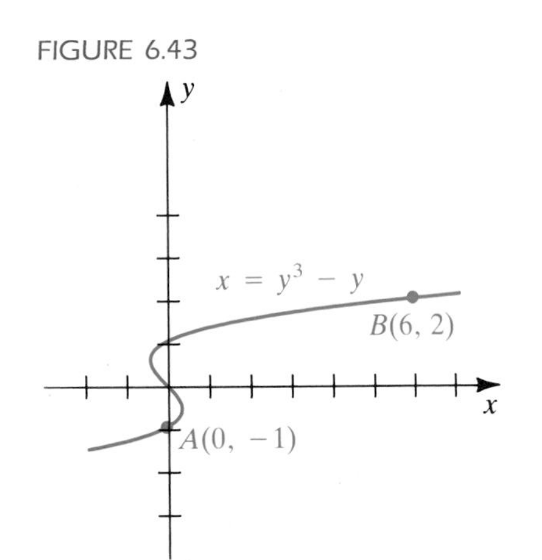

(a) Since the equation is not of the form $y = f(x)$, Definition (6.14) cannot be applied directly. However, if we write $x = y^3 - y$, then we can employ (6.15) with $g(y) = y^3 - y$. The graph of the equation is sketched in Figure 6.43. Using (6.15) with $c = -1$ and $d = 2$ yields

$$L_{-1}^{2} = \int_{-1}^{2} \sqrt{1 + (3y^2 - 1)^2}\, dy$$
$$= \int_{-1}^{2} \sqrt{9y^4 - 6y^2 + 2}\, dy.$$

(b) To use Simpson's rule, we let $f(y) = \sqrt{9y^4 - 6y^2 + 2}$ and arrange our work as we did in Section 5.6, obtaining the following table.

k	y_k	$f(y_k)$	m	$mf(y_k)$
0	−1.0	2.2361	1	2.2361
1	−0.5	1.0308	4	4.1232
2	0.0	1.4142	2	2.8284
3	0.5	1.0308	4	4.1232
4	1.0	2.2361	2	4.4722
5	1.5	5.8363	4	23.3452
6	2.0	11.0454	1	11.0454

The sum of the numbers in the last column is 52.1737. Applying Simpson's rule with $a = -1$, $b = 2$, and $n = 6$ gives us

$$\int_{-1}^{2} \sqrt{9y^4 - 6y^2 + 2}\, dy \approx \frac{2 - (-1)}{3(6)}(52.1737) \approx 8.7.$$

A function f is **piecewise smooth** on its domain if the graph of f can be decomposed into a finite number of parts, each of which is the graph of a smooth function. The arc length of the graph is defined as the sum of the arc lengths of the individual graphs.

To avoid any misunderstanding in the following discussion, we shall denote the variable of integration by t. In this case the arc length formula

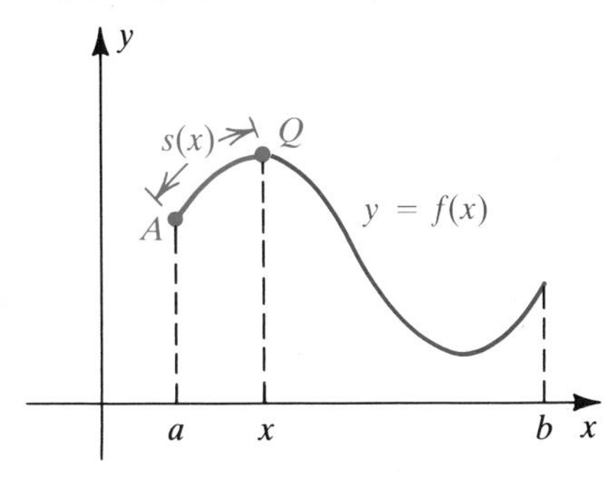

FIGURE 6.44

in Definition (6.14) is written

$$L_a^b = \int_a^b \sqrt{1 + [f'(t)]^2}\, dt.$$

If f is smooth on $[a, b]$, then f is smooth on $[a, x]$ for every number x in $[a, b]$, and the length of the graph from the point $A(a, f(a))$ to the point $Q(x, f(x))$ is

$$L_a^x = \int_a^x \sqrt{1 + [f'(t)]^2}\, dt.$$

If we change the notation and use the symbol $s(x)$ in place of L_a^x, then s may be regarded as a function with domain $[a, b]$, since to each x in $[a, b]$ there corresponds a unique number $s(x)$. As shown in Figure 6.44, $s(x)$ is the length of arc of the graph of f from $A(a, f(a))$ to $Q(x, f(x))$. We shall call s the *arc length function* for the graph of f, as in the next definition.

Definition **(6.16)**

Let f be smooth on $[a, b]$. The **arc length function** s for the graph of f on $[a, b]$ is defined by

$$s(x) = \int_a^x \sqrt{1 + [f'(t)]^2}\, dt$$

for $a \le x \le b$.

If s is the arc length function, the differential $ds = s'(x)\, dx$ is called the **differential of arc length**. The next theorem specifies formulas for finding ds.

Theorem **(6.17)**

Let f be smooth on $[a, b]$, and let s be the arc length function for the graph of $y = f(x)$ on $[a, b]$. If dx and dy are the differentials of x and y, then

(i) $ds = \sqrt{1 + [f'(x)]^2}\, dx$

(ii) $(ds)^2 = (dx)^2 + (dy)^2$

PROOF By Definition (6.16) and Theorem (5.35),

$$s'(x) = D_x\,[s(x)] = D_x\left[\int_a^x \sqrt{1 + [f'(t)]^2}\, dt\right] = \sqrt{1 + [f'(x)]^2}$$

Applying Definition (3.28) yields

$$ds = s'(x)\, dx = \sqrt{1 + [f'(x)]^2}\, dx.$$

This proves (i).

To prove (ii), we square both sides of (i), obtaining

$$\begin{aligned}(ds)^2 &= \{1 + [f'(x)]^2\}(dx)^2 \\ &= (dx)^2 + [f'(x)\, dx]^2 \\ &= (dx)^2 + (dy)^2.\end{aligned}$$

The last equality follows from Definition (3.28). ■

Theorem (6.17)(ii) has an interesting and useful geometric interpretation. Consider $y = f(x)$, and let Δx be an increment of x. Let Δy denote the change in y, and let Δs denote the change in arc length corresponding to Δx. Typical increments are illustrated in Figure 6.45, where l is the tangent line at (x, y) (compare with Figure 3.27). Since $(ds)^2 = (dx)^2 + (dy)^2$, we may regard $|ds|$ as the length of the hypotenuse of a right triangle that has sides $|dx|$ and $|dy|$, as illustrated in the figure. Note that if Δx is small, then ds may be used to approximate the increment Δs of arc length.

FIGURE 6.45

(i)

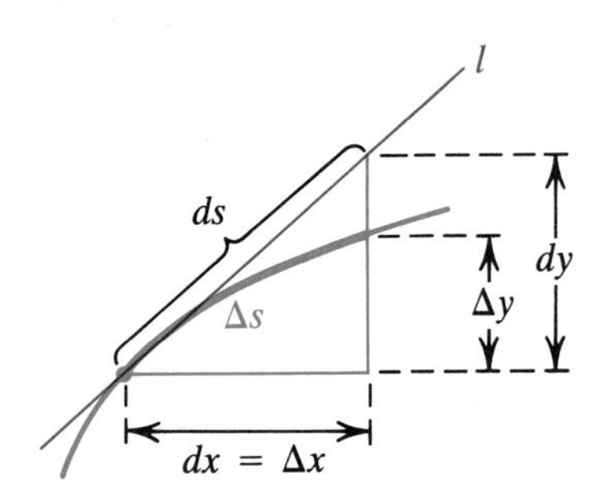

(ii)

(iii)

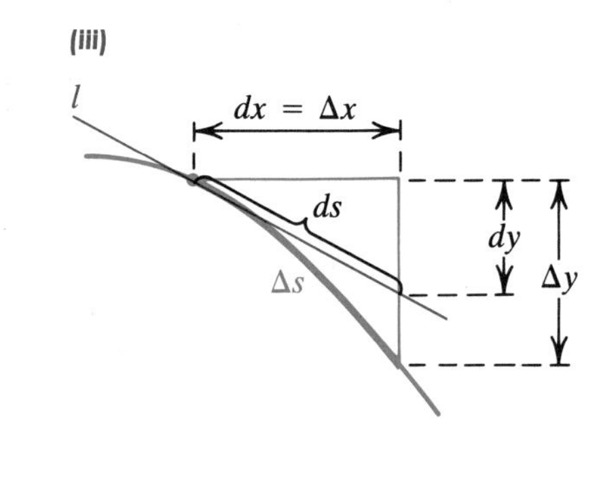

EXAMPLE 3 Use differentials to approximate the arc length of the graph of $y = x^3 + 2x$ from $A(1, 3)$ to $B(1.2, 4.128)$.

SOLUTION If we let $f(x) = x^3 + 2x$, then, by Theorem (6.17)(i),

$$ds = \sqrt{1 + (3x^2 + 2)^2}\, dx.$$

An approximation may be obtained by letting $x = 1$ and $dx = 0.2$:

$$ds = \sqrt{1 + 5^2}\,(0.2) = \sqrt{26}(0.2) \approx 1.02$$

FIGURE 6.46

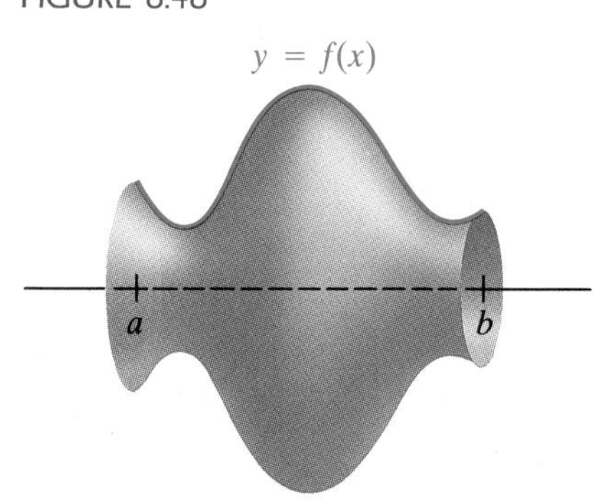

Let f be a function that is nonnegative throughout a closed interval $[a, b]$. If the graph of f is revolved about the x-axis, a **surface of revolution** is generated (see Figure 6.46). For example, if $f(x) = \sqrt{r^2 - x^2}$ for a positive constant r, the graph of f on $[-r, r]$ is the upper half of the circle $x^2 + y^2 = r^2$, and a revolution about the x-axis produces a sphere of radius r having surface area $4\pi r^2$.

FIGURE 6.47

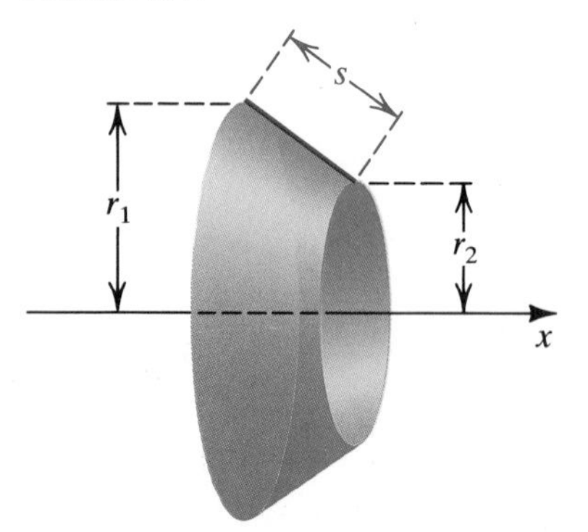

If the graph of f is the line segment shown in Figure 6.47, then the surface generated is a frustum of a cone having base radii r_1 and r_2 and slant height s. It can be shown that the surface area is

$$\pi(r_1 + r_2)s = 2\pi\left(\frac{r_1 + r_2}{2}\right)s.$$

You may remember this formula as follows.

Surface area S of a frustum of a cone **(6.18)**

$$S = 2\pi \text{ (average radius)(slant height)}$$

We shall use this fact in the following discussion.

Let f be a smooth function that is nonnegative on $[a, b]$, and consider the surface generated by revolving the graph of f about the x-axis (see Figure 6.46). We wish to find a formula for the area S of this surface. Let P be a partition of $[a, b]$ determined by $a = x_0, x_1, \ldots, x_n = b$, and, for each k, let Q_k denote the point $(x_k, f(x_k))$ on the graph of f (see Figure 6.48). If the norm $\|P\|$ is close to zero, then the broken line l_P obtained by connecting Q_{k-1} to Q_k for each k is an approximation to the graph of f, and hence the area of the surface generated by revolving l_P about the x-axis should approximate S. The line segment $Q_{k-1}Q_k$ generates a frustum of a cone having base radii $f(x_{k-1})$ and $f(x_k)$ and slant height $d(Q_{k-1}, Q_k)$. By (6.18), its surface area is

$$2\pi \frac{f(x_{k-1}) + f(x_k)}{2} d(Q_{k-1}, Q_k).$$

FIGURE 6.48

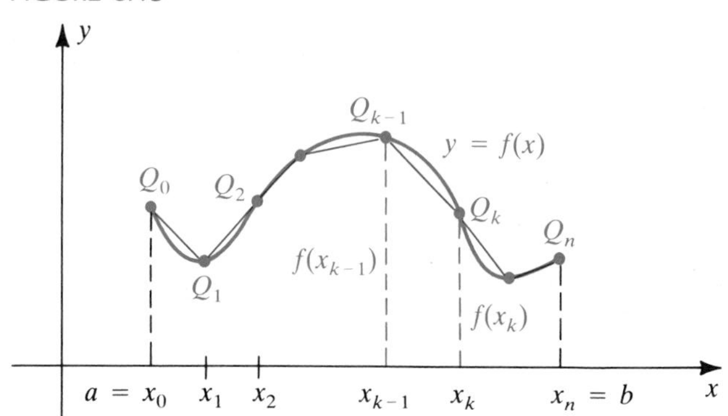

Summing terms of this form from $k = 1$ to $k = n$ gives us the area S_P of the surface generated by the broken line l_P. If we use the expression for $d(Q_{k-1}, Q_k)$ on page 334, then

$$S_P = \sum_{k=1}^{n} 2\pi \frac{f(x_{k-1}) + f(x_k)}{2} \sqrt{1 + [f'(w_k)]^2}\, \Delta x_k,$$

where $x_{k-1} < w_k < x_k$. We define the area S of the surface of revolution as

$$S = \lim_{\|P\| \to 0} S_P.$$

From the form of S_P, it is reasonable to expect that the limit is given by

$$\int_a^b 2\pi \frac{f(x) + f(x)}{2} \sqrt{1 + [f'(x)]^2}\, dx = \int_a^b 2\pi f(x)\sqrt{1 + [f'(x)]^2}\, dx.$$

The proof of this fact requires results from advanced calculus and is omitted. The following definition summarizes our discussion.

Definition **(6.19)**

If f is smooth and $f(x) \geq 0$ on $[a, b]$, then the **area** S of the surface generated by revolving the graph of f about the x-axis is

$$S = \int_a^b 2\pi f(x)\sqrt{1 + [f'(x)]^2}\, dx.$$

If f is negative for some x in $[a, b]$, then the following extension of Definition (6.19) can be used to find the surface area S:

$$S = \int_a^b 2\pi\,|f(x)|\sqrt{1 + [f'(x)]^2}\, dx$$

FIGURE 6.49

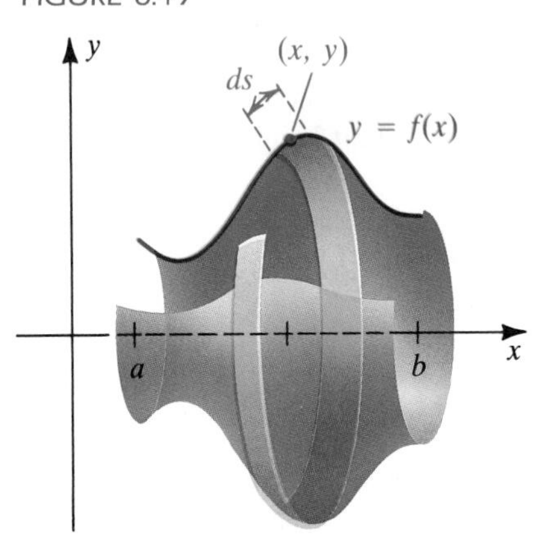

We can use (6.18) to remember the formula for S in Definition (6.19). As in Figure 6.49, let (x, y) denote an arbitrary point on the graph of f and, as in Theorem (6.17)(i), consider the differential of arc length

$$ds = \sqrt{1 + [f'(x)]^2}\, dx.$$

Next, regard ds as the slant height of the frustum of a cone that has average radius $y = f(x)$ (see Figure 6.49). Applying (6.18), the surface area of this frustum is given by

$$2\pi f(x)\, ds = 2\pi y\, ds.$$

As with our work in Sections 6.1 through 6.3, applying $\int_a^b$ may be regarded as taking a limit of sums of these areas of frustums. Thus,

$$S = \int_a^b 2\pi f(x)\, ds = \int_a^b 2\pi y\, ds.$$

FIGURE 6.50

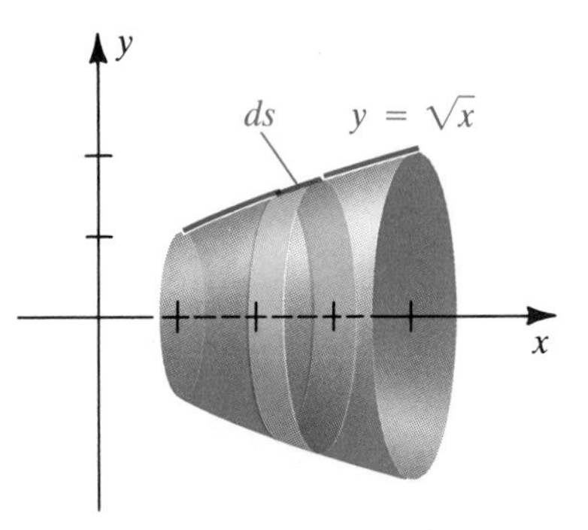

EXAMPLE 4 The graph of $y = \sqrt{x}$ from $(1, 1)$ to $(4, 2)$ is revolved about the x-axis. Find the area of the resulting surface.

SOLUTION The surface is illustrated in Figure 6.50. Using Definition (6.19) or the previous discussion, we have

$$\begin{aligned} S &= \int_1^4 2\pi y\, ds \\ &= \int_1^4 2\pi x^{1/2}\sqrt{1 + \left(\frac{1}{2x^{1/2}}\right)^2}\, dx \\ &= \int_1^4 2\pi x^{1/2}\sqrt{\frac{4x+1}{4x}}\, dx = \pi\int_1^4 \sqrt{4x+1}\, dx \\ &= \frac{\pi}{6}\left[(4x+1)^{3/2}\right]_1^4 = \frac{\pi}{6}(17^{3/2} - 5^{3/2}) \approx 30.85 \text{ square units.} \end{aligned}$$

FIGURE 6.51

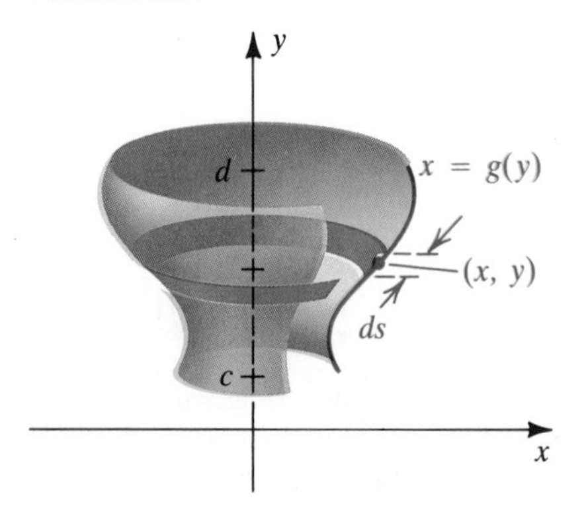

If we interchange the roles of x and y in the preceding discussion, then a formula analogous to (6.19) can be obtained for integration with respect to y. Thus, if $x = g(y)$ and g is smooth and nonnegative on $[c, d]$, then the area S of the surface generated by revolving the graph of g about the y-axis (see Figure 6.51) is

$$S = \int_c^d 2\pi g(y)\sqrt{1 + [g'(y)]^2}\, dy.$$

EXERCISES 6.5

Exer. 1–4: Set up an integral that can be used to find the arc length of the graph from A to B by integrating with respect to (a) x and (b) y.

1

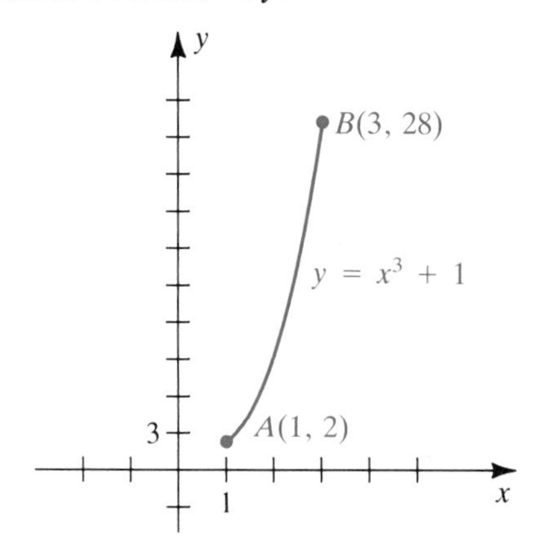

2

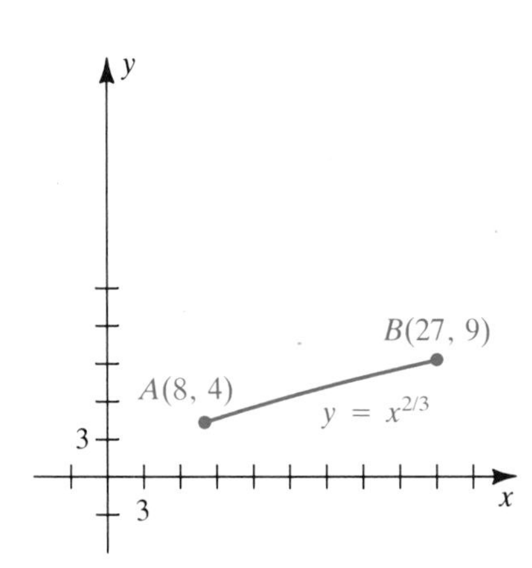

3

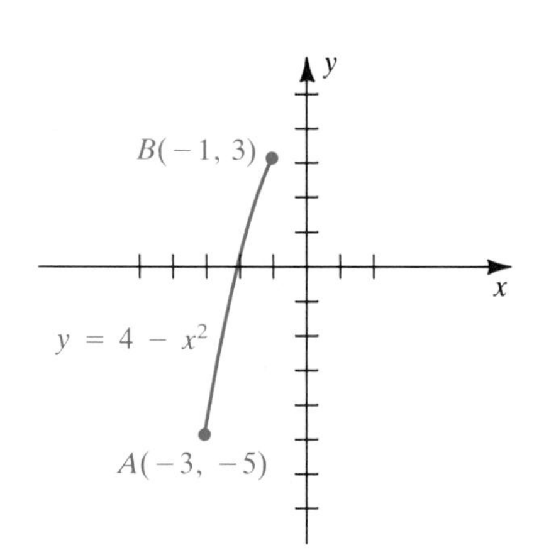

4

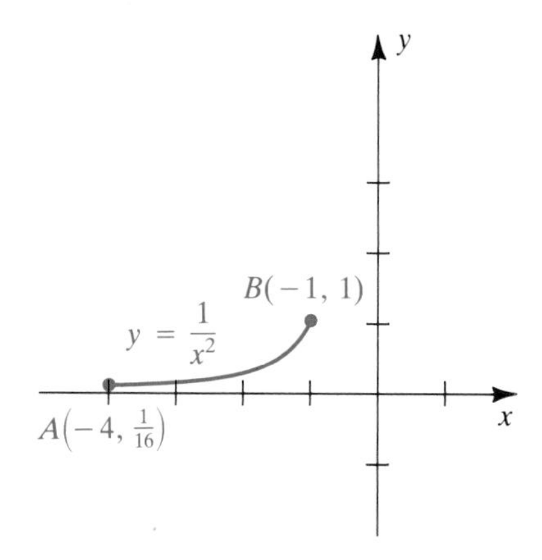

Exer. 5–12: Find the arc length of the graph of the equation from A to B.

5 $y = \frac{2}{3}x^{2/3}$; $A(1, \frac{2}{3})$, $B(8, \frac{8}{3})$

6 $(y + 1)^2 = (x - 4)^3$; $A(5, 0)$, $B(8, 7)$

7 $y = 5 - \sqrt{x^3}$; $A(1, 4)$, $B(4, -3)$

8 $y = 6\sqrt[3]{x^2} + 1$; $A(-1, 7)$, $B(-8, 25)$

9 $y = \dfrac{x^3}{12} + \dfrac{1}{x}$; $A(1, \frac{13}{12})$, $B(2, \frac{7}{6})$

10 $y + \dfrac{1}{4x} + \dfrac{x^3}{3} = 0$; $A(2, \frac{67}{24})$, $B(3, \frac{109}{12})$

11 $30xy^3 - y^8 = 15$; $A(\frac{8}{15}, 1)$, $B(\frac{271}{240}, 2)$

12 $x = \dfrac{y^4}{16} + \dfrac{1}{2y^2}$; $A(\frac{9}{8}, -2)$, $B(\frac{9}{16}, -1)$

Exer. 13–14: Set up an integral for finding the arc length of the graph of the equation from A to B.

13 $2y^3 - 7y + 2x = 8$; $A(3, 2)$, $B(4, 0)$

14 $11x - 4x^3 - 7y = -7$; $A(1, 2)$, $B(0, 1)$

15 Find the arc length of the graph of $x^{2/3} + y^{2/3} = 1$. (*Hint:* Use symmetry with respect to the line $y = x$.)

16 Find the arc length of the graph of $y = \dfrac{3x^8 + 5}{30x^3}$ from $(1, \frac{4}{5})$ to $(2, \frac{773}{240})$.

Exer. 17–18: (a) Find $s(x)$, where s is an arc length function for the graph of f. (b) If x increases from 1 to 1.1, find Δs and ds.

17 $f(x) = \sqrt[3]{x^2}$

18 $f(x) = \sqrt{x^3}$

Exer. 19–20: Use differentials to approximate the arc length of the graph of the equation from A to B.

19 $y = x^2$; $A(2, 4)$, $B(2.1, 4.41)$

20 $y + x^3 = 0$; $A(1, -1)$, $B(1.1, -1.331)$

Exer. 21–22: Use differentials to approximate the arc length of the graph of the equation between the points with x-coordinates a and b.

21 $y = \cos x;\quad a = \pi/6,\quad b = 31\pi/180$

22 $y = \sin x;\quad a = 0,\quad b = \pi/90$

c **Exer. 23–26: Use Simpson's rule with $n = 4$ to approximate the arc length of the graph of the equation from A to B. (Round the answer to two decimal places.)**

23 $y = x^2 + x + 3;\quad A(-2, 5),\quad B(2, 9)$

24 $y = x^3;\quad A(0, 0),\quad B(2, 8)$

25 $y = \sin x;\quad A(0, 0),\quad B(\pi/2, 1)$

26 $y = \tan x;\quad A(0, 0),\quad B(\pi/4, 1)$

c 27 (a) Approximate the arc length of the graph of $f(x) = \sin x$ from $(0, 0)$ to $(\pi, 0)$ by using $\sum_{k=1}^{4} d_k(Q_{k-1}, Q_k)$, where Q_k is the point $(\frac{1}{4}\pi k, f(\frac{1}{4}\pi k))$.

(b) If n is any positive integer, how does $\sum_{k=1}^{n} d_k(Q_{k-1}, Q_k)$ compare to the exact arc length?

c 28 (a) Set up an integral for the arc length in Exercise 27(a).

(b) Use the trapezoidal rule with $n = 4$ to approximate the integral in (a).

Exer. 29–32: The graph of the equation from A to B is revolved about the x-axis. Find the area of the resulting surface.

29 $4x = y^2;\quad A(0, 0),\quad B(1, 2)$

30 $y = x^3;\quad A(1, 1),\quad B(2, 8)$

31 $8y = 2x^4 + x^{-2};\quad A(1, \frac{3}{8}),\quad B(2, \frac{129}{32})$

32 $y = 2\sqrt{x + 1};\quad A(0, 2),\quad B(3, 4)$

Exer. 33–34: The graph of the equation from A to B is revolved about the y-axis. Find the area of the resulting surface.

33 $y = 2\sqrt[3]{x};\quad A(1, 2),\quad B(8, 4)$

34 $x = 4\sqrt{y};\quad A(4, 1),\quad B(12, 9)$

Exer. 35–36: If the smaller arc of the circle $x^2 + y^2 = 25$ between the points $(-3, 4)$ and $(3, 4)$ is revolved about the given axis, find the area of the resulting surface.

35 The y-axis

36 The x-axis

Exer. 37–39: Use a definite integral to derive a formula for the surface area of the indicated solid.

37 A right circular cone of altitude h and base radius r

38 A spherical segment of altitude h in a sphere of radius r

39 A sphere of radius r

40 Show that the area of the surface of a sphere of radius a between two parallel planes depends only on the distance between the planes. (*Hint:* Use Exercise 38.)

41 If the graph in Figure 6.49 is revolved about the y-axis, show that the area of the resulting surface is given by

$$\int_a^b 2\pi x\sqrt{1 + [f'(x)]^2}\, dx.$$

42 Use Exercise 41 to find the area of the surface generated by revolving the graph of $y = 3\sqrt[3]{x}$ from $A(1, 3)$ to $B(8, 6)$ about the y-axis.

c 43 The graph of $f(x) = 1 - x^3$ from $(0, 1)$ to $(1, 0)$ is revolved about the x-axis. Approximate the area of the resulting surface by using

$$\sum_{k=1}^{4} 2\pi \frac{f(x_{k-1}) + f(x_k)}{2} d(Q_{k-1}, Q_k),$$

where Q_k is the point $(\frac{1}{4}k, f(\frac{1}{4}k))$.

c 44 (a) Set up an integral for the area of the surface generated in Exercise 43.

(b) Use Simpson's rule with $n = 4$ to approximate the integral in (a).

6.6 WORK

The concept of force may be considered as a push or a pull on an object. For example, a force is needed to push or pull furniture across a floor, to lift an object off the ground, to stretch or compress a spring, or to move a charged particle through an electromagnetic field.

If an object weighs 10 pounds, then by definition the force required to lift it (or hold it off the ground) is 10 pounds. A force of this type is a **constant force**, since its magnitude does not change while it is applied to the object.

The concept of *work* is used when a force acts through a distance. The following definition covers the simplest case, in which the object moves along a line in the same direction as the applied force.

Definition **(6.20)**

> If a constant force F acts on an object, moving it a distance d in the direction of the force, the **work** W done is
>
> $$W = Fd.$$

The following table lists units of force and work in the British system and the International System (abbreviated SI, for the French *Système International*). In SI units, 1 Newton is the force required to impart an acceleration of 1 m/sec^2 to a mass of 1 kilogram.

System	Unit of force	Unit of distance	Unit of work
British	pound (lb)	foot (ft) inch (in.)	foot-pound (ft-lb) inch-pound (in.-lb)
International (SI)	Newton (N)	meter (m)	Newton-meter (N-m)

A Newton-meter is also called a *joule* (J). It can be shown that

$$1 \text{ N} \approx 0.225 \text{ lb} \quad \text{and} \quad 1 \text{ N-m} \approx 0.74 \text{ ft-lb}.$$

For simplicity, in examples and most exercises we will use the British system, in which the magnitude of the force is the same as the weight, in pounds, of the object. In using SI units it is often necessary to consider a gravitational constant a (9.81 m/sec^2) and use Newton's second law of motion, $F = ma$, to change a mass m (in kilograms) to a force F (in Newtons).

FIGURE 6.52

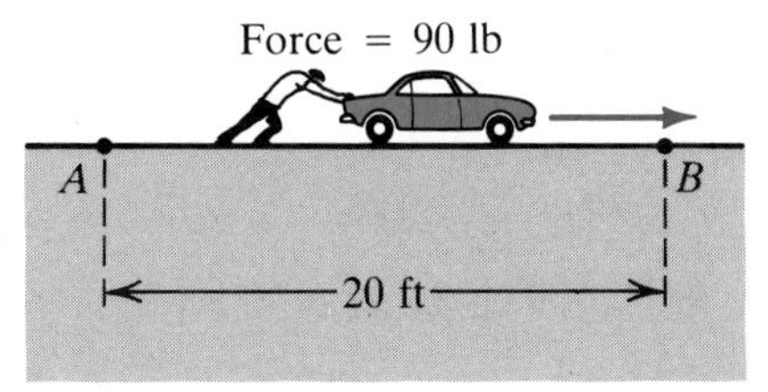

EXAMPLE 1 Find the work done in pushing an automobile a distance of 20 feet along a level road while exerting a constant force of 90 pounds.

SOLUTION The problem is illustrated in Figure 6.52. Since the constant force is $F = 90$ lb and the distance the automobile moves is $d = 20$ ft, it follows from Definition (6.20) that the work done is

$$W = (90)(20) = 1800 \text{ ft-lb}.$$

Anyone who has pushed an automobile (or some other object) is aware that the force applied is seldom constant. Thus, if an automobile is stalled, a larger force may be required to get it moving than to keep it in motion. The force may also vary because of friction, since part of the road may be smooth and another part rough. A force that is not constant is a **variable force**. We shall next develop a method for determining the work done by a variable force in moving an object rectilinearly in the same direction as the force.

Suppose a force moves an object along the x-axis from $x = a$ to $x = b$ and that the force at x is given by $f(x)$, where f is continuous on $[a, b]$. (The phrase *force at* x means the force acting at the point with coordinate x.) As shown in Figure 6.53, we begin by considering a partition P of $[a, b]$ determined by

$$a = x_0, x_1, x_2, \ldots, x_n = b \quad \text{with} \quad \Delta x_k = x_k - x_{k-1}.$$

FIGURE 6.53

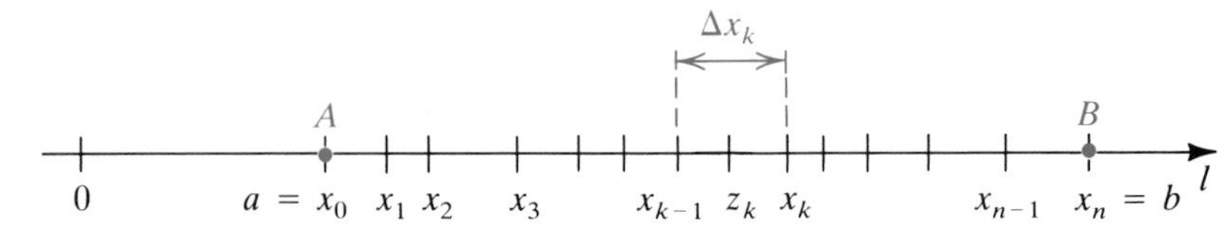

If ΔW_k is the **increment of work**—that is, the amount of work done from x_{k-1} to x_k—then the work W done from a to b is the sum

$$W = \Delta W_1 + \Delta W_2 + \cdots + \Delta W_n = \sum_{k=1}^{n} \Delta W_k.$$

To approximate ΔW_k, choose any number z_k in $[x_{k-1}, x_k]$ and consider the force $f(z_k)$ at z_k. If the norm $\|P\|$ is small, then intuitively we know that the function values change very little on $[x_{k-1}, x_k]$; that is, f is *almost constant* on this interval. Applying Definition (6.20) gives us

$$\Delta W_k \approx f(z_k)\, \Delta x_k$$

and hence

$$W = \sum_{k=1}^{n} \Delta W_k \approx \sum_{k=1}^{n} f(z_k)\, \Delta x_k.$$

Since this approximation should improve as $\|P\| \to 0$, we define W as a limit of such sums. This limit leads to a definite integral.

Definition **(6.21)**

If $f(x)$ is the force at x and if f is continuous on $[a, b]$, then the **work** W done in moving an object along the x-axis from $x = a$ to $x = b$ is

$$W = \lim_{\|P\| \to 0} \sum_k f(z_k)\, \Delta x_k = \int_a^b f(x)\, dx.$$

An analogous definition can be stated for an interval on the y-axis by replacing x with y throughout our discussion.

Definition (6.21) can be used to find the work done in stretching or compressing a spring. To solve problems of this type, it is necessary to use the following law from physics:

Hooke's Law: The force $f(x)$ required to stretch a spring x units beyond its natural length is given by $f(x) = kx$, where k is a constant called the **spring constant**.

EXAMPLE 2 A force of 9 pounds is required to stretch a spring from its natural length of 6 inches to a length of 8 inches. Find the work done in stretching the spring

(a) from its natural length to a length of 10 inches

(b) from a length of 7 inches to a length of 9 inches

SOLUTION

FIGURE 6.54

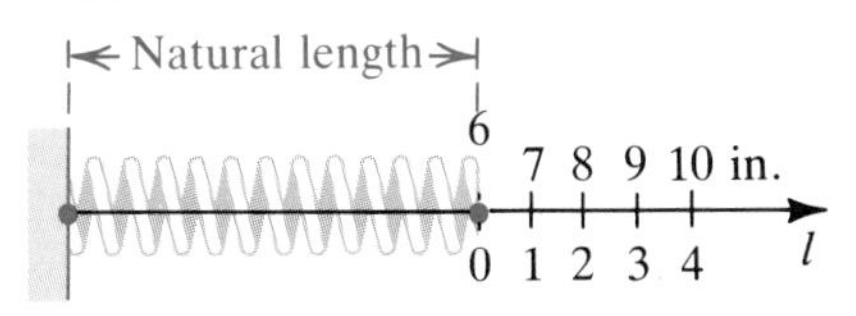

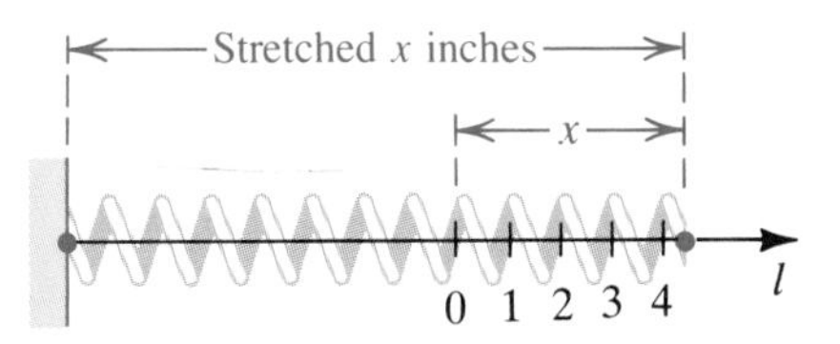

(a) Let us introduce an x-axis as shown in Figure 6.54, with one end of the spring attached to a point to the left of the origin and the end to be pulled located at the origin. According to Hooke's law, the force $f(x)$ required to stretch the spring x units beyond its natural length is $f(x) = kx$ for some constant k. Since a 9-pound force is required to stretch the spring 2 inches beyond its natural length, we have $f(2) = 9$. We let $x = 2$ in $f(x) = kx$:

$$9 = k \cdot 2, \quad \text{or} \quad k = \tfrac{9}{2}$$

Consequently, for this spring, Hooke's law has the form

$$f(x) = \tfrac{9}{2}x.$$

Applying Definition (6.21) with $a = 0$ and $b = 4$, we can determine the work done in stretching the spring 4 inches:

$$W = \int_0^4 \frac{9}{2}x\,dx = \frac{9}{2}\left[\frac{x^2}{2}\right]_0^4 = 36 \text{ in.-lb}$$

(b) We again use the force $f(x) = \tfrac{9}{2}x$ obtained in part (a). By Definition (6.21), the work done in stretching the spring from $x = 1$ to $x = 3$ is

$$W = \int_1^3 \frac{9}{2}x\,dx = \frac{9}{2}\left[\frac{x^2}{2}\right]_1^3 = 18 \text{ in.-lb.}$$

FIGURE 6.55

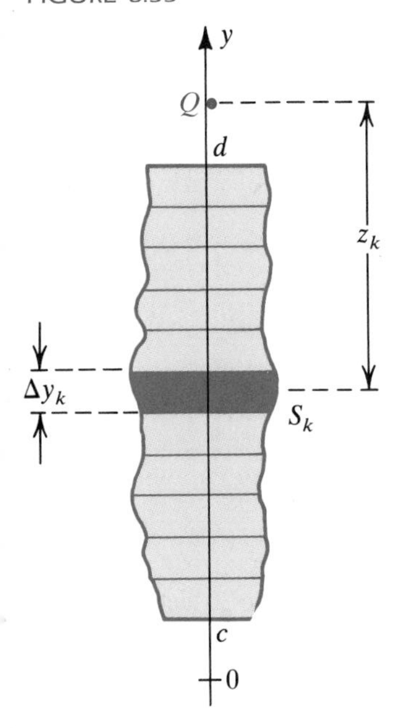

In some applications we wish to determine the work done in pumping out a tank containing a fluid or in lifting an object, such as a chain or a cable, that extends vertically between two points. A general situation is illustrated in Figure 6.55, which shows a solid that extends along the y-axis from $y = c$ to $y = d$. We wish to vertically lift all particles contained in the solid to the level of point Q. Let us consider a partition P of $[c, d]$ and imagine slicing the solid by means of planes perpendicular to the y-axis at each number y_k in the partition. As shown in the figure, $\Delta y_k = y_k - y_{k-1}$, and S_k represents the kth slice. We next introduce the following notation:

$$z_k = \text{the (approximate) distance } S_k \text{ is lifted}$$

$$\Delta F_k = \text{the (approximate) force required to lift } S_k$$

If ΔW_k is the work done in lifting S_k, then, by Definition (6.20),

$$\Delta W_k \approx \Delta F_k \cdot z_k = z_k \cdot \Delta F_k.$$

We define the work W done in lifting the entire solid as a limit of sums.

Definition **(6.22)**

$$W = \lim_{\|P\| \to 0} \sum_k z_k \cdot \Delta F_k$$

The limit leads to a definite integral. Notice the difference between this type of problem and that in our earlier discussion. To obtain (6.21), we considered *distance increments* Δx_k and the force $f(z_k)$ that acts through Δx_k. In the present situation we consider *force increments* ΔF_k and the distance z_k through which ΔF_k acts. The next two examples illustrate this technique. As in preceding sections, we shall use dy to represent a typical increment Δy_k and y to denote a number in $[c, d]$.

FIGURE 6.56

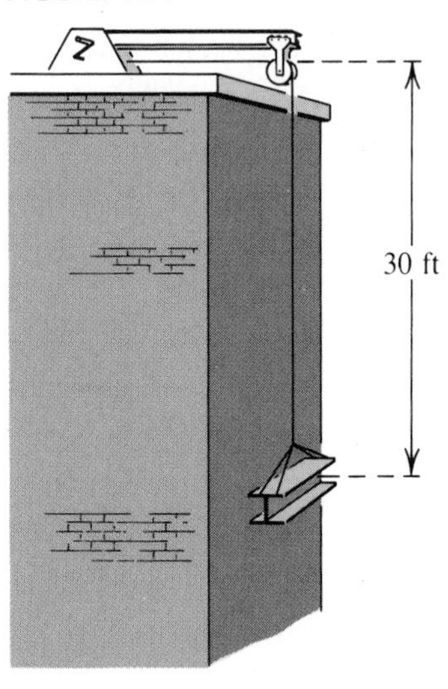

EXAMPLE 3 A uniform cable 30 feet long and weighing 60 pounds hangs vertically from a pulley system at the top of a building, as shown in Figure 6.56. A steel beam that weighs 500 pounds is attached to the end of the cable; find the work required to pull it to the top.

SOLUTION Let W_B denote the work required to pull the beam to the top, and let W_C denote the work required for the cable. Since the beam weighs 500 pounds and must move through a distance of 30 feet, we have, by Definition (6.20),

$$W_B = 500 \cdot 30 = 15{,}000 \text{ ft-lb}.$$

The work required to pull the cable to the top may be found by the method used to obtain (6.22). Consider a y-axis with the lower end of the cable at the origin and the upper end at $y = 30$, as in Figure 6.57. Let dy denote an increment of length of the cable. Since each foot of cable weighs $60/30 = 2$ lb, the weight of the increment (and hence the force required to lift it) is $2\,dy$. If y denotes the distance from O to a point in the increment, then we have the following:

FIGURE 6.57

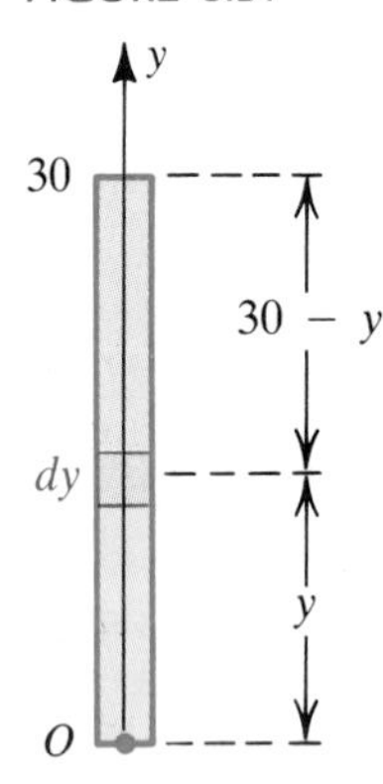

increment of force: $2\,dy$

distance lifted: $30 - y$

increment of work: $(30 - y)2\,dy$

Applying $\int_0^{30}$ takes a limit of sums of the increments of work. Hence

$$\begin{aligned} W_C &= \int_0^{30} (30 - y)2\,dy \\ &= 2\Big[30y - \tfrac{1}{2}y^2\Big]_0^{30} \\ &= 900 \text{ ft-lb}. \end{aligned}$$

The total work required is

$$W = W_B + W_C = 15{,}000 + 900 = 15{,}900 \text{ ft-lb}.$$

EXAMPLE 4 A right circular conical tank of altitude 20 feet and radius of base 5 feet has its vertex at ground level and axis vertical. If the tank is full of water weighing 62.5 lb/ft^3, find the work done in pumping all the water over the top of the tank.

SOLUTION We begin by introducing a coordinate system, as shown in Figure 6.58. The cone intersects the xy-plane along the line of slope 4

FIGURE 6.58

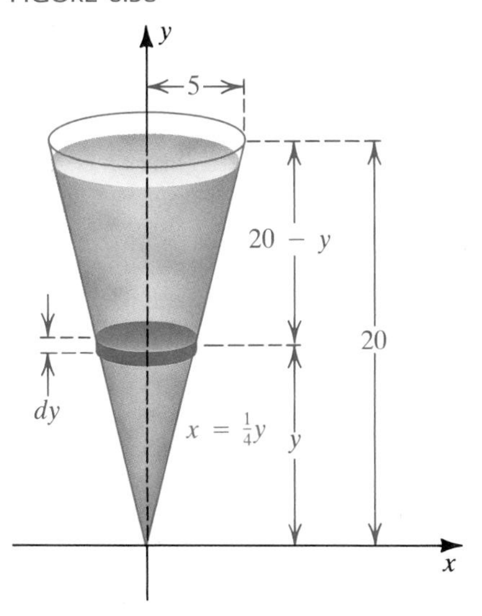

through the origin. An equation for this line is

$$y = 4x, \quad \text{or} \quad x = \tfrac{1}{4}y.$$

Let us imagine subdividing the water into slices, using planes perpendicular to the y-axis, from $y = 0$ to $y = 20$. If dy represents the width of a typical slice, then its volume may be approximated by the circular disk shown in Figure 6.58. As we did in our work with volumes of revolution in Section 6.2, we obtain

$$\text{volume of disk} = \pi x^2\, dy = \pi(\tfrac{1}{4}y)^2\, dy.$$

Since water weighs 62.5 lb/ft^3, the weight of the disk, and hence the force required to lift it, is $62.5\pi(\frac{1}{4}y)^2\, dy$. Thus, we have

$$\begin{aligned} \text{increment of force:} &\quad 62.5\pi(\tfrac{1}{16}y^2)\, dy \\ \text{distance lifted:} &\quad 20 - y \\ \text{increment of work:} &\quad (20 - y)62.5\pi(\tfrac{1}{16}y^2)\, dy \end{aligned}$$

Applying $\int_0^{20}$ takes a limit of sums of the increments of work. Hence

$$\begin{aligned} W &= \int_0^{20} (20 - y)62.5\pi(\tfrac{1}{16}y^2)\, dy \\ &= \frac{62.5}{16}\pi \int_0^{20} (20y^2 - y^3)\, dy \\ &= \frac{62.5}{16}\pi \left[20\left(\frac{y^3}{3}\right) - \frac{y^4}{4} \right]_0^{20} \\ &= \frac{62.5}{16}\pi \left(\frac{40{,}000}{3}\right) \approx 163{,}625 \text{ ft-lb.} \end{aligned}$$

The next example is another illustration of how work may be calculated by means of a limit of sums—that is, by a definite integral.

EXAMPLE 5 A confined gas has pressure p (lb/in.^{2}) and volume v (in.^{3}). If the gas expands from $v = a$ to $v = b$, show that the work done (in.-lb) is given by

$$W = \int_a^b p\, dv.$$

SOLUTION Since the work done is independent of the shape of the container, we may assume that the gas is enclosed in a right circular cylinder of radius r and that the expansion takes place against a piston head, as illustrated in Figure 6.59 on the following page. As in the figure, let dv denote the change in volume that corresponds to a change of h inches in the position of the piston head. Thus,

$$dv = \pi r^2 h, \quad \text{or} \quad h = \frac{1}{\pi r^2}\, dv.$$

If p denotes the pressure at some point in the volume increment shown in Figure 6.59, then the force against the piston head is the product of p and the area πr^2 of the piston head. Thus, we have the following for the

FIGURE 6.59

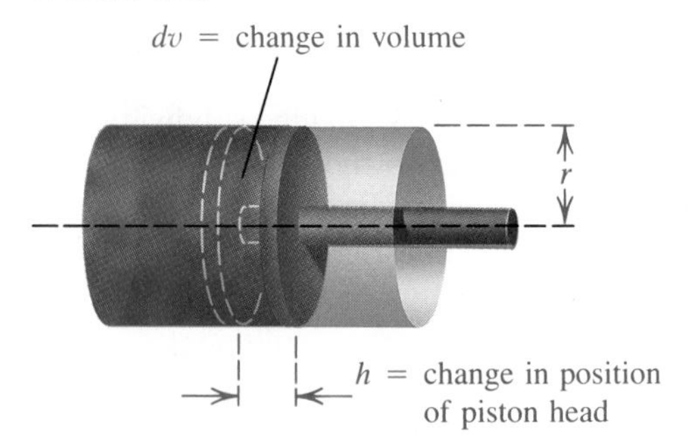

indicated volume increment: dv

force against piston head: $p(\pi r^2)$

distance piston head moves: h

increment of work: $(p\pi r^2)h = (p\pi r^2)\dfrac{1}{\pi r^2}\,dv = p\,dv$

Applying $\int_a^b$ to the increments of work gives us the work done as the gas expands from $v = a$ to $v = b$:

$$W = \int_a^b p\,dv$$

EXERCISES 6.6

1 A 400-pound gorilla climbs a vertical tree 15 feet high. Find the work done if the gorilla reaches the top in

(a) 10 seconds (b) 5 seconds

2 Find the work done in lifting an 80-pound sandbag a height of 4 feet.

3 A spring of natural length 10 inches stretches 1.5 inches under a weight of 8 pounds. Find the work done in stretching the spring

(a) from its natural length to a length of 14 inches

(b) from a length of 11 inches to a length of 13 inches

4 A force of 25 pounds is required to compress a spring of natural length 0.80 foot to a length of 0.75 foot. Find the work done in compressing the spring from its natural length to a length of 0.70 foot.

5 If a spring is 12 inches long, compare the work W_1 done in stretching it from 12 inches to 13 inches with the work W_2 done in stretching it from 13 inches to 14 inches.

6 It requires 60 in.-lb of work to stretch a certain spring from a length of 6 inches to 7 inches and another 120 in.-lb of work to stretch it from 7 inches to 8 inches. Find the spring constant and the natural length of the spring.

7 A freight elevator weighing 3000 pounds is supported by a 12-foot-long cable that weighs 14 pounds per linear foot. Approximate the work required to lift the elevator 9 feet by winding the cable onto a winch.

8 A construction worker pulls a 50-pound motor from ground level to the top of a 60-foot-high building using a rope that weighs $\frac{1}{4}$ lb/ft. Find the work done.

9 A bucket containing water is lifted vertically at a constant rate of 1.5 ft/sec by means of a rope of negligible weight. As the bucket rises, water leaks out at the rate of 0.25 lb/sec. If the bucket weighs 4 pounds when empty and if it contained 20 pounds of water at the instant that the lifting began, determine the work done in raising the bucket 12 feet.

10 In Exercise 9, find the work required to raise the bucket until half the water has leaked out.

11 A fishtank has a rectangular base of width 2 feet and length 4 feet, and rectangular sides of height 3 feet. If the tank is filled with water weighing 62.5 lb/ft^3, find the work required to pump all the water over the top of the tank.

12 Generalize Example 4 of this section to the case of a conical tank of altitude h feet and radius of base a feet that is filled with a liquid weighing ρ lb/ft^3.

13 A vertical cylindrical tank of diameter 3 feet and height 6 feet is full of water. Find the work required to pump all the water

(a) over the top of the tank

(b) through a pipe that rises to a height of 4 feet above the top of the tank

14 Work Exercise 13 if the tank is only half-full of water.

15 The ends of an 8-foot-long water trough are equilateral triangles having sides of length 2 feet. If the trough is full of water, find the work required to pump all of it over the top.

16 A cistern has the shape of the lower half of a sphere of radius 5 feet. If the cistern is full of water, find the work required to pump all the water to a point 4 feet above the top of the cistern.

17 Refer to Example 5 in this section. The volume and pressure of a certain gas vary in accordance with the law $pv^{1.2} = 115$, where the units of measurement are inches and pounds. Find the work done if the gas expands from 32 in.^{3} to 40 in.^{3}

18 Refer to Example 5. The pressure and volume of a quantity of enclosed steam are related by the formula $pv^{1.14} = c$, where c is a constant. If the initial pressure and volume are p_0 and v_0, respectively, find a formula for the work done if the steam expands to twice its volume.

19 Newton's law of gravitation states that the force F of attraction between two particles having masses m_1 and m_2 is given by $F = Gm_1m_2/s^2$, where G is a gravitational constant and s is the distance between the particles. If the mass m_1 of the earth is regarded as concentrated at the center of the earth and a rocket of mass m_2 is on the surface (a distance 4000 miles from the center), find a general formula for the work done in firing the rocket vertically upward to an altitude h (see figure).

EXERCISE 19

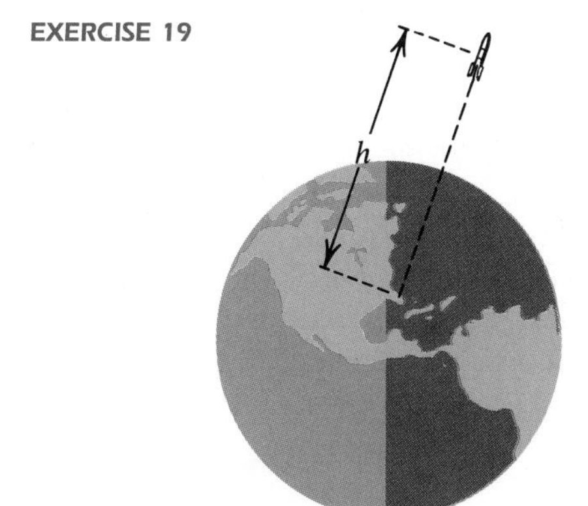

20 In the study of electricity, the formula $F = kq/r^2$, where k is a constant, is used to find the force (in Newtons) with which a positive charge Q of strength q units repels a unit positive charge located r meters from Q. Find the work done in moving a unit charge from a point d centimeters from Q to a point $\frac{1}{2}d$ centimeters from Q.

c **Exer. 21–22: Suppose the table was obtained experimentally for a force $f(x)$ acting at the point with coordinate x on a coordinate line. Use the trapezoidal rule to approximate the work done on the interval $[a, b]$, where a and b are the smallest and largest values of x, respectively.**

21

x (ft)	0	0.5	1.0	1.5	2.0	2.5
$f(x)$ (lb)	7.4	8.1	8.4	7.8	6.3	7.1

x (ft)	3.0	3.5	4.0	4.5	5.0
$f(x)$ (lb)	5.9	6.8	7.0	8.0	9.2

22

x (m)	1	2	3	4	5
$f(x)$ (N)	125	120	130	146	165

x (m)	6	7	8	9
$f(x)$ (N)	157	150	143	140

23 The force (in Newtons) with which two electrons repel each other is inversely proportional to the square of the distance (in meters) between them.

(a) If one electron is held fixed at the point (5, 0), find the work done in moving a second electron along the x-axis from the origin to the point (3, 0).

(b) If two electrons are held fixed at the points (5, 0) and (−5, 0), respectively, find the work done in moving a third electron from the origin to (3, 0).

24 If the force function is constant, show that Definition (6.21) reduces to Definition (6.20).

6.7 MOMENTS AND CENTERS OF MASS

In this section we consider some topics involving the mass of an object. The terms *mass* and *weight* are sometimes confused with each other. Weight is determined by the force of gravity. For example, the weight of an object on the moon is approximately one-sixth its weight on earth, because the force of gravity is weaker. However, the mass is the same. Newton used the term *mass* synonymously with *quantity of matter* and related it to force by his *second law of motion*, $F = ma$, where F denotes the force acting on an object of mass m that has acceleration a. In the British system, we often approximate a by 32 ft/sec^2 and use the **slug** as the unit of mass. In SI units, $a \approx 9.81$ m/sec^2, and the kilogram is the unit of mass. It can be shown that

$$1 \text{ slug} \approx 14.6 \text{ kg} \quad \text{and} \quad 1 \text{ kg} \approx 0.07 \text{ slug}.$$

In applications we usually assume that the mass of an object is concentrated at a point, and we refer to the object as a **point-mass**, regardless of its size. For example, using the earth as a frame of reference, we may regard a human being, an automobile, or a building as a point-mass.

FIGURE 6.60

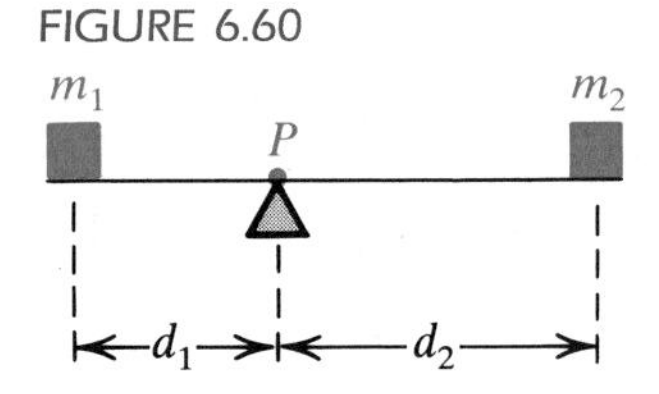

In an elementary physics experiment we consider two point-masses m_1 and m_2 attached to the ends of a thin rod, as illustrated in Figure 6.60, and then locate the point P at which a fulcrum should be placed so that the rod balances. (This situation is similar to balancing a seesaw with a person sitting at each end.) If the distances from m_1 and m_2 to P are d_1 and d_2, respectively, then it can be shown experimentally that P is the balance point if

$$m_1 d_1 = m_2 d_2.$$

FIGURE 6.61

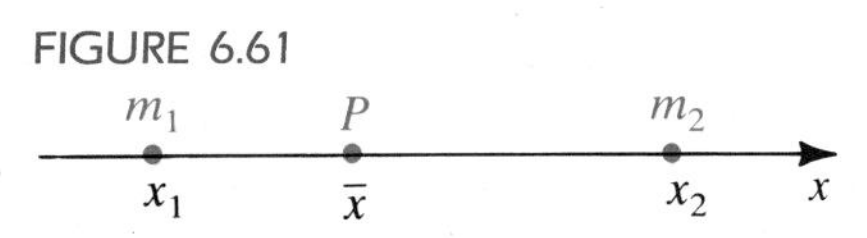

In order to generalize this concept, let us introduce an x-axis, as illustrated in Figure 6.61, with m_1 and m_2 located at points with coordinates x_1 and x_2. If the coordinate of the balance point P is $\bar{x}$, then using the formula $m_1 d_1 = m_2 d_2$ yields

$$m_1(\bar{x} - x_1) = m_2(x_2 - \bar{x})$$

$$m_1\bar{x} + m_2\bar{x} = m_1 x_1 + m_2 x_2$$

$$\bar{x} = \frac{m_1 x_1 + m_2 x_2}{m_1 + m_2}.$$

This gives us a formula for locating the balance point P.

If a mass m is located at a point on the axis with coordinate x, then the product mx is called the *moment M_0 of the mass about the origin*. Our formula for $\bar{x}$ states that to find the coordinate of the balance point, we may divide the sum of the moments about the origin by the total mass. The point with coordinate $\bar{x}$ is called the *center of mass* (or *center of gravity*) of the two point-masses. The next definition extends this discussion to many point-masses located on an axis, as shown in Figure 6.62.

FIGURE 6.62

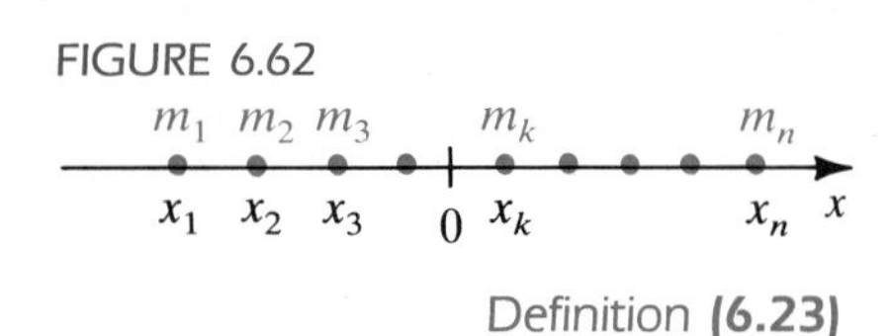

Definition **(6.23)**

Let S denote a system of point-masses $m_1, m_2, \ldots, m_n$ located at $x_1, x_2, \ldots, x_n$ on a coordinate line, and let $m = \sum_{k=1}^{n} m_k$ denote the total mass.

(i) The **moment of S about the origin** is $M_0 = \sum_{k=1}^{n} m_k x_k$.

(ii) The **center of mass** (or **center of gravity**) of S is given by $\bar{x} = \dfrac{M_0}{m}$.

The point with coordinate $\bar{x}$ is the balance point of the system S in the same sense as in our seesaw illustration.

EXAMPLE 1 Three point-masses of 40, 60, and 100 kilograms are located at -2, 3, and 7, respectively, on an x-axis. Find the center of mass.

SOLUTION If we denote the three masses by m_1, m_2, and m_3, we have the situation illustrated in Figure 6.63, with $x_1 = -2$, $x_2 = 3$, and $x_3 = 7$. Applying Definition (6.23) gives us the coordinate $\bar{x}$ of the center of mass:

FIGURE 6.63

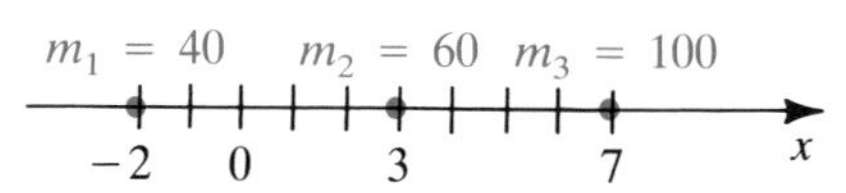

$$\bar{x} = \frac{40(-2) + 60(3) + 100(7)}{40 + 60 + 100} = \frac{800}{200} = 4$$

FIGURE 6.64

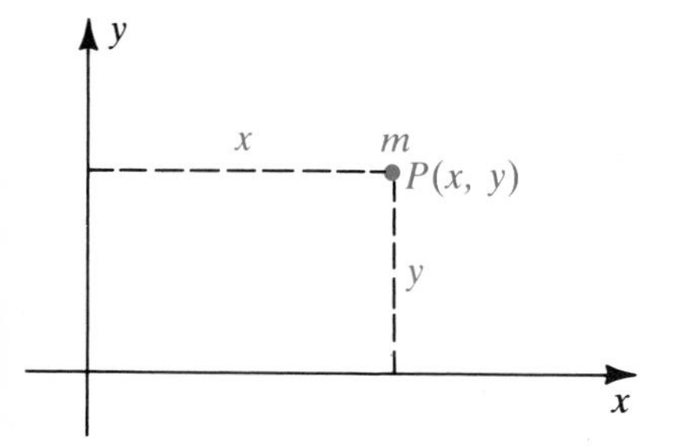

Let us next consider a point-mass m located at $P(x, y)$ in a coordinate plane (see Figure 6.64). We define the moments M_x and M_y of m about the coordinate axes as follows:

$$\textit{moment about the x-axis:} \quad M_x = my$$

$$\textit{moment about the y-axis:} \quad M_y = mx$$

In words, to find M_x we multiply m by the y-coordinate of P, and to find M_y we multiply m by the x-coordinate. To find M_x and M_y for a *system* of point-masses, we add the individual moments, as in (i) and (ii) of the next definition.

Definition **(6.24)**

Let S denote a system of point-masses $m_1, m_2, \ldots, m_n$ located at (x_1, y_1), $(x_2, y_2), \ldots, (x_n, y_n)$ in a coordinate plane, and let $m = \sum_{k=1}^{n} m_k$ denote the total mass.

(i) The **moment of S about the x-axis** is $M_x = \sum_{k=1}^{n} m_k y_k$.

(ii) The **moment of S about the y-axis** is $M_y = \sum_{k=1}^{n} m_k x_k$.

(iii) The **center of mass** (or **center of gravity**) of S is the point $(\bar{x}, \bar{y})$ such that

$$\bar{x} = \frac{M_y}{m}, \quad \bar{y} = \frac{M_x}{m}.$$

From (iii) of this definition,

$$m\bar{x} = M_y \quad \text{and} \quad m\bar{y} = M_x.$$

Since $m\bar{x}$ and $m\bar{y}$ are the moments about the y-axis and x-axis, respectively, of a single point-mass m located at $(\bar{x}, \bar{y})$, we may interpret the center of mass as the point at which the total mass can be concentrated to obtain the moments M_y and M_x of S.

FIGURE 6.65

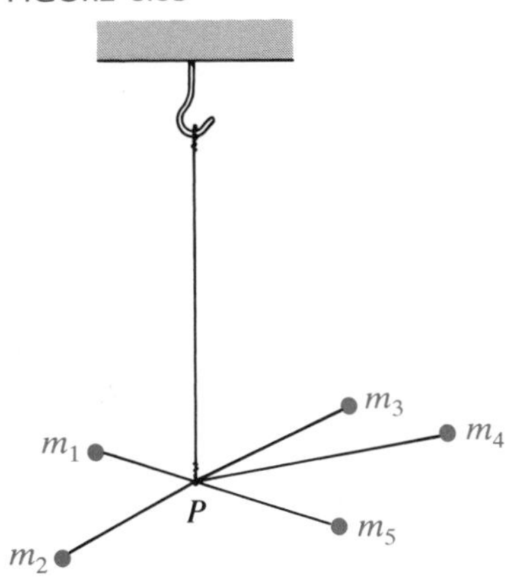

We might think of the n point-masses in (6.24) as being fastened to the center of mass P by weightless rods, as spokes of a wheel are attached to the center of the wheel. The system S would balance if supported by a cord attached to P, as illustrated in Figure 6.65. The appearance would be similar to that of a mobile having all its objects in the same horizontal plane.

EXAMPLE 2 Point-masses of 4, 8, 3, and 2 kilograms are located at $(-2, 3)$, $(2, -6)$, $(7, -3)$, and $(5, 1)$, respectively. Find M_x, M_y, and the center of mass of the system.

FIGURE 6.66

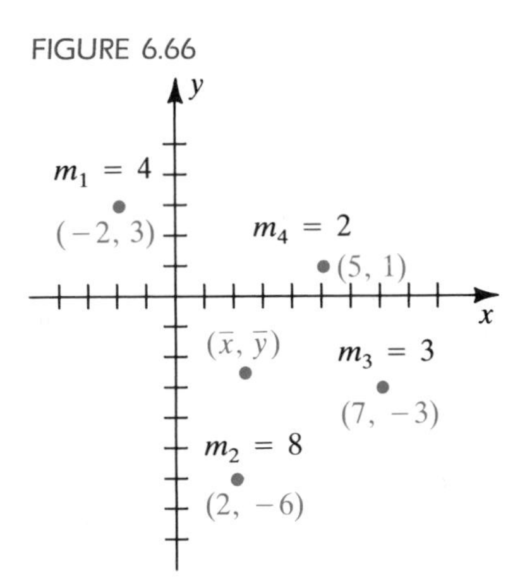

SOLUTION The masses are illustrated in Figure 6.66, in which we have also anticipated the position of $(\bar{x}, \bar{y})$. Applying Definition (6.24) gives us

$$M_x = (4)(3) + (8)(-6) + (3)(-3) + (2)(1) = -43$$

$$M_y = (4)(-2) + (8)(2) + (3)(7) + (2)(5) = 39.$$

Since $m = 4 + 8 + 3 + 2 = 17$,

$$\bar{x} = \frac{M_y}{m} = \frac{39}{17} \approx 2.3 \quad \text{and} \quad \bar{y} = \frac{M_x}{m} = -\frac{43}{17} \approx -2.5.$$

Thus, the center of mass is $(\frac{39}{17}, -\frac{43}{17})$.

Later in the text we shall consider solid objects that are **homogeneous** in the sense that the mass is uniformly distributed throughout the solid. In physics, the **density** ρ (rho) of a homogeneous solid of mass m and volume V is defined by $\rho = m/V$. Thus, *density is mass per unit volume*. The SI unit for density is kg/m^3; however, g/cm^3 is also used. The British unit is lb/ft^3 or lb/in.^3

In this section we shall restrict our discussion to homogeneous **laminas** (thin flat plates) that have **area density** (mass per unit area) ρ. Area density is measured in kg/m^2, lb/ft^2, and so on. If the area of one face of a lamina is A and the area density is ρ, then its mass m is given by $m = \rho A$. We wish to define the center of mass P such that if the tip of a sharp pencil were placed at P, as illustrated in Figure 6.67, the lamina would balance in

FIGURE 6.67

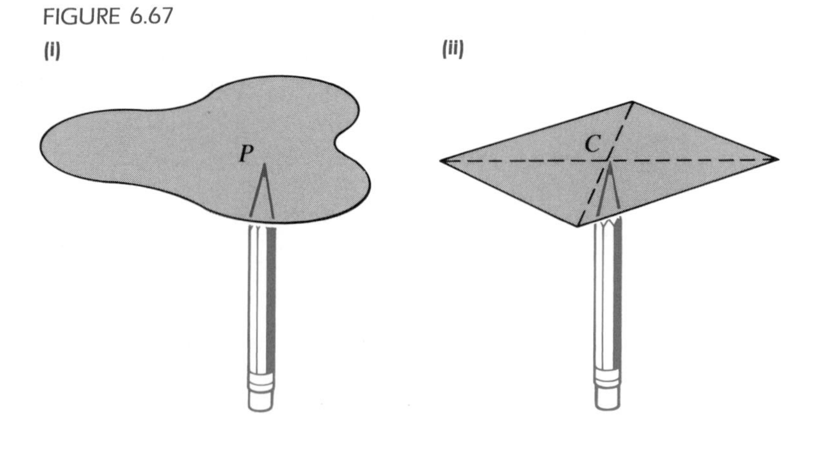

a horizontal position. As in (ii) of the figure, we shall assume that *the center of mass of a rectangular lamina is the point C at which the diagonals intersect*. We call C the *center* of the rectangle. Thus, for problems involving mass, we may assume that a rectangular lamina is a point-mass located at the center of the rectangle. This assumption is the key to our definition of the center of mass of a lamina.

Consider a lamina that has area density ρ and the shape of the R_x region in Figure 6.68. Since we have had ample experience using limits of Riemann sums for definitions in Sections 6.1 through 6.6, let us proceed directly to the method of representing the width of the rectangle in the figure by dx (instead of Δx_k), obtaining

$$\text{area of rectangle:} \quad [f(x) - g(x)]\,dx.$$

FIGURE 6.68

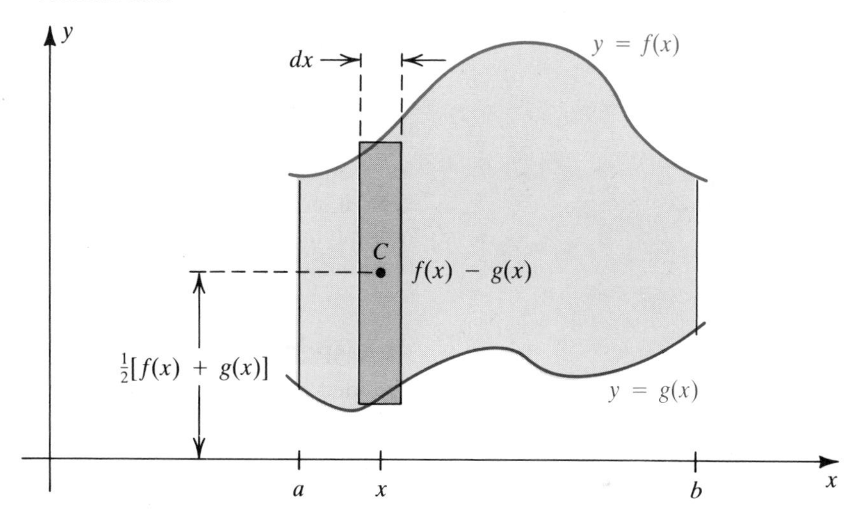

Since the area density of the lamina is ρ, we may write

$$\text{mass of rectangular lamina:} \quad \rho[f(x) - g(x)]\,dx.$$

If, as in previous sections, we regard $\int_a^b$ as an operator that takes limits of sums, we arrive at the following definition for the mass m of the lamina:

$$m = \int_a^b \rho[f(x) - g(x)]\,dx$$

We next assume that the rectangular lamina in Figure 6.68 is a point-mass located at the center C of the rectangle. Since, by the midpoint formula (1.5), the distance from the x-axis to C is $\frac{1}{2}[f(x) + g(x)]$, we obtain the following result for the rectangular lamina:

$$\textit{moment about the x-axis:} \quad \tfrac{1}{2}[f(x) + g(x)] \cdot \rho[f(x) - g(x)]\,dx$$

Similarly, since the distance from the y-axis to C is x,

$$\textit{moment about the y-axis:} \quad x \cdot \rho[f(x) - g(x)]\,dx.$$

Taking limits of sums by applying $\int_a^b$ leads to the next definition.

Definition (6.25)

Let a lamina L of area density ρ have the shape of the R_x region in Figure 6.68.

(i) The **mass** of L is $m = \int_a^b \rho[f(x) - g(x)]\,dx$.

(ii) The **moments of L about the x-axis and y-axis** are

$$M_x = \int_a^b \tfrac{1}{2}[f(x) + g(x)] \cdot \rho[f(x) - g(x)]\,dx$$

and $$M_y = \int_a^b x \cdot \rho[f(x) - g(x)]\,dx.$$

(iii) The **center of mass** (or **center of gravity**) of L is the point $(\bar{x}, \bar{y})$ such that

$$\bar{x} = \frac{M_y}{m} \quad \text{and} \quad \bar{y} = \frac{M_x}{m}.$$

An analogous definition can be stated if L has the shape of an R_y region and the integrations are with respect to y. We could also obtain formulas for moments with respect to lines other than the x-axis or y-axis; however, it is advisable to remember the *technique* for finding moments—multiplying a mass by a distance from an axis—instead of memorizing formulas that cover all possible cases.

EXAMPLE 3 A lamina of area density ρ has the shape of the region bounded by the graphs of $y = x^2 + 1$, $x = 0$, $x = 1$, and $y = 0$. Find the center of mass.

FIGURE 6.69

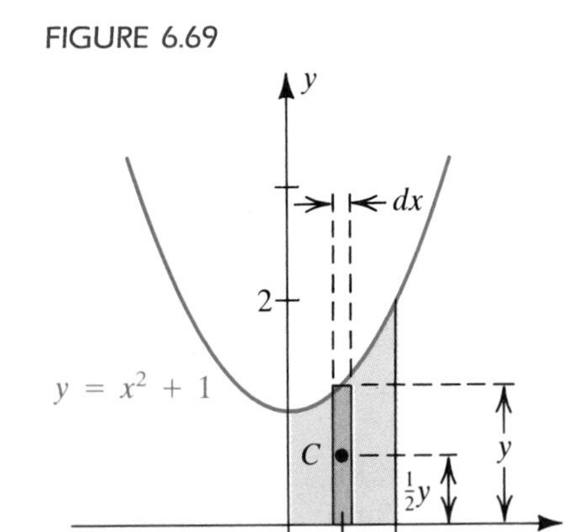

SOLUTION The region and a typical rectangle of width dx and height y are sketched in Figure 6.69. As indicated in the figure, the distance from the x-axis to the center C of the rectangle is $\frac{1}{2}y$, and the distance from the y-axis to C is x. Hence, *for the rectangular lamina*, we have the following:

$$\begin{aligned} \text{mass: } & \rho y\,dx = \rho(x^2 + 1)\,dx \\ \text{moment about } x\text{-axis: } & \tfrac{1}{2}y \cdot \rho y\,dx = \tfrac{1}{2}\rho(x^2 + 1)^2\,dx \\ \text{moment about } y\text{-axis: } & x \cdot \rho y\,dx = \rho x(x^2 + 1)\,dx \end{aligned}$$

We now take a limit of sums of these expressions by applying the operator $\int_0^1$:

$$m = \int_0^1 \rho(x^2 + 1)\,dx = \rho\left[\tfrac{1}{3}x^3 + x\right]_0^1 = \tfrac{4}{3}\rho$$

$$\begin{aligned} M_x &= \int_0^1 \tfrac{1}{2}\rho(x^2 + 1)^2\,dx = \tfrac{1}{2}\rho \int_0^1 (x^4 + 2x^2 + 1)\,dx \\ &= \tfrac{1}{2}\rho\left[\tfrac{1}{5}x^5 + \tfrac{2}{3}x^3 + x\right]_0^1 = \tfrac{14}{15}\rho \end{aligned}$$

$$\begin{aligned} M_y &= \int_0^1 \rho x(x^2 + 1)\,dx = \rho \int_0^1 (x^3 + x)\,dx \\ &= \rho\left[\tfrac{1}{4}x^4 + \tfrac{1}{2}x^2\right]_0^1 = \tfrac{3}{4}\rho \end{aligned}$$

To find the center of mass $(\bar{x}, \bar{y})$, we use (iii) of Definition (6.25):

$$\bar{x} = \frac{M_y}{m} = \frac{\frac{3}{4}\rho}{\frac{4}{3}\rho} = \frac{9}{16} \quad \text{and} \quad \bar{y} = \frac{M_x}{m} = \frac{\frac{14}{15}\rho}{\frac{4}{3}\rho} = \frac{7}{10}$$

When we found $(\bar{x}, \bar{y})$ in Example 3, the constant ρ in the numerator and denominator canceled. This will always be the case for a homogeneous lamina. Hence the center of mass is independent of the area density ρ; that is, $\bar{x}$ and $\bar{y}$ depend only on the shape of the lamina. For this reason, the point $(\bar{x}, \bar{y})$ is sometimes referred to as the center of mass of a *region* in the plane, or as the **centroid** of the region. We can obtain formulas for moments of centroids by letting $\rho = 1$ and $m = A$ (the area of the region) in our previous work.

EXAMPLE 4 Find the centroid of the region bounded by the graphs of $y = 6 - x^2$ and $y = 3 - 2x$.

FIGURE 6.70

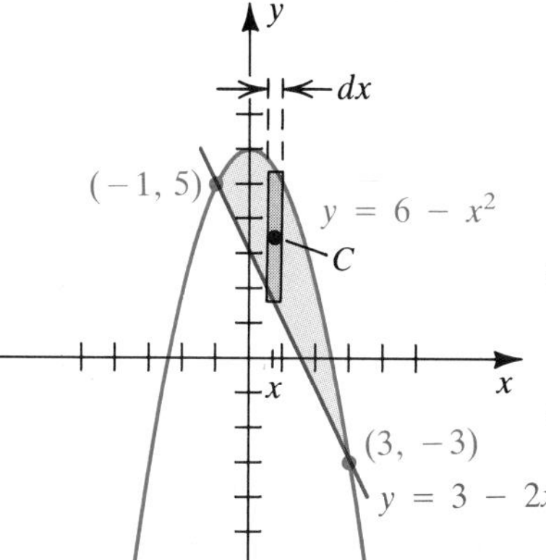

SOLUTION The region is the same as that considered in Example 2 of Section 6.1 and is resketched in Figure 6.70. To find the moments and the centroid, we take $\rho = 1$ and $m = A$. Referring to the typical rectangle with center C shown in Figure 6.70, we obtain the following:

area of rectangle: $[(6 - x^2) - (3 - 2x)]\,dx$

distance from x-axis to C: $\frac{1}{2}[(6 - x^2) + (3 - 2x)]$

moment about x-axis: $\frac{1}{2}[(6 - x^2) + (3 - 2x)] \cdot [(6 - x^2) - (3 - 2x)]\,dx$

distance from y-axis to C: x

moment about y-axis: $x[(6 - x^2) - (3 - 2x)]\,dx$

We now take a limit of sums by applying the operator $\int_{-1}^{3}$:

$$\begin{aligned} M_x &= \int_{-1}^{3} \tfrac{1}{2}[(6 - x^2) + (3 - 2x)] \cdot [(6 - x^2) - (3 - 2x)]\,dx \\ &= \tfrac{1}{2}\int_{-1}^{3} [(6 - x^2)^2 - (3 - 2x)^2]\,dx \\ &= \tfrac{1}{2}\int_{-1}^{3} (x^4 - 16x^2 + 12x + 27)\,dx = \tfrac{416}{15} \end{aligned}$$

$$\begin{aligned} M_y &= \int_{-1}^{3} x[(6 - x^2) - (3 - 2x)]\,dx \\ &= \int_{-1}^{3} (3x + 2x^2 - x^3)\,dx = \tfrac{32}{3} \end{aligned}$$

Using $A = \frac{32}{3}$ and (iii) of Definition (6.25), we determine the centroid:

$$\bar{x} = \frac{M_y}{m} = \frac{\frac{32}{3}}{\frac{32}{3}} = 1 \quad \text{and} \quad \bar{y} = \frac{M_x}{m} = \frac{\frac{416}{15}}{\frac{32}{3}} = \frac{13}{5}$$

We could have found the centroid by using Definition (6.25) with $f(x) = 6 - x^2$, $g(x) = 3 - 2x$, $a = -1$, and $b = 3$, but that would merely teach you how to substitute and not how to think.

If a homogeneous lamina has the shape of a region that has an axis of symmetry, then the center of mass must lie on that axis. This fact is used in the next example.

EXAMPLE 5 Find the centroid of the semicircular region bounded by the x-axis and the graph of $y = \sqrt{a^2 - x^2}$ with $a > 0$.

FIGURE 6.71

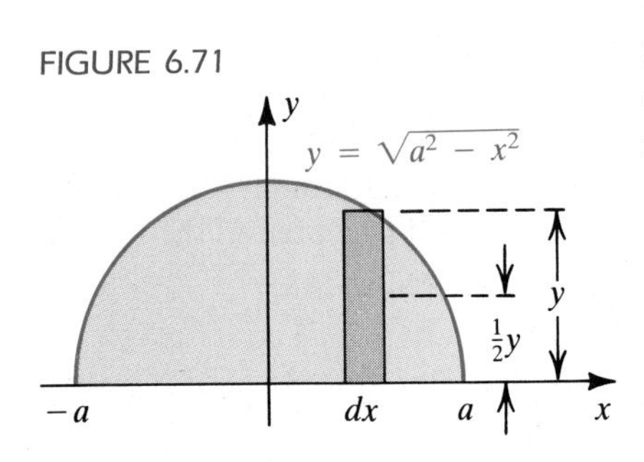

SOLUTION The region is sketched in Figure 6.71. By symmetry, the centroid is on the y-axis; that is, $\bar{x} = 0$. Hence we need find only $\bar{y}$. Referring to the rectangle in the figure and using $\rho = 1$ gives us the following result.

$$\text{moment about } x\text{-axis: } \tfrac{1}{2}y \cdot y\,dx = \tfrac{1}{2}y^2\,dx = \tfrac{1}{2}(a^2 - x^2)\,dx$$

We now take a limit of sums by applying the operator $\int_{-a}^{a}$:

$$M_x = \int_{-a}^{a} \tfrac{1}{2}(a^2 - x^2)\,dx = 2\int_0^a \tfrac{1}{2}(a^2 - x^2)\,dx$$
$$= \left[a^2x - \tfrac{1}{3}x^3\right]_0^a = \tfrac{2}{3}a^3$$

Using $m = A = \frac{1}{2}\pi a^2$ gives us

$$\bar{y} = \frac{M_x}{m} = \frac{\frac{2}{3}a^3}{\frac{1}{2}\pi a^2} = \frac{4a}{3\pi} \approx 0.42a.$$

Thus, the centroid is the point $\left(0, \dfrac{4}{3\pi}a\right)$.

We shall conclude this section by stating a useful theorem about solids of revolution. To illustrate a special case of the theorem, consider an R_x region R of the type shown in Figure 6.68. Using $\rho = 1$ and $m = A$ (the area of R), we find that the moment of R about the y-axis is given by

$$M_y = \int_a^b x[f(x) - g(x)]\,dx.$$

If R is revolved about the y-axis, then using cylindrical shells, we find that the volume V of the resulting solid is given by

$$V = \int_a^b 2\pi x[f(x) - g(x)]\,dx.$$

Comparing these two equations, we see that

$$M_y = \frac{V}{2\pi}.$$

If $(\bar{x}, \bar{y})$ is the centroid of R, then, by Definition (6.25)(iii),

$$\bar{x} = \frac{M_y}{m} = \frac{(V/2\pi)}{A} = \frac{V}{2\pi A}$$

and hence

$$V = 2\pi\bar{x}A.$$

Since $\bar{x}$ is the distance from the y-axis to the centroid of R, the last formula states that the volume V of the solid of revolution may be found by multiplying the area A of R by the distance $2\pi\bar{x}$ that the centroid travels when R is revolved once about the y-axis. A similar statement is true if R is revolved about the x-axis. In Chapter 17 we shall prove the following more general theorem, named after the mathematician Pappus of Alexandria (ca. 300 A.D.).

Theorem of Pappus **(6.26)**

> Let R be a region in a plane that lies entirely on one side of a line l in the plane. If R is revolved once about l, the volume of the resulting solid is the product of the area of R and the distance traveled by the centroid of R.

FIGURE 6.72

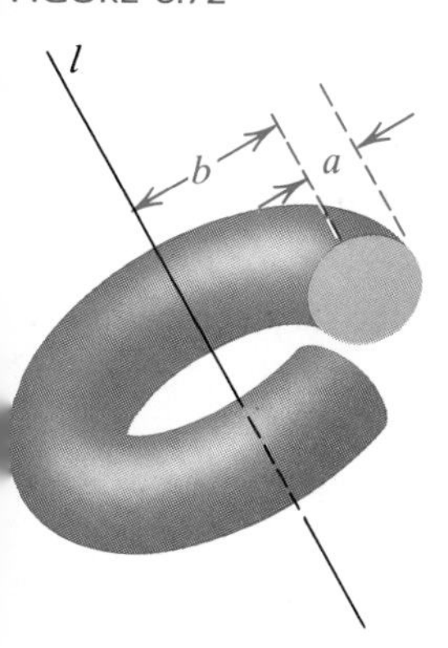

EXAMPLE 6 The region bounded by a circle of radius a is revolved about a line l, in the plane of the circle, that is a distance b from the center of the circle, where $b > a$ (see Figure 6.72). Find the volume V of the resulting solid. (The surface of this doughnut-shaped solid is called a **torus**.)

SOLUTION The region bounded by the circle has area πa^2, and the distance traveled by the centroid is $2\pi b$. Hence, by the theorem of Pappus,

$$V = (2\pi b)(\pi a^2) = 2\pi^2 a^2 b.$$

EXERCISES 6.7

Exer. 1–2: The table lists point-masses (in kilograms) and their coordinates (in meters) on an x-axis. Find m, M_0, and the center of mass.

1

mass	100	80	70
coordinate	−3	2	4

2

mass	50	100	50
coordinate	−10	2	3

Exer. 3–4: The table lists point-masses (in kilograms) and their locations (in meters) in an xy-plane. Find m, M_x, M_y, and the center of mass of the system.

3

mass	2	7	5
location	(4, −1)	(−2, 0)	(−8, −5)

4

mass	10	3	4	1	8
location	(−5, −2)	(3, 7)	(0, −3)	(−8, −3)	(0, 0)

Exer. 5–14: Sketch the region bounded by the graphs of the equations, and find m, M_x, M_y, and the centroid.

5 $y = x^3$, $y = 0$, $x = 1$

6 $y = \sqrt{x}$, $y = 0$, $x = 9$

7 $y = 4 - x^2$, $y = 0$

8 $2x + 3y = 6$, $y = 0$, $x = 0$

9 $y^2 = x$, $2y = x$

10 $y = x^2$, $y = x^3$

11 $y = 1 - x^2$, $x - y = 1$

12 $y = x^2$, $x + y = 2$

13 $x = y^2$, $x - y = 2$

14 $x = 9 - y^2$, $x + y = 3$

15 Find the centroid of the region in the first quadrant bounded by the circle $x^2 + y^2 = a^2$ and the coordinate axes.

16 Let R be the region in the first quadrant bounded by part of the parabola $y^2 = cx$ with $c > 0$, the x-axis, and the vertical line through the point (a, b) on the parabola, as shown in the figure on the following page. Find the centroid of R.

EXERCISE 16

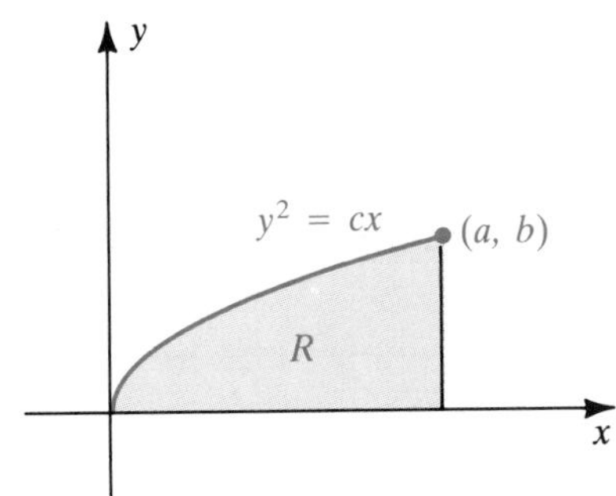

17 A region has the shape of a square of side $2a$ surmounted by a semicircle of radius a. Find the centroid. (*Hint:* Use Example 5 and the fact that the moment of the region is the sum of the moments of the square and the semicircle.)

18 Let the points P, Q, R, and S have coordinates $(-b, 0)$, $(-a, 0)$, $(a, 0)$, and $(b, 0)$, respectively, with $0 < a < b$. Find the centroid of the region bounded by the graphs of $y = \sqrt{b^2 - x^2}$, $y = \sqrt{a^2 - x^2}$, and the line segments PQ and RS. (*Hint:* Use Example 5.)

19 Prove that the centroid of a triangle coincides with the intersection of the medians. (*Hint:* Take the vertices at the points $(0, 0)$, (a, b), and $(0, c)$, with a, b, and c positive.)

20 A region has the shape of a square of side a surmounted by an equilateral triangle of side a. Find the centroid. (*Hint:* See Exercise 19 and the hint given for Exercise 17.)

Exer. 21–24: Use the theorem of Pappus.

21 Let R be the rectangular region with vertices $(1, 2)$, $(2, 1)$, $(5, 4)$, and $(4, 5)$. Find the volume of the solid generated by revolving R about the y-axis.

22 Let R be the triangular region with vertices $(1, 1)$, $(2, 2)$, and $(3, 1)$. Find the volume of the solid generated by revolving R about the y-axis.

23 Find the centroid of the region in the first quadrant bounded by the graph of $y = \sqrt{a^2 - x^2}$ and the coordinate axes.

24 Find the centroid of the triangular region with vertices $O(0, 0)$, $A(0, a)$, and $B(b, 0)$ for positive numbers a and b.

c **25** A lamina of area density ρ has the shape of the region bounded by the graphs of $f(x) = \sqrt{|\cos x|}$ and $g(x) = x^2$. Graph f and g on the same coordinate axes.

(a) Set up an integral that can be used to approximate the mass of the lamina.

(b) Use Simpson's rule with $n = 4$ to approximate the integral in (a).

c **26** Use Simpson's rule with $n = 4$ to approximate the centroid of the region bounded by the graphs of $y = 0$, $y = (\sin x)/x$, $x = 1$, and $x = 2$.

6.8 OTHER APPLICATIONS

It should be evident from our work in this chapter that if a quantity can be approximated by a sum of many terms, then it is a candidate for representation as a definite integral. The main requirement is that as the number of terms increases, the sums approach a limit. In this section we consider several miscellaneous applications of the definite integral. Let us begin with the force exerted by a liquid on a submerged object.

In physics, the **pressure** p at a depth h in a fluid is defined as the weight of fluid contained in a column that has a cross-sectional area of one square unit and an altitude h. Pressure may also be regarded as the force per unit area exerted by the fluid. If a fluid has density ρ, then the pressure p at depth h is given by

$$p = \rho h.$$

The following illustration is for water, with $\rho = 62.5 \text{ lb/ft}^3$.

ILLUSTRATION

DENSITY ρ (lb/ft^3)	DEPTH h (ft)	PRESSURE $p = \rho h$ (lb/ft^2)
62.5	2	125
62.5	4	250
62.5	6	375

FIGURE 6.73

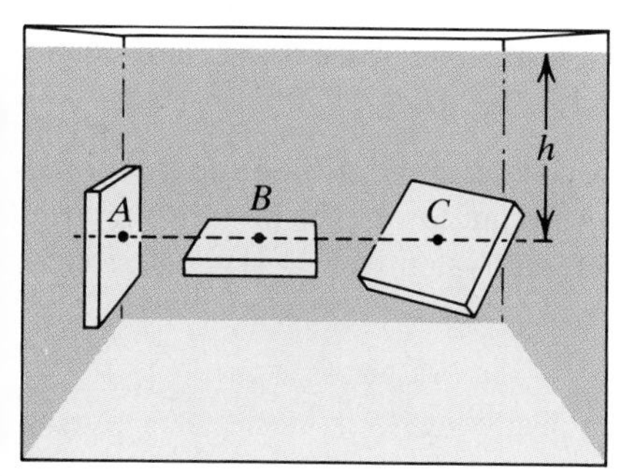

Pascal's principle in physics states that the pressure at a depth h in a fluid is the same in every direction. Thus, if a flat plate is submerged in a fluid, then the pressure on one side of the plate at a point that is h units below the surface is ρh, regardless of whether the plate is submerged vertically, horizontally, or obliquely (see Figure 6.73, where the pressure at points A, B, and C is ρh).

If a rectangular tank, such as a fish aquarium, is filled with water (see Figure 6.74), the total force exerted by the water on the base may be calculated as follows:

$$\text{force on base} = (\text{pressure at base}) \cdot (\text{area of base})$$

FIGURE 6.74

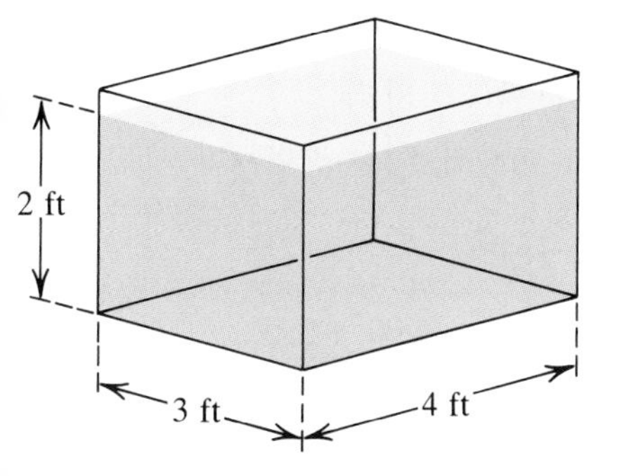

For the tank in Figure 6.74, we use $\rho = 62.5$ lb/ft^3 and $h = 2$ ft to obtain

$$\text{force on base} = (125\ \text{lb/ft}^2) \cdot (12\ \text{ft}^2) = 1500\ \text{lb}.$$

This corresponds to 12 columns of water, each having cross-sectional area 1 ft^2 and each weighing 125 pounds.

It is more complicated to find the force exerted on one of the sides of the aquarium, since the pressure is not constant there but increases as the depth increases. Instead of investigating this particular problem, let us consider the following more general situation.

Suppose a flat plate is submerged in a fluid of density ρ such that the face of the plate is perpendicular to the surface of the fluid. Let us introduce a coordinate system as shown in Figure 6.75, where the width of the plate extends over the interval $[c, d]$ on the y-axis. Assume that for each y in $[c, d]$, the corresponding depth of the fluid is $h(y)$ and the length of the plate is $L(y)$, where h and L are continuous functions.

FIGURE 6.75

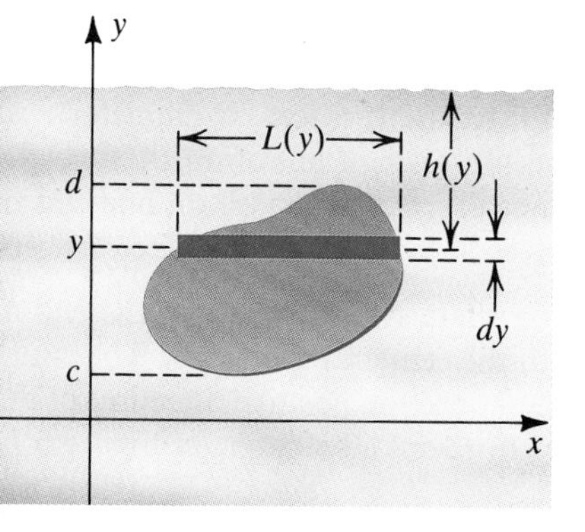

We shall employ our standard technique of considering a typical horizontal rectangle of width dy and length $L(y)$, as illustrated in Figure 6.75. If dy is small, then the pressure at any point in the rectangle is approximately $\rho h(y)$. Thus, the force on one side of the rectangle can be approximated by

$$\text{force on rectangle} \approx (\text{pressure}) \cdot (\text{area of rectangle}),$$

or

$$\text{force on rectangle} \approx \rho h(y) \cdot L(y)\, dy.$$

Taking a limit of sums of these forces by applying the operator $\int_c^d$ leads to the following definition.

Definition (6.27)

> The **force F exerted by a fluid** of constant density ρ on one side of a submerged region of the type illustrated in Figure 6.75 is
>
> $$F = \int_c^d \rho h(y) \cdot L(y)\, dy.$$

If a more complicated region is divided into subregions of the type illustrated in Figure 6.75, we apply Definition (6.27) to each subregion and add the resulting forces. The coordinate system may be introduced in various ways: In Example 2 we choose the x-axis along the surface of the liquid and the positive direction of the y-axis downward.

EXAMPLE 1 The ends of a water trough 8 feet long have the shape of isosceles trapezoids of lower base 4 feet, upper base 6 feet, and height 4 feet. Find the total force on one end if the trough is full of water.

FIGURE 6.76

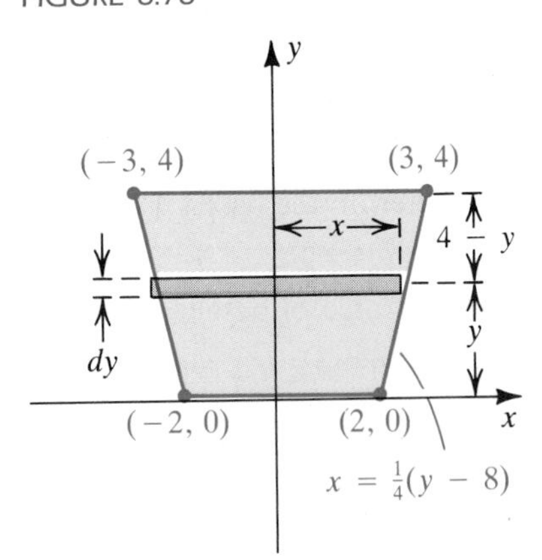

SOLUTION Figure 6.76 illustrates one end of the trough superimposed on a rectangular coordinate system. An equation of the line through the points (2, 0) and (3, 4) is $y = 4x - 8$, or, equivalently, $x = \frac{1}{4}(y + 8)$. Referring to Figure 6.76 gives us the following, for a horizontal rectangle of width dy:

$$\begin{aligned} \text{length:}&\quad 2x = 2 \cdot \tfrac{1}{4}(y + 8) = \tfrac{1}{2}(y + 8) \\ \text{area:}&\quad \tfrac{1}{2}(y + 8)\, dy \\ \text{depth:}&\quad 4 - y \\ \text{pressure:}&\quad 62.5(4 - y) \\ \text{force:}&\quad 62.5(4 - y) \cdot \tfrac{1}{2}(y + 8)\, dy \end{aligned}$$

Taking a limit of sums by applying the operator $\int_0^4$, we obtain, as in Definition (6.27),

$$\begin{aligned} F &= \int_0^4 62.5(4 - y) \cdot \tfrac{1}{2}(y + 8)\, dy \\ &= 31.25 \int_0^4 (32 - 4y - y^2)\, dy = \frac{7000}{3} \approx 2333 \text{ lb.} \end{aligned}$$

In the preceding example, the *length* of the water trough was irrelevant when we considered the force on one end. The same is true for the oil tank in the next example.

EXAMPLE 2 A cylindrical oil storage tank 6 feet in diameter and 10 feet long is lying on its side. If the tank is half full of oil that weighs 58 lb/ft^3, set up an integral for the force exerted by the oil on one end of the tank.

FIGURE 6.77

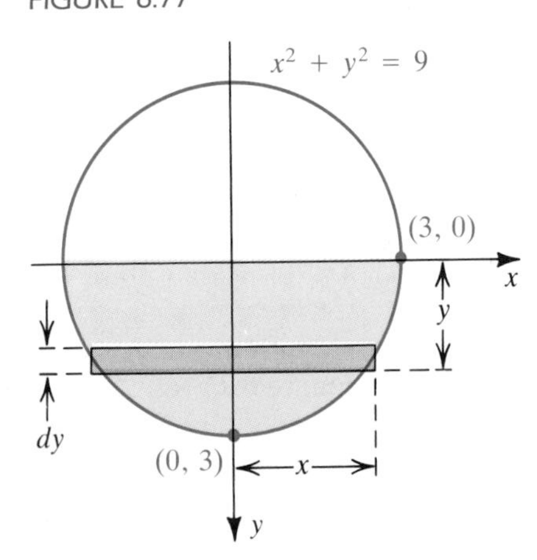

SOLUTION Let us introduce a coordinate system such that the end of the tank is a circle of radius 3 feet with the center at the origin. The equation of the circle is $x^2 + y^2 = 9$. If we choose the positive direction of the y-axis *downward*, then referring to the horizontal rectangle in Figure 6.77 gives us the following:

$$\begin{aligned} \text{length:}&\quad 2x = 2\sqrt{9 - y^2} \\ \text{area:}&\quad 2\sqrt{9 - y^2}\, dy \\ \text{depth:}&\quad y \\ \text{pressure:}&\quad 58y \\ \text{force:}&\quad 58y \cdot 2\sqrt{9 - y^2}\, dy \end{aligned}$$

Taking a limit of sums by applying $\int_0^3$ we obtain

$$F = \int_0^3 116y\sqrt{9 - y^2}\, dy.$$

Evaluating the integral by using the method of substitution would give us

$$F = 1044 \text{ lb}.$$

FIGURE 6.78

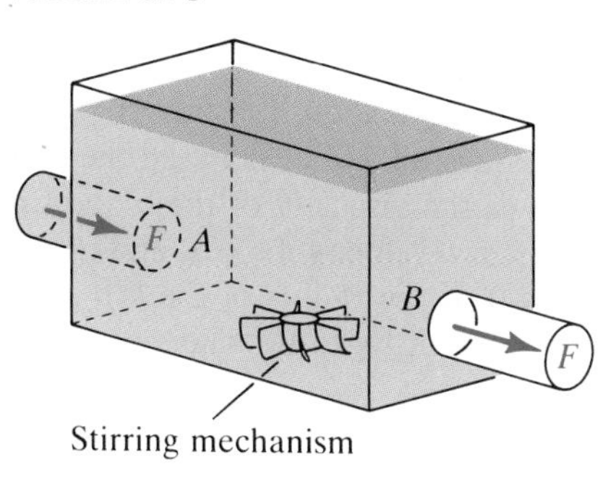

Definite integrals can be applied to dye-dilution or tracer methods used in physiological tests and elsewhere. One example involves the measurement of cardiac output—that is, the rate at which blood flows through the aorta. A simple model for tracer experiments is sketched in Figure 6.78, where a liquid (or gas) flows into a tank at A and exits at B, with a constant flow rate F (in L/sec). Suppose that at time $t = 0$, Q_0 grams of tracer (or dye) are introduced into the tank at A and that a stirring mechanism thoroughly mixes the solution at all times. The concentration $c(t)$ (in g/L) of tracer at time t is monitored at B. Thus the amount of tracer passing B at time t is given by

$$(\text{flow rate}) \cdot (\text{concentration}) = F \cdot c(t) \text{ g/sec}.$$

If the amount of tracer in the tank at time t is $Q(t)$, where Q is a differentiable function, then the rate of change $Q'(t)$ of Q is given by

$$Q'(t) = -F \cdot c(t)$$

(the negative sign indicates that Q is decreasing).

If T is a time at which all the tracer has left the tank, then $Q(T) = 0$ and, by the fundamental theorem of calculus,

$$\int_0^T Q'(t)\, dt = Q(t)\Big]_0^T = Q(T) - Q(0)$$
$$= 0 - Q_0 = -Q_0.$$

We may also write

$$\int_0^T Q'(t)\, dt = \int_0^T [-F \cdot c(t)]\, dt = -F \int_0^T c(t)\, dt.$$

Equating the two forms for the integral gives us the following formula.

Flow concentration formula **(6.28)**

$$Q_0 = F \int_0^T c(t)\, dt$$

Usually an explicit form for $c(t)$ will not be known, but, instead, a table of function values will be given. By employing numerical integration, we may find an approximation to the flow rate F (see Exercises 11 and 12).

Let us next consider another aspect of the flow of liquids. If a liquid flows through a cylindrical tube and if the velocity is a constant v_0, then the volume of liquid passing a fixed point per unit time is given by $v_0 A$, where A is the area of a cross section of the tube (see Figure 6.79).

FIGURE 6.79

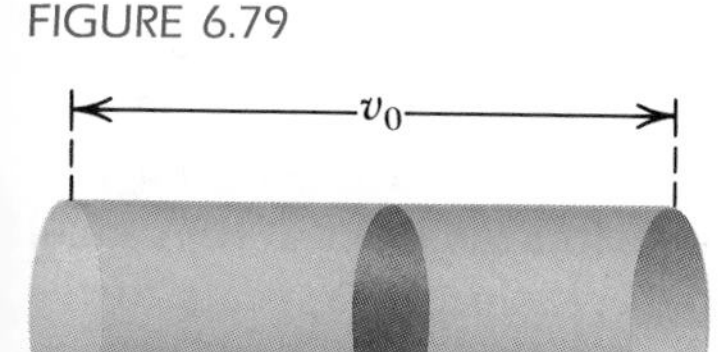

A more complicated formula is required to study the flow of blood in an arteriole. In this case the flow is in layers, as illustrated in Figure 6.80, on the following page. In the layer closest to the wall of the arteriole, the blood tends to stick to the wall, and its velocity may be considered zero. The velocity increases as the layers approach the center of the arteriole.

For computational purposes, we may regard the blood flow as consisting of thin cylindrical shells that slide along, with the outer shell fixed and the velocity of the shells increasing as the radii of the shells decrease (see Figure 6.80). If the velocity in each shell is considered constant, then from the theory of liquids in motion, the velocity $v(r)$ in a shell having average radius r is

FIGURE 6.80

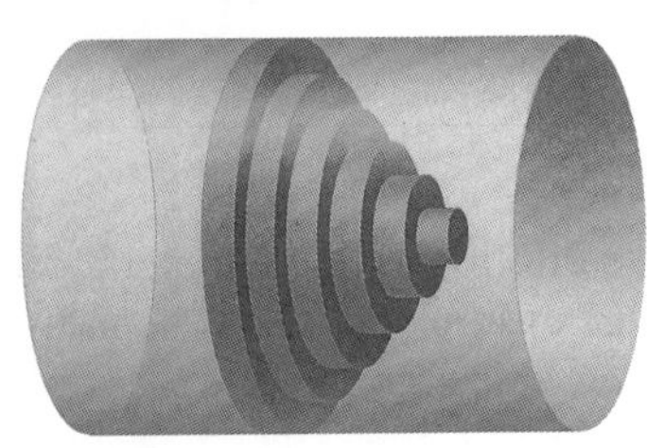

$$v(r) = \frac{P}{4\nu l}(R^2 - r^2),$$

where R is the radius of the arteriole (in centimeters), l is the length of the arteriole (in centimeters), P is the pressure difference between the two ends of the arteriole (in dyn/cm^2), and ν is the viscosity of the blood (in dyn-sec/cm^2). Note that the formula gives zero velocity if $r = R$ and maximum velocity $PR^2/(4\nu l)$ as r approaches 0. If the radius of the kth shell is r_k and the thickness of the shell is Δr_k, then, by (6.10), the volume of blood in this shell is

$$2\pi r_k v(r_k)\,\Delta r_k = \frac{2\pi r_k P}{4\nu l}(R^2 - r_k^2)\,\Delta r_k.$$

If there are n shells, then the total flow in the arteriole per unit time may be approximated by

$$\sum_{k=1}^{n} \frac{2\pi r_k P}{4\nu l}(R^2 - r_k^2)\,\Delta r_k.$$

To estimate the total flow F (the volume of blood per unit time), we consider the limit of these sums as n increases without bound. This leads to the following definite integral:

$$\begin{aligned} F &= \int_0^R \frac{2\pi r P}{4\nu l}(R^2 - r^2)\,dr \\ &= \frac{2\pi P}{4\nu l}\int_0^R (R^2 r - r^3)\,dr \\ &= \frac{\pi P}{2\nu l}\left[\frac{1}{2}R^2 r^2 - \frac{1}{4}r^4\right]_0^R \\ &= \frac{\pi P R^4}{8\nu l}\ \text{cm}^3 \end{aligned}$$

This formula for F is not exact, because the thickness of the shells cannot be made arbitrarily small. The lower limit is the width of a red blood cell, or approximately 2×10^{-4} centimeter. We may assume, however, that the formula gives a reasonable estimate. It is interesting to observe that a small change in the radius of an arteriole produces a large change in the flow, since F is directly proportional to the fourth power of R. A small change in pressure difference has a lesser effect, since P appears to the first power.

In many types of employment, a worker must perform the same assignment repeatedly. For example, a bicycle shop employee may be asked to assemble new bicycles. As more and more bicycles are assembled, the time required for each assembly should decrease until a certain minimum assembly time is reached. Another example of this process of learning by repetition is that of a data processor who must keyboard information from written forms into a computer system. The time required to process each entry should decrease as the number of entries increases. As a final illustration, the time required for a person to trace a path through a maze should improve with practice.

Let us consider a general situation in which a certain task is to be repeated many times. Suppose experience has shown that the time required to perform the task for the kth time can be approximated by $f(k)$ for a continuous decreasing function f on a suitable interval. The total time required to perform the task n times is given by the sum

FIGURE 6.81

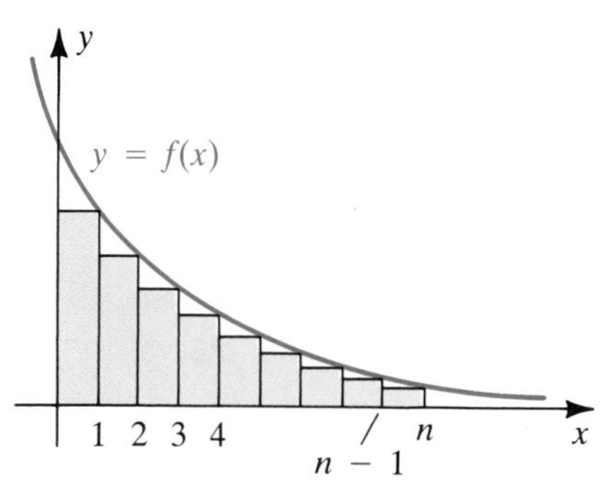

$$\sum_{k=1}^{n} f(k) = f(1) + f(2) + \cdots + f(n).$$

If we consider the graph of f, then, as illustrated in Figure 6.81, the preceding sum equals the area of the pictured inscribed rectangular polygon and, therefore, may be approximated by the definite integral $\int_0^n f(x)\,dx$. Evidently, the approximation will be close to the actual sum if f decreases slowly on $[0, n]$. If f changes rapidly per unit change in x, then an integral should not be used as an approximation.

EXAMPLE 3 A company that conducts polls via telephone interviews finds that the time required by an employee to complete one interview depends on the number of interviews that the employee has completed previously. Suppose it is estimated that, for a certain survey, the number of minutes required to complete the kth interview is given by $f(k) = 6(1 + k)^{-1/5}$ for $0 \le k \le 500$. Use a definite integral to approximate the time required for an employee to complete 100 interviews and 200 interviews. If an interviewer receives \$4.80 per hour, estimate how much more expensive it is to have two employees each conduct 100 interviews than to have one employee conduct 200 interviews.

SOLUTION From the preceding discussion, the time required for 100 interviews is approximately

$$\int_0^{100} 6(1 + x)^{-1/5}\,dx = 6 \cdot \tfrac{5}{4}(1 + x)^{4/5}\Big]_0^{100} \approx 293.5 \text{ min.}$$

The time required for 200 interviews is approximately

$$\int_0^{200} 6(1 + x)^{-1/5}\,dx \approx 514.4 \text{ min.}$$

Since an interviewer receives \$0.08 per minute, the cost for one employee to conduct 200 interviews is roughly (\$0.08)(514.4), or \$41.15. If two employees each conduct 100 interviews, the cost is about 2(\$0.08)(293.5), or \$46.96, which is \$5.81 more than the cost of one employee. Note, however, that the time saved in using two people is approximately 221 minutes.

Using a computer, we have

$$\sum_{k=1}^{100} 6(1+k)^{-1/5} \approx 291.75$$

and

$$\sum_{k=1}^{200} 6(1+k)^{-1/5} \approx 512.57.$$

Hence the results obtained by integration (the area under the graph of f) are roughly 2 minutes more than the value of the corresponding sum (the area of the inscribed rectangular polygon).

In economics, the process that a corporation uses to increase its accumulated wealth is called **capital formation**. If the amount K of capital at time t can be approximated by $K = f(t)$ for a differentiable function f, the rate of change of K with respect to t is called the **net investment flow**. Hence, if I denotes the investment flow, then

$$I = \frac{dK}{dt} = f'(t).$$

Conversely, if I is given by $g(t)$ for a function g that is continuous on an interval $[a, b]$, then the increase in capital over this time interval is

$$\int_a^b g(t)\,dt = f(b) - f(a).$$

EXAMPLE 4 Suppose a corporation wishes to have its net investment flow approximated by $g(t) = t^{1/3}$ for t in years and $g(t)$ in millions of dollars per year. If $t = 0$ corresponds to the present time, estimate the amount of capital formation over the next eight years.

SOLUTION From the preceding discussion, the increase in capital over the next eight years is

$$\int_0^8 g(t)\,dt = \int_0^8 t^{1/3}\,dt = \tfrac{3}{4}t^{4/3}\Big]_0^8 = 12.$$

Consequently, the amount of capital formation is \$12,000,000.

FIGURE 6.82

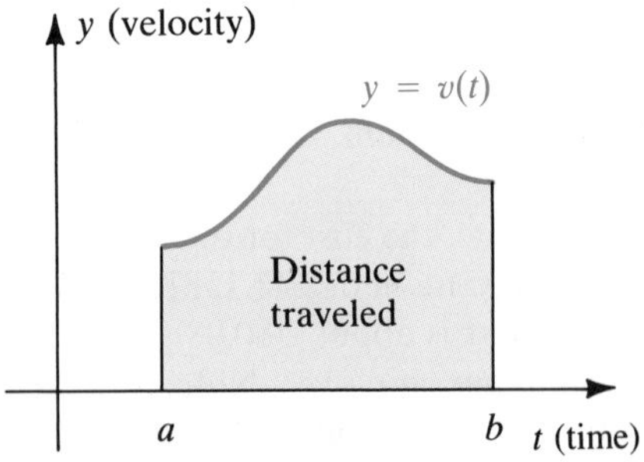

Any quantity that can be interpreted as an area of a region in a plane may be investigated by means of a definite integral. (See, for example, the discussion of hysteresis at the end of Section 6.1.) Conversely, definite integrals allow us to represent physical quantities as areas. In the following illustrations, a quantity is *numerically equal* to an area of a region; that is, we *disregard units of measurement*, such as centimeter, foot-pound, and so on.

Suppose $v(t)$ is the velocity, at time t, of an object that is moving on a coordinate line. If s is the position function, then $s'(t) = v(t)$ and

$$\int_a^b v(t)\,dt = \int_a^b s'(t)\,dt = s(t)\Big]_a^b = s(b) - s(a).$$

FIGURE 6.83

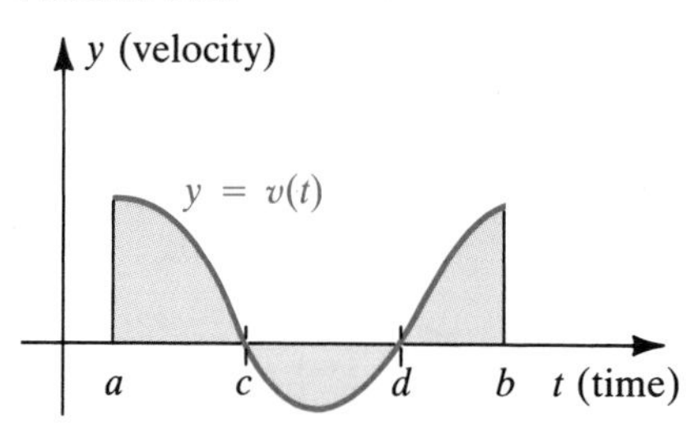

If $v(t) > 0$ throughout the time interval $[a, b]$, this tells us that the area under the graph of the function v from a to b represents the distance the object travels, as illustrated in Figure 6.82. This observation is useful to an engineer or physicist, who may not have an explicit form for $v(t)$ but merely a graph (or table) indicating the velocity at various times. The distance traveled may then be estimated by approximating the area under the graph.

If $v(t) < 0$ at certain times in $[a, b]$, the graph of v may resemble that in Figure 6.83. The figure indicates that the object moved in the negative direction from $t = c$ to $t = d$. The distance it traveled during that time is given by $\int_c^d |v(t)|\, dt$. It follows that $\int_a^b |v(t)|\, dt$ is the *total* distance traveled in $[a, b]$, whether $v(t)$ is positive or negative.

FIGURE 6.84

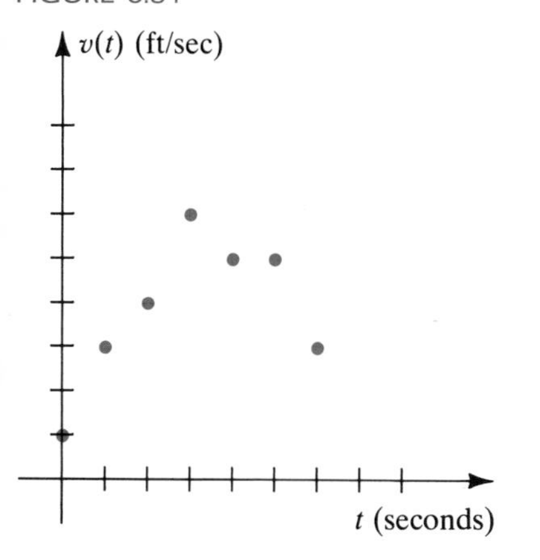

EXAMPLE 5 As an object moves along a straight path, its velocity $v(t)$ (in ft/sec) at time t is recorded each second for 6 seconds. The results are given in the following table.

t	0	1	2	3	4	5	6
$v(t)$	1	3	4	6	5	5	3

Approximate the distance traveled by the object.

SOLUTION The points $(t, v(t))$ are plotted in Figure 6.84. If we assume that v is a continuous function, then, as in the preceding discussion, the distance traveled during the time interval $[0, 6]$ is $\int_0^6 v(t)\, dt$. Let us approximate this definite integral by means of Simpson's rule, with $n = 6$:

$$\int_0^6 v(t)\, dt \approx \frac{6-0}{3 \cdot 6}[v(0) + 4v(1) + 2v(2) + 4v(3) + 2v(4) + 4v(5) + v(6)]$$
$$= \tfrac{1}{3}[1 + 4 \cdot 3 + 2 \cdot 4 + 4 \cdot 6 + 2 \cdot 5 + 4 \cdot 5 + 3] = 26 \text{ ft}$$

FIGURE 6.85

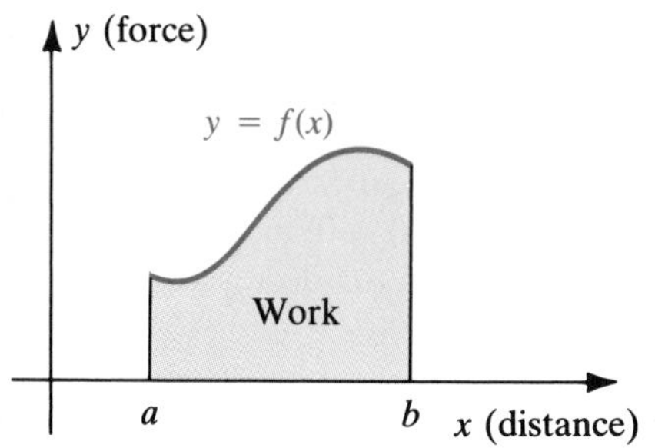

In (6.21) we defined the work W done by a variable force $f(x)$ that acts along a coordinate line from $x = a$ to $x = b$ by $W = \int_a^b f(x)\, dx$. Suppose $f(x) \geq 0$ throughout $[a, b]$. If we sketch the graph of f as illustrated in Figure 6.85, then the work W is numerically equal to the area under the graph from a to b.

FIGURE 6.86

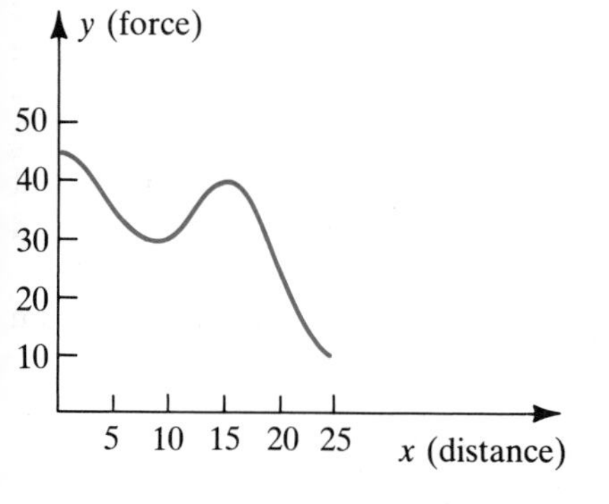

EXAMPLE 6 An engineer obtains the graph in Figure 6.86, which shows the force (in pounds) acting on a small cart as it moves 25 feet along horizontal ground. Estimate the work done.

SOLUTION If we assume that the force is a continuous function f for $0 \leq x \leq 25$, the work done is

$$W = \int_0^{25} f(x)\, dx.$$

We do not have an explicit form for $f(x)$; however, we may estimate function values from the graph and approximate W by means of numerical integration.

Let us apply the trapezoidal rule with $a = 0$, $b = 25$, and $n = 5$. Referring to the graph to estimate function values gives us the following table.

k	x_k	$f(x_k)$	m	$mf(x_k)$
0	0	45	1	45
1	5	35	2	70
2	10	30	2	60
3	15	40	2	80
4	20	25	2	50
5	25	10	1	10

The sum of the numbers in the last column is 315. Since

$$(b - a)/(2n) = (25 - 0)/10 = 2.5,$$

it follows from (5.36) that

$$W = \int_0^{25} f(x)\, dx \approx 2.5(315) \approx 790 \text{ ft-lb}.$$

For greater accuracy we could use a larger value of n or Simpson's rule.

Suppose that the amount of a physical entity, such as oil, water, electric power, money supply, bacteria count, or blood flow, is increasing or decreasing in some manner, and that $R(t)$ is the rate at which it is changing at time t. If $Q(t)$ is the amount of the entity present at time t and if Q is differentiable, then $Q'(t) = R(t)$. If $R(t) > 0$ (or $R(t) < 0$) in a time interval $[a, b]$, then the amount that the entity increases (or decreases) between $t = a$ and $t = b$ is

$$Q(b) - Q(a) = \int_a^b Q'(t)\, dt = \int_a^b R(t)\, dt.$$

This number may be represented as the area of the region in a ty-plane bounded by the graphs of R, $t = a$, $t = b$, and $y = 0$.

EXAMPLE 7 Starting at 9:00 A.M., oil is pumped into a storage tank at a rate of $(150t^{1/2} + 25)$ gal/hr, for time t (in hours) after 9:00 A.M. How many gallons will have been pumped into the tank at 1:00 P.M.?

SOLUTION Letting $R(t) = 150t^{1/2} + 25$ in the preceding discussion, we obtain the following:

$$\int_0^4 (150t^{1/2} + 25)\, dt = \Big[100t^{3/2} + 25t\Big]_0^4 = 900 \text{ gal}$$

We have given only a few illustrations of the use of definite integrals. The interested reader may find many more in books on the physical and biological sciences, economics, and business, and even such areas as political science and sociology.

EXERCISES 6.8

1 A glass aquarium tank is 3 feet long and has square ends of width 1 foot. If the tank is filled with water, find the force exerted by the water on

(a) one end (b) one side

2 If one of the square ends of the tank in Exercise 1 is divided into two parts by means of a diagonal, find the force exerted on each part.

3 The ends of a water trough 6 feet long have the shape of isosceles triangles with equal sides of length 2 feet and the third side of length $2\sqrt{3}$ feet at the top of the trough. Find the force exerted by the water on one end of the trough if the trough is

(a) full of water (b) half full of water

4 The ends of a water trough have the shape of the region bounded by the graphs of $y = x^2$ and $y = 4$, with x and y measured in feet. If the trough is full of water, find the force on one end.

5 A cylindrical oil storage tank 4 feet in diameter and 5 feet long is lying on its side. If the tank is half full of oil weighing 60 lb/ft^3, find the force exerted by the oil on one end of the tank.

6 A rectangular gate in a dam is 5 feet long and 3 feet high. If the gate is vertical, with the top of the gate parallel to the surface of the water and 6 feet below it, find the force of the water against the gate.

7 A plate having the shape of an isosceles trapezoid with upper base 4 feet long and lower base 8 feet long is submerged vertically in water such that the bases are parallel to the surface. If the distances from the surface of the water to the lower and upper bases are 10 feet and 6 feet, respectively, find the force exerted by the water on one side of the plate.

8 A circular plate of radius 2 feet is submerged vertically in water. If the distance from the surface of the water to the center of the plate is 6 feet, find the force exerted by the water on one side of the plate.

9 A rectangular plate 3 feet wide and 6 feet long is submerged vertically in oil weighing 50 lb/ft^3, with its short side parallel to, and 2 feet below, the surface.

(a) Find the total force exerted on one side of the plate.

(b) If the plate is divided into two parts by means of a diagonal, find the force exerted on each part.

c 10 A flat, irregularly shaped plate is submerged vertically in water (see figure). Measurements of its width, taken at successive depths at intervals of 0.5 foot, are compiled in the following table.

Water depth (ft)	1	1.5	2	2.5	3	3.5	4
Width of plate (ft)	0	2	3	5.5	4.5	3.5	0

Estimate the force of the water on one side of the plate by using, with $n = 6$,

(a) the trapezoidal rule (b) Simpson's rule

EXERCISE 10

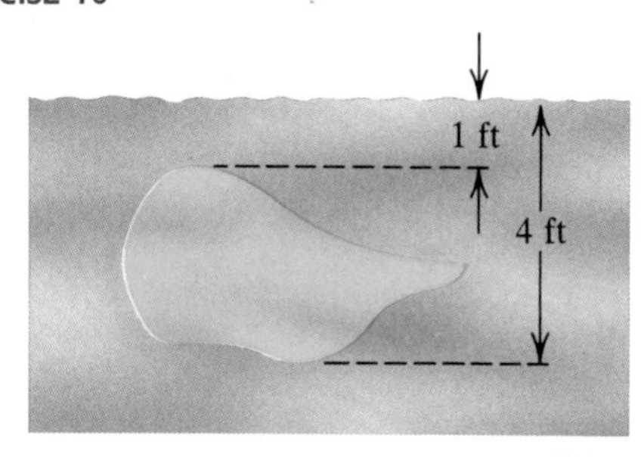

c 11 Refer to (6.28). To estimate cardiac output F (the number of liters of blood per minute that the heart pumps through the aorta), a 5-milligram dose of the tracer indocyanine-green is injected into a pulmonary artery and dye concentration measurements $c(t)$ are taken every minute from a peripheral artery near the aorta. The results are given in the following table. Use Simpson's rule, with $n = 12$, to estimate the cardiac output.

t (min)	$c(t)$ (mg/L)
0	0
1	0
2	0.15
3	0.48
4	0.86
5	0.72
6	0.48
7	0.26
8	0.15
9	0.09
10	0.05
11	0.01
12	0

c 12 Refer to (6.28). Suppose 1200 kilograms of sodium dichromate are mixed into a river at point A and sodium dichromate samples are taken every 30 seconds at a point B downstream. The concentration $c(t)$ at time t is recorded in the following table. Use the trapezoidal rule, with $n = 12$, to estimate the river flow rate F.

t (sec)	$c(t)$ (mg/L or g/m^3)
0	0
30	2.14
60	3.89
90	5.81
120	8.95
150	7.31
180	6.15
210	4.89
240	2.98
270	1.42
300	0.89
330	0.29
360	0

13 A manufacturer estimates that the time required for a worker to assemble a certain item depends on the number of this item the worker has previously assembled. If the time (in minutes) required to assemble the kth item is given by $f(k) = 20(k + 1)^{-0.4} + 3$, use a definite integral to approximate the time, to the nearest minute, required to assemble

(a) 1 item (b) 4 items (c) 8 items (d) 16 items

14 The number of minutes needed for a person to trace a path through a certain maze without error is estimated to be $f(k) = 5k^{-1/2}$, where k is the number of trials previously completed. Use a definite integral to approximate the time required to complete 10 trials.

15 A data processor keyboards registration data for college students from written forms to electronic files. The number of minutes required to process the kth registration is estimated to be approximately $f(k) = 6(1 + k)^{-1/3}$. Use a definite integral to estimate the time required for

(a) one person to keyboard 600 registrations

(b) two people to keyboard 300 each

16 If, in Example 4, the rate of investment is approximated by $g(t) = 2t(3t + 1)$, with $g(t)$ in thousands of dollars, use a definite integral to approximate the amount of capital formation over the intervals $[0, 5]$ and $[5, 10]$.

Exer. 17–18: Use a definite integral to approximate the sum, and round your answer to the nearest integer.

17 $\displaystyle\sum_{k=1}^{100} k(k^2 + 1)^{-1/4}$

18 $\displaystyle\sum_{k=1}^{200} 5k(k^2 + 10)^{-1/3}$

c 19 The velocity (in mi/hr) of an automobile as it traveled along a freeway over a 12-minute interval is indicated in the figure. Use the trapezoidal rule to approximate the distance traveled to the nearest mile.

EXERCISE 19

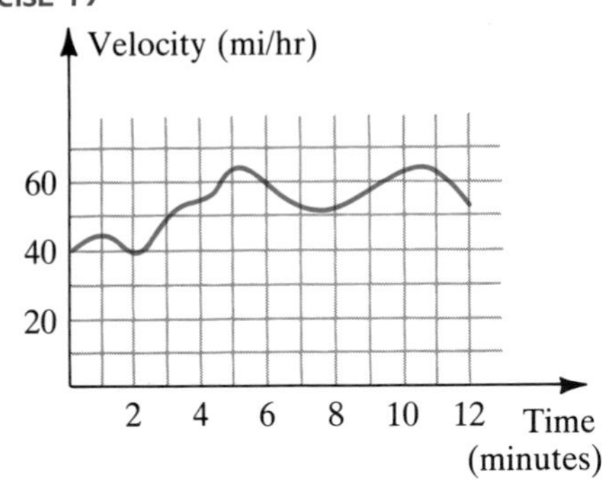

c 20 The acceleration (in ft/sec^2) of an automobile over a period of 8 seconds is indicated in the figure. Use the trapezoidal rule to approximate the net change in velocity in this time period.

EXERCISE 20

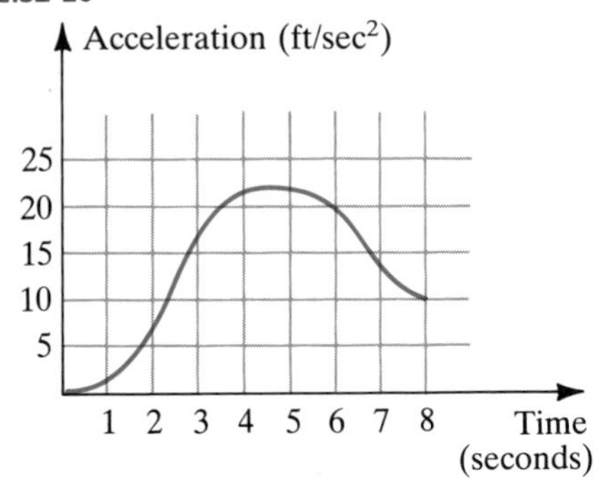

21 The following table was obtained by recording the force $f(x)$ (in Newtons) acting on a particle as it moved 6 meters along a coordinate line from $x = 1$ to $x = 7$. Estimate the work done using

(a) the trapezoidal rule, with $n = 6$

(b) Simpson's rule, with $n = 6$

x	1	2	3	4	5	6	7
$f(x)$	20	23	25	22	26	30	28

22 A bicyclist pedals directly up a hill, recording the velocity $v(t)$ (in ft/sec) at the end of every two seconds. Referring to the results recorded in the following table, use the trapezoidal rule to approximate the distance traveled.

t	0	2	4	6	8	10
$v(t)$	24	22	16	10	2	0

23 A motorboat uses gasoline at the rate of $t\sqrt{9 - t^2}$ gal/hr. If the motor is started at $t = 0$, how much gasoline is used in 2 hours?

24 The population of a city has increased since 1985 at a rate of $1.5 + 0.3\sqrt{t} + 0.006t^2$ thousand people per year, where t is the number of years after 1985. Assuming that this rate continues and that the population was 50,000 in 1985, estimate the population in 1994.

25 A simple thermocouple, in which heat is transformed into electrical energy, is shown in the figure. To determine the total charge Q (in coulombs) transferred to the copper wire, current readings (in amperes) are recorded every $\frac{1}{2}$ second, and the results are shown in the following table.

t (sec)	0	0.5	1.0	1.5	2.0	2.5	3.0
I (amp)	0	0.2	0.6	0.7	0.8	0.5	0.2

Use the fact that $I = dQ/dt$ and the trapezoidal rule, with $n = 6$, to estimate the total charge transferred to the copper wire during the first three seconds.

EXERCISE 25

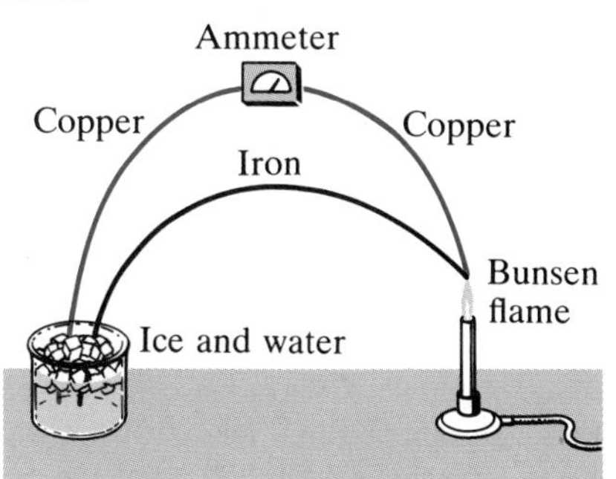

26 Suppose $\rho(x)$ is the density (in cm/km) of ozone in the atmosphere at a height of x kilometers above the ground. For example, if $\rho(6) = 0.0052$, then at a height of 6 kilometers there is effectively a thickness of 0.0052 centimeter of ozone for each kilometer of atmosphere. If ρ is a continuous function, the thickness of the ozone layer between heights a and b can be found by evaluating $\int_a^b \rho(x)\,dx$. Values for $\rho(x)$ found experimentally are shown in the following table.

x (km)	$\rho(x)$ (spring)	$\rho(x)$ (autumn)
0	0.0034	0.0038
6	0.0052	0.0043
12	0.0124	0.0076
18	0.0132	0.0104
24	0.0136	0.0109
30	0.0084	0.0072
36	0.0034	0.0034
42	0.0017	0.0016

(a) Use the trapezoidal rule to estimate the thickness of the ozone layer between the altitudes of 6 and 42 kilometers during both spring and autumn.

(b) Work part (a) using Simpson's rule.

c **27** Radon gas can pose a serious health hazard if inhaled. If $V(t)$ is the volume of air (in cm^3) in an adult's lungs at time t (in minutes), then the rate of change of V can often be approximated by $V'(t) = 12{,}450\pi \sin(30\pi t)$. Inhaling and exhaling correspond to $V'(t) > 0$ and $V'(t) < 0$, respectively. Suppose an adult lives in a home that has a radioactive energy concentration due to radon of 4.1×10^{-12} joule/cm^3.

(a) Approximate the volume of air inhaled by the adult with each breath.

(b) If inhaling more than 0.02 joule of radioactive energy in one year is considered hazardous, is it safe for the adult to remain at home?

28 A stationary exercise bicycle is programmed so that it can be set for different intensity levels L and workout times T. It displays the elapsed time t (in minutes), for $0 \le t \le T$, and the number of calories $C(t)$ that are being burned per minute at time t, where

$$C(t) = 5 + 3L - 6\frac{L}{T}\left|t - \frac{1}{2}T\right|.$$

Suppose an individual exercises 16 minutes, with $L = 3$ for $0 \le t \le 8$ and with $L = 2$ for $8 \le t \le 16$. Find the total number of calories burned during the workout.

29 The rate of growth R (in cm/yr) of an average boy who is t years old is shown in the following table for $10 \le t \le 15$.

t (yr)	10	11	12	13	14	15
R (cm/yr)	5.3	5.2	4.9	6.5	9.3	7.0

Use the trapezoidal rule, with $n = 5$, to approximate the number of centimeters the boy grows between his tenth and fifteenth birthdays.

30 To determine the number of zooplankton in a portion of an ocean 80 meters deep, marine biologists take samples at successive depths of 10 meters, obtaining the following table, where $\rho(x)$ is the density (in number/m^3) of zooplankton at a depth of x meters.

x	0	10	20	30	40	50	60	70	80
$\rho(x)$	0	10	25	30	20	15	10	5	0

Use Simpson's rule, with $n = 8$, to estimate the total number of zooplankton in a *water column* (a column of water) having a cross section 1 meter square extending from the surface to the ocean floor.

6.9 REVIEW EXERCISES

Exer. 1–2: Sketch the region bounded by the graphs of the equations, and find the area by integrating with respect to (a) x and (b) y.

1 $y = -x^2, \quad y = x^2 - 8$

2 $y^2 = 4 - x, \quad x + 2y = 1$

Exer. 3–4: Find the area of the region bounded by the graphs of the equations.

3 $x = y^2, \quad x + y = 1$

4 $y = -x^3, \quad y = \sqrt{x}, \quad 7x + 3y = 10$

5 Find the area of the region between the graphs of the equations $y = \cos \frac{1}{2}x$ and $y = \sin x$, from $x = \pi/3$ to $x = \pi$.

6 The region bounded by the graph of $y = \sqrt{1 + \cos 2x}$ and the x-axis, from $x = 0$ to $x = \pi/2$, is revolved about the x-axis. Find the volume of the resulting solid.

Exer. 7–10: Sketch the region R bounded by the graphs of the equations, and find the volume of the solid generated by revolving R about the indicated axis.

7 $y = \sqrt{4x + 1}, \quad y = 0, \quad x = 0, \quad x = 2;$ x-axis

8 $y = x^4, \quad y = 0, \quad x = 1;$ y-axis

9 $y = x^3 + 1, \quad x = 0, \quad y = 2;$ y-axis

10 $y = \sqrt[3]{x}, \quad y = \sqrt{x};$ x-axis

Exer. 11–12: The region bounded by the x-axis and the graph of the given equation, from $x = 0$ to $x = b$, is revolved about the y-axis. Find the volume of the resulting solid.

11 $y = \cos x^2; \quad b = \sqrt{\pi/2}$

12 $y = x \sin x^3; \quad b = 1$

13 Find the volume of the solid generated by revolving the region bounded by the graphs of $y = 4x^2$ and $4x + y = 8$ about

(a) the x-axis (b) $x = 1$ (c) $y = 16$

14 Find the volume of the solid generated by revolving the region bounded by the graphs of $y = x^3$, $x = 2$, and $y = 0$ about

(a) the x-axis (b) the y-axis (c) $x = 2$

(d) $x = 3$ (e) $y = 8$ (f) $y = -1$

15 Find the arc length of the graph of $(x + 3)^2 = 8(y - 1)^3$ from $A(-2, \frac{3}{2})$ to $B(5, 3)$.

16 A solid has for its base the region in the xy-plane bounded by the graphs of $y^2 = 4x$ and $x = 4$. Find the volume of the solid if every cross section by a plane perpendicular to the x-axis is an isosceles right triangle with one of its equal sides on the base of the solid.

17 An above-ground swimming pool has the shape of a right circular cylinder of diameter 12 feet and height 5 feet. If the depth of the water in the pool is 4 feet, find the work required to empty the pool by pumping the water out over the top.

18 As a bucket is raised a distance of 30 feet from the bottom of a well, water leaks out at a uniform rate. Find the work done if the bucket originally contains 24 pounds of water and one-third leaks out. Assume that the weight of the empty bucket is 4 pounds, and disregard the weight of the rope.

19 A square plate of side 4 feet is submerged vertically in water such that one of the diagonals is parallel to the surface. If the distance from the surface of the water to the center of the plate is 6 feet, find the force exerted by the water on one side of the plate.

20 Use differentials to approximate the arc length of the graph of $y = 2 \sin \frac{1}{3}x$ between the points with x-coordinates π and $91\pi/90$.

Exer. 21–22: Sketch the region bounded by the graphs of the equations, and find m, M_x, M_y, and the centroid.

21 $y = x^3 + 1, \quad x + y = -1, \quad x = 1$

22 $y = x^2 + 1, \quad y = x, \quad x = -1, \quad x = 2$

23 The graph of the equation $12y = 4x^3 + (3/x)$ from $A(1, \frac{7}{12})$ to $B(2, \frac{67}{24})$ is revolved about the x-axis. Find the area of the resulting surface.

24 The shape of a reflector in a searchlight is obtained by revolving a parabola about its axis. If, as shown in the figure, the reflector is 4 feet across at the opening and 1 foot deep, find its surface area.

EXERCISE 24

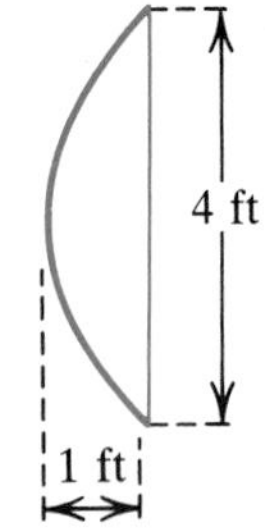

25 The velocity $v(t)$ of a rocket that is traveling directly upward is given in the following table. Use the trapezoidal rule to approximate the distance the rocket travels from $t = 0$ to $t = 5$.

t (sec)	0	1	2	3	4	5
$v(t)$ (ft/sec)	100	120	150	190	240	300

26 An electrician suspects that a meter showing the total consumption Q in kilowatt hours (kwh) of electricity is not functioning properly. To check the accuracy, the electrician measures the consumption rate R directly every 10 minutes, obtaining the results in the following table.

t (min)	0	10	20	30
R (kwh/min)	1.31	1.43	1.45	1.39

t (min)	40	50	60
R (kwh/min)	1.36	1.47	1.29

(a) Use Simpson's rule to estimate the total consumption during this one-hour period.

(b) If the meter read 48,792 kwh at the beginning of the experiment and 48,953 kwh at the end, what should the electrician conclude?

27 Interpret $\int_0^1 2\pi x^4\,dx$ in the following ways:

(a) as the area of a region in the xy-plane

(b) as the volume of a solid obtained by revolving a region in the xy-plane about
  (i) the x-axis
  (ii) the y-axis

(c) as the work done by a force

28 Let R be the semicircular region in the xy-plane with endpoints of its diameter at $(4, 0)$ and $(10, 0)$. Use the theorem of Pappus to find the volume of the solid obtained by revolving R about the y-axis.

CHAPTER 7

LOGARITHMIC AND EXPONENTIAL FUNCTIONS

INTRODUCTION

In precalculus mathematics we sketch graphs of equations such as $y = a^x$ for $a > 0$ without defining a^x if x is irrational. Instead, we *assume* that real numbers such as a^π and $a^{\sqrt{3}}$ exist, and that the graph rises if $a > 1$ or falls if $0 < a < 1$. We *further* assume that the laws of exponents are true for all real exponents. Next, we define the logarithm $\log_a x$ using our *undefined* exponential expressions and "prove" properties of logarithms by applying the *unproved* laws of exponents! Although this development is acceptable in elementary algebra, it is unsatisfactory in calculus, where the standards of mathematical rigor are higher.

Our approach in this chapter is to employ a definite integral to introduce the *natural logarithmic function*. We then use this function to define the *natural exponential function*. Finally, we give precise meanings to a^x and $\log_a x$. You may think this is an awkward procedure; however, it is the simplest way to treat these topics *rigorously*. Moreover, our approach enables us to establish results on continuity, derivatives, and integrals in a simple manner and to *prove* the laws of exponents that are assumed in precalculus mathematics.

Many applications of logarithmic and exponential functions are included throughout the chapter. Since these functions are inverses of each other, we shall begin by discussing the general concept of inverse functions.

7.1 INVERSE FUNCTIONS

A function f may have the same value for different numbers in its domain. For example, if $f(x) = x^2$, then $f(2) = 4 = f(-2)$ but $2 \neq -2$. In order to define the *inverse of a function*, it is essential that different numbers in the domain *always* give different values of f. Such functions are called *one-to-one functions*.

Definition (7.1)

A function f with domain D and range R is a **one-to-one function** if whenever $a \neq b$ in D, then $f(a) \neq f(b)$ in R.

FIGURE 7.1

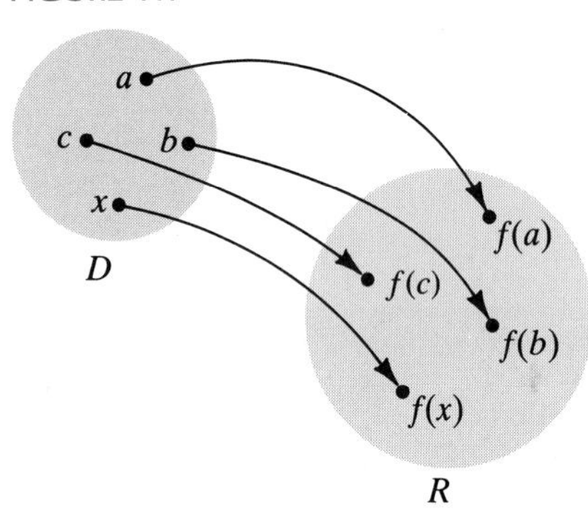

FIGURE 7.2

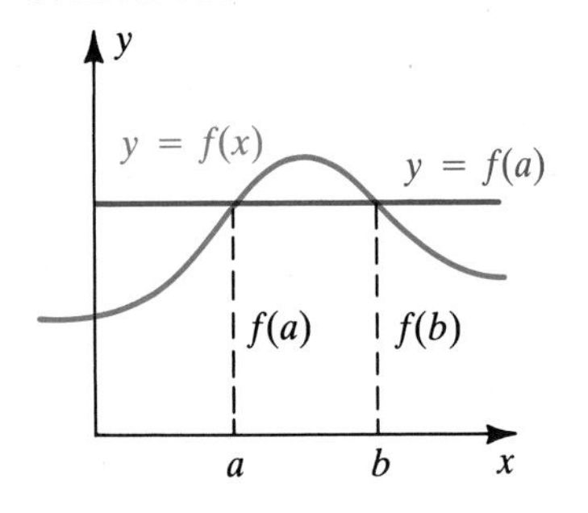

The diagram in Figure 7.1 illustrates a one-to-one function, because each function value in the range R corresponds to *exactly one* element in the domain D. The function whose graph is illustrated in Figure 7.2 is not one-to-one, because $a \neq b$ but $f(a) = f(b)$. Note that the horizontal line $y = f(a)$ (or $y = f(b)$) intersects the graph in more than one point. Thus, *if any horizontal line intersects the graph of a function f in more than one point, then f is not one-to-one*. Every increasing function is one-to-one, because if $a < b$, then $f(a) < f(b)$, and if $b < a$, then $f(b) < f(a)$. Thus, if $a \neq b$, then $f(a) \neq f(b)$. Similarly, every decreasing function is one-to-one.

If f is a one-to-one function with domain D and range R, then for each number y in R, there is *exactly one* number x in D such that $y = f(x)$, as illustrated by the arrow in Figure 7.3(i). Since x is *unique*, we may define a function g from R to D by means of the rule $x = g(y)$. As in Figure 7.3(ii), *g reverses the correspondence given by f*. We call g the *inverse function* of f, as in the following definition.

FIGURE 7.3

(i)

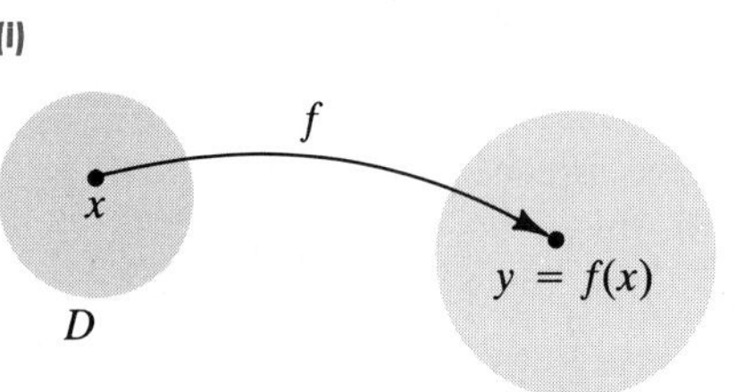

(ii)

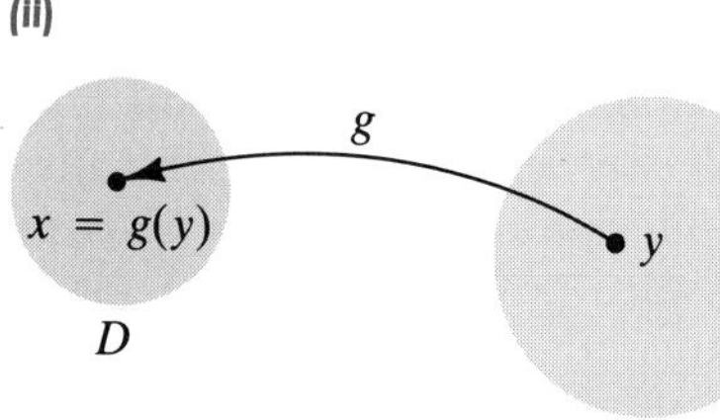

Definition (7.2)

Let f be one-to-one function with domain D and range R. A function g with domain R and range D is the **inverse function** of f, provided the following condition is true for every x in D and every y in R:

$$y = f(x) \quad \text{if and only if} \quad x = g(y)$$

The following theorem can be used to verify that a function g is the inverse of f.

Theorem **(7.3)**

> Let f be a one-to-one function with domain D and range R. If g is a function with domain R and range D, then g is the inverse function of f if and only if both of the following conditions are true:
>
> **(i)** $g(f(x)) = x$ for every x in D
>
> **(ii)** $f(g(y)) = y$ for every y in R

PROOF Let us first prove that if g is the inverse function of f, then conditions (i) and (ii) are true. By the definition of an inverse function,

$$y = f(x) \quad \text{if and only if} \quad x = g(y)$$

for every x in D and every y in R. If we substitute $f(x)$ for y in the equation $x = g(y)$, we obtain condition (i): $x = g(f(x))$. Similarly, if we substitute $g(y)$ for x in the equation $y = f(x)$, we obtain condition (ii): $y = f(g(y))$. Thus, if g is the inverse function of f, then conditions (i) and (ii) are true.

Conversely, let g be a function with domain R and range D, and suppose that conditions (i) and (ii) are true. To show that g is the inverse function of f, we must prove that

$$y = f(x) \quad \text{if and only if} \quad x = g(y)$$

for every x in D and every y in R.

First suppose that $y = f(x)$. Since (i) is true, $g(f(x)) = x$; that is, $g(y) = x$. This shows that if $y = f(x)$, then $x = g(y)$.

Next suppose that $x = g(y)$. Since (ii) is true, $f(g(y)) = y$; that is, $f(x) = y$. This shows that if $x = g(y)$, then $y = f(x)$, which completes the proof. ■

A one-to-one function f can have only one inverse function. Conditions (i) and (ii) of Theorem (7.3) imply that if g is the inverse function of f, then f is the inverse function of g. We say that *f and g are inverse functions of each other*.

If a function f has an inverse function g, we often denote g by f^{-1}. The -1 used in this notation should not be mistaken for an exponent; that is, $f^{-1}(y)$ *does not mean* $1/[f(y)]$. The reciprocal $1/[f(y)]$ may be denoted by $[f(y)]^{-1}$. It is important to remember the following relationships.

Domains and ranges of f and f^{-1} **(7.4)**

> domain of f^{-1} = range of f
>
> range of f^{-1} = domain of f

When we discuss functions, we often let x denote an arbitrary number in the domain. Thus, for the inverse function f^{-1}, we may consider $f^{-1}(x)$, *where x is in the domain of f^{-1}*. In this case the two conditions in Theorem (7.3) are written as follows:

(i) $f^{-1}(f(x)) = x$ for every x in the domain of f

(ii) $f(f^{-1}(x)) = x$ for every x in the domain of f^{-1}

In some cases we can find the inverse of a one-to-one function by solving the equation $y = f(x)$ for x in terms of y, obtaining an equation of the form $x = g(y)$. If the two conditions $g(f(x)) = x$ and $f(g(x)) = x$ are true for every x in the domains of f and g, respectively, then g is the required inverse function f^{-1}. The following guidelines summarize this procedure. In guideline 2, in anticipation of finding f^{-1}, we shall write $x = f^{-1}(y)$ instead of $x = g(y)$.

Guidelines for finding f^{-1} in simple cases **(7.5)**

1. Verify that f is a one-to-one function (or that f is increasing or is decreasing) throughout its domain.
2. Solve the equation $y = f(x)$ for x in terms of y, obtaining an equation of the form $x = f^{-1}(y)$.
3. Verify the two conditions

$$f^{-1}(f(x)) = x \quad \text{and} \quad f(f^{-1}(x)) = x$$

for every x in the domains of f and f^{-1}, respectively.

The success of this method depends on the nature of the equation $y = f(x)$, since we must be able to solve for x in terms of y. For this reason, we include *simple cases* in the title of the guidelines.

EXAMPLE 1 Let $f(x) = 3x - 5$. Find the inverse function of f.

SOLUTION We shall follow the three guidelines. First, we note that the graph of the linear function f is a line of slope 3. Since f is increasing throughout $\mathbb{R}$, f is one-to-one, and hence the inverse function f^{-1} exists. Moreover, since the domain and range of f are $\mathbb{R}$, the same is true for f^{-1}.

As in guideline 2, we consider the equation

$$y = 3x - 5$$

and then solve for x in terms of y, obtaining

$$x = \frac{y + 5}{3}.$$

We now let

$$f^{-1}(y) = \frac{y + 5}{3}.$$

Since the symbol used for the variable is immaterial, we may also write

$$f^{-1}(x) = \frac{x + 5}{3}.$$

We next verify the conditions (i) $f^{-1}(f(x)) = x$ and (ii) $f(f^{-1}(x)) = x$:

(i)
$$\begin{aligned} f^{-1}(f(x)) &= f^{-1}(3x - 5) && \text{(definition of } f\text{)} \\ &= \frac{(3x - 5) + 5}{3} && \text{(definition of } f^{-1}\text{)} \\ &= x && \text{(simplifying)} \end{aligned}$$

(ii) $f(f^{-1}(x)) = f\left(\dfrac{x+5}{3}\right)$ (definition of f^{-1})

$= 3\left(\dfrac{x+5}{3}\right) - 5$ (definition of f)

$= x$ (simplifying)

Thus, by Theorem (7.3), the inverse function of f is given by $f^{-1}(x) = (x+5)/3$.

EXAMPLE 2 Let $f(x) = x^2 - 3$ for $x \geq 0$. Find the inverse function of f.

FIGURE 7.4

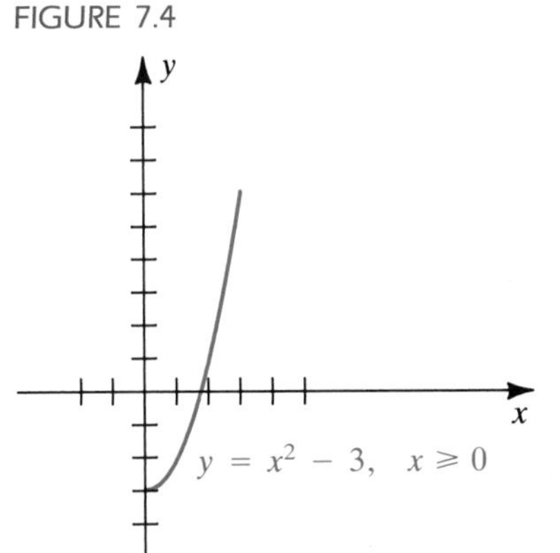

SOLUTION The graph of f is sketched in Figure 7.4. The domain of f is $[0, \infty)$, and the range is $[-3, \infty)$. Since f is increasing, it is one-to-one and hence has an inverse function f^{-1} that has domain $[-3, \infty)$ and range $[0, \infty)$.

As in guideline 2, we consider the equation

$$y = x^2 - 3$$

and solve for x, obtaining

$$x = \pm\sqrt{y+3}.$$

Since x is nonnegative, we reject $x = -\sqrt{y+3}$ and let

$$f^{-1}(y) = \sqrt{y+3}, \quad \text{or, equivalently,} \quad f^{-1}(x) = \sqrt{x+3}.$$

Finally, we verify that (i) $f^{-1}(f(x)) = x$ for x in $[0, \infty)$ and (ii) $f(f^{-1}(x)) = x$ for x in $[-3, \infty)$:

(i) $f^{-1}(f(x)) = f^{-1}(x^2 - 3) = \sqrt{(x^2 - 3) + 3} = \sqrt{x^2} = x \quad \text{if} \quad x \geq 0$

(ii) $f(f^{-1}(x)) = f(\sqrt{x+3}) = (\sqrt{x+3})^2 - 3 = (x+3) - 3 = x \quad \text{if} \quad x \geq -3$

Thus, the inverse function is given by $f^{-1}(x) = \sqrt{x+3}$ for $x \geq -3$.

FIGURE 7.5

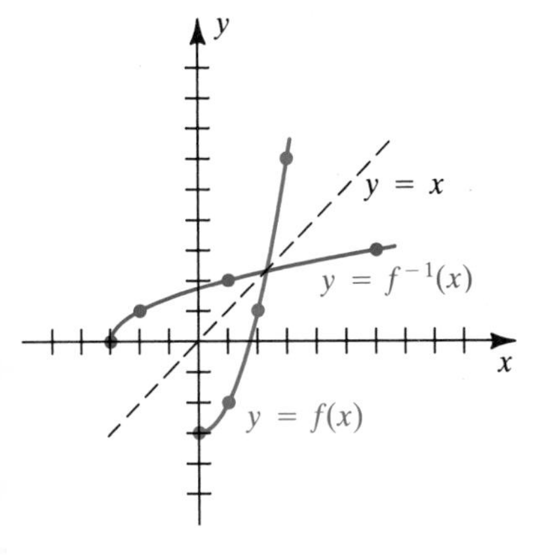

There is an interesting relationship between the graph of a function f and the graph of its inverse function f^{-1}. We first note that $b = f(a)$ is equivalent to $a = f^{-1}(b)$. These equations imply that *the point (a, b) is on the graph of f if and only if the point (b, a) is on the graph of f^{-1}.*

As an illustration, in Example 2 we found that the functions f and f^{-1} given by

$$f(x) = x^2 - 3 \quad \text{and} \quad f^{-1}(x) = \sqrt{x+3}$$

are inverse functions of each other, provided that x is suitably restricted. Some points on the graph of f are $(0, -3)$, $(1, -2)$, $(2, 1)$, and $(3, 6)$. Corresponding points on the graph of f^{-1} are $(-3, 0)$, $(-2, 1)$, $(1, 2)$, and $(6, 3)$. The graphs of f and f^{-1} are sketched on the same coordinate plane in Figure 7.5. If the page is folded along the line $y = x$ that bisects quadrants I and III (as indicated by the dashes in the figure), then the graphs of f and f^{-1} coincide. The two graphs are *reflections* of each other through the line $y = x$. This is typical of the graph of every function f that has an inverse function f^{-1} (see Exercise 14).

FIGURE 7.6

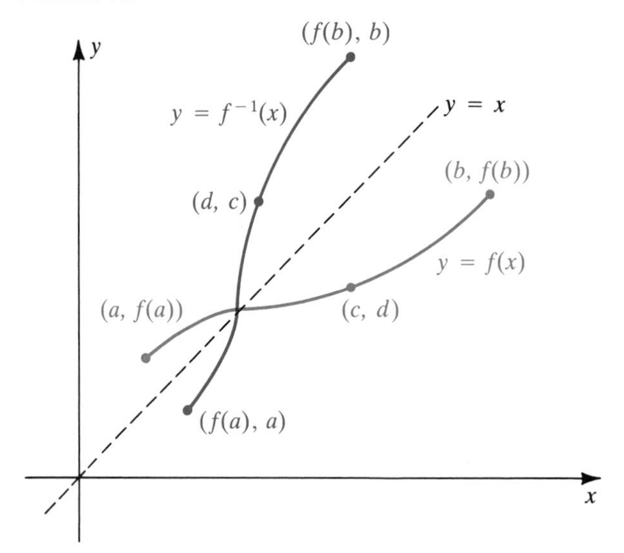

Figure 7.6 illustrates the graphs of an arbitrary one-to-one function f and its inverse function f^{-1}. As indicated in the figure, (c, d) is on the graph of f if and only if (d, c) is on the graph of f^{-1}. Thus, if we restrict the domain of f to the interval $[a, b]$, then the domain of f^{-1} is restricted to $[f(a), f(b)]$. If f is continuous, then the graph of f has no breaks or holes, and hence the same is true for the (reflected) graph of f^{-1}. Thus, we see intuitively that if f is continuous on $[a, b]$, then f^{-1} is continuous on $[f(a), f(b)]$. We can also show that if f is increasing, then so is f^{-1}. These facts are stated in the next theorem. A proof is given in Appendix II.

Theorem (7.6)

If f is continuous and increasing on $[a, b]$, then f has an inverse function f^{-1} that is continuous and increasing on $[f(a), f(b)]$.

We can also prove the analogous result obtained by replacing the word *increasing* in Theorem (7.6) by *decreasing*.

The next theorem provides a method for finding the derivative of an inverse function.

Theorem (7.7)

If a differentiable function f has an inverse function $g = f^{-1}$ and if $f'(g(c)) \neq 0$, then g is differentiable at c and

$$g'(c) = \frac{1}{f'(g(c))}.$$

PROOF By Definition (3.6),

$$g'(c) = \lim_{x \to c} \frac{g(x) - g(c)}{x - c}.$$

Let $y = g(x)$ and $a = g(c)$. Since f and g are inverse functions of each other,

$$g(x) = y \quad \text{if and only if} \quad f(y) = x$$

and

$$g(c) = a \quad \text{if and only if} \quad f(a) = c.$$

Since f is differentiable, it is continuous and hence, by Theorem (7.6), so is the inverse function $g = f^{-1}$. Thus, if $x \to c$, then $g(x) \to g(c)$; that is, $y \to a$. If $y \to a$, then $f(y) \to f(a)$. Thus, we may write

$$\begin{aligned} g'(c) &= \lim_{x \to c} \frac{g(x) - g(c)}{x - c} \\ &= \lim_{y \to a} \frac{y - a}{f(y) - f(a)} \\ &= \lim_{y \to a} \frac{1}{\dfrac{f(y) - f(a)}{y - a}} \\ &= \frac{1}{\displaystyle\lim_{y \to a} \frac{f(y) - f(a)}{y - a}} \\ &= \frac{1}{f'(a)} = \frac{1}{f'(g(c))}. \end{aligned}$$

It is convenient to restate Theorem (7.7) as follows.

Corollary **(7.8)**

If g is the inverse function of a differentiable function f and if $f'(g(x)) \neq 0$, then

$$g'(x) = \frac{1}{f'(g(x))}.$$

EXAMPLE 3 If $f(x) = x^3 + 2x - 1$, prove that f has an inverse function g, and find the slope of the tangent line to the graph of g at the point $P(2, 1)$.

SOLUTION Since $f'(x) = 3x^2 + 2 > 0$ for every x, f is increasing and hence is one-to-one. Thus, f has an inverse function g. Since $f(1) = 2$, it follows that $g(2) = 1$, and consequently the point $P(2, 1)$ is on the graph of g. It would be difficult to find g using Guidelines (7.5), because we would have to solve the equation $y = x^3 + 2x - 1$ for x in terms of y. However, even if we cannot find g explicitly, we *can* find the slope $g'(2)$

of the tangent line to the graph of g at $P(2, 1)$. Thus, by Theorem (7.7),

$$g'(2) = \frac{1}{f'(g(2))} = \frac{1}{f(1)} = \frac{1}{5}.$$

An easy way to remember Corollary (7.8) is to let $y = f(x)$. If g is the inverse function of f, then $g(y) = g(f(x)) = x$. From (7.8),

$$g'(y) = \frac{1}{f'(g(y))} = \frac{1}{f'(x)},$$

or, in differential notation,

$$\frac{dx}{dy} = \frac{1}{\left(\frac{dy}{dx}\right)}.$$

This shows that, in a sense, the derivative of the inverse function g is the reciprocal of the derivative of f. A disadvantage of using the last two formulas is that neither is stated in terms of the independent variable for the inverse function. To illustrate, in Example 3 let $y = x^3 + 2x - 1$ and $x = g(y)$. Then

$$\frac{dx}{dy} = \frac{1}{dy/dx} = \frac{1}{3x^2 + 2};$$

that is,

$$g'(y) = \frac{1}{3x^2 + 2} = \frac{1}{3(g(y))^2 + 2}.$$

This may also be written in the form

$$g'(x) = \frac{1}{3(g(x))^2 + 2}.$$

Consequently, to find $g'(x)$ it is necessary to know $g(x)$, just as in Corollary (7.8).

EXERCISES 7.1

Exer. 1–12: Find $f^{-1}(x)$.

1 $f(x) = 3x + 5$

2 $f(x) = 7 - 2x$

3 $f(x) = \dfrac{1}{3x - 2}$

4 $f(x) = \dfrac{1}{x + 3}$

5 $f(x) = \dfrac{3x + 2}{2x - 5}$

6 $f(x) = \dfrac{4x}{x - 2}$

7 $f(x) = 2 - 3x^2, \quad x \le 0$

8 $f(x) = 5x^2 + 2, \quad x \ge 0$

9 $f(x) = \sqrt{3 - x}$

10 $f(x) = \sqrt{4 - x^2}, \quad 0 \le x \le 2$

11 $f(x) = \sqrt[3]{x} + 1$

12 $f(x) = (x^3 + 1)^5$

13 (a) Prove that the linear function defined by $f(x) = ax + b$ with $a \ne 0$ has an inverse function, and find $f^{-1}(x)$.

(b) Does a constant function have an inverse? Explain.

14 Show that the graph of f^{-1} is the reflection of the graph of f through the line $y = x$ by verifying the following conditions:

(i) If $P(a, b)$ is on the graph of f, then $Q(b, a)$ is on the graph of f^{-1}.

(ii) The midpoint of line segment PQ is on the line $y = x$.

(iii) The line PQ is perpendicular to the line $y = x$.

Exer. 15–18: The graph of a one-to-one function f is shown in the figure. (a) Use a reflection to sketch the graph of f^{-1}. (b) Find the domain and range of f. (c) Find the domain and range of f^{-1}.

15

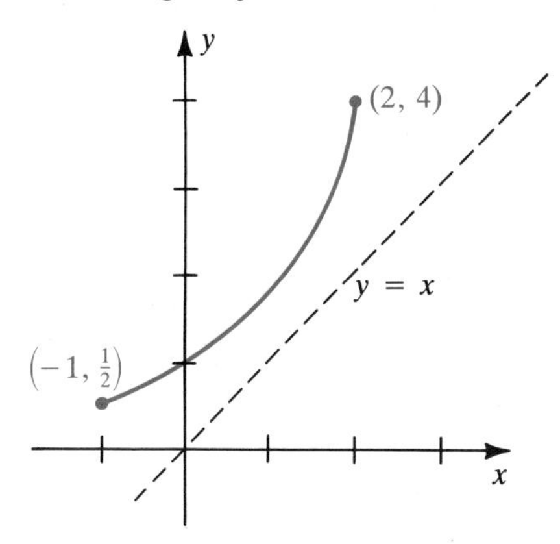

16

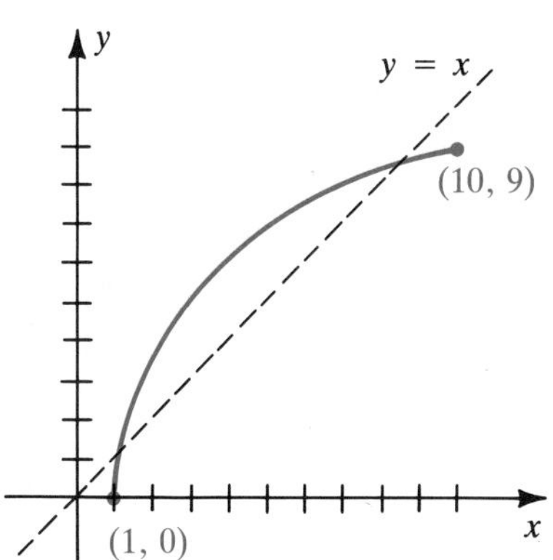

17

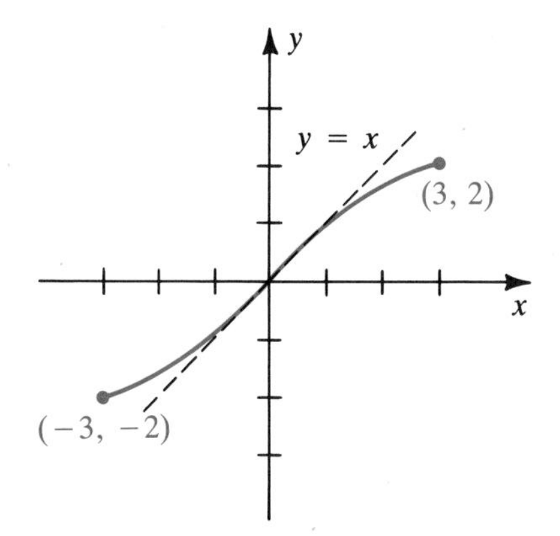

18

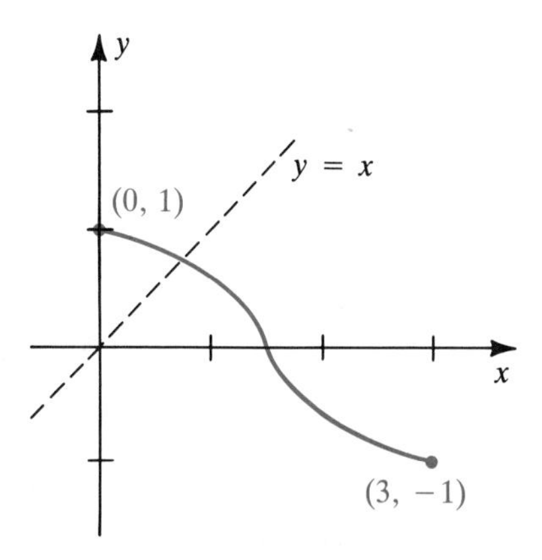

c **Exer. 19–20: Graph f on the given interval. (a) Estimate the largest interval $[a, b]$ with $a < 0 < b$ on which f is one-to-one. (b) If g is the function with domain $[a, b]$ such that $g(x) = f(x)$ for $a \leq x \leq b$, estimate the domain and range of g^{-1}.**

19 $f(x) = 2.1x^3 - 2.98x^2 - 2.11x + 3;\quad [-1, 2]$

20 $f(x) = \sin(\sin(1.1x));\quad [-2, 2]$

Exer. 21–26: (a) Prove that f has an inverse function. (b) State the domain of f^{-1}. (c) Use Corollary (7.8) to find $D_x f^{-1}(x)$.

21 $f(x) = \sqrt{2x + 3}$

22 $f(x) = \sqrt[3]{5x + 2}$

23 $f(x) = 4 - x^2,\quad x \geq 0$

24 $f(x) = x^2 - 4x + 5,\quad x \geq 2$

25 $f(x) = \dfrac{1}{x},\quad x \neq 0$

26 $f(x) = \sqrt{9 - x^2},\quad 0 \leq x \leq 3$

Exer. 27–32: (a) Use f' to prove that f has an inverse function. (b) Find the slope of the tangent line at the point P on the graph of f^{-1}.

27 $f(x) = x^5 + 3x^3 + 2x - 1;\quad P(5, 1)$

28 $f(x) = 2 - x - x^3;\quad P(-8, 2)$

29 $f(x) = -2x + (8/x^3),\ x > 0;\quad P(-3, 2)$

30 $f(x) = 4x^5 - (1/x^3),\ x > 0;\quad P(3, 1)$

31 $f(x) = x^3 + 4x - 1;\quad P(15, 2)$

32 $f(x) = x^5 + x;\quad P(2, 1)$

7.2 THE NATURAL LOGARITHMIC FUNCTION

In the chapter introduction we explained why it is necessary to take a different approach to logarithmic and exponential functions than that used in precalculus mathematics. At first you may think it strange to define a logarithmic function as a definite integral; however, later you will see

FIGURE 7.7

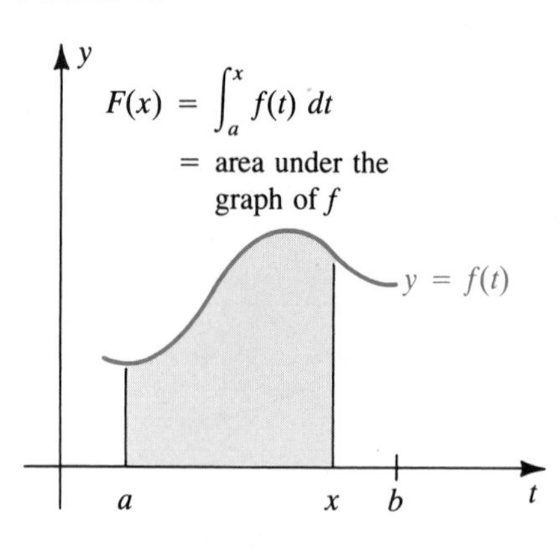

that the function we obtain obeys the familiar laws of logarithms considered in precalculus courses.

Let f be a function that is continuous on a closed interval $[a, b]$. As in the proof of Part I of the fundamental theorem of calculus (5.30), we can define a function F by

$$F(x) = \int_a^x f(t)\,dt$$

for x in $[a, b]$. If $f(t) \geq 0$ throughout $[a, b]$, then $F(x)$ is the area under the graph of f from a to x, as illustrated in Figure 7.7. For the special case $f(t) = t^n$, where n is a rational number and $n \neq -1$, we can find an explicit form for F. Thus, by the power rule for integrals,

$$F(x) = \int_a^x t^n\,dt = \left[\frac{t^{n+1}}{n+1}\right]_a^x$$

$$= \frac{1}{n+1}(x^{n+1} - a^{n+1}) \quad \text{if} \quad n \neq -1.$$

As indicated, we cannot use $t^{-1} = 1/t$ for the integrand, since $1/(n+1)$ is undefined if $n = -1$. Up to this point in our work we have been unable to determine an antiderivative of $1/x$—that is, a function F such that $F'(x) = 1/x$. The next definition will remedy this situation.

Definition **(7.9)**

> The **natural logarithmic function**, denoted by **ln**, is defined by
>
> $$\ln x = \int_1^x \frac{1}{t}\,dt$$
>
> for every $x > 0$.

The expression $\ln x$ (read *ell-en of* x) is called the **natural logarithm of** $\boldsymbol{x}$. We use this terminology because, as we shall see, ln has the same properties as the logarithmic functions studied in precalculus courses. The restriction $x > 0$ is necessary because if $x \leq 0$, the integrand $1/t$ has an infinite discontinuity between x and 1 and hence $\int_1^x (1/t)\,dt$ does not exist.

If $x > 1$, the definite integral $\int_1^x (1/t)\,dt$ may be interpreted as the area of the region the graph of $y = 1/t$ from $t = 1$ to $t = x$ (see Figure 7.8(i)).

FIGURE 7.8

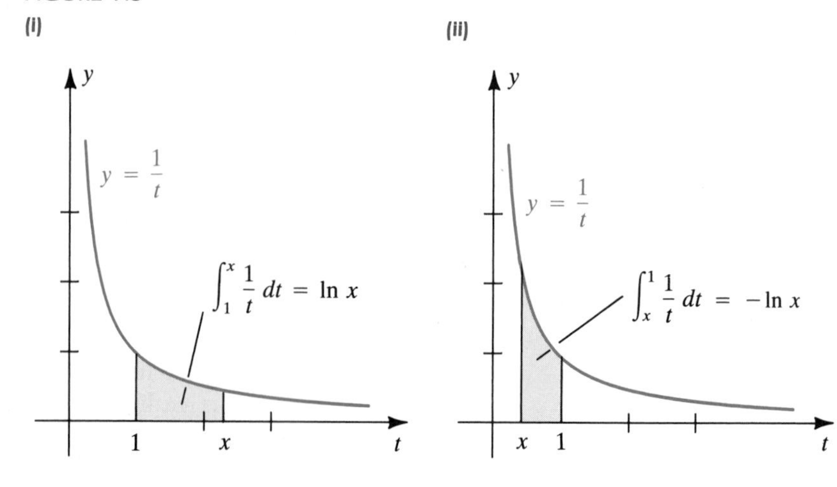

If $0 < x < 1$, then, since

$$\int_1^x \frac{1}{t}\,dt = -\int_x^1 \frac{1}{t}\,dt,$$

the integral is the *negative* of the area of the region under the graph of $y = 1/t$ from $t = x$ to $t = 1$ (see Figure 7.8(ii)). This shows that *$\ln x$ is negative for $0 < x < 1$ and positive for $x > 1$.* Also note that, by Definition (5.18),

$$\ln 1 = \int_1^1 \frac{1}{t}\,dt = 0.$$

Applying Theorem (5.35) yields

$$D_x \int_1^x \frac{1}{t}\,dt = \frac{1}{x}$$

for every $x > 0$. Substituting $\ln x$ for $\int_1^x (1/t)\,dt$ gives us the following theorem.

Theorem **(7.10)**

$$D_x \ln x = \frac{1}{x}$$

By Theorem (7.10), $\ln x$ *is an antiderivative of* $1/x$. Since $\ln x$ is differentiable and its derivative $1/x$ is positive for every $x > 0$, it follows from Theorems (3.11) and (4.13) that *the natural logarithmic function is continuous and increasing throughout its domain.* Also note that

$$D_x^2 \ln x = D_x\,(D_x \ln x) = D_x\left(\frac{1}{x}\right) = -\frac{1}{x^2},$$

which is negative for every $x > 0$. Hence, by (4.16), the graph of the natural logarithmic function is concave downward on $(0, \infty)$.

Let us sketch the graph of $y = \ln x$. If $0 < x < 1$, then $\ln x < 0$ and the graph is below the x-axis. If $x > 1$, the graph is above the x-axis. Since $\ln 1 = 0$, the x-intercept is 1. We may approximate y-coordinates of points on the graph by applying the trapezoidal rule or Simpson's rule. If $x = 2$, then, by Example 2 in Section 5.7,

$$\ln 2 = \int_1^2 \frac{1}{t}\,dt \approx 0.693.$$

We will show in Theorem (7.12) that if $a > 0$, then $\ln a^r = r \ln a$ for every rational number r. Using this result yields the following:

$$\ln 4 = \ln 2^2 = 2 \ln 2 \approx 2(0.693) \approx 1.386$$

$$\ln 8 = \ln 2^3 = 3 \ln 2 \approx 2.079$$

$$\ln \tfrac{1}{2} = \ln 2^{-1} = -\ln 2 \approx -0.693$$

$$\ln \tfrac{1}{4} = \ln 2^{-2} = -2 \ln 2 \approx -1.386$$

$$\ln \tfrac{1}{8} = \ln 2^{-3} = -3 \ln 2 \approx -2.079$$

FIGURE 7.9

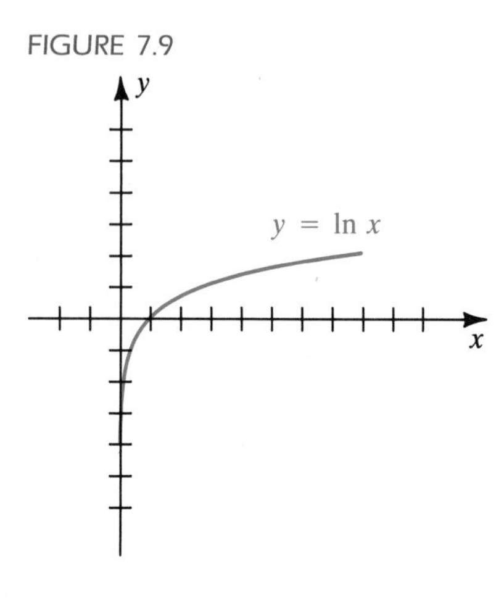

Table C in Appendix III provides a list of natural logarithms of many other numbers, correct to three decimal places. Calculators may also be used to estimate values of ln.

Plotting the points that correspond to the y-coordinates we have calculated and using the fact that ln is continuous and increasing gives us the sketch in Figure 7.9.

At the end of this section we prove that

$$\lim_{x \to \infty} \ln x = \infty \quad \text{and} \quad \lim_{x \to 0^+} \ln x = -\infty.$$

The first of these results tells us that $y = \ln x$ increases without bound as $x \to \infty$. Note, however, that the *rate of change* of y with respect to x is very small if x is large. For example, if $x = 10^6$, then

$$\left.\frac{dy}{dx}\right]_{10^6} = \left.\frac{1}{x}\right]_{10^6} = \frac{1}{10^6} = 0.000001.$$

Thus, the tangent line is *almost* horizontal at the point on the graph with x-coordinate 10^6, and hence the graph is very flat near that point. The fact that $\lim_{x \to 0^+} \ln x = -\infty$ tells us that the line $x = 0$ (the y-axis) is a vertical asymptote for the graph (see Figure 7.9).

The next result generalizes Theorem (7.10).

Theorem **(7.11)**

If $u = g(x)$ and g is differentiable, then

(i) $D_x \ln u = \dfrac{1}{u} D_x u \quad \text{if} \quad g(x) > 0$

(ii) $D_x \ln |u| = \dfrac{1}{u} D_x u \quad \text{if} \quad g(x) \neq 0$

PROOF

(i) If we let $y = \ln u$ and $u = g(x)$, then, by the chain rule and Theorem (7.10),

$$D_x \ln u = \frac{dy}{dx} = \frac{dy}{du}\frac{du}{dx} = \frac{1}{u} D_x u.$$

(ii) If $u > 0$, then $|u| = u$ and, by part (i),

$$D_x \ln |u| = D_x \ln u = \frac{1}{u} D_x u.$$

If $u < 0$, then $|u| = -u > 0$ and, by part (i),

$$D_x \ln |u| = D_x \ln (-u) = \frac{1}{-u} D_x (-u)$$

$$= -\frac{1}{u}(-1) D_x u = \frac{1}{u} D_x u. \quad ■$$

In examples and exercises, if a function is defined in terms of the natural logarithmic function, its domain will not usually be stated explicitly. Instead *we shall tacitly assume that x is restricted to values for which the logarithmic expression has meaning*. Thus, in Example 1, we assume $x^2 - 6 > 0$; that is, $|x| > \sqrt{6}$. In Example 2, we assume $x + 1 > 0$.

EXAMPLE 1 If $f(x) = \ln (x^2 - 6)$, find $f'(x)$.

SOLUTION Letting $u = x^2 - 6$ in Theorem (7.11)(i) yields

$$f'(x) = D_x \ln (x^2 - 6) = \frac{1}{x^2 - 6} D_x (x^2 - 6) = \frac{2x}{x^2 - 6}.$$

EXAMPLE 2 If $y = \ln \sqrt{x + 1}$, find $\dfrac{dy}{dx}$.

SOLUTION Letting $u = \sqrt{x + 1}$ in Theorem (7.11)(i) gives us

$$\begin{aligned}\frac{dy}{dx} &= \frac{d}{dx} \ln \sqrt{x + 1} \\ &= \frac{1}{\sqrt{x + 1}} \frac{d}{dx} \sqrt{x + 1} = \frac{1}{\sqrt{x + 1}} \cdot \frac{1}{2} (x + 1)^{-1/2} \\ &= \frac{1}{\sqrt{x + 1}} \cdot \frac{1}{2} \frac{1}{\sqrt{x + 1}} = \frac{1}{2(x + 1)}.\end{aligned}$$

EXAMPLE 3 If $f(x) = \ln |4 + 5x - 2x^3|$, find $f'(x)$.

SOLUTION Using Theorem (7.11)(ii) with $u = 4 + 5x - 2x^3$ yields

$$\begin{aligned}f'(x) &= D_x \ln |4 + 5x - 2x^3| \\ &= \frac{1}{4 + 5x - 2x^3} D_x (4 + 5x - 2x^3) = \frac{5 - 6x^2}{4 + 5x - 2x^3}.\end{aligned}$$

The next result states that natural logarithms satisfy the laws of logarithms studied in precalculus mathematics courses.

Laws of natural logarithms **(7.12)**

> If $p > 0$ and $q > 0$, then
>
> **(i)** $\ln pq = \ln p + \ln q$
>
> **(ii)** $\ln \dfrac{p}{q} = \ln p - \ln q$
>
> **(iii)** $\ln p^r = r \ln p$ for every rational number r

PROOF

(i) If $p > 0$, then using Theorem (7.11) with $u = px$ gives us

$$D_x \ln px = \frac{1}{px} D_x (px) = \frac{1}{px} p = \frac{1}{x}.$$

Thus, $\ln px$ and $\ln x$ are both antiderivatives of $1/x$, and hence, by Theorem (5.2),

$$\ln px = \ln x + C$$

for some constant C. Letting $x = 1$, we obtain

$$\ln p = \ln 1 + C.$$

Since $\ln 1 = 0$, we see that $C = \ln p$, and therefore

$$\ln px = \ln x + \ln p.$$

Substituting q for x in the last equation gives us

$$\ln pq = \ln q + \ln p,$$

which is what we wished to prove.

(ii) Using the formula $\ln p + \ln q = \ln pq$ with $p = 1/q$, we see that

$$\ln \frac{1}{q} + \ln q = \ln \left(\frac{1}{q} \cdot q\right) = \ln 1 = 0$$

and hence $$\ln \frac{1}{q} = -\ln q.$$

Consequently,

$$\ln \frac{p}{q} = \ln \left(p \cdot \frac{1}{q}\right) = \ln p + \ln \frac{1}{q} = \ln p - \ln q.$$

(iii) If r is a rational number and $x > 0$, then, by Theorem (7.11) with $u = x^r$,

$$D_x (\ln x^r) = \frac{1}{x^r} D_x (x^r) = \frac{1}{x^r} rx^{r-1} = r\left(\frac{1}{x}\right) = \frac{r}{x}.$$

By Theorems (3.18)(iv) and (7.7), we may also write

$$D_x (r \ln x) = r\, D_x (\ln x) = r\left(\frac{1}{x}\right) = \frac{r}{x}.$$

Since $\ln x^r$ and $r \ln x$ are both antiderivatives of r/x, it follows from Theorem (5.2) that

$$\ln x^r = r \ln x + C$$

for some constant C. If we let $x = 1$ in the last formula, we obtain

$$\ln 1 = r \ln 1 + C.$$

Since $\ln 1 = 0$, this implies that $C = 0$ and, therefore,

$$\ln x^r = r \ln x.$$

In Section 7.5 we shall extend this law to irrational exponents. ■

As shown in the following illustration, sometimes it is convenient to use laws of natural logarithms *before* differentiating.

ILLUSTRATION

$f(x)$	$f(x)$ AFTER USING LAWS OF LOGARITHMS	$f'(x)$
$\ln [(x+2)(3x-5)]$	$\ln (x+2) + \ln (3x-5)$	$\dfrac{1}{x+2} + \dfrac{1}{3x-5} \cdot 3 = \dfrac{6x+1}{(x+2)(3x-5)}$
$\ln \dfrac{x+2}{3x-5}$	$\ln (x+2) - \ln (3x-5)$	$\dfrac{1}{x+2} - \dfrac{1}{3x-5} \cdot 3 = \dfrac{-11}{(x+2)(3x-5)}$
$\ln (x^2+1)^5$	$5 \ln (x^2+1)$	$5 \cdot \dfrac{1}{x^2+1} \cdot 2x = \dfrac{10x}{x^2+1}$
$\ln \sqrt{x+1}$	$\dfrac{1}{2} \ln (x+1)$	$\dfrac{1}{2} \cdot \dfrac{1}{x+1} = \dfrac{1}{2(x+1)}$

An advantage of using laws of logarithms before differentiating may be seen by comparing the method of finding $D_x \ln \sqrt{x+1}$ in the preceding illustration with the solution of Example 2.

In the next two examples we apply laws of logarithms to complicated expressions before differentiating.

EXAMPLE 4 If $f(x) = \ln [\sqrt{6x-1}(4x+5)^3]$, find $f'(x)$.

SOLUTION We first write $\sqrt{6x-1} = (6x-1)^{1/2}$ and then use laws of logarithms (i) and (iii):

$$\begin{aligned} f(x) &= \ln [(6x-1)^{1/2}(4x+5)^3] \\ &= \ln (6x-1)^{1/2} + \ln (4x+5)^3 \\ &= \tfrac{1}{2} \ln (6x-1) + 3 \ln (4x+5) \end{aligned}$$

By Theorem (7.11),

$$\begin{aligned} f'(x) &= \left(\frac{1}{2} \cdot \frac{1}{6x-1} \cdot 6\right) + \left(3 \cdot \frac{1}{4x+5} \cdot 4\right) \\ &= \frac{3}{6x-1} + \frac{12}{4x+5} \\ &= \frac{84x+3}{(6x-1)(4x+5)}. \end{aligned}$$

EXAMPLE 5 If $y = \ln \sqrt[3]{\dfrac{x^2-1}{x^2+1}}$, find $\dfrac{dy}{dx}$.

SOLUTION We first use laws of logarithms to change the form of y as follows:

$$\begin{aligned} y &= \ln \left(\frac{x^2-1}{x^2+1}\right)^{1/3} = \frac{1}{3} \ln \left(\frac{x^2-1}{x^2+1}\right) \\ &= \tfrac{1}{3}[\ln (x^2-1) - \ln (x^2+1)] \end{aligned}$$

Next we use Theorem (7.11) to obtain

$$\begin{aligned}\frac{dy}{dx} &= \frac{1}{3}\left(\frac{1}{x^2-1}\cdot 2x - \frac{1}{x^2+1}\cdot 2x\right)\\ &= \frac{2x}{3}\left(\frac{1}{x^2-1} - \frac{1}{x^2+1}\right)\\ &= \frac{2x}{3}\left[\frac{2}{(x^2-1)(x^2+1)}\right] = \frac{4x}{3(x^2-1)(x^2+1)}.\end{aligned}$$

Given $y = f(x)$, we may sometimes find $D_x\, y$ by **logarithmic differentiation**. This method is especially useful if $f(x)$ involves complicated products, quotients, or powers. In the following guidelines it is assumed that $f(x) > 0$; however, we shall show that the same steps can be used if $f(x) < 0$.

Guidelines for logarithmic differentiation **(7.13)**

1	$y = f(x)$	(given)
2	$\ln y = \ln f(x)$	(take natural logarithms and simplify)
3	$D_x [\ln y] = D_x [\ln f(x)]$	(differentiate implicitly)
4	$\frac{1}{y} D_x\, y = D_x [\ln f(x)]$	(by Theorem (7.11))
5	$D_x\, y = f(x)\, D_x [\ln f(x)]$	(multiply by $y = f(x)$)

Of course, to complete the solution we must differentiate $\ln f(x)$ at some stage after guideline 3. If $f(x) < 0$ for some x, then guideline 2 is invalid, since $\ln f(x)$ is undefined. In this event we can replace guideline 1 by $|y| = |f(x)|$ and take natural logarithms, obtaining $\ln |y| = \ln |f(x)|$. If we now differentiate implicitly and use Theorem (7.11)(ii), we again arrive at guideline 4. Thus, negative values of $f(x)$ do not change the outcome, and we are not concerned whether $f(x)$ is positive or negative. The method should not be used to find $f'(a)$ if $f(a) = 0$, since $\ln 0$ is undefined.

EXAMPLE 6 If $y = \dfrac{(5x-4)^3}{\sqrt{2x+1}}$, use logarithmic differentiation to find $D_x\, y$.

SOLUTION As in guideline 2, we begin by taking the natural logarithm of each side, obtaining

$$\begin{aligned}\ln y &= \ln (5x-4)^3 - \ln \sqrt{2x+1}\\ &= 3 \ln (5x-4) - \tfrac{1}{2} \ln (2x+1).\end{aligned}$$

Applying guidelines 3 and 4, we differentiate implicitly with respect to x and then use Theorem (7.8) to obtain

$$\begin{aligned}\frac{1}{y} D_x\, y &= \left(3\cdot\frac{1}{5x-4}\cdot 5\right) - \left(\frac{1}{2}\cdot\frac{1}{2x+1}\cdot 2\right)\\ &= \frac{25x+19}{(5x-4)(2x+1)}.\end{aligned}$$

Finally, as in guideline 5, we multiply both sides of the last equation by y (that is, by $(5x - 4)^3/\sqrt{2x + 1}$) to get

$$D_x\, y = \frac{25x + 19}{(5x - 4)(2x + 1)} \cdot \frac{(5x - 4)^3}{\sqrt{2x + 1}}$$
$$= \frac{(25x + 19)(5x - 4)^2}{(2x + 1)^{3/2}}.$$

We could check this result by applying the quotient rule to y.

An application of natural logarithms to growth processes is given in the next example. Many additional applied problems involving ln appear in other examples and exercises of this chapter.

EXAMPLE 7 The *Count model* is an empirically based formula that can be used to predict the height of a preschooler. If $h(x)$ denotes the height (in centimeters) at age x (in years) for $\frac{1}{4} \leq x \leq 6$, then $h(x)$ can be approximated by

$$h(x) = 70.228 + 5.104x + 9.222 \ln x.$$

(a) Predict the height and rate of growth when a child reaches age 2.

(b) When is the rate of growth largest?

SOLUTION

(a) The height at age 2 is approximately

$$h(2) = 70.228 + 5.104(2) + 9.222 \ln 2 \approx 86.8 \text{ cm}.$$

The rate of change of h with respect to x is

$$h'(x) = 5.104 + 9.222\left(\frac{1}{x}\right).$$

Letting $x = 2$ gives us

$$h'(2) = 5.104 + 9.222(\tfrac{1}{2}) = 9.715.$$

Hence the rate of growth on the child's second birthday is about 9.7 cm/yr.

(b) To determine the maximum value of the rate of growth $h'(x)$, we first find the critical numbers of h'. Differentiating $h'(x)$, we obtain

$$h''(x) = 9.222\left(-\frac{1}{x^2}\right) = -\frac{9.222}{x^2}.$$

Since $h''(x)$ is negative for every x in $[\frac{1}{4}, 6]$, h' has no critical numbers in $(\frac{1}{4}, 6)$. It follows from Theorem (4.13) that h' is decreasing on $[\frac{1}{4}, 6]$. Consequently, the maximum value of $h'(x)$ occurs at $x = \frac{1}{4}$; that is, the rate of growth is largest at the age of 3 months.

We shall conclude this section by investigating $\ln x$ as $x \to \infty$ and as $x \to 0^+$. If $x > 1$, we may interpret the integral $\int_1^x (1/t)\, dt = \ln x$ as the

FIGURE 7.10

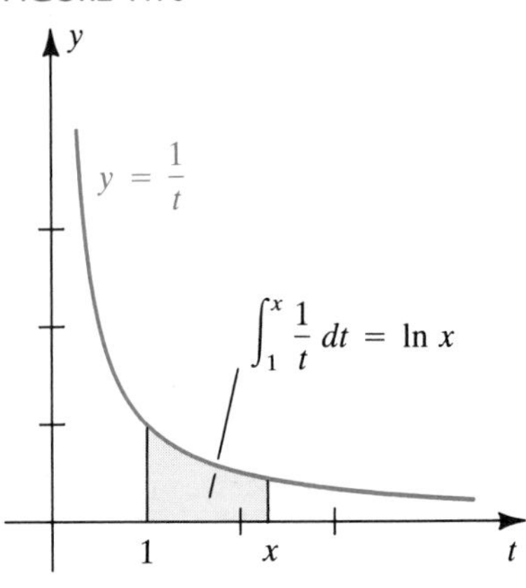

FIGURE 7.11

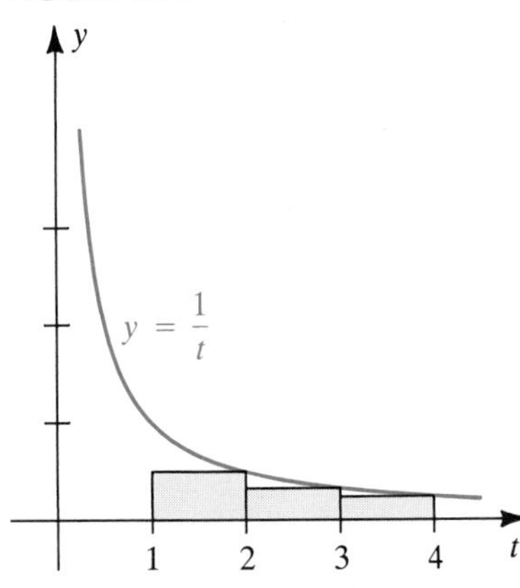

area of the region shown in Figure 7.10. The sum of the areas of the three rectangles shown in Figure 7.11 is

$$\tfrac{1}{2} + \tfrac{1}{3} + \tfrac{1}{4} = \tfrac{13}{12}.$$

Since the area under the graph of $y = 1/t$ from $t = 1$ to $t = 4$ is ln 4, we see that

$$\ln 4 > \tfrac{13}{12} > 1.$$

It follows that if M is any positive rational number, then

$$M \ln 4 > M, \quad \text{or} \quad \ln 4^M > M.$$

If $x > 4^M$, then since ln is an increasing function,

$$\ln x > \ln 4^M > M.$$

This proves that $\ln x$ can be made as large as desired by choosing x sufficiently large; that is,

$$\lim_{x\to\infty} \ln x = \infty.$$

To investigate the case $x \to 0^+$, we first note that

$$\ln \frac{1}{x} = \ln 1 - \ln x = 0 - \ln x = -\ln x.$$

Hence $$\lim_{x\to 0^+} \ln x = \lim_{x\to 0^+}\left(-\ln \frac{1}{x}\right).$$

As x approaches zero through positive values, $1/x$ increases without bound and, therefore, so does $\ln(1/x)$. Consequently, $-\ln(1/x)$ *decreases* without bound; that is,

$$\lim_{x\to 0^+} \ln x = -\infty.$$

EXERCISES 7.2

Exer. 1–34: Find $f'(x)$ if $f(x)$ is the given expression.

1 $\ln(9x + 4)$
2 $\ln(x^4 + 1)$
3 $\ln(3x^2 - 2x + 1)$
4 $\ln(4x^3 - x^2 + 2)$
5 $\ln|3 - 2x|$
6 $\ln|4 - 3x|$
7 $\ln|2 - 3x|^5$
8 $\ln|5x^2 - 1|^3$
9 $\ln\sqrt{7 - 2x^3}$
10 $\ln\sqrt[3]{6x + 7}$
11 $x \ln x$
12 $\ln(\ln x)$
13 $\ln\sqrt{x} + \sqrt{\ln x}$
14 $\ln x^3 + (\ln x)^3$
15 $\dfrac{1}{\ln x} + \ln\dfrac{1}{x}$
16 $\dfrac{x^2}{\ln x}$
17 $\ln[(5x - 7)^4(2x + 3)^3]$
18 $\ln[\sqrt[3]{4x - 5}\,(3x + 8)^2]$
19 $\ln\dfrac{\sqrt{x^2 + 1}}{(9x - 4)^2}$
20 $\ln\dfrac{x^2(2x - 1)^3}{(x + 5)^2}$
21 $\ln\sqrt{\dfrac{x^2 - 1}{x^2 + 1}}$
22 $\ln\sqrt{\dfrac{4 + x^2}{4 - x^2}}$
23 $\ln(x + \sqrt{x^2 - 1})$
24 $\ln(x + \sqrt{x^2 + 1})$
25 $\ln\cos 2x$
26 $\cos(\ln 2x)$
27 $\ln\tan^3 3x$
28 $\ln\cot(x^2)$
29 $\ln\ln\sec 2x$
30 $\ln\csc^2 4x$
31 $\ln|\sec x|$
32 $\ln|\sin x|$
33 $\ln|\sec x + \tan x|$
34 $\ln|\csc x - \cot x|$

Exer. 35–38: Use implicit differentiation to find y'.

35 $3y - x^2 + \ln xy = 2$
36 $y^2 + \ln(x/y) - 4x = -3$
37 $x\ln y - y\ln x = 1$
38 $y^3 + x^2\ln y = 5x + 3$

Exer. 39–44: Use logarithmic differentiation to find dy/dx.

39 $y = (5x + 2)^3(6x + 1)^2$

40 $y = (x + 1)^2(x + 2)^3(x + 3)^4$

41 $y = \sqrt{4x + 7}(x - 5)^3$

42 $y = \sqrt{(3x^2 + 2)\sqrt{6x - 7}}$

43 $y = \dfrac{(x^2 + 3)^5}{\sqrt{x + 1}}$

44 $y = \dfrac{(x^2 + 3)^{2/3}(3x - 4)^4}{\sqrt{x}}$

45 Find an equation of the tangent line to the graph of $y = x^2 + \ln(2x - 5)$ at the point $P(3, 9)$.

46 Find an equation of the tangent line to the graph of $y = x + \ln x$ that is perpendicular to the line whose equation is $2x + 6y = 5$.

47 Shown in the figure is a graph of $y = 5 \ln x - \frac{1}{2}x$. Find the coordinates of the highest point, and show that the graph is concave downward for $x > 0$.

EXERCISE 47

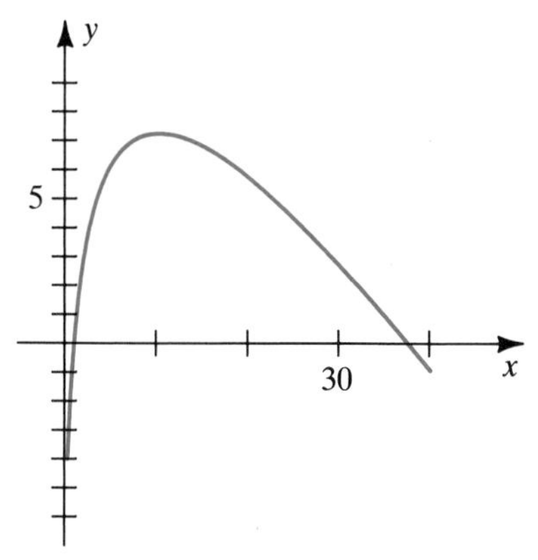

48 Shown in the figure is a graph of $y = \ln(x^2 + 1)$. Find the points of inflection.

EXERCISE 48

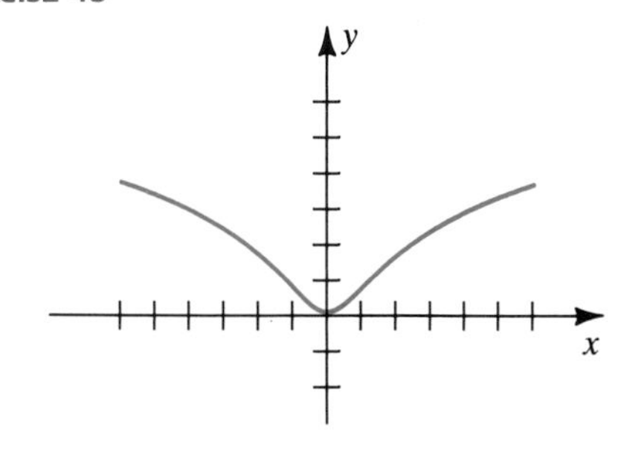

49 An approximation to the age T (in years) of a female blue whale can be obtained from a length measurement L (in feet) using the formula $T = -2.57 \ln[(87 - L)/63]$. A blue whale has been spotted by a research vessel, and her length is estimated to be 80 feet. If the maximum error in estimating L is ± 2 feet, use differentials to approximate the maximum error in T.

50 The *Ehrenberg relation*, $\ln W = \ln 2.4 + 0.0184h$, is an empirically based formula relating the height h (in centimeters) to the weight W (in kilograms) for children aged 5 through 13. The formula, with minor changes in the constants, has been verified in many different countries. Find the relationship between the rates of change dW/dt and dh/dt, for time t (in years).

51 A rocket of mass m_1 is filled with fuel of mass m_2, which will be burned at a constant rate of b kg/sec. If the fuel is expelled from the rocket at a constant rate, the distance $s(t)$ (in meters) that the rocket has traveled after t seconds is

$$s(t) = ct + \frac{c}{b}(m_1 + m_2 - bt) \ln\left(\frac{m_1 + m_2 - bt}{m_1 + m_2}\right)$$

for some constant $c > 0$.

(a) Find the initial velocity and initial acceleration of the rocket.

(b) Burnout occurs when $t = m_2/b$. Find the velocity and acceleration at burnout.

52 One method of estimating the thickness of the ozone layer is to use the formula $\ln(I/I_0) = -\beta T$, where I_0 is the intensity of a particular wavelength of light from the sun before it reaches the atmosphere, I is the intensity of the same wavelength after passing through a layer of ozone T centimeters thick, and β is the absorption coefficient for that wavelength. Suppose that for a wavelength of 3055×10^{-8} centimeter with $\beta \approx 2.7$, I_0/I is measured as 2.3.

(a) Approximate the thickness of the ozone layer to the nearest 0.01 centimeter.

(b) If the maximum error in the measured value of I_0/I is ± 0.1, use differentials to approximate the maximum error in the approximation obtained in (a).

53 Describe the difference between the graphs of $y = \ln(x^2)$ and $y = 2 \ln x$.

54 Sketch the graphs of

(a) $y = \ln|x|$ (b) $y = |\ln x|$

c 55 Use Newton's method to approximate the real root of $\ln x + x = 0$ to three decimal places.

56 Show that $x > \ln x$ for every $x > 0$.

c 57 Let $f(x) = \ln(x^4)$ and $g(x) = \sqrt{x}$.

(a) Is $f(x) \geq g(x)$ on $[1, 2]$?

(b) Is $f(x) \geq g(x)$ on $[3, 64]$?

(c) Is there a positive integer M such that $f(x) \geq g(x)$ on $[M, \infty)$? (*Hint:* Estimate $\lim_{x\to\infty} [f(x)/g(x)]$.)

c 58 (a) Use the trapezoidal rule, with $n = 4$, to approximate $\int_1^2 \ln(x^2)\,dx$.

(b) Estimate the error in (a) using (5.37).

7.3 THE NATURAL EXPONENTIAL FUNCTION

In Section 7.2 we saw that

$$\lim_{x \to \infty} \ln x = \infty \quad \text{and} \quad \lim_{x \to 0^+} \ln x = -\infty.$$

These facts are used in the proof of the following result.

Theorem **(7.14)**

> To every real number x there corresponds exactly one positive real number y such that $\ln y = x$.

PROOF First note that if $x = 0$, then $\ln 1 = 0$. Moreover, since ln is an increasing function, 1 is the only value of y such that $\ln y = 0$.

If x is positive, then we may choose a number b such that

$$\ln 1 < x < \ln b.$$

Since ln is continuous, it takes on every value between ln 1 and ln b (see the intermediate value theorem (2.26)). Thus, there is a number y between 1 and b such that $\ln y = x$. Since ln is an increasing function, there is only one such number.

Finally, if x is negative, then there is a number $b > 0$ such that

$$\ln b < x < \ln 1,$$

and, as before, there is exactly one number y between b and 1 such that $\ln y = x$. ■

It follows from Theorem (7.14) that the range of the natural logarithms is ℝ. Since ln is an increasing function, it is one-to-one and therefore has an inverse function, to which we give the following special name.

Definition **(7.15)**

> The **natural exponential function**, denoted by **exp**, is the inverse of the natural logarithmic function.

The reason for the term *exponential* in this definition will become clear shortly. Since exp is the inverse of ln, its domain is ℝ and its range is $(0, \infty)$. Moreover, as in (7.2),

$$y = \exp x \quad \text{if and only if} \quad x = \ln y,$$

where x is any real number and $y > 0$. By Theorem (7.3), we may also write

$$\ln(\exp x) = x \quad \text{and} \quad \exp(\ln y) = y.$$

FIGURE 7.12

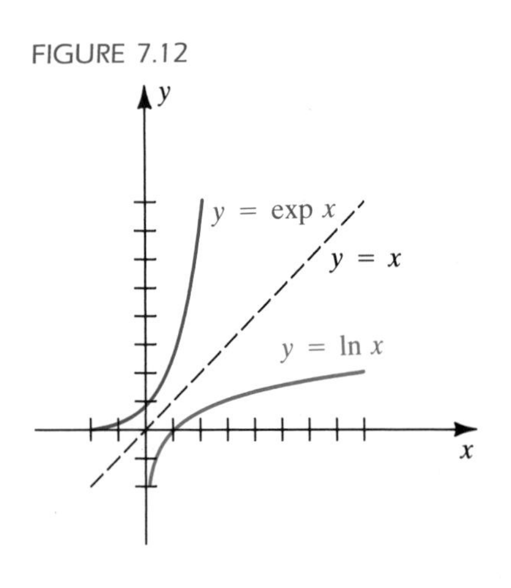

As we observed in Section 7.1, if two functions are inverses of each other, then their graphs are reflections through the line $y = x$. Hence the graph of $y = \exp x$ can be obtained by reflecting the graph of $y = \ln x$ through this line, as illustrated in Figure 7.12. Note that

$$\lim_{x \to \infty} \exp x = \infty \quad \text{and} \quad \lim_{x \to -\infty} \exp x = 0.$$

By Theorem (7.14), there exists exactly one positive real number whose natural logarithm is 1. This number is denoted by e. The great Swiss mathematician Leonhard Euler (1707–1783) was among the first to study its properties extensively.

Definition of e **(7.16)**

The letter e denotes the positive real number such that $\ln e = 1$.

Several values of ln were calculated in Section 7.2. We can show, by means of the trapezoidal rule, that

$$\int_1^{2.7} \frac{1}{t}\,dt < 1 < \int_1^{2.8} \frac{1}{t}\,dt.$$

(See Exercise 25 in Section 5.7.) Applying Definitions (7.9) and (7.16) yields

$$\ln 2.7 < \ln e < \ln 2.8$$

and hence

$$2.7 < e < 2.8.$$

Later, in Theorem (7.32), we show that

$$e = \lim_{h \to 0} (1 + h)^{1/h}.$$

This limit formula can be used to approximate e to any degree of accuracy. In Section 7.5 we shall obtain the following approximation to five decimal places.

Approximation to e **(7.17)**

$$e \approx 2.71828$$

It can be shown that e is an irrational number.

If r is any *rational* number, then

$$\ln e^r = r \ln e = r(1) = r.$$

The formula $\ln e^r = r$ may be used to motivate a definition of e^x for every *real* number x. Specifically, we shall *define* e^x as the real number y such that $\ln y = x$. The following statement is a convenient way to remember this definition.

Definition of e^x **(7.18)**

If x is any real number, then

$$e^x = y \quad \text{if and only if} \quad \ln y = x.$$

Since exp is the inverse function of ln,

$$\exp x = y \quad \text{if and only if} \quad \ln y = x.$$

Comparing this relationship with Definition (7.18), we see that

$$e^x = \exp x \quad \text{for every } x.$$

This is the reason for calling exp an *exponential* function and referring to it as the **exponential function with base** e. The graph of $y = e^x$ is the same as that of $y = \exp x$, illustrated in Figure 7.12. Hereafter *we shall use* e^x *instead of* $\exp x$ *to denote values of the natural exponential function.*

The fact that $\ln(\exp x) = x$ for every x and $\exp(\ln x) = x$ for every $x > 0$ may now be written as follows:

Theorem **(7.19)**

(i) $\ln e^x = x$ for every x

(ii) $e^{\ln x} = x$ for every $x > 0$

Some special cases of this theorem are given in the following illustration.

ILLUSTRATION

- $\ln e^5 = 5$
- $e^{\ln 5} = 5$
- $e^{3 \ln x} = (e^{\ln x})^3 = x^3$
- $\ln e^{\sqrt{x+1}} = \sqrt{x+1}$
- $e^{\ln \sqrt{x+1}} = \sqrt{x+1}$
- $e^{k \ln x} = (e^{\ln x})^k = x^k$

A brief table of values of e^x and e^{-x} is given in Table B of Appendix III. To approximate values of e^x with a scientific calculator, we employ either an $\boxed{e^x}$ key or the succession $\boxed{\text{INV}}$ $\boxed{\ln x}$. Still another method is to use the $\boxed{y^x}$ key with $y = e \approx 2.71828$.

The next theorem states that the laws of exponents are true for powers of e.

Theorem **(7.20)**

If p and q are real numbers and r is a rational number, then

(i) $e^p e^q = e^{p+q}$ **(ii)** $\dfrac{e^p}{e^q} = e^{p-q}$ **(iii)** $(e^p)^r = e^{pr}$

PROOF Using Theorems (7.12) and (7.19), we obtain

$$\ln e^p e^q = \ln e^p + \ln e^q = p + q = \ln e^{p+q}.$$

Since the natural logarithmic function is one-to-one,

$$e^p e^q = e^{p+q}.$$

This proves (i). The proofs for (ii) and (iii) are similar. We show in Section 7.5 that (iii) is also true if r is irrational. ■

By Theorem (7.7), the inverse function of a differentiable function is differentiable, and hence $D_x e^x$ exists. The next theorem states that e^x *is its own derivative.*

Theorem **(7.21)**

$$D_x e^x = e^x$$

PROOF By (i) of Theorem (7.19),

$$\ln e^x = x.$$

Differentiating each side of this equation and using Theorem (7.11)(i) with $u = e^x$ gives us the following:

$$D_x\,(\ln e^x) = D_x\,(x)$$

$$\frac{1}{e^x}\,D_x\,e^x = 1$$

$$D_x\,e^x = e^x \quad ■$$

EXAMPLE 1 If $f(x) = x^2e^x$, find $f'(x)$.

SOLUTION By the product rule and Theorem 7.21,

$$\begin{aligned} f'(x) &= x^2(D_x\,e^x) + e^x(D_x\,x^2) \\ &= x^2e^x + e^x(2x) = xe^x(x + 2). \end{aligned}$$

The next result is a generalization of Theorem (7.21).

Theorem **(7.22)**

> If $u = g(x)$ and g is differentiable, then
>
> $$D_x\,e^u = e^u\,D_x\,u.$$

PROOF Letting $y = e^u$ with $u = g(x)$, and using the chain rule and Theorem (7.21), we have

$$D_x\,e^u = \frac{dy}{dx} = \frac{dy}{du}\frac{du}{dx} = e^u\,D_x\,u. \quad ■$$

If $u = x$, then Theorem (7.22) reduces to (7.21).

EXAMPLE 2 If $y = e^{\sqrt{x^2+1}}$, find dy/dx.

SOLUTION By Theorem (7.22),

$$\begin{aligned} \frac{dy}{dx} = \frac{d}{dx}\,e^{\sqrt{x^2+1}} &= e^{\sqrt{x^2+1}}\,\frac{d}{dx}\,\sqrt{x^2+1} \\ &= e^{\sqrt{x^2+1}}\,\frac{d}{dx}\,(x^2+1)^{1/2} \\ &= e^{\sqrt{x^2+1}}\,(\tfrac{1}{2})(x^2+1)^{-1/2}(2x) \\ &= e^{\sqrt{x^2+1}} \cdot \frac{x}{\sqrt{x^2+1}} \\ &= \frac{xe^{\sqrt{x^2+1}}}{\sqrt{x^2+1}}. \end{aligned}$$

EXAMPLE 3 The function f defined by $f(x) = e^{-x^2/2}$ occurs in the branch of mathematics called *probability*. Find the local extrema of f, discuss concavity, find the points of inflection, and sketch the graph of f.

SOLUTION By Theorem (7.22),

$$f'(x) = e^{-x^2/2}\, D_x\left(-\frac{x^2}{2}\right) = e^{-x^2/2}\left(-\frac{2x}{2}\right) = -xe^{-x^2/2}.$$

Since $e^{-x^2/2}$ is always positive, the only critical number of f is 0. If $x < 0$, then $f'(x) > 0$, and if $x > 0$, then $f'(x) < 0$. It follows from the first derivative test that f has a local maximum at 0. The maximum value is $f(0) = e^{-0} = 1$.

Applying the product rule to $f'(x)$ yields

$$\begin{aligned} f''(x) &= -x\, D_x\, e^{-x^2/2} + e^{-x^2/2}\, D_x(-x) \\ &= -xe^{-x^2/2}(-2x/2) - e^{-x^2/2} \\ &= e^{-x^2/2}(x^2 - 1), \end{aligned}$$

and hence the second derivative is zero at -1 and 1. If $-1 < x < 1$, then $f''(x) < 0$ and, by (4.16), the graph of f is concave downward in the open interval $(-1, 1)$. If $x < -1$ or $x > 1$, then $f''(x) > 0$ and, therefore, the graph is concave upward throughout the infinite intervals $(-\infty, -1)$ and $(1, \infty)$. Consequently, $P(-1, e^{-1/2})$ and $Q(1, e^{-1/2})$ are points of inflection. From the expression

$$f(x) = \frac{1}{e^{x^2/2}}$$

it is evident that as x increases numerically, $f(x)$ approaches 0. We can prove that $\lim_{x\to\infty} f(x) = 0$ and $\lim_{x\to-\infty} f(x) = 0$; that is, the x-axis is a horizontal asymptote. The graph of f is sketched in Figure 7.13.

FIGURE 7.13

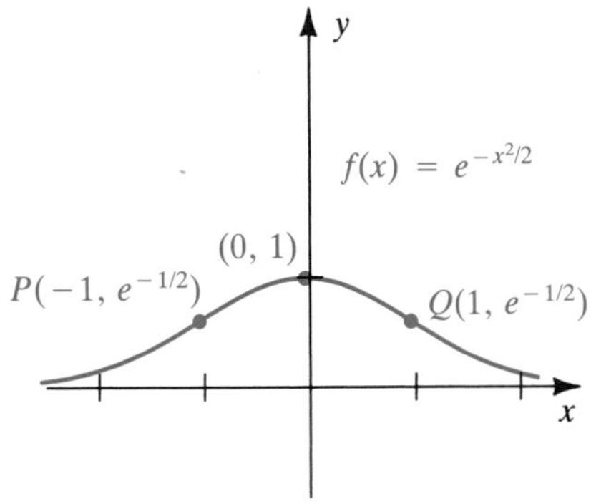

Exponential functions play an important role in the field of *radiotherapy*, the treatment of tumors by radiation. The fraction of cells in a tumor that survive a treatment, called the *surviving fraction*, depends not only on the energy and nature of the radiation, but also on the depth, size, and characteristics of the tumor itself. The exposure to radiation may be thought of as a number of potentially damaging events, where only one *hit* is required to kill a tumor cell. Suppose that each cell has exactly one *target* that must be hit. If k denotes the average target size of a tumor cell and if x is the number of damaging events (the *dose*), then the surviving fraction $f(x)$ is given by

$$f(x) = e^{-kx}.$$

This is called the *one-target–one hit surviving fraction*.

Suppose next that each cell has n targets and that hitting each target once results in the death of a cell. In this case, the *n target–one hit surviving fraction* is given by

$$f(x) = 1 - (1 - e^{-kx})^n.$$

In the next example we examine the case where $n = 2$.

EXAMPLE 4 If each cell of a tumor has two targets, then the two target–one hit surviving fraction is given by

$$f(x) = 1 - (1 - e^{-kx})^2,$$

where k is the average size of a cell. Analyze the graph of f to determine what effect increasing the dosage x has on decreasing the surviving fraction of tumor cells.

SOLUTION First note that if $x = 0$, then $f(0) = 1$; that is, if there is no dose, then all cells survive. Differentiating, we obtain

$$\begin{aligned} f'(x) &= 0 - 2(1 - e^{-kx})\, D_x\,(1 - e^{-kx}) \\ &= -2(1 - e^{-kx})(ke^{-kx}) \\ &= -2ke^{-kx}(1 - e^{-kx}). \end{aligned}$$

Since $f'(x) < 0$ for every $x > 0$ and $f'(0) = 0$, the function f is decreasing and the graph has a horizontal tangent line at the point (0, 1). We may verify that the second derivative is

$$f''(x) = 2k^2e^{-kx}(1 - 2e^{-kx}).$$

We see that $f''(x) = 0$ if $1 - 2e^{-kx} = 0$ (that is, if $e^{-kx} = \frac{1}{2}$, or, equivalently, $-kx = \ln \frac{1}{2} = -\ln 2$). This gives us

$$x = \frac{1}{k} \ln 2.$$

It can be verified that if $0 \le x < (1/k) \ln 2$, then $f''(x) < 0$, and hence the graph is concave downward. If $x > (1/k) \ln 2$, then $f''(x) > 0$, and the graph is concave upward. This implies that there is a point of inflection with x-coordinate $(1/k) \ln 2$. The y-coordinate of this point is

$$\begin{aligned} f\left(\frac{1}{k} \ln 2\right) &= 1 - (1 - e^{-\ln 2})^2 \\ &= 1 - (1 - \tfrac{1}{2})^2 = \tfrac{3}{4}. \end{aligned}$$

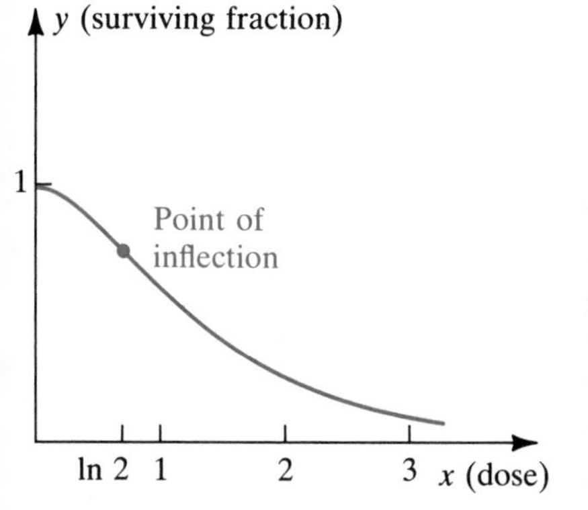

FIGURE 7.14
Surviving fraction of tumor cells after a radiation treatment

The graph is sketched in Figure 7.14 for the case $k = 1$. The *shoulder* on the curve near the point (0, 1) represents the threshold nature of the treatment; that is, a small dose results in very little tumor elimination. Note that if x is large, then an increase in dosage has little effect on the surviving fraction. To determine the ideal dose that should be administered to a given patient, specialists in radiation therapy must also take into account the number of healthy cells that are killed during a treatment.

EXERCISES 7.3

Exer. 1–30: Find $f'(x)$ if $f(x)$ equals the given expression.

1 e^{-5x}

2 e^{3x}

3 e^{3x^2}

4 e^{1-x^3}

5 $\sqrt{1 + e^{2x}}$

6 $1/(e^x + 1)$

7 $e^{\sqrt{x+1}}$

8 xe^{-x}

9 x^2e^{-2x}

10 $\sqrt{e^{2x} + 2x}$

11 $e^x/(x^2 + 1)$

12 $x/e^{(x^2)}$

13 $(e^{4x} - 5)^3$

14 $(e^{3x} - e^{-3x})^4$

15 $e^{1/x} + (1/e^x)$

16 $e^{\sqrt{x}} + \sqrt{e^x}$

17 $\dfrac{e^x - e^{-x}}{e^x + e^{-x}}$

18 $e^{x \ln x}$

19 $e^{-2x} \ln x$

20 $\ln e^x$

21 $\sin e^{5x}$

22 $e^{\sin 5x}$

23 $\ln \cos e^{-x}$

24 $e^{-3x} \cos 3x$

25 $e^{3x} \tan \sqrt{x}$

26 $\sec e^{-2x}$

27 $\sec^2 (e^{-4x})$

28 $e^{-x} \tan^2 x$

29 $xe^{\cot x}$

30 $\ln (\csc e^{3x})$

Exer. 31–34: Use implicit differentiation to find y'.

31 $e^{xy} - x^3 + 3y^2 = 11$

32 $xe^y + 2x - \ln (y + 1) = 3$

33 $e^x \cot y = xe^{2y}$

34 $e^x \cos y = xe^y$

35 Find an equation of the tangent line to the graph of $y = (x - 1)e^x + 3 \ln x + 2$ at the point $P(1, 2)$.

36 Find an equation of the tangent line to the graph of $y = x - e^{-x}$ that is parallel to the line $6x - 2y = 7$.

Exer. 37–42: Find the local extrema of f. Determine where f is increasing or is decreasing, discuss concavity, find the points of inflection, and sketch the graph of f.

37 $f(x) = xe^x$

38 $f(x) = x^2e^{-2x}$

39 $f(x) = e^{1/x}$

40 $f(x) = xe^{-x}$

41 $f(x) = x \ln x$

42 $f(x) = (1 - \ln x)^2$

43 A radioactive substance decays according to the formula $q(t) = q_0e^{-ct}$, where q_0 is the initial amount of the substance, c is a positive constant, and $q(t)$ is the amount remaining after time t. Show that the rate at which the substance decays is proportional to $q(t)$.

44 The current $I(t)$ at time t in an electrical circuit is given by $I(t) = I_0e^{-Rt/L}$, where R is the resistance, L is the inductance, and I_0 is the current at time $t = 0$. Show that the rate of change of the current at any time t is proportional to $I(t)$.

45 If a drug is injected into the bloodstream, then its concentration t minutes later is given by

$$C(t) = \frac{k}{a - b}(e^{-bt} - e^{-at})$$

for positive constants a, b, and k.

(a) At what time does the maximum concentration occur?

(b) What can be said about the concentration after a long period of time?

46 If a beam of light that has intensity k is projected vertically downward into water, then its intensity $I(x)$ at a depth of x meters is $I(x) = ke^{-1.4x}$.

(a) At what rate is the intensity changing with respect to depth at 1 meter? 5 meters? 10 meters?

(b) At what depth is the intensity one-half its value at the surface? One-tenth its value?

47 The *Jenss model* is generally regarded as the most accurate formula for predicting the height of a preschooler. If $h(x)$ denotes the height (in centimeters) at age x (in years) for $\frac{1}{4} \le x \le 6$, then $h(x)$ can be approximated by

$$h(x) = 79.041 + 6.39x - e^{3.261 - 0.993x}.$$

(Compare with Example 7 of Section 7.2.)

(a) Predict the height and rate of growth when a child reaches the age of 1.

(b) When is the rate of growth largest, and when is it smallest?

48 For a population of female African elephants, the weight $W(t)$ (in kilograms) at age t (in years) may be approximated by a *von Bertanlanffy growth function* W such that

$$W(t) = 2600(1 - 0.51e^{-0.075t})^3.$$

(a) Approximate the weight and the rate of growth of a newborn.

(b) Assuming that an adult female weighs 1800 kilograms, estimate her age and her rate of growth at present.

(c) Find and interpret $\lim_{t\to\infty} W(t)$.

(d) Show that the rate of growth is largest between the ages of 5 years and 6 years.

49 Gamma distributions, which are important in traffic control studies and probability theory, are determined by $f(x) = cx^ne^{-ax}$ for $x > 0$, a positive integer n, a positive constant a, and $c = a^{n+1}/n!$. Shown in the figure are graphs corresponding to $a = 1$ for $n = 2$, 3, and 4.

(a) Show that f has exactly one local maximum.

(b) If $n = 4$, determine where $f(x)$ is increasing most rapidly.

EXERCISE 49

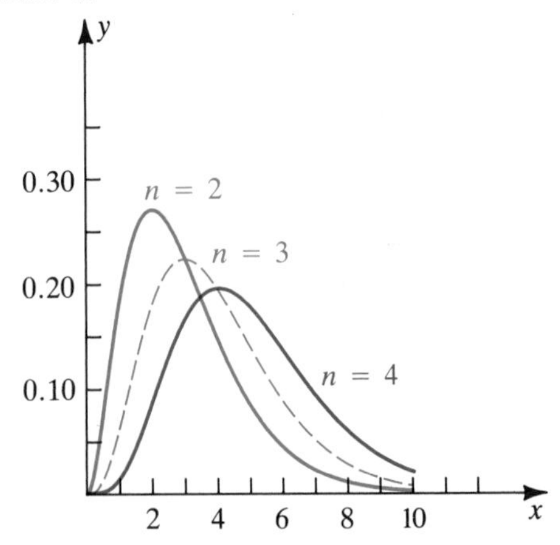

50 The relative number of gas molecules in a container that travel at a velocity of v cm/sec can be computed by means of the *Maxwell-Boltzmann speed distribution*,

$F(v) = cv^2 e^{-mv^2/(2kT)}$, where T is the temperature (in °K), m is the mass of a molecule, and c and k are positive constants. Show that the maximum value of F occurs when $v = \sqrt{2kT/m}$.

51 An *urban density model* is a formula that relates the population density (in number/mi^2) to the distance r (in miles) from the center of the city. The formula $D = ae^{-br+cr^2}$, where a, b, and c are positive constants, has been found to be appropriate for certain cities. Determine the shape of the graph for $r \geq 0$.

52 The effect of light on the rate of photosynthesis can be described by

$$f(x) = x^a e^{(a/b)(1-x^b)}$$

for $x > 0$ and positive constants a and b.

(a) Show that f has a maximum at $x = 1$.

(b) Conclude that if $x_0 > 0$ and $y_0 > 0$, then $g(x) = y_0 f(x/x_0)$ has a maximum $g(x_0) = y_0$.

53 The rate R at which a tumor grows is related to its size x by the equation $R = rx \ln (K/x)$, where r and K are positive constants. Show that the tumor is growing most rapidly when $x = e^{-1}K$.

54 If p denotes the selling price (in dollars) of a commodity and x is the corresponding demand (in number sold per day), then the relationship between p and x may be given by $p = p_0 e^{-ax}$ for positive constants p_0 and a. Suppose $p = 300e^{-0.02x}$. Find the selling price that will maximize daily revenues (see page 228).

55 In statistics the probability density function for the normal distribution is defined by

$$f(x) = \frac{1}{\sigma\sqrt{2\pi}} e^{-z^2/2} \quad \text{with} \quad z = \frac{x-\mu}{\sigma}$$

for real numbers μ and $\sigma > 0$ (μ is the *mean* and σ^2 is the *variance* of the distribution). Find the local extrema of f, and determine where f is increasing or is decreasing. Discuss concavity, find points of inflection, find $\lim_{x\to\infty} f(x)$ and $\lim_{x\to-\infty} f(x)$, and sketch the graph of f (see Example 3).

c 56 The integral $\int_a^b e^{-x^2}\, dx$ has applications in statistics. Use the trapezoidal rule, with $n = 10$, to approximate this integral if $a = 0$ and $b = 1$.

57 Nerve impulses in the human body travel along nerve fibers that consist of an *axon*, which transports the impulse, and an insulating coating surrounding the axon, called the *myelin sheath* (see figure). The nerve fiber is similar to an insulated cylindrical cable, for which the velocity v of an impulse is given by $v = -k(r/R)^2 \ln (r/R)$, where r is the radius of the cable and R is the insulation radius. Find the value of r/R that maximizes v. (In most nerve fibers $r/R \approx 0.6$.)

EXERCISE 57

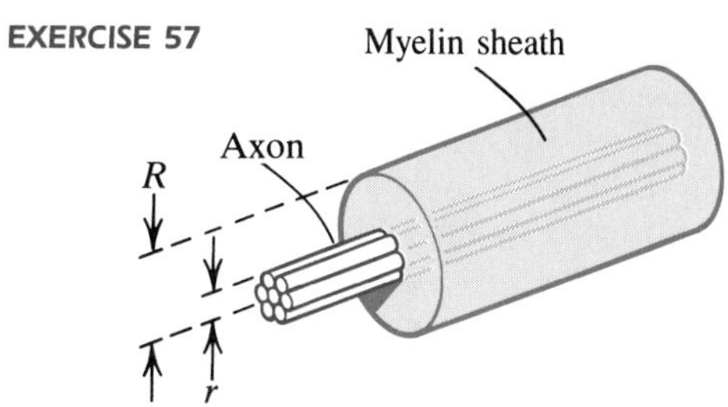

58 Sketch the graph of the *three target–one hit surviving fraction* (with $k = 1$) given by $f(x) = 1 - (1 - e^{-x})^3$ (see Example 4).

c **Exer. 59–60: Use Newton's method to approximate the real root of the equation to two decimal places.**

59 $e^{-x} = x$

60 $x \ln x = 1$

7.4 INTEGRATION

We may use differentiation formulas for ln to obtain formulas for integration. In particular, by Theorem (7.11),

$$D_x \ln |g(x)| = \frac{1}{g(x)} g'(x),$$

which gives us the integration formula

$$\int \frac{1}{g(x)} g'(x)\, dx = \ln |g(x)| + C.$$

This is restated in the next theorem in terms of the variable u.

Theorem (7.23)

If $u = g(x) \neq 0$ and g is differentiable, then

$$\int \frac{1}{u}\,du = \ln|u| + C.$$

Of course, if $u > 0$, then the absolute value sign may be deleted. A special case of Theorem (7.23) is

$$\int \frac{1}{x}\,dx = \ln|x| + C.$$

EXAMPLE 1 Evaluate $\int \frac{x}{3x^2 - 5}\,dx$.

SOLUTION Rewriting the integral as

$$\int \frac{x}{3x^2 - 5}\,dx = \int \frac{1}{3x^2 - 5}\,x\,dx$$

suggests that we use Theorem (7.23) with $u = 3x^2 - 5$. Thus, we make the substitution

$$u = 3x^2 - 5, \qquad du = 6x\,dx.$$

Introducing a factor 6 in the integrand and using Theorem (7.23) yields

$$\begin{aligned}\int \frac{x}{3x^2 - 5}\,dx &= \frac{1}{6}\int \frac{1}{3x^2 - 5}\,6x\,dx = \frac{1}{6}\int \frac{1}{u}\,du \\ &= \tfrac{1}{6}\ln|u| + C = \tfrac{1}{6}\ln|3x^2 - 5| + C.\end{aligned}$$

Another technique is to replace the expression $x\,dx$ in the integral by $\frac{1}{6}\,du$ and then integrate.

EXAMPLE 2 Evaluate $\int_2^4 \frac{1}{9 - 2x}\,dx$.

SOLUTION Since $1/(9 - 2x)$ is continuous on $[2, 4]$, the definite integral exists. One method of evaluation consists of using an indefinite integral to find an antiderivative of $1/(9 - 2x)$. We let

$$u = 9 - 2x, \qquad du = -2\,dx$$

and proceed as follows:

$$\begin{aligned}\int \frac{1}{9 - 2x}\,dx &= -\frac{1}{2}\int \frac{1}{9 - 2x}(-2)\,dx \\ &= -\frac{1}{2}\int \frac{1}{u}\,du = -\frac{1}{2}\ln|u| + C \\ &= -\tfrac{1}{2}\ln|9 - 2x| + C\end{aligned}$$

Applying the fundamental theorem of calculus yields

$$\int_2^4 \frac{1}{9-2x}\,dx = -\frac{1}{2}\Big[\ln|9-2x|\Big]_2^4$$
$$= -\tfrac{1}{2}(\ln 1 - \ln 5) = \tfrac{1}{2}\ln 5.$$

Another method is to use the same substitution in the *definite* integral and change the limits of integration. Since $u = 9 - 2x$, we obtain the following:

(i) If $x = 2$, then $u = 5$.

(ii) If $x = 4$, then $u = 1$.

Thus,

$$\int_2^4 \frac{1}{9-2x}\,dx = -\frac{1}{2}\int_2^4 \frac{1}{9-2x}(-2)\,dx$$
$$= -\frac{1}{2}\int_5^1 \frac{1}{u}\,du = -\frac{1}{2}\Big[\ln|u|\Big]_5^1$$
$$= -\tfrac{1}{2}(\ln 1 - \ln 5) = \tfrac{1}{2}\ln 5.$$

EXAMPLE 3 Evaluate $\displaystyle\int \frac{\sqrt{\ln x}}{x}\,dx$.

SOLUTION Two possible substitutions are $u = \sqrt{\ln x}$ and $u = \ln x$. If we use

$$u = \ln x, \qquad du = \frac{1}{x}\,dx,$$

then

$$\int \frac{\sqrt{\ln x}}{x}\,dx = \int \sqrt{\ln x}\cdot\frac{1}{x}\,dx = \int u^{1/2}\,du = \frac{u^{3/2}}{3/2} + C$$
$$= \tfrac{2}{3}(\ln x)^{3/2} + C.$$

The substitution $u = \sqrt{\ln x}$ could also be used; however, the algebraic manipulations would be somewhat more involved.

The derivative formula $D_x\, e^{g(x)} = e^{g(x)}g'(x)$ gives us the following integration formula for the natural exponential function:

$$\int e^{g(x)}g'(x)\,dx = e^{g(x)} + C$$

This is restated in the next theorem in terms of the variable u.

Theorem **(7.24)**

> If $u = g(x)$ and g is differentiable, then
>
> $$\int e^u\,du = e^u + C.$$

As a special case of Theorem (7.24), if $u = x$, then

$$\int e^x \, dx = e^x + C.$$

EXAMPLE 4 Evaluate

(a) $\int \frac{e^{3/x}}{x^2} \, dx$ **(b)** $\int_1^2 \frac{e^{3/x}}{x^2} \, dx$

SOLUTION

(a) Rewriting the integral as

$$\int \frac{e^{3/x}}{x^2} \, dx = \int e^{3/x} \frac{1}{x^2} \, dx$$

suggests that we use Theorem (7.24) with $u = 3/x$. Thus, we make the substitution

$$u = \frac{3}{x}, \qquad du = -\frac{3}{x^2} \, dx.$$

The integrand may be written in the form of (7.24) by introducing the factor -3. Doing this and compensating by multiplying the integral by $-\frac{1}{3}$, we obtain

$$\begin{aligned} \int \frac{e^{3/x}}{x^2} \, dx &= -\frac{1}{3} \int e^{3/x} \left(-\frac{3}{x^2}\right) dx \\ &= -\tfrac{1}{3} \int e^u \, du \\ &= -\tfrac{1}{3} e^u + C \\ &= -\tfrac{1}{3} e^{3/x} + C. \end{aligned}$$

(b) Using the antiderivative found in (a) and applying the fundamental theorem of calculus yields

$$\begin{aligned} \int_1^2 \frac{e^{3/x}}{x^2} \, dx &= -\frac{1}{3} \Big[e^{3/x} \Big]_1^2 \\ &= -\tfrac{1}{3}(e^{3/2} - e^3) \approx 5.2. \end{aligned}$$

We can also evaluate the integral by using the method of substitution. As in (a), we let $u = 3/x$, $du = (-3/x^2) \, dx$, and we note that if $x = 1$, then $u = 3$, and if $x = 2$, then $u = \frac{3}{2}$. Consequently,

$$\begin{aligned} \int_1^2 \frac{e^{3/x}}{x^2} \, dx &= -\frac{1}{3} \int_1^2 e^{3/x} \left(-\frac{3}{x^2}\right) dx \\ &= -\tfrac{1}{3} \int_3^{3/2} e^u \, du \\ &= -\tfrac{1}{3} \Big[e^u \Big]_3^{3/2} = -\tfrac{1}{3}(e^{3/2} - e^3) \approx 5.2. \end{aligned}$$

The integral $\int e^{ax} \, dx$, with $a \neq 0$, occurs frequently. We can show that

$$\int e^{ax} \, dx = \frac{1}{a} e^{ax} + C$$

either by using Theorem (7.24) or by showing that $(1/a)e^{ax}$ is an antiderivative of e^{ax}.

ILLUSTRATION

- $\int e^{3x}\,dx = \frac{1}{3}e^{3x} + C$
- $\int e^{-5x}\,dx = -\frac{1}{5}e^{-5x} + C$
- $\int e^{-x}\,dx = -e^{-x} + C$

In the next example we solve a differential equation that contains exponential expressions.

EXAMPLE 5 Solve the differential equation

$$\frac{dy}{dx} = 3e^{2x} + 6e^{-3x}$$

subject to the initial condition $y = 4$ if $x = 0$.

SOLUTION As in Example 6 of Section 5.1, we may multiply both sides of the equation by dx and then integrate as follows:

$$dy = (3e^{2x} + 6e^{-3x})\,dx$$

$$\int dy = \int (3e^{2x} + 6e^{-3x})\,dx = 3\int e^{2x}\,dx + 6\int e^{-3x}\,dx$$

$$y = 3(\tfrac{1}{2})e^{2x} + 6(-\tfrac{1}{3})e^{-3x} + C$$

$$= \tfrac{3}{2}e^{2x} - 2e^{-3x} + C$$

Using the initial condition $y = 4$ if $x = 0$ gives us

$$4 = \tfrac{3}{2}e^{0} - 2e^{0} + C = \tfrac{3}{2} - 2 + C.$$

Hence $C = 4 - \frac{3}{2} + 2 = \frac{9}{2}$, and the solution of the differential equation is

$$y = \tfrac{3}{2}e^{2x} - 2e^{-3x} + \tfrac{9}{2}.$$

EXAMPLE 6 Find the area of the region bounded by the graphs of the equations $y = e^x$, $y = \sqrt{x}$, $x = 0$, and $x = 1$.

FIGURE 7.15

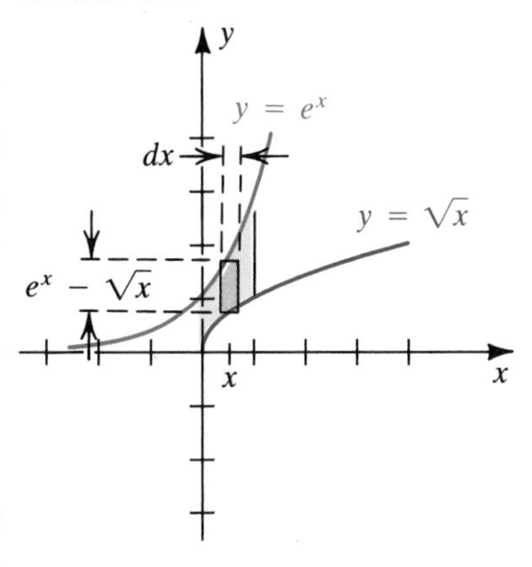

SOLUTION The region and a typical rectangle of the type considered in Chapter 6 are shown in Figure 7.15. As usual, we list the following:

width of rectangle: dx

length of rectangle: $e^x - \sqrt{x}$

area of rectangle: $(e^x - \sqrt{x})\,dx$

We next take a limit of sums of these rectangular areas by applying the operator $\int_0^1$:

$$\int_0^1 (e^x - \sqrt{x})\,dx = \int_0^1 (e^x - x^{1/2})\,dx$$

$$= \left[e^x - \tfrac{2}{3}x^{3/2}\right]_0^1 = e - \tfrac{5}{3} \approx 1.05$$

In Chapter 5 we obtained integration formulas for the sine and cosine functions. We were unable to consider the remaining four trigonometric functions at that time because, as indicated in the next theorem, their integrals are logarithmic functions. In the theorem we assume that $u = g(x)$, with g differentiable whenever the function is defined.

Theorem **(7.25)**

(i) $\int \tan u\, du = -\ln|\cos u| + C$

(ii) $\int \cot u\, du = \ln|\sin u| + C$

(iii) $\int \sec u\, du = \ln|\sec u + \tan u| + C$

(iv) $\int \csc u\, du = \ln|\csc u - \cot u| + C$

PROOF It is sufficient to consider the case $u = x$, since the formulas for $u = g(x)$ then follow from Theorem (5.7).

To find $\int \tan x\, dx$, we first use a trigonometric identity to express $\tan x$ in terms of $\sin x$ and $\cos x$ as follows:

$$\int \tan x\, dx = \int \frac{\sin x}{\cos x}\, dx = \int \frac{1}{\cos x} \sin x\, dx$$

The form of the integrand on the right suggests that we make the substitution

$$v = \cos x, \qquad dv = -\sin x\, dx.$$

This gives us

$$\int \tan x\, dx = -\int \frac{1}{v}\, dv.$$

If $\cos x \neq 0$, then, by Theorem (7.11)(ii),

$$\int \tan x\, dx = -\ln|v| + C = -\ln|\cos x| + C.$$

A formula for $\int \cot x\, dx$ may be obtained in similar fashion by first writing $\cot x = (\cos x)/(\sin x)$.

To find a formula for $\int \sec x\, dx$, we begin as follows:

$$\begin{aligned}\int \sec x\, dx &= \int \sec x \frac{\sec x + \tan x}{\sec x + \tan x}\, dx \\ &= \int \frac{\sec^2 x + \sec x \tan x}{\sec x + \tan x}\, dx \\ &= \int \frac{1}{\sec x + \tan x}(\sec x \tan x + \sec^2 x)\, dx\end{aligned}$$

Using the substitution

$$v = \sec x + \tan x, \qquad dv = (\sec x \tan x + \sec^2 x)\, dx$$

gives us

$$\int \sec x\,dx = \int \frac{1}{v}\,dv$$
$$= \ln|v| + C$$
$$= \ln|\sec x + \tan x| + C.$$

A similar proof can be given for (iv). ■

If we use $\cos u = 1/\sec u$, $\sin u = 1/\csc u$, and $\ln(1/v) = -\ln v$, then formulas (i) and (ii) of Theorem (7.25) can be written as follows:

$$\int \tan u\,du = \ln|\sec u| + C$$

$$\int \cot u\,du = -\ln|\csc u| + C$$

EXAMPLE 7 Evaluate $\int x \cot x^2\,dx$.

SOLUTION To obtain the form $\int \cot u\,du$, we make the substitution

$$u = x^2, \qquad du = 2x\,dx.$$

We next introduce the factor 2 in the integrand as follows:

$$\int x \cot x^2\,dx = \tfrac{1}{2}\int (\cot x^2)2x\,dx$$

Since $u = x^2$ and $du = 2x\,dx$,

$$\int x \cot x^2\,dx = \tfrac{1}{2}\int \cot u\,du = \tfrac{1}{2}\ln|\sin u| + C$$
$$= \tfrac{1}{2}\ln|\sin x^2| + C.$$

EXAMPLE 8 Evaluate $\int_0^{\pi/2} \tan \frac{x}{2}\,dx$.

SOLUTION We make the substitution

$$u = \frac{x}{2}, \qquad du = \frac{1}{2}\,dx$$

and note that $u = 0$ if $x = 0$, and $u = \pi/4$ if $x = \pi/2$. Thus,

$$\int_0^{\pi/2} \tan\frac{x}{2}\,dx = 2\int_0^{\pi/2} \tan\frac{x}{2}\cdot\frac{1}{2}\,dx$$
$$= 2\int_0^{\pi/4} \tan u\,du = 2\Big[\ln \sec u\Big]_0^{\pi/4}.$$

In this case we may drop the absolute value sign given in Theorem (7.25)(iii), since $\sec u$ is positive if u is between 0 and $\pi/4$. Since $\ln \sec(\pi/4) = \ln\sqrt{2} = \tfrac{1}{2}\ln 2$ and $\ln \sec 0 = \ln 1 = 0$, it follows that

$$\int_0^{\pi/2} \tan\frac{x}{2}\,dx = 2\cdot\frac{1}{2}\ln 2 = \ln 2 \approx 0.69.$$

EXAMPLE 9 Evaluate $\int e^{2x} \sec e^{2x}\, dx$.

SOLUTION We let

$$u = e^{2x}, \qquad du = 2e^{2x}\, dx$$

and proceed as follows:

$$\begin{aligned}\int e^{2x} \sec e^{2x}\, dx &= \tfrac{1}{2}\int (\sec e^{2x})2e^{2x}\, dx\\ &= \tfrac{1}{2}\int \sec u\, du\\ &= \tfrac{1}{2}\ln|\sec u + \tan u| + C\\ &= \tfrac{1}{2}\ln|\sec e^{2x} + \tan e^{2x}| + C\end{aligned}$$

EXAMPLE 10 Evaluate $\int (\csc x - 1)^2\, dx$.

SOLUTION

$$\begin{aligned}\int (\csc x - 1)^2\, dx &= \int (\csc^2 x - 2\csc x + 1)\, dx\\ &= \int \csc^2 x\, dx - 2\int \csc x\, dx + \int dx\\ &= -\cot x - 2\ln|\csc x - \cot x| + x + C.\end{aligned}$$

We shall discuss additional methods for integrating trigonometric expressions in Chapter 9.

EXERCISES 7.4

Exer. 1–36: Evaluate the integral.

1 (a) $\int \frac{1}{2x+7}\, dx$ (b) $\int_{-2}^{1} \frac{1}{2x+7}\, dx$

2 (a) $\int \frac{1}{4-5x}\, dx$ (b) $\int_{-1}^{0} \frac{1}{4-5x}\, dx$

3 (a) $\int \frac{4x}{x^2-9}\, dx$ (b) $\int_{1}^{2} \frac{4x}{x^2-9}\, dx$

4 (a) $\int \frac{3x}{x^2+4}\, dx$ (b) $\int_{1}^{2} \frac{3x}{x^2+4}\, dx$

5 (a) $\int e^{-4x}\, dx$ (b) $\int_{1}^{3} e^{-4x}\, dx$

6 (a) $\int x^2 e^{3x^3}\, dx$ (b) $\int_{1}^{2} x^2 e^{3x^3}\, dx$

7 (a) $\int \tan 2x\, dx$ (b) $\int_{0}^{\pi/8} \tan 2x\, dx$

8 (a) $\int \cot \frac{1}{3}x\, dx$ (b) $\int_{3\pi/2}^{9\pi/4} \cot \frac{1}{3}x\, dx$

9 (a) $\int \csc \frac{1}{2}x\, dx$ (b) $\int_{\pi}^{5\pi/3} \csc \frac{1}{2}x\, dx$

10 (a) $\int \sec 3x\, dx$ (b) $\int_{0}^{\pi/12} \sec 3x\, dx$

11 $\int \frac{x-2}{x^2-4x+9}\, dx$

12 $\int \frac{x^3}{x^4-5}\, dx$

13 $\int \frac{(x+2)^2}{x}\, dx$

14 $\int \frac{(2+\ln x)^{10}}{x}\, dx$

15 $\int \frac{\ln x}{x}\, dx$

16 $\int \frac{1}{x(\ln x)^2}\, dx$

17 $\int (x + e^{5x})\, dx$

18 $\int \frac{e^{\sqrt{x}}}{\sqrt{x}}\, dx$

19 $\int \frac{3 \sin x}{1 + 2 \cos x}\, dx$

20 $\int \frac{\sec^2 x}{1 + \tan x}\, dx$

21 $\int \frac{(e^x + 1)^2}{e^x}\, dx$

22 $\int \frac{e^x}{(e^x + 1)^2}\, dx$

23 $\int \frac{e^x - e^{-x}}{e^x + e^{-x}}\, dx$

24 $\int \frac{e^x}{e^x + 1}\, dx$

25 $\int \frac{\cot \sqrt[3]{x}}{\sqrt[3]{x^2}}\, dx$

26 $\int e^x(1 + \tan e^x)\, dx$

27 $\int \frac{1}{\cos 2x}\, dx$

28 $\int (x + \csc 8x)\, dx$

29 $\int \frac{\tan e^{-3x}}{e^{3x}}\, dx$

30 $\int e^{\cos x} \sin x\, dx$

31 $\int \frac{\cos^2 x}{\sin x}\, dx$

32 $\int \frac{\tan^2 2x}{\sec 2x}\, dx$

33 $\int \frac{\cos x \sin x}{\cos^2 x - 1}\, dx$

34 $\int (\tan 3x + \sec 3x)\, dx$

35 $\int (1 + \sec x)^2\, dx$

36 $\int \csc x\,(1 - \csc x)\, dx$

Exer. 37–38: Find the area of the region bounded by the graphs of the given equations.

37 $y = e^{2x}, \quad y = 0, \quad x = 0, \quad x = \ln 3$

38 $y = 2 \tan x, \quad y = 0, \quad x = 0, \quad x = \pi/4$

Exer. 39–40: Find the volume of the solid generated if the region bounded by the graphs of the equations is revolved about the indicated axis.

39 $y = e^{-x^2}, \quad x = 0, \quad x = 1, \quad y = 0;$ y-axis

40 $y = \sec x, \quad x = -\pi/3, \quad x = \pi/3, \quad y = 0;$ x-axis

Exer. 41–44: Solve the differential equation subject to the given conditions.

41 $y' = 4e^{2x} + 3e^{-2x}; \quad y = 4$ if $x = 0$

42 $y' = 3e^{4x} - 8e^{-2x}; \quad y = -2$ if $x = 0$

43 $y'' = 3e^{-x}; \quad y = -1$ and $y' = 1$ if $x = 0$

44 $y'' = 6e^{2x}; \quad y = -3$ and $y' = 2$ if $x = 0$

Exer. 45–46: A nonnegative function f defined on a closed interval $[a, b]$ is called a *probability density function* if $\int_a^b f(x)\, dx = 1$. Determine c so that the resulting function is a probability density function.

45 $f(x) = \frac{cx}{x^2 + 4}$ for $0 \le x \le 3$

46 $f(x) = cxe^{-x^2}$ for $0 \le x \le 10$

47 A culture of bacteria is growing at a rate of $3e^{0.2t}$ per hour, with t in hours and $0 \le t \le 20$.

(a) How many new bacteria will be in the culture after the first five hours?

(b) How many new bacteria are introduced in the sixth through the fourteenth hours?

(c) For approximately what value of t will the culture contain 150 new bacteria?

48 If a savings bond is purchased for \$500 with interest compounded continuously at 7% per year, then after t years the bond will be worth $500e^{0.07t}$ dollars.

(a) Approximately when will the bond be worth \$1000?

(b) Approximately when will the value of the bond be growing at a rate of \$50 per year?

49 The specific heat c of a metal such as silver is constant at temperatures T above 200 °K. If the temperature of the metal increases from T_1 to T_2, the area under the curve $y = c/T$ from T_1 to T_2 is called the *change in entropy* ΔS, a measurement of the increased molecular disorder of the system. Express ΔS in terms of T_1 and T_2.

50 The 1952 earthquake in Assam had a magnitude of 8.7 on the Richter scale—the largest ever recorded. (The October 1989 San Francisco earthquake had a magnitude of 7.1.) Seismologists have determined that if the largest earthquake in a given year has magnitude R, then the energy E (in joules) released by all earthquakes in that year can be estimated by using the formula

$$E = 9.13 \times 10^{12} \int_0^R e^{1.25x}\, dx.$$

Find E if $R = 8$.

51 In a circuit containing a 12-volt battery, a resistor, and a capacitor, the current $I(t)$ at time t is predicted to be $I(t) = 10e^{-4t}$ amperes. If $Q(t)$ is the charge (in coulombs) on the capacitor, then $I = dQ/dt$.

(a) If $Q(0) = 0$, find $Q(t)$.

(b) Find the charge on the capacitor after a long period of time.

52 A country that presently has coal reserves of 50 million tons used 6.5 million tons last year. Based on population projections, the rate of consumption R (in million tons/year) is expected to increase according to the formula $R = 6.5e^{0.02t}$, where t is the time in years. If the country uses only its own resources, estimate how many years the coal reserves will last.

53 A very small spherical particle (on the order of 5 microns in diameter) is projected into still air with an initial velocity of v_0 m/sec, but its velocity decreases because of drag forces. Its velocity after t seconds is given by $v(t) = v_0 e^{-t/k}$ for some positive constant k.

(a) Express the distance that the particle travels as a function of t.

(b) The *stopping distance* is the distance traveled by the particle before it comes to rest. Express the stopping distance in terms of v_0 and k.

54 If the temperature remains constant, the pressure p and volume v of an expanding gas are related by the equation $pv = k$ for some constant k. Show that the work done if the gas expands from v_0 to v_1 is $k \ln (v_1/v_0)$. (*Hint:* See Example 5 of Section 6.6.)

c **Exer. 55–56: If T_2 and T_4 are approximations of a definite integral obtained by using the trapezoidal rule with $n = 2$ and $n = 4$, respectively, then it can be shown that $R = \frac{1}{3}(4T_4 - T_2)$ is usually a better approximation. Find T_2, T_4, T_{10}, and R for the given integral, and decide whether R or T_{10} is the better approximation.**

55 $\int_0^1 e^{-x^2}\, dx \approx 0.746824$

56 $\int_1^2 (\ln x)^2\, dx \approx 0.188317$

7.5 GENERAL EXPONENTIAL AND LOGARITHMIC FUNCTIONS

Throughout this section a will denote a positive real number. Let us begin by defining a^x for every real number x. If the exponent is a *rational* number r, then applying Theorems (7.19)(ii) and (7.12)(iii) yields

$$a^r = e^{\ln a^r} = e^{r \ln a}.$$

This formula is the motivation for the following definition of a^x.

Definition of a^x **(7.26)**

$$a^x = e^{x \ln a}$$

for every $a > 0$ and every real number x

ILLUSTRATION

- $2^\pi = e^{\pi \ln 2} \approx e^{2.18} \approx 8.8$
- $(\frac{1}{2})^{\sqrt{3}} = e^{\sqrt{3} \ln (1/2)} \approx e^{-1.20} \approx 0.3$

If $f(x) = a^x$, then f is the **exponential function with base a**. Since e^x is positive for every x, so is a^x. To approximate values of a^x, we may use a calculator or refer to tables of logarithmic and exponential functions.

It is now possible to prove that the law of logarithms stated in Theorem (7.12)(iii) is also true for irrational exponents. Thus, if u is any *real* number, then, by Definition (7.26) and Theorem (7.19)(i),

$$\ln a^u = \ln e^{u \ln a} = u \ln a.$$

The next theorem states that properties of rational exponents from elementary algebra are also true for real exponents.

Laws of exponents **(7.27)**

Let $a > 0$ and $b > 0$. If u and v are any real numbers, then

$$a^u a^v = a^{u+v} \qquad (a^u)^v = a^{uv} \qquad (ab)^u = a^u b^u$$

$$\frac{a^u}{a^v} = a^{u-v} \qquad \left(\frac{a}{b}\right)^u = \frac{a^u}{b^u}.$$

PROOF To show that $a^u a^v = a^{u+v}$, we use Definition (7.26) and Theorem (7.20)(i) as follows:

$$\begin{aligned} a^u a^v &= e^{u \ln a} e^{v \ln a} \\ &= e^{u \ln a + v \ln a} \\ &= e^{(u+v) \ln a} \\ &= a^{u+v} \end{aligned}$$

To prove that $(a^u)^v = a^{uv}$, we first use Definition (7.26) with a^u in place of a and $v = x$ to write

$$(a^u)^v = e^{v \ln a^u}.$$

Using the fact that $\ln a^u = u \ln a$ and then applying Definition (7.26), we obtain

$$(a^u)^v = e^{vu \ln a} = a^{vu} = a^{uv}.$$

The proofs of the remaining laws are similar. ■

As usual, in part (ii) of the next theorem, $u = g(x)$, where g is differentiable.

Theorem (7.28)

(i) $D_x\, a^x = a^x \ln a$ **(ii)** $D_x\, a^u = (a^u \ln a)\, D_x\, u$

PROOF Applying Definition (7.26) and Theorem (7.22), we obtain

$$D_x\, a^x = D_x\, e^{x \ln a} = e^{x \ln a}\, D_x\, (x \ln a) = e^{x \ln a}(\ln a).$$

Since $e^{x \ln a} = a^x$, this gives us formula (i):

$$D_x\, a^x = a^x \ln a$$

Formula (ii) follows from the chain rule. ■

Note that if $a = e$, then Theorem (7.28)(i) reduces to (7.21), since $\ln e = 1$.

ILLUSTRATION

- $D_x\, 3^x = 3^x \ln 3$
- $D_x\, 10^x = 10^x \ln 10$
- $D_x\, 3^{\sqrt{x}} = (3^{\sqrt{x}} \ln 3)\, D_x \sqrt{x} = (3^{\sqrt{x}} \ln 3)\left(\dfrac{1}{2\sqrt{x}}\right) = \dfrac{3^{\sqrt{x}} \ln 3}{2\sqrt{x}}$
- $D_x\, 10^{\sin x} = (10^{\sin x} \ln 10)\, D_x \sin x = (10^{\sin x} \ln 10) \cos x$

FIGURE 7.16

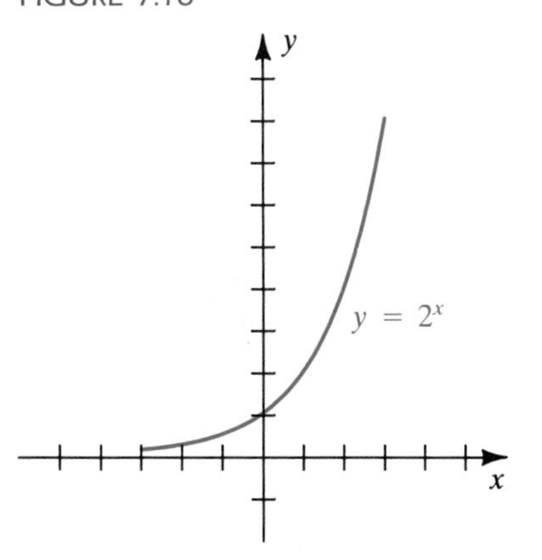

If $a > 1$, then $\ln a > 0$ and, therefore, $D_x\, a^x = a^x \ln a > 0$. Hence a^x is increasing on the interval $(-\infty, \infty)$ if $a > 1$.

If $0 < a < 1$, then $\ln a < 0$ and $D_x\, a^x = a^x \ln a < 0$. Thus, a^x is decreasing for every x if $0 < a < 1$.

The graphs of $y = 2^x$ and $y = (\frac{1}{2})^x = 2^{-x}$ are sketched in Figures 7.16 and 7.17 (on the following page). The graph of $y = a^x$ has the general shape illustrated in Figure 7.16 or 7.17 if $a > 1$ or $0 < a < 1$, respectively.

FIGURE 7.17

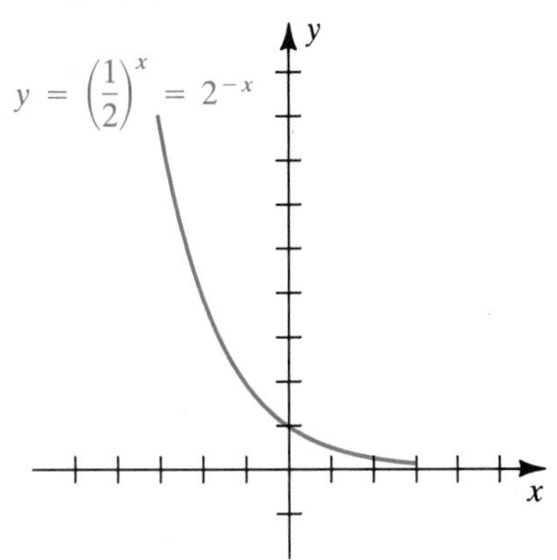

If $u = g(x)$, *it is important to distinguish between expressions of the form* a^u *and* u^a. To differentiate a^u, we use (7.28); for u^a the power rule must be employed, as illustrated in the next example.

EXAMPLE 1 Find y' if $y = (x^2 + 1)^{10} + 10^{x^2+1}$.

SOLUTION Using the power rule for functions and Theorem (7.28), we obtain

$$\begin{aligned} y' &= 10(x^2 + 1)^9(2x) + (10^{x^2+1} \ln 10)(2x) \\ &= 20x[(x^2 + 1)^9 + 10^{x^2} \ln 10]. \end{aligned}$$

The integration formula in (i) of the next theorem may be verified by showing that the integrand is the derivative of the expression on the right side of the equation. Formula (ii) follows from Theorem (7.28)(ii), where $u = g(x)$.

Theorem **(7.29)**

(i) $\int a^x \, dx = \left(\frac{1}{\ln a}\right) a^x + C$ **(ii)** $\int a^u \, du = \left(\frac{1}{\ln a}\right) a^u + C$

EXAMPLE 2 Evaluate

(a) $\int 3^x \, dx$ **(b)** $\int x3^{(x^2)} \, dx$

SOLUTION

(a) Using (i) of Theorem (7.29) yields

$$\int 3^x \, dx = \left(\frac{1}{\ln 3}\right) 3^x + C.$$

(b) To use (ii) of Theorem (7.29), we make the substitution

$$u = x^2, \qquad du = 2x \, dx$$

and proceed as follows:

$$\begin{aligned} \int x3^{(x^2)} \, dx &= \tfrac{1}{2} \int 3^{(x^2)}(2x) \, dx = \tfrac{1}{2} \int 3^u \, du \\ &= \frac{1}{2}\left(\frac{1}{\ln 3}\right) 3^u + C = \left(\frac{1}{2 \ln 3}\right) 3^{(x^2)} + C \end{aligned}$$

EXAMPLE 3 An important problem in oceanography is determining the light intensity at different ocean depths. The *Beer-Lambert law* states that at a depth x (in meters), the light intensity $I(x)$ (in calories/cm^2/sec) is given by $I(x) = I_0 a^x$, where I_0 and a are positive constants.

(a) What is the light intensity at the surface?

(b) Find the rate of change of the light intensity with respect to depth at a depth x.

(c) If $a = 0.4$ and $I_0 = 10$, find the average light intensity between the surface and a depth of x meters.

(d) Show that $I(x) = I_0 e^{kx}$ for some constant k.

SOLUTION

(a) At the surface, $x = 0$ and

$$I(0) = I_0 a^0 = I_0.$$

Hence the light intensity at the surface is I_0.

(b) The rate of change of $I(x)$ with respect to x is $I'(x)$. Thus,

$$I'(x) = I_0(a^x \ln a) = (\ln a)(I_0 a^x) = (\ln a)I(x).$$

Hence the rate of change $I'(x)$ at depth x is directly proportional to $I(x)$, and the constant of proportionality is $\ln a$.

(c) If $I(x) = 10(0.4)^x$, then, by Definition (5.29), the average value of I on the interval $[0, 5]$ is

$$\begin{aligned} I_{av} &= \frac{1}{5-0}\int_0^5 10(0.4)^x\, dx = 2\int_0^5 (0.4)^x\, dx \\ &= 2\left[\frac{1}{\ln(0.4)}(0.4)^x\right]_0^5 = \frac{2}{\ln(0.4)}[(0.4)^5 - (0.4)^0] \\ &= \frac{-1.97952}{\ln(0.4)} \approx 2.16. \end{aligned}$$

(d) Using Definition (7.26) yields

$$I(x) = I_0 a^x = I_0 e^{x \ln a} = I_0 e^{kx},$$

where $k = \ln a$.

If $a \neq 1$ and $f(x) = a^x$, then f is a one-to-one function. Its inverse function is denoted by $\log_a$ and is called the **logarithmic function with base a**. Another way of stating this relationship is as follows.

Definition of $\log_a x$ **(7.30)**

$$y = \log_a x \quad \text{if and only if} \quad x = a^y$$

The expression $\log_a x$ is called **the logarithm of x with base a**. In this terminology, natural logarithms are logarithms with base e; that is,

$$\ln x = \log_e x.$$

Laws of logarithms similar to Theorem (7.12) are true for logarithms with base a.

To obtain the relationship between $\log_a$ and ln, consider $y = \log_a x$, or, equivalently, $x = a^y$. Taking the natural logarithm of both sides of the last equation gives us $\ln x = y \ln a$, or $y = (\ln x)/(\ln a)$. This proves that

$$\log_a x = \frac{\ln x}{\ln a}.$$

Differentiating both sides of the last equation leads to (i) of the next theorem. Using the chain rule and generalizing to absolute values as in Theorem (7.11) gives us (ii), where $u = g(x)$.

Theorem **(7.31)**

(i) $D_x \log_a x = D_x\left(\dfrac{\ln x}{\ln a}\right) = \dfrac{1}{\ln a}\cdot\dfrac{1}{x}$

(ii) $D_x \log_a |u| = D_x\left(\dfrac{\ln |u|}{\ln a}\right) = \dfrac{1}{\ln a}\cdot\dfrac{1}{u} D_x u$

ILLUSTRATION

- $D_x \log_2 x = D_x\left(\dfrac{\ln x}{\ln 2}\right) = \dfrac{1}{\ln 2}\cdot\dfrac{1}{x} = \dfrac{1}{(\ln 2)x}$
- $D_x \log_2 |x^2-9| = D_x\left(\dfrac{\ln |x^2-9|}{\ln 2}\right) = \dfrac{1}{\ln 2}\cdot\dfrac{1}{x^2-9}\cdot 2x = \dfrac{2x}{(\ln 2)(x^2-9)}$

Logarithms with base 10 are useful for certain applications (see Exercises 48–51 and 53). We refer to such logarithms as **common logarithms** and use the symbol **log *x*** as an abbreviation for $\log_{10} x$. This notation is used in the next example.

EXAMPLE 4 If $f(x) = \log \sqrt[3]{(2x+5)^2}$, find $f'(x)$.

SOLUTION We first write $f(x) = \log (2x+5)^{2/3}$. The law $\log u^r = r \log u$ is true only if $u > 0$; however, since $(2x+5)^{2/3} = |2x+5|^{2/3}$, we may proceed as follows:

$$\begin{aligned} f(x) &= \log (2x+5)^{2/3} \\ &= \log |2x+5|^{2/3} \\ &= \tfrac{2}{3} \log |2x+5| \\ &= \frac{2}{3}\,\frac{\ln |2x+5|}{\ln 10} \end{aligned}$$

Differentiating yields

$$f'(x) = \frac{2}{3}\cdot\frac{1}{\ln 10}\cdot\frac{1}{2x+5}(2) = \frac{4}{3(2x+5)\ln 10}.$$

Now that we have defined irrational exponents, we may consider the **general power function** f given by $f(x) = x^c$ for any real number c. If c is irrational, then, by definition, the domain of f is the set of positive real numbers. Using Definition (7.26) and Theorems (7.22) and (7.11)(i), we have

$$\begin{aligned} D_x x^c &= D_x e^{c \ln x} = e^{c \ln x} D_x (c \ln x) \\ &= e^{c \ln x}\left(\frac{c}{x}\right) = x^c\left(\frac{c}{x}\right) = cx^{c-1}. \end{aligned}$$

This proves that the power rule is true for irrational as well as rational exponents. The power rule for functions may also be extended to irrational exponents.

ILLUSTRATION

- $D_x\,(x^{\sqrt{2}}) = \sqrt{2}x^{\sqrt{2}-1}$
- $D_x\,(1 + e^{2x})^{\pi} = \pi(1 + e^{2x})^{\pi-1}\,D_x\,(1 + e^{2x})$
  $= \pi(1 + e^{2x})^{\pi-1}(2e^{2x}) = 2\pi e^{2x}(1 + e^{2x})^{\pi-1}$

EXAMPLE 5 If $y = x^x$ and $x > 0$, find $D_x\,y$.

SOLUTION Since the exponent in x^x is a variable, the power rule may not be used. Similarly, Theorem (7.28) is not applicable, since the base a is not a fixed real number. However, by Definition (7.26), $x^x = e^{x\ln x}$ for every $x > 0$, and hence

$$\begin{aligned} D_x\,(x^x) &= D_x\,(e^{x\ln x}) \\ &= e^{x\ln x}\,D_x\,(x\ln x) \\ &= e^{x\ln x}\left[x\left(\frac{1}{x}\right) + (1)\ln x\right] \\ &= x^x(1 + \ln x). \end{aligned}$$

Another way of solving this problem is to use the method of logarithmic differentiation introduced in the preceding section. In this case we take the natural logarithm of both sides of the equation $y = x^x$ and then differentiate implicitly as follows:

$$\begin{aligned} \ln y &= \ln x^x = x\ln x \\ D_x\,(\ln y) &= D_x\,(x\ln x) \\ \frac{1}{y}D_x\,y &= 1 + \ln x \\ D_x\,y &= y(1 + \ln x) = x^x(1 + \ln x) \end{aligned}$$

We shall conclude this section by expressing the number e as a limit.

Theorem **(7.32)**

(i) $\lim_{h\to 0}(1 + h)^{1/h} = e$ **(ii)** $\lim_{n\to\infty}\left(1 + \frac{1}{n}\right)^n = e$

PROOF Applying the definition of derivative (3.5) to $f(x) = \ln x$ and using laws of logarithms yields

$$\begin{aligned} f'(x) &= \lim_{h\to 0}\frac{\ln(x + h) - \ln x}{h} = \lim_{h\to 0}\frac{1}{h}\ln\frac{x + h}{x} \\ &= \lim_{h\to 0}\frac{1}{h}\ln\left(1 + \frac{h}{x}\right) = \lim_{h\to 0}\ln\left(1 + \frac{h}{x}\right)^{1/h}. \end{aligned}$$

Since $f'(x) = 1/x$, we have, for $x = 1$,

$$1 = \lim_{h \to 0} \ln (1 + h)^{1/h}.$$

We next observe, from Theorem (7.19), that

$$(1 + h)^{1/h} = e^{\ln (1+h)^{1/h}}.$$

Since the natural exponential function is continuous at 1, it follows from Theorem (2.25) that

$$\lim_{h \to 0} (1 + h)^{1/h} = \lim_{h \to 0} \left[e^{\ln (1+h)^{1/h}}\right]$$

$$= e^{\left[\lim_{h \to 0} \ln (1+h)^{1/h}\right]} = e^1 = e.$$

This establishes part (i) of the theorem. The limit in part (ii) may be obtained by introducing the change of variable $n = 1/h$ with $h > 0$. ■

The formulas in Theorem (7.32) are sometimes used to *define* the number e. You may find it instructive to calculate $(1 + h)^{1/h}$ for numerically small values of h. Some approximate values are given in the following table.

h	$(1 + h)^{1/h}$	h	$(1 + h)^{1/h}$
0.01	2.704814	−0.01	2.731999
0.001	2.716924	−0.001	2.719642
0.0001	2.718146	−0.0001	2.718418
0.00001	2.718268	−0.00001	2.718295
0.000001	2.718280	−0.000001	2.718283

To five decimal places, $e \approx 2.71828$.

EXERCISES 7.5

Exer. 1–26: Find $f'(x)$ if $f(x)$ is the given expression.

1 7^x
2 5^{-x}
3 8^{x^2+1}
4 $9^{\sqrt{x}}$
5 $\log (x^4 + 3x^2 + 1)$
6 $\log_3 |6x - 7|$
7 5^{3x-4}
8 3^{2-x^2}
9 $(x^2 + 1)10^{1/x}$
10 $(10^x + 10^{-x})^{10}$
11 $\log (3x^2 + 2)^5$
12 $\log \sqrt{x^2 + 1}$
13 $\log_5 \left|\dfrac{6x + 4}{2x - 3}\right|$
14 $\log \left|\dfrac{1 - x^2}{2 - 5x^3}\right|$
15 $\log \ln x$
16 $\ln \log x$
17 $x^e + e^x$
18 $x^\pi \pi^x$
19 $(x + 1)^x$
20 x^{4+x^2}
21 $2^{\sin^2 x}$
22 $4^{\sec 3x}$
23 $x^{\tan x}$
24 $(\cos 2x)^x$
25 (a) e^e (b) x^5 (c) $x^{\sqrt{5}}$ (d) $(\sqrt{5})^x$ (e) $x^{(x^2)}$
26 (a) π^π (b) x^4 (c) x^π (d) π^x (e) x^{2x}

Exer. 27–42: Evaluate the integral.

27 (a) $\int 7^x \, dx$ (b) $\int_{-2}^{1} 7^x \, dx$
28 (a) $\int 3^x \, dx$ (b) $\int_{-1}^{0} 3^x \, dx$
29 (a) $\int 5^{-2x} \, dx$ (b) $\int_{1}^{2} 5^{-2x} \, dx$
30 (a) $\int 2^{3x-1} \, dx$ (b) $\int_{-1}^{1} 2^{3x-1} \, dx$
31 $\int 10^{3x} \, dx$
32 $\int 5^{-5x} \, dx$
33 $\int x(3^{-x^2}) \, dx$
34 $\int \dfrac{(2^x + 1)^2}{2^x} \, dx$

35 $\int \frac{2^x}{2^x + 1}\,dx$

36 $\int \frac{3^x}{\sqrt{3^x + 4}}\,dx$

37 $\int \frac{1}{x \log x}\,dx$

38 $\int \frac{10^{\sqrt{x}}}{\sqrt{x}}\,dx$

39 $\int 3^{\cos x} \sin x\,dx$

40 $\int \frac{5^{\tan x}}{\cos^2 x}\,dx$

41 (a) $\int \pi^\pi\,dx$ (b) $\int x^4\,dx$

(c) $\int x^\pi\,dx$ (d) $\int \pi^x\,dx$

42 (a) $\int e^e\,dx$ (b) $\int x^5\,dx$

(c) $\int x^{\sqrt{5}}\,dx$ (d) $\int (\sqrt{5})^x\,dx$

43 Find the area of the region bounded by the graphs of $y = 2^x$, $x + y = 1$, and $x = 1$.

44 The region under the graph of $y = 3^{-x}$ from $x = 1$ to $x = 2$ is revolved about the x-axis. Find the volume of the resulting solid.

45 An economist predicts that the buying power $B(t)$ of a dollar t years from now will decrease according to the formula $B(t) = (0.95)^t$.

(a) At approximately what rate will the buying power be decreasing two years from now?

(b) Estimate the *average buying power* of the dollar over the next two years.

46 When a person takes a 100-milligram tablet of an asthma drug orally, the rate R at which the drug enters the bloodstream is predicted to be $R = 5(0.95)^t$ mg/min. If the bloodstream does not contain any trace of the drug when the tablet is taken, determine the number of minutes needed for 50 milligrams to enter the bloodstream.

47 One thousand trout, each one year old, are introduced into a large pond. The number still alive after t years is predicted to be $N(t) = 1000(0.9)^t$.

(a) Approximate the death rate dN/dt at times $t = 1$ and $t = 5$. At what rate is the population decreasing when $N = 500$?

(b) The weight $W(t)$ (in pounds) of an individual trout is expected to increase according to the formula $W(t) = 0.2 + 1.5t$. After approximately how many years is the total number of pounds of trout in the pond a maximum?

48 The vapor pressure P (in psi), a measure of the volatility of a liquid, is related to its temperature T (in °F) by the *Antoine equation*: $\log P = a + [b/(c + T)]$, for constants a, b, and c. Vapor pressure increases rapidly with an increase in temperature. Find conditions on a, b, and c that guarantee that P is an increasing function of T.

49 Chemists use a number denoted by pH to describe quantitatively the acidity or basicity of solutions. By definition, $\text{pH} = -\log [\text{H}^+]$, where $[\text{H}^+]$ is the hydrogen ion concentration in moles per liter. For a certain brand of vinegar it is estimated (with a maximum percentage error of $\pm 0.5\%$) that $[\text{H}^+] \approx 6.3 \times 10^{-3}$. Calculate the pH and use differentials to estimate the maximum percentage error in the calculation.

50 The magnitude R (on the Richter scale) of an earthquake of intensity I may be found by means of the formula $R = \log (I/I_0)$, where I_0 is a certain minimum intensity. Suppose the intensity of an earthquake is estimated to be 100 times I_0. If the maximum percentage error in the estimate is $\pm 1\%$, use differentials to approximate the maximum percentage error in the calculated value of R.

51 Let $R(x)$ be the reaction of a subject to a stimulus of strength x. For example, if the stimulus x is *saltiness* (in grams of salt/liter), $R(x)$ may be the subject's estimate of how salty the solution tasted on a scale from 0 to 10. A function that has been proposed to relate R to x is given by the *Weber-Fechner formula*: $R = a \log (x/x_0)$, where a is a positive constant.

(a) Show that $R = 0$ for the *threshold stimulus* $x = x_0$.

(b) The derivative $S = dR/dx$ is the *sensitivity* at stimulus level x and measures the ability to detect small changes in stimulus level. Show that S is inversely proportional to x, and compare $S(x)$ to $S(2x)$.

52 The loudness of sound, as experienced by the human ear, is based on intensity level. A formula used for finding the intensity level α that corresponds to a sound intensity I is $\alpha = 10 \log (I/I_0)$ decibels, where I_0 is a special value of I agreed to be the weakest sound that can be detected by the ear under certain conditions. Find the rate of change of α with respect to I if

(a) I is 10 times as great as I_0

(b) I is 1000 times as great as I_0

(c) I is 10,000 times as great as I_0 (This is the intensity level of the average voice.)

53 If a principal of P dollars is invested in a savings account for t years and the yearly interest rate r (expressed as a decimal) is compounded n times per year, then the amount A in the account after t years is given by the *compound interest formula*: $A = P[1 + (r/n)]^{nt}$.

(a) Let $h = r/n$ and show that

$$\ln A = \ln P + rt \ln (1 + h)^{1/h}.$$

(b) Let $n \to \infty$ and use the expression in part (a) to establish the formula $A = Pe^{rt}$ for interest *compounded continuously*.

54 Establish Theorem (7.32)(ii) by using the limit in part (i) and the change of variable $n = 1/h$.

55 Prove that $\lim_{n\to\infty} [1 + (x/n)]^n = e^x$ by letting $h = x/n$ and using Theorem (7.32)(i).

c 56 By letting $h = 0.1$, 0.01, and 0.001, predict which of the following expressions gives the best approximation of e for small values of h:

$$(1 + h)^{1/h}, \quad (1 + h + h^2)^{1/h}, \quad (1 + h + \tfrac{1}{2}h^2)^{1/h}$$

c 57 Graph, on the same coordinate axes, $y = 2^{-x}$ and $y = \log_2 x$.

(a) Estimate the x-coordinate of the point of intersection of the graphs.

(b) If the region R bounded by the graphs and the line $x = 1$ is revolved about the x-axis, set up an integral that can be used to approximate the volume of the resulting solid.

(c) Use Simpson's rule, with $n = 4$, to approximate the integral in (b).

7.6 LAWS OF GROWTH AND DECAY

Suppose that a physical quantity varies with time and that the magnitude of the quantity at time t is given by $q(t)$, where q is differentiable and $q(t) > 0$ for every t. The derivative $q'(t)$ is the rate of change of $q(t)$ with respect to time. In many applications this rate of change is directly proportional to the magnitude of the quantity at time t; that is,

$$q'(t) = cq(t)$$

for some constant c. The number of bacteria in certain cultures behaves in this way. If the number of bacteria $q(t)$ is small, then the rate of increase $q'(t)$ is small; however, as the number of bacteria increases, the *rate of increase* also increases. The decay of a radioactive substance obeys a similar law: as the amount of matter decreases, the rate of decay—that is, the amount of radiation—also decreases. As a final illustration, suppose an electrical condenser is allowed to discharge. If the charge on the condenser is large at the outset, the rate of discharge is also large, but as the charge weakens, the condenser discharges less rapidly.

In applied problems the equation $q'(t) = cq(t)$ is often expressed in terms of differentials. Thus, if $y = q(t)$, we may write

$$\frac{dy}{dt} = cy, \quad \text{or} \quad dy = cy\,dt.$$

Dividing both sides of the last equation by y, we obtain

$$\frac{1}{y}\,dy = c\,dt.$$

Since it is possible to **separate the variables** y and t—in the sense that they can be placed on opposite sides of the equals sign—the differential equation $dy/dt = cy$ is a **separable differential equation**. We will study such equations in more detail later in the text and will show that solutions can be found by integrating both sides of the "separated" equation $(1/y)\,dy = c\,dt$. Thus,

$$\int \frac{1}{y}\,dy = \int c\,dt$$

and, assuming $y > 0$,

$$\ln y = ct + d$$

for some constant d. It follows that

$$y = e^{ct+d} = e^d e^{ct}.$$

If y_0 denotes the initial value of y (that is, the value corresponding to $t = 0$), then letting $t = 0$ in the last equation gives us

$$y_0 = e^d e^0 = e^d,$$

and hence the solution $y = e^d e^{ct}$ may be written

$$y = y_0 e^{ct}.$$

We have proved the following theorem.

Theorem **(7.33)**

> Let y be a differentiable function of t such that $y > 0$ for every t, and let y_0 be the value of y at $t = 0$. If $dy/dt = cy$ for some constant c, then
>
> $$y = y_0 e^{ct}.$$

The preceding theorem states that *if the rate of change of $y = q(t)$ with respect to t is directly proportional to y, then y may be expressed in terms of an exponential function.* If y increases with t, the formula $y = y_0 e^{ct}$ is a **law of growth**, and if y decreases, it is a **law of decay**.

EXAMPLE 1 The number of bacteria in a culture increases from 600 to 1800 in two hours. Assuming that the rate of increase is directly proportional to the number of bacteria present, find

(a) a formula for the number of bacteria at time t

(b) the number of bacteria at the end of four hours

SOLUTION

(a) Let $y = q(t)$ denote the number of bacteria after t hours. Thus, $y_0 = q(0) = 600$ and $q(2) = 1800$. By hypothesis,

$$\frac{dy}{dt} = cy.$$

Following exactly the same steps used in the proof of Theorem (7.33), we obtain

$$y = y_0 e^{ct} = 600e^{ct}.$$

Since $y = 1800$ when $t = 2$, we obtain the following equivalent equations:

$$1800 = 600e^{2c}, \quad 3 = e^{2c}, \quad e^c = 3^{1/2}.$$

Substituting for e^c in $y = 600e^{ct}$ gives us

$$y = 600(3^{1/2})^t, \quad \text{or} \quad y = 600(3)^{t/2}.$$

(b) Letting $t = 4$ in $y = 600(3)^{t/2}$ yields

$$y = 600(3)^{4/2} = 600(9) = 5400.$$

EXAMPLE 2 Radium decays exponentially and has a half-life of approximately 1600 years; that is, given any quantity, one-half of it will disintegrate in 1600 years.

(a) Find a formula for the amount y remaining from 50 milligrams of pure radium after t years.

(b) When will the amount remaining be 20 milligrams?

SOLUTION

(a) If we let $y = q(t)$, then

$$y_0 = q(0) = 50 \quad \text{and} \quad q(1600) = \tfrac{1}{2}(50) = 25.$$

Since $dy/dt = cy$ for some c, it follows from Theorem (7.33) that

$$y = 50e^{ct}.$$

Since $y = 25$ when $t = 1600$,

$$25 = 50e^{1600c}, \quad \text{or} \quad e^{1600c} = \tfrac{1}{2}.$$

Hence

$$e^c = (\tfrac{1}{2})^{1/1600} = 2^{-1/1600}.$$

Substituting for e^c in $y = 50e^{ct}$ gives us

$$y = 50(2^{-1/1600})^t, \quad \text{or} \quad y = 50(2)^{-t/1600}.$$

(b) Using $y = 50(2)^{-t/1600}$, we see that the value of t at which $y = 20$ is a solution of the equation

$$20 = 50(2)^{-t/1600}, \quad \text{or} \quad 2^{t/1600} = \tfrac{5}{2}.$$

Taking the natural logarithm of each side, we obtain

$$\frac{t}{1600} \ln 2 = \ln \frac{5}{2},$$

or

$$t = \frac{1600 \ln \frac{5}{2}}{\ln 2} \approx 2115 \text{ yr}.$$

EXAMPLE 3 Newton's law of cooling states that the rate at which an object cools is directly proportional to the difference in temperature between the object and the surrounding medium. If an object cools from 125 °F to 100 °F in half an hour when surrounded by air at a temperature of 75 °F, find its temperature at the end of the next half hour.

SOLUTION Let y denote the temperature of the object after t hours of cooling. Since the temperature of the surrounding medium is 75°, the difference in temperature is $y - 75$, and hence, by Newton's law of cooling,

$$\frac{dy}{dt} = c(y - 75)$$

for some constant c. We separate variables and integrate as follows:

$$\frac{1}{y-75}\,dy = c\,dt$$

$$\int \frac{1}{y-75}\,dy = \int c\,dt$$

$$\ln(y-75) = ct + b$$

for some constant b. The last equation is equivalent to

$$y - 75 = e^{ct+b} = e^b e^{ct}.$$

Since $y = 125$ when $t = 0$,

$$125 - 75 = e^b e^0 = e^b, \quad \text{or} \quad e^b = 50.$$

Hence

$$y - 75 = 50e^{ct}, \quad \text{or} \quad y = 50e^{ct} + 75.$$

Using the fact that $y = 100$ when $t = \frac{1}{2}$ leads to the following equivalent equations:

$$100 = 50e^{c/2} + 75, \quad e^{c/2} = \tfrac{25}{50} = \tfrac{1}{2}, \quad e^c = \tfrac{1}{4}$$

Substituting $\frac{1}{4}$ for e^c in $y = 50e^{ct} + 75$ gives us a formula for the temperature after t hours:

$$y = 50(\tfrac{1}{4})^t + 75$$

In particular, if $t = 1$,

$$y = 50(\tfrac{1}{4}) + 75 = 87.5\,^\circ\text{F}.$$

In biology, a function G is sometimes used as follows to estimate the size of a quantity at time t:

$$G(t) = ke^{(-Ae^{-Bt})}$$

for positive constants k, A, and B. The function G is called a **Gompertz growth function**. It is always positive and increasing, but has a limit as t increases without bound. The graph of G is called a **Gompertz growth curve**.

EXAMPLE 4 Discuss and sketch the graph of the Gompertz growth function G.

SOLUTION We first observe that the y-intercept is $G(0) = ke^{-A}$ and that $G(t) > 0$ for every t. Differentiating twice, we obtain

$$\begin{aligned} G'(t) &= ke^{(-Ae^{-Bt})}\,D_t\,(-Ae^{-Bt}) \\ &= ABke^{(-Bt-Ae^{-Bt})} \end{aligned}$$

$$\begin{aligned} G''(t) &= ABke^{(-Bt-Ae^{-Bt})}\,D_t\,(-Bt - Ae^{-Bt}) \\ &= ABk(-B + ABe^{-Bt})e^{-Bt-Ae^{-Bt}} \end{aligned}$$

FIGURE 7.18

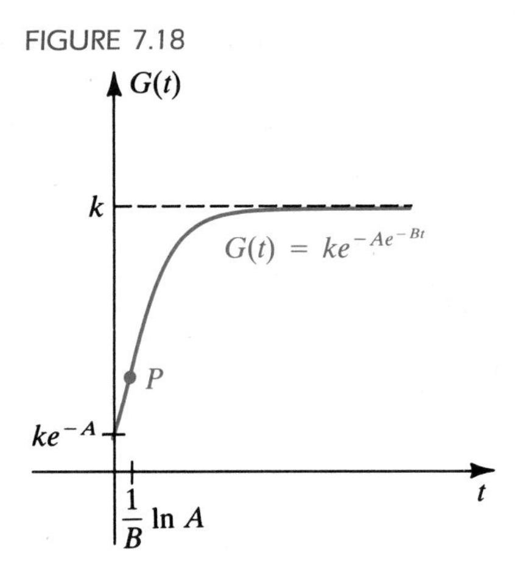

Since $G'(t) > 0$ for every t, the function G is increasing on $[0, \infty)$. The second derivative $G''(t)$ is zero if

$$-B + ABe^{-Bt} = 0 \quad \text{or} \quad e^{Bt} = A.$$

Solving the last equation for t gives us $t = (1/B) \ln A$, which is a critical number for the function G'. We leave it as an exercise to show that at this time the rate of growth G' has a maximum value Bk/e. We can also show that

$$\lim_{t \to \infty} G'(t) = 0 \quad \text{and} \quad \lim_{t \to \infty} G(t) = k.$$

Hence, as t increases without bound, the rate of growth approaches 0 and the graph of G has a horizontal asymptote $y = k$. A typical graph is sketched in Figure 7.18. The point P on the graph, corresponding to $t = (1/B) \ln A$, is a point of inflection, and the concavity changes from upward to downward at P.

In the next example we consider a physical quantity that increases to a maximum value and then decreases asymptotically to 0.

EXAMPLE 5 When uranium disintegrates into lead, one step in the process is the radioactive decay of radium into radon gas. Radon gas enters homes by diffusing through the soil into basements, where it presents a health hazard if inhaled. If a quantity Q of radium is present initially, then the amount of radon gas present after t years is given by

$$A(t) = \frac{c_1 Q}{c_2 - c_1}(e^{-c_1 t} - e^{-c_2 t}),$$

where $c_1 = \frac{1}{1600} \ln 2$ and $c_2 = \frac{1}{0.0105} \ln 2$ are the *decay constants* for radium and radon gas, respectively.

(a) Find the amount of radon gas present initially and after a long period of time.

(b) When is the amount of radon gas greatest?

SOLUTION

(a) The initial amount of radon gas is

$$A(0) = \frac{c_1 Q}{c_2 - c_1}(e^0 - e^0) = 0.$$

If we let t increase without bound, then

$$\begin{aligned}\lim_{t \to \infty} A(t) &= \frac{c_1 Q}{c_2 - c_1} \lim_{t \to \infty} (e^{-c_1 t} - e^{-c_2 t}) \\ &= \frac{c_1 Q}{c_2 - c_1}(0 - 0) = 0.\end{aligned}$$

Hence, over a long period of time, the amount of radon gas decreases to 0.

(b) To find the critical numbers of A, we differentiate, obtaining

$$A'(t) = \frac{c_1 Q}{c_2 - c_1}(-c_1 e^{-c_1 t} + c_2 e^{-c_2 t}).$$

Thus, $A'(t) = 0$ if

$$c_1 e^{-c_1 t} = c_2 e^{-c_2 t}, \quad \text{or} \quad e^{(c_2 - c_1)t} = \frac{c_2}{c_1}.$$

It follows that

$$(c_2 - c_1)t = \ln \frac{c_2}{c_1} = \ln c_2 - \ln c_1,$$

or

$$t = \frac{\ln c_2 - \ln c_1}{c_2 - c_1}.$$

This value of t yields the maximum value of A. Substituting the given values for c_1 and c_2 gives us the approximation

$$t \approx 0.181 \text{ year} \approx 66 \text{ days}.$$

EXERCISES 7.6

1 The number of bacteria in a culture increases from 5000 to 15,000 in 10 hours. Assuming that the rate of increase is proportional to the number of bacteria present, find a formula for the number of bacteria in the culture at any time t. Estimate the number at the end of 20 hours. When will the number be 50,000?

2 The polonium isotope ^{210}Po has a half-life of approximately 140 days. If a sample weighs 20 milligrams initially, how much remains after t days? Approximately how much will be left after two weeks?

3 If the temperature is constant, then the rate of change of barometric pressure p with respect to altitude h is proportional to p. If $p = 30$ in. at sea level and $p = 29$ in. when $h = 1000$ ft, find the pressure at an altitude of 5000 feet.

4 The population of a city is increasing at the rate of 5% per year. If the present population is 500,000 and the rate of increase is proportional to the number of people, what will the population be in 10 years?

5 Agronomists use the assumption that a quarter acre of land is required to provide food for one person and estimate that there are 10 billion acres of tillable land in the world. Hence a maximum population of 40 billion people can be sustained if no other food source is available. The world population at the beginning of 1980 was approximately 4.5 billion. Assuming that the population increases at a rate of 2% per year and the rate of increase is proportional to the number of people, when will the maximum population be reached?

6 A metal plate that has been heated cools from 180 °F to 150 °F in 20 minutes when surrounded by air at a temperature of 60 °F. Use Newton's law of cooling (see Example 3) to approximate its temperature at the end of one hour of cooling. When will the temperature be 100 °F?

7 An outdoor thermometer registers a temperature of 40 °F. Five minutes after it is brought into a room where the temperature is 70 °F, the thermometer registers 60 °F. When will it register 65 °F?

8 The rate at which salt dissolves in water is directly proportional to the amount that remains undissolved. If 10 pounds of salt are placed in a container of water and 4 pounds dissolve in 20 minutes, how long will it take for two more pounds to dissolve?

9 According to Kirchhoff's first law for electrical circuits, $V = RI + L(dI/dt)$, where the constants V, R, and L denote the electromotive force, the resistance, and the inductance, respectively, and I denotes the current at time t. If the electromotive force is terminated at time $t = 0$ and if the current is I_0 at the instant of removal, prove that $I = I_0 e^{-Rt/L}$.

10 A physicist finds that an unknown radioactive substance registers 2000 counts per minute on a Geiger counter. Ten days later the substance registers 1500 counts per minute. Approximate its half-life.

11 The air pressure P (in atmospheres) at an elevation of z meters above sea level is a solution of the differential equation $dP/dz = -9.81\rho(z)$, where $\rho(z)$ is the density of air at elevation z. Assuming that air obeys the ideal gas law, this differential equation can be rewritten as $dP/dz = -0.0342P/T$, where T is the temperature (in °K) at elevation z. If $T = 288 - 0.01z$ and if the pressure is 1 atmosphere at sea level, express P as a function of z.

12 During the first month of growth for crops such as maize, cotton, and soybeans, the rate of growth (in grams/day) is proportional to the present weight W. For a species of cotton, $dW/dt = 0.21W$. Predict the weight of a plant at the end of the month ($t = 30$) if the plant weighs 70 milligrams at the beginning of the month.

13 Radioactive strontium-90, ^{90}Sr, with a half-life of 29 years, can cause bone cancer in humans. The substance is carried by acid rain, soaks into the ground, and is passed through the food chain. The radioactivity level in a particular field is estimated to be 2.5 times the safe level S. For approximately how many years will this field be contaminated?

14 The radioactive tracer ^{51}Cr, with a half-life of 27.8 days, can be used in medical testing to locate the position of a placenta in a pregnant woman. Often the tracer must be ordered from a medical supply lab. If 35 units are needed for a test and delivery from the lab requires two days, estimate the minimum number of units that should be ordered.

15 Veterinarians use sodium pentobarbital to anesthetize animals. Suppose that to anesthetize a dog, 30 milligrams are required for each kilogram of body weight. If sodium pentobarbital is eliminated exponentially from the bloodstream and half is eliminated in four hours, approximate the single dose that will anesthetize a 20-kilogram dog for 45 minutes.

16 In the study of lung physiology, the following differential equation is used to describe the transport of a substance across a capillary wall:

$$\frac{dh}{dt} = -\frac{V}{Q}\left(\frac{h}{k+h}\right),$$

where h is the hormone concentration in the bloodstream, t is time, V is the maximum transport rate, Q is the volume of the capillary, and k is a constant that measures the affinity between the hormones and enzymes that assist with the transport process. Find the general solution of the differential equation.

17 A space probe is shot upward from the earth. If air resistance is disregarded, a differential equation for the velocity after burnout is $v\,(dv/dy) = -ky^{-2}$, where y is the distance from the center of the earth and k is a positive constant. If y_0 is the distance from the center of the earth at burnout and v_0 is the corresponding velocity, express v as a function of y.

18 At high temperatures, nitrogen dioxide, NO_2, decomposes into NO and O_2. If $y(t)$ is the concentration of NO_2 (in moles per liter), then, at 600 °K, $y(t)$ changes according to the *second-order reaction law* $dy/dt = -0.05y^2$ for time t in seconds. Express y in terms of t and the initial concentration y_0.

19 The technique of carbon-14 dating is used to determine the age of archeological or geological specimens. This method is based on the fact that the unstable isotope carbon-14 (^{14}C) is present in the CO_2 in the atmosphere. Plants take in carbon from the atmosphere; when they die, the ^{14}C that has accumulated begins to decay, with a half-life of approximately 5700 years. By measuring the amount of ^{14}C that remains in a specimen, it is possible to approximate when the organism died. Suppose that a bone fossil contains 20% as much ^{14}C as an equal amount of carbon in present-day bone. Approximate the age of the bone.

20 Refer to Exercise 19. The hydrogen isotope ^3_1H, which has a half-life of 12.3 years, is produced in the atmosphere by cosmic rays and is brought to earth by rain. If the wood siding of an old house contains 10% as much ^3_1H as the siding on a similar new house, approximate the age of the old house.

21 The earth's atmosphere absorbs approximately 32% of the sun's incoming radiation. The earth also emits radiation (mostly in the form of heat), and the atmosphere absorbs approximately 93% of this outgoing radiation. This difference in absorption of incoming and outgoing radiation by the atmosphere is called the *greenhouse effect*. Changes in this balance will affect the earth's climate. Suppose I_0 is the intensity of the sun's radiation and I is the intensity of the radiation after traveling a distance x through the atmosphere. If $\rho(h)$ is the density of the atmosphere at height h, then the *optical thickness* is $f(x) = k\int_0^x \rho(h)\,dh$, where k is an absorption constant, and I is given by $I = I_0e^{-f(x)}$. Show that $dI/dx = -k\rho(x)I$.

22 Certain learning processes may be illustrated by the graph of $f(x) = a + b(1 - e^{-cx})$ for positive constants a, b, and c. Suppose a manufacturer estimates that a new employee can produce five items the first day on the job. As the employee becomes more proficient, the daily production increases until a certain maximum production is reached. Suppose that on the nth day on the job, the number $f(n)$ of items produced is approximated by the formula $f(n) = 3 + 20(1 - e^{-0.1n})$.

(a) Estimate the number of items produced on the fifth day, the ninth day, the twenty-fourth day, and the thirtieth day.

(b) Sketch the graph of f from $n = 0$ to $n = 30$. (Graphs of this type are called *learning curves* and are used frequently in education and psychology.)

(c) What happens as n increases without bound?

23 A spherical cell has volume V and surface area S. A simple model for cell growth before mitosis assumes that the rate of growth dV/dt is proportional to the surface area of the cell. Show that $dV/dt = kV^{2/3}$ for some $k > 0$, and express V as a function of t.

24 In Theorem (7.33) we assumed that the rate of change of a quantity $q(t)$ at time t is directly proportional to $q(t)$. Find $q(t)$ if its rate of change is directly proportional to $[q(t)]^2$.

25 Refer to Example 4.

(a) Verify that Bk/e is a maximum value for G'.

(b) Show that $\lim_{t\to\infty} G'(t) = 0$ and $\lim_{t\to\infty} G(t) = k$.

(c) Sketch the graph of G if $k = 10$, $A = 2$, and $B = 1$.

c 26 Graph the Gompertz growth function G on the interval $[0, 5]$ for $k = 1.1$, $A = 3.2$, and $B = 1.1$.

7.7 REVIEW EXERCISES

Exer. 1–2: Find $f^{-1}(x)$.

1 $f(x) = 10 - 15x$

2 $f(x) = 9 - 2x^2, \quad x \le 0$

Exer. 3–4: Show that the function f has an inverse function, and find $[D_x f^{-1}(x)]_{x=a}$ for the given number a.

3 $f(x) = 2x^3 - 8x + 5, \quad -1 \le x \le 1; \quad a = 5$

4 $f(x) = e^{3x} + 2e^x - 5, \quad x \ge 0; \quad a = -2$

Exer. 5–38: Find $f'(x)$ if $f(x)$ is the given expression.

5 $\ln |4 - 5x^3|^5$

6 $\ln |x^2 - 7|^3$

7 $(1 - 2x) \ln |1 - 2x|$

8 $\log \left|\dfrac{2 - 9x}{1 - x^2}\right|$

9 $\ln \dfrac{(3x + 2)^4\sqrt{6x - 5}}{8x - 7}$

10 $\ln \sqrt[4]{\dfrac{x}{3x + 5}}$

11 $\dfrac{1}{\ln (2x^2 + 3)}$

12 $\dfrac{\ln x}{e^{2x} + 1}$

13 $\dfrac{x}{\ln x}$

14 $\dfrac{\ln x}{x}$

15 $e^{\ln (x^2 + 1)}$

16 $\ln e^{\sqrt{x}}$

17 $\ln (e^{4x} + 9)$

18 $4^{\sqrt{2x+3}}$

19 $10^x \log x$

20 $5^{3x} + (3x)^5$

21 $\sqrt{\ln \sqrt{x}}$

22 $(1 + \sqrt{x})^e$

23 $x^2e^{-x^2}$

24 $\dfrac{2^{-3x}}{x^3 + 4}$

25 $\sqrt{e^{3x} + e^{-3x}}$

26 $(x^2 + 1)^{2x}$

27 $10^{\ln x}$

28 $7^{\ln |x|}$

29 $x^{\ln x}$

30 $(\ln x)^{\ln x}$

31 $\ln |\tan x - \sec x|$

32 $\ln \csc \sqrt{x}$

33 $\csc e^{-2x} \cot e^{-2x}$

34 $x^2e^{\tan 2x}$

35 $\ln \cos^4 4x$

36 $3^{\sin 3x}$

37 $(\sin x)^{\cos x}$

38 $\dfrac{1}{\sin^2 e^{-3x}}$

Exer. 39–40: Use implicit differentiation to find y'.

39 $1 + xy = e^{xy}$

40 $\ln (x+y)+x^2-2y^3=1$

Exer. 41–42: Use logarithmic differentiation to find dy/dx.

41 $y = (x + 2)^{4/3}(x - 3)^{3/2}$

42 $y = \sqrt[3]{(3x - 1)\sqrt{2x + 5}}$

Exer. 43–78: Evaluate the integral.

43 (a) $\displaystyle\int \frac{1}{\sqrt{x}\, e^{\sqrt{x}}}\, dx$ (b) $\displaystyle\int_1^4 \frac{1}{\sqrt{x}\, e^{\sqrt{x}}}\, dx$

44 (a) $\displaystyle\int e^{-3x+2}\, dx$ (b) $\displaystyle\int_0^1 e^{-3x+2}\, dx$

45 (a) $\displaystyle\int x4^{-x^2}\, dx$ (b) $\displaystyle\int_0^1 x4^{-x^2}\, dx$

46 (a) $\displaystyle\int \frac{x^2 + 1}{x^3 + 3x}\, dx$ (b) $\displaystyle\int_1^2 \frac{x^2 + 1}{x^3 + 3x}\, dx$

47 $\displaystyle\int x \tan x^2\, dx$

48 $\displaystyle\int \cot\left(x + \frac{\pi}{6}\right) dx$

49 $\displaystyle\int x^e\, dx$

50 $\displaystyle\int \frac{1}{7 - 5x}\, dx$

51 $\displaystyle\int \frac{1}{x - x\ln x}\, dx$

52 $\displaystyle\int \frac{1}{x\ln x}\, dx$

53 $\displaystyle\int \frac{(1 + e^x)^2}{e^{2x}}\, dx$

54 $\displaystyle\int \frac{(e^{2x} + e^{3x})^2}{e^{5x}}\, dx$

55 $\displaystyle\int \frac{x^2}{x + 2}\, dx$

56 $\displaystyle\int \frac{x^2 + 1}{x + 1}\, dx$

57 $\displaystyle\int \frac{e^{4/x^2}}{x^3}\, dx$

58 $\displaystyle\int \frac{e^{1/x}}{x^2}\, dx$

59 $\displaystyle\int \frac{x}{x^4 + 2x^2 + 1}\, dx$

60 $\displaystyle\int \frac{5x^3}{x^4 + 1}\, dx$

61 $\int \frac{e^x}{1 + e^x}\,dx$

62 $\int (1 + e^{-3x})^2\,dx$

63 $\int 5^x e^x\,dx$

64 $\int x10^{(x^2)}\,dx$

65 $\int \frac{1}{x\sqrt{\log x}}\,dx$

66 $\int 7^x\sqrt{1 + 7^x}\,dx$

67 $\int e^{-x}\sin e^{-x}\,dx$

68 $\int \tan x\, e^{\sec x}\sec x\,dx$

69 $\int \frac{\csc^2 x}{1 + \cot x}\,dx$

70 $\int \frac{\cos x + \sin x}{\sin x - \cos x}\,dx$

71 $\int \frac{\cos 2x}{1 - 2\sin 2x}\,dx$

72 $\int 3^x(3 + \sin 3^x)\,dx$

73 $\int e^x \tan e^x\,dx$

74 $\int \frac{\sec (1/x)}{x^2}\,dx$

75 $\int (\csc 3x + 1)^2\,dx$

76 $\int \cos 2x \csc 2x\,dx$

77 $\int (\cot 9x + \csc 9x)\,dx$

78 $\int \frac{\sin x + 1}{\cos x}\,dx$

79 Solve the differential equation $y'' = -e^{-3x}$ subject to the conditions $y = -1$ and $y' = 2$ if $x = 0$.

80 In *seasonal population growth*, the population $q(t)$ at time t (in years) increases during the spring and summer but decreases during the fall and winter. A differential equation that is sometimes used to describe this type of growth is $q'(t)/q(t) = k \sin 2\pi t$, where $k > 0$ and $t = 0$ corresponds to the first day of spring.

(a) Show that the population $q(t)$ is seasonal.

(b) If $q_0 = q(0)$, find a formula for $q(t)$.

81 A particle moves on a coordinate line with an acceleration at time t of $e^{t/2}$ cm/sec^2. At $t = 0$ the particle is at the origin and its velocity is 6 cm/sec. How far does it travel during the time interval $[0, 4]$?

82 Find the local extrema of $f(x) = x^2 \ln x$ for $x > 0$. Discuss concavity, find the points of inflection, and sketch the graph of f.

83 Find an equation of the tangent line to the graph of the equation $y = xe^{1/x^3} + \ln|2 - x^2|$ at the point $P(1, e)$.

84 Find the area of the region bounded by the graphs of the equations $y = e^{2x}$, $y = x/(x^2 + 1)$, $x = 0$, and $x = 1$.

85 The region bounded by the graphs of $y = e^{4x}$, $x = -2$, $x = -3$, and $y = 0$ is revolved about the x-axis. Find the volume of the resulting solid.

86 The 1980 population estimate for India was 651 million, and the population has been increasing at a rate of about 2% per year, with the rate of increase proportional to the number of people. If t denotes the time (in years) after 1980, find a formula for $N(t)$, the population (in millions) at time t. Assuming that this rapid growth rate continues, estimate the population and the rate of population growth in the year 2000.

87 A radioactive substance has a half-life of 5 days. How long will it take for an amount A to disintegrate to the extent that only 1% of A remains?

88 The carbon-14 dating equation $T = -8310 \ln x$ is used to predict the age T (in years) of a fossil in terms of the percentage $100x$ of carbon still present in the specimen (see Exercise 19, Section 7.6).

(a) If $x = 0.04$, estimate the age of the fossil to the nearest 1000 years.

(b) If the maximum error in estimating x in part (a) is ± 0.005, use differentials to approximate the maximum error in T.

89 The rate at which sugar dissolves in water is proportional to the amount that remains undissolved. Suppose that 10 pounds of sugar are placed in a container of water at 1:00 P.M., and one-half is dissolved at 4:00 P.M.

(a) How long will it take two more pounds to dissolve?

(b) How much of the 10 pounds will be dissolved at 8:00 P.M.?

90 According to Newton's law of cooling, the rate at which an object cools is directly proportional to the difference in temperature between the object and its surrounding medium. If $f(t)$ denotes the temperature at time t, show that $f(t) = T + [f(0) - T]e^{-kt}$, where T is the temperature of the surrounding medium and k is a positive constant.

91 The bacterium *E. coli* undergoes cell division approximately every 20 minutes. Starting with 100,000 cells, determine the number of cells after 2 hours.

92 The differential equation $p\,dv + cv\,dp = 0$ describes the adiabatic change of state of air for pressure p, volume v, and a constant c. Solve for p as a function of v.

CHAPTER

8

INVERSE TRIGONOMETRIC AND HYPERBOLIC FUNCTIONS

INTRODUCTION

The chapter begins with a review of the inverse trigonometric functions that are studied in trigonometry courses. We next apply methods of calculus to obtain formulas for derivatives and integrals. The fact that function values may be regarded as angles allows us to consider applications such as measuring the rate of change in the angle of elevation as an observer tracks an object in flight, finding the rate at which a searchlight is rotating, and determining an angle that minimizes energy loss as blood flows through a blood vessel.

The hyperbolic functions, defined in Section 8.3, are used in the physical sciences and engineering to describe the shape of a flexible cable that is supported at each end, to find the velocity of an object in a resisting medium such as air or water, and to study the diffusion of radon gas through a basement wall.

The chapter closes with a discussion of the inverse hyperbolic functions. These functions are used primarily for evaluating certain types of integrals.

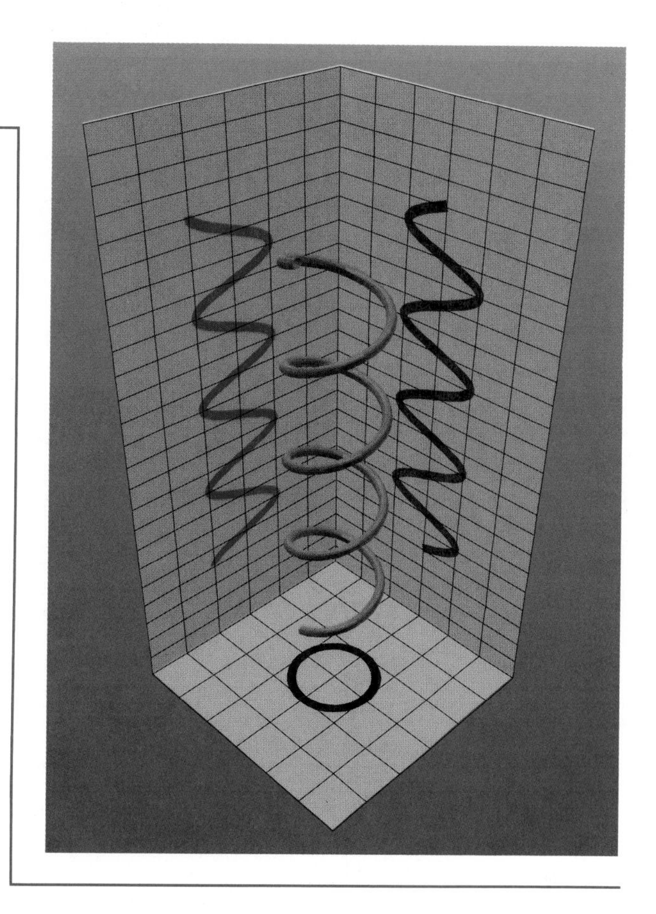

8.1 INVERSE TRIGONOMETRIC FUNCTIONS

Since the trigonometric functions are not one-to-one, they do not have inverse functions (see Section 7.1). By restricting their domains, however, we may obtain one-to-one functions that have the same values as the trigonometric functions and that *do* have inverses over these restricted domains.

FIGURE 8.1

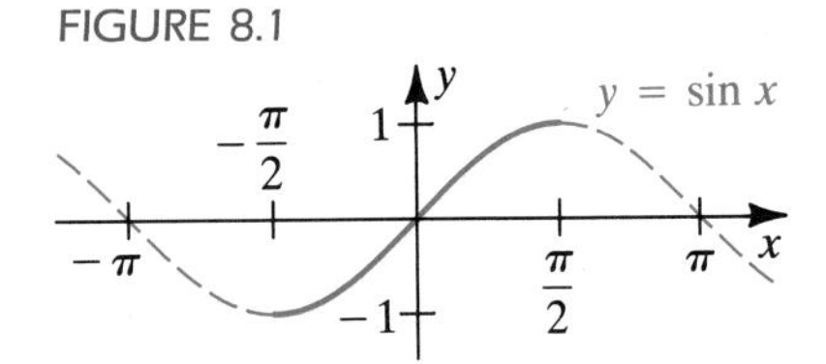

Let us first consider the graph of the sine function, whose domain is $\mathbb{R}$ and range is the closed interval $[-1, 1]$ (see Figure 8.1). The sine function is not one-to-one, since a horizontal line such as $y = \frac{1}{2}$ intersects the graph in more than one point. Thus, numbers such as $\pi/6$, $5\pi/6$, and $-7\pi/6$ yield the same function value, $\frac{1}{2}$. If we restrict the domain to $[-\pi/2, \pi/2]$, then, as illustrated by the solid portion of the graph in Figure 8.1, we obtain an increasing function that takes on every value of the sine function once and only once. This new function, with domain $[-\pi/2, \pi/2]$ and range $[-1, 1]$, is continuous and increasing and hence, by Theorem (7.6), has an inverse function that is continuous and increasing. The inverse function has domain $[-1, 1]$ and range $[-\pi/2, \pi/2]$. This leads to the following definition.

Definition **(8.1)**

> The **inverse sine function**, denoted $\sin^{-1}$, is defined by
>
> $$y = \sin^{-1} x \quad \text{if and only if} \quad x = \sin y$$
>
> for $-1 \le x \le 1$ and $-\pi/2 \le y \le \pi/2$.

The inverse sine function is also called the **arcsine function**, and $\arcsin x$ is often used in place of $\sin^{-1} x$. The -1 in $\sin^{-1}$ is not to be regarded as an exponent, but rather as a means of denoting this inverse function. The notation $y = \sin^{-1} x$ may be read *y is the inverse sine of x*. The equation $x = \sin y$ in the definition allows us to regard y as an angle, and hence $y = \sin^{-1} x$ may also be read *y is the angle whose sine is x*. Observe that, by Definition (8.1),

$$-\frac{\pi}{2} \le \sin^{-1} x \le \frac{\pi}{2}, \quad \text{or} \quad -\frac{\pi}{2} \le \arcsin x \le \frac{\pi}{2}.$$

ILLUSTRATION

FIGURE 8.2

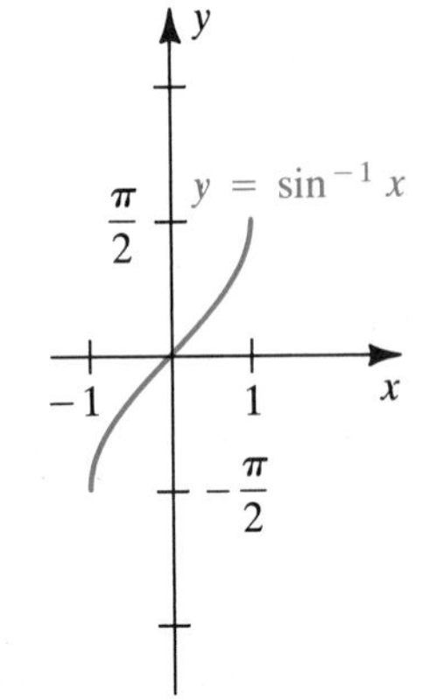

- If $y = \sin^{-1} \frac{1}{2}$, then $\sin y = \frac{1}{2}$ and $-\frac{\pi}{2} \le y \le \frac{\pi}{2}$. Hence $y = \frac{\pi}{6}$.
- If $y = \arcsin\left(-\frac{1}{2}\right)$, then $\sin y = -\frac{1}{2}$ and $-\frac{\pi}{2} \le y \le \frac{\pi}{2}$. Hence $y = -\frac{\pi}{6}$.

Using the method we introduced in Section 7.1 for sketching the graph of an inverse function, we can sketch the graph of $y = \sin^{-1} x$ by reflecting the solid portion of Figure 8.1 through the line $y = x$. This gives us Figure 8.2. We could also use the equation $x = \sin y$ with $-\pi/2 \le y \le \pi/2$ to find points on the graph.

The relationships $f(f^{-1}(x)) = x$ and $f^{-1}(f(x)) = x$ that hold for any inverse function f^{-1} give us the following properties.

Properties of $\sin^{-1}$ **(8.2)**

(i) $\sin(\sin^{-1} x) = \sin(\arcsin x) = x$ if $-1 \le x \le 1$

(ii) $\sin^{-1}(\sin x) = \arcsin(\sin x) = x$ if $-\dfrac{\pi}{2} \le x \le \dfrac{\pi}{2}$

ILLUSTRATION

- $\sin\left(\sin^{-1}\dfrac{1}{2}\right) = \dfrac{1}{2}$ since $-1 < \dfrac{1}{2} < 1$
- $\arcsin\left(\sin\dfrac{\pi}{4}\right) = \dfrac{\pi}{4}$ since $-\dfrac{\pi}{2} < \dfrac{\pi}{4} < \dfrac{\pi}{2}$
- $\sin^{-1}\left(\sin\dfrac{2\pi}{3}\right) = \sin^{-1}\left(\dfrac{\sqrt{3}}{2}\right) = \dfrac{\pi}{3}$

Be careful when using (8.2). In the third part of the preceding illustration, $2\pi/3$ is *not* between $-\pi/2$ and $\pi/2$, and hence we cannot use (ii) of (8.2). Instead, we use properties of special angles (see Section 1.3) to first evaluate $\sin(2\pi/3)$ and then find $\sin^{-1}(\sqrt{3}/2)$.

FIGURE 8.3

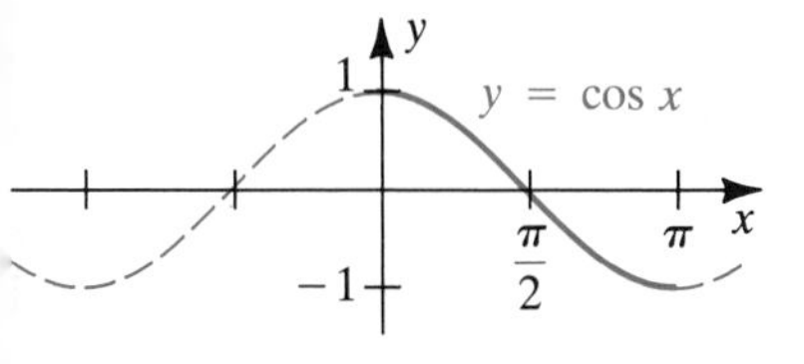

We may use the other five trigonometric functions to define inverse trigonometric functions. If the domain of the cosine function is restricted to the interval $[0, \pi]$ (see the solid portion of Figure 8.3), we obtain a one-to-one continuous decreasing function that has a continuous decreasing inverse function. This leads to the next definition.

Definition **(8.3)**

The **inverse cosine function**, denoted $\cos^{-1}$, is defined by

$$y = \cos^{-1} x \quad \text{if and only if} \quad x = \cos y$$

for $-1 \le x \le 1$ and $0 \le y \le \pi$.

The domain of the inverse cosine function is $[-1, 1]$, and the range is $[0, \pi]$. The notation $y = \cos^{-1} x$ may be read *y is the inverse cosine of x* or *y is the angle whose cosine is x*. The inverse cosine function is also called the **arccosine function**, and the notation $\arccos x$ is used interchangeably with $\cos^{-1} x$.

ILLUSTRATION

- If $y = \cos^{-1}\dfrac{1}{2}$, then $\cos y = \dfrac{1}{2}$ and $0 \le y \le \pi$. Hence $y = \dfrac{\pi}{3}$.
- If $y = \arccos\left(-\dfrac{1}{2}\right)$, then $\cos y = -\dfrac{1}{2}$ and $0 \le y \le \pi$. Hence $y = \dfrac{2\pi}{3}$.

The graph of the inverse cosine function may be found by reflecting the solid portion of Figure 8.3 through the line $y = x$. This gives us the sketch in Figure 8.4. We could also use the equation $x = \cos y$ with $0 \le y \le \pi$ to find points on the graph.

FIGURE 8.4

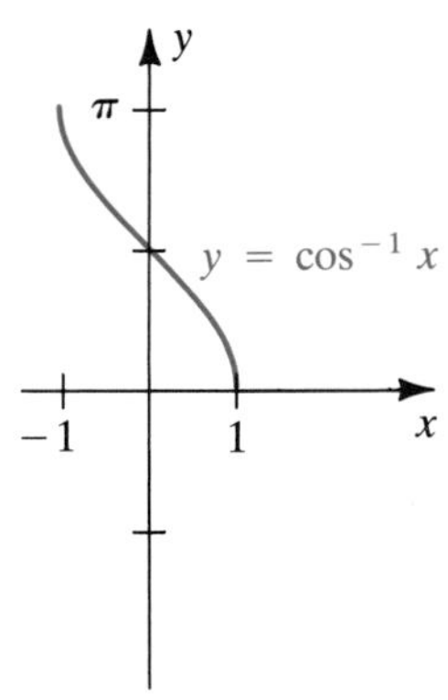

Since cos and $\cos^{-1}$ are inverse functions of each other, we obtain the following properties.

Properties of $\cos^{-1}$ **(8.4)**

(i) $\cos(\cos^{-1} x) = \cos(\arccos x) = x \quad \text{if} \quad -1 \le x \le 1$

(ii) $\cos^{-1}(\cos x) = \arccos(\cos x) = x \quad \text{if} \quad 0 \le x \le \pi$

ILLUSTRATION

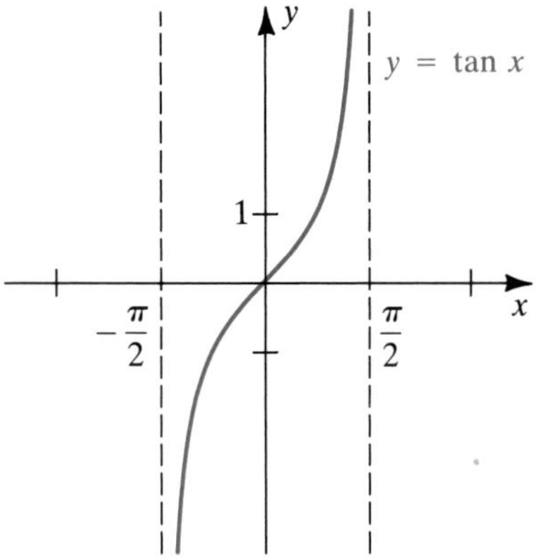

- $\cos\left[\cos^{-1}\left(-\frac{1}{2}\right)\right] = -\frac{1}{2} \quad \text{since} \quad -1 < -\frac{1}{2} < 1$
- $\arccos\left(\cos\frac{2\pi}{3}\right) = \frac{2\pi}{3} \quad \text{since} \quad 0 < \frac{2\pi}{3} < \pi$
- $\cos^{-1}\left[\cos\left(-\frac{\pi}{4}\right)\right] = \cos^{-1}\left(\frac{\sqrt{2}}{2}\right) = \frac{\pi}{4}$

FIGURE 8.5
$y = \tan x, \; -\pi/2 \le x \le \pi/2$

Note that in the third part of the preceding illustration, $-\pi/4$ is *not* between 0 and π, and hence we cannot use (ii) of (8.4). Instead, we first evaluate $\cos(-\pi/4)$ and then find $\cos^{-1}(\sqrt{2}/2)$.

If we restrict the domain of the tangent function to the open interval $(-\pi/2, \pi/2)$, we obtain a continuous increasing function (see Figure 8.5). We use this *new* function to define the *inverse tangent function.*

Definition **(8.5)**

The **inverse tangent function**, or **arctangent function**, denoted by $\tan^{-1}$, or arctan, is defined by

$$y = \tan^{-1} x = \arctan x \quad \text{if and only if} \quad x = \tan y$$

for every x and $-\pi/2 < y < \pi/2$.

The domain of the arctangent function is $\mathbb{R}$, and the range is the open interval $(-\pi/2. \pi/2)$. We can obtain the graph of $y = \tan^{-1} x$ in Figure 8.6 by reflecting tne graph in Figure 8.5 through the line $y = x$.

FIGURE 8.6

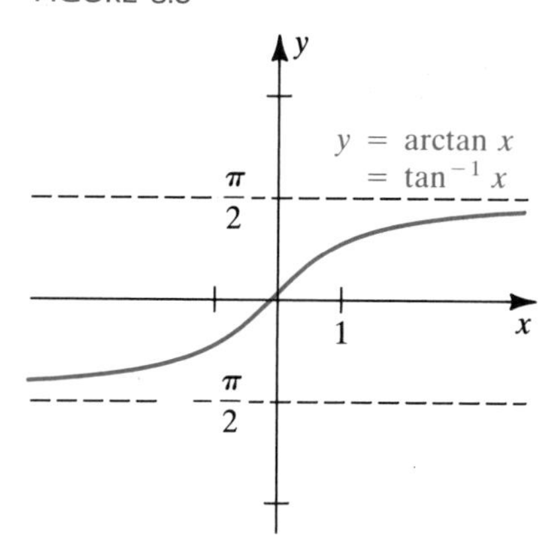

As with $\sin^{-1}$ and $\cos^{-1}$, we have the following.

Properties of $\tan^{-1}$ **(8.6)**

> **(i)** $\tan(\tan^{-1} x) = \tan(\arctan x) = x$ for every x
>
> **(ii)** $\tan^{-1}(\tan x) = \arctan(\tan x) = x$ if $-\dfrac{\pi}{2} < x < \dfrac{\pi}{2}$

ILLUSTRATION

- If $y = \arctan(-1)$, then $\tan y = -1$ and $-\dfrac{\pi}{2} < y < \dfrac{\pi}{2}$.

  Hence $y = -\dfrac{\pi}{4}$.

- $\tan(\tan^{-1} 1000) = 1000$ by (8.6)(i)

- $\tan^{-1}\left(\tan \dfrac{\pi}{4}\right) = \dfrac{\pi}{4}$ since $-\dfrac{\pi}{2} < \dfrac{\pi}{4} < \dfrac{\pi}{2}$

- $\arctan(\tan \pi) = \arctan 0 = 0$

EXAMPLE 1 Find the exact value of $\sec(\arctan \frac{2}{3})$.

FIGURE 8.7

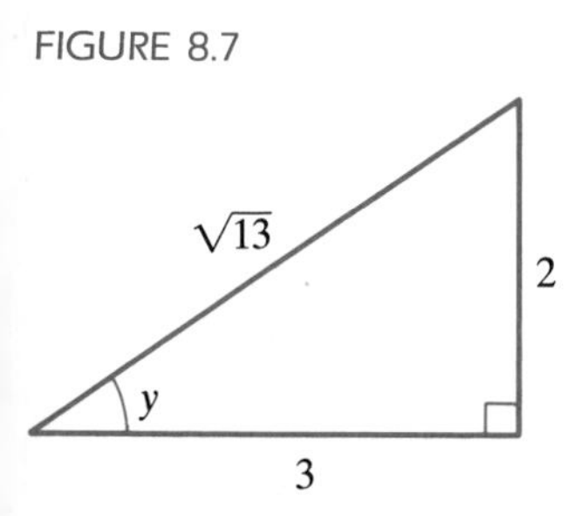

SOLUTION If we let $y = \arctan \frac{2}{3}$, then $\tan y = \frac{2}{3}$. We wish to find $\sec y$. Since $-\pi/2 < \arctan x < \pi/2$ for every x and $\tan y > 0$, it follows that $0 < y < \pi/2$. Thus, we may regard y as the radian measure of an angle of a right triangle such that $\tan y = \frac{2}{3}$, as illustrated in Figure 8.7. By the Pythagorean theorem, the hypotenuse is $\sqrt{3^2 + 2^2} = \sqrt{13}$. Referring to the triangle, we obtain

$$\sec\left(\arctan \frac{2}{3}\right) = \sec y = \frac{\sqrt{13}}{3}.$$

If we consider the graph of $y = \sec x$, there are many ways to restrict x so that we obtain a one-to-one function that takes on every value of the secant function. There is no universal agreement on how this should be done. It is convenient to restrict x to the intervals $[0, \pi/2)$ and $[\pi, 3\pi/2)$, as indicated by the solid portion of the graph of $y = \sec x$ in Figure 8.8, instead of to the "more natural" intervals $[0, \pi/2)$ and $(\pi/2, \pi]$, because the differentiation formula for the inverse secant is simpler. We show in the next section that $D_x \sec^{-1} x = 1/(x\sqrt{x^2 - 1})$. Thus, the slope of the tangent line to the graph of $y = \sec^{-1} x$ is negative if $x < -1$ or positive if $x > 1$. For the more natural intervals, the slope is always positive, and we would have $D_x \sec^{-1} x = 1/(|x|\sqrt{x^2 - 1})$.

FIGURE 8.8
$y = \sec x$

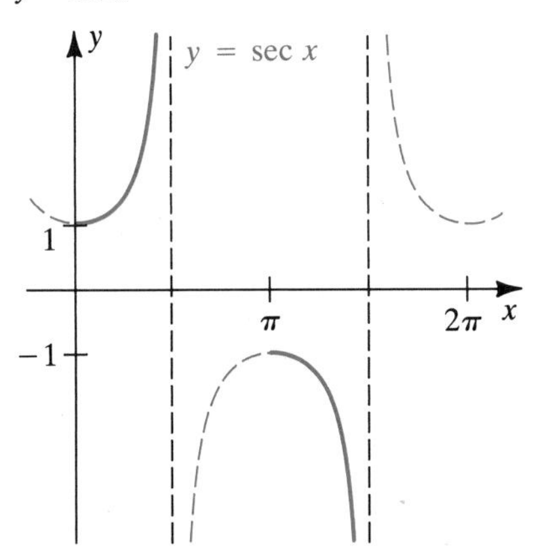

Definition (8.7)

The **inverse secant function**, or **arcsecant function**, denoted by $\sec^{-1}$, or arcsec, is defined by

$$y = \sec^{-1} x = \operatorname{arcsec} x \quad \text{if and only if} \quad x = \sec y$$

for $|x| \geq 1$ and y in $[0, \pi/2)$ or in $[\pi, 3\pi/2)$.

FIGURE 8.9
$y = \sec^{-1} x$

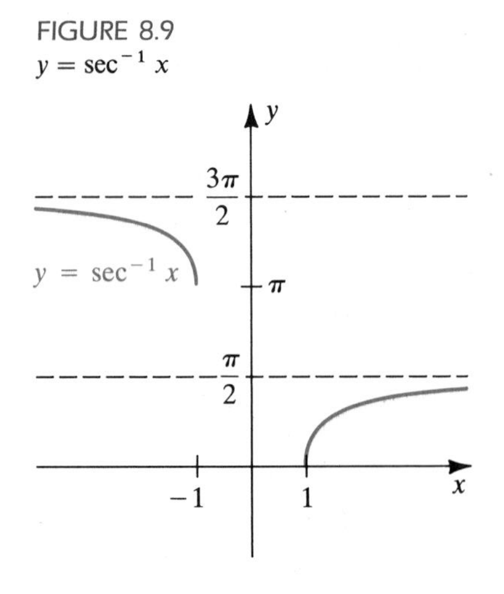

The graph of $y = \sec^{-1} x$ is sketched in Figure 8.9.

The inverse cotangent function, $\cot^{-1}$, and inverse cosecant function, $\csc^{-1}$, can be defined in similar fashion (see Exercises 31–32).

The following examples illustrate some of the manipulations that can be carried out with inverse trigonometric functions.

EXAMPLE 2 Find the exact value of $\sin(\arctan \frac{1}{2} - \arccos \frac{4}{5})$.

SOLUTION If we let

$$u = \arctan \tfrac{1}{2} \quad \text{and} \quad v = \arccos \tfrac{4}{5},$$

then

$$\tan u = \tfrac{1}{2} \quad \text{and} \quad \cos v = \tfrac{4}{5}.$$

We wish to find $\sin(u - v)$. Since u and v are in the interval $(0, \pi/2)$, they can be considered as the radian measures of positive acute angles, and

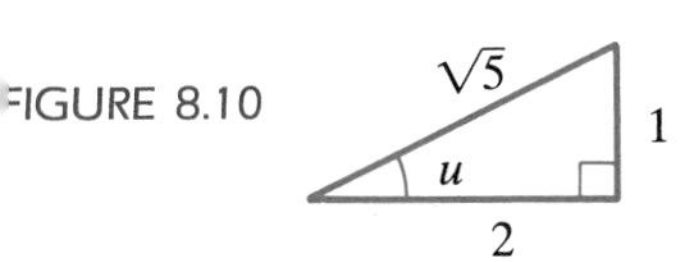

FIGURE 8.10

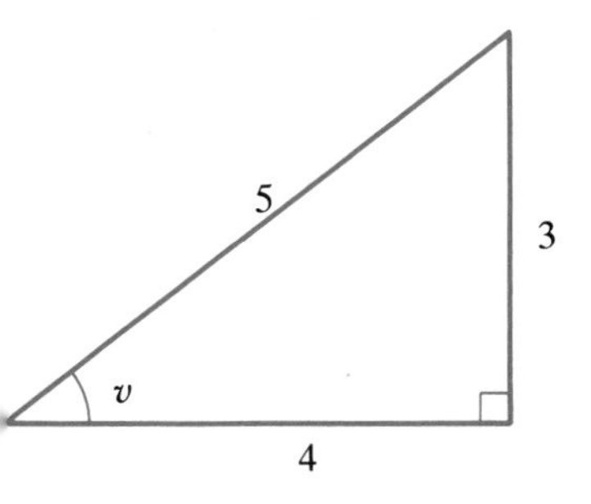

we may refer to the right triangles in Figure 8.10. This gives us

$$\sin u = \frac{1}{\sqrt{5}}, \qquad \cos u = \frac{2}{\sqrt{5}}, \qquad \sin v = \frac{3}{5}, \qquad \cos v = \frac{4}{5}.$$

Using the subtraction formula for the sine function, we obtain

$$\begin{aligned} \sin(u - v) &= \sin u \cos v - \cos u \sin v \\ &= \frac{1}{\sqrt{5}}\frac{4}{5} - \frac{2}{\sqrt{5}}\frac{3}{5} \\ &= -\frac{2}{5\sqrt{5}} = -\frac{2\sqrt{5}}{25}. \end{aligned}$$

EXAMPLE 3 If $-1 \le x \le 1$, rewrite $\cos(\sin^{-1} x)$ as an algebraic expression in x.

SOLUTION Let

$$y = \sin^{-1} x, \quad \text{or, equivalently,} \quad \sin y = x.$$

We wish to express $\cos y$ in terms of x. Since $-\pi/2 \le y \le \pi/2$, it follows that $\cos y \ge 0$, and hence

$$\cos y = \sqrt{1 - \sin^2 y} = \sqrt{1 - x^2}.$$

Consequently $$\cos(\sin^{-1} x) = \sqrt{1 - x^2}.$$

$y = \sin^{-1} x$
FIGURE 8.11

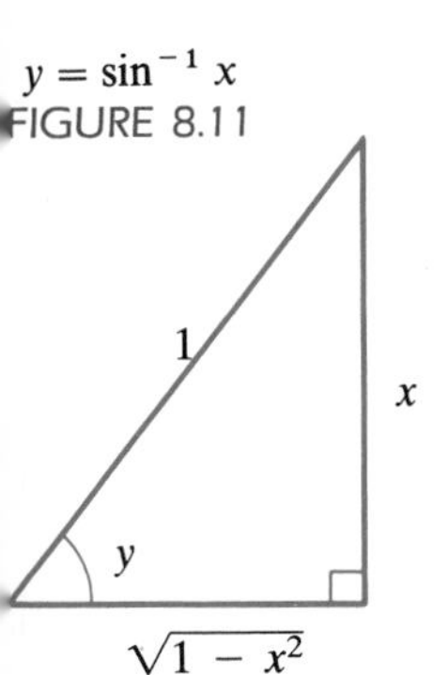

The last identity can also be seen geometrically if $0 < x < 1$. In this case $0 < y < \pi/2$, and we may regard y as the radian measure of an angle of a right triangle such that $\sin y = x$, as illustrated in Figure 8.11. (The side of length $\sqrt{1 - x^2}$ is found by using the Pythagorean theorem.) Referring to the triangle, we have

$$\cos(\sin^{-1} x) = \cos y = \frac{\sqrt{1 - x^2}}{1} = \sqrt{1 - x^2}.$$

EXAMPLE 4 Find the solutions of $5 \sin^2 t + 3 \sin t - 1 = 0$ that are in the interval $[-\pi/2, \pi/2]$.

SOLUTION The equation may be regarded as a quadratic equation in $\sin t$. Applying the quadratic formula yields

$$\sin t = \frac{-3 \pm \sqrt{9 + 20}}{10} = \frac{-3 \pm \sqrt{29}}{10}.$$

Using the definition of the inverse sine function, we obtain the following solutions:

$$t = \sin^{-1} \tfrac{1}{10}(-3 + \sqrt{29}) \approx 0.2408$$

$$t = \sin^{-1} \tfrac{1}{10}(-3 - \sqrt{29}) \approx -0.9946$$

EXERCISES 8.1

Exer. 1–18: Find the exact value of the expression, whenever it is defined.

1 (a) $\sin^{-1}\left(-\frac{\sqrt{2}}{2}\right)$ (b) $\cos^{-1}\left(-\frac{1}{2}\right)$
(c) $\tan^{-1}(-\sqrt{3})$

2 (a) $\sin^{-1}\left(-\frac{1}{2}\right)$ (b) $\cos^{-1}\left(-\frac{\sqrt{2}}{2}\right)$
(c) $\tan^{-1}(-1)$

3 (a) $\arcsin \frac{\sqrt{3}}{2}$ (b) $\arccos \frac{\sqrt{2}}{2}$
(c) $\arctan \frac{1}{\sqrt{3}}$

4 (a) $\arcsin 0$ (b) $\arccos(-1)$
(c) $\arctan 0$

5 (a) $\sin^{-1}\frac{\pi}{3}$ (b) $\cos^{-1}\frac{\pi}{2}$
(c) $\tan^{-1} 1$

6 (a) $\arcsin \frac{\pi}{2}$ (b) $\arccos \frac{\pi}{3}$
(c) $\arctan\left(-\frac{\sqrt{3}}{3}\right)$

7 (a) $\sin\left[\arcsin\left(-\frac{3}{10}\right)\right]$ (b) $\cos\left(\arccos \frac{1}{2}\right)$
(c) $\tan(\arctan 14)$

8 (a) $\sin\left(\sin^{-1}\frac{2}{3}\right)$ (b) $\cos\left[\cos^{-1}\left(-\frac{1}{5}\right)\right]$
(c) $\tan[\tan^{-1}(-9)]$

9 (a) $\sin^{-1}\left(\sin \frac{\pi}{3}\right)$ (b) $\cos^{-1}\left(\cos \frac{5\pi}{6}\right)$
(c) $\tan^{-1}\left[\tan\left(-\frac{\pi}{6}\right)\right]$

10 (a) $\arcsin\left[\sin\left(-\frac{\pi}{2}\right)\right]$ (b) $\arccos(\cos 0)$
(c) $\arctan\left(\tan \frac{\pi}{4}\right)$

11 (a) $\arcsin\left(\sin \frac{5\pi}{4}\right)$ (b) $\arccos\left(\cos \frac{5\pi}{4}\right)$
(c) $\arctan\left(\tan \frac{7\pi}{4}\right)$

12 (a) $\sin^{-1}\left(\sin \frac{2\pi}{3}\right)$ (b) $\cos^{-1}\left(\cos \frac{4\pi}{3}\right)$
(c) $\tan^{-1}\left(\tan \frac{7\pi}{6}\right)$

13 (a) $\sin\left[\cos^{-1}\left(-\frac{1}{2}\right)\right]$ (b) $\cos(\tan^{-1} 1)$
(c) $\tan[\sin^{-1}(-1)]$

14 (a) $\sin(\tan^{-1}\sqrt{3})$ (b) $\cos(\sin^{-1} 1)$
(c) $\tan(\cos^{-1} 0)$

15 (a) $\cot\left(\sin^{-1}\frac{2}{3}\right)$ (b) $\sec\left[\tan^{-1}\left(-\frac{3}{5}\right)\right]$
(c) $\csc\left[\cos^{-1}\left(-\frac{1}{4}\right)\right]$

16 (a) $\cot\left[\sin^{-1}\left(-\frac{2}{5}\right)\right]$ (b) $\sec\left(\tan^{-1}\frac{7}{4}\right)$
(c) $\csc\left(\cos^{-1}\frac{1}{5}\right)$

17 (a) $\sin\left(\arcsin \frac{1}{2} + \arccos 0\right)$
(b) $\cos\left[\arctan\left(-\frac{3}{4}\right) - \arcsin \frac{4}{5}\right]$
(c) $\tan\left(\arctan \frac{4}{3} + \arccos \frac{8}{17}\right)$

18 (a) $\sin\left[\sin^{-1}\frac{5}{13} - \cos^{-1}\left(-\frac{3}{5}\right)\right]$
(b) $\cos\left(\sin^{-1}\frac{4}{5} + \tan^{-1}\frac{3}{4}\right)$
(c) $\tan\left[\cos^{-1}\frac{1}{2} - \sin^{-1}\left(-\frac{1}{2}\right)\right]$

Exer. 19–22: Rewrite as an algebraic expression in x for $x > 0$.

19 $\sin(\tan^{-1} x)$ 20 $\tan(\arccos x)$

21 $\sec\left(\sin^{-1}\frac{x}{3}\right)$ 22 $\cot\left(\sin^{-1}\frac{1}{x}\right)$

Exer. 23–30: Sketch the graph of the equation.

23 $y = \sin^{-1} 2x$ 24 $y = \frac{1}{2}\sin^{-1} x$

25 $y = \cos^{-1}\frac{1}{2}x$ 26 $y = 2\cos^{-1} x$

27 $y = 2\tan^{-1} x$ 28 $y = \tan^{-1} 2x$

29 $y = \sin(\arccos x)$ 30 $y = \sin(\sin^{-1} x)$

31 (a) Define $\cot^{-1}$ by restricting the domain of the cotangent function to the interval $(0, \pi)$.
(b) Sketch the graph of $y = \cot^{-1} x$.

32 (a) Define $\csc^{-1}$ by restricting the domain of the cosecant function to $[-\pi/2, 0) \cup (0, \pi/2]$.

(b) Sketch the graph of $y = \csc^{-1} x$.

Exer. 33–36: (a) Use inverse trigonometric functions to find the solutions of the equation that are in the given interval. (b) Approximate the solutions to four decimal places.

33 $2 \tan^2 t + 9 \tan t + 3 = 0$; $(-\pi/2, \pi/2)$

34 $3 \sin^2 t + 7 \sin t + 3 = 0$; $[-\pi/2, \pi/2]$

35 $15 \cos^4 x - 14 \cos^2 x + 3 = 0$; $[0, \pi]$

36 $3 \tan^4 \theta - 19 \tan^2 \theta + 2 = 0$; $(-\pi/2, \pi/2)$

37 As shown in the figure, a sailboat is following a straight-line course l. The shortest distance from a tracking station T to the course is d miles. As the boat sails, the tracking station records its distance k from T and its direction θ with respect to T. Angle α specifies the direction of the sailboat.

(a) Express α in terms of d, k, and θ.

(b) Estimate α to the nearest degree if $d = 50$ mi and $k = 210$ mi and $\theta = 53.4°$.

EXERCISE 37

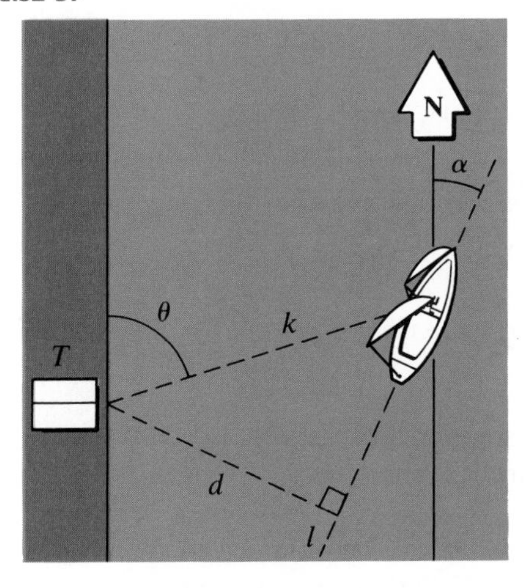

38 An art critic whose eye level is 6 feet above the floor views a painting that is 10 feet in height and is mounted 4 feet above the floor, as shown in the figure.

(a) If the critic is standing x feet from the wall, express the viewing angle θ in terms of x.

(b) Use the addition formula for the tangent to show that $\theta = \tan^{-1}\left(\dfrac{10x}{x^2 - 16}\right)$.

(c) For what value of x is $\theta = 45°$?

EXERCISE 38

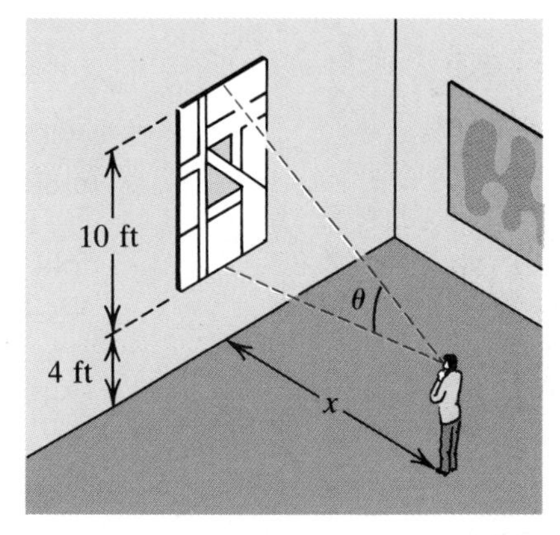

39 The only inverse trigonometric function available in some computer languages is $\tan^{-1}$. In BASIC, this function is denoted by ATN(X). Express the following in terms of $\tan^{-1}$.

(a) $\sin^{-1} x$ for $|x| < 1$

(b) $\cos^{-1} x$ for $|x| < 1$ and $x \neq 0$

40 If $-1 \le x \le 1$, is it always possible to find $\sin^{-1}(\sin^{-1} x)$ by pressing the calculator key sequence INV SIN twice? If not, determine the permissible values of x.

8.2 DERIVATIVES AND INTEGRALS

In this section we shall concentrate on the inverse sine, cosine, tangent, and secant functions. Formulas for their derivatives and for integrals that result in inverse trigonometric functions are listed in the next two theorems, with $u = g(x)$ differentiable and x restricted to values for which the indicated expressions have meaning. You may find it surprising to learn that although we used trigonometric functions to define inverse trigonometric functions, their derivatives are *algebraic* functions.

Theorem **(8.8)**

(i) $D_x \sin^{-1} u = \dfrac{1}{\sqrt{1 - u^2}} D_x u$

(ii) $D_x \cos^{-1} u = -\dfrac{1}{\sqrt{1 - u^2}} D_x u$

(iii) $D_x \tan^{-1} u = \dfrac{1}{1 + u^2} D_x u$

(iv) $D_x \sec^{-1} u = \dfrac{1}{u\sqrt{u^2 - 1}} D_x u$

PROOF We shall consider only the special case $u = x$, since the formulas for $u = g(x)$ may then be obtained by applying the chain rule.

If we let $f(x) = \sin x$ and $g(x) = \sin^{-1} x$ in Theorem (7.7), then it follows that the inverse sine function g is differentiable if $|x| < 1$. We shall use implicit differentiation to find $g'(x)$. First note that the equations

$$y = \sin^{-1} x \quad \text{and} \quad \sin y = x$$

are equivalent if $-1 < x < 1$ and $-\pi/2 < y < \pi/2$. Differentiating $\sin y = x$ implicitly, we have

$$\cos y \, D_x y = 1$$

and hence
$$D_x \sin^{-1} x = D_x y = \frac{1}{\cos y}.$$

Since $-\pi/2 < y < \pi/2$, $\cos y$ is positive and, therefore,

$$\cos y = \sqrt{1 - \sin^2 y} = \sqrt{1 - x^2}.$$

Thus,
$$D_x \sin^{-1} x = \frac{1}{\sqrt{1 - x^2}}$$

for $|x| < 1$. The inverse sine function is not differentiable at ± 1. This fact is evident from Figure 8.2, since vertical tangent lines occur at the endpoints of the graph.

The formula for $D_x \cos^{-1} x$ can be obtained in similar fashion.

It follows from Theorem (7.7) that the inverse tangent function is differentiable at every real number. Let us consider the equivalent equations

$$y = \tan^{-1} x \quad \text{and} \quad \tan y = x$$

for $-\pi/2 < y < \pi/2$. Differentiating $\tan y = x$ implicitly, we have

$$\sec^2 y \, D_x y = 1.$$

Consequently,
$$D_x \tan^{-1} x = D_x y = \frac{1}{\sec^2 y}.$$

Using the fact that $\sec^2 y = 1 + \tan^2 y = 1 + x^2$ gives us

$$D_x \tan^{-1} x = \frac{1}{1 + x^2}.$$

Finally, consider the equivalent equations

$$y = \sec^{-1} x \quad \text{and} \quad \sec y = x$$

for y in either $(0, \pi/2)$ or $(\pi, 3\pi/2)$. Differentiating $\sec y = x$ implicitly yields

$$\sec y \tan y \, D_x y = 1.$$

Since $0 < y < \pi/2$ or $\pi < y < 3\pi/2$, it follows that $\sec y \tan y \neq 0$ and, hence,

$$D_x \sec^{-1} x = D_x y = \frac{1}{\sec y \tan y}.$$

Using the fact that $\tan y = \sqrt{\sec^2 y - 1} = \sqrt{x^2 - 1}$, we obtain

$$D_x \sec^{-1} x = \frac{1}{x\sqrt{x^2 - 1}}$$

for $|x| > 1$. The inverse secant function is not differentiable at $x = \pm 1$. Note that the graph has vertical tangent lines at the points with these x-coordinates (see Figure 8.9).

ILLUSTRATION

$f(x)$	$f'(x)$
$\sin^{-1} 3x$	$\dfrac{1}{\sqrt{1-(3x)^2}} D_x(3x) = \dfrac{3}{\sqrt{1-9x^2}}$
$\arccos(\ln x)$	$-\dfrac{1}{\sqrt{1-(\ln x)^2}} D_x \ln x = -\dfrac{1}{x\sqrt{1-(\ln x)^2}}$
$\tan^{-1} e^{2x}$	$\dfrac{1}{1+(e^{2x})^2} D_x e^{2x} = \dfrac{2e^{2x}}{1+e^{4x}}$
$\operatorname{arcsec}(x^2)$	$\dfrac{1}{x^2\sqrt{(x^2)^2-1}} D_x(x^2) = \dfrac{2}{x\sqrt{x^4-1}}$

FIGURE 8.12

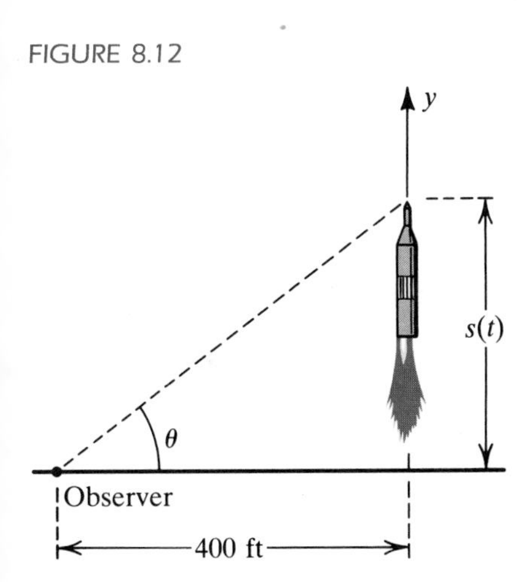

EXAMPLE 1 A rocket is fired directly upward with initial velocity 0 and burns fuel at a rate that produces a constant acceleration of 50 ft/sec^2 for $0 \leq t \leq 5$, with time t in seconds. As illustrated in Figure 8.12, an observer 400 feet from the launching pad visually follows the flight of the rocket.

(a) Express the angle of elevation θ of the rocket as a function of t.

(b) The observer perceives the rocket to be rising fastest when $d\theta/dt$ is largest. (Of course, this is an illusion, since the velocity is steadily increasing.) Determine the height of the rocket at the moment of perceived maximum velocity.

SOLUTION

(a) Let $s(t)$ denote the height of the rocket at time t (see Figure 8.12). The fact that the acceleration is always 50 gives us the differential equation

$$s''(t) = 50,$$

subject to the initial conditions $s'(0) = 0$ and $s(0) = 0$. Integrating with respect to t, we obtain

$$\int s''(t)\,dt = \int 50\,dt$$

$$s'(t) = 50t + C$$

for some constant C. Substituting $t = 0$ and using $s'(0) = 0$ gives us $0 = 50(0) + C$, or $C = 0$. Hence

$$s'(t) = 50t.$$

Integrating again, we have

$$\int s'(t)\,dt = \int 50t\,dt$$

$$s(t) = 25t^2 + D$$

for some constant D. If we substitute $t = 0$ and use $s(0) = 0$, we obtain $0 = 25(0) + D$, or $D = 0$. Hence

$$s(t) = 25t^2.$$

Referring to Figure 8.12, with $s(t) = 25t^2$, we find

$$\tan\theta = \frac{25t^2}{400} = \frac{t^2}{16}, \quad \text{or} \quad \theta = \arctan\frac{t^2}{16}.$$

(b) By Theorem (8.8), the rate of change of θ with respect to t is

$$\frac{d\theta}{dt} = \frac{1}{1 + (t^2/16)^2}\left(\frac{2t}{16}\right) = \frac{32t}{256 + t^4}.$$

Since we wish to find the maximum value of $d\theta/dt$, we begin by finding the critical numbers of $d\theta/dt$. Using the quotient rule, we obtain

$$\frac{d}{dt}\left(\frac{d\theta}{dt}\right) = \frac{d^2\theta}{dt^2} = \frac{(256 + t^4)(32) - 32t(4t^3)}{(256 + t^4)^2} = \frac{32(256 - 3t^4)}{(256 + t^4)^2}.$$

Considering $d^2\theta/dt^2 = 0$ gives us the critical number $t = \sqrt[4]{256/3}$. It follows from the first (or second) derivative test that $d\theta/dt$ has a maximum value at $t = \sqrt[4]{256/3} \approx 3.04$ sec. The height of the rocket at this time is

$$s(\sqrt[4]{256/3}) = 25(\sqrt[4]{256/3})^2 = 25\sqrt{256/3} \approx 230.9 \text{ ft.}$$

We may use differentiation formulas (i), (ii), and (iv) of Theorem (8.7) to obtain the following integration formulas:

(1) $$\int \frac{1}{\sqrt{1 - u^2}}\,du = \sin^{-1} u + C$$

(2) $$\int \frac{1}{1 + u^2}\,du = \tan^{-1} u + C$$

(3) $$\int \frac{1}{u\sqrt{u^2 - 1}}\,du = \sec^{-1} u + C$$

These formulas can be generalized as follows for $a > 0$.

Theorem **(8.9)**

(i) $$\int \frac{1}{\sqrt{a^2 - u^2}}\,du = \sin^{-1}\frac{u}{a} + C$$

(ii) $$\int \frac{1}{a^2 + u^2}\,du = \frac{1}{a}\tan^{-1}\frac{u}{a} + C$$

(iii) $$\int \frac{1}{u\sqrt{u^2 - a^2}}\,du = \frac{1}{a}\sec^{-1}\frac{u}{a} + C$$

PROOF Let us prove (ii). As usual, it is sufficient to consider the case $u = x$. We begin by writing

$$\int \frac{1}{a^2 + x^2}\,dx = \frac{1}{a^2}\int \frac{1}{1 + (x/a)^2}\,dx.$$

Next we make the substitution $v = x/a$, $dv = (1/a)\,dx$. Introducing the factor $1/a$ in the integrand, compensating by multiplying the integral by a, and using formula (2), preceding this theorem, gives us the following:

$$\begin{aligned}
\int \frac{1}{a^2 + x^2}\,dx &= \frac{1}{a}\int \frac{1}{1 + (x/a)^2}\cdot\frac{1}{a}\,dx \\
&= \frac{1}{a}\int \frac{1}{1 + v^2}\,dv \\
&= \frac{1}{a}\tan^{-1} v + C \\
&= \frac{1}{a}\tan^{-1}\frac{x}{a} + C
\end{aligned}$$

The remaining formulas may be proved in similar fashion. ■

EXAMPLE 2 Evaluate $\displaystyle\int \frac{e^{2x}}{\sqrt{1 - e^{4x}}}\,dx.$

SOLUTION The integral may be written as in the first formula of Theorem (8.9) by letting $a = 1$ and using the substitution

$$u = e^{2x}, \qquad du = 2e^{2x}\,dx.$$

We introduce a factor 2 in the integrand and proceed as follows:

$$\begin{aligned}
\int \frac{e^{2x}}{\sqrt{1 - e^{4x}}}\,dx &= \frac{1}{2}\int \frac{1}{\sqrt{1 - (e^{2x})^2}}\,2e^{2x}\,dx \\
&= \frac{1}{2}\int \frac{1}{\sqrt{1 - u^2}}\,du \\
&= \tfrac{1}{2}\sin^{-1} u + C \\
&= \tfrac{1}{2}\sin^{-1} e^{2x} + C
\end{aligned}$$

EXAMPLE 3 Evaluate $\int \frac{x^2}{5 + x^6}\,dx$.

SOLUTION The integral may be written as in the second formula of Theorem (8.9) by letting $a^2 = 5$ and using the substitution

$$u = x^3, \qquad du = 3x^2\,dx.$$

We introduce a factor 3 in the integrand and proceed as follows:

$$\begin{aligned}\int \frac{x^2}{5 + x^6}\,dx &= \frac{1}{3}\int \frac{1}{5 + (x^3)^2}\,3x^2\,dx \\ &= \frac{1}{3}\int \frac{1}{(\sqrt{5})^2 + u^2}\,du \\ &= \frac{1}{3}\cdot\frac{1}{\sqrt{5}}\tan^{-1}\frac{u}{\sqrt{5}} + C \\ &= \frac{\sqrt{5}}{15}\tan^{-1}\frac{x^3}{\sqrt{5}} + C\end{aligned}$$

EXAMPLE 4 Evaluate $\int \frac{1}{x\sqrt{x^4 - 9}}\,dx$.

SOLUTION The integral may be written as in Theorem (8.9)(iii) by letting $a^2 = 9$ and using the substitution

$$u = x^2, \qquad du = 2x\,dx.$$

We introduce $2x$ in the integrand by multiplying numerator and denominator by $2x$ and then proceed as follows:

$$\begin{aligned}\int \frac{1}{x\sqrt{x^4 - 9}}\,dx &= \int \frac{1}{2x\cdot x\sqrt{(x^2)^2 - 3^2}}\,2x\,dx \\ &= \frac{1}{2}\int \frac{1}{u\sqrt{u^2 - 3^2}}\,du \\ &= \frac{1}{2}\cdot\frac{1}{3}\sec^{-1}\frac{u}{3} + C \\ &= \frac{1}{6}\sec^{-1}\frac{x^2}{3} + C\end{aligned}$$

EXERCISES 8.2

Exer. 1–26: Find $f'(x)$ if $f(x)$ is the given expression.

1 $\sin^{-1}\sqrt{x}$

2 $\sin^{-1}\frac{1}{3}x$

3 $\tan^{-1}(3x - 5)$

4 $\tan^{-1}(x^2)$

5 $e^{-x}\,\mathrm{arcsec}\,e^{-x}$

6 $\sqrt{\mathrm{arcsec}\,3x}$

7 $x^2\arctan(x^2)$

8 $\tan^{-1}\sin 2x$

9 $\sec^{-1}\sqrt{x^2 - 1}$

10 $x^2\sec^{-1}5x$

11 $\frac{1}{\sin^{-1}x}$

12 $\arcsin\ln x$

13 $(1 + \cos^{-1}3x)^3$

14 $\cos^{-1}\cos e^x$

15 $\ln\arctan(x^2)$

16 $\arctan\frac{x + 1}{x - 1}$

17 $\cos(x^{-1}) + (\cos x)^{-1} + \cos^{-1} x$

18 $x \arccos \sqrt{4x+1}$

19 $3^{\arcsin(x^3)}$

20 $\left(\dfrac{1}{x} - \arcsin \dfrac{1}{x}\right)^4$

21 $\dfrac{\arctan x}{x^2+1}$

22 $\dfrac{e^{2x}}{\sin^{-1} 5x}$

23 $\sqrt{x}\sec^{-1}\sqrt{x}$

24 $(\sin 2x)(\sin^{-1} 2x)$

25 $(\tan x)^{\arctan x}$

26 $(\tan^{-1} 4x)e^{\tan^{-1} 4x}$

Exer. 27–28: Find y'.

27 $x^2 + x\sin^{-1} y = ye^x$

28 $\ln(x+y) = \tan^{-1} xy$

Exer. 29–44: Evaluate the integral.

29 (a) $\displaystyle\int \frac{1}{x^2+16}\,dx$ (b) $\displaystyle\int_0^4 \frac{1}{x^2+16}\,dx$

30 (a) $\displaystyle\int \frac{e^x}{1+e^{2x}}\,dx$ (b) $\displaystyle\int_0^1 \frac{e^x}{1+e^{2x}}\,dx$

31 (a) $\displaystyle\int \frac{x}{\sqrt{1-x^4}}\,dx$ (b) $\displaystyle\int_0^{\sqrt{2}/2} \frac{x}{\sqrt{1-x^4}}\,dx$

32 (a) $\displaystyle\int \frac{1}{x\sqrt{x^2-1}}\,dx$ (b) $\displaystyle\int_{2/\sqrt{3}}^2 \frac{1}{x\sqrt{x^2-1}}\,dx$

33 $\displaystyle\int \frac{\sin x}{\cos^2 x + 1}\,dx$

34 $\displaystyle\int \frac{\cos x}{\sqrt{9-\sin^2 x}}\,dx$

35 $\displaystyle\int \frac{1}{\sqrt{x}(1+x)}\,dx$

36 $\displaystyle\int \frac{1}{e^x\sqrt{1-e^{-2x}}}\,dx$

37 $\displaystyle\int \frac{e^x}{\sqrt{16-e^{2x}}}\,dx$

38 $\displaystyle\int \frac{\sec x \tan x}{1+\sec^2 x}\,dx$

39 $\displaystyle\int \frac{1}{x\sqrt{x^6-4}}\,dx$

40 $\displaystyle\int \frac{x}{\sqrt{36-x^2}}\,dx$

41 $\displaystyle\int \frac{x}{x^2+9}\,dx$

42 $\displaystyle\int \frac{1}{x\sqrt{x-1}}\,dx$

43 $\displaystyle\int \frac{1}{\sqrt{e^{2x}-25}}\,dx$

44 $\displaystyle\int \frac{e^x}{\sqrt{4-e^x}}\,dx$

45 The floor of a storage shed has the shape of a right triangle. The sides opposite and adjacent to an acute angle θ of the triangle are measured as 10 feet and 7 feet, respectively, with a possible error of ± 0.5 inch in the 10-foot measurement. Use the differential of an inverse trigonometric function to approximate the error in the calculated value of θ.

46 Use differentials to approximate the arc length of the graph of $y = \tan^{-1} x$ from $A(0, 0)$ to $B(0.1, \tan^{-1} 0.1)$.

47 An airplane at a constant altitude of 5 miles and a speed of 500 mi/hr is flying in a direction away from an observer on the ground. Use inverse trigonometric functions to find the rate at which the angle of elevation is changing when the airplane flies over a point 2 miles from the observer.

48 A searchlight located $\frac{1}{8}$ mile from the nearest point P on a straight road is trained on an automobile traveling on the road at a rate of 50 mi/hr. Use inverse trigonometric functions to find the rate at which the searchlight is rotating when the car is $\frac{1}{4}$ mile from P.

49 A billboard 20 feet high is located on top of a building, with its lower edge 60 feet above the level of a viewer's eye. Use inverse trigonometric functions to find how far from a point directly below the sign a viewer should stand to maximize the angle between the lines of sight of the top and bottom of the billboard (see Example 8 of Section 4.5).

50 The velocity, at time t, of a point moving on a coordinate line is $(1+t^2)^{-1}$ ft/sec. If the point is at the origin at $t = 0$, find its position at the instant that the acceleration and the velocity have the same absolute value.

51 A missile is fired vertically from a point that is 5 miles from a tracking station and at the same elevation. For the first 20 seconds of flight, its angle of elevation changes at a constant rate of 2° per second. Use inverse trigonometric functions to find the velocity of the missile when the angle of elevation is 30°.

52 Blood flowing through a blood vessel causes a loss of energy due to friction. According to *Poiseuille's law*, this energy loss E is given by $E = kl/r^4$, where r is the radius of the blood vessel, l is the length, and k is a constant. Suppose a blood vessel of radius r_2 and length l_2 branches off, at an angle θ, from a blood vessel of radius r_1 and length l_1, as illustrated in the figure, where the white arrows indicate the direction of blood flow. The energy loss is then the sum of the individual energy losses; that is,

$$E = \frac{kl_1}{r_1^4} + \frac{kl_2}{r_2^4}.$$

Express l_1 and l_2 in terms of a, b, and θ, and find the angle that minimizes the energy loss.

EXERCISE 52

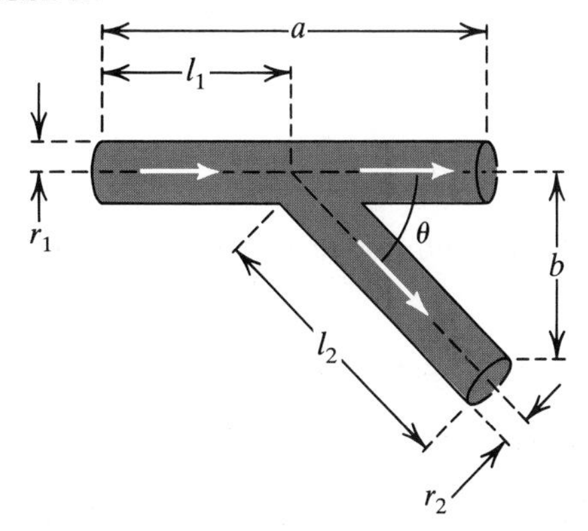

c 53 Use Simpson's rule, with $n = 4$, to approximate the arc length of the graph of $y = \arcsin x$ from $A(0, 0)$ to $B(1/2, \pi/6)$.

c 54 The graph of $y = 4 \arctan (x^2)$ from $A(0, 0)$ to $B(1, \pi)$ is revolved about the x-axis. Use the trapezoidal rule, with $n = 8$, to approximate the area of the resulting surface.

8.3 HYPERBOLIC FUNCTIONS

The exponential expressions

$$\frac{e^x - e^{-x}}{2} \quad \text{and} \quad \frac{e^x + e^{-x}}{2}$$

occur in advanced applications of calculus. Their properties are similar in many ways to those of $\sin x$ and $\cos x$. Later in our discussion, we shall see why they are called the *hyperbolic sine* and the *hyperbolic cosine* of x.

Definition **(8.10)**

The **hyperbolic sine function**, denoted by **sinh**, and the **hyperbolic cosine function**, denoted by **cosh**, are defined by

$$\sinh x = \frac{e^x - e^{-x}}{2} \quad \text{and} \quad \cosh x = \frac{e^x + e^{-x}}{2}$$

for every real number x.

We pronounce $\sinh x$ and $\cosh x$ as *sinch* x and *kosh* x, respectively.

The graph of $y = \cosh x$ may be found by **addition of y-coordinates.** Noting that $\cosh x = \frac{1}{2}e^x + \frac{1}{2}e^{-x}$, we first sketch the graphs of $y = \frac{1}{2}e^x$ and $y = \frac{1}{2}e^{-x}$ on the same coordinate plane, as shown with dashes in Figure 8.13. We then add the y-coordinates of points on these graphs to obtain the graph of $y = \cosh x$. Note that the range of cosh is $[1, \infty)$.

FIGURE 8.13 FIGURE 8.14

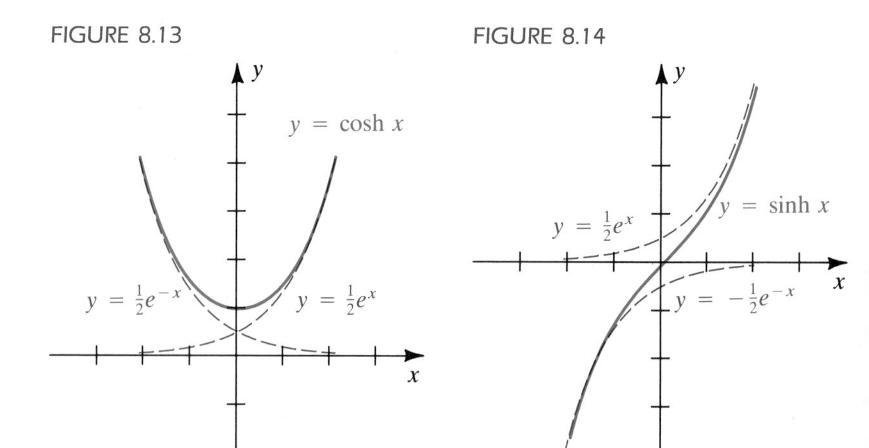

We may find the graph of $y = \sinh x$ by adding y-coordinates of the graphs of $y = \frac{1}{2}e^x$ and $y = -\frac{1}{2}e^{-x}$, as shown in Figure 8.14.

Some scientific calculators have keys that can be used to find values of sinh and cosh directly. We can also substitute numbers for x in Definition (8.10), as in the following illustration.

ILLUSTRATION

■ $\sinh 2 = \dfrac{e^2 - e^{-2}}{2} \approx 3.63$ ■ $\cosh 0.5 = \dfrac{e^{0.5} + e^{-0.5}}{2} \approx 1.13$

FIGURE 8.15

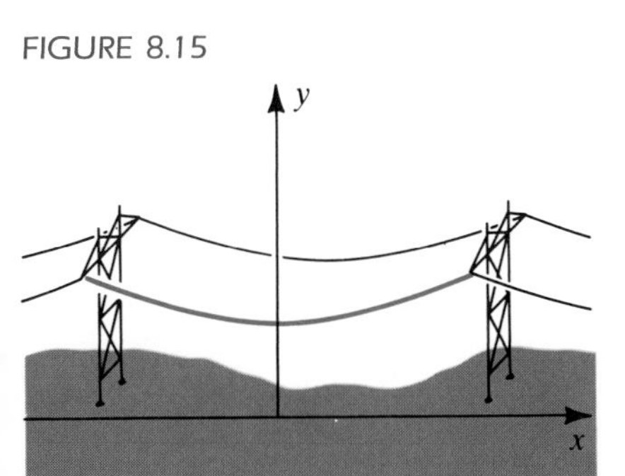

The hyperbolic cosine function can be used to describe the shape of a uniform flexible cable, or chain, whose ends are supported from the same height. As illustrated in Figure 8.15, telephone or power lines may be strung between poles in this manner. The shape of the cable appears to be a parabola, but is actually a **catenary** (after the Latin word for *chain*). If we introduce a coordinate system, as in Figure 8.15 it can be shown that an equation corresponding to the shape of the cable is $y = a \cosh (x/a)$ for some real number a.

The hyperbolic cosine function also occurs in the analysis of motion in a resisting medium. If an object is dropped from a given height and if air resistance is disregarded, then the distance y that it falls in t seconds is $y = \frac{1}{2}gt^2$, where g is a gravitational constant. However, air resistance cannot always be disregarded. As the velocity of the object increases, air resistance may significantly affect its motion. For example, if the air resistance is directly proportional to the square of the velocity, then the distance y that the object falls in t seconds is given by

$$y = A \ln (\cosh Bt)$$

for constants A and B. Another application is given in Example 2 of this section.

Many identities similar to those for trigonometric functions hold for the hyperbolic sine and cosine functions. For example, if $\cosh^2 x$ and $\sinh^2 x$ denote $(\cosh x)^2$ and $(\sinh x)^2$, respectively, we have the following identity.

Theorem **(8.11)**

$$\cosh^2 x - \sinh^2 x = 1$$

PROOF By Definition (8.10),

$$\begin{aligned}\cosh^2 x - \sinh^2 x &= \left(\frac{e^x + e^{-x}}{2}\right)^2 - \left(\frac{e^x - e^{-x}}{2}\right)^2 \\ &= \frac{e^{2x} + 2 + e^{-2x}}{4} - \frac{e^{2x} - 2 + e^{-2x}}{4} \\ &= \frac{e^{2x} + 2 + e^{-2x} - e^{2x} + 2 - e^{-2x}}{4} \\ &= \tfrac{4}{4} = 1. \quad ■\end{aligned}$$

Theorem (8.11) is analogous to the trigonometric identity $\cos^2 x + \sin^2 x = 1$. Other hyperbolic identities are stated in the exercises. To verify an identity, it is sufficient to express the hyperbolic functions in terms of exponential functions and show that one side of the equation can be transformed into the other as illustrated in the proof of Theorem (8.11). The hyperbolic identities are similar to (but not always the same as) certain trigonometric identities—differences usually involve signs of terms.

FIGURE 8.16
$x^2 + y^2 = 1$

FIGURE 8.17
$x^2 - y^2 = 1$

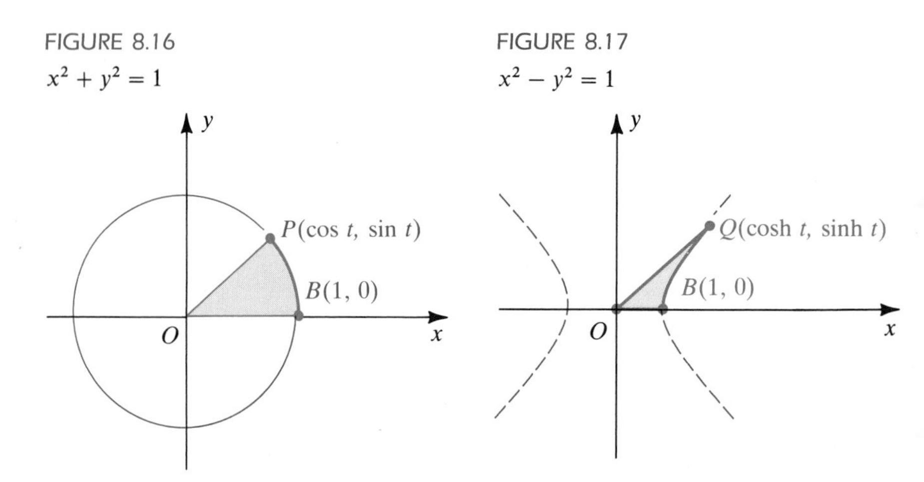

If t is a real number, there is an interesting geometric relationship between the points $P(\cos t, \sin t)$ and $Q(\cosh t, \sinh t)$ in a coordinate plane. Let us consider the graphs of $x^2 + y^2 = 1$ and $x^2 - y^2 = 1$, sketched in Figures 8.16 and 8.17. The graph in Figure 8.16 is the unit circle with center at the origin. The graph in Figure 8.17 is a *hyperbola*. (Hyperbolas and their properties will be discussed in detail in Chapter 12.) Note first that since $\cos^2 t + \sin^2 t = 1$, the point $P(\cos t, \sin t)$ is on the circle $x^2 + y^2 = 1$. Next, by Theorem (8.11), $\cosh^2 t - \sinh^2 t = 1$, and hence the point $Q(\cosh t, \sinh t)$ is on the hyperbola $x^2 - y^2 = 1$. These are the reasons for referring to cos and sin as *circular* functions and to cosh and sinh as *hyperbolic* functions.

The graphs in Figures 8.16 and 8.17 are related in another way. If $0 < t < \pi/2$, then t is the radian measure of angle POB, shown in Figure 8.16. By Theorem (1.15), the area A of the shaded circular sector is $A = \frac{1}{2}(1)^2 t = \frac{1}{2}t$, and hence $t = 2A$. Similarly, if $Q(\cosh t, \sinh t)$ is the point in Figure 8.17, then $t = 2A$ for the area A of the shaded hyperbolic sector (see Exercise 47).

The impressive analogies between the trigonometric and hyperbolic sine and cosine functions motivate us to define hyperbolic functions that correspond to the four remaining trigonometric functions. The **hyperbolic tangent**, **hyperbolic cotangent**, **hyperbolic secant**, and **hyperbolic cosecant functions**, denoted by tanh, coth, sech, and csch, respectively, are defined as follows.

Definition **(8.12)**

(i) $$\tanh x = \frac{\sinh x}{\cosh x} = \frac{e^x - e^{-x}}{e^x + e^{-x}}$$

(ii) $$\coth x = \frac{\cosh x}{\sinh x} = \frac{e^x + e^{-x}}{e^x - e^{-x}}, \quad x \neq 0$$

(iii) $$\operatorname{sech} x = \frac{1}{\cosh x} = \frac{2}{e^x + e^{-x}}$$

(iv) $$\operatorname{csch} x = \frac{1}{\sinh x} = \frac{2}{e^x - e^{-x}}, \quad x \neq 0$$

We pronounce the four function values in the preceding definition as *tansh* x, *cotansh* x, *setch* x, and *cosetch* x. Their graphs are sketched in Figure 8.18.

FIGURE 8.18

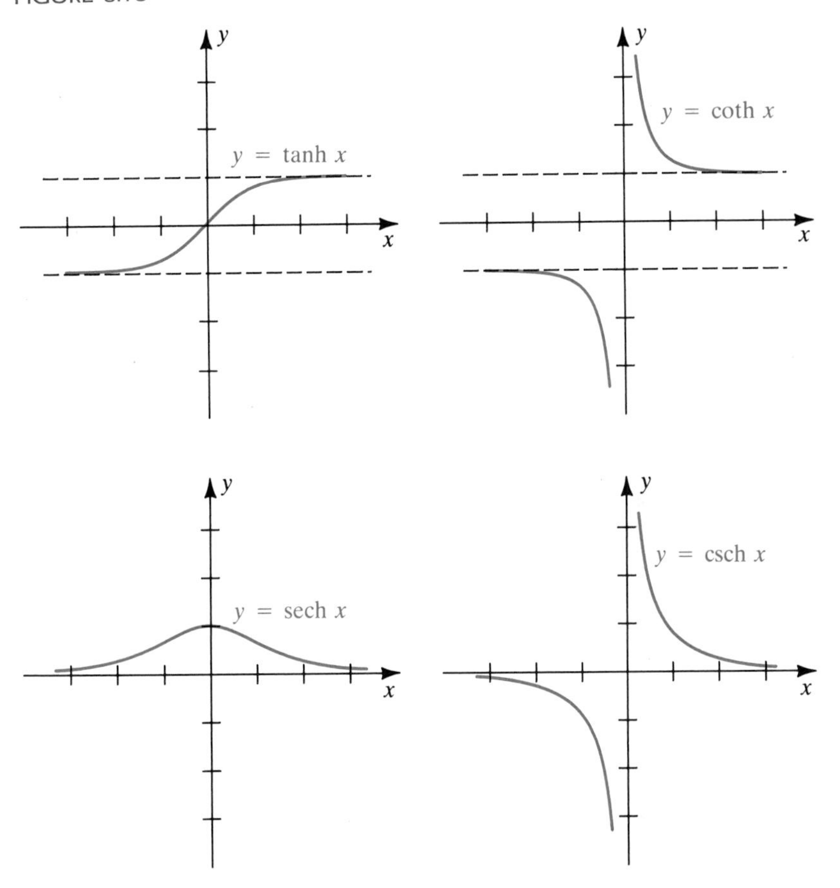

If we divide both sides of the identity $\cosh^2 x - \sinh^2 x = 1$ (see (8.11)) by $\cosh^2 x$, we obtain

$$\frac{\cosh^2 x}{\cosh^2 x} - \frac{\sinh^2 x}{\cosh^2 x} = \frac{1}{\cosh^2 x}.$$

Using the definitions of tanh x and sech x gives us (i) of the next theorem. Formula (ii) may be obtained by dividing both sides of (8.11) by $\sinh^2 x$.

Theorem **(8.13)**

(i) $1 - \tanh^2 x = \operatorname{sech}^2 x$ **(ii)** $\coth^2 x - 1 = \operatorname{csch}^2 x$

Note the similarities and differences between (8.13) and the analogous trigonometric identities.

Derivative formulas for the hyperbolic functions are listed in the next theorem, where $u = g(x)$ and g is differentiable.

Theorem **(8.14)**

(i) $D_x \sinh u = \cosh u \, D_x u$

(ii) $D_x \cosh u = \sinh u \, D_x u$

(iii) $D_x \tanh u = \operatorname{sech}^2 u \, D_x u$

(iv) $D_x \coth u = -\operatorname{csch}^2 u \, D_x u$

(v) $D_x \operatorname{sech} u = -\operatorname{sech} u \tanh u \, D_x u$

(vi) $D_x \operatorname{csch} u = -\operatorname{csch} u \coth u \, D_x u$

PROOF As usual, we consider only the case $u = x$. Since $D_x e^x = e^x$ and $D_x e^{-x} = -e^{-x}$,

$$D_x \sinh x = D_x\left(\frac{e^x - e^{-x}}{2}\right) = \frac{e^x + e^{-x}}{2} = \cosh x$$

and $$D_x \cosh x = D_x\left(\frac{e^x + e^{-x}}{2}\right) = \frac{e^x - e^{-x}}{2} = \sinh x.$$

To differentiate $\tanh x$, we apply the quotient rule as follows:

$$\begin{aligned} D_x \tanh x &= D_x \frac{\sinh x}{\cosh x} \\ &= \frac{\cosh x \, D_x \sinh x - \sinh x \, D_x \cosh x}{\cosh^2 x} \\ &= \frac{\cosh^2 x - \sinh^2 x}{\cosh^2 x} \\ &= \frac{1}{\cosh^2 x} = \operatorname{sech}^2 x \end{aligned}$$

The remaining formulas can be proved in similar fashion. ■

EXAMPLE 1 If $f(x) = \cosh(x^2 + 1)$, find $f'(x)$.

SOLUTION Applying Theorem (8.14)(i), with $u = x^2 + 1$, we obtain

$$\begin{aligned} f'(x) &= \sinh(x^2 + 1) \cdot D_x(x^2 + 1) \\ &= 2x \sinh(x^2 + 1). \end{aligned}$$

FIGURE 8.19

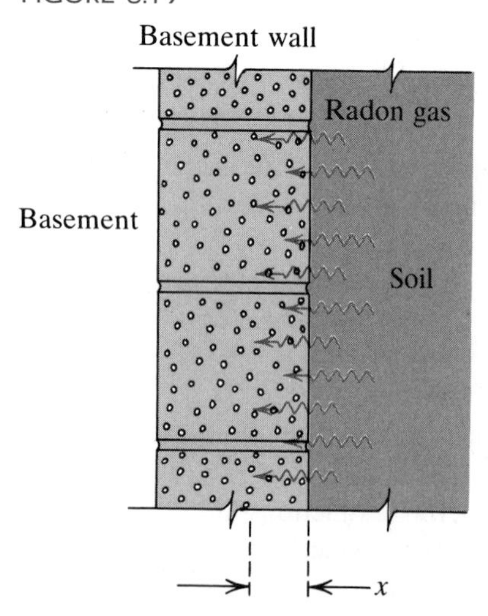

EXAMPLE 2 Radon gas can readily diffuse through solid materials such as brick and cement. If the direction of diffusion in a basement wall is perpendicular to the surface, as illustrated in Figure 8.19, then the radon concentration $f(x)$ (in joules/cm^3) in the air-filled pores within the wall at a distance x from the outside surface can be approximated by

$$f(x) = A \sinh(qx) + B \cosh(qx) + k,$$

where the constant q depends on the porosity of the wall, the half-life of radon, and a diffusion coefficient; the constant k is the maximum radon concentration in the air-filled pores; and A and B are constants that depend on initial conditions. Show that $y = f(x)$ is a solution of the *diffusion equation*

$$\frac{d^2y}{dx^2} - q^2y + q^2k = 0.$$

SOLUTION Differentiating $y = f(x)$ twice gives us

$$\frac{dy}{dx} = qA \cosh(qx) + qB \sinh(qx)$$

and

$$\frac{d^2y}{dx^2} = q^2A \sinh(qx) + q^2B \cosh(qx).$$

Since $y = A \sinh(qx) + B \cosh(qx) + k$, we have

$$q^2y = q^2A \sinh(qx) + q^2B \cosh(qx) + q^2k.$$

Subtracting the expressions for d^2y/dx^2 and q^2y yields

$$\frac{d^2y}{dx^2} - q^2y = -q^2k$$

and hence

$$\frac{d^2y}{dx^2} - q^2y + q^2k = 0.$$

The integration formulas that correspond to the derivative formulas in Theorem (8.14) are as follows.

Theorem **(8.15)**

(i) $\int \sinh u\, du = \cosh u + C$

(ii) $\int \cosh u\, du = \sinh u + C$

(iii) $\int \operatorname{sech}^2 u\, du = \tanh u + C$

(iv) $\int \operatorname{csch}^2 u\, du = -\coth u + C$

(v) $\int \operatorname{sech} u \tanh u\, du = -\operatorname{sech} u + C$

(vi) $\int \operatorname{csch} u \coth u\, du = -\operatorname{csch} u + C$

EXAMPLE 3 Evaluate $\int x^2 \sinh x^3\, dx$.

SOLUTION If we let $u = x^3$, then $du = 3x^2\, dx$ and

$$\begin{aligned}\int x^2 \sinh x^3\, dx &= \tfrac{1}{3}\int (\sinh x^3)3x^2\, dx\\ &= \tfrac{1}{3}\int \sinh u\, du = \tfrac{1}{3}\cosh u + C\\ &= \tfrac{1}{3}\cosh x^3 + C.\end{aligned}$$

EXERCISES 8.3

c **Exer. 1–2: Approximate to four decimal places.**

1 (a) $\sinh 4$ (b) $\cosh \ln 4$ (c) $\tanh(-3)$
(d) $\coth 10$ (e) $\operatorname{sech} 2$ (f) $\operatorname{csch}(-1)$

2 (a) $\sinh \ln 4$ (b) $\cosh 4$ (c) $\tanh 3$
(d) $\coth(-10)$ (e) $\operatorname{sech}(-2)$ (f) $\operatorname{csch} 1$

Exer. 3–26: Find $f'(x)$ if $f(x)$ is the given expression.

3 $\sinh 5x$ 4 $\sinh(x^2+1)$ 5 $\cosh(x^3)$

6 $\cosh^3 x$ 7 $\sqrt{x}\tanh\sqrt{x}$ 8 $\arctan\tanh x$

9 $\coth\dfrac{1}{x}$ 10 $\dfrac{\coth x}{\cot x}$ 11 $\dfrac{\operatorname{sech}(x^2)}{x^2+1}$

12 $\sqrt{\operatorname{sech} 5x}$ 13 $\operatorname{csch}^2 6x$ 14 $x\operatorname{csch} e^{4x}$

15 $\ln\sinh 2x$

16 $\sinh^2 3x$

17 $\cosh\sqrt{4x^2+3}$

18 $\dfrac{1+\cosh x}{1-\cosh x}$ 19 $\dfrac{1}{\tanh x+1}$ 20 $\ln|\tanh x|$

21 $\coth\ln x$ 22 $\coth^3 2x$ 23 $e^{3x}\operatorname{sech} x$

24 $\frac{1}{2}\operatorname{sech}(x^2+1)$

25 $\tan^{-1}(\operatorname{csch} x)$

26 $\operatorname{csch}\ln x$

Exer. 27–42: Evaluate the integral.

27 $\int x^2\cosh(x^3)\,dx$ 28 $\int \dfrac{1}{\operatorname{sech} 7x}\,dx$

29 $\int \dfrac{\sinh\sqrt{x}}{\sqrt{x}}\,dx$ 30 $\int x\sinh(2x^2)\,dx$

31 $\int \dfrac{1}{\cosh^2 3x}\,dx$ 32 $\int \operatorname{sech}^2(5x)\,dx$

33 $\int \operatorname{csch}^2(\frac{1}{2}x)\,dx$ 34 $\int (\sinh 4x)^{-2}\,dx$

35 $\int \tanh 3x\operatorname{sech} 3x\,dx$ 36 $\int \sinh x\operatorname{sech}^2 x\,dx$

37 $\int \cosh x\operatorname{csch}^2 x\,dx$ 38 $\int \coth 6x\operatorname{csch} 6x\,dx$

39 $\int \coth x\,dx$ 40 $\int \tanh x\,dx$

41 $\int \sinh x\cosh x\,dx$ 42 $\int \operatorname{sech} x\,dx$

43 Find the area of the region bounded by the graphs of $y=\sinh 3x$, $y=0$, and $x=1$.

44 Find the arc length of the graph of $y=\cosh x$ from $(0, 1)$ to $(1, \cosh 1)$.

45 Find the points on the graph of $y=\sinh x$ at which the tangent line has slope 2.

46 The region bounded by the graphs of $y=\cosh x$, $x=-1$, $x=1$, and $y=0$ is revolved about the x-axis. Find the volume of the resulting solid.

47 If A is the region shown in Figure 8.17, prove that $t=2A$.

48 Sketch the graph of $x^2-y^2=1$ and show that as t varies, the point $P(\cosh t, \sinh t)$ traces the part of the graph in quadrants I and IV.

49 The Gateway Arch in St. Louis has the shape of an inverted catenary (see figure). Rising 630 feet at its center and stretching 630 feet across its base, the shape of the arch can be approximated by

$$y=-127.7\cosh(x/127.7)+757.7$$

for $-315\le x\le 315$.

(a) Approximate the total open area under the arch.

(b) Approximate the total length of the arch.

EXERCISE 49

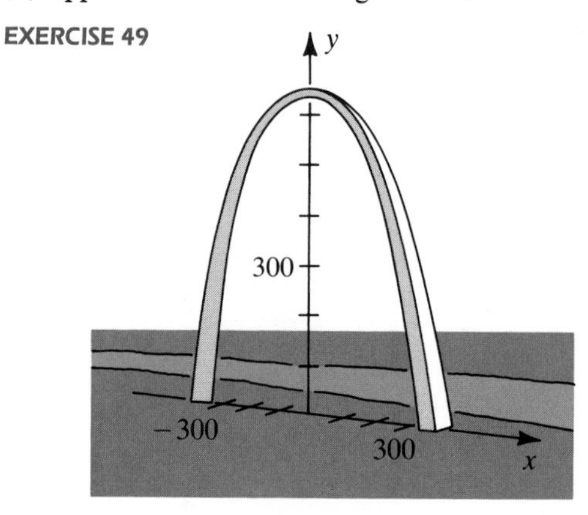

50 If a steel ball of mass m is released into water and the force of resistance is directly proportional to the square of the velocity, then the distance y the ball travels in t seconds is given by

$$y=km\ln\cosh\left(\sqrt{\frac{g}{km}}\,t\right),$$

where g is a gravitational constant and $k>0$. Show that y is a solution of the differential equation

$$m\frac{d^2y}{dt^2}+\frac{1}{k}\left(\frac{dy}{dt}\right)^2=mg.$$

51 If a wave of length L is traveling across water of depth h (see figure on next page), the velocity v, or *celerity*, of the wave is related to L and h by the formula

$$v^2=\frac{gL}{2\pi}\tanh\frac{2\pi h}{L},$$

where g is a gravitational constant.

(a) Find $\lim_{h\to\infty} v^2$ and conclude that $v \approx \sqrt{gL/(2\pi)}$ in deep water.

(b) If $x \approx 0$ and f is a continuous function, then, by the mean value theorem (4.12), $f(x) - f(0) \approx f'(0)\,x$. Use this fact to show that $v \approx \sqrt{gh}$ if $h/L \approx 0$. Conclude that wave velocity is independent of wave length in shallow water.

EXERCISE 51

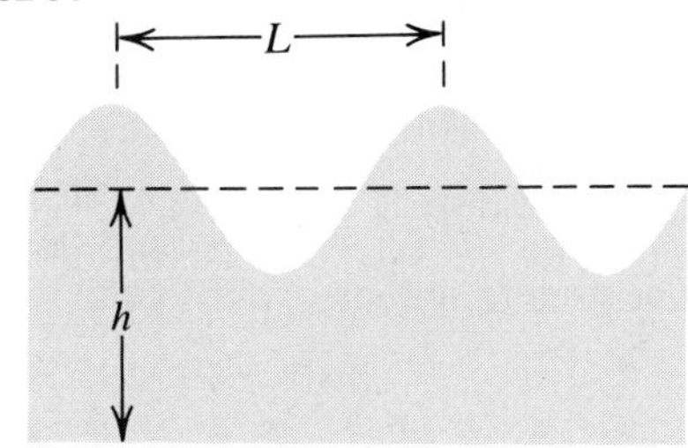

52 A soap bubble formed by two parallel concentric rings is shown in the figure. If the rings are not too far apart, it can be shown that the function f whose graph generates this surface of revolution is a solution of the differential equation $yy'' = 1 + (y')^2$, where $y = f(x)$. If A and B are positive constants, show that $y = A \cosh Bx$ is a solution if and only if $AB = 1$. Conclude that the graph is a catenary.

EXERCISE 52

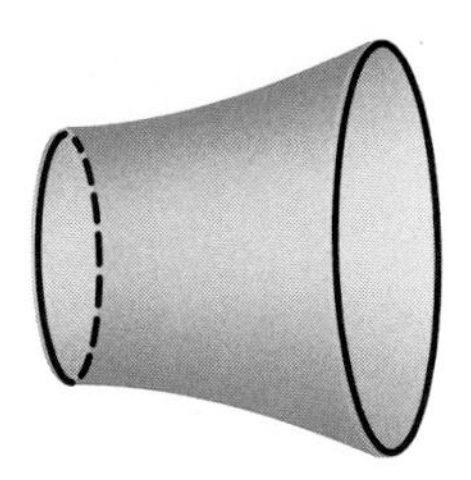

c 53 Graph, on the same coordinate axes, $y = \tanh x$ and $y = \operatorname{sech}^2 x$ for $0 \le x \le 2$.

(a) Estimate the x-coordinate a of the point of intersection of the graphs.

(b) Use Newton's method to approximate a to three decimal places.

c 54 Graph, on the same coordinate axes, $y = \cosh^2 x$ and $y = 2$.

(a) Set up integrals for estimating the centroid of the region R bounded by the graphs.

(b) Use Simpson's rule, with $n = 4$, to approximate the coordinates of the centroid of R.

Exer. 55–72: Verify the identity.

55 $\cosh x + \sinh x = e^x$

56 $\cosh x - \sinh x = e^{-x}$

57 $\sinh(-x) = -\sinh x$

58 $\cosh(-x) = \cosh x$

59 $\sinh(x + y) = \sinh x \cosh y + \cosh x \sinh y$

60 $\cosh(x + y) = \cosh x \cosh y + \sinh x \sinh y$

61 $\sinh(x - y) = \sinh x \cosh y - \cosh x \sinh y$

62 $\cosh(x - y) = \cosh x \cosh y - \sinh x \sinh y$

63 $\tanh(x + y) = \dfrac{\tanh x + \tanh y}{1 + \tanh x \tanh y}$

64 $\tanh(x - y) = \dfrac{\tanh x - \tanh y}{1 - \tanh x \tanh y}$

65 $\sinh 2x = 2 \sinh x \cosh x$

66 $\cosh 2x = \cosh^2 x + \sinh^2 x$

67 $\sinh^2 \dfrac{x}{2} = \dfrac{\cosh x - 1}{2}$

68 $\cosh^2 \dfrac{x}{2} = \dfrac{\cosh x + 1}{2}$

69 $\tanh 2x = \dfrac{2 \tanh x}{1 + \tanh^2 x}$

70 $\tanh \dfrac{x}{2} = \dfrac{\sinh x}{1 + \cosh x}$

71 $(\cosh x + \sinh x)^n = \cosh nx + \sinh nx$ for every positive integer n (*Hint:* Use Exercise 55.)

72 $(\cosh x - \sinh x)^n = \cosh nx - \sinh nx$ for every positive integer n

8.4 INVERSE HYPERBOLIC FUNCTIONS

The hyperbolic sine function is continuous and increasing for every x and hence, by Theorem (7.6), has a continuous, increasing inverse function, denoted by $\sinh^{-1}$. Since $\sinh x$ is defined in terms of e^x, we might expect that $\sinh^{-1}$ can be expressed in terms of the inverse, ln, of the natural exponential function. The first formula of the next theorem shows that this is the case.

Theorem (8.16)

(i) $\sinh^{-1} x = \ln(x + \sqrt{x^2 + 1})$

(ii) $\cosh^{-1} x = \ln(x + \sqrt{x^2 - 1}), \quad x \geq 1$

(iii) $\tanh^{-1} x = \dfrac{1}{2} \ln \dfrac{1 + x}{1 - x}, \quad |x| < 1$

(iv) $\operatorname{sech}^{-1} x = \ln \dfrac{1 + \sqrt{1 - x^2}}{x}, \quad 0 < x \leq 1$

PROOF To prove (i), we begin by noting that

$$y = \sinh^{-1} x \quad \text{if and only if} \quad x = \sinh y.$$

The equation $x = \sinh y$ can be used to find an explicit form for $\sinh^{-1} x$. Thus, if

$$x = \sinh y = \frac{e^y - e^{-y}}{2},$$

then

$$e^y - 2x - e^{-y} = 0.$$

Multiplying both sides by e^y, we obtain

$$e^{2y} - 2xe^y - 1 = 0.$$

Applying the quadratic formula yields

$$e^y = \frac{2x \pm \sqrt{4x^2 + 4}}{2}, \quad \text{or} \quad e^y = x \pm \sqrt{x^2 + 1}.$$

Since $x - \sqrt{x^2 + 1} < 0$ and e^y is never negative, we must have

$$e^y = x + \sqrt{x^2 + 1}.$$

The equivalent logarithmic form is

$$y = \ln(x + \sqrt{x^2 + 1});$$

that is,

$$\sinh^{-1} x = \ln(x + \sqrt{x^2 + 1}).$$

Formulas (ii)–(iv) are obtained in similar fashion. As with trigonometric functions, some inverse functions exist only if the domain is restricted. For example, if the domain of cosh is restricted to the set of nonnegative real numbers, then the resulting function is continuous and increasing, and its inverse function $\cosh^{-1}$ is defined by

$$y = \cosh^{-1} x \quad \text{if and only if} \quad \cosh y = x, \quad y \geq 0.$$

Employing the process used for $\sinh^{-1} x$ leads us to (ii).
Similarly,

$$y = \tanh^{-1} x \quad \text{if and only if} \quad \tanh y = x \quad \text{for} \quad |x| < 1.$$

Using Definition (8.12), we may write $\tanh y = x$ as

$$\frac{e^y - e^{-y}}{e^y + e^{-y}} = x.$$

Solving for y gives us (iii).

Finally, if we restrict the domain of sech to nonnegative numbers, the result is a one-to-one function, and we define

$$y = \operatorname{sech}^{-1} x \quad \text{if and only if} \quad \operatorname{sech} y = x, \quad y \geq 0.$$

Again, introducing the exponential form leads to (iv). ■

In the next theorem $u = g(x)$, where g is differentiable and x is suitably restricted.

Theorem **(8.17)**

(i) $D_x \sinh^{-1} u = \dfrac{1}{\sqrt{u^2 + 1}} D_x u$

(ii) $D_x \cosh^{-1} u = \dfrac{1}{\sqrt{u^2 - 1}} D_x u, \quad u > 1$

(iii) $D_x \tanh^{-1} u = \dfrac{1}{1 - u^2} D_x u, \quad |u| < 1$

(iv) $D_x \operatorname{sech}^{-1} u = \dfrac{-1}{u\sqrt{1 - u^2}} D_x u, \quad 0 < u < 1$

PROOF By Theorem (8.16)(i),

$$\begin{aligned} D_x \sinh^{-1} x &= D_x \ln (x + \sqrt{x^2 + 1}) \\ &= \frac{1}{x + \sqrt{x^2 + 1}} \left(1 + \frac{x}{\sqrt{x^2 + 1}}\right) \\ &= \frac{\sqrt{x^2 + 1} + x}{(x + \sqrt{x^2 + 1})\sqrt{x^2 + 1}} \\ &= \frac{1}{\sqrt{x^2 + 1}}. \end{aligned}$$

This formula can be extended to $D_x \sinh^{-1} u$ by applying the chain rule. The remaining formulas can be proved in similar fashion. ■

EXAMPLE 1 If $y = \sinh^{-1} (\tan x)$, find dy/dx.

SOLUTION Using Theorem (8.17)(i) with $u = \tan x$, we have

$$\begin{aligned} \frac{dy}{dx} &= \frac{1}{\sqrt{\tan^2 x + 1}} \frac{d}{dx} \tan x = \frac{1}{\sqrt{\sec^2 x}} \sec^2 x \\ &= \frac{1}{|\sec x|} |\sec x|^2 = |\sec x|. \end{aligned}$$

The following theorem may be verified by differentiating the right-hand side of each formula.

Theorem **(8.18)**

(i) $\displaystyle\int \frac{1}{\sqrt{a^2+u^2}}\,du = \sinh^{-1}\frac{u}{a} + C, \quad a > 0$

(ii) $\displaystyle\int \frac{1}{\sqrt{u^2-a^2}}\,du = \cosh^{-1}\frac{u}{a} + C, \quad 0 < a < u$

(iii) $\displaystyle\int \frac{1}{a^2-u^2}\,du = \frac{1}{a}\tanh^{-1}\frac{u}{a} + C, \quad |u| < a$

(iv) $\displaystyle\int \frac{1}{u\sqrt{a^2-u^2}}\,du = -\frac{1}{a}\operatorname{sech}^{-1}\frac{|u|}{a} + C, \quad 0 < |u| < a$

If we use Theorem (8.16), then each of the integration formulas in the preceding theorem can be expressed in terms of the natural logarithmic function. To illustrate,

$$\begin{aligned}\int \frac{1}{\sqrt{a^2+u^2}}\,du &= \sinh^{-1}\frac{u}{a} + C\\ &= \ln\left(\frac{u}{a} + \sqrt{\left(\frac{u}{a}\right)^2 + 1}\right) + C.\end{aligned}$$

We can show that if $a > 0$, then the last formula can be written as

$$\int \frac{1}{\sqrt{a^2+u^2}}\,du = \ln\,(u + \sqrt{a^2+u^2}) + D,$$

where D is a constant. In Section 9.3 we shall discuss another method for evaluating the integrals in Theorem (8.18).

EXAMPLE 2 Evaluate $\displaystyle\int \frac{1}{\sqrt{25+9x^2}}\,dx$.

SOLUTION We may express the integral as in Theorem (8.18)(i), by using the substitution

$$u = 3x, \qquad du = 3\,dx.$$

Since du contains the factor 3, we adjust the integrand by multiplying by 3 and then compensate by multiplying the integral by $\frac{1}{3}$ before substituting:

$$\begin{aligned}\int \frac{1}{\sqrt{25+9x^2}}\,dx &= \frac{1}{3}\int \frac{1}{\sqrt{5^2+(3x)^2}}\,3\,dx\\ &= \frac{1}{3}\int \frac{1}{\sqrt{5^2+u^2}}\,du\\ &= \frac{1}{3}\sinh^{-1}\frac{u}{5} + C\\ &= \frac{1}{3}\sinh^{-1}\frac{3x}{5} + C\end{aligned}$$

EXAMPLE 3 Evaluate $\int \frac{e^x}{16 - e^{2x}}\,dx$.

SOLUTION Substituting $u = e^x$, $du = e^x\,dx$ and applying Theorem (8.18)(iii) with $a = 4$, we have

$$\begin{aligned}\int \frac{e^x}{16 - e^{2x}}\,dx &= \int \frac{1}{4^2 - (e^x)^2}\,e^x\,dx \\ &= \int \frac{1}{4^2 - u^2}\,du \\ &= \frac{1}{4}\tanh^{-1}\frac{u}{4} + C \\ &= \frac{1}{4}\tanh^{-1}\frac{e^x}{4} + C\end{aligned}$$

for $|u| < a$ (that is, $e^x < 4$).

EXERCISES 8.4

Exer. 1–2: Approximate to four decimal places.

1 (a) $\sinh^{-1} 1$ (b) $\cosh^{-1} 2$
(c) $\tanh^{-1}(-\frac{1}{2})$ (d) $\text{sech}^{-1}\frac{1}{2}$

2 (a) $\sinh^{-1}(-2)$ (b) $\cosh^{-1} 5$
(c) $\tanh^{-1}\frac{1}{3}$ (d) $\text{sech}^{-1}\frac{3}{5}$

Exer. 3–18: Find $f'(x)$ if $f(x)$ is the given expression.

3 $\sinh^{-1} 5x$ 4 $\sinh^{-1} e^x$

5 $\cosh^{-1}\sqrt{x}$ 6 $\sqrt{\cosh^{-1} x}$

7 $\tanh^{-1}(-4x)$ 8 $\tanh^{-1}\sin 3x$

9 $\text{sech}^{-1} x^2$ 10 $\text{sech}^{-1}\sqrt{1-x}$

11 $x\sinh^{-1}\frac{1}{x}$ 12 $\frac{1}{\sinh^{-1} x^2}$

13 $\ln\cosh^{-1} 4x$ 14 $\cosh^{-1}\ln 4x$

15 $\tanh^{-1}(x+1)$ 16 $\tanh^{-1} x^3$

17 $\text{sech}^{-1}\sqrt{x}$ 18 $(\text{sech}^{-1} x)^{-1}$

Exer. 19–26: Evaluate the integral.

19 $\int \frac{1}{\sqrt{81 + 16x^2}}\,dx$ 20 $\int \frac{1}{\sqrt{16x^2 - 9}}\,dx$

21 $\int \frac{1}{49 - 4x^2}\,dx$ 22 $\int \frac{\sin x}{\sqrt{1 + \cos^2 x}}\,dx$

23 $\int \frac{e^x}{\sqrt{e^{2x} - 16}}\,dx$ 24 $\int \frac{2}{5 - 3x^2}\,dx$

25 $\int \frac{1}{x\sqrt{9 - x^4}}\,dx$ 26 $\int \frac{1}{\sqrt{5 - e^{2x}}}\,dx$

27 A point moves along the line $x = 1$ in a coordinate plane with a velocity that is directly proportional to its distance from the origin. If the initial position of the point is $(1, 0)$ and the initial velocity is 3 ft/sec, express the y-coordinate of the point as a function of time t (in seconds).

28 The rectangular coordinate system shown in the figure illustrates the problem of a dog seeking its master. The dog, initially at the point $(1, 0)$, sees its master at the point $(0, 0)$. The master proceeds up the y-axis at a constant speed, and the dog runs directly toward its master at all times. If the speed of the dog is twice that of the master, it can be shown that the path of the dog is given by $y = f(x)$, where y is a solution of the differential equation $2xy'' = \sqrt{1 + (y')^2}$. Solve this equation by first letting $z = dy/dx$ and solving $2xz' = \sqrt{1 + z^2}$, obtaining $z = \frac{1}{2}[\sqrt{x} - (1/\sqrt{x})]$. Finally, solve $y' = \frac{1}{2}[\sqrt{x} - (1/\sqrt{x})]$.

EXERCISE 28

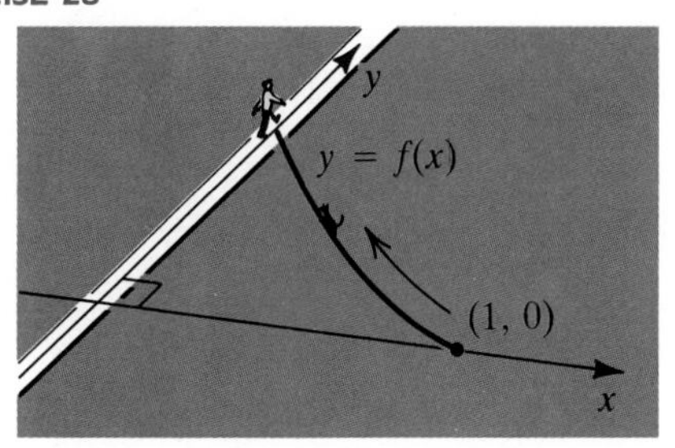

Exer. 29–32: Sketch the graph of the equation.

29 $y = \sinh^{-1} x$

30 $y = \cosh^{-1} x$

31 $y = \tanh^{-1} x$

32 $y = \operatorname{sech}^{-1} x$

Exer. 33–38: Verify the formula.

33 $D_x \cosh^{-1} u = \dfrac{1}{\sqrt{u^2 - 1}} D_x u, \quad u > 1$

34 $D_x \tanh^{-1} u = \dfrac{1}{1 - u^2} D_x u, \quad |u| < 1$

35 $D_x \operatorname{sech}^{-1} u = -\dfrac{1}{u\sqrt{1 - u^2}} D_x u, \quad 0 < u < 1$

36 $\displaystyle\int \frac{1}{\sqrt{u^2 - a^2}}\, du = \cosh^{-1} \frac{u}{a} + C, \quad 0 < a < u$

37 $\displaystyle\int \frac{1}{a^2 - u^2}\, du = \frac{1}{a} \tanh^{-1} \frac{u}{a} + C, \quad |u| < a$

38 $\displaystyle\int \frac{1}{u\sqrt{a^2 - u^2}}\, du = -\frac{1}{a} \operatorname{sech}^{-1} \frac{|u|}{a} + C, \quad 0 < |u| < a$

Exer. 39–41: Derive the formula (see Theorem (8.16)).

39 $\cosh^{-1} x = \ln (x + \sqrt{x^2 - 1}), \quad x \geq 1$

40 $\tanh^{-1} x = \dfrac{1}{2} \ln \dfrac{1 + x}{1 - x}, \quad |x| < 1$

41 $\operatorname{sech}^{-1} x = \ln \dfrac{1 + \sqrt{1 - x^2}}{x}, \quad 0 < x \leq 1$

8.5 REVIEW EXERCISES

Exer. 1–24: Find $f'(x)$ if $f(x)$ is the given expression.

1 $\arctan \sqrt{x - 1}$

2 $\tan^{-1} (\ln 3x)$

3 $x^2 \operatorname{arcsec} (x^2)$

4 $\dfrac{1}{\cos^{-1} x}$

5 $2^{\arctan 2x}$

6 $(1 + \operatorname{arcsec} 2x)^{\sqrt{2}}$

7 $\ln \tan^{-1} (x^2)$

8 $\dfrac{1 - x^2}{\arccos x}$

9 $\sin^{-1} \sqrt{1 - x^2}$

10 $\sqrt{\sin^{-1} (1 - x^2)}$

11 $(\tan x + \tan^{-1} x)^4$

12 $\tan^{-1} \sqrt{\tan 2x}$

13 $\tan^{-1} (\tan^{-1} x)$

14 $e^{4x} \sec^{-1} e^{4x}$

15 $\cosh e^{-5x}$

16 $\dfrac{\ln \sinh x}{x}$

17 $e^{-x} \sinh e^{-x}$

18 $e^{x \cosh x}$

19 $\dfrac{\sinh x}{\cosh x - \sinh x}$

20 $\ln \tanh (5x + 1)$

21 $\sinh^{-1} (x^2)$

22 $\cosh^{-1} \tan x$

23 $\tanh^{-1} (\tanh \sqrt[3]{x})$

24 $\dfrac{1}{x} \tanh \dfrac{1}{x}$

Exer. 25–40: Evaluate the integral.

25 $\displaystyle\int \frac{1}{4 + 9x^2}\, dx$

26 $\displaystyle\int \frac{x}{4 + 9x^2}\, dx$

27 $\displaystyle\int \frac{e^{2x}}{\sqrt{1 - e^{2x}}}\, dx$

28 $\displaystyle\int \frac{e^x}{\sqrt{1 - e^{2x}}}\, dx$

29 $\displaystyle\int \frac{x}{\operatorname{sech} (x^2)}\, dx$

30 $\displaystyle\int \frac{1}{x\sqrt{x^4 - 1}}\, dx$

31 $\displaystyle\int_{-1/2}^{1/2} \frac{1}{\sqrt{1 - x^2}}\, dx$

32 $\displaystyle\int_0^{\pi/2} \frac{\cos x}{1 + \sin^2 x}\, dx$

33 $\displaystyle\int \frac{\sinh (\ln x)}{x}\, dx$

34 $\displaystyle\int \operatorname{sech}^2 (1 - 2x)\, dx$

35 $\displaystyle\int \frac{1}{\sqrt{9 - 4x^2}}\, dx$

36 $\displaystyle\int \frac{x}{\sqrt{9 - 4x^2}}\, dx$

37 $\displaystyle\int \frac{1}{x\sqrt{9 - 4x^2}}\, dx$

38 $\displaystyle\int \frac{1}{x\sqrt{4x^2 - 9}}\, dx$

39 $\displaystyle\int \frac{x}{\sqrt{25x^2 + 36}}\, dx$

40 $\displaystyle\int \frac{1}{\sqrt{25x^2 + 36}}\, dx$

41 Find the points on the graph of $y = \sin^{-1} 3x$ at which the tangent line is parallel to the line through $A(2, -3)$ and $B(4, 7)$.

42 Find the points of inflection and discuss the concavity of the graph of $y = x \sin^{-1} x$.

43 Find the local extrema of $f(x) = 8 \sec x + \csc x$ on the interval $(0, \pi/2)$, and describe where $f(x)$ is increasing or is decreasing on that interval.

44 Find the area of the region bounded by the graphs of $y = x/(x^4 + 1)$, $x = 1$, and $y = 0$.

45 Damped oscillations are oscillations of decreasing magnitude that occur when frictional forces are considered. Shown in the figure on the next page is a graph of the damped oscillations given by $f(x) = e^{-x/2} \sin 2x$.

(a) Find the x-coordinates of the extrema of f for $0 \leq x \leq 2\pi$.

(b) Approximate the x-coordinates in part (a) to two decimal places.

EXERCISE 45

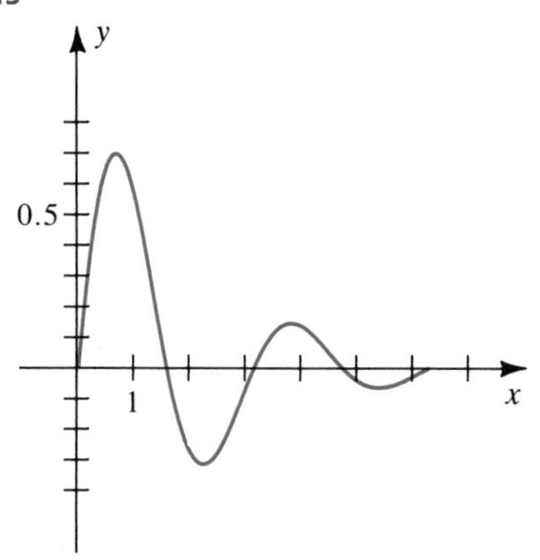

46 Find the arc length of the graph of $y = \ln \tanh \frac{1}{2}x$ from $x = 1$ to $x = 2$.

47 A balloon is released from level ground, 500 meters away from a person who observes its vertical ascent. If the balloon rises at a constant rate of 2 m/sec, use inverse trigonometric functions to find the rate at which the angle of elevation of the observer's line of sight is changing at the instant the balloon is at a height of 100 meters. (Disregard the observer's height.)

48 A square picture with sides 2 feet long is hung on a wall with the base 6 feet above the floor. A person whose eye level is 5 feet above the floor approaches the picture at a rate of 2 ft/sec. If θ is the angle between the line of sight and the top and bottom of the picture, find

(a) the rate at which θ is changing when the person is 8 feet from the wall

(b) the distance from the wall at which θ has its maximum value

49 A stuntman jumps from a hot-air balloon that is hovering at a constant altitude, 100 feet above a lake. A movie camera on shore, 200 feet from a point directly below the balloon, follows the stuntman's descent (see figure). At what rate is the angle of elevation θ of the camera changing 2 seconds after the stuntman jumps? (Disregard the height of the camera.)

EXERCISE 49

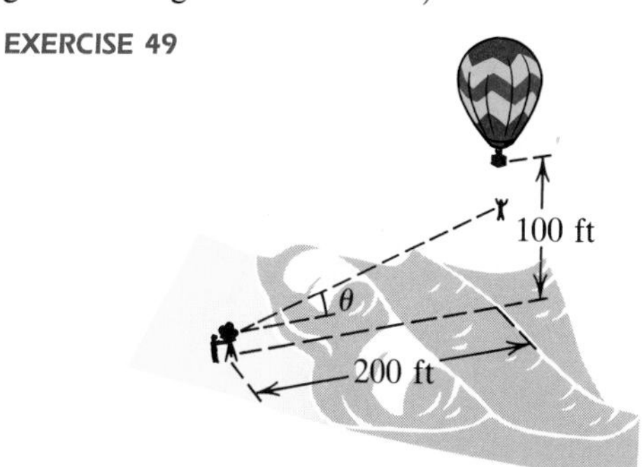

50 A person on a small island I, which is k miles from the closest point A on a straight shoreline, wishes to reach a camp that is d miles downshore from A by swimming to some point P on shore and then walking the rest of the way (see figure). Suppose the person burns c_1 calories per mile while swimming and c_2 calories per mile while walking, where $c_1 > c_2$.

(a) Find a formula for the total number c of calories burned in completing the trip.

(b) For what angle AIP does c have a minimum value?

EXERCISE 50

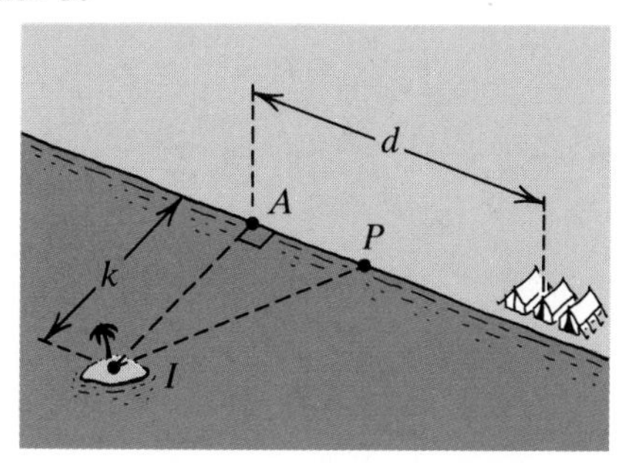

CHAPTER

9

TECHNIQUES OF INTEGRATION

INTRODUCTION

In previous chapters we obtained formulas for evaluating various types of integrals. Many are listed on the inside front cover of this text. We also discussed the *method of substitution*, which is used to change a complicated integral into one that can be readily evaluated. In this chapter we consider additional ways to simplify integrals. Foremost among these is *integration by parts*, which we discuss in the first section. This powerful device allows us to obtain indefinite integrals of $\ln x$, $\tan^{-1} x$, and other important transcendental expressions. In later sections we develop techniques for simplifying integrals that contain powers of trigonometric functions, radicals, and rational expressions.

The use of a table of integrals is explained in Section 9.7. Such tables are always incomplete, and it is sometimes necessary to use skills obtained in previous sections *before* consulting a table. The same can be said for computer programs that are designed to evaluate various (but not all) indefinite integrals.

For applications involving *definite* integrals, it is often unnecessary to find an antiderivative and apply the fundamental theorem of calculus, because the trapezoidal rule or Simpson's rule can be used to obtain numerical approximations. In such cases either a computer or a programmable calculator is invaluable, since it can usually arrive at an approximation in a matter of seconds.

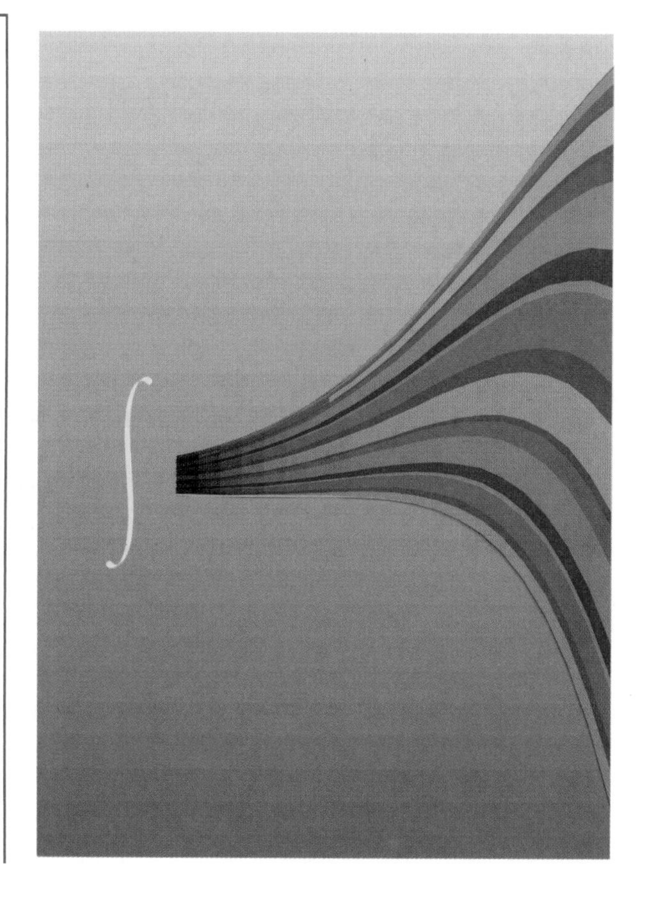

9.1 INTEGRATION BY PARTS

Up to this stage of our work we have been unable to evaluate integrals such as the following:

$$\int \ln x\, dx, \quad \int xe^x\, dx, \quad \int x^2 \sin x\, dx, \quad \int \tan^{-1} x\, dx$$

The next formula will enable us to evaluate not only these, but also many other types of integrals.

Integration by parts formula **(9.1)**

If $u = f(x)$ and $v = g(x)$ and if f' and g' are continuous, then

$$\int u\, dv = uv - \int v\, du.$$

PROOF By the product rule,

$$D_x [f(x)g(x)] = f(x)g'(x) + g(x)f'(x),$$

or, equivalently,

$$f(x)g'(x) = D_x [f(x)g(x)] - g(x)f'(x).$$

Integrating both sides of the previous equation gives us

$$\int f(x)g'(x)\, dx = \int D_x [f(x)g(x)]\, dx - \int g(x)f'(x)\, dx.$$

By Theorem (5.5)(i), the first integral on the right side equals $f(x)g(x) + C$. Since another constant of integration is obtained from the second integral, we may omit C in the formula; that is,

$$\int f(x)g'(x)\, dx = f(x)g(x) - \int g(x)f'(x)\, dx.$$

Since $dv = g'(x)\, dx$ and $du = f'(x)\, dx$, we may write the preceding formula as in (9.1). ■

When applying Formula (9.1) to an integral, we begin by letting one part of the integrand correspond to dv. The expression we choose for dv must include the differential dx. After selecting dv, we designate the remaining part of the integrand by u and then find du. Since this process involves splitting the integrand into two parts, the use of (9.1) is referred to as **integrating by parts**. A proper choice for dv is crucial. We usually *let dv equal the most complicated part of the integrand that can be readily integrated*. The following examples illustrate this method of integration.

EXAMPLE 1 Evaluate $\int xe^{2x}\, dx$.

SOLUTION The following list contains all possible choices for dv:

$$dx, \quad x\, dx, \quad e^{2x}\, dx, \quad xe^{2x}\, dx$$

The most complicated of these expressions that can be readily integrated is $e^{2x}\, dx$. Thus, we let

$$dv = e^{2x}\, dx.$$

The remaining part of the integrand is u—that is, $u = x$. To find v, we integrate dv, obtaining $v = \frac{1}{2}e^{2x}$. Note that a constant of integration is not added at this stage of the solution. (In Exercise 51 you are asked to prove that if a constant *is* added to v, the same final result is obtained.) If $u = x$, then $du = dx$. For ease of reference let us display these expressions as follows:

$$dv = e^{2x}\,dx \qquad u = x$$
$$v = \tfrac{1}{2}e^{2x} \qquad du = dx$$

Substituting these expressions in Formula (9.1)—that is, *integrating by parts*—we obtain

$$\int xe^{2x}\,dx = x(\tfrac{1}{2}e^{2x}) - \int \tfrac{1}{2}e^{2x}\,dx.$$

We may find the integral on the right side as in Section 7.4. This gives us

$$\int xe^{2x}\,dx = \tfrac{1}{2}xe^{2x} - \tfrac{1}{4}e^{2x} + C.$$

It takes considerable practice to become proficient in making a suitable choice for dv. To illustrate, if we had chosen $dv = x\,dx$ in Example 1, then it would have been necessary to let $u = e^{2x}$, giving us

$$dv = x\,dx \qquad u = e^{2x}$$
$$v = \tfrac{1}{2}x^2 \qquad du = 2e^{2x}\,dx.$$

Integrating by parts, we obtain

$$\int xe^{2x}\,dx = \tfrac{1}{2}x^2e^{2x} - \int x^2e^{2x}\,dx.$$

Since the exponent associated with x has increased, the integral on the right is more complicated than the given integral. This indicates that we have made an incorrect choice for dv.

EXAMPLE 2 Evaluate

(a) $\int x \sec^2 x\,dx$ **(b)** $\int_0^{\pi/3} x \sec^2 x\,dx$

SOLUTION

(a) The possible choices for dv are

$$dx, \quad x\,dx, \quad \sec x\,dx, \quad x\sec x\,dx, \quad \sec^2 x\,dx, \quad x\sec^2 x\,dx.$$

The most complicated of these expressions that can be readily integrated is $\sec^2 x\,dx$. Thus, we let

$$dv = \sec^2 x\,dx \qquad u = x$$
$$v = \tan x \qquad du = dx.$$

Integrating by parts gives us

$$\begin{aligned}\int x\sec^2 x\,dx &= x\tan x - \int \tan x\,dx\\ &= x\tan x + \ln|\cos x| + C.\end{aligned}$$

(b) The indefinite integral obtained in part (a) is an antiderivative of $x \sec^2 x$. Using the fundamental theorem of calculus (and dropping the constant of integration C), we obtain

$$\begin{aligned}\int_0^{\pi/3} x \sec^2 x\, dx &= \Big[x \tan x + \ln|\cos x|\Big]_0^{\pi/3} \\ &= \left(\frac{\pi}{3}\tan\frac{\pi}{3} + \ln\left|\cos\frac{\pi}{3}\right|\right) - (0 + \ln 1) \\ &= \left(\frac{\pi}{3}\sqrt{3} + \ln\frac{1}{2}\right) - (0 + 0) \\ &= \frac{\pi}{3}\sqrt{3} - \ln 2 \approx 1.12.\end{aligned}$$

If, in Example 2, we had chosen $dv = x\, dx$ and $u = \sec^2 x$, then the integration by parts formula (9.1) would have led to a more complicated integral. (Verify this fact.)

In the next example we use integration by parts to find an antiderivative of the natural logarithmic function.

EXAMPLE 3 Evaluate $\int \ln x\, dx$.

SOLUTION Let

$$dv = dx \qquad u = \ln x$$
$$v = x \qquad du = \frac{1}{x}\, dx$$

and integrate by parts as follows:

$$\begin{aligned}\int \ln x\, dx &= (\ln x)x - \int (x)\frac{1}{x}\, dx \\ &= x \ln x - \int dx \\ &= x \ln x - x + C\end{aligned}$$

Sometimes it is necessary to use integration by parts more than once in the same problem. This is illustrated in the next example.

EXAMPLE 4 Evaluate $\int x^2 e^{2x}\, dx$.

SOLUTION Let

$$dv = e^{2x}\, dx \qquad u = x^2$$
$$v = \tfrac{1}{2}e^{2x} \qquad du = 2x\, dx$$

and integrate by parts as follows:

$$\int x^2 e^{2x}\,dx = x^2(\tfrac{1}{2}e^{2x}) - \int (\tfrac{1}{2}e^{2x})2x\,dx$$
$$= \tfrac{1}{2}x^2 e^{2x} - \int x e^{2x}\,dx$$

To evaluate the integral on the right side of the last equation, we must again integrate by parts. Proceeding exactly as in Example 1 leads to

$$\int x^2 e^{2x}\,dx = \tfrac{1}{2}x^2 e^{2x} - \tfrac{1}{2}x e^{2x} + \tfrac{1}{4}e^{2x} + C.$$

The following example illustrates another device for evaluating an integral by means of two applications of the integration by parts formula.

EXAMPLE 5 Evaluate $\int e^x \cos x\,dx$.

SOLUTION We could either let $dv = \cos x\,dx$ or let $dv = e^x\,dx$, since each of these expressions is readily integrable. Let us choose

$$dv = \cos x\,dx \qquad u = e^x$$
$$v = \sin x \qquad du = e^x\,dx$$

and integrate by parts as follows:

$$\int e^x \cos x\,dx = e^x \sin x - \int (\sin x)e^x\,dx$$

$$\int e^x \cos x\,dx = e^x \sin x - \int e^x \sin x\,dx \tag{1}$$

We next apply integration by parts to the integral on the right side of equation (1). Since we chose a trigonometric form for dv in the first integration by parts, we shall also choose a trigonometric form for the second. Letting

$$dv = \sin x\,dx \qquad u = e^x$$
$$v = -\cos x \qquad du = e^x\,dx$$

and integrating by parts, we have

$$\int e^x \sin x\,dx = e^x(-\cos x) - \int (-\cos x)e^x\,dx$$

$$\int e^x \sin x\,dx = -e^x \cos x + \int e^x \cos x\,dx. \tag{2}$$

If we now use equation (2) to substitute on the right side of equation (1), we obtain

$$\int e^x \cos x\,dx = e^x \sin x - \left[-e^x \cos x + \int e^x \cos x\,dx\right],$$

or $$\int e^x \cos x\,dx = e^x \sin x + e^x \cos x - \int e^x \cos x\,dx.$$

Adding $\int e^x \cos x\,dx$ to both sides of the last equation gives us

$$2\int e^x \cos x\,dx = e^x(\sin x + \cos x).$$

Finally, dividing both sides by 2 and adding the constant of integration yields

$$\int e^x \cos x \, dx = \tfrac{1}{2}e^x (\sin x + \cos x) + C.$$

We could have evaluated the given integral by using $dv = e^x \, dx$ for both the first and second applications of the integration by parts formula.

We must choose substitutions carefully when evaluating an integral of the type given in Example 5. To illustrate, suppose that in the evaluation of the integral on the right in equation (1) of the solution we had used

$$dv = e^x \, dx \qquad u = \sin x$$

$$v = e^x \qquad du = \cos x \, dx.$$

Integration by parts then leads to

$$\begin{aligned}\int e^x \sin x \, dx &= (\sin x)\, e^x - \int e^x \cos x \, dx \\ &= e^x \sin x - \int e^x \cos x \, dx.\end{aligned}$$

If we now substitute in (1), we obtain

$$\int e^x \cos x \, dx = e^x \sin x - \left[e^x \sin x - \int e^x \cos x \, dx\right],$$

which reduces to

$$\int e^x \cos x \, dx = \int e^x \cos x \, dx.$$

Although this is a true statement, it is not an evaluation of the given integral.

EXAMPLE 6 Evaluate $\int \sec^3 x \, dx$.

SOLUTION The possible choices for dv are

$$dx, \quad \sec x \, dx, \quad \sec^2 x \, dx, \quad \sec^3 x \, dx.$$

The most complicated of these expressions that can be readily integrated is $\sec^2 x \, dx$. Thus, we let

$$dv = \sec^2 x \, dx \qquad u = \sec x$$

$$v = \tan x \qquad du = \sec x \tan x \, dx$$

and integrate by parts as follows:

$$\int \sec^3 x \, dx = \sec x \tan x - \int \sec x \tan^2 x \, dx$$

Instead of applying another integration by parts, let us change the form of the integral on the right by using the identity $1 + \tan^2 x = \sec^2 x$. This gives us

$$\int \sec^3 x \, dx = \sec x \tan x - \int \sec x \, (\sec^2 x - 1) \, dx,$$

or $$\int \sec^3 x\, dx = \sec x \tan x - \int \sec^3 x\, dx + \int \sec x\, dx.$$

Adding $\int \sec^3 x\, dx$ to both sides of the last equation gives us

$$2 \int \sec^3 x\, dx = \sec x \tan x + \int \sec x\, dx.$$

If we now evaluate $\int \sec x\, dx$ and divide both sides of the resulting equation by 2 (and then add the constant of integration), we obtain

$$\int \sec^3 x\, dx = \tfrac{1}{2} \sec x \tan x + \tfrac{1}{2} \ln |\sec x + \tan x| + C.$$

Integration by parts may sometimes be employed to obtain **reduction formulas** for integrals. We can use such formulas to write an integral involving powers of an expression in terms of integrals that involve lower powers of the expression.

EXAMPLE 7 Find a reduction formula for $\int \sin^n x\, dx$.

SOLUTION Let

$$dv = \sin x\, dx \qquad u = \sin^{n-1} x$$
$$v = -\cos x \qquad du = (n-1) \sin^{n-2} x \cos x\, dx$$

and integrate by parts as follows:

$$\int \sin^n x\, dx = -\cos x \sin^{n-1} x + (n-1) \int \sin^{n-2} x \cos^2 x\, dx$$

Since $\cos^2 x = 1 - \sin^2 x$, we may write

$$\int \sin^n x\, dx = -\cos x \sin^{n-1} x + (n-1) \int \sin^{n-2} x\, dx - (n-1) \int \sin^n x\, dx.$$

Consequently,

$$\int \sin^n x\, dx + (n-1) \int \sin^n x\, dx = -\cos x \sin^{n-1} x + (n-1) \int \sin^{n-2} x\, dx.$$

The left side of the last equation reduces to $n \int \sin^n x\, dx$. Dividing both sides by n, we obtain

$$\int \sin^n x\, dx = -\frac{1}{n} \cos x \sin^{n-1} x + \frac{n-1}{n} \int \sin^{n-2} x\, dx.$$

EXAMPLE 8 Use the reduction formula in Example 7 to evaluate $\int \sin^4 x\, dx$.

SOLUTION Using the formula with $n = 4$ gives us

$$\int \sin^4 x\, dx = -\tfrac{1}{4} \cos x \sin^3 x + \tfrac{3}{4} \int \sin^2 x\, dx.$$

Applying the reduction formula, with $n = 2$, to the integral on the right, we have

$$\int \sin^2 x\, dx = -\tfrac{1}{2}\cos x \sin x + \tfrac{1}{2}\int dx$$
$$= -\tfrac{1}{2}\cos x \sin x + \tfrac{1}{2}x + C.$$

Consequently,

$$\int \sin^4 x\, dx = -\tfrac{1}{4}\cos x \sin^3 x - \tfrac{3}{8}\cos x \sin x + \tfrac{3}{8}x + D$$

with $D = \tfrac{3}{4}C$.

It should be evident that by repeated applications of the formula in Example 7 we can find $\int \sin^n x\, dx$ for any positive integer n, because these reductions end with either $\int \sin x\, dx$ or $\int dx$, and each of these can be evaluated easily.

EXERCISES 9.1

Exer. 1–38: Evaluate the integral.

1 $\int xe^{-x}\, dx$

2 $\int x \sin x\, dx$

3 $\int x^2 e^{3x}\, dx$

4 $\int x^2 \sin 4x\, dx$

5 $\int x \cos 5x\, dx$

6 $\int xe^{-2x}\, dx$

7 $\int x \sec x \tan x\, dx$

8 $\int x \csc^2 3x\, dx$

9 $\int x^2 \cos x\, dx$

10 $\int x^3 e^{-x}\, dx$

11 $\int \tan^{-1} x\, dx$

12 $\int \sin^{-1} x\, dx$

13 $\int \sqrt{x} \ln x\, dx$

14 $\int x^2 \ln x\, dx$

15 $\int x \csc^2 x\, dx$

16 $\int x \tan^{-1} x\, dx$

17 $\int e^{-x} \sin x\, dx$

18 $\int e^{3x} \cos 2x\, dx$

19 $\int \sin x \ln \cos x\, dx$

20 $\int_0^1 x^3 e^{-x^2}\, dx$

21 $\int \csc^3 x\, dx$

22 $\int \sec^5 x\, dx$

23 $\int_0^1 \frac{x^3}{\sqrt{x^2+1}}\, dx$

24 $\int \sin \ln x\, dx$

25 $\int_0^{\pi/2} x \sin 2x\, dx$

26 $\int x \sec^2 5x\, dx$

27 $\int x(2x+3)^{99}\, dx$

28 $\int \frac{x^5}{\sqrt{1-x^3}}\, dx$

29 $\int e^{4x} \sin 5x\, dx$

30 $\int x^3 \cos (x^2)\, dx$

31 $\int (\ln x)^2\, dx$

32 $\int x\, 2^x\, dx$

33 $\int x^3 \sinh x\, dx$

34 $\int (x+4) \cosh 4x\, dx$

35 $\int \cos \sqrt{x}\, dx$

36 $\int \tan^{-1} 3x\, dx$

37 $\int \cos^{-1} x\, dx$

38 $\int (x+1)^{10}(x+2)\, dx$

Exer. 39–42: Use integration by parts to derive the reduction formula.

39 $\int x^m e^x\, dx = x^m e^x - m \int x^{m-1} e^x\, dx$

40 $\int x^m \sin x\, dx = -x^m \cos x + m \int x^{m-1} \cos x\, dx$

41 $\int (\ln x)^m\, dx = x (\ln x)^m - m \int (\ln x)^{m-1}\, dx$

42 $\int \sec^m x\, dx = \frac{\sec^{m-2} x \tan x}{m-1} + \frac{m-2}{m-1}\int \sec^{m-2} x\, dx$ for $m \neq 1$.

43 Use Exercise 39 to evaluate $\int x^5 e^x\, dx$.

44 Use Exercise 41 to evaluate $\int (\ln x)^4\, dx$.

45 If $f(x) = \sin \sqrt{x}$, find the area of the region under the graph of f from $x = 0$ to $x = \pi^2$.

46 The region between the graph of $y = x\sqrt{\sin x}$ and the x-axis from $x = 0$ to $x = \pi/2$ is revolved about the x-axis. Find the volume of the resulting solid.

47 The region bounded by the graphs of $y = \ln x$, $y = 0$, and $x = e$ is revolved about the y-axis. Find the volume of the resulting solid.

48 Suppose the force $f(x)$ acting at the point with coordinate x on a coordinate line l is given by $f(x) = x^5\sqrt{x^3+1}$. Find the work done in moving an object from $x = 0$ to $x = 1$.

49 Find the centroid of the region bounded by the graphs of the equations $y = e^x$, $y = 0$, $x = 0$, and $x = \ln 3$.

50 The velocity (at time t) of a point moving along a coordinate line is t/e^{2t} ft/sec. If the point is at the origin at $t = 0$, find its position at time t.

51 When applying the integration by parts formula (9.1), show that if, after choosing dv, we use $v + C$ in place of v, the same result is obtained.

52 In Section 6.3 the discussion of finding volumes by means of cylindrical shells was incomplete because we did not show that the same result is obtained if the disk method is also applicable. Use integration by parts to prove that if f is differentiable and either $f'(x) > 0$ on $[a, b]$ or $f'(x) < 0$ on $[a, b]$, and if V is the volume of the solid obtained by revolving the region bounded by the graphs of f, $x = a$, and $x = b$ about the x-axis, then the same value of V is obtained using either the disk method or the shell method. (*Hint:* Let g be the inverse function of f, and use integration by parts on $\int_a^b \pi[f(x)]^2\,dx$.)

53 Discuss the following use of Formula (9.1): Given $\int (1/x)\,dx$, let $dv = dx$ and $u = 1/x$ so that $v = x$ and $du = (-1/x^2)\,dx$. Hence

$$\int \frac{1}{x}\,dx = \left(\frac{1}{x}\right)x - \int x\left(-\frac{1}{x^2}\right)dx,$$

or

$$\int \frac{1}{x}\,dx = 1 + \int \frac{1}{x}\,dx.$$

Consequently, $0 = 1$.

54 If $u = f(x)$ and $v = g(x)$, prove that the analogue of Formula (9.1) for definite integrals is

$$\int_a^b u\,dv = \Big[uv\Big]_a^b - \int_a^b v\,du$$

for values a and b of x.

9.2 TRIGONOMETRIC INTEGRALS

In Example 7 of Section 9.1 we obtained a reduction formula for $\int \sin^n x\,dx$. Integrals of this type may also be found without using integration by parts. If n is an odd positive integer, we begin by writing

$$\int \sin^n x\,dx = \int \sin^{n-1} x \sin x\,dx.$$

Since the integer $n - 1$ is even, we may then use the trigonometric identity $\sin^2 x = 1 - \cos^2 x$ to obtain a form that is easy to integrate, as illustrated in the following example.

EXAMPLE 1 Evaluate $\int \sin^5 x\,dx$.

SOLUTION As in the preceding discussion, we write

$$\begin{aligned}\int \sin^5 x\,dx &= \int \sin^4 x \sin x\,dx\\ &= \int (\sin^2 x)^2 \sin x\,dx\\ &= \int (1 - \cos^2 x)^2 \sin x\,dx\\ &= \int (1 - 2\cos^2 x + \cos^4 x)\sin x\,dx.\end{aligned}$$

If we substitute

$$u = \cos x, \quad du = -\sin x\, dx,$$

we obtain

$$\begin{aligned}\int \sin^5 x\, dx &= -\int (1 - 2\cos^2 x + \cos^4 x)(-\sin x)\, dx \\ &= -\int (1 - 2u^2 + u^4)\, du \\ &= -u + \tfrac{2}{3}u^3 - \tfrac{1}{5}u^5 + C \\ &= -\cos x + \tfrac{2}{3}\cos^3 x - \tfrac{1}{5}\cos^5 x + C.\end{aligned}$$

Similarly, for odd powers of $\cos x$ we write

$$\int \cos^n x\, dx = \int \cos^{n-1} x \cos x\, dx$$

and use the fact that $\cos^2 x = 1 - \sin^2 x$ to obtain an integrable form.

If the integrand is $\sin^n x$ or $\cos^n x$ and n is *even*, then the half-angle formula

$$\sin^2 x = \frac{1 - \cos 2x}{2} \quad \text{or} \quad \cos^2 x = \frac{1 + \cos 2x}{2}$$

may be used to simplify the integrand.

EXAMPLE 2 Evaluate $\int \cos^2 x\, dx$.

SOLUTION Using a half-angle formula, we have

$$\begin{aligned}\int \cos^2 x\, dx &= \tfrac{1}{2}\int (1 + \cos 2x)\, dx \\ &= \tfrac{1}{2}x + \tfrac{1}{4}\sin 2x + C.\end{aligned}$$

EXAMPLE 3 Evaluate $\int \sin^4 x\, dx$.

SOLUTION

$$\begin{aligned}\int \sin^4 x\, dx &= \int (\sin^2 x)^2\, dx \\ &= \int \left(\frac{1 - \cos 2x}{2}\right)^2 dx \\ &= \tfrac{1}{4}\int (1 - 2\cos 2x + \cos^2 2x)\, dx\end{aligned}$$

We appy a half-angle formula again and write

$$\cos^2 2x = \tfrac{1}{2}(1 + \cos 4x) = \tfrac{1}{2} + \tfrac{1}{2}\cos 4x.$$

Substituting in the last integral and simplifying gives us

$$\int \sin^4 x\,dx = \tfrac{1}{4}\int\left(\tfrac{3}{2} - 2\cos 2x + \tfrac{1}{2}\cos 4x\right)dx$$
$$= \tfrac{3}{8}x - \tfrac{1}{4}\sin 2x + \tfrac{1}{32}\sin 4x + C.$$

Integrals involving only products of $\sin x$ and $\cos x$ may be evaluated using the following guidelines.

Guidelines for evaluating $\int \sin^m x \cos^n x\,dx$ **(9.2)**

1 **If m is an odd integer:** Write the integral as

$$\int \sin^m x \cos^n x\,dx = \int \sin^{m-1} x \cos^n x \sin x\,dx$$

and express $\sin^{m-1} x$ in terms of $\cos x$ by using the trigonometric identity $\sin^2 x = 1 - \cos^2 x$. Make the substitution

$$u = \cos x, \quad du = -\sin x\,dx$$

and evaluate the resulting integral.

2 **If n is an odd integer:** Write the integral as

$$\int \sin^m x \cos^n x\,dx = \int \sin^m x \cos^{n-1} x \cos x\,dx$$

and express $\cos^{n-1} x$ in terms of $\sin x$ by using the trigonometric identity $\cos^2 x = 1 - \sin^2 x$. Make the substitution

$$u = \sin x, \quad du = \cos x\,dx$$

and evaluate the resulting integral.

3 **If m and n are even:** Use half-angle formulas for $\sin^2 x$ and $\cos^2 x$ to reduce the exponents by one-half.

EXAMPLE 4 Evaluate $\int \cos^3 x \sin^4 x\,dx$.

SOLUTION By guideline 2 of (9.2),

$$\int \cos^3 x \sin^4 x\,dx = \int \cos^2 x \sin^4 x \cos x\,dx$$
$$= \int (1 - \sin^2 x)\sin^4 x \cos x\,dx.$$

If we let $u = \sin x$, then $du = \cos x\,dx$, and the integral may be written

$$\int \cos^3 x \sin^4 x\,dx = \int (1 - u^2)u^4\,du = \int (u^4 - u^6)\,du$$
$$= \tfrac{1}{5}u^5 - \tfrac{1}{7}u^7 + C$$
$$= \tfrac{1}{5}\sin^5 x - \tfrac{1}{7}\sin^7 x + C.$$

The following guidelines are analogous to those in (9.2) for integrands of the form $\tan^m x \sec^n x$.

Guidelines for evaluating $\int \tan^m x \sec^n x\, dx$ **(9.3)**

1 **If m is an odd integer:** Write the integral as

$$\int \tan^m x \sec^n x\, dx = \int \tan^{m-1} x \sec^{n-1} x \sec x \tan x\, dx$$

and express $\tan^{m-1} x$ in terms of $\sec x$ by using the trigonometric identity $\tan^2 x = \sec^2 x - 1$. Make the substitution

$$u = \sec x, \quad du = \sec x \tan x\, dx$$

and evaluate the resulting integral.

2 **If n is an even integer:** Write the integral as

$$\int \tan^m x \sec^n x\, dx = \int \tan^m x \sec^{n-2} x \sec^2 x\, dx$$

and express $\sec^{n-2} x$ in terms of $\tan x$ by using the trigonometric identity $\sec^2 x = 1 + \tan^2 x$. Make the substitution

$$u = \tan x, \quad du = \sec^2 x\, dx$$

and evaluate the resulting integral.

3 **If m is even and n is odd:** There is no standard method of evaluation. Possibly use integration by parts.

EXAMPLE 5 Evaluate $\int \tan^3 x \sec^5 x\, dx$.

SOLUTION By guideline 1 of (9.3),

$$\int \tan^3 x \sec^5 x\, dx = \int \tan^2 x \sec^4 x\, (\sec x \tan x)\, dx$$
$$= \int (\sec^2 x - 1) \sec^4 x\, (\sec x \tan x)\, dx.$$

Substituting $u = \sec x$ and $du = \sec x \tan x\, dx$, we obtain

$$\int \tan^3 x \sec^5 x\, dx = \int (u^2 - 1)u^4\, du$$
$$= \int (u^6 - u^4)\, du$$
$$= \tfrac{1}{7}u^7 - \tfrac{1}{5}u^5 + C$$
$$= \tfrac{1}{7} \sec^7 x - \tfrac{1}{5} \sec^5 x + C.$$

EXAMPLE 6 Evaluate $\int \tan^2 x \sec^4 x\, dx$.

SOLUTION By guideline 2 of (9.3),

$$\int \tan^2 x \sec^4 x\, dx = \int \tan^2 x \sec^2 x \sec^2 x\, dx$$
$$= \int \tan^2 x\, (\tan^2 x + 1) \sec^2 x\, dx.$$

If we let $u = \tan x$, then $du = \sec^2 x\, dx$, and

$$\int \tan^2 x \sec^4 x\, dx = \int u^2(u^2 + 1)\, du$$
$$= \int (u^4 + u^2)\, du$$
$$= \tfrac{1}{5}u^5 + \tfrac{1}{3}u^3 + C$$
$$= \tfrac{1}{5} \tan^2 x + \tfrac{1}{3} \tan^3 x + C.$$

Integrals of the form $\int \cot^m x \csc^n x\, dx$ may be evaluated in similar fashion.

Finally, if an integrand has one of the forms $\cos mx \cos nx$, $\sin mx \sin nx$, or $\sin mx \cos nx$, we use a product-to-sum formula to help evaluate the integral, as illustrated in the next example.

EXAMPLE 7 Evaluate $\int \cos 5x \cos 3x\, dx$.

SOLUTION Using the product-to-sum formula for $\cos u \cos v$, we obtain

$$\begin{aligned}\int \cos 5x \cos 3x\, dx &= \int \tfrac{1}{2}(\cos 8x + \cos 2x)\, dx \\ &= \tfrac{1}{16}\sin 8x + \tfrac{1}{4}\sin 2x + C.\end{aligned}$$

EXERCISES 9.2

Exer. 1–30: Evaluate the integral.

1 $\int \cos^3 x\, dx$

2 $\int \sin^2 2x\, dx$

3 $\int \sin^2 x \cos^2 x\, dx$

4 $\int \cos^7 x\, dx$

5 $\int \sin^3 x \cos^2 x\, dx$

6 $\int \sin^5 x \cos^3 x\, dx$

7 $\int \sin^6 x\, dx$

8 $\int \sin^4 x \cos^2 x\, dx$

9 $\int \tan^3 x \sec^4 x\, dx$

10 $\int \sec^6 x\, dx$

11 $\int \tan^3 x \sec^3 x\, dx$

12 $\int \tan^5 x \sec x\, dx$

13 $\int \tan^6 x\, dx$

14 $\int \cot^4 x\, dx$

15 $\int \sqrt{\sin x} \cos^3 x\, dx$

16 $\int \frac{\cos^3 x}{\sqrt{\sin x}}\, dx$

17 $\int (\tan x + \cot x)^2\, dx$

18 $\int \cot^3 x \csc^3 x\, dx$

19 $\int_0^{\pi/4} \sin^3 x\, dx$

20 $\int_0^1 \tan^2 (\tfrac{1}{4}\pi x)\, dx$

21 $\int \sin 5x \sin 3x\, dx$

22 $\int_0^{\pi/4} \cos x \cos 5x\, dx$

23 $\int_0^{\pi/2} \sin 3x \cos 2x\, dx$

24 $\int \sin 4x \cos 3x\, dx$

25 $\int \csc^4 x \cot^4 x\, dx$

26 $\int (1 + \sqrt{\cos x})^2 \sin x\, dx$

27 $\int \frac{\cos x}{2 - \sin x}\, dx$

28 $\int \frac{\tan^2 x - 1}{\sec^2 x}\, dx$

29 $\int \frac{\sec^2 x}{(1 + \tan x)^2}\, dx$

30 $\int \frac{\sec x}{\cot^5 x}\, dx$

31 The region bounded by the x-axis and the graph of $y = \cos^2 x$ from $x = 0$ to $x = 2\pi$ is revolved about the x-axis. Find the volume of the resulting solid.

32 The region between the graphs of $y = \tan^2 x$ and $y = 0$ from $x = 0$ to $x = \pi/4$ is revolved about the x-axis. Find the volume of the resulting solid.

33 The velocity (at time t) of a point moving on a coordinate line is $\cos^2 \pi t$ ft/sec. How far does the point travel in 5 seconds?

34 The acceleration (at time t) of a point moving on a coordinate line is $\sin^2 t \cos t$ ft/sec^2. At $t = 0$ the point is at the origin and its velocity is 10 ft/sec. Find its position at time t.

35 (a) Prove that if m and n are positive integers,

$$\int \sin mx \sin nx\, dx = \begin{cases} \dfrac{\sin (m-n)x}{2(m-n)} - \dfrac{\sin (m+n)x}{2(m+n)} + C & \text{if } m \neq n \\[2ex] \dfrac{x}{2} - \dfrac{\sin 2mx}{4m} + C & \text{if } m = n \end{cases}$$

(b) Obtain formulas similar to that in part (a) for

$$\int \sin mx \cos nx\, dx$$

and $$\int \cos mx \cos nx\, dx.$$

36 (a) Use part (a) of Exercise 35 to prove that

$$\int_{-\pi}^{\pi} \sin mx \sin nx\, dx = \begin{cases} 0 & \text{if } m \neq n \\ \pi & \text{if } m = n \end{cases}$$

(b) Find

(i) $\int_{-\pi}^{\pi} \sin mx \cos nx\, dx$

(ii) $\int_{-\pi}^{\pi} \cos mx \cos nx\, dx$

9.3 TRIGONOMETRIC SUBSTITUTIONS

In Example 1 of Section 1.3 we showed how to change the expression $\sqrt{a^2 - x^2}$, with $a > 0$, into a trigonometric expression without radicals, by using the *trigonometric substitution* $x = a \sin \theta$. We can use a similar procedure for $\sqrt{a^2 + x^2}$ and $\sqrt{x^2 - a^2}$. This technique is useful for eliminating radicals from certain types of integrands. The substitutions are listed in the following table.

Trigonometric substitutions **(9.4)**

EXPRESSION IN INTEGRAND	TRIGONOMETRIC SUBSTITUTION
$\sqrt{a^2 - x^2}$	$x = a \sin \theta$
$\sqrt{a^2 + x^2}$	$x = a \tan \theta$
$\sqrt{x^2 - a^2}$	$x = a \sec \theta$

When making a trigonometric substitution we shall assume that θ is in the range of the corresponding inverse trigonometric function. Thus, for the substitution $x = a \sin \theta$, we have $-\pi/2 \leq \theta \leq \pi/2$. In this case, $\cos \theta \geq 0$ and

$$\begin{aligned}\sqrt{a^2 - x^2} &= \sqrt{a^2 - a^2 \sin^2 \theta} \\ &= \sqrt{a^2(1 - \sin^2 \theta)} \\ &= \sqrt{a^2 \cos^2 \theta} \\ &= a \cos \theta.\end{aligned}$$

If $\sqrt{a^2 - x^2}$ occurs in a denominator, we add the restriction $|x| \neq a$, or, equivalently, $-\pi/2 < \theta < \pi/2$.

EXAMPLE 1 Evaluate $\displaystyle\int \frac{1}{x^2\sqrt{16 - x^2}}\, dx.$

SOLUTION The integrand contains $\sqrt{16 - x^2}$, which is of the form $\sqrt{a^2 - x^2}$ with $a = 4$. Hence, by (9.4), we let

$$x = 4 \sin \theta \quad \text{for} \quad -\pi/2 < \theta < \pi/2.$$

It follows that

$$\sqrt{16 - x^2} = \sqrt{16 - 16 \sin^2 \theta} = 4\sqrt{1 - \sin^2 \theta} = 4\sqrt{\cos^2 \theta} = 4 \cos \theta.$$

Since $x = 4 \sin \theta$, we have $dx = 4 \cos \theta\, d\theta$. Substituting in the given integral yields

$$\begin{aligned}\int \frac{1}{x^2\sqrt{16 - x^2}}\, dx &= \int \frac{1}{(16 \sin^2 \theta)4 \cos \theta}\, 4 \cos \theta\, d\theta \\ &= \frac{1}{16} \int \frac{1}{\sin^2 \theta}\, d\theta \\ &= \tfrac{1}{16} \int \csc^2 \theta\, d\theta \\ &= -\tfrac{1}{16} \cot \theta + C.\end{aligned}$$

FIGURE 9.1
$\sin \theta = \dfrac{x}{4}$

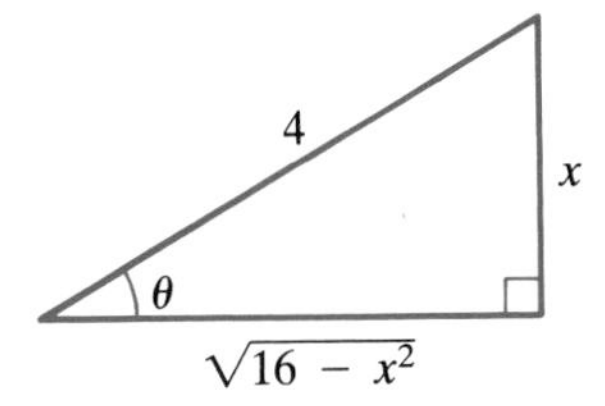

We must now return to the original variable of integration, x. Since $\theta = \arcsin(x/4)$, we could write $-\frac{1}{16}\cot\theta$ as $-\frac{1}{16}\cot\arcsin(x/4)$, but this is a cumbersome expression. Since the integrand contains $\sqrt{16-x^2}$, it is preferable that the evaluated form also contain this radical. There is a simple geometric method for ensuring that it does. If $0 < \theta < \pi/2$ and $\sin\theta = x/4$, we may interpret θ as an acute angle of a right triangle having opposite side and hypotenuse of lengths x and 4, respectively (see Figure 9.1). By the Pythagorean theorem, the length of the adjacent side is $\sqrt{16-x^2}$. Referring to the triangle, we find

$$\cot\theta = \frac{\sqrt{16-x^2}}{x}.$$

It can be shown that the last formula is also true if $-\pi/2 < \theta < 0$. Thus, Figure 9.1 may be used if θ is either positive or negative.

Substituting $\sqrt{16-x^2}/x$ for $\cot\theta$ in our integral evaluation gives us

$$\begin{aligned}\int \frac{1}{x^2\sqrt{16-x^2}}\,dx &= -\frac{1}{16}\cdot\frac{\sqrt{16-x^2}}{x} + C\\ &= -\frac{\sqrt{16-x^2}}{16x} + C.\end{aligned}$$

If an integrand contains $\sqrt{a^2+x^2}$ for $a > 0$, then, by (9.4), we use the substitution $x = a\tan\theta$ to eliminate the radical. When using this substitution we assume that θ is in the range of the inverse tangent function; that is, $-\pi/2 < \theta < \pi/2$. In this case, $\sec\theta > 0$ and

FIGURE 9.2
$\tan\theta = \dfrac{x}{a}$

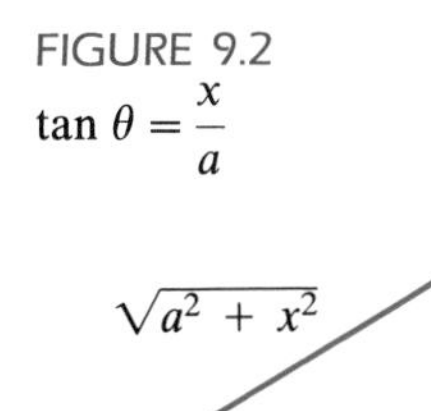

$$\begin{aligned}\sqrt{a^2+x^2} &= \sqrt{a^2 + a^2\tan^2\theta}\\ &= \sqrt{a^2(1+\tan^2\theta)}\\ &= \sqrt{a^2\sec^2\theta}\\ &= a\sec\theta.\end{aligned}$$

After substituting and evaluating the resulting trigonometric integral, it is necessary to return to the variable x. We can do this by using the formula $\tan\theta = x/a$ and referring to the right triangle in Figure 9.2.

EXAMPLE 2 Evaluate $\displaystyle\int \frac{1}{\sqrt{4+x^2}}\,dx$.

SOLUTION The denominator of the integrand has the form $\sqrt{a^2+x^2}$ with $a = 2$. Hence, by (9.4), we make the substitution

$$x = 2\tan\theta, \quad dx = 2\sec^2\theta\,d\theta.$$

Consequently

$$\sqrt{4+x^2} = \sqrt{4+4\tan^2\theta} = 2\sqrt{1+\tan^2\theta} = 2\sqrt{\sec^2\theta} = 2\sec\theta$$

and
$$\int \frac{1}{\sqrt{4+x^2}}\,dx = \int \frac{1}{2\sec\theta}\,2\sec^2\theta\,d\theta$$
$$= \int \sec\theta\,d\theta$$
$$= \ln|\sec\theta + \tan\theta| + C.$$

FIGURE 9.3
$\tan\theta = \dfrac{x}{2}$

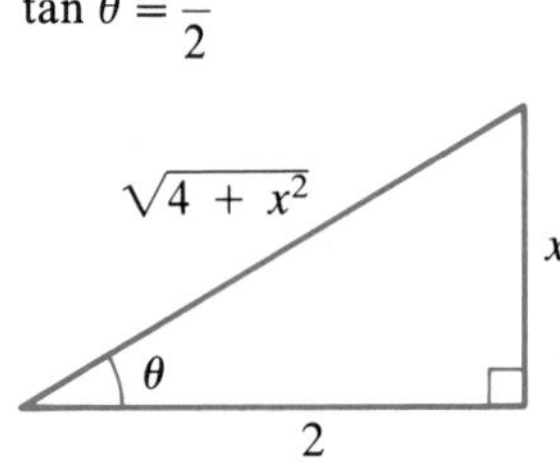

Using $\tan\theta = x/2$, we sketch the triangle in Figure 9.3, from which we obtain

$$\sec\theta = \frac{\sqrt{4+x^2}}{2}.$$

Hence

$$\int \frac{1}{\sqrt{4+x^2}}\,dx = \ln\left|\frac{\sqrt{4+x^2}}{2} + \frac{x}{2}\right| + C.$$

The expression on the right may be written

$$\ln\left|\frac{\sqrt{4+x^2}+x}{2}\right| + C = \ln\left|\sqrt{4+x^2}+x\right| - \ln 2 + C.$$

Since $\sqrt{4+x^2} + x > 0$ for every x, the absolute value sign is unnecessary. If we also let $D = -\ln 2 + C$, then

$$\int \frac{1}{\sqrt{4+x^2}}\,dx = \ln\left(\sqrt{4+x^2}+x\right) + D.$$

If an integrand contains $\sqrt{x^2 - a^2}$, then using (9.4) we substitute $x = a\sec\theta$, where θ is chosen in the range of the inverse secant function; that is, either $0 \le \theta < \pi/2$ or $\pi \le \theta < 3\pi/2$. In this case, $\tan\theta \ge 0$ and

$$\sqrt{x^2-a^2} = \sqrt{a^2\sec^2\theta - a^2}$$
$$= \sqrt{a^2(\sec^2\theta - 1)}$$
$$= \sqrt{a^2\tan^2\theta}$$
$$= a\tan\theta.$$

FIGURE 9.4
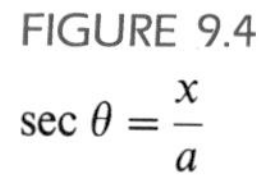
$\sec\theta = \dfrac{x}{a}$

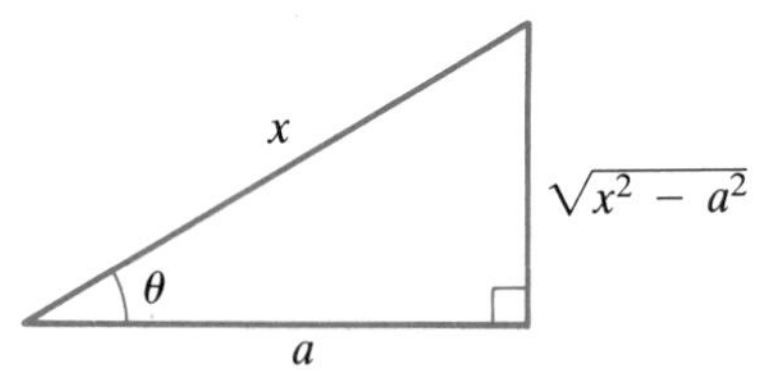

Since

$$\sec\theta = \frac{x}{a},$$

we may refer to the triangle in Figure 9.4 when changing from the variable θ to the variable x.

EXAMPLE 3 Evaluate $\displaystyle\int \frac{\sqrt{x^2-9}}{x}\,dx$.

SOLUTION The integrand contains $\sqrt{x^2-9}$, which is of the form $\sqrt{x^2-a^2}$ with $a = 3$. Referring to (9.4), we substitute as follows:

$$x = 3\sec\theta, \quad dx = 3\sec\theta\tan\theta\,d\theta$$

Consequently

$$\sqrt{x^2-9} = \sqrt{9\sec^2\theta - 9} = 3\sqrt{\sec^2\theta - 1} = 3\sqrt{\tan^2\theta} = 3\tan\theta$$

and

$$\begin{aligned}\int \frac{\sqrt{x^2-9}}{x}\,dx &= \int \frac{3\tan\theta}{3\sec\theta}\,3\sec\theta\tan\theta\,d\theta \\ &= 3\int \tan^2\theta\,d\theta \\ &= 3\int(\sec^2\theta - 1)\,d\theta = 3\int \sec^2\theta\,d\theta - 3\int d\theta \\ &= 3\tan\theta - 3\theta + C.\end{aligned}$$

FIGURE 9.5

$\sec\theta = \dfrac{x}{3}$

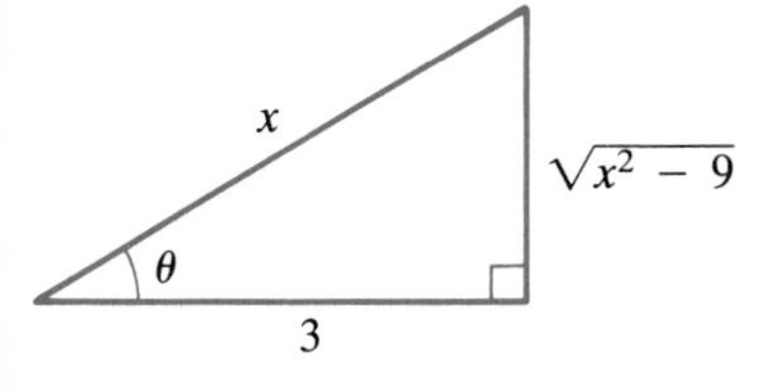

Since $\sec\theta = x/3$, we may refer to the right triangle in Figure 9.5. Using $\tan\theta = \sqrt{x^2-9}/3$ and $\theta = \sec^{-1}(x/3)$, we obtain

$$\begin{aligned}\int \frac{\sqrt{x^2-9}}{x}\,dx &= 3\frac{\sqrt{x^2-9}}{3} - 3\sec^{-1}\left(\frac{x}{3}\right) + C \\ &= \sqrt{x^2-9} - 3\sec^{-1}\left(\frac{x}{3}\right) + C.\end{aligned}$$

As shown in the next example, we can use trigonometric substitutions to evaluate certain integrals that involve $(a^2 - x^2)^n$, $(a^2 + x^2)^n$, or $(x^2 - a^2)^n$, in cases other than $n = \frac{1}{2}$.

EXAMPLE 4 Evaluate $\displaystyle\int \frac{(1-x^2)^{3/2}}{x^6}\,dx.$

SOLUTION The integrand contains the expression $1 - x^2$, which is of the form $a^2 - x^2$ with $a = 1$. Using (9.4), we substitute

$$x = \sin\theta, \quad dx = \cos\theta\,d\theta.$$

Thus, $1 - x^2 = 1 - \sin^2\theta = \cos^2\theta$, and

$$\begin{aligned}\int \frac{(1-x^2)^{3/2}}{x^6}\,dx &= \int \frac{(\cos^2\theta)^{3/2}}{\sin^6\theta}\cos\theta\,d\theta \\ &= \int \frac{\cos^4\theta}{\sin^6\theta}\,d\theta = \int \frac{\cos^4\theta}{\sin^4\theta}\cdot\frac{1}{\sin^2\theta}\,d\theta \\ &= \int \cot^4\theta\csc^2\theta\,d\theta \\ &= -\tfrac{1}{5}\cot^5\theta + C.\end{aligned}$$

FIGURE 9.6

$\sin\theta = x$

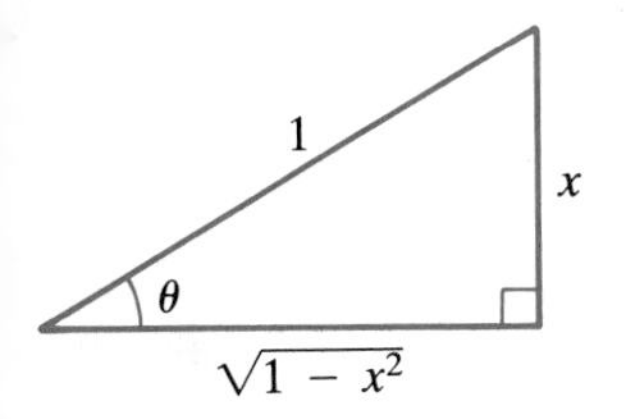

To return to the variable x, we note that $\sin\theta = x = x/1$ and refer to the right triangle in Figure 9.6, obtaining $\cot\theta = \sqrt{1-x^2}/x$. Hence

$$\begin{aligned}\int \frac{(1-x^2)^{3/2}}{x^6}\,dx &= -\frac{1}{5}\left(\frac{\sqrt{1-x^2}}{x}\right)^5 + C \\ &= -\frac{(1-x^2)^{5/2}}{5x^5} + C.\end{aligned}$$

Although we now have additional integration techniques available, it is a good idea to keep earlier methods in mind. For example, the integral $\int (x/\sqrt{9+x^2})\,dx$ could be evaluated by means of the trigonometric substitution $x = 3\tan\theta$. However, it is simpler to use the algebraic substitution $u = 9 + x^2$ and $du = 2x\,dx$, for in this event the integral takes on the form $\frac{1}{2}\int u^{-1/2}\,du$, which is readily integrated by means of the power rule. The following exercises include integrals that can be evaluated using simpler techniques than trigonometric substitutions.

EXERCISES 9.3

Exer. 1–22: Evaluate the integral.

1 $\displaystyle\int \frac{1}{x\sqrt{4-x^2}}\,dx$

2 $\displaystyle\int \frac{\sqrt{4-x^2}}{x^2}\,dx$

3 $\displaystyle\int \frac{1}{x\sqrt{9+x^2}}\,dx$

4 $\displaystyle\int \frac{1}{x^2\sqrt{x^2+9}}\,dx$

5 $\displaystyle\int \frac{1}{x^2\sqrt{x^2-25}}\,dx$

6 $\displaystyle\int \frac{1}{x^3\sqrt{x^2-25}}\,dx$

7 $\displaystyle\int \frac{x}{\sqrt{4-x^2}}\,dx$

8 $\displaystyle\int \frac{x}{x^2+9}\,dx$

9 $\displaystyle\int \frac{1}{(x^2-1)^{3/2}}\,dx$

10 $\displaystyle\int \frac{1}{\sqrt{4x^2-25}}\,dx$

11 $\displaystyle\int \frac{1}{(36+x^2)^2}\,dx$

12 $\displaystyle\int \frac{1}{(16-x^2)^{5/2}}\,dx$

13 $\displaystyle\int \frac{1}{\sqrt{9-x^2}}\,dx$

14 $\displaystyle\int \frac{1}{49+x^2}\,dx$

15 $\displaystyle\int \frac{x}{(16-x^2)^2}\,dx$

16 $\displaystyle\int x\sqrt{x^2-9}\,dx$

17 $\displaystyle\int \frac{x^3}{\sqrt{9x^2+49}}\,dx$

18 $\displaystyle\int \frac{1}{x\sqrt{25x^2+16}}\,dx$

19 $\displaystyle\int \frac{1}{x^4\sqrt{x^2-3}}\,dx$

20 $\displaystyle\int \frac{x^2}{(1-9x^2)^{3/2}}\,dx$

21 $\displaystyle\int \frac{(4+x^2)^2}{x^3}\,dx$

22 $\displaystyle\int \frac{3x-5}{\sqrt{1-x^2}}\,dx$

23 The region bounded by the graphs of $y = x(x^2+25)^{-1/2}$, $y = 0$, and $x = 5$ is revolved about the y-axis. Find the volume of the resulting solid.

24 Find the area of the region bounded by the graph of $y = x^3(10-x^2)^{-1/2}$, the x-axis, and the line $x = 1$.

Exer. 25–26: Solve the differential equation subject to the given initial condition.

25 $x\,dy = \sqrt{x^2-16}\,dx;\quad y = 0 \text{ if } x = 4$

26 $\sqrt{1-x^2}\,dy = x^3\,dx;\quad y = 0 \text{ if } x = 0$

Exer. 27–32: Use a trigonometric substitution to derive the formula. (See Formulas 21, 27, 31, 36, 41, and 44 in Appendix IV.)

27 $\displaystyle\int \sqrt{a^2+u^2}\,du = \frac{u}{2}\sqrt{a^2+u^2} + \frac{a^2}{2}\ln\left|u+\sqrt{a^2+u^2}\right| + C$

28 $\displaystyle\int \frac{1}{u\sqrt{a^2+u^2}}\,du = -\frac{1}{a}\ln\left|\frac{\sqrt{a^2+u^2}+a}{u}\right| + C$

29 $\displaystyle\int u^2\sqrt{a^2-u^2}\,du = \frac{u}{8}(2u^2-a^2)\sqrt{a^2-u^2} + \frac{a^4}{8}\sin^{-1}\frac{u}{a} + C$

30 $\displaystyle\int \frac{1}{u^2\sqrt{a^2-u^2}}\,du = -\frac{1}{a^2u}\sqrt{a^2-u^2} + C$

31 $\displaystyle\int \frac{\sqrt{u^2-a^2}}{u}\,du = \sqrt{u^2-a^2} - a\cos^{-1}\frac{a}{u} + C$

32 $\displaystyle\int \frac{u^2}{\sqrt{u^2-a^2}}\,du = \frac{u}{2}\sqrt{u^2-a^2} + \frac{a^2}{2}\ln\left|u+\sqrt{u^2-a^2}\right| + C$

9.4 INTEGRALS OF RATIONAL FUNCTIONS

Recall that if q is a rational function, then $q(x) = f(x)/g(x)$, where $f(x)$ and $g(x)$ are polynomials. In this section we shall state rules for evaluating $\int q(x)\,dx$.

Let us consider the specific case $q(x) = 2/(x^2 - 1)$. It is easy to verify that

$$\frac{1}{x-1} + \frac{-1}{x+1} = \frac{2}{x^2-1}.$$

The expression on the left side of the equation is called the *partial fraction decomposition* of $2/(x^2 - 1)$. To find $\int q(x)\,dx$, we integrate each of the fractions that make up the decomposition, obtaining

$$\begin{aligned}\int \frac{2}{x^2-1}\,dx &= \int \frac{1}{x-1}\,dx + \int \frac{-1}{x+1}\,dx \\ &= \ln|x-1| - \ln|x+1| + C \\ &= \ln\left|\frac{x-1}{x+1}\right| + C.\end{aligned}$$

It is theoretically possible to write *any* rational expression $f(x)/g(x)$ as a sum of rational expressions whose denominators involve powers of polynomials of degree not greater than two. Specifically, if $f(x)$ and $g(x)$ are polynomials *and the degree of $f(x)$ is less than the degree of $g(x)$*, then it can be proved that

$$\frac{f(x)}{g(x)} = F_1 + F_2 + \cdots + F_r$$

such that each term F_k of the sum has one of the forms

$$\frac{A}{(ax+b)^n} \quad \text{or} \quad \frac{Ax+B}{(ax^2+bx+c)^n}$$

for real numbers A and B and a nonnegative integer n, where $ax^2 + bx + c$ is **irreducible** in the sense that this quadratic polynomial has no real zeros (that is, $b^2 - 4ac < 0$). In this case, $ax^2 + bx + c$ cannot be expressed as a product of two first-degree polynomials with real coefficients.

The sum $F_1 + F_2 + \cdots + F_r$ is the **partial fraction decomposition** of $f(x)/g(x)$, and each F_k is a **partial fraction**. We shall not prove this algebraic result but shall, instead, state guidelines for obtaining the decomposition.

The guidelines for finding the partial fraction decomposition of $f(x)/g(x)$ should be used only if $f(x)$ has lower degree than $g(x)$. If this is not the case, then we may use long division to arrive at the proper form. For example, given

$$\frac{x^3 - 6x^2 + 5x - 3}{x^2 - 1},$$

we obtain, by long division,

$$\frac{x^3 - 6x^2 + 5x - 3}{x^2 - 1} = x - 6 + \frac{6x - 9}{x^2 - 1}.$$

We then find the partial fraction decomposition for $(6x - 9)/(x^2 - 1)$.

Guidelines for partial fraction decompositions of $f(x)/g(x)$ **(9.5)**

1. If the degree of $f(x)$ is not lower than the degree of $g(x)$, use long division to obtain the proper form.
2. Express $g(x)$ as a product of linear factors $ax + b$ or irreducible quadratic factors $ax^2 + bx + c$, and collect repeated factors so that $g(x)$ is a product of *different* factors of the form $(ax + b)^n$ or $(ax^2 + bx + c)^n$ for a nonnegative integer n.
3. Apply the following rules.

Rule a For each factor $(ax + b)^n$ with $n \geq 1$, the partial fraction decomposition contains a sum of n partial fractions of the form

$$\frac{A_1}{ax + b} + \frac{A_2}{(ax + b)^2} + \cdots + \frac{A_n}{(ax + b)^n},$$

where each numerator A_k is a real number.

Rule b For each factor $(ax^2 + bx + c)^n$ with $n \geq 1$ and with $ax^2 + bx + c$ irreducible, the partial fraction decomposition contains a sum of n partial fractions of the form

$$\frac{A_1x + B_1}{ax^2 + bx + c} + \frac{A_2x + B_2}{(ax^2 + bx + c)^2} + \cdots + \frac{A_nx + B_n}{(ax^2 + bx + c)^n},$$

where each A_k and B_k is a real number.

EXAMPLE 1 Evaluate $\displaystyle\int \frac{4x^2 + 13x - 9}{x^3 + 2x^2 - 3x}\,dx$.

SOLUTION We may factor the denominator of the integrand as follows:

$$x^3 + 2x^2 - 3x = x(x^2 + 2x - 3) = x(x + 3)(x - 1)$$

Each factor has the form stated in Rule a of (9.5), with $m = 1$. Thus, to the factor x there corresponds a partial fraction of the form A/x. Similarly, to the factors $x + 3$ and $x - 1$ there correspond partial fractions $B/(x + 3)$ and $C/(x - 1)$, respectively. Therefore the partial fraction decomposition has the form

$$\frac{4x^2 + 13x - 9}{x(x + 3)(x - 1)} = \frac{A}{x} + \frac{B}{x + 3} + \frac{C}{x - 1}.$$

Multiplying by the lowest common denominator gives us

$$(*) \qquad 4x^2 + 13x - 9 = A(x + 3)(x - 1) + Bx(x - 1) + Cx(x + 3).$$

In a case such as this, in which the factors are all linear and nonrepeated, the values for A, B, and C can be found by substituting values for x that make the various factors zero. If we let $x = 0$ in $(*)$, then

$$-9 = -3A, \quad \text{or} \quad A = 3.$$

Letting $x = 1$ in $(*)$ gives us

$$8 = 4C, \quad \text{or} \quad C = 2.$$

Finally, if $x = -3$ in $(*)$, then

$$-12 = 12B, \quad \text{or} \quad B = -1.$$

The partial fraction decomposition is, therefore,

$$\frac{4x^2 + 13x - 9}{x(x + 3)(x - 1)} = \frac{3}{x} + \frac{-1}{x + 3} + \frac{2}{x - 1}.$$

Integrating and letting K denote the sum of the constants of integration, we have

$$\begin{aligned}\int \frac{4x^2 + 13x - 9}{x(x + 3)(x - 1)}\,dx &= \int \frac{3}{x}\,dx + \int \frac{-1}{x + 3}\,dx + \int \frac{2}{x - 1}\,dx \\ &= 3\ln|x| - \ln|x + 3| + 2\ln|x - 1| + K \\ &= \ln|x^3| - \ln|x + 3| + \ln|x - 1|^2 + K \\ &= \ln\left|\frac{x^3(x - 1)^2}{x + 3}\right| + K.\end{aligned}$$

Another technique for finding A, B, and C is to expand the right-hand side of (*) and collect like powers of x as follows:

$$4x^2 + 13x - 9 = (A + B + C)x^2 + (2A - B + 3C)x - 3A$$

We now use the fact that if two polynomials are equal, then coefficients of like powers of x are the same. It is convenient to arrange our work in the following way, which we call **comparing coefficients of x**.

$$\begin{aligned} &\textit{coefficients of } x^2\text{:} && A + B + C = 4 \\ &\textit{coefficients of } x\text{:} && 2A - B + 3C = 13 \\ &\textit{constant terms}\text{:} && -3A = -9 \end{aligned}$$

We may show that the solution of this system of equations is $A = 3$, $B = -1$, and $C = 2$.

EXAMPLE 2 Evaluate $\displaystyle\int \frac{3x^3 - 18x^2 + 29x - 4}{(x + 1)(x - 2)^3}\,dx.$

SOLUTION By Rule a of (9.5), there is a partial fraction of the form $A/(x + 1)$ corresponding to the factor $x + 1$ in the denominator of the integrand. For the factor $(x - 2)^3$ we apply Rule a (with $m = 3$), obtaining a sum of three partial fractions $B/(x - 2)$, $C/(x - 2)^2$, and $D/(x - 2)^3$. Consequently, the partial fraction decomposition has the form

$$\frac{3x^3 - 18x^2 + 29x - 4}{(x + 1)(x - 2)^3} = \frac{A}{x + 1} + \frac{B}{x - 2} + \frac{C}{(x - 2)^2} + \frac{D}{(x - 2)^3}.$$

Multiplying both sides by $(x + 1)(x - 2)^3$ gives us

$$(*) \qquad \begin{aligned} 3x^3 - 18x^2 + 29x - 4 = {} & A(x - 2)^3 + B(x + 1)(x - 2)^2 \\ & + C(x + 1)(x - 2) + D(x + 1). \end{aligned}$$

Two of the unknown constants may be determined easily. If we let $x = 2$ in (*), we obtain

$$6 = 3D, \quad \text{or} \quad D = 2.$$

Similarly, letting $x = -1$ in $(*)$ yields

$$-54 = -27A, \quad \text{or} \quad A = 2.$$

The remaining constants may be found by comparing coefficients. Examining the right-hand side of $(*)$, we see that the coefficient of x^3 is $A + B$. This must equal the coefficient of x^3 on the left. Thus, by comparison,

$$\textit{coefficients of } x^3\text{:} \quad 3 = A + B.$$

Since $A = 2$, it follows that $B = 1$.

Finally, we compare the constant terms in $(*)$ by letting $x = 0$. This gives us the following:

$$\textit{constant terms:} \quad -4 = -8A + 4B - 2C + D$$

Substituting the values we have found for A, B, and D into the preceding equation yields

$$-4 = -16 + 4 - 2C + 2,$$

which has the solution $C = -3$. The partial fraction decomposition is, therefore,

$$\frac{3x^3 - 18x^2 + 29x - 4}{(x + 1)(x - 2)^3} = \frac{2}{x + 1} + \frac{1}{x - 2} + \frac{-3}{(x - 2)^2} + \frac{2}{(x - 2)^3}.$$

To find the given integral, we integrate each of the partial fractions on the right side of the last equation, obtaining

$$2 \ln |x + 1| + \ln |x - 2| + \frac{3}{x - 2} - \frac{1}{(x - 2)^2} + K$$

with K the sum of the four constants of integration. This may be written in the form

$$\ln [(x + 1)^2 |x - 2|] + \frac{3}{x - 2} - \frac{1}{(x - 2)^2} + K.$$

EXAMPLE 3 Evaluate $\displaystyle\int \frac{x^2 - x - 21}{2x^3 - x^2 + 8x - 4}\, dx.$

SOLUTION The denominator may be factored by grouping as follows:

$$2x^3 - x^2 + 8x - 4 = x^2(2x - 1) + 4(2x - 1) = (x^2 + 4)(2x - 1)$$

Applying Rule b of (9.5) to the irreducible quadratic factor $x^2 + 4$, we see that one of the partial fractions has the form $(Ax + B)/(x^2 + 4)$. By Rule a, there is also a partial fraction $C/(2x - 1)$ corresponding to the factor $2x - 1$. Consequently,

$$\frac{x^2 - x - 21}{2x^3 - x^2 + 8x - 4} = \frac{Ax + B}{x^2 + 4} + \frac{C}{2x - 1}.$$

As in previous examples, this leads to

$$(*) \qquad x^2 - x - 21 = (Ax + B)(2x - 1) + C(x^2 + 4).$$

We can find one constant easily. Substituting $x = \frac{1}{2}$ in $(*)$ gives us

$$-\tfrac{85}{4} = \tfrac{17}{4}C, \quad \text{or} \quad C = -5.$$

The remaining constants may be found by comparing coefficients of x in $(*)$:

$$\begin{aligned} \textit{coefficients of } x^2\text{:} \quad & 1 = 2A + C \\ \textit{coefficients of } x\text{:} \quad & -1 = -A + 2B \\ \textit{constant terms:} \quad & -21 = -B + 4C \end{aligned}$$

Since $C = -5$, it follows from $1 = 2A + C$ that $A = 3$. Similarly, using the coefficients of x with $A = 3$ gives us $-1 = -3 + 2B$, or $B = 1$. Thus the partial fraction decomposition of the integrand is

$$\begin{aligned} \frac{x^2 - x - 21}{2x^3 - x^2 + 8x - 4} &= \frac{3x + 1}{x^2 + 4} + \frac{-5}{2x - 1} \\ &= \frac{3x}{x^2 + 4} + \frac{1}{x^2 + 4} - \frac{5}{2x - 1}. \end{aligned}$$

The given integral may now be found by integrating the right side of the last equation. This gives us

$$\frac{3}{2}\ln(x^2 + 4) + \frac{1}{2}\tan^{-1}\frac{x}{2} - \frac{5}{2}\ln|2x - 1| + K.$$

EXAMPLE 4 Evaluate $\displaystyle\int \frac{5x^3 - 3x^2 + 7x - 3}{(x^2 + 1)^2}\,dx.$

SOLUTION Applying Rule b of (9.5), with $n = 2$, yields

$$\frac{5x^3 - 3x^2 + 7x - 3}{(x^2 + 1)^2} = \frac{Ax + B}{x^2 + 1} + \frac{Cx + D}{(x^2 + 1)^2}.$$

Multiplying by the lcd $(x^2 + 1)^2$ gives us

$$\begin{aligned} 5x^3 - 3x^2 + 7x - 3 &= (Ax + B)(x^2 + 1) + Cx + D \\ 5x^3 - 3x^2 + 7x - 3 &= Ax^3 + Bx^2 + (A + C)x + (B + D). \end{aligned}$$

We next compare coefficients as follows:

$$\begin{aligned} \textit{coefficients of } x^3\text{:} \quad & 5 = A \\ \textit{coefficients of } x^2\text{:} \quad & -3 = B \\ \textit{coefficients of } x\text{:} \quad & 7 = A + C \\ \textit{constant terms:} \quad & -3 = B + D \end{aligned}$$

This gives us $A = 5$, $B = -3$, $C = 7 - A = 2$, and $D = -3 - B = 0$. Therefore

$$\frac{5x^3 - 3x^2 + 7x - 3}{(x^2 + 1)^2} = \frac{5x - 3}{x^2 + 1} + \frac{2x}{(x^2 + 1)^2} = \frac{5x}{x^2 + 1} - \frac{3}{x^2 + 1} + \frac{2x}{(x^2 + 1)^2}.$$

Integrating yields

$$\int \frac{5x^3 - 3x^2 + 7x - 3}{(x^2 + 1)^2}\,dx = \frac{5}{2}\ln(x^2 + 1) - 3\tan^{-1} x - \frac{1}{x^2 + 1} + K.$$

EXERCISES 9.4

Exer. 1–32: Evaluate the integral.

1 $\int \frac{5x - 12}{x(x - 4)}\,dx$

2 $\int \frac{x + 34}{(x - 6)(x + 2)}\,dx$

3 $\int \frac{37 - 11x}{(x + 1)(x - 2)(x - 3)}\,dx$

4 $\int \frac{4x^2 + 54x + 134}{(x - 1)(x + 5)(x + 3)}\,dx$

5 $\int \frac{6x - 11}{(x - 1)^2}\,dx$

6 $\int \frac{-19x^2 + 50x - 25}{x^2(3x - 5)}\,dx$

7 $\int \frac{x + 16}{x^2 + 2x - 8}\,dx$

8 $\int \frac{11x + 2}{2x^2 - 5x - 3}\,dx$

9 $\int \frac{5x^2 - 10x - 8}{x^3 - 4x}\,dx$

10 $\int \frac{4x^2 - 5x - 15}{x^3 - 4x^2 - 5x}\,dx$

11 $\int \frac{2x^2 - 25x - 33}{(x + 1)^2(x - 5)}\,dx$

12 $\int \frac{2x^2 - 12x + 4}{x^3 - 4x^2}\,dx$

13 $\int \frac{9x^4 + 17x^3 + 3x^2 - 8x + 3}{x^5 + 3x^4}\,dx$

14 $\int \frac{5x^2 + 30x + 43}{(x + 3)^3}\,dx$

15 $\int \frac{x^3 + 6x^2 + 3x + 16}{x^3 + 4x}\,dx$

16 $\int \frac{2x^2 + 7x}{x^2 + 6x + 9}\,dx$

17 $\int \frac{5x^2 + 11x + 17}{x^3 + 5x^2 + 4x + 20}\,dx$

18 $\int \frac{4x^3 - 3x^2 + 6x - 27}{x^4 + 9x^2}\,dx$

19 $\int \frac{x^2 + 3x + 1}{x^4 + 5x^2 + 4}\,dx$

20 $\int \frac{4x}{(x^2 + 1)^3}\,dx$

21 $\int \frac{2x^3 + 10x}{(x^2 + 1)^2}\,dx$

22 $\int \frac{x^4 + 2x^2 + 4x + 1}{(x^2 + 1)^3}\,dx$

23 $\int \frac{x^3 + 3x - 2}{x^2 - x}\,dx$

24 $\int \frac{x^4 + 2x^2 + 3}{x^3 - 4x}\,dx$

25 $\int \frac{x^6 - x^3 + 1}{x^4 + 9x^2}\,dx$

26 $\int \frac{x^5}{(x^2 + 4)^2}\,dx$

27 $\int \frac{2x^3 - 5x^2 + 46x + 98}{(x^2 + x - 12)^2}\,dx$

28 $\int \frac{-2x^4 - 3x^3 - 3x^2 + 3x + 1}{x^2(x + 1)^3}\,dx$

29 $\int \frac{4x^3 + 2x^2 - 5x - 18}{(x - 4)(x + 1)^3}\,dx$

30 $\int \frac{10x^2 + 9x + 1}{2x^3 + 3x^2 + x}\,dx$

31 $\int \frac{x^3 + 3x^2 + 3x + 63}{(x^2 - 9)^2}\,dx$

32 $\int \frac{x^5 - x^4 - 2x^3 + 4x^2 - 15x + 5}{(x^2 + 1)^2(x^2 + 4)}\,dx$

Exer. 33–36: Use partial fractions to evaluate the integral (see Formulas 19, 49, 50, and 52 of the table of integrals in Appendix IV).

33 $\int \frac{1}{a^2 - u^2}\,du$

34 $\int \frac{1}{u(a + bu)}\,du$

35 $\int \frac{1}{u^2(a + bu)}\,du$

36 $\int \frac{1}{u(a + bu)^2}\,du$

37 If $f(x) = x/(x^2 - 2x - 3)$, find the area of the region under the graph of f from $x = 0$ to $x = 2$.

38 The region bounded by the graphs of $y = 1/(x-1)(4-x)$, $y = 0$, $x = 2$, and $x = 3$ is revolved about the y-axis. Find the volume of the resulting solid.

39 If the region described in Exercise 38 is revolved about the x-axis, find the volume of the resulting solid.

40 In the *law of logistic growth*, it is assumed that at time t, the rate of growth $f'(t)$ of a quantity $f(t)$ is given by $f'(t) = Af(t)[B - f(t)]$, where A and B are constants. If $f(0) = C$, show that

$$f(t) = \frac{BC}{C + (B - C)e^{-ABt}}.$$

41 As an alternative to partial fractions, show that an integral of the form

$$\int \frac{1}{ax^2 + bx}\,dx$$

may be evaluated by writing it as

$$\int \frac{(1/x^2)}{a + (b/x)}\,dx$$

and using the substitution $u = a + (b/x)$.

42 Generalize Exercise 41 to integrals of the form

$$\int \frac{1}{ax^n + bx}\,dx.$$

43 Suppose $g(x) = (x - c_1)(x - c_2)\cdots(x - c_n)$ for a positive integer n and distinct real numbers $c_1, c_2, \ldots, c_n$. If $f(x)$ is a polynomial of degree less than n, show that

$$\frac{f(x)}{g(x)} = \frac{A_1}{x - c_1} + \frac{A_2}{x - c_2} + \cdots + \frac{A_n}{x - c_n}$$

with $A_k = f(c_k)/g'(c_k)$ for $k = 1, 2, \ldots, n$. (This is a method for finding the partial fraction decomposition if the denominator can be factored into distinct linear factors.)

44 Use Exercise 43 to find the partial fraction decomposition of

$$\frac{2x^4 - x^3 - 3x^2 + 5x + 7}{x^5 - 5x^3 + 4x}.$$

9.5 INTEGRALS INVOLVING QUADRATIC EXPRESSIONS

Partial fraction decompositions may lead to integrands containing an irreducible quadratic expression $ax^2 + bx + c$. If $b \neq 0$, it is sometimes necessary to complete the square as follows:

$$\begin{aligned} ax^2 + bx + c &= a\left(x^2 + \frac{b}{a}x\right) + c \\ &= a\left(x + \frac{b}{2a}\right)^2 + c - \frac{b^2}{4a} \end{aligned}$$

The substitution $u = x + b/(2a)$ may then lead to an integrable form.

EXAMPLE 1 Evaluate $\int \frac{2x - 1}{x^2 - 6x + 13}\,dx.$

SOLUTION Note that the quadratic expression $x^2 - 6x + 13$ is irreducible, since $b^2 - 4ac = -16 < 0$. We complete the square as follows:

$$\begin{aligned} x^2 - 6x + 13 &= (x^2 - 6x \qquad) + 13 \\ &= (x^2 - 6x + 9) + 13 - 9 = (x - 3)^2 + 4 \end{aligned}$$

Thus,

$$\int \frac{2x - 1}{x^2 - 6x + 13}\,dx = \int \frac{2x - 1}{(x - 3)^2 + 4}\,dx.$$

We now make the substitution

$$u = x - 3, \quad x = u + 3, \quad dx = du.$$

Thus,

$$\begin{aligned}\int \frac{2x-1}{x^2-6x+13}\,dx &= \int \frac{2(u+3)-1}{u^2+4}\,du \\ &= \int \frac{2u+5}{u^2+4}\,du \\ &= \int \frac{2u}{u^2+4}\,du + 5\int \frac{1}{u^2+4}\,du \\ &= \ln(u^2+4) + \frac{5}{2}\tan^{-1}\frac{u}{2} + C \\ &= \ln(x^2-6x+13) + \frac{5}{2}\tan^{-1}\frac{x-3}{2} + C.\end{aligned}$$

We may also employ the technique of completing the square if a quadratic expression appears under a radical sign.

EXAMPLE 2 Evaluate $\displaystyle\int \frac{1}{\sqrt{8+2x-x^2}}\,dx.$

SOLUTION We complete the square for the quadratic expression $8 + 2x - x^2$ as follows:

$$\begin{aligned}8 + 2x - x^2 = 8 - (x^2 - 2x) &= 8 + 1 - (x^2 - 2x + 1) \\ &= 9 - (x-1)^2\end{aligned}$$

Thus,

$$\int \frac{1}{\sqrt{8+2x-x^2}}\,dx = \int \frac{1}{\sqrt{9-(x-1)^2}}\,dx.$$

Using the substitution

$$u = x - 1, \quad du = dx$$

yields

$$\begin{aligned}\int \frac{1}{\sqrt{8+2x-x^2}}\,dx &= \int \frac{1}{\sqrt{9-(x-1)^2}}\,dx \\ &= \int \frac{1}{\sqrt{9-u^2}}\,du \\ &= \sin^{-1}\frac{u}{3} + C \\ &= \sin^{-1}\frac{x-1}{3} + C.\end{aligned}$$

In the next example we make a trigonometric substitution after completing the square.

EXAMPLE 3 Evaluate $\int \frac{1}{\sqrt{x^2 + 8x + 25}}\,dx.$

SOLUTION We complete the square for the quadratic expression as follows:

$$\begin{aligned} x^2 + 8x + 25 &= (x^2 + 8x \qquad) + 25 \\ &= (x^2 + 8x + 16) + 25 - 16 \\ &= (x + 4)^2 + 9 \end{aligned}$$

Thus,

$$\int \frac{1}{\sqrt{x^2 + 8x + 25}}\,dx = \int \frac{1}{\sqrt{(x+4)^2 + 9}}\,dx.$$

If we make the trigonometric substitution

$$x + 4 = 3\tan\theta, \quad dx = 3\sec^2\theta\,d\theta,$$

then

$$\sqrt{(x+4)^2 + 9} = \sqrt{9\tan^2\theta + 9} = 3\sqrt{\tan^2\theta + 1} = 3\sec\theta$$

and

$$\begin{aligned} \int \frac{1}{\sqrt{x^2 + 8x + 25}}\,dx &= \int \frac{1}{3\sec\theta}\,3\sec^2\theta\,d\theta \\ &= \int \sec\theta\,d\theta \\ &= \ln|\sec\theta + \tan\theta| + C. \end{aligned}$$

FIGURE 9.7

$\tan\theta = \frac{x+4}{3}$

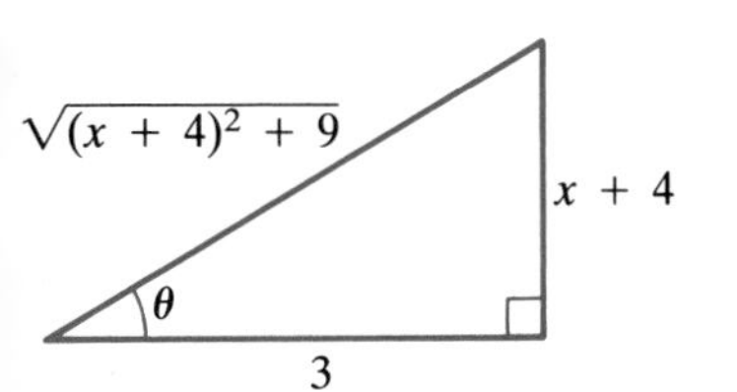

To return to the variable x, we use the triangle in Figure 9.7, obtaining

$$\begin{aligned} \int \frac{1}{\sqrt{x^2 + 8x + 25}}\,dx &= \ln\left|\frac{\sqrt{x^2 + 8x + 25}}{3} + \frac{x+4}{3}\right| + C \\ &= \ln\left|\sqrt{x^2 + 8x + 25} + x + 4\right| - \ln|3| + C \\ &= \ln\left|\sqrt{x^2 + 8x + 25} + x + 4\right| + K \end{aligned}$$

with $K = C - \ln 3$.

EXERCISES 9.5

Exer. 1–18: Evaluate the integral.

1 $\int \frac{1}{(x+1)^2 + 4}\,dx$

2 $\int \frac{1}{\sqrt{16 - (x-3)^2}}\,dx$

3 $\int \frac{1}{x^2 - 4x + 8}\,dx$

4 $\int \frac{1}{x^2 - 2x + 2}\,dx$

5 $\int \frac{1}{\sqrt{4x - x^2}}\,dx$

6 $\int \frac{1}{\sqrt{7 + 6x - x^2}}\,dx$

7 $\int \frac{2x + 3}{\sqrt{9 - 8x - x^2}}\,dx$

8 $\int \frac{x + 5}{9x^2 + 6x + 17}\,dx$

9 $\int \frac{1}{(x^2 + 4x + 5)^2}\,dx$

10 $\int \frac{1}{(x^2 - 6x + 34)^{3/2}}\,dx$

11 $\int \frac{1}{(x^2 + 6x + 13)^{3/2}}\,dx$

12 $\int \sqrt{x(6 - x)}\,dx$

13 $\int \frac{1}{2x^2 - 3x + 9}\,dx$

14 $\int \frac{2x}{(x^2 + 2x + 5)^2}\,dx$

15 $\int \frac{e^x}{e^{2x} + 3e^x + 2}\,dx$

16 $\int \sqrt{x^2 + 10x}\,dx$

17 $\int_2^3 \frac{x^2 - 4x + 6}{x^2 - 4x + 5}\,dx$

18 $\int_0^1 \frac{x - 1}{x^2 + x + 1}\,dx$

19 Find the area of the region bounded by the graphs of $y = 1/(x^2 + 4x + 29)$, $y = 0$, $x = -2$, and $x = 3$.

20 The region bounded by the graph of $y = 1/(x^2 + 2x + 10)$, the coordinate axes, and the line $x = 2$ is revolved about the x-axis. Find the volume of the resulting solid.

9.6 MISCELLANEOUS SUBSTITUTIONS

In this section we shall consider substitutions that are useful for evaluating certain types of integrals. The first example illustrates that if an integral contains an expression of the form $\sqrt[n]{f(x)}$, then one of the substitutions $u = \sqrt[n]{f(x)}$ or $u = f(x)$ may simplify the evaluation.

EXAMPLE 1 Evaluate $\int \frac{x^3}{\sqrt[3]{x^2 + 4}}\,dx$.

SOLUTION 1 The substitution $u = \sqrt[3]{x^2 + 4}$ leads to the following equivalent equations:

$$u = \sqrt[3]{x^2 + 4}, \quad u^3 = x^2 + 4, \quad x^2 = u^3 - 4$$

Taking the differential of each side of the last equation, we obtain

$$2x\,dx = 3u^2\,du, \quad \text{or} \quad x\,dx = \tfrac{3}{2}u^2\,du.$$

We now substitute as follows:

$$\begin{aligned}
\int \frac{x^3}{\sqrt[3]{x^2 + 4}}\,dx &= \int \frac{x^2}{\sqrt[3]{x^2 + 4}} \cdot x\,dx \\
&= \int \frac{u^3 - 4}{u} \cdot \frac{3}{2}u^2\,du = \frac{3}{2}\int (u^4 - 4u)\,du \\
&= \tfrac{3}{2}(\tfrac{1}{5}u^5 - 2u^2) + C = \tfrac{3}{10}u^2(u^3 - 10) + C \\
&= \tfrac{3}{10}(x^2 + 4)^{2/3}(x^2 - 6) + C
\end{aligned}$$

SOLUTION 2 If we substitute u for the expression *underneath* the radical, then

$$u = x^2 + 4, \quad \text{or} \quad x^2 = u - 4$$

and

$$2x\,dx = du, \quad \text{or} \quad x\,dx = \tfrac{1}{2}\,du.$$

In this case we may write

$$\begin{aligned}
\int \frac{x^3}{\sqrt[3]{x^2 + 4}}\,dx &= \int \frac{x^2}{\sqrt[3]{x^2 + 4}} \cdot x\,dx \\
&= \int \frac{u - 4}{u^{1/3}} \cdot \frac{1}{2}\,du = \frac{1}{2}\int (u^{2/3} - 4u^{-1/3})\,du \\
&= \tfrac{1}{2}(\tfrac{3}{5}u^{5/3} - 6u^{2/3}) + C = \tfrac{3}{10}u^{2/3}(u - 10) + C \\
&= \tfrac{3}{10}(x^2 + 4)^{2/3}(x^2 - 6) + C.
\end{aligned}$$

EXAMPLE 2 Evaluate $\int \frac{1}{\sqrt{x} + \sqrt[3]{x}}\,dx$.

SOLUTION To obtain a substitution that will eliminate the two radicals $\sqrt{x} = x^{1/2}$ and $\sqrt[3]{x} = x^{1/3}$, we use $u = x^{1/n}$, where n is the least common denominator of $\frac{1}{2}$ and $\frac{1}{3}$. Thus, we let

$$u = x^{1/6}, \quad \text{or, equivalently,} \quad x = u^6.$$

Hence

$$dx = 6u^5\,du, \quad x^{1/2} = (u^6)^{1/2} = u^3, \quad x^{1/3} = (u^6)^{1/3} = u^2$$

and, therefore,

$$\int \frac{1}{\sqrt{x} + \sqrt[3]{x}}\,dx = \int \frac{1}{u^3 + u^2}\,6u^5\,du = 6\int \frac{u^3}{u+1}\,du.$$

By long division,

$$\frac{u^3}{u+1} = u^2 - u + 1 - \frac{1}{u+1}.$$

Consequently,

$$\begin{aligned}\int \frac{1}{\sqrt{x} + \sqrt[3]{x}}\,dx &= 6\int \left(u^2 - u + 1 - \frac{1}{u+1}\right) du \\ &= 6(\tfrac{1}{3}u^3 - \tfrac{1}{2}u^2 + u - \ln|u+1|) + C \\ &= 2\sqrt{x} - 3\sqrt[3]{x} + 6\sqrt[6]{x} - 6\ln(\sqrt[6]{x} + 1) + C.\end{aligned}$$

If an integrand is a rational expression in $\sin x$ and $\cos x$, then the substitution

$$u = \tan\frac{x}{2} \quad \text{for} \quad -\pi < x < \pi$$

will transform the integrand into a rational (algebraic) expression in u. To prove this, first note that

$$\cos\frac{x}{2} = \frac{1}{\sec(x/2)} = \frac{1}{\sqrt{1 + \tan^2(x/2)}} = \frac{1}{\sqrt{1+u^2}}$$

$$\sin\frac{x}{2} = \tan\frac{x}{2}\cos\frac{x}{2} = u\frac{1}{\sqrt{1+u^2}}.$$

Consequently,

$$\sin x = 2\sin\frac{x}{2}\cos\frac{x}{2} = \frac{2u}{1+u^2}$$

$$\cos x = 1 - 2\sin^2\frac{x}{2} = 1 - \frac{2u^2}{1+u^2} = \frac{1-u^2}{1+u^2}.$$

Moreover, since $x/2 = \tan^{-1} u$, we have $x = 2\tan^{-1} u$ and, therefore,

$$dx = \frac{2}{1+u^2}\,du.$$

The following theorem summarizes this discussion.

Theorem (9.6)

If an integrand is a rational expression in $\sin x$ and $\cos x$, the following substitutions will produce a rational expression in u:

$$\sin x = \frac{2u}{1+u^2}, \quad \cos x = \frac{1-u^2}{1+u^2}, \quad dx = \frac{2}{1+u^2}\,du,$$

where $u = \tan \dfrac{x}{2}$.

EXAMPLE 3 Evaluate $\displaystyle\int \frac{1}{4\sin x - 3\cos x}\,dx$.

SOLUTION Applying Theorem (9.6) and simplifying the integrand yields

$$\begin{aligned}\int \frac{1}{4\sin x - 3\cos x}\,dx &= \int \frac{1}{4\left(\dfrac{2u}{1+u^2}\right) - 3\left(\dfrac{1-u^2}{1+u^2}\right)} \cdot \frac{2}{1+u^2}\,du \\ &= \int \frac{2}{8u - 3(1-u^2)}\,du \\ &= 2\int \frac{1}{3u^2+8u-3}\,du.\end{aligned}$$

Using partial fractions, we have

$$\frac{1}{3u^2+8u-3} = \frac{1}{10}\left(\frac{3}{3u-1} - \frac{1}{u+3}\right)$$

and hence

$$\begin{aligned}\int \frac{1}{4\sin x - 3\cos x}\,dx &= \frac{1}{5}\int\left(\frac{3}{3u-1} - \frac{1}{u+3}\right)du \\ &= \frac{1}{5}(\ln|3u-1| - \ln|u+3|) + C \\ &= \frac{1}{5}\ln\left|\frac{3u-1}{u+3}\right| + C \\ &= \frac{1}{5}\ln\left|\frac{3\tan(x/2)-1}{\tan(x/2)+3}\right| + C.\end{aligned}$$

Theorem (9.6) may be used for any integrand that is a rational expression in $\sin x$ and $\cos x$. However, it is important to also consider simpler substitutions, as illustrated in the next example.

EXAMPLE 4 Evaluate $\displaystyle\int \frac{\cos x}{1+\sin^2 x}\,dx$.

SOLUTION We could use the formulas in Theorem (9.6) to change the integrand into a rational expression in u. The following substitution is simpler:

$$u = \sin x, \quad du = \cos x\, dx$$

Thus,

$$\begin{aligned}\int \frac{\cos x}{1 + \sin^2 x}\, dx &= \int \frac{1}{1 + u^2}\, du \\ &= \arctan u + C \\ &= \arctan \sin x + C.\end{aligned}$$

EXERCISES 9.6

Exer. 1–26: Evaluate the integral.

1 $\int x \sqrt[3]{x + 9}\, dx$

2 $\int x^2 \sqrt{2x + 1}\, dx$

3 $\int \frac{x}{\sqrt[5]{3x + 2}}\, dx$

4 $\int \frac{5x}{(x + 3)^{2/3}}\, dx$

5 $\int_4^9 \frac{1}{\sqrt{x} + 4}\, dx$

6 $\int_0^{25} \frac{1}{\sqrt{4 + \sqrt{x}}}\, dx$

7 $\int \frac{\sqrt{x}}{1 + \sqrt[3]{x}}\, dx$

8 $\int \frac{1}{\sqrt[4]{x} + \sqrt[3]{x}}\, dx$

9 $\int \frac{1}{(x + 1)\sqrt{x - 2}}\, dx$

10 $\int_0^4 \frac{2x + 3}{\sqrt{1 + 2x}}\, dx$

11 $\int \frac{x + 1}{(x + 4)^{1/3}}\, dx$

12 $\int \frac{x^{1/3} + 1}{x^{1/3} - 1}\, dx$

13 $\int e^{3x} \sqrt{1 + e^x}\, dx$

14 $\int \frac{e^{2x}}{\sqrt[3]{1 + e^x}}\, dx$

15 $\int \frac{e^{2x}}{e^x + 4}\, dx$

16 $\int \frac{\sin 2x}{\sqrt{1 + \sin x}}\, dx$

17 $\int \sin \sqrt{x + 4}\, dx$

18 $\int \sqrt{x}\, e^{\sqrt{x}}\, dx$

19 $\int_2^3 \frac{x}{(x - 1)^6}\, dx$

20 $\int \frac{x^2}{(3x + 4)^{10}}\, dx$

21 $\int \frac{\sin x}{\cos x(\cos x - 1)}\, dx$

22 $\int \frac{\cos x}{\sin^2 x - \sin x - 2}\, dx$

23 $\int \frac{e^x}{e^{2x} - 1}\, dx$

24 $\int \frac{1}{e^x + e^{-x}}\, dx$

25 $\int \frac{\sin 2x}{\sin^2 x - 2 \sin x - 8}\, dx$

26 $\int \frac{\sin x}{5 \cos x + \cos^2 x}\, dx$

Exer. 27–32: Use Theorem (9.6) to evaluate the integral.

27 $\int \frac{1}{2 + \sin x}\, dx$

28 $\int \frac{1}{3 + 2 \cos x}\, dx$

29 $\int \frac{1}{1 + \sin x + \cos x}\, dx$

30 $\int \frac{1}{\tan x + \sin x}\, dx$

31 $\int \frac{\sec x}{4 - 3 \tan x}\, dx$

32 $\int \frac{1}{\sin x - \sqrt{3} \cos x}\, dx$

Exer. 33–34: Use Theorem (9.6) to derive the formula.

33 $\int \sec x\, dx = \ln \left| \frac{1 + \tan \frac{1}{2}x}{1 - \tan \frac{1}{2}x} \right| + C$

34 $\int \csc x\, dx = \frac{1}{2} \ln \left(\frac{1 - \cos x}{1 + \cos x} \right) + C$

9.7 TABLES OF INTEGRALS

Mathematicians and scientists who use integrals in their work sometimes refer to tables of integrals. Many of the formulas contained in these tables may be obtained by methods we have studied. In general, tables of integrals should be used only after gaining experience with standard methods of integration. For complicated integrals it is often necessary to make substitutions or to use partial fractions, integration by parts, or other techniques to obtain integrands to which the table is applicable.

The following examples illustrate the use of several formulas stated in the brief table of integrals in Appendix IV. To guard against errors introduced when using the table, you should always check answers by differentiation.

EXAMPLE 1 Evaluate $\int x^3 \cos x\, dx$.

SOLUTION We first use reduction Formula 85 in the table of integrals with $n = 3$ and $u = x$, obtaining

$$\int x^3 \cos x\, dx = x^3 \sin x - 3 \int x^2 \sin x\, dx.$$

Next we apply Formula 84 with $n = 2$, and then Formula 83, obtaining

$$\begin{aligned}\int x^2 \sin x\, dx &= -x^2 \cos x + 2 \int x \cos x\, dx \\ &= -x^2 \cos x + 2(\cos x + x \sin x) + C.\end{aligned}$$

Substitution in the first expression gives us

$$\int x^3 \cos x\, dx = x^3 \sin x + 3x^2 \cos x - 6 \cos x - 6x \sin x + C.$$

EXAMPLE 2 Evaluate $\int \frac{1}{x^2\sqrt{3 + 5x^2}}\, dx$ for $x > 0$.

SOLUTION The integrand suggests that we use that part of the table dealing with the form $\sqrt{a^2 + u^2}$. Specifically, Formula 28 states that

$$\int \frac{du}{u^2\sqrt{a^2 + u^2}} = -\frac{\sqrt{a^2 + u^2}}{a^2 u} + C.$$

(In tables, the differential du is placed in the numerator instead of to the right of the integrand.) To use this formula, we must adjust the given integral so that it matches *exactly* with the formula. If we let

$$a^2 = 3 \quad \text{and} \quad u^2 = 5x^2,$$

then the expression underneath the radical is taken care of; however, we also need

(i) u^2 to the left of the radical

(ii) du in the numerator

We can obtain (i) by writing the integral as

$$5 \int \frac{1}{5x^2\sqrt{3 + 5x^2}}\, dx.$$

For (ii) we note that

$$u = \sqrt{5}x \quad \text{and} \quad du = \sqrt{5}\, dx$$

and write the preceding integral as

$$5 \cdot \frac{1}{\sqrt{5}} \int \frac{1}{5x^2\sqrt{3 + 5x^2}} \sqrt{5}\, dx.$$

The last integral matches exactly with that in Formula 28, and hence

$$\begin{aligned} \int \frac{1}{x^2\sqrt{3 + 5x^2}}\, dx &= \sqrt{5}\left[-\frac{\sqrt{3 + 5x^2}}{3(\sqrt{5}x)}\right] + C \\ &= -\frac{\sqrt{3 + 5x^2}}{3x} + C. \end{aligned}$$

As illustrated in the next example, it may be necessary to make a substitution of some type before a table can be used to help evaluate an integral.

EXAMPLE 3 Evaluate $\displaystyle\int \frac{\sin 2x}{\sqrt{3 - 5\cos x}}\, dx.$

SOLUTION Let us begin by rewriting the integral:

$$\int \frac{\sin 2x}{\sqrt{3 - 5\cos x}}\, dx = \int \frac{2\sin x \cos x}{\sqrt{3 - 5\cos x}}\, dx$$

Since no formulas in the table have this form, we consider making the substitution $u = \cos x$. In this case $du = -\sin x\, dx$ and the integral may be written

$$\begin{aligned} 2\int \frac{\sin x \cos x}{\sqrt{3 - 5\cos x}}\, dx &= -2\int \frac{\cos x}{\sqrt{3 - 5\cos x}}(-\sin x)\, dx \\ &= -2\int \frac{u}{\sqrt{3 - 5u}}\, du. \end{aligned}$$

Referring to the table of integrals, we see that Formula 55 is

$$\int \frac{u\, du}{\sqrt{a + bu}} = \frac{2}{3b^2}(bu - 2a)\sqrt{a + bu}.$$

Using this result with $a = 3$ and $b = -5$ gives us

$$-2\int \frac{u}{\sqrt{3 - 5u}}\, du = -2\left(\frac{2}{75}\right)(-5u - 6)\sqrt{3 - 5u} + C.$$

Finally, since $u = \cos x$, we obtain

$$\int \frac{\sin 2x}{\sqrt{3 - 5\cos x}}\, dx = \frac{4}{75}(5\cos x + 6)\sqrt{3 - 5\cos x} + C.$$

We have discussed various methods for evaluating indefinite integrals; however, the types of integrals we have considered constitute only a small percentage of those that occur in applications. The following are examples of indefinite integrals for which antiderivatives of the integrands cannot be expressed in terms of a finite number of algebraic or transcendental functions:

$$\int \sqrt[3]{x^2 + 4x - 1}\, dx, \quad \int \sqrt{3 \cos^2 x + 1}\, dx, \quad \int e^{-x^2}\, dx$$

In Chapter 11 we shall consider methods involving *infinite* sums that are sometimes useful in evaluating such integrals.

EXERCISES 9.7

Exer. 1–30: Use the table of integrals in Appendix IV to evaluate the integral.

1 $\int \frac{\sqrt{4 + 9x^2}}{x}\, dx$

2 $\int \frac{1}{x\sqrt{2 + 3x^2}}\, dx$

3 $\int (16 - x^2)^{3/2}\, dx$

4 $\int x^2 \sqrt{4x^2 - 16}\, dx$

5 $\int x\sqrt{2 - 3x}\, dx$

6 $\int x^2 \sqrt{5 + 2x}\, dx$

7 $\int \sin^6 3x\, dx$

8 $\int x \cos^5 (x^2)\, dx$

9 $\int \csc^4 x\, dx$

10 $\int \sin 5x \cos 3x\, dx$

11 $\int x \sin^{-1} x\, dx$

12 $\int x^2 \tan^{-1} x\, dx$

13 $\int e^{-3x} \sin 2x\, dx$

14 $\int x^5 \ln x\, dx$

15 $\int \frac{\sqrt{5x - 9x^2}}{x}\, dx$

16 $\int \frac{1}{x\sqrt{3x - 2x^2}}\, dx$

17 $\int \frac{x}{5x^4 - 3}\, dx$

18 $\int \cos x \sqrt{\sin^2 x - \frac{1}{4}}\, dx$

19 $\int e^{2x} \cos^{-1} e^x\, dx$

20 $\int \sin^2 x \cos^3 x\, dx$

21 $\int x^3 \sqrt{2 + x}\, dx$

22 $\int \frac{7x^3}{\sqrt{2 - x}}\, dx$

23 $\int \frac{\sin 2x}{4 + 9 \sin x}\, dx$

24 $\int \frac{\tan x}{\sqrt{4 + 3 \sec x}}\, dx$

25 $\int \frac{\sqrt{9 + 2x}}{x}\, dx$

26 $\int \sqrt{8x^3 - 3x^2}\, dx$

27 $\int \frac{1}{x(4 + \sqrt[3]{x})}\, dx$

28 $\int \frac{1}{2x^{3/2} + 5x^2}\, dx$

29 $\int \sqrt{16 - \sec^2 x} \tan x\, dx$

30 $\int \frac{\cot x}{\sqrt{4 - \csc^2 x}}\, dx$

9.8 REVIEW EXERCISES

Exer. 1–100: Evaluate the integral.

1 $\int x \sin^{-1} x\, dx$

2 $\int \sec^3 (3x)\, dx$

3 $\int_0^1 \ln (1 + x)\, dx$

4 $\int_0^1 e^{\sqrt{x}}\, dx$

5 $\int \cos^3 2x \sin^2 2x\, dx$

6 $\int \cos^4 x\, dx$

7 $\int \tan x \sec^5 x\, dx$

8 $\int \tan x \sec^6 x\, dx$

9 $\int \frac{1}{(x^2 + 25)^{3/2}}\, dx$

10 $\int \frac{1}{x^2\sqrt{16 - x^2}}\, dx$

11 $\int \frac{\sqrt{4 - x^2}}{x}\, dx$

12 $\int \frac{x}{(x^2 + 1)^2}\, dx$

13 $\int \frac{x^3 + 1}{x(x - 1)^3}\, dx$

14 $\int \frac{1}{x + x^3}\, dx$

15 $\int \frac{x^3 - 20x^2 - 63x - 198}{x^4 - 81}\, dx$

16 $\int \frac{x - 1}{(x + 2)^5}\, dx$

17 $\int \frac{x}{\sqrt{4 + 4x - x^2}}\, dx$

18 $\int \frac{x}{x^2 + 6x + 13}\, dx$

19 $\int \frac{\sqrt[3]{x + 8}}{x}\, dx$

20 $\int \frac{\sin x}{2 \cos x + 3}\, dx$

21 $\int e^{2x} \sin 3x\, dx$

22 $\int \cos (\ln x)\, dx$

23 $\int \sin^3 x \cos^3 x\, dx$

24 $\int \cot^2 3x\, dx$

25 $\int \frac{x}{\sqrt{4 - x^2}}\, dx$

26 $\int \frac{1}{x\sqrt{9x^2 + 4}}\, dx$

27 $\int \frac{x^5 - x^3 + 1}{x^3 + 2x^2}\,dx$

28 $\int \frac{x^3}{x^3 - 3x^2 + 9x - 27}\,dx$

29 $\int \frac{1}{x^{3/2} + x^{1/2}}\,dx$

30 $\int \frac{2x + 1}{(x + 5)^{100}}\,dx$

31 $\int e^x \sec e^x\,dx$

32 $\int x \tan x^2\,dx$

33 $\int x^2 \sin 5x\,dx$

34 $\int \sin 2x \cos x\,dx$

35 $\int \sin^3 x \cos^{1/2} x\,dx$

36 $\int \sin 3x \cot 3x\,dx$

37 $\int e^x \sqrt{1 + e^x}\,dx$

38 $\int x(4x^2 + 25)^{-1/2}\,dx$

39 $\int \frac{x^2}{\sqrt{4x^2 + 25}}\,dx$

40 $\int \frac{3x + 2}{x^2 + 8x + 25}\,dx$

41 $\int \sec^2 x \tan^2 x\,dx$

42 $\int \sin^2 x \cos^5 x\,dx$

43 $\int x \cot x \csc x\,dx$

44 $\int (1 + \csc 2x)^2\,dx$

45 $\int x^2(8 - x^3)^{1/3}\,dx$

46 $\int x\,(\ln x)^2\,dx$

47 $\int \sqrt{x} \sin \sqrt{x}\,dx$

48 $\int x\sqrt{5 - 3x}\,dx$

49 $\int \frac{e^{3x}}{1 + e^x}\,dx$

50 $\int \frac{e^{2x}}{4 + e^{4x}}\,dx$

51 $\int \frac{x^2 - 4x + 3}{\sqrt{x}}\,dx$

52 $\int \frac{\cos^3 x}{\sqrt{1 + \sin x}}\,dx$

53 $\int \frac{x^3}{\sqrt{16 - x^2}}\,dx$

54 $\int \frac{x}{25 - 9x^2}\,dx$

55 $\int \frac{1 - 2x}{x^2 + 12x + 35}\,dx$

56 $\int \frac{7}{x^2 - 6x + 18}\,dx$

57 $\int \tan^{-1} 5x\,dx$

58 $\int \sin^4 3x\,dx$

59 $\int \frac{e^{\tan x}}{\cos^2 x}\,dx$

60 $\int \frac{x}{\csc 5x^2}\,dx$

61 $\int \frac{1}{\sqrt{7 + 5x^2}}\,dx$

62 $\int \frac{2x + 3}{x^2 + 4}\,dx$

63 $\int \cot^6 x\,dx$

64 $\int \cot^5 x \csc x\,dx$

65 $\int x^3\sqrt{x^2 - 25}\,dx$

66 $\int (\sin x)10^{\cos x}\,dx$

67 $\int (x^2 - \operatorname{sech}^2 4x)\,dx$

68 $\int x \cosh x\,dx$

69 $\int x^2 e^{-4x}\,dx$

70 $\int x^5\sqrt{x^3 + 1}\,dx$

71 $\int \frac{3}{\sqrt{11 - 10x - x^2}}\,dx$

72 $\int \frac{12x^3 + 7x}{x^4}\,dx$

73 $\int \tan 7x \cos 7x\,dx$

74 $\int e^{1 + \ln 5x}\,dx$

75 $\int \frac{4x^2 - 12x - 10}{(x - 2)(x^2 - 4x + 3)}\,dx$

76 $\int \frac{1}{x^4\sqrt{16 - x^2}}\,dx$

77 $\int (x^3 + 1)\cos x\,dx$

78 $\int (x - 3)^2(x + 1)\,dx$

79 $\int \frac{\sqrt{9 - 4x^2}}{x^2}\,dx$

80 $\int \frac{4x^3 - 15x^2 - 6x + 81}{x^4 - 18x^2 + 81}\,dx$

81 $\int (5 - \cot 3x)^2\,dx$

82 $\int x(x^2 + 5)^{3/4}\,dx$

83 $\int \frac{1}{x(\sqrt{x} + \sqrt[4]{x})}\,dx$

84 $\int \frac{x}{\cos^2 4x}\,dx$

85 $\int \frac{\sin x}{\sqrt{1 + \cos x}}\,dx$

86 $\int \frac{4x^2 - 6x + 4}{(x^2 + 4)(x - 2)}\,dx$

87 $\int \frac{x^2}{(25 + x^2)^2}\,dx$

88 $\int \sin^4 x \cos^3 x\,dx$

89 $\int \tan^3 x \sec x\,dx$

90 $\int \frac{x}{\sqrt{4 + 9x^2}}\,dx$

91 $\int \frac{2x^3 + 4x^2 + 10x + 13}{x^4 + 9x^2 + 20}\,dx$

92 $\int \frac{\sin x}{(1 + \cos x)^3}\,dx$

93 $\int \frac{(x^2 - 2)^2}{x}\,dx$

94 $\int \cot^2 x \csc x\,dx$

95 $\int x^{3/2} \ln x\,dx$

96 $\int \frac{x}{\sqrt[3]{x} - 1}\,dx$

97 $\int \frac{x^2}{\sqrt[3]{2x + 3}}\,dx$

98 $\int \frac{1 - \sin x}{\cot x}\,dx$

99 $\int x^3 e^{(x^2)}\,dx$

100 $\int (x + 2)^2(x + 1)^{10}\,dx$

CHAPTER

10

INDETERMINATE FORMS AND IMPROPER INTEGRALS

INTRODUCTION

The first important limit we considered in Chapter 3 was the derivative formula

$$f'(a) = \lim_{x \to a} \frac{f(x) - f(a)}{x - a}.$$

If f is continuous at $x = a$, then taking the limit of the numerator and denominator separately gives us

$$f'(a) = \frac{f(a) - f(a)}{a - a} = \frac{0}{0},$$

an undefined expression. However, we know that derivatives are not always undefined. You may recall that to arrive at each rule for finding derivatives we used an algebraic or trigonometric simplification, which was sometimes accompanied by an ingenious manipulation or geometric argument. In this chapter we introduce techniques that allow us to proceed in a more direct manner when considering similar problems about limits. The most important result we shall discuss is *L'Hôpital's rule*, used for investigating limits of quotients in which both numerator and denominator approach 0 or both approach ∞ or $-\infty$. Other so-called *indeterminate forms* are considered in Section 10.2. In the last two sections we study definite integrals that have discontinuous integrands or infinite limits of integration.

The topics discussed in this chapter have many mathematical and physical applications. Our most important uses for them will occur in the next chapter, when we discuss infinite series.

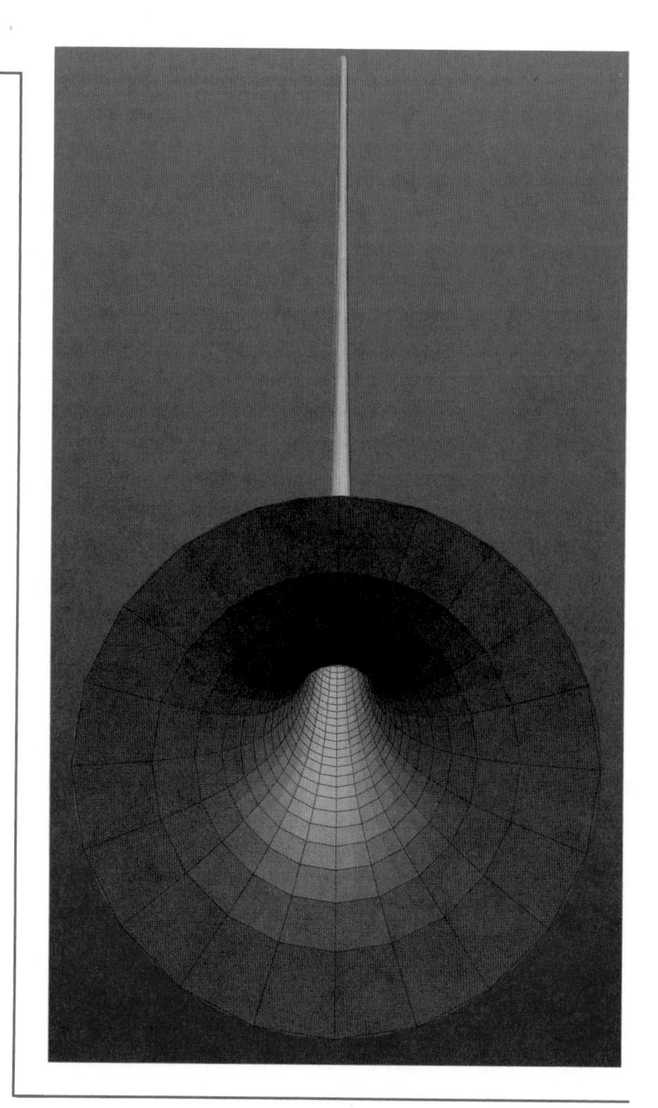

10.1 THE INDETERMINATE FORMS 0/0 AND ∞/∞

In Chapter 2 we considered limits of quotients such as

$$\lim_{x\to 3} \frac{x^2 - 9}{x - 3} \quad \text{and} \quad \lim_{x\to 0} \frac{\sin x}{x}.$$

In each case, taking the limits of the numerator and denominator gives us the undefined expression 0/0. We say that the indicated quotients have the **indeterminate form 0/0** at $x = 3$ and at $x = 0$, respectively. We previously used algebraic, geometric, and trigonometric methods to calculate such limits. In this section we develop another technique that employs the derivatives of the numerator and denominator of the quotient. We also consider the **indeterminate form** ∞/∞, where both the numerator and the denominator approach ∞ or $-\infty$. The following table displays general definitions of the forms we shall discuss.

Indeterminate form	Limit form: $\lim_{x\to c} \frac{f(x)}{g(x)}$
$\frac{0}{0}$	$\lim_{x\to c} f(x) = 0$ and $\lim_{x\to c} g(x) = 0$
$\frac{\infty}{\infty}$	$\lim_{x\to c} f(x) = \infty$ or $-\infty$ and $\lim_{x\to c} g(x) = \infty$ or $-\infty$

The main tool for investigating these indeterminate forms is *L'Hôpital's rule*. The proof of this rule makes use of the following formula, which bears the name of the French mathematician Augustin Cauchy (1789–1857).

Cauchy's formula **(10.1)**

If f and g are continuous on $[a, b]$ and differentiable on (a, b) and if $g'(x) \neq 0$ for every x in (a, b), then there is a number w in (a, b) such that

$$\frac{f(b) - f(a)}{g(b) - g(a)} = \frac{f'(w)}{g'(w)}.$$

PROOF We first note that $g(b) - g(a) \neq 0$, because otherwise $g(a) = g(b)$ and, by Rolle's theorem (4.10), there is a number c in (a, b) such that $g'(c) = 0$, contrary to our assumption about g'.

Let us introduce a new function h as follows:

$$h(x) = [f(b) - f(a)]g(x) - [g(b) - g(a)]f(x)$$

for every x in $[a, b]$. It follows that h is continuous on $[a, b]$ and differentiable on (a, b) and that $h(a) = h(b)$. By Rolle's theorem, there is a number w in (a, b) such that $h'(w) = 0$; that is,

$$[f(b) - f(a)]g'(w) - [g(b) - g(a)]f'(w) = 0.$$

This is equivalent to Cauchy's formula. ■

Cauchy's formula is a generalization of the mean value theorem (4.12), for if we let $g(x) = x$ in (10.1), we obtain

$$\frac{f(b) - f(a)}{b - a} = \frac{f'(w)}{1} = f'(w).$$

The next result is the main theorem on indeterminate forms.

L'Hôpital's rule* (10.2)

> Suppose f and g are differentiable on an open interval (a, b) containing c, except possibly at c itself. If $f(x)/g(x)$ has the indeterminate form 0/0 or ∞/∞ at $x = c$ and if $g'(x) \neq 0$ for $x \neq c$, then
>
> $$\lim_{x \to c} \frac{f(x)}{g(x)} = \lim_{x \to c} \frac{f'(x)}{g'(x)},$$
>
> provided either
>
> $$\lim_{x \to c} \frac{f'(x)}{g'(x)} \text{ exists} \quad \text{or} \quad \lim_{x \to c} \frac{f'(x)}{g'(x)} = \infty.$$

*G. L'Hôpital (1661–1704) was a French nobleman who published the first calculus book. The rule appeared in that book; however, it was actually discovered by his teacher, the Swiss mathematician Johann Bernoulli (1667–1748), who communicated the result to L'Hôpital in 1694.

PROOF Suppose $f(x)/g(x)$ has the indeterminate form 0/0 at $x = c$ and $\lim_{x \to c} [f'(x)/g'(x)] = L$ for some number L. We wish to prove that $\lim_{x \to c} [f(x)/g(x)] = L$. Let us introduce two functions F and G as follows:

$$F(x) = f(x) \quad \text{if } x \neq c \quad \text{and} \quad F(c) = 0$$

$$G(x) = g(x) \quad \text{if } x \neq c \quad \text{and} \quad G(c) = 0$$

Since

$$\lim_{x \to c} F(x) = \lim_{x \to c} f(x) = 0 = F(c),$$

the function F is continuous at c and hence is continuous *throughout* the interval (a, b). Similarly, G is continuous on (a, b). Moreover, at every $x \neq c$ we have $F'(x) = f'(x)$ and $G'(x) = g'(x)$. It follows from Cauchy's formula, applied either to the interval $[c, x]$ or to $[x, c]$, that there is a number w between c and x such that

$$\frac{F(x) - F(c)}{G(x) - G(c)} = \frac{F'(w)}{G'(w)} = \frac{f'(w)}{g'(w)}.$$

FIGURE 10.1

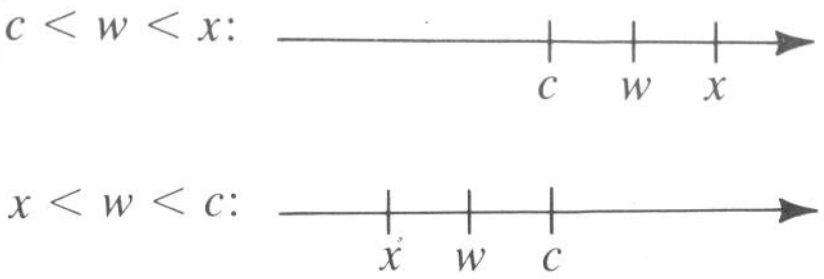

$c < w < x$:

$x < w < c$:

Using the fact that $F(x) = f(x)$, $G(x) = g(x)$, and $F(c) = G(c) = 0$ gives us

$$\frac{f(x)}{g(x)} = \frac{f'(w)}{g'(w)}.$$

Since w is always between c and x (see Figure 10.1), it follows that

$$\lim_{x \to c} \frac{f(x)}{g(x)} = \lim_{x \to c} \frac{f'(w)}{g'(w)} = \lim_{w \to c} \frac{f'(w)}{g'(w)} = L,$$

which is what we wished to prove.

A similar argument may be given if $\lim_{x \to c} [f'(x)/g'(x)] = \infty$. The proof for the indeterminate form ∞/∞ is more difficult and may be found in texts on advanced calculus. ■

L'Hôpital's rule is sometimes used incorrectly, by applying the quotient rule to $f(x)/g(x)$. Note that (10.2) states that the derivatives of $f(x)$ and $g(x)$ are taken *separately*, after which the limit of $f'(x)/g'(x)$ is investigated.

EXAMPLE 1 Find $\lim_{x\to 0} \dfrac{\cos x + 2x - 1}{3x}$.

SOLUTION Both the numerator and the denominator have the limit 0 as $x \to 0$. Hence the quotient has the indeterminate form 0/0 at $x = 0$. By L'Hôpital's rule (10.2),

$$\lim_{x\to 0} \frac{\cos x + 2x - 1}{3x} = \lim_{x\to 0} \frac{-\sin x + 2}{3},$$

provided the limit on the right exists or equals ∞. Since

$$\lim_{x\to 0} \frac{-\sin x + 2}{3} = \frac{2}{3},$$

it follows that

$$\lim_{x\to 0} \frac{\cos x + 2x - 1}{3x} = \frac{2}{3}.$$

Sometimes it is necessary to employ L'Hôpital's rule several times in the same problem, as illustrated in the next example.

EXAMPLE 2 Find $\lim_{x\to 0} \dfrac{e^x + e^{-x} - 2}{1 - \cos 2x}$.

SOLUTION The given quotient has the indeterminate form 0/0. By L'Hôpital's rule,

$$\lim_{x\to 0} \frac{e^x + e^{-x} - 2}{1 - \cos 2x} = \lim_{x\to 0} \frac{e^x - e^{-x}}{2 \sin 2x},$$

provided the second limit exists. Because the last quotient has the indeterminate form 0/0, we apply L'Hôpital's rule a second time, obtaining

$$\lim_{x\to 0} \frac{e^x - e^{-x}}{2 \sin 2x} = \lim_{x\to 0} \frac{e^x + e^{-x}}{4 \cos 2x} = \frac{2}{4} = \frac{1}{2}.$$

It follows that the given limit exists and equals $\frac{1}{2}$.

L'Hôpital's rule is also valid for one-sided limits, as illustrated in the following example.

EXAMPLE 3 Find $\lim_{x\to(\pi/2)^-} \dfrac{4\tan x}{1+\sec x}$.

SOLUTION The indeterminate form is ∞/∞. By L'Hôpital's rule,

$$\lim_{x\to(\pi/2)^-} \frac{4\tan x}{1+\sec x} = \lim_{x\to(\pi/2)^-} \frac{4\sec^2 x}{\sec x \tan x} = \lim_{x\to(\pi/2)^-} \frac{4\sec x}{\tan x}.$$

The last quotient again has the indeterminate form ∞/∞ at $x = \pi/2$; however, additional applications of L'Hôpital's rule always produce the form ∞/∞ (verify this fact). In this case the limit may be found by using trigonometric identities to change the quotient as follows:

$$\frac{4\sec x}{\tan x} = \frac{4/\cos x}{\sin x/\cos x} = \frac{4}{\sin x}$$

Consequently

$$\lim_{x\to(\pi/2)^-} \frac{4\tan x}{1+\sec x} = \lim_{x\to(\pi/2)^-} \frac{4}{\sin x} = \frac{4}{1} = 4.$$

Another form of L'Hôpital's rule can be proved for $x \to \infty$ or $x \to -\infty$. Let us give a partial proof of this fact. Suppose

$$\lim_{x\to\infty} f(x) = \lim_{x\to\infty} g(x) = 0.$$

If we let $u = 1/x$ and apply L'Hôpital's rule,

$$\lim_{x\to\infty} \frac{f(x)}{g(x)} = \lim_{u\to0^+} \frac{f(1/u)}{g(1/u)} = \lim_{u\to0^+} \frac{D_u f(1/u)}{D_u g(1/u)}.$$

By the chain rule,

$$D_u f(1/u) = f'(1/u)(-1/u^2) \quad \text{and} \quad D_u g(1/u) = g'(1/u)(-1/u^2).$$

Substituting in the last limit and simplifying, we obtain

$$\lim_{x\to\infty} \frac{f(x)}{g(x)} = \lim_{u\to0^+} \frac{f'(1/u)}{g'(1/u)} = \lim_{x\to\infty} \frac{f'(x)}{g'(x)}.$$

We shall also refer to this result as L'Hôpital's rule. The next two examples illustrate the application of the rule to the form ∞/∞.

EXAMPLE 4 Find $\lim_{x\to\infty} \dfrac{\ln x}{\sqrt{x}}$.

SOLUTION The indeterminate form is ∞/∞. By L'Hôpital's rule,

$$\lim_{x\to\infty} \frac{\ln x}{\sqrt{x}} = \lim_{x\to\infty} \frac{1/x}{1/(2\sqrt{x})}.$$

The last expression has the indeterminate form 0/0. However, further applications of L'Hôpital's rule would again lead to 0/0 (verify this fact). If, instead, we simplify the expression algebraically, we can find the limit as follows:

$$\lim_{x\to\infty} \frac{1/x}{1/(2\sqrt{x})} = \lim_{x\to\infty} \frac{2\sqrt{x}}{x} = \lim_{x\to\infty} \frac{2}{\sqrt{x}} = 0$$

EXAMPLE 5 Find $\lim_{x\to\infty} \frac{e^{3x}}{x^2}$, if it exists.

SOLUTION The indeterminate form is ∞/∞. We apply L'Hôpital's rule:

$$\lim_{x\to\infty} \frac{e^{3x}}{x^2} = \lim_{x\to\infty} \frac{3e^{3x}}{2x}$$

The last quotient has the indeterminate form ∞/∞, so we apply L'Hôpital's rule a second time, obtaining

$$\lim_{x\to\infty} \frac{3e^{3x}}{2x} = \lim_{x\to\infty} \frac{9e^{3x}}{2} = \infty.$$

Thus, e^{3x}/x^2 has no limit, increasing without bound as $x \to \infty$.

It is extremely important to verify that a given quotient has the indeterminate form 0/0 or ∞/∞ before using L'Hôpital's rule. If we apply the rule to a form that is not indeterminate, we may obtain an incorrect conclusion, as illustrated in the next example.

EXAMPLE 6 Find $\lim_{x\to 0} \frac{e^x + e^{-x}}{x^2}$, if it exists.

SOLUTION The quotient does *not* have either of the indeterminate forms, 0/0 or ∞/∞, at $x = 0$. To investigate the limit, we write

$$\lim_{x\to 0} \frac{e^x + e^{-x}}{x^2} = \lim_{x\to 0} (e^x + e^{-x})\left(\frac{1}{x^2}\right).$$

Since

$$\lim_{x\to 0} (e^x + e^{-x}) = 2 \quad \text{and} \quad \lim_{x\to 0} \frac{1}{x^2} = \infty,$$

it follows that

$$\lim_{x\to 0} \frac{e^x + e^{-x}}{x^2} = \infty.$$

If we had overlooked the fact that the quotient does not have the indeterminate form 0/0 or ∞/∞ at $x = 0$ and had (incorrectly) applied L'Hôpital's rule, we would have obtained

$$\lim_{x\to 0} \frac{e^x + e^{-x}}{x^2} = \lim_{x\to 0} \frac{e^x - e^{-x}}{2x}.$$

Since the last quotient has the indeterminate form 0/0, we might have applied L'Hôpital's rule, obtaining

$$\lim_{x\to 0}\frac{e^x - e^{-x}}{2x} = \lim_{x\to 0}\frac{e^x + e^{-x}}{2} = \frac{1+1}{2} = 1.$$

This would have given us the (wrong) conclusion that the given limit exists and equals 1.

The next example illustrates an application of an indeterminate form in the analysis of an electrical circuit.

FIGURE 10.2

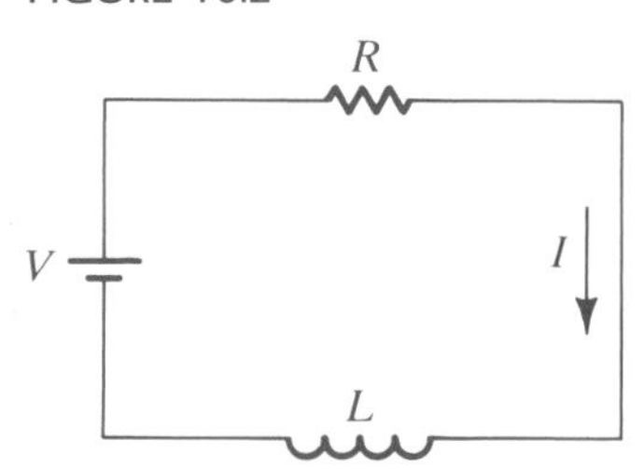

EXAMPLE 7 The schematic diagram in Figure 10.2 illustrates an electrical circuit consisting of an electromotive force V, a resistor R, and an inductor L. The current I at time t is given by

$$I = \frac{V}{R}(1 - e^{-Rt/L}).$$

When the voltage is first applied (at $t = 0$), the inductor opposes the rate of increase of current and I is small; however, as t increases, I approaches V/R.

(a) If L is the only independent variable, find $\lim_{L\to 0^+} I$.

(b) If R is the only independent variable, find $\lim_{R\to 0^+} I$.

SOLUTION

(a) If we consider V, R, and t as constants and L as a variable, then the expression for I is not indeterminate at $L = 0$. Using standard limit theorems, we obtain

$$\begin{aligned}\lim_{L\to 0^+} I &= \lim_{L\to 0^+}\frac{V}{R}(1 - e^{-Rt/L}) \\ &= \frac{V}{R}\left(1 - \lim_{L\to 0^+} e^{-Rt/L}\right) \\ &= \frac{V}{R}(1-0) = \frac{V}{R}.\end{aligned}$$

Thus, if $L \approx 0$, then the current can be approximated by Ohm's law $I = V/R$.

(b) If V, L, and t are constant and if R is a variable, then I has the indeterminate form 0/0 at $R = 0$. Applying L'Hôpital's rule, we have

$$\begin{aligned}\lim_{R\to 0^+} I &= V \lim_{R\to 0^+}\frac{1 - e^{-Rt/L}}{R} \\ &= V \lim_{R\to 0^+}\frac{0 - e^{-Rt/L}(-t/L)}{1} \\ &= V\,[0 - (1)(-t/L)] = \frac{V}{L}t.\end{aligned}$$

This may be interpreted as follows. As $R \to 0^+$, the current I is directly proportional to the time t, with the constant of proportionality V/L. Thus, at $t = 1$ the current is V/L, at $t = 2$ it is $(V/L)(2)$, at $t = 3$ it is $(V/L)(3)$, and so on.

EXERCISES 10.1

Exer. 1–52: Find the limit, if it exists.

1 $\lim_{x\to 0} \dfrac{\sin x}{2x}$

2 $\lim_{x\to 0} \dfrac{5x}{\tan x}$

3 $\lim_{x\to 5} \dfrac{\sqrt{x-1}-2}{x^2-25}$

4 $\lim_{x\to 4} \dfrac{x-4}{\sqrt[3]{x+4}-2}$

5 $\lim_{x\to 2} \dfrac{2x^2-5x+2}{5x^2-7x-6}$

6 $\lim_{x\to -3} \dfrac{x^2+2x-3}{2x^2+3x-9}$

7 $\lim_{x\to 1} \dfrac{x^3-3x+2}{x^2-2x+1}$

8 $\lim_{x\to 2} \dfrac{x^2-5x+6}{2x^2-x-7}$

9 $\lim_{x\to 0} \dfrac{\sin x - x}{\tan x - x}$

10 $\lim_{x\to 0} \dfrac{\sin x}{x - \tan x}$

11 $\lim_{x\to 0} \dfrac{x+1-e^x}{x^2}$

12 $\lim_{x\to 0^+} \dfrac{x+1-e^x}{x^3}$

13 $\lim_{x\to 0} \dfrac{x - \sin x}{x^3}$

14 $\lim_{x\to \pi/2} \dfrac{1-\sin x}{\cos x}$

15 $\lim_{x\to \pi/2} \dfrac{1+\sin x}{\cos^2 x}$

16 $\lim_{x\to 0^+} \dfrac{\cos x}{x}$

17 $\lim_{x\to (\pi/2)^-} \dfrac{2+\sec x}{3\tan x}$

18 $\lim_{x\to 0^+} \dfrac{\ln x}{\cot x}$

19 $\lim_{x\to \infty} \dfrac{x^2}{\ln x}$

20 $\lim_{x\to \infty} \dfrac{\ln x}{x^2}$

21 $\lim_{x\to 0^+} \dfrac{\ln \sin x}{\ln \sin 2x}$

22 $\lim_{x\to 0} \dfrac{2x}{\tan^{-1} x}$

23 $\lim_{x\to 0} \dfrac{e^x - e^{-x} - 2\sin x}{x \sin x}$

24 $\lim_{x\to 2} \dfrac{\ln (x-1)}{x-2}$

25 $\lim_{x\to 0} \dfrac{x\cos x + e^{-x}}{x^2}$

26 $\lim_{x\to 0} \dfrac{2e^x - 3x - e^{-x}}{x^2}$

27 $\lim_{x\to \infty} \dfrac{2x^2+3x+1}{5x^2+x+4}$

28 $\lim_{x\to \infty} \dfrac{x^3+x+1}{3x^3+4}$

29 $\lim_{x\to \infty} \dfrac{x\ln x}{x+\ln x}$

30 $\lim_{x\to \infty} \dfrac{e^{3x}}{\ln x}$

31 $\lim_{x\to \infty} \dfrac{x^n}{e^x}, n > 0$

32 $\lim_{x\to \infty} \dfrac{e^x}{x^n}, n > 0$

33 $\lim_{x\to 2^+} \dfrac{\ln (x-1)}{(x-2)^2}$

34 $\lim_{x\to 0} \dfrac{\sin^2 x + 2\cos x - 2}{\cos^2 x - x\sin x - 1}$

35 $\lim_{x\to 0} \dfrac{\sin^{-1} 2x}{\sin^{-1} x}$

36 $\lim_{x\to \infty} \dfrac{\ln (\ln x)}{\ln x}$

37 $\lim_{x\to 0} \dfrac{\tan x - \sin x}{x^3 \tan x}$

38 $\lim_{x\to 1} \dfrac{2x^3-5x^2+6x-3}{x^3-2x^2+x-1}$

39 $\lim_{x\to -\infty} \dfrac{3-3^x}{5-5^x}$

40 $\lim_{x\to 0} \dfrac{2-e^x-e^{-x}}{1-\cos^2 x}$

41 $\lim_{x\to 1} \dfrac{x^4-x^3-3x^2+5x-2}{x^4-5x^3+9x^2-7x+2}$

42 $\lim_{x\to 1} \dfrac{x^4+x^3-3x^2-x+2}{x^4-5x^3+9x^2-7x+2}$

43 $\lim_{x\to 0} \dfrac{x - \tan^{-1} x}{x \sin x}$

44 $\lim_{x\to \infty} \dfrac{e^{-x}}{1+e^{-x}}$

45 $\lim_{x\to \infty} \dfrac{x^{3/2}+5x-4}{x\ln x}$

46 $\lim_{x\to 0} \dfrac{x\sin^{-1} x}{x - \sin x}$

47 $\lim_{x\to \infty} \dfrac{\sqrt{x^2+1}}{\tan^{-1} x}$

48 $\lim_{x\to (\pi/2)^-} \dfrac{\tan x}{\cot 2x}$

49 $\lim_{x\to \infty} \dfrac{2e^{3x}+\ln x}{e^{3x}+x^2}$

50 $\lim_{x\to 0^+} \dfrac{e^{-1/x}}{x}$

51 $\lim_{x\to \infty} \dfrac{x - \cos x}{x}$

52 $\lim_{x\to \infty} \dfrac{x+\cosh x}{x^2+1}$

c **Exer. 53–54: Predict the limit after substituting the indicated values of x for $k = 1, 2, 3$, and 4.**

53 $\lim_{x\to 0^+} \dfrac{\ln (\tan x + \cos x)}{\sqrt{\ln (x^2+1)}}$; $x = 10^{-k}$

54 $\lim_{x\to 0} \dfrac{\tan^2 (\sin^{-1} x)}{1 - \cos [\ln (1+x)]}$; $x = \pm 10^{-k}$

55 An object of mass m is released from a hot-air balloon. If the force of resistance due to air is directly proportional to the velocity $v(t)$ of the object at time t, then it can be shown that

$$v(t) = (mg/k)(1 - e^{-(k/m)t}),$$

where $k > 0$ and g is a gravitational constant. Find $\lim_{k \to 0^+} v(t)$.

56 If a steel ball of mass m is released into water and the force of resistance is directly proportional to the square of the velocity, then the distance $s(t)$ that the ball travels in time t is given by

$$s(t) = (m/k) \ln \cosh (\sqrt{gk/m}\, t),$$

where $k > 0$ and g is a gravitational constant. Find $\lim_{k \to 0^+} s(t)$.

57 Refer to Definition (4.22) for simple harmonic motion. The following is an example of the phenomenon of resonance. A weight of mass m is attached to a spring suspended from a support. The weight is set in motion by moving the support up and down according to the formula $h = A \cos \omega t$, where A and ω are positive constants and t is time. If frictional forces are negligible, then the displacement s of the weight from its initial position at time t is given by

$$s = \frac{A\omega^2}{\omega_0^2 - \omega^2}(\cos \omega t - \cos \omega_0 t),$$

with $\omega_0 = \sqrt{k/m}$ for some constant k and with $\omega \neq \omega_0$. Find $\lim_{\omega \to \omega_0} s$, and show that the resulting oscillations increase in magnitude.

58 The logistic model for population growth predicts the size $y(t)$ of a population at time t by means of the formula $y(t) = K/(1 + ce^{-rt})$, where r and K are positive constants and $c = [K - y(0)]/y(0)$. Ecologists call K the *carrying capacity* and interpret it as the maximum number of individuals that the environment can sustain. Find $\lim_{t \to \infty} y(t)$ and $\lim_{K \to \infty} y(t)$, and discuss the graphical significance of these limits.

59 The *sine integral* $Si(x) = \int_0^x [(\sin u)/u]\, du$ is a special function in applied mathematics. Find

(a) $\lim_{x \to 0} \dfrac{Si(x)}{x}$ (b) $\lim_{x \to 0} \dfrac{Si(x) - x}{x^3}$

60 The *Fresnel cosine integral* $C(x) = \int_0^x \cos u^2\, du$ is used in the analysis of the diffraction of light. Find

(a) $\lim_{x \to 0} \dfrac{C(x)}{x}$ (b) $\lim_{x \to 0} \dfrac{C(x) - x}{x^5}$

c 61 (a) Refer to Exercise 60. Use Simpson's rule, with $n = 4$, to approximate $C(x)$ for $x = \frac{1}{4}, \frac{1}{2}, \frac{3}{4}$, and 1.

(b) Graph C on $[0, 1]$ using the values found in (a).

c 62 Refer to Exercise 61. Let R be the region under the graph of C from $x = 0$ to $x = 1$, and let V be the volume of the solid obtained by revolving R about the x-axis. Approximate V by using Simpson's rule with $n = 4$.

63 Let $x > 0$. If $n \neq -1$, then $\int_1^x t^n\, dt = [t^{n+1}/(n + 1)]_1^x$. Show that

$$\lim_{n \to -1} \int_1^x t^n\, dt = \int_1^x t^{-1}\, dt.$$

64 Find $\lim_{x \to \infty} f(x)/g(x)$ if

$$f(x) = \int_0^x e^{(t^2)}\, dt \quad \text{and} \quad g(x) = e^{(x^2)}.$$

10.2 OTHER INDETERMINATE FORMS

In the preceding section we discussed limits of quotients that have the indeterminate forms $0/0$ or ∞/∞. Products may lead to the indeterminate form $0 \cdot \infty$, as defined in the following table.

Indeterminate form	**Limit form: $\lim_{x \to c} [f(x)\, g(x)]$**
$0 \cdot \infty$	$\lim_{x \to c} f(x) = 0$ and $\lim_{x \to c} g(x) = \infty$ or $-\infty$

In exercises we shall also consider the indeterminate form $0 \cdot \infty$ for the case $x \to \infty$ or $x \to -\infty$. The following guidelines may be used.

Guidelines for investigating $\lim_{x \to c} [f(x)\, g(x)]$ for the form $0 \cdot \infty$ **(10.3)**

1 Write $f(x)\, g(x)$ as

$$\frac{f(x)}{1/g(x)} \quad \text{or} \quad \frac{g(x)}{1/f(x)}.$$

2 Apply L'Hôpital's rule (10.2) to the resulting indeterminate form $0/0$ or ∞/∞.

The choice in guideline 1 is not arbitrary. The following example shows that using $f(x)/[1/g(x)]$ gives us the limit whereas using $g(x)/[1/f(x)]$ leads to a more complicated expression.

EXAMPLE 1 Find $\lim_{x \to 0^+} x^2 \ln x$.

SOLUTION The indeterminate form is $0 \cdot \infty$. Applying guideline 1 of (10.3), we write

$$x^2 \ln x = \frac{\ln x}{1/x^2}.$$

Because the quotient on the right has the indeterminate form ∞/∞ at $x = 0$, we may apply L'Hôpital's rule:

$$\lim_{x \to 0^+} x^2 \ln x = \lim_{x \to 0^+} \frac{\ln x}{1/x^2} = \lim_{x \to 0^+} \frac{1/x}{-2/x^3}$$

The last quotient has the indeterminate form ∞/∞; however, further applications of L'Hôpital's rule would again lead to ∞/∞. In this case we simplify the quotient algebraically and find the limit as follows:

$$\lim_{x \to 0^+} \frac{1/x}{-2/x^3} = \lim_{x \to 0^+} \frac{x^3}{-2x} = \lim_{x \to 0^+} \frac{x^2}{-2} = 0$$

If, in applying guideline 1, we had rewritten the given expression as

$$x^2 \ln x = \frac{x^2}{1/\ln x} = \frac{x^2}{(\ln x)^{-1}},$$

then the resulting indeterminate form would have been 0/0. By L'Hôpital's rule,

$$\begin{aligned} \lim_{x \to 0^+} x^2 \ln x &= \lim_{x \to 0^+} \frac{x^2}{(\ln x)^{-1}} \\ &= \lim_{x \to 0^+} \frac{2x}{-(\ln x)^{-2}(1/x)} \\ &= \lim_{x \to 0^+} [-2x^2(\ln x)^2]. \end{aligned}$$

The expression $-2x^2(\ln x)^2$ is more complicated than $x^2 \ln x$, so this choice in guideline 1 does *not* give us the limit.

EXAMPLE 2 Find $\lim_{x \to (\pi/2)^-} (2x - \pi) \sec x$.

SOLUTION The indeterminate form is $0 \cdot \infty$. Using guideline 1 of (10.3), we begin by writing

$$(2x - \pi) \sec x = \frac{2x - \pi}{1/\sec x} = \frac{2x - \pi}{\cos x}.$$

Because the last expression has the indeterminate form 0/0 at $x = \pi/2$, L'Hôpital's rule may be applied as follows:

$$\lim_{x \to (\pi/2)^-} \frac{2x - \pi}{\cos x} = \lim_{x \to (\pi/2)^-} \frac{2}{-\sin x} = \frac{2}{-1} = -2$$

The indeterminate forms defined in the next table may occur in investigating limits involving exponential expressions.

Indeterminate form	Limit form: $\lim_{x \to c} f(x)^{g(x)}$
0^0	$\lim_{x \to c} f(x) = 0$ and $\lim_{x \to c} g(x) = 0$
∞^0	$\lim_{x \to c} f(x) = \infty$ or $-\infty$ and $\lim_{x \to c} g(x) = 0$
1^∞	$\lim_{x \to c} f(x) = 1$ and $\lim_{x \to c} g(x) = \infty$ or $-\infty$

In exercises we shall also consider cases in which $x \to \infty$ or $x \to -\infty$. One method for investigating these forms is to consider

$$y = f(x)^{g(x)}$$

and take the natural logarithm of both sides, obtaining

$$\ln y = \ln f(x)^{g(x)} = g(x) \ln f(x).$$

If the indeterminate form for y is 0^0 or ∞^0, then the indeterminate form for $\ln y$ is $0 \cdot \infty$, which may be handled using earlier methods. Similarly, if y has the form 1^∞, then the indeterminate form for $\ln y$ is $\infty \cdot 0$. It follows that

$$\text{if} \quad \lim_{x \to c} \ln y = \ln \left(\lim_{x \to c} y \right) = L, \quad \text{then} \quad \lim_{x \to c} y = e^L;$$

that is,

$$\lim_{x \to c} f(x)^{g(x)} = e^L.$$

This procedure may be summarized as follows.

Guidelines for investigating $\lim_{x \to c} f(x)^{g(x)}$ for the forms 0^0, 1^∞, and ∞^0 (10.4)

1. Let $y = f(x)^{g(x)}$.
2. Take natural logarithms in guideline 1:
$$\ln y = \ln f(x)^{g(x)} = g(x) \ln f(x)$$
3. Investigate $\lim_{x \to c} \ln y = \lim_{x \to c} [g(x) \ln f(x)]$ and conclude the following:
   (a) If $\lim_{x \to c} \ln y = L$, then $\lim_{x \to c} y = e^L$.
   (b) If $\lim_{x \to c} \ln y = \infty$, then $\lim_{x \to c} y = \infty$.
   (c) If $\lim_{x \to c} \ln y = -\infty$, then $\lim_{x \to c} y = 0$.

A common error is to stop after showing $\lim_{x \to c} \ln y = L$ and conclude that the given expression has the limit L. Remember that *we wish to find the limit of* y. Thus, if $\ln y$ has the limit L, then y has the limit e^L. The guidelines may also be used if $x \to \infty$ or if $x \to -\infty$ or for one-sided limits.

EXAMPLE 3 Find $\lim_{x \to 0^+} (1 + 3x)^{1/(2x)}$.

SOLUTION The indeterminate form is 1^∞. Employing Guidelines (10.4), we proceed as follows:

Guideline 1 $$y = (1 + 3x)^{1/(2x)}$$

Guideline 2 $$\ln y = \frac{1}{2x} \ln (1 + 3x) = \frac{\ln (1 + 3x)}{2x}$$

Guideline 3 The last expression has the indeterminate form 0/0 at $x = 0$, so we apply L'Hôpital's rule:

$$\lim_{x \to 0^+} \ln y = \lim_{x \to 0^+} \frac{\ln (1 + 3x)}{2x} = \lim_{x \to 0^+} \frac{3/(1 + 3x)}{2} = \frac{3}{2}$$

Consequently we arrive at the following:

$$\lim_{x \to 0^+} (1 + 3x)^{1/(2x)} = \lim_{x \to 0^+} y = e^{3/2}$$

The final indeterminate form we shall consider is defined in the following table.

Indeterminate form	Limit form: $\lim_{x \to c} [f(x) - g(x)]$
$\infty - \infty$	$\lim_{x \to c} f(x) = \infty$ and $\lim_{x \to c} g(x) = \infty$

When investigating $\infty - \infty$, we try to change the form of $f(x) - g(x)$ to a quotient or product and then apply L'Hôpital's rule or some other method of evaluation, as illustrated in the next example.

EXAMPLE 4 Find $\lim_{x \to 0^+} \left(\frac{1}{e^x - 1} - \frac{1}{x} \right)$.

SOLUTION The form is $\infty - \infty$; however, if the difference is written as a single fraction, then

$$\lim_{x \to 0^+} \left(\frac{1}{e^x - 1} - \frac{1}{x} \right) = \lim_{x \to 0^+} \frac{x - e^x + 1}{xe^x - x}.$$

This gives us the indeterminate form 0/0. It is necessary to apply L'Hôpital's rule twice, since the first application leads to the indeterminate form 0/0. Thus,

$$\lim_{x\to 0^+} \frac{x - e^x + 1}{xe^x - x} = \lim_{x\to 0^+} \frac{1 - e^x}{xe^x + e^x - 1}$$
$$= \lim_{x\to 0^+} \frac{-e^x}{xe^x + 2e^x} = -\frac{1}{2}.$$

FIGURE 10.3

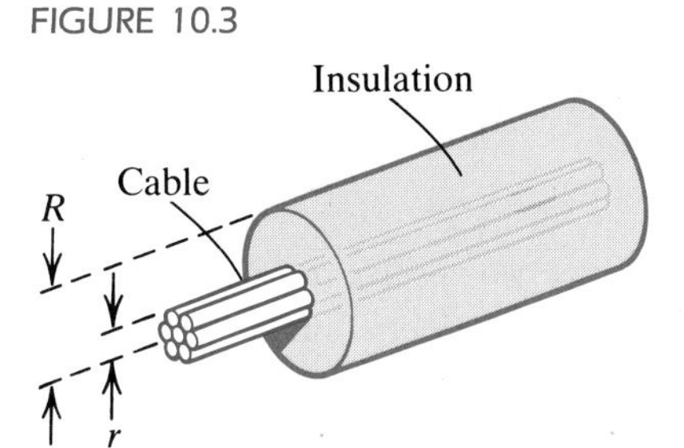

EXAMPLE 5 The velocity v of an electrical impulse in an insulated cable is given by

$$v = -k\left(\frac{r}{R}\right)^2 \ln\left(\frac{r}{R}\right),$$

where k is a positive constant, r is the radius of the cable, and R is the distance from the center of the cable to the outside of the insulation, as shown in Figure 10.3. Find

(a) $\lim_{R\to r^+} v$ **(b)** $\lim_{r\to 0^+} v$

SOLUTION

(a) The limit notation implies that r is fixed and R is a variable. In this case the expression for v is not indeterminate, and

$$\lim_{R\to r^+} v = -k \lim_{R\to r^+} \left(\frac{r}{R}\right)^2 \ln\left(\frac{r}{R}\right) = -k(1)^2 \ln 1 = -k(0) = 0.$$

(b) If R is fixed and r is a variable, then the expression for v has the indeterminate form $0 \cdot \infty$ at $r = 0$, and we first change the form of the expression algebraically, as follows:

$$\lim_{r\to 0^+} v = -k \lim_{r\to 0^+} \frac{\ln (r/R)}{(r/R)^{-2}} = -kR^2 \lim_{r\to 0^+} \frac{\ln r - \ln R}{r^{-2}}$$

The last quotient has the indeterminate form ∞/∞ at $r = 0$, so we may apply L'Hôpital's rule, obtaining

$$\lim_{r\to 0^+} v = -kR^2 \lim_{r\to 0^+} \frac{(1/r) - 0}{-2r^{-3}}$$
$$= -kR^2 \lim_{r\to 0^+} \left(\frac{r^2}{-2}\right) = -kR^2(0) = 0.$$

EXERCISES 10.2

Exer. 1–42: Find the limit, if it exists.

1 $\lim_{x\to 0^+} x \ln x$

2 $\lim_{x\to(\pi/2)^-} \tan x \ln \sin x$

3 $\lim_{x\to\infty} (x^2 - 1)e^{-x^2}$

4 $\lim_{x\to\infty} x(e^{1/x} - 1)$

5 $\lim_{x\to 0} e^{-x} \sin x$

6 $\lim_{x\to -\infty} x \tan^{-1} x$

7 $\lim_{x\to 0^+} \sin x \ln \sin x$

8 $\lim_{x\to\infty} x\left(\frac{\pi}{2} - \tan^{-1} x\right)$

9 $\lim_{x\to\infty} x \sin \frac{1}{x}$

10 $\lim_{x\to\infty} e^{-x} \ln x$

11 $\lim_{x\to 0} x \sec^2 x$

12 $\lim_{x\to 0} (\cos x)^{x+1}$

13 $\lim_{x\to\infty}\left(1+\frac{1}{x}\right)^{5x}$

14 $\lim_{x\to 0^+}(e^x+3x)^{1/x}$

15 $\lim_{x\to 0^+}(e^x-1)^x$

16 $\lim_{x\to 0^+} x^x$

17 $\lim_{x\to\infty} x^{1/x}$

18 $\lim_{x\to(\pi/2)^-}(\tan x)^{\cos x}$

19 $\lim_{x\to(\pi/2)^-}(\tan x)^x$

20 $\lim_{x\to 2^+}(x-2)^x$

21 $\lim_{x\to 0^+}(2x+1)^{\cot x}$

22 $\lim_{x\to 0^+}(1+3x)^{\csc x}$

23 $\lim_{x\to\infty}\left(\frac{x^2}{x-1}-\frac{x^2}{x+1}\right)$

24 $\lim_{x\to 1^+}\left(\frac{1}{x-1}-\frac{1}{\ln x}\right)$

25 $\lim_{x\to 0^-}\left(\frac{1}{x}-\frac{1}{\sin x}\right)$

26 $\lim_{x\to(\pi/2)^-}(\sec x-\tan x)$

27 $\lim_{x\to 1^-}(1-x)^{\ln x}$

28 $\lim_{x\to\infty}(1+e^x)^{e^{-x}}$

29 $\lim_{x\to 0^+}\left(\frac{1}{\sqrt{x^2+1}}-\frac{1}{x}\right)$

30 $\lim_{x\to 0}(\cot^2 x-\csc^2 x)$

31 $\lim_{x\to 0^+}\cot 2x\tan^{-1} x$

32 $\lim_{x\to\infty} x^3\, 2^{-x}$

33 $\lim_{x\to 0}(\cot^2 x-e^{-x})$

34 $\lim_{x\to\infty}(\sqrt{x^2+4}-\tan^{-1} x)$

35 $\lim_{x\to(\pi/2)^-}(1+\cos x)^{\tan x}$

36 $\lim_{x\to 0^+}(1+ax)^{b/x}$

37 $\lim_{x\to -3^-}\left(\frac{x}{x^2+2x-3}-\frac{4}{x+3}\right)$

38 $\lim_{x\to\infty}(\sqrt{x^4+5x^2+3}-x^2)$

39 $\lim_{x\to 0^+}(x+\cos 2x)^{\csc 3x}$

40 $\lim_{x\to(\pi/2)^-}\sec x\cos 3x$

41 $\lim_{x\to\infty}(\sinh x-x)$

42 $\lim_{x\to\infty}[\ln(4x+3)-\ln(3x+4)]$

c **Exer. 43–44: Graph f on the given interval and use the graph to estimate $\lim_{x\to 0} f(x)$.**

43 $f(x)=(x\tan x)^{(x^2)}$; $[-1,1]$

44 $f(x)=\left[\frac{\ln(x+1)}{\tan x}\right]^{1/x}$; $[-0.5,0.5]$

Exer. 45–46: (a) Find the local extrema and discuss the behavior of $f(x)$ near $x=0$. (b) Find horizontal asymptotes, if they exist. (c) Sketch the graph of f for $x>0$.

45 $f(x)=x^{1/x}$

46 $f(x)=x^x$

47 The *geometric mean* of two positive real numbers a and b is defined as $\sqrt{ab}$. Use L'Hôpital's rule to prove that

$$\sqrt{ab}=\lim_{x\to\infty}\left(\frac{a^{1/x}+b^{1/x}}{2}\right)^x.$$

48 If a sum of money P is invested at an interest rate of $100r$ percent per year, compounded m times per year, then the principal at the end of t years is given by $P(1+rm^{-1})^{mt}$. If we regard m as a real number and let m increase without bound, then the interest is said to be *compounded continuously*. Use L'Hôpital's rule to show that in this case the principal after t years is Pe^{rt}.

49 Refer to Exercise 55 of Section 10.1. In the velocity formula

$$v(t)=(mg/k)(1-e^{-(k/m)t}),$$

m represents the mass of the falling object. Find $\lim_{m\to\infty} v(t)$ and conclude that $v(t)$ is approximately proportional to time t if the mass is very large.

10.3 INTEGRALS WITH INFINITE LIMITS OF INTEGRATION

FIGURE 10.4
$\int_a^t f(x)\,dx$

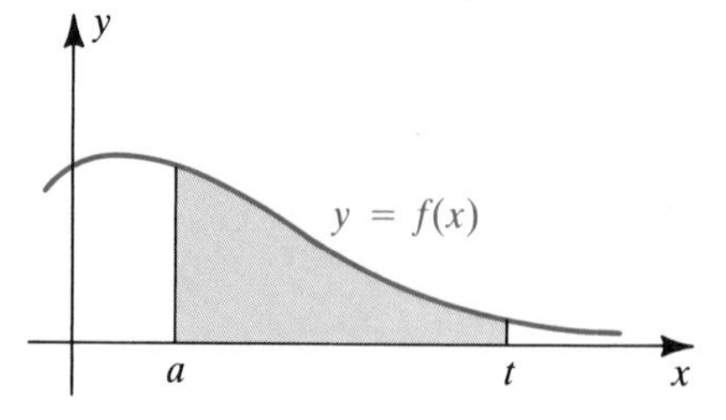

Suppose a function f is continuous and nonnegative on an infinite interval $[a,\infty)$ and $\lim_{x\to\infty} f(x)=0$. If $t>a$, then the area $A(t)$ under the graph of f from a to t, as illustrated in Figure 10.4, is

$$A(t)=\int_a^t f(x)\,dx.$$

If $\lim_{t\to\infty} A(t)$ exists, then the limit may be interpreted as the area of the region that lies under the graph of f, over the x-axis, and to the right of $x=a$, as illustrated in Figure 10.5. The symbol $\int_a^\infty f(x)\,dx$ is used to denote this number. If $\lim_{t\to\infty} A(t)=\infty$, we cannot assign an area to this (unbounded) region.

Part (i) of the next definition generalizes the preceding remarks to the case where $f(x)$ may be negative for some x in $[a,\infty)$.

Definition (10.5)

(i) If f is continuous on $[a, \infty)$, then

$$\int_a^{\infty} f(x)\,dx = \lim_{t\to\infty} \int_a^t f(x)\,dx,$$

provided the limit exists.

(ii) If f is continuous on $(-\infty, a]$, then

$$\int_{-\infty}^{a} f(x)\,dx = \lim_{t\to-\infty} \int_t^a f(x)\,dx,$$

provided the limit exists.

FIGURE 10.5

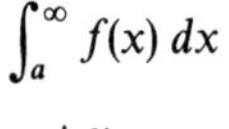

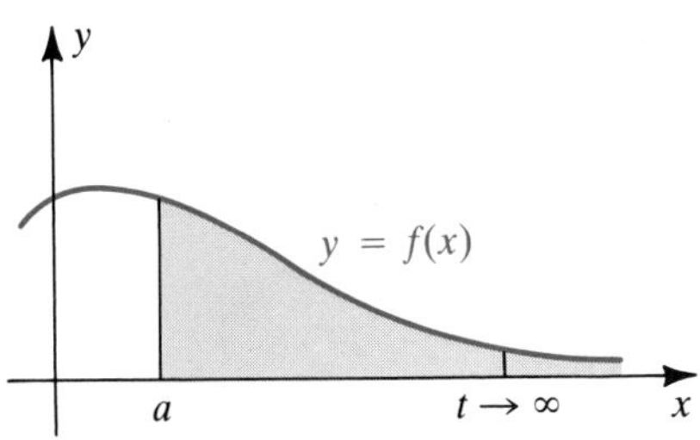

FIGURE 10.6

$\int_{-\infty}^{a} f(x)\,dx$

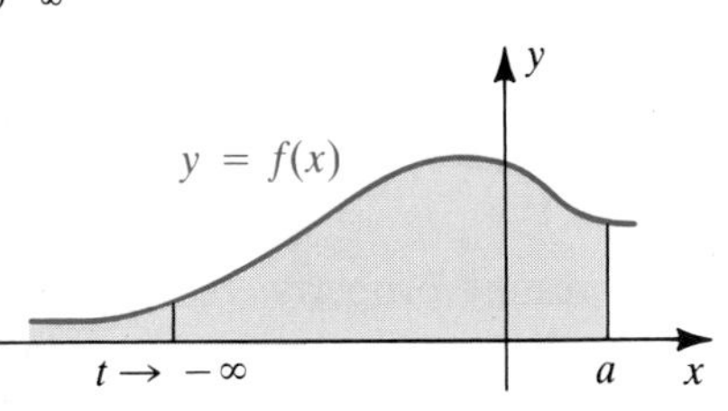

If $f(x) \geq 0$ for every x, then the limit in Definition (10.5)(ii) may be regarded as the area under the graph of f, over the x-axis, and to the *left* of $x = a$ (see Figure 10.6).

The expressions in Definition (10.5) are **improper integrals**. They differ from definite integrals in that one of the limits of integration is not a real number. An improper integral is said to **converge** if the limit exists, and the limit is the **value** of the improper integral. If the limit does not exist, the improper integral **diverges**.

Definition (10.5) is useful in many applications. In Example 4 we shall use an improper integral to calculate the work required to project an object from the surface of the earth to a point outside of the earth's gravitational field. Another important application occurs in the investigation of infinite series.

EXAMPLE 1 Determine whether the integral converges or diverges, and if it converges, find its value.

(a) $\displaystyle\int_2^{\infty} \frac{1}{(x-1)^2}\,dx$ **(b)** $\displaystyle\int_2^{\infty} \frac{1}{x-1}\,dx$

SOLUTION

(a) By Definition (10.5)(i),

$$\int_2^{\infty} \frac{1}{(x-1)^2}\,dx = \lim_{t\to\infty} \int_2^t \frac{1}{(x-1)^2}\,dx = \lim_{t\to\infty} \left[\frac{-1}{x-1}\right]_2^t$$

$$= \lim_{t\to\infty} \left(\frac{-1}{t-1} + \frac{1}{2-1}\right) = 0 + 1 = 1.$$

Thus, the integral converges and has the value 1.

(b) By Definition (10.5)(i),

$$\int_2^{\infty} \frac{1}{x-1}\,dx = \lim_{t\to\infty} \int_2^t \frac{1}{x-1}\,dx$$

$$= \lim_{t\to\infty} \Big[\ln(x-1)\Big]_2^t$$

$$= \lim_{t\to\infty} [\ln(t-1) - \ln(2-1)]$$

$$= \lim_{t\to\infty} \ln(t-1) = \infty.$$

Since the limit does not exist, the improper integral diverges.

The graphs of the two functions given by the integrands in Example 1, together with the (unbounded) regions that lie under the graphs for $x \geq 2$, are sketched in Figures 10.7 and 10.8. Note that although the graphs have the same general shape for $x \geq 2$, we may assign an area to the region under the graph shown in Figure 10.7, but not to that shown in Figure 10.8.

FIGURE 10.7 FIGURE 10.8

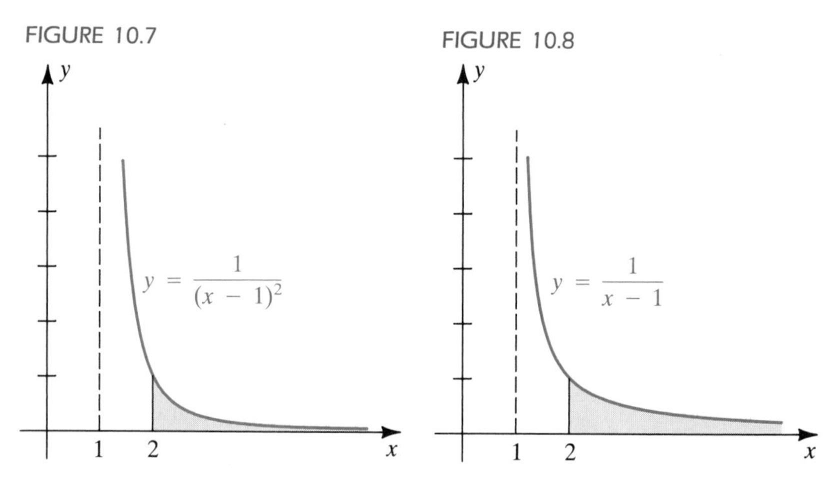

FIGURE 10.9

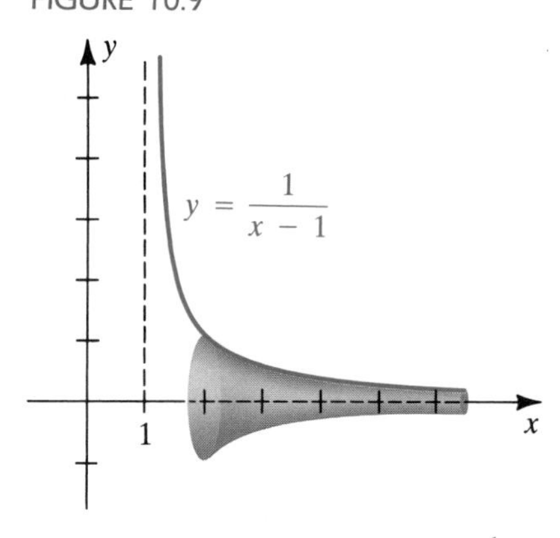

The graph in Figure 10.8 has an interesting property. If the region under the graph of $y = 1/(x - 1)$ is revolved about the x-axis, we obtain the (unbounded) solid of revolution shown in Figure 10.9. The improper integral

$$\int_2^{\infty} \pi \frac{1}{(x-1)^2}\, dx$$

may be regarded as the volume of this solid. By (a) of Example 1, the value of this improper integral is $\pi \cdot 1$, or π. This gives us the curious fact that although we cannot assign an area to the region in Figure 10.8, the volume of the solid of revolution generated by the region (see Figure 10.9) is finite. (A similar situation is described in Exercise 35.)

FIGURE 10.10

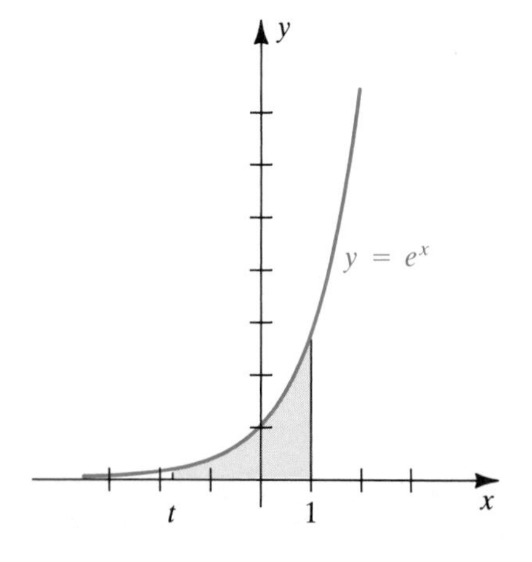

EXAMPLE 2 Assign an area to the region that lies under the graph of $y = e^x$, over the x-axis, and to the left of $x = 1$.

SOLUTION The region bounded by the graphs of $y = e^x$, $y = 0$, $x = 1$, and $x = t$, for $t < 1$, is sketched in Figure 10.10. The area of the *unbounded* region to the left of $x = 1$ is

$$\begin{aligned}\int_{-\infty}^{1} e^x\, dx &= \lim_{t \to -\infty} \int_t^1 e^x\, dx = \lim_{t \to -\infty} \Big[e^x\Big]_t^1 \\ &= \lim_{t \to -\infty} (e - e^t) = e - 0 = e.\end{aligned}$$

An improper integral may have *two* infinite limits of integration, as in the following definition.

Definition **(10.6)**

Let f be continuous for every x. If a is any real number, then

$$\int_{-\infty}^{\infty} f(x)\,dx = \int_{-\infty}^{a} f(x)\,dx + \int_{a}^{\infty} f(x)\,dx,$$

provided both of the improper integrals on the right converge.

If either of the integrals on the right in (10.6) diverges, then $\int_{-\infty}^{\infty} f(x)\,dx$ is said to **diverge**. It can be shown that (10.6) does not depend on the choice of the real number a. It can also be shown that $\int_{-\infty}^{\infty} f(x)\,dx$ is not necessarily the same as $\lim_{t\to\infty} \int_{-t}^{t} f(x)\,dx$ (consider $f(x) = x$).

EXAMPLE 3

(a) Evaluate $\displaystyle\int_{-\infty}^{\infty} \frac{1}{1+x^2}\,dx.$

(b) Sketch the graph of $f(x) = \dfrac{1}{1+x^2}$ and interpret the integral in (a) as an area.

SOLUTION **(a)** Using Definition (10.6), with $a = 0$, yields

$$\int_{-\infty}^{\infty} \frac{1}{1+x^2}\,dx = \int_{-\infty}^{0} \frac{1}{1+x^2}\,dx + \int_{0}^{\infty} \frac{1}{1+x^2}\,dx.$$

Next, applying Definition (10.5)(i), we have

$$\int_{0}^{\infty} \frac{1}{1+x^2}\,dx = \lim_{t\to\infty} \int_{0}^{t} \frac{1}{1+x^2}\,dx = \lim_{t\to\infty} \Big[\arctan x\Big]_0^t$$

$$= \lim_{t\to\infty} (\arctan t - \arctan 0) = \frac{\pi}{2} - 0 = \frac{\pi}{2}.$$

Similarly, we may show, by using (10.5)(ii), that

$$\int_{-\infty}^{0} \frac{1}{1+x^2}\,dx = \frac{\pi}{2}.$$

Consequently the given improper integral converges and has the value $(\pi/2) + (\pi/2) = \pi$.

(b) The graph of $y = 1/(1+x^2)$ is sketched in Figure 10.11. As in our previous discussion, the unbounded region that lies under the graph and above the x-axis may be assigned an area of π square units.

FIGURE 10.11

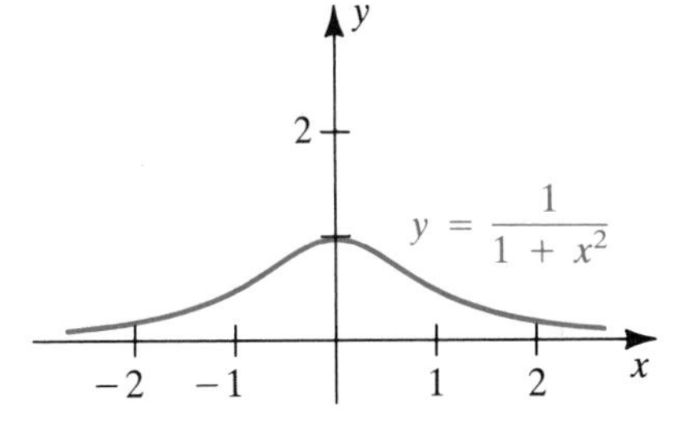

Let us conclude this section with a physical application of an improper integral. If a and b are the coordinates of two points A and B on a coordinate line l (see Figure 10.12) and if $f(x)$ is the force acting at the point P with coordinate x, then, by Definition (6.21), the work done as P moves from A to B is given by

$$W = \int_{a}^{b} f(x)\,dx.$$

FIGURE 10.12

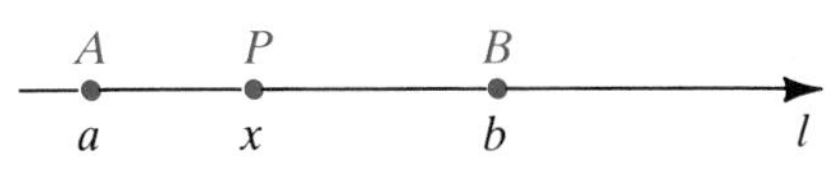

In similar fashion, the improper integral $\int_{a}^{\infty} f(x)\,dx$ may be used to define the work done as P moves indefinitely to the right (in applications, we use the terminology *P moves to infinity*). For example, if $f(x)$ is the force of attraction between a particle fixed at point A and a (movable) particle at P and if $c > a$, then $\int_c^\infty f(x)\,dx$ represents the work required to move P from the point with coordinate c to infinity.

FIGURE 10.13

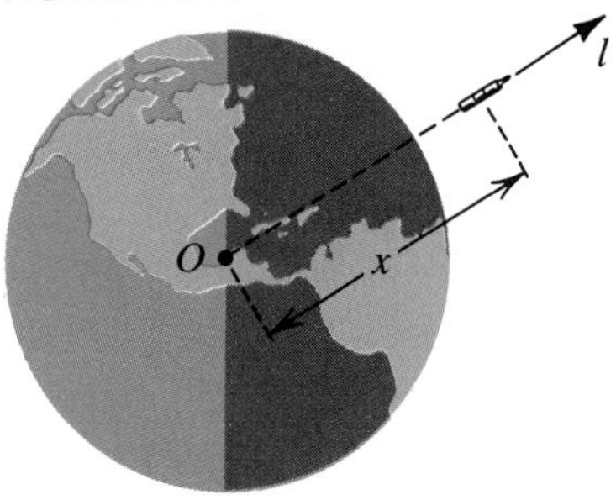

EXAMPLE 4 Let l be a coordinate line with origin O at the center of the earth, as shown in Figure 10.13. The gravitational force exerted at a point on l that is a distance x from O is given by $f(x) = k/x^2$, for some constant k. Using 4000 miles for the radius of the earth, find the work required to project an object weighing 100 pounds along l, from the surface to a point outside of the earth's gravitational field.

SOLUTION Theoretically, there is *always* a gravitational force $f(x)$ acting on the object; however, we may think of projecting the object from the surface to infinity. From the preceding discussion we wish to find

$$W = \int_{4000}^{\infty} f(x)\,dx.$$

By definition, $f(x) = k/x^2$ is the weight of an object that is a distance x from O, and hence

$$100 = f(4000) = \frac{k}{(4000)^2},$$

or, equivalently,

$$k = 100(4000)^2 = 10^2 \cdot 16 \cdot 10^6 = 16 \cdot 10^8.$$

Thus,

$$f(x) = (16 \cdot 10^8)\frac{1}{x^2}$$

and the required work is

$$\begin{aligned} W &= \int_{4000}^{\infty} (16 \cdot 10^8)\frac{1}{x^2}\,dx = 16 \cdot 10^8 \lim_{t\to\infty} \int_{4000}^{t} \frac{1}{x^2}\,dx \\ &= 16 \cdot 10^8 \lim_{t\to\infty} \left[-\frac{1}{x}\right]_{4000}^{t} = 16 \cdot 10^8 \lim_{t\to\infty}\left(-\frac{1}{t} + \frac{1}{4000}\right) \\ &= \frac{16 \cdot 10^8}{4000} = 4 \cdot 10^5 \text{ mi-lb}. \end{aligned}$$

In terms of foot-pounds,

$$W = 5280 \cdot 4 \cdot 10^5 \approx (2.1)10^9 \text{ ft-lb},$$

or approximately 2 trillion ft-lb.

EXERCISES 10.3

Exer. 1–24: Determine whether the integral converges or diverges, and if it converges, find its value.

1 $\displaystyle\int_1^\infty \frac{1}{x^{4/3}}\,dx$

2 $\displaystyle\int_{-\infty}^{0} \frac{1}{(x-1)^3}\,dx$

3 $\displaystyle\int_1^\infty \frac{1}{x^{3/4}}\,dx$

4 $\displaystyle\int_0^\infty \frac{x}{1+x^2}\,dx$

5 $\displaystyle\int_{-\infty}^{2} \frac{1}{5-2x}\,dx$

6 $\displaystyle\int_{-\infty}^{\infty} \frac{x}{x^4+9}\,dx$

7 $\int_0^\infty e^{-2x}\,dx$

8 $\int_{-\infty}^0 e^x\,dx$

9 $\int_{-\infty}^{-1} \frac{1}{x^3}\,dx$

10 $\int_0^\infty \frac{1}{\sqrt[3]{x+1}}\,dx$

11 $\int_{-\infty}^0 \frac{1}{(x-8)^{2/3}}\,dx$

12 $\int_1^\infty \frac{x}{(1+x^2)^2}\,dx$

13 $\int_0^\infty \frac{\cos x}{1+\sin^2 x}\,dx$

14 $\int_{-\infty}^2 \frac{1}{x^2+4}\,dx$

15 $\int_{-\infty}^\infty xe^{-x^2}\,dx$

16 $\int_{-\infty}^\infty \cos^2 x\,dx$

17 $\int_1^\infty \frac{\ln x}{x}\,dx$

18 $\int_3^\infty \frac{1}{x^2-1}\,dx$

19 $\int_0^\infty \cos x\,dx$

20 $\int_{-\infty}^{\pi/2} \sin 2x\,dx$

21 $\int_{-\infty}^\infty \operatorname{sech} x\,dx$

22 $\int_0^\infty xe^{-x}\,dx$

23 $\int_{-\infty}^0 \frac{1}{x^2-3x+2}\,dx$

24 $\int_4^\infty \frac{x+18}{x^2+x-12}\,dx$

Exer. 25–28: If f and g are continuous functions and $0 \le f(x) \le g(x)$ for every x in $[a, \infty)$, then the following *comparison tests for improper integrals* are true:

(I) If $\int_a^\infty g(x)\,dx$ converges, then $\int_a^\infty f(x)\,dx$ converges.

(II) If $\int_a^\infty f(x)\,dx$ diverges, then $\int_a^\infty g(x)\,dx$ diverges.

Determine whether the first integral converges by comparing it with the second integral.

25 $\int_1^\infty \frac{1}{1+x^4}\,dx;\quad \int_1^\infty \frac{1}{x^4}\,dx$

26 $\int_2^\infty \frac{1}{\sqrt[3]{x^2-1}}\,dx;\quad \int_2^\infty \frac{1}{\sqrt[3]{x^2}}\,dx$

27 $\int_2^\infty \frac{1}{\ln x}\,dx;\quad \int_2^\infty \frac{1}{x}\,dx$

28 $\int_1^\infty e^{-x^2}\,dx;\quad \int_1^\infty e^{-x}\,dx$

Exer. 29–32: Assign, if possible, a value to (a) the area of the region R and (b) the volume of the solid obtained by revolving R about the x-axis.

29 $R = \{(x, y): x \ge 1, 0 \le y \le 1/x\}$

30 $R = \{(x, y): x \ge 1, 0 \le y \le 1/\sqrt{x}\}$

31 $R = \{(x, y): x \ge 4, 0 \le y \le x^{-3/2}\}$

32 $R = \{(x, y): x \ge 8, 0 \le y \le x^{-2/3}\}$

33 The unbounded region to the right of the y-axis and between the graphs of $y = e^{-x^2}$ and $y = 0$ is revolved about the y-axis. Show that a volume can be assigned to the resulting unbounded solid, and find the volume.

34 The graph of $y = e^{-x}$ for $x \ge 0$ is revolved about the x-axis. Show that an area can be assigned to the resulting unbounded surface, and find the area.

35 The solid of revolution known as *Gabriel's horn* is generated by rotating the region under the graph of $y = 1/x$ for $x \ge 1$ about the x-axis (see figure).

(a) Show that Gabriel's horn has a finite volume of π cubic units.

(b) Is a finite volume obtained if the graph is rotated about the y-axis?

(c) Show that the surface area of Gabriel's horn is given by $\int_1^\infty 2\pi(1/x)\sqrt{1+(1/x^4)}\,dx$. Use a comparison test (see Exercises 25–28) with $f(x) = 2\pi/x$ to establish that this integral diverges. Thus, we cannot assign an area to the surface, even though the volume of the horn is finite.

EXERCISE 35

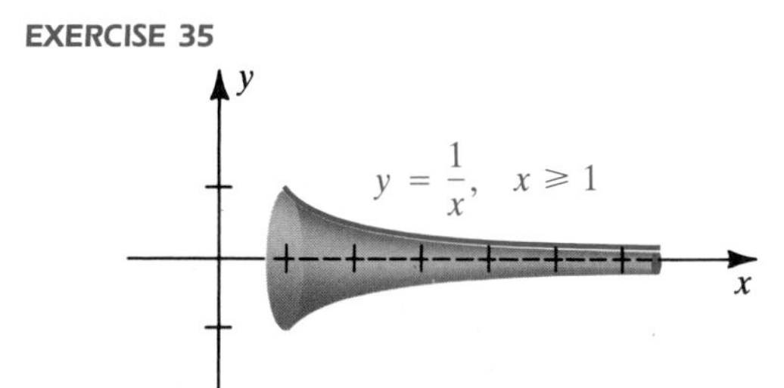

36 A spacecraft carries a fuel supply of mass m. As a conservation measure, the captain decides to burn fuel at a rate of $R(t) = mke^{-kt}$ g/sec, for some positive constant k.

(a) What does the improper integral $\int_0^\infty R(t)\,dt$ represent?

(b) When will the spacecraft run out of fuel?

37 The force (in joules) with which two electrons repel one another is inversely proportional to the square of the distance (in meters) between them. If, in Figure 10.12, one electron is fixed at A, find the work done if another electron is repelled along l from a point B, which is 1 meter from A, to infinity.

38 An electric dipole consists of opposite charges separated by a small distance d. Suppose that charges of $+q$ and $-q$ units are located on a coordinate line l at $\frac{1}{2}d$ and $-\frac{1}{2}d$, respectively (see figure). By Coulomb's law, the net force acting on a unit charge of -1 unit at $x > \frac{1}{2}d$ is given by

$$f(x) = \frac{-kq}{(x-\frac{1}{2}d)^2} + \frac{kq}{(x+\frac{1}{2}d)^2}$$

for some positive constant k. If $a > \frac{1}{2}d$, find the work done in moving the unit charge along l from a to infinity.

EXERCISE 38

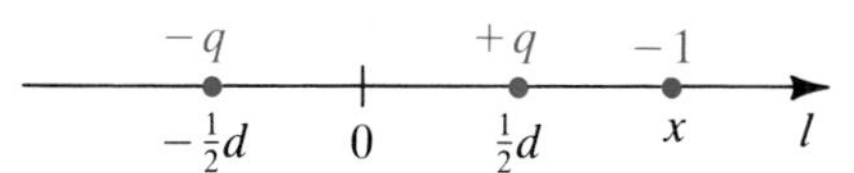

39 The reliability $R(t)$ of a product is the probability that it will not require repair for at least t years. To design a warranty guarantee, a manufacturer must know the average time of service before first repair of a product. This is given by the improper integral $\int_0^\infty (-t)R'(t)\,dt$.

(a) For many high-quality products, $R(t)$ has the form e^{-kt} for some positive constant k. Find an expression in terms of k for the average time of service before repair.

(b) Is it possible to manufacture a product for which $R(t) = 1/(t+1)$?

40 A sum of money is deposited into an account that pays interest at 8% per year, compounded continuously (see Exercise 48 of Section 10.2). Starting T years from now, money will be withdrawn at the *capital flow rate* of $f(t)$ dollars per year, continuing indefinitely. For future income to be generated at this rate, the minimum amount A that must be deposited, or the *present value of the capital flow*, is given by the improper integral $A = \int_T^\infty f(t)e^{-0.08t}\,dt$. Find A if the income desired 20 years from now is

(a) 12,000 dollars per year

(b) $12{,}000e^{0.04t}$ dollars per year

41 (a) Use integration by parts to establish the formula

$$\int_0^\infty x^2 e^{-ax^2}\,dx = \frac{1}{2a^{3/2}}\int_0^\infty e^{-u^2}\,du.$$

It can be shown that the value of this integral is $\sqrt{\pi}/2$.

(b) The relative number of gas molecules in a container that travel at a speed of v cm/sec can be found by using the *Maxwell-Boltzmann speed distribution* F:

$$F(v) = cv^2 e^{-mv^2/(2kT)},$$

where T is the temperature (in °K), m is the mass of a molecule, and c and k are positive constants. The constant c must be selected so that $\int_0^\infty F(v)\,dv = 1$. Use part (a) to express c in terms of k, T, and m.

42 The *Fourier transform* is useful for solving certain differential equations. The *Fourier cosine transform* of a function f is defined by

$$F_C[f(x)] = \int_0^\infty f(x)\cos sx\,dx$$

for every real number s for which the improper integral converges. Find $F_C[e^{-ax}]$ for $a > 0$.

Exer. 43–48: In the theory of differential equations, if f is a function, then the *Laplace transform* L of $f(x)$ is defined by

$$L[f(x)] = \int_0^\infty e^{-sx}f(x)\,dx$$

for every real number s for which the improper integral converges. Find $L[f(x)]$ if $f(x)$ is the given expression.

43 1 44 x 45 $\cos x$

46 $\sin x$ 47 e^{ax} 48 $\sin ax$

49 The *gamma function* Γ is defined by $\Gamma(n) = \int_0^\infty x^{n-1}e^{-x}\,dx$ for every positive real number n.

(a) Find $\Gamma(1)$, $\Gamma(2)$, and $\Gamma(3)$.

(b) Prove that $\Gamma(n+1) = n\Gamma(n)$.

(c) Use mathematical induction to prove that if n is any positive integer, then $\Gamma(n+1) = n!$. (This shows that factorials are special values of the gamma function.)

50 Refer to Exercise 49. Functions given by $f(x) = cx^k e^{-ax}$ with $x > 0$ are called *gamma distributions* and play an important role in probability theory. The constant c must be selected so that $\int_0^\infty f(x)\,dx = 1$. Express c in terms of the positive constants k and a and the gamma function Γ.

c **Exer. 51–52: Approximate the improper integral by making the substitution $u = 1/x$ and then using Simpson's rule with $n = 4$.**

51 $\displaystyle\int_2^\infty \frac{1}{\sqrt{x^4+x}}\,dx$ 52 $\displaystyle\int_{-\infty}^{-10} \frac{\sqrt{|x|}}{x^3+1}\,dx$

10.4 INTEGRALS WITH DISCONTINUOUS INTEGRANDS

If a function f is continuous on a closed interval $[a, b]$, then, by Theorem (5.20), the definite integral $\int_a^b f(x)\,dx$ exists. If f has an infinite discontinuity at some number in the interval, it may still be possible to assign a value to the integral. Suppose, for example, that f is continuous and nonnegative on the half-open interval $[a, b)$ and $\lim_{x\to b^-} f(x) = \infty$. If $a < t < b$, then the area $A(t)$ under the graph of f from a to t (see Figure 10.14, on the next page) is

$$A(t) = \int_a^t f(x)\,dx.$$

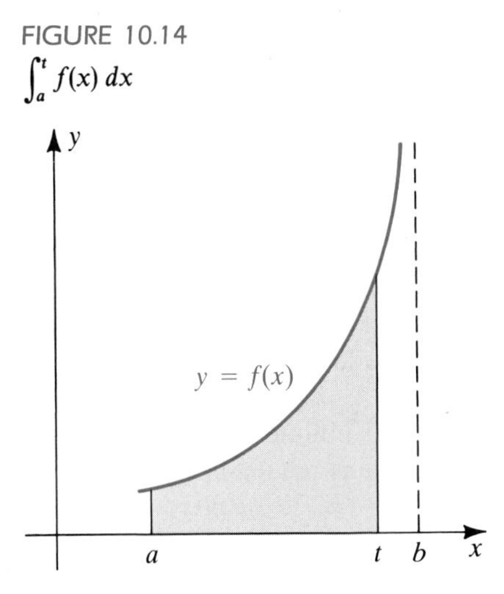

FIGURE 10.14
$\int_a^t f(x)\,dx$

If $\lim_{t\to b^-} A(t)$ exists, then the limit may be interpreted as the area of the unbounded region that lies under the graph of f, over the x-axis, and between $x = a$ and $x = b$. We shall denote this number by $\int_a^b f(x)\,dx$.

For the situation illustrated in Figure 10.15, $\lim_{x\to a^+} f(x) = \infty$, and we define $\int_a^b f(x)\,dx$ as the limit of $\int_t^b f(x)\,dx$ as $t \to a^+$.

FIGURE 10.15
$\int_t^b f(x)\,dx$

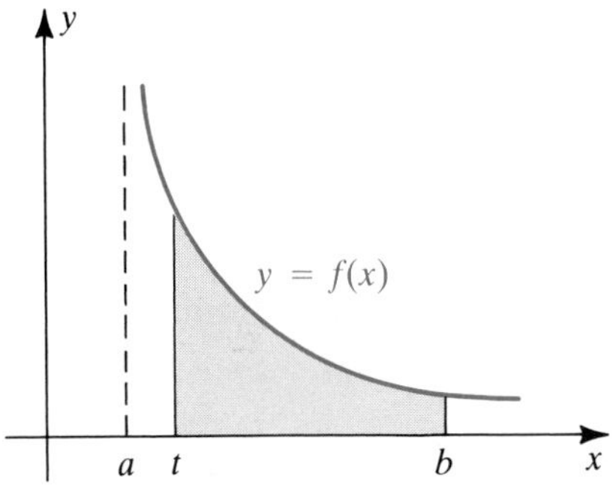

These remarks are the motivation for the following definition.

Definition **(10.7)**

(I) If f is continuous on $[a, b)$ and discontinuous at b, then

$$\int_a^b f(x)\,dx = \lim_{t\to b^-} \int_a^t f(x)\,dx,$$

provided the limit exists.

(II) If f is continuous on $(a, b]$ and discontinuous at a, then

$$\int_a^b f(x)\,dx = \lim_{t\to a^+} \int_t^b f(x)\,dx,$$

provided the limit exists.

As in the preceding section, the integrals defined in (10.7) are referred to as *improper integrals* and they *converge* if the limits exist. The limits are called the *values* of the improper integrals. If the limits do not exist, the improper integrals *diverge*.

Another type of improper integral is defined as follows.

Definition **(10.8)**

If f has a discontinuity at a number c in the open interval (a, b) but is continuous elsewhere on $[a, b]$, then

$$\int_a^b f(x)\,dx = \int_a^c f(x)\,dx + \int_c^b f(x)\,dx,$$

provided *both* of the improper integrals on the right converge. If both converge, then the value of the improper integral $\int_a^b f(x)\,dx$ is the sum of the two values.

FIGURE 10.16

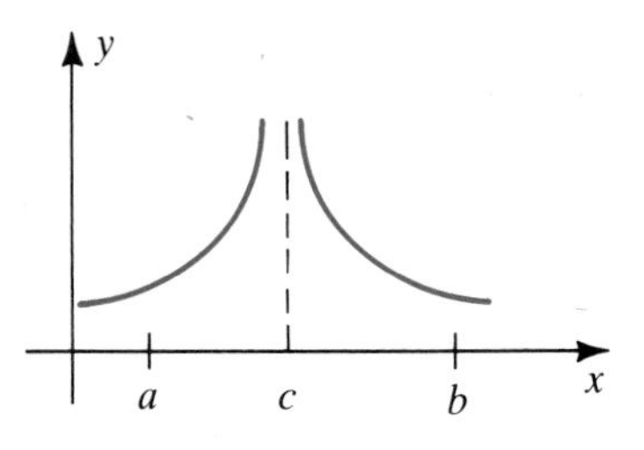

FIGURE 10.17

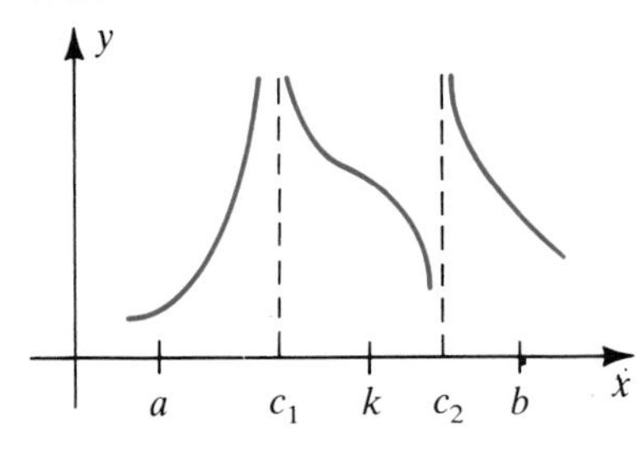

The graph of a function satisfying the conditions of Definition (10.8) is sketched in Figure 10.16.

A definition similar to (10.8) is used if f has any finite number of discontinuities in (a, b). For example, suppose f has discontinuities at c_1 and c_2, with $c_1 < c_2$, but is continuous elsewhere on $[a, b]$. One possibility is illustrated in Figure 10.17. In this case we choose a number k between c_1 and c_2 and express $\int_a^b f(x)\,dx$ as a sum of four improper integrals over the intervals $[a, c_1]$, $[c_1, k]$, $[k, c_2]$, and $[c_2, b]$, respectively. By definition, $\int_a^b f(x)\,dx$ converges if and only if each of the four improper integrals in the sum converges. We can show that this definition is independent of the number k.

Finally, if f is continuous on (a, b) but has infinite discontinuities at a and b, then we again define $\int_a^b f(x)\,dx$ by means of (10.8).

EXAMPLE 1 Evaluate $\displaystyle\int_0^3 \frac{1}{\sqrt{3-x}}\,dx$.

SOLUTION Since the integrand has an infinite discontinuity at $x = 3$, we apply Definition (10.7)(i) as follows:

$$\begin{aligned}\int_0^3 \frac{1}{\sqrt{3-x}}\,dx &= \lim_{t\to 3^-}\int_0^t \frac{1}{\sqrt{3-x}}\,dx\\ &= \lim_{t\to 3^-}\Big[-2\sqrt{3-x}\Big]_0^t\\ &= \lim_{t\to 3^-}\left(-2\sqrt{3-t}+2\sqrt{3}\right)\\ &= 0 + 2\sqrt{3} = 2\sqrt{3}\end{aligned}$$

EXAMPLE 2 Determine whether the improper integral $\displaystyle\int_0^1 \frac{1}{x}\,dx$ converges or diverges.

SOLUTION The integrand is undefined at $x = 0$. Applying (10.7)(ii) gives us

$$\begin{aligned}\int_0^1 \frac{1}{x}\,dx = \lim_{t\to 0^+}\int_t^1 \frac{1}{x}\,dx &= \lim_{t\to 0^+}\Big[\ln x\Big]_t^1\\ &= \lim_{t\to 0^+}(0-\ln t) = \infty.\end{aligned}$$

Since the limit does not exist, the improper integral diverges.

EXAMPLE 3 Determine whether the improper integral $\displaystyle\int_0^4 \frac{1}{(x-3)^2}\,dx$ converges or diverges.

SOLUTION The integrand is undefined at $x = 3$. Since this number is in the interval $(0, 4)$, we use Definition (10.8) with $c = 3$:

$$\int_0^4 \frac{1}{(x-3)^2}\,dx = \int_0^3 \frac{1}{(x-3)^2}\,dx + \int_3^4 \frac{1}{(x-3)^2}\,dx$$

For the integral on the left to converge, *both* integrals on the right must converge. Equivalently, the integral on the left diverges if either of the integrals on the right diverges. Applying Definition (10.7)(i) to the first integral on the right gives us

$$\begin{aligned}\int_0^3 \frac{1}{(x-3)^2}\,dx &= \lim_{t\to 3^-} \int_0^t \frac{1}{(x-3)^2}\,dx \\ &= \lim_{t\to 3^-} \left[\frac{-1}{x-3}\right]_0^t \\ &= \lim_{t\to 3^-} \left(\frac{-1}{t-3} - \frac{1}{3}\right) = \infty.\end{aligned}$$

Thus, the given improper integral diverges.

It is important to note that the fundamental theorem of calculus cannot be applied to the integral in Example 3, since the function given by the integrand is not continuous on $[0, 4]$. If we had (incorrectly) applied the fundamental theorem, we would have obtained

$$\left[\frac{-1}{x-3}\right]_0^4 = -1 - \frac{1}{3} = -\frac{4}{3}.$$

This result is obviously incorrect, since the integrand is never negative.

EXAMPLE 4 Evaluate $\displaystyle\int_{-2}^{7} \frac{1}{(x+1)^{2/3}}\,dx.$

SOLUTION The integrand is undefined at $x = -1$, which is in the interval $(-2, 7)$. Hence we apply Definition (10.8), with $c = -1$:

$$\int_{-2}^{7} \frac{1}{(x+1)^{2/3}}\,dx = \int_{-2}^{-1} \frac{1}{(x+1)^{2/3}}\,dx + \int_{-1}^{7} \frac{1}{(x+1)^{2/3}}\,dx$$

We next investigate each of the integrals on the right-hand side of this equation. Using (10.7)(i) with $b = -1$ gives us

$$\begin{aligned}\int_{-2}^{-1} \frac{1}{(x+1)^{2/3}}\,dx &= \lim_{t\to -1^-} \int_{-2}^{t} \frac{1}{(x+1)^{2/3}}\,dx \\ &= \lim_{t\to -1^-} \Big[3(x+1)^{1/3}\Big]_{-2}^{t} \\ &= 3 \lim_{t\to -1^-} \left[(t+1)^{1/3} - (-1)^{1/3}\right] \\ &= 3(0+1) = 3.\end{aligned}$$

Similarly, using (10.7)(ii) with $a = -1$ yields

$$\begin{aligned}
\int_{-1}^{7} \frac{1}{(x+1)^{2/3}}\,dx &= \lim_{t\to -1^+} \int_t^7 \frac{1}{(x+1)^{2/3}}\,dx \\
&= \lim_{t\to -1^+} \Big[3(x+1)^{1/3}\Big]_t^7 \\
&= 3 \lim_{t\to -1^+} \left[(8)^{1/3} - (t+1)^{1/3}\right] \\
&= 3(2-0) = 6.
\end{aligned}$$

Since both integrals converge, the given integral converges and has the value $3 + 6 = 9$.

An improper integral may have both a discontinuity in the integrand and an infinite limit of integration. Integrals of this type may be investigated by expressing them as sums of improper integrals, each of which has *one* of the forms previously defined. As an illustration, since the integrand of $\int_0^\infty (1/\sqrt{x})\,dx$ is discontinuous at $x = 0$, we choose any number greater than 0—say 1—and write

$$\int_0^\infty \frac{1}{\sqrt{x}}\,dx = \int_0^1 \frac{1}{\sqrt{x}}\,dx + \int_1^\infty \frac{1}{\sqrt{x}}\,dx.$$

We can show that the first integral on the right-hand side of the equation converges and the second diverges. Hence (by definition) the given integral diverges.

Improper integrals of the types considered in this section arise in physical applications. Figure 10.18 is a schematic drawing of a spring with an attached weight that is oscillating between points with coordinates $-c$ and c on a coordinate line y (the y-axis has been positioned at the right for clarity). The **period** T is the time required for one complete oscillation—that is, *twice* the time required for the weight to cover the interval $[-c, c]$. The next example illustrates how an improper integral results when we derive a formula for T.

FIGURE 10.18

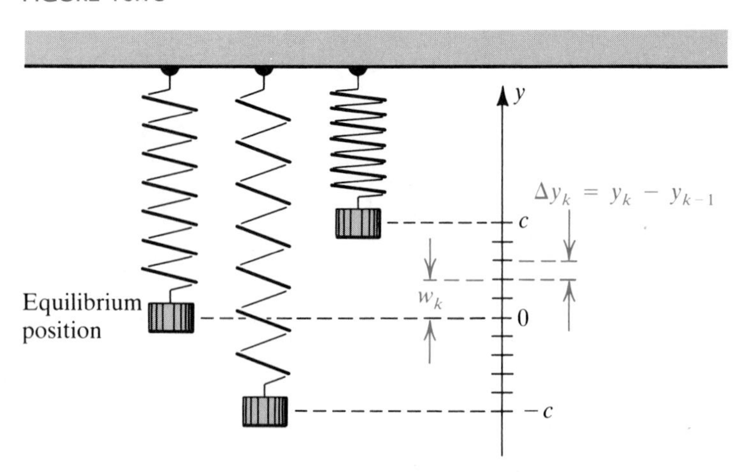

EXAMPLE 5 Let $v(y)$ denote the velocity of the weight in Figure 10.18 when it is at the point with coordinate y in $[-c, c]$. Show that the period T is given by

$$T = 2\int_{-c}^{c} \frac{1}{v(y)}\,dy.$$

SOLUTION Let us partition $[-c, c]$ in the usual way, and let $\Delta y_k = y_k - y_{k-1}$ denote the distance the weight travels during the time interval Δt_k. If w_k is any number in the subinterval $[y_{k-1}, y_k]$, then $v(w_k)$ is the velocity of the weight when it is at the point with coordinate w_k. If the norm of the partition is small and if we assume v is a continuous function, then the distance Δy_k may be approximated by the product $v(w_k)\,\Delta t_k$; that is,

$$\Delta y_k \approx v(w_k)\,\Delta t_k.$$

Hence the time required for the weight to cover the distance Δy_k may be approximated by

$$\Delta t_k \approx \frac{1}{v(w_k)}\,\Delta y_k$$

and, therefore,

$$T = 2\sum_k \Delta t_k \approx 2\sum_k \frac{1}{v(w_k)}\,\Delta y_k.$$

By considering the limit of the sums on the right and using the definition of definite integral, we conclude that

$$T = 2\int_{-c}^{c} \frac{1}{v(y)}\,dy.$$

Note that $v(c) = 0$ and $v(-c) = 0$, so the integral is improper.

EXERCISES 10.4

Exer. 1–30: Determine whether the integral converges or diverges, and if it converges, find its value.

1 $\int_0^8 \frac{1}{\sqrt[3]{x}}\,dx$

2 $\int_0^9 \frac{1}{\sqrt{x}}\,dx$

3 $\int_{-3}^1 \frac{1}{x^2}\,dx$

4 $\int_{-2}^{-1} \frac{1}{(x+2)^{5/4}}\,dx$

5 $\int_0^{\pi/2} \sec^2 x\,dx$

6 $\int_0^1 \frac{e^{\sqrt{x}}}{\sqrt{x}}\,dx$

7 $\int_0^4 \frac{1}{(4-x)^{3/2}}\,dx$

8 $\int_0^{-1} \frac{1}{\sqrt[3]{x+1}}\,dx$

9 $\int_0^4 \frac{1}{(4-x)^{2/3}}\,dx$

10 $\int_1^2 \frac{x}{x^2-1}\,dx$

11 $\int_{-2}^2 \frac{1}{(x+1)^3}\,dx$

12 $\int_{-1}^1 x^{-4/3}\,dx$

13 $\int_{-2}^0 \frac{1}{\sqrt{4-x^2}}\,dx.$

14 $\int_{-2}^0 \frac{x}{\sqrt{4-x^2}}\,dx$

15 $\int_{-1}^2 \frac{1}{x}\,dx$

16 $\int_0^4 \frac{1}{x^2-x-2}\,dx$

17 $\int_0^1 x \ln x\,dx$

18 $\int_0^{\pi/2} \tan^2 x\,dx$

19 $\int_0^{\pi/2} \tan x\,dx$

20 $\int_0^{\pi/2} \frac{1}{1-\cos x}\,dx$

21 $\int_2^4 \frac{x-2}{x^2-5x+4}\,dx$

22 $\int_{1/e}^e \frac{1}{x(\ln x)^2}\,dx$

23 $\int_{-1}^2 \frac{1}{x^2}\cos\frac{1}{x}\,dx$

24 $\int_0^{\pi} \sec x\,dx$

25 $\int_0^{\pi} \frac{\cos x}{\sqrt{1 - \sin x}}\, dx$

26 $\int_0^9 \frac{x}{\sqrt[3]{x-1}}\, dx$

27 $\int_0^4 \frac{1}{x^2 - 4x + 3}\, dx$

28 $\int_{-1}^3 \frac{x}{\sqrt[3]{x^2-1}}\, dx$

29 $\int_0^{\infty} \frac{1}{(x-4)^2}\, dx$

30 $\int_{-\infty}^0 \frac{1}{x+2}\, dx$

Exer. 31–34: Suppose that f and g are continuous and $0 \le f(x) \le g(x)$ for every x in $(a, b]$. If f and g are discontinuous at $x = a$, then the following *comparison tests* can be proved:

(I) If $\int_a^b g(x)\, dx$ converges, then $\int_a^b f(x)\, dx$ converges.

(II) If $\int_a^b f(x)\, dx$ diverges, then $\int_a^b g(x)\, dx$ diverges.

Analogous tests may be stated for continuity on $[a, b)$ with a discontinuity at $x = b$. Determine whether the first integral converges or diverges by comparing it with the second integral.

31 $\int_0^{\pi} \frac{\sin x}{\sqrt{x}}\, dx$; $\int_0^{\pi} \frac{1}{\sqrt{x}}\, dx$

32 $\int_0^{\pi/4} \frac{\sec x}{x^3}\, dx$; $\int_0^{\pi/4} \frac{1}{x^3}\, dx$

33 $\int_0^2 \frac{\cosh x}{(x-2)^2}\, dx$; $\int_0^2 \frac{1}{(x-2)^2}\, dx$

34 $\int_0^1 \frac{e^{-x}}{x^{2/3}}\, dx$; $\int_0^1 \frac{1}{x^{2/3}}\, dx$

Exer. 35–36: Find all real values of n for which the integral converges.

35 $\int_0^1 x^n\, dx$

36 $\int_0^1 x^n \ln x\, dx$

Exer. 37–40: Assign, if possible, a value to (a) the area of the region R and (b) the volume of the solid obtained by revolving R about the x-axis.

37 $R = \{(x, y)\colon 0 \le x \le 1, 0 \le y \le 1/\sqrt{x}\}$

38 $R = \{(x, y)\colon 0 \le x \le 1, 0 \le y \le 1/\sqrt[3]{x}\}$

39 $R = \{(x, y)\colon -4 \le x \le 4, 0 \le y \le 1/(x+4)\}$

40 $R = \{(x, y)\colon 1 \le x \le 2, 0 \le y \le 1/(x-1)\}$

c 41 Approximate $\int_0^1 \frac{\cos x}{\sqrt{x}}\, dx$ by making the substitution $u = \sqrt{x}$ and then using the trapezoidal rule with $n = 4$.

c 42 Approximate $\int_0^1 \frac{\sin x}{x}\, dx$ by removing the discontinuity at $x = 0$ and then using Simpson's rule with $n = 4$.

43 Refer to Example 5. If the weight in Figure 10.18 has mass m and if the spring obeys Hooke's law (with spring constant $k > 0$), then, in the absence of frictional forces, the velocity v of the weight is a solution of the differential equation

$$mv\frac{dv}{dy} + ky = 0.$$

(a) Use separation of variables (see Section 7.6) to show that $v^2 = (k/m)(c^2 - y^2)$. (*Hint:* Recall from Example 5 that $v(c) = v(-c) = 0$.)

(b) Find the period T of the oscillation.

44 A simple pendulum consists of a bob of mass m attached to a string of length L (see figure). If we assume that the string is weightless and that no other frictional forces are present, then the angular velocity $v = d\theta/dt$ is a solution of the differential equation

$$v\frac{dv}{d\theta} + \frac{g}{L}\sin\theta = 0,$$

where g is a gravitational constant.

(a) If $v = 0$ at $\theta = \pm\theta_0$, use separation of variables to show that

$$v^2 = \frac{2g}{L}(\cos\theta - \cos\theta_0).$$

(b) The period T of the pendulum is twice the amount of time needed for θ to change from $-\theta_0$ to θ_0. Show that T is given by the improper integral

$$T = 2\sqrt{\frac{2L}{g}}\int_0^{\theta_0} \frac{1}{\sqrt{\cos\theta - \cos\theta_0}}\, d\theta.$$

EXERCISE 44

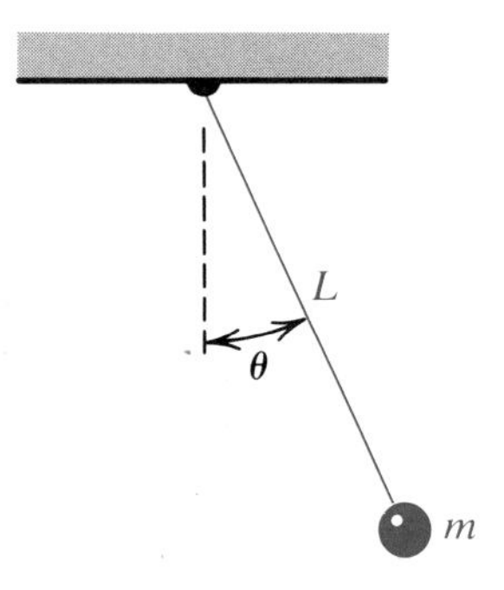

45 When a dose of y_0 milligrams of a drug is injected directly into the bloodstream, the average length of time T that a molecule remains in the bloodstream is given by the formula $T = (1/y_0)\int_0^{y_0} t\, dy$ for the time t at which exactly y milligrams is still present.

(a) If $y = y_0 e^{-kt}$ for some positive constant k, explain why the integral for T is improper.

(b) If τ is the half-life of the drug in the bloodstream, show that $T = \tau/\ln 2$.

46 In fishery science, the collection of fish that results from one annual reproduction is referred to as a *cohort*. The number N of fish still alive after t years is usually given by an exponential function. For North Sea haddock with initial size of a cohort N_0, $N = N_0 e^{-0.2t}$. The average life expectancy T (in years) of a fish in a cohort is given by $T = (1/N_0) \int_0^{N_0} t\, dN$ for the time t when precisely N fish are still alive.

(a) Find the value of T for North Sea haddock.

(b) Is it possible to have a species such that $N = N_0/(1 + kN_0 t)$ for some positive constant k? If so, compute T for such a species.

10.5 REVIEW EXERCISES

Exer. 1–16: Find the limit, if it exists.

1 $\lim_{x\to 0} \dfrac{\ln(2-x)}{1+e^{2x}}$

2 $\lim_{x\to 0} \dfrac{\sin 2x - \tan 2x}{x^2}$

3 $\lim_{x\to\infty} \dfrac{x^2+2x+3}{\ln(x+1)}$

4 $\lim_{x\to 0} \dfrac{\tan^{-1} x}{\sin^{-1} x}$

5 $\lim_{x\to 0} \dfrac{e^{2x} - e^{-2x} - 4x}{x^3}$

6 $\lim_{x\to(\pi/2)^-} \dfrac{\tan x}{\sec x}$

7 $\lim_{x\to\infty} \dfrac{x^e}{e^x}$

8 $\lim_{x\to(\pi/2)^-} \cos x \ln \cos x$

9 $\lim_{x\to\infty} (1 - 2e^{1/x})x$

10 $\lim_{x\to 0^+} \tan^{-1} x \csc x$

11 $\lim_{x\to 0} (1 + 8x^2)^{1/x^2}$

12 $\lim_{x\to 1^+} (\ln x)^{x-1}$

13 $\lim_{x\to\infty} (e^x + 1)^{1/x}$

14 $\lim_{x\to 0^+} \left(\dfrac{1}{\tan x} - \dfrac{1}{x}\right)$

15 $\lim_{x\to\infty} \dfrac{\sqrt{x^2+1}}{x}$

16 $\lim_{x\to\infty} \dfrac{3^x + 2x}{x^3+1}$

Exer. 17–28: Determine whether the integral converges or diverges, and if it converges, find its value.

17 $\int_4^\infty \dfrac{1}{\sqrt{x}}\, dx$

18 $\int_4^\infty \dfrac{1}{x\sqrt{x}}\, dx$

19 $\int_{-\infty}^0 \dfrac{1}{x+2}\, dx$

20 $\int_0^\infty \sin x\, dx$

21 $\int_{-8}^1 \dfrac{1}{\sqrt[3]{x}}\, dx$

22 $\int_{-4}^0 \dfrac{1}{x+4}\, dx$

23 $\int_0^2 \dfrac{x}{(x^2-1)^2}\, dx$

24 $\int_1^2 \dfrac{1}{x\sqrt{x^2-1}}\, dx$

25 $\int_{-\infty}^\infty \dfrac{1}{e^x + e^{-x}}\, dx$

26 $\int_{-\infty}^0 xe^x\, dx$

27 $\int_0^1 \dfrac{\ln x}{x}\, dx$

28 $\int_0^{\pi/2} \csc x\, dx$

c **Exer. 29–30: Approximate the improper integral by making the substitution $u = 1/x$ and then using Simpson's rule with $n = 4$.**

29 $\int_1^\infty e^{-x^2} dx$

30 $\int_1^\infty e^{-x} \sin\sqrt{x}\, dx$

31 Find $\lim_{x\to\infty} f(x)/g(x)$ if $f(x) = \int_1^x (\sin t)^{2/3}\, dt$ and $g(x) = x^2$.

32 Gauss' error integral $\operatorname{erf}(x) = (2/\sqrt{\pi}) \int_0^x e^{-u^2}\, du$ is used in probability theory. It has the special property $\lim_{x\to\infty} \operatorname{erf}(x) = 1$. Find $\lim_{x\to\infty} e^{(x^2)}[1 - \operatorname{erf}(x)]$.

CHAPTER 11

INFINITE SERIES

INTRODUCTION

If x is a real number, we usually find $\sin x$, e^x, $\ln x$, $\tan^{-1} x$, and values of other transcendental functions by using a calculator or a table. A more fundamental problem is that of determining *how* calculators compute these numbers or *how* a table is constructed. In Section 11.6 we consider how *infinite series* (infinite sums of a certain type) can be used to find function values. Specifically, if a function f satisfies certain conditions, we develop techniques for *representing* $f(x)$ as an infinite series whose terms contain powers of a variable x. Substituting a number c for x and then finding (or approximating) the resulting infinite sum gives us the value (or an approximation) of $f(c)$. This is essentially what a calculator does when it computes function values.

This new way of representing functions is the most important reason for developing the theory in the first five sections. Infinite series representations for $\sin x$, e^x, and other expressions allow us to consider problems that cannot be solved using finite methods. For example, if x is suitably restricted, we can evaluate integrals such as $\int \sin \sqrt{x}\, dx$ and $\int e^{-x^2}\, dx$—something we could not do in Chapter 9. As another application, in Chapter 19 we use infinite series to extend the definitions of $\sin x$, e^x, and other expressions to the case where x is a *complex number* $a + bi$, with a and b real and $i^2 = -1$.

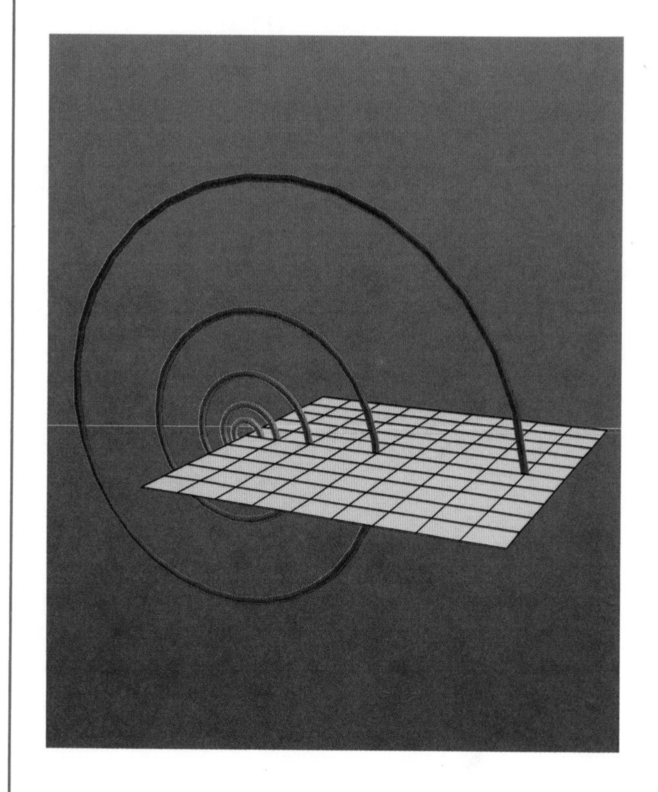

11.1 SEQUENCES

An arbitrary *infinite sequence* is often denoted as follows.

Sequence notation **(11.1)**

$$a_1, a_2, a_3, \ldots, a_n, \ldots$$

For convenience we usually refer to infinite sequences as *sequences*. We may regard (11.1) as a collection of real numbers that is in one-to-one correspondence with the positive integers. Each number a_k is a **term** of the sequence. The sequence is *ordered* in the sense that there is a **first term** a_1, a **second term** a_2, and, if n denotes an arbitrary positive integer, an ***n*th term** a_n.

We may also define a sequence as a function. Recall that a function f is a correspondence that associates with each number x in the domain exactly one number $f(x)$ in the range. If we restrict the domain to the positive integers 1, 2, 3, . . . , we obtain a sequence.

Definition **(11.2)**

A **sequence** is a function f whose domain is the set of positive integers.

In this text the range of a sequence will be a set of real numbers. If a function f is a sequence, then to each positive integer k there corresponds a real number $f(k)$. The numbers in the range of f may be denoted by

$$f(1), f(2), f(3), \ldots, f(n), \ldots$$

The three dots at the end indicate that the sequence does not terminate.

Note that Definition (11.2) leads to the subscript form (11.1) if we let $a_k = f(k)$ for each positive integer k. Conversely, given (11.1), we can obtain the function f in (11.2) by letting $f(k) = a_k$ for each k.

If we regard a sequence as a function f, then we may consider its graph in an xy-plane. Since the domain of f is the set of positive integers, the only points on the graph are

$$(1, a_1), (2, a_2), (3, a_3), \ldots, (n, a_n), \ldots,$$

where a_n is the nth term of the sequence (see Figure 11.1). We sometimes use the graph of a sequence to illustrate the behavior of the nth term a_n as n increases without bound.

FIGURE 11.1 Graph of a sequence

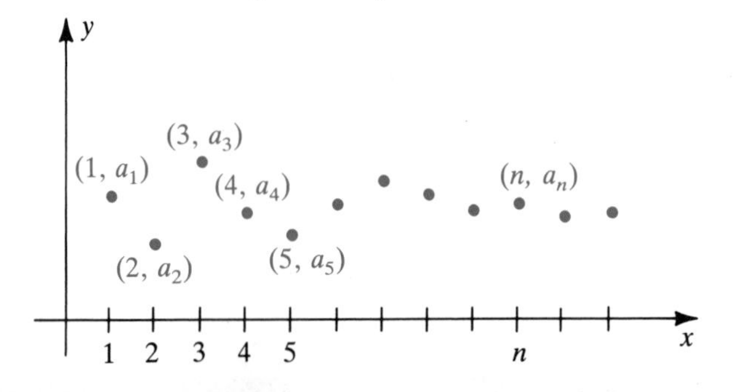

Another notation for a sequence with nth term a_n is $\{a_n\}$. For example, the sequence $\{2^n\}$ has nth term 2^n. Using the notation in (11.1), we write this sequence as follows:

$$2^1, 2^2, 2^3, \ldots, 2^n, \ldots$$

By Definition (11.2), the sequence $\{2^n\}$ is the function f with $f(n) = 2^n$ for every positive integer n.

EXAMPLE 1 List the first four terms and the tenth term of each sequence:

(a) $\left\{\dfrac{n}{n+1}\right\}$ **(b)** $\{2 + (0.1)^n\}$ **(c)** $\left\{(-1)^{n+1}\dfrac{n^2}{3n-1}\right\}$ **(d)** $\{4\}$

SOLUTION To find the first four terms, we substitute, successively, $n = 1, 2, 3,$ and 4 in the formula for a_n. The tenth term is found by substituting 10 for n. Doing this and simplifying gives us the following:

SEQUENCE	nTH TERM a_n	FIRST FOUR TERMS	TENTH TERM
(a) $\left\{\dfrac{n}{n+1}\right\}$	$\dfrac{n}{n+1}$	$\dfrac{1}{2}, \dfrac{2}{3}, \dfrac{3}{4}, \dfrac{4}{5}$	$\dfrac{10}{11}$
(b) $\{2 + (0.1)^n\}$	$2 + (0.1)^n$	2.1, 2.01, 2.001, 2.0001	2.0000000001
(c) $\left\{(-1)^{n+1}\dfrac{n^2}{3n-1}\right\}$	$(-1)^{n+1}\dfrac{n^2}{3n-1}$	$\dfrac{1}{2}, -\dfrac{4}{5}, \dfrac{9}{8}, -\dfrac{16}{11}$	$-\dfrac{100}{29}$
(d) $\{4\}$	4	4, 4, 4, 4	4

For some sequences we state the first term a_1, together with a rule for obtaining any term a_{k+1} from the preceding term a_k whenever $k \geq 1$. We call this a **recursive definition**, and the sequence is said to be defined **recursively**.

EXAMPLE 2 Find the first four terms and the nth term of the sequence defined recursively as follows:

$$a_1 = 3 \quad \text{and} \quad a_{k+1} = 2a_k \quad \text{for } k \geq 1$$

SOLUTION The sequence is defined recursively, since the first term is given, as well as a rule for finding a_{k+1} whenever a_k is known. Thus, the first four terms of the sequence are

$$a_1 = 3$$

$$a_2 = 2a_1 = 2 \cdot 3 = 6$$

$$a_3 = 2a_2 = 2 \cdot 2 \cdot 3 = 2^2 \cdot 3 = 12$$

$$a_4 = 2a_3 = 2 \cdot 2 \cdot 2 \cdot 3 = 2^3 \cdot 3 = 24.$$

We have written the terms as products to gain insight into the nature of the nth term. Continuing, we obtain $a_5 = 2^4 \cdot 3$ and $a_6 = 2^5 \cdot 3$; it appears that $a_n = 2^{n-1} \cdot 3$. We can use mathematical induction to prove that this guess is correct. Using the notation in (11.1), we write the sequence as

$$3, 2 \cdot 3, 2^2 \cdot 3, 2^3 \cdot 3, \ldots, 2^{n-1} \cdot 3, \ldots.$$

A sequence $\{a_n\}$ may have the property that as n increases, a_n gets very close to some real number L; that is, $|a_n - L| \approx 0$ if n is large. As an illustration, suppose that

$$a_n = 2 + (-\tfrac{1}{2})^n.$$

The first few terms of the sequence $\{a_n\}$ are

$$2 - \tfrac{1}{2},\ 2 + \tfrac{1}{4},\ 2 - \tfrac{1}{8},\ 2 + \tfrac{1}{16},\ 2 - \tfrac{1}{32},\ 2 + \tfrac{1}{64},\ \ldots,$$

or, equivalently,

$$1.5, 2.25, 1.875, 2.0625, 1.96875, 2.015625, \ldots.$$

It appears that the terms get closer to 2 as n increases. Note that for every positive integer n,

$$|a_n - 2| = |2 + (-\tfrac{1}{2})^n - 2| = |(-\tfrac{1}{2})^n| = (\tfrac{1}{2})^n = \frac{1}{2^n}.$$

The number $1/2^n$, and hence $|a_n - 2|$, *can be made arbitrarily close to* 0 *by choosing n sufficiently large*. According to the next definition, the sequence *has the limit* 2, or *converges to* 2, and we write

$$\lim_{n \to \infty} [2 + (-\tfrac{1}{2})^n] = 2.$$

This type of limit is almost the same as $\lim_{x \to \infty} f(x) = L$, given in Chapter 2. The only difference is that if $f(n) = a_n$, the domain of f is the set of positive integers and not an infinite interval of real numbers. As in Definition (2.16), but using a_n instead of $f(x)$, we state the following.

Definition **(11.3)**

A sequence $\{a_n\}$ **has the limit L**, or **converges to L**, denoted by either

$$\lim_{n \to \infty} a_n = L \quad \text{or} \quad a_n \to L \text{ as } n \to \infty,$$

if for every $\epsilon > 0$ there exists a positive number N such that

$$|a_n - L| < \epsilon \quad \text{whenever} \quad n > N.$$

If such a number L does not exist, the sequence **has no limit**, or **diverges**.

A graphical interpretation similar to that shown for the limit of a function in Figure 2.32 can be given for the limit of a sequence. The only difference is that the x-coordinate of each point on the graph is a positive integer. Figure 11.2 is the graph of a sequence $\{a_n\}$ for a specific case in which $\lim_{n \to \infty} a_n = L$. Note that for any $\epsilon > 0$, the points (n, a_n) lie between the lines $y = L \pm \epsilon$, provided n is sufficiently large. Of course, the approach to L may vary from that illustrated in the figure (see, for example, Figures 11.3 and 11.6).

FIGURE 11.2

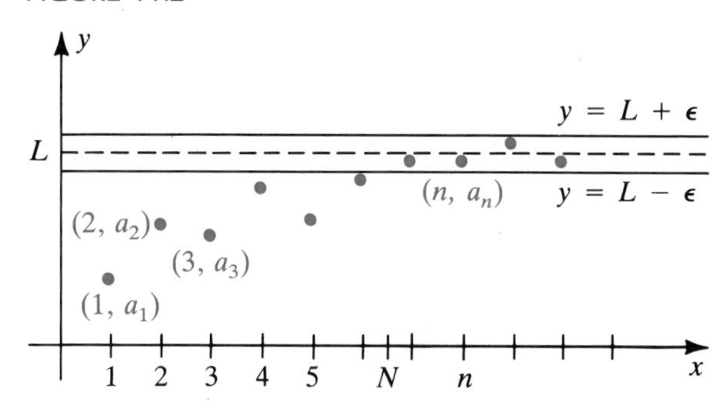

If we can make a_n as large as desired by choosing n sufficiently large, then the sequence $\{a_n\}$ diverges, but we still use the limit notation and write $\lim_{n\to\infty} a_n = \infty$. A more precise way of specifying this follows.

Definition **(11.4)**

> The notation
>
> $$\lim_{n\to\infty} a_n = \infty$$
>
> means that for every positive real number P there exists a number N such that $a_n > P$ whenever $n > N$.

As was the case for functions in Section 2.4, $\lim_{n\to\infty} a_n = \infty$ does *not* mean that the limit exists, but rather that the number a_n increases without bound as n increases. Similarly, $\lim_{n\to\infty} a_n = -\infty$ means that a_n decreases without bound as n increases.

The next theorem is important because it allows us to use results from Chapter 2 to investigate convergence or divergence of sequences. The proof follows from Definitions (11.3) and (2.16).

Theorem **(11.5)**

> Let $\{a_n\}$ be a sequence, let $f(n) = a_n$, and suppose that $f(x)$ exists for every real number $x \geq 1$.
>
> **(i)** If $\lim_{x\to\infty} f(x) = L$, then $\lim_{n\to\infty} f(n) = L$.
>
> **(ii)** If $\lim_{x\to\infty} f(x) = \infty$ (or $-\infty$), then $\lim_{n\to\infty} f(n) = \infty$ (or $-\infty$).

The following example illustrates the use of Theorem (11.5).

EXAMPLE 3 If $a_n = 1 + \dfrac{1}{n}$, determine whether $\{a_n\}$ converges or diverges.

SOLUTION We let $f(n) = 1 + \dfrac{1}{n}$ and consider

$$f(x) = 1 + \frac{1}{x} \quad \text{for every real number } x \geq 1.$$

From our work in Section 2.4,

$$\lim_{x\to\infty} f(x) = \lim_{x\to\infty}\left(1+\frac{1}{x}\right) = \lim_{x\to\infty} 1 + \lim_{x\to\infty}\frac{1}{x} = 1 + 0 = 1.$$

Hence, by Theorem (11.5),

$$\lim_{n\to\infty}\left(1+\frac{1}{n}\right) = 1.$$

Thus, the sequence $\{a_n\}$ converges to 1.

The difference between

$$\lim_{x\to\infty}\left(1+\frac{1}{x}\right) = 1 \quad\text{and}\quad \lim_{n\to\infty}\left(1+\frac{1}{n}\right) = 1$$

is illustrated in Figure 11.3. Note that for $1 + (1/x)$, the function f is continuous if $x \geq 1$, and the graph has a horizontal asymptote $y = 1$. For $1 + (1/n)$ we consider only the points whose x-coordinates are positive integers.

FIGURE 11.3

(i) $\lim_{x\to\infty}\left(1+\frac{1}{x}\right) = 1$

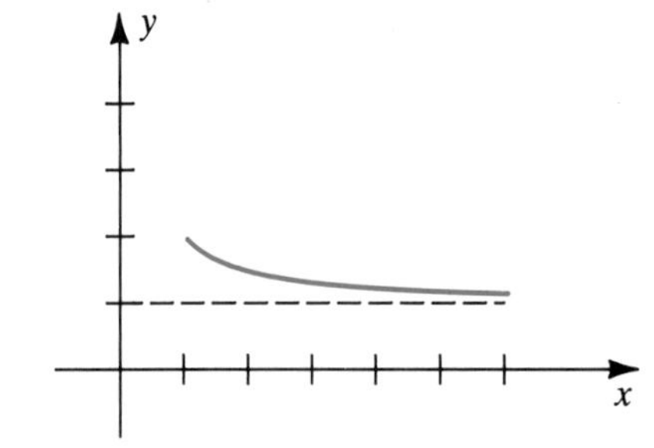

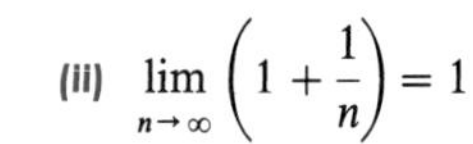

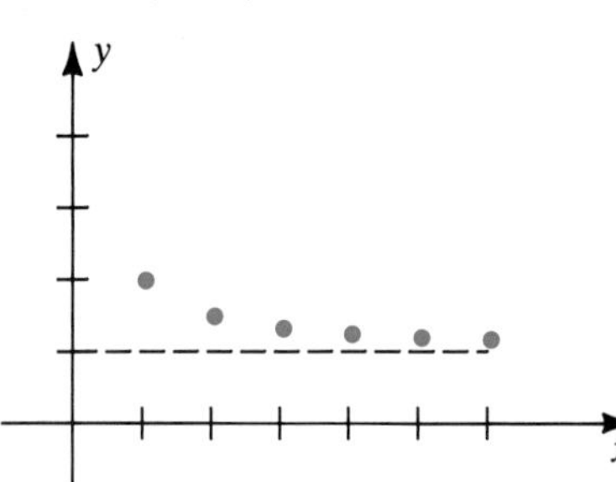

EXAMPLE 4 Determine whether the sequence converges or diverges:
(a) $\{\frac{1}{4}n^2 - 1\}$ **(b)** $\{(-1)^{n-1}\}$

FIGURE 11.4

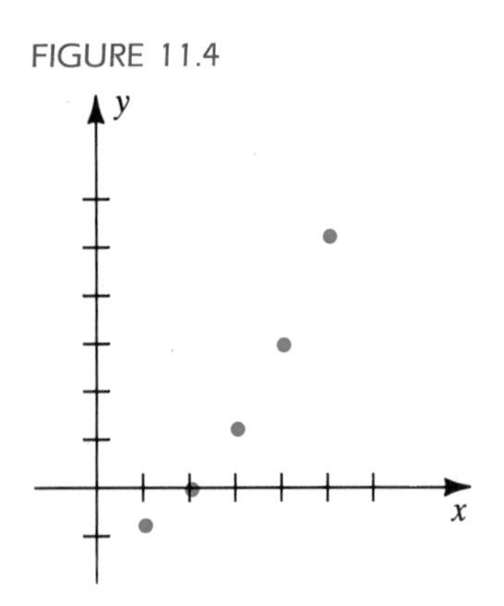

SOLUTION

(a) If we let $f(x) = \frac{1}{4}x^2 - 1$, then $f(x)$ exists for every $x \geq 1$ and

$$\lim_{x\to\infty}\left(\tfrac{1}{4}x^2 - 1\right) = \infty.$$

Hence, by Theorem (11.5),

$$\lim_{n\to\infty}\left(\tfrac{1}{4}n^2 - 1\right) = \infty.$$

Since the limit does not exist, the sequence diverges. The graph in Figure 11.4 illustrates the manner in which the sequence diverges.

FIGURE 11.5

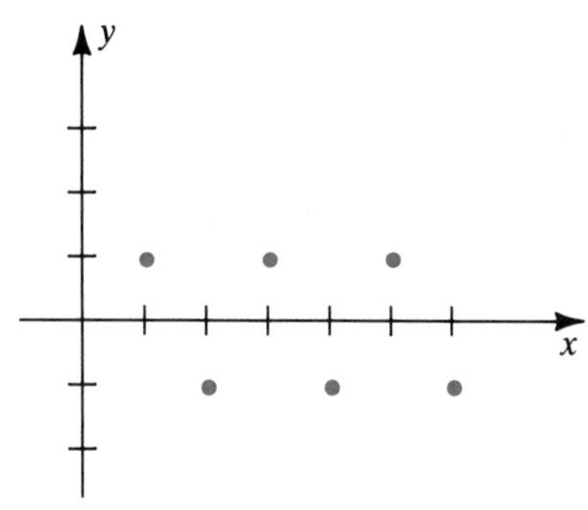

(b) Letting $n = 1, 2, 3, \ldots$, we see that the terms of $(-1)^{n-1}$ oscillate between 1 and -1 as follows:

$$1, -1, 1, -1, 1, -1, \ldots$$

This is illustrated graphically in Figure 11.5. Thus, $\lim_{n\to\infty}(-1)^{n-1}$ does not exist, so the sequence diverges.

The next example shows how we may use L'Hôpital's rule (10.2) to find limits of certain sequences.

EXAMPLE 5 Determine whether the sequence $\left\{\dfrac{5n}{e^{2n}}\right\}$ converges or diverges.

SOLUTION Let $f(x) = 5x/e^{2x}$ for every real number x. Since f takes on the indeterminate form ∞/∞ as $x \to \infty$, we may use L'Hôpital's rule, obtaining

$$\lim_{x\to\infty} \frac{5x}{e^{2x}} = \lim_{x\to\infty} \frac{5}{2e^{2x}} = 0.$$

Hence, by Theorem (11.5), $\lim_{n\to\infty} (5n/e^{2n}) = 0$. Thus, the sequence converges to 0.

The proof of the next theorem illustrates the use of Definition (11.3).

Theorem **(11.6)**

(i) $\lim\limits_{n\to\infty} r^n = 0 \quad \text{if} \quad |r| < 1$

(ii) $\lim\limits_{n\to\infty} |r^n| = \infty \quad \text{if} \quad |r| > 1$

PROOF If $r = 0$, it follows trivially that the limit is 0. Let us assume that $0 < |r| < 1$. To prove (i) by means of Definition (11.3), we must show that for every $\epsilon > 0$ there exists a positive number N such that

$$\text{if} \quad n > N, \quad \text{then} \quad |r^n - 0| < \epsilon.$$

The inequality $|r^n - 0| < \epsilon$ is equivalent to each inequality in the following list:

$$|r|^n < \epsilon, \quad \ln |r|^n < \ln \epsilon, \quad n \ln |r| < \ln \epsilon, \quad n > \frac{\ln \epsilon}{\ln |r|}$$

The final inequality sign is reversed because $\ln |r|$ is negative if $0 < |r| < 1$. The last inequality in the list provides a clue to the choice of N. Let us consider the two cases $\epsilon < 1$ and $\epsilon \geq 1$ separately. If $\epsilon < 1$, then $\ln \epsilon < 0$ and we let $N = \ln \epsilon/\ln |r| > 0$. In this event, if $n > N$, then the last inequality in the list is true and hence so is the first, which is what we wished to prove. If $\epsilon \geq 1$, then $\ln \epsilon \geq 0$ and hence $\ln \epsilon/\ln |r| \leq 0$. In this case, if N is *any* positive number, then whenever $n > N$ the last inequality in the list is again true.

To prove (ii), let $|r| > 1$ and consider any positive real number P. The following inequalities are equivalent:

$$|r|^n > P, \quad \ln |r|^n > \ln P, \quad n \ln |r| > \ln P, \quad n > \frac{\ln P}{\ln |r|}$$

If we choose $N = \ln P/\ln |r|$, then whenever $n > N$ the last inequality is true and hence so is the first; that is, $|r|^n > P$. By Definition (11.4), this means that $\lim_{n\to\infty} |r|^n = \infty$. ■

EXAMPLE 6 List the first four terms of the sequence, and determine whether the sequence converges or diverges:

(a) $\{(-\frac{2}{3})^n\}$ **(b)** $\{(1.01)^n\}$

SOLUTION

(a) The first four terms of $\{(-\frac{2}{3})^n\}$ are

$$-\tfrac{2}{3}, \tfrac{4}{9}, -\tfrac{8}{27}, \tfrac{16}{81}.$$

If we let $r = -\frac{2}{3}$, then, by Theorem (11.6)(i), with $|r| = \frac{2}{3} < 1$,

$$\lim_{n\to\infty} (-\tfrac{2}{3})^n = 0.$$

Hence the sequence converges to 0.

(b) The first four terms of $\{(1.01)^n\}$ are

$$1.01, 1.0201, 1.030301, 1.04060401.$$

If we let $r = 1.01$, then, by Theorem (11.6)(ii),

$$\lim_{n\to\infty} (1.01)^n = \infty.$$

Since the limit does not exist, the sequence diverges.

Limit theorems that are analogous to those stated in Chapter 2 for sums, differences, products, and quotients of functions can be established for sequences. For example, if $\{a_n\}$ and $\{b_n\}$ are convergent sequences, then

$$\lim_{n\to\infty} (a_n + b_n) = \lim_{n\to\infty} a_n + \lim_{n\to\infty} b_n,$$

$$\lim_{n\to\infty} (a_n b_n) = \left(\lim_{n\to\infty} a_n\right)\left(\lim_{n\to\infty} b_n\right),$$

and so on.

If $a_n = c$ for every n, the sequence $\{a_n\}$ is $c, c, \ldots, c, \ldots$ and

$$\lim_{n\to\infty} c = c.$$

Similarly, if c is a real number and k is a positive rational number, then, as in Theorem (2.18),

$$\lim_{n\to\infty} \frac{c}{n^k} = 0.$$

EXAMPLE 7 Find the limit of the sequence $\left\{\dfrac{2n^2}{5n^2 - 3}\right\}$.

SOLUTION We wish to find $\lim_{n\to\infty} a_n$, where $a_n = 2n^2/(5n^2 - 3)$. Dividing both numerator and denominator of a_n by n^2 and applying limit theorems, we obtain

$$\lim_{n\to\infty} \frac{2n^2}{5n^2 - 3} = \lim_{n\to\infty} \frac{2}{5 - (3/n^2)} = \frac{\lim_{n\to\infty} 2}{\lim_{n\to\infty} [5 - (3/n^2)]}$$

$$= \frac{2}{\lim_{n\to\infty} 5 - \lim_{n\to\infty} (3/n^2)} = \frac{2}{5 - 0} = \frac{2}{5}.$$

Hence the sequence has the limit $\frac{2}{5}$. We can also prove this by applying L'Hôpital's rule to $2x^2/(5x^2 - 3)$.

The next theorem, which is similar to Theorem (2.15), states that if the terms of a sequence are always sandwiched between corresponding terms of two sequences that have the same limit L, then the given sequence also has the limit L.

Sandwich theorem for sequences **(11.7)**

If $\{a_n\}$, $\{b_n\}$, and $\{c_n\}$ are sequences and $a_n \le b_n \le c_n$ for every n and if

$$\lim_{n\to\infty} a_n = L = \lim_{n\to\infty} c_n,$$

then

$$\lim_{n\to\infty} b_n = L.$$

EXAMPLE 8 Find the limit of the sequence $\left\{\dfrac{\cos^2 n}{3^n}\right\}$.

SOLUTION Since $0 < \cos^2 n < 1$ for every positive integer n,

$$0 < \frac{\cos^2 n}{3^n} < \frac{1}{3^n}.$$

Applying Theorem (11.6)(i) with $r = \frac{1}{3}$, we have

$$\lim_{n\to\infty} \frac{1}{3^n} = \lim_{n\to\infty} \left(\frac{1}{3}\right)^n = 0.$$

Moreover, $\lim_{n\to\infty} 0 = 0$. It follows from the sandwich theorem (11.7), with $a_n = 0$, $b_n = (\cos^2 n)/3^n$, and $c_n = (\frac{1}{3})^n$, that

$$\lim_{n\to\infty} \frac{\cos^2 n}{3^n} = 0.$$

Hence the limit of the sequence is 0.

The next theorem can be proved using Definition (11.3).

Theorem **(11.8)**

> Let $\{a_n\}$ be a sequence. If $\lim_{n\to\infty} |a_n| = 0$, then $\lim_{n\to\infty} a_n = 0$.

EXAMPLE 9 Suppose the nth term of a sequence is

$$a_n = (-1)^{n+1}\frac{1}{n}.$$

Prove that $\lim_{n\to\infty} a_n = 0$.

SOLUTION The terms of the sequence are alternately positive and negative. For example, the first seven terms are

$$1, -\tfrac{1}{2}, \tfrac{1}{3}, -\tfrac{1}{4}, \tfrac{1}{5}, -\tfrac{1}{6}, \tfrac{1}{7}.$$

Since $$\lim_{n\to\infty} |a_n| = \lim_{n\to\infty} \frac{1}{n} = 0,$$

it follows from Theorem (11.8) that $\lim_{n\to\infty} a_n = 0$.

A sequence is **monotonic** if successive terms are nondecreasing:

$$a_1 \le a_2 \le \cdots \le a_n \le \cdots;$$

or if they are nonincreasing:

$$a_1 \ge a_2 \ge \cdots \ge a_n \ge \cdots.$$

A sequence is **bounded** if there is a positive real number M such that $|a_k| \le M$ for every k. To illustrate, the sequence

$$\frac{1}{2}, \frac{2}{3}, \frac{3}{4}, \frac{4}{5}, \ldots, \frac{n}{n+1}, \ldots$$

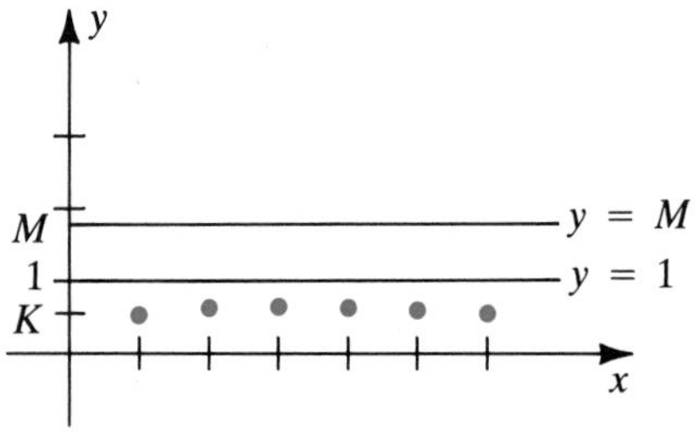

is both monotonic (the terms are increasing) and bounded (since $k/(k+1) < 1$ for every k). The graph of the sequence is illustrated in Figure 11.6. Note that any number $M \ge 1$ is a bound for the sequence; however, if $K < 1$, then K is not a bound, since $K < k/(k+1)$ when k is sufficiently large.

FIGURE 11.6 $\left\{\frac{n}{n+1}\right\}$

The next theorem is fundamental for later developments.

Theorem **(11.9)**

> A bounded, monotonic sequence has a limit.

To prove Theorem (11.9), it is necessary to use an important property of real numbers. Let us first state several definitions. If S is a nonempty set of real numbers, then a real number u is an **upper bound** of S if $x \le u$ for every x in S. A number v is a **least upper bound** of S if v is an upper bound and no number less than v is an upper bound of S. Thus, *the least upper bound is the smallest real number that is greater than or equal to every number in S.* To illustrate, if S is the open interval (a, b), then any number greater than b is an upper bound of S; however, the least upper bound of S is unique and equals b.

The following statement is an axiom for the real number system.

Completeness property **(11.10)**

> If a nonempty set S of real numbers has an upper bound, then S has a least upper bound.

PROOF OF THEOREM (11.9) Let $\{a_n\}$ be a bounded, monotonic sequence with nondecreasing terms. Thus,

$$a_1 \leq a_2 \leq \cdots \leq a_n \leq \cdots,$$

and there is a number M such that $a_k \leq M$ for every positive integer k. Since M is an upper bound for the set S of all numbers in the sequence, it follows from the completeness property that S has a least upper bound L such that $L \leq M$ (see Figure 11.7).

FIGURE 11.7

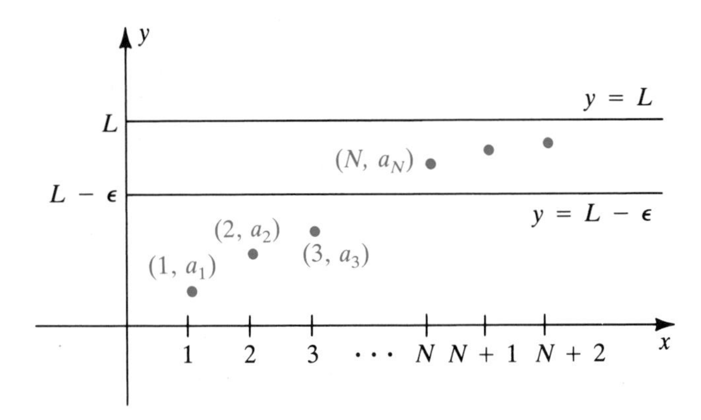

If $\epsilon > 0$, then $L - \epsilon$ is not an upper bound of S, and hence at least one term of $\{a_n\}$ is greater than $L - \epsilon$; that is,

$$L - \epsilon < a_N \quad \text{for some positive integer } N,$$

as shown in Figure 11.7. Since the terms of $\{a_n\}$ are nondecreasing,

$$a_N \leq a_{N+1} \leq a_{N+2} \leq \cdots$$

and, therefore,

$$L - \epsilon < a_n \quad \text{for every } n \geq N.$$

It follows that if $n > N$, then

$$0 \leq L - a_n < \epsilon \quad \text{or} \quad |L - a_n| < \epsilon.$$

By Definition (11.3), this means that

$$\lim_{n \to \infty} a_n = L \leq M;$$

that is, $\{a_n\}$ has a limit.

We may obtain the proof for a sequence $\{a_n\}$ of nonincreasing terms in a similar fashion or by considering the sequence $\{-a_n\}$. ■

As an illustration of Theorem (11.9), the monotonic sequence $\{n/(n+1)\}$ illustrated in Figure 11.6 has the least upper bound (and limit) 1.

There are many applications of sequences. In particular, sequences may be applied to the investigation of the time course of an S → I → S epidemic.* Suppose that physicians issue daily reports indicating the number of persons who have become infected with the disease and those who have been cured. We shall label the reporting days as $1, 2, \ldots, n, \ldots$ and let N denote the total population. In addition, let

*The notation S → I → S is an abbreviation for *Susceptible* → *Infected* → *Susceptible* and signifies that an infected person who becomes cured is not immune to the disease, but may contract it again. Examples of such diseases are gonorrhea and strep throat.

$$I_n = \text{number of persons who have the disease on day } n$$

$$F_n = \text{number of newly infected persons on day } n$$

$$C_n = \text{number of persons cured on day } n.$$

It follows that for every $n \geq 1$,

$$I_{n+1} = I_n + F_{n+1} - C_{n+1}.$$

Suppose health officials decide that the number of new cases on a given day is directly proportional to the product of the number ill and the number not infected on the previous day. (This is known as the *law of mass action* and is typical of a population of students on a college campus.) Moreover, suppose that the number cured each day is directly proportional to the number ill the previous day. Hence

$$F_{n+1} = aI_n(N - I_n) \quad \text{and} \quad C_{n+1} = bI_n,$$

where a and b are positive constants that can be approximated from early data. Substituting in the preceding formula for I_{n+1}, we have

$$I_{n+1} = I_n + aI_n(N - I_n) - bI_n.$$

In the early stages of an epidemic, I_n will be very small compared to N, and from the point of view of public health, it is better to *overestimate* the number ill than to underestimate and be unprepared for the spread of the disease. With this in mind, we drop the term $-a_nI_n^2$ in the formula for I_{n+1} and investigate the early dynamics of the epidemic by examining the equation

$$I_{n+1} = I_n + aNI_n - bI_n = (1 + aN - b)I_n.$$

If we let $r = 1 + aN - b$, then $I_{n+1} = rI_n$ and, therefore,

$$I_2 = rI_1, \quad I_3 = rI_2 = r^2I_1, \quad I_4 = rI_3 = r^3I_1, \quad \ldots, \quad I_n = r^{n-1}I_1, \quad \ldots$$

This gives us the following sequence of numbers of infected individuals:

$$I_1, rI_1, r^2I_1, \ldots, r^{n-1}I_1, \ldots$$

The number $r = 1 + aN - b$ is of critical import. If $r > 1$, then, by Theorem (11.6)(ii), $\lim_{n\to\infty} I_n = \infty$ and an epidemic is in progress. In this case, when n is large, I_n is no longer small compared to N, and the formula for I_{n+1} becomes invalid. If $r < 1$, then, by Theorem (11.6)(i), $\lim_{n\to\infty} I_n = 0$ and health officials need not be concerned. The case $r = 1$ results in the constant sequence $I_1, I_1, \ldots, I_1, \ldots$.

EXERCISES 11.1

Exer. 1–16: The expression is the nth term a_n of a sequence $\{a_n\}$. Find the first four terms and $\lim_{n\to\infty} a_n$, if it exists.

1 $\dfrac{n}{3n+2}$

2 $\dfrac{6n-5}{5n+1}$

3 $\dfrac{7-4n^2}{3+2n^2}$

4 $\dfrac{4}{8-7n}$

5 -5

6 $\sqrt{2}$

7 $\dfrac{(2n-1)(3n+1)}{n^3+1}$

8 $8n+1$

9 $\dfrac{2}{\sqrt{n^2+9}}$

10 $\dfrac{100n}{n^{3/2}+4}$

11 $(-1)^{n+1}\dfrac{3n}{n^2+4n+5}$

12 $(-1)^{n+1}\dfrac{\sqrt{n}}{n+1}$

13 $1+(0.1)^n$

14 $1-\dfrac{1}{2^n}$

15 $1+(-1)^{n+1}$

16 $\dfrac{n+1}{\sqrt{n}}$

Exer. 17–42: Determine whether the sequence converges or diverges, and if it converges, find the limit.

17 $\{6(-\frac{5}{6})^n\}$

18 $\{8-(\frac{7}{8})^n\}$

19 $\{\arctan n\}$

20 $\left\{\dfrac{\tan^{-1} n}{n}\right\}$

21 $\{1000-n\}$

22 $\left\{\dfrac{(1.0001)^n}{1000}\right\}$

23 $\left\{(-1)^n\dfrac{\ln n}{n}\right\}$

24 $\left\{\dfrac{n^2}{\ln(n+1)}\right\}$

25 $\left\{\dfrac{4n^4+1}{2n^2-1}\right\}$

26 $\left\{\dfrac{\cos n}{n}\right\}$

27 $\left\{\dfrac{e^n}{n^4}\right\}$

28 $\{e^{-n}\ln n\}$

29 $\left\{\left(1+\dfrac{1}{n}\right)^n\right\}$

30 $\{(-1)^n n^3 3^{-n}\}$

31 $\{2^{-n}\sin n\}$

32 $\left\{\dfrac{4n^3+5n+1}{2n^3-n^2+5}\right\}$

33 $\left\{\dfrac{n^2}{2n-1}-\dfrac{n^2}{2n+1}\right\}$

34 $\left\{n\sin\dfrac{1}{n}\right\}$

35 $\{\cos \pi n\}$

36 $\{4+\sin\frac{1}{2}\pi n\}$

37 $\{n^{1/n}\}$

38 $\left\{\dfrac{n^2}{2^n}\right\}$

39 $\left\{\dfrac{n^{-10}}{\sec n}\right\}$

40 $\left\{(-1)^n\dfrac{n^2}{1+n^2}\right\}$

41 $\{\sqrt{n+1}-\sqrt{n}\}$

42 $\{\sqrt{n^2+n}-n\}$

43 A stable population of 35,000 birds lives on three islands. Each year 10% of the population on island A migrates to island B, 20% of the population on island B migrates to island C, and 5% of the population on island C migrates to island A. Let A_n, B_n, and C_n denote the numbers of birds on islands A, B, and C, respectively, in year n before migration takes place.

(a) Show that

$$A_{n+1}=0.9A_n+0.05C_n$$
$$B_{n+1}=0.1A_n+0.80B_n$$

and

$$C_{n+1}=0.95C_n+0.20B_n.$$

(b) Assuming that $\lim_{n\to\infty} A_n$, $\lim_{n\to\infty} B_n$, and $\lim_{n\to\infty} C_n$ exist, approximate the number of birds on each island after many years.

44 A bobcat population is classified by age as kittens (less than one year old) and adults (at least one year old). All adult females, including those born the prior year, have a litter each June, with an average litter size of three kittens. The survival rate of kittens is 50%, while that of adults is $66\frac{2}{3}$% per year. Let K_n be the number of newborn kittens in June of the nth year, let A_n be the number of adults, and assume that the ratio of males to females is always 1.

(a) Show that $K_{n+1}=\frac{3}{2}A_{n+1}$ and $A_{n+1}=\frac{2}{3}A_n+\frac{1}{2}K_n$.

(b) Conclude that $A_{n+1}=\frac{17}{12}A_n$ and $K_{n+1}=\frac{17}{12}K_n$, and that $A_n=(\frac{17}{12})^{n-1}A_1$ and $K_n=(\frac{17}{12})^{n-1}K_1$. What can you conclude about the population?

c 45 Terms of the sequence defined recursively by $a_1=5$ and $a_{k+1}=\sqrt{a_k}$ may be generated by entering 5 and pressing $\boxed{\sqrt{x}}$ repeatedly.

(a) Describe what happens to the terms of the sequence as k increases.

(b) Show that $a_n=5^{1/2^n}$, and find $\lim_{n\to\infty} a_n$.

c 46 If a sequence is generated by entering a number and pressing $\boxed{1/x}$ repeatedly, under what conditions does the sequence have a limit?

c 47 Terms of the sequence defined recursively by $a_1=1$ and $a_{k+1}=\cos a_k$ may be generated by entering 1 (in radian mode) and pressing $\boxed{\text{COS}}$ repeatedly.

(a) Describe what happens to terms of the sequence as k increases.

(b) Assuming that $\lim_{n\to\infty} a_n=L$, prove that $L=\cos L$. (*Hint:* $\lim_{n\to\infty} a_{n+1}=L$.)

c 48 A sequence $\{x_n\}$ is defined recursively by the formula $x_{k+1}=x_k-\tan x_k$.

(a) If $x_1=3$, approximate the first five terms of the sequence. Predict $\lim_{n\to\infty} x_n$.

(b) If $x_1=6$, approximate the first five terms of the sequence. Predict $\lim_{n\to\infty} x_n$.

(c) Assuming that $\lim_{n\to\infty} x_n=L$, prove that $L=\pi n$ for some integer n.

c **49** Approximations to $\sqrt{N}$ may be generated from the sequence defined recursively by

$$x_1 = \frac{N}{2}, \qquad x_{k+1} = \frac{1}{2}\left(x_k + \frac{N}{x_k}\right).$$

(a) Approximate x_2, x_3, x_4, x_5, x_6 if $N = 10$.

(b) Assuming that $\lim_{n\to\infty} x_n = L$, prove that $L = \sqrt{N}$.

c **50** The famous *Fibonacci sequence* is defined recursively by $a_{k+1} = a_k + a_{k-1}$ with $a_1 = a_2 = 1$.

(a) Find the first ten terms of the sequence.

(b) The terms of the sequence $r_k = a_{k+1}/a_k$ give approximations to τ, the *golden ratio*. Approximate the first ten terms of this sequence.

(c) Assuming that $\lim_{n\to\infty} r_n = \tau$, prove that

$$\tau = \tfrac{1}{2}(1 + \sqrt{5}).$$

c **Exer. 51–52: If f is differentiable, then a sequence $\{a_n\}$ defined recursively by $a_{k+1} = f(a_k)$, for $k \geq 1$, will converge for any a_1 if the derivative f' is continuous and $|f'(x)| \leq B < 1$ for some positive constant B.**

(a) For the given f, verify that the sequence $\{a_n\}$ converges for any a_1 by finding a suitable B.

(b) Approximate, to two decimal places, $\lim_{n\to\infty} a_n$, if $a_1 = 1$ and also if $a_1 = -100$.

51 $f(a_k) = \frac{1}{4}\sin a_k \cos a_k + 1$ **52** $f(a_k) = \dfrac{a_k^2}{a_k^2 + 1} + 2$

11.2 CONVERGENT OR DIVERGENT SERIES

We may use sequences to define expressions of the form

$$0.6 + 0.06 + 0.006 + 0.0006 + 0.00006 + \cdots,$$

where the three dots indicate that the sum continues indefinitely. In Definition (11.11) we call such an expression an *infinite series*. Since only finite sums may be added algebraically, we must *define* what is meant by this "infinite sum." As we shall see, the key to the definition is to consider the *sequence of partial sums* $\{S_n\}$, *where* S_k *is the sum of the first k numbers of the infinite series*. For the preceding illustration,

$$S_1 = 0.6$$
$$S_2 = 0.6 + 0.06 = 0.66$$
$$S_3 = 0.6 + 0.06 + 0.006 = 0.666$$
$$S_4 = 0.6 + 0.06 + 0.006 + 0.0006 = 0.6666$$

and so on. Thus, the sequence of partial sums $\{S_n\}$ may be written

$$0.6,\ 0.66,\ 0.666,\ 0.6666,\ 0.66666, \ldots.$$

It will follow from Theorem (11.15) that

$$S_n \to \tfrac{2}{3} \quad \text{as} \quad n \to \infty.$$

From an intuitive point of view, the more numbers of the infinite series that we add, the closer the sum gets to $\frac{2}{3}$. Thus, we write

$$\tfrac{2}{3} = 0.6 + 0.06 + 0.006 + 0.0006 + \cdots$$

and call $\frac{2}{3}$ the *sum* of the infinite series.

With this special case in mind, let us introduce terminology that will be used throughout the remainder of this chapter. In the following definition, we assume that $a_1, a_2, \ldots, a_n, \ldots$ are the terms of some sequence.

Definition (11.11)

An **infinite series** (or simply a **series**) is an expression of the form

$$a_1 + a_2 + \cdots + a_n + \cdots,$$

or, in summation notation,

$$\sum_{n=1}^{\infty} a_n, \quad \text{or} \quad \sum a_n.$$

Each number a_k is a **term** of the series, and a_n is the ***n*th term**.

Sometimes there is confusion between the concept of a series and that of a sequence. Remember that a series is an expression that represents an *infinite sum* of numbers. A sequence is a collection of numbers that are in one-to-one correspondence with the positive integers. The sequence of partial sums in the next definition is a special type of sequence that we obtain by using the terms of a series.

As in the special case introduced at the beginning of this section, we define the *sequence of partial sums* of a series as follows.

Definition (11.12)

(i) The ***k*th partial sum** S_k of the series $\sum a_n$ is

$$S_k = a_1 + a_2 + \cdots + a_k.$$

(ii) The **sequence of partial sums** of the series $\sum a_n$ is

$$S_1, S_2, S_3, \ldots, S_n, \ldots.$$

By Definition (11.12)(i),

$$S_1 = a_1$$
$$S_2 = a_1 + a_2$$
$$S_3 = a_1 + a_2 + a_3$$
$$S_4 = a_1 + a_2 + a_3 + a_4.$$

To calculate S_5, S_6, S_7, and so on, we add more terms of the series. Thus, S_{1000} is the sum of the first one thousand terms of $\sum a_n$. If the sequence $\{S_n\}$ has a limit S, we call S the *sum* of the series $\sum a_n$, as in the next definition.

Definition (11.13)

A series $\sum a_n$ is **convergent** (or **converges**) if its sequence of partial sums $\{S_n\}$ converges—that is, if

$$\lim_{n\to\infty} S_n = S \quad \text{for some real number } S.$$

The limit S is the **sum** of the series $\sum a_n$, and we write

$$S = a_1 + a_2 + \cdots + a_n + \cdots.$$

The series $\sum a_n$ is **divergent** (or **diverges**) if $\{S_n\}$ diverges. A divergent series has no sum.

For most series it is very difficult to find a formula for S_n. However, as we shall see in later sections, it may be possible to establish the convergence or divergence of a series using other methods. In the remainder of this section we shall consider several important series for which we *can* find a formula for S_n.

EXAMPLE 1 Given the series

$$\frac{1}{1\cdot 2}+\frac{1}{2\cdot 3}+\frac{1}{3\cdot 4}+\cdots+\frac{1}{n(n+1)}+\cdots,$$

(a) find S_1, S_2, S_3, S_4, S_5, and S_6

(b) find S_n

(c) show that the series converges and find its sum

SOLUTION

(a) By Definition (11.12), the first six partial sums are as follows:

$$S_1=\frac{1}{1\cdot 2}=\frac{1}{2}$$

$$S_2=\frac{1}{1\cdot 2}+\frac{1}{2\cdot 3}=\frac{2}{3}$$

$$S_3=\frac{1}{1\cdot 2}+\frac{1}{2\cdot 3}+\frac{1}{3\cdot 4}=\frac{3}{4}$$

$$S_4=\frac{1}{1\cdot 2}+\frac{1}{2\cdot 3}+\frac{1}{3\cdot 4}+\frac{1}{4\cdot 5}=\frac{4}{5}$$

$$S_5=S_4+a_5=\frac{4}{5}+\frac{1}{5\cdot 6}=\frac{5}{6}$$

$$S_6=S_5+a_6=\frac{5}{6}+\frac{1}{6\cdot 7}=\frac{6}{7}$$

(b) To find S_n, we shall write the terms of the series in a different way. Using partial fractions, we can show that

$$a_n=\frac{1}{n(n+1)}=\frac{1}{n}-\frac{1}{n+1}.$$

Consequently the nth partial sum of the series may be written

$$\begin{aligned}S_n&=a_1+a_2+a_3+\cdots+a_n\\&=\left(1-\frac{1}{2}\right)+\left(\frac{1}{2}-\frac{1}{3}\right)+\left(\frac{1}{3}-\frac{1}{4}\right)+\cdots+\left(\frac{1}{n}-\frac{1}{n+1}\right).\end{aligned}$$

Regrouping, we see that all numbers except the first and last cancel, and hence

$$S_n=1-\frac{1}{n+1}=\frac{n}{n+1}.$$

(c) Using the formula for S_n obtained in part (b), we obtain

$$\lim_{n\to\infty}S_n=\lim_{n\to\infty}\frac{n}{n+1}=1.$$

Thus, the series converges and has the sum 1. As in Definition (11.13), we may write

$$1 = \frac{1}{1 \cdot 2} + \frac{1}{2 \cdot 3} + \frac{1}{3 \cdot 4} + \cdots + \frac{1}{n(n+1)} + \cdots.$$

The series $\sum 1/[n(n+1)]$ of Example 1 is called a **telescoping series**, since writing S_n as shown in (b) of the solution causes the terms to *telescope* to $1 - [1/(n+1)]$.

EXAMPLE 2 Given the series

$$\sum_{n=1}^{\infty} (-1)^{n-1} = 1 + (-1) + 1 + (-1) + \cdots + (-1)^{n-1} + \cdots,$$

(a) find S_1, S_2, S_3, S_4, S_5, and S_6

(b) find S_n

(c) show that the series diverges

SOLUTION

(a) By Definition (11.12),

$$S_1 = 1, \quad S_2 = 0, \quad S_3 = 1, \quad S_4 = 0, \quad S_5 = 1, \quad \text{and} \quad S_6 = 0.$$

(b) We can write S_n as follows:

$$S_n = \begin{cases} 1 & \text{if } n \text{ is odd} \\ 0 & \text{if } n \text{ is even} \end{cases}$$

(c) Since the sequence of partial sums $\{S_n\}$ oscillates between 1 and 0, it follows that $\lim_{n \to \infty} S_n$ does not exist. Hence the series diverges.

EXAMPLE 3 Prove that the following series is divergent:

$$1 + \frac{1}{2} + \frac{1}{3} + \frac{1}{4} + \cdots + \frac{1}{n} + \cdots$$

SOLUTION Let us group the terms of the series as follows:

$$1 + \tfrac{1}{2} + (\tfrac{1}{3} + \tfrac{1}{4}) + (\tfrac{1}{5} + \tfrac{1}{6} + \tfrac{1}{7} + \tfrac{1}{8}) + (\tfrac{1}{9} + \cdots + \tfrac{1}{16}) + (\tfrac{1}{17} + \cdots + \tfrac{1}{32}) + \cdots$$

Note that each group contains twice the number of terms as the preceding group. Moreover, since increasing the denominator *decreases* the value of a fraction, we have the following:

$$\tfrac{1}{3} + \tfrac{1}{4} > \tfrac{1}{4} + \tfrac{1}{4} = \tfrac{1}{2}$$

$$\tfrac{1}{5} + \tfrac{1}{6} + \tfrac{1}{7} + \tfrac{1}{8} > \tfrac{1}{8} + \tfrac{1}{8} + \tfrac{1}{8} + \tfrac{1}{8} = \tfrac{1}{2}$$

$$\tfrac{1}{9} + \tfrac{1}{10} + \cdots + \tfrac{1}{16} > \tfrac{1}{16} + \tfrac{1}{16} + \cdots + \tfrac{1}{16} = \tfrac{1}{2}$$

$$\tfrac{1}{17} + \tfrac{1}{18} + \cdots + \tfrac{1}{32} > \tfrac{1}{32} + \tfrac{1}{32} + \cdots + \tfrac{1}{32} = \tfrac{1}{2}$$

Since the sum of the terms within each set of parentheses is greater than $\frac{1}{2}$, we obtain the following inequalities.

$$S_4 > 1 + \tfrac{1}{2} + \tfrac{1}{2} > 3(\tfrac{1}{2})$$
$$S_8 > 1 + \tfrac{1}{2} + \tfrac{1}{2} + \tfrac{1}{2} > 4(\tfrac{1}{2})$$
$$S_{16} > 1 + \tfrac{1}{2} + \tfrac{1}{2} + \tfrac{1}{2} + \tfrac{1}{2} > 5(\tfrac{1}{2})$$
$$S_{32} > 1 + \tfrac{1}{2} + \tfrac{1}{2} + \tfrac{1}{2} + \tfrac{1}{2} + \tfrac{1}{2} > 6(\tfrac{1}{2})$$

It can be shown, by mathematical induction, that

$$S_{2^k} > (k+1)(\tfrac{1}{2}) \quad \text{for every positive integer } k.$$

It follows that S_n can be made as large as desired by taking n sufficiently large; that is, $\lim_{n\to\infty} S_n = \infty$. Since $\{S_n\}$ diverges, the given series diverges.

The series in Example 3 will be useful in later developments. It is given the following special name.

Definition **(11.14)**

The **harmonic series** is the divergent series

$$1 + \frac{1}{2} + \frac{1}{3} + \cdots + \frac{1}{n} + \cdots.$$

In the next section we shall give another proof of the divergence of the harmonic series.

Certain types of series occur frequently in solutions of applied problems. One of the most important is the **geometric series**

$$a + ar + ar^2 + \cdots + ar^{n-1} + \cdots,$$

where a and r are real numbers, with $a \neq 0$.

Theorem **(11.15)**

Let $a \neq 0$. The geometric series

$$a + ar + ar^2 + \cdots + ar^{n-1} + \cdots$$

(i) converges and has the sum $S = \dfrac{a}{1-r}$ if $|r| < 1$

(ii) diverges if $|r| \geq 1$

PROOF If $r = 1$, then $S_n = a + a + \cdots + a = na$ and the series diverges, since $\lim_{n\to\infty} S_n$ does not exist.

If $r = -1$, then $S_k = a$ if k is odd and $S_k = 0$ if k is even. Since the sequence of partial sums oscillates between a and 0, the series diverges.

If $r \neq 1$, then

$$S_n = a + ar + ar^2 + \cdots + ar^{n-1}$$

and

$$rS_n = ar + ar^2 + ar^3 + \cdots + ar^n.$$

Subtracting corresponding sides of these equations, we obtain

$$(1 - r)S_n = a - ar^n.$$

Dividing both sides by $1 - r$ gives us

$$S_n = \frac{a}{1 - r} - \frac{ar^n}{1 - r}.$$

Consequently

$$\begin{aligned}\lim_{n\to\infty} S_n &= \lim_{n\to\infty} \left(\frac{a}{1 - r} - \frac{ar^n}{1 - r}\right)\\ &= \lim_{n\to\infty} \frac{a}{1 - r} - \lim_{n\to\infty} \frac{ar^n}{1 - r}\\ &= \frac{a}{1 - r} - \frac{a}{1 - r} \lim_{n\to\infty} r^n.\end{aligned}$$

If $|r| < 1$, then $\lim_{r\to\infty} r^n = 0$, by Theorem (11.6)(i), and hence

$$\lim_{n\to\infty} S_n = \frac{a}{1 - r} = S.$$

If $|r| > 1$, then $\lim_{n\to\infty} r^n$ does not exist, by Theorem (11.6)(ii), and hence $\lim_{n\to\infty} S_n$ does not exist. In this case the series diverges. ■

EXAMPLE 4 Prove that the following series converges, and find its sum:

$$0.6 + 0.06 + 0.006 + \cdots + \frac{6}{10^n} + \cdots$$

SOLUTION This is the series considered at the beginning of this section. It is geometric with $a = 0.6$ and $r = 0.1$. Since $|r| < 1$, we conclude from Theorem (11.15)(i) that the series converges and has the sum

$$S = \frac{a}{1 - r} = \frac{0.6}{1 - 0.1} = \frac{0.6}{0.9} = \frac{2}{3}.$$

Thus, $$\frac{2}{3} = 0.6 + 0.06 + 0.006 + \cdots + \frac{6}{10^n} + \cdots.$$

This justifies the nonterminating decimal notation $\frac{2}{3} = 0.66666\ldots$.

EXAMPLE 5 Prove that the following series converges, and find its sum:

$$2 + \frac{2}{3} + \frac{2}{3^2} + \cdots + \frac{2}{3^{n-1}} + \cdots$$

SOLUTION The series converges, since it is geometric with $r = \frac{1}{3} < 1$. By Theorem (11.15)(i), the sum is

$$S = \frac{a}{1 - r} = \frac{2}{1 - \frac{1}{3}} = \frac{2}{\frac{2}{3}} = 3.$$

Theorem (11.16)

> If a series $\sum a_n$ is convergent, then $\lim_{n\to\infty} a_n = 0$.

PROOF The nth term a_n of the series can be expressed as

$$a_n = S_n - S_{n-1}.$$

If S is the sum of the series $\sum a_n$, then we know $\lim_{n\to\infty} S_n = S$ and also $\lim_{n\to\infty} S_{n-1} = S$. Hence

$$\lim_{n\to\infty} a_n = \lim_{n\to\infty} (S_n - S_{n-1}) = \lim_{n\to\infty} S_n - \lim_{n\to\infty} S_{n-1} = S - S = 0.$$

The preceding theorem states that *if* a series converges, *then* its nth term a_n has the limit 0 as $n \to \infty$. The converse is false—that is, *if* $\lim_{n\to\infty} a_n = 0$, *it does not necessarily follow* that the series $\sum a_n$ is convergent. The harmonic series (11.14) is an illustration of a divergent series $\sum a_n$ for which $\lim_{n\to\infty} a_n = 0$. Consequently, to establish convergence of a series, *it is not enough* to prove that $\lim_{n\to\infty} a_n = 0$, since that may be true for divergent as well as for convergent series.

The next result is a corollary of Theorem (11.16) and the preceding remarks.

nth-term test (11.17)

> **(i)** If $\lim_{n\to\infty} a_n \neq 0$, then the series $\sum a_n$ is divergent.
>
> **(ii)** If $\lim_{n\to\infty} a_n = 0$, then further investigation is necessary to determine whether the series $\sum a_n$ is convergent or divergent.

The next illustration shows how to apply the nth-term test to a series.

ILLUSTRATION

SERIES	nTH-TERM TEST	CONCLUSION
$\sum_{n=1}^{\infty} \frac{n}{2n+1}$	$\lim_{n\to\infty} \frac{n}{2n+1} = \frac{1}{2} \neq 0$	Diverges, by (11.17)(i)
$\sum_{n=1}^{\infty} \frac{1}{n^2}$	$\lim_{n\to\infty} \frac{1}{n^2} = 0$	Further investigation is necessary, by (11.17)(ii)
$\sum_{n=1}^{\infty} \frac{1}{\sqrt{n}}$	$\lim_{n\to\infty} \frac{1}{\sqrt{n}} = 0$	Further investigation is necessary, by (11.17)(ii)
$\sum_{n=1}^{\infty} \frac{e^n}{n}$	$\lim_{n\to\infty} \frac{e^n}{n} = \infty$	Diverges, by (11.17)(i)

We shall see in the next section that the second series in the illustration converges and that the third series diverges.

The next theorem states that if corresponding terms of two series are identical after a certain term, then both series converge or both series diverge.

Theorem **(11.18)**

If $\sum a_n$ and $\sum b_n$ are series such that $a_j = b_j$ for every $j > k$, where k is a positive integer, then both series converge or both series diverge.

PROOF By hypothesis, we may write the following:

$$\sum a_n = a_1 + a_2 + \cdots + a_k + a_{k+1} + \cdots + a_n + \cdots$$
$$\sum b_n = b_1 + b_2 + \cdots + b_k + a_{k+1} + \cdots + a_n + \cdots$$

Let S_n and T_n denote the nth partial sums of $\sum a_n$ and $\sum b_n$, respectively. It follows that if $n \geq k$, then

$$S_n - S_k = T_n - T_k,$$

or

$$S_n = T_n + (S_k - T_k).$$

Consequently

$$\lim_{n\to\infty} S_n = \lim_{n\to\infty} T_n + (S_k - T_k),$$

and hence either both of the limits exist or both do not exist. This gives us the desired conclusion. If both series converge, then their sums differ by $S_k - T_k$. ■

Theorem (11.18) implies that changing a finite number of terms of a series has no effect on its convergence or divergence (although it does change the sum of a convergent series). In particular, if we replace the first k terms of $\sum a_n$ by 0, convergence is unaffected. It follows that the series

$$a_{k+1} + a_{k+2} + \cdots + a_n + \cdots$$

converges or diverges if $\sum a_n$ converges or diverges, respectively. The series $a_{k+1} + a_{k+2} + \cdots$ is obtained from $\sum a_n$ by **deleting the first** $\boldsymbol{k}$ **terms**.

Let us state this result for reference as follows.

Theorem **(11.19)**

For any positive integer k, the series

$$\sum_{n=1}^{\infty} a_n = a_1 + a_2 + \cdots \quad \text{and} \quad \sum_{n=k+1}^{\infty} a_n = a_{k+1} + a_{k+2} + \cdots$$

either both converge or both diverge.

EXAMPLE 6 Show that the following series converges:

$$\frac{1}{3\cdot 4} + \frac{1}{4\cdot 5} + \cdots + \frac{1}{(n+2)(n+3)} + \cdots$$

SOLUTION The series can be obtained by deleting the first two terms of the convergent telescoping series of Example 1. Hence, by Theorem (11.19), the given series converges.

The proof of the next theorem follows directly from Definition (11.13).

Theorem **(11.20)**

If $\sum a_n$ and $\sum b_n$ are convergent series with sums A and B, respectively, then

(i) $\sum (a_n + b_n)$ converges and has sum $A + B$

(ii) $\sum ca_n$ converges and has sum cA for every real number c

(iii) $\sum (a_n - b_n)$ converges and has sum $A - B$

It is also easy to show that if $\sum a_n$ diverges, then so does $\sum ca_n$ for every $c \neq 0$.

EXAMPLE 7 Prove that the following series converges, and find its sum:

$$\sum_{n=1}^{\infty} \left[\frac{7}{n(n+1)} + \frac{2}{3^{n-1}} \right]$$

SOLUTION The telescoping series $\sum 1/[n(n+1)]$ was considered in Example 1, where we found that it converges and has the sum 1. Using Theorem (11.20)(ii) with $c = 7$ and $a_n = 1/[n(n+1)]$, we see that the series $\sum 7/[n(n+1)]$ converges and has the sum $7(1) = 7$.

The geometric series $\sum 2/3^{n-1}$ converges and has the sum 3 (see Example 5). Hence, by Theorem (11.20)(i), the given series converges and has the sum $7 + 3 = 10$.

Theorem **(11.21)**

If $\sum a_n$ is a convergent series and $\sum b_n$ is divergent, then $\sum (a_n + b_n)$ is divergent.

PROOF As in the statement of the theorem, let $\sum a_n$ be convergent and $\sum b_n$ be divergent. We shall give an indirect proof—that is, we shall assume that the conclusion of the theorem is *false* and arrive at a contradiction. Thus, *suppose* that $\sum (a_n + b_n)$ is convergent. Applying Theorem (11.20)(iii), we find that the series

$$\sum [(a_n + b_n) - a_n] = \sum b_n$$

is convergent. This contradicts the fact that $\sum b_n$ is divergent, and hence our supposition is false; that is, $\sum (a_n + b_n)$ is divergent. ■

EXAMPLE 8 Determine the convergence or divergence of the series

$$\sum_{n=1}^{\infty} \left(\frac{1}{5^n} + \frac{1}{n} \right).$$

SOLUTION Since $\sum (1/5^n)$ is a convergent geometric series and $\sum (1/n)$ is the divergent harmonic series, the given series diverges, by Theorem (11.21).

We may apply infinite series to the S → I → S epidemic discussed at the end of Section 11.1. Suppose that instead of I_n (the number ill on day n) we are interested in the *total number* S_n of individuals who have been ill at some time between the first and nth days. As in our earlier discussion, let us overestimate S_n by approximating the number F_{n+1} of new cases on day $n + 1$ by aNI_n. Thus,

$$\begin{aligned} S_n &= I_1 + F_2 + F_3 + F_4 + \cdots + F_n \\ &= I_1 + aNI_1 + aNI_2 + aNI_3 + \cdots + aNI_{n-1}. \end{aligned}$$

Recalling that $I_n = r^{n-1}I_1$, with $r = 1 + aN - b$, we obtain

$$\begin{aligned} S_n &= I_1 + aNI_1 + aNrI_1 + aNr^2I_1 + \cdots + aNr^{n-2}I_1 \\ &= I_1 + aNI_1(1 + r + r^2 + \cdots + r^{n-2}). \end{aligned}$$

As in the proof of Theorem (11.15), this may be written

$$S_n = I_1 + aNI_1\left(\frac{1}{1-r} - \frac{r^{n-1}}{1-r}\right).$$

If $r < 1$, then

$$\begin{aligned} \lim_{n\to\infty} S_n &= I_1 + aNI_1\left(\frac{1}{1-r}\right) \\ &= I_1\left(1 + \frac{aN}{1-r}\right) \\ &= I_1\left(1 + \frac{aN}{b-aN}\right) \\ &= I_1\left(\frac{b}{b-aN}\right). \end{aligned}$$

If a and b are approximated from early data, this result enables health officials to determine an upper bound for the total number of individuals who will be ill at some stage of the epidemic.

EXERCISES 11.2

Exer. 1–6: Use the method of Example 1 to find (a) S_1, S_2, and S_3; (b) S_n; and (c) the sum of the series, if it converges.

1 $\displaystyle\sum_{n=1}^{\infty} \frac{-2}{(2n+5)(2n+3)}$

2 $\displaystyle\sum_{n=1}^{\infty} \frac{5}{(5n+2)(5n+7)}$

3 $\displaystyle\sum_{n=1}^{\infty} \frac{1}{4n^2-1}$

4 $\displaystyle\sum_{n=1}^{\infty} \frac{-1}{9n^2+3n-2}$

5 $\displaystyle\sum_{n=1}^{\infty} \ln\frac{n}{n+1}$

6 $\displaystyle\sum_{n=1}^{\infty} \frac{1}{\sqrt{n+1}+\sqrt{n}}$ (*Hint:* Rationalize the denominator.)

Exer. 7–16: Use Theorem (11.15) to determine whether the geometric series converges or diverges; if it converges, find its sum.

7 $3 + \frac{3}{4} + \cdots + \frac{3}{4^{n-1}} + \cdots$

8 $3 + \frac{3}{(-4)} + \cdots + \frac{3}{(-4)^{n-1}} + \cdots$

9 $1 + \left(\frac{-1}{\sqrt{5}}\right) + \cdots + \left(\frac{-1}{\sqrt{5}}\right)^{n-1} + \cdots$

10 $1 + \left(\frac{e}{3}\right) + \cdots + \left(\frac{e}{3}\right)^{n-1} + \cdots$

11 $0.37 + 0.0037 + \cdots + \frac{37}{(100)^n} + \cdots$

12 $0.628 + 0.000628 + \cdots + \dfrac{628}{(1000)^n} + \cdots$

13 $\sum_{n=1}^{\infty} 2^{-n}3^{n-1}$

14 $\sum_{n=1}^{\infty} (-5)^{n-1}4^{-n}$

15 $\sum_{n=1}^{\infty} (-1)^{n-1}$

16 $\sum_{n=1}^{\infty} (\sqrt{2})^{n-1}$

Exer. 17–20: Use Theorem (11.15) to find all values of x for which the series converges, and find the sum of the series.

17 $1 - x + x^2 - x^3 + \cdots + (-1)^n x^n + \cdots$

18 $1 + x^2 + x^4 + \cdots + x^{2n} + \cdots$

19 $\dfrac{1}{2} + \dfrac{(x-3)}{4} + \dfrac{(x-3)^2}{8} + \cdots + \dfrac{(x-3)^n}{2^{n+1}} + \cdots$

20 $3 + (x-1) + \dfrac{(x-1)^2}{3} + \cdots + \dfrac{(x-1)^n}{3^{n-1}} + \cdots$

Exer. 21–24: The overbar indicates that the digits underneath repeat indefinitely. Express the repeating decimal as a series, and find the rational number it represents.

21 $0.\overline{23}$

22 $5.\overline{146}$

23 $3.2\overline{394}$

24 $2.\overline{71828}$

Exer. 25–32: Use Example 1 or 3 and Theorem (11.19) or (11.20) to determine whether the series converges or diverges.

25 $\dfrac{1}{4 \cdot 5} + \dfrac{1}{5 \cdot 6} + \cdots + \dfrac{1}{(n+3)(n+4)} + \cdots$

26 $\dfrac{1}{10 \cdot 11} + \dfrac{1}{11 \cdot 12} + \cdots + \dfrac{1}{(n+9)(n+10)} + \cdots$

27 $\dfrac{5}{1 \cdot 2} + \dfrac{5}{2 \cdot 3} + \cdots + \dfrac{5}{n(n+1)} + \cdots$

28 $\dfrac{-1}{1 \cdot 2} + \dfrac{-1}{2 \cdot 3} + \cdots + \dfrac{-1}{n(n+1)} + \cdots$

29 $\dfrac{1}{4} + \dfrac{1}{5} + \cdots + \dfrac{1}{n+3} + \cdots$

30 $6^{-1} + 7^{-1} + \cdots + (n+5)^{-1} + \cdots$

31 $3 + \dfrac{3}{2} + \cdots + \dfrac{3}{n} + \cdots$

32 $-4 - 2 - \dfrac{4}{3} - \cdots - \dfrac{4}{n} - \cdots$

Exer. 33–40: Use the nth-term test (11.17) to determine whether the series diverges or needs further investigation.

33 $\sum_{n=1}^{\infty} \dfrac{3n}{5n-1}$

34 $\sum_{n=1}^{\infty} \dfrac{1}{1 + (0.3)^n}$

35 $\sum_{n=1}^{\infty} \dfrac{1}{n^2 + 3}$

36 $\sum_{n=1}^{\infty} \dfrac{1}{e^n + 1}$

37 $\sum_{n=1}^{\infty} \dfrac{1}{\sqrt[n]{e}}$

38 $\sum_{n=1}^{\infty} n \sin \dfrac{1}{n}$

39 $\sum_{n=1}^{\infty} \dfrac{n}{\ln (n+1)}$

40 $\sum_{n=1}^{\infty} \ln \left(\dfrac{2n}{7n-5} \right)$

Exer. 41–48: Use known convergent or divergent series, together with Theorem (11.20) or (11.21), to determine whether the series is convergent or divergent; if it converges, find its sum.

41 $\sum_{n=3}^{\infty} \left[\left(\dfrac{1}{4}\right)^n + \left(\dfrac{3}{4}\right)^n \right]$

42 $\sum_{n=1}^{\infty} \left[\left(\dfrac{3}{2}\right)^n + \left(\dfrac{2}{3}\right)^n \right]$

43 $\sum_{n=1}^{\infty} (2^{-n} - 2^{-3n})$

44 $\sum_{n=1}^{\infty} \left(\dfrac{1}{3^n} - \dfrac{1}{4^n} \right)$

45 $\sum_{n=1}^{\infty} \left[\dfrac{1}{8^n} + \dfrac{1}{n(n+1)} \right]$

46 $\sum_{n=1}^{\infty} \left[\dfrac{1}{n(n+1)} - \dfrac{4}{n} \right]$

47 $\sum_{n=1}^{\infty} \left(\dfrac{5}{n+2} - \dfrac{5}{n+3} \right)$

48 $\sum_{n=1}^{\infty} \left(\dfrac{1}{n+1} - \dfrac{1}{n} \right)$

c **Exer. 49–50: For the given convergent series, (a) approximate S_1, S_2, and S_3 to five decimal places and (b) approximate the sum of the series to three decimal places.**

49 $\sum_{n=1}^{\infty} \dfrac{\sin n}{4^n}$

50 $\sum_{n=1}^{\infty} \dfrac{\sqrt{n}}{e^{(n^2)}}$

c 51 Let S_n be the nth partial sum of the harmonic series. If $M = 3$, use Example 3 to find a positive integer m such that $S_m \geq M$, and approximate S_m to two decimal places.

c 52 Work Exercise 51 if $M = 8$.

53 Prove or disprove: If $\sum a_n$ and $\sum b_n$ both diverge, then $\sum (a_n + b_n)$ diverges.

54 What is wrong with the following "proof" that the divergent geometric series $\sum_{n=1}^{\infty} (-1)^{n+1}$ has the sum 0? (See Example 2.)

$$\sum_{n=1}^{\infty} (-1)^{n+1}$$
$$= [1 + (-1)] + [1 + (-1)] + [1 + (-1)] + \cdots$$
$$= 0 + 0 + 0 + \cdots = 0$$

55 A rubber ball is dropped from a height of 10 meters. If it rebounds approximately one-half the distance after each fall, use a geometric series to approximate the total distance the ball travels before coming to rest.

56 The bob of a pendulum swings through an arc 24 centimeters long on its first swing. If each successive swing is approximately five-sixths the length of the preceding swing, use a geometric series to approximate the total distance the bob travels before coming to rest.

57 If a dosage of Q units of a certain drug is administered to an individual, then the amount remaining in the bloodstream at the end of t minutes is given by Qe^{-ct}, where $c > 0$. Suppose this same dosage is given at successive T-minute intervals.

(a) Show that the amount $A(k)$ of the drug in the bloodstream immediately after the kth dose is given by $A(k) = \sum_{n=0}^{k-1} Qe^{-ncT}$.

(b) Find an upper bound for the amount of the drug in the bloodstream after any number of doses.

(c) Find the smallest time between doses that will ensure that $A(k)$ does not exceed a certain level M for $M > Q$.

58 Suppose that each dollar introduced into the economy recirculates as follows: 85% of the original dollar is spent, then 85% of that \$0.85 is spent, and so on. Find the economic impact (the total amount spent) if \$1,000,000 is introduced into the economy.

59 In a pest eradication program, N sterilized male flies are released into the general population each day, and 90% of these flies will survive a given day.

(a) Show that the number of sterilized flies in the population after n days is $N + (0.9)N + \cdots + (0.9)^{n-1}N$.

(b) If the *long-range* goal of the program is to keep 20,000 sterilized males in the population, how many such flies should be released each day?

60 A certain drug has a half-life in the bloodstream of about 2 hours. Doses of K milligrams will be administered every four hours, with K still to be determined.

(a) Show that the number of milligrams of drug in the bloodstream after the nth dose has been administered is $K + \frac{1}{4}K + \cdots + (\frac{1}{4})^{n-1}K$, and that this sum is approximately $\frac{4}{3}K$ for large values of n.

(b) If more than 500 milligrams of the drug in the bloodstream is considered to be a dangerous level, find the largest possible dose that can be given repeatedly over a long period of time.

(c) Refer to Exercise 57. If the dose K is 50 milligrams, how frequently can the drug be safely administered?

61 The first figure shows some terms of a sequence of squares $S_1, S_2, \ldots, S_k, \ldots$. Let a_k, A_k, and P_k denote the side, area, and perimeter, respectively, of the square S_k. The square S_{k+1} is constructed from S_k by connecting four points on S_k, with each point a distance of $\frac{1}{4}a_k$ from a vertex, as shown in the second figure.

(a) Find a relationship between a_{k+1} and a_k.

(b) Find a_n, A_n, and P_n.

(c) Calculate $\sum_{n=1}^{\infty} P_n$ and $\sum_{n=1}^{\infty} A_n$.

EXERCISE 61

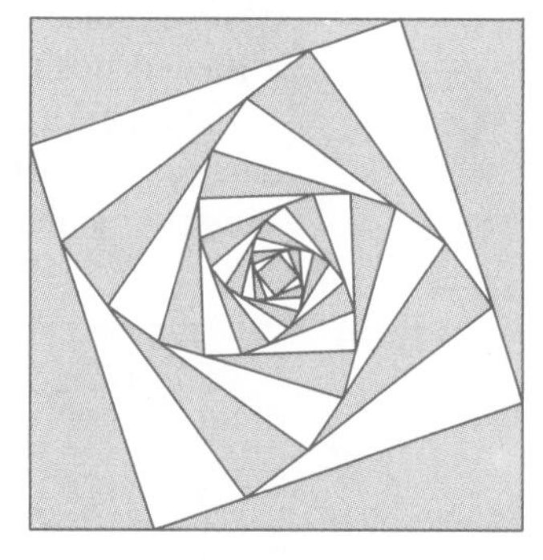

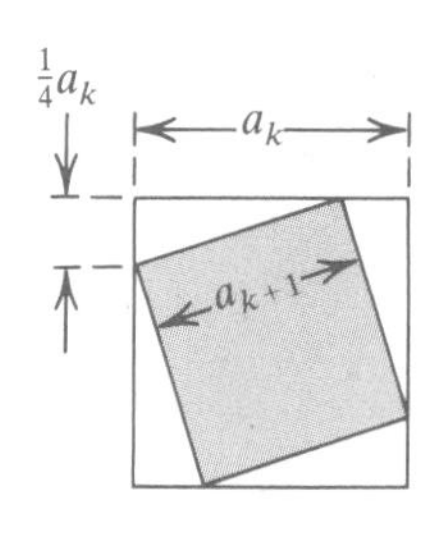

62 The figure shows several terms of a sequence consisting of alternating circles and squares. Each circle is inscribed in a square, and each square (excluding the largest) is inscribed in a circle. Let S_n denote the area of the nth square and C_n the area of the nth circle.

(a) Find relationships between S_n and C_n and between C_n and S_{n+1}.

(b) What portion of the largest square is shaded in the figure?

EXERCISE 62

11.3 POSITIVE-TERM SERIES

In the preceding section we established the convergence or divergence of several series by finding a formula for the nth partial sum S_n and then determining whether or not $\lim_{n\to\infty} S_n$ exists. Unfortunately, except in special cases such as a geometric series or a telescoping series, it is often impossible to find an explicit formula for S_n. However, we can develop tests for convergence or divergence of a series $\sum a_n$ that employ the nth term a_n. *These tests will not give us the sum S of the series*, but instead will tell us only whether the sum *exists*. This is sufficient in most applications, because knowing that the sum exists, we can usually approximate it to any degree of accuracy by adding a sufficient number of terms of the series.

In this section we shall consider only **positive-term series**—that is, series $\sum a_n$ such that $a_n > 0$ for every n. Although this approach may appear to be very specialized, positive-term series are the foundation for all of our future work with series. As we shall see later, the convergence or divergence of an *arbitrary* series can often be determined from that of a related positive-term series.

The next theorem shows that to establish convergence or divergence of a positive-term series, it is sufficient to determine whether the sequence of partial sums $\{S_n\}$ is bounded.

Theorem **(11.22)**

If $\sum a_n$ is a positive-term series and if there exists a number M such that

$$S_n = a_1 + a_2 + \cdots + a_n < M$$

for every n, then the series converges and has a sum $S \le M$. If no such M exists, the series diverges.

PROOF If $\{S_n\}$ is the sequence of partial sums of the positive-term series $\sum a_n$, then

$$S_1 < S_2 < \cdots < S_n < \cdots$$

and therefore $\{S_n\}$ is monotonic. If there exists a number M such that $S_n < M$ for every n, then $\{S_n\}$ is bounded monotonic. As in the proof of Theorem (11.9),

$$\lim_{n\to\infty} S_n = S \le M$$

for some S, and hence the series converges. If no such M exists, then $\lim_{n\to\infty} S_n = \infty$ and the series diverges. ■

We may use the nth term a_n of a series $\sum a_n$ to define a function f such that $f(n) = a_n$ for every positive integer n. In some cases, if we replace n with x, we obtain a function that is defined for every *real* number $x \ge 1$. For example,

$$\text{given } \sum_{n=1}^{\infty} \frac{1}{n^2}, \quad \text{let} \quad f(n) = \frac{1}{n^2}.$$

Replacing n with x, we obtain $f(x) = 1/x^2$, which gives us the desired function f. Note that

$$\sum_{n=1}^{\infty} \frac{1}{n^2} = \sum_{n=1}^{\infty} f(n) = f(1) + f(2) + \cdots + f(n) + \cdots.$$

The next result shows that if a function f obtained in this way satisfies certain conditions, then we may use the improper integral $\int_1^\infty f(x)\,dx$ to test the series $\sum_{n=1}^\infty f(n)$ for convergence or divergence.

Integral test **(11.23)**

> If $\sum a_n$ is a series, let $f(n) = a_n$ and let f be the function obtained by replacing n with x. If f is positive-valued, continuous, and decreasing for every real number $x \ge 1$, then the series $\sum a_n$
>
> **(i)** converges if $\int_1^\infty f(x)\,dx$ converges
>
> **(ii)** diverges if $\int_1^\infty f(x)\,dx$ diverges

FIGURE 11.8

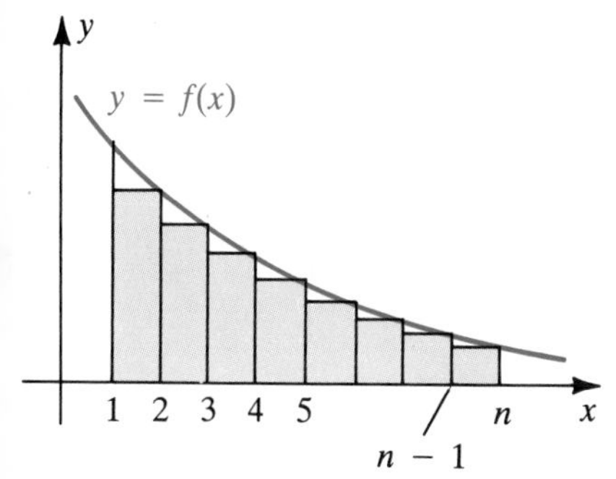

FIGURE 11.9

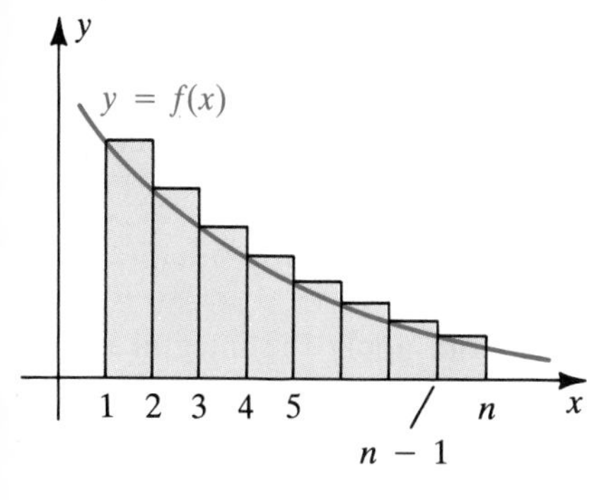

PROOF As in the hypotheses, we let $f(n) = a_n$ and consider $f(x)$ for every real number $x \ge 1$. A typical graph of this positive-valued, continuous, decreasing function is sketched in Figure 11.8. If n is a positive integer greater than 1, the area of the inscribed rectangular polygon illustrated in Figure 11.8 is

$$\sum_{k=2}^{n} f(k) = f(2) + f(3) + \cdots + f(n).$$

Similarly, the area of the circumscribed rectangular polygon illustrated in Figure 11.9 is

$$\sum_{k=1}^{n-1} f(k) = f(1) + f(2) + \cdots + f(n-1).$$

Since $\int_1^n f(x)\,dx$ is the area under the graph of f from 1 to n,

$$\sum_{k=2}^{n} f(k) \le \int_1^n f(x)\,dx \le \sum_{k=1}^{n-1} f(k).$$

If S_n denotes the nth partial sum of the series $f(1) + f(2) + \cdots + f(n) + \cdots$, then this inequality may be written

$$S_n - f(1) \le \int_1^n f(x)\,dx \le S_{n-1}.$$

The preceding inequality implies that if the integral $\int_1^\infty f(x)\,dx$ converges and equals $K > 0$, then

$$S_n - f(1) \le K, \quad \text{or} \quad S_n \le K + f(1)$$

for every positive integer n. Hence, by Theorem (11.22), the series $\sum f(n)$ converges.

If the improper integral diverges, then

$$\lim_{n\to\infty} \int_1^n f(x)\,dx = \infty,$$

and since $\int_1^n f(x)\,dx \le S_{n-1}$, we also have $\lim_{n\to\infty} S_{n-1} = \infty$; that is, the series $\sum f(n)$ diverges. ■

In using the integral test (11.23) it is necessary to consider

$$\int_1^{\infty} f(x)\,dx = \lim_{t\to\infty} \int_1^t f(x)\,dx.$$

Thus, we must integrate $f(x)$ and then take a limit. If $f(x)$ is not readily integrable, a different test for convergence or divergence should be employed.

EXAMPLE 1 Use the integral test (11.23) to prove that the harmonic series

$$1 + \frac{1}{2} + \frac{1}{3} + \cdots + \frac{1}{n} + \cdots$$

diverges (see Example 3 of Section 11.2).

SOLUTION Since $a_n = 1/n$, we let $f(n) = 1/n$. Replacing n by x gives us $f(x) = 1/x$. Because f is positive-valued, continuous, and decreasing for $x \geq 1$, we can apply the integral test (11.23):

$$\begin{aligned}\int_1^{\infty} \frac{1}{x}\,dx &= \lim_{t\to\infty} \int_1^t \frac{1}{x}\,dx = \lim_{t\to\infty} \Big[\ln x\Big]_1^t \\ &= \lim_{t\to\infty} [\ln t - \ln 1] = \infty\end{aligned}$$

The series diverges, by (11.23)(ii).

EXAMPLE 2 Determine whether the infinite series $\sum ne^{-n^2}$ converges or diverges.

SOLUTION Since $a_n = ne^{-n^2}$, we let $f(n) = ne^{-n^2}$ and consider $f(x) = xe^{-x^2}$. If $x \geq 1$, f is positive-valued and continuous. The first derivative may be used to determine whether f is decreasing. Since

$$f'(x) = e^{-x^2} - 2x^2e^{-x^2} = e^{-x^2}(1 - 2x^2) < 0,$$

f is decreasing on $[1, \infty)$. We may therefore apply the integral test as follows:

$$\begin{aligned}\int_1^{\infty} xe^{-x^2}\,dx &= \lim_{t\to\infty} \int_1^t xe^{-x^2}\,dx = \lim_{t\to\infty} \Big[(-\tfrac{1}{2})e^{-x^2}\Big]_1^t \\ &= \left(-\frac{1}{2}\right) \lim_{t\to\infty} \left[\frac{1}{e^{(t^2)}} - \frac{1}{e}\right] = \frac{1}{2e}\end{aligned}$$

Hence the series converges, by (11.23)(i).

In Example 2 we proved that the series $\sum ne^{-n^2}$ converges and therefore has a sum S. However, *we have not found the numerical value of* S. The number $1/(2e)$ in the solution is the value of an improper integral, not the sum of the series. If desired, we could *approximate* S by using a partial sum S_n, with n sufficiently large. (See Exercise 59.)

An integral test may also be used if the function f satisfies the conditions of (11.23) for every $x \geq k$ for some positive integer k. In this case we merely replace the integral in (11.23) by $\int_k^\infty f(x)\,dx$. This corresponds to deleting the first $k - 1$ terms of the series.

The following series, which is a generalization of the harmonic series (11.14), will be useful when we apply comparison tests later in this section.

Definition **(11.24)**

A **p-series**, or a **hyperharmonic series**, is a series of the form

$$\sum_{n=1}^{\infty} \frac{1}{n^p} = 1 + \frac{1}{2^p} + \frac{1}{3^p} + \cdots + \frac{1}{n^p} + \cdots,$$

where p is a positive real number.

Note that if $p = 1$ in (11.24), we obtain the harmonic series. The following theorem provides information about convergence or divergence of p-series.

Theorem **(11.25)**

The p-series $\sum_{n=1}^{\infty} \frac{1}{n^p}$

(i) converges if $p > 1$

(ii) diverges if $p \leq 1$

PROOF The special case $p = 1$ is the divergent harmonic series. Suppose that p is a positive real number and $p \neq 1$. We shall employ the integral test (11.23), letting $f(n) = 1/n^p$ and considering $f(x) = 1/x^p = x^{-p}$. The function f is positive-valued and continuous for $x \geq 1$. Moreover, for these values of x we see that $f'(x) = -px^{-p-1} < 0$, and hence f is decreasing. Thus, f satisfies the conditions stated in the integral test (11.23), and we consider

$$\begin{aligned}\int_1^\infty \frac{1}{x^p}\,dx &= \lim_{t\to\infty} \int_1^t x^{-p}\,dx = \lim_{t\to\infty}\left[\frac{x^{1-p}}{1-p}\right]_1^t \\ &= \frac{1}{1-p}\lim_{t\to\infty}(t^{1-p} - 1).\end{aligned}$$

If $p > 1$, then $p - 1 > 0$ and the last expression may be written

$$\frac{1}{1-p}\lim_{t\to\infty}\left(\frac{1}{t^{p-1}} - 1\right) = \frac{1}{1-p}(0-1) = \frac{1}{p-1}.$$

Thus, by (11.23)(i), the p-series converges if $p > 1$.

If $0 < p < 1$, then $1 - p > 0$ and

$$\frac{1}{1-p}\lim_{t\to\infty}(t^{1-p} - 1) = \infty.$$

Hence, by (11.23)(ii), the p-series diverges.

If $p \leq 0$, then $\lim_{n\to\infty}(1/n^p) \neq 0$ and, by (11.17)(i), the series diverges. ■

The following illustration contains some special p-series.

ILLUSTRATION

p-SERIES	VALUE OF p	CONCLUSION
$\sum_{n=1}^{\infty} \frac{1}{n^2} = 1 + \frac{1}{2^2} + \frac{1}{3^2} + \cdots$	$p = 2$	Converges, by (11.25)(i), since $2 > 1$
$\sum_{n=1}^{\infty} \frac{1}{\sqrt{n}} = 1 + \frac{1}{\sqrt{2}} + \frac{1}{\sqrt{3}} + \cdots$	$p = \frac{1}{2}$	Diverges, by (11.25)(ii), since $\frac{1}{2} < 1$
$\sum_{n=1}^{\infty} \frac{1}{n^{3/2}} = 1 + \frac{1}{2^{3/2}} + \frac{1}{3^{3/2}} + \cdots$	$p = \frac{3}{2}$	Converges, by (11.25)(i), since $\frac{3}{2} > 1$
$\sum_{n=1}^{\infty} \frac{1}{\sqrt[3]{n}} = 1 + \frac{1}{\sqrt[3]{2}} + \frac{1}{\sqrt[3]{3}} + \cdots$	$p = \frac{1}{3}$	Diverges, by (11.25)(ii), since $\frac{1}{3} < 1$

The next theorem allows us to use known convergent (divergent) series to establish the convergence (divergence) of other series.

Basic comparison tests (11.26)

Let $\sum a_n$ and $\sum b_n$ be positive-term series.

(i) If $\sum b_n$ converges and $a_n \leq b_n$ for every positive integer n, then $\sum a_n$ converges.

(ii) If $\sum b_n$ diverges and $a_n \geq b_n$ for every positive integer n, then $\sum a_n$ diverges.

PROOF Let S_n and T_n denote the nth partial sums of $\sum a_n$ and $\sum b_n$, respectively. Suppose $\sum b_n$ converges and has the sum T. If $a_n \leq b_n$ for every n, then $S_n \leq T_n < T$ and hence, by Theorem (11.22), $\sum a_n$ converges. This proves part (i).

To prove (ii), suppose $\sum b_n$ diverges and $a_n \geq b_n$ for every n. Then $S_n \geq T_n$, and since T_n increases without bound as n becomes infinite, so does S_n. Consequently $\sum a_n$ diverges. ■

The convergence or divergence of a series is not affected by deleting a finite number of terms, so the condition $a_n \leq b_n$ or $a_n \geq b_n$ of (11.26) is only required from the kth term on, for some positive integer k.

A series $\sum d_n$ is said to **dominate** a series $\sum c_n$ if $c_n \leq d_n$ for every positive integer n. In this terminology, (11.26)(i) states that *a positive-term series that is dominated by a convergent series is also convergent*. Part (ii) states that *a series that dominates a divergent positive-term series is also divergent*.

EXAMPLE 3 Determine whether the series converges or diverges:

(a) $\sum_{n=1}^{\infty} \frac{1}{2 + 5^n}$ **(b)** $\sum_{n=2}^{\infty} \frac{3}{\sqrt{n} - 1}$

SOLUTION

(a) For every $n \geq 1$,

$$\frac{1}{2 + 5^n} < \frac{1}{5^n} = \left(\frac{1}{5}\right)^n.$$

Since $\sum (1/5)^n$ is a convergent geometric series, the given series converges, by the basic comparison test (11.26)(i).

(b) The p-series $\sum 1/\sqrt{n}$ diverges, and hence so does the series obtained by disregarding the first term $1/\sqrt{1}$. If $n \geq 2$, then

$$\frac{1}{\sqrt{n} - 1} > \frac{1}{\sqrt{n}} \quad \text{and hence} \quad \frac{3}{\sqrt{n} - 1} > \frac{1}{\sqrt{n}}.$$

It follows from the basic comparison test (11.26)(ii) that the given series diverges.

When we use a basic comparison test, we must first decide on a suitable series $\sum b_n$ and then prove that either $a_n \leq b_n$ or $a_n \geq b_n$ for every n greater than some positive integer k. This proof can be very difficult if a_n is a complicated expression. The following comparison test is often easier to apply, because after deciding on $\sum b_n$, we need only take a limit of the quotient a_n/b_n as $n \to \infty$.

Limit comparison test **(11.27)**

> Let $\sum a_n$ and $\sum b_n$ be positive-term series. If
>
> $$\lim_{n \to \infty} \frac{a_n}{b_n} = c > 0,$$
>
> then either both series converge or both diverge.

PROOF If $\lim_{n \to \infty} (a_n/b_n) = c > 0$, then a_n/b_n is close to c if n is large. Hence there exists a number N such that

$$\frac{c}{2} < \frac{a_n}{b_n} < \frac{3c}{2} \quad \text{whenever} \quad n > N$$

FIGURE 11.10

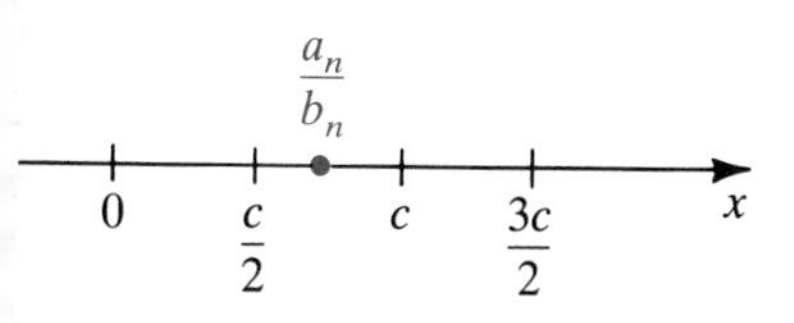

(see Figure 11.10). This is equivalent to

$$\frac{c}{2} b_n < a_n < \frac{3c}{2} b_n \quad \text{whenever} \quad n > N.$$

If the series $\sum a_n$ converges, then $\sum (c/2)b_n$ also converges, because it is dominated by $\sum a_n$. Applying (11.20)(ii), we find that the series

$$\sum b_n = \sum \left(\frac{2}{c}\right)\left(\frac{c}{2}\right) b_n$$

converges.

Conversely, if $\sum b_n$ converges, then so does $\sum a_n$, since it is dominated by the convergent series $\sum (3c/2)b_n$. We have proved that $\sum a_n$ converges if and only if $\sum b_n$ converges. Consequently $\sum a_n$ diverges if and only if $\sum b_n$ diverges. ■

If, in (11.27), the limit equals 0 or ∞, it may be possible to determine whether the series $\sum a_n$ converges or diverges by using the comparison test stated in Exercise 51 or 52, respectively.

To find a suitable series $\sum b_n$ to use in the limit comparison test (11.27) when a_n is a quotient, a good procedure is to *delete all terms in the numerator and denominator of a_n except those that have the greatest effect on the magnitude*. We may also replace any constant factor c by 1, since $\sum b_n$ and $\sum cb_n$ either both converge or both diverge (see Theorem 11.20). The next illustration demonstrates this procedure for several series $\sum a_n$.

ILLUSTRATION

a_n	DELETING TERMS OF LEAST MAGNITUDE	CHOICE OF b_n IN (11.27)
$\dfrac{3n+1}{4n^3+n^2-2}$	$\dfrac{3n}{4n^3}=\dfrac{3}{4n^2}$	$\dfrac{1}{n^2}$
$\dfrac{5}{\sqrt{n^2+2n+7}}$	$\dfrac{5}{\sqrt{n^2}}=\dfrac{5}{n}$	$\dfrac{1}{n}$
$\dfrac{\sqrt[3]{n^2+4}}{6n^2-n-1}$	$\dfrac{\sqrt[3]{n^2}}{6n^2}=\dfrac{n^{2/3}}{6n^2}=\dfrac{1}{6n^{4/3}}$	$\dfrac{1}{n^{4/3}}$

EXAMPLE 4 Determine whether the series converges or diverges:

(a) $\displaystyle\sum_{n=1}^{\infty}\frac{1}{\sqrt[3]{n^2+1}}$ **(b)** $\displaystyle\sum_{n=1}^{\infty}\frac{3n^2+5n}{2^n(n^2+1)}$

SOLUTION

(a) The nth term of the series is

$$a_n=\frac{1}{\sqrt[3]{n^2+1}}.$$

If we delete the number 1 in the radicand, we obtain $b_n=1/\sqrt[3]{n^2}$, which is the nth term of a divergent p-series, with $p=\frac{2}{3}$. Applying the limit comparison test (11.27) gives us the following:

$$\lim_{n\to\infty}\frac{a_n}{b_n}=\lim_{n\to\infty}\frac{\sqrt[3]{n^2}}{\sqrt[3]{n^2+1}}=\lim_{n\to\infty}\sqrt[3]{\frac{n^2}{n^2+1}}=1>0$$

Since $\sum b_n$ diverges, so does $\sum a_n$.

It is important to note that we cannot use $b_n=1/\sqrt[3]{n^2}$ with the basic comparison test (11.26), because $a_n<b_n$ instead of $a_n\ge b_n$.

(b) The nth term of the series is

$$a_n=\frac{3n^2+5n}{2^n n^2+2^n}.$$

Deleting the terms of least magnitude in the numerator and denominator, we obtain

$$\frac{3n^2}{2^n n^2} = \frac{3}{2^n},$$

and hence we choose $b_n = 1/2^n$. Applying the limit comparison test (11.27) gives us

$$\lim_{n\to\infty} \frac{a_n}{b_n} = \lim_{x\to\infty} \frac{3n^2 + 5n}{2^n(n^2 + 1)} \cdot \frac{2^n}{1} = \lim_{x\to\infty} \frac{3n^2 + 5n}{n^2 + 1} = 3 > 0.$$

Since, by Theorem (11.15)(i), $\sum b_n$ is a convergent geometric series (with $r = \frac{1}{2} < 1$), the series $\sum a_n$ is also convergent.

EXAMPLE 5 Let $a_n = \dfrac{8n + \sqrt{n}}{5 + n^2 + n^{7/2}}$. Determine whether $\sum a_n$ converges or diverges.

SOLUTION To find a suitable comparison series $\sum b_n$, we delete all but the highest powers of n in the numerator and denominator, obtaining

$$\frac{8n}{n^{7/2}} = \frac{8}{n^{5/2}}.$$

Applying the limit comparison test (11.27), with $b_n = 1/n^{5/2}$, we find

$$\begin{aligned}\lim_{n\to\infty} \frac{a_n}{b_n} &= \lim_{n\to\infty} \frac{8n + n^{1/2}}{5 + n^2 + n^{7/2}} \cdot \frac{n^{5/2}}{1}\\ &= \lim_{n\to\infty} \frac{8n^{7/2} + n^3}{5 + n^2 + n^{7/2}} = 8 > 0.\end{aligned}$$

Since $\sum b_n$ is a convergent p-series with $p = \frac{5}{2} > 1$, it follows from (11.27) that $\sum a_n$ is also convergent.

We shall conclude this section with several general remarks about positive-term series. Suppose $\sum a_n$ is a positive-term series and the terms are grouped in some manner, such as

$$(a_1 + a_2) + a_3 + (a_4 + a_5 + a_6 + a_7) + \cdots.$$

If we denote the last series by $\sum b_n$, so that

$$b_1 = a_1 + a_2, \quad b_2 = a_3, \quad b_3 = a_4 + a_5 + a_6 + a_7, \quad \ldots,$$

then any partial sum of the series $\sum b_n$ is also a partial sum of $\sum a_n$. It follows that if $\sum a_n$ converges, then $\sum b_n$ converges and has the same sum. A similar argument may be used for any grouping of the terms of $\sum a_n$. Thus, *if a positive-term series converges, then the series obtained by grouping the terms in any manner also converges and has the same sum.* We cannot make a similar statement about arbitrary divergent series. For example, the terms of the divergent series $\sum (-1)^n$ may be grouped to produce a convergent series (see Exercise 54 of Section 11.2).

Next, suppose that a convergent positive-term series $\sum a_n$ has the sum S and that a new series $\sum b_n$ is formed by rearranging the terms in some way. For example, $\sum b_n$ could be the series

$$a_2 + a_8 + a_1 + a_5 + a_7 + a_3 + \cdots.$$

If T_n is the nth partial sum of $\sum b_n$, then it is a sum of terms of $\sum a_n$. If m is the largest subscript associated with the terms a_k in T_n, then $T_n \le S_m < S$. Consequently $T_n < S$ for every n. Applying Theorem (11.22), we find that $\sum b_n$ converges and has a sum $T \le S$. The preceding proof is independent of the particular rearrangement of terms. We may also regard the series $\sum a_n$ as having been obtained by rearranging the terms of $\sum b_n$ and hence, by the same argument, $S \le T$. We have proved that *if the terms of a convergent positive-term series $\sum a_n$ are rearranged in any manner, then the resulting series converges and has the same sum.*

EXERCISES 11.3

Exer. 1–12: (a) Show that the function f determined by the nth term of the series satisfies the hypotheses of the integral test. (b) Use the integral test to determine whether the series converges or diverges.

1 $\displaystyle\sum_{n=1}^{\infty} \frac{1}{(3+2n)^2}$

2 $\displaystyle\sum_{n=1}^{\infty} \frac{1}{(4+n)^{3/2}}$

3 $\displaystyle\sum_{n=1}^{\infty} \frac{1}{4n+7}$

4 $\displaystyle\sum_{n=1}^{\infty} \frac{n}{n^2+1}$

5 $\displaystyle\sum_{n=1}^{\infty} n^2 e^{-n^3}$

6 $\displaystyle\sum_{n=3}^{\infty} \frac{1}{n(2n-5)}$

7 $\displaystyle\sum_{n=3}^{\infty} \frac{\ln n}{n}$

8 $\displaystyle\sum_{n=2}^{\infty} \frac{1}{n(\ln n)^2}$

9 $\displaystyle\sum_{n=2}^{\infty} \frac{1}{n\sqrt{n^2-1}}$

10 $\displaystyle\sum_{n=4}^{\infty} \left(\frac{1}{n-3} - \frac{1}{n}\right)$

11 $\displaystyle\sum_{n=1}^{\infty} \frac{\arctan n}{1+n^2}$

12 $\displaystyle\sum_{n=1}^{\infty} \frac{1}{1+16n^2}$

Exer. 13–20: Use a basic comparison test to determine whether the series converges or diverges.

13 $\displaystyle\sum_{n=1}^{\infty} \frac{1}{n^4+n^2+1}$

14 $\displaystyle\sum_{n=1}^{\infty} \frac{\sqrt{n}}{n^2+1}$

15 $\displaystyle\sum_{n=1}^{\infty} \frac{1}{n3^n}$

16 $\displaystyle\sum_{n=1}^{\infty} \frac{2+\cos n}{n^2}$

17 $\displaystyle\sum_{n=1}^{\infty} \frac{\arctan n}{n}$

18 $\displaystyle\sum_{n=1}^{\infty} \frac{\operatorname{arcsec} n}{(0.5)^n}$

19 $\displaystyle\sum_{n=1}^{\infty} \frac{1}{n^n}$

20 $\displaystyle\sum_{n=1}^{\infty} \frac{1}{n!}$

Exer. 21–28: Use the limit comparison test to determine whether the series converges or diverges.

21 $\displaystyle\sum_{n=1}^{\infty} \frac{\sqrt{n}}{n+4}$

22 $\displaystyle\sum_{n=1}^{\infty} \frac{2}{3+\sqrt{n}}$

23 $\displaystyle\sum_{n=2}^{\infty} \frac{1}{\sqrt{4n^3-5n}}$

24 $\displaystyle\sum_{n=1}^{\infty} \frac{1}{\sqrt{n(n+1)(n+2)}}$

25 $\displaystyle\sum_{n=1}^{\infty} \frac{8n^2-7}{e^n(n+1)^2}$

26 $\displaystyle\sum_{n=1}^{\infty} \frac{3n+5}{n2^n}$

27 $\displaystyle\sum_{n=1}^{\infty} \frac{1}{\sqrt{n+9}}$

28 $\displaystyle\sum_{n=1}^{\infty} \frac{n^2}{n^3+1}$

Exer. 29–46: Determine whether the series converges or diverges.

29 $\displaystyle\sum_{n=1}^{\infty} \frac{2n+n^2}{n^3+1}$

30 $\displaystyle\sum_{n=1}^{\infty} \frac{n^5+4n^3+1}{2n^8+n^4+2}$

31 $\displaystyle\sum_{n=1}^{\infty} \frac{1+2^n}{1+3^n}$

32 $\displaystyle\sum_{n=4}^{\infty} \frac{3n}{2n^2-7}$

33 $\displaystyle\sum_{n=1}^{\infty} \frac{1}{\sqrt[3]{5n^2+1}}$

34 $\displaystyle\sum_{n=1}^{\infty} \frac{\ln n}{n^4}$

35 $\displaystyle\sum_{n=1}^{\infty} \frac{1}{\sqrt[3]{2n+1}}$

36 $\displaystyle\sum_{n=1}^{\infty} \frac{n+\ln n}{n^3+n+1}$

37 $\displaystyle\sum_{n=1}^{\infty} ne^{-n}$

38 $\displaystyle\sum_{n=1}^{\infty} \frac{1}{n(n+1)(n+2)}$

39 $\displaystyle\sum_{n=1}^{\infty} \sin\frac{1}{n^2}$

40 $\displaystyle\sum_{n=1}^{\infty} \tan\frac{1}{n}$

41 $\displaystyle\sum_{n=1}^{\infty} \frac{(2n+1)^3}{(n^3+1)^2}$

42 $\displaystyle\sum_{n=1}^{\infty} \frac{n+\ln n}{n^2+1}$

43 $\sum_{n=1}^{\infty} \frac{n^2 + 2^n}{n + 3^n}$

44 $\sum_{n=1}^{\infty} \ln\left(1 + \frac{1}{2^n}\right)$

45 $\sum_{n=1}^{\infty} \frac{\ln n}{n^3}$

46 $\sum_{n=1}^{\infty} \frac{\sin n + 2^n}{n + 5^n}$

Exer. 47–48: Find every real number k for which the series converges.

47 $\sum_{n=2}^{\infty} \frac{1}{n^k \ln n}$

48 $\sum_{n=2}^{\infty} \frac{1}{n (\ln n)^k}$

49 (a) Use the proof of the integral test (11.23) to show that, for every positive integer $n > 1$,

$$\ln (n + 1) < 1 + \frac{1}{2} + \frac{1}{3} + \cdots + \frac{1}{n} < 1 + \ln n.$$

(b) Estimate the number of terms of the harmonic series that should be added so that $S_n > 100$.

50 Consider the hypothetical problem illustrated in the figure: Starting with a ball of radius 1 foot, a person stacks balls vertically such that if r_k is the radius of the kth ball, then $r_{n+1} = r_n\sqrt{n/(n + 1)}$ for each positive integer n.

(a) Show that the height of the stack can be made arbitrarily large.

(b) If the balls are made of a material that weighs 1 lb/ft^3, show that the total weight of the stack is always less than 4π pounds.

EXERCISE 50

51 Suppose $\sum a_n$ and $\sum b_n$ are positive-term series. Prove that if $\lim_{n\to\infty} (a_n/b_n) = 0$ and $\sum b_n$ converges, then $\sum a_n$ converges. (This is not necessarily true for series that contain negative terms.)

52 Prove that if $\lim_{n\to\infty} (a_n/b_n) = \infty$ and $\sum b_n$ diverges, then $\sum a_n$ diverges.

53 Let $\sum a_n$ be a convergent, positive-term series. Let $f(n) = a_n$, and suppose f is continuous and decreasing for $x \geq N$ for some integer N. Prove that the error in approximating the sum of the given series by $\sum_{n=1}^{N} a_n$ is less than $\int_N^\infty f(x)\,dx$.

Exer. 54–56: Use Exercise 53 to estimate the smallest number of terms that can be added to approximate the sum of the series with an error less than E.

54 $\sum_{n=1}^{\infty} \frac{1}{n^2}$, $E = 0.001$

55 $\sum_{n=1}^{\infty} \frac{1}{n^3}$, $E = 0.01$

56 $\sum_{n=2}^{\infty} \frac{1}{n (\ln n)^2}$, $E = 0.05$

57 Prove that if a positive-term series $\sum a_n$ converges, then $\sum (1/a_n)$ diverges.

58 Prove that if a positive-term series $\sum a_n$ converges, then $\sum \sqrt{a_n a_{n+1}}$ converges. (*Hint:* First show that the following is true: $\sqrt{a_n a_{n+1}} \leq (a_n + a_{n+1})/2$.)

c **Exer. 59–60: Approximate the sum of the given series to three decimal places. (Use Exercise 53 to justify the accuracy of your answer.)**

59 $\sum_{n=1}^{\infty} ne^{-n^2}$

60 $\sum_{n=1}^{\infty} \frac{1}{n^4}$

c 61 Graph, on the same coordinate axes, $y = x$ and $y = \ln (x^k)$ for $k = 1, 2, 3$ and $1 \leq x \leq 20$, and then use the graphs to predict whether the series $\sum_{n=1}^{\infty} \frac{1}{\ln (n^k)}$ converges or diverges for $k = 1, 2$, and 3.

c 62 Graph, on the same coordinate axes, $y = x$ and $y = (\ln x)^k$ for $k = 1, 2, 3$ and $1 \leq x \leq 200$, and then use the graphs to predict whether the series $\sum_{n=1}^{\infty} \frac{1}{(\ln n)^k}$ converges or diverges for $k = 1, 2$, and 3.

11.4 THE RATIO AND ROOT TESTS

For the integral test to be applied to a positive-term series $\sum a_n$ with $a_n = f(n)$, the terms must be decreasing and we must be able to integrate $f(x)$. This often rules out series that involve factorials and other complicated expressions. We shall now introduce two tests that can be used to help determine convergence or divergence when other tests are not applicable. Unfortunately, as indicated by part (iii) of both tests, they are inconclusive for certain series.

Ratio test (11.28)

> Let $\sum a_n$ be a positive-term series, and suppose
>
> $$\lim_{n\to\infty} \frac{a_{n+1}}{a_n} = L.$$
>
> **(i)** If $L < 1$, the series is convergent.
>
> **(ii)** If $L > 1$ or $\lim_{n\to\infty} \dfrac{a_{n+1}}{a_n} = \infty$, the series is divergent.
>
> **(iii)** If $L = 1$, apply a different test; the series may be convergent or divergent.

PROOF

(i) Suppose $\lim_{n\to\infty} (a_{n+1}/a_n) = L < 1$. Let r be any number such that $0 \le L < r < 1$. Since a_{n+1}/a_n is close to L if n is large, there exists an integer N such that whenever $n \ge N$,

$$\frac{a_{n+1}}{a_n} < r \quad \text{or} \quad a_{n+1} < a_n r.$$

Substituting $N, N+1, N+2, \ldots$ for n, we obtain

$$a_{N+1} < a_N r$$
$$a_{N+2} < a_{N+1} r < a_N r^2$$
$$a_{N+3} < a_{N+2} r < a_N r^3$$

and, in general,

$$a_{N+m} < a_N r^m \quad \text{whenever} \quad m > 0.$$

It follows from the basic comparison test (11.26)(i) that the series

$$a_{N+1} + a_{N+2} + \cdots + a_{N+m} + \cdots$$

converges, since its terms are less than the corresponding terms of the convergent geometric series

$$a_N r + a_N r^2 + \cdots + a_N r^n + \cdots.$$

Since convergence or divergence is unaffected by discarding a finite number of terms (see Theorem (11.19)), the series $\sum_{n=1}^{\infty} a_n$ also converges.

(ii) Suppose $\lim_{n\to\infty} (a_{n+1}/a_n) = L > 1$. If r is a real number such that $L > r > 1$, then there exists an integer N such that

$$\frac{a_{n+1}}{a_n} > r > 1 \quad \text{whenever} \quad n \ge N.$$

Consequently $a_{n+1} > a_n$ if $n \ge N$. Thus, $\lim_{n\to\infty} a_n \ne 0$ and, by the nth-term test (11.17)(i), the series $\sum a_n$ diverges.

The proof for $\lim_{n\to\infty} (a_{n+1}/a_n) = \infty$ is similar and is left as an exercise.

(iii) The ratio test is inconclusive if

$$\lim_{n\to\infty} \frac{a_{n+1}}{a_n} = 1,$$

for it is easy to verify that the limit is 1 for both the convergent series $\sum (1/n^2)$ and the divergent series $\sum (1/n)$. Consequently, *if the limit is 1 then a different test must be employed.* ■

EXAMPLE 1 Determine whether the series is convergent or divergent.

(a) $\sum_{n=1}^{\infty} \frac{3^n}{n!}$ **(b)** $\sum_{n=1}^{\infty} \frac{3^n}{n^2}$

SOLUTION

(a) Applying the ratio test, we have

$$\begin{aligned}\lim_{n\to\infty} \frac{a_{n+1}}{a_n} &= \lim_{n\to\infty} \left(a_{n+1} \cdot \frac{1}{a_n}\right) \\ &= \lim_{n\to\infty} \frac{3^{n+1}}{(n+1)!} \cdot \frac{n!}{3^n} \\ &= \lim_{n\to\infty} \frac{3}{n+1} = 0.\end{aligned}$$

Since $0 < 1$, the series is convergent.

(b) Applying the ratio test, we obtain

$$\begin{aligned}\lim_{n\to\infty} \frac{a_{n+1}}{a_n} &= \lim_{n\to\infty} \frac{3^{n+1}}{(n+1)^2} \cdot \frac{n^2}{3^n} \\ &= \lim_{n\to\infty} \frac{3n^2}{n^2+2n+1} = 3.\end{aligned}$$

Since $3 > 1$, the series diverges, by (11.28)(ii).

EXAMPLE 2 Determine the convergence or divergence of $\sum_{n=1}^{\infty} \frac{n^n}{n!}$.

SOLUTION Applying the ratio test gives us the following:

$$\begin{aligned}\lim_{n\to\infty} \frac{a_{n+1}}{a_n} &= \lim_{n\to\infty} \frac{(n+1)^{n+1}}{(n+1)!} \cdot \frac{n!}{n^n} \\ &= \lim_{n\to\infty} \frac{(n+1)^{n+1}}{(n+1)} \cdot \frac{1}{n^n} \\ &= \lim_{n\to\infty} \frac{(n+1)^n}{n^n} = \lim_{n\to\infty} \left(\frac{n+1}{n}\right)^n \\ &= \lim_{n\to\infty} \left(1 + \frac{1}{n}\right)^n = e\end{aligned}$$

The last equality is a consequence of Theorem (7.32)(ii). Since $e > 1$, the series diverges.

If $\sum a_n$ is a series such that $\lim_{n\to\infty} (a_{n+1}/a_n) = 1$, we must use a different test (see (iii) of (11.28)). The next illustration contains several series of this type and suggestions on how to show convergence or divergence.

ILLUSTRATION

SERIES $\sum a_n$	$\lim_{n\to\infty} \frac{a_{n+1}}{a_n}$	SUGGESTION
$\sum_{n=1}^{\infty} \frac{2n^2+3n+4}{5n^5-7n^3+n}$	1	Show convergence by using the limit comparison test (11.27) with $b_n = 1/n^3$.
$\sum_{n=1}^{\infty} \frac{2n+1}{\sqrt{n^3+5n+3}}$	1	Show divergence by using the limit comparison test (11.27) with $b_n = 1/\sqrt{n}$.
$\sum_{n=1}^{\infty} \frac{\ln n}{n}$	1	Show divergence by using the integral test (11.23).

The following test is often useful if a_n contains powers of n.

Root test **(11.29)**

Let $\sum a_n$ be a positive-term series, and suppose

$$\lim_{n\to\infty} \sqrt[n]{a_n} = L.$$

(i) If $L < 1$, the series is convergent.

(ii) If $L > 1$ or $\lim_{n\to\infty} \sqrt[n]{a_n} = \infty$, the series is divergent.

(iii) If $L = 1$, apply a different test; the series may be convergent or divergent.

PROOF If $L < 1$, let us consider any number r such that $0 \le L < r < 1$. By the definition of limit, there exists a positive integer N such that if $n \ge N$, then

$$\sqrt[n]{a_n} < r \quad \text{or} \quad a_n < r^n.$$

Since $0 < r < 1$, $\sum_{n=N}^{\infty} r^n$ is a convergent geometric series, and hence, by the basic comparison test (11.26), $\sum_{n=N}^{\infty} a_n$ converges. Consequently $\sum_{n=1}^{\infty} a_n$ converges. This proves (i). The remainder of the proof is similar to that used for the ratio test.

EXAMPLE 3 Determine the convergence or divergence of $\sum_{n=1}^{\infty} \frac{2^{3n+1}}{n^n}$.

SOLUTION Applying the root test yields

$$\lim_{n\to\infty} \sqrt[n]{\frac{2^{3n+1}}{n^n}} = \lim_{n\to\infty} \left(\frac{2^{3n+1}}{n^n}\right)^{1/n}$$

$$= \lim_{n\to\infty} \frac{2^{3+(1/n)}}{n} = 0.$$

Since $0 < 1$, the series converges. We could have applied the ratio test; however, the process of evaluating the limit would have been more complicated.

EXERCISES 11.4

Exer. 1–10: Find $\lim_{n\to\infty} \frac{a_{n+1}}{a_n}$, and use the ratio test to determine if the series converges or diverges or if the test is inconclusive.

1 $\sum_{n=1}^{\infty} \frac{3n+1}{2^n}$

2 $\sum_{n=1}^{\infty} \frac{3^n}{n^2+4}$

3 $\sum_{n=1}^{\infty} \frac{5^n}{n(3^{n+1})}$

4 $\sum_{n=1}^{\infty} \frac{2^{n-1}}{5^n(n+1)}$

5 $\sum_{n=1}^{\infty} \frac{100^n}{n!}$

6 $\sum_{n=1}^{\infty} \frac{n^{10}+10}{n!}$

7 $\sum_{n=1}^{\infty} \frac{n+3}{n^2+2n+5}$

8 $\sum_{n=1}^{\infty} \frac{3n}{\sqrt{n^3+1}}$

9 $\sum_{n=1}^{\infty} \frac{n!}{e^n}$

10 $\sum_{n=1}^{\infty} \frac{n!}{(n+1)^5}$

Exer. 11–18: Find $\lim_{n\to\infty} \sqrt[n]{a_n}$, and use the root test to determine if the series converges or diverges or if the test is inconclusive.

11 $\sum_{n=1}^{\infty} \frac{1}{n^n}$

12 $\sum_{n=1}^{\infty} \frac{(\ln n)^n}{n^{n/2}}$

13 $\sum_{n=1}^{\infty} \frac{2^n}{n^2}$

14 $\sum_{n=2}^{\infty} \frac{5^{n+1}}{(\ln n)^n}$

15 $\sum_{n=1}^{\infty} \frac{n}{3^n}$

16 $\sum_{n=1}^{\infty} \frac{n^{10}}{10^n}$

17 $\sum_{n=1}^{\infty} \left(\frac{n}{2n+1}\right)^n$

18 $\sum_{n=2}^{\infty} \left(\frac{n}{\ln n}\right)^n$

Exer. 19–40: Determine whether the series converges or diverges.

19 $\sum_{n=1}^{\infty} \frac{\sqrt{n}}{n^2+1}$

20 $\sum_{n=1}^{\infty} \frac{\sqrt{n}}{3n+4}$

21 $\sum_{n=1}^{\infty} \frac{99^n(n^5+2)}{n^2 10^{2n}}$

22 $\sum_{n=1}^{\infty} \frac{n3^{2n}}{5^{n-1}}$

23 $\sum_{n=1}^{\infty} \frac{2}{n^3+e^n}$

24 $\sum_{n=1}^{\infty} \frac{n+1}{n^3+1}$

25 $\sum_{n=1}^{\infty} \left(\frac{2}{n}\right)^n n!$

26 $\sum_{n=1}^{\infty} \frac{n!}{n^n}$

27 $\sum_{n=1}^{\infty} \frac{n^n}{10^{n+1}}$

28 $\sum_{n=1}^{\infty} \frac{10+2^n}{n!}$

29 $\sum_{n=1}^{\infty} \frac{(n!)^2}{(2n)!}$

30 $\sum_{n=1}^{\infty} \frac{(2n)!}{2^n}$

31 $\sum_{n=2}^{\infty} \frac{1}{n\sqrt[3]{\ln n}}$

32 $\sum_{n=1}^{\infty} \frac{(2n)^n}{(5n+3n^{-1})^n}$

33 $\sum_{n=1}^{\infty} \frac{\ln n}{(1.01)^n}$

34 $\sum_{n=1}^{\infty} 3^{1/n}$

35 $\sum_{n=1}^{\infty} n \tan \frac{1}{n}$

36 $\sum_{n=1}^{\infty} \frac{\arctan n}{n^2}$

37 $\sum_{n=1}^{\infty} \left(1+\frac{1}{n}\right)^n$

38 $\sum_{n=2}^{\infty} \frac{1}{(\ln n)^n}$

39 $1 + \frac{1\cdot 3}{2!} + \frac{1\cdot 3\cdot 5}{3!} + \cdots + \frac{1\cdot 3\cdot 5\cdot \cdots \cdot (2n-1)}{n!} + \cdots$

40 $\frac{1}{2} + \frac{1\cdot 4}{2\cdot 4} + \frac{1\cdot 4\cdot 7}{2\cdot 4\cdot 6} + \cdots + \frac{1\cdot 4\cdot 7\cdot \cdots \cdot (3n-2)}{2\cdot 4\cdot 6\cdot \cdots \cdot (2n)} + \cdots$

11.5 ALTERNATING SERIES AND ABSOLUTE CONVERGENCE

The tests for convergence that we have discussed thus far can be applied only to positive-term series. We shall next consider infinite series that contain both positive and negative terms. One of the simplest, and most useful, series of this type is an **alternating series**, in which the terms are alternately positive and negative. It is customary to express an alternating series in one of the forms

$$a_1 - a_2 + a_3 - a_4 + \cdots + (-1)^{n-1}a_n + \cdots$$

or

$$-a_1 + a_2 - a_3 + a_4 - \cdots + (-1)^n a_n + \cdots$$

with $a_k > 0$ for every k. The next theorem provides the main test for convergence of these series. For convenience we shall consider $\sum_{n=1}^{\infty} (-1)^{n-1}a_n$. A similar proof holds for $\sum_{n=1}^{\infty} (-1)^n a_n$.

Alternating series test **(11.30)**

The alternating series

$$\sum_{n=1}^{\infty} (-1)^{n-1} a_n = a_1 - a_2 + a_3 - a_4 + \cdots + (-1)^{n-1} a_n + \cdots$$

is convergent if the following two conditions are satisfied:

(i) $a_k \geq a_{k+1} > 0$ for every k

(ii) $\lim_{n \to \infty} a_n = 0$

PROOF By condition (i), we may write

$$a_1 \geq a_2 \geq a_3 \geq a_4 \geq a_5 \geq \cdots \geq a_k \geq a_{k+1} \geq \cdots.$$

Let us consider the partial sums

$$S_2, S_4, S_6, \ldots, S_{2n}, \ldots$$

which contain an even number of terms of the series. Since

$$S_{2n} = (a_1 - a_2) + (a_3 - a_4) + \cdots + (a_{2n-1} - a_{2n})$$

and $a_k - a_{k+1} \geq 0$ for every k, we see that

$$0 \leq S_2 \leq S_4 \leq \cdots \leq S_{2n} \leq \cdots;$$

that is, $\{S_{2n}\}$ is a monotonic sequence. This fact is also evident from Figure 11.11, where we have used a coordinate line l to represent the following four partial sums of the series:

$$S_1 = a_1, \quad S_2 = a_1 - a_2, \quad S_3 = a_1 - a_2 + a_3, \quad S_4 = a_1 - a_2 + a_3 - a_4$$

FIGURE 11.11

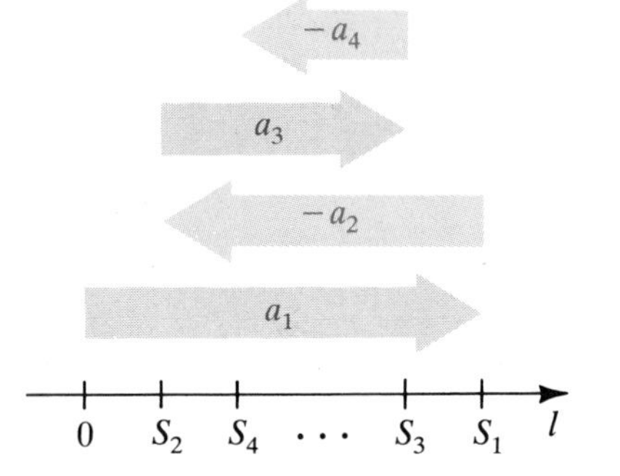

You may find it instructive to locate the points on l that correspond to S_5 and S_6.

Referring to Figure 11.11, we see that $S_{2n} \leq a_1$ for every positive integer n. This may also be proved algebraically by observing that

$$S_{2n} = a_1 - (a_2 - a_3) - (a_4 - a_5) - \cdots - (a_{2n-2} - a_{2n-1}) - a_{2n} \leq a_1.$$

Thus, $\{S_{2n}\}$ is a *bounded* monotonic sequence. As in the proof of Theorem (11.9),

$$\lim_{n \to \infty} S_{2n} = S \leq a_1$$

for some number S.

If we next consider a partial sum S_{2n+1} having an *odd* number of terms of the series, then $S_{2n+1} = S_{2n} + a_{2n+1}$ and, since $\lim_{n \to \infty} a_{2n+1} = 0$,

$$\lim_{n \to \infty} S_{2n+1} = \lim_{n \to \infty} S_{2n} = S.$$

Because both the sequence of even partial sums and the sequence of odd partial sums have the same limit S, it follows that

$$\lim_{n \to \infty} S_n = S \leq a_1;$$

that is, the series converges. ■

EXAMPLE 1 Determine whether the alternating series converges or diverges.

(a) $\sum_{n=1}^{\infty} (-1)^{n-1} \frac{2n}{4n^2 - 3}$ **(b)** $\sum_{n=1}^{\infty} (-1)^{n-1} \frac{2n}{4n - 3}$

SOLUTION

(a) Let

$$a_n = f(n) = \frac{2n}{4n^2 - 3}.$$

To apply the alternating series test (11.30), we must show that

(i) $a_k \geq a_{k+1}$ for every positive integer k

(ii) $\lim_{n \to \infty} a_n = 0$

There are several ways to prove (i). One method is to show that $f(x) = 2x/(4x^2 - 3)$ is decreasing for $x \geq 1$. By the quotient rule,

$$\begin{aligned} f'(x) &= \frac{(4x^2 - 3)(2) - (2x)(8x)}{(4x^2 - 3)^2} \\ &= \frac{-8x^2 - 6}{(4x^2 - 3)^2} < 0. \end{aligned}$$

By Theorem (4.13), $f(x)$ is decreasing and, therefore, $f(k) \geq f(k + 1)$; that is, $a_k \geq a_{k+1}$ for every positive integer k.

We can also prove (i) directly, by proving that $a_k - a_{k+1} \geq 0$. Thus, if $a_n = 2n/(4n^2 - 3)$, then for every positive integer k,

$$\begin{aligned} a_k - a_{k+1} &= \frac{2k}{4k^2 - 3} - \frac{2(k + 1)}{4(k + 1)^2 - 3} \\ &= \frac{8k^2 + 8k + 6}{(4k^2 - 3)(4k^2 + 8k + 1)} \geq 0. \end{aligned}$$

Still another technique for proving that $a_k \geq a_{k+1}$ is to show that $a_{k+1}/a_k \leq 1$.

To prove (ii), we see that

$$\lim_{n \to \infty} a_n = \lim_{n \to \infty} \frac{2n}{4n^2 - 3} = 0.$$

Thus, the alternating series converges.

(b) We can show that $a_k \geq a_{k+1}$ for every k; however,

$$\lim_{n \to \infty} a_n = \lim_{n \to \infty} \frac{2n}{4n - 3} = \frac{1}{2} \neq 0,$$

and hence the series diverges, by the nth-term test (11.17)(i).

The alternating series test (11.30) may be used if condition (i) holds for $k > m$ for some positive integer m, because this corresponds to deleting the first m terms of the series.

If a series converges, then the nth partial sum S_n can be used to approximate the sum S of the series. In many cases it is difficult to determine the accuracy of the approximation. However, for an *alternating series*, the next theorem provides a simple way of estimating the error that is involved.

Theorem (11.31)

Let $\sum_{n=1}^{\infty} (-1)^{n-1} a_n$ be an alternating series that satisfies conditions (i) and (ii) of the alternating series test. If S is the sum of the series and S_n is a partial sum, then

$$|S - S_n| \le a_{n+1};$$

that is, the error involved in approximating S by S_n is less than or equal to a_{n+1}.

PROOF The series obtained by deleting the first n terms of $\sum (-1)^{n-1} a_n$, namely

$$(-1)^n a_{n+1} + (-1)^{n+1} a_{n+2} + (-1)^{n+2} a_{n+3} + \cdots,$$

also satisfies the conditions of (11.30) and therefore has a sum R_n. Thus,

$$S - S_n = R_n = (-1)^n (a_{n+1} - a_{n+2} + a_{n+3} - \cdots)$$

and

$$|R_n| = a_{n+1} - a_{n+2} + a_{n+3} - \cdots.$$

Employing the same argument used in the proof of the alternating series test, we see that $|R_n| \le a_{n+1}$. Consequently

$$E = |S - S_n| = |R_n| \le a_{n+1},$$

which is what we wished to prove. ■

In the next example we use Theorem (11.31) to approximate the sum of an alternating series. In order to discuss the accuracy of an approximation, we must first agree on what is meant by one-decimal-place accuracy, two-decimal-place accuracy, and so on. Let us adopt the following convention. If E is the error in an approximation, then the approximation will be considered accurate to k decimal places if $|E| < 0.5 \times 10^{-k}$. For example, we have

1-decimal-place accuracy if $|E| < 0.5 \times 10^{-1} = 0.05$

2-decimal-place accuracy if $|E| < 0.5 \times 10^{-2} = 0.005$

3-decimal-place accuracy if $|E| < 0.5 \times 10^{-3} = 0.0005$.

EXAMPLE 2 Prove that the series

$$1 - \frac{1}{3!} + \frac{1}{5!} - \cdots + (-1)^{n-1} \frac{1}{(2n-1)!} + \cdots$$

is convergent, and approximate its sum S to five decimal places.

SOLUTION The nth term $a_n = 1/(2n - 1)!$ has the limit 0 as $n \to \infty$, and $a_k > a_{k+1}$ for every positive integer k. Hence the series converges, by the alternating series test. If we use S_n to approximate S, then, by Theorem (11.31), the error involved is less than or equal to $a_{n+1} = 1/(2n + 1)!$. Calculating several values of a_{n+1}, we find that for $n = 4$,

$$a_5 = \frac{1}{9!} \approx 0.0000028 < 0.000005.$$

Hence the partial sum S_4 approximates S to five decimal places. Since

$$\begin{aligned} S_4 &= 1 - \frac{1}{3!} + \frac{1}{5!} - \frac{1}{7!} \\ &= 1 - \frac{1}{6} + \frac{1}{120} - \frac{1}{5040} \approx 0.841468, \end{aligned}$$

we have $S \approx 0.84147$.

It will follow from (11.48)(a) that the sum of the series is sin 1, and hence $\sin 1 \approx 0.84147$.

The following concept is useful in investigating a series that contains both positive and negative terms but is not alternating. It allows us to use tests for positive-term series to establish convergence for other types of series (see Theorem 11.34).

Definition **(11.32)**

> A series $\sum a_n$ is **absolutely convergent** if the series
>
> $$\sum |a_n| = |a_1| + |a_2| + \cdots + |a_n| + \cdots$$
>
> is convergent.

Note that if $\sum a_n$ is a positive-term series, then $|a_n| = a_n$, and in this case absolute convergence is the same as convergence.

EXAMPLE 3 Prove that the following alternating series is absolutely convergent:

$$1 - \frac{1}{2^2} + \frac{1}{3^2} - \frac{1}{4^2} + \cdots + (-1)^n \frac{1}{n^2} + \cdots$$

SOLUTION Taking the absolute value of each term gives us

$$1 + \frac{1}{2^2} + \frac{1}{3^2} + \frac{1}{4^2} + \cdots + \frac{1}{n^2} + \cdots,$$

which is a convergent p-series. Hence, by Definition (11.32), the alternating series is absolutely convergent.

EXAMPLE 4 The **alternating harmonic series** is

$$\sum_{n=1}^{\infty} (-1)^{n-1}\frac{1}{n} = 1 - \frac{1}{2} + \frac{1}{3} - \frac{1}{4} + \cdots + (-1)^{n-1}\frac{1}{n} + \cdots.$$

Show that this series is

(a) convergent **(b)** not absolutely convergent

SOLUTION

(a) Conditions (i) and (ii) of the alternating series test (11.30) are satisfied, because

$$\frac{1}{k} > \frac{1}{k+1} \quad \text{for every } k \quad \text{and} \quad \lim_{n\to\infty} \frac{1}{n} = 0.$$

Hence the alternating harmonic series is convergent.

(b) To examine the series for absolute convergence, we apply Definition (11.32) and consider

$$\sum_{n=1}^{\infty} \left|(-1)^{n-1}\frac{1}{n}\right| = 1 + \frac{1}{2} + \frac{1}{3} + \frac{1}{4} + \cdots + \frac{1}{n} + \cdots.$$

This series is the divergent harmonic series (see Example 3 of Section 11.2). Hence, by Definition (11.32), the alternating harmonic series is not absolutely convergent.

Series that are convergent but not absolutely convergent, such as the alternating harmonic series in Example 4, are given a special name, as indicated in the next definition.

Definition **(11.33)**

A series $\sum a_n$ is **conditionally convergent** if $\sum a_n$ is convergent and $\sum |a_n|$ is divergent.

The following theorem tells us that absolute convergence implies convergence.

Theorem **(11.34)**

If a series $\sum a_n$ is absolutely convergent, then $\sum a_n$ is convergent.

PROOF If we let $b_n = a_n + |a_n|$ and use the property $-|a_n| \le a_n \le |a_n|$, then

$$0 \le a_n + |a_n| \le 2|a_n|, \quad \text{or} \quad 0 \le b_n \le 2|a_n|.$$

If $\sum a_n$ is absolutely convergent, then $\sum |a_n|$ is convergent and hence, by Theorem (11.20)(ii), $\sum 2|a_n|$ is convergent. If we apply the basic comparison test (11.26), it follows that $\sum b_n$ is convergent. By (11.20)(iii), $\sum (b_n - |a_n|)$ is convergent. Since $b_n - |a_n| = a_n$, the proof is complete. ■

EXAMPLE 5 Let $\sum a_n$ be the series

$$\frac{1}{2} + \frac{1}{2^2} - \frac{1}{2^3} - \frac{1}{2^4} + \frac{1}{2^5} + \frac{1}{2^6} - \frac{1}{2^7} - \frac{1}{2^8} + \cdots,$$

where the signs of the terms vary in pairs as indicated and where $|a_n| = 1/2^n$. Determine whether $\sum a_n$ converges or diverges.

SOLUTION The series is neither alternating nor geometric nor positive-term, so none of the earlier tests can be applied. Let us consider the series of absolute values:

$$\sum |a_n| = \frac{1}{2} + \frac{1}{2^2} + \frac{1}{2^3} + \frac{1}{2^4} + \cdots + \frac{1}{2^n} + \cdots$$

This is a geometric series with $r = \frac{1}{2}$, and since $\frac{1}{2} < 1$, it is convergent, by Theorem (11.15)(i). Thus, the given series is absolutely convergent and hence, by Theorem (11.34), it is convergent.

EXAMPLE 6 Determine whether the following series is convergent or divergent:

$$\sin 1 + \frac{\sin 2}{2^2} + \frac{\sin 3}{3^2} + \cdots + \frac{\sin n}{n^2} + \cdots$$

SOLUTION The series contains both positive and negative terms, but it is not an alternating series, because, for example, the first three terms are positive and the next three are negative. The series of absolute values is

$$\sum_{n=1}^{\infty} \left| \frac{\sin n}{n^2} \right| = \sum_{n=1}^{\infty} \frac{|\sin n|}{n^2}.$$

Since
$$\frac{|\sin n|}{n^2} < \frac{1}{n^2},$$

the series of absolute values $\sum |(\sin n)/n^2|$ is dominated by the convergent p-series $\sum (1/n^2)$ and hence is convergent. Thus, the given series is absolutely convergent and therefore is convergent, by Theorem (11.34).

We see from the preceding discussion that an arbitrary series may be classified in exactly *one* of the following ways:

(i) absolutely convergent
(ii) conditionally convergent
(iii) divergent

Of course, for positive-term series we need only determine convergence or divergence.

The following form of the ratio test may be used to investigate absolute convergence.

Ratio test for absolute convergence (11.35)

> Let $\sum a_n$ be a series of nonzero terms, and suppose
>
> $$\lim_{n\to\infty}\left|\frac{a_{n+1}}{a_n}\right| = L.$$
>
> **(i)** If $L < 1$, the series is absolutely convergent.
>
> **(ii)** If $L > 1$ or $\lim_{n\to\infty}\left|\frac{a_{n+1}}{a_n}\right| = \infty$, the series is divergent.
>
> **(iii)** If $L = 1$, apply a different test; the series may be absolutely convergent, conditionally convergent, or divergent.

The proof is similar to that of (11.28). Note that for positive-term series the two ratio tests are identical.

We can also state a root test for absolute convergence. The statement is the same as that of (11.29), except that we replace $\sqrt{a_n}$ with $\sqrt{|a_n|}$.

EXAMPLE 6 Determine whether the following series is absolutely convergent, conditionally convergent, or divergent:

$$\sum_{n=1}^{\infty} (-1)^n \frac{n^2+4}{2^n}$$

SOLUTION Using the ratio test (11.35), we obtain

$$\lim_{n\to\infty}\left|\frac{a_{n+1}}{a_n}\right| = \lim_{n\to\infty}\left|\frac{(n+1)^2+4}{2^{n+1}}\cdot\frac{2^n}{n^2+4}\right|$$

$$= \lim_{n\to\infty}\frac{1}{2}\left(\frac{n^2+2n+5}{n^2+4}\right) = \frac{1}{2}(1) = \frac{1}{2} < 1.$$

Hence, by (11.35)(i), the series is absolutely convergent.

It can be proved that if a series $\sum a_n$ is absolutely convergent and if the terms are rearranged in any manner, then the resulting series converges and has the same sum as the given series. This is not true for conditionally convergent series. If $\sum a_n$ is conditionally convergent, then by suitably rearranging terms, we can obtain either a divergent series or a series that converges and has any desired sum S.*

*See, for example, R. C. Buck, *Advanced Calculus*, Third Edition (New York: McGraw-Hill, 1978), pp. 238–239.

We now have a variety of tests that can be used to investigate a series for convergence or divergence. Considerable skill is needed to determine which test is best suited for a particular series. This skill can be obtained by working many exercises involving different types of series. The following summary may be helpful in deciding which test to apply; however, some series cannot be investigated by any of these tests. In those cases it may be necessary to use results from advanced mathematics courses.

Summary of convergence and divergence tests for series

TEST	SERIES	CONVERGENCE OR DIVERGENCE	COMMENTS
nth-term	$\sum a_n$	Diverges if $\lim_{n\to\infty} a_n \neq 0$	Inconclusive if $\lim_{n\to\infty} a_n = 0$
Geometric series	$\sum_{n=1}^{\infty} ar^{n-1}$	(i) Converges with sum $S = \dfrac{a}{1-r}$ if $\lvert r\rvert < 1$ (ii) Diverges if $\lvert r\rvert \geq 1$	Useful for comparison tests if the nth term a_n of a series is *similar* to ar^{n-1}
p-series	$\sum_{n=1}^{\infty} \dfrac{1}{n^p}$	(i) Converges if $p > 1$ (ii) Diverges if $p \leq 1$	Useful for comparison tests if the nth term a_n of a series is *similar* to $1/n^p$
Integral	$\sum_{n=1}^{\infty} a_n$ $a_n = f(n)$	(i) Converges if $\int_1^{\infty} f(x)\,dx$ converges (ii) Diverges if $\int_1^{\infty} f(x)\,dx$ diverges	The function f obtained from $a_n = f(n)$ must be continuous, positive, decreasing, and readily integrable.
Comparison	$\sum a_n, \sum b_n$ $a_n > 0, b_n > 0$	(i) If $\sum b_n$ converges and $a_n \leq b_n$ for every n, then $\sum a_n$ converges. (ii) If $\sum b_n$ diverges and $a_n \geq b_n$ for every n, then $\sum a_n$ diverges. (iii) If $\lim_{n\to\infty} (a_n/b_n) = c > 0$, then both series converge or both diverge.	The comparison series $\sum b_n$ is often a geometric series or a p-series. To find b_n in (iii), consider only the terms of a_n that have the greatest effect on the magnitude.
Ratio	$\sum a_n$	If $\lim_{n\to\infty} \left\lvert \dfrac{a_{n+1}}{a_n} \right\rvert = L$ (or ∞), the series (i) converges (absolutely) if $L < 1$ (ii) diverges if $L > 1$ (or ∞)	Inconclusive if $L = 1$ Useful if a_n involves factorials or nth powers If $a_n > 0$ for every n, the absolute value sign may be disregarded.
Root	$\sum a_n$	If $\lim_{n\to\infty} \sqrt[n]{\lvert a_n\rvert} = L$ (or ∞), the series (i) converges (absolutely) if $L < 1$ (ii) diverges if $L > 1$ (or ∞)	Inconclusive if $L = 1$ Useful if a_n involves nth powers If $a_n > 0$ for every n, the absolute value sign may be disregarded.
Alternating series	$\sum (-1)^n a_n$ $a_n > 0$	Converges if $a_k \geq a_{k+1}$ for every k and $\lim_{n\to\infty} a_n = 0$	Applicable only to an alternating series
$\sum \lvert a_n\rvert$	$\sum a_n$	If $\sum \lvert a_n\rvert$ converges, then $\sum a_n$ converges.	Useful for series that contain both positive and negative terms

EXERCISES 11.5

Exer. 1–4: Determine whether the series (a) satisfies conditions (i) and (ii) of the alternating series test (11.30) and (b) converges or diverges.

1 $\displaystyle\sum_{n=1}^{\infty} (-1)^{n-1} \frac{1}{n^2+7}$

2 $\displaystyle\sum_{n=1}^{\infty} (-1)^{n-1} n5^{-n}$

3 $\displaystyle\sum_{n=1}^{\infty} (-1)^n (1 + e^{-n})$

4 $\displaystyle\sum_{n=1}^{\infty} (-1)^n \frac{e^{2n}+1}{e^{2n}-1}$

Exer. 5–32: Determine whether the series is absolutely convergent, conditionally convergent, or divergent.

5 $\displaystyle\sum_{n=1}^{\infty} (-1)^{n-1} \frac{1}{\sqrt{2n+1}}$

6 $\displaystyle\sum_{n=1}^{\infty} (-1)^{n-1} \frac{1}{n^{2/3}}$

7 $\displaystyle\sum_{n=1}^{\infty} (-1)^{n+1} \frac{1}{\ln(n+1)}$

8 $\displaystyle\sum_{n=1}^{\infty} (-1)^{n+1} \frac{n}{n^2+4}$

9 $\sum_{n=2}^{\infty} (-1)^n \frac{n}{\ln n}$

10 $\sum_{n=1}^{\infty} (-1)^n \frac{\ln n}{n}$

11 $\sum_{n=1}^{\infty} (-1)^n \frac{5}{n^3+1}$

12 $\sum_{n=1}^{\infty} (-1)^n e^{-n}$

13 $\sum_{n=1}^{\infty} \frac{(-10)^n}{n!}$

14 $\sum_{n=1}^{\infty} \frac{n!}{(-5)^n}$

15 $\sum_{n=1}^{\infty} (-1)^n \frac{n^2+3}{(2n-5)^2}$

16 $\sum_{n=1}^{\infty} \frac{\sin\sqrt{n}}{\sqrt{n^3+4}}$

17 $\sum_{n=1}^{\infty} (-1)^{n-1} \frac{\sqrt[3]{n}}{n+1}$

18 $\sum_{n=1}^{\infty} (-1)^n \frac{(n+1)^2}{n^5+1}$

19 $\sum_{n=1}^{\infty} \frac{\cos \frac{1}{6}\pi n}{n^2}$

20 $\sum_{n=1}^{\infty} (-1)^n \frac{\ln n}{(1.5)^n}$

21 $\sum_{n=1}^{\infty} (-1)^n n \sin \frac{1}{n}$

22 $\sum_{n=1}^{\infty} (-1)^n \frac{\arctan n}{n^2}$

23 $\sum_{n=2}^{\infty} (-1)^n \frac{1}{n\sqrt{\ln n}}$

24 $\sum_{n=1}^{\infty} (-1)^n \frac{2^{1/n}}{n!}$

25 $\sum_{n=1}^{\infty} \frac{n^n}{(-5)^n}$

26 $\sum_{n=1}^{\infty} \frac{(n^2+1)^n}{(-n)^n}$

27 $\sum_{n=1}^{\infty} (-1)^n \frac{1+4^n}{1+3^n}$

28 $\sum_{n=1}^{\infty} (-1)^n \frac{n^4}{e^n}$

29 $\sum_{n=1}^{\infty} (-1)^n \frac{\cos \pi n}{n}$

30 $\sum_{n=1}^{\infty} \frac{1}{n} \sin \frac{(2n-1)\pi}{2}$

31 $\sum_{n=1}^{\infty} (-1)^n \frac{1}{(n-4)^2+5}$

32 $\sum_{n=1}^{\infty} (-1)^n \frac{\ln n}{\sqrt[3]{n}}$

c **Exer. 33–38: Approximate the sum of each series to three decimal places.**

33 $\sum_{n=0}^{\infty} (-1)^n \frac{1}{n!}$

34 $\sum_{n=0}^{\infty} (-1)^{n+1} \frac{1}{(2n)!}$

35 $\sum_{n=1}^{\infty} (-1)^{n-1} \frac{1}{n^3}$

36 $\sum_{n=1}^{\infty} (-1)^{n-1} \frac{1}{n^5}$

37 $\sum_{n=1}^{\infty} (-1)^{n-1} \frac{n+1}{5^n}$

38 $\sum_{n=1}^{\infty} (-1)^n \frac{1}{n}\left(\frac{1}{2}\right)^n$

c **Exer. 39–42: Use Theorem (11.31) to find a positive integer n such that S_n approximates the sum of the series to four decimal places.**

39 $\sum_{n=1}^{\infty} (-1)^n \frac{1}{n^2}$

40 $\sum_{n=1}^{\infty} (-1)^n \frac{1}{\sqrt{n}}$

41 $\sum_{n=1}^{\infty} (-1)^n \frac{1}{n^n}$

42 $\sum_{n=1}^{\infty} (-1)^n \frac{1}{n^3+1}$

Exer. 43–44: Show that the alternating series converges for every positive integer k.

43 $\sum_{n=1}^{\infty} (-1)^n \frac{(\ln n)^k}{n}$

44 $\sum_{n=1}^{\infty} (-1)^n \frac{1}{\sqrt[k]{n}}$

45 If $\sum a_n$ and $\sum b_n$ are both convergent series, is $\sum a_n b_n$ convergent? Explain.

46 If $\sum a_n$ and $\sum b_n$ are both divergent series, is $\sum a_n b_n$ divergent? Explain.

11.6 POWER SERIES

As stated in the chapter introduction, the most important reason for developing the theory in the previous sections is to represent functions as *power series*—that is, as series whose terms contain powers of a variable x. To illustrate, if we use the formula $S = a/(1-r)$ for the sum of a geometric series (see Theorem (11.15)(i)), we obtain

$$1 + x + x^2 + \cdots + x^n + \cdots = \frac{1}{1-x},$$

provided $|x| < 1$. If we let $f(x) = 1/(1-x)$ with $|x| < 1$, then

$$f(x) = 1 + x + x^2 + \cdots + x^n + \cdots.$$

We say that $f(x)$ is *represented* by this power series. To find a function value $f(c)$, we can let $x = c$ and find the sum of a series. For example,

$$f\left(\frac{1}{2}\right) = 1 + \frac{1}{2} + \left(\frac{1}{2}\right)^2 + \cdots + \left(\frac{1}{2}\right)^n + \cdots = \frac{1}{1-\frac{1}{2}} = 2.$$

Later we shall apply other techniques to express many different types of functions as series.

The following definition may be considered as a generalization of the notion of a polynomial to an infinite series.

Definition (11.36)

> Let x be a variable. A **power series in x** is a series of the form
>
> $$\sum_{n=0}^{\infty} a_n x^n = a_0 + a_1 x + a_2 x^2 + \cdots + a_n x^n + \cdots,$$
>
> where each a_k is a real number.

If a number c is substituted for x in the power series $\sum_{n=0}^{\infty} a_n x^n$, we obtain

$$\sum_{n=0}^{\infty} a_n c^n = a_0 + a_1 c + a_2 c^2 + \cdots + a_n c^n + \cdots.$$

This series of constant terms may then be tested for convergence or divergence. To simplify the nth term, we assume that $x^0 = 1$, even if $x = 0$. The main objective of this section is to determine all values of x for which a power series converges. Every power series in x converges if $x = 0$, since

$$a_0 + a_1(0) + a_2(0)^2 + \cdots + a_n(0)^n + \cdots = a_0.$$

To find other values of x that produce convergent series, we often employ the ratio test for absolute convergence (11.35), as illustrated in the following examples.

EXAMPLE 1 Find all values of x for which the following power series is absolutely convergent:

$$1 + \frac{1}{5}x + \frac{2}{5^2}x^2 + \cdots + \frac{n}{5^n}x^n + \cdots$$

SOLUTION If we let

$$u_n = \frac{n}{5^n}x^n = \frac{nx^n}{5^n},$$

then

$$\lim_{n\to\infty}\left|\frac{u_{n+1}}{u_n}\right| = \lim_{n\to\infty}\left|\frac{(n+1)x^{n+1}}{5^{n+1}} \cdot \frac{5^n}{nx^n}\right|$$

$$= \lim_{n\to\infty}\left|\frac{(n+1)x}{5n}\right| = \lim_{n\to\infty}\left(\frac{n+1}{5n}\right)|x| = \tfrac{1}{5}|x|.$$

By the ratio test (11.35), with $L = \frac{1}{5}|x|$, the series is absolutely convergent if the following equivalent inequalities are true:

$$\tfrac{1}{5}|x| < 1, \quad |x| < 5, \quad -5 < x < 5$$

The series diverges if $\frac{1}{5}|x| > 1$—that is, if $x > 5$ or $x < -5$.

If $\frac{1}{5}|x| = 1$, the ratio test is inconclusive, and hence the numbers 5 and -5 require special consideration. Substituting 5 for x in the power series, we obtain

$$1 + 1 + 2 + 3 + \cdots + n + \cdots,$$

which is divergent, by the nth-term test (11.17), because $\lim_{n \to \infty} a_n \neq 0$. If we let $x = -5$, we obtain

$$1 - 1 + 2 - 3 + \cdots + (-1)^n n + \cdots,$$

which is also divergent, by the nth-term test. Consequently the power series is absolutely convergent for every x in the open interval $(-5, 5)$ and diverges elsewhere.

EXAMPLE 2 Find all values of x for which the following power series is absolutely convergent:

$$1 + \frac{1}{1!}x + \frac{1}{2!}x^2 + \cdots + \frac{1}{n!}x^n + \cdots$$

SOLUTION We shall employ the same technique as was used in Example 1. If we let

$$u_n = \frac{1}{n!}x^n = \frac{x^n}{n!},$$

then

$$\lim_{n \to \infty} \left| \frac{u_{n+1}}{u_n} \right| = \lim_{n \to \infty} \left| \frac{x^{n+1}}{(n+1)!} \cdot \frac{n!}{x^n} \right|$$

$$= \lim_{n \to \infty} \left| \frac{x}{n+1} \right| = \lim_{n \to \infty} \frac{1}{n+1} |x| = 0.$$

The limit 0 is less than 1 for every value of x, and hence, from the ratio test (11.35), the power series is absolutely convergent for *every* real number x.

EXAMPLE 3 Find all values of x for which $\sum n!x^n$ is convergent.

SOLUTION Let $u_n = n!x^n$. If $x \neq 0$, then

$$\lim_{n \to \infty} \left| \frac{u_{n+1}}{u_n} \right| = \lim_{n \to \infty} \left| \frac{(n+1)!x^{n+1}}{n!x^n} \right|$$

$$= \lim_{n \to \infty} |(n+1)x| = \lim_{n \to \infty} (n+1)|x| = \infty$$

and, by the ratio test (11.35), the series diverges. Hence the power series is convergent only if $x = 0$.

Theorem (11.38) will show that the solutions of the preceding examples are typical in the sense that if a power series converges for nonzero values of x, then either it is absolutely convergent for every real number or it is absolutely convergent throughout some open interval $(-r, r)$ and diverges outside of the closed interval $[-r, r]$. The proof of this fact depends on the next theorem.

Theorem **(11.37)**

(i) If a power series $\sum a_n x^n$ converges for a nonzero number c, then it is absolutely convergent whenever $|x| < |c|$.

(ii) If a power series $\sum a_n x^n$ diverges for a nonzero number d, then it diverges whenever $|x| > |d|$.

PROOF If $\sum a_n c^n$ converges and $c \neq 0$, then, by Theorem (11.16), $\lim_{n\to\infty} a_n c^n = 0$. Employing Definition (11.3) with $\epsilon = 1$, we know that there is a positive integer N such that

$$|a_n c^n| < 1 \quad \text{whenever} \quad n \geq N.$$

Consequently $$|a_n x^n| = \left|\frac{a_n c^n x^n}{c^n}\right| = |a_n c^n| \left|\frac{x}{c}\right|^n < \left|\frac{x}{c}\right|^n,$$

provided $n \geq N$. If $|x| < |c|$, then $|x/c| < 1$ and $\sum |x/c|^n$ is a convergent geometric series. Hence, by the basic comparison test (11.26), the series obtained by deleting the first N terms of $\sum |a_n x^n|$ is convergent. It follows that the series $\sum |a_n x^n|$ is also convergent, which proves (i).

To prove (ii), suppose the series diverges for $x = d \neq 0$. If the series converges for some number c_1 with $|c_1| > |d|$, then, by (i), it converges whenever $|x| < |c_1|$. In particular, the series converges for $x = d$, contrary to our supposition. Hence the series diverges whenever $|x| > |d|$. ■

We may now prove the following.

Theorem **(11.38)**

If $\sum a_n x^n$ is a power series, then exactly one of the following is true:

(i) The series converges only if $x = 0$.

(ii) The series is absolutely convergent for every x.

(iii) There is a number $r > 0$ such that the series is absolutely convergent if x is in the open interval $(-r, r)$ and divergent if $x < -r$ or $x > r$.

PROOF If neither (i) nor (ii) is true, then there exist nonzero numbers c and d such that the series converges if $x = c$ and diverges if $x = d$. Let S denote the set of all real numbers for which the series is absolutely convergent. By Theorem (11.37), the series diverges if $|x| > |d|$, and hence every number in S is less than $|d|$. By the completeness property (11.10), S has a least upper bound r. It follows that the series is absolutely convergent if $|x| < r$ and diverges if $|x| > r$. ■

FIGURE 11.12
$\sum a_n x^n$ with radius of convergence r

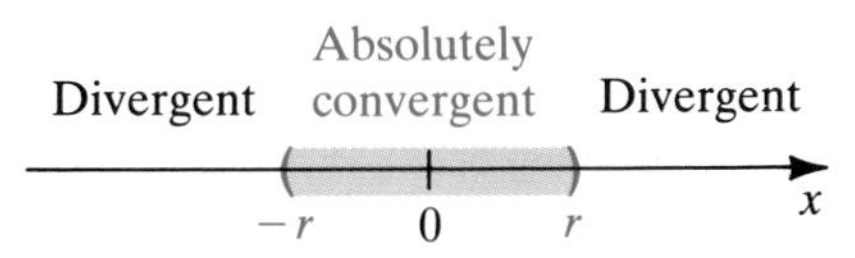

Case (iii) of Theorem (11.38) is illustrated graphically in Figure 11.12. The number r is called the **radius of convergence** of the series. Either convergence or divergence may occur at $-r$ or r, depending on the nature of the series.

FIGURE 11.13
Intervals of convergence

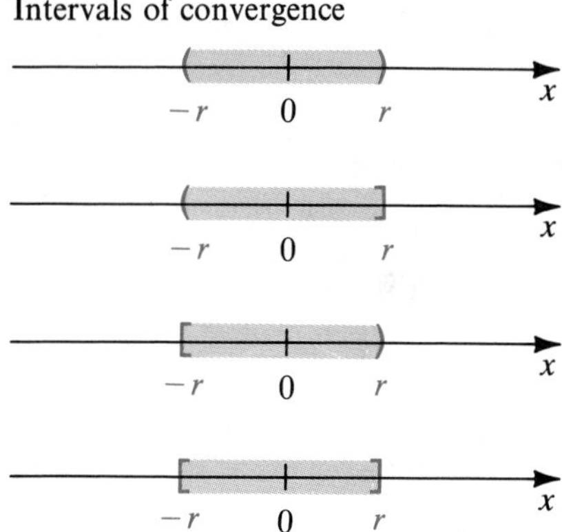

The totality of numbers for which a power series converges is called its **interval of convergence**. If the radius of convergence r is positive, then the interval of convergence is one of the following (see Figure 11.13):

$$(-r, r), \quad (-r, r], \quad [-r, r), \quad [-r, r]$$

To determine which of these intervals occurs, we must conduct separate investigations for the numbers $x = r$ and $x = -r$.

In (i) or (ii) of Theorem (11.38), the radius of convergence is denoted by 0 or ∞, respectively. In Example 1 of this section, the interval of convergence is $(-5, 5)$ and the radius of convergence is 5. In Example 2, the interval of convergence is $(-\infty, \infty)$ and we write $r = \infty$. In Example 3, $r = 0$. The next example illustrates the case of a half-open interval of convergence.

EXAMPLE 4 Find the interval of convergence of the power series $\sum_{n=1}^{\infty} \dfrac{1}{\sqrt{n}} x^n$.

SOLUTION Note that in this example the coefficient of x^0 is 0, and the summation begins with $n = 1$. We let $u_n = x^n/\sqrt{n}$ and consider

$$\lim_{n\to\infty} \left|\frac{u_{n+1}}{u_n}\right| = \lim_{n\to\infty} \left|\frac{x^{n+1}}{\sqrt{n+1}} \cdot \frac{\sqrt{n}}{x^n}\right| = \lim_{n\to\infty} \left|\frac{\sqrt{n}}{\sqrt{n+1}}\, x\right|$$

$$= \lim_{n\to\infty} \sqrt{\frac{n}{n+1}}\,|x| = (1)|x| = |x|.$$

It follows from the ratio test (11.35) that the power series is absolutely convergent if $|x| < 1$—that is, if x is in the open interval $(-1, 1)$. The series diverges if $x > 1$ or $x < -1$. The numbers 1 and -1 must be investigated separately by substitution in the power series.

If we substitute $x = 1$, we obtain

$$\sum_{n=1}^{\infty} \frac{1}{\sqrt{n}}(1)^n = 1 + \frac{1}{\sqrt{2}} + \frac{1}{\sqrt{3}} + \cdots + \frac{1}{\sqrt{n}} + \cdots,$$

which is a divergent p-series, with $p = \frac{1}{2}$. If we substitute $x = -1$, we obtain

$$\sum_{n=1}^{\infty} \frac{1}{\sqrt{n}}(-1)^n = -1 + \frac{1}{\sqrt{2}} - \frac{1}{\sqrt{3}} + \cdots + \frac{(-1)^n}{\sqrt{n}} + \cdots,$$

which converges, by the alternating series test. Hence the power series converges if $-1 \le x < 1$. The interval of convergence $[-1, 1)$ is sketched in Figure 11.14.

FIGURE 11.14
Interval of convergence of $\sum \dfrac{1}{\sqrt{n}} x^n$

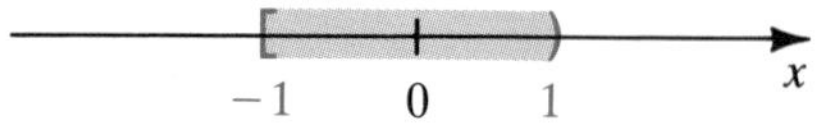

We shall also consider the following more general types of power series.

Definition **(11.39)**

Let c be a real number and x a variable. A **power series in $x - c$** is a series of the form

$$\sum_{n=0}^{\infty} a_n(x-c)^n = a_0 + a_1(x-c) + a_2(x-c)^2 + \cdots + a_n(x-c)^n + \cdots,$$

where each a_k is a real number.

To simplify the nth term in (11.39), we assume that $(x-c)^0 = 1$ even if $x = c$. As in the proof of Theorem (11.38), but with x replaced by $x - c$, exactly one of the following cases is true:

(i) The series converges only if $x - c = 0$—that is, if $x = c$.

(ii) The series is absolutely convergent for every x.

(iii) There is a number $r > 0$ such that the series is absolutely convergent if x is in the open interval $(c - r, c + r)$ and divergent if $x < c - r$ or $x > c + r$.

FIGURE 11.15
$\sum a_n(x-c)^n$ with radius of convergence r

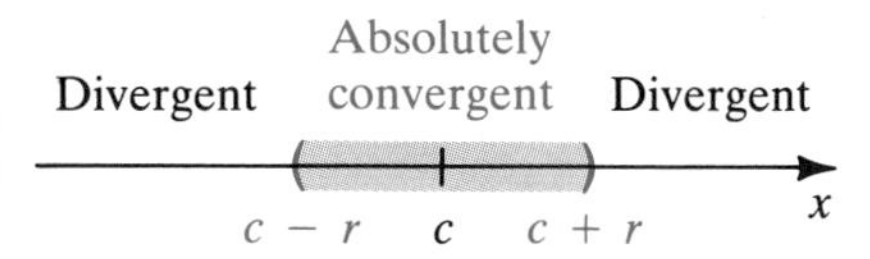

Case (iii) is illustrated in Figure 11.15. The endpoints $c - r$ and $c + r$ of the interval must be investigated separately. As before, the totality of numbers for which the series converges is called the *interval of convergence*, and r is the *radius of convergence*.

EXAMPLE 5 Find the interval of convergence of the series

$$1 - \frac{1}{2}(x-3) + \frac{1}{3}(x-3)^2 + \cdots + (-1)^n \frac{1}{n+1}(x-3)^n + \cdots.$$

SOLUTION If we let

$$u_n = (-1)^n \frac{(x-3)^n}{n+1},$$

then

$$\begin{aligned}\lim_{n\to\infty}\left|\frac{u_{n+1}}{u_n}\right| &= \lim_{n\to\infty}\left|\frac{(x-3)^{n+1}}{n+2}\cdot\frac{n+1}{(x-3)^n}\right| \\ &= \lim_{n\to\infty}\left|\frac{n+1}{n+2}(x-3)\right| \\ &= \lim_{n\to\infty}\left(\frac{n+1}{n+2}\right)|x-3| \\ &= (1)|x-3| = |x-3|.\end{aligned}$$

By the ratio test, the series is absolutely convergent if $|x-3| < 1$—that is, if

$$-1 < x - 3 < 1, \quad \text{or} \quad 2 < x < 4.$$

Thus, the series is absolutely convergent for every x in the open interval $(2, 4)$. The series diverges if $x < 2$ or $x > 4$. The numbers 2 and 4 each require separate investigation.

If we substitute $x = 4$ in the series, we obtain

$$1 - \frac{1}{2} + \frac{1}{3} - \cdots + (-1)^n \frac{1}{n+1} + \cdots,$$

which converges, by the alternating series test. Substituting $x = 2$ gives us

$$1 + \frac{1}{2} + \frac{1}{3} + \cdots + \frac{1}{n+1} + \cdots,$$

which is the divergent harmonic series. Hence the interval of convergence is $(2, 4]$, as illustrated in Figure 11.16.

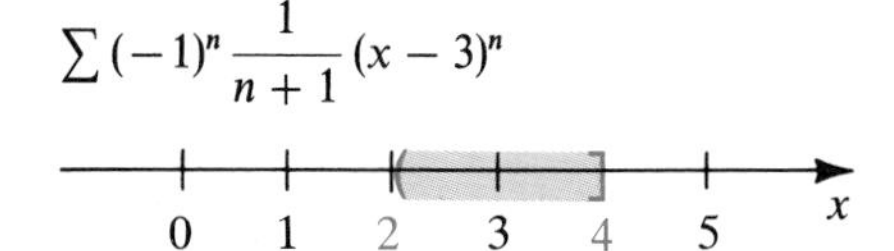

FIGURE 11.16
Interval of convergence of
$\sum (-1)^n \frac{1}{n+1}(x-3)^n$

EXERCISES 11.6

Exer. 1–30: Find the interval of convergence of the power series.

1 $\sum_{n=0}^{\infty} \frac{1}{n+4} x^n$

2 $\sum_{n=0}^{\infty} \frac{1}{n^2+4} x^n$

3 $\sum_{n=0}^{\infty} \frac{n^2}{2^n} x^n$

4 $\sum_{n=1}^{\infty} \frac{(-3)^n}{n} x^{n+1}$

5 $\sum_{n=1}^{\infty} (-1)^{n-1} \frac{1}{\sqrt{n}} x^n$

6 $\sum_{n=1}^{\infty} \frac{1}{\ln(n+1)} x^n$

7 $\sum_{n=2}^{\infty} \frac{n}{n^2+1} x^n$

8 $\sum_{n=1}^{\infty} \frac{1}{4^n \sqrt{n}} x^n$

9 $\sum_{n=2}^{\infty} \frac{\ln n}{n^3} x^n$

10 $\sum_{n=0}^{\infty} \frac{10^{n+1}}{3^{2n}} x^n$

11 $\sum_{n=0}^{\infty} \frac{n+1}{10^n} (x-4)^n$

12 $\sum_{n=1}^{\infty} \frac{1}{n(n+1)} (x-2)^n$

13 $\sum_{n=0}^{\infty} \frac{n!}{100^n} x^n$

14 $\sum_{n=0}^{\infty} \frac{(3n)!}{(2n)!} x^n$

15 $\sum_{n=0}^{\infty} \frac{1}{(-4)^n} x^{2n+1}$

16 $\sum_{n=1}^{\infty} (-1)^{n-1} \frac{1}{\sqrt[3]{n}3^n} x^n$

17 $\sum_{n=0}^{\infty} \frac{2^n}{(2n)!} x^{2n}$

18 $\sum_{n=0}^{\infty} \frac{10^n}{n!} x^n$

19 $\sum_{n=0}^{\infty} \frac{3^{2n}}{n+1} (x-2)^n$

20 $\sum_{n=1}^{\infty} \frac{1}{n5^n} (x-5)^n$

21 $\sum_{n=0}^{\infty} \frac{n^2}{2^{3n}} (x+4)^n$

22 $\sum_{n=0}^{\infty} \frac{1}{2n+1} (x+3)^n$

23 $\sum_{n=1}^{\infty} (-1)^n \frac{n^n}{n+1} (x-3)^n$

24 $\sum_{n=1}^{\infty} (-1)^n \frac{n!}{n^3} (x+2)^n$

25 $\sum_{n=1}^{\infty} \frac{\ln n}{e^n} (x-e)^n$

26 $\sum_{n=0}^{\infty} \frac{n}{3^{2n-1}} (x-1)^{2n}$

27 $\sum_{n=1}^{\infty} (-1)^n \frac{1}{n6^n} (2x-1)^n$

28 $\sum_{n=0}^{\infty} \frac{1}{\sqrt{3n+4}} (3x+4)^n$

29 $\sum_{n=1}^{\infty} (-1)^n \frac{3^n}{n!} (x-4)^n$

30 $\sum_{n=1}^{\infty} (-1)^n \frac{e^{n+1}}{n^n} (x-1)^n$

Exer. 31–34: Find the radius of convergence of the power series.

31 $\sum_{n=1}^{\infty} (-1)^n \frac{1 \cdot 3 \cdot 5 \cdot \cdots \cdot (2n-1)}{3 \cdot 6 \cdot 9 \cdot \cdots \cdot (3n)} x^n$

32 $\sum_{n=1}^{\infty} \frac{2 \cdot 4 \cdot 6 \cdot \cdots \cdot (2n)}{4 \cdot 7 \cdot 10 \cdot \cdots \cdot (3n+1)} x^n$

33 $\sum_{n=1}^{\infty} \frac{n^n}{n!} x^n$

34 $\sum_{n=0}^{\infty} \frac{(n+1)!}{10^n} (x-5)^n$

Exer. 35–36: Find the radius of convergence of the power series for positive integers c and d.

35 $\sum_{n=0}^{\infty} \frac{(n+c)!}{n!(n+d)!} x^n$

36 $\sum_{n=0}^{\infty} \frac{(cn)!}{(n!)^c} x^n$

37 *Bessel functions* are useful in the analysis of problems that involve oscillations. If α is a positive integer, the Bessel function $J_\alpha(x)$ *of the first kind of order* α is defined by the power series

$$J_\alpha(x) = \sum_{n=0}^{\infty} \frac{(-1)^n}{n!(n+\alpha)!} \left(\frac{x}{2}\right)^{2n+\alpha}.$$

Show that this power series is convergent for every real number.

38 Refer to Exercise 37. The sixth-degree polynomial

$$1 - \frac{x^2}{4} + \frac{x^4}{64} - \frac{x^6}{2304}$$

is sometimes used to approximate the Bessel function $J_0(x)$ of the first kind of order zero for $0 \le x \le 1$. Show that the error E involved in this approximation is less than 0.00001.

c **Exer. 39–40: Refer to Exercise 37. For the given α, find the first four terms of the series for $J_\alpha(x)$ and graph J_α on the given interval.**

39 $\alpha = 0$; $[0, 2]$

40 $\alpha = 1$; $[0, 4]$

41 If $\lim_{n\to\infty} |a_{n+1}/a_n| = k$ and $k \neq 0$, prove that the radius of convergence of $\sum a_n x^n$ is $1/k$.

42 If $\lim_{n\to\infty} \sqrt[n]{|a_n|} = k$ and $k \neq 0$, prove that the radius of convergence of $\sum a_n x^n$ is $1/k$.

43 If $\sum a_n x^n$ has radius of convergence r, prove that $\sum a_n x^{2n}$ has radius of convergence $\sqrt{r}$.

44 If $\sum a_n$ is absolutely convergent, prove that $\sum a_n x^n$ is absolutely convergent for every x in the interval $[-1, 1]$.

45 If the interval of convergence of $\sum a_n x^n$ is $(-r, r]$, prove that the series is conditionally convergent at r.

46 If $\sum a_n x^n$ is absolutely convergent at one endpoint of its interval of convergence, prove that it is also absolutely convergent at the other endpoint.

11.7 POWER SERIES REPRESENTATIONS OF FUNCTIONS

A power series $\sum a_n x^n$ determines a function f whose domain is the interval of convergence of the series. Specifically, for each x in this interval we let $f(x)$ equal the sum of the series; that is,

$$f(x) = a_0 + a_1 x + a_2 x^2 + \cdots + a_n x^n + \cdots.$$

If a function f is defined in this way, we say that $\sum a_n x^n$ is a **power series representation for $f(x)$** (or *of* $f(x)$). We also use the phrase **f is represented by the power series.**

Numerical computations using power series provide the basis for the design of calculators and the construction of mathematical tables. In addition to this use, power series representations for functions have far-reaching consequences in advanced mathematics and applications. The proof of Theorem (11.41) will show that e^x may be represented as follows:

$$e^x = 1 + x + \frac{x^2}{2!} + \frac{x^3}{3!} + \cdots + \frac{x^n}{n!} + \cdots$$

This will allow us to consider e^x as a *series* instead of as the inverse of the natural logarithmic function. As we shall see, algebraic manipulations, differentiation, and integration can be performed by using the series for e^x, instead of previous methods. The same will be true for trigonometric, inverse trigonometric, logarithmic, and hyperbolic functions. In the next example we consider a power series representation for a simple algebraic function.

EXAMPLE 1 Find a function f that is represented by the power series

$$1 - x + x^2 - x^3 + \cdots + (-1)^n x^n + \cdots.$$

SOLUTION If $|x| < 1$, then, by Theorem (11.15)(i), the given geometric series converges and has the sum

$$S = \frac{a}{1-r} = \frac{1}{1-(-x)} = \frac{1}{1+x}.$$

Hence we may write

$$\frac{1}{1+x} = 1 - x + x^2 - x^3 + \cdots + (-1)^n x^n + \cdots.$$

This is a power series representation for $f(x) = 1/(1 + x)$ on the interval $(-1, 1)$.

If a function f is represented by a power series in x, then

$$f(x) = \sum_{n=0}^{\infty} a_n x^n$$

for every x in the interval of convergence of the series. Since a polynomial in x is a *finite* sum of terms of the form $a_n x^n$, it may not be surprising that f has properties similar to those for polynomial functions. In particular, in the next theorem (stated without proof) we see that f has a derivative f' whose power series representation can be found by differentiating each term of the series for f. Similarly, definite integrals of $f(x)$ may be obtained by integrating each term of the series $\sum a_n x^n$. In the statement of the theorem, note that for the nth term $a_n x^n$ of the series, we have

$$D_x(a_n x^n) = n a_n x^{n-1} \quad \text{and} \quad \int_0^x a_n t^n \, dt = \left[a_n \frac{t^{n+1}}{n+1} \right]_0^x = a_n \frac{x^{n+1}}{n+1}.$$

Theorem (11.40)

Suppose a power series $\sum a_n x^n$ has a radius of convergence $r > 0$, and let f be defined by

$$f(x) = \sum_{n=0}^{\infty} a_n x^n = a_0 + a_1 x + a_2 x^2 + a_3 x^3 + \cdots + a_n x^n + \cdots$$

for every x in the interval of convergence. If $-r < x < r$, then

(i) $f'(x) = a_1 + 2a_2 x + 3a_3 x^2 + \cdots + n a_n x^{n-1} + \cdots = \sum_{n=1}^{\infty} n a_n x^{n-1}$

(ii) $\int_0^x f(t)\, dt = a_0 x + a_1 \frac{x^2}{2} + a_2 \frac{x^3}{3} + \cdots + a_n \frac{x^{n+1}}{n+1} + \cdots$

$$= \sum_{n=0}^{\infty} \frac{a_n}{n+1} x^{n+1}$$

The series obtained by differentiation in (i) or integration in (ii) of Theorem (11.40) has the same radius of convergence as $\sum a_n x^n$. However, convergence at the endpoints $x = r$ and $x = -r$ of the interval may change. As usual, these numbers require separate investigation.

As a corollary of Theorem (11.40)(i), *a function that is represented by a power series in an interval* $(-r, r)$ *is continuous throughout* $(-r, r)$ (see Theorem (3.11)). Similar results are true for functions represented by power series of the form $\sum a_n(x - c)^n$.

EXAMPLE 2 Use a power series representation for $1/(1 + x)$ to obtain a power series representation for

$$\frac{1}{(1+x)^2} \quad \text{if} \quad |x| < 1.$$

SOLUTION From Example 1,

$$\frac{1}{1+x} = 1 - x + x^2 - x^3 + \cdots + (-1)^n x^n + \cdots \quad \text{if} \quad |x| < 1.$$

If we differentiate each term of this series, then, by Theorem (11.40)(i),

$$-\frac{1}{(1+x)^2} = -1 + 2x - 3x^2 - \cdots + (-1)^n n x^{n-1} + \cdots.$$

By Theorem (11.20)(ii), we may multiply both sides by -1, obtaining

$$\frac{1}{(1+x)^2} = 1 - 2x + 3x^2 + \cdots + (-1)^{n+1} n x^{n-1} + \cdots$$

if $|x| < 1$.

EXAMPLE 3 Find a power series representation for $\ln(1+x)$ if $|x| < 1$.

SOLUTION If $|x| < 1$, then

$$\begin{aligned}\ln(1+x) &= \int_0^x \frac{1}{1+t}\,dt \\ &= \int_0^x [1 - t + t^2 - \cdots + (-1)^n t^n + \cdots]\,dt,\end{aligned}$$

where the last equality follows from Example 1. By Theorem (11.40)(ii), we may integrate each term of the series as follows:

$$\begin{aligned}\ln(1+x) &= \int_0^x 1\,dt - \int_0^x t\,dt + \int_0^x t^2\,dt - \cdots + (-1)^n \int_0^x t^n\,dt + \cdots \\ &= \Big[t\Big]_0^x - \left[\frac{t^2}{2}\right]_0^x + \left[\frac{t^3}{3}\right]_0^x - \cdots + (-1)^n\left[\frac{t^{n+1}}{n+1}\right]_0^x + \cdots\end{aligned}$$

Hence

$$\ln(1+x) = x - \frac{x^2}{2} + \frac{x^3}{3} - \cdots + (-1)^n \frac{x^{n+1}}{n+1} + \cdots$$

if $|x| < 1$.

EXAMPLE 4 Use the results of Example 3 to calculate $\ln(1.1)$ to five decimal places.

SOLUTION In Example 3 we found a series representation for $\ln(1+x)$ if $|x| < 1$. Substituting 0.1 for x in that series gives us the alternating series

$$\begin{aligned}\ln(1.1) &= 0.1 - \frac{(0.1)^2}{2} + \frac{(0.1)^3}{3} - \frac{(0.1)^4}{4} + \frac{(0.1)^5}{5} - \cdots \\ &\approx 0.1 - 0.005 + 0.000333 - 0.000025 + 0.000002 - \cdots.\end{aligned}$$

If we sum the first four terms on the right and round off to five decimal places, we obtain

$$\ln(1.1) \approx 0.09531.$$

By Theorem (11.31), the error E is less than or equal to the absolute value 0.000002 of the fifth term of the series, and therefore the number 0.09531 is accurate to five decimal places.

EXAMPLE 5 Find a power series representation for $\arctan x$.

SOLUTION We first observe that

$$\arctan x = \int_0^x \frac{1}{1+t^2}\,dt.$$

Next we note that if $|t| < 1$, then, by Theorem (11.15)(i) with $a = 1$ and $r = -t^2$,

$$\frac{1}{1+t^2} = 1 - t^2 + t^4 - \cdots + (-1)^n t^{2n} + \cdots.$$

By Theorem (11.40)(ii), we may integrate each term of the series from 0 to x to obtain

$$\arctan x = x - \frac{x^3}{3} + \frac{x^5}{5} - \cdots + (-1)^n \frac{x^{2n+1}}{2n+1} + \cdots,$$

provided $|x| < 1$. It can be proved that this series representation is also valid if $|x| = 1$.

In the next theorem we find a power series representation for e^x.

Theorem (11.41)

If x is any real number,

$$e^x = 1 + x + \frac{x^2}{2!} + \frac{x^3}{3!} + \cdots + \frac{x^n}{n!} + \cdots.$$

PROOF We considered the indicated power series in Example 2 of the preceding section and found that it is absolutely convergent for every real number x. If we let f denote the function represented by the series, then

$$f(x) = \sum_{n=0}^{\infty} \frac{x^n}{n!}.$$

Applying Theorem (11.40)(i) gives us

$$f'(x) = \sum_{n=1}^{\infty} \frac{nx^{n-1}}{n!} = \sum_{n=1}^{\infty} \frac{x^{n-1}}{(n-1)!}$$

$$= 1 + x + \frac{x^2}{2!} + \frac{x^3}{3!} + \cdots + \frac{x^n}{n!} + \cdots;$$

that is,

$$f'(x) = f(x) \text{ for every } x.$$

If, in Theorem (7.33), we let $y = f(t)$, $t = x$, and $c = 1$, we obtain

$$f(x) = f(0)e^x.$$

However,

$$f(0) = 1 + 0 + \frac{0^2}{2!} + \cdots + \frac{0^n}{n!} + \cdots = 1$$

and hence $$f(x) = e^x,$$

which is what we wished to prove. ■

Note that Theorem (11.41) allows us to express the number e as the sum of a convergent positive-term series, namely

$$e = 1 + 1 + \frac{1}{2!} + \frac{1}{3!} + \cdots + \frac{1}{n!} + \cdots.$$

We can use a power series representation for a function to obtain representations for other related functions by making algebraic substitutions. Thus, by Theorem (11.41), if x is any real number,

$$e^x = 1 + x + \frac{x^2}{2!} + \frac{x^3}{3!} + \cdots + \frac{x^n}{n!} + \cdots.$$

To obtain a power series representation for e^{-x}, we need only substitute $-x$ for x:

$$e^{-x} = 1 + (-x) + \frac{(-x)^2}{2!} + \frac{(-x)^3}{3!} + \cdots + \frac{(-x)^n}{n!} + \cdots,$$

or $$e^{-x} = 1 - x + \frac{x^2}{2!} - \frac{x^3}{3!} + \cdots + (-1)^n \frac{x^n}{n!} + \cdots$$

By Theorem (11.20)(i), we may add corresponding terms of the series for e^x and e^{-x}, obtaining

$$e^x + e^{-x} = 2 + 2 \cdot \frac{x^2}{2!} + 2 \cdot \frac{x^4}{4!} + \cdots + 2 \cdot \frac{x^{2n}}{(2n)!} + \cdots.$$

(Note that odd powers of x cancel.) We can now find a power series for $\cosh x = \frac{1}{2}(e^x + e^{-x})$ by multiplying each term of the last series by $\frac{1}{2}$ (see Theorem (11.20)(ii)). Thus,

$$\cosh x = 1 + \frac{x^2}{2!} + \frac{x^4}{4!} + \cdots + \frac{x^{2n}}{(2n)!} + \cdots.$$

We could find a power series representation for $\sinh x$ by using $\frac{1}{2}(e^x - e^{-x})$ or by differentiating each term of the series for $\cosh x$. It is left as an exercise to show that

$$\sinh x = x + \frac{x^3}{3!} + \frac{x^5}{5!} + \cdots + \frac{x^{2n+1}}{(2n+1)!} + \cdots.$$

EXAMPLE 6 Find a power series representation for xe^{-2x}.

SOLUTION First we substitute $-2x$ for x in Theorem (11.41):

$$e^{-2x} = 1 + (-2x) + \frac{(-2x)^2}{2!} + \frac{(-2x)^3}{3!} + \cdots + \frac{(-2x)^n}{n!} + \cdots,$$

or $$e^{-2x} = 1 - 2x + (2^2)\frac{x^2}{2!} - (2^3)\frac{x^3}{3!} + \cdots + (-2)^n \frac{x^n}{n!} + \cdots$$

Multiplying both sides by x gives us

$$xe^{-2x} = x - 2x^2 + (2^2)\frac{x^3}{2!} - (2^3)\frac{x^4}{3!} + \cdots + (-2)^n \frac{x^{n+1}}{n!} + \cdots,$$

which may be written as

$$xe^{-2x} = \sum_{n=0}^{\infty} (-2)^n \frac{x^{n+1}}{n!}.$$

EXAMPLE 7 Approximate $\int_0^{0.1} e^{-x^2}\,dx$.

SOLUTION We cannot use the fundamental theorem of calculus to evaluate the integral, because we do not know of an antiderivative for e^{-x^2}. Although we could use the trapezoidal rule or Simpson's rule, the following method is simpler and, in addition, produces a high degree of accuracy using a sum of only two terms. Letting $x = -t^2$ in Theorem (11.41), we obtain

$$e^{-t^2} = 1 - t^2 + \frac{t^4}{2!} - \cdots + \frac{(-1)^n t^{2n}}{n!} + \cdots$$

for every t. Applying Theorem (11.40)(ii) yields

$$\begin{aligned}\int_0^{0.1} e^{-x^2}\,dx &= \int_0^{0.1} e^{-t^2}\,dt\\ &= \Big[t\Big]_0^{0.1} - \left[\frac{t^3}{3}\right]_0^{0.1} + \left[\frac{t^5}{10}\right]_0^{0.1} - \cdots\\ &= 0.1 - \frac{(0.1)^3}{3} + \frac{(0.1)^5}{10} - \cdots.\end{aligned}$$

If we use the first two terms to approximate the sum of this convergent alternating series, then, by Theorem (11.31), the error is less than the third term $(0.1)^5/10 = 0.000001$. Hence

$$\int_0^{0.1} e^{-x^2}\,dx \approx 0.1 - \frac{0.001}{3} \approx 0.09967,$$

which is accurate to five decimal places.

The method used in Example 7 is accurate because the numbers in the interval $[0, 0.1]$ are close to 0. The method would be much less accurate (for the same number of terms of the series) if, for example, the limits of integration were 3 and 3.1.

Thus far, the methods we have used to obtain power series representations of functions are *indirect* in the sense that we started with known series and then differentiated or integrated. In the next section we shall discuss a *direct* method that can be used to find power series representations for a large variety of functions.

EXERCISES 11.7

Exer. 1–4: (a) Find a power series representation for $f(x)$. (b) Use Theorem (11.40) to find power series representations for $f'(x)$ and $\int_0^x f(t)\,dt$.

1 $f(x) = \dfrac{1}{1-3x}, \quad |x| < \frac{1}{3}$

2 $f(x) = \dfrac{1}{1+5x}, \quad |x| < \frac{1}{5}$

3 $f(x) = \dfrac{1}{2+7x}, \quad |x| < \frac{2}{7}$

4 $f(x) = \dfrac{1}{3-2x}, \quad |x| < \frac{3}{2}$

Exer. 5–10: Find a power series in x that has the given sum, and specify the radius of convergence. (*Hint:* Use (11.15), (11.40), or long division, as necessary.)

5 $\dfrac{x^2}{1-x^2}$ 6 $\dfrac{x}{1-x^4}$ 7 $\dfrac{x}{2-3x}$

8 $\dfrac{x^3}{4-x^3}$ 9 $\dfrac{x^2+1}{x-1}$ 10 $\dfrac{x^2-3}{x-2}$

11 (a) Prove that

$$\ln(1-x) = -\sum_{n=1}^{\infty} \frac{x^n}{n} \quad \text{if} \quad |x| < 1.$$

(b) Use the series in (a) to approximate ln (1.2) to three decimal places, and compare the approximation with that obtained using a calculator.

12 Use the first three terms of the series in Exercise 11(a) to approximate ln (0.9), and compare the approximation with that obtained using a calculator.

13 Use Example 5 to prove that

$$\frac{\pi}{6} = \frac{1}{\sqrt{3}} \sum_{n=0}^{\infty} (-1)^n \frac{1}{3^n(2n+1)}.$$

14 (a) Use the first five terms of the series in Example 5 to approximate $\pi/4$.

(b) Estimate the error in the approximation obtained in (a).

Exer. 15–26: Use a power series representation obtained in this section to find a power series representation for $f(x)$.

15 $f(x) = xe^{3x}$ 16 $f(x) = x^2e^{(x^2)}$

17 $f(x) = x^3e^{-x}$ 18 $f(x) = xe^{-3x}$

19 $f(x) = x^2 \ln(1+x^2), \quad |x| < 1$

20 $f(x) = x \ln(1-x), \quad |x| < 1$

21 $f(x) = \arctan \sqrt{x}, \quad |x| < 1$

22 $f(x) = x^4 \arctan(x^4), \quad |x| < 1$

23 $f(x) = \sinh(-5x)$ 24 $f(x) = \sinh(x^2)$

25 $f(x) = x^2 \cosh(x^3)$ 26 $f(x) = \cosh(-2x)$

c **Exer. 27–32: Use an infinite series to approximate the integral to four decimal places.**

27 $\displaystyle\int_0^{1/3} \frac{1}{1+x^6}\,dx$ 28 $\displaystyle\int_0^{1/2} \arctan x^2\,dx$

29 $\displaystyle\int_{0.1}^{0.2} \frac{\arctan x}{x}\,dx$ 30 $\displaystyle\int_0^{0.2} \frac{x^3}{1+x^5}\,dx$

31 $\displaystyle\int_0^1 e^{-x^2/10}\,dx$ 32 $\displaystyle\int_0^{0.5} e^{-x^3}\,dx$

33 Use the power series representation for $(1-x^2)^{-1}$ to find a power series representation for $2x(1-x^2)^{-2}$.

34 Use the method of Example 3 to find a power series representation for ln (3 + 2x).

35 Refer to Exercise 37 of Section 11.6. Use Theorem (11.40) to prove the following.

(a) If $J_0(x)$ and $J_1(x)$ are Bessel functions of the first kind of orders 0 and 1, respectively, then

$$D_x\, J_0(x) = -J_1(x).$$

(b) If $J_2(x)$ and $J_3(x)$ are Bessel functions of the first kind of orders 2 and 3, respectively, then

$$\int x^3 J_2(x)\,dx = x^3 J_3(x) + C.$$

36 Light is absorbed by rods and cones in the retina of the eye. The number of photons absorbed by a photoreceptor during a given flash of light is governed by the *Poisson distribution.* More precisely, the probability p_n that a photoreceptor absorbs exactly n photons is given by the formula $p_n = e^{-\lambda}\lambda^n/n!$ for some $\lambda > 0$.

(a) Show that $\sum_{n=0}^{\infty} p_n = 1$.

(b) Sight usually occurs when two or more photons are absorbed by a photoreceptor. Show that the probability that this will occur is $1 - e^{-\lambda}(\lambda + 1)$.

Exer. 37–38: Find a power series representation for $f(x)$. (If the integrand is denoted by $g(t)$, assume that the value of $g(0)$ is $\lim_{t\to 0} g(t)$.)

37 $f(x) = \displaystyle\int_0^x \frac{\ln(1-t)}{t}\,dt$ 38 $f(x) = \displaystyle\int_0^x \frac{e^t - 1}{t}\,dt$

11.8 MACLAURIN AND TAYLOR SERIES

In the preceding section we considered power series representations for several special functions, including those where $f(x)$ has the form

$$\frac{1}{1+x}, \quad \ln(1+x), \quad \arctan x, \quad e^x, \quad \text{or} \quad \cosh x,$$

provided x is suitably restricted. We now wish to consider the following two general questions:

Question 1: If a function f has a power series representation

$$f(x) = \sum_{n=0}^{\infty} a_n x^n \quad \text{or} \quad f(x) = \sum_{n=0}^{\infty} a_n (x-c)^n,$$

what is the form of a_n?

Question 2: What conditions are sufficient for a function f to have a power series representation?

Let us begin with question 1. Suppose

$$f(x) = \sum_{n=0}^{\infty} a_n x^n = a_0 + a_1 x + a_2 x^2 + a_3 x^3 + a_4 x^4 + \cdots$$

and the radius of convergence of the series is $r > 0$. By Theorem (11.40)(i), a power series representation for $f'(x)$ may be obtained by differentiating each term of the series for $f(x)$. We may then find a series for $f''(x)$ by differentiating the terms of the series for $f'(x)$. Series for $f'''(x)$, $f^{(4)}(x)$, and so on, can be found in similar fashion. Thus,

$$f'(x) = a_1 + 2a_2 x + 3a_3 x^2 + 4a_4 x^3 + \cdots = \sum_{n=1}^{\infty} n a_n x^{n-1}$$

$$f''(x) = 2a_2 + (3 \cdot 2)a_3 x + (4 \cdot 3)a_4 x^2 + \cdots = \sum_{n=2}^{\infty} n(n-1)a_n x^{n-2}$$

$$f'''(x) = (3 \cdot 2)a_3 + (4 \cdot 3 \cdot 2)a_4 x + \cdots = \sum_{n=3}^{\infty} n(n-1)(n-2)a_n x^{n-3},$$

and for every positive integer k,

$$f^{(k)}(x) = \sum_{n=k}^{\infty} n(n-1)(n-2)\cdots(n-k+1)a_n x^{n-k}.$$

Each series obtained by differentiation has the same radius of convergence r as the series for $f(x)$. Substituting 0 for x in each of these series representations, we obtain

$$f(0) = a_0, \quad f'(0) = a_1, \quad f''(0) = 2a_2, \quad f'''(0) = (3 \cdot 2)a_3,$$

and for every positive integer k,

$$f^{(k)}(0) = k(k-1)(k-2)\cdots(1)a_k.$$

If we let $k = n$, then

$$f^{(n)}(0) = n!\, a_n.$$

Solving the previous equations for $a_0, a_1, a_2, \ldots$, we see that

$$a_0 = f(0), \quad a_1 = f'(0), \quad a_2 = \frac{f''(0)}{2}, \quad a_3 = \frac{f'''(0)}{3 \cdot 2},$$

and, in general,

$$a_n = \frac{f^{(n)}(0)}{n!}$$

We have proved that the power series for $f(x)$ has the form stated in the next theorem. It is called a *Maclaurin series for* $f(x)$—named after the Scottish mathematician Colin Maclaurin (1698–1746).

Maclaurin series for $f(x)$ **(11.42)**

If a function f has a power series representation

$$f(x) = \sum_{n=0}^{\infty} a_n x^n$$

with radius of convergence $r > 0$, then $f^{(k)}(0)$ exists for every positive integer k and $a_n = f^{(n)}(0)/n!$. Thus,

$$f(x) = f(0) + f'(0)x + \frac{f''(0)}{2!}x^2 + \cdots + \frac{f^{(n)}(0)}{n!}x^n + \cdots.$$

Employing the type of proof used for (11.42) gives us the next theorem. If $c \neq 0$, we call the series a *Taylor series for* $f(x)$ *at* c—named after the English mathematician Brook Taylor (1685–1731).

Taylor series for $f(x)$ **(11.43)**

If a function f has a power series representation

$$f(x) = \sum_{n=0}^{\infty} a_n (x - c)^n$$

with radius of convergence $r > 0$, then $f^{(k)}(c)$ exists for every positive integer k and $a_n = f^{(n)}(c)/n!$. Thus,

$$f(x) = f(c) + f'(c)(x - c) + \frac{f''(c)}{2!}(x - c)^2 + \cdots$$
$$+ \frac{f^{(n)}(c)}{n!}(x - c)^n + \cdots.$$

Note that the special Taylor series with $c = 0$ is the Maclaurin series (11.42). If we use the convention $f^{(0)}(c) = f(c)$, then the Maclaurin and Taylor series for f may be written in the following summation forms:

$$f(x) = \sum_{n=0}^{\infty} \frac{f^{(n)}(0)}{n!} x^n \quad \text{and} \quad f(x) = \sum_{n=0}^{\infty} \frac{f^{(n)}(c)}{n!} (x - c)^n$$

EXAMPLE 1 By Theorem (11.41), e^x has the following power series representation:

$$e^x = 1 + x + \frac{1}{2!}x^2 + \frac{1}{3!}x^3 + \cdots + \frac{1}{n!}x^n + \cdots$$

Verify that this is a Maclaurin series.

SOLUTION If $f(x) = e^x$, then the nth derivative of f is $f^{(n)}(x) = e^x$ and

$$f^{(n)}(0) = e^0 = 1 \quad \text{for } n = 0, 1, 2, \ldots.$$

Hence the Maclaurin series (11.42) is

$$\sum_{n=0}^{\infty} \frac{f^{(n)}(0)}{n!} x = \sum_{n=0}^{\infty} \frac{1}{n!} x^n = 1 + x + \frac{1}{2!}x^2 + \cdots + \frac{1}{n!}x^n + \cdots,$$

which is the same as the given series.

Theorems (11.42) and (11.43) imply that *if* a function f is represented by a power series in x or in $x - c$, then the series *must* be a Maclaurin or Taylor series, respectively. However, the theorems do *not* answer question 2 posed at the beginning of this section: What conditions on a function guarantee that a power series representation *exists*? We shall next obtain such conditions for any series in $x - c$ (including $c = 0$). Let us begin with the following definition.

Definition **(11.44)**

> Let c be a real number and let f be a function that has n derivatives at c: $f'(c), f''(c), \ldots, f^{(n)}(c)$. The ***n*th-degree Taylor polynomial $P_n(x)$** of f at c is
>
> $$P_n(x) = f(c) + f'(c)(x - c) + \frac{f''(c)}{2!}(x - c)^2 + \cdots + \frac{f^{(n)}(c)}{n!}(x - c)^n.$$

In summation notation,

$$P_n(x) = \sum_{k=0}^{n} \frac{f^{(k)}(c)}{k!}(x - c)^k.$$

If $c = 0$ in (11.44), we call $P_n(x)$ the ***n*th-degree Maclaurin polynomial of f**. Note that $P_n(x)$ in (11.44) is the $(n + 1)$st partial sum of the Taylor series (11.43). If we let $c = 0$, then $P_n(x)$ is the $(n + 1)$st partial sum of the Maclaurin series (11.42). The next result will lead to an answer to question 2.

Taylor's formula with remainder **(11.45)**

> Let f have $n + 1$ derivatives throughout an interval containing c. If x is any number in the interval that is different from c, then there is a number z between c and x such that
>
> $$f(x) = P_n(x) + R_n(x), \quad \text{where} \quad R_n(x) = \frac{f^{(n+1)}(z)}{(n + 1)!}(x - c)^{n+1}.$$

PROOF If x is any number in the interval that is different from c, let us *define* $R_n(x)$ as follows:

$$R_n(x) = f(x) - P_n(x)$$

This equation may be rewritten as

$$f(x) = P_n(x) + R_n(x).$$

All we need to show is that for a suitable number z, $R_n(x)$ has the form stated in the conclusion of the theorem.

If t is any number in the interval, let g be the function defined by

$$g(t) = f(x) - \left[f(t) + f'(t)(x - t) + \frac{f''(t)}{2!}(x - t)^2 + \cdots + \frac{f^{(n)}(t)}{n!}(x - t)^n\right] - R_n(x)\frac{(x - t)^{n+1}}{(x - c)^{n+1}}.$$

If we differentiate each side of the equation *with respect to t* (regarding x as a constant), then many terms on the right-hand side cancel. You may verify that

$$g'(t) = -\frac{f^{(n+1)}(t)}{n!}(x - t)^n + R_n(x) \cdot (n + 1)\frac{(x - t)^n}{(x - c)^{n+1}}.$$

By referring to the formula for $g(t)$, we can verify that $g(x) = 0$. We also see that

$$\begin{aligned} g(c) &= f(x) - [P_n(x)] - R_n(x)\frac{(x - c)^{n+1}}{(x - c)^{n+1}} \\ &= f(x) - P_n(x) - R_n(x) \\ &= f(x) - [P_n(x) + R_n(x)] \\ &= f(x) - f(x) = 0. \end{aligned}$$

Hence, by Rolle's theorem, there is a number z between c and x such that $g'(z) = 0$; that is,

$$-\frac{f^{(n+1)}(z)}{n!}(x - z)^n + R_n(x) \cdot (n + 1)\frac{(x - z)^n}{(x - c)^{n+1}} = 0.$$

Solving for $R_n(x)$, we obtain

$$R_n(x) = \frac{f^{(n+1)}(z)}{(n + 1)!}(x - c)^{n+1},$$

which is what we wished to prove. ■

If $c = 0$, we refer to (11.45) as **Maclaurin's formula with remainder.** The expression $R_n(x)$ obtained in Theorem (11.45) is called the **Taylor remainder of f at c**. If $c = 0$, $R_n(x)$ is the **Maclaurin remainder of f**. In the next theorem we shall use the Taylor remainder to obtain sufficient conditions for the existence of power series representations for a function f.

Theorem **(11.46)**

Let f have derivatives of all orders throughout an interval containing c, and let $R_n(x)$ be the Taylor remainder of f at c. If

$$\lim_{n \to \infty} R_n(x) = 0$$

for every x in the interval, then $f(x)$ is represented by the Taylor series for $f(x)$ at c.

PROOF The Taylor polynomial $P_n(x)$ is the $(n+1)$st term for the sequence of partial sums of the Taylor series for $f(x)$ at c. By Theorem (11.45), $P_n(x) = f(x) - R_n(x)$, and hence

$$\lim_{n\to\infty} P_n(x) = \lim_{n\to\infty} f(x) - \lim_{n\to\infty} R_n(x) = f(x) - 0 = f(x).$$

Thus, the sequence of partial sums converges to $f(x)$, which proves the theorem. ■

In Example 2 of Section 11.6 we proved that the power series $\sum x^n/n!$ is absolutely convergent for every real number x. Since the nth term of a convergent series must approach 0 as $n \to \infty$ (see Theorem (11.16)), we obtain the following resulting.

Theorem **(11.47)**

If x is any real number,

$$\lim_{n\to\infty} \frac{|x|^n}{n!} = 0.$$

We shall use Theorem (11.47) in the solution of the following example.

EXAMPLE 2 Find the Maclaurin series for $\sin x$, and prove that it represents $\sin x$ for every real number x.

SOLUTION Let us arrange our work as follows:

$$\begin{aligned} f(x) &= \sin x & f(0) &= 0 \\ f'(x) &= \cos x & f'(0) &= 1 \\ f''(x) &= -\sin x & f''(0) &= 0 \\ f'''(x) &= -\cos x & f'''(0) &= -1 \end{aligned}$$

Successive derivatives follow this pattern. Substitution in (11.45) gives us the following Maclaurin series:

$$\sin x = x - \frac{x^3}{3!} + \frac{x^5}{5!} - \frac{x^7}{7!} + \cdots + (-1)^n \frac{x^{2n+1}}{(2n+1)!} + \cdots$$

At this stage all we know is that *if* $\sin x$ is represented by a power series in x, then it is given by the preceding series. To prove that $\sin x$ *is* actually represented by this Maclaurin series, let us employ Theorem (11.46) with $c = 0$. If n is a positive integer, then either

$$|f^{(n+1)}(x)| = |\cos x| \quad \text{or} \quad |f^{(n+1)}(x)| = |\sin x|.$$

Hence $|f^{(n+1)}(z)| \le 1$ for every number z. Using the formula for $R_n(x)$ in Theorem (11.45), with $c = 0$, we obtain

$$|R_n(x)| = \frac{|f^{(n+1)}(z)|}{(n+1)!} |x|^{n+1} \le \frac{|x|^{n+1}}{(n+1)!}.$$

It follows from Theorem (11.47) and the sandwich theorem (11.7) that $\lim_{n\to\infty} |R_n(x)| = 0$. Consequently $\lim_{n\to\infty} R_n(x) = 0$, and the Maclaurin series representation for $\sin x$ is true for every x.

EXAMPLE 3 Find the Maclaurin series for $\cos x$.

SOLUTION We could proceed directly, as in Example 2; however, let us obtain the series for $\cos x$ by differentiating the series for $\sin x$ obtained in Example 2. This gives us

$$\cos x = 1 - \frac{x^2}{2!} + \frac{x^4}{4!} - \frac{x^6}{6!} + \cdots + (-1)^n \frac{x^{2n}}{(2n)!} + \cdots.$$

The Maclaurin series for e^x was obtained in Theorem (11.41) by an indirect technique (see also Example 1 of this section). We shall next give a direct derivation of this important formula.

EXAMPLE 4 Find a Maclaurin series that represents e^x for every real number x.

SOLUTION If $f(x) = e^x$, then $f^{(k)}(x) = e^x$ for every positive integer k. Hence $f^{(k)}(0) = 1$, and substitution in (11.42) gives us

$$e^x = 1 + x + \frac{x^2}{2!} + \frac{x^3}{3!} + \cdots + \frac{x^n}{n!} + \cdots.$$

As in the solution of Example 2, we now employ Theorem (11.46) to prove that this power series representation of e^x is true for every real number x. Using the formula for $R_n(x)$ with $c = 0$, we obtain

$$R_n(x) = \frac{f^{(n+1)}(z)}{(n+1)!} x^{n+1} = \frac{e^z}{(n+1)!} x^{n+1},$$

where z is a number between 0 and x. If $0 < x$, then $e^z < e^x$, since the natural exponential function is increasing, and hence for every positive integer n,

$$0 < R_n(x) < \frac{e^x}{(n+1)!} x^{n+1}.$$

By Theorem (11.47),

$$\lim_{n\to\infty} \frac{e^x}{(n+1)!} x^{n+1} = e^x \lim_{n\to\infty} \frac{x^{n+1}}{(n+1)!} = 0,$$

and by the sandwich theorem (11.7),

$$\lim_{n\to\infty} R_n(x) = 0.$$

If $x < 0$, then $z < 0$, and hence $e^z < e^0 = 1$. Consequently

$$0 < |R_n(x)| < \left|\frac{x^{n+1}}{(n+1)!}\right|,$$

and we again see that $R_n(x)$ has the limit 0 as $n \to \infty$. It follows from Theorem (11.46) that the power series representation for e^x is valid for all nonzero x. Finally, note that if $x = 0$, then the series reduces to $e^0 = 1$.

EXAMPLE 5 Find the Taylor series for $\sin x$ in powers of $x - (\pi/6)$.

SOLUTION The derivatives of $f(x) = \sin x$ are listed in Example 2. If we evaluate them at $c = \pi/6$, we obtain

$$f\left(\frac{\pi}{6}\right) = \frac{1}{2}, \quad f'\left(\frac{\pi}{6}\right) = \frac{\sqrt{3}}{2}, \quad f''\left(\frac{\pi}{6}\right) = -\frac{1}{2}, \quad f'''\left(\frac{\pi}{6}\right) = -\frac{\sqrt{3}}{2},$$

and this pattern of four numbers repeats itself indefinitely. Substitution in (11.45) gives us

$$\sin x = \frac{1}{2} + \frac{\sqrt{3}}{2}\left(x - \frac{\pi}{6}\right) - \frac{1}{2(2!)}\left(x - \frac{\pi}{6}\right)^2 - \frac{\sqrt{3}}{2(3!)}\left(x - \frac{\pi}{6}\right)^3 + \cdots.$$

The nth term u_n of this series is given by

$$u_n = \begin{cases} (-1)^{n/2}\dfrac{1}{2(n!)}\left(x - \dfrac{\pi}{6}\right)^n & \text{if } n = 0, 2, 4, 6, \ldots \\[2ex] (-1)^{(n-1)/2}\dfrac{\sqrt{3}}{2(n!)}\left(x - \dfrac{\pi}{6}\right)^n & \text{if } n = 1, 3, 5, 7, \ldots \end{cases}$$

The proof that the series represents $\sin x$ for every x is similar to that given in Example 2 and is therefore omitted.

The next example brings out the fact that a function f may have derivatives of all orders at some number c, but may not have a Taylor series representation at that number. This shows that an additional condition, such as $\lim_{n\to\infty} R_n(x) = 0$, is required to guarantee the existence of a Taylor series.

EXAMPLE 6 Let f be the function defined by

$$f(x) = \begin{cases} e^{-1/x^2} & \text{if } x \neq 0 \\ 0 & \text{if } x = 0 \end{cases}$$

Show that $f(x)$ cannot be represented by a Maclaurin series.

SOLUTION By Definition (3.6), the derivative of f at 0 is

$$f'(0) = \lim_{x\to 0}\frac{f(x) - f(0)}{x - 0} = \lim_{x\to 0}\frac{e^{-1/x^2}}{x} = \lim_{x\to 0}\frac{(1/x)}{e^{1/x^2}}.$$

The last expression has the indeterminate form ∞/∞. Applying L'Hôpital's rule (10.2), we see that

$$f'(0) = \lim_{x\to 0}\frac{(-1/x^2)}{(-2/x^3)e^{1/x^2}} = \lim_{x\to 0}\frac{x}{2e^{1/x^2}} = 0.$$

It can be proved that $f''(0) = 0$, $f'''(0) = 0$, and, in general, $f^{(n)}(0) = 0$ for every positive integer n. According to Theorem (11.42), *if* $f(x)$ has a Maclaurin series representation, then it is given by

$$f(x) = 0 + 0x + \frac{0}{2!}x^2 + \cdots + \frac{0}{n!}x^n + \cdots,$$

which implies that $f(x) = 0$ throughout an interval containing 0. However, this contradicts the definition of f. Consequently $f(x)$ does not have a Maclaurin series representation.

As a by-product of Example 6, it follows from Theorem (11.46) that for the given function f, $\lim_{n\to\infty} R_n(x) \neq 0$ at $c = 0$.

We next list, for reference, Maclaurin series that have been obtained in examples in this section and Section 11.7. These series are important because of their uses in advanced mathematics and applications.

Important Maclaurin series **(11.48)**

MACLAURIN SERIES	INTERVAL OF CONVERGENCE
(a) $\sin x = x - \frac{x^3}{3!} + \frac{x^5}{5!} - \frac{x^7}{7!} + \cdots + (-1)^n \frac{x^{2n+1}}{(2n+1)!} + \cdots$	$(-\infty, \infty)$
(b) $\cos x = 1 - \frac{x^2}{2!} + \frac{x^4}{4!} - \frac{x^6}{6!} + \cdots + (-1)^n \frac{x^{2n}}{(2n)!} + \cdots$	$(-\infty, \infty)$
(c) $e^x = 1 + x + \frac{x^2}{2!} + \frac{x^3}{3!} + \cdots + \frac{x^n}{n!} + \cdots$	$(-\infty, \infty)$
(d) $\ln(1 + x) = x - \frac{x^2}{2} + \frac{x^3}{3} - \frac{x^4}{4} + \cdots + (-1)^n \frac{x^{n+1}}{n+1} + \cdots$	$(-1, 1]$
(e) $\tan^{-1} x = x - \frac{x^3}{3} + \frac{x^5}{5} - \frac{x^7}{7} + \cdots + (-1)^n \frac{x^{2n+1}}{2n+1} + \cdots$	$[-1, 1]$
(f) $\sinh x = x + \frac{x^3}{3!} + \frac{x^5}{5!} + \frac{x^7}{7!} + \cdots + \frac{x^{2n+1}}{(2n+1)!} + \cdots$	$(-\infty, \infty)$
(g) $\cosh x = 1 + \frac{x^2}{2!} + \frac{x^4}{4!} + \cdots + \frac{x^{2n}}{(2n)!} + \cdots$	$(-\infty, \infty)$

We can use Maclaurin or Taylor series to approximate values of functions and definite integrals, as illustrated in the next two examples.

EXAMPLE 7 Use the first two nonzero terms of a Maclaurin series to approximate the following, and estimate the error in the approximation:
(a) $\sin(0.1)$ **(b)** $\sin x$ for any nonzero real number x in $[-1, 1]$

SOLUTION

(a) Letting $x = 0.1$ in the Maclaurin series for $\sin x$ (see (11.48)(a)) yields

$$\sin(0.1) = 0.1 - \frac{0.001}{6} + \frac{0.00001}{120} - \cdots.$$

By Theorem (11.31), the error involved in approximating $\sin(0.1)$ by using the first two terms of this alternating series is less than the third term, 0.00001/120 ≈ 0.0000000.8. To six decimal places,

$$\sin(0.1) \approx 0.1 - \frac{0.001}{6} \approx 0.099833.$$

(b) Using the first two terms of (11.48)(a) gives us the approximation formula

$$\sin x \approx x - \frac{x^3}{6}.$$

By Theorem (11.31), the error involved in using this formula for a real number x is less than $|x^5|/5!$.

EXAMPLE 8 Approximate $\int_0^1 \sin(x^2)\,dx$ to four decimal places.

SOLUTION Substituting x^2 for x in (11.48)(a) gives us

$$\sin(x^2) = x^2 - \frac{x^6}{3!} + \frac{x^{10}}{5!} - \frac{x^{14}}{7!} + \cdots.$$

Integrating each term of this series, we obtain

$$\int_0^1 \sin(x^2)\,dx = \frac{1}{3} - \frac{1}{42} + \frac{1}{1320} - \frac{1}{75{,}600} + \cdots.$$

Summing the first three terms yields

$$\int_0^1 \sin(x^2)\,dx \approx 0.31028.$$

By Theorem (11.31), the error is less than $\frac{1}{75{,}600} \approx 0.00001$.

Note that in the preceding example we achieved accuracy to four decimal places by summing only *three* terms of the integrated series for $\sin(x^2)$. To obtain this degree of accuracy by means of the trapezoidal rule or Simpson's rule, it would be necessary to use a large value of n for the interval $[0, 1]$. However, if the interval were $[10, 11]$, the efficiency of each method would be quite different. This brings out an important point. For numerical applications, in addition to analyzing a given problem, we should also strive to find the most efficient method for computing the answer.

To obtain a Taylor or Maclaurin series representation for a function f, it is necessary to find a general formula for $f^{(n)}(x)$ and, in addition, to investigate $\lim_{n\to\infty} R_n(x)$. Because of this, our examples have been restricted to expressions such as $\sin x$, $\cos x$, and e^x. The method cannot be used if, for example, $f(x)$ equals $\tan x$ or $\sin^{-1} x$, because $f^{(n)}(x)$ becomes very complicated as n increases. Most of the exercises that follow are based on functions whose nth derivatives can be determined easily or on series representations that we have already established. In more complicated cases we shall restrict our attention to only the first few terms of a Taylor or Maclaurin series representation.

EXERCISES 11.8

Exer. 1–6: If $f(x) = \sum_{n=0}^{\infty} a_n x^n$, find a_n by using the formula for a_n in (11.42).

1 $f(x) = e^{3x}$

2 $f(x) = e^{-2x}$

3 $f(x) = \sin 2x$

4 $f(x) = \cos 3x$

5 $f(x) = \dfrac{1}{1+3x}$

6 $f(x) = \dfrac{1}{1-2x}$

7 Let $f(x) = \cos x$.

(a) Use the method of Example 2 to prove that $\lim_{n\to\infty} R_n(x) = 0$.

(b) Use (11.42) to find a Maclaurin series for $f(x)$.

8 Let $f(x) = e^{-x}$.

(a) Use the method of Example 4 to prove that $\lim_{n\to\infty} R_n(x) = 0$.

(b) Use (11.42) to find a Maclaurin series for $f(x)$.

Exer. 9–14: Use a Maclaurin series obtained in this section to obtain a Maclaruin series for $f(x)$.

9 $f(x) = x \sin 3x$

10 $f(x) = x^2 \sin x$

11 $f(x) = \cos(-2x)$

12 $f(x) = \cos(x^2)$

13 $f(x) = \cos^2 x$ (*Hint:* Use a half-angle formula.)

14 $f(x) = \sin^2 x$

Exer. 15–16: Find a Maclaurin series for $f(x)$. (Do not verify that $\lim_{n\to\infty} R_n(x) = 0$.)

15 $f(x) = 10^x$

16 $f(x) = \ln(3 + x)$

Exer. 17–20: Find a Taylor series for $f(x)$ at c. (Do not verify that $\lim_{n\to\infty} R_n(x) = 0$.)

17 $f(x) = \sin x$; $c = \pi/4$

18 $f(x) = \cos x$; $c = \pi/3$

19 $f(x) = 1/x$; $c = 2$

20 $f(x) = e^x$; $c = -3$

21 Find a series representation for e^{2x} in powers of $x + 1$.

22 Find a series representation of $\ln x$ in powers of $x - 1$.

Exer. 23–28: Find the first three terms of the Taylor series for $f(x)$ at c.

23 $f(x) = \sec x$; $c = \pi/3$

24 $f(x) = \tan x$; $c = \pi/4$

25 $f(x) = \sin^{-1} x$; $c = \frac{1}{2}$

26 $f(x) = \tan^{-1} x$; $c = 1$

27 $f(x) = xe^x$; $c = -1$

28 $f(x) = \csc x$; $c = 2\pi/3$

Exer. 29–38: Use the first two nonzero terms of a Maclaurin series to approximate the number, and estimate the error in the approximation.

29 $\dfrac{1}{\sqrt{e}}$

30 $\dfrac{1}{e}$

31 $\cos 3°$

32 $\sin 1°$

33 $\tan^{-1} 0.1$

34 $\ln 1.5$

35 $\int_0^1 e^{-x^2}\, dx$

36 $\int_0^{1/2} x \cos(x^3)\, dx$

37 $\int_0^{0.5} \cos(x^2)\, dx$

38 $\int_0^{0.1} \tan^{-1}(x^2)\, dx$

c **Exer. 39–42: Approximate the improper integral to four decimal places. (Assume that if the integrand is $f(x)$, then $f(0) = \lim_{x\to 0} f(x)$.)**

39 $\int_0^1 \dfrac{1 - \cos x}{x^2}\, dx$

40 $\int_0^1 \dfrac{\sin x}{x}\, dx$

41 $\int_0^{1/2} \dfrac{\ln(1 + x)}{x}\, dx$

42 $\int_0^1 \dfrac{1 - e^{-x}}{x}\, dx$

c **Exer. 43–44: (a) Let $g(x)$ be the sum of the first two nonzero terms of the Maclaurin series for $f(x)$. Use $g(x)$ to approximate $\int_0^1 f(x)\, dx$ and $\int_1^2 f(x)\, dx$. (b) Graph, on the same coordinate axes, f and g for $0 \le x \le 2$, and then use the graphs to compare the accuracy of the approximations in (a).**

43 $f(x) = \sin(x^2)$

44 $f(x) = \sinh x$

45 Use (11.48)(d) to find the Maclaurin series for

$$f(x) = \ln \frac{1 + x}{1 - x}.$$

c 46 Use the first five terms of the series in Exercise 45 to calculate ln 2, and compare your answer to the value obtained using a calculator or table.

47 (a) Use (11.48)(e) with $x = 1$ to represent π as the sum of an infinite series.

(b) What accuracy is obtained by using the first five terms of the series to approximate π?

(c) Approximately how many terms of the series are required to obtain four-decimal-place accuracy for π?

48 (a) Use the identity

$$\tan^{-1}\frac{1}{2} + \tan^{-1}\frac{1}{3} = \frac{\pi}{4}$$

to express π as the sum of two infinite series.

(b) Use the first five terms of each series in (a) to approximate π, and compare the result with that obtained in Exercise 47.

49 In planning a highway across a desert, a surveyor must make compensations for the curvature of the earth when measuring differences in elevation (see figure).

(a) If s is the length of the highway and R is the radius of the earth, show that the correction C is given by $C = R[\sec(s/R) - 1]$.

(b) Use the Maclaurin series for $\sec x$ to show that C is approximately $s^2/(2R) + (5s^4)/(24R^3)$.

(c) The average radius of the earth in 3959 miles. Estimate the correction, to the nearest 0.1 foot, for a stretch of highway 5 miles long.

EXERCISE 49

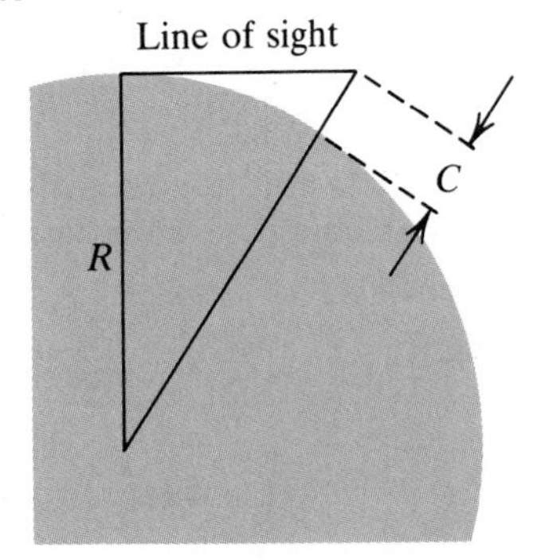

50 The velocity v of a water wave is related to its length L and the depth h of the water by

$$v^2 = \frac{gL}{2\pi} \tanh \frac{2\pi h}{L},$$

where g is a gravitational constant.

(a) Show that $\tanh x \approx x - \frac{1}{3}x^3$ if $x \approx 0$.

(b) Use the approximation $\tanh x \approx x$ to show that $v^2 \approx gh$ if h/L is small.

(c) Use part (a) and the fact that the Maclaurin series for $\tanh x$ is an alternating series to show that if $L > 20h$, then the error involved in using $v^2 \approx gh$ is less than $0.002gL$.

51 If a large downward force of P pounds is applied to a cantilever column of length L at a point x units to the right of center (see figure), the column will buckle. The horizontal deflection δ can be expressed as

$$\delta = x(\sec kL - 1) \quad \text{with} \quad k = \sqrt{P/R},$$

where R is a constant called the *flexural rigidity* of the material and $0 \leq kL < \pi/2$. Use Exercise 49(b) to show that $\delta \approx \frac{1}{2}PxL^2/R$ if P is small.

EXERCISE 51

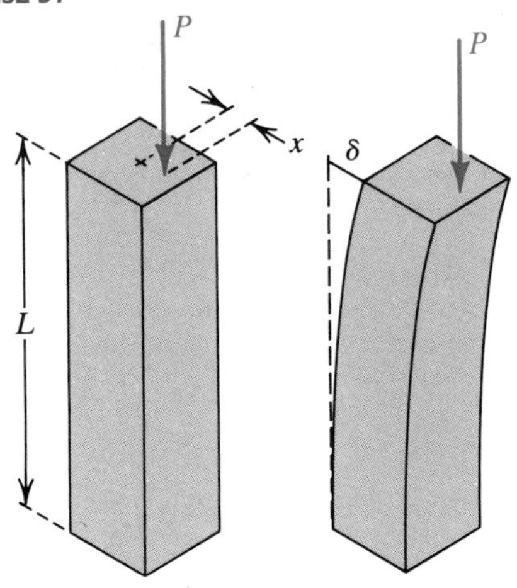

52 Show that $\cos x \approx 1 - \frac{1}{2}x^2 + \frac{1}{24}x^4 - \frac{1}{720}x^6$ is accurate to five decimal places if $0 \leq x \leq \pi/4$.

11.9 APPLICATIONS OF TAYLOR POLYNOMIALS

In Example 7(b) of the preceding section, we used the first two nonzero terms of the Maclaurin series for $\sin x$ to obtain the approximation formula

$$\sin x \approx x - \frac{x^3}{6}.$$

By Definition (11.44), the expression on the right-hand side of this formula is the third-degree Taylor polynomial $P_3(x)$ of $\sin x$ at $c = 0$. Thus, we could write

$$\sin x \approx P_3(x).$$

Using additional terms of the Maclaurin series for $\sin x$ would give us other approximation formulas. To illustrate,

$$\sin x \approx x - \frac{x^3}{3!} + \frac{x^5}{5!} = P_5(x).$$

By Theorem (11.31), the error involved in using this formula is less than $|x^7|/7!$. Thus, the approximation is very accurate if x is close to 0.

We can use this procedure for any function f that has a sufficient number of derivatives. Specifically, if f satisfies the hypotheses of Taylor's formula with remainder (11.45), then

$$f(x) = P_n(x) + R_n(x),$$

where $P_n(x)$ is the nth-degree Taylor polynomial of f at c and $R_n(x)$ is the Taylor remainder. If $\lim_{n\to\infty} R_n(x) = 0$, then, as n increases, $P_n(x) \to f(x)$; hence the approximation formula $f(x) \approx P_n(x)$ improves as n gets larger. Thus, we can approximate values of many different transcendental functions by using *polynomial* functions. This is a very important fact, because polynomial functions are the simplest functions to use for calculations—their values can be found by employing only additions and multiplications of real numbers.

FIGURE 11.17

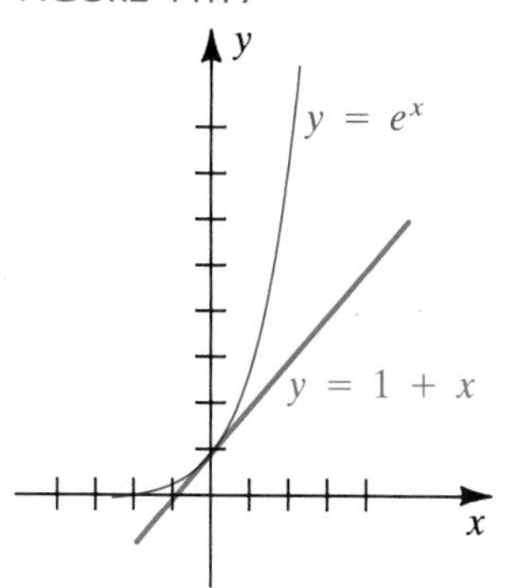

As another illustration, consider the exponential function given by $f(x) = e^x$. From (11.48)(c), the Maclaurin series for e^x is

$$e^x = 1 + x + \frac{x^2}{2!} + \frac{x^3}{3!} + \cdots + \frac{x^n}{n!} + \cdots.$$

If we approximate e^x by means of Taylor polynomials (with $c = 0$), we obtain

$$e^x \approx P_1(x) = 1 + x$$

$$e^x \approx P_2(x) = 1 + x + \frac{x^2}{2}$$

$$e^x \approx P_3(x) = 1 + x + \frac{x^2}{2} + \frac{x^3}{6}$$

FIGURE 11.18

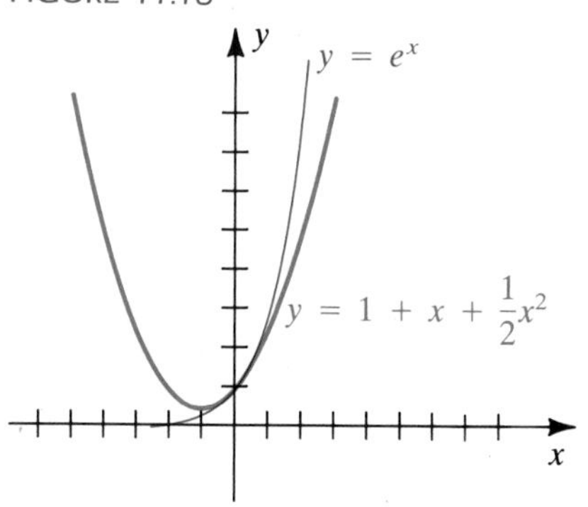

and so on. These approximation formulas are accurate only if x is close to 0. To approximate e^x for larger values of x, we may use Taylor polynomials with $c \neq 0$.

The accuracy of the preceding three approximation formulas for e^x is illustrated by the graphs of the functions P_1, P_2, and P_3 in Figures 11.17–11.19. It is of interest to note that the graph of $y = P_1(x) = 1 + x$ in Figure 11.17 is the tangent line to the graph of $y = e^x$ at the point (0, 1). You may verify that the parabola $y = P_2(x) = 1 + x + \frac{1}{2}x^2$ in Figure 11.18 has the same tangent line and the same concavity as the graph of $y = e^x$ at (0, 1). The graph of $y = P_3(x)$ in Figure 11.19 has the same tangent line and concavity at (0, 1) and also the same *rate of change of concavity*, since $P_3'''(0) = D_x^3\, e^x]_{x=0}$. In general, we can show that for any positive integer n, $P_n^{(n)}(0) = D_x^n\, e^x]_{x=0}$. As n increases without bound, the graph of the equation $y = P_n(x)$ more closely resembles the graph of $y = e^x$.

FIGURE 11.19

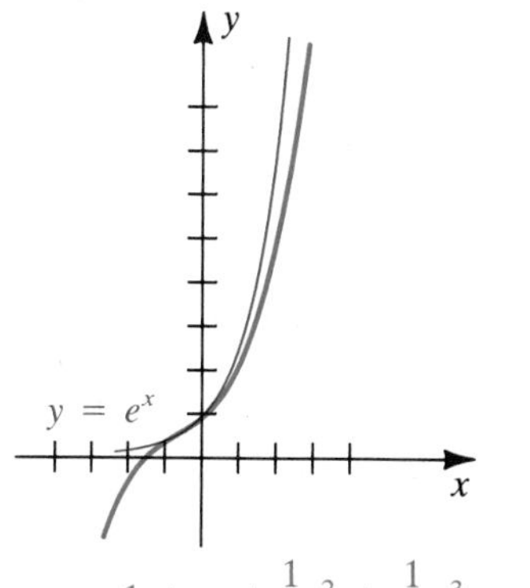

The table on the following page indicates the accuracy of the approximation formula $e^x \approx P_3(x)$ for several values of x, approximated to the nearest hundredth. If more accuracy were desired, we could use any larger positive integer n to obtain

$$e^x \approx P_n(x) = 1 + x + \frac{x^2}{2!} + \frac{x^3}{3!} + \cdots + \frac{x^n}{n!}.$$

We can use this remarkably simple formula to approximate e^x to any degree of accuracy.

x	-1.5	-1.0	-0.5	0	0.5	1.0	1.5
e^x	0.22	0.37	0.61	1	1.65	2.72	4.48
$P_3(x)$	0.06	0.33	0.60	1	1.65	2.67	4.19

In the remainder of this section we shall use Taylor polynomials to approximate values of functions that satisfy the hypotheses of Taylor's formula with remainder (11.45). Using the conclusion $f(x) = P_n(x) + R_n(x)$ of that theorem, we see that the error involved in approximating $f(x)$ by $P_n(x)$ is

$$|f(x) - P_n(x)| = |R_n(x)|.$$

The complete statement of this result is given in the next theorem.

Theorem (11.49)

Let f have $n + 1$ derivatives throughout an interval containing c. If x is any number in the interval and $x \neq c$, then the error in approximating $f(x)$ by the nth-degree Taylor polynomial of f at c,

$$P_n(x) = f(c) + f'(c)(x - c) + \frac{f''(c)}{2!}(x - c)^2 + \cdots + \frac{f^{(n)}(c)}{n!}(x - c)^n,$$

is $|R_n(x)|$, where

$$R_n(x) = \frac{f^{(n+1)}(z)}{(n+1)!}(x - c)^{n+1}$$

and z is the number between c and x given by (11.45).

In the next two examples, Taylor polynomials are used to approximate function values. If we are interested in k-decimal-place accuracy in the approximation of a sum, we often approximate each term of the sum to $k + 1$ decimal places and then round off the final result to k decimal places. In certain cases this may fail to produce the required degree of accuracy; however, it is customary to proceed in this way for elementary approximations. More precise techniques may be found in texts on *numerical analysis*.

EXAMPLE 1 Let $f(x) = \ln x$.

(a) Find $P_3(x)$ and $R_3(x)$ at $c = 1$.

(b) Approximate ln 1.1 to four decimal places by means of $P_3(1.1)$, and use $R_3(1.1)$ to estimate the error in this approximation.

SOLUTION

(a) As in Theorem (11.49), the general Taylor polynomial $P_3(x)$ and Taylor remainder $R_3(x)$ at $c = 1$ are

$$P_3(x) = f(1) + f'(1)(x - 1) + \frac{f''(1)}{2!}(x - 1)^2 + \frac{f'''(1)}{3!}(x - 1)^3$$

and

$$R_3(x) = \frac{f^{(4)}(z)}{4!}(x - 1)^4,$$

where z is a number between 1 and x. Thus, we need the first four derivatives of f. It is convenient to arrange our work as follows:

$$\begin{array}{ll} f(x) = \ln x & f(1) = 0 \\ f'(x) = x^{-1} & f'(1) = 1 \\ f''(x) = -x^{-2} & f''(1) = -1 \\ f'''(x) = 2x^{-3} & f'''(1) = 2 \\ f^{(4)}(x) = -6x^{-4} & f^{(4)}(z) = -6z^{-4} \end{array}$$

Substituting in $P_3(x)$ and $R_3(x)$, we obtain

$$P_3(x) = 0 + 1(x - 1) - \frac{1}{2!}(x - 1)^2 + \frac{2}{3!}(x - 1)^3$$

and $$R_3(x) = \frac{-6z^{-4}}{4!}(x - 1)^4 = -\frac{1}{4z^4}(x - 1)^4,$$

where z is between 1 and x.

(b) From part (a),

$$\ln 1.1 \approx P_3(1.1) = 0.1 - \tfrac{1}{2}(0.1)^2 + \tfrac{1}{3}(0.1)^3,$$

or $$\ln 1.1 \approx 0.0953.$$

To estimate the error in this approximation, we consider

$$|R_3(1.1)| = \left| -\frac{(0.1)^4}{4z^4} \right|, \quad \text{where} \quad 1 < z < 1.1.$$

Since $z > 1$, $1/z < 1$ and therefore $1/z^4 < 1$. Consequently

$$|R_3(1.1)| = \left| -\frac{(0.1)^4}{4z^4} \right| < \left| -\frac{0.0001}{4} \right| = 0.000025.$$

Because $0.000025 < 0.00005$, it follows from Theorem (11.49) that the approximation $\ln 1.1 \approx 0.0953$ is accurate to four decimal places.

If we wish to approximate a function value $f(x)$ for some x, it is desirable to choose the number c in Theorem (11.49) such that the remainder $R_n(x)$ is very close to 0 when n is relatively small (say, $n = 3$ or $n = 4$). This will be true if we choose c close to x. In addition, we should choose c so that values of the first $n + 1$ derivatives of f at c are easy to calculate. This was done in Example 1, where to approximate $\ln x$ for $x = 1.1$ we selected $c = 1$. The next example provides another illustration of a suitable choice of c.

EXAMPLE 2 Use a Taylor polynomial to approximate $\cos 61°$, and estimate the accuracy of the approximation.

SOLUTION We wish to approximate $f(x) = \cos x$ if $x = 61°$. Let us begin by observing that $61°$ is close to $60°$, or $\pi/3$ radians, and that it is easy to calculate values of trigonometric functions at $\pi/3$. This suggests that we choose $c = \pi/3$ in (11.49). The choice of n will depend on the degree of accuracy we wish to attain. Let us try $n = 2$. In this case, the first three derivatives of f are required and we arrange our work as follows:

$$\begin{aligned} f(x) &= \cos x & f\left(\frac{\pi}{3}\right) &= \frac{1}{2} \\ f'(x) &= -\sin x & f'\left(\frac{\pi}{3}\right) &= -\frac{\sqrt{3}}{2} \\ f''(x) &= -\cos x & f''\left(\frac{\pi}{3}\right) &= -\frac{1}{2} \\ f'''(x) &= \sin x & f'''(z) &= \sin z \end{aligned}$$

As in Theorem (11.49), the second-degree Taylor polynomial of f at $c = \pi/3$ is

$$P_2(x) = \frac{1}{2} - \frac{\sqrt{3}}{2}\left(x - \frac{\pi}{3}\right) - \frac{\frac{1}{2}}{2!}\left(x - \frac{\pi}{3}\right)^2.$$

Since x represents a real number, we must convert 61° to radian measure before substituting into $P_2(x)$. Writing

$$61° = 60° + 1° = \frac{\pi}{3} + \frac{\pi}{180}$$

and substituting, we obtain

$$P_2\left(\frac{\pi}{3} + \frac{\pi}{180}\right) = \frac{1}{2} - \left(\frac{\sqrt{3}}{2}\right)\left(\frac{\pi}{180}\right) - \frac{1}{4}\left(\frac{\pi}{180}\right)^2 \approx 0.48481,$$

and hence

$$\cos 61° \approx 0.48481.$$

To estimate the accuracy of this approximation, we consider

$$|R_2(x)| = \left|\frac{f'''(z)}{3!}\left(x - \frac{\pi}{3}\right)^3\right| = \left|\frac{\sin z}{3!}\left(x - \frac{\pi}{3}\right)^3\right|,$$

where z is between $\pi/3$ and x. Substituting $x = (\pi/3) + (\pi/180)$ and using the fact that $|\sin z| \le 1$, we obtain

$$\left|R_2\left(\frac{\pi}{3} + \frac{\pi}{180}\right)\right| = \left|\frac{\sin z}{3!}\left(\frac{\pi}{180}\right)^3\right| \le \left|\frac{1}{3!}\left(\frac{\pi}{180}\right)^3\right| \le 0.000001.$$

Thus, by Theorem (11.49), the approximation $\cos 61° \approx 0.48481$ is accurate to five decimal places. If we desire greater accuracy, we must find a value of n such that the maximum value of $|R_n[(\pi/3) + (\pi/180)]|$ is within the desired range.

EXAMPLE 3 If $f(x) = e^x$, use the Taylor polynomial $P_9(x)$ of f at $c = 0$ to approximate e, and estimate the error in the approximation.

SOLUTION For every positive integer k, $f^{(k)}(x) = e^x$, and hence $f^{(k)}(0) = e^0 = 1$. Thus, using $n = 9$ and $c = 0$ in Theorem (11.49) yields

$$P_9(x) = 1 + x + \frac{x^2}{2!} + \frac{x^3}{3!} + \cdots + \frac{x^9}{9!},$$

and therefore

$$e \approx P_9(1) = 1 + 1 + \frac{1}{2!} + \frac{1}{3!} + \cdots + \frac{1}{9!}.$$

This gives us $e \approx 2.71828153$.

To estimate the error, we consider

$$R_9(x) = \frac{e^z}{10!}x^{10}.$$

If $x = 1$, then $0 < z < 1$. Using results about e^x from Chapter 7, we have $e^z < e^1 < 3$, and

$$|R_9(1)| = \left|\frac{e^z}{10!}(1)\right| < \frac{3}{10!} < 0.000001.$$

Hence the approximation $e \approx 2.71828$ is accurate to five decimal places.

EXERCISES 11.9

Exer. 1–4: (a) Find the Maclaurin polynomials $P_1(x)$, $P_2(x)$, and $P_3(x)$ for $f(x)$. (b) Sketch the graphs of P_1, P_2, P_3, and f on the same coordinate plane. (c) Approximate $f(a)$ to four decimal places by means of $P_3(a)$, and use $R_3(a)$ to estimate the error in this approximation.

1 $f(x) = \sin x$; $a = 0.05$

2 $f(x) = \cos x$; $a = 0.2$

3 $f(x) = \ln(x + 1)$; $a = 0.9$

4 $f(x) = \tan^{-1} x$; $a = 0.1$

Exer. 5–6: Graph, on the same coordinate plane, f, P_1, P_3, and P_5 for $-3 \le x \le 3$.

5 $f(x) = \sinh x$

6 $f(x) = \cosh x$

Exer. 7–18: Find Taylor's formula with remainder (11.45) for the given $f(x)$, c, and n.

7 $f(x) = \sin x$, $c = \pi/2$, $n = 3$

8 $f(x) = \cos x$, $c = \pi/4$, $n = 3$

9 $f(x) = \sqrt{x}$, $c = 4$, $n = 3$

10 $f(x) = e^{-x}$, $c = 1$, $n = 3$

11 $f(x) = \tan x$, $c = \pi/4$, $n = 2$

12 $f(x) = 1/(x - 1)^2$, $c = 2$, $n = 5$

13 $f(x) = 1/x$, $c = -2$, $n = 5$

14 $f(x) = \sqrt[3]{x}$, $c = -8$, $n = 3$

15 $f(x) = \tan^{-1} x$, $c = 1$, $n = 2$

16 $f(x) = \ln \sin x$, $c = \pi/6$, $n = 3$

17 $f(x) = xe^x$, $c = -1$, $n = 4$

18 $f(x) = \log x$, $c = 10$, $n = 2$

Exer. 19–30: Find Maclaurin's formula with remainder for the given $f(x)$ and n.

19 $f(x) = \ln(x + 1)$, $n = 4$

20 $f(x) = \sin x$, $n = 7$

21 $f(x) = \cos x$, $n = 8$

22 $f(x) = \tan^{-1} x$, $n = 3$

23 $f(x) = e^{2x}$, $n = 5$

24 $f(x) = \sec x$, $n = 3$

25 $f(x) = 1/(x - 1)^2$, $n = 5$

26 $f(x) = \sqrt{4 - x}$, $n = 3$

27 $f(x) = \arcsin x$, $n = 2$

28 $f(x) = e^{-x^2}$, $n = 3$

29 $f(x) = 2x^4 - 5x^3$, $n = 4$ and $n = 5$

30 $f(x) = \cosh x$, $n = 4$ and $n = 5$

c **Exer. 31–34: Approximate the number to four decimal places by using the indicated exercise and the fact that $\pi/180 \approx 0.0175$. Prove that your answer is correct by showing that $|R_n(x)| < 0.5 \times 10^{-4}$.**

31 $\sin 89°$ (Exercise 7)

32 $\cos 47°$ (Exercise 8)

33 $\sqrt{4.03}$ (Exercise 9)

34 $e^{-1.02}$ (Exercise 10)

c **Exer. 35–40: Approximate the number by using the indicated exercise, and estimate the error in the approximation by means of $R_n(x)$.**

35 $-1/(2.2)$ (Exercise 13)

36 $\sqrt[3]{-8.5}$ (Exercise 14)

37 $\ln 1.25$ (Exercise 19)

38 $\sin 0.1$ (Exercise 20)

39 $\cos 30°$ (Exercise 21)

40 $\log 10.01$ (Exercise 18)

Exer. 41–46: Use Maclaurin's formula with remainder to establish the approximation formula, and state, in terms of decimal places, the accuracy of the approximation if $|x| \leq 0.1$.

41 $\cos x \approx 1 - \dfrac{x^2}{2}$

42 $\sqrt[3]{1+x} \approx 1 + \dfrac{1}{3}x$

43 $e^x \approx 1 + x + \dfrac{x^2}{2}$

44 $\sin x \approx x - \dfrac{x^3}{6}$

45 $\ln(1+x) \approx x - \dfrac{x^2}{2} + \dfrac{x^3}{3}$

46 $\cosh x \approx 1 + \dfrac{x^2}{2}$

47 Let $P_n(x)$ be the nth-degree Maclaurin polynomial. If $f(x)$ is a polynomial of degree n, prove that $f(x) = P_n(x)$.

11.10 THE BINOMIAL SERIES

The binomial theorem states that if k is a positive integer, then for all numbers a and b,

$$(a+b)^k = a^k + ka^{k-1}b + \frac{k(k-1)}{2!}a^{k-2}b^2 + \cdots + \frac{k(k-1)\cdots(k-n+1)}{n!}a^{k-n}b^n + \cdots + b^k.$$

If we let $a = 1$ and $b = x$, then

$$(1+x)^k = 1 + kx + \frac{k(k-1)}{2!}x^2 + \cdots + \frac{k(k-1)\cdots(k-n+1)}{n!}x^n + \cdots + x^k.$$

If k is not a positive integer (or 0), it is useful to study the power series $\sum a_n x^n$ with $a_0 = 1$ and $a_n = k(k-1)\cdots(k-n+1)/n!$ for $n \geq 1$. This infinite series has the form

$$1 + kx + \frac{k(k-1)}{2!}x^2 + \cdots + \frac{k(k-1)\cdots(k-n+1)}{n!}x^n + \cdots$$

and is called the **binomial series**. If k is a nonnegative integer, the series reduces to the finite sum given in the binomial theorem. Otherwise, the series does not terminate. Using the formula for a_n, we can show that

$$\lim_{n\to\infty}\left|\frac{a_{n+1}x^{n+1}}{a_n x^n}\right| = \lim_{n\to\infty}\left|\frac{k-n}{n+1}\right||x| = |x|.$$

Hence, by the ratio test (11.35), the series is absolutely convergent if $|x| < 1$ and is divergent if $|x| > 1$. Thus, the binomial series represents a function f such that

$$f(x) = 1 + \sum_{n=1}^{\infty}\frac{k(k-1)\cdots(k-n+1)}{n!}x^n \quad \text{if} \quad |x| < 1.$$

We have already noted that if k is a nonnegative integer, then $f(x) = (1+x)^k$. We shall now prove that the same is true for *every* real number k. Differentiating each term of the binomial series gives us

$$f'(x) = k + k(k-1)x + \cdots + \frac{nk(k-1)\cdots(k-n+1)}{n!}x^{n-1} + \cdots$$

and therefore

$$xf'(x) = kx + k(k-1)x^2 + \cdots + \frac{nk(k-1)\cdots(k-n+1)}{n!}x^n + \cdots.$$

If we add corresponding terms of the preceding two power series, then the coefficient of x^n is

$$\frac{(n+1)k(k-1)\cdots(k-n)}{(n+1)!} + \frac{nk(k-1)\cdots(k-n+1)}{n!},$$

which simplifies to

$$[(k-n)+n]\frac{k(k-1)\cdots(k-n+1)}{n!} = ka_n.$$

Consequently

$$f'(x) + xf'(x) = \sum_{n=0}^{\infty} ka_n x^n = kf(x),$$

or, equivalently,

$$f'(x)(1+x) - kf(x) = 0.$$

If we define the function g by $g(x) = f(x)/(1+x)^k$, then

$$\begin{aligned} g'(x) &= \frac{(1+x)^k f'(x) - f(x)k(1+x)^{k-1}}{(1+x)^{2k}} \\ &= \frac{(1+x)f'(x) - kf(x)}{(1+x)^{k+1}} = 0. \end{aligned}$$

It follows that $g(x) = c$ for some constant c; that is,

$$\frac{f(x)}{(1+x)^k} = c.$$

Since $f(0) = 1$, we see that $c = 1$ and hence $f(x) = (1+x)^k$, which is what we wished to prove. The next statement summarizes this discussion.

Binomial series **(11.50)**

If $|x| < 1$, then for every real number k,

$$\begin{aligned} (1+x)^k = 1 + kx &+ \frac{k(k-1)}{2!}x^2 + \cdots \\ &+ \frac{k(k-1)\cdots(k-n+1)}{n!}x^n + \cdots \end{aligned}$$

EXAMPLE 1 Find a power series representation for $\sqrt[3]{1+x}$.

SOLUTION Using (11.50) with $k = \frac{1}{3}$, we obtain

$$\begin{aligned} \sqrt[3]{1+x} = 1 + \frac{1}{3}x &+ \frac{\frac{1}{3}(\frac{1}{3}-1)}{2!}x^2 + \frac{\frac{1}{3}(\frac{1}{3}-1)(\frac{1}{3}-2)}{3!}x^3 + \cdots \\ &+ \frac{\frac{1}{3}(\frac{1}{3}-1)\cdots(\frac{1}{3}-n+1)}{n!}x^n + \cdots, \end{aligned}$$

which may be written

$$\sqrt[3]{1+x} = 1 + \frac{1}{3}x - \frac{2}{3^2 \cdot 2!}x^2 + \frac{1 \cdot 2 \cdot 5}{3^3 \cdot 3!}x^3 + \cdots$$
$$+ (-1)^{n+1}\frac{1 \cdot 2 \cdots (3n-4)}{3^n \cdot n!}x^n + \cdots$$

for $|x| < 1$. The formula for the nth term of this series is valid provided $n \geq 2$.

EXAMPLE 2 Find a power series representation for $\sqrt[3]{1+x^4}$.

SOLUTION The power series can be obtained by substituting x^4 for x in the series of Example 1. Hence, if $|x| < 1$, then

$$\sqrt[3]{1+x^4} = 1 + \frac{1}{3}x^4 - \frac{2}{3^2 \cdot 2!}x^8 + \cdots$$
$$+ (-1)^{n+1}\frac{1 \cdot 2 \cdots (3n-4)}{3^n \cdot n!}x^{4n} + \cdots.$$

EXAMPLE 3 Approximate $\int_0^{0.3} \sqrt[3]{1+x^4}\,dx$.

SOLUTION Integrating the terms of the series obtained in Example 2 gives us

$$\int_0^{0.3} \sqrt[3]{1+x^4}\,dx = 0.3 + 0.000162 - 0.000000243 + \cdots.$$

Consequently the integral may be approximated by 0.300162, which is accurate to six decimal places, since the error is less than 0.000000243. (Why?)

The binomial series can be used to obtain polynomial approximation formulas for $(1+x)^k$. To illustrate, if $|x| < 1$, then from Example 1,

$$\sqrt[3]{1+x} \approx 1 + \tfrac{1}{3}x.$$

Since the series is alternating and satisfies (11.30) from the second term onward, the error involved in this approximation is less than the third term, $\frac{1}{9}x^2$.

EXERCISES 11.10

Exer. 1–12: Find a power series representation for the expression, and state the radius of convergence.

1 (a) $\sqrt{1+x}$ (b) $\sqrt{1-x^3}$

2 (a) $\dfrac{1}{\sqrt[3]{1+x}}$ (b) $\dfrac{1}{\sqrt[3]{1-x^2}}$

3 $(1+x)^{-2/3}$

4 $(1+x)^{1/4}$

5 $(1-x)^{3/5}$

6 $(1-x)^{2/3}$

7 $(1+x)^{-2}$

8 $(1+x)^{-4}$

9 $(1+x)^{-3}$

10 $x(1+2x)^{-2}$

11 $\sqrt[3]{8+x}$ (*Hint:* Consider $2\sqrt[3]{1+\frac{1}{8}x}$.)

12 $(4+x)^{3/2}$

Exer. 13–14: (a) Obtain a power series representation for $f(x)$ by using the given relationship. (b) Find the radius of convergence.

13 $f(x) = \sin^{-1} x;\quad \sin^{-1} x = \int_0^x (1/\sqrt{1 - t^2})\, dt$

14 $f(x) = \sinh^{-1} x;\quad \sinh^{-1} x = \int_0^x (1/\sqrt{1 + t^2})\, dt$

Exer. 15–20: Approximate the integral to three decimal places, using the indicated exercise.

15 $\int_0^{1/2} \sqrt{1 + x^3}\, dx$ (Exercise 1)

16 $\int_0^{1/2} \frac{1}{\sqrt[3]{1 + x^2}}\, dx$ (Exercise 2)

17 $\int_0^{0.2} (1 - x^2)^{3/5}\, dx$ (Exercise 5)

18 $\int_0^{0.4} (1 - x^3)^{2/3}\, dx$ (Exercise 6)

19 $\int_0^{0.3} \frac{1}{(1 + x^3)^2}\, dx$ (Exercise 7)

20 $\int_0^{0.1} \frac{1}{(1 + 5x^2)^4}\, dx$ (Exercise 8)

c **Exer. 21–22: For the given k, graph $f(x) = (1 + x)^k$ and $g(x) = 1 + kx$ on the same coordinate plane for $-1 \le x \le 1$. Use the graphs to estimate the largest closed interval on which $|f(x) - g(x)| \le \frac{1}{10}$.**

21 $k = \frac{1}{2}$

22 $k = \frac{5}{2}$

23 Refer to Exercise 44 of Section 10.4. The formula for the period T of a pendulum of length L, initially displaced from equilibrium through an angle of θ_0 radians, is given by the improper integral

$$T = 2\sqrt{\frac{2L}{g}} \int_0^{\theta_0} \frac{1}{\sqrt{\cos\theta - \cos\theta_0}}\, d\theta.$$

By making the substitution $\sin u = (1/k)\sin\frac{1}{2}\theta$, with $k = \sin\frac{1}{2}\theta_0$, it can be shown that

$$T = 4\sqrt{\frac{L}{g}} \int_0^{\pi/2} \frac{1}{\sqrt{1 - k^2\sin^2 u}}\, du.$$

(a) Use the binomial series for $(1 - x)^{-1/2}$ to show that

$$T \approx 2\pi\sqrt{\frac{L}{g}}\left(1 + \frac{1}{4}k^2\right).$$

(b) Approximate T if $\theta_0 = \pi/6$.

11.11 REVIEW EXERCISES

Exer. 1–6: Determine whether the sequence converges or diverges, and if it converges, find the limit.

1 $\left\{\frac{\ln(n^2 + 1)}{n}\right\}$

2 $\{100(0.99)^n\}$

3 $\left\{\frac{10^n}{n^{10}}\right\}$

4 $\left\{\frac{1}{n} + (-2)^n\right\}$

5 $\left\{\frac{n}{\sqrt{n + 4}} - \frac{n}{\sqrt{n + 9}}\right\}$

6 $\left\{\left(1 + \frac{2}{n}\right)^{2n}\right\}$

Exer. 7–32: If the series is positive-term, determine whether it is convergent or divergent; if the series contains negative terms, determine whether it is absolutely convergent, conditionally convergent, or divergent.

7 $\sum_{n=1}^{\infty} \frac{1}{\sqrt[3]{n(n + 1)(n + 2)}}$

8 $\sum_{n=0}^{\infty} \frac{(2n + 3)^2}{(n + 1)^3}$

9 $\sum_{n=1}^{\infty} \left(-\frac{2}{3}\right)^{n-1}$

10 $\sum_{n=0}^{\infty} \frac{1}{2 + (\frac{1}{2})^n}$

11 $\sum_{n=1}^{\infty} \frac{3^{2n+1}}{n5^{n-1}}$

12 $\sum_{n=1}^{\infty} \frac{1}{3^n + 2}$

13 $\sum_{n=1}^{\infty} \frac{n!}{\ln(n + 1)}$

14 $\sum_{n=1}^{\infty} \frac{n^2 - 1}{n^2 + 1}$

15 $\sum_{n=1}^{\infty} (n^2 + 9)(-2)^{1-n}$

16 $\sum_{n=1}^{\infty} \frac{n + \cos n}{n^3 + 1}$

17 $\sum_{n=1}^{\infty} \frac{e^n}{n^e}$

18 $\sum_{n=1}^{\infty} (-1)^{n-1} \frac{n}{n^2 + 1}$

19 $\sum_{n=1}^{\infty} (-1)^n \frac{1}{\sqrt[3]{n}}$

20 $\sum_{n=2}^{\infty} (-1)^n \frac{(0.9)^n}{\ln n}$

21 $\sum_{n=1}^{\infty} \frac{\sin \frac{5\pi}{3} n}{n^{5\pi/3}}$

22 $\sum_{n=2}^{\infty} (-1)^n \frac{\sqrt[3]{n} - 1}{n^2 - 1}$

23 $\sum_{n=1}^{\infty} (-1)^{n-1} \frac{\sqrt{n}}{n + 1}$

24 $\sum_{n=1}^{\infty} (-1)^n \frac{2n + 3}{n!}$

25 $\sum_{n=1}^{\infty} \frac{1 - \cos n}{n^2}$

26 $\frac{2}{1!} - \frac{2 \cdot 4}{2!} + \cdots + (-1)^{n-1} \frac{2 \cdot 4 \cdot \cdots \cdot (2n)}{n!} + \cdots$

27 $\sum_{n=1}^{\infty} \frac{(2n)^n}{n^{2n}}$

28 $\sum_{n=1}^{\infty} \frac{3^{n-1}}{n^2 + 9}$

29 $\sum_{n=1}^{\infty} \frac{e^{2n}}{(2n - 1)!}$

30 $\sum_{n=1}^{\infty} \left(\frac{1}{3^n} - \frac{5}{\sqrt{n}}\right)$

31 $\sum_{n=2}^{\infty} (-1)^n \frac{\sqrt{\ln n}}{n}$

32 $\sum_{n=1}^{\infty} \frac{\tan^{-1} n}{\sqrt{1 + n^2}}$

Exer. 33–38: Use the integral test (11.23) to determine the convergence or divergence of the series.

33 $\sum_{n=1}^{\infty} \frac{1}{(3n+2)^3}$

34 $\sum_{n=2}^{\infty} \frac{n}{\sqrt{n^2-1}}$

35 $\sum_{n=1}^{\infty} n^{-2}e^{1/n}$

36 $\sum_{n=2}^{\infty} \frac{1}{n(\ln n)^3}$

37 $\sum_{n=1}^{\infty} \frac{10}{\sqrt[3]{n+8}}$

38 $\sum_{n=5}^{\infty} \frac{1}{n^2-4n}$

c **Exer. 39–40: Approximate the sum of the series to three decimal places.**

39 $\sum_{n=1}^{\infty} (-1)^{n-1} \frac{1}{(2n+1)!}$

40 $\sum_{n=1}^{\infty} (-1)^{n-1} \frac{1}{n^2(n^2+1)}$

Exer. 41–44: Find the interval of convergence of the series.

41 $\sum_{n=0}^{\infty} \frac{n+1}{(-3)^n} x^n$

42 $\sum_{n=0}^{\infty} (-1)^n \frac{4^{2n}}{\sqrt{n+1}} x^n$

43 $\sum_{n=1}^{\infty} \frac{1}{n \cdot 2^n} (x+10)^n$

44 $\sum_{n=2}^{\infty} \frac{1}{n(\ln n)^2} (x-1)^n$

Exer. 45–46: Find the radius of convergence of the series.

45 $\sum_{n=0}^{\infty} \frac{(2n)!}{(n!)^2} x^n$

46 $\sum_{n=0}^{\infty} \frac{1}{(n+5)!} (x+5)^n$

Exer. 47–52: Find the Maclaurin series for $f(x)$, and state the radius of convergence.

47 $f(x) = \begin{cases} \dfrac{1-\cos x}{x} & \text{if } x \neq 0 \\ 0 & \text{if } x = 0 \end{cases}$

48 $f(x) = xe^{-2x}$

49 $f(x) = \sin x \cos x$

50 $f(x) = \ln(2+x)$

51 $f(x) = (1+x)^{2/3}$

52 $f(x) = \dfrac{1}{\sqrt{1-x^2}}$

53 Find a series representation for e^{-x} in powers of $x+2$.

54 Find a series representation for $\cos x$ in powers of $x - (\pi/2)$.

55 Find a series representation for $\sqrt{x}$ in powers of $x-4$.

c **Exer. 56–60: Use an infinite series to approximate the number to three decimal places.**

56 $1/\sqrt[3]{e}$

57 $\int_0^1 x^2 e^{-x^2}\, dx$

58 $\int_0^1 f(x)\, dx$ with $f(x) = (\sin x)/\sqrt{x}$ if $x \neq 0$ and $f(0) = 0$

59 $\sqrt[5]{1.01}$

60 $e^{-0.25}$

Exer. 61–62: Find Taylor's formula with remainder for the given $f(x)$, c, and n.

61 $f(x) = \ln \cos x,\quad c = \pi/6,\quad n = 3$

62 $f(x) = \sqrt{x-1},\quad c = 2,\quad n = 4$

Exer. 63–64: Find Maclaurin's formula with remainder for the given $f(x)$ and n.

63 $f(x) = e^{-x^2},\quad n = 3$

64 $f(x) = \dfrac{1}{1-x},\quad n = 6$

c 65 Use Taylor's formula with remainder to approximate $\cos 43°$ to four decimal places.

66 Use Taylor's formula with remainder to show that the approximation formula

$$\sin x \approx x - \tfrac{1}{6}x^3 + \tfrac{1}{120}x^5$$

is accurate to four decimal places for $0 \leq x \leq \pi/4$.

CHAPTER

12

TOPICS FROM ANALYTIC GEOMETRY

INTRODUCTION

The *conic sections*, also called *conics*, can be obtained by intersecting a double-napped right circular cone with a plane. By varying the position of the plane, we obtain an *ellipse*, a *parabola*, or a *hyperbola*, as illustrated in the computer art on this page. *Degenerate conics* are obtained if the plane intersects the cone in only one point or along either one or two lines that lie on the cone.

Conic sections were studied extensively by the ancient Greeks, who discovered properties that enable us to state their definitions in terms of points and lines. A remarkable fact about conic sections is that although they were studied thousands of years ago, they are far from obsolete. They are important for present-day investigations in outer space and for the study of the behavior of atomic particles.

Much of the material in this chapter may be review, because it also appears in precalculus textbooks. However, methods of calculus are required to solve problems that include finding equations of tangent lines to conics, calculating areas and volumes of regions determined by conics, and proving reflective properties used in the design of telescopes and searchlight reflectors.

12.1 PARABOLAS

In this section we define a *parabola* and derive equations for parabolas that have either vertical or horizontal axes.

Definition **(12.1)**

> A **parabola** is the set of all points in a plane equidistant from a fixed point F (the **focus**) and a fixed line l (the **directrix**) in the plane.

FIGURE 12.1

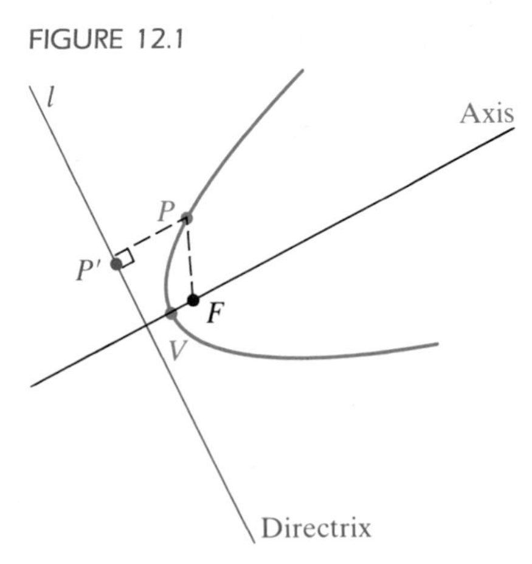

We shall assume that F is not on l, for this would result in a line. If P is a point in the plane and P' is the point on l determined by a line through P that is perpendicular to l (see Figure 12.1), then, by Definition (12.1), P is on the parabola if and only if the distances $d(P, F)$ and $d(P, P')$ are equal. The **axis** of the parabola is the line through F that is perpendicular to the directrix. The **vertex** of the parabola is the point V on the axis halfway from F to l.

To obtain a simple equation for a parabola, place the y-axis along the axis of the parabola, with the origin at the vertex V, as shown in Figure 12.2. In this case, the focus F has coordinates $(0, p)$ for some real number $p \neq 0$, and the equation of the directrix is $y = -p$. (The figure shows the case $p > 0$.) By the distance formula, a point $P(x, y)$ is on the parabola if and only if

$$\sqrt{(x - 0)^2 + (y - p)^2} = \sqrt{(x - x)^2 + (y + p)^2}.$$

FIGURE 12.2

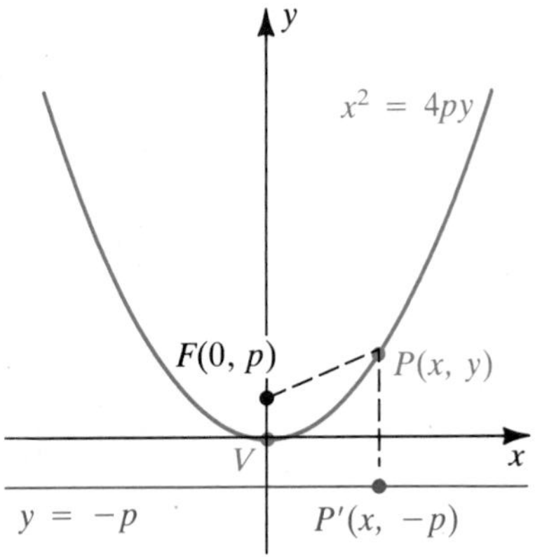

We square both sides and simplify:

$$(x - 0)^2 + (y - p)^2 = (y + p)^2$$
$$x^2 + y^2 - 2py + p^2 = y^2 + 2py + p^2$$
$$x^2 = 4py$$

An equivalent equation is $y = \dfrac{1}{4p}x^2$.

We have shown that the coordinates of every point (x, y) on the parabola satisfy $x^2 = 4py$. Conversely, if (x, y) is a solution of $x^2 = 4py$, then by reversing the previous steps we see that the point (x, y) is on the parabola. If $p > 0$, the parabola opens upward, as in Figure 12.2. If $p < 0$, the parabola opens downward. The graph is symmetric with respect to the y-axis, since substitution of $-x$ for x does not change the equation $x^2 = 4py$.

If we interchange the roles of x and y, we obtain $y^2 = 4px$, or, equivalently, $x = y^2/(4p)$. This is an equation of a parabola with vertex at the origin, focus $F(p, 0)$, and opening right if $p > 0$ or left if $p < 0$.

For convenience we often refer to *the parabola* $x^2 = 4py$ (or $y^2 = 4px$) instead of *the parabola with equation* $x^2 = 4py$ (or $y^2 = 4px$).

The next table summarizes our discussion.

Parabolas with vertex $V(0, 0)$ **(12.2)**

EQUATION	GRAPH FOR $p > 0$	GRAPH FOR $p < 0$
$x^2 = 4py$ $y = \frac{1}{4p}x^2$	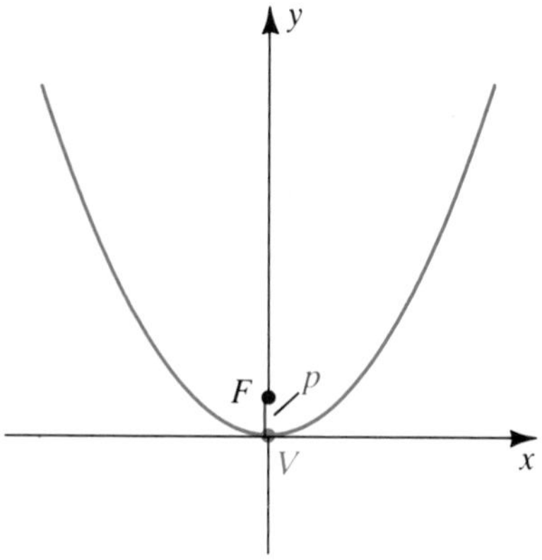	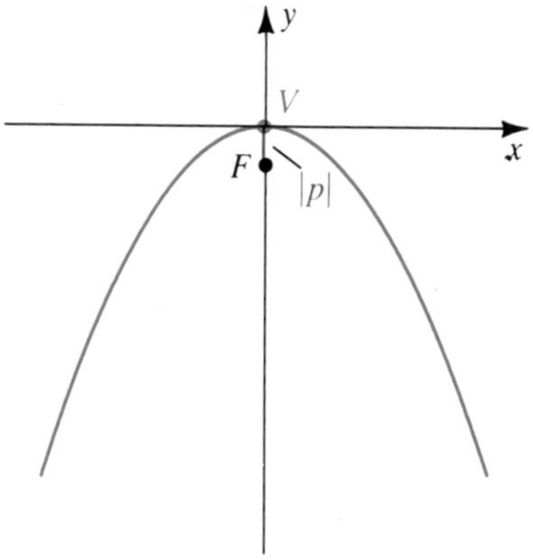
$y^2 = 4px$ $x = \frac{1}{4p}y^2$	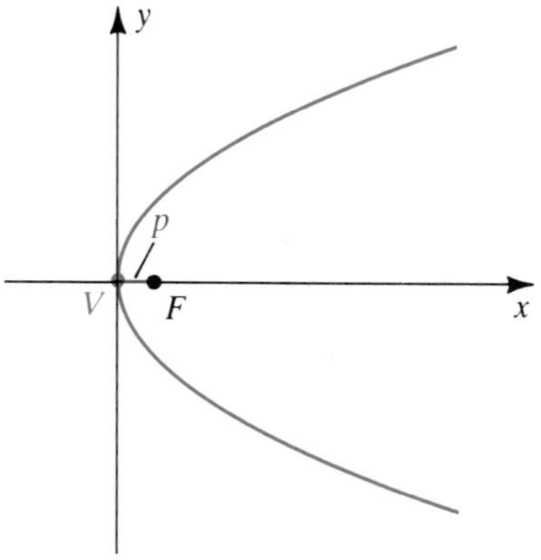	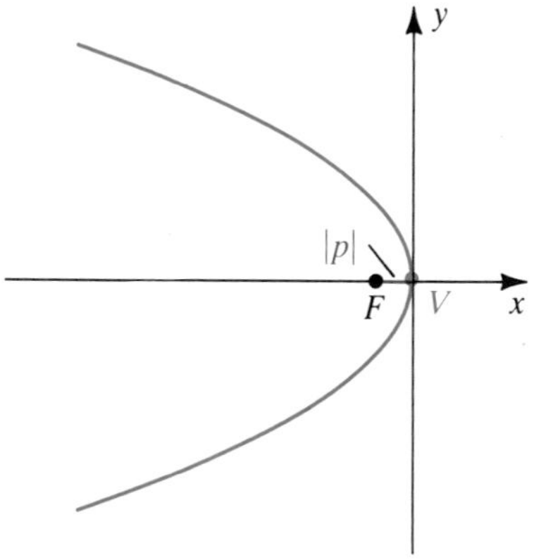

We see from (12.2) that for any nonzero real number a, the graph of $y = ax^2$ or $x = ay^2$ is a parabola with vertex $V(0, 0)$. Moreover, $a = 1/(4p)$, or, equivalently, $p = 1/(4a)$, where $|p|$ is the distance between the focus F and vertex V. To find the directrix l, recall that l is also a distance $|p|$ from V.

FIGURE 12.3

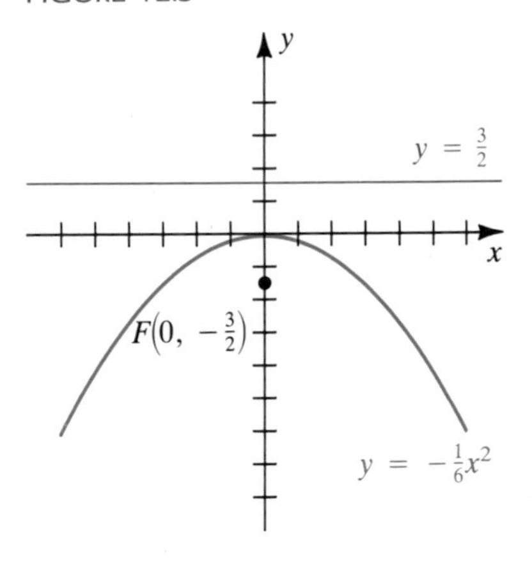

EXAMPLE 1 Find the focus and directrix of the parabola $y = -\frac{1}{6}x^2$, and sketch its graph.

SOLUTION The equation has the form $y = ax^2$ with $a = -\frac{1}{6}$. As in (12.2), $a = 1/(4p)$, or

$$p = \frac{1}{4a} = \frac{1}{4(-\frac{1}{6})} = -\frac{3}{2}.$$

The parabola opens downward and has focus $F(0, -\frac{3}{2})$, as illustrated in Figure 12.3. The directrix is the horizontal line $y = \frac{3}{2}$, which is a distance $\frac{3}{2}$ above V, as shown in the figure.

EXAMPLE 2

(a) Find an equation of a parabola that has vertex at the origin, opens right, and passes through the point $P(7, -3)$.

(b) Find the focus.

FIGURE 12.4

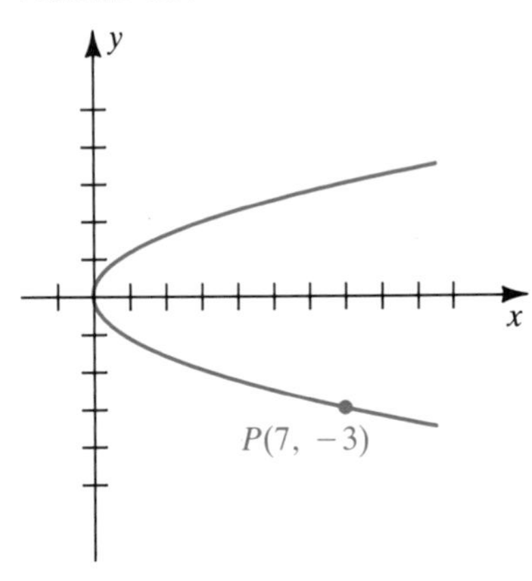

SOLUTION

(a) The parabola is sketched in Figure 12.4. By (12.2), the equation of the parabola has the form $x = ay^2$ for some number a. If $P(7, -3)$ is on the graph, then

$$7 = a(-3)^2, \quad \text{or} \quad a = \tfrac{7}{9}.$$

Hence an equation for the parabola is $x = \frac{7}{9}y^2$.

(b) The focus is a distance p to the right of the vertex, where

$$p = \frac{1}{4a} = \frac{1}{4(\frac{7}{9})} = \frac{9}{28}.$$

Thus, the focus has coordinates $(\frac{9}{28}, 0)$.

To extend our discussion to a parabola with vertex *not* at the origin, we use a **translation of axes**, as illustrated in Figure 12.5, where the x- and y-axes are shifted to positions—denoted by x' and y'—that are par-

FIGURE 12.5

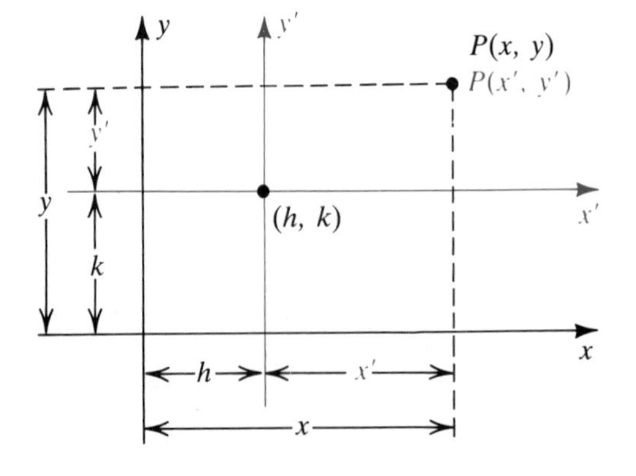

allel to their original positions. Every point P in the plane then has two different ordered pair representations: $P(x, y)$ in the xy-system and $P(x', y')$ in the $x'y'$-system. If the origin of the new $x'y'$-system has coordinates (h, k) in the xy-plane, as illustrated in Figure 12.5, we see that

$$x = x' + h \quad \text{and} \quad y = y' + k.$$

These formulas are true for all values of h and k. Equivalent formulas are

$$x' = x - h \quad \text{and} \quad y' = y - k.$$

This gives us the following.

Translation of axes formulas **(12.3)**

If (x, y) are the coordinates of a point P in an xy-plane and if (x', y') are the coordinates of P in an $x'y'$-plane with origin at the point (h, k) of the xy-plane, then

(i) $x = x' + h, \quad y = y' + k$

(ii) $x' = x - h, \quad y' = y - k$

If, in the xy-plane, a certain collection of points is the graph of an equation in x and y, then to find an equation in x' and y' that has the same graph in the $x'y'$-plane, we substitute $x' + h$ for x and $y' + k$ for y. Conversely, if a set of points in the $x'y'$-plane is the graph of an equation in x' and y', then to find the corresponding equation in x and y, we substitute $x - h$ for x' and $y - k$ for y'.

The following discussion illustrates (12.3). We know that

$$(x')^2 = 4py'$$

is an equation of a parabola with vertex at the origin O' of the $x'y'$-plane. Using translation of axes formulas (ii), we see that

$$(x - h)^2 = 4p(y - k)$$

is an equation of the same parabola in the xy-plane with vertex $V(h, k)$. The focus is $F(h, k + p)$, and the directrix is $y = k - p$.

Squaring the left side of $(x - h)^2 = 4p(y - k)$ and simplifying leads to an equation of the form

$$y = ax^2 + bx + c,$$

where a, b, and c are real numbers. Conversely, if $a \neq 0$, then the graph of $y = ax^2 + bx + c$ is a parabola with a vertical axis. As with $y = ax^2$, we can show that $a = 1/(4p)$, or $p = 1/(4a)$.

Similarly, if we begin with $(y')^2 = 4px'$ and use (12.3), we obtain

$$(y - k)^2 = 4p(x - h),$$

which is an equation of a parabola in the xy-plane with vertex $V(h, k)$, opening right if $p > 0$ or left if $p < 0$. This equation may be written in the form $x = ay^2 + by + c$, where $a = 1/(4p)$, or $p = 1/(4a)$.

The next table summarizes this discussion. For convenience, we have taken $V(h, k)$ in the first quadrant.

Parabolas with vertex $V(h, k)$ **(12.4)**

EQUATION	GRAPH FOR $p > 0$	GRAPH FOR $p < 0$
$(x - h)^2 = 4p(y - k)$ $y = ax^2 + bx + c$ $p = \dfrac{1}{4a}$	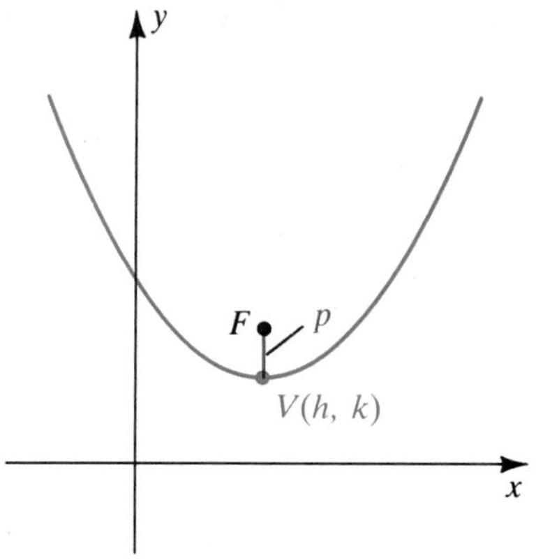 y, F, p, $V(h, k)$, x	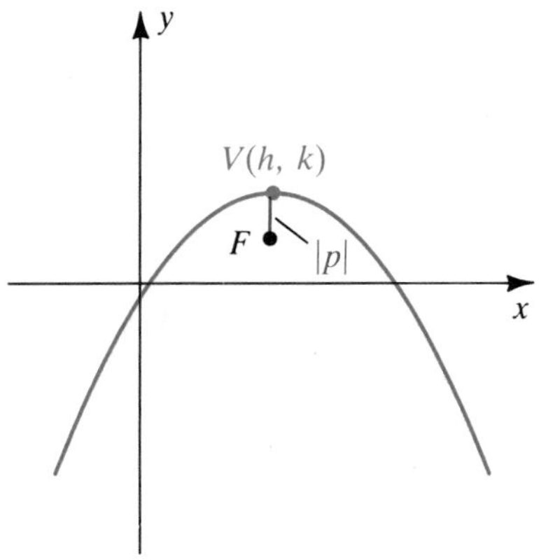 y, $V(h, k)$, F, $\|p\|$, x
$(y - k)^2 = 4p(x - h)$ $x = ay^2 + by + c$ $p = \dfrac{1}{4a}$	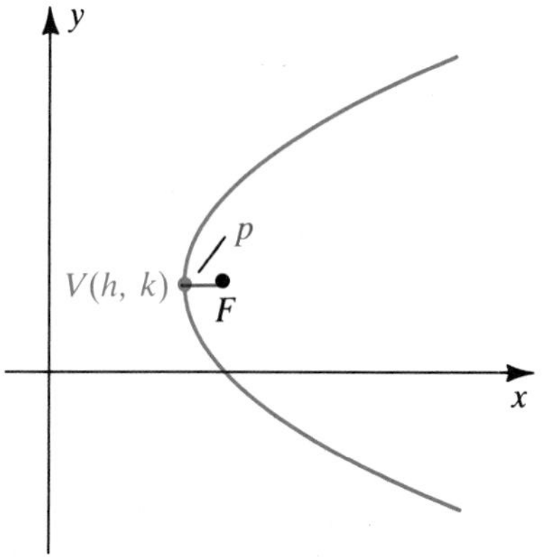 y, p, $V(h, k)$, F, x	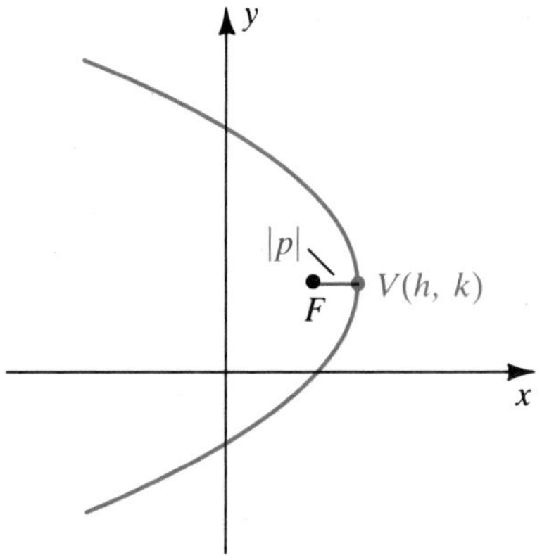 y, $\|p\|$, $V(h, k)$, F, x

If a is any nonzero real number, then, by (12.4), the graph of the equation $y = ax^2 + bx + c$ is a parabola with vertical axis. We could find the vertex $V(h, k)$ algebraically, by completing squares and changing the equation to the form $(x - h)^2 = 4p(y - k)$. Another method for finding V is to use a derivative. Since the tangent line at V is horizontal, we can find the x-coordinate h of V by differentiating $y = ax^2 + bx + c$ and then solving the equation $dy/dx = 0$ for x, as follows:

$$\frac{dy}{dx} = 2ax + b = 0, \quad \text{or} \quad x = -\frac{b}{2a}$$

The y-coordinate k of V can be found by substituting $-b/(2a)$ for x in $y = ax^2 + bx + c$. The coordinates of the focus F and an equation of the directrix l may then be determined by using $p = 1/(4a)$.

The second derivative $d^2y/dx^2 = 2a$ tells us that the parabola is concave upward if $a > 0$ or downward if $a < 0$.

EXAMPLE 3 Discuss and sketch the graph of $y = 2x^2 - 6x + 4$.

SOLUTION By (12.4), the graph is a parabola with vertical axis. As noted in the preceding discussion, the vertex V is the point at which the tangent line is horizontal. Thus, we consider

$$\frac{dy}{dx} = 4x - 6 = 0, \quad \text{or} \quad x = \frac{6}{4} = \frac{3}{2}.$$

Letting $x = \frac{3}{2}$ in the given equation, we obtain

$$y = 2(\tfrac{3}{2})^2 - 6(\tfrac{3}{2}) + 4 = -\tfrac{1}{2}.$$

Hence the vertex is $V(\frac{3}{2}, -\frac{1}{2})$.

The coefficient of x^2 is $a = 2$, so

$$p = \frac{1}{4a} = \frac{1}{4(2)} = \frac{1}{8}.$$

Since the parabola opens upward, the focus F is a distance $p = \frac{1}{8}$ above V. This gives us

$$F(\tfrac{3}{2}, -\tfrac{1}{2} + \tfrac{1}{8}) = F(\tfrac{3}{2}, -\tfrac{3}{8}).$$

The directrix is the horizontal line l that is a distance $p = \frac{1}{8}$ below V. Therefore an equation for l is

$$y = -\tfrac{1}{2} - \tfrac{1}{8}, \quad \text{or} \quad y = -\tfrac{5}{8}.$$

The graph is sketched in Figure 12.6. Note that the y-intercept is 4 and the x-intercepts (the solutions of $2x^2 - 6x + 4 = 0$) are 1 and 2.

FIGURE 12.6

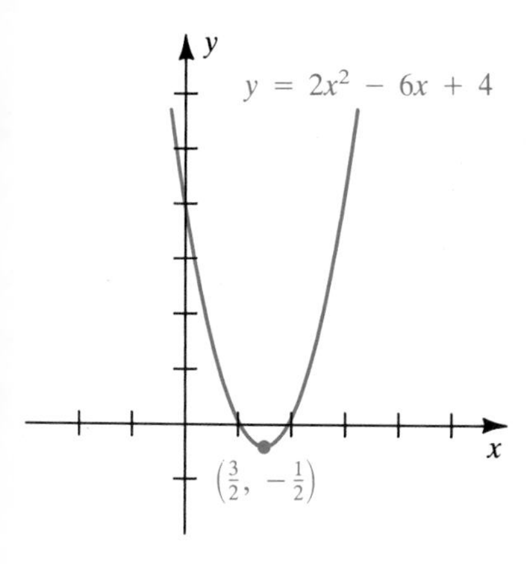

For equations of the form $x = ay^2 + by + c$, we can find the y-coordinate k of the vertex $V(h, k)$ by solving the equation $dx/dy = 0$ for y. The x-coordinate h of V can then be found by substituting k for y in the equation. In this case the parabola opens right or left, and V is the point at which the tangent line is *vertical*, since $dx/dy = 1/(dy/dx)$.

EXAMPLE 4 Discuss and sketch the graph of $x = \frac{1}{2}y^2 + 4y + 11$.

SOLUTION By (12.4), the graph is a parabola with horizontal axis. To find the y-coordinate of the vertex V (the point at which the tangent line is vertical), we consider

$$\frac{dx}{dy} = y + 4 = 0, \quad \text{or} \quad y = -4.$$

Substituting -4 for y in the given equation, we obtain

$$x = \tfrac{1}{2}(-4)^2 + 4(-4) + 11 = 3.$$

Hence the vertex is $V(3, -4)$.

The coefficient of y^2 is $a = \frac{1}{2}$; thus,

$$p = \frac{1}{4a} = \frac{1}{4(\frac{1}{2})} = \frac{1}{2}.$$

FIGURE 12.7

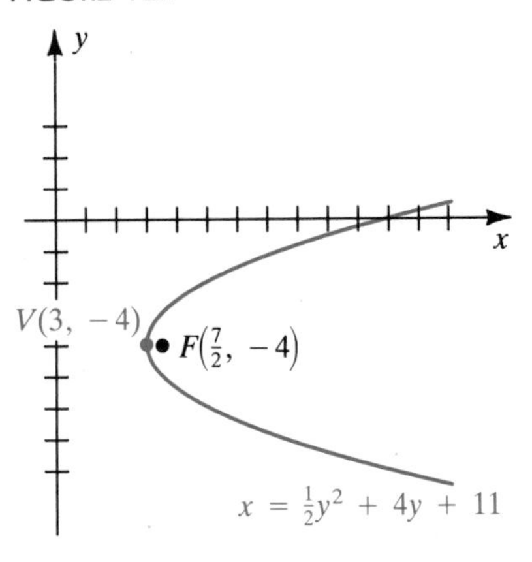

The parabola opens right, so the focus F is a distance $\frac{1}{2}$ to the right of V. This gives us

$$F(3 + \tfrac{1}{2}, -4) = F(\tfrac{7}{2}, -4).$$

The directrix is the vertical line l, a distance $\frac{1}{2}$ to the left of V. Therefore an equation for l is

$$x = 3 - \tfrac{1}{2}, \quad \text{or} \quad x = \tfrac{5}{2}.$$

The parabola is sketched in Figure 12.7.

FIGURE 12.8

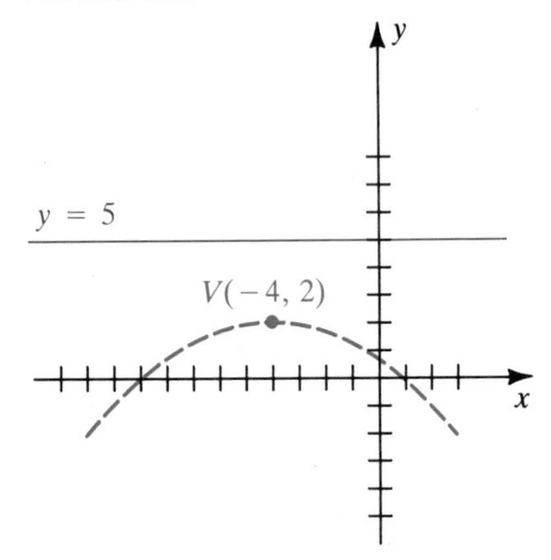

EXAMPLE 5 Find an equation of the parabola with vertex $V(-4, 2)$ and directrix $y = 5$.

SOLUTION The vertex and directrix are shown in Figure 12.8. The dashes indicate a possible shape for the parabola. It follows that an equation of the parabola is

$$(x - h)^2 = 4p(y - k),$$

with $h = -4$, $k = 2$, and $p = -3$. This gives us

$$(x + 4)^2 = -12(y - 2).$$

This equation can be expressed in the form $y = ax^2 + bx + c$, as follows:

$$\begin{aligned} x^2 + 8x + 16 &= -12y + 24 \\ 12y &= -x^2 - 8x + 8 \\ y &= -\tfrac{1}{12}x^2 - \tfrac{2}{3}x + \tfrac{2}{3} \end{aligned}$$

EXAMPLE 6 Traffic engineers are designing a stretch of highway that will connect a horizontal highway to one having a 20% grade (slope $\frac{1}{5}$), as illustrated by the side view in Figure 12.9. The smooth transition is to take place over a horizontal distance of 800 feet, with a parabolic-shaped highway connecting points A and B.

(a) Find an equation of the parabola with vertex A that has a tangent line of slope 0 at A and $\frac{1}{5}$ at B.

FIGURE 12.9

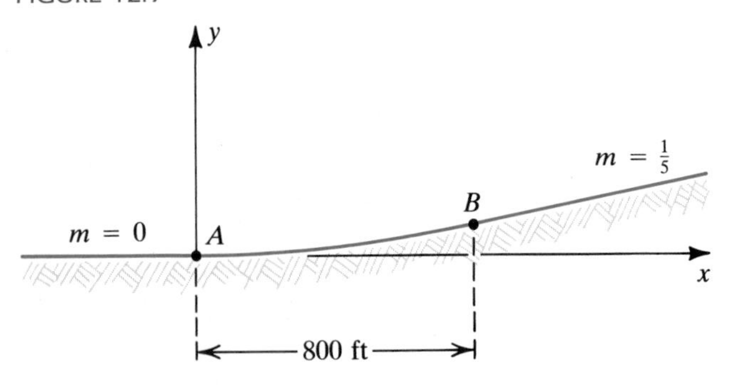

(b) Find the difference in elevation between B and the horizontal portion of the highway.

(c) Set up an integral for the length of the highway between A and B.

(d) Evaluate the integral in (c).

SOLUTION

(a) If we introduce an xy-plane, as in Figure 12.9, then a parabola with vertex $(0, 0)$ and opening upward has the equation $y = ax^2$ for some number a. The slope m of the tangent line at any point $P(x, y)$ on the parabola is

$$m = \frac{dy}{dx} = 2ax.$$

Thus, the tangent line has slope 0 at A (the origin). The transition at B is smooth if $m = \frac{1}{5}$ when $x = 800$; that is, if

$$\tfrac{1}{5} = 2a(800), \quad \text{or} \quad a = \tfrac{1}{8000}.$$

Hence an equation of the parabola is $y = \frac{1}{8000}x^2$.

(b) The difference in elevation between B and the horizontal portion of the highway is the y-coordinate of B. Thus, we let $x = 800$ in the equation found in (a), obtaining

$$y = \tfrac{1}{8000}(800)^2 = 80 \text{ ft}.$$

(c) By (6.14), the length (in feet) of the parabolic-shaped highway between A and B is

$$L = \int_0^{800} \sqrt{1 + \left(\frac{dy}{dx}\right)^2}\, dx = \int_0^{800} \sqrt{1 + \left(\frac{x}{4000}\right)^2}\, dx.$$

(d) The value of the integral in (c) can be found by using either the trigonometric substitution $x/4000 = \tan\theta$ followed by integration by parts or the table of integrals in Appendix IV. For an approximate value we can use the trapezoidal rule or Simpson's rule. Using Formula 21 from the table of integrals,

$$\int \sqrt{a^2 + u^2}\, du = \frac{u}{2}\sqrt{a^2 + u^2} + \frac{a^2}{2} \ln\left|u + \sqrt{a^2 + u^2}\right| + C,$$

and the substitution

$$u = \tfrac{1}{4000}x, \quad du = \tfrac{1}{4000}\,dx$$

and noting that $u = \frac{1}{5}$ when $x = 800$, we obtain the following:

$$\begin{aligned} L &= 4000 \int_0^{1/5} \sqrt{1 + u^2}\, du \\ &= 4000\left[\frac{u}{2}\sqrt{1 + u^2} + \frac{1}{2}\ln\left|u + \sqrt{1 + u^2}\right|\right]_0^{1/5} \\ &= 80\sqrt{26} + 2000 \ln\left(\frac{1 + \sqrt{26}}{5}\right) \\ &\approx 805 \text{ ft} \end{aligned}$$

It is of interest to note that $d(A, B) \approx 804$ ft.

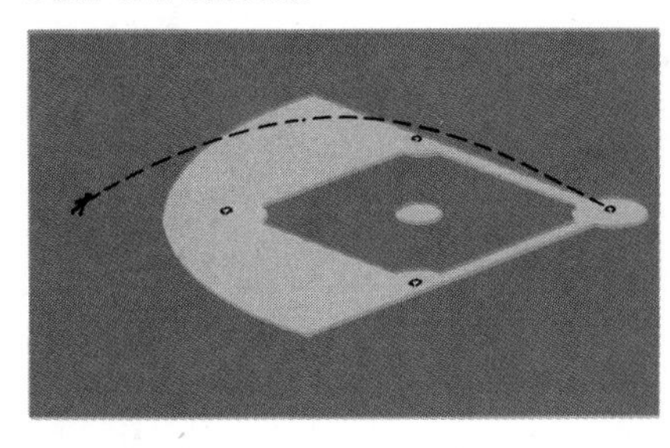
FIGURE 12.10
Path of a baseball

There are many applications of parabolas. If, as shown in Figure 12.10, a baseball player throws a ball in a nonvertical direction and if gravity is the only force acting on the ball (that is, air resistance and other outside factors are negligible), then the path of the ball (or any projected object) is parabolic. We can use this fact (which will be proved in Chapter 15) to determine where the ball (or object) will land, to find its maximum height, and so on.

The shapes of cables in certain types of suspension bridges are parabolic. However, as we pointed out in Section 8.3, for a *freely* hanging cable, the curve is a catenary, not a parabola.

An important property is associated with tangent lines to parabolas. Suppose l is the tangent line at a point $P(x_1, y_1)$ on the parabola $y^2 = 4px$, and let F be the focus. As in Figure 12.11, let α denote the angle between l and the line segment FP, and let β denote the angle between l and the horizontal half-line with endpoint P. In Exercise 42 you are asked to prove that $\alpha = \beta$. This *reflective property* has many applications. For example, the shape of the mirror in a searchlight is obtained by revolving a parabola about its axis. If a light source is placed at F, then, by a law of physics (*the angle of reflection equals the angle of incidence*), a beam of light will be reflected along a line parallel to the axis (see Figure 12.12(i)). The same principle is employed in the construction of mirrors for telescopes or solar ovens—a beam of light coming toward the parabolic mirror, parallel to the axis, will be reflected into the focus (see Figure 12.12(ii)). Antennas for radar systems, radio telescopes, and field microphones used at football games also make use of this property.

FIGURE 12.11
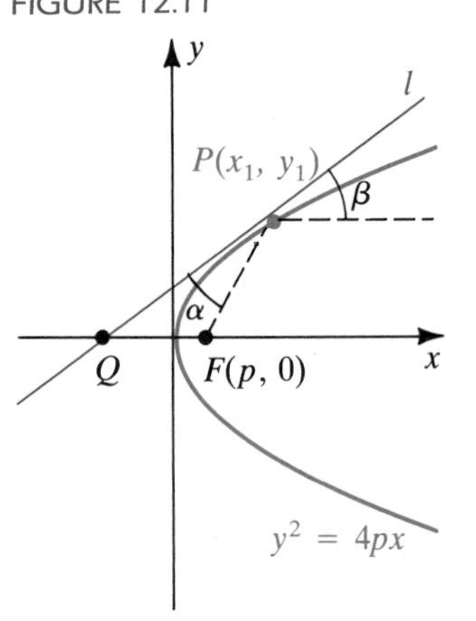

FIGURE 12.12

(i) Searchlight mirror

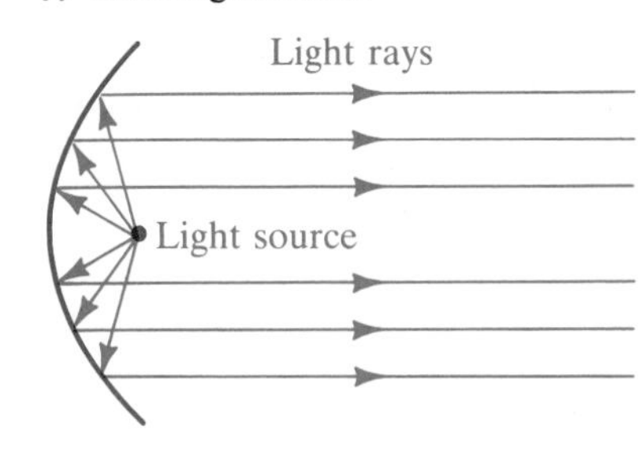

(ii) Telescope mirror

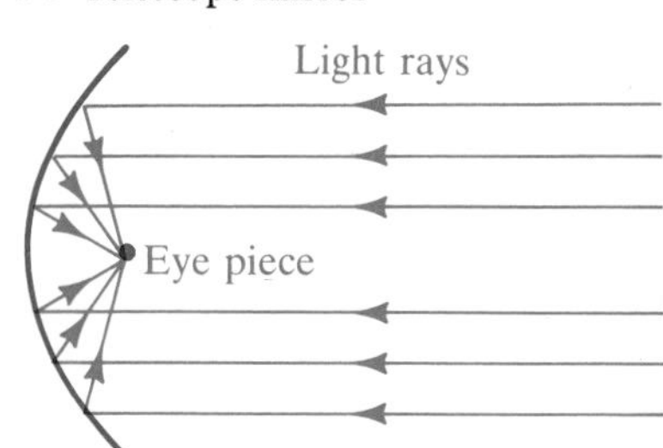

EXERCISES 12.1

Exer. 1–6: Find the vertex, focus, and directrix of the parabola. Sketch its graph, showing the focus and the directrix.

1 $y = -\frac{1}{12}x^2$

2 $x = 2y^2$

3 $2y^2 = -3x$

4 $x^2 = -3y$

5 $y = 8x^2$

6 $y^2 = -100x$

Exer. 7–16: Find the vertex and focus of the parabola. Sketch its graph, showing the focus.

7 $y = x^2 - 4x + 2$

8 $y = 8x^2 + 16x + 10$

9 $y^2 - 12 = 12x$

10 $y^2 - 20y + 100 = 6x$

11 $y^2 - 4y - 2x - 4 = 0$

12 $y^2 + 14y + 4x + 45 = 0$

13 $4x^2 + 40x + y + 106 = 0$

14 $y = 40x - 97 - 4x^2$

15 $x^2 + 20y = 10$

16 $4x^2 + 4x + 4y + 1 = 0$

Exer. 17–24: Find an equation of the parabola that satisfies the given conditions.

17 focus $F(2, 0)$, directrix $x = -2$

18 focus $F(0, -4)$, directrix $y = 4$

19 vertex $V(3, -5)$, directrix $x = 2$

20 vertex $V(-2, 3)$, directrix $y = 5$

21 vertex $V(-1, 0)$, focus $F(-4, 0)$

22 vertex $V(1, -2)$, focus $F(1, 0)$

23 vertex at the origin, symmetric to the y-axis, and passing through the point $(2, -3)$

24 vertex $V(-3, 5)$, axis parallel to the x-axis, and passing through the point $(5, 9)$

25 A searchlight reflector is designed so that every cross section containing its axis of symmetry is a parabola with the light source at the focus. Where is the focus if the reflector is 3 feet across at the opening and 1 foot deep?

26 Find an equation of the parabola that has a vertical axis and passes through $A(2, 5)$, $B(-2, -3)$, and $C(1, 6)$.

27 Let R be the region bounded by the parabola $x^2 = 4y$ and the line l through the focus that is perpendicular to the axis of the parabola.

(a) Find the area of R.

(b) If R is revolved about the y-axis, find the volume of the resulting solid.

(c) If R is revolved about the x-axis, find the volume of the resulting solid.

28 Work (a)–(c) of Exercise 27 if R is the region bounded by the graphs of $y^2 = 2x - 6$ and $x = 5$.

29 A *paraboloid of revolution* is formed by revolving a parabola about its axis. Paraboloids are the basic shape for a wide variety of collectors and reflectors. Shown in the figure is a (finite) paraboloid of altitude h and radius of base r.

(a) The *focal length* of the paraboloid is the distance p between the vertex and the focus of the parabola. Express p in terms of r and h.

(b) Find the volume of the paraboloid.

EXERCISE 29

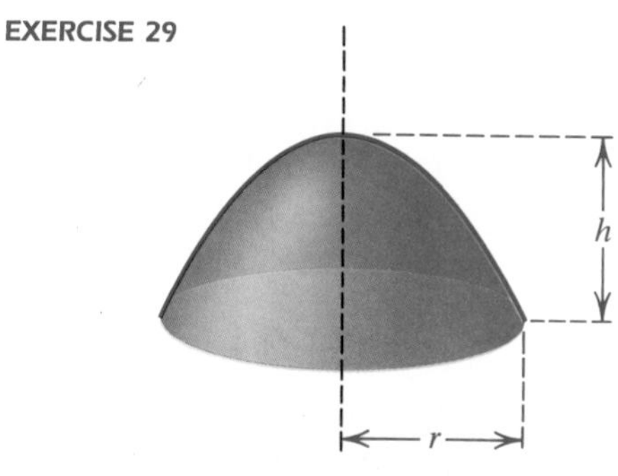

30 Refer to Exercise 29. A radio telescope has the shape of a paraboloid of revolution with focal length p and diameter of base $2a$.

(a) Show that the surface area S available for collecting radio waves is

$$S = \frac{8\pi p^2}{3}\left[\left(1 + \frac{a^2}{4p^2}\right)^{3/2} - 1\right].$$

(b) One of the largest radio telescopes, located in Jodrell Bank, Cheshire, England, has diameter 250 feet and focal length 50 feet. Approximate S to the nearest square foot.

31 One section of a suspension bridge has its weight uniformly distributed between twin towers that are 400 feet apart and that rise 90 feet above the horizontal roadway. A cable strung between the tops of the towers has the shape of a parabola, with center point 10 feet above the roadway. Suppose coordinate axes are introduced, as shown in the figure.

(a) Find an equation for the parabola.

(b) Set up an integral whose value is the length of the cable.

(c) If nine equispaced vertical cables are used to support the parabolic cable, find the total length of these supports.

EXERCISE 31

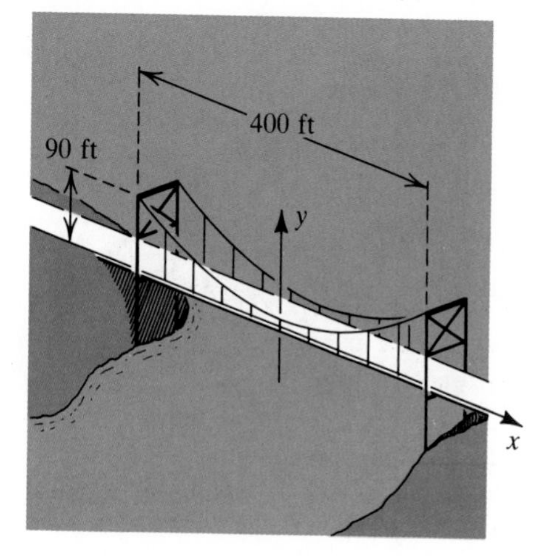

32 Let R be the region bounded by the parabola $x = ay^2$ and the line l through the focus that is perpendicular to the axis of the parabola. Find the area of the curved surface obtained by revolving R about the x-axis.

33 A parabolic arch has a center height of k feet, as shown in the figure. Prove that the height of the largest rectangle that can fit under the arch is $\frac{2}{3}k$ feet.

EXERCISE 33

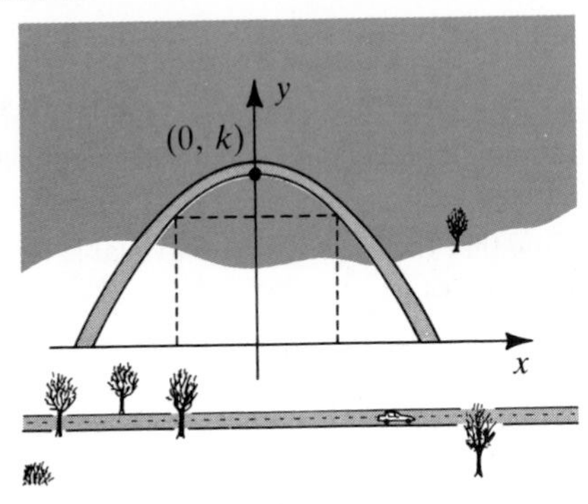

34 Prove that the point on a parabola that is closest to the focus is the vertex.

35 A child throws a ball at an angle of 45° from the edge of a plateau, above a hill that has slope $-\frac{3}{4}$, as shown in the figure.

(a) If the ball lands 50 feet down the hill, find an equation for its parabolic path. (Disregard the height of the child.)

(b) What is the maximum height of the ball *off the ground*?

EXERCISE 35

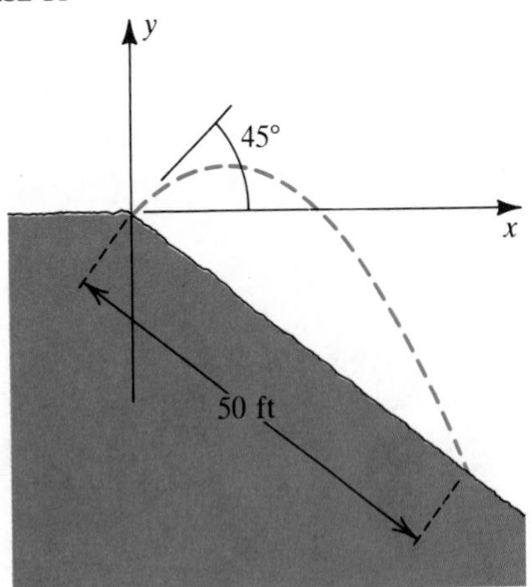

36 A cylindrical container, partially filled with mercury, is rotated about its axis at the rate of ω radians/sec (see figure). From physics, the function f, whose graph generates the inside surface of the mercury, is a solution of the differential equation $y' = (\omega^2/g)x$, where $g = 32 \text{ ft/sec}^2$.

(a) Show that $y = \frac{1}{64}\omega^2 x^2 + f(0)$.

(b) For what value of ω is the focus of the parabola in (a) 2 feet from the vertex?

EXERCISE 36

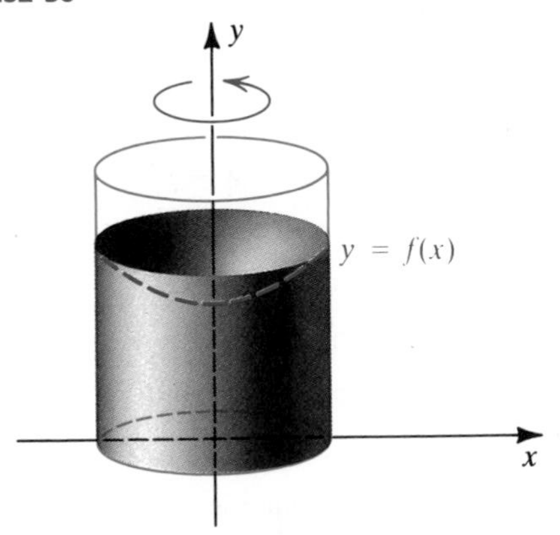

37 The parabola $y^2 = 4p(x + p)$ has its focus at the origin and axis along the x-axis. By assigning different values to p, a family of *confocal parabolas* is obtained, as shown in the figure. Families of this type occur in the study of electricity and magnetism.

(a) Show that there are exactly two parabolas in the family that pass through a given point $P(x_1, y_1)$ if $y_1 \neq 0$.

(b) Show that the two parabolas in (a) are mutually orthogonal—that is, the tangent lines at $P(x_1, y_1)$ are perpendicular.

EXERCISE 37

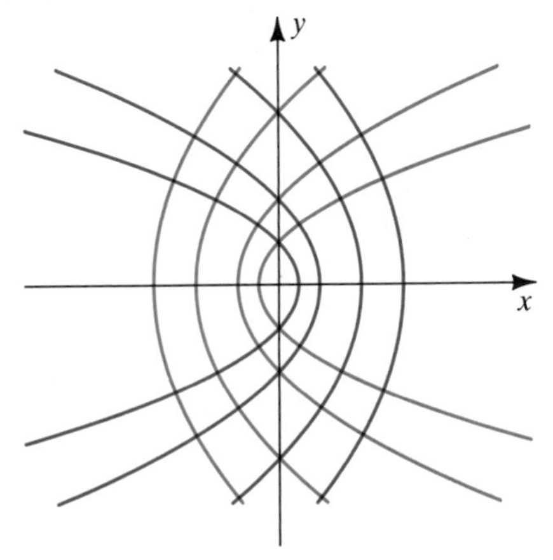

38 A *focal chord* is a line segment that passes through the focus of a parabola and has its endpoints on the parabola. If AB is a focal chord, prove that the tangent lines

at A and B

(a) are perpendicular

(b) intersect on the directrix

Exer. 39–40: Refer to Exercise 37. Graph, on the same coordinate axes, the confocal parabolas $x^2 = 4p(y + p)$ for the given values of p. Estimate the points of intersection and verify that the tangent lines at these points are perpendicular.

39 $p = 1.3, -1$

40 $p = 0.2, -2.1$

41 Let m and b be nonzero real numbers.

(a) If the line $y = mx + b$ intersects the parabola $y^2 = 4px$ in only one point, show that $p = mb$.

(b) Show that the slope of the tangent line to $y^2 = 4px$ at $P(x_1, y_1)$ is $y_1/(2x_1)$.

42 Establish the reflective property of the parabola. (*Hint:* Show that $d(Q, F) = d(F, P)$ in Figure 12.11, and use the result in Exercise 41.)

43 Suppose the tangent line l to a parabola at P intersects the directrix at Q, as shown in the figure. If F is the focus, prove that angle PFQ is a right angle.

EXERCISE 43

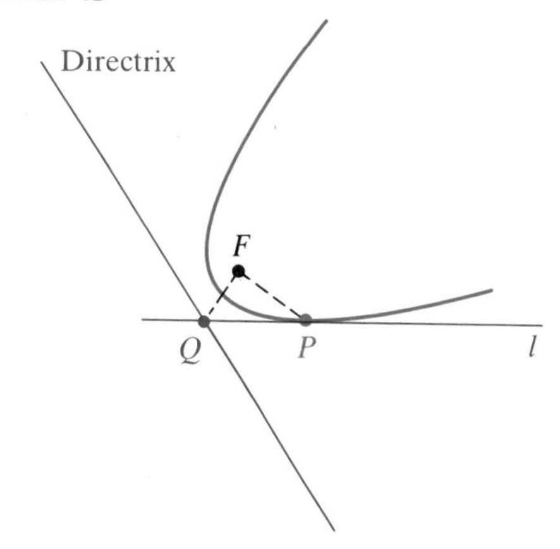

12.2 ELLIPSES

An ellipse may be defined as follows. (*Foci* is the plural of *focus*.)

Definition (12.5)

An **ellipse** is the set of all points in a plane, the sum of whose distances from two fixed points (the **foci**) in the plane is constant.

FIGURE 12.13

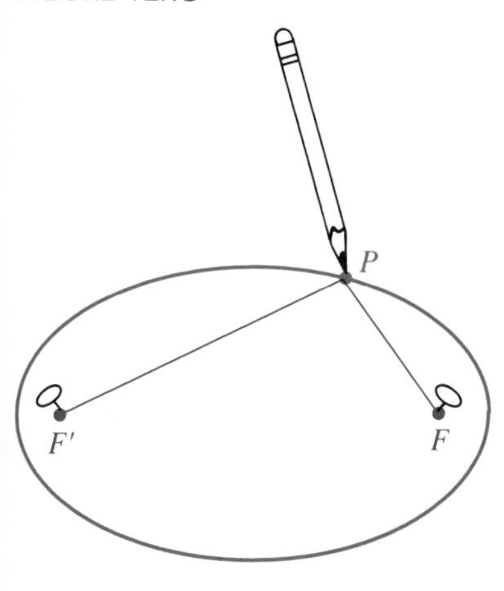

We can construct an ellipse on paper as follows: Insert two pushpins in the paper at any points F and F', and fasten the ends of a piece of string to the pins. After looping the string around a pencil and drawing it tight, as at point P in Figure 12.13, move the pencil, keeping the string tight. The sum of the distances $d(F, P)$ and $d(F', P)$ is the length of the string and hence is constant; thus, the pencil will trace out an ellipse with foci at F and F'. By changing the positions of F and F' while keeping the length of the string fixed, we can vary the shape of the ellipse considerably. If F and F' are far apart so that $d(F, F')$ is almost the same as the length of the string, the ellipse is flat. If $d(F, F')$ is close to zero, the ellipse is almost circular. If $F = F'$, we obtain a circle with center F.

By introducing suitable coordinate systems, we may derive simple equations for ellipses. Let us choose the x-axis as the line through the two foci F and F', with the origin at the midpoint of the segment $F'F$. This midpoint is the **center** of the ellipse. If F has coordinates $(c, 0)$ with $c > 0$, then, as in Figure 12.14, on the following page, F' has coordinates $(-c, 0)$. Hence the distance between F and F' is $2c$. The constant sum of the distances of P from F and F' will be denoted by $2a$. To obtain points that are not on the x-axis, we must have $2a > 2c$—that is, $a > c$. By

FIGURE 12.14

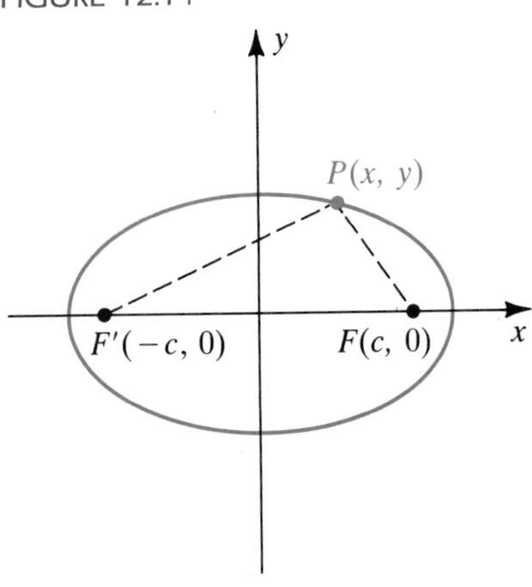

definition, $P(x, y)$ is on the ellipse if and only if the following equivalent equations are true:

$$d(P, F) + d(P, F') = 2a$$

$$\sqrt{(x - c)^2 + (y - 0)^2} + \sqrt{(x + c)^2 + (y - 0)^2} = 2a$$

$$\sqrt{(x - c)^2 + y^2} = 2a - \sqrt{(x + c)^2 + y^2}$$

Squaring both sides of the last equation gives us

$$x^2 - 2cx + c^2 + y^2 = 4a^2 - 4a\sqrt{(x + c)^2 + y^2} + x^2 + 2cx + c^2 + y^2,$$

or $a\sqrt{(x + c)^2 + y^2} = a^2 + cx.$

Squaring both sides again yields

$$a^2(x^2 + 2cx + c^2 + y^2) = a^4 + 2a^2cx + c^2x^2,$$

or
$$x^2(a^2 - c^2) + a^2y^2 = a^2(a^2 - c^2).$$

Dividing both sides by $a^2(a^2 - c^2)$, we obtain

$$\frac{x^2}{a^2} + \frac{y^2}{a^2 - c^2} = 1.$$

Recalling that $a > c$ and therefore $a^2 - c^2 > 0$, we let

$$b = \sqrt{a^2 - c^2}, \quad \text{or} \quad b^2 = a^2 - c^2.$$

This gives us

$$\frac{x^2}{a^2} + \frac{y^2}{b^2} = 1.$$

Since $c > 0$ and $b^2 = a^2 - c^2$, it follows that $a^2 > b^2$ and hence $a > b$.

We have shown that the coordinates of every point (x, y) on the ellipse in Figure 12.15 satisfy the equation $(x^2/a^2) + (y^2/b^2) = 1$. Conversely, if

FIGURE 12.15

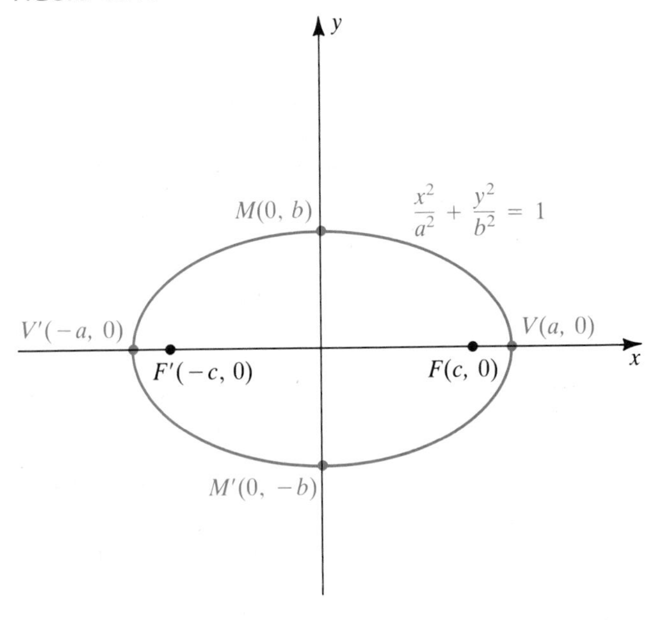

(x, y) is a solution of this equation, then by reversing the preceding steps we see that the point (x, y) is on the ellipse.

We may find the x-intercepts of the ellipse by letting $y = 0$ in the equation. Doing so gives us $x^2/a^2 = 1$, or $x^2 = a^2$, and consequently the x-intercepts are a and $-a$. The corresponding points $V(a, 0)$ and $V'(-a, 0)$ on the graph are the **vertices** of the ellipse (see Figure 12.15). The line segment $V'V$ is the **major axis**. Similarly, letting $x = 0$ in the equation, we obtain $y^2/b^2 = 1$, or $y^2 = b^2$. Hence the y-intercepts are b and $-b$. The segment between $M'(0, -b)$ and $M(0, b)$ is the **minor axis** of the ellipse. The major axis is always longer than the minor axis, since $a > b$.

Applying tests for symmetry, we see that the ellipse is symmetric with respect to the x-axis, the y-axis, and the origin.

The preceding discussion may be summarized as follows.

Theorem **(12.6)**

The graph of the equation

$$\frac{x^2}{a^2} + \frac{y^2}{b^2} = 1$$

for $a^2 > b^2$ is an ellipse with vertices $(\pm a, 0)$. The endpoints of the minor axis are $(0, \pm b)$. The foci are $(\pm c, 0)$, where $c^2 = a^2 - b^2$.

FIGURE 12.16

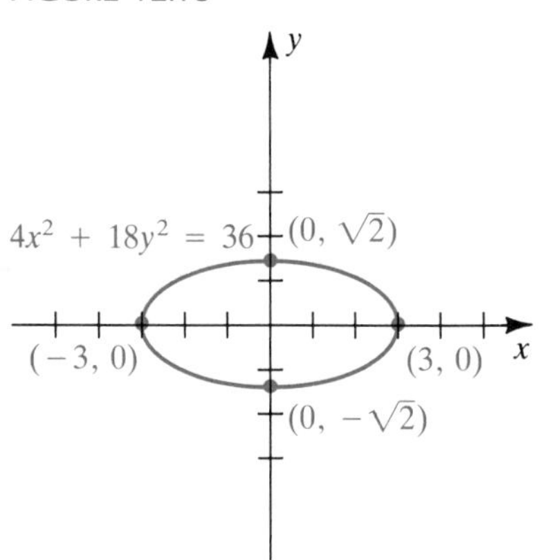

EXAMPLE 1 Discuss and sketch the graph of $4x^2 + 18y^2 = 36$.

SOLUTION To obtain the form in Theorem (12.6), we divide both sides of the equation by 36 and simplify. This leads to

$$\frac{x^2}{9} + \frac{y^2}{2} = 1,$$

which is in the proper form, with $a^2 = 9$ and $b^2 = 2$. Thus, $a = 3$ and $b = \sqrt{2}$; hence the endpoints of the major axis are $(\pm 3, 0)$, and the endpoints of the minor axis are $(0, \pm\sqrt{2})$. Since

$$c^2 = a^2 - b^2 = 9 - 2 = 7, \quad \text{or} \quad c = \sqrt{7},$$

the foci are $(\pm\sqrt{7}, 0)$. The graph is sketched in Figure 12.16.

EXAMPLE 2 Find an equation of the ellipse with vertices $(\pm 4, 0)$ and foci $(\pm 2, 0)$.

SOLUTION In the notation of Theorem (12.6), $a = 4$ and $c = 2$. Since $c^2 = a^2 - b^2$, we see that $b^2 = a^2 - c^2 = 16 - 4 = 12$. This gives us

$$\frac{x^2}{16} + \frac{y^2}{12} = 1.$$

We sometimes choose the major axis of the ellipse along the y-axis. If the foci are $(0, \pm c)$, then by the same type of argument used previously, we obtain the following.

Theorem (12.7)

> The graph of the equation
>
> $$\frac{x^2}{b^2} + \frac{y^2}{a^2} = 1$$
>
> for $a^2 > b^2$ is an ellipse with vertices $(0, \pm a)$. The endpoints of the minor axis are $(\pm b, 0)$. The foci are $(0, \pm c)$, where $c^2 = a^2 - b^2$.

FIGURE 12.17

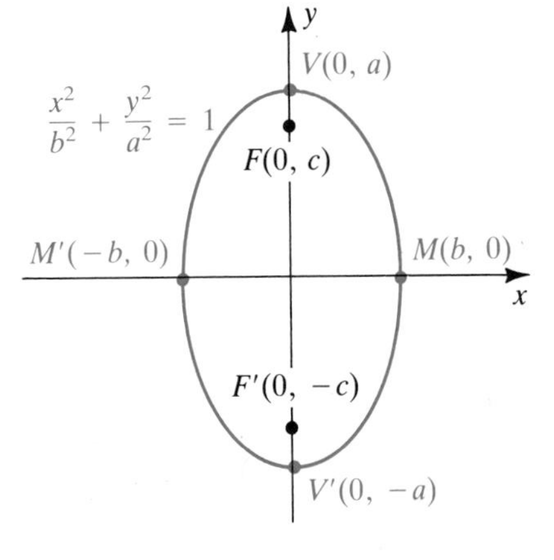

A typical graph is sketched in Figure 12.17.

The preceding discussion shows that an equation of an ellipse with center at the origin and foci on a coordinate axis can always be written in the form

$$\frac{x^2}{p} + \frac{y^2}{q} = 1, \quad \text{or} \quad qx^2 + py^2 = pq,$$

with p and q positive and $p \neq q$. If $p > q$, the major axis is on the x-axis, and if $q > p$, the major axis is on the y-axis. It is unnecessary to memorize these facts, because in any given problem the major axis can be determined by examining the x- and y-intercepts.

EXAMPLE 3 Sketch the graph of $9x^2 + 4y^2 = 25$.

FIGURE 12.18

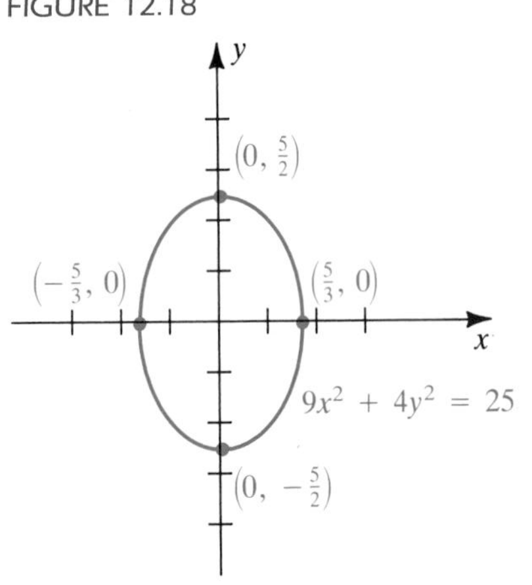

SOLUTION The graph is an ellipse with center at the origin and foci on one of the coordinate axes. To find x-intercepts, we let $y = 0$, obtaining

$$9x^2 = 25, \quad \text{or} \quad x = \pm\tfrac{5}{3}.$$

Similarly, to find the y-intercepts, we let $x = 0$, obtaining

$$4y^2 = 25, \quad \text{or} \quad y = \pm\tfrac{5}{2}.$$

This enables us to sketch the ellipse (see Figure 12.18). Since $\frac{5}{3} < \frac{5}{2}$, the major axis is on the y-axis.

To find the foci, we first calculate

$$c^2 = a^2 - b^2 = (\tfrac{5}{2})^2 - (\tfrac{5}{3})^2 = \tfrac{125}{36}.$$

Thus, $c = 5\sqrt{5}/6$ and the foci are $(0, \pm 5\sqrt{5}/6)$.

We can use the translation of axes formulas (12.3) to extend our work to an ellipse with center at any point $C(h, k)$ in the xy-plane. For example, since the graph of

$$\frac{(x')^2}{a^2} + \frac{(y')^2}{b^2} = 1$$

is an ellipse with center at O' in an $x'y'$-plane (see Figure 12.19, on the next page), its equation relative to the xy-coordinate system is

$$\frac{(x-h)^2}{a^2} + \frac{(y-k)^2}{b^2} = 1.$$

FIGURE 12.19

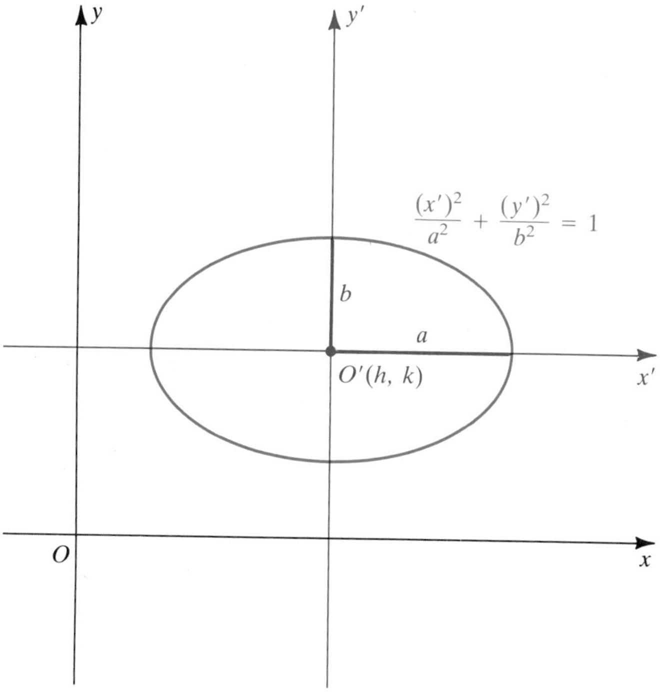

Squaring the indicated terms in the last equation and simplifying gives us an equation of the form

$$Ax^2 + Cy^2 + Dx + Ey + F = 0,$$

where the coefficients are real numbers and both A and C are positive. Conversely, if we start with such an equation, then by completing squares we can obtain a form that displays the center of the ellipse and the lengths of the major and minor axes. This technique is illustrated in the next example.

EXAMPLE 4 Discuss and sketch the graph of the equation

$$16x^2 + 9y^2 + 64x - 18y - 71 = 0.$$

SOLUTION We begin by writing the equation in the form

$$16(x^2 + 4x \qquad) + 9(y^2 - 2y \qquad) = 71.$$

Next, we complete the squares for the expressions within parentheses:

$$16(x^2 + 4x + 4) + 9(y^2 - 2y + 1) = 71 + 64 + 9$$

By adding 4 to the expression within the first parentheses we have added 64 to the left-hand side of the equation, and hence we must compensate by adding 64 to the right-hand side. Similarly, by adding 1 to the expression within the second parentheses we have added 9 to the left side, and consequently we must also add 9 to the right side. The last equation may be written

$$16(x + 2)^2 + 9(y - 1)^2 = 144.$$

Dividing by 144, we obtain

$$\frac{(x+2)^2}{9} + \frac{(y-1)^2}{16} = 1,$$

which is of the form

$$\frac{(x')^2}{9} + \frac{(y')^2}{16} = 1,$$

FIGURE 12.20

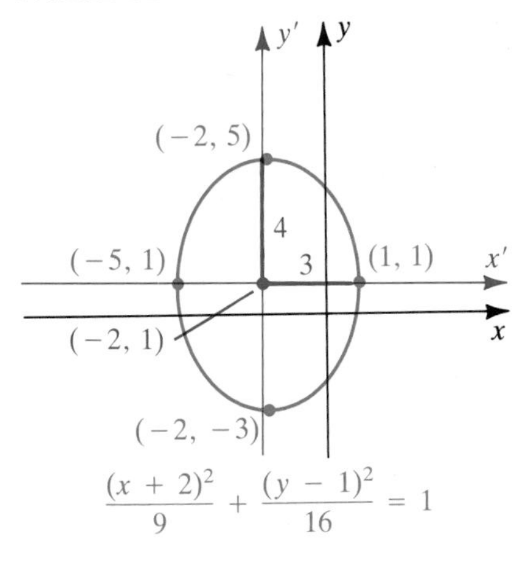

with $x' = x + 2$ and $y' = y - 1$. This corresponds to letting $h = -2$ and $k = 1$ in the translation of axes formulas.

The graph of the equation $(x')^2/9 + (y')^2/16 = 1$ is an ellipse with center at the origin O' in the $x'y'$-plane and major axis on the y'-axis. It follows that the graph of the given equation is an ellipse with center $C(-2, 1)$ in the xy-plane and major axis on the vertical line $x = -2$. Using $a = 4$ and $b = 3$ gives us the ellipse in Figure 12.20.

To find the foci, we first calculate

$$c^2 = a^2 - b^2 = 16 - 9 = 7.$$

The distance from the center of the ellipse to the foci is $c = \sqrt{7}$. Since the center is $(-2, 1)$, the foci are $(-2, 1 \pm \sqrt{7})$.

In the next two examples we apply methods of calculus to solve several problems involving ellipses.

EXAMPLE 5

(a) Use implicit differentiation to find a formula for the slope of the tangent line at the point $P(x_1, y_1)$ on the ellipse

$$\frac{x^2}{a^2} + \frac{y^2}{b^2} = 1,$$

where $a > b > 0$.

(b) Apply the formula found in (a) to determine where the tangent line is horizontal or vertical.

(c) Prove that an equation of the tangent line at $P(x_1, y_1)$ is

$$\frac{x_1 x}{a^2} + \frac{y_1 y}{b^2} = 1.$$

SOLUTION

(a) Differentiating the equation of the ellipse implicitly, we obtain

$$\frac{2x}{a^2} + \frac{2yy'}{b^2} = 0, \quad \text{or} \quad y' = -\frac{b^2 x}{a^2 y}.$$

Hence the slope m of the tangent line at $P(x_1, y_1)$ is $m = -(b^2 x_1)/(a^2 y_1)$.

(b) The tangent line is horizontal if $b^2 x = 0$, or $x = 0$. This gives us the endpoints $(0, \pm b)$ of the minor axis.

The tangent line is vertical if $a^2y = 0$, or $y = 0$. This gives us the vertices $(\pm a, 0)$ of the ellipse.

(c) Using the formula for m in (a), we find that an equation of the tangent line at $P(x_1, y_1)$ is

$$y - y_1 = -\frac{b^2x_1}{a^2y_1}(x - x_1).$$

We may change the form of this equation as follows:

$$a^2y_1y - a^2y_1^2 = -b^2x_1x + b^2x_1^2$$

$$b^2x_1x + a^2y_1y = b^2x_1^2 + a^2y_1^2$$

Dividing both sides by a^2b^2 gives us

$$\frac{x_1x}{a^2} + \frac{y_1y}{b^2} = \frac{x_1^2}{a^2} + \frac{y_1^2}{b^2}.$$

Since $P(x_1, y_1)$ is on the ellipse, (x_1, y_1) is a solution of the equation in (a), and hence the right-hand side of the last equation equals 1. This completes the proof.

EXAMPLE 6 Find the area of the region bounded by an ellipse whose major and minor axes have lengths $2a$ and $2b$, respectively.

SOLUTION By Theorem (12.6), we see that an equation for the ellipse is $(x^2/a^2) + (y^2/b^2) = 1$. Solving for y gives us

$$y = \pm\frac{b}{a}\sqrt{a^2 - x^2}.$$

The graph of the ellipse has the general shape shown in Figure 12.16, and hence, by symmetry, it is sufficient to find the area of the region in the first quadrant and multiply the result by 4. By (5.19),

$$A = 4\int_0^a y\,dx = 4\frac{b}{a}\int_0^a \sqrt{a^2 - x^2}\,dx.$$

If we make the trigonometric substitution $x = a \sin \theta$, then

$$\sqrt{a^2 - x^2} = a \cos \theta \quad \text{and} \quad dx = a \cos \theta\,d\theta.$$

Since the values of θ that correspond to $x = 0$ and $x = a$ are $\theta = 0$ and $\theta = \pi/2$, respectively, we obtain

$$A = 4\frac{b}{a}\int_0^{\pi/2} a^2 \cos^2 \theta\,d\theta = 4ab\int_0^{\pi/2} \frac{1 + \cos 2\theta}{2}\,d\theta$$

$$= 2ab\left[\theta + \tfrac{1}{2}\sin 2\theta\right]_0^{\pi/2} = 2ab[\pi/2] = \pi ab.$$

As a special case, if $b = a$, the ellipse is a circle and $A = \pi a^2$.

Ellipses can be very flat or almost circular. To obtain information about the *roundness* of an ellipse, we sometimes use the *eccentricity e* (not to be confused with the base of the natural logarithms). Eccentricity is defined as follows, where a, b, and c have the same meanings as before.

Definition **(12.8)**

> The **eccentricity** e of an ellipse is
> $$e = \frac{c}{a} = \frac{\sqrt{a^2 - b^2}}{a}.$$

Consider the ellipse $(x^2/a^2) + (y^2/b^2) = 1$, and suppose that the length $2a$ of the major axis is fixed and the length $2b$ of the minor axis is variable. Since $\sqrt{a^2 - b^2} < a$, we see that $0 < e < 1$. If $e \approx 1$, then $\sqrt{a^2 - b^2} \approx a$ and hence $b \approx 0$. In this case the ellipse is very flat. If $e \approx 0$, then $\sqrt{a^2 - b^2} \approx 0$ and $a \approx b$. In this case the ellipse is almost circular.

After many years of analyzing an enormous amount of empirical data, the German astronomer Johannes Kepler (1571–1630) formulated three laws that describe the motion of planets about the sun. (These laws will be proved in Chapter 15.) Kepler's first law states that the orbit of each planet in the solar system is an ellipse with the sun at one focus. Most of these orbits are almost circular, and hence their corresponding eccentricities are close to 0. To illustrate, for Earth, $e \approx 0.017$; for Mars, $e \approx 0.093$; and for Uranus, $e \approx 0.046$. The orbits of Mercury and Pluto are less circular, with eccentricities of 0.206 and 0.249, respectively.

Many comets have elliptical orbits with the sun at a focus. In this case the eccentricity e is close to 1, and the ellipse is very flat. As in the next example, we use the **astronomical unit** (AU)—that is, the average distance from the earth to the sun—to specify large distances. (1 AU $\approx$ 93,000,000 miles.)

EXAMPLE 7 Halley's comet has an elliptical orbit with eccentricity $e = 0.967$. The closest that Halley's comet comes to the sun is 0.587 AU. Approximate the maximum distance of the comet from the sun, to the nearest 0.1 AU.

FIGURE 12.21

Halley's comet

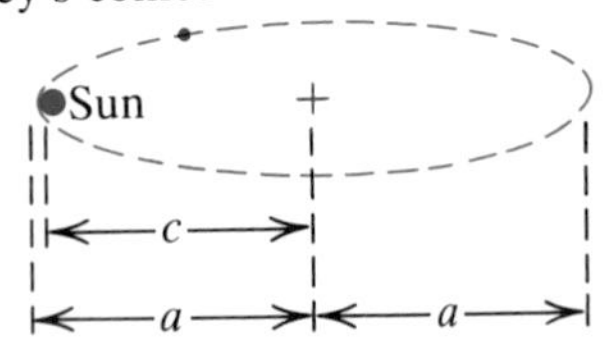

SOLUTION Figure 12.21 illustrates the orbit of the comet, where c is the distance from the center of the ellipse to a focus (the sun) and $2a$ is the length of the major axis.

Since $a - c$ is the minimum distance between the sun and the comet, we have (in AU)

$$a - c = 0.587, \quad \text{or} \quad a = c + 0.587.$$

Since $e = c/a = 0.967$,

$$c = 0.967a = 0.967(c + 0.587)$$
$$c \approx 0.967c + 0.568.$$

Thus,

$$0.033c \approx 0.568 \quad \text{and} \quad c \approx \frac{0.568}{0.033} \approx 17.2.$$

Consequently

$$a = c + 0.587$$
$$a \approx 17.2 + 0.587 \approx 17.8,$$

and the maximum distance between the sun and the comet is

$$a + c \approx 17.8 + 17.2, \quad \text{or} \quad a + c \approx 35.0 \text{ AU}.$$

FIGURE 12.22

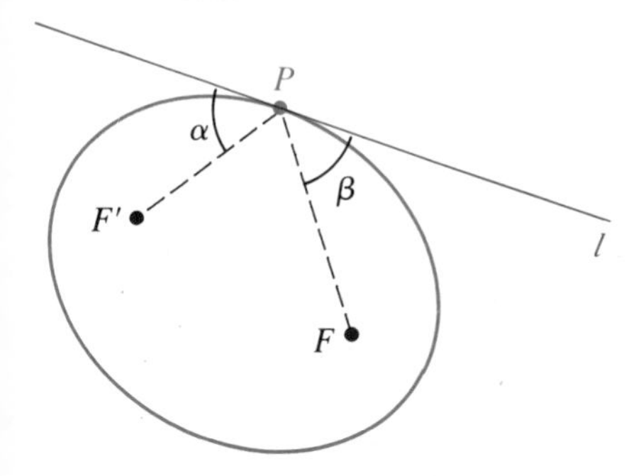

An ellipse has a reflective property analogous to that of the parabola discussed at the end of Section 12.1. To illustrate, let l denote the tangent line at a point P on an ellipse with foci F and F', as shown in Figure 12.22. If α is the angle between $F'P$ and l and if β is the angle between FP and l, it can be shown that $\alpha = \beta$. Thus, if a ray of light or sound emanates from one focus, it is reflected to the other focus. This property is used in the design of certain types of optical equipment. It is also evident in *whispering galleries*—that is, rooms with elliptically shaped ceilings, in which a person who whispers at one focus can be heard at the other focus. Examples of whispering galleries may be found in the Rotunda of the Capitol Building in Washington, D.C., and in the Mormon Tabernacle in Salt Lake City.

EXERCISES 12.2

Exer. 1–14: Find the vertices and foci of the ellipse. Sketch its graph, showing the foci.

1 $\dfrac{x^2}{9} + \dfrac{y^2}{4} = 1$

2 $\dfrac{x^2}{25} + \dfrac{y^2}{16} = 1$

3 $4x^2 + y^2 = 16$

4 $y^2 + 9x^2 = 9$

5 $5x^2 + 2y^2 = 10$

6 $\frac{1}{2}x^2 + 2y^2 = 8$

7 $4x^2 + 25y^2 = 1$

8 $10y^2 + x^2 = 5$

9 $4x^2 + 9y^2 - 32x - 36y + 64 = 0$

10 $x^2 + 2y^2 + 2x - 20y + 43 = 0$

11 $9x^2 + 16y^2 + 54x - 32y - 47 = 0$

12 $4x^2 + 9y^2 + 24x + 18y + 9 = 0$

13 $25x^2 + 4y^2 - 250x - 16y + 541 = 0$

14 $4x^2 + y^2 = 2y$

Exer. 15–24: Find an equation for the ellipse that has its center at the origin and satisfies the given conditions.

15 vertices $V(\pm 8, 0)$, foci $F(\pm 5, 0)$

16 vertices $V(0, \pm 7)$, foci $F(0, \pm 2)$

17 vertices $V(0, \pm 5)$, minor axis of length 3

18 foci $F(\pm 3, 0)$, minor axis of length 2

19 vertices $V(0, \pm 6)$, passing through $(3, 2)$

20 passing through $(2, 3)$ and $(6, 1)$

21 eccentricity $\frac{3}{4}$, vertices $V(0, \pm 4)$

22 eccentricity $\frac{1}{2}$, vertices on the x-axis, passing through $(1, 3)$

23 x-intercepts ± 2, y-intercepts $\pm \frac{1}{3}$

24 x-intercepts $\pm \frac{1}{2}$, y-intercepts ± 4

25 The arch of a bridge is semielliptical, with major axis horizontal. The base of the arch is 30 feet across, and the highest part is 10 feet above the horizontal roadway, as shown in the figure. Find the height of the arch 6 feet from the center of the base.

EXERCISE 25

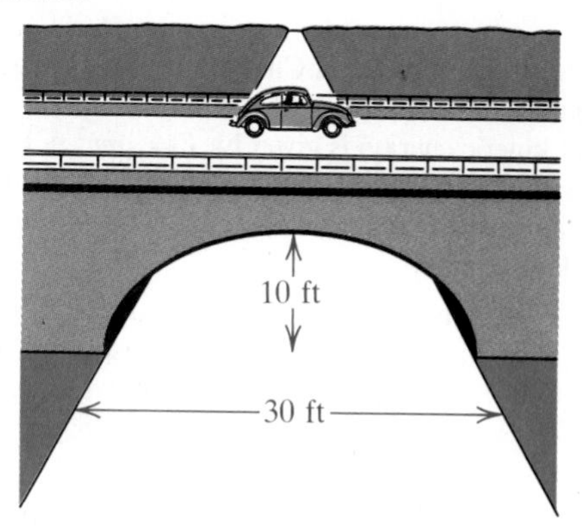

26 Assume that the length of the major axis of the earth's orbit is 186,000,000 miles and the eccentricity is 0.017. Find, to the nearest 1000 miles, the maximum and minimum distances between the earth and the sun.

27 A line segment of length $a + b$ moves with its endpoints A and B on the coordinate axes, as illustrated in the figure. Prove that the point P traces an ellipse.

EXERCISE 27

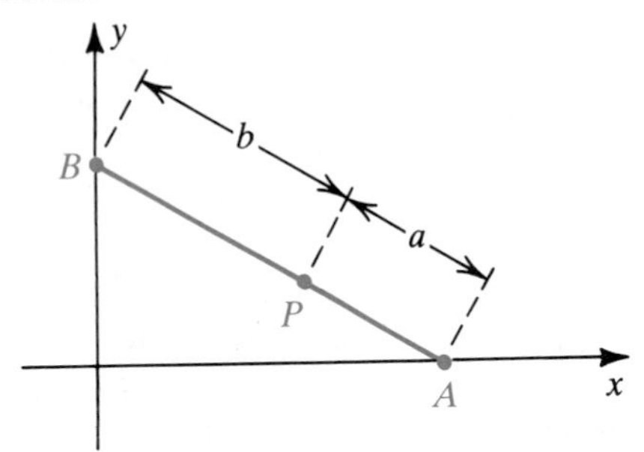

28 From a point P on the circle $x^2 + y^2 = 4$, a line segment PQ is drawn perpendicular to the diameter AB, and the midpoint M is found (see figure). Find an equation of the collection of all such midpoints M, and sketch the graph.

EXERCISE 28

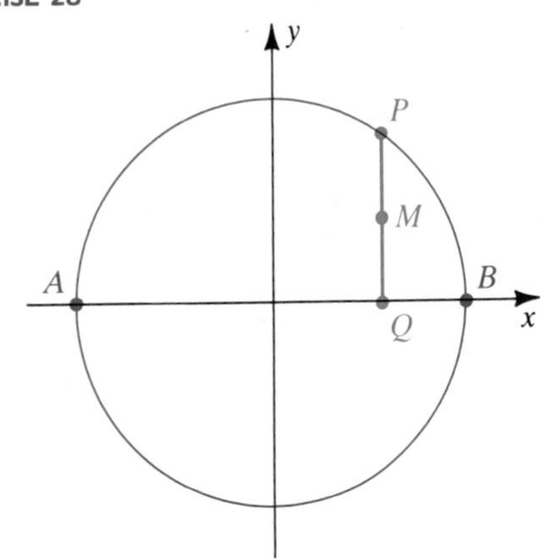

29 Refer to Figure 4.56 in Chapter 4. In a spring-mass system, the total energy E (the sum of the potential energy and kinetic energy) is given by $E = \frac{1}{2}mv^2 + \frac{1}{2}kx^2$, where x is the displacement of the mass m from equilibrium, v is the velocity of the mass, and k is the spring constant in Hooke's law.

(a) For a given initial velocity and displacement, the total energy of the system is constant. Show that v and x are related by an equation whose graph is an ellipse, and find the lengths of the major and minor axes.

(b) Find a relationship between the area A of the ellipse in (a) and the total energy E.

30 An ellipse has a vertex at the origin and foci $F_1(p, 0)$ and $F_2(p + 2c, 0)$, as shown in the figure. If the focus at F_1 is fixed and (x, y) is on the ellipse, show that $\lim_{c \to \infty} y^2 = 4px$. (Thus, a parabola may be considered as an ellipse with "one focus at infinity.")

EXERCISE 30

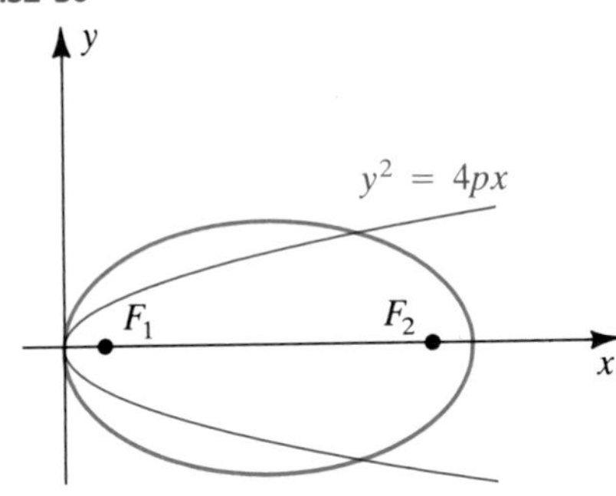

31 If tangent lines to the ellipse $9x^2 + 4y^2 = 36$ intersect the y-axis at the point $(0, 6)$, find the points of tangency.

32 (a) Show that the circumference C of the ellipse with the equation $(x^2/a^2) + (y^2/b^2) = 1$ is given by

$$C = 4a \int_0^{\pi/2} \sqrt{1 - e^2 \sin^2 \theta}\, d\theta,$$

where e is the eccentricity. (This is an *elliptic integral*, which cannot be evaluated using the methods of Chapter 9.)

(b) The planet Mercury travels in an elliptical orbit with $e = 0.206$ and $a = 0.387$ AU. Use part (a) and Simpson's rule, with $n = 10$, to approximate the length of the orbit.

(c) Find the maximum and minimum distances between Mercury and the sun.

Exer. 33–34: Find equations of the tangent line and normal line to the ellipse at the point P.

33 $5x^2 + 4y^2 = 56$; $P(-2, 3)$

34 $9x^2 + 4y^2 = 72$; $P(2, 3)$

35 Find the volume of the solid obtained by revolving the region bounded by the ellipse $b^2x^2 + a^2y^2 = a^2b^2$ about the x-axis.

36 Work Exercise 35 with the region revolved about the y-axis.

37 The base of a solid is a region bounded by an ellipse with major and minor axes of lengths 16 and 9, respectively. Find the volume of the solid if every cross section by a plane perpendicular to the major axis has the shape of a square.

38 Work Exercise 37 with the cross section having the shape of an equilateral triangle

39 A common model for human limbs is the *elliptical frustum* shown in the figure on the next page, where cross sections perpendicular to the axis of the frustum are elliptical and have the same eccentricity. For human limbs, the eccentricity typically varies from 0.6 to values near 1. If $k = a_1/b_1 = a_2/b_2$ and if L is the length of the limb, show that the volume V is given by the equation $V = (\frac{1}{3}\pi L/k)(a_1^2 + a_1a_2 + a_2^2)$.

EXERCISE 39

40 The base of a right elliptic cone has major and minor axes of lengths $2a$ and $2b$, respectively. Find the volume if the altitude of the cone is h.

41 The shape of the earth's surface can be approximated by revolving the ellipse $(x^2/a^2) + (y^2/b^2) = 1$, with $a = 6378$ km and $b = 6356$ km, about the x-axis. Approximate the surface area of the earth to the nearest 10^6 km^2. (*Hint:* Use (6.19) with $f(x) = \sqrt{b^2 - (b^2/a^2)x^2}$, and make the substitution $u = (b/a)x$.)

42 Find the dimensions of the rectangle of maximum area that can be inscribed in an ellipse of semiaxes a and b if two sides of the rectangle are parallel to the major axis.

43 Prove that if a normal line to a point $P(x_1, y_1)$ on an ellipse passes through the center of the ellipse, then the ellipse is a circle.

44 Establish the reflective property of the ellipse by showing that $\alpha = \beta$ in Figure 12.22.

c **Exer. 45–46: Graph, on the same coordinate axes, the given ellipses. (a) Estimate their points of intersection. (b) Set up an integral that can be used to approximate the area of the region bounded by and inside both ellipses.**

45 $\dfrac{x^2}{2.9} + \dfrac{y^2}{2.1} = 1;\quad \dfrac{x^2}{4.3} + \dfrac{(y - 2.1)^2}{4.9} = 1$

46 $\dfrac{x^2}{3.9} + \dfrac{y^2}{2.4} = 1;\quad \dfrac{(x + 1.9)^2}{4.1} + \dfrac{y^2}{2.5} = 1$

12.3 HYPERBOLAS

The definition of a hyperbola is similar to that of an ellipse. The only change is that instead of using the *sum* of distances from two fixed points, we use the *difference*.

Definition (12.9)

> A **hyperbola** is the set of all points in a plane, the difference of whose distances from two fixed points in the plane (the **foci**) is a positive constant.

FIGURE 12.23

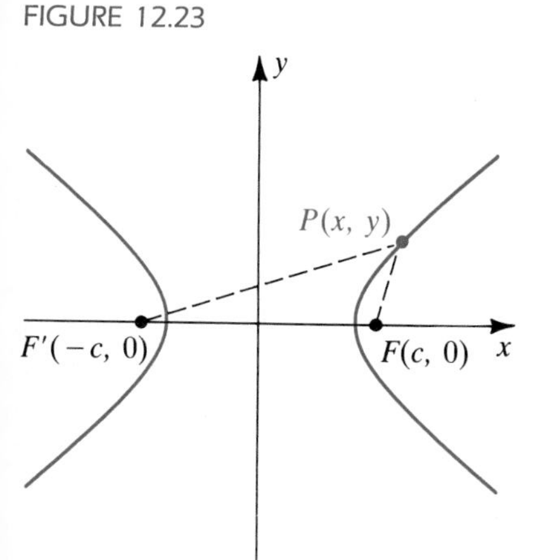

To find a simple equation for a hyperbola, we choose a coordinate system with foci at $F(c, 0)$ and $F'(-c, 0)$ and denote the (constant) distance by $2a$. Referring to Figure 12.23, we see that a point $P(x, y)$ is on the hyperbola if and only if either one of the following is true:

$$d(P, F') - d(P, F) = 2a$$

or

$$d(P, F) - d(P, F') = 2a$$

For hyperbolas (unlike ellipses), we need $a < c$ to obtain points on the hyperbola that are not on the x-axis, for if P is such a point, then from Figure 12.23 we see that

$$d(P, F) < d(F', F) + d(P, F'),$$

because the length of one side of a triangle is always less than the sum of the lengths of the other two sides. Similarly,

$$d(P, F') < d(F', F) + d(P, F).$$

Equivalent forms for the previous two inequalities are

$$d(P, F) - d(P, F') < d(F', F)$$

and

$$d(P, F') - d(P, F) < d(F', F).$$

Since the differences on the left-hand sides of these inequalities both equal $2a$ and since $d(F', F) = 2c$, the last two inequalities imply that $2a < 2c$, or $a < c$.

The equations $d(P, F') - d(P, F) = 2a$ and $d(P, F) - d(P, F') = 2a$ may be replaced by the single equation

$$|d(P, F) - d(P, F')| = 2a.$$

Using the distance formula to find $d(P, F)$ and $d(P, F')$, we obtain an equation of the hyperbola:

$$\left|\sqrt{(x - c)^2 + (y - 0)^2} - \sqrt{(x + c)^2 + (y - 0)^2}\right| = 2a$$

Employing the type of simplification procedure that we used to derive an equation for an ellipse, we can rewrite the preceding equation as

$$\frac{x^2}{a^2} - \frac{y^2}{c^2 - a^2} = 1.$$

If, in this equation, we let

$$b^2 = c^2 - a^2 \quad \text{with} \quad b > 0,$$

we obtain

$$\frac{x^2}{a^2} - \frac{y^2}{b^2} = 1.$$

We have shown that the coordinates of every point (x, y) on the hyperbola in Figure 12.23 satisfy the last equation. Conversely, if (x, y) is a solution of that equation, then by reversing the preceding steps we see that the point (x, y) is on the hyperbola.

By tests for symmetry, the hyperbola is symmetric with respect to both axes and the origin. The x-intercepts are $\pm a$. The corresponding points $V(a, 0)$ and $V'(-a, 0)$ are the **vertices**, and the line segment $V'V$ is the **transverse axis** of the hyperbola. The origin is the **center** of the hyperbola. The graph has no y-intercept, since the equation $-y^2/b^2 = 1$ has no solution.

The preceding discussion may be summarized as follows.

Theorem **(12.10)**

The graph of the equation

$$\frac{x^2}{a^2} - \frac{y^2}{b^2} = 1$$

is a hyperbola with vertices $(\pm a, 0)$. The foci are $(\pm c, 0)$, where $c^2 = a^2 + b^2$.

If we solve the equation $(x^2/a^2) - (y^2/b^2) = 1$ for y, we obtain

$$y = \pm\frac{b}{a}\sqrt{x^2 - a^2}.$$

There are no points (x, y) on the graph if $x^2 - a^2 < 0$, or, equivalently, $-a < x < a$. There *are* points $P(x, y)$ on the graph if $x \geq a$ or $x \leq -a$.

The line $y = (b/a)x$ is an asymptote for the hyperbola because the vertical distance $p(x)$ between the point $P(x, y)$ on the hyperbola and the corresponding point $P'(x, y_1)$ on the line approaches 0 as x increases without bound. To prove this, we note that if $x > 0$, then

$$p(x) = \frac{b}{a}x - \frac{b}{a}\sqrt{x^2 - a^2} = \frac{b}{a}(x - \sqrt{x^2 - a^2}).$$

We can show, by rationalizing the numerator, that

$$\frac{x - \sqrt{x^2 - a^2}}{1} = \frac{a^2}{x + \sqrt{x^2 - a^2}}.$$

Hence $$\lim_{x \to \infty} p(x) = \lim_{x \to \infty} \frac{b}{a} \frac{a^2}{x + \sqrt{x^2 - a^2}} = 0.$$

Similarly, $\lim_{x \to -\infty} p(x) = 0$. We can also show that the line $y = (-b/a)x$ is an asymptote for the hyperbola in Theorem 12.10.

The asymptotes $y = \pm(b/a)x$ serve as excellent guides for sketching the graph. A convenient way to sketch the asymptotes is to first plot the vertices $V(a, 0)$, $V'(-a, 0)$ and the points $W(0, b)$, $W'(0, -b)$ (see Figure 12.24). The line segment $W'W$ of length $2b$ is the **conjugate axis** of the hyperbola. If horizontal and vertical lines are drawn through the endpoints of the conjugate and transverse axes, respectively, then the

FIGURE 12.24
$\frac{x^2}{a^2} - \frac{y^2}{b^2} = 1$

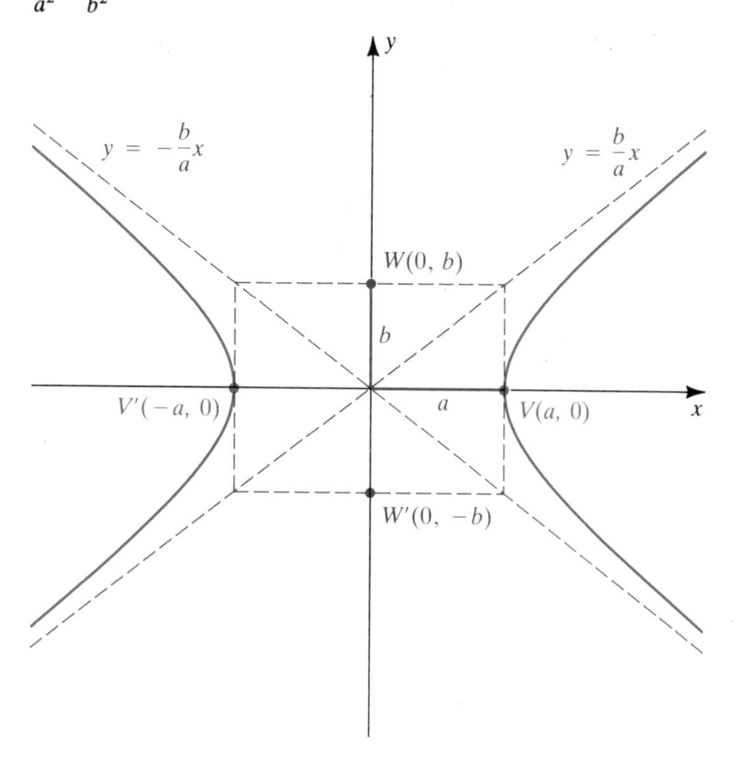

diagonals of the resulting rectangle have slopes b/a and $-b/a$. Hence, by extending these diagonals we obtain lines with equations $y = (\pm b/a)x$. The hyperbola is then sketched as in Figure 12.24, using the asymptotes as guides. The two curves that make up the hyperbola are the **branches** of the hyperbola.

EXAMPLE 1 Discuss and sketch the graph of $9x^2 - 4y^2 = 36$.

FIGURE 12.25
$9x^2 - 4y^2 = 36$

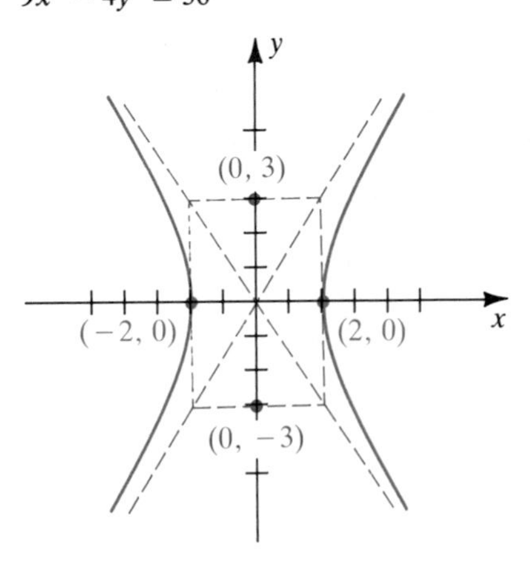

SOLUTION Dividing both sides by 36 gives us

$$\frac{x^2}{4} - \frac{y^2}{9} = 1,$$

which is of the form stated in Theorem (12.10), with $a^2 = 4$ and $b^2 = 9$. Hence $a = 2$ and $b = 3$. The vertices $(\pm 2, 0)$ and the endpoints $(0, \pm 3)$ of the conjugate axis determine a rectangle whose diagonals (extended) give us the asymptotes. The graph of the equation is sketched in Figure 12.25.

The equations of the asymptotes, $y = \pm\frac{3}{2}x$, can be found by referring to the graph or to the equations $y = \pm(b/a)x$.

To find the foci, we calculate

$$c^2 = a^2 + b^2 = 4 + 9 = 13.$$

Thus, $c = \sqrt{13}$ and the foci are $(\pm\sqrt{13}, 0)$.

The preceding example indicates that for hyperbolas it is not always true that $a > b$, as is the case for ellipses. Indeed, we may have $a < b$, $a > b$, or $a = b$.

EXAMPLE 2 A hyperbola has vertices $(\pm 3, 0)$ and passes through the point $P(5, 2)$. Find its equation, foci, and asymptotes.

SOLUTION We begin by sketching a hyperbola with vertices $(\pm 3, 0)$ that passes through the point $P(5, 2)$, as in Figure 12.26.

An equation of the hyperbola has the form

$$\frac{x^2}{3^2} - \frac{y^2}{b^2} = 1.$$

FIGURE 12.26

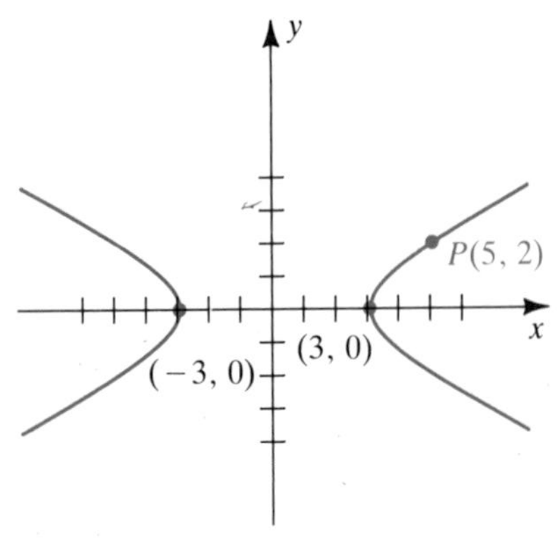

Since $P(5, 2)$ is on the hyperbola, the x- and y-coordinates satisfy the last equation; that is,

$$\frac{25}{9} - \frac{4}{b^2} = 1.$$

Solving for b^2 gives us $b^2 = \frac{9}{4}$, and hence the desired equation is

$$\frac{x^2}{9} - \frac{y^2}{\frac{9}{4}} = 1,$$

or, equivalently, $$x^2 - 4y^2 = 9.$$

To find the foci, we first calculate

$$c^2 = a^2 + b^2 = 9 + \tfrac{9}{4} = \tfrac{45}{4}.$$

Hence $c = \sqrt{\tfrac{45}{4}} = \tfrac{3}{2}\sqrt{5}$, and the foci are $(\pm\tfrac{3}{2}\sqrt{5}, 0)$.

The general equations of the asymptotes are $y = \pm(b/a)x$. Substituting $a = 3$ and $b = \tfrac{3}{2}$ gives us $y = \pm\tfrac{1}{2}x$.

If the foci of a hyperbola are the points $(0, \pm c)$ on the y-axis, then by the same type of argument used previously, we obtain the following theorem.

Theorem **(12.11)**

> The graph of the equation
>
> $$\frac{y^2}{a^2} - \frac{x^2}{b^2} = 1$$
>
> is a hyperbola with vertices $(0, \pm a)$. The foci are $(0, \pm c)$, where $c^2 = a^2 + b^2$.

For the hyperbola in the preceding theorem, the endpoints of the conjugate axis are $W(b, 0)$ and $W'(-b, 0)$. We find the asymptotes as before, by using the diagonals of the rectangle determined by these points, the vertices, and lines parallel to the coordinate axes. The graph is sketched in Figure 12.27. The equations of the asymptotes are $y = \pm(a/b)x$. Note

FIGURE 12.27
$\dfrac{y^2}{a^2} - \dfrac{x^2}{b^2} = 1$

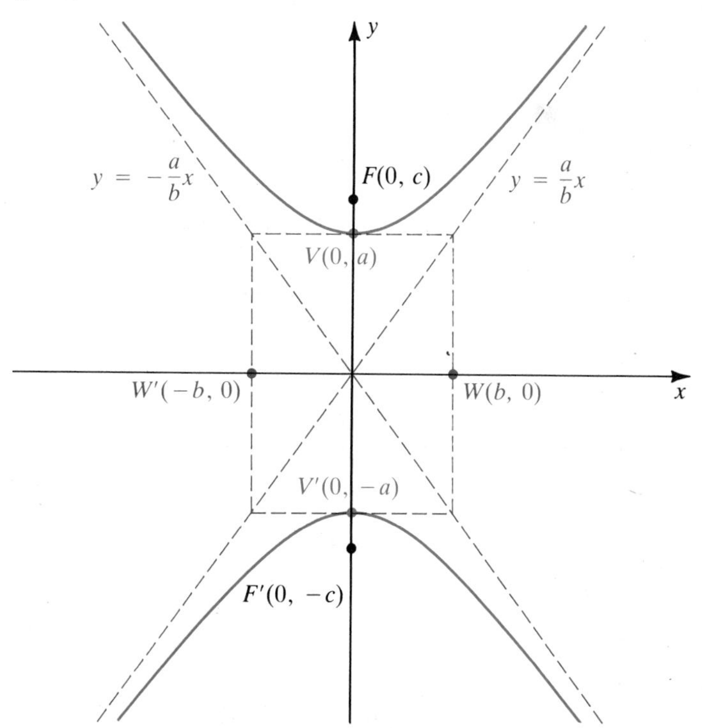

the difference between these equations and the equations $y = \pm(b/a)x$ for the asymptotes of the hyperbola considered first in this section.

FIGURE 12.28
$4y^2 - 2x^2 = 1$

EXAMPLE 3 Discuss and sketch the graph of $4y^2 - 2x^2 = 1$.

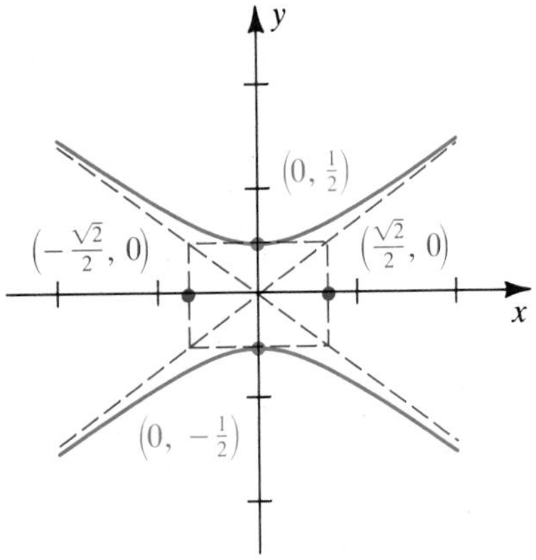

SOLUTION We may obtain the form in the last theorem by writing the equation as

$$\frac{y^2}{\frac{1}{4}} - \frac{x^2}{\frac{1}{2}} = 1.$$

Thus, $\quad a^2 = \frac{1}{4}, \quad b^2 = \frac{1}{2}, \quad c^2 = a^2 + b^2 = \frac{3}{4},$

and consequently

$$a = \frac{1}{2}, \qquad b = \frac{\sqrt{2}}{2}, \qquad c = \frac{\sqrt{3}}{2}.$$

The vertices are $(0, \pm\frac{1}{2})$, the foci are $(0, \pm\sqrt{3}/2)$, and the endpoints of the conjugate axes are $(\pm\sqrt{2}/2, 0)$. The graph is sketched in Figure 12.28.

To find equations of the asymptotes, we can use $y = \pm(a/b)x$, obtaining $y = \pm(\sqrt{2}/2)x$.

As was the case for ellipses, we may use translations of axes to generalize our work. The following example illustrates this technique.

EXAMPLE 4 Discuss and sketch the graph of the equation

$$9x^2 - 4y^2 - 54x - 16y + 29 = 0.$$

SOLUTION We arrange our work as follows:

$$9(x^2 - 6x \qquad) - 4(y^2 + 4y \qquad) = -29$$

$$9(x^2 - 6x + 9) - 4(y^2 + 4y + 4) = -29 + 81 - 16$$

$$9(x - 3)^2 - 4(y + 2)^2 = 36$$

$$\frac{(x - 3)^2}{4} - \frac{(y + 2)^2}{9} = 1$$

This equation is of the form

$$\frac{(x')^2}{4} - \frac{(y')^2}{9} = 1,$$

with $x' = x - 3$ and $y' = y + 2$. Thus, we translate the x- and y-axes to the new origin $C(3, -2)$. The graph is a hyperbola with vertices on the x'-axis (the line $y = -2$) and

$$a^2 = 4, \qquad b^2 = 9, \qquad c^2 = a^2 + b^2 = 13.$$

Hence

$$a = 2, \qquad b = 3, \qquad c = \sqrt{13}.$$

FIGURE 12.29

$$\frac{(x-3)^2}{4} - \frac{(y+2)^2}{9} = 1$$

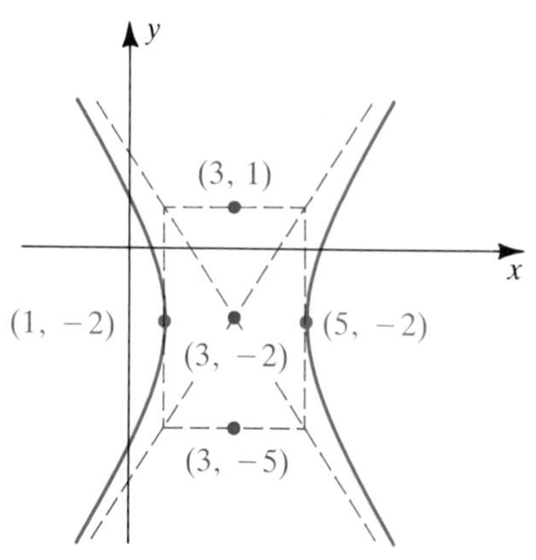

As illustrated in Figure 12.29, the vertices are $(3 \pm 2, -2)$—that is, $(5, -2)$ and $(1, -2)$. The endpoints of the conjugate axis are $(3, -2 \pm 3)$—that is, $(3, 1)$ and $(3, -5)$. The foci are $(3 \pm \sqrt{13}, -2)$, and equations of the asymptotes are

$$y + 2 = \pm\tfrac{3}{2}(x - 3).$$

The results of the last three sections indicate that the graph of every equation of the form

$$Ax^2 + Cy^2 + Dx + Ey + F = 0$$

is a conic, except for certain degenerate cases in which a point, one or two lines, or no graph is obtained. Although we have considered only special examples, our methods are perfectly general. If A and C are equal and not 0, then the graph, when it exists, is a circle or, in exceptional cases, a point. If A and C are unequal but have the same sign, then by completing squares and properly translating axes, we obtain an equation whose graph, when it exists, is an ellipse (or a point). If A and C have opposite signs, an equation of a hyperbola is obtained or possibly, in the degenerate case, two intersecting straight lines. If either A or C (but not both) is 0, the graph is a parabola or, in certain cases, a pair of parallel lines.

We shall conclude this section with several applications of hyperbolas.

EXAMPLE 5 Coast Guard station A is 200 miles directly east of another station B. A ship is sailing on a line parallel to, and 50 miles north of, the line through stations A and B. Radio signals are sent out from A and B at the rate of 980 ft/μsec (microsecond). If, at 1:00 P.M., the signal from B reaches the ship 400 μsec after the signal from A, locate the position of the ship at that time.

SOLUTION Let us introduce a coordinate system, as shown in Figure 12.30(i), on the following page, with the stations at points A and B on the x-axis and the ship at P on the line $y = 50$. Since at 1:00 P.M. it takes 400 μsec longer for the signal to arrive from B than from A, the difference $d_1 - d_2$ in the indicated distances at that time is

$$d_1 - d_2 = (980)(400) = 392{,}000 \text{ ft}.$$

Dividing by 5280 (ft/mi) gives us

$$d_1 - d_2 = \frac{392{,}000}{5280} = 74.2424\ldots \text{ mi}.$$

At 1:00 P.M., point P is on the right branch of a hyperbola whose equation is $(x^2/a^2) - (y^2/b^2) = 1$ (see Figure 12.30(ii)), consisting of all points whose difference in distances from the foci B and A is $d_1 - d_2$. In our derivation of the equation $(x^2/a^2) - (y^2/b^2) = 1$ we let $d_1 - d_2 = 2a$; it follows that

FIGURE 12.30

(i)

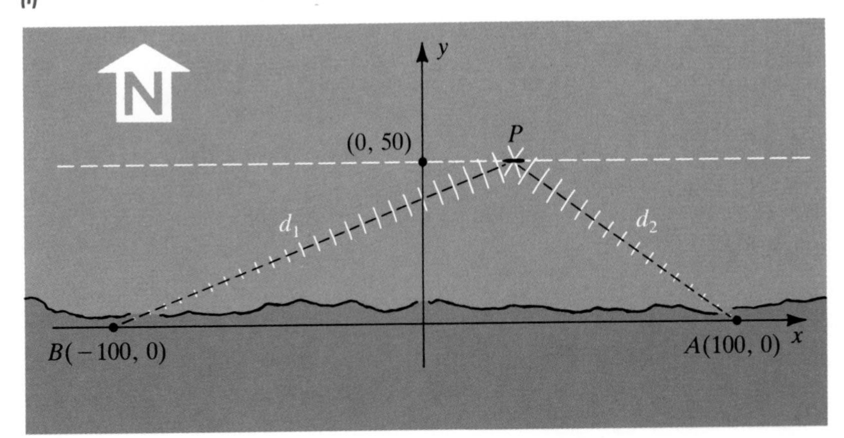

(ii)

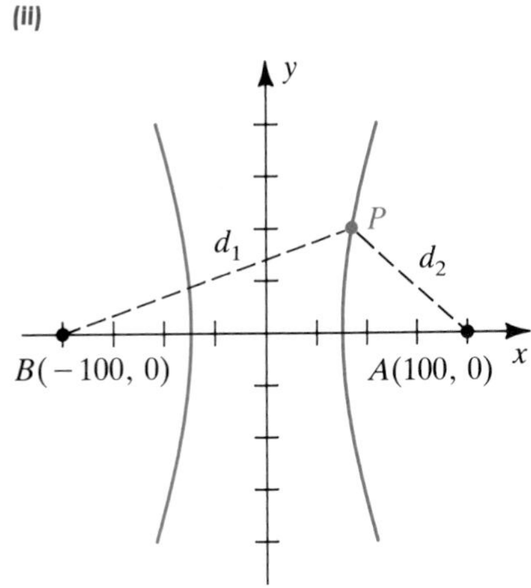

in the present situation

$$a = \frac{74.2424\ldots}{2} = 37.1212\ldots \quad \text{and} \quad a^2 \approx 1378.$$

Since the distance c from the origin to either focus is 100,

$$b^2 = c^2 - a^2 \approx 10{,}000 - 1378, \quad \text{or} \quad b^2 \approx 8622.$$

Hence an (approximate) equation for the hyperbola that has foci A and B and passes through P is

$$\frac{x^2}{1378} - \frac{y^2}{8622} = 1.$$

If we let $y = 50$ (the y-coordinate of P), we obtain

$$\frac{x^2}{1378} - \frac{2500}{8622} = 1.$$

Solving for x gives us $x \approx 42.16$. Rounding off to the nearest mile, we find that the coordinates of P are approximately (42, 50).

FIGURE 12.31

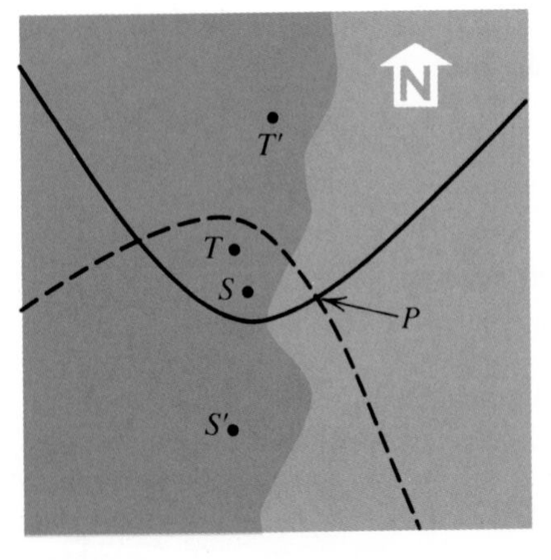

An extension of the method used in Example 5 is the basis for the navigational system LORAN (for *Long Range Navigation*). This system involves two pairs of radio transmitters, such as those located at T, T' and S, S' in Figure 12.31. Suppose that signals sent out by the transmitters at T and T' reach a radio receiver in a ship located at some point P. The difference in the times of arrival of the signals can be used to determine the difference in the distances of P from T and T'. Thus, P lies on one branch of a hyperbola with foci at T and T'. Repeating this process for the other pair of transmitters, we see that P also lies on one branch of a hyperbola with foci at S and S'. The intersection of these two branches determines the position of P.

In the next example we use a definite integral to find the volume of a solid of revolution obtained from a region that has a hyperbola as part of its boundary.

EXAMPLE 6 The region bounded by the hyperbola with equation $(x^2/a^2) - (y^2/b^2) = 1$ and a vertical line through a focus is revolved about the x-axis. Find the volume V of the resulting solid.

SOLUTION The region and the solid of revolution are sketched in Figure 12.32(i). The point $F(c, 0)$, with $c = \sqrt{a^2 + b^2}$, is a focus of the hyperbola. As in Figure 12.32(ii), we consider the disk generated by a vertical rectangle of height y and width dx. The volume of the disk is $\pi y^2\,dx$.

FIGURE 12.32

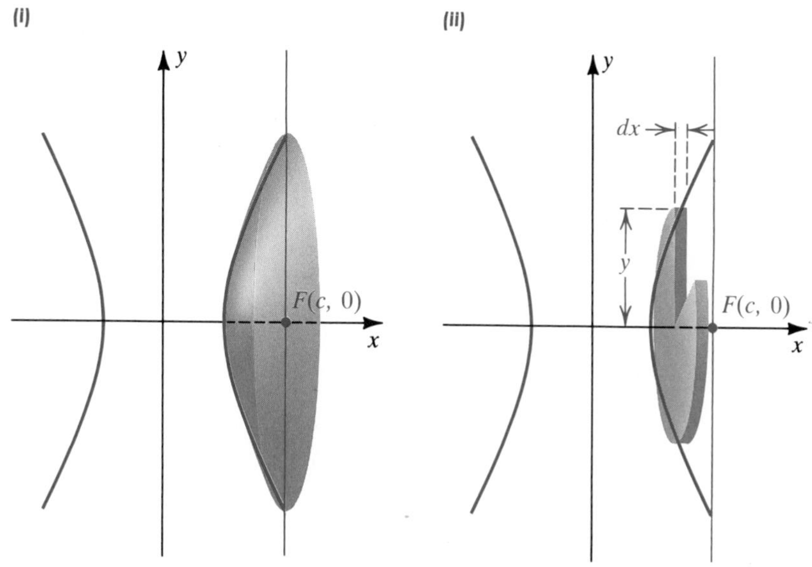

Applying the limit of sums operator $\int_a^c$, we obtain

$$
\begin{aligned}
V = \int_a^c \pi y^2\,dx &= \pi \int_a^c b^2\left(\frac{x^2}{a^2} - 1\right) dx \\
&= \pi b^2\left[\frac{x^3}{3a^2} - x\right]_a^c \\
&= \frac{\pi b^2}{3a^2}\left[x(x^2 - 3a^2)\right]_a^c \\
&= \frac{\pi b^2}{3a^2}\left[c(c^2 - 3a^2) - a(a^2 - 3a^2)\right] \\
&= \frac{\pi b^2}{3a^2}\left[\sqrt{a^2 + b^2}(b^2 - 2a^2) + 2a^3\right].
\end{aligned}
$$

FIGURE 12.33

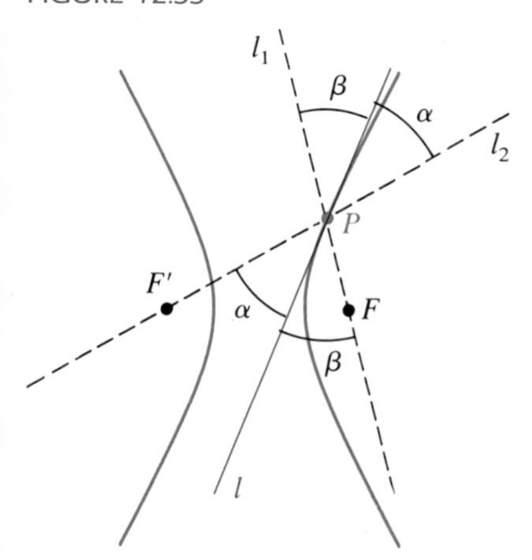

The surface of the solid in Example 6 is part of the *hyperboloid* generated by revolving the entire hyperbola about the x-axis. We shall study hyperboloids in detail in Chapter 14.

A hyperbola has a reflective property analogous to that of the ellipse discussed at the end of Section 12.2. To illustrate, let l denote the tangent line at a point P on a hyperbola with foci F and F', as shown in Figure 12.33. If α is the acute angle between $F'P$ and l and if β is the acute angle between FP and l, then $\alpha = \beta$. If a ray of light is directed along the line l_1 toward F, it will be reflected back at P along the line l_2 toward F'. This property is used in the design of telescopes of the Cassegrain type (see Exercise 43).

EXERCISES 12.3

Exer. 1–18: Find the vertices and foci of the hyperbola. Sketch its graph, showing the asymptotes and the foci.

1 $\dfrac{x^2}{9} - \dfrac{y^2}{4} = 1$

2 $\dfrac{y^2}{49} - \dfrac{x^2}{16} = 1$

3 $\dfrac{y^2}{9} - \dfrac{x^2}{4} = 1$

4 $\dfrac{x^2}{49} - \dfrac{y^2}{16} = 1$

5 $y^2 - 4x^2 = 16$

6 $x^2 - 2y^2 = 8$

7 $x^2 - y^2 = 1$

8 $y^2 - 16x^2 = 1$

9 $x^2 - 5y^2 = 25$

10 $4y^2 - 4x^2 = 1$

11 $3x^2 - y^2 = -3$

12 $16x^2 - 36y^2 = 1$

13 $25x^2 - 16y^2 + 250x + 32y + 109 = 0$

14 $y^2 - 4x^2 - 12y - 16x + 16 = 0$

15 $4y^2 - x^2 + 40y - 4x + 60 = 0$

16 $25x^2 - 9y^2 + 100x - 54y + 10 = 0$

17 $9y^2 - x^2 - 36y + 12x - 36 = 0$

18 $4x^2 - y^2 + 32x - 8y + 49 = 0$

Exer. 19–28: Find an equation for the hyperbola that has its center at the origin and satisfies the given conditions.

19 foci $F(0, \pm 4)$, vertices $V(0, \pm 1)$

20 foci $F(\pm 8, 0)$, vertices $V(\pm 5, 0)$

21 foci $F(\pm 5, 0)$, vertices $V(\pm 3, 0)$

22 foci $F(0, \pm 3)$, vertices $V(0, \pm 2)$

23 foci $F(0, \pm 5)$, conjugate axis of length 4

24 vertices $V(\pm 4, 0)$, passing through $(8, 2)$

25 vertices $V(\pm 3, 0)$, asymptotes $y = \pm 2x$

26 foci $F(0, \pm 10)$, asymptotes $y = \pm \frac{1}{3}x$

27 x-intercepts ± 5, asymptotes $y = \pm 2x$

28 y-intercepts ± 2, asymptotes $y = \pm \frac{1}{4}x$

29 The graphs of the equations

$$\frac{x^2}{a^2} - \frac{y^2}{b^2} = 1 \quad \text{and} \quad \frac{x^2}{a^2} - \frac{y^2}{b^2} = -1$$

are called *conjugate hyperbolas*. Sketch the graphs of both equations on the same coordinate plane, with $a = 5$ and $b = 3$. Describe the relationship between the two graphs.

30 In 1911, the physicist Ernest Rutherford (1871–1937) discovered that when alpha particles are shot toward the nucleus of an atom, they are eventually repulsed away from the nucleus along hyperbolic paths. The figure illustrates the path of a particle that starts toward the origin along the line $y = \frac{1}{2}x$ and comes within 3 units of the nucleus. Find an equation of the path.

EXERCISE 30

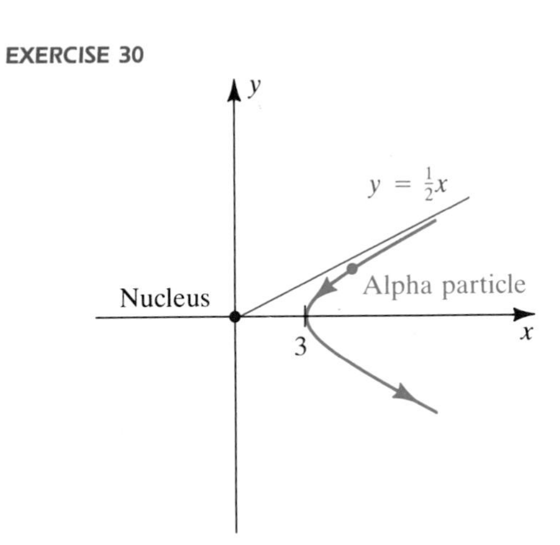

31 A cruise ship is traveling a course that is 100 miles from, and parallel to, a straight shoreline. The ship sends out a distress signal, which is received by two Coast Guard stations A and B, located 200 miles apart, as shown in the figure. By measuring the difference in signal reception times, officials determine that the ship is 160 miles closer to B than to A. Where is the ship?

EXERCISE 31

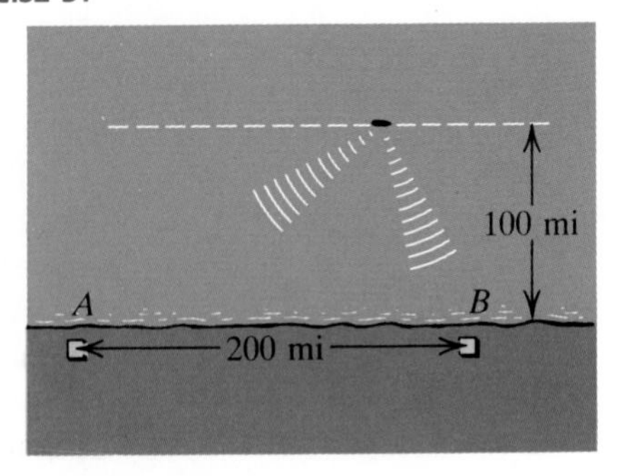

32 See Example 5 of Section 12.2. Prove that an equation of the tangent line to the graph of the hyperbola $(x^2/a^2) - (y^2/b^2) = 1$ at the point $P(x_1, y_1)$ is

$$\frac{x_1 x}{a^2} - \frac{y_1 y}{b^2} = 1.$$

33 If tangent lines to the hyperbola $9x^2 - y^2 = 36$ intersect the y-axis at the point $(0, 6)$, find the points of tangency.

34 Find an equation of a line through $P(2, -1)$ that is tangent to the hyperbola $x^2 - 4y^2 = 16$.

Exer. 35–36: Find equations of the tangent line and normal line to the hyperbola at the point P.

35 $2x^2 - 5y^2 = 3$; $P(-2, 1)$

36 $3y^2 - 2x^2 = 40$; $P(2, -4)$

Exer. 37–38: Let R be the region bounded by the hyperbola with equation $b^2x^2 - a^2y^2 = a^2b^2$ and a vertical line through a focus.

37 Find the area of the region R.

38 Find the volume of the solid obtained by revolving R about the y-axis.

39 Let R be the region bounded by the right branch of the hyperbola $x^2 - y^2 = 8$ and the vertical line through the focus. Find the area of the curved surface of the solid obtained by revolving R about the x-axis.

40 Show that the vertex is the point on a branch of a hyperbola that is closest to the focus associated with that branch.

41 Some comets travel along hyperbolic paths with the sun at a focus, as illustrated in the figure. If an equation of the path of the comet is $12x^2 + 24x - 4y^2 + 9 = 0$, approximately how close (in AU) does the comet come to the sun? (*Hint:* See Exercise 40.)

EXERCISE 41

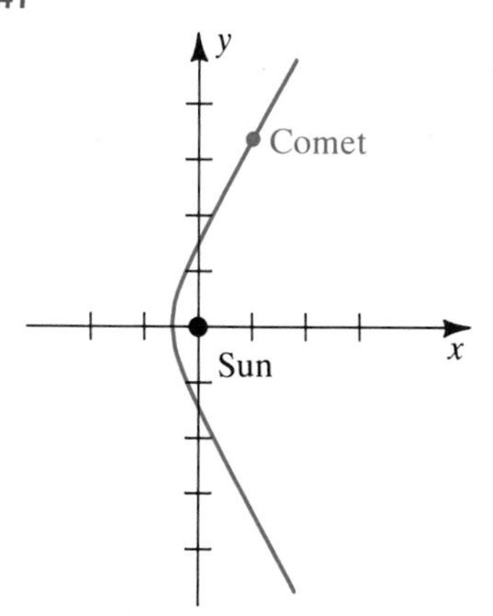

42 Establish the reflective property of the hyperbola by showing that $\alpha = \beta$ in Figure 12.33.

43 The Cassegrain telescope design (dating to 1672) makes use of the reflective properties of both the parabola and the hyperbola. Shown in the figure is a (split) parabolic mirror, with focus at F_1 and axis along the line l, and a second hyperbolic mirror, with one focus also at F_1 and transverse axis along l. Where do incoming light waves parallel to the common axis finally collect?

EXERCISE 43

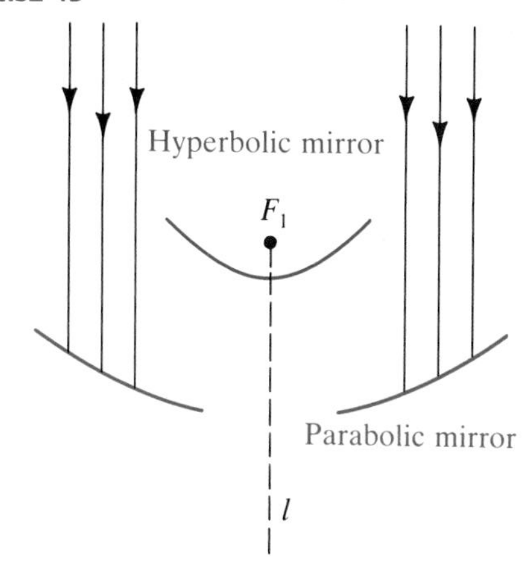

44 Let l denote the tangent line at a point P on a hyperbola (see figure). If l intersects the asymptotes at Q and R, prove that P is the midpoint of QR.

EXERCISE 44

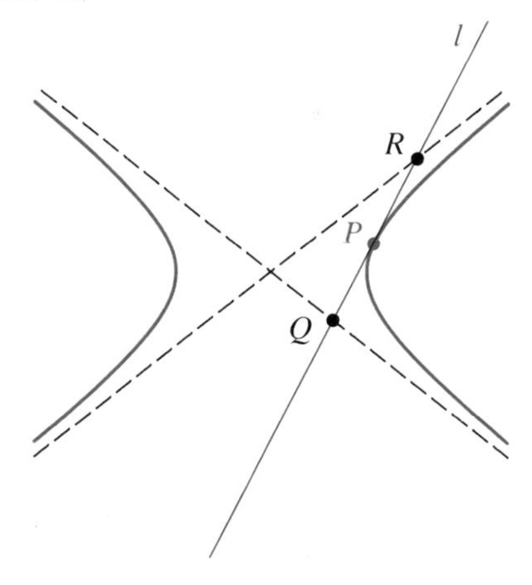

c **Exer. 45–46: Graph, on the same coordinate axes, the given hyperbolas. (a) Estimate their first-quadrant point of intersection. (b) Set up an integral that can be used to approximate the area of the region in the first quadrant bounded by the hyperbolas and a coordinate axis.**

45 $\dfrac{(y - 0.1)^2}{1.6} - \dfrac{(x + 0.2)^2}{0.5} = 1$; $\dfrac{(y - 0.5)^2}{2.7} - \dfrac{(x - 0.1)^2}{5.3} = 1$

46 $\dfrac{(x - 0.1)^2}{0.12} - \dfrac{y^2}{0.1} = 1$; $\dfrac{x^2}{0.9} - \dfrac{(y - 0.3)^2}{2.1} = 1$

12.4 ROTATION OF AXES

We may obtain the $x'y'$-plane used in a translation of axes by moving the origin O of the xy-plane to a new position $C(h, k)$ without changing the positive directions of the axes or the units of length. We shall next introduce a new coordinate plane obtained by keeping the origin O fixed and rotating the x- and y-axes about O to another position, denoted by x' and y'. A transformation of this type is a **rotation of axes**.

FIGURE 12.34

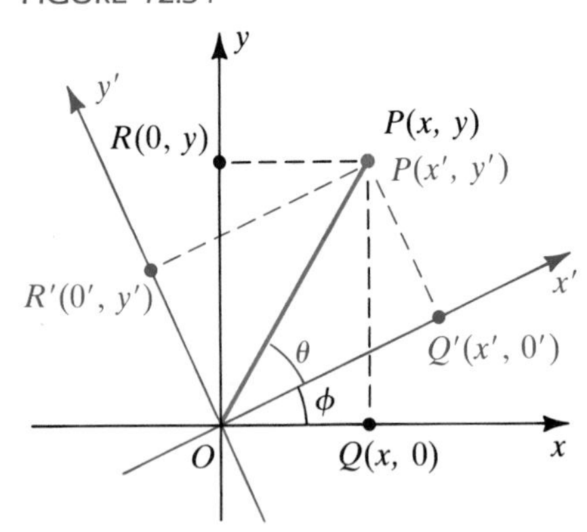

Consider the rotation of axes in Figure 12.34, and let ϕ denote the acute angle through which the positive x-axis must be rotated in order to coincide with the positive x'-axis. If (x, y) are the coordinates of a point P relative to the xy-plane, then (x', y') will denote its coordinates relative to the new $x'y'$-plane.

Let the projections of P on the various axes be denoted as in Figure 12.34, and let θ denote angle POQ'. If $p = d(O, P)$, then

$$x' = p\cos\theta, \qquad y' = p\sin\theta$$

$$x = p\cos(\theta + \phi), \qquad y = p\sin(\theta + \phi).$$

Applying the addition formulas for the sine and cosine, we see that

$$x = p\cos\theta\cos\phi - p\sin\theta\sin\phi$$

$$y = p\sin\theta\cos\phi + p\cos\theta\sin\phi.$$

Using the fact that $x' = p\cos\theta$ and $y' = p\sin\theta$ gives us (i) of the next theorem. The formulas in (ii) may be obtained from (i) by solving for x' and y'.

Rotation of axes formulas **(12.12)**

If the x- and y-axes are rotated about the origin O, through an acute angle ϕ, then the coordinates (x, y) and (x', y') of a point P in the xy- and $x'y'$-planes are related as follows:

(i) $x = x'\cos\phi - y'\sin\phi, \quad y = x'\sin\phi + y'\cos\phi$

(ii) $x' = x\cos\phi + y\sin\phi, \quad y' = -x\sin\phi + y\cos\phi$

FIGURE 12.35
$y = 1/x$

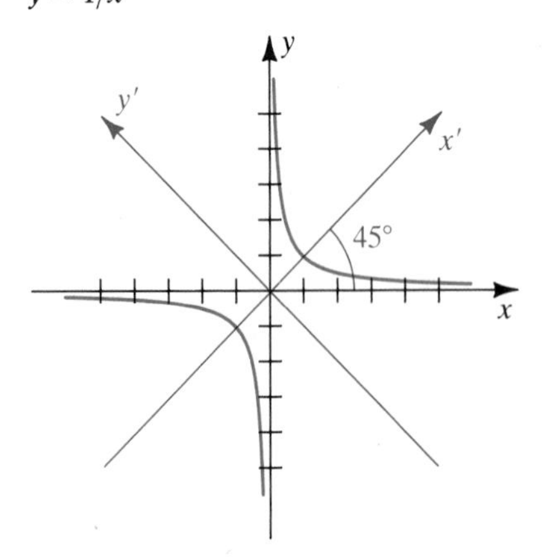

EXAMPLE 1 The graph of $xy = 1$, or, equivalently, $y = 1/x$, is sketched in Figure 12.35. If the coordinate axes are rotated through an angle of 45°, find an equation of the graph relative to the new $x'y'$-plane.

SOLUTION We let $\phi = 45°$ in rotation of axes formulas (12.12)(i):

$$x = x'\left(\frac{\sqrt{2}}{2}\right) - y'\left(\frac{\sqrt{2}}{2}\right) = \frac{\sqrt{2}}{2}(x' - y')$$

$$y = x'\left(\frac{\sqrt{2}}{2}\right) + y'\left(\frac{\sqrt{2}}{2}\right) = \frac{\sqrt{2}}{2}(x' + y')$$

Substituting for x and y in the equation $xy = 1$ gives us

$$\frac{\sqrt{2}}{2}(x' - y')\cdot\frac{\sqrt{2}}{2}(x' + y') = 1.$$

This reduces to
$$\frac{(x')^2}{2} - \frac{(y')^2}{2} = 1,$$
which is an equation of a hyperbola with vertices $(\pm\sqrt{2}, 0)$ on the x'-axis. Note that the asymptotes for the hyperbola have equations $y' = \pm x'$ in the new system. These correspond to the original x- and y-axes.

Example 1 illustrates a method for eliminating a term of an equation that contains the product xy. This method can be used to transform any equation of the form

$$Ax^2 + Bxy + Cy^2 + Dx + Ey + F = 0,$$

where $B \neq 0$, into an equation in x' and y' that contains no $x'y'$ term. Let us prove that this may always be done. If we rotate the axes through an angle ϕ, then using rotation of axes formulas (12.12)(i) to substitute for x and y gives us

$$\begin{aligned}&A(x' \cos\phi - y' \sin\phi)^2 + B(x' \cos\phi - y' \sin\phi)(x' \sin\phi + y' \cos\phi)\\&+ C(x' \sin\phi + y' \cos\phi)^2 + D(x' \cos\phi - y' \sin\phi)\\&+ E(x' \sin\phi + y' \cos\phi) + F\\&\qquad = 0.\end{aligned}$$

By performing the multiplications and rearranging terms, we may write this equation in the form

$$A'(x')^2 + B'x'y' + C'(y')^2 + D'x' + E'y' + F' = 0$$

with

$$\begin{aligned}A' &= A\cos^2\phi + B\cos\phi\sin\phi + C\sin^2\phi\\B' &= 2(C - A)\sin\phi\cos\phi + B(\cos^2\phi - \sin^2\phi)\\C' &= A\sin^2\phi - B\sin\phi\cos\phi + C\cos^2\phi\\D' &= D\cos\phi + E\sin\phi\\E' &= -D\sin\phi + E\cos\phi\\F' &= F.\end{aligned}$$

To eliminate the $x'y'$ term, we must select ϕ such that $B' = 0$; that is,

$$2(C - A)\sin\phi\cos\phi + B(\cos^2\phi - \sin^2\phi) = 0.$$

Using double-angle formulas, we may write this equation as

$$(C - A)\sin 2\phi + B\cos 2\phi = 0,$$

which is equivalent to

$$\cot 2\phi = \frac{A - C}{B}.$$

This proves the next result.

Theorem (12.13)

To eliminate the xy-term from the equation

$$Ax^2 + Bxy + Cy^2 + Dx + Ey + F = 0,$$

where $B \neq 0$, choose an angle ϕ such that

$$\cot 2\phi = \frac{A - C}{B} \quad \text{with} \quad 0^\circ < 2\phi < 180^\circ$$

and use the rotation of axes formulas.

The graph of any equation in x and y of the type displayed in the preceding theorem is a conic, except for certain degenerate cases.

When using the preceding theorem, note that $\sin 2\phi > 0$, since $0^\circ < 2\phi < 180^\circ$. Moreover, because $\cot 2\phi = \cos 2\phi/\sin 2\phi$, the signs of $\cot 2\phi$ and $\cos 2\phi$ are always the same.

EXAMPLE 2 Discuss and sketch the graph of the equation

$$41x^2 - 24xy + 34y^2 - 25 = 0.$$

SOLUTION Use the notation of Theorem (12.13):

$$A = 41, \qquad B = -24, \qquad C = 34$$

$$\cot 2\phi = \frac{41 - 34}{-24} = -\frac{7}{24}$$

Since $\cot 2\phi$ is negative, we choose 2ϕ such that $90^\circ < 2\phi < 180^\circ$, and consequently $\cos 2\phi = -\frac{7}{25}$. We now use the half-angle formulas to obtain

$$\sin \phi = \sqrt{\frac{1 - \cos 2\phi}{2}} = \sqrt{\frac{1 - (-\frac{7}{25})}{2}} = \frac{4}{5}$$

$$\cos \phi = \sqrt{\frac{1 + \cos 2\phi}{2}} = \sqrt{\frac{1 + (-\frac{7}{25})}{2}} = \frac{3}{5}.$$

FIGURE 12.36

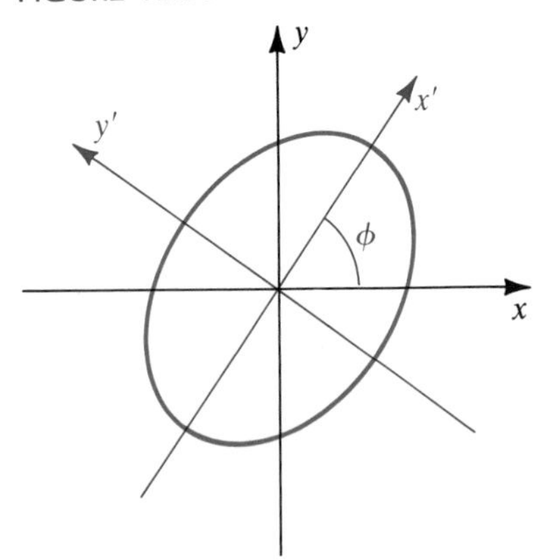

Thus, the desired rotation of axes formulas are

$$x = \tfrac{3}{5}x' - \tfrac{4}{5}y' \quad \text{and} \quad y = \tfrac{4}{5}x' + \tfrac{3}{5}y'.$$

After substituting for x and y in the given equation and simplifying, we obtain the equation

$$(x')^2 + 2(y')^2 = 1.$$

The graph therefore is an ellipse with vertices at $(\pm 1, 0)$ on the x'-axis. Since $\tan \phi = \sin \phi/\cos \phi = (\frac{4}{5})/(\frac{3}{5}) = \frac{4}{3}$, we obtain $\phi = \tan^{-1}(\frac{4}{3})$. To the nearest minute, $\phi \approx 53^\circ 8'$. The graph is sketched in Figure 12.36.

The next theorem states rules that we can apply to identify the type of conic *before* rotating the axes.

Identification theorem (12.14)

The graph of the equation

$$Ax^2 + Bxy + Cy^2 + Dx + Ey + F = 0$$

is either a conic or a degenerate conic. If the graph is a conic, then it is

(i) a parabola if $B^2 - 4AC = 0$

(ii) an ellipse if $B^2 - 4AC < 0$

(iii) a hyperbola if $B^2 - 4AC > 0$

PROOF If the x and y axes are rotated through an angle ϕ, using the rotation of axes formulas gives us

$$A'(x')^2 + B'x'y' + C'(y')^2 + D'x' + E'y' + F' = 0.$$

Using the formulas for A', B', and C' on page 635, we can show that

$$(B')^2 - 4A'C' = B^2 - 4AC.$$

For a suitable rotation of axes, we obtain $B' = 0$ and

$$A'(x')^2 + C'(y')^2 + D'x' + E'y' + F' = 0.$$

Except for degenerate cases, the graph of this equation is an ellipse if $A'C' > 0$ (A' and C' have the same sign), a hyperbola if $A'C' < 0$ (A' and C' have opposite signs), or a parabola if $A'C' = 0$ (either $A' = 0$ or $C' = 0$). However, if $B' = 0$, then $B^2 - 4AC = -4A'C'$, and hence the graph is an ellipse if $B^2 - 4AC < 0$, a hyperbola if $B^2 - 4AC > 0$, or a parabola if $B^2 - 4AC = 0$. ■

The expression $B^2 - 4AC$ is called the **discriminant** of the equation in the identification theorem (12.14). We say that this discriminant is **invariant** under a rotation of axes, because it is unchanged by any such rotation.

EXAMPLE 3 Use the identification theorem (12.14) to determine if the graph of the equation

$$41x^2 - 24xy + 34y^2 - 25 = 0$$

is a parabola, an ellipse, or a hyperbola.

SOLUTION The equation was considered in Example 2, where we performed a rotation of axes. Since $A = 41$, $B = -24$, and $C = 34$, the discriminant is

$$B^2 - 4AC = 576 - 4(41)(34) = -5000 < 0.$$

Hence, by the identification theorem, the graph is an ellipse.

In some cases, after eliminating the xy term, it may be necessary to translate the axes of the $x'y'$-coordinate system to obtain the graph, as illustrated in the next example.

EXAMPLE 4 Discuss and sketch the graph of the equation

$$x^2 + 2\sqrt{3}xy + 3y^2 + 8\sqrt{3}x - 8y + 32 = 0.$$

SOLUTION Using $A = 1$, $B = 2\sqrt{3}$, and $C = 3$, we see that

$$B^2 - 4AC = 12 - 12 = 0.$$

By the identification theorem (12.14), the graph is a parabola.

To apply a rotation of axes, we calculate

$$\cot 2\phi = \frac{A - C}{B} = \frac{1 - 3}{2\sqrt{3}} = -\frac{1}{\sqrt{3}}.$$

Hence $2\phi = 120°$, $\phi = 60°$, and

$$\sin \phi = \frac{\sqrt{3}}{2}, \qquad \cos \phi = \frac{1}{2}.$$

The rotation of axes formulas (12.12)(i) are as follows:

FIGURE 12.37

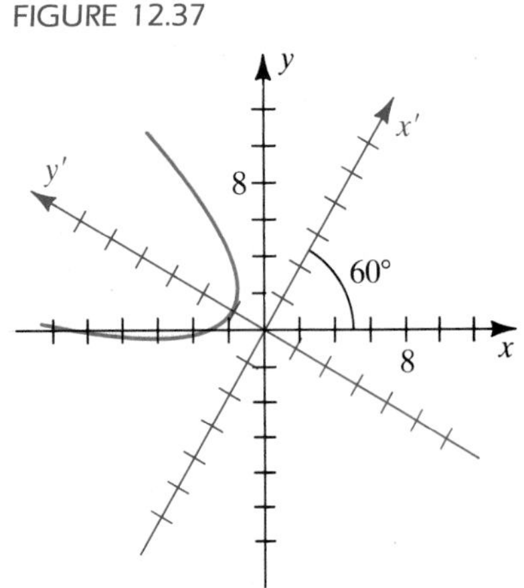

$$x = \frac{1}{2}x' - \frac{\sqrt{3}}{2}y' = \frac{1}{2}(x' - \sqrt{3}y')$$

$$y = \frac{\sqrt{3}}{2}x' + \frac{1}{2}y' = \frac{1}{2}(\sqrt{3}x' + y')$$

Substituting for x and y in the given equation and simplifying leads to

$$4(x')^2 - 16y' + 32 = 0,$$

or, equivalently,
$$(x')^2 = 4(y' - 2).$$

The parabola is sketched in Figure 12.37, where each tic represents two units. Note that the vertex is at the point $(0, 2)$ in the $x'y'$-plane, and the graph is symmetric with respect to the y'-axis.

EXERCISES 12.4

Exer. 1–13: (a) Use the identification theorem (12.14) to determine whether the graph of the equation is a parabola, an ellipse, or a hyperbola. (b) Use a suitable rotation of axes to find an equation for the graph in an $x'y'$-plane, and sketch the graph, labeling vertices.

1 $x^2 - 2xy + y^2 - 2\sqrt{2}x - 2\sqrt{2}y = 0$

2 $x^2 - 2xy + y^2 + 4x + 4y = 0$

3 $5x^2 - 8xy + 5y^2 = 9$

4 $x^2 - xy + y^2 = 3$

5 $11x^2 + 10\sqrt{3}xy + y^2 = 4$

6 $7x^2 - 48xy - 7y^2 = 225$

7 $16x^2 - 24xy + 9y^2 - 60x - 80y + 100 = 0$

8 $x^2 + 4xy + 4y^2 + 6\sqrt{5}x - 18\sqrt{5}y + 45 = 0$

9 $40x^2 - 36xy + 25y^2 - 8\sqrt{13}x - 12\sqrt{13}y = 0$

10 $18x^2 - 48xy + 82y^2 + 6\sqrt{10}x + 2\sqrt{10}y - 80 = 0$

11 $5x^2 + 6\sqrt{3}xy - y^2 + 8x - 8\sqrt{3}y - 12 = 0$

12 $15x^2 + 20xy - 4\sqrt{5}x + 8\sqrt{5}y - 100 = 0$

13 $32x^2 - 72xy + 53y^2 = 80$

c **Exer. 14–16: Graph the equation.**

14 $1.1x^2 - 1.3xy + y^2 - 2.9x - 1.9y = 0$

15 $2.1x^2 - 4xy + 1.5y^2 - 4x + y - 1 = 0$

16 $3.2x^2 - 4\sqrt{2}xy + 2.5y^2 + 2.1y + 3x - 2.1 = 0$

12.5 REVIEW EXERCISES

Exer. 1–16: Find the vertices and foci of the conic, and sketch its graph.

1 $y^2 = 64x$

2 $y = 8x^2 + 32x + 33$

3 $9y^2 = 144 - 16x^2$

4 $9y^2 = 144 + 16x^2$

5 $x^2 - y^2 - 4 = 0$

6 $25x^2 + 36y^2 = 1$

7 $25y = 100 - x^2$

8 $3x^2 + 4y^2 - 18x + 8y + 19 = 0$

9 $x^2 - 9y^2 + 8x + 90y - 210 = 0$

10 $x = 2y^2 + 8y + 3$

11 $4x^2 + 9y^2 + 24x - 36y + 36 = 0$

12 $4x^2 - y^2 - 40x - 8y + 88 = 0$

13 $y^2 - 8x + 8y + 32 = 0$

14 $4x^2 + y^2 - 24x + 4y + 36 = 0$

15 $x^2 - 9y^2 + 8x + 7 = 0$

16 $y^2 - 2x^2 + 6y + 8x - 3 = 0$

Exer. 17–26: Find an equation for the conic that satisfies the given conditions.

17 hyperbola with vertices $V(0, \pm 7)$ and endpoints of conjugate axis $(\pm 3, 0)$

18 parabola, with focus $F(-4, 0)$ and directrix $x = 4$

19 parabola, with focus $F(0, -10)$ and directrix $y = 10$

20 parabola, with vertex at the origin, symmetric to the x-axis, and passing through the point $(5, -1)$

21 ellipse, with vertices $V(0, \pm 10)$ and foci $F(0, \pm 5)$

22 hyperbola, with foci $F(\pm 10, 0)$ and vertices $V(\pm 5, 0)$

23 hyperbola, with vertices $V(0, \pm 6)$ and asymptotes $y = \pm 9x$

24 ellipse, with foci $F(\pm 2, 0)$ and passing through the point $(2, \sqrt{2})$

25 ellipse, with eccentricity $\frac{2}{3}$ and endpoints of minor axis $(\pm 5, 0)$

26 ellipse, with eccentricity $\frac{3}{4}$ and foci $F(\pm 12, 0)$

27 A bridge is to be constructed across a river that is 200 feet wide. The arch of the bridge is to be semielliptical and must be constructed so that a ship up to 50 feet wide and 30 feet high can pass safely through the arch, as shown in the figure.

(a) Find an equation for the arch.

(b) Approximate the height of the arch in the middle of the bridge.

EXERCISE 27

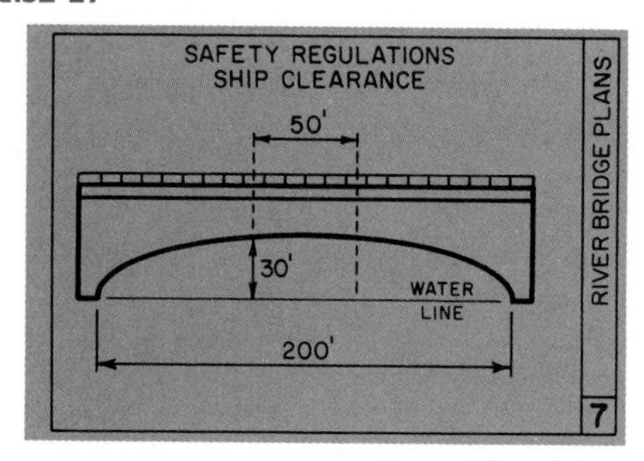

28 A point $P(x, y)$ is the same distance from $(4, 0)$ as it is from the circle $x^2 + y^2 = 4$, as illustrated in the figure. Show that the collection of all such points forms a branch of a hyperbola, and sketch its graph.

EXERCISE 28

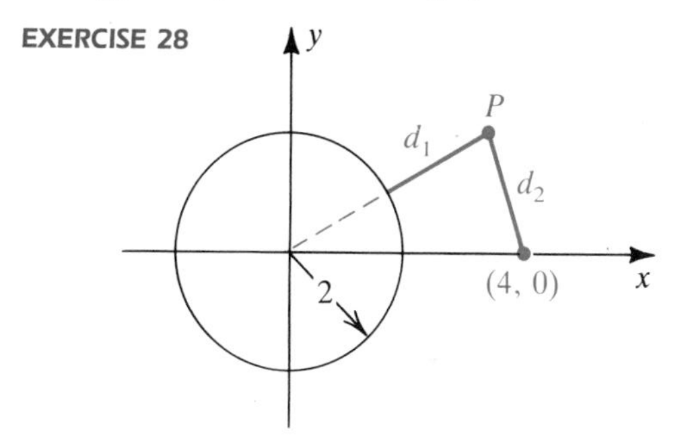

29 Find equations of the tangent line and normal line to the hyperbola $4x^2 - 9y^2 - 8x + 6y - 36 = 0$ at the point $P(-3, 2)$.

30 Tangent lines to the parabola $y = 2x^2 + 3x + 1$ pass through the point $P(2, -1)$. Find the x-coordinates of the points of tangency.

31 Prove that there is exactly one line of a given slope m that is tangent to the parabola $x^2 = 4py$, and show that its equation is $y = mx - pm^2$.

32 Consider the ellipse $px^2 + qy^2 = pq$, with $p > 0$ and $q > 0$. Prove that if m is any real number, there are exactly two lines of slope m that are tangent to the ellipse, and show that their equations are $y = mx \pm \sqrt{p + qm^2}$.

33 Let R be the region bounded by a parabola and the line through the focus that is perpendicular to the axis, and let p be the distance from the vertex V to the focus F, as shown in the figure on the following page. Find the area of R.

EXERCISE 33

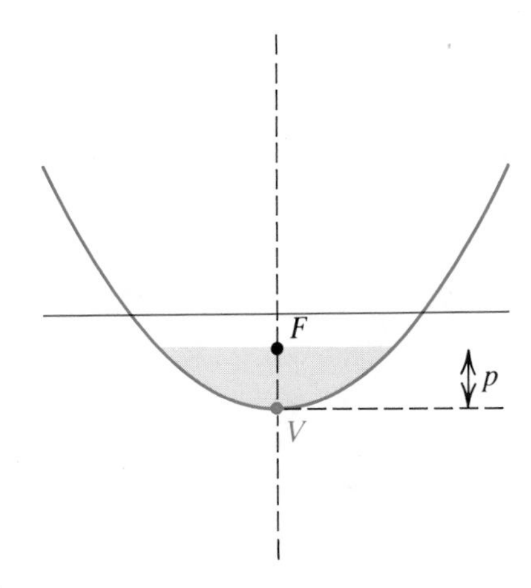

34 Let R be the region that is bounded by the parabola $9y = 5x^2 - 39x + 90$ and an asymptote of the hyperbola $9y^2 - 4x^2 = 36$.

(a) Find the area of R.

(b) Find the volume of the solid obtained by revolving R about the y-axis.

35 An ellipse having axes of lengths 8 and 4 is revolved about its major axis. Find the volume of the resulting solid.

36 A solid has, for its base, the region in the xy-plane bounded by the graph of $x^2 + y^2 = 16$. Find the volume of the solid if every cross section by a plane perpendicular to the x-axis is half of an ellipse with semi-axis of length 5.

37 Find the centroid of the region in the xy-plane that is bounded by the hyperbola $b^2y^2 - a^2x^2 = a^2b^2$ and a horizontal line $y = c$ through a focus.

38 Use the discriminant to identify the graph of each equation. (Do not sketch the graph.)

(a) $2x^2 - 3xy + 4y^2 + 6x - 2y - 6 = 0$

(b) $3x^2 + 2xy - y^2 - 2x + y + 4 = 0$

(c) $x^2 - 6xy + 9y^2 + x - 3y + 5 = 0$

Exer. 39–40: Use a suitable rotation of axes to find an equation for the graph in an $x'y'$-plane, and sketch the graph, labeling vertices.

39 $x^2 - 8xy + 16y^2 - 12\sqrt{17}x - 3\sqrt{17}y = 0$

40 $8x^2 + 12xy + 17y^2 - 16\sqrt{5}x - 12\sqrt{5}y = 0$

CHAPTER 13

PLANE CURVES AND POLAR COORDINATES

INTRODUCTION

The concept of *curve* is more general than that of the graph of a function, since a curve may cross itself in figure-eight style, be closed (as are circles and ellipses), or spiral around a fixed point. In fact, some curves studied in advanced mathematics pass through every point in a coordinate plane!

The curves discussed in this chapter lie in an xy-plane, and each has the property that the coordinates x and y of an arbitrary point P on the curve can be expressed as functions of a variable t, called a *parameter*. The reason for choosing the letter t is that in many applications this variable denotes time and P represents a moving object that has position (x, y) at time t. In later chapters we use such representations to define velocity, acceleration, and other concepts associated with motion.

In Sections 13.3 and 13.4 we discuss polar coordinates and use definite integrals to find areas enclosed by graphs of polar equations. Our methods are analogous to those developed in Chapter 6. The principal difference is that we consider limits of sums of circular sectors instead of vertical or horizontal rectangles.

The chapter closes with a unified description of conics in terms of polar equations. Such equations are indispensable in analyzing orbits of planets, satellites, and atomic particles.

13.1 PLANE CURVES

If f is a continuous function, the graph of the equation $y = f(x)$ is often called a *plane curve*. However, this definition is restrictive, because it excludes many useful graphs. The following definition is more general.

Definition **(13.1)**

A **plane curve** is a set C of ordered pairs $(f(t), g(t))$, where f and g are continuous functions on an interval I.

For simplicity, we often refer to a plane curve as a **curve**. The **graph** of C in Definition (13.1) consists of all points $P(t) = (f(t), g(t))$ in an xy-plane, for t in I. We shall use the term *curve* interchangeably with *graph of a curve*. We sometimes regard the point $P(t)$ as tracing the curve C as t varies through the interval I.

The graphs of several curves are sketched in Figure 13.1, where I is a closed interval $[a, b]$. In (i) of the figure, $P(a) \neq P(b)$, and $P(a)$ and $P(b)$ are called the **endpoints** of C. The curve in (i) intersects itself; that is, two different values of t produce the same point. If $P(a) = P(b)$, as in Figure 13.1(ii), then C is a **closed curve**. If $P(a) = P(b)$ and C does not intersect itself at any other point, as in (iii), then C is a **simple closed curve**.

FIGURE 13.1

(i) Curve

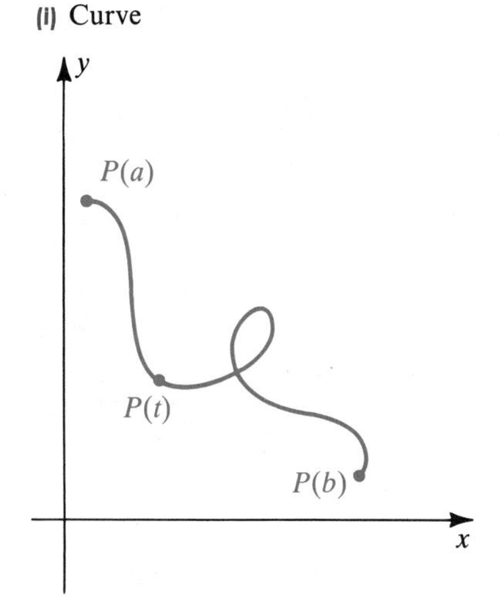

(ii) Closed curve

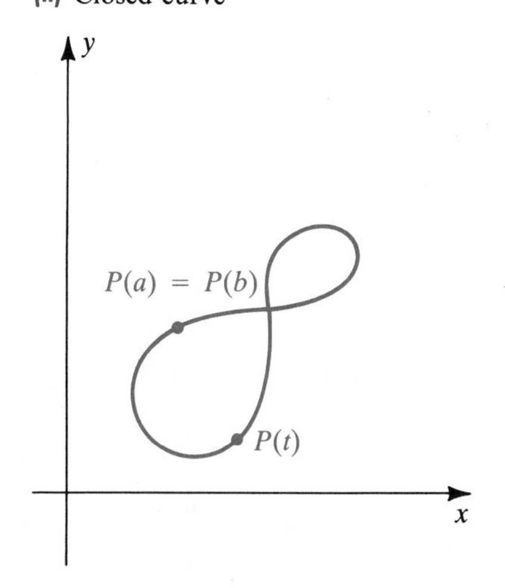

(iii) Simple closed curve

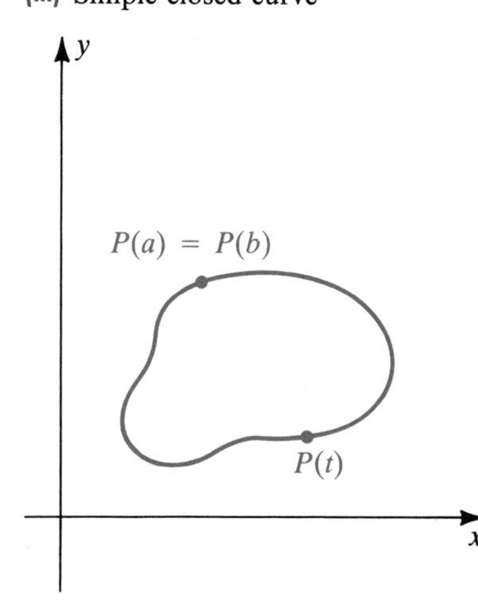

A convenient way to represent curves is given in the next definition.

Definition **(13.2)**

Let C be the curve consisting of all ordered pairs $(f(t), g(t))$, where f and g are continuous on an interval I. The equations

$$x = f(t), \quad y = g(t),$$

for t in I, are **parametric equations** for C with **parameter** t.

The curve C in this definition is referred to as a **parametrized curve**, and the parametric equations are a **parametrization** for C. We often use the notation

$$x = f(t), \quad y = g(t); \quad t \text{ in } I$$

to indicate the domain I of f and g. Sometimes it may be possible to eliminate the parameter and obtain a familiar equation in x and y for C. In simple cases we may sketch a graph of a parametrized curve by plotting points and connecting them in the order of increasing t, as illustrated in the next example.

EXAMPLE 1 Sketch the graph of the curve C that has the parametrization

$$x = 2t, \quad y = t^2 - 1; \quad -1 \le t \le 2.$$

SOLUTION We use the parametric equations to tabulate coordinates of points $P(x, y)$ on C as follows.

t	-1	$-\frac{1}{2}$	0	$\frac{1}{2}$	1	$\frac{3}{2}$	2
x	-2	-1	0	1	2	3	4
y	0	$-\frac{3}{4}$	-1	$-\frac{3}{4}$	0	$\frac{5}{4}$	3

FIGURE 13.2
$x = 2t, y = t^2 - 1; -1 \le t \le 2$

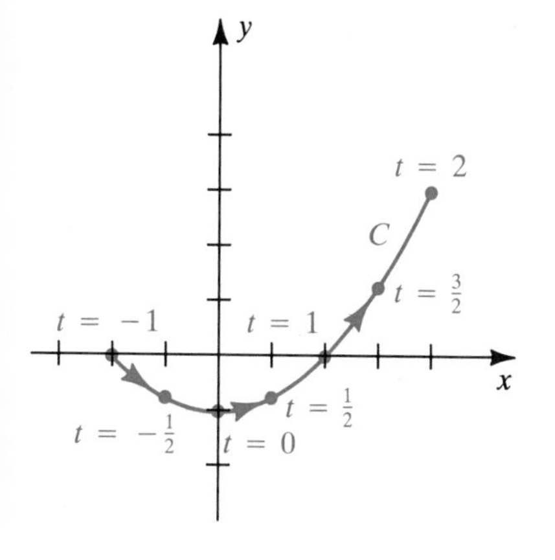

Plotting points leads to the sketch in Figure 13.2. The arrowheads on the graph indicate the direction in which $P(x, y)$ traces the curve as t *increases* from -1 to 2.

We may obtain a clearer description of the graph by eliminating the parameter. Solving the first parametric equation for t, we obtain $t = \frac{1}{2}x$. Substituting this expression for t in the second equation gives us

$$y = (\tfrac{1}{2}x)^2 - 1.$$

The graph of this equation in x and y is a parabola symmetric with respect to the y-axis with vertex $(0, -1)$. However, since $x = 2t$ and $-1 \le t \le 2$, we see that $-2 \le x \le 4$ for points (x, y) on C, and hence C is that part of the parabola between the points $(-2, 0)$ and $(4, 3)$ shown in Figure 13.2.

As indicated by the arrowheads in Figure 13.2, the point $P(x, y)$ traces the curve C from *left to right* as t increases. The parametric equations

$$x = -2t, \quad y = t^2 - 1; \quad -2 \le t \le 1$$

give us the same graph; however, as t increases, $P(x, y)$ traces the curve from *right to left*. For other parametrizations, the point $P(x, y)$ may oscillate back and forth as t increases.

The **orientation** of a parametrized curve C is the direction determined by *increasing* values of the parameter. We often indicate an orientation by placing arrowheads on C as in Figure 13.2. If $P(x, y)$ moves back and forth as t increases, we may place arrows *alongside* of C. As we have observed, a curve may have different orientations, depending on the parametrization.

The next example demonstrates that it is sometimes useful to eliminate the parameter *before* plotting points.

EXAMPLE 2 A point moves in a plane such that its position $P(x, y)$ at time t is given by

$$x = a \cos t, \quad y = a \sin t; \quad t \text{ in } \mathbb{R},$$

where $a > 0$. Describe the motion of the point.

SOLUTION We may eliminate the parameter by rewriting the parametric equations as

$$\frac{x}{a} = \cos t, \quad \frac{y}{a} = \sin t$$

and using the identity $\cos^2 t + \sin^2 t = 1$ to obtain

$$\left(\frac{x}{a}\right)^2 + \left(\frac{y}{a}\right)^2 = 1,$$

or

$$x^2 + y^2 = a^2.$$

FIGURE 13.3
$x = a \cos t$, $y = a \sin t$; t in $\mathbb{R}$

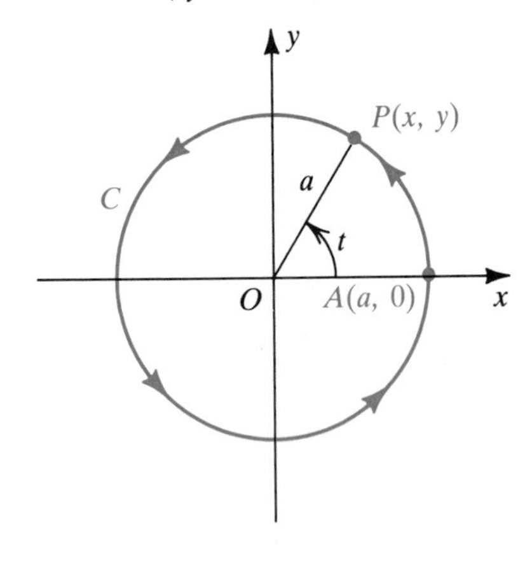

This shows that the point $P(x, y)$ moves on the circle C of radius a with center at the origin (see Figure 13.3). The point is at $A(a, 0)$ when $t = 0$, at $(0, a)$ when $t = \pi/2$, at $(-a, 0)$ when $t = \pi$, at $(0, -a)$ when $t = 3\pi/2$, and back at $A(a, 0)$ when $t = 2\pi$. Thus, P moves around C in a counterclockwise direction, making one revolution every 2π units of time. The orientation of C is indicated by the arrowheads in the figure.

Note that in this example we may interpret t geometrically as the radian measure of the angle generated by the line segment OP.

EXAMPLE 3 Sketch the graph of the curve C that has the parametrization

$$x = -2 + t^2, \quad y = 1 + 2t^2; \quad t \text{ in } \mathbb{R}$$

and indicate the orientation.

SOLUTION To eliminate the parameter, we use the first equation to obtain $t^2 = x + 2$ and then substitute for t^2 in the second equation. Thus,

$$y = 1 + 2(x + 2).$$

FIGURE 13.4
(i) (ii)

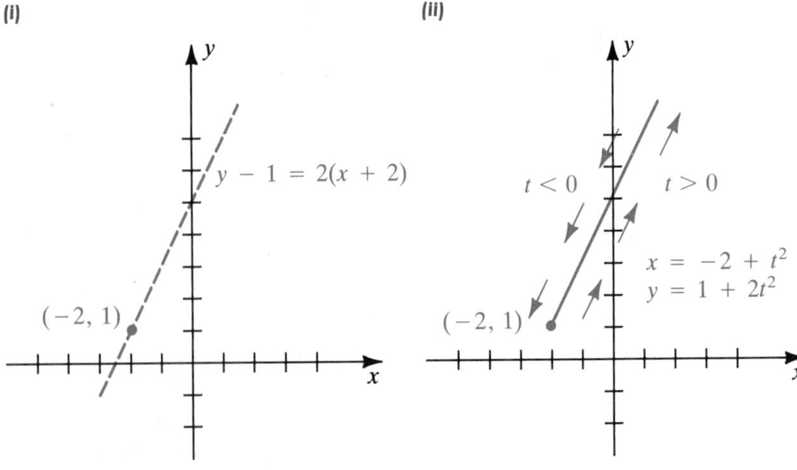

This is an equation of the line of slope 2 through the point $(-2, 1)$, as indicated by the dashes in Figure 13.4(i). However, since $t^2 \geq 0$, we see from the parametric equations for C that

$$x = -2 + t^2 \geq -2 \quad \text{and} \quad y = 1 + 2t^2 \geq 1.$$

Thus, the graph of C is that part of the line to the right of $(-2, 1)$ (the point corresponding to $t = 0$), as shown in Figure 13.4(ii). The orientation is indicated by the arrows alongside of C. As t increases in the interval $(-\infty, 0]$, $P(x, y)$ moves down the curve toward the point $(-2, 1)$. As t increases in $[0, \infty)$, $P(x, y)$ moves up the curve away from $(-2, 1)$.

If a curve C is described by an equation $y = f(x)$ for a continuous function f, then an easy way to obtain parametric equations for C is to let

$$x = t, \quad y = f(t),$$

where t is in the domain of f. For example, if $y = x^3$, then parametric equations are

$$x = t, \quad y = t^3; \quad t \text{ in } \mathbb{R}.$$

We can use many different substitutions for x, provided that as t varies through some interval, x takes on every value in the domain of f. Thus, the graph of $y = x^3$ is also given by

$$x = t^{1/3}, \quad y = t; \quad t \text{ in } \mathbb{R}.$$

Note, however, that the parametric equations

$$x = \sin t, \quad y = \sin^3 t; \quad t \text{ in } \mathbb{R}$$

give only that part of the graph of $y = x^3$ between the points $(-1, -1)$ and $(1, 1)$.

EXAMPLE 4 Find three parametrizations for the line of slope m through the point (x_1, y_1).

SOLUTION By the point-slope form, an equation for the line is

$$y - y_1 = m(x - x_1).$$

If we let $x = t$, then $y - y_1 = m(t - x_1)$ and we obtain the parametrization

$$x = t, \quad y = y_1 + m(t - x_1); \quad t \text{ in } \mathbb{R}.$$

We obtain another parametrization for the line if we let $x - x_1 = t$. In this case $y - y_1 = mt$, and we have

$$x = x_1 + t, \quad y = y_1 + mt; \quad t \text{ in } \mathbb{R}.$$

As a third illustration, if we let $x - x_1 = \tan t$, then

$$x = x_1 + \tan t, \quad y = y_1 + m \tan t; \quad -\frac{\pi}{2} < t < \frac{\pi}{2}.$$

There are many other parametrizations for the line.

Parametric equations of the form

$$x = a \sin \omega_1 t, \quad y = b \cos \omega_2 t; \quad t \geq 0,$$

where a, b, ω_1, and ω_2 are constants, occur in electrical theory. The variables x and y usually represent voltages or currents at time t. The resulting curve is often difficult to sketch; however, using an oscilloscope and imposing voltages or currents on the input terminals, we can represent the graph, a **Lissajous figure**, on the screen of the oscilloscope. Computers are also useful in obtaining these complicated graphs.

FIGURE 13.5
$x = \sin 2t$, $y = \cos t$; $0 \leq t \leq 2\pi$

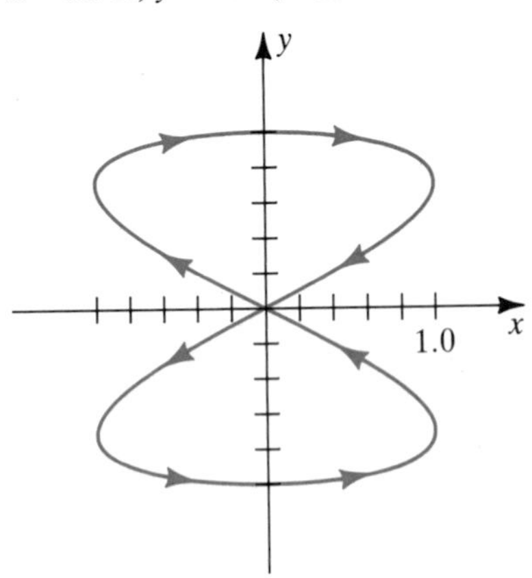

EXAMPLE 5 A computer-generated graph of the Lissajous figure

$$x = \sin 2t, \quad y = \cos t; \quad 0 \leq t \leq 2\pi$$

is shown in Figure 13.5, with the arrowheads indicating the orientation. Verify the orientation and find an equation in x and y for the curve.

SOLUTION Referring to the parametric equations, we see that as t increases from 0 to $\pi/2$, the point $P(x, y)$ starts at (0, 1) and traces the part of the curve in quadrant I (in a generally clockwise direction). As t increases from $\pi/2$ to π, $P(x, y)$ traces the part in quadrant III (in a counterclockwise direction). For $\pi \leq t \leq 3\pi/2$, we obtain the part in quadrant IV; and $3\pi/2 \leq t \leq 2\pi$ gives us the part in quadrant II.

We may find an equation in x and y for the curve by employing trigonometric identities and algebraic manipulations. Writing $x = 2 \sin t \cos t$ and squaring, we have

$$x^2 = 4 \sin^2 t \cos^2 t,$$

or

$$x^2 = 4(1 - \cos^2 t) \cos^2 t.$$

Using $y = \cos t$ gives us

$$x^2 = 4(1 - y^2)y^2.$$

To express y in terms of x, let us rewrite the last equation as

$$4y^4 - 4y^2 + x^2 = 0$$

and use the quadratic formula to solve for y^2 as follows:

$$y^2 = \frac{4 \pm \sqrt{16 - 16x^2}}{8} = \frac{1 \pm \sqrt{1 - x^2}}{2}$$

Taking square roots, we obtain

$$y = \pm\sqrt{\frac{1 \pm \sqrt{1 - x^2}}{2}}.$$

These complicated equations should indicate the advantage of expressing the curve in parametric form.

A curve C is **smooth** if it has a parametrization $x = f(t)$, $y = g(t)$ on an interval I such that the derivatives f' and g' are continuous and not simultaneously zero, except possibly at endpoints of I. A curve C is **piecewise smooth** if the interval I can be partitioned into closed subintervals with

C smooth on each subinterval. The graph of a smooth curve has no corners or cusps. The curves given in Examples 1–5 are smooth. The curve in the next example is piecewise smooth.

EXAMPLE 6 The curve traced by a fixed point P on the circumference of a circle as the circle rolls along a line in a plane is called a **cycloid**. Find parametric equations for a cycloid and determine the intervals on which it is smooth.

SOLUTION Suppose the circle has radius a and that it rolls along (and above) the x-axis in the positive direction. If one position of P is the origin, then Figure 13.6 displays part of the curve and a possible position of the circle.

FIGURE 13.6

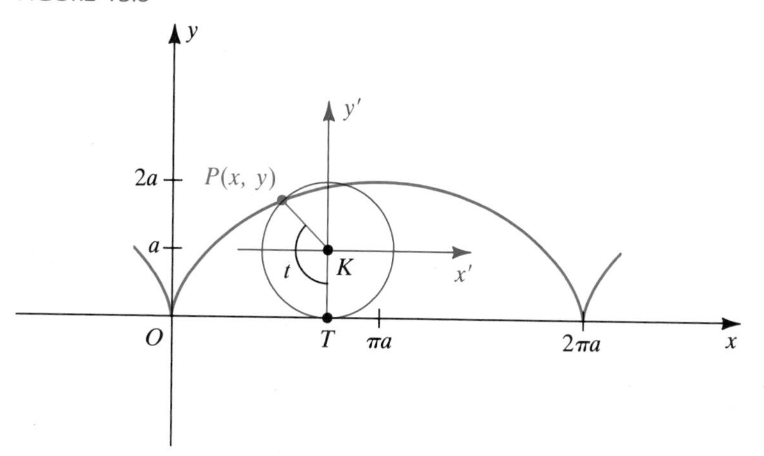

Let K denote the center of the circle and T the point of tangency with the x-axis. We introduce, as a parameter t, the radian measure of angle TKP. The distance the circle has rolled is $d(O, T) = at$. Consequently the coordinates of K are (at, a). If we consider an $x'y'$-coordinate system with origin at $K(at, a)$ and if $P(x', y')$ denotes the point P relative to this system, then, by the translation of axes formulas with $h = at$ and $k = a$,

$$x = at + x', \quad y = a + y'.$$

FIGURE 13.7

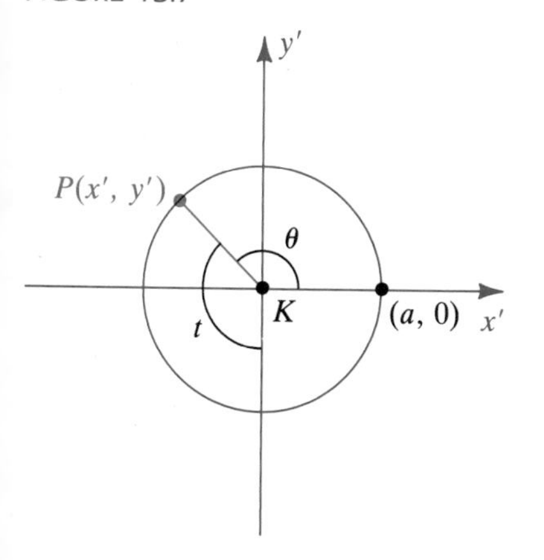

If, as in Figure 13.7, θ denotes an angle in standard position on the $x'y'$-plane, then $\theta = (3\pi/2) - t$. Hence

$$x' = a \cos \theta = a \cos \left(\frac{3\pi}{2} - t\right) = -a \sin t$$

$$y' = a \sin \theta = a \sin \left(\frac{3\pi}{2} - t\right) = -a \cos t,$$

and substitution in $x = at + x'$, $y = a + y'$ gives us parametric equations for the cycloid:

$$x = a(t - \sin t), \quad y = a(1 - \cos t); \quad t \text{ in } \mathbb{R}.$$

Differentiating the parametric equations of the cycloid yields

$$\frac{dx}{dt} = a(1 - \cos t), \quad \frac{dy}{dt} = a \sin t.$$

These derivatives are continuous for every t, but are simultaneously 0 at $t = 2\pi n$ for every integer n. The points corresponding to $t = 2\pi n$ are the x-intercepts of the graph, and the cycloid has a cusp at each such point (see Figure 13.6). The graph is piecewise smooth, since it is smooth on the t-interval $[2\pi n, 2\pi(n + 1)]$ for every integer n.

FIGURE 13.8

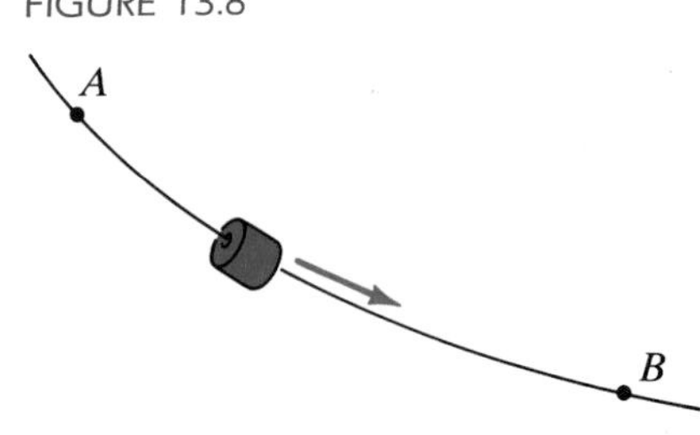

If $a < 0$, then the graph of $x = a(t - \sin t)$, $y = a(1 - \cos t)$ is the inverted cycloid that results if the circle of Example 6 rolls *below* the x-axis. This curve has a number of important physical properties. To illustrate, suppose a thin wire passes through two fixed points A and B, as shown in Figure 13.8, and that the shape of the wire can be changed by bending it in any manner. Suppose further that a bead is allowed to slide along the wire and the only force acting on the bead is gravity. We now ask which of all the possible paths will allow the bead to slide from A to B in the least amount of time. It is natural to believe that the desired path is the straight line segment from A to B; however, this is not the correct answer. The path that requires the least time coincides with the graph of an inverted cycloid with A at the origin. Because the velocity of the bead increases more rapidly along the cycloid than along the line through A and B, the bead reaches B more rapidly, even though the distance is greater.

There is another interesting property of this **curve of least descent**. Suppose that A is the origin and B is the point with x-coordinate $\pi|a|$—that is, the lowest point on the cycloid in the first arc to the right of A. If the bead is released at *any* point between A and B, it can be shown that the time required for it to reach B is always the same.

Variations of the cycloid occur in applications. For example, if a motorcycle wheel rolls along a straight road, then the curve traced by a fixed point on one of the spokes is a cycloidlike curve. In this case the curve does not have corners or cusps, nor does it intersect the road (the x-axis) as does the graph of a cycloid. If the wheel of a train rolls along a railroad track, then the curve traced by a fixed point on the circumference of the wheel (which extends below the track) contains loops at regular intervals. Other cycloids are defined in Exercises 33 and 34.

EXERCISES 13.1

Exer. 1–24: (a) Find an equation in x and y whose graph contains the points on the curve C. (b) Sketch the graph of C and indicate the orientation.

1 $x = t - 2$, $y = 2t + 3$; $0 \le t \le 5$

2 $x = 1 - 2t$, $y = 1 + t$; $-1 \le t \le 4$

3 $x = t^2 + 1$, $y = t^2 - 1$; $-2 \le t \le 2$

4 $x = t^3 + 1$, $y = t^3 - 1$; $-2 \le t \le 2$

5 $x = 4t^2 - 5$, $y = 2t + 3$; t in $\mathbb{R}$

6 $x = t^3$, $y = t^2$; t in $\mathbb{R}$

7 $x = e^t$, $y = e^{-2t}$; t in $\mathbb{R}$

8 $x = \sqrt{t}$, $y = 3t + 4$; $t \ge 0$

9 $x = 2 \sin t$, $y = 3 \cos t$; $0 \le t \le 2\pi$

10 $x = \cos t - 2$, $y = \sin t + 3$; $0 \le t \le 2\pi$

11 $x = \sec t,\quad y = \tan t;\quad -\pi/2 < t < \pi/2$

12 $x = \cos 2t,\quad y = \sin t;\quad -\pi \le t \le \pi$

13 $x = t^2,\quad y = 2 \ln t;\quad t > 0$

14 $x = \cos^3 t,\quad y = \sin^3 t;\quad 0 \le t \le 2\pi$

15 $x = \sin t,\quad y = \csc t;\quad 0 < t \le \pi/2$

16 $x = e^t,\quad y = e^{-t};\quad t \text{ in } \mathbb{R}$

17 $x = \cosh t,\quad y = \sinh t;\quad t \text{ in } \mathbb{R}$

18 $x = 3 \cosh t,\quad y = 2 \sinh t;\quad t \text{ in } \mathbb{R}$

19 $x = t,\quad y = \sqrt{t^2 - 1};\quad |t| \ge 1$

20 $x = -2\sqrt{1 - t^2},\quad y = t;\quad |t| \le 1$

21 $x = t,\quad y = \sqrt{t^2 - 2t + 1};\quad 0 \le t \le 4$

22 $x = 2t,\quad y = 8t^3;\quad -1 \le t \le 1$

23 $x = (t + 1)^3,\quad y = (t + 2)^2;\quad 0 \le t \le 2$

24 $x = \tan t,\quad y = 1;\quad -\pi/2 < t < \pi/2$

Exer. 25–26: Curves C_1, C_2, C_3, and C_4 are given parametrically, for t in $\mathbb{R}$. Sketch their graphs and indicate orientations.

25 C_1: $x = t^2,\quad y = t$
C_2: $x = t^4,\quad y = t^2$
C_3: $x = \sin^2 t,\quad y = \sin t$
C_4: $x = e^{2t},\quad y = -e^t$

26 C_1: $x = t,\quad y = 1 - t$
C_2: $x = 1 - t^2,\quad y = t^2$
C_3: $x = \cos^2 t,\quad y = \sin^2 t$
C_4: $x = \ln t - t,\quad y = 1 + t - \ln t;\quad t > 0$

Exer. 27–28: The parametric equations specify the position of a moving point $P(x, y)$ at time t. Sketch the graph and indicate the motion of P as t increases.

27 (a) $x = \cos t,\quad y = \sin t;\quad 0 \le t \le \pi$
(b) $x = \sin t,\quad y = \cos t;\quad 0 \le t \le \pi$
(c) $x = t,\quad y = \sqrt{1 - t^2};\quad -1 \le t \le 1$

28 (a) $x = t^2,\quad y = 1 - t^2;\quad 0 \le t \le 1$
(b) $x = 1 - \ln t,\quad y = \ln t;\quad 1 \le t \le e$
(c) $x = \cos^2 t,\quad y = \sin^2 t;\quad 0 \le t \le 2\pi$

29 Show that

$$x = a \cos t + h,\quad y = b \sin t + k;\quad 0 \le t \le 2\pi$$

are parametric equations of an ellipse with center (h, k) and axes of lengths $2a$ and $2b$.

30 Show that

$$x = a \sec t + h,\quad y = b \tan t + k;$$

$$-\frac{\pi}{2} < t < \frac{3\pi}{2} \quad \text{and} \quad t \ne \frac{\pi}{2}$$

are parametric equations of a hyperbola with center (h, k), transverse axis of length $2a$, and conjugate axis of length $2b$. Determine the values of t for each branch.

31 If $P_1(x_1, y_1)$ and $P_2(x_2, y_2)$ are distinct points, show that

$$x = (x_2 - x_1)t + x_1,\quad y = (y_2 - y_1)t + y_1;\quad t \text{ in } \mathbb{R}$$

are parametric equations for the line l through P_1 and P_2.

32 Describe the difference between the graph of the hyperbola $(x^2/a^2) - (y^2/b^2) = 1$ and the graph of

$$x = a \cosh t,\quad y = b \sinh t;\quad t \text{ in } \mathbb{R}.$$

(*Hint:* Use Theorem (8.11).)

33 A circle C of radius b rolls on the outside of the circle $x^2 + y^2 = a^2$, and $b < a$. Let P be a fixed point on C, and let the initial position of P be $A(a, 0)$, as shown in the figure. If the parameter t is the angle from the positive x-axis to the line segment from O to the center of C, show that parametric equations for the curve traced by P (an *epicycloid*) are

$$x = (a + b) \cos t - b \cos\left(\frac{a + b}{b} t\right),$$

$$y = (a + b) \sin t - b \sin\left(\frac{a + b}{b} t\right);\quad 0 \le t \le 2\pi.$$

EXERCISE 33

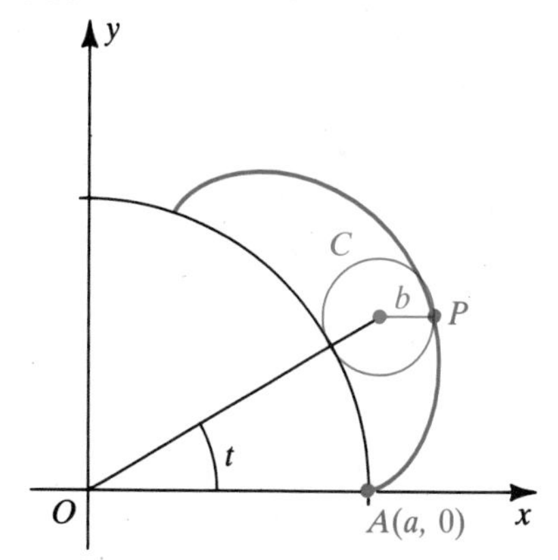

34 If the circle C of Exercise 33 rolls on the inside of the second circle (see the figure on the following page), then the curve traced by P is a *hypocycloid*.

(a) Show that parametric equations for this curve are

$$x = (a - b) \cos t + b \cos\left(\frac{a - b}{b} t\right),$$

$$y = (a - b) \sin t - b \sin\left(\frac{a - b}{b} t\right);\quad 0 \le t \le 2\pi.$$

(b) If $b = \frac{1}{4}a$, show that $x = a \cos^3 t$, $y = a \sin^3 t$ and sketch the graph.

EXERCISE 34

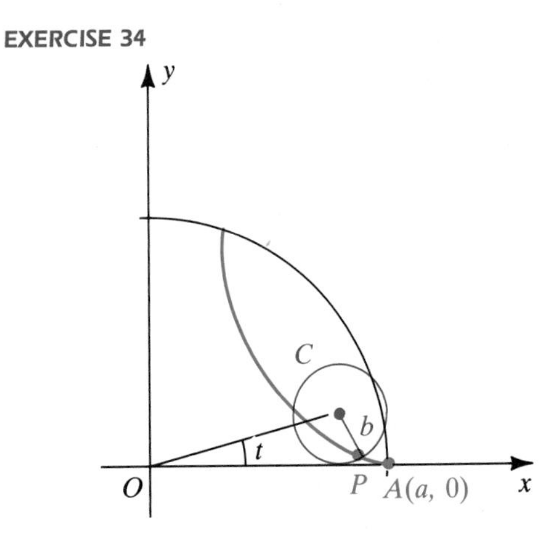

35 If $b = \frac{1}{3}a$ in Exercise 33, find parametric equations for the epicycloid and sketch the graph.

36 The radius of circle B is one-third that of circle A. How many revolutions will circle B make as it rolls around circle A until it reaches its starting point? (*Hint:* Use Exercise 35.)

37 If a string is unwound from around a circle of radius a and is kept tight in the plane of the circle, then a fixed point P on the string will trace a curve called the *involute of the circle*. Let the circle be chosen as in the figure. If the parameter t is the measure of the indicated angle and the initial position of P is $A(a, 0)$, show that parametric equations for the involute are

$$x = a(\cos t + t \sin t), \quad y = a(\sin t - t \cos t).$$

EXERCISE 37

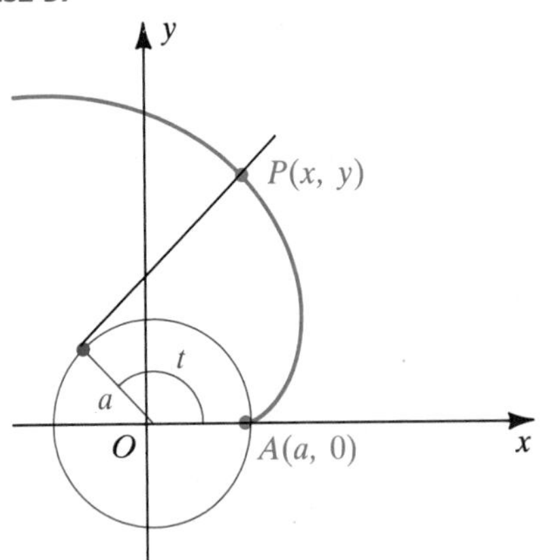

38 Generalize the cycloid of Example 6 to the case where P is any point on a fixed line through the center C of the circle. If $b = d(C, P)$, show that

$$x = at - b \sin t, \quad y = a - b \cos t.$$

Sketch a typical graph if $b < a$ (a *curtate cycloid*) and if $b > a$ (a *prolate cycloid*). The term *trochoid* is sometimes used for either of these curves.

39 Refer to Example 5.

(a) Describe the Lissajous figure given by $f(t) = a \sin \omega t$ and $g(t) = b \cos \omega t$ for $t \geq 0$ and $a \neq b$.

(b) Suppose $f(t) = a \sin \omega_1 t$ and $g(t) = b \sin \omega_2 t$, where ω_1 and ω_2 are positive rational numbers, and write ω_2/ω_1 as m/n for positive integers m and n. Show that if $p = 2\pi n/\omega_1$, then $f(t + p) = f(t)$ and $g(t + p) = g(t)$. Conclude that the curve retraces itself every p units of time.

40 Shown in the figure is the Lissajous figure given by

$$x = 2 \sin 3t, \quad y = 3 \sin 1.5t; \quad t \geq 0.$$

(a) Find the period of the figure—that is, the length of the smallest t-interval that traces the curve.

(b) Find the maximum distance from the origin to a point on the graph.

EXERCISE 40

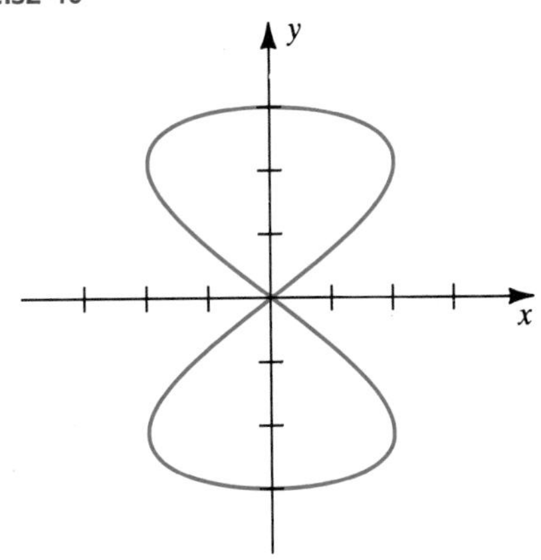

c Exer. 41–44: Graph the curve.

41 $x = 3 \sin^5 t, \quad y = 3 \cos^5 t; \quad 0 \leq t \leq 2\pi$

42 $x = 8 \cos t - 2 \cos 4t, \quad y = 8 \sin t - 2 \sin 4t; \quad 0 \leq t \leq 2\pi$

43 $x = 3t - 2 \sin t, \quad y = 3 - 2 \cos t; \quad -8 \leq t \leq 8$

44 $x = 2t - 3 \sin t, \quad y = 2 - 3 \cos t; \quad -8 \leq t \leq 8$

Exer. 45–48: Graph the given curves on the same coordinate axes and describe the shape of the resulting figure.

c 45 C_1: $x = 2 \sin 3t, \quad y = 3 \cos 2t; \quad -\pi/2 \leq t \leq \pi/2$
C_2: $x = \frac{1}{4} \cos t + \frac{3}{4}, \quad y = \frac{1}{4} \sin t + \frac{3}{2}; \quad 0 \leq t \leq 2\pi$
C_3: $x = \frac{1}{4} \cos t - \frac{3}{4}, \quad y = \frac{1}{4} \sin t + \frac{3}{2}; \quad 0 \leq t \leq 2\pi$
C_4: $x = \frac{3}{4} \cos t, \quad y = \frac{1}{4} \sin t; \quad 0 \leq t \leq 2\pi$
C_5: $x = \frac{1}{4} \cos t, \quad y = \frac{1}{8} \sin t + \frac{3}{4}; \quad \pi \leq t \leq 2\pi$

46 C_1: $x = \frac{3}{2} \cos t + 1, \quad y = \sin t - 1; \quad -\pi/2 \leq t \leq \pi/2$
C_2: $x = \frac{3}{2} \cos t + 1, \quad y = \sin t + 1; \quad -\pi/2 \leq t \leq \pi/2$
C_3: $x = 1, \quad y = 2 \tan t; \quad -\pi/4 \leq t \leq \pi/4$

47 C_1: $x = \tan t$, $y = 3 \tan t$; $0 \le t \le \pi/4$
C_2: $x = 1 + \tan t$, $y = 3 - 3 \tan t$; $0 \le t \le \pi/4$
C_3: $x = \frac{1}{2} + \tan t$, $y = \frac{3}{2}$; $0 \le t \le \pi/4$

48 C_1: $x = 1 + \cos t$, $y = 1 + \sin t$; $\pi/3 \le t \le 2\pi$
C_2: $x = 1 + \tan t$, $y = 1$; $0 \le t \le \pi/4$

13.2 TANGENT LINES AND ARC LENGTH

The curve C given parametrically by

$$x = 2t, \quad y = t^2 - 1; \quad -1 \le t \le 2$$

can also be represented by an equation of the form $y = k(x)$, where k is a function defined on a suitable interval. In Example 1 of the preceding section, we eliminated the parameter t, obtaining

$$y = k(x) = \tfrac{1}{4}x^2 - 1 \quad \text{for} \quad -2 \le x \le 4.$$

The slope of the tangent line at any point $P(x, y)$ on C is

$$k'(x) = \tfrac{1}{2}x, \quad \text{or} \quad k'(x) = \tfrac{1}{2}(2t) = t.$$

Since it is often difficult to eliminate a parameter, we shall next derive a formula that can be used to find the slope directly from the parametric equations.

Theorem **(13.3)**

If a smooth curve C is given parametrically by $x = f(t)$, $y = g(t)$, then the slope dy/dx of the tangent line to C at $P(x, y)$ is

$$\frac{dy}{dx} = \frac{dy/dt}{dx/dt}, \quad \text{provided} \quad \frac{dx}{dt} \neq 0.$$

PROOF If $dx/dt \neq 0$ at $x = c$, then, since f is continuous at c, $dx/dt > 0$ or $dx/dt < 0$ throughout an interval $[a, b]$, with $a < c < b$ (see Theorem (2.27)). Applying Theorem (7.6) or the analogous result for decreasing functions, we know that f has an inverse function f^{-1}, and we may consider $t = f^{-1}(x)$ for x in $[f(a), f(b)]$. Applying the chain rule to $y = g(t)$ and $t = f^{-1}(x)$, we obtain

$$\frac{dy}{dx} = \frac{dy}{dt}\frac{dt}{dx} = \frac{dy/dt}{dx/dt},$$

where the last equality follows from Corollary (7.8). ■

EXAMPLE 1 Let C be the curve with parametrization

$$x = 2t, \quad y = t^2 - 1; \quad -1 \le t \le 2.$$

Find the slopes of the tangent line and normal line to C at $P(x, y)$.

SOLUTION The curve C was considered in Example 1 of the preceding section (see Figure 13.2). Using Theorem (13.3) with $x = 2t$ and $y = t^2 - 1$, we find that the slope of the tangent line at $P(x, y)$ is

$$\frac{dy}{dx} = \frac{dy/dt}{dx/dt} = \frac{2t}{2} = t.$$

This result agrees with that of the discussion at the beginning of this section, where we used the form $y = k(x)$ to show that $m = \frac{1}{2}x = t$.

The slope of the normal line is the negative reciprocal $-1/t$, provided $t \neq 0$.

EXAMPLE 2 Let C be the curve with parametrization

$$x = t^3 - 3t, \quad y = t^2 - 5t - 1; \quad t \text{ in } \mathbb{R}.$$

(a) Find an equation of the tangent line to C at the point corresponding to $t = 2$.

(b) For what values of t is the tangent line horizontal or vertical?

SOLUTION

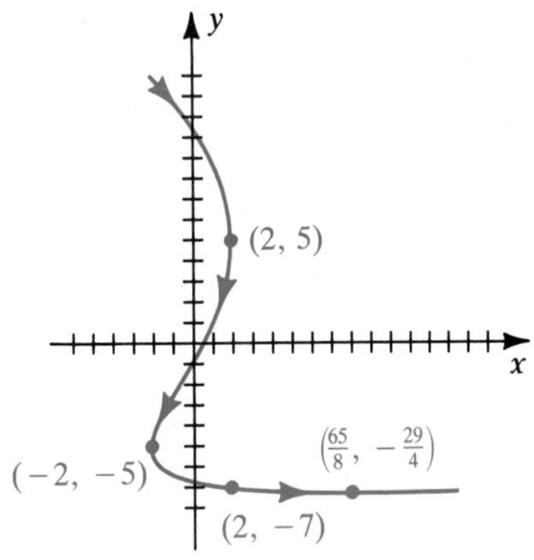

FIGURE 13.9
$x = t^3 - 3t$, $y = t^2 - 5t - 1$; t in $\mathbb{R}$

(a) A portion of the graph of C is sketched in Figure 13.9, where we have also plotted several points and indicated the orientation. Using the parametric equations for C, we find that the point corresponding to $t = 2$ is $(2, -7)$. By Theorem (13.3),

$$\frac{dy}{dx} = \frac{dy/dt}{dx/dt} = \frac{2t - 5}{3t^2 - 3}.$$

The slope m of the tangent line at $(2, -7)$ is

$$m = \left.\frac{dy}{dx}\right]_{t=2} = \frac{2(2) - 5}{3(2^2) - 3} = -\frac{1}{9}.$$

Applying the point-slope form, we obtain an equation of the tangent line:

$$y + 7 = -\tfrac{1}{9}(x - 2), \quad \text{or} \quad x + 9y = -61$$

(b) The tangent line is horizontal if $dy/dx = 0$—that is, if $2t - 5 = 0$, or $t = \frac{5}{2}$. The corresponding point on C is $(\frac{65}{8}, -\frac{29}{4})$, as shown in Figure 13.9.

The tangent line is vertical if $3t^2 - 3 = 0$. Thus, there are vertical tangent lines at the points corresponding to $t = 1$ and $t = -1$—that is, at $(-2, 5)$ and $(2, 5)$.

If a curve C is parametrized by $x = f(t)$, $y = g(t)$ and if y' is a differentiable function of t, we can find d^2y/dx^2 by applying Theorem (13.3) to y' as follows.

Second derivative in parametric form **(13.4)**

$$\frac{d^2y}{dx^2} = \frac{d}{dx}(y') = \frac{dy'/dt}{dx/dt}$$

It is important to observe that

$$\frac{d^2y}{dx^2} \neq \frac{d^2y/dt^2}{d^2x/dt^2}.$$

EXAMPLE 3 Let C be the curve with parametrization

$$x = e^{-t}, \quad y = e^{2t}; \quad t \text{ in } \mathbb{R}.$$

(a) Sketch the graph of C and indicate the orientation.

(b) Use (13.3) and (13.4) to find dy/dx and d^2y/dx^2.

(c) Find a function k that has the same graph as C, and use $k'(x)$ and $k''(x)$ to check the answers to (b).

(d) Discuss the concavity of C.

SOLUTION

(a) To help us sketch the graph, let us first eliminate the parameter. Using $x = e^{-t} = 1/e^t$, we see that $e^t = 1/x$. Substituting in $y = e^{2t} = (e^t)^2$ gives us

$$y = \left(\frac{1}{x}\right)^2 = \frac{1}{x^2}.$$

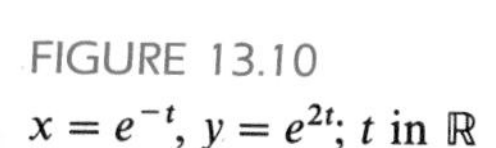
FIGURE 13.10
$x = e^{-t}$, $y = e^{2t}$; t in $\mathbb{R}$

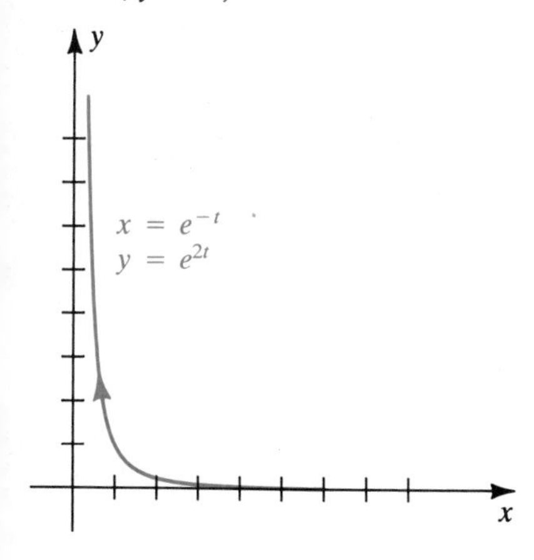

Remembering that $x = e^{-t} > 0$ leads to the graph in Figure 13.10. Note that the point $(1, 1)$ corresponds to $t = 0$. If t increases in $(-\infty, 0]$, the point $P(x, y)$ approaches $(1, 1)$ from the right as indicated by the arrowhead. If t increases in $[0, \infty)$, $P(x, y)$ moves up the curve, approaching the y-axis.

(b) By (13.3) and (13.4),

$$y' = \frac{dy}{dx} = \frac{dy/dt}{dx/dt} = \frac{2e^{2t}}{-e^{-t}} = -2e^{3t}$$

$$\frac{d^2y}{dx^2} = \frac{dy'}{dx} = \frac{dy'/dt}{dx/dt} = \frac{-6e^{3t}}{-e^{-t}} = 6e^{4t}.$$

(c) From part (a), a function k that has the same graph as C is given by

$$k(x) = \frac{1}{x^2} = x^{-2} \quad \text{for} \quad x > 0.$$

Differentiating twice yields

$$k'(x) = -2x^{-3} = -2(e^{-t})^{-3} = -2e^{3t}$$

$$k''(x) = 6x^{-4} = 6(e^{-t})^{-4} = 6e^{4t},$$

which is in agreement with part (b).

(d) Since $d^2y/dx^2 = 6e^{4t} > 0$ for every t, the curve C is concave upward at every point.

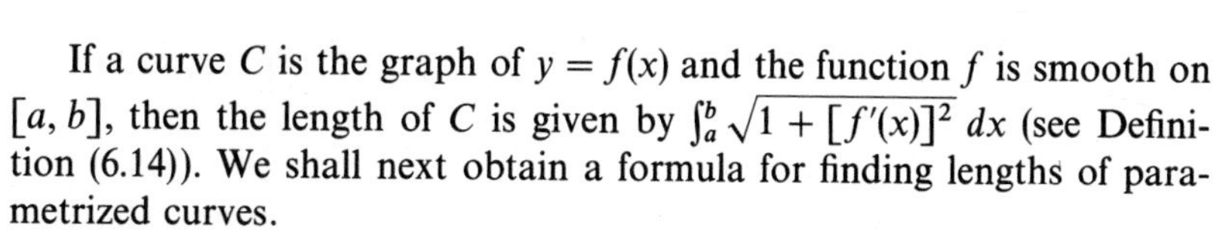
If a curve C is the graph of $y = f(x)$ and the function f is smooth on $[a, b]$, then the length of C is given by $\int_a^b \sqrt{1 + [f'(x)]^2}\, dx$ (see Definition (6.14)). We shall next obtain a formula for finding lengths of parametrized curves.

Suppose a smooth curve C is given parametrically by

$$x = f(t), \quad y = g(t); \quad a \le t \le b.$$

Furthermore, suppose C does not intersect itself—that is, different values of t between a and b determine different points on C. Consider a partition

FIGURE 13.11

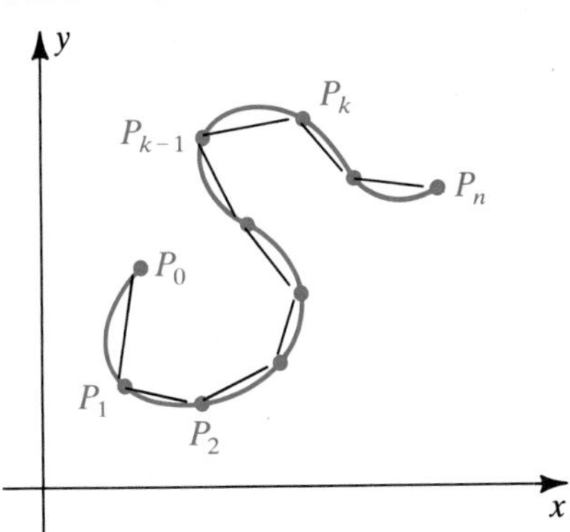

P of $[a, b]$ given by $a = t_0 < t_1 < t_2 < \cdots < t_n = b$. Let $\Delta t_k = t_k - t_{k-1}$ and let $P_k = (f(t_k), g(t_k))$ be the point on C that corresponds to t_k. If $d(P_{k-1}, P_k)$ is the length of the line segment $P_{k-1}P_k$, then the length L_P of the broken line in Figure 13.11 is

$$L_P = \sum_{k=1}^{n} d(P_{k-1}, P_k).$$

As in Section 6.5, we define

$$L = \lim_{\|P\| \to 0} L_P$$

and call L the **length of C** from P_0 to P_n if for every $\epsilon > 0$ there exists a $\delta > 0$ such that $|L_P - L| < \epsilon$ for every partition P with $\|P\| < \delta$.

By the distance formula,

$$d(P_{k-1}, P_k) = \sqrt{[f(t_k) - f(t_{k-1})]^2 + [g(t_k) - g(t_{k-1})]^2}.$$

By the mean value theorem (4.12), there exist numbers w_k and z_k in the open interval (t_{k-1}, t_k) such that

$$f(t_k) - f(t_{k-1}) = f'(w_k)\,\Delta t_k$$

$$g(t_k) - g(t_{k-1}) = g'(z_k)\,\Delta t_k.$$

Substituting these in the formula for $d(P_{k-1}, P_k)$ and removing the common factor $(\Delta t_k)^2$ from the radicand gives us

$$d(P_{k-1}, P_k) = \sqrt{[f'(w_k)]^2 + [g'(z_k)]^2}\,\Delta t_k.$$

Consequently

$$L = \lim_{\|P\| \to 0} L_P = \lim_{\|P\| \to 0} \sum_{k=1}^{n} \sqrt{[f'(w_k)]^2 + [g'(z_k)]^2}\,\Delta t_k,$$

provided the limit exists. If $w_k = z_k$ for every k, then the sums are Riemann sums for the function defined by $\sqrt{[f'(t)]^2 + [g'(t)]^2}$. The limit of these sums is

$$L = \int_a^b \sqrt{[f'(t)]^2 + [g'(t)]^2}\,dt.$$

The limit exists even if $w_k \neq z_k$; however, the proof requires advanced methods and is omitted. The next theorem summarizes this discussion.

Theorem **(13.5)**

If a smooth curve C is given parametrically by $x = f(t)$, $y = g(t)$; $a \le t \le b$, and if C does not intersect itself, except possibly for $t = a$ and $t = b$, then the length L of C is

$$L = \int_a^b \sqrt{[f'(t)]^2 + [g'(t)]^2}\,dt = \int_a^b \sqrt{\left(\frac{dx}{dt}\right)^2 + \left(\frac{dy}{dt}\right)^2}\,dt.$$

The integral formula in Theorem (13.5) is not necessarily true if C intersects itself. For example, if C has the parametrization $x = \cos t$, $y = \sin t$; $0 \le t \le 4\pi$, then the graph is a unit circle with center at the origin. If t varies from 0 to 4π, the circle is traced twice and hence intersects itself infinitely many times. If we use Theorem (13.5) with $a = 0$ and $b = 4\pi$, we obtain the incorrect value 4π for the length of C. The correct value 2π can be obtained by using the t-interval $[0, 2\pi]$. Note that in this case the curve intersects itself only at the points corresponding to $t = 0$ and $t = 2\pi$, which is allowable by the theorem.

If a curve C is given by $y = k(x)$, with k' continuous on $[a, b]$, then parametric equations for C are

$$x = t, \quad y = k(t); \quad a \le t \le b.$$

In this case

$$\frac{dx}{dt} = 1, \quad \frac{dy}{dt} = k'(t) = k'(x), \quad dt = dx,$$

and from Theorem (13.5)

$$L = \int_a^b \sqrt{1 + [k'(x)]^2}\, dx.$$

This is in agreement with the arc length formula given in Definition (6.14).

EXAMPLE 4 Find the length of one arch of the cycloid that has the parametrization

$$x = t - \sin t, \quad y = 1 - \cos t; \quad t \text{ in } \mathbb{R}.$$

SOLUTION The graph has the shape illustrated in Figure 13.6. The radius a of the circle is 1. One arch is obtained if t varies from 0 to 2π. Applying Theorem (13.5) yields

$$\begin{aligned} L &= \int_0^{2\pi} \sqrt{(1 - \cos t)^2 + (\sin t)^2}\, dt \\ &= \int_0^{2\pi} \sqrt{1 - 2\cos^2 t + \cos^2 t + \sin^2 t}\, dt. \end{aligned}$$

Since $\cos^2 t + \sin^2 t = 1$, the integrand reduces to

$$\sqrt{2 - 2\cos t} = \sqrt{2}\sqrt{1 - \cos t}.$$

Thus,

$$L = \int_0^{2\pi} \sqrt{2}\sqrt{1 - \cos t}\, dt.$$

By a half-angle formula, $\sin^2 \frac{1}{2}t = \frac{1}{2}(1 - \cos t)$, or, equivalently,

$$1 - \cos t = 2\sin^2 \tfrac{1}{2}t.$$

Hence

$$\sqrt{1 - \cos t} = \sqrt{2\sin^2 \tfrac{1}{2}t} = \sqrt{2}\,|\sin \tfrac{1}{2}t|.$$

The absolute value sign may be deleted, since if $0 \le t \le 2\pi$, then $0 \le \frac{1}{2}t \le \pi$ and hence $\sin \frac{1}{2}t \ge 0$. Consequently

$$\begin{aligned} L &= \int_0^{2\pi} \sqrt{2}\sqrt{2}\sin \tfrac{1}{2}t\, dt = 2\int_0^{2\pi} \sin \tfrac{1}{2}t\, dt \\ &= -4\Big[\cos \tfrac{1}{2}t\Big]_0^{2\pi} = -4(-1 - 1) = 8. \end{aligned}$$

To remember Theorem (13.5), recall that if ds is the differential of arc length, then, by Theorem (6.17),

$$(ds)^2 = (dx)^2 + (dy)^2.$$

Assuming that ds and dt are positive, we have the following.

Parametric differential of arc length **(13.6)**

$$ds = \sqrt{(dx)^2 + (dy)^2} = \sqrt{\left(\frac{dx}{dt}\right)^2 + \left(\frac{dy}{dt}\right)^2}\, dt$$

Using (13.6), we can rewrite the formula for arc length in Theorem (13.5) as

$$L = \int_{t=a}^{t=b} ds.$$

The limits of integration specify that the independent variable is t, not s.

If a function f is smooth and nonnegative for $a \le x \le b$, then, by Definition (6.19), the area S of the surface that is generated by revolving the graph of $y = f(x)$ about the x-axis (see Figure 13.12) is given by

FIGURE 13.12

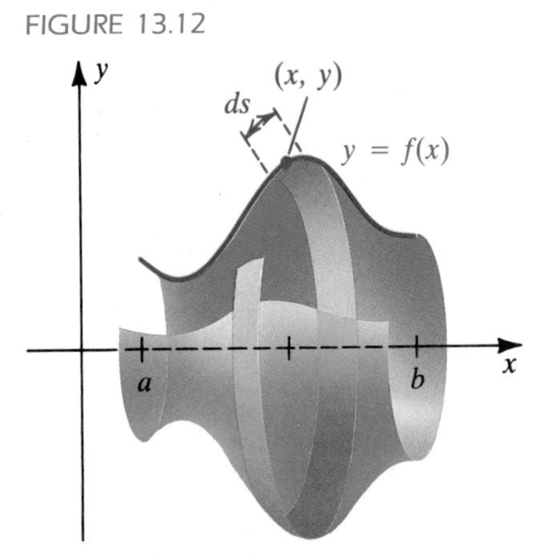

$$S = \int_{x=a}^{x=b} 2\pi y\, ds,$$

where $ds = \sqrt{1 + [f'(x)]^2}\, dx$. We can regard $2\pi y\, ds$ as the surface area of a frustum of a cone of slant height ds and average radius y (see (6.18)).

If a curve C is given parametrically by $x = f(t)$, $y = g(t)$; $a \le t \le b$ and if $g(t) \ge 0$ throughout $[a, b]$, we can use an argument similar to that given in Section 6.5 to show that the area of the surface generated by revolving C about the y-axis is $S = \int_{t=a}^{t=b} 2\pi y\, ds$, where ds is the parametric differential of arc length (13.6). Let us state this for reference as follows.

Theorem **(13.7)**

Let a smooth curve C be given by $x = f(t)$, $y = g(t)$; $a \le t \le b$, and suppose C does not intersect itself, except possibly at the point corresponding to $t = a$ and $t = b$. If $g(t) \ge 0$ throughout $[a, b]$, then the area S of the surface of revolution obtained by revolving C about the x-axis is

$$S = \int_{t=a}^{t=b} 2\pi y\, ds = \int_a^b 2\pi g(t) \sqrt{\left(\frac{dx}{dt}\right)^2 + \left(\frac{dy}{dt}\right)^2}\, dt.$$

FIGURE 13.13

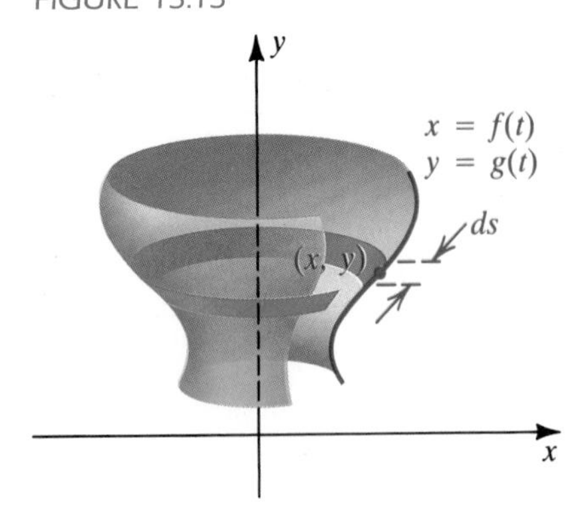

The formula for S in Theorem (13.7) can be extended to the case in which $y = g(t)$ is negative for some t in $[a, b]$ by replacing the variable y that precedes ds by $|y|$.

If the curve C in Theorem (13.7) is revolved about the y-axis and if $x = f(t) \ge 0$ for $a \le t \le b$ (see Figure 13.13), then

$$S = \int_{t=a}^{t=b} 2\pi x\, ds = \int_a^b 2\pi f(t) \sqrt{\left(\frac{dx}{dt}\right)^2 + \left(\frac{dy}{dt}\right)^2}\, dt.$$

In this case we may regard $2\pi x\, ds$ as the surface area of a frustum of a cone of slant height ds and average radius x.

EXAMPLE 5 Verify that the surface area of a sphere of radius a is $4\pi a^2$.

SOLUTION If C is the upper half of the circle $x^2 + y^2 = a^2$, then the spherical surface may be obtained by revolving C about the x-axis. Parametric equations for C are

$$x = a \cos t, \quad y = a \sin t; \quad 0 \le t \le \pi.$$

Applying Theorem (13.7) and using the identity $\sin^2 t + \cos^2 t = 1$, we have

$$S = \int_0^\pi 2\pi a \sin t \sqrt{a^2 \sin^2 t + a^2 \cos^2 t}\, dt = 2\pi a^2 \int_0^\pi \sin t\, dt$$
$$= -2\pi a^2 \Big[\cos t\Big]_0^\pi = -2\pi a^2[-1 - 1] = 4\pi a^2.$$

EXERCISES 13.2

Exer. 1–8: Find the slopes of the tangent line and the normal line at the point on the curve that corresponds to $t = 1$.

1 $x = t^2 + 1, \quad y = t^2 - 1; \quad -2 \le t \le 2$

2 $x = t^3 + 1, \quad y = t^3 - 1; \quad -2 \le t \le 2$

3 $x = 4t^2 - 5, \quad y = 2t + 3; \quad t \text{ in } \mathbb{R}$

4 $x = t^3, \quad y = t^2; \quad t \text{ in } \mathbb{R}$

5 $x = e^t, \quad y = e^{-2t}; \quad t \text{ in } \mathbb{R}$

6 $x = \sqrt{t}, \quad y = 3t + 4; \quad t \ge 0$

7 $x = 2 \sin t, \quad y = 3 \cos t; \quad 0 \le t \le 2\pi$

8 $x = \cos t - 2, \quad y = \sin t + 3; \quad 0 \le t \le 2\pi$

Exer. 9–10: Let C be the curve with the given parametrization, for t in $\mathbb{R}$. Find the points on C at which the slope of the tangent line is m.

9 $x = -t^3, \quad y = -6t^2 - 18t; \quad m = 2$

10 $x = t^2 + t, \quad y = 5t^2 - 3; \quad m = 4$

Exer. 11–18: (a) Find the points on the curve C at which the tangent line is either horizontal or vertical. (b) Find d^2y/dx^2. (c) Sketch the graph of C.

11 $x = 4t^2, \quad y = t^3 - 12t; \quad t \text{ in } \mathbb{R}$

12 $x = t^3 - 4t, \quad y = t^2 - 4; \quad t \text{ in } \mathbb{R}$

13 $x = t^3 + 1, \quad y = t^2 - 2t; \quad t \text{ in } \mathbb{R}$

14 $x = 12t - t^3, \quad y = t^2 - 5t; \quad t \text{ in } \mathbb{R}$

15 $x = 3t^2 - 6t, \quad y = \sqrt{t}; \quad t \ge 0$

16 $x = \sqrt[3]{t}, \quad y = \sqrt[3]{t} - t; \quad t \text{ in } \mathbb{R}$

17 $x = \cos^3 t, \quad y = \sin^3 t; \quad 0 \le t \le 2\pi$

18 $x = \cosh t, \quad y = \sinh t; \quad t \text{ in } \mathbb{R}$

Exer. 19–20: Shown is a Lissajous figure (see Example 5, Section 13.1). Determine where the tangent line is horizontal or vertical.

19 $x = 4 \sin 2t, \quad y = 2 \cos 3t$

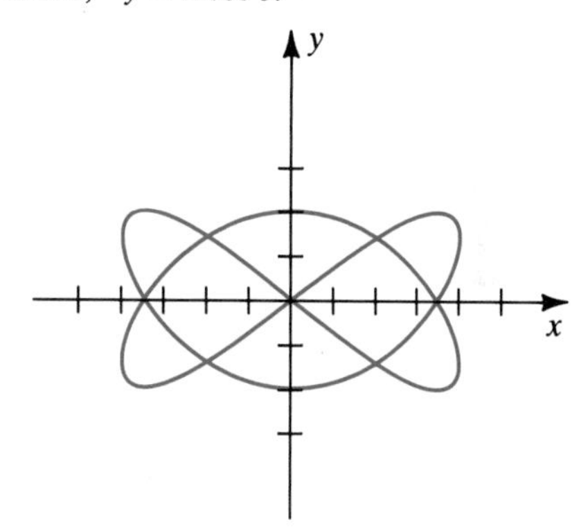

20 $x = 5 \sin \frac{1}{3}t, \quad y = 4 \sin 2t$

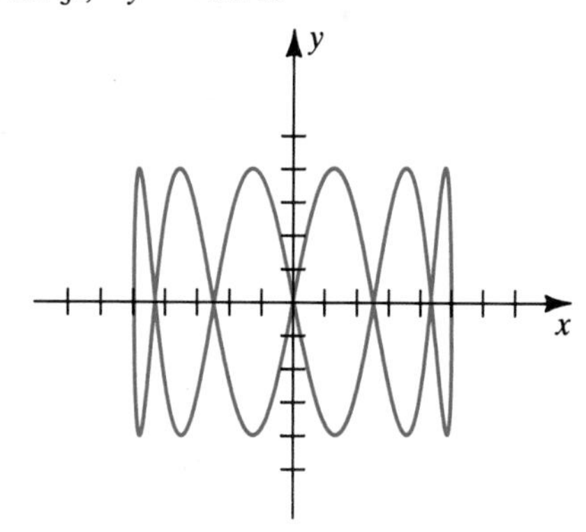

Exer. 21–26: Find the length of the curve.

21 $x = 5t^2, \quad y = 2t^3; \quad 0 \le t \le 1$

22 $x = 3t, \quad y = 2t^{3/2}; \quad 0 \le t \le 4$

23 $x = e^t \cos t, \quad y = e^t \sin t; \quad 0 \le t \le \pi/2$

24 $x = \cos 2t, \quad y = \sin^2 t; \quad 0 \le t \le \pi$

25 $x = t \cos t - \sin t, \quad y = t \sin t + \cos t; \quad 0 \le t \le \pi/2$

26 $x = \cos^3 t, \quad y = \sin^3 t; \quad 0 \le t \le \pi/2$

c **Exer. 27–28: Use Simpson's rule, with $n = 6$, to approximate the length of the curve.**

27 $x = 2\cos t,\quad y = 3\sin t;\quad 0 \le t \le 2\pi$

28 $x = 4t^3 - t,\quad y = 2t^2;\quad 0 \le t \le 1$

Exer. 29–34: Find the area of the surface generated by revolving the curve about the x-axis.

29 $x = t^2,\quad y = 2t;\quad 0 \le t \le 4$

30 $x = 4t,\quad y = t^3;\quad 1 \le t \le 2$

31 $x = t^2,\quad y = t - \frac{1}{3}t^3;\quad 0 \le t \le 1$

32 $x = 4t^2 + 1,\quad y = 3 - 2t;\quad -2 \le t \le 0$

33 $x = t - \sin t,\quad y = 1 - \cos t;\quad 0 \le t \le 2\pi$

34 $x = t,\quad y = \frac{1}{3}t^3 + \frac{1}{4}t^{-1};\quad 1 \le t \le 2$

Exer. 35–38: Find the area of the surface generated by revolving the curve about the y-axis.

35 $x = 4t^{1/2},\quad y = \frac{1}{2}t^2 + t^{-1};\quad 1 \le t \le 4$

36 $x = 3t,\quad y = t + 1;\quad 0 \le t \le 5$

37 $x = e^t \sin t,\quad y = e^t \cos t;\quad 0 \le t \le \pi/2$

38 $x = 3t^2,\quad y = 2t^3;\quad 0 \le t \le 1$

c **Exer. 39–40: Use Simpson's rule, with $n = 4$, to approximate the area of the surface generated by revolving the curve about the given axis.**

39 $x = \cos(t^2),\quad y = \sin^2 t;\quad 0 \le t \le 1;$ the x-axis

40 $x = t^2 + 2t,\quad y = t^4;\quad 0 \le t \le 1;$ the y-axis

13.3 POLAR COORDINATES

In a rectangular coordinate system, the ordered pair (a, b) denotes the point whose directed distances from the x- and y-axes are b and a, respectively. Another method for representing points is to use *polar coordinates*. We begin with a fixed point O (the **origin**, or **pole**) and a directed half-line (the **polar axis**) with endpoint O. Next we consider any point P in the plane different from O. If, as illustrated in Figure 13.14, $r = d(O, P)$ and θ denotes the measure of any angle determined by the polar axis and OP, then r and θ are **polar coordinates** of P, and the symbols (r, θ) or $P(r, \theta)$ are used to denote P. As usual, θ is considered positive if the angle is generated by a counterclockwise rotation of the polar axis and negative if the rotation is clockwise. Either radian or degree measure may be used for θ.

FIGURE 13.14

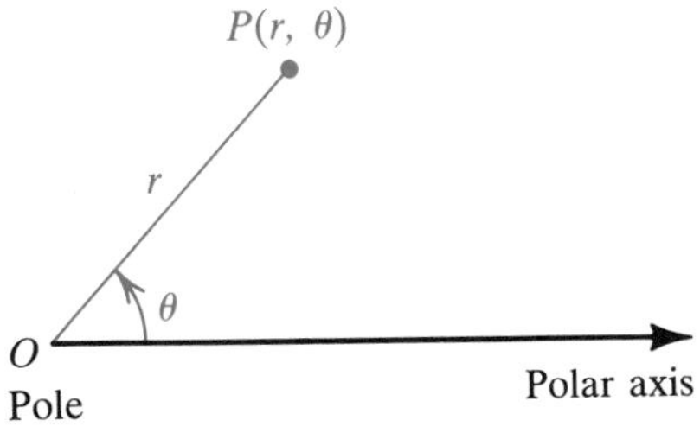

P(r, θ)
r
θ
O
Pole
Polar axis

The polar coordinates of a point are not unique. For example $(3, \pi/4)$, $(3, 9\pi/4)$, and $(3, -7\pi/4)$ all represent the same point (see Figure 13.15). We shall also allow r to be negative. In this case, instead of measuring $|r|$ units along the terminal side of the angle θ, we measure along the half-line with endpoint O that has direction *opposite* that of the terminal side. The points corresponding to the pairs $(-3, 5\pi/4)$ and $(-3, -3\pi/4)$ are also plotted in Figure 13.15.

FIGURE 13.15

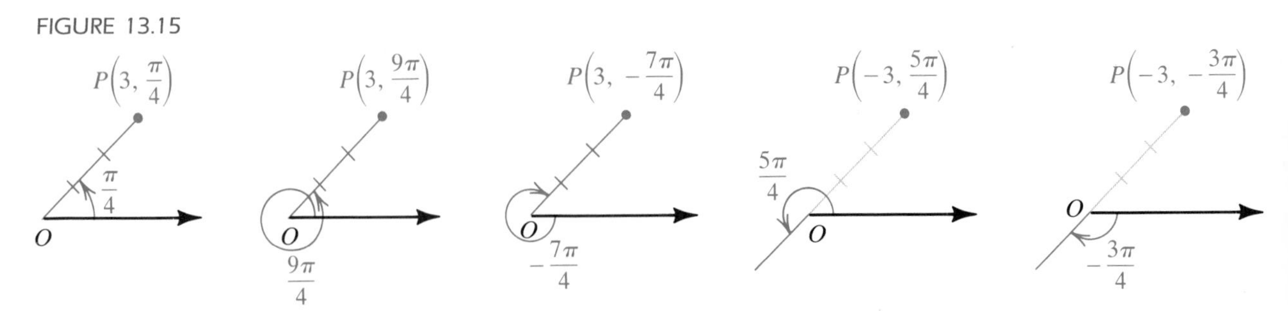

We agree that the pole O has polar coordinates $(0, \theta)$ for *any* θ. An assignment of ordered pairs of the form (r, θ) to points in a plane is a **polar coordinate system**, and the plane is an **$r\theta$-plane**.

FIGURE 13.16

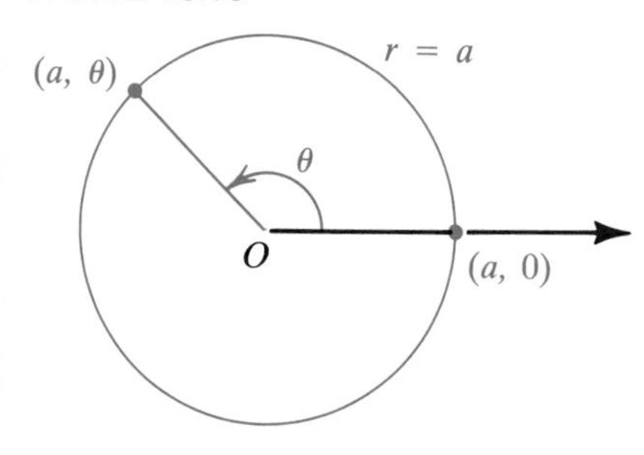

FIGURE 13.17

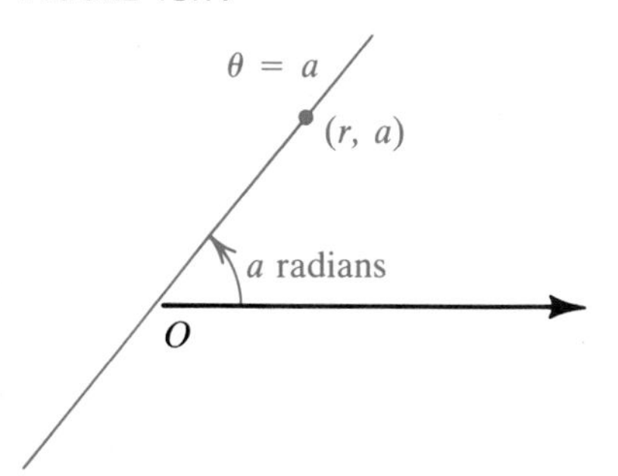

A **polar equation** is an equation in r and θ. A **solution** of a polar equation is an ordered pair (a, b) that leads to equality if a is substituted for r and b for θ. The **graph** of a polar equation is the set of all points (in an $r\theta$-plane) that correspond to the solutions.

The simplest polar equations are $r = a$ and $\theta = a$, where a is a nonzero real number. Since the solutions of the polar equation $r = a$ are of the form (a, θ) for *any* angle θ, it follows that the graph is a circle of radius $|a|$ with center at the pole. A graph for $a > 0$ is sketched in Figure 13.16. The same graph is obtained for $r = -a$.

The solutions of the polar equation $\theta = a$ are of the form (r, a) for *any* real number r. Since the (angle) coordinate a is constant, the graph is a line through the origin, as illustrated in Figure 13.17 for the case $0 < a < \pi/2$.

In the following examples we obtain the graphs of polar equations by plotting points. As you proceed through this section, you should try to recognize forms of polar equations so that you will be able to sketch their graphs by plotting few, if any, points.

EXAMPLE 1 Sketch the graph of the polar equation $r = 4 \sin \theta$.

SOLUTION The following table displays some solutions of the equation. We have included a third row in the table that contains one-decimal-place approximations to r.

θ	0	$\frac{\pi}{6}$	$\frac{\pi}{4}$	$\frac{\pi}{3}$	$\frac{\pi}{2}$	$\frac{2\pi}{3}$	$\frac{3\pi}{4}$	$\frac{5\pi}{6}$	π
r	0	2	$2\sqrt{2}$	$2\sqrt{3}$	4	$2\sqrt{3}$	$2\sqrt{2}$	2	0
r (approx.)	0	2	2.8	3.4	4	3.4	2.8	2	0

FIGURE 13.18

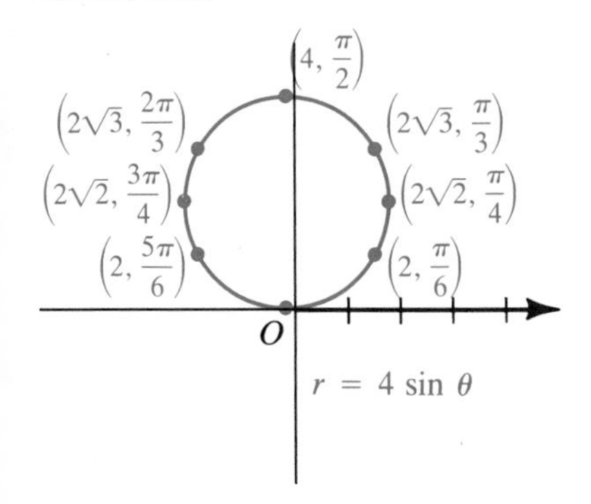

The points in an $r\theta$-plane that correspond to the pairs in the table appear to lie on a circle of radius 2, and we draw the graph accordingly (see Figure 13.18). As an aid to plotting points, we have extended the polar axis in the negative direction and introduced a vertical line through the pole.

The proof that the graph of $r = 4 \sin \theta$ is a circle is given in Example 6. Additional points obtained by letting θ vary from π to 2π lie on the same circle. For example, the solution $(-2, 7\pi/6)$ gives us the same point as $(2, \pi/6)$; the point corresponding to $(-2\sqrt{2}, 5\pi/4)$ is the same as that obtained from $(2\sqrt{2}, \pi/4)$; and so on. If we let θ increase through all real numbers, we obtain the same points again and again because of the periodicity of the sine function.

EXAMPLE 2 Sketch the graph of the polar equation $r = 2 + 2 \cos \theta$.

SOLUTION Since the cosine function decreases from 1 to -1 as θ varies from 0 to π, it follows that r decreases from 4 to 0 in this θ-interval.

The following table exhibits some solutions of $r = 2 + 2\cos\theta$, together with one-decimal-place approximations to r.

FIGURE 13.19
$r = 2 + 2\cos\theta$

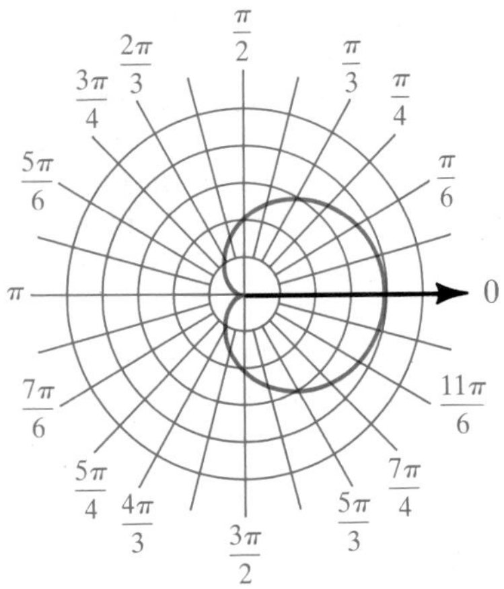

θ	0	$\frac{\pi}{6}$	$\frac{\pi}{4}$	$\frac{\pi}{3}$	$\frac{\pi}{2}$	$\frac{2\pi}{3}$	$\frac{3\pi}{4}$	$\frac{5\pi}{6}$	π
r	4	$2+\sqrt{3}$	$2+\sqrt{2}$	3	2	1	$2-\sqrt{2}$	$2-\sqrt{3}$	0
r (approx.)	4	3.7	3.4	3	2	1	0.6	0.3	0

Plotting points in an $r\theta$-plane leads to the upper half of the graph sketched in Figure 13.19. (We have used polar coordinate graph paper, which displays lines through O at various angles and concentric circles with centers at the pole.)

If θ increases from π to 2π, then $\cos\theta$ increases from -1 to 1 and, consequently, r increases from 0 to 4. Plotting points for $\pi \le \theta \le 2\pi$ gives us the lower half of the graph.

The same graph may be obtained by taking other intervals of length 2π for θ.

The heart-shaped graph in Example 2 is a **cardioid**. In general, the graph of any of the following polar equations, with $a \ne 0$, is a cardioid:

$$r = a(1 + \cos\theta) \qquad r = a(1 + \sin\theta)$$
$$r = a(1 - \cos\theta) \qquad r = a(1 - \sin\theta)$$

If a and b are not zero, then the graphs of the following polar equations are **limaçons**:

$$r = a + b\cos\theta \qquad r = a + b\sin\theta$$

Note that the special limaçons in which $|a| = |b|$ are cardioids. Some limaçons contain a loop, as shown in the next example.

FIGURE 13.20

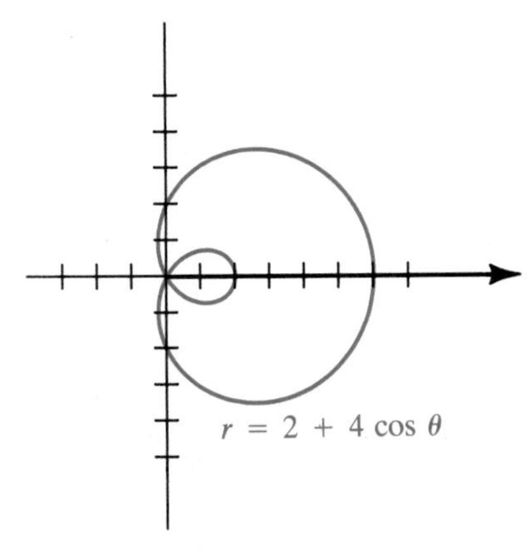

EXAMPLE 3 Sketch the graph of the polar equation $r = 2 + 4\cos\theta$.

SOLUTION Coordinates of some points in an $r\theta$-plane that correspond to $0 \le \theta \le \pi$ are listed in the following table.

θ	0	$\frac{\pi}{6}$	$\frac{\pi}{4}$	$\frac{\pi}{3}$	$\frac{\pi}{2}$	$\frac{2\pi}{3}$	$\frac{3\pi}{4}$	$\frac{5\pi}{6}$	π
r	6	$2+2\sqrt{3}$	$2+2\sqrt{2}$	4	2	0	$2-2\sqrt{2}$	$2-2\sqrt{3}$	-2
r (approx.)	6	5.4	4.8	4	2	0	-0.8	-1.4	-2

Note that $r = 0$ at $\theta = 2\pi/3$. The values of r are negative if $2\pi/3 < \theta \le \pi$, and this leads to the lower half of the small loop in Figure 13.20. Letting θ range from π to 2π gives us the upper half of the small loop and the lower half of the large loop.

EXAMPLE 4 Sketch the graph of the polar equation $r = a \sin 2\theta$ for $a > 0$.

SOLUTION Instead of tabulating solutions, let us reason as follows. If θ increases from 0 or $\pi/4$, then 2θ varies from 0 to $\pi/2$ and hence $\sin 2\theta$ increases from 0 to 1. It follows that r increases from 0 to a in the θ-interval $[0, \pi/4]$. If we next let θ increase from $\pi/4$ to $\pi/2$, then 2θ changes from $\pi/2$ to π and hence $\sin 2\theta$ decreases from 1 to 0. Thus, r decreases from a to 0 in the θ-interval $[\pi/4, \pi/2]$. The corresponding points on the graph constitute the first-quadrant loop illustrated in Figure 13.21. Note that the point $P(r, \theta)$ traces the loop in a *counterclockwise* direction (indicated by the arrows) as θ increases from 0 to $\pi/2$.

FIGURE 13.21

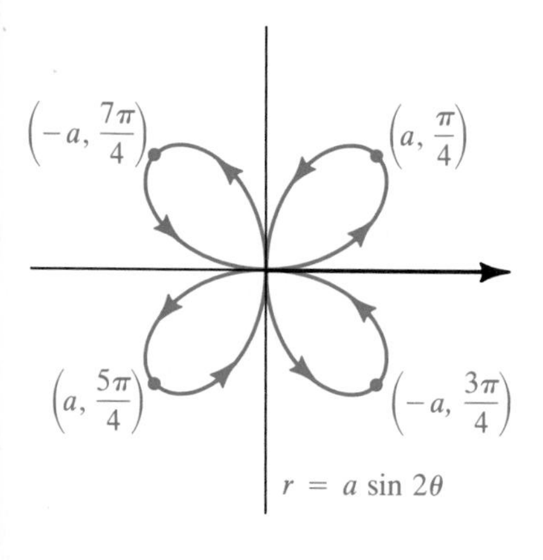

If $\pi/2 \le \theta \le \pi$, then $\pi \le 2\theta \le 2\pi$ and, therefore, $r = a \sin 2\theta \le 0$. Thus, *if* $\pi/2 < \theta < \pi$, *then* r *is negative and the points* $P(r, \theta)$ *are in the fourth quadrant*. If θ increases from $\pi/2$ to π, then we can show, by plotting points, that $P(r, \theta)$ traces (in a counterclockwise direction) the loop shown in the fourth quadrant.

Similarly, for $\pi \le \theta \le 3\pi/2$ we get the loop in the third quadrant, and for $3\pi/2 \le \theta \le 2\pi$ we get the loop in the second quadrant. Both loops are traced in a counterclockwise direction as θ increases. You should verify these facts by plotting some points with, say, $a = 1$. In Figure 13.21 we have plotted only those points on the graph that correspond to the largest numerical values of r.

The graph in Example 4 is a **four-leafed rose**. In general, a polar equation of the form

$$r = a \sin n\theta \quad \text{or} \quad r = a \cos n\theta$$

for any positive integer n greater than 1 and any nonzero real number a has a graph that consists of a number of loops through the origin. If n is even, there are $2n$ loops, and if n is odd, there are n loops (see Exercises 15–18).

The graph of the polar equation $r = a\theta$ for any nonzero real number a is a **spiral of Archimedes**. The case $a = 1$ is considered in the next example.

FIGURE 13.22

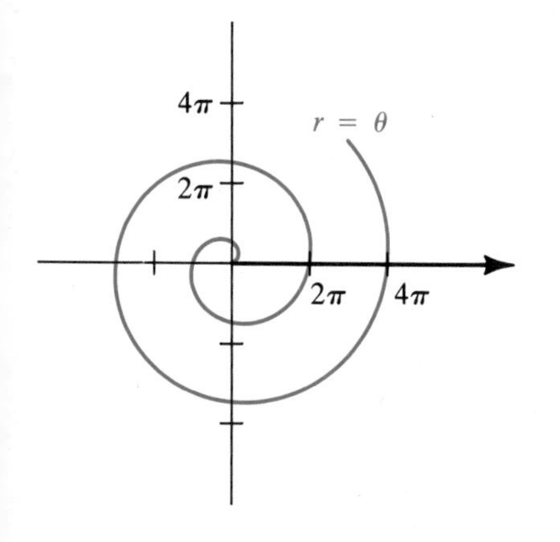

EXAMPLE 5 Sketch the graph of the polar equation $r = \theta$ for $\theta \ge 0$.

SOLUTION The graph consists of all points that have polar coordinates of the form (c, c) for every real number $c \ge 0$. Thus, the graph contains the points $(0, 0)$, $(\pi/2, \pi/2)$, (π, π), and so on. As θ increases, r increases at the same rate, and the spiral winds around the origin in a counterclockwise direction, intersecting the polar axis at $0, 2\pi, 4\pi, \ldots,$ as illustrated in Figure 13.22.

If θ is allowed to be negative, then as θ decreases through negative values, the resulting spiral winds around the origin and is the symmetric image, with respect to the vertical axis, of the curve sketched in Figure 13.22.

Let us next superimpose an xy-plane on an $r\theta$-plane so that the positive x-axis coincides with the polar axis. Any point P in the plane may then be assigned rectangular coordinates (x, y) or polar coordinates (r, θ). If $r > 0$, we have a situation similar to that illustrated in Figure 13.23(i). If $r < 0$, we have that shown in (ii) of the figure, where, for later purposes, we have also plotted the point P' having polar coordinates $(|r|, \theta)$ and rectangular coordinates $(-x, -y)$.

FIGURE 13.23

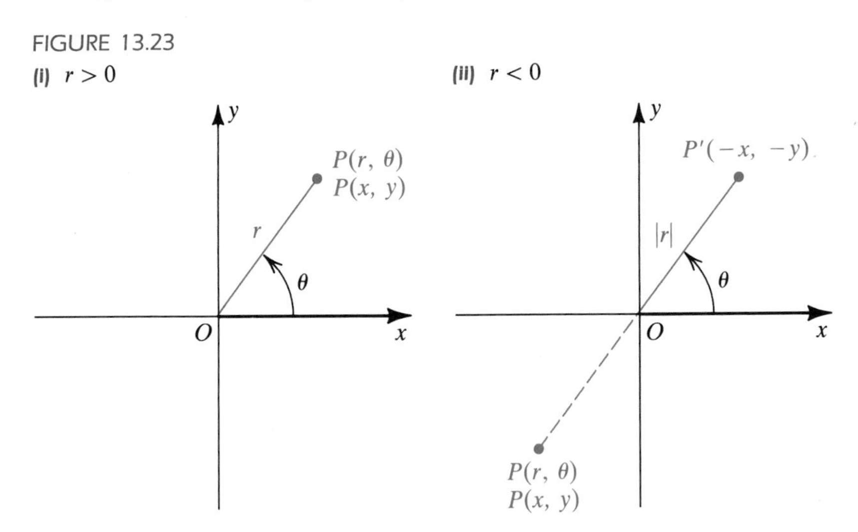

The following result specifies relationships between (x, y) and (r, θ), where it is assumed that the positive x-axis coincides with the polar axis.

Relationships between rectangular and polar coordinates **(13.8)**

The rectangular coordinates (x, y) and polar coordinates (r, θ) of a point P are related as follows:

(i) $x = r\cos\theta, \quad y = r\sin\theta$

(ii) $r^2 = x^2 + y^2, \quad \tan\theta = \dfrac{y}{x} \quad \text{if} \quad x \neq 0$

PROOF Although we have pictured θ as an acute angle in Figure 13.23, the discussion that follows is valid for all angles. If $r > 0$ as in Figure 13.23(i), then $\cos\theta = x/r$, $\sin\theta = y/r$, and hence

$$x = r\cos\theta, \quad y = r\sin\theta.$$

If $r < 0$, then $|r| = -r$, and from Figure 13.23(ii) we see that

$$\cos\theta = \frac{-x}{|r|} = \frac{-x}{-r} = \frac{x}{r}, \quad \sin\theta = \frac{-y}{|r|} = \frac{-y}{-r} = \frac{y}{r}.$$

Multiplication by r gives us relationship (i), and therefore these formulas hold if r is either positive or negative. If $r = 0$, then the point is the pole and we again see that the formulas in (i) are true.

The formulas in (ii) follow readily from Figure 13.23. ■

We may use the preceding result to change from one system of coordinates to the other. A more important use is for transforming a polar equation to an equation in x and y, and vice versa. This is illustrated in the next three examples.

EXAMPLE 6 Find an equation in x and y that has the same graph as the polar equation $r = a \sin \theta$, with $a \neq 0$. Sketch the graph.

SOLUTION From (13.8)(i), a relationship between $\sin \theta$ and y is given by $y = r \sin \theta$. To introduce this expression into the equation $r = a \sin \theta$, we multiply both sides by r, obtaining

$$r^2 = ar \sin \theta.$$

FIGURE 13.24

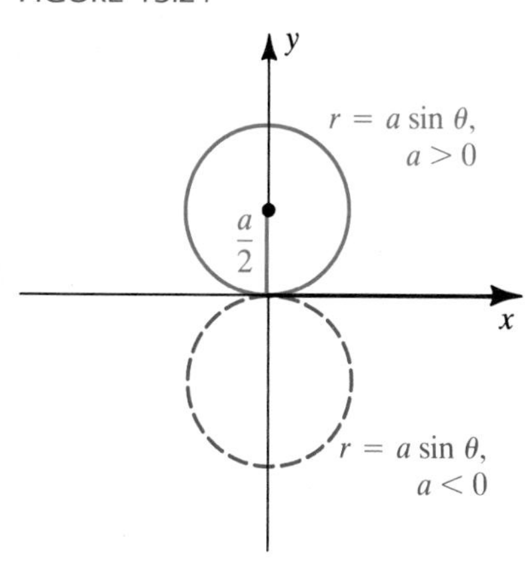

Next, using $r^2 = x^2 + y^2$ and $y = r \sin \theta$, we have

$$x^2 + y^2 = ay,$$

or

$$x^2 + y^2 - ay = 0.$$

Completing the square in y gives us

$$x^2 + y^2 - ay + \left(\frac{a}{2}\right)^2 = \left(\frac{a}{2}\right)^2,$$

or

$$x^2 + \left(y - \frac{a}{2}\right)^2 = \left(\frac{a}{2}\right)^2.$$

In the xy-plane, the graph of the last equation is a circle with center $(0, a/2)$ and radius $|a|/2$, as illustrated in Figure 13.24 for the case $a > 0$ (the solid circle) and $a < 0$ (the dashed circle).

FIGURE 13.25

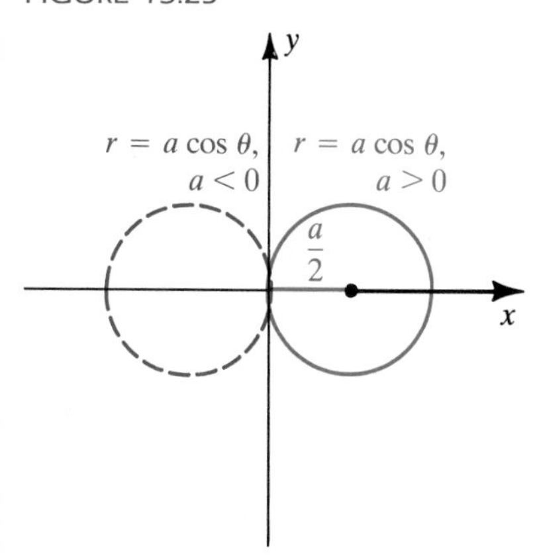

Using the same method as in the preceding example, we can show that the graph of $r = a \cos \theta$, with $a \neq 0$, is a circle of radius $a/2$ of the type illustrated in Figure 13.25.

EXAMPLE 7 Find a polar equation for the hyperbola $x^2 - y^2 = 16$.

SOLUTION Using the formulas $x = r \cos \theta$ and $y = r \sin \theta$, we obtain the following polar equations:

$$(r \cos \theta)^2 - (r \sin \theta)^2 = 16$$

$$r^2 \cos^2 \theta - r^2 \sin^2 \theta = 16$$

$$r^2 (\cos^2 \theta - \sin^2 \theta) = 16$$

$$r^2 \cos 2\theta = 16$$

$$r^2 = \frac{16}{\cos 2\theta}$$

$$r^2 = 16 \sec 2\theta$$

The division by $\cos 2\theta$ is allowable because $\cos 2\theta \neq 0$. (Note that if $\cos 2\theta = 0$, then $r^2 \cos 2\theta \neq 16$.)

EXAMPLE 8 Find a polar equation of an arbitrary line.

SOLUTION Every line in an xy-coordinate plane is the graph of a linear equation $ax + by = c$. Using the formulas $x = r\cos\theta$ and $y = r\sin\theta$ gives us the following equivalent polar equations:

$$ar\cos\theta + br\sin\theta = c$$

$$r(a\cos\theta + b\sin\theta) = c$$

$$r = \frac{c}{a\cos\theta + b\sin\theta}$$

If we superimpose an xy-plane on an $r\theta$-plane, then the graph of a polar equation may be symmetric with respect to the x-axis (the polar axis), the y-axis (the line $\theta = \pi/2$), or the origin (the pole). Some typical symmetries are illustrated in Figure 13.26. This leads to the next result.

FIGURE 13.26 Symmetries of graphs of polar equations

(i) Polar axis

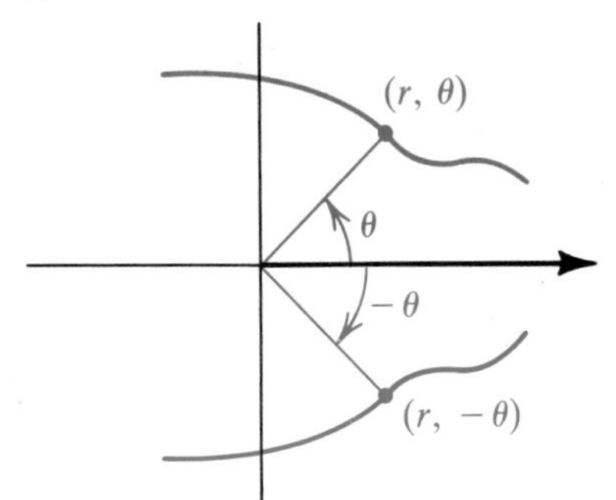

(ii) Line $\theta = \pi/2$

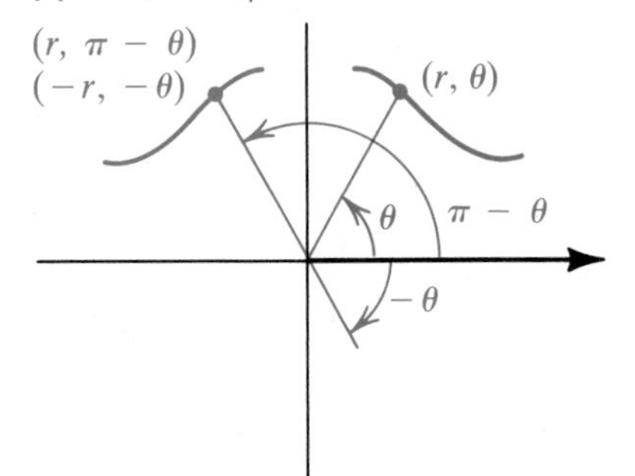

(iii) Pole

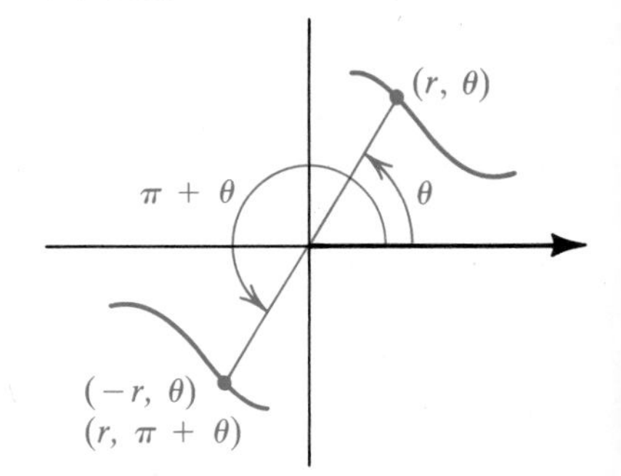

Tests for symmetry **(13.9)**

> **(i)** The graph of $r = f(\theta)$ is symmetric with respect to the polar axis if substitution of $-\theta$ for θ leads to an equivalent equation.
>
> **(ii)** The graph of $r = f(\theta)$ is symmetric with respect to the vertical line $\theta = \pi/2$ if substitution of either (a) $\pi - \theta$ for θ or (b) $-r$ for r and $-\theta$ for θ leads to an equivalent equation.
>
> **(iii)** The graph of $r = f(\theta)$ is symmetric with respect to the pole if substitution of either (a) $-r$ for r or (b) $\pi + \theta$ for θ leads to an equivalent equation.

To illustrate, since $\cos(-\theta) = \cos\theta$, the graph of the polar equation $r = 2 + 4\cos\theta$ in Example 3 is symmetric with respect to the polar axis, by test (i). Since $\sin(\pi - \theta) = \sin\theta$, the graph in Example 1 is symmetric with respect to the line $\theta = \pi/2$, by test (ii). The graph in Example 4 is symmetric to the polar axis, the line $\theta = \pi/2$ and the pole. Other tests for symmetry may be stated; however, those we have listed are among the easiest to apply.

Unlike the graph of an equation in x and y, the graph of a polar equation $r = f(\theta)$ can be symmetric with respect to the polar axis, the line

FIGURE 13.27

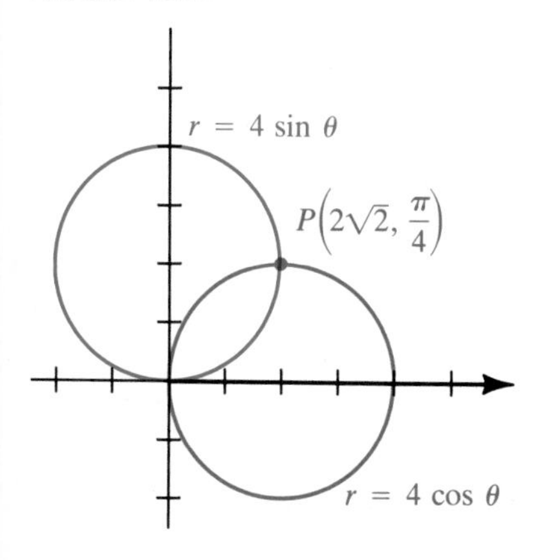

$\theta = \pi/2$, or the pole *without* satisfying one of the preceding tests for symmetry. This is true because of the many different ways of specifying a point in polar coordinates.

Another difference between rectangular and polar coordinate systems is that the points of intersection of two graphs cannot always be found by solving the polar equations simultaneously. To illustrate, from Example 1, the graph of $r = 4 \sin \theta$ is a circle of diameter 4 with center at $(2, \pi/2)$ (see Figure 13.27). Similarly, the graph of $r = 4 \cos \theta$ a is circle of diameter 4 with center at $(2, 0)$ on the polar axis. Referring to Figure 13.27, we see that the coordinates of the point of intersection $P(2\sqrt{2}, \pi/4)$ in quadrant I satisfy both equations; however, the origin O, which is on each circle, *cannot* be found by solving the equations simultaneously. Thus, in searching for points of intersection of polar graphs, it is sometimes necessary to refer to the graphs themselves, *in addition* to solving the two equations simultaneously. An alternative method is to use different (equivalent) equations for the graphs.

Tangent lines to graphs of polar equations may be found by means of the next theorem.

Theorem **(13.10)**

The slope m of the tangent line to the graph of $r = f(\theta)$ at the point $P(r, \theta)$ is

$$m = \frac{\dfrac{dr}{d\theta} \sin \theta + r \cos \theta}{\dfrac{dr}{d\theta} \cos \theta - r \sin \theta}.$$

PROOF If (x, y) are the rectangular coordinates of $P(r, \theta)$, then, by Theorem (13.8),

$$x = r \cos \theta = f(\theta) \cos \theta$$
$$y = r \sin \theta = f(\theta) \sin \theta.$$

These may be considered as parametric equations for the graph with parameter θ. Applying Theorem (13.3), we find that the slope of the tangent line at (x, y) is

$$\frac{dy}{dx} = \frac{dy/d\theta}{dx/d\theta} = \frac{f(\theta) \cos \theta + f'(\theta) \sin \theta}{f(\theta)(-\sin \theta) + f'(\theta) \cos \theta}$$
$$= \frac{f'(\theta) \sin \theta + f(\theta) \cos \theta}{f'(\theta) \cos \theta - f(\theta) \sin \theta}.$$

This is equivalent to the formula in the statement of the theorem. ■

Horizontal tangent lines occur if the numerator in the formula for m is 0 and the denominator is not 0. Vertical tangent lines occur if the denominator is 0 and the numerator is not 0. The case 0/0 requires further investigation.

To find the slopes of the tangent lines at the pole, we must determine the values of θ for which $r = f(\theta) = 0$. For such values (and with $r = 0$

and $dr/d\theta \neq 0$), the formula in Theorem (13.10) reduces to $m = \tan\theta$. These remarks are illustrated in the next example.

EXAMPLE 9 For the cardioid $r = 2 + 2\cos\theta$ with $0 \le \theta < 2\pi$, find

(a) the slope of the tangent line at $\theta = \pi/6$

(b) the points at which the tangent line is horizontal

(c) the points at which the tangent line is vertical

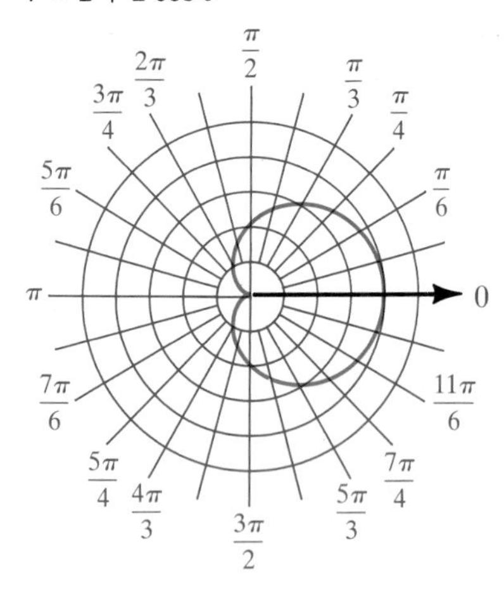

FIGURE 13.28
$r = 2 + 2\cos\theta$

SOLUTION **(a)** The graph of $r = 2 + 2\cos\theta$ was considered in Example 2 and is resketched in Figure 13.28. Applying Theorem (13.10), we find that the slope m of the tangent line is

$$\begin{aligned} m &= \frac{(-2\sin\theta)\sin\theta + (2 + 2\cos\theta)\cos\theta}{(-2\sin\theta)\cos\theta - (2 + 2\cos\theta)\sin\theta} \\ &= \frac{2(\cos^2\theta - \sin^2\theta) + 2\cos\theta}{-2(2\sin\theta\cos\theta) - 2\sin\theta} \\ &= -\frac{\cos 2\theta + \cos\theta}{\sin 2\theta + \sin\theta}. \end{aligned}$$

At $\theta = \pi/6$ (that is, at the point $(2 + \sqrt{3}, \pi/6)$),

$$m = -\frac{\cos(\pi/3) + \cos(\pi/6)}{\sin(\pi/3) + \sin(\pi/6)} = -\frac{(1/2) + (\sqrt{3}/2)}{(\sqrt{3}/2) + (1/2)} = -1.$$

(b) To find horizontal tangents, we let

$$\cos 2\theta + \cos\theta = 0.$$

This equation may be written as

$$2\cos^2\theta - 1 + \cos\theta = 0,$$

or

$$(2\cos\theta - 1)(\cos\theta + 1) = 0.$$

From $\cos\theta = \frac{1}{2}$ we obtain $\theta = \pi/3$ and $\theta = 5\pi/3$. The corresponding points are $(3, \pi/3)$ and $(3, 5\pi/3)$.

Using $\cos\theta = -1$ gives us $\theta = \pi$. The denominator in the formula for m is 0 at $\theta = \pi$, and hence further investigation is required. If $\theta = \pi$, then $r = 0$ and the formula for m in (13.10) reduces to $m = \tan\theta$. Thus, the slope at $(0, \pi)$ is $m = \tan\pi = 0$, and therefore the tangent line is horizontal at the pole.

(c) To find vertical tangent lines, we let

$$\sin 2\theta + \sin\theta = 0.$$

Equivalent equations are

$$2\sin\theta\cos\theta + \sin\theta = 0$$

and

$$\sin\theta\,(2\cos\theta + 1) = 0.$$

Letting $\sin\theta = 0$ and $\cos\theta = -\frac{1}{2}$ leads to the following values of θ: 0, π, $2\pi/3$, and $4\pi/3$. We found, in part (b), that π gives us a horizontal tangent. The remaining values result in the points $(4, 0)$, $(1, 2\pi/3)$, and $(1, 4\pi/3)$, at which the graph has vertical tangent lines.

EXERCISES 13.3

Exer. 1–26: Sketch the graph of the polar equation.

1 $r = 5$

2 $r = -2$

3 $\theta = -\pi/6$

4 $\theta = \pi/4$

5 $r = 3 \cos \theta$

6 $r = -2 \sin \theta$

7 $r = 4 - 4 \sin \theta$

8 $r = -6(1 + \cos \theta)$

9 $r = 2 + 4 \sin \theta$

10 $r = 1 + 2 \cos \theta$

11 $r = 2 - \cos \theta$

12 $r = 5 + 3 \sin \theta$

13 $r = 4 \csc \theta$

14 $r = -3 \sec \theta$

15 $r = 8 \cos 3\theta$

16 $r = 2 \sin 4\theta$

17 $r = 3 \sin 2\theta$

18 $r = 8 \cos 5\theta$

19 $r^2 = 4 \cos 2\theta$ (lemniscate)

20 $r^2 = -16 \sin 2\theta$

21 $r = e^{\theta}, \quad \theta \geq 0$ (logarithmic spiral)

22 $r = 6 \sin^2 (\theta/2)$

23 $r = 2\theta, \quad \theta \geq 0$

24 $r\theta = 1, \quad \theta > 0$ (spiral)

25 $r = 2 + 2 \sec \theta$ (conchoid)

26 $r = 1 - \csc \theta$

Exer. 27–36: Find a polar equation that has the same graph as the equation in x and y.

27 $x = -3$

28 $y = 2$

29 $x^2 + y^2 = 16$

30 $x^2 = 8y$

31 $2y = -x$

32 $y = 6x$

33 $y^2 - x^2 = 4$

34 $xy = 8$

35 $(x^2 + y^2) \tan^{-1} (y/x) = ay, \quad a > 0$ (cochleoid, or *Oui-ja board curve*)

36 $x^3 + y^3 - 3axy = 0$ (Folium of Descartes)

Exer. 37–50: Find an equation in x and y that has the same graph as the polar equation and use it to help sketch the graph in an $r\theta$-plane.

37 $r \cos \theta = 5$

38 $r \sin \theta = -2$

39 $r = -3 \csc \theta$

40 $r = 4 \sec \theta$

41 $r^2 \cos 2\theta = 1$

42 $r^2 \sin 2\theta = 4$

43 $r(\sin \theta - 2 \cos \theta) = 6$

44 $r(3 \cos \theta - 4 \sin \theta) = 12$

45 $r(\sin \theta + r \cos^2 \theta) = 1$

46 $r(r \sin^2 \theta - \cos \theta) = 3$

47 $r = 8 \sin \theta - 2 \cos \theta$

48 $r = 2 \cos \theta - 4 \sin \theta$

49 $r = \tan \theta$

50 $r = 6 \cot \theta$

Exer. 51–60: Find the slope of the tangent line to the graph of the polar equation at the point corresponding to the given value of θ.

51 $r = 2 \cos \theta; \quad \theta = \pi/3$

52 $r = -2 \sin \theta; \quad \theta = \pi/6$

53 $r = 4(1 - \sin \theta); \quad \theta = 0$

54 $r = 1 + 2 \cos \theta; \quad \theta = \pi/2$

55 $r = 8 \cos 3\theta; \quad \theta = \pi/4$

56 $r = 2 \sin 4\theta; \quad \theta = \pi/4$

57 $r^2 = 4 \cos 2\theta; \quad \theta = \pi/6$

58 $r^2 = -2 \sin 2\theta; \quad \theta = 3\pi/4$

59 $r = 2^{\theta}; \quad \theta = \pi$

60 $r\theta = 1; \quad \theta = 2\pi$

61 If $P_1(r_1, \theta_1)$ and $P_2(r_2, \theta_2)$ are points in an $r\theta$-plane, use the law of cosines to prove that

$$[d(P_1, P_2)]^2 = r_1^2 + r_2^2 - 2r_1 r_2 \cos (\theta_2 - \theta_1).$$

62 If a and b are nonzero real numbers, prove that the graph of $r = a \sin \theta + b \cos \theta$ is a circle, and find its center and radius.

63 If the graphs of the polar equations $r = f(\theta)$ and $r = g(\theta)$ intersect at $P(r, \theta)$, prove that the tangent lines at P are perpendicular if and only if

$$f'(\theta)g'(\theta) + f(\theta)g(\theta) = 0.$$

(The graphs are said to be *orthogonal* at P.)

64 Use Exercise 63 to prove that the graphs of each pair of equations are orthogonal at their point of intersection:

(a) $r = a \sin \theta, \quad r = a \cos \theta$ (b) $r = a\theta, \quad r\theta = a$

65 If $\cos \theta \neq 0$, show that the slope of the tangent line to the graph of $r = f(\theta)$ is

$$m = \frac{(dr/d\theta) \tan \theta + r}{(dr/d\theta) - r \tan \theta}.$$

66 A logarithmic spiral has a polar equation of the form $r = ae^{b\theta}$ for nonzero constants a and b (see Exercise 21). A famous *four bugs problem* illustrates such a curve. Four bugs A, B, C, and D are placed at the four corners of a square. The center of the square corresponds to the pole. The bugs begin to crawl simultaneously—bug A crawls toward B, B toward C, C toward D, and D toward A, as shown in the figure on the following page. Assume that all bugs crawl at the same rate, that they move directly toward the next bug at all times, and that they approach one another but never meet. (The bugs are infinitely small!) At any instant, the positions of the bugs are the vertices of a square, which shrinks and rotates toward the center of the original square as the bugs continue to crawl. If the position of bug A has polar coordinates (r, θ), then the position of B has coordinates $(r, \theta + \pi/2)$.

EXERCISE 66

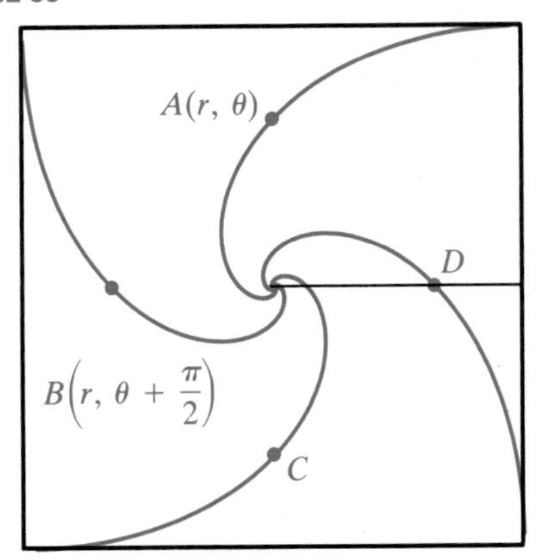

(a) Show that the line through A and B has slope

$$\frac{\sin\theta - \cos\theta}{\sin\theta + \cos\theta}.$$

(b) The line through A and B is tangent to the path of bug A. Use the formula in Exercise 65 to conclude that $dr/d\theta = -r$.

(c) Prove that the path of bug A is a logarithmic spiral. (*Hint:* Solve the differential equation in (b) by separating variables.)

c **Exer. 67–68: Graph the polar equation for the given values of θ, and use the graph to determine symmetries.**

67 $r = 2\sin^2\theta\tan^2\theta;\quad -\pi/3 \le \theta \le \pi/3$

68 $r = \dfrac{4}{1 + \sin^2\theta};\qquad 0 \le \theta \le 2\pi$

c **Exer. 69–70: Graph the polar equations on the same coordinate plane, and estimate the points of intersection of the graphs.**

69 $r = 8\cos 3\theta,\quad r = 4 - 2.5\cos\theta$

70 $r = 2\sin^2\theta,\quad r = \frac{3}{4}(\theta + \cos^2\theta)$

13.4 INTEGRALS IN POLAR COORDINATES

FIGURE 13.29

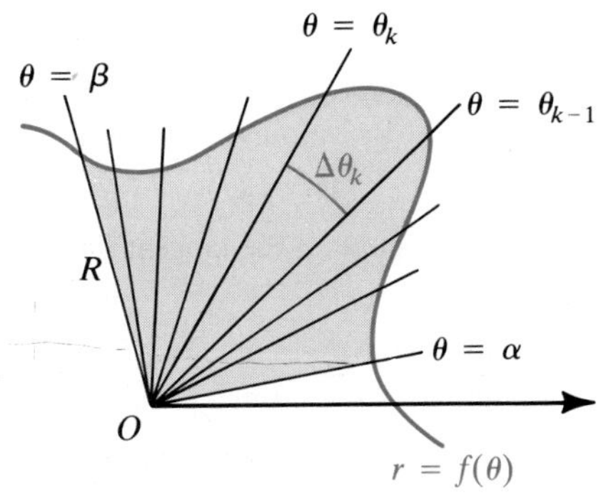

The areas of certain regions bounded by graphs of polar equations can be found by using limits of sums of areas of circular sectors. We shall call a region R in the $r\theta$-plane an **R_θ region** (for integration with respect to θ) if R is bounded by lines $\theta = \alpha$ and $\theta = \beta$ for $0 \le \alpha < \beta \le 2\pi$ and by the graph of a polar equation $r = f(\theta)$, where f is continuous and $f(\theta) \ge 0$ on $[\alpha, \beta]$. An R_θ region is illustrated in Figure 13.29.

Let P denote a partition of $[\alpha, \beta]$ determined by

$$\alpha = \theta_0 < \theta_1 < \theta_2 < \cdots < \theta_n = \beta$$

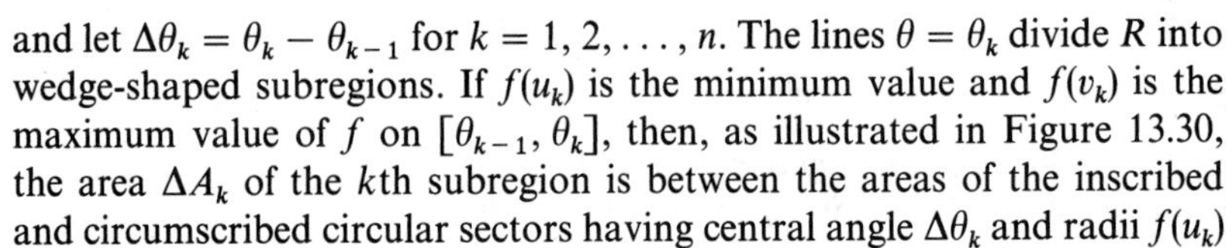

and let $\Delta\theta_k = \theta_k - \theta_{k-1}$ for $k = 1, 2, \ldots, n$. The lines $\theta = \theta_k$ divide R into wedge-shaped subregions. If $f(u_k)$ is the minimum value and $f(v_k)$ is the maximum value of f on $[\theta_{k-1}, \theta_k]$, then, as illustrated in Figure 13.30, the area ΔA_k of the kth subregion is between the areas of the inscribed and circumscribed circular sectors having central angle $\Delta\theta_k$ and radii $f(u_k)$

FIGURE 13.30

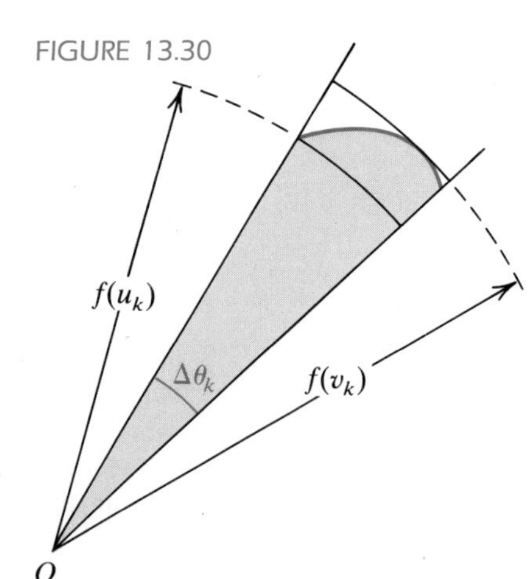

FIGURE 13.31

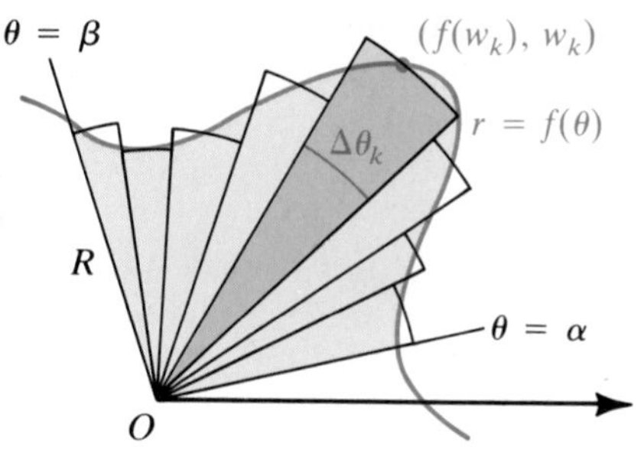

and $f(v_k)$, respectively. Hence, by Theorem (1.15),

$$\tfrac{1}{2}[f(u_k)]^2\,\Delta\theta_k \le \Delta A_k \le \tfrac{1}{2}[f(v_k)]^2\,\Delta\theta_k.$$

Summing from $k = 1$ to $k = n$ and using the fact that the sum of the ΔA_k is the area A of R, we obtain

$$\sum_{k=1}^{n} \tfrac{1}{2}[f(u_k)]^2\,\Delta\theta_k \le A \le \sum_{k=1}^{n} \tfrac{1}{2}[f(v_k)]^2\,\Delta\theta_k.$$

The limits of the sums, as the norm $\|P\|$ of the subdivision approaches zero, both equal the integral $\int_\alpha^\beta \frac{1}{2}[f(\theta)]^2\,d\theta$. This gives us the following result.

Theorem **(13.11)**

> If f is continuous and $f(\theta) \ge 0$ on $[\alpha, \beta]$, where $0 \le \alpha < \beta \le 2\pi$, then the area A of the region bounded by the graphs of $r = f(\theta)$, $\theta = \alpha$, and $\theta = \beta$ is
>
> $$A = \int_\alpha^\beta \tfrac{1}{2}[f(\theta)]^2\,d\theta = \int_\alpha^\beta \tfrac{1}{2}r^2\,d\theta.$$

FIGURE 13.32

The integral in Theorem (13.11) may be interpreted as a limit of sums by writing

$$A = \int_\alpha^\beta \tfrac{1}{2}[f(\theta)]^2\,d\theta = \lim_{\|P\|\to 0} \sum_{k=1}^{n} \tfrac{1}{2}[f(w_k)]^2\,\Delta\theta_k$$

for *any* number w_k in the subinterval $[\theta_{k-1}, \theta_k]$ of $[\alpha, \beta]$. Figure 13.31 is a geometric illustration of a typical Riemann sum.

The following guidelines may be useful for remembering this limit of sums formula (see Figure 13.32).

Guidelines for finding the area of an R_θ region **(13.12)**

1. Sketch the region, labeling the graph of $r = f(\theta)$. Find the smallest value $\theta = \alpha$ and the largest value $\theta = \beta$ for points (r, θ) in the region.
2. Sketch a typical circular sector and label its central angle $d\theta$.
3. Express the area of the sector in guideline 2 as $\frac{1}{2}r^2\,d\theta$.
4. Apply the limit of sums operator $\int_\alpha^\beta$ to the expression in guideline 3 and evaluate the integral.

EXAMPLE 1 Find the area of the region bounded by the cardioid $r = 2 + 2\cos\theta$.

FIGURE 13.33

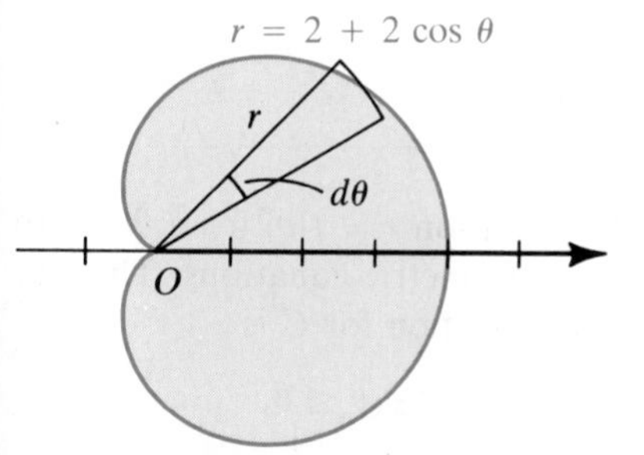

SOLUTION Following guideline 1, we first sketch the region as in Figure 13.33. The cardioid is obtained by letting θ vary from 0 to 2π; however, using symmetry we may find the area of the top half and multiply by 2. Thus, we use $\alpha = 0$ and $\beta = \pi$ for the smallest and largest values of θ. As in guideline 2, we sketch a typical circular sector and label its central angle $d\theta$. To apply guideline 3, we refer to the figure, obtaining the following:

radius of circular sector: $r = 2 + 2\cos\theta$

area of sector: $\frac{1}{2}r^2\,d\theta = \frac{1}{2}(2 + 2\cos\theta)^2\,d\theta$

We next use guideline 4, with $\alpha = 0$ and $\beta = \pi$, remembering that applying $\int_0^\pi$ to the expression $\frac{1}{2}(2 + 2\cos\theta)^2\, d\theta$ represents taking a limit of sums of areas of circular sectors, *sweeping out* the region by letting θ vary from 0 to π. Thus,

$$A = 2\int_0^\pi \tfrac{1}{2}(2 + 2\cos\theta)^2\, d\theta$$
$$= \int_0^\pi (4 + 8\cos\theta + 4\cos^2\theta)\, d\theta.$$

Using the fact that $\cos^2\theta = \frac{1}{2}(1 + \cos 2\theta)$ yields

$$A = \int_0^\pi (6 + 8\cos\theta + 2\cos 2\theta)\, d\theta$$
$$= \Big[6\theta + 8\sin\theta + \sin 2\theta\Big]_0^\pi = 6\pi.$$

We could also have found the area by using $\alpha = 0$ and $\beta = 2\pi$.

FIGURE 13.34

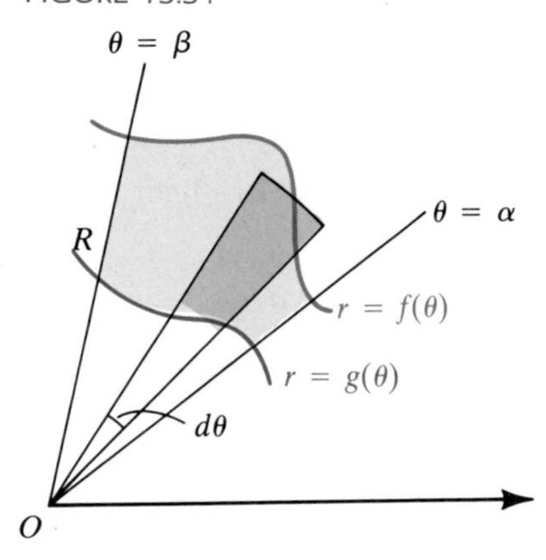

A region R between the graphs of two polar equations $r = f(\theta)$ and $r = g(\theta)$ and the lines $\theta = \alpha$ and $\theta = \beta$ is sketched in Figure 13.34. We may find the area A of R by subtracting the area of the *inner* region bounded by $r = g(\theta)$ from the area of the *outer* region bounded by $r = f(\theta)$ as follows:

$$A = \int_\alpha^\beta \tfrac{1}{2}[f(\theta)]^2\, d\theta - \int_\alpha^\beta \tfrac{1}{2}[g(\theta)]^2\, d\theta$$

We use this technique in the next example.

EXAMPLE 2 Find the area A of the region R that is inside the cardioid $r = 2 + 2\cos\theta$ and outside the circle $r = 3$.

FIGURE 13.35

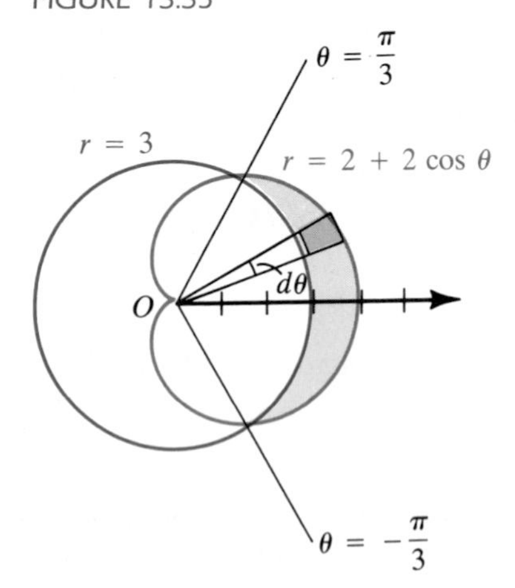

SOLUTION Figure 13.35 shows the region R and circular sectors that extend from the pole to the graphs of the two polar equations. The points of intersection $(3, -\pi/3)$ and $(3, \pi/3)$ can be found by solving the equations simultaneously. Since the angles α and β in Guidelines 13.12 are nonnegative, we shall find the area of the top half of R (using $\alpha = 0$ and $\beta = \pi/3$) and then double the result. Subtracting the area of the inner region (bounded by $r = 3$) from the area of the outer region (bounded by $r = 2 + 2\cos\theta$), we obtain

$$A = 2\left[\int_0^{\pi/3} \tfrac{1}{2}(2 + 2\cos\theta)^2\, d\theta - \int_0^{\pi/3} \tfrac{1}{2}(3)^2\, d\theta\right]$$
$$= \int_0^{\pi/3} (4\cos^2\theta + 8\cos\theta - 5)\, d\theta$$

As in Example 1, the integral may be evaluated by using the substitution $\cos^2\theta = \frac{1}{2}(1 + \cos 2\theta)$. It can be shown that

$$A = \tfrac{9}{2}\sqrt{3} - \pi \approx 4.65.$$

If a curve C is the graph of a polar equation $r = f(\theta)$ from $\theta = \alpha$ to $\theta = \beta$, we can find its length L by using parametric equations. Thus, as in the proof of Theorem (13.10), a parametrization for C is

$$x = f(\theta)\cos\theta, \quad y = f(\theta)\sin\theta; \quad \alpha \le \theta \le \beta.$$

Differentiating with respect to θ, we obtain

$$\frac{dx}{d\theta} = -f(\theta)\sin\theta + f'(\theta)\cos\theta$$

$$\frac{dy}{d\theta} = f(\theta)\cos\theta + f'(\theta)\sin\theta.$$

Using the trigonometric identity $\sin^2\theta + \cos^2\theta = 1$, we can show that

$$\left(\frac{dx}{d\theta}\right)^2 + \left(\frac{dy}{d\theta}\right)^2 = [f(\theta)]^2 + [f'(\theta)]^2.$$

Substitution in Theorem (13.5) with $t = \theta$, $a = \alpha$, and $b = \beta$ gives us

$$L = \int_\alpha^\beta \sqrt{[f(\theta)]^2 + [f'(\theta)]^2}\, d\theta = \int_\alpha^\beta \sqrt{r^2 + \left(\frac{dr}{d\theta}\right)^2}\, d\theta.$$

As an aid to remembering this formula, we may use the differential of arc length $ds = \sqrt{(dx)^2 + (dy)^2}$ in (13.6). The preceding manipulations give us the following.

Differential of arc length in polar coordinates **(13.13)**

$$ds = \sqrt{r^2 + \left(\frac{dr}{d\theta}\right)^2}\, d\theta$$

We may now write the formula for L as

$$L = \int_{\theta=\alpha}^{\theta=\beta} ds.$$

The limits of integration specify that the independent variable is θ, not s.

EXAMPLE 3 Find the length of the cardioid $r = 1 + \cos\theta$.

SOLUTION The cardioid is sketched in Figure 13.36. Making use of symmetry, we shall find the length of the upper half and double the result. Applying (13.13), we have

FIGURE 13.36

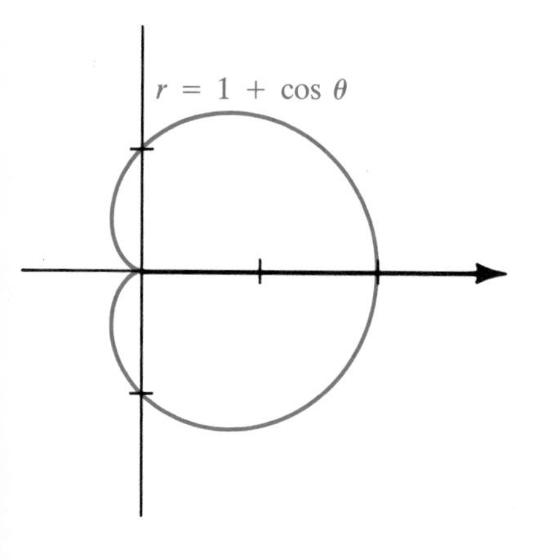

$$\begin{aligned} ds &= \sqrt{(1+\cos\theta)^2 + (-\sin\theta)^2}\, d\theta \\ &= \sqrt{1 + 2\cos\theta + \cos^2\theta + \sin^2\theta}\, d\theta \\ &= \sqrt{2 + 2\cos\theta}\, d\theta \\ &= \sqrt{2}\sqrt{1+\cos\theta}\, d\theta. \end{aligned}$$

Hence

$$L = 2\int_{\theta=0}^{\theta=\pi} ds = 2\int_0^\pi \sqrt{2}\sqrt{1+\cos\theta}\, d\theta.$$

The last integral may be evaluated by employing the trigonometric identity $\cos^2\frac{1}{2}\theta = \frac{1}{2}(1+\cos\theta)$, or, equivalently, $1 + \cos\theta = 2\cos^2\frac{1}{2}\theta$. Thus,

$$\begin{aligned} L &= 2\sqrt{2}\int_0^\pi \sqrt{2\cos^2\tfrac{1}{2}\theta}\, d\theta \\ &= 4\int_0^\pi \cos\tfrac{1}{2}\theta\, d\theta \\ &= 8\Big[\sin\tfrac{1}{2}\theta\Big]_0^\pi = 8. \end{aligned}$$

FIGURE 13.37

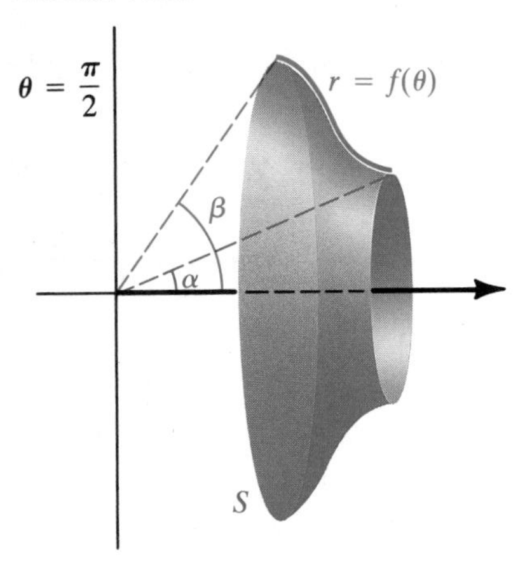

In the solution to Example 3, it was legitimate to replace $\sqrt{\cos^2 \frac{1}{2}\theta}$ by $\cos \frac{1}{2}\theta$, because if $0 \le \theta \le \pi$, then $0 \le \frac{1}{2}\theta \le \pi/2$, and hence $\cos \frac{1}{2}\theta$ is *positive* on $[0, \pi]$. If we had *not* used symmetry, but had written L as $\int_0^{2\pi} \sqrt{r^2 + (dr/d\theta)^2}\, d\theta$, this simplification would not have been valid. Generally, in determining areas or arc lengths that involve polar coordinates, it is a good idea to use any symmetries that exist.

Let C be the graph of a polar equation $r = f(\theta)$ for $\alpha \le \theta \le \beta$. Let us obtain a formula for the area S of the surface generated by revolving C about the polar axis, as illustrated in Figure 13.37. Since parametric equations for C are

$$x = f(\theta) \cos \theta, \quad y = f(\theta) \sin \theta; \quad \alpha \le \theta \le \beta,$$

we may find S by using Theorem (13.7) with $\theta = t$. This gives us the following result, where the arc length differential ds is given by (13.13).

Surfaces of revolution in polar coordinates **(13.14)**

About the polar axis: $S = \int_{\theta=\alpha}^{\theta=\beta} 2\pi y\, ds = \int_{\theta=\alpha}^{\theta=\beta} 2\pi r \sin \theta\, ds$

About the line $\theta = \pi/2$: $S = \int_{\theta=\alpha}^{\theta=\beta} 2\pi x\, ds = \int_{\theta=\alpha}^{\theta=\beta} 2\pi r \cos \theta\, ds$

When applying (13.14), *we must choose α and β so that the surface does not retrace itself when C is revolved*, as would be the case if the circle $r = \cos \theta$, with $0 \le \theta \le \pi$, were revolved about the polar axis.

EXAMPLE 4 The part of the spiral $r = e^{\theta/2}$ from $\theta = 0$ to $\theta = \pi$ is revolved about the polar axis. Find the area of the resulting surface.

FIGURE 13.38

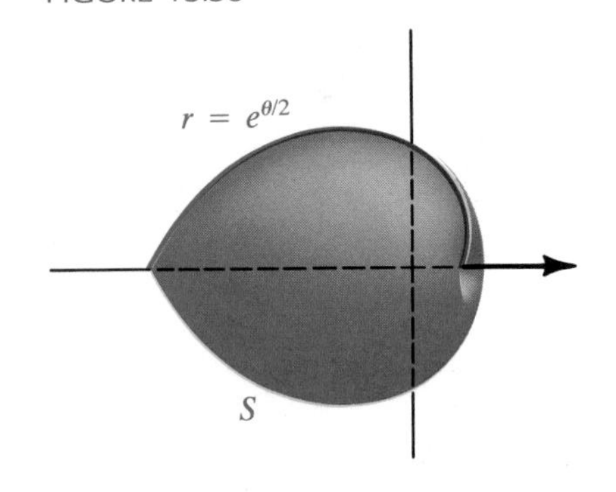

SOLUTION The surface is illustrated in Figure 13.38. By (13.13), the polar differential of arc length in polar coordinates is

$$\begin{aligned} ds &= \sqrt{(e^{\theta/2})^2 + (\tfrac{1}{2}e^{\theta/2})^2}\, d\theta \\ &= \sqrt{\frac{5}{4}e^{\theta}}\, d\theta = \frac{\sqrt{5}}{2} e^{\theta/2}\, d\theta. \end{aligned}$$

Hence, by (13.14),

$$\begin{aligned} S &= \int_{\theta=0}^{\theta=\pi} 2\pi y\, ds = \int_{\theta=0}^{\theta=\pi} 2\pi r \sin \theta\, ds \\ &= \int_0^{\pi} 2\pi e^{\theta/2} \sin \theta \left(\frac{\sqrt{5}}{2} e^{\theta/2}\right) d\theta \\ &= \sqrt{5}\pi \int_0^{\pi} e^{\theta} \sin \theta\, d\theta. \end{aligned}$$

Using integration by parts or Formula 98 in the table of integrals (see Appendix IV), we have

$$S = \frac{\sqrt{5}\pi}{2}\Big[e^{\theta}(\sin \theta - \cos \theta)\Big]_0^{\pi} = \frac{\sqrt{5}\pi}{2}(e^{\pi} + 1) \approx 84.8.$$

EXERCISES 13.4

Exer. 1–6: Find the area of the region bounded by the graph of the polar equation.

1 $r = 2\cos\theta$

2 $r = 5\sin\theta$

3 $r = 1 - \cos\theta$

4 $r = 6 - 6\sin\theta$

5 $r = \sin 2\theta$

6 $r^2 = 9\cos 2\theta$

Exer. 7–8: Find the area of region R.

7 $R = \{(r, \theta): 0 \le \theta \le \pi/2, 0 \le r \le e^{\theta}\}$

8 $R = \{(r, \theta): 0 \le \theta \le \pi, 0 \le r \le 2\theta\}$

Exer. 9–12: Find the area of the region bounded by one loop of the graph of the polar equation.

9 $r^2 = 4\cos 2\theta$

10 $r = 2\cos 3\theta$

11 $r = 3\cos 5\theta$

12 $r = \sin 6\theta$

Exer. 13–16: Set up integrals in polar coordinates that can be used to find the area of the region shown in the figure.

13

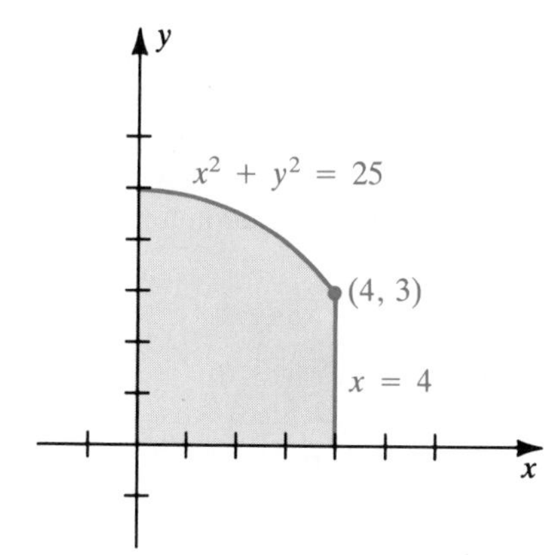

14

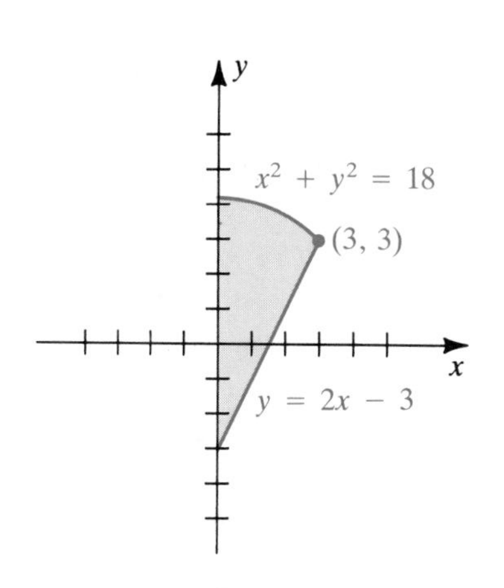

15

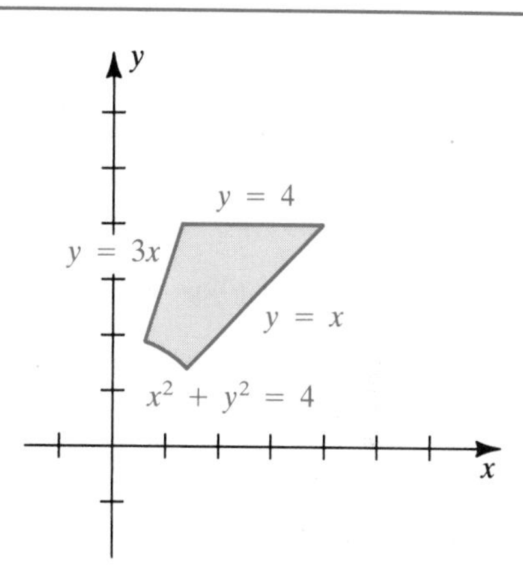

16

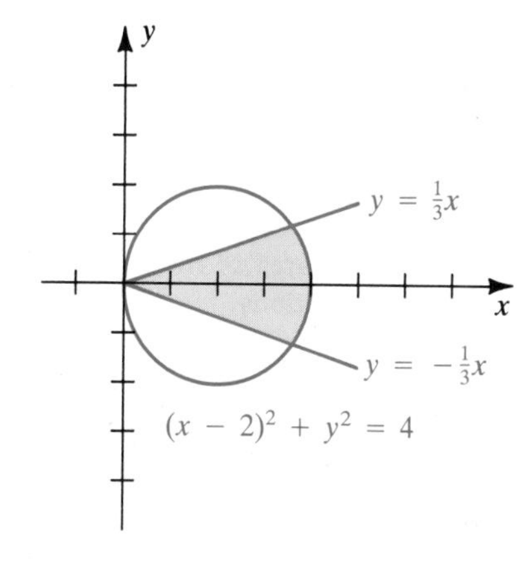

Exer. 17–18: Set up integrals in polar coordinates that can be used to find the area of (a) the blue region and (b) the green region.

17

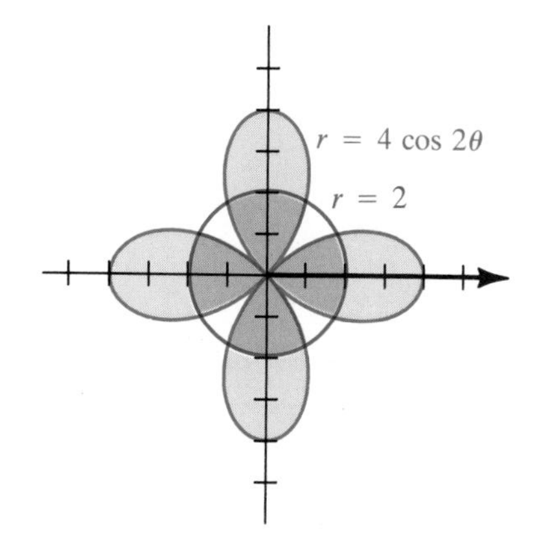

18

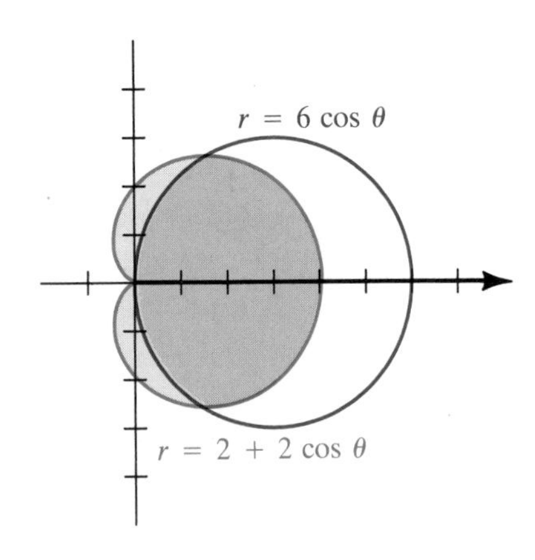

Exer. 19–22: Find the area of the region that is outside the graph of the first equation and inside the graph of the second equation.

19 $r = 2 + 2 \cos \theta, \quad r = 3$

20 $r = 2, \quad r = 4 \cos \theta$

21 $r = 2, \quad r^2 = 8 \sin 2\theta$

22 $r = 1 - \sin \theta, \quad r = 3 \sin \theta$

Exer. 23–26: Find the area of the region that is inside the graphs of *both* equations.

23 $r = \sin \theta, \quad r = \sqrt{3} \cos \theta$

24 $r = 2(1 + \sin \theta), \quad r = 1$

25 $r = 1 + \sin \theta, \quad r = 5 \sin \theta$

26 $r^2 = 4 \cos 2\theta, \quad r = 1$

Exer. 27–32: Find the length of the curve.

27 $r = e^{-\theta}$ from $\theta = 0$ to $\theta = 2\pi$

28 $r = \theta$ from $\theta = 0$ to $\theta = 4\pi$

29 $r = \cos^2 \frac{1}{2}\theta$ from $\theta = 0$ to $\theta = \pi$

30 $r = 2^\theta$ from $\theta = 0$ to $\theta = \pi$

31 $r = \sin^3 \frac{1}{3}\theta$

32 $r = 2 - 2 \cos \theta$

c **Exer. 33–34: Use Simpson's rule, with $n = 4$, to approximate the length of the curve.**

33 $r = \theta + \cos \theta$ from $\theta = 0$ to $\theta = \pi/2$

34 $r = \sin \theta + \cos^2 \theta$ from $\theta = 0$ to $\theta = \pi$

Exer. 35–38: Find the area of the surface generated by revolving the graph of the equation about the polar axis.

35 $r = 2 + 2 \cos \theta$

36 $r^2 = 4 \cos 2\theta$

37 $r = 2a \sin \theta$

38 $r = 2a \cos \theta$

c **Exer. 39–40: Use the trapezoidal rule, with $n = 4$, to approximate the area of the surface generated by revolving the graph of the polar equation about the line $\theta = \pi/2$. (Use symmetry when setting up the integral.)**

39 $r = \sin^2 \theta$

40 $r = \cos^2 \theta$

41 A *torus* is the surface generated by revolving a circle about a nonintersecting line in its plane. Use polar coordinates to find the surface area of the torus generated by revolving the circle $x^2 + y^2 = a^2$ about the line $x = b$, where $0 < a < b$.

42 Let OP be the ray from the pole to the point $P(r, \theta)$ on the spiral $r = a\theta$, where $a > 0$. If the ray makes two revolutions (starting from $\theta = 0$), find the area of the region swept out in the second revolution that was not swept out in the first revolution (see figure).

EXERCISE 42

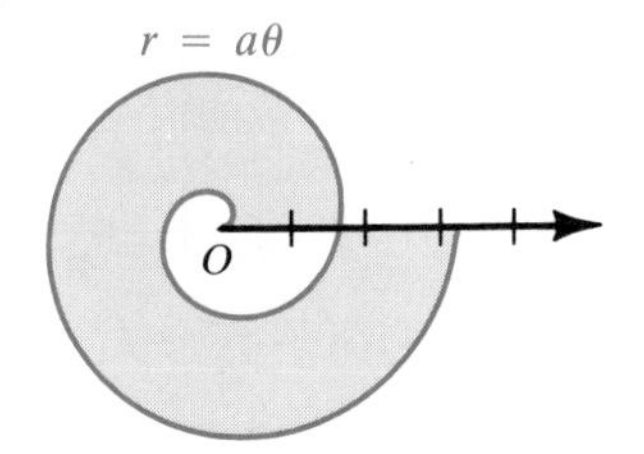

43 The part of the spiral $r = e^{-\theta}$ from $\theta = 0$ to $\theta = \pi/2$ is revolved about the line $\theta = \pi/2$. Find the area of the resulting surface.

13.5 POLAR EQUATIONS OF CONICS

The following theorem combines the definitions of parabola, ellipse, and hyperbola into a unified description of the conic sections. The constant e in the statement of the theorem is the **eccentricity** of the conic. The point

F is a **focus** of the conic, and the line l is a **directrix**. Possible positions of F and l are illustrated in Figure 13.39.

FIGURE 13.39

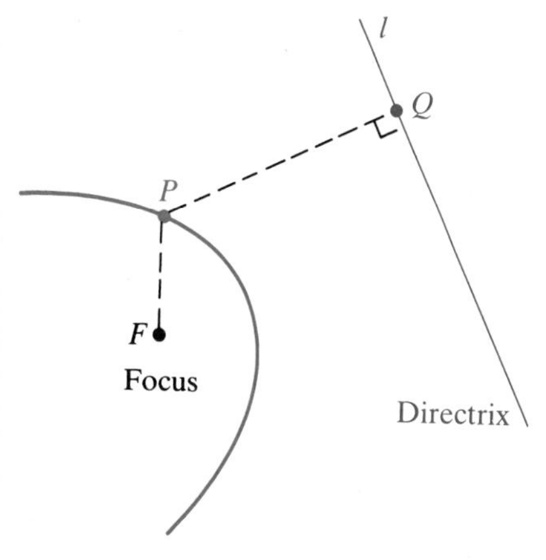

Theorem **(13.15)**

> Let F be a fixed point and l a fixed line in a plane. The set of all points P in the plane, such that the ratio $d(P, F)/d(P, Q)$ is a positive constant e with $d(P, Q)$ the distance from P to l, is a conic section. The conic is a parabola if $e = 1$, an ellipse if $0 < e < 1$, and a hyperbola if $e > 1$.

PROOF If $e = 1$, then $d(P, F) = d(P, Q)$ and, by definition, the resulting conic is a parabola with focus F and directrix l.

Suppose next that $0 < e < 1$. It is convenient to introduce a polar coordinate system in the plane with F as the pole and l perpendicular to the polar axis at the point $D(d, 0)$, with $d > 0$, as illustrated in Figure 13.40. If $P(r, \theta)$ is a point in the plane such that $d(P, F)/d(P, Q) = e < 1$, then P lies to the left of l. Let C be the projection of P on the polar axis. Since

FIGURE 13.40

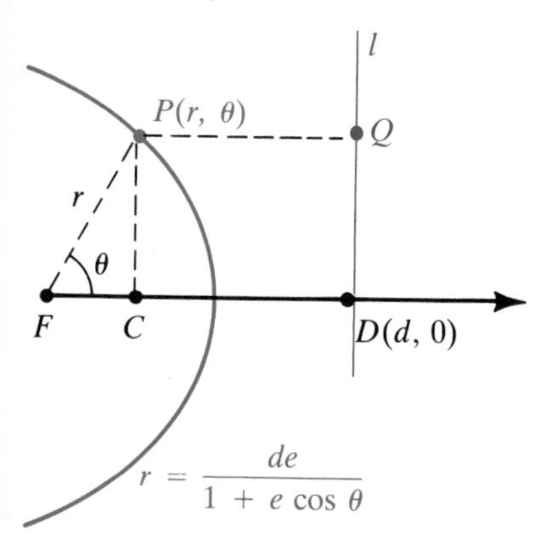

$$d(P, F) = r \quad \text{and} \quad d(P, Q) = d - r\cos\theta,$$

it follows that P satisfies the condition in the theorem if and only if the following are true:

$$\frac{r}{d - r\cos\theta} = e$$

$$r = de - er\cos\theta$$

$$r(1 + e\cos\theta) = de$$

$$r = \frac{de}{1 + e\cos\theta}$$

The same equations are obtained if $e = 1$; however, there is no point (r, θ) on the graph if $1 + \cos\theta = 0$.

An equation in x and y corresponding to $r = de - er\cos\theta$ is

$$\sqrt{x^2 + y^2} = de - ex.$$

Squaring both sides and rearranging terms leads to

$$(1 - e^2)x^2 + 2de^2x + y^2 = d^2e^2.$$

Completing the square and simplifying, we obtain

$$\left(x + \frac{de^2}{1 - e^2}\right)^2 + \frac{y^2}{1 - e^2} = \frac{d^2e^2}{(1 - e^2)^2}.$$

Finally, dividing both sides by $d^2e^2/(1 - e^2)^2$ gives us an equation of the form

$$\frac{(x - h)^2}{a^2} + \frac{y^2}{b^2} = 1,$$

with $h = -de^2/(1 - e^2)$. Consequently the graph is an ellipse with center at the point $(h, 0)$ on the x-axis and with

$$a^2 = \frac{d^2e^2}{(1 - e^2)^2}, \qquad b^2 = \frac{d^2e^2}{1 - e^2}.$$

Since

$$c^2 = a^2 - b^2 = \frac{d^2e^4}{(1 - e^2)^2},$$

we obtain $c = de^2/(1 - e^2)$ and hence $|h| = c$. This proves that F is a focus of the ellipse. It also follows that $e = c/a$. A similar proof may be given for the case $e > 1$. ■

FIGURE 13.41

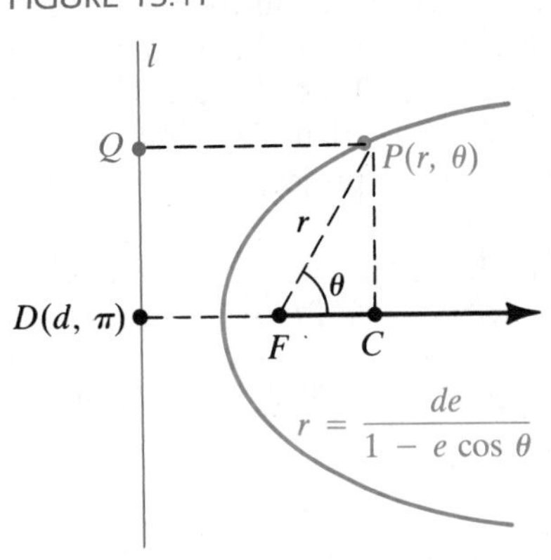

We also can show that every conic that is not degenerate may be described by means of the statement in Theorem (13.15). This gives us a formulation of conic sections that is equivalent to the approach used previously. Since the theorem includes all three types of conics, it is sometimes regarded as a definition for the conic sections.

If we had chosen the focus F to the *right* of the directrix, as illustrated in Figure 13.41 (with $d > 0$), then the equation $r = de/(1 - e \cos \theta)$ would have resulted. (Note the minus sign in place of the plus sign.) Other sign changes occur if d is allowed to be negative.

If we had taken l *parallel* to the polar axis through one of the points $(d, \pi/2)$ or $(d, 3\pi/2)$, as illustrated in Figure 13.42, then the corresponding equations would have contained $\sin \theta$ instead of $\cos \theta$.

FIGURE 13.42

(i)

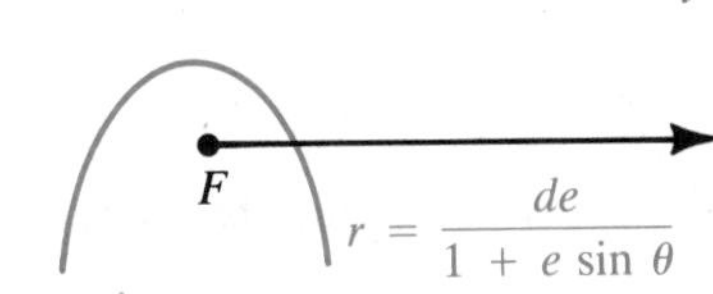

(ii)

$r = \dfrac{de}{1 - e \sin \theta}$ F l

The following theorem summarizes our discussion.

Theorem (13.16)

> A polar equation that has one of the four forms
>
> $$r = \frac{de}{1 \pm e \cos \theta}, \qquad r = \frac{de}{1 \pm e \sin \theta}$$
>
> is a conic section. The conic is a parabola if $e = 1$, an ellipse if $0 < e < 1$, or a hyperbola if $e > 1$.

EXAMPLE 1 Describe and sketch the graph of the polar equation

$$r = \frac{10}{3 + 2 \cos \theta}.$$

SOLUTION We first divide numerator and denominator of the fraction by 3:

$$r = \frac{\frac{10}{3}}{1 + \frac{2}{3}\cos\theta}$$

FIGURE 13.43

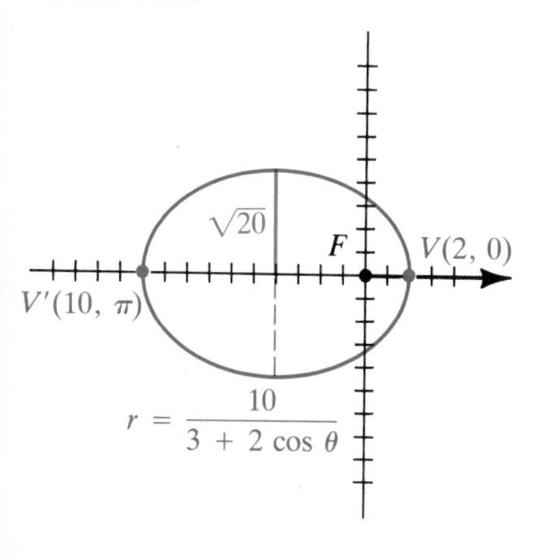

This equation has one of the forms in Theorem (13.16), with $e = \frac{2}{3}$. Thus, the graph is an ellipse with focus F at the pole and major axis along the polar axis. We find the endpoints of the major axis by letting $\theta = 0$ and $\theta = \pi$. This gives us $V(2, 0)$ and $V'(10, \pi)$. Hence

$$2a = d(V', V) = 12, \quad \text{or} \quad a = 6.$$

The center of the ellipse is the midpoint $(4, \pi)$ of the segment $V'V$. Using the fact that $e = c/a$, we obtain

$$c = ae = 6(\tfrac{2}{3}) = 4.$$

Hence $$b^2 = a^2 - c^2 = 36 - 16 = 20.$$

Thus, $b = \sqrt{20}$. The graph is sketched in Figure 13.43. For reference, we have superimposed an xy-coordinate system on the polar system.

EXAMPLE 2 Describe and sketch the graph of the polar equation $r = \dfrac{10}{2 + 3\sin\theta}$.

SOLUTION To express the equation in the proper form, we divide numerator and denominator of the fraction by 2:

$$r = \frac{5}{1 + \frac{3}{2}\sin\theta}$$

Thus, $e = \frac{3}{2}$, and, by Theorem (13.16), the graph is a hyperbola with a focus at the pole. The expression $\sin\theta$ tells us that the transverse axis of the hyperbola is perpendicular to the polar axis. To find the vertices, we let $\theta = \pi/2$ and $\theta = 3\pi/2$ in the given equation. This gives us the points $V(2, \pi/2)$, $V'(-10, 3\pi/2)$, and hence

$$2a = d(V, V') = 8, \quad \text{or} \quad a = 4.$$

FIGURE 13.44

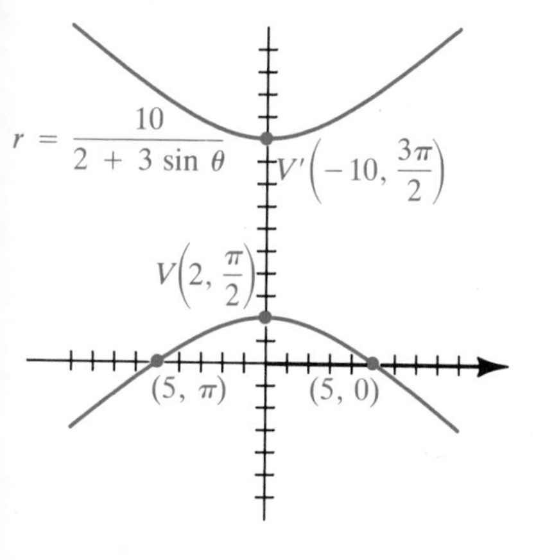

The points $(5, 0)$ and $(5, \pi)$ on the graph can be used to sketch the lower branch of the hyperbola. The upper branch is obtained by symmetry, as illustrated in Figure 13.44. If we desire more accuracy or additional information, we calculate

$$c = ae = 4(\tfrac{3}{2}) = 6$$

and $$b^2 = c^2 - a^2 = 36 - 16 = 20.$$

Asymptotes may then be constructed in the usual way.

EXAMPLE 3 Sketch the graph of the polar equation $r = \dfrac{15}{4 - 4\cos\theta}$.

SOLUTION To obtain the proper form, we divide numerator and denominator by 4:

$$r = \frac{\frac{15}{4}}{1 - \cos\theta}$$

Consequently $e = 1$, and, by Theorem (13.16), the graph is a parabola with focus at the pole. We may obtain a sketch by plotting the points that correspond to the x- and y-intercepts. These are indicated in the following table.

FIGURE 13.45

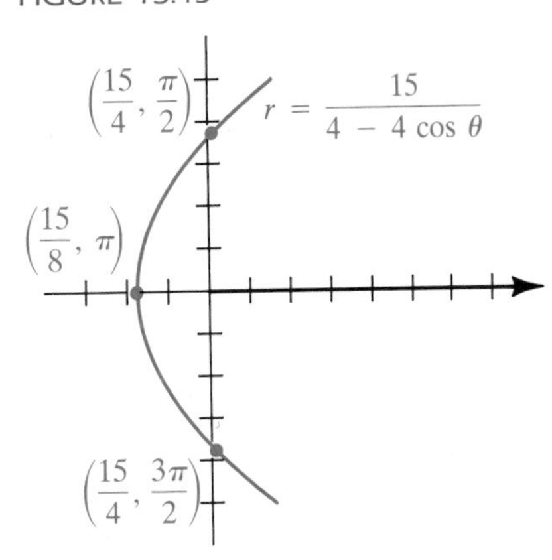

θ	0	$\frac{\pi}{2}$	π	$\frac{3\pi}{2}$
r	undefined	$\frac{15}{4}$	$\frac{15}{8}$	$\frac{15}{4}$

Note that there is no point on the graph corresponding to $\theta = 0$, since the denominator $1 - \cos\theta$ is 0 for that value. Plotting the three points and using the fact that the graph is a parabola with focus at the pole gives us the sketch in Figure 13.45.

If we desire only a rough sketch of a conic, then the technique employed in Example 3 is recommended. To use this method, we plot (if possible) points corresponding to $\theta = 0$, $\pi/2$, π, and $3\pi/2$. These points, together with the type of conic (obtained from the value of e), readily lead to the sketch.

EXAMPLE 4 Find an equation in x and y that has the same graph as the polar equation

$$r = \frac{15}{4 - 4\sin\theta}.$$

SOLUTION We multiply both sides of the polar equation by the lcd, $4 - 4\sin\theta$, and then use relationships between rectangular and polar coordinates as follows:

$$r(4 - 4\sin\theta) = 15$$
$$4r = 15 + 4r\sin\theta$$
$$4(\pm\sqrt{x^2 + y^2}) = 15 + 4y$$

Squaring both sides and simplifying yields

$$16(x^2 + y^2) = 225 + 120y + 16y^2$$
$$16x^2 = 120y + 225.$$

EXAMPLE 5 Find a polar equation of the conic with a focus at the pole, eccentricity $e = \frac{1}{2}$, and directrix $r = -3 \sec \theta$.

SOLUTION The equation $r = -3 \sec \theta$ of the directrix may be written $r \cos \theta = -3$, which is equivalent to $x = -3$ in a rectangular coordinate system. This gives us the situation illustrated in Figure 13.41, with $d = 3$. Hence a polar equation has the form

$$r = \frac{de}{1 - e \cos \theta}.$$

We now substitute $d = 3$ and $e = \frac{1}{2}$:

$$r = \frac{3(\frac{1}{2})}{1 - \frac{1}{2} \cos \theta}$$

$$r = \frac{3}{2 - \cos \theta}$$

EXERCISES 13.5

Exer. 1–10: (a) Find the eccentricity and classify the conic. (b) Sketch the graph and label the vertices.

1 $r = \dfrac{12}{6 + 2 \sin \theta}$

2 $r = \dfrac{12}{6 - 2 \sin \theta}$

3 $r = \dfrac{12}{2 - 6 \cos \theta}$

4 $r = \dfrac{12}{2 + 6 \cos \theta}$

5 $r = \dfrac{3}{2 + 2 \cos \theta}$

6 $r = \dfrac{3}{2 - 2 \sin \theta}$

7 $r = \dfrac{4}{\cos \theta - 2}$

8 $r = \dfrac{4 \sec \theta}{2 \sec \theta - 1}$

9 $r = \dfrac{6 \csc \theta}{2 \csc \theta + 3}$

10 $r = \csc \theta \, (\csc \theta - \cot \theta)$

Exer. 11–20: Find equations in x and y for the polar equations in Exercises 1–10.

Exer. 21–26: Find a polar equation of the conic with focus at the pole and the given eccentricity and equation of the directrix.

21 $e = \frac{1}{3}$; $r = 2 \sec \theta$

22 $e = \frac{2}{5}$; $r = 4 \csc \theta$

23 $e = 4$; $r = -3 \csc \theta$

24 $e = 3$; $r = -4 \sec \theta$

25 $e = 1$; $r \cos \theta = 5$

26 $e = 1$; $r \sin \theta = -2$

27 Find a polar equation of the parabola with focus at the pole and vertex $V(4, \pi/2)$.

28 An ellipse has a focus at the pole, center $C(3, \pi/2)$, and vertex $V(1, 3\pi/2)$.

(a) Find the eccentricity.

(b) Find a polar equation for the ellipse.

Exer. 29–32: Find the slope of the tangent line to the conic at the point corresponding to the given value of θ.

29 $r = \dfrac{12}{6 + 2 \sin \theta}$; $\theta = \dfrac{\pi}{6}$

30 $r = \dfrac{12}{6 - 2 \sin \theta}$; $\theta = 0$

31 $r = \dfrac{12}{2 - 6 \cos \theta}$; $\theta = \dfrac{\pi}{2}$

32 $r = \dfrac{12}{2 + 6 \cos \theta}$; $\theta = \dfrac{\pi}{3}$

Exer. 33–36: Set up an integral in polar coordinates that can be used to find the area of the region bounded by the graphs of the equations.

33 $r = 2 \sec \theta$; $\theta = \pi/6$, $\theta = \pi/3$

34 $r = \csc \theta \cot \theta$; $\theta = \pi/6$, $\theta = \pi/4$

35 $r(1 - \cos \theta) = 4$; $\theta = \pi/4$

36 $r(1 + \sin \theta) = 2$; $\theta = \pi/3$

37 Kepler's first law states that planets travel in elliptical orbits with the sun at a focus. To find an equation of an orbit, place the pole O at the center of the sun and the polar axis along the major axis of the ellipse (see figure).

(a) Show that an equation of the orbit is

$$r = \frac{(1 - e^2)a}{1 - e \cos \theta},$$

where e is the eccentricity and $2a$ is the length of the major axis.

(b) The *perihelion distance* r_{per} and *aphelion distance* r_{aph} are defined as the minimum and maximum distances, respectively, of a planet from the sun. Show that $r_{per} = a(1 - e)$ and $r_{aph} = a(1 + e)$.

EXERCISE 37

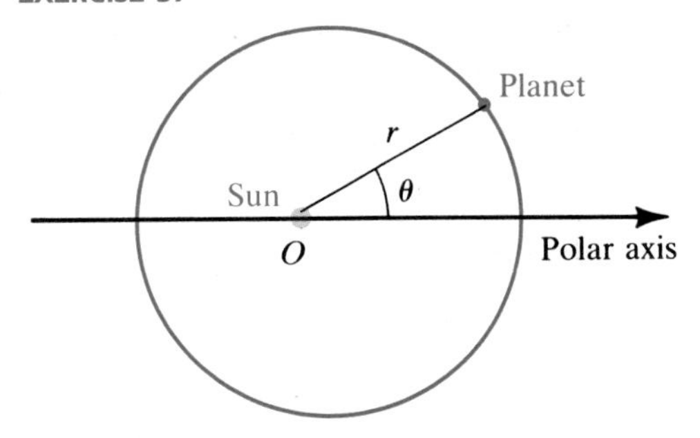

38 Refer to Exercise 37. The planet Pluto travels in an elliptical orbit of eccentricity 0.249. If the perihelion distance is 29.62 AU, find a polar equation for the orbit and estimate the aphelion distance.

39 (a) Use Theorem (9.6) to show that

$$\int_{-\pi}^{\pi} \frac{1}{1 - e\cos\theta}\,d\theta = \frac{2\pi}{\sqrt{1 - e^2}}.$$

(b) Use Definition (5.29) to find the average distance of a planet from the sun.

40 Refer to Exercises 37 and 39. Eros, with an average distance from the sun of 1.46 AU, is an asteroid in the solar system. If Eros has an elliptical orbit of eccentricity 0.223, find a polar equation that approximates the orbit and estimate how close Eros comes to the sun.

13.6 REVIEW EXERCISES

Exer. 1–4: (a) Find an equation in x and y whose graph contains the points on the curve C. (b) Sketch the graph of C and indicate the orientation.

1 $x = \dfrac{1}{t} + 1, \quad y = \dfrac{2}{t} - t; \quad 0 < t \le 4$

2 $x = \cos^2 t - 2, \quad y = \sin t + 1; \quad 0 \le t \le 2\pi$

3 $x = \sqrt{t}, \quad y = 2^{-t}; \quad t \ge 0$

4 $x = 3\cos t + 2, \quad y = -3\sin t - 1; \quad 0 \le t \le 2\pi$

Exer. 5–6: Sketch the graphs of C_1, C_2, C_3, and C_4, and indicate their orientations.

5 C_1: $x = t, \quad y = \sqrt{16 - t^2}; \quad -4 \le t \le 4$
C_2: $x = -\sqrt{16 - t}, \quad y = -\sqrt{t}; \quad 0 \le t \le 16$
C_3: $x = 4\cos t, \quad y = 4\sin t; \quad 0 \le t \le 2\pi$
C_4: $x = e^t, \quad y = -\sqrt{16 - e^{2t}}; \quad t \le \ln 4$

6 C_1: $x = t^2, \quad y = t^3; \quad t$ in $\mathbb{R}$
C_2: $x = t^4, \quad y = t^6; \quad t$ in $\mathbb{R}$
C_3: $x = e^{2t}, \quad y = e^{3t}; \quad t$ in $\mathbb{R}$
C_4: $x = 1 - \sin^2 t, \quad y = \cos^3 t; \quad t$ in $\mathbb{R}$

Exer. 7–8: Let C be the given parametrized curve. (a) Express dy/dx in terms of t. (b) Find the values of t that correspond to horizontal or vertical tangent lines to the graph of C. (c) Express d^2y/dx^2 in terms of t.

7 $x = t^2, \quad y = 2t^3 + 4t - 1; \quad t$ in $\mathbb{R}$

8 $x = t - 2\sin t, \quad y = 1 - 2\cos t; \quad t$ in $\mathbb{R}$

Exer. 9–26: Sketch the graph of the polar equation.

9 $r = -4\sin\theta$

10 $r = 10\cos\theta$

11 $r = 6 - 3\cos\theta$

12 $r = 3 + 2\cos\theta$

13 $r^2 = 9\sin 2\theta$

14 $r^2 = -4\sin 2\theta$

15 $r = 3\sin 5\theta$

16 $r = 2\sin 3\theta$

17 $2r = \theta$

18 $r = e^{-\theta}, \quad \theta \ge 0$

19 $r = 8\sec\theta$

20 $r(3\cos\theta - 2\sin\theta) = 6$

21 $r = 4 - 4\cos\theta$

22 $r = 4\cos^2 \frac{1}{2}\theta$

23 $r = 6 - r\cos\theta$

24 $r = 6\cos 2\theta$

25 $r = \dfrac{8}{3 + \cos\theta}$

26 $r = \dfrac{8}{1 - 3\sin\theta}$

Exer. 27–32: Find a polar equation that has the same graph as the given equation.

27 $y^2 = 4x$

28 $x^2 + y^2 - 3x + 4y = 0$

29 $2x - 3y = 8$

30 $x^2 + y^2 = 2xy$

31 $y^2 = x^2 - 2x$

32 $x^2 = y^2 + 3y$

Exer. 33–38: Find an equation in x and y that has the same graph as the polar equation.

33 $r^2 = \tan\theta$

34 $r = 2\cos\theta + 3\sin\theta$

35 $r^2 = 4\sin 2\theta$

36 $r^2 = \sec 2\theta$

37 $\theta = \sqrt{3}$

38 $r = -6$

Exer. 39–40: Find the slope of the tangent line to the graph of the polar equation at the point corresponding to the given value of θ.

39 $r = \dfrac{3}{2 + 2\cos\theta}; \quad \theta = \pi/2$

40 $r = e^{3\theta}; \quad \theta = \pi/4$

41 Find the area of the region bounded by one loop of $r^2 = 4\sin 2\theta$.

42 Find the area of the region that is inside the graph of $r = 3 + 2 \sin \theta$ and outside the graph of $r = 4$.

43 The position (x, y) of a moving point at time t is given by $x = 2 \sin t$, $y = \sin^2 t$. Find the distance the point travels from $t = 0$ to $t = \pi/2$.

44 Find the length of the spiral $r = 1/\theta$ from $\theta = 1$ to $\theta = 2$.

45 The curve with parametrization $x = 2t^2 + 1$, $y = 4t - 3$; $0 \leq t \leq 1$ is revolved about the y-axis. Find the area of the resulting surface.

46 The arc of the spiral $r = e^\theta$ from $\theta = 0$ to $\theta = 1$ is revolved about the line $\theta = \pi/2$. Find the area of the resulting surface.

47 Find the area of the surface generated by revolving the lemniscate $r^2 = a^2 \cos 2\theta$ about the polar axis.

48 A line segment of fixed length has endpoints A and B on the y-axis and x-axis, respectively. A fixed point P on AB is selected with $d(A, P) = a$ and $d(B, P) = b$ (see figure). If A and B may slide freely along their respective axes, a curve C is traced by P. If t is the radian measure of angle ABO, find parametric equations for C with parameter t and describe C.

EXERCISE 48

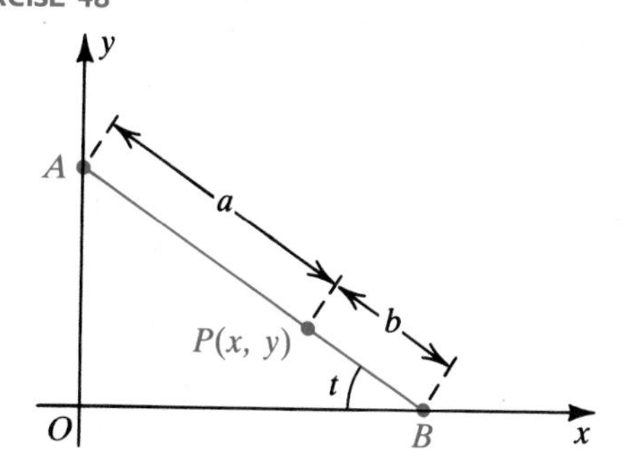

CHAPTER

14

VECTORS AND SURFACES

INTRODUCTION

A vector has two different natures, one geometric and the other algebraic. Both aspects are useful in applications. Generally, geometry helps us develop an intuitive understanding of vector concepts; however, rigorous proofs require algebraic techniques.

There are different ways to introduce vectors so that they can be applied in either two or three dimensions. In one approach we define vectors in three dimensions and then *specialize* to vectors in a coordinate plane. Another procedure is to define vectors in a coordinate plane and then *generalize* to three dimensions. We shall use the latter method, beginning with the familiar xy-plane in Section 14.1 and then proceeding to an xyz-coordinate system in Section 14.2. An advantage of this approach is that we can plot (and visualize) initial and terminal points of vectors more easily in the plane than in three dimensions. Moreover, proofs of properties of vector addition and scalar multiples are shorter, because there is one less vector component to consider. Since the corresponding proofs in three dimensions follow exactly the same steps, it is unnecessary to reprove these properties in Section 14.2.

In Sections 14.5 and 14.6 we discuss topics from solid analytic geometry that include lines, planes, and surfaces. Our main tool for sketching a surface is its trace on a plane. A good background in finding traces is helpful in analyzing the computer-generated graphs of surfaces considered in Chapter 16.

14.1 VECTORS IN TWO DIMENSIONS

Quantities such as area, volume, arc length, temperature, and time have magnitude only and can be completely characterized by a single real number (with an appropriate unit of measurement such as in.^{2}, ft^3, cm, deg, or sec). A quantity of this type is a **scalar quantity**, and the corresponding real number is a **scalar**. A concept such as velocity or force has both magnitude and direction and is often represented by a **directed line segment**—that is, a line segment to which a direction has been assigned. Another name for a directed line segment is a **vector**.

FIGURE 14.1

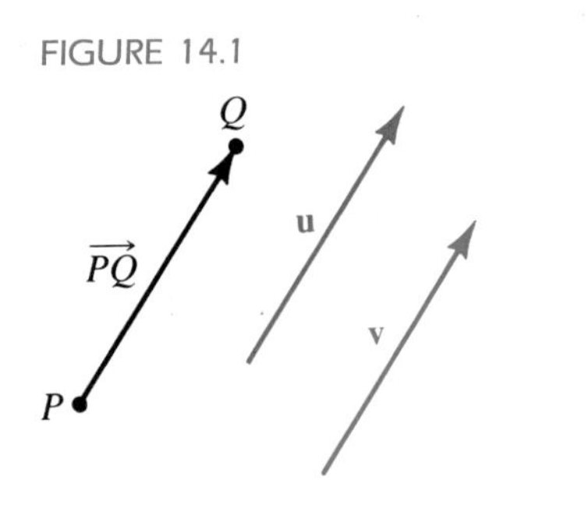

As shown in Figure 14.1, we use $\overrightarrow{PQ}$ to denote the vector with **initial point** P and **terminal point** Q and indicate the direction of the vector by placing an arrowhead at Q. The **magnitude** of $\overrightarrow{PQ}$ is the length of the segment PQ and is denoted by $\|\overrightarrow{PQ}\|$. As in the figure, we use boldface letters such as **u** and **v** to denote vectors whose endpoints are not specified. In handwritten work, a notation such as $\vec{u}$ or $\vec{v}$ is often used.

Vectors that have the same magnitude and direction are said to be **equivalent**. In calculus, a vector is determined only by its magnitude and direction, not by its location. Thus, we regard equivalent vectors, such as those in Figure 14.1, as **equal** and write

$$\mathbf{u} = \overrightarrow{PQ}, \quad \mathbf{v} = \overrightarrow{PQ}, \quad \text{and} \quad \mathbf{u} = \mathbf{v}.$$

FIGURE 14.2

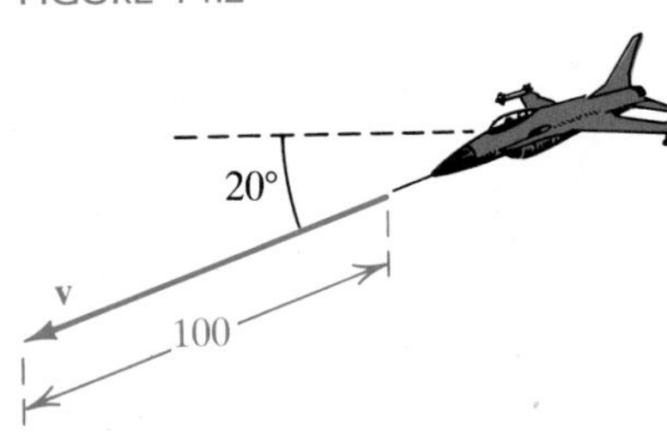

Thus, *a vector may be translated from one location to another, provided neither the magnitude nor the direction is changed.*

We can represent many physical concepts by vectors. To illustrate, suppose an airplane is descending at a constant speed of 100 mi/hr and the line of flight makes an angle of 20° with the horizontal. Both of these facts are represented by the vector **v** of magnitude 100 in Figure 14.2. The vector **v** is a **velocity vector**.

FIGURE 14.3

A vector that represents a pull or push of some type is a **force vector**. The force exerted when a person holds a 5-pound weight is illustrated by the vector **F** of magnitude 5 in Figure 14.3. This force has the same magnitude as the force exerted on the weight by gravity, but acts in the opposite direction. As a result, there is no movement upward or downward.

We sometimes use $\overrightarrow{AB}$ to represent the path of a point (or particle) as it moves along the line segment from A to B. We then refer to $\overrightarrow{AB}$ as a **displacement** of the point (or particle). As in Figure 14.4, a displacement $\overrightarrow{AB}$ followed by a displacement $\overrightarrow{BC}$ leads to the same point as the single displacement $\overrightarrow{AC}$. By definition, the vector $\overrightarrow{AC}$ is the **sum** of $\overrightarrow{AB}$ and $\overrightarrow{BC}$, and we write

$$\overrightarrow{AC} = \overrightarrow{AB} + \overrightarrow{BC}.$$

FIGURE 14.4
$\overrightarrow{AC} = \overrightarrow{AB} + \overrightarrow{BC}$

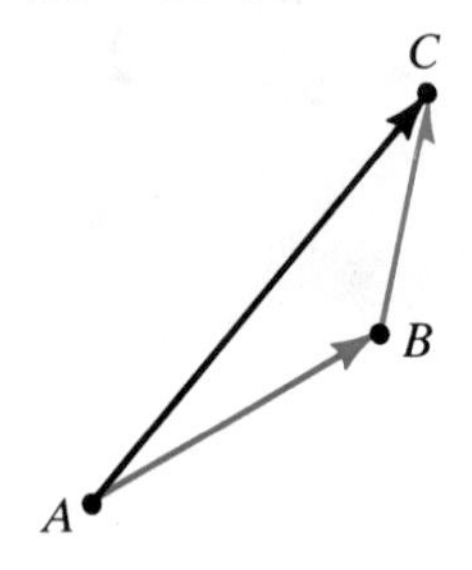

Since vectors may be translated from one location to another, *any* two vectors may be added by placing the initial point of the second vector on the terminal point of the first and then drawing the line segment from the initial point of the first to the terminal point of the second, as in Figure 14.4.

FIGURE 14.5
$\overrightarrow{PS} = \overrightarrow{PQ} + \overrightarrow{PR}$

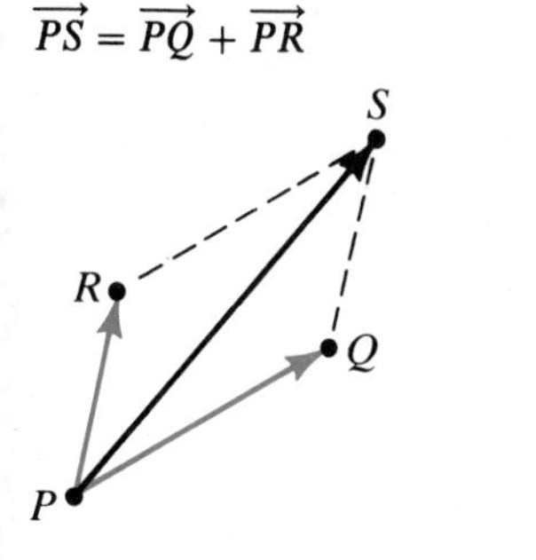

FIGURE 14.6

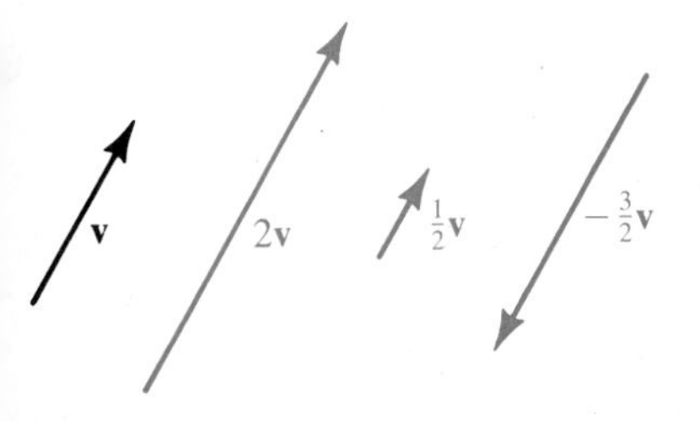

FIGURE 14.7

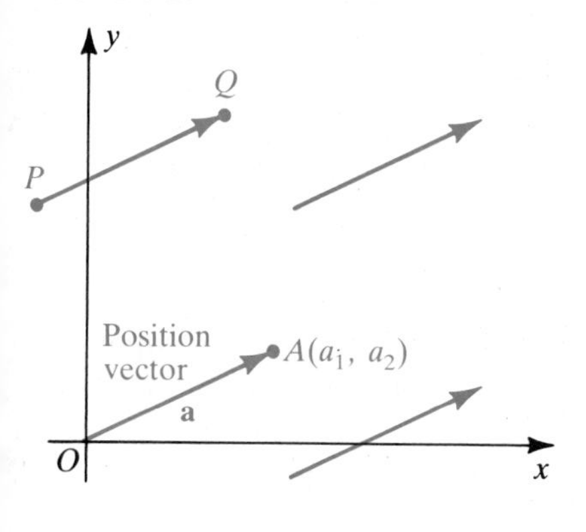

Another way to find the sum is to choose vectors $\overrightarrow{PQ}$ and $\overrightarrow{PR}$ that are equal to $\overrightarrow{AB}$ and $\overrightarrow{BC}$, respectively, and have the same initial point P, as shown in Figure 14.5. If we construct parallelogram $RPQS$, then since $\overrightarrow{PR} = \overrightarrow{QS}$, it follows that $\overrightarrow{PS} = \overrightarrow{PQ} + \overrightarrow{PR}$. If $\overrightarrow{PQ}$ and $\overrightarrow{PR}$ are two forces acting at P, then $\overrightarrow{PS}$ is the **resultant force**—that is, the single force that produces the same effect as the two combined forces.

If c is a scalar and $\mathbf{v}$ is a vector, then $c\mathbf{v}$ is defined as a vector whose magnitude is $|c|$ times the magnitude $\|\mathbf{v}\|$ of $\mathbf{v}$ and whose direction is either the same as that of $\mathbf{v}$ (if $c > 0$) or opposite that of $\mathbf{v}$ (if $c < 0$). Illustrations are given in Figure 14.6. We refer to $c\mathbf{v}$ as a **scalar multiple** of $\mathbf{v}$.

Throughout the remainder of this section we shall restrict our discussion to the special case in which all vectors lie in an xy-plane. If $\overrightarrow{PQ}$ is such a vector, then, as indicated in Figure 14.7, there are many vectors that are equivalent to $\overrightarrow{PQ}$; however, there is precisely *one* equivalent vector $\mathbf{a} = \overrightarrow{OA}$ with initial point at the origin. This vector $\overrightarrow{OA}$ is designated as the **position vector** for $\overrightarrow{PQ}$. Thus, *each vector determines a unique ordered pair of real numbers*, the coordinates (a_1, a_2) of the terminal point of its position vector. Conversely, every ordered pair (a_1, a_2) determines the vector $\overrightarrow{OA}$, where A has coordinates (a_1, a_2). Thus, *there is a one-to-one correspondence between vectors in an xy-plane and ordered pairs of real numbers*. This correspondence allows us to interpret a vector as both a directed line segment *and* an ordered pair. To avoid confusion with the notation for open intervals or points, we use the symbol $\langle a_1, a_2 \rangle$ for an ordered pair that represents a vector, and we denote it by a boldface letter, such as $\mathbf{a} = \langle a_1, a_2 \rangle$. The numbers a_1 and a_2 are the **components** of the vector $\langle a_1, a_2 \rangle$. Two vectors $\langle a_1, a_2 \rangle$ and $\langle b_1, b_2 \rangle$ are **equal** if and only if $a_1 = b_1$ and $a_2 = b_2$. If A is the point (a_1, a_2) as in Figure 14.7, we call $\overrightarrow{OA}$ the **position vector** for $\langle a_1, a_2 \rangle$, or for the *point A*.

The preceding discussion shows that vectors have two different natures, one geometric and the other algebraic. Often we do not distinguish between the two. It should always be clear from our discussion whether we are referring to ordered pairs or directed line segments. We shall use the symbol V_2 for the set of all vectors $\langle x, y \rangle$, where x and y are real numbers.

The **magnitude** $\|\mathbf{a}\|$ of the vector $\mathbf{a} = \langle a_1, a_2 \rangle$ is, by definition, the length of its position vector $\overrightarrow{OA}$, as illustrated in Figure 14.8. Another way of stating this is given in the following definition.

FIGURE 14.8

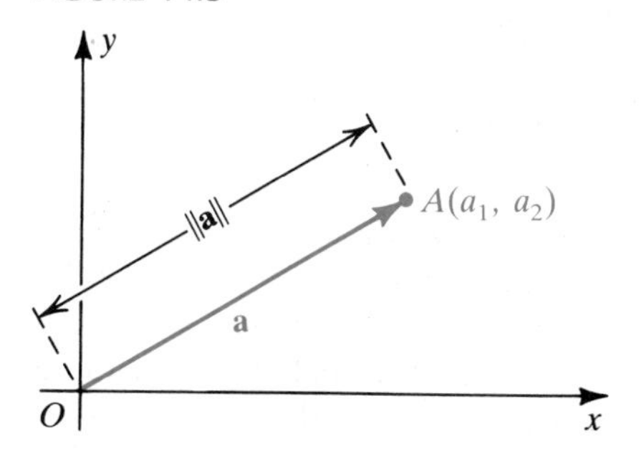

Definition **(14.1)**

The **magnitude** $\|\mathbf{a}\|$ of the vector $\mathbf{a} = \langle a_1, a_2 \rangle$ is

$$\|\mathbf{a}\| = \|\langle a_1, a_2 \rangle\| = \sqrt{a_1^2 + a_2^2}.$$

FIGURE 14.9

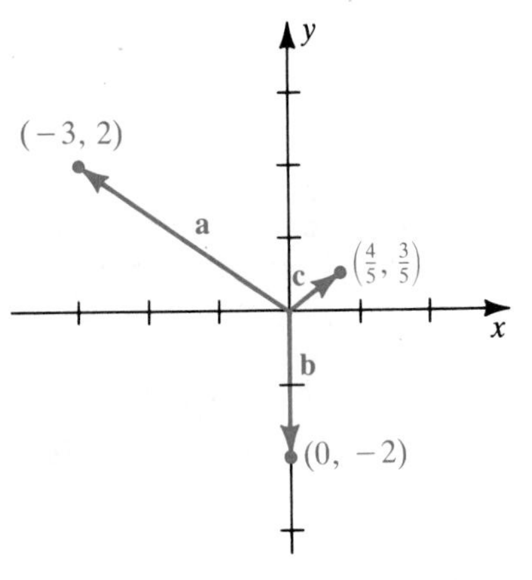

EXAMPLE 1 Sketch the position vectors for

$$\mathbf{a} = \langle -3, 2 \rangle, \quad \mathbf{b} = \langle 0, -2 \rangle, \quad \mathbf{c} = \langle \tfrac{4}{5}, \tfrac{3}{5} \rangle$$

and find the magnitude of each vector.

SOLUTION The position vectors are sketched in Figure 14.9. By Definition (14.1),

$$\|\mathbf{a}\| = \|\langle -3, 2 \rangle\| = \sqrt{(-3)^2 + 2^2} = \sqrt{13}$$
$$\|\mathbf{b}\| = \|\langle 0, -2 \rangle\| = \sqrt{0^2 + (-2)^2} = \sqrt{4} = 2$$
$$\|\mathbf{c}\| = \|\langle \tfrac{4}{5}, \tfrac{3}{5} \rangle\| = \sqrt{(\tfrac{4}{5})^2 + (\tfrac{3}{5})^2} = \sqrt{\tfrac{25}{25}} = 1.$$

FIGURE 14.10

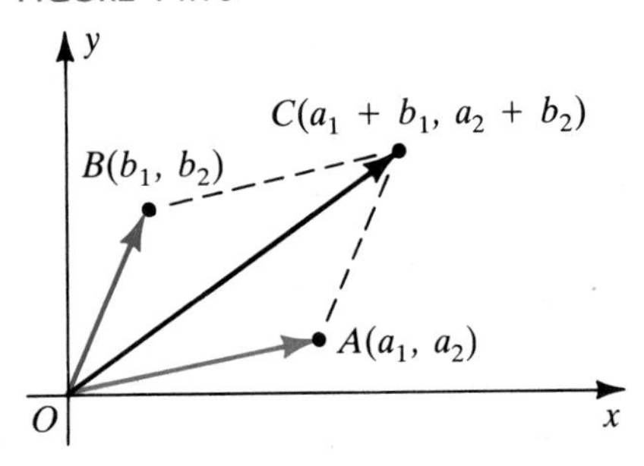

Let $\mathbf{a} = \langle a_1, a_2 \rangle$ and $\mathbf{b} = \langle b_1, b_2 \rangle$ be vectors in V_2, and let $\overrightarrow{OA}$ and $\overrightarrow{OB}$ be their position vectors, as illustrated in Figure 14.10. If $\overrightarrow{OC}$ is the position vector for $\mathbf{c} = \langle a_1 + b_1, a_2 + b_2 \rangle$, then it can be shown, using slopes, that O, A, C, and B are vertices of a parallelogram; that is,

$$\overrightarrow{OA} + \overrightarrow{OB} = \overrightarrow{OC}.$$

This motivates the following definition of addition in V_2.

Addition of vectors **(14.2)**

$$\langle a_1, a_2 \rangle + \langle b_1, b_2 \rangle = \langle a_1 + b_1, a_2 + b_2 \rangle$$

Note that to add two vectors of V_2 we add corresponding components.

It can also be shown that if c is a scalar, then the ordered pair determined by $c\overrightarrow{OA}$ is (ca_1, ca_2), as illustrated in Figure 14.11 for $c > 0$. This suggests the following definition of scalar multiples in V_2.

FIGURE 14.11

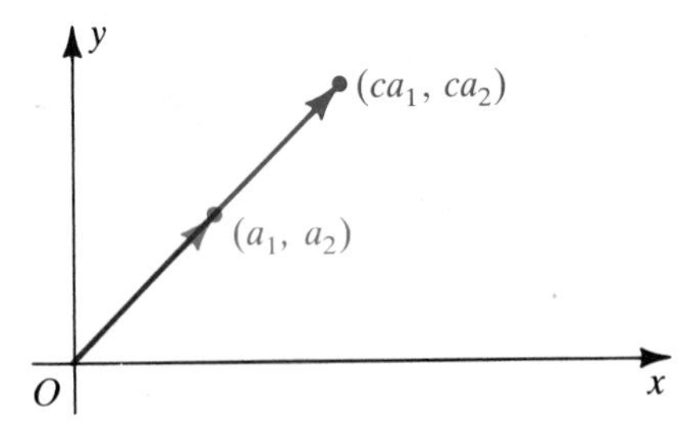

Scalar multiple of a vector **(14.3)**

$$c\langle a_1, a_2 \rangle = \langle ca_1, ca_2 \rangle$$

Thus, to find a scalar multiple of a vector in V_2, we multiply each component by the scalar.

FIGURE 14.12

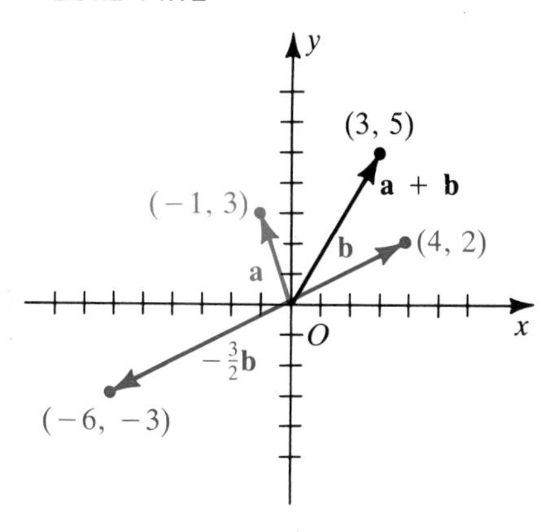

EXAMPLE 2 Let $\mathbf{a} = \langle -1, 3\rangle$ and $\mathbf{b} = \langle 4, 2\rangle$.

(a) Find $\mathbf{a} + \mathbf{b}$, $-\frac{3}{2}\mathbf{b}$, and $2\mathbf{a} + 3\mathbf{b}$.

(b) Sketch position vectors for $\mathbf{a}$, $\mathbf{b}$, $\mathbf{a} + \mathbf{b}$, and $-\frac{3}{2}\mathbf{b}$.

SOLUTION

(a) Applying Definitions (14.2) and (14.3), we have the following:

$$\mathbf{a} + \mathbf{b} = \langle -1, 3\rangle + \langle 4, 2\rangle = \langle -1 + 4, 3 + 2\rangle = \langle 3, 5\rangle$$

$$-\tfrac{3}{2}\mathbf{b} = -\tfrac{3}{2}\langle 4, 2\rangle = \langle -\tfrac{3}{2}(4), -\tfrac{3}{2}(2)\rangle = \langle -6, -3\rangle$$

$$2\mathbf{a} + 3\mathbf{b} = 2\langle -1, 3\rangle + 3\langle 4, 2\rangle = \langle -2, 6\rangle + \langle 12, 6\rangle = \langle 10, 12\rangle$$

(b) Position vectors for $\mathbf{a}$, $\mathbf{b}$, $\mathbf{a} + \mathbf{b}$, and $-\frac{3}{2}\mathbf{b}$ are sketched in Figure 14.12.

The **zero vector 0** and the **negative** $-\mathbf{a}$ of a vector $\mathbf{a} = \langle a_1, a_2\rangle$ are defined as follows.

Definition **(14.4)**

$$\mathbf{0} = \langle 0, 0\rangle \qquad -\mathbf{a} = -\langle a_1, a_2\rangle = \langle -a_1, -a_2\rangle$$

In the next theorem we state some important properties of addition and scalar multiples of vectors for any vectors $\mathbf{a}$, $\mathbf{b}$, and $\mathbf{c}$ in V_2 and scalars c and d. There should be no difficulty in remembering these properties, since they resemble familiar laws of real numbers.

Theorem **(14.5)**

(i) $\mathbf{a} + \mathbf{b} = \mathbf{b} + \mathbf{a}$

(ii) $\mathbf{a} + (\mathbf{b} + \mathbf{c}) = (\mathbf{a} + \mathbf{b}) + \mathbf{c}$

(iii) $\mathbf{a} + \mathbf{0} = \mathbf{a}$

(iv) $\mathbf{a} + (-\mathbf{a}) = \mathbf{0}$

(v) $c(\mathbf{a} + \mathbf{b}) = c\mathbf{a} + c\mathbf{b}$

(vi) $(c + d)\mathbf{a} = c\mathbf{a} + d\mathbf{a}$

(vii) $(cd)\mathbf{a} = c(d\mathbf{a}) = d(c\mathbf{a})$

(viii) $1\mathbf{a} = \mathbf{a}$

(ix) $0\mathbf{a} = \mathbf{0} = c\mathbf{0}$

PROOF Let $\mathbf{a} = \langle a_1, a_2\rangle$ and $\mathbf{b} = \langle b_1, b_2\rangle$. To prove (i), we note that

$$\mathbf{a} + \mathbf{b} = \langle a_1 + b_1, a_2 + b_2\rangle = \langle b_1 + a_1, b_2 + a_2\rangle = \mathbf{b} + \mathbf{a}.$$

The proof of (v) is as follows:

$$\begin{aligned} c(\mathbf{a} + \mathbf{b}) &= c\langle a_1 + b_1, a_2 + b_2\rangle \\ &= \langle ca_1 + cb_1, ca_2 + cb_2\rangle \\ &= \langle ca_1, ca_2\rangle + \langle cb_1, cb_2\rangle \\ &= c\mathbf{a} + c\mathbf{b} \end{aligned}$$

Proofs of the remaining properties are similar and are left as exercises.

The operation of **subtraction** of vectors in V_2, denoted by $-$, is defined as follows.

Definition **(14.6)**

Let $\mathbf{a} = \langle a_1, a_2 \rangle$ and $\mathbf{b} = \langle b_1, b_2 \rangle$. The **difference** $\mathbf{a} - \mathbf{b}$ is

$$\mathbf{a} - \mathbf{b} = \mathbf{a} + (-\mathbf{b}).$$

Using Definitions (14.6), (14.4), and (14.2), we see that

$$\begin{aligned}\mathbf{a} - \mathbf{b} &= \mathbf{a} + (-\mathbf{b}) = \langle a_1, a_2 \rangle + \langle -b_1, -b_2 \rangle \\ &= \langle a_1 - b_1, a_2 - b_2 \rangle.\end{aligned}$$

Thus, to find $\mathbf{a} - \mathbf{b}$, we merely subtract the components of $\mathbf{b}$ from the corresponding components of $\mathbf{a}$.

EXAMPLE 3 If $\mathbf{a} = \langle 5, -4 \rangle$ and $\mathbf{b} = \langle -3, 2 \rangle$, find $\mathbf{a} - \mathbf{b}$ and $2\mathbf{a} - 3\mathbf{b}$.

SOLUTION

$$\mathbf{a} - \mathbf{b} = \langle 5, -4 \rangle - \langle -3, 2 \rangle = \langle 5 - (-3), -4 - 2 \rangle = \langle 8, -6 \rangle$$

$$2\mathbf{a} - 3\mathbf{b} = 2\langle 5, -4 \rangle - 3\langle -3, 2 \rangle = \langle 10, -8 \rangle - \langle -9, 6 \rangle = \langle 19, -14 \rangle$$

FIGURE 14.13

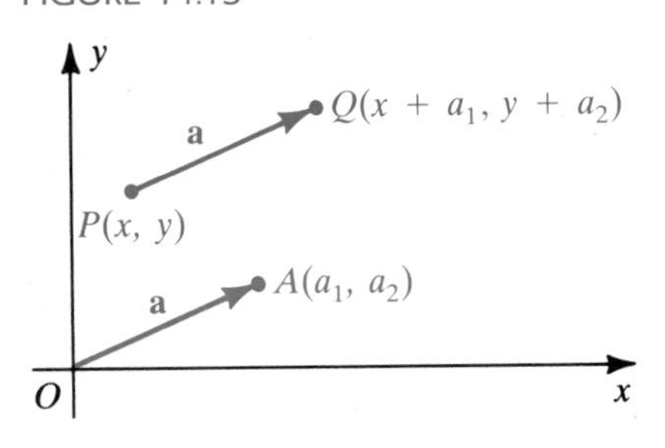

We sometimes represent a nonzero vector $\mathbf{a} = \langle a_1, a_2 \rangle$ of V_2 by a directed line segment $\overrightarrow{PQ}$ with *any* initial point $P(x, y)$ and terminal point $Q(x + a_1, y + a_2)$, as illustrated in Figure 14.13. Note that the position vectors for $\overrightarrow{PQ}$ and $\mathbf{a}$ are both $\overrightarrow{OA}$, and that $\|\overrightarrow{PQ}\| = \|\mathbf{a}\|$. We say that $\overrightarrow{PQ}$ *corresponds to* $\mathbf{a}$, or that $\mathbf{a}$ corresponds to $\overrightarrow{PQ}$. Strictly speaking, $\overrightarrow{PQ}$ *represents* $\mathbf{a}$; however, we often refer to $\overrightarrow{PQ}$ as the *vector* $\mathbf{a}$. The zero vector will be represented by any point in the plane.

We shall make frequent use of the next theorem.

Theorem **(14.7)**

If $P_1(x_1, y_1)$ and $P_2(x_2, y_2)$ are any points, the vector $\mathbf{a}$ in V_2 that corresponds to $\overrightarrow{P_1P_2}$ is

$$\mathbf{a} = \langle x_2 - x_1, y_2 - y_1 \rangle.$$

FIGURE 14.14

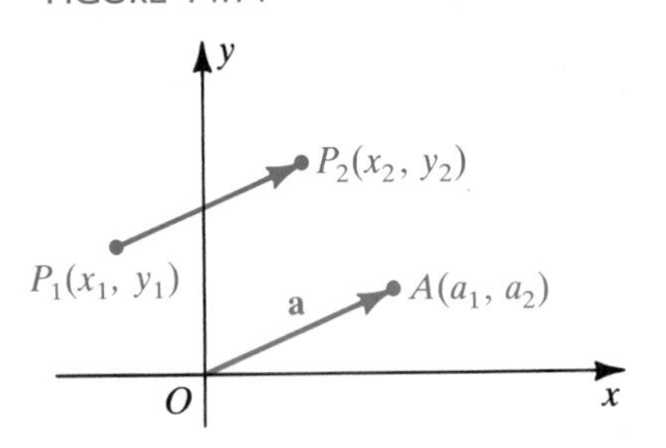

PROOF The vector $\overrightarrow{P_1P_2}$ and the position vector for $\mathbf{a} = \langle a_1, a_2 \rangle$ are illustrated in Figure 14.14. Since $\overrightarrow{P_1P_2}$ corresponds to $\mathbf{a}$, we see that

$$x_2 = x_1 + a_1 \quad \text{and} \quad y_2 = y_1 + a_2.$$

Solving for a_1 and a_2 gives us $\mathbf{a} = \langle x_2 - x_1, y_2 - y_1 \rangle$. ■

EXAMPLE 4 Given the points $P(-2, 3)$ and $Q(4, 5)$,

(a) sketch $\overrightarrow{PQ}$ and $\overrightarrow{QP}$

(b) find vectors $\mathbf{a}$ and $\mathbf{b}$ in V_2 that correspond to $\overrightarrow{PQ}$ and $\overrightarrow{QP}$

(c) sketch the position vectors for $\mathbf{a}$ and $\mathbf{b}$

SOLUTION

(a) The vectors $\overrightarrow{PQ}$ and $\overrightarrow{QP}$ are sketched in Figure 14.15.

(b) By Theorem (14.7),

$$\mathbf{a} = \langle 4 - (-2), 5 - 3 \rangle = \langle 6, 2 \rangle$$
$$\mathbf{b} = \langle -2 - 4, 3 - 5 \rangle = \langle -6, -2 \rangle = -\langle 6, 2 \rangle.$$

(c) The position vectors $\overrightarrow{OA}$ and $\overrightarrow{OB}$ for $\mathbf{a}$ and $\mathbf{b}$ are sketched in Figure 14.15.

FIGURE 14.15

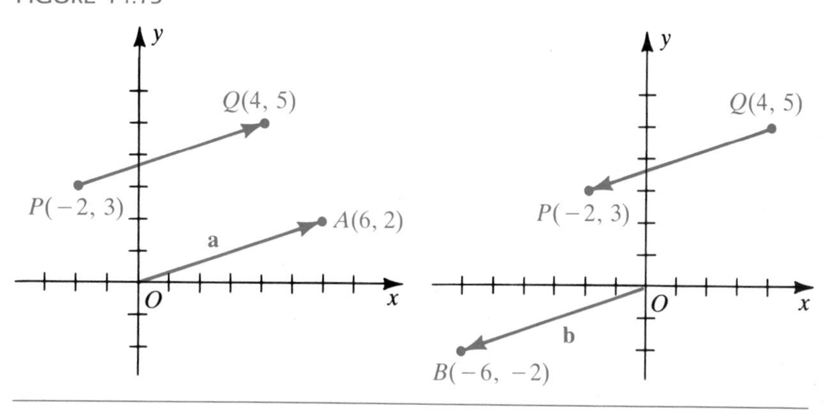

FIGURE 14.16

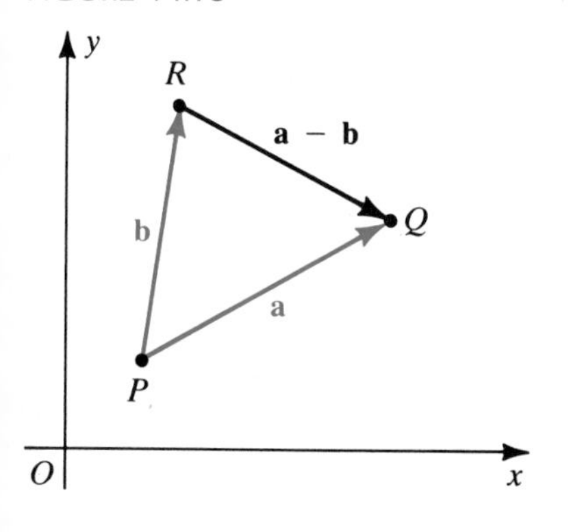

If $\mathbf{a}$ and $\mathbf{b}$ are arbitrary vectors, then

$$\mathbf{b} + (\mathbf{a} - \mathbf{b}) = \mathbf{a};$$

that is, $\mathbf{a} - \mathbf{b}$ *is the vector that, when added to* $\mathbf{b}$, *gives us* $\mathbf{a}$. If we represent $\mathbf{a}$ and $\mathbf{b}$ by vectors $\overrightarrow{PQ}$ and $\overrightarrow{PR}$ *with the same initial point*, as in Figure 14.16, then $\overrightarrow{RQ}$ represents $\mathbf{a} - \mathbf{b}$.

At the beginning of this section we used the symbol $c\overrightarrow{AB}$ to denote a vector whose direction is the same as that of $\overrightarrow{AB}$ if $c > 0$ or opposite that of $\overrightarrow{AB}$ if $c < 0$. The next definition is the analogous concept for vectors in V_2.

Definition **(14.8)**

Nonzero vectors $\mathbf{a}$ and $\mathbf{b}$ in V_2 have

(i) the **same direction** if $\mathbf{b} = c\mathbf{a}$ for some scalar $c > 0$

(ii) the **opposite direction** if $\mathbf{b} = c\mathbf{a}$ for some scalar $c < 0$

FIGURE 14.17

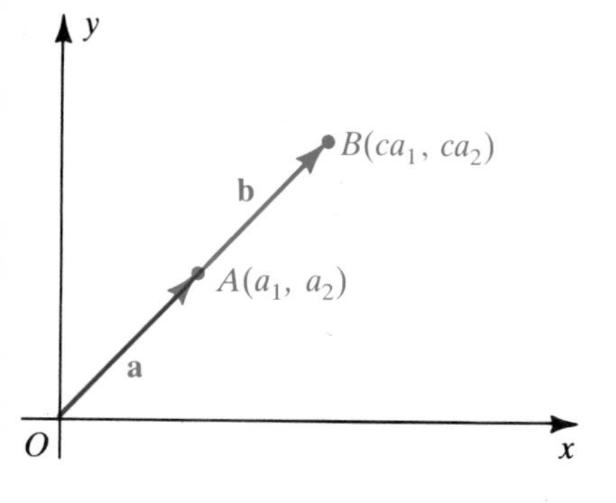

Let $\mathbf{a} = \langle a_1, a_2 \rangle$ and $\mathbf{b} = \langle b_1, b_2 \rangle$. If $\mathbf{b} = c\mathbf{a}$, then $b_1 = ca_1$ and $b_2 = ca_2$, and the position vectors $\overrightarrow{OA}$ and $\overrightarrow{OB}$ for $\mathbf{a}$ and $\mathbf{b}$ lie on the same line through the origin, as illustrated in Figure 14.17. Moreover, if $c > 0$, then A and B are in the same quadrant (or on the same positive or negative coordinate axis). If $c < 0$ and A is not on a coordinate axis, then A and B are in diagonally opposite quadrants.

We shall assume that the zero vector $\mathbf{0}$ has *every* direction. Do not think of this peculiar property as something that can be proved. It is merely a definition of convenience, which will help us state certain results about a vector $\mathbf{a}$ without adding the condition $\mathbf{a} \neq \mathbf{0}$. Remember that $\mathbf{0}$

is also special in another way. It is the only vector that has zero magnitude!

By definition, two vectors **a** and **b** in V_2 are **parallel** if and only if $\mathbf{b} = c\mathbf{a}$ for some scalar c. Thus, nonzero vectors **a** and **b** are parallel if they have the same or opposite direction. Note that the zero vector is parallel to *every* vector **a**, since we can write $\mathbf{0} = 0\mathbf{a}$.

The next theorem specifies the relationship between the magnitudes of **a** and $c\mathbf{a}$.

Theorem **(14.9)**

If **a** is a vector and c is a scalar, then $\|c\mathbf{a}\| = |c|\|\mathbf{a}\|$.

PROOF If $\mathbf{a} = \langle a_1, a_2\rangle$, then $c\mathbf{a} = \langle ca_1, ca_2\rangle$. Using Definition (14.1) and properties of real numbers gives us

$$\begin{aligned}\|c\mathbf{a}\| &= \sqrt{(ca_1)^2 + (ca_2)^2} = \sqrt{c^2(a_1^2 + a_2^2)} \\ &= \sqrt{c^2}\sqrt{a_1^2 + a_2^2} = |c|\|\mathbf{a}\|.\end{aligned}$$

Theorem (14.9) implies that the length of a directed line segment that represents $c\mathbf{a}$ is $|c|$ times the length of a directed line segment that represents **a**. This agrees with the geometric interpretation discussed at the beginning of this section and illustrated in Figure 14.6.

The vectors **i** and **j** defined next will be useful in future developments.

Definition **(14.10)**

$$\mathbf{i} = \langle 1, 0\rangle, \qquad \mathbf{j} = \langle 0, 1\rangle$$

Both **i** and **j** have magnitude 1 since, by Definition (14.1),

$$\|\mathbf{i}\| = \sqrt{1^2 + 0^2} = 1 \quad \text{and} \quad \|\mathbf{j}\| = \sqrt{0^2 + 1^2} = 1.$$

We may use **i** and **j** to obtain another way of denoting vectors in V_2, as indicated by the next result.

i, j form for vectors **(14.11)**

If $\mathbf{a} = \langle a_1, a_2\rangle$, then $\mathbf{a} = a_1\mathbf{i} + a_2\mathbf{j}$.

PROOF Using Definitions (14.1) and (14.2), we have

$$\langle a_1, a_2\rangle = \langle a_1, 0\rangle + \langle 0, a_2\rangle = a_1\langle 1, 0\rangle + a_2\langle 0, 1\rangle = a_1\mathbf{i} + a_2\mathbf{j}.$$

The formula $\mathbf{a} = a_1\mathbf{i} + a_2\mathbf{j}$ for the vector $\mathbf{a} = \langle a_1, a_2\rangle$ has an interesting geometric interpretation. Position vectors for **i**, **j**, and **a** are sketched in Figure 14.18(i). Since **i** and **j** both have magnitude 1, we see from Theorem (14.9) that

$$\|a_1\mathbf{i}\| = |a_1|\|\mathbf{i}\| = |a_1| \quad \text{and} \quad \|a_2\mathbf{j}\| = |a_2|\|\mathbf{j}\| = |a_2|.$$

Hence, as in Figure 14.18(ii), $a_1\mathbf{i}$ and $a_2\mathbf{j}$ may be represented by horizontal and vertical vectors of magnitudes $|a_1|$ and $|a_2|$, respectively. The

FIGURE 14.18

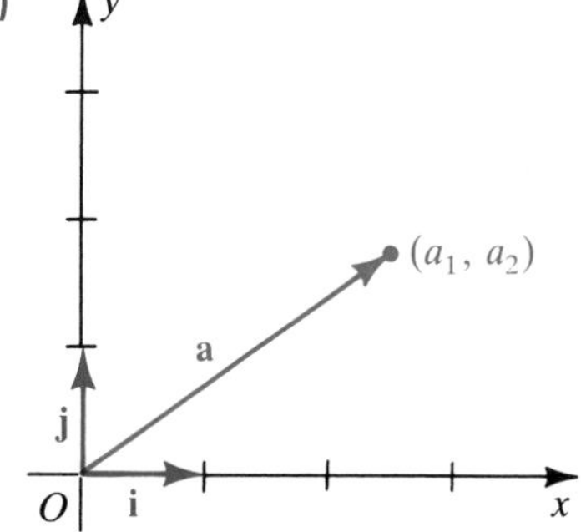

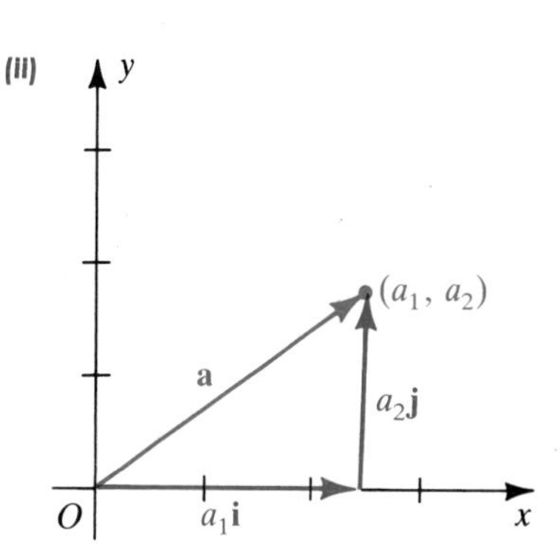

position vector for **a** may be regarded as the sum of these vectors. For this reason a_1 is called the **horizontal component** and a_2 the **vertical component** of the vector **a**.

The vector sum $a_1\mathbf{i} + a_2\mathbf{j}$ is called a **linear combination** of **i** and **j**. Note that if $\mathbf{b} = \langle b_1, b_2 \rangle = b_1\mathbf{i} + b_2\mathbf{j}$ and if c is a scalar,

$$(a_1\mathbf{i} + a_2\mathbf{j}) + (b_1\mathbf{i} + b_2\mathbf{j}) = (a_1 + b_1)\mathbf{i} + (a_2 + b_2)\mathbf{j}$$

$$(a_1\mathbf{i} + a_2\mathbf{j}) - (b_1\mathbf{i} + b_2\mathbf{j}) = (a_1 - b_1)\mathbf{i} + (a_2 - b_2)\mathbf{j}$$

$$c(a_1\mathbf{i} + a_2\mathbf{j}) = (ca_1)\mathbf{i} + (ca_2)\mathbf{j}.$$

The preceding formulas show that we may regard linear combinations of **i** and **j** as ordinary algebraic sums.

EXAMPLE 5 If $\mathbf{a} = 5\mathbf{i} + \mathbf{j}$ and $\mathbf{b} = 4\mathbf{i} - 7\mathbf{j}$, express $3\mathbf{a} - 2\mathbf{b}$ as a linear combination of **i** and **j**.

SOLUTION

$$\begin{aligned} 3\mathbf{a} - 2\mathbf{b} &= 3(5\mathbf{i} + \mathbf{j}) - 2(4\mathbf{i} - 7\mathbf{j}) \\ &= (15\mathbf{i} + 3\mathbf{j}) - (8\mathbf{i} - 14\mathbf{j}) \\ &= 7\mathbf{i} + 17\mathbf{j} \end{aligned}$$

A **unit vector** is a vector of magnitude 1. The vectors **i** and **j** are unit vectors.

Theorem **(14.12)**

> If $\mathbf{a} \neq \mathbf{0}$, then the unit vector **u** that has the same direction as **a** is
>
> $$\mathbf{u} = \frac{1}{\|\mathbf{a}\|}\mathbf{a}.$$

PROOF If we let $c = 1/\|\mathbf{a}\|$, then $c > 0$ and, by Definition (14.8), the vector $\mathbf{u} = c\mathbf{a}$ has the same direction as **a**. Moreover, by Theorem (14.9),

$$\|\mathbf{u}\| = \|c\mathbf{a}\| = |c|\,\|\mathbf{a}\| = \frac{1}{\|\mathbf{a}\|}\|\mathbf{a}\| = 1. \quad \blacksquare$$

EXAMPLE 6 Find the unit vector $\mathbf{u}$ that has the same direction as $3\mathbf{i} - 4\mathbf{j}$.

SOLUTION By Definition (14.1), $\|\mathbf{a}\| = \sqrt{9 + 16} = 5$. Hence, from Theorem (14.12),

$$\mathbf{u} = \tfrac{1}{5}(3\mathbf{i} - 4\mathbf{j}) = \tfrac{3}{5}\mathbf{i} - \tfrac{4}{5}\mathbf{j}.$$

Vectors have many applications in physics and engineering. We shall conclude this section with an example from the field of aerodynamics.

EXAMPLE 7 In air navigation, directions are specified by measuring from the north in a *clockwise* direction. Suppose an airplane with an airspeed of 200 mi/hr is flying in the direction 60°, and a wind is blowing directly from the west at 40 mi/hr. As in Figure 14.19, these velocities may be represented by vectors $\mathbf{v}$ and $\mathbf{w}$ of magnitudes 200 and 40, respectively. The direction of the resultant $\mathbf{v} + \mathbf{w}$ is the **true course** of the airplane relative to the ground, and the magnitude $\|\mathbf{v} + \mathbf{w}\|$ is the **ground speed** of the airplane. Approximate the ground speed to the nearest mile per hour and the true course to the nearest degree.

FIGURE 14.19

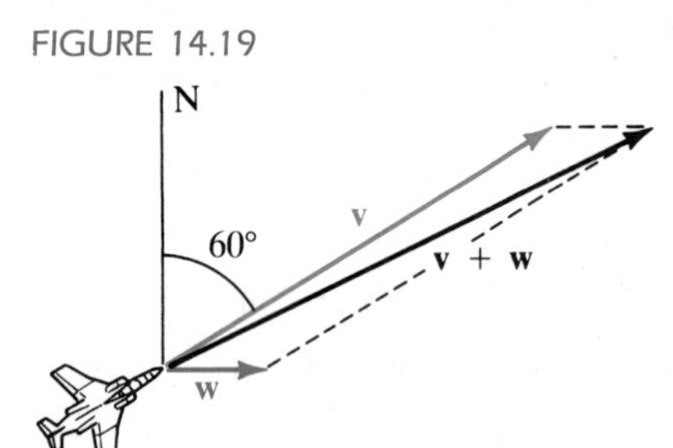

SOLUTION As in Figure 14.20, let us consider an xy-plane with the airplane at the origin, the y-axis on the north-south line, and $\mathbf{v} = \langle v_1, v_2 \rangle$. Since $d(O, A) = \|\mathbf{v}\| = 200$, it follows that

$$v_1 = 200 \cos 30° = 100\sqrt{3} \quad \text{and} \quad v_2 = 200 \sin 30° = 100.$$

FIGURE 14.20

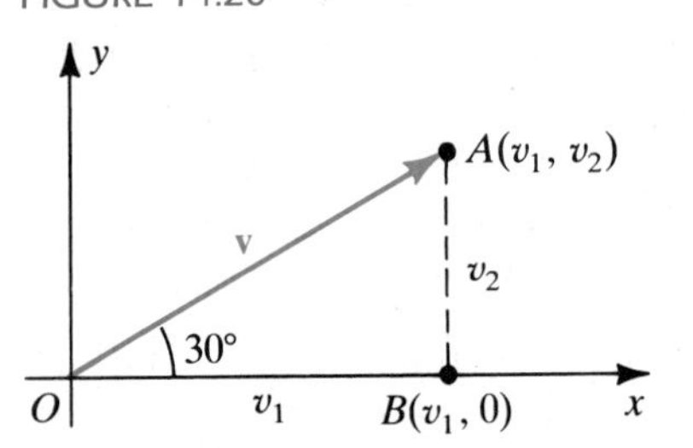

Hence

$$\mathbf{v} = \langle 100\sqrt{3}, 100 \rangle \quad \text{and} \quad \mathbf{w} = \langle 40, 0 \rangle.$$

The resultant is

$$\mathbf{v} + \mathbf{w} = \langle 100\sqrt{3} + 40, 100 \rangle,$$

and the ground speed is

$$\|\mathbf{v} + \mathbf{w}\| = \sqrt{(100\sqrt{3} + 40)^2 + (100)^2} \approx 235 \text{ mi/hr}.$$

If θ is the angle between $\mathbf{v} + \mathbf{w}$ and the positive x-axis, then

$$\tan\theta = \frac{100}{100\sqrt{3} + 40} \approx 0.469 \quad \text{and} \quad \theta \approx \tan^{-1}(0.469) \approx 25°.$$

Hence the true course of the airplane is approximately $90° - 25° = 65°$.

EXERCISES 14.1

Exer. 1–6: Sketch the position vector of a and find $\|\mathbf{a}\|$.

1 $\mathbf{a} = \langle 2, 5 \rangle$

2 $\mathbf{a} = \langle -4, -7 \rangle$

3 $\mathbf{a} = \langle -5, 0 \rangle$

4 $\mathbf{a} = -18\mathbf{j}$

5 $\mathbf{a} = -4\mathbf{i} + 5\mathbf{j}$

6 $\mathbf{a} = 2\mathbf{i} - 3\mathbf{j}$

Exer. 7–12: Find $\mathbf{a} + \mathbf{b}$, $\mathbf{a} - \mathbf{b}$, $2\mathbf{a}$, $-3\mathbf{b}$, and $4\mathbf{a} - 5\mathbf{b}$.

7 $\mathbf{a} = \langle 2, -3 \rangle$, $\mathbf{b} = \langle 1, 4 \rangle$

8 $\mathbf{a} = \langle -2, -5 \rangle$, $\mathbf{b} = \langle -3, 4 \rangle$

9 $\mathbf{a} = -\langle 7, -2 \rangle$, $\mathbf{b} = 4\langle -2, 1 \rangle$

10 $\mathbf{a} = 2\langle 1, 5\rangle$, $\mathbf{b} = -3\langle -1, -4\rangle$

11 $\mathbf{a} = 3\mathbf{i} + 2\mathbf{j}$, $\mathbf{b} = -\mathbf{i} + 5\mathbf{j}$

12 $\mathbf{a} = -5\mathbf{i} + 2\mathbf{j}$, $\mathbf{b} = \mathbf{i} - 3\mathbf{j}$

Exer. 13–18: Use components to express the sum or difference as a scalar multiple of one of the vectors a, b, c, d, e, or f shown in the figure.

EXERCISES 13–18

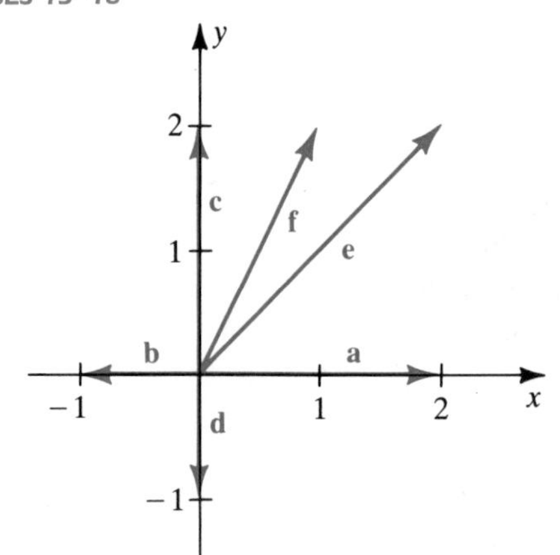

13 $\mathbf{a} + \mathbf{b}$ 14 $\mathbf{c} - \mathbf{d}$ 15 $\mathbf{b} + \mathbf{e}$

16 $\mathbf{f} - \mathbf{b}$ 17 $\mathbf{b} + \mathbf{d}$ 18 $\mathbf{e} + \mathbf{c}$

Exer. 19–24: Find the vector a in V_2 that corresponds to $\overrightarrow{PQ}$. Sketch $\overrightarrow{PQ}$ and the position vector for a.

19 $P(1, -4)$, $Q(5, 3)$

20 $P(7, -3)$, $Q(-2, 4)$

21 $P(2, 5)$, $Q(-4, 5)$

22 $P(-4, 6)$, $Q(-4, -2)$

23 $P(-3, -1)$, $Q(6, -4)$

24 $P(2, 3)$, $Q(-6, 0)$

Exer. 25–28: Find a unit vector that has (a) the same direction as a and (b) the opposite direction of a.

25 $\mathbf{a} = -8\mathbf{i} + 15\mathbf{j}$ 26 $\mathbf{a} = 5\mathbf{i} - 3\mathbf{j}$

27 $\mathbf{a} = \langle 2, -5\rangle$ 28 $\mathbf{a} = \langle 0, 6\rangle$

29 Find a vector that has the same direction as $\langle -6, 3\rangle$ and

(a) twice the magnitude

(b) one-half the magnitude

30 Find a vector that has the opposite direction of $8\mathbf{i} - 5\mathbf{j}$ and

(a) three times the magnitude

(b) one-third the magnitude

31 Find a vector of magnitude 6 that has the same direction as $\mathbf{a} = 4\mathbf{i} - 7\mathbf{j}$.

32 Find a vector of magnitude 4 that has the opposite direction of $\mathbf{a} = \langle 2, -5\rangle$.

Exer. 33–34: Find all real numbers c such that (a) $\|c\mathbf{a}\| = 3$, (b) $\|c\mathbf{a}\| = -3$, and (c) $\|c\mathbf{a}\| = 0$.

33 $\mathbf{a} = 3\mathbf{i} - 4\mathbf{j}$ 34 $\mathbf{a} = \langle -5, 12\rangle$

c **Exer. 35–38: Approximate the horizontal and vertical components of the vector that is described.**

35 A quarterback releases the football with a velocity of 50 ft/sec at an angle of 35° with the horizontal.

36 A child pulls a sled through the snow by exerting a force of 20 pounds at an angle of 40° with the horizontal.

37 The biceps muscle, in supporting the forearm and a weight held in the hand, exerts a force of 20 pounds. As shown in the figure, the muscle makes an angle of 108° with the forearm.

EXERCISE 37

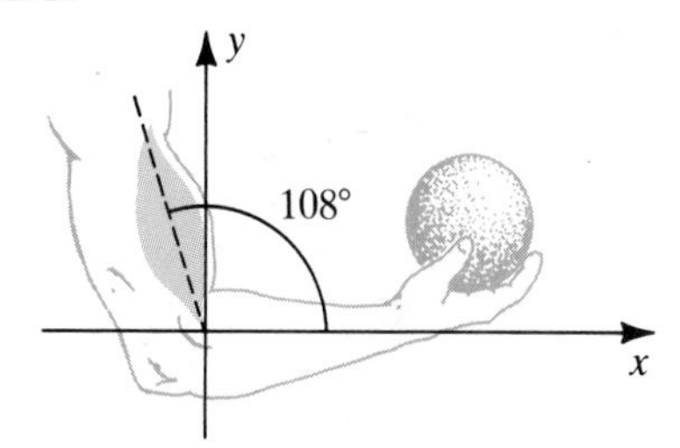

38 A jet airplane approaches a runway at an angle of 7.5° with the horizontal, traveling at a velocity of 160 mi/hr.

39 An airplane is flying in the direction 150° with an airspeed of 300 mi/hr, and the wind is blowing at 30 mi/hr in the direction 60°. Approximate the true course and the ground speed of the airplane.

40 An airplane pilot wishes to maintain a true course in the direction 240° with a ground speed of 400 mi/hr when the wind is blowing directly north at 50 mi/hr. Find the required airspeed and compass heading.

41 Two tugboats are towing a large ship into port, as shown in the figure. The larger tug exerts a force of 4000

EXERCISE 41

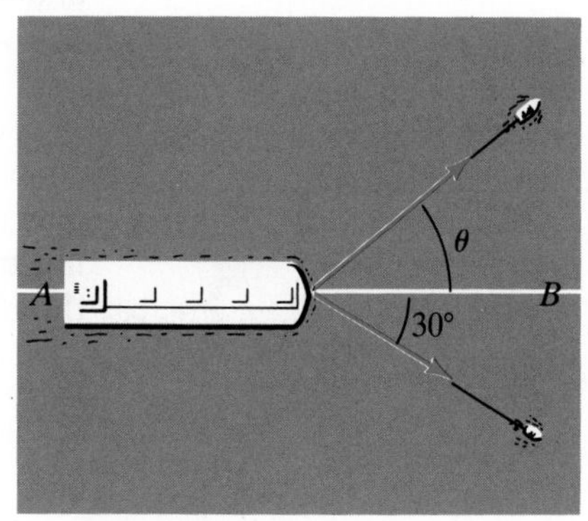

pounds on its cable, and the smaller tug exerts a force of 3200 pounds on its cable. If the ship is to travel in a straight line from A to B, find the angle θ that the larger tug must make with the line segment AB.

c 42 Shown in the figure is an apparatus used to simulate gravity conditions on other planets. A rope is attached to an astronaut who maneuvers on an inclined plane that makes an angle of θ degrees with the horizontal.

(a) If the astronaut weighs 160 pounds, find the x- and y-components of this downward force (see figure for axes).

(b) The y-component in part (a) is the weight of the astronaut relative to the inclined plane. The astronaut would weigh 27 pounds on the moon and 60 pounds on Mars. Approximate each angle θ (to the nearest $0.01°$) so that the inclined-plane apparatus will simulate walking on these surfaces.

EXERCISE 42

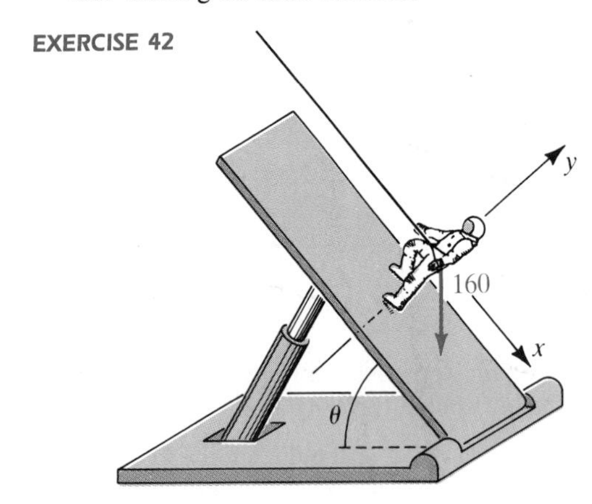

43 Find scalars p and q such that

$$p\langle 3, -1\rangle + q\langle 4, 3\rangle = \langle -6, -11\rangle.$$

44 If $\mathbf{a} = \langle a_1, a_2\rangle$ and $\mathbf{b} = \langle b_1, b_2\rangle$ are any nonzero, nonparallel vectors and if $\mathbf{c}$ is any other vector, prove that there exist scalars p and q such that $\mathbf{c} = p\mathbf{a} + q\mathbf{b}$. Interpret this fact geometrically.

45 Under what conditions is $\|\mathbf{a} + \mathbf{b}\| = \|\mathbf{a}\| + \|\mathbf{b}\|$?

46 (a) Let $\mathbf{a} = \langle a_1, a_2\rangle$ be any nonzero vector, and let $\overrightarrow{OA}$ be the position vector for $\mathbf{a}$. If θ is the smallest nonnegative angle from the positive x-axis to $\overrightarrow{OA}$, show that $\mathbf{a} = \|\mathbf{a}\|(\cos\theta\mathbf{i} + \sin\theta\mathbf{j})$.

(b) Show that every unit vector in V_2 can be expressed in the form $\cos\theta\mathbf{i} + \sin\theta\mathbf{j}$ for some θ.

47 Let $\mathbf{r}_0 = \langle x_0, y_0\rangle$, $\mathbf{r} = \langle x, y\rangle$, and $c > 0$. Describe the set of all points $P(x, y)$ such that $\|\mathbf{r} - \mathbf{r}_0\| = c$.

48 Let $\mathbf{r}_0 = \langle x_0, y_0\rangle$, $\mathbf{r} = \langle x, y\rangle$, and $\mathbf{a} = \langle a_1, a_2\rangle \neq \mathbf{0}$. Describe the set of all points $P(x, y)$ such that $\mathbf{r} - \mathbf{r}_0$ is a scalar multiple of $\mathbf{a}$.

Exer. 49–56: Prove the given property if $\mathbf{a} = \langle a_1, a_2\rangle$, $\mathbf{b} = \langle b_1, b_2\rangle$, $\mathbf{c} = \langle c_1, c_2\rangle$, and p and q are real numbers.

49 $\mathbf{a} + (\mathbf{b} + \mathbf{c}) = (\mathbf{a} + \mathbf{b}) + \mathbf{c}$

50 $\mathbf{a} + \mathbf{0} = \mathbf{a}$

51 $0\mathbf{a} = \mathbf{0} = p\mathbf{0}$

52 $1\mathbf{a} = \mathbf{a}$

53 $(p + q)\mathbf{a} = p\mathbf{a} + q\mathbf{a}$

54 $p(\mathbf{a} - \mathbf{b}) = p\mathbf{a} - p\mathbf{b}$

55 If $p\mathbf{a} = \mathbf{0}$ and $p \neq 0$, then $\mathbf{a} = \mathbf{0}$.

56 If $p\mathbf{a} = \mathbf{0}$ and $\mathbf{a} \neq \mathbf{0}$, then $p = 0$.

14.2 VECTORS IN THREE DIMENSIONS

FIGURE 14.21

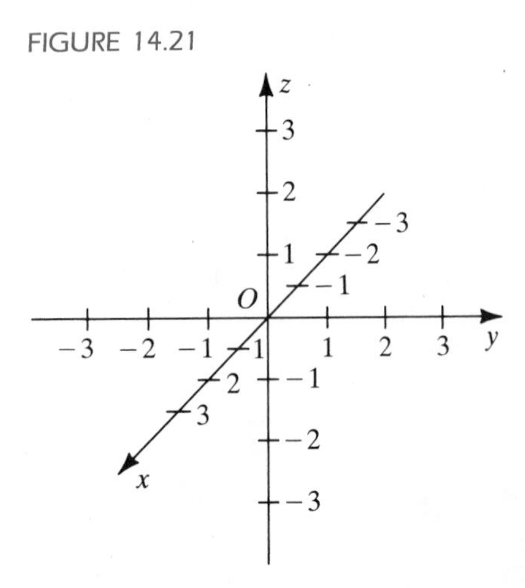

To discuss vectors in space, we use *ordered triples* and a *three-dimensional coordinate system*. An **ordered triple** (a, b, c) is a set $\{a, b, c\}$ of three numbers in which a is considered the first number of the set, b the second, and c the third. Two ordered triples (a_1, a_2, a_3) and (b_1, b_2, b_3) are **equal**, and we write $(a_1, a_2, a_3) = (b_1, b_2, b_3)$, if and only if $a_1 = b_1$, $a_2 = b_2$, and $a_3 = b_3$. The totality of ordered triples is denoted by $\mathbb{R}^3$.

To obtain a rectangular coordinate system in three dimensions, we choose a fixed point O (the **origin**) and consider three mutually perpendicular coordinate lines (the x-, y-, and z-axes) with common origin O, as illustrated in Figure 14.21. We may regard the y- and z-axes as lying in the plane of the paper and the x-axis as projecting out from the paper. The three coordinate lines determine three coordinate planes: the ***xy*-plane**, the ***yz*-plane**, and the ***xz*-plane** shown in Figure 14.22.

The three axes determine a **right-handed** coordinate system. (If the x- and y-axes are interchanged, the coordinate system is **left-handed**.) The

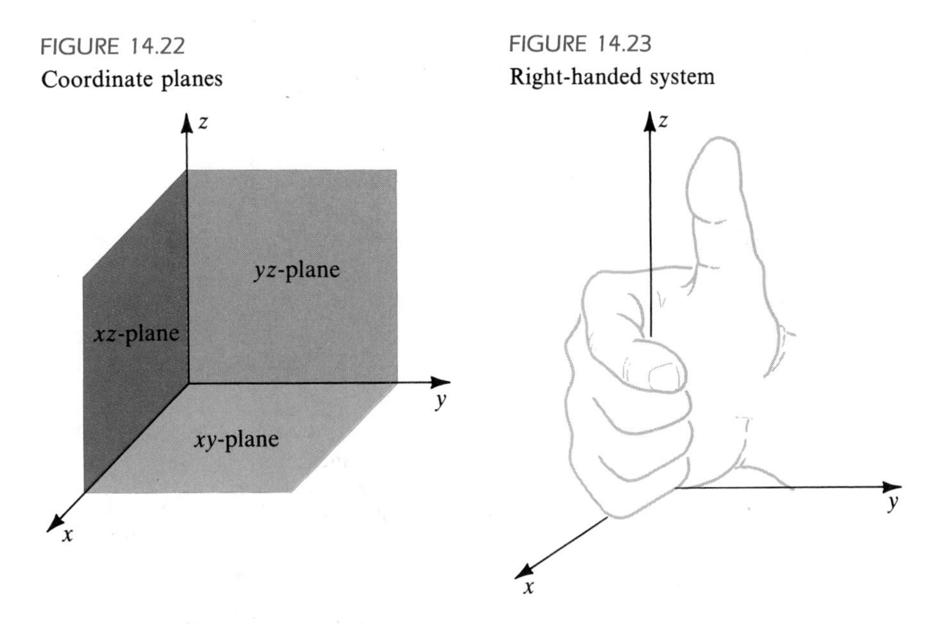

FIGURE 14.22
Coordinate planes

FIGURE 14.23
Right-handed system

terminology *right-handed* may be justified as follows. If, as in Figure 14.23, the fingers of the right hand are curled in the direction of a 90° rotation of axes in the xy-plane (so that the positive x-axis is transformed into the positive y-axis), then the extended thumb points in the direction of the positive z-axis. In this text we usually use right-handed coordinate systems.

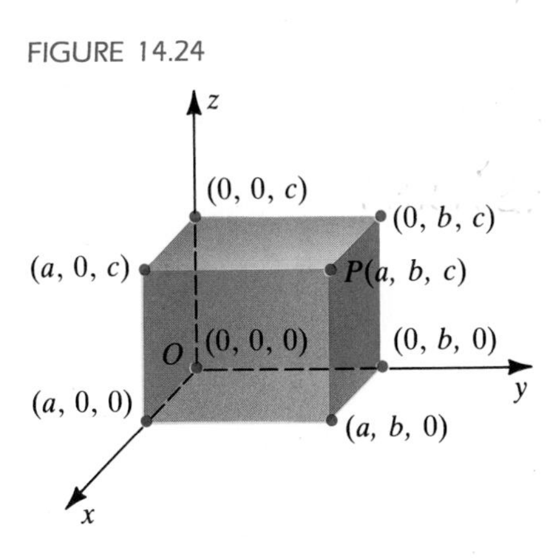

FIGURE 14.24

If P is a point, then the (perpendicular) projection of P on the x-axis has coordinate a, the **x-coordinate** of P. Similarly, the coordinates b and c of the projections of P on the y- and z-axes, respectively, are the **y-coordinate** and **z-coordinate** of P. We use $P(a, b, c)$ or (a, b, c) to denote the point P with coordinates a, b, and c. If P is not on a coordinate plane, then the three planes through P that are parallel to the coordinate planes, together with the coordinate planes, determine a rectangular box of the type illustrated in Figure 14.24.

The concept of plotting points in three dimensions is similar to that in two dimensions. Several points are plotted in Figure 14.25. To plot $(3, 4, -2)$, we first locate $(3, 4, 0)$ on the xy-plane and then move *down* 2 units. To plot $(-2, 3, 4)$, we first locate $(-2, 3, 0)$ on the xy-plane and then move *up* 4 units.

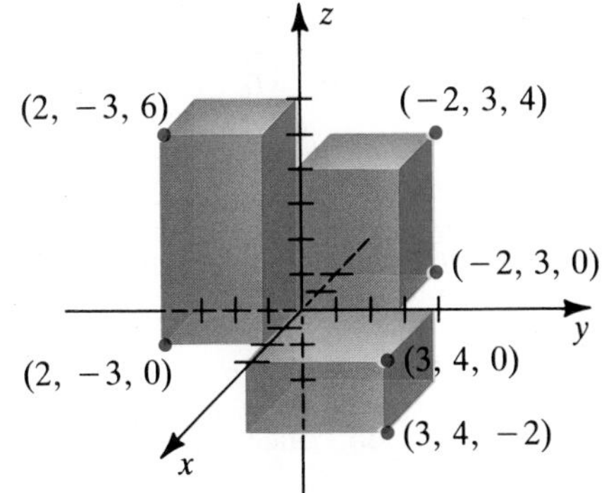

FIGURE 14.25

The one-to-one correspondence between points in space and ordered triples of real numbers is a **rectangular coordinate system in three dimensions**, or an ***xyz*-coordinate system**. The three coordinate planes partition space into eight parts, called **octants**. The part consisting of all points $P(a, b, c)$ with a, b, and c positive is the **first octant**. The remaining octants are usually not numbered.

Let us derive a formula for the distance $d(P_1, P_2)$ between two points P_1 and P_2. If P_1 and P_2 are on a line parallel to the z-axis, as in Figure 14.26 on the following page, and if their projections on the z-axis are $A_1(0, 0, z_1)$ and $A_2(0, 0, z_2)$, respectively, then $d(P_1, P_2) = d(A_1, A_2) = |z_2 - z_1|$. Similar formulas hold if the line through P_1 and P_2 is parallel to the x- or y-axis.

FIGURE 14.26

FIGURE 14.27

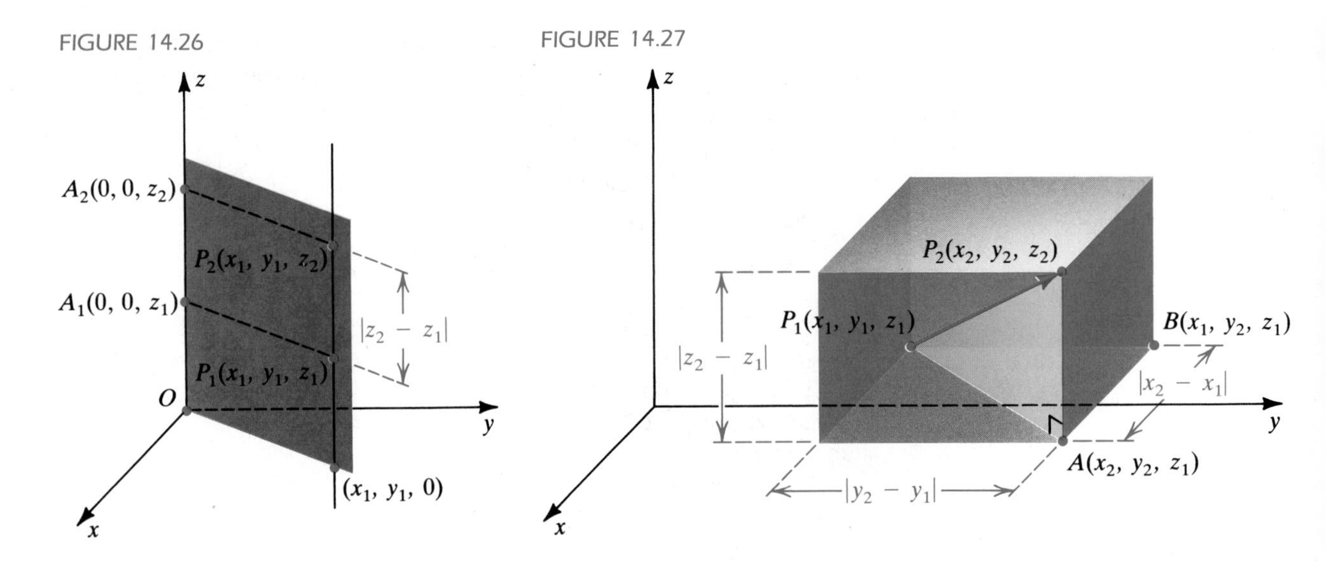

To find the distance between two points P_1 and P_2 that are not parallel to an axis, we consider a situation of the type illustrated in Figure 14.27. Triangle P_1AP_2 is a right triangle, and hence, by the Pythagorean theorem,

$$d(P_1, P_2) = \sqrt{[d(P_1, A)]^2 + [d(A, P_2)]^2}.$$

Since P_1 and A are in a plane parallel to the xy-plane, the distance formula (1.4) gives us $[d(P_1, A)]^2 = (x_2 - x_1)^2 + (y_2 - y_1)^2$. From our previous remarks, $[d(A, P_2)]^2 = (z_2 - z_1)^2$. Substituting in the preceding formula for $d(P_1, P_2)$, we obtain the following.

Distance formula **(14.13)**

> The distance between $P_1(x_1, y_1, z_1)$ and $P_2(x_2, y_2, z_2)$ is
> $$d(P_1, P_2) = \sqrt{(x_2 - x_1)^2 + (y_2 - y_1)^2 + (z_2 - z_1)^2}.$$

If P_1 and P_2 are on the xy-plane so that $z_1 = z_2 = 0$, then (14.13) reduces to the two-dimensional distance formula (1.4).

EXAMPLE 1 Find the distance between $A(-1, -3, 1)$ and $B(3, 4, -2)$.

SOLUTION Using the distance formula (14.13), we have

$$d(A, B) = \sqrt{(3 + 1)^2 + (4 + 3)^2 + (-2 - 1)^2}$$
$$= \sqrt{16 + 49 + 9} = \sqrt{74}.$$

FIGURE 14.28

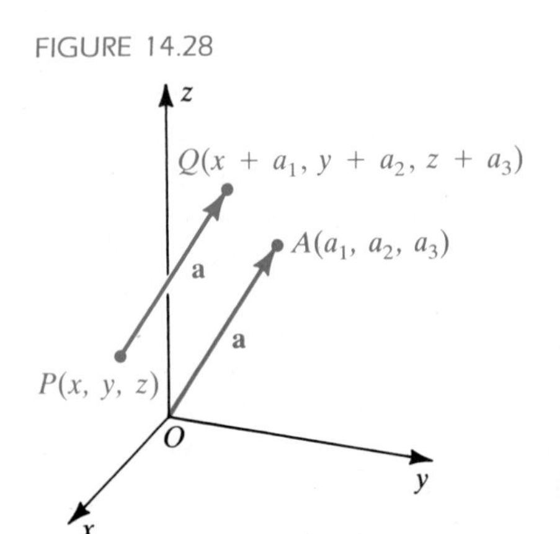

To discuss vectors in three dimensions, we include a third component and use symbols of the form

$$\mathbf{a} = \langle a_1, a_2, a_3 \rangle, \qquad \mathbf{b} = \langle b_1, b_2, b_3 \rangle.$$

The set of all such vectors is denoted by V_3. Each concept discussed in the preceding section for V_2 readily extends to V_3. For example, if A is the point (a_1, a_2, a_3), we refer to the position vector $\overrightarrow{OA}$ for **a**, illustrated in Figure 14.28, which also shows another vector $\overrightarrow{PQ}$ that corresponds to

a. As in Section 14.1, we seldom distinguish between the geometric and algebraic natures of vectors. Magnitude of a vector, vector addition, and scalar multiples of vectors are defined as before by simply including an additional component as follows:

$$\|\mathbf{a}\| = \sqrt{a_1^2 + a_2^2 + a_3^2}$$

$$\mathbf{a} + \mathbf{b} = \langle a_1 + b_1, a_2 + b_2, a_3 + b_3 \rangle$$

$$c\mathbf{a} = \langle ca_1, ca_2, ca_3 \rangle$$

The zero vector in V_3 is $\mathbf{0} = \langle 0, 0, 0 \rangle$, and the negative of $\mathbf{a}$ is $-\mathbf{a} = \langle -a_1, -a_2, -a_3 \rangle$. The theorems and definitions stated in (14.5)–(14.9) extend to V_3, the only change being the inclusion of a third component for each vector. The following examples contain specific illustrations of these concepts.

EXAMPLE 2 If $\mathbf{a} = \langle 2, 5, -3 \rangle$ and $\mathbf{b} = \langle -4, 1, 7 \rangle$, find $\mathbf{a} + \mathbf{b}$, $2\mathbf{a} - 3\mathbf{b}$, and $\|\mathbf{a}\|$.

SOLUTION

$$\mathbf{a} + \mathbf{b} = \langle 2, 5, -3 \rangle + \langle -4, 1, 7 \rangle = \langle -2, 6, 4 \rangle$$

$$\begin{aligned} 2\mathbf{a} - 3\mathbf{b} &= 2\langle 2, 5, -3 \rangle - 3\langle -4, 1, 7 \rangle = \langle 4, 10, -6 \rangle - \langle -12, 3, 21 \rangle \\ &= \langle 16, 7, -27 \rangle \end{aligned}$$

$$\|\mathbf{a}\| = \sqrt{(2)^2 + (5)^2 + (-3)^2} = \sqrt{38}$$

EXAMPLE 3 If $\mathbf{a} = \langle 15, -6, 24 \rangle$, $\mathbf{b} = \langle 5, -2, 8 \rangle$, and $\mathbf{c} = \langle -\frac{15}{2}, 3, -12 \rangle$, show that $\mathbf{a}$ and $\mathbf{b}$ have the same direction and that $\mathbf{a}$ and $\mathbf{c}$ have opposite directions.

SOLUTION We see, by inspection, that

$$\mathbf{a} = 3\mathbf{b}, \quad \text{or} \quad \mathbf{b} = \tfrac{1}{3}\mathbf{a}.$$

Since the scalar 3 (or $\frac{1}{3}$) is positive, $\mathbf{a}$ and $\mathbf{b}$ have the same direction.

Again, by inspection,

$$\mathbf{a} = -2\mathbf{c}, \quad \text{or} \quad \mathbf{c} = -\tfrac{1}{2}\mathbf{a}.$$

Since the scalar -2 (or $-\frac{1}{2}$) is negative, $\mathbf{a}$ and $\mathbf{c}$ have opposite directions.

EXAMPLE 4 Given the points $P_1(5, 6, -2)$ and $P_2(-3, 8, 7)$, find the vector $\mathbf{a}$ in V_3 that corresponds to $\overrightarrow{P_1P_2}$.

SOLUTION By the analogue of Theorem (14.7) in V_3,

$$\mathbf{a} = \langle -3 - 5, 8 - 6, 7 + 2 \rangle = \langle -8, 2, 9 \rangle.$$

A vector $\mathbf{a}$ in V_3 is a **unit vector** if $\|\mathbf{a}\| = 1$. The special unit vectors

$$\mathbf{i} = \langle 1, 0, 0 \rangle, \qquad \mathbf{j} = \langle 0, 1, 0 \rangle, \qquad \mathbf{k} = \langle 0, 0, 1 \rangle$$

FIGURE 14.29

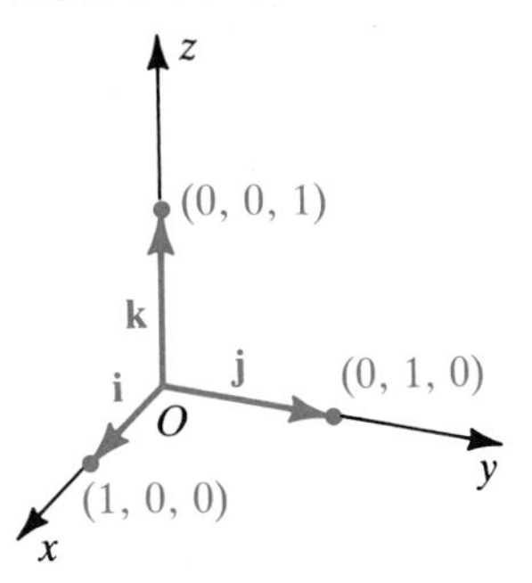

are important, since any vector $\mathbf{a} = \langle a_1, a_2, a_3 \rangle$ can be expressed as a linear combination of **i**, **j**, and **k**. Specifically,

$$\mathbf{a} = \langle a_1, a_2, a_3 \rangle = a_1\mathbf{i} + a_2\mathbf{j} + a_3\mathbf{k}.$$

Position vectors for **i**, **j**, and **k** are sketched in Figure 14.29. Figure 14.30 illustrates how the position vector for $\mathbf{a} = \langle a_1, a_2, a_3 \rangle$ may be regarded as the sum of three vectors corresponding to $a_1\mathbf{i}$, $a_2\mathbf{j}$, and $a_3\mathbf{k}$. It is often convenient to regard V_2 as a subset of V_3 by identifying the vector $\langle a_1, a_2 \rangle$ with $\langle a_1, a_2, 0 \rangle$. Thus, the unit vectors $\mathbf{i} = \langle 1, 0 \rangle$ and $\mathbf{j} = \langle 0, 1 \rangle$ are essentially the same as $\mathbf{i} = \langle 1, 0, 0 \rangle$ and $\mathbf{j} = \langle 0, 1, 0 \rangle$.

FIGURE 14.30

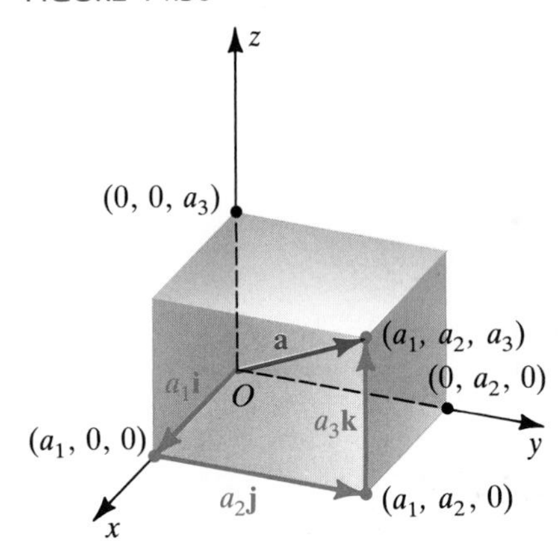

EXAMPLE 5 Express $\mathbf{a} = \langle 3, -4, 2 \rangle$ in terms of **i**, **j**, and **k**, and find the unit vector **u** that has the same direction as **a**.

SOLUTION We may write

$$\mathbf{a} = \langle 3, -4, 2 \rangle = 3\mathbf{i} - 4\mathbf{j} + 2\mathbf{k}.$$

The magnitude of **a** is

$$\|\mathbf{a}\| = \sqrt{3^2 + (-4)^2 + 2^2} = \sqrt{29}.$$

As in Theorem (14.12), a unit vector **u** having the same direction as **a** is

$$\mathbf{u} = \frac{1}{\|\mathbf{a}\|}\mathbf{a} = \frac{1}{\sqrt{29}}(3\mathbf{i} - 4\mathbf{j} + 2\mathbf{k}) = \frac{3}{\sqrt{29}}\mathbf{i} - \frac{4}{\sqrt{29}}\mathbf{j} + \frac{2}{\sqrt{29}}\mathbf{k}.$$

We shall conclude this section by using vectors to derive several useful formulas. The first may be stated as follows.

Midpoint formula **(14.14)**

> The midpoint of the line segment from $P_1(x_1, y_1, z_1)$ to $P_2(x_2, y_2, z_2)$ is
>
> $$\left(\frac{x_1 + x_2}{2}, \frac{y_1 + y_2}{2}, \frac{z_1 + z_2}{2}\right).$$

FIGURE 14.31

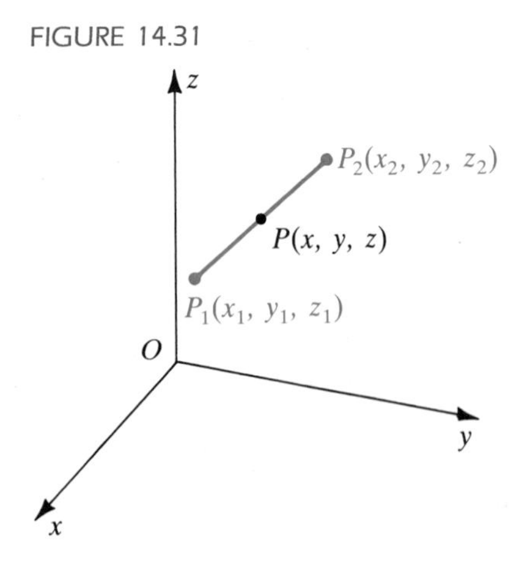

PROOF Let $P(x, y, z)$ be the midpoint of the line segment P_1P_2 (see Figure 14.31). The fact that P is halfway from P_1 to P_2 may be expressed in vector form as follows:

$$\overrightarrow{P_1P} = \tfrac{1}{2}\overrightarrow{P_1P_2},$$

or $$\langle x - x_1, y - y_1, z - z_1 \rangle = \tfrac{1}{2}\langle x_2 - x_1, y_2 - y_1, z_2 - z_1 \rangle$$

Equating components, we have

$$x - x_1 = \tfrac{1}{2}(x_2 - x_1), \qquad y - y_1 = \tfrac{1}{2}(y_2 - y_1), \qquad z - z_1 = \tfrac{1}{2}(z_2 - z_1).$$

Solving these three equations for x, y, and z gives us the following coordinates of the midpoint P:

$$x = \tfrac{1}{2}(x_1 + x_2), \qquad y = \tfrac{1}{2}(y_1 + y_2), \qquad z = \tfrac{1}{2}(z_1 + z_2) \quad ■$$

EXAMPLE 6 Find the midpoint of the line segment from $(2, -3, 6)$ to $(3, 4, -2)$.

SOLUTION The points are plotted in Figure 14.25. By Theorem (14.14), the midpoint is

$$\left(\frac{2+3}{2}, \frac{-3+4}{2}, \frac{6+(-2)}{2}\right) = \left(\frac{5}{2}, \frac{1}{2}, 2\right).$$

FIGURE 14.32

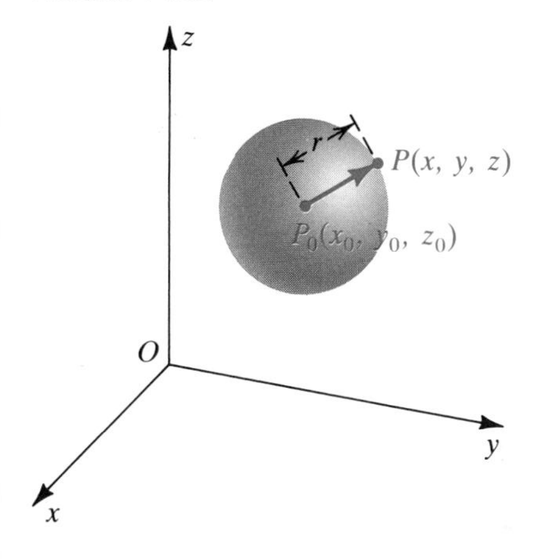

The **graph of an equation** in three variables x, y, and z is the set of all points $P(a, b, c)$ in a rectangular coordinate system such that the ordered triple (a, b, c) is a solution of the equation; that is, equality is obtained when a, b, and c are substituted for x, y, and z, respectively. The graph of such an equation is a **surface**. We can easily derive an equation for a sphere of radius r with center at $P_0(x_0, y_0, z_0)$. As illustrated in Figure 14.32, a point $P(x, y, z)$ is on the sphere if and only if $\|\overrightarrow{P_0P}\| = r$. An equivalent statement is

$$\sqrt{(x-x_0)^2 + (y-y_0)^2 + (z-z_0)^2} = r.$$

Squaring both sides of the last equation gives us the following theorem.

Theorem **(14.15)**

An equation of a sphere of radius r and center $P_0(x_0, y_0, z_0)$ is

$$(x-x_0)^2 + (y-y_0)^2 + (z-z_0)^2 = r^2.$$

If we square the expressions in (14.15) and simplify, then an equation of a sphere may be written in the form

$$x^2 + y^2 + z^2 + ax + by + cz + d = 0,$$

where a, b, c, and d are real numbers. Conversely, if we begin with an equation of this form and if the graph exists, then by completing squares we can obtain the form in (14.15), and hence the graph is a sphere or a point.

EXAMPLE 7 Discuss the graph of the equation

$$x^2 + y^2 + z^2 - 6x + 8y + 4z + 4 = 0.$$

SOLUTION We complete squares as follows:

$$(x^2 - 6x) + (y^2 + 8y) + (z^2 + 4z) = -4$$

$$(x^2 - 6x + 9) + (y^2 + 8y + 16) + (z^2 + 4z + 4) = -4 + 9 + 16 + 4$$

$$(x-3)^2 + (y+4)^2 + (z+2)^2 = 25$$

Comparing the last equation with (14.15), we see that the graph is a sphere of radius 5 with center $(3, -4, -2)$.

EXERCISES 14.2

Exer. 1–6: Plot A and B and find (a) $d(A, B)$, (b) the midpoint of AB, and (c) the vector in V_3 that corresponds to $\overrightarrow{AB}$.

1 $A(2, 4, -5)$, $B(4, -2, 3)$

2 $A(1, -2, 7)$, $B(2, 4, -1)$

3 $A(-4, 0, 1)$, $B(3, -2, 1)$

4 $A(0, 5, -4)$, $B(1, 1, 0)$

5 $A(1, 0, 0)$, $B(0, 1, 1)$

6 $A(0, 0, 0)$, $B(-8, -1, 4)$

Exer. 7–12: Find (a) $\mathbf{a} + \mathbf{b}$, (b) $\mathbf{a} - \mathbf{b}$, (c) $5\mathbf{a} - 4\mathbf{b}$, (d) $\|\mathbf{a}\|$, and (e) $\|-3\mathbf{a}\|$.

7 $\mathbf{a} = \langle -2, 6, 1\rangle$, $\mathbf{b} = \langle 3, -3, -1\rangle$

8 $\mathbf{a} = \langle 1, 2, -3\rangle$, $\mathbf{b} = \langle -4, 0, 1\rangle$

9 $\mathbf{a} = 3\mathbf{i} - 4\mathbf{j} + 2\mathbf{k}$, $\mathbf{b} = \mathbf{i} + 2\mathbf{j} - 5\mathbf{k}$

10 $\mathbf{a} = 2\mathbf{i} - \mathbf{j} + 4\mathbf{k}$, $\mathbf{b} = \mathbf{i} - \mathbf{k}$

11 $\mathbf{a} = \mathbf{i} + \mathbf{j}$, $\mathbf{b} = -\mathbf{j} + \mathbf{k}$

12 $\mathbf{a} = 2\mathbf{i}$, $\mathbf{b} = 3\mathbf{k}$

Exer. 13–14: Sketch position vectors for $\mathbf{a}$, $\mathbf{b}$, $2\mathbf{a}$, $-3\mathbf{b}$, $\mathbf{a} + \mathbf{b}$, and $\mathbf{a} - \mathbf{b}$.

13 $\mathbf{a} = \langle 2, 3, 4\rangle$, $\mathbf{b} = \langle 1, -2, 2\rangle$

14 $\mathbf{a} = -\mathbf{i} + 2\mathbf{j} + 3\mathbf{k}$, $\mathbf{b} = -2\mathbf{j} + \mathbf{k}$

Exer. 15–16: Find the unit vector that has the same direction as a.

15 $\mathbf{a} = 2\langle -2, 5, -1\rangle$

16 $\mathbf{a} = 3\mathbf{i} - 7\mathbf{j} + 2\mathbf{k}$

Exer. 17–18: Find the vector that has (a) the same direction as a and twice the magnitude of a, (b) the opposite direction of a and one-third the magnitude of a, and (c) the same direction as a and magnitude 2.

17 $\mathbf{a} = 14\mathbf{i} - 15\mathbf{j} + 6\mathbf{k}$

18 $\mathbf{a} = \langle -6, -3, 6\rangle$

Exer. 19–22: Find an equation of the sphere with center C and radius r.

19 $C(3, -1, 2)$; $r = 3$

20 $C(4, -5, 1)$; $r = 5$

21 $C(-5, 0, 1)$; $r = \frac{1}{2}$

22 $C(0, -3, -6)$; $r = \sqrt{3}$

Exer. 23–24: Find an equation of the sphere with center C that is tangent to (a) the xy-plane, (b) the xz-plane, and (c) the yz-plane.

23 $C(-2, 4, -6)$

24 $C(3, -1, 2)$

Exer. 25–26: Find an equation of the sphere that has endpoints of a diameter at A and B.

25 $A(1, 4, -2)$, $B(-7, 1, 2)$

26 $A(4, -3, 4)$, $B(0, 3, -10)$

27 Find an equation of the sphere that has radius 1 and center in the first octant and is tangent to each coordinate plane.

28 Find an equation of the sphere that has center $(2, 3, -1)$ and contains the point $(1, 7, -9)$.

Exer. 29–34: Find the center and radius of the sphere.

29 $x^2 + y^2 + z^2 + 4x - 2y + 2z + 2 = 0$

30 $x^2 + y^2 + z^2 - 6x - 10y + 6z + 34 = 0$

31 $x^2 + y^2 + z^2 - 8x + 8z + 16 = 0$

32 $4x^2 + 4y^2 + 4z^2 - 4x + 8y - 3 = 0$

33 $x^2 + y^2 + z^2 + 4y = 0$

34 $x^2 + y^2 + z^2 - z = 0$

Exer. 35–42: Describe the region R in a three-dimensional coordinate system.

35 $R = \{(x, y, z): x^2 + y^2 + z^2 \le 1\}$

36 $R = \{(x, y, z): x^2 + y^2 + z^2 > 1\}$

37 $R = \{(x, y, z): |x| \le 1, |y| \le 2, |z| \le 3\}$

38 $R = \{(x, y, z): |x| \ge 1, |y| \ge 2, |z| \ge 3\}$

39 $R = \{(x, y, z): x^2 + y^2 \le 25, |z| \le 3\}$

40 $R = \{(x, y, z): (x^2/4) + (y^2/9) \ge 1\}$

41 $R = \{(x, y, z): xyz \ne 0\}$

42 $R = \{(x, y, z): 4 < x^2 + y^2 + z^2 < 9\}$

43 Show that the line segments P_1P_1', P_2P_2', and P_3P_3' (see figure) that join the midpoints of the opposite edges of a

EXERCISE 43

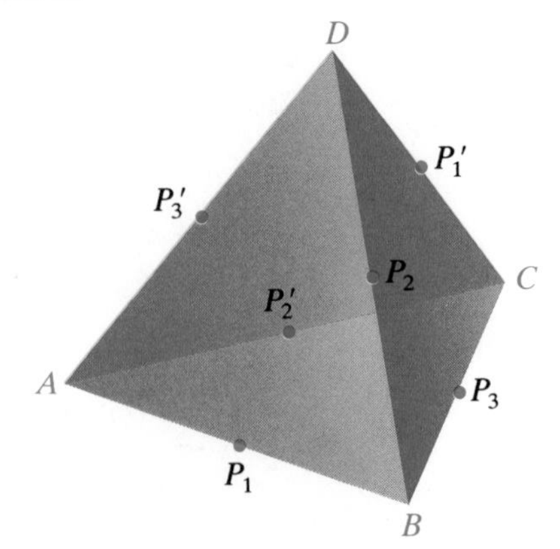

tetrahedron intersect at a point P, and each is bisected by that point. (*Hint:* Place vertex D on the z-axis and triangle ABC in the xy-plane.)

44 A cube has edge of length k. The centers of the six faces of the cube are the vertices of an octahedron, as shown in the figure.

(a) Find the coordinates of each vertex of the octahedron.

(b) Find the length of each edge of the octahedron in terms of k.

EXERCISE 44

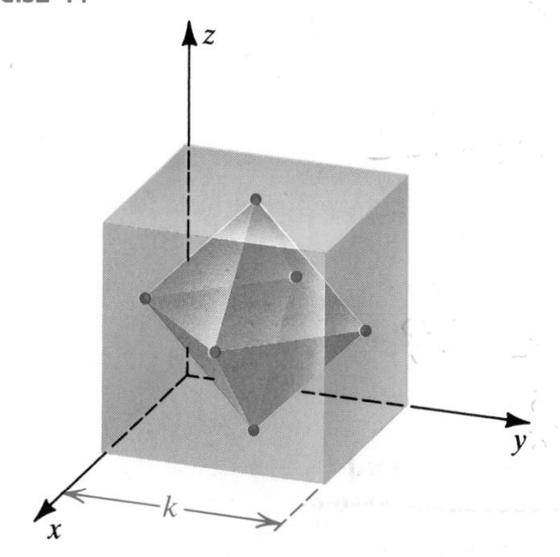

45 *Coulomb's law* states that the magnitude of the force of attraction between two oppositely charged particles is directly proportional to the product of the magnitudes q_1 and q_2 of the charges and inversely proportional to the square of the distance d between them. Show that if a particle with charge $+q$ is fixed at point A and a particle of charge -1 is placed at B (see figure), then the force of attraction $\mathbf{F}$ at A is given by

$$\mathbf{F} = \frac{kq}{\|\overrightarrow{BA}\|^3}\overrightarrow{BA} \quad \text{for some positive constant } k.$$

EXERCISE 45

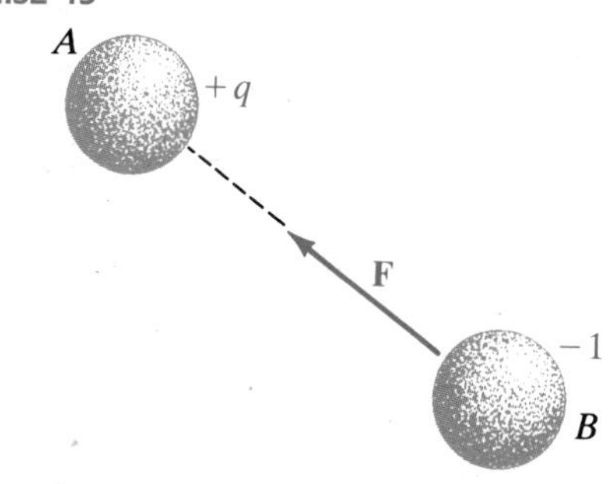

46 Refer to Exercise 45. Particles of charge $+q$ are placed and held fixed at points (1, 0, 0), (0, 1, 0), and (0, 0, 1). A charge of -1 is then placed at $P(x, y, z)$.

(a) If $\mathbf{v} = \overrightarrow{OP}$, show that the net force $\mathbf{F}$ on the negatively charged particle is given by

$$\mathbf{F} = kq\left(\frac{\mathbf{v}-\mathbf{i}}{\|\mathbf{v}-\mathbf{i}\|^3} + \frac{\mathbf{v}-\mathbf{j}}{\|\mathbf{v}-\mathbf{j}\|^3} + \frac{\mathbf{v}-\mathbf{k}}{\|\mathbf{v}-\mathbf{k}\|^3}\right).$$

(b) The negatively charged particle is to be placed at a point $P(x, y, z)$, equidistant from the three positive charges, such that the net force acting on the particle is $\mathbf{0}$. Find the coordinates of P.

14.3 THE DOT PRODUCT

Two useful concepts that involve vectors $\mathbf{a}$ and $\mathbf{b}$ are the *dot product*, which is a scalar, and the *vector product*, which is a vector. In this section we define the dot product. The vector product will be discussed in the next section.

The **dot product** $\mathbf{a} \cdot \mathbf{b}$ of two vectors may be defined either geometrically or algebraically. We shall use the algebraic approach for two reasons:

1. It can be readily generalized for use in advanced mathematics courses such as *linear algebra*, where vectors in four or more dimensions are studied.

2. It allows us to give easier proofs of properties of dot products.

We shall consider geometric and physical interpretations of $\mathbf{a} \cdot \mathbf{b}$ later in the section.

Definition (14.16)

The **dot product** $\mathbf{a} \cdot \mathbf{b}$ of $\mathbf{a} = \langle a_1, a_2, a_3 \rangle$ and $\mathbf{b} = \langle b_1, b_2, b_3 \rangle$ is

$$\mathbf{a} \cdot \mathbf{b} = a_1b_1 + a_2b_2 + a_3b_3.$$

The symbol $\mathbf{a} \cdot \mathbf{b}$ is read **a** *dot* **b**. The dot product is also referred to as the **scalar product**, or **inner product**. It is important to note that $\mathbf{a} \cdot \mathbf{b}$ is a scalar, not a vector. If $\mathbf{a} = \langle a_1, a_2 \rangle$ and $\mathbf{b} = \langle b_1, b_2 \rangle$ are vectors in V_2, then we define their dot product as $\mathbf{a} \cdot \mathbf{b} = a_1b_1 + a_2b_2$.

EXAMPLE 1 Find $\mathbf{a} \cdot \mathbf{b}$ if

(a) $\mathbf{a} = \langle 2, 4, -3 \rangle$, $\mathbf{b} = \langle -1, 5, 2 \rangle$

(b) $\mathbf{a} = 3\mathbf{i} - 2\mathbf{j} + \mathbf{k}$, $\mathbf{b} = 4\mathbf{i} + 5\mathbf{j} - 2\mathbf{k}$

SOLUTION

(a) $\langle 2, 4, -3 \rangle \cdot \langle -1, 5, 2 \rangle = (2)(-1) + (4)(5) + (-3)(2) = 12$

(b) $(3\mathbf{i} - 2\mathbf{j} + \mathbf{k}) \cdot (4\mathbf{i} + 5\mathbf{j} - 2\mathbf{k}) = (3)(4) + (-2)(5) + (1)(-2) = 0$

Some properties of the dot product are listed next for any vectors **a**, **b**, **c** and a scalar c.

Properties of the dot product **(14.17)**

(i) $\mathbf{a} \cdot \mathbf{a} = \|\mathbf{a}\|^2$

(ii) $\mathbf{a} \cdot \mathbf{b} = \mathbf{b} \cdot \mathbf{a}$

(iii) $\mathbf{a} \cdot (\mathbf{b} + \mathbf{c}) = \mathbf{a} \cdot \mathbf{b} + \mathbf{a} \cdot \mathbf{c}$

(iv) $(c\mathbf{a}) \cdot \mathbf{b} = c(\mathbf{a} \cdot \mathbf{b}) = \mathbf{a} \cdot (c\mathbf{b})$

(v) $\mathbf{0} \cdot \mathbf{a} = 0$

PROOF We shall prove (iii) and leave proofs of the remaining properties as exercises. If $\mathbf{a} = \langle a_1, a_2, a_3 \rangle$, $\mathbf{b} = \langle b_1, b_2, b_3 \rangle$, and $\mathbf{c} = \langle c_1, c_2, c_3 \rangle$, then

$$\begin{aligned}
\mathbf{a} \cdot (\mathbf{b} + \mathbf{c}) &= \langle a_1, a_2, a_3 \rangle \cdot \langle b_1 + c_1, b_2 + c_2, b_3 + c_3 \rangle \\
&= a_1(b_1 + c_1) + a_2(b_2 + c_2) + a_3(b_3 + c_3) \\
&= a_1b_1 + a_1c_1 + a_2b_2 + a_2c_2 + a_3b_3 + a_3c_3 \\
&= (a_1b_1 + a_2b_2 + a_3b_3) + (a_1c_1 + a_2c_2 + a_3c_3) \\
&= \mathbf{a} \cdot \mathbf{b} + \mathbf{a} \cdot \mathbf{c}.
\end{aligned}$$

We next show that there is a close relationship between dot products and the angle between two vectors.

Angle between **a** and **b** **(14.18)**

Let **a** and **b** be nonzero vectors.

(i) If **b** is not a scalar multiple of **a** and if $\overrightarrow{OA}$ and $\overrightarrow{OB}$ are the position vectors for **a** and **b**, respectively, then **the angle θ between a and b** (or between $\overrightarrow{OA}$ and $\overrightarrow{OB}$) is angle AOB of the triangle determined by points A, O, and B (see Figure 14.33).

(ii) If $\mathbf{b} = c\mathbf{a}$ for some scalar c (that is, if **a** and **b** are parallel), then $\theta = 0$ if $c > 0$ and $\theta = \pi$ if $c < 0$.

The vectors **a** and **b** are **orthogonal**, or **perpendicular**, if $\theta = \pi/2$. We shall assume that the zero vector **0** is parallel and orthogonal to every vector **a**. (Recall that **0** has *every* direction.)

Theorem **(14.19)**

> If θ is the angle between nonzero vectors **a** and **b**, then
> $$\mathbf{a} \cdot \mathbf{b} = \|\mathbf{a}\| \|\mathbf{b}\| \cos \theta.$$

FIGURE 14.33

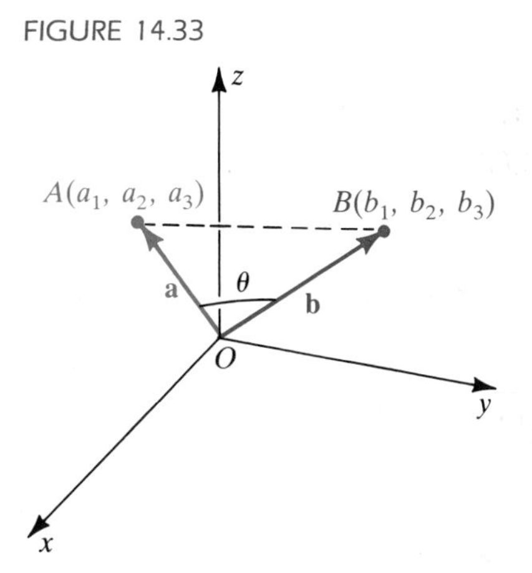

PROOF If $\mathbf{b} \neq c\mathbf{a}$, we have the situation illustrated in Figure 14.33. Applying the law of cosines to triangle AOB yields

$$\|\overrightarrow{AB}\|^2 = \|\mathbf{a}\|^2 + \|\mathbf{b}\|^2 - 2\|\mathbf{a}\|\|\mathbf{b}\| \cos \theta.$$

Consequently

$$(b_1 - a_1)^2 + (b_2 - a_2)^2 + (b_3 - a_3)^2 \\ = (a_1^2 + a_2^2 + a_3^2) + (b_1^2 + b_2^2 + b_3^2) - 2\|\mathbf{a}\|\|\mathbf{b}\| \cos \theta,$$

which simplifies to

$$-2a_1b_1 - 2a_2b_2 - 2a_3b_3 = -2\|\mathbf{a}\|\|\mathbf{b}\| \cos \theta.$$

Dividing both sides of the last equation by -2 gives us the desired conclusion.

If $\mathbf{b} = c\mathbf{a}$, then, by properties (iv) and (i) of Theorem (14.17),

$$\mathbf{a} \cdot \mathbf{b} = \mathbf{a} \cdot (c\mathbf{a}) = c(\mathbf{a} \cdot \mathbf{a}) = c\|\mathbf{a}\|^2.$$

Also, $$\|\mathbf{a}\|\|\mathbf{b}\| \cos \theta = \|\mathbf{a}\|\|c\mathbf{a}\| \cos \theta = |c|\|\mathbf{a}\|^2 \cos \theta.$$

If $c > 0$, then $|c| = c$, $\theta = 0$, and $|c|\|\mathbf{a}\|^2 \cos \theta$ reduces to $c\|\mathbf{a}\|^2$. Thus, $\mathbf{a} \cdot \mathbf{b} = \|\mathbf{a}\|\|\mathbf{b}\| \cos \theta$. If $c < 0$, then $|c| = -c$, $\theta = \pi$, and again $|c|\|\mathbf{a}\|^2 \cos \theta$ reduces to $c\|\mathbf{a}\|^2$. This completes the proof of the theorem. ■

In words, Theorem (14.19) may be stated as follows: *The dot product of two vectors is equal to the product of their magnitudes and the cosine of the angle between the vectors.* In some texts this statement is used as the definition of the dot product, because it has an interesting geometric interpretation that is useful in applications. Two disadvantages of making it the definition are that it is difficult to visualize vectors in four or more dimensions and that it is cumbersome to use it to prove some of the properties of the dot product.

Dividing both sides of the formula in Theorem (14.19) by $\|\mathbf{a}\|\|\mathbf{b}\|$ gives us the following.

Corollary **(14.20)**

> If θ is the angle between nonzero vectors **a** and **b**, then
> $$\cos \theta = \frac{\mathbf{a} \cdot \mathbf{b}}{\|\mathbf{a}\|\|\mathbf{b}\|}.$$

EXAMPLE 2 Find the angle between the vectors $\mathbf{a} = \langle 4, -3, 1\rangle$ and $\mathbf{b} = \langle -1, -2, 2\rangle$.

SOLUTION Applying Corollary (14.20), we have

$$\cos\theta = \frac{\mathbf{a}\cdot\mathbf{b}}{\|\mathbf{a}\|\|\mathbf{b}\|} = \frac{(4)(-1)+(-3)(-2)+(1)(2)}{\sqrt{16+9+1}\sqrt{1+4+4}}$$
$$= \frac{4}{3\sqrt{26}} = \frac{4\sqrt{26}}{78} = \frac{2\sqrt{26}}{39}.$$

Hence $\qquad \theta = \arccos(2\sqrt{26}/39) \approx \arccos(0.2615).$

Using a calculator or table, we obtain

$$\theta \approx 74.84^\circ \approx 1.31 \text{ radians.}$$

The next theorem follows from Theorem (14.19).

Theorem **(14.21)**

> Two vectors $\mathbf{a}$ and $\mathbf{b}$ are orthogonal if and only if $\mathbf{a}\cdot\mathbf{b} = 0$.

EXAMPLE 3 Prove that the vectors are orthogonal:
(a) $\mathbf{i}, \mathbf{j}$ **(b)** $3\mathbf{i} - 7\mathbf{j} + 2\mathbf{k},\ 10\mathbf{i} + 4\mathbf{j} - \mathbf{k}$

SOLUTION The proof follows from Theorem (14.21).

(a) $\mathbf{i}\cdot\mathbf{j} = \langle 1, 0, 0\rangle\cdot\langle 0, 1, 0\rangle = (1)(0) + (0)(1) + (0)(0) = 0$

(b) $(3\mathbf{i} - 7\mathbf{j} + 2\mathbf{k})\cdot(10\mathbf{i} + 4\mathbf{j} - \mathbf{k}) = 30 - 28 - 2 = 0$

Each of the next two results about magnitudes is true for all vectors $\mathbf{a}$ and $\mathbf{b}$.

Cauchy-Schwarz inequality **(14.22)**

$$|\mathbf{a}\cdot\mathbf{b}| \le \|\mathbf{a}\|\|\mathbf{b}\|$$

PROOF The result is trivial if either $\mathbf{a}$ or $\mathbf{b}$ is $\mathbf{0}$. If $\mathbf{a}$ and $\mathbf{b}$ are nonzero vectors, then, by Theorem (14.19), $|\mathbf{a}\cdot\mathbf{b}| = \|\mathbf{a}\|\|\mathbf{b}\||\cos\theta|$, where θ is the angle between $\mathbf{a}$ and $\mathbf{b}$. Since $|\cos\theta| \le 1$, $|\mathbf{a}\cdot\mathbf{b}| \le \|\mathbf{a}\|\|\mathbf{b}\|$. ■

FIGURE 14.34

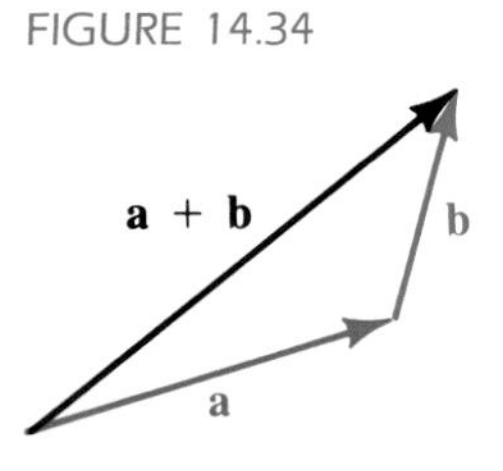

If we represent $\mathbf{a}$, $\mathbf{b}$, and $\mathbf{a} + \mathbf{b}$ geometrically, as in Figure 14.34, and use the fact that the length $\|\mathbf{a} + \mathbf{b}\|$ of one side of the triangle cannot exceed the sum of the lengths of the other two sides, we obtain the following result.

Triangle inequality **(14.23)**

$$\|\mathbf{a} + \mathbf{b}\| \le \|\mathbf{a}\| + \|\mathbf{b}\|$$

Let us give an alternative *algebraic* proof of the triangle inequality, using the dot product.

ALGEBRAIC PROOF OF (14.23) Using properties of the dot product, we obtain

$$\|\mathbf{a}+\mathbf{b}\|^2 = (\mathbf{a}+\mathbf{b})\cdot(\mathbf{a}+\mathbf{b}) = \mathbf{a}\cdot\mathbf{a} + 2(\mathbf{a}\cdot\mathbf{b}) + \mathbf{b}\cdot\mathbf{b}$$
$$= \|\mathbf{a}\|^2 + 2(\mathbf{a}\cdot\mathbf{b}) + \|\mathbf{b}\|^2.$$

Since $\mathbf{a}\cdot\mathbf{b} \le |\mathbf{a}\cdot\mathbf{b}| \le \|\mathbf{a}\|\|\mathbf{b}\|$ (see (14.22)),

$$\|\mathbf{a}+\mathbf{b}\|^2 \le |\mathbf{a}\|^2 + 2\|\mathbf{a}\|\|\mathbf{b}\| + \|\mathbf{b}\|^2 = (\|\mathbf{a}\| + \|\mathbf{b}\|)^2.$$

Taking square roots, we have

$$\|\mathbf{a}+\mathbf{b}\| \le \|\mathbf{a}\| + \|\mathbf{b}\|. \quad ■$$

FIGURE 14.35

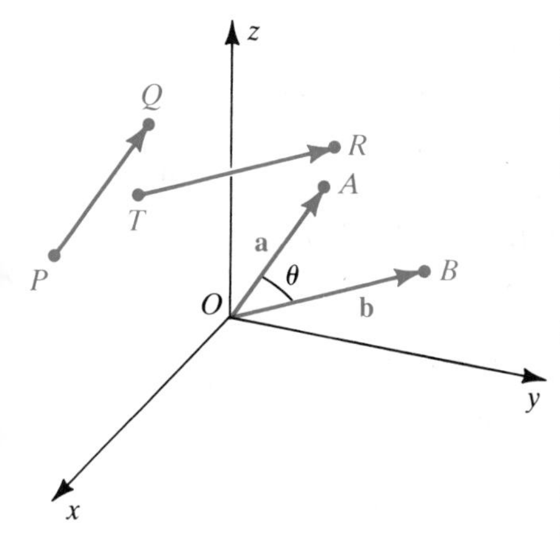

The angle between vectors in V_3 was defined in (14.18). We define the angle between two vectors (directed line segments) $\overrightarrow{PQ}$ and $\overrightarrow{TR}$ as the angle θ between their corresponding vectors **a** and **b** in V_3. A typical angle is illustrated in Figure 14.35, where $\overrightarrow{OA}$ and $\overrightarrow{OB}$ are position vectors for **a** and **b**, respectively. If $\theta = \pi/2$, then $\overrightarrow{PQ}$ and $\overrightarrow{TR}$ are **orthogonal**.

The **dot product** of $\overrightarrow{PQ}$ and $\overrightarrow{TR}$ in Figure 14.35 is defined by

$$\overrightarrow{PQ}\cdot\overrightarrow{TR} = \mathbf{a}\cdot\mathbf{b} = \|\mathbf{a}\|\|\mathbf{b}\|\cos\theta.$$

Since $\|\overrightarrow{PQ}\| = \|\mathbf{a}\|$ and $\|\overrightarrow{TR}\| = \|\mathbf{b}\|$,

$$\overrightarrow{PQ}\cdot\overrightarrow{TR} = \|\overrightarrow{PQ}\|\|\overrightarrow{TR}\|\cos\theta.$$

If $\overrightarrow{PQ}$ and $\overrightarrow{PR}$ have the same initial point and if S is the projection of Q on the line through P and R (see Figure 14.36), then the scalar $\|\overrightarrow{PQ}\|\cos\theta$ is called the **component of $\overrightarrow{PQ}$ along $\overrightarrow{PR}$**, abbreviated $\mathbf{comp}_{\overrightarrow{PR}}\,\overrightarrow{PQ}$.

FIGURE 14.36
$\text{comp}_{\overrightarrow{PR}}\,\overrightarrow{PQ}$

(i)

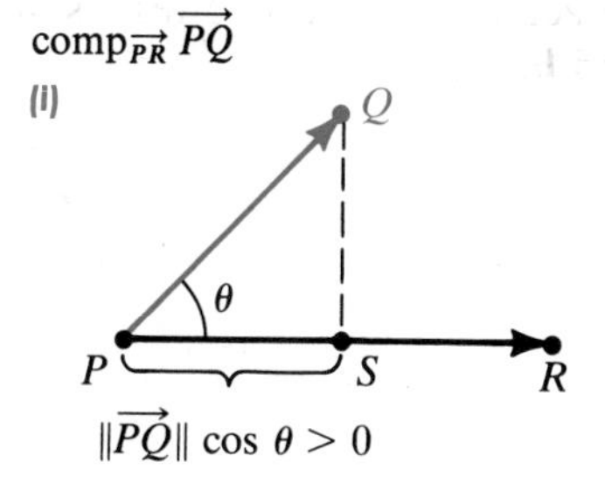

(ii)

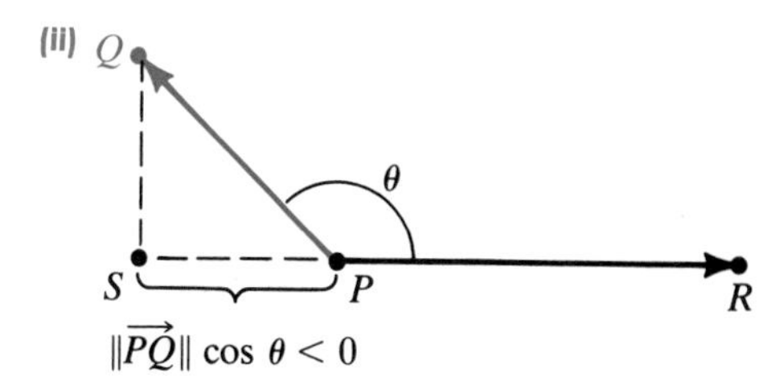

Note that the scalar $\|\overrightarrow{PQ}\|\cos\theta$ is positive if $0 \le \theta < \pi/2$ or negative if $\pi/2 < \theta \le \pi$. If $\theta = \pi/2$, the component is 0. It follows that

$$\text{comp}_{\overrightarrow{PR}}\,\overrightarrow{PQ} = \|\overrightarrow{PQ}\|\cos\theta$$
$$= \|\overrightarrow{PQ}\|\frac{\overrightarrow{PQ}\cdot\overrightarrow{PR}}{\|\overrightarrow{PQ}\|\|\overrightarrow{PR}\|}$$
$$= \overrightarrow{PQ}\cdot\frac{1}{\|\overrightarrow{PR}\|}\overrightarrow{PR}.$$

One way to remember this fact is to use the following statement.

Theorem (14.24)

> The component of $\overrightarrow{PQ}$ along $\overrightarrow{PR}$ equals the dot product of $\overrightarrow{PQ}$ with the unit vector that has the same direction as $\overrightarrow{PR}$.

We use Theorem (14.24) to define the component of a vector **a** along a vector **b** in V_3 as follows.

Definition (14.25)

> Let **a** and **b** be vectors in V_3 with $\mathbf{b} \neq \mathbf{0}$. The **component of a along b**, denoted by $\text{comp}_\mathbf{b}\, \mathbf{a}$, is
>
> $$\text{comp}_\mathbf{b}\, \mathbf{a} = \mathbf{a} \cdot \frac{1}{\|\mathbf{b}\|}\mathbf{b}.$$

If $\mathbf{a} = a_1\mathbf{i} + a_2\mathbf{j} + a_3\mathbf{k}$, then, by Definition (14.25),

$$\text{comp}_\mathbf{i}\, \mathbf{a} = \mathbf{a} \cdot \mathbf{i} = a_1, \quad \text{comp}_\mathbf{j}\, \mathbf{a} = \mathbf{a} \cdot \mathbf{j} = a_2, \quad \text{and} \quad \text{comp}_\mathbf{k}\, \mathbf{a} = \mathbf{a} \cdot \mathbf{k} = a_3.$$

Thus, the components of **a** along **i**, **j**, and **k** are the same as the components a_1, a_2, and a_3 of **a**.

EXAMPLE 4 If $\mathbf{a} = 4\mathbf{i} - \mathbf{j} + 5\mathbf{k}$ and $\mathbf{b} = 6\mathbf{i} + 3\mathbf{j} - 2\mathbf{k}$, find

(a) $\text{comp}_\mathbf{b}\, \mathbf{a}$ **(b)** $\text{comp}_\mathbf{a}\, \mathbf{b}$

SOLUTION

(a) Using Definition (14.25), we have

$$\begin{aligned} \text{comp}_\mathbf{b}\, \mathbf{a} = \mathbf{a} \cdot \frac{1}{\|\mathbf{b}\|}\mathbf{b} &= (4\mathbf{i} - \mathbf{j} + 5\mathbf{k}) \cdot \frac{1}{7}(6\mathbf{i} + 3\mathbf{j} - 2\mathbf{k}) \\ &= \frac{24 - 3 - 10}{7} = \frac{11}{7} \approx 1.6. \end{aligned}$$

(b) Interchanging the roles of **a** and **b** in (14.25) yields

$$\begin{aligned} \text{comp}_\mathbf{a}\, \mathbf{b} = \mathbf{b} \cdot \frac{1}{\|\mathbf{a}\|}\mathbf{a} &= (6\mathbf{i} + 3\mathbf{j} - 2\mathbf{k}) \cdot \frac{1}{\sqrt{42}}(4\mathbf{i} - \mathbf{j} + 5\mathbf{k}) \\ &= \frac{24 - 3 - 10}{\sqrt{42}} = \frac{11}{\sqrt{42}} \approx 1.7. \end{aligned}$$

We conclude this section with an important physical interpretation for the dot product. Recall from Definition (6.20) that the work done if a constant force F is exerted through a distance d is given by $W = Fd$. This formula is very restrictive, since it can be used only if the force is applied along the line of motion. More generally, suppose a vector $\overrightarrow{PQ}$ represents a force, and its point of application moves along a vector $\overrightarrow{PR}$. This is illustrated in Figure 14.37, where a force $\overrightarrow{PQ}$ is used to pull an object along a level path from P to R. Since

FIGURE 14.37

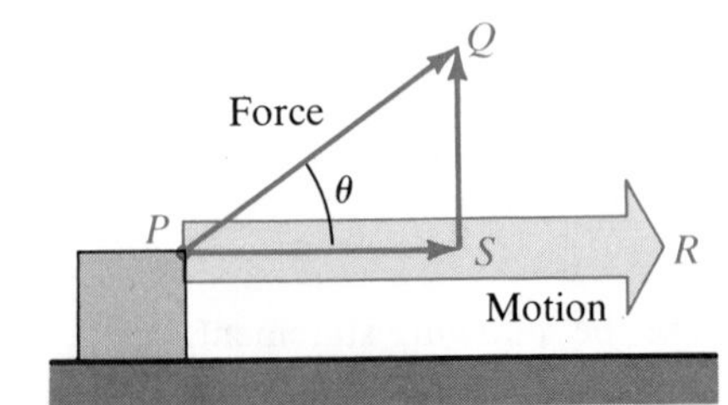

$$\overrightarrow{PQ} = \overrightarrow{PS} + \overrightarrow{SQ},$$

$\overrightarrow{SQ}$ does not contribute to the horizontal movement, and hence the motion from P to R is caused by $\overrightarrow{PS}$ alone. Applying (6.20) with $F = \|\overrightarrow{PS}\|$ and

$d = \|\overrightarrow{PR}\|$, we find that the work W is

$$\begin{aligned} W &= (\|\overrightarrow{PS}\|)\|\overrightarrow{PR}\| \\ &= (\|\overrightarrow{PQ}\| \cos \theta)\|\overrightarrow{PR}\| = \overrightarrow{PQ} \cdot \overrightarrow{PR}. \end{aligned}$$

This is the motivation for the following definition.

Definition **(14.26)**

> The **work done by a constant force** $\overrightarrow{PQ}$ as its point of application moves along the vector $\overrightarrow{PR}$ is $\overrightarrow{PQ} \cdot \overrightarrow{PR}$.

EXAMPLE 5 The magnitude and direction of a constant force are given by $\mathbf{a} = 5\mathbf{i} + 2\mathbf{j} + 6\mathbf{k}$. Find the work done if the point of application of the force moves from $P(1, -1, 2)$ to $R(4, 3, -1)$.

SOLUTION The vector in V_3 that corresponds to $\overrightarrow{PR}$ is

$$\mathbf{b} = \langle 4 - 1, 3 - (-1), -1 - 2 \rangle = \langle 3, 4, -3 \rangle.$$

If $\overrightarrow{PQ}$ corresponds to $\mathbf{a}$, then, by Definition (14.26), the work done is

$$\overrightarrow{PQ} \cdot \overrightarrow{PR} = \mathbf{a} \cdot \mathbf{b} = \langle 5, 2, 6 \rangle \cdot \langle 3, 4, -3 \rangle = 15 + 8 - 18 = 5.$$

If, for example, distance is in feet and the magnitude of the force is in pounds, then the work done is 5 ft-lb. If distance is in meters and force is in Newtons, then the work done is 5 joules.

EXAMPLE 6 A small cart weighing 100 pounds is pushed up an incline that makes an angle of 30° with the horizontal, as shown in Figure 14.38(i). Find the work done against gravity in pushing the cart a distance of 80 feet.

SOLUTION We shall treat this as a two-dimensional problem and introduce an xy-coordinate system as shown in Figure 14.38(ii). The vector $\overrightarrow{PQ}$ represents the force of gravity acting vertically downward with a magnitude of 100 pounds. The corresponding vector $\mathbf{F}$ in V_2 is $0\mathbf{i} - 100\mathbf{j}$. The

FIGURE 14.38

(i)

30°

(ii)

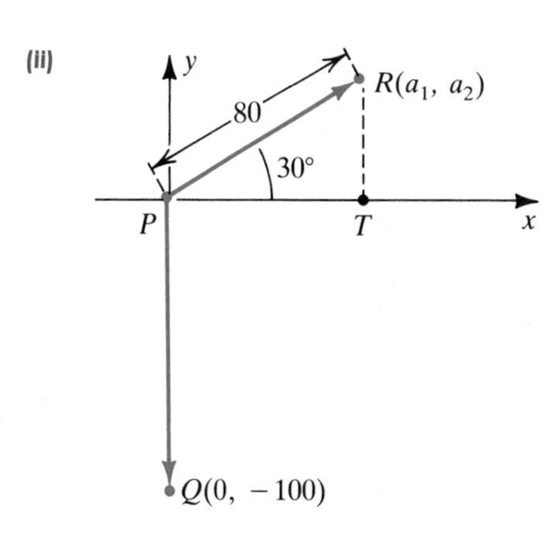

point of application of this force moves along the vector $\overrightarrow{PR}$ of magnitude 80. If $\overrightarrow{PR}$ corresponds to $\mathbf{a} = a_1\mathbf{i} + a_2\mathbf{j}$, then referring to triangle PTR, we see that

$$a_1 = 80 \cos 30° = 40\sqrt{3}, \qquad a_2 = 80 \sin 30° = 40$$

and hence

$$\mathbf{a} = 40\sqrt{3}\mathbf{i} + 40\mathbf{j}.$$

The work done *by* gravity is $\overrightarrow{PQ} \cdot \overrightarrow{PR}$, or $\mathbf{F} \cdot \mathbf{a}$. Applying the definition of dot product yields

$$\mathbf{F} \cdot \mathbf{a} = (0\mathbf{i} - 100\mathbf{j}) \cdot (40\sqrt{3}\mathbf{i} + 40\mathbf{j}) = 0 - 4000 = -4000 \text{ ft-lb}.$$

Thus, the work done *against* gravity is

$$-\mathbf{F} \cdot \mathbf{a} = 4000 \text{ ft-lb}.$$

EXERCISES 14.3

Exer. 1–10: Given $\mathbf{a} = \langle -2, 3, 1\rangle$, $\mathbf{b} = \langle 7, 4, 5\rangle$, and $\mathbf{c} = \langle 1, -5, 2\rangle$, find the number.

1 $\mathbf{a} \cdot \mathbf{b}$

2 $\mathbf{b} \cdot \mathbf{c}$

3 (a) $\mathbf{a} \cdot (\mathbf{b} + \mathbf{c})$ (b) $\mathbf{a} \cdot \mathbf{b} + \mathbf{a} \cdot \mathbf{c}$

4 (a) $(\mathbf{a} - \mathbf{c}) \cdot \mathbf{b}$ (b) $\mathbf{a} \cdot \mathbf{b} - \mathbf{c} \cdot \mathbf{b}$

5 $(2\mathbf{a} + \mathbf{b}) \cdot 3\mathbf{c}$

6 $(\mathbf{a} - \mathbf{b}) \cdot (\mathbf{b} + \mathbf{c})$

7 $\text{comp}_{\mathbf{c}}\, \mathbf{b}$

8 $\text{comp}_{\mathbf{b}}\, \mathbf{c}$

9 $\text{comp}_{\mathbf{b}}\, (\mathbf{a} + \mathbf{c})$

10 $\text{comp}_{\mathbf{c}}\, \mathbf{c}$

Exer. 11–14: Find the angle between a and b.

11 $\mathbf{a} = -4\mathbf{i} + 8\mathbf{j} - 3\mathbf{k}$, $\mathbf{b} = 2\mathbf{i} + \mathbf{j} + \mathbf{k}$

12 $\mathbf{a} = \mathbf{i} - 7\mathbf{j} + 4\mathbf{k}$, $\mathbf{b} = 5\mathbf{i} - \mathbf{k}$

13 $\mathbf{a} = \langle -2, -3, 0\rangle$, $\mathbf{b} = \langle -6, 0, 4\rangle$

14 $\mathbf{a} = \langle 3, -5, -1\rangle$, $\mathbf{b} = \langle 2, 1, -3\rangle$

Exer. 15–16: Show that a and b are orthogonal.

15 $\mathbf{a} = 3\mathbf{i} - 2\mathbf{j} + \mathbf{k}$, $\mathbf{b} = 4\mathbf{i} + 5\mathbf{j} - 2\mathbf{k}$

16 $\mathbf{a} = \langle 4, -1, -2\rangle$, $\mathbf{b} = \langle 2, -2, 5\rangle$

Exer. 17–18: Find all values of c such that a and b are orthogonal.

17 $\mathbf{a} = \langle c, -2, 3\rangle$, $\mathbf{b} = \langle c, c, -5\rangle$

18 $\mathbf{a} = 4\mathbf{i} + 2\mathbf{j} + c\mathbf{k}$, $\mathbf{b} = \mathbf{i} + 22\mathbf{j} - 3c\mathbf{k}$

Exer. 19–24: Given points $P(3, -2, -1)$, $Q(1, 5, 4)$, $R(2, 0, -6)$, and $S(-4, 1, 5)$, find the indicated quantity.

19 $\overrightarrow{PQ} \cdot \overrightarrow{RS}$

20 $\overrightarrow{QS} \cdot \overrightarrow{RP}$

21 The angle between $\overrightarrow{PQ}$ and $\overrightarrow{RS}$

22 The angle between $\overrightarrow{QS}$ and $\overrightarrow{RP}$

23 The component of $\overrightarrow{PS}$ along $\overrightarrow{QR}$

24 The component of $\overrightarrow{QR}$ along $\overrightarrow{PS}$

Exer. 25–26: If the vector a represents a constant force, find the work done when its point of application moves along the line segment from P to Q.

25 $\mathbf{a} = -\mathbf{i} + 5\mathbf{j} - 3\mathbf{k}$; $P(4, 0, -7)$, $Q(2, 4, 0)$

26 $\mathbf{a} = \langle 8, 0, -4\rangle$; $P(-1, 2, 5)$, $Q(4, 1, 0)$

27 A constant force of magnitude 4 pounds has the same direction as the vector $\mathbf{a} = \mathbf{i} + \mathbf{j} + \mathbf{k}$. If distance is measured in feet, find the work done if the point of application moves along the y-axis from $(0, 2, 0)$ to $(0, -1, 0)$.

28 A constant force of magnitude 5 Newtons has the same direction as the positive z-axis. If distance is measured in meters, find the work done if the point of application moves along a line from the origin to the point $P(1, 2, 3)$.

29 A person pulls a wagon along level ground by exerting a force of 20 pounds on a handle that makes an angle of 30° with the horizontal (see figure). Find the work done in pulling the wagon 100 feet.

EXERCISE 29

30 Refer to Exercise 29. Find the work done if the wagon is pulled, with the same force, 100 feet up an incline that makes an angle of 30° with the horizontal (see figure).

EXERCISE 30

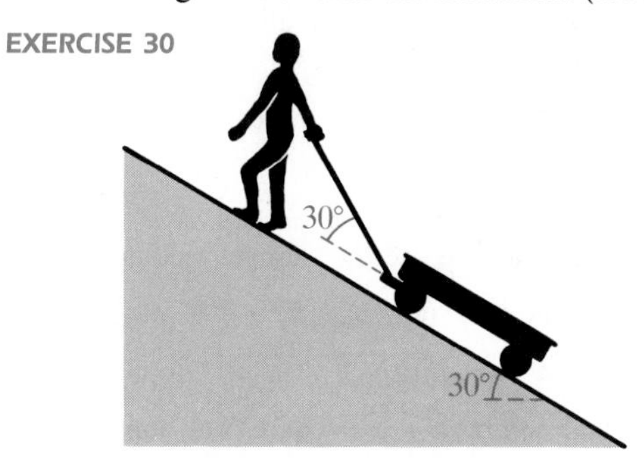

31 If AB is a diameter of a sphere with center O and radius r and if P is a third point on the sphere, use vectors to show that APB is a right triangle. (*Hint:* Let $\mathbf{v}_1 = \overrightarrow{OA}$, $\mathbf{v}_2 = \overrightarrow{OP}$, and write $\overrightarrow{PA}$ and $\overrightarrow{PB}$ in terms of $\mathbf{v}_1$ and $\mathbf{v}_2$.)

32 A rectangular box has length a, width b, and height c (see figure). If P is the center of the box, use vectors to find an expression for angle APB in terms of a, b, and c.

EXERCISE 32

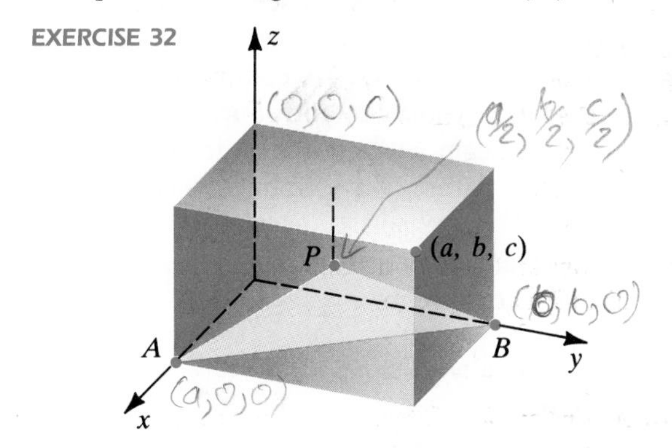

33 Refer to Exercise 32. In the mineral sphalerite, each zinc atom is surrounded by four sulphur atoms, which form a tetrahedron with the zinc atom at its center (see figure). The *bond angle* θ is the angle formed by the S-Zn-S combination. Use vectors to show that the *tetrahedral angle* θ is approximately 109.5°.

EXERCISE 33

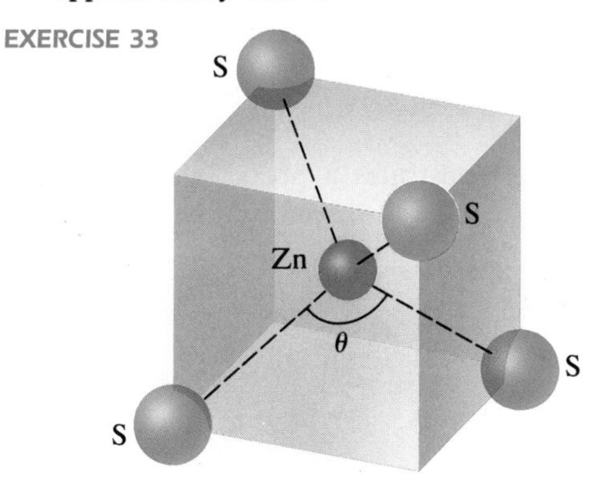

34 Given a sequence A-B-C-D of four bonded atoms, the angle between the plane formed by A, B, and C and the plane formed by B, C, and D is called the *torsion angle* θ of the bond. This torsion angle is used to explain the stability of molecular structures. If segment BC is placed along the z-axis (see figure), how can θ be computed in terms of the components of vectors $\overrightarrow{BA}$ and $\overrightarrow{CD}$?

EXERCISE 34

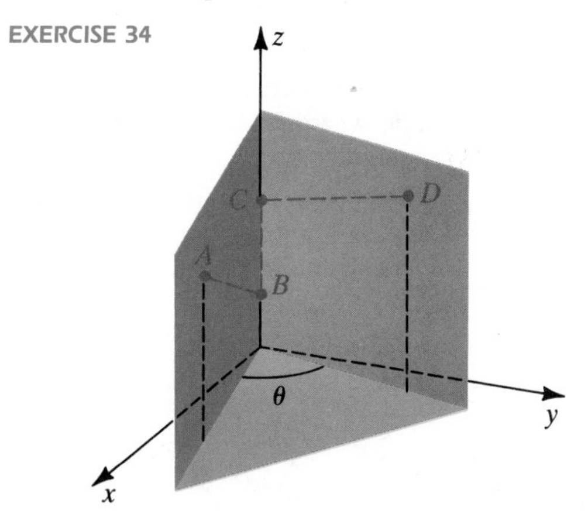

35 The *direction angles* of a nonzero vector $\mathbf{a} = \langle a_1, a_2, a_3 \rangle$ are defined as the angles α, β, and γ between the vectors $\mathbf{i}$, $\mathbf{j}$, and $\mathbf{k}$, respectively, and the vector $\mathbf{a}$. The *direction cosines* of $\mathbf{a}$ are $\cos\alpha$, $\cos\beta$, and $\cos\gamma$. Prove the following:

(a) $\cos\alpha = \dfrac{a_1}{\|\mathbf{a}\|}$, $\cos\beta = \dfrac{a_2}{\|\mathbf{a}\|}$, $\cos\gamma = \dfrac{a_3}{\|\mathbf{a}\|}$

(b) $\cos^2\alpha + \cos^2\beta + \cos^2\gamma = 1$

36 Refer to Exercise 35.

(a) Find the direction cosines of $\mathbf{a} = \langle -2, 1, 5 \rangle$.

(b) Find the direction angles and direction cosines of $\mathbf{i}$, $\mathbf{j}$, and $\mathbf{k}$.

(c) Find two unit vectors that satisfy the condition

$$\cos\alpha = \cos\beta = \cos\gamma.$$

37 Three nonzero numbers l, m, and n are *direction numbers* of a nonzero vector $\mathbf{a}$ if they are proportional to the direction cosines—that is, if there exists a positive number k such that

$$l = k\cos\alpha, \qquad m = k\cos\beta, \qquad n = k\cos\gamma.$$

If $d = (l^2 + m^2 + n^2)^{1/2}$, prove that $\cos\alpha = l/d$, $\cos\beta = m/d$, $\cos\gamma = n/d$.

38 Refer to Exercise 37. If l_1, m_1, n_1 and l_2, m_2, n_2 are direction numbers of $\mathbf{a}$ and $\mathbf{b}$, respectively, prove that

(a) $\mathbf{a}$ and $\mathbf{b}$ are orthogonal if and only if

$$l_1l_2 + m_1m_2 + n_1n_2 = 0$$

(b) $\mathbf{a}$ and $\mathbf{b}$ are parallel if and only if there is a number k such that $l_1 = kl_2$, $m_1 = km_2$, and $n_1 = kn_2$

Exer. 39–42: Prove the given property of vectors if $\mathbf{a} = \langle a_1, a_2, a_3 \rangle$, $\mathbf{b} = \langle b_1, b_2, b_3 \rangle$, and c is a scalar.

39 $\mathbf{a} \cdot \mathbf{a} = \|\mathbf{a}\|^2$

40 $\mathbf{a} \cdot \mathbf{b} = \mathbf{b} \cdot \mathbf{a}$

41 $(c\mathbf{a}) \cdot \mathbf{b} = c(\mathbf{a} \cdot \mathbf{b}) = \mathbf{a} \cdot (c\mathbf{b})$

42 $\mathbf{0} \cdot \mathbf{a} = 0$

Exer. 43–52: Solve *without using components* for the vectors.

43 Let $\mathbf{a}$ and $\mathbf{b}$ be nonzero, nonparallel vectors such that $c\mathbf{a} + d\mathbf{b} = p\mathbf{a} + q\mathbf{b}$, where c, d, p, and q are scalars. Prove that $c = p$ and $d = q$.

44 Prove that if $\mathbf{c}$ is orthogonal to both $\mathbf{a}$ and $\mathbf{b}$, then $\mathbf{c}$ is orthogonal to $p\mathbf{a} + q\mathbf{b}$ for all scalars p and q.

45 Prove that $\|\mathbf{a} + \mathbf{b}\|^2 = \|\mathbf{a}\|^2 + 2(\mathbf{a} \cdot \mathbf{b}) + \|\mathbf{b}\|^2$.

46 Prove that $\|\mathbf{a} + \mathbf{b}\|^2 + \|\mathbf{a} - \mathbf{b}\|^2 = 2(\|\mathbf{a}\|^2 + \|\mathbf{b}\|^2)$.

47 In the Cauchy-Schwarz inequality, under what conditions is $|\mathbf{a} \cdot \mathbf{b}| = \|\mathbf{a}\| \|\mathbf{b}\|$?

48 In the triangle inequality, under what conditions is $\|\mathbf{a} + \mathbf{b}\| = \|\mathbf{a}\| + \|\mathbf{b}\|$?

49 Prove that $\|\mathbf{a} - \mathbf{b}\| \geq \|\mathbf{a}\| - \|\mathbf{b}\|$. (*Hint:* Consider $\mathbf{a} = \mathbf{b} + (\mathbf{a} - \mathbf{b})$ and use the triangle inequality.)

50 Prove that $(\mathbf{a} + \mathbf{b}) \cdot (\mathbf{a} - \mathbf{b}) = \mathbf{a} \cdot \mathbf{a} - \mathbf{b} \cdot \mathbf{b}$.

51 Prove that $\mathbf{a} \cdot \mathbf{b} = \frac{1}{4}(\|\mathbf{a} + \mathbf{b}\|^2 - \|\mathbf{a} - \mathbf{b}\|^2)$.

52 Prove that $\text{comp}_{\mathbf{c}}\,(\mathbf{a} + \mathbf{b}) = \text{comp}_{\mathbf{c}}\,\mathbf{a} + \text{comp}_{\mathbf{c}}\,\mathbf{b}$.

14.4 THE VECTOR PRODUCT

In this section we introduce the *vector product* (or *cross product*) $\mathbf{a} \times \mathbf{b}$ of two vectors $\mathbf{a}$ and $\mathbf{b}$ in V_3. Unlike the dot product $\mathbf{a} \cdot \mathbf{b}$, which is a scalar, the cross product is a vector. As we shall see, the vector $\mathbf{a} \times \mathbf{b}$ is orthogonal to both $\mathbf{a}$ and $\mathbf{b}$. It is used in physics and engineering to describe rotational effects produced by forces. We could define $\mathbf{a} \times \mathbf{b}$ geometrically and then obtain an $\mathbf{i}$, $\mathbf{j}$, $\mathbf{k}$-form by introducing a rectangular coordinate system; however, we begin with an algebraic definition. This approach disguises the geometric nature of $\mathbf{a} \times \mathbf{b}$, but it leads to simpler proofs of properties.

It is convenient to use *determinants* when working with vector products. A **determinant of order 2** is defined by

$$\begin{vmatrix} a_1 & a_2 \\ b_1 & b_2 \end{vmatrix} = a_1 b_2 - a_2 b_1,$$

where all letters represent real numbers.

ILLUSTRATION

- $\begin{vmatrix} 2 & -3 \\ 4 & 5 \end{vmatrix} = (2)(5) - (-3)(4) = 10 + 12 = 22$
- $\begin{vmatrix} 4 & 5 \\ 2 & -3 \end{vmatrix} = (4)(-3) - (5)(2) = -12 - 10 = -22$

The preceding illustration demonstrates a property that is true for determinants of all orders: *interchanging two rows changes the sign of a determinant.*

A **determinant of order 3** is given by

$$\begin{vmatrix} c_1 & c_2 & c_3 \\ a_1 & a_2 & a_3 \\ b_1 & b_2 & b_3 \end{vmatrix} = \begin{vmatrix} a_2 & a_3 \\ b_2 & b_3 \end{vmatrix} c_1 - \begin{vmatrix} a_1 & a_3 \\ b_1 & b_3 \end{vmatrix} c_2 + \begin{vmatrix} a_1 & a_2 \\ b_1 & b_2 \end{vmatrix} c_3.$$

This is sometimes called the *expansion of the determinant by the first row*. The value can be found by evaluating the second-order determinants on the right side of the equation.

ILLUSTRATION

$$\begin{vmatrix} 2 & -1 & 3 \\ -2 & 5 & 1 \\ 1 & 2 & -4 \end{vmatrix} = \begin{vmatrix} 5 & 1 \\ 2 & -4 \end{vmatrix}(2) - \begin{vmatrix} -2 & 1 \\ 1 & -4 \end{vmatrix}(-1) + \begin{vmatrix} -2 & 5 \\ 1 & 2 \end{vmatrix}(3)$$

$$= (-20 - 2)(2) - (8 - 1)(-1) + (-4 - 5)(3)$$

$$= -44 + 7 - 27 = -64$$

Definition **(14.27)**

The **vector product** (or **cross product**) $\mathbf{a} \times \mathbf{b}$ of $\mathbf{a} = \langle a_1, a_2, a_3 \rangle$ and $\mathbf{b} = \langle b_1, b_2, b_3 \rangle$ is

$$\mathbf{a} \times \mathbf{b} = \begin{vmatrix} a_2 & a_3 \\ b_2 & b_3 \end{vmatrix}\mathbf{i} - \begin{vmatrix} a_1 & a_3 \\ b_1 & b_3 \end{vmatrix}\mathbf{j} + \begin{vmatrix} a_1 & a_2 \\ b_1 & b_2 \end{vmatrix}\mathbf{k}.$$

The symbol $\mathbf{a} \times \mathbf{b}$ is read **a** *cross* **b**. Note that the formula for $\mathbf{a} \times \mathbf{b}$ can be obtained by replacing c_1, c_2, c_3 in our definition of a determinant of order 3 by the unit vectors **i**, **j**, **k**. This suggests the following notation for the formula in Definition (14.27).

Notation for vector products **(14.28)**

$$\mathbf{a} \times \mathbf{b} = \begin{vmatrix} \mathbf{i} & \mathbf{j} & \mathbf{k} \\ a_1 & a_2 & a_3 \\ b_1 & b_2 & b_3 \end{vmatrix}$$

The symbol on the right side in (14.28) is *not* a determinant, since the first row contains vectors instead of scalars. However, determinant notation is useful for remembering the more cumbersome formula in Definition (14.27). With this distinction in mind, we shall use (14.28) to find vector products, as in the following example.

EXAMPLE 1 Find $\mathbf{a} \times \mathbf{b}$ if $\mathbf{a} = \langle 2, -1, 6 \rangle$ and $\mathbf{b} = \langle -3, 5, 1 \rangle$.

SOLUTION We proceed as follows:

$$\mathbf{a} \times \mathbf{b} = \begin{vmatrix} \mathbf{i} & \mathbf{j} & \mathbf{k} \\ 2 & -1 & 6 \\ -3 & 5 & 1 \end{vmatrix}$$

$$= \begin{vmatrix} -1 & 6 \\ 5 & 1 \end{vmatrix}\mathbf{i} - \begin{vmatrix} 2 & 6 \\ -3 & 1 \end{vmatrix}\mathbf{j} + \begin{vmatrix} 2 & -1 \\ -3 & 5 \end{vmatrix}\mathbf{k}$$

$$= (-1 - 30)\mathbf{i} - (2 + 18)\mathbf{j} + (10 - 3)\mathbf{k}$$

$$= -31\mathbf{i} - 20\mathbf{j} + 7\mathbf{k}$$

If $\mathbf{a}$ is any vector in V_3, then

$$\mathbf{a} \times \mathbf{0} = \mathbf{0} = \mathbf{0} \times \mathbf{a},$$

for if one of the vectors in Definition (14.27) is $\mathbf{0}$, then each determinant has a row of zeros and hence is 0 (verify this fact). It is also easy to show that $\mathbf{a} \times \mathbf{a} = \mathbf{0}$ for every $\mathbf{a}$.

The next theorem brings out an important property of vector products.

Theorem (14.29) The vector $\mathbf{a} \times \mathbf{b}$ is orthogonal to both $\mathbf{a}$ and $\mathbf{b}$.

PROOF By Theorem (14.21), it is sufficient to show that

$$(\mathbf{a} \times \mathbf{b}) \cdot \mathbf{a} = 0 \quad \text{and} \quad (\mathbf{a} \times \mathbf{b}) \cdot \mathbf{b} = 0.$$

Taking the dot product of $\mathbf{a} \times \mathbf{b}$ in (14.27) with $\mathbf{a} = \langle a_1, a_2, a_3 \rangle$ gives us

$$\begin{aligned}(\mathbf{a} \times \mathbf{b}) \cdot \mathbf{a} &= \begin{vmatrix} a_2 & a_3 \\ b_2 & b_3 \end{vmatrix} a_1 - \begin{vmatrix} a_1 & a_3 \\ b_1 & b_3 \end{vmatrix} a_2 + \begin{vmatrix} a_1 & a_2 \\ b_1 & b_2 \end{vmatrix} a_3 \\ &= (a_2b_3 - a_3b_2)a_1 - (a_1b_3 - a_3b_1)a_2 + (a_1b_2 - a_2b_1)a_3 \\ &= a_2b_3a_1 - a_3b_2a_1 - a_1b_3a_2 + a_3b_1a_2 + a_1b_2a_3 - a_2b_1a_3 \\ &= 0.\end{aligned}$$

Hence $\mathbf{a} \times \mathbf{b}$ is orthogonal to $\mathbf{a}$. The proof that $(\mathbf{a} \times \mathbf{b}) \cdot \mathbf{b} = 0$ is similar. ■

Let us interpret Theorem (14.29) geometrically. As illustrated in Figure 14.39(i), if nonzero vectors $\mathbf{a}$ and $\mathbf{b}$ correspond to vectors $\overrightarrow{PQ}$ and $\overrightarrow{PR}$ with the same initial point P, then $\mathbf{a} \times \mathbf{b}$ corresponds to a vector $\overrightarrow{PS}$ that is orthogonal to both $\overrightarrow{PQ}$ and $\overrightarrow{PR}$. Hence $\overrightarrow{PS}$ is perpendicular to the plane determined by P, Q, and R. We shall write

$$\overrightarrow{PS} = \overrightarrow{PQ} \times \overrightarrow{PR}.$$

FIGURE 14.39

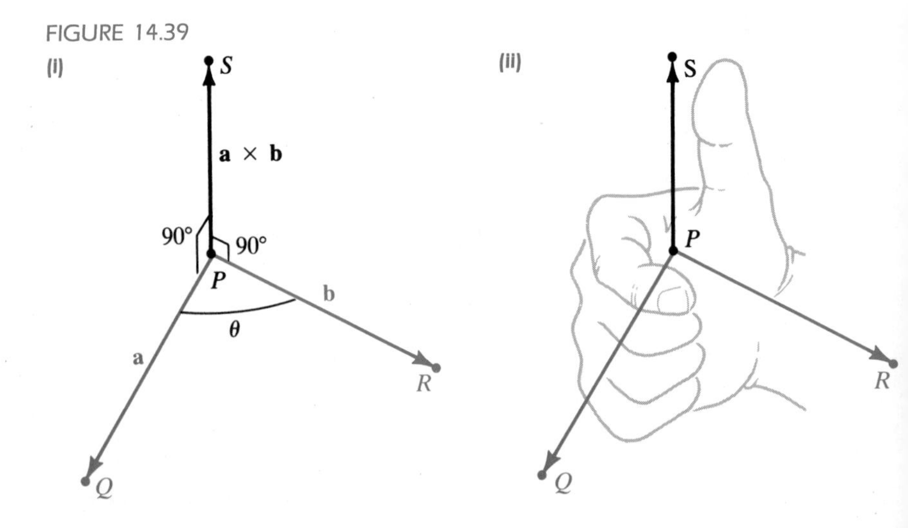

The direction of $\overrightarrow{PS}$ may be obtained using the right-hand rule illustrated in Figure 14.39(ii). Thus, if θ is the angle between $\overrightarrow{PQ}$ and $\overrightarrow{PR}$ and if the fingers of the right hand are curled such that a rotation through θ will transform $\overrightarrow{PQ}$ into a vector that has the same direction as $\overrightarrow{PR}$, then the extended thumb points in the direction of $\overrightarrow{PQ} \times \overrightarrow{PR}$.

The following result describes the magnitude of $\mathbf{a} \times \mathbf{b}$.

Theorem **(14.30)**

If θ is the angle between nonzero vectors $\mathbf{a}$ and $\mathbf{b}$, then

$$\|\mathbf{a} \times \mathbf{b}\| = \|\mathbf{a}\|\|\mathbf{b}\| \sin \theta.$$

PROOF By Theorem (14.17)(i) and Definition (14.27),

$$\begin{aligned}
\|\mathbf{a} \times \mathbf{b}\|^2 &= (\mathbf{a} \times \mathbf{b}) \cdot (\mathbf{a} \times \mathbf{b}) \\
&= \begin{vmatrix} a_2 & a_3 \\ b_2 & b_3 \end{vmatrix}^2 + \begin{vmatrix} a_1 & a_3 \\ b_1 & b_3 \end{vmatrix}^2 + \begin{vmatrix} a_1 & a_2 \\ b_1 & b_2 \end{vmatrix}^2 \\
&= (a_2b_3 - a_3b_2)^2 + (a_1b_3 - a_3b_1)^2 + (a_1b_2 - a_2b_1)^2 \\
&= a_2^2b_3^2 - 2a_2a_3b_2b_3 + a_3^2b_2^2 + a_1^2b_3^2 - 2a_1a_3b_1b_3 \\
&\qquad + a_3^2b_1^2 + a_1^2b_2^2 - 2a_1a_2b_1b_2 + a_2^2b_1^2 \\
&= (a_1^2 + a_2^2 + a_3^2)(b_1^2 + b_2^2 + b_3^2) - (a_1b_1 + a_2b_2 + a_3b_3)^2.
\end{aligned}$$

The last equality may be verified by multiplying the indicated expressions. In vector form, the preceding identity is

$$\|\mathbf{a} \times \mathbf{b}\|^2 = (\|\mathbf{a}\|\|\mathbf{b}\|)^2 - (\mathbf{a} \cdot \mathbf{b})^2,$$

or, since $\mathbf{a} \cdot \mathbf{b} = \|\mathbf{a}\|\|\mathbf{b}\| \cos \theta$,

$$\begin{aligned}
\|\mathbf{a} \times \mathbf{b}\|^2 &= (\|\mathbf{a}\|\|\mathbf{b}\|)^2 - (\|\mathbf{a}\|\|\mathbf{b}\|)^2 \cos^2 \theta \\
&= (\|\mathbf{a}\|\|\mathbf{b}\|)^2(1 - \cos^2 \theta).
\end{aligned}$$

Hence $$\|\mathbf{a} \times \mathbf{b}\|^2 = (\|\mathbf{a}\|\|\mathbf{b}\|)^2 \sin^2 \theta.$$

Taking square roots and observing that $\sin \theta \geq 0$ (since $0 \leq \theta \leq \pi$), we obtain

$$\|\mathbf{a} \times \mathbf{b}\| = \|\mathbf{a}\|\|\mathbf{b}\| \sin \theta. \quad ■$$

Corollary **(14.31)**

Two vectors $\mathbf{a}$ and $\mathbf{b}$ are parallel if and only if $\mathbf{a} \times \mathbf{b} = \mathbf{0}$.

PROOF Suppose $\mathbf{a}$ and $\mathbf{b}$ are nonzero vectors. If θ is the angle between $\mathbf{a}$ and $\mathbf{b}$, then the vectors are parallel if and only if $\theta = 0$ or $\theta = \pi$, or, equivalently, $\sin \theta = 0$. By Theorem (14.30), the last statement is equivalent to $\mathbf{a} \times \mathbf{b} = \mathbf{0}$. If either $\mathbf{a}$ or $\mathbf{b}$ is the zero vector, the proof is trivial, since $\mathbf{0}$ is parallel to every vector. ■

To interpret $\|\mathbf{a} \times \mathbf{b}\|$ geometrically, let us represent $\mathbf{a}$ and $\mathbf{b}$ by vectors $\overrightarrow{PQ}$ and $\overrightarrow{PR}$ having the same initial point P. As in Figure 14.40, let S be

FIGURE 14.40

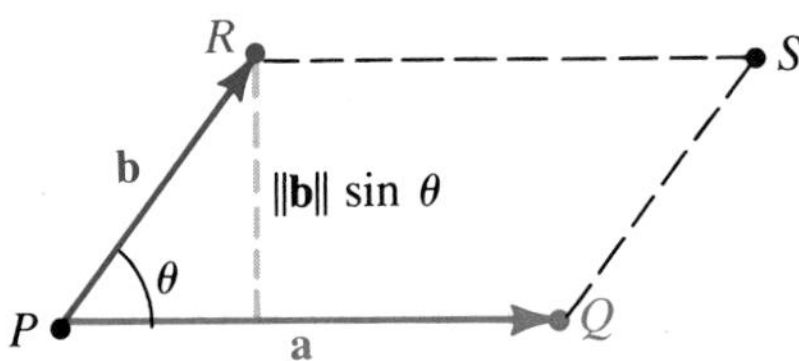

the point such that segments PQ and PR are adjacent sides of a parallelogram with vertices P, Q, R, and S. An altitude of the parallelogram is $\|\mathbf{b}\| \sin \theta$, and hence its area is $\|\mathbf{a}\|\|\mathbf{b}\| \sin \theta$. Comparing this with Theorem (14.30), we see that *the magnitude of the vector product* $\mathbf{a} \times \mathbf{b}$ *equals the area of the parallelogram determined by* $\mathbf{a}$ *and* $\mathbf{b}$.

EXAMPLE 2 Find the area of the triangle determined by $P(4, -3, 1)$, $Q(6, -4, 7)$, and $R(1, 2, 2)$.

SOLUTION The vectors in V_3 corresponding to $\overrightarrow{PQ}$ and $\overrightarrow{PR}$ are given by

$$\mathbf{a} = \langle 6 - 4, -4 - (-3), 7 - 1 \rangle = \langle 2, -1, 6 \rangle$$

$$\mathbf{b} = \langle 1 - 4, 2 - (-3), 2 - 1 \rangle = \langle -3, 5, 1 \rangle.$$

From Example 1,

$$\mathbf{a} \times \mathbf{b} = -31\mathbf{i} - 20\mathbf{j} + 7\mathbf{k}.$$

Hence the area of the parallelogram with adjacent sides PQ and PR is

$$\|\mathbf{a} \times \mathbf{b}\| = \sqrt{961 + 400 + 49} = \sqrt{1410}.$$

It follows that the area of the triangle is $\frac{1}{2}\sqrt{1410} \approx 18.8$.

The vector products of the special unit vectors $\mathbf{i}$, $\mathbf{j}$, and $\mathbf{k}$ are of interest. For example, using Definition (14.27) with $\mathbf{a} = \mathbf{i} = \langle 1, 0, 0 \rangle$ and $\mathbf{b} = \mathbf{j} = \langle 0, 1, 0 \rangle$, we have

$$\mathbf{i} \times \mathbf{j} = \begin{vmatrix} 0 & 0 \\ 1 & 0 \end{vmatrix} \mathbf{i} - \begin{vmatrix} 1 & 0 \\ 0 & 0 \end{vmatrix} \mathbf{j} + \begin{vmatrix} 1 & 0 \\ 0 & 1 \end{vmatrix} \mathbf{k} = \mathbf{k}.$$

We can prove the following in similar fashion.

Properties of **i, j, k (14.32)**

$$\mathbf{i} \times \mathbf{j} = \mathbf{k} \qquad \mathbf{j} \times \mathbf{k} = \mathbf{i} \qquad \mathbf{k} \times \mathbf{i} = \mathbf{j}$$

$$\mathbf{j} \times \mathbf{i} = -\mathbf{k} \qquad \mathbf{k} \times \mathbf{j} = -\mathbf{i} \qquad \mathbf{i} \times \mathbf{k} = -\mathbf{j}$$

$$\mathbf{i} \times \mathbf{i} = \mathbf{j} \times \mathbf{j} = \mathbf{k} \times \mathbf{k} = \mathbf{0}$$

FIGURE 14.41

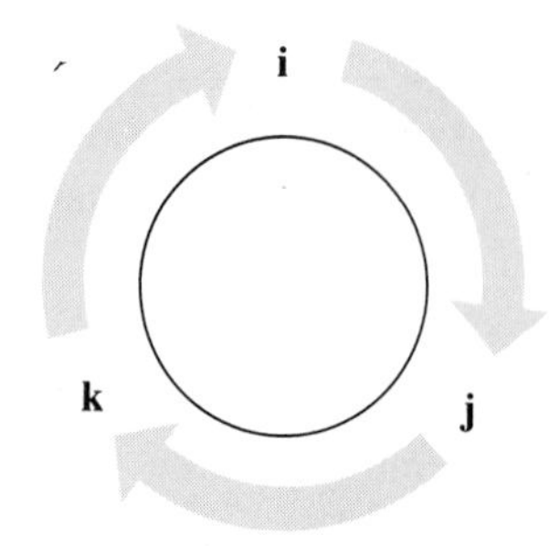

Figure 14.41 is a simple mnemonic device for remembering six of the formulas in (14.32). If we place the symbols $\mathbf{i}$, $\mathbf{j}$, $\mathbf{k}$ around a circle as shown and proceed in the *clockwise* direction (as denoted by the arrows), then the vector product of any two successive vectors is the next vector. To illustrate, $\mathbf{i} \times \mathbf{j} = \mathbf{k}$. For the *counterclockwise* direction, take the *negative* of the next vector, as in $\mathbf{j} \times \mathbf{i} = -\mathbf{k}$.

The fact that $\mathbf{i} \times \mathbf{j} \neq \mathbf{j} \times \mathbf{i}$ shows that *the vector product is not commutative*. Moreover, *the associative law does not hold*, since, for example,

$$\mathbf{i} \times (\mathbf{j} \times \mathbf{j}) = \mathbf{i} \times \mathbf{0} = \mathbf{0},$$

but

$$(\mathbf{i} \times \mathbf{j}) \times \mathbf{j} = \mathbf{k} \times \mathbf{j} = -\mathbf{i}.$$

The following properties are true for any vectors **a**, **b**, and **c** and a scalar m.

Properties of the vector product **(14.33)**

(i) $\mathbf{b} \times \mathbf{a} = -(\mathbf{a} \times \mathbf{b})$

(ii) $(m\mathbf{a}) \times \mathbf{b} = m(\mathbf{a} \times \mathbf{b}) = \mathbf{a} \times (m\mathbf{b})$

(iii) $\mathbf{a} \times (\mathbf{b} + \mathbf{c}) = (\mathbf{a} \times \mathbf{b}) + (\mathbf{a} \times \mathbf{c})$

(iv) $(\mathbf{a} + \mathbf{b}) \times \mathbf{c} = (\mathbf{a} \times \mathbf{c}) + (\mathbf{b} \times \mathbf{c})$

(v) $(\mathbf{a} \times \mathbf{b}) \cdot \mathbf{c} = \mathbf{a} \cdot (\mathbf{b} \times \mathbf{c})$

(vi) $\mathbf{a} \times (\mathbf{b} \times \mathbf{c}) = (\mathbf{a} \cdot \mathbf{c})\mathbf{b} - (\mathbf{a} \cdot \mathbf{b})\mathbf{c}$

PROOF Each property may be established by straightforward (but sometimes lengthy) applications of Definition (14.27). To prove (i), we let $\mathbf{a} = \langle a_1, a_2, a_3 \rangle$ and $\mathbf{b} = \langle b_1, b_2, b_3 \rangle$ and find

$$\mathbf{b} \times \mathbf{a} = \begin{vmatrix} b_2 & b_3 \\ a_2 & a_3 \end{vmatrix} \mathbf{i} - \begin{vmatrix} b_1 & b_3 \\ a_1 & a_3 \end{vmatrix} \mathbf{j} + \begin{vmatrix} b_1 & b_2 \\ a_1 & a_2 \end{vmatrix} \mathbf{k}.$$

Since interchanging two rows of a determinant changes its sign, we obtain

$$\begin{aligned} \mathbf{b} \times \mathbf{a} &= -\begin{vmatrix} a_2 & a_3 \\ b_2 & b_3 \end{vmatrix} \mathbf{i} + \begin{vmatrix} a_1 & a_3 \\ b_1 & b_3 \end{vmatrix} \mathbf{j} - \begin{vmatrix} a_1 & a_2 \\ b_1 & b_2 \end{vmatrix} \mathbf{k} \\ &= -(\mathbf{a} \times \mathbf{b}). \end{aligned}$$

To prove (iii), consider $\mathbf{c} = \langle c_1, c_2, c_3 \rangle$. The **i** component of $\mathbf{a} \times (\mathbf{b} + \mathbf{c})$ is

$$\begin{aligned} \begin{vmatrix} a_2 & a_3 \\ b_1 + c_2 & b_3 + c_3 \end{vmatrix} &= a_2(b_3 + c_3) - a_3(b_2 + c_2) \\ &= (a_2b_3 - a_3b_2) + (a_2c_3 - a_3c_2) \\ &= \begin{vmatrix} a_2 & a_3 \\ b_2 & b_3 \end{vmatrix} + \begin{vmatrix} a_2 & a_3 \\ c_2 & c_3 \end{vmatrix}, \end{aligned}$$

which is equal to the **i** component of $(\mathbf{a} \times \mathbf{b}) + (\mathbf{a} \times \mathbf{c})$. A similar calculation can be used to prove that the **j** and **k** components of $\mathbf{a} \times (\mathbf{b} + \mathbf{c})$ are the same as those of $(\mathbf{a} \times \mathbf{b}) + (\mathbf{a} \times \mathbf{c})$. This establishes (iii). The proofs of the remaining properties are left as exercises. ■

Formula (vi) of Theorem (14.33) is called the **triple vector product** of **a**, **b**, and **c**.

EXAMPLE 3 Find a formula for the distance d from a point R to a line l.

FIGURE 14.42

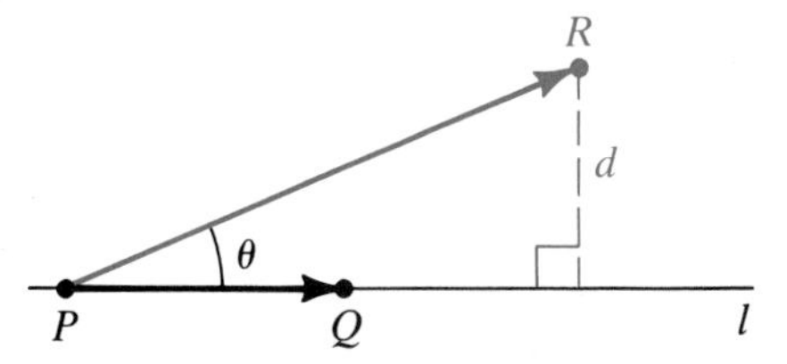

SOLUTION Let P and Q be points on l, as shown in Figure 14.42, and let θ be the angle between $\overrightarrow{PQ}$ and $\overrightarrow{PR}$. Since $d = \|\overrightarrow{PR}\| \sin\theta$, we obtain

$$\|\overrightarrow{PQ} \times \overrightarrow{PR}\| = \|\overrightarrow{PQ}\|\,\|\overrightarrow{PR}\| \sin\theta = \|\overrightarrow{PQ}\|\,d.$$

Hence

$$d = \frac{\|\overrightarrow{PQ} \times \overrightarrow{PR}\|}{\|\overrightarrow{PQ}\|}.$$

FIGURE 14.43

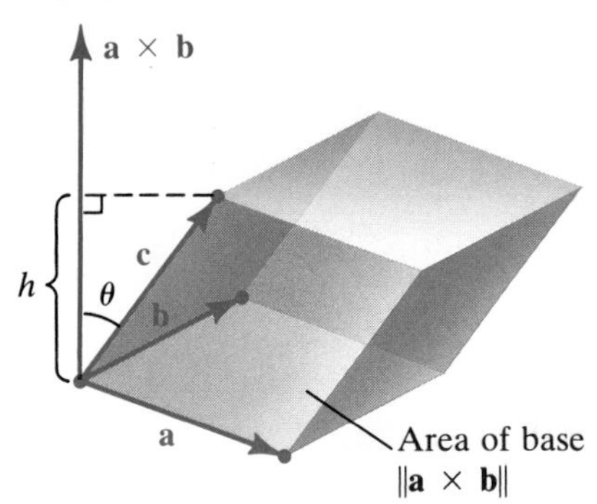

EXAMPLE 4 Suppose **a**, **b**, and **c** represent adjacent sides of the oblique box in Figure 14.43. Show that $|(\mathbf{a} \times \mathbf{b}) \cdot \mathbf{c}|$ is the volume of the box.

SOLUTION The area of the base of the box is $\|\mathbf{a} \times \mathbf{b}\|$. Let θ be the angle between **c** and $\mathbf{a} \times \mathbf{b}$. Since $\mathbf{a} \times \mathbf{b}$ is perpendicular to the base, the altitude h of the box is given by $h = \|\mathbf{c}\| \, |\cos \theta|$. We must use the absolute value $|\cos \theta|$, since θ may be an obtuse angle. Thus, the volume V of the box is

$$\begin{aligned} V &= (\text{area of base})(\text{altitude}) \\ &= \|\mathbf{a} \times \mathbf{b}\| \, \|\mathbf{c}\| \, |\cos \theta| \\ &= |(\mathbf{a} \times \mathbf{b}) \cdot \mathbf{c}|, \end{aligned}$$

where the last equality follows from Theorem (14.19).

The number $(\mathbf{a} \times \mathbf{b}) \cdot \mathbf{c}$ considered in Example 4 is called the **triple scalar product** of **a**, **b**, and **c**. This number can be expressed as a determinant of order 3 that involves the components of the three vectors (see Exercise 21).

FIGURE 14.44

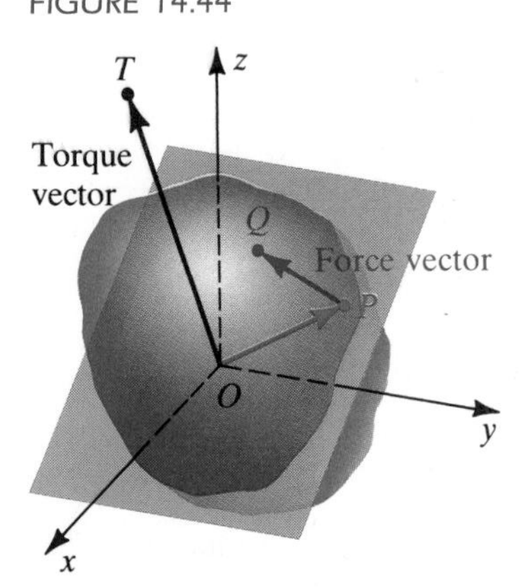

Vector products are used to study rotational effects produced by forces. Suppose a force $\overrightarrow{PQ}$ acts at a point P in an object, as illustrated in Figure 14.44, where we have superimposed an xyz-coordinate system. The force tends to make the object rotate about a line through O that is perpendicular to the plane determined by $\overrightarrow{PQ}$ and the position vector $\overrightarrow{OP}$ of P. The **torque vector** $\overrightarrow{OT}$ for the force (also called **the moment of** $\overrightarrow{PQ}$ **about** O) is defined by

$$\overrightarrow{OT} = \overrightarrow{OP} \times \overrightarrow{PQ}.$$

In physics or mechanics, $\overrightarrow{OT}$ specifies both the magnitude and the direction of the rotational effect induced by the force $\overrightarrow{PQ}$. If $\overrightarrow{OP}$ represents the handle of a wrench and $\overrightarrow{PQ}$ is the force applied to the end of the handle, then $\overrightarrow{OT}$ points in the direction that a (right-handed) bolt moves when the wrench is turned.

Vector products are also useful in electromagnetic field theory. To illustrate, let us consider an **electric dipole**—that is, two point charges $+q$ and $-q$ that are always a constant distance d apart. If **d** is the vector shown in Figure 14.45(i), then $\|\mathbf{d}\| = d$ and the **electric dipole moment p** is defined as $\mathbf{p} = q\mathbf{d}$. Suppose, as illustrated in Figure 14.45(ii), the electric dipole is placed in an electromagnetic field where every force vector **E** has the same magnitude and direction. The electric dipole will tend to

FIGURE 14.45

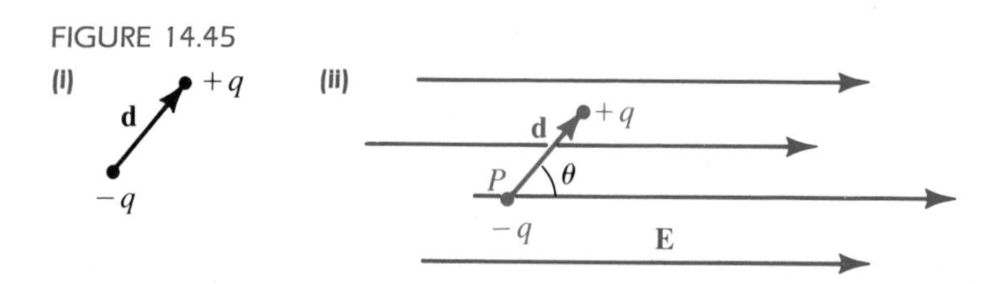

rotate until it is aligned with the direction of the field; that is, the angle θ between $\mathbf{d}$ and $\mathbf{E}$ will approach 0. The torque $\mathbf{T}$ caused by $\mathbf{E}$ at P is given by $\mathbf{T} = \mathbf{p} \times \mathbf{E}$, and the work done by $\mathbf{E}$ in the alignment process is $W = \mathbf{p} \cdot \mathbf{E}$.

We shall consider other interpretations of the vector product in Chapter 15.

EXERCISES 14.4

Exer. 1–10: Find a × b.

1 $\mathbf{a} = \langle 1, -2, 3\rangle$, $\mathbf{b} = \langle 2, 1, -4\rangle$

2 $\mathbf{a} = \langle -5, 1, -1\rangle$, $\mathbf{b} = \langle 3, 6, -2\rangle$

3 $\mathbf{a} = \langle 0, 1, 2\rangle$, $\mathbf{b} = \langle 1, 2, 0\rangle$

4 $\mathbf{a} = \langle 0, 0, 4\rangle$, $\mathbf{b} = \langle -7, 1, 0\rangle$

5 $\mathbf{a} = 5\mathbf{i} - 6\mathbf{j} - \mathbf{k}$, $\mathbf{b} = 3\mathbf{i} + \mathbf{k}$

6 $\mathbf{a} = 2\mathbf{i} + \mathbf{j}$, $\mathbf{b} = -5\mathbf{j} + 2\mathbf{k}$

7 $\mathbf{a} = -3\mathbf{i} + \mathbf{j} + 2\mathbf{k}$, $\mathbf{b} = 9\mathbf{i} - 3\mathbf{j} - 6\mathbf{k}$

8 $\mathbf{a} = 3\mathbf{i} - \mathbf{j} + 8\mathbf{k}$, $\mathbf{b} = 5\mathbf{j}$

9 $\mathbf{a} = 4\mathbf{i} - 6\mathbf{j} + 2\mathbf{k}$, $\mathbf{b} = -2\mathbf{i} + 3\mathbf{j} - \mathbf{k}$

10 $\mathbf{a} = 3\mathbf{i}$, $\mathbf{b} = 4\mathbf{k}$

Exer. 11–12: Use the vector product to show that a and b are parallel.

11 $\mathbf{a} = \langle -6, -10, 4\rangle$, $\mathbf{b} = \langle 3, 5, -2\rangle$

12 $\mathbf{a} = 2\mathbf{i} - \mathbf{j} + 4\mathbf{k}$, $\mathbf{b} = -6\mathbf{i} + 3\mathbf{j} - 12\mathbf{k}$

Exer. 13–14: Let $\mathbf{a} = \langle 2, 0, -1\rangle$, $\mathbf{b} = \langle -3, 1, 0\rangle$, and $\mathbf{c} = \langle 1, -2, 4\rangle$.

13 Find $\mathbf{a} \times (\mathbf{b} \times \mathbf{c})$ and $(\mathbf{a} \times \mathbf{b}) \times \mathbf{c}$.

14 Find $\mathbf{a} \times (\mathbf{b} - \mathbf{c})$ and $(\mathbf{a} \times \mathbf{b}) - (\mathbf{a} \times \mathbf{c})$.

Exer. 15–18: (a) Find a vector perpendicular to the plane determined by P, Q, and R. (b) Find the area of the triangle PQR.

15 $P(1, -1, 2)$, $Q(0, 3, -1)$, $R(3, -4, 1)$

16 $P(-3, 0, 5)$, $Q(2, -1, -3)$, $R(4, 1, -1)$

17 $P(4, 0, 0)$, $Q(0, 5, 0)$, $R(0, 0, 2)$

18 $P(-1, 2, 0)$, $Q(0, 2, -3)$, $R(5, 0, 1)$

Exer. 19–20: Refer to Example 3. Find the distance from P to the line through Q and R.

19 $P(3, 1, -2)$, $Q(2, 5, 1)$, $R(-1, 4, 2)$

20 $P(-2, 5, 1)$, $Q(3, -1, 4)$, $R(1, 6, -3)$

21 If $\mathbf{a} = \langle a_1, a_2, a_3\rangle$, $\mathbf{b} = \langle b_1, b_2, b_3\rangle$, $\mathbf{c} = \langle c_1, c_2, c_3\rangle$, prove that

$$\mathbf{a} \cdot (\mathbf{b} \times \mathbf{c}) = (\mathbf{a} \times \mathbf{b}) \cdot \mathbf{c} = \begin{vmatrix} a_1 & a_2 & a_3 \\ b_1 & b_2 & b_3 \\ c_1 & c_2 & c_3 \end{vmatrix}$$

Exer. 22–23: Use Example 4 and Exercise 21 to find the volume of the box having adjacent sides AB, AC, and AD.

22 $A(0, 0, 0)$, $B(1, -1, 2)$, $C(0, 3, -1)$, $D(3, -4, 1)$

23 $A(2, 1, -1)$, $B(3, 0, 2)$, $C(4, -2, 1)$, $D(5, -3, 0)$

24 If $\mathbf{a}$, $\mathbf{b}$, and $\mathbf{c}$ are represented by vectors with a common initial point, show that $\mathbf{a} \cdot (\mathbf{b} \times \mathbf{c}) = 0$ if and only if the vectors are coplanar.

25 Prove that $(\mathbf{a} \times \mathbf{b}) \cdot \mathbf{b} = 0$ for all vectors $\mathbf{a}$ and $\mathbf{b}$.

26 If $\mathbf{a} \times \mathbf{b} = \mathbf{a} \times \mathbf{c}$ and $\mathbf{a} \neq \mathbf{0}$, does it follow that $\mathbf{b} = \mathbf{c}$? Explain.

27 Let $\mathbf{a} \neq \mathbf{0}$. If $\mathbf{a} \times \mathbf{b} = \mathbf{a} \times \mathbf{c}$ *and* $\mathbf{a} \cdot \mathbf{b} = \mathbf{a} \cdot \mathbf{c}$, prove that $\mathbf{b} = \mathbf{c}$.

Exer. 28–31: Prove the given property if $\mathbf{a} = \langle a_1, a_2, a_3\rangle$, $\mathbf{b} = \langle b_1, b_2, b_3\rangle$, $\mathbf{c} = \langle c_1, c_2, c_3\rangle$, and m is a scalar.

28 $(m\mathbf{a}) \times \mathbf{b} = m(\mathbf{a} \times \mathbf{b}) = \mathbf{a} \times (m\mathbf{b})$

29 $(\mathbf{a} + \mathbf{b}) \times \mathbf{c} = (\mathbf{a} \times \mathbf{c}) + (\mathbf{b} \times \mathbf{c})$

30 $(\mathbf{a} \times \mathbf{b}) \cdot \mathbf{c} = \mathbf{a} \cdot (\mathbf{b} \times \mathbf{c})$

31 $\mathbf{a} \times (\mathbf{b} \times \mathbf{c}) = (\mathbf{a} \cdot \mathbf{c})\mathbf{b} - (\mathbf{a} \cdot \mathbf{b})\mathbf{c}$

Exer. 32–37: Verify *without using components* for the vectors.

32 $(\mathbf{a} + \mathbf{b}) \times (\mathbf{a} - \mathbf{b}) = 2(\mathbf{b} \times \mathbf{a})$

33 $\mathbf{a} \times (\mathbf{b} \times \mathbf{c}) + \mathbf{b} \times (\mathbf{c} \times \mathbf{a}) + \mathbf{c} \times (\mathbf{a} \times \mathbf{b}) = \mathbf{0}$
(*Hint:* Use (vi) of Theorem (14.33).)

34 $(\mathbf{a} \times \mathbf{b}) \times \mathbf{c} = (\mathbf{a} \cdot \mathbf{c})\mathbf{b} - (\mathbf{b} \cdot \mathbf{c})\mathbf{a}$

35 $(\mathbf{a} \times \mathbf{b}) \cdot (\mathbf{c} \times \mathbf{d}) = \begin{vmatrix} \mathbf{a} \cdot \mathbf{c} & \mathbf{b} \cdot \mathbf{c} \\ \mathbf{a} \cdot \mathbf{d} & \mathbf{b} \cdot \mathbf{d} \end{vmatrix}$

36 $(\mathbf{a} \times \mathbf{b}) \times (\mathbf{c} \times \mathbf{d}) = (\mathbf{a} \times \mathbf{b} \cdot \mathbf{d})\mathbf{c} - (\mathbf{a} \times \mathbf{b} \cdot \mathbf{c})\mathbf{d}$

37 $(\mathbf{a} \times \mathbf{b}) \cdot (\mathbf{b} \times \mathbf{c}) \times (\mathbf{c} \times \mathbf{a}) = (\mathbf{a} \cdot \mathbf{b} \times \mathbf{c})^2$

14.5 LINES AND PLANES

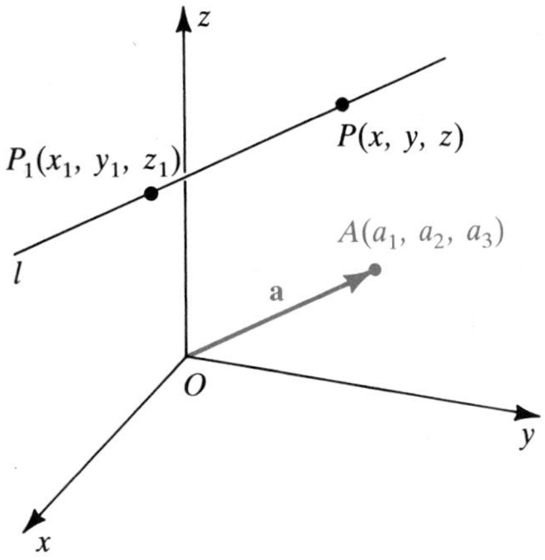

In this section we describe lines and planes by using the vector concepts of *parallel* and *orthogonal*, respectively. We will assume that all lines and planes lie in an xyz-coordinate system.

FIGURE 14.46

Let $\mathbf{a} = \langle a_1, a_2, a_3 \rangle$ be a nonzero vector in V_3, let $P_1(x_1, y_1, z_1)$ be any point, and let $\overrightarrow{OA}$ be the position vector for $\mathbf{a}$. As illustrated in Figure 14.46, we describe **the line l through $P_1(x_1, y_1, z_1)$ parallel to a** as the set of all points $P(x, y, z)$ such that $\overrightarrow{P_1P}$ is parallel to $\overrightarrow{OA}$; that is,

$$\overrightarrow{P_1P} = t\,\overrightarrow{OA} \quad \text{for some scalar } t.$$

In terms of vectors in V_3,

$$\langle x - x_1, y - y_1, z - z_1 \rangle = t\langle a_1, a_2, a_3 \rangle = \langle a_1t, a_2t, a_3t \rangle.$$

Equating components and solving for x, y, and z gives us

$$x = x_1 + a_1t, \quad y = y_1 + a_2t, \quad z = z_1 + a_3t,$$

where t is a real number. These are parametric equations for the line l, with parameter t. The points $P(x, y, z)$ on l are obtained by letting t take on all real values. We have proved the following.

Theorem **(14.34)**

Parametric equations for the line through $P_1(x_1, y_1, z_1)$ parallel to $\mathbf{a} = \langle a_1, a_2, a_3 \rangle$ are

$$x = x_1 + a_1t, \quad y = y_1 + a_2t, \quad z = z_1 + a_3t; \quad t \text{ in } \mathbb{R}.$$

Note that the coefficients of t in (14.34) are the components of the vector **a**. We may obtain the same line by using any nonzero vector that is parallel to **a**, for in this case $\mathbf{b} = c\mathbf{a} = \langle ca_1, ca_2, ca_3 \rangle$ and parametric equations for the line are

$$x = x_1 + (ca_1)v, \quad y = y_1 + (ca_2)v, \quad z = z_1 + (ca_3)v; \quad v \text{ in } \mathbb{R}.$$

These equations determine the same line, since the point given by t can be obtained by letting $v = t/c$.

EXAMPLE 1

(a) Find parametric equations for the line l through $P(5, -2, 4)$ that is parallel to $\mathbf{a} = \langle \frac{1}{2}, 2, -\frac{2}{3} \rangle$.

(b) Where does l intersect the xy-plane?

(c) Sketch the position vector for **a** and the line l.

SOLUTION

(a) To avoid fractions, we shall use the vector $\mathbf{b} = 6\mathbf{a} = \langle 3, 12, -4 \rangle$ instead of **a**. Applying Theorem (14.34), we obtain the following parametric equations for l:

$$x = 5 + 3t, \quad y = -2 + 12t, \quad z = 4 - 4t; \quad t \text{ in } \mathbb{R}$$

FIGURE 14.47

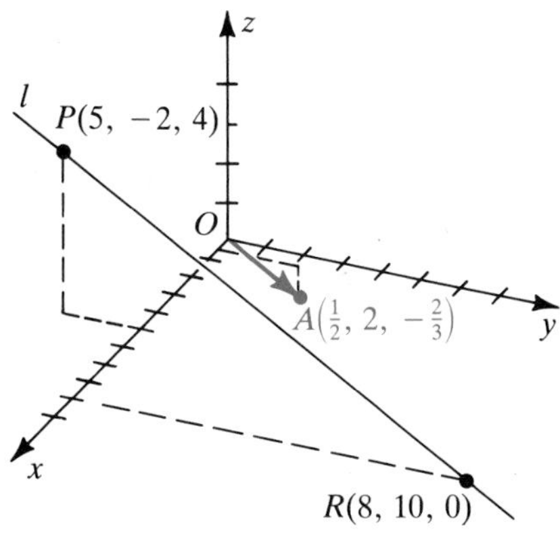

(b) The line intersects the xy-plane at the point $R(x, y, z)$ if $z = 4 - 4t = 0$ (that is, if $t = 1$). Letting $t = 1$ in the parametric equations from part (a), we find that the x- and y-coordinates of R are

$$x = 5 + 3(1) = 8 \quad \text{and} \quad y = -2 + 12(1) = 10.$$

Hence R is the point with coordinates $(8, 10, 0)$.

(c) The position vector $\overrightarrow{OA}$ for $\mathbf{a}$ is shown in Figure 14.47. The line l may be sketched by using the given point $P(5, -2, 4)$ and the point $R(8, 10, 0)$ found in part (b).

FIGURE 14.48

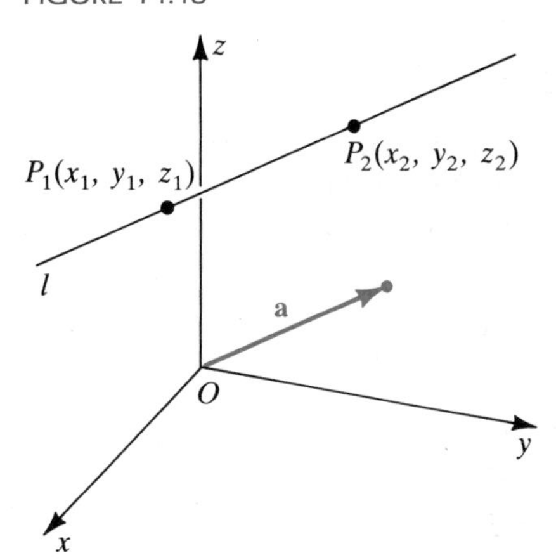

To find parametric equations for the line through two arbitrary points $P_1(x_1, y_1, z_1)$ and $P_2(x_2, y_2, z_2)$, as illustrated in Figure 14.48, we use the vector $\mathbf{a}$ that corresponds to $\overrightarrow{P_1P_2}$—that is,

$$\mathbf{a} = \langle x_2 - x_1, y_2 - y_1, z_2 - z_1 \rangle.$$

Substituting in Theorem (14.34), we obtain the parametric equations

$$x = x_1 + (x_2 - x_1)t, \quad y = y_1 + (y_2 - y_1)t, \quad z = z_1 + (z_2 - z_1)t; \quad t \text{ in } \mathbb{R}.$$

Note that $t = 0$ gives us the point P_1, $t = \frac{1}{2}$ the midpoint of P_1P_2, and $t = 1$ the point P_2. Generally, as t varies from 0 to 1, the point $P(x, y, z)$ traces the line segment P_1P_2. Other values of t give different points on l.

EXAMPLE 2 Find parametric equations for the line l through $P_1(3, 1, -2)$ and $P_2(-2, 7, -4)$.

SOLUTION The vector $\mathbf{a}$ in V_3 corresponding to $\overrightarrow{P_1P_2}$ is

$$\mathbf{a} = \langle -2 - 3, 7 - 1, -4 + 2 \rangle = \langle -5, 6, -2 \rangle.$$

Applying Theorem (14.34), we find that parametric equations for l are

$$x = 3 - 5t, \quad y = 1 + 6t, \quad z = -2 - 2t; \quad t \text{ in } \mathbb{R}.$$

EXAMPLE 3 Let lines l_1 and l_2 have the respective parametrizations

$$x = -2 + 3t, \quad y = 5 - 4t, \quad z = 1 + 2t; \qquad t \text{ in } \mathbb{R}$$
$$x = 1 - v, \qquad y = 3 + 2v, \quad z = -4 - 3v; \quad v \text{ in } \mathbb{R}.$$

Determine whether l_1 and l_2 intersect, and if so, find the point of intersection.

SOLUTION The lines intersect if there are values of t (for l_1) and v (for l_2) that give us the same point $P(x, y, z)$. Hence we consider the following system of three equations in two variables:

$$\begin{cases} -2 + 3t = 1 - v \\ 5 - 4t = 3 + 2v \\ 1 + 2t = -4 - 3v \end{cases} \quad \text{or, equivalently,} \quad \begin{cases} 3t + v = 3 \\ -4t - 2v = -2 \\ 2t + 3v = -5 \end{cases}$$

Solving the first two equations of the system simultaneously, we obtain $t = 2$ and $v = -3$. Substituting into the third equation, $2t + 3v = -5$, we obtain $2(2) + 3(-3) = -5$, or $-5 = -5$. Since $t = 2$ and $v = -3$ is a solution of each of the three equations, the lines intersect. The point of intersection may be found by letting $t = 2$ in the parametrization for l_1 or by letting $v = -3$ in the parametrization for l_2. In either case we obtain the point $P(4, -3, 5)$.

It is important to note that if the solution $t = 2$, and $v = -3$ of the first two equations in the system had *not* satisfied the third equation, then the lines would not have intersected.

In the next definition we use the angle between two vectors (see (14.18)) to define the angle between two lines. We also use vectors to define parallel lines and orthogonal lines.

Definition **(14.35)**

Let l_1 and l_2 be lines that are parallel to the vectors $\mathbf{a}$ and $\mathbf{b}$, and let θ be the angle between $\mathbf{a}$ and $\mathbf{b}$ (see Figure 14.49).

(i) The **angles between l_1 and l_2** are θ and $\pi - \theta$.

(ii) The lines l_1 and l_2 are **parallel** if and only if $\mathbf{a}$ and $\mathbf{b}$ are parallel—that is, $\mathbf{b} = c\mathbf{a}$ for some scalar c.

(iii) The lines l_1 and l_2 are **orthogonal** if and only if $\mathbf{a}$ and $\mathbf{b}$ are orthogonal—that is, $\mathbf{a} \cdot \mathbf{b} = 0$.

FIGURE 14.49

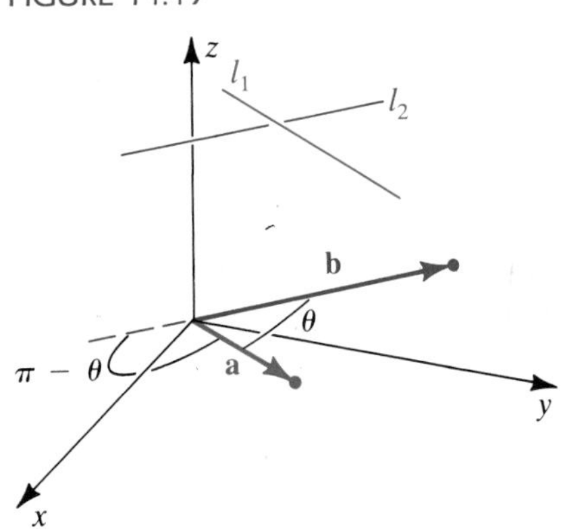

Note that the angle between l_1 and l_2 is defined for either intersecting or nonintersecting lines.

EXAMPLE 4 Let lines l_1 and l_2 have the respective parametrizations for t in $\mathbb{R}$:

$$x = 1 + 2t, \quad y = 3 - 4t, \quad z = -2 + t$$

$$x = -5 - t, \quad y = 2 - 3t, \quad z = 4 + 3t$$

Find the angles between l_1 and l_2.

SOLUTION Referring to the coefficients of t, we see that l_1 and l_2 are parallel, respectively, to

$$\mathbf{a} = \langle 2, -4, 1\rangle \quad \text{and} \quad \mathbf{b} = \langle -1, -3, 3\rangle.$$

If θ is the angle between $\mathbf{a}$ and $\mathbf{b}$, then, by (14.18),

$$\cos\theta = \frac{\mathbf{a}\cdot\mathbf{b}}{\|\mathbf{a}\|\|\mathbf{b}\|} = \frac{(2)(-1) + (-4)(-3) + (1)(3)}{\sqrt{4 + 16 + 1}\sqrt{1 + 9 + 9}} = \frac{13}{\sqrt{399}}.$$

Hence, by Definition (14.35), the angles between l_1 and l_2 are given by

$$\theta = \cos^{-1}\frac{13}{\sqrt{399}} \approx 0.86 \text{ radian} \approx 49°$$

$$\pi - \theta \approx 2.28 \text{ radians} \approx 131°.$$

As illustrated in Figure 14.50(i), if $P_1(x_1, y_1, z_1)$ is a point on a line l, then **the plane through P_1 with normal line l** is the set of points on all lines l' that are orthogonal to l at P_1. To obtain an equivalent definition using vectors, choose another point P_2 on l and consider the vector $\overrightarrow{P_1P_2}$ as in Figure 14.50(ii). **The plane through P_1 with normal vector $\overrightarrow{P_1P_2}$** is defined as the set of all points $P(x, y, z)$ such that $\overrightarrow{P_1P}$ is orthogonal to $\overrightarrow{P_1P_2}$.

FIGURE 14.50

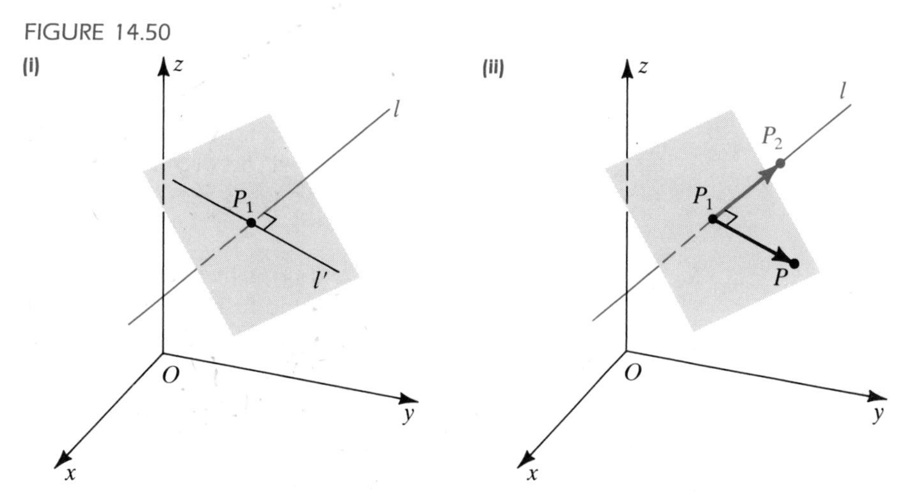

By Theorem (14.21), P is on the plane if and only if $\overrightarrow{P_1P_2} \cdot \overrightarrow{P_1P} = 0$. If $\mathbf{a} = \langle a_1, a_2, a_3 \rangle$ corresponds to $\overrightarrow{P_1P_2}$, this is equivalent to

$$\langle a_1, a_2, a_3 \rangle \cdot \langle x - x_1, y - y_1, z - z_1 \rangle = 0.$$

Applying the definition of dot product (14.16) gives us the next result.

Theorem **(14.36)**

An equation of the plane through $P_1(x_1, y_1, z_1)$ with normal vector $\mathbf{a} = \langle a_1, a_2, a_3 \rangle$ is

$$a_1(x - x_1) + a_2(y - y_1) + a_3(z - z_1) = 0.$$

EXAMPLE 5 Find an equation of the plane through the point $(5, -2, 4)$ with normal vector $\mathbf{a} = \langle 1, 2, 3 \rangle$.

SOLUTION Using Theorem (14.36), we obtain

$$1(x - 5) + 2(y + 2) + 3(z - 4) = 0,$$

or

$$x + 2y + 3z - 13 = 0.$$

EXAMPLE 6 Find an equation of the plane determined by the points $P(4, -3, 1)$, $Q(6, -4, 7)$, and $R(1, 2, 2)$.

SOLUTION The points P, Q, and R determine a plane that contains the triangle shown in Figure 14.51, on the following page. Vectors $\mathbf{a}$ and $\mathbf{b}$

FIGURE 14.51

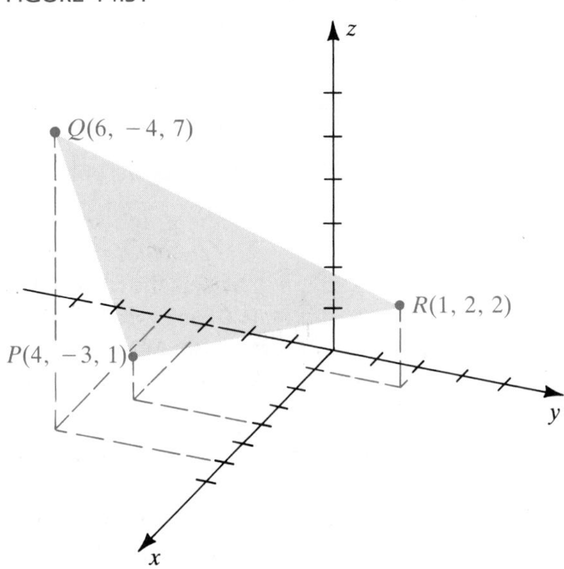

corresponding to $\overrightarrow{PQ}$ and $\overrightarrow{PR}$ are

$$\mathbf{a} = \langle 2, -1, 6 \rangle \quad \text{and} \quad \mathbf{b} = \langle -3, 5, 1 \rangle.$$

The vector $\mathbf{a} \times \mathbf{b}$ is normal to the plane determined by P, Q, and R. From Example 1 of Section 14.4,

$$\mathbf{a} \times \mathbf{b} = -31\mathbf{i} - 20\mathbf{j} + 7\mathbf{k}.$$

Using Theorem (14.36) with $P_1 = P(4, -3, 1)$ and the normal vector $\mathbf{a} \times \mathbf{b}$ gives us the equation

$$-31(x - 4) - 20(y + 3) + 7(z - 1) = 0,$$

or

$$-31x - 20y + 7z + 57 = 0.$$

The equation of the plane in Theorem (14.36) may be written in the form

$$ax + by + cz + d = 0,$$

where $a = a_1$, $b = b_1$, $c = c_1$, and $d = -a_1x_1 - a_2y_1 - a_3z_1$. Conversely, given $ax + by + cz + d = 0$, with a, b, and c not all zero, we may choose numbers x_1, y_1, and z_1 such that $ax_1 + by_1 + cz_1 + d = 0$. It follows that $d = -ax_1 - by_1 - cz_1$, and hence

$$ax + by + cz - ax_1 - by_1 - cz_1 = 0,$$

or

$$a(x - x_1) + b(y - y_1) + c(z - z_1) = 0.$$

According to Theorem (14.36), the graph of the last equation is a plane through $P(x_1, y_1, z_1)$ with normal vector $\langle a, b, c \rangle$. An equation of the form $ax + by + cz + d = 0$, with a, b, and c not all zero, is called a **linear equation in three variables** x, y, and z. We have proved the following theorem.

Theorem (14.37)

The graph of every linear equation $ax + by + cz + d = 0$ is a plane with normal vector $\langle a, b, c \rangle$.

For simplicity we often use the phrase *the plane* $ax + by + cz + d = 0$ instead of the more accurate statement *the plane that has equation* $ax + by + cz + d = 0$.

To sketch the graph of a linear equation, we often find, if possible, the **trace** of the graph in each coordinate plane—that is, the line in which the graph intersects the coordinate plane. To find the trace in the xy-plane, we substitute 0 for z, since this will lead to all points of the graph that lie on the xy-plane. Similarly, to find the trace in the yz-plane or the xz-plane, we let $x = 0$ or $y = 0$, respectively, in the equation $ax + by + cz + d = 0$.

FIGURE 14.52

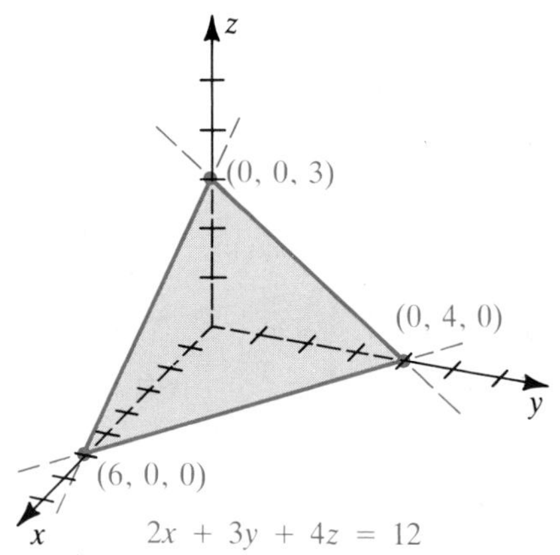

EXAMPLE 7 Sketch the graph of the equation $2x + 3y + 4z = 12$.

SOLUTION We can easily find three points on the plane—namely, the points of intersection of the plane with the coordinate axes. Substituting 0 for both y and z in the equation, we obtain $2x = 12$, or $x = 6$. Thus, the point $(6, 0, 0)$ is on the graph. As in two dimensions, 6 is called the *x-intercept* of the graph. Similarly, substitution of 0 for x and z gives us the *y-intercept* 4, and hence the point $(0, 4, 0)$ is on the graph. The point $(0, 0, 3)$ (or *z-intercept* 3) is obtained by substituting 0 for x and y. The trace in the xy-plane is found by substituting 0 for z in the given equation. This leads to $2x + 3y = 12$, which has as its graph in the xy-plane a line with x-intercept 6 and y-intercept 4. This trace and the traces of the graph in the xz- and yz-planes are illustrated in Figure 14.52.

Definition (14.38)

Two planes with normal vectors **a** and **b** are

(i) **parallel** if **a** and **b** are parallel

(ii) **orthogonal** if **a** and **b** are orthogonal

EXAMPLE 8 Prove that the planes $2x - 3y - z - 5 = 0$ and $-6x + 9y + 3z + 2 = 0$ are parallel.

SOLUTION By Theorem (14.37), the planes have normal vectors $\mathbf{a} = \langle 2, -3, -1 \rangle$ and $\mathbf{b} = \langle -6, 9, 3 \rangle$. Since $\mathbf{b} = -3\mathbf{a}$, the vectors **a** and **b** are parallel and hence, by Definition (14.38), so are the planes.

EXAMPLE 9 Find an equation of the plane through $P(5, -2, 4)$ that is parallel to the plane $3x + y - 6z + 8 = 0$.

SOLUTION By Theorem (14.37), the plane $3x + y - 6z + 8 = 0$ has a normal vector $\mathbf{a} = \langle 3, 1, -6 \rangle$. Hence an equation for a parallel plane has the form

$$3x + y - 6z + d = 0$$

for some real number d. If $P(5, -2, 4)$ is on this plane, then its coordinates must satisfy the equation; that is,

$$3(5) + 1(-2) - 6(4) + d = 0, \quad \text{or} \quad d = 11.$$

This gives us the equation $3x + y - 6z + 11 = 0$.

The vector $\mathbf{i} = \langle 1, 0, 0 \rangle$ is a normal vector for the yz-plane. A plane that has an equation of the form $x - x_1 = 0$ (or $x = x_1$) also has normal vector $\mathbf{i}$ and hence is parallel to the yz-plane (and orthogonal to both the xy- and xz-planes). The portion of the graph of $x = a$ that lies in the first octant is sketched in Figure 14.53(i). Similarly, the graph of $y = b$ is a plane parallel to the xz-plane with y-intercept b, and the graph of $z = c$ is a plane parallel to the xy-plane with z-intercept c (see (ii) and (iii) of Figure 14.53).

FIGURE 14.53

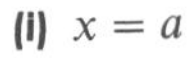

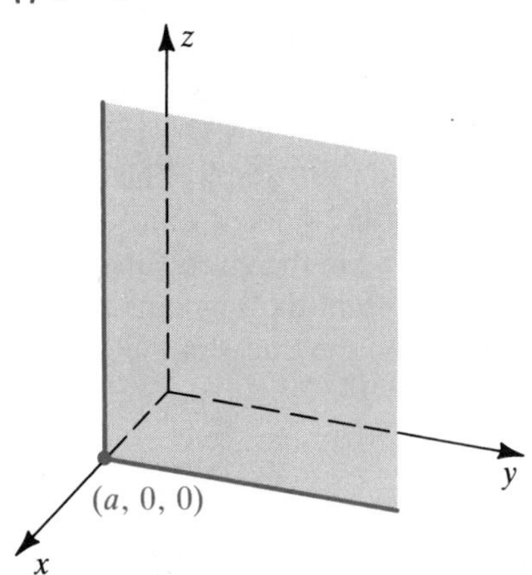

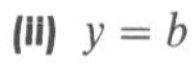

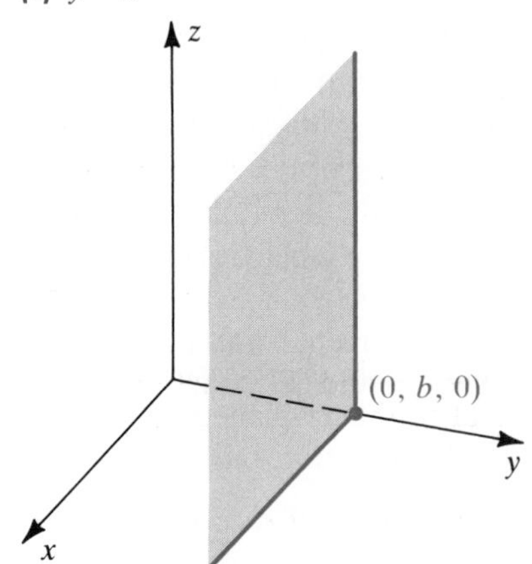

(iii) $z = c$

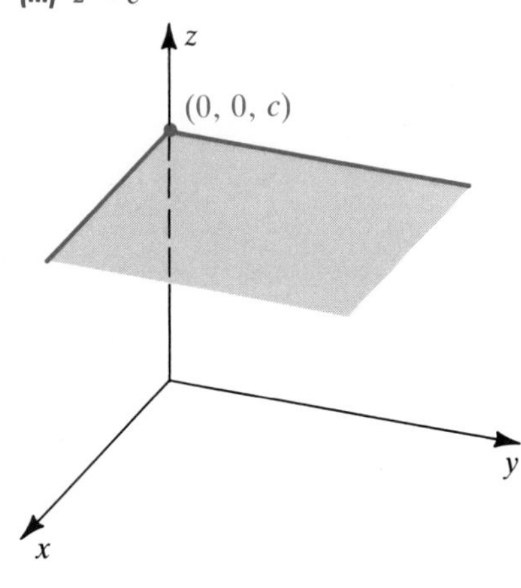

A plane with an equation of the form $by + cz + d = 0$ has normal vector $\mathbf{a} = \langle 0, b, c \rangle$ and is orthogonal to the yz-plane, since $\mathbf{a} \cdot \mathbf{i} = 0$. Similarly, graphs of $ax + by + d = 0$ and $ax + cz + d = 0$ are planes that are orthogonal to the xy-plane and xz-plane, respectively.

FIGURE 14.54

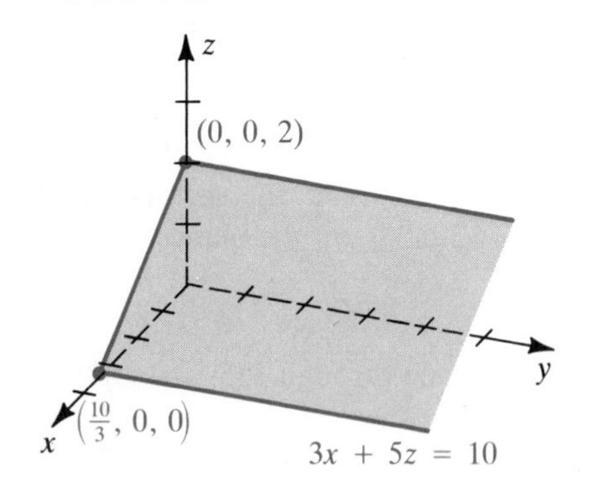

EXAMPLE 10 Sketch the plane $3x + 5z = 10$.

SOLUTION The plane is orthogonal to the xz-plane and has x-intercept $\frac{10}{3}$ and z-intercept 2. Note that the trace in the yz-plane has equation $5z = 10$ and hence is a line parallel to the y-axis with z-intercept 2. Similarly, the trace in the xy-plane has equation $3x = 10$ and is a line parallel to the y-axis with x-intercept $\frac{10}{3}$. A portion of the graph showing traces in the three coordinate planes is sketched in Figure 14.54.

Lines may be described as intersections of planes. If a line l is given parametrically as in Theorem (14.34) and if a_1, a_2, a_3 are different from zero, we may solve each equation for t, obtaining

$$t = \frac{x - x_1}{a_1}, \quad t = \frac{y - y_1}{a_2}, \quad t = \frac{z - z_1}{a_3}.$$

It follows that a point $P(x, y, z)$ is on l if and only if the following equations, called *the symmetric form for* l, are satisfied.

Symmetric form for a line **(14.39)**

$$\frac{x - x_1}{a_1} = \frac{y - y_1}{a_2} = \frac{z - z_1}{a_3}$$

The symmetric form for a line is not unique, since in (14.39) we may use any three numbers b_1, b_2, b_3 that are proportional to a_1, a_2, a_3 or any point on l other than (x_1, y_1, z_1).

If, in (14.39), we take the indicated expressions in pairs, say

$$\frac{x - x_1}{a_1} = \frac{y - y_1}{a_2} \quad \text{and} \quad \frac{x - x_1}{a_1} = \frac{z - z_1}{a_3},$$

we obtain a description of l as an intersection of two planes, the first orthogonal to the xy-plane and the second orthogonal to the xz-plane. If one of the numbers a_1, a_2, or a_3 is zero, we cannot solve each equation in (14.34) for t. For example, if $a_3 = 0$ and $a_1 a_2 \neq 0$, then the third equation reduces to $z = z_1$, and a symmetric form may be written as

$$\frac{x - x_1}{a_1} = \frac{y - y_1}{a_2}, \quad z = z_1,$$

which again expresses l as an intersection of two planes, the first orthogonal to the xy-plane and the second, $z = z_1$, parallel to the xy-plane. A similar situation exists if $a_1 = 0$ or $a_2 = 0$.

EXAMPLE 11 Find a symmetric form for the line through $P_1(3, 1, -2)$ and $P_2(-2, 7, -4)$.

SOLUTION As in Example 2, a parametric representation for the line is

$$x = 3 - 5t, \quad y = 1 + 6t, \quad z = -2 - 2t; \quad t \text{ in } \mathbb{R}.$$

Solving each equation for t and equating the results, we obtain the symmetric form

$$\frac{x - 3}{-5} = \frac{y - 1}{6} = \frac{z + 2}{-2}.$$

EXAMPLE 12 Let l be the line of intersection of the two planes

$$2x - y + 4z = 4 \quad \text{and} \quad x + 3y - 2z = 1.$$

Find parametric equations for l.

SOLUTION A point $P(x, y, z)$ is on l if and only if P is on each of the planes—that is, if and only if (x, y, z) is a solution of the system of equations

$$\begin{cases} 2x - y + 4z = 4 \\ x + 3y - 2z = 1 \end{cases}$$

or, equivalently,

$$\begin{cases} 2x - y = 4 - 4z \\ x + 3y = 1 + 2z \end{cases}$$

Let us regard this as a system of two equations in x and y. By eliminating y from the system, we can express x in terms of z. Similarly, eliminating x from the system leads to a formula for y in terms of z. This procedure gives us the equivalent system

$$\begin{cases} x = \frac{13}{7} - \frac{10}{7}z \\ y = -\frac{2}{7} + \frac{8}{7}z \end{cases}$$

By letting z take on every real value t, we obtain the solutions of the system—that is, the points on l. Thus, parametric equations for l are

$$x = \tfrac{13}{7} - \tfrac{10}{7}t, \quad y = -\tfrac{2}{7} + \tfrac{8}{7}t, \quad z = t; \quad t \text{ in } \mathbb{R}.$$

In the next example we use vector methods to find the distance from a point to a plane.

EXAMPLE 13 Find a formula for the distance h from a point $P_0(x_0, y_0, z_0)$ to the plane $ax + by + cz + d = 0$.

FIGURE 14.55

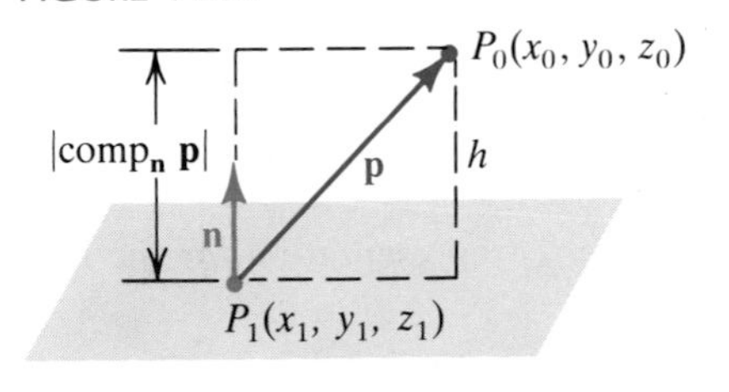

SOLUTION As in Figure 14.55, let $P_1(x_1, y_1, z_1)$ be any point in the plane, and let $\mathbf{n}$ be a normal vector to the plane. The vector $\mathbf{p}$ corresponding to $\overrightarrow{P_1P_0}$ is

$$\mathbf{p} = \langle x_0 - x_1, y_0 - y_1, z_0 - z_1 \rangle.$$

Referring to the figure, we see that

$$h = |\text{comp}_{\mathbf{n}}\, \mathbf{p}| = \left| \mathbf{p} \cdot \frac{1}{\|\mathbf{n}\|} \mathbf{n} \right|.$$

Since $\langle a, b, c \rangle$ is a normal vector to the plane, we may let

$$\frac{1}{\|\mathbf{n}\|}\mathbf{n} = \frac{1}{\sqrt{a^2 + b^2 + c^2}} \langle a, b, c \rangle.$$

Consequently

$$\begin{aligned} h = \left| \mathbf{p} \cdot \frac{1}{\|\mathbf{n}\|} \mathbf{n} \right| &= \left| \frac{a(x_0 - x_1) + b(y_0 - y_1) + c(z_0 - z_1)}{\sqrt{a^2 + b^2 + c^2}} \right| \\ &= \frac{|(ax_0 + by_0 + cz_0) + (-ax_1 - by_1 - cz_1)|}{\sqrt{a^2 + b^2 + c^2}}. \end{aligned}$$

Since P_1 is on the plane, (x_1, y_1, z_1) is a solution of $ax + by + cz + d = 0$, and hence $ax_1 + by_1 + cz_1 + d = 0$, or $d = -ax_1 - by_1 - cz_1$. Thus, the preceding formula for h may be written

$$h = \frac{|ax_0 + by_0 + cz_0 + d|}{\sqrt{a^2 + b^2 + c^2}}.$$

Two lines are **skew** if they are not parallel and do not intersect. In the next example we find a formula for the shortest distance between skew lines.

EXAMPLE 14 Find a formula for the shortest distance d between two skew lines l_1 and l_2.

FIGURE 14.56

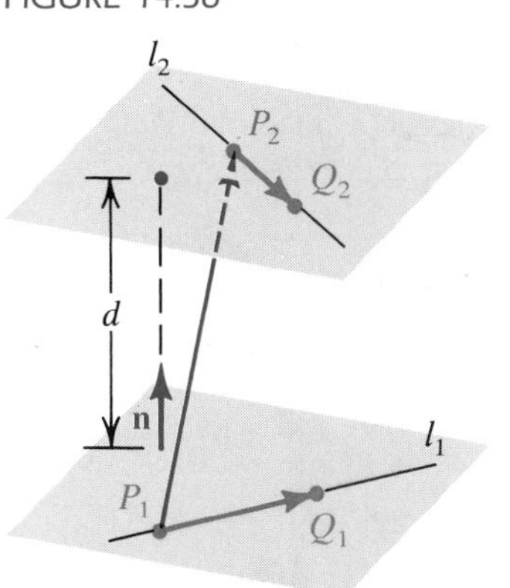

SOLUTION Choose points P_1, Q_1 on l_1 and P_2, Q_2 on l_2, as illustrated in Figure 14.56. By Theorem (14.29), $\overrightarrow{P_1Q_1} \times \overrightarrow{P_2Q_2}$ is orthogonal to both $\overrightarrow{P_1Q_1}$ and $\overrightarrow{P_2Q_2}$, and hence a *unit* vector $\mathbf{n}$ orthogonal to both $\overrightarrow{P_1Q_1}$ and $\overrightarrow{P_2Q_2}$ is

$$\mathbf{n} = \frac{1}{\|\overrightarrow{P_1Q_1} \times \overrightarrow{P_2Q_2}\|}(\overrightarrow{P_1Q_1} \times \overrightarrow{P_2Q_2}).$$

Let us consider planes through P_1 and P_2, respectively, each having normal vector $\mathbf{n}$. These planes are parallel and contain l_1 and l_2, respectively. The distance d between the planes is measured along a line parallel to the common normal $\mathbf{n}$, as shown in the figure. It follows that d is the shortest distance between l_1 and l_2. Moreover,

$$\begin{aligned} d &= |\text{comp}_{\mathbf{n}}\, \overrightarrow{P_1P_2}| = |\mathbf{n} \cdot \overrightarrow{P_1P_2}| \\ &= \frac{1}{\|\overrightarrow{P_1Q_1} \times \overrightarrow{P_2Q_2}\|} |(\overrightarrow{P_1Q_1} \times \overrightarrow{P_2Q_2}) \cdot \overrightarrow{P_1P_2}|. \end{aligned}$$

EXERCISES 14.5

Exer. 1–4: Find parametric equations for the line through P parallel to a.

1 $P(4, 2, -3)$; $\mathbf{a} = \langle \frac{1}{3}, 2, \frac{1}{2} \rangle$

2 $P(5, 0, -2)$; $\mathbf{a} = \langle -1, -4, 1 \rangle$

3 $P(0, 0, 0)$; $\mathbf{a} = \mathbf{j}$

4 $P(1, 2, 3)$; $\mathbf{a} = \mathbf{i} + 2\mathbf{j} + 3\mathbf{k}$

Exer. 5–8: Find parametric equations for the line through P_1 and P_2. Determine (if possible) the points at which the line intersects each of the coordinate planes.

5 $P_1(5, -2, 4)$, $P_2(2, 6, 1)$

6 $P_1(-3, 1, -1)$, $P_2(7, 11, -8)$

7 $P_1(2, 0, 5)$, $P_2(-6, 0, 3)$

8 $P_1(2, -2, 4)$, $P_2(2, -2, -3)$

9 If l has parametric equations $x = 5 - 3t$, $y = -2 + t$, $z = 1 + 9t$, find parametric equations for the line through $P(-6, 4, -3)$ that is parallel to l.

10 Find parametric equations for the line through the point $P(4, -1, 0)$ that is parallel to the line through the points $P_1(-3, 9, -2)$ and $P_2(5, 7, -3)$.

Exer. 11–14: Determine whether the two lines intersect, and if so, find the point of intersection.

11 $x = 1 + 2t, \quad y = 1 - 4t, \quad z = 5 - t$

$x = 4 - v, \quad y = -1 + 6v, \quad z = 4 + v$

12 $x = 1 - 6t, \quad y = 3 + 2t, \quad z = 1 - 2t$

$x = 2 + 2v, \quad y = 6 + v, \quad z = 2 + v$

13 $x = 3 + t, \quad y = 2 - 4t, \quad z = t$

$x = 4 - v, \quad y = 3 + v, \quad z = -2 + 3v$

14 $x = 2 - 5t, \quad y = 6 + 2t, \quad z = -3 - 2t$

$x = 4 - 3v, \quad y = 7 + 5v, \quad z = 1 + 4v$

Exer. 15–18: Equations for two lines l_1 and l_2 are given. Find the angles between l_1 and l_2.

15 $x = 7 - 2t, \quad y = 4 + 3t, \quad z = 5t$

$x = -1 + 4t, \quad y = 3 + 4t, \quad z = 1 + t$

16 $x = 5 + 3t, \quad y = 4 - t, \quad z = 3 + 2t$

$x = -t, \quad y = 1 - 2t, \quad z = 3 + t$

17 $\dfrac{x-1}{-3} = \dfrac{y+2}{8} = \dfrac{z}{-3}; \quad \dfrac{x+2}{10} = \dfrac{y}{10} = \dfrac{z-4}{-7}$

18 $\dfrac{x}{3} = \dfrac{y-2}{3} = z - 1; \quad \dfrac{x+5}{4} = \dfrac{y-1}{-3} = \dfrac{z+7}{-9}$

Exer. 19–26: Find an equation of the plane that satisfies the stated conditions.

19 Through $P(6, -7, 4)$ and parallel to

(a) the xy-plane (b) the yz-plane (c) the xz-plane

20 Through $P(-2, 5, -8)$ with normal vector

(a) $\mathbf{i}$ (b) $\mathbf{j}$ (c) $\mathbf{k}$

21 Through $P(-11, 4, -2)$ with normal vector $\mathbf{a} = 6\mathbf{i} - 5\mathbf{j} - \mathbf{k}$

22 Through $P(4, 2, -9)$ with normal vector $\overrightarrow{OP}$

23 Through $P(2, 5, -6)$ and parallel to the plane $3x - y + 2z = 10$

24 Through the origin and parallel to the plane $x - 6y + 4z = 7$

25 Through $P(-4, 1, 6)$ and having the same trace in the xz-plane as the plane $x + 4y - 5z = 8$

26 Through the origin and the points $P(0, 2, 5)$ and $Q(1, 4, 0)$

Exer. 27–28: Find an equation of the plane through P, Q, and R.

27 $P(1, 1, 3), \quad Q(-1, 3, 2), \quad R(1, -1, 2)$

28 $P(3, 2, 1), \quad Q(-1, 1, -2), \quad R(3, -4, 1)$

Exer. 29–36: Sketch the graph of the equation in an xyz-coordinate system.

29 (a) $x = 3$ (b) $y = -2$ (c) $z = 5$

30 (a) $x = -4$ (b) $y = 0$ (c) $z = -\frac{2}{3}$

31 $2x + y - 6 = 0$

32 $3x - 2z - 24 = 0$

33 $2y - 3z - 9 = 0$

34 $5x + y - 4z + 20 = 0$

35 $2x - y + 5z + 10 = 0$

36 $x + y + z = 0$

Exer. 37–40: Find an equation of the plane.

37

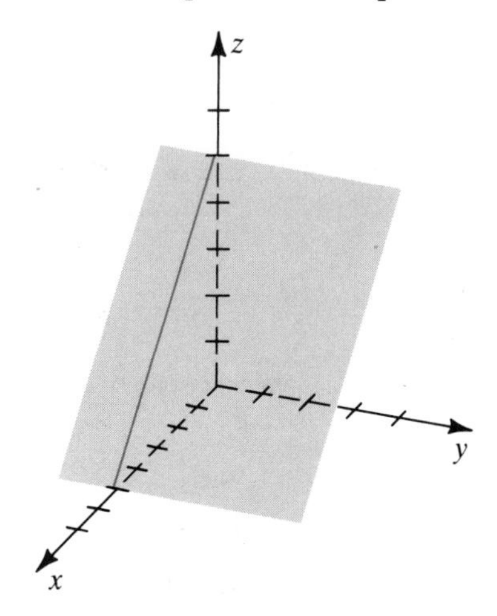

38

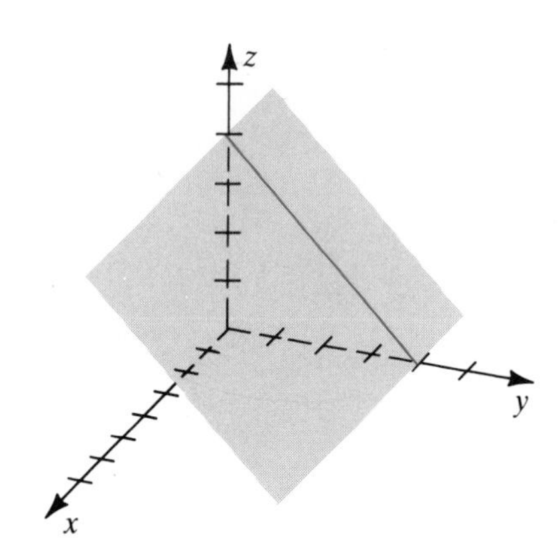

39

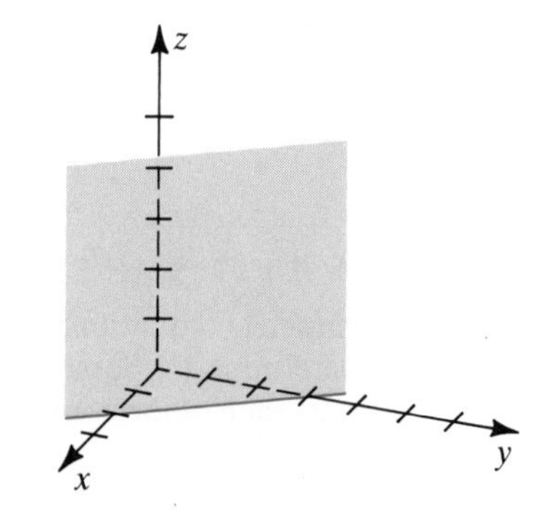

40

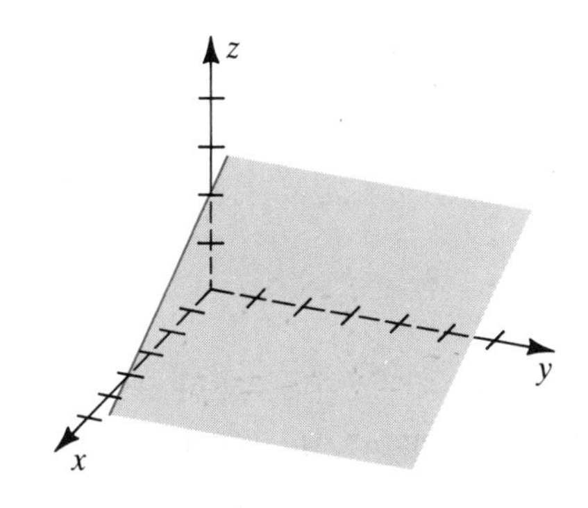

Exer. 41–42: Find an equation of the plane through P that is parallel to the given plane.

41 $P(1, 2, -3)$; $4x - y + 3z - 7 = 0$

42 $P(3, -2, 4)$; $-2x + 3y - z + 5 = 0$

Exer. 43–46: Find a symmetric form for the line through P_1 and P_2.

43 $P_1(5, -2, 4)$, $P_2(2, 6, 1)$

44 $P_1(-3, 1, -1)$, $P_2(7, 11, -8)$

45 $P_1(4, 2, -3)$, $P_2(-3, 2, 5)$

46 $P_1(5, -7, 4)$, $P_2(-2, -1, 4)$

Exer. 47–50: Find parametric equations for the line of intersection of the two planes.

47 $x + 2y - 9z = 7$, $2x - 3y + 17z = 0$

48 $2x + 5y + 16z = 13$, $-x - 2y - 6z = -5$

49 $-2x + 3y + 9z = 12$, $x - 2y - 5z = -8$

50 $5x - y - 12z = 15$, $2x + 3y + 2z = 6$

Exer. 51–52: Refer to Example 13. Find the distance from P to the plane.

51 $P(1, -1, 2)$; $3x - 7y + z - 5 = 0$

52 $P(3, 1, -2)$; $2x + 4y - 5z + 1 = 0$

Exer. 53–54: Show that the two planes are parallel and find the distance between the planes.

53 $4x - 2y + 6z = 3$, $-6x + 3y - 9z = 4$

54 $3x + 12y - 6z = -2$, $5x + 20y - 10z = 7$

Exer. 55–56: Refer to Example 14. Let l_1 be the line through A and B, and let l_2 be the line through C and D. Find the shortest distance between l_1 and l_2.

55 $A(1, -2, 3)$, $B(2, 0, 5)$; $C(4, 1, -1)$, $D(-2, 3, 4)$

56 $A(1, 3, 0)$, $B(0, 4, 5)$; $C(-2, -1, 2)$, $D(5, 1, 0)$

Exer. 57–58: Find an equation of the plane that contains the point P and the line.

57 $P(5, 0, 2)$; $x = 3t + 1$, $y = -2t + 4$, $z = t - 3$

58 $P(4, -3, 0)$; $x = t + 5$, $y = 2t - 1$, $z = -t + 7$

Exer. 59–60: Use a dot product to find the distance from A to the line through B and C.

59 $A(2, -6, 1)$; $B(3, 4, -2)$, $C(7, -1, 5)$

60 $A(1, 5, 0)$; $B(-2, 1, -4)$, $C(0, -3, 2)$

Exer. 61–62: Find the distance from the point P to the line.

61 $P(2, 1, -2)$; $x = 3 - 2t$, $y = -4 + 3t$, $z = 1 + 2t$

62 $P(3, 1, -1)$; $x = 1 + 4t$, $y = 3 - t$, $z = 3t$

Exer. 63–64: If a plane has nonzero x-, y-, and z-intercepts a, b, and c, respectively, then its *intercept form* is

$$\frac{x}{a} + \frac{y}{b} + \frac{z}{c} = 1.$$

Find the intercept form for the given plane.

63 $10x - 15y + 6z = 30$

64 $12x + 15y - 20z = 60$

Exer. 65–66: Find an equation for the plane of the form $Ax + By + Cz = D$.

65

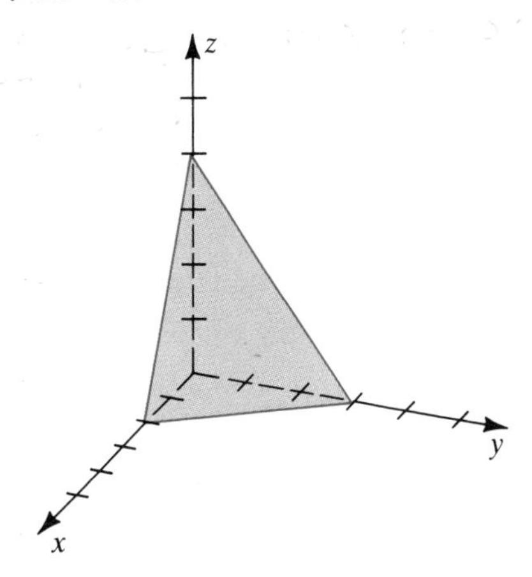

66

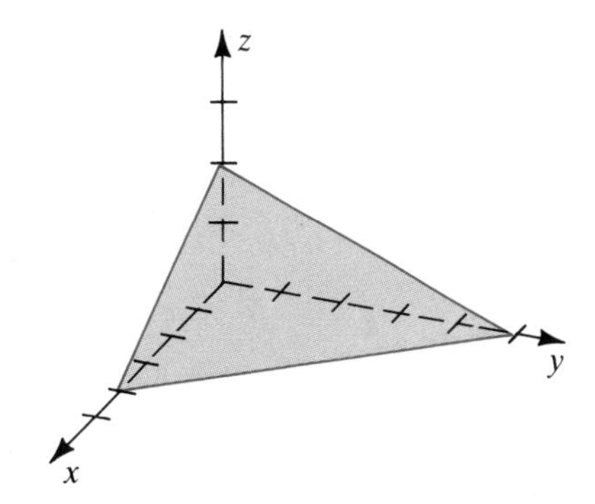

c **Exer. 67–68: Graph f and g on the same coordinate plane for $-2 \le x \le 2$. (a) Estimate the coordinates of their point P of intersection. (b) Approximate the angles between the tangent lines to the graphs at P.**

67 $f(x) = \sin(x^2)$, $g(x) = \cos x - x$

68 $f(x) = 1 - 3x + x^3$, $g(x) = x^5 + \frac{1}{2}$

14.6 SURFACES

To sketch a surface with pencil and paper, we usually choose the coordinate axes as in Figure 14.21, regarding the y- and z-axes as lying in the plane of the paper and the x-axis as projecting out from the paper. This technique is illustrated in Figures 14.57–14.67 and the charts in this section. A disadvantage of choosing coordinate axes in this way is that when a specific equation is graphed, the shape of the resulting surface may seem distorted. For example, circular cross-sections may appear to be elliptical and vice versa. For this reason, graphs in three dimensions that are illustrated later in the text are computer-generated, with axes and units of distance chosen to provide an undistorted view of a surface. In such an *axonometric system*, none of the coordinate planes coincides with the plane of the paper. After you have studied this section carefully and learned how to identify a surface from its equation, you should strive to become proficient in using any choice of coordinate axes.

In the preceding section we defined the *trace* of the graph of a linear equation in a coordinate plane. More generally, if S is any surface (that is, the graph of an equation in x, y, and z), then the **trace** of S in a plane is the intersection of S and the plane. To sketch a surface, we make considerable use of traces. Of special importance are traces in the coordinate planes. These three traces are the **xy-trace**, the **yz-trace**, and the **xz-trace**. It may also be convenient to find traces in planes that are *parallel* to the coordinate planes. The following charts show how to find equations of traces from an equation of a surface. The indicated sketch of the trace is only one possibility. In typical problems the trace may be a conic section, a line, or some other curve. Sometimes there is *no* trace; that is, the surface may not intersect a given plane.

Trace	To find equation of trace	Sketch of trace
xy-trace	Let $z = 0$	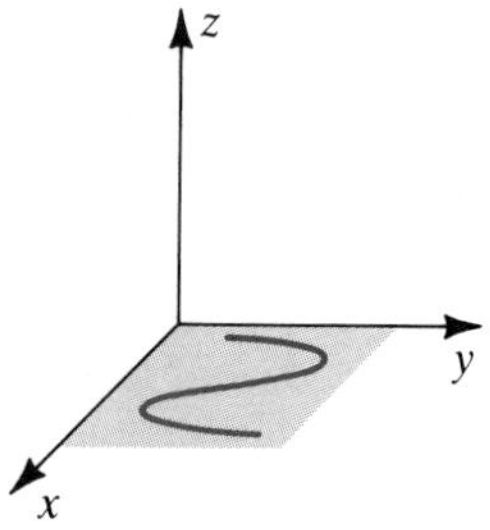 z, y, x

Trace	To find equation of trace	Sketch of trace
yz-trace	Let $x = 0$	z, y, x
xz-trace	Let $y = 0$	z, y, x
On $z = k$	Let $z = k$	z, $z = k$, y, x
On $x = k$	Let $x = k$	z, $x = k$, y, x
On $y = k$	Let $y = k$	z, y, $y = k$, x

EXAMPLE 1 Sketch the graph of $z = x^2 + y^2$.

SOLUTION The graph is a surface in an xyz-coordinate system. The following chart indicates pertinent traces on planes.

Trace	Equation of trace	Description of trace	Sketch of trace
xy-trace	$0 = x^2 + y^2$	Origin	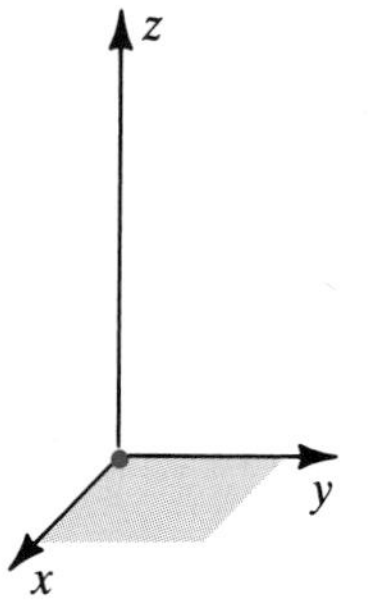
yz-trace	$z = y^2$	Parabola	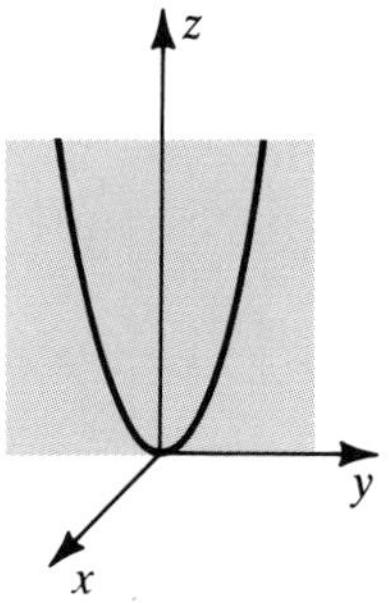
xz-trace	$z = x^2$	Parabola	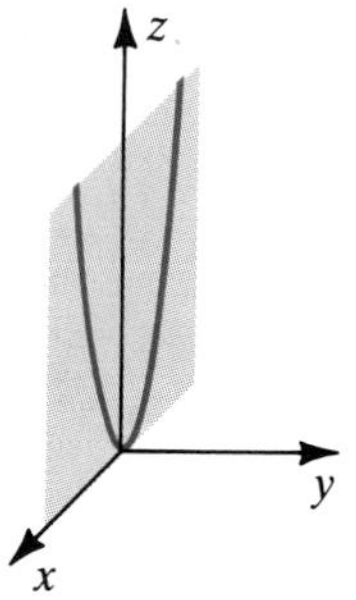
On $z = k$	$k = x^2 + y^2$, or $x^2 + y^2 = k$	Circle, point, or no graph	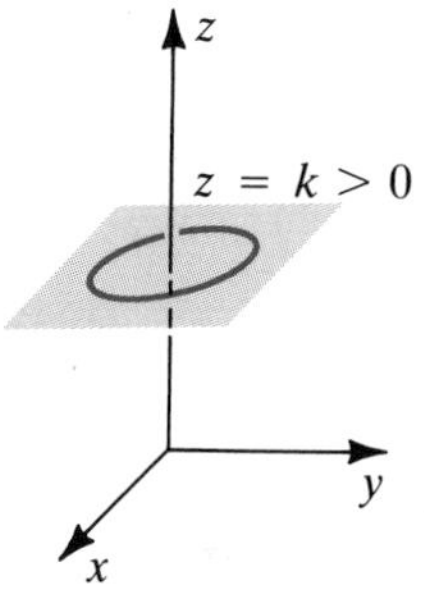 $z = k > 0$ No graph if $k < 0$

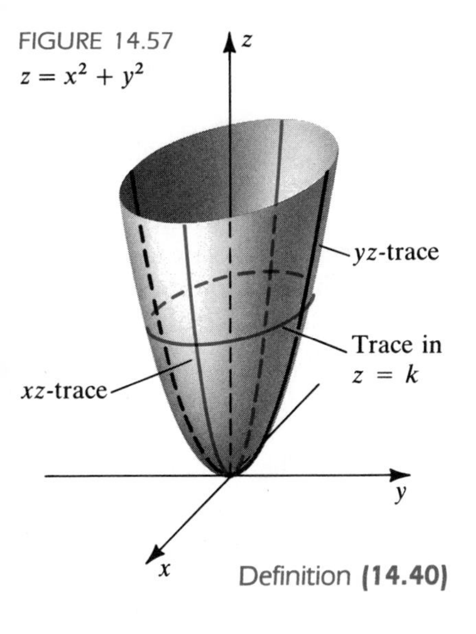

FIGURE 14.57
$z = x^2 + y^2$

We next sketch these traces using only *one* set of coordinate axes. Noting that the radius $\sqrt{k}$ of the circular trace on $z = k$ increases as z increases leads to the sketch in Figure 14.57.

Traces on the planes $x = k$ and $y = k$ are parabolas. To illustrate, the trace on the plane $y = 1$ is the parabola $z = x^2 + 1$, and the trace on the plane $x = 2$ is the parabola $z = y^2 + 4$. We shall not sketch these, since the circular traces on $z = k$ are sufficient for indicating the graph. The surface in this example may be regarded as having been generated by revolving the graph of the parabola $z = y^2$ in the yz-plane about the z-axis. This surface is called a **circular paraboloid**, or **paraboloid of revolution**.

You may have previously regarded a cylinder as an object that has the shape of a tin can with circular cross sections. The next definition indicates that cylinders may have a variety of shapes.

Definition **(14.40)**

> Let C be a curve in a plane, and let l be a line that is not in a parallel plane. The set of points on all lines that are parallel to l and intersect C is a **cylinder**.

Two examples of cylinders are sketched in Figure 14.58. The curve C in the plane is called a **directrix** for the cylinder, and each line through C parallel to l is a **ruling** of the cylinder. The cylinder in Figure 14.58(i) is a **right circular cylinder**, obtained if C is a circle in a plane and l is a line perpendicular to the plane. Although we have cut off the cylinder in the figure, the rulings extend indefinitely. As illustrated in (ii) of the figure, the *directrix C is not necessarily a closed curve*.

FIGURE 14.58

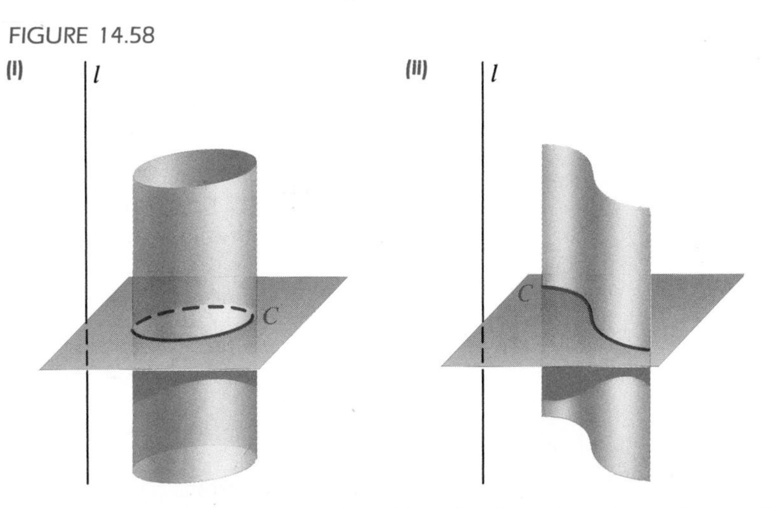

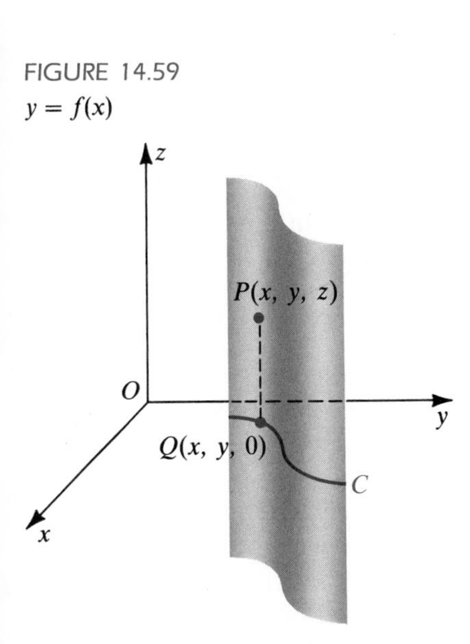

FIGURE 14.59
$y = f(x)$

Let us consider a cylinder in which the directrix C is on the xy-plane and has equation $y = f(x)$ for some function f. Suppose also that the rulings are parallel to the z-axis. As in Figure 14.59, a point $P(x, y, z)$ is on the cylinder if and only if $Q(x, y, 0)$ is on C—that is, if and only if the first two coordinates x and y of P satisfy the equation $y = f(x)$. Thus, an equation of the cylinder is $y = f(x)$, which is the same as the equation of the directrix in the xy-plane.

FIGURE 14.60
$\frac{x^2}{4} + \frac{y^2}{9} = 1$

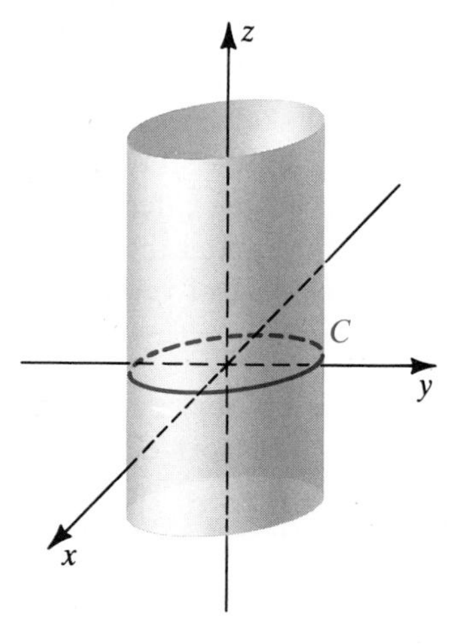

EXAMPLE 2 Sketch the graph of $\frac{x^2}{4} + \frac{y^2}{9} = 1$ in three dimensions.

SOLUTION From our previous remarks, the graph is a cylinder with rulings parallel to the z-axis. We begin by sketching the graph of the equation $(x^2/4) + (y^2/9) = 1$ in the xy-plane. This ellipse is a directrix C for the cylinder. All traces in planes parallel to the xy-plane are ellipses congruent to this directrix. When sketching a cylinder we usually draw at least two such traces, as indicated in Figure 14.60. Since the directrix is an ellipse, we call this surface an **elliptic cylinder**.

The graph of an equation that contains only the variables y and z is a cylinder with rulings parallel to the x-axis and whose trace (directrix) in the yz-plane is the graph of the given equation. Similarly, the graph of an equation that does not contain the variable y is a cylinder with rulings parallel to the y-axis and whose directrix is the graph of the given equation in the xz-plane.

FIGURE 14.61
$y^2 = 9 - z$

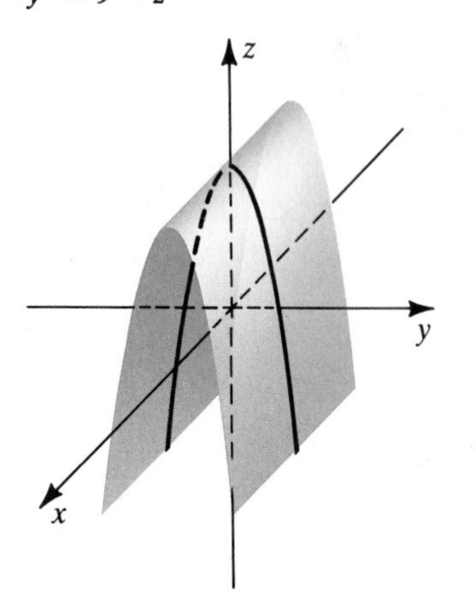

EXAMPLE 3 Sketch the graphs of the following equations in three dimensions:

(a) $y^2 = 9 - z$ **(b)** $z = \sin x$

SOLUTION

(a) The graph of $y^2 = 9 - z$ is a cylinder with rulings parallel to the x-axis. A directrix in the yz-plane is the parabola $y^2 = 9 - z$. Part of the graph (a **parabolic cylinder**) is sketched in Figure 14.61.

(b) The graph of $z = \sin x$ is a cylinder with rulings parallel to the y-axis and whose directrix in the xz-plane is the sine curve $z = \sin x$. A portion of the graph is sketched in Figure 14.62.

FIGURE 14.62
$z = \sin x$

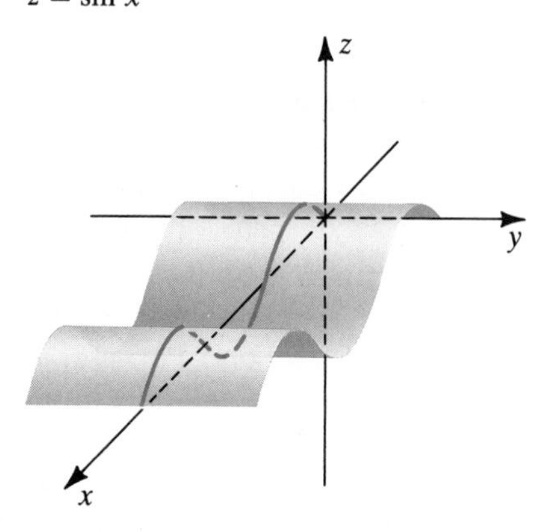

In Chapter 12 we saw that, in two dimensions, the graph of any second-degree equation in x and y,

$$Ax^2 + Bxy + Cy^2 + Dx + Ey + F = 0,$$

is a conic section (except for degenerate cases). In three dimensions, the graph of a second-degree equation in x, y, and z,

$$Ax^2 + By^2 + Cz^2 + Dxy + Exz + Fyz + Gx + Hy + Iz + J = 0,$$

is a **quadric surface** (except for degenerate cases). For simplicity, we shall limit our discussion to the case in which the coefficients D, E, F, G, H, and I are all zero. More general equations can be reduced to this case by using translations and possible rotations of axes.

There are three types of quadric surfaces: *ellipsoids*, *hyperboloids*, and *paraboloids*. The names are derived from the fact that traces in planes parallel to coordinate planes are usually ellipses, hyperbolas, and parabolas, respectively. In the remainder of this section, unless noted otherwise, a, b, and c denote positive real numbers. The graph of the following equation is an ellipsoid. The chart that follows indicates traces of this surface in the coordinate planes.

Ellipsoid **(14.41)**

$$\frac{x^2}{a^2} + \frac{y^2}{b^2} + \frac{z^2}{c^2} = 1$$

Trace	Equation of trace	Description of trace	Sketch of trace
xy-trace	$\frac{x^2}{a^2} + \frac{y^2}{b^2} = 1$	Ellipse	
yz-trace	$\frac{y^2}{b^2} + \frac{z^2}{c^2} = 1$	Ellipse	
xz-trace	$\frac{x^2}{a^2} + \frac{z^2}{c^2} = 1$	Ellipse	

FIGURE 14.63
$\frac{x^2}{a^2} + \frac{y^2}{b^2} + \frac{z^2}{c^2} = 1$

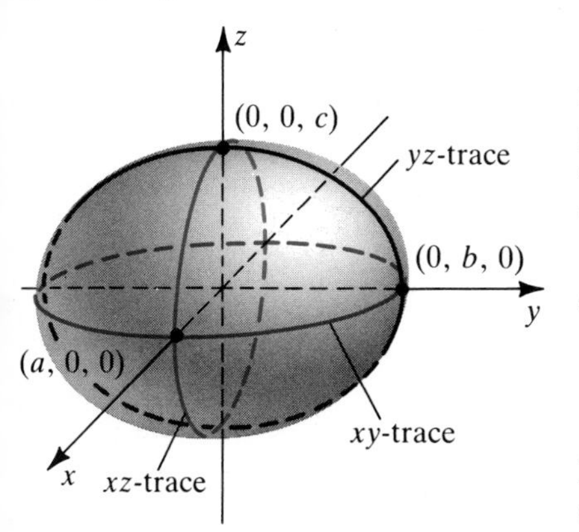

We next sketch these traces using only *one* set of coordinate axes, as illustrated in Figure 14.63. After you become familiar with quadric surfaces, these three traces will suffice to sketch an accurate graph. However, to complete our discussion, let us consider traces on planes parallel to the coordinate planes.

To find the trace on a plane $z = k$ parallel to the xy-plane, we let $z = k$ in the equation of the ellipsoid, obtaining

$$\frac{x^2}{a^2} + \frac{y^2}{b^2} = 1 - \frac{k^2}{c^2}.$$

If $|k| > c$, then $1 - (k^2/c^2) < 0$ and there is no graph. Thus, the graph of (14.41) lies between the planes $z = -k$ and $z = k$. If $|k| < c$, then

FIGURE 14.64

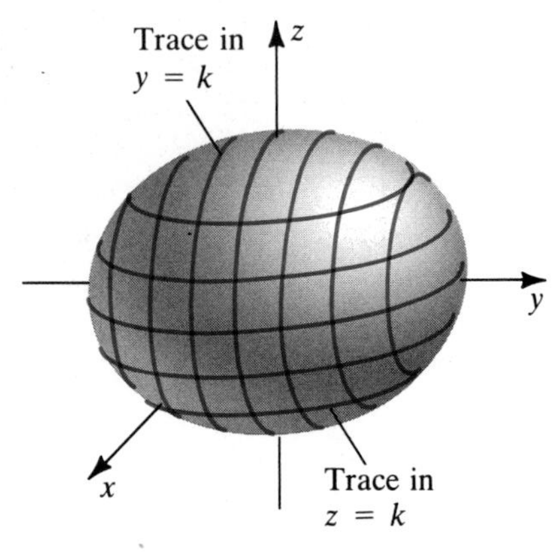

$1 - (k^2/c^2) > 0$ and hence the trace in the plane $z = k$ is an ellipse, as illustrated in Figure 14.64. Viewable portions of other traces in planes parallel to the xy-plane are also shown in the figure.

The trace on a plane $y = k$ parallel to the xz-plane is

$$\frac{x^2}{a^2} + \frac{z^2}{c^2} = 1 - \frac{k^2}{b^2}.$$

Viewable portions of several of these traces (ellipses) for $-b < k < b$ are shown in Figure 14.64.

Finally, letting $x = k$ for $-a < k < a$, we see that the trace in a plane parallel to the yz-plane is an ellipse. (These ellipses are not shown in Figure 14.64).

Note that if $a = b = c$, then the graph of (14.41) is a sphere of radius a with center at the origin.

The graph of the following equation is a *hyperboloid of one sheet*. The chart that follows indicates traces in the coordinate planes.

Hyperboloid of one sheet **(14.42)**

$$\frac{x^2}{a^2} + \frac{y^2}{b^2} - \frac{z^2}{c^2} = 1$$

Trace	Equation of trace	Description of trace	Sketch of trace
xy-trace	$\frac{x^2}{a^2} + \frac{y^2}{b^2} = 1$	Ellipse	z, $(0, b, 0)$, y, $(a, 0, 0)$, x
yz-trace	$\frac{y^2}{b^2} - \frac{z^2}{c^2} = 1$	Hyperbola	z, $(0, b, 0)$, y, x
xz-trace	$\frac{x^2}{a^2} - \frac{z^2}{c^2} = 1$	Hyperbola	z, $(a, 0, 0)$, y, x

FIGURE 14.65
$$\frac{x^2}{a^2}+\frac{y^2}{b^2}-\frac{z^2}{c^2}=1$$

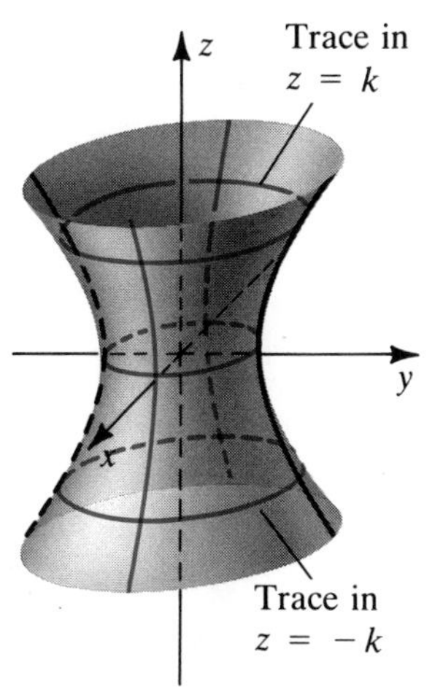

These traces are sketched on one set of coordinate axes in Figure 14.65. The z-axis is the **axis of the hyperboloid**.

The trace in a plane $z = k$ parallel to the xy-plane is given by

$$\frac{x^2}{a^2}+\frac{y^2}{b^2}=1+\frac{k^2}{c^2}$$

and hence is an ellipse. As k increases through positive values, the lengths of the axes of the ellipse increase.

Traces in planes $x = k$ or $y = k$ (that is, in planes parallel to the yz-plane or xz-plane, respectively) are hyperbolas (verify this fact).

The graphs of

$$\frac{x^2}{a^2}-\frac{y^2}{b^2}+\frac{z^2}{c^2}=1 \quad \text{and} \quad -\frac{x^2}{a^2}+\frac{y^2}{b^2}+\frac{z^2}{c^2}=1$$

are also hyperboloids of one sheet; however, in the first case the axis of the hyperboloid is the y-axis, and in the second case the axis of the hyperboloid coincides with the x-axis. Thus, the term that is negative in these equations indicates the axis of the hyperboloid.

The following is a different type of hyperboloid.

Hyperboloid of two sheets **(14.43)**

$$-\frac{x^2}{a^2}-\frac{y^2}{b^2}+\frac{z^2}{c^2}=1$$

Trace	Equation of trace	Description of trace	Sketch of trace
xy-trace	$-\frac{x^2}{a^2}-\frac{y^2}{b^2}=1$	None	No graph
yz-trace	$-\frac{y^2}{b^2}+\frac{z^2}{c^2}=1$	Hyperbola	
xz-trace	$-\frac{x^2}{a^2}+\frac{z^2}{c^2}=1$	Hyperbola	

FIGURE 14.66
$$-\frac{x^2}{a^2}-\frac{y^2}{b^2}+\frac{z^2}{c^2}=1$$

These traces are sketched on one set of axes in Figure 14.66, where we have also shown traces on the planes $z = k$ and $z = -k$ for $k > c$.

(Verify that these traces are ellipses if $a \neq b$.) Traces in planes parallel to the yz-plane or xz-plane are hyperbolas. The z-axis is the axis of the hyperboloid.

By using minus signs on different terms, we can obtain a hyperboloid of two sheets whose axis is the x-axis or the y-axis. (Which terms should be negative?)

A (double-napped) *cone* may be thought of as the degenerate hyperboloid obtained by replacing the number 1 in (14.42) or (14.43) by 0. This gives us the following equation.

Cone **(14.44)**

$$\frac{x^2}{a^2} + \frac{y^2}{b^2} - \frac{z^2}{c^2} = 0$$

As indicated in the following chart, the yz-trace and the xz-trace are the lines $z = \pm(c/b)y$ and $z = \pm(c/a)x$, respectively.

Trace	Equation of trace	Description of trace	Sketch of trace
xy-trace	$\frac{x^2}{a^2} + \frac{y^2}{b^2} = 0$	Origin	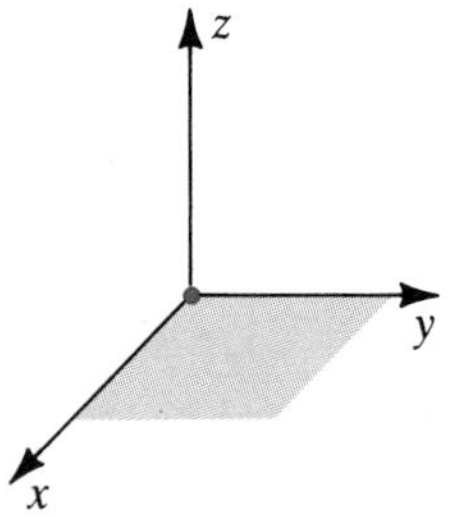
yz-trace	$\frac{y^2}{b^2} - \frac{z^2}{c^2} = 0$	Two intersecting lines	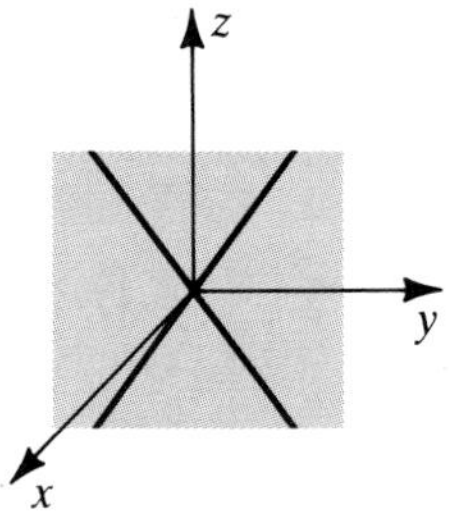
xz-trace	$\frac{x^2}{a^2} - \frac{z^2}{c^2} = 0$	Two intersecting lines	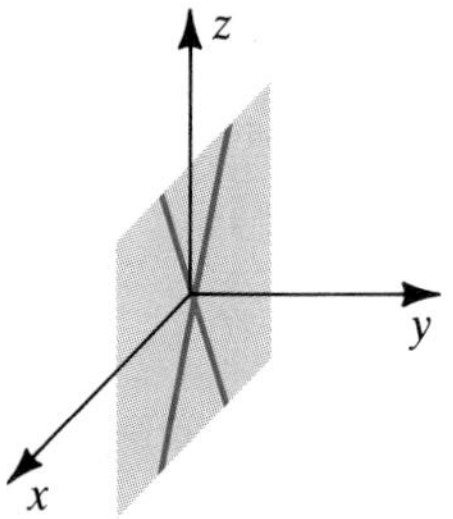

The trace in a plane $z = k$ parallel to the xy-plane has the equation

$$\frac{x^2}{a^2} + \frac{y^2}{b^2} = \frac{k^2}{c^2}$$

and hence is an ellipse. Traces in planes parallel to the other coordinate axes are hyperbolas. The graph is sketched in Figure 14.67. The z-axis is the axis of the cone.

FIGURE 14.67
$\frac{x^2}{a^2} + \frac{y^2}{b^2} - \frac{z^2}{c^2} = 0$

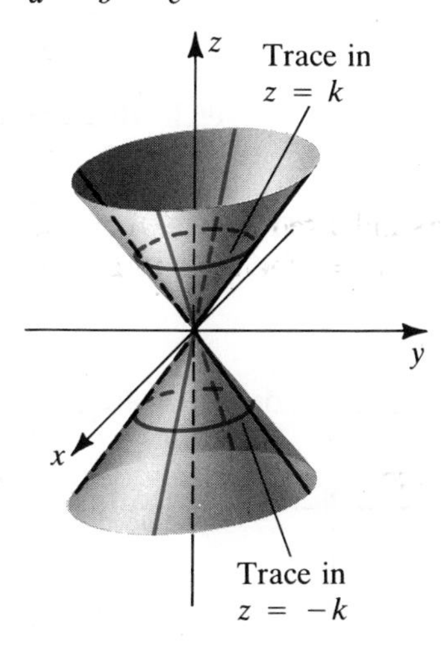

The graph of the next equation, for c either positive or negative, is a *paraboloid*.

Paraboloid **(14.45)**

$$\frac{x^2}{a^2} + \frac{y^2}{b^2} = cz$$

Example 1 is the special case of (14.45) with $a = b = c = 1$. If $c > 0$, then the graph of (14.45) is similar to that shown in Figure 14.57, except that if $a \neq b$, then traces in planes $z = k$ parallel to the xy-plane are ellipses instead of circles. If $c < 0$, then the paraboloid *opens downward*. The z-axis is the **axis of the paraboloid**. The graphs of the equations

$$\frac{x^2}{a^2} + \frac{z^2}{b^2} = cy \quad \text{and} \quad \frac{y^2}{a^2} + \frac{z^2}{b^2} = cx$$

are paraboloids whose axes are the y-axis and x-axis, respectively. If we change the sign of one of the terms on the left in (14.45), we obtain a *hyperbolic paraboloid*.

Hyperbolic paraboloid **(14.46)**

$$\frac{y^2}{a^2} - \frac{x^2}{b^2} = cz$$

FIGURE 14.68
$\frac{y^2}{a^2} - \frac{x^2}{b^2} = cz$

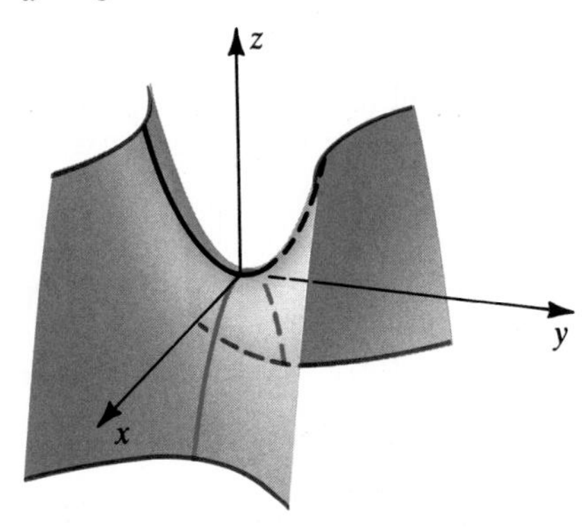

A typical sketch of this saddle-shaped surface for the case $c > 0$ is shown in Figure 14.68. Variations are obtained by interchanging x, y, and z in (14.46). A hyperbolic paraboloid is the most difficult quadric surface to visualize, and it takes considerable practice to become proficient at sketching the graph. Traces of the type shown in Figure 14.68 can be helpful. Note that the trace in the xy-plane has equation

$$\frac{y^2}{a^2} - \frac{x^2}{b^2} = 0, \quad \text{or} \quad y = \pm\frac{a}{b}x,$$

and hence is a pair of intersecting lines through the origin. This trace is not shown in Figure 14.68.

The next two examples are special cases of quadric surfaces.

EXAMPLE 4 Sketch the graph of $16x^2 - 9y^2 + 36z^2 = 144$, and identify the surface.

SOLUTION Dividing both sides of the equation by 144 leads to

$$\frac{x^2}{9} - \frac{y^2}{16} + \frac{z^2}{4} = 1.$$

Traces in coordinate planes are as follows.

Trace	Equation of trace	Description of trace
xy-plane	$\frac{x^2}{9} - \frac{y^2}{16} = 1$	Hyperbola
yz-plane	$\frac{z^2}{4} - \frac{y^2}{16} = 1$	Hyperbola
xz-plane	$\frac{x^2}{9} + \frac{z^2}{4} = 1$	Ellipse

FIGURE 14.69
$16x^2 - 9y^2 + 36z^2 = 144$

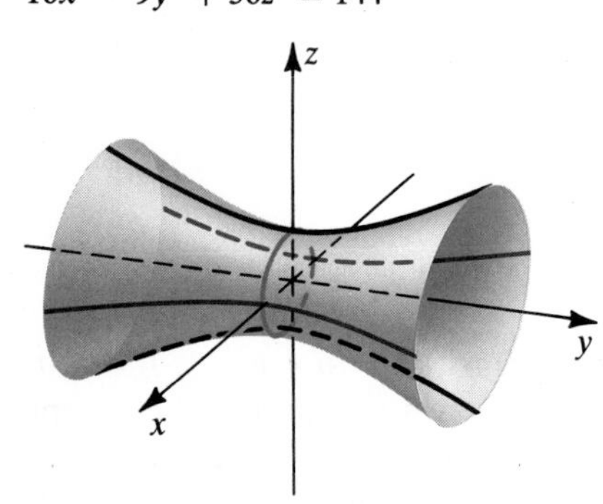

The graph, a hyperboloid of one sheet with the y-axis as its axis, is sketched in Figure 14.69. Traces in planes parallel to the xz-plane are ellipses, and traces in planes parallel to the xy- or yz-planes are hyperbolas.

EXAMPLE 5 Sketch the graph of $y^2 + 4z^2 = x$, and identify the surface.

SOLUTION Traces are as follows.

Trace	Equation of trace	Description of trace
xy-plane	$y^2 = x$	Parabola
yz-plane	$y^2 + 4z^2 = 0$	Origin
xz-plane	$4z^2 = x$	Parabola

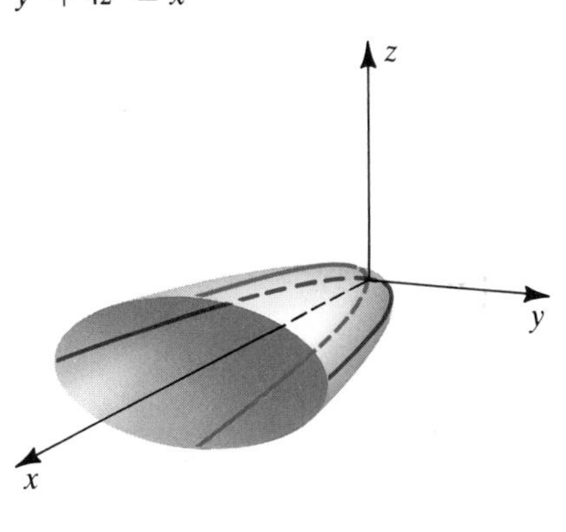

FIGURE 14.70
$y^2 + 4z^2 = x$

The trace in a plane $x = k$ parallel to the yz-plane has equation $y^2 + 4z^2 = k$, which is an ellipse if $k > 0$. Traces in planes parallel to the xz- or xy-planes are parabolas. The surface, a paraboloid having the x-axis as its axis, is sketched in Figure 14.70.

A *surface of revolution* is obtained by revolving a plane curve C about a line (the axis of revolution) in the plane. In the following discussion, C will always lie in a coordinate plane and the axis of revolution will be one of the coordinate axes. We use the symbol $f(x, y)$ for an expression in the variables x and y. In this case $f(a, b)$ denotes the number obtained by substituting a for x and b for y. This notation will be discussed further in Chapter 16.

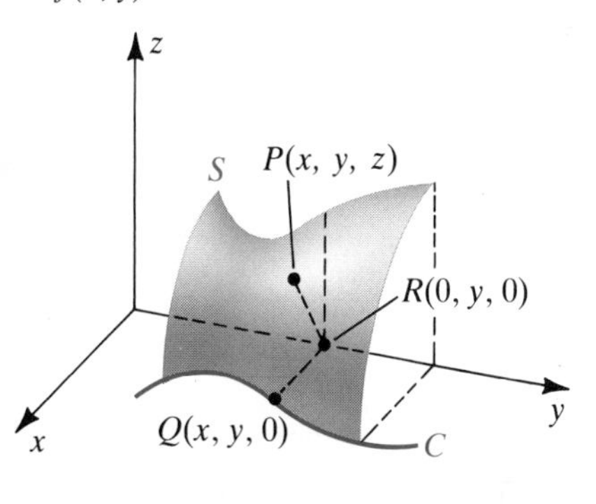

FIGURE 14.71
C: $f(x, y) = 0$

The graph of the equation $f(x, y) = 0$ in the xy-plane is a curve C. (We are interested here only in the graph in the xy-plane and not in the three-dimensional graph, which is a cylinder.) Suppose, for simplicity, that x and y are nonnegative for all points (x, y) on C, and let S denote the surface obtained by revolving C about the y-axis, as illustrated in Figure 14.71. A point $P(x, y, z)$ is on S if and only if $Q(x_1, y, 0)$ is on C and $x_1 = \sqrt{x^2 + z^2}$. Consequently, $P(x, y, z)$ is on S if and only if $f(\sqrt{x^2 + z^2}, y) = 0$. Thus, to find an equation for S, we replace the variable x in the equation for C by $\sqrt{x^2 + z^2}$. Similarly, if the graph of $f(x, y) = 0$ is revolved about the x-axis, then an equation for the resulting surface may be found by replacing y by $\sqrt{y^2 + z^2}$. For some curves that contain points (x, y) with x or y negative, the preceding discussion can be extended by substituting $\pm\sqrt{x^2 + z^2}$ for x or $\pm\sqrt{y^2 + z^2}$ for y. If, as in the next example, x or y occurs only in even powers, then this distinction is unnecessary, since the radical disappears when the equation is simplified.

EXAMPLE 6 The graph of $9x^2 + 4y^2 = 36$ is revolved about the y-axis. Find an equation for the resulting surface.

SOLUTION We find an equation for the surface by substituting $x^2 + z^2$ for x^2. This gives us

$$9(x^2 + z^2) + 4y^2 = 36.$$

The surface is an *ellipsoid of revolution*. If we divide both sides by 36 and rearrange terms, we obtain

$$\frac{x^2}{4} + \frac{y^2}{9} + \frac{z^2}{4} = 1,$$

which is of the form in (14.41).

A similar discussion can be given for curves that lie in the yz-plane or the xz-plane. For example, if a suitable curve C in the xz-plane is revolved about the z-axis, then we may find an equation for the resulting surface by replacing x by $\sqrt{x^2 + y^2}$. If C is revolved about the x-axis, we replace z by $\sqrt{y^2 + z^2}$.

Finally, note that equations for surfaces of revolution are characterized by the fact that two of the variables occur in combinations such as $x^2 + y^2$, $y^2 + z^2$, or $x^2 + z^2$.

EXERCISES 14.6

Exer. 1–8: Sketch the graph of the cylinder in an xyz-coordinate system.

1 $x^2 + y^2 = 9$

2 $y^2 + z^2 = 16$

3 $4y^2 + 9z^2 = 36$

4 $x^2 + 5z^2 = 25$

5 $x^2 = 9z$

6 $x^2 - 4y = 0$

7 $y^2 - x^2 = 16$

8 $xz = 1$

Exer. 9–20: Match each graph with one of the equations.

A. $\dfrac{x^2}{25} + \dfrac{y^2}{9} + \dfrac{z^2}{4} = 1$

B. $x = z^2 + \dfrac{y^2}{4}$

C. $y^2 + z^2 - x^2 = 1$

D. $\dfrac{x^2}{4} + \dfrac{y^2}{9} - \dfrac{z^2}{4} = 0$

E. $z = \dfrac{x^2}{9} - \dfrac{y^2}{4}$

F. $z^2 - \dfrac{x^2}{4} - y^2 = 1$

G. $\dfrac{z^2}{9} + \dfrac{y^2}{4} - \dfrac{x^2}{4} = 0$

H. $\dfrac{x^2}{4} - y^2 - z^2 = 1$

I. $y = \dfrac{x^2}{4} - \dfrac{z^2}{9}$

J. $x^2 + \dfrac{y^2}{4} + \dfrac{z^2}{16} = 1$

K. $z = \dfrac{x^2}{9} + y^2$

L. $\dfrac{x^2}{4} + \dfrac{y^2}{16} + \dfrac{z^2}{9} = 1$

M. $y = \dfrac{z^2}{9} - \dfrac{x^2}{4}$

N. $y = \dfrac{x^2}{4} + \dfrac{z^2}{4}$

O. $z^2 + \dfrac{x^2}{4} - y^2 = 1$

P. $\dfrac{x^2}{4} + \dfrac{z^2}{9} - \dfrac{y^2}{4} = 0$

Q. $y^2 - \dfrac{x^2}{4} - z^2 = 1$

R. $x^2 + \dfrac{y^2}{4} - z^2 = 1$

9

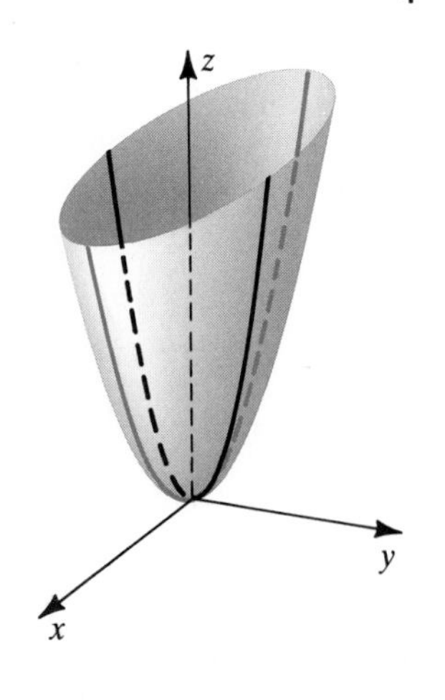

10

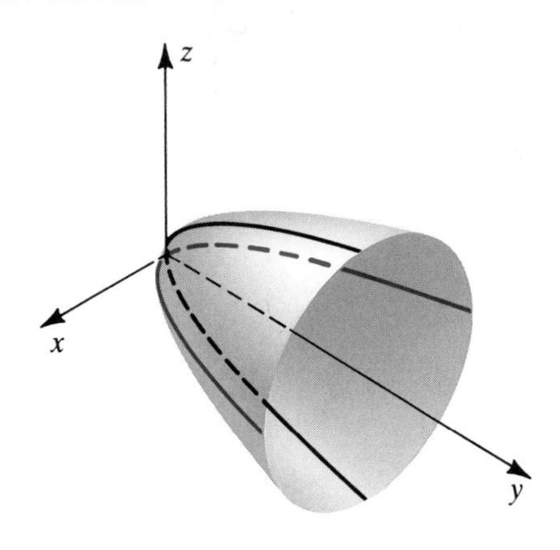

11

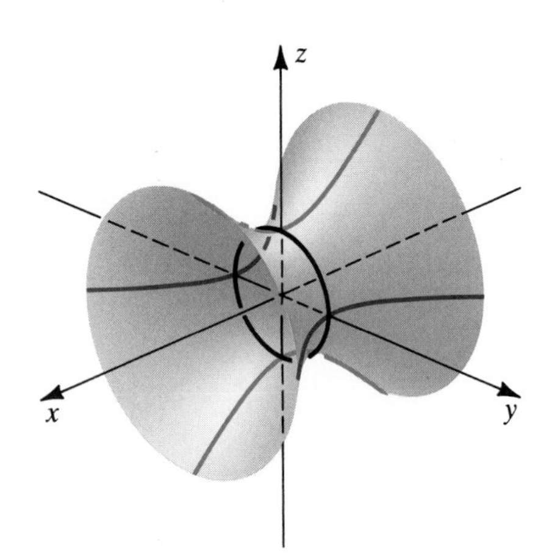

12

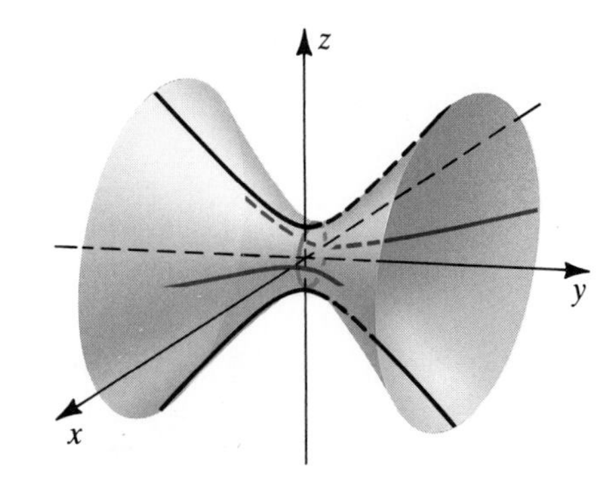

13

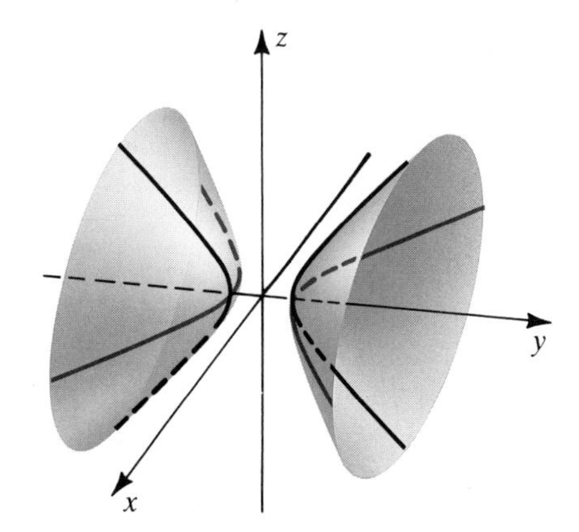

14

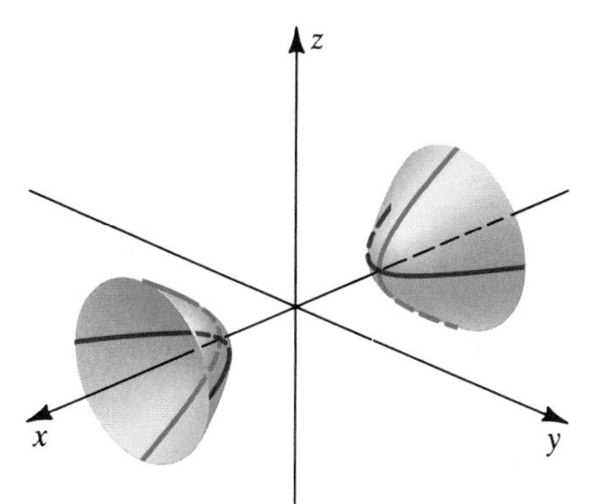

15

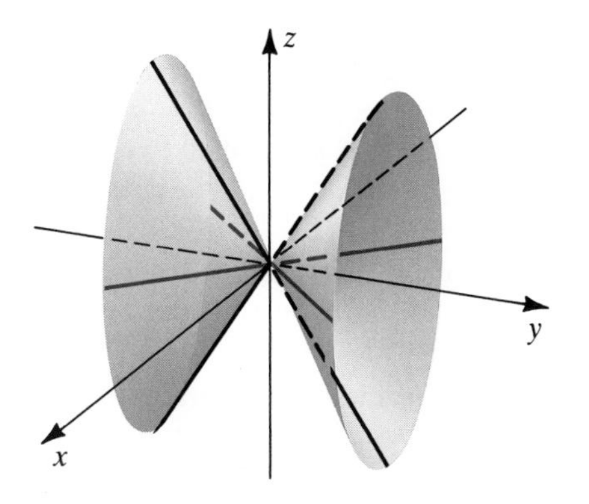

16

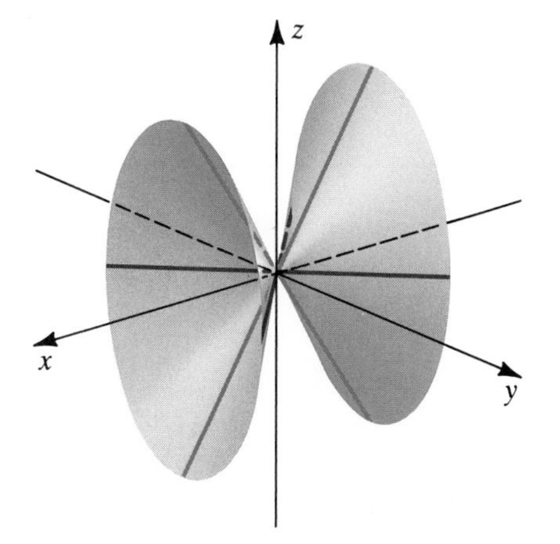

17

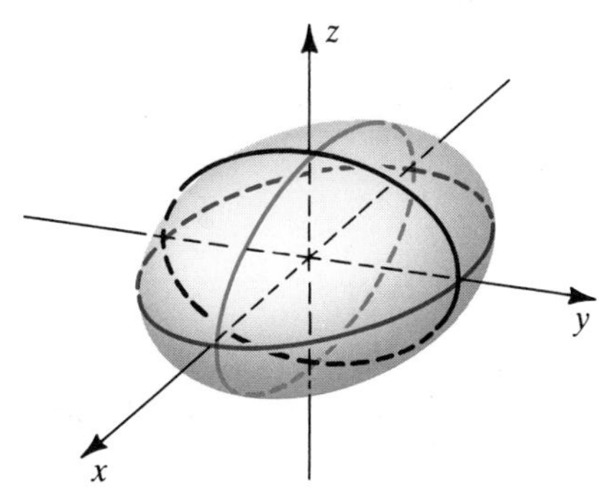

18

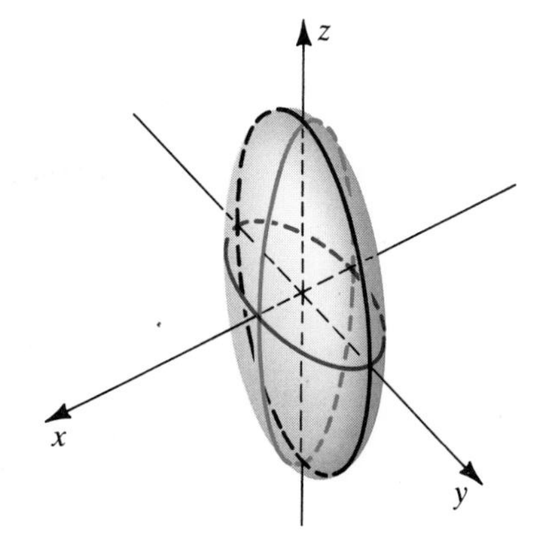

19

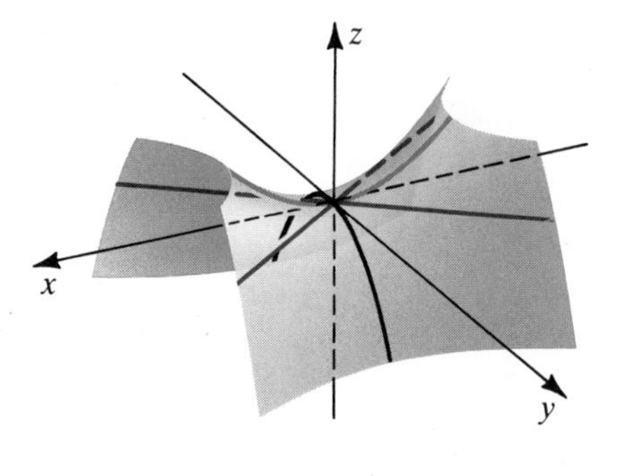

20

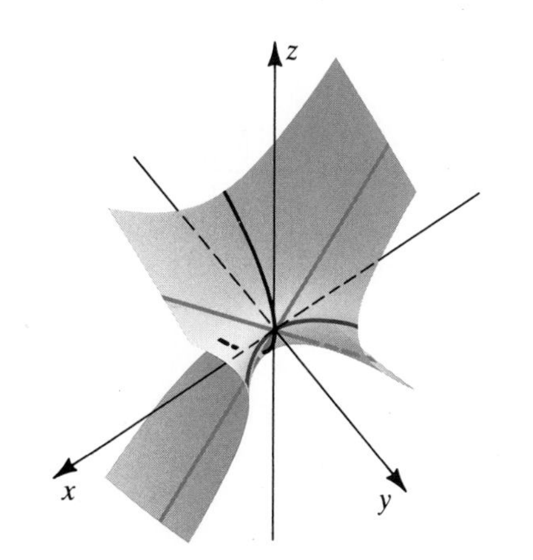

Exer. 21–32: Sketch the graph of the quadric surface.

Ellipsoids

21 $\dfrac{x^2}{4}+\dfrac{y^2}{9}+\dfrac{z^2}{16}=1$

22 $x^2+\dfrac{y^2}{9}+\dfrac{z^2}{4}=1$

Hyperboloids of one sheet

23 (a) $\dfrac{x^2}{4}+y^2-z^2=1$ (b) $x^2+\dfrac{z^2}{4}-y^2=1$

24 (a) $z^2+x^2-y^2=1$ (b) $y^2+\dfrac{z^2}{4}-x^2=1$

Hyperboloids of two sheets

25 (a) $x^2-\dfrac{y^2}{4}-z^2=1$ (b) $\dfrac{z^2}{4}-y^2-x^2=1$

26 (a) $z^2-\dfrac{x^2}{4}-\dfrac{y^2}{4}=1$ (b) $\dfrac{y^2}{4}-x^2-\dfrac{z^2}{9}=1$

Cones

27 (a) $\dfrac{x^2}{9}+\dfrac{y^2}{4}=\dfrac{z^2}{4}$ (b) $\dfrac{x^2}{4}-y^2+\dfrac{z^2}{9}=0$

28 (a) $\dfrac{x^2}{25}+\dfrac{y^2}{9}-z^2=0$ (b) $x^2=4y^2+z^2$

Paraboloids

29 (a) $y=\dfrac{x^2}{4}+\dfrac{z^2}{9}$ (b) $x^2+\dfrac{y^2}{4}-z=0$

30 (a) $z=x^2+\dfrac{y^2}{9}$ (b) $\dfrac{z^2}{25}+\dfrac{y^2}{9}-x=0$

Hyperbolic paraboloids

31 (a) $z=x^2-y^2$ (b) $z=y^2-x^2$

32 (a) $z=\dfrac{y^2}{9}-\dfrac{x^2}{4}$ (b) $z=\dfrac{x^2}{4}-\dfrac{y^2}{9}$

Exer. 33–46: Sketch the graph of the equation in an *xyz*-coordinate system, and identify the surface.

33 $16x^2-4y^2-z^2+1=0$

34 $8x^2+4y^2+z^2=16$

35 $36x=9y^2+z^2$

36 $16x^2+100y^2-25z^2=400$

37 $x^2-16y^2=4z^2$

38 $3x^2-4y^2-z^2=12$

39 $9x^2+4y^2+z^2=36$

40 $16y=x^2+4z^2$

41 $z=e^y$

42 $x^2+(y-2)^2=1$

43 $4x-3y=12$

44 $2x+4y+3z=12$

45 $y^2-9x^2-z^2-9=0$

46 $36x^2-16y^2+9z^2=0$

c **Exer. 47–50: Graph the surface.**

47 $z=\frac{1}{5}y^2-3|x|$

48 $z=x^2+3xy+4y^2$

49 $z=xy+x^2$

50 $z=\dfrac{y^2}{16}-\dfrac{xy}{15}-\dfrac{x^2}{9}$

Exer. 51–56: Find an equation of the surface obtained by revolving the graph of the equation about the indicated axis.

51 $x^2+4y^2=16$; y-axis

52 $y^2=4x$; x-axis

53 $z=4-y^2$; z-axis

54 $z=e^{-y^2}$; y-axis

55 $z^2-x^2=1$; x-axis

56 $xz=1$; z-axis

57 Although we often use a sphere as a model of the earth, a more precise relationship is needed for surveying the earth's surface. The *Clarke ellipsoid* (1866), with equation $(x^2/a^2)+(y^2/b^2)+(z^2/c^2)=1$ for $a=b=6378.2064$ km and $c=6356.5838$ km, is used to establish the geographic positions of control points in the U.S. national geodetic network.

(a) Explain briefly the difference between the Clarke ellipsoid and the usual spherical representation of the earth's surface.

(b) Curves of equal latitude are traces in a plane of the form $z=k$. Describe these curves.

(c) Curves of equal longitude (or *meridians*) are traces in a plane of the form $y=mx$. Describe these curves.

14.7 REVIEW EXERCISES

Exer. 1–20: If $\mathbf{a}=3\mathbf{i}-\mathbf{j}-4\mathbf{k}$, $\mathbf{b}=2\mathbf{i}+5\mathbf{j}-2\mathbf{k}$, and $\mathbf{c}=-\mathbf{i}+6\mathbf{k}$, find the vector or scalar.

1 $3\mathbf{a}-2\mathbf{b}$

2 $\mathbf{a}\cdot(\mathbf{b}-\mathbf{c})$

3 $\|-3\mathbf{b}\|$

4 $\|\mathbf{b}+\mathbf{c}\|$

5 $\|\mathbf{a}\|^2$

6 $\|\mathbf{a}\times\mathbf{a}\|$

7 The angle between **a** and **c**

8 The angles between **a** and the unit vectors **i**, **j**, and **k**

9 A unit vector having the same direction as **a**

10 A vector having the opposite direction of **a** and twice the magnitude of **b**

11 $\mathbf{a}\times\mathbf{b}$

12 $(\mathbf{b}+\mathbf{c})\times\mathbf{a}$

13 $\text{comp}_{\mathbf{b}}\,\mathbf{a}$

14 $\text{comp}_{\mathbf{a}}\,(\mathbf{b}\times\mathbf{c})$

15 $(2\mathbf{a})\cdot(3\mathbf{a})$

16 $(2\mathbf{a})\times(3\mathbf{a})$

17 $(\mathbf{a}\times\mathbf{c})+(\mathbf{c}\times\mathbf{a})$

18 $(\mathbf{a}\times\mathbf{c})\times(\mathbf{c}\times\mathbf{a})$

19 The volume of the box determined by **a**, **b**, and **c**

20 The area of the triangle determined by **a** and **b**

21 Show that $\mathbf{a} = \langle 4, 3, -5 \rangle$ and $\mathbf{b} = \langle -1, 3, 1 \rangle$ are orthogonal.

22 A constant force of magnitude 18 pounds has the same direction as the vector $\mathbf{a} = 4\mathbf{i} + 7\mathbf{j} + 4\mathbf{k}$. If the distance is measured in feet, find the work done if the point of application of the force moves along the line segment from $P(1, 1, 1)$ to $Q(3, 5, 4)$.

23 Given the points $A(5, -3, 2)$ and $B(-1, -4, 3)$, find the following:

(a) $d(A, B)$

(b) the midpoint of AB

(c) an equation of the sphere with center B and tangent to the xz-plane

(d) an equation of the plane through B parallel to the xz-plane

(e) parametric equations for the line through A and B

(f) an equation of the plane through A with normal vector $\overrightarrow{AB}$

24 Find an equation of the plane through $A(0, 4, 9)$ and $B(0, -3, 7)$ that is perpendicular to the yz-plane.

25 Find an equation of the plane that has x-intercept 5, y-intercept -2, and z-intercept 6.

26 Find a symmetric form for the line $x = 3 + 4t$, $y = 2 - t$, $z = 2t$.

27 Find parametric equations for the line of intersection of the planes $2x + y + 4z = 8$ and $x + 3y - z = -1$.

28 Find an equation of the plane through $P(1, 3, -2)$ that is parallel to the plane $5x - 4y + 3z = 7$.

29 Find an equation of the plane through $P(4, 1, 2)$ that has the same yz-trace as the plane $2x + 3y - 4z = 11$.

30 Find an equation of the cylinder that has rulings parallel to the z-axis and that has, for its directrix, the circle in the xy-plane with center $C(4, -3, 0)$ and radius 5.

31 Find an equation of the ellipsoid with center O, x-intercept 8, y-intercept 3, and z-intercept 1.

32 Find an equation of the surface obtained by revolving the graph of the equation $z = x$ about the z-axis.

33 Given the points $P(2, -1, 1)$, $Q(-3, 2, 0)$, and $R(4, -5, 3)$, find the following:

(a) a unit normal vector to the plane determined by P, Q, and R

(b) an equation of the plane through P, Q, and R

(c) parametric equations for a line through P that is parallel to the line through Q and R

(d) $\overrightarrow{QP} \cdot \overrightarrow{QR}$

(e) the angle between $\overrightarrow{QP}$ and $\overrightarrow{QR}$

(f) the area of triangle PQR

(g) the distance from R to the line through P and Q

34 Find the angles between the two lines

$$\frac{x-3}{2} = \frac{y+1}{-4} = \frac{z-5}{8} \quad \text{and} \quad \frac{x+1}{7} = \frac{6-y}{2} = \frac{2z+7}{-4}.$$

35 Find parametric equations for each of the lines in Exercise 34.

36 Determine whether the following lines intersect, and if so, find the point of intersection:

$$x = 2 + t, \quad y = 1 + t, \quad z = 4 + 7t$$
$$x = -4 + 5v, \quad y = 2 - 2v, \quad z = 1 - 4v$$

37 Find the angles between the lines in Exercise 36.

38 Suppose a line l has parametric equations $x = 2t + 1$, $y = -t + 3$, $z = 5t$. Find the distance from $A(3, 1, -1)$ to l.

Exer. 39–50: Sketch the graph of the equation.

39 $x^2 + y^2 + z^2 - 14x + 6y - 8z + 10 = 0$

40 $4y - 3z - 15 = 0$

41 $3x - 5y + 2z = 10$

42 $y = z^2 + 1$

43 $9x^2 + 4z^2 = 36$

44 $x^2 + 4y + 9z^2 = 0$

45 $z^2 - 4x^2 = 9 - 4y^2$

46 $2x^2 + 4z^2 - y^2 = 0$

47 $z^2 - 4x^2 - y^2 = 4$

48 $x^2 + 2y^2 + 4z^2 = 16$

49 $x^2 - 4y^2 = 4z$

50 $z = 9 - x^2 - y^2$

CHAPTER

15

VECTOR-VALUED FUNCTIONS

INTRODUCTION

The functions considered up to this point in our work have values that are real numbers. In this chapter we introduce functions whose values are vectors. An example of such a *vector-valued function* is the velocity, at time t, of an object moving through space. We discuss this application and other aspects of motion, such as acceleration and centripetal force, in Section 15.3.

The concepts of limits, derivatives, and integrals of vector-valued functions are defined by applying earlier methods to components of vectors. This approach enables us to readily obtain properties that are analogous to those obtained for functions in Chapters 3 and 5.

If a vector-valued function is continuous, then the terminal points of the vectors in its range determine a curve in three dimensions, and, conversely, to each curve there corresponds a continuous vector-valued function. This one-to-one correspondence is useful in applications involving a point moving along a curve. In particular, the derivative of a vector-valued function gives us *tangent vectors* to a curve.

In Sections 15.4 and 15.5 we use vector-valued functions to define *curvature*—a concept for determining how much a curve bends, or changes its shape. The final section contains an astonishing use of vectors in the proofs of Kepler's laws. The proofs are difficult and require many properties of vectors. This material should be read carefully, the goal being to appreciate the power of vector methods.

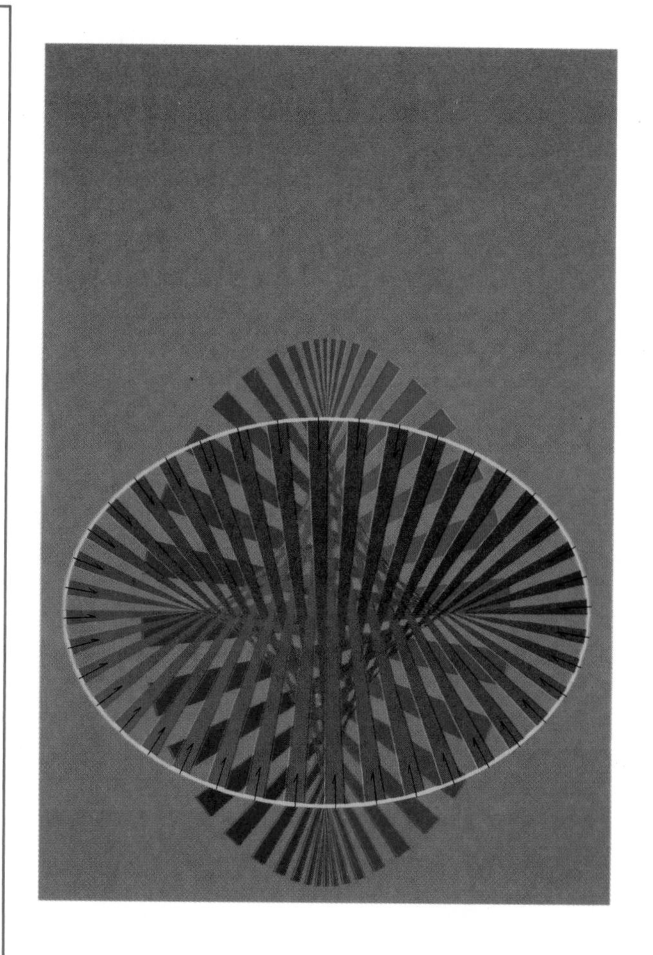

15.1 VECTOR-VALUED FUNCTIONS AND SPACE CURVES

By Definition (1.10), a *function* is a correspondence that assigns to each element of its domain D exactly one element of its range E. If the function values are real numbers, we sometimes refer to the function as a **real-valued function**, or as a **scalar function**. We next consider *vector-valued functions*, denoted by $\mathbf{r}$ (or by $\vec{r}$), for which the function values are *vectors*.

Definition **(15.1)**

> Let D be a set of real numbers. A **vector-valued function r** with domain D is a correspondence that assigns to each number t in D exactly one vector $\mathbf{r}(t)$ in V_3.

FIGURE 15.1

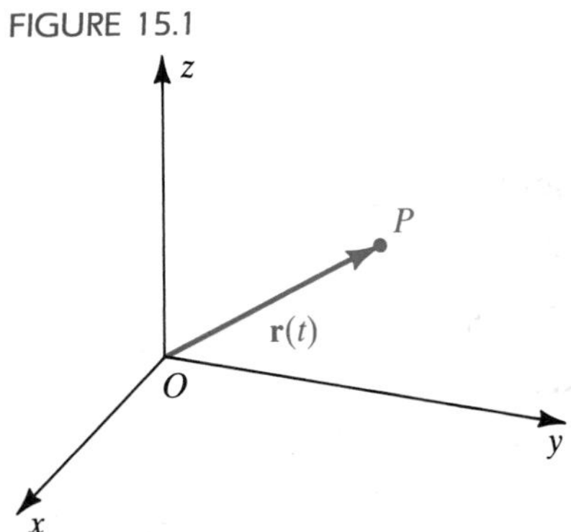

The range of $\mathbf{r}$ in Definition (15.1) consists of all possible vectors $\mathbf{r}(t)$, for t in D. In Figure 15.1, the domain D is a closed interval; however, D could be *any* set of real numbers. We also have sketched the position vector for $\mathbf{r}(t)$. In the future we shall not make a distinction between the vector $\mathbf{r}(t)$ in V_3 and its position vector $\overrightarrow{OP}$. Thus, to *sketch* $\mathbf{r}(t)$ will mean to sketch $\overrightarrow{OP}$. The terminal point P of $\overrightarrow{OP}$ will be referred to as the **endpoint** of $\mathbf{r}(t)$.

Since for each t in D the three components of $\mathbf{r}(t)$ are uniquely determined, we may write

$$\mathbf{r}(t) = f(t)\mathbf{i} + g(t)\mathbf{j} + h(t)\mathbf{k},$$

where f, g, and h are *scalar* functions with domain D. Conversely, a formula of this type for $\mathbf{r}(t)$ determines a vector-valued function $\mathbf{r}$. This gives us the following.

Theorem **(15.2)**

> If D is a set of real numbers, then $\mathbf{r}$ is a vector-valued function with domain D if and only if there are scalar functions f, g, and h such that
>
> $$\mathbf{r}(t) = f(t)\mathbf{i} + g(t)\mathbf{j} + h(t)\mathbf{k}$$
>
> for every t in D.

FIGURE 15.2

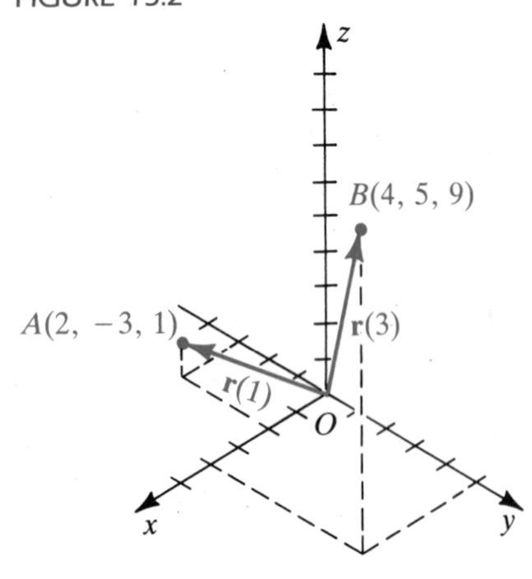

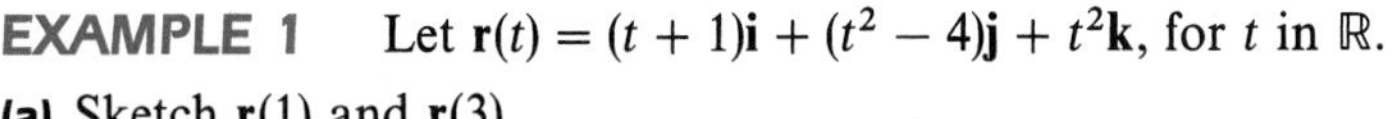
EXAMPLE 1 Let $\mathbf{r}(t) = (t + 1)\mathbf{i} + (t^2 - 4)\mathbf{j} + t^2\mathbf{k}$, for t in $\mathbb{R}$.

(a) Sketch $\mathbf{r}(1)$ and $\mathbf{r}(3)$.

(b) Find the vectors $\mathbf{r}(t)$ that are in a coordinate plane.

SOLUTION

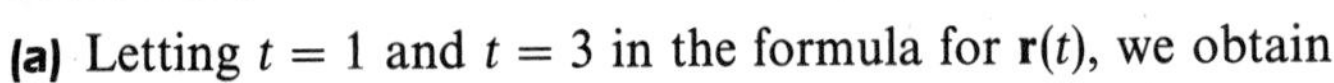
(a) Letting $t = 1$ and $t = 3$ in the formula for $\mathbf{r}(t)$, we obtain

$$\mathbf{r}(1) = 2\mathbf{i} - 3\mathbf{j} + \mathbf{k} \quad \text{and} \quad \mathbf{r}(3) = 4\mathbf{i} + 5\mathbf{j} + 9\mathbf{k}.$$

These vectors are sketched in Figure 15.2.

(b) The following chart displays the vectors $\mathbf{r}(t)$ that are in a coordinate plane.

Plane containing $\mathbf{r}(t)$	Zero component of $\mathbf{r}(t)$	Value of t	$\mathbf{r}(t)$
xy-plane	$\mathbf{k}$-component	$t = 0$	$\mathbf{r}(0) = \mathbf{i} - 4\mathbf{j}$
yz-plane	$\mathbf{i}$-component	$t = -1$	$\mathbf{r}(-1) = -3\mathbf{j} + \mathbf{k}$
xz-plane	$\mathbf{j}$-component	$t = \pm 2$	$\mathbf{r}(2) = 3\mathbf{i} + 4\mathbf{k}$ $\mathbf{r}(-2) = -\mathbf{i} + 4\mathbf{k}$

EXAMPLE 2 Let $\mathbf{r}(t) = (9 - 4t)\mathbf{i} + (-4 + 6t)\mathbf{j} + (3 + 3t)\mathbf{k}$, for t in $\mathbb{R}$.

(a) Sketch $\mathbf{r}(0)$, $\mathbf{r}(1)$, and $\mathbf{r}(2)$.

(b) Give a graphical description of the set of all endpoints of $\mathbf{r}(t)$.

SOLUTION

(a) Let $t = 0$, 1, and 2 in the formula for $\mathbf{r}(t)$, as follows:

$$\mathbf{r}(0) = 9\mathbf{i} - 4\mathbf{j} + 3\mathbf{k}$$

$$\mathbf{r}(1) = 5\mathbf{i} + 2\mathbf{j} + 6\mathbf{k}$$

$$\mathbf{r}(2) = \mathbf{i} + 8\mathbf{j} + 9\mathbf{k}$$

These are the vectors $\overrightarrow{OA}$, $\overrightarrow{OB}$, and $\overrightarrow{OC}$ sketched in Figure 15.3.

FIGURE 15.3

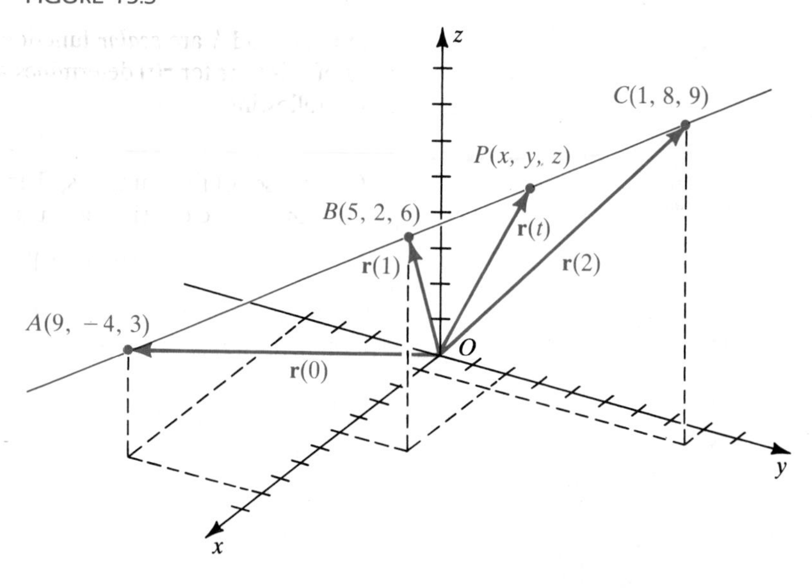

(b) If $P(x, y, z)$ is the endpoint of $\mathbf{r}(t)$, as indicated in Figure 15.3, then

$$x = 9 - 4t, \quad y = -4 + 6t, \quad z = 3 + 3t; \quad t \text{ in } \mathbb{R}.$$

By Theorem (14.34), these are parametric equations for the line l through $A(9, -4, 3)$ parallel to the vector $\langle -4, 6, 3 \rangle$. Since $\langle -4, 6, 3 \rangle$ corresponds to $\overrightarrow{AB}$, we see that the endpoints of $\mathbf{r}(t)$ coincide with the line l.

The endpoints of $\mathbf{r}(t)$ in Example 2 determine a line l in an xyz-coordinate system. In general, the endpoints of

$$\mathbf{r}(t) = f(t)\mathbf{i} + g(t)\mathbf{j} + h(t)\mathbf{k},$$

where the scalar functions f, g, and h are continuous on an interval I, determine a **space curve** (or simply a **curve**)—that is, a set C of ordered triples $(f(t), g(t), h(t))$. The **graph** of C consists of all points $P(f(t), g(t), h(t))$ in an xyz-coordinate system that correspond to the ordered triples. (As in two dimensions, *curve* and *graph of a curve* are used interchangeably.) The equations

$$x = f(t), \quad y = g(t), \quad z = h(t); \quad t \text{ in } I$$

are **parametric equations** for C. We call C a **parametrized curve** and the equations a **parametrization** for C. The **orientation** of C is the direction determined by increasing values of t.

The notions of **endpoints, closed curve**, and **simple closed curve** are defined exactly as for plane curves (see Section 13.1). A curve C is **smooth** if it has a **smooth parametrization**: $x = f(t)$, $y = g(t)$, $z = h(t)$, with f', g', and h' continuous and not simultaneously zero (that is, not all zero for some t), except possibly at endpoints. A smooth curve has no sharp corners, cusps, or breaks. A **piecewise-smooth curve** is defined as in Section 13.1.

FIGURE 15.4

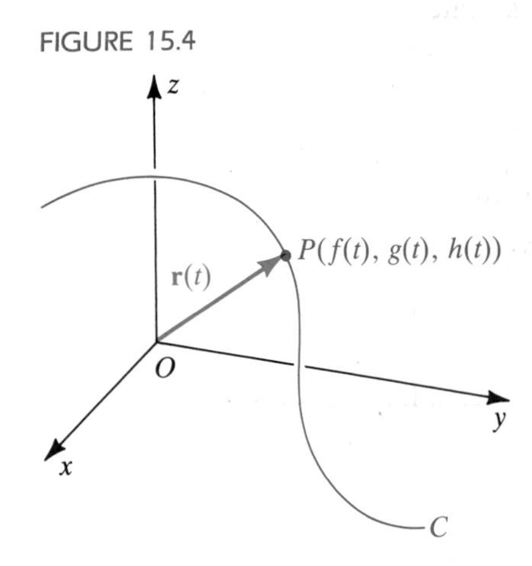

If $\mathbf{r}(t) = f(t)\mathbf{i} + g(t)\mathbf{j} + h(t)\mathbf{k}$, with f, g, and h continuous on an interval I, then, as illustrated in Figure 15.4, as t varies through I, the endpoint of $\mathbf{r}(t)$ traces the curve C with parametrization

$$x = f(t), \quad y = g(t), \quad z = h(t); \quad t \text{ in } I.$$

We refer to C as **the curve determined by** $\mathbf{r}(t)$. In Section 15.3 we shall use this vector representation of C, with t as time, to describe the motion of a point P as it moves in space.

A **twisted cubic** is a curve that has a parametrization

$$x = at, \quad y = bt^2, \quad z = ct^3,$$

where a, b, and c are nonzero constants. The special case $a = 1$, $b = 1$, and $c = 1$ is the curve determined by $\mathbf{r}(t)$ in the next example.

FIGURE 15.5

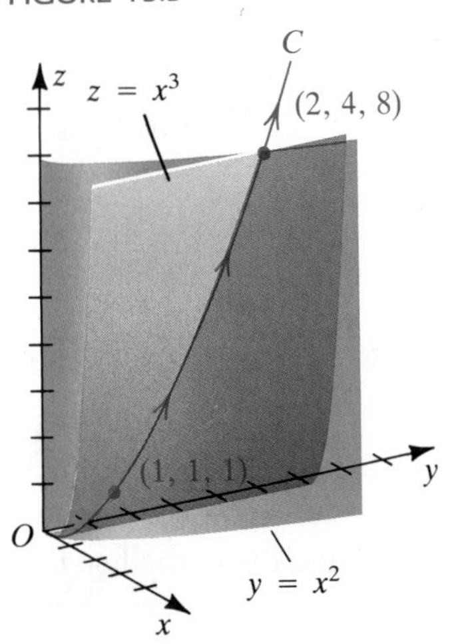

EXAMPLE 3 Let $\mathbf{r}(t) = t\mathbf{i} + t^2\mathbf{j} + t^3\mathbf{k}$ for $t \geq 0$. Sketch the curve C determined by $\mathbf{r}(t)$, and indicate the orientation.

SOLUTION Parametric equations for C are

$$x = t, \quad y = t^2, \quad z = t^3; \quad t \geq 0.$$

Since x, y, and z are nonnegative, C lies in the first octant.

To help us visualize the graph, let us eliminate the parameter from the first two equations, obtaining $y = x^2$. This implies that every point $P(x, y, z)$ on C is also on the parabolic cylinder $y = x^2$.

If we eliminate the parameter from $x = t$ and $z = t^3$, we obtain $z = x^3$, which is an equation of a cylinder with rulings parallel to the y-axis. The curve C is the intersection of the two cylinders $y = x^2$ and $z = x^3$, as shown in Figure 15.5.

At $t = 0$, the endpoint $P(x, y, z)$ of the position vector for $\mathbf{r}(t)$ is $(0, 0, 0)$. As t increases, the endpoint traces the curve C, passing through the points $(1, 1, 1)$, $(2, 4, 8)$, $(3, 9, 27)$, and so on. The arrowheads in the figure indicate the orientation of C. Note that y increases more rapidly than x, and z more rapidly than y. Thus, at $t = 10$, the endpoint of the position vector is (10, 100, 1000).

FIGURE 15.6

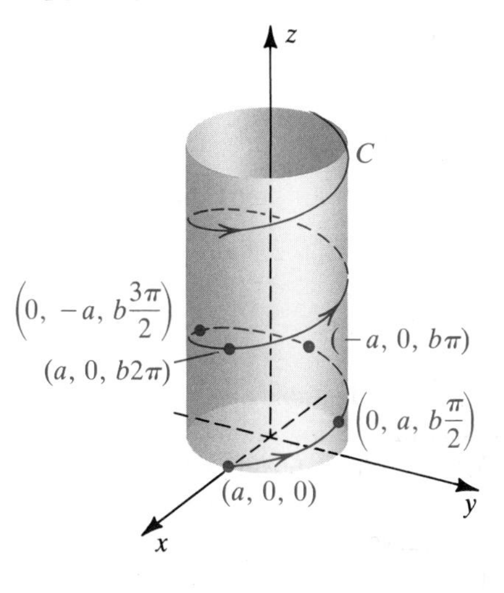

EXAMPLE 4 Let $\mathbf{r}(t) = a \cos t\mathbf{i} + a \sin t\mathbf{j} + bt\mathbf{k}$ for $t \geq 0$ and positive constants a and b. Sketch the curve C determined by $\mathbf{r}(t)$, and indicate the orientation.

SOLUTION Parametric equations for C are

$$x = a \cos t, \quad y = a \sin t, \quad z = bt; \quad t \geq 0.$$

If we eliminate the parameter in the first two equations (see Example 2 of Section 13.1), we obtain

$$x^2 + y^2 = a^2.$$

Thus, every point $P(x, y, z)$ on C is on the circular cylinder $x^2 + y^2 = a^2$. If t varies from 0 to 2π, the point P starts at $(a, 0, 0)$ and moves upward, while making one revolution around the cylinder, as illustrated in Figure 15.6. Additional t-intervals of length 2π result in similar portions of the curve. The orientation of C is indicated by arrowheads.

The curve C in Figure 15.6 is called a **circular helix**.

If we regard V_2 as the subset of V_3 consisting of all vectors whose third component is 0, then $\mathbf{r}(t)$ in (15.2) may be written

$$\mathbf{r}(t) = f(t)\mathbf{i} + g(t)\mathbf{j}.$$

In this case the curve determined by $\mathbf{r}(t)$ is in the xy-plane.

EXAMPLE 5 Let $\mathbf{r}(t) = 2t\mathbf{i} + (8 - 2t^2)\mathbf{j}$ for $-2 \leq t \leq 2$.

(a) Sketch the curve C determined by $\mathbf{r}(t)$, and indicate the orientation.

(b) Sketch $\mathbf{r}(t)$ for $t = -2, -1, 0, \frac{3}{2}, 2$.

SOLUTION

(a) The curve C is in the xy-plane and has the parametrization

$$x = 2t, \quad y = 8 - 2t^2; \quad -2 \leq t \leq 2.$$

Eliminating the parameter, we see that C is on the parabola

$$y = 8 - 2(\tfrac{1}{2}x)^2, \quad \text{or} \quad y = 8 - \tfrac{1}{2}x^2.$$

The values $-2 \leq t \leq 2$ give us the part of the parabola between $(-4, 0)$ and $(4, 0)$, sketched in Figure 15.7, where the arrowheads on C indicate the orientation.

FIGURE 15.7

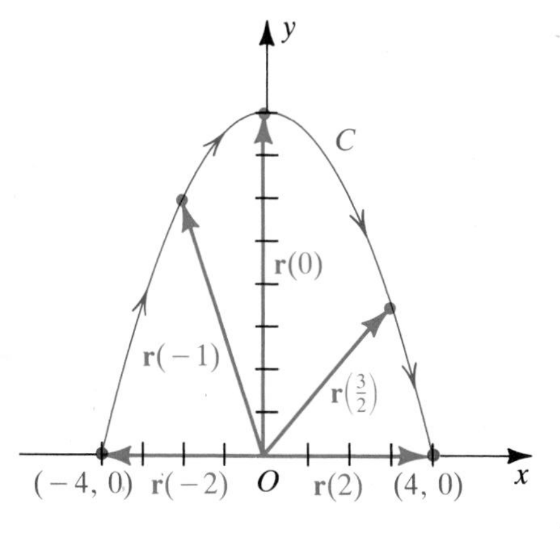

(b) Letting $t = -2, -1, 0, \frac{3}{2}, 2$ in the formula for $\mathbf{r}(t)$, we obtain the following:

$$\mathbf{r}(-2) = -4\mathbf{i} \qquad \mathbf{r}(-1) = -2\mathbf{i} + 6\mathbf{j} \qquad \mathbf{r}(0) = 8\mathbf{j}$$

$$\mathbf{r}(\tfrac{3}{2}) = 3\mathbf{i} + \tfrac{7}{2}\mathbf{j} \qquad \mathbf{r}(2) = 4\mathbf{i}$$

These vectors are sketched in Figure 15.7.

EXAMPLE 6 Let $\mathbf{r}(t) = (\frac{1}{2}t \cos t)\mathbf{i} + (\frac{1}{2}t \sin t)\mathbf{j}$ for $0 \le t \le 3\pi$.

(a) Sketch the curve determined by $\mathbf{r}(t)$, and indicate the orientation.

(b) Sketch $\mathbf{r}(6)$.

SOLUTION

(a) The curve C lies in the xy-plane and has parametric equations

$$x = \tfrac{1}{2}t \cos t, \quad y = \tfrac{1}{2}t \sin t; \quad 0 \le t \le 3\pi.$$

We may assume that t is the radian measure of an angle θ and write

$$x = \tfrac{1}{2}\theta \cos \theta, \quad y = \tfrac{1}{2}\theta \sin \theta; \quad 0 \le \theta \le 3\pi.$$

Moreover, since

$$\frac{y}{x} = \frac{\frac{1}{2}\theta \sin \theta}{\frac{1}{2}\theta \cos \theta} = \tan \theta,$$

it follows from Theorem (13.8) that we can use θ to express (x, y) in polar coordinates. We next note that

$$\begin{aligned} x^2 + y^2 &= \tfrac{1}{4}\theta^2 \cos^2 \theta + \tfrac{1}{4}\theta^2 \sin^2 \theta \\ &= \tfrac{1}{4}\theta^2 (\cos^2 \theta + \sin^2 \theta) = \tfrac{1}{4}\theta^2. \end{aligned}$$

FIGURE 15.8

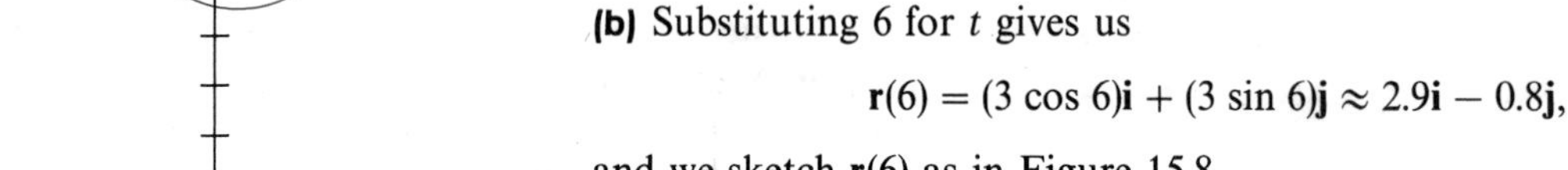

Changing $x^2 + y^2$ to polar coordinates gives us

$$r^2 = \tfrac{1}{4}\theta^2, \quad \text{or} \quad r = \tfrac{1}{2}\theta.$$

(Note that $r \ne -\frac{1}{2}\theta$, because $r = \lVert \mathbf{r}(t) \rVert \ge 0$.)

The graph of the polar equation $r = \frac{1}{2}\theta$ for $0 \le \theta \le 3\pi$ is the portion of the spiral of Archimedes illustrated in Figure 15.8 (see Example 5, Section 13.3). The orientation is in the counterclockwise direction, as indicated by the arrowheads.

(b) Substituting 6 for t gives us

$$\mathbf{r}(6) = (3 \cos 6)\mathbf{i} + (3 \sin 6)\mathbf{j} \approx 2.9\mathbf{i} - 0.8\mathbf{j},$$

and we sketch $\mathbf{r}(6)$ as in Figure 15.8.

We may define the **length of a curve** C in three dimensions exactly as we did for plane curves in Section 13.1.

The next result is analogous to Theorem (13.5).

Theorem **(15.3)**

If a curve C has a smooth parametrization

$$x = f(t), \quad y = g(t), \quad z = h(t); \quad a \le t \le b$$

and if C does not intersect itself, except possibly for $t = a$ and $t = b$, then the length L of C is

$$\begin{aligned} L &= \int_a^b \sqrt{[f'(t)]^2 + [g'(t)]^2 + [h'(t)]^2}\, dt \\ &= \int_a^b \sqrt{\left(\frac{dx}{dt}\right)^2 + \left(\frac{dy}{dt}\right)^2 + \left(\frac{dz}{dt}\right)^2}\, dt. \end{aligned}$$

EXAMPLE 7 Find the length L of the circular helix

$$x = a\cos t, \quad y = a\sin t, \quad z = bt; \quad 0 \le t \le 2\pi.$$

SOLUTION The helix is that part of the curve in Figure 15.6 between the points $(a, 0, 0)$ and $(a, 0, 2\pi b)$. Using Theorem (15.3), we obtain

$$\begin{aligned} L &= \int_0^{2\pi} \sqrt{(-a\sin t)^2 + (a\cos t)^2 + (b)^2}\, dt \\ &= \int_0^{2\pi} \sqrt{a^2\sin^2 t + a^2\cos^2 t + b^2}\, dt \\ &= \int_0^{2\pi} \sqrt{a^2 + b^2}\, dt \\ &= \left[\sqrt{a^2+b^2}\, t\right]_0^{2\pi} = 2\pi\sqrt{a^2+b^2}. \end{aligned}$$

EXERCISES 15.1

Exer. 1–8: (a) Sketch the two vectors listed after the formula for r(*t*). (b) Sketch, on the same plane, the curve *C* determined by r(*t*), and indicate the orientation for the given values of *t*.

1 $\mathbf{r}(t) = 3t\mathbf{i} + (1 - 9t^2)\mathbf{j}$, $\quad \mathbf{r}(0)$, $\mathbf{r}(1)$; $\quad t$ in $\mathbb{R}$

2 $\mathbf{r}(t) = (1 - t^3)\mathbf{i} + t\mathbf{j}$, $\quad \mathbf{r}(1)$, $\mathbf{r}(2)$; $\quad t \ge 0$

3 $\mathbf{r}(t) = (t^3 - 1)\mathbf{i} + (t^2 + 2)\mathbf{j}$, $\quad \mathbf{r}(1)$, $\mathbf{r}(2)$; $\quad -2 \le t \le 2$

4 $\mathbf{r}(t) = (2 + \cos t)\mathbf{i} - (3 - \sin t)\mathbf{j}$, $\quad \mathbf{r}(\pi/2)$, $\mathbf{r}(\pi)$; $\quad 0 \le t \le 2\pi$

5 $\mathbf{r}(t) = (3 + t)\mathbf{i} + (2 - t)\mathbf{j} + (1 + 2t)\mathbf{k}$, $\quad \mathbf{r}(-1)$, $\mathbf{r}(0)$; $\quad t \ge -1$

6 $\mathbf{r}(t) = t\mathbf{i} - 3\sin t\mathbf{j} + 3\cos t\mathbf{k}$, $\quad \mathbf{r}(0)$, $\mathbf{r}(\pi/2)$; $\quad t \ge 0$

7 $\mathbf{r}(t) = t\mathbf{i} + 4\cos t\mathbf{j} + 9\sin t\mathbf{k}$, $\quad \mathbf{r}(0)$, $\mathbf{r}(\pi/2)$; $\quad t \ge 0$

8 $\mathbf{r}(t) = \tan t\mathbf{i} + \sec t\mathbf{j} + 2\mathbf{k}$, $\quad \mathbf{r}(0)$, $\mathbf{r}(\pi/4)$; $\quad -\pi/2 < t < \pi/2$

Exer. 9–18: Sketch the curve *C* determined by r(*t*), and indicate the orientation.

9 $\mathbf{r}(t) = e^t\cos t\mathbf{i} + e^t\sin t\mathbf{j}$; $\quad 0 \le t \le \pi$

10 $\mathbf{r}(t) = 2\cosh t\mathbf{i} + 3\sinh t\mathbf{j}$; $\quad t$ in $\mathbb{R}$

11 $\mathbf{r}(t) = t\mathbf{i} + 2t^2\mathbf{j} + 3t^3\mathbf{k}$; $\quad t$ in $\mathbb{R}$

12 $\mathbf{r}(t) = t^3\mathbf{i} + t^2\mathbf{j} + t\mathbf{k}$; $\quad 0 \le t \le 4$

13 $\mathbf{r}(t) = (t^2 + 1)\mathbf{i} + t\mathbf{j} + 3\mathbf{k}$; $\quad t$ in $\mathbb{R}$

14 $\mathbf{r}(t) = 6\sin t\mathbf{i} + 4\mathbf{j} + 25\cos t\mathbf{k}$; $\quad -2\pi \le t \le 2\pi$

15 $\mathbf{r}(t) = t\mathbf{i} + t\mathbf{j} + \sin t\mathbf{k}$; $\quad t$ in $\mathbb{R}$

16 $\mathbf{r}(t) = t\mathbf{i} + 2t\mathbf{j} + e^t\mathbf{k}$; $\quad t$ in $\mathbb{R}$

c 17 $\mathbf{r}(t) = 3\sin(t^2)\mathbf{i} + (4 - t^{3/2})\mathbf{j}$; $\quad 0 \le t \le 5$

c 18 $\mathbf{r}(t) = e^{\sin 3t}\mathbf{i} + e^{-\cos t}\mathbf{j}$; $\quad 0 \le t \le 2\pi$

Exer. 19–24: Find the length of the parametrized curve.

19 $x = 5t$, $\quad y = 4t^2$, $\quad z = 3t^2$; $\quad 0 \le t \le 2$

20 $x = t^2$, $\quad y = t\sin t$, $\quad z = t\cos t$; $\quad 0 \le t \le 1$

21 $x = e^t\cos t$, $\quad y = e^t$, $\quad z = e^t\sin t$; $\quad 0 \le t \le 2\pi$

22 $x = 2t$, $\quad y = 4\sin 3t$, $\quad z = 4\cos 3t$; $\quad 0 \le t \le 2\pi$

23 $x = 3t^2$, $\quad y = t^3$, $\quad z = 6t$; $\quad 0 \le t \le 1$

24 $x = 1 - 2t^2$, $\quad y = 4t$, $\quad z = 3 + 2t^2$; $\quad 0 \le t \le 2$

25 A *concho-spiral* is a curve C that has a parametrization $x = ae^{\mu t}\cos t$, $y = ae^{\mu t}\sin t$, $z = be^{\mu t}$; $t \ge 0$, where a, b, and μ are constants.

(a) Show that C lies on the cone $a^2z^2 = b^2(x^2 + y^2)$.

(b) Sketch C for $a = b = 4$ and $\mu = -1$.

(c) Find the length of C corresponding to the t-interval $[0, \infty)$.

26 A curve C has the parametrization

$$x = a\sin t\sin\alpha, \quad y = b\sin t\cos\alpha, \quad z = c\cos t; \quad t \ge 0,$$

where a, b, c, and α are positive constants.

(a) Show that C lies on the ellipsoid

$$\frac{x^2}{a^2} + \frac{y^2}{b^2} + \frac{z^2}{c^2} = 1.$$

(b) Show that C also lies on a plane that contains the z-axis.

(c) Describe the curve C.

27 (a) Show that a twisted cubic $x = at$, $y = bt^2$, $z = ct^3$; $t \ge 0$ intersects a given plane in at most three points.

(b) Find the length of the twisted cubic $x = 6t$, $y = 3t^2$, $z = t^3$ between the points corresponding to $t = 0$ and $t = 1$.

28 A rectangle can be made into a cylinder by joining together two opposite and parallel edges. Shown on the left in the figure is a rectangle $ABCD$ of width 2π. Edge AD is joined to edge BC, and point A is then positioned at $(1, 0, 0)$ to form part of the cylinder $x^2 + y^2 = 1$, sketched on the right.

(a) If l is the line segment from A to another point P in the rectangle and m is the slope of l, show that, as a curve on the cylinder, l has a parametrization $x = \cos t$, $y = \sin t$, $z = mt$.

(b) Use part (a) to show that the curve on the cylinder with shortest length from $(1, 0, 0)$ to another point P is a helix (see Example 4).

EXERCISE 28

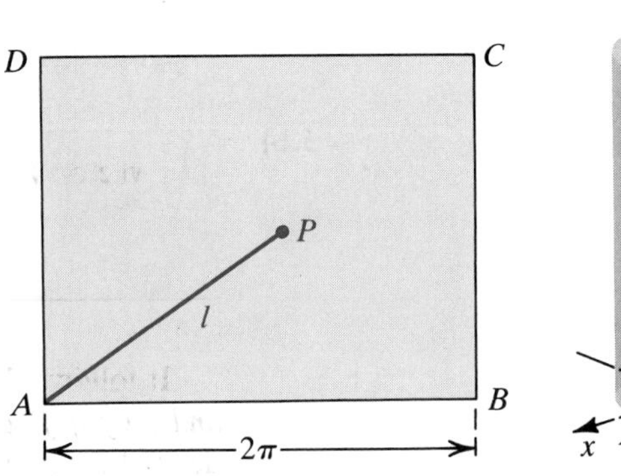

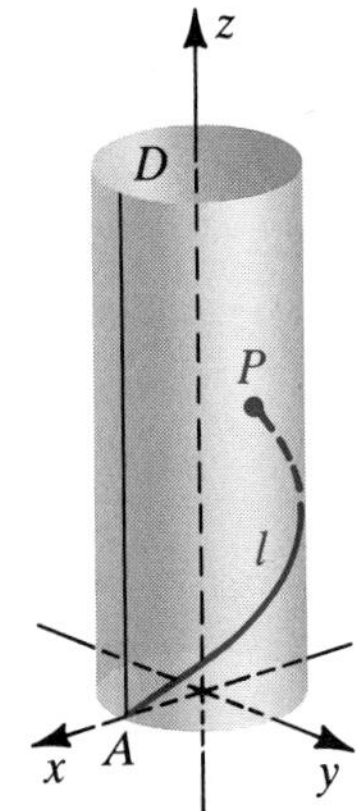

15.2 LIMITS, DERIVATIVES, AND INTEGRALS

We could define the limit of a vector-valued function $\mathbf{r}$ by using an ϵ-δ approach similar to that used in Definition (2.4) (see Exercise 41); however, since $\mathbf{r}(t)$ may be expressed in terms of $\mathbf{i}$, $\mathbf{j}$, and $\mathbf{k}$, where the components are scalar functions f, g, and h, it is simpler to use the following definition.

Definition **(15.4)**

Let $\mathbf{r}(t) = f(t)\mathbf{i} + g(t)\mathbf{j} + h(t)\mathbf{k}$. The **limit of r**$(t)$ as t approaches a is

$$\lim_{t\to a} \mathbf{r}(t) = \left[\lim_{t\to a} f(t)\right]\mathbf{i} + \left[\lim_{t\to a} g(t)\right]\mathbf{j} + \left[\lim_{t\to a} h(t)\right]\mathbf{k},$$

provided f, g, and h have limits as t approaches a.

Thus, to find $\lim_{t\to a} \mathbf{r}(t)$, we take the limit of each component of $\mathbf{r}(t)$. We may state a definition similar to (15.4) for one-sided limits.

ILLUSTRATION

$$\lim_{t\to 2} (t^2\mathbf{i} + 3t\mathbf{j} + 5\mathbf{k}) = \left[\lim_{t\to 2} t^2\right]\mathbf{i} + \left[\lim_{t\to 2} 3t\right]\mathbf{j} + \left[\lim_{t\to 2} 5\right]\mathbf{k}$$
$$= 4\mathbf{i} + 6\mathbf{j} + 5\mathbf{k}$$

FIGURE 15.9

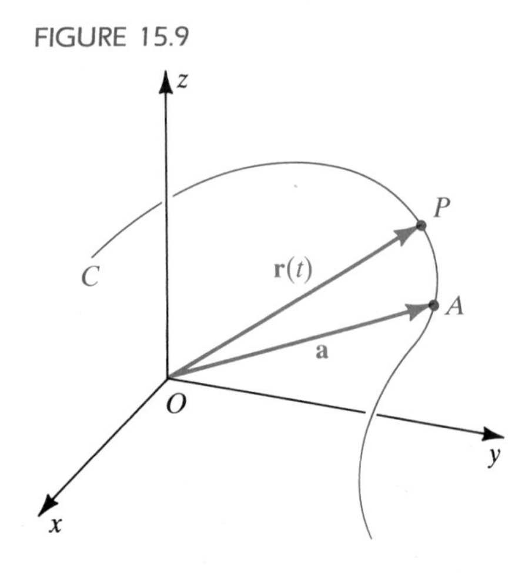

If, in (15.4),

$$\lim_{t\to a} f(t) = a_1, \quad \lim_{t\to a} g(t) = a_2, \quad \text{and} \quad \lim_{t\to a} h(t) = a_3,$$

then

$$\lim_{t\to a} \mathbf{r}(t) = a_1\mathbf{i} + a_2\mathbf{j} + a_3\mathbf{k}.$$

Letting $\mathbf{a} = a_1\mathbf{i} + a_2\mathbf{j} + a_3\mathbf{k}$ gives us the situation illustrated in Figure 15.9, where C is the curve determined by $\mathbf{r}(t)$ and $\overrightarrow{OA}$ is the position vector for **a**. As t approaches a, the endpoint P of $\mathbf{r}(t)$ approaches A; that is, $\mathbf{r}(t)$ approaches $\overrightarrow{OA}$.

Continuity of a vector function **r** is defined in the same manner as for a scalar function.

Definition **(15.5)**

A vector-valued function **r** is **continuous** at a if

$$\lim_{t \to a} \mathbf{r}(t) = \mathbf{r}(a).$$

It follows that if $\mathbf{r}(t) = f(t)\mathbf{i} + g(t)\mathbf{j} + h(t)\mathbf{k}$, then **r** *is continuous at a if and only if f, g, and h are continuous at a.* Continuity on an interval is defined in the usual way.

Definition **(15.6)**

Let **r** be a vector-valued function. The **derivative** of **r** is the vector-valued function $\mathbf{r}'$ defined by

$$\mathbf{r}'(t) = \lim_{\Delta t \to 0} \frac{1}{\Delta t} [\mathbf{r}(t + \Delta t) - \mathbf{r}(t)]$$

for every t such that the limit exists.

If we write

$$\frac{1}{\Delta t} [\mathbf{r}(t + \Delta t) - \mathbf{r}(t)] = \frac{\mathbf{r}(t + \Delta t) - \mathbf{r}(t)}{\Delta t},$$

then Definition (15.6) takes on the familiar form for derivatives of scalar functions, introduced in Chapter 3.

The next theorem states that to find $\mathbf{r}'(t)$ we differentiate each component of $\mathbf{r}(t)$.

Theorem **(15.7)**

If $\mathbf{r}(t) = f(t)\mathbf{i} + g(t)\mathbf{j} + h(t)\mathbf{k}$ and f, g, and h are differentiable, then

$$\mathbf{r}'(t) = f'(t)\mathbf{i} + g'(t)\mathbf{j} + h'(t)\mathbf{k}.$$

PROOF By Definition (15.6),

$$\begin{aligned}
\mathbf{r}'(t) &= \lim_{\Delta t \to 0} \frac{\mathbf{r}(t + \Delta t) - \mathbf{r}(t)}{\Delta t} \\
&= \lim_{\Delta t \to 0} \frac{[f(t + \Delta t)\mathbf{i} + g(t + \Delta t)\mathbf{j} + h(t + \Delta t)\mathbf{k}] - [f(t)\mathbf{i} + g(t)\mathbf{j} + h(t)\mathbf{k}]}{\Delta t} \\
&= \lim_{\Delta t \to 0} \left[\frac{f(t + \Delta t) - f(t)}{\Delta t}\mathbf{i} + \frac{g(t + \Delta t) - g(t)}{\Delta t}\mathbf{j} + \frac{h(t + \Delta t) - h(t)}{\Delta t}\mathbf{k} \right].
\end{aligned}$$

Taking the limit of each component gives us the conclusion of the theorem. ■

If $\mathbf{r}'(t)$ exists, we say that $\mathbf{r}$ is **differentiable** at t. We also denote derivatives as follows:

$$\mathbf{r}'(t) = D_t\,\mathbf{r}(t) = \frac{d}{dt}\,\mathbf{r}(t)$$

Higher derivatives may be obtained in similar fashion. For example, if f, g, and h have second derivatives, then

$$\mathbf{r}''(t) = f''(t)\mathbf{i} + g''(t)\mathbf{j} + h''(t)\mathbf{k}.$$

EXAMPLE 1 Let $\mathbf{r}(t) = (\ln t)\mathbf{i} + e^{-3t}\mathbf{j} + t^2\mathbf{k}$.

(a) Find the domain of $\mathbf{r}$ and determine where $\mathbf{r}$ is continuous.

(b) Find $\mathbf{r}'(t)$ and $\mathbf{r}''(t)$.

SOLUTION

(a) Since $\ln t$ is undefined if $t \leq 0$, the domain of $\mathbf{r}$ is the set of positive real numbers. Moreover, $\mathbf{r}$ is continuous throughout its domain, since each component determines a continuous function.

(b) Applying Theorem (15.7), we have

$$\mathbf{r}'(t) = \frac{1}{t}\mathbf{i} - 3e^{-3t}\mathbf{j} + 2t\mathbf{k}$$

$$\mathbf{r}''(t) = -\frac{1}{t^2}\mathbf{i} + 9e^{-3t}\mathbf{j} + 2\mathbf{k}.$$

The next theorem lists differentiation formulas for vector-valued functions. Note the similarity to those for scalar functions.

Theorem **(15.8)**

If $\mathbf{u}$ and $\mathbf{v}$ are differentiable vector-valued functions and c is a scalar, then

(i) $D_t\,[\mathbf{u}(t) + \mathbf{v}(t)] = \mathbf{u}'(t) + \mathbf{v}'(t)$

(ii) $D_t\,[c\mathbf{u}(t)] = c\mathbf{u}'(t)$

(iii) $D_t\,[\mathbf{u}(t) \cdot \mathbf{v}(t)] = \mathbf{u}(t) \cdot \mathbf{v}'(t) + \mathbf{u}'(t) \cdot \mathbf{v}(t)$

(iv) $D_t\,[\mathbf{u}(t) \times \mathbf{v}(t)] = \mathbf{u}(t) \times \mathbf{v}'(t) + \mathbf{u}'(t) \times \mathbf{v}(t)$

PROOF We shall prove (iii) and leave the other parts as exercises. Let

$$\mathbf{u}(t) = f_1(t)\mathbf{i} + f_2(t)\mathbf{j} + f_3(t)\mathbf{k}$$

$$\mathbf{v}(t) = g_1(t)\mathbf{i} + g_2(t)\mathbf{j} + g_3(t)\mathbf{k},$$

where each scalar function f_k and g_k is a differentiable function of t. By the definition of dot product,

$$\mathbf{u}(t) \cdot \mathbf{v}(t) = f_1(t)g_1(t) + f_2(t)g_2(t) + f_3(t)g_3(t) = \sum_{k=1}^{3} f_k(t)g_k(t).$$

Consequently

$$\begin{aligned} D_t\,[\mathbf{u}(t)\cdot\mathbf{v}(t)] &= D_t \sum_{k=1}^{3} f_k(t)g_k(t) = \sum_{k=1}^{3} D_t\,[f_k(t)g_k(t)] \\ &= \sum_{k=1}^{3} [f_k(t)g_k'(t) + f_k'(t)g_k(t)] \\ &= \sum_{k=1}^{3} f_k(t)g_k'(t) + \sum_{k=1}^{3} f_k'(t)g_k(t) \\ &= \mathbf{u}(t)\cdot\mathbf{v}'(t) + \mathbf{u}'(t)\cdot\mathbf{v}(t). \end{aligned}$$

To obtain a geometric interpretation for the derivative $\mathbf{r}'(t)$ in Definition (15.6), consider

$$\mathbf{r}(t) = f(t)\mathbf{i} + g(t)\mathbf{j} + h(t)\mathbf{k},$$

where f, g, and h are differentiable (scalar) functions, and let C be the curve determined by $\mathbf{r}(t)$. As shown in Figure 15.10(i), if $\overrightarrow{OP}$ and $\overrightarrow{OQ}$ correspond to $\mathbf{r}(t)$ and $\mathbf{r}(t+\Delta t)$, respectively, then

$$\overrightarrow{PQ} = \overrightarrow{OQ} - \overrightarrow{OP} \quad \text{corresponds to} \quad \mathbf{r}(t+\Delta t) - \mathbf{r}(t).$$

Hence, for any nonzero real number Δt,

$$\frac{1}{\Delta t}\overrightarrow{PQ} \quad \text{corresponds to} \quad \frac{1}{\Delta t}[\mathbf{r}(t+\Delta t) - \mathbf{r}(t)].$$

If $\Delta t > 0$, then $(1/\Delta t)\overrightarrow{PQ}$ is a vector $\overrightarrow{PR}$ having the same direction as $\overrightarrow{PQ}$. Moreover, if $0 < \Delta t < 1$, then $(1/\Delta t) > 1$ and $\|\overrightarrow{PR}\| > \|\overrightarrow{PQ}\|$, as illustrated in Figure 15.10(ii).

FIGURE 15.10

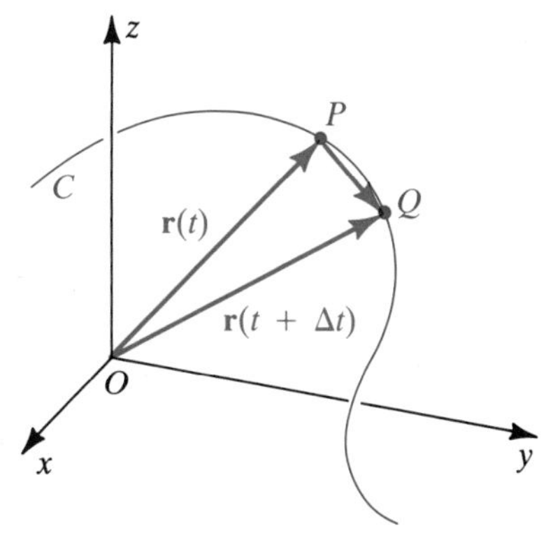

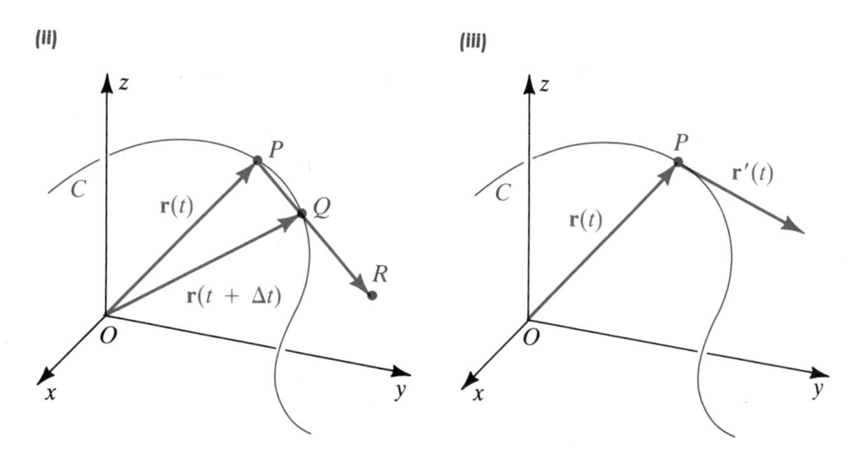

If we let $\Delta t \to 0^+$, then Q approaches P along C, and since $\overrightarrow{PR}$ lies on the secant line through P and Q, the vector $\overrightarrow{PR}$ should approach a vector that lies on the tangent line to C at P. A similar discussion may be given if $\Delta t < 0$. For this reason we refer to $\mathbf{r}'(t)$, or any of its geometric representations, as a **tangent vector** to C at P. *We shall always represent* $\mathbf{r}'(t)$ *by a vector with initial point P on the curve C*, as shown in Figure 15.10(iii). *The vector* $\mathbf{r}'(t)$ *points in the direction determined by the orientation of C.* By definition, the **tangent line** to C at P is the line through P that is parallel to $\mathbf{r}'(t)$.

FIGURE 15.11

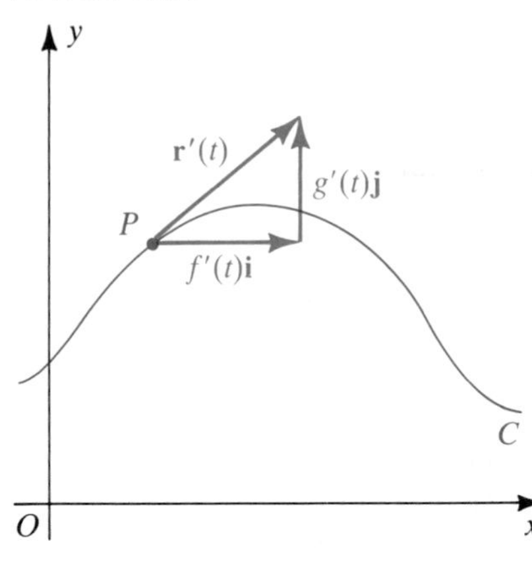

For the two-dimensional case $\mathbf{r}(t) = f(t)\mathbf{i} + g(t)\mathbf{j}$, the tangent vector is

$$\mathbf{r}'(t) = f'(t)\mathbf{i} + g'(t)\mathbf{j}.$$

If $f'(t) \neq 0$, then the slope of the line through P parallel to $\mathbf{r}'(t)$ is $g'(t)/f'(t)$ (see Figure 15.11). This is in agreement with the formula in Theorem (13.3) obtained for tangent lines to plane curves.

EXAMPLE 2 Let $\mathbf{r}(t) = 2t\mathbf{i} + (t^2 - 4)\mathbf{j}$ for $-2 \le t \le 3$.

(a) Sketch the curve C determined by $\mathbf{r}(t)$, and indicate the orientation.

(b) Find $\mathbf{r}'(t)$ and sketch $\mathbf{r}(1)$ and $\mathbf{r}'(1)$.

SOLUTION

(a) Since $\mathbf{r}(t)$ is in V_2, we shall use an xy-plane. Parametric equations for C are

$$x = 2t, \quad y = t^2 - 4; \quad -2 \le t \le 3.$$

Eliminating the parameter, we see that C is on the parabola

$$y = (\tfrac{1}{2}x)^2 - 4 = \tfrac{1}{4}x^2 - 4.$$

FIGURE 15.12

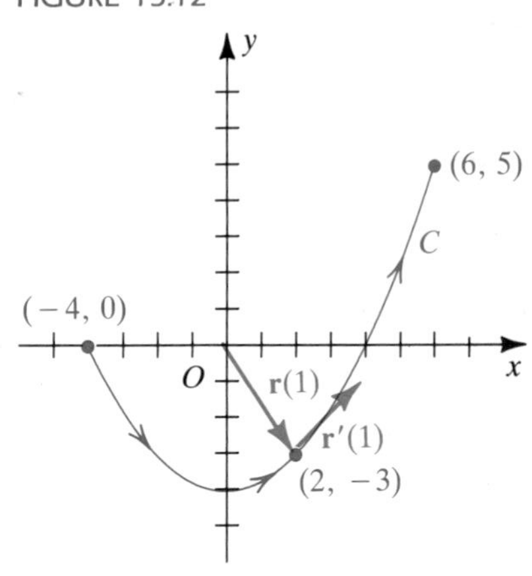

The values $-2 \le t \le 3$ give us the part of the parabola between $(-4, 0)$ and $(6, 5)$ sketched in Figure 15.12, where the orientation is from left to right.

(b) Differentiating $\mathbf{r}(t) = 2t\mathbf{i} + (t^2 - 4)\mathbf{j}$ yields

$$\mathbf{r}'(t) = 2\mathbf{i} + 2t\mathbf{j}.$$

Letting $t = 1$ in the formulas for $\mathbf{r}(t)$ and $\mathbf{r}'(t)$, we obtain

$$\mathbf{r}(1) = 2\mathbf{i} - 3\mathbf{j}, \quad \mathbf{r}'(1) = 2\mathbf{i} + 2\mathbf{j}.$$

As mentioned in our discussion, we always represent $\mathbf{r}'(t)$ graphically by taking the initial point as the point on C that corresponds to t. Thus, in Figure 15.12, the initial point of $\mathbf{r}'(1)$ is the point $P(2, -3)$ that corresponds to $t = 1$. Since $\mathbf{r}'(1) = 2\mathbf{i} + 2\mathbf{j}$, the terminal point of $\mathbf{r}'(1)$ is $(2 + 2, (-3) + 2)$, or $(4, -1)$. Note that $\mathbf{r}'(1)$ points in the direction determined by the orientation of C.

EXAMPLE 3 Let C be the curve with parametric equations

$$x = t, \quad y = t^2, \quad z = t^3; \quad t \ge 0.$$

Find parametric equations for the tangent line to C at the point corresponding to $t = 2$.

SOLUTION The curve C is determined by

$$\mathbf{r}(t) = t\mathbf{i} + t^2\mathbf{j} + t^3\mathbf{k} \quad \text{for} \quad t \ge 0$$

and is the twisted cubic discussed in Example 3 of Section 15.1 (see Figure 15.5). From our previous discussion, a tangent vector to C at the point corresponding to t is

$$\mathbf{r}'(t) = \mathbf{i} + 2t\mathbf{j} + 3t^2\mathbf{k}.$$

FIGURE 15.13

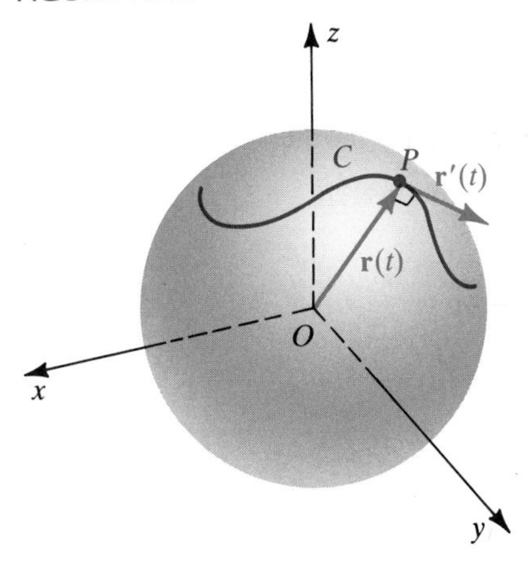

In particular, the point on C corresponding to $t = 2$ is $(2, 4, 8)$, and a tangent vector is

$$\mathbf{r}'(2) = \mathbf{i} + 4\mathbf{j} + 12\mathbf{k}.$$

By Theorem (14.34), parametric equations for the corresponding tangent line are

$$x = 2 + t, \quad y = 4 + 4t, \quad z = 8 + 12t; \quad t \text{ in } \mathbb{R}.$$

If $\mathbf{r}$ is differentiable and if the magnitude of $\mathbf{r}(t)$ is constant, say $\|\mathbf{r}(t)\| = c$ for some c, then the curve C determined by $\mathbf{r}(t)$ lies on a sphere of radius c with center at O, as illustrated in Figure 15.13. Our geometric intuition tells us that the tangent vector $\mathbf{r}'(t)$ at any point P on C is orthogonal to the vector $\mathbf{r}(t)$ from O to P. We shall prove this fact next.

Theorem **(15.9)**

> If $\mathbf{r}$ is differentiable and $\|\mathbf{r}(t)\|$ is constant, then $\mathbf{r}'(t)$ is orthogonal to $\mathbf{r}(t)$ for every t in the domain of $\mathbf{r}'$.

PROOF If $\|\mathbf{r}(t)\| = c$ for some constant c, then

$$\mathbf{r}(t) \cdot \mathbf{r}(t) = \|\mathbf{r}(t)\|^2 = c^2$$

and, therefore,

$$D_t\,[\mathbf{r}(t) \cdot \mathbf{r}(t)] = D_t\,c^2 = 0.$$

Using Theorem (15.8)(iii), we obtain

$$\mathbf{r}(t) \cdot \mathbf{r}'(t) + \mathbf{r}'(t) \cdot \mathbf{r}(t) = 0,$$

or

$$2[\mathbf{r}(t) \cdot \mathbf{r}'(t)] = 0.$$

Hence, by Theorem (14.21), $\mathbf{r}(t)$ and $\mathbf{r}'(t)$ are orthogonal. ■

Definite integrals of vector-valued functions may be defined as follows.

Definition **(15.10)**

> Let $\mathbf{r}(t) = f(t)\mathbf{i} + g(t)\mathbf{j} + h(t)\mathbf{k}$, where f, g, and h are integrable on $[a, b]$. The **definite integral of r from *a* to *b*** is
>
> $$\int_a^b \mathbf{r}(t)\,dt = \left[\int_a^b f(t)\,dt\right]\mathbf{i} + \left[\int_a^b g(t)\,dt\right]\mathbf{j} + \left[\int_a^b h(t)\,dt\right]\mathbf{k}.$$

If $\mathbf{R}'(t) = \mathbf{r}(t)$, then $\mathbf{R}(t)$ is an **antiderivative** of $\mathbf{r}(t)$. The next result is analogous to the fundamental theorem of calculus.

Theorem **(15.11)**

> If $\mathbf{R}(t)$ is an antiderivative of $\mathbf{r}(t)$ on $[a, b]$, then
>
> $$\int_a^b \mathbf{r}(t)\,dt = \mathbf{R}(t)\Big]_a^b = \mathbf{R}(b) - \mathbf{R}(a).$$

EXAMPLE 4 Find $\int_0^2 \mathbf{r}(t)\,dt$ if $\mathbf{r}(t) = 12t^3\mathbf{i} + 4e^{2t}\mathbf{j} + (t+1)^{-1}\mathbf{k}$.

SOLUTION Finding an antiderivative for each component of $\mathbf{r}(t)$, we obtain

$$\mathbf{R}(t) = 3t^4\mathbf{i} + 2e^{2t}\mathbf{j} + \ln(t+1)\mathbf{k}.$$

Since $\mathbf{R}'(t) = \mathbf{r}(t)$, it follows from Theorem (15.11) that

$$\begin{aligned}\int_0^2 \mathbf{r}(t)\,dt &= \mathbf{R}(2) - \mathbf{R}(0)\\ &= (48\mathbf{i} + 2e^4\mathbf{j} + \ln 3\mathbf{k}) - (0\mathbf{i} + 2\mathbf{j} + 0\mathbf{k})\\ &= 48\mathbf{i} + 2(e^4 - 1)\mathbf{j} + \ln 3\mathbf{k}.\end{aligned}$$

The theory of indefinite integrals of vector-valued functions is similar to that developed for real-valued functions in Chapter 5. The proofs of theorems require only minor modifications of those given earlier and are thus omitted. If $\mathbf{R}(t)$ is an antiderivative of $\mathbf{r}(t)$, then every antiderivative has the form $\mathbf{R}(t) + \mathbf{c}$ for some (constant) vector $\mathbf{c}$, and we write

$$\int \mathbf{r}(t)\,dt = \mathbf{R}(t) + \mathbf{c} \quad \text{where} \quad \mathbf{R}'(t) = \mathbf{r}(t).$$

EXERCISES 15.2

Exer. 1–8: (a) Find the domain of r. (b) Find $\mathbf{r}'(t)$ and $\mathbf{r}''(t)$.

1 $\mathbf{r}(t) = \sqrt{t-1}\,\mathbf{i} + \sqrt{2-t}\,\mathbf{j}$

2 $\mathbf{r}(t) = \dfrac{1}{t}\mathbf{i} + \sin 3t\mathbf{j}$

3 $\mathbf{r}(t) = \tan t\mathbf{i} + (t^2 + 8t)\mathbf{j}$

4 $\mathbf{r}(t) = e^{(t^2)}\mathbf{i} + \sin^{-1} t\mathbf{j}$

5 $\mathbf{r}(t) = t^2\mathbf{i} + \tan t\mathbf{j} + 3\mathbf{k}$

6 $\mathbf{r}(t) = \sqrt[3]{t}\,\mathbf{i} + \dfrac{1}{t}\mathbf{j} + e^{-t}\mathbf{k}$

7 $\mathbf{r}(t) = \sqrt{t}\,\mathbf{i} + e^{2t}\mathbf{j} + t\mathbf{k}$

8 $\mathbf{r}(t) = \ln(1-t)\mathbf{i} + \sin t\mathbf{j} + t^2\mathbf{k}$

Exer. 9–16: (a) Sketch the curve in the xy-plane determined by $\mathbf{r}(t)$ and indicate the orientation. (b) Find $\mathbf{r}'(t)$ and sketch $\mathbf{r}(t)$ and $\mathbf{r}'(t)$ for the indicated value of t.

9 $\mathbf{r}(t) = -\frac{1}{4}t^4\mathbf{i} + t^2\mathbf{j}$; $t = 2$

10 $\mathbf{r}(t) = e^{2t}\mathbf{i} + e^{-4t}\mathbf{j}$; $t = 0$

11 $\mathbf{r}(t) = 4\cos t\mathbf{i} + 2\sin t\mathbf{j}$; $t = 3\pi/4$

12 $\mathbf{r}(t) = 2\sec t\mathbf{i} + 3\tan t\mathbf{j}$; $t = \pi/4$, $|t| < \pi/2$

13 $\mathbf{r}(t) = t^3\mathbf{i} + t^{-3}\mathbf{j}$; $t = 1$, $t > 0$

14 $\mathbf{r}(t) = t^2\mathbf{i} + t^3\mathbf{j}$; $t = -1$

15 $\mathbf{r}(t) = (2t-1)\mathbf{i} + (4-t)\mathbf{j}$; $t = 3$

16 $\mathbf{r}(t) = 5\mathbf{i} + t^3\mathbf{j}$; $t = 2$

Exer. 17–20: A curve C is given parametrically. Find parametric equations for the tangent line to C at P.

17 $x = 2t^3 - 1$, $y = -5t^2 + 3$, $z = 8t + 2$; $P(1, -2, 10)$

18 $x = 4\sqrt{t}$, $y = t^2 - 10$, $z = 4/t$; $P(8, 6, 1)$

19 $x = e^t$, $y = te^t$, $z = t^2 + 4$; $P(1, 0, 4)$

20 $x = t\sin t$, $y = t\cos t$, $z = t$; $P(\pi/2, 0, \pi/2)$

Exer. 21–22: A curve C is given parametrically. Find two unit tangent vectors to C at P.

21 $x = e^{2t}$, $y = e^{-t}$, $z = t^2 + 4$; $P(1, 1, 4)$

22 $x = \sin t + 2$, $y = \cos t$, $z = t$; $P(2, 1, 0)$

23 Refer to Exercise 25, Section 15.1. Show that the concho-spiral has the special property that the angle between $\mathbf{k}$ and the tangent vector $\mathbf{r}'(t)$ is a constant.

24 The *general helix* is a curve whose tangent vector makes a constant angle with a fixed unit vector $\mathbf{u}$. Show that the curve with parametrization $x = 3t - t^3$, $y = 3t^2$, $z = 3t + t^3$; t in $\mathbb{R}$ is a general helix by finding an appropriate vector $\mathbf{u}$.

25 A point P moves along a curve C in such a way that the position vector $\mathbf{r}(t)$ of P is equal to the tangent vector $\mathbf{r}'(t)$ for every t. Find parametric equations for C, and describe the graph.

26 A point moves along a curve C in such a way that the position vector $\mathbf{r}(t)$ and the tangent vector $\mathbf{r}'(t)$ are always orthogonal. Prove that C lies on a sphere with center at the origin. (*Hint:* Show that $D_t \|\mathbf{r}(t)\|^2 = 0$.)

Exer. 27–30: Evaluate the integral.

27 $\int_0^2 (6t^2\mathbf{i} - 4t\mathbf{j} + 3\mathbf{k})\, dt$

28 $\int_{-1}^1 (-5t\mathbf{i} + 8t^3\mathbf{j} - 3t^2\mathbf{k})\, dt$

29 $\int_0^{\pi/4} (\sin t\mathbf{i} - \cos t\mathbf{j} + \tan t\mathbf{k})\, dt$

30 $\int_0^1 [te^{(t^2)}\mathbf{i} + \sqrt{t}\mathbf{j} + (t^2+1)^{-1}\mathbf{k}]\, dt$

Exer. 31–34: Find $\mathbf{r}(t)$ subject to the given conditions.

31 $\mathbf{r}'(t) = t^2\mathbf{i} + (6t+1)\mathbf{j} + 8t^3\mathbf{k}$, $\quad \mathbf{r}(0) = 2\mathbf{i} - 3\mathbf{j} + \mathbf{k}$

32 $\mathbf{r}'(t) = 2\mathbf{i} - 4t^3\mathbf{j} + 6\sqrt{t}\mathbf{k}$, $\quad \mathbf{r}(0) = \mathbf{i} + 5\mathbf{j} + 3\mathbf{k}$

33 $\mathbf{r}''(t) = 6t\mathbf{i} - 12t^2\mathbf{j} + \mathbf{k}$, $\quad \mathbf{r}'(0) = \mathbf{i} + 2\mathbf{j} - 3\mathbf{k}$, $\quad \mathbf{r}(0) = 7\mathbf{i} + \mathbf{k}$

34 $\mathbf{r}''(t) = 6t\mathbf{i} + 3\mathbf{j}$, $\quad \mathbf{r}'(0) = 4\mathbf{i} - \mathbf{j} + \mathbf{k}$, $\quad \mathbf{r}(0) = 5\mathbf{j}$

Exer. 35–36: If a curve C has a tangent vector a at a point P, then the *normal plane* to C at P is the plane through P with normal vector a. Find an equation of the normal plane to the given curve at P.

35 $x = e^t$, $\quad y = te^t$, $\quad z = t^2 + 4$; $\quad P(1, 0, 4)$

36 $x = t \sin t$, $\quad y = t \cos t$, $\quad z = t$; $\quad P(\pi/2, 0, \pi/2)$

Exer. 37–38: Find $D_t\,[\mathbf{u}(t) \cdot \mathbf{v}(t)]$ and $D_t\,[\mathbf{u}(t) \times \mathbf{v}(t)]$.

37 $\mathbf{u}(t) = t\mathbf{i} + t^2\mathbf{j} + t^3\mathbf{k}$, $\quad \mathbf{v}(t) = \sin t\mathbf{i} + \cos t\mathbf{j} + 2 \sin t\mathbf{k}$

38 $\mathbf{u}(t) = 2t\mathbf{i} + 6t\mathbf{j} + t^2\mathbf{k}$, $\quad \mathbf{v}(t) = e^{-t}\mathbf{i} - e^{-t}\mathbf{j} + \mathbf{k}$

39 If $\mathbf{u}$ and $\mathbf{v}$ are vector-valued functions that have limits as $t \to a$, prove the following:

(a) $\lim_{t\to a} [\mathbf{u}(t) + \mathbf{v}(t)] = \lim_{t\to a} \mathbf{u}(t) + \lim_{t\to a} \mathbf{v}(t)$

(b) $\lim_{t\to a} [\mathbf{u}(t) \cdot \mathbf{v}(t)] = \lim_{t\to a} \mathbf{u}(t) \cdot \lim_{t\to a} \mathbf{v}(t)$

(c) $\lim_{t\to a} c\mathbf{u}(t) = c \lim_{t\to a} \mathbf{u}(t)$, where c is a scalar

40 If a scalar function f and a vector-valued function $\mathbf{u}$ have limits as $t \to a$, prove that

$$\lim_{t\to a} f(t)\mathbf{u}(t) = \left[\lim_{t\to a} f(t)\right]\left[\lim_{t\to a} \mathbf{u}(t)\right].$$

41 Prove that $\lim_{t\to a} \mathbf{u}(t) = \mathbf{b}$ if and only if for every $\epsilon > 0$ there is a $\delta > 0$ such that $\|\mathbf{u}(t) - \mathbf{b}\| < \epsilon$ whenever $0 < |t - a| < \delta$. Give a graphical description of this result.

42 If $\mathbf{u}$ and $\mathbf{v}$ have limits as $t \to a$, prove that

$$\lim_{t\to a} [\mathbf{u}(t) \times \mathbf{v}(t)] = \left[\lim_{t\to a} \mathbf{u}(t)\right] \times \left[\lim_{t\to a} \mathbf{v}(t)\right].$$

Exer. 43–44: If u and v are differentiable, prove the stated rule for derivatives.

43 $D_t\,[\mathbf{u}(t) + \mathbf{v}(t)] = \mathbf{u}'(t) + \mathbf{v}'(t)$

44 $D_t\,[\mathbf{u}(t) \times \mathbf{v}(t)] = \mathbf{u}(t) \times \mathbf{v}'(t) + \mathbf{u}'(t) \times \mathbf{v}(t)$

45 If f and $\mathbf{u}$ are differentiable, prove that

$$D_t\,[f(t)\mathbf{u}(t)] = f(t)\mathbf{u}'(t) + f'(t)\mathbf{u}(t).$$

46 If f and $\mathbf{u}$ are differentiable with suitably restricted domains, prove the chain rule

$$D_t\,\mathbf{u}(f(t)) = f'(t)\mathbf{u}'(f(t)).$$

47 If $\mathbf{u}$, $\mathbf{v}$, and $\mathbf{w}$ are differentiable, prove that

$$D_t\,[\mathbf{u}(t) \cdot \mathbf{v}(t) \times \mathbf{w}(t)] = [\mathbf{u}'(t) \cdot \mathbf{v}(t) \times \mathbf{w}(t)] + [\mathbf{u}(t) \cdot \mathbf{v}'(t) \times \mathbf{w}(t)] + [\mathbf{u}(t) \cdot \mathbf{v}(t) \times \mathbf{w}'(t)].$$

48 If $\mathbf{u}'(t)$ and $\mathbf{u}''(t)$ exist, prove that

$$D_t\,[\mathbf{u}(t) \times \mathbf{u}'(t)] = \mathbf{u}(t) \times \mathbf{u}''(t).$$

49 If $\mathbf{u}$ and $\mathbf{v}$ are integrable on $[a, b]$ and if c is a scalar, prove the following:

(a) $\int_a^b [\mathbf{u}(t) + \mathbf{v}(t)]\, dt = \int_a^b \mathbf{u}(t)\, dt + \int_a^b \mathbf{v}(t)\, dt$

(b) $\int_a^b c\mathbf{u}(t)\, dt = c \int_a^b \mathbf{u}(t)\, dt$

50 If $\mathbf{u}$ is integrable on $[a, b]$ and $\mathbf{c}$ is in V_3, prove that

$$\int_a^b \mathbf{c} \cdot \mathbf{u}(t)\, dt = \mathbf{c} \cdot \int_a^b \mathbf{u}(t)\, dt.$$

15.3 MOTION

Motion often takes place in a plane. For example, although the earth moves through space, its orbit lies in a plane. (This will be proved in Section 15.6.) To study the motion of a point P in a coordinate plane, it is essential to know its position (x, y) at every instant. As usual, for *objects* in motion we assume that the mass is concentrated at P. Suppose the coordinates of P are given by the parametric equations

$$x = f(t), \quad y = g(t),$$

where t is in some interval I. If we let

$$\mathbf{r}(t) = f(t)\mathbf{i} + g(t)\mathbf{j},$$

then as t varies through I, the endpoint of $\mathbf{r}(t)$ traces the path C of the point. We refer to $\mathbf{r}(t)$ as the **position vector of P**. As in Figure 15.14, we represent

FIGURE 15.14

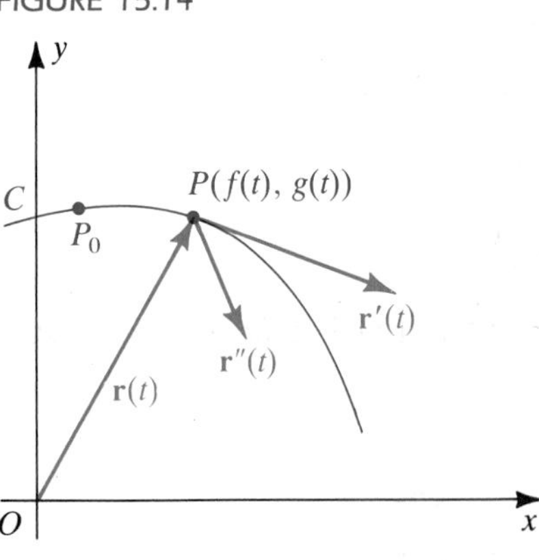

$$\mathbf{r}'(t) = f'(t)\mathbf{i} + g'(t)\mathbf{j}$$

as a tangent vector to C with initial point P. The vector $\mathbf{r}'(t)$ points in the direction of increasing values of t and has magnitude

$$\|\mathbf{r}'(t)\| = \sqrt{[f'(t)]^2 + [g'(t)]^2}.$$

Let t_0 be any number in I, and let P_0 be the point on C that corresponds to t_0 (see Figure 15.14). If C is smooth, then, by Theorem (13.5), the arc length $s(t)$ of C from P_0 to P is

$$s(t) = \int_{t_0}^{t} \sqrt{[f'(t)]^2 + [g'(t)]^2}\, dt = \int_{t_0}^{t} \|\mathbf{r}'(t)\|\, dt.$$

Applying Theorem (5.35), we obtain

$$D_t\,[s(t)] = D_t \int_{t_0}^{t} \|\mathbf{r}'(t)\|\, dt = \|\mathbf{r}'(t)\|.$$

Thus, $\|\mathbf{r}'(t)\|$ *is the rate of change of arc length with respect to time*. For this reason we refer to $\|\mathbf{r}'(t)\|$ as the *speed* of the point. The tangent vector $\mathbf{r}'(t)$ is defined as the *velocity* of the point P at time t, and the vector $\mathbf{r}''(t)$ is the *acceleration* of P. As with $\mathbf{r}'(t)$, we shall represent $\mathbf{r}''(t)$ graphically by a vector with initial point P. In most cases $\mathbf{r}''(t)$ is directed toward the concave side of C, as illustrated in Figure 15.14.

The next definition summarizes this discussion and introduces the symbols $\mathbf{v}$ and $\mathbf{a}$ for velocity and acceleration.

Definition **(15.12)**

Let the position vector for a point $P(x, y)$ moving in an xy-plane be

$$\mathbf{r}(t) = x\mathbf{i} + y\mathbf{j} = f(t)\mathbf{i} + g(t)\mathbf{j},$$

where t is time and f and g have first and second derivatives. The velocity, speed, and acceleration of P at time t are as follows:

$$\textbf{Velocity:}\quad \mathbf{v}(t) = \mathbf{r}'(t) = \frac{dx}{dt}\mathbf{i} + \frac{dy}{dt}\mathbf{j}$$

$$\textbf{Speed:}\quad v(t) = \|\mathbf{v}(t)\| = \|\mathbf{r}'(t)\| = \sqrt{\left(\frac{dx}{dt}\right)^2 + \left(\frac{dy}{dt}\right)^2}$$

$$\textbf{Acceleration:}\quad \mathbf{a}(t) = \mathbf{v}'(t) = \mathbf{r}''(t) = \frac{d^2x}{dt^2}\mathbf{i} + \frac{d^2y}{dt^2}\mathbf{j}$$

EXAMPLE 1 The position vector of a point P moving in an xy-plane is

$$\mathbf{r}(t) = (t^2 + t)\mathbf{i} + t^3\mathbf{j} \quad \text{for} \quad 0 \le t \le 2.$$

(a) Find the velocity and acceleration of P at time t.

(b) Sketch the path C of the point, together with $\mathbf{v}(1)$ and $\mathbf{a}(1)$.

SOLUTION

(a) The velocity and acceleration are

$$\mathbf{v}(t) = \mathbf{r}'(t) = (2t + 1)\mathbf{i} + 3t^2\mathbf{j} \quad \text{and} \quad \mathbf{a}(t) = \mathbf{r}''(t) = 2\mathbf{i} + 6t\mathbf{j}.$$

(b) Parametric equations for C (obtained from $\mathbf{r}(t)$) are

$$x = t^2 + t, \quad y = t^3; \quad 0 \le t \le 2.$$

Coordinates of several points (x, y) on C are listed in the following table:

t	0	0.5	1	1.5	2
x	0	0.75	2	3.75	6
y	0	0.125	1	3.375	8

FIGURE 15.15

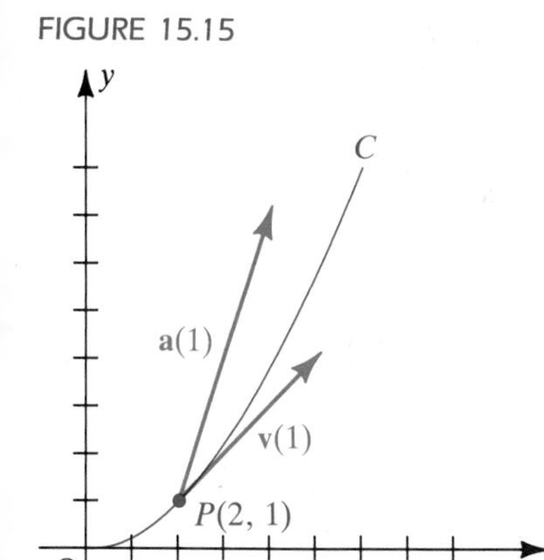

Plotting points and using the continuity of the scalar components of $\mathbf{r}(t)$ gives us the sketch of C in Figure 15.15.

At $t = 1$, the point is at $P(2, 1)$ with

$$\mathbf{v}(1) = \mathbf{r}'(1) = 3\mathbf{i} + 3\mathbf{j} \quad \text{and} \quad \mathbf{a}(1) = \mathbf{r}''(1) = 2\mathbf{i} + 6\mathbf{j}.$$

These vectors are sketched in Figure 15.15.

EXAMPLE 2 Show that if a point P moves around a circle of radius k at a constant speed v, then the acceleration vector has constant magnitude v^2/k and is directed from P toward the center of the circle.

FIGURE 15.16

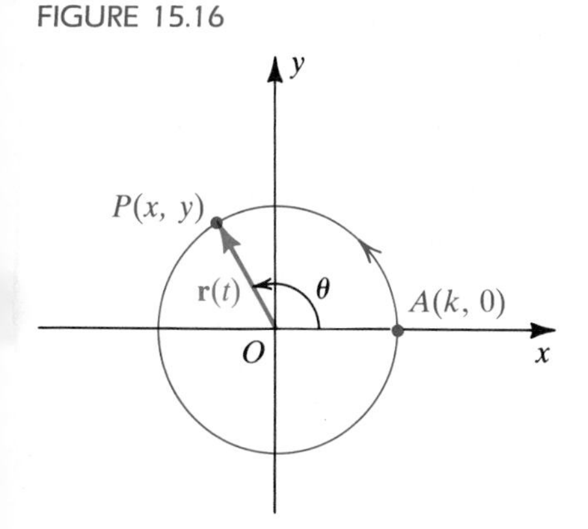

SOLUTION Let us assume that the center of the circle is the origin O in an xy-plane, and the orientation is counterclockwise. Suppose that at time $t = 0$ the point P is at $A(k, 0)$ and that θ is the angle generated by $\overrightarrow{OP}$ after t units of time (see Figure 15.16). Since P moves around the circle at a constant speed, the rate of change of θ with respect to t (the **angular speed**) is a constant ω. Thus,

$$\frac{d\theta}{dt} = \omega, \quad \text{or} \quad d\theta = \omega\, dt.$$

Integration gives us $\int d\theta = \int \omega\, dt$, or

$$\theta = \omega t + c$$

for some constant c. Since $\theta = 0$ when $t = 0$, we see that $c = 0$; that is, $\theta = \omega t$. Thus, the coordinates (x, y) of P are

$$x = k \cos \omega t, \quad y = k \sin \omega t,$$

and the position vector of P is

$$\mathbf{r}(t) = k \cos \omega t\mathbf{i} + k \sin \omega t\mathbf{j}.$$

Consequently

$$\mathbf{v}(t) = \mathbf{r}'(t) = -\omega k \sin \omega t\mathbf{i} + \omega k \cos \omega t\mathbf{j}$$
$$\mathbf{a}(t) = \mathbf{r}''(t) = -\omega^2 k \cos \omega t\mathbf{i} - \omega^2 k \sin \omega t\mathbf{j},$$

and hence

$$\mathbf{a}(t) = -\omega^2(k \cos \omega t\mathbf{i} + k \sin \omega t\mathbf{j}) = -\omega^2\mathbf{r}(t).$$

FIGURE 15.17

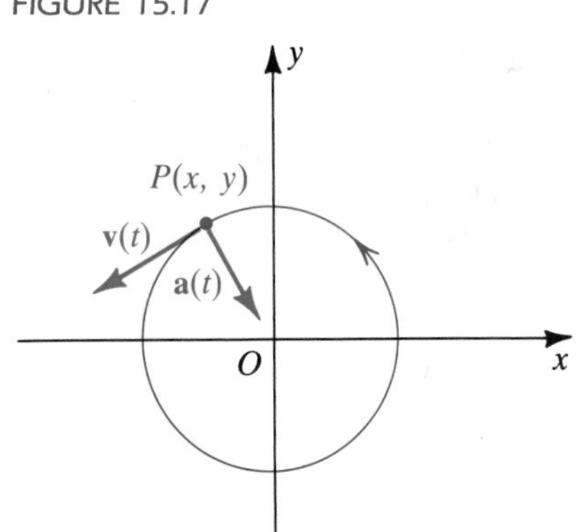

This shows that the direction of the acceleration vector $\mathbf{a}(t)$ is opposite that of $\mathbf{r}(t)$, and hence $\mathbf{a}(t)$ is directed from P toward O, as illustrated in Figure 15.17, which also shows the velocity vector $\mathbf{v}(t)$.

The magnitude of $\mathbf{a}(t)$ is

$$\|\mathbf{a}(t)\| = \|-\omega^2\mathbf{r}(t)\| = |-\omega^2|\,\|\mathbf{r}(t)\| = \omega^2 k,$$

which is a constant. Moreover,

$$v = \|\mathbf{r}'(t)\| = \sqrt{(-\omega k)^2 \sin^2 \omega t + (\omega k)^2 \cos^2 \omega t} = \sqrt{\omega^2 k^2} = \omega k,$$

and hence $\omega = v/k$. Substitution in the formula $\|\mathbf{a}(t)\| = \omega^2 k$ gives us the constant magnitude

$$\|\mathbf{a}(t)\| = \frac{v^2}{k}.$$

The acceleration vector $\mathbf{a}(t)$ in Example 2 is called a **centripetal acceleration vector**, and the force that produces $\mathbf{a}(t)$ is a **centripetal force**. Note that the magnitude $\|\mathbf{a}(t)\| = v^2/k$ will increase if we either increase v or decrease k, a fact that is evident to anyone who has attached an object to a string or rope and twirled it about in a circular path.

Our discussion of motion may be extended to a point P moving in a three-dimensional coordinate system. Suppose the coordinates of P at time t are $(f(t), g(t), h(t))$, where f, g, and h are defined on an interval I. The position vector of P is

$$\mathbf{r}(t) = f(t)\mathbf{i} + g(t)\mathbf{j} + h(t)\mathbf{k}.$$

As t varies, the endpoint P of $\mathbf{r}(t)$ traces the path. As in Definition (15.12), the derivative $\mathbf{r}'(t)$, if it exists, is the **velocity** $\mathbf{v}(t)$ at time t and is a tangent vector to C at P. The vector $\mathbf{a}(t) = \mathbf{r}''(t)$ is the **acceleration** at time t. The **speed** $v(t)$ is the magnitude $\|\mathbf{v}(t)\|$ of the velocity and, as in two dimensions, is equal to the rate of change of arc length with respect to time.

EXAMPLE 3 The position vector of a point P is $\mathbf{r}(t) = 2t\mathbf{i} + 3t^2\mathbf{j} + t^3\mathbf{k}$ for $0 \le t \le 2$. Find the velocity and acceleration of P at time t. Sketch the path of P and show $\mathbf{v}(1)$ and $\mathbf{a}(1)$.

FIGURE 15.18

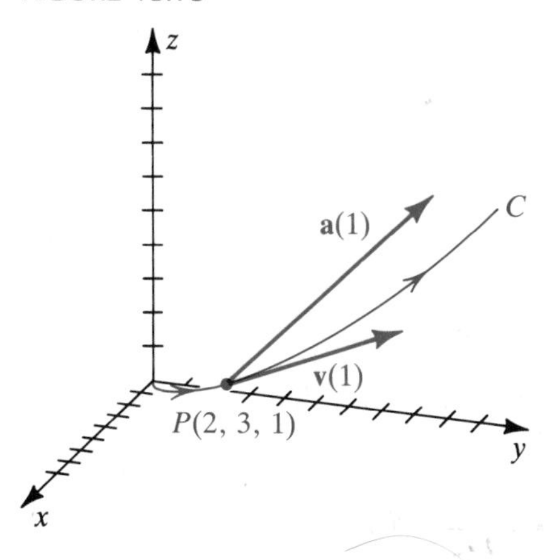

SOLUTION The velocity and acceleration are

$$\mathbf{v}(t) = \mathbf{r}'(t) = 2\mathbf{i} + 6t\mathbf{j} + 3t^2\mathbf{k} \quad \text{and} \quad \mathbf{a}(t) = \mathbf{r}''(t) = 6\mathbf{j} + 6t\mathbf{k}.$$

At $t = 1$, the point is at $P(2, 3, 1)$ and

$$\mathbf{v}(1) = 2\mathbf{i} + 6\mathbf{j} + 3\mathbf{k} \quad \text{and} \quad \mathbf{a}(1) = 6\mathbf{j} + 6\mathbf{k}.$$

These vectors and the path of the point are sketched in Figure 15.18. The path is a twisted cubic (see Example 3 and Exercise 27 of Section 15.1).

In the discussion of torque at the end of Section 14.4, we noted that the cross product of vectors is used in the investigation of rotational motion. The following example is another illustration of this fact.

FIGURE 15.19

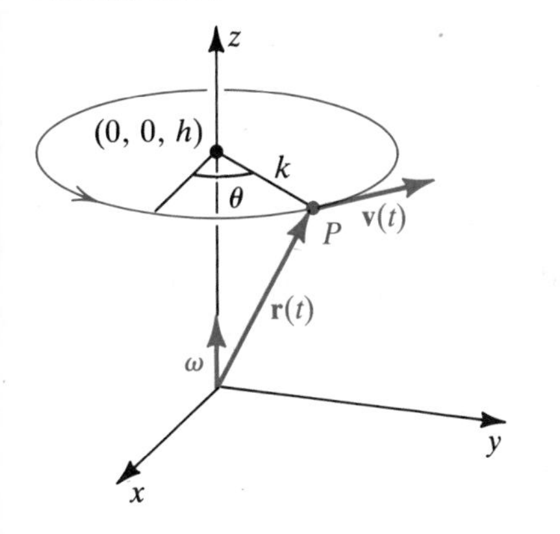

EXAMPLE 4 A point P is rotating about the z-axis on a circle of radius k that lies in the plane $z = h$, as illustrated in Figure 15.19. The angular speed $d\theta/dt$ is a constant ω. The vector $\boldsymbol{\omega} = \omega\mathbf{k}$ directed along the z-axis and having magnitude ω is the **angular velocity** of P. Show that the velocity $\mathbf{v}(t)$ of P is the cross product of $\boldsymbol{\omega}$ and the position vector $\mathbf{r}(t)$ for P.

SOLUTION Extending Example 2 to three dimensions, we find that the motion of P is given by

$$\mathbf{r}(t) = k \cos \omega t\mathbf{i} + k \sin \omega t\mathbf{j} + h\mathbf{k}.$$

Using the definition of cross product, we have

$$\begin{aligned} \boldsymbol{\omega} \times \mathbf{r}(t) &= \begin{vmatrix} \mathbf{i} & \mathbf{j} & \mathbf{k} \\ 0 & 0 & \omega \\ k\cos \omega t & k \sin \omega t & h \end{vmatrix} \\ &= -\omega k \sin \omega t\mathbf{i} + \omega k \cos \omega t\mathbf{j} = \mathbf{r}'(t) = \mathbf{v}(t). \end{aligned}$$

The following law is used frequently in the investigation of objects in motion.

Newton's second law of motion **(15.13)**

> The force $\mathbf{F}$ acting on an object of constant mass m is related to the acceleration $\mathbf{a}$ of the object as follows:
>
> $$\mathbf{F} = m\mathbf{a}$$

In the next example we shall use (15.13) to determine the position of an object that is moving near the surface of the earth under the influence of gravity alone; that is, air resistance and other forces that could affect acceleration are negligible. We shall also assume that the ground is level and the curvature of the earth is not a factor in determining the path of the object.

EXAMPLE 5 A projectile is fired, with an initial velocity $\mathbf{v}_0$, from a point h_0 feet above ground. If the only force acting on the projectile is caused by the gravitational acceleration $\mathbf{g}$, determine its position after t seconds.

FIGURE 15.20

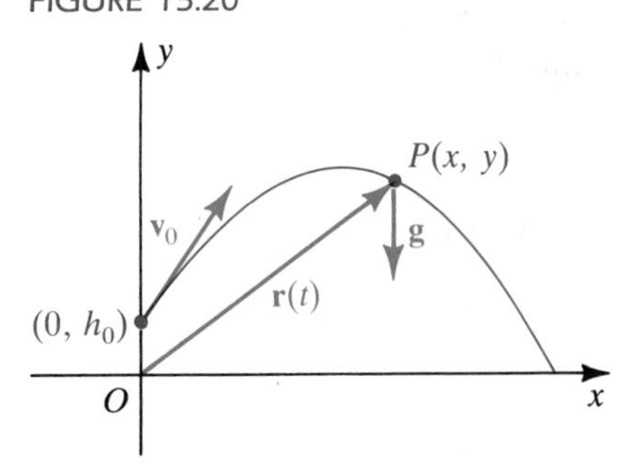

SOLUTION Let us introduce an xy-plane, as in Figure 15.20, where $(0, h_0)$ is the point from which the projectile is fired. Let $P(x, y)$ denote the position of the projectile after t seconds, and let $\mathbf{r}(t)$ be the position vector for P. Since $\mathbf{g}$ acts in the downward direction,

$$\mathbf{g} = -g\mathbf{j} \quad \text{and} \quad \|\mathbf{g}\| = g \approx 32 \text{ ft/sec}^2.$$

By Newton's second law (15.13), $\mathbf{F} = m\mathbf{a}$, where $\mathbf{a}$ is the acceleration of the projectile. In the present situation, $m\mathbf{a} = m\mathbf{g}$, or $\mathbf{a} = \mathbf{g}$, which leads to the vector differential equation

$$\mathbf{r}''(t) = \mathbf{g}.$$

Indefinite integration gives us

$$\mathbf{r}'(t) = t\mathbf{g} + \mathbf{c}$$

for some constant vector $\mathbf{c}$. Since $\mathbf{r}'(t)$ is the velocity at time t,

$$\mathbf{v}_0 = \mathbf{r}'(0) = \mathbf{c}$$

and hence $$\mathbf{r}'(t) = t\mathbf{g} + \mathbf{v}_0.$$

By indefinite integration,

$$\mathbf{r}(t) = \tfrac{1}{2}t^2\mathbf{g} + t\mathbf{v}_0 + \mathbf{d}$$

for some constant vector $\mathbf{d}$. Since $\mathbf{r}(0) = h_0\mathbf{j}$ (see Figure 15.20), it follows that $\mathbf{d} = h_0\mathbf{j}$. Consequently

$$\mathbf{r}(t) = \tfrac{1}{2}t^2\mathbf{g} + t\mathbf{v}_0 + h_0\mathbf{j}.$$

Since $\mathbf{g} = -g\mathbf{j}$, this may also be written

$$\mathbf{r}(t) = (-\tfrac{1}{2}t^2 g + h_0)\mathbf{j} + t\mathbf{v}_0.$$

EXAMPLE 6 Suppose, in Example 5, that the projectile is fired from O and the initial velocity vector $\mathbf{v}_0$ makes an angle α with the horizontal.

(a) Find parametric equations and an equation in x and y for the path of the projectile.

(b) Find the range of the projectile—that is, the horizontal distance it travels before hitting the ground. What is the maximum range?

(c) Find the maximum altitude of the projectile.

FIGURE 15.21

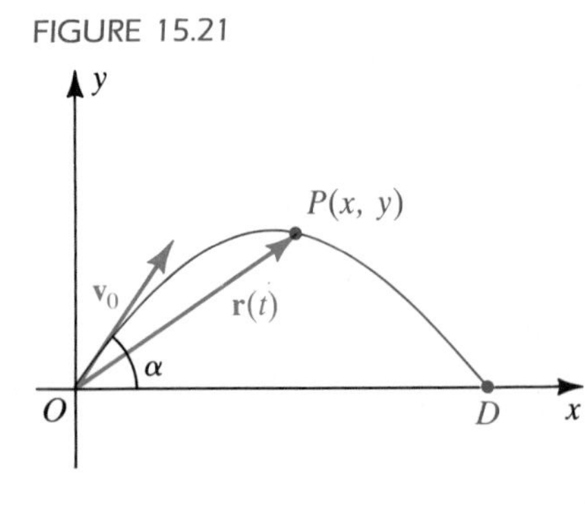

SOLUTION

(a) This special case of Example 5 is illustrated in Figure 15.21. A unit vector $\mathbf{u}$ having the same direction as $\mathbf{v}_0$ is

$$\mathbf{u} = \cos\alpha\mathbf{i} + \sin\alpha\mathbf{j}.$$

Thus, if $\| \mathbf{v}_0 \| = v_0$, then

$$\mathbf{v}_0 = v_0\mathbf{u} = (v_0 \cos \alpha)\mathbf{i} + (v_0 \sin \alpha)\mathbf{j}.$$

Using the formula for $\mathbf{r}(t)$ obtained in Example 5 with $h_0 = 0$, we obtain

$$\begin{aligned}\mathbf{r}(t) &= (-\tfrac{1}{2}gt^2)\mathbf{j} + t[(v_0 \cos \alpha)\mathbf{i} + (v_0 \sin \alpha)\mathbf{j}] \\ &= t(v_0 \cos \alpha)\mathbf{i} + (-\tfrac{1}{2}gt^2 + tv_0 \sin \alpha)\mathbf{j}.\end{aligned}$$

Hence parametric equations for the path of the projectile are

$$x = (v_0 \cos \alpha)t, \quad y = -\tfrac{1}{2}gt^2 + (v_0 \sin \alpha)t.$$

Eliminating the parameter gives us the following equation in rectangular coordinates:

$$y = \frac{-g}{2v_0^2 \cos^2 \alpha} x^2 + (\tan \alpha)x$$

This shows that the path of the projectile is parabolic.

(b) To find the range of the projectile, we must find the point D in Figure 15.21 at which it hits the ground. Since the y-coordinate of D is 0, we let $y = 0$ in the second parametric equation of part (a) and factor, obtaining

$$t(-\tfrac{1}{2}gt + v_0 \sin \alpha) = 0.$$

Thus, either

$$t = 0 \quad \text{or} \quad -\tfrac{1}{2}gt + v_0 \sin \alpha = 0.$$

It follows that the projectile is at O if $t = 0$ and at D if

$$\tfrac{1}{2}gt = v_0 \sin \alpha, \quad \text{or} \quad t = \frac{2v_0 \sin \alpha}{g}.$$

Using the second parametric equation of part (a), we can find the distance $d(O, D)$ from O to D:

$$d(O, D) = (v_0 \cos \alpha)\left(\frac{2v_0 \sin \alpha}{g}\right) = \frac{v_0^2 \sin 2\alpha}{g}$$

In particular, the range will have its maximum value v_0^2/g if $\sin 2\alpha = 1$, or $\alpha = 45°$.

(c) Differentiating $\mathbf{r}(t)$ (see (a)), we obtain the velocity vector

$$\mathbf{v}(t) = \mathbf{r}'(t) = (v_0 \cos \alpha)\mathbf{i} + (-gt + v_0 \sin \alpha)\mathbf{j}.$$

At the maximum altitude, this tangent vector to the path of the projectile is horizontal; that is, the $\mathbf{j}$-component is zero. This gives us $t = (v_0 \sin \alpha)/g$. We can also obtain this by taking one-half the time required for the projectile to reach D. Substituting this value of t into the parametric equation for y found in part (a), we obtain

$$h = -\frac{1}{2}g\left(\frac{v_0 \sin \alpha}{g}\right)^2 + (v_0 \sin \alpha)\left(\frac{v_0 \sin \alpha}{g}\right) = \frac{v_0^2 \sin^2 \alpha}{2g}.$$

EXERCISES 15.3

Exer. 1–8: Let r(*t*) be the position vector of a moving point *P*. Sketch the path *C* of *P* together with v(*t*) and a(*t*) for the given value of *t*.

1 $\mathbf{r}(t) = 2t\mathbf{i} + (4t^2 + 1)\mathbf{j}$; $t = 1$

2 $\mathbf{r}(t) = (4 - 9t^2)\mathbf{i} + 3t\mathbf{j}$; $t = 1$

3 $\mathbf{r}(t) = \sin t\mathbf{i} + 4\cos 2t\mathbf{j}$; $t = \pi/6$

4 $\mathbf{r}(t) = \cos^2 t\mathbf{i} + 2\sin t\mathbf{j}$; $t = 3\pi/4$

5 $\mathbf{r}(t) = \cos t\mathbf{i} + \sin t\mathbf{j} + t\mathbf{k}$; $t = \pi/2$

6 $\mathbf{r}(t) = 4\sin t\mathbf{i} + 2t\mathbf{j} + 9\cos t\mathbf{k}$; $t = 3\pi/4$

7 $\mathbf{r}(t) = t^2\mathbf{i} + t\mathbf{j} + 2t\mathbf{k}$; $t = 1$

8 $\mathbf{r}(t) = t^2\mathbf{i} + t^3\mathbf{j} + t\mathbf{k}$; $t = 2$

Exer. 9–16: If r(*t*) is the position vector of a moving point *P*, find its velocity, acceleration, and speed at the given time *t*.

9 $\mathbf{r}(t) = \dfrac{2}{t}\mathbf{i} + \dfrac{3}{t+1}\mathbf{j}$; $t = 2$

10 $\mathbf{r}(t) = \sqrt{t}\mathbf{i} + (1 + \sqrt{t})\mathbf{j}$; $t = 4$

11 $\mathbf{r}(t) = e^{2t}\mathbf{i} + e^{-t}\mathbf{j}$; $t = 0$

12 $\mathbf{r}(t) = 2t\mathbf{i} + e^{-t^2}\mathbf{j}$; $t = 1$

13 $\mathbf{r}(t) = e^t(\cos t\mathbf{i} + \sin t\mathbf{j} + \mathbf{k})$; $t = \pi/2$

14 $\mathbf{r}(t) = t(\cos t\mathbf{i} + \sin t\mathbf{j} + t\mathbf{k})$; $t = \pi/2$

15 $\mathbf{r}(t) = (1 + t)\mathbf{i} + 2t\mathbf{j} + (2 + 3t)\mathbf{k}$; $t = 2$

16 $\mathbf{r}(t) = 2t\mathbf{i} + \mathbf{j} + 9t^2\mathbf{k}$; $t = 2$

17 If a point moves at a constant speed, prove that the velocity and acceleration vectors are orthogonal. (*Hint:* See Theorem (15.9).)

18 If the acceleration of a moving point is always **0**, prove that the motion is along a line.

Exer. 19–20: Solve, using the results of Example 2 and 4000 miles for the radius of the earth.

19 A space shuttle is in a circular orbit 150 miles above the surface of the earth. Approximate

(a) its speed

(b) the time required for one revolution

20 An earth satellite is in a circular orbit. If the time required for one revolution is 88 minutes, approximate the satellite's altitude.

Exer. 21–24: Solve, using the results of Example 5.

21 A projectile is fired from level ground with an initial speed of 1500 ft/sec and angle of elevation 30°. Find

(a) the velocity at time t

(b) the maximum altitude

(c) the range

(d) the speed at which the projectile strikes the ground

22 Work Exercise 21 if the angle of elevation is 60°.

23 A baseball player throws a ball a distance of 250 feet. If the ball is released at an angle of 45° with the horizontal, find its initial speed.

24 A projectile is fired horizontally with a velocity of 1800 ft/sec from an altitude of 1000 feet above level ground. When and where does it strike the ground?

25 To test ability to withstand G-forces, an astronaut is placed at the end of a centrifuge device (see figure) that rotates at an angular velocity ω. If the arm is 30 feet in length, find the number of revolutions per second that will result in an acceleration that is eight times that of gravity **g**. (Use $\|\mathbf{g}\| = 32$ ft/sec^2.)

EXERCISE 25

30 ft

26 The orbits of Earth, Venus, and Neptune are nearly circular. Given the information in the table, estimate the (average) speed of each planet to the nearest 0.1 km/sec.

Planet	Period (days)	Distance from Sun (10^6 km)
Earth	365.3	149.6
Venus	224.7	108.2
Neptune	60,188	4498

27 A satellite moves in a circular orbit about the earth at a distance of d miles from the earth's surface. The magnitude of the force of attraction **F** between the satellite and the earth is

$$\|\mathbf{F}\| = G\frac{mM}{(R + d)^2},$$

where m is the mass of the satellite, M is the mass of the earth, R is the radius of the earth, and G is a gravitational constant. Use the results of Example 2 to estab-

lish *Kepler's third law* for circular orbits:

$$T^2 = \frac{4\pi^2}{GM}(R + d)^3,$$

where T is the period of the satellite. (*Hint:* $\|\mathbf{F}\|$ is also given by $m\|\mathbf{r}''(t)\|$.)

28 Refer to Exercise 27. If the period of a satellite is measured in days, Kepler's third law for circular orbits may be written $T^2 = 0.00346[1 + (d/R)]^3$.

(a) A satellite is moving in a circular orbit that is 1000 miles above the earth's surface. Assuming that the radius of the earth is 3959 miles, estimate the period of the satellite's orbit to the nearest 0.01 hour.

(b) In a *geosynchronous orbit* a satellite is always located in the same position relative to the earth; that is, the period of the satellite is one day. Estimate, to the nearest mile, the distance of such a satellite from the earth's surface. (Information of this type is needed for positioning communication satellites.)

Exer. 29–30: Solve, using the results of Example 6.

29 A major-league pitcher releases a ball at a point 6 feet above the ground and 58 feet from home plate at a speed of 100 mi/hr. If gravity had no effect, the ball would travel along a line and cross home plate 4 feet off the ground, as shown in the figure. Find the drop d caused by gravity.

EXERCISE 29

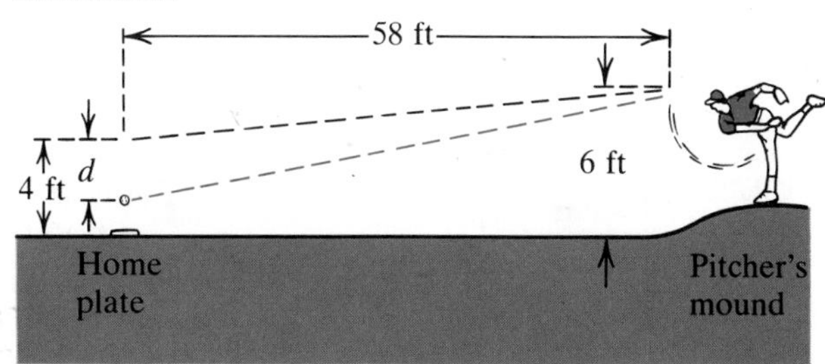

30 A quarterback on a football team throws a pass, releasing the ball at an angle of 30° with the horizontal. Approximate the velocity at which the football must be released to reach a receiver 150 feet downfield. (Neglect air resistance.)

31 Vertical wind shear in the lowest 300 feet of the atmosphere is of great importance to aircraft during takeoffs and landings. Vertical wind shear is defined as $D_h\,\mathbf{v}$, where $\mathbf{v}$ is the wind velocity and h is the height above the ground. During strong wind gusts at a certain airport, the wind velocity (in mi/hr) for altitudes h between 0 and 200 feet is estimated to be

$$\mathbf{v} = (12 + 0.006h^{3/2})\mathbf{i} + (10 + 0.005h^{3/2})\mathbf{j}.$$

Calculate the magnitude of the vertical wind shear 150 feet above ground.

15.4 CURVATURE

As a point moves along a curve C, it may change direction rapidly or slowly, depending on whether C bends sharply or gradually. To measure the rate at which C bends, or changes shape, we use the notion of *curvature*. Let us begin by introducing certain unit tangent and normal vectors to curves that will be useful in our discussion of this concept. If $\mathbf{r}$ is a vector-valued function and the curve C determined by $\mathbf{r}(t)$ is smooth, then we know (from our previous work) that $\mathbf{r}'(t)$ is a tangent vector to C that points in a direction determined by the orientation of C. If $\mathbf{r}'(t) \neq \mathbf{0}$, then the *unit tangent vector* $\mathbf{T}(t)$ to C is defined as follows.

Unit tangent vector **(15.14)**

$$\mathbf{T}(t) = \frac{1}{\|\mathbf{r}'(t)\|}\mathbf{r}'(t)$$

Since $\|\mathbf{T}(t)\| = 1$ for every t, we see from Theorem (15.9) that if $\mathbf{T}$ is differentiable, then $\mathbf{T}'(t)$ is orthogonal to $\mathbf{T}(t)$. The *principal unit normal vector* $\mathbf{N}(t)$ to C is defined as the *unit* vector having the same direction as $\mathbf{T}'(t)$, as follows.

Principal unit normal vector (15.15)

$$\mathbf{N}(t) = \frac{1}{\|\mathbf{T}'(t)\|}\mathbf{T}'(t)$$

FIGURE 15.22

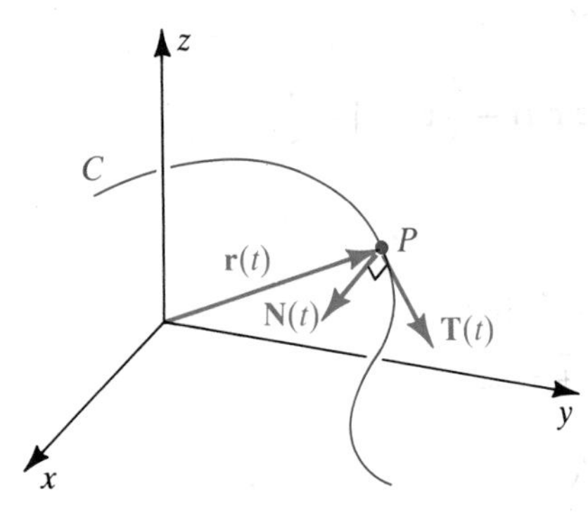

Formulas (15.14) and (15.15) may be applied to either plane curves or space curves.

When we represent $\mathbf{T}(t)$ or $\mathbf{N}(t)$ by directed line segments, we shall take the initial point at the point P on C corresponding to t, as illustrated in Figure 15.22, where we have used the right angle symbol ⌉ to emphasize that $\mathbf{N}(t)$ and $\mathbf{T}(t)$ are orthogonal. Since the tangent vector $\mathbf{r}'(t)$ points in the direction of increasing values of t, so does $\mathbf{T}(t)$.

EXAMPLE 1 Let C be the plane curve determined by $\mathbf{r}(t) = t^2\mathbf{i} + t\mathbf{j}$.

(a) Find the unit tangent and normal vectors $\mathbf{T}(t)$ and $\mathbf{N}(t)$.

(b) Sketch C, $\mathbf{T}(1)$, and $\mathbf{N}(1)$.

SOLUTION

(a) By (15.14) with $\mathbf{r}'(t) = 2t\mathbf{i} + \mathbf{j}$,

$$\mathbf{T}(t) = \frac{1}{(4t^2+1)^{1/2}}(2t\mathbf{i} + \mathbf{j}) = \frac{2t}{(4t^2+1)^{1/2}}\mathbf{i} + \frac{1}{(4t^2+1)^{1/2}}\mathbf{j}.$$

Differentiating the components of $\mathbf{T}(t)$ gives us

$$\mathbf{T}'(t) = \frac{2}{(4t^2+1)^{3/2}}\mathbf{i} - \frac{4t}{(4t^2+1)^{3/2}}\mathbf{j} = \frac{2}{(4t^2+1)^{3/2}}(\mathbf{i} - 2t\mathbf{j}).$$

It is easy to verify that

$$\|\mathbf{T}'(t)\| = \frac{2}{4t^2+1}.$$

Applying (15.15) and simplifying, we obtain

$$\mathbf{N}(t) = \frac{1}{(4t^2+1)^{1/2}}(\mathbf{i} - 2t\mathbf{j}).$$

FIGURE 15.23

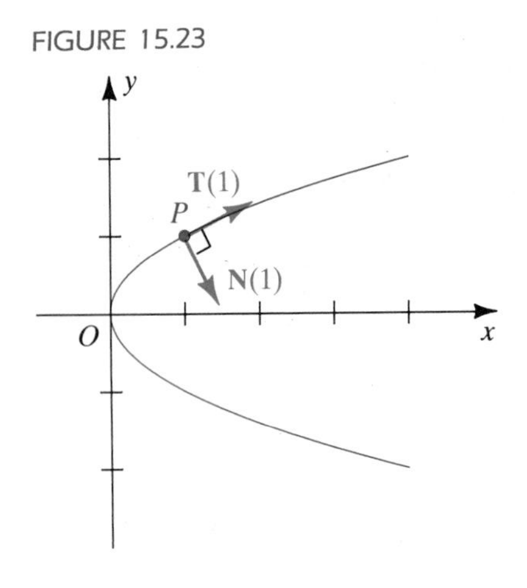

(b) The point P corresponding to $t = 1$ is $(1, 1)$. By substitution,

$$\mathbf{T}(1) = \frac{1}{\sqrt{5}}(2\mathbf{i} + \mathbf{j}) \quad \text{and} \quad \mathbf{N}(1) = \frac{1}{\sqrt{5}}(\mathbf{i} - 2\mathbf{j}).$$

To check the orthogonality of $\mathbf{T}(1)$ and $\mathbf{N}(1)$, we can verify the fact that $\mathbf{T}(1) \cdot \mathbf{N}(1) = 0$. These vectors and the curve (a parabola) are sketched in Figure 15.23.

EXAMPLE 2 Let C be determined by

$$\mathbf{r}(t) = 4\cos t\,\mathbf{i} + 4\sin t\,\mathbf{j} + 3t\mathbf{k} \quad \text{for} \quad t \geq 0.$$

Sketch C and find $\mathbf{T}(t)$ and $\mathbf{N}(t)$.

FIGURE 15.24

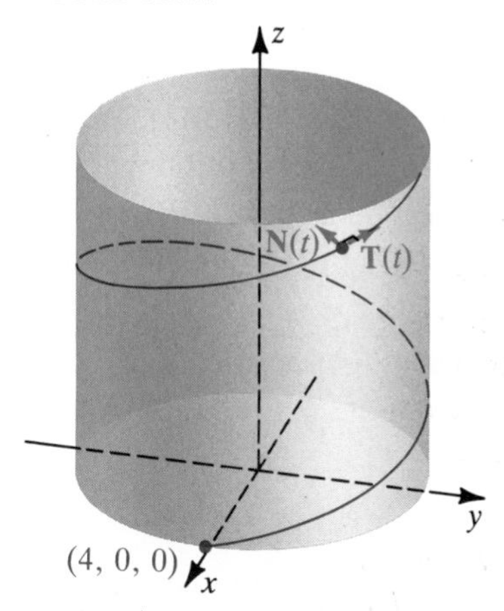

SOLUTION The curve is the circular helix of radius 4 shown in Figure 15.24 (see Example 4, Section 15.1). Differentiating $\mathbf{r}(t)$ and applying Definition (15.14) gives us the following:

$$\mathbf{r}'(t) = -4 \sin t\mathbf{i} + 4 \cos t\mathbf{j} + 3\mathbf{k}$$

$$\|\mathbf{r}'(t)\| = \sqrt{16 \sin^2 t + 16 \cos^2 t + 9} = \sqrt{16 + 9} = \sqrt{25} = 5$$

$$\mathbf{T}(t) = \frac{1}{\|\mathbf{r}'(t)\|}\mathbf{r}'(t) = -\frac{4}{5} \sin t\mathbf{i} + \frac{4}{5} \cos t\mathbf{j} + \frac{3}{5}\mathbf{k}$$

Differentiating $\mathbf{T}(t)$ and using Definition (15.15), we obtain

$$\mathbf{T}'(t) = -\tfrac{4}{5} \cos t\mathbf{i} - \tfrac{4}{5} \sin t\mathbf{j}$$

$$\|\mathbf{T}'(t)\| = \sqrt{\tfrac{16}{25} \cos^2 t + \tfrac{16}{25} \sin^2 t} = \sqrt{\tfrac{16}{25}} = \tfrac{4}{5}$$

$$\mathbf{N}(t) = \frac{1}{\|\mathbf{T}'(t)\|}\mathbf{T}'(t) = -\cos t\mathbf{i} - \sin t\mathbf{j}.$$

To check the orthogonality of $\mathbf{T}(t)$ and $\mathbf{N}(t)$, we can verify the fact that $\mathbf{T}(t) \cdot \mathbf{N}(t) = 0$.

Typical unit vectors $\mathbf{T}(t)$ and $\mathbf{N}(t)$ are shown in Figure 15.24. Note that the principal unit normal $\mathbf{N}(t)$ to the circular helix is always parallel to the xy-plane and points toward the z-axis.

To define curvature, we shall begin with plane curves. Space curves will be considered later in this section.

FIGURE 15.25

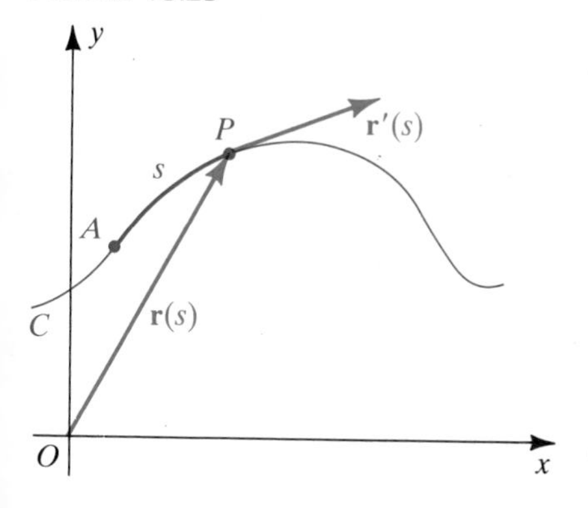

A smooth plane curve C has many different parametrizations. Sometimes it is convenient to use arc length measured along C as a parameter. As illustrated in Figure 15.25, let A be a fixed point on C and let s denote the length of arc $\widehat{AP}$ from A to an arbitrary point $P(x, y)$ on C. Suppose C is given parametrically by

$$x = f(s), \quad y = g(s); \quad s = \text{length of } \widehat{AP}.$$

Thus, to each value of s there corresponds a point $P(f(s), g(s))$ on C that is s units from A, measured along C. The orientation of C is determined by increasing values of s. We shall call s an *arc length parameter* for the curve C.

The next definition summarizes these remarks.

Definition **(15.16)**

A variable s is an **arc length parameter** for a plane curve C if there is a parametrization

$$x = f(s), \quad y = g(s)$$

such that s is the length of C from a fixed point A on the curve to the point $P(x, y)$. The parametric equations are called an **arc length parametrization** for C.

EXAMPLE 3 Find an arc length parametrization for the circle $x^2 + y^2 = k^2$, where $k > 0$.

FIGURE 15.26

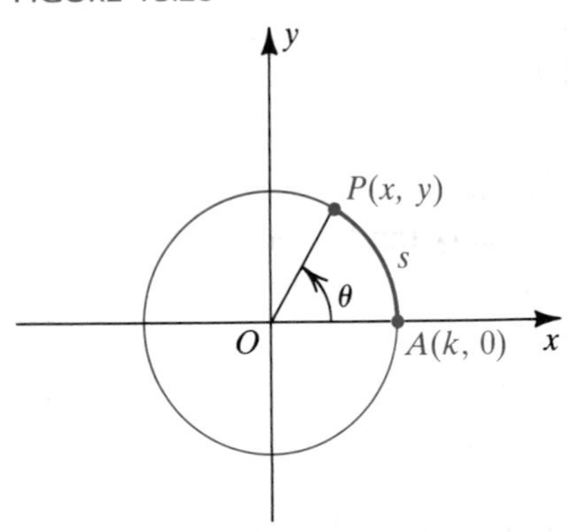

SOLUTION Let $A(k, 0)$ be the fixed point in Definition (15.16), and let $P(x, y)$ be any point on the circle. Let s denote the length of the circular arc $\widehat{AP}$, measured in the counterclockwise direction from A to P (see Figure 15.26). If θ is the radian measure of angle AOP, then parametric equations for the circle are

$$x = k\cos\theta, \quad y = k\sin\theta; \quad 0 \le \theta \le 2\pi.$$

By (1.14),

$$s = k\theta, \quad \text{or} \quad \theta = \frac{s}{k}.$$

Since $0 \le s \le 2\pi k$ corresponds to $0 \le \theta \le 2\pi$, an arc length parametrization for the circle is

$$x = k\cos\frac{s}{k}, \quad y = k\sin\frac{s}{k}; \quad 0 \le s \le 2\pi k.$$

By using a different fixed point we could obtain other arc length parametrizations.

FIGURE 15.27

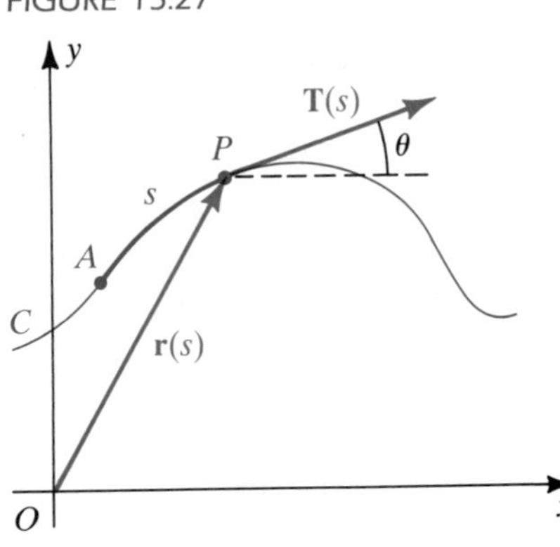

A general method for finding arc length parametrizations for curves is outlined in the instructions for Exercises 49–52. It is a very tedious process to say the least. However, as we shall see, there is great value in working with such parametrizations.

If a curve C has an arc length parametrization as in Definition (15.16), then the position vector for the point $P(x, y)$ on C is

$$\mathbf{r}(s) = x\mathbf{i} + y\mathbf{j} = f(s)\mathbf{i} + g(s)\mathbf{j}.$$

Differentiating with respect to the arc length parameter s gives us a tangent vector to C,

$$\mathbf{r}'(s) = \frac{dx}{ds}\mathbf{i} + \frac{dy}{ds}\mathbf{j} = f'(s)\mathbf{i} + g'(s)\mathbf{j}.$$

The magnitude of $\mathbf{r}'(s)$ is

$$\|\mathbf{r}'(s)\| = \sqrt{\left(\frac{dx}{ds}\right)^2 + \left(\frac{dy}{ds}\right)^2} = \sqrt{\left(\frac{ds}{ds}\right)^2} = 1,$$

FIGURE 15.28

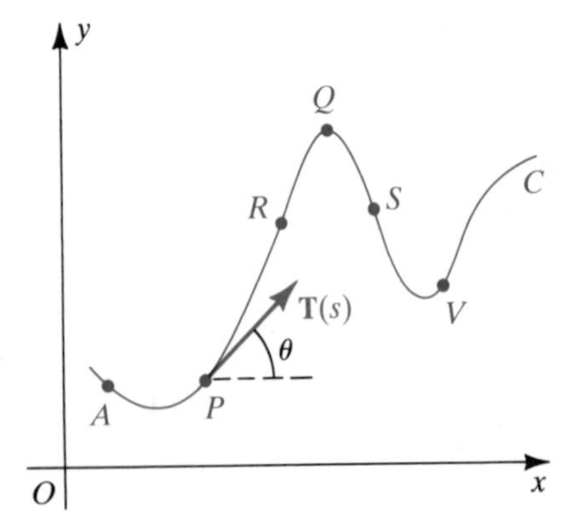

where we have used (13.6). Thus, *for an arc length parametrization*, $\mathbf{r}'(s)$ *is a unit tangent vector* to C at P. As in (15.14), with $t = s$, we shall denote this unit tangent vector by $\mathbf{T}(s)$.

For each value of s, let θ denote the angle between $\mathbf{T}(s)$ and $\mathbf{i}$, as illustrated in Figure 15.27. Note that *θ is a function of s*, since $\mathbf{T}(s)$ is a function of s. We may use the rate of change $d\theta/ds$ of θ with respect to s to measure how much the curve C bends at various points. To illustrate, suppose a point P moves along the curve C in Figure 15.28. At point R the curve bends gradually, and the angle θ between $\mathbf{T}(s)$ and $\mathbf{i}$ changes slowly; that is, $|d\theta/ds|$ is small. At point Q the curve bends sharply, and $|d\theta/ds|$ is large. We use these observations to motivate the next definition.

Definition (15.17)

Let a smooth plane curve C have an arc length parametrization $x = f(s)$, $y = g(s)$, and let θ be the angle between the unit tangent vector $\mathbf{T}(s)$ and $\mathbf{i}$. The **curvature** K of C at the point $P(x, y)$ is

$$K = \left|\frac{d\theta}{ds}\right|.$$

For the curve in Figure 15.28, the curvature is relatively small at points R and S, and the curvature is large at Q and V. The next two examples are illustrations of Definition (15.17).

FIGURE 15.29

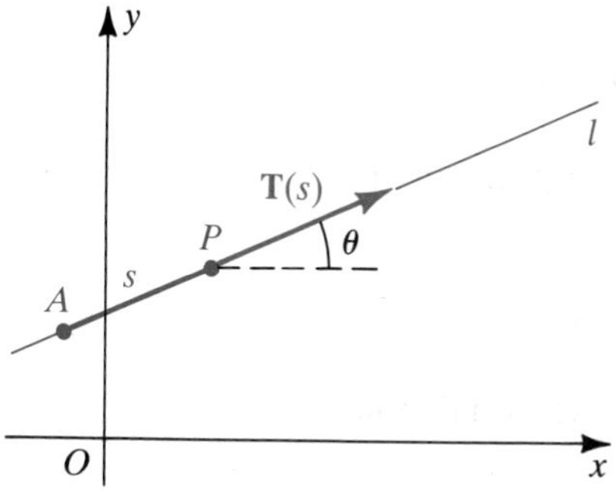

EXAMPLE 4 Prove that the curvature of a line l is 0 at every point on l.

SOLUTION As in Figure 15.29, let A be a fixed point on l, and let $s = d(A, P)$. Since the unit tangent vector $\mathbf{T}(s)$ to l always lies on l, the angle θ is the same for every s; that is, θ is constant. Hence

$$K = \left|\frac{d\theta}{ds}\right| = |0| = 0.$$

EXAMPLE 5 Prove that the curvature at every point on a circle of radius k is $1/k$.

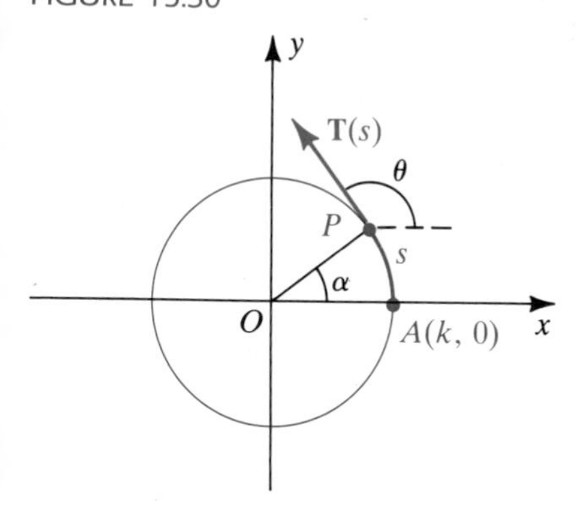

FIGURE 15.30

SOLUTION As in Figure 15.30, we assume that the circle has its center at O, that $A(k, 0)$ is the fixed point, and that the point P is in the first quadrant. Let s be the length of the arc $\overset{\frown}{AP}$. If α is the radian measure of angle POA, then

$$s = k\alpha, \quad \text{or} \quad \alpha = \frac{s}{k}.$$

Referring to Figure 15.30 and using the fact that $\mathbf{T}(s)$ is orthogonal to $\overrightarrow{OP}$, we see that

$$\theta = \alpha + \frac{\pi}{2} = \frac{s}{k} + \frac{\pi}{2}.$$

Differentiating yields

$$\frac{d\theta}{ds} = \frac{d}{ds}\left(\frac{s}{k} + \frac{\pi}{2}\right) = \frac{1}{k} + 0 = \frac{1}{k}.$$

Hence

$$K = \left|\frac{d\theta}{ds}\right| = \frac{1}{k}.$$

Note from Example 5 that as the radius k of a circle increases, its curvature $K = 1/k$ decreases. If we let k increase without bound, K approaches 0, which is the curvature of a line. As the radius k approaches 0, the curvature K increases without bound.

FIGURE 15.31

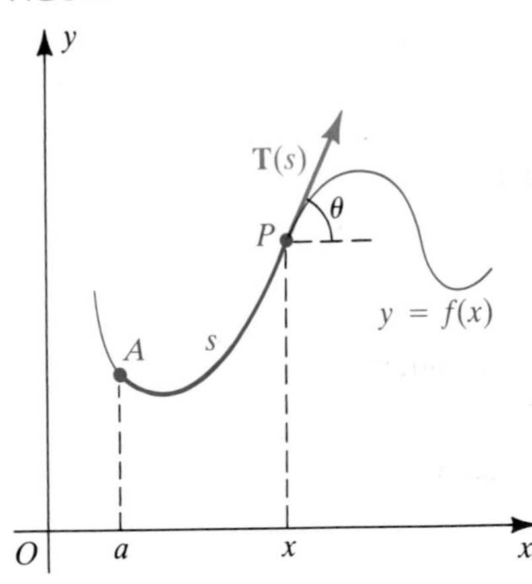

Let us next derive a formula that can be used to find the curvature K at a point (x, y) on the graph of $y = f(x)$, where f' is continuous on some interval. Let $\mathbf{T}(s)$ and θ be defined as in (15.17) (see Figure 15.31). Since y' is the slope of the tangent line at P, we see that

(1) $$\tan\theta = y', \quad \text{or} \quad \theta = \tan^{-1} y'.$$

From Definition (6.16), the arc length function s may be defined by

(2) $$s(x) = \int_a^x \sqrt{1 + (y')^2}\, dx,$$

where a is the x-coordinate of a fixed point A on C. If y'' exists, then, by the chain rule,

$$\frac{d\theta}{dx} = \frac{d\theta}{ds}\frac{ds}{dx}, \quad \text{or} \quad \frac{d\theta}{ds} = \frac{d\theta/dx}{ds/dx}.$$

Hence, by Definition (15.17),

(3) $$K = \left|\frac{d\theta}{ds}\right| = \left|\frac{d\theta/dx}{ds/dx}\right|.$$

Referring to equations (1) and (2), we have

$$\frac{d\theta}{dx} = \frac{d}{dx}\tan^{-1} y' = \frac{y''}{1 + (y')^2},$$

$$\frac{ds}{dx} = \frac{d}{dx}\int_a^x \sqrt{1 + (y')^2}\, dx = \sqrt{1 + (y')^2}.$$

Substitution in (3) gives us the following theorem.

Theorem **(15.18)**

> If a smooth curve C is the graph of $y = f(x)$, then the curvature K at $P(x, y)$ is
>
> $$K = \frac{|y''|}{[1 + (y')^2]^{3/2}}.$$

EXAMPLE 6 Sketch the graph of $y = 1 - x^2$, and find the curvature at the points (x, y), $(0, 1)$, $(1, 0)$, and $(2, -3)$.

FIGURE 15.32

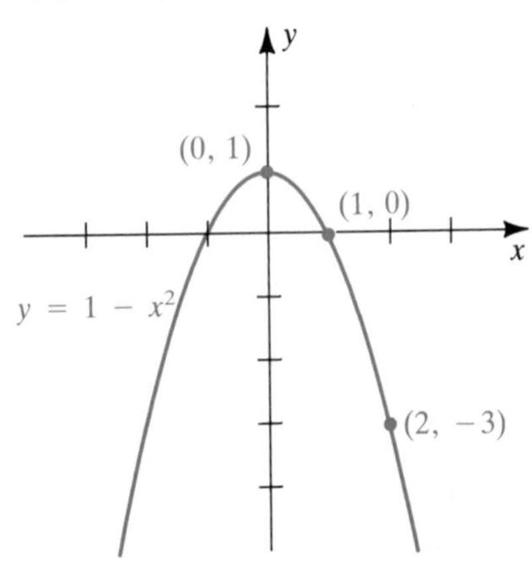

SOLUTION The graph (a parabola) is sketched in Figure 15.32. Since $y' = -2x$ and $y'' = -2$, we have, from Theorem (15.18),

$$K = \frac{2}{(1 + 4x^2)^{3/2}}.$$

Letting $x = 0$, 1, and 2, we obtain the following table.

Point on C	$(0, 1)$	$(1, 0)$	$(2, -3)$
Curvature K	2	$\frac{2}{5^{3/2}} \approx 0.18$	$\frac{2}{17^{3/2}} \approx 0.03$

Note that the maximum curvature occurs at (0, 1) and as x increases (or decreases) without bound, the curvature at $P(x, y)$ approaches 0.

Let us next derive a formula for finding K if C is described in terms of *any* parameter t. Suppose C is given parametrically by

$$x = f(t), \quad y = g(t)$$

and that f'' and g'' exist for every t in some interval I. As in the discussion preceding Theorem (15.18),

$$K = \left|\frac{d\theta}{ds}\right| = \left|\frac{d\theta/dt}{ds/dt}\right|.$$

Since the slope of the tangent line at $P(x, y)$ is $g'(t)/f'(t)$ (see Theorem (13.3)),

$$\tan\theta = \frac{g'(t)}{f'(t)}, \quad \text{or} \quad \theta = \tan^{-1}\frac{g'(t)}{f'(t)},$$

provided $f'(t) \neq 0$. Hence

$$\frac{d\theta}{dt} = \frac{1}{1 + [g'(t)/f'(t)]^2}\,\frac{f'(t)g''(t) - g'(t)f''(t)}{[f'(t)]^2} = \frac{f'(t)g''(t) - g'(t)f''(t)}{[f'(t)]^2 + [g'(t)]^2}.$$

Moreover, by (13.6),

$$\left|\frac{ds}{dt}\right| = \sqrt{[f'(t)]^2 + [g'(t)]^2}.$$

Substituting in the formula $K = |(d\theta/dt)/(ds/dt)|$ gives us the following.

Theorem (15.19)

If a plane curve C has a parametrization $x = f(t)$, $y = g(t)$ and if f'' and g'' exist, then the curvature K at $P(x, y)$ is

$$K = \frac{|f'(t)g''(t) - g'(t)f''(t)|}{[(f'(t))^2 + (g'(t))^2]^{3/2}}.$$

In the next example we use Theorem (15.19) to rework Example 5.

EXAMPLE 7 Prove that the curvature at every point on a circle of radius k is $1/k$.

SOLUTION Parametric equations for a circle of radius k with center at the origin are

$$x = k\cos t = f(t), \quad y = k\sin t = g(t); \quad 0 \le t \le 2\pi.$$

Differentiating gives us

$$f'(t) = -k\sin t \qquad f''(t) = -k\cos t$$
$$g'(t) = k\cos t \qquad g''(t) = -k\sin t.$$

Substituting in (15.19), we obtain

$$K = \frac{|(-k\sin t)(-k\sin t) - (k\cos t)(-k\cos t)|}{[(-k\sin t)^2 + (k\cos t)^2]^{3/2}}$$

$$= \frac{k^2\sin^2 t + k^2\cos^2 t}{(k^2\sin^2 t + k^2\cos^2 t)^{3/2}}$$

$$= \frac{k^2}{(k^2)^{3/2}} = \frac{k^2}{k^3} = \frac{1}{k}.$$

If the curvature K at a point P on a curve C is not 0, then the circle of radius $\rho = 1/K$ whose center lies on the concave side of C and that has the same tangent line at P as C is the **circle of curvature** for P. Its radius ρ and center are the **radius of curvature** and **center of curvature**, respectively, for P. According to Examples 5 and 7, the curvature of the circle of curvature is $1/\rho$, or K, and hence is the same as the curvature of C. For this reason *the circle of curvature may be thought of as the circle that best coincides with C at P*. A circle of curvature is shown in Figure 15.33.

FIGURE 15.33

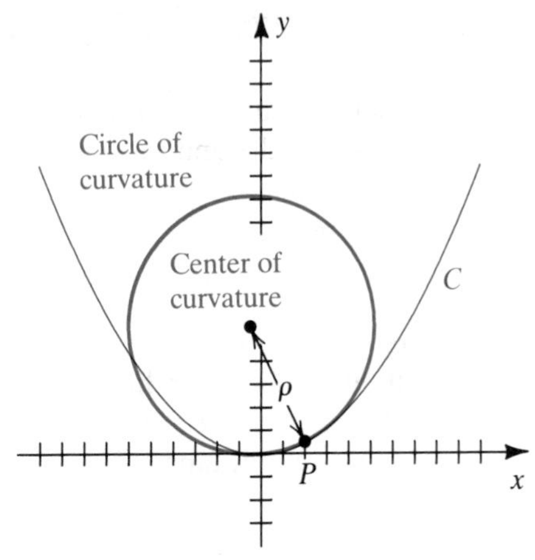

EXAMPLE 8 A curve C has the parametrization $x = t^2$, $y = t^3$; t in $\mathbb{R}$. Find the curvature at the point P corresponding to $t = \frac{1}{2}$. Sketch the graph of C and the circle of curvature for P.

SOLUTION Letting $f(t) = t^2$, $g(t) = t^3$ and differentiating, we have

$$f'(t) = 2t, \quad f''(t) = 2, \qquad g'(t) = 3t^2, \quad g''(t) = 6t.$$

Substituting in Theorem (15.19) gives us

$$K = \frac{|(2t)(6t) - (3t^2)(2)|}{[(2t)^2 + (3t^2)^2]^{3/2}} = \frac{6t^2}{(4t^2 + 9t^4)^{3/2}}.$$

If $t = \frac{1}{2}$, then

$$K = \frac{6/4}{(25/16)^{3/2}} = \frac{96}{125} \approx 0.768.$$

FIGURE 15.34

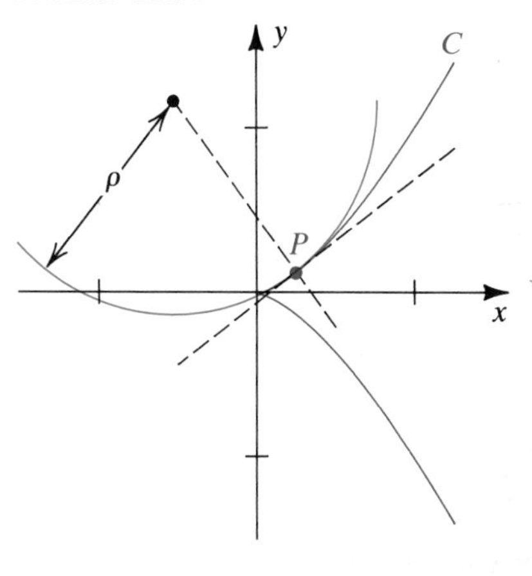

The point corresponding to $t = \frac{1}{2}$ has coordinates $(\frac{1}{4}, \frac{1}{8})$, and the radius of curvature ρ at that point is $1/K$, or $\frac{125}{96} \approx 1.3$. The graph of C and the circle of curvature are sketched in Figure 15.34. Note that the curvature at the origin does not exist, since K is undefined if $t = 0$.

The definition of curvature $K = |d\theta/ds|$ for plane curves has no immediate analogue in three dimensions, because the unit tangent vector $\mathbf{T}(s)$ cannot be specified in terms of a single angle θ. Thus, it is necessary to use a different approach for space curves. Of course, we then must show that if we specialize to vectors in a plane, the new definition of curvature agrees with (15.17).

To find a clue to a suitable definition, we first observe that in two dimensions the unit tangent vector $\mathbf{T}(s)$ can be written

$$\mathbf{T}(s) = \cos\theta\mathbf{i} + \sin\theta\mathbf{j},$$

where θ is the angle between $\mathbf{T}(s)$ and $\mathbf{i}$ (see Figure 15.28). Regarding θ as a function of s and differentiating yields

$$\mathbf{T}'(s) = \left(-\sin\theta\frac{d\theta}{ds}\right)\mathbf{i} + \left(\cos\theta\frac{d\theta}{ds}\right)\mathbf{j} = \frac{d\theta}{ds}(-\sin\theta\mathbf{i} + \cos\theta\mathbf{j}).$$

Hence $$\|\mathbf{T}'(s)\| = \left|\frac{d\theta}{ds}\right| \|-\sin\theta\mathbf{i} + \cos\theta\mathbf{j}\| = \left|\frac{d\theta}{ds}\right| = K.$$

We shall use this fact to define curvature in three dimensions. Our plan is to describe the unit tangent vector $\mathbf{T}(s)$ *without* referring to the angle θ and then *define* K as $\|\mathbf{T}'(s)\|$.

Suppose a curve C has an arc length parametrization

$$x = f(s), \quad y = g(s), \quad z = h(s)$$

where f'', g'', and h'' exist. As illustrated in Figure 15.35, s is the arc length along C from a fixed point A to the point $P(f(s), g(s), h(s))$. As in two dimensions, if $\mathbf{r}(s) = f(s)\mathbf{i} + g(s)\mathbf{j} + h(s)\mathbf{k}$ is the position vector for P, then $\mathbf{r}'(s)$ is a unit tangent vector to C at P, denoted by $\mathbf{T}(s)$. Since $\|\mathbf{T}(s)\|$

FIGURE 15.35

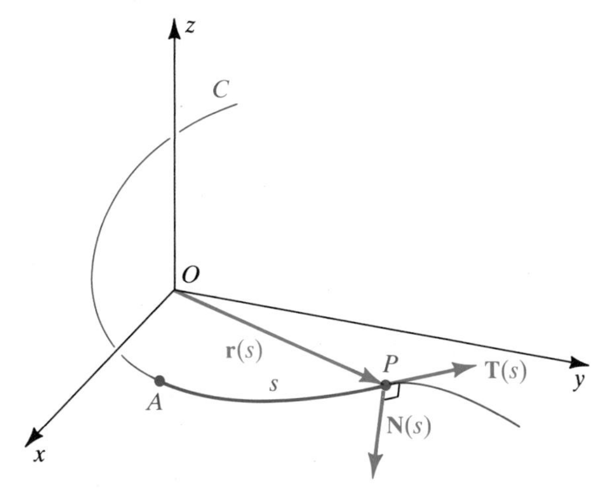

is a constant, it follows that $\mathbf{T}'(s)$ is orthogonal to $\mathbf{T}(s)$ (see Theorem (15.9)). If $\mathbf{T}'(s) \neq \mathbf{0}$, we let

$$\mathbf{N}(s) = \frac{1}{\|\mathbf{T}'(s)\|}\mathbf{T}'(s).$$

The vector $\mathbf{N}(s)$ is a unit vector orthogonal to $\mathbf{T}(s)$ and is referred to as the **principal unit normal vector** to C at the point P. These concepts are illustrated in Figure 15.35.

Having obtained a suitable unit tangent vector $\mathbf{T}(s)$ and assuming that $\mathbf{T}'(s)$ exists, we follow our plan and define curvature in three dimensions as follows.

Definition (15.20)

Let a smooth space curve C have an arc length parametrization

$$x = f(s), \quad y = g(s), \quad z = h(s).$$

Let $\mathbf{r}(s) = f(s)\mathbf{i} + g(s)\mathbf{j} + h(s)\mathbf{k}$ and let $\mathbf{T}(s) = \mathbf{r}'(s)$. The **curvature** K of C at the point $P(x, y, z)$ is

$$K = \|\mathbf{T}'(s)\|.$$

As we have already proved, the preceding definition reduces to (15.17) if C is a plane curve. Note that

$$\mathbf{N}(s) = \frac{1}{K}\mathbf{T}'(s), \quad \text{or} \quad \mathbf{T}'(s) = K\mathbf{N}(s).$$

The formula for K in Definition (15.20) is usually cumbersome to apply to specific problems. In the next section we shall derive a more practical formula that can be used to find curvature.

EXERCISES 15.4

Exer. 1–6: (a) Find the unit tangent and normal vectors $\mathbf{T}(t)$ and $\mathbf{N}(t)$ for the curve C determined by $\mathbf{r}(t)$. (b) Sketch the graph of C, and show $\mathbf{T}(t)$ and $\mathbf{N}(t)$ for the given value of t.

1 $\mathbf{r}(t) = t\mathbf{i} - \frac{1}{2}t^2\mathbf{j}$; $t = 1$

2 $\mathbf{r}(t) = -t^2\mathbf{i} + 2t\mathbf{j}$; $t = 1$

3 $\mathbf{r}(t) = t^3\mathbf{i} + 3t\mathbf{j}$; $t = 1$

4 $\mathbf{r}(t) = (4 + \cos t)\mathbf{i} - (3 - \sin t)\mathbf{j}$; $t = \pi/6$

5 $\mathbf{r}(t) = 2\sin t\mathbf{i} + 3\mathbf{j} + 2\cos t\mathbf{k}$; $t = \pi/4$

6 $\mathbf{r}(t) = t\mathbf{i} + \frac{1}{2}t^2\mathbf{j} + t^2\mathbf{k}$; $t = 1$

Exer. 7–18: Find the curvature of the curve at P.

7 $y = 2 - x^3$; $P(1, 1)$

8 $y = x^4$; $P(1, 1)$

9 $y = e^{(x^2)}$; $P(0, 1)$

10 $y = \ln(x - 1)$; $P(2, 0)$

11 $y = \cos 2x$; $P(0, 1)$

12 $y = \sec x$; $P(\pi/3, 2)$

13 $x = t - 1$, $y = \sqrt{t}$; $P(3, 2)$

14 $x = t + 1$, $y = t^2 + 4t + 3$; $P(1, 3)$

15 $x = t - t^2$, $y = 1 - t^3$; $P(0, 1)$

16 $x = t - \sin t$, $y = 1 - \cos t$; $P(\pi/2 - 1, 1)$

17 $x = 2\sin t$, $y = 3\cos t$; $P(1, \frac{3}{2}\sqrt{3})$

18 $x = \cos^3 t$, $y = \sin^3 t$; $P(\frac{1}{4}\sqrt{2}, \frac{1}{4}\sqrt{2})$

Exer. 19–22: For the given curve and point P, (a) find the radius of curvature, (b) find the center of curvature, and (c) sketch the graph and the circle of curvature for P.

19 $y = \sin x$; $P(\pi/2, 1)$

20 $y = \sec x$; $P(0, 1)$

21 $y = e^x$; $P(0, 1)$

22 $xy = 1$; $P(1, 1)$

Exer. 23–28: Find the points on the given curve at which the curvature is a maximum.

23 $y = e^{-x}$

24 $y = \cosh x$

25 $9x^2 + 4y^2 = 36$

26 $9x^2 - 4y^2 = 36$

27 $y = \ln x$

28 $y = \sin x$

Exer. 29–32: Find the points on the graph of the equation at which the curvature is 0.

29 $y = x^4 - 12x^2$

30 $y = \tan x$

31 $y = \sinh x$

32 $y = e^{-x^2}$

33 Suppose that a curve C is the graph of a polar equation $r = f(\theta)$. If $r' = dr/d\theta$ and $r'' = d^2r/d\theta^2$, show that the curvature K at $P(r, \theta)$ is

$$K = \frac{|2(r')^2 - rr'' + r^2|}{[(r')^2 + r^2]^{3/2}}.$$

(*Hint:* Use $x = r\cos\theta$ and $y = r\sin\theta$ to express C in parametric form.)

Exer. 34–36: Use the formula in Exercise 33 to find the curvature of the polar curve at $P(r, \theta)$.

34 $r = a(1 - \cos\theta)$; $0 < \theta < 2\pi$

35 $r = \sin 2\theta$; $0 < \theta < 2\pi$

36 $r = e^{a\theta}$

37 Let $P(x, y)$ be a point on the graph of $y = f(x)$ at which $K \neq 0$. If (h, k) is the center of curvature for P, show that

$$h = x - \frac{y'[1 + (y')^2]}{y''}, \quad k = y + \frac{[1 + (y')^2]}{y''}.$$

Exer. 38–42: Use the formulas in Exercise 37 to find the center of curvature for the point P on the graph of the equation. (Refer to Exercises 7–11.)

38 $y = 2 - x^3$; $P(1, 1)$

39 $y = x^4$; $P(1, 1)$

40 $y = e^{(x^2)}$; $P(0, 1)$

41 $y = \ln(x - 1)$; $P(2, 0)$

42 $y = \cos 2x$; $P(0, 1)$

43 The path of a highway and exit ramp are superimposed on a rectangular coordinate system such that the highway coincides with the x-axis. The exit ramp begins at the origin O. After following the graph of $y = -\frac{1}{27}x^3$ from O to the point $P(3, -1)$, the path follows along the arc of a circle, as shown in the figure. If $K(x)$ is the curvature of the exit ramp at (x, y), find the center of the circular arc that makes the curvature at $P(3, -1)$ continuous at $x = 3$.

EXERCISE 43

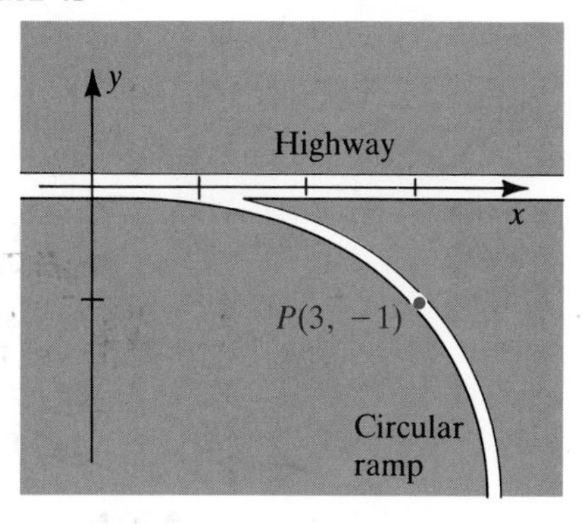

44 Use the equation $y = mx + b$ to prove that the curvature at every point on a line is 0 (see Example 4).

45 Prove that the maximum curvature of a parabola is at the vertex.

46 Prove that the maximum and minimum curvatures of an ellipse are at the ends of the major and minor axes, respectively.

47 Prove that the maximum curvature of a hyperbola is at the ends of the transverse axis.

48 Prove that lines and circles are the only *plane* curves that have a constant curvature. (Exercise 17 of Section 15.5 shows that this result is not true for space curves.)

Exer. 49–52: If the curve C in Figure 15.22 has a smooth parametrization $x = f(t)$, $y = g(t)$, then, by Theorem (13.5), the relationship between t and the arc length parameter s is given by

$$s = \int_a^t \sqrt{[f'(u)]^2 + [g'(u)]^2}\, du,$$

where a is the value of t corresponding to the fixed point A. Use this relationship to express the given curve in terms of the arc length parameter s if the fixed point A corresponds to $t = 0$. (*Hint:* First evaluate the integral to find the relationship between t and s, and then substitute for t in the parametric equations.)

49 $x = 4t - 3$, $y = 3t + 5$; $t \geq 0$

50 $x = 3t^2$, $y = 2t^3$; $t \geq 0$

51 $x = 4\cos t$, $y = 4\sin t$; $0 \leq t \leq 2\pi$

52 $x = e^t\cos t$, $y = e^t\sin t$; $t \geq 0$

53 Prove that if k is a nonnegative function that is nonzero and continuous on an interval $[0, a]$, then there is a plane curve C such that k represents the curvature of C as a function of arc length. (*Hint:* If s is in $[0, a]$, define

$$h(s) = \int_0^s k(t)\,dt, \quad x = f(s) = \int_0^s \cos h(t)\,dt,$$

$$\text{and} \quad y = g(s) = \int_0^s \sin h(t)\,dt.)$$

54 Use Exercise 53 to work Exercise 48.

15.5 TANGENTIAL AND NORMAL COMPONENTS OF ACCELERATION

In this section we use the concept of curvature to help analyze the motion of objects.

Theorem (15.21)

Let $\mathbf{r}(t)$ be the position vector of a moving point P at time t, and let s be an arc length parameter for the curve C determined by $\mathbf{r}(t)$. If $\mathbf{T}(s)$ and $\mathbf{N}(s)$ are the unit tangent and principal unit normal vectors and if K is the curvature of C at the point corresponding to s, then the velocity and acceleration of P may be written as follows:

$$\mathbf{v}(t) = \frac{ds}{dt}\mathbf{T}(s)$$

$$\mathbf{a}(t) = \frac{d^2s}{dt^2}\mathbf{T}(s) + K\left(\frac{ds}{dt}\right)^2\mathbf{N}(s)$$

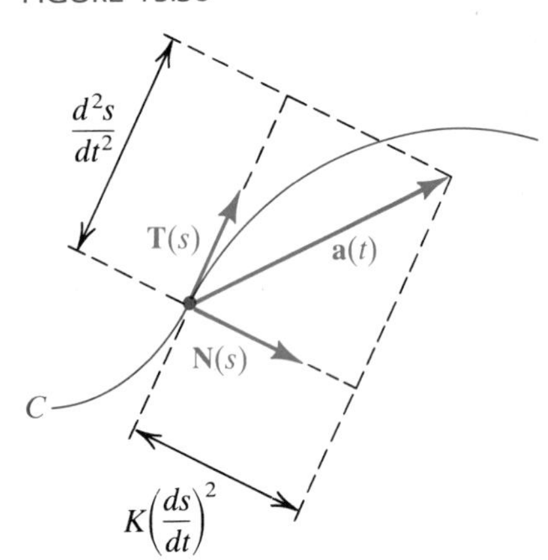

FIGURE 15.36

PROOF The sketch in Figure 15.36 is a geometric interpretation of the formula for $\mathbf{a}(t)$ in the statement of the theorem. Recall that the unit tangent vector $\mathbf{T}(s)$ to C is given by

$$\mathbf{T}(s) = \frac{1}{\|\mathbf{r}'(t)\|}\mathbf{r}'(t),$$

and hence $\quad \mathbf{r}'(t) = \|\mathbf{r}'(t)\|\,\mathbf{T}(s).$

Since $\mathbf{r}'(t) = \mathbf{v}(t)$ and $\|\mathbf{r}'(t)\| = \|\mathbf{v}(t)\| = ds/dt$, this gives us the velocity formula

$$\mathbf{v}(t) = \frac{ds}{dt}\mathbf{T}(s).$$

Differentiating the preceding equation with respect to t and using the chain rules in Exercises 45 and 46 of Section 15.2, we obtain

$$\begin{aligned}\mathbf{a}(t) = \mathbf{v}'(t) &= \frac{d^2s}{dt^2}\mathbf{T}(s) + \frac{ds}{dt}\frac{d}{dt}\mathbf{T}(s)\\ &= \frac{d^2s}{dt^2}\mathbf{T}(s) + \frac{ds}{dt}\frac{ds}{dt}\mathbf{T}'(s).\end{aligned}$$

From the remark following Definition (15.20), $\mathbf{T}'(s) = K\mathbf{N}(s)$. Thus, the last formula may be written

$$\mathbf{a}(t) = \frac{d^2s}{dt^2}\mathbf{T}(s) + K\left(\frac{ds}{dt}\right)^2\mathbf{N}(s). \quad ■$$

If we denote the speed $\|\mathbf{v}(t)\| = ds/dt$ of P by v, then the formulas in Theorem (15.21) are

$$\mathbf{v}(t) = v\mathbf{T}(s) \quad \text{and} \quad \mathbf{a}(t) = \frac{dv}{dt}\mathbf{T}(s) + Kv^2\mathbf{N}(s).$$

The second formula expresses the acceleration $\mathbf{a}(t)$ in terms of a *tangential component* $a_{\mathbf{T}}$ and a *normal component* $a_{\mathbf{N}}$; that is,

$$\mathbf{a}(t) = a_{\mathbf{T}}\mathbf{T}(s) + a_{\mathbf{N}}\mathbf{N}(s),$$

where $$a_{\mathbf{T}} = \frac{dv}{dt} = \frac{d^2s}{dt^2} \quad \text{and} \quad a_{\mathbf{N}} = Kv^2 = K\left(\frac{ds}{dt}\right)^2.$$

Tangential and normal components of acceleration are useful in the investigation of the forces that act on a moving object as it travels along a curve C. Note that the normal component $a_{\mathbf{N}} = Kv^2$ depends only on the speed of P and the curvature of C. If the speed or curvature is large, then the normal component of acceleration is large. This result, obtained theoretically, proves the well-known fact that an automobile driver should slow down when attempting to negotiate a sharp turn.

We can also derive formulas for the tangential and normal components of acceleration that depend only on $\mathbf{r}(t)$. Taking the dot product of $\mathbf{v}(t)$ and $\mathbf{a}(t)$ in Theorem (15.21), we obtain

$$\begin{aligned}\mathbf{v}(t)\cdot\mathbf{a}(t) &= \left[\frac{ds}{dt}\mathbf{T}(s)\right]\cdot\left[\frac{d^2s}{dt^2}\mathbf{T}(s) + K\left(\frac{ds}{dt}\right)^2\mathbf{N}(s)\right] \\ &= \left(\frac{ds}{dt}\right)\left(\frac{d^2s}{dt^2}\right)[\mathbf{T}(s)\cdot\mathbf{T}(s)] + K\left(\frac{ds}{dt}\right)^3[\mathbf{T}(s)\cdot\mathbf{N}(s)].\end{aligned}$$

Since $\mathbf{T}(s)$ has magnitude 1 and is orthogonal to $\mathbf{N}(s)$, this formula reduces to

$$\mathbf{v}(t)\cdot\mathbf{a}(t) = \left(\frac{ds}{dt}\right)\left(\frac{d^2s}{dt^2}\right),$$

or, in terms of $\mathbf{r}$,

$$\mathbf{r}'(t)\cdot\mathbf{r}''(t) = \|\mathbf{r}'(t)\|\frac{d^2s}{dt^2} = \|\mathbf{r}'(t)\|\,a_{\mathbf{T}}.$$

If $\mathbf{r}'(t) \neq 0$, this gives us the following formula.

Tangential component of acceleration **(15.22)**

$$a_{\mathbf{T}} = \frac{d^2s}{dt^2} = \frac{\mathbf{r}'(t)\cdot\mathbf{r}''(t)}{\|\mathbf{r}'(t)\|}$$

If, instead of the dot product, we take the *cross* product of $\mathbf{a}(t)$ and $\mathbf{v}(t)$, we obtain

$$\mathbf{v}(t)\times\mathbf{a}(t) = \left(\frac{ds}{dt}\right)\left(\frac{d^2s}{dt^2}\right)[\mathbf{T}(s)\times\mathbf{T}(s)] + K\left(\frac{ds}{dt}\right)^3[\mathbf{T}(s)\times\mathbf{N}(s)].$$

Since the cross product of a vector with itself is $\mathbf{0}$, $\mathbf{T}(s)\times\mathbf{T}(s) = \mathbf{0}$. Since $\mathbf{T}(s)$ and $\mathbf{N}(s)$ are orthogonal unit vectors, $\|\mathbf{T}(s)\times\mathbf{N}(s)\| = 1$ (see (14.30)). Hence the formula for $\mathbf{v}(t)\times\mathbf{a}(t)$ gives us

$$\|\mathbf{v}(t)\times\mathbf{a}(t)\| = K\left(\frac{ds}{dt}\right)^3$$

and, therefore,

$$a_{\mathbf{N}} = K\left(\frac{ds}{dt}\right)^2 = \frac{\|\mathbf{v}(t)\times\mathbf{a}(t)\|}{ds/dt}.$$

Rewriting this in terms of $\mathbf{r}$, we obtain the following.

Normal component of acceleration **(15.23)**

$$a_{\mathbf{N}} = K\left(\frac{ds}{dt}\right)^2 = \frac{\|\mathbf{r}'(t) \times \mathbf{r}''(t)\|}{\|\mathbf{r}'(t)\|}$$

EXAMPLE 1 The position vector of a moving point at time t is $\mathbf{r}(t) = t\mathbf{i} + t^2\mathbf{j} + t^3\mathbf{k}$ for $1 \le t \le 4$.

(a) Find the tangential and normal components of acceleration at time t.

(b) Find two-decimal-place approximations for $a_{\mathbf{N}}$, $a_{\mathbf{T}}$, and $\|\mathbf{v}(t)\|$ at times $t = 1, 2, 3$, and 4, and describe the motion of the point.

SOLUTION

(a) Differentiating yields

$$\mathbf{v}(t) = \mathbf{r}'(t) = \mathbf{i} + 2t\mathbf{j} + 3t^2\mathbf{k}$$

$$\mathbf{a}(t) = \mathbf{r}''(t) = 2\mathbf{j} + 6t\mathbf{k}$$

$$\|\mathbf{v}(t)\| = \|\mathbf{r}'(t)\| = (1 + 4t^2 + 9t^4)^{1/2}.$$

By (15.22),

$$a_{\mathbf{T}} = \frac{\mathbf{r}'(t) \cdot \mathbf{r}''(t)}{\|\mathbf{r}'(t)\|} = \frac{4t + 18t^3}{(1 + 4t^2 + 9t^4)^{1/2}}.$$

To find $a_{\mathbf{N}}$, we first calculate

$$\mathbf{r}'(t) \times \mathbf{r}''(t) = \begin{vmatrix} \mathbf{i} & \mathbf{j} & \mathbf{k} \\ 1 & 2t & 3t^2 \\ 0 & 2 & 6t \end{vmatrix} = 6t^2\mathbf{i} - 6t\mathbf{j} + 2\mathbf{k}.$$

Applying (15.23) gives us

$$a_{\mathbf{N}} = \frac{\|\mathbf{r}'(t) \times \mathbf{r}''(t)\|}{\|\mathbf{r}'(t)\|} = \frac{(36t^4 + 36t^2 + 4)^{1/2}}{(1 + 4t^2 + 9t^4)^{1/2}} = 2\left(\frac{9t^4 + 9t^2 + 1}{9t^4 + 4t^2 + 1}\right)^{1/2}.$$

(b) The path of the point—that is, the curve C determined by $\mathbf{r}(t)$—is the portion of the twisted cubic discussed in Example 3 of Section 15.1 that lies between (1, 1, 1) and (4, 16, 64) (see Figure 15.5). Substitution in the formulas obtained in part (a) gives us the following table of approximations.

t	1	2	3	4
Position of P	(1, 1, 1)	(2, 4, 8)	(3, 9, 27)	(4, 16, 64)
$a_{\mathbf{N}}$	2.33	2.12	2.06	2.03
$a_{\mathbf{T}}$	5.88	11.98	17.99	24.00
$\|\mathbf{v}(t)\|$	3.74	12.69	27.68	48.67

Thus, as t increases from 1 to 4, the point moves along C from (1, 1, 1) to (4, 16, 64), gaining speed rapidly. The normal component of acceleration $a_{\mathbf{N}}$ approaches 2 (note that $\lim_{t\to\infty} a_{\mathbf{N}} = 2$). We see from the table that the tangential component $a_{\mathbf{T}}$ increases at approximately 6 units per unit change in time.

We can obtain an alternative formula for the normal component of acceleration. To simplify the notation, let us denote $\mathbf{T}(s)$, $\mathbf{N}(s)$, and $\mathbf{a}(t)$ in Theorem (15.21) by $\mathbf{T}$, $\mathbf{N}$, and $\mathbf{a}$, respectively. Thus,

$$\mathbf{a} = a_{\mathbf{T}}\mathbf{T} + a_{\mathbf{N}}\mathbf{N}.$$

Since $\mathbf{T}$ and $\mathbf{N}$ are mutually orthogonal unit vectors, $\mathbf{T} \cdot \mathbf{N} = 0$, $\mathbf{T} \cdot \mathbf{T} = 1$, and $\mathbf{N} \cdot \mathbf{N} = 1$. Consequently

$$\begin{aligned} \|\mathbf{a}\|^2 &= \mathbf{a} \cdot \mathbf{a} = (a_{\mathbf{T}}\mathbf{T} + a_{\mathbf{N}}\mathbf{N}) \cdot (a_{\mathbf{T}}\mathbf{T} + a_{\mathbf{N}}\mathbf{N}) \\ &= a_{\mathbf{T}}^2(\mathbf{T} \cdot \mathbf{T}) + 2a_{\mathbf{N}}a_{\mathbf{T}}(\mathbf{N} \cdot \mathbf{T}) + a_{\mathbf{N}}^2(\mathbf{N} \cdot \mathbf{N}) \\ &= a_{\mathbf{T}}^2 + a_{\mathbf{N}}^2. \end{aligned}$$

This relationship can also be obtained by applying the Pythagorean theorem to one of the right triangles with hypotenuse $\|\mathbf{a}(t)\|$ in Figure 15.36. Solving for $a_{\mathbf{N}}$ gives us the following.

Theorem **(15.24)**

$$a_{\mathbf{N}} = \sqrt{\|\mathbf{a}\|^2 - a_{\mathbf{T}}^2}$$

Theorem (15.24) should be used if the normal component is difficult to find by means of (15.23).

EXAMPLE 2 Find $a_{\mathbf{N}}$ in Example 1 by using Theorem (15.24).

SOLUTION As in the solution to Example 1, we obtain

$$a_{\mathbf{T}} = \frac{4t + 18t^3}{(1 + 4t^2 + 9t^4)^{1/2}} \quad \text{and} \quad \mathbf{a} = \mathbf{r}''(t) = 2\mathbf{j} + 6t\mathbf{k}.$$

Applying (15.24) gives us

$$\begin{aligned} a_{\mathbf{N}} &= \sqrt{\|\mathbf{a}\|^2 - a_{\mathbf{T}}^2} \\ &= \sqrt{(4 + 36t^2) - \frac{(4t + 18t^3)^2}{1 + 4t^2 + 9t^4}}. \end{aligned}$$

This simplifies to the expression for $a_{\mathbf{N}}$ obtained in Example 1.

EXAMPLE 3 If a point P moves around a circle of radius k with a constant speed v, find the tangential and normal components of acceleration.

SOLUTION Since $v = ds/dt$ is a constant, the tangential component $a_{\mathbf{T}} = d^2s/dt^2$ is 0.

By Example 5 of the previous section, the curvature of the circle is $1/k$. Hence the normal component of acceleration (see (15.23)) is $(1/k)v^2$. This shows that the acceleration is a vector of constant magnitude v^2/k directed from P toward the center of the circle, as was shown in Example 2 of Section 15.3.

We may use (15.23) to obtain a formula for curvature of a space curve C. If C has the parametrization

$$x = f(t), \quad y = g(t), \quad z = h(t),$$

then C is determined by the position vector

$$\mathbf{r}(t) = f(t)\mathbf{i} + g(t)\mathbf{j} + h(t)\mathbf{k}.$$

If we solve the equation in (15.23) for K and use $ds/dt = \|\mathbf{r}'(t)\|$, we obtain the next theorem.

Theorem (15.25)

Let a space curve C have the parametrization $x = f(t)$, $y = g(t)$, $z = h(t)$, where f'', g'', and h'' exist. The curvature K at the point $P(x, y, z)$ on C is

$$K = \frac{\|\mathbf{r}'(t) \times \mathbf{r}''(t)\|}{\|\mathbf{r}'(t)\|^3} = a_{\mathbf{N}} \frac{1}{\|\mathbf{r}'(t)\|^2}.$$

This formula may also be used for plane curves (see Exercise 16).

EXAMPLE 4

(a) Find the curvature K of the twisted cubic $x = t$, $y = t^2$, $z = t^3$ at the point (x, y, z).

(b) Find four-decimal-place approximations for K at the points corresponding to $t = 1, 2, 3$, and 4.

SOLUTION

(a) If we let

$$\mathbf{r}(t) = t\mathbf{i} + t^2\mathbf{j} + t^3\mathbf{k},$$

then the curve is the same as that considered in Example 1. Substituting the expressions obtained there for $\mathbf{r}'(t)$ and $\mathbf{r}'(t) \times \mathbf{r}''(t)$ into the formula for K in Theorem (15.25) yields

$$K = \frac{2(9t^4 + 9t^2 + 1)^{1/2}}{(9t^4 + 4t^2 + 1)^{3/2}}.$$

We could also find K by substituting for $a_{\mathbf{N}}$ and $\|\mathbf{r}'(t)\|$ in Theorem (15.25).

(b) Substituting $t = 1, 2, 3$, and 4 into the formula for K obtained in part (a), we obtain the following approximations for K (compare with the table on page 782).

t	1	2	3	4
(x, y, z)	(1, 1, 1)	(2, 4, 8)	(3, 9, 27)	(4, 16, 64)
K	0.1664	0.0132	0.0027	0.0009

Using limit theorems, we can show that $\lim_{t\to\infty} K = 0$; that is, the curvature of the curve approaches that of a line as t increases.

EXERCISES 15.5

Exer. 1–8: Find general formulas for the tangential and normal components of acceleration and for the curvature of the curve C determined by $\mathbf{r}(t)$.

1 $\mathbf{r}(t) = t^2\mathbf{i} + (3t + 2)\mathbf{j}$

2 $\mathbf{r}(t) = (2t^2 - 1)\mathbf{i} + 5t\mathbf{j}$

3 $\mathbf{r}(t) = 3t\mathbf{i} + t^3\mathbf{j} + 3t^2\mathbf{k}$

4 $\mathbf{r}(t) = 4t\mathbf{i} + t^2\mathbf{j} + 2t^2\mathbf{k}$

5 $\mathbf{r}(t) = t(\cos t\mathbf{i} + \sin t\mathbf{j})$

6 $\mathbf{r}(t) = \cosh t\mathbf{i} + \sinh t\mathbf{j}$

7 $\mathbf{r}(t) = 4\cos t\mathbf{i} + 9\sin t\mathbf{j} + t\mathbf{k}$

8 $\mathbf{r}(t) = e^t(\sin t\mathbf{i} + \cos t\mathbf{j} + \mathbf{k})$

9 A point moves along the parabola $y = x^2$ such that the horizontal component of velocity is always 3. Find the tangential and normal components of acceleration at $P(1, 1)$.

10 Work Exercise 9 if the point moves along the graph of $y = 2x^3 - x$.

11 Prove that if a point moves along a curve C with a constant speed, then the acceleration is always normal to C.

12 Use Theorem (15.25) to prove that if a point moves through space with an acceleration that is always $\mathbf{0}$, then the motion is on a line.

13 If a point P moves along a curve C with a constant speed, show that the magnitude of the acceleration is directly proportional to the curvature of the curve.

14 If, in Exercise 13, a second point Q moves along C with a speed twice that of P, show that the magnitude of the acceleration of Q is four times greater than that of P.

15 Show that if a point moves along the graph of $y = f(x)$ for $a \le x \le b$, then the normal component of acceleration is 0 at a point of inflection.

16 If a plane curve is given parametrically by $x = f(t)$, $y = g(t)$ and if f'' and g'' exist, use Theorem (15.25) to prove that the curvature at the point $P(x, y)$ is given by Theorem (15.19).

17 Show that the curvature at every point on the circular helix $x = a\cos t$, $y = a\sin t$, $z = bt$, where $a > 0$, is given by $K = a/(a^2 + b^2)$.

18 An *elliptic helix* has parametric equations $x = a\cos t$, $y = b\sin t$, $z = ct$, where a, b, and c are positive real numbers and $a \ne b$. Find the curvature at (x, y, z).

15.6 KEPLER'S LAWS

It is fitting to conclude this chapter with a display of the power and beauty of vector methods when applied to the derivation of three classical physical laws. The discussion in this section is not simple, for it is not a simple problem that we intend to consider. There is no exercise set at the end of this section. The reason is that we are not interested in numerical calculations involving the laws to be developed. Your objective should be to carefully read and understand each step of the discussion. Proceed slowly. You will gain considerable insight into vector methods by studying the material that follows.

After many years of analyzing an enormous amount of empirical data, the German astronomer Johannes Kepler (1571–1630) formulated three laws that describe the motion of planets about the sun. These laws may be stated as follows.

Kepler's laws **(15.26)**

First Law: The orbit of each planet is an ellipse with the sun at one focus.

Second Law: The vector from the sun to a moving planet sweeps out area at a constant rate.

Third Law: If the time required for a planet to travel once around its elliptical orbit is T and if the major axis of the ellipse is $2a$, then $T^2 = ka^3$ for some constant k.

Approximately 50 years later, Sir Isaac Newton (1642–1727) proved that Kepler's laws were consequences of Newton's law of universal gravitation and second law of motion. The achievements of both men were monumental, because these laws clarified all astronomical observations that had been made up to that time.

In this section we shall prove Kepler's laws through the use of vectors. Since the force of gravity that the sun exerts on a planet far exceeds that exerted by other celestial bodies, we shall neglect all other forces acting on a planet. From this point of view we have only two objects to consider: the sun and a planet revolving around it.

FIGURE 15.37

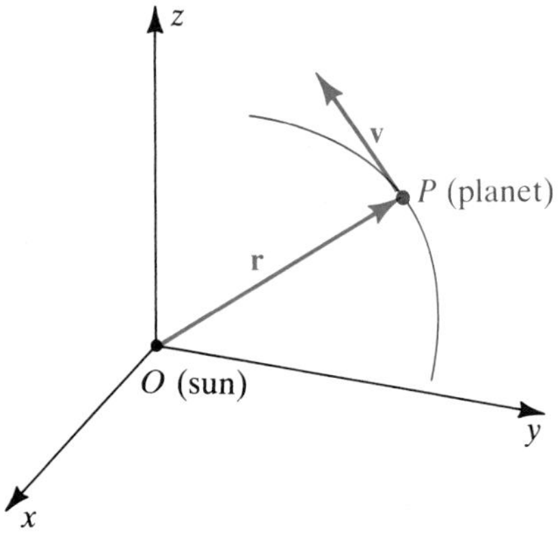

It is convenient to introduce a coordinate system with the center of mass of the sun at the origin, O, as illustrated in Figure 15.37. The point P represents the center of mass of the planet. To simplify the notation, we shall denote the position vector of P by $\mathbf{r}$ instead of $\mathbf{r}(t)$ and use $\mathbf{v}$ and $\mathbf{a}$ to denote the velocity $\mathbf{r}'(t)$ and acceleration $\mathbf{r}''(t)$, respectively.

Before proving Kepler's laws, let us show that the motion of the planet takes place in one plane. If we let $r = \|\mathbf{r}\|$, then $\mathbf{u} = (1/r)\mathbf{r}$ is a unit vector having the same direction as $\mathbf{r}$. According to Newton's law of gravitation, the force $\mathbf{F}$ of gravitational attraction on the planet is given by

$$\mathbf{F} = -G\frac{Mm}{r^2}\mathbf{u},$$

where M is the mass of the sun, m is the mass of the planet, and G is a gravitational constant. Newton's second law of motion (15.13) states that

$$\mathbf{F} = m\mathbf{a}.$$

If we equate these two expressions for $\mathbf{F}$ and solve for $\mathbf{a}$, we obtain

(1) $$\mathbf{a} = -\frac{GM}{r^2}\mathbf{u}.$$

This shows that $\mathbf{a}$ is parallel to $\mathbf{r} = r\mathbf{u}$ and hence $\mathbf{r} \times \mathbf{a} = \mathbf{0}$. In addition, since $\mathbf{v} \times \mathbf{v} = \mathbf{0}$, we see that

$$\begin{aligned}\frac{d}{dt}(\mathbf{r} \times \mathbf{v}) &= \mathbf{r} \times \frac{d\mathbf{v}}{dt} + \frac{d\mathbf{r}}{dt} \times \mathbf{v}\\ &= \mathbf{r} \times \mathbf{a} + \mathbf{v} \times \mathbf{v} = \mathbf{0}.\end{aligned}$$

It follows that

(2) $$\mathbf{r} \times \mathbf{v} = \mathbf{c}$$

FIGURE 15.38

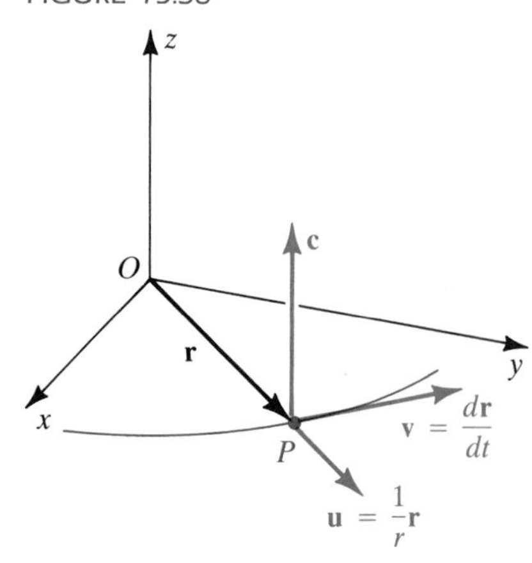

for a constant vector $\mathbf{c}$. The vector $\mathbf{c}$ will play an important role in the proofs of Kepler's laws.

Since $\mathbf{r} \times \mathbf{v} = \mathbf{c}$, the vector $\mathbf{r}$ is orthogonal to $\mathbf{c}$ for every value of t. This implies that the curve traced by P lies in one plane; that is, *the orbit of the planet is a plane curve*, as illustrated in Figure 15.38.

Let us now prove Kepler's first law. We may assume that the motion of the planet takes place in the xy-plane. In this case the vector $\mathbf{c}$ is perpendicular to the xy-plane, and we may assume that $\mathbf{c}$ has the same direction as the positive z-axis, as illustrated in Figure 15.38.

Since $\mathbf{r} = r\mathbf{u}$, we see that

$$\mathbf{v} = \frac{d\mathbf{r}}{dt} = r\frac{d\mathbf{u}}{dt} + \frac{dr}{dt}\mathbf{u}.$$

Substitution in $\mathbf{c} = \mathbf{r} \times \mathbf{v}$ and use of properties of vector products gives us

$$\begin{aligned}\mathbf{c} &= r\mathbf{u} \times \left(r\frac{d\mathbf{u}}{dt} + \frac{dr}{dt}\mathbf{u}\right) \\ &= r^2\left(\mathbf{u} \times \frac{d\mathbf{u}}{dt}\right) + r\frac{dr}{dt}(\mathbf{u} \times \mathbf{u}).\end{aligned}$$

Since $\mathbf{u} \times \mathbf{u} = \mathbf{0}$, this reduces to

(3) $$\mathbf{c} = r^2\left(\mathbf{u} \times \frac{d\mathbf{u}}{dt}\right).$$

Using (3) and (1) together with (ii) and (vi) of Theorem (14.33), we see that

$$\begin{aligned}\mathbf{a} \times \mathbf{c} &= \left(-\frac{GM}{r^2}\mathbf{u}\right) \times \left[r^2\left(\mathbf{u} \times \frac{d\mathbf{u}}{dt}\right)\right] \\ &= -GM\left[\mathbf{u} \times \left(\mathbf{u} \times \frac{d\mathbf{u}}{dt}\right)\right] \\ &= -GM\left[\left(\mathbf{u} \cdot \frac{d\mathbf{u}}{dt}\right)\mathbf{u} - (\mathbf{u} \cdot \mathbf{u})\frac{d\mathbf{u}}{dt}\right].\end{aligned}$$

Since $\|\mathbf{u}\| = 1$, it follows from Theorem (15.9) that $\mathbf{u} \cdot (d\mathbf{u}/dt) = 0$. In addition, $\mathbf{u} \cdot \mathbf{u} = \|\mathbf{u}\|^2 = 1$, and hence the last formula for $\mathbf{a} \times \mathbf{c}$ reduces to

$$\mathbf{a} \times \mathbf{c} = GM\frac{d\mathbf{u}}{dt} = \frac{d}{dt}(GM\mathbf{u}).$$

We may also write

$$\mathbf{a} \times \mathbf{c} = \frac{d\mathbf{v}}{dt} \times \mathbf{c} = \frac{d}{dt}(\mathbf{v} \times \mathbf{c})$$

and, consequently,

$$\frac{d}{dt}(\mathbf{v} \times \mathbf{c}) = \frac{d}{dt}(GM\mathbf{u}).$$

Integrating both sides of this equation gives us

(4) $$\mathbf{v} \times \mathbf{c} = GM\mathbf{u} + \mathbf{b},$$

where $\mathbf{b}$ is a constant vector.

The vector $\mathbf{v} \times \mathbf{c}$ is orthogonal to $\mathbf{c}$ and therefore is in the xy-plane. Since $\mathbf{u}$ is also in the xy-plane, it follows from (4) that $\mathbf{b}$ is in the xy-plane.

Up to this point, our proof has been independent of the positions of the x- and y-axes. Let us now choose a coordinate system such that the positive x-axis has the direction of the constant vector $\mathbf{b}$, as illustrated in Figure 15.39.

FIGURE 15.39

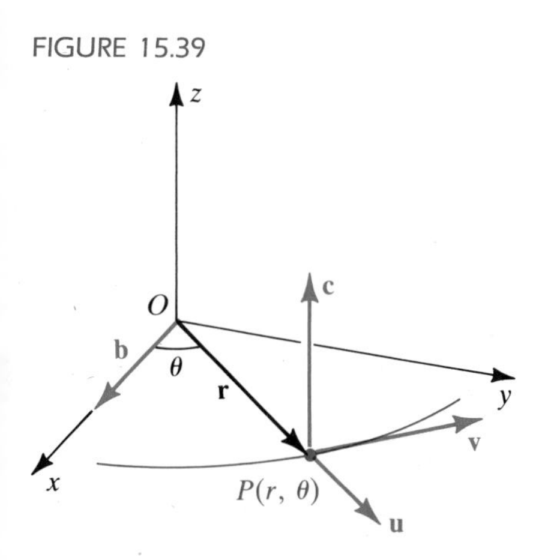

Let (r, θ) be polar coordinates for the point P, with $r = \|\mathbf{r}\|$. It follows that

$$\mathbf{u} \cdot \mathbf{b} = \|\mathbf{u}\| \|\mathbf{b}\| \cos \theta = b \cos \theta,$$

where $b = \|\mathbf{b}\|$. If we let $c = \|\mathbf{c}\|$, then using (2), together with properties of the dot and vector products, and also (4), yields

$$\begin{aligned} c^2 &= \mathbf{c} \cdot \mathbf{c} = (\mathbf{r} \times \mathbf{v}) \cdot \mathbf{c} = \mathbf{r} \cdot (\mathbf{v} \times \mathbf{c}) \\ &= (r\mathbf{u}) \cdot (GM\mathbf{u} + \mathbf{b}) \\ &= rGM(\mathbf{u} \cdot \mathbf{u}) + r(\mathbf{u} \cdot \mathbf{b}) \\ &= rGM + rb \cos \theta. \end{aligned}$$

Solving the last equation for r gives us

$$r = \frac{c^2}{GM + b \cos \theta}.$$

Dividing numerator and denominator of this fraction by GM, we obtain

$$\text{(5)} \qquad r = \frac{p}{1 + e \cos \theta}$$

with $p = c^2/(GM)$ and $e = b/(GM)$. From Theorem (13.16), the graph of this polar equation is a conic with eccentricity e and focus at the origin. Since the orbit is a closed curve, it follows that $0 < e < 1$ and that the conic is an ellipse. This completes the proof of Kepler's first law.

Let us next prove Kepler's second law. We may assume that the orbit of the planet is an ellipse in the xy-plane. Let $r = f(\theta)$ be a polar equation of the orbit, with the center of the sun at the focus O. Let P_0 denote the position of the planet at time t_0 and P its position at any time $t \geq t_0$. As illustrated in Figure 15.40, θ_0 and θ will denote the angles measured from the positive x-axis to $\overrightarrow{OP_0}$ and $\overrightarrow{OP}$, respectively.

FIGURE 15.40

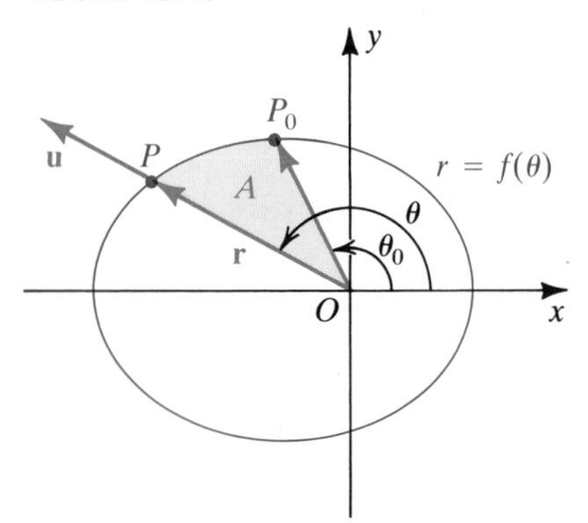

By Theorem (13.11), the area A swept out by $\overrightarrow{OP}$ in the time interval $[t_0, t]$ is

$$A = \int_{\theta_0}^{\theta} \tfrac{1}{2}r^2 \, d\theta,$$

and hence

$$\frac{dA}{d\theta} = \frac{d}{d\theta} \int_{\theta_0}^{\theta} \tfrac{1}{2}r^2 \, d\theta = \tfrac{1}{2}r^2.$$

Using this fact and the chain rule gives us

$$\text{(6)} \qquad \frac{dA}{dt} = \frac{dA}{d\theta}\frac{d\theta}{dt} = \tfrac{1}{2}r^2 \frac{d\theta}{dt}.$$

Next we observe that since $\mathbf{r} = r \cos \theta \mathbf{i} + r \sin \theta \mathbf{j} + 0\mathbf{k}$, the unit vector $\mathbf{u} = (1/r)\mathbf{r}$ may be expressed in the form

$$\mathbf{u} = \cos \theta \mathbf{i} + \sin \theta \mathbf{j} + 0\mathbf{k}.$$

Consequently

$$\frac{d\mathbf{u}}{dt} = -\sin \theta \frac{d\theta}{dt}\mathbf{i} + \cos \theta \frac{d\theta}{dt}\mathbf{j} + 0\mathbf{k}.$$

A direct calculation may be used to show that

$$\mathbf{u} \times \frac{d\mathbf{u}}{dt} = \frac{d\theta}{dt}\mathbf{k}.$$

If $\mathbf{c}$ is the vector obtained in the proof of Kepler's first law, then, by (3) and the last equation,

$$\mathbf{c} = r^2\left(\mathbf{u} \times \frac{d\mathbf{u}}{dt}\right) = r^2\frac{d\theta}{dt}\mathbf{k}$$

and hence

(7) $$c = |\mathbf{c}| = r^2\frac{d\theta}{dt}.$$

Combining (6) and (7), we see that

(8) $$\frac{dA}{dt} = \tfrac{1}{2}c;$$

that is, the rate at which A is swept out by $\overrightarrow{OP}$ is a constant. This establishes Kepler's second law.

To prove Kepler's third law, we shall retain the notation used in the proofs of the first two laws. In particular, we assume that a polar equation of the planetary orbit is given by

$$r = \frac{p}{1 + e\cos\theta}$$

with $p = c^2/(GM)$ and $e = b/(GM)$.

Let T denote the time required for the planet to make one complete revolution about the sun. By (8), the area swept out in the time interval $[0, T]$ is given by

$$A = \int_0^T \frac{dA}{dt}\,dt = \int_0^T \tfrac{1}{2}c\,dt = \tfrac{1}{2}cT.$$

This also equals the area of the plane region bounded by the ellipse. However, in Example 6 of Section 12.2 we found that $A = \pi ab$ for an ellipse whose major and minor axes have lengths $2a$ and $2b$, respectively. Consequently

$$\tfrac{1}{2}cT = \pi ab, \quad \text{or} \quad T = \frac{2\pi ab}{c}.$$

By (12.8),

$$e = \frac{\sqrt{a^2 - b^2}}{a}$$

and hence

$$a^2e^2 = a^2 - b^2, \quad \text{or} \quad b^2 = a^2(1 - e^2).$$

Thus, $$T^2 = \frac{4\pi^2a^2b^2}{c^2} = \frac{4\pi^2a^4(1 - e^2)}{c^2}.$$

From the proof of Theorem (13.15) (where we used $p = de$) we know that

$$a^2 = \frac{p^2}{(1-e^2)^2}, \quad \text{or} \quad a = \frac{p}{1-e^2},$$

and hence

$$T^2 = \frac{4\pi^2 a^4}{c^2}\left(\frac{p}{a}\right).$$

Since $p = c^2/(GM)$, this reduces to

$$T^2 = \frac{4\pi^2}{GM}a^3 = ka^3$$

with $k = 4\pi^2/(GM)$. This completes the proof.

Remember that in our proofs of Kepler's laws we assumed that the only gravitational force acting on a planet was that of the sun. If forces exerted by other planets are taken into account, then irregularities in the elliptical orbits may occur. Observed irregularities in the motion of Uranus led both the British astronomer J. Adams (1819–1892) and the French astronomer U. Leverrier (1811–1877) to predict the presence of an unknown planet that was causing the irregularities. On the basis of their predictions, this planet, later named Neptune, was first observed with a telescope by the German astronomer J. Galle in 1846.

15.7 REVIEW EXERCISES

Exer. 1–2: Let $\mathbf{r}(t)$ be the position vector of a moving point P. Sketch the path C of P together with $\mathbf{v}(t)$ and $\mathbf{a}(t)$ for the given value of t.

1 $\mathbf{r}(t) = \tan t\mathbf{i} - \sec t\mathbf{j}$; $\quad t = \pi/4, \quad |t| < \pi/2$

2 $\mathbf{r}(t) = t^2\mathbf{i} + 3\cos t\mathbf{j} + 5\sin t\mathbf{k}$; $\quad t = \pi$

Exer. 3–4: The position of a point moving in a coordinate plane is given by $\mathbf{r}(t)$. Find its velocity, acceleration, and speed at time t.

3 $\mathbf{r}(t) = t^2\mathbf{i} + (4t^2 - t^4)\mathbf{j}$

4 $\mathbf{r}(t) = (2 - \sin t)\mathbf{i} + (1 - \cos t)\mathbf{j}$

5 Let C be the curve determined by

$$\mathbf{r}(t) = (e^t \sin t)\mathbf{i} + (e^t \cos t)\mathbf{j} + e^t\mathbf{k} \quad \text{for} \quad 0 \le t \le 1.$$

(a) Find a unit tangent vector to C at the point corresponding to $t = 0$.

(b) Find an equation of the tangent line to C at the point corresponding to $t = 0$.

6 Find the length of the curve C in Exercise 5.

Exer. 7–8: Let

$$\mathbf{u}(t) = t^2\mathbf{i} + 6t\mathbf{j} + t\mathbf{k} \quad \text{and} \quad \mathbf{v}(t) = t\mathbf{i} - 5t\mathbf{j} + 4t^2\mathbf{k}.$$

7 Find the values of t for which $\mathbf{u}(t)$ and $\mathbf{v}'(t)$ are orthogonal.

8 Find $D_t[\mathbf{u}(t) \times \mathbf{v}(t)]$ and $D_t[\mathbf{u}(t) \cdot \mathbf{v}(t)]$.

9 The position vector of a point at time t is

$$\mathbf{r}(t) = 3t\mathbf{i} + t^3\mathbf{j} + t^4\mathbf{k} \quad \text{for} \quad t \text{ in } \mathbb{R}.$$

Find the velocity, acceleration, and speed at t and at $t = 1$.

Exer. 10–11: Evaluate the integral.

10 $\int (\sin 3t\mathbf{i} + e^{-2t}\mathbf{j} + \cos t\mathbf{k})\,dt$

11 $\int_0^1 (4t\mathbf{i} + t^3\mathbf{j} - \mathbf{k})\,dt$

12 Find $\mathbf{u}(t)$ if $\mathbf{u}'(t) = e^{-t}\mathbf{i} - 4\sin 2t\mathbf{j} + 3\sqrt{t}\,\mathbf{k}$ and $\mathbf{u}(0) = -\mathbf{i} + 2\mathbf{j}$.

13 A point P is moving with an acceleration $12t\mathbf{j} + 5\mathbf{k}$ at time t. If, at $t = 1$, the coordinates of P are $(-1, 3, \frac{3}{2})$ and its velocity is $3\mathbf{i} - 2\mathbf{j} + 4\mathbf{k}$, find its position vector $\mathbf{r}(t)$.

Exer. 14–15: Verify the identity *without using components*.

14 $D_t\|\mathbf{u}(t)\|^2 = 2\mathbf{u}(t) \cdot \mathbf{u}'(t)$

15 $D_t[\mathbf{u}(t) \cdot \mathbf{u}'(t) \times \mathbf{u}''(t)] = \mathbf{u}(t) \cdot \mathbf{u}'(t) \times \mathbf{u}'''(t)$

Exer. 16–18: Find the curvature of the curve at the point P.

16 $y = xe^x$; $\quad P(0, 0)$

17 $x = \dfrac{1}{1+t}, \quad y = \dfrac{1}{1-t}; \quad P(\frac{2}{3}, 2)$

18 $x = 2t^2, \quad y = t^4, \quad z = 4t; \quad P(x, y, z)$

19 Find the x-coordinates of the points on the graph of $y = x^3 - 3x$ at which the curvature is a maximum.

20 If C is the graph of $y = \cosh x$, find an equation of the circle of curvature for the point $P(0, 1)$.

21 Find the radius of curvature for the point $P(2, \pi)$ on the graph of the polar equation $r = 2 + \sin \theta$.

22 Let C be the curve determined by $\mathbf{r}(t) = (t^2 + 1)\mathbf{i} + 4t\mathbf{j}$, for t in $\mathbb{R}$.

(a) Find the unit tangent and normal vectors $\mathbf{T}(t)$ and $\mathbf{N}(t)$.

(b) Sketch the graph of C and show $\mathbf{T}(1)$ and $\mathbf{N}(1)$.

Exer. 23–24: If r(t) is the position vector for a point, find the tangential and normal components of acceleration at time t.

23 $\mathbf{r}(t) = \sin 2t\mathbf{i} + \cos t\mathbf{j}$

24 $\mathbf{r}(t) = 3t\mathbf{i} + t^3\mathbf{j} + t\mathbf{k}$

CHAPTER 16

PARTIAL DIFFERENTIATION

INTRODUCTION

A manufacturer may find that the production cost C of a commodity depends on the quality of the material used, the hourly wages of employees, the type of machinery needed, maintenance charges, and overhead. We say that C is a *function of five variables*, because it depends on five different quantities. In this chapter we study such scalar functions, starting with the case of two independent variables and then proceeding to any number of variables.

First we show how to graphically represent functions of two or three variables. Next we generalize the single-variable concepts of limits, continuous functions, derivatives, and differentials. (Integrals of functions of several variables are defined in the next chapter.) Among the applications considered in examples and exercises are finding the rate at which the temperature of a solid changes, examining the rate at which a pollutant emitted from a smokestack dissipates, measuring the intensity of winds inside a hurricane, designing computer chips, and investigating relationships between the market price of a commodity and consumer demand.

In Section 16.6 we introduce an important tool for analyzing values of a function of several variables—the *gradient*. We shall use gradients to find the rate at which a scalar quantity such as temperature or pressure changes as a point moves in any specified direction. Another important use is the determination of extreme values of functions.

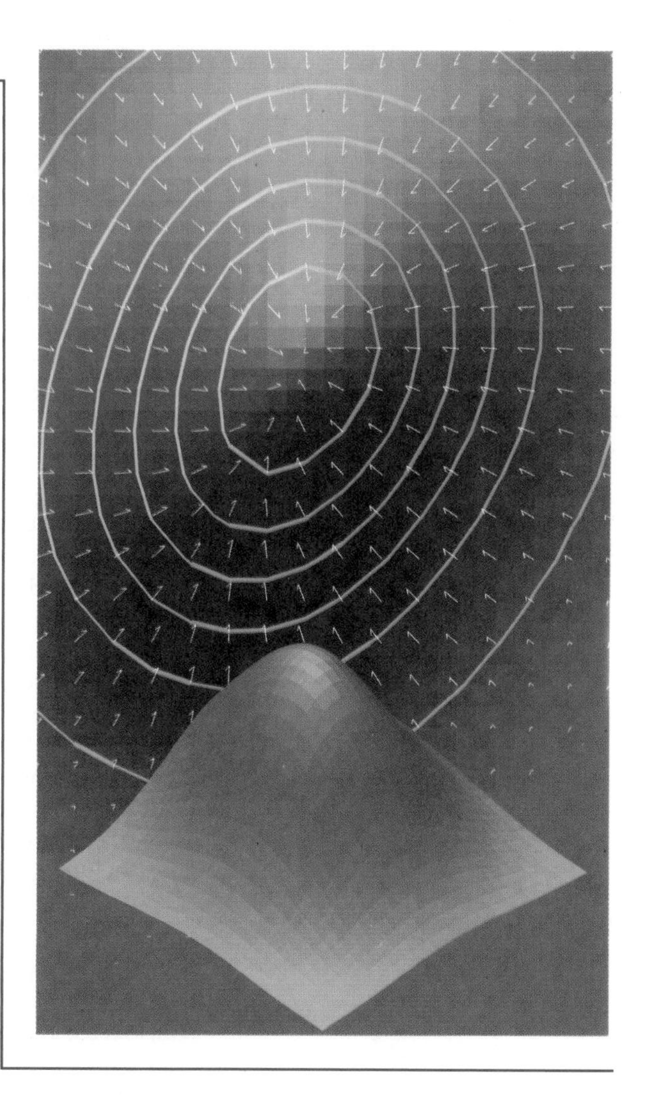

16.1 FUNCTIONS OF SEVERAL VARIABLES

The functions discussed in previous chapters involved only one independent variable. Such functions have many applications; however, in some problems *several* independent variables occur.

ILLUSTRATION

- The area of a rectangle depends on *two* quantities, length and width.
- If an object is located in space, the temperature at a point P in the object may depend on *three* rectangular coordinates, x, y, and z of P.
- If the temperature of an object in space changes with time t, then *four* variables, x, y, z, and t, are involved.

In this section we define functions of several variables and discuss some of their properties. Recall that a function f is a correspondence that assigns to each element in its domain D exactly one element in its range E. If D is a subset of $\mathbb{R}^2$ (the set of all ordered pairs (x, y) of real numbers) and E is a subset of $\mathbb{R}$, then f is a *function of two (real) variables*. Another way of stating this is as follows.

Definition **(16.1)**

> Let D be a set of ordered pairs of real numbers. A **function f of two variables** is a correspondence that assigns to each pair (x, y) in D exactly one real number, denoted by $f(x, y)$. The set D is the **domain** of f. The **range** of f consists of all real numbers $f(x, y)$, where (x, y) is in D.

FIGURE 16.1

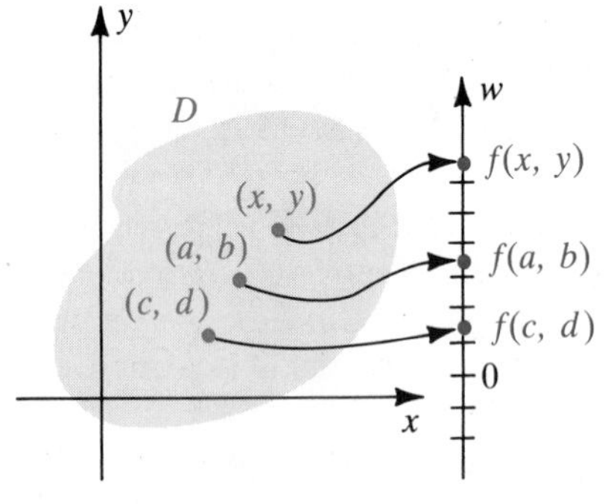

We may represent the domain D in Definition (16.1) by points in an xy-plane and the range by points on a real line, say a w-axis, as illustrated in Figure 16.1. Several curved arrows are drawn from ordered pairs in D to the corresponding numbers in the range. As an application, suppose a flat metal plate has the shape of D. To each point (x, y) on the plate there corresponds a temperature $f(x, y)$ that can be recorded on a thermometer, represented by the w-axis. We could also regard D as the surface of a lake and let $f(x, y)$ denote the depth of the water under the point (x, y). There are many other physical interpretations for Definition (16.1).

We often use an expression in x and y to specify $f(x, y)$ and assume that the domain is the set of all pairs (x, y) for which the given expression is meaningful. We then call f **a function of x and y**.

EXAMPLE 1 Let $f(x, y) = \dfrac{xy - 5}{2\sqrt{y - x^2}}$.

(a) Sketch the domain D of f.

(b) Represent the numbers $f(2, 5)$, $f(1, 2)$, and $f(-1, 2)$ on a w-axis, as in Figure 16.1.

SOLUTION

(a) Referring to the radical $\sqrt{y - x^2}$ in the denominator of $f(x, y)$, we see that the domain D is the set of all pairs (x, y) such that $y - x^2$ is positive; that is, $y > x^2$. The graph of D is the set of all points that lie above the parabola $y = x^2$, as shown in Figure 16.2.

FIGURE 16.2

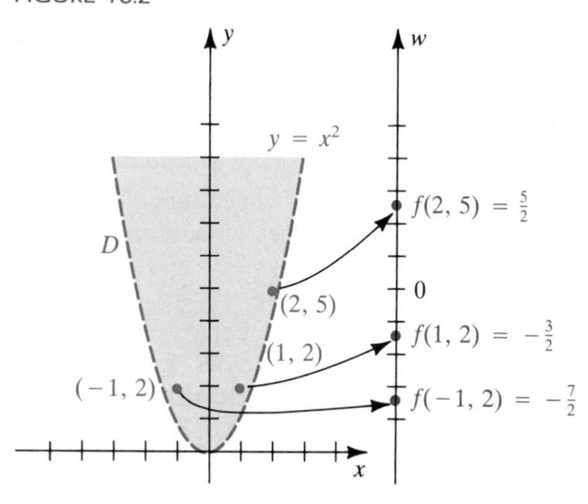

(b) By substitution,

$$f(2, 5) = \frac{(2)(5) - 5}{2\sqrt{5 - 4}} = \frac{5}{2}.$$

Similarly, $f(1, 2) = -\frac{3}{2}$ and $f(-1, 2) = -\frac{7}{2}$. These function values are indicated on the w-axis in Figure 16.2.

Formulas may be used to define functions of two variables. For example, the formula $V = \pi r^2 h$ expresses the volume V of a right circular cylinder as a function of the altitude h and base radius r. The symbols r and h are referred to as **independent variables**, and V is the **dependent variable**.

A function f of three (real) variables is defined as in (16.1), except that the domain D is a subset of $\mathbb{R}^3$. In this case, to each ordered triple (x, y, z) in D there is assigned exactly one real number $f(x, y, z)$. If we represent D by a region in three dimensions, as illustrated in Figure 16.3, then to each point (x, y, z) in D there corresponds a unique point on the w-axis with coordinate $f(x, y, z)$. As an application, we could regard D as a solid object and let $f(x, y, z)$ be the temperature at (x, y, z).

FIGURE 16.3

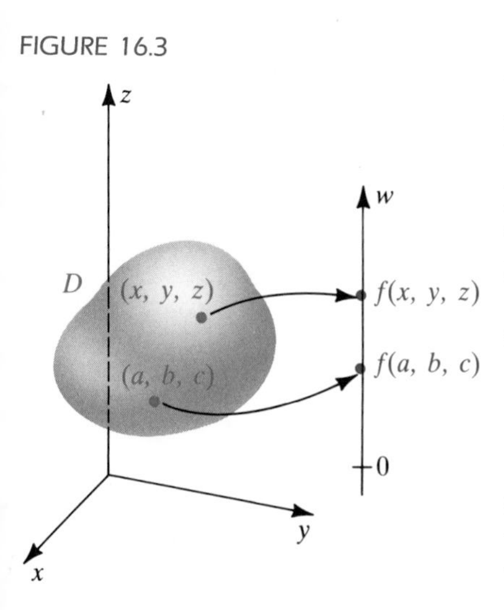

Functions of three variables are often defined by expressions. For example,

$$f(x, y, z) = \frac{xe^y + \sqrt{z^2 + 1}}{xy \sin z}$$

determines a function of x, y, and z. Formulas, such as $V = lwh$ for the volume of a rectangular box of dimensions l, w, and h, may also be used to define functions of three variables.

FIGURE 16.4

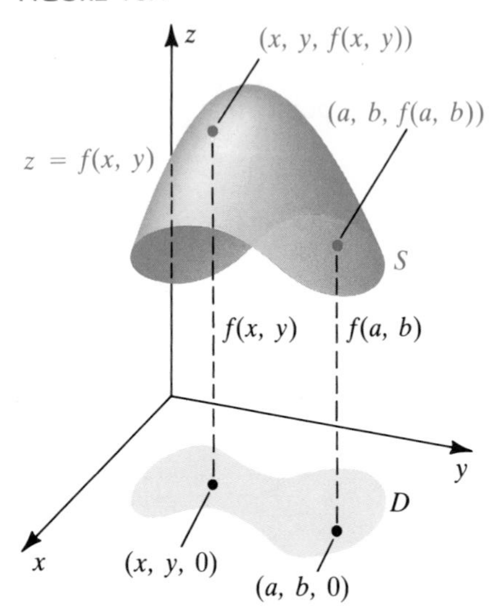

Let us return to a function f of two variables x and y. The **graph** of f is, by definition, the graph of the equation $z = f(x, y)$ in an xyz-coordinate system and, hence, is usually a surface S of some type. If we represent the domain D by a region in the xy-plane, then the pair (x, y) in D is represented by the point $(x, y, 0)$. Function values $f(x, y)$ are the (signed) distances from the xy-plane to S, as illustrated in Figure 16.4.

EXAMPLE 2 Let f be the function with domain D given by

$$f(x, y) = 9 - x^2 - y^2 \quad \text{and} \quad D = \{(x, y): x^2 + y^2 \leq 9\}.$$

Sketch the graph of f and show the traces on the planes $z = 0, z = 2, z = 4, z = 6$, and $z = 8$.

SOLUTION The domain D may be represented by all points within and on the circle $x^2 + y^2 = 9$ in the xy-plane. The graph of f is the portion of the graph of

$$z = 9 - x^2 - y^2$$

(a paraboloid) that lies above or on the xy-plane (see Figure 16.5).

To find the trace on the plane $z = k$, we consider

$$k = 9 - x^2 - y^2, \quad \text{or} \quad x^2 + y^2 = 9 - k.$$

Letting $k = 0, 2, 4, 6$, and 8, we obtain the circles of radii 3, $\sqrt{7}$, $\sqrt{5}$, $\sqrt{3}$, and 1, respectively, shown in Figure 16.5.

FIGURE 16.5
Circular traces on the planes $z = k$

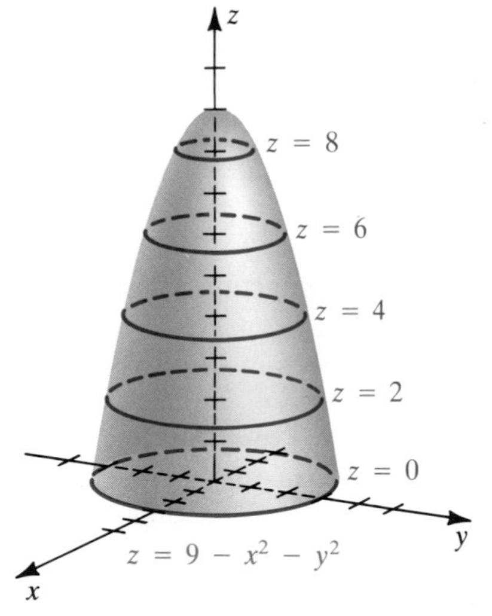

Let f be a function of two variables, and consider the trace of the graph of f on the plane $z = k$, as illustrated in Figure 16.6. If we project this trace onto the xy-plane, we obtain a curve C with equation $f(x, y) = k$. Note that if a point $(x, y, 0)$ moves along C, the corresponding function values $f(x, y)$ always equal k. We call C a **level curve** of f.

EXAMPLE 3 Sketch some level curves of the function f in Example 2.

SOLUTION The level curves are graphs, in the xy-plane, of equations of the form $f(x, y) = k$—that is,

$$9 - x^2 - y^2 = k, \quad \text{or} \quad x^2 + y^2 = 9 - k.$$

These are circles, provided $0 \leq k < 9$. Figure 16.7 shows level curves corresponding to $k = 0, 2, 4, 6$, and 8. These level curves are the projections, in the xy-plane, of the circular traces shown in Figure 16.5.

FIGURE 16.6

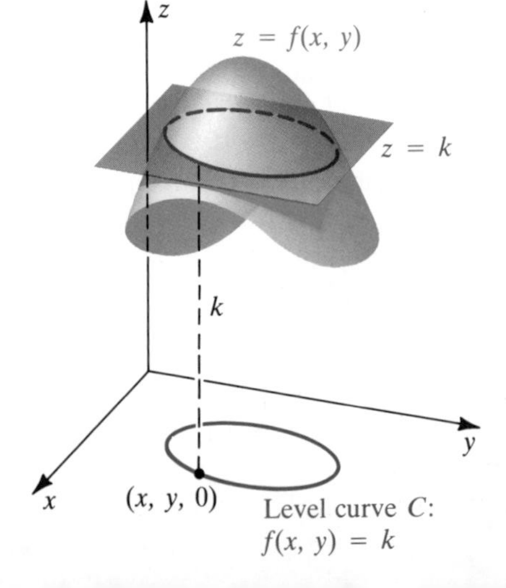

If f is a function of two variables and we sketch the level curves $f(x, y) = k$ for equispaced values of k, such as $k = 0, 2, 4, 6$, and 8 in Example 3, then the nearness of successive curves gives us information about the steepness of the graph of f. In Figure 16.7 the level curves corresponding to $k = 0$ and $k = 2$ are closer than those corresponding to $k = 6$ and $k = 8$, which indicates that the surface shown in Figure 16.5 is steeper at points near the xy-plane than at points farther away.

FIGURE 16.7
Level curves: $9 - x^2 - y^2 = k$

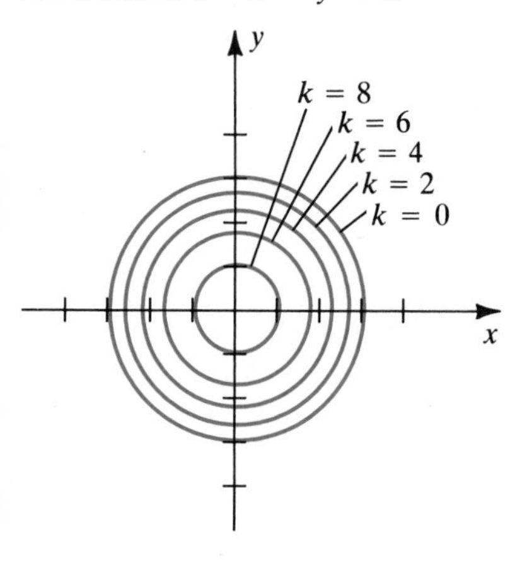

EXAMPLE 4 If $f(x, y) = y^2 - x^2$, sketch some level curves of f.

SOLUTION The graph of f is the hyperbolic paraboloid $z = y^2 - x^2$ obtained by letting $a = b = c = 1$ in (14.46). The graph is sketched in Figure 16.8.

FIGURE 16.8

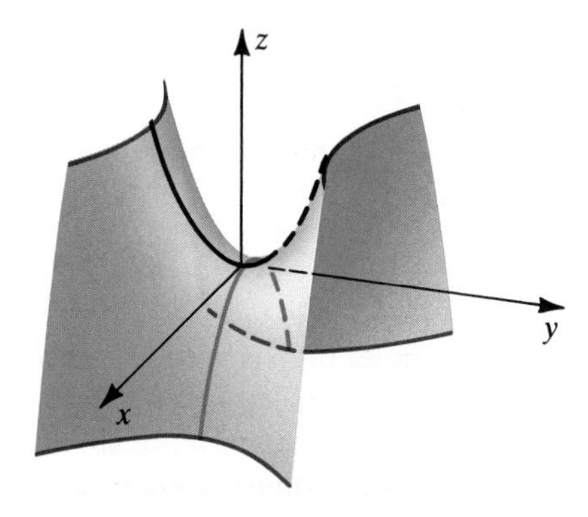

The level curves in the xy-plane are the graphs of the equations $f(x, y) = k$—that is, of

$$y^2 - x^2 = k, \quad \text{for } k \text{ in } \mathbb{R}.$$

These curves are described in the following table.

Value of k	Equation of level curve C	Description of C in xy-plane
$k > 0$	$y^2 - x^2 = k$	Hyperbola with vertices $(0, \pm\sqrt{k})$
$k = 0$	$y^2 - x^2 = 0$, or $y = \pm x$	Two lines of slopes ± 1 through the origin
$k < 0$	$x^2 - y^2 = k$, or $y^2 - x^2 = -k$	Hyperbola with vertices $(\pm\sqrt{-k}, 0)$

FIGURE 16.9
Level curves: $y^2 - x^2 = k$

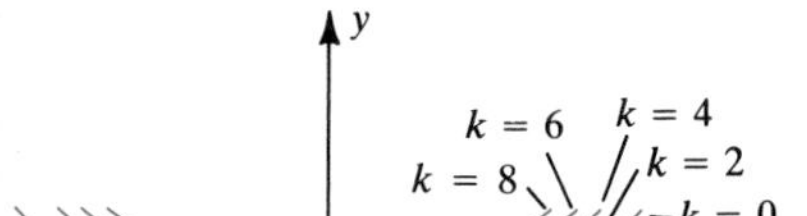

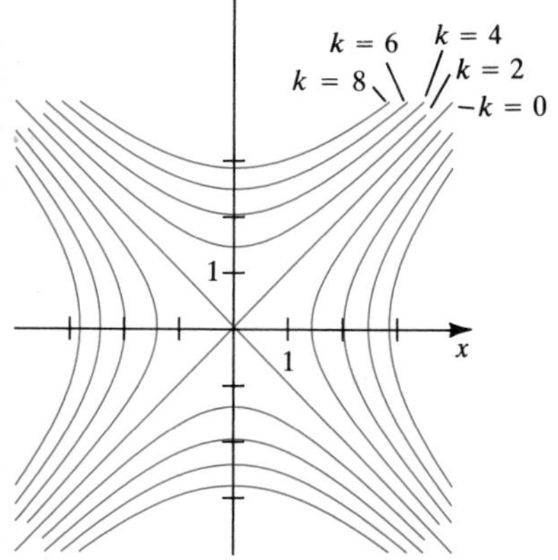

The level curves corresponding to $k = 0, \pm 2, \pm 4, \pm 6$, and ± 8 are sketched in Figure 16.9.

Level curves are often used in making **topographic**, or **contour**, **maps**. For example, suppose $f(x, y)$ denotes the elevation (in feet) at a point (x, y) of latitude x and longitude y. On the hill pictured in Figure 16.10 (on the following page), we have sketched curves (in three dimensions) corresponding to elevations of 0, 100, 200, 300, and 400 feet. We may regard these curves as having been obtained by slicing the hill with planes parallel to the base. A person walking along one of these curves would always remain at the same elevation. The (two-dimensional) level curves corresponding to the same elevations are shown in Figure 16.11. They represent the view obtained by looking down on the hill from an airplane.

FIGURE 16.10
Mountain elevations

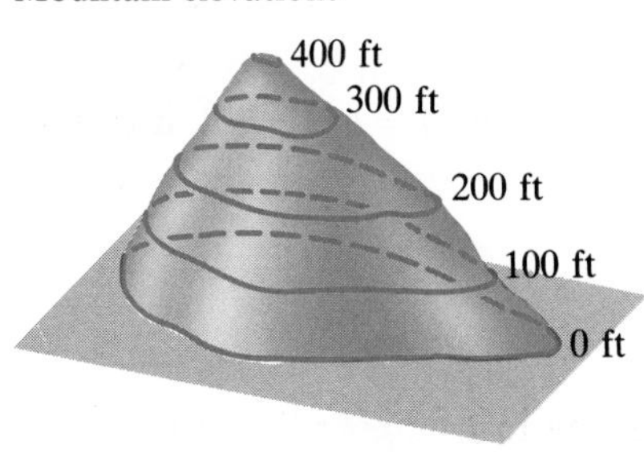

FIGURE 16.11
Topographic map of mountain

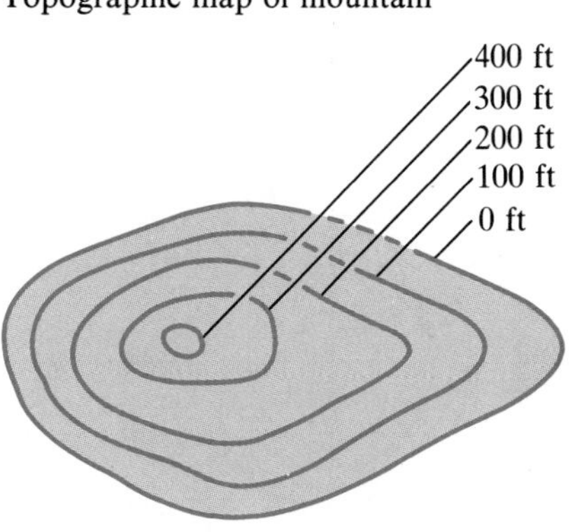

FIGURE 16.12
Depth of water in a lake

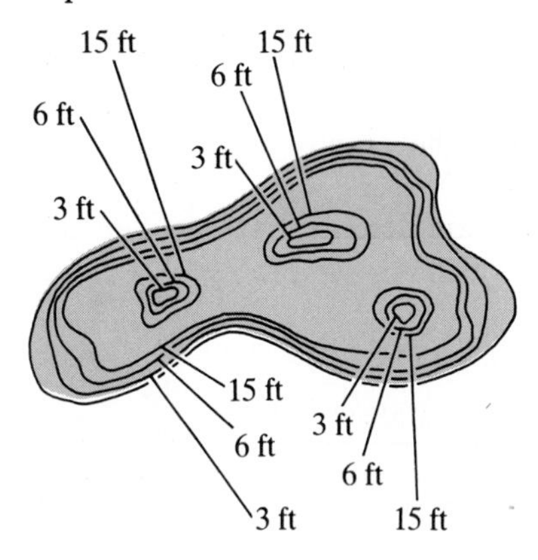

A similar map is used to indicate the depth of the water in a lake. One example is sketched in Figure 16.12, where $f(x, y)$ is the depth under the point (x, y). What parts of the lake should be avoided by water skiers?

As another illustration of level curves, Figure 16.13 shows a weather map of the United States, where $f(x, y)$ denotes the high temperature at (x, y) during a certain day. On the level curves, called **isothermal curves** or **isotherms**, temperature is constant. A different weather map could be drawn in which $f(x, y)$ represented the barometric pressure at (x, y). In this case the level curves would be called **isobars**.

If f is a function of three variables x, y, and z, then by definition the **level surfaces** of f are the graphs of $f(x, y, z) = k$ for suitable values of k. If we let $k = w_0$, w_1, and w_2, the resulting graphs are surfaces S_0, S_1, and S_2, as illustrated in Figure 16.14. As a point (x, y, z) moves along one of these surfaces, $f(x, y, z)$ does not change. If $f(x, y, z)$ is the temperature at (x, y, z), the level surfaces are **isothermal surfaces**, and the temperature is constant on each surface. If $f(x, y, z)$ represents electrical potential, the

FIGURE 16.13
Isothermal curves

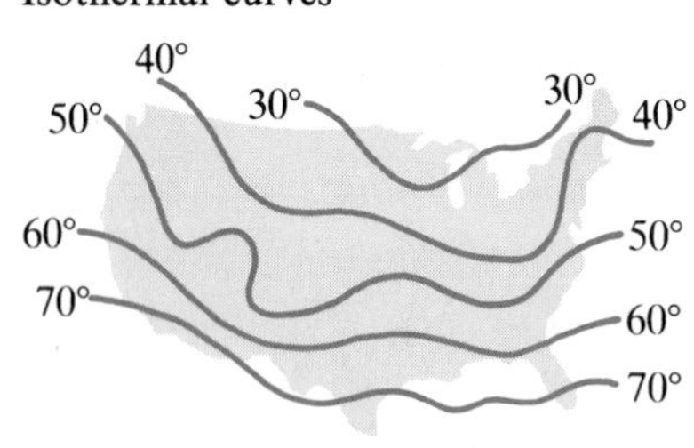

FIGURE 16.14
Level surfaces of $f(x, y, z)$

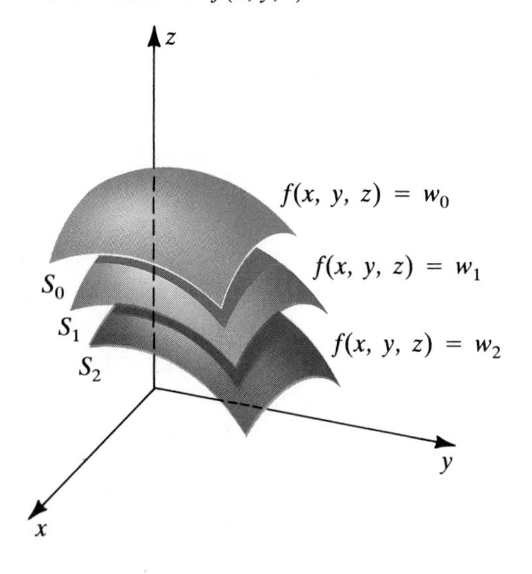

level surfaces are **equipotential surfaces**, and the voltage does not change if (x, y, z) remains on such a surface.

FIGURE 16.15
Level surfaces: $z - \sqrt{x^2 + y^2} = k$

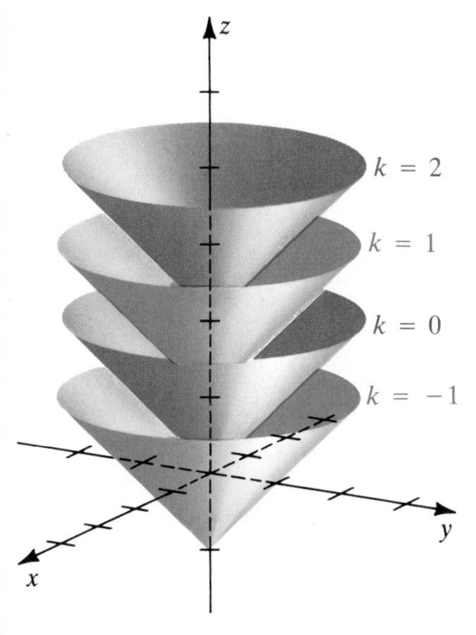

EXAMPLE 5 If $f(x, y, z) = z - \sqrt{x^2 + y^2}$, sketch some level surfaces of f.

SOLUTION The level surfaces are graphs of $f(x, y, z) = k$—that is, of

$$z - \sqrt{x^2 + y^2} = k, \quad \text{or} \quad z = \sqrt{x^2 + y^2} + k,$$

where k is any real number. For each k we obtain a right circular cone whose axis is along the z-axis. The cases $k = -1, 0, 1$, and 2 are sketched in Figure 16.15.

EXAMPLE 6 If $f(x, y, z) = x^2 - y^2 + z^2$, describe the level surfaces of f.

SOLUTION The level surfaces are graphs of equations of the form $f(x, y, z) = k$—that is, of

$$x^2 - y^2 + z^2 = k,$$

where k is a real number. These surfaces are described in the following table.

Value of k	Equation of level surface	Description of surface
$k > 0$	$x^2 - y^2 + z^2 = k$	Hyperboloid of one sheet with axis along the y-axis
$k = 0$	$x^2 - y^2 + z^2 = 0$	Cone with axis along the y-axis
$k < 0$	$-x^2 + y^2 - z^2 = -k$	Hyperboloid of two sheets with axis along the y-axis

A level surface of each type is sketched in Figure 16.16.

FIGURE 16.16

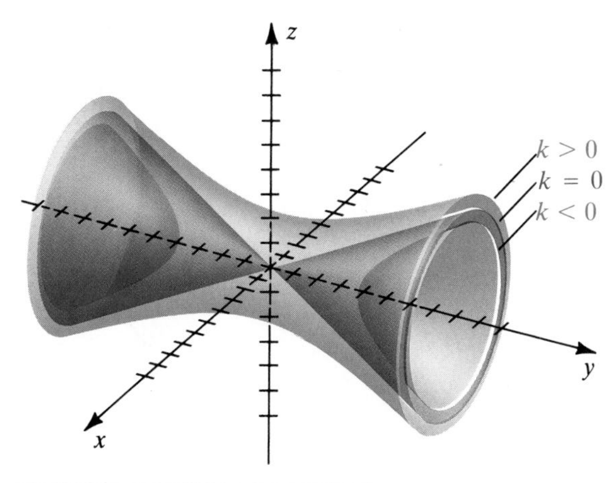

FIGURE 16.17

(i) $z = -(x^2 + y^{2/3})$

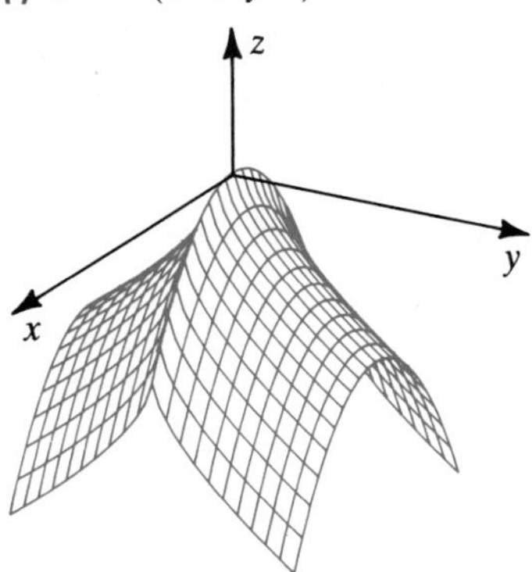

(ii) Level curves: $k = -(x^2 + y^{2/3})$

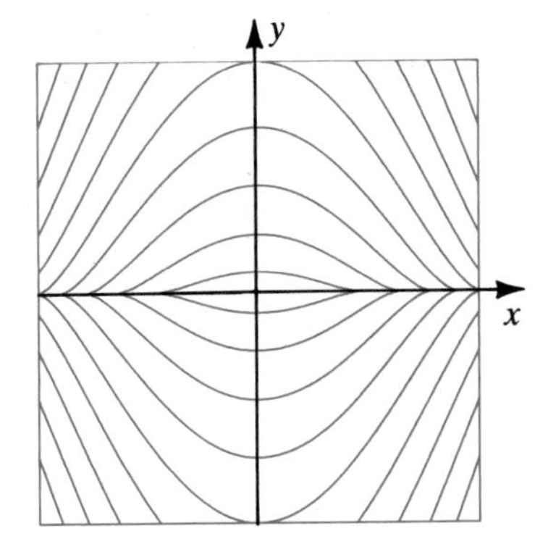

(iii) $z = \dfrac{1}{9x^2 + y^2}$

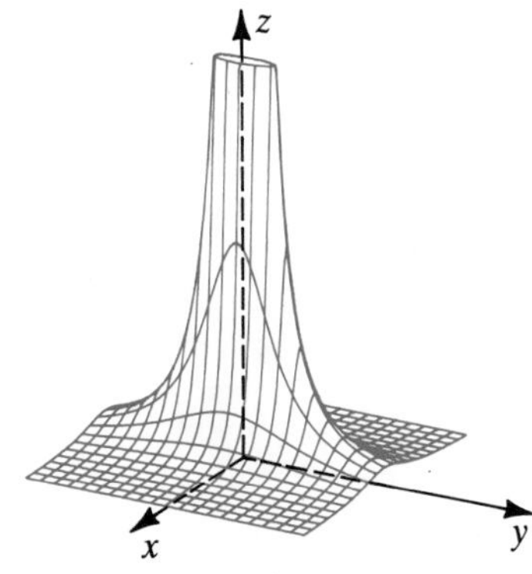

(iv) $z = x^3 - 2xy^2 - x^2 - y^2$

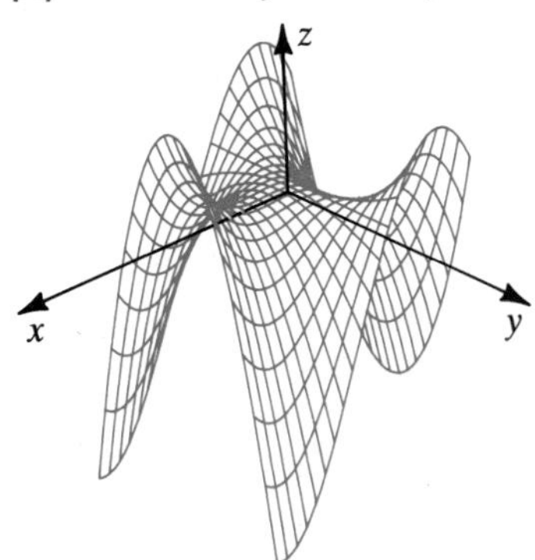

(v) Level curves: $k = x^3 - 2xy^2 - x^2 - y^2$

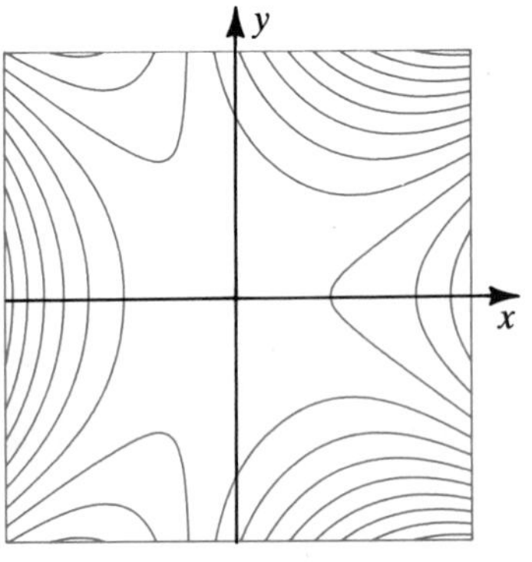

(vi) $z = 8e^{-(x^2+y^2)/4}$

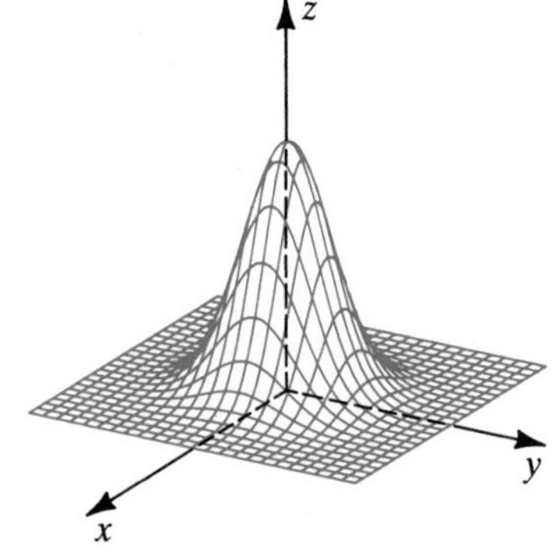

(vii) $z = \cos x + \cos y$

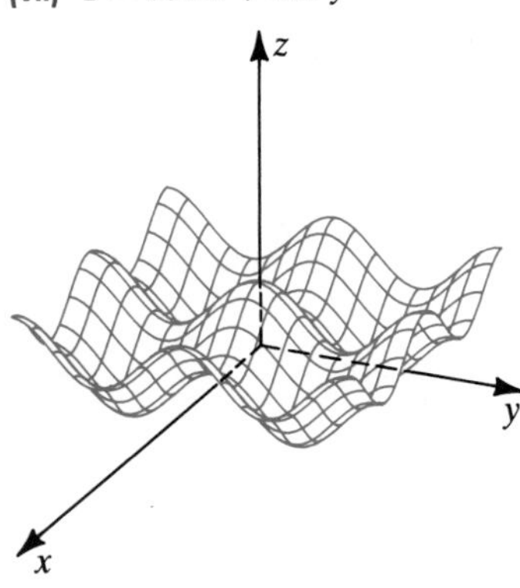

(viii) $z = \dfrac{\frac{1}{2}\cos(2x^2 + y^2)}{1 + 2x^2 + y^2}$

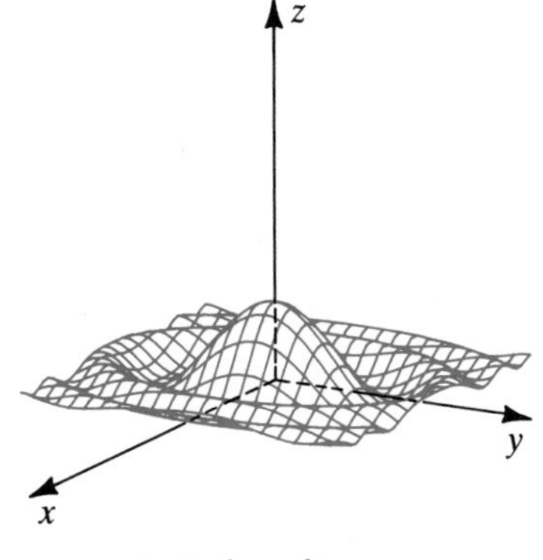

(ix) $z = \dfrac{1 - 2\cos(x^2 + y^2)}{x^2 + y^2}$

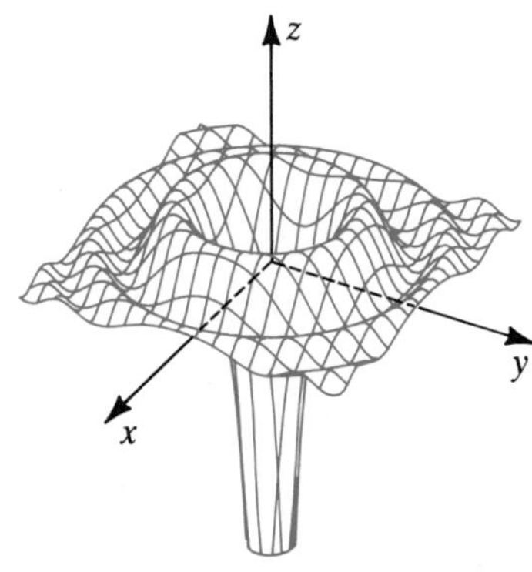

(x) $z = \cos x + 3\cos(3x + y)$

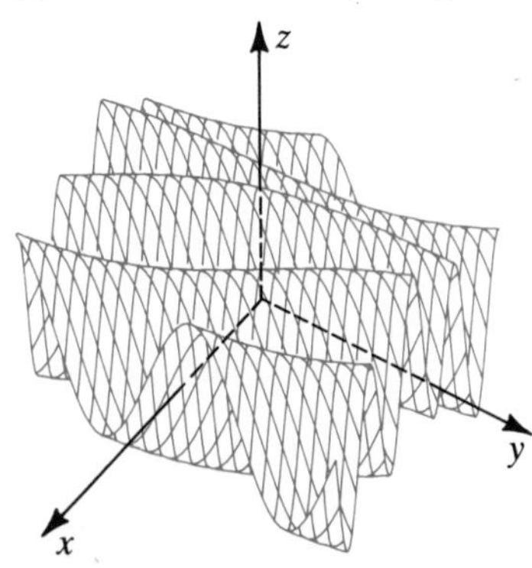

(xi) $z = \ln(2x^2 + y^2)$

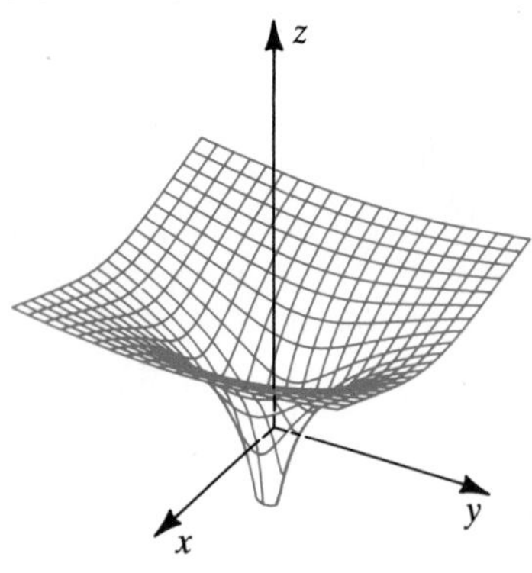

(xii)
$$z = \begin{cases} \sqrt{1 - y^2} & \text{if } |x| \ge 1 \text{ and } |y| \le 1, \text{ or} \\ & \text{if } |x| \le 1 \text{ and } |y| \le |x| \\ \sqrt{1 - x^2} & \text{if } |y| \ge 1 \text{ and } |x| \le 1, \text{ or} \\ & \text{if } |y| \le 1 \text{ and } |x| \le |y| \\ 0 & \text{if } |x| \ge 1 \text{ and } |y| \ge 1 \end{cases}$$

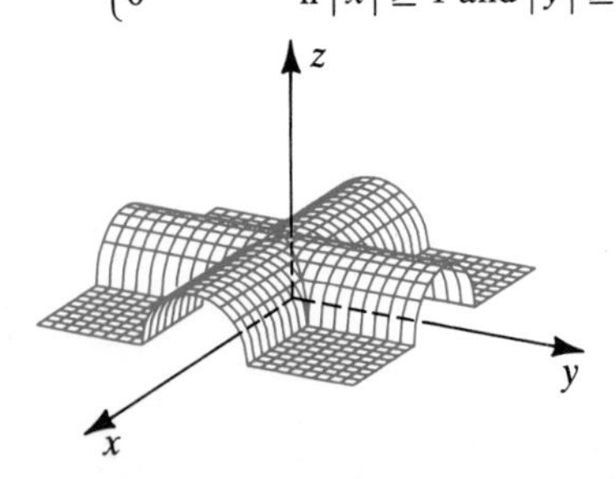

Computers can be programmed to show level curves and traces of surfaces on various planes and to represent surfaces from various perspectives. Sketches of this type are referred to as *computer graphics*, or *computer-generated graphs*. Several illustrations obtained by means of a computer are shown in Figure 16.17 on the opposite page and in Exercises 29–34 on the following pages.

EXERCISES 16.1

Exer. 1–6: Describe the domain of f, and find the indicated function values.

1 $f(x, y) = 2x - y^2$; $\quad f(-2, 5),\quad f(5, -2),\quad f(0, -2)$

2 $f(x, y) = \dfrac{y + 2}{x}$; $\quad f(3, 1),\quad f(1, 3),\quad f(2, 0)$

3 $f(u, v) = \dfrac{uv}{u - 2v}$; $\quad f(2, 3),\quad f(-1, 4),\quad f(0, 1)$

4 $f(r, s) = \sqrt{1 - r} - e^{r/s}$; $\quad f(1, 1),\quad f(0, 4),\quad f(-3, 3)$

5 $f(x, y, z) = \sqrt{25 - x^2 - y^2 - z^2}$; $\quad f(1, -2, 2),\quad f(-3, 0, 2)$

6 $f(x, y, z) = 2 + \tan x + y \sin z$; $\quad f(\pi/4, 4, \pi/6),\quad f(0, 0, 0)$

Exer. 7–14: Sketch the graph of f.

7 $f(x, y) = \sqrt{1 - x^2 - y^2}$

8 $f(x, y) = 4 - x^2 - 4y^2$

9 $f(x, y) = x^2 + y^2 - 1$

10 $f(x, y) = \frac{1}{6}\sqrt{9x^2 + 4y^2}$

11 $f(x, y) = 6 - 2x - 3y$

12 $f(x, y) = \sqrt{72 + 4x^2 - 9y^2}$

13 $f(x, y) = \sqrt{y^2 - 4x^2 - 16}$

14 $f(x, y) = \sqrt{x^2 + 4y^2 + 25}$

Exer. 15–20: Sketch the level curves of f for the given values of k.

15 $f(x, y) = y^2 - x^2$; $\quad k = -4, 0, 9$

16 $f(x, y) = 3x - 2y$; $\quad k = -4, 0, 6$

17 $f(x, y) = x^2 - y$; $\quad k = -2, 0, 3$

18 $f(x, y) = xy$; $\quad k = -4, 1, 4$

19 $f(x, y) = (x - 2)^2 + (y + 3)^2$; $\quad k = 1, 4, 9$

20 $f(x, y) = 4x^2 + y^2$; $\quad k = 4, 9, 16$

Exer. 21–22: Find an equation of the level curve of f that contains the point P.

21 $f(x, y) = y \arctan x$; $\quad P(1, 4)$

22 $f(x, y) = (2x + y^2)e^{xy}$; $\quad P(0, 2)$

Exer. 23–24: Find an equation of the level surface of f that contains the point P.

23 $f(x, y, z) = x^2 + 4y^2 - z^2$; $\quad P(2, -1, 3)$

24 $f(x, y, z) = z^2y + x$; $\quad P(1, 4, -2)$

Exer. 25–28: Describe the level curves, whenever they exist, for the surface in the indicated figure if (a) $k > 0$, (b) $k = 0$, and (c) $k < 0$.

25 Figure 16.17(iii)

26 Figure 16.17(xi)

27 Figure 16.17(vi)

28 Figure 16.17(xii)

Exer. 29–34: Correctly match one of the graphs shown in (a)–(f) to the system of level curves of the function given by $z = f(x, y)$.

(a)

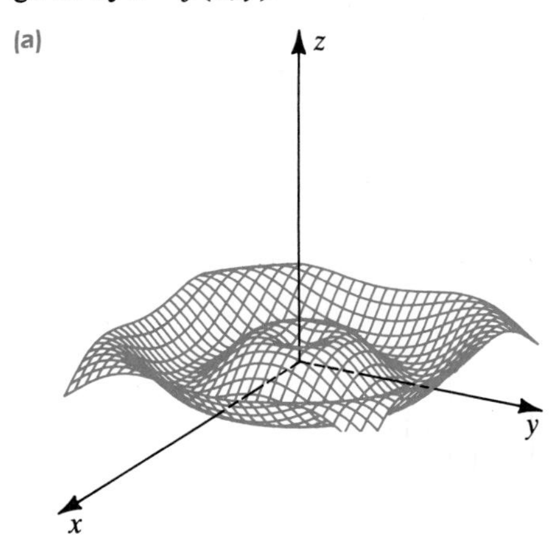

(b)

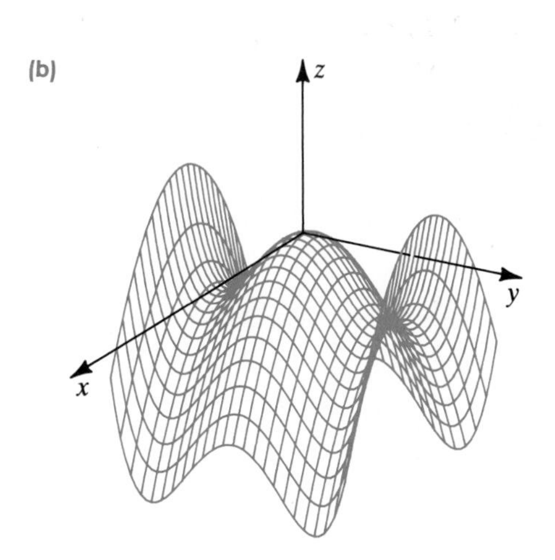

(c)

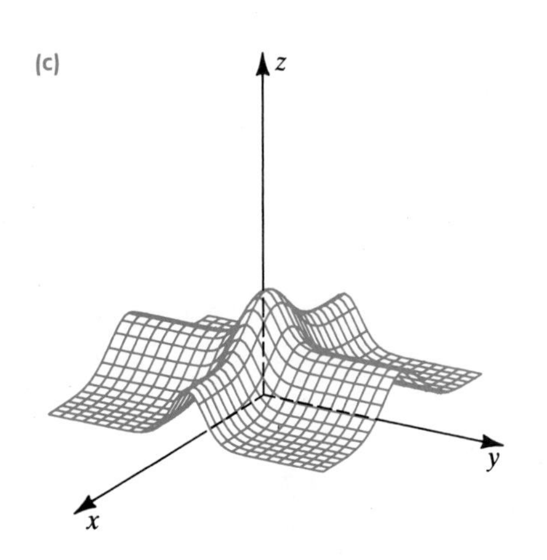

(d)

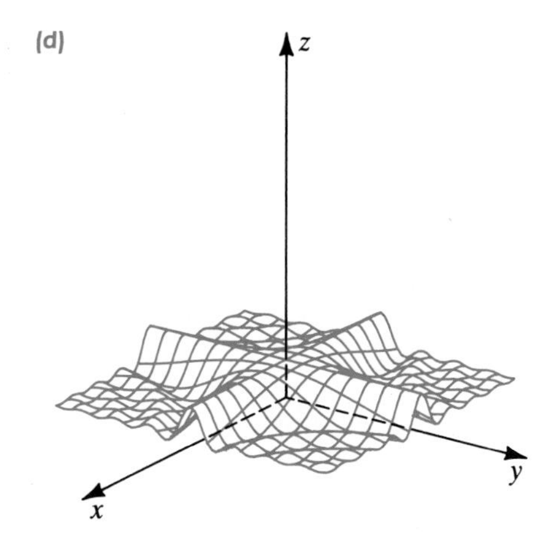

(e)

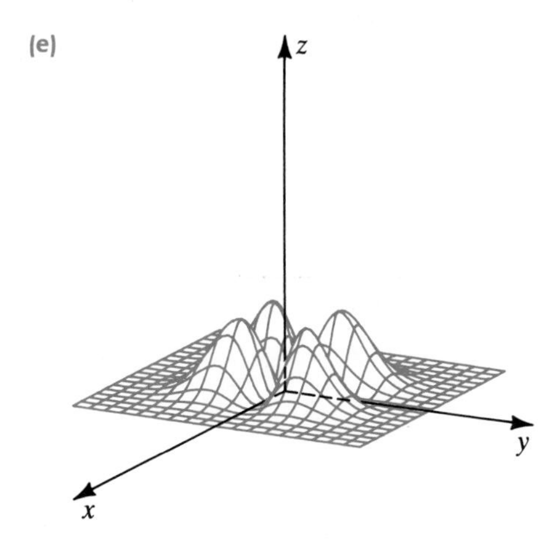

(f)

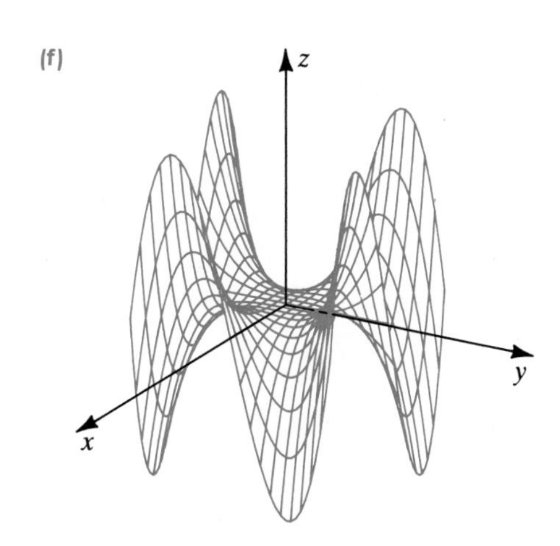

29 $z = \dfrac{\sin xy}{xy}$

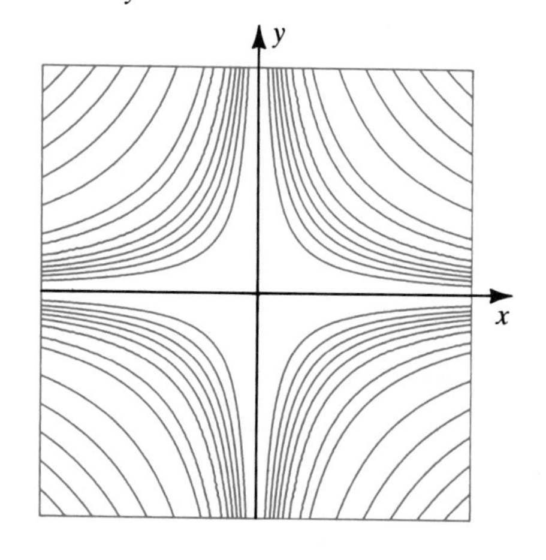

30 $z = \dfrac{15x^2y^2e^{-x^2-y^2}}{x^2+y^2}$

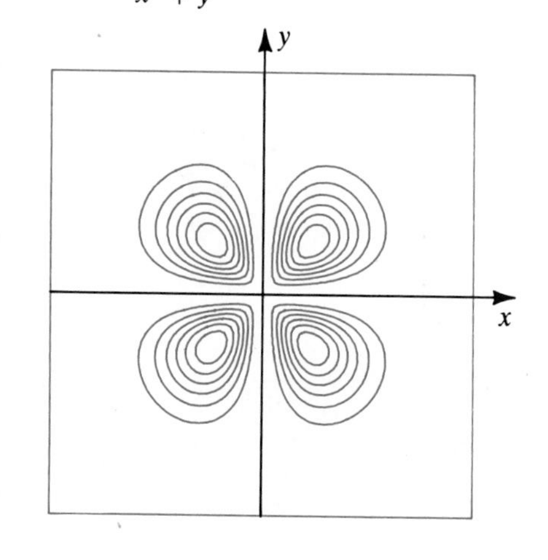

31 $z = \sin\sqrt{x^2+y^2}$;
$-7 \le x \le 7,\ -7 \le y \le 7$

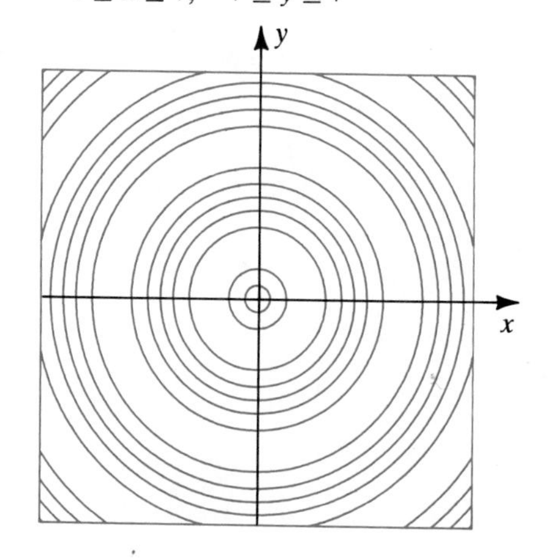

32 $z = e^{-x^2} + e^{-4y^2}$;
$-2 \le x \le 2,\ -2 \le y \le 2$

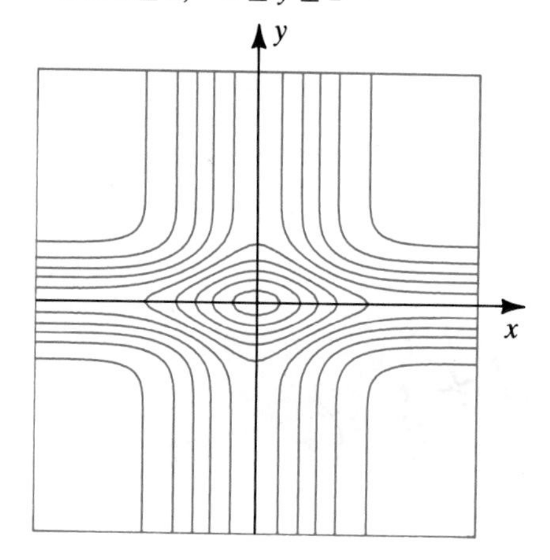

33 $z = \dfrac{xy^3 - x^3y}{2}$;
$-2 \le x \le 2,\ -2 \le y \le 2$

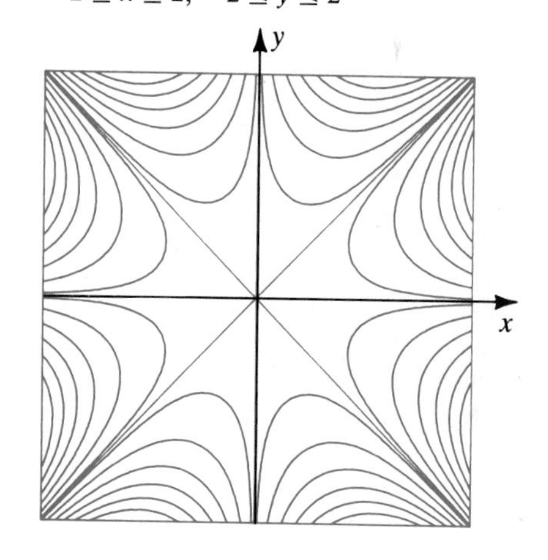

34 $z = y^4 - 8y^2 - 4x^2$;
$-3 \le x \le 3,\ -3 \le y \le 3$

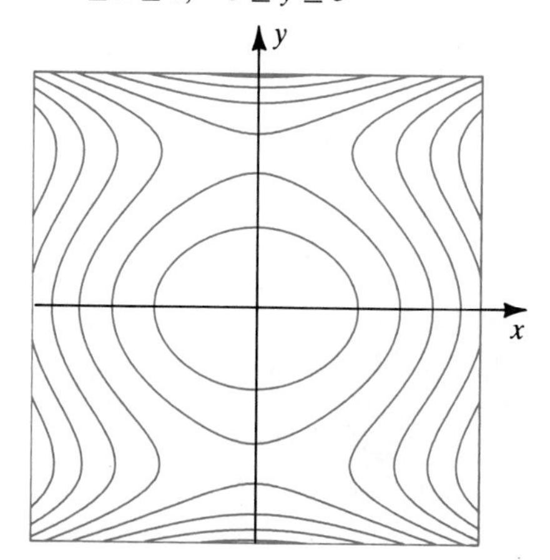

c **Exer. 35–36: Graph, on the same coordinate plane, the level curves for $k = 1, 2$, and 3, using polar coordinates.**

35 $f(x, y) = x^4 + 8x^2y^2 + y^4$

36 $f(x, y) = 3x^4 + x^2y^2 + y^4$

Exer. 37–42: Describe the level surfaces of f for the given values of k.

37 $f(x, y, z) = x^2 + y^2 + z^2$; $k = -1, 0, 4$

38 $f(x, y, z) = z + x^2 + 4y^2$; $k = -1, 0, 2$

39 $f(x, y, z) = x + 2y + 3z$; $k = -6, 6, 12$

40 $f(x, y, z) = x^2 + y^2 - z^2$; $k = -1, 0, 1$

41 $f(x, y, z) = x^2 + y^2$; $k = -1, 0, 4$

42 $f(x, y, z) = z$; $k = -2, 0, 3$

43 A flat metal plate is situated in an xy-plane such that the temperature T (in °C) at the point (x, y) is inversely proportional to the distance from the origin.

(a) Describe the isotherms.

(b) If the temperature at the point $P(4, 3)$ is 40 °C, find an equation of the isotherm for a temperature of 20 °C.

44 If the voltage V at the point $P(x, y, z)$ is given by $V = 6/(x^2 + 4y^2 + 9z^2)^{1/2}$,

(a) describe the equipotential surfaces

(b) find an equation of the equipotential surface $V = 120$

45 According to Newton's law of universal gravitation, if a particle of mass m_0 is at the origin of an xyz-coordinate system, then the magnitude F of the force exerted on a particle of mass m located at the point (x, y, z) is given by

$$F = \frac{Gm_0m}{x^2 + y^2 + z^2},$$

where G is the universal gravitational constant. How many independent variables are present? If m_0 and m are constant, describe the level surfaces of the resulting function of x, y, and z. What is the physical significance of these level surfaces?

46 According to the *ideal gas law*, the pressure P, volume V, and temperature T of a confined gas are related by the formula $PV = kT$ for a constant k. Express P as a function of V and T, and describe the level curves associated with this function. What is the physical significance of these level curves?

47 The power P generated by a wind rotor is proportional to the product of the area A swept out by the blades and the third power of the wind velocity v.

(a) Express P as a function of A and v.

(b) Describe the level curves of P and explain their physical significance.

(c) When the diameter of the circular area swept out by the blades is 10 feet and the wind velocity is 20 mi/hr, then $P = 3000$ watts. Find an equation of the level curve $P = 4000$.

48 If x is wind velocity (in m/sec) and y is temperature (in °C), then the *windchill factor* F (in (kcal/m^2)/hr) is given by $F = (33 - y)(10\sqrt{x} - x + 10.5)$.

(a) Find the velocities and temperatures for which the windchill factor is 0. (Assume that $0 \le x \le 50$ and $-50 \le y \le 50$.)

(b) If $F \ge 1400$, frostbite will occur on exposed human skin. Sketch the graph of the level curve $F = 1400$.

49 Shown in the figure are level curves for the surface area of the human body (in tenths of a square meter) as a function of weight x (in kilograms) and height y (in centimeters). Use this system of level curves to estimate your own body surface area. Compare this estimate to the more exact estimate given by the *DuBois and DuBois formula*

$$S = 0.007184x^{0.425}y^{0.725}.$$

EXERCISE 49

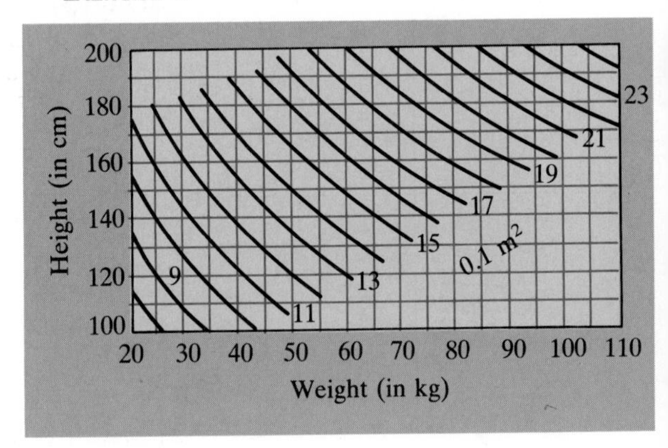

50 The upper boundary of the semicircular region shown in the figure is kept at a temperature of 10 °C, while the lower boundary is maintained at 0 °C. The long-range, or *steady state*, temperature T at a point (x, y) inside the region is given by

$$T(x, y) = \frac{20}{\pi}\tan^{-1}\left(\frac{2y}{1 - x^2 - y^2}\right).$$

EXERCISE 50

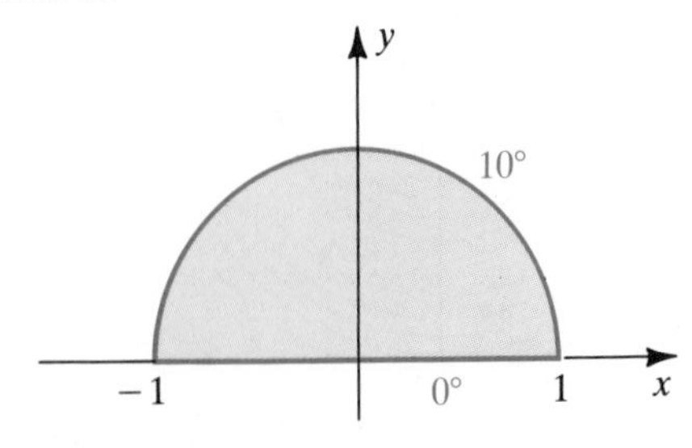

Show that the isotherms are arcs of circles that have centers on the negative y-axis and pass through the points $(-1, 0)$ and $(1, 0)$. Sketch the isotherm corresponding to a temperature of 5 °C.

51 A garbage incineration plant will be built to service two cities. Each city would like to maximize its distance from the plant, but for economic reasons the sum of the distances from each city to the plant cannot exceed M miles. Show that the level curves for the location of the plant are ellipses.

52 The atmospheric pressure near ground level in a certain region is given by

$$p(x, y) = ax^2 + by^2 + c,$$

where a, b, and c are positive constants.

(a) Describe the isobars in this region for pressures greater than c.

(b) Is this a region of high or low pressure?

16.2 LIMITS AND CONTINUITY

If f is a function of two variables, we may wish to consider changes in the function values $f(x, y)$ as (x, y) varies through the domain D of f. As a physical illustration, suppose a flat metal plate has the shape of the region D in Figure 16.18. To each point (x, y) on the plate there corresponds a temperature $f(x, y)$, which is recorded on a thermometer represented by the w-axis. As the point (x, y) moves on the plate, the temperature may increase, decrease, or remain constant; therefore, the point on the w-axis that corresponds to $f(x, y)$ will move in a positive direction, move in a negative direction, or remain fixed, respectively. If the temperature $f(x, y)$ gets closer to a fixed value L as (x, y) gets closer to a fixed point (a, b), we use the following notation.

Limit notation **(16.2)**

$$\lim_{(x,y)\to(a,b)} f(x, y) = L, \quad \text{or} \quad f(x, y) \to L \text{ as } (x, y) \to (a, b)$$

This may be read *the limit of* $f(x, y)$ *as* (x, y) *approaches* (a, b) *is* L.

To make (16.2) mathematically precise, let us proceed as follows. For any $\epsilon > 0$, consider the open interval $(L - \epsilon, L + \epsilon)$ on the w-axis, as illustrated in Figure 16.19. If (16.2) is true, then there is a $\delta > 0$ such that for every point (x, y) inside the circle of radius δ with center (a, b), except possibly (a, b) itself, the function value $f(x, y)$ is in the interval $(L - \epsilon, L + \epsilon)$.

FIGURE 16.18

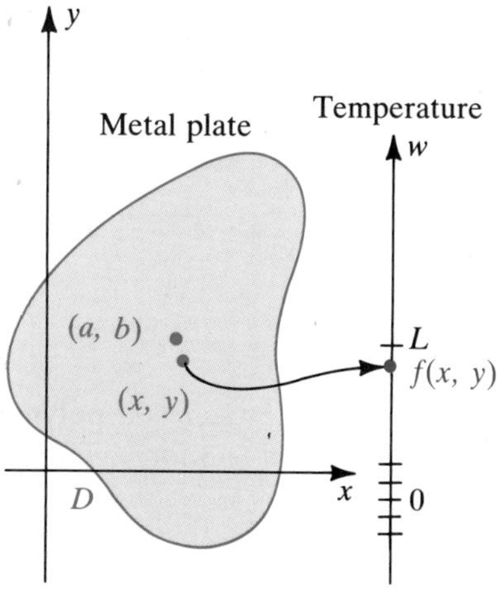

FIGURE 16.19

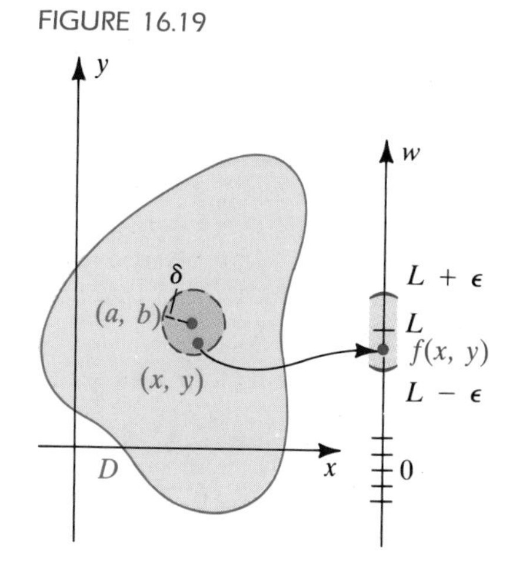

This is equivalent to the following statement:

$$\text{If} \quad 0 < \sqrt{(x-a)^2 + (y-b)^2} < \delta, \quad \text{then} \quad |f(x, y) - L| < \epsilon.$$

Thus, we have the following.

Definition of limit (16.3)

Let a function f of two variables be defined throughout the interior of a circle with center (a, b), except possibly at (a, b) itself. The statement

$$\lim_{(x,y)\to(a,b)} f(x, y) = L$$

means that for every $\epsilon > 0$ there is a $\delta > 0$ such that

$$\text{if} \quad 0 < \sqrt{(x-a)^2 + (y-b)^2} < \delta, \quad \text{then} \quad |f(x, y) - L| < \epsilon.$$

FIGURE 16.20

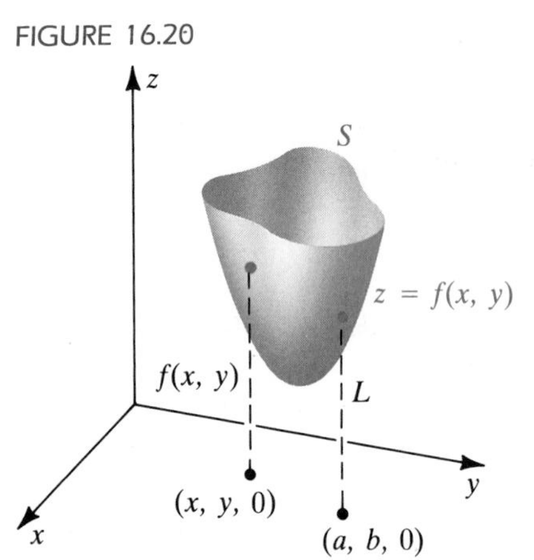

Let us consider the graph S of f illustrated in Figure 16.20. Intuitively we know that Definition (16.3) means that as the point $(x, y, 0)$ approaches $(a, b, 0)$ in the xy-plane, the corresponding point $(x, y, f(x, y))$ on S approaches (a, b, L) (which may or may not be on S). We can show that *if the limit L exists, it is unique.*

If f and g are functions of two variables, then $f + g$, $f - g$, fg, and f/g are defined in the usual way and Theorem (2.8) concerning limits of sums, products, and quotients can be extended. For example, if f and g have limits as (x, y) approaches (a, b), then

$$\lim_{(x,y)\to(a,b)} [f(x, y) + g(x, y)] = \lim_{(x,y)\to(a,b)} f(x, y) + \lim_{(x,y)\to(a,b)} g(x, y),$$

$$\lim_{(x,y)\to(a,b)} \frac{f(x, y)}{g(x, y)} = \frac{\lim_{(x,y)\to(a,b)} f(x, y)}{\lim_{(x,y)\to(a,b)} g(x, y)} \quad \text{if} \quad \lim_{(x,y)\to(a,b)} g(x, y) \neq 0,$$

$$\lim_{(x,y)\to(a,b)} \sqrt{f(x, y)} = \sqrt{\lim_{(x,y)\to(a,b)} f(x, y)} \quad \text{if} \quad \lim_{(x,y)\to(a,b)} f(x, y) > 0,$$

and so on.

A function f of two variables is a **polynomial function** if $f(x, y)$ can be expressed as a sum of terms of the form $cx^m y^n$, where c is a real number and m and n are nonnegative integers. A **rational function** is a quotient of two polynomial functions. As for single-variable functions, limits of polynomial and rational functions in two variables may be found by substituting for x and y.

EXAMPLE 1 Find

(a) $\displaystyle\lim_{(x,y)\to(2,-3)} (x^3 - 4xy^2 + 5y - 7)$ **(b)** $\displaystyle\lim_{(x,y)\to(3,4)} \frac{x^2 - y^2}{\sqrt{x^2 + y^2}}$

SOLUTION

(a) Since $x^3 - 4xy^2 + 5y - 7$ is a polynomial, we may find the limit by substituting 2 for x and -3 for y. Thus,

$$\begin{aligned}\lim_{(x,y)\to(2,-3)} (x^3 - 4xy^2 + 5y - 7) &= 2^3 - 4(2)(-3)^2 + 5(-3) - 7 \\ &= 8 - 72 - 15 - 7 = -86.\end{aligned}$$

(b) We may proceed as follows:

$$\lim_{(x,y)\to(3,4)} \frac{x^2 - y^2}{\sqrt{x^2 + y^2}} = \frac{\lim_{(x,y)\to(3,4)} (x^2 - y^2)}{\lim_{(x,y)\to(3,4)} \sqrt{x^2 + y^2}} = \frac{9 - 16}{\sqrt{\lim_{(x,y)\to(3,4)} (x^2 + y^2)}}$$

$$= \frac{-7}{\sqrt{9 + 16}} = -\frac{7}{5}$$

FIGURE 16.21

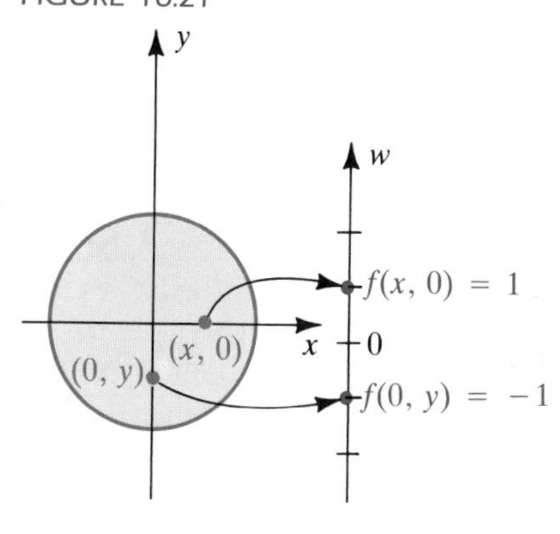

EXAMPLE 2 Show that $\lim_{(x,y)\to(0,0)} \frac{x^2 - y^2}{x^2 + y^2}$ does not exist.

SOLUTION Let $f(x, y) = \frac{x^2 - y^2}{x^2 + y^2}.$

The function f is rational; however, we cannot substitute 0 for both x and y, since that would lead to a zero denominator.

If we consider any point $(x, 0)$ on the x-axis, then

$$f(x, 0) = \frac{x^2 - 0}{x^2 + 0} = 1, \quad \text{provided} \quad x \neq 0.$$

FIGURE 16.22

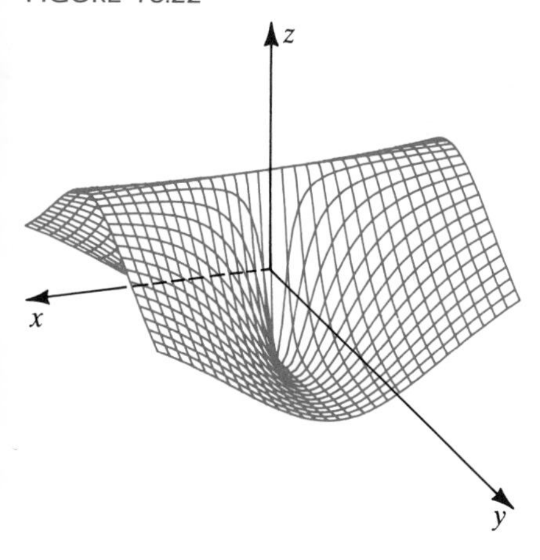

For any point $(0, y)$ on the y-axis, we have

$$f(0, y) = \frac{0 - y^2}{0 + y^2} = -1, \quad \text{provided} \quad y \neq 0.$$

Consequently, as illustrated in Figure 16.21, *every* circle with center (0, 0) contains points at which the value of f is 1 and points at which the value of f is -1. It follows that the limit does not exist, for if we take $\epsilon = 1$ in Definition (16.3), there is no open interval $(L - \epsilon, L + \epsilon)$ on the w-axis containing both 1 and -1. Hence it is impossible to find a $\delta > 0$ that satisfies the conditions of the definition.

Figure 16.22 is a computer-generated graph of f. Note that for $x \neq 0$ and $y \neq 0$, the graph contains all points $(x, 0, 1)$ and $(0, y, -1)$, as we have already observed. We can show that the entire graph lies between the planes $z = -1$ and $z = 1$.

FIGURE 16.23

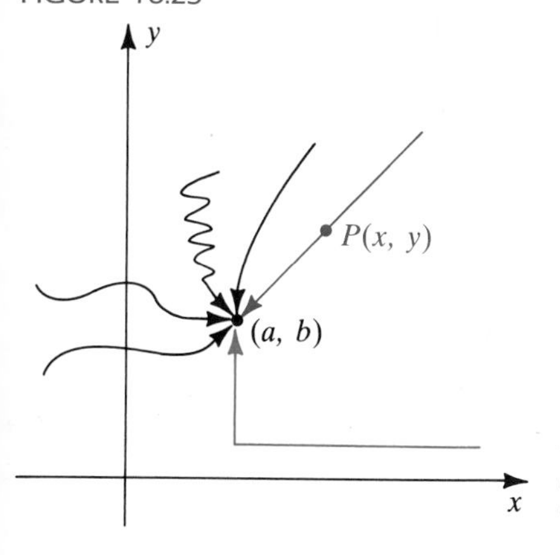

In Chapter 2, for a function of one variable with a jump discontinuity at $x = a$, we proved that $\lim_{x\to a} f(x)$ does not exist by showing that $\lim_{x\to a^-} f(x)$ and $\lim_{x\to a^+} f(x)$ are not equal. When considering such one-sided limits, we may regard the point on the x-axis with coordinate x as approaching the point with coordinate a either from the left or from the right, respectively. The analogous situation for functions of two variables is more complicated, since in a coordinate plane there are an infinite number of different curves, or **paths**, along which (x, y) can approach (a, b), as illustrated in Figure 16.23. However, if the limit in Definition (16.3) exists, then $f(x, y)$ must have the limit L, regardless of the path taken. This illustrates the following rule for investigating limits.

Two-path rule **(16.4)**

> If two different paths to a point $P(a, b)$ produce two different limiting values for f, then $\lim_{(x,y)\to(a,b)} f(x, y)$ does not exist.

We shall next rework Example 2 using (16.4).

EXAMPLE 3 Show that $\lim_{(x,y)\to(0,0)} \frac{x^2 - y^2}{x^2 + y^2}$ does not exist.

SOLUTION If $P(x, y)$ approaches $(0, 0)$ along the x-axis (see Figure 16.24(i)), the y-coordinate of P is always zero and the expression $(x^2 - y^2)/(x^2 + y^2)$ reduces to x^2/x^2, or 1. Hence the limiting value along this path is 1.

If $P(x, y)$ approaches $(0, 0)$ along the y-axis (see Figure 16.24(ii)), the x-coordinate of P is 0 and $(x^2 - y^2)/(x^2 + y^2)$ reduces to $-y^2/y^2$, or -1. Since two different values are obtained, the limit does not exist, by the two-path rule (16.4).

We could, of course, have chosen other paths to the origin $(0, 0)$. For example, if we let $P(x, y)$ approach $(0, 0)$ along the line $y = 2x$ (see Figure 16.24(iii)), then

$$\frac{x^2 - y^2}{x^2 + y^2} = \frac{x^2 - (2x)^2}{x^2 + (2x)^2} = \frac{-3x^2}{5x^2} = -\frac{3}{5}$$

and the limiting value along the line $y = 2x$ is $-\frac{3}{5}$.

FIGURE 16.24

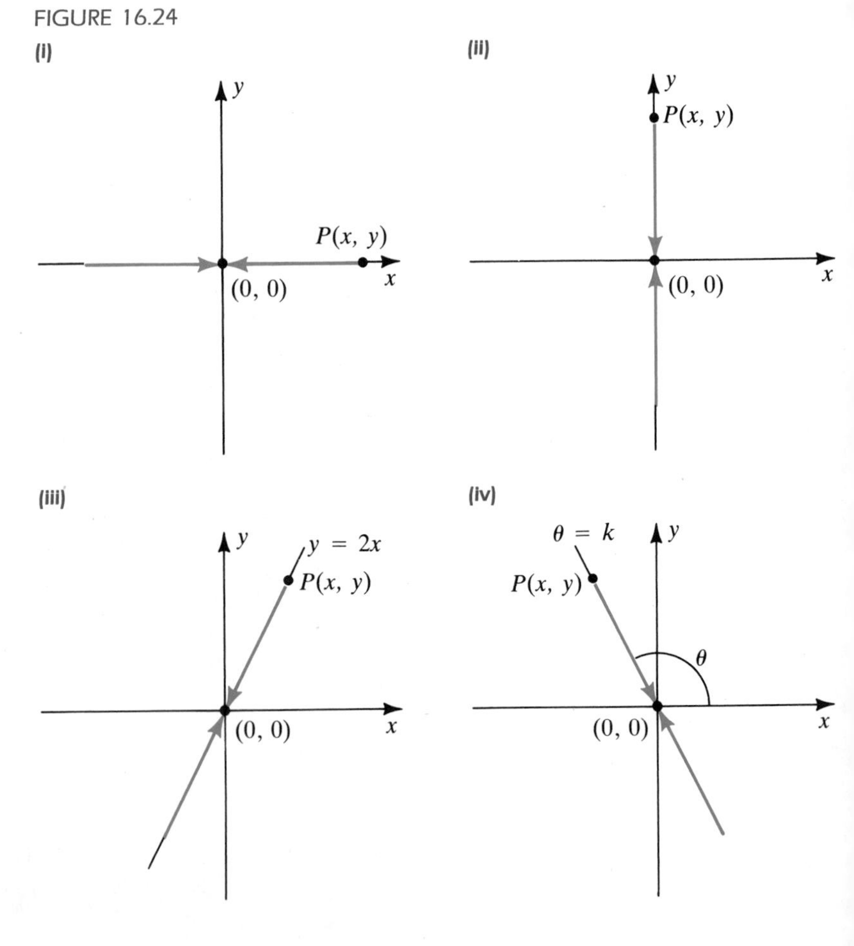

As an alternative solution, we could change the expression in x and y to polar coordinates, as follows:

$$\frac{x^2 - y^2}{x^2 + y^2} = \frac{r^2 \cos^2 \theta - r^2 \sin^2 \theta}{r^2} = \cos^2 \theta - \sin^2 \theta = \cos 2\theta$$

As $P(x, y) \to (0, 0)$ along any ray $\theta = k$ (see Figure 16.24(iv)), the function value is always $\cos 2k$, and hence $f(x, y)$ would have this limiting value. By choosing suitable values of k, we could make $f(x, y)$ approach any value between -1 and 1. Hence, by (16.4), the limit does not exist.

EXAMPLE 4 Show that $\displaystyle\lim_{(x,y)\to(0,0)} \frac{x^2 y}{x^4 + y^2}$ does not exist.

SOLUTION If we let $P(x, y)$ approach $(0, 0)$ along any line $y = mx$ that passes through the origin (see Figure 16.25(i)), we see that if $m \neq 0$,

$$\lim_{(x,y)\to(0,0)} \frac{x^2 y}{x^4 + y^2} = \lim_{(x,y)\to(0,0)} \frac{x^2(mx)}{x^4 + (mx)^2} = \lim_{(x,y)\to(0,0)} \frac{mx^3}{x^4 + m^2x^2}$$

$$= \lim_{(x,y)\to(0,0)} \frac{mx}{x^2 + m^2} = \frac{0}{0 + m^2} = 0.$$

FIGURE 16.25

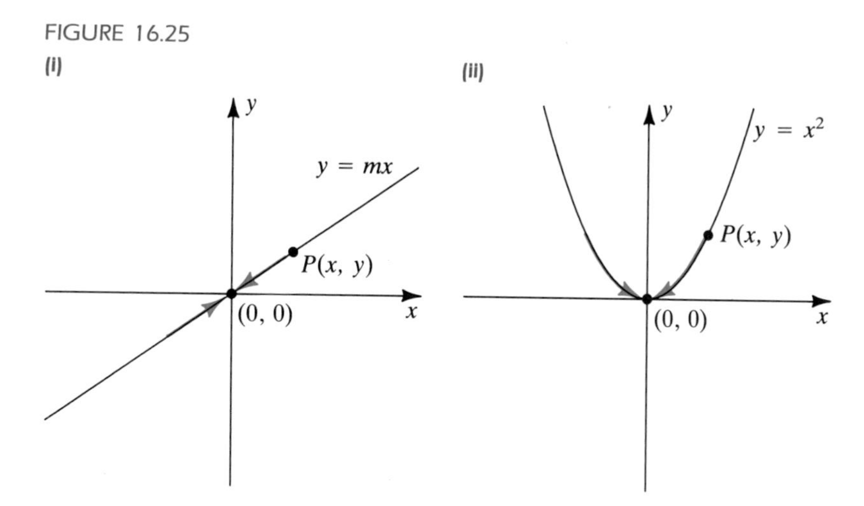

It is tempting to conclude that the limit is 0 as (x, y) approaches $(0, 0)$; however, if we let $P(x, y)$ approach $(0, 0)$ along the parabola $y = x^2$ (see Figure 16.25(ii)), then

$$\lim_{(x,y)\to(0,0)} \frac{x^2 y}{x^4 + y^2} = \lim_{(x,y)\to(0,0)} \frac{x^2(x^2)}{x^4 + (x^2)^2}$$

$$= \lim_{(x,y)\to(0,0)} \frac{x^4}{2x^4} = \lim_{(x,y)\to(0,0)} \frac{1}{2} = \frac{1}{2}.$$

Therefore not every path to $(0, 0)$ leads to the same limiting value, and hence, by the two-path rule (16.4), the limit does not exist.

If, in the preceding solution, we had obtained the limit 0 using the path $y = x^2$, we *still* could not have concluded that the limit is 0 as the point $P(x, y) \to (0, 0)$, because there could be some *other* path that yields a nonzero number. Remember that the two-path rule cannot be used to prove that a limit exists—only that a limit does *not* exist.

When investigating properties of a function of two variables, we often restrict the domain to points (x, y) that lie in certain types of regions of the xy-plane. In particular, an **open disk** consists of all points that lie *inside* a circle. A **closed disk** contains both the points inside and the points *on* the circle. A point (a, b) is an **interior point** of a region R if there is an open disk with center (a, b) that lies completely within R. An interior point is illustrated in Figure 16.26(i). A point (a, b) is a **boundary point** of R if every disk with center (a, b) contains points that are in R and points that are not in R, as illustrated in Figure 16.26(ii). The concept of an interior (or boundary) point (a, b, c) in a *three*-dimensional region R is defined in similar fashion using spherical regions instead of disks.

FIGURE 16.26

(i) Interior point of R (ii) Boundary point of R

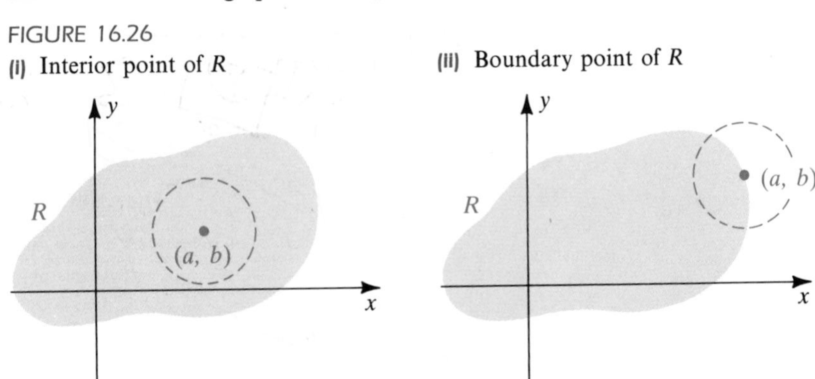

A region is **closed** if it contains all of its boundary points. A region is **open** if it contains *none* of its boundary points—that is, every point of the region is an interior point. A region that contains some, but not all, of its boundary points is neither open nor closed. These concepts are analogous to those of closed, open, and half-open intervals of real numbers.

Suppose a function f of two variables is defined for every (x, y) in a region R, except possibly at (a, b). If (a, b) is an interior point, then Definition (16.3) can be used to investigate the limit of f as (x, y) approaches (a, b). *If* (a, b) *is a boundary point, we shall use Definition* (16.3) *with the added restriction that* (x, y) *must be both in R and inside the circle of radius* δ. Theorems on limits may then be extended to boundary points.

The definition of continuity of a function f of two variables is analogous to that for a function of one variable (see Definition (2.20)).

Definition **(16.5)**

A function f of two variables is **continuous** at an interior point (a, b) of its domain if

$$\lim_{(x,y)\to(a,b)} f(x, y) = f(a, b).$$

In order to define continuity at a boundary point (a, b), we add the restriction that (x, y) is in the domain D of f. Intuitively we know that

$f(x, y) \to f(a, b)$ only if $(x, y) \to (a, b)$ along every path C that lies entirely within D. The function f is **continuous on its domain D** if it is continuous at every pair (a, b) in D. If f is continuous on D, then a small change in (x, y) produces a small change in $f(x, y)$. Referring to the graph S of f (see Figure 16.20), we note that if (x, y) is close to (a, b), then the point $(x, y, f(x, y))$ on S is close to $(a, b, f(a, b))$. Thus, there are no holes or vertical steps in the graph of a continuous function of two variables.

We may prove theorems on continuity for functions of two variables that are analogous to those for functions of one variable. Thus, sums, products, and quotients of continuous functions are continuous, provided no zero denominators occur for the case of quotients. In particular, polynomial functions are continuous everywhere, and rational functions are continuous everywhere except at points where the denominator is zero.

The preceding discussion on limits and continuity can be extended to functions of three or more variables. For example, if f is a function of three variables, we have the following definition.

Definition **(16.6)**

Let a function f of three variables be defined throughout the interior of a sphere with center (a, b, c), except possibly at (a, b, c) itself. The statement

$$\lim_{(x,y,z)\to(a,b,c)} f(x, y, z) = L$$

means that for every $\epsilon > 0$ there is a $\delta > 0$ such that if

$$0 < \sqrt{(x-a)^2 + (y-b)^2 + (z-c)^2} < \delta,$$

then $$|f(x, y, z) - L| < \epsilon.$$

FIGURE 16.27

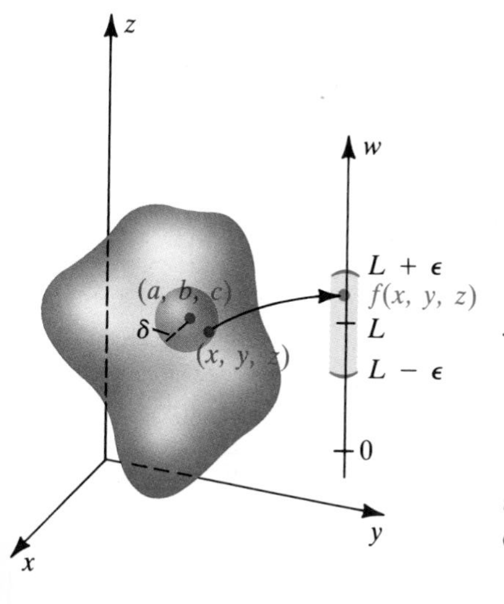

We may give a graphical interpretation for Definition (16.6) that is similar to the one illustrated in Figure 16.19. The difference is that in place of the *circle* in the xy-plane, we use a *sphere* of radius δ in an xyz-coordinate system. Specifically, for any $\epsilon > 0$, consider the open interval $(L - \epsilon, L + \epsilon)$ on the w-axis, as in Figure 16.27. If (16.6) is true, then there is a $\delta > 0$ such that for every point (x, y, z) within the sphere of radius δ with center (a, b, c), except possibly for (a, b, c) itself, the function value $f(x, y, z)$ is in the interval $(L - \epsilon, L + \epsilon)$.

The limit definition (16.6) may be extended to a boundary point (a, b, c) of a region R by adding the restriction that (x, y, z) is both in R and inside the sphere of radius δ.

The two-path rule (16.4) can be extended to functions of three variables. In the next example we use this rule to show that a limit does not exist.

EXAMPLE 5 Show that $\displaystyle\lim_{(x,y,z)\to(1,2,-1)} \frac{(x + y + 3z)^3}{(x-1)(y-2)(z+1)}$ does not exist.

SOLUTION Suppose we let $P(x, y, z)$ approach $(1, 2, -1)$ along the line l that is parallel to the vector $\langle a, b, c\rangle$, where a, b, and c are real numbers. By (14.34), parametric equations for l are

$$x = 1 + at, \quad y = 2 + bt, \quad z = -1 + ct; \quad t \text{ in } \mathbb{R}.$$

Hence, for any $P(x, y, z)$ on l different from $(1, 2, -1)$,

$$\frac{(x+y+3z)^3}{(x-1)(y-2)(z+1)} = \frac{(at+bt+3ct)^3}{(at)(bt)(ct)} = \frac{(a+b+3c)^3t^3}{abct^3} = \frac{(a+b+3c)^3}{abc}.$$

By assigning different values to a, b, and c—that is, by approaching $(1, 2, -1)$ along different lines—we get different limits. Hence, by the two-path rule, the given limit does not exist.

A function f of three variables is **continuous** at an interior point (a, b, c) of a region if

$$\lim_{(x,y,z)\to(a,b,c)} f(x, y, z) = f(a, b, c).$$

Continuity at a boundary point is defined by using the extension of Definition (16.6) to boundary points.

We may extend the definition of limits to functions of four or more variables; however, in these cases we do not give geometric interpretations.

In Section 16.5 we shall discuss composite functions of several variables. As a simple illustration, suppose that f is a function of two variables x and y and that g is a function of one variable t. A function h of two variables may be obtained by substituting $f(x, y)$ for t—that is, by letting $h(x, y) = g(f(x, y))$—provided the range of f is in the domain of g. This is illustrated in the next example.

EXAMPLE 6 Express $g(f(x, y))$ in terms of x and y, and find the domain of the resulting composite function.

(a) $f(x, y) = xe^y, \quad g(t) = 3t^2 + t + 1$

(b) $f(x, y) = y - 4x^2, \quad g(t) = \sin\sqrt{t}$

SOLUTION In each case we substitute $f(x, y)$ for t in the expression for $g(t)$:

(a) $g(f(x, y)) = g(xe^y) = 3x^2e^{2y} + xe^y + 1$

(b) $g(f(x, y)) = g(y - 4x^2) = \sin\sqrt{y - 4x^2}$

The domain in part (a) is $\mathbb{R}^2$, and the domain in part (b) consists of every ordered pair (x, y) such that $y \geq 4x^2$.

The following analogue of Theorem (2.24) may be proved for functions of the type illustrated in Example 6.

Theorem **(16.7)**

> If a function f of two variables is continuous at (a, b) and a function g of one variable is continuous at $f(a, b)$, then the function h defined by $h(x, y) = g(f(x, y))$ is continuous at (a, b).

Theorem (16.7) allows us to establish the continuity of composite functions of several variables, as illustrated in the next example.

EXAMPLE 7 If $h(x, y) = e^{x^2+5xy+y^3}$, show that h is continuous at every pair (a, b).

SOLUTION If we let $f(x, y) = x^2 + 5xy + y^3$ and $g(t) = e^t$, it follows that $h(x, y) = g(f(x, y))$. Since f is a polynomial function, it is continuous at every pair (a, b). Moreover, g is continuous at every $t = f(a, b)$. Thus, by Theorem (16.7), h is continuous at (a, b).

EXERCISES 16.2

Exer. 1–10: Find the limit.

1 $\lim_{(x,y)\to(0,0)} \dfrac{x^2-2}{3+xy}$

2 $\lim_{(x,y)\to(2,1)} \dfrac{4+x}{2-y}$

3 $\lim_{(x,y)\to(\pi/2,1)} \dfrac{y+1}{2-\cos x}$

4 $\lim_{(x,y)\to(-1,3)} \dfrac{y^2+x}{(x-1)(y+2)}$

5 $\lim_{(x,y)\to(0,0)} \dfrac{x^4-y^4}{x^2+y^2}$

6 $\lim_{(x,y)\to(1,2)} \dfrac{xy-y}{x^2-x+2xy-2y}$

7 $\lim_{(x,y)\to(0,0)} \dfrac{3x^3-2x^2y+3y^2x-2y^3}{x^2+y^2}$

8 $\lim_{(x,y)\to(0,0)} \dfrac{x^3-x^2y+xy^2-y^3}{x^2+y^2}$

9 $\lim_{(x,y,z)\to(2,3,1)} \dfrac{y^2-4y+3}{x^2z(y-3)}$

10 $\lim_{(x,y,z)\to(2,1,2)} \dfrac{x^2-z^2}{x^3-z^3}$

Exer. 11–20: Show that the limit does not exist.

11 $\lim_{(x,y)\to(0,0)} \dfrac{2x^2-y^2}{x^2+2y^2}$

12 $\lim_{(x,y)\to(0,0)} \dfrac{x^2-2xy+5y^2}{3x^2+4y^2}$

13 $\lim_{(x,y)\to(1,2)} \dfrac{xy-2x-y+2}{x^2+y^2-2x-4y+5}$

14 $\lim_{(x,y)\to(2,1)} \dfrac{x^2-4x+4}{xy-2y-x+2}$

15 $\lim_{(x,y)\to(0,0)} \dfrac{3x^2y^3}{2y^5-2x^5}$

16 $\lim_{(x,y)\to(0,0)} \dfrac{3xy}{5x^4+2y^4}$

17 $\lim_{(x,y,z)\to(0,0,0)} \dfrac{xy+yz+xz}{x^2+y^2+z^2}$

18 $\lim_{(x,y,z)\to(0,0,0)} \dfrac{2x^2+3y^2+z^2}{x^2+y^2+z^2}$

19 $\lim_{(x,y,z)\to(0,0,0)} \dfrac{x^3+y^3+z^3}{xyz}$

20 $\lim_{(x,y,z)\to(2,1,0)} \dfrac{(x+y+z-3)^5}{z^3(x-2)(y-1)}$

Exer. 21–24: Use polar coordinates to find the limit, if it exists.

21 $\lim_{(x,y)\to(0,0)} \dfrac{xy^2}{x^2+y^2}$

22 $\lim_{(x,y)\to(0,0)} \dfrac{x^3-y^3}{x^2+y^2}$

23 $\lim_{(x,y)\to(0,0)} \dfrac{x^2+y^2}{\sin(x^2+y^2)}$

24 $\lim_{(x,y)\to(0,0)} \dfrac{\sinh(x^2+y^2)}{x^2+y^2}$

Exer. 25–28: Describe the set of all points in the *xy*-plane at which *f* is continuous.

25 $f(x, y) = \ln(x+y-1)$

26 $f(x, y) = \dfrac{xy}{x^2-y^2}$

27 $f(x, y) = \sqrt{x}e^{\sqrt{1-y^2}}$

28 $f(x, y) = \sqrt{25-x^2-y^2}$

Exer. 29–32: Describe the set of all points in an *xyz*-coordinate system at which *f* is continuous.

29 $f(x, y, z) = \dfrac{1}{x^2+y^2-z^2}$

30 $f(x, y, z) = \sqrt{xy}\tan z$

31 $f(x, y, z) = \sqrt{x-2}\ln(yz)$

32 $f(x, y, z) = \sqrt{4-x^2-y^2-z^2}$

Exer. 33–34: The graph of f is shown in the figure. Use polar coordinates to investigate $\lim_{(x,y)\to(0,0)} f(x, y)$.

33 $f(x, y) = \dfrac{x^2 + y^2}{\ln(x^2 + y^2)}$

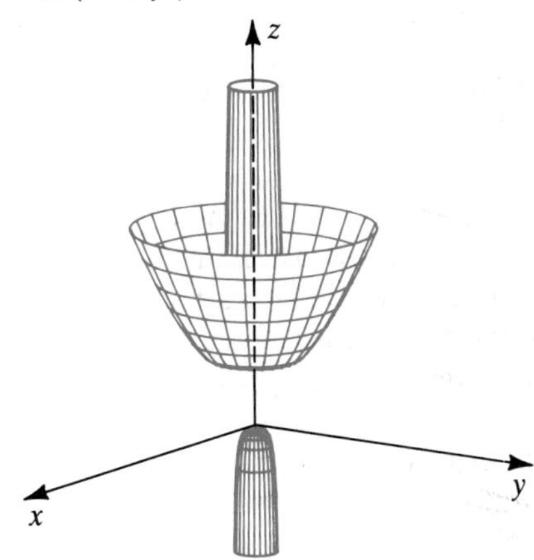

34 $f(x, y) = \dfrac{\sin(2x^2 + y^2)}{x^2 + y^2}$

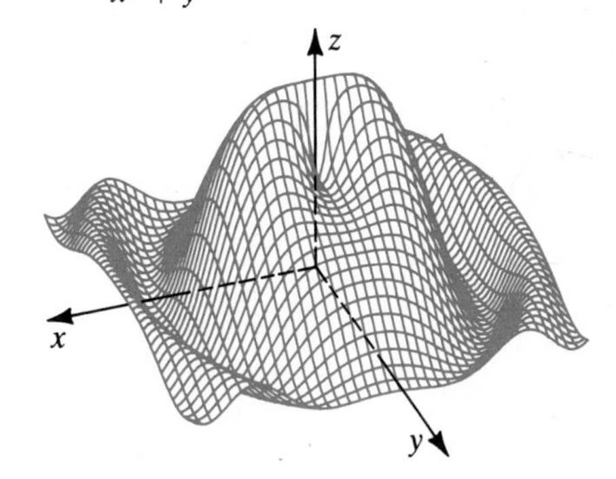

Exer. 35–38: Find $h(x, y) = g(f(x, y))$ and use Theorem (16.7) to determine where h is continuous.

35 $f(x, y) = x^2 - y^2$; $\quad g(t) = (t^2 - 4)/t$

36 $f(x, y) = 3x + 2y - 4$; $\quad g(t) = \ln(t + 5)$

37 $f(x, y) = x + \tan y$; $\quad g(z) = z^2 + 1$

38 $f(x, y) = y \ln x$; $\quad g(w) = e^w$

39 If $f(x, y) = x^2 + 2y$, $g(t) = e^t$, and $h(t) = t^2 - 3t$, find $g(f(x, y))$, $h(f(x, y))$, and $f(g(t), h(t))$.

40 If $f(x, y, z) = 2x + ye^z$ and $g(t) = t^2$, find $g(f(x, y, z))$.

41 If $f(u, v) = uv - 3u + v$, $g(x, y) = x - 2y$, and $k(x, y) = 2x + y$, find $f(g(x, y), k(x, y))$.

42 If $f(x, y) = 2x + y$, find $f(f(x, y), f(x, y))$.

43 Extend Definition (16.3) to functions of four variables.

44 Prove that if f is a continuous function of two variables and $f(a, b) > 0$, then there is a circle C in the xy-plane with center (a, b) such that $f(x, y) > 0$ for every pair (x, y) that is in the domain of f and within C.

45 Prove, directly from Definition (16.3), that

(a) $\lim_{(x,y)\to(a,b)} x = a$ $\quad$ (b) $\lim_{(x,y)\to(a,b)} y = b$

46 If $\lim_{(x,y)\to(a,b)} f(x, y) = L$ and c is any real number, prove, directly from Definition (16.3), that

$$\lim_{(x,y)\to(a,b)} cf(x, y) = cL.$$

16.3 PARTIAL DERIVATIVES

In Chapter 3 we defined the derivative $f'(x)$ of a function of *one* variable as

$$f'(x) = \lim_{h\to 0} \frac{f(x + h) - f(x)}{h}.$$

We may interpret this formula as follows. First change the independent variable x by an amount h; then divide the corresponding change in f, $f(x + h) - f(x)$, by h; and finally, let h approach 0. An analogous procedure can be applied to functions of several variables. Given $f(x, y)$, first change *one* of the variables, say x, by an amount h; then divide the corresponding change in f, $f(x + h, y) - f(x, y)$, by h; and finally, let h approach 0. This leads to the concept of the *partial derivative $f_x(x, y)$ of f with respect to x*, defined in (16.8). A similar process is used for the variable y.

Definition **(16.8)**

Let f be a function of two variables. The **first partial derivatives of f with respect to x and y** are the functions f_x and f_y such that

$$f_x(x, y) = \lim_{h \to 0} \frac{f(x + h, y) - f(x, y)}{h}$$

$$f_y(x, y) = \lim_{h \to 0} \frac{f(x, y + h) - f(x, y)}{h}.$$

In Definition (16.8), x and y are fixed (but arbitrary) and h is the only variable; hence we use the notation for limits of functions of one variable instead of the $(x, y) \to (a, b)$ notation introduced in the previous section. We can find partial derivatives without using limits, as follows. If we let $y = b$ and define a function g of one variable by $g(x) = f(x, b)$, then $g'(x) = f_x(x, b) = f_x(x, y)$. Thus, *to find $f_x(x, y)$, we regard y as a constant and differentiate $f(x, y)$ with respect to x*. Similarly, *to find $f_y(x, y)$, we regard the variable x as a constant and differentiate $f(x, y)$ with respect to y*. This technique is illustrated in Example 1.

Other common notations for partial derivatives are listed in the next box.

Notations for partial derivatives **(16.9)**

If $w = f(x, y)$, then

$$f_x = \frac{\partial f}{\partial x}, \qquad f_y = \frac{\partial f}{\partial y}$$

$$f_x(x, y) = \frac{\partial}{\partial x} f(x, y) = \frac{\partial w}{\partial x} = w_x$$

$$f_y(x, y) = \frac{\partial}{\partial y} f(x, y) = \frac{\partial w}{\partial y} = w_y.$$

For brevity we often speak of $\partial f/\partial x$ or $\partial f/\partial y$ as *the partial of f with respect to x* or *to y*, respectively.

EXAMPLE 1 If $f(x, y) = x^3y^2 - 2x^2y + 3x$, find

(a) $f_x(x, y)$ and $f_y(x, y)$ **(b)** $f_x(2, -1)$ and $f_y(2, -1)$

SOLUTION

(a) We regard y as a constant and differentiate with respect to x:

$$f_x(x, y) = 3x^2y^2 - 4xy + 3$$

We regard x as a constant and differentiate with respect to y:

$$f_y(x, y) = 2x^3y - 2x^2$$

(b) Substituting $x = 2$ and $y = -1$ in part (a), we have

$$f_x(2, -1) = 3(2)^2(-1)^2 - 4(2)(-1) + 3 = 23,$$

$$f_y(2, -1) = 2(2)^3(-1) - 2(2)^2 = -24.$$

Formulas similar to those for functions of one variable are true for partial derivatives. For example, if $u = f(x, y)$ and $v = g(x, y)$, then a product rule and quotient rule for partial derivatives are

$$\frac{\partial}{\partial x}(uv) = u\frac{\partial v}{\partial x} + v\frac{\partial u}{\partial x}, \qquad \frac{\partial}{\partial x}\left(\frac{u}{v}\right) = \frac{v\dfrac{\partial u}{\partial x} - u\dfrac{\partial v}{\partial x}}{v^2}$$

or, in subscript notation,

$$(fg)_x = fg_x + gf_x, \qquad \left(\frac{f}{g}\right)_x = \frac{gf_x - fg_x}{g^2}.$$

A power rule for partial differentiation is

$$\frac{\partial}{\partial x}(u^n) = nu^{n-1}\frac{\partial u}{\partial x},$$

where n is a real number. Similarly,

$$\frac{\partial}{\partial x}\cos u = -\sin u\frac{\partial u}{\partial x}, \qquad \frac{\partial}{\partial x}e^u = e^u\frac{\partial u}{\partial x},$$

and so on.

EXAMPLE 2 Find $\dfrac{\partial w}{\partial y}$ if $w = xy^2e^{xy}$.

SOLUTION We write $w = (xy^2)(e^{xy})$ and use a product rule:

$$\begin{aligned}\frac{\partial w}{\partial y} &= (xy^2)\frac{\partial}{\partial y}(e^{xy}) + (e^{xy})\frac{\partial}{\partial y}(xy^2)\\ &= xy^2(xe^{xy}) + e^{xy}(2xy) = xy(xy+2)e^{xy}\end{aligned}$$

FIGURE 16.28

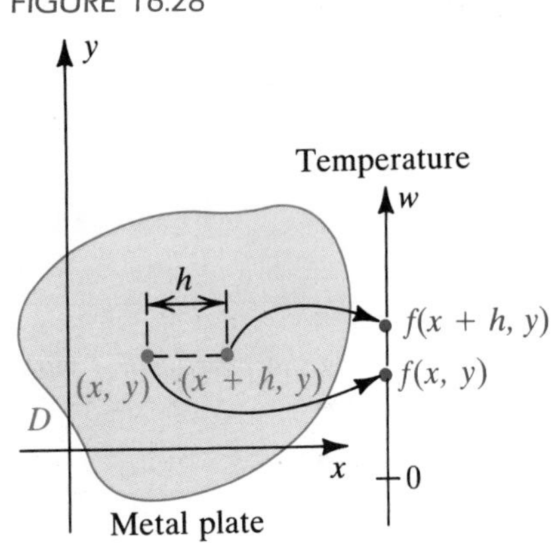

Let us apply Definition (16.8) to the physical illustration given at the beginning of the previous section: $f(x, y)$ is the temperature at the point (x, y) of a flat metal plate in an xy-plane. Note that the points (x, y) and $(x + h, y)$ are on a horizontal line, as indicated in Figure 16.28. If a point moves horizontally from (x, y) to $(x + h, y)$, then the difference $f(x + h, y) - f(x, y)$ is the net change in temperature, and we have

$$\text{Average change in temperature: } \quad \frac{f(x+h, y) - f(x, y)}{h}$$

FIGURE 16.29

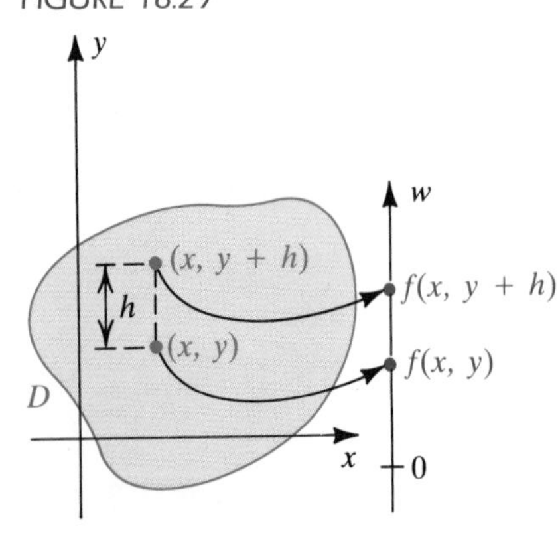

For example, if the temperature change is 2 degrees and the distance h is 4, then the average change in temperature from (x, y) to $(x + h, y)$ is $\frac{2}{4}$, or $\frac{1}{2}$. Thus, *on the average*, the temperature changes at a rate of $\frac{1}{2}$ degree per unit change in distance. Taking the limit of the average change as h approaches 0, we see that $f_x(x, y)$ *is the (instantaneous) rate of change of temperature with respect to distance as the point* (x, y) *moves in a horizontal direction.*

Similarly, if a point moves on a vertical line from (x, y) to $(x, y + h)$, as shown in Figure 16.29, we obtain

$$\text{Average change in temperature: } \quad \frac{f(x, y+h) - f(x, y)}{h}$$

Letting h approach 0, we see that *$f_y(x, y)$ is the (instantaneous) rate of change of temperature as the point (x, y) moves in a vertical direction.*

As another application, suppose D represents the surface of a lake and $f(x, y)$ is the depth of the water under the point (x, y) on the surface. In this case $f_x(x, y)$ is the rate at which the depth changes as a point moves away from (x, y) parallel to the x-axis. Similarly, $f_y(x, y)$ is the rate of change of the depth in the direction of the y-axis.

FIGURE 16.30

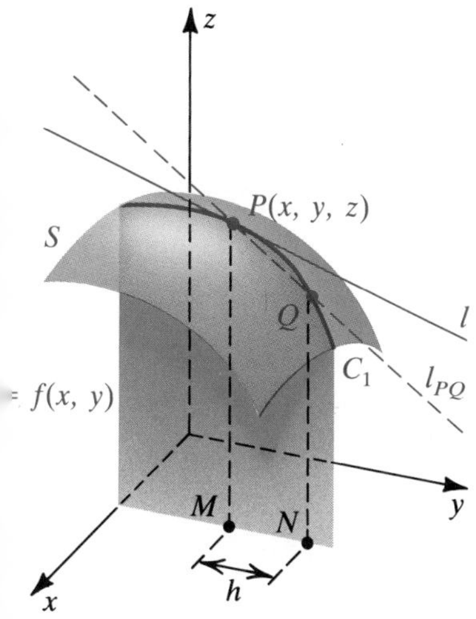

We shall next use the graph S of f to obtain a geometric interpretation of Definition (16.8). To illustrate $f_y(x, y)$, consider the points $M(x, y, 0)$ and $N(x, y + h, 0)$. The plane parallel to the yz-plane and passing through M and N intersects S in a curve C_1 (see Figure 16.30). Let P and Q be the points on C_1 that have projections M and N on the xy-plane.

If we introduce a rectangular coordinate system in the plane containing M, N, P, and Q, then the slope m_{PQ} of the secant line l_{PQ} through P and Q is

$$m_{PQ} = \frac{f(x, y + h) - f(x, y)}{h}.$$

The limit of m_{PQ} as h approaches 0—that is, the partial derivative $f_y(x, y)$—is the slope of the tangent line l to C_1 at P.

Similarly, if C_2 is the trace of S on the plane parallel to the xz-plane and passing through M, then $f_x(x, y)$ is the slope of the tangent line to C_2 at P.

The following theorem summarizes this discussion.

Theorem (16.10)

Let S be the graph of $z = f(x, y)$, and let $P(a, b, f(a, b))$ be a point on S at which f_x and f_y exist. Let C_1 and C_2 be the traces of S on the planes $x = a$ and $y = b$, respectively, and let l_1 and l_2 be the tangent lines to C_1 and C_2 at P (see Figure 16.31).

(i) The slope of l_1 in the plane $x = a$ is $f_y(a, b)$.

(ii) The slope of l_2 in the plane $y = b$ is $f_x(a, b)$.

FIGURE 16.31

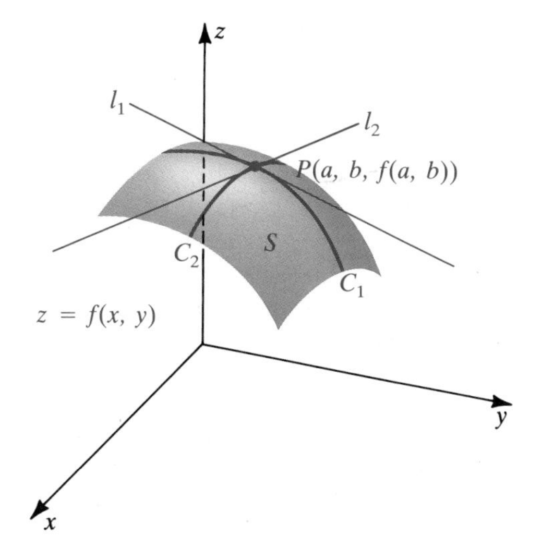

First partial derivatives of functions of three or more variables are defined in the same manner as in Definition (16.8). Specifically, all variables except one are regarded as constant, and we differentiate with respect to the remaining variable. Given $f(x, y, z)$, we may find f_x, f_y, and f_z (or, equivalently, $\partial f/\partial x$, $\partial f/\partial y$, and $\partial f/\partial z$). For example,

$$f_z(x, y, z) = \lim_{h \to 0} \frac{f(x, y, z + h) - f(x, y, z)}{h}.$$

As an application, let $f(x, y, z)$ be the temperature at the point $P(x, y, z)$. The partial derivative $f_z(x, y, z)$ is the instantaneous rate of change of temperature with respect to distance along a line through P that is parallel to the z-axis. The partial derivatives $f_x(x, y, z)$ and $f_y(x, y, z)$ give the rates of change in the directions of the x- and y-axes, respectively.

EXAMPLE 3 If $w = x^2y^3 \sin z + e^{xz}$, find $\partial w/\partial x$, $\partial w/\partial y$, and $\partial w/\partial z$.

SOLUTION We proceed as follows:

Regard y and z as constant: $\dfrac{\partial w}{\partial x} = 2xy^3 \sin z + ze^{xz}$

Regard x and z as constant: $\dfrac{\partial w}{\partial y} = 3x^2y^2 \sin z$

Regard x and y as constant: $\dfrac{\partial w}{\partial z} = x^2y^3 \cos z + xe^{xz}$

If f is a function of two variables x and y, then f_x and f_y are also functions of two variables, and we may consider *their* first partial derivatives. These are called the **second partial derivatives of f** and are denoted as follows.

Second partial derivatives **(16.11)**

$$\frac{\partial}{\partial x} f_x = (f_x)_x = f_{xx} = \frac{\partial}{\partial x}\left(\frac{\partial f}{\partial x}\right) = \frac{\partial^2 f}{\partial x^2}$$

$$\frac{\partial}{\partial y} f_x = (f_x)_y = f_{xy} = \frac{\partial}{\partial y}\left(\frac{\partial f}{\partial x}\right) = \frac{\partial^2 f}{\partial y\, \partial x}$$

$$\frac{\partial}{\partial x} f_y = (f_y)_x = f_{yx} = \frac{\partial}{\partial x}\left(\frac{\partial f}{\partial y}\right) = \frac{\partial^2 f}{\partial x\, \partial y}$$

$$\frac{\partial}{\partial y} f_y = (f_y)_y = f_{yy} = \frac{\partial}{\partial y}\left(\frac{\partial f}{\partial y}\right) = \frac{\partial^2 f}{\partial y^2}$$

If $w = f(x, y)$, we write

$$\frac{\partial^2}{\partial x^2} f(x, y) = f_{xx}(x, y) = \frac{\partial^2 w}{\partial x^2} = w_{xx},$$

$$\frac{\partial^2}{\partial y\, \partial x} f(x, y) = f_{xy}(x, y) = \frac{\partial^2 w}{\partial y\, \partial x} = w_{xy},$$

and so on.

Note that f_{xy} means that we differentiate first with respect to x and then with respect to y, and f_{yx} means that we differentiate first with respect to y and then with respect to x. For the "∂" notation, the reverse order is used; that is, for $\partial^2 f/\partial x\, \partial y$, we differentiate first with respect to y and then with respect to x.

We shall refer to f_{xy} and f_{yx} as the **mixed second partial derivatives of f** (or simply *mixed partials of f*). The next theorem states that, under suitable conditions, these mixed partials are equal; that is, the order of differentiation is immaterial. The proof may be found in texts on advanced calculus.

Theorem **(16.12)**

> Let f be a function of two variables x and y. If f, f_x, f_y, f_{xy}, and f_{yx} are continuous on an open region R, then $f_{xy} = f_{yx}$ throughout R.

The hypothesis of Theorem (16.12) is satisfied for most functions encountered in calculus and its applications. Similarly, if $w = f(x, y, z)$ and f has continuous first and second partial derivatives, then the following equalities hold for mixed partials:

$$\frac{\partial^2 w}{\partial y\, \partial x} = \frac{\partial^2 w}{\partial x\, \partial y}, \qquad \frac{\partial^2 w}{\partial z\, \partial x} = \frac{\partial^2 w}{\partial x\, \partial z}, \qquad \frac{\partial^2 w}{\partial z\, \partial y} = \frac{\partial^2 w}{\partial y\, \partial z}$$

EXAMPLE 4 Find the second partial derivatives of f if

$$f(x, y) = x^3y^2 - 2x^2y + 3x.$$

SOLUTION This function was considered in Example 1, where we obtained

$$f_x(x, y) = 3x^2y^2 - 4xy + 3 \quad \text{and} \quad f_y(x, y) = 2x^3y - 2x^2.$$

Hence the second partial derivatives are as follows:

$$f_{xx}(x, y) = \frac{\partial}{\partial x} f_x(x, y) = \frac{\partial}{\partial x}(3x^2y^2 - 4xy + 3) = 6xy^2 - 4y$$

$$f_{xy}(x, y) = \frac{\partial}{\partial y} f_x(x, y) = \frac{\partial}{\partial y}(3x^2y^2 - 4xy + 3) = 6x^2y - 4x$$

$$f_{yx}(x, y) = \frac{\partial}{\partial x} f_y(x, y) = \frac{\partial}{\partial x}(2x^3y - 2x^2) = 6x^2y - 4x$$

$$f_{yy}(x, y) = \frac{\partial}{\partial y} f_y(x, y) = \frac{\partial}{\partial y}(2x^3y - 2x^2) = 2x^3$$

Third and higher partial derivatives are defined in similar fashion. For example,

$$\frac{\partial}{\partial x} f_{xx} = f_{xxx} = \frac{\partial}{\partial x}\left(\frac{\partial^2 f}{\partial x^2}\right) = \frac{\partial^3 f}{\partial x^3},$$

$$\frac{\partial}{\partial x} f_{xy} = f_{xyx} = \frac{\partial}{\partial x}\left(\frac{\partial^2 f}{\partial y\, \partial x}\right) = \frac{\partial^3 f}{\partial x\, \partial y\, \partial x},$$

and so on. If first, second, and third partial derivatives are continuous, then the order of differentiation is immaterial; that is,

$$f_{xyx} = f_{yxx} = f_{xxy} \quad \text{and} \quad f_{yxy} = f_{xyy} = f_{yyx}.$$

Of course, letters other than x and y may be used. If f is a function of r and s, then symbols such as

$$f_r(r, s), \quad f_s(r, s), \quad f_{rs}, \quad \frac{\partial f}{\partial r}, \quad \frac{\partial^2 f}{\partial r^2}$$

are employed for partial derivatives.

Similar notations and results apply to partials of functions of more than two variables.

EXERCISES 16.3

Exer. 1–18: Find the first partial derivatives of f.

1 $f(x, y) = 2x^4y^3 - xy^2 + 3y + 1$

2 $f(x, y) = (x^3 - y^2)^5$

3 $f(r, s) = \sqrt{r^2 + s^2}$

4 $f(s, t) = \dfrac{t}{s} - \dfrac{s}{t}$

5 $f(x, y) = xe^y + y \sin x$

6 $f(x, y) = e^x \ln xy$

7 $f(t, v) = \ln \sqrt{\dfrac{t + v}{t - v}}$

8 $f(u, w) = \arctan \dfrac{u}{w}$

9 $f(x, y) = x \cos \dfrac{x}{y}$

10 $f(x, y) = \sqrt{4x^2 - y^2} \sec x$

11 $f(x, y, z) = 3x^2z + xy^2$

12 $f(x, y, z) = x^2y^3z^4 + 2x - 5yz$

13 $f(r, s, t) = r^2e^{2s} \cos t$

14 $f(x, y, t) = \dfrac{x^2 - t^2}{1 + \sin 3y}$

15 $f(x, y, z) = xe^z - ye^x + ze^{-y}$

16 $f(r, s, v, p) = r^3 \tan s + \sqrt{s}e^{(v^2)} - v \cos 2p$

17 $f(q, v, w) = \sin^{-1} \sqrt{qv} + \sin vw$

18 $f(x, y, z) = xyze^{xyz}$

Exer. 19–24: Verify that $w_{xy} = w_{yx}$.

19 $w = xy^4 - 2x^2y^3 + 4x^2 - 3y$

20 $w = \dfrac{x^2}{x + y}$

21 $w = x^3e^{-2y} + y^{-2} \cos x$

22 $w = y^2e^{(x^2)} + \dfrac{1}{x^2y^3}$

23 $w = x^2 \cosh \dfrac{z}{y}$

24 $w = \sqrt{x^2 + y^2 + z^2}$

25 If $w = 3x^2y^3z + 2xy^4z^2 - yz$, find w_{xyz}.

26 If $w = u^4vt^2 - 3uv^2t^3$, find w_{tut}.

27 If $u = v \sec rt$, find u_{rvr}.

28 If $v = y \ln (x^2 + z^4)$, find v_{zzy}.

29 If $w = \sin xyz$, find $\dfrac{\partial^3 w}{\partial z\, \partial y\, \partial x}$.

30 If $w = \dfrac{x^2}{y^2 + z^2}$, find $\dfrac{\partial^3 w}{\partial z\, \partial y^2}$.

31 If $w = r^4s^3t - 3s^2e^{rt}$, verify that $w_{rrs} = w_{rsr} = w_{srr}$.

32 If $w = \tan uv + 2 \ln (u + v)$, verify that $w_{uvv} = w_{vuv} = w_{vvu}$.

Exer. 33–36: A function f of x and y is *harmonic* if

$$\frac{\partial^2 f}{\partial x^2} + \frac{\partial^2 f}{\partial y^2} = 0$$

throughout the domain of f. Prove that the given function is harmonic.

33 $f(x, y) = \ln \sqrt{x^2 + y^2}$

34 $f(x, y) = \arctan \dfrac{y}{x}$

35 $f(x, y) = \cos x \sinh y + \sin x \cosh y$

36 $f(x, y) = e^{-x} \cos y + e^{-y} \cos x$

37 If $w = \cos (x - y) + \ln (x + y)$, show that $\dfrac{\partial^2 w}{\partial x^2} - \dfrac{\partial^2 w}{\partial y^2} = 0$.

38 If $w = (y - 2x)^3 - \sqrt{y - 2x}$, show that $w_{xx} - 4w_{yy} = 0$.

39 If $w = e^{-c^2t} \sin cx$, show that $w_{xx} = w_t$ for every real number c.

40 The ideal gas law may be stated as $PV = knT$, where n is the number of moles of gas, V is the volume, T is the temperature, P is the pressure, and k is a constant. Show

that

$$\frac{\partial V}{\partial T}\frac{\partial T}{\partial P}\frac{\partial P}{\partial V} = -1.$$

Exer. 41–42: Show that v satisfies the *wave equation*

$$\frac{\partial^2 v}{\partial t^2} = a^2 \frac{\partial^2 v}{\partial x^2}.$$

41 $v = (\sin akt)(\sin kx)$

42 $v = (x - at)^4 + \cos(x + at)$

Exer. 43–46: Show that the functions u and v satisfy the *Cauchy-Riemann equations* $u_x = v_y$ and $u_y = -v_x$.

43 $u(x, y) = x^2 - y^2$; $v(x, y) = 2xy$

44 $u(x, y) = \dfrac{y}{x^2 + y^2}$; $v(x, y) = \dfrac{x}{x^2 + y^2}$

45 $u(x, y) = e^x \cos y$; $v(x, y) = e^x \sin y$

46 $u(x, y) = \cos x \cosh y + \sin x \sinh y$;
$v(x, y) = \cos x \cosh y - \sin x \sinh y$

47 List all possible second partial derivatives of $w = f(x, y, z)$.

48 If $w = f(x, y, z, t, v)$, define w_t as a limit.

49 A flat metal plate lies on an xy-plane such that the temperature T at (x, y) is given by $T = 10(x^2 + y^2)^2$, where T is in degrees and x and y are in centimeters. Find the instantaneous rate of change of T with respect to distance at $(1, 2)$ in the direction of

(a) the x-axis (b) the y-axis

50 The surface of a certain lake is represented by a region D in an xy-plane such that the depth under the point corresponding to (x, y) is given by $f(x, y) = 300 - 2x^2 - 3y^2$, where x, y, and $f(x, y)$ are in feet. If a water skier is in the water at the point $(4, 9)$, find the instantaneous rate at which the depth changes in the direction of

(a) the x-axis (b) the y-axis

51 Suppose the electrical potential V at the point (x, y, z) is given by $V = 100/(x^2 + y^2 + z^2)$, where V is in volts and x, y, and z are in inches. Find the instantaneous rate of change of V with respect to distance at $(2, -1, 1)$ in the direction of

(a) the x-axis (b) the y-axis (c) the z-axis

52 An object is situated in a rectangular coordinate system such that the temperature T at the point $P(x, y, z)$ is given by $T = 4x^2 - y^2 + 16z^2$, where T is in degrees and x, y, and z are in centimeters. Find the instantaneous rate of change of T with respect to distance at the point $P(4, -2, 1)$ in the direction of

(a) the x-axis (b) the y-axis (c) the z-axis

53 When a pollutant such as nitric oxide is emitted from a smokestack of height h meters, the long-range concentration $C(x, y)$ (in $\mu g/m^3$) of the pollutant at a point x kilometers from the smokestack and at a height of y meters (see figure) can often be represented by

$$C(x, y) = \frac{a}{x^2}\left[e^{-b(y-h)^2/x^2} + e^{-b(y+h)^2/x^2}\right],$$

where a and b are positive constants that depend on atmospheric conditions and the pollution emission rate. Suppose that

$$C(x, y) = \frac{200}{x^2}\left[e^{-0.02(y-10)^2/x^2} + e^{-0.02(y+10)^2/x^2}\right].$$

Compute and interpret $\partial C/\partial x$ and $\partial C/\partial y$ at the point $(2, 5)$.

EXERCISE 53

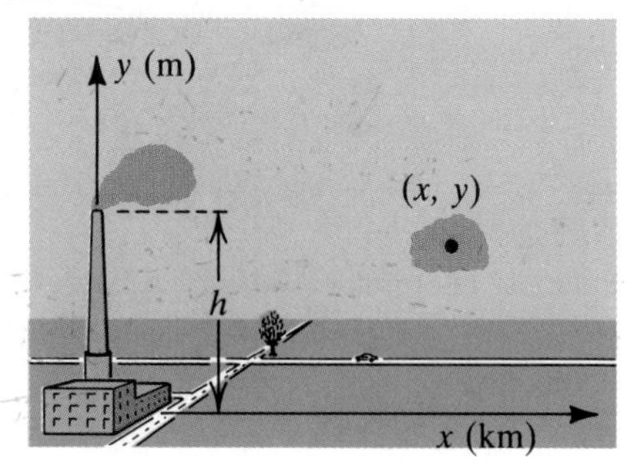

54 The analysis of certain electrical circuits involves the formula $I = V/\sqrt{R^2 + L^2\omega^2}$, where I is the current, V the voltage, R the resistance, L the inductance, and ω a positive constant. Find and interpret $\partial I/\partial R$ and $\partial I/\partial L$.

55 Most computers have only one processor that can be used for computations. Modern supercomputers, however, have anywhere from two to several thousand processors. A multiprocessor supercomputer is compared with a uniprocessor computer in terms of speedup. The *speedup* S is the number of times faster that a given computation can be accomplished with a multiprocessor than with a uniprocessor. A formula used to determine S is *Amdahl's law*,

$$S(p, q) = \frac{p}{q + p(1 - q)},$$

where p is the number of processors and q is the fraction of the computation that can be performed using all available processors in parallel—that is, using them in such a way that data are processed concurrently by separate units. The ideal situation, *complete parallelism*, occurs when $q = 1$.

(a) If $q = 0.8$, find the speedup when $p = 10$, 100, and 1000. Show that the speedup S cannot exceed 5, regardless of the number of processors available.

(b) Find the instantaneous rate of change of S with respect to q.

(c) What is the rate of change in (b) if there is complete parallelism, and how does the number of processors affect this rate of change?

56 Refer to Exercise 55. The efficiency E of a multiprocessor computation can be calculated using the equation

$$E = \frac{S(p, q)}{p} = \frac{1}{q + p(1 - q)}.$$

Show that if $0 \leq q < 1$, E is a decreasing function of p and therefore, without complete parallelism, increasing the number of processors does not increase the efficiency of the computation.

57 In the study of frost penetration in highway engineering, the temperature T at time t hours and depth x feet can be approximated by

$$T = T_0 e^{-\lambda x} \sin(\omega t - \lambda x),$$

where T_0, ω, and λ are constants. The period of $\sin(\omega t - \lambda x)$ is 24 hours.

(a) Find and interpret $\partial T/\partial t$ and $\partial T/\partial x$.

(b) Show that T satisfies the one-dimensional heat equation

$$\frac{\partial T}{\partial t} = k \frac{\partial^2 T}{\partial x^2},$$

where k is a constant.

58 Show that any function given by

$$w = (\sin ax)(\cos by)e^{-\sqrt{a^2 + b^2}\, z}$$

satisfies Laplace's equation in three dimensions:

$$\frac{\partial^2 w}{\partial x^2} + \frac{\partial^2 w}{\partial y^2} + \frac{\partial^2 w}{\partial z^2} = 0$$

59 The vital capacity V of the lungs is the largest volume (in milliliters) that can be exhaled after a maximum inhalation of air. For a typical male x years old and y centimeters tall, V may be approximated by the formula $V = 27.63y - 0.112xy$. Compute and interpret

(a) $\dfrac{\partial V}{\partial x}$ (b) $\dfrac{\partial V}{\partial y}$

60 On a clear day, the intensity of sunlight $I(x, t)$ (in foot-candles) at t hours after sunrise and at ocean depth x (in meters) can be approximated by

$$I(x, t) = I_0 e^{-kx} \sin^3(\pi t/D),$$

where I_0 is the intensity at midday, D is the length of the day (in hours), and k is a positive constant. If $I_0 = 1000$, $D = 12$, and $k = 0.10$, compute and interpret $\partial I/\partial t$ and $\partial I/\partial x$ when $t = 6$ and $x = 5$.

61 In economics the *price elasticity of demand* for a commodity indicates the responsiveness of consumers to a change in the market price of the commodity. Suppose n commodities $C_1, C_2, \ldots, C_n$ have prices $p_1, p_2, \ldots, p_n$, respectively, and consumer demand for C_k is a function q_k of $p_1, p_2, \ldots, p_n$. The price elasticity of C_k is the function e_k defined by

$$e_k = \frac{p_k}{q_k} \frac{\partial q_k}{\partial p_k}.$$

If, for each k,

$$q_k = b_k p_1^{-a_{k1}} p_2^{-a_{k2}} \cdots p_n^{-a_{kn}},$$

where $a_{k1}, a_{k2}, \ldots, a_{kn}$ and b_k are nonnegative constants, show that e_k is a constant function.

62 Refer to Exercise 61. Commodity C_k is said to be *independent* of commodity C_j if a change in price p_k does not affect demand q_j. This is equivalent to the condition $\partial q_j/\partial p_k = 0$. If q_k has the form in Exercise 61, show that C_k is independent of C_j if and only if $a_{jk} = 0$.

Exer. 63–64: Use Theorem (16.10).

63 Let C be the trace of the paraboloid $z = 9 - x^2 - y^2$ on the plane $x = 1$. Find parametric equations of the tangent line l to C at the point $P(1, 2, 4)$. Sketch the paraboloid, C, and l.

64 Let C be the trace of the graph of $z = \sqrt{36 - 9x^2 - 4y^2}$ on the plane $y = 2$. Find parametric equations of the tangent line l to C at the point $P(1, 2, \sqrt{11})$. Sketch the surface, C, and l.

c **Exer. 65–66: Use the following formulas with $h = 0.01$ to approximate $f_x(0.5, 0.2)$ and $f_y(0.5, 0.2)$, and compare the results with the values obtained from $f_x(x, y)$ and $f_y(x, y)$.**

$$f_x(x, y) \approx \frac{f(x - 2h, y) - 4f(x - h, y) + 3f(x, y)}{2h}$$

$$f_y(x, y) \approx \frac{f(x, y - 2h) - 4f(x, y - h) + 3f(x, y)}{2h}$$

65 $f(x, y) = y^2 \sin(xy)$

66 $f(x, y) = xy^3 + 4x^3y^2$

c **Exer. 67–68: Use the following formulas with $h = 0.01$ to approximate $f_{xx}(0.6, 0.8)$ and $f_{yy}(0.6, 0.8)$.**

$$f_{xx}(x, y) \approx \frac{f(x + h, y) - 2f(x, y) + f(x - h, y)}{h^2}$$

$$f_{yy}(x, y) \approx \frac{f(x, y + h) - 2f(x, y) + f(x, y - h)}{h^2}$$

67 $f(x, y) = \sec^2[\tan(xy^2)] \sin(xy)$

68 $f(x, y) = \dfrac{x^3 + xy^2}{\tan(xy) + 4x^2y^3}$

16.4 INCREMENTS AND DIFFERENTIALS

If f is a function of two variables x and y, then the symbols Δx and Δy will denote increments of x and y. Note that Δy is an increment of the *independent* variable y and is not the same as that defined in (3.26) for a *dependent* variable y. In terms of this increment notation, Definition (16.8) may be written

$$f_x(x, y) = \lim_{\Delta x \to 0} \frac{f(x + \Delta x, y) - f(x, y)}{\Delta x}$$

$$f_y(x, y) = \lim_{\Delta y \to 0} \frac{f(x, y + \Delta y) - f(x, y)}{\Delta y}.$$

The increment of the *dependent* variable $w = f(x, y)$ is defined as follows.

Definition **(16.13)**

Let $w = f(x, y)$, and let Δx and Δy be increments of x and y, respectively. The **increment Δw of $w = f(x, y)$** is

$$\Delta w = f(x + \Delta x, y + \Delta y) - f(x, y).$$

The increment Δw represents the change in the function value if (x, y) *changes to* $(x + \Delta x, y + \Delta y)$. If we regard $f(x, y)$ as the temperature at the point (x, y) in a metal plate D, as shown in Figure 16.32, then Δw is the net change in temperature in going from (x, y) to $(x + \Delta x, y + \Delta y)$.

FIGURE 16.32

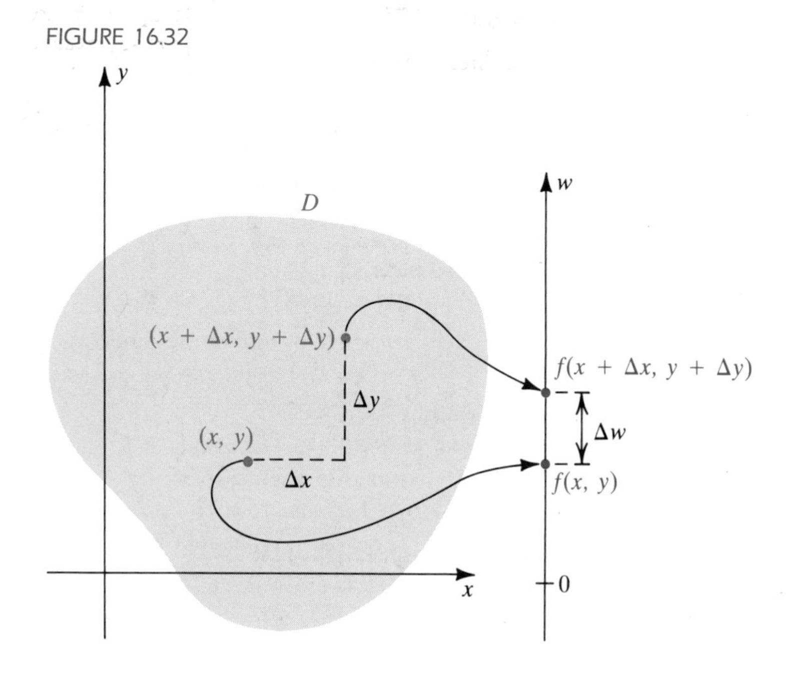

EXAMPLE 1 Let $w = f(x, y) = 3x^2 - xy$.

(a) If Δx and Δy are increments of x and y, find Δw.

(b) Use Δw to calculate the change in $f(x, y)$ if (x, y) changes from $(1, 2)$ to $(1.01, 1.98)$.

FIGURE 16.33

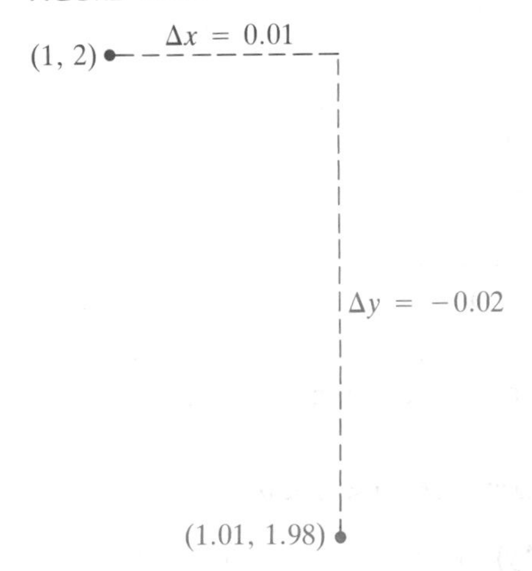

SOLUTION

(a) Using Definition (16.13), we have

$$\begin{aligned}\Delta w &= f(x + \Delta x, y + \Delta y) - f(x, y)\\ &= [3(x + \Delta x)^2 - (x + \Delta x)(y + \Delta y)] - (3x^2 - xy)\\ &= [3x^2 + 6x\,\Delta x + 3(\Delta x)^2 - (xy + x\,\Delta y + y\,\Delta x + \Delta x\,\Delta y)] - 3x^2 + xy\\ &= 6x\,\Delta x + 3(\Delta x)^2 - x\,\Delta y - y\,\Delta x - \Delta x\,\Delta y.\end{aligned}$$

(b) If (x, y) changes from $(1, 2)$ to $(1.01, 1.98)$, we have the situation shown in Figure 16.33. Substituting $x = 1$, $y = 2$, $\Delta x = 0.01$, and $\Delta y = -0.02$ into the formula for Δw gives us

$$\begin{aligned}\Delta w &= 6(1)(0.01) + 3(0.01)^2 - (1)(-0.02) - 2(0.01) - (0.01)(-0.02)\\ &= 0.0605.\end{aligned}$$

This number can also be found by calculating $f(1.01, 1.98) - f(1, 2)$.

The formula for Δw in Definition (16.13) is used only for finding differences in function values between two points. It is not well suited for establishing other results about the variation of $f(x, y)$. The following theorem provides a formula that is more useful for applications. To simplify the statement, we shall use the abbreviations ϵ_1 and ϵ_2 for certain functions of Δx and Δy, as well as for their *values* $\epsilon_1(\Delta x, \Delta y)$ and $\epsilon_2(\Delta x, \Delta y)$ at $(\Delta x, \Delta y)$.

Theorem **(16.14)**

Let $w = f(x, y)$, where the function f is defined on a rectangular region $R = \{(x, y): a < x < b, c < y < d\}$. Suppose f_x and f_y exist throughout R and are continuous at the point (x_0, y_0) in R. If $(x_0 + \Delta x, y_0 + \Delta y)$ is in R and

$$\Delta w = f(x_0 + \Delta x, y_0 + \Delta y) - f(x_0, y_0),$$

then

$$\Delta w = f_x(x_0, y_0)\,\Delta x + f_y(x_0, y_0)\,\Delta y + \epsilon_1\,\Delta x + \epsilon_2\,\Delta y,$$

where ϵ_1 and ϵ_2 are functions of Δx and Δy that have the limit 0 as $(\Delta x, \Delta y) \to (0, 0)$.

PROOF The graph of R and the points we wish to consider are illustrated in Figure 16.34.

Let us rewrite the increment Δw by subtracting and adding $f(x_0, y_0 + \Delta y)$ as follows:

(1)
$$\begin{aligned}\Delta w &= f(x_0 + \Delta x, y_0 + \Delta y) - f(x_0, y_0)\\ &= [f(x_0 + \Delta x, y_0 + \Delta y) - f(x_0, y_0 + \Delta y)]\\ &\qquad + [f(x_0, y_0 + \Delta y) - f(x_0, y_0)].\end{aligned}$$

FIGURE 16.34

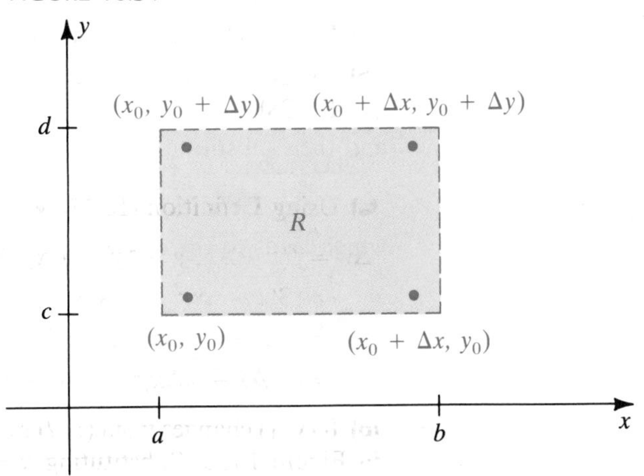

We next consider Δy as a constant and define a function g of *one* variable x by letting

$$g(x) = f(x, y_0 + \Delta y) \quad \text{for} \quad a < x < b.$$

It follows that $g'(x) = f_x(x, y_0 + \Delta y)$ for $a < x < b$. Applying the mean value theorem (4.12) to g on the interval $[x_0, x_0 + \Delta x]$, we obtain

$$g(x_0 + \Delta x) - g(x_0) = g'(u)\,\Delta x,$$

where u is between x_0 and $x_0 + \Delta x$. Since $g(x) = f(x, y_0 + \Delta y)$ and $g'(x) = f_x(x, y_0 + \Delta y)$, the last equation can be written

(2) $$f(x_0 + \Delta x, y_0 + \Delta y) - f(x_0, y_0 + \Delta y) = f_x(u, y_0 + \Delta y)\,\Delta x.$$

Similarly, if we let $h(y) = f(x_0, y)$ for $c < y < d$, then $h'(y) = f_y(x_0, y)$. Applying the mean value theorem to the function h on the interval $[y_0, y_0 + \Delta y]$, we obtain

$$h(y_0 + \Delta y) - h(y_0) = h'(v)\,\Delta y,$$

where v is between y_0 and $y_0 + \Delta y$, or, equivalently,

(3) $$f(x_0, y_0 + \Delta y) - f(x_0, y_0) = f_y(x_0, v)\,\Delta y.$$

Substituting (2) and (3) into the expression for Δw in (1) gives us

(4) $$\Delta w = f_x(u, y_0 + \Delta y)\,\Delta x + f_y(x_0, v)\,\Delta y.$$

FIGURE 16.35

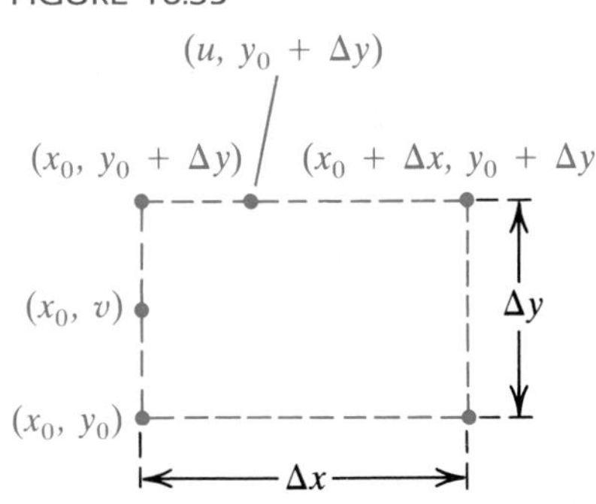

Typical points corresponding to $(u, y_0 + \Delta y)$ and (x_0, v) are shown in Figure 16.35.

Let us define ϵ_1 and ϵ_2 as follows:

$$\epsilon_1 = f_x(u, y_0 + \Delta y) - f_x(x_0, y_0)$$
$$\epsilon_2 = f_y(x_0, v) - f_y(x_0, y_0)$$

Using the fact that f_x and f_y are continuous and noting from Figure 16.35 that $u \to x_0$ as $\Delta x \to 0$ and $v \to y_0$ as $\Delta y \to 0$, we can show that ϵ_1 and ϵ_2 have the limit 0 as $(\Delta x, \Delta y) \to (0, 0)$. If we write the preceding equations for ϵ_1 and ϵ_2 in the form

$$f_x(u, y_0 + \Delta y) = f_x(x_0, y_0) + \epsilon_1$$
$$f_y(x_0, v) = f_y(x_0, y_0) + \epsilon_2$$

and then substitute into (4), the expression for Δw, we see that

$$\Delta w = [f_x(x_0, y_0) + \epsilon_1]\,\Delta x + [f_y(x_0, y_0) + \epsilon_2]\,\Delta y,$$

which leads to the conclusion of the theorem. ■

EXAMPLE 2 If $w = 3x^2 - xy$, find expressions for ϵ_1 and ϵ_2 that satisfy the conclusion of Theorem (16.14) with $(x_0, y_0) = (x, y)$.

SOLUTION From the solution of Example 1,

$$\Delta w = 6x\,\Delta x + 3(\Delta x)^2 - x\,\Delta y - y\,\Delta x - \Delta x\,\Delta y,$$

or, equivalently,

$$\Delta w = (6x - y)\,\Delta x + (-x)\,\Delta y + (3\,\Delta x)(\Delta x) + (-\Delta x)\,\Delta y.$$

This is of the form given in Theorem (16.14) with

$$f_x(x, y) = 6x - y, \quad f_y(x, y) = -x, \quad \epsilon_1 = 3\,\Delta x, \quad \epsilon_2 = -\Delta x.$$

Note that ϵ_1 and ϵ_2 are not unique, since we may also write

$$\Delta w = (6x - y)\,\Delta x + (-x)\,\Delta y + (3\,\Delta x - \Delta y)\,\Delta x + (0)\,\Delta y,$$

in which case $\epsilon_1 = 3\,\Delta x - \Delta y$ and $\epsilon_2 = 0$.

In Chapter 3 we defined differentials for functions of one variable. Recall from (3.27) that if $u = f(x)$ and Δx is an increment of x, then $dx = \Delta x$ and $du = f'(x)\,dx$. The next definition extends this concept to functions of two variables.

Definition (16.15)

Let $w = f(x, y)$, and let Δx and Δy be increments of x and y, respectively.

(i) The **differentials** $\boldsymbol{dx}$ **and** $\boldsymbol{dy}$ of the independent variables x and y are

$$dx = \Delta x \quad \text{and} \quad dy = \Delta y.$$

(ii) The **differential** $\boldsymbol{dw}$ of the dependent variable w is

$$dw = f_x(x, y)\,dx + f_y(x, y)\,dy = \frac{\partial w}{\partial x}\,dx + \frac{\partial w}{\partial y}\,dy.$$

If f satisfies the hypotheses of Theorem (16.14), then using the conclusion of that theorem, with the pair (x_0, y_0) replaced by (x, y), we see that

$$\Delta w = dw + \epsilon_1\,\Delta x + \epsilon_2\,\Delta y.$$

Hence

$$\Delta w - dw = \epsilon_1\,\Delta x + \epsilon_2\,\Delta y,$$

where both ϵ_1 and ϵ_2 approach 0 as $(\Delta x, \Delta y) \to (0, 0)$. It follows that if Δx and Δy are small, then $\Delta w - dw \approx 0$; that is, $dw \approx \Delta w$. We may use this fact to approximate the change in w corresponding to small changes in x and y.

EXAMPLE 3 If $w = 3x^2 - xy$, find dw and use it to approximate the change in w if (x, y) changes from $(1, 2)$ to $(1.01, 1.98)$. How does this compare with the exact change in w?

SOLUTION This is the same function considered in Examples 1 and 2. Applying Definition (16.15) yields

$$\begin{aligned} dw &= \frac{\partial w}{\partial x}\,dx + \frac{\partial w}{\partial y}\,dy \\ &= (6x - y)\,dx + (-x)\,dy. \end{aligned}$$

Substituting $x = 1$, $y = 2$, $dx = \Delta x = 0.01$, and $dy = \Delta y = -0.02$, we obtain

$$dw = (6 - 2)(0.01) + (-1)(-0.02) = 0.06.$$

In Example 1 we showed that $\Delta w = 0.0605$. Hence the error involved in using dw is 0.0005.

EXAMPLE 4 The radius and altitude of a right circular cylinder are measured as 3 inches and 8 inches, respectively, with a possible error in measurement of ± 0.05 inch. Use differentials to approximate the maximum error in the calculated volume of the cylinder.

FIGURE 16.36

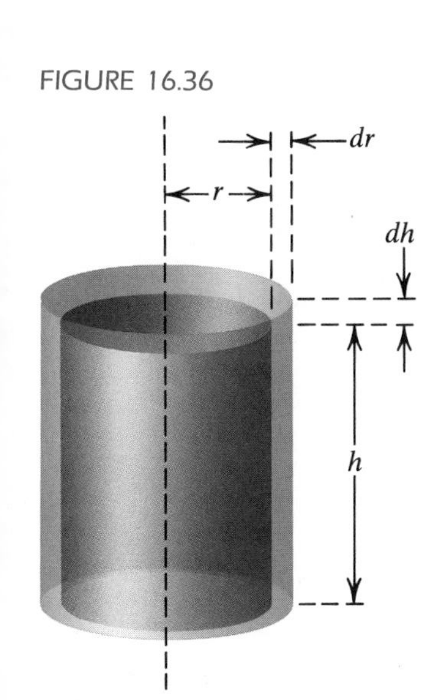

SOLUTION The method is analogous to that used for functions of one variable. Thus, we begin by considering the *general* formula $V = \pi r^2 h$ for the volume V of a cylinder of radius r and altitude h. We regard r and h as measured values, with maximum errors in measurement dr and dh, respectively. The case where both dr and dh are positive is illustrated in Figure 16.36. Of course, the error in one or both measurements could be negative. The error in the calculated volume is the change in V that corresponds to dr and dh. Using differentials, we have, by our previous remarks and Definition (16.15)(ii),

$$\Delta V \approx dV = \frac{\partial V}{\partial r}\,dr + \frac{\partial V}{\partial h}\,dh = 2\pi rh\,dr + \pi r^2\,dh.$$

Finally, we substitute specific values for the variables. Letting $r = 3$, $h = 8$, and $dr = dh = \pm 0.05$ gives us the following approximation to the maximum error:

$$dV = 48\pi(\pm 0.05) + 9\pi(\pm 0.05) = \pm 2.85\pi \approx \pm 8.95 \text{ in.}^3$$

Recall from Definition (3.32) that if a measurement w has an error (approximated by) dw, then the *average error* is approximately dw/w. The *percentage error* is the average error multiplied by 100%. In the next example we estimate the percentage error caused by errors in two related measurements.

EXAMPLE 5 A weather balloon is released at sea level, a distance R from the eye of a hurricane, and its altitude increases as it moves toward the eye. The altitude h that the balloon will attain at the eye may be approximated by

$$h = \pi^2 g \frac{R^4}{C^2},$$

where g is a gravitational constant and C is a meteorological measurement called the *circulation* of the wind velocity. (Circulation is closely related to the intensity and direction of the winds inside the hurricane.) If the maximum percentage errors in the measurements of R and h are $\pm 2\%$ and $\pm 5\%$, respectively, estimate the maximum percentage error in the calculation of C.

SOLUTION Solving the given formula for C, we obtain

$$C = \pi g^{1/2} \frac{R^2}{h^{1/2}}.$$

If the errors in measurement of R and h are dR and dh, then the error in C may be approximated by

$$\begin{aligned} dC &= \frac{\partial C}{\partial R}\, dR + \frac{\partial C}{\partial h}\, dh \\ &= \pi g^{1/2} \frac{2R}{h^{1/2}}\, dR - \pi g^{1/2} \frac{1}{2} \frac{R^2}{h^{3/2}}\, dh. \end{aligned}$$

We are interested in percentage errors, so we divide both sides of this equation by $C = \pi g^{1/2} R^2/h^{1/2}$, obtaining

$$\frac{dC}{C} = 2\frac{dR}{R} - \frac{1}{2}\frac{dh}{h}.$$

The maximum percentage error will occur when dR and dh have opposite signs. Thus, if $dR/R = 2\%$ and $dh/h = -5\%$,

$$\frac{dC}{C} = 2(2\%) - \frac{1}{2}(-5\%) = 6.5\%.$$

If $dR/R = -2\%$ and $dh/h = 5\%$,

$$\frac{dC}{C} = 2(-2\%) - \frac{1}{2}(5\%) = -6.5\%.$$

For a function of one variable, the term *differentiable* means that the derivative exists. For functions of two variables, we use the condition stated in the next definition, which is based on the formula for Δw in Theorem (16.14).

Definition **(16.16)**

Let $w = f(x, y)$. The function f is **differentiable** at (x_0, y_0) if Δw can be expressed in the form

$$\Delta w = f_x(x_0, y_0)\,\Delta x + f_y(x_0, y_0)\,\Delta y + \epsilon_1\,\Delta x + \epsilon_2\,\Delta y,$$

where ϵ_1 and ϵ_2 are functions of Δx and Δy that have the limit 0 as $(\Delta x, \Delta y) \to (0, 0)$.

A function f of two variables is said to be **differentiable on a region *R*** if it is differentiable at every point of R. The next theorem follows directly from Theorem (16.14) and Definition (16.16). The term *rectangular region* means a region of the type described in (16.14).

Theorem **(16.17)**

If $w = f(x, y)$ and if f_x and f_y are continuous on a rectangular region R, then f is differentiable on R.

The following result shows that a differentiable function is continuous, and it is one of the reasons for using Definition (16.16) as the definition of differentiability.

Theorem **(16.18)**

If a function f of two variables is differentiable at (x_0, y_0), then f is continuous at (x_0, y_0).

PROOF Let Δx and Δy be increments. We can write Δw in the following two ways, using (16.13) and (16.16):

$$\Delta w = f(x_0 + \Delta x, y_0 + \Delta y) - f(x_0, y_0)$$

$$\Delta w = [f_x(x_0, y_0) + \epsilon_1]\,\Delta x + [f_y(x_0, y_0) + \epsilon_2]\,\Delta y$$

If we equate these expressions and let $x = x_0 + \Delta x$, $y = y_0 + \Delta x$, then

$$f(x, y) - f(x_0, y_0) = [f_x(x_0, y_0) + \epsilon_1](x - x_0) + [f_y(x_0, y_0) + \epsilon_2](y - y_0).$$

It follows that

$$\lim_{(x,y)\to(x_0,y_0)} [f(x, y) - f(x_0, y_0)] = 0,$$

or, equivalently,

$$\lim_{(x,y)\to(x_0,y_0)} f(x, y) = f(x_0, y_0).$$

Hence f is continuous at (x_0, y_0). ■

If f_x and f_y are continuous, then, by Theorem (16.17), f is differentiable. This gives us the following corollary to Theorem (16.18).

Corollary **(16.19)**

If f is a function of two variables and if f_x and f_y are continuous on a rectangular region R, then f is continuous on R.

We can show by means of examples that the mere *existence* of f_x and f_y is not enough to ensure continuity of f (see Exercise 41). This is different from the single-variable case, where the existence of f' implies continuity of f.

The preceding discussion can be extended to functions of more than two variables. For example, suppose $w = f(x, y, z)$, where f is defined on a suitable region R (such as a rectangular box) and f_x, f_y, f_z exist in R and are continuous at (x, y, z). If x, y, and z are given increments Δx, Δy, and Δz, respectively, then the corresponding increment

$$\Delta w = f(x + \Delta x, y + \Delta y, z + \Delta z) - f(x, y, z)$$

can be written in the form

$$\begin{aligned}\Delta w = f_x(x, y, z)\,\Delta x + f_y(x, y, z)\,\Delta y + f_z(x, y, z)\,\Delta z \\ + \epsilon_1\,\Delta x + \epsilon_2\,\Delta y + \epsilon_3\,\Delta z,\end{aligned}$$

where ϵ_1, ϵ_2, and ϵ_3 are functions of Δx, Δy, and Δz that have the limit 0 as $(\Delta x, \Delta y, \Delta z) \to (0, 0, 0)$. Results that are analogous to (16.17)–(16.19) may be established for functions of three variables.

The following definition is an extension of (16.15) to functions of three variables.

Definition **(16.20)**

Let $w = f(x, y, z)$, and let Δx, Δy, and Δz be increments of x, y, and z, respectively.

(i) The **differentials** dx, dy, and dz of the independent variables x, y, and z are

$$dx = \Delta x, \quad dy = \Delta y, \quad dz = \Delta z.$$

(ii) The **differential** dw of the dependent variable w is

$$dw = \frac{\partial w}{\partial x}\,dx + \frac{\partial w}{\partial y}\,dy + \frac{\partial w}{\partial z}\,dz.$$

We can use dw to approximate Δw if the increments of x, y, and z are small. Differentiability is defined as in (16.16). The extension to four or more variables is made in similar fashion.

EXAMPLE 6 Suppose the dimensions (in inches) of a rectangular box change from 9, 6, and 4 to 9.02, 5.97, and 4.01, respectively

(a) Use differentials to approximate the change in volume.

(b) Find the exact change in volume.

SOLUTION

(a) The method of solution is similar to that used for the right circular cylinder in Example 4. Thus, we begin with the general formula $V = xyz$ for the volume of a rectangular box of measured dimensions x, y, and z. We next regard dx, dy, and dz as errors in measurement. The error in the calculated volume is then

$$\begin{aligned}\Delta V \approx dV &= \frac{\partial V}{\partial x}\,dx + \frac{\partial V}{\partial y}\,dy + \frac{\partial V}{\partial z}\,dz \\ &= yz\,dx + xz\,dy + xy\,dz.\end{aligned}$$

Finally, we substitute the given values $x = 9$, $y = 6$, $z = 4$, $dx = 0.02$, $dy = -0.03$, and $dz = 0.01$, obtaining

$$\begin{aligned} dV &= 24(0.02) + 36(-0.03) + 54(0.01) \\ &= 0.48 - 1.08 + 0.54 = -0.06. \end{aligned}$$

Thus, the volume decreases by approximately 0.06 in.^{3}

(b) The exact change in volume is

$$\Delta V = (9.02)(5.97)(4.01) - (9)(6)(4) = -0.063906 \text{ in.}^3$$

In precalculus we study methods of solving systems of linear equations, such as

$$\begin{cases} 2x - 3y = 4 \\ 5x + 2y = 9 \end{cases} \quad \text{or} \quad \begin{cases} x + 3y - z = 4 \\ 2x - y + 3z = 1 \\ x + 2y - z = -5 \end{cases}$$

In applications it is often necessary to solve a system of *nonlinear* equations, such as

FIGURE 16.37

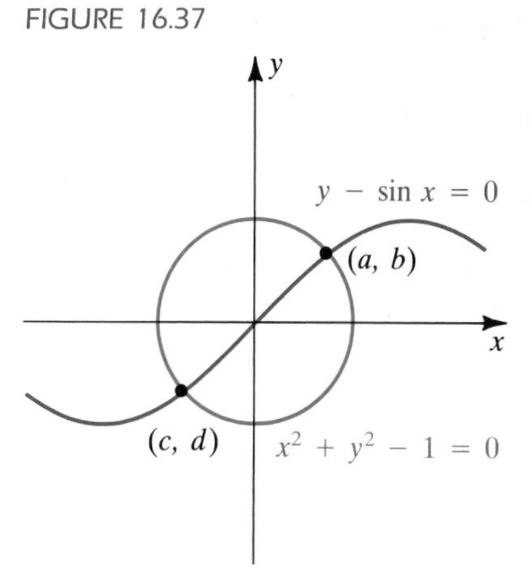

$$\begin{cases} x^2 + y^2 - 1 = 0 \\ y - \sin x = 0 \end{cases}$$

Since we cannot find exact solutions of this system using algebraic techniques, we must develop methods for *approximating* solutions. We could graph both equations on the same coordinate axes, as in Figure 16.37, and estimate the points of intersection (a, b) and (c, d); however, for most applications we need more accuracy. We shall next discuss a method based on differentials that can be used to approximate solutions to any degree of accuracy.

Let us consider a general system of two equations in x and y:

$$\begin{cases} f(x, y) = 0 \\ g(x, y) = 0 \end{cases}$$

FIGURE 16.38

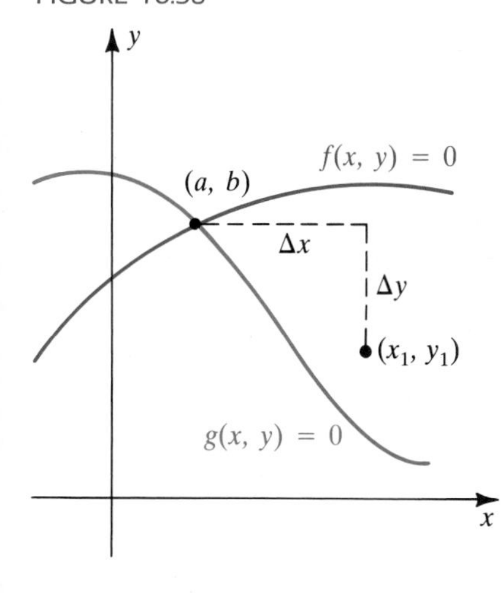

where f and g are functions of two variables. (In the previous discussion $f(x, y) = x^2 + y^2 - 1$ and $g(x, y) = y - \sin x$.) We wish to approximate a point of intersection (a, b) of the graphs of the two equations (see Figure 16.38). Let us begin with a *first approximation* (x_1, y_1), where, as indicated in the figure, there are increments Δx and Δy such that

$$a = x_1 + \Delta x, \quad b = y_1 + \Delta y.$$

Let Δf and Δg be the increments of f and g corresponding to the change from (x_1, y_1) to (a, b). Since $f(a, b) = 0$ and $g(a, b) = 0$, we see that

$$\begin{aligned} \Delta f &= f(a, b) - f(x_1, y_1) = -f(x_1, y_1) \\ \Delta g &= g(a, b) - g(x_1, y_1) = -g(x_1, y_1). \end{aligned}$$

Thus, if df and dg are the differentials of f and g evaluated at (x_1, y_1), then

$$df \approx -f(x_1, y_1) \quad \text{and} \quad dg \approx -g(x_1, y_1).$$

Using the form for df and dg given by Definition (16.15)(ii) gives us the following system of *approximate equalities* in the unknowns Δx and Δy:

$$\text{(1)} \qquad \begin{cases} f_x(x_1, y_1)\,\Delta x + f_y(x_1, y_1)\,\Delta y \approx -f(x_1, y_1) \\ g_x(x_1, y_1)\,\Delta x + g_y(x_1, y_1)\,\Delta y \approx -g(x_1, y_1) \end{cases}$$

In practice this system is often written in *matrix form*, as follows, where all functions are evaluated at (x_1, y_1):

$$\text{(2)} \qquad \begin{bmatrix} f_x & f_y \\ g_x & g_y \end{bmatrix} \begin{bmatrix} \Delta x \\ \Delta y \end{bmatrix} \approx \begin{bmatrix} -f \\ -g \end{bmatrix}$$

We solve the approximate equalities in (1) for Δx and Δy by treating $\approx$ as $=$, which gives us *approximations* for Δx and Δy. Although $(x_2, y_2) = (x_1 + \Delta x, y_1 + \Delta y)$ will not usually be the *exact* solution (a, b), it is often a better approximation than (x_1, y_1). The process can be repeated using the approximation (x_2, y_2) instead of (x_1, y_1) in (1) to obtain a third approximation, (x_3, y_3). We continue the process until $|\Delta x| \le \epsilon$ and $|\Delta y| \le \epsilon$ for some positive error tolerance $\epsilon > 0$. The convergence of x_k and y_k to a and b, respectively, is usually rapid, and the number of correct digits approximately doubles with each step. This method of successive approximations is known as **Newton's method** for a system of two equations in two unknowns. It can be extended to systems of n equations in n unknowns. This is one of the most widely used methods for approximating solutions of systems of nonlinear equations.

EXAMPLE 7 Use Newton's method to approximate the solutions of the following system:

$$\begin{cases} x^2 + y^2 - 1 = 0 \\ y - \sin x = 0 \end{cases}$$

SOLUTION The graphs of the two equations are sketched in Figure 16.37. We shall take $(x_1, y_1) = (0.5, 0.5)$ as a first approximation for the solution corresponding to the point of intersection in the first quadrant. Letting $f(x, y) = x^2 + y^2 - 1$ and $g(x, y) = y - \sin x$, we obtain

$$f_x(x, y) = 2x, \quad f_y(x, y) = 2y, \quad g_x(x, y) = -\cos x, \quad g_y(x, y) = 1.$$

We now let $x_1 = 0.5$ and $y_1 = 0.5$ in (1) and solve for (the approximate values) Δx and Δy. We then take, as a second approximation,

$$(x_2, y_2) = (x_1 + \Delta x, y_1 + \Delta y).$$

Using (x_2, y_2) in place of (x_1, y_1) in (1), we find (approximate values for) Δx_1 and Δy_1 and take, as a third approximation,

$$(x_3, y_3) = (x_2 + \Delta x_1, y_2 + \Delta y_1).$$

The process continues until we reach the desired accuracy. The following table lists eight-decimal-place approximations to the values used in equations (1). Note that if $k = 5$, we have $|\Delta x| \le \epsilon$ and $|\Delta y| \le \epsilon$ with $\epsilon = 0.000000005$. Hence the solution (x_5, y_5) is accurate to eight decimal places.

k	x_k	y_k	$f(x_k, y_k)$	$g(x_k, y_k)$
1	0.5	0.5	−0.5	0.02057446
2	0.77725783	0.72274217	0.12648598	0.02141484
3	0.74030063	0.67498279	0.00364679	0.00047290
4	0.73908622	0.67361333	0.00000335	0.00000050
5	0.73908513	0.67361203	0.00000000	0.00000000

$f_x(x_k, y_k)$	$f_y(x_k, y_k)$	$g_x(x_k, y_k)$	$g_y(x_k, y_k)$	Δx_k	Δy_k
1	1	−0.87758256	1	0.27725783	0.22274217
1.55451565	1.44548435	−0.71283938	1	−0.03695719	−0.04775939
1.48060127	1.34996558	−0.73826581	1	−0.00121442	−0.00136946
1.47817243	1.34722665	−0.73908440	1	−0.00000108	−0.00000130
1.47817027	1.34722406	−0.73908513	1	0.00000000	0.00000000

From the table,

$$(x_5, y_5) \approx (0.73908513, 0.67361203).$$

An approximation to the solution corresponding to the point in the third quadrant in Figure 16.37 is, by symmetry, $(-x_5, -y_5)$.

EXERCISES 16.4

Exer. 1–2: If Δx and Δy are increments of x and y, find (a) Δw, (b) dw, and (c) $dw - \Delta w$.

1 $w = 5y^2 - xy$

2 $w = xy - y^2 + 3x$

Exer. 3–6: Find expressions for ϵ_1 and ϵ_2 that satisfy Definition (16.16).

3 $f(x, y) = 4y^2 - 3xy + 2x$

4 $f(x, y) = (2x - y)^2$

5 $f(x, y) = x^3 + y^3$

6 $f(x, y) = 2x^2 - xy^2 + 3y$

Exer. 7–18: Find dw.

7 $w = x^3 - x^2y + 3y^2$

8 $w = 5x^2 + 4y - 3xy^3$

9 $w = x^2 \sin y + 2y^{3/2}$

10 $w = ye^{-2x} - 3x^4$

11 $w = x^2e^{xy} + (1/y^2)$

12 $w = \ln(x^2 + y^2) + x \tan^{-1} y$

13 $w = x^2 \ln(y^2 + z^2)$

14 $w = x^2y^3z + e^{-2z}$

15 $w = \dfrac{xyz}{x + y + z}$

16 $w = x^2e^{yz} + y \ln z$

17 $w = x^2z + 4yt^3 - xz^2t$

18 $w = x^2y^3zt^{-1}v^4$

Exer. 19–22: Use differentials to approximate the change in f if the independent variables change as indicated.

19 $f(x, y) = x^2 - 3x^3y^2 + 4x - 2y^3 + 6$; $(-2, 3)$ to $(-2.02, 3.01)$

20 $f(x, y) = x^2 - 2xy + 3y$; $(1, 2)$ to $(1.03, 1.99)$

21 $f(x, y, z) = x^2z^3 - 3yz^2 + x^{-3} + 2y^{1/2}z$; $(1, 4, 2)$ to $(1.02, 3.97, 1.96)$

22 $f(x, y, z) = xy + xz + yz$; $(-1, 2, 3)$ to $(-0.98, 1.99, 3.03)$

23 The dimensions of a closed rectangular box are measured as 3 feet, 4 feet, and 5 feet, with a possible error of $\pm\frac{1}{16}$ inch in each measurement. Use differentials to approximate the maximum error in the calculated value of

(a) the surface area (b) the volume

24 The two shortest sides of a right triangle are measured as 3 centimeters and 4 centimeters, respectively, with a possible error of ± 0.02 centimeter in each measurement. Use differentials to approximate the maximum error in the calculated value of

(a) the hypotenuse

(b) the area of the triangle

25 The *withdrawal resistance* of a nail indicates its holding strength in wood. An empirical formula used for bright, common nails is $P = 15{,}700S^{5/2}RD$, where P is the maximum withdrawal resistance in pounds, S is the specific gravity of the wood at 12% moisture content, R is the radius of the nail (in inches), and D is the depth (in inches) that the nail has penetrated the wood. A 6d bright, common nail of length 2 inches and diameter 0.113 inch is driven completely into a piece of Douglas fir that has a specific gravity of 0.54.

(a) Approximate the maximum withdrawal resistance. (In applications only one-sixth of this maximum is considered safe for extended periods of time.)

(b) When nails are manufactured, R and D can vary by $\pm 2\%$, and the specific gravity of different samples of Douglas fir can vary by $\pm 3\%$. Approximate the maximum percentage error in the calculated value of P.

26 The total resistance R of three resistances R_1, R_2, and R_3 connected in parallel is given by

$$\frac{1}{R} = \frac{1}{R_1} + \frac{1}{R_2} + \frac{1}{R_3}.$$

If measurements of R_1, R_2, and R_3 are 100, 200, and 400 ohms, respectively, with a maximum error of $\pm 1\%$ in each measurement, approximate the maximum error in the calculated value of R.

27 The specific gravity of an object more dense than water is given by $s = A/(A - W)$, where A and W are the weights (in pounds) of the object in air and water, respectively. If measurements are $A = 12$ lb and $W = 5$ lb, with maximum errors of $\pm\frac{1}{2}$ ounce in air and ± 1 ounce in water, what is the maximum error in the calculated value of s?

28 The pressure P, volume V, and temperature T (in °K) of a confined gas are related by the ideal gas law $PV = kT$, where k is a constant. If $P = 0.5$ lb/in.^{2} when $V = 64$ in.^{3} and $T = 350$ °K, approximate the change in P if V and T change to 70 in.^{3} and 345 °K, respectively.

29 Suppose that when the specific gravity formula $s = A/(A - W)$ is used (see Exercise 27), there are percentage errors of $\pm 2\%$ and $\pm 4\%$ in the measurements of A and W, respectively. Express the maximum percentage error in the calculated value of s as a function of A and W.

30 Suppose that when the ideal gas law $PV = kT$ is used (see Exercise 28), there are percentage errors of $\pm 0.8\%$ and $\pm 0.5\%$ in the measurements of T and P, respectively. Approximate the maximum percentage error in the calculated value of V.

31 The electrical resistance R of a wire is directly proportional to its length and inversely proportional to the square of its diameter. If the length is measured with a possible error of $\pm 1\%$ and the diameter is measured with a possible error of $\pm 3\%$, what is the maximum percentage error in the calculated value of R?

32 The flow of blood through an arteriole is given by $F = \pi PR^4/(8vl)$, where l is the length of the arteriole, R is the radius, P is the pressure difference between the two ends, and v is the viscosity of the blood (see Section 6.8). Suppose that v and l are constant. Use differentials to approximate the percentage change in the blood flow if the radius decreases by 2% and the pressure increases by 3%.

33 The temperature T at the point $P(x, y, z)$ in an xyz-coordinate system is given by $T = 8(2x^2 + 4y^2 + 9z^2)^{1/2}$. Use differentials to approximate the temperature difference between the points (6, 3, 2) and (6.1, 3.3, 1.98).

34 Approximate the change in area of an isosceles triangle if each of the two equal sides increases from 100 to 101 and the angle between them decreases from 120° to 119°.

35 If a mountaintop is viewed from the point P shown in the figure, the angle of elevation is α. From a point Q that is a distance x units closer to the mountain, the angle of elevation is β. From trigonometry, the height h of the mountain is given by

$$h = \frac{x}{\cot \alpha - \cot \beta}.$$

A surveyor measures α and β to an accuracy of 30″ (approximately 0.000145 radian). Suppose $\alpha = 15°$, $\beta = 20°$, and $x = 2000$ ft. Use differentials to estimate, to the nearest 0.1 foot, how accurate the length measurement must be so that the maximum error in the calculated value of h is no greater than ± 10 feet.

EXERCISE 35

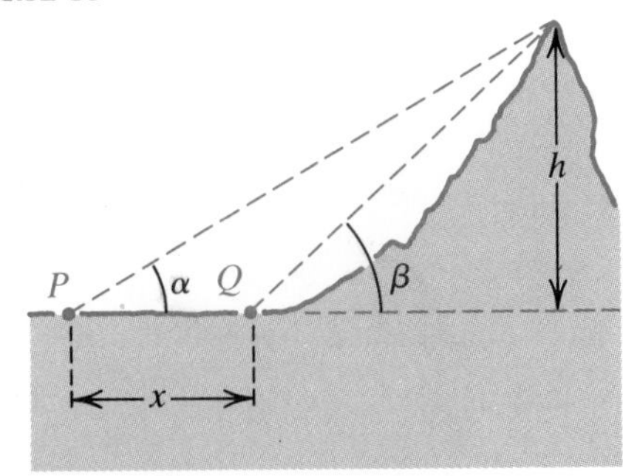

36 If a drug is taken orally, the time T at which the largest amount of drug is in the bloodstream can be calculated using the half-life x of the drug in the stomach and the half-life y of the drug in the bloodstream. For many common drugs (such as penicillin), T is given by

$$T = \frac{xy(\ln x - \ln y)}{(x - y)\ln 2}.$$

For a certain drug, $x = 30$ min and $y = 1$ hr. If the maximum error in estimating each half-life is $\pm 10\%$, find the maximum error in the calculated value of T.

37 Assume the cylinder described in Example 4 has a closed top and bottom. Use differentials to approximate the maximum error in the calculated total surface area.

38 Use differentials to approximate the change in surface area of the box described in Example 6. What is the exact change?

Exer. 39–40: Prove that f is differentiable throughout its domain.

39 $f(x, y) = \dfrac{x^2 - y^2}{x^2 + y^2}$

40 $f(x, y, z) = \dfrac{x + y + z}{x^2 + y^2 + z^2}$

41 Let $f(x, y) = \begin{cases} \dfrac{xy}{x^2 + y^2} & \text{if } (x, y) \neq (0, 0) \\ 0 & \text{if } (x, y) = (0, 0) \end{cases}$

(a) Prove that $f_x(0, 0)$ and $f_y(0, 0)$ exist. (*Hint:* Use Definition (16.8).)

(b) Prove that f is not continuous at $(0, 0)$.

(c) Prove that f is not differentiable at $(0, 0)$.

42 Let $f(x, y, z) = \begin{cases} \dfrac{xyz}{x^3 + y^3 + z^3} & \text{if } (x, y, z) \neq (0, 0, 0) \\ 0 & \text{if } (x, y, z) = (0, 0, 0) \end{cases}$

(a) Prove that f_x, f_y, and f_z exist at $(0, 0, 0)$.

(b) Prove that f is not differentiable at $(0, 0, 0)$.

c **Exer. 43–44: Refer to Example 7. Use Newton's method to approximate a solution of the system of equations to four decimal places using the given first approximation.**

43 $\begin{cases} \dfrac{x^2}{4} + \dfrac{y^2}{9} = 1 \\ \dfrac{(x-1)^2}{10} + \dfrac{(y+1)^2}{5} = 1 \end{cases}$ $\quad (x_1, y_1) = (2, 1)$

44 $\begin{cases} \dfrac{x^2}{9} + \dfrac{y^2}{4} = 1 \\ \dfrac{(x-1)^2}{2} - \dfrac{(y-1)^2}{3} = 1 \end{cases}$ $\quad (x_1, y_1) = (-1.5, -1.5)$

45 Derive Newton's method for a system of three equations,

$$\begin{cases} f(x, y, z) = 0 \\ g(x, y, z) = 0 \\ h(x, y, z) = 0 \end{cases}$$

where f, g, and h are functions of three variables. Express the solution in a matrix form similar to (2) on page 832.

16.5 CHAIN RULES

If f and g are functions of *one* variable such that

$$w = f(u) \quad \text{and} \quad u = g(x),$$

then the composite function of f and g is given by

$$w = f(g(x)).$$

Applying chain rule (3.33), we may find the derivative of w with respect to x as follows:

$$\frac{dw}{dx} = \frac{dw}{du}\frac{du}{dx}.$$

In this section we shall extend this formula to functions of several variables.

Let f, g, and h be functions of two variables such that

$$w = f(u, v), \quad \text{with} \quad u = g(x, y), \quad v = h(x, y).$$

If, for each pair (x, y) in a subset D of $\mathbb{R}^2$, the corresponding pair (u, v) is in the domain of f, then

$$w = f(g(x, y), h(x, y))$$

defines w as a (composite) function of x and y with domain D. For example, if

$$w = u^2 + u \sin v, \quad \text{with} \quad u = xe^{2y},\ v = xy,$$

then $$w = x^2e^{4y} + xe^{2y} \sin xy.$$

The next theorem provides formulas for expressing $\partial w/\partial x$ and $\partial w/\partial y$ in terms of the first partial derivatives of the functions g, h, and f. In the statement of the theorem we have assumed that domains are chosen so that the composite function is defined on a suitable domain D. Each of the formulas stated in Theorem (16.21) is referred to as a *chain rule*.

Chain rules (16.21)

If $w = f(u, v)$, with $u = g(x, y)$, $v = h(x, y)$, and if f, g, and h are differentiable, then

$$\frac{\partial w}{\partial x} = \frac{\partial w}{\partial u}\frac{\partial u}{\partial x} + \frac{\partial w}{\partial v}\frac{\partial v}{\partial x}$$

$$\frac{\partial w}{\partial y} = \frac{\partial w}{\partial u}\frac{\partial u}{\partial y} + \frac{\partial w}{\partial v}\frac{\partial v}{\partial y}.$$

PROOF If x is given an increment Δx and y is held constant (that is, $\Delta y = 0$), we obtain the following increments of u and v:

$$\text{(1)} \qquad \begin{aligned} \Delta u &= g(x + \Delta x, y) - g(x, y) \\ \Delta v &= h(x + \Delta x, y) - h(x, y) \end{aligned}$$

These in turn lead to the following increment of w:

$$\Delta w = f(u + \Delta u, v + \Delta v) - f(u, v)$$

Since f is differentiable, we have, by Definition (16.16),

$$\text{(2)} \qquad \Delta w = \frac{\partial w}{\partial u}\Delta u + \frac{\partial w}{\partial v}\Delta v + \epsilon_1 \Delta u + \epsilon_2 \Delta v,$$

where ϵ_1 and ϵ_2 are functions of Δu and Δv that have the limit 0 as $(\Delta u, \Delta v) \to (0, 0)$. Moreover, we may assume that both ϵ_1 and ϵ_2 equal 0 if $(\Delta u, \Delta v) = (0, 0)$, because if they do not, they can be replaced by functions μ_1 and μ_2 that have this property and are equal to ϵ_1 and ϵ_2 elsewhere. With this agreement, the functions ϵ_1 and ϵ_2 in (2) are continuous at $(0, 0)$. Dividing both sides of equation (2) by Δx gives us

$$\text{(3)} \qquad \frac{\Delta w}{\Delta x} = \frac{\partial w}{\partial u}\frac{\Delta u}{\Delta x} + \frac{\partial w}{\partial v}\frac{\Delta v}{\Delta x} + \epsilon_1 \frac{\Delta u}{\Delta x} + \epsilon_2 \frac{\Delta v}{\Delta x}.$$

If we regard w as a function of x and y, then

$$\lim_{\Delta x \to 0} \frac{\Delta w}{\Delta x} = \frac{\partial w}{\partial x}.$$

Also, from equations (1),

$$\lim_{\Delta x \to 0} \frac{\Delta u}{\Delta x} = \frac{\partial u}{\partial x} \quad \text{and} \quad \lim_{\Delta x \to 0} \frac{\Delta v}{\Delta x} = \frac{\partial v}{\partial x}.$$

If Δx approaches 0, we see from (1) that Δu and Δv also approach 0 and hence so do ϵ_1 and ϵ_2. Consequently, if we take the limit in equation (3) as $\Delta x \to 0$, we obtain

$$\frac{\partial w}{\partial x} = \frac{\partial w}{\partial u}\frac{\partial u}{\partial x} + \frac{\partial w}{\partial v}\frac{\partial v}{\partial x}.$$

The second formula in the statement of the theorem is established in similar fashion. ■

FIGURE 16.39

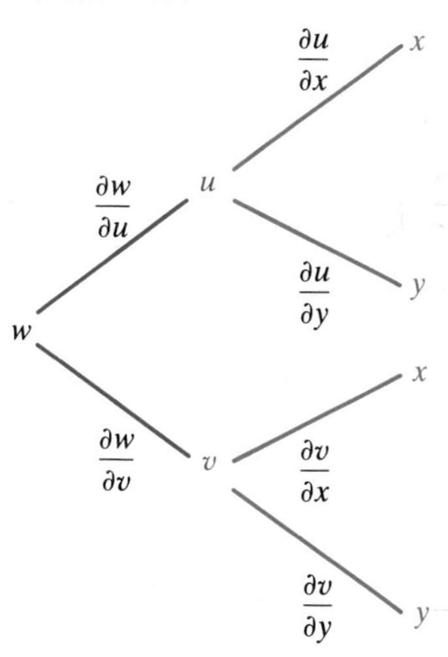

The *tree diagram* in Figure 16.39 is a device for remembering chain rules (16.21). We first draw *branches* (line segments) from w to u and v, to indicate that w is a function of those two variables. Since u is a function of x and y, we next draw branches from u to x and y. Similarly, branches are drawn from v to x and y. In this diagram we have also indicated the various partial derivatives that are involved for the given variables.

To find $\partial w/\partial x$ by means of the diagram, we consider all pairs of branches (paths) that lead from w to x. There are two possibilities:

$$w \xrightarrow{\frac{\partial w}{\partial u}} u \xrightarrow{\frac{\partial u}{\partial x}} x \quad \text{and} \quad w \xrightarrow{\frac{\partial w}{\partial v}} v \xrightarrow{\frac{\partial v}{\partial x}} x$$

We next add the products of the corresponding pairs of partial derivatives, to obtain

$$\frac{\partial w}{\partial x} = \frac{\partial w}{\partial u}\frac{\partial u}{\partial x} + \frac{\partial w}{\partial v}\frac{\partial v}{\partial x}.$$

The formula for $\partial w/\partial y$ may be found in a like manner, using the pairs of branches that lead from w to y.

FIGURE 16.40

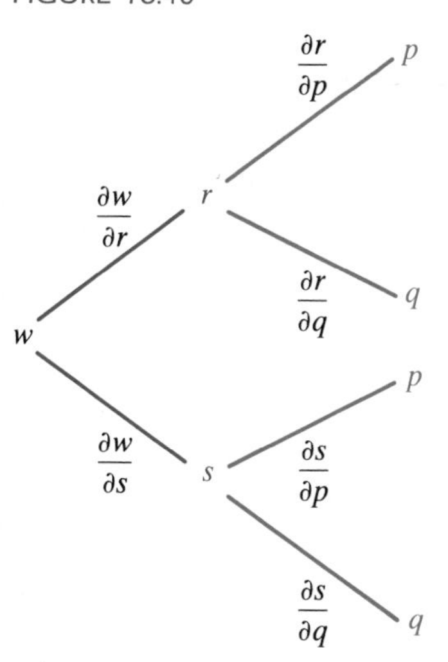

EXAMPLE 1 Use a chain rule to find $\partial w/\partial p$ and $\partial w/\partial q$ if

$$w = r^3 + s^2, \quad \text{with} \quad r = pq^2, \quad s = p^2 \sin q.$$

SOLUTION Note that w is a (composite) function of p and q. We could substitute for r and s, obtaining

$$w = (pq^2)^3 + (p^2 \sin q)^2,$$

and then find $\partial w/\partial p$ and $\partial w/\partial q$ directly; however, we wish to illustrate chain rules. Moreover, in some cases such substitutions lead to very complicated expressions. The procedure here is analogous to procedures used in computer programming languages to express a complex computation as a succession of simpler ones.

Since w is a function of r and s and both r and s are functions of p and q, we begin by constructing the tree diagram in Figure 16.40. Referring to the diagram, we take the products of all pairs of partial derivatives that lead from w to p, obtaining

$$\frac{\partial w}{\partial p} = \frac{\partial w}{\partial r}\frac{\partial r}{\partial p} + \frac{\partial w}{\partial s}\frac{\partial s}{\partial p}$$
$$= (3r^2)(q^2) + (2s)(2p \sin q).$$

We now substitute $r = pq^2$ and $s = p^2 \sin q$:

$$\frac{\partial w}{\partial p} = 3(pq^2)^2(q^2) + (2p^2 \sin q)(2p \sin q)$$
$$= 3p^2q^6 + 4p^3 \sin^2 q$$

To find $\partial w/\partial q$, we again refer to the diagram in Figure 16.40:

$$\frac{\partial w}{\partial q} = \frac{\partial w}{\partial r}\frac{\partial r}{\partial q} + \frac{\partial w}{\partial s}\frac{\partial s}{\partial q}$$
$$= (3r^2)(2pq) + (2s)(p^2 \cos q)$$

Substituting for r and s, we obtain

$$\frac{\partial w}{\partial q} = 3(pq^2)^2(2pq) + 2(p^2 \sin q)(p^2 \cos q)$$
$$= 6p^3q^5 + 2p^4 \sin q \cos q.$$

FIGURE 16.41

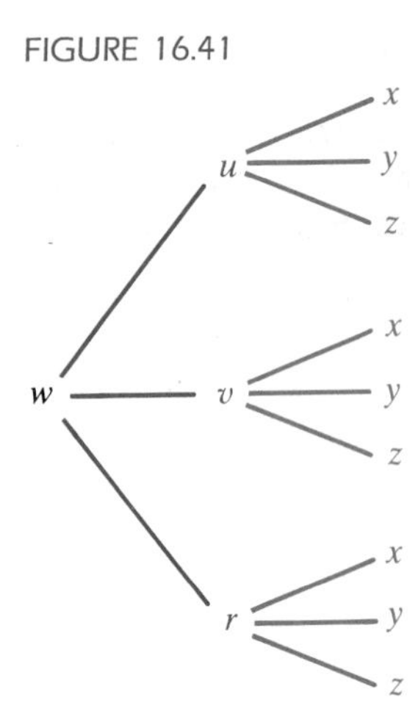

FIGURE 16.42

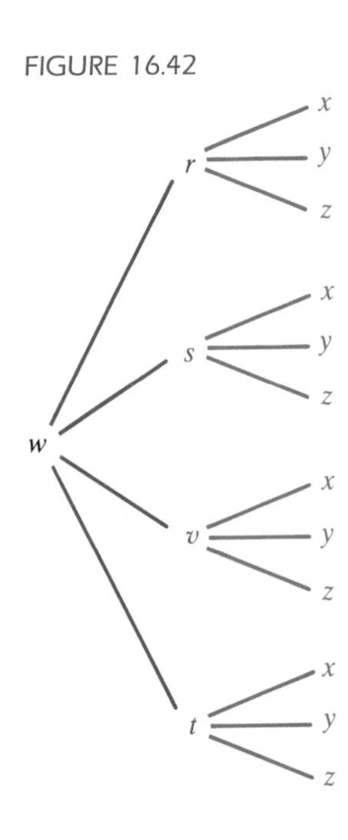

Note that after applying a chain rule in Example 1 we substituted for r and s, thereby expressing $\partial w/\partial p$ and $\partial w/\partial q$ in terms of p and q. This was done to emphasize the fact that w is a (composite) function of the *two* variables p and q.

We can apply chain rules to composite functions of any number of variables and construct tree diagrams to help formulate these rules. Hereafter, we shall not attach partial derivative symbols to the branches. It will be understood that if a branch leads from a variable w to another variable r, as in Figure 16.40, then the partial derivative is $\partial w/\partial r$. However, if w is a function of only *one* variable r, then we write dw/dr instead of $\partial w/\partial r$.

As an illustration, suppose w is a function of u, v, and r, and that u, v, and r are each functions of x, y, and z. This is indicated by the diagram in Figure 16.41. If we wish to find $\partial w/\partial y$, we take the pairs of products of the partials that lead from w to y and add, obtaining

$$\frac{\partial w}{\partial y} = \frac{\partial w}{\partial u}\frac{\partial u}{\partial y} + \frac{\partial w}{\partial v}\frac{\partial v}{\partial y} + \frac{\partial w}{\partial r}\frac{\partial r}{\partial y}.$$

EXAMPLE 2 Use a chain rule to find $\partial w/\partial z$ if

$$w = r^2 + sv + t^3, \quad \text{with} \quad r = x^2 + y^2 + z^2, \quad s = xyz, \quad v = xe^y, \quad t = yz^2.$$

SOLUTION Note that w is a function of r, s, v, and t and that each of these four variables is a function of x, y, and z. An appropriate tree diagram is shown in Figure 16.42. We wish to find $\partial w/\partial z$, so we trace all branches that lead from w to z. This gives us

$$\begin{aligned}\frac{\partial w}{\partial z} &= \frac{\partial w}{\partial r}\frac{\partial r}{\partial z} + \frac{\partial w}{\partial s}\frac{\partial s}{\partial z} + \frac{\partial w}{\partial v}\frac{\partial v}{\partial z} + \frac{\partial w}{\partial t}\frac{\partial t}{\partial z} \\ &= (2r)(2z) + v(xy) + s(0) + (3t^2)(2yz) \\ &= 4z(x^2 + y^2 + z^2) + xe^y(xy) + 0 + 3(yz^2)^2(2yz) \\ &= 4z(x^2 + y^2 + z^2) + x^2ye^y + 6y^3z^5.\end{aligned}$$

If w is a function of several variables, each of which is a function of only *one* variable, say t, then w is a function of the one variable t and we may consider dw/dt. A chain rule may be applied, as in the next example.

EXAMPLE 3 Use a chain rule to find dw/dt if

$$w = x^2 + yz, \quad \text{with} \quad x = 3t^2 + 1, \quad y = 2t - 4, \quad z = t^3.$$

FIGURE 16.43

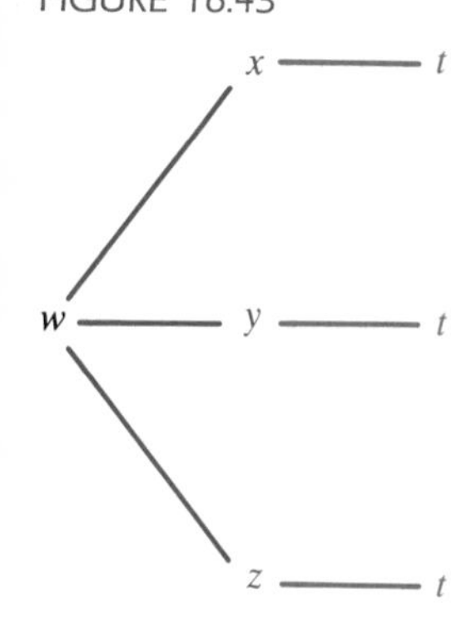

SOLUTION To apply a chain rule, we construct the tree diagram in Figure 16.43. The pairs of branches that lead from w to t give us the following, where we use d/dt for differentiation with respect to the *single* variable t:

$$\begin{aligned}\frac{dw}{dt} &= \frac{\partial w}{\partial x}\frac{dx}{dt} + \frac{\partial w}{\partial y}\frac{dy}{dt} + \frac{\partial w}{\partial z}\frac{dz}{dt} \\ &= (2x)(6t) + z(2) + y(3t^2) \\ &= 2(3t^2 + 1)6t + t^3(2) + (2t - 4)3t^2 \\ &= 44t^3 - 12t^2 + 12t\end{aligned}$$

The problem could also be solved without a chain rule by writing

$$w = (3t^2 + 1)^2 + (2t - 4)t^3$$

and then finding dw/dt by single-variable methods.

Chain rules are useful for solving related rate problems, as illustrated in the next example.

EXAMPLE 4 A simple electrical circuit consists of a resistor R and an electromotive force V. At a certain instant V is 80 volts and is increasing at a rate of 5 volts/min, while R is 40 ohms and is decreasing at a rate of 2 ohms/min. Use Ohm's law, $I = V/R$, and a chain rule to find the rate at which the current I (in amperes) is changing.

FIGURE 16.44

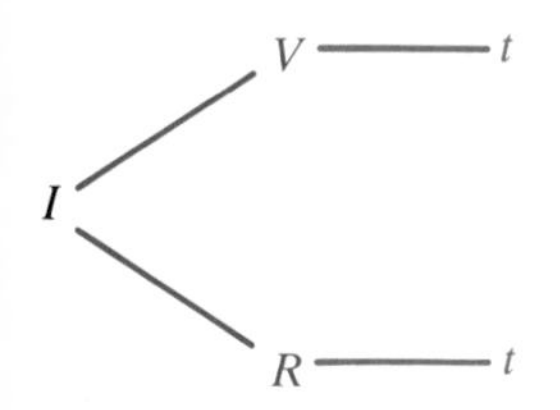

SOLUTION Since I is a function of V and R and both V and R are functions of time t (in minutes), we consider the tree diagram in Figure 16.44. Applying a chain rule gives us

$$\begin{aligned}\frac{dI}{dt} &= \frac{\partial I}{\partial V}\frac{dV}{dt} + \frac{\partial I}{\partial R}\frac{dR}{dt} \\ &= \left(\frac{1}{R}\right)\frac{dV}{dt} + \left(-\frac{V}{R^2}\right)\frac{dR}{dt}.\end{aligned}$$

Substituting

$$V = 80, \quad \frac{dV}{dt} = 5, \quad R = 40, \quad \text{and} \quad \frac{dR}{dt} = -2,$$

we obtain

$$\frac{dI}{dt} = \left(\frac{1}{40}\right)(5) + \left(-\frac{80}{1600}\right)(-2) = \frac{9}{40} = 0.225 \text{ amp/min.}$$

Partial derivatives can be used to find derivatives of functions that are determined implicitly. Suppose, as in Section 3.7, an equation $F(x, y) = 0$ determines a differentiable function f such that $y = f(x)$; that is, $F(x, f(x)) = 0$ for every x in the domain D of f. Let us introduce the following composite function F:

$$w = F(u, y), \quad \text{where} \quad u = x \quad \text{and} \quad y = f(x)$$

FIGURE 16.45

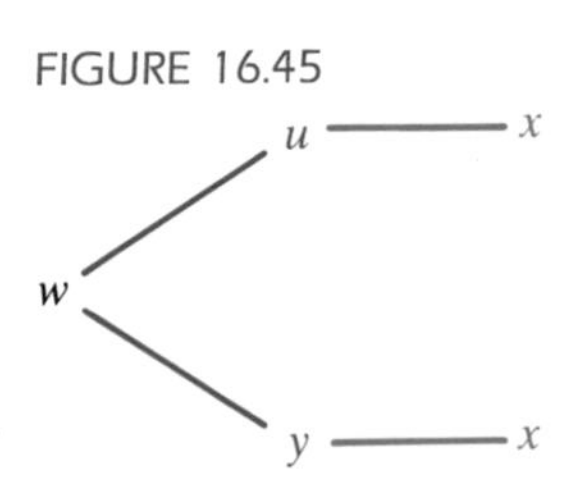

This leads to the tree diagram in Figure 16.45. Using a chain rule and the fact that u and y are functions of *one* variable x yields

$$\frac{dw}{dx} = \frac{\partial w}{\partial u}\frac{du}{dx} + \frac{\partial w}{\partial y}\frac{dy}{dx}.$$

Since $w = F(x, f(x)) = 0$ for every x, it follows that $dw/dx = 0$. Moreover, since $u = x$ and $y = f(x)$,

$$\frac{du}{dx} = 1 \quad \text{and} \quad \frac{dy}{dx} = f'(x).$$

Substituting in the preceding chain rule formula for dw/dx, we obtain

$$0 = \frac{\partial w}{\partial u}(1) + \frac{\partial w}{\partial y} f'(x).$$

If $\partial w/\partial y \neq 0$, then (since $u = x$)

$$f'(x) = -\frac{\partial w/\partial u}{\partial w/\partial y} = -\frac{\partial w/\partial x}{\partial w/\partial y} = -\frac{F_x(x, y)}{F_y(x, y)}.$$

We may summarize the preceding discussion as follows.

Theorem **(16.22)**

If an equation $F(x, y) = 0$ determines, implicitly, a differentiable function f of one variable x such that $y = f(x)$, then

$$\frac{dy}{dx} = -\frac{F_x(x, y)}{F_y(x, y)}.$$

EXAMPLE 5 Find dy/dx if $y = f(x)$ is determined implicitly by

$$y^4 + 3y - 4x^3 - 5x - 1 = 0.$$

SOLUTION If $F(x, y)$ is the expression on the left side of the equation, then, by Theorem (16.22),

$$\frac{dy}{dx} = -\frac{-12x^2 - 5}{4y^3 + 3} = \frac{12x^2 + 5}{4y^3 + 3}.$$

Compare this solution with that of Example 2 in Section 3.7, which was obtained using single-variable methods.

Given the equation

$$x^2 - 4y^3 + 2z - 7 = 0,$$

we can solve for z, obtaining

$$z = \tfrac{1}{2}(-x^2 + 4y^3 + 7),$$

which is of the form

$$z = f(x, y).$$

In analogy with the single-variable case, we say that the function f of two variables x and y is determined *implicitly* by the given equation. The next theorem gives us formulas for finding f_x and f_y, or, equivalently, $\partial z/\partial x$ and $\partial z/\partial y$, without actually solving the equation for z.

Theorem **(16.23)**

> If an equation $F(x, y, z) = 0$ determines an implicit differentiable function f of two variables x and y such that $z = f(x, y)$ for every (x, y) in the domain of f, then
>
> $$\frac{\partial z}{\partial x} = -\frac{F_x(x, y, z)}{F_z(x, y, z)}, \qquad \frac{\partial z}{\partial y} = -\frac{F_y(x, y, z)}{F_z(x, y, z)}.$$

PROOF The statement $F(x, y, z) = 0$ *determines a function f such that* $z = f(x, y)$ means that $F(x, y, f(x, y)) = 0$ for every (x, y) in the domain of f. Consider the composite function F of x and y defined as follows:

FIGURE 16.46

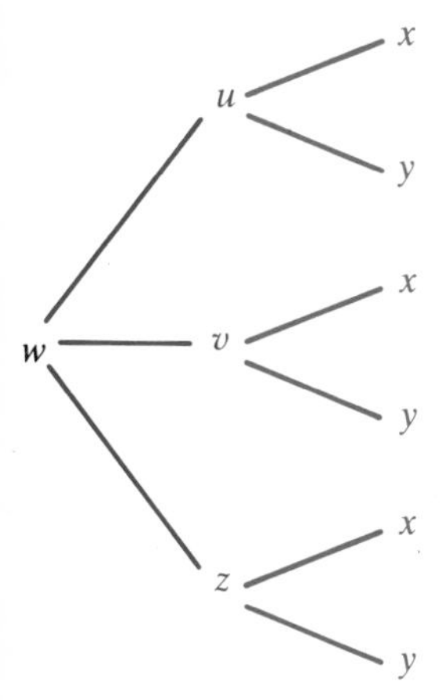

$$w = F(u, v, z), \quad \text{where} \quad u = x, \quad v = y, \quad z = f(x, y).$$

Note that u and v are functions of x and y, since we can write $u = x + (0 \cdot y)$ and $v = y + (0 \cdot x)$. Referring to the diagram in Figure 16.46 and considering the branches that lead from w to x, we obtain

$$\frac{\partial w}{\partial x} = \frac{\partial w}{\partial u}\frac{\partial u}{\partial x} + \frac{\partial w}{\partial v}\frac{\partial v}{\partial x} + \frac{\partial w}{\partial z}\frac{\partial z}{\partial x}.$$

Since $w = F(x, y, f(x, y)) = 0$ for every x and every y, it follows that $\partial w/\partial x = 0$. Moreover, since $\partial u/\partial x = 1$ and $\partial v/\partial x = 0$, our chain rule formula for $\partial w/\partial x$ may be written

$$0 = \frac{\partial w}{\partial x}(1) + \frac{\partial w}{\partial y}(0) + \frac{\partial w}{\partial z}\frac{\partial z}{\partial x},$$

and if $\partial w/\partial z \neq 0$,

$$\frac{\partial z}{\partial x} = -\frac{\partial w/\partial x}{\partial w/\partial z} = -\frac{F_x(x, y, z)}{F_z(x, y, z)}.$$

The formula for $\partial z/\partial y$ may be obtained in similar fashion. ■

EXAMPLE 6 Find $\partial z/\partial x$ and $\partial z/\partial y$ if $z = f(x, y)$ is determined implicitly by

$$x^2z^2 + xy^2 - z^3 + 4yz - 5 = 0.$$

SOLUTION If we let $F(x, y, z)$ denote the expression on the left of the given equation, then, by Theorem (16.23),

$$\frac{\partial z}{\partial x} = -\frac{2xz^2 + y^2}{2x^2z - 3z^2 + 4y}$$

$$\frac{\partial z}{\partial y} = -\frac{2xy + 4z}{2x^2z - 3z^2 + 4y}.$$

EXERCISES 16.5

Use a chain rule in Exercises 1–14.

Exer. 1–2: Find $\partial w/\partial x$ and $\partial w/\partial y$.

1 $w = u \sin v$; $u = x^2 + y^2$, $v = xy$

2 $w = uv + v^2$; $u = x \sin y$, $v = y \sin x$

Exer. 3–4: Find $\partial w/\partial r$ and $\partial w/\partial s$.

3 $w = u^2 + 2uv$; $u = r \ln s$, $v = 2r + s$

4 $w = e^{tv}$; $t = r + s$, $v = rs$

Exer. 5–6: Find $\partial z/\partial x$ and $\partial z/\partial y$.

5 $z = r^3 + s + v^2$; $r = xe^y$, $s = ye^x$, $v = x^2y$

6 $z = pq + qw$; $p = 2x - y$, $q = x - 2y$, $w = -2x + 2y$

Exer. 7–8: Find $\partial r/\partial u$, $\partial r/\partial v$, and $\partial r/\partial t$.

7 $r = x \ln y$; $x = 3u + vt$, $y = uvt$

8 $r = w^2 \cos z$; $w = u^2vt$, $z = ut^2$

9 If $p = u^2 + 3v^2 - 4w^2$, where $u = x - 3y + 2r - s$, $v = 2x + y - r + 2s$, and $w = -x + 2y + r + s$, find $\partial p/\partial r$.

10 Find $\partial s/\partial y$ if $s = tr + ue^v$, where $t = xy^2z$, $r = x^2yz$, $u = xyz^2$, and $v = xyz$.

Exer. 11–14: Find dw/dt.

11 $w = x^3 - y^3$; $x = \dfrac{1}{t+1}$, $y = \dfrac{t}{t+1}$

12 $w = \ln(u + v)$; $u = e^{-2t}$, $v = t^3 - t^2$

13 $w = r^2 - s \tan v$; $r = \sin^2 t$, $s = \cos t$, $v = 4t$

14 $w = x^2y^3z^4$; $x = 2t + 1$, $y = 3t - 2$, $z = 5t + 4$

Exer. 15–18: Use partial derivatives to find dy/dx if $y = f(x)$ is determined implicitly by the given equation.

15 $2x^3 + x^2y + y^3 = 1$

16 $x^4 + 2x^2y^2 - 3xy^3 + 2x = 0$

17 $6x + \sqrt{xy} = 3y - 4$

18 $x^{2/3} + y^{2/3} = 4$

Exer. 19–22: Find $\partial z/\partial x$ and $\partial z/\partial y$ if $z = f(x, y)$ is determined implicitly by the given equation.

19 $2xz^3 - 3yz^2 + x^2y^2 + 4z = 0$

20 $xz^2 + 2x^2y - 4y^2z + 3y - 2 = 0$

21 $xe^{yz} - 2ye^{xz} + 3ze^{xy} = 1$

22 $yx^2 + z^2 + \cos xyz = 4$

Exer. 23–32: Use a chain rule.

23 The radius r and altitude h of a right circular cylinder are increasing at rates of 0.01 in./min and 0.02 in./min, respectively.

(a) Find the rate at which the volume is increasing at the time when $r = 4$ in. and $h = 7$ in.

(b) At what rate is the curved surface area changing at this time?

24 The equal sides and the included angle of an isosceles triangle are increasing at rates of 0.1 ft/hr and 2°/hr, respectively. Find the rate at which the area of the triangle is increasing at the time when the length of each of the equal sides is 20 feet and the included angle is 60°.

25 The pressure P, volume V, and temperature T of a confined gas are related by the ideal gas law $PV = kT$, where k is a constant. If P and V are changing at the rates dP/dt and dV/dt, respectively, find a formula for dT/dt.

26 If the base radius r and altitude h of a right circular cylinder are changing at the rates dr/dt and dh/dt, respectively, find a formula for dV/dt, where V is the volume of the cylinder.

27 A certain gas obeys the ideal gas law $PV = 8T$. Suppose that the gas is being heated at a rate of 2°/min and the pressure is increasing at a rate of $\frac{1}{2}$ (lb/in.^{2})/min. If, at a certain instant, the temperature is 200° and the pressure is 10 lb/in.^{2}, find the rate at which the volume is changing.

28 Sand is leaking out of a hole in a container at a rate of 6 in.^{3}/min. As it leaks out, it forms a pile in the shape of a right circular cone whose base radius is increasing at a rate of $\frac{1}{4}$ in./min. If, at the instant that 40 in.^{3} has leaked out, the radius is 5 inches, find the rate at which the height of the pile is increasing.

29 At age 2 years, a typical boy is 86 centimeters tall, weighs 13 kilograms, and is growing at the rate of 9 cm/year and 2 kg/year. Use the DuBois and DuBois surface area formula, $S = 0.007184x^{0.425}y^{0.725}$ for weight x and height y, to estimate the rate at which the body surface area is growing (see Exercise 49 of Section 16.1).

30 When the size of the molecules and their forces of attraction are taken into account, the pressure P, volume V, and temperature T of a mole of confined gas are related by the *van der Waals equation*,

$$\left(P + \frac{a}{V^2}\right)(V - b) = kT,$$

where a, b, and k are positive constants. If t is time, find a formula for dT/dt in terms of dP/dt, dV/dt, P, and V.

31 If n resistances $R_1, R_2, \ldots, R_n$ are connected in parallel, then the total resistance R is given by

$$\frac{1}{R} = \sum_{k=1}^{n} \frac{1}{R_k}.$$

Prove that for $k = 1, 2, \ldots, n$,

$$\frac{\partial R}{\partial R_k} = \left(\frac{R}{R_k}\right)^2.$$

32 A function f of two variables is *homogeneous of degree* n if $f(tx, ty) = t^n f(x, y)$ for every t such that (tx, ty) is in the domain of f. Show that, for such functions, $xf_x(x, y) + yf_y(x, y) = nf(x, y)$. (*Hint:* Differentiate $f(tx, ty)$ with respect to t.)

Exer. 33–36: Refer to Exercise 32. Find the degree n of the homogeneous function f and verify the formula

$$xf_x(x, y) + yf_y(x, y) = nf(x, y).$$

33 $f(x, y) = 2x^3 + 3x^2y + y^3$

34 $f(x, y) = \dfrac{x^3y}{x^2 + y^2}$

35 $f(x, y) = \arctan \dfrac{y}{x}$

36 $f(x, y) = xye^{y/x}$

37 If $w = f(x, y)$, where $x = r \cos \theta$ and $y = r \sin \theta$, show that

$$\left(\frac{\partial w}{\partial x}\right)^2 + \left(\frac{\partial w}{\partial y}\right)^2 = \left(\frac{\partial w}{\partial r}\right)^2 + \frac{1}{r^2}\left(\frac{\partial w}{\partial \theta}\right)^2.$$

38 If $w = f(x, y)$, where $x = e^r \cos \theta$ and $y = e^r \sin \theta$, show that

$$\frac{\partial^2 w}{\partial x^2} + \frac{\partial^2 w}{\partial y^2} = e^{-2r}\left(\frac{\partial^2 w}{\partial r^2} + \frac{\partial^2 w}{\partial \theta^2}\right).$$

39 If $w = f(x, y)$, where $x = r \cos \theta$ and $y = r \sin \theta$, show that

$$\frac{\partial^2 w}{\partial x^2} + \frac{\partial^2 w}{\partial y^2} = \frac{\partial^2 w}{\partial r^2} + \frac{1}{r^2}\frac{\partial^2 w}{\partial \theta^2} + \frac{1}{r}\frac{\partial w}{\partial r}.$$

40 If $v = f(x - at) + g(x + at)$ and f and g have second partial derivatives, show that v satisfies the wave equation

$$\frac{\partial^2 v}{\partial t^2} = a^2 \frac{\partial^2 v}{\partial x^2}.$$

(Compare with Exercise 42 of Section 16.3.)

41 If $w = \cos(x + y) + \cos(x - y)$, show, without using addition formulas, that $w_{xx} - w_{yy} = 0$.

42 If $w = f(x^2 + y^2)$, show that $y(\partial w/\partial x) - x(\partial w/\partial y) = 0$. (*Hint:* Let $u = x^2 + y^2$.)

43 If $w = f(u, v)$, where $u = g(x, y)$ and $v = k(x, y)$, show that

$$\frac{\partial^2 w}{\partial x^2} = \frac{\partial^2 w}{\partial u^2}\left(\frac{\partial u}{\partial x}\right)^2 + \left(\frac{\partial^2 w}{\partial v\, \partial u} + \frac{\partial^2 w}{\partial u\, \partial v}\right)\frac{\partial u}{\partial x}\frac{\partial v}{\partial x} + \frac{\partial^2 w}{\partial v^2}\left(\frac{\partial v}{\partial x}\right)^2 + \frac{\partial w}{\partial u}\frac{\partial^2 u}{\partial x^2} + \frac{\partial w}{\partial v}\frac{\partial^2 v}{\partial x^2}.$$

44 For w, u, and v as given in Exercise 43, show that

$$\frac{\partial^2 w}{\partial y\, \partial x} = \frac{\partial^2 w}{\partial u^2}\frac{\partial u}{\partial x}\frac{\partial u}{\partial y} + \frac{\partial^2 w}{\partial v\, \partial u}\frac{\partial u}{\partial x}\frac{\partial v}{\partial y} + \frac{\partial^2 w}{\partial u\, \partial v}\frac{\partial u}{\partial y}\frac{\partial v}{\partial x} + \frac{\partial^2 w}{\partial v^2}\frac{\partial v}{\partial x}\frac{\partial v}{\partial y} + \frac{\partial w}{\partial u}\frac{\partial^2 u}{\partial y\, \partial x} + \frac{\partial w}{\partial v}\frac{\partial^2 v}{\partial y\, \partial x}.$$

45 Prove the following mean value theorem for a function f of two variables x and y:
If f has first partial derivatives that are continuous on a rectangular region $R = \{(x, y): a < x < b, c < y < d\}$ and if $A(x_1, y_1)$ and $B(x_2, y_2)$ are points in R, then there

is a point $P(x^*, y^*)$ on the line segment AB such that

$$f(x_2, y_2) - f(x_1, y_1) = f_x(x^*, y^*)(x_2 - x_1) + f_y(x^*, y^*)(y_2 - y_1).$$

46 Use Exercise 45 to prove the following:
If $f_x(x, y) = 0$ and $f_y(x, y) = 0$ for every (x, y) in a rectangular region R, then $f(x, y)$ is constant on R.

47 Extend Exercise 45 to a function of three variables.

48 Extend Exercise 46 to a function of three variables.

49 Suppose that $u = f(x, y)$ and $v = g(x, y)$ satisfy the Cauchy-Riemann equations $u_x = v_y$ and $u_y = -v_x$. If $x = r\cos\theta$ and $y = r\sin\theta$, show that

$$\frac{\partial u}{\partial r} = \frac{1}{r}\frac{\partial v}{\partial \theta} \quad \text{and} \quad \frac{\partial v}{\partial r} = -\frac{1}{r}\frac{\partial u}{\partial \theta}.$$

50 If $u = y/(x^2 + y^2)$ and $v = x/(x^2 + y^2)$, verify the formulas for $\partial u/\partial r$ and $\partial v/\partial r$ in Exercise 49 directly, by substituting $r\cos\theta$ for x and $r\sin\theta$ for y and then differentiating.

16.6 DIRECTIONAL DERIVATIVES

In Section 16.3 we discussed the fact that if $f(x, y)$ is the temperature of a flat metal plate at the point $P(x, y)$ in an xy-plane, then the partial derivatives $f_x(x, y)$ and $f_y(x, y)$ give us the instantaneous rates of change of temperature with respect to distance in the horizontal and vertical directions, respectively (see Figures 16.28 and 16.29). In this section we generalize this fact to the rate of change of $f(x, y)$ in *any* direction.

Let $\mathbf{u} = u_1\mathbf{i} + u_2\mathbf{j}$ be a unit vector. If we represent $\mathbf{u}$ by a vector with initial point $P(x, y)$, as in Figure 16.47(i), then the terminal point has coordinates $(x + u_1, y + u_2)$. We wish to define the rate of change of $f(x, y)$ with respect to distance in the direction determined by $\mathbf{u}$.

Consider the line l through P parallel to $\mathbf{u}$, and let Q be any point on l, as in Figure 16.47(ii). The vector $\overrightarrow{PQ}$ corresponds to a scalar multiple $s\mathbf{u} = (su_1)\mathbf{i} + (su_2)\mathbf{j}$ for some s. Hence the coordinates of Q are $(x + su_1, y + su_2)$. Since $\mathbf{u}$ is a unit vector,

$$\|\overrightarrow{PQ}\| = \|s\mathbf{u}\| = |s|\,\|\mathbf{u}\| = |s|.$$

Thus, s is the (signed) distance from P, measured along l. If $s > 0$, then $s\mathbf{u}$ has the same direction as $\mathbf{u}$. If $s < 0$, then $s\mathbf{u}$ has the opposite direction.

If the position of a point changes from P to Q, then the increment Δw of $w = f(x, y)$ is

$$\Delta w = f(x + su_1, y + su_2) - f(x, y).$$

The *average rate of change* in $f(x, y)$ is

$$\frac{\Delta w}{s} = \frac{f(x + su_1, y + su_2) - f(x, y)}{s}.$$

FIGURE 16.47

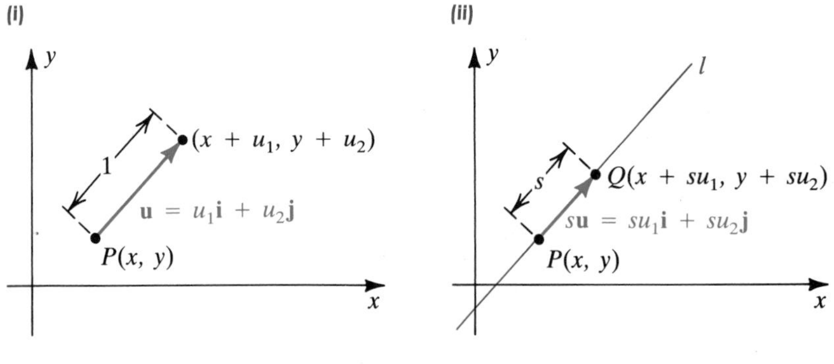

(i)

(ii)

To illustrate, suppose $w = f(x, y)$ is the temperature at (x, y). If the temperature at P is 50° and the temperature at Q is 51.5°, then $\Delta w = 1.5°$. If $\|\overrightarrow{PQ}\| = 3$ cm, then the average rate of change in temperature as a point moves from P to Q is

$$\frac{\Delta w}{s} = 0.5° \text{ per cm.}$$

As in our previous work with derivatives, to find the *instantaneous* rate of change of $f(x, y)$ at P in the direction determined by $\mathbf{u}$, we consider the limit of the average rate of change $\Delta w/s$ as $s \to 0$. This is the motivation for the following definition.

Definition **(16.24)**

Let $w = f(x, y)$, and let $\mathbf{u} = u_1\mathbf{i} + u_2\mathbf{j}$ be a unit vector. The **directional derivative of f at $P(x, y)$ in the direction of $\mathbf{u}$**, denoted by $D_{\mathbf{u}} f(x, y)$, is

$$D_{\mathbf{u}} f(x, y) = \lim_{s \to 0} \frac{f(x + su_1, y + su_2) - f(x, y)}{s}.$$

If $\mathbf{a}$ is any vector that has the same direction as $\mathbf{u}$, we shall also refer to $D_{\mathbf{u}} f(x, y)$ as the **directional derivative of f in the direction of $\mathbf{a}$.**

The following theorem provides a formula for finding directional derivatives.

Theorem **(16.25)**

If f is a differentiable function of two variables and $\mathbf{u} = u_1\mathbf{i} + u_2\mathbf{j}$ is a unit vector, then

$$D_{\mathbf{u}} f(x, y) = f_x(x, y)u_1 + f_y(x, y)u_2.$$

PROOF Consider x, y, u_1, and u_2 as fixed (but arbitrary), and let g be the function of *one* variable s defined by

$$g(s) = f(x + su_1, y + su_2).$$

From Definitions (3.6) and (16.24),

$$\begin{aligned} g'(0) &= \lim_{s \to 0} \frac{g(s) - g(0)}{s - 0} \\ &= \lim_{s \to 0} \frac{f(x + su_1, y + su_2) - f(x, y)}{s} \\ &= D_{\mathbf{u}} f(x, y). \end{aligned}$$

We next consider g as a composite function, with

$$w = g(s) = f(r, v) \quad \text{and} \quad r = x + su_1, \quad v = y + su_2.$$

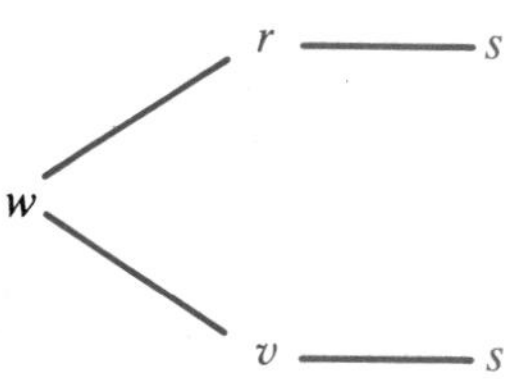

FIGURE 16.48

As indicated by the tree diagram in Figure 16.48, w is a function of r and v, and r and v are functions of *one* variable s. Applying a chain rule, we obtain

$$\frac{dw}{ds} = \frac{\partial w}{\partial r}\frac{dr}{ds} + \frac{\partial w}{\partial v}\frac{dv}{ds}.$$

Using the definitions of $w = g(s)$, r, and v, we may rewrite this as

$$g'(s) = f_r(r, v)u_1 + f_v(r, v)u_2.$$

If we let $s = 0$, then $r = x$, $v = y$, and, by the first part of the proof, $g'(0) = D_{\mathbf{u}} f(x, y)$. Hence

$$D_{\mathbf{u}} f(x, y) = f_x(x, y)u_1 + f_y(x, y)u_2. \quad ■$$

The first partial derivatives of f are special cases of the directional derivative $D_{\mathbf{u}} f(x, y)$. Thus, if we let $\mathbf{u} = \mathbf{i}$, then $u_1 = 1$, $u_2 = 0$, and the formula in (16.25) becomes

$$D_{\mathbf{i}} f(x, y) = f_x(x, y).$$

If $\mathbf{u} = \mathbf{j}$, then $u_1 = 0$, $u_2 = 1$, and we obtain

$$D_{\mathbf{j}} f(x, y) = f_y(x, y).$$

EXAMPLE 1 Let $f(x, y) = x^3y^2$.

(a) Find the directional derivative of f at the point $P(-1, 2)$ in the direction of the vector $\mathbf{a} = 4\mathbf{i} - 3\mathbf{j}$.

(b) Discuss the significance of part (a) if $f(x, y)$ is the temperature at (x, y).

SOLUTION

FIGURE 16.49

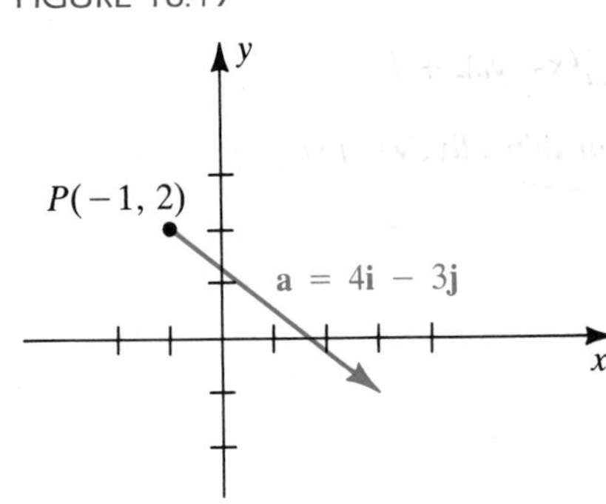

(a) The vector $\mathbf{a} = 4\mathbf{i} - 3\mathbf{j}$ with initial point $P(-1, 2)$ is represented in Figure 16.49. We wish to find $D_{\mathbf{u}} f(-1, 2)$ for the *unit* vector $\mathbf{u}$ that has the direction of $\mathbf{a}$. By Theorem (14.12),

$$\mathbf{u} = \frac{1}{\|\mathbf{a}\|}\mathbf{a} = \frac{1}{5}(4\mathbf{i} - 3\mathbf{j}) = \frac{4}{5}\mathbf{i} - \frac{3}{5}\mathbf{j}.$$

Since $\qquad f_x(x, y) = 3x^2y^2 \quad \text{and} \quad f_y(x, y) = 2x^3y,$

it follows from (16.25) that

$$D_{\mathbf{u}} f(x, y) = 3x^2y^2(\tfrac{4}{5}) + 2x^3y(-\tfrac{3}{5}).$$

Hence, at $P(-1, 2)$,

$$D_{\mathbf{u}} f(-1, 2) = 3(-1)^2(2)^2(\tfrac{4}{5}) + 2(-1)^3(2)(-\tfrac{3}{5}) = 12.$$

(b) If $f(x, y)$ is the temperature (in degrees) at (x, y), then $D_{\mathbf{u}} f(-1, 2) = 12$ tells us that if a point moves in the direction of $\mathbf{u}$, the temperature at P is increasing at the rate of 12° per unit change in distance. It is of interest to compare this value with $f_x(-1, 2) = 12$ and $f_y(-1, 2) = -4$, which are the rates of change in the horizontal and vertical directions, respectively.

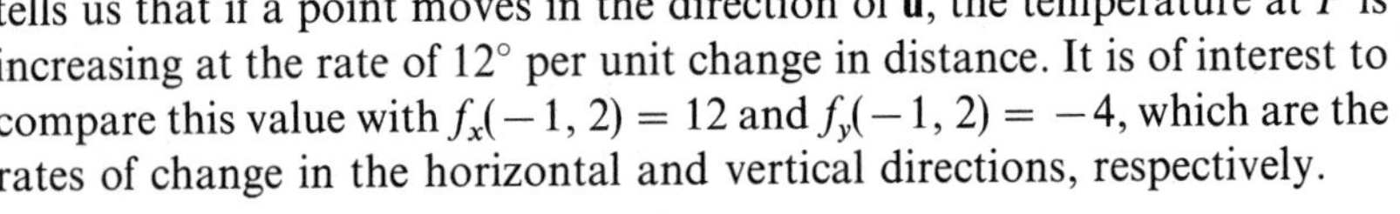

We may use Theorem (16.25) to express a directional derivative as a dot product of two vectors as follows:

$$D_{\mathbf{u}} f(x, y) = [f_x(x, y)\mathbf{i} + f_y(x, y)\mathbf{j}] \cdot [u_1\mathbf{i} + u_2\mathbf{j}]$$

The vector in the first bracket, whose components are the first partial derivatives of $f(x, y)$, is very important. It is denoted by $\nabla f(x, y)$ (read *del* $f(x, y)$) and is given the following special name.

Definition (16.26)

> Let f be a function of two variables. The **gradient** of f (or of $f(x, y)$) is the vector function given by
>
> $$\nabla f(x, y) = f_x(x, y)\mathbf{i} + f_y(x, y)\mathbf{j}.$$

The gradient $\nabla f(x, y)$ is sometimes denoted by **grad** $f(x, y)$. From the previous discussion we have the following.

Directional derivative (gradient form) (16.27)

$$D_{\mathbf{u}} f(x, y) = \nabla f(x, y) \cdot \mathbf{u}$$

Thus, *to find the directional derivative of f in the direction of the unit vector* $\mathbf{u}$, *we may dot the gradient of f with* $\mathbf{u}$. Henceforth, we shall use this formula instead of Theorem (16.25) to find $D_{\mathbf{u}} f(x, y)$.

The symbol ∇ (called **del**) is a *vector differential operator* and is defined by

$$\nabla = \mathbf{i}\frac{\partial}{\partial x} + \mathbf{j}\frac{\partial}{\partial y}.$$

Its properties are similar to those of the operator d/dx (see Exercises 35–40). Operating on $f(x, y)$, it produces the gradient vector $\nabla f(x, y)$.

If $P_0(x_0, y_0)$ is a specific point in the xy-plane, we sometimes denote the gradient vector at P_0 by $\nabla f]_{P_0}$. Thus,

$$\nabla f]_{P_0} = \nabla f(x_0, y_0) = f_x(x_0, y_0)\mathbf{i} + f_y(x_0, y_0)\mathbf{j}.$$

When we represent the vector $\nabla f]_{P_0}$ graphically, we always take the initial point at P_0, as illustrated in Figure 16.50.

FIGURE 16.50

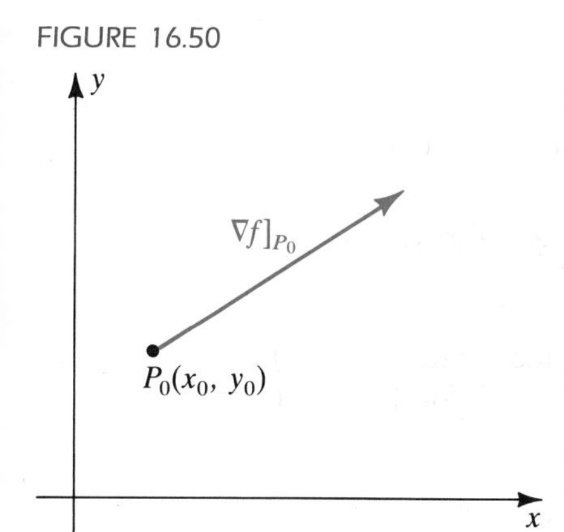

EXAMPLE 2 Let $f(x, y) = x^2 - 4xy$.

(a) Find the gradient of f at the point $P(1, 2)$, and sketch the vector $\nabla f]_P$.

(b) Use the gradient to find the directional derivative of f at $P(1, 2)$ in the direction from $P(1, 2)$ to $Q(2, 5)$.

SOLUTION

(a) By Definition (16.26),

$$\nabla f(x, y) = (2x - 4y)\mathbf{i} - 4x\mathbf{j}.$$

At $P(1, 2)$,

$$\nabla f]_P = \nabla f(1, 2) = (2 - 8)\mathbf{i} - 4\mathbf{j} = -6\mathbf{i} - 4\mathbf{j}.$$

This vector is sketched in Figure 16.51, with initial point $P(1, 2)$.

(b) If we let $\mathbf{a} = \overrightarrow{PQ}$, then

$$\mathbf{a} = \langle 2 - 1, 5 - 2 \rangle = \langle 1, 3 \rangle = \mathbf{i} + 3\mathbf{j}$$

(see Figure 16.51). A unit vector having the direction of $\overrightarrow{PQ}$ is

$$\mathbf{u} = \frac{1}{\|\mathbf{a}\|}\mathbf{a} = \frac{1}{\sqrt{10}}(\mathbf{i} + 3\mathbf{j}).$$

FIGURE 16.51

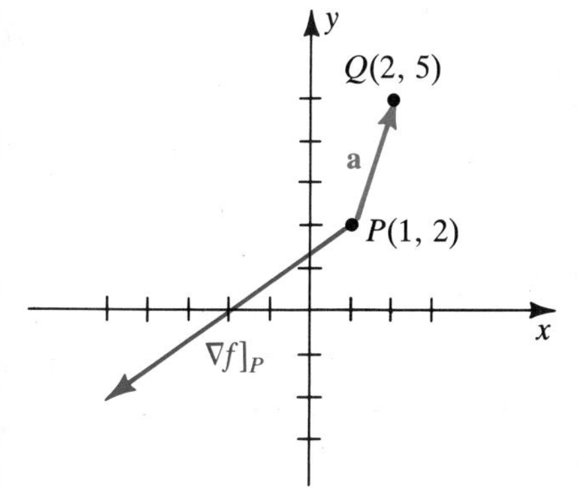

Applying (16.27), we have

$$\begin{aligned} D_{\mathbf{u}} f(1, 2) &= \nabla f(1, 2) \cdot \mathbf{u} \\ &= (-6\mathbf{i} - 4\mathbf{j}) \cdot \frac{1}{\sqrt{10}}(\mathbf{i} + 3\mathbf{j}) \\ &= \frac{1}{\sqrt{10}}(-6 - 12) = -\frac{18}{\sqrt{10}} \approx -5.7. \end{aligned}$$

Let $P(x, y)$ be a fixed point, and consider the directional derivative $D_{\mathbf{u}} f(x, y)$ as $\mathbf{u} = \langle u_1, u_2 \rangle$ varies. For a given unit vector $\mathbf{u}$, the directional derivative may be positive (that is, $f(x, y)$ may increase) or negative ($f(x, y)$ may decrease) or 0. In many applications it is important to find the direction in which $f(x, y)$ increases most rapidly and the maximum rate of change. The next theorem provides this information.

Gradient theorem **(16.28)**

> Let f be a function of two variables that is differentiable at the point $P(x, y)$.
>
> **(i)** The maximum value of $D_{\mathbf{u}} f(x, y)$ at $P(x, y)$ is $\|\nabla f(x, y)\|$.
>
> **(ii)** The maximum rate of increase of $f(x, y)$ at $P(x, y)$ occurs in the direction of $\nabla f(x, y)$.

FIGURE 16.52

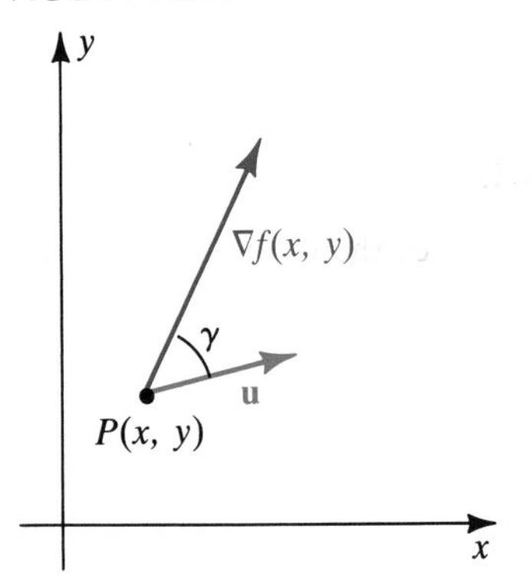

PROOF

(i) We shall consider the point $P(x, y)$ and the vector $\nabla f(x, y)$ as fixed (but arbitrary) and the unit vector $\mathbf{u}$ as variable (see Figure 16.52). Let γ be the angle between $\mathbf{u}$ and $\nabla f(x, y)$. By (16.27) and Theorem (14.22),

$$\begin{aligned} D_{\mathbf{u}} f(x, y) &= \nabla f(x, y) \cdot \mathbf{u} \\ &= \|\nabla f(x, y)\| \, \|\mathbf{u}\| \cos \gamma \\ &= \|\nabla f(x, y)\| \cos \gamma. \end{aligned}$$

Since $-1 \le \cos \gamma \le 1$, the maximum value of $D_{\mathbf{u}} f(x, y)$ occurs if $\cos \gamma = 1$; in this case $D_{\mathbf{u}} f(x, y) = \|\nabla f(x, y)\|$.

(ii) The directional derivative $D_{\mathbf{u}} f(x, y)$ is the rate of change of $f(x, y)$ with respect to distance at $P(x, y)$ in the direction determined by $\mathbf{u}$. As in part (i), this rate of change has its maximum value if $\cos \gamma = 1$—that is, if $\gamma = 0$. However, if $\gamma = 0$, then $\mathbf{u}$ has the same direction as $\nabla f(x, y)$. ■

FIGURE 16.53

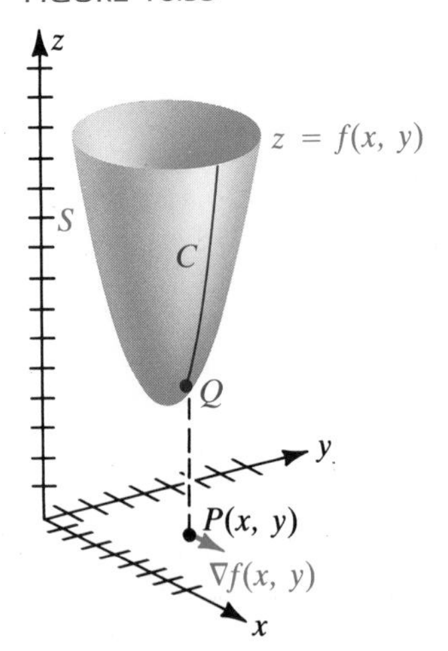

The gradient theorem (16.28) has an interesting and useful geometric interpretation. Let S denote the graph of $z = f(x, y)$ in an xyz-coordinate system. As illustrated in Figure 16.53, the point $P(x, y)$ and the vector $\nabla f(x, y)$ are in the xy-plane, and $Q(x, y, f(x, y))$ is the point on S corresponding to P. If a (variable) point in the xy-plane moves through P in the direction of $\nabla f(x, y)$, then, by (16.28)(ii), the corresponding point on S moves up a curve C of *steepest ascent* at the point Q on the graph of f. In Section 16.7 we discuss another geometric aspect of $\nabla f(x, y)$, which involves the level curves of f.

Referring to the proof of the gradient theorem (16.28), we see that the *minimum* value of $D_{\mathbf{u}}\, f(x, y)$ occurs if $\cos \gamma = -1$. In this case $\gamma = \pi$ (or $\gamma = 180^\circ$) and $D_{\mathbf{u}}\, f(x, y) = -\|\nabla f(x, y)\|$, which is the maximum rate of *decrease* of $f(x, y)$. This gives us the following corollary to Theorem (16.28).

Corollary **(16.29)**

> Let f be a function of two variables that is differentiable at the point $P(x, y)$.
>
> **(i)** The minimum value of $D_{\mathbf{u}}\, f(x, y)$ at $P(x, y)$ is $-\|\nabla f(x, y)\|$.
>
> **(ii)** The minimum rate of increase (or maximum rate of decrease) of $f(x, y)$ at $P(x, y)$ occurs in the direction of $-\nabla f(x, y)$.

FIGURE 16.54

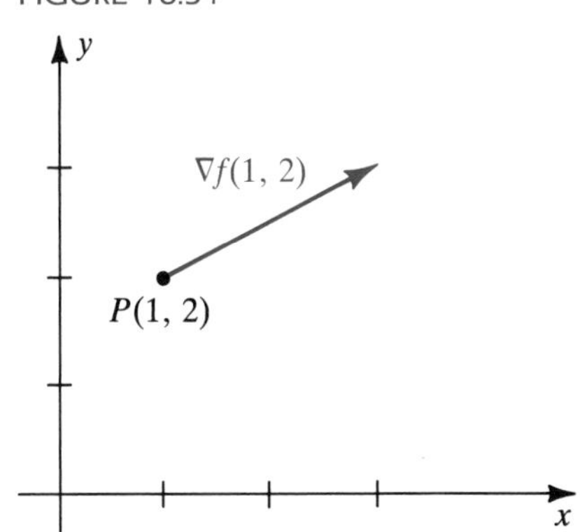

EXAMPLE 3 Let $f(x, y) = 2 + x^2 + \frac{1}{4}y^2$.

(a) Find the direction in which $f(x, y)$ increases most rapidly at the point $P(1, 2)$, and find the maximum rate of increase of f at P.

(b) Interpret (a) using the graph of f.

SOLUTION

(a) The gradient of f is

$$\nabla f(x, y) = 2x\mathbf{i} + \tfrac{1}{2}y\mathbf{j}.$$

At $P(1, 2)$,

$$\nabla f(1, 2) = 2\mathbf{i} + \mathbf{j}.$$

This vector, sketched in Figure 16.54, determines the direction in which $f(x, y)$ increases most rapidly at $P(1, 2)$.

The maximum rate of increase of f at $P(1, 2)$ is

$$\|\nabla f(1, 2)\| = \|-2\mathbf{i} - \mathbf{j}\| = \sqrt{2^2 + 1^2} = \sqrt{5} \approx 2.2.$$

FIGURE 16.55

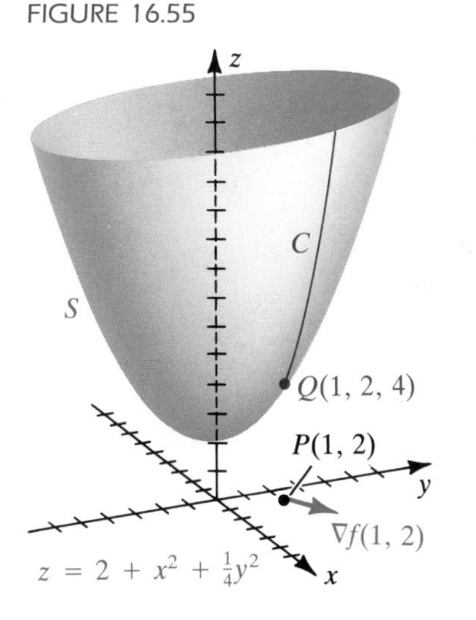

(b) The graph S of f, an elliptic paraboloid, is sketched in Figure 16.55. The point $P(1, 2)$ and the vector $\nabla f(1, 2)$ are shown in the xy-plane. The point on S corresponding to P is $Q(1, 2, 4)$. If a (variable) point in the xy-plane moves through P in the direction of $\nabla f(1, 2)$, the corresponding point on the graph moves along a curve C of steepest ascent on the paraboloid.

The directional derivative of a function f of three variables is defined in the same manner as in Definition (16.24). Specifically, if $\mathbf{u} = u_1\mathbf{i} + u_2\mathbf{j} + u_3\mathbf{k}$ is a unit vector, we have the following.

Directional derivative of $f(x, y, z)$ **(16.30)**

$$D_{\mathbf{u}}\, f(x, y, z) = \lim_{s \to 0} \frac{f(x + su_1, y + su_2, z + su_3) - f(x, y, z)}{s}$$

As in the two-variable case, $D_{\mathbf{u}}\, f(x, y, z)$ is the rate of change of f with respect to distance at $P(x, y, z)$ in the direction of $\mathbf{u}$. The **gradient** of f (or of $f(x, y, z)$), denoted by $\nabla f(x, y, z)$ or **grad** $f(x, y, z)$, is defined as follows.

Gradient of $f(x, y, z)$ **(16.31)**

$$\nabla f(x, y, z) = f_x(x, y, z)\mathbf{i} + f_y(x, y, z)\mathbf{j} + f_z(x, y, z)\mathbf{k}$$

As in two dimensions, the gradient at a specific point $P_0(x_0, y_0, z_0)$ is denoted by $\nabla f]_{P_0}$.

The next result is the three-dimensional version of Theorem (16.25).

Theorem **(16.32)**

If f is a differentiable function of three variables and $\mathbf{u} = u_1\mathbf{i} + u_2\mathbf{j} + u_3\mathbf{k}$ is a unit vector, then

$$\begin{aligned} D_{\mathbf{u}} f(x, y, z) &= \nabla f(x, y, z) \cdot \mathbf{u} \\ &= f_x(x, y, z)u_1 + f_y(x, y, z)u_2 + f_z(x, y, z)u_3. \end{aligned}$$

As in Theorem (16.28), of all possible directional derivatives $D_{\mathbf{u}} f(x, y, z)$ at the point $P(x, y, z)$, the one in the direction of $\nabla f(x, y, z)$ has the largest value, and that maximum value is $\|\nabla f(x, y, z)\|$.

EXAMPLE 4 Suppose an xyz-coordinate system is located in space such that the temperature T at the point (x, y, z) is given by the formula $T = 100/(x^2 + y^2 + z^2)$.

(a) Find the rate of change of T with respect to distance at the point $P(1, 3, -2)$ in the direction of the vector $\mathbf{a} = \mathbf{i} - \mathbf{j} + \mathbf{k}$.

(b) In what direction from P does T increase most rapidly? What is the maximum rate of change of T at P?

SOLUTION

(a) By Definition (16.31), the gradient of $T = 100/(x^2 + y^2 + z^2)$ is

$$\nabla T = \frac{\partial T}{\partial x}\mathbf{i} + \frac{\partial T}{\partial y}\mathbf{j} + \frac{\partial T}{\partial z}\mathbf{k}.$$

Since

$$\frac{\partial T}{\partial x} = \frac{-200x}{(x^2 + y^2 + z^2)^2}, \quad \frac{\partial T}{\partial y} = \frac{-200y}{(x^2 + y^2 + z^2)^2}, \quad \frac{\partial T}{\partial z} = \frac{-200z}{(x^2 + y^2 + z^2)^2},$$

we obtain

$$\nabla T = \frac{-200}{(x^2 + y^2 + z^2)^2}(x\mathbf{i} + y\mathbf{j} + z\mathbf{k}).$$

Hence

$$\nabla T]_P = \frac{-200}{196}(\mathbf{i} + 3\mathbf{j} - 2\mathbf{k}).$$

A unit vector $\mathbf{u}$ having the same direction as $\mathbf{a} = \mathbf{i} - \mathbf{j} + \mathbf{k}$ is

$$\mathbf{u} = \frac{1}{\sqrt{3}}(\mathbf{i} - \mathbf{j} + \mathbf{k}).$$

By Theorem (16.32), the rate of change of T at P in the direction of $\mathbf{a}$ is

$$D_{\mathbf{u}}\, T]_P = \nabla T]_P \cdot \mathbf{u} = \frac{-200}{196}\frac{(1-3-2)}{\sqrt{3}} = \frac{200}{49\sqrt{3}} \approx 2.4.$$

If, for example, T is measured in degrees and distance is in inches, then T is increasing at a rate of 2.4 degrees per inch at P in the direction of $\mathbf{a}$.

(b) The maximum rate of change of T at P occurs in the direction of the gradient—that is, in the direction of the vector $-\mathbf{i} - 3\mathbf{j} + 2\mathbf{k}$. The maximum rate of change equals the magnitude of the gradient; that is,

$$\|\nabla T]_P\| = \frac{200}{196}\sqrt{1+9+4} \approx 3.8.$$

In this section we used the gradient of a function f primarily for finding directional derivatives. Later in the text other important aspects will be considered. We shall conclude this section with an application to thermodynamics.

Let $T = f(x, y)$ be the steady state temperature at the point (x, y) of a two-dimensional region R. (*Steady state* means that T is independent of time.) *Fourier's law of heat conduction* asserts that the rate q at which heat flows past a point $P(x, y)$ on the boundary of R is directly proportional to the component of the temperature gradient ∇T along the unit (outer) normal vector $\mathbf{n}$ to the boundary at P (see Figure 16.56). Thus,

FIGURE 16.56

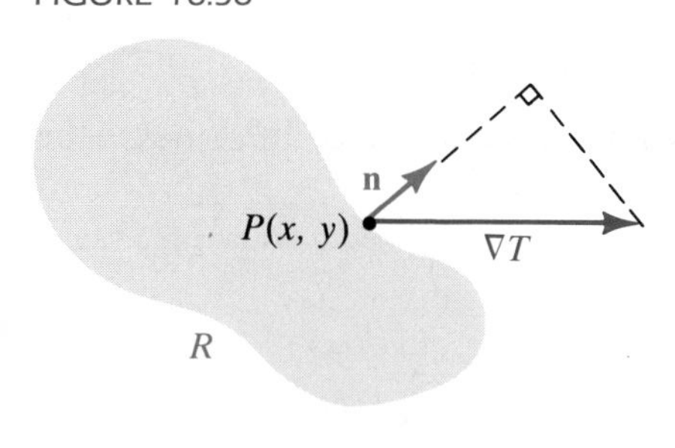

$$q = k\ \text{comp}_{\mathbf{n}}\ \nabla T = k\,\nabla T \cdot \mathbf{n}$$

for some scalar k (see Definition (14.25)). If $\nabla T \cdot \mathbf{n} = 0$, the component in the direction of the normal will be 0 (∇T and $\mathbf{n}$ are orthogonal), and hence no heat will escape from the region at P. The boundary is then said to be *insulated* at the point P. The region is *insulated along part of a boundary* if it is insulated at every point on that part. Analogous statements can be made for three-dimensional regions.

FIGURE 16.57

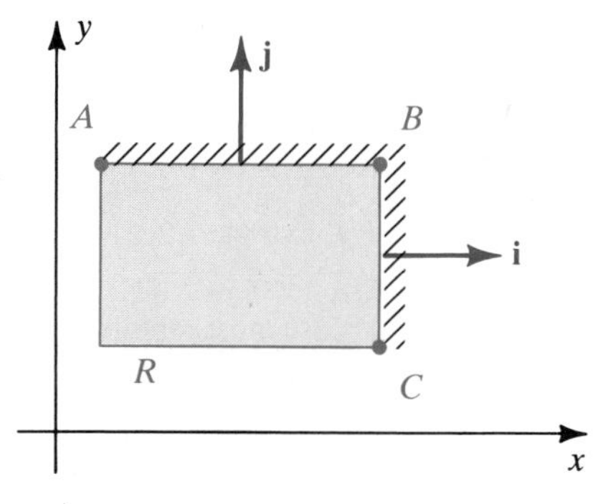

EXAMPLE 5 Let R be the rectangular region in the xy-plane shown in Figure 16.57, and let $T = f(x, y)$ be the steady state temperature at (x, y). Find conditions on T that make the following insulated:

(a) edge BC **(b)** edge AB

SOLUTION

(a) We may use the vector $\mathbf{i}$ for the unit normal $\mathbf{n}$ to edge BC. Thus, BC is insulated if and only if, at each point P on BC,

$$\nabla T \cdot \mathbf{i} = 0, \quad \text{or} \quad \left(\frac{\partial T}{\partial x}\mathbf{i} + \frac{\partial T}{\partial y}\mathbf{j}\right) \cdot \mathbf{i} = \frac{\partial T}{\partial x} = 0.$$

This means that the rate of change of T in the horizontal direction is 0.

(b) A normal to edge AB is $\mathbf{j}$. In this case AB is insulated if and only if

$$\nabla T \cdot \mathbf{j} = 0 \quad \text{or, equivalently,} \quad \frac{\partial T}{\partial y} = 0;$$

that is, the rate of change of T in the vertical direction is 0.

EXERCISES 16.6

Exer. 1–6: Find the gradient of f at P.

1 $f(x, y) = \sqrt{x^2 + y^2}$; $P(-4, 3)$

2 $f(x, y) = 7y - 5x$; $P(2, 6)$

3 $f(x, y) = e^{3x} \tan y$; $P(0, \pi/4)$

4 $f(x, y) = x \ln (x - y)$; $P(5, 4)$

5 $f(x, y, z) = yz^3 - 2x^2$; $P(2, -3, 1)$

6 $f(x, y, z) = xy^2e^z$; $P(2, -1, 0)$

Exer. 7–20: Find the directional derivative of f at the point P in the indicated direction.

7 $f(x, y) = x^2 - 5xy + 3y^2$; $P(3, -1)$, $\mathbf{u} = (\sqrt{2}/2)(\mathbf{i} + \mathbf{j})$

8 $f(x, y) = x^3 - 3x^2y - y^3$; $P(1, -2)$, $\mathbf{u} = \frac{1}{2}(-\mathbf{i} + \sqrt{3}\mathbf{j})$

9 $f(x, y) = \arctan \dfrac{y}{x}$; $P(4, -4)$, $\mathbf{a} = 2\mathbf{i} - 3\mathbf{j}$

10 $f(x, y) = x^2 \ln y$; $P(5, 1)$, $\mathbf{a} = -\mathbf{i} + 4\mathbf{j}$

11 $f(x, y) = \sqrt{9x^2 - 4y^2 - 1}$; $P(3, -2)$, $\mathbf{a} = \mathbf{i} + 5\mathbf{j}$

12 $f(x, y) = \dfrac{x - y}{x + y}$; $P(2, -1)$, $\mathbf{a} = 3\mathbf{i} + 4\mathbf{j}$

13 $f(x, y) = x \cos^2 y$; $P(2, \pi/4)$, $\mathbf{a} = \langle 5, 1 \rangle$

14 $f(x, y) = xe^{3y}$; $P(4, 0)$, $\mathbf{a} = \langle -1, 3 \rangle$

15 $f(x, y, z) = xy^3z^2$; $P(2, -1, 4)$, $\mathbf{a} = \mathbf{i} + 2\mathbf{j} - 3\mathbf{k}$

16 $f(x, y, z) = x^2 + 3yz + 4xy$; $P(1, 0, -5)$, $\mathbf{a} = 2\mathbf{i} - 3\mathbf{j} + \mathbf{k}$

17 $f(x, y, z) = z^2e^{xy}$; $P(-1, 2, 3)$, $\mathbf{a} = 3\mathbf{i} + \mathbf{j} - 5\mathbf{k}$

18 $f(x, y, z) = \sqrt{xy} \sin z$; $P(4, 9, \pi/4)$, $\mathbf{a} = 2\mathbf{i} + 3\mathbf{j} - 2\mathbf{k}$

19 $f(x, y, z) = (x + y)(y + z)$; $P(5, 7, 1)$, $\mathbf{a} = \langle -3, 0, 1 \rangle$

20 $f(x, y, z) = z^2 \tan^{-1} (x + y)$; $P(0, 0, 4)$, $\mathbf{a} = \langle 6, 0, 1 \rangle$

Exer. 21–24: (a) Find the directional derivative of f at P in the direction from P to Q. (b) Find a unit vector in the direction in which f increases most rapidly at P, and find the rate of change of f in that direction. (c) Find a unit vector in the direction in which f decreases most rapidly at P, and find the rate of change of f in that direction.

21 $f(x, y) = x^2e^{-2y}$; $P(2, 0)$, $Q(-3, 1)$

22 $f(x, y) = \sin (2x - y)$; $P(-\pi/3, \pi/6)$, $Q(0, 0)$

23 $f(x, y, z) = \sqrt{x^2 + y^2 + z^2}$; $P(-2, 3, 1)$, $Q(0, -5, 4)$

24 $f(x, y, z) = \dfrac{x}{y} - \dfrac{y}{z}$; $P(0, -1, 2)$ $Q(3, 1, -4)$

25 A metal plate is located in an xy-plane such that the temperature T at (x, y) is inversely proportional to the distance from the origin, and the temperature at $P(3, 4)$ is 100 °F.

(a) Find the rate of change of T at P in the direction of $\mathbf{i} + \mathbf{j}$.

(b) In what direction does T increase most rapidly at P?

(c) In what direction does T decrease most rapidly at P?

(d) In what direction is the rate of change 0?

26 The surface of a lake is represented by a region D in the xy-plane such that the depth (in feet) under the point (x, y) is $f(x, y) = 300 - 2x^2 - 3y^2$.

(a) In what direction should a boat at $P(4, 9)$ sail in order for the depth of the water to decrease most rapidly?

(b) In what direction does the depth remain the same?

27 The electrical potential V at (x, y, z) is

$$V = x^2 + 4y^2 + 9z^2.$$

(a) Find the rate of change of V at $P(2, -1, 3)$ in the direction from P to the origin.

(b) Find the direction that produces the maximum rate of change of V at P.

(c) What is the maximum rate of change at P?

28 The temperature T at (x, y, z) is given by

$$T = 4x^2 - y^2 + 16z^2.$$

(a) Find the rate of change of T at $P(4, -2, 1)$ in the direction of $2\mathbf{i} + 6\mathbf{j} - 3\mathbf{k}$.

(b) In what direction does T increase most rapidly at P?

(c) What is this maximum rate of change?

(d) In what direction does T decrease most rapidly at P?

(e) What is this rate of change?

Exer. 29–30: Refer to the discussion that precedes Example 5. In each case, T is the temperature at (x, y).

29 Shown in the figure is a semicircular region R.

(a) Use polar coordinates to show that the upper boundary $\widehat{AB}$ is insulated if and only if $\partial T/\partial r = 0$. (*Hint:*

Show that if $T = f(x, y)$, with $x = r\cos\theta$ and $y = r\sin\theta$, then $\partial T/\partial r = (\partial T/\partial x)\cos\theta + (\partial T/\partial y)\sin\theta$.)

(b) Interpret $\partial T/\partial r$ as a rate of change of T.

EXERCISE 29

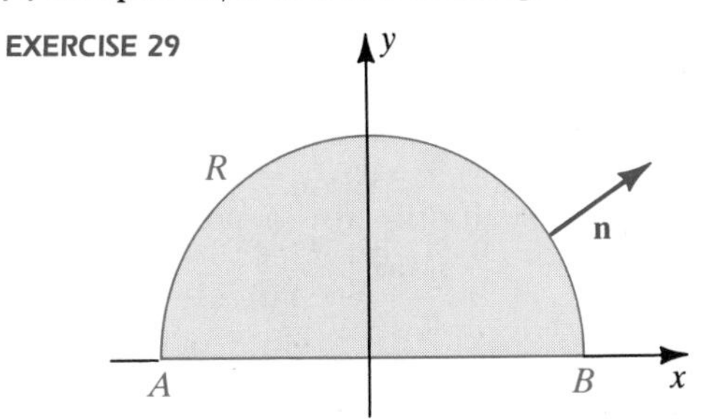

30 Shown in the figure is a circular sector whose boundary AB is insulated.

(a) Use polar coordinates to show that the insulation condition is equivalent to $\partial T/\partial\theta = 0$ for every point on the segment AB.

(b) Interpret $\partial T/\partial\theta$ as a rate of change of T.

EXERCISE 30

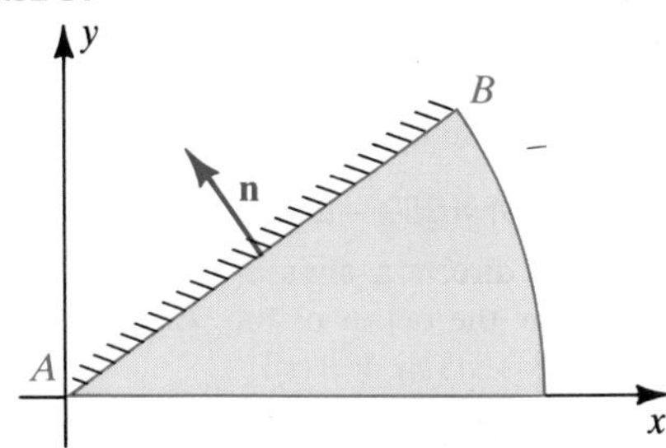

31 In some applications it may be difficult to directly calculate the gradient $\nabla f(x, y)$. The following approximations are sometimes used for the components, where $h \approx 0$:

$$f_x(x, y) \approx \frac{f(x+h, y) - f(x-h, y)}{2h}$$

$$f_y(x, y) \approx \frac{f(x, y+h) - f(x, y-h)}{2h}$$

(a) Show that these approximations improve as $h \to 0$.

(b) If $f(x, y) = x^3/(1+y)$, approximate $\nabla f(1, 2)$ using $h = 0.01$ and compare the approximation with the exact result.

32 An analysis of the temperature of each component is crucial to the design of a computer chip. Suppose that for a chip to operate properly, the temperature of each component must not exceed 78°F. If a component is likely to become too warm, engineers will usually place it in a cool portion of the chip. The designing of chips is aided by computer simulation in which temperature gradients are analyzed. A computer simulation for a new chip has resulted in the temperature grid (in °F) shown in the figure.

(a) If $T(x, y)$ is the temperature at (x, y), use Exercise 31 with $h = 1$ to approximate $\nabla T(3, 3)$.

(b) Estimate the direction of maximum heat transfer at $(3, 3)$.

(c) Estimate the instantaneous rate of change of T in the direction of $\mathbf{a} = -\mathbf{i} + 2\mathbf{j}$ at $(3, 3)$.

EXERCISE 32

Temperature grid (°F)

y (mm) \ x (mm)	0	1	2	3	4
6	62	62	65	63	61
5	62	67	69	65	64
4	63	70	70	69	67
3	65	66	72	74	76
2	61	67	73	80	75
1	60	60	71	76	72
0	60	60	63	65	69

c **Exer. 33–34: Extend the approximation formulas in Exercise 31 to the first partial derivatives of $f(x, y, z)$, and then use $h = 0.01$ to approximate the directional derivative of f at $P_0(1, 1, 1)$ in the direction of u.**

33 $f(x, y, z) = \dfrac{x^2 \sin y \tan z}{x^3 + y^2 z^4}$; $\quad \mathbf{u} = \dfrac{1}{\sqrt{3}}(\mathbf{i} + \mathbf{j} + \mathbf{k})$

34 $f(x, y, z) = \dfrac{x^4 + 3x^2z^2}{4x^2y^2 + \cosh(yz)}$; $\quad \mathbf{u} = \dfrac{1}{\sqrt{6}}(2\mathbf{i} - \mathbf{j} + \mathbf{k})$

Exer. 35–40: If $u = f(x, y)$, $v = g(x, y)$, and f and g are differentiable, prove the identity.

35 $\nabla(cu) = c\,\nabla u$ for a constant c

36 $\nabla(u + v) = \nabla u + \nabla v$

37 $\nabla(uv) = u\,\nabla v + v\,\nabla u$

38 $\nabla\left(\dfrac{u}{v}\right) = \dfrac{v\,\nabla u - u\,\nabla v}{v^2}$ with $v \neq 0$

39 $\nabla u^n = nu^{n-1}\,\nabla u$ for every real number n

40 If $w = h(u)$, then $\nabla w = \dfrac{dw}{du}\nabla u$.

41 Let $\mathbf{u}$ be a unit vector and let θ be the angle, measured in the counterclockwise direction, from the positive x-axis to the position vector corresponding to $\mathbf{u}$.

(a) Show that

$$D_{\mathbf{u}} f(x, y) = f_x(x, y)\cos\theta + f_y(x, y)\sin\theta.$$

(b) If $f(x, y) = x^2 + 2xy - y^2$ and $\theta = 5\pi/6$, find $D_{\mathbf{u}} f(2, -3)$.

42 Refer to Exercise 41. If $f(x, y) = (xy + y^2)^4$ and $\theta = \pi/3$, find $D_{\mathbf{u}} f(2, -1)$.

43 If f, f_x, and f_y are continuous and $\nabla f(x, y) = \mathbf{0}$ on a rectangular region $R = \{(x, y): a < x < b,\ c < y < d\}$, prove that $f(x, y)$ is constant on R (see Exercise 46 of Section 16.5).

44 Suppose $w = f(x, y)$, $x = g(t)$, $y = h(t)$, and all functions are differentiable. If $\mathbf{r}(t) = x\mathbf{i} + y\mathbf{j}$, prove that

$$\frac{dw}{dt} = \nabla w \cdot \mathbf{r}'(t).$$

16.7 TANGENT PLANES AND NORMAL LINES

FIGURE 16.58

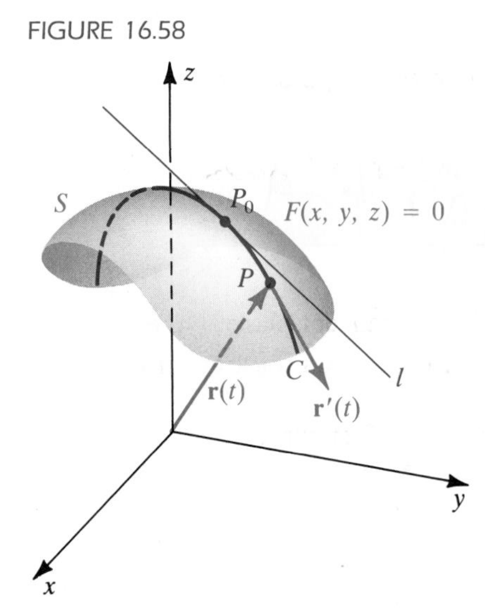

Suppose a surface S is the graph of an equation $F(x, y, z) = 0$ and F has continuous first partial derivatives. Let $P_0(x_0, y_0, z_0)$ be a point on S at which F_x, F_y, and F_z are not all zero. A **tangent line** to S at P_0 is, by definition, a tangent line l to any curve C that lies on S and contains P_0 (see Figure 16.58). If C has the parametrization

$$x = f(t), \quad y = g(t), \quad z = h(t)$$

for t in some interval I and if $\mathbf{r}(t)$ is the position vector of $P(x, y, z)$, then

$$\mathbf{r}(t) = f(t)\mathbf{i} + g(t)\mathbf{j} + h(t)\mathbf{k}.$$

Hence

$$\mathbf{r}'(t) = f'(t)\mathbf{i} + g'(t)\mathbf{j} + h'(t)\mathbf{k}$$

is a tangent vector to C at $P(x, y, z)$, as indicated in Figure 16.58.

For each t, the point $(f(t), g(t), h(t))$ on C is also on S, and therefore

$$F(f(t), g(t), h(t)) = 0.$$

If we let

$$w = F(x, y, z), \quad \text{with} \quad x = f(t), \quad y = g(t), \quad z = h(t),$$

then using a chain rule and the fact that $w = 0$ for every t, we have

$$\frac{dw}{dt} = \frac{\partial w}{\partial x}\frac{dx}{dt} + \frac{\partial w}{\partial y}\frac{dy}{dt} + \frac{\partial w}{\partial z}\frac{dz}{dt} = 0.$$

Thus, for every point $P(x, y, z)$ on C,

$$F_x(x, y, z)f'(t) + F_y(x, y, z)g'(t) + F_z(x, y, z)h'(t) = 0,$$

or, equivalently,

$$\nabla F(x, y, z) \cdot \mathbf{r}'(t) = 0.$$

FIGURE 16.59

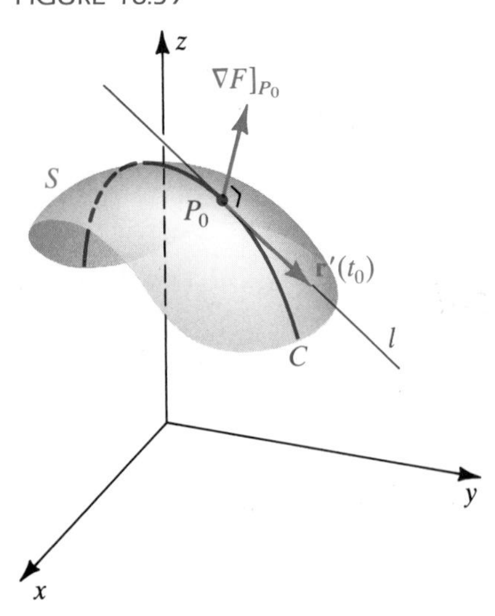

In particular, if $P_0(x_0, y_0, z_0)$ corresponds to $t = t_0$, then

$$\nabla F]_{P_0} \cdot \mathbf{r}'(t_0) = \nabla F(x_0, y_0, z_0) \cdot \mathbf{r}'(t_0) = 0.$$

Since $\mathbf{r}'(t_0)$ is a tangent vector to C at P_0, this implies that *the vector* $\nabla F]_{P_0}$ *is orthogonal to every tangent line l to S at P_0* (see Figure 16.59).

The plane through P_0 with normal vector $\nabla F]_{P_0}$ is the **tangent plane to S at P_0**. We have shown that *every tangent line l to S at P_0 lies in the tangent plane at P_0*. The following theorem summarizes our discussion.

Theorem **(16.33)**

Suppose $F(x, y, z)$ has continuous first partial derivatives and S is the graph of $F(x, y, z) = 0$. If P_0 is a point on S and if F_x, F_y, and F_z are not all 0 at P_0, then the vector $\nabla F]_{P_0}$ is normal to the tangent plane to S at P_0.

We shall refer to $\nabla F]_{P_0}$ in Theorem (16.33) as a vector that is *normal to the surface* S at P_0. Applying Theorem (14.36) gives us the following corollary, where we assume that F_x, F_y, and F_z are not all 0 at P_0.

Corollary **(16.34)**

An equation for the tangent plane to the graph of $F(x, y, z) = 0$ at the point $P_0(x_0, y_0, z_0)$ is

$$F_x(x_0, y_0, z_0)(x - x_0) + F_y(x_0, y_0, z_0)(y - y_0) + F_z(x_0, y_0, z_0)(z - z_0) = 0.$$

EXAMPLE 1 Find an equation for the tangent plane to the ellipsoid $\frac{3}{4}x^2 + 3y^2 + z^2 = 12$ at the point $P_0(2, 1, \sqrt{6})$, and illustrate graphically.

SOLUTION To use Corollary (16.34), we first express the equation of the surface in the form $F(x, y, z) = 0$ by letting

$$F(x, y, z) = \tfrac{3}{4}x^2 + 3y^2 + z^2 - 12 = 0.$$

The partial derivatives of F are

$$F_x(x, y, z) = \tfrac{3}{2}x, \quad F_y(x, y, z) = 6y, \quad F_z(x, y, z) = 2z,$$

and hence, at $P_0(2, 1, \sqrt{6})$,

$$F_x(2, 1, \sqrt{6}) = 3, \quad F_y(2, 1, \sqrt{6}) = 6, \quad F_z(2, 1, \sqrt{6}) = 2\sqrt{6}.$$

We now apply Corollary (16.34), obtaining the equation

$$3(x - 2) + 6(y - 1) + 2\sqrt{6}(z - \sqrt{6}) = 0,$$

or, equivalently,

$$3x + 6y + 2\sqrt{6}z = 24.$$

FIGURE 16.60

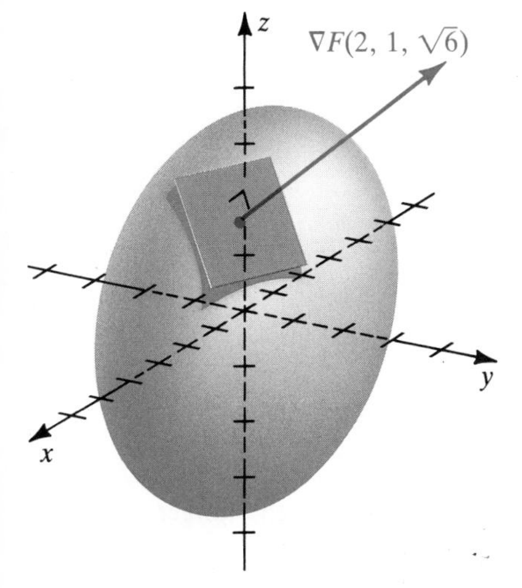

The ellipsoid has x-, y-, and z-intercepts ± 4, ± 2, and $\pm\sqrt{12}$, respectively, and is sketched in Figure 16.60. A normal vector to the tangent plane at the point $P_0(2, 1, \sqrt{6})$ is

$$\nabla F]_{P_0} = \nabla F(2, 1, \sqrt{6}) = 3\mathbf{i} + 6\mathbf{j} + 2\sqrt{6}\mathbf{k},$$

as shown in the figure.

If $z = f(x, y)$ is an equation for S and we let $F(x, y, z) = f(x, y) - z$, then the equation in Corollary (16.34) takes on the form

$$f_x(x_0, y_0)(x - x_0) + f_y(x_0, y_0)(y - y_0) + (-1)(z - z_0) = 0.$$

This gives us the following.

Theorem **(16.35)**

> An equation for the tangent plane to the graph of $z = f(x, y)$ at the point (x_0, y_0, z_0) is
>
> $$z - z_0 = f_x(x_0, y_0)(x - x_0) + f_y(x_0, y_0)(y - y_0).$$

The line perpendicular to the tangent plane at a point $P_0(x_0, y_0, z_0)$ on a surface S is a **normal line** to S at P_0. If S is the graph of $F(x, y, z) = 0$, then the normal line is parallel to the vector $\nabla F(x_0, y_0, z_0)$.

EXAMPLE 2 Find an equation of the normal line to the ellipsoid $\frac{3}{4}x^2 + 3y^2 + z^2 = 12$ at the point $P_0(2, 1, \sqrt{6})$.

SOLUTION This surface is the same as that considered in Example 1. Consequently the vector

$$\nabla F(2, 1, \sqrt{6}) = 3\mathbf{i} + 6\mathbf{j} + 2\sqrt{6}\mathbf{k},$$

shown in Figure 16.60, is parallel to the normal line. Using Theorem (14.34), we obtain the following parametric equations for the normal line:

$$x = 2 + 3t, \quad y = 1 + 6t, \quad z = \sqrt{6} + 2\sqrt{6}t; \quad t \text{ in } \mathbb{R}$$

EXAMPLE 3 Let P_0 be the point $(3, -4, 2)$ on the hyperboloid $16x^2 - 9y^2 + 36z^2 = 144$. Find equations for the tangent plane and the normal line at P_0.

SOLUTION The hyperboloid sketched in Figure 16.61 is the same as that sketched in Figure 14.69. Letting

$$F(x, y, z) = 16x^2 - 9y^2 + 36z^2 - 144,$$

we obtain

$$\nabla F(x, y, z) = 32x\mathbf{i} - 18y\mathbf{j} + 72z\mathbf{k}.$$

Hence a normal vector to the hyperboloid at $P_0(3, -4, 2)$ is

$$\nabla F]_{P_0} = 96\mathbf{i} + 72\mathbf{j} + 144\mathbf{k}.$$

Any scalar multiple of $\nabla F]_{P_0}$ is also a normal vector. In particular, since

$$\nabla F]_{P_0} = 24(4\mathbf{i} + 3\mathbf{j} + 6\mathbf{k}),$$

we may use $\mathbf{a} = 4\mathbf{i} + 3\mathbf{j} + 6\mathbf{k}$. Thus, an equation for the tangent plane at $P_0(3, -4, 2)$ is

$$4(x - 3) + 3(y + 4) + 6(z - 2) = 0,$$

FIGURE 16.61
$16x^2 - 9y^2 + 36z^2 = 144$

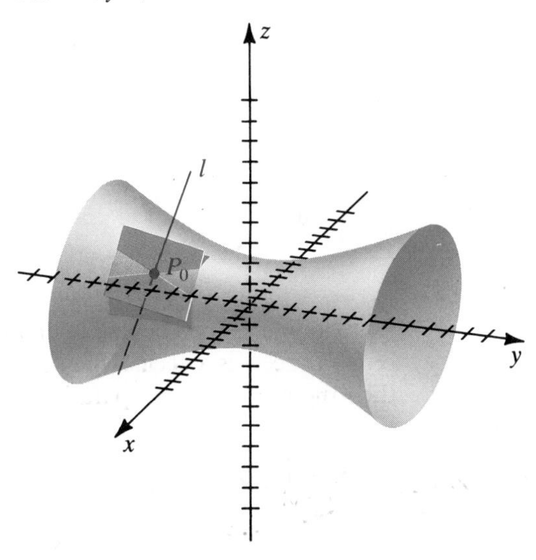

or, equivalently,

$$4x + 3y + 6z = 12.$$

This plane has x-, y-, and z-intercepts 3, 4, and 2, respectively (see Figure 16.61). Note that unlike the one in Example 1, the tangent plane at P_0 intersects the surface at many points.

Using **a** as a normal vector leads to the following parametric equations for the normal line l in the figure:

$$x = 3 + 4t, \quad y = -4 + 3t, \quad z = 2 + 6t; \quad t \text{ in } \mathbb{R}$$

Suppose S is the graph of the equation $z = f(x, y)$. The tangent plane to S at $P_0(x_0, y_0, z_0)$ may be used to obtain a geometric interpretation for the differential

$$dz = f_x(x_0, y_0)\,\Delta x + f_y(x_0, y_0)\,\Delta y,$$

which was defined in (16.15). Since

$$\Delta z = f(x_0 + \Delta x, y_0 + \Delta y) - f(x_0, y_0),$$

the point $P(x_0 + \Delta x, y_0 + \Delta y, z_0 + \Delta z)$ is on S (see Figure 16.62).

FIGURE 16.62

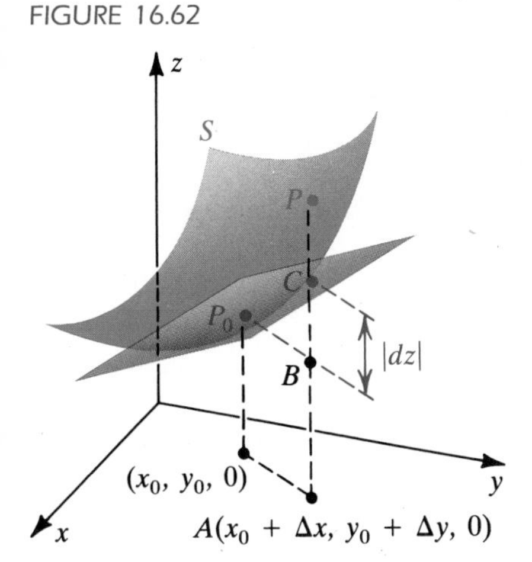

Let $C(x_0 + \Delta x, y_0 + \Delta y, z)$ be on the tangent plane, and consider the points $A(x_0 + \Delta x, y_0 + \Delta y, 0)$ and $B(x_0 + \Delta x, y_0 + \Delta y, z_0)$ shown in Figure 16.62. Since C is on the tangent plane, its coordinates satisfy the equation in Theorem (16.35); that is,

$$\begin{aligned} z - z_0 &= f_x(x_0, y_0)(x_0 + \Delta x - x_0) + f_y(x_0, y_0)(y_0 + \Delta y - y_0) \\ &= f_x(x_0, y_0)\,\Delta x + f_y(x_0, y_0)\,\Delta y = dz. \end{aligned}$$

In terms of distances, we have shown that $|dz| = \|\overrightarrow{BC}\|$; that is, $|dz|$ *is the distance from B to the point on the tangent plane that lies above (or below) B*. This is analogous to the single-variable case $y = f(x)$ discussed in Section 3.5, where we interpreted dy as a distance to a point on a tangent line to the graph of f (see Figure 3.27). It follows that when dz is used as an approximation to the change in z, we assume that the surface S almost coincides with its tangent plane at points close to P_0.

FIGURE 16.63
Isothermal surfaces

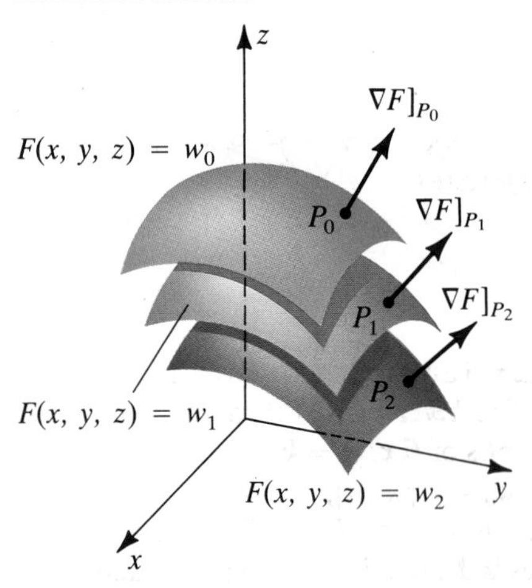

As an application of the discussion in this section, suppose a solid, liquid, or gas is situated in a three-dimensional coordinate system such that the temperature at the point $P(x, y, z)$ is $w = F(x, y, z)$. If $P_0(x_0, y_0, z_0)$ is a fixed point, then the graph of the equation

$$F(x, y, z) = F(x_0, y_0, z_0)$$

is the level (isothermal) surface that passes through P_0. For every point on this surface, the temperature is $w_0 = F(x_0, y_0, z_0)$. It is convenient to visualize all possible level surfaces that can be obtained by choosing different points. Several surfaces through P_0, P_1, and P_2 corresponding to temperatures w_0, w_1, and w_2 are sketched in Figure 16.63. By Theorem (16.33), $\nabla F]_{P_0}$, $\nabla F]_{P_1}$, and $\nabla F]_{P_2}$ are normal vectors to their corresponding surfaces at the points P_0, P_1, and P_2, respectively.

In Section 16.6 we saw that the maximum rate of change of $F(x, y, z)$ at $P_0(x_0, y_0, z_0)$ is in the direction of $\nabla F]_{P_0}$. From the preceding remarks, this maximum rate of change takes place in a direction that is normal to the level surface of F that contains P_0.

The following theorem summarizes these remarks.

Theorem (16.36)

> Let a function F of three variables be differentiable at $P_0(x_0, y_0, z_0)$, and let S be the level surface of F containing P_0. If $\nabla F(x_0, y_0, z_0) \neq \mathbf{0}$, then this gradient vector is normal to S at P_0. Thus, the direction for the maximum rate of change of $F(x, y, z)$ at P_0 is normal to S.

The preceding theorem is illustrated in Figure 16.63, where the maximum rate of change of temperature takes place in directions normal to the isothermal surfaces. Another illustration is found in electrical theory, where $F(x, y, z)$ is the potential at (x, y, z). If a particle moves on one of the level (equipotential) surfaces, the potential remains constant. The potential changes most rapidly when a particle moves in a direction that is normal to an equipotential surface.

FIGURE 16.64

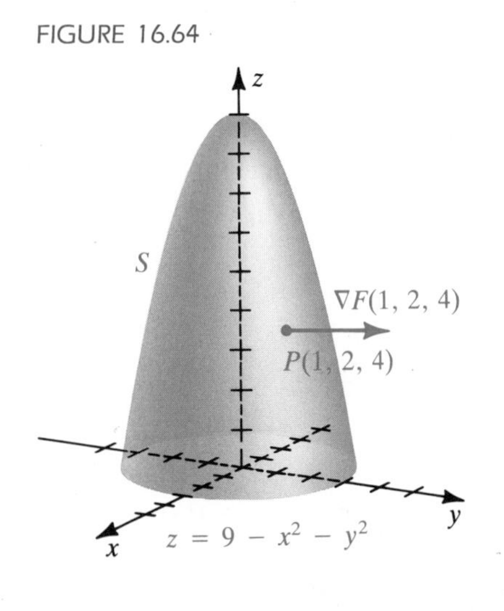

EXAMPLE 4 If $F(x, y, z) = x^2 + y^2 + z$, sketch the level surface of F that passes through $P(1, 2, 4)$ and sketch $\nabla F(1, 2, 4)$.

SOLUTION The level surfaces of F are graphs of equations of the form $F(x, y, z) = k$, where k is a constant. Since $F(1, 2, 4) = 1^2 + 2^2 + 4 = 9$, the level surface S that passes through $P(1, 2, 4)$ is the graph of the equation $F(x, y, z) = 9$—that is, of

$$x^2 + y^2 + z = 9, \quad \text{or} \quad z = 9 - x^2 - y^2.$$

The graph of this level surface, a paraboloid of revolution, is sketched in Figure 16.64.

The gradient of F is

$$\nabla F(x, y, z) = 2x\mathbf{i} + 2y\mathbf{j} + \mathbf{k}.$$

At $P(1, 2, 4)$,

$$\nabla F(1, 2, 4) = 2(1)\mathbf{i} + 2(2)\mathbf{j} + \mathbf{k} = 2\mathbf{i} + 4\mathbf{j} + \mathbf{k}.$$

The vector $\nabla F(1, 2, 4)$ is sketched in Figure 16.64. By Theorem (16.36), $\nabla F(1, 2, 4)$ is orthogonal to the level surface S at P.

The following result for functions of *two* variables is analogous to Theorem (16.36).

Theorem **(16.37)**

> Let a function f of two variables be differentiable at $P_0(x_0, y_0)$, and let C be the level curve of f that contains P_0. If $\nabla f(x_0, y_0) \neq \mathbf{0}$, then this gradient vector is orthogonal to C at P_0. Thus, the direction for the maximum rate of change of $f(x, y)$ at P_0 is orthogonal to C.

FIGURE 16.65

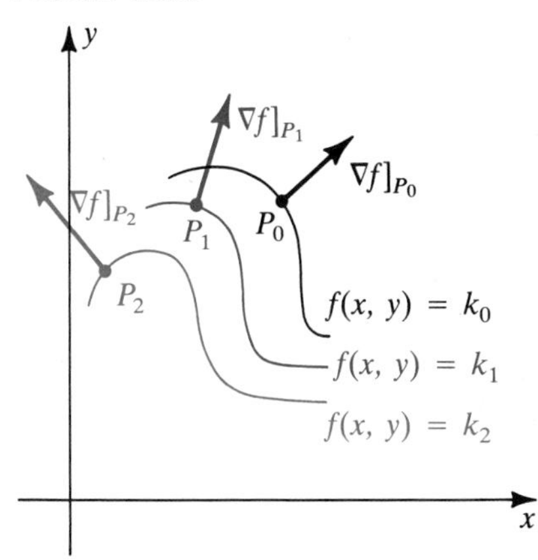

Theorem (16.37) is illustrated graphically in Figure 16.65 for an arbitrary function f of two variables. The level curves through $P_0(x_0, y_0)$, $P_1(x_1, y_1)$, and $P_2(x_2, y_2)$ are the graphs of $f(x, y) = k_0$, $f(x, y) = k_1$, and $f(x, y) = k_2$, respectively, where $k_0 = f(x_0, y_0)$, $k_1 = f(x_1, y_1)$, and $k_2 = f(x_2, y_2)$. By Theorem (16.37), the vectors $\nabla f]_{P_0}$, $\nabla f]_{P_1}$, and $\nabla f]_{P_2}$ are orthogonal to the level curves, as indicated in the figure. If each level curve C is an isobar on a weather map, then the barometric pressure does not change as a point moves along C. The vector $\nabla f]_P$ at a point P on an isobar points in the direction in which the barometric pressure changes most rapidly.

As a specific illustration of Theorem (16.37), if $f(x, y) = 9 - x^2 - y^2$, then the level curves are given by $9 - x^2 - y^2 = k$, where k is a real number (see Example 3 of Section 16.1). Some of these level curves (circles) are shown in Figure 16.66. By Theorem 16.37, the maximum (or minimum) rate of change of $f(x, y)$ occurs if (x, y) moves in a direction orthogonal to these circles, and hence along lines through the origin. This corresponds to movement of the point $(x, y, f(x, y))$ up (or down) the steepest part of the graph of f sketched in Figure 16.64.

FIGURE 16.66
Level curves: $9 - x^2 - y^2 = k$

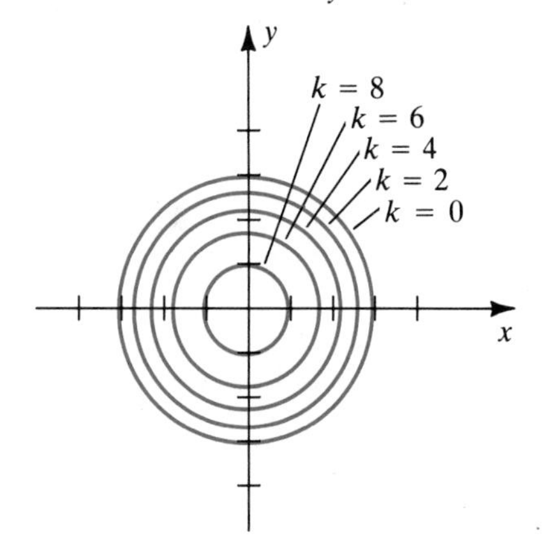

FIGURE 16.67

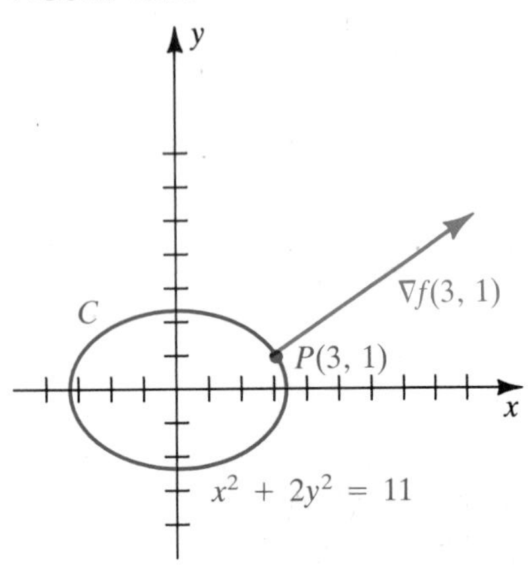

FIGURE 16.68

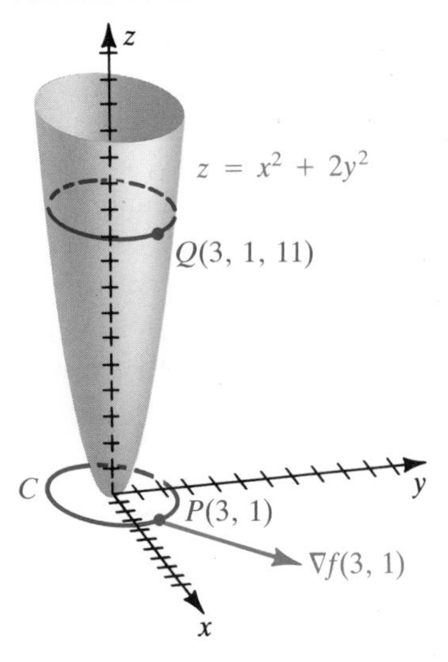

EXAMPLE 5 Let $f(x, y) = x^2 + 2y^2$.

(a) Sketch the level curve C of f that passes through the point $P(3, 1)$, and sketch $\nabla f(3, 1)$.

(b) Discuss the significance of part (a) in terms of the graph of f.

SOLUTION

(a) Since $f(3, 1) = 3^2 + 2(1)^2 = 9 + 2 = 11$, the level curve of f that passes through the point $P(3, 1)$ is given by $f(x, y) = 11$—that is, by

$$x^2 + 2y^2 = 11.$$

The graph is the ellipse sketched in Figure 16.67. The gradient of f is

$$\nabla f(x, y) = 2x\mathbf{i} + 4y\mathbf{j},$$

and hence $$\nabla f(3, 1) = 2(3)\mathbf{i} + 4(1)\mathbf{j} = 6\mathbf{i} + 4\mathbf{j}.$$

The vector $\nabla f(3, 1)$ is shown in Figure 16.67. By Theorem (16.37), this vector is orthogonal to the level curve C.

(b) The graph of f (that is, of $z = x^2 + 2y^2$) is an elliptic paraboloid (see Figure 16.68). The level curve $x^2 + 2y^2 = 11$ in the xy-plane corresponds to the trace of the paraboloid on the plane $z = 11$. The maximum rate of change of $f(x, y)$ occurs if the point (x, y) in the xy-plane moves in the direction of $\nabla f(3, 1)$ at $P(3, 1)$. This corresponds to movement of the point $(x, y, f(x, y))$ up the steepest part of the paraboloid at $Q(3, 1, 11)$.

Tangent planes can be used to obtain successive approximations to a solution of a system of two nonlinear equations, $f(x, y) = 0$, $g(x, y) = 0$, from a first approximation (x_1, y_1) by performing the following steps.

Step 1 Use Theorem (16.35) to find equations of the tangent planes to the graphs of f and g at the points $(x_1, y_1, f(x_1, y_1))$ and $(x_1, y_1, g(x_1, y_1))$.

Step 2 Find the trace in the xy-plane of each tangent plane in Step 1.

Step 3 Find the point of intersection (x_2, y_2) of the traces found in Step 2.

Step 4 Take (x_2, y_2) as a second approximation and repeat Steps 1–3.

This process is a geometric description of Newton's method, discussed at the end of Section 16.4.

EXERCISES 16.7

Exer. 1–10: Find equations for the tangent plane and the normal line to the graph of the equation at the point P.

1 $4x^2 - y^2 + 3z^2 = 10$; $P(2, -3, 1)$

2 $9x^2 - 4y^2 - 25z^2 = 40$; $P(4, 1, -2)$

3 $z = 4x^2 + 9y^2$; $P(-2, -1, 25)$

4 $z = 4x^2 - y^2$; $P(5, -8, 36)$

5 $xy + 2yz - xz^2 + 10 = 0$; $P(-5, 5, 1)$

6 $x^3 - 2xy + z^3 + 7y + 6 = 0$; $P(1, 4, -3)$

7 $z = 2e^{-x} \cos y$; $P(0, \pi/3, 1)$

8 $z = \ln xy$; $P(\frac{1}{2}, 2, 0)$

9 $x = \ln \dfrac{y}{2z}$; $P(0, 2, 1)$

10 $xyz - 4xz^3 + y^3 = 10$; $P(-1, 2, 1)$

Exer. 11–14: Sketch both the level curve C of f that contains P and $\nabla f]_P$.

11 $f(x, y) = y^2 - x^2$; $P(2, 1)$

12 $f(x, y) = 3x - 2y$; $P(-2, 1)$

13 $f(x, y) = x^2 - y$; $P(-3, 5)$

14 $f(x, y) = xy$; $P(3, 2)$

Exer. 15–20: Sketch both the level surface S of F that contains P and $\nabla F]_P$.

15 $F(x, y, z) = x^2 + y^2 + z^2$; $P(1, 5, 2)$

16 $F(x, y, z) = z - x^2 - y^2$; $P(2, -2, 1)$

17 $F(x, y, z) = x + 2y + 3z$; $P(3, 4, 1)$

18 $F(x, y, z) = x^2 + y^2 - z^2$; $P(3, -1, 1)$

19 $F(x, y, z) = x^2 + y^2$; $P(2, 0, 3)$

20 $F(x, y, z) = z$; $P(2, 3, 4)$

Exer. 21–24: Prove that an equation of the tangent plane to the given quadric surface at the point $P_0(x_0, y_0, z_0)$ may be written in the indicated form.

21 $\dfrac{x^2}{a^2} + \dfrac{y^2}{b^2} + \dfrac{z^2}{c^2} = 1$; $\dfrac{xx_0}{a^2} + \dfrac{yy_0}{b^2} + \dfrac{zz_0}{c^2} = 1$

22 $\dfrac{x^2}{a^2} - \dfrac{y^2}{b^2} + \dfrac{z^2}{c^2} = 1$; $\dfrac{xx_0}{a^2} - \dfrac{yy_0}{b^2} + \dfrac{zz_0}{c^2} = 1$

23 $\dfrac{x^2}{a^2} - \dfrac{y^2}{b^2} - \dfrac{z^2}{c^2} = 1$; $\dfrac{xx_0}{a^2} - \dfrac{yy_0}{b^2} - \dfrac{zz_0}{c^2} = 1$

24 $\dfrac{x^2}{a^2} + \dfrac{y^2}{b^2} = cz$; $\dfrac{2xx_0}{a^2} + \dfrac{2yy_0}{b^2} = c(z + z_0)$

25 Find the points on the hyperboloid of two sheets $x^2 - 2y^2 - 4z^2 = 16$ at which the tangent plane is parallel to the plane $4x - 2y + 4z = 5$.

26 Show that the sum of the squares of the x-, y-, and z-intercepts of every tangent plane to the graph of the equation $x^{2/3} + y^{2/3} + z^{2/3} = a^{2/3}$ is the constant a^2.

27 Prove that every normal line to a sphere passes through the center of the sphere.

28 Find the points on the paraboloid $z = 4x^2 + 9y^2$ at which the normal line is parallel to the line through $P(-2, 4, 3)$ and $Q(5, -1, 2)$.

29 Two surfaces are said to be orthogonal at a point of intersection $P(x, y, z)$ if their normal lines at P are orthogonal. Show that the graphs of $F(x, y, z) = 0$ and $G(x, y, z) = 0$ (where F and G have partial derivatives) are orthogonal at P if and only if

$$F_xG_x + F_yG_y + F_zG_z = 0.$$

30 Refer to Exercise 29. Prove that the sphere with equation $x^2 + y^2 + z^2 = a^2$ and the cone $x^2 + y^2 - z^2 = 0$ are orthogonal at every point of intersection.

c **Exer. 31–32: For the given system of equations and first approximate solution (x_1, y_1), use Steps 1–3 stated at the end of this section to find a second approximation (x_2, y_2).**

31 $x^2 - y^3 = 0$, $x^2 + y^2 - 3 = 0$; $(1.3, 1.1)$

32 $\sin x - \cos y = 0$, $x^2 - y^2 - 1.3 = 0$; $(1.2, 0.3)$

16.8 EXTREMA OF FUNCTIONS OF SEVERAL VARIABLES

In Chapter 4 we discussed local and absolute extrema for functions of one variable. In this section we extend these concepts to functions of several variables.

A function f of two variables has a **local maximum** at (a, b) if there is an open disk R containing (a, b) such that $f(x, y) \le f(a, b)$ for every (x, y) in R. The local maxima correspond to the high points on the graph S of f, as illustrated in Figure 16.69. Similarly, the function f has a **local minimum** at (c, d) if there is an open disk R containing (c, d) such that $f(x, y) \ge f(c, d)$ for every (x, y) in R. The local minima correspond to the low points on the graph of f, as illustrated in Figure 16.70.

A region in the xy-plane is **bounded** if it is a subregion of a closed disk. If f is continuous on a closed and bounded region R, then f has a **maximum** $f(a, b)$ and **minimum** $f(c, d)$ for some (a, b) and (c, d) in R; that is,

$$f(c, d) \le f(x, y) \le f(a, b)$$

for every (x, y) in R. The proof may be found in texts on advanced calculus.

FIGURE 16.69
Local maximum $f(a, b)$

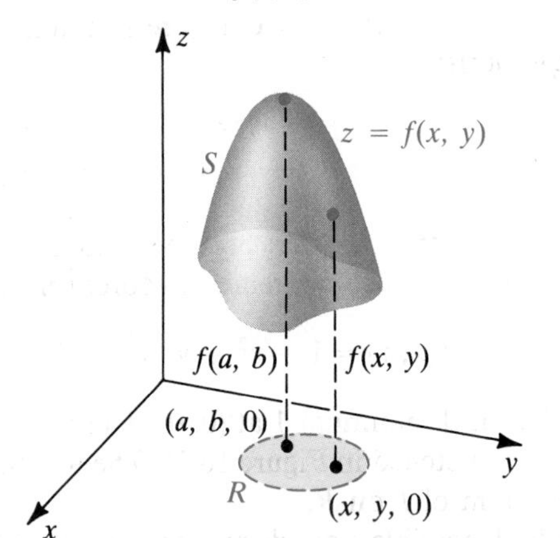

FIGURE 16.70
Local minimum $f(c, d)$

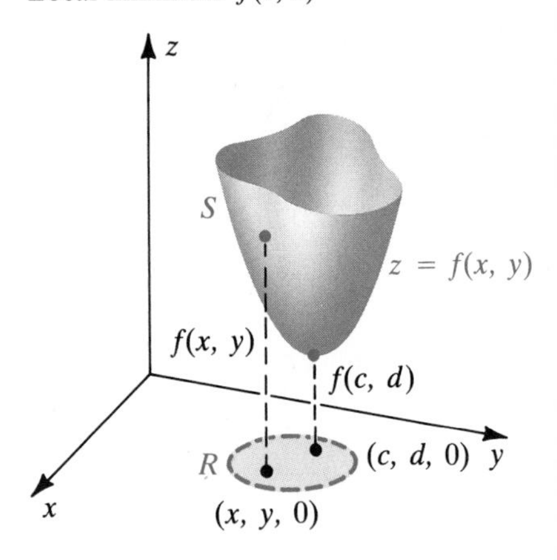

The local maxima and minima are the **local extrema** of f. The **extrema** also include the maximum and minimum (if they exist). If f has continuous first partial derivatives at (x_0, y_0) and if $f(x_0, y_0)$ is a local extrema of f, then the tangent plane to the graph of $z = f(x, y)$ at (x_0, y_0, z_0) is parallel to the xy-plane, and hence its equation is $z = z_0$. It follows from Theorem (16.35) that $f_x(x_0, y_0) = 0$ and $f_y(x_0, y_0) = 0$. Therefore the pairs that yield local extrema must be solutions of *both*

$$f_x(x, y) = 0 \quad \text{and} \quad f_y(x, y) = 0.$$

As was the case for functions of one variable, local extrema can also occur at (x, y) if either $f_x(x, y)$ or $f_y(x, y)$ does not exist. All of these pairs are crucial for finding local extrema, so we shall give them a special name.

Definition **(16.38)**

> Let f be a function of two variables. A pair (a, b) is a **critical point** of f if either
>
> **(i)** $f_x(a, b) = 0$ and $f_y(a, b) = 0$, or
>
> **(ii)** $f_x(a, b)$ or $f_y(a, b)$ does not exist.

When searching for local extrema of a function, we usually begin by finding the critical points. We then test each pair in some way to determine if it yields a local maximum or minimum.

A maximum or minimum of a function of two variables may occur at a boundary point of its domain R. The investigation of such **boundary extrema** usually requires a separate procedure, as was the case with endpoint extrema for functions of one variable. If there are no boundary points—for example, when R is the entire xy-plane or an open disk—then there can be no boundary extrema.

EXAMPLE 1 Let $f(x, y) = 1 + x^2 + y^2$, with $x^2 + y^2 \leq 4$. Find the extrema of f.

FIGURE 16.71

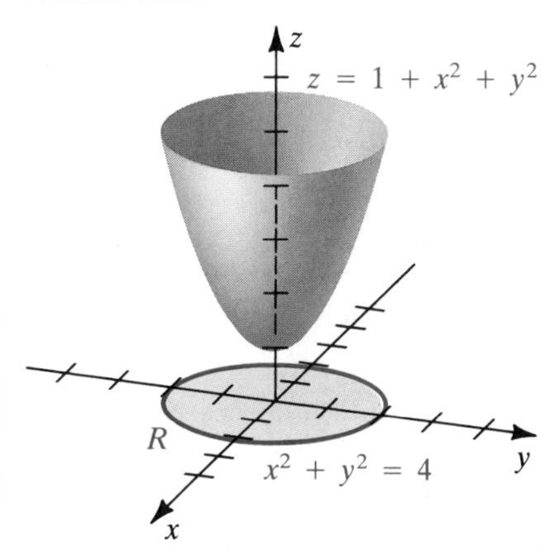

SOLUTION The restriction $x^2 + y^2 \leq 4$ corresponds to the closed disk R of radius 2 and center O in the xy-plane (see Figure 16.71). By Definition (16.38)(i), the critical points are solutions of the following system of two equations:

$$\begin{cases} f_x(x, y) = 0 \\ f_y(x, y) = 0 \end{cases} \quad \text{or} \quad \begin{cases} 2x = 0 \\ 2y = 0 \end{cases}$$

The only pair that satisfies both equations is (0, 0); hence $f(0, 0) = 1$ is the only possible local extremum. Moreover, because

$$f(x, y) = 1 + x^2 + y^2 > 1 \quad \text{if} \quad (x, y) \neq (0, 0),$$

f has the local minimum 1 at (0, 0). This fact may also be seen from the graph of f, sketched in Figure 16.71. The function value $f(0, 0) = 1$ is also the minimum of f on R.

To find possible boundary extrema, we investigate points (a, b) that lie on the boundary of R. Referring to the figure, we see that any such point (for example, (0, 2)) leads to the maximum $f(0, 2) = 5$.

As illustrated in the next example, not every critical point may lead to an extremum.

EXAMPLE 2 If $f(x, y) = y^2 - x^2$ and the domain of f is $\mathbb{R}^2$, find the extrema of f.

FIGURE 16.72

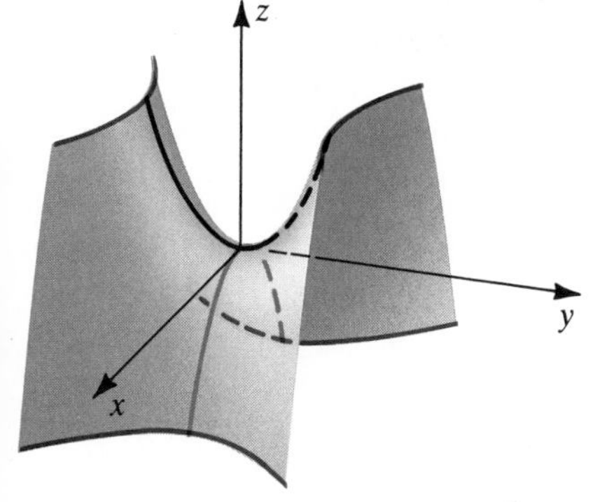

SOLUTION By Definition (16.38)(i), the critical points are solutions of the following system of two equations:

$$\begin{cases} f_x(x, y) = 0 \\ f_y(x, y) = 0 \end{cases} \quad \text{or} \quad \begin{cases} -2x = 0 \\ 2y = 0 \end{cases}$$

As in Example 1, the only possible local extremum is $f(0, 0) = 0$. However, if $y \neq 0$, then $f(0, y) = y^2 > 0$, and if $x \neq 0$, then $f(x, 0) = -x^2 < 0$. Thus, every open disk in the xy-plane containing (0, 0) contains pairs at which function values are greater than $f(0, 0)$ and also pairs at which values are less than $f(0, 0)$. Consequently f has no local extrema. This fact is also evident from the graph of f, a hyperbolic paraboloid, sketched in Figure 16.72.

To determine the extrema of a more complicated function f of two variables, it is convenient to use the following function D, called the *discriminant* of f.

Definition **(16.39)**

Let f be a function of two variables that has continuous second partial derivatives. The **discriminant** D of f is given by

$$D(x, y) = f_{xx}(x, y)f_{yy}(x, y) - [f_{xy}(x, y)]^2.$$

One way to remember the formula for the discriminant is to use the following determinant, with all functions evaluated at (x, y):

$$D = \begin{vmatrix} f_{xx} & f_{xy} \\ f_{yx} & f_{yy} \end{vmatrix} = f_{xx}f_{yy} - (f_{xy})^2$$

Note that we have used the fact that $f_{yx} = f_{xy}$.

The following test is stated without proof. It is the counterpart of the second derivative test for functions of one variable.

Test for local extrema **(16.40)**

Let f be a function of two variables that has continuous second partial derivatives throughout an open disk R containing (a, b). If $f_x(a, b) = f_y(a, b) = 0$ and $D(a, b) > 0$, then $f(a, b)$ is

(i) a local maximum of f if $f_{xx}(a, b) < 0$

(ii) a local minimum of f if $f_{xx}(a, b) > 0$

Note that if $D(a, b) > 0$, then $f_{xx}(a, b)f_{yy}(a, b) > [f_{xy}(a, b)]^2$ and hence $f_{xx}(a, b)f_{yy}(a, b)$ is positive; that is, $f_{xx}(a, b)$ and $f_{yy}(a, b)$ have the same sign. Thus, we may replace $f_{xx}(a, b)$ in (i) and (ii) of the preceding test by $f_{yy}(a, b)$.

A point $P(a, b, f(a, b))$ on the graph of f is a **saddle point**, or a *saddle point for* f, if $f_x(a, b) = f_y(a, b) = 0$ and if there is an open disk R containing (a, b) such that $f(x, y) > f(a, b)$ for some points (x, y) in R and $f(x, y) < f(a, b)$ for other points. Note that the graph of the hyperbolic paraboloid in Figure 16.72 has the saddle point $(0, 0, 0)$. The following theorem is proved in advanced calculus texts.

Theorem **(16.41)**

Let f have continuous second partial derivatives throughout an open disk R containing (a, b). If $f_x(a, b) = f_y(a, b) = 0$ and $D(a, b) < 0$, then the point $P(a, b, f(a, b))$ is a saddle point on the graph of f.

We *cannot* use (16.40) to determine whether $f(a, b)$ is a local extrema if $D(a, b) = 0$ or if either $f_x(a, b)$ or $f_y(a, b)$ does not exist. In these cases we must analyze the variation of $f(x, y)$ near (a, b) by referring to the definition of $f(x, y)$ or the graph of f. To illustrate, if $f(x, y) = -(x^2 + y^{2/3})$, then the partial derivative $f_y(x, y) = -\frac{2}{3}y^{-1/3}$ does not exist if $y = 0$. Hence *every* point $(a, 0)$ on the x-axis is a critical point of f. From the graph of f in Figure 16.17(i) it is evident that $f(0, 0) = 0$ is both a local maximum and a maximum of f in $\mathbb{R}^2$. We can show this algebraically by observing that if $(x, y) \neq (0, 0)$, then $f(x, y) < 0$. Since many such exceptions to (16.40) require complicated analyses, we will not consider examples or exercises of this type.

EXAMPLE 3 If $f(x, y) = x^2 - 4xy + y^3 + 4y$, find the local extrema and saddle points of f.

SOLUTION The first partial derivatives of f are

$$f_x(x, y) = 2x - 4y \quad \text{and} \quad f_y(x, y) = -4x + 3y^2 + 4.$$

Since f_x and f_y exist for every (x, y), the only critical points are the solutions of the following system of two equations in two unknowns:

$$\begin{cases} 2x - 4y = 0 \\ -4x + 3y^2 + 4 = 0 \end{cases}$$

Solving this system, we find that f has the two critical points $(4, 2)$ and $(\frac{4}{3}, \frac{2}{3})$. The second partial derivatives of f are

$$f_{xx}(x, y) = 2, \quad f_{xy}(x, y) = -4, \quad f_{yy}(x, y) = 6y.$$

Hence the discriminant D is given by

$$D(x, y) = (2)(6y) - (-4)^2 = 12y - 16.$$

The remainder of the solution is arranged in the following table.

Critical point	Value of the discriminant	Value of f_{xx}	Conclusion
$(4, 2)$	$D(4, 2) = 8 > 0$	$f_{xx}(4, 2) = 2 > 0$	$f(4, 2) = 0$ is a local minimum by (16.40)(ii)
$(\frac{4}{3}, \frac{2}{3})$	$D(\frac{4}{3}, \frac{2}{3}) = -8 < 0$	irrelevant	$(\frac{4}{3}, \frac{2}{3}, f(\frac{4}{3}, \frac{2}{3}))$ is a saddle point by (16.41)

EXAMPLE 4 If $f(x, y) = x^2 - 4xy + y^3 + 4y$, find the extrema of f on the triangular region R that has vertices $(-1, -1)$, $(7, -1)$, and $(7, 7)$.

FIGURE 16.73

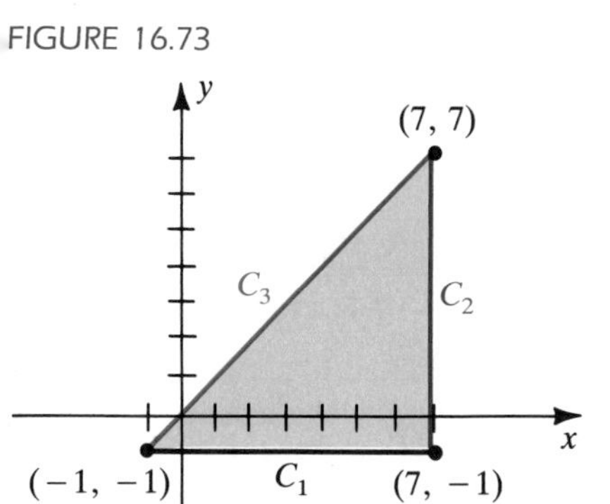

SOLUTION The boundary of R consists of the line segments C_1, C_2, and C_3, as shown in Figure 16.73. In Example 3 we found that f has a local minimum 0 at the point $(4, 2)$, which is within R. Thus, we need only check for boundary extrema.

On C_1 we have $y = -1$, so values of f are given by

$$f(x, -1) = x^2 + 4x - 1 - 4 = x^2 + 4x - 5.$$

This determines a function of one variable whose domain is the interval $[-1, 7]$. The first derivative is $2x + 4$, which equals 0 at $x = -2$, a number outside the interval $[-1, 7]$. Thus, there is no local extremum of $f(x, -1)$ on $[-1, 7]$. Since $f(x, -1)$ is increasing throughout this interval, we obtain the endpoint extrema

$$f(-1, -1) = -8 \quad \text{and} \quad f(7, -1) = 72.$$

On C_2 we have $x = 7$, and values of f are given by

$$f(7, y) = 49 - 28y + y^3 + 4y = y^3 - 24y + 49$$

for $-1 \le y \le 7$. The first derviative of this function of y is $3y^2 - 24$, and hence there is a critical number if $3y^2 = 24$, or $y = \sqrt{8} = 2\sqrt{2}$. Using the second derivative $6y$ of $f(7, y)$, we see that $f(7, 2\sqrt{2}) = 49 - 32\sqrt{2} \approx 3.7$ is a local minimum for f *on the segment* C_2. However, it is not a minimum for f on R, since $f(-1, -1) = -8 < 3.7$. The values of f at the endpoints of C_2 are

$$f(7, -1) = 72 \quad \text{and} \quad f(7, 7) = 224.$$

Finally, on C_3 we have $y = x$, and values of f are given by

$$f(x, x) = x^2 - 4x^2 + x^3 + 4x = x^3 - 3x^2 + 4x.$$

The first derivative $3x^2 - 6x + 4$ has no real roots, and hence there are no critical numbers for $f(x, x)$. We have already calculated the endpoint values $f(-1, -1) = -8$ and $f(7, 7) = 224$. These, therefore, are the minimum and maximum values for f on the triangular region R.

EXAMPLE 5 If $f(x, y) = \frac{1}{3}x^3 + \frac{4}{3}y^3 - x^2 - 3x - 4y - 3$, find the local extrema and saddle points of f.

SOLUTION The first partial derivatives of f are

$$f_x(x, y) = x^2 - 2x - 3 \quad \text{and} \quad f_y(x, y) = 4y^2 - 4.$$

Since f_x and f_y exist for every (x, y), the only critical points are the solutions of the following system:

$$\begin{cases} x^2 - 2x - 3 = 0 \\ 4y^2 - 4 = 0 \end{cases}$$

Solving this system, we obtain the four critical points $(3, 1)$, $(3, -1)$, $(-1, 1)$, and $(-1, -1)$. The second partial derivatives of f are

$$f_{xx}(x, y) = 2x - 2, \quad f_{xy}(x, y) = 0, \quad f_{yy}(x, y) = 8y.$$

Hence the discriminant D is given by

$$D(x, y) = (2x - 2)(8y) - 0^2 = 16y(x - 1).$$

Let us arrange our work in the following table.

Critical point	Value of the discriminant	Value of f_{xx}	Conclusion
$(3, 1)$	$D(3, 1) = 32 > 0$	$f_{xx}(3, 1) = 4 > 0$	$f(3, 1) = -\frac{44}{3}$ is a local minimum by (16.40)(ii)
$(3, -1)$	$D(3, -1) = -32 < 0$	irrelevant	$(3, -1, f(3, -1))$ is a saddle point by (16.41)
$(-1, 1)$	$D(-1, 1) = -32 < 0$	irrelevant	$(-1, 1, f(-1, 1))$ is a saddle point by (16.41)
$(-1, -1)$	$D(-1, -1) = 32 > 0$	$f_{xx}(-1, -1) = -4 < 0$	$f(-1, -1) = \frac{4}{3}$ is a local maximum by (16.40)(i)

EXAMPLE 6 A rectangular box with no top is to be constructed to have a volume $V = 12\ \text{ft}^3$. The cost per square foot of the material to be used is \$4 for the bottom, \$3 for two of the opposite sides, and \$2 for the remaining pair of opposite sides. Find the dimensions of the box that will minimize the cost.

FIGURE 16.74

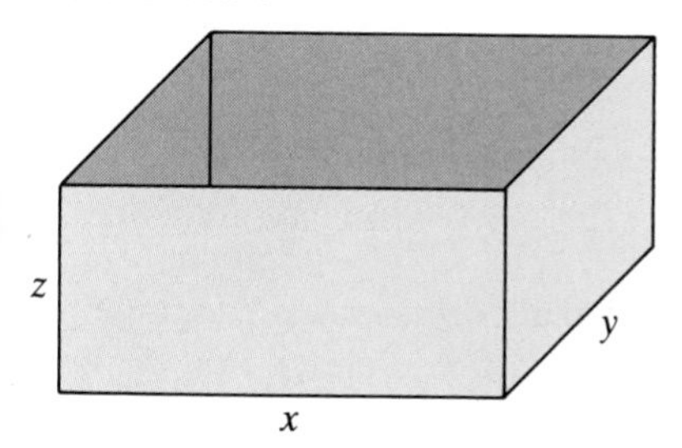

SOLUTION As in Figure 16.74, let x and y denote the dimensions (in feet) of the base and z the altitude (in feet). Since the area of the base is xy and there are two sides of area xz and two sides of area yz, the cost C (in dollars) of the material is

$$C = 4xy + 3(2xz) + 2(2yz),$$

where x, y, and z are positive. Since $V = xyz = 12$, it follows that $z = 12/(xy)$. Substituting for z in the formula for C and simplifying, we obtain

$$C = 4xy + \frac{72}{y} + \frac{48}{x}.$$

There are no boundary points, since $x > 0$ and $y > 0$ for every (x, y). Hence there are no boundary extrema. To determine possible local extrema, we solve the following system of two equations:

$$C_x = 4y - \frac{48}{x^2} = 0, \quad C_y = 4x - \frac{72}{y^2} = 0.$$

Equivalent equations are

$$y = \frac{12}{x^2} \quad \text{and} \quad xy^2 = 18.$$

Substituting $y = 12/x^2$ into the second equation gives us

$$x\left(\frac{144}{x^4}\right) = 18, \quad \text{or} \quad 144 = 18x^3.$$

Hence $x^3 = 8$, or $x = 2$. Since $y = 12/x^2$, the corresponding value of y is $\frac{12}{4}$, or 3. Using Theorem (16.40), we can show that these values of x and y determine a minimum value of C. Finally, using $z = 12/(xy)$, we obtain $z = 12/(2 \cdot 3) = 2$. Thus, the minimum cost occurs if the dimensions of the base are 3 feet by 2 feet and the altitude is 2 feet. The longer sides should be made out of the $2 material and the shorter sides out of the $3 material.

EXAMPLE 7 In the design of modern computers, it is necessary to consider the speed of light (approximately 186,000 mi/sec) when positioning electrical components, so that they can communicate with one another as rapidly as possible. Suppose that three components are located in a coordinate plane at $P_1(x_1, y_1)$, $P_2(x_2, y_2)$, and $P_3(x_3, y_3)$ and that each angle of triangle $P_1P_2P_3$ is less than 120°. A fourth component, which will communicate equally with the other three, is to be placed at a point $P(x, y)$ that will minimize the total delay time between the transmission and reception of signals (see Figure 16.75). If P is positioned properly, find the angles between the following pairs of vectors:

$$\overrightarrow{P_1P}, \overrightarrow{P_2P}; \quad \overrightarrow{P_1P}, \overrightarrow{P_3P}; \quad \overrightarrow{P_2P}, \overrightarrow{P_3P}$$

SOLUTION Let d_1, d_2, and d_3 be the distances between P and P_1, P and P_2, and P and P_3, respectively. Thus,

$$d_k = \sqrt{(x - x_k)^2 + (y - y_k)^2} \quad \text{for} \quad k = 1, 2, 3.$$

FIGURE 16.75

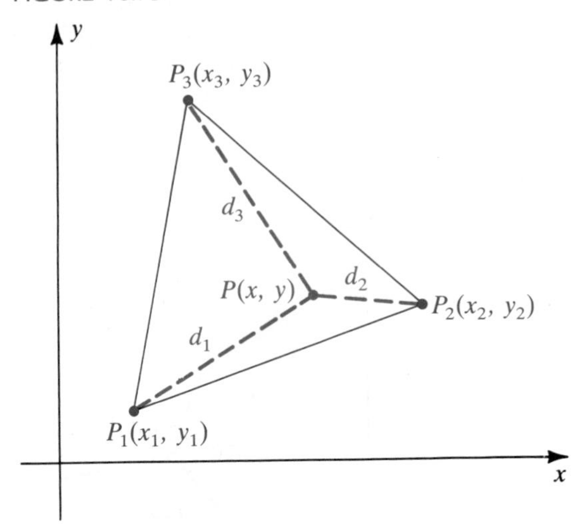

The minimal delay time between the transmission and the reception of signals will occur if (x, y) gives the minimal value of

$$f(x, y) = d_1 + d_2 + d_3.$$

To find the critical points of f, we consider the following system of equations:

$$\begin{cases} f_x(x, y) = \dfrac{x - x_1}{d_1} + \dfrac{x - x_2}{d_2} + \dfrac{x - x_3}{d_3} = 0 \\ f_y(x, y) = \dfrac{y - y_1}{d_1} + \dfrac{y - y_2}{d_2} + \dfrac{y - y_3}{d_3} = 0 \end{cases}$$

If we transpose the last term in each equation to the right-hand side, square each side, and add, we obtain

$$\left(\frac{x - x_1}{d_1} + \frac{x - x_2}{d_2}\right)^2 + \left(\frac{y - y_1}{d_1} + \frac{y - y_2}{d_2}\right)^2 = \frac{(x - x_3)^2 + (y - y_3)^2}{d_3^2} = 1.$$

You should verify that this simplifies to

$$(*) \qquad \frac{(x - x_1)(x - x_2) + (y - y_1)(y - y_2)}{d_1 d_2} = -\frac{1}{2}.$$

Let θ be the angle between the vectors

$$\overrightarrow{P_1P} = (x - x_1)\mathbf{i} + (y - y_1)\mathbf{j}$$

and

$$\overrightarrow{P_2P} = (x - x_2)\mathbf{i} + (y - y_2)\mathbf{j}.$$

Applying Corollary (14.20) and using $(*)$, we have

$$\cos\theta = \frac{\overrightarrow{P_1P} \cdot \overrightarrow{P_2P}}{\|\overrightarrow{P_1P}\|\|\overrightarrow{P_2P}\|} = -\frac{1}{2}.$$

Hence $\theta = \arccos\left(-\frac{1}{2}\right) = 120°$. By symmetry, the angles between $\overrightarrow{P_1P}$ and $\overrightarrow{P_3P}$ and between $\overrightarrow{P_2P}$ and $\overrightarrow{P_3P}$ must also equal 120°, as illustrated in Figure 16.76.

FIGURE 16.76

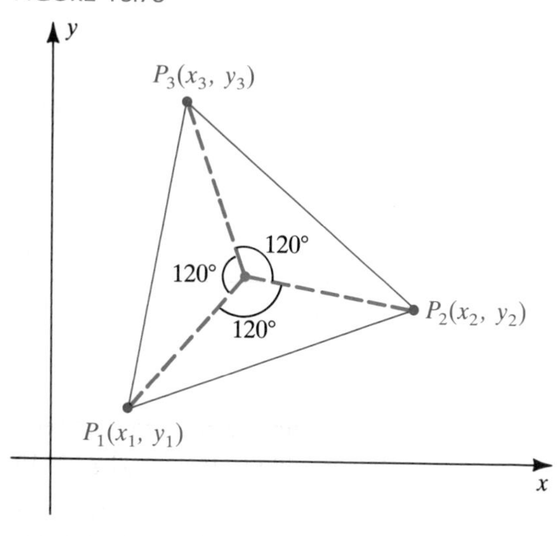

The discussion in this section can be generalized to functions of more than two variables. For example, given $f(x, y, z)$, we define local maxima and minima in a manner analogous to that used for the two-variable case. If f has first partial derivatives, then a local extremum can occur only at a point where f_x, f_y, and f_z are simultaneously 0. It is difficult to obtain tests for determining whether such a point corresponds to a maximum, a minimum, or neither. However, in applications we can often determine this by analyzing the physical nature of the problem.

EXERCISES 16.8

Exer. 1–20: Find the extrema and saddle points of f.

1 $f(x, y) = -x^2 - 4x - y^2 + 2y - 1$

2 $f(x, y) = x^2 - 2x + y^2 - 6y + 12$

3 $f(x, y) = x^2 + 4y^2 - x + 2y$

4 $f(x, y) = 5 + 4x - 2x^2 + 3y - y^2$

5 $f(x, y) = x^2 + 2xy + 3y^2$

6 $f(x, y) = x^2 - 3xy - y^2 + 2y - 6x$

7 $f(x, y) = x^3 + 3xy - y^3$

8 $f(x, y) = x^2 + xy$

9 $f(x, y) = \frac{1}{2}x^2 + 2xy - \frac{1}{2}y^2 + x - 8y$

10 $f(x, y) = -2x^2 - 2xy - \frac{3}{2}y^2 - 14x - 5y$

11 $f(x, y) = \frac{1}{3}x^3 - \frac{2}{3}y^3 + \frac{1}{2}x^2 - 6x + 32y + 4$

12 $f(x, y) = \frac{1}{3}x^3 + \frac{1}{3}y^3 - \frac{3}{2}x^2 - 4y$

13 $f(x, y) = \frac{1}{2}x^4 - 2x^3 + 4xy + y^2$

14 $f(x, y) = \frac{1}{3}x^3 + 4xy - 9x - y^2$

15 $f(x, y) = x^4 + y^3 + 32x - 9y$

16 $f(x, y) = -\frac{1}{3}x^3 + xy + \frac{1}{2}y^2 - 12y$

17 $f(x, y) = e^x \sin y$

18 $f(x, y) = x \sin y$

19 $f(x, y) = \dfrac{4y + x^2y^2 + 8x}{xy}$

20 $f(x, y) = \dfrac{x}{x + y}$

21 Shown in the figure is a graph of

$$f(x, y) = (x^2 + 3y^2)e^{-(x^2+y^2)}.$$

Show that there are five critical points, and find the extrema of f.

EXERCISE 21

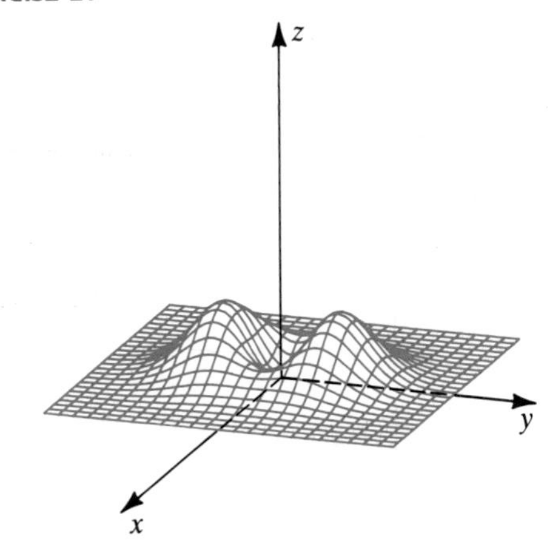

22 Shown in the figure is a graph of

$$f(x, y) = xy^2e^{-(x^2+y^2)/4}.$$

(a) Show that there are an infinite number of critical points.

(b) Find the coordinates of the four critical points shown in the figure.

EXERCISE 22

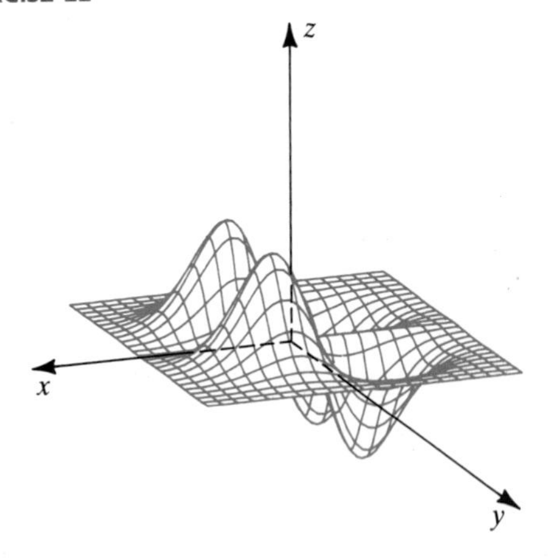

Exer. 23–28: Find the maximum and minimum values of f on R. (Refer to Exercises 3–8 for local extrema.)

23 $f(x, y) = x^2 + 4y^2 - x + 2y$;
the region R bounded by the ellipse $x^2 + 4y^2 = 1$

24 $f(x, y) = 5 + 4x - 2x^2 + 3y - y^2$;
the triangular region R bounded by the lines $y = x$, $y = -x$, and $y = 2$

25 $f(x, y) = x^2 + 2xy + 3y^2$;
$R = \{(x, y): -2 \le x \le 4, -1 \le y \le 3\}$

26 $f(x, y) = x^2 - 3xy - y^2 + 2y - 6x$;
$R = \{(x, y): |x| \le 3, |y| \le 2\}$

27 $f(x, y) = x^3 + 3xy - y^3$;
the triangular region R with vertices $(1, 2)$, $(1, -2)$, and $(-1, -2)$

28 $f(x, y) = x^2 + xy$;
the region R bounded by the graphs of $y = x^2$ and $y = 9$

29 Find the shortest distance from the point $P(2, 1, -1)$ to the plane $4x - 3y + z = 5$.

30 Find the shortest distance between the parallel planes $2x + 3y - z = 2$ and $2x + 3y - z = 4$.

31 Find the points on the graph of $xy^3z^2 = 16$ that are closest to the origin.

32 Find three positive real numbers whose sum is 1000 and whose product is a maximum.

33 If an open rectangular box is to have a fixed volume V, what relative dimensions will make the surface area a minimum?

34 If an open rectangular box is to have a fixed surface area A, what relative dimensions will make the volume a maximum?

35 Find the dimensions of the rectangular box of maximum volume with faces parallel to the coordinate planes that can be inscribed in the ellipsoid

$$16x^2 + 4y^2 + 9z^2 = 144.$$

36 Generalize Exercise 35 to any ellipsoid

$$\frac{x^2}{a^2} + \frac{y^2}{b^2} + \frac{z^2}{c^2} = 1.$$

37 Find the dimensions of the rectangular box of maximum volume that has three of its faces in the coordinate planes, one vertex at the origin, and another vertex in the first octant on the plane $4x + 3y + z = 12$.

38 Generalize Exercise 37 to any plane

$$\frac{x}{a} + \frac{y}{b} + \frac{z}{c} = 1,$$

where a, b, and c are positive real numbers.

39 A company plans to manufacture closed rectangular boxes that have a volume of 8 ft^3. Find the dimensions that will minimize the cost if the material for the top and bottom costs twice as much as the material for the sides.

40 A window has the shape of a rectangle surmounted by an isosceles triangle, as illustrated in the figure on the next page. If the perimeter of the window is 12 feet, what values of x, y, and θ will maximize the total area?

EXERCISE 40

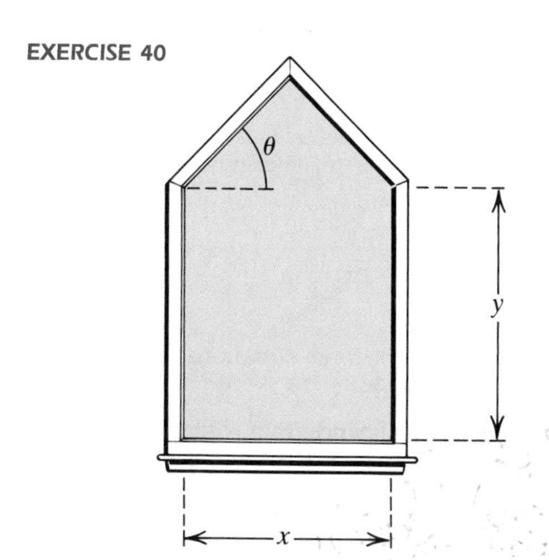

41 The U.S. Postal Service will not accept a rectangular box if the sum of its length and girth (the perimeter of a cross section that is perpendicular to the length) is more than 108 inches. Find the dimensions of the box of maximum volume that can be mailed.

42 Find a vector in three dimensions having magnitude 8 such that the sum of its components is as large as possible.

Exer. 43–44: Refer to Example 7.

43 Three electrical components of a computer are located at $P_1(0, 0)$, $P_2(4, 0)$, and $P_3(0, 4)$. Locate the position of a fourth component so that the signal delay time is minimal.

44 Show that if triangle $P_1P_2P_3$ contains an angle that is greater than or equal to 120°, then P cannot be located as in Figure 16.76.

45 In scientific experiments, corresponding values of two quantities x and y are often tabulated as follows:

x values	x_1	x_2	$\cdots$	x_n
y values	y_1	y_2	$\cdots$	y_n

Plotting the points (x_k, y_k) may lead the investigator to conjecture that x and y are related *linearly*; that is, $y = mx + b$ for some m and b. Thus, it is desirable to find a line l having that equation which best fits the data, as illustrated in the figure. Statisticians call l a *linear regression line*.

One technique for finding l is to employ the *method of least squares*. To use this method, consider, for each k, the *vertical deviation* $d_k = y_k - (mx_k + b)$ of the point (x_k, y_k) from the line $y = mx + b$ (see figure). Values of m and b are then determined that minimize the sum of the squares $\sum_{k=1}^{n} d_k^2$ (the squares d_k^2 are used because some of the d_k may be negative). Substituting for d_k produces the following function f of m and b:

$$f(m, b) = \sum_{k=1}^{n} (y_k - mx_k - b)^2$$

Show that the line $y = mx + b$ of best fit occurs if

$$\left(\sum_{k=1}^{n} x_k\right)m + nb = \sum_{k=1}^{n} y_k$$

and
$$\left(\sum_{k=1}^{n} x_k^2\right)m + \left(\sum_{k=1}^{n} x_k\right)b = \sum_{k=1}^{n} x_k y_k.$$

Thus, the line can be found by solving this system of two equations for the two unknowns m and b.

EXERCISE 45

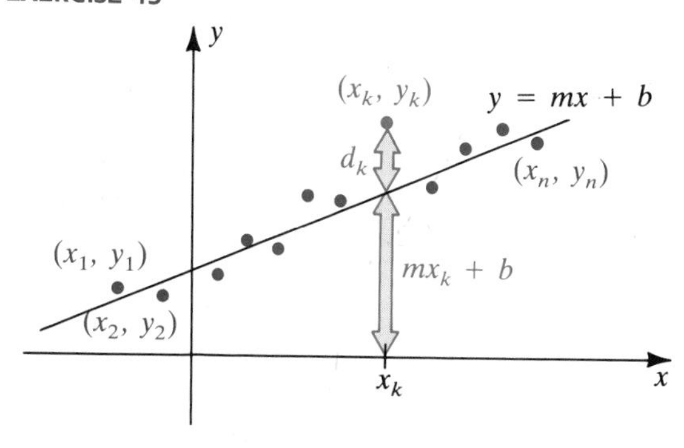

46 Given the equations in Exercise 45, show that the sum $\sum_{k=1}^{n} d_k$ of the deviations is 0. (This means that the positive and negative deviations cancel one another, and it is one reason for using $\sum_{k=1}^{n} d_k^2$ in the method of least squares.)

Exer. 47–48. Use the method of least squares (see Exercise 45) to find a line $y = mx + b$ that best fits the given data.

47

x values	1	4	7
y values	3	5	6

48

x values	1	4	6	8
y values	1	3	2	4

c 49 The following table lists the relationship between semester averages and scores on the final examination for ten students in a mathematics class.

Semester average	40	55	62	68	72	76	80	86	90	94
Final examination	30	45	65	72	60	82	76	92	88	98

Fit these data to a line, and use the line to estimate the final examination grade of a student with an average of 70.

C **50** In studying the stress-strain diagram of an elastic material (see page 311), an engineer finds that part of the curve appears to be linear. Experimental values are listed in the following table.

Stress (lb)	2	2.2	2.4	2.6	2.8	3.0
Strain (in.)	0.10	0.30	0.40	0.60	0.70	0.90

Fit these data to a straight line, and estimate the strain when the stress is 2.5 pounds.

51 Shown in the figure are the relative positions of three towns, A, B, and C. City planners want to use the least-squares criterion to decide where to construct a new high school that will serve all three communities. They will construct the school about a point $P(x, y)$ at which the sum of the squares of the distances from towns A, B, and C is a minimum. Find the relative position of the construction site.

EXERCISE 51

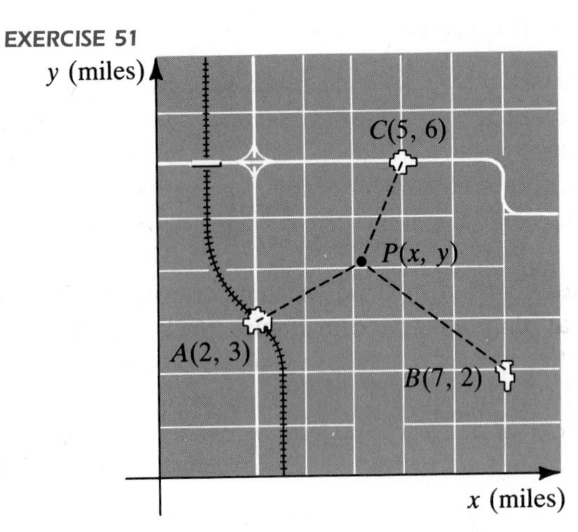

52 Generalize Exercise 51 to the case of n towns at positions $Q_1(x_1, y_1), Q_2(x_2, y_2), \ldots, Q_n(x_n, y_n)$.

53 Exercise 45 may be generalized to solve the problem of finding the plane $z = ax + by + c$ that best fits data $(x_1, y_1, z_1), (x_2, y_2, z_2), \ldots, (x_n, y_n, z_n)$. The method of least squares attempts to determine values a, b, and c such that

$$f(a, b, c) = \sum_{k=1}^{n} (z_k - ax_k - by_k - c)^2$$

is minimized.

(a) Find a system of three equations for the three unknowns a, b, and c.

(b) Find the plane of best fit that passes through $(0, 0, 0)$, $(0, 1, 0)$, $(0, 0, 1)$, and $(1, 1, 2)$.

(*Hint:* Solve the system of three equations $f_a = 0, f_b = 0, f_c = 0$.)

54 Three alleles (alternative forms of a gene) A, B, and O determine the four human blood types: A (AA or AO), B (BB or BO), O (OO), and AB. The *Hardy-Weinberg law* asserts that the proportion of individuals in a population who carry two different alleles is given by the formula

$$P = 2pq + 2pr + 2rq,$$

where p, q, and r are the proportions of alleles A, B, and O, respectively, in the population. Show that P must be less than or equal to $\frac{2}{3}$. (*Hint:* $p \geq 0$, $q \geq 0$, $r \geq 0$, and $p + q + r = 1$.)

C **Exer. 55–56: Estimate the critical points of f on $R = \{(x, y): |x| \leq 1.5 \text{ and } |y| \leq 1.5\}$ by graphing $f_x(x, y) = 0$ and $f_y(x, y) = 0$ on the same coordinate plane.**

55 $f(x, y) = x^3 \sin x - xy + 4y^2 + y$

56 $f(x, y) = xy - \arctan x - y^{5/4}$

16.9 LAGRANGE MULTIPLIERS

In many applications we must find the extrema of a function f of several variables when the variables are restricted in some manner. As an illustration, suppose we wish to find the volume of the largest rectangular box with faces parallel to the coordinate planes that can be inscribed in the ellipsoid $16x^2 + 4y^2 + 9z^2 = 144$. Note that, by symmetry, it is sufficient to examine the part in the first octant illustrated in Figure 16.77. If $P(x, y, z)$ is the vertex shown in the figure, then the volume V of the entire box is $V = 8xyz$. We must find the maximum value of V subject to the **constraint** (or **side condition**)

$$16x^2 + 4y^2 + 9z^2 - 144 = 0.$$

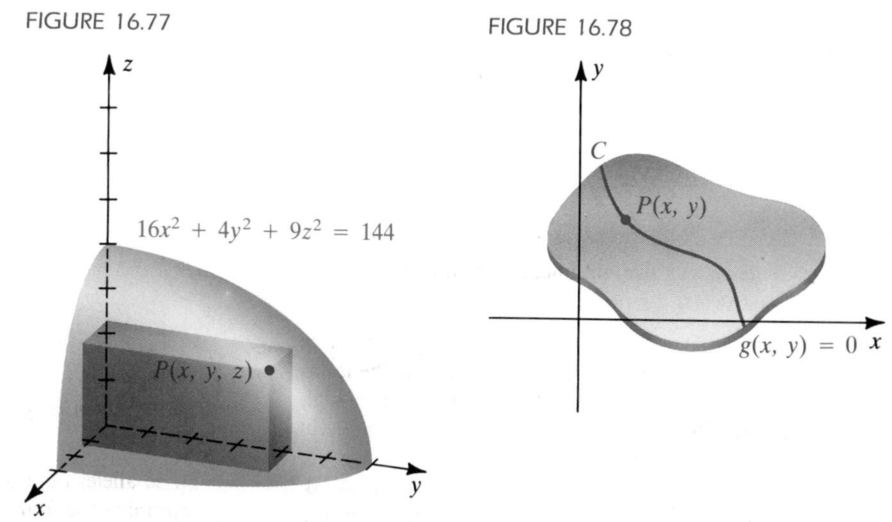

Solving this equation for z and substituting in the formula for V, we obtain

$$V = (8xy)\tfrac{1}{3}\sqrt{144 - 16x^2 - 4y^2}.$$

We may now find the extrema by (16.40); however, this method is cumbersome because of the manipulations involved in finding partial derivatives and critical points. Another disadvantage of this technique is that for similar problems it may be impossible to solve for z. For these reasons it is sometimes simpler to employ the method of *Lagrange multipliers* discussed in this section.*

*This method was invented by the French mathematician Joseph-Louis Lagrange (1736–1813).

As another illustration, let $f(x, y)$ be the temperature at the point $P(x, y)$ on the flat metal plate illustrated in Figure 16.78, and let C be a curve that has the equation $g(x, y) = 0$. Suppose we wish to find the points on C at which the temperature attains its largest or smallest value. This amounts to finding the extrema of $f(x, y)$ *subject to the constraint* $g(x, y) = 0$. One technique for accomplishing this is to use the following result.

Lagrange's theorem (16.42)

> Suppose f and g are functions of two variables that have continuous first partial derivatives, and that $\nabla g \neq \mathbf{0}$ throughout a region of the xy-plane. If f has an extremum $f(x_0, y_0)$ subject to the constraint $g(x, y) = 0$, then there is a real number λ such that
>
> $$\nabla f(x_0, y_0) = \lambda \nabla g(x_0, y_0).$$

FIGURE 16.79

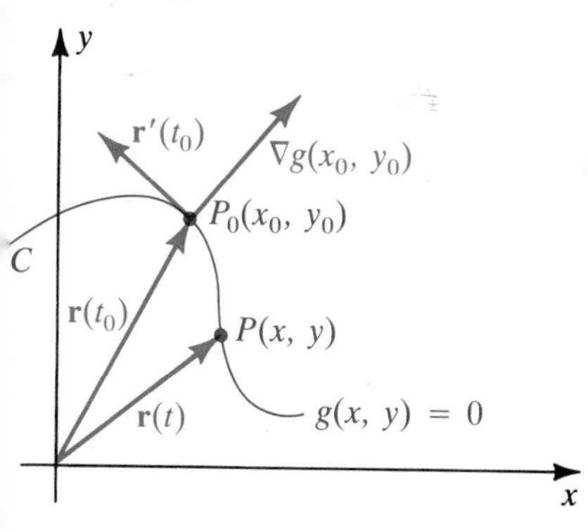

PROOF The graph of $g(x, y) = 0$ is a curve C in the xy-plane. It can be shown that, under the given conditions, C has a smooth parametrization

$$x = h(t), \quad y = k(t)$$

for t in some interval I. Let

$$\mathbf{r}(t) = x\mathbf{i} + y\mathbf{j} = h(t)\mathbf{i} + k(t)\mathbf{j}$$

be the position vector of the point $P(x, y)$ on C (see Figure 16.79), and let the point $P_0(x_0, y_0)$ at which f has an extremum correspond to $t = t_0$; that is,

$$\mathbf{r}(t_0) = x_0\mathbf{i} + y_0\mathbf{j} = h(t_0)\mathbf{i} + k(t_0)\mathbf{j}.$$

If we define a function F of one variable t by

$$F(t) = f(h(t), k(t)),$$

then, as t varies, we obtain function values $f(x, y)$ that correspond to (x, y) on C; that is, $f(x, y)$ is subject to the constraint $g(x, y) = 0$. Since $f(x_0, y_0)$ is an extremum of f under these conditions, it follows that $F(t_0) = f(h(t_0), k(t_0))$ is an extremum of $F(t)$. Thus, $F'(t_0) = 0$. If we regard F as a composite function, then, by a chain rule,

$$F'(t) = f_x(x, y)\frac{dx}{dt} + f_y(x, y)\frac{dy}{dt} = f_x(x, y)h'(t) + f_y(x, y)k'(t).$$

Letting $t = t_0$, we have

$$0 = F'(t_0) = f_x(x_0, y_0)h'(t_0) + f_y(x_0, y_0)k'(t_0) = \nabla f(x_0, y_0) \cdot \mathbf{r}'(t_0).$$

This shows that the vector $\nabla f(x_0, y_0)$ is orthogonal to the tangent vector $\mathbf{r}'(t_0)$ to C (see Theorem (16.37)). However, $\nabla g(x_0, y_0)$ is also orthogonal to $\mathbf{r}'(t_0)$, because C is a level curve for g. Since $\nabla f(x_0, y_0)$ and $\nabla g(x_0, y_0)$ are orthogonal to the same vector, they are parallel; that is, $\nabla f(x_0, y_0) = \lambda \nabla g(x_0, y_0)$ for some λ. ■

The number λ in the preceding theorem is called a **Lagrange multiplier**.

The equation $\nabla f(x_0, y_0) = \lambda \nabla g(x_0, y_0)$ in Theorem (16.42) may be written

$$f_x(x_0, y_0)\mathbf{i} + f_y(x_0, y_0)\mathbf{j} = \lambda g_x(x_0, y_0)\mathbf{i} + \lambda g_y(x_0, y_0)\mathbf{j}.$$

Equating the **i** and **j** components and using the constraint $g(x, y) = 0$ leads to the following non-vector restatement of Lagrange's theorem, in which it is assumed that f and g have continuous first partial derivatives and $g_x^2 + g_y^2 \neq 0$—that is, g_x and g_y are not simultaneously zero.

Corollary **(16.43)**

The points at which a function f of two variables has relative extrema subject to the constraint $g(x, y) = 0$ are included among the points (x, y) determined by the first two coordinates of the solutions (x, y, λ) of the system of equations

$$\begin{cases} f_x(x, y) = \lambda g_x(x, y) \\ f_y(x, y) = \lambda g_y(x, y) \\ g(x, y) = 0 \end{cases}$$

When using Corollary (16.43), we first find all solutions

$$(x_1, y_1, \lambda_1), \quad (x_2, y_2, \lambda_2), \quad (x_3, y_3, \lambda_3), \quad \ldots$$

of the indicated system of equations. (Sometimes there is only *one* solution.) The points at which the relative extrema of f occur are among

$$(x_1, y_1), \quad (x_2, y_2), \quad (x_3, y_3), \quad \ldots.$$

Thus, the Lagrange multipliers $\lambda_1, \lambda_2, \lambda_3, \ldots$ are discarded after the solutions (x_k, y_k, λ_k) are found. Each point (x_k, y_k) is then examined to determine whether $f(x_k, y_k)$ is a relative maxima, a relative minima, or neither.

There is a complicated test (proved in advanced mathematics) for distinguishing between maxima and minima; however, it is usually easier to use graphical or physical considerations to make the distinction. In solutions to problems we do not usually give rigorous proofs that our results yield extrema.

EXAMPLE 1 Find the extrema of $f(x, y) = xy$ if (x, y) is restricted to the ellipse $4x^2 + y^2 = 4$.

SOLUTION In this example the constraint is $g(x, y) = 4x^2 + y^2 - 4 = 0$. Setting $\nabla f(x, y) = \lambda \nabla g(x, y)$ (see (16.42)), we obtain

$$y\mathbf{i} + x\mathbf{j} = \lambda(8x\mathbf{i} + 2y\mathbf{j}).$$

Thus, the system of equations in Corollary (16.43) is

$$\begin{cases} y = 8x\lambda \\ x = 2y\lambda \\ 4x^2 + y^2 - 4 = 0 \end{cases}$$

There are many ways to find the solutions of this system. One method is to let $y = 8x\lambda$ in $x = 2y\lambda$, obtaining

$$x = 2(8x\lambda)\lambda = 16x\lambda^2.$$

An equivalent equation is

$$x - 16x\lambda^2 = 0, \quad \text{or} \quad x(1 - 16\lambda^2) = 0.$$

Therefore either $x = 0$ or $\lambda = \pm\frac{1}{4}$.

If $\lambda = \pm\frac{1}{4}$, then

$$y = 8x\lambda = 8x(\pm\tfrac{1}{4}), \quad \text{or} \quad y = \pm 2x.$$

Substituting for y in the equation $4x^2 + y^2 - 4 = 0$, we obtain

$$4x^2 + (\pm 2x)^2 - 4 = 0, \quad 8x^2 = 4, \quad \text{or} \quad x = \pm\frac{1}{\sqrt{2}} = \pm\frac{\sqrt{2}}{2}.$$

The corresponding values of y are

$$y = \pm 2x = \pm 2\left(\pm\frac{\sqrt{2}}{2}\right) = \pm\sqrt{2}.$$

Discarding the Lagrange multipliers gives us the four points $(\sqrt{2}/2, \pm\sqrt{2})$ and $(-\sqrt{2}/2, \pm\sqrt{2})$.

If $x = 0$, then substituting into the equation $4x^2 + y^2 - 4 = 0$ gives us $y^2 - 4 = 0$, or $y = \pm 2$. Thus, the points $(0, \pm 2)$ may also determine extrema of f.

The value of f at each of the points we have found is listed in the following table.

(x, y)	$(0, 2)$	$(0, -2)$	$(\sqrt{2}/2, \sqrt{2})$	$(\sqrt{2}/2, -\sqrt{2})$	$(-\sqrt{2}/2, \sqrt{2})$	$(-\sqrt{2}/2, -\sqrt{2})$
$f(x, y)$	0	0	1	-1	-1	1

Thus, $f(x, y)$ takes on a maximum value of 1 at either $(\sqrt{2}/2, \sqrt{2})$ or $(-\sqrt{2}/2, -\sqrt{2})$ and a minimum value of -1 at $(\sqrt{2}/2, -\sqrt{2})$ or $(-\sqrt{2}/2, \sqrt{2})$. These facts are indicated on the ellipse shown in Figure 16.80.

FIGURE 16.80

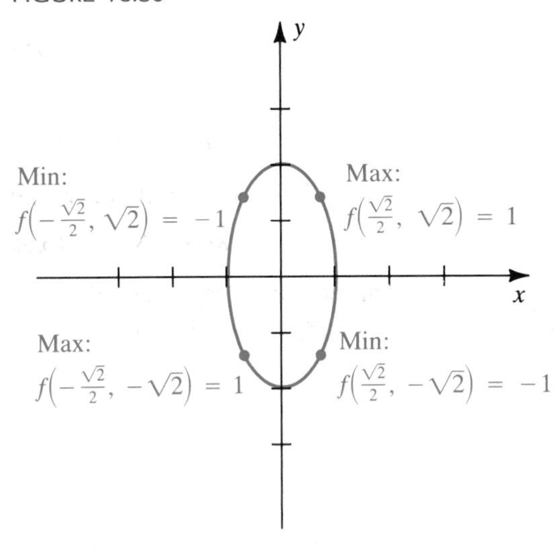

Lagrange's theorem can be extended to functions of more than two variables. The next result is one of many such extensions. We can state similar theorems for functions of any number of variables.

Extension of Lagrange's theorem **(16.44)**

Suppose that f and g are functions of three variables having continuous first partial derivatives and that $\nabla g \neq 0$ throughout a region in an xyz-coordinate system. If f has an extremum $f(x_0, y_0, z_0)$ subject to the constraint $g(x, y, z) = 0$, then there is a real number λ such that

$$\nabla f(x_0, y_0, z_0) = \lambda \nabla g(x_0, y_0, z_0).$$

The following is an analogue of Corollary (16.43).

Corollary **(16.45)**

The points at which a function f of three variables has relative extrema subject to the constraint $g(x, y, z) = 0$ are included among the points (x, y, z) determined by the first three coordinates of the solutions (x, y, z, λ) of the system of equations

$$\begin{cases} f_x(x, y, z) = \lambda g_x(x, y, z) \\ f_y(x, y, z) = \lambda g_y(x, y, z) \\ f_z(x, y, z) = \lambda g_z(x, y, z) \\ g(x, y, z) = 0 \end{cases}$$

The next example was discussed briefly at the beginning of this section.

EXAMPLE 2 Find the volume of the largest rectangular box with faces parallel to the coordinate planes that can be inscribed in the ellipsoid $16x^2 + 4y^2 + 9z^2 = 144$.

SOLUTION A typical position for the box is illustrated in Figure 16.77. We wish to maximize the volume

$$V = f(x, y, z) = 8xyz$$

subject to the constraint

$$g(x, y, z) = 16x^2 + 4y^2 + 9z^2 - 144 = 0.$$

Setting $\nabla f(x, y, z) = \lambda \nabla g(x, y, z)$ (see (16.44)), we obtain

$$8yz\mathbf{i} + 8xz\mathbf{j} + 8xy\mathbf{k} = \lambda(32x\mathbf{i} + 8y\mathbf{j} + 18z\mathbf{k}).$$

This equation, together with the constraint $g(x, y, z) = 0$, gives us the following system of four equations (see Corollary (16.45)):

$$\begin{cases} 8yz = 32x\lambda \\ 8xz = 8y\lambda \\ 8xy = 18z\lambda \\ 16x^2 + 4y^2 + 9z^2 - 144 = 0 \end{cases}$$

Multiplying the first equation by x, the second by y, and the third by z and adding gives us

$$24xyz = 32x^2\lambda + 8y^2\lambda + 18z^2\lambda = 2\lambda(16x^2 + 4y^2 + 9z^2).$$

This result, coupled with the constraint $16x^2 + 4y^2 + 9z^2 = 144$, implies that

$$24xyz = 2\lambda(144), \quad \text{or} \quad xyz = 12\lambda.$$

The last equation may be used to find x, y, and z. For example, multiplying both sides of the equation $8yz = 32x\lambda$ by x and using $xyz = 12\lambda$, we obtain

$$\begin{aligned} 8xyz &= 32x^2\lambda \\ 8(12\lambda) &= 32x^2\lambda \\ 96\lambda - 32x^2\lambda &= 0 \\ 32\lambda(3 - x^2) &= 0. \end{aligned}$$

Consequently either $\lambda = 0$ or $x = \sqrt{3}$. We may reject $\lambda = 0$, since this leads to $xyz = 0$ and, hence, $V = 8xyz = 0$. Thus, the only possibility is $x = \sqrt{3}$.

Similarly, multiplying both sides of the equation $8xz = 8y\lambda$ by y leads to

$$\begin{aligned} 8xyz &= 8y^2\lambda \\ 8(12\lambda) &= 8y^2\lambda \\ 96\lambda - 8y^2\lambda &= 0 \\ 8\lambda(12 - y^2) &= 0 \end{aligned}$$

and thus to $y = \sqrt{12} = 2\sqrt{3}$.

Finally, multiplying the equation $8xy = 18z\lambda$ by z and using the same technique, we obtain $z = 4/\sqrt{3}$. It follows that the desired volume is

$$V = 8xyz = 8(\sqrt{3})(2\sqrt{3})(4/\sqrt{3}) = 64\sqrt{3} \approx 111.$$

EXAMPLE 3 If $f(x, y, z) = 4x^2 + y^2 + 5z^2$, find the point on the plane $2x + 3y + 4z = 12$ at which $f(x, y, z)$ has its least value.

SOLUTION We shall examine $f(x, y, z) = 4x^2 + y^2 + 5z^2$ subject to the constraint $g(x, y, z) = 2x + 3y + 4z - 12 = 0$. As in Example 2 we consider $\nabla f(x, y, z) = \lambda g(x, y, z)$, that is,

$$\nabla(4x^2 + y^2 + 5z^2) = \lambda\nabla(2x + 3y + 4z - 12),$$

obtaining

$$8x\mathbf{i} + 2y\mathbf{j} + 10z\mathbf{k} = \lambda(2\mathbf{i} + 3\mathbf{j} + 4\mathbf{k}).$$

Equating components and using $g(x, y, z) = 0$ leads to the following system of equations (see Corollary (16.45)):

$$\begin{cases} 8x = 2\lambda \\ 2y = 3\lambda \\ 10z = 4\lambda \\ 2x + 3y + 4z - 12 = 0 \end{cases}$$

Solving the first three equations for λ gives us

$$\lambda = 4x = \tfrac{2}{3}y = \tfrac{5}{2}z.$$

These conditions imply that

$$y = 6x \quad \text{and} \quad z = \tfrac{8}{5}x.$$

Substituting into the constraint equation, $2x + 3y + 4z - 12 = 0$, we obtain

$$2x + 18x + \tfrac{32}{5}x - 12 = 0,$$

or $x = \frac{5}{11}$. Hence $y = 6(\frac{5}{11}) = \frac{30}{11}$ and $z = (\frac{8}{5})(\frac{5}{11}) = \frac{8}{11}$. Since there is only one critical point, it follows that the minimum value occurs at that point, $(\frac{5}{11}, \frac{30}{11}, \frac{8}{11})$.

Some applications may involve more than one constraint. In particular, consider the problem of finding the extrema of $f(x, y, z)$ subject to the *two* constraints

$$g(x, y, z) = 0 \quad \text{and} \quad h(x, y, z) = 0.$$

If f has an extremum subject to these constraints, then the following condition must be satisfied for some real numbers λ and μ:

$$\nabla f(x, y, z) = \lambda\nabla g(x, y, z) + \mu\nabla h(x, y, z)$$

By equating components and using the constraints, we obtain a system of five equations in the five unknowns x, y, z, λ, and μ. A specific illustration of this situation is given in the next example. This method can also be extended to functions of more than three variables and to more than two constraints.

EXAMPLE 4 Let C denote the first-octant arc of the curve in which the paraboloid $2z = 16 - x^2 - y^2$ and the plane $x + y = 4$ intersect. Find the points on C that are closest to and farthest from the origin. Find the minimum and maximum distances from the origin to C.

FIGURE 16.81

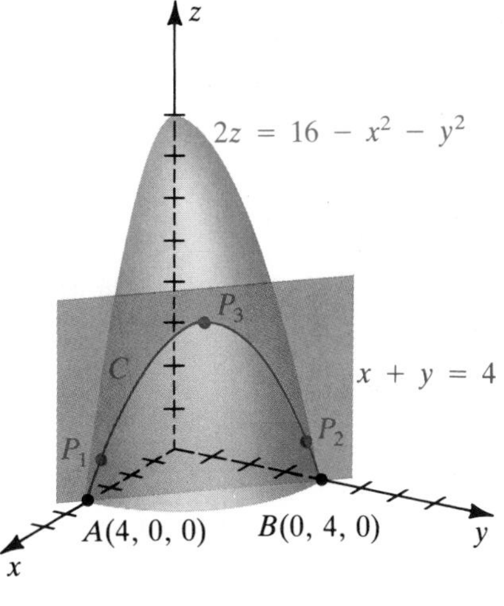

SOLUTION The curve C is sketched in Figure 16.81. If $P(x, y, z)$ is an arbitrary point on C, then we wish to find the largest and smallest values of $d(O, P) = \sqrt{x^2 + y^2 + z^2}$. These values may be found by determining the triples (x, y, z) that give the extrema of the radicand

$$f(x, y, z) = x^2 + y^2 + z^2$$

subject to the two constraints

$$g(x, y, z) = x^2 + y^2 + 2z - 16 = 0$$

$$h(x, y, z) = x + y - 4 = 0.$$

As in the discussion preceding this example, consider

$$\nabla(x^2 + y^2 + z^2) = \lambda\nabla(x^2 + y^2 + 2z - 16) + \mu\nabla(x + y - 4).$$

Thus,

$$\begin{aligned} 2x\mathbf{i} + 2y\mathbf{j} + 2z\mathbf{k} &= \lambda(2x\mathbf{i} + 2y\mathbf{j} + 2\mathbf{k}) + \mu(\mathbf{i} + \mathbf{j}) \\ &= (2x\lambda + \mu)\mathbf{i} + (2y\lambda + \mu)\mathbf{j} + (2\lambda)\mathbf{k}. \end{aligned}$$

Equating components and using the two constraints gives us the following system of five equations:

$$\begin{cases} 2x = 2x\lambda + \mu \\ 2y = 2y\lambda + \mu \\ 2z = 2\lambda \\ x^2 + y^2 + 2z - 16 = 0 \\ x + y - 4 = 0 \end{cases}$$

Subtracting the second equation from the first, we obtain the following equivalent equations:

$$2x - 2y = (2x\lambda + \mu) - (2y\lambda + \mu) = 2x\lambda - 2y\lambda$$

$$2(x - y) - 2\lambda(x - y) = 0$$

$$2(x - y)(1 - \lambda) = 0$$

Consequently either $\lambda = 1$ or $x = y$.

If $\lambda = 1$, we have $2z = 2\lambda = 2(1)$, or $z = 1$. The first constraint, that is, $x^2 + y^2 + 2z - 16 = 0$, then gives us $x^2 + y^2 - 14 = 0$. Solving this equation simultaneously with $x + y - 4 = 0$, we find that either

$$x = 2 + \sqrt{3}, \quad y = 2 - \sqrt{3} \quad \text{or} \quad x = 2 - \sqrt{3}, \quad y = 2 + \sqrt{3}.$$

Thus, points on C that may lead to extrema are

$$P_1(2 + \sqrt{3}, 2 - \sqrt{3}, 1) \quad \text{and} \quad P_2(2 - \sqrt{3}, 2 + \sqrt{3}, 1).$$

The corresponding distances from O are

$$d(O, P_1) = \sqrt{15} = d(O, P_2).$$

If $x = y$, then, using the constraint $x + y - 4 = 0$, we obtain the equivalent equations $x + x - 4 = 0$, $2x = 4$, or $x = 2$. This gives us $P_3(2, 2, 4)$ and $d(O, P_3) = 2\sqrt{6}$.

Referring to Figure 16.81, we may now make the following observations. As a point moves continuously along C from $A(4, 0, 0)$ to $B(0, 4, 0)$, its distance from the origin starts at $d(O, A) = 4$, decreases to a minimum value $\sqrt{15}$ at P_1, and then increases to a maximum value $2\sqrt{6}$ at P_3. The distance then decreases to $\sqrt{15}$ at P_2 and again increases to 4 at B.

As a check on the solution, note that parametric equations for C are

$$x = 4 - t, \quad y = t, \quad z = 4t - t^2; \quad 0 \le t \le 4.$$

In this case

$$f(x, y, z) = (4 - t)^2 + t^2 + (4t - t^2)^2,$$

and the extrema of f may be found using single-variable methods. It can be verified that the same points are obtained.

EXERCISES 16.9

Exer. 1–10: Use Lagrange multipliers to find the extrema of f subject to the stated constraints.

1 $f(x, y) = y^2 - 4xy + 4x^2$; $\quad x^2 + y^2 = 1$

2 $f(x, y) = 2x^2 + xy - y^2 + y$; $\quad 2x + 3y = 1$

3 $f(x, y, z) = x + y + z$; $\quad x^2 + y^2 + z^2 = 25$

4 $f(x, y, z) = x^2 + y^2 + z^2$; $\quad x + y + z = 25$

5 $f(x, y, z) = x^2 + y^2 + z^2$; $\quad x - y + z = 1$

6 $f(x, y, z) = x + 2y - 3z$; $\quad z = 4x^2 + y^2$

7 $f(x, y, z) = x^2 + y^2 + z^2$; $\quad x - y = 1, \quad y^2 - z^2 = 1$

8 $f(x, y, z) = z - x^2 - y^2$; $\quad x + y + z = 1, \quad x^2 + y^2 = 4$

9 $f(x, y, z, t) = xyzt$; $\quad x - z = 2, \quad y^2 + t = 4$

10 $f(x, y, z, t) = x^2 + y^2 + z^2 + t^2$; $\quad 3x + 4y = 5, \quad z + t = 2$

11 Find the point on the sphere $x^2 + y^2 + z^2 = 9$ that is closest to the point $(2, 3, 4)$.

12 Find the point on the line of intersection of the planes $x + 3y - 2z = 11$ and $2x - y + z = 3$ that is closest to the origin.

13 A closed rectangular box having a volume of 2 ft^3 is to be constructed. If the cost per square foot of the material for the sides, bottom, and top is \$1.00, \$2.00, and \$1.50, respectively, find the dimensions that will minimize the cost.

14 Prove that a closed rectangular box of fixed volume and minimal surface area is a cube.

15 Find the volume of the largest rectangular box that has three of its vertices on the positive x-, y-, and z-axes, respectively, and a fourth vertex on the plane $2x + 3y + 4z = 12$.

16 Find the dimensions of the rectangular box of maximum volume that has three of its faces in the coordinate planes, one vertex at the origin, and another vertex in the first octant on the plane $2x + 3y + 5z = 90$.

17 A container with a closed top and fixed surface area is to be constructed in the shape of a right circular cylinder. Find the relative dimensions that maximize the volume.

18 Find the dimensions of the rectangular box of maximum volume, with faces parallel to the coordinate planes, that can be inscribed in the ellipsoid $4x^2 + 4y^2 + z^2 = 36$.

19 Prove that the triangle of maximum area and fixed perimeter p is equilateral. (*Hint:* If the sides are x, y, z and if $s = \frac{1}{2}p$, then the area A is given by Heron's formula, $A = \sqrt{s(s - x)(s - y)(s - z)}$.)

20 Prove that the product of the sines of the angles of a triangle is greatest when the triangle is equilateral.

21 The strength of a rectangular beam varies as the product of its width and the square of its depth. Find the dimensions of the strongest rectangular beam that can be cut from a cylindrical log whose cross sections are elliptical with major and minor axes of lengths 24 inches and 16 inches, respectively.

22 If x units of capital and y units of labor are required to manufacture $f(x, y)$ units of a certain commodity, the *Cobb-Douglas production function* is defined by $f(x, y) = kx^a y^b$, where k is a constant and a and b are positive numbers such that $a + b = 1$. Suppose that $f(x, y) = x^{1/5}y^{4/5}$ and that each unit of capital costs C dollars and each unit of labor costs L dollars. If the total amount available for these costs is M dollars, so $xC + yL = M$, how many units of capital and labor will maximize production?

c **Exer. 23–24: Use Lagrange multipliers and graphs to estimate the extrema of $f(x, y)$ subject to the constraint $g(x, y) = 0$.**

23 $f(x, y) = y - \cos x + 2x$; $\quad g(x, y) = x^2 + 2y^2 - 1$

24 $f(x, y) = \frac{1}{5}x^5 + \frac{1}{3}y^3$; $\quad g(x, y) = x^2 + y^2 - 1$

16.10 REVIEW EXERCISES

Exer. 1–4: Describe the domain of f and the level curve or surface through P.

1 $f(x, y) = \sqrt{36 - 4x^2 + 9y^2}$; $\quad P(3, 4)$

2 $f(x, y) = \ln xy$; $\quad P(2, 3)$

3 $f(x, y, z) = (z^2 - x^2 - y^2)^{-3/2}$; $\quad P(0, 0, 1)$

4 $f(x, y, z) = \dfrac{\sec z}{x - y}$; $\quad P(5, 3, 0)$

Exer. 5–8: Find the limit or show that it does not exist.

5 $\displaystyle\lim_{(x,y)\to(0,0)} \frac{3xy + 5}{y^2 + 4}$

6 $\displaystyle\lim_{(x,y,z)\to(1,3,-2)} \frac{z^2 + z - 2}{xyz + 2xy}$

7 $\displaystyle\lim_{(x,y)\to(0,0)} \left(\frac{x^2 - y^2}{x^2 + y^2}\right)^2$

8 $\displaystyle\lim_{(x,y)\to(0,0)} \frac{x^2y^2}{x^4 + 2y^4}$

9 Use polar coordinates to help describe the level curves of

$$f(x, y) = \frac{xy}{(x^2 + y^2)^{3/2}},$$

and investigate $\displaystyle\lim_{(x,y)\to(0,0)} f(x, y)$.

10 The graph of $f(x, y) = x^2y^2/(x^2 + y^2)^2$ shown in the figure indicates that f has first partial derivatives at $(0, 0)$ but is not continuous there. Verify each of these facts.

EXERCISE 10

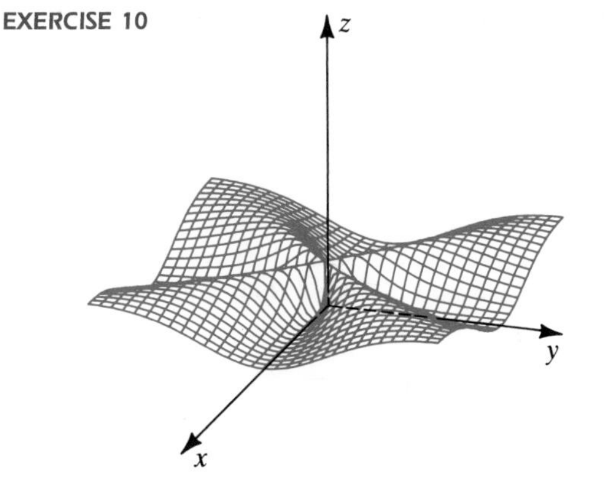

Exer. 11–16: Find the first partial derivatives of f.

11 $f(x, y) = x^3 \cos y - y^2 + 4x$

12 $f(r, s) = r^2 e^{rs}$

13 $f(x, y, z) = \dfrac{x^2 + y^2}{y^2 + z^2}$

14 $f(u, v, t) = u \ln \dfrac{v}{t}$

15 $f(x, y, z, t) = x^2 z\sqrt{2y + t}$

16 $f(v, w) = v^2 \cos w + w^2 \cos v$

Exer. 17–18: Find the second partial derivatives of f.

17 $f(x, y) = x^3y^2 - 3xy^3 + x^4 - 3y + 2$

18 $f(x, y, z) = x^2 e^{y^2 - z^2}$

19 If $u = (x^2 + y^2 + z^2)^{-1/2}$, prove that

$$\frac{\partial^2 u}{\partial x^2} + \frac{\partial^2 u}{\partial y^2} + \frac{\partial^2 u}{\partial z^2} = 0.$$

20 Find dw if

(a) $w = y^3 \tan^{-1} x^2 + 2x - y$ $\quad$ (b) $w = x^2 \sin yz$

21 (a) Find Δw and dw if $w = x^2 + 3xy - y^2$.

(b) If (x, y) changes from $(-1, 2)$ to $(-1.1, 2.1)$, use Δw to find the exact change in w, and use dw to find an approximate change in w.

22 Suppose that when Ohm's law, $R = V/I$, is used, there are percentage errors of $\pm 3\%$ and $\pm 2\%$ in the measurements of V and I, respectively. Use differentials to estimate the maximum percentage error in the calculated value of R.

Exer. 23–24: Prove that f is differentiable throughout its domain.

23 $f(x, y) = \dfrac{xy}{y^2 - x^2}$

24 $f(x, y, z) = \dfrac{xyz}{x - y}$

Exer. 25–27: Use a chain rule.

25 Find $\partial s/\partial x$ and $\partial s/\partial y$ if $s = uv + vw - uw$, $u = 2x + 3y$, $v = 4x - y$, and $w = -x + 2y$.

26 Find $\partial z/\partial r$, $\partial z/\partial s$, and $\partial z/\partial t$ if $z = ye^x$, $x = r + st$, and $y = 2r + 3s - t$.

27 Find dw/dt if $w = x \sin yz$, $x = 3e^{-t}$, $y = t^2$, and $z = 3t$.

28 Use partial derivatives to find dy/dx if $y = f(x)$ is determined implicitly by

$$x^3 - 4xy^3 - 3y + x - 2 = 0.$$

29 Find $\partial z/\partial x$ and $\partial z/\partial y$ if $z = f(x, y)$ is determined implicitly by $x^2y + z \cos y - xz^3 = 0$.

30 **(a)** Find the directional derivative of the function $f(x, y) = 3x^2 - y^2 + 5xy$ at $P(2, -1)$ in the direction of $\mathbf{a} = -3\mathbf{i} - 4\mathbf{j}$.

(b) Find the maximum rate of increase of $f(x, y)$ at P.

31 Suppose that the temperature at the point (x, y, z) is given by $T(x, y, z) = 3x^2 + 2y^2 - 4z$.

(a) Find the instantaneous rate of change of T at the point $P(-1, -3, 2)$ in the direction from P to the point $Q(-4, 1, -2)$.

(b) Find the maximum rate of change of T at P.

32 An *urban density model* is a formula that relates the population density (in number of people per square mile) to the distance from the center of the city. Suppose a rectangular coordinate system is introduced with the center of the city at the origin, and let $Q(x, y)$ denote the population density near the point $P(x, y)$. Decide the direction in which the population density increases most rapidly at $P(x, y)$ for the given density function Q, where a, b, and c are positive constants.

(a) $Q(x, y) = ae^{-b\sqrt{x^2+y^2}}$

(b) $Q(x, y) = ae^{-b\sqrt{x^2+y^2}+c(x^2+y^2)}$

33 Find equations of the tangent plane and normal line to the graph of $7z = 4x^2 - 2y^2$ at $P(-2, -1, 2)$.

34 A curve C has parametrization $x = t$, $y = t^2$, $z = t^3$; t in $\mathbb{R}$. If $f(x, y, z) = y^2 + xz$ and $\mathbf{u}$ is a unit tangent vector to C at $P(2, 4, 8)$, find $D_{\mathbf{u}} f(2, 4, 8)$.

35 Show that every plane tangent to the following cone passes through the origin:

$$\frac{x^2}{a^2} - \frac{y^2}{b^2} + \frac{z^2}{c^2} = 0$$

36 If $f(x, y) = \frac{1}{4}x^2 + \frac{1}{25}y^2$, sketch the level curve of f that contains $P(0, 5)$, and sketch $\nabla f]_P$.

37 If $F(x, y, z) = z + 4x^2 + 9y^2$, sketch the level surface of F that contains $P(1, 0, 0)$, and sketch $\nabla F]_P$.

Exer. 38–39: Find the extrema of f.

38 $f(x, y) = \dfrac{4}{x} + \dfrac{2}{y} - xy$

39 $f(x, y) = x^2 + 3y - y^3$

40 The material for the bottom of a rectangular box costs twice as much per square inch as the material for the sides and top. If the volume is fixed, find the relative dimensions that minimize the cost.

Exer. 41–42: Use Lagrange multipliers to find the extrema of f subject to the stated constraints.

41 $f(x, y, z) = xyz$;
$$x^2 + 4y^2 + 2z^2 = 8$$

42 $f(x, y, z) = 4x^2 + y^2 + z^2$;
$$2x - y + z = 4, \qquad x + 2y - z = 1$$

43 Find the point on the paraboloid

$$z = \frac{x^2}{25} + \frac{y^2}{4}$$

that is closest to the point $(0, 5, 0)$.

44 A grain elevator hopper has the shape of a right circular cone of radius 2 feet surmounted by a right circular cylinder, as shown in the figure. If the total volume is 100 ft^3, find the altitudes h and k of the cylinder and cone, respectively, that will minimize the curved surface area.

EXERCISE 44

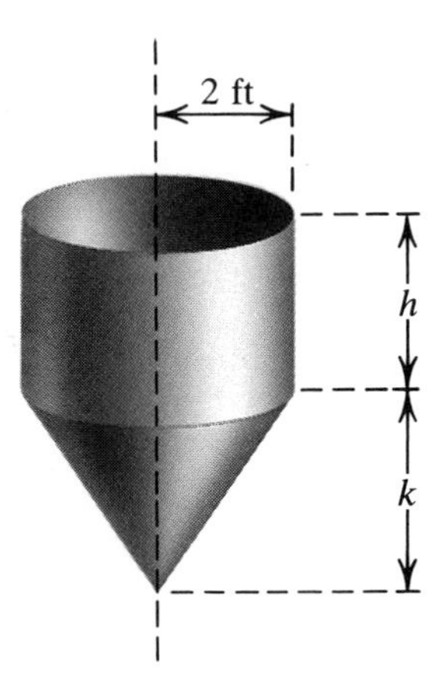

45 A *pipeline processor* in a supercomputer is like an assembly line designed to compute a large number of similar arithmetic operations in stages. Once the pipeline is full, it will compute one arithmetic operation in each clock period. Pipelines usually have less than 10 stages. The *performance-cost ratio* P for a pipeline processor is given by

$$P = \frac{s}{(T + sk_0)(C + sk_1)},$$

where s is the number of stages, T is the time required to process the operation on a nonpipeline computer, C is the total cost of the pipeline, and k_0 and k_1 are constants that depend on the hardware used. For the optimal design $\partial P/\partial s = 0$. Determine the number of stages in the optimal design.

46 A thin slab of metal having the shape of a square lamina with vertices on a coordinate plane at points (0, 0), (1, 0), (1, 1), and (0, 1) has each of its four edges inserted in ice at 0 °C. The initial temperature T at $P(x, y)$ is given by

$$T = 20 \sin \pi x \sin \pi y.$$

(a) Find the point at which the initial temperature is greatest.

(b) If $U(x, y, t)$ is the temperature at $P(x, y)$ at time t, then it can be shown that U satisfies the two-dimensional heat equation

$$\frac{\partial U}{\partial t} - k\left(\frac{\partial^2 U}{\partial x^2} + \frac{\partial^2 U}{\partial y^2}\right) = 0,$$

where k is a constant that depends on the thermal conductivity and specific heat of the slab. Show that the following is a solution of this heat equation:

$$U(x, y, t) = 20e^{-2k\pi^2 t} \sin \pi x \sin \pi y$$

(c) Describe the temperature of the slab over a long period of time.

CHAPTER

17

MULTIPLE INTEGRALS

INTRODUCTION

Multiple integrals are defined using functions of several variables and limits of sums in a manner analogous to that used to define the definite integral $\int_a^b f(x)\,dx$ in Chapter 5. The principal difference is that instead of beginning with a partition of the interval $[a, b]$, we subdivide a region R of the xy-plane or an xyz-coordinate system. Evaluation theorems are based on equations of curves and surfaces that form the boundary of R, and hence they are more complicated than the fundamental theorem of calculus for definite integrals.

Multiple integrals enable us to consider applications involving *nonhomogeneous* solids and surfaces that are more general than those obtained by revolving a plane region or curve about a line. In particular, we may find moments and centers of gravity of irregularly shaped solids in which the density is not constant—something we were unable to do earlier in the text.

In Sections 17.7 and 17.8 we define cylindrical and spherical coordinate systems in three dimensions. These systems are useful for investigating properties of solids determined by quadric surfaces—especially cylinders, spheres, and cones.

The chapter closes by introducing the concept of *Jacobian* and applying it to the problem of changing variables in multiple integrals.

17.1 DOUBLE INTEGRALS

In Chapter 5 we defined the definite integral $\int_a^b f(x)\,dx$ of a function f of one variable. We can also consider integrals of functions of several variables, called *double integrals*, *triple integrals*, *surface integrals*, and *line integrals*. Each integral is defined in a similar manner. The principal difference is the domain of the integrand.

Recall that $\int_a^b f(x)\,dx$ may be defined by applying the following four steps. (Step 1 and Step 2 are illustrated in Figure 17.1.)

FIGURE 17.1

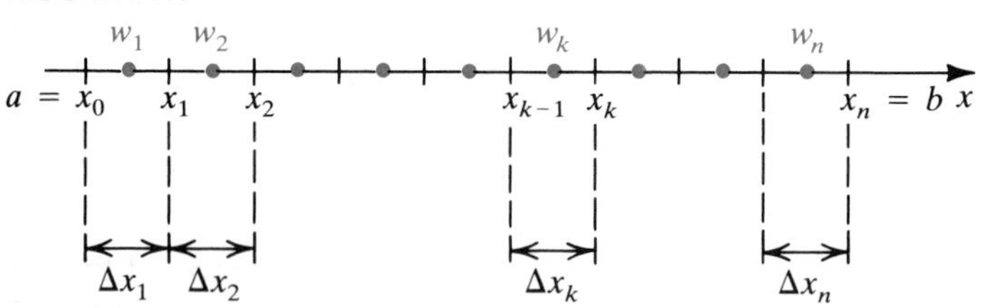

Step 1 Partition $[a, b]$ by choosing $a = x_0 < x_1 < x_2 < \cdots < x_n = b$.

Step 2 For each k, select any number w_k in the subinterval $[x_{k-1}, x_k]$.

Step 3 Form the Riemann sum $\sum_k f(w_k)\,\Delta x_k$, where $\Delta x_k = x_k - x_{k-1}$.

Step 4 If $\|P\|$ is the norm of the partition (the largest Δx_k), then

$$\int_a^b f(x)\,dx = \lim_{\|P\|\to 0} \sum_k f(w_k)\,\Delta x_k.$$

FIGURE 17.2

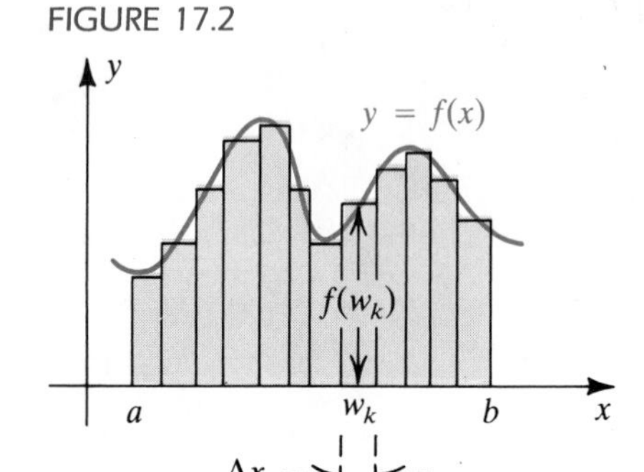

If f is nonnegative on $[a, b]$, then the Riemann sum $\sum_k f(w_k)\,\Delta x_k$ in Step 3 is a sum of areas of rectangles of the type illustrated in Figure 17.2. This sum approximates the area of the region that lies under the graph of f from $x = a$ to $x = b$. If the limit of Riemann sums exists as $\|P\| \to 0$ (Step 4), we obtain the definite integral $\int_a^b f(x)\,dx$, whose value is the (exact) area under the graph.

Let us now consider a function f of *two* variables such that $f(x, y)$ exists throughout a region R of the xy-plane. We shall define the double integral $\iint_R f(x, y)\,dA$ by using a four-step process similar to that used for $\int_a^b f(x)\,dx$. Throughout this chapter R will denote a region that can be subdivided into a finite number of R_x and R_y regions, as defined in Section 6.1. Every such region R is contained in a closed rectangular region W. If W is divided into smaller rectangles by means of a grid of horizontal and vertical lines, then the collection of all closed rectangular subregions that lie completely *within* R is an **inner partition** P of R. (This partitioning corresponds to Step 1 of the four-step process.)

FIGURE 17.3
Inner partition of R

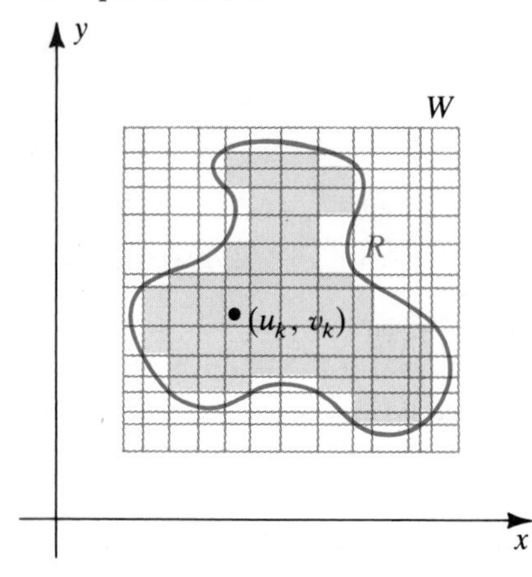

The blue rectangular regions in Figure 17.3 illustrate an inner partition of R. If we label them $R_1, R_2, \ldots, R_n$, then this inner partition is denoted by $\{R_k\}$. The length of the longest diagonal of all the R_k is the **norm $\|P\|$ of the partition P**. The symbol ΔA_k denotes the area of R_k. For each k, we choose any point (u_k, v_k) in R_k, as shown in Figure 17.3. (This point selection corresponds to Step 2.) Riemann sums (Step 3) are defined as follows.

Definition **(17.1)**

Let f be a function of two variables that is defined on a region R, and let $P = \{R_k\}$ be an inner partition of R. A **Riemann sum** of f for P is any sum of the form

$$\sum_k f(u_k, v_k)\,\Delta A_k,$$

where (u_k, v_k) is a point in R_k and ΔA_k is the area of R_k. The summation extends over all the subregions $R_1, R_2, \ldots, R_n$ of P.

Finally, we consider a limit of Riemann sums as $\|P\| \to 0$ (Step 4). It can be shown that if f is continuous on R, the Riemann sums in Definition (17.1) approach a real number L as $\|P\| \to 0$, regardless of which points (u_k, v_k) are chosen in the subregions R_k. The number L is the *double integral* $\iint_R f(x, y)\,dA$. The mathematically precise definition of a limit of Riemann sums may be stated as follows.

Definition **(17.2)**

Let f be a function of two variables that is defined on a region R, and let L be a real number. The statement

$$\lim_{\|P\|\to 0} \sum_k f(u_k, v_k)\,\Delta A_k = L$$

means that for every $\epsilon > 0$ there is a $\delta > 0$ such that if $P = \{R_k\}$ is an inner partition of R with $\|P\| < \delta$, then

$$\left| \sum_k f(u_k, v_k)\,\Delta A_k - L \right| < \epsilon$$

for every choice of (u_k, v_k) in R_k.

Definition (17.2) states that *we can make every Riemann sum of f as close as desired to L by choosing an inner partition of sufficiently small norm* $\|P\|$.

The double integral of f may be defined as follows.

Definition **(17.3)**

Let f be a function of two variables that is defined on a region R. The **double integral of f over R**, denoted by $\iint_R f(x, y)\,dA$, is

$$\iint_R f(x, y)\,dA = \lim_{\|P\|\to 0} \sum_k f(u_k, v_k)\,\Delta A_k,$$

provided the limit exists.

If the double integral of f over R exists, then f is said to be **integrable over R**. It is proved in more advanced texts that if f is continuous on R, then f is integrable over R.

There is a useful geometric interpretation for Riemann sums and double integrals if f is continuous and $f(x, y) \geq 0$ throughout R. Let S denote the graph of f, and let Q denote the solid that lies under S and over R, as illustrated in Figure 17.4. If $P_k(u_k, v_k, 0)$ is a point in the subregion R_k of an inner partition P of R, then $f(u_k, v_k)$ is the distance from the xy-plane to the point B_k on S, directly above P_k. The product $f(u_k, v_k)\,\Delta A_k$

FIGURE 17.4

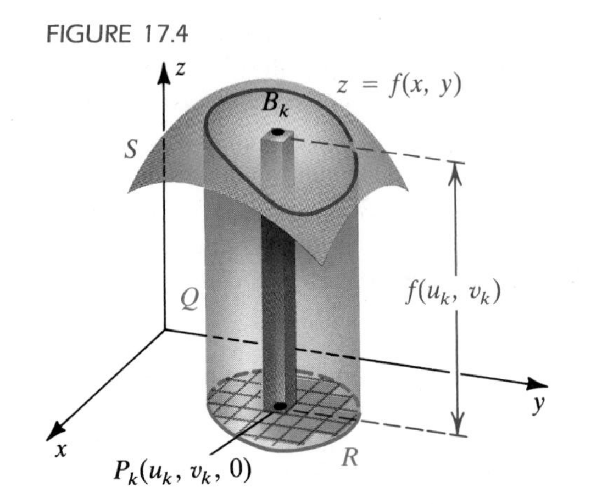

FIGURE 17.5

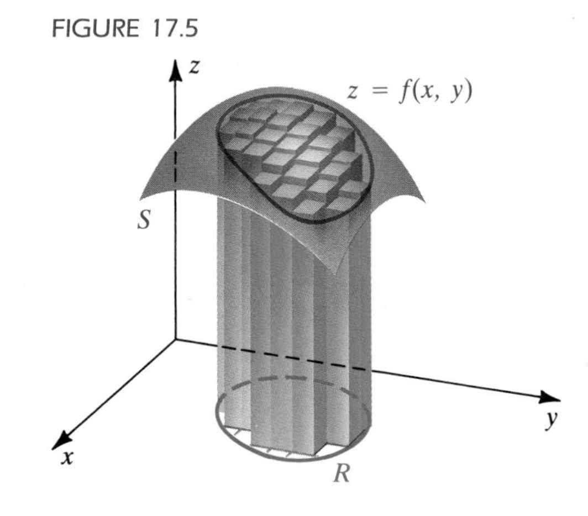

is the volume of the prism with a rectangular base of area ΔA_k illustrated in Figure 17.4. The sum of all such volumes of prisms (see Figure 17.5) is an approximation to the volume V of Q. Since this approximation to the volume improves as $\|P\|$ approaches zero, we define V as the limit of sums of the numbers $f(u_k, v_k)\,\Delta A_k$. Applying (17.3) gives us the following.

Definition **(17.4)**

> Let f be a continuous function of two variables such that $f(x, y) \geq 0$ for every (x, y) in a region R. The **volume** V of the solid that lies under the graph of $z = f(x, y)$ and over R is
>
> $$V = \iint_R f(x, y)\, dA.$$

FIGURE 17.6
$R = R_1 \cup R_2$

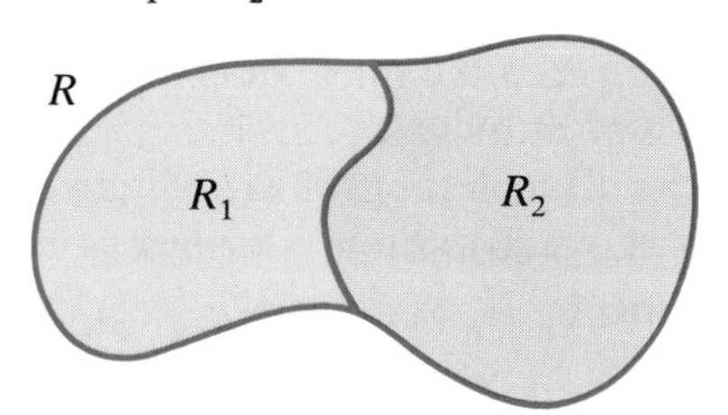

If $f(x, y) \leq 0$ throughout R, the double integral of f over R is the *negative* of the volume of the solid that lies *over* the graph of f and *under* the region R.

In the next theorem we list, without proof, some properties of double integrals that correspond to those given for definite integrals in Chapter 5. It is assumed that all regions and functions are suitably restricted so that the indicated integrals exist. The term *nonoverlapping*, in (iii) of the theorem, means that the regions R_1 and R_2 have at most only boundary points in common, as illustrated in Figure 17.6.

Theorem **(17.5)**

> **(i)** $\iint_R cf(x, y)\, dA = c \iint_R f(x, y)\, dA$ for every real number c
>
> **(ii)** $\iint_R [f(x, y) + g(x, y)]\, dA = \iint_R f(x, y)\, dA + \iint_R g(x, y)\, dA$
>
> **(iii)** If R is the union of two nonoverlapping regions R_1 and R_2,
>
> $$\iint_R f(x, y)\, dA = \iint_{R_1} f(x, y)\, dA + \iint_{R_2} f(x, y)\, dA.$$
>
> **(iv)** If $f(x, y) \geq 0$ throughout R, then $\iint_R f(x, y)\, dA \geq 0$.

FIGURE 17.7

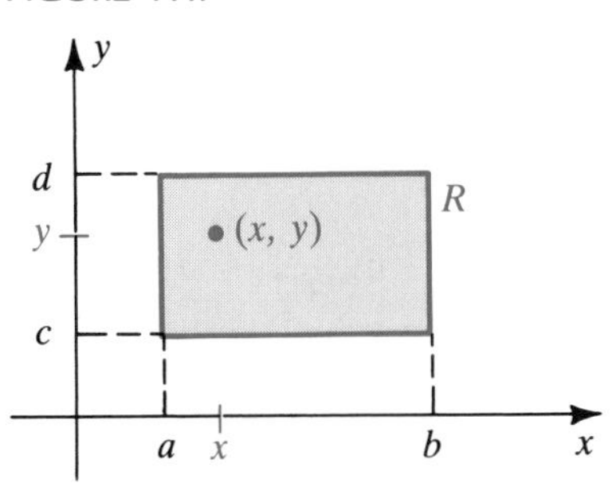

Except for elementary cases, it is virtually impossible to find the value of a double integral $\iint_R f(x, y)\,dA$ directly from Definition (17.3). If R is an R_x or R_y region, however, the double integral can be evaluated by using two successive integrals, each involving only one independent variable.

Let us begin with a simple case, a function f that is continuous on a closed rectangular region R of the type illustrated in Figure 17.7. It is shown in advanced calculus that the double integral $\iint_R f(x, y)\,dA$ can be evaluated by using an **iterated integral** of the following type:

$$\int_a^b \left[\int_c^d f(x, y)\,dy\right] dx$$

As indicated by the brackets, we first perform a **partial integration** *with respect to* y, regarding x as a constant. Substituting the limits of integration c and d for y in the usual way, we obtain an expression in x, which is then integrated from a to b. We can also use the following iterated integral:

$$\int_c^d \left[\int_a^b f(x, y)\,dx\right] dy$$

In this case we first perform a partial integration *with respect to* x, regarding y as a constant. After substituting the limits of integration a and b for x, we integrate the resulting expression in y from c to d.

The notation for iterated integrals is usually shortened by omitting the brackets, as in the following definition.

Definition **(17.6)**

(i) $$\int_a^b \int_c^d f(x, y)\,dy\,dx = \int_a^b \left[\int_c^d f(x, y)\,dy\right] dx$$

(ii) $$\int_c^d \int_a^b f(x, y)\,dx\,dy = \int_c^d \left[\int_a^b f(x, y)\,dx\right] dy$$

Note that the *first* differential to the *right* of the integrand $f(x, y)$ in (17.6) specifies the variable in the *first* partial integration. The first integral sign to the *left* of $f(x, y)$ specifies the limits of integration for that variable. Thus, when evaluating an iterated integral, work with the *innermost* integral first.

EXAMPLE 1 Evaluate $\int_1^4 \int_{-1}^2 (2x + 6x^2y)\,dy\,dx$.

SOLUTION By Definition (17.6)(i), the integral equals

$$\begin{aligned}
\int_1^4 \left[\int_{-1}^2 (2x + 6x^2y)\,dy\right] dx &= \int_1^4 \left[2xy + 6x^2\left(\frac{y^2}{2}\right)\right]_{-1}^2 dx \\
&= \int_1^4 [(4x + 12x^2) - (-2x + 3x^2)]\,dx \\
&= \int_1^4 (6x + 9x^2)\,dx \\
&= \Big[3x^2 + 3x^3\Big]_1^4 = 234.
\end{aligned}$$

EXAMPLE 2 Evaluate $\int_{-1}^{2}\int_{1}^{4}(2x+6x^2y)\,dx\,dy$.

SOLUTION By Definition (17.6)(ii), the integral equals

$$\begin{aligned}\int_{-1}^{2}\left[\int_{1}^{4}(2x+6x^2y)\,dx\right]dy &= \int_{-1}^{2}\left[2\left(\frac{x^2}{2}\right)+6\left(\frac{x^3}{3}\right)y\right]_{1}^{4}dy\\ &= \int_{-1}^{2}[(16+128y)-(1+2y)]\,dy\\ &= \int_{-1}^{2}(126y+15)\,dy\\ &= \Big[63y^2+15y\Big]_{-1}^{2} = 234.\end{aligned}$$

The fact that the iterated integrals in Examples 1 and 2 are equal is no accident. If f is continuous, then the two iterated integrals in (17.6) are *always* equal. We say that the *order of integration* is immaterial.

An iterated double integral may be defined over an R_x or R_y region of the type shown in Figure 17.8 as follows.

FIGURE 17.8

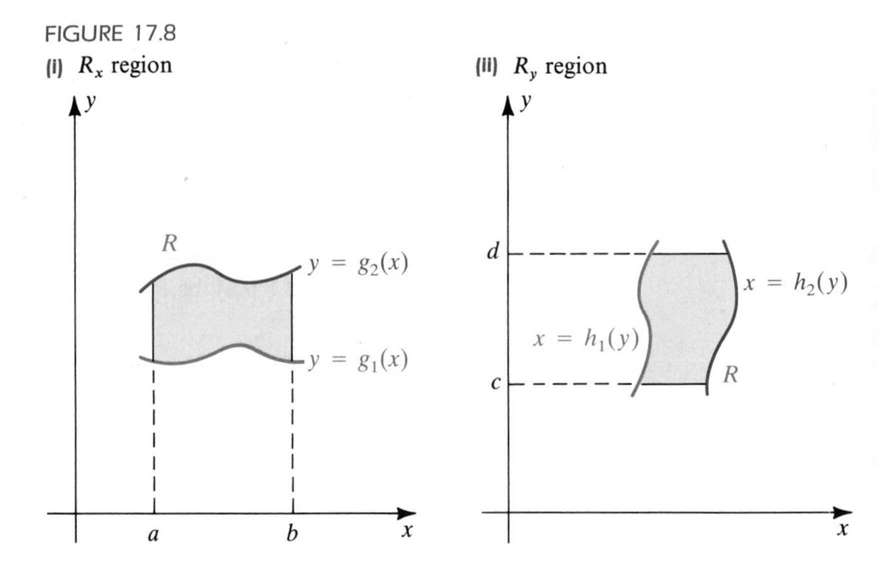

Definition of iterated integrals **(17.7)**

(i) $\int_a^b\int_{g_1(x)}^{g_2(x)} f(x, y)\,dy\,dx = \int_a^b\left[\int_{g_1(x)}^{g_2(x)} f(x, y)\,dy\right]dx$

(ii) $\int_c^d\int_{h_1(y)}^{h_2(y)} f(x, y)\,dx\,dy = \int_c^d\left[\int_{h_1(y)}^{h_2(y)} f(x, y)\,dx\right]dy$

In Definition (17.7)(i), we first perform a partial integration with respect to y and then substitute $g_2(x)$ and $g_1(x)$ for y in the usual way. The resulting expression in x is then integrated from a to b. In (ii) of the definition, we first integrate with respect to x and then, after substituting $h_2(y)$ and $h_1(y)$ for x, integrate the resulting expression with respect to y from c to d. Thus, as in Definition (17.6) for rectangular regions, we work with the *innermost* integral first.

EXAMPLE 3 Evaluate $\int_0^2 \int_{x^2}^{2x} (x^3 + 4y)\, dy\, dx$.

SOLUTION By Definition (17.7)(i), the integral equals

$$\int_0^2 \left[\int_{x^2}^{2x} (x^3 + 4y)\, dy\right] dx = \int_0^2 \left[x^3 y + 4\left(\frac{y^2}{2}\right)\right]_{x^2}^{2x} dx$$
$$= \int_0^2 [(2x^4 + 8x^2) - (x^5 + 2x^4)]\, dx$$
$$= \left[\tfrac{8}{3}x^3 - \tfrac{1}{6}x^6\right]_0^2 = \tfrac{32}{3}.$$

EXAMPLE 4 Evaluate $\int_1^3 \int_{\pi/6}^{y^2} 2y \cos x\, dx\, dy$.

SOLUTION By Definition (17.7)(ii), the integral equals

$$\int_1^3 \left[\int_{\pi/6}^{y^2} 2y \cos x\, dx\right] dy = \int_1^3 2y\left[\sin x\right]_{\pi/6}^{y^2} dy$$
$$= \int_1^3 2y(\sin y^2 - \tfrac{1}{2})\, dy$$
$$= \int_1^3 (2y \sin y^2 - y)\, dy$$
$$= \left[-\cos y^2 - \tfrac{1}{2}y^2\right]_1^3$$
$$= (-\cos 9 - \tfrac{9}{2}) - (-\cos 1 - \tfrac{1}{2}) \approx -2.55.$$

The next theorem states that if the region R is an R_x or R_y region, then $\iint_R f(x, y)\, dA$ may be evaluated by means of an iterated integral.

Evaluation theorem for double integrals **(17.8)**

(i) Let R be the R_x region shown in Figure 17.8(i). If f is continuous on R, then

$$\iint_R f(x, y)\, dA = \int_a^b \int_{g_1(x)}^{g_2(x)} f(x, y)\, dy\, dx.$$

(ii) Let R be the R_y region shown in Figure 17.8(ii). If f is continuous on R, then

$$\iint_R f(x, y)\, dA = \int_c^d \int_{h_1(y)}^{h_2(y)} f(x, y)\, dx\, dy.$$

For more complicated regions, we divide R into R_x or R_y subregions, apply (17.8) to each, and add the values of the resulting integrals.

A proof of Theorem (17.8) may be found in texts on advanced calculus. The following intuitive discussion makes the result plausible, at least for nonnegative-valued functions. Suppose $f(x, y) \geq 0$ throughout the R_x region R illustrated in Figure 17.8(i). Let S denote the graph of f, Q the solid that lies under S and over R, and V the volume of Q. Consider the plane that is parallel to the yz-plane and intersects the x-axis at $(x, 0, 0)$, with $a \leq x \leq b$, and let C be the trace of S on this plane (see Figure 17.9).

FIGURE 17.9

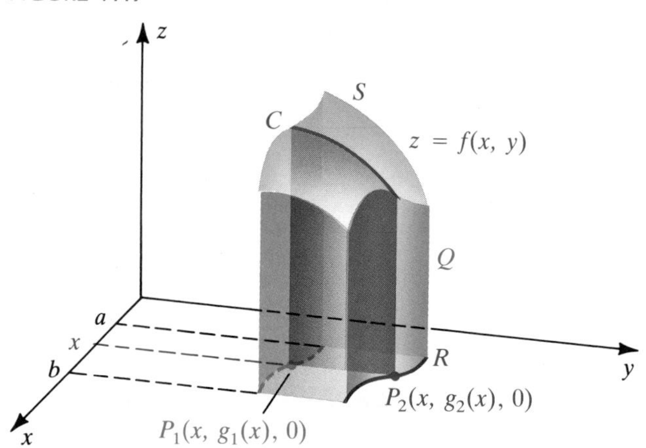

This plane intersects the boundaries of R—$y = g_1(x)$ and $y = g_2(x)$—at $P_1(x, g_1(x), 0)$ and $P_2(x, g_2(x), 0)$.

From Chapter 6 we know that the area $A(x)$ of this portion of the plane is

$$A(x) = \int_{g_1(x)}^{g_2(x)} f(x, y)\, dy.$$

Since $A(x)$ is the area of a typical cross section of Q, it follows from (6.13) that

$$V = \int_a^b A(x)\, dx = \int_a^b \int_{g_1(x)}^{g_2(x)} f(x, y)\, dy\, dx.$$

Using the fact that V is also given by Definition (17.4), we obtain the evaluation formula in part (i) of Theorem (17.8). A similar discussion can be given for part (ii).

Before using Theorem (17.8) to evaluate a double integral, *it is important to sketch the region R* and determine its boundaries. The following examples illustrate the use of Theorem (17.8).

EXAMPLE 5 Let R be the region in the xy-plane bounded by the graphs of $y = x^2$ and $y = 2x$. Evaluate $\iint_R (x^3 + 4y)\, dA$ using

(a) Theorem (17.8)(i) **(b)** Theorem (17.8)(ii)

FIGURE 17.10

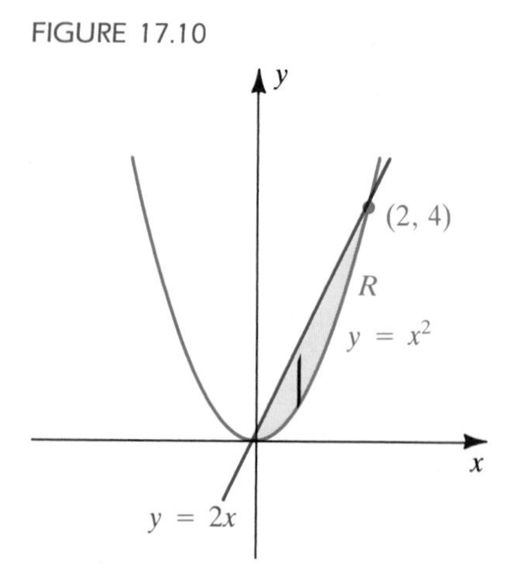

SOLUTION

(a) The region R is sketched in Figure 17.10. Note that R is both an R_x region and an R_y region. Let us regard R as an R_x region having lower boundary $y = x^2$ and upper boundary $y = 2x$, with $0 \le x \le 2$. We have drawn a vertical line segment between these boundaries to indicate that the first integration is with respect to y (from the lower boundary to the upper boundary). By Theorem (17.8)(i),

$$\iint_R (x^3 + 4y)\, dA = \int_0^2 \int_{x^2}^{2x} (x^3 + 4y)\, dy\, dx.$$

From Example 3 we know that the last integral equals $\frac{32}{3}$.

FIGURE 17.11

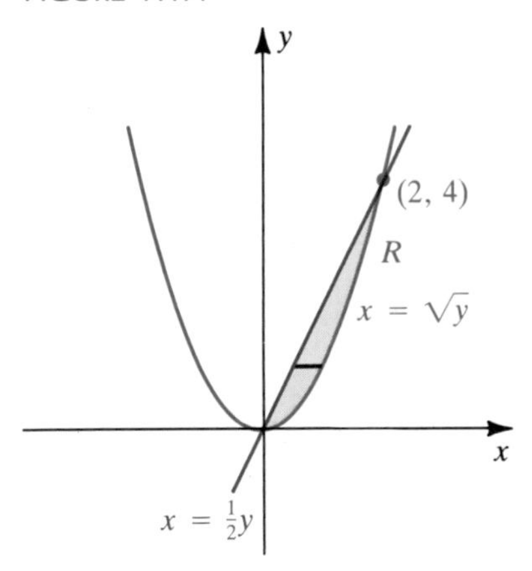

(b) To use Theorem (17.8)(ii), we regard R as an R_y region and solve each of the given equations for x in terms of y, obtaining

$$x = \tfrac{1}{2}y \quad \text{and} \quad x = \sqrt{y}, \quad \text{with} \quad 0 \le y \le 4.$$

The horizontal line segment in Figure 17.11 extends from the left boundary to the right boundary, indicating that the first integration is with respect to x. By Theorem (17.8)(ii),

$$\iint_R f(x, y)\, dA = \int_0^4 \int_{y/2}^{\sqrt{y}} (x^3 + 4y)\, dx\, dy$$

$$= \int_0^4 \left[\tfrac{1}{4}x^4 + 4yx\right]_{y/2}^{\sqrt{y}} dy$$

$$= \int_0^4 \left[\left(\tfrac{1}{4}y^2 + 4y^{3/2}\right) - \left(\tfrac{1}{64}y^4 + 2y^2\right)\right] dy = \tfrac{32}{3}.$$

EXAMPLE 6 Let R be the region bounded by the graphs of the equations $y = \sqrt{x}$, $y = \sqrt{3x - 18}$, and $y = 0$. If f is an arbitrary continuous function on R, express the double integral $\iint_R f(x, y)\, dA$ in terms of iterated integrals using only

(a) Theorem (17.8)(i) **(b)** Theorem (17.8)(ii)

SOLUTION The graphs of $y = \sqrt{x}$ and $y = \sqrt{3x - 18}$ are the top halves of the parabolas $y^2 = x$ and $y^2 = 3x - 18$. The region R is sketched in Figures 17.12 and 17.13.

(a) If we wish to use only Theorem 17.8(i), then it is necessary to employ two iterated integrals, because if $0 \le x \le 6$, the lower boundary of the region is the graph of $y = 0$, and if $6 \le x \le 9$, the lower boundary is the graph of $y = \sqrt{3x - 18}$. (The vertical line segments in Figure 17.12 extend from the lower boundary to the upper boundary of R.)

FIGURE 17.12

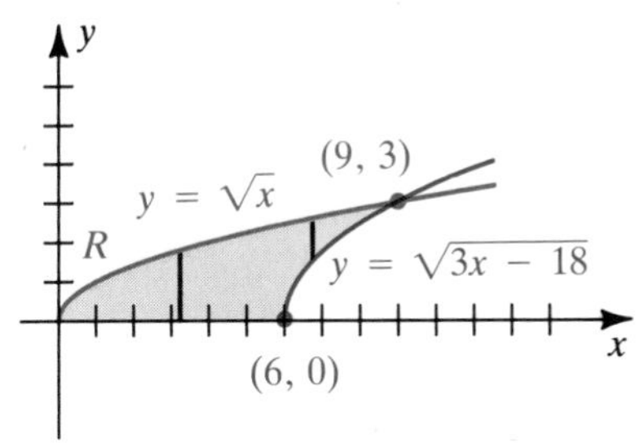

If R_1 denotes the part of the region R that lies between $x = 0$ and $x = 6$ and if R_2 denotes the part between $x = 6$ and $x = 9$, then both R_1 and R_2 are R_x regions. Hence

$$\iint_R f(x, y)\, dA = \iint_{R_1} f(x, y)\, dA + \iint_{R_2} f(x, y)\, dA$$

$$= \int_0^6 \int_0^{\sqrt{x}} f(x, y)\, dy\, dx + \int_6^9 \int_{\sqrt{3x-18}}^{\sqrt{x}} f(x, y)\, dy\, dx.$$

(b) To use Theorem (17.8)(ii), we must solve each of the given equations for x in terms of y, obtaining

$$x = y^2 \quad \text{and} \quad x = \frac{y^2 + 18}{3} = \frac{1}{3}y^2 + 6, \quad \text{with} \quad 0 \le y \le 3.$$

FIGURE 17.13

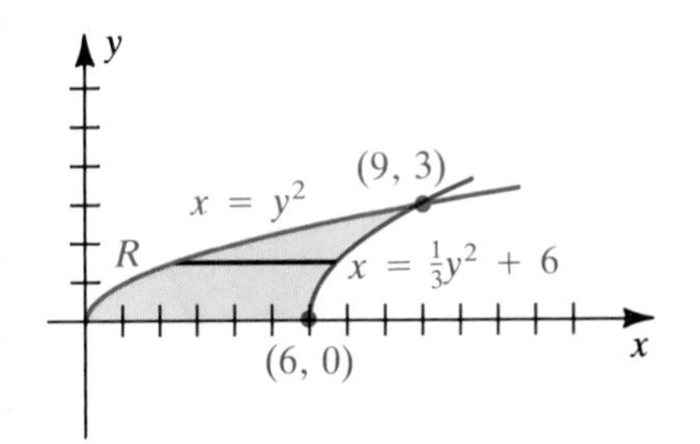

The horizontal line segment in Figure 17.13 extends from the left boundary to the right boundary. Only one iterated integral is required in this case, since R is an R_y region. Thus,

$$\iint_R f(x, y)\, dA = \int_0^3 \int_{y^2}^{(y^2/3)+6} f(x, y)\, dx\, dy.$$

Examples 5 and 6 indicate how certain double integrals may be evaluated by using either (i) or (ii) of Theorem (17.8). Generally, the choice of the *order of integration dy dx* or *dx dy* depends on the form of $f(x, y)$ and the region R. Sometimes it is extremely difficult, or even impossible, to evaluate a given iterated double integral. However, by *reversing the order of integration* from $dy\,dx$ to $dx\,dy$, or vice versa, it may be possible to find an equivalent iterated double integral that can be easily evaluated. This technique is illustrated in the next example.

EXAMPLE 7 Given $\int_0^4 \int_{\sqrt{y}}^2 y \cos x^5 \, dx \, dy$, reverse the order of integration and evaluate the resulting integral.

FIGURE 17.14

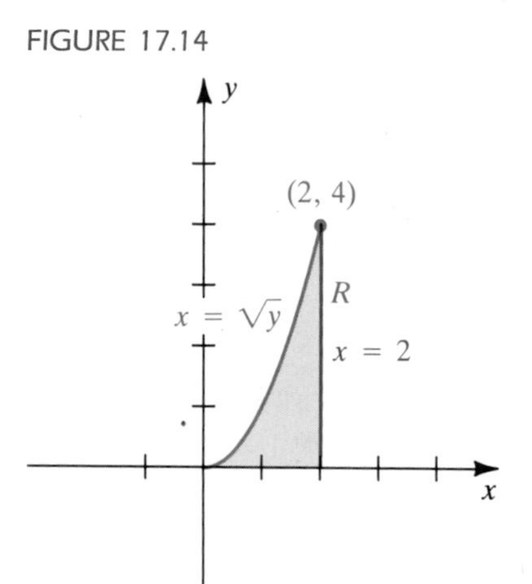

SOLUTION The given order of integration, $dx\,dy$, indicates that the region is an R_y region. As illustrated in Figure 17.14, the left and right boundaries are the graphs of $x = \sqrt{y}$ and $x = 2$, respectively, with $0 \le y \le 4$.

Note that R is also an R_x region whose lower and upper boundaries are given by $y = 0$ and $y = x^2$, respectively, with $0 \le x \le 2$. Hence, by Theorem (17.8)(i), the integral can be evaluated as follows:

$$\begin{aligned}
\int_0^4 \int_{\sqrt{y}}^2 y \cos x^5 \, dx \, dy &= \iint_R y \cos x^5 \, dA = \int_0^2 \int_0^{x^2} y \cos x^5 \, dy \, dx \\
&= \int_0^2 \left[\frac{y^2}{2} \cos x^5 \right]_0^{x^2} dx = \int_0^2 \frac{x^4}{2} \cos x^5 \, dx \\
&= \tfrac{1}{10} \int_0^2 (\cos x^5)(5x^4) \, dx \\
&= \tfrac{1}{10} \left[\sin x^5 \right]_0^2 = \tfrac{1}{10} \sin 32 \approx 0.055
\end{aligned}$$

EXERCISES 17.1

Exer. 1–10: Determine whether the region is R_x, R_y, or neither.

1

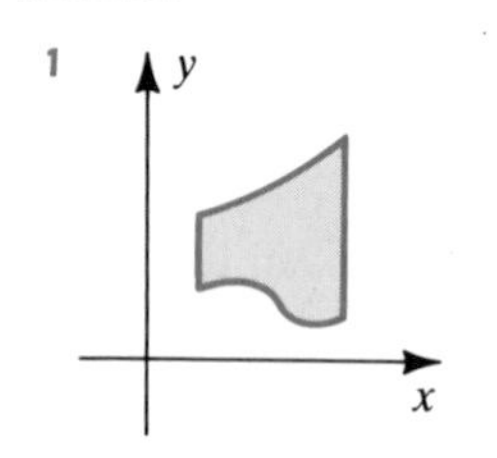

2

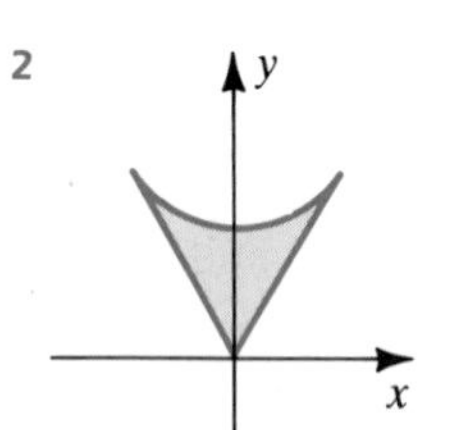

3

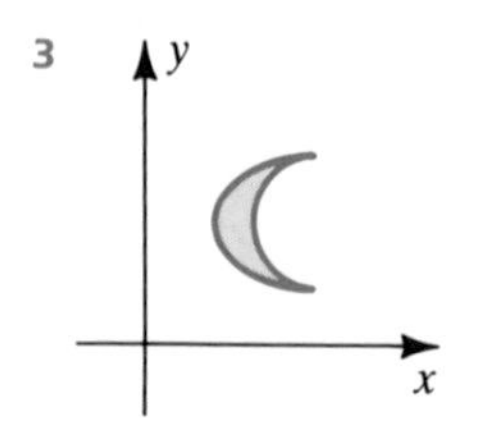

4

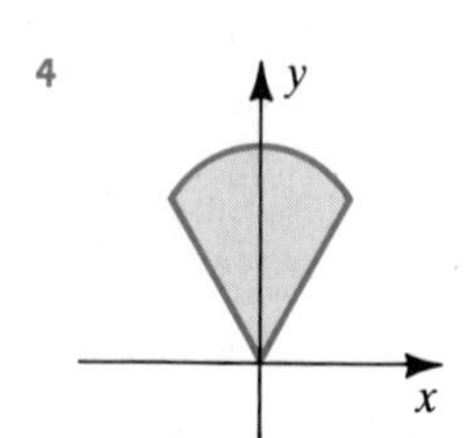

5

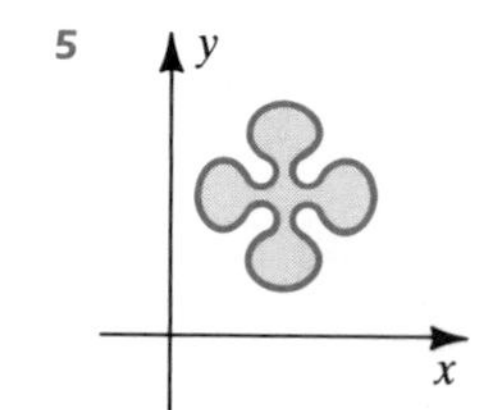

6

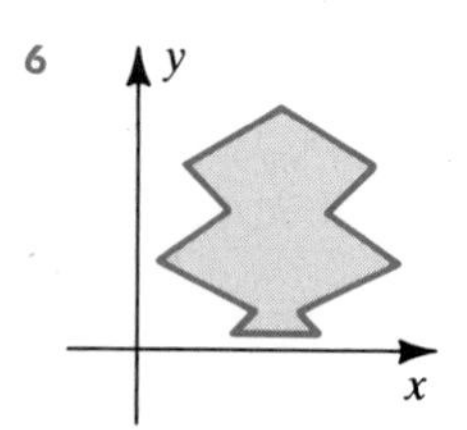

7

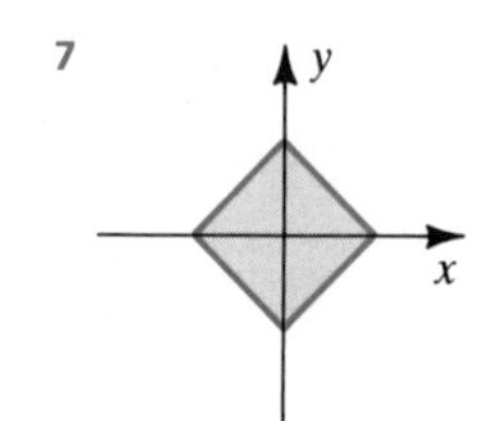

8

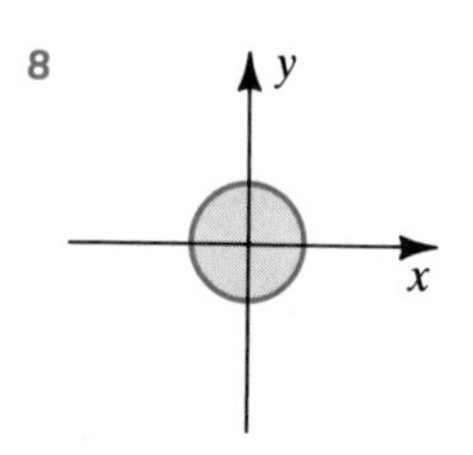

9

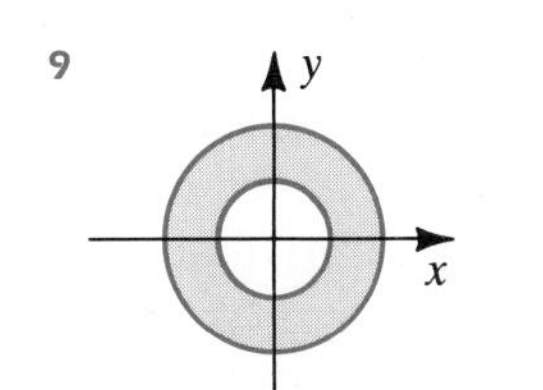

10

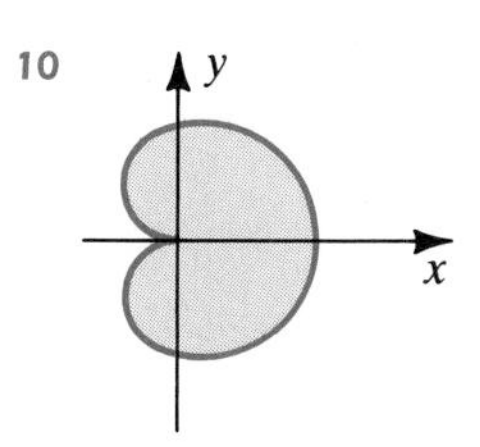

11 Suppose $f(x, y) = 4x + 2y + 1$ and R is the region in the first quadrant bounded by the coordinate axes and the graph of $y = 4 - \frac{1}{4}x^2$. Let $\{R_k\}$, with $k = 1, 2, \ldots, 7$, be the inner partition of R shown in the figure. Calculate the Riemann sum $\sum_k f(u_k, v_k)\,\Delta A_k$ if, for each k, (u_k, v_k) is the point

(a) in the lower left corner of R_k

(b) in the upper right corner of R_k

(c) at the center of R_k

EXERCISE 11

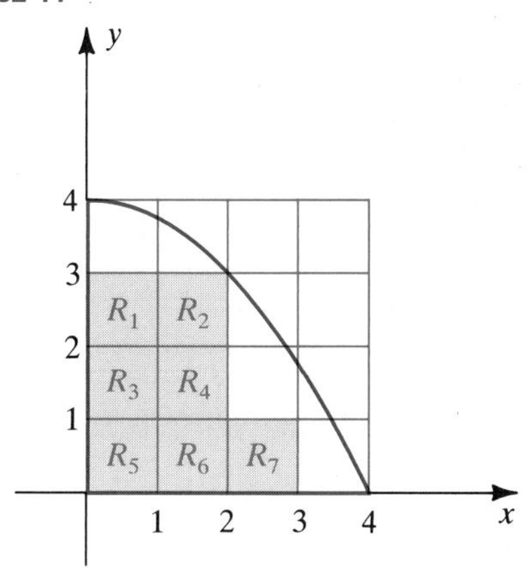

12 Suppose $f(x, y) = xy$ and R is the region bounded by the trapezoid with vertices $(0, 0)$, $(4, 4)$, $(8, 4)$, and $(12, 0)$. Let P be the inner partition of R determined by vertical lines with x-intercepts 0, 2, 4, 6, 8, 10, and 12 and by horizontal lines with y-intercepts 0, 2, and 4. Calculate the Riemann sums as in (a)–(c) of Exercise 11.

Exer. 13–20: Evaluate the iterated integral.

13 $\int_1^2 \int_{-1}^2 (12xy^2 - 8x^3)\,dy\,dx$

14 $\int_0^3 \int_{-2}^{-1} (4xy^3 + y)\,dx\,dy$

15 $\int_1^2 \int_{1-x}^{\sqrt{x}} x^2 y\,dy\,dx$

16 $\int_{-1}^1 \int_{x^3}^{x+1} (3x + 2y)\,dy\,dx$

17 $\int_0^2 \int_{y^2}^{2y} (4x - y)\,dx\,dy$

18 $\int_0^1 \int_{-y-1}^{y-1} (x^2 + y^2)\,dx\,dy$

19 $\int_1^2 \int_{x^3}^{x} e^{y/x}\,dy\,dx$

20 $\int_0^{\pi/6} \int_0^{\pi/2} (x \cos y - y \cos x)\,dy\,dx$

Exer. 21–26: Sketch the region R bounded by the graphs of the given equations. If $f(x, y)$ is an arbitrary continuous function, express $\iint_R f(x, y)\,dA$ as an iterated integral by using (a) only Theorem (17.8)(i) and (b) only Theorem (17.8)(ii).

21 $y = \sqrt{x}$, $x = 4$, $y = 0$

22 $y = \sqrt{x}$, $x = 0$, $y = 2$

23 $y = x^3$, $x = 0$, $y = 8$

24 $y = x^3$, $x = 2$, $y = 0$

25 $y = \sqrt{x}$, $y = x^3$

26 $y = \sqrt{1 - x^2}$, $y = 0$

Exer. 27–32: Express the double integral over the indicated region R as an iterated integral, and find its value.

27 $\iint_R (y + 2x)\,dA$; the rectangular region with vertices $(-1, -1)$, $(2, -1)$, $(2, 4)$, $(-1, 4)$

28 $\iint_R (x - y)\,dA$; the triangular region with vertices $(2, 9)$, $(2, 1)$, $(-2, 1)$

29 $\iint_R xy^2\,dA$; the triangular region with vertices $(0, 0)$, $(3, 1)$, $(-2, 1)$

30 $\iint_R (y + 1)\,dA$; the region between the graphs of $y = \sin x$ and $y = \cos x$ from $x = 0$ to $x = \pi/4$

31 $\iint_R x^3 \cos xy\,dA$; the region bounded by the graphs of $y = x^2$, $y = 0$, and $x = 2$

32 $\iint_R e^{x/y}\,dA$; the region bounded by the graphs of $y = 2x$, $y = -x$, and $y = 4$

Exer. 33–38: Sketch the region R bounded by the graphs of the given equations. If f is an arbitrary continuous function on R, express $\iint_R f(x, y)\,dA$ as a sum of two iterated integrals of the type used in (a) Theorem (17.8)(i) and (b) Theorem (17.8)(ii).

33 $8y = x^3$, $y - x = 4$, $4x + y = 9$

34 $x = 2\sqrt{y}$, $\sqrt{3}x = \sqrt{y}$, $y = 2x + 5$

35 $x = \sqrt{3 - y}, \quad y = 2x, \quad x + y + 3 = 0$

36 $x + 2y = 5, \quad x - y = 2, \quad 2x + y = -2$

37 $y = e^x, \quad y = \ln x, \quad x + y = 1, \quad x + y = 1 + e$

38 $y = \sin x, \quad \pi y = 2x$

Exer. 39–44: Sketch the region of integration for the iterated integral.

39 $\int_{-1}^{2} \int_{-\sqrt{4-x^2}}^{4-x^2} f(x, y)\, dy\, dx$

40 $\int_{-1}^{2} \int_{x^2-4}^{x-2} f(x, y)\, dy\, dx$

41 $\int_{0}^{1} \int_{\sqrt{y}}^{\sqrt[3]{y}} f(x, y)\, dx\, dy$

42 $\int_{-2}^{-1} \int_{3y}^{2y} f(x, y)\, dx\, dy$

43 $\int_{-3}^{1} \int_{\arctan x}^{e^x} f(x, y)\, dy\, dx$

44 $\int_{\pi}^{2\pi} \int_{\sin y}^{\ln y} f(x, y)\, dx\, dy$

Exer. 45–50: Reverse the order of integration, and evaluate the resulting integral.

45 $\int_{0}^{1} \int_{2x}^{2} e^{(y^2)}\, dy\, dx$

46 $\int_{0}^{9} \int_{\sqrt{y}}^{3} \sin x^3\, dx\, dy$

47 $\int_{0}^{2} \int_{y^2}^{4} y \cos x^2\, dx\, dy$

48 $\int_{1}^{e} \int_{0}^{\ln x} y\, dy\, dx$

49 $\int_{0}^{8} \int_{\sqrt[3]{y}}^{2} \frac{y}{\sqrt{16 + x^7}}\, dx\, dy$

50 $\int_{0}^{2} \int_{x}^{2} y^4 \cos(xy^2)\, dy\, dx$

c **Exer. 51–52: Numerical integration can be used to approximate $\int_a^b \int_c^d f(x, y)\, dx\, dy$ by first letting $G(y) = \int_c^d f(x, y)\, dx$ and then applying the trapezoidal rule or Simpson's rule to $\int_a^b G(y)\, dy$. To find each value $G(k)$ needed for these rules, approximate $\int_c^d f(x, k)\, dx$. Use this method and the trapezoidal rule, with $n = 3$, to approximate the given double integral.**

51 $\int_{0}^{1} \int_{0}^{1} e^{x^2 y^2}\, dx\, dy$

52 $\int_{0}^{1} \int_{0}^{1} \sin(x^2 + y^3)\, dx\, dy$

c **Exer. 53–54: Use the method described for Exercises 51–52 and Simpson's rule, with $n = 2$, to approximate the double integral.**

53 $\int_{0}^{1} \int_{0}^{1} \cos(x^2 e^y)\, dx\, dy$

54 $\int_{0}^{1} \int_{-1/3}^{1/3} \frac{1}{1 + x^4 + y^4}\, dx\, dy$

17.2 AREA AND VOLUME

In Section 17.1 we saw that if $f(x, y) \geq 0$ and f is continuous, then the volume V of the solid that lies under the graph of $z = f(x, y)$ and over an R_x region in the xy-plane is given by

$$V = \int_a^b A(x)\, dx = \int_a^b \int_{g_1(x)}^{g_2(x)} f(x, y)\, dy\, dx,$$

where $A(x)$ is the area of a typical cross section of the solid (see Figure 17.9). In this section we discuss a method for interpreting this formula for V as a limit of (double) sums. By the volume-by-cross-section formula (6.13),

$$V = \int_a^b A(x)\, dx = \lim_{\|P'\| \to 0} \sum_k A(u_k)\, \Delta x_k,$$

where P' is a partition of the interval $[a, b]$, u_k is any number in the kth subinterval $[x_{k-1}, x_k]$ of P', and $\Delta x_k = x_k - x_{k-1}$. As illustrated in Figure 17.15, $A(u_k)\, \Delta x_k$ is the volume of a lamina L_k with face parallel to the yz-plane and rectangular base of width Δx_k in the xy-plane. Thus, *the volume V is a limit of a sum of volumes of such laminas.*

The *iterated* integral $V = \int_a^b \int_{g_1(x)}^{g_2(x)} f(x, y)\, dy\, dx$ may also be expressed in terms of limits of sums. First we apply the definition of the definite integral as follows. If x is in $[a, b]$, then

$$A(x) = \int_{g_1(x)}^{g_2(x)} f(x, y)\, dy = \lim_{\|P''\| \to 0} \sum_j f(x, v_j)\, \Delta y_j,$$

FIGURE 17.15

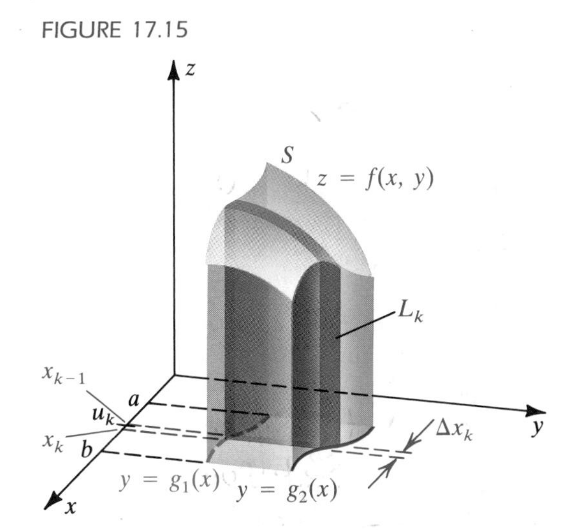

FIGURE 17.16

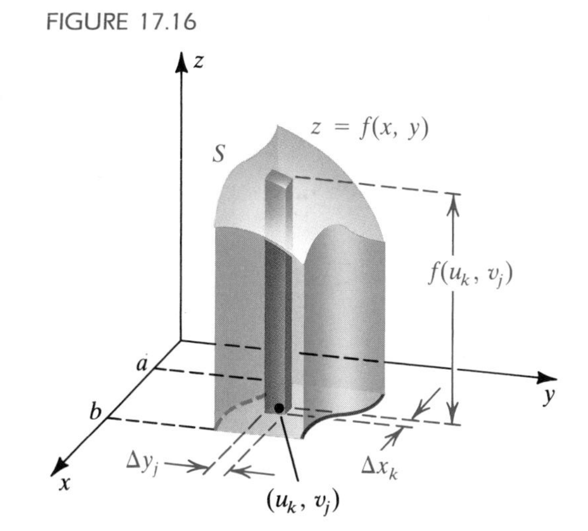

where P'' is a partition of the y-interval $[g_1(x), g_2(x)]$, v_j is any number in the jth subinterval $[y_{j-1}, y_j]$ of P'', and $\Delta y_j = y_j - y_{j-1}$. Hence, for each u_k in $[a, b]$,

$$A(u_k) = \lim_{\|P''\| \to 0} \sum_j f(u_k, v_j)\, \Delta y_j.$$

Consequently

$$V = \int_a^b A(x)\, dx = \lim_{\|P'\| \to 0} \sum_k A(u_k)\, \Delta x_k$$

$$= \lim_{\|P'\| \to 0} \sum_k \left[\lim_{\|P''\| \to 0} \sum_j f(u_k, v_j)\, \Delta y_j \right] \Delta x_k.$$

If we disregard the limits in this formula, we obtain a **double sum**:

$$\sum_k \sum_j f(u_k, v_j)\, \Delta y_j\, \Delta x_k$$

Referring to Figure 17.16, we see that $f(u_k, v_j)\, \Delta y_j\, \Delta x_k$ is the volume of a prism having base area $\Delta y_j\, \Delta x_k$ and altitude $f(u_k, v_j)$. The double sum may be interpreted as follows. *Holding x fixed*, we sum the volumes of the prisms in the direction of the y-axis, thereby obtaining the volume of the lamina illustrated in Figure 17.15. The volumes of all such laminas are then added by summing in the direction of the x-axis. In a certain sense, the iterated integral does this and, at the same time, takes limits as the bases of the prisms shrink to points. This interpretation is often useful in visualizing applications of the double integral. To emphasize this point of view, we sometimes express the volume as follows, where $\|P\|$ is the length of the longest diagonal of the bases of the prisms.

Volume as a limit of double sums **(17.9)**

$$V = \lim_{\|P\| \to 0} \sum_k \sum_j f(u_k, v_j)\, \Delta y_j\, \Delta x_k = \int_a^b \int_{g_1(x)}^{g_2(x)} f(x, y)\, dy\, dx$$

Areas can also be considered as limits of double sums. We begin by denoting the double integral $\iint_R 1\, dA$ by $\iint_R dA$. If R is the R_x region in

FIGURE 17.17

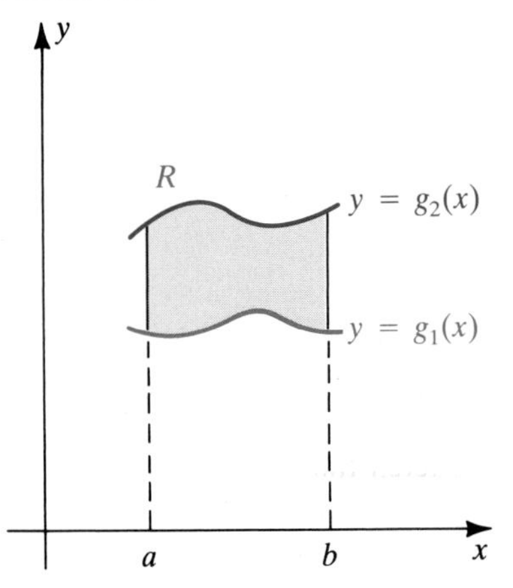

Figure 17.17, then, by Theorem (17.8)(i),

$$\iint_R dA = \int_a^b \int_{g_1(x)}^{g_2(x)} 1\,dy\,dx = \int_a^b \Big[y\Big]_{g_1(x)}^{g_2(x)} dx$$
$$= \int_a^b [g_2(x) - g_1(x)]\,dx,$$

which, according to Theorem (6.1), equals the area A of R. A similar formula holds if R is an R_y region. These facts are also evident from the definition of the double integral, for if $f(x, y) = 1$ throughout R, then the Riemann sum in Definition (17.3) is a sum of areas of rectangles in an inner partition P of R. As the norm $\|P\|$ approaches zero, these rectangles cover more of R and the limit equals the area of R.

If an iterated integral is used to find an area, it may be regarded as a limit of double sums in a manner similar to that used for volumes. Specifically, as in (17.9), with $f(x, y) = 1$, we have the following result.

Area as a limit of double sums **(17.10)**

$$A = \iint_R dA = \int_a^b \int_{g_1(x)}^{g_2(x)} dy\,dx = \lim_{\|P\|\to 0} \sum_k \sum_j \Delta y_j\,\Delta x_k$$

As for functions of one variable (see Guidelines (6.3)), we may employ the following *intuitive* method for remembering the limit of sums formula (17.10).

Guidelines for finding the area of an R_x region by means of a double integral **(17.11)**

1. Sketch the region, as in Figure 17.18(i), and show a typical rectangle of dimensions dx and dy.
2. *Holding x fixed*, regard $\int_{g_1(x)}^{g_2(x)}$ as an operator that sums the area elements $dy\,dx$ in the same direction as the y-axis, from the lower boundary to the upper boundary. The expression $[\int_{g_1(x)}^{g_2(x)} dy]\,dx$ represents the area of the vertical rectangle in Figure 17.18(ii).
3. Apply the operator $\int_a^b$ to $[\int_{g_1(x)}^{g_2(x)} dy]\,dx$, thereby taking a limit of sums of areas of the vertical rectangles in guideline 2, from $x = a$ to $x = b$.

FIGURE 17.18

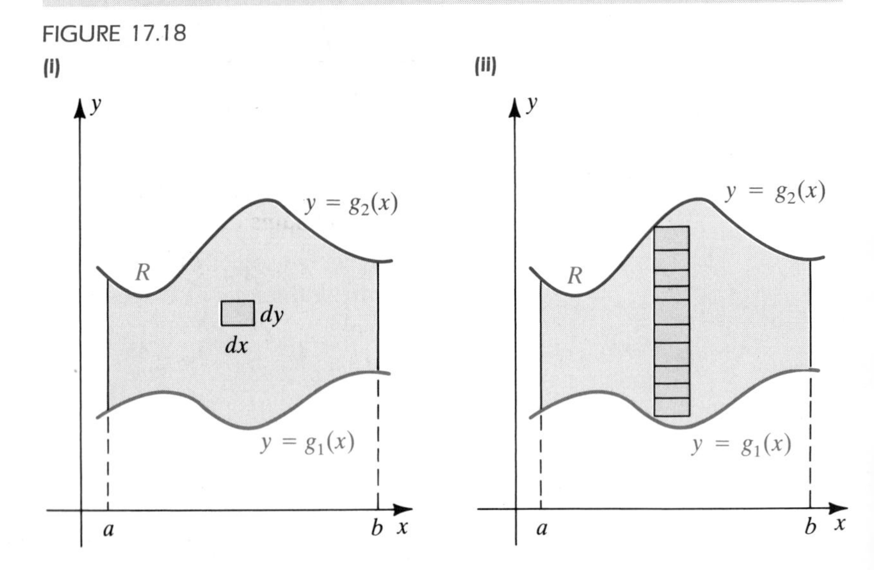

It is important to remember that when Guidelines (17.11) are applied to an R_x region, the first integration is with respect to y and the second with respect to x. We can, of course, obtain the area of the region in Figure 17.18 by using the methods of Section 6.1. The reason for using a double integral is to gain practice in working with limits of double sums. This same technique will be used later for problems that *cannot* be solved using single-variable methods.

EXAMPLE 1 Find the area A of the region in the xy-plane bounded by the graphs of $2y = 16 - x^2$ and $x + 2y = 4$.

SOLUTION Following guideline 1, we sketch the region and a typical rectangle of area $dy\,dx$, as in Figure 17.19(i). We solve the equations for y in terms of x, and we label the boundaries of the region $y = 2 - \frac{1}{2}x$ and $y = 8 - \frac{1}{2}x^2$.

As in guideline 2, the partial integration

$$\int_{2-(x/2)}^{8-(x^2/2)} dy\,dx = \left[\int_{2-(x/2)}^{8-(x^2/2)} dy\right] dx$$

represents the sum of the area elements $dy\,dx$ from the lower boundary to the upper boundary. This gives us the area of the vertical rectangle in Figure 17.19(ii).

FIGURE 17.19

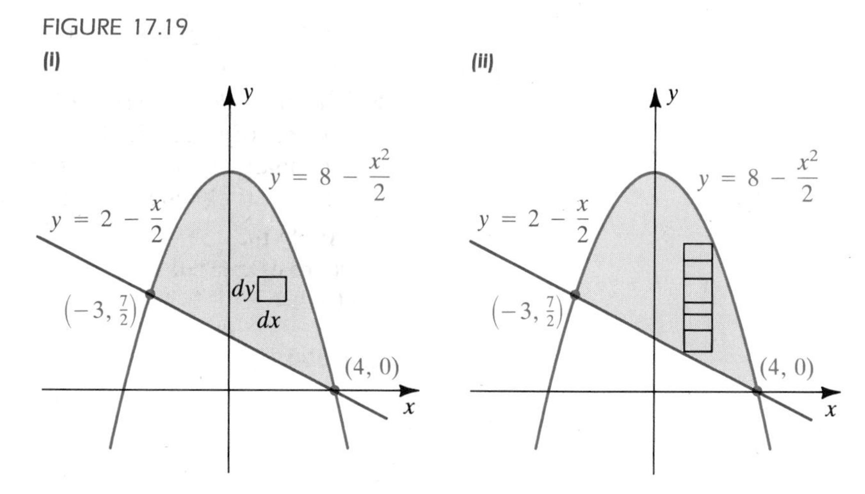

Finally, using guideline 3, we apply the operator $\int_{-3}^{4}$, thereby taking a limit of sums of these vertical rectangles from $x = -3$ to $x = 4$:

$$A = \int_{-3}^{4} \int_{2-(x/2)}^{8-(x^2/2)} dy\,dx = \int_{-3}^{4} \Big[y\Big]_{2-(x/2)}^{8-(x^2/2)} dx$$

$$= \int_{-3}^{4} \left[\left(8 - \frac{x^2}{2}\right) - \left(2 - \frac{x}{2}\right)\right] dx$$

$$= \left[6x - \frac{x^3}{6} + \frac{x^2}{4}\right]_{-3}^{4} = \frac{343}{12} \approx 28.6$$

We next state guidelines similar to (17.11) for R_y regions.

Guidelines for finding the area of an R_y region by means of a double integral **(17.12)**

1. Sketch the region, as in Figure 17.20(i), and show a typical rectangle of dimensions dx and dy.
2. *Holding y fixed*, regard $\int_{h_1(y)}^{h_2(y)}$ as an operator that sums the area elements $dx\,dy$ in the same direction as the x-axis, from the left boundary to the right boundary. The expression $[\int_{h_1(y)}^{h_2(y)} dx]\,dy$ represents the area of the horizontal rectangle in Figure 17.20(ii).
3. Apply the operator $\int_c^d$ to $[\int_{h_1(y)}^{h_2(y)} dx]\,dy$, thereby taking a limit of sums of areas of the horizontal rectangles in guideline 2, from $y = c$ to $y = d$.

FIGURE 17.20

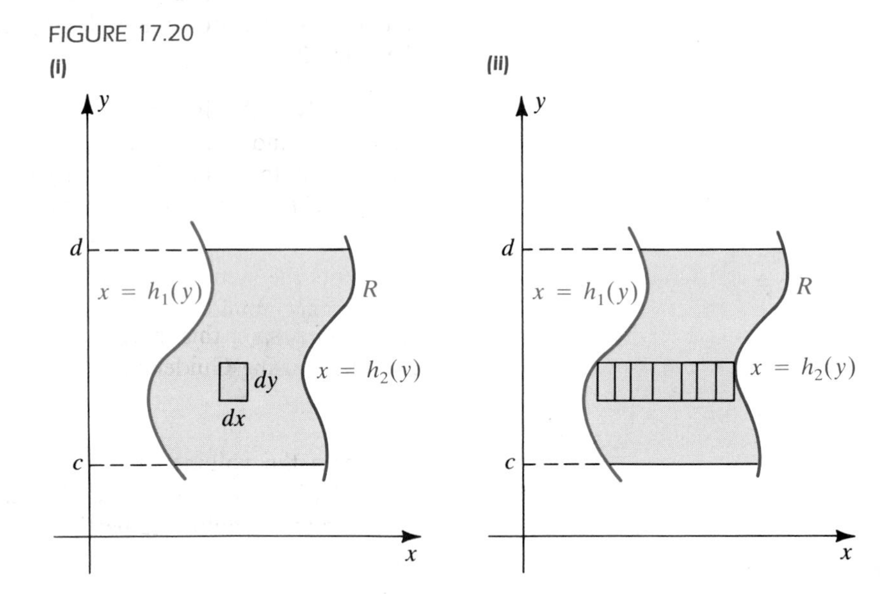

Note that when Guidelines (17.12) are applied to an R_y region, the first integration is with respect to x and the second with respect to y.

EXAMPLE 2 Find the area A of the region in the xy-plane bounded by the graphs of $x = y^3$, $x + y = 2$, and $y = 0$.

SOLUTION The region is sketched in Figure 17.21(i), together with a rectangle of area $dx\,dy$. As in guideline 2 of (17.12), the partial integration

$$\int_{y^3}^{2-y} dx\,dy = \left[\int_{y^3}^{2-y} dx\right] dy$$

represents the sum of the area elements $dx\,dy$ from the left boundary to the right boundary. This gives us the area of the horizontal rectangle in Figure 17.21(ii).

Following guideline 3, we apply the operator $\int_0^1$, thereby taking a limit of sums of these horizontal rectangles from $y = 0$ to $y = 1$:

$$A = \iint_R dA = \int_0^1 \int_{y^3}^{2-y} dx\,dy = \int_0^1 \Big[x\Big]_{y^3}^{2-y} dy$$

$$= \int_0^1 (2 - y - y^3)\,dy = \left[2y - \frac{y^2}{2} - \frac{y^4}{4}\right]_0^1 = \frac{5}{4}$$

FIGURE 17.21

(i)

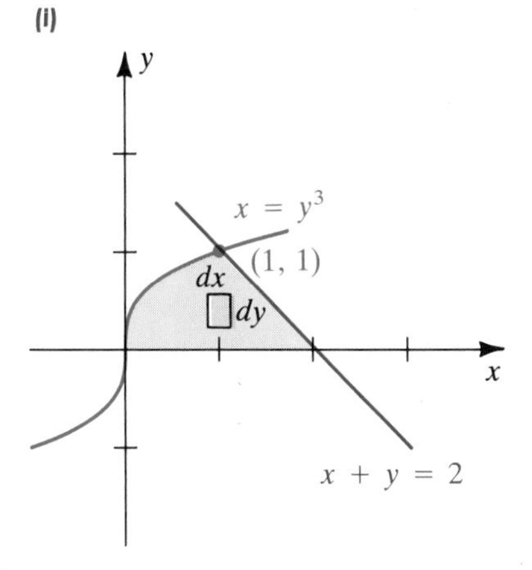

(ii)

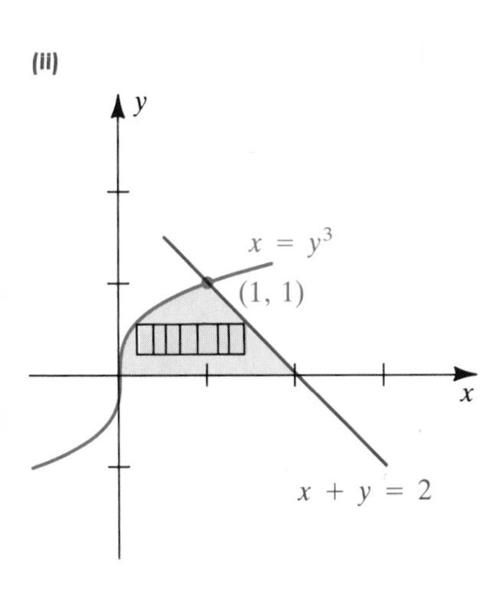

The area can also be found by employing Theorem (17.8)(i); however, in this case it is necessary to divide R into two parts by means of a vertical line through (1, 1). We then have

$$A = \int_0^1 \int_0^{\sqrt[3]{x}} dy\,dx + \int_1^2 \int_0^{2-x} dy\,dx.$$

In the next two examples we find volumes by means of (17.9).

EXAMPLE 3 Find the volume V of the solid in the first octant bounded by the coordinate planes, the paraboloid $z = x^2 + y^2 + 1$, and the plane $2x + y = 2$.

SOLUTION As illustrated in Figure 17.22(i), the solid lies under the paraboloid and over the triangular region R in the xy-plane bounded by the coordinate axes and the line $y = 2 - 2x$.

By Definition (17.4), with $f(x, y) = x^2 + y^2 + 1$,

$$V = \iint_R (x^2 + y^2 + 1)\,dA.$$

We may interpret this integral as a limit of double sums, as in (17.9), and evaluate it using Guidelines (17.11). If we represent dA by $dy\,dx$, then

$$(x^2 + y^2 + 1)\,dy\,dx$$

represents the volume of the prism illustrated in Figure 17.22(i). Figure 17.22(ii) shows the region R in the xy-plane and a rectangle corresponding to the following partial integration with respect to y (holding x fixed):

$$\int_0^{2-x} (x^2 + y^2 + 1)\,dy\,dx = \left[\int_0^{2-x} (x^2 + y^2 + 1)\,dy\right] dx$$

FIGURE 17.22

(i) (ii)

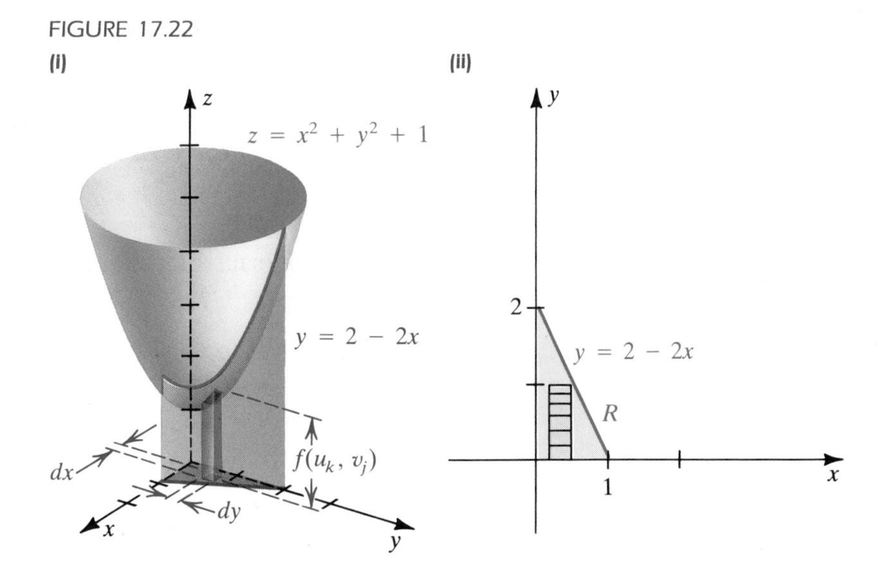

This formula represents the sum of the prism volumes $(x^2 + y^2 + 1)\,dy\,dx$ *in the direction parallel to the y-axis* and gives us the volume of a lamina whose face is parallel to the yz-plane. Finally, we apply the operator $\int_0^1$; that is, we take a limit of sums of these laminar volumes from $x = 0$ to $x = 1$. Thus,

$$\begin{aligned} V &= \int_0^1 \left[\int_0^{2-2x} (x^2 + y^2 + 1)\,dy\right] dx = \int_0^1 \left[x^2 y + \tfrac{1}{3}y^3 + y\right]_0^{2-2x} dx \\ &= \int_0^1 \left(-\tfrac{14}{3}x^3 + 10x^2 - 10x + \tfrac{14}{3}\right) dx \\ &= \left[-\tfrac{7}{6}x^4 + \tfrac{10}{3}x^3 - 5x^2 + \tfrac{14}{3}x\right]_0^1 = \tfrac{11}{6}. \end{aligned}$$

We may also find V by integrating first with respect to x, in which case the iterated integral has the form

$$V = \int_0^2 \int_0^{(2-y)/2} (x^2 + y^2 + 1)\,dx\,dy.$$

EXAMPLE 4 Find the volume V of the solid that lies in the first octant and is bounded by the three coordinate planes and the cylinders $x^2 + y^2 = 9$ and $y^2 + z^2 = 9$.

SOLUTION The cylinder $y^2 + z^2 = 9$ has radius 3 and axis on the x-axis, with the part in the first octant above the xy-plane, as illustrated in Figure 17.23(i). The cylinder $x^2 + y^2 = 9$ has radius 3 and axis on the z-axis, with the part in the first octant intersecting the xy-plane in a quarter-circle. Thus, the solid lies *under* the graph of $z = \sqrt{9 - y^2}$ and *over* the region R shown in Figure 17.23(ii). By Definition (17.4),

$$V = \iint_R \sqrt{9 - y^2}\,dA.$$

FIGURE 17.23

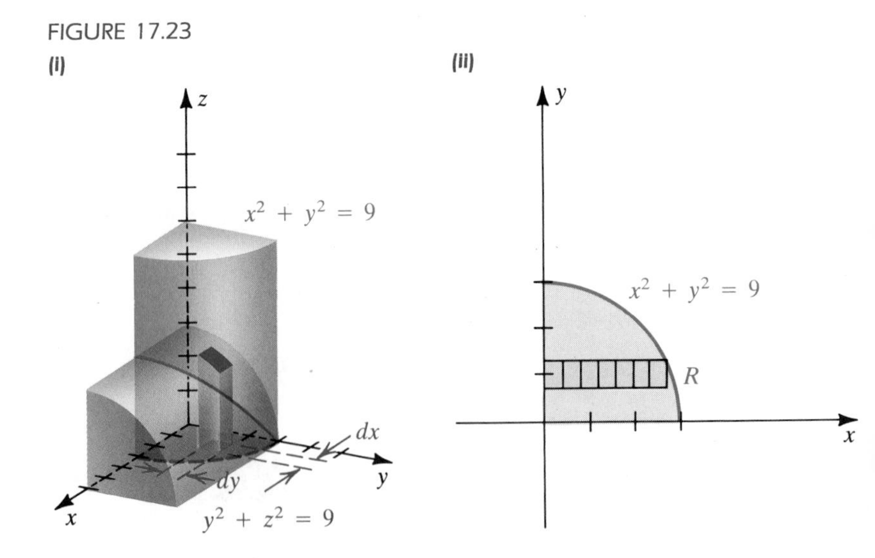

We may regard the integral as a limit of double sums of volumes of prisms of the type shown in Figure 17.23(i). We first sum in the direction of the x-axis from $x = 0$ to $x = \sqrt{9 - y^2}$. (Figure 17.23(ii) shows a strip in the xy-plane corresponding to this first summation.) This sum gives us the volume of a lamina whose face is parallel to the xz-plane. We then sum these laminar volumes in the direction of the y-axis from $y = 0$ to $y = 3$. Using an iterated integral, we have

$$\begin{aligned} V &= \int_0^3 \int_0^{\sqrt{9-y^2}} \sqrt{9 - y^2}\, dx\, dy \\ &= \int_0^3 \sqrt{9 - y^2} \Big[x\Big]_0^{\sqrt{9-y^2}} dy \\ &= \int_0^3 (9 - y^2)\, dy \\ &= \Big[9y - \tfrac{1}{3}y^3\Big]_0^3 = 18. \end{aligned}$$

The double integral could also be evaluated by integrating first with respect to y; however, this order of integration is much more complicated.

EXERCISES 17.2

Exer. 1–4: Set up an iterated double integral that can be used to find the area of the region.

1

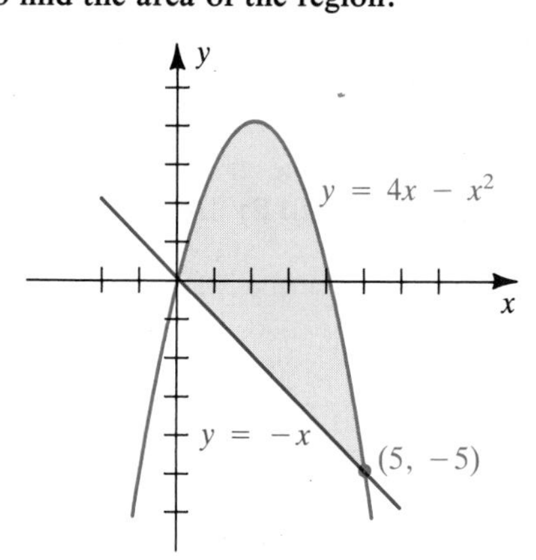

2

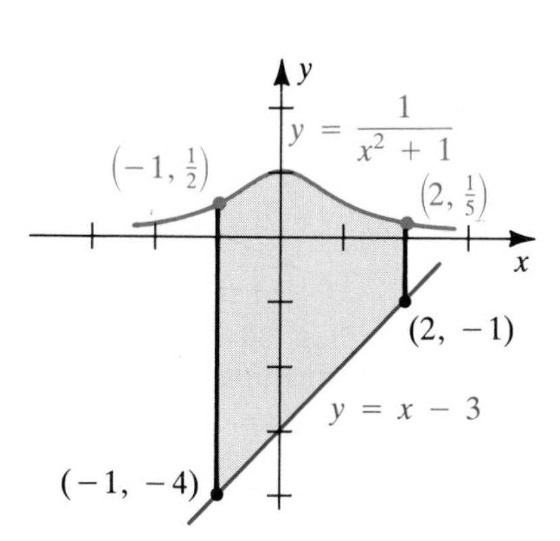

3

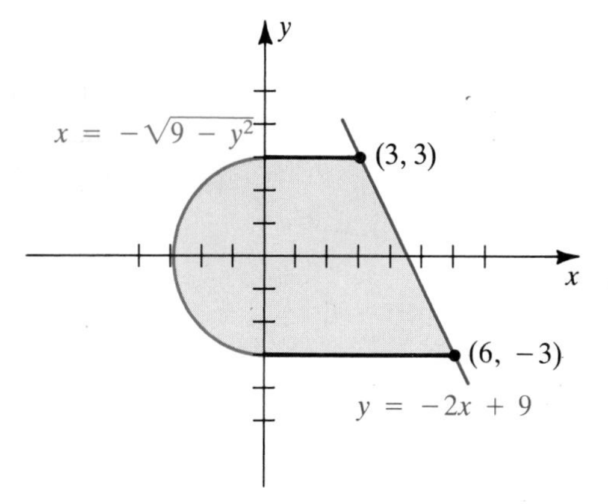

4

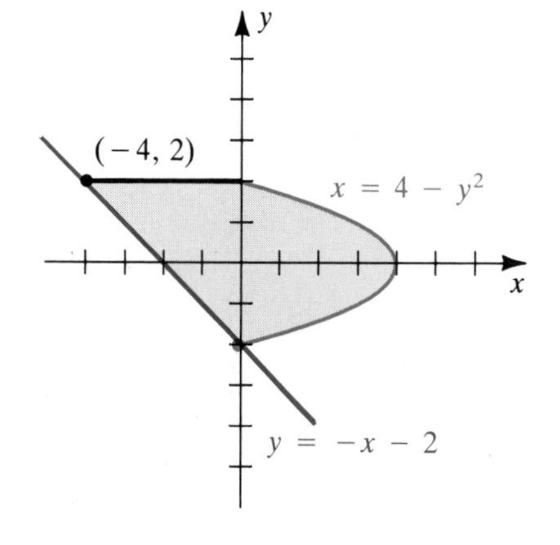

Exer. 5–12: Sketch the region bounded by the graphs of the equations, and find its area by using one or more double integrals.

5 $y = 1/x^2$, $y = -x^2$, $x = 1$, $x = 2$

6 $y = \sqrt{x}$, $y = -x$, $x = 1$, $x = 4$

7 $y^2 = -x$, $x - y = 4$, $y = -1$, $y = 2$

8 $x = y^2$, $y - x = 2$, $y = -2$, $y = 3$

9 $y = x$, $y = 3x$, $x + y = 4$

10 $x - y = -1$, $7x - y = 17$, $2x + y = -2$

11 $y = e^x$, $y = \sin x$, $x = -\pi$, $x = \pi$

12 $y = x^2$, $y = 1/(x^2 + 1)$

Exer. 13–16: Set up an iterated double integral that can be used to find the volume of the solid.

13

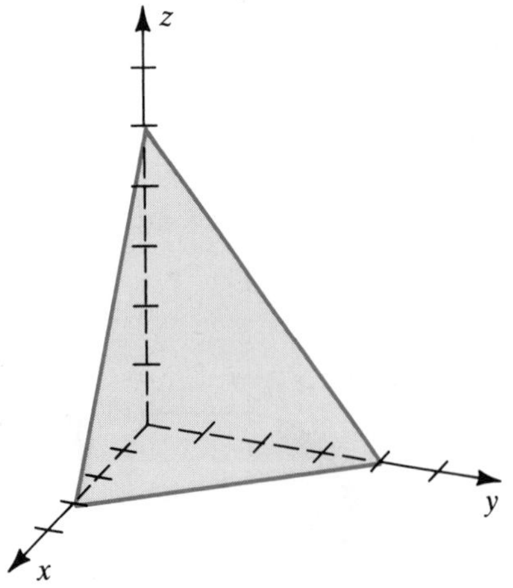

14

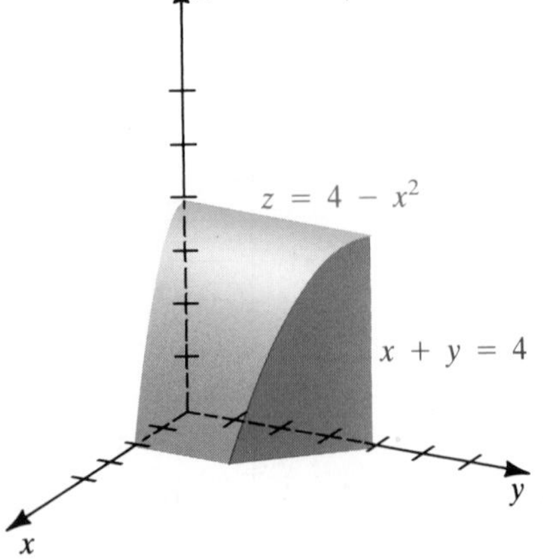

$z = 4 - x^2$

$x + y = 4$

15

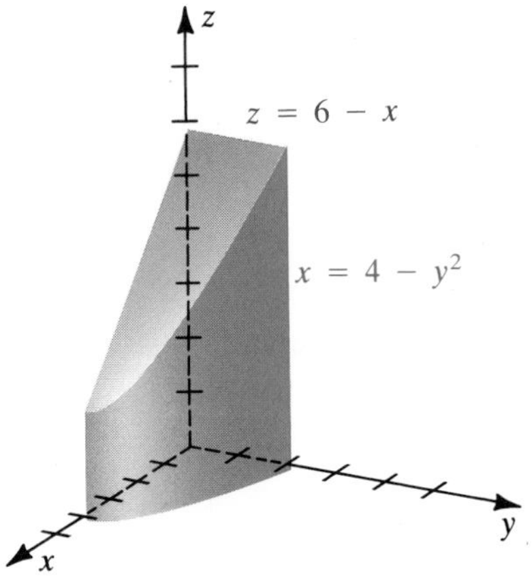

$z = 6 - x$

$x = 4 - y^2$

16

$x^2 + y^2 + z^2 = 25$

$x + y = 4$

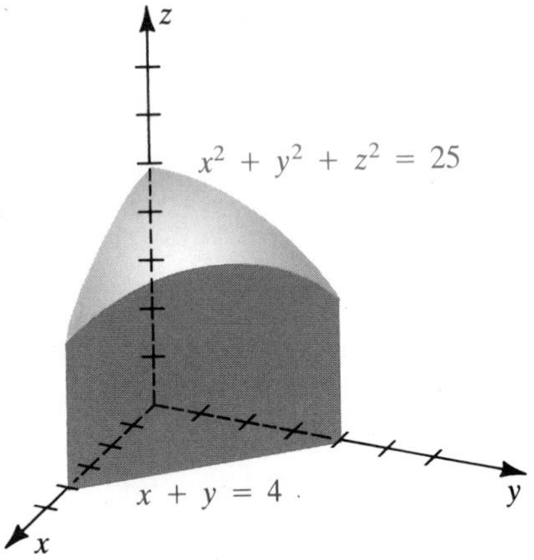

Exer. 17–20: The iterated double integral represents the volume of a solid under a surface S and over a region R in the xy-plane. Describe S and sketch R.

17 $\int_0^4 \int_{-1}^2 3\, dy\, dx$

18 $\int_0^1 \int_{3-x}^{3-x^2} \sqrt{25 - x^2 - y^2}\, dy\, dx$

19 $\int_{-2}^1 \int_{x-1}^{1-x^2} (x^2 + y^2)\, dy\, dx$

20 $\int_0^4 \int_0^{\sqrt{y}} \sqrt{x^2 + y^2}\, dx\, dy$

Exer. 21–22: Find the volume of the solid that lies under the graph of the equation and over the region in the xy-plane bounded by the polygon with the given vertices.

21 $z = 4x^2 + y^2$; (0, 0), (0, 1), (2, 0), (2, 1)

22 $z = x^2 + 4y^2$; (0, 0), (1, 0), (1, 2)

Exer. 23–30: Sketch the solid in the first octant bounded by the graphs of the equations, and find its volume.

23 $x^2 + z^2 = 9$, $y = 2x$, $y = 0$, $z = 0$

24 $z = 4 - x^2$, $x + y = 2$, $x = 0$, $y = 0$, $z = 0$

25 $2x + y + z = 4$, $x = 0$, $y = 0$, $z = 0$

26 $y^2 = z, \quad y = x, \quad x = 4, \quad z = 0$

27 $z = x^2 + y^2, \quad y = 4 - x^2, \quad x = 0, \quad y = 0, \quad z = 0$

28 $z = y^3, \quad y = x^3, \quad x = 0, \quad z = 0, \quad y = 1$

29 $z = x^3, \quad x = 4y^2, \quad 16y = x^2, \quad z = 0$

30 $x^2 + y^2 = 16, \quad x = z, \quad y = 0, \quad z = 0$

Exer. 31–32: Find the volume of the solid bounded by the graphs of the equations.

31 $z = x^2 + 4, \quad y = 4 - x^2, \quad x + y = 2, \quad z = 0$

32 $y = x^3, \quad y = x^4, \quad z - x - y = 4, \quad z = 0$

c **Exer. 33–34: Refer to (17.9). Approximate the volume $V = \int_0^1 \int_0^{1/2} f(x, y)\, dy\, dx$ for the given f by using**

$$\sum_{k=1}^{4} \sum_{j=1}^{2} f(u_k, v_j)\, \Delta y_j\, \Delta x_k$$

with $\Delta x_k = \frac{1}{4}$, $\Delta y_j = \frac{1}{4}$, $u_k = \frac{1}{4}k - \frac{1}{8}$, and $v_j = \frac{1}{4}j - \frac{1}{8}$ for $k = 1, 2, 3, 4$ and $j = 1, 2$.

33 $f(x, y) = \sin\,[\cos\,(xy)]$

34 $f(x, y) = \sqrt{x^4 + y^4}$

17.3 DOUBLE INTEGRALS IN POLAR COORDINATES

FIGURE 17.24

$\Delta A = \bar{r}\, \Delta r\, \Delta\theta$

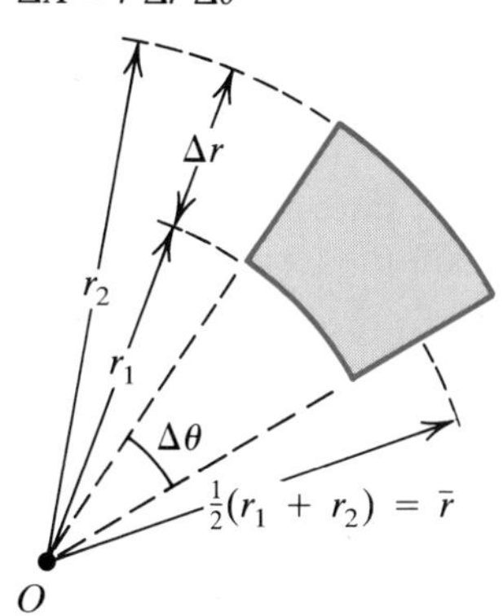

In this section we shall discuss how to evaluate a double integral over a region R that is bounded by graphs of polar equations. Let us first consider the *elementary polar region* in Figure 17.24, bounded by arcs of circles of radii r_1 and r_2, with centers at the origin, and by two rays from the origin. If $\Delta\theta$ is the radian measure of the angle between the rays and if $\Delta r = r_2 - r_1$, then the area ΔA of the region is

$$\Delta A = \tfrac{1}{2}r_2^2\, \Delta\theta - \tfrac{1}{2}r_1^2\, \Delta\theta.$$

This formula may also be written

$$\Delta A = \tfrac{1}{2}(r_2^2 - r_1^2)\, \Delta\theta = \tfrac{1}{2}(r_2 + r_1)(r_2 - r_1)\, \Delta\theta.$$

If we denote the **average radius** $\frac{1}{2}(r_2 + r_1)$ by $\bar{r}$, then

$$\Delta A = \bar{r}\, \Delta r\, \Delta\theta.$$

Next, consider the region R of the type illustrated in Figure 17.25(i), bounded by two rays that make positive angles α and β with the polar axis and by the graphs of two polar equations $r = g_1(\theta)$ and $r = g_2(\theta)$, where g_1 and g_2 are continuous functions and $g_1(\theta) \le g_2(\theta)$ for $\alpha \le \theta \le \beta$. If R is subdivided by means of circular arcs and rays as shown in Figure 17.25(ii),

FIGURE 17.25

(i)

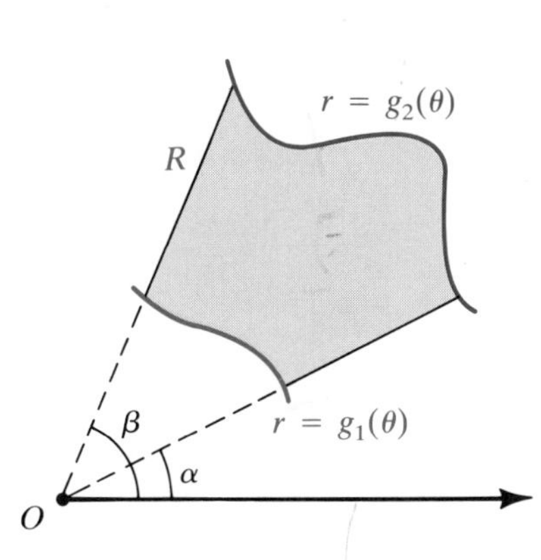

(ii)

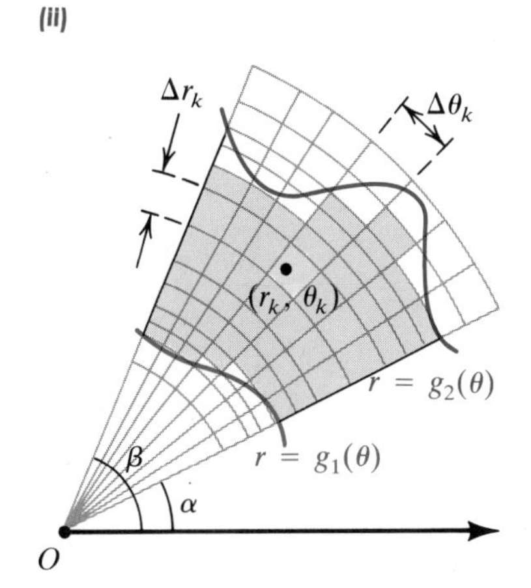

then the collection of elementary polar regions $R_1, R_2, \ldots, R_n$ that lie completely within R is called an **inner polar partition** P of R. The norm $\|P\|$ of the partition is the length of the longest diagonal of the R_k. If we choose a point (r_k, θ_k) in R_k such that r_k is the average radius, then

$$\Delta A_k = r_k \, \Delta r_k \, \Delta\theta_k.$$

If f is a continuous function of the polar variables r and θ, then the following can be proved.

Evaluation theorem **(17.13)**

$$\lim_{\|P\|\to 0} \sum_k f(r_k, \theta_k) r_k \, \Delta r_k \, \Delta\theta_k = \iint_R f(r, \theta)\, dA = \int_\alpha^\beta \int_{g_1(\theta)}^{g_2(\theta)} f(r, \theta) r \, dr \, d\theta$$

Note that the integrand on the right in (17.13) is the product of $f(r, \theta)$ and r. This is due to the fact that ΔA_k equals $r_k \, \Delta r_k \, \Delta\theta_k$.

We may regard the iterated integral in (17.13) as a limit of double sums. We *first hold* θ *fixed* and sum along the wedge-shaped region shown in Figure 17.26, from the graph of g_1 to the graph of g_2. For the second summation we *sweep out* the region by letting θ vary from α to β. The iterated integral denotes the limit of these sums as $\|P\| \to 0$.

If $f(r, \theta) = 1$ throughout R, then the integral in (17.13) equals the area of R. This also follows from our work in Chapter 13, since after performing the partial integration with respect to r, we obtain the formula in Theorem (13.11).

FIGURE 17.26

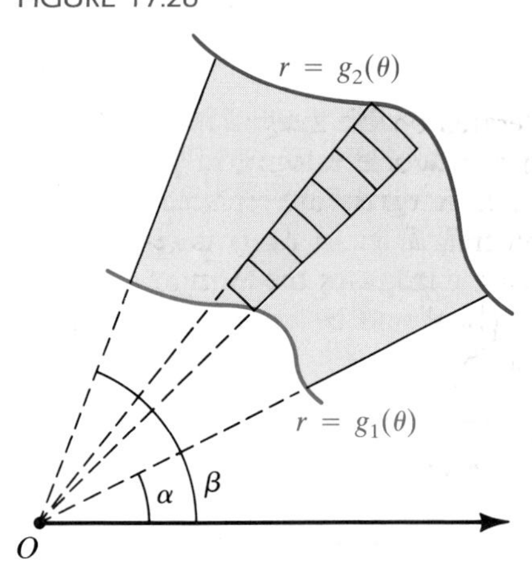

EXAMPLE 1 Find the area of the region R that lies outside the circle $r = a$ and inside the circle $r = 2a \sin\theta$.

SOLUTION The region is sketched in Figure 17.27, together with a typical wedge of elementary polar regions obtained by summing from the inner boundary $r = a$ to the outer boundary $r = 2a \sin\theta$. We then sweep out the region by letting θ vary from $\pi/6$ to $5\pi/6$. (A common error is to let θ vary from 0 to π. Why is this incorrect?)

From Theorem (17.13) with $f(r, \theta) = 1$,

$$A = \iint_R dA = \int_{\pi/6}^{5\pi/6} \int_a^{2a \sin\theta} r \, dr \, d\theta.$$

FIGURE 17.27

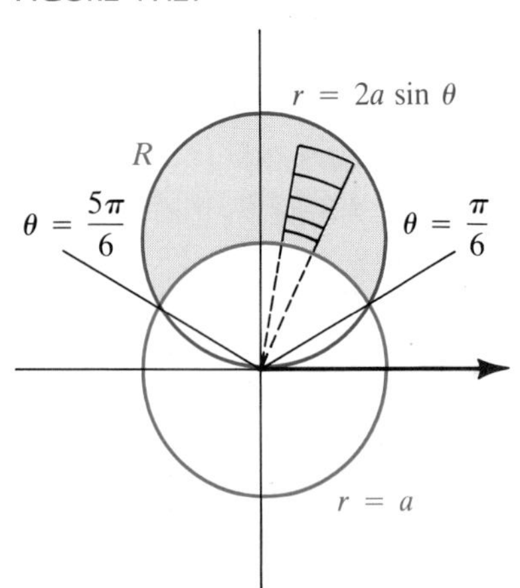

To simplify the evaluation, we may use the symmetry of R with respect to the y-axis. In this case we let θ vary from $\pi/6$ to $\pi/2$ and double the integral. Thus,

$$\begin{aligned}
A = \iint_R dA &= 2 \int_{\pi/6}^{\pi/2} \int_a^{2a \sin\theta} r \, dr \, d\theta \\
&= 2 \int_{\pi/6}^{\pi/2} \left[\tfrac{1}{2} r^2\right]_a^{2a \sin\theta} d\theta = \int_{\pi/6}^{\pi/2} (4a^2 \sin^2\theta - a^2) \, d\theta \\
&= a^2 \int_{\pi/6}^{\pi/2} \left(4 \frac{1 - \cos 2\theta}{2} - 1\right) d\theta = a^2 \int_{\pi/6}^{\pi/2} (1 - 2\cos 2\theta) \, d\theta \\
&= a^2 \Big[\theta - \sin 2\theta\Big]_{\pi/6}^{\pi/2} = a^2 \left[\left(\frac{\pi}{2} - 0\right) - \left(\frac{\pi}{6} - \frac{\sqrt{3}}{2}\right)\right] \\
&= a^2 \left(\frac{\pi}{3} + \frac{\sqrt{3}}{2}\right) \approx 1.9a^2.
\end{aligned}$$

EXAMPLE 2 Find the area of the region R bounded by one loop of the lemniscate $r^2 = a^2 \sin 2\theta$, where $a > 0$.

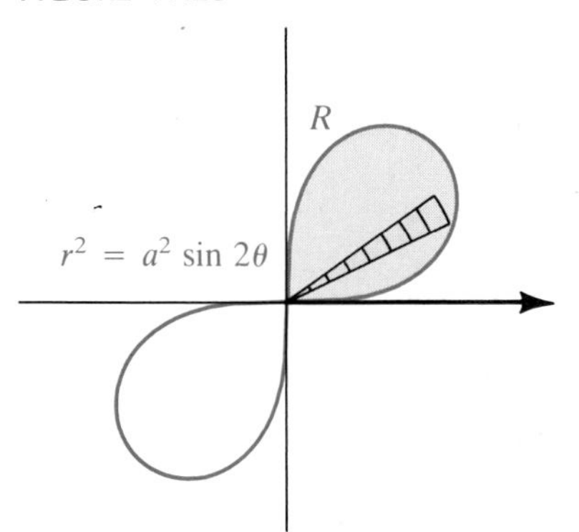

FIGURE 17.28

SOLUTION The lemniscate is sketched in Figure 17.28, together with a wedge of elementary polar regions obtained by summing from the origin ($r = 0$) to the boundary $r^2 = a^2 \sin 2\theta$ of R in the first quadrant. We then sweep out the loop by letting θ vary from 0 to $\pi/2$. (Note that r is *undefined* if $\pi/2 < \theta < \pi$ or $3\pi/2 < \theta < 2\pi$.)

By Theorem (17.13),

$$\begin{aligned} A = \iint_R dA &= \int_0^{\pi/2} \int_0^{a\sqrt{\sin 2\theta}} r \, dr \, d\theta \\ &= \int_0^{\pi/2} \left[\tfrac{1}{2} r^2\right]_0^{a\sqrt{\sin 2\theta}} d\theta = \tfrac{1}{2} \int_0^{\pi/2} a^2 \sin 2\theta \, d\theta \\ &= -\tfrac{1}{4} a^2 \Big[\cos 2\theta\Big]_0^{\pi/2} = -\tfrac{1}{4} a^2(-1 - 1) = \tfrac{1}{2} a^2. \end{aligned}$$

Under suitable conditions, an iterated double integral in rectangular coordinates can be transformed into a double integral in polar coordinates. First, the variables x and y in the integrand are replaced by $r \cos \theta$ and $r \sin \theta$. Next, in the iterated integral, $dy\,dx$ or $dx\,dy$ is replaced by $r\,dr\,d\theta$ or $r\,d\theta\,dr$. The following formula indicates the form of a typical integrand. The double integral sign $\iint_R$ should be replaced by suitable *iterated* integral signs and $dA = dy\,dx$ by $r\,dr\,d\theta$.

Change of variables formula **(17.14)**

$$\iint_R f(x, y) \, dy \, dx = \iint_R f(r \cos \theta, r \sin \theta) \, r \, dr \, d\theta$$

If the integrand $f(x, y)$ of a double integral $\iint_R f(x, y)\,dA$ contains the expression $x^2 + y^2$ or if the region R involves circular arcs with centers at the origin, then the introduction of polar coordinates often leads to a simple evaluation, because $x^2 + y^2 = r^2$ and the circular arcs have equations of the form $r = k$. This is illustrated in the following example.

EXAMPLE 3 Use polar coordinates to evaluate

$$\int_{-a}^{a} \int_0^{\sqrt{a^2 - x^2}} (x^2 + y^2)^{3/2} \, dy \, dx.$$

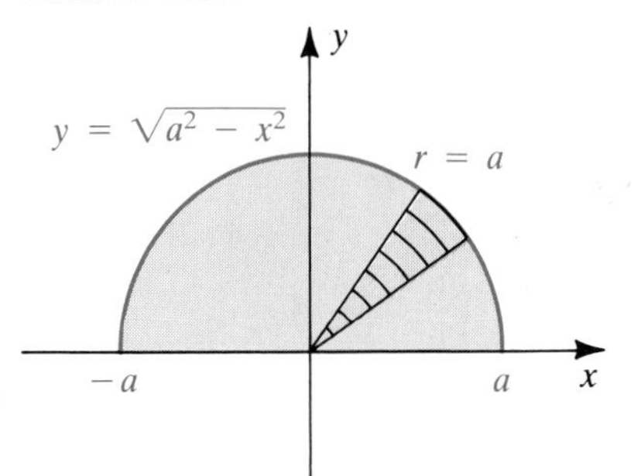

FIGURE 17.29

SOLUTION The region of integration is bounded by the graphs of $y = 0$ (the x-axis) and $y = \sqrt{a^2 - x^2}$ (a semicircle), as shown in Figure 17.29. We first replace $x^2 + y^2$ in the integrand by r^2 and $dy\,dx$ by $r\,dr\,d\theta$. Next we change the limits to polar coordinates. Referring to Figure 17.29, we find that

$$\begin{aligned} \int_{-a}^{a} \int_0^{\sqrt{a^2 - x^2}} (x^2 + y^2)^{3/2} \, dy \, dx &= \int_0^{\pi} \int_0^{a} r^3 \, r \, dr \, d\theta \\ &= \int_0^{\pi} \left[\tfrac{1}{5} r^5\right]_0^a d\theta = \tfrac{1}{5} a^5 \int_0^{\pi} d\theta \\ &= \tfrac{1}{5} a^5 \Big[\theta\Big]_0^{\pi} = \tfrac{1}{5} \pi a^5. \end{aligned}$$

Polar coordinates can also be used for double integrals if R is a region of the type illustrated in Figure 17.30(i). In this case R is bounded by the arcs of two circles of radii a and b and by the graphs of two polar equations $\theta = h_1(r)$ and $\theta = h_2(r)$, where h_1 and h_2 are continuous functions and $h_1(r) \le h_2(r)$ for $a \le r \le b$. If f is a function of r and θ and if f is continuous on R, then the limit in (17.13) exists and the evaluation formula is given in the next theorem.

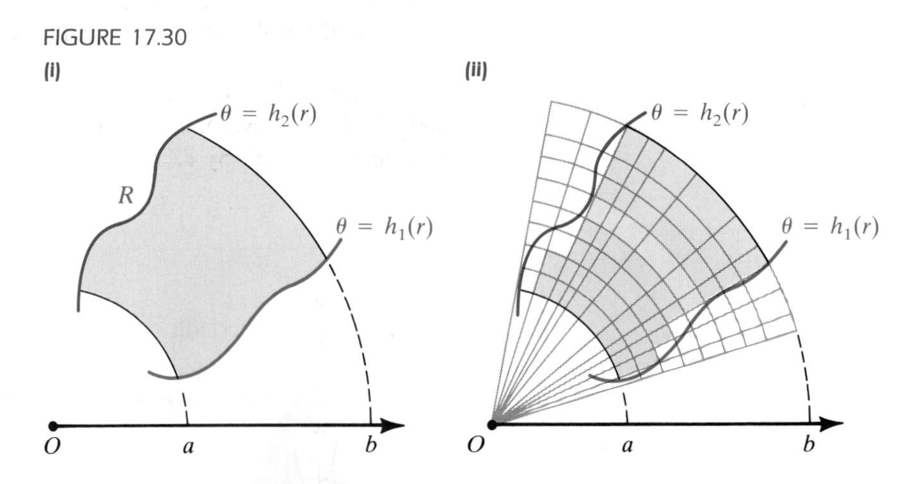

FIGURE 17.30

Evaluation theorem **(17.15)**

$$\iint_R f(r, \theta)\, dA = \int_a^b \int_{h_1(r)}^{h_2(r)} f(r, \theta) r \, d\theta \, dr$$

The iterated integral in Theorem (17.15) may be interpreted in terms of limits of double sums. We first hold r fixed and sum along a circular arc, illustrated by the pink region in Figure 17.30(ii). We next sweep out R by summing terms that correspond to such ring-shaped regions from $r = a$ to $r = b$.

EXAMPLE 4 Find the area of the smaller of the regions bounded by the polar axis, the graphs of $r = 1$ and $r = 2$, and the part of the spiral $r\theta = 1$ from $\theta = \frac{1}{2}$ to $\theta = 1$.

FIGURE 17.31

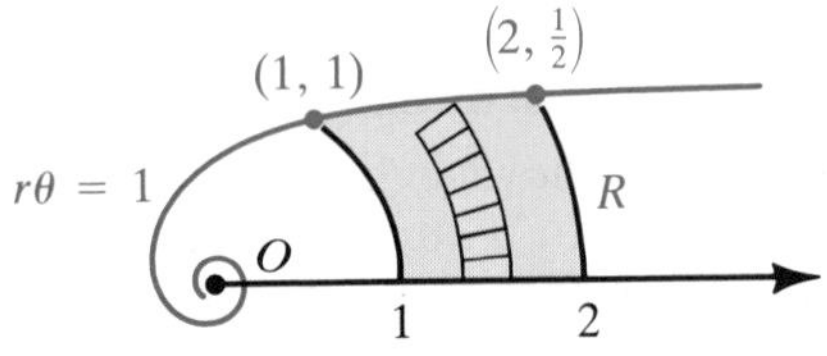

$(1, 1)$ $\left(2, \frac{1}{2}\right)$ $r\theta = 1$ O R 1 2

SOLUTION The region R, together with a ring of elementary polar regions, is sketched in Figure 17.31. We apply Theorem (17.15) with $f(r, \theta) = 1$:

$$\begin{aligned} A &= \iint_R dA = \int_1^2 \int_0^{1/r} r \, d\theta \, dr \\ &= \int_1^2 r \Big[\theta\Big]_0^{1/r} dr = \int_1^2 r\left(\frac{1}{r}\right) dr \\ &= \int_1^2 dr = \Big[r\Big]_1^2 = 2 - 1 = 1 \end{aligned}$$

As in the preceding section, if $f(x, y) \ge 0$, then the volume V of the solid that lies under the graph of $z = f(x, y)$ and over a region R in the

xy-plane is given by

$$V = \iint_R f(x, y)\,dA.$$

Sometimes it is convenient to evaluate this double integral by using polar coordinates, as in the next example.

EXAMPLE 5 Find the volume V of the solid that is bounded by the paraboloid $z = 4 - x^2 - y^2$ and the xy-plane.

SOLUTION The first-octant portion of the solid is sketched in Figure 17.32(i). By symmetry, we may find the volume of this portion and multiply the result by 4.

FIGURE 17.32

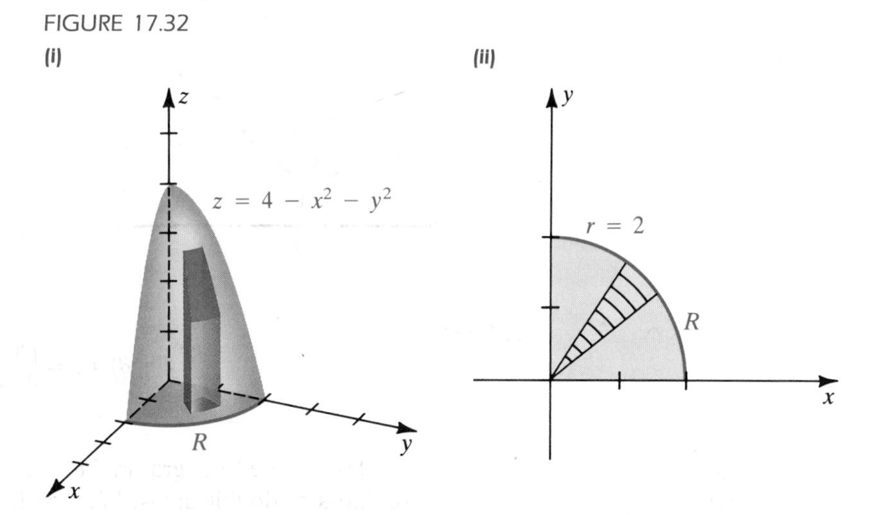

The region R in the xy-plane is bounded by the coordinate axes and one-fourth of the circle with polar equation $r = 2$, sketched in Figure 17.32(ii), where we have also shown a wedge of elementary polar regions. The base of the prism in Figure 17.32(i) corresponds to *one* elementary polar region. We can obtain the volume of the solid by using $r^2 = x^2 + y^2$ and the change of variables formula (17.14):

$$\begin{aligned} V &= 4 \iint_R (4 - x^2 - y^2)\,dA \\ &= 4 \int_0^{\pi/2} \int_0^2 (4 - r^2)\,r\,dr\,d\theta \\ &= 4 \int_0^{\pi/2} \left[2r^2 - \tfrac{1}{4}r^4\right]_0^2 d\theta = 4 \int_0^{\pi/2} 4\,d\theta = 16\Big[\theta\Big]_0^{\pi/2} = 8\pi \end{aligned}$$

The same problem, stated in terms of rectangular coordinates, leads to the following double integral:

$$V = 4 \iint_R (4 - x^2 - y^2)\,dA = 4 \int_0^2 \int_0^{\sqrt{4-x^2}} (4 - x^2 - y^2)\,dy\,dx$$

Evaluating the last integral should thoroughly convince anyone of the advantage of using polar coordinates in certain problems.

EXERCISES 17.3

Exer. 1–6: Express the area of the region as an iterated double integral in polar coordinates, using symmetry whenever possible.

1

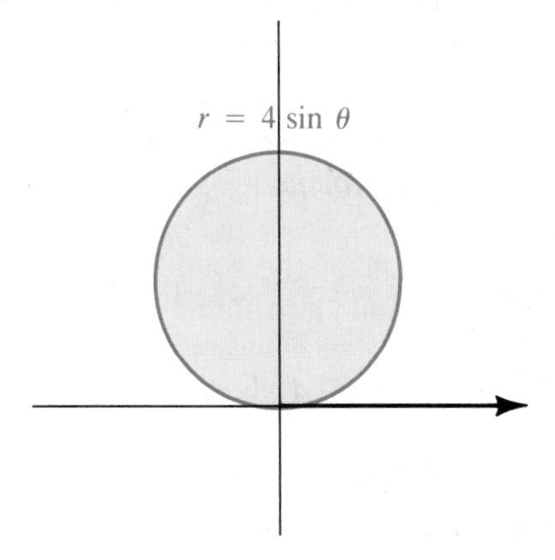

2

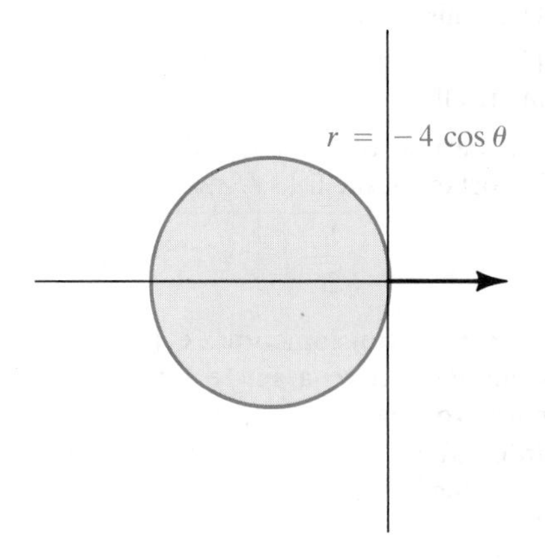

3

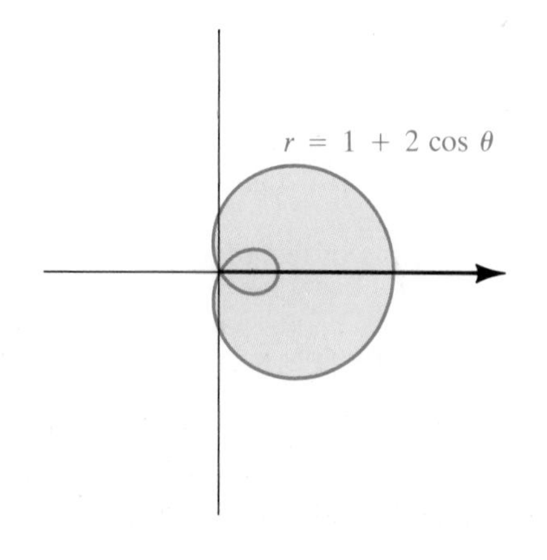

4

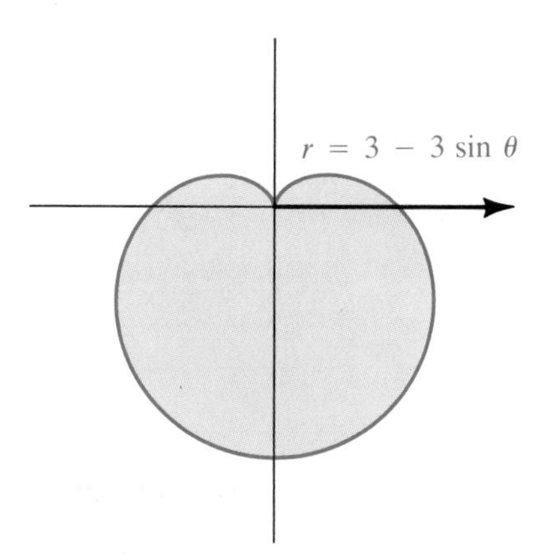

5

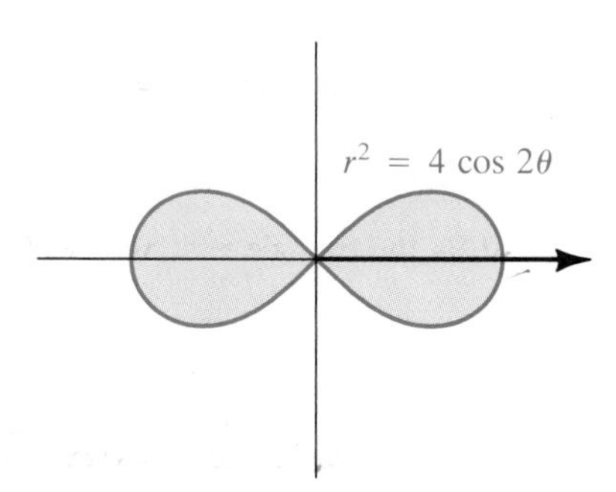

6

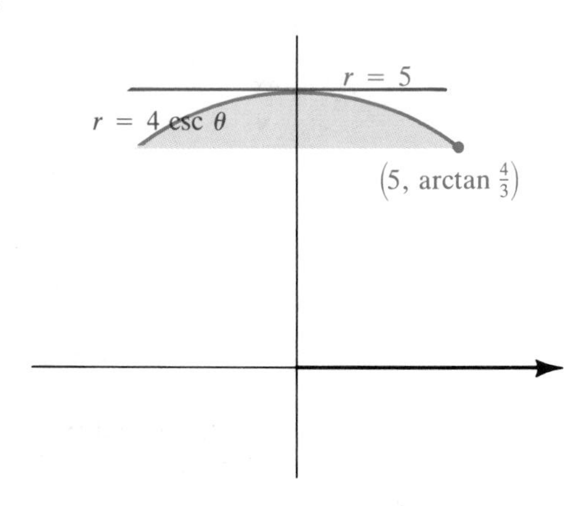

Exer. 7–12: Use a double integral to find the area of the region that has the indicated shape.

7 One loop of $r = 4 \sin 3\theta$

8 One loop of $r = 2 \cos 4\theta$

9 Inside $r = 2 - 2 \cos \theta$ and outside $r = 3$

10 Bounded by $r = 3 + 2 \sin \theta$

11 One loop of $r^2 = 9 \cos 2\theta$

12 Inside $r = 3 \sin \theta$ and outside $r = 1 + \sin \theta$

Exer. 13–24: Use polar coordinates to evaluate the integral.

13 $\iint\limits_R (x^2+y^2)^{3/2}\,dA$;
R is bounded by the circle $x^2+y^2=4$.

14 $\iint\limits_R x^2(x^2+y^2)^3\,dA$;
R is bounded by the semicircle $y=\sqrt{1-x^2}$ and the x-axis.

15 $\iint\limits_R \dfrac{x^2}{x^2+y^2}\,dA$;
R is the annular region bounded by $x^2+y^2=a^2$ and $x^2+y^2=b^2$ with $0<a<b$.

16 $\iint\limits_R (x+y)\,dA$;
R is bounded by the circle $x^2+y^2=2y$.

17 $\iint\limits_R \sqrt{x^2+y^2}\,dA$;
R is bounded by the triangle with vertices $(0, 0)$, $(3, 0)$, $(3, 3)$.

18 $\iint\limits_R \sqrt{x^2+y^2}\,dA$;
R is bounded by the semicircle $y=\sqrt{2x-x^2}$ and the line $y=x$.

19 $\int_{-a}^{a}\int_{0}^{\sqrt{a^2-x^2}} e^{-(x^2+y^2)}\,dy\,dx$

20 $\int_{0}^{a}\int_{0}^{\sqrt{a^2-x^2}} (x^2+y^2)^{3/2}\,dy\,dx$

21 $\int_{1}^{2}\int_{0}^{x} \dfrac{1}{\sqrt{x^2+y^2}}\,dy\,dx$

22 $\int_{0}^{1}\int_{0}^{\sqrt{1-x^2}} e^{\sqrt{x^2+y^2}}\,dy\,dx$

23 $\int_{0}^{2}\int_{0}^{\sqrt{4-y^2}} \cos(x^2+y^2)\,dx\,dy$

24 $\int_{0}^{2}\int_{-\sqrt{2y-y^2}}^{\sqrt{2y-y^2}} x\,dx\,dy$

Exer. 25–30: Use polar coordinates to find the volume of the solid that has the shape of Q.

25 Q is the region inside the sphere $x^2+y^2+z^2=25$ and outside the cylinder $x^2+y^2=9$.

26 Q is cut out of the ellipsoid $4x^2+4y^2+z^2=16$ by the cylinder $x^2+y^2=1$.

27 Q is bounded by the cone $z^2=x^2+y^2$ and the cylinder $x^2+y^2=2x$.

28 Q is bounded by the paraboloid $z=4x^2+4y^2$, the cylinder $x^2+y^2=3y$, and the plane $z=0$.

29 Q is the largest region that is inside both the sphere $x^2+y^2+z^2=16$ and the cylinder $x^2+y^2=4y$.

30 Q is bounded by the paraboloid $z=9-x^2-y^2$ and the plane $z=5$.

31 Let R_a be the region bounded by the circle $x^2+y^2=a^2$. If we define

$$\int_{-\infty}^{\infty}\int_{-\infty}^{\infty} e^{-(x^2+y^2)}\,dx\,dy=\lim_{a\to\infty}\iint\limits_{R_a} e^{-(x^2+y^2)}\,dA,$$

evaluate the improper double integral.

32 **(a)** It can be proved, using advanced methods, that

$$\int_{-\infty}^{\infty}\int_{-\infty}^{\infty} e^{-(x^2+y^2)}\,dx\,dy=\int_{-\infty}^{\infty} e^{-x^2}\,dx\int_{-\infty}^{\infty} e^{-y^2}\,dy.$$

Use this fact and Exercise 31 to show that $\int_{-\infty}^{\infty} e^{-x^2}\,dx=\sqrt{\pi}$. Interpret this result geometrically.

(b) Use part (a) to prove the following result, which is important in the field of statistics:

$$\frac{1}{\sqrt{2\pi}}\int_{-\infty}^{\infty} e^{-x^2/2}\,dx=1$$

c **Exer. 33–34: We can sometimes evaluate a double integral by changing it to a single integral. (a) Use polar coordinates to change the given double integral to a single integral involving only the variable r. (b) Use Simpson's rule, with $n=4$, to approximate the single integral.**

33 $\iint\limits_R \sqrt{1+(x^2+y^2)^2}\,dA$;
R is the region in the first quadrant bounded by the circle $x^2+y^2=4$ and the coordinate axes.

34 $\iint\limits_R \sin\sqrt[3]{x^2+y^2}\,dA$;
R is the region bounded by the semicircle $x=\sqrt{1-y^2}$ and the y-axis.

17.4 SURFACE AREA

In Section 6.5 we obtained formulas for the area of a surface of revolution. We now consider a method for finding areas of more general surfaces. Suppose that $f(x, y)\ge 0$ throughout a region R in the xy-plane and that

FIGURE 17.33

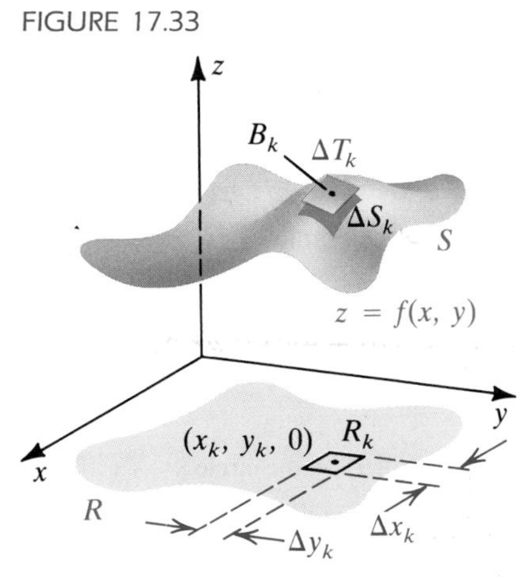

f has continuous first partial derivatives on R. Let S denote the portion of the graph of f whose projection on the xy-plane is R, as illustrated in Figure 17.33. To simplify the discussion, we assume that no normal vector to S is parallel to the xy-plane. We wish to define the area A of S and find a formula that can be used to calculate A.

Let $P = \{R_k\}$ be an inner partition of R, and let the dimensions of the rectangle R_k be Δx_k and Δy_k. For each k, choose any point $(x_k, y_k, 0)$ in R_k and let $B_k(x_k, y_k, f(x_k, y_k))$ be the corresponding point on S. Next consider the tangent plane to S at B_k (see Figure 17.33), and let ΔT_k and ΔS_k be the areas of the regions on the tangent plane and on S, respectively, obtained by projecting R_k vertically upward. If the norm $\|P\|$ of the partition is small, then ΔT_k is an approximation to ΔS_k, and $\sum_k \Delta T_k$ is an approximation to the area A of S. This approximation should improve as $\|P\|$ approaches 0, so we arrive at the following.

Definition **(17.16)**

$$A = \lim_{\|P\|\to 0} \sum_k \Delta T_k$$

To find an integral formula for A, let us choose $(x_k, y_k, 0)$ as the corner of R_k closest to the origin. As in the isolated view of ΔT_k shown in Figure 17.34, consider the vectors **a** and **b** with initial point $B_k(x_k, y_k, f(x_k, y_k))$, which are tangent to the traces of S on the planes $y = y_k$ and $x = x_k$, respectively. By Theorem (16.10), the slopes of the lines determined by **a** and **b** in these planes are $f_x(x_k, y_k)$ and $f_y(x_k, y_k)$, respectively. It follows that

FIGURE 17.34

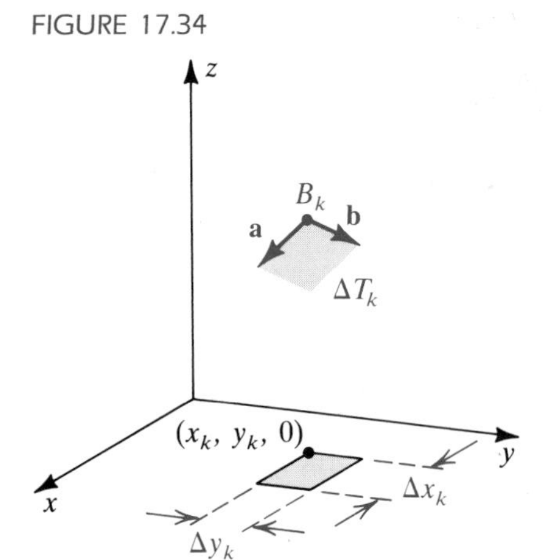

$$\mathbf{a} = \Delta x_k\mathbf{i} + f_x(x_k, y_k)\,\Delta x_k\mathbf{k}$$
$$\mathbf{b} = \Delta y_k\mathbf{j} + f_y(x_k, y_k)\,\Delta y_k\mathbf{k}.$$

From Section 14.4, the area ΔT_k of the parallelogram determined by **a** and **b** is $\|\mathbf{a} \times \mathbf{b}\|$. Using

$$\mathbf{a} \times \mathbf{b} = \begin{vmatrix} \mathbf{i} & \mathbf{j} & \mathbf{k} \\ \Delta x_k & 0 & f_x(x_k, y_k)\,\Delta x_k \\ 0 & \Delta y_k & f_y(x_k, y_k)\,\Delta y_k \end{vmatrix}$$

gives us

$$\mathbf{a} \times \mathbf{b} = -f_x(x_k, y_k)\,\Delta x_k\,\Delta y_k\mathbf{i} - f_y(x_k, y_k)\,\Delta x_k\,\Delta y_k\mathbf{j} + \Delta x_k\,\Delta y_k\mathbf{k}.$$

Consequently

$$\Delta T_k = \|\mathbf{a} \times \mathbf{b}\| = \sqrt{[f_x(x_k, y_k)]^2 + [f_y(x_k, y_k)]^2 + 1}\;\Delta x_k\,\Delta y_k,$$

where $\Delta x_k\,\Delta y_k = \Delta A_k$. If, as in (17.16), we take the limit of sums of the ΔT_k and apply the definition of the double integral, we obtain the following.

Integral for surface area **(17.17)**

$$A = \iint_R \sqrt{[f_x(x, y)]^2 + [f_y(x, y)]^2 + 1}\; dA$$

The formula in (17.17) may also be used if $f(x, y) \le 0$ on R.

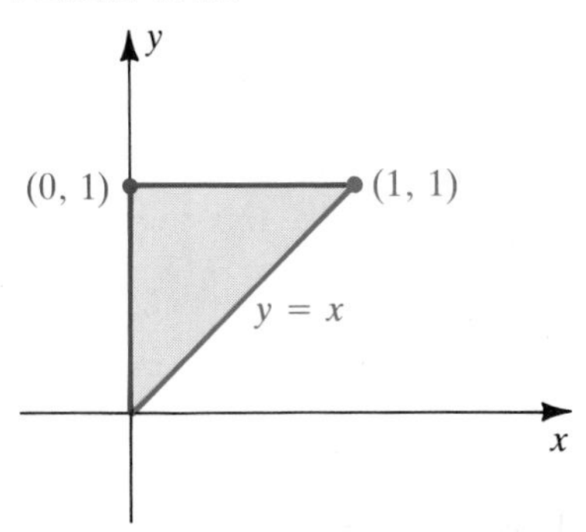

FIGURE 17.35

EXAMPLE 1 Let R be the triangular region in the xy-plane with vertices (0, 0, 0), (0, 1, 0), and (1, 1, 0). Find the surface area of that portion of the graph of $z = 3x + y^2$ that lies over R.

SOLUTION The region R in the xy-plane is bounded by the graphs of $y = x$, $x = 0$, and $y = 1$, as shown in Figure 17.35. Letting $f(x, y) = 3x + y^2$ and applying (17.17) gives us

$$\begin{aligned} A &= \iint_R \sqrt{3^2 + (2y)^2 + 1}\, dA = \int_0^1 \int_0^y (10 + 4y^2)^{1/2}\, dx\, dy \\ &= \int_0^1 (10 + 4y^2)^{1/2} \Big[x\Big]_0^y dy = \int_0^1 (10 + 4y^2)^{1/2} y\, dy \\ &= \left[\frac{1}{12}(10 + 4y^2)^{3/2}\right]_0^1 = \frac{14^{3/2} - 10^{3/2}}{12} \approx 1.7 \text{ square units.} \end{aligned}$$

EXAMPLE 2 Find the surface area of the paraboloid $z = 4 - x^2 - y^2$ for $z \geq 0$.

SOLUTION The paraboloid is sketched in Figure 17.36(i). Applying (17.17) with $f(x, y) = 4 - x^2 - y^2$, we have

$$\begin{aligned} A &= \iint_R \sqrt{(-2x)^2 + (-2y)^2 + 1}\, dA \\ &= \iint_R \sqrt{4x^2 + 4y^2 + 1}\, dA, \end{aligned}$$

where R is the region in the xy-plane bounded by the circle $x^2 + y^2 = 4$. By symmetry, we can find the surface area corresponding to the first-quadrant portion of R shown in Figure 17.36(ii) and then multiply this result by 4. Using an iterated integral yields

$$A = 4 \int_0^2 \int_0^{\sqrt{4 - x^2}} \sqrt{4x^2 + 4y^2 + 1}\, dy\, dx.$$

FIGURE 17.36

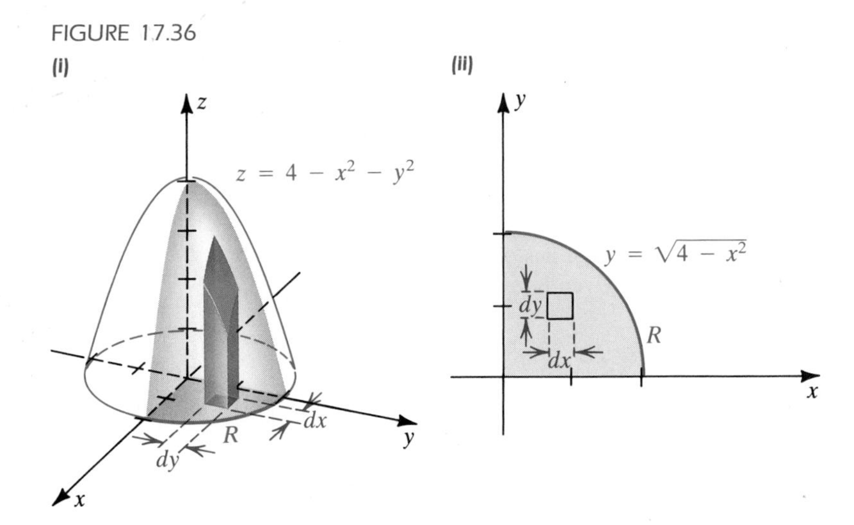

It would be very tedious to evaluate this integral, so we change to polar coordinates and proceed as follows:

$$\begin{aligned} A &= 4\int_0^{\pi/2}\int_0^2 \sqrt{4r^2+1}\, r\, dr\, d\theta \\ &= 4\int_0^{\pi/2}\left[\tfrac{1}{12}(4r^2+1)^{3/2}\right]_0^2 d\theta \\ &= 4\int_0^{\pi/2}\tfrac{1}{12}(17^{3/2}-1)\, d\theta \\ &= \frac{1}{3}(17^{3/2}-1)\Big[\theta\Big]_0^{\pi/2} = \frac{\pi}{6}(17^{3/2}-1) \approx 36.2 \end{aligned}$$

Formulas similar to (17.17) may be stated if the surface S has suitable projections on the yz- or xz-planes. Thus, if S is the graph of an equation $y = g(x, z)$ and if the projection on the xz-plane is R_1, then

$$A = \iint_{R_1} \sqrt{[g_x(x, z)]^2 + [g_z(x, z)]^2 + 1}\, dA.$$

A formula of this type may also be stated if S is given by $x = h(y, z)$.

EXERCISES 17.4

Exer. 1–4: Set up an iterated double integral that can be used to find the surface area of the portion of the graph of the equation that lies over the region R in the xy-plane that has the given boundary. Use symmetry whenever possible.

1 $x^2 + y^2 + z^2 = 4$;
the square with vertices $(1, 1), (1, -1), (-1, 1), (-1, -1)$

2 $x^2 - y^2 + z^2 = 1$;
the square with vertices $(0, 1), (1, 0), (-1, 0), (0, -1)$

3 $36z^2 = 16x^2 + 9y^2 + 144$;
the circle with center at the origin and radius 3

4 $9y^2 = 225z^2 + 25x^2$;
the triangle with vertices $(0, 0), (3, 5), (-3, 5)$

Exer. 5–6: Find the surface area of the portion of the graph of the equation that lies over the region R in the xy-plane that has the given boundary.

5 $z = y + \frac{1}{2}x^2$;
the square with vertices $(0, 0), (1, 0), (1, 1), (0, 1)$

6 $z = y^2$;
the triangle with vertices $(0, 0), (0, 2), (2, 2)$

7 A portion of the plane $(x/a) + (y/b) + (z/c) = 1$ is cut out by the cylinder $x^2 + y^2 = k^2$, where a, b, c, and k are positive. Find the area of that portion.

8 Find the surface area of the first-octant portion of the cylinder $y^2 + z^2 = 9$ that lies inside the cylinder $x^2 + y^2 = 9$ (see Figure 17.23).

Exer. 9–12: Find the area of the surface S.

9 S is the part of the paraboloid $z = x^2 + y^2$ cut off by the plane $z = 1$.

10 S is the part of the plane $z = y + 1$ that is inside the cylinder $x^2 + y^2 = 1$.

11 S is the part of the sphere $x^2 + y^2 + z^2 = a^2$ that is inside the cylinder $x^2 + y^2 = ay$.

12 S is the first-octant part of the hyperbolic paraboloid $z = x^2 - y^2$ that is inside the cylinder $x^2 + y^2 = 1$.

13 A dome-shaped camping tent is designed to have a circular floor of radius 5 feet and a roof described by the graph of $z = 7 - \frac{7}{25}(x^2 + y^2)$ for $z \geq 0$. Approximate the number of square feet of material needed to construct the tent. (Do not consider wasted or overlapping material.)

14 A Van de Graaff generator can be used to produce millions of volts of static electricity and large sparks. Part of the generator consists of a spherical metal shell with a circular hole cut out of the bottom. The total charge C (in coulombs) generated is given by $C = \delta A$, where A

is the area of the metal surface and δ is the area charge density on the surface. Suppose the sphere can be described by $x^2 + y^2 + z^2 = 1$, and the hole is obtained by intersecting the cylinder $x^2 + y^2 = \frac{1}{16}$ with the sphere for $z < 0$, where x, y, and z are in meters. Approximate C if $\delta = 1.2 \times 10^{-5}$ coulomb/m^2. (This value of C corresponds to a voltage potential of approximately 1.35 million volts.)

c **Exer. 15–16: Refer to Exercises 51–52 of Section 17.1. Use the trapezoidal rule, with $n = 2$, to approximate the surface area of the portion of the graph of f that lies over the square region R with the given vertices in the xy-plane.**

15 $f(x, y) = \frac{1}{3}x^3 + \cos y$; $(0, 0), (0, 1), (1, 1), (1, 0)$

16 $f(x, y) = x^3 y$; $(\frac{1}{4}, \frac{1}{4}), (\frac{1}{4}, \frac{3}{4}), (\frac{3}{4}, \frac{1}{4}), (\frac{3}{4}, \frac{3}{4})$

17.5 TRIPLE INTEGRALS

Triple integrals for a function f of *three* variables x, y, and z can be defined using a four-step process similar to that given in Section 17.1 for functions of two variables. The simplest case occurs if f is continuous throughout a box-shaped region Q in three dimensions of the following type:

$$Q = \{(x, y, z)\colon a \le x \le b,\ c \le y \le d,\ m \le z \le n\}$$

A possible position for Q is illustrated in Figure 17.37(i). If Q is divided into subregions $Q_1, Q_2, \ldots, Q_n$ by means of planes parallel to the three coordinate planes (see Figure 17.37(ii)), then the collection $\{Q_k\}$ is a **partition** P of Q. The **norm** $\|P\|$ of the partition is the length of the longest diagonal of all the Q_k. As shown in Figure 17.37(iii), if Δx_k, Δy_k, and Δz_k are the dimensions of Q_k, then its volume ΔV_k is

$$\Delta V_k = \Delta x_k\, \Delta y_k\, \Delta z_k.$$

FIGURE 17.37

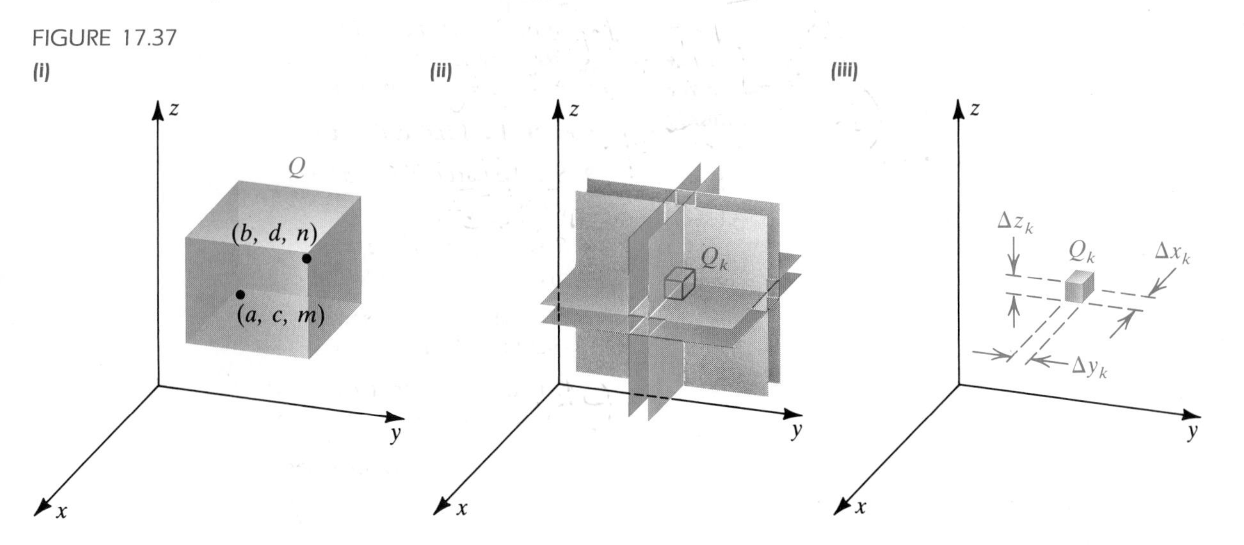

A **Riemann sum** of f for P is

$$\sum_k f(u_k, v_k, w_k)\, \Delta V_k,$$

where (u_k, v_k, w_k) is any point in Q_k. A limit of Riemann sums, as $\|P\| \to 0$, is defined as it was for functions of two variables. If the limit exists, it is called the **triple integral of f over Q** and is denoted by $\iiint_Q f(x, y, z)\, dV$. Using limit notation, we have the following.

Triple integral of f over Q **(17.18)**

$$\iiint_Q f(x, y, z)\, dV = \lim_{\|P\|\to 0} \sum_k f(u_k, v_k, w_k)\, \Delta V_k$$

It can be shown that if Q is the region in Figure 17.37(i), then

$$\iiint_Q f(x, y, z)\, dV = \int_m^n \int_c^d \int_a^b f(x, y, z)\, dx\, dy\, dz.$$

The iterated integral on the right is evaluated by beginning with the *innermost* integral and working outward. Thus, the first integration is with respect to x (with y and z fixed), the second is with respect to y (with z fixed), and the third is with respect to z. There are five other iterated integrals that equal the triple integral of f over Q. For example, if we integrate in the order y, z, x, then

$$\iiint_Q f(x, y, z)\, dV = \int_a^b \int_m^n \int_c^d f(x, y, z)\, dy\, dz\, dx.$$

It can be shown that for the box-shaped region Q in Figure 17.37(i), the order of integration is immaterial.

EXAMPLE 1 Evaluate $\iiint_Q (xy^2 + yz^3)\, dV$ if

$$Q = \{(x, y, z)\colon -1 \le x \le 1, 3 \le y \le 4, 0 \le z \le 2\}.$$

SOLUTION Of the six possible iterated integrals, we shall use the following:

$$\begin{aligned}
\int_3^4 \int_{-1}^1 \int_0^2 (xy^2 + yz^3)\, dz\, dx\, dy &= \int_3^4 \int_{-1}^1 \left[\int_0^2 (xy^2 + yz^3)\, dz\right] dx\, dy \\
&= \int_3^4 \int_{-1}^1 \left[xy^2 z + y\left(\frac{z^4}{4}\right)\right]_0^2 dx\, dy \\
&= \int_3^4 \int_{-1}^1 (2xy^2 + 4y)\, dx\, dy \\
&= \int_3^4 \left[2\left(\frac{x^2}{2}\right)y^2 + 4yx\right]_{-1}^1 dy \\
&= \int_3^4 [(y^2 + 4y) - (y^2 - 4y)]\, dy \\
&= \int_3^4 8y\, dy = 8\left[\frac{y^2}{2}\right]_3^4 = 28
\end{aligned}$$

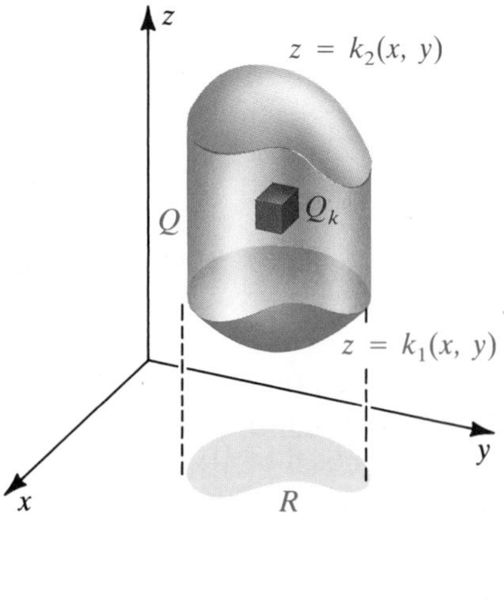

Triple integrals may be defined over a region more complicated than a rectangular box. Suppose, for example, that R is a region in the xy-plane that can be divided into R_x and R_y regions and that Q is the region in three dimensions defined by

$$Q = \{(x, y, z)\colon (x, y) \text{ is in } R \text{ and } k_1(x, y) \le z \le k_2(x, y)\},$$

where k_1 and k_2 are functions that have continuous first partial derivatives throughout R. The region Q lies between the graphs of $z = k_1(x, y)$ and $z = k_2(x, y)$ and over or under the region R (see Figure 17.38). If Q is sub-

FIGURE 17.38

divided by means of planes parallel to the three coordinate planes, then the resulting (small) box-shaped regions $Q_1, Q_2, \ldots, Q_n$ that lie completely within Q form an **inner partition** P of Q. A typical element Q_k of an inner partition of Q is shown in Figure 17.38.

A **Riemann sum** of f for P is any sum of the form $\sum_k f(u_k, v_k, w_k)\,\Delta V_k$, where (u_k, v_k, w_k) is an arbitrary point in Q_k and ΔV_k is the volume of Q_k. The triple integral of f over Q is again defined as the limit (17.18). If f is continuous throughout Q, then the following can be proved.

Evaluation theorem **(17.19)**

$$\iiint_Q f(x, y, z)\, dV = \iint_R \left[\int_{k_1(x,y)}^{k_2(x,y)} f(x, y, z)\, dz \right] dA$$

The notation on the right in (17.19) means that after first integrating with respect to z, we evaluate the resulting double integral over the region R in the xy-plane using the methods of Section 17.1. Thus, for an R_x region such as that in Figure 17.8(i),

$$\iiint_Q f(x, y, z)\, dV = \int_a^b \int_{g_1(x)}^{g_2(x)} \int_{k_1(x,y)}^{k_2(x,y)} f(x, y, z)\, dz\, dy\, dx.$$

The symbol on the right-hand side of this equation is an **iterated triple integral**. It is evaluated by means of partial integrations of $f(x, y, z)$ in the order z, y, x, with the indicated limits substituted in the usual way. Similarly, for an R_y region such as that in Figure 17.8(ii),

$$\iiint_Q f(x, y, z)\, dV = \int_c^d \int_{h_1(y)}^{h_2(y)} \int_{k_1(x,y)}^{k_2(x,y)} f(x, y, z)\, dz\, dx\, dy.$$

FIGURE 17.39

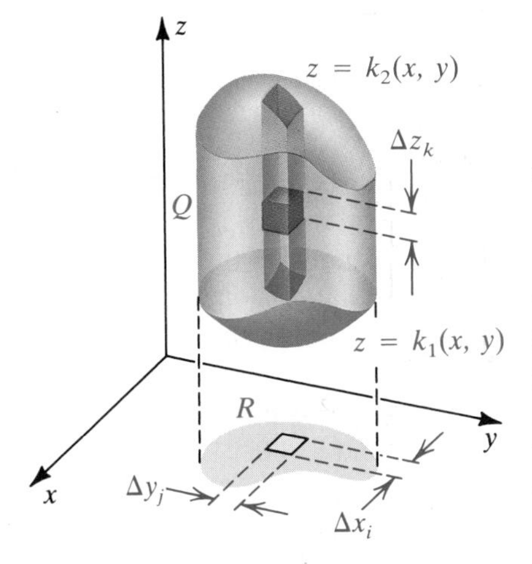

It is convenient to regard either of the last two integrals as a limit of a *triple sum*,

$$\sum_i \sum_j \sum_k f(u_i, v_j, w_k)\, \Delta z_k\, \Delta y_j\, \Delta x_i,$$

where the first summation (with respect to k) corresponds to a column of (small) boxes, in the direction parallel to the z-axis, from the lower surface (with equation $z = k_1(x, y)$) to the upper surface (with equation $z = k_2(x, y)$), as illustrated in Figure 17.39. The remaining two summations take place over the region R in the xy-plane, as in our work with double integrals in Section 17.1.

EXAMPLE 2 Express $\iiint_Q f(x, y, z)\, dV$ as an iterated integral if Q is the region in the first octant bounded by the coordinate planes, the paraboloid $z = 2 + x^2 + \frac{1}{4}y^2$, and the cylinder $x^2 + y^2 = 1$.

SOLUTION As illustrated in Figure 17.40(i), Q lies under the paraboloid and over the xy-plane. The region R in the xy-plane is bounded by the coordinate axes and the graph of $y = \sqrt{1 - x^2}$. Also shown in the figure is a column that corresponds to a first summation over the boxes of volumes $\Delta z_k\, \Delta y_j\, \Delta x_i$ in the direction of the z-axis. Since the column extends from the xy-plane to the paraboloid, the lower limit for the integration with respect to z is $z = 0$ and the upper limit is $z = 2 + x^2 + \frac{1}{4}y^2$. The second and third integrations are taken over the region R in the xy-plane

FIGURE 17.40

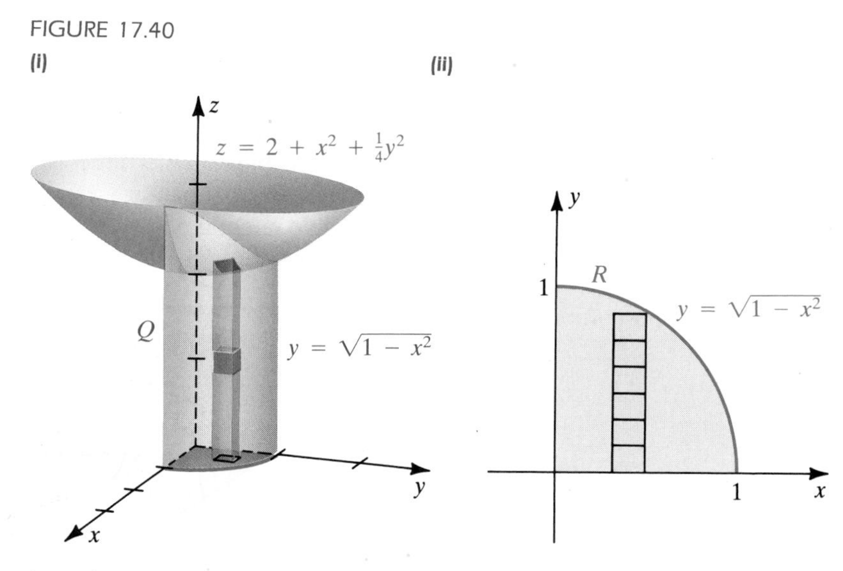

(see Figure 17.40(ii)). Thus, the integral in (17.19) has the form

$$\int_0^1 \int_0^{\sqrt{1-x^2}} \int_0^{2+x^2+(1/4)y^2} f(x, y, z)\, dz\, dy\, dx.$$

If $f(x, y, z) = 1$ throughout a region Q, then the triple integral of f over Q is written $\iiint_Q dV$ and its value is the volume of Q. In Example 2 the volume V of the region Q in Figure 17.40(i) is given by

$$V = \int_0^1 \int_0^{\sqrt{1-x^2}} \int_0^{2+x^2+(1/4)y^2} dz\, dy\, dx.$$

We could, of course, obtain the volume using a double integral, as in Section 17.2. The reason for using a triple integral here and in the following examples is to gain practice in working with limits of triple sums. This same technique will be used later for problems that *cannot* be solved using double integrals.

EXAMPLE 3 Find the volume V of the solid that is bounded by the cylinder $y = x^2$ and by the planes $y + z = 4$ and $z = 0$.

SOLUTION The solid is sketched in Figure 17.41(i), together with a column corresponding to a first summation in the direction of the z-axis. Note that the column extends from $z = 0$ to $z = 4 - y$. The region R in the xy-plane is shown in Figure 17.41(ii), together with a rectangle corresponding to the first (of a double) integration with respect to y. Applying (17.19) with $f(x, y, z) = 1$, we have

$$\begin{aligned} V &= \int_{-2}^2 \int_{x^2}^4 \int_0^{4-y} dz\, dy\, dx = \int_{-2}^2 \int_{x^2}^4 \Big[z\Big]_0^{4-y} dy\, dx \\ &= \int_{-2}^2 \int_{x^2}^4 (4 - y)\, dy\, dx = \int_{-2}^2 \Big[4y - \tfrac{1}{2}y^2\Big]_{x^2}^4 dx \\ &= \int_{-2}^2 (8 - 4x^2 + \tfrac{1}{2}x^4)\, dx = \Big[8x - \tfrac{4}{3}x^3 + \tfrac{1}{10}x^5\Big]_{-2}^2 \\ &= \tfrac{256}{15} \approx 17. \end{aligned}$$

FIGURE 17.41

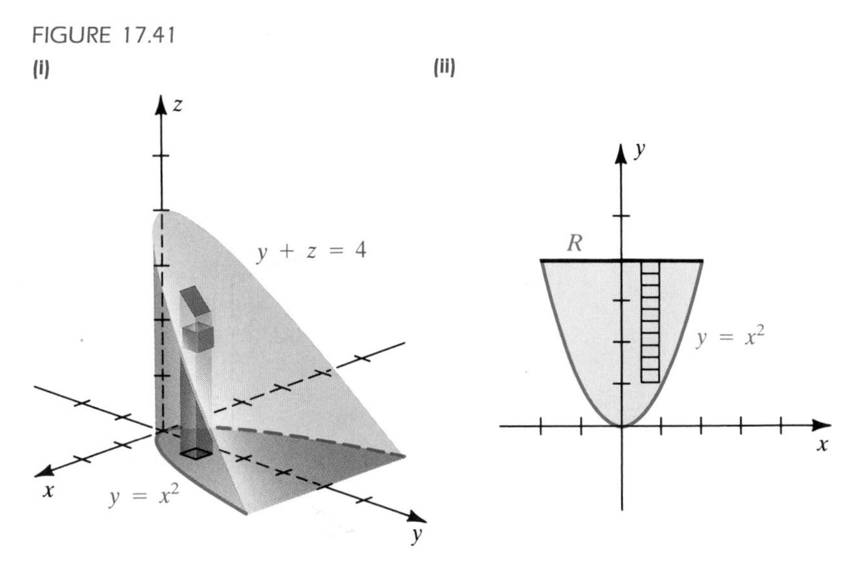

If we use the order $dx\,dy$ for the double integral, then the rectangle in Figure 17.41(ii) will be horizontal and the formula for V will become

$$V = \int_0^4 \int_{-\sqrt{y}}^{\sqrt{y}} \int_0^{4-y} dz\,dx\,dy.$$

Some triple integrals may be evaluated by means of an iterated triple integral in which the first integration is with respect to y. Thus, consider

$$Q = \{(x, y, z): a \le x \le b, h_1(x) \le z \le h_2(x), k_1(x, z) \le y \le k_2(x, z)\},$$

where h_1 and h_2 are functions that are continuous on $[a, b]$ and k_1 and k_2 are functions that have continuous first partial derivatives on the region R in the xz-plane shown in Figure 17.42. Note that Q lies between the graphs of $y = k_1(x, z)$ and $y = k_2(x, z)$. The projection R of Q onto the xz-plane is an R_x region. In this case we have the following.

FIGURE 17.42

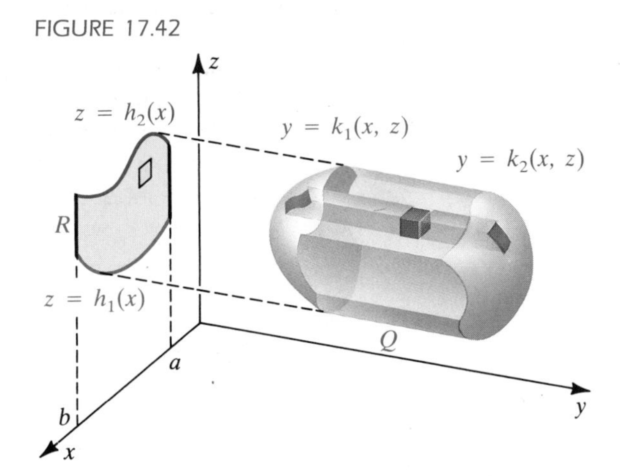

Evaluation theorem **(17.20)**

$$\iiint_Q f(x, y, z)\, dV = \int_a^b \int_{h_1(x)}^{h_2(x)} \int_{k_1(x,z)}^{k_2(x,z)} f(x, y, z)\, dy\, dz\, dx$$

The iterated integral in (17.20) may be interpreted as a limit of triple sums obtained by first summing over a row of small boxes in the direction parallel to the y-axis from the left surface (with equation $y = k_1(x, z)$) to the right surface (with equation $y = k_2(x, z)$), as indicated in Figure 17.42. The resulting double integral is then evaluated over the region R in the xz-plane, as illustrated in the next example.

EXAMPLE 4 Find the volume of the region Q bounded by the graphs of $z = 3x^2$, $z = 4 - x^2$, $y = 0$, and $z + y = 6$.

SOLUTION As shown in Figure 17.43(i), Q lies under the cylinder $z = 4 - x^2$, over the cylinder $z = 3x^2$, to the right of the xz-plane, and to the left of the plane $z + y = 6$. Hence Q is a region of the type illustrated in Figure 17.42 with $k_1(x, z) = 0$ and $k_2(x, z) = 6 - z$. The region R in the xz-plane is sketched in Figure 17.43(ii). Applying (17.20), we have

$$\begin{aligned} V = \iiint_Q dV &= \int_{-1}^{1} \int_{3x^2}^{4-x^2} \int_0^{6-z} dy\, dz\, dx = \int_{-1}^{1} \int_{3x^2}^{4-x^2} \Big[y \Big]_0^{6-z} dz\, dx \\ &= \int_{-1}^{1} \int_{3x^2}^{4-x^2} (6 - z)\, dz\, dx = \int_{-1}^{1} \Big[6z - \tfrac{1}{2}z^2 \Big]_{3x^2}^{4-x^2} dx \\ &= \int_{-1}^{1} (16 - 20x^2 + 4x^4)\, dx = \tfrac{304}{15} \approx 20.3. \end{aligned}$$

FIGURE 17.43

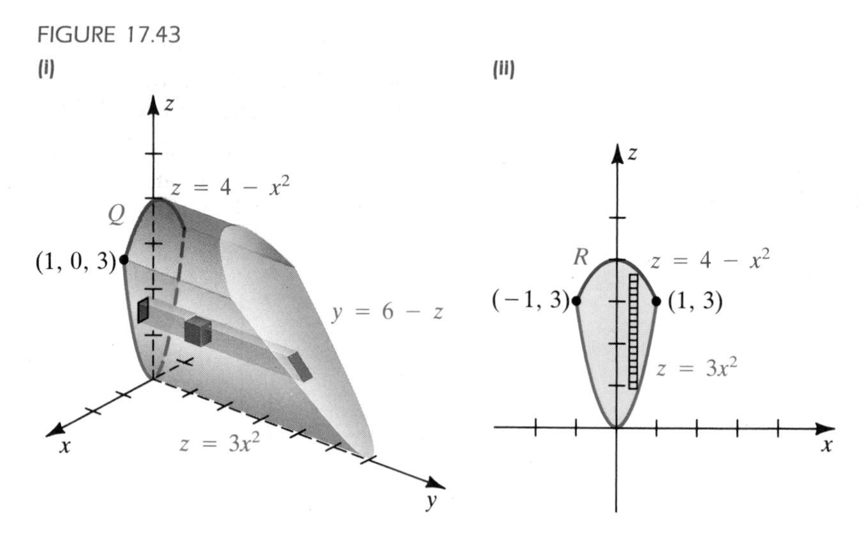

If we used a different order of integration, it would be necessary to use several triple integrals. (Do you see why this is true?)

Finally, if Q is a region of the type illustrated in Figure 17.44, where the functions p_1 and p_2 have continuous first partial derivatives on a

FIGURE 17.44

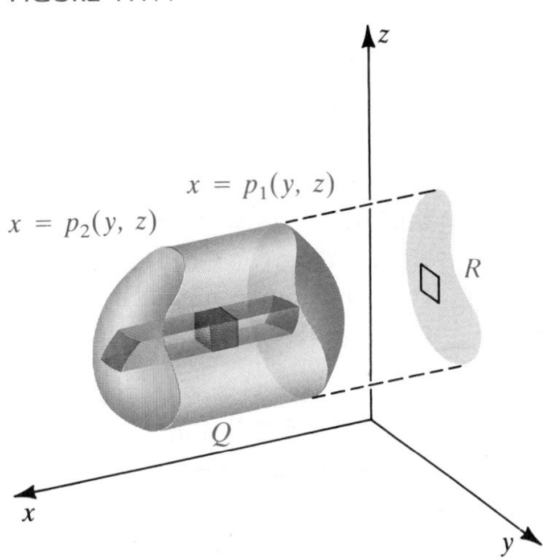

suitable region R in the yz-plane, then

$$\iiint_Q f(x, y, z)\, dV = \iint_R \left[\int_{p_1(y,z)}^{p_2(y,z)} f(x, y, z)\, dx\right] dA.$$

In the final iterated double integral, dA will be replaced by $dz\, dy$ or $dy\, dz$.

EXAMPLE 5 In Example 3 we considered the solid bounded by the cylinder $y = x^2$ and the planes $y + z = 4$ and $z = 0$. Set up a triple integral for the volume of this solid in which the first integration is with respect to x.

SOLUTION The solid is resketched in Figure 17.45(i). A first integration with respect to x corresponds to a summation along a column of small boxes in the direction parallel to the x-axis, extending from the graph of $x = -\sqrt{y}$ to the graph of $x = \sqrt{y}$. The region R in the yz-plane lies between these graphs and is bounded by the y- and z-axes and the line $y + z = 4$, as shown in Figure 17.45(ii).

FIGURE 17.45

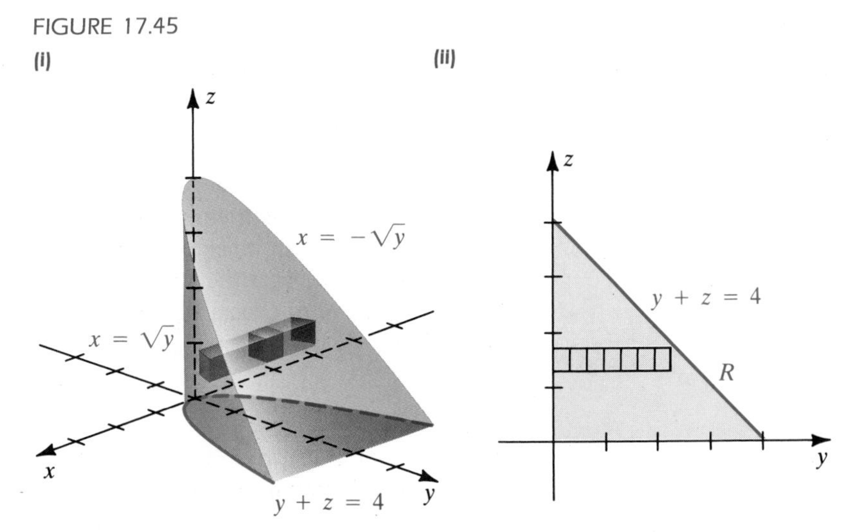

If the second integration is with respect to y, as indicated by the row of rectangles in Figure 17.45(ii), then

$$V = \int_0^4 \int_0^{4-z} \int_{-\sqrt{y}}^{\sqrt{y}} dx\, dy\, dz.$$

If we take the second integration with respect to z instead of y, then the iterated integral is

$$V = \int_0^4 \int_0^{4-y} \int_{-\sqrt{y}}^{\sqrt{y}} dx\, dz\, dy.$$

In the next section we shall use triple integrals to find the *center of mass* of a solid. We conclude this section by discussing the preliminary application of finding the mass of a solid.

If a solid has mass m and volume V and if the mass is uniformly distributed throughout the solid, we say that the solid is **homogeneous**. The **mass density** δ is defined by

$$\delta = \frac{m}{V}, \quad \text{or} \quad m = \delta V.$$

FIGURE 17.46

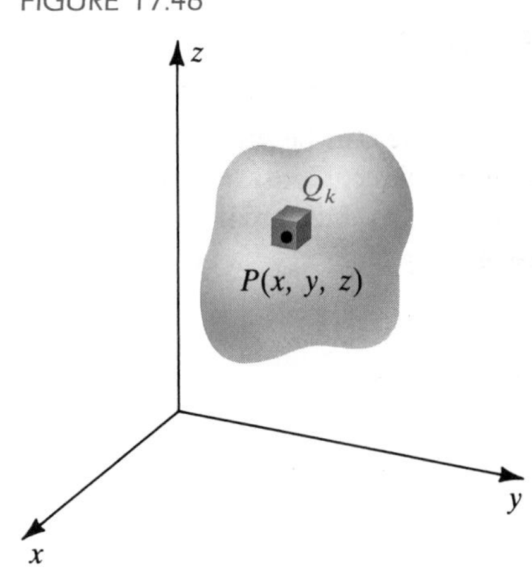

Thus, δ is mass per unit volume. If, for example, m is in kilograms and V in cubic meters, then δ is in kg/m^3.

Let us next consider a **nonhomogeneous solid**, in which the mass density is not the same throughout. For instance, an object might consist of different metals, such as copper, iron, and gold, formed into one solid. We begin by introducing a coordinate system as in Figure 17.46, where the solid has the shape of the region Q. To define the mass density $\delta(x, y, z)$ at a point $P(x, y, z)$, consider a box-shaped subregion Q_k that contains P and has volume ΔV_k and mass Δm_k (see Figure 17.46). If the longest diagonal $\|\Delta V_k\|$ of the box is small, we expect $\Delta m_k/\Delta V_k$ to be an approximation to $\delta(x, y, z)$. This motivates the following definition.

Definition of mass density **(17.21)**

$$\delta(x, y, z) = \lim_{\|\Delta V_k\| \to 0} \frac{\Delta m_k}{\Delta V_k}$$

If the limit in (17.21) exists and $\|\Delta V_k\| \approx 0$, then

$$\Delta m_k \approx \delta(x, y, z)\, \Delta V_k.$$

In some cases we may know the density $\delta(x, y, z)$ and wish to find the mass. If δ is a continuous function and Q is a suitable region, then we consider an inner partition $\{Q_k\}$ of Q, choose a point (x_k, y_k, z_k) in each Q_k, and form the Riemann sum $\sum_k \delta(x_k, y_k, z_k)\, \Delta V_k$. We may define the mass m of Q as a limit of such sums. This gives us the following.

Mass of a solid **(17.22)**

$$m = \iiint_Q \delta(x, y, z)\, dV$$

EXAMPLE 6 A solid has the shape of a right circular cylinder of base radius a and height h. Find the mass if the density at a point P is directly proportional to the distance from one of the bases to P.

FIGURE 17.47

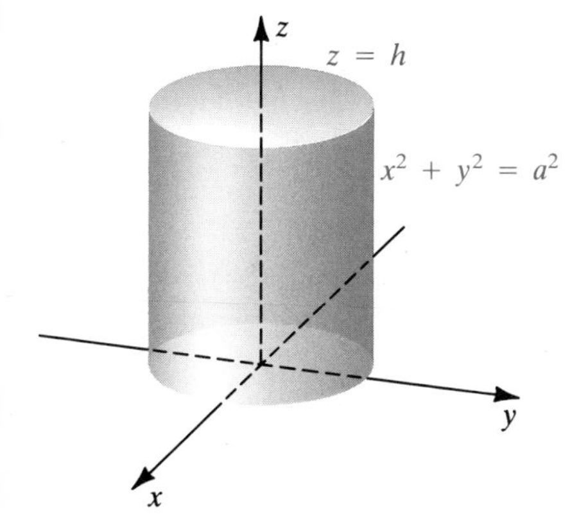

SOLUTION If we introduce a coordinate system as in Figure 17.47, the solid is bounded by the graphs of $x^2 + y^2 = a^2$, $z = 0$, and $z = h$. By hypothesis, the density at (x, y, z) is $\delta(x, y, z) = kz$ for some constant k. From the form of δ and the symmetry of the solid we may calculate m for the first-octant portion and multiply by 4. Using (17.22), we have

$$\begin{aligned} m &= 4\int_0^a \int_0^{\sqrt{a^2-x^2}} \int_0^h kz\,dz\,dy\,dx \\ &= 4k\int_0^a \int_0^{\sqrt{a^2-x^2}} \left[\tfrac{1}{2}z^2\right]_0^h dy\,dx \\ &= 4k\int_0^a \int_0^{\sqrt{a^2-x^2}} \tfrac{1}{2}h^2\,dy\,dx \\ &= 4k\cdot\tfrac{1}{2}h^2 \int_0^a \Big[y\Big]_0^{\sqrt{a^2-x^2}} dx \\ &= 2kh^2 \int_0^a \sqrt{a^2-x^2}\,dx \\ &= 2kh^2\cdot\tfrac{1}{4}\pi a^2 = \tfrac{1}{2}\pi k h^2 a^2. \end{aligned}$$

A discussion similar to the preceding one can be given for a lamina that has the shape of a region R in the xy-plane. Consider a rectangular subregion R_k that contains the point P and has area ΔA_k (see Figure 17.48). In this case the **area mass density** at $P(x, y)$ is defined by

$$\delta(x, y) = \lim_{\|\Delta A_k\|\to 0} \frac{\Delta m_k}{\Delta A_k}.$$

FIGURE 17.48

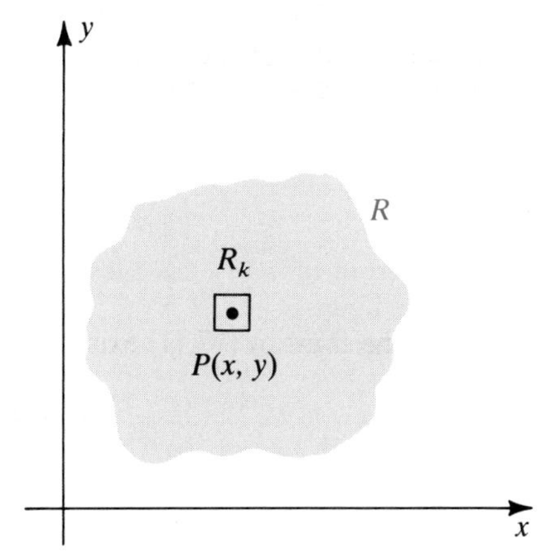

The units for $\delta(x, y)$ are *mass per unit area*, such as kg/m^2. We can state a formula analogous to (17.22) for a lamina as follows.

Mass of a lamina **(17.23)**

$$m = \iint_R \delta(x, y)\,dA$$

EXAMPLE 7 A lamina has the shape of an isosceles right triangle with equal sides of length a. Find the mass if the area mass density at a point P is directly proportional to the square of the distance from P to the vertex that is opposite the hypotenuse.

FIGURE 17.49

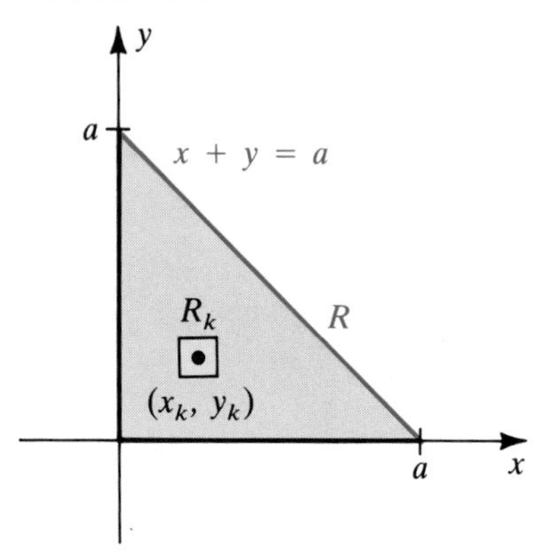

SOLUTION It is convenient to introduce a coordinate system, as in Figure 17.49, with the vertex at the origin and the hypotenuse of the triangle along the line $x + y = a$. Also shown in the figure is a typical rectangle R_k (of area ΔA_k) of an inner partition with a point (x_k, y_k) in the rectangle.

By hypothesis, the area mass density at (x, y) is $\delta(x, y) = k(x^2 + y^2)$ for some constant k. Applying Definition (17.23) and Theorem (17.8), we find that the mass is

$$m = \iint_R k(x^2 + y^2)\, dA = \int_0^a \int_0^{a-x} k(x^2 + y^2)\, dy\, dx$$

$$= k \int_0^a \left[x^2 y + \tfrac{1}{3}y^3\right]_0^{a-x} dx = k \int_0^a \left[x^2(a - x) + \tfrac{1}{3}(a - x)^3\right] dx.$$

The last integral equals $\frac{1}{6}ka^4$.

EXERCISES 17.5

Exer. 1–6: Evaluate the iterated integral.

1 $\int_0^3 \int_{-1}^0 \int_1^2 (x + 2y + 4z)\, dx\, dy\, dz$

2 $\int_0^1 \int_{-1}^2 \int_1^3 (6x^2 z + 5xy^2)\, dz\, dx\, dy$

3 $\int_0^1 \int_{x+1}^{2x} \int_z^{x+z} x\, dy\, dz\, dx$

4 $\int_1^2 \int_0^{z^2} \int_{x+z}^{x-z} z\, dy\, dx\, dz$

5 $\int_{-1}^2 \int_1^{x^2} \int_0^{x+y} 2x^2 y\, dz\, dy\, dx$

6 $\int_2^3 \int_0^{3y} \int_1^{yz} (2x + y + z)\, dx\, dz\, dy$

Exer. 7–10: If f is an arbitrary continuous function of three variables and Q is the region shown in the figure, express $\iiint_Q f(x, y, z)\, dV$ as an iterated triple integral in six different ways.

7 

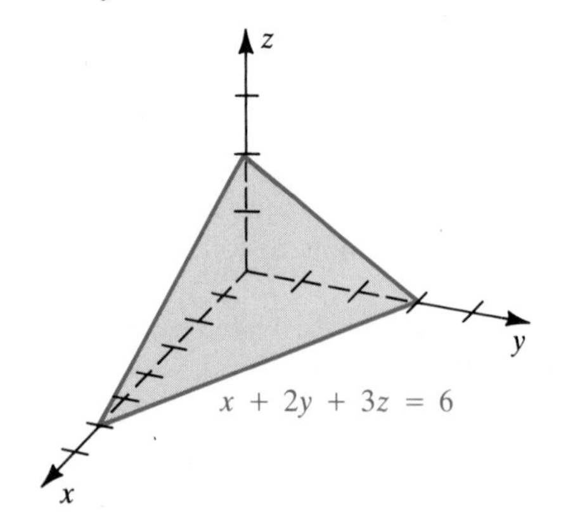

8

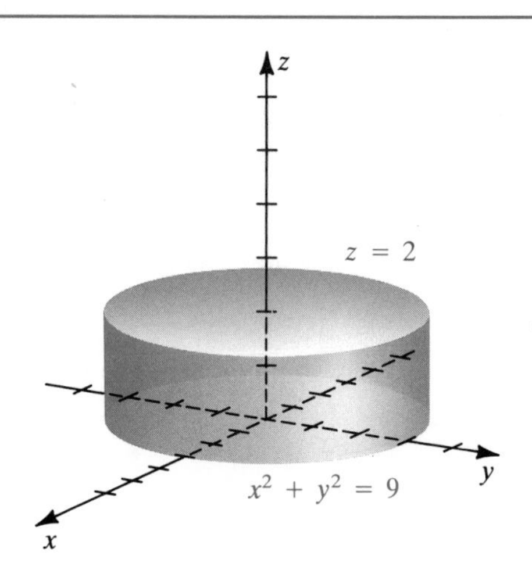

9

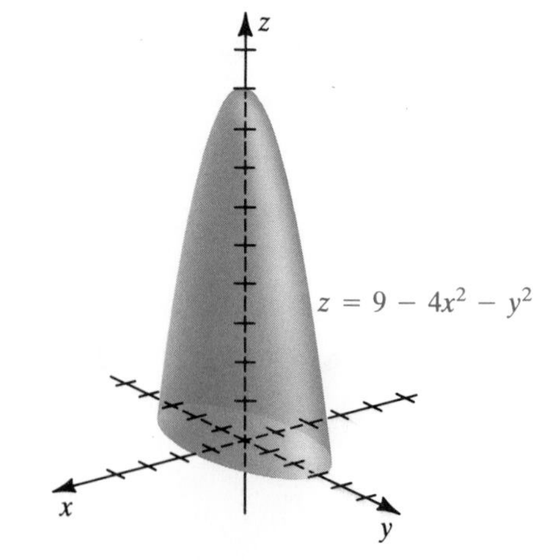

10

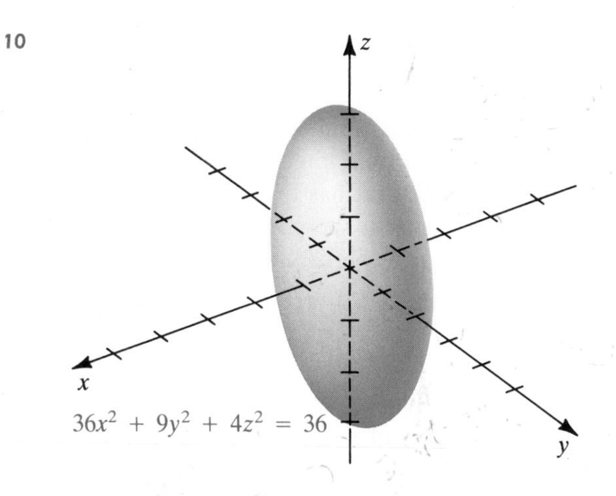

Exer. 11–20: Sketch the region bounded by the graphs of the equations, and use a triple integral to find its volume.

11 $z + x^2 = 4, \quad y + z = 4, \quad y = 0, \quad z = 0$

12 $x^2 + z^2 = 4, \quad y^2 + z^2 = 4$

13 $y = 2 - z^2, \quad y = z^2, \quad x + z = 4, \quad x = 0$

14 $z = 4y^2, \quad z = 2, \quad x = 2, \quad x = 0$

15 $y^2 + z^2 = 1, \quad x + y + z = 2, \quad x = 0$

16 $z = x^2 + y^2, \quad y + z = 2$

17 $z = 9 - x^2, \quad z = 0, \quad y = -1, \quad y = 2$

18 $z = e^{x+y}, \quad y = 3x, \quad x = 2, \quad y = 0, \quad z = 0$

19 $z = x^2, \quad z = x^3, \quad y = z^2, \quad y = 0$

20 $y = x^2 + z^2, \quad z = x^2, \quad z = 4, \quad y = 0$

Exer. 21–22: Let a, b, and c be positive numbers.

21 Find the volume of the tetrahedron bounded by the coordinate planes and the plane

$$\frac{x}{a} + \frac{y}{b} + \frac{z}{c} = 1.$$

22 Find the volume of the solid bounded by the ellipsoid

$$\frac{x^2}{a^2} + \frac{y^2}{b^2} + \frac{z^2}{c^2} = 1.$$

Exer. 23–28: The iterated integral represents the volume of a region Q in an xyz-coordinate system. Describe Q.

23 $\int_0^1 \int_{\sqrt{1-z}}^{\sqrt{4-z}} \int_2^3 dx\, dy\, dz$

24 $\int_0^1 \int_{z^3}^{\sqrt{z}} \int_0^{4-x} dy\, dx\, dz$

25 $\int_0^2 \int_{x^2}^{2x} \int_0^{x+y} dz\, dy\, dx$

26 $\int_0^1 \int_x^{3x} \int_0^{xy} dz\, dy\, dx$

27 $\int_1^2 \int_{-\sqrt{z}}^{\sqrt{z}} \int_{-\sqrt{z-x^2}}^{\sqrt{z-x^2}} dy\, dx\, dz$

28 $\int_1^4 \int_{-z}^{z} \int_{-\sqrt{z^2-y^2}}^{\sqrt{z^2-y^2}} dx\, dy\, dz$

Exer. 29–30: A lamina having area mass density $\delta(x, y)$ has the shape of the region bounded by the graphs of the equations. Set up an iterated double integral that can be used to find the mass of the lamina.

29 $\delta(x, y) = y^2; \quad y = e^{-x}, \quad x = 0, \quad x = 1, \quad y = 0$

30 $\delta(x, y) = x^2 + y^2; \quad xy^2 = 1, \quad x = 0, \quad y = 1, \quad y = 2$

Exer. 31–32: A solid having density $\delta(x, y, z)$ has the shape of the region bounded by the graphs of the equations. Set up an iterated triple integral that can be used to find the mass of the solid.

31 $\delta(x, y, z) = x^2 + y^2; \; x + 2y + z = 4, \quad x = 0, \; y = 0, \; z = 0$

32 $\delta(x, y, z) = z + 1, \quad z = 4 - x^2 - y^2, \; z = 0$

33 Near sea level, the density δ of the earth's atmosphere at a height of z meters can be approximated by $\delta = 1.225 - 0.000113z$ kg/m^3. Approximate the mass of a region of the atmosphere that has the shape of a cube with edge of length 1 kilometer and one face on the surface of the earth.

c 34 For an altitude z up to 30,000 meters, the density δ (in kg/m^3) of the earth's atmosphere can be approximated by

$$\delta = 1.2 - (1.096 \times 10^{-4})z + (3.42 \times 10^{-9})z^2 - (3.6 \times 10^{-14})z^3.$$

The average pressure P (in newtons) on a region of the earth's surface of area A caused by a volume Q of the atmosphere can be estimated from

$$P = \frac{1}{A} \iiint_Q \delta g \, dV,$$

where $g = 9.80$ m/sec^2 is a gravitational constant. Estimate the pressure exerted by a column of air 30,000 meters high on one square meter of the earth's surface. (Standard atmospheric pressure on one square meter is 101,325 newtons.)

c **Exer. 35–36: A cubical solid bounded by the coordinate planes and the planes $x = 1$, $y = 1$, and $z = 1$ has the given density $\delta(x, y, z)$. Approximate its mass by evaluating**

$$\sum_{i=1}^{n} \sum_{j=1}^{n} \sum_{k=1}^{n} \delta(u_i, v_j, w_k)\, \Delta z_k\, \Delta y_j\, \Delta x_i$$

for the indicated value of n, where each increment equals $1/n$ and $u_i = (i - \frac{1}{2})/n$, $v_j = (j - \frac{1}{2})/n$, and $w_k = (k - \frac{1}{2})/n$.

35 $\delta(x, y, z) = \sqrt{x^3 + y^3 + z^3}; \quad n = 2$

36 $\delta(x, y, z) = \tan(xy \cos z); \quad n = 4$

17.6 MOMENTS AND CENTER OF MASS

Moments and the center of mass of a homogeneous lamina were considered in Section 6.7. We can use double integrals to extend these concepts to a *nonhomogeneous* lamina L that has the shape of a region R in the xy-plane. If the area mass density at the point (x, y) is $\delta(x, y)$ and δ is a continuous function on R, then, by (17.23), the mass m of L is given by

$$m = \iint_R \delta(x, y)\, dA.$$

FIGURE 17.50

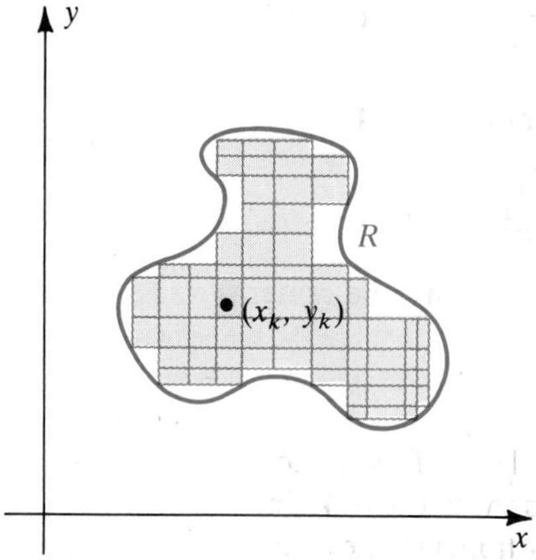

Consider an inner partition $P = \{R_k\}$ of R, and for each k, let (x_k, y_k) be any point in R_k (see Figure 17.50). Since δ is continuous, a small change in (x, y) produces a small change in the density $\delta(x, y)$; that is, δ is *almost constant* on R_k. Hence, if $\|P\| \approx 0$, the mass Δm_k that corresponds to R_k may be approximated by $\delta(x_k, y_k)\,\Delta A_k$, where ΔA_k is the area of R_k. If we assume that the mass Δm_k is concentrated at (x_k, y_k), then the moment with respect to the x-axis of this element of L is the product $y_k\delta(x_k, y_k)\,\Delta A_k$. Since the sum of the moments $\sum_k y_k\delta(x_k, y_k)\,\Delta A_k$ should approximate the moment of the *lamina*, we define the moment M_x of L with respect to the x-axis as follows:

$$M_x = \lim_{\|P\| \to 0} \sum_k y_k\delta(x_k, y_k)\,\Delta A_k = \iint_R y\delta(x, y)\, dA$$

Similarly, the moment M_y of L with respect to the y-axis is

$$M_y = \lim_{\|P\| \to 0} \sum_k x_k\delta(x_k, y_k)\,\Delta A_k = \iint_R x\delta(x, y)\, dA.$$

As in Definition (6.24), we define the center of mass (or center of gravity) of the lamina as the point $(\bar{x}, \bar{y})$ such that $\bar{x} = M_y/m$, $\bar{y} = M_x/m$. This gives us the following.

Definition (17.24)

Let L be a lamina that has the shape of a region R in the xy-plane. If the area mass density at (x, y) is $\delta(x, y)$ and if δ is continuous on R, then the **mass** m, the **moments** M_x and M_y, and the **center of mass** $(\bar{x}, \bar{y})$ are

(i) $m = \iint_R \delta(x, y)\, dA$

(ii) $M_x = \iint_R y\delta(x, y)\, dA, \qquad M_y = \iint_R x\delta(x, y)\, dA$

(iii) $\bar{x} = \dfrac{M_y}{m} = \dfrac{\iint_R x\delta(x, y)\, dA}{\iint_R \delta(x, y)\, dA}, \qquad \bar{y} = \dfrac{M_x}{m} = \dfrac{\iint_R y\delta(x, y)\, dA}{\iint_R \delta(x, y)\, dA}$

If L is homogeneous, then the area mass density $\delta(x, y)$ is a constant and can be canceled in (17.24)(iii). Thus, as in Section 6.7, the center of mass of a homogeneous lamina depends only on its shape, and we refer to $(\bar{x}, \bar{y})$ as the **centroid** of the region R.

EXAMPLE 1 A lamina has the shape of an isosceles right triangle with equal sides of length a. The area mass density at a point P is directly proportional to the square of the distance from P to the vertex that is opposite the hypotenuse. Find the center of mass.

FIGURE 17.51

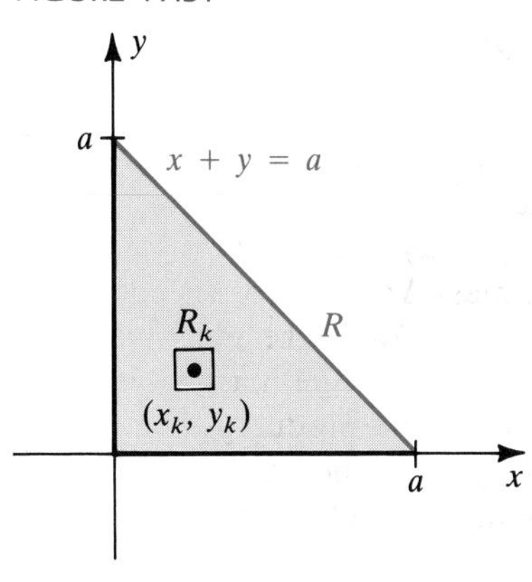

SOLUTION The lamina is the same as that considered in Example 7 of Section 17.5, where we placed the triangle as in Figure 17.51. Using the area mass density $\delta(x, y) = k(x^2 + y^2)$, we found that $m = \frac{1}{6}ka^4$. By Definition (17.24)(ii),

$$M_y = \iint_R xk(x^2 + y^2)\, dA = \int_0^a \int_0^{a-x} xk(x^2 + y^2)\, dy\, dx,$$

which is equal to $\frac{1}{15}ka^5$. From Definition (17.24)(iii),

$$\bar{x} = \frac{\frac{1}{15}ka^5}{\frac{1}{6}ka^4} = \frac{2}{5}a.$$

Similarly, $M_x = \frac{1}{15}ka^5$ and $\bar{y} = \frac{2}{5}a$. Thus, the center of mass (the *balance point*) of the lamina is $(\frac{2}{5}a, \frac{2}{5}a)$.

EXAMPLE 2 A lamina has the shape of the region R in the xy-plane bounded by the parabola $x = y^2$ and the line $x = 4$. The area mass density at the point $P(x, y)$ is directly proportional to the distance from the y-axis to P. Find the center of mass.

FIGURE 17.52

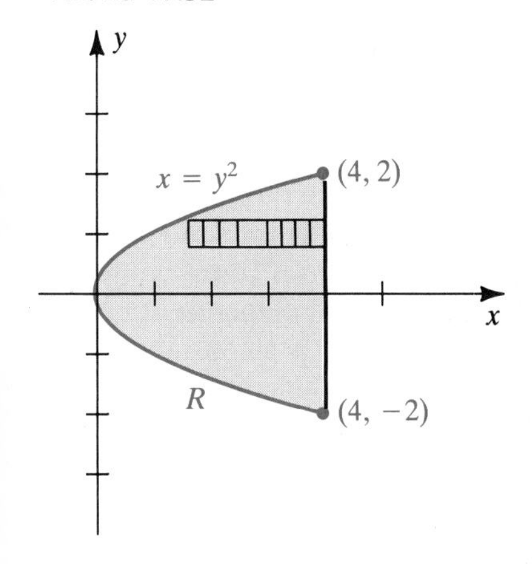

SOLUTION The region is sketched in Figure 17.52. By hypothesis, the area mass density at (x, y) is $\delta(x, y) = kx$ for some constant k. It follows from the form of δ and the symmetry of the region that the center of mass is on the x-axis; that is, $\bar{y} = 0$.

Using Definition (17.24)(i) and then integrating first with respect to x, as indicated by the row of rectangles in Figure 17.52, we have

$$m = \iint_R kx\, dA = k \int_{-2}^{2} \int_{y^2}^{4} x\, dx\, dy$$

$$= k \int_{-2}^{2} \left[\tfrac{1}{2}x^2\right]_{y^2}^{4} dy = \tfrac{1}{2}k \int_{-2}^{2} (16 - y^4)\, dy = \tfrac{128}{5}k.$$

By Definition (17.24)(ii),

$$M_y = \iint_R x(kx)\, dA = k \int_{-2}^{2} \int_{y^2}^{4} x^2\, dx\, dy$$

$$= k \int_{-2}^{2} \left[\tfrac{1}{3}x^3\right]_{y^2}^{4} dy = \tfrac{1}{3}k \int_{-2}^{2} (64 - y^6)\, dy = \tfrac{512}{7}k.$$

Consequently

$$\bar{x} = \frac{M_y}{m} = \frac{512k}{7} \cdot \frac{5}{128k} = \frac{20}{7} \approx 2.86.$$

Hence the center of mass is $(\frac{20}{7}, 0)$.

The moments M_x and M_y in Definition (17.24) are also called the *first moments* of L with respect to the coordinate axes. If we use the *squares* of the distances from the coordinate axes, we obtain the *second moments*, or **moments of inertia**, I_x and I_y with respect to the x-axis and y-axis, respectively. The sum $I_O = I_x + I_y$ is the **polar moment of inertia**, or the **moment of inertia with respect to the origin**. In the following formula, I_O can be obtained as a limit of a sum, using the *square* of the distance from the origin to (x_k, y_k) in Figure 17.50.

Moments of inertia of a lamina **(17.25)**

$$I_x = \lim_{\|P\|\to 0} \sum_k y_k^2\delta(x_k, y_k)\,\Delta A_k = \iint_R y^2\delta(x, y)\,dA$$

$$I_y = \lim_{\|P\|\to 0} \sum_k x_k^2\delta(x_k, y_k)\,\Delta A_k = \iint_R x^2\delta(x, y)\,dA$$

$$I_O = \lim_{\|P\|\to 0} \sum_k (x_k^2 + y_k^2)\delta(x_k, y_k)\,\Delta A_k = \iint_R (x^2 + y^2)\delta(x, y)\,dA$$

FIGURE 17.53

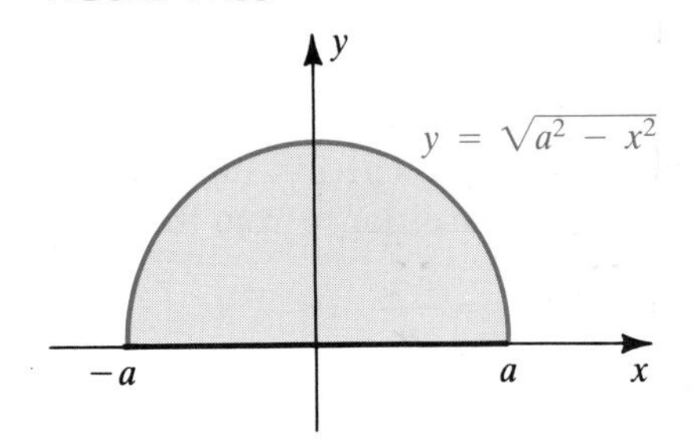

EXAMPLE 3 A lamina has the semicircular shape illustrated in Figure 17.53. The area mass density is directly proportional to the distance from the x-axis. Find the moment of inertia with respect to the x-axis.

SOLUTION By hypothesis, the area mass density at (x, y) is $\delta(x, y) = ky$. Applying (17.25), we find that the moment of inertia with respect to the x-axis is

$$I_x = \int_{-a}^{a}\int_0^{\sqrt{a^2-x^2}} y^2(ky)\,dy\,dx = k\int_{-a}^{a}\left[\tfrac{1}{4}y^4\right]_0^{\sqrt{a^2-x^2}} dx$$
$$= \tfrac{1}{4}k\int_{-a}^{a}(a^4 - 2a^2x^2 + x^4)\,dx = \tfrac{4}{15}ka^5.$$

FIGURE 17.54

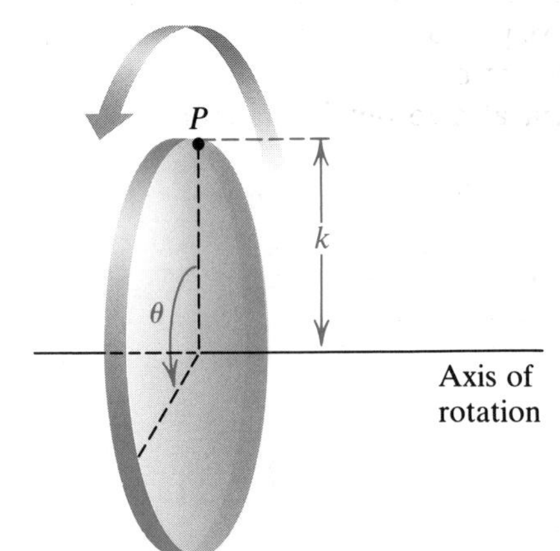

Moments of inertia are useful in problems that involve rotation of an object about a fixed axis, such as rotation of a wheel (or disk) about an axle (see Figure 17.54). If a particle P on the wheel has mass m and is a distance k from the axis of rotation, then the moment of inertia I of P with respect to the axis is mk^2. If the angular speed $d\theta/dt$ is a constant ω, then the speed v of the particle is $k\omega$. From physics, the **kinetic energy** K.E. of P is

$$\text{K.E.} = \tfrac{1}{2}mv^2.$$

Since $v = k\omega$,

$$\text{K.E.} = \tfrac{1}{2}mk^2\omega^2 = \tfrac{1}{2}I\omega^2.$$

If we represent the wheel by a disk and introduce a limit of sums, then the same formula may be derived for the kinetic energy of the wheel. The formula can also be extended to laminas having noncircular shapes. The kinetic energy of a rotating object tells an engineer or physicist the amount of work necessary to bring the object to rest. Since K.E. $= \tfrac{1}{2}I\omega^2$, *the kinetic energy is directly proportional to the moment of inertia*. For a fixed ω, the

FIGURE 17.55

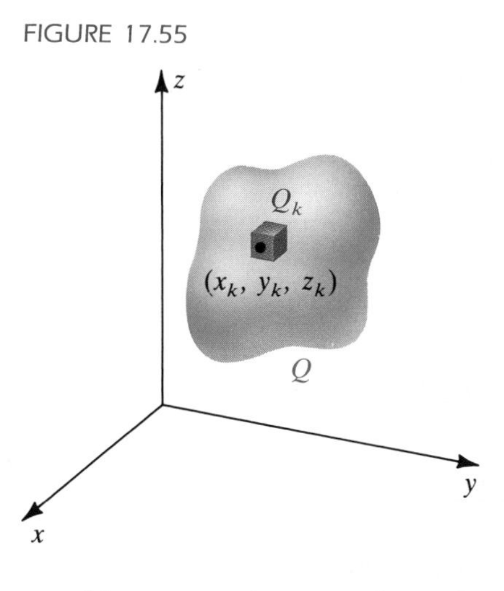

larger the moment of inertia, the larger the amount of work required to stop the rotation.

Let us next consider a solid that has the shape of a three-dimensional region Q. Suppose that the mass density (mass per unit volume) at (x, y, z) is $\delta(x, y, z)$ and that δ is continuous throughout Q. Let $\{Q_k\}$ be an inner partition of Q, and let ΔV_k be the volume of Q_k. If (x_k, y_k, z_k) is a point in Q_k (see Figure 17.55), then the corresponding mass is approximately $\delta(x_k, y_k, z_k)\,\Delta V_k$. By (17.22), the mass of the solid is the limit of such sums—that is, $\iiint_Q \delta(x, y, z)\,dV$.

The *moment with respect to the xy-plane* of the portion of the solid that corresponds to Q_k is approximated by $z_k\delta(x_k, y_k, z_k)\,\Delta V_k$ (see Figure 17.55). Summing and taking a limit gives us the **moment M_{xy} of the solid with respect to the xy-plane** (see (17.26)(ii)). The **moments M_{xz} and M_{yz} with respect to the xz-plane and yz-plane**, respectively, are obtained in similar fashion. The **center of mass** $(\bar{x}, \bar{y}, \bar{z})$ of the solid is defined in (17.26)(iii).

Moments and center of mass in three dimensions **(17.26)**

(i) $$m = \iiint_Q \delta(x, y, z)\,dV$$

(ii) $$M_{xy} = \iiint_Q z\delta(x, y, z)\,dV$$

$$M_{xz} = \iiint_Q y\delta(x, y, z)\,dV$$

$$M_{yz} = \iiint_Q x\delta(x, y, z)\,dV$$

(iii) $$\bar{x} = \frac{M_{yz}}{m}; \quad \bar{y} = \frac{M_{xz}}{m}; \quad \bar{z} = \frac{M_{xy}}{m}$$

From (iii) of (17.26),

$$m\bar{x} = M_{yz}, \quad m\bar{y} = M_{xz}, \quad m\bar{z} = M_{xy}.$$

Since $m\bar{x}$, $m\bar{y}$, and $m\bar{z}$ are the moments with respect to the yz-plane, the xz-plane, and the xy-plane, respectively, of a single point mass located at $(\bar{x}, \bar{y}, \bar{z})$, we may interpret the center of mass of a solid as the point at which the total mass can be concentrated to obtain the moments M_{yz}, M_{xz}, and M_{xy} of the solid.

If the solid is homogeneous, the mass density δ is a constant and hence δ can be canceled after substitution in (17.26)(iii). Consequently the center of mass of a homogeneous solid depends only on the shape of Q. As in two dimensions, the corresponding point for geometric solids is called the **centroid** of the solid. To find the centroid, we let $\delta(x, y, z) = 1$ in (17.26).

FIGURE 17.56

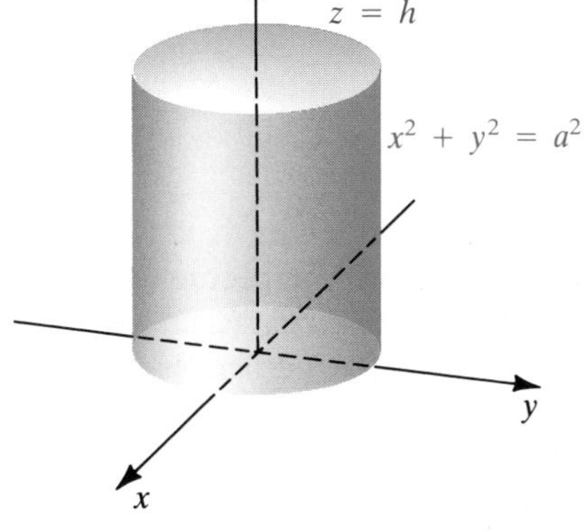

EXAMPLE 4 A solid has the shape of a right circular cylinder of base radius a and height h. The density at a point P is directly proportional to the distance from one of the bases to P. Find the center of mass.

SOLUTION In Example 6 of Section 17.5 we positioned the solid as shown in Figure 17.56. Using $\delta(x, y, z) = kz$, we found that $m = \frac{1}{2}\pi k h^2 a^2$.

The center of mass is on the z-axis, so it is sufficient to find $\bar{z} = M_{xy}/m$. Moreover, because of the form of δ and the symmetry of the solid, we may calculate M_{xy} for the first-octant portion and multiply by 4. Using (17.26), we have

$$\begin{aligned} M_{xy} &= 4 \int_0^a \int_0^{\sqrt{a^2-x^2}} \int_0^h z(kz)\,dz\,dy\,dx \\ &= 4k \int_0^a \int_0^{\sqrt{a^2-x^2}} \left[\frac{z^3}{3}\right]_0^h dy\,dx \\ &= 4k \int_0^a \int_0^{\sqrt{a^2-x^2}} \frac{h^3}{3}\,dy\,dx \\ &= \tfrac{4}{3}kh^3 \int_0^a \Big[y\Big]_0^{\sqrt{a^2-x^2}} dx \\ &= \tfrac{4}{3}kh^3 \int_0^a \sqrt{a^2-x^2}\,dx \\ &= \tfrac{4}{3}kh^3 \cdot \tfrac{1}{4}\pi a^2 = \tfrac{1}{3}\pi kh^3a^2 \end{aligned}$$

Finally, we calculate the center of mass:

$$\bar{z} = \frac{M_{xy}}{m} = \frac{\frac{1}{3}\pi kh^3a^2}{\frac{1}{2}\pi kh^2a^2} = \frac{2}{3}h$$

Hence the center of mass is on the axis of the cylinder, two-thirds of the way from the lower base.

FIGURE 17.57

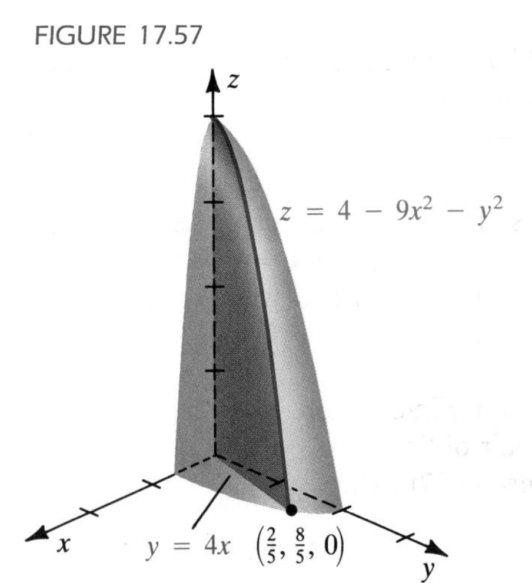

EXAMPLE 5 A solid has the shape of the region in the first octant bounded by the paraboloid $z = 4 - 9x^2 - y^2$ and the planes $y = 4x$, $z = 0$, and $y = 0$. The density at $P(x, y, z)$ is proportional to the distance from the origin to P. Set up iterated integrals that can be used to find $\bar{x}$.

SOLUTION The region is sketched in Figure 17.57. The density at (x, y, z) is $\delta(x, y, z) = k(x^2 + y^2 + z^2)^{1/2}$ for some k. Using (17.26) gives us

$$m = \int_0^{8/5} \int_{y/4}^{\sqrt{4-y^2}/3} \int_0^{4-9x^2-y^2} k(x^2 + y^2 + z^2)^{1/2}\,dz\,dx\,dy$$

$$M_{yz} = \int_0^{8/5} \int_{y/4}^{\sqrt{4-y^2}/3} \int_0^{4-9x^2-y^2} xk(x^2 + y^2 + z^2)^{1/2}\,dz\,dx\,dy$$

$$\bar{x} = \frac{M_{yz}}{m}.$$

If a particle of mass m is at the point (x, y, z), then its distance to the z-axis is $(x^2 + y^2)^{1/2}$ and its **moment of inertia I_z with respect to the z-axis** is defined as $(x^2 + y^2)m$. Similarly, **moments of inertia I_x and I_y with respect to the x-axis and y-axis** are $(y^2 + z^2)m$ and $(x^2 + z^2)m$, respectively. For solids having the shape of Q in Figure 17.55, we employ limits of sums in the usual way to obtain the following.

Moments of inertia of solids **(17.27)**

$$I_z = \iiint_Q (x^2 + y^2)\delta(x, y, z)\, dV$$

$$I_x = \iiint_Q (y^2 + z^2)\delta(x, y, z)\, dV$$

$$I_y = \iiint_Q (x^2 + z^2)\delta(x, y, z)\, dV$$

EXAMPLE 6 Find the moment of inertia, with respect to the axis of symmetry, of the cylindrical solid described in Example 4.

SOLUTION The solid is sketched in Figure 17.56. Employing (17.27) with $\delta(x, y, z) = kz$ and making use of symmetry and the fact that δ is a function of z alone, we have

$$\begin{aligned} I_z &= 4\int_0^a \int_0^{\sqrt{a^2-x^2}} \int_0^h (x^2 + y^2)kz\, dz\, dy\, dx \\ &= 4k \int_0^a \int_0^{\sqrt{a^2-x^2}} (x^2 + y^2)\left[\frac{z^2}{2}\right]_0^h dy\, dx \\ &= 2kh^2 \int_0^a \int_0^{\sqrt{a^2-x^2}} (x^2 + y^2)\, dy\, dx \\ &= 2kh^2 \int_0^a \left[x^2 y + \tfrac{1}{3}y^3\right]_0^{\sqrt{a^2-x^2}} dx \\ &= 2kh^2 \int_0^a \left[x^2\sqrt{a^2 - x^2} + \tfrac{1}{3}\sqrt{(a^2 - x^2)^3}\right] dx. \end{aligned}$$

The last integral may be evaluated by using a trigonometric substitution or a table of integrals. This results in $I_z = \frac{1}{4}\pi k h^2 a^4$.

EXAMPLE 7 A homogeneous solid has the shape of the region Q bounded by the cone $x^2 - y^2 + z^2 = 0$ and the plane $y = 3$. Set up a triple integral that can be used to find its moment of inertia with respect to the y-axis.

FIGURE 17.58

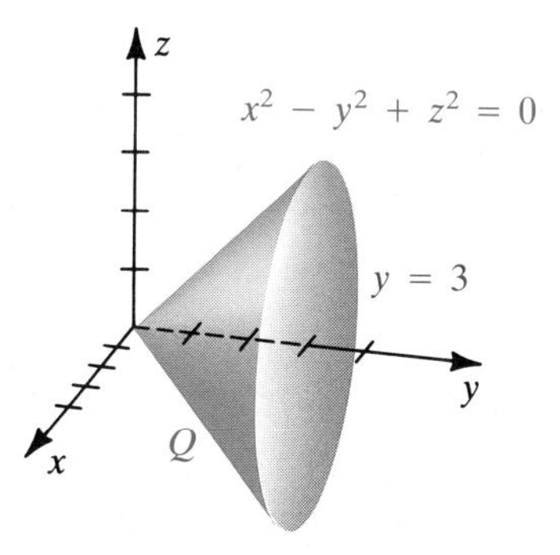

SOLUTION The region is sketched in Figure 17.58. Note that the trace of the cone on the xy-plane is the pair of lines $x = \pm y$. If we denote the constant density by k, then applying (17.27) yields

$$I_y = \int_0^3 \int_{-y}^{y} \int_{-\sqrt{y^2-x^2}}^{\sqrt{y^2-x^2}} (x^2 + z^2)k\, dz\, dx\, dy.$$

We could also find I_y by multiplying the moment of inertia of that part of the solid that lies in the first octant by 4. Thus,

$$I_y = 4\int_0^3 \int_0^y \int_0^{\sqrt{y^2-x^2}} (x^2 + z^2)k\, dz\, dx\, dy.$$

We shall conclude this section with a proof of the following theorem, which was stated without proof in (6.26).

Theorem of Pappus (17.28)

> Let R be a region in a plane that lies entirely on one side of a line l in the plane. If R is revolved once about l, the volume of the resulting solid is the product of the area of R and the distance traveled by the centroid of R.

FIGURE 17.59

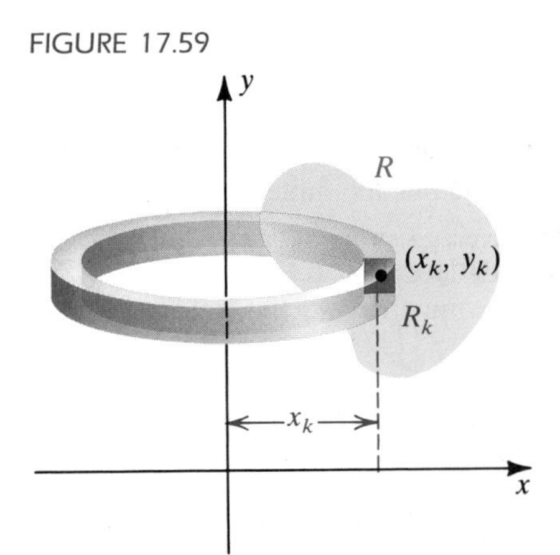

PROOF We may assume that l is the y-axis and R is a region in the first quadrant, as shown in Figure 17.59. Let $\{R_k\}$ be an inner partition P of R, and let ΔA_k denote the area of R_k. For each k, choose (x_k, y_k) as the center of the rectangle R_k.

If R is revolved about the y-axis, then, as shown in Figure 17.59, R_k generates a cylindrical shell having volume $2\pi x_k \, \Delta A_k$. Thus, the volume V of the solid generated by R is

$$V = \lim_{\|P\| \to 0} \sum_k 2\pi x_k \, \Delta A_k = \iint_R 2\pi x \, dA.$$

If we apply Definition (17.24) to centroids, by taking $\delta(x, y) = 1$ and $m = A$, then

$$V = 2\pi \iint_R x \, dA = 2\pi M_y = 2\pi \bar{x} A.$$

Thus, we may find the volume V by multiplying the area A of R by the distance $2\pi\bar{x}$ traveled by the centroid when R is revolved once about the y-axis. ■

EXERCISES 17.6

Exer. 1–8: Find the mass and the center of mass of the lamina that has the shape of the region bounded by the graphs of the given equations and has the indicated area mass density.

1 $y = \sqrt{x}, \quad x = 9, \quad y = 0; \quad \delta(x, y) = x + y$

2 $y = \sqrt[3]{x}, \quad x = 8, \quad y = 0; \quad \delta(x, y) = y^2$

3 $y = x^2, \quad y = 4;$
density at $P(x, y)$ is directly proportional to the distance from the y-axis to P

4 $y = x^3, \quad y = 2x;$
density at $P(x, y)$ is directly proportional to the distance from the x-axis to P

5 $y = e^{-x^2}, \quad y = 0, \quad x = -1, \quad x = 1; \quad \delta(x, y) = |xy|$

6 $y = \sin x, \quad y = 0, \quad x = 0, \quad x = \pi; \quad \delta(x, y) = y$

7 $y = \sec x, \quad x = -\pi/4, \quad x = \pi/4, \quad y = \frac{1}{2}; \quad \delta(x, y) = 4$

8 $y = \ln x, \quad y = 0, \quad x = 2; \quad \delta(x, y) = 1/x$

9 Find I_x, I_y, and I_O for the lamina of Exercise 1.

10 Find I_x, I_y, and I_O for the lamina of Exercise 2.

11 Find I_x, I_y, and I_O for the lamina of Exercise 3.

12 Find I_x, I_y, and I_O for the lamina of Exercise 4.

13 A homogeneous lamina has the shape of a square of side a. Find the moment of inertia with respect to

(a) a side (b) a diagonal (c) the center of mass

14 A homogeneous lamina has the shape of an equilateral triangle of side a. Find the moment of inertia with respect to

(a) an altitude (b) a side (c) a vertex

15 If a solid of mass m has moment of inertia I with respect to a line, then the *radius of gyration* is, by definition, the number d such that $I = md^2$. This formula implies that the radius of gyration is the distance from the line at which all the mass could be concentrated without changing the moment of inertia of the solid. Find the radius of gyration in (a) of Exercise 13.

16 Refer to Exercise 15. Find the radius of gyration in (a) of Exercise 14.

Exer. 17–18: Find the center of mass of Q.

17 The density at a point P in a cubical solid Q of edge a is directly proportional to the square of the distance from P to a fixed corner of the cube.

18 Let Q be the tetrahedron bounded by the coordinate planes and the plane $2x + 5y + z = 10$. The density at the point $P(x, y, z)$ is directly proportional to the distance from the xz-plane to P.

Exer. 19–20: Set up iterated integrals that can be used to find the center of mass of the solid that has the shape of the region Q and density $\delta(x, y, z)$.

19 Q is bounded by the paraboloid $x = y^2 + 4z^2$ and the plane $x = 4$; $\delta(x, y, z) = x^2 + z^2$.

20 Q is bounded by the hyperboloid $y^2 - x^2 - z^2 = 1$ and the plane $y = 2$; $\delta(x, y, z) = x^2y^2z^2$.

Exer. 21–22: Set up iterated integrals that can be used to find the centroid of the solid shown in the figure.

21

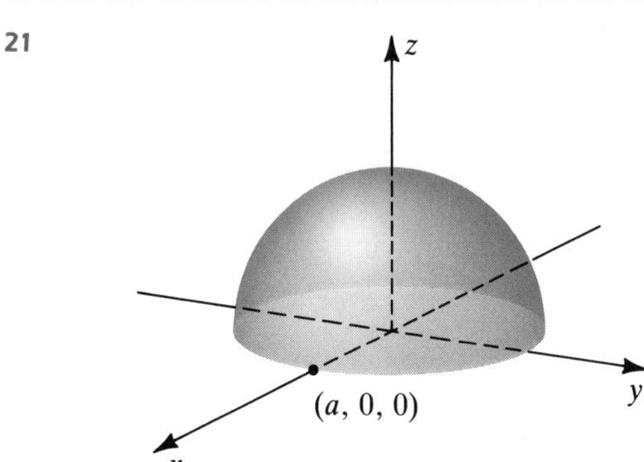

22

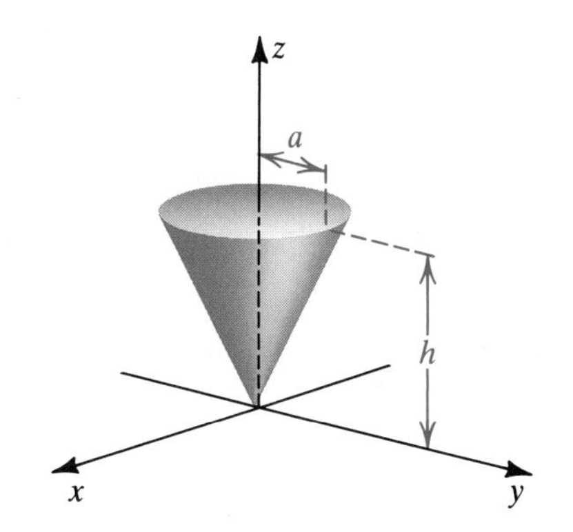

23 Let Q be the solid in the first octant bounded by the coordinate planes and the graphs of $z = 9 - x^2$ and $2x + y = 6$.

(a) Set up iterated integrals that can be used to find the centroid.

(b) Find the centroid.

24 Set up iterated integrals that can be used to find the centroid of the solid bounded by the graphs of $z = x^2$, $y = x^2$, $y = x^3$, and $z = 0$.

Exer. 25–28: Set up an iterated integral that can be used to find the moment of inertia with respect to the z-axis of the indicated solid.

25 The sphere of radius a with center at the origin; $\delta(x, y, z) = x^2 + y^2 + z^2$

26 The cone bounded by the graphs of $x^2 + 9y^2 - z^2 = 0$ and $z = 36$; $\delta(x, y, z) = x^2 + y^2$

27 The homogeneous tetrahedron bounded by the coordinate planes and the plane $(x/a) + (y/b) + (z/c) = 1$ for positive numbers a, b, and c

28 The homogeneous solid bounded by the ellipsoid $(x^2/a^2) + (y^2/b^2) + (z^2/c^2) = 1$

c **Exer. 29–30: The average value of $f(x, y)$ on a region R having area A is defined by**

$$f_{\text{av}} = \frac{1}{A} \iint_R f(x, y)\, dA.$$

If points $(x_1, y_1), \ldots, (x_n, y_n)$ are uniformly distributed throughout R, then

$$f_{\text{av}} \approx \frac{1}{n} \sum_{k=1}^{n} f(x_k, y_k)$$

and

$$\iint_R f(x, y)\, dA = A \cdot f_{\text{av}} \approx \frac{A}{n} \sum_{k=1}^{n} f(x_k, y_k).$$

If the (x_k, y_k) are generated using random numbers, this technique is called a *Monte Carlo method*. (If your calculator cannot generate random numbers, choose ten points that are evenly distributed throughout R.) Use this method, with $n = 10$, to approximate the mass of a lamina that has the given area mass density $\delta(x, y)$ and the shape of R.

29 $\delta(x, y) = \ln [3 + \sin (xy)]$;
R is the square region with vertices $(0, 0)$, $(0, 1)$, $(1, 1)$, and $(1, 0)$.

30 $\delta(x, y) = 2 + \cos \sqrt[4]{x^2 + y^2}$;
R is the region in the first quadrant bounded by the circle $x^2 + y^2 = 1$ and the coordinate axes. (Note that when (x_k, y_k) is generated, if $x_k^2 + y_k^2 > 1$ then (x_k, y_k) is not in R and should not be used.)

c **Exer. 31–32: Extend the Monte Carlo method in Exercises 29–30 to $f(x, y, z)$ and a region Q having volume V, and use this extension, with $n = 10$, to approximate the mass of a solid that has the given mass density $\delta(x, y, z)$ and the shape of Q.**

31 $\delta(x, y, z) = 1 + \sin \sqrt{xyz}$;
Q is the cubical region in the first octant bounded by the coordinate planes and the planes $x = 1$, $y = 1$, and $z = 1$.

32 $\delta(x, y, z) = \ln (2 + \sqrt[3]{x^2 + y^2 + z^2})$;
Q is the region in the first octant bounded by the coordinate planes and the sphere $x^2 + y^2 + z^2 = 1$.

17.7 CYLINDRICAL COORDINATES

FIGURE 17.60

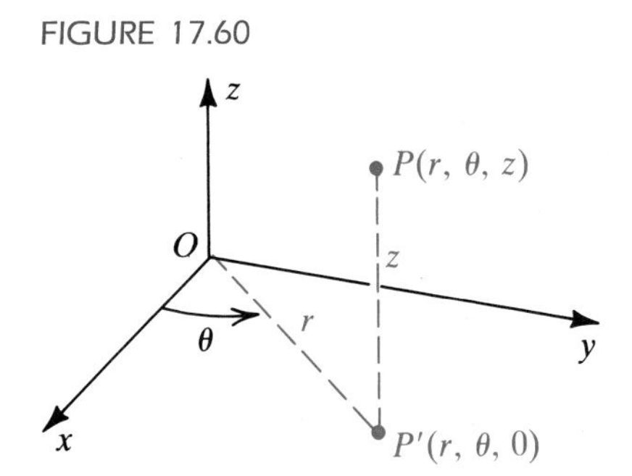

It is easy to extend the system of polar coordinates to three dimensions. We merely represent a point P by an ordered triple (r, θ, z), where z is the usual (third) rectangular coordinate of P and r and θ are polar coordinates for the projection P' of P onto the xy-plane, as illustrated in Figure 17.60. We call (r, θ, z) **cylindrical coordinates** for P. Our principal use for cylindrical coordinates will be to simplify certain types of multiple integrals.

Since polar coordinates are used in the xy-plane, the formulas in Theorem (13.8) give relationships between rectangular coordinates and cylindrical coordinates. Let us restate these formulas as follows.

Theorem **(17.29)**

The rectangular coordinates (x, y, z) and the cylindrical coordinates (r, θ, z) of a point P are related as follows:

$$x = r\cos\theta, \qquad y = r\sin\theta, \qquad \tan\theta = \frac{y}{x},$$

$$r^2 = x^2 + y^2, \qquad z = z$$

If $r_0 > 0$, then the graph of the equation $r = r_0$, or, equivalently, $x^2 + y^2 = r_0^2$, is a circular cylinder of radius r_0 with axis along the z-axis. If θ_0 and z_0 are numbers, then the graph of $\theta = \theta_0$ is a plane containing the z-axis and the graph of $z = z_0$ is a plane perpendicular to the z-axis. Typical graphs are sketched in Figure 17.61.

FIGURE 17.61

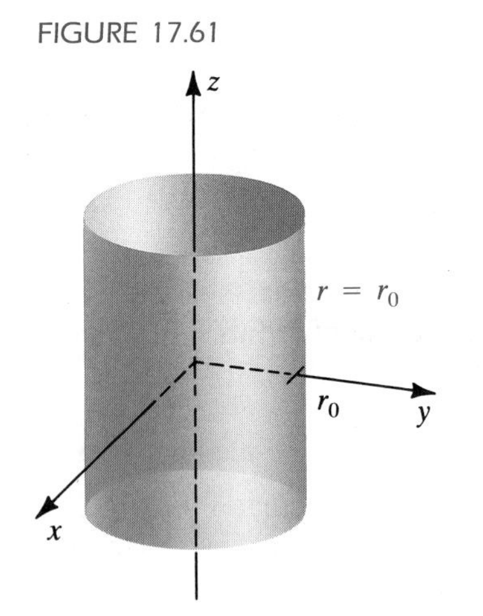

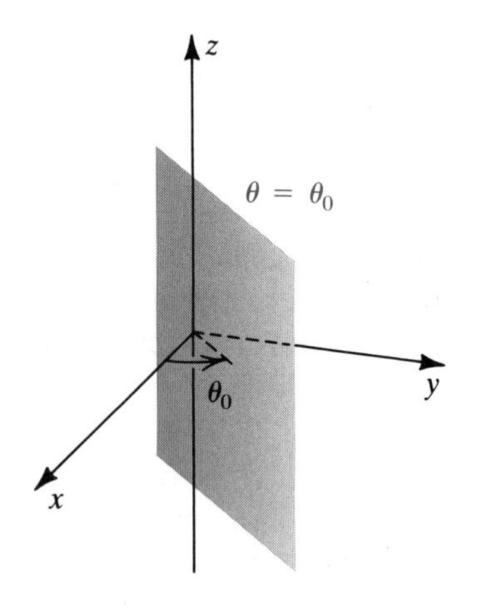

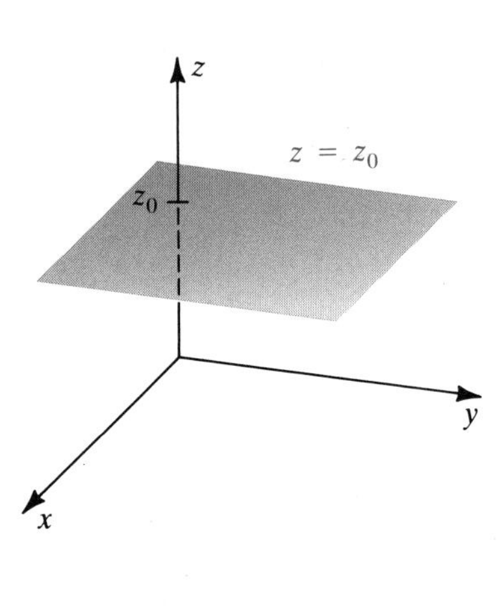

EXAMPLE 1 Change the equation $z^2 = x^2 + y^2$ to cylindrical coordinates, and sketch its graph.

SOLUTION Using Theorem (17.29), we obtain

$$z^2 = r^2, \quad \text{or} \quad z = \pm r.$$

FIGURE 17.62

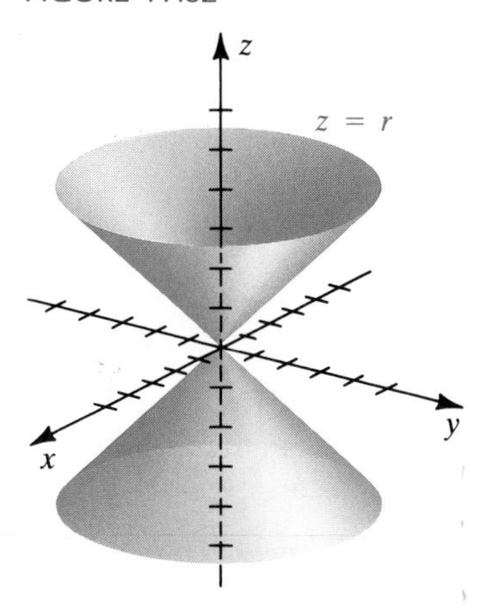

Since r can be positive or negative, $z = \pm r$ is equivalent to $z = r$. The graph is a circular cone with axis along the z-axis, as illustrated in Figure 17.62.

EXAMPLE 2 Change the equation to rectangular coordinates, and sketch its graph in an xyz-coordinate system.

(a) $z = 4r^2$ **(b)** $r = 4 \sin \theta$

SOLUTION

(a) Applying Theorem (17.29), we have

$$z = 4(x^2 + y^2), \quad \text{or} \quad z = 4x^2 + 4y^2.$$

The graph is the paraboloid of revolution sketched in Figure 17.63.

FIGURE 17.63

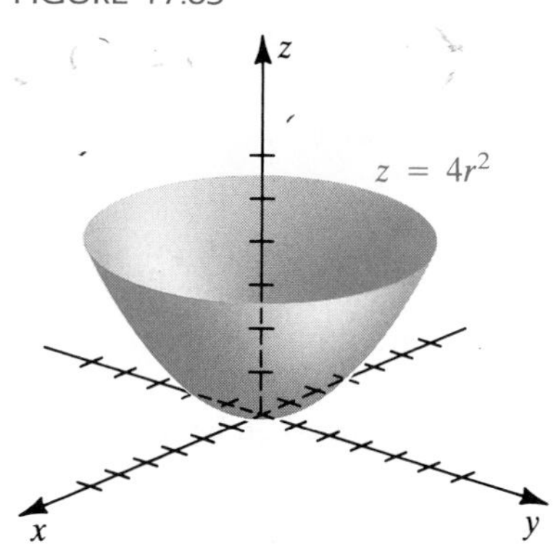

FIGURE 17.64

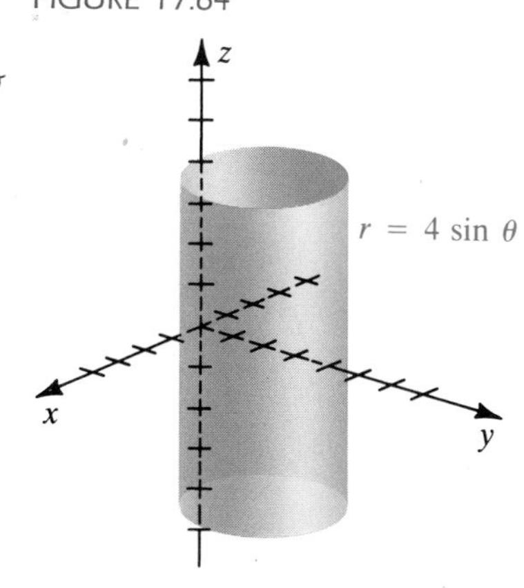

(b) Multiplying both sides of $r = 4 \sin \theta$ by r gives us $r^2 = 4r \sin \theta$. Using (17.29), we obtain $x^2 + y^2 = 4y$, or, equivalently,

$$x^2 + (y - 2)^2 = 4.$$

The graph is a cylinder with rulings parallel to the z-axis, as shown in Figure 17.64. A directrix of the cylinder is a circle of radius 2 in the xy-plane whose center in rectangular coordinates is (0, 2, 0).

Sometimes it is convenient to use cylindrical coordinates to evaluate triple integrals. The simplest case occurs when a function f of r, θ, and z is continuous throughout a region of the form

$$Q = \{(r, \theta, z): a \le r \le b, c \le \theta \le d, m \le z \le n\}.$$

We divide Q into subregions $Q_1, Q_2, \ldots, Q_n$ that have the same shape as Q by using graphs of equations of the form $r = a_k$, $\theta = c_k$, and $z = m_k$, where a_k, c_k, and m_k are in the intervals $[a, b]$, $[c, d]$, and $[m, n]$, respectively. If $a_k \neq 0$, these graphs are circular cylinders, planes containing the z-axis, and planes parallel to the xy-plane, respectively (see Figure 17.61).

A typical subregion Q_k is sketched in Figure 17.65(i). If $\overline{r_k}$ is the *average radius* of the base of Q_k, as shown in Figure 17.65(ii), then the volume ΔV_k of Q_k is the area of the base $\overline{r_k}\,\Delta r_k\,\Delta\theta_k$ (see Section 17.3) times the altitude Δz_k:

$$\Delta V_k = \overline{r_k}\,\Delta r_k\,\Delta\theta_k\,\Delta z_k$$

FIGURE 17.65

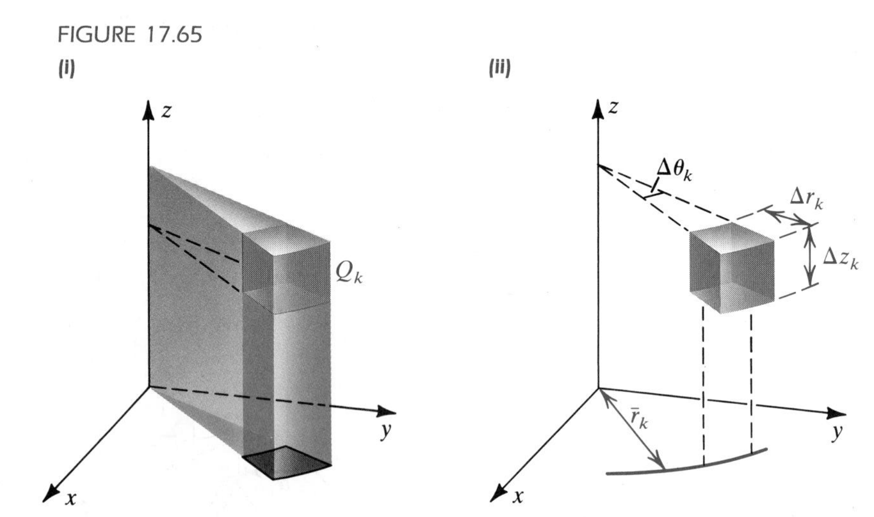

If (r_k, θ_k, z_k) is any point in Q_k and $\|P\|$ is the length of the longest diagonal of the Q_k, then the triple integral of f over Q is defined as

$$\iiint_Q f(r, \theta, z)\,dV = \lim_{\|P\|\to 0} \sum_k f(r_k, \theta_k, z_k)\,\Delta V_k.$$

It can be proved that

$$\iiint_Q f(r, \theta, z)\,dV = \int_m^n \int_c^d \int_a^b f(r, \theta, z) r\,dr\,d\theta\,dz.$$

There are five other possible orders of integration for this iterated integral.

Triple integrals in cylindrical coordinates may be defined over regions more complicated than the one just discussed by employing inner partitions. Thus, if R is a polar region of the type discussed in Section 17.5 and if

$$Q = \{(r, \theta, z): (r, \theta) \text{ is in } R \text{ and } k_1(r, \theta) \le z \le k_2(r, \theta)\},$$

where k_1 and k_2 are functions that have continuous first partial derivatives throughout R, then it can be shown that

$$\iiint_Q f(r, \theta, z)\,dV = \iint_R \left[\int_{k_1(r,\theta)}^{k_2(r,\theta)} f(r, \theta, z)\,dz\right] dA.$$

FIGURE 17.66

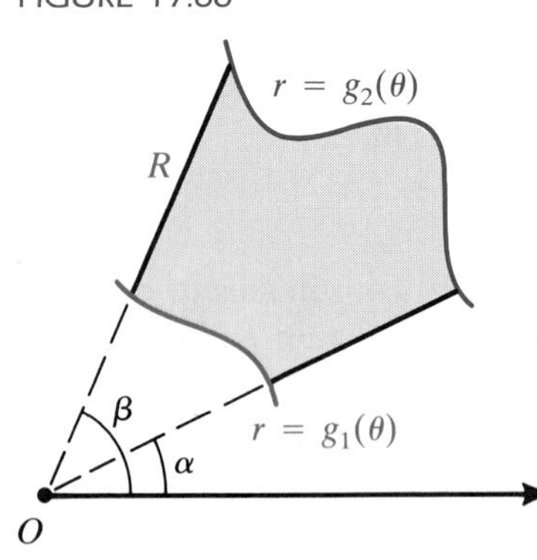

In particular, if R is a region of the type illustrated in Figure 17.66, we may evaluate the triple integral of f over Q as follows.

Evaluation theorem (cylindrical coordinates) **(17.30)**

$$\iiint_Q f(r, \theta, z)\,dV = \int_\alpha^\beta \int_{g_1(\theta)}^{g_2(\theta)} \int_{k_1(r,\theta)}^{k_2(r,\theta)} f(r, \theta, z) r\,dz\,dr\,d\theta$$

The iterated integral in (17.30) may be interpreted as a limit of triple sums. The first summation takes place along a column of the subregions Q_k from the lower surface (with equation $z = k_1(r, \theta)$) to the upper surface (with equation $z = k_2(r, \theta)$) (see Figure 17.67). The second summation extends over a wedge of these columns, with θ fixed and r allowed to vary. At this stage we have summed over a slice of Q, as illustrated in Figure 17.67. Finally, we sweep out Q by letting θ vary from α to β.

FIGURE 17.67

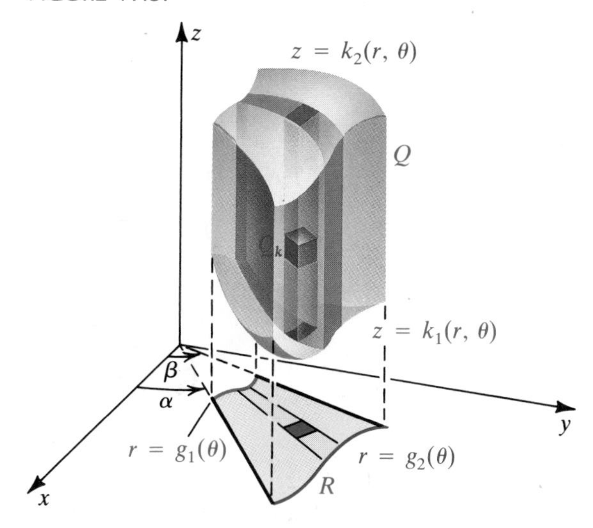

The following examples illustrate the use of (17.30).

EXAMPLE 3 Find the centroid of a hemispherical solid Q of radius a.

SOLUTION If we introduce an xyz-coordinate system, as in Figure 17.68, then an equation of the hemisphere is $z = \sqrt{a^2 - x^2 - y^2}$, or, in cylindrical coordinates, $z = \sqrt{a^2 - r^2}$. By symmetry, the centroid is on the z-axis, and hence we need only find $\bar{z} = M_{xy}/V$, where the volume V is given by $V = \frac{1}{2} \cdot \frac{4}{3}\pi a^3 = \frac{2}{3}\pi a^3$.

FIGURE 17.68

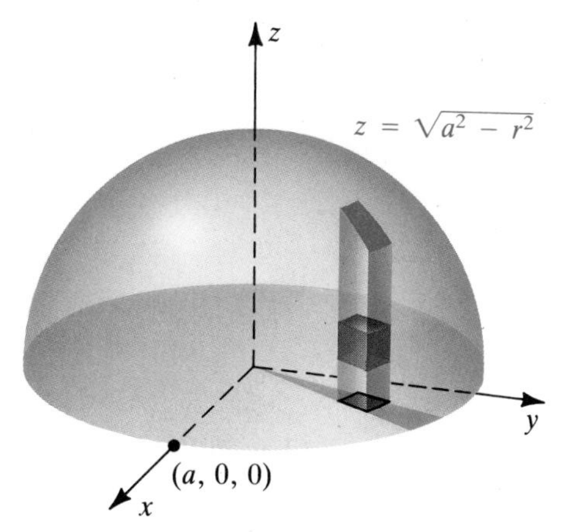

We use (17.26)(ii) with $\delta(x, y, z) = 1$, obtaining

$$M_{xy} = \iiint_Q z\, dV.$$

Applying the evaluation theorem (17.30), we note that the first integration (with respect to z) is from $z = 0$ (the xy-plane) to the hemisphere $z = \sqrt{a^2 - r^2}$. This partial integration represents the volume of the column shown in Figure 17.68. The remaining integrations with respect to r and θ are over the region bounded by the circle $r = a$. Thus,

$$\begin{aligned} M_{xy} &= \int_0^{2\pi} \int_0^a \int_0^{\sqrt{a^2 - r^2}} zr\, dz\, dr\, d\theta \\ &= \int_0^{2\pi} \int_0^a \left[\frac{z^2}{2}\right]_0^{\sqrt{a^2 - r^2}} r\, dr\, d\theta \\ &= \tfrac{1}{2} \int_0^{2\pi} \int_0^a (a^2 - r^2) r\, dr\, d\theta \\ &= \tfrac{1}{2} \int_0^{2\pi} \int_0^a (a^2 r - r^3)\, dr\, d\theta \\ &= \frac{1}{2} \int_0^{2\pi} \left[a^2\left(\frac{r^2}{2}\right) - \frac{r^4}{4}\right]_0^a d\theta \\ &= \frac{a^4}{8} \int_0^{2\pi} d\theta = \frac{a^4}{8}\Big[\theta\Big]_0^{2\pi} = \frac{1}{4}\pi a^4. \end{aligned}$$

Hence

$$\bar{z} = \frac{M_{xy}}{V} = \frac{\frac{1}{4}\pi a^4}{\frac{2}{3}\pi a^3} = \frac{3}{8}a.$$

EXAMPLE 4 A solid Q is bounded by the cone $z = \sqrt{x^2 + y^2}$ and the plane $z = 2$. The density at $P(x, y, z)$ is directly proportional to the square of the distance from the origin to P. Find its mass.

SOLUTION The solid is sketched in Figure 17.69. An equation of the cone in cylindrical coordinates is $z = r$, with $r \ge 0$. Note that if $z = 2$, then $2 = \sqrt{x^2 + y^2}$, or $4 = x^2 + y^2$. Therefore the top of the cone projects

FIGURE 17.69

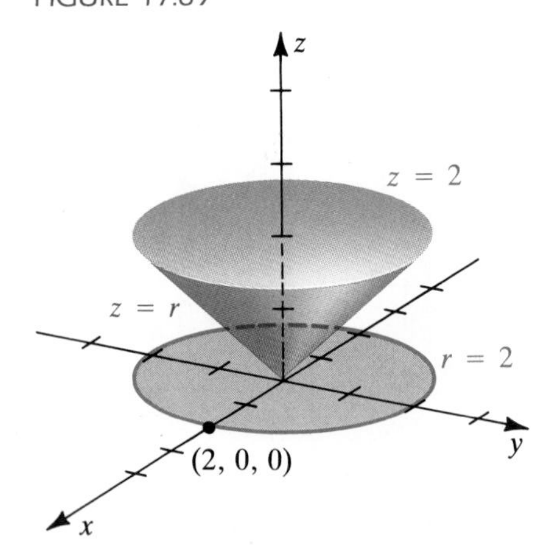

onto the circle $x^2 + y^2 = 4$, or $r = 2$, in the xy-plane. The density at (x, y, z) is given by

$$\delta = k(x^2 + y^2 + z^2), \quad \text{or} \quad \delta = k(r^2 + z^2).$$

By (17.26) and (17.30),

$$m = \iiint_Q \delta\, dV = \int_0^{2\pi} \int_0^2 \int_r^2 k(r^2 + z^2) r\, dz\, dr\, d\theta.$$

Note that the first integration (with respect to z) is from the cone $z = r$ (the lower surface) to the plane $z = 2$ (the upper surface). The remaining integrations with respect to r and θ are over the circular region bounded by $r = 2$. Integrating, we obtain

$$\begin{aligned}
m &= k \int_0^{2\pi} \int_0^2 \left[r^2 z + \frac{z^3}{3} \right]_r^2 r\, dr\, d\theta \\
&= k \int_0^{2\pi} \int_0^2 \left[\left(2r^2 + \frac{8}{3} \right) - \left(r^3 + \frac{r^3}{3} \right) \right] r\, dr\, d\theta \\
&= k \int_0^{2\pi} \int_0^2 (2r^3 + \tfrac{8}{3}r - \tfrac{4}{3}r^4)\, dr\, d\theta \\
&= k \int_0^{2\pi} \left[2\frac{r^4}{4} + \frac{8}{3}\frac{r^2}{2} - \frac{4}{3}\frac{r^5}{5} \right]_0^2 d\theta \\
&= k \int_0^{2\pi} \tfrac{24}{5}\, d\theta = \tfrac{24}{5}k\Big[\theta\Big]_0^{2\pi} = \tfrac{48}{5}\pi k.
\end{aligned}$$

EXAMPLE 5 A solid has the shape of the region Q that lies inside the cylinder $r = a$, within the sphere $r^2 + z^2 = 4a^2$, and above the xy-plane. The density at a point P is directly proportional to the distance from the xy-plane to P. Find the mass and the moment of inertia I_z of the solid.

SOLUTION The region Q is sketched in Figure 17.70, together with a column that corresponds to a first summation in the direction parallel to the z-axis. Since the density at the point P is given by kz for some constant k, we may find the mass m by applying (17.26)(i), with $\delta(x, y, z) = kz$

FIGURE 17.70

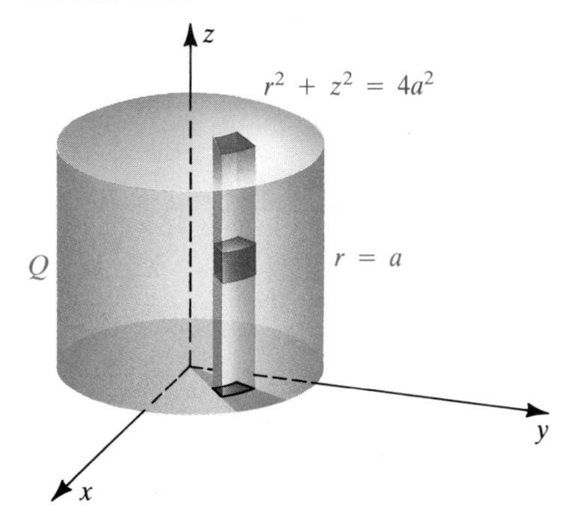

and $dV = r\,dz\,dr\,d\theta$. Thus,

$$m = \int_0^{2\pi}\int_0^a\int_0^{\sqrt{4a^2-r^2}} (kz)r\,dz\,dr\,d\theta = k\int_0^{2\pi}\int_0^a \left[\frac{z^2}{2}\right]_0^{\sqrt{4a^2-r^2}} r\,dr\,d\theta$$

$$= \frac{1}{2}k\int_0^{2\pi}\int_0^a (4a^2r - r^3)\,dr\,d\theta = \frac{1}{2}k\int_0^{2\pi}\left[2a^2r^2 - \frac{r^4}{4}\right]_0^a d\theta$$

$$= \tfrac{7}{8}a^4k\int_0^{2\pi} d\theta = \tfrac{7}{8}a^4k\Big[\theta\Big]_0^{2\pi} = \tfrac{7}{4}a^4\pi k.$$

To find I_z, we employ (17.27) and use cylindrical coordinates, obtaining

$$I_z = \int_0^{2\pi}\int_0^a\int_0^{\sqrt{4a^2-r^2}} r^2(kz)r\,dz\,dr\,d\theta = k\int_0^{2\pi}\int_0^a\left[\frac{z^2}{2}\right]_0^{\sqrt{4a^2-r^2}} r^3\,dr\,d\theta$$

$$= \tfrac{1}{2}k\int_0^{2\pi}\int_0^a (4a^2r^3 - r^5)\,dr\,d\theta = \tfrac{5}{6}a^6\pi k.$$

EXAMPLE 6 A glass chamber is to be used in experiments to test the effects of exposure to radon gas. The chamber has the shape of the paraboloid $z = 25 - x^2 - y^2$, with $z \geq 0$ and units in centimeters. If the density of radioactive energy is to be 6×10^{-11} joule/cm^3, calculate the total amount of radioactive energy that is required in the chamber.

SOLUTION The shape of the first-quadrant part of the chamber is similar to that of the surface in Figure 17.36. In this example we let $\delta(x, y, z)$ denote the *energy density* at (x, y, z) instead of the mass density. If dV is the volume of a subregion dQ within the chamber and (x, y, z) is a point in dQ, then the radioactive energy within dQ may be approximated by $\delta(x, y, z)\,dV$. Taking a limit of triple sums with $\delta(x, y, z) = \delta = 6 \times 10^{-11}$, we find that the total radioactive energy E is given by

$$E = \iiint_V \delta\,dV.$$

Employing cylindrical coordinates yields

$$\begin{aligned} E &= \delta\int_0^{2\pi}\int_0^5\int_0^{25-r^2} r\,dz\,dr\,d\theta \\ &= \delta\int_0^{2\pi}\int_0^5 r\Big[z\Big]_0^{25-r^2} dr\,d\theta \\ &= \delta\int_0^{2\pi}\int_0^5 (25r - r^3)\,dr\,d\theta \\ &= \tfrac{625}{2}\pi\delta \approx 5.89 \times 10^{-8} \text{ joule.}\end{aligned}$$

EXERCISES 17.7

Exer. 1–14: Describe the graph of the equation in three dimensions.

1 (a) $r = 4$ (b) $\theta = -\pi/2$ (c) $z = 1$

2 (a) $r = -3$ (b) $\theta = \pi/4$ (c) $z = -2$

3 $r = -3\sec\theta$ 4 $r = -\csc\theta$ 5 $z = 4r^2$

6 $z = 4 - r^2$ 7 $r = 6\sin\theta$ 8 $r\sec\theta = 4$

9 $z = 2r$ 10 $3z = r$ 11 $r^2 = 9 - z^2$

12 $r^2 + z^2 = 16$ 13 $r = 2\csc\theta\cot\theta$ 14 $r = \tan\theta\sec\theta$

Exer. 15–24: Change the equation to cylindrical coordinates.

15 $x^2 + y^2 + z^2 = 4$ 16 $x^2 + y^2 = 4z$

17 $3x + y - 4z = 12$ 18 $y = x$

19 $x^2 = 4 - y^2$

20 $x^2 + (y - 2)^2 = 4$

21 $x^2 - 4z^2 + y^2 = 0$

22 $x^2 - y^2 - z^2 = 1$

23 $y^2 + z^2 = 9$

24 $x^2 + z^2 = 9$

Exer. 25–28: Let f be an arbitrary continuous function of r, θ, and z, and let Q be the region shown in the figure. Set up an iterated triple integral in cylindrical coordinates for $\iiint_Q f(r, \theta, z)\, dV$.

25

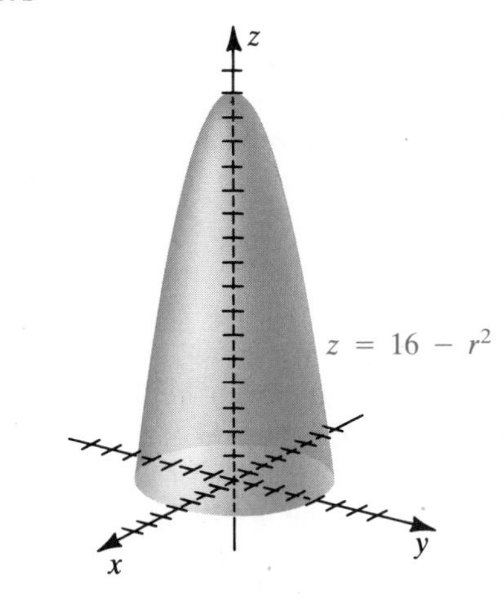

26

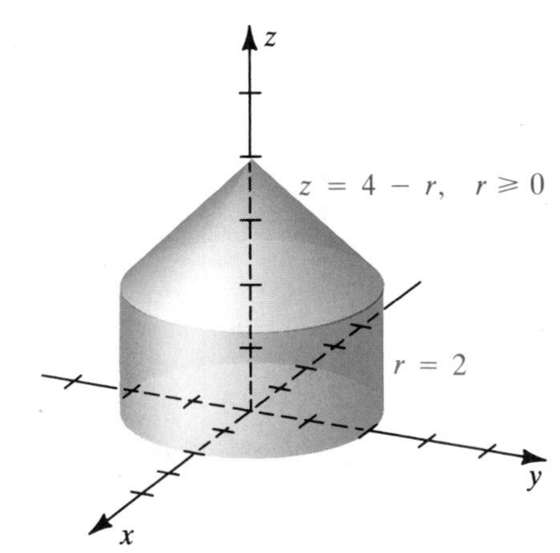

27

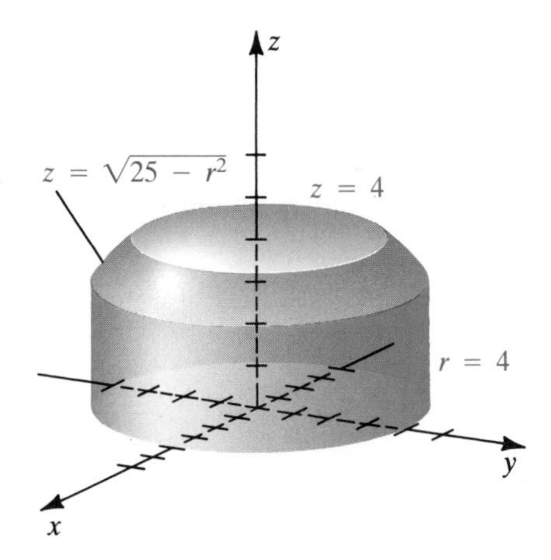

28

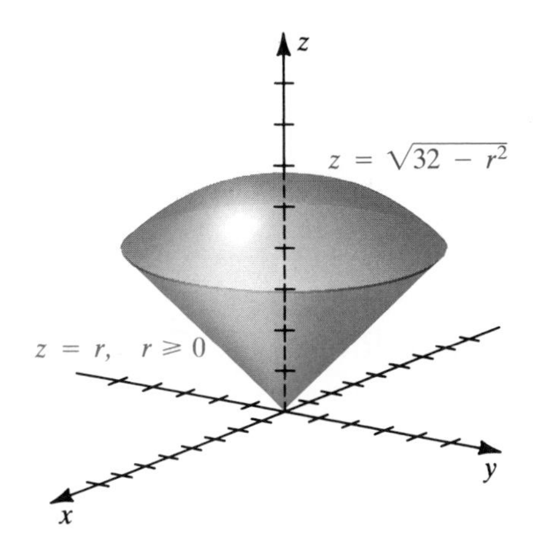

Exer. 29–38: Use cylindrical coordinates.

29 A solid is bounded by the paraboloid $z = x^2 + y^2$, the cylinder $x^2 + y^2 = 4$, and the xy-plane. Find

(a) its volume (b) its centroid

30 A solid is bounded by the cone $z = \sqrt{x^2 + y^2}$, the cylinder $x^2 + y^2 = 4$, and the xy-plane. Find

(a) its volume (b) its centroid

31 A homogeneous right circular cylinder of density δ has altitude h and radius of base a. Find its moment of inertia with respect to

(a) the axis of the cylinder (b) a diameter of the base

32 A homogeneous solid of density δ is bounded by the cone $z = \sqrt{x^2 + y^2}$ and the paraboloid $z = x^2 + y^2$. Find

(a) its center of mass

(b) its moment of inertia with respect to the z-axis

33 A spherical solid has radius a, and the density at $P(x, y, z)$ is directly proportional to the distance from P to a fixed line l through the center of the solid. Find its mass.

34 A solid is bounded by the cone $z = \sqrt{x^2 + y^2}$ and the plane $z = 4$, and the density at $P(x, y, z)$ is directly proportional to the distance from the z-axis to P. Find its mass.

35 Find the moment of inertia with respect to l for the solid described in Exercise 33.

36 Find the moment of inertia with respect to the z-axis for the solid described in Exercise 34.

c 37 For an altitude z up to 10,000 meters, the density δ (in kg/m^3) of the earth's atmosphere can be approximated by

$$\delta = 1.2 - (1.05 \times 10^{-4})z + (2.6 + 10^{-9})z^2.$$

Estimate the mass of a column of air 10 kilometers high that has a circular base of radius 3 meters.

38 A solid is cut out of the sphere $x^2 + y^2 + z^2 = 4$ by the cylinder $x^2 + y^2 = 2y$. The density at $P(x, y, z)$ is directly proportional to the distance from the xy-plane to P. Find

(a) its mass (b) its center of mass

Exer. 39–40: Use cylindrical coordinates to evaluate the integral.

39 $\displaystyle\int_0^1 \int_0^{\sqrt{1-y^2}} \int_0^{\sqrt{4-x^2-y^2}} z\, dz\, dx\, dy$

40 $\displaystyle\int_0^2 \int_{-\sqrt{2x-x^2}}^{\sqrt{2x-x^2}} \int_0^{x^2+y^2} \sqrt{x^2 + y^2}\, dz\, dy\, dx$

17.8 SPHERICAL COORDINATES

FIGURE 17.71

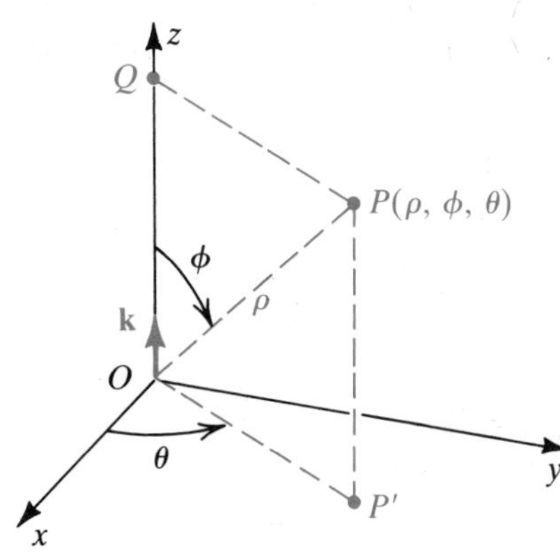

In this section we discuss the **spherical coordinate system** in three dimensions. The *spherical coordinates* of a point P are given by an ordered triple (ρ, ϕ, θ), where, as shown in Figure 17.71, $\rho = \|\overrightarrow{OP}\|$, ϕ is the angle between $\overrightarrow{OP}$ and the vector $\mathbf{k}$, and θ is a polar angle associated with the projection P' of P onto the xy-plane. The point Q is the projection of P onto the z-axis. Note that $\rho \geq 0$ and $0 \leq \phi \leq \pi$.

If $\rho_0 > 0$, the graph of $\rho = \rho_0$ is a sphere of radius ρ_0 with center O (see Figure 17.72(i)). The graph of $\rho = 0$ consists of one point, the origin.

If $0 < \phi_0 < \pi$, the graph of $\phi = \phi_0$ is a half-cone with vertex at O (see Figure 17.72(ii)). The graph of $\phi = 0$ is the nonnegative part of the z-axis, and the graph of $\phi = \pi$ is the nonpositive part.

The graph of $\theta = \theta_0$ is a half-plane containing the z-axis (see Figure 17.72(iii)).

FIGURE 17.72

(i)

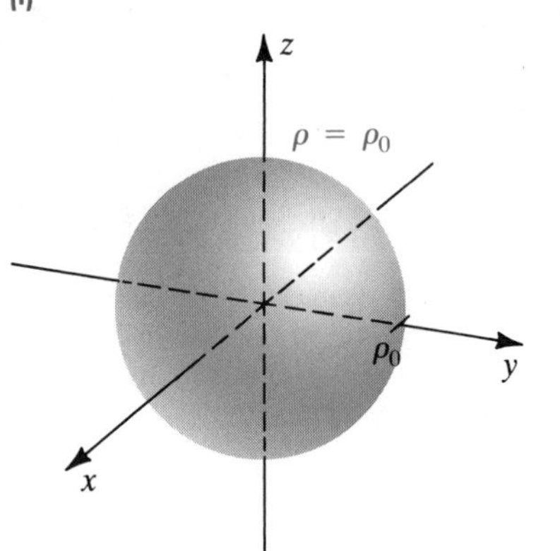

(ii) (iii)

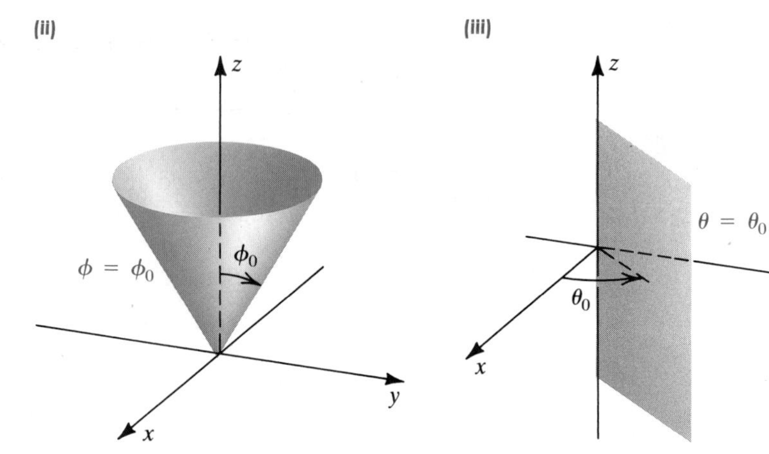

The relationship between spherical coordinates (ρ, ϕ, θ) and rectangular coordinates (x, y, z) of a point P can be found by referring to Figure 17.71. Using the fact that

$$x = \|\overrightarrow{OP'}\| \cos\theta \quad \text{and} \quad y = \|\overrightarrow{OP'}\| \sin\theta$$

together with

$$\|\overrightarrow{OP'}\| = \|\overrightarrow{QP}\| = \rho \sin\phi \quad \text{and} \quad \|\overrightarrow{OQ}\| = \rho\cos\phi$$

gives us (i) of the next theorem. Part (ii) is a consequence of the distance formula (14.13).

Theorem **(17.31)**

> The rectangular coordinates (x, y, z) and the spherical coordinates (ρ, ϕ, θ) of a point P are related as follows:
>
> **(i)** $x = \rho \sin \phi \cos \theta, \quad y = \rho \sin \phi \sin \theta, \quad z = \rho \cos \phi$
>
> **(ii)** $\rho^2 = x^2 + y^2 + z^2$

FIGURE 17.73

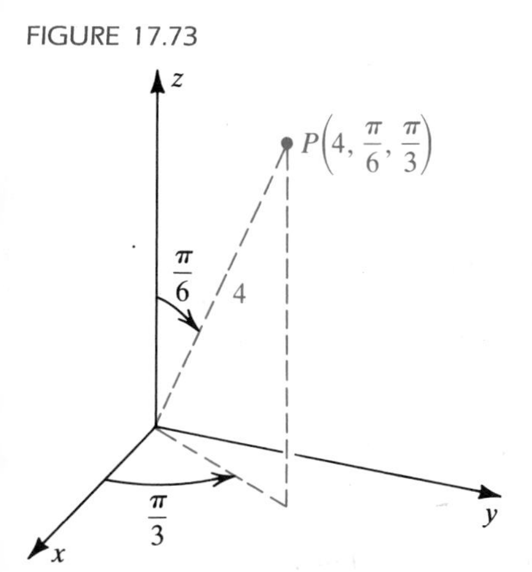

EXAMPLE 1 If a point P has spherical coordinates $(4, \pi/6, \pi/3)$, find rectangular and cylindrical coordinates for P.

SOLUTION The point P is plotted in Figure 17.73. Using Theorem (17.31)(i), with $\rho = 4$, $\phi = \pi/6$, and $\theta = \pi/3$, we obtain the rectangular coordinates:

$$x = 4 \sin \frac{\pi}{6} \cos \frac{\pi}{3} = 4\left(\frac{1}{2}\right)\left(\frac{1}{2}\right) = 1$$

$$y = 4 \sin \frac{\pi}{6} \sin \frac{\pi}{3} = 4\left(\frac{1}{2}\right)\left(\frac{\sqrt{3}}{2}\right) = \sqrt{3}$$

$$z = 4 \cos \frac{\pi}{6} = 4\left(\frac{\sqrt{3}}{2}\right) = 2\sqrt{3}$$

To find cylindrical coordinates for P, we first note that

$$r^2 = x^2 + y^2 = 1 + 3 = 4,$$

and hence $r = \pm 2$. Thus, cylindrical coordinates (r, θ, z) for P are $(2, \pi/3, 2\sqrt{3})$.

EXAMPLE 2 Find an equation in spherical coordinates whose graph is the paraboloid $z = x^2 + y^2$.

SOLUTION Substituting from (17.31)(i), we obtain

$$\begin{aligned}\rho \cos \phi &= \rho^2 \sin^2 \phi \cos^2 \theta + \rho^2 \sin^2 \phi \sin^2 \theta \\ &= \rho^2 \sin^2 \phi(\cos^2 \theta + \sin^2 \theta) \\ &= \rho^2 \sin^2 \phi,\end{aligned}$$

or, equivalently,

$$\rho(\cos \phi - \rho \sin^2 \phi) = 0.$$

Hence either

$$\rho = 0 \quad \text{or} \quad \rho \sin^2 \phi = \cos \phi.$$

The graph of $\rho = 0$ is the origin, which is also on the graph of $\rho \sin^2 \phi = \cos \phi$. Hence $z = x^2 + y^2$ is equivalent to $\rho \sin^2 \phi = \cos \phi$.

We cannot have $\sin \phi = 0$ in the equation $\rho \sin^2 \phi = \cos \phi$, because in this case $\phi = 0$ or $\phi = \pi$, and hence $\cos \phi = 1$ or $\cos \phi = -1$, which gives us the absurdity $0 = 1$ or $0 = -1$. Therefore, we may divide both sides of the equation $\rho \sin^2 \phi = \cos \phi$ by $\sin^2 \phi$, obtaining

$$\rho = \frac{\cos \phi}{\sin^2 \phi} = \frac{\cos \phi}{\sin \phi} \frac{1}{\sin \phi},$$

or, equivalently,

$$\rho = \cot \phi \csc \phi.$$

EXAMPLE 3 Change the equation $\rho = 2 \sin \phi \cos \theta$ to rectangular coordinates, and describe its graph.

SOLUTION Since we plan to use the first formula in (17.31)(i), we multiply both sides of the equation by ρ, obtaining

$$\rho^2 = 2\rho \sin \phi \cos \theta.$$

Applying (17.31) gives us the following equivalent equation in x, y, and z:

$$x^2 + y^2 + z^2 = 2x$$

$$(x - 1)^2 + y^2 + z^2 = 1$$

By (14.15), the graph is a sphere with radius 1 whose center, in rectangular coordinates, is $(1, 0, 0)$.

Sometimes it is convenient to use spherical coordinates to evaluate triple integrals. The simplest case occurs when a function f of ρ, ϕ, and θ is continuous throughout a region of the form

$$Q = \{(\rho, \phi, \theta): a \le \rho \le b, c \le \phi \le d, m \le \theta \le n\}.$$

We divide Q into subregions $Q_1, Q_2, \ldots, Q_n$ that have the same shape as Q by using graphs of equations $\rho = a_k$, $\phi = c_k$, and $\theta = m_k$, where a_k, c_k, and m_k are numbers in the intervals $[a, b]$, $[c, d]$, and $[m, n]$, respectively. These graphs are spheres with centers at the origin, cones with axes along the z-axis, and half-planes containing the z-axis, respectively (see Figure 17.72).

A typical subregion Q_k is sketched in Figure 17.74(i). The increment $\Delta\rho_k$ is the distance between two successive spheres, $\Delta\phi_k$ is the change in the vertex angle between successive cones, and $\Delta\theta_k$ is the angle between two successive half-planes. If, as shown in Figure 17.74(ii), P has spherical coordinates $(\rho_k, \phi_k, \theta_k)$ and P' is the projection of P onto the z-axis, then, from trigonometry, the length of segment $P'P$ is $\rho_k \sin \phi_k$ and hence, by Theorem (1.14), the length of arc $\widehat{PS}$ is $\rho_k \sin \phi_k \, \Delta\theta_k$. The distance from O to P is ρ_k, so the length of arc $\widehat{PR}$ is $\rho_k \, \Delta\phi_k$. The length of segment PQ is $\Delta\rho_k$. We shall approximate the volume ΔV_k of Q_k by regarding it as a rectangular box with these three dimensions. Thus,

$$\Delta V_k \approx (\rho_k \sin \phi_k \, \Delta\theta_k)(\rho_k \, \Delta\phi_k) \, \Delta\rho_k$$

or

$$\Delta V_k \approx \rho_k^2 \sin \phi_k \, \Delta\rho_k \, \Delta\phi_k \, \Delta\theta_k.$$

We define the triple integral of f over Q as a limit of Riemann sums $\sum_k f(\rho_k, \phi_k, \theta_k) \, \Delta V_k$. The following theorem can be proved.

FIGURE 17.74

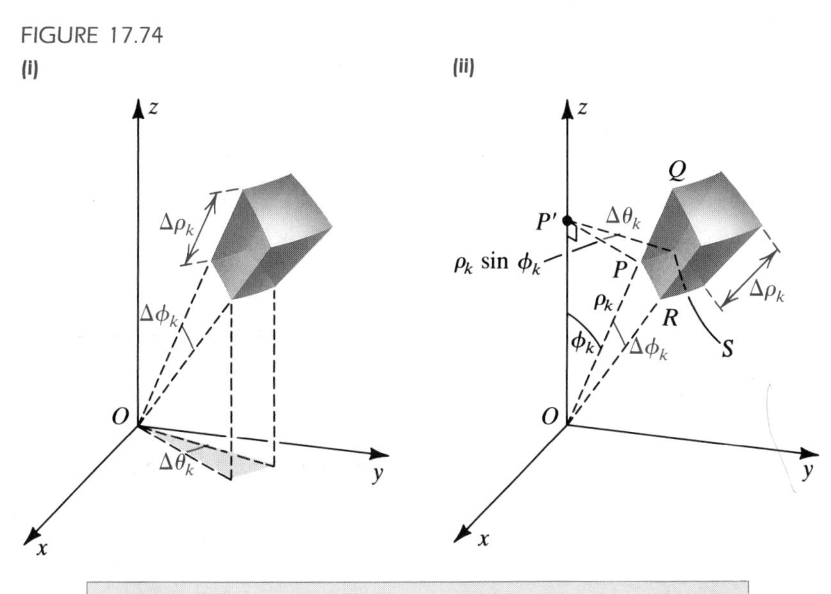

Evaluation theorem (spherical coordinates) **(17.32)**

$$\iiint_Q f(\rho, \phi, \theta)\, dV = \int_m^n \int_c^d \int_a^b f(\rho, \phi, \theta)\rho^2 \sin\phi \, d\rho\, d\phi\, d\theta$$

There are five other possible orders of integration for this iterated integral.

Spherical coordinates may also be employed over regions more complicated than the one just discussed by using inner partitions. In such cases the limits of integration in the iterated integral must be suitably adjusted.

In the following two examples we reconsider Examples 3 and 4 of the preceding section, which were solved using cylindrical coordinates.

EXAMPLE 4 Find the centroid of a hemispherical solid Q of radius a.

FIGURE 17.75

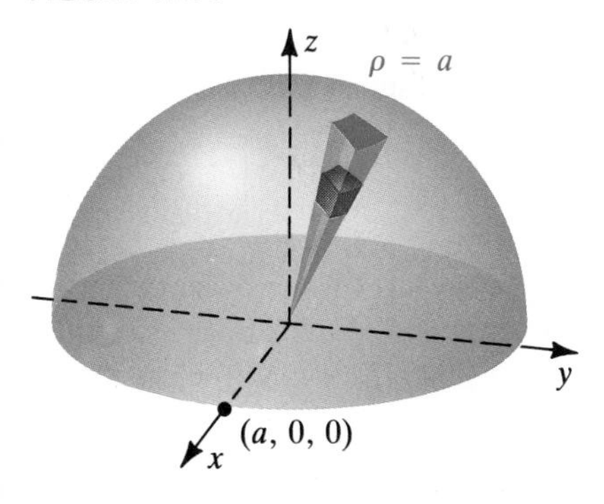

SOLUTION The solid is resketched in Figure 17.75, together with a spherical coordinates element of volume and a column indicating a first (partial) integration with respect to ρ. As in Example 3 of Section 17.7, it is sufficient to find $\bar{z}$. Using spherical coordinates, we have

$$\begin{aligned} M_{xy} &= \iiint_Q z\, dV \\ &= \int_0^{2\pi}\int_0^{\pi/2}\int_0^a (\rho\cos\phi)\rho^2 \sin\phi\, d\rho\, d\phi\, d\theta \\ &= \int_0^{2\pi}\int_0^{\pi/2}\left[\frac{\rho^4}{4}\right]_0^a \sin\phi\cos\phi\, d\phi\, d\theta \\ &= \tfrac{1}{4}a^4 \int_0^{2\pi}\int_0^{\pi/2} \sin\phi\cos\phi\, d\phi\, d\theta \\ &= \tfrac{1}{4}a^4 \int_0^{2\pi}\left[\tfrac{1}{2}\sin^2\phi\right]_0^{\pi/2} d\theta \\ &= \tfrac{1}{8}a^4\Big[\theta\Big]_0^{2\pi} = \tfrac{1}{4}\pi a^4. \end{aligned}$$

Hence
$$\bar{z} = \frac{M_{xy}}{V} = \frac{\frac{1}{4}\pi a^4}{\frac{2}{3}\pi a^3} = \frac{3}{8}a.$$

EXAMPLE 5 A solid Q is bounded by the cone $z = \sqrt{x^2 + y^2}$ and the plane $z = 2$. The density at $P(x, y, z)$ is directly proportional to the square of the distance from the origin to P. Find its mass.

FIGURE 17.76

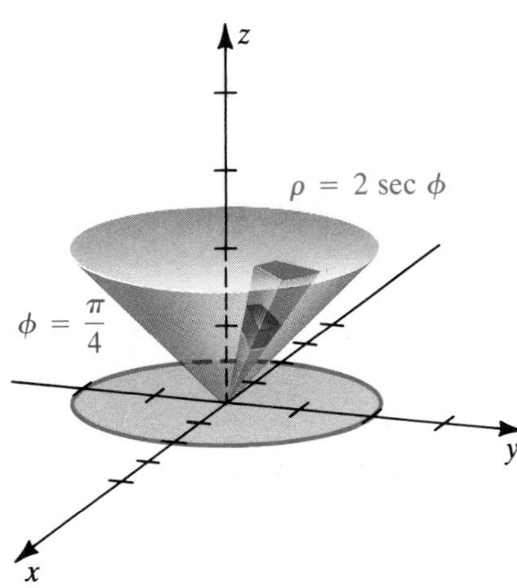

SOLUTION The solid is resketched in Figure 17.76, together with a spherical coordinates element of volume and a column indicating a first (partial) integration with respect to ρ. An equation of the plane $z = 2$ in spherical coordinates is $\rho \cos \phi = 2$, or $\rho = 2 \sec \phi$. The density at (x, y, z) is given by

$$\delta = k(x^2 + y^2 + z^2) = k\rho^2$$

for some constant k. Hence

$$\begin{aligned}
m &= \iiint_Q \delta \, dV \\
&= k \int_0^{2\pi} \int_0^{\pi/4} \int_0^{2 \sec \phi} \rho^2 \cdot \rho^2 \sin \phi \, d\rho \, d\phi \, d\theta \\
&= k \int_0^{2\pi} \int_0^{\pi/4} \left[\frac{\rho^5}{5} \right]_0^{2 \sec \phi} \sin \phi \, d\phi \, d\theta \\
&= k \int_0^{2\pi} \int_0^{\pi/4} \tfrac{32}{5} \sec^5 \phi \sin \phi \, d\phi \, d\theta \\
&= \tfrac{32}{5}k \int_0^{2\pi} \int_0^{\pi/4} \sec^3 \phi \sec \phi \tan \phi \, d\phi \, d\theta \\
&= \tfrac{32}{5}k \int_0^{2\pi} \left[\tfrac{1}{4} \sec^4 \phi \right]_0^{\pi/4} d\theta \\
&= \tfrac{8}{5}k \int_0^{2\pi} (4 - 1) \, d\theta = \tfrac{48}{5}\pi k.
\end{aligned}$$

EXAMPLE 6 Find the volume and the centroid of the region Q that is bounded above by the sphere $\rho = a$ and below by the cone $\phi = c$, where $0 < c < \pi/2$.

FIGURE 17.77

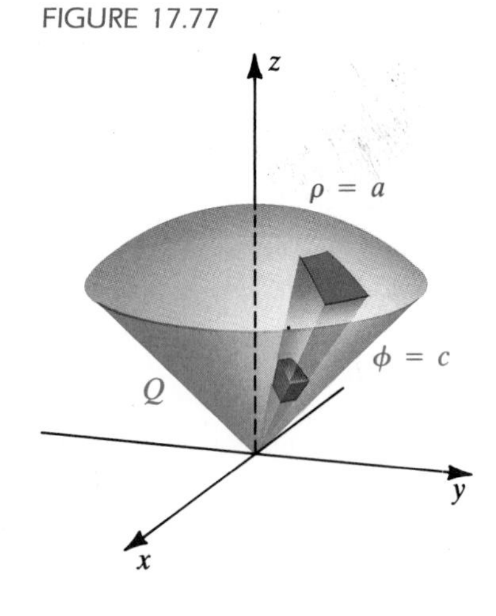

SOLUTION The region Q is sketched in Figure 17.77, together with a spherical coordinates element of volume and a column indicating a first (partial) integration with respect to ρ. The volume V is

$$\begin{aligned}
V &= \iiint_Q dV = \int_0^{2\pi} \int_0^c \int_0^a \rho^2 \sin \phi \, d\rho \, d\phi \, d\theta = \int_0^{2\pi} \int_0^c \left[\frac{\rho^3}{3} \right]_0^a \sin \phi \, d\phi \, d\theta \\
&= \tfrac{1}{3}a^3 \int_0^{2\pi} \int_0^c \sin \phi \, d\phi \, d\theta = \tfrac{1}{3}a^3 \int_0^{2\pi} \Big[-\cos \phi \Big]_0^c d\theta \\
&= \tfrac{1}{3}a^3 \int_0^{2\pi} (1 - \cos c) \, d\theta = \tfrac{2}{3}\pi a^3 (1 - \cos c).
\end{aligned}$$

By symmetry, the centroid is on the z-axis, so we need only find $\bar{z} = M_{xy}/V$. Using (17.26)(ii) with $\delta(x, y, z) = 1$, we find that

$$\begin{aligned}
M_{xy} &= \iiint_Q z \, dV = \int_0^{2\pi} \int_0^c \int_0^a (\rho \cos \phi) \rho^2 \sin \phi \, d\rho \, d\phi \, d\theta \\
&= \int_0^{2\pi} \int_0^c \left[\frac{\rho^4}{4} \right]_0^a \sin \phi \cos \phi \, d\phi \, d\theta \\
&= \tfrac{1}{4}a^4 \int_0^{2\pi} \int_0^c \sin \phi \cos \phi \, d\phi \, d\theta = \tfrac{1}{4}a^4 \int_0^{2\pi} \left[\tfrac{1}{2} \sin^2 \phi \right]_0^c d\theta \\
&= \tfrac{1}{8}a^4 \sin^2 c \int_0^{2\pi} d\theta = \tfrac{1}{4}\pi a^4 \sin^2 c.
\end{aligned}$$

The centroid, in rectangular coordinates, is $(0, 0, \bar{z})$ with

$$\bar{z} = \frac{M_{xy}}{V} = \frac{3}{8}a(1 + \cos c).$$

EXERCISES 17.8

Exer. 1–2: Change the spherical coordinates to (a) rectangular coordinates and (b) cylindrical coordinates.

1 $(4, \pi/6, \pi/2)$

2 $(1, 3\pi/4, 2\pi/3)$

Exer. 3–4: Change the rectangular coordinates to (a) spherical coordinates and (b) cylindrical coordinates.

3 $(1, 1, -2\sqrt{2})$

4 $(1, \sqrt{3}, 0)$

Exer. 5–20: Describe the graph of the equation in three dimensions.

5 (a) $\rho = 3$ (b) $\phi = \pi/6$ (c) $\theta = \pi/3$

6 (a) $\rho = 5$ (b) $\phi = 2\pi/3$ (c) $\theta = \pi/4$

7 $\rho = 4 \cos \phi$

8 $\rho \sec \phi = 6$

9 $\rho \cos \phi = 3$

10 $\rho = 4 \sec \phi$

11 $\rho = 6 \sin \phi \cos \theta$

12 $\rho = 8 \sin \phi \sin \theta$

13 $\rho = 5 \csc \phi \csc \theta$

14 $\rho = 2 \csc \phi \sec \theta$

15 $\rho = 5 \csc \phi$

16 $\rho \sin \phi = 3$

17 $\tan \phi = 2$

18 $\tan \theta = 4$

19 $\rho = 6 \cot \phi \csc \phi$

20 $\rho^2 - 3\rho + 2 = 0$

Exer. 21–30: Change the equation to spherical coordinates.

21 $x^2 + y^2 + z^2 = 4$

22 $x^2 + y^2 = 4z$

23 $3x + y - 4z = 12$

24 $y = x$

25 $x^2 = 4 - y^2$

26 $x^2 + (y - 2)^2 = 4$

27 $x^2 - 4z^2 + y^2 = 0$

28 $x^2 - y^2 - z^2 = 1$

29 $y^2 + z^2 = 9$

30 $x^2 + z^2 = 9$

Exer. 31–38: Use spherical coordinates.

31 Find the mass and the center of mass of a solid hemisphere of radius a if the density at a point P is directly proportional to the distance from the center of the base to P.

32 Find the volume and the centroid of the solid bounded by the cone $z = \sqrt{x^2 + y^2}$, the cylinder $x^2 + y^2 = 4$, and the xy-plane.

33 Find the moment of inertia with respect to the axis of symmetry for the hemisphere in Exercise 31.

34 Find the moment of inertia with respect to a diameter of the base of a homogeneous solid hemisphere of radius a.

35 Find the volume of the solid that lies above the cone $z^2 = x^2 + y^2$ and inside the sphere $x^2 + y^2 + z^2 = 4z$.

36 Find the volume of the solid that lies outside the cone $z^2 = x^2 + y^2$ and inside the sphere $x^2 + y^2 + z^2 = 1$.

37 Find the mass of the solid that lies outside the sphere $x^2 + y^2 + z^2 = 1$ and inside the sphere $x^2 + y^2 + z^2 = 2$ if the density at a point P is directly proportional to the square of the distance from the center of the spheres to P.

c 38 If the earth is assumed to be spherical with a radius of 6370 kilometers, the density δ (in kg/m^3) of the atmosphere a distance ρ meters from the center of the earth can be approximated by

$$\delta = 619.09 - (9.7 \times 10^{-5})\rho$$

for $6{,}370{,}000 \le \rho \le 6{,}373{,}000$.

(a) Estimate the mass of the atmosphere between ground level and an altitude of 3 kilometers.

(b) The atmosphere extends beyond an altitude of 100 kilometers and has a total mass of approximately 5.1×10^{18} kilograms. What percentage of the mass is in the lowest 3 kilometers of the atmosphere?

Exer. 39–40: Evaluate the integral by changing to spherical coordinates.

39 $\displaystyle\int_{-2}^{2} \int_{-\sqrt{4-x^2}}^{\sqrt{4-x^2}} \int_{\sqrt{x^2+y^2}}^{\sqrt{8-x^2-y^2}} (x^2 + y^2 + z^2)\, dz\, dy\, dx$

40 $\displaystyle\int_{0}^{\sqrt{2}} \int_{y}^{\sqrt{4-y^2}} \int_{0}^{\sqrt{4-x^2-y^2}} \sqrt{x^2 + y^2 + z^2}\, dz\, dx\, dy$

41 The relationship between spherical and rectangular coordinates is important in the design of joints for robot arms. The movement caused by three joints is illustrated in the figure on the following page: a hinge joint rotates the arm about the z-axis through an angle θ in the xy-plane, a sliding joint extends and contracts the length L of the arm, and another hinge joint tilts the arm up and down through the (spherical coordinate) angle ϕ.

(a) If the joints are currently set at $\theta = 120°$, $\phi = 135°$, and $L = 12$ in., find the rectangular coordinates of the joint P in the hand of the arm.

(b) Describe the changes in θ, ϕ, and L that are necessary to pick up an object located at the point with rectangular coordinates $(-8, -8, 8)$.

EXERCISE 41

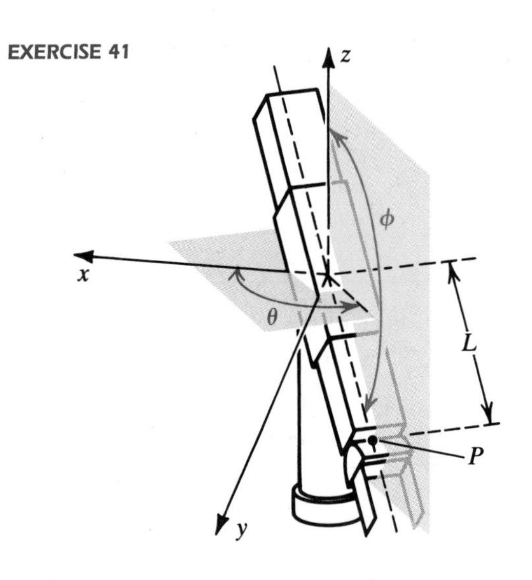

42 Small movements in the joints of robot arms, called *differential motions*, are used to describe the change in position of the hand when joint settings are not exact. Suppose that the joint settings θ, ϕ, and L for the arm in Exercise 41 have errors $\Delta\theta$, $\Delta\phi$, and ΔL.

(a) Use differentials to find formulas for the corresponding errors Δx, Δy, and Δz in the rectangular coordinates of P.

(b) Use the formulas found in (a) to estimate Δx, Δy, and Δz if $\theta = 135°$, $\phi = 45°$, $L = 12$ in., $\Delta\theta = -1°$, $\Delta\phi = 0.5°$, and $\Delta L = 0.1$.

17.9 CHANGE OF VARIABLES AND JACOBIANS

FIGURE 17.78

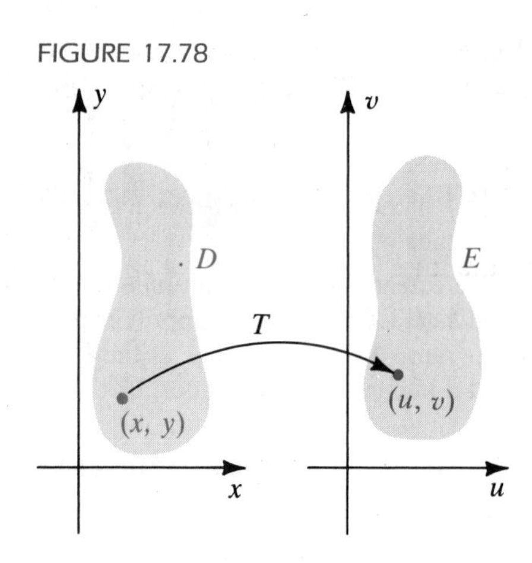

In Section 17.3 we discussed how to change a double integral in rectangular coordinates to a double integral in polar coordinates. In the preceding two sections we considered triple integrals in cylindrical and spherical coordinates. We now introduce a more general method for changing variables in multiple integrals. This method is closely connected to *transformations* (or *correspondences*) from one rectangular coordinate system to another.

Let us begin by considering a function T whose domain D is a region in the xy-plane and whose range E is a region in a uv-plane. As illustrated in Figure 17.78, to each point (x, y) in D there corresponds exactly one point (u, v) in E such that $T(x, y) = (u, v)$. We call T a **transformation of coordinates** from the xy-plane to the uv-plane. Since each pair (u, v) is uniquely determined by (x, y), both u and v are functions of x and y. Thus, we have the following *transformation formulas*, where f and g are functions that have the same domain D as T.

Transformation of coordinates formulas **(17.33)**

$$u = f(x, y), \quad v = g(x, y); \quad (x, y) \text{ in } D, \quad (u, v) \text{ in } E$$

Given the transformation of coordinates (17.33), let us partition a region in the uv-plane by means of vertical lines $u = c_1$, $u = c_2$, $u = c_3, \ldots$ and horizontal lines $v = d_1$, $v = d_2$, $v = d_3, \ldots$, as illustrated in Figure 17.79(i). The corresponding level curves for f and g in the xy-plane are graphs of

$$u = f(x, y) = c_j \quad \text{and} \quad v = g(x, y) = d_k$$

for $j = 1, 2, 3, \ldots$ and $k = 1, 2, 3, \ldots$. These curves determine a *curvilinear partition* of a region in the xy-plane, as illustrated in Figure 17.79(ii). We refer to the curves $u = f(x, y) = c_j$ and $v = g(x, y) = d_k$ in the xy-plane as ***u*-curves** and ***v*-curves**, respectively. Of course, the types of curves obtained depend on the nature of the functions f and g. The next example illustrates a case for which each u-curve and v-curve is a line.

FIGURE 17.79

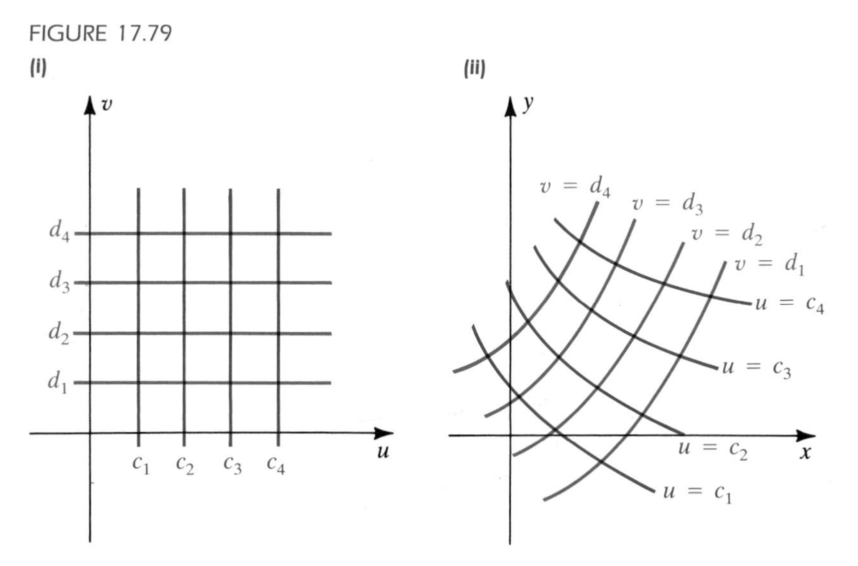

EXAMPLE 1 Let T be the transformation of coordinates from the xy-plane to the uv-plane determined by

$$u = x + 2y, \quad v = x - 2y.$$

Sketch, in the uv-plane, the vertical lines $u = 2$, $u = 4$, $u = 6$, $u = 8$ and the horizontal lines $v = -1$, $v = 1$, $v = 3$, $v = 5$. Sketch the corresponding u-curves and v-curves in the xy-plane.

SOLUTION The vertical and horizontal lines in the uv-plane are sketched in Figure 17.80(i).

The u-curves in the xy-plane are the lines

$$x + 2y = 2, \quad x + 2y = 4, \quad x + 2y = 6, \quad x + 2y = 8,$$

and the v-curves are the lines

$$x - 2y = -1, \quad x - 2y = 1, \quad x - 2y = 3, \quad x - 2y = 5.$$

These lines are sketched in Figure 17.80(ii).

FIGURE 17.80

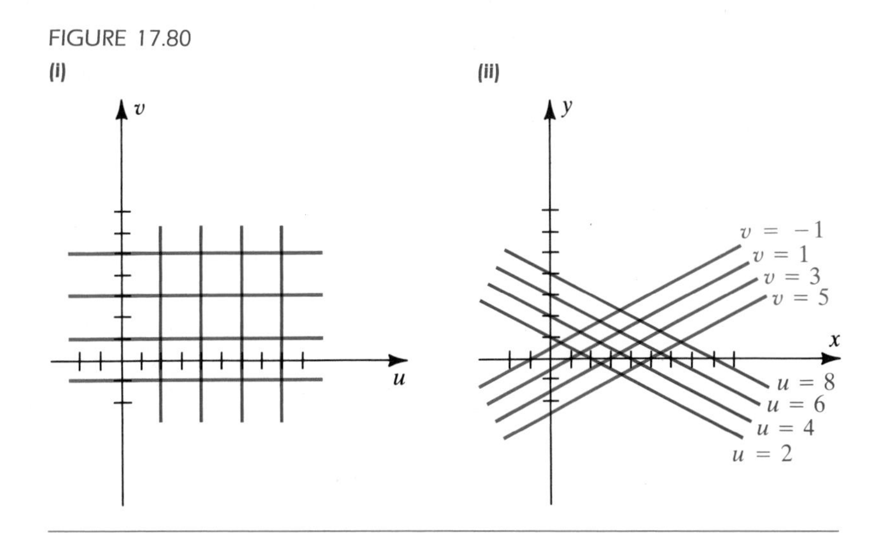

The transformation T in Example 1 is one-to-one; that is, if $(x_1, y_1) \neq (x_2, y_2)$ in the xy-plane, then $T(x_1, y_1) \neq T(x_2, y_2)$ in the uv-plane. In general, if T is a one-to-one transformation of coordinates, then by reversing the correspondence we obtain a transformation T^{-1} from the uv-plane to the xy-plane called the **inverse** of T. We may specify T^{-1} by means of equations of the form

$$x = F(u, v), \quad y = G(u, v)$$

for certain functions F and G. If a one-to-one transformation T is illustrated by an arrow, as in Figure 17.78, then the inverse T^{-1} can be illustrated by *reversing* the arrow, as in Figure 17.81. Note that $T^{-1}(T(x, y)) = (x, y)$ and $T(T^{-1}(u, v)) = (u, v)$ for every (x, y) in D and every (u, v) in E.

FIGURE 17.81

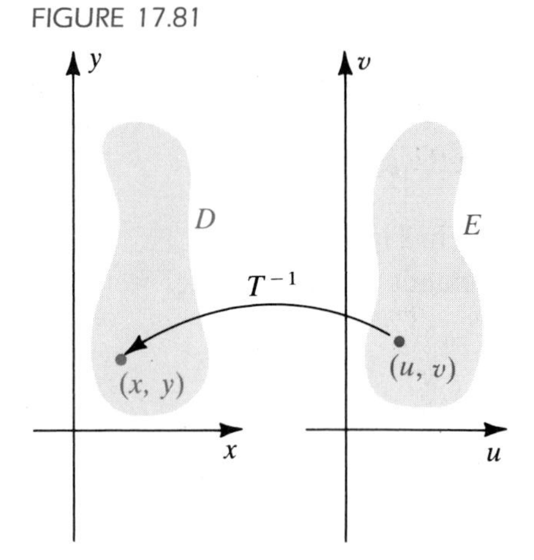

EXAMPLE 2

(a) Find the inverse of the transformation T in Example 1.

(b) Find the curve in the uv-plane that T^{-1} transforms onto the ellipse $x^2 + 4y^2 = 1$.

SOLUTION

(a) The transformation T is given by

$$u = x + 2y, \quad v = x - 2y.$$

If we add corresponding sides of these equations, we obtain $u + v = 2x$. Subtracting corresponding sides leads to $u - v = 4y$. Thus, T^{-1} is given by

$$x = \tfrac{1}{2}(u + v), \quad y = \tfrac{1}{4}(u - v).$$

(b) Since x and y are related by means of the last two equations in part (a), the points (u, v) corresponding to $x^2 + 4y^2 = 1$ must satisfy the equation

$$[\tfrac{1}{2}(u + v)]^2 + 4[\tfrac{1}{4}(u - v)]^2 = 1.$$

This simplifies to

$$u^2 + v^2 = 2.$$

FIGURE 17.82

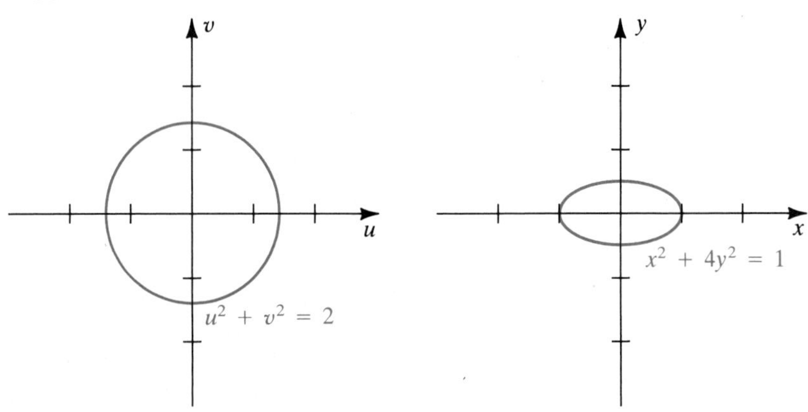

It follows that the circle of radius $\sqrt{2}$ with center at the origin in the uv-plane is transformed onto the ellipse $x^2 + 4y^2 = 1$ by T^{-1} (see Figure 17.82).

EXAMPLE 3 If $P(x, y)$ is any point with $x \neq 0$ in the xy-plane, let

$$r = \sqrt{x^2 + y^2}, \quad \theta = \arctan \frac{y}{x}$$

and let $P(0, b)$ correspond to $r = b$, $\theta = \pi/2$. Thus, r and θ are polar coordinates for P, and the preceding equalities determine a transformation of coordinates T from the xy-plane to the $r\theta$-plane. Sketch, in the $r\theta$-plane, the graphs of $r = 1$, $r = 2$, $r = 3$ and $\theta = -\frac{1}{2}$, $\theta = \frac{1}{2}$, $\theta = \frac{3}{2}$. Sketch the corresponding r-curves and θ-curves in the xy-plane.

SOLUTION The graphs are sketched in Figure 17.83.

FIGURE 17.83

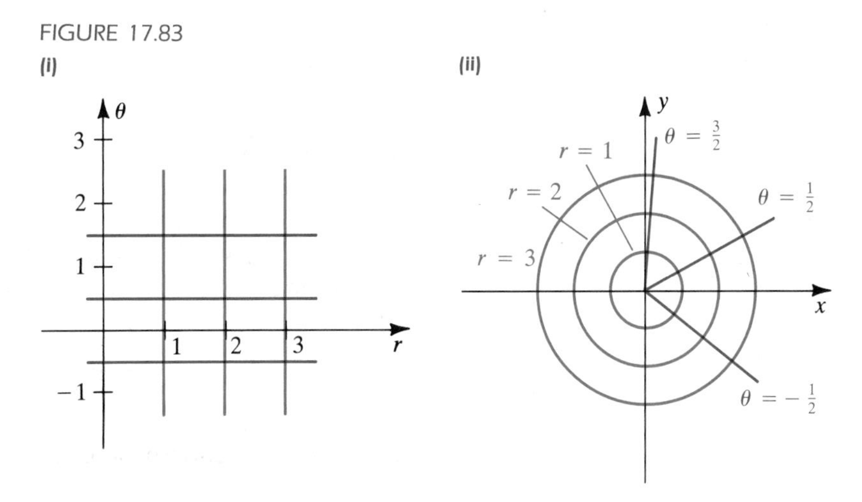

In Section 5.6 we discussed how to change variables in a definite integral $\int_a^b f(x)\,dx$. Under suitable conditions, if we substitute $x = g(u)$, then $dx = g'(u)\,du$ and

$$\int_a^b f(x)\,dx = \int_c^d f(g(u))g'(u)\,du,$$

where $a = g(c)$ and $b = g(d)$ (see Theorem (5.33)). We shall next obtain a formula for changing variables in a double integral.

Consider $\iint_R F(x, y)\,dA$, where R is a region in the xy-plane, and suppose we make the substitution

$$x = f(u, v), \quad y = g(u, v),$$

where f and g are functions that have continuous second partial derivatives. These equations define a transformation of coordinates W from the uv-plane to the xy-plane. After the substitution for x and y, the integrand becomes a function of u and v. One of our objectives is to find a region S in the uv-plane that is transformed onto R by W, as illustrated in Figure 17.84, such that

$$\iint_R F(x, y)\, dA = \iint_S F(f(u, v), g(u, v))\, dA.$$

FIGURE 17.84

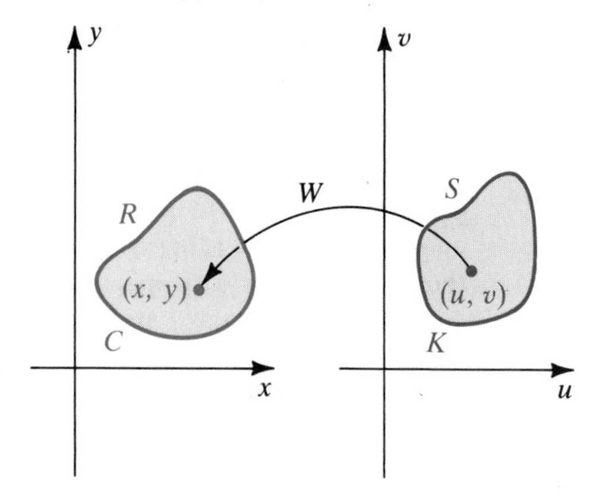

For a suitable region S to exist, it is necessary to place restrictions on the region R and on the integrand $F(x, y)$. We shall assume that R consists of all points that are either inside or on a piecewise-smooth simple closed curve C, and that F has continuous first partial derivatives throughout an open region containing R. The **positive direction along C** is such that as a point P traces C, the region R is always on the left. For the **negative direction**, R is on the right as P traces C. We shall also require that W transform a region S of the uv-plane in a one-to-one manner onto R, and that S be bounded by a piecewise-smooth simple closed curve K that W transforms onto C. Our final requirement is that, as (u, v) traces K once in the positive direction, the corresponding point (x, y) trace C once in either the positive or the negative direction. It is possible to weaken these conditions; however, that is beyond the scope of our work.

The function of u and v introduced in the next definition will be used in the change of variable process. It is named after the German mathematician C. G. Jacobi (1804–1851).

Definition **(17.34)**

If $x = f(u, v)$ and $y = g(u, v)$, then the **Jacobian** of x and y with respect to u and v, denoted by $\partial(x, y)/\partial(u, v)$, is

$$\frac{\partial(x, y)}{\partial(u, v)} = \begin{vmatrix} \dfrac{\partial x}{\partial u} & \dfrac{\partial x}{\partial v} \\[2ex] \dfrac{\partial y}{\partial u} & \dfrac{\partial y}{\partial v} \end{vmatrix} = \frac{\partial x}{\partial u}\frac{\partial y}{\partial v} - \frac{\partial y}{\partial u}\frac{\partial x}{\partial v}.$$

In the next theorem *all symbols have the same meanings as before, and we assume that the regions and functions satisfy the conditions we have discussed*. In the statement of the theorem, the notations $dx\, dy$ and $du\, dv$ are used in place of dA so that there is no misunderstanding about the region of integration.

Theorem (17.35)

If $x = f(u, v)$, $y = g(u, v)$ is a transformation of coordinates, then

$$\iint_R F(x, y)\, dx\, dy = \pm \iint_S F(f(u, v), g(u, v)) \frac{\partial(x, y)}{\partial(u, v)}\, du\, dv.$$

As (u, v) traces the boundary K of S once in the positive direction, the corresponding point (x, y) traces the boundary C of R once in either the positive direction, in which case the plus sign is chosen, or the negative direction, in which case the minus sign is chosen.

A proof of Theorem (17.35), based on Green's theorem (18.19), is given in Appendix II.

If we let $F(x, y) = 1$ for every (x, y) in R, then the integral on the left in (17.35) is the area A of R and we obtain

$$A = \pm \iint_S \frac{\partial(x, y)}{\partial(u, v)}\, du\, dv.$$

If $\partial(x, y)/\partial(u, v) > 0$ throughout S, then the double integral on the right side of this equation is positive, and hence the + sign should be used. This shows that if the Jacobian is always positive, then as the point (u, v) traces the boundary K of S once in the positive direction, the point (x, y) traces the boundary C of R once in the positive direction. If the Jacobian is always negative, then (x, y) traces C once in the negative direction. This gives us the following corollary to Theorem (17.35).

Corollary (17.36)

If $x = f(u, v)$, $y = g(u, v)$ is a transformation of coordinates and if the Jacobian $\partial(x, y)/\partial(u, v)$ does not change sign in S, then

$$\iint_R F(x, y)\, dx\, dy = \iint_S F(f(u, v), g(u, v)) \left| \frac{\partial(x, y)}{\partial(u, v)} \right| du\, dv.$$

EXAMPLE 4 Evaluate

$$\iint_R e^{(y-x)/(y+x)}\, dx\, dy,$$

where R is the region in the xy-plane bounded by the trapezoid with vertices $(0, 1)$, $(0, 2)$, $(2, 0)$, and $(1, 0)$.

SOLUTION The fact that $y - x$ and $y + x$ appear in the integrand suggests the following substitution:

$$u = y - x, \quad v = y + x$$

These equations define a transformation T from the xy-plane to the uv-plane. To apply Theorem (17.35), we must use the inverse transformation T^{-1}. To find T^{-1}, we solve the preceding equations for x and y in terms of u and v, obtaining

$$x = \tfrac{1}{2}(-u + v), \quad y = \tfrac{1}{2}(u + v).$$

By (17.34), the Jacobian is

$$\frac{\partial(x, y)}{\partial(u, v)} = \begin{vmatrix} -\frac{1}{2} & \frac{1}{2} \\ \frac{1}{2} & \frac{1}{2} \end{vmatrix} = -\frac{1}{4} - \frac{1}{4} = -\frac{1}{2}.$$

The region R is sketched in Figure 17.85(i). To determine the region S in the uv-plane that transforms onto R, we note that the sides of R lie on the lines

$$x = 0, \quad x + y = 2, \quad y = 0, \quad x + y = 1.$$

Substituting $x = \frac{1}{2}(-u + v)$ and $y = \frac{1}{2}(u + v)$, we see that the corresponding curves in the uv-plane are, respectively,

$$v = u, \quad v = 2, \quad v = -u, \quad v = 1.$$

These lines form the boundary of the trapezoidal region S shown in Figure 17.85(ii).

FIGURE 17.85

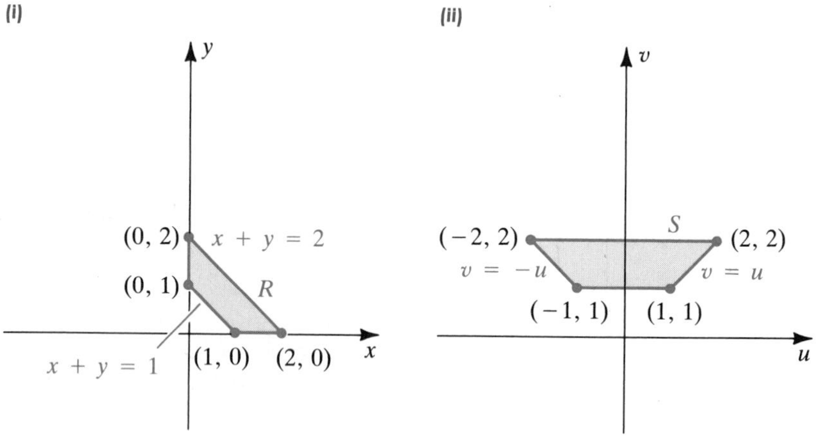

(i)

y; (0, 2); x + y = 2; (0, 1); R; x + y = 1; (1, 0); (2, 0); x

(ii)

v; S; (−2, 2); (2, 2); v = −u; v = u; (−1, 1); (1, 1); u

If the point (u, v) moves along the line $v = u$ from $(1, 1)$ to $(2, 2)$, then the corresponding point (x, y) moves along the line $x = 0$ from $(0, 1)$ to $(0, 2)$. Similarly, if (u, v) moves along the line $v = 2$ from $(2, 2)$ to $(-2, 2)$, then (x, y) moves along the line $x + y = 2$ from $(0, 2)$ to $(2, 0)$. Proceeding in this manner, we see that if (u, v) traces the boundary of S once in the *positive* direction, then (x, y) traces the boundary of R once in the *negative* direction. It can also be shown, using inequalities, that interior points of S correspond to interior points of R. Applying Theorem (17.35) yields

$$\begin{aligned}\iint_R e^{(y-x)/(y+x)}\,dx\,dy &= -\iint_S e^{u/v}\left(-\tfrac{1}{2}\right)du\,dv \\ &= \tfrac{1}{2}\int_1^2 \int_{-v}^{v} e^{u/v}\,du\,dv \\ &= \tfrac{1}{2}\int_1^2 \Big[ve^{u/v}\Big]_{-v}^{v} dv = \tfrac{1}{2}\int_1^2 v(e - e^{-1})\,dv \\ &= \tfrac{1}{2}(e - e^{-1})\Big[\tfrac{1}{2}v^2\Big]_1^2 = \tfrac{3}{4}(e - e^{-1}) \approx 1.763.\end{aligned}$$

FIGURE 17.86

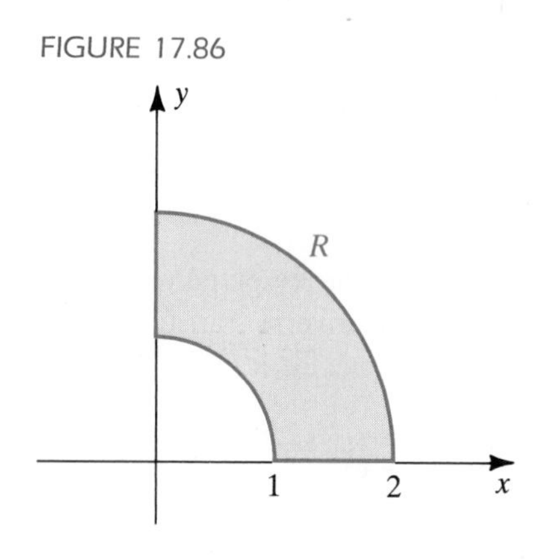

EXAMPLE 5 Evaluate $\iint_R e^{-(x^2+y^2)}\,dx\,dy$ for the first-quadrant region R in the xy-plane between the circular arcs shown in Figure 17.86.

SOLUTION This integral could be evaluated by changing to polar coordinates as we did in Section 17.3; however, our objective is to demonstrate the use of Theorem (17.35). The polar substitution

$$x = r\cos\theta, \quad y = r\sin\theta$$

determines a transformation from the $r\theta$-plane to the xy-plane, whose Jacobian is

$$\frac{\partial(x, y)}{\partial(r, \theta)} = \begin{vmatrix} \cos\theta & -r\sin\theta \\ \sin\theta & r\cos\theta \end{vmatrix} = r.$$

FIGURE 17.87

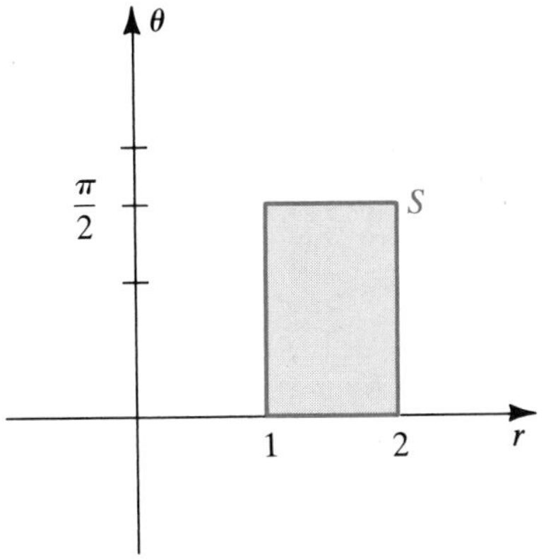

We can verify that the rectangular region S in the $r\theta$-plane bounded by $r = 1, r = 2, \theta = 0$, and $\theta = \pi/2$ (see Figure 17.87) corresponds to R under the transformation. Moreover, as (r, θ) traces the boundary of S once in the positive direction, the corresponding point (x, y) traces R once in the positive direction. Hence, from Theorem (17.35),

$$\begin{aligned} \iint_R e^{-(x^2+y^2)}\,dx\,dy &= \iint_S e^{-r^2} r\,dr\,d\theta \\ &= \int_0^{\pi/2}\int_1^2 e^{-r^2} r\,dr\,d\theta = \int_0^{\pi/2}\left[-\tfrac{1}{2}e^{-r^2}\right]_1^2 d\theta \\ &= -\tfrac{1}{2}(e^{-4} - e^{-1})\int_0^{\pi/2} d\theta = -\tfrac{1}{4}\pi(e^{-4} - e^{-1}) \approx 0.275. \end{aligned}$$

Definition (17.34) can be extended to functions of three variables as follows.

Definition (17.37)

If $x = f(u, v, w)$, $y = g(u, v, w)$, and $z = h(u, v, w)$, then the **Jacobian** of x, y, and z with respect to u, v, and w, denoted by $\partial(x, y, z)/\partial(u, v, w)$, is defined by

$$\frac{\partial(x, y, z)}{\partial(u, v, w)} = \begin{vmatrix} \dfrac{\partial x}{\partial u} & \dfrac{\partial x}{\partial v} & \dfrac{\partial x}{\partial w} \\ \dfrac{\partial y}{\partial u} & \dfrac{\partial y}{\partial v} & \dfrac{\partial y}{\partial w} \\ \dfrac{\partial z}{\partial u} & \dfrac{\partial z}{\partial v} & \dfrac{\partial z}{\partial w} \end{vmatrix}$$

The following theorem is a three-dimensional analogue of Corollary (17.36).

Theorem (17.38)

If $x = f(u, v, w)$, $y = g(u, v, w)$, $z = h(u, v, w)$ is a transformation of coordinates from a region S in a uvw-coordinate system onto a region R in an xyz-coordinate system and if the Jacobian $\partial(x, y, z)/\partial(u, v, w)$ does not change sign in S, then

$$\iiint_R F(x, y, z)\,dx\,dy\,dz = \iiint_S G(u, v, w)\left|\frac{\partial(x, y, z)}{\partial(u, v, w)}\right| du\,dv\,dw,$$

where $G(u, v, w)$ is the expression obtained by substituting for x, y, and z in $F(x, y, z)$.

To illustrate Definition (17.37), if we use the spherical coordinate formulas (17.31), then

$$x = \rho \sin \phi \cos \theta, \quad y = \rho \sin \phi \sin \theta, \quad z = \rho \cos \phi.$$

In Exercise 33 you are asked to show that

$$\frac{\partial(x, y, z)}{\partial(\rho, \phi, \theta)} = \rho^2 \sin \phi.$$

Since $\rho^2 \sin \phi \geq 0$, the conclusion of Theorem (17.38) takes on the form

$$\iiint_R F(x, y, z)\, dx\, dy\, dz = \iiint_S G(\rho, \phi, \theta)\rho^2 \sin \phi\, d\rho\, d\phi\, d\theta.$$

This is in agreement with formula (17.32), which was obtained in an intuitive manner.

We conclude this section with an applied problem that involves a change of variables in a double integral.

EXAMPLE 6 *Elliptic sinusoids* are good approximations to the shapes of many lake beds. Suppose that the boundary of a lake has the shape of the ellipse $(x^2/a^2) + (y^2/b^2) = 1$, where a and b are positive. If the maximum depth of the water is h_M, then an elliptic sinusoid is given by

$$f(x, y) = -h_M \cos\left(\frac{\pi}{2}\sqrt{\frac{x^2}{a^2} + \frac{y^2}{b^2}}\right),$$

where $(x^2/a^2) + (y^2/b^2) \leq 1$ (see Figure 17.88). Find the volume V and the average depth h_{av} of the water in the lake.

FIGURE 17.88

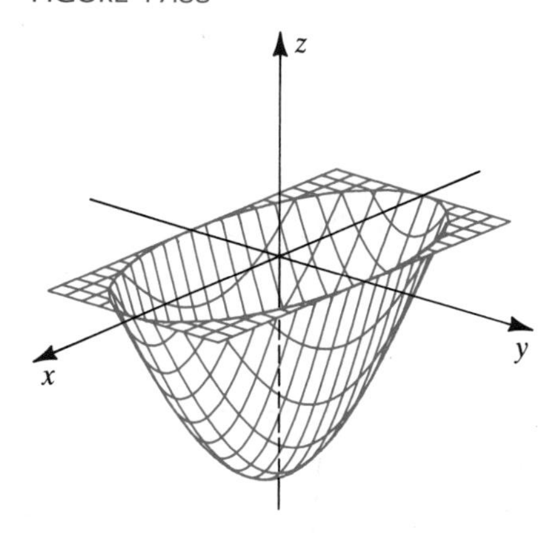

SOLUTION If R is the elliptical region that corresponds to the surface of the lake (see Figure 17.89(i)), then since the water depth at (x, y) is $|f(x, y)|$,

$$V = \iint_R |f(x, y)|\, dA = h_M \iint_R \cos\left(\frac{\pi}{2}\sqrt{\frac{x^2}{a^2} + \frac{y^2}{b^2}}\right) dA.$$

The form of the integrand suggests the substitution

$$x = au, \quad y = bv.$$

FIGURE 17.89

(i)

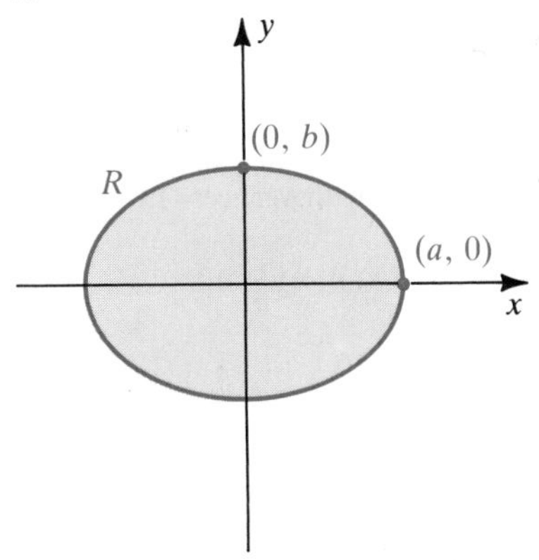

(ii)

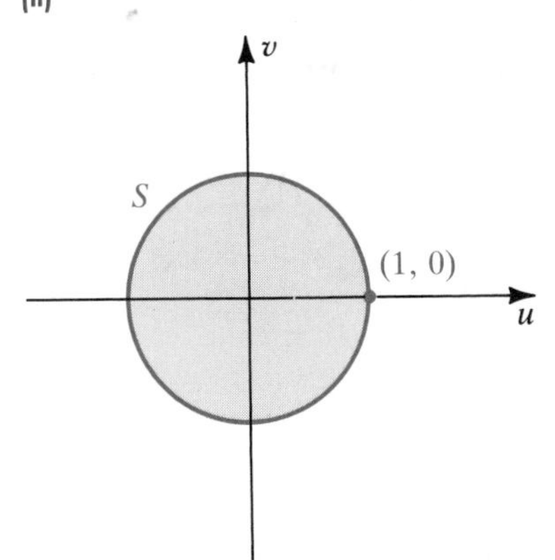

In this case the Jacobian is $\partial(x, y)/\partial(u, v) = ab > 0$, and the region S that corresponds to R is the circular disk $u^2 + v^2 \leq 1$ (see Figure 17.89(ii)). Applying Corollary (17.36) gives us

$$V = abh_M \iint_S \cos\left(\frac{\pi}{2}\sqrt{u^2 + v^2}\right) du\, dv.$$

We may evaluate the last integral by means of the polar coordinate substitution $u = r\cos\theta$, $v = r\sin\theta$. This gives us

$$V = abh_M \int_0^{2\pi}\int_0^1 \cos\left(\frac{\pi}{2}r\right) r\, dr\, d\theta.$$

Using integration by parts, or Formula 83 of the table of integrals, we have

$$\begin{aligned} V &= abh_M \int_0^{2\pi}\left[\frac{4}{\pi^2}\cos\left(\frac{\pi}{2}r\right) + \frac{2}{\pi}r\sin\left(\frac{\pi}{2}r\right)\right]_0^1 d\theta \\ &= abh_M \int_0^{2\pi}\left(\frac{2}{\pi} - \frac{4}{\pi^2}\right) d\theta \\ &= abh_M\left(\frac{2}{\pi} - \frac{4}{\pi^2}\right)2\pi \approx 1.4535abh_M. \end{aligned}$$

This formula can be used to estimate lake volume from the three measurements a, b, and h_M.

The area of the elliptical region R is πab. Applying an analogue of Definition (5.29) to double integrals, we obtain the average depth:

$$h_{av} = \frac{1}{\pi ab}\iint_R |f(x, y)|\, dA.$$

Hence
$$h_{av} \approx \frac{1}{\pi ab}(1.4535)abh_M \approx 0.4627h_M.$$

A study of 107 lakes worldwide yielded an average value of h_{av}/h_M of 0.467.

EXERCISES 17.9

Exer. 1–8: Let T be the transformation from the xy-plane to the uv-plane determined by the given formulas. (a) Describe the u-curves and the v-curves. (b) Find formulas $x = F(u, v)$, $y = G(u, v)$ that specify T^{-1}.

1 $u = 3x$, $v = 5y$

2 $u = \frac{1}{2}y$, $v = \frac{1}{3}x$

3 $u = x - y$, $v = 2x + 3y$

4 $u = -5x + 4y$, $v = 2x - 3y$

5 $u = x^3$, $v = x + y$

6 $u = x + 1$, $v = 2 - y^3$

7 $u = e^x$, $v = e^y$

8 $u = e^{2y}$, $v = e^{-3x}$

Exer. 9–12: Let T be the indicated transformation of coordinates (see Exercises 1–4). Find the curve in the uv-plane that corresponds to the curve in the xy-plane specified in (a) and (b).

9 $u = 3x$, $v = 5y$

(a) the rectangle with vertices (0, 0), (0, 1), (2, 1), (2, 0)

(b) the circle $x^2 + y^2 = 1$

10 $u = \frac{1}{2}y, \quad v = \frac{1}{3}x$

(a) the triangle with vertices $(0, 0), (3, 6), (9, 4)$

(b) the line $3x - 2y = 4$

11 $u = x - y, \quad v = 2x + 3y$

(a) the triangle with vertices $(0, 0), (0, 1), (2, 0)$

(b) the line $x + 2y = 1$

12 $u = -5x + 4y, \quad v = 2x - 3y$

(a) the square with vertices $(0, 0), (1, -1), (2, 0), (1, 1)$

(b) the circle $x^2 + y^2 = 1$

Exer. 13–16: Find the Jacobian $\dfrac{\partial(x, y)}{\partial(u, v)}$.

13 $x = u^2 - v^2, \quad y = 2uv$

14 $x = e^u \sin v, \quad y = e^u \cos v$

15 $x = ve^{-2u}, \quad y = u^2 e^{-v}$

16 $x = \dfrac{u}{u^2 + v^2}, \quad y = \dfrac{v}{u^2 + v^2}$

Exer. 17–18: Find the Jacobian $\dfrac{\partial(x, y, z)}{\partial(u, v, w)}$.

17 $x = 2u + 3v - w, \quad y = v - 5w, \quad z = u + 4w$

18 $x = u^2 + vw, \quad y = 2v + u^2 w, \quad z = uvw$

Exer. 19–22: Express the integral as an iterated double integral over a region S in the uv-plane by making the indicated change of variables.

19 $\displaystyle\iint_R (y - x)\, dx\, dy$

boundary of R: $y = 2x$, $y = 0$, $x = 2$
change of variables: $x = u + v$, $y = 2v$

20 $\displaystyle\iint_R (3x - 4y)\, dx\, dy$

boundary of R: $y = 3x$, $y = \frac{1}{2}x$, $x = 4$
change of variables: $x = u - 2v$, $y = 3u - v$

21 $\displaystyle\iint_R (\tfrac{1}{4}x^2 + \tfrac{1}{9}y^2)\, dx\, dy$

boundary of R: $\frac{1}{4}x^2 + \frac{1}{9}y^2 = 1$
change of variables: $x = 2u$, $y = 3v$

22 $\displaystyle\iint_R xy\, dx\, dy$

boundary of R: $y = 2\sqrt{1 - x}$, $x = 0$, $y = 0$
change of variables: $x = u^2 - v^2$, $y = 2uv$

Exer. 23–28: Evaluate the integral by making the indicated change of variables.

23 $\displaystyle\iint_R (x - y)^2 \cos^2 (x + y)\, dx\, dy$

boundary of R: the square with vertices $(0, 1), (1, 2), (2, 1), (1, 0)$
change of variables: $u = x - y$, $v = x + y$

24 $\displaystyle\iint_R \sin \frac{y - x}{y + x}\, dx\, dy$

boundary of R: the trapezoid with vertices $(1, 1), (2, 2), (4, 0), (2, 0)$
change of variables: $u = y - x$, $v = y + x$

25 $\displaystyle\iint_R (x^2 + 2y^2)\, dx\, dy$

boundary of R: $xy = 1$, $xy = 2$, $y = |x|$, $y = 2x$
change of variables: $x = u/v$, $y = v$

26 $\displaystyle\iint_R \frac{1}{(4x - 4y + 1)^2}\, dx\, dy$

boundary of R: $x = \sqrt{-y}$, $x = y$, $x = 1$
change of variables: $x = u + v$, $y = v - u^2$

27 $\displaystyle\iint_R \frac{2y + x}{y - 2x}\, dx\, dy$

boundary of R: the trapezoid with vertices $(-1, 0), (-2, 0), (0, 4), (0, 2)$
change of variables: $u = y - 2x$, $v = 2y + x$

28 $\displaystyle\iint_R (\sqrt{x - 2y} + \tfrac{1}{4}y^2)\, dx\, dy$

boundary of R: the triangle with vertices $(0, 0), (4, 0), (4, 2)$
change of variables: $u = \frac{1}{2}y$, $v = x - 2y$

29 Because of rotation, the earth is not perfectly spherical but is slightly flattened at the poles, with a polar radius of 6356 kilometers and equatorial radius of 6378 kilometers. As a result, the shape of the earth's surface can be approximated by the ellipsoid

$$\frac{x^2}{a^2} + \frac{y^2}{b^2} + \frac{z^2}{c^2} = 1,$$

with $a = b = 6378$ and $c = 6356$. Estimate the volume of the earth. (*Hint:* Let $x = au$, $y = bv$, and $z = cw$.)

30 A balloon has the shape of the sphere $x^2 + y^2 + z^2 = 9$, where x, y, and z are in meters. The balloon is slowly heated at a constant pressure of one atmosphere, that is, 1.01×10^5 N/m^2, and expands until it has the shape of the ellipsoid $(x^2/10) + (y^2/16) + (z^2/12) = 1$. For an ideal gas, the work W done by the gas is given by

$$W = \iiint_{V_1}^{V_2} P\, dV,$$

where P is the pressure and V_1 and V_2 are the initial and final volumes. Use Exercise 29 to find W.

31 Refer to Exercise 41 of Section 17.8. It is very difficult to find formulas that describe the final position of the hand if five or six joints of a robotic arm are moved. The process can be simplified by using the three *Euler angles* ϕ, θ, and ψ for rotations of coordinate axes. Referring to the figure, first rotate the xy-plane about the z-axis through the angle ϕ, obtaining the new x''- and y''-axes. Second, rotate the $x''z$-plane about the y''-axis through the angle θ, obtaining the new x'''- and z'-axes. Third, rotate the $x'''y''$-plane about the z'-axis through the angle ψ, obtaining the new x'- and y'-axes. If the coordinates of point P in the hand are (x, y, z) in the xyz-system, then its coordinates (x', y', z') in the transformed system are given by

$$\begin{aligned} x' &= (\cos\phi\cos\theta\cos\psi - \sin\phi\sin\psi)x \\ &\quad + (-\cos\phi\cos\theta\sin\psi - \sin\phi\cos\psi)y \\ &\quad + (\cos\phi\sin\theta)z \end{aligned}$$

$$\begin{aligned} y' &= (\sin\phi\cos\theta\cos\psi + \cos\phi\sin\psi)x \\ &\quad + (-\sin\phi\cos\theta\sin\psi + \cos\phi\cos\psi)y \\ &\quad + (\sin\phi\sin\theta)z \end{aligned}$$

$$z' = (-\sin\theta\cos\psi)x + (\sin\theta\sin\psi)y + (\cos\theta)z.$$

(a) A robot's hand is positioned at $P(\sqrt{2}, 0, 1)$ and then rotated through $\phi = 45°$. Find the new location of the hand.

(b) The hand is next rotated through $\theta = 90°$. Find the new location of the hand.

(c) Finally, the hand is rotated through $\psi = 90°$. Find the final location of the hand.

EXERCISE 31

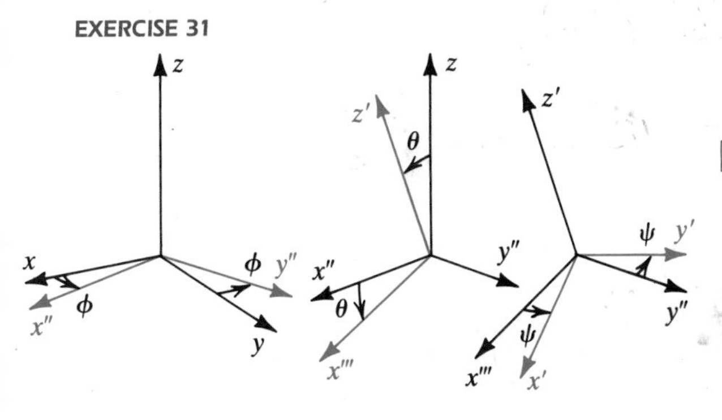

32 If a robot arm has several joints, it may be possible to position its hand in a specific location in more than one way by rotating the joints differently. There will be exactly one choice for the rotation if and only if the Jacobian of the transformation of coordinates is nonzero.

(a) Find the Jacobian $\partial(x', y', z')/\partial(x, y, z)$ in Exercise 31 when $\phi = 45°$, $\theta = 90°$, and $\psi = 90°$.

(b) Can the hand be positioned in the same location by using different values of ϕ, θ, and ψ?

33 Verify that, for the spherical coordinate transformation,

$$\frac{\partial(x, y, z)}{\partial(\rho, \phi, \theta)} = \rho^2 \sin\phi.$$

34 Use (17.38) to derive a formula for

$$\iiint_R F(x, y, z)\, dx\, dy\, dz$$

for a transformation from rectangular to cylindrical coordinates.

35 If the transformation of coordinates $x = f(u, v)$, $y = g(u, v)$ is one-to-one, show that

$$\frac{\partial(x, y)}{\partial(u, v)} \frac{\partial(u, v)}{\partial(x, y)} = 1.$$

Hint: Use the following property of determinants:

$$\begin{vmatrix} a & b \\ c & d \end{vmatrix} \begin{vmatrix} p & q \\ r & s \end{vmatrix} = \begin{vmatrix} ap + br & aq + bs \\ cp + dr & cq + ds \end{vmatrix}$$

36 Given the transformation $x = f(u, v)$, $y = g(u, v)$ and the transformation $u = h(r, s)$, $v = k(r, s)$, show that

$$\frac{\partial(x, y)}{\partial(u, v)} \frac{\partial(u, v)}{\partial(r, s)} = \frac{\partial(x, y)}{\partial(r, s)}.$$

(*Hint:* Use the hint given in Exercise 35.)

c **Exer. 37–38: Let T be the indicated transformation from the uv-plane to the xy-plane. Graph the points (u, v) such that $\partial(x, y)/\partial(u, v) = 0$ for the indicated values of u. (This graph shows where T is not one-to-one.)**

37 $x = \frac{1}{4}v^4 + u^2$, $\quad y = u - \cos v$; $\quad -\pi \le u \le \pi$

38 $x = \frac{1}{20}u^4 - 10\cos u + \frac{1}{2}v^2$, $\quad y = u + v$; $\quad -5 \le u \le 5$

17.10 REVIEW EXERCISES

Exer. 1–6: Evaluate the iterated integral.

1 $\int_{-1}^{0} \int_{x+1}^{x^3} (x^2 - 2y)\, dy\, dx$

2 $\int_{1}^{2} \int_{1}^{y^2} \frac{1}{y}\, dx\, dy$

3 $\int_{0}^{3} \int_{r}^{r^2+1} r\, d\theta\, dr$

4 $\int_{2}^{0} \int_{0}^{z^2} \int_{x}^{z} (x + z)\, dy\, dx\, dz$

5 $\int_{0}^{2} \int_{\sqrt{y}}^{1} \int_{z^2}^{y} xy^2z^3\, dx\, dz\, dy$

6 $\int_0^{\pi/2} \int_0^{\pi/4} \int_0^{a \cos \phi} \rho^2 \sin \phi \, d\rho \, d\phi \, d\theta$

Exer. 7–10: If $f(x, y)$ is an arbitrary continuous function, express $\iint_R f(x, y) \, dA$ as an interated double integral, where R is the region bounded by the graphs of the equations.

7 $x^2 - y^2 = 4, \quad x = 4$

8 $x^2 - y^2 = 4, \quad y = 4, \quad y = 0$

9 $y^2 = 4 + x, \quad y^2 = 4 - x$

10 $y = -x^2 + 4, \quad y = 3x^2$

Exer. 11–14: Sketch the region of integration for the iterated integral.

11 $\int_{-1}^{1} \int_{e^y}^{y^3} dx \, dy$

12 $\int_{-1}^{0} \int_{x}^{-x^2} dy \, dx$

13 $\int_0^{\pi} \int_0^{2+4\cos\theta} r \, dr \, d\theta$

14 $\int_0^2 \int_{-\sqrt{y^2+1}}^{\sqrt{y^2+1}} \int_0^{\sqrt{1-x^2+y^2}} dz \, dx \, dy$

Exer. 15–16: Reverse the order of integration and evaluate the resulting integral.

15 $\int_0^3 \int_{y^2}^{9} y e^{-x^2} \, dx \, dy$

16 $\int_0^1 \int_x^{\sqrt{x}} e^{x/y} \, dy \, dx$

Exer. 17–18: Find the volume of the solid that is under the graph of the equation and over the region R in the xy-plane.

17 $z = xy^2$; R is the rectangle with vertices $(1, 1)$, $(2, 1)$, $(1, 3)$, $(2, 3)$.

18 $z = 7 - x^2 - y^2$; R is the circle $x^2 + y^2 = 4$.

19 Find the surface area of the part of the cone $z = \sqrt{x^2 + y^2}$ that is inside the cylinder $x^2 + y^2 = 4x$.

20 Find the surface area of the part of the hyperboloid $x^2 + y^2 - z^2 = 1$ that lies over the circle $x^2 + y^2 = 1$.

21 Use a double integral to find the area of the region bounded by the polar axis and the graphs of $r = e^{\theta}$ and $r = 2$ from $\theta = 0$ to $\theta = \ln 2$.

22 Use polar coordinates to evaluate

$$\int_{-a}^{0} \int_{-\sqrt{a^2-x^2}}^{0} \sqrt{x^2 + y^2} \, dy \, dx.$$

23 Use spherical coordinates to evaluate

$$\int_0^4 \int_0^{\sqrt{16-x^2}} \int_{\sqrt{x^2+y^2}}^{\sqrt{32-x^2-y^2}} \sqrt{x^2 + y^2 + z^2} \, dz \, dy \, dx.$$

24 Let f be an arbitrary continuous function of x, y, and z, and let Q be the region bounded by the paraboloid $y = x^2 + 4z^2$ and the plane $y = 4$. Express the triple integral $\iiint_Q f(x, y, z) \, dV$ as an iterated triple integral in six different ways.

Exer. 25–26: Find the mass and the center of mass of the lamina that has the shape of the region bounded by the graphs of the equations and has the indicated density.

25 $y = x$, $y = 2x$, $x = 3$; area mass density at $P(x, y)$ is directly proportional to the distance from the y-axis to P.

26 $y^2 = x$, $x = 4$; area mass density at $P(x, y)$ is directly proportional to the distance from the line $x = -1$ to P.

27 A lamina has the shape of the region that is inside the limacon $r = 2 + \sin \theta$ and outside the circle $r = 1$. The area mass density at $P(r, \theta)$ is inversely proportional to the distance from the pole to P. Find the mass.

28 A lamina has the shape of the region bounded by the graphs of $y = x^2$ and $y = x^3$, and the area mass density at $P(x, y)$ is directly proportional to the distance from the y-axis to P. Find I_x, I_y, and I_O.

29 A lamina has the shape of a right triangle with sides of lengths a, b, and $\sqrt{a^2 + b^2}$. The area mass density is directly proportional to the distance from the side of length a. Find the moment of inertia with respect to a line containing the side of length a.

30 A lamina has the shape of the region between concentric circles of radii a and b, with $a < b$. The area mass density at a point P is directly proportional to the distance from the center of the circles to P. Use polar coordinates to find the moment of inertia with respect to a line through the center.

31 Use triple integrals to find the volume and centroid of the solid bounded by the graphs of the cylinder $z = x^2$ and the planes $y = 0$ and $y + z = 4$.

32 A solid is bounded by the paraboloid $z = 9x^2 + y^2$ and the plane $z = 9$. The density at $P(x, y, z)$ is inversely proportional to the square of the distance from $(0, 0, -1)$ to P. Set up an iterated triple integral that can be used to find the moment of inertia with respect to the z-axis.

33 A solid is bounded by the hyperboloid $x^2 - y^2 + z^2 = 1$ and the planes $y = 0$ and $y = 4$. The density at $P(x, y, z)$ is directly proportional to the distance from the y-axis to P. Set up an iterated triple integral that can be used to find the moment of inertia with respect to the y-axis.

34 A homogeneous solid of density δ is bounded by the paraboloid $z = 9 - x^2 - y^2$, the interior of the cylinder $x^2 + y^2 = 4$, and the xy-plane. Use cylindrical coordinates to find

(a) the mass (b) the center of mass

(c) the moment of inertia with respect to the z-axis

35 A spherical solid has radius a, and the density at any point is directly proportional to its distance from the center of the sphere. Use spherical coordinates to find the mass.

Exer. 36–37: Complete the table.

	Rectangular coordinates	Cylindrical coordinates	Spherical coordinates
36	$(2, -2, 1)$	______	______
37	______	______	$(12, \pi/6, 3\pi/4)$

Exer. 38–43: Change the equation to rectangular coordinates, and describe its graph in three dimensions.

38 $r = -\csc\theta$

39 $z + 3r^2 = 9$

40 $z = \frac{1}{4}r$

41 $\rho \sin\phi = 4$

42 $\rho \sin\phi \cos\theta = 1$

43 $\rho^2 - 3\rho = 0$

Exer. 44–47: Change the equation to (a) cylindrical coordinates and (b) spherical coordinates.

44 $x^2 + y^2 = 1$

45 $z = x^2 - y^2$

46 $x^2 + y^2 + z^2 - 2z = 0$

47 $2x + y - 3z = 4$

Exer. 48–49: Express the area of the region in terms of one or more iterated double integrals that involve only the order of integration (a) $dy\,dx$, (b) $dx\,dy$, and (c) $dr\,d\theta$.

48

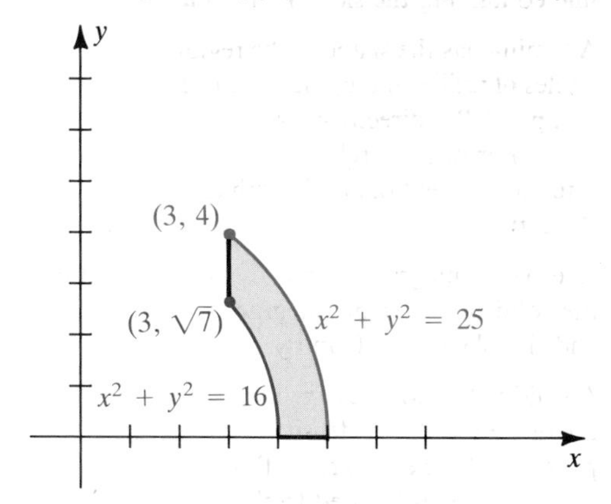

49

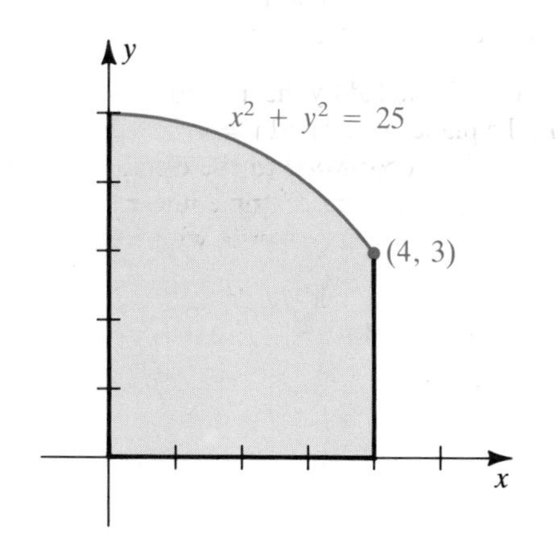

Exer. 50–51: Express the area of the region in terms of one or more iterated triple integrals in (a) rectangular coordinates, (b) cylindrical coordinates, and (c) spherical coordinates.

50

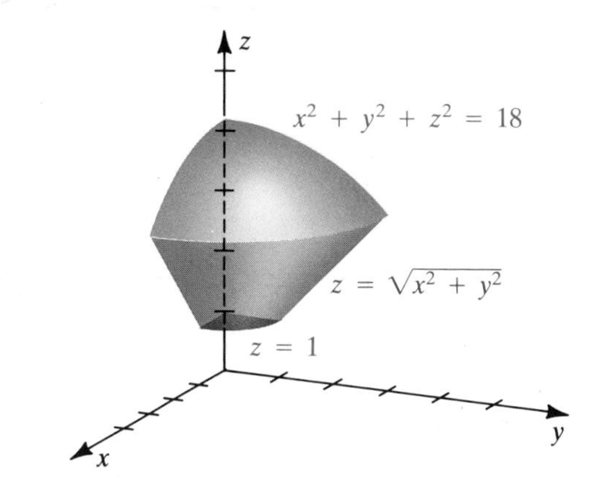

51

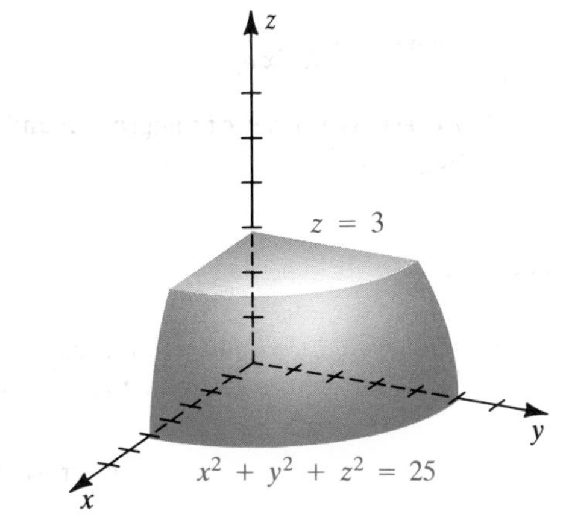

52 If a transformation T of the xy-plane to the uv-plane is given by $u = 2x + 5y$, $v = 3x - 4y$, find

(a) the u-curves (b) the v-curves

(c) T^{-1} (d) $\dfrac{\partial(x, y)}{\partial(u, v)}$

(e) the curve in the uv-plane corresponding to the line $ax + by + c = 0$

(f) the curve in the uv-plane corresponding to the circle $x^2 + y^2 = a^2$

53 Evaluate the integral

$$\int_0^1 \int_y^{2-y} e^{(x-y)/(x+y)}\,dx\,dy$$

by means of the change of variables $u = x - y$, $v = x + y$.

CHAPTER 18

VECTOR CALCULUS

INTRODUCTION

In this chapter we extend previous methods to new concepts and obtain results that have many uses in the sciences. We first discuss *vector fields*. The principal applications involve *velocity fields* and *force fields* of either solids, liquids, or gases—so named because with each particle in the substance we associate a velocity vector or force vector. Of primary importance are *conservative* vector fields, because they include the gravitational and electromagnetic fields that occur throughout the universe.

Line integrals, which are generalizations of the definite integral, enable us to find the work done as a particle moves through a force field. *Surface integrals* are extensions of double integrals and can be applied to problems about *flux*—the rate of flow of energy, fluids, and gases across a surface. Line and surface integrals can be used to interpret the *divergence* and *curl* of a vector field. Divergence measures the rate at which a physical entity, such as heat, is absorbed or generated near a point. Curl is used to describe rotational aspects of a vector field, such as the direction and magnitude of wind circulation inside a tornado.

The final sections contain discussions of the *divergence theorem* and *Stokes' theorem*, which relate divergence to flux and work to curl, respectively. These two outstanding results have far-reaching consequences in science and engineering.

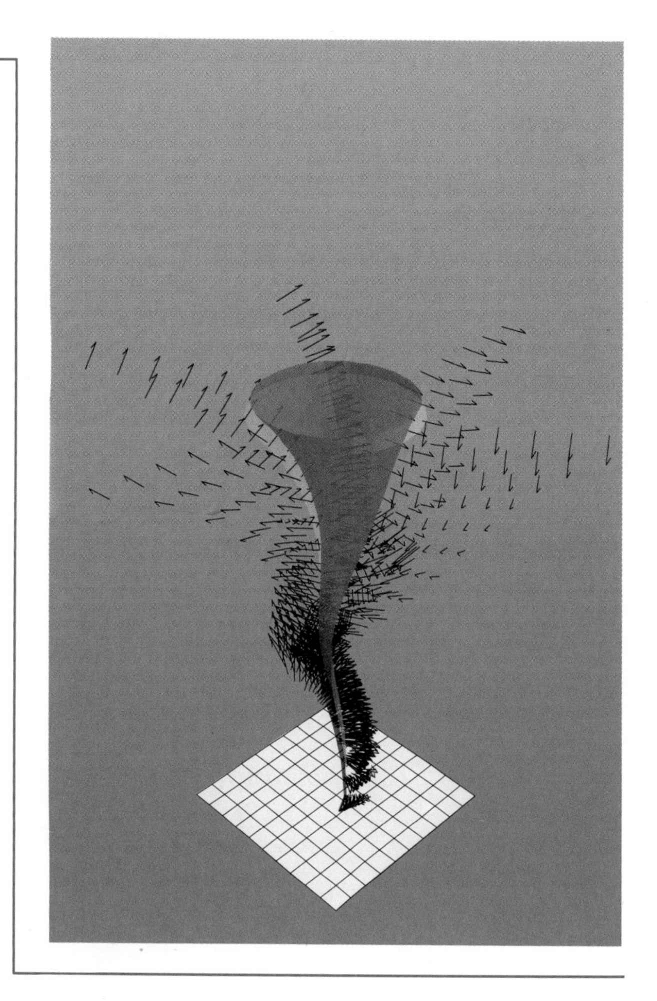

18.1 VECTOR FIELDS

FIGURE 18.1

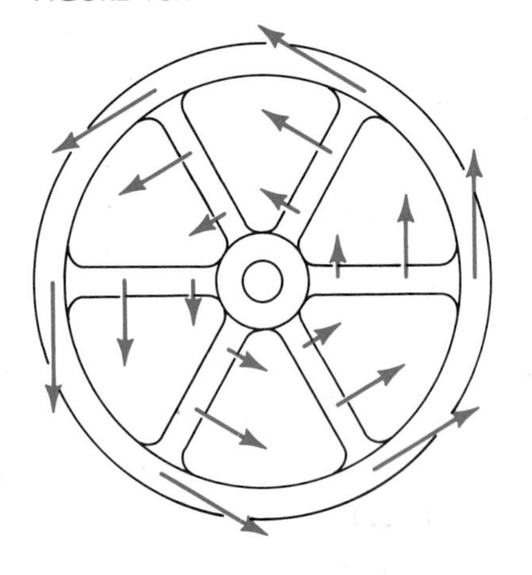

FIGURE 18.2

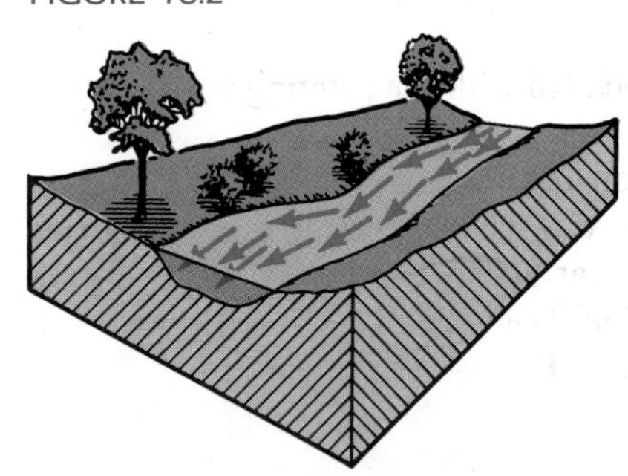

If to each point K in a region there is assigned exactly one vector having initial point K, then the collection of all such vectors is a **vector field**. Figure 18.1 illustrates a vector field determined by a wheel rotating about an axle. To each point on the wheel there corresponds a velocity vector. A vector field of this type is a **velocity field**. A velocity field may also be determined by the flow of water or of wind. Figure 18.2 displays velocity vectors associated with fluid particles moving in a stream. Sketches of this type show only a few vectors of the vector field. Remember that a vector is assigned to *every* point in the region.

A **force field** is a vector field in which a force vector is assigned to each point. Force fields are common in the study of mechanics and electricity. In our illustrations we have assumed that the vectors are independent of time. Such **steady vector fields** are the only types we shall consider in this chapter.

Given a vector field, let us introduce an xyz-coordinate system and denote the vector assigned to the point $K(x, y, z)$ by $\mathbf{F}(x, y, z)$ (see Figure 18.3). Since the components of $\mathbf{F}(x, y, z)$ depend on the coordinates x, y, and z of K, we may write

$$\mathbf{F}(x, y, z) = M(x, y, z)\mathbf{i} + N(x, y, z)\mathbf{j} + P(x, y, z)\mathbf{k},$$

FIGURE 18.3

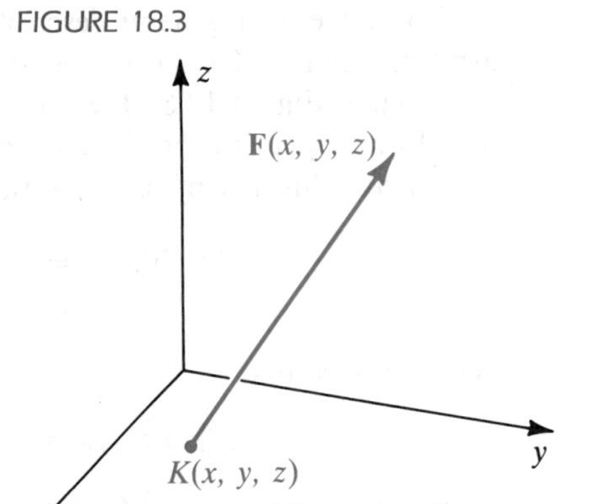

FIGURE 18.4

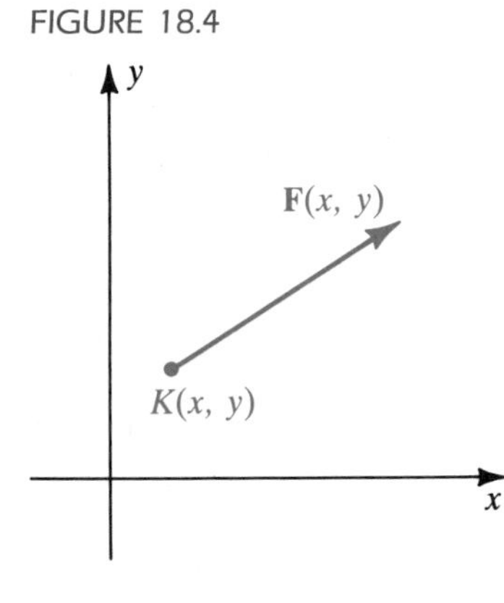

where M, N, and P are scalar functions. Conversely, every such equation determines a vector field. Thus, a vector field may be regarded as a function $\mathbf{F}$ of the type stated in the following definition.

Definition **(18.1)**

> A **vector field in three dimensions** is a function $\mathbf{F}$ whose domain D is a subset of $\mathbb{R}^3$ and whose range is a subset of V_3. If (x, y, z) is in D, then
>
> $$\mathbf{F}(x, y, z) = M(x, y, z)\mathbf{i} + N(x, y, z)\mathbf{j} + P(x, y, z)\mathbf{k},$$
>
> where M, N, and P are scalar functions.

As a special case of (18.1), if the domain D corresponds to a region in the xy-plane and if the range is a subset of V_2 (see Figure 18.4), then a

vector field F in two dimensions is given by

$$\mathbf{F}(x, y) = M(x, y)\mathbf{i} + N(x, y)\mathbf{j},$$

where M and N are scalar functions. We sometimes describe a vector field by sketching vectors that correspond to $\mathbf{F}(x, y, z)$ or to $\mathbf{F}(x, y)$ for typical points (x, y, z) or (x, y), as illustrated in the following example.

FIGURE 18.5

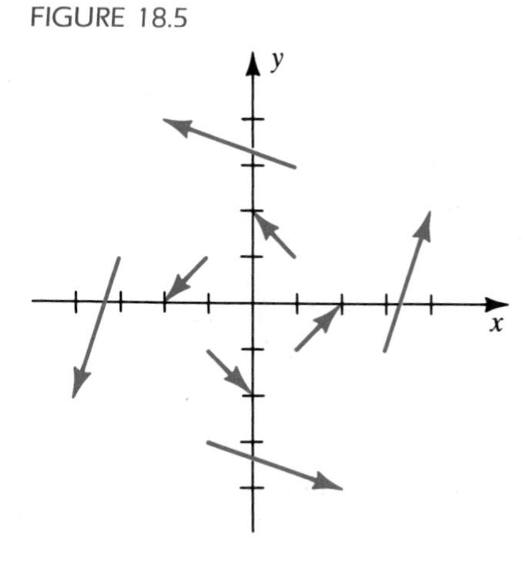

EXAMPLE 1 Describe the vector field $\mathbf{F}$ if $\mathbf{F}(x, y) = -y\mathbf{i} + x\mathbf{j}$.

SOLUTION The vectors $\mathbf{F}(x, y)$ associated with several points (x, y) are listed in the following table and are sketched in Figure 18.5.

(x, y)	$\mathbf{F}(x, y)$
$(1, 1)$	$-\mathbf{i} + \mathbf{j}$
$(-1, 1)$	$-\mathbf{i} - \mathbf{j}$
$(-1, -1)$	$\mathbf{i} - \mathbf{j}$
$(1, -1)$	$\mathbf{i} + \mathbf{j}$

(x, y)	$\mathbf{F}(x, y)$
$(1, 3)$	$-3\mathbf{i} + \mathbf{j}$
$(-3, 1)$	$-\mathbf{i} - 3\mathbf{j}$
$(-1, -3)$	$3\mathbf{i} - \mathbf{j}$
$(3, -1)$	$\mathbf{i} + 3\mathbf{j}$

The vectors are similar to those associated with the rotating wheel in Figure 18.1.

FIGURE 18.6

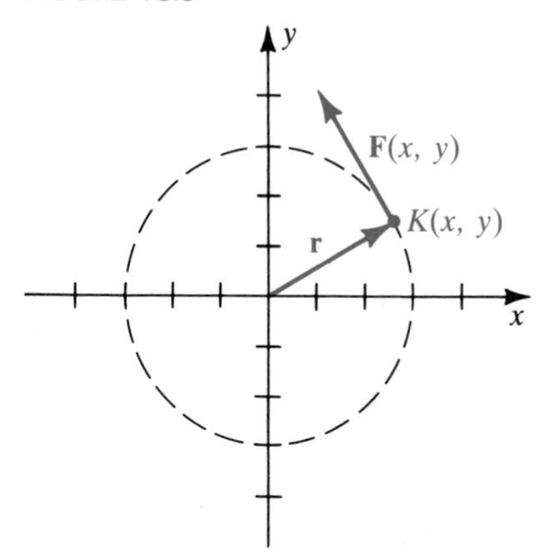

To arrive at a general description of the vector field $\mathbf{F}$, consider an arbitrary point $K(x, y)$ and let $\mathbf{r} = x\mathbf{i} + y\mathbf{j}$ be the position vector for $K(x, y)$ (see Figure 18.6). It appears that $\mathbf{F}(x, y)$ is orthogonal to $\mathbf{r}$ and is therefore tangent to the circle of radius $\|\mathbf{r}\|$ with center at the origin. We may prove this fact by showing that $\mathbf{r} \cdot \mathbf{F}(x, y) = 0$, as follows:

$$\begin{aligned} \mathbf{r} \cdot \mathbf{F}(x, y) &= (x\mathbf{i} + y\mathbf{j}) \cdot (-y\mathbf{i} + x\mathbf{j}) \\ &= -xy + yx = 0 \end{aligned}$$

We also note that

$$\|\mathbf{F}(x, y)\| = \sqrt{y^2 + x^2} = \|\mathbf{r}\|.$$

Hence the magnitude of $\mathbf{F}(x, y)$ equals the radius of the circle. This implies that as the point $K(x, y)$ moves away from the origin, the magnitude of $\mathbf{F}(x, y)$ increases, as is the case for the rotating wheel in Figure 18.1.

The next definition introduces one of the most important vector fields that occurs in the physical sciences.

Definition **(18.2)**

> Let $\mathbf{r} = x\mathbf{i} + y\mathbf{j} + z\mathbf{k}$ be the position vector for (x, y, z), and let $\mathbf{u} = (1/\|\mathbf{r}\|)\mathbf{r}$ denote the unit vector that has the same direction as $\mathbf{r}$. A vector field $\mathbf{F}$ is an **inverse square field** if
>
> $$\mathbf{F}(x, y, z) = \frac{c}{\|\mathbf{r}\|^2}\mathbf{u} = \frac{c}{\|\mathbf{r}\|^3}\mathbf{r},$$
>
> where c is a scalar.

EXAMPLE 2 Describe the inverse square field $\mathbf{F}(x, y, z)$ of (18.2).

SOLUTION Since $\mathbf{r} = x\mathbf{i} + y\mathbf{j} + z\mathbf{k}$,

$$\mathbf{F}(x, y, z) = \frac{c}{\|\mathbf{r}\|^3}\mathbf{r} = \frac{c}{(x^2 + y^2 + z^2)^{3/2}}(x\mathbf{i} + y\mathbf{j} + z\mathbf{k}).$$

We may write the last expression as in (18.1); however, it is simpler to analyze the vectors in the field by using $\mathbf{r}$. If $c < 0$, then $\mathbf{F}(x, y, z)$ is a negative scalar multiple of $\mathbf{r}$, and hence the direction of $\mathbf{F}(x, y, z)$ is toward the origin O. Moreover, since

$$\|\mathbf{F}(x, y, z)\| = \frac{|c|}{\|\mathbf{r}\|^3}\|\mathbf{r}\| = \frac{|c|}{\|\mathbf{r}\|^2},$$

the magnitude of $\mathbf{F}(x, y, z)$ is inversely proportional to the square of the distance from O to the point (x, y, z). This means that as the point $K(x, y, z)$ moves away from the origin, the length of the vector $\mathbf{F}(x, y, z)$ decreases. Some typical vectors in $\mathbf{F}$ are sketched in Figure 18.7(i) for $c < 0$. If $c > 0$, the vectors point *away* from the origin, as in Figure 18.7(ii).

FIGURE 18.7
Inverse square vector fields

(i) $c < 0$

(ii) $c > 0$

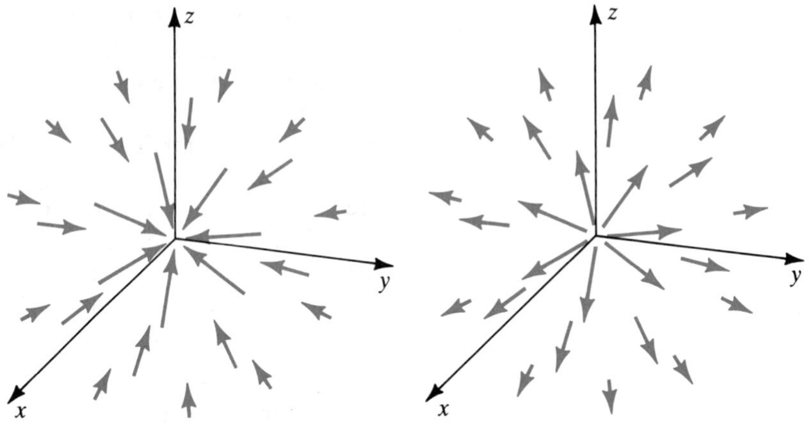

The force of gravity determines inverse square vector fields. According to **Newton's law of universal gravitation**, if a particle of mass m_0 is located at the origin of an xyz-coordinate system, then the force exerted on a particle of mass m located at $K(x, y, z)$ is

$$\mathbf{F}(x, y, z) = -G\frac{m_0 m}{\|\mathbf{r}\|^2}\mathbf{u},$$

where G is a gravitational constant, $\mathbf{r}$ is the position vector for the point K, and $\mathbf{u} = (1/\|\mathbf{r}\|)\mathbf{r}$ is a unit vector.

Inverse square fields also occur in electrical theory. **Coulomb's law** states that if a point charge of Q coulombs is at the origin, then the force $\mathbf{F}(x, y, z)$ it exerts on a point charge of q coulombs located at $K(x, y, z)$

is

$$\mathbf{F}(x, y, z) = c\frac{Qq}{\|\mathbf{r}\|^2}\mathbf{u},$$

where c is a constant and $\mathbf{r}$ and $\mathbf{u}$ are as in Definition (18.2). Note that Coulomb's law has the same form as Newton's law of universal gravitation.

If f is a function of three variables, then, as in (16.31), the *gradient* of $f(x, y, z)$ is the vector field defined as follows:

$$\nabla f(x, y, z) = f_x(x, y, z)\mathbf{i} + f_y(x, y, z)\mathbf{j} + f_z(x, y, z)\mathbf{k}$$

Vector fields that are gradients of scalar functions are given the following special name.

Definition **(18.3)**

A vector field $\mathbf{F}$ is **conservative** if

$$\mathbf{F}(x, y, z) = \nabla f(x, y, z)$$

for some scalar function f.

FIGURE 18.8

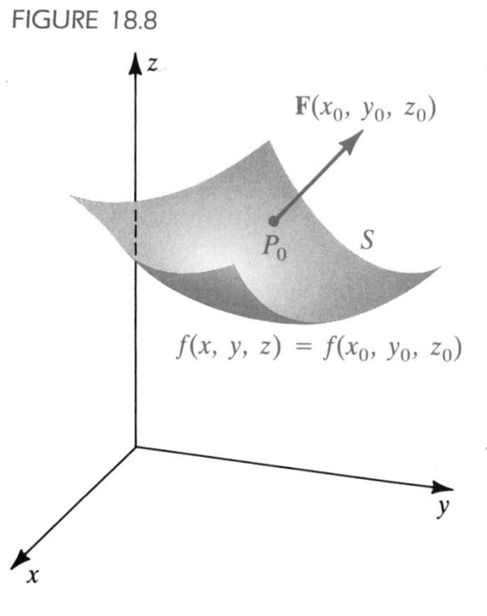

If $\mathbf{F}$ is conservative, then the function f in (18.3) is called a **potential function** for $\mathbf{F}$ and $f(x, y, z)$ is the **potential** at the point (x, y, z). Recall from Theorem (16.36) that if (x_0, y_0, z_0) is in the domain of $\mathbf{F} = \nabla f$, then the gradient vector $\nabla f(x_0, y_0, z_0) = \mathbf{F}(x_0, y_0, z_0)$ is normal to the level surface S of f that contains $P_0(x_0, y_0, z_0)$. The surface S is the graph of the equation $f(x, y, z) = f(x_0, y_0, z_0)$. Thus, *every vector* $\mathbf{F}(x_0, y_0, z_0)$ *in a conservative vector field is normal to the level surface of a potential function* f *for* $\mathbf{F}$ *that contains* $P_0(x_0, y_0, z_0)$. A typical case is illustrated in Figure 18.8. In Section 16.6 we proved that $\|\mathbf{F}(x_0, y_0, z_0)\|$ is the maximum rate of change of the potential at P_0. In applications where $f(x, y, z)$ is the temperature (or potential) at (x, y, z), this maximum rate of change of f is normal to the isothermal (or equipotential) surface containing P_0.

The next theorem is important in applications.

Theorem **(18.4)**

Every inverse square vector field is conservative.

PROOF If $\mathbf{F}$ is an inverse square field, then, as in the solution to Example 2,

$$\mathbf{F}(x, y, z) = \frac{cx}{(x^2 + y^2 + z^2)^{3/2}}\mathbf{i} + \frac{cy}{(x^2 + y^2 + z^2)^{3/2}}\mathbf{j} + \frac{cz}{(x^2 + y^2 + z^2)^{3/2}}\mathbf{k},$$

where c is a constant. By Definition (18.3), if $\mathbf{F}$ is conservative, then $\mathbf{F}(x, y, z) = \nabla f(x, y, z)$ for some scalar function f, and the components of $\mathbf{F}$ are $f_x(x, y, z)$, $f_y(x, y, z)$, and $f_z(x, y, z)$, respectively. Partial integrations of these components with respect to x, y, and z, respectively, *suggest* that

$$f(x, y, z) = \frac{-c}{(x^2 + y^2 + z^2)^{1/2}}.$$

Calculation of the partial derivatives of f proves that this guess is correct. This gives us the following:

$$\mathbf{F}(x, y, z) = \nabla f(x, y, z) = \nabla\left(\frac{-c}{r}\right),$$

where $r = \|\mathbf{r}\| = (x^2 + y^2 + z^2)^{1/2}$. ■

The level surfaces for the potential function f of the inverse square vector field $\mathbf{F}$ in the proof of Theorem (18.4) are graphs of equations of the form

$$\frac{-c}{(x^2 + y^2 + z^2)^{1/2}} = k,$$

where $k < 0$. Squaring both sides and taking reciprocals gives us

$$x^2 + y^2 + z^2 = \frac{c^2}{k^2}.$$

Thus, the level surfaces are spheres with centers at the origin O. The vectors $\mathbf{F}(x, y, z)$ are orthogonal to these spheres and hence are directed toward O or away from O, as in Figure 18.7.

In physics, the potential function of a conservative vector field $\mathbf{F}$ is defined as the function p such that $\mathbf{F}(x, y, z) = -\nabla p(x, y, z)$. In this case, letting $p = -f$ in the proof of Theorem (18.4), we obtain $\mathbf{F}(x, y, z) = \nabla(c/r)$. We shall say more about conservative vector fields later in this chapter. In particular, in Section 18.3 we discuss a method for finding $f(x, y, z)$ when given $\mathbf{F}(x, y, z) = \nabla f(x, y, z)$.

The vector differential operator ∇ in three dimensions is

$$\nabla = \mathbf{i}\frac{\partial}{\partial x} + \mathbf{j}\frac{\partial}{\partial y} + \mathbf{k}\frac{\partial}{\partial z}.$$

From (16.31), if ∇ operates on a *scalar* function f, it produces the gradient of f,

$$\text{grad } f = \nabla f = \frac{\partial f}{\partial x}\mathbf{i} + \frac{\partial f}{\partial y}\mathbf{j} + \frac{\partial f}{\partial z}\mathbf{k}.$$

We shall next use ∇ as an operator on a *vector* field to define another vector field called the *curl* of $\mathbf{F}$, which we denote by curl $\mathbf{F}$ or $\nabla \times \mathbf{F}$.

Definition **(18.5)**

Let $\mathbf{F}(x, y, z) = M(x, y, z)\mathbf{i} + N(x, y, z)\mathbf{j} + P(x, y, z)\mathbf{k}$, where M, N, and P have partial derivatives in some region. The **curl** of $\mathbf{F}$ is given by

$$\text{curl } \mathbf{F} = \nabla \times \mathbf{F} = \left(\frac{\partial P}{\partial y} - \frac{\partial N}{\partial z}\right)\mathbf{i} + \left(\frac{\partial M}{\partial z} - \frac{\partial P}{\partial x}\right)\mathbf{j} + \left(\frac{\partial N}{\partial x} - \frac{\partial M}{\partial y}\right)\mathbf{k}.$$

We shall also use the symbol curl $\mathbf{F}$ or $\nabla \times \mathbf{F}$ to denote the vector curl $\mathbf{F}(x, y, z)$ or $\nabla \times \mathbf{F}(x, y, z)$ associated with (x, y, z). The formula for curl $\mathbf{F}$ can be written in the following *determinant form.*

Determinant notation for curl $\mathbf{F}$ **(18.6)**

$$\operatorname{curl}\mathbf{F} = \nabla \times \mathbf{F} = \begin{vmatrix} \mathbf{i} & \mathbf{j} & \mathbf{k} \\ \dfrac{\partial}{\partial x} & \dfrac{\partial}{\partial y} & \dfrac{\partial}{\partial z} \\ M & N & P \end{vmatrix}$$

The expression on the right in (18.6) is not actually a determinant, since the first row is made up of vectors, the second row of partial derivative operators, and the third row of scalar functions; however, it is an extremely useful device for remembering the cumbersome formula given in Definition (18.5).

EXAMPLE 3 If $\mathbf{F}(x, y, z) = xy^2z^4\mathbf{i} + (2x^2y + z)\mathbf{j} + y^3z^2\mathbf{k}$, find $\nabla \times \mathbf{F}$.

SOLUTION Applying (18.6), we have

$$\nabla \times \mathbf{F} = \begin{vmatrix} \mathbf{i} & \mathbf{j} & \mathbf{k} \\ \dfrac{\partial}{\partial x} & \dfrac{\partial}{\partial y} & \dfrac{\partial}{\partial z} \\ xy^2z^4 & (2x^2y + z) & y^3z^2 \end{vmatrix}$$

$$= (3y^2z^2 - 1)\mathbf{i} + 4xy^2z^3\mathbf{j} + (4xy - 2xyz^4)\mathbf{k}.$$

FIGURE 18.9

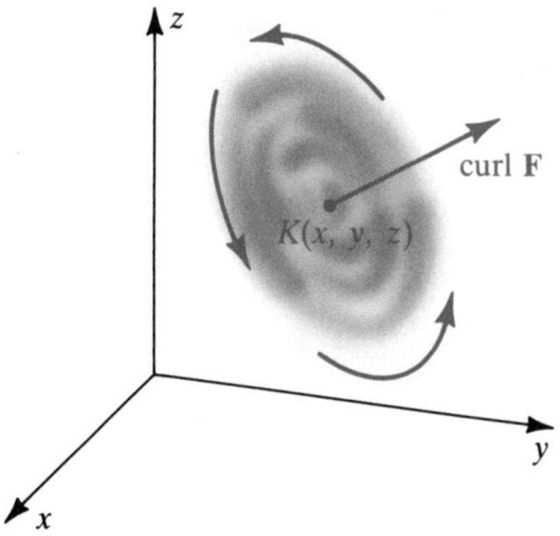

In Section 18.7 we shall see that if $\mathbf{F}$ is the velocity field of a fluid or gas that is moving through an xyz-coordinate system, then curl $\mathbf{F}$ provides information about rotational aspects of the motion. If we consider a point $K(x, y, z)$ at which the fluid or gas is rotating, or swirling (or *curl*ing), then curl $\mathbf{F}$ lies on the axis of rotation (see Figure 18.9) and can be used to describe rotational properties of the field.

The operator ∇ also can be used to obtain a *scalar* function from a vector field $\mathbf{F}$ as follows.

Definition **(18.7)**

Let $\mathbf{F}(x, y, z) = M(x, y, z)\mathbf{i} + N(x, y, z)\mathbf{j} + P(x, y, z)\mathbf{k}$, where M, N, and P have partial derivatives in some region. The **divergence** of $\mathbf{F}$, denoted by div $\mathbf{F}$ or $\nabla \cdot \mathbf{F}$, is given by

$$\operatorname{div}\mathbf{F} = \nabla \cdot \mathbf{F} = \frac{\partial M}{\partial x} + \frac{\partial N}{\partial y} + \frac{\partial P}{\partial z}.$$

We use the symbol $\nabla \cdot \mathbf{F}$ for divergence because the formula may be obtained by taking what *appears* to be the dot product of ∇ and $\mathbf{F}$, as follows:

$$\nabla \cdot \mathbf{F} = \left(\frac{\partial}{\partial x}\mathbf{i} + \frac{\partial}{\partial y}\mathbf{j} + \frac{\partial}{\partial z}\mathbf{k}\right) \cdot (M\mathbf{i} + N\mathbf{j} + P\mathbf{k})$$

$$= \frac{\partial}{\partial x}(M) + \frac{\partial}{\partial y}(N) + \frac{\partial}{\partial z}(P)$$

As in the next example, we shall also use the symbol $\nabla \cdot \mathbf{F}$ to denote the function value $(\nabla \cdot \mathbf{F})(x, y, z)$. The value of $\nabla \cdot \mathbf{F}$, or div $\mathbf{F}$, at a particular point K is denoted by $[\nabla \cdot \mathbf{F}]_K$ or $[\operatorname{div}\mathbf{F}]_K$.

EXAMPLE 4 If $\mathbf{F}(x, y, z) = xy^2z^4\mathbf{i} + (2x^2y + z)\mathbf{j} + y^3z^2\mathbf{k}$, find $\nabla \cdot \mathbf{F}$.

SOLUTION By Definition (18.7),

$$\begin{aligned}\nabla \cdot \mathbf{F} &= \frac{\partial}{\partial x}(xy^2z^4) + \frac{\partial}{\partial y}(2x^2y + z) + \frac{\partial}{\partial z}(y^3z^2) \\ &= y^2z^4 + 2x^2 + 2y^3z.\end{aligned}$$

In Section 18.6 we shall see that if $\mathbf{F}$ is the velocity field of a fluid or gas, then div $\mathbf{F}$ provides information about the flow of mass. If K is a point and $[\text{div } \mathbf{F}]_K < 0$, there is more mass flowing toward the point than away from it and we say there is a **sink** at K. If $[\text{div } \mathbf{F}]_K > 0$, more mass is flowing away from K than toward K and there is a **source** at K. If $[\text{div } \mathbf{F}]_K = 0$, as is true for incompressible fluids, there is neither a sink nor a source at K. Similarly, if $\mathbf{F}$ represents the flow of heat and if $[\nabla \cdot \mathbf{F}]_K > 0$, then there is a source of heat at K (or heat is *leaving* K and therefore the temperature at K is decreasing). If $[\nabla \cdot \mathbf{F}]_K < 0$, then heat is being *absorbed* at K (or the temperature at K is increasing).

Various algebraic properties can be established for *div* and *curl*. A property that is analogous to the product rule for derivatives is considered in the next example. Other properties are stated in Exercises 23–26.

EXAMPLE 5 Let f be a scalar function and $\mathbf{F}$ a vector function. If partial derivatives exist, show that

$$\nabla \cdot (f\mathbf{F}) = f(\nabla \cdot \mathbf{F}) + (\nabla f) \cdot \mathbf{F}.$$

SOLUTION If we write $\mathbf{F} = M\mathbf{i} + N\mathbf{j} + P\mathbf{k}$, where M, N, and P are functions of x, y, and z, then

$$f\mathbf{F} = fM\mathbf{i} + fN\mathbf{j} + fP\mathbf{k}.$$

Applying Definition (18.7) yields

$$\begin{aligned}\nabla \cdot (f\mathbf{F}) &= \frac{\partial}{\partial x}(fM) + \frac{\partial}{\partial y}(fN) + \frac{\partial}{\partial z}(fP) \\ &= f\frac{\partial M}{\partial x} + \frac{\partial f}{\partial x}M + f\frac{\partial N}{\partial y} + \frac{\partial f}{\partial y}N + f\frac{\partial P}{\partial z} + \frac{\partial f}{\partial z}P.\end{aligned}$$

Rearranging terms gives us

$$\begin{aligned}\nabla \cdot (f\mathbf{F}) &= f\left(\frac{\partial M}{\partial x} + \frac{\partial N}{\partial y} + \frac{\partial P}{\partial z}\right) + \left(\frac{\partial f}{\partial x}M + \frac{\partial f}{\partial y}N + \frac{\partial f}{\partial z}P\right) \\ &= f(\nabla \cdot \mathbf{F}) + (\nabla f) \cdot \mathbf{F}.\end{aligned}$$

Finally, if a vector field $\mathbf{F}$ is described as in (18.1), then we may define limits, continuity, partial derivatives, and multiple integrals by using the components of $\mathbf{F}(x, y, z)$ as we did for vector-valued functions in Chapter 15. For example, if we wish to differentiate or integrate $\mathbf{F}(x, y, z)$, we differentiate or integrate each component. The usual theorems may be established: **F** is continuous if and only if the component functions M, N, and P are continuous; **F** has partial derivatives if and only if the same is true for M, N, and P; and so on.

EXERCISES 18.1

Exer. 1–10: Sketch a sufficient number of vectors to illustrate the pattern of the vectors in the field F.

1 $\mathbf{F}(x, y) = x\mathbf{i} - y\mathbf{j}$

2 $\mathbf{F}(x, y) = -x\mathbf{i} + y\mathbf{j}$

3 $\mathbf{F}(x, y) = 2x\mathbf{i} + 3y\mathbf{j}$

4 $\mathbf{F}(x, y) = 3\mathbf{i} + x\mathbf{j}$

5 $\mathbf{F}(x, y) = (x^2 + y^2)^{-1/2}(x\mathbf{i} + y\mathbf{j})$

6 $\mathbf{F}(x, y, z) = x\mathbf{i} + z\mathbf{k}$

7 $\mathbf{F}(x, y, z) = -x\mathbf{i} - y\mathbf{j} - z\mathbf{k}$

8 $\mathbf{F}(x, y, z) = x\mathbf{i} + y\mathbf{j} + z\mathbf{k}$

9 $\mathbf{F}(x, y, z) = \mathbf{i} + \mathbf{j} + \mathbf{k}$

10 $\mathbf{F}(x, y, z) = 2\mathbf{k}$

Exer. 11–14: Find a conservative vector field that has the given potential.

11 $f(x, y, z) = x^2 - 3y^2 + 4z^2$

12 $f(x, y, z) = \sin(x^2 + y^2 + z^2)$

13 $f(x, y) = \arctan(xy)$

14 $f(x, y) = y^2e^{-3x}$

Exer. 15–18: Find $\nabla \times \mathbf{F}$ and $\nabla \cdot \mathbf{F}$.

15 $\mathbf{F}(x, y, z) = x^2z\mathbf{i} + y^2x\mathbf{j} + (y + 2z)\mathbf{k}$

16 $\mathbf{F}(x, y, z) = (3x + y)\mathbf{i} + xy^2z\mathbf{j} + xz^2\mathbf{k}$

17 $\mathbf{F}(x, y, z) = 3xyz^2\mathbf{i} + y^2 \sin z\mathbf{j} + xe^{2z}\mathbf{k}$

18 $\mathbf{F}(x, y, z) = x^3 \ln z\mathbf{i} + xe^{-y}\mathbf{j} - (y^2 + 2z)\mathbf{k}$

19 If $\mathbf{r} = x\mathbf{i} + y\mathbf{j} + z\mathbf{k}$, prove that

(a) $\nabla \cdot \mathbf{r} = 3$ (b) $\nabla \times \mathbf{r} = \mathbf{0}$ (c) $\nabla \|\mathbf{r}\| = \mathbf{r}/\|\mathbf{r}\|$

20 If $\mathbf{r} = x\mathbf{i} + y\mathbf{j} + z\mathbf{k}$ and $\mathbf{a}$ is a constant vector, prove that

(a) curl $(\mathbf{a} \times \mathbf{r}) = 2\mathbf{a}$ (b) div $(\mathbf{a} \times \mathbf{r}) = 0$

21 Prove that the curl and divergence of an inverse square vector field are $\mathbf{0}$ and 0, respectively.

22 If a vector field $\mathbf{F}$ is conservative and has continuous partial derivatives, prove that curl $\mathbf{F} = \mathbf{0}$.

Exer. 23–26: Verify the identity.

23 $\nabla \times (\mathbf{F} + \mathbf{G}) = \nabla \times \mathbf{F} + \nabla \times \mathbf{G}$

24 $\nabla \cdot (\mathbf{F} + \mathbf{G}) = \nabla \cdot \mathbf{F} + \nabla \cdot \mathbf{G}$

25 $\nabla \times (f\mathbf{F}) = f(\nabla \times \mathbf{F}) + (\nabla f) \times \mathbf{F}$

26 $\nabla \cdot (\mathbf{F} \times \mathbf{G}) = (\nabla \times \mathbf{F}) \cdot \mathbf{G} - (\nabla \times \mathbf{G}) \cdot \mathbf{F}$

Exer. 27–30: If f and F have continuous second partial derivatives and a is a constant vector, verify the identity.

27 curl grad $f = \mathbf{0}$

28 div curl $\mathbf{F} = 0$

29 curl (grad f + curl $\mathbf{F}$) = curl curl $\mathbf{F}$

30 curl $\mathbf{a} = \mathbf{0}$

31 Let $\mathbf{r} = x\mathbf{i} + y\mathbf{j} + z\mathbf{k}$ and $r = \|\mathbf{r}\|$. If $\mathbf{F}(x, y, z) = (c/r^k)\mathbf{r}$, where c and k are constants with $k > 0$, prove that curl $\mathbf{F} = \mathbf{0}$. (*Hint:* Use Exercise 25.)

32 The differential operator $\nabla^2 = \nabla \cdot \nabla$ is defined by

$$\nabla^2 = \frac{\partial^2}{\partial x^2} + \frac{\partial^2}{\partial y^2} + \frac{\partial^2}{\partial z^2}.$$

If ∇^2 operates on $f(x, y, z)$, it produces a scalar function called the *Laplacian* of f, given by

$$\nabla^2 f = \frac{\partial^2 f}{\partial x^2} + \frac{\partial^2 f}{\partial y^2} + \frac{\partial^2 f}{\partial z^2}.$$

If f and g are scalar functions that have second partial derivatives, prove that

(a) $\nabla \cdot (\nabla f) = \nabla^2 f$

(b) $\nabla^2(fg) = f\nabla^2 g + g\nabla^2 f + 2\nabla f \cdot \nabla g$

Exer. 33–34: Using the notation of Exercise 32, prove that the function satisfies *Laplace's equation* $\nabla^2 f = 0$. (Functions of this type are *harmonic* and are important in physical applications.)

33 $f(x, y, z) = (x^2 + y^2 + z^2)^{-1/2}$

34 $f(x, y, z) = ax^2 + by^2 + cz^2$ for $a + b + c = 0$

35 If $\mathbf{F}$ is given by (18.1), define

$$\lim_{(x,y,z)\to(x_0,y_0,z_0)} \mathbf{F}(x, y, z) = \mathbf{a}$$

in a manner analogous to that of Definition (15.4). Give an ϵ-δ definition for this limit. What is the geometric significance of the limit?

36 If $\mathbf{F}$ is given by (18.1), define the notion of *continuity* at (x_0, y_0, z_0) in a manner similar to that of Definition (15.5). What is the geometric significance of a continuous vector function?

c **Exer. 37–38: Extend the approximation formulas in Exercises 67–68 of Section 16.3 to $f(x, y, z)$, and then use $h = 0.05$ to approximate $\nabla^2 f(0.3, 0.5, 0.2)$.**

37 $f(x, y, z) = \dfrac{\tan^2 \sqrt{xyz}}{\cos z \sec^2 (x^2y^2)}$

38 $f(x, y, z) = \dfrac{x^2 + xy^2z + z^3}{4x^3 + 2y^2z + 4xz^2}$

18.2 LINE INTEGRALS

To define $\int_a^b f(x)\,dx$ in Chapter 5, we began by dividing the interval $[a, b]$ into subintervals of lengths $\Delta x_1, \Delta x_2, \ldots, \Delta x_n$. We then chose any number w_k in each subinterval and considered the limit of Riemann sums $\sum_k f(w_k)\,\Delta x_k$ as $\Delta x_k \to 0$. We may use a similar process to define *line integrals* of functions of several variables along (or over) curves in two or three dimensions.

Recall that a plane curve C is *smooth* if it has a parametrization

$$x = g(t), \quad y = h(t); \quad a \le t \le b$$

such that g' and h' are continuous and not simultaneously 0 on $[a, b]$. For space curves we also consider a third function k of the same type with $z = k(t)$. The **positive direction**, or *orientation*, of C is the direction determined by increasing values of t. The curve C is *piecewise smooth* if $[a, b]$ can be partitioned into closed subintervals such that C is smooth for each subinterval.

Suppose f is a function of two variables x and y that is continuous on a region D containing a smooth curve C, with parametrization $x = g(t)$, $y = h(t)$; $a \le t \le b$. We shall define three different integrals of f along C. Let us begin by partitioning the parameter interval $[a, b]$, choosing

$$a = t_0 < t_1 < t_2 < \cdots < t_n = b.$$

The norm of this partition (that is, the length of the largest subinterval $[t_{k-1}, t_k]$) will be denoted by $\|P\|$. If $P_k(x_k, y_k)$ is the point on C corresponding to t_k, then the points $P_0, P_1, P_2, \ldots, P_n$ divide C into n parts $\widehat{P_{k-1}P_k}$, as illustrated in Figure 18.10. Let

$$\Delta x_k = x_k - x_{k-1}, \quad \Delta y_k = y_k - y_{k-1}, \quad \Delta s_k = \text{length of } \widehat{P_{k-1}P_k}.$$

For each k, let $Q_k(u_k, v_k)$ be a point in $\widehat{P_{k-1}P_k}$ obtained by choosing some number in $[t_{k-1}, t_k]$ (see Figure 18.10). We now consider the three sums

$$\sum_k f(u_k, v_k)\,\Delta s_k, \quad \sum_k f(u_k, v_k)\,\Delta x_k, \quad \sum_k f(u_k, v_k)\,\Delta y_k.$$

FIGURE 18.10

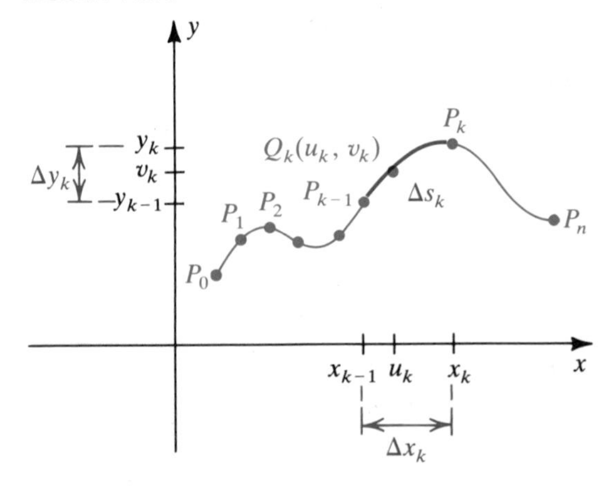

If the limits of these sums exist as $\|P\| \to 0$, they are the **line integrals of f along** (or **over**) **C** with respect to s, x, and y, respectively, and are denoted as follows.

Line integrals in two dimensions **(18.8)**

$$\int_C f(x, y)\, ds = \lim_{\|P\| \to 0} \sum_k f(u_k, v_k)\, \Delta s_k$$

$$\int_C f(x, y)\, dx = \lim_{\|P\| \to 0} \sum_k f(u_k, v_k)\, \Delta x_k$$

$$\int_C f(x, y)\, dy = \lim_{\|P\| \to 0} \sum_k f(u_k, v_k)\, \Delta y_k$$

The term *line integral* is a misnomer, since C is not necessarily a line. It would be more descriptive to use *curve integral*.

If f is continuous on D, then the limits in (18.8) exist and are the same for every parametrization of C (provided the same orientation is used). Moreover, the integrals can be evaluated by substituting $x = g(t)$, $y = h(t)$ from the parametrization of C and replacing the differentials by

$$ds = \sqrt{(dx)^2 + (dy)^2} = \sqrt{[g'(t)]^2 + [h'(t)]^2}\, dt$$

$$dx = g'(t)\, dt, \quad dy = h'(t)\, dt.$$

Note that the formula for ds is the differential of arc length from (13.6). Let us state these facts for reference as follows.

Evaluation theorem for line integrals **(18.9)**

If a smooth curve C is given by $x = g(t)$, $y = h(t)$; $a \le t \le b$ and if $f(x, y)$ is continuous on a region D containing C, then

(i) $\displaystyle\int_C f(x, y)\, ds = \int_a^b f(g(t), h(t))\sqrt{[g'(t)]^2 + [h'(t)]^2}\, dt$

(ii) $\displaystyle\int_C f(x, y)\, dx = \int_a^b f(g(t), h(t))g'(t)\, dt$

(iii) $\displaystyle\int_C f(x, y)\, dy = \int_a^b f(g(t), h(t))h'(t)\, dt$

The preceding discussion can be extended to more complicated curves. In particular, suppose C is a piecewise-smooth curve that can be expressed as the union of a finite number of smooth curves $C_1, C_2, \ldots, C_n$, where the terminal point of C_k is the initial point of C_{k+1} for $k = 1, 2, \ldots, n - 1$. In this case the line integral of f along C is defined as the sum of the line integrals along the individual curves.

We may establish properties of line integrals that are similar to those obtained for definite integrals in Chapter 5. For example, reversing the direction of integration changes the sign of the integral; the integral of a sum of two functions is the sum of the integrals of the individual functions; and so on.

FIGURE 18.11

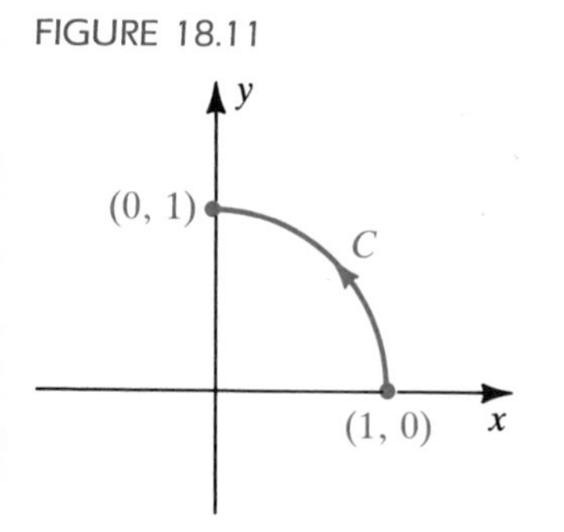

EXAMPLE 1 Evaluate $\int_C xy^2\, ds$ if C has the parametrization

$$x = \cos t, \quad y = \sin t; \quad 0 \le t \le \pi/2.$$

SOLUTION The curve C is the first-quadrant part of the unit circle with center at the origin, as shown in Figure 18.11, where the arrowhead indicates the orientation of C and the direction of integration. Applying Theorem (18.9)(i) gives us

$$\int_C xy^2\,ds = \int_0^{\pi/2} \cos t \sin^2 t\sqrt{\sin^2 t + \cos^2 t}\,dt$$
$$= \int_0^{\pi/2} \sin^2 t \cos t\,dt = \left[\tfrac{1}{3}\sin^3 t\right]_0^{\pi/2} = \tfrac{1}{3}.$$

EXAMPLE 2 Evaluate $\int_C xy^2\,dx$ and $\int_C xy^2\,dy$ if C is the portion of the parabola $y = x^2$ from $(0, 0)$ to $(2, 4)$.

FIGURE 18.12

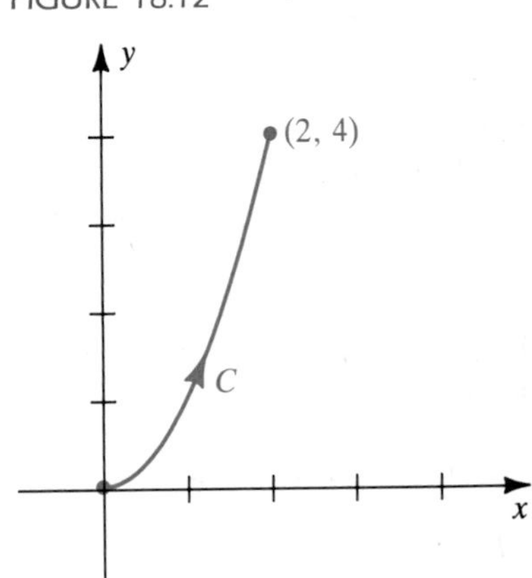

SOLUTION The graph of $y = x^2$ from $(0, 0)$ to $(2, 4)$ is sketched in Figure 18.12. Parametric equations for C are

$$x = t, \quad y = t^2; \quad 0 \le t \le 2.$$

The differentials are

$$dx = dt, \quad dy = 2t\,dt.$$

Applying Theorem (18.9)(ii) and (iii), we have

$$\int_C xy^2\,dx = \int_0^2 t(t^2)^2\,dt = \int_0^2 t^5\,dt = \left[\tfrac{1}{6}t^6\right]_0^2 = \tfrac{32}{3}$$
$$\int_C xy^2\,dy = \int_0^2 t(t^2)^2\,2t\,dt = \int_0^2 2t^6\,dt = \left[\tfrac{2}{7}t^7\right]_0^2 = \tfrac{256}{7}.$$

If, as in Example 2, C is the graph of an equation $y = g(x)$ such that $a \le x \le b$, then parametric equations for C are

$$x = t, \quad y = g(t); \quad a \le t \le b.$$

The line integrals (ii) and (iii) of Theorem (18.9) may then be evaluated as follows:

$$\int_C f(x, y)\,dx = \int_a^b f(t, g(t))\,dt = \int_a^b f(x, g(x))\,dx$$
$$\int_C f(x, y)\,dy = \int_a^b f(t, g(t))g'(t)\,dt = \int_a^b f(x, g(x))g'(x)\,dx$$

This shows that *for curves given in the form $y = g(x)$ with $a \le x \le b$, we may bypass the parametric equations by substituting $y = g(x)$, $dy = g'(x)\,dx$ and then using a and b for the limits of integration.* To illustrate, in Example 2, where $y = x^2$, we could have written

$$\int_C xy^2\,dx = \int_0^2 x(x^4)\,dx = \int_0^2 x^5\,dx = \tfrac{32}{3}$$
$$\int_C xy^2\,dy = \int_0^2 x(x^4)2x\,dx = \int_0^2 2x^6\,dx = \tfrac{256}{7}.$$

A physical application of the line integral $\int_C f(x, y)\,ds$ may be obtained by regarding the curve as a thin wire of variable density. If the wire is represented by the curve C in Figure 18.10 and the **linear mass density** (the mass per unit length) at the point (x, y) is given by $\delta(x, y)$, then $\delta(u_k, v_k)\,\Delta s_k$ is an approximation to the mass Δm_k of the part of the wire between P_{k-1} and P_k. The sum

$$\sum_k \Delta m_k = \sum_k \delta(u_k, v_k)\,\Delta s_k$$

is an approximation to the total mass m of the wire. To define m, we take the limit of these sums, obtaining the following:

Mass of a wire **(18.10)**

$$m = \int_C \delta(x, y)\, ds$$

EXAMPLE 3 A thin wire is bent into the shape of a semicircle of radius a. If the linear mass density at a point P is directly proportional to its distance from the line through the endpoints, find the mass of the wire.

FIGURE 18.13

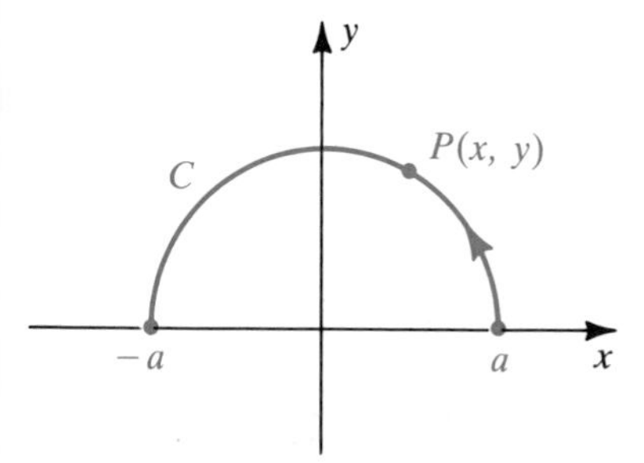

SOLUTION Let us introduce a coordinate system such that the shape of the wire coincides with the upper half of a circle of radius a with center at the origin (see Figure 18.13). Parametric equations for C are

$$x = a \cos t, \quad y = a \sin t; \quad 0 \le t \le \pi.$$

By hypothesis, the linear mass density at $P(x, y)$ is given by $\delta(x, y) = ky$, where k is a constant. By (18.10), the mass of the wire is

$$\begin{aligned} m = \int_C (ky)\, ds &= \int_0^\pi (ka \sin t)\sqrt{a^2 \sin^2 t + a^2 \cos^2 t}\, dt \\ &= \int_0^\pi (ka \sin t)a\, dt \\ &= ka^2 \int_0^\pi \sin t\, dt \\ &= ka^2 \Big[-\cos t\Big]_0^\pi = 2ka^2. \end{aligned}$$

In applications involving work, line integrals often occur in the combination

$$\int_C M(x, y)\, dx + \int_C N(x, y)\, dy,$$

where the functions M and N are continuous on a domain D containing C. This sum is usually abbreviated as

$$\int_C M(x, y)\, dx + N(x, y)\, dy.$$

FIGURE 18.14

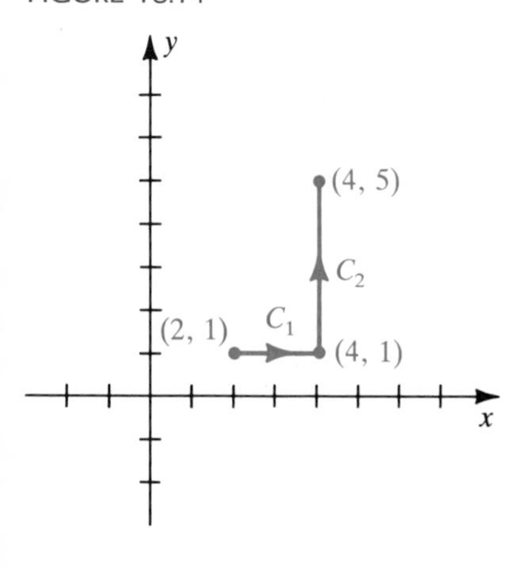

EXAMPLE 4 Evaluate $\int_C xy\, dx + x^2\, dy$ if

(a) C consists of line segments from (2, 1) to (4, 1) and from (4, 1) to (4, 5)

(b) C is the line segment from (2, 1) to (4, 5)

(c) parametric equations for C are $x = 3t - 1$, $y = 3t^2 - 2t$; $1 \le t \le \frac{5}{3}$

SOLUTION

(a) If C is subdivided into two parts C_1 and C_2, as shown in Figure 18.14, then parametric equations for these curves are

$$C_1: \quad x = t, \quad y = 1; \quad 2 \le t \le 4$$

$$C_2: \quad x = 4, \quad y = t; \quad 1 \le t \le 5.$$

FIGURE 18.15

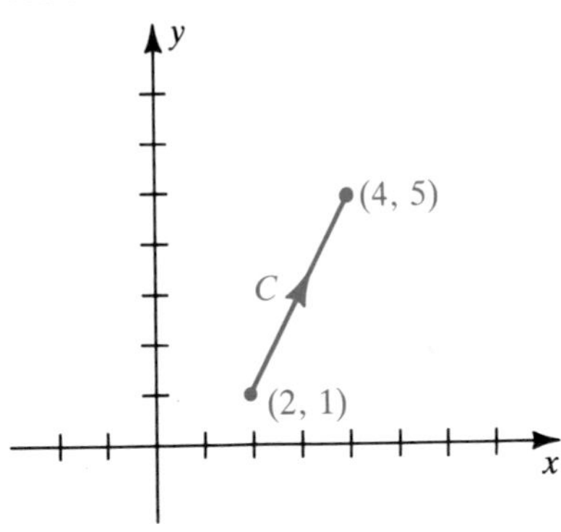

The line integral along C may be expressed as a sum of two line integrals, the first along C_1 and the second along C_2. On C_1 we have $dy = 0$, $dx = dt$, and hence

$$\int_{C_1} xy\,dx + x^2\,dy = \int_2^4 t(1)\,dt + t^2(0) = \left[\tfrac{1}{2}t^2\right]_2^4 = 6.$$

On C_2 we have $dx = 0$, $dy = dt$, and therefore

$$\int_{C_2} xy\,dx + x^2\,dy = \int_1^5 4t(0) + 16\,dt = 16\Big[t\Big]_1^5 = 64.$$

Consequently the line integral along C equals $6 + 64$, or 70.

(b) The graph of C is sketched in Figure 18.15. An equation for C is $y = 2x - 3$ with $2 \le x \le 4$. In this case $dy = 2\,dx$ and

$$\begin{aligned}\int_C xy\,dx + x^2\,dy &= \int_2^4 x(2x - 3)\,dx + x^2(2)\,dx \\ &= \int_2^4 (4x^2 - 3x)\,dx = \tfrac{170}{3}.\end{aligned}$$

FIGURE 18.16

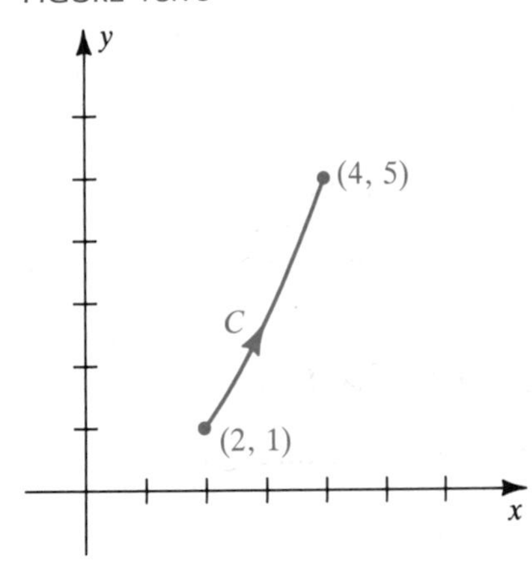

(c) The graph of C is part of a parabola (see Figure 18.16). In this case, using the parametric equations $x = 3t - 1$, $y = 3t^2 - 2t$; $1 \le t \le \frac{5}{3}$, we obtain $dx = 3\,dt$, $dy = (6t - 2)\,dt$, and the line integral equals

$$\int_1^{5/3} (3t - 1)(3t^2 - 2t)3\,dt + (3t - 1)^2(6t - 2)\,dt.$$

We can show that the value is 58.

Another method of solution is to use the equation $y = \frac{1}{3}(x^2 - 1)$ for the parabola, with $2 \le x \le 4$. We then obtain the integral

$$\begin{aligned}\int_C xy\,dx + x^2\,dy &= \int_2^4 x \cdot \tfrac{1}{3}(x^2 - 1)\,dx + x^2(\tfrac{2}{3}x)\,dx \\ &= \int_2^4 (x^3 - \tfrac{1}{3}x)\,dx = \left[\tfrac{1}{4}x^4 - \tfrac{1}{6}x^2\right]_2^4 = 58.\end{aligned}$$

In the preceding example we obtained three different values for the line integral along three different paths from (2, 1) to (4, 5). In Section 18.3 we consider line integrals that have the same value along every curve joining two points A and B. For such integrals we use the phrase *independent of path.*

If a smooth curve C in three dimensions has the parametrization

$$x = g(t), \quad y = h(t), \quad z = k(t); \quad a \le t \le b,$$

then line integrals of a function f of *three* variables are defined in a manner similar to that used for integrals of a function of two variables. In this case, instead of using (x_k, y_k) and (u_k, v_k) as the coordinates of P_k and Q_k on C as in Figure 18.10, we use (x_k, y_k, z_k) and (u_k, v_k, w_k), respectively (see Figure 18.17). We now have

$$\int_C f(x, y, z)\,ds = \lim_{\|P\| \to 0} \sum_k f(u_k, v_k, w_k)\,\Delta s_k.$$

This integral may be evaluated by using the formula

$$\int_a^b f(g(t), h(t), k(t))\sqrt{[g'(t)]^2 + [h'(t)]^2 + [k'(t)]^2}\,dt.$$

FIGURE 18.17

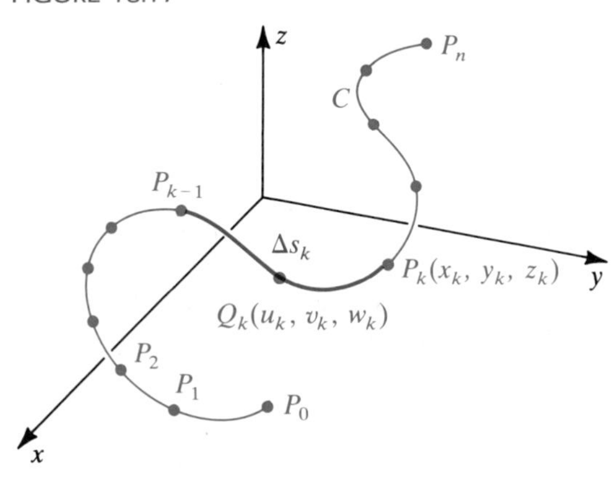

If a wire has the shape of C and if $f(x, y, z)$ is the linear mass density at (x, y, z), then the value of this integral is the total mass of the wire.

In addition to line integrals $\int_C f(x, y, z)\,dx$ and $\int_C f(x, y, z)\,dy$, a function of three variables also has a line integral with respect to z, given by

$$\int_C f(x, y, z)\,dz = \lim_{\|P\|\to 0} \sum_k f(u_k, v_k, w_k)\,\Delta z_k,$$

where $\Delta z_k = z_k - z_{k-1}$. As in two dimensions, these line integrals often occur as sums, which we abbreviate as

$$\int_C M(x, y, z)\,dx + N(x, y, z)\,dy + P(x, y, z)\,dz,$$

FIGURE 18.18

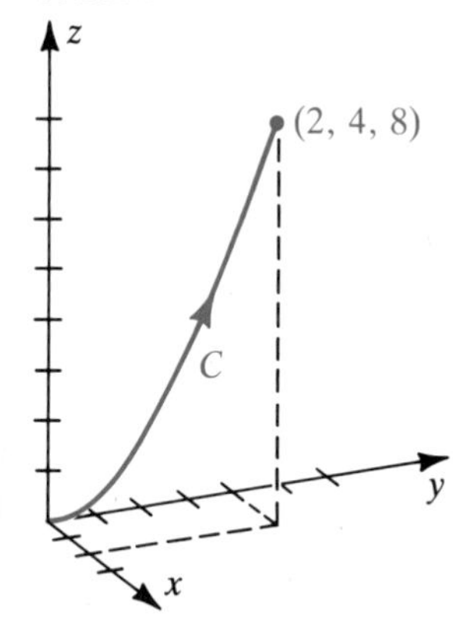

where the functions M, N, and P are continuous throughout a region containing C. If C is given parametrically, then this line integral may be evaluated by substituting for x, y, and z in the same manner as in the two-variable case.

EXAMPLE 5 Evaluate $\int_C yz\,dx + xz\,dy + xy\,dz$ if C is given by

$$x = t, \quad y = t^2, \quad z = t^3; \quad 0 \le t \le 2.$$

SOLUTION The curve C (a twisted cubic) is sketched in Figure 18.18. We substitute for x, y, and z and use $dx = dt$, $dy = 2t\,dt$, $dz = 3t^2\,dt$:

$$\int_0^2 t^5\,dt + 2t^5\,dt + 3t^5\,dt = \int_0^2 6t^5\,dt = \Big[t^6\Big]_0^2 = 64$$

One of the most important physical applications of line integrals involves force fields. Suppose that the force acting at the point (x, y, z) is

$$\mathbf{F}(x, y, z) = M(x, y, z)\mathbf{i} + N(x, y, z)\mathbf{j} + P(x, y, z)\mathbf{k},$$

where M, N, and P are continuous functions. We shall formulate a definition for the work done as the point of application of $\mathbf{F}(x, y, z)$ moves along a smooth curve C that has a parametrization

$$x = g(t), \quad y = h(t), \quad z = k(t); \quad a \le t \le b.$$

FIGURE 18.19

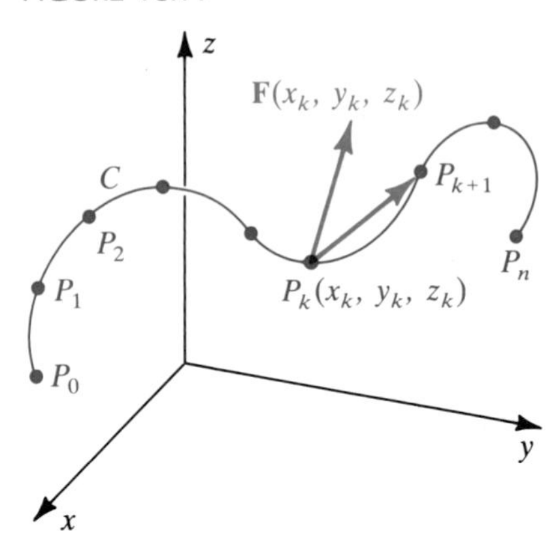

We assume that the motion is in the direction determined by *increasing* values of t.

Let us subdivide C by means of points $P_0, P_1, P_2, \ldots, P_n$, where P_k has coordinates (x_k, y_k, z_k) (see Figure 18.19). If the norm $\|P\|$ is small, then P_k is close to P_{k+1} for each k. Hence the work done by $\mathbf{F}(x, y, z)$ from P_k to P_{k+1} can be approximated by the work ΔW_k done by the *constant* force $\mathbf{F}(x_k, y_k, z_k)$ as its point of application moves along $\overrightarrow{P_kP_{k+1}}$.

If $\Delta x_k = x_{k+1} - x_k$, $\Delta y_k = y_{k+1} - y_k$, and $\Delta z_k = z_{k+1} - z_k$, then $\overrightarrow{P_kP_{k+1}}$ corresponds to the vector $\Delta x_k\mathbf{i} + \Delta y_k\mathbf{j} + \Delta z_k\mathbf{k}$ (see Figure 18.20). By (14.26), the work ΔW_k done by $\mathbf{F}(x_k, y_k, z_k)$ along $\overrightarrow{P_kP_{k+1}}$ is

$$\begin{aligned}\Delta W_k &= \mathbf{F}(x_k, y_k, z_k) \cdot (\Delta x_k\mathbf{i} + \Delta y_k\mathbf{j} + \Delta z_k\mathbf{k}) \\ &= M(x_k, y_k, z_k)\,\Delta x_k + N(x_k, y_k, z_k)\,\Delta y_k + P(x_k, y_k, z_k)\,\Delta z_k.\end{aligned}$$

The **work W done by F along C** is defined as follows.

Definition of work **(18.11)**

$$\begin{aligned}W &= \lim_{\|P\|\to 0} \sum_k \Delta W_k \\ &= \int_C M(x, y, z)\,dx + N(x, y, z)\,dy + P(x, y, z)\,dz\end{aligned}$$

FIGURE 18.20

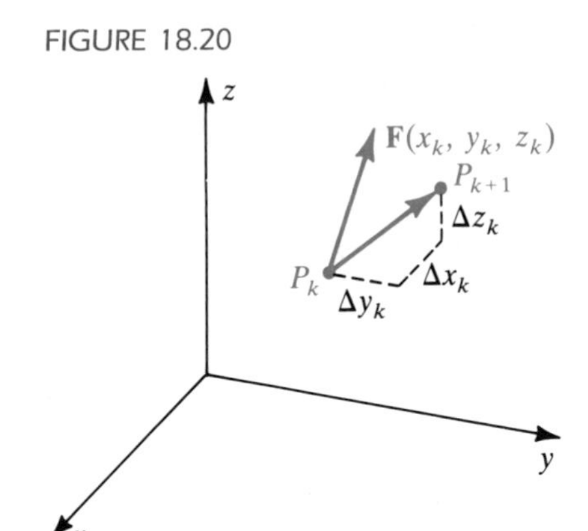

For applications, the integral in (18.11) is usually expressed in vector form. If we let

$$\mathbf{r}(t) = x\mathbf{i} + y\mathbf{j} + z\mathbf{k},$$

where $x = g(t)$, $y = h(t)$, and $z = k(t)$, then $\mathbf{r}(t)$ is the position vector for the point $Q(x, y, z)$ on C. If s is an arc length parameter for C, then, as in Section 15.4, a unit tangent vector $\mathbf{T}(s)$ to C at Q is

$$\mathbf{T}(s) = \frac{d}{ds}\mathbf{r}(t) = \frac{dx}{ds}\mathbf{i} + \frac{dy}{ds}\mathbf{j} + \frac{dz}{ds}\mathbf{k}.$$

The vectors $\mathbf{r}(t)$, $\mathbf{F}(x, y, z)$, and $\mathbf{T}(s)$ are illustrated in Figure 18.21.

The **tangential component of F at Q** (see Figure 18.21) is

FIGURE 18.21

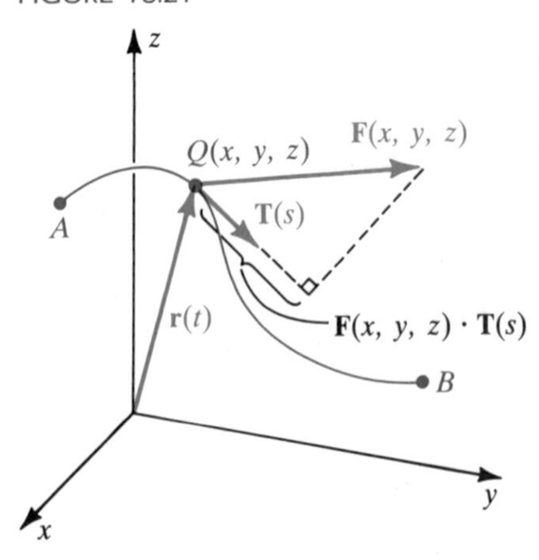

$$\mathbf{F}(x, y, z) \cdot \mathbf{T}(s) = M(x, y, z)\frac{dx}{ds} + N(x, y, z)\frac{dy}{ds} + P(x, y, z)\frac{dz}{ds}.$$

Note that this component is positive if the angle between $\mathbf{F}(x, y, z)$ and $\mathbf{T}(s)$ is acute and negative if this angle is obtuse. Formula (18.11) may now be rewritten as follows:

$$W = \int_C \mathbf{F}(x, y, z) \cdot \mathbf{T}(s)\,ds$$

Thus, *the work done as the point of application of* $\mathbf{F}(x, y, z)$ *moves along* C *equals the line integral with respect to s of the tangential component of* $\mathbf{F}$ *along* C. To simplify the notation, we shall denote $\mathbf{F}(x, y, z)$ by $\mathbf{F}$ and $\mathbf{T}(s)$ by $\mathbf{T}$ and let

$$d\mathbf{r} = dx\mathbf{i} + dy\mathbf{j} + dz\mathbf{k} = \mathbf{T}\,ds.$$

Our discussion may be summarized as follows.

Definition (18.12)

> Let C be a smooth space curve, let $\mathbf{T}$ be a unit tangent vector to C at (x, y, z), and let $\mathbf{F}$ be the force acting at (x, y, z). The **work W done by $\mathbf{F}$ along C** is
>
> $$W = \int_C \mathbf{F} \cdot \mathbf{T}\, ds = \int_C \mathbf{F} \cdot d\mathbf{r},$$
>
> where $\mathbf{r} = x\mathbf{i} + y\mathbf{j} + z\mathbf{k}$.

Intuitively, we may regard $\mathbf{F} \cdot d\mathbf{r}$ in Definition (18.12) as the work done as the point of application of $\mathbf{F}$ moves along the tangent vector $d\mathbf{r}$ to C. The integral sign represents the limit of sums of the elements $\mathbf{F} \cdot d\mathbf{r}$ of work.

EXAMPLE 6 If an inverse force field $\mathbf{F}$ is given by

$$\mathbf{F}(x, y, z) = \frac{k}{\|\mathbf{r}\|^3}\mathbf{r},$$

where k is a constant, find the work done by $\mathbf{F}$ as its point of application moves along the x-axis from $A(1, 0, 0)$ to $B(2, 0, 0)$.

SOLUTION Let C denote the line segment from A to B. Parametric equations for C are $x = t$, $y = 0$, $z = 0$; $1 \le t \le 2$. As in Example 2 of Section 18.1,

$$\mathbf{F}(x, y, z) = \frac{k}{(x^2 + y^2 + z^2)^{3/2}}(x\mathbf{i} + y\mathbf{j} + z\mathbf{k}).$$

Applying Definition (18.12), we have

$$W = \int_C \mathbf{F} \cdot d\mathbf{r} = \int_C \frac{k}{(x^2 + y^2 + z^2)^{3/2}}(x\, dx + y\, dy + z\, dz).$$

Substituting for x, y, and z from the parametric equations for C gives us

$$W = \int_1^2 \frac{k}{(t^2)^{3/2}}\, t\, dt = \int_1^2 \frac{k}{t^2}\, dt = \left[-\frac{k}{t}\right]_1^2 = \frac{k}{2}.$$

The units for W depend on those for distance and $\|\mathbf{F}(x, y, z)\|$.

If the preceding discussion is restricted to two dimensions, then a force field may be expressed in the form

$$\mathbf{F}(x, y) = M(x, y)\mathbf{i} + N(x, y)\mathbf{j}.$$

If C is a finite piecewise-smooth plane curve, then the work done as the point of application of $\mathbf{F}(x, y)$ moves along C is

$$W = \int_C M(x, y)\, dx + N(x, y)\, dy = \int_C \mathbf{F} \cdot d\mathbf{r},$$

where $\mathbf{r} = x\mathbf{i} + y\mathbf{j}$.

EXAMPLE 7 Let C be the part of the parabola $y = x^2$ between $(0, 0)$ and $(3, 9)$. If $\mathbf{F}(x, y) = -y\mathbf{i} + x\mathbf{j}$ is a force acting at (x, y), find the work done by $\mathbf{F}$ along C from

(a) $(0, 0)$ to $(3, 9)$ **(b)** $(3, 9)$ to $(0, 0)$

FIGURE 18.22

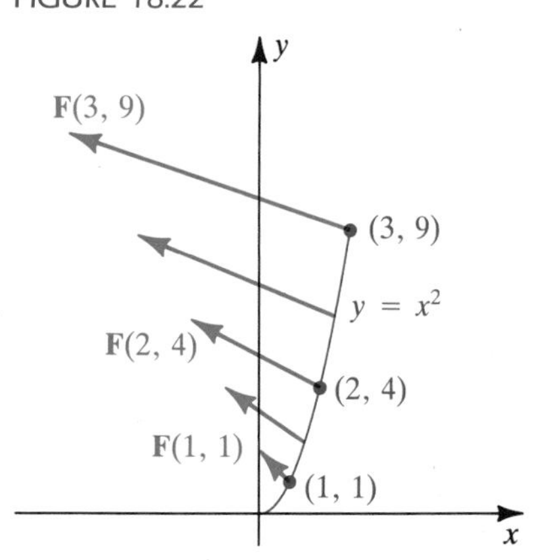

SOLUTION The vector field $\mathbf{F}$ was discussed in Example 1 of Section 18.1. Some typical force vectors acting along C are shown in Figure 18.22. For example,

$$\mathbf{F}(1, 1) = -\mathbf{i} + \mathbf{j}, \quad \mathbf{F}(2, 4) = -4\mathbf{i} + 2\mathbf{j}, \quad \mathbf{F}(3, 9) = -9\mathbf{i} + 3\mathbf{j}.$$

Note the change in direction and magnitude of $\mathbf{F}(x, y)$ between $(0, 0)$ and $(3, 9)$.

(a) Using the parametrization $x = t$, $y = t^2$; $0 \le t \le 3$, we obtain

$$W = \int_C \mathbf{F} \cdot d\mathbf{r} = \int_C -y\,dx + x\,dy$$

$$= \int_0^3 -t^2\,dt + t(2t)\,dt = \int_0^3 t^2\,dt = \left[\frac{t^3}{3}\right]_0^3 = 9.$$

If, for example, $\|\mathbf{F}\|$ is in pounds and s is in feet, then $W = 9$ ft-lb.

(b) We could use the parametrization $x = -t$, $y = t^2$; $-3 \le t \le 0$ for C and evaluate $\int_C \mathbf{F} \cdot d\mathbf{r}$; however, it is simpler to reverse the limits of integration in (a). This gives us $W = -9$.

The significance of the sign difference between 9 and -9 in solutions (a) and (b) of Example 7 may be interpreted as follows. In (a), imagine a particle moving through the force field $\mathbf{F}$ along the path $y = x^2$ from $(0, 0)$ to $(3, 9)$. The particle moves in a northeasterly direction, and the force vectors act in a northwesterly direction. The angle θ between $\mathbf{F}(x, y)$ and a tangent vector to the path (in the direction of the motion) is always acute, and hence $\mathbf{F}$ *assists* the movement of the particle. Thus, the work done by $\mathbf{F}$ is positive.

In (b), the particle moves from $(3, 9)$ to $(0, 0)$ in a southwesterly direction. The angle θ is always obtuse, and hence $\mathbf{F}$ *hinders* the movement of the particle. Thus, the work done by $\mathbf{F}$ is negative.

EXERCISES 18.2

Exer. 1–2: Evaluate the line integrals $\int_C f(x, y)\,ds$, $\int_C f(x, y)\,dx$, and $\int_C f(x, y)\,dy$ if C has the given parametrization.

1 $f(x, y) = x^3 + y$, $x = 3t$, $y = t^3$; $0 \le t \le 1$

2 $f(x, y) = xy^{2/5}$, $x = \frac{1}{2}t$, $y = t^{5/2}$; $0 \le t \le 1$

Exer. 3–6: Evaluate the line integral along C.

3 $\int_C 6x^2y\,dx + xy\,dy$;
C is the graph of $y = x^3 + 1$ from $(-1, 0)$ to $(1, 2)$.

4 $\int_C y\,dx + (x + y)\,dy$;
C is the graph of $y = x^2 + 2x$ from $(0, 0)$ to $(2, 8)$.

5 $\int_C (x - y)\,dx + x\,dy$;
C is the graph of $y^2 = x$ from $(4, -2)$ to $(4, 2)$.

6 $\int_C xy\,dx + x^2y^3\,dy$;
C is the graph of $x = y^3$ from $(0, 0)$ to $(1, 1)$.

7 Evaluate $\int_C xy\,dx + (x + y)\,dy$ along each curve C from $(0, 0)$ to $(1, 3)$.

(a)

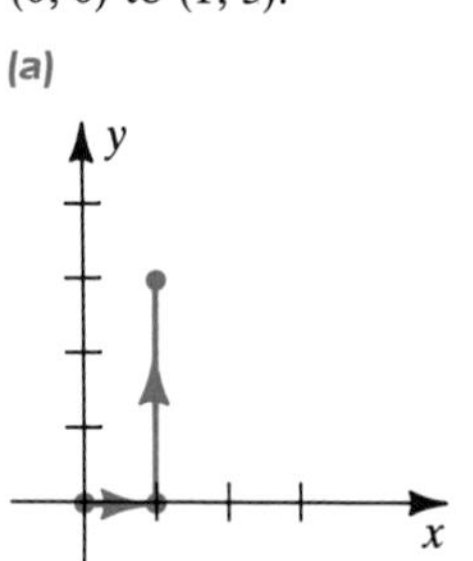

(b)

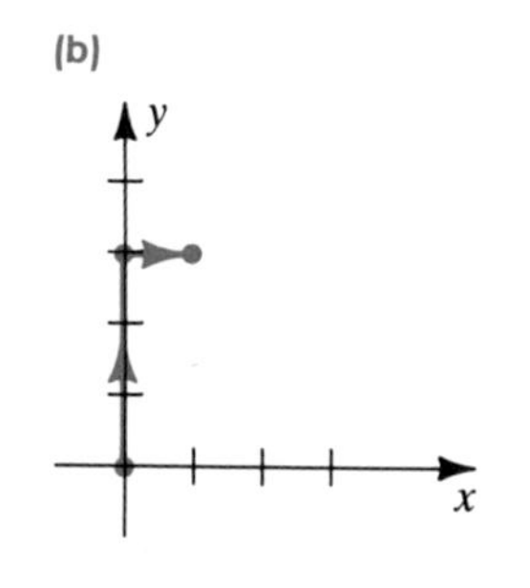

(c)

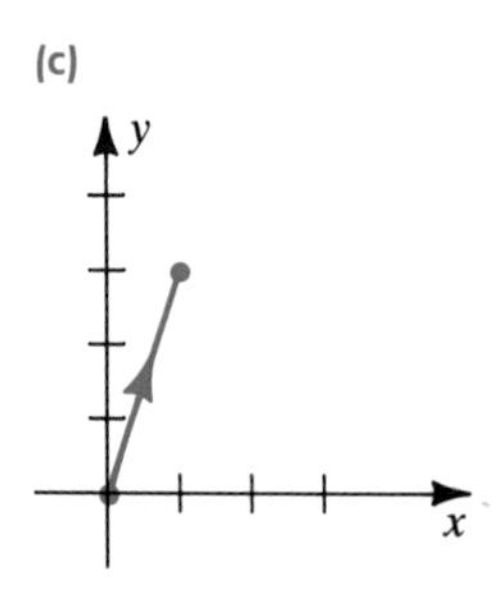

(d)

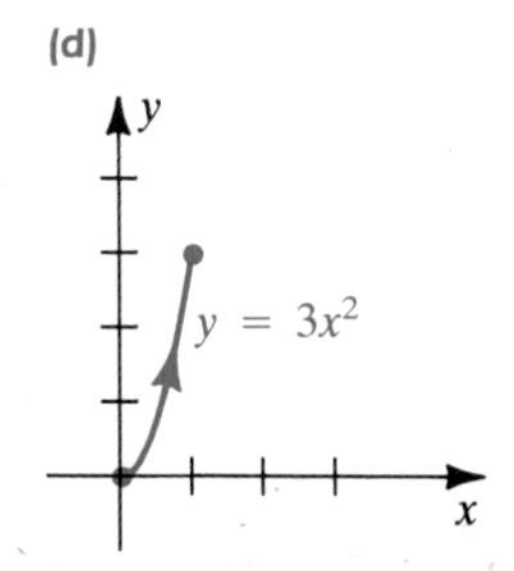

8 Evaluate $\int_C (x^2 + y^2)\,dx + 2x\,dy$ along each curve C from $(1, 2)$ to $(-2, 8)$.

(a)

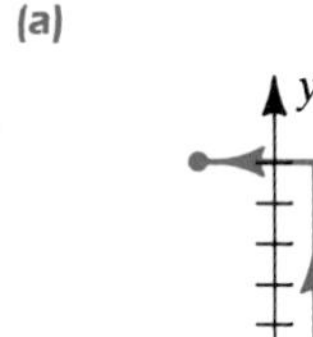

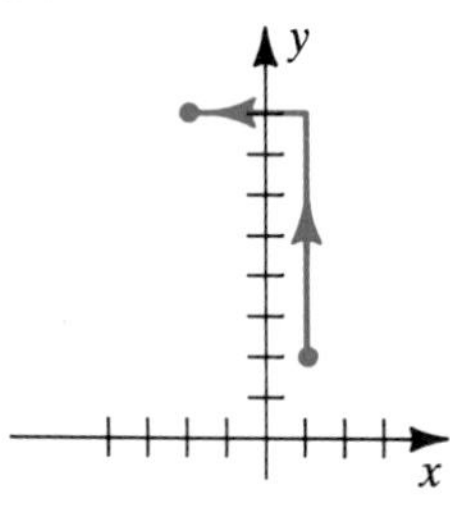

(b)

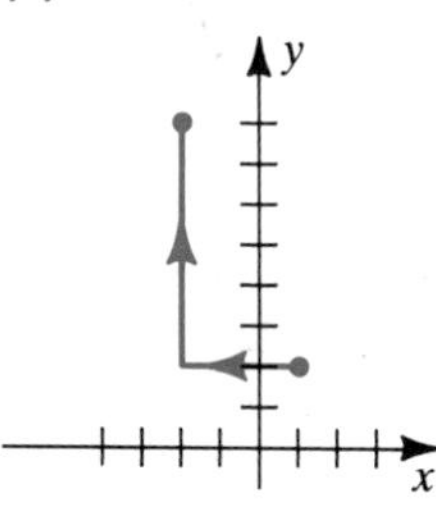

(c)

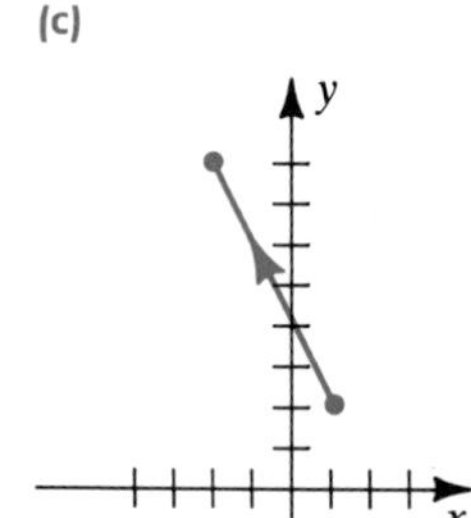

(d)

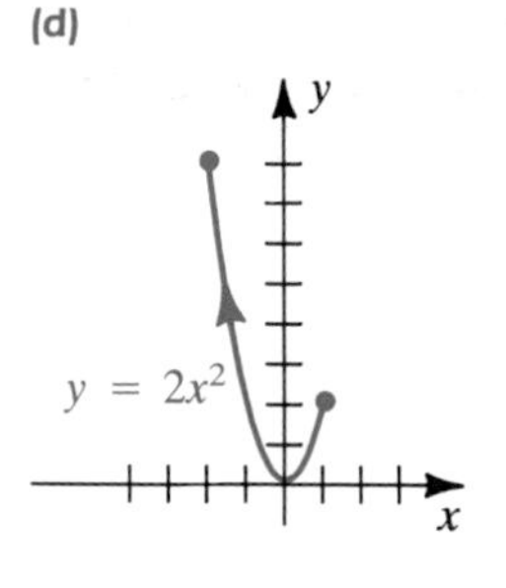

9 Evaluate $\int_C xz\,dx + (y + z)\,dy + x\,dz$ if C is the graph of $x = e^t$, $y = e^{-t}$, $z = e^{2t}$; $0 \le t \le 1$.

10 Evaluate $\int_C y\,dx + z\,dy + x\,dz$ if C is the graph of $x = \sin t$, $y = 2\sin t$, $z = \sin^2 t$; $0 \le t \le \pi/2$.

11 Evaluate

$$\int_C (x + y + z)\,dx + (x - 2y + 3z)\,dy + (2x + y - z)\,dz,$$

where C is the curve from $(0, 0, 0)$ to $(2, 3, 4)$, if

(a) C consists of three line segments, the first parallel to the x-axis, the second parallel to the y-axis, and the third parallel to the z-axis

(b) C consists of three line segments, the first parallel to the z-axis, the second parallel to the x-axis, and the third parallel to the y-axis

(c) C is a line segment

12 Evaluate $\int_C (x - y)\,dx + (y - z)\,dy + x\,dz$ if C is the curve from $(1, -2, 3)$ to $(-4, 5, 2)$ of the type described in (a)–(c) of Exercise 11.

13 Evaluate $\int_C xyz\,ds$ if C is the line segment from $(0, 0, 0)$ to $(1, 2, 3)$.

14 Evaluate $\int_C (xy + z)\,ds$ if C is the helix $x = a\cos t$, $y = a\sin t$, $z = bt$; $0 \le t \le 2\pi$.

15 If the force at (x, y) is $\mathbf{F}(x, y) = xy^2\mathbf{i} + x^2y\mathbf{j}$, find the work done by $\mathbf{F}$ along the curves in (a)–(d) of Exercise 7.

16 If the force at (x, y) is $\mathbf{F}(x, y) = (2x + y)\mathbf{i} + (x + 2y)\mathbf{j}$, find the work done by $\mathbf{F}$ along the curves in (a)–(d) of Exercise 8.

17 The force acting at a point (x, y) in a coordinate plane is $\mathbf{F}(x, y) = (4/\|\mathbf{r}\|^3)\mathbf{r}$, where $\mathbf{r} = x\mathbf{i} + y\mathbf{j}$. Find the work done by $\mathbf{F}$ along the upper half of the circle $x^2 + y^2 = a^2$ from $(-a, 0)$ to $(a, 0)$.

18 The force at a point (x, y) in a coordinate plane is given by $\mathbf{F}(x, y) = (x^2 + y^2)\mathbf{i} + xy\mathbf{j}$. Find the work done by $\mathbf{F}(x, y)$ along the graph of $y = x^3$ from $(0, 0)$ to $(2, 8)$.

19 The force at a point (x, y, z) in three dimensions is given by $\mathbf{F}(x, y, z) = y\mathbf{i} + z\mathbf{j} + x\mathbf{k}$. Find the work done by $\mathbf{F}(x, y, z)$ along the twisted cubic $x = t$, $y = t^2$, $z = t^3$ from $(0, 0, 0)$ to $(2, 4, 8)$.

20 Work Exercise 19 if $\mathbf{F}(x, y, z) = e^x\mathbf{i} + e^y\mathbf{j} + e^z\mathbf{k}$.

21 If an object moves through a force field $\mathbf{F}$ such that at each point (x, y, z) its velocity vector is orthogonal to $\mathbf{F}(x, y, z)$, show that the work done by $\mathbf{F}$ on the object is 0.

22 If a constant force $\mathbf{c}$ acts on a moving object as it travels once around a circle, show that the work done by $\mathbf{c}$ on the object is 0.

23 If a thin wire has the shape of a plane curve C and if the linear mass density at (x, y) is $\delta(x, y)$, then the *moments with respect to the x- and y-axes* are defined by

$$M_x = \int_C y\delta(x, y)\,ds \quad \text{and} \quad M_y = \int_C x\delta(x, y)\,ds.$$

Use limits of sums to show that these are natural definitions for M_x and M_y. The center of mass $(\bar{x}, \bar{y})$ of the wire is defined by $\bar{x} = M_y/m$ and $\bar{y} = M_x/m$, where m is the mass of the wire. Find the center of mass of the wire in Example 3.

24 Refer to Exercise 23. A thin wire is situated in a coordinate plane such that its shape coincides with the part of the parabola $y = 4 - x^2$ between $(-2, 0)$ and $(2, 0)$. The density at the point (x, y) is directly proportional to its distance from the y-axis. Find the mass and center of mass.

25 Extend the definitions of moments and center of mass of a wire to three dimensions (see Exercise 23).

26 A wire of constant density is bent into the shape of the helix $x = a \cos t$, $y = a \sin t$, $z = bt$; $0 \le t \le 3\pi$. Find the mass and center of mass of the wire (see Exercise 25).

27 If a thin wire of variable density is represented by a curve C in the xy-plane, define the moments of inertia I_x and I_y with respect to the x- and y-axes, respectively. Find I_x and I_y for the wire in Example 3.

28 Find the moments of inertia I_x and I_y for the wire in Exercise 24.

29 If a thin wire of variable density is represented by a curve in three dimensions, define the moments of inertia with respect to the x-, y-, and z-axes.

30 Given the wire in Exercise 26, find the moment of inertia with respect to the z-axis (see also Exercise 29).

c 31 If C is the graph of $y = x^4$ from $(0, 0)$ to $(1, 1)$, approximate $\int_C \sin(xy)\, dx$ by using

$$\sum_{k=1}^{10} \sin(u_k v_k)\, \Delta x_k$$

with $\Delta x_k = \frac{1}{10}$ and $u_k = \frac{1}{10}k - \frac{1}{20}$.

c 32 If C is the graph of $y = 3x + 4$ from $(1, 7)$ to $(2, 10)$, approximate $\int_C \ln \sqrt{xy}\, dy$ by using

$$\sum_{k=1}^{10} \ln \sqrt{u_k v_k}\, \Delta y_k$$

with $\Delta y_k = \frac{3}{10}$ and $u_k = \frac{19}{20} + \frac{1}{10}k$.

18.3 INDEPENDENCE OF PATH

A piecewise-smooth curve with endpoints A and B is sometimes called a **path** from A to B. We now obtain conditions under which a line integral is **independent of path** in a region in the sense that if A and B are arbitrary points, then the same value is obtained for *every* path in that region from A to B. The results will be established for line integrals in two dimensions. Proofs for the three-dimensional case are similar and are omitted.

If the line integral $\int_C f(x, y)\, ds$ is independent of path, we may denote it by $\int_A^B f(x, y)\, ds$, since the value of the integral depends only on the endpoints A and B of the curve C. A similar notation is used for $\int_C f(x, y)\, dx$, $\int_C f(x, y)\, dy$, and line integrals in three dimensions.

Throughout this section we assume that all regions are **connected**; that is, any two points in a region can be joined by a piecewise-smooth curve that lies in the region. The next theorem is a fundamental result about independence of path.

Theorem **(18.13)**

If $\mathbf{F}(x, y) = M(x, y)\mathbf{i} + N(x, y)\mathbf{j}$ is continuous on an open connected region D, then the line integral $\int_C \mathbf{F} \cdot d\mathbf{r}$ is independent of path if and only if $\mathbf{F}$ is conservative; that is, $\mathbf{F}(x, y) = \nabla f(x, y)$ for some scalar function f.

PROOF Suppose the integral is independent of path in D. If (x_0, y_0) is a fixed point in D, let f be defined by

$$f(x, y) = \int_{(x_0, y_0)}^{(x, y)} \mathbf{F} \cdot d\mathbf{r}$$

for every point (x, y) in D. We shall show that $\mathbf{F}(x, y) = \nabla f(x, y)$.

Since the integral is independent of path, f depends only on x and y, not on the path C from (x_0, y_0) to (x, y). Choose a circle in D with center (x, y), and let (x_1, y) be a point within the circle such that $x_1 \ne x$. As

FIGURE 18.23

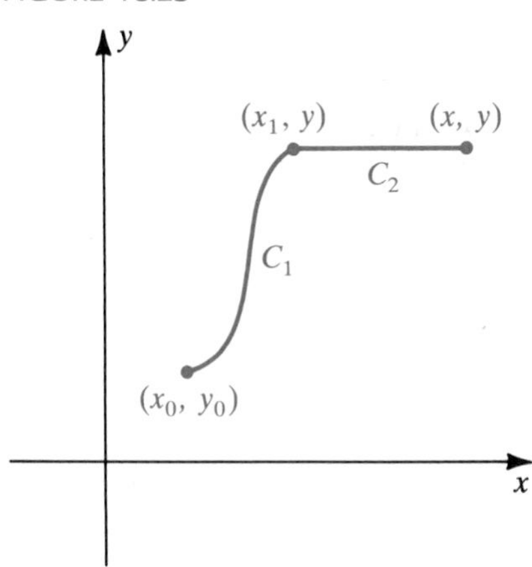

illustrated in Figure 18.23, if C_1 is any path from (x_0, y_0) to (x_1, y) and C_2 is the horizontal segment from (x_1, y) to (x, y), then

$$f(x, y) = \int_{C_1} \mathbf{F} \cdot d\mathbf{r} + \int_{C_2} \mathbf{F} \cdot d\mathbf{r} = \int_{(x_0,y_0)}^{(x_1,y)} \mathbf{F} \cdot d\mathbf{r} + \int_{(x_1,y)}^{(x,y)} \mathbf{F} \cdot d\mathbf{r}.$$

Since the first integral does not depend on x,

$$\frac{\partial}{\partial x} f(x, y) = 0 + \frac{\partial}{\partial x} \int_{(x_1,y)}^{(x,y)} \mathbf{F} \cdot d\mathbf{r}.$$

If we write $\mathbf{F} \cdot d\mathbf{r} = M(x, y)\,dx + N(x, y)\,dy$ and use the fact that $dy = 0$ on C_2 (see Figure 18.23), we obtain

$$\frac{\partial}{\partial x} f(x, y) = \frac{\partial}{\partial x} \int_{(x_1,y)}^{(x,y)} M(x, y)\,dx.$$

Since y is fixed in this partial differentiation, we may regard the integrand as a function of one variable x. Applying (5.35) gives us

$$\frac{\partial}{\partial x} f(x, y) = M(x, y).$$

FIGURE 18.24

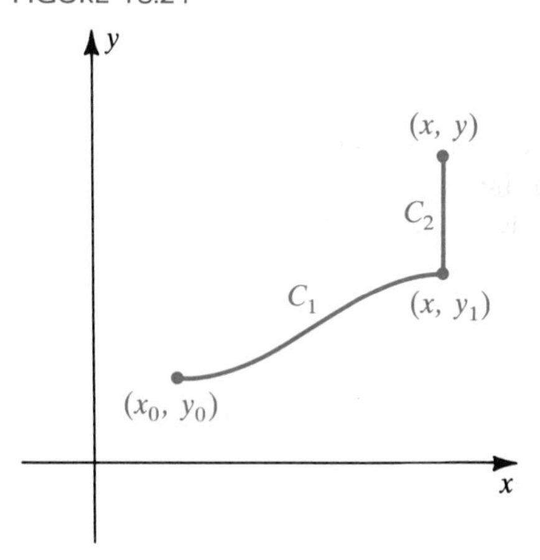

Similarly, if we choose the path shown in Figure 18.24 and differentiate with respect to y, we obtain

$$\frac{\partial}{\partial y} f(x, y) = N(x, y).$$

This proves that $\nabla f(x, y) = M(x, y)\mathbf{i} + N(x, y)\mathbf{j} = \mathbf{F}(x, y)$, which is what we wished to show.

Conversely, if there exists a function f such that $\mathbf{F}(x, y) = \nabla f(x, y)$, then

$$M(x, y)\mathbf{i} + N(x, y)\mathbf{j} = f_x(x, y)\mathbf{i} + f_y(x, y)\mathbf{j}$$

Hence $\qquad M(x, y) = f_x(x, y) \quad \text{and} \quad N(x, y) = f_y(x, y).$

If $A(x_1, y_1)$ and $B(x_2, y_2)$ are points in the region D and if C is any piecewise-smooth curve with endpoints A and B (see Figure 18.25), then

FIGURE 18.25

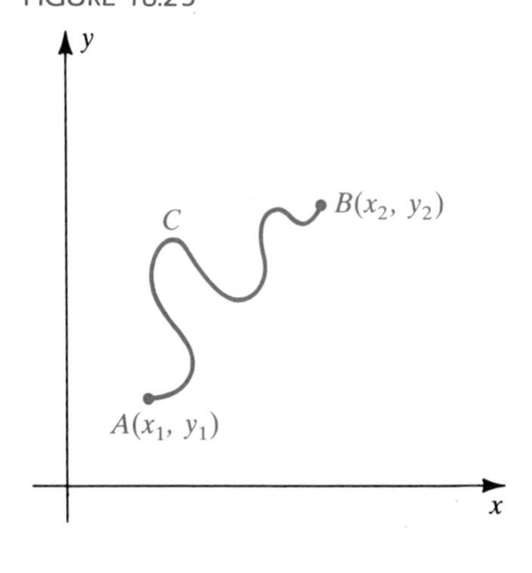

$$\begin{aligned}\int_C \mathbf{F} \cdot d\mathbf{r} &= \int_C M(x, y)\,dx + N(x, y)\,dy \\ &= \int_C f_x(x, y)\,dx + f_y(x, y)\,dy.\end{aligned}$$

If C has the smooth parametrization $x = g(t)$, $y = h(t)$; $t_1 \le t \le t_2$, then, by Theorem (18.9),

$$\int_C \mathbf{F} \cdot d\mathbf{r} = \int_{t_1}^{t_2} [f_x(g(t), h(t))g'(t) + f_y(g(t), h(t))h'(t)]\,dt.$$

Applying a chain rule and the fundamental theorem of calculus, we have

$$\begin{aligned}\int_C \mathbf{F} \cdot d\mathbf{r} &= \int_{t_1}^{t_2} \frac{d}{dt} [f(g(t), h(t))]\,dt \\ &= f(g(t_2), h(t_2)) - f(g(t_1), h(t_1)) \\ &= f(x_2, y_2) - f(x_1, y_1). \\ &= \Big[f(x, y)\Big]_{(x_1,y_1)}^{(x_2,y_2)}.\end{aligned}$$

where the last bracket symbol is analogous to that used for functions of one variable. Thus, the line integral depends only on the coordinates of A and B, not on the path C; that is, $\int_C \mathbf{F} \cdot d\mathbf{r}$ is independent of path. The proof can be extended to piecewise-smooth curves by subdividing C into a finite number of smooth curves. ■

The proof of Theorem (18.13) contains a method for evaluating line integrals that are independent of path. We may state this result, which is analogous to the fundamental theorem of calculus, as follows.

Theorem (18.14)

Let $\mathbf{F}(x, y) = M(x, y)\mathbf{i} + N(x, y)\mathbf{j}$ be continuous on an open connected region D, and let C be a piecewise-smooth curve in D with endpoints $A(x_1, y_1)$ and $B(x_2, y_2)$. If $\mathbf{F}(x, y) = \nabla f(x, y)$, then

$$\int_C M(x, y)\,dx + N(x, y)\,dy = \int_{(x_1,y_1)}^{(x_2,y_2)} \mathbf{F} \cdot d\mathbf{r} = \Big[f(x, y)\Big]_{(x_1,y_1)}^{(x_2,y_2)}.$$

If a line integral $\int_C \mathbf{F} \cdot d\mathbf{r}$ is independent of path, then using Theorem (18.14) with $(x_1, y_1) = (x_2, y_2)$ we see that $\int_C \mathbf{F} \cdot d\mathbf{r} = 0$ *for every simple closed curve* C.

EXAMPLE 1 Let $\mathbf{F}(x, y) = (2x + y^3)\mathbf{i} + (3xy^2 + 4)\mathbf{j}$.

(a) Show that $\int_C \mathbf{F} \cdot d\mathbf{r}$ is independent of path.

(b) Evaluate $\int_{(0,1)}^{(2,3)} \mathbf{F} \cdot d\mathbf{r}$.

SOLUTION

(a) By Theorem (18.13), the line integral is independent of path if and only if there exists a differentiable function f of x and y such that $\nabla f(x, y) = \mathbf{F}(x, y)$; that is,

$$f_x(x, y)\mathbf{i} + f_y(x, y)\mathbf{j} = (2x + y^3)\mathbf{i} + (3xy^2 + 4)\mathbf{j},$$

or, equivalently,

$$(*) \qquad f_x(x, y) = 2x + y^3 \quad \text{and} \quad f_y(x, y) = 3xy^2 + 4.$$

If we (partially) integrate $f_x(x, y) = 2x + y^3$ with respect to x, we obtain

$$f(x, y) = x^2 + xy^3 + g(y),$$

where g is a function of y alone. (We must use $g(y)$ instead of a constant of integration in order to obtain the most *general* expression for $f(x, y)$ such that $f_x(x, y) = 2x + y^3$.)

We next differentiate $f(x, y) = x^2 + xy^3 + g(y)$ with respect to y and compare the result with the expression $f_y(x, y) = 3xy^2 + 4$ in $(*)$. Thus,

$$f_y(x, y) = 0 + 3xy^2 + g'(y) = 3xy^2 + 4$$

and hence

$$g'(y) = 4.$$

Integrating with respect to y, we obtain

$$g(y) = 4y + c,$$

where c is a constant of integration. Thus,

$$f(x, y) = x^2 + xy^3 + 4y + c$$

gives us a function f such that $\nabla f(x, y) = \mathbf{F}(x, y)$.

(b) Since any potential function may be used to evaluate the integral, we apply Theorem (18.14) with $f(x, y) = x^2 + xy^3 + 4y$, obtaining

$$\begin{aligned}\int_{(0,1)}^{(2,3)} \mathbf{F} \cdot d\mathbf{r} &= \Big[f(x, y)\Big]_{(0,1)}^{(2,3)} \\ &= \Big[x^2 + xy^3 + 4y\Big]_{(0,1)}^{(2,3)} \\ &= (4 + 54 + 12) - 4 = 66.\end{aligned}$$

The following result is an application of the preceding discussion to conservative vector fields.

Theorem **(18.15)**

> If $\mathbf{F}$ is a conservative force field in two dimensions, then the work done by $\mathbf{F}$ along any path C from $A(x_1, y_1)$ to $B(x_2, y_2)$ is equal to the difference in potentials between A and B.

PROOF If $\mathbf{F}$ is conservative, then $\mathbf{F}(x, y) = \nabla f(x, y)$ for some potential function f. By Definition (18.12) and Theorem (18.14), the work W done along any path from $A(x_1, y_1)$ to $B(x_2, y_2)$ is

$$W = \int_{(x_1,y_1)}^{(x_2,y_2)} \mathbf{F} \cdot d\mathbf{r} = f(x_2, y_2) - f(x_1, y_1),$$

which is the difference in the potentials between A and B. ■

FIGURE 18.26

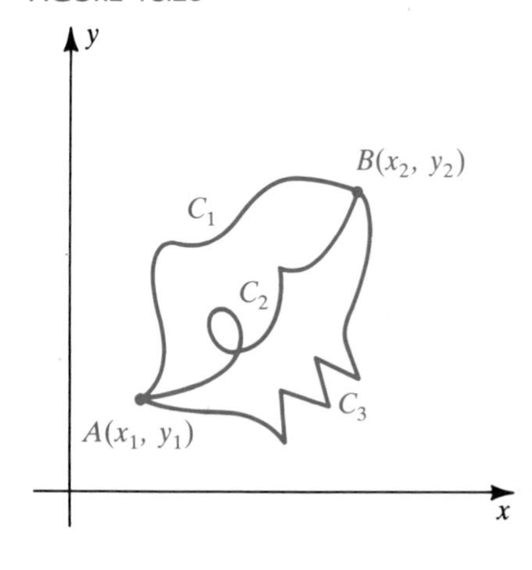

The preceding theorem is illustrated in Figure 18.26 for piecewise-smooth curves C_1, C_2, and C_3. The same amount of work is done no matter which path is taken from A to B. If C is a closed curve—that is, if $A = B$—then the difference in potentials is 0, and hence the work done in traversing C is 0. A converse of this result is also true—namely, if $\int_C \mathbf{F} \cdot d\mathbf{r} = 0$ for every simple closed curve C, then the line integral is independent of path and hence the field is conservative. These facts are important in applications, since many of the vector fields that occur in nature are inverse square fields and hence are conservative (see Theorem (18.4)). In physical terms, if a unit particle moves completely around a closed curve in a conservative force field, then the work done is 0. We will discuss conservative fields in more detail at the end of this section.

Theorems (18.14) and (18.15) can be extended to a vector field in three dimensions. The next example is an illustration of these results.

EXAMPLE 2 Let $\mathbf{F}$ be the gravitational field of a particle of mass m_0 located at the origin of an xyz-coordinate system. Find the work done by $\mathbf{F}$ as a particle of mass m moves from $A(2, 3, 4)$ to $B(1, 0, 0)$.

SOLUTION From page 966, the force exerted on a particle of mass m located at $K(x, y, z)$ is

$$\mathbf{F}(x, y, z) = -G\frac{m_0 m}{\|\mathbf{r}\|^3}\mathbf{r},$$

where $\mathbf{r} = x\mathbf{i} + y\mathbf{j} + z\mathbf{k}$. As in the proof of Theorem (18.4), $\mathbf{F}(x, y, z) = \nabla f(x, y, z)$, where

$$f(x, y, z) = \frac{Gm_0m}{(x^2 + y^2 + z^2)^{1/2}}.$$

Hence, by the extension of Theorems (18.14) and (18.15) to three dimensions,

$$W = \int_A^B \mathbf{F} \cdot d\mathbf{r} = \left[\frac{Gm_0m}{(x^2 + y^2 + z^2)^{1/2}}\right]_{(2,3,4)}^{(1,0,0)} = Gm_0m\left(1 - \frac{1}{\sqrt{29}}\right).$$

If the integral $\int_C M(x, y)\,dx + N(x, y)\,dy$ is independent of path, then, by Theorem (18.13), there is a function f such that

$$M = \frac{\partial f}{\partial x} \quad \text{and} \quad N = \frac{\partial f}{\partial y}.$$

Consequently, $$\frac{\partial M}{\partial y} = \frac{\partial^2 f}{\partial y\, \partial x} \quad \text{and} \quad \frac{\partial N}{\partial x} = \frac{\partial^2 f}{\partial x\, \partial y}.$$

If M and N have continuous first partial derivatives, then f has continuous *second* partial derivatives, and hence the order of differentiation is immaterial; that is,

$$\frac{\partial M}{\partial y} = \frac{\partial N}{\partial x}.$$

The converse of this result is false, unless we place additional restrictions on the domain D of $\mathbf{F}(x, y)$. In particular, if D is a **simply connected region** in the sense that every simple closed curve C in D encloses only points in D (there are no *holes* in the region), then the condition $\partial M/\partial y = \partial N/\partial x$ implies that the line integral $\int_C M(x, y)\,dx + N(x, y)\,dy$ is independent of path. (A proof may be found in books on advanced calculus.) Our discussion can be summarized as follows.

Theorem **(18.16)**

If $M(x, y)$ and $N(x, y)$ have continuous first partial derivatives on a simply connected region D, then the line integral

$$\int_C M(x, y)\,dx + N(x, y)\,dy$$

is independent of path in D if and only if

$$\frac{\partial M}{\partial y} = \frac{\partial N}{\partial x}.$$

EXAMPLE 3 Show that the line integral

$$\int_C (e^{3y} - y^2 \sin x)\,dx + (3xe^{3y} + 2y \cos x)\,dy$$

is independent of path in a simply connected region.

SOLUTION Letting

$$M = e^{3y} - y^2 \sin x \quad \text{and} \quad N = 3xe^{3y} + 2y \cos x,$$

we see that

$$\frac{\partial M}{\partial y} = 3e^{3y} - 2y \sin x = \frac{\partial N}{\partial x}.$$

Hence, by Theorem (18.16), the line integral is independent of path.

EXAMPLE 4 Determine whether $\int_C x^2 y\, dx + 3xy^2\, dy$ is independent of path.

SOLUTION If we let $M = x^2 y$ and $N = 3xy^2$, then

$$\frac{\partial M}{\partial y} = x^2 \quad \text{and} \quad \frac{\partial N}{\partial x} = 3y^2.$$

Since $\partial M/\partial y \neq \partial N/\partial x$, the integral is not independent of path, by Theorem (18.16).

In Example 1 we determined a potential function f for a two-dimensional conservative vector field. The solution of the next example illustrates a similar technique that can be used in three dimensions.

EXAMPLE 5 Let $\mathbf{F}(x, y, z) = y^2 \cos x\mathbf{i} + (2y \sin x + e^{2z})\mathbf{j} + 2ye^{2z}\mathbf{k}$.

(a) Show that $\int_C \mathbf{F} \cdot d\mathbf{r}$ is independent of path, and find a potential function f for $\mathbf{F}$.

(b) If $\mathbf{F}$ is a force field, find the work done by $\mathbf{F}$ along any curve C from $(0, 1, \frac{1}{2})$ to $(\pi/2, 3, 2)$.

SOLUTION

(a) The integral is independent of path if there exists a differentiable function f of x, y, and z such that $\nabla f(x, y, z) = \mathbf{F}(x, y, z)$; that is,

(1)
$$\begin{aligned} f_x(x, y, z) &= y^2 \cos x \\ f_y(x, y, z) &= 2y \sin x + e^{2z} \\ f_z(x, y, z) &= 2ye^{2z}. \end{aligned}$$

If we (partially) integrate $f_x(x, y, z)$ with respect to x, we see that

(2)
$$f(x, y, z) = y^2 \sin x + g(y, z)$$

for some function g of y and z. (We must use $g(y, z)$ in order to obtain the *most general* expression $f(x, y, z)$ such that $f_x(x, y, z) = y^2 \cos x$.)

Differentiating $f(x, y, z)$ in (2) with respect to y and then comparing the result with the formula for $f_y(x, y, z)$ in (1), we obtain

$$f_y(x, y, z) = 2y \sin x + g_y(y, z) = 2y \sin x + e^{2z}$$

and hence

$$g_y(y, z) = e^{2z}.$$

A partial integration of $g_y(y, z)$ with respect to y gives us

$$g(y, z) = ye^{2z} + h(z),$$

where h is a function of the one variable z. Using formula (2) yields

$$f(x, y, z) = y^2 \sin x + ye^{2z} + h(z).$$

Differentiating the preceding expression with respect to z and then comparing the result with the formula for $f_z(x, y, z)$ in (1), we have

$$f_z(x, y, z) = 2ye^{2z} + h'(z) = 2ye^{2z}.$$

It follows that $h'(z) = 0$ or $h(z) = c$ for some constant c. Thus,

$$f(x, y, z) = y^2 \sin x + ye^{2z} + c$$

gives us a potential function f for **F**.

(b) Using the extension of Theorem (18.15) to vector fields in three dimensions, we find that the work done by **F** along any curve C from $(0, 1, \frac{1}{2})$ to $(\pi/2, 3, 2)$ is

$$\begin{aligned}\int_C \mathbf{F} \cdot d\mathbf{r} &= \Big[y^2 \sin x + ye^{2z}\Big]_{(0,1,1/2)}^{(\pi/2,3,2)} \\ &= (9 + 3e^4) - (0 + e) = 9 + 3e^4 - e \approx 170.\end{aligned}$$

We shall conclude this section by considering some applications of independence of path. The next definition is used in physics.

Definition **(18.17)**

Let $\mathbf{F}(x, y, z)$ be a conservative vector force field with potential function f. The **potential energy** $p(x, y, z)$ of a particle at the point (x, y, z) is

$$p(x, y, z) = -f(x, y, z).$$

Since the potential energy is the *negative* of the potential $f(x, y, z)$,

$$\mathbf{F}(x, y, z) = -\nabla p(x, y, z).$$

If A and B are two points, then, as in Theorem (18.15), the work done by **F** along a smooth curve C with endpoints A and B is

$$W = \int_A^B \mathbf{F} \cdot d\mathbf{r} = \Big[-p(x, y, z)\Big]_A^B = p(A) - p(B),$$

where $p(A)$ and $p(B)$ denote the potential energy at A and B, respectively. Thus, the work W is the difference in the potential energies at A and B. In particular, if B is a point at which the potential is 0, then $W = p(A)$. This is in agreement with the classical physical description of potential energy as the type of energy that a body has by virtue of its position.

If a particle has mass m and speed v, then its *kinetic energy* is $\frac{1}{2}mv^2$. Thus, kinetic energy is the energy a body has by virtue of its speed and mass.

Let us establish the following fundamental physical law, which is the principal reason for referring to certain vector fields as *conservative*.

Law of conservation of energy **(18.18)**

> If a particle moves from one point to another in a conservative vector force field, then the sum of the potential and kinetic energies remains constant; that is, the total energy does not change.

FIGURE 18.27

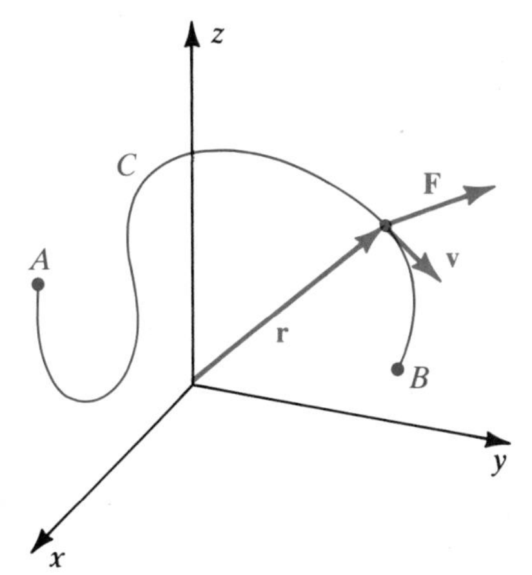

PROOF Let $\mathbf{F}$ be a conservative vector field, and suppose that a particle moves from A to B along a piecewise-smooth curve C such that its position at time t is given by $x = g(t)$, $y = h(t)$, $z = k(t)$; $a \le t \le b$. If $\mathbf{r} = x\mathbf{i} + y\mathbf{j} + z\mathbf{k}$ is the position vector for the particle (see Figure 18.27), then the velocity and acceleration at time t are

$$\mathbf{v} = \frac{d\mathbf{r}}{dt} \quad \text{and} \quad \mathbf{a} = \frac{d\mathbf{v}}{dt},$$

where we have used the abbreviations $\mathbf{v}$ and $\mathbf{a}$ for $\mathbf{v}(t)$ and $\mathbf{a}(t)$, respectively. The speed of the particle is $v(t) = v = \|\mathbf{v}\|$. The work done by $\mathbf{F}(x, y, z)$ along C is

$$W = \int_A^B \mathbf{F} \cdot d\mathbf{r} = \int_a^b \left(\mathbf{F} \cdot \frac{d\mathbf{r}}{dt}\right) dt = \int_a^b (\mathbf{F} \cdot \mathbf{v})\, dt.$$

By Newton's second law of motion (15.13),

$$\mathbf{F} = m\mathbf{a} = m\frac{d\mathbf{v}}{dt}$$

and hence

$$\begin{aligned}
W &= \int_a^b \left(m\frac{d\mathbf{v}}{dt} \cdot \mathbf{v}\right) dt \\
&= m\int_a^b \left(\frac{d\mathbf{v}}{dt} \cdot \mathbf{v}\right) dt = m\int_a^b \frac{1}{2}\frac{d}{dt}(\mathbf{v} \cdot \mathbf{v})\, dt \\
&= \frac{1}{2}m\int_a^b \frac{d}{dt}(v^2)\, dt = \frac{1}{2}m\Big[v^2\Big]_a^b \\
&= \frac{1}{2}m[v(b)]^2 - \frac{1}{2}m[v(a)]^2.
\end{aligned}$$

Since the kinetic energy K of a particle of mass m and velocity v is $\frac{1}{2}mv^2$, the last equality may be written

$$W = K(B) - K(A),$$

where $K(A)$ and $K(B)$ denote the kinetic energy of the particle at A and B, respectively. From the discussion following Definition (18.17), $W = p(A) - p(B)$, and hence

$$p(A) - p(B) = K(B) - K(A),$$

or

$$p(A) + K(A) = p(B) + K(B).$$

The last formula states that if a particle moves from one point to another in a conservative vector field, then the sum of the potential and kinetic energies is constant. ■

EXERCISES 18.3

Exer. 1–10: Show that $\int_C \mathbf{F} \cdot d\mathbf{r}$ is independent of path by finding a potential function f for F.

1 $\mathbf{F}(x, y) = (3x^2y + 2)\mathbf{i} + (x^3 + 4y^3)\mathbf{j}$

2 $\mathbf{F}(x, y) = (6xy^2 + 2y)\mathbf{i} + (6x^2y + 2x)\mathbf{j}$

3 $\mathbf{F}(x, y) = (2x \sin y + 4e^x)\mathbf{i} + x^2 \cos y\mathbf{j}$

4 $\mathbf{F}(x, y) = (2xe^{2y} + 4y^3)\mathbf{i} + (2x^2e^{2y} + 12xy^2)\mathbf{j}$

5 $\mathbf{F}(x, y) = -2y^3 \sin x\mathbf{i} + (6y^2 \cos x + 5)\mathbf{j}$

6 $\mathbf{F}(x, y) = (5y^3 + 4y^3 \sec^2 x)\mathbf{i} + (15xy^2 + 12y^2 \tan x)\mathbf{j}$

7 $\mathbf{F}(x, y, z) = 8xz\mathbf{i} + (1 - 6yz^3)\mathbf{j} + (4x^2 - 9y^2z^2)\mathbf{k}$

8 $\mathbf{F}(x, y, z) = (y + z)\mathbf{i} + (x + z)\mathbf{j} + (x + y)\mathbf{k}$

9 $\mathbf{F}(x, y, z) = (y \sec^2 x - ze^x)\mathbf{i} + \tan x\mathbf{j} - e^x\mathbf{k}$

10 $\mathbf{F}(x, y, z) = 2x \sin z\mathbf{i} + 2y \cos z\mathbf{j} + (x^2 \cos z - y^2 \sin z)\mathbf{k}$

Exer. 11–14: Show that the line integral is independent of path, and find its value.

11 $\int_{(-1,2)}^{(3,1)} (y^2 + 2xy)\, dx + (x^2 + 2xy)\, dy$

12 $\int_{(0,0)}^{(1,\pi/2)} e^x \sin y\, dx + e^x \cos y\, dy$

13 $\int_{(1,0,2)}^{(-2,1,3)} (6xy^3 + 2z^2)\, dx + 9x^2y^2\, dy + (4xz + 1)\, dz$

14 $\int_{(4,0,3)}^{(-1,1,2)} (yz + 1)\, dx + (xz + 1)\, dy + (xy + 1)\, dz$

Exer. 15–18: Use Theorem (18.16) to show that $\int_C \mathbf{F} \cdot d\mathbf{r}$ is not independent of path.

15 $\mathbf{F}(x, y) = 4xy^3\mathbf{i} + 2xy^3\mathbf{j}$

16 $\mathbf{F}(x, y) = (6x^2 - 2xy^2)\mathbf{i} + (2x^2y + 5)\mathbf{j}$

17 $\mathbf{F}(x, y) = e^x\mathbf{i} + (3 - e^x \sin y)\mathbf{j}$

18 $\mathbf{F}(x, y) = y^3 \cos x\mathbf{i} - 3y^2 \sin x\mathbf{j}$

19 If $\int_C M(x, y, z)\, dx + N(x, y, z)\, dy + P(x, y, z)\, dz$ is independent of path and M, N, and P have continuous first partial derivatives, prove that

$$\frac{\partial M}{\partial y} = \frac{\partial N}{\partial x}, \quad \frac{\partial M}{\partial z} = \frac{\partial P}{\partial x}, \quad \frac{\partial N}{\partial z} = \frac{\partial P}{\partial y}.$$

Exer. 20–22: Use Exercise 19 to show that the line integral is not independent of path.

20 $\int_C 5y\, dx + 5x\, dy + yz^2\, dz$

21 $\int_C 2xy\, dx + (x^2 + z^2)\, dy + yz\, dz$

22 $\int_C e^y \cos z\, dx + xe^y \cos z\, dy + xe^y \sin z\, dz$

23 Suppose a force $\mathbf{F}(x, y, z)$ is directed toward the origin with a magnitude that is inversely proportional to the distance from the origin. Prove that $\mathbf{F}$ is conservative by finding a potential function f for $\mathbf{F}$.

24 Suppose a force $\mathbf{F}(x, y, z)$ is directed away from the origin with a magnitude that is directly proportional to the distance from the origin. Prove that $\mathbf{F}$ is conservative by finding a potential function f for $\mathbf{F}$.

25 If $\mathbf{F}$ is a constant force, prove that the work done along any curve with endpoints P and Q is $\mathbf{F} \cdot \overrightarrow{PQ}$.

26 If $\mathbf{F}(x, y, z) = g(x)\mathbf{i} + h(y)\mathbf{j} + k(z)\mathbf{k}$, where g, h, and k are continuous functions, show that $\mathbf{F}$ is conservative.

27 Let $\mathbf{F}(x, y) = M(x, y)\mathbf{i} + N(x, y)\mathbf{j}$, where

$$M(x, y) = \frac{-y}{x^2 + y^2} \quad \text{and} \quad N(x, y) = \frac{x}{x^2 + y^2}.$$

Show that $\partial M/\partial y = \partial N/\partial x$ for every (x, y) in the domain D of $\mathbf{F}$ but that $\int_C \mathbf{F} \cdot d\mathbf{r}$ is not independent of path in D. (*Hint:* Consider two different semicircles with endpoints $(-1, 0)$ and $(1, 0)$.) Why doesn't this contradict Theorem (18.16)?

28 Suppose that a satellite of mass m is orbiting the earth at a constant altitude of h miles and that it completes one revolution every k minutes. If the radius of the earth is taken as 4000 miles, find the work done by the gravitational field of the earth during any time interval.

29 Let $\mathbf{F}$ be an inverse square field such that $\mathbf{F}(x, y, z) = (c/\|\mathbf{r}\|^3)\mathbf{r}$ for $\mathbf{r} = x\mathbf{i} + y\mathbf{j} + z\mathbf{k}$, where c is a constant. Let P_1 and P_2 be any points whose distances from the origin are d_1 and d_2, respectively. Express, in terms of d_1 and d_2, the work done by $\mathbf{F}$ along any piecewise-smooth curve joining P_1 to P_2.

30 Suppose that as a particle moves through a conservative vector field, its potential energy is decreasing at a rate of k units per second. At what rate is its kinetic energy changing?

18.4 GREEN'S THEOREM

Let C be a smooth plane curve with parametrization $x = g(t)$, $y = h(t)$; $a \le t \le b$. Recall that if $A = (g(a), h(a)) = (g(b), h(b)) = B$, then C is a smooth *closed* curve. If C does not cross itself between A and B, it is *simple*. Some common examples of smooth simple closed curves are circles and ellipses. A *piecewise-smooth simple closed curve* is a finite union of smooth curves C_k such that as t varies from a to b, the point $P(t)$ obtained from parametrizations of the C_k traces C exactly once, with the exception that $P(a) = P(b)$. A curve of this type forms the boundary of a region R in the xy-plane, and, by definition, the **orientation**, or **positive direction**, along C is such that R is on the left as $P(t)$ traces C, as illustrated by the arrowheads in Figure 18.28. The notation

FIGURE 18.28

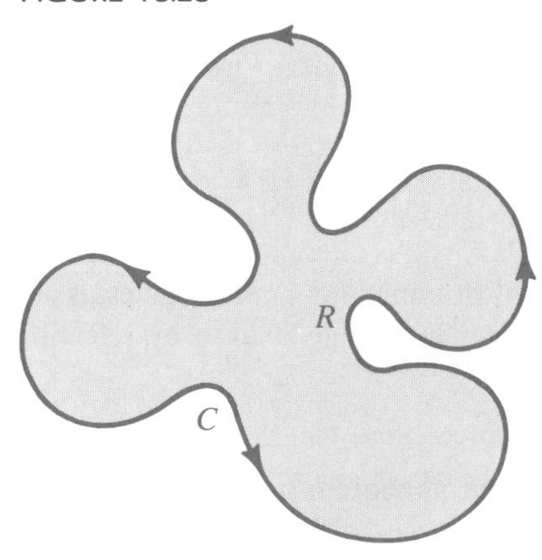

$$\oint_C M(x, y)\,dx + N(x, y)\,dy$$

denotes a line integral along a simple closed curve C once in the positive direction.

The next theorem, named after the English mathematical physicist George Green (1793–1841), specifies a relationship between the line integral around C and a double integral over R. To simplify the statement, the symbols M, N, $\partial N/\partial x$, and $\partial M/\partial y$ are used in the integrands to denote the values of these functions at (x, y).

Green's theorem **(18.19)**

Let C be a piecewise-smooth simple closed curve, and let R be the region consisting of C and its interior. If M and N are continuous functions that have continuous first partial derivatives throughout an open region D containing R, then

$$\oint_C M\,dx + N\,dy = \iint_R \left(\frac{\partial N}{\partial x} - \frac{\partial M}{\partial y}\right) dA.$$

PROOF FOR A SPECIAL CASE We shall prove the theorem if R is *both* an R_x region and an R_y region of the following types:

$$R = \{(x, y): a \le x \le b,\ g_1(x) \le y \le g_2(x)\}$$

$$R = \{(x, y): c \le y \le d,\ h_1(y) \le x \le h_2(y)\},$$

FIGURE 18.29

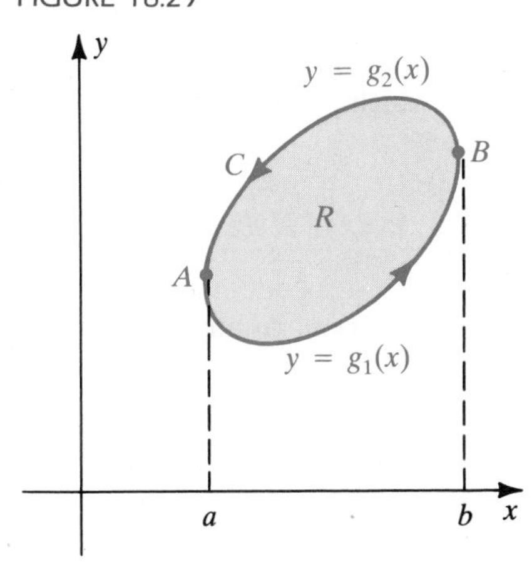

where g_1, g_2, h_1, and h_2 are smooth functions. It is sufficient to show that each of the following is true, since addition of these integrals produces the desired conclusion:

(1) $$\oint_C M\,dx = -\iint_R \frac{\partial M}{\partial y}\,dA$$

(2) $$\oint_C N\,dy = \iint_R \frac{\partial N}{\partial x}\,dA$$

To prove (1), we refer to Figure 18.29 and note that C consists of two curves C_1 and C_2 that have equations $y = g_1(x)$ and $y = g_2(x)$, respectively.

The line integral $\oint_C M\,dx$ may be written

$$\begin{aligned}\oint_C M\,dx &= \int_{C_1} M(x, y)\,dx + \int_{C_2} M(x, y)\,dx \\ &= \int_a^b M(x, g_1(x))\,dx + \int_b^a M(x, g_2(x))\,dx \\ &= \int_a^b M(x, g_1(x))\,dx - \int_a^b M(x, g_2(x))\,dx.\end{aligned}$$

FIGURE 18.30

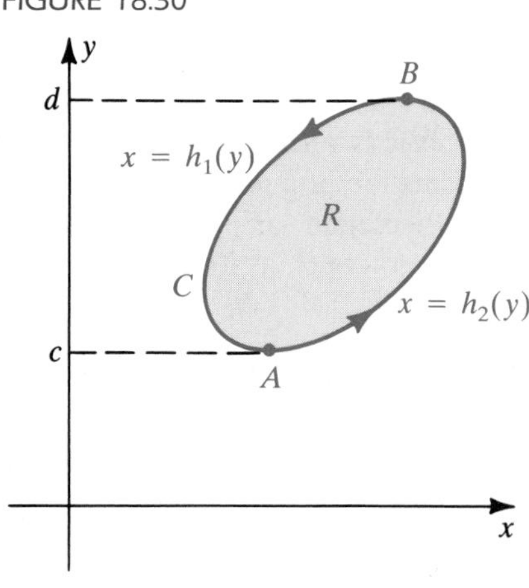

We next apply Theorem (17.8)(i) to the double integral $\iint_R (\partial M/\partial y)\,dA$, obtaining

$$\begin{aligned}\iint_R \frac{\partial M}{\partial y}\,dA &= \int_a^b \int_{g_1(x)}^{g_2(x)} \frac{\partial M}{\partial y}\,dy\,dx \\ &= \int_a^b \Big[M(x, y)\Big]_{g_1(x)}^{g_2(x)}\,dx \\ &= \int_a^b [M(x, g_2(x)) - M(x, g_1(x))]\,dx.\end{aligned}$$

Comparing the last expression with that obtained for $\oint_C M\,dx$ gives us (1).

The formula in (2) may be established in similar fashion by referring to Figure 18.30. ■

Although we shall not prove it, Green's theorem may be extended to a region R where part of the boundary consists of horizontal or vertical line segments. The theorem may then be further extended to the case where R is a finite union of such regions. For example, if $R = R_1 \cup R_2$ and if the boundary of R_1 is $C_1 \cup C_1'$ and the boundary of R_2 is $C_2 \cup C_2'$, as illustrated in Figure 18.31, then

FIGURE 18.31

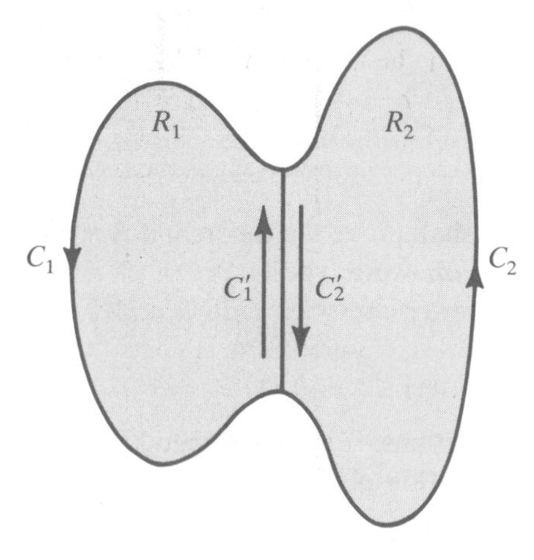

$$\iint_{R_1} \left(\frac{\partial N}{\partial x} - \frac{\partial M}{\partial y}\right) dy\,dx = \oint_{C_1 \cup C_1'} M\,dx + N\,dy$$

$$\iint_{R_2} \left(\frac{\partial N}{\partial x} - \frac{\partial M}{\partial y}\right) dy\,dx = \oint_{C_2 \cup C_2'} M\,dx + N\,dy.$$

The sum of the two double integrals equals a double integral over R. We next observe that a line integral along C_1' in the direction indicated in Figure 18.31 is the negative of that along C_2', since the curve is the same but the direction is opposite. Hence the sum of the two line integrals reduces to a line integral along $C_1 \cup C_2$, which is the boundary C of R. Thus,

$$\iint_R \left(\frac{\partial N}{\partial x} - \frac{\partial M}{\partial y}\right) dy\,dx = \oint_C M\,dx + N\,dy.$$

We may now proceed, by mathematical induction, to any finite union. The proof of Green's theorem for the most general case is beyond the scope of this book.

EXAMPLE 1 Use Green's theorem to evaluate $\oint_C 5xy\,dx + x^3\,dy$, where C is the closed curve consisting of the graphs of $y = x^2$ and $y = 2x$ between the points $(0, 0)$ and $(2, 4)$.

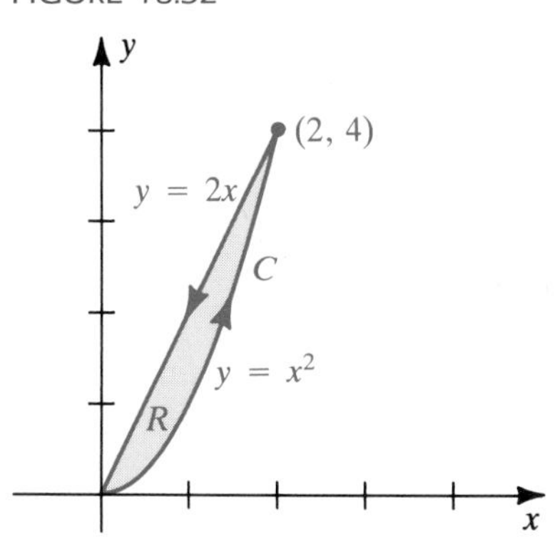

FIGURE 18.32

SOLUTION The region R bounded by C is shown in Figure 18.32. Applying Green's theorem, with $M(x, y) = 5xy$ and $N(x, y) = x^3$, gives us

$$\begin{aligned}\oint_C 5xy\,dx + x^3\,dy &= \iint_R \left[\frac{\partial}{\partial x}(x^3) - \frac{\partial}{\partial y}(5xy)\right] dA \\ &= \int_0^2 \int_{x^2}^{2x} (3x^2 - 5x)\,dy\,dx \\ &= \int_0^2 \Big[3x^2y - 5xy\Big]_{x^2}^{2x} dx \\ &= \int_0^2 (11x^3 - 10x^2 - 3x^4)\,dx = -\tfrac{28}{15}.\end{aligned}$$

The line integral can also be evaluated directly.

EXAMPLE 2 Use Green's theorem to evaluate the line integral

$$\oint_C 2xy\,dx + (x^2 + y^2)\,dy$$

if C is the ellipse $4x^2 + 9y^2 = 36$.

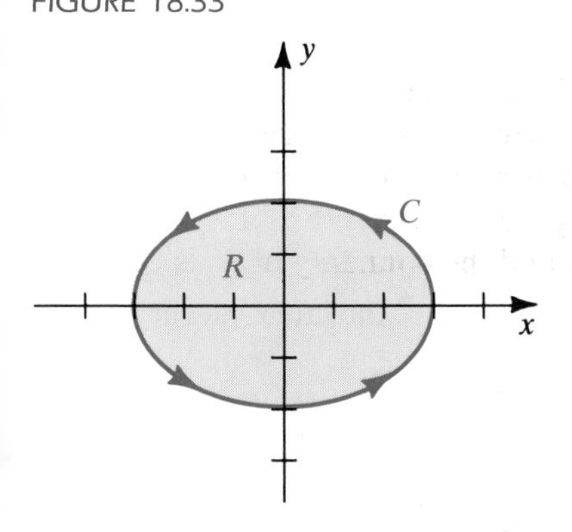

FIGURE 18.33

SOLUTION The region R bounded by C is shown in Figure 18.33. Applying Green's theorem, with $M(x, y) = 2xy$ and $N(x, y) = x^2 + y^2$, we have

$$\begin{aligned}\oint_C 2xy\,dx + (x^2 + y^2)\,dy &= \iint_R (2x - 2x)\,dA \\ &= \iint_R 0\,dA = 0.\end{aligned}$$

Note that we did not use the curve C in the evaluation. We can show that the line integral is independent of path and hence is zero for *every* simple closed curve C (see the remark following Theorem (18.14)).

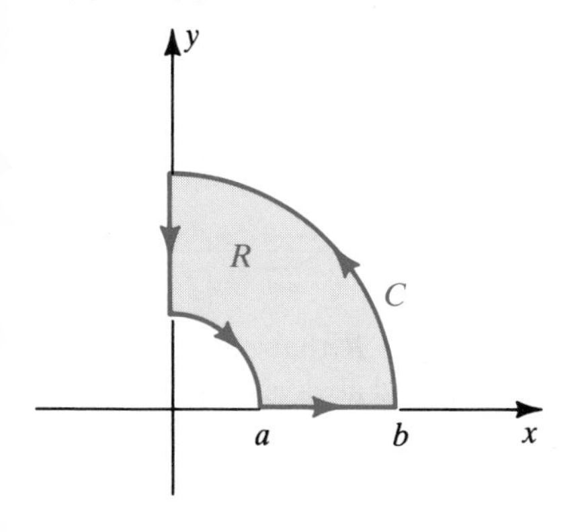

FIGURE 18.34

EXAMPLE 3 Evaluate $\oint_C (4 + e^{\sqrt{x}})\,dx + (\sin y + 3x^2)\,dy$ if C is the boundary of the region R between quarter-circles of radii a and b and segments on the x- and y-axes, as shown in Figure 18.34.

SOLUTION If we apply Green's theorem and then change the double integral to polar coordinates, the line integral equals

$$\begin{aligned}\iint_R (6x - 0)\,dA &= \int_0^{\pi/2} \int_a^b (6r\cos\theta)r\,dr\,d\theta \\ &= 6\int_0^{\pi/2} \cos\theta\Big[\tfrac{1}{3}r^3\Big]_a^b d\theta = 2(b^3 - a^3)\int_0^{\pi/2} \cos\theta\,d\theta \\ &= 2(b^3 - a^3)\Big[\sin\theta\Big]_0^{\pi/2} = 2(b^3 - a^3).\end{aligned}$$

A deeper appreciation of Green's theorem can be gained by attempting to evaluate the line integral in this example directly.

Green's theorem may be used to derive a formula for finding the area A of a region R that is bounded by a piecewise-smooth simple closed curve C. If we let $M = 0$ and $N = x$ in (18.19), we obtain

$$A = \iint_R dA = \oint_C x\,dy.$$

However, if we let $M = -y$ and $N = 0$, the result is

$$A = \iint_R dA = -\oint_C y\,dx.$$

We may combine these two formulas for A by adding both sides of the equations and dividing by 2. This gives us the third formula in the next theorem.

Theorem **(18.20)**

If a region R in the xy-plane is bounded by a piecewise-smooth simple closed curve C, then the area A of R is

$$A = \oint_C x\,dy = -\oint_C y\,dx = \tfrac{1}{2}\oint_C x\,dy - y\,dx.$$

Although the first two formulas in (18.20) appear to be easier to apply, for certain curves the third formula leads to a simpler integration. If you doubt this, try to find the area in Exercise 20 by using either $\oint_C x\,dy$ or $-\oint_C y\,dx$.

EXAMPLE 4 Use Theorem (18.20) to find the area of the ellipse $\dfrac{x^2}{a^2} + \dfrac{y^2}{b^2} = 1$.

SOLUTION Parametric equations for the ellipse C are $x = a\cos t$, $y = b\sin t$; $0 \le t \le 2\pi$. Applying the third formula in Theorem (18.20) yields

$$\begin{aligned} A &= \tfrac{1}{2}\oint_C x\,dy - y\,dx \\ &= \tfrac{1}{2}\int_0^{2\pi} (a\cos t)(b\cos t)\,dt - (b\sin t)(-a\sin t)\,dt \\ &= \tfrac{1}{2}\int_0^{2\pi} ab(\cos^2 t + \sin^2 t)\,dt \\ &= \tfrac{1}{2}ab\int_0^{2\pi} dt = \tfrac{1}{2}ab(2\pi) = \pi ab. \end{aligned}$$

FIGURE 18.35

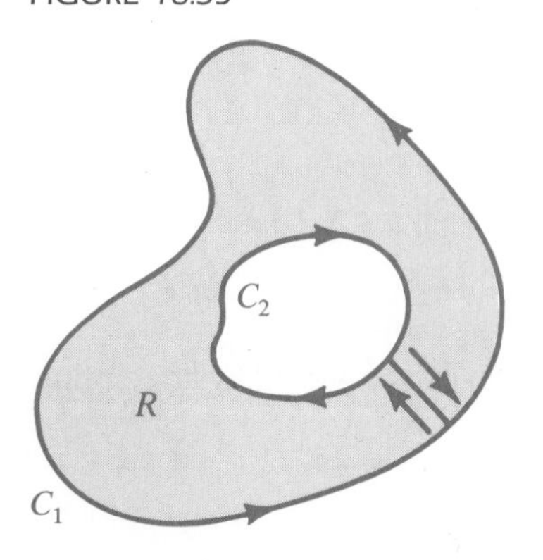

Green's theorem can be extended to a region R that contains holes, provided we integrate over the *entire* boundary and always keep the region R to the left of C. This is illustrated by the region in Figure 18.35, where the double integral over R equals the sum of the line integrals along C_1 and C_2 in the indicated directions. The proof consists of making a slit in R as illustrated in the figure, then noting that the sum of two line integrals in opposite directions along the same curve is zero. A similar argument can be used if the region has several holes. We shall use this observation in the solution of the next example.

EXAMPLE 5 Let C_1 and C_2 be two nonintersecting piecewise-smooth simple closed curves, each having the origin O as an interior point. If

$$M = \frac{-y}{x^2 + y^2}, \quad N = \frac{x}{x^2 + y^2},$$

prove that

$$\oint_{C_1} M\,dx + N\,dy = \oint_{C_2} M\,dx + N\,dy.$$

SOLUTION If R denotes the region between C_1 and C_2, then we have a situation similar to that illustrated in Figure 18.35, with O inside C_2. By the remarks preceding this example, Green's theorem gives us

$$\oint_{C_1} M\,dx + N\,dy + \ointclockwise_{C_2} M\,dx + N\,dy = \iint_R \left(\frac{\partial N}{\partial x} - \frac{\partial M}{\partial y}\right) dA,$$

where we have used the symbol $\circlearrowright$ to indicate the positive direction along C_2 *with respect to* R. Since

$$\frac{\partial N}{\partial x} = \frac{(x^2 + y^2)(1) - x(2x)}{(x^2 + y^2)^2} = \frac{y^2 - x^2}{(x^2 + y^2)^2} = \frac{\partial M}{\partial y},$$

the double integral over R is zero. Consequently

$$\oint_{C_1} M\,dx + N\,dy = -\ointclockwise_{C_2} M\,dx + N\,dy.$$

Since the positive direction along C_2 *with respect to the region in the interior of* C_2 is opposite to that indicated in the preceding integral, we obtain

$$\oint_{C_1} M\,dx + N\,dy = \oint_{C_2} M\,dx + N\,dy,$$

which is what we wished to prove.

EXAMPLE 6 If $\mathbf{F}$ is defined by $\mathbf{F}(x, y) = \dfrac{1}{x^2 + y^2}(-y\mathbf{i} + x\mathbf{j})$, prove that $\oint_C \mathbf{F} \cdot d\mathbf{r} = 2\pi$ for every piecewise-smooth simple closed curve C having the origin in its interior.

FIGURE 18.36

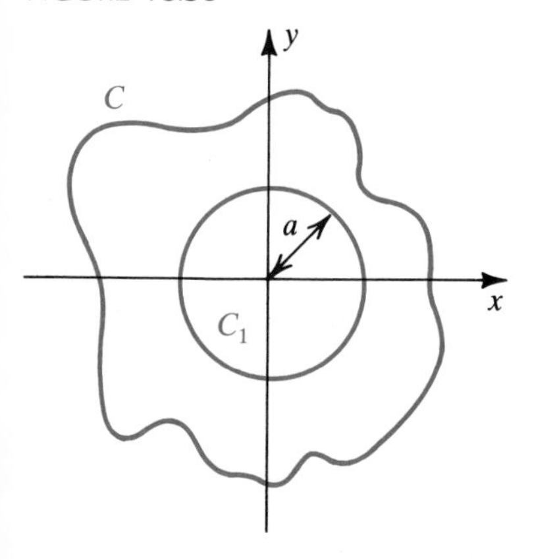

SOLUTION If we let $\mathbf{F}(x, y) = M\mathbf{i} + N\mathbf{j}$, then M and N are the same as in Example 5. Hence, if we choose a circle C_1 of radius a with center at the origin that lies entirely within C (see Figure 18.36), then, by Example 5,

$$\oint_C \mathbf{F} \cdot d\mathbf{r} = \oint_{C_1} \mathbf{F} \cdot d\mathbf{r}.$$

Since parametric equations for C_1 are

$$x = a \cos t, \quad y = a \sin t; \quad 0 \le t \le 2\pi,$$

we obtain

$$\begin{aligned}
\oint_C \mathbf{F} \cdot d\mathbf{r} &= \oint_{C_1} \frac{-y}{x^2 + y^2}\,dx + \frac{x}{x^2 + y^2}\,dy \\
&= \int_0^{2\pi} \frac{-a \sin t}{a^2}(-a \sin t)\,dt + \frac{a \cos t}{a^2}(a \cos t)\,dt \\
&= \int_0^{2\pi} (\sin^2 t + \cos^2 t)\,dt = \int_0^{2\pi} dt = 2\pi.
\end{aligned}$$

The conclusion of Green's theorem (18.19) may be expressed in vector form. First note that if $\mathbf{F}$ is a vector field in two dimensions, then we may write

$$\mathbf{F}(x, y) = M\mathbf{i} + N\mathbf{j} + 0\mathbf{k},$$

where $M = M(x, y)$ and $N = N(x, y)$. The curl of $\mathbf{F}$ is

$$\nabla \times \mathbf{F} = \begin{vmatrix} \mathbf{i} & \mathbf{j} & \mathbf{k} \\ \dfrac{\partial}{\partial x} & \dfrac{\partial}{\partial y} & \dfrac{\partial}{\partial z} \\ M & N & 0 \end{vmatrix} = 0\mathbf{i} + 0\mathbf{j} + \left(\frac{\partial N}{\partial x} - \frac{\partial M}{\partial y}\right)\mathbf{k}.$$

The coefficient of $\mathbf{k}$ has the same form as the integrand of the double integral in Green's theorem (18.19). Let s be an arc length parameter for C, and consider the unit tangent vector

$$\mathbf{T} = \frac{dx}{ds}\mathbf{i} + \frac{dy}{ds}\mathbf{j} + \frac{dz}{ds}\mathbf{k}.$$

Using this notation, we may write Green's theorem as follows.

Vector form of Green's theorem **(18.21)**

$$\oint_C \mathbf{F} \cdot \mathbf{T}\, ds = \iint_R (\nabla \times \mathbf{F}) \cdot \mathbf{k}\, dA$$

Since $(\nabla \times \mathbf{F}) \cdot \mathbf{k}$ is the component of curl $\mathbf{F}$ in the direction parallel to the z-axis, we refer to it as the **normal component** (to R) of curl $\mathbf{F}$. In words, (18.21) may be phrased: *The line integral of the tangential component of* $\mathbf{F}$ *taken along C once in the positive direction is equal to the double integral over R of the normal component of* curl $\mathbf{F}$. The three-dimensional analogue of this result is *Stokes' theorem*, which will be discussed in Section 18.7. We will also discuss physical interpretations of curl $\mathbf{F}$ at that time.

Green's theorem is important in the area of mathematics known as *complex variables*, a subject that is fundamental for many applications in the physical sciences and engineering.

EXERCISES 18.4

Exer. 1–14: Use Green's theorem to evaluate the line integral.

1 $\oint_C (x^2 + y)\, dx + (xy^2)\, dy$;
C is the closed curve determined by $y^2 = x$ and $y = -x$ with $0 \le x \le 1$.

2 $\oint_C (x + y^2)\, dx + (1 + x^2)\, dy$;
C is the closed curve determined by $y = x^3$ and $y = x^2$ with $0 \le x \le 1$.

3 $\oint_C x^2y^2\, dx + (x^2 - y^2)\, dy$;
C is the square with vertices $(0, 0)$, $(1, 0)$, $(1, 1)$, $(0, 1)$.

4 $\oint_C \sqrt{y}\, dx + \sqrt{x}\, dy$;
C is the triangle with vertices $(1, 1)$, $(3, 1)$, $(2, 2)$.

5 $\oint_C xy\, dx + (y + x)\, dy$;
C is the circle $x^2 + y^2 = 1$.

6 $\oint_C y^2\, dx + x^2\, dy$;
C is the boundary of the region bounded by the semicircle $y = \sqrt{4 - x^2}$ and the x-axis.

7 $\oint_C xy\, dx + \sin y\, dy$;
C is the triangle with vertices $(1, 1)$, $(2, 2)$, $(3, 0)$.

8 $\oint_C \tan^{-1} x\, dx + 3x\, dy$;
C is the rectangle with vertices $(1, 0)$, $(0, 1)$, $(2, 3)$, $(3, 2)$.

9 $\oint_C y^2(1 + x^2)^{-1}\, dx + 2y \tan^{-1} x\, dy$;
C is the hypocycloid $x^{2/3} + y^{2/3} = 1$.

10 $\oint_C (x^2 + y^2)\, dx + 2xy\, dy$;
C is the boundary of the region bounded by the graphs of $y = \sqrt{x}$, $y = 0$, and $x = 4$.

11 $\oint_C (x^4 + 4)\, dx + xy\, dy$;
C is the cardioid $r = 1 + \cos\theta$.

12 $\oint_C xy\, dx + (x^2 + y^2)\, dy$;
C is the first-quadrant loop of $r = \sin 2\theta$.

13 $\oint_C (x + y)\, dx + (y + x^2)\, dy$;
C is the boundary of the region between the circles $x^2 + y^2 = 1$ and $x^2 + y^2 = 4$.

14 $\oint_C (1 - x^2 y)\, dx + \sin y\, dy$;
C is the boundary of the region that lies inside the square with vertices $(\pm 2, \pm 2)$ and outside the square with vertices $(\pm 1, \pm 1)$.

Exer. 15–18: Use Theorem (18.20) to find the area of the region bounded by the graphs of the equations.

15 $y = 4x^2$, $y = 16x$ 16 $y = x^3$, $y^2 = x$

17 $y^2 - x^2 = 5$, $y = 3$ 18 $xy = 2$, $x + y = 3$

Exer. 19–20: Use Theorem (18.20) to find the area of the region bounded by the curve C.

19 C is the hypocycloid $x = a\cos^3 t$, $y = a \sin^3 t$; $0 \le t \le 2\pi$.

20 C is the piecewise-smooth curve consisting of the folium of Descartes

$$x = \frac{3t}{t^3 + 1}, \quad y = \frac{3t^2}{t^3 + 1}; \quad 0 \le t \le 1$$

and the line segment from $(\frac{3}{2}, \frac{3}{2})$ to $(0, 0)$.

21 If $\mathbf{F}(x, y)$ is a two-dimensional vector field and $\int_A^B \mathbf{F} \cdot d\mathbf{r}$ is independent of path in a region D, use Green's theorem to prove that $\oint_C \mathbf{F} \cdot d\mathbf{r} = 0$ for every piecewise-smooth simple closed curve in D.

22 If f and g are differentiable functions of one variable, prove that $\oint_C f(x)\, dx + g(y)\, dy = 0$ for every piecewise-smooth simple closed curve C.

23 Let $M = y/(x^2 + y^2)$ and $N = -x/(x^2 + y^2)$. If R is the region bounded by the unit circle C with center at the origin, show that

$$\oint_C M\, dx + N\, dy \neq \iint_R \left(\frac{\partial N}{\partial x} - \frac{\partial M}{\partial y}\right) dA.$$

Explain why Green's theorem is not true in this instance.

24 Suppose $\mathbf{F}(x, y)$ has continuous first partial derivatives in a simply connected region R. If $\operatorname{curl} \mathbf{F} = \mathbf{0}$ in R, prove that $\oint_C \mathbf{F} \cdot d\mathbf{r} = 0$ for every piecewise-smooth simple closed curve C in R.

25 Let R be the region bounded by a piecewise-smooth simple closed curve C in the xy-plane. If the area of R is A, use Green's theorem to prove that the centroid $(\bar{x}, \bar{y})$ is

$$\bar{x} = \frac{1}{2A} \oint_C x^2\, dy, \quad \bar{y} = -\frac{1}{2A} \oint_C y^2\, dx.$$

26 Suppose a homogeneous lamina of density k has the shape of a region in the xy-plane that is bounded by a piecewise-smooth simple closed curve C. Prove that the moments of inertia with respect to the x- and y-axes are

$$I_x = -\frac{k}{3} \oint_C y^3\, dx, \quad I_y = \frac{k}{3} \oint_C x^3\, dy.$$

27 Use Exercise 25 to find the centroid of a semicircular region of radius a.

28 Use Exercise 26 to find the moment of inertia of a homogeneous circular disk of radius a with respect to a diameter.

18.5 SURFACE INTEGRALS

Line integrals are evaluated along curves. Double and triple integrals are defined on regions in two and three dimensions, respectively. We may also consider an integral of a function over a surface. For simplicity we shall restrict our discussion to surfaces that are graphs of rather simple equations, and our demonstrations will be at an intuitive level. Rigorous treatments may be found in texts on advanced calculus.

FIGURE 18.37

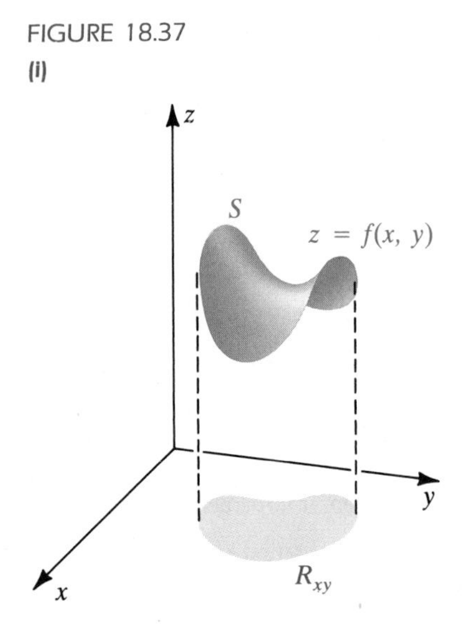

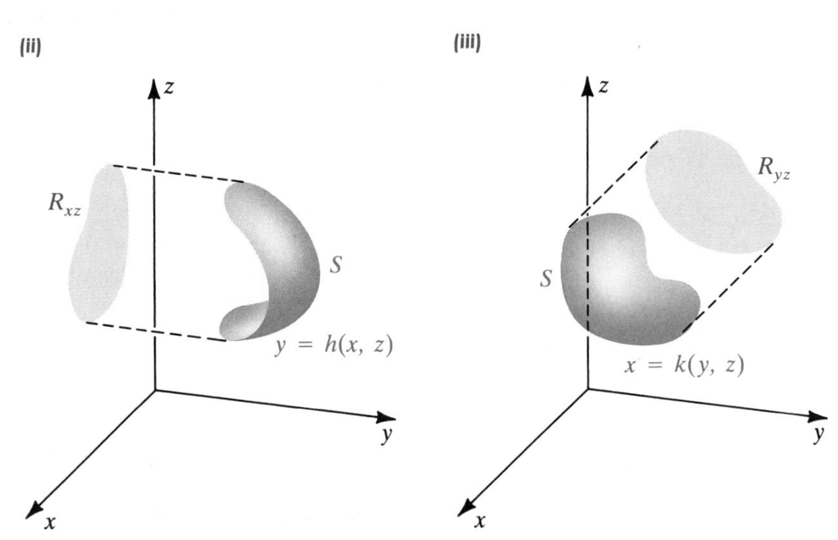

If the projection of a surface S on a coordinate plane is a region of a type considered for double integrals, we say that S has a **regular projection** on the coordinate plane. If S has a regular projection on the xy-plane, the xz-plane, or the yz-plane, we sometimes denote the region by R_{xy}, R_{xz}, or R_{yz}, respectively. For such projections we assume that S is the graph of an equation of the form $z = f(x, y)$, $y = h(x, z)$, or $x = k(y, z)$, respectively, as shown in Figure 18.37. Moreover, we assume that the function f, g, or k has continuous first partial derivatives on its respective region.

FIGURE 18.38

Let us first consider the case of the graph S of $z = f(x, y)$ (see Figure 18.37(i)). In (17.16) we defined the area A of S. A similar technique may be used to define an integral of $g(x, y, z)$ over the surface S, if the function g is continuous throughout a region containing S. We use the notation of Section 17.2. Thus, as illustrated in Figure 18.38, ΔS_k and ΔT_k will denote areas of portions of the surface S and the tangent plane at $B_k(x_k, y_k, z_k)$, respectively, that project onto the rectangle R_k of an inner partition of R_{xy}. We evaluate g at B_k for each k and form the sum $\sum_k g(x_k, y_k, z_k)\,\Delta T_k$. As in the following definition, the **surface integral** $\iint_S g(x, y, z)\,dS$ **of g over S** is the limit of such sums as the norms of the partitions approach 0.

Surface integral **(18.22)**

$$\iint_S g(x, y, z)\,dS = \lim_{\|P\|\to 0} \sum_k g(x_k, y_k, z_k)\,\Delta T_k$$

If S is the union of several surfaces of the proper types, then the surface integral is defined as the sum of the individual surface integrals. If $g(x, y, z) = 1$ for every (x, y, z), then (18.22) reduces to (17.17) and hence the surface integral gives us the area of the surface S.

A physical application of surface integrals may be obtained by considering a thin metal sheet, such as tinfoil, that has the shape of S. If the area mass density at (x, y, z) is $g(x, y, z)$, then (18.22) is the mass of the sheet. The center of gravity and moments of inertia may be obtained by employing the methods used for solids in Chapter 17 (see Exercise 21).

In a manner similar to the way formula (17.17) for surface area was developed, the surface integral (18.22) may be evaluated by means of formula (i) of the next theorem. Formulas (ii) and (iii) are used for surfaces of the type illustrated in Figure 18.37(ii) and (iii).

Evaluation theorem for surface integrals **(18.23)**

(i) $$\iint_S g(x, y, z)\, dS = \iint_{R_{xy}} g(x, y, f(x, y))\sqrt{[f_x(x, y)]^2 + [f_y(x, y)]^2 + 1}\, dA$$

(ii) $$\iint_S g(x, y, z)\, dS = \iint_{R_{xz}} g(x, h(x, z), z)\sqrt{[h_x(x, z)]^2 + [h_z(x, z)]^2 + 1}\, dA$$

(iii) $$\iint_S g(x, y, z)\, dS = \iint_{R_{yz}} g(k(y, z), y, z)\sqrt{[(k_y(y, z)]^2 + [k_z(y, z)]^2 + 1}\, dA$$

EXAMPLE 1 Evaluate $\iint_S x^2 z\, dS$ if S is the portion of the cone $z^2 = x^2 + y^2$ that lies between the planes $z = 1$ and $z = 4$.

FIGURE 18.39

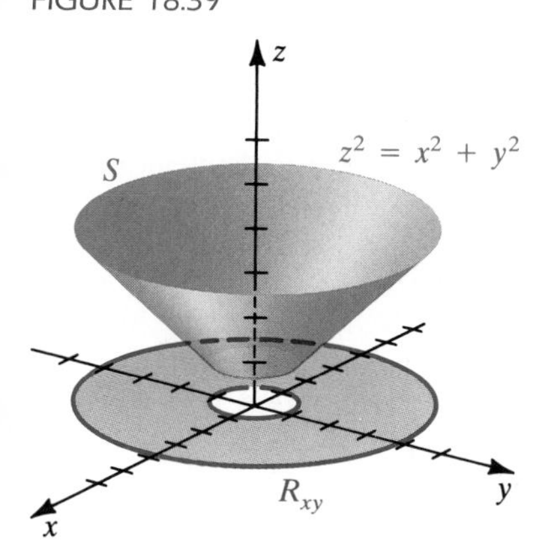

SOLUTION As shown in Figure 18.39, the projection R_{xy} of S onto the xy-plane is the annular region bounded by circles of radii 1 and 4 with centers at the origin. If we write the equation for S in the form

$$z = (x^2 + y^2)^{1/2} = f(x, y),$$

then $$f_x(x, y) = \frac{x}{(x^2 + y^2)^{1/2}} \quad \text{and} \quad f_y(x, y) = \frac{y}{(x^2 + y^2)^{1/2}}.$$

Applying (18.23)(i) and noting that the radical reduces to $\sqrt{2}$, we obtain

$$\iint_S x^2 z\, dS = \iint_{R_{xy}} x^2(x^2 + y^2)^{1/2}\sqrt{2}\, dA.$$

We use polar coordinates to evaluate the double integral as follows:

$$\begin{aligned}
\iint_{R_{xy}} x^2(x^2 + y^2)^{1/2}\sqrt{2}\, dA &= \int_0^{2\pi}\int_1^4 (r^2\cos^2\theta) r\sqrt{2} r\, dr\, d\theta \\
&= \sqrt{2}\int_0^{2\pi} \cos^2\theta \left[\frac{r^5}{5}\right]_1^4 d\theta \\
&= \frac{1023\sqrt{2}}{5}\int_0^{2\pi} \frac{1 + \cos 2\theta}{2}\, d\theta \\
&= \frac{1023\sqrt{2}}{10}\left[\theta + \frac{1}{2}\sin 2\theta\right]_0^{2\pi} \\
&= \frac{1023\sqrt{2}\pi}{5} \approx 909
\end{aligned}$$

FIGURE 18.40

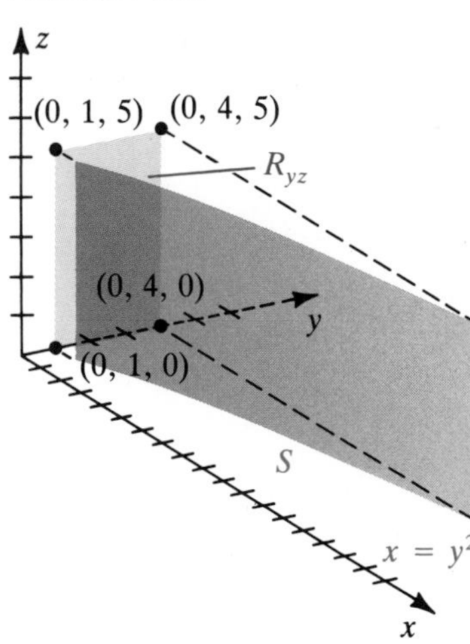

EXAMPLE 2 Evaluate $\iint_S (xz/y)\, dS$ if S is the portion of the cylinder $x = y^2$ that lies in the first octant between the planes $z = 0$, $z = 5$, $y = 1$, and $y = 4$.

SOLUTION The surface S is sketched in Figure 18.40. The projection R_{yz} of S on the yz-plane is the rectangle with vertices $(0, 1, 0)$, $(0, 4, 0)$, $(0, 4, 5)$, and $(0, 1, 5)$. Thus, by Theorem (18.23)(iii) with $k(y, z) = y^2$,

$$\begin{aligned}\iint_S \frac{xz}{y}\, dS &= \int_1^4 \int_0^5 \frac{y^2 z}{y} \sqrt{(2y)^2 + 0^2 + 1}\; dz\, dy \\ &= \int_1^4 \int_0^5 y\sqrt{4y^2 + 1}\; z\, dz\, dy = \int_1^4 y\sqrt{4y^2+1}\left[\frac{z^2}{2}\right]_0^5 dy \\ &= \tfrac{25}{2}\int_1^4 y\sqrt{4y^2+1}\; dy = \tfrac{25}{24}\Big[(4y^2+1)^{3/2}\Big]_1^4 \\ &= \tfrac{25}{24}(65^{3/2} - 5^{3/2}) \approx 534.2.\end{aligned}$$

The formulas in Theorem (18.23) are stated under the assumption that the functions f, h, and k have continuous first partial derivatives on R_{xy}, R_{xz}, and R_{yz}, respectively. In certain cases we can remove these restrictions by using an improper integral, as in the next example.

EXAMPLE 3 Evaluate $\iint_S (z + y)\, dS$ if S is the part of the graph of $z = \sqrt{1 - x^2}$ in the first octant between the xz-plane and the plane $y = 3$.

FIGURE 18.41

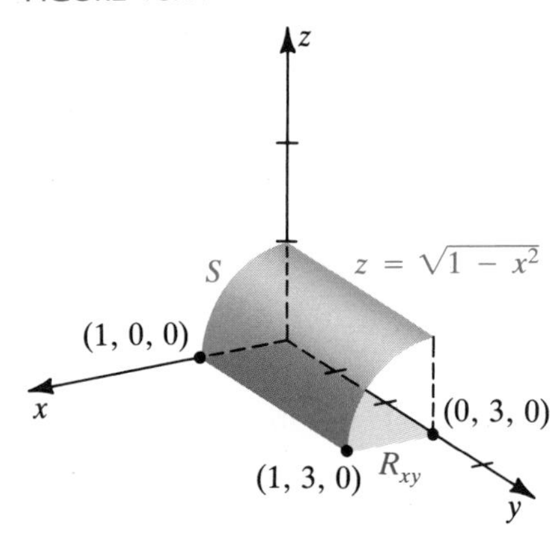

SOLUTION The surface S is the first-octant portion of the cylinder $x^2 + z^2 = 1$ from $y = 0$ to $y = 3$. The graph is sketched in Figure 18.41. The projection R_{xy} of S onto the xy-plane is the rectangular region with vertices $(0, 0, 0)$, $(1, 0, 0)$, $(1, 3, 0)$, and $(0, 3, 0)$. If we wish to use Theorem (18.23)(i) with $f(x, y) = \sqrt{1 - x^2}$, then

$$\sqrt{[f_x(x, y)]^2 + [f_y(x, y)]^2 + 1} = \sqrt{\left(\frac{-x}{\sqrt{1 - x^2}}\right)^2 + (0)^2 + 1} = \frac{1}{\sqrt{1 - x^2}}.$$

Note that $f_x(x, y)$ is undefined at $x = 1$ and hence is not continuous throughout R_{xy}. However, we may apply Theorem (18.23)(i) with $0 \le x < 1$ and then use an improper integral, as follows:

$$\begin{aligned}\iint_S (z + y)\, dS &= \iint_{R_{xy}} (\sqrt{1 - x^2} + y)\frac{1}{\sqrt{1 - x^2}}\, dA \\ &= \int_0^1 \int_0^3 \left(1 + \frac{y}{\sqrt{1 - x^2}}\right) dy\, dx \\ &= \int_0^1 \left[y + \frac{1}{2}\frac{y^2}{\sqrt{1 - x^2}}\right]_0^3 dx = \int_0^1 \left[3 + \frac{1}{2}\frac{9}{\sqrt{1 - x^2}}\right] dx \\ &= \lim_{t \to 1^-} \int_0^t \left[3 + \frac{9}{2}\frac{1}{\sqrt{1 - x^2}}\right] dx \\ &= \lim_{t \to 1^-} (3t + \tfrac{9}{2}\arcsin t) = 3 + \tfrac{9}{4}\pi \approx 10.1\end{aligned}$$

In Section 6.8 we discussed how to find the force exerted on one face of a lamina when it is submerged in a fluid (see (6.27)). We can use surface integrals to solve similar problems for non-laminar surfaces, as illustrated in the next example.

EXAMPLE 4 A paper cup has the shape of a right circular cone of altitude 6 inches and radius of base 3 inches. If the cup is full of water weighing 62.5 lb/ft^3, find the total force exerted by the water on the inside surface of the cup.

FIGURE 18.42

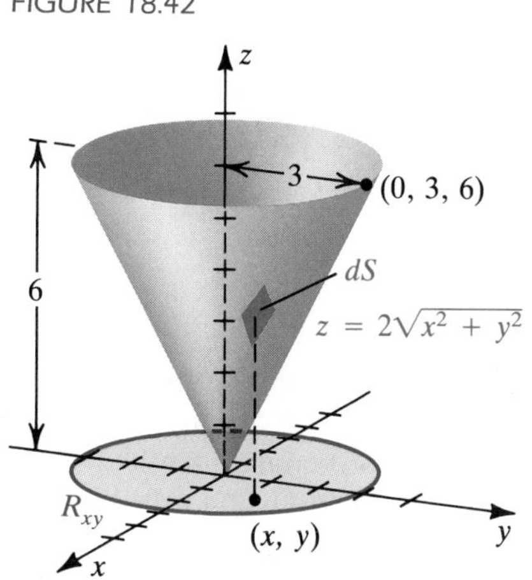

SOLUTION If we position the cup in an xyz-coordinate system as illustrated in Figure 18.42 (where the units are inches), then an equation of the cone has the form $z = k\sqrt{x^2 + y^2}$ for some constant k. Since the point $(0, 3, 6)$ is on the cone, $6 = k\sqrt{0^2 + 3^2}$, or $k = 2$. Hence $z = 2\sqrt{x^2 + y^2}$.

Let dS represent an element of surface area of the cone that is (approximately) a distance z from the xy-plane. The corresponding depth of the water in the cup is $6 - z$. Changing the density of the water to lb/in.^{3}, we obtain $(62.5/12^3)$ lb/in.^{3} Thus, the pressure at a depth of $6 - z$ inches is $(62.5/12^3)(6 - z)$ lb/in.^{2}, and the force exerted by the fluid on this portion of the cup is approximately

$$\frac{62.5}{12^3}(6 - z)\, dS \text{ lb}.$$

Taking a limit of sums, we find that the total force exerted by the water on the inside surface S of the cone is given by

$$F = \frac{62.5}{12^3} \iint_S (6 - z)\, dS.$$

Note that the region R_{xy} that lies under the cone is bounded by the circle $x^2 + y^2 = 9$. Applying (18.23)(i) with $f(x, y) = 2\sqrt{x^2 + y^2} = z$, we can verify that

$$F = \frac{62.5}{12^3} \iint_{R_{xy}} (6 - z)\sqrt{5}\, dA.$$

Using polar coordinates, we have

$$\begin{aligned} F &= \frac{62.5}{12^3} \int_0^{2\pi} \int_0^3 (6 - 2r)\sqrt{5}\, r\, dr\, d\theta \\ &= \frac{125\sqrt{5}}{192}\pi \text{ lb} \approx 4.57 \text{ lb}. \end{aligned}$$

The most important applications of surface integrals occur when working with vector fields. Let $\mathbf{F}$ be a vector field such that

$$\mathbf{F}(x, y, z) = M(x, y, z)\mathbf{i} + N(x, y, z)\mathbf{j} + P(x, y, z)\mathbf{k},$$

where M, N, and P are continuous (scalar) functions. In Section 18.2 we considered the following line integral for a piecewise-smooth curve C:

$$\int_C \mathbf{F} \cdot \mathbf{T}\, ds,$$

where s is an arc length parameter for C. As in Figure 18.43, $\mathbf{T} = d\mathbf{r}/ds$ is a unit tangent vector to C at the point (x, y, z), where $\mathbf{r} = x\mathbf{i} + y\mathbf{j} + z\mathbf{k}$ is the position vector of (x, y, z). If $\mathbf{F}$ is a force field, then, from (18.12), the value of this line integral is the work done by $\mathbf{F}$ along C.

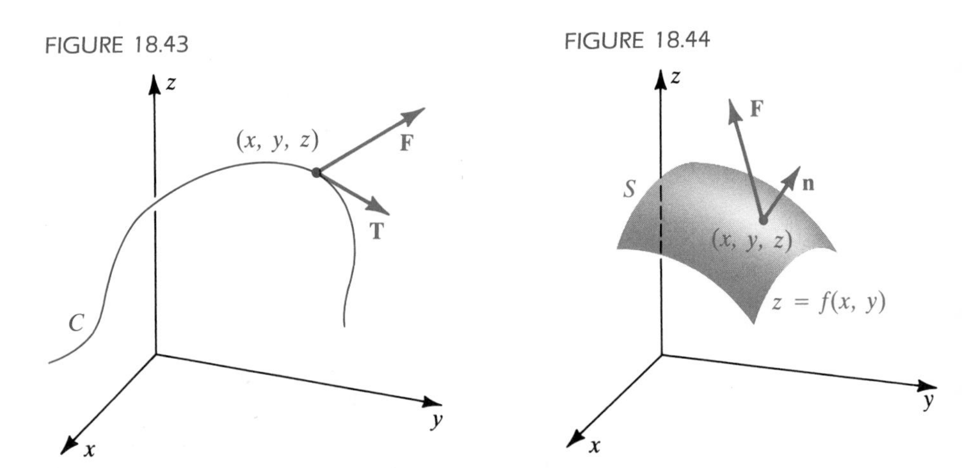

FIGURE 18.43

FIGURE 18.44

Let us next consider a surface S and a unit normal vector $\mathbf{n}$ to S at the point (x, y, z), as illustrated in Figure 18.44. If the components of $\mathbf{n}$ are continuous functions of x, y, and z, then $\mathbf{F} \cdot \mathbf{n}$ is a continuous (scalar) function and we may consider the following surface integral over S, which (for reasons we will discuss at the end of this section) is called the **flux integral** of $\mathbf{F}$ over S.

Flux integral of $\mathbf{F}$ over S **(18.24)**

$$\iint_S \mathbf{F} \cdot \mathbf{n}\, dS$$

To evaluate (18.24), we must be more precise about properties of the unit vector $\mathbf{n}$. If the surface S is the graph of an equation $z = f(x, y)$ and if we let $g(x, y, z) = z - f(x, y)$, then S is also the graph of the equation $g(x, y, z) = 0$. Since the gradient $\nabla g(x, y, z)$ is a normal vector to the graph of $g(x, y, z) = 0$ at the point (x, y, z), a *unit* normal vector $\mathbf{n}$ can be obtained as follows:

$$\mathbf{n} = \frac{\nabla g(x, y, z)}{\|\nabla g(x, y, z)\|} = \frac{-f_x(x, y)\mathbf{i} - f_y(x, y)\mathbf{j} + \mathbf{k}}{\sqrt{[f_x(x, y)]^2 + [f_y(x, y)]^2 + 1}}$$

Similar formulas can be stated if S is given by $y = h(x, z)$ or by $x = k(y, z)$. Note that for the case $z = f(x, y)$, the unit vector $\mathbf{n}$ is an **upper normal** to S, since the $\mathbf{k}$ component is positive. A **lower normal** can be obtained by using $-\mathbf{n}$.

In the remainder of this chapter, we shall assume that every surface S is **oriented** (or **orientable**) in the sense that a unit normal vector $\mathbf{n}$ exists at every nonboundary point (x, y, z), and that the components of $\mathbf{n}$ are continuous functions of x, y, and z. We say that $\mathbf{n}$ *varies continuously* over the surface S. We shall also assume that S has two sides, such as the *top side* and *bottom side* of the graph of $z = f(x, y)$ in Figure 18.44. For a *closed* surface S such as a sphere, we may consider the *outside* and *inside*

FIGURE 18.45

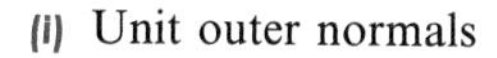

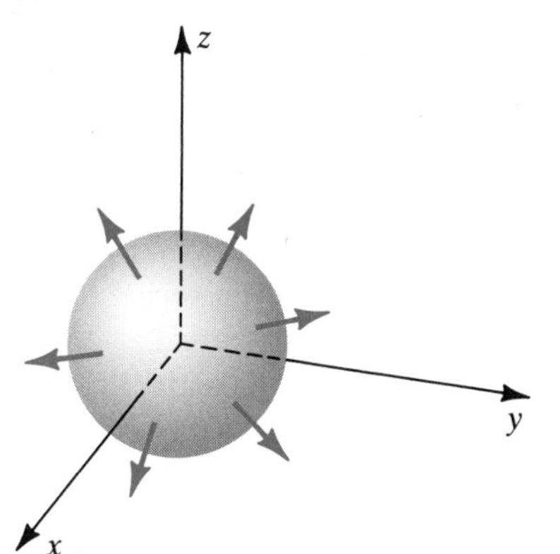

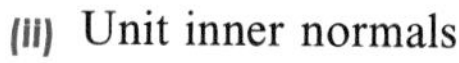

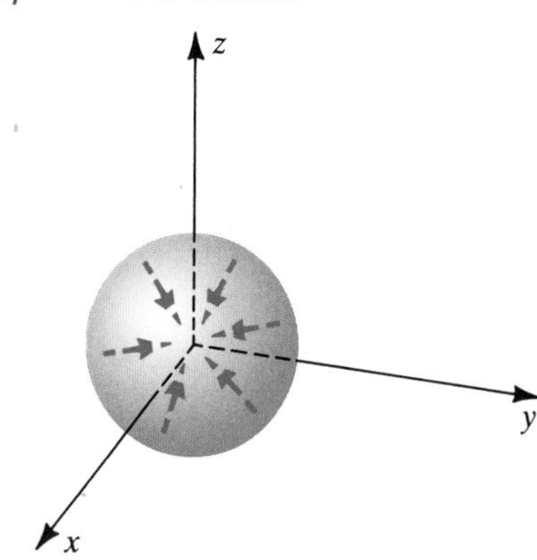

FIGURE 18.46
Möbius strip

FIGURE 18.47

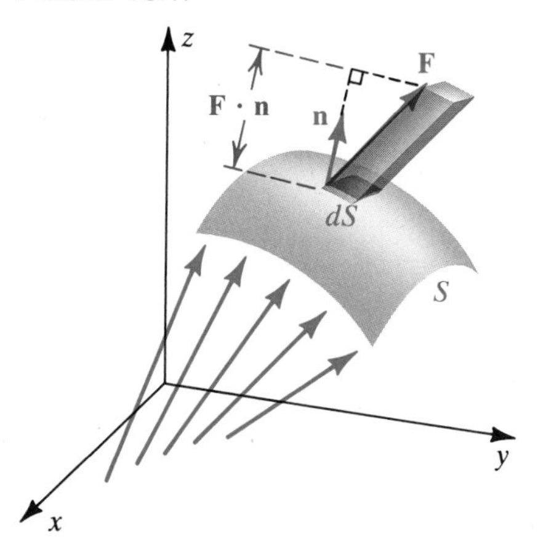

of S. We then refer to **n** as the **unit outer normal** or the **unit inner normal** (see Figure 18.45).

Our restrictions on S rule out *one-sided surfaces*, such as a *Möbius strip*, named after the German mathematician A. F. Möbius (1790–1868). A Möbius strip can be constructed from a long rectangular strip of paper by making a half twist in the paper and taping the short edges together, producing the surface illustrated in Figure 18.46. Imagine a painter who starts at some point and attempts to paint only one side of the original strip. The painter will eventually return to the same point, having painted the entire strip without crossing any edge. This one-sided surface is not orientable.

To interpret the physical significance of the flux integral (18.24), let us regard the surface S as a thin membrane through which a fluid can pass. Suppose S is submerged in a fluid having a velocity field $\mathbf{F}(x, y, z)$. Some typical vectors of the field are shown in the lower part of Figure 18.47. Each vector of the field is the velocity of a fluid particle (or molecule).

Let dS represent a small element of area of S. If $\mathbf{F}$ is continuous, then it is almost constant on dS, and the amount of fluid crossing dS per unit of time may be approximated by the volume of a prism of base area dS and altitude $\mathbf{F} \cdot \mathbf{n}$, as illustrated in Figure 18.47. If dV denotes the volume of the prism in the figure, then $dV = \mathbf{F} \cdot \mathbf{n}\, dS$. Since dV represents the amount of fluid crossing dS per unit time, the flux integral is a limit of sums of elements of volume, and hence $\iint_S \mathbf{F} \cdot \mathbf{n}\, dS$ *is the volume of fluid crossing S per unit time*. This quantity is called the *flux* of $\mathbf{F}$ *through* (or *over*) S. If $\mathbf{F}$ and S satisfy the conditions we have imposed, then we have the following.

Definition **(18.25)**

The **flux** of a vector field $\mathbf{F}$ through (or over) a surface S is

$$\iint_S \mathbf{F} \cdot \mathbf{n}\, dS.$$

If the fluid in the previous discussion has density $\delta = \delta(x, y, z)$, *then the value of the flux integral* $\iint_S \delta\mathbf{F} \cdot \mathbf{n}\, dS$ *is the mass of the fluid crossing* S.

The vector field $\mathbf{F}$ in Definition (18.25) may also represent steady state heat flow, a uniformly expanding gas, or fields encountered in the theory of electricity. In these cases flux is a measure of the amount of heat crossing S, the flow of gas through S, or magnetic or electric flux through S, respectively.

If, for a closed surface such as a sphere, we choose $\mathbf{n}$ as the unit *outer normal*, then the flux measures the *net outward flow* per unit time. If the flux integral is positive, then the flow out of S exceeds the flow into S, and we say there is a **source** of $\mathbf{F}$ within S. If the integral is negative, then the flow into S exceeds the flow out of S, and there is a **sink** within S. If the integral is 0, then the flow into and the flow out of S are equal; that is, the sources and sinks balance one another.

EXAMPLE 5 Let S be the part of the graph of $z = 9 - x^2 - y^2$ with $z \geq 0$. If $\mathbf{F}(x, y, z) = 3x\mathbf{i} + 3y\mathbf{j} + z\mathbf{k}$, find the flux of $\mathbf{F}$ through S.

FIGURE 18.48

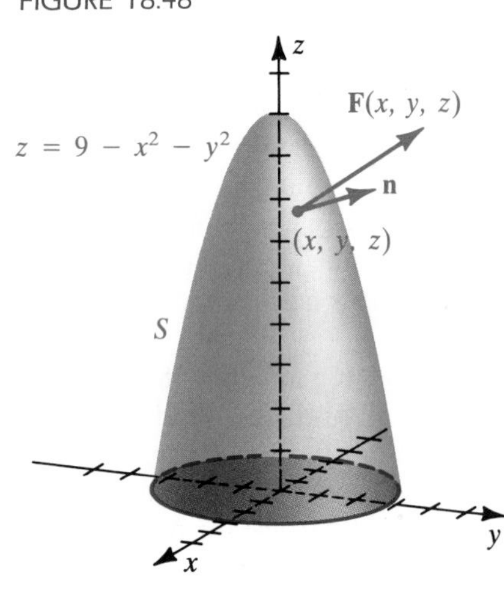

SOLUTION The graph is sketched in Figure 18.48, together with a typical unit upper normal $\mathbf{n}$ and the vector $\mathbf{F}(x, y, z)$. As in our discussion following (18.24), to find a formula for $\mathbf{n}$ we consider

$$g(x, y, z) = z - (9 - x^2 - y^2) = z - 9 + x^2 + y^2$$

and let

$$\mathbf{n} = \frac{\nabla g(x, y, z)}{\|\nabla g(x, y, z)\|} = \frac{2x\mathbf{i} + 2y\mathbf{j} + \mathbf{k}}{\sqrt{4x^2 + 4y^2 + 1}}.$$

By (18.25), the flux of $\mathbf{F}$ through S is

$$\iint_S \mathbf{F} \cdot \mathbf{n}\, dS = \iint_S \frac{6x^2 + 6y^2 + z}{\sqrt{4x^2 + 4y^2 + 1}}\, dS.$$

Applying Theorem (18.23)(i) gives us

$$\iint_S \mathbf{F} \cdot \mathbf{n}\, dS = \iint_{R_{xy}} \frac{6x^2 + 6y^2 + 9 - x^2 - y^2}{\sqrt{4x^2 + 4y^2 + 1}} \sqrt{4x^2 + 4y^2 + 1}\, dA$$

$$= \iint_{R_{xy}} (5x^2 + 5y^2 + 9)\, dA,$$

where R_{xy} is the circular region in the xy-plane bounded by the graph of $x^2 + y^2 = 9$. Changing to polar coordinates, we have

$$\iint_S \mathbf{F} \cdot \mathbf{n}\, dS = \int_0^{2\pi} \int_0^3 (5r^2 + 9)r\, dr\, d\theta = \frac{567\pi}{2} \approx 890.6.$$

If, for example, $\mathbf{F}$ is the velocity field of an expanding gas and $\|\mathbf{F}\|$ is measured in m/sec, then the unit of flux is m^3/sec. Hence the flow of gas across the surface S in Figure 18.48 is approximately 890.6 m^3/sec.

In the next section we discuss a result, called the *divergence theorem*, that will enable us to calculate the flux over a closed surface by using a triple integral.

EXERCISES 18.5

Exer. 1–4: Evaluate $\iint_S g(x, y, z)\, dS$.

1 $g(x, y, z) = x^2$;
S is the upper half of the sphere $x^2 + y^2 + z^2 = a^2$.

2 $g(x, y, z) = x^2 + y^2 + z^2$;
S is the part of the plane $z = y + 4$ that is inside the cylinder $x^2 + y^2 = 4$.

3 $g(x, y, z) = x + y$;
S is the first-octant portion of the plane $2x + 3y + z = 6$.

4 $g(x, y, z) = (x^2 + y^2 + 1)^{1/2}$;
S is the portion of the paraboloid $2z = x^2 + y^2$ that lies inside the cylinder $x^2 + y^2 = 2y$.

Exer. 5–8: Express the surface integral as an iterated double integral by using a projection of S on (a) the yz-plane and (b) the xz-plane.

5 $\iint_S xy^2z^3\, dS$;
S is the first-octant portion of the plane $2x + 3y + 4z = 12$.

6 $\iint_S (xz + 2y)\, dS$;
S is the portion of the graph of $y = x^3$ between the planes $y = 0$, $y = 8$, $z = 2$, and $z = 0$.

7 $\iint_S (x^2 - 2y + z)\, dS$;
S is the portion of the graph of $4x + y = 8$ bounded by the coordinate planes and the plane $z = 6$.

8 $\iint_S (x^2 + y^2 + z^2)\, dS$;
S is the first-octant portion of the graph of $x^2 + y^2 = 4$ bounded by the coordinate planes and the plane $x + z = 2$.

9 Interpret $\iint_S g(x, y, z)\, dS$ geometrically if f is the constant function $g(x, y, z) = c$ with $c > 0$ and S has a regular projection on the xy-plane.

10 Show that a double integral $\iint_R f(x, y)\, dA$ is a special case of a surface integral.

Exer. 11–14: Find $\iint_S \mathbf{F} \cdot \mathbf{n}\, dS$ if $\mathbf{n}$ is a unit upper normal to S.

11 $\mathbf{F} = x\mathbf{i} + y\mathbf{j} + z\mathbf{k}$;
S is the upper half of the sphere $x^2 + y^2 + z^2 = a^2$.

12 $\mathbf{F} = x\mathbf{i} - y\mathbf{j}$;
S is the first-octant portion of the sphere $x^2 + y^2 + z^2 = a^2$.

13 $\mathbf{F} = 2\mathbf{i} + 5\mathbf{j} + 3\mathbf{k}$;
S is the portion of the cone $z = (x^2 + y^2)^{1/2}$ that is inside the cylinder $x^2 + y^2 = 1$.

14 $\mathbf{F} = x\mathbf{i} + y\mathbf{j} + z\mathbf{k}$;
S is the portion of the plane $3x + 2y + z = 12$ cut out by the planes $x = 0$, $y = 0$, $x = 1$, and $y = 2$.

Exer. 15–16: Find the flux of $\mathbf{F}$ through S.

15 $\mathbf{F}(x, y, z) = x\mathbf{i} + y\mathbf{j} + z\mathbf{k}$;
S is the first-octant portion of the plane $2x + 3y + z = 6$.

16 $\mathbf{F}(x, y, z) = (x^2 + z)\mathbf{i} + y^2z\mathbf{j} + (x^2 + y^2 + z)\mathbf{k}$;
S is the first-octant portion of the paraboloid $z = x^2 + y^2$ that is cut off by the plane $z = 4$.

Exer. 17–18: Find the flux of $\mathbf{F}$ over the closed surface S. (Use the *outer* normal to S.)

17 $\mathbf{F}(x, y, z) = (x + y)\mathbf{i} + z\mathbf{j} + xz\mathbf{k}$;
S is the surface of the cube having vertices $(\pm 1, \pm 1, \pm 1)$.

18 $\mathbf{F}(x, y, z) = x\mathbf{i} - y\mathbf{j} + z\mathbf{k}$;
S is the surface of the solid bounded by the graphs of $z = x^2 + y^2$ and $z = 4$.

19 If S is given by $x = k(y, z)$ and $\mathbf{F} = M\mathbf{i} + N\mathbf{j} + P\mathbf{k}$, show that

$$\iint_S \mathbf{F} \cdot \mathbf{n}\, dS = \iint_{R_{yz}} [M - Nk_y(y, z) - Pk_z(y, z)]\, dy\, dz.$$

20 By Coulomb's law, if a point charge of q coulombs is located at the origin, then the force exerted on a unit charge at (x, y, z) is given by $\mathbf{F}(x, y, z) = (cq/\|\mathbf{r}\|^3)\mathbf{r}$, where $\mathbf{r} = x\mathbf{i} + y\mathbf{j} + z\mathbf{k}$. If S is any sphere with center at the origin, show that the flux of $\mathbf{F}$ through S is $4\pi cq$.

21 If a thin metal sheet T has the shape of a surface S and has area mass density $\delta(x, y, z)$, then the *moments* of T with respect to the coordinate planes are defined by

$$M_{xy} = \iint_S z\delta(x, y, z)\, dS$$

$$M_{xz} = \iint_S y\delta(x, y, z)\, dS$$

$$M_{yz} = \iint_S x\delta(x, y, z)\, dS.$$

Use limits of sums to show that these are natural definitions. The *center of mass* $(\bar{x}, \bar{y}, \bar{z})$ is defined by

$$\bar{x}m = M_{yz}, \quad \bar{y}m = M_{xz}, \quad \bar{z}m = M_{xy},$$

where m is the mass of T. Suppose a metal funnel has the shape of the surface S described in Example 1 (see Figure 18.39). If the unit of length is the centimeter and the area mass density at the point (x, y, z) is z^2 gm/cm^2, find

(a) the mass and the center of mass of the funnel

(b) the moment of inertia of the funnel with respect to the z-axis

22 Refer to Exercise 21. A thin metal sheet of constant density k has the shape of the hemisphere $z = \sqrt{a^2 - x^2 - y^2}$. Find its center of mass.

18.6 THE DIVERGENCE THEOREM

One of the most important theorems of vector calculus is the *divergence theorem*. It is often called *Gauss' theorem* in honor of Carl Friedrich Gauss (1777–1855), whom many consider the greatest mathematician of all time. The theorem specifies the flux of a vector field over a **closed surface** S that is the boundary of a region Q in three dimensions. For example, S could be a sphere, an ellipsoid, a cube, or a tetrahedron. The divergence theorem can be proved for very complicated regions; however, the proof would take us into the field of advanced calculus. Instead, *throughout this section* we shall assume that Q is a region over which triple integrals can be evaluated using methods developed in Chapter 17. We will also assume that surface integrals can be evaluated over S. In this section $\mathbf{n}$ will denote a unit *outer* normal to S, as illustrated by some typical vectors in Figure 18.49.

FIGURE 18.49

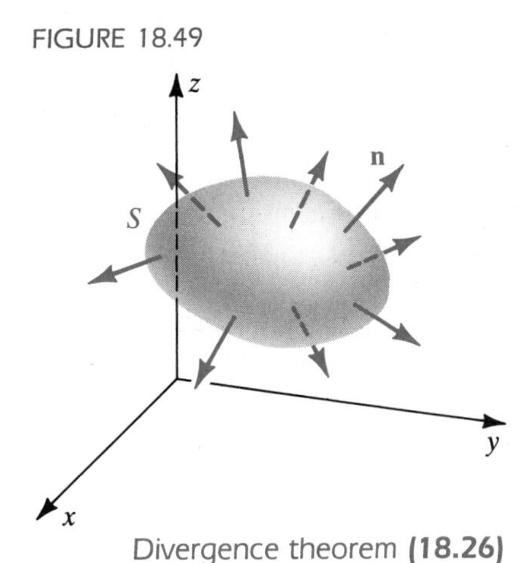

Under the restrictions we have placed on Q and S, the divergence theorem may be stated as follows.

Divergence theorem **(18.26)**

> Let Q be a region in three dimensions bounded by a closed surface S, and let $\mathbf{n}$ denote the unit outer normal to S at (x, y, z). If $\mathbf{F}$ is a vector function that has continuous partial derivatives on Q, then
>
> $$\iint_S \mathbf{F} \cdot \mathbf{n}\, dS = \iiint_Q \nabla \cdot \mathbf{F}\, dV;$$
>
> that is, *the flux of* $\mathbf{F}$ *over* S *equals the triple integral of the divergence of* $\mathbf{F}$ *over* Q.

PROOF If $\mathbf{F}(x, y, z) = M(x, y, z)\mathbf{i} + N(x, y, z)\mathbf{j} + P(x, y, z)\mathbf{k}$, let us simplify the notation by writing $\mathbf{F} = M\mathbf{i} + N\mathbf{j} + P\mathbf{k}$. Using properties of the dot product and the definition of $\mathbf{F}$, we may write the conclusion of the theorem as

$$\iint_S (M\mathbf{i} \cdot \mathbf{n} + N\mathbf{j} \cdot \mathbf{n} + P\mathbf{k} \cdot \mathbf{n})\, dS = \iiint_Q \left(\frac{\partial M}{\partial x} + \frac{\partial N}{\partial y} + \frac{\partial P}{\partial x}\right) dV.$$

To prove this equality, it is sufficient to show that

$$\iint_S M\mathbf{i} \cdot \mathbf{n}\, dS = \iiint_Q \frac{\partial M}{\partial x}\, dV$$

$$\iint_S N\mathbf{j} \cdot \mathbf{n}\, dS = \iiint_Q \frac{\partial N}{\partial y}\, dV$$

$$\iint_S P\mathbf{k} \cdot \mathbf{n}\, dS = \iiint_Q \frac{\partial P}{\partial z}\, dV.$$

FIGURE 18.50

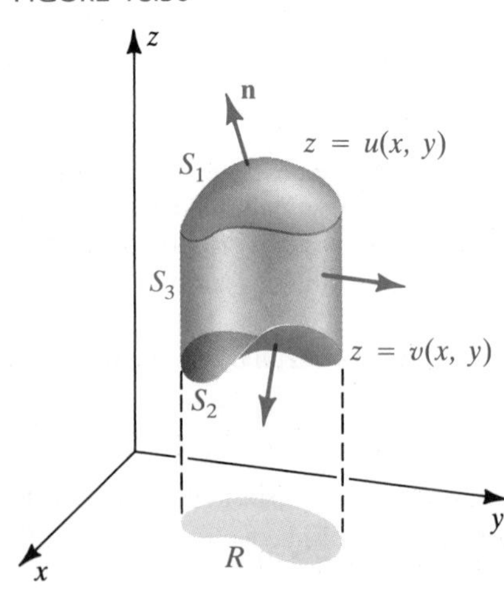

Since the proofs of all three formulas are similar, we shall consider only the third. Moreover, the proof will be restricted to the case, illustrated in Figure 18.50, where S is the surface of a region Q that lies between the graphs of $z = v(x, y)$ and $z = u(x, y)$ and over a suitable region R in the xy-plane. We shall denote the upper surface by S_1, the lower surface by S_2, and the lateral surface by S_3. Some typical unit outer normal vectors are shown in the figure.

On S_3, the $\mathbf{k}$ component of $\mathbf{n}$ is 0, and hence $\mathbf{k} \cdot \mathbf{n} = 0$. Thus, the flux integral over S_3 is 0, and we may write

$$\iint_S P\mathbf{k} \cdot \mathbf{n}\, dS = \iint_{S_1} P\mathbf{k} \cdot \mathbf{n}\, dS + \iint_{S_2} P\mathbf{k} \cdot \mathbf{n}\, dS.$$

As in Section 18.5, to find a unit (upper) normal to S_1, we let

$$g_1(x, y, z) = z - u(x, y)$$

and calculate

$$\mathbf{n} = \frac{\nabla g_1(x, y, z)}{\|\nabla g_1(x, y, z)\|} = \frac{-u_x(x, y)\mathbf{i} - u_y(x, y)\mathbf{j} + \mathbf{k}}{\sqrt{[u_x(x, y)]^2 + [u_y(x, y)]^2 + 1}}.$$

Hence $\mathbf{k} \cdot \mathbf{n} = 1/\sqrt{[u_x(x, y)]^2 + [u_y(x, y)]^2 + 1}$. If we apply Theorem (18.23)(i) with $R = R_{xy}$ and $u(x, y) = f(x, y)$, the radicals cancel and we obtain

$$\iint_{S_1} P\mathbf{k} \cdot \mathbf{n}\, dS = \iint_R P(x, y, u(x, y))\, dA.$$

On S_2, we let

$$g_2(x, y, z) = z - v(x, y)$$

and use the *lower* unit normal $\mathbf{n}$ given by

$$\mathbf{n} = -\frac{\nabla g_2(x, y, z)}{\|\nabla g_2(x, y, z)\|} = \frac{v_x(x, y)\mathbf{i} + v_y(x, y)\mathbf{j} - \mathbf{k}}{\sqrt{[v_x(x, y)]^2 + [v_y(x, y)]^2 + 1}}.$$

Again applying (18.23)(i) gives us the flux integral over S_2,

$$\iint_{S_2} P\mathbf{k} \cdot \mathbf{n}\, dS = -\iint_R P(x, y, v(x, y))\, dA.$$

Adding the flux integrals over S_1 and S_2 yields

$$\begin{aligned}\iint_S P\mathbf{k} \cdot \mathbf{n}\, dS &= \iint_R [P(x, y, u(x, y)) - P(x, y, v(x, y))]\, dA \\ &= \iint_R \left[\int_{v(x,y)}^{u(x,y)} \frac{\partial P}{\partial z}\, dz\right] dA = \iiint_Q \frac{\partial P}{\partial z}\, dV,\end{aligned}$$

which is what we wished to prove. Our proof can be extended to finite unions of regions of the type we have considered. The proofs of the formulas for $\iint_S M\mathbf{i} \cdot \mathbf{n}\, dS$ and $\iint_S N\mathbf{j} \cdot \mathbf{n}\, dS$ are similar, provided Q and S are suitably restricted. ■

FIGURE 18.51

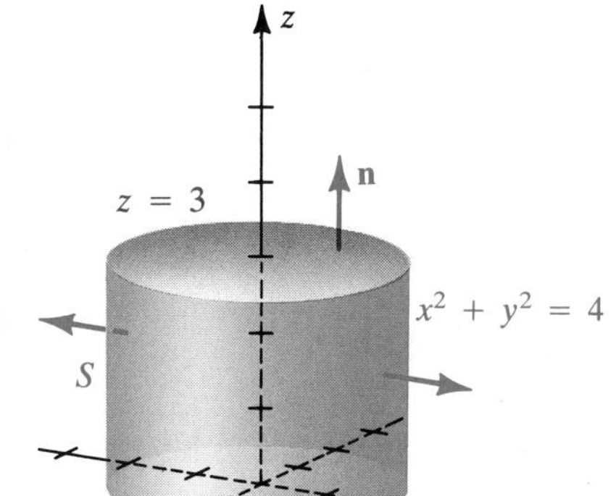

EXAMPLE 1 Let Q be the region bounded by the circular cylinder $x^2 + y^2 = 4$ and the planes $z = 0$ and $z = 3$. Let S denote the surface of Q. If $\mathbf{F}(x, y, z) = x^3\mathbf{i} + y^3\mathbf{j} + z^3\mathbf{k}$, use the divergence theorem to find $\iint_S \mathbf{F} \cdot \mathbf{n}\, dS$.

SOLUTION The surface S and some typical positions of $\mathbf{n}$ are sketched in Figure 18.51. Since

$$\nabla \cdot \mathbf{F} = 3x^2 + 3y^2 + 3z^2 = 3(x^2 + y^2 + z^2),$$

we have, by the divergence theorem (18.26),

$$\iint_S \mathbf{F} \cdot \mathbf{n}\, dS = 3 \iiint_Q (x^2 + y^2 + z^2)\, dV.$$

If we use cylindrical coordinates to evaluate the triple integral, then

$$\iint_S \mathbf{F}\cdot\mathbf{n}\,dS = 3\int_0^{2\pi}\int_0^2\int_0^3 (r^2+z^2)r\,dz\,dr\,d\theta$$

$$= 3\int_0^{2\pi}\int_0^2 \Big[r^2z+\tfrac{1}{3}z^3\Big]_0^3 r\,dr\,d\theta = 3\int_0^{2\pi}\int_0^2 (3r^2+9)r\,dr\,d\theta$$

$$= 3\int_0^{2\pi}\Big[\tfrac{3}{4}r^4+\tfrac{9}{2}r^2\Big]_0^2 d\theta = 3\int_0^{2\pi} 30\,d\theta = 90\Big[\theta\Big]_0^{2\pi} = 180\pi.$$

EXAMPLE 2 Let Q be the region bounded by the cylinder $z = 4 - x^2$, the plane $y + z = 5$, and the xy- and xz-planes. Let S be the surface of Q. If $\mathbf{F}(x, y, z) = (x^3 + \sin z)\mathbf{i} + (x^2y + \cos z)\mathbf{j} + e^{x^2+y^2}\mathbf{k}$, find $\iint_S \mathbf{F}\cdot\mathbf{n}\,dS$.

FIGURE 18.52

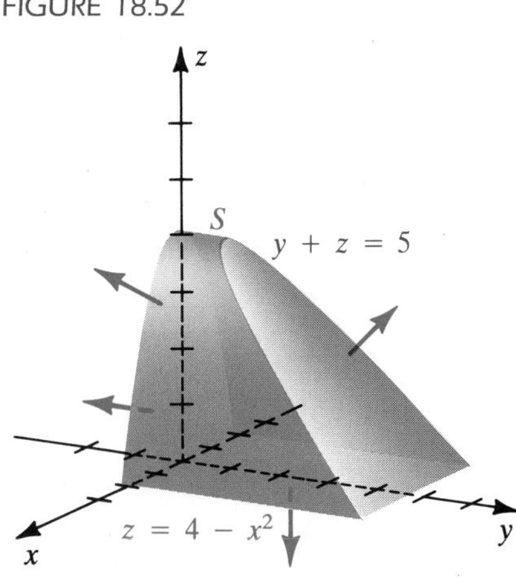

SOLUTION The region Q is sketched in Figure 18.52 with several unit outer normal vectors. It would be an incredibly difficult job to evaluate the surface integral directly; however, using the divergence theorem we can obtain the value from the triple integral

$$\iiint_Q (3x^2 + x^2)\,dV = \iiint_Q 4x^2\,dV.$$

Referring to Figure 18.52 and using (17.20), we see that this integral equals

$$\int_{-2}^{2}\int_0^{4-x^2}\int_0^{5-z} 4x^2\,dy\,dz\,dx = \int_{-2}^{2}\int_0^{4-x^2} 4x^2(5-z)\,dz\,dx$$

$$= \int_{-2}^{2}\Big[20x^2z - 2x^2z^2\Big]_0^{4-x^2} dx$$

$$= \int_{-2}^{2}(48x^2 - 4x^4 - 2x^6)\,dx$$

$$= \tfrac{4608}{35} \approx 131.7.$$

We may use the divergence theorem to obtain a physical interpretation for the divergence of a vector field. Let us begin by recalling from the mean value theorem for definite integrals (5.28) that if a function f of one variable is continuous on a closed interval $[a, b]$, then there is a number c in (a, b) such that

$$\int_a^b f(x)\,dx = f(c)L,$$

where $L = b - a$ is the length of $[a, b]$. Analogous results may be proved for double and triple integrals. Thus, if a function f of three variables is continuous throughout a spherical region Q, then there is a point $A(x_1, y_1, z_1)$ in the interior of Q such that

$$\iiint_Q f(x, y, z)\,dV = [f(x, y, z)]_A\,V,$$

where V is the volume of Q and $[f(x, y, z)]_A = f(x_1, y_1, z_1)$. Let us call this result the **mean value theorem for triple integrals**. It follows that if $\mathbf{F}$ is a continuous vector function, then

$$\iiint_Q \nabla\cdot\mathbf{F}\,dV = [\nabla\cdot\mathbf{F}]_A\,V,$$

or, equivalently,

$$[\nabla \cdot \mathbf{F}]_A = \frac{1}{V} \iiint_Q \nabla \cdot \mathbf{F}\, dV.$$

Applying the divergence theorem gives us

FIGURE 18.53

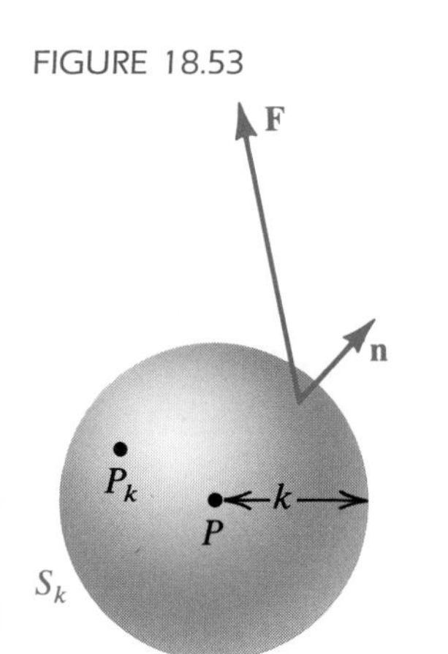

$$[\nabla \cdot \mathbf{F}]_A = \frac{1}{V} \iint_S \mathbf{F} \cdot \mathbf{n}\, dS,$$

where S is the surface of Q. The quantity on the right may be regarded as *the flux of* **F** *per unit volume over the sphere.*

Next, let P be an arbitrary point, and suppose **F** is continuous throughout a region containing P in its interior. Let S_k be the surface of a sphere of radius k with center at P. From the previous discussion, for each k there is a point P_k within S_k such that

$$[\nabla \cdot \mathbf{F}]_{P_k} = \frac{1}{V_k} \iint_{S_k} \mathbf{F} \cdot \mathbf{n}\, dS,$$

where V_k is the volume of the sphere (see Figure 18.53).

If we let $k \to 0$, then $P_k \to P$ and $[\nabla \cdot \mathbf{F}]_{P_k} \to [\nabla \cdot \mathbf{F}]_P = [\text{div } \mathbf{F}]_P$. This gives us the following.

Divergence as a limit **(18.27)**

$$[\text{div } \mathbf{F}]_P = \lim_{k \to 0} \frac{1}{V_k} \iint_{S_k} \mathbf{F} \cdot \mathbf{n}\, dS$$

Thus, the *divergence of* **F** *at P is the limiting value of the flux per unit volume over a sphere with center P, as the radius of the sphere approaches* 0. In particular, if **F** represents the velocity of a fluid, then $[\text{div } \mathbf{F}]_P$ can be interpreted as the rate of loss or gain of fluid per unit volume at P per unit time. If $\delta = \delta(x, y, z)$ is the density at (x, y, z), then $[\text{div } \delta\mathbf{F}]_P$ is the change in *mass* per unit volume per unit time. It follows from the remarks at the end of Section 18.5 that there is a source or sink at P if $[\text{div } \mathbf{F}]_P > 0$ or $[\text{div } \mathbf{F}]_P < 0$, respectively. If the fluid is incompressible and no source or sink is present, then there can be no gain or loss within the volume element V_k and hence, at every point P, div $\mathbf{F} = 0$. The limit form for divergence in (18.27) may also be applied to physical concepts such as magnetic or electric flux, since these entities have many characteristics that are similar to those of fluids.

FIGURE 18.54

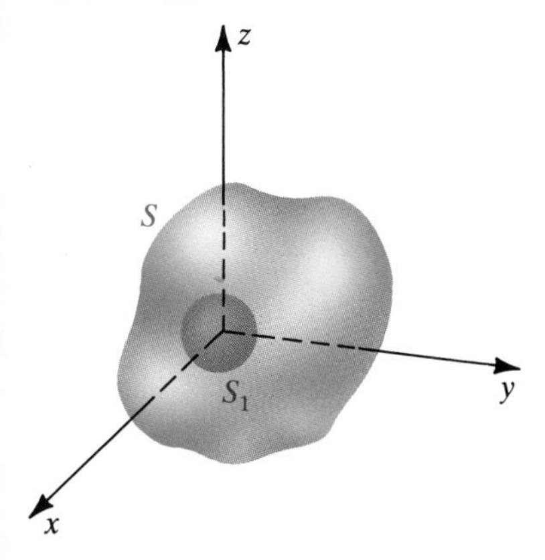

EXAMPLE 3 Suppose a closed surface S forms the boundary of a region Q and the origin O is an interior point of Q. If an inverse square field is given by $\mathbf{F} = (q/r^3)\mathbf{r}$, where q is a constant, $\mathbf{r} = x\mathbf{i} + y\mathbf{j} + z\mathbf{k}$, and $r = \|\mathbf{r}\|$, prove that the flux of **F** over S is $4\pi q$ regardless of the shape of Q.

SOLUTION Since **F** is not continuous at O, the divergence theorem cannot be applied directly; however, we may proceed as follows. Let S_1 be a sphere of radius a and center O that lies completely within S (see Figure 18.54), and let Q_1 denote the region that lies *outside* S_1 and *inside*

S. Since $\mathbf{F}$ is continuous throughout Q_1, we may apply the divergence theorem, obtaining

$$\iiint_{Q_1} \nabla \cdot \mathbf{F}\, dV = \iint_S \mathbf{F} \cdot \mathbf{n}\, dS + \iint_{S_1} \mathbf{F} \cdot \mathbf{n}\, dS.$$

We can show that if $\mathbf{F}$ is an inverse square field, then $\nabla \cdot \mathbf{F} = 0$ (see Exercise 21 of Section 18.1) and therefore

$$\iint_S \mathbf{F} \cdot \mathbf{n}\, dS = -\iint_{S_1} \mathbf{F} \cdot \mathbf{n}\, dS.$$

Since the unit normal $\mathbf{n}$ on S_1 is an *outer* normal, it points *away* from the region Q_1 that has S_1 as part of its boundary. Hence on S_1 the unit normal $\mathbf{n}$ points *toward* the origin. Thus, $\mathbf{n} = (-1/r)\mathbf{r}$ with $r = a$, and

$$\begin{aligned}\iint_S \mathbf{F} \cdot \mathbf{n}\, dS &= -\iint_{S_1} \left(\frac{q}{r^3}\right)\mathbf{r} \cdot \left(-\frac{1}{r}\mathbf{r}\right) dS \\ &= \iint_{S_1} \frac{q}{r^4}(\mathbf{r} \cdot \mathbf{r})\, dS = \iint_{S_1} \frac{q}{r^2}\, dS \\ &= \frac{q}{a^2} \iint_{S_1} dS = \frac{q}{a^2}(4\pi a^2) = 4\pi q.\end{aligned}$$

Example 3 has an important application in the theory of electricity. By Coulomb's law, if a point charge of q coulombs is located at the origin, then the force exerted on a unit charge at (x, y, z) is $\mathbf{F}(x, y, z) = (cq/\|\mathbf{r}\|^3)\mathbf{r}$, where $\mathbf{r} = x\mathbf{i} + y\mathbf{j} + z\mathbf{k}$ and c is a constant. It follows that the *electric flux* of $\mathbf{F}$ over any closed surface containing the origin is $4\pi cq$. Thus, the electric flux is independent of the size or shape of S and depends only on q. It is not difficult to extend this to the case of more than one point charge within S. This result, called *Gauss' law*, has far-reaching consequences in the study of electric fields.

EXERCISES 18.6

Exer. 1–4: Use the divergence theorem (18.26) to find $\iint_S \mathbf{F} \cdot \mathbf{n}\, dS$.

1 $\mathbf{F} = y \sin x\mathbf{i} + y^2z\mathbf{j} + (x + 3z)\mathbf{k}$;
S is the surface of the region bounded by the planes $x = \pm 1$, $y = \pm 1$, $z = \pm 1$.

2 $\mathbf{F} = y^3e^z\mathbf{i} - xy\mathbf{j} + x \arctan y\mathbf{k}$;
S is the surface of the region bounded by the coordinate planes and the plane $x + y + z = 1$.

3 $\mathbf{F} = (x^2 + \sin yz)\mathbf{i} + (y - xe^{-z})\mathbf{j} + z^2\mathbf{k}$;
S is the surface of the region bounded by the cylinder $x^2 + y^2 = 4$ and the planes $x + z = 2$ and $z = 0$.

4 $\mathbf{F} = 2xy\mathbf{i} + z \cosh x\mathbf{j} + (z^2 + y \sin^{-1} x)\mathbf{k}$;
S is the surface of the region bounded by the paraboloid $z = x^2 + y^2$ and the plane $z = 1$.

Exer. 5–10: Use the divergence theorem (18.26) to find the flux of F through S.

5 $\mathbf{F}(x, y, z) = yz\mathbf{i} + xz\mathbf{j} + xy\mathbf{k}$;
S is the graph of $x^{2/3} + y^{2/3} + z^{2/3} = 1$.

6 $\mathbf{F}(x, y, z) = (x^2 + z^2)\mathbf{i} + (y^2 - 2xy)\mathbf{j} + (4z - 2yz)\mathbf{k}$;
S is the surface of the region bounded by the cone $x = \sqrt{y^2 + z^2}$ and the plane $x = 9$.

7 $\mathbf{F}(x, y, z) = 3x\mathbf{i} + xz\mathbf{j} + z^2\mathbf{k}$;
S is the surface of the region bounded by the paraboloid $z = 4 - x^2 - y^2$ and the xy-plane.

8 $\mathbf{F}(x, y, z) = xy^2\mathbf{i} + yz^2\mathbf{j} + zx^2\mathbf{k}$;
S is the surface of the region that lies between the cylinders $x^2 + y^2 = 4$ and $x^2 + y^2 = 9$ and between the planes $z = -1$ and $z = 2$.

9 $\mathbf{F}(x, y, z) = 2xz\mathbf{i} + xyz\mathbf{j} + yz\mathbf{k}$;
S is the surface of the region bounded by the coordinate planes and the planes $x + 2z = 4$ and $y = 2$.

10 $\mathbf{F}(x, y, z) = x^3\mathbf{i} + y^3\mathbf{j} + z^3\mathbf{k}$;
S is the surface of the region that is inside both the cone $z = \sqrt{x^2 + y^2}$ and the sphere $x^2 + y^2 + z^2 = 25$.

Exer. 11–14: Verify the divergence theorem (18.26) by evaluating both the surface integral and the triple integral.

11 $\mathbf{F} = x\mathbf{i} + y\mathbf{j} + z\mathbf{k}$;
S is the sphere $x^2 + y^2 + z^2 = a^2$.

12 $\mathbf{F} = x^2\mathbf{i} + y^2\mathbf{j} + z^2\mathbf{k}$;
S is the surface of the cube bounded by the coordinate planes and the planes $x = a$, $y = a$, $z = a$ with $a > 0$.

13 $\mathbf{F} = (x + z)\mathbf{i} + (y + z)\mathbf{j} + (x + y)\mathbf{k}$;
S is the surface of the region
$Q = \{(x, y, z): 0 \le y^2 + z^2 \le 1, 0 \le x \le 2\}$.

14 $\mathbf{F} = \|\mathbf{r}\|^2\mathbf{r}$ for $\mathbf{r} = x\mathbf{i} + y\mathbf{j} + z\mathbf{k}$;
S is the sphere $x^2 + y^2 + z^2 = a^2$.

15 If $\iint_S \mathbf{F} \cdot \mathbf{n}\, dS = 0$ for every closed surface of the type considered in the divergence theorem, prove that $\operatorname{div} \mathbf{F} = 0$.

16 Use the divergence theorem to prove that if a scalar function f has continuous second partial derivatives, then

$$\iiint_Q \nabla^2 f\, dV = \iint_S D_{\mathbf{n}} f\, dS,$$

where $D_{\mathbf{n}} f$ is the directional derivative of f in the direction of an outer unit normal $\mathbf{n}$ to S.

Exer. 17–18: Assume that S and Q satisfy the conditions of the divergence theorem and that f and g are scalar functions that have continuous second partial derivatives. Prove the identity.

17 $$\iiint_Q (f\nabla^2 g + \nabla f \cdot \nabla g)\, dV = \iint_S (f\nabla g) \cdot \mathbf{n}\, dS$$

(*Hint:* Let $\mathbf{F} = f\nabla g$ in (18.26).)

18 $$\iiint_Q (f\nabla^2 g - g\nabla^2 f)\, dV = \iint_S (f\nabla g - g\nabla f) \cdot \mathbf{n}\, dS$$

(*Hint:* Use the identity in Exercise 17 together with that obtained by interchanging f and g.)

Exer. 19–22: Assume that S and Q satisfy the conditions of the divergence theorem.

19 If $\mathbf{F}$ is a conservative vector field with potential function f and if $\operatorname{div} \mathbf{F} = 0$, prove that $\iiint_Q \mathbf{F} \cdot \mathbf{F}\, dV = \iint_S f\mathbf{F} \cdot \mathbf{n}\, dS$.

20 If $\mathbf{r} = x\mathbf{i} + y\mathbf{j} + z\mathbf{k}$ and V is the volume of Q, prove that $V = \frac{1}{3}\iint_S \mathbf{r} \cdot \mathbf{n}\, dS$.

21 If $\mathbf{F}$ has continuous second partial derivatives, prove that $\iint_S \operatorname{curl} \mathbf{F} \cdot \mathbf{n}\, dS = 0$.

22 If $\mathbf{a}$ is a constant vector, prove that $\iint_S \mathbf{a} \cdot \mathbf{n}\, dS = 0$.

Exer. 23–24: A surface (or triple) integral of a vector function is defined as the sum of the surface (or triple) integrals of each component of the function. Using this fact and assuming that S and Q satisfy the conditions of the divergence theorem, establish the result.

23 $$\iint_S \mathbf{F} \times \mathbf{n}\, dS = -\iiint_Q \nabla \times \mathbf{F}\, dV$$

(*Hint:* Apply the divergence theorem and Exercise 26 of Section 18.1 to $\mathbf{c} \times \mathbf{F}$, where $\mathbf{c}$ is an arbitrary constant vector.)

24 $$\iint_S f\mathbf{n}\, dS = \iiint_Q \nabla f\, dV$$

(*Hint:* Apply the divergence theorem to $f\mathbf{c}$, where $\mathbf{c}$ is an arbitrary constant vector.)

25 If $\mathbf{F}$ is orthogonal to S at each point (x, y, z), prove that $\iiint_Q \operatorname{curl} \mathbf{F}\, dV = \mathbf{0}$.

26 If $\mathbf{r}$ is the position vector for (x, y, z) and $r = \|\mathbf{r}\|$, prove that $\iiint_Q \mathbf{r}\, dV = \frac{1}{2}\iint_S r^2\mathbf{n}\, dS$.

27 An empty cylindrical tank of diameter 2 feet and length 10 feet is immersed in water having a density of 62.5 lb/ft^3. The *buoyancy force* on the tank is given by $-\iint_S p\mathbf{n}\, dS$, where $\mathbf{n}$ is the unit outer normal to the surface S of the tank and p is the pressure exerted by the water on S. Use the formula

$$\iint_S p\mathbf{n}\, dS = \iiint_Q \nabla p\, dV$$

(see Exercise 24) to calculate the buoyancy force. (*Hint:* If the surface of the water coincides with the xy-plane, then $p = -62.5z$, where z is the depth of the water.)

18.7 STOKES' THEOREM

In (18.21) we stated the following vector form for Green's theorem:

$$\oint_C \mathbf{F} \cdot \mathbf{T}\, ds = \iint_R (\operatorname{curl} \mathbf{F}) \cdot \mathbf{k}\, dA,$$

where the plane curve C is the boundary of R. This result may be extended to a piecewise-smooth simple closed curve C in three dimensions that forms the boundary of a surface S. The sketch in Figure 18.55 illustrates a case in which S is the graph of $z = f(x, y)$, f has continuous first partial derivatives, and the projection C_1 of C onto the xy-plane is a curve that bounds a region R of the type considered in Green's theorem. In the figure, $\mathbf{n}$ is a unit *upper* normal to S. We take the positive direction along C as that which corresponds to the positive direction along C_1. The vector $\mathbf{T}$ is a unit tangent vector to C that points in the positive direction. If $\mathbf{F}$ is a vector field whose components have continuous partial derivatives in a region containing S, then we have the following theorem, named after the English mathematical physicist George G. Stokes (1819–1903).

Stokes' theorem **(18.28)**

$$\oint_C \mathbf{F}\cdot\mathbf{T}\,ds = \iint_S (\text{curl }\mathbf{F})\cdot\mathbf{n}\,dS$$

FIGURE 18.55

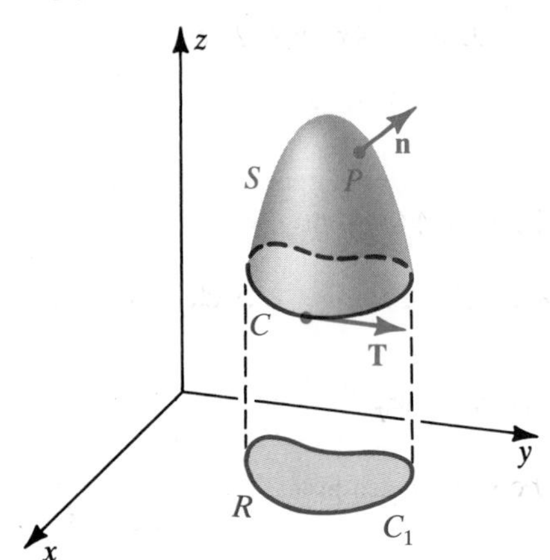

Stokes' theorem may be stated: *The line integral of the tangential component of* $\mathbf{F}$ *taken along* C *once in the positive direction equals the surface integral of the normal component of curl* $\mathbf{F}$ *over* S. If $\mathbf{F}$ is a force field, the theorem states that the work done by $\mathbf{F}$ along C equals the flux of curl $\mathbf{F}$ over S. The line integral in (18.28) may also be written $\oint_C \mathbf{F}\cdot d\mathbf{r}$, where $\mathbf{r}$ is the position vector of the point (x, y, z) on C. To consider situations more general than the one pictured in Figure 18.55, we must consider an *oriented* surface S and define a positive direction along C in a suitable manner. (A proof of Stokes' theorem may be found in more advanced texts.)

EXAMPLE 1 Let S be the part of the paraboloid $z = 9 - x^2 - y^2$ with $z \geq 0$, and let C be the trace of S on the xy-plane. Verify Stokes' theorem (18.28) for the vector field $\mathbf{F} = 3z\mathbf{i} + 4x\mathbf{j} + 2y\mathbf{k}$.

SOLUTION We wish to show that the two integrals in Theorem (18.28) have the same value.

The surface is the same as that considered in Example 5 of Section 18.5 (see Figure 18.48), where we found that

$$\mathbf{n} = \frac{2x\mathbf{i} + 2y\mathbf{j} + \mathbf{k}}{\sqrt{4x^2 + 4y^2 + 1}}.$$

By (18.6),

$$\text{curl }\mathbf{F} = \begin{vmatrix} \mathbf{i} & \mathbf{j} & \mathbf{k} \\ \dfrac{\partial}{\partial x} & \dfrac{\partial}{\partial y} & \dfrac{\partial}{\partial z} \\ 3z & 4x & 2y \end{vmatrix} = 2\mathbf{i} + 3\mathbf{j} + 4\mathbf{k}.$$

Consequently

$$\iint_S (\text{curl }\mathbf{F})\cdot\mathbf{n}\,dS = \iint_S \frac{4x + 6y + 4}{\sqrt{4x^2 + 4y^2 + 1}}\,dS.$$

Using (18.23)(i) to evaluate this surface integral, we have

$$\iint_S (\text{curl }\mathbf{F})\cdot\mathbf{n}\,dS = \iint_R (4x + 6y + 4)\,dA,$$

where R is the region in the xy-plane bounded by the circle of radius 3 with center at the origin. Changing to polar coordinates, we obtain

$$\iint_S (\text{curl } \mathbf{F}) \cdot \mathbf{n} \, dS = \int_0^{2\pi} \int_0^3 (4r \cos\theta + 6r \sin\theta + 4) r \, dr \, d\theta$$

$$= \int_0^{2\pi} \int_0^3 [(4 \cos\theta + 6 \sin\theta) r^2 + 4r] \, dr \, d\theta$$

$$= \int_0^{2\pi} \left[(4 \cos\theta + 6 \sin\theta)\tfrac{1}{3}r^3 + 2r^2\right]_0^3 d\theta$$

$$= \int_0^{2\pi} (36 \cos\theta + 54 \sin\theta + 18) \, d\theta$$

$$= \left[36 \sin\theta - 54 \cos\theta + 18\theta\right]_0^{2\pi} = 36\pi.$$

The line integral in Stokes' theorem (18.28) may be written

$$\oint_C \mathbf{F} \cdot \mathbf{T} \, ds = \oint_C \mathbf{F} \cdot d\mathbf{r} = \oint_C 3z \, dx + 4x \, dy + 2y \, dz,$$

where C is the circle $x^2 + y^2 = 9$ in the xy-plane. Since $z = 0$ on C, this line integral reduces to

$$\oint_C \mathbf{F} \cdot d\mathbf{r} = \oint_C 4x \, dy = 4 \oint_C x \, dy.$$

By Theorem (18.20), $\oint_C x \, dy$ is the area of the region (a circle of radius 3) bounded by C, and hence

$$\oint \mathbf{F} \cdot d\mathbf{r} = 4(9\pi) = 36\pi,$$

which is the same as the value of the surface integral.

FIGURE 18.56

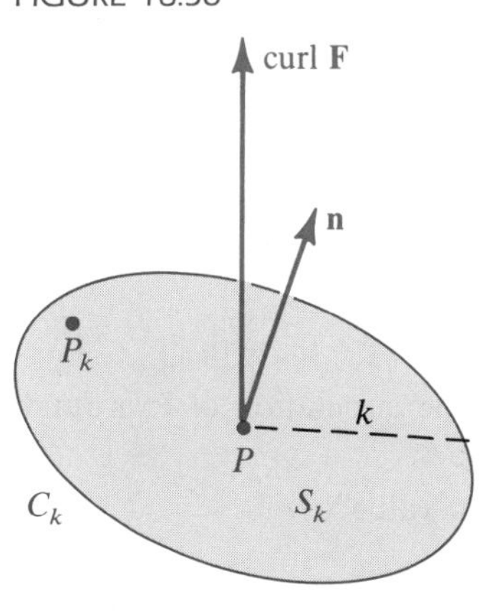

We may use Stokes' theorem to obtain a physical interpretation for curl $\mathbf{F}$. If P is any point, let S_k be a circular disk of radius k with center at P and let C_k denote the boundary of S_k (see Figure 18.56). Applying Stokes' theorem and a mean value theorem for double integrals leads to

$$\oint_{C_k} \mathbf{F} \cdot \mathbf{T} \, ds = \iint_{S_k} (\text{curl } \mathbf{F}) \cdot \mathbf{n} \, dS = [(\text{curl } \mathbf{F}) \cdot \mathbf{n}]_{P_k} (\pi k^2),$$

where P_k is some interior point of S_k and πk^2 is the area of S_k. Thus,

$$[(\text{curl } \mathbf{F}) \cdot \mathbf{n}]_{P_k} = \frac{1}{\pi k^2} \oint_{C_k} \mathbf{F} \cdot \mathbf{T} \, ds.$$

If we let $k \to 0$, then $P_k \to P$ and we obtain the following.

(Curl F) · n as a limit (18.29)

$$[(\text{curl } \mathbf{F}) \cdot \mathbf{n}]_P = \lim_{k \to 0} \frac{1}{\pi k^2} \oint_{C_k} \mathbf{F} \cdot \mathbf{T} \, ds$$

If $\mathbf{F}$ is the velocity field of a fluid, then the line integral $\oint_C \mathbf{F} \cdot \mathbf{T} \, ds$ is the **circulation around C**. It measures the average tendency of the fluid to move, or *circulate*, around the curve. We see from (18.29) that $[(\text{curl } \mathbf{F}) \cdot \mathbf{n}]_P$ provides information about the motion of the fluid around the circumference of a circular disk that is perpendicular to $\mathbf{n}$, as the disk shrinks to a point. Since $[(\text{curl } \mathbf{F}) \cdot \mathbf{n}]_P$ has its maximum value when $\mathbf{n}$ is

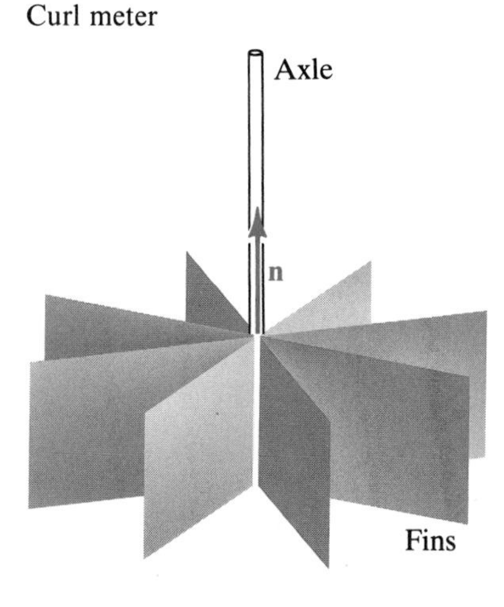

FIGURE 18.57
Curl meter

parallel to curl **F**, the direction of curl **F** at P is that for which the circulation around the boundary of a disk perpendicular to curl **F** will have its maximum value as the disk shrinks to a point.

To visualize these ideas, consider a miniature (imaginary) paddlewheel, or a **curl meter**, of the type shown in Figure 18.57. A velocity field acting on the fins may cause the wheel to turn on its axle. If **n** is a unit vector directed along the axle, then the curl meter will rotate most rapidly when **n** is parallel to curl **F**.

The scalar (curl **F**) · **n** is sometimes called the **rotation** of **F** about **n** and is denoted by rot **F**, since it measures the tendency of the vector field to *rotate* about P. If rot $\mathbf{F} > 0$ throughout a region containing the curl meter in Figure 18.57, the fins rotate in a counterclockwise direction (a direction determined by the right-hand rule with thumb along **n**). If rot $\mathbf{F} < 0$, the fins rotate in a clockwise direction. If rot $\mathbf{F} = 0$, as is the case for curl $\mathbf{F} = \mathbf{0}$, then the circulation around every closed curve C is 0 and the fins do not rotate. For this reason we define an **irrotational vector field F** as one for which curl $\mathbf{F} = \mathbf{0}$.

EXAMPLE 2 If $\mathbf{F}(x, y) = a\mathbf{i} + b\mathbf{j} + c\mathbf{k}$ and a, b, and c are constants, discuss the rotational properties of the field **F**.

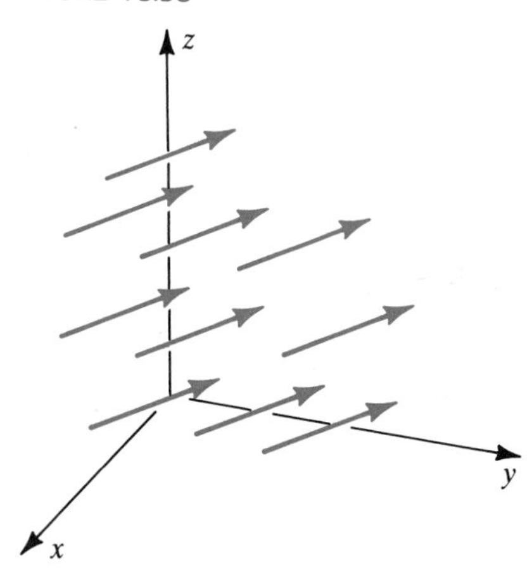

FIGURE 18.58

SOLUTION Every vector of the field **F** has the same length and direction, as illustrated in Figure 18.58, where a, b, and c are positive. If we inserted a curl meter into this vector field at any point, the fins would not rotate. Note that

$$\text{curl } \mathbf{F} = \nabla \times \mathbf{F} = \begin{vmatrix} \mathbf{i} & \mathbf{j} & \mathbf{k} \\ \dfrac{\partial}{\partial x} & \dfrac{\partial}{\partial y} & \dfrac{\partial}{\partial z} \\ a & b & c \end{vmatrix} = \mathbf{0}.$$

Hence **F** is irrotational.

EXAMPLE 3 Let $\mathbf{F}(x, y, z) = 3(y^2 + 1)^{-1}\mathbf{i} + 0\mathbf{j} + 0\mathbf{k}$ with $|y| \le 4$.

(a) Describe the vector field **F** and discuss the circulation of **F** around several circles C_k that lie in the xy-plane.

(b) Where does (curl **F**) · **n** have its maximum value?

SOLUTION

(a) If we restrict our attention to the xy-plane, the vectors in **F** form the pattern shown in Figure 18.59. (Verify this fact.) The same pattern is repeated in any plane parallel to the xy-plane.

Let us regard **F** as the velocity field of a fluid that is flowing in a culvert. Note that the speed of the fluid is greatest along the x-axis and decreases as the point (x, y) approaches either of the lines $y = \pm 4$. If we consider the circle C_1, then evidently $\oint_{C_1} \mathbf{F} \cdot \mathbf{T}\, ds > 0$, since as C_1 is traversed in the positive direction, the tangential component $\mathbf{F} \cdot \mathbf{T}$ is large and positive on the lower half of C_1 and negative (but close to 0) on the

FIGURE 18.59

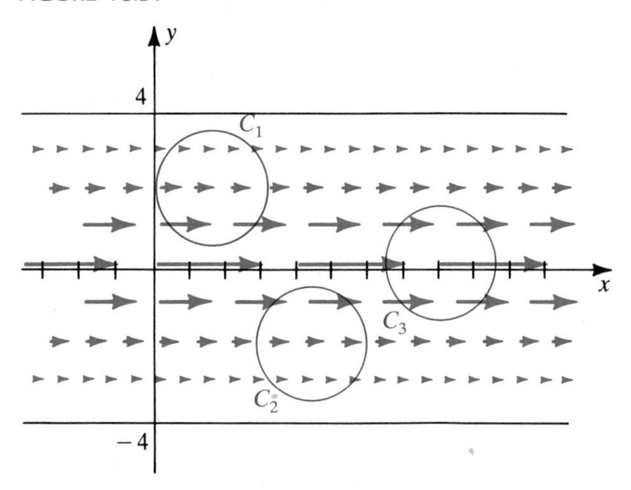

upper half. If a curl meter were inserted at the center of C_1, with its axle perpendicular to the xy-plane, the fins would rotate in the counterclockwise direction.

Again considering values of $\mathbf{F} \cdot \mathbf{T}$, we see that $\oint_{C_2} \mathbf{F} \cdot \mathbf{T}\, ds < 0$ and $\oint_{C_3} \mathbf{F} \cdot \mathbf{T}\, ds = 0$. A curl meter for C_2 would rotate in a clockwise direction, and that for C_3 would not rotate at all.

(b) Applying (18.6) gives us

$$\text{curl}\ \mathbf{F} = \frac{6y}{(y^2+1)^2}\,\mathbf{k}.$$

Since the unit normal $\mathbf{n}$ is $\mathbf{k}$,

$$(\text{curl}\ \mathbf{F}) \cdot \mathbf{n} = (\text{curl}\ \mathbf{F}) \cdot \mathbf{k} = \frac{6y}{(y^2+1)^2}.$$

Observe that $(\text{curl}\ \mathbf{F}) \cdot \mathbf{n}$ is positive for points above the x-axis, negative for points below the x-axis, and 0 for points on the x-axis. Differentiating with respect to y yields

$$D_y[(\text{curl}\ \mathbf{F}) \cdot \mathbf{n}] = \frac{6(1-3y^2)}{(y^2+1)^3},$$

and hence critical numbers for $(\text{curl}\ \mathbf{F}) \cdot \mathbf{n}$ are $y = \pm 1/\sqrt{3} \approx \pm 0.577$. We see from the first derivative test that these lead to maximum values. Thus, the maximum values for $(\text{curl}\ \mathbf{F}) \cdot \mathbf{n}$ occur at any point on either of the horizontal lines $y = \pm 1/\sqrt{3}$.

To demonstrate another connection between curl $\mathbf{F}$ and rotational aspects of motion, consider a fluid rotating uniformly about the z-axis, as if it were a rigid body. If we concentrate on a single fluid particle at the point $P(x, y, z)$ and if x, y, and z are functions of time t, then we have a situation similar to that illustrated in Figure 18.60, where $\mathbf{r}(t) = x\mathbf{i} + y\mathbf{j} + z\mathbf{k}$ is the position vector of P and $\boldsymbol{\omega} = \omega_1\mathbf{i} + \omega_2\mathbf{j} + \omega_3\mathbf{k}$ is the (constant) angular velocity. Referring to Figure 18.60, we see that

FIGURE 18.60

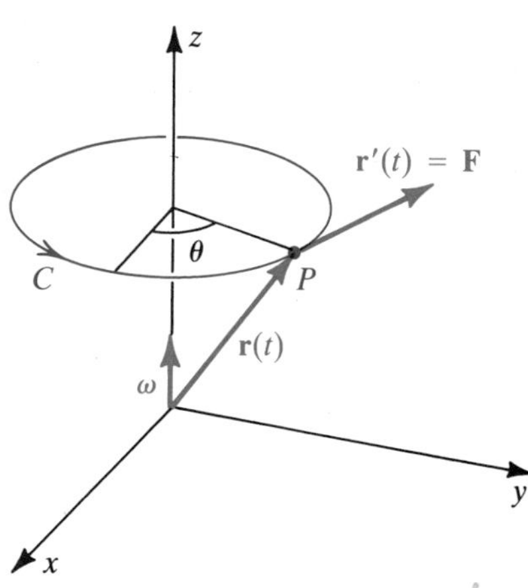

$d\theta/dt = \|\boldsymbol{\omega}\|$. If we denote the velocity $\mathbf{r}'(t)$ of P by $\mathbf{F}$, then, as in Example 4 of Section 15.3,

$$\mathbf{F} = \boldsymbol{\omega} \times \mathbf{r} = (\omega_2 z - \omega_3 y)\mathbf{i} + (\omega_3 x - \omega_1 z)\mathbf{j} + (\omega_1 y - \omega_2 x)\mathbf{k}.$$

By (18.6),

$$\text{curl } \mathbf{F} = \begin{vmatrix} \mathbf{i} & \mathbf{j} & \mathbf{k} \\ \dfrac{\partial}{\partial x} & \dfrac{\partial}{\partial y} & \dfrac{\partial}{\partial z} \\ \omega_2 z - \omega_3 y & \omega_3 x - \omega_1 z & \omega_1 y - \omega_2 x \end{vmatrix}$$

Using the fact that $\boldsymbol{\omega}$ is a constant vector, we may verify that

$$\text{curl } \mathbf{F} = 2\omega_1\mathbf{i} + 2\omega_2\mathbf{j} + 2\omega_3\mathbf{k} = 2\boldsymbol{\omega}.$$

This shows that the magnitude of curl **F** is twice the angular velocity, and the direction of curl **F** is along the axis of rotation.

We can apply the preceding discussion to the velocity field **F** of the wind inside the core of a tornado. (We often call **F** the *tangential wind velocity*.) The study of the circulation around a curve C inside the core is important in the analysis of a tornado. In the next example we calculate the circulation if C is a circle.

EXAMPLE 4 The core of a tornado rotates as if it were a rigid body, in the manner illustrated in Figure 18.60, where the angular wind speed $\omega = d\theta/dt$ is constant. If C is a circle of radius a in a cross section of the core, as shown in the figure, the magnitude of the tangential wind velocity $\mathbf{v}$ is $\|\mathbf{v}\| = \omega a$. Find the circulation of $\mathbf{v}$ around C by

(a) using Stokes' theorem **(b)** evaluating $\int_C \mathbf{v} \cdot \mathbf{T}\, ds$

SOLUTION

(a) Applying Stokes' theorem, we find that the circulation is

$$\int_C \mathbf{v} \cdot \mathbf{T}\, ds = \iint_S (\text{curl } \mathbf{v}) \cdot \mathbf{n}\, dS,$$

where S is the disk bounded by C and where $\mathbf{n} = \mathbf{k}$. From the discussion preceding this example,

$$\text{curl } \mathbf{v} = 2\boldsymbol{\omega} = 2\omega\mathbf{k}.$$

Hence

$$\begin{aligned} \int_C \mathbf{v} \cdot \mathbf{T}\, ds &= \iint_S (2\omega\mathbf{k}) \cdot \mathbf{k}\, dS \\ &= 2\omega \iint_S dS \\ &= 2\omega\pi a^2. \end{aligned}$$

This shows that as the radius a of C increases, the tendency of the winds to move around C also increases.

(b) Since **T** is a unit tangent vector to C having the same direction as $\mathbf{v}$,

$$\mathbf{v} \cdot \mathbf{T} = \|\mathbf{v}\|\,\|\mathbf{T}\| = \|\mathbf{v}\|$$

(see Theorem (14.19)). Hence

$$\int_C \mathbf{v} \cdot \mathbf{T}\, ds = \int_C \|\mathbf{v}\|\, ds = \omega a \int_C ds$$
$$= \omega a(2\pi a) = 2\omega\pi a^2.$$

EXAMPLE 5 For points outside the core, a tornado does not rotate as if it were a rigid body. Let r_0 be the radius of the core and ω the (constant) angular wind speed inside the core, as in Example 4. If C is a cross-sectional circle of radius $r > r_0$, as illustrated in Figure 18.60, then the magnitude of the tangential wind velocity $\mathbf{v}$ on C is (approximately) $\|\mathbf{v}\| = \omega r_0^2/r$.

(a) Show that the circulation around C is the same for every $r > r_0$.

(b) Show that curl $\mathbf{v} = \mathbf{0}$ outside the core.

SOLUTION

(a) The circulation around every circle C of radius $r > r_0$ is

$$\int_C \mathbf{v} \cdot \mathbf{T}\, ds = \int_C \|\mathbf{v}\|\, ds$$
$$= \frac{\omega r_0^2}{r} \int_C ds$$
$$= \frac{\omega r_0^2}{r} 2\pi r = 2\pi\omega r_0^2.$$

FIGURE 18.61

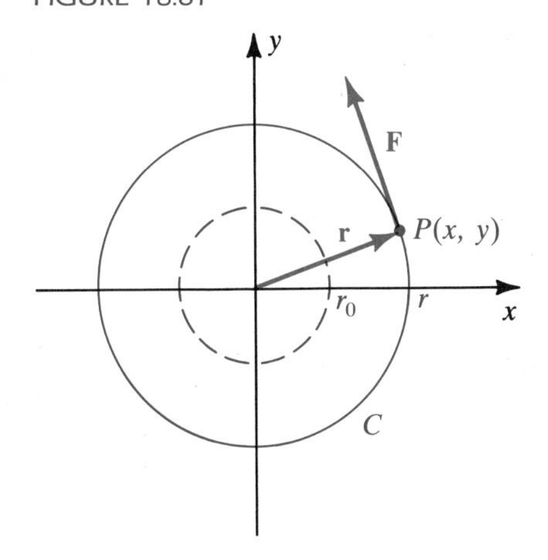

(b) A cross-sectional view of the tornado is shown in Figure 18.61, where $\mathbf{r} = x\mathbf{i} + y\mathbf{j}$ is the position vector of a point $P(x, y)$ on C and $\|\mathbf{r}\| = r$. In Example 1 of Section 18.1 we proved that $\mathbf{F}(x, y) = -y\mathbf{i} + x\mathbf{j}$ is orthogonal to $\mathbf{r}$ at every point (x, y), as illustrated in the figure. Hence a *unit* vector $\mathbf{u}$ in the direction of $\mathbf{F}$ is

$$\mathbf{u} = \frac{1}{\sqrt{x^2 + y^2}}(-y\mathbf{i} + x\mathbf{j}).$$

Since the magnitude of $\mathbf{v}$ is $\omega r_0^2/r$ and $r = \sqrt{x^2 + y^2}$,

$$\mathbf{v} = \frac{\omega r_0^2}{r}\mathbf{u} = \frac{\omega r_0^2}{x^2 + y^2}(-y\mathbf{i} + x\mathbf{j}).$$

Applying Definition (18.6), we obtain

$$\text{curl } \mathbf{v} = \nabla \times \mathbf{v} = \mathbf{0}.$$

In our discussion of line integrals that are independent of path, we introduced the notion of a simply connected region in the plane. This concept can be extended to three dimensions; however, the usual definition requires properties of curves and surfaces studied in more advanced courses. For our purposes a region D in three dimensions will be **simply connected** if every simple closed curve C in D is the boundary of a surface S in D that satisfies the conditions of Stokes' theorem. In regions of this type, any simple closed curve can be continuously shrunk to a point in D without crossing the boundary of D. For example, the region inside of a sphere or a rectangular box is simply connected. The region inside of a torus (a doughnut-shaped surface) is not simply connected. Using this restriction, we can establish the following result.

Theorem **(18.30)**

> If $\mathbf{F}(x, y, z)$ has continuous partial derivatives throughout a simply connected region D, then curl $\mathbf{F} = \mathbf{0}$ in D if and only if $\oint_C \mathbf{F} \cdot d\mathbf{r} = 0$ for every simple closed curve C in D.

PROOF If curl $\mathbf{F} = \mathbf{0}$, then, by Stokes' theorem (18.28),

$$\oint_C \mathbf{F} \cdot d\mathbf{r} = \iint_S (\text{curl } \mathbf{F}) \cdot \mathbf{n}\, dS = 0.$$

Conversely, suppose $\oint_C \mathbf{F} \cdot d\mathbf{r} = 0$ for every simple closed curve C. If, at some point P, curl $\mathbf{F} \neq \mathbf{0}$, then, by continuity, there is a subregion of D containing P throughout which curl $\mathbf{F} \neq \mathbf{0}$. If, in this subregion, we choose a circular disk of the type illustrated in Figure 18.56 with $\mathbf{n}$ parallel to curl $\mathbf{F}$, then

$$\oint_{C_k} \mathbf{F} \cdot d\mathbf{r} = \iint_{S_k} (\text{curl } \mathbf{F}) \cdot \mathbf{n}\, dS > 0,$$

a contradiction. Consequently curl $\mathbf{F} = \mathbf{0}$ throughout D. ■

If $\mathbf{F}$ is continuous on a suitable region, then the condition $\oint_C \mathbf{F} \cdot d\mathbf{r} = 0$ for every simple closed curve is equivalent to independence of path. Moreover, as in Theorem (18.13), independence of path is equivalent to $\mathbf{F} = \nabla f$ for some scalar function f (that is, $\mathbf{F}$ is conservative). Combining these facts with Theorem (18.30), we obtain the following.

Theorem **(18.31)**

> If $\mathbf{F}(x, y, z)$ has continuous first partial derivatives throughout a simply connected region D, then the following conditions are equivalent:
>
> **(i)** $\mathbf{F}$ is conservative; that is, $\mathbf{F} = \nabla f$ for some scalar function f.
>
> **(ii)** $\int_C \mathbf{F} \cdot d\mathbf{r}$ is independent of path in D.
>
> **(iii)** $\oint_C \mathbf{F} \cdot d\mathbf{r} = 0$ for every simple closed curve C in D.
>
> **(iv)** $\mathbf{F}$ is irrotational; that is, curl $\mathbf{F} = \mathbf{0}$.

EXAMPLE 6 Prove that if $\mathbf{F}(x, y, z) = (3x^2 + y^2)\mathbf{i} + 2xy\mathbf{j} - 3z^2\mathbf{k}$, then $\mathbf{F}$ is conservative.

SOLUTION The function $\mathbf{F}$ has continuous first partial derivatives for every (x, y, z). Since

$$\text{curl } \mathbf{F} = \begin{vmatrix} \mathbf{i} & \mathbf{j} & \mathbf{k} \\ \dfrac{\partial}{\partial x} & \dfrac{\partial}{\partial y} & \dfrac{\partial}{\partial z} \\ 3x^2 + y^2 & 2xy & -3z^2 \end{vmatrix}$$

$$= (0 - 0)\mathbf{i} + (0 - 0)\mathbf{j} + (2y - 2y)\mathbf{k} = \mathbf{0},$$

$\mathbf{F}$ is conservative, by Theorem (18.31)(iv).

EXERCISES 18.7

Exer. 1–4: Verify Stokes' theorem (18.28) for F and S.

1 $\mathbf{F} = y^2\mathbf{i} + z^2\mathbf{j} + x^2\mathbf{k}$;
S is the first-octant portion of the plane $x + y + z = 1$.

2 $\mathbf{F} = 2y\mathbf{i} - z\mathbf{j} + 3\mathbf{k}$;
S is the portion of the paraboloid $z = 4 - x^2 - y^2$ that lies inside the cylinder $x^2 + y^2 = 1$.

3 $\mathbf{F} = z\mathbf{i} + x\mathbf{j} + y\mathbf{k}$;
S is the hemisphere $z = (a^2 - x^2 - y^2)^{1/2}$.

4 $\mathbf{F} = x^2\mathbf{i} + y^2\mathbf{j} + z^2\mathbf{k}$;
S is the part of the cone $z = (x^2 + y^2)^{1/2}$ cut off by the plane $z = 1$.

5 If $\mathbf{F} = (3z - \sin x)\mathbf{i} + (x^2 + e^y)\mathbf{j} + (y^3 - \cos z)\mathbf{k}$, use Stokes' theorem to evaluate $\oint_C \mathbf{F} \cdot d\mathbf{r}$, where C is the curve given by $x = \cos t$, $y = \sin t$, $z = 1$; $0 \le t \le 2\pi$.

6 If $\mathbf{F} = yz\mathbf{i} + xy\mathbf{j} + xz\mathbf{k}$ and C is the square with vertices $(0, 0, 2)$, $(1, 0, 2)$, $(1, 1, 2)$, and $(0, 1, 2)$, use Stokes' theorem to evaluate $\oint_C \mathbf{F} \cdot d\mathbf{r}$.

7 If $\mathbf{F} = 2y\mathbf{i} + e^z\mathbf{j} - \arctan x\mathbf{k}$ and S is the portion of the paraboloid $z = 4 - x^2 - y^2$ cut off by the xy-plane, use Stokes' theorem to evaluate $\iint_S \text{curl } \mathbf{F} \cdot \mathbf{n}\, dS$.

8 If $\mathbf{F} = xy^2\mathbf{i} + yz^2\mathbf{j} + zx^2\mathbf{k}$ and S is the region bounded by the triangle with vertices $(1, 0, 0)$, $(0, 2, 0)$, and $(0, 0, 3)$, use Stokes' theorem to evaluate $\iint_S \text{curl } \mathbf{F} \cdot \mathbf{n}\, dS$.

Exer. 9–12: Sketch some of the field vectors F(x, y, z), and discuss the rotational properties of F by using both a curl meter and curl F. Where does (curl F) · n have its maximum value?

9 $\mathbf{F}(x, y, z) = (y^2 - 2y)\mathbf{i} + 0\mathbf{j} + 0\mathbf{k}$; $0 \le y \le 2$

10 $\mathbf{F}(x, y, z) = 0\mathbf{i} + \sin x\mathbf{j} + 0\mathbf{k}$; $0 \le x \le \pi$

11 $\mathbf{F}(x, y, z) = -y\mathbf{i} + x\mathbf{j} + 0\mathbf{k}$ (see Example 1 of Section 18.1 and the discussion following Example 3, page 1015)

12 The inverse square field of Definition (18.2)

13 If $\mathbf{F}(x, y, z) = y\mathbf{i} + (x + e^z)\mathbf{j} + (1 + ye^z)\mathbf{k}$, prove that $\mathbf{F}$ is irrotational.

14 A vector field $\mathbf{F}$ is a *central force field* if $\mathbf{F} = f(r)\,\mathbf{r}$, where $\mathbf{r} = x\mathbf{i} + y\mathbf{j} + z\mathbf{k}$, $r = \|\mathbf{r}\|$, and f is a scalar function. If f is differentiable, show that $\mathbf{F}$ is irrotational.

15 Let $\mathbf{n}$ denote the unit outer normal at any point P on the surface of a sphere S. If $\mathbf{F}$ has continuous first partial derivatives within and on S, prove that $\iint_S \text{curl } \mathbf{F} \cdot \mathbf{n}\, dS = 0$ by using

(a) the divergence theorem (b) Stokes' theorem

16 If S and C satisfy the conditions of Stokes' theorem and if $\mathbf{F}$ is a constant vector function, use (18.28) to prove that $\oint_C \mathbf{F} \cdot \mathbf{T}\, ds = 0$.

Exer. 17–19: Establish the identity under the assumption that C and S satisfy the conditions of Stokes' theorem.

17 $\oint_C f\nabla g \cdot d\mathbf{r} = \iint_S (\nabla f \times \nabla g) \cdot \mathbf{n}\, dS$, where f and g are scalar functions

18 $\oint_C \mathbf{a} \times \mathbf{r} \cdot d\mathbf{r} = 2\mathbf{a} \cdot \iint_S \mathbf{n}\, dS$, where $\mathbf{a}$ is a constant vector

19 $\oint_C f\, d\mathbf{r} = \iint_S \mathbf{n} \times \nabla f\, dS$, where f is a scalar function

20 If M, N, and P are functions of x, y, and z that have continuous first partial derivatives on a simply connected region, prove that

$$\int_C M(x, y, z)\, dx + N(x, y, z)\, dy + P(x, y, z)\, dz$$

is independent of path if and only if

$$\frac{\partial M}{\partial y} = \frac{\partial N}{\partial x}, \quad \frac{\partial M}{\partial z} = \frac{\partial P}{\partial x}, \quad \frac{\partial N}{\partial z} = \frac{\partial P}{\partial y}.$$

18.8 REVIEW EXERCISES

Exer. 1–4: Sketch some typical vectors in the vector field F.

1 $\mathbf{F}(x, y) = 2x\mathbf{i} + y\mathbf{j}$

2 $\mathbf{F}(x, y, z) = x\mathbf{i} + y\mathbf{j} + \mathbf{k}$

3 $\mathbf{F}(x, y, z) = -\mathbf{k}$

4 $\mathbf{F}(x, y, z) = \nabla(x^2 + y^2 + z^2)^{-1/2}$

5 Find a conservative vector field in two dimensions that has the potential function $f(x, y) = y^2 \tan x$.

6 Find a conservative vector field in three dimensions that has the potential function $f(x, y, z) = \ln(x + y + z)$.

Exer. 7–10: Evaluate $\int_C y^2\, dx + xy\, dy$, where C is the curve from $(1, 0)$ to $(-1, 4)$ shown in the figure.

7 8

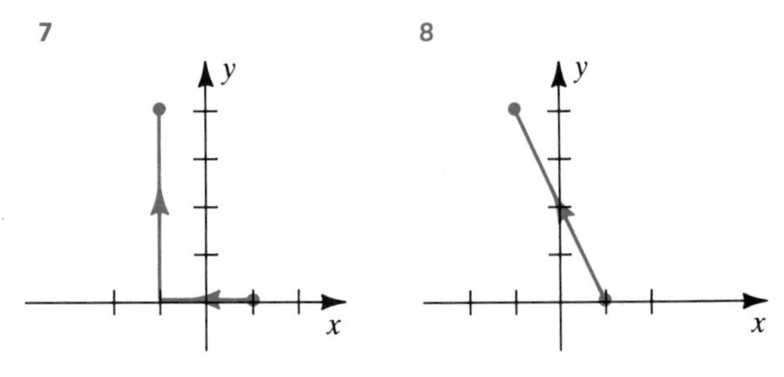

9 10

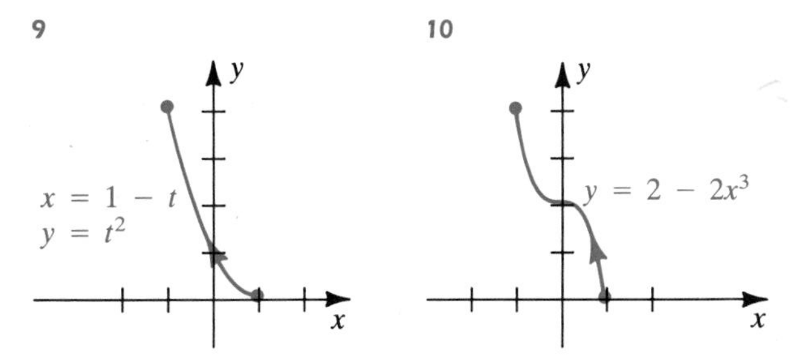

11 Evaluate $\int_C xy\, ds$; C is the graph of $y = x^4$ from $(-1, 1)$ to $(2, 16)$.

Exer. 12–14: Evaluate

$$\int_C x\, dx + (x + y)\, dy + (x + y + z)\, dz$$

for the given curve C from $A(0, 0, 0)$ to $B(2, 4, 8)$.

12 C consists of three line segments, the first parallel to the z-axis, the second parallel to the x-axis, and the third parallel to the y-axis.

13 C consists of the line segment from A to B.

14 C has the parametrization $x = t$, $y = t^2$, $z = t^3$.

15 If $\mathbf{F}(x, y) = (x + y)\mathbf{i} + (x - y)\mathbf{j}$ and C is given by $x = \cos t$, $y = \sin t$; $-\pi \le t \le 0$, evaluate $\int_C \mathbf{F} \cdot d\mathbf{r}$.

16 The force $\mathbf{F}$ at a point (x, y, z) in three dimensions is given by $\mathbf{F}(x, y, z) = xy\mathbf{i} + y^2z\mathbf{j} + xz^2\mathbf{k}$. If C is the square with vertices $A_1(1, 1, 1)$, $A_2(-1, 1, 1)$, $A_3(-1, -1, 1)$, and $A_4(1, -1, 1)$, find the work done by $\mathbf{F}$ as its point of application moves around C once in the direction determined by increasing subscripts on the A_k.

Exer. 17–18: Prove that the line integral is independent of path, and find its value.

17 $\int_{(1,-1)}^{(2,3)} (x + y)\, dx + (x + y)\, dy$

18 $\int_{(0,0,0)}^{(2,1,3)} (8x^3 + z^2)\, dx - 3z\, dy + (2xz - 3y)\, dz$

19 If $\mathbf{F}(x, y, z) = 2xe^{2y}\mathbf{i} + 2(x^2e^{2y} + y \cot z)\mathbf{j} - y^2 \csc^2 z\mathbf{k}$, prove that $\int_C \mathbf{F} \cdot d\mathbf{r}$ is independent of path, and find a potential function f for $\mathbf{F}$.

Exer. 20–22: Use Green's theorem to evaluate the line integral

$$\oint_C xy\, dx + (x^2 + y^2)\, dy$$

if C is the given curve.

20 C is the closed curve determined by $y = x^2$ and $y - x = 2$ between $(-1, 1)$ and $(2, 4)$.

21 C is the triangle with vertices $(0, 0)$, $(1, 0)$, $(0, 1)$.

22 C is the circle $x^2 + y^2 - 2x = 0$.

23 If $\mathbf{F}(x, y, z) = x^3z^4\mathbf{i} + xyz^2\mathbf{j} + x^2y^2\mathbf{k}$, find div $\mathbf{F}$ and curl $\mathbf{F}$.

24 If f and g are scalar functions of three variables, prove that $\nabla \cdot (f\nabla g) = f\nabla^2 g + \nabla f \cdot \nabla g$.

25 Evaluate $\iint_S xyz\, dS$ if S is the part of the plane $z = x + y$ that lies over the triangular region in the xy-plane having vertices $(0, 0, 0)$, $(1, 0, 0)$, and $(0, 2, 0)$.

26 Evaluate $\iint_S x^2z^2\, dS$ if S is the top half of the cylinder $y^2 + z^2 = 4$ between $x = 0$ and $x = 1$.

27 Let Q be the region bounded by the cylinder $x^2 + y^2 = 1$ and the planes $z = 0$ and $z = 1$. If $\mathbf{F} = x^3\mathbf{i} + y^3\mathbf{j} + z^3\mathbf{k}$, use the divergence theorem to find $\iint_S \mathbf{F} \cdot \mathbf{n}\, dS$, where S is the surface of Q and $\mathbf{n}$ is the unit outer normal to S.

28 Verify the divergence theorem if $\mathbf{F} = 2x\mathbf{i} + y\mathbf{j} - z\mathbf{k}$ and S is the surface of the rectangular box bounded by the planes $x = \pm 1$, $y = \pm 2$, $z = \pm 3$.

Exer. 19–20: Verify Stokes' theorem for F and S.

29 $\mathbf{F} = y^2\mathbf{i} + 2x\mathbf{j} + 5y\mathbf{k}$;
S is the hemisphere $z = (4 - x^2 - y^2)^{1/2}$.

30 $\mathbf{F} = (x + y)\mathbf{i} + (y + z)\mathbf{j} + (x + z)\mathbf{k}$;
S is the region bounded by the triangle with vertices $(1, 0, 0)$, $(0, 1, 0)$, $(0, 0, 1)$.

CHAPTER

19

DIFFERENTIAL EQUATIONS

INTRODUCTION

Differential equations are used to describe and help solve complex problems about motion, growth, vibrations, electricity and magnetism, aerodynamics, thermodynamics, hydrodynamics, nuclear energy, and every other type of physical phenomenon that involves rates of change of variable quantities.

Earlier in the text we solved certain types of differential equations by separating variables and then using indefinite integration. Section 19.1 contains additional examples and exercises involving such *separable* equations. In later sections we discuss several other varieties that can be solved systematically using standard techniques and formulas. Although the latter differential equations are more general than separable equations, they are still very specialized. In modern applications that involve placing satellites into orbit or sending a spacecraft to distant parts of the solar system, the differential equations are very complicated and supercomputers are used to obtain approximations to solutions.

The material in this chapter is not intended to be a treatise on the subject, but rather to serve as an introduction to this vast and important branch of mathematics. There are separate courses and books devoted entirely to the study of differential equations.

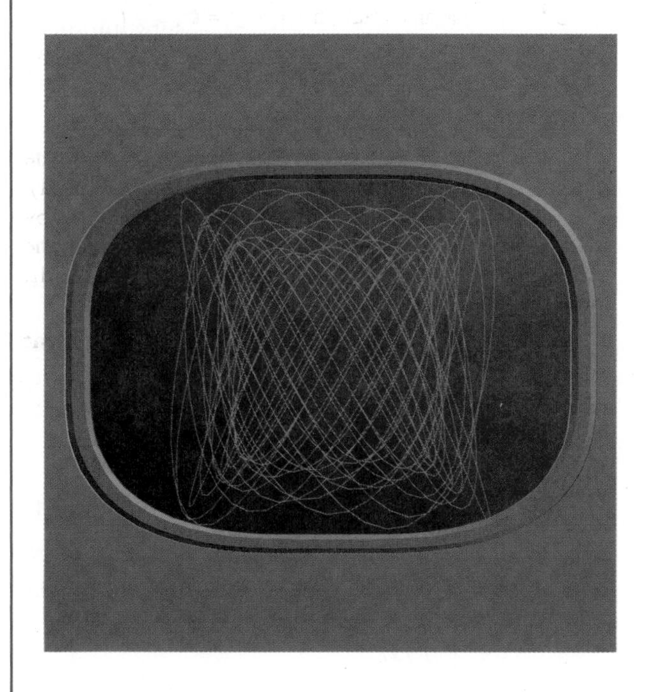

19.1 SEPARABLE DIFFERENTIAL EQUATIONS

If y is a function of x and if n is a positive integer, then an equation that involves $x, y, y', y'', \ldots, y^{(n)}$ is called an **ordinary differential equation of order *n*.**

ILLUSTRATION

DIFFERENTIAL EQUATION	ORDER
$y' = 2x$	1
$\dfrac{d^2y}{dx^2} + x^2\left(\dfrac{dy}{dx}\right)^3 - 15y = 0$	2
$(y''')^4 - x^2(y'')^5 + 4xy = xe^x$	3
$\left(\dfrac{d^4y}{dx^4}\right)^2 - 1 = x^3\dfrac{dy}{dx}$	4

We considered differential equations briefly in Sections 5.1 and 7.6. Recall that a function f (or $f(x)$) is a **solution** of a differential equation if substitution of $f(x)$ for y results in an identity for every x in some interval. For example, the differential equation

$$y' = 6x^2 - 5$$

has the solution

$$f(x) = 2x^3 - 5x + C$$

for every real number C, since substitution of $f(x)$ for y leads to the identity $6x^2 - 5 = 6x^2 - 5$. We call $f(x) = 2x^3 - 5x + C$ the **general solution** of $y' = 6x^2 - 5$, since every solution is of this form. A **particular solution** is obtained by assigning C a specific value. To illustrate, if $C = 0$, we obtain the particular solution $y = 2x^3 - 5x$. Sometimes **initial conditions** are stated that determine a particular solution, as illustrated in the following example.

FIGURE 19.1
$y = x^2 + C$

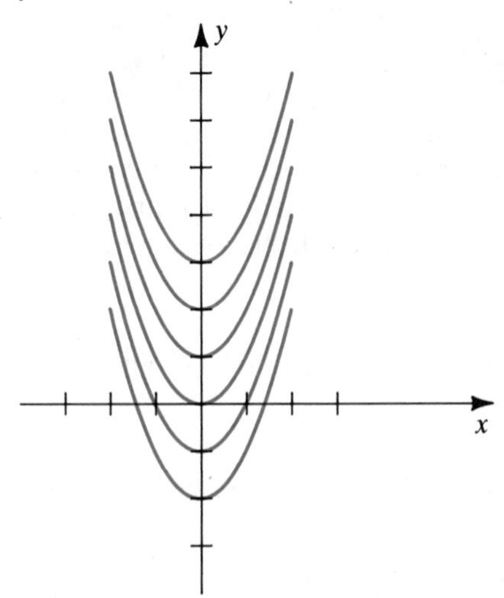

EXAMPLE 1 Given the differential equation $y' = 2x$,

(a) find the general solution and illustrate it graphically

(b) find the particular solution that satisfies the condition $y = 3$ if $x = 0$

SOLUTION

(a) We use indefinite integration as follows:

$$y' = 2x$$

$$\int y'\,dx = \int 2x\,dx$$

$$y = x^2 + C$$

This is the general solution of the differential equation. Particular solutions are obtained by assigning values to C. This leads to the family of parabolas illustrated in Figure 19.1.

(b) If $y = 3$ when $x = 0$, then substitution in the general solution $y = x^2 + C$ gives us $3 = 0 + C$, or $C = 3$. Hence the particular solution is $y = x^2 + 3$. The graph is the parabola in Figure 19.1 with y-intercept 3.

Differential equations of the type given in the next example occur in the analysis of vibrations. We will consider these in detail in Sections 19.3–19.5.

EXAMPLE 2 Show that the differential equation $y'' - 25y = 0$ has the solution

$$y = C_1e^{5x} + C_2e^{-5x}$$

for all real numbers C_1 and C_2.

SOLUTION Differentiating y twice, we have

$$y' = 5C_1e^{5x} - 5C_2e^{-5x}$$
$$y'' = 25C_1e^{5x} + 25C_2e^{-5x}.$$

Substituting for y and y'' in the differential equation $y'' - 25y = 0$ gives us

$$(25C_1e^{5x} + 25C_2e^{-5x}) - 25(C_1e^{5x} + C_2e^{-5x}) = 0,$$

or

$$(25C_1e^{5x} - 25C_1e^{5x}) + (25C_2e^{-5x} - 25C_2e^{-5x}) = 0.$$

Since the left side is 0 for every x, it follows that $y = C_1e^{5x} + C_2e^{-5x}$ is a solution of $y'' - 25y = 0$.

The solution $y = C_1e^{5x} + C_2e^{-5x}$ in Example 2 is the general solution of $y'' - 25y = 0$. Observe that the differential equation is of order 2 and the general solution contains two arbitrary constants (called **parameters**) C_1 and C_2. The precise definition of a general solution involves the concept of **independent parameters** and is left for more advanced courses. General solutions of nth-order differential equations contain n independent parameters $C_1, C_2, \ldots, C_n$. A particular solution is obtained by assigning a specific value to each parameter. Some differential equations have solutions that are not special cases of the general solution. Such **singular solutions** will not be discussed in this text.

The solutions in Examples 1 and 2 express y *explicitly* in terms of x. Solutions of certain differential equations are stated *implicitly*, and we use implicit differentiation to check such solutions, as illustrated in the following example.

EXAMPLE 3 Show that $x^3 + x^2y - 2y^3 = C$ is an implicit solution of

$$(x^2 - 6y^2)y' + 3x^2 + 2xy = 0.$$

SOLUTION If the first equation is satisfied by $y = f(x)$, then differentiating implicitly yields

$$3x^2 + 2xy + x^2y' - 6y^2y' = 0,$$

or
$$(x^2 - 6y^2)y' + 3x^2 + 2xy = 0.$$

Thus, $f(x)$ is a solution of the differential equation.

One of the simplest types of differential equations is

$$M(x) + N(y)y' = 0, \quad \text{or} \quad M(x) + N(y)\frac{dy}{dx} = 0,$$

where M and N are continuous functions. If $y = f(x)$ is a solution, then

$$M(x) + N(f(x))f'(x) = 0.$$

If $f'(x)$ is continuous, then indefinite integration leads to

$$\int M(x)\,dx + \int N(f(x))f'(x)\,dx = C,$$

or
$$\int M(x)\,dx + \int N(y)\,dy = C.$$

The last equation is an (implicit) solution of the differential equation. The differential equation $M(x) + N(y)y' = 0$ is **separable**, since the variables x and y may be separated as we have indicated.

An easy way to remember the method of separating the variables is to change the equation

$$M(x) + N(y)\frac{dy}{dx} = 0$$

to the *differential form*

$$M(x)\,dx + N(y)\,dy = 0$$

and then integrate each term.

EXAMPLE 4 Solve the differential equation $y^4e^{2x} + \dfrac{dy}{dx} = 0.$

SOLUTION We first express the equation in differential form,

$$y^4e^{2x}\,dx + dy = 0.$$

If $y \neq 0$, we may separate the variables by dividing by y^4, as follows:

$$e^{2x}\,dx + \frac{1}{y^4}\,dy = 0$$

Integrating each term, we obtain the (implicit) solution

$$\frac{1}{2}e^{2x} - \frac{1}{3y^3} = C$$

If we multiply both sides by 6, then another form of the solution is

$$3e^{2x} - \frac{2}{y^3} = K$$

with $K = 6C$. If an explicit form is desired, we may solve for y, obtaining

$$y = \left(\frac{2}{3e^{2x} - K}\right)^{1/3}$$

We have assumed that $y \neq 0$; however, $y = 0$ *is* a solution of the differential equation. In future examples we will not point out such singular solutions.

EXAMPLE 5 Solve the differential equation $2y + (xy + 3x)\frac{dy}{dx} = 0$, where $x \neq 0$.

SOLUTION We first express the equation in differential form as follows:

$$2y\,dx + (xy + 3x)\,dy = 0$$

$$2y\,dx + x(y + 3)\,dy = 0$$

The variables may be separated by dividing both sides of the last equation by xy. This gives us

$$\frac{2}{x}\,dx + \left(\frac{y+3}{y}\right)dy = 0,$$

or

$$\frac{2}{x}\,dx + \left(1 + \frac{3}{y}\right)dy = 0.$$

Integrating each term leads to the following equivalent equations:

$$2\ln|x| + y + 3\ln|y| = c$$

$$\ln|x|^2 + \ln|y|^3 = c - y$$

$$\ln|x|^2|y|^3 = c - y$$

Changing to exponential form, we have

$$|x|^2|y|^3 = e^{c-y} = e^c e^{-y} = ke^{-y},$$

where $k = e^c$. We can show, by differentiation, that

$$x^2y^3 = ke^{-y}, \quad \text{or} \quad x^2y^3e^y = k,$$

is an implicit solution of the differential equation.

EXAMPLE 6 Solve the differential equation

$$(1 + y^2) + (1 + x^2)\frac{dy}{dx} = 0.$$

SOLUTION We first express the equation in differential form and then separate the variables, obtaining

$$(1 + y^2)\,dx + (1 + x^2)\,dy = 0$$

$$\frac{1}{1 + x^2}\,dx + \frac{1}{1 + y^2}\,dy = 0.$$

Integrating each term, we obtain the (implicit) solution

$$\tan^{-1} x + \tan^{-1} y = C.$$

To find an explicit solution, we may solve for y as follows:

$$\tan^{-1} y = C - \tan^{-1} x$$

$$y = \tan (C - \tan^{-1} x)$$

The form of this solution may be changed by using the subtraction formula for the tangent function. Thus,

$$y = \frac{\tan C - \tan (\tan^{-1} x)}{1 + \tan C \tan (\tan^{-1} x)}.$$

If we let $k = \tan C$ and use the fact that $\tan (\tan^{-1} x) = x$, this may be written

$$y = \frac{k - x}{1 + kx}.$$

FIGURE 19.2
$y = 2x + b$; $y = -\frac{1}{2}x + c$

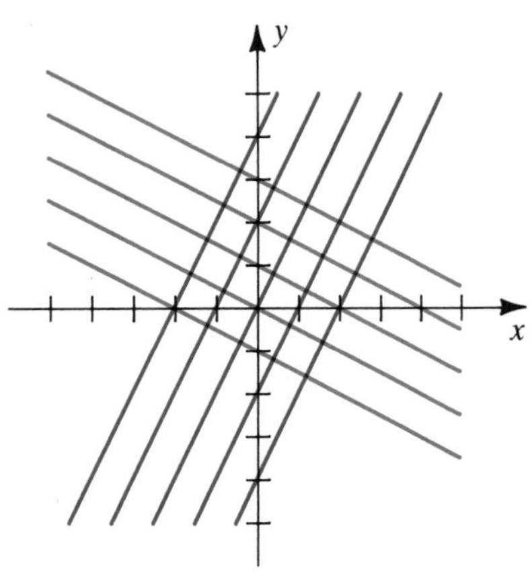

FIGURE 19.3
$y = mx$; $x^2 + y^2 = a^2$

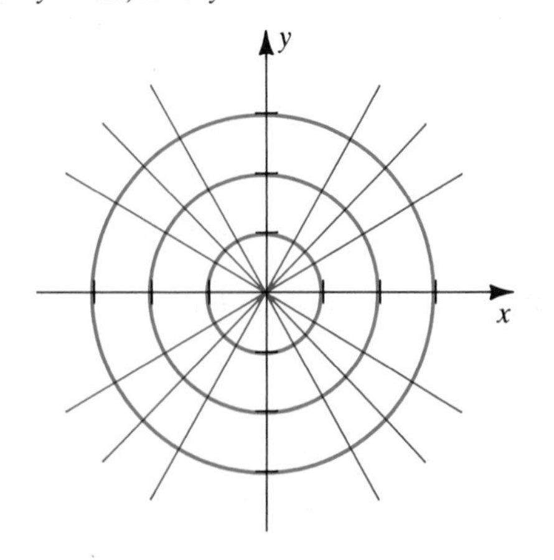

An **orthogonal trajectory** of a family of curves is a curve that intersects each curve of the family orthogonally. We shall restrict our discussion to curves in a coordinate plane. To illustrate, for the family $y = 2x + b$ of all lines of slope 2, every line $y = -\frac{1}{2}x + c$ of slope $-\frac{1}{2}$ is an orthogonal trajectory. Several curves from each family are sketched in Figure 19.2. We say that two such families of curves are **mutually orthogonal**. As another example, the family $y = mx$ of all lines through the origin and the family $x^2 + y^2 = a^2$ of all concentric circles with centers at the origin are mutually orthogonal (see Figure 19.3).

Pairs of mutually orthogonal families of curves occur frequently in applications. In the theory of electricity and magnetism, the lines of force associated with a given field are orthogonal trajectories of the corresponding equipotential curves. Similarly, the stream lines studied in aerodynamics and hydrodynamics are orthogonal trajectories of the *velocity-equipotential* curves. As a final illustration, in the study of thermodynamics, the flow of heat across a plane surface is orthogonal to the isothermal curves.

The next example illustrates a technique for finding the orthogonal trajectories of a family of curves.

EXAMPLE 7 Find the orthogonal trajectories of the family of ellipses $x^2 + 3y^2 = c$, and sketch several members of each family.

SOLUTION Differentiating the given equation implicitly, we obtain

$$2x + 6yy' = 0, \quad \text{or} \quad y' = -\frac{x}{3y}.$$

Hence the slope of the tangent line at any point (x, y) on each of the ellipses is $y' = -x/(3y)$. If dy/dx is the slope of the tangent line on a corresponding orthogonal trajectory, then it must equal the negative reciprocal of y'. This gives us the following differential equation for the family of ortho-

FIGURE 19.4
$x^2 + 3y^2 = c;\ y = kx^3$

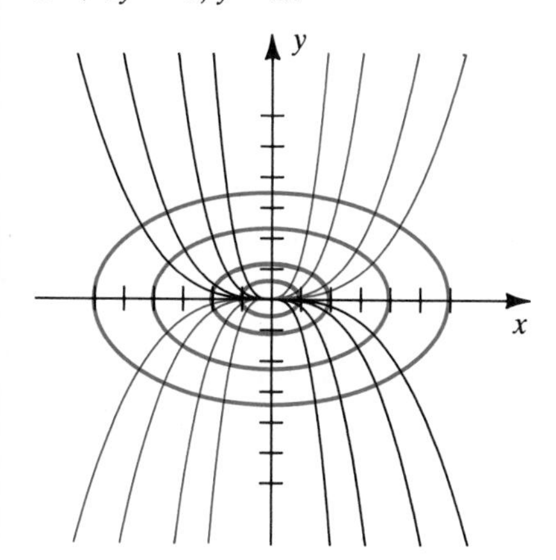

gonal trajectories:

$$\frac{dy}{dx} = \frac{3y}{x}$$

Separating the variables, we have

$$\frac{dy}{y} = 3\,\frac{dx}{x}.$$

Integrating and writing the constant of integration as $\ln|k|$ gives us

$$\ln|y| = 3\ln|x| + \ln|k| = \ln|kx^3|.$$

It follows that $y = kx^3$ is an equation for the family of orthogonal trajectories. We have sketched several members of the family of ellipses (in blue) and corresponding orthogonal trajectories (in red or in black) in Figure 19.4.

EXERCISES 19.1

Exer. 1–4: (a) Find the general solution of the differential equation, and illustrate it graphically. (b) Find the particular solution that satisfies the condition $y = 2$ when $x = 0$.

1 $y' = 3x^2$

2 $y' = x - 1$

3 $y' = \dfrac{-x}{\sqrt{4 - x^2}}$

4 $y' = 3$

Exer. 5–10: Prove that y is a solution of the differential equation.

5 $y'' - 3y' + 2y = 0;\quad y = C_1e^x + C_2e^{2x}$

6 $y' + 3y = 0;\quad y = Ce^{-3x}$

7 $2xy^3 + 3x^2y^2\dfrac{dy}{dx} = 0;\quad y = Cx^{-2/3}$

8 $x^3y''' + x^2y'' - 3xy' - 3y = 0;\quad y = Cx^3$

9 $(x - 2y)\dfrac{dy}{dx} + 2x + y = 0;\quad y^2 - x^2 - xy = C$

10 $y\dfrac{dy}{dx} = x;\quad x^2 - y^2 = C$

Exer. 11–26: Solve the differential equation.

11 $\sec x\,dy - 2y\,dx = 0$

12 $x^2\,dy - \csc 2y\,dx = 0$

13 $x\,dy - y\,dx = 0$

14 $(4 + y^2)\,dx + (9 + x^2)\,dy = 0$

15 $3y\,dx + (xy + 5x)\,dy = 0$

16 $(xy - 4x)\,dx + (x^2y + y)\,dy = 0$

17 $y' = x - 1 + xy - y$

18 $(y + yx^2)\,dy + (x + xy^2)\,dx = 0$

19 $e^{x+2y}\,dx - e^{2x-y}\,dy = 0$

20 $\cos x\,dy - y\,dx = 0$

21 $y(1 + x^3)y' + x^2(1 + y^2) = 0$

22 $x^2y' - yx^2 = y$

23 $x\tan y - y'\sec x = 0$

24 $xy + y'e^{(x^2)}\ln y = 0$

25 $e^y\sin x\,dx - \cos^2 x\,dy = 0$

26 $\sin y\cos x\,dx + (1 + \sin^2 x)\,dy = 0$

Exer. 27–34: Find the particular solution of the differential equation that satisfies the given condition.

27 $2y^2y' = 3y - y';\quad y = 1$ when $x = 3$

28 $\sqrt{x}\,y' - \sqrt{y} = x\sqrt{y};\quad y = 4$ when $x = 9$

29 $x\,dy - (2x + 1)e^{-y}\,dx = 0;\quad y = 2$ when $x = 1$

30 $\sec 2y\,dx - \cos^2 x\,dy = 0;\quad y = \pi/6$ when $x = \pi/4$

31 $(xy + x)\,dx + \sqrt{4 + x^2}\,dy = 0;\quad y = 1$ when $x = 0$

32 $x\,dy - \sqrt{1 - y^2}\,dx = 0;\quad y = \frac{1}{2}$ when $x = 1$

33 $\cot x\,dy - (1 + y^2)\,dx = 0;\quad y = 1$ when $x = 0$

34 $\csc y\,dx - e^x\,dy = 0;\quad y = 0$ when $x = 0$

Exer. 35–40: Find the orthogonal trajectories of the family of curves. Describe the graphs.

35 $x^2 - y^2 = c$

36 $xy = c$

37 $y^2 = cx$

38 $y = cx^2$

39 $y^2 = cx^3$

40 $y = ce^{-x}$

c Exer. 41–42: If a differential equation of the form $y' = f(x, y)$ has initial condition $y_0 = A$ at $x = a$ and if $b > a$, then its solution at $x = b$ can be approximated by using the following steps, called *Euler's method.*

Step 1 Let $a = x_0 < x_1 < \cdots < x_n = b$, where n is an integer and $x_{k+1} - x_k = (b - a)/n = h$ for $k = 0, 1, 2, \ldots, n - 1$.

Step 2 Let $y_0 = A$.

Step 3 Let $y_{k+1} = y_k + hf(x_k, y_k)$.

Step 4 If $h \approx 0$, then $y \approx y_n$ at $x = b$.

(a) For the given differential equation with initial condition $y_0 = 1$ at $a = 0$, use Euler's method, with $n = 8$, to approximate y at $x = 1$. (b) Solve the differential equation and find the exact value of y at $x = 1$.

41 $y' = xy$

42 $y' = 1/y$

c Exer. 43–44: A more accurate method than that used in Exercises 41–42 is the *improved Euler's method* (or *Heun's method*), in which the formula in Step 3 is replaced by

$$y_{k+1} = y_k + \tfrac{1}{2}h[f(x_k, y_k) + f(x_k + h, y_k + hf(x_k, y_k))].$$

Work the following exercises using this method.

43 Part (a) of Exercise 41

44 Part (a) of Exercise 42

19.2 FIRST-ORDER LINEAR DIFFERENTIAL EQUATIONS

The following type of differential equation occurs frequently in the study of physical phenomena.

Definition (19.1)

A **first-order linear differential equation** is an equation of the form

$$y' + P(x)y = Q(x),$$

where P and Q are continuous functions.

If, in Definition (19.1), $Q(x) = 0$ for every x, we may separate the variables and then integrate as follows (provided $y \neq 0$):

$$\frac{dy}{dx} + P(x)y = 0$$

$$\frac{1}{y}\frac{dy}{dx} = -P(x)$$

$$\frac{1}{y}\,dy = -P(x)\,dx$$

$$\ln|y| = -\int P(x)\,dx + \ln|C|$$

We have expressed the constant of integration as $\ln|C|$ in order to change the form of the last equation as follows:

$$\ln|y| - \ln|C| = -\int P(x)\,dx$$

$$\ln\left|\frac{y}{C}\right| = -\int P(x)\,dx$$

$$\frac{y}{C} = e^{-\int P(x)\,dx}$$

$$ye^{\int P(x)\,dx} = C$$

We next observe that, by the product rule,

$$\begin{aligned} D_x\,(ye^{\int P(x)\,dx}) &= (D_x\,y)e^{\int P(x)\,dx} + y(D_x\,e^{\int P(x)\,dx}) \\ &= y'e^{\int P(x)\,dx} + yP(x)e^{\int P(x)\,dx} \\ &= (y' + P(x)y)e^{\int P(x)\,dx}. \end{aligned}$$

Consequently, if we multiply both sides of $y' + P(x)y = Q(x)$ by $e^{\int P(x)\,dx}$, then the resulting equation may be written

$$D_x\,(ye^{\int P(x)\,dx}) = Q(x)e^{\int P(x)\,dx}.$$

Integrating both sides gives us the following implicit solution of the first-order differential equation in Definition (19.1):

$$ye^{\int P(x)\,dx} = \int Q(x)e^{\int P(x)\,dx}\,dx + K$$

for a constant K. Solving this equation for y leads to an explicit solution. The expression $e^{\int P(x)\,dx}$ is an **integrating factor** of the differential equation. We have proved the following result.

Theorem (19.2)

> The first-order linear differential equation $y' + P(x)y = Q(x)$ may be transformed into a separable differential equation by multiplying both sides by the integrating factor $e^{\int P(x)\,dx}$.

EXAMPLE 1 Solve the differential equation $\dfrac{dy}{dx} - 3x^2y = x^2$.

SOLUTION The differential equation has the form in Theorem (19.1) with $P(x) = -3x^2$ and $Q(x) = x^2$. By Theorem (19.2), an integrating factor is

$$e^{\int -3x^2\,dx} = e^{-x^3}$$

There is no need to introduce a constant of integration, since $e^{-x^3+c} = e^ce^{-x^3}$, which differs from e^{-x^3} by a constant factor e^c. Multiplying both sides of the given differential equation by the integrating factor e^{-x^3}, we obtain

$$e^{-x^3}\frac{dy}{dx} - 3x^2e^{-x^3}y = x^2e^{-x^3},$$

or, equivalently, $$D_x\,(e^{-x^3}y) = x^2e^{-x^3}$$

Integrating both sides of the last equation gives us

$$e^{-x^3}y = \int x^2e^{-x^3}\,dx = -\tfrac{1}{3}e^{-x^3} + C.$$

Finally, multiplying by e^{x^3} gives us the explicit solution

$$y = -\tfrac{1}{3} + Ce^{(x^3)}.$$

EXAMPLE 2 Solve the differential equation $x^2y' + 5xy + 3x^5 = 0$, where $x \neq 0$.

SOLUTION To find an integrating factor, we begin by expressing the differential equation in the standardized form (19.2), where y' has the coefficient 1. Thus, dividing both sides by x^2, we obtain

$$y' + \frac{5}{x}y = -3x^3,$$

which has the form given in (19.1) with $P(x) = 5/x$. By Theorem (19.2), an integrating factor is

$$e^{\int P(x)\,dx} = e^{5\ln|x|} = e^{\ln|x|^5} = |x|^5.$$

If $x > 0$, then $|x|^5 = x^5$. If $x < 0$, then $|x|^5 = -x^5$. In either case, multiplying both sides of the standardized form by $|x|^5$ yields

$$x^5y' + 5x^4y = -3x^8,$$

or, equivalently,
$$D_x(x^5y) = -3x^8.$$

Integrating both sides of the last equation gives us the solution

$$x^5y = \int -3x^8\,dx = -\frac{x^9}{3} + C,$$

or
$$y = -\frac{x^4}{3} + \frac{C}{x^5}.$$

EXAMPLE 3 Solve the differential equation

$$y' + y\tan x = \sec x + 2x\cos x.$$

SOLUTION The equation is a first-order linear differential equation. By Theorem (19.2), with $P(x) = \tan x$, an integrating factor is

$$e^{\int \tan x\,dx} = e^{\ln|\sec x|} = |\sec x|.$$

Multiplying both sides of the differential equation by $|\sec x|$ and discarding the absolute value sign gives us

$$y'\sec x + y\sec x\tan x = \sec^2 x + 2x\cos x\sec x,$$

or
$$D_x(y\sec x) = \sec^2 x + 2x.$$

Integrating both sides yields the (implicit) solution

$$y\sec x = \tan x + x^2 + C.$$

Finally, multiplying both sides of the last equation by $1/\sec x = \cos x$, we obtain the explicit solution

$$y = \sin x + (x^2 + C)\cos x.$$

We have previously used indefinite integration to derive laws of motion for falling bodies, assuming that air resistance could be disregarded (see Example 8 in Section 5.1). This is a valid assumption for small, slowly moving objects; however, in many cases air resistance must be taken into account. This frictional force often increases as the speed of the object increases. In the following example we shall derive the law of motion for a falling body under the assumption that the resistance due to the air is directly proportional to the speed of the body.

EXAMPLE 4 An object of mass m is released from a hot-air balloon. Find the distance it falls in t seconds if the force of resistance due to the air is directly proportional to the speed of the object.

FIGURE 19.5

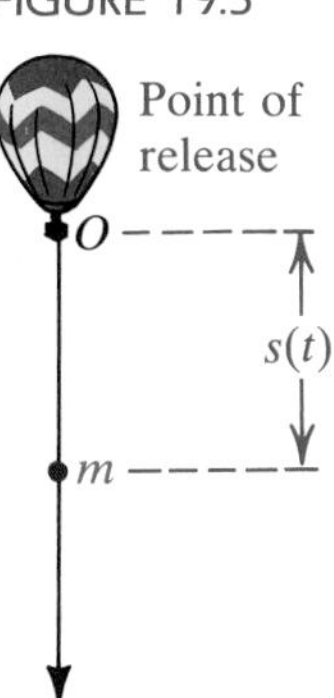

SOLUTION Let us introduce a vertical axis with positive direction downward and origin at the point of release, as illustrated in Figure 19.5. We wish to find the distance $s(t)$ from the origin to the object at time t. The speed of the object is $v = s'(t)$, and the magnitude of the acceleration is $a = dv/dt = s''(t)$. If g is the gravitational constant, then the object is attracted toward the earth with a force of magnitude mg. By hypothesis, the force of resistance due to the air is kv for some constant k, and this force is directed opposite to the motion. It follows that the downward force F on the object is $mg - kv$. Since Newton's second law of motion states that $F = ma = m(dv/dt)$, we arrive at the following differential equation:

$$m\frac{dv}{dt} = mg - kv,$$

or, equivalently,

$$\frac{dv}{dt} + \frac{k}{m}v = g.$$

If we denote the constant k/m by c, this equation may be written

$$\frac{dv}{dt} + cv = g,$$

which is a first-order differential equation with t as the independent variable. By (19.2), an integrating factor is

$$e^{\int c\,dt} = e^{ct}.$$

Multiplying both sides of the last differential equation by e^{ct} gives us

$$e^{ct}\frac{dv}{dt} + ce^{ct}v = ge^{ct},$$

or, equivalently,

$$D_t\,(ve^{ct}) = ge^{ct}.$$

Integrating both sides, we have

$$ve^{ct} = \frac{g}{c}e^{ct} + K, \quad \text{or} \quad v = \frac{g}{c} + Ke^{-ct},$$

where K is a constant.

If we let $t = 0$, then $v = 0$ and hence

$$0 = \frac{g}{c} + K, \quad \text{or} \quad K = -\frac{g}{c}.$$

Consequently

$$v = \frac{g}{c} - \frac{g}{c} e^{-ct}.$$

Integrating both sides of this equation with respect to t and using the fact that $v = s'(t)$, we see that

$$s(t) = \frac{g}{c} t + \frac{g}{c^2} e^{-ct} + E.$$

We may find the constant of integration E by letting $t = 0$. Since $s(0) = 0$,

$$0 = 0 + \frac{g}{c^2} + E, \quad \text{or} \quad E = -\frac{g}{c^2}.$$

Thus, the distance the object falls in t seconds is

$$\begin{aligned} s(t) &= \frac{g}{c} t + \frac{g}{c^2} e^{-ct} - \frac{g}{c^2} \\ &= \frac{g}{c^2} (ct + e^{-ct} - 1) \\ &= \frac{gm^2}{k^2} \left(\frac{k}{m} t + e^{-(k/m)t} - 1 \right) \end{aligned}$$

It is interesting to compare this formula for $s(t)$ with that obtained when the air resistance is disregarded. In the latter case the differential equation $m(dv/dt) = mg - kv$ reduces to $dv/dt = g$, and hence $s'(t) = v = gt$. Integrating both sides leads to the much simpler formula $s(t) = \frac{1}{2}gt^2$.

FIGURE 19.6

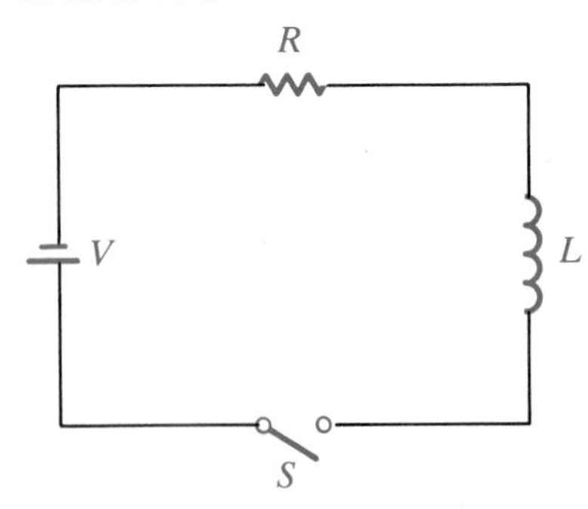

EXAMPLE 5 A simple electrical circuit consists of a resistance R and an inductance L connected in series, as illustrated schematically in Figure 19.6, with a constant electromotive force V. If the switch S is closed at $t = 0$, then it follows from one of Kirchhoff's rules for electrical circuits that if $t > 0$, the current I satisfies the differential equation

$$L \frac{dI}{dt} + RI = V.$$

Express I as a function of t.

SOLUTION The differential equation may be written

$$\frac{dI}{dt} + \frac{R}{L} I = \frac{V}{L},$$

which is a first-order linear differential equation. Applying Theorem (19.2), we multiply both sides by the integrating factor $e^{\int (R/L)\, dt} = e^{(R/L)t}$, obtaining

$$e^{(R/L)t} \frac{dI}{dt} + \frac{R}{L} e^{(R/L)t} I = \frac{V}{L} e^{(R/L)t},$$

or

$$D_t \left(I e^{(R/L)t} \right) = \frac{V}{L} e^{(R/L)t}.$$

Integrating with respect to t yields

$$Ie^{(R/L)t} = \int \frac{V}{L} e^{(R/L)t}\, dt = \frac{V}{R} e^{(R/L)t} + C.$$

Since $I = 0$ when $t = 0$, it follows that $C = -V/R$. Substituting for C leads to

$$Ie^{(R/L)t} = \frac{V}{R} e^{(R/L)t} - \frac{V}{R}.$$

Finally, multiplying both sides by $e^{-(R/L)t}$ gives us

$$I = \frac{V}{R} [1 - e^{-(R/L)t}].$$

Observe that as t increases without bound, I approaches V/R, which is the current predicted by Ohm's law when no inductance is present.

EXERCISES 19.2

Exer. 1–22: Solve the differential equation.

1 $y' + 2y = e^{2x}$

2 $y' - 3y = 2$

3 $xy' - 3y = x^5$

4 $y' + y \cot x = \csc x$

5 $xy' + y + x = e^x$

6 $xy' + (1 + x)y = 5$

7 $x^2\, dy + (2xy - e^x)\, dx = 0$

8 $x^2\, dy + (x - 3xy + 1)\, dx = 0$

9 $y' + y \cot x = 4x^2 \csc x$

10 $y' + y \tan x = \sin x$

11 $(y \sin x - 2)\, dx + \cos x\, dy = 0$

12 $(x^2y - 1)\, dx + x^3\, dy = 0$

13 $(x^2 \cos x + y)\, dx - x\, dy = 0$

14 $y' + y = \sin x$

15 $xy' + (2 + 3x)y = xe^{-3x}$

16 $(x + 4)y' + 5y = x^2 + 8x + 16$

17 $x^{-1}y' + 2y = 3$

18 $y' - 5y = e^{5x}$

19 $\tan x\, dy + (y - \sin x)\, dx = 0$

20 $\cos x\, dy - (y \sin x + e^{-x})\, dx = 0$

21 $y' + 3x^2y = x^2 + e^{-x^3}$

22 $y' + y \tan x = \cos^3 x$

Exer. 23–26: Find the particular solution of the differential equation that satisfies the given condition.

23 $xy' - y = x^2 + x$; $y = 2$ when $x = 1$

24 $y' + 2y = e^{-3x}$; $y = 2$ when $x = 0$

25 $xy' + y + xy = e^{-x}$; $y = 0$ when $x = 1$

26 $y' + 2xy - e^{-x^2} = x$; $y = 1$ when $x = 0$

Exer. 27–28: Solve the differential equation by (a) using an integrating factor and (b) separating the variables.

27 The differential equation

$$R\frac{dQ}{dt} + \frac{Q}{C} = V$$

describes the charge Q on a capacitor with capacitance C during a charging process involving a resistance R and electromotive force V. If the charge is 0 when $t = 0$, express Q as a function of t.

28 The differential equation

$$R\frac{dI}{dt} + \frac{I}{C} = \frac{dV}{dt}$$

describes an electrical circuit consisting of an electromotive force V with a resistance R and capacitance C connected in series. If V is constant and if $I = I_0$ when $t = 0$, express I as a function of t.

29 At time $t = 0$, a tank contains K pounds of salt dissolved in 80 gallons of water. Suppose that water containing $\frac{1}{3}$ pound of salt per gallon is being added to the tank at a rate of 6 gal/min, and that the well-stirred solution is being drained from the tank at the same rate. Find a formula for the amount $f(t)$ of salt in the tank at time t.

30 An object of mass m is moving on a coordinate line subject to a force given by $F(t) = e^{-t}$, where t is time. The motion is resisted by a frictional force that is numerically equal to twice the speed of the object. If $v = 0$ at $t = 0$, find a formula for v at any time $t > 0$.

31 A *learning curve* is used to describe the rate at which a skill is acquired. Suppose a manufacturer estimates that a new employee will produce A items the first day on the job, and that as the employee's proficiency increases, items will be produced more rapidly until the employee produces a maximum of M items per day. Let $f(t)$ denote the number produced on day t, where $t \geq 1$. Suppose that the rate of production $f'(t)$ is proportional to $M - f(t)$.

(a) Find a formula for $f(t)$.

(b) If $M = 30$, $f(1) = 5$, and $f(2) = 8$, estimate the number of items produced on day 20.

32 A room having dimensions 10 ft × 15 ft × 8 ft originally contains 0.001% carbon monoxide (CO). Suppose that at time $t = 0$, fumes containing 5% CO begin entering the room at a rate of 0.12 ft^3/min, and that the well-circulated mixture is eliminated from the room at the same rate.

(a) Find a formula for the volume $f(t)$ of CO in the room at time t.

(b) The minimum level of CO considered to be hazardous to health is 0.015%. After approximately how many minutes will the room contain this level of CO?

33 The von Bertalanffy model for the growth of an animal assumes that there is an upper bound of y_L centimeters on length. If y is the length at age t years, the rate of growth is assumed to be proportional to the length yet to be achieved. Find the general solution of the resulting differential equation.

34 A cell is in a liquid containing a solute, such as potassium, of constant concentration c_0. If $c(t)$ is the concentration of the solute inside the cell at time t, then *Fick's principle* for passive diffusion across the cell membrane asserts that the rate at which the concentration changes is proportional to the *concentration gradient* $c_0 - c(t)$. Find $c(t)$ if $c(0) = 0$.

35 Many common drugs (such as forms of penicillin) are eliminated from the bloodstream at a rate that is proportional to the amount y still present.

(a) If y_0 milligrams are injected directly into the bloodstream, show that $y = y_0 e^{-kt}$ for some $k > 0$.

(b) If the drug is fed intravenously into the bloodstream at a rate of I mg/min, then $y' = -ky + I$. If $y = 0$ at $t = 0$, express y as a function of t and find $\lim_{t \to \infty} y$.

(c) If the half-life of the drug in the bloodstream is 2 hours, estimate the infusion rate that will result in a long-term amount of 100 milligrams in the bloodstream.

36 Shown in the figure is a two-tank system in which water flows in and out of the tanks at the rate of 5 gal/min. Each tank contains 50 gallons of water.

(a) If 1 pound of dye is thoroughly mixed into the water in Tank 1 and if $x(t)$ denotes the amount of dye in the tank after t minutes, show that $x'(t) = -0.1x(t)$ and that $x(t) = e^{-0.1t}$.

(b) From part (a), the dye is entering Tank 2 at the rate of $5 \cdot \frac{1}{50}x(t) = 0.1e^{-0.1t}$ lb/min. Show that if $y(t)$ is the amount of dye in Tank 2 after t minutes, then $y'(t) = -0.1y(t) + 0.1e^{-0.1t}$. Assuming that $y(0) = 0$, find $y(t)$.

(c) What is the maximum amount of dye in Tank 2?

(d) Rework part (b) if the volume of Tank 2 is 40 gallons.

EXERCISE 36

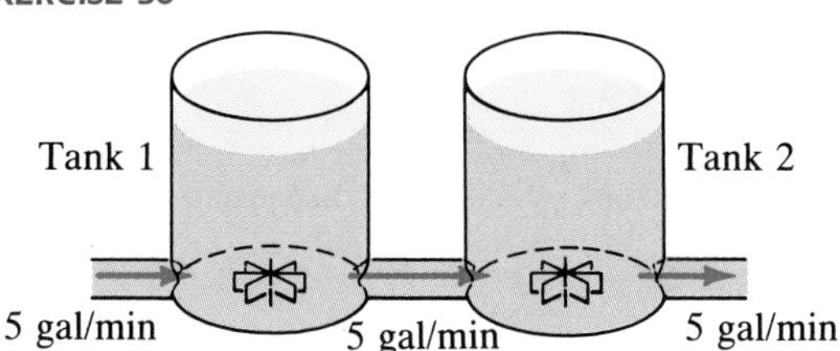

C **Exer. 37–38: Refer to Exercises 41–44 of Section 19.1. Show that numerical methods can fail to provide accurate approximations by first verifying that the differential equation $y' = 10y - 11e^{-x}$ with initial condition $y = 1$ at $x = 0$ has the solution $y = e^{-x}$ for every x and then approximating y at $x = 1$ using the indicated method with $n = 8$.**

37 Euler's method

38 Improved Euler's method

C **Exer. 39–40: Another numerical technique for approximating solutions of differential equations is the classical *Runge-Kutta method*, obtained by replacing the formula in Step 3 of the improved Euler's method (Exercises 41–42 of Section 19.1) by**

$$y_{k+1} = y_k + \tfrac{1}{6}h(K_1 + 2K_2 + 2K_3 + K_4),$$

where

$$K_1 = f(x_k, y_k)$$
$$K_2 = f(x_k + \tfrac{1}{2}h,\ y_k + \tfrac{1}{2}hK_1)$$
$$K_3 = f(x_k + \tfrac{1}{2}h,\ y_k + \tfrac{1}{2}hK_2)$$
$$K_4 = f(x_k + h,\ y_k + hK_3).$$

For the given differential equation with initial condition $y_0 = 0.1$ at $x = 0$, use this method, with $n = 4$, to approximate y at $x = \frac{1}{2}$.

39 $y' = x^2 + y^2$

40 $y' = \sin(x + y^2)$

19.3 SECOND-ORDER LINEAR DIFFERENTIAL EQUATIONS

The following definition is a generalization of (19.1).

Definition **(19.3)**

> A **linear differential equation of order *n*** is an equation of the form
>
> $$y^{(n)} + f_1(x)y^{(n-1)} + \cdots + f_{n-1}(x)y' + f_n(x)y = k(x),$$
>
> where $f_1, f_2, \ldots, f_n$ and k are functions of one variable that have the same domain. If $k(x) = 0$ for every x, the equation is **homogeneous**. If $k(x) \neq 0$ for some x, the equation is **nonhomogeneous**.

A thorough analysis of (19.3) may be found in textbooks on differential equations. We shall restrict our work to second-order equations in which f_1 and f_2 are constant functions. In this section we shall consider the homogeneous case. Nonhomogeneous equations will be discussed in the next section. Applications are considered in Section 19.5.

The general second-order homogeneous linear differential equation with constant coefficients has the form

$$y'' + by' + cy = 0,$$

where b and c are constants. Before attempting to find particular solutions, let us establish the following result.

Theorem **(19.4)**

> If $y = f(x)$ and $y = g(x)$ are solutions of $y'' + by' + cy = 0$, then
>
> $$y = C_1f(x) + C_2g(x)$$
>
> is a solution for all real numbers C_1 and C_2.

PROOF Since $f(x)$ and $g(x)$ are solutions of $y'' + by' + cy = 0$,

$$f''(x) + bf'(x) + cf(x) = 0$$

and

$$g''(x) + bg'(x) + cg(x) = 0.$$

If we multiply the first of these equations by C_1, the second by C_2, and add, the result is

$$[C_1f''(x) + C_2g''(x)] + b[C_1f'(x) + C_2g'(x)] + c[C_1f(x) + C_2g(x)] = 0.$$

Thus, $C_1f(x) + C_2g(x)$ is a solution. ■

We can show that if the solutions f and g in Theorem (19.4) have the property that $f(x) \neq Cg(x)$ for every real number C and if $g(x)$ is not identically 0, then $y = C_1f(x) + C_2g(x)$ is the general solution of the differential equation $y'' + by' + cy = 0$. Thus, to determine the general solution, it is sufficient to find two such functions f and g and employ (19.4).

In our search for a solution of $y'' + by' + cy = 0$, we shall use $y = e^{mx}$ as a trial solution. Since $y' = me^{mx}$ and $y'' = m^2e^{mx}$, it follows that $y = e^{mx}$ is a solution if and only if

$$m^2e^{mx} + bme^{mx} + ce^{mx} = 0$$

or, since $e^{mx} \neq 0$, if and only if

$$m^2 + bm + c = 0.$$

The last equation is very important in finding solutions of $y'' + by' + cy = 0$ and is given the following special name.

Definition (19.5)

The **auxiliary equation** of the differential equation $y'' + by' + cy = 0$ is $m^2 + bm + c = 0$.

Note that the auxiliary equation can be obtained from the differential equation by replacing y'' by m^2, y' by m, and y by 1. Since the auxiliary equation is quadratic, it either has unequal real roots m_1 and m_2, a double real root m, or two complex conjugate roots. The next theorem is a consequence of the remark following the proof of Theorem (19.4).

Theorem (19.6)

If the roots m_1, m_2 of the auxiliary equation are real and unequal, then the general solution of $y'' + by' + cy = 0$ is

$$y = C_1 e^{m_1 x} + C_2 e^{m_2 x}.$$

EXAMPLE 1 Solve the differential equation $y'' - 3y' - 10y = 0$.

SOLUTION The auxiliary equation is $m^2 - 3m - 10 = 0$, or, equivalently, $(m - 5)(m + 2) = 0$. Since the roots $m_1 = 5$ and $m_2 = -2$ are real and unequal, it follows from Theorem (19.6) that the general solution is

$$y = C_1 e^{5x} + C_2 e^{-2x}.$$

Theorem (19.7)

If the auxiliary equation has a double root m, then the general solution of $y'' + by' + cy = 0$ is

$$y = C_1 e^{mx} + C_2 x e^{mx}.$$

PROOF By the quadratic formula, the roots of $m^2 + bm + c = 0$ are $m = (-b \pm \sqrt{b^2 - 4c})/2$. If $b^2 - 4c = 0$, then m is a double root and we obtain $m = -b/2$, or $2m + b = 0$. Since m satisfies the auxiliary equation, $y = e^{mx}$ is a solution of the differential equation. According to the remark following the proof of Theorem (19.4), it is sufficient to show that $y = xe^{mx}$ is also a solution. Substitution of xe^{mx} for y in the equation $y'' + by' + cy = 0$ gives us

$$\begin{aligned}(2me^{mx} + m^2xe^{mx}) + b(mxe^{mx} + e^{mx}) + cxe^{mx} \\ &= (m^2 + bm + c)xe^{mx} + (2m + b)e^{mx} \\ &= 0xe^{mx} + 0e^{mx} = 0,\end{aligned}$$

which is what we wished to show. ■

EXAMPLE 2 Solve the differential equation $y'' - 6y' + 9y = 0$.

SOLUTION The auxiliary equation $m^2 - 6m + 9 = 0$, or, equivalently, $(m - 3)^2 = 0$, has a double root 3. Hence, by Theorem (19.7), the general solution is

$$y = C_1e^{3x} + C_2xe^{3x} = e^{3x}(C_1 + C_2x).$$

We may also consider second-order differential equations of the form

$$ay'' + by' + cy = 0$$

with $a \neq 1$. It is possible to obtain the form stated in Theorems (19.6) and (19.7) by dividing both sides by a; however, it is usually simpler to employ the auxiliary equation

$$am^2 + bm + c = 0,$$

as illustrated in the next example.

EXAMPLE 3 Solve the differential equation $6y'' - 7y' + 2y = 0$.

SOLUTION The auxiliary equation is $6m^2 - 7m + 2 = 0$, or, equivalently, $(2m - 1)(3m - 2) = 0$. Hence the roots are $m_1 = \frac{1}{2}$ and $m_2 = \frac{2}{3}$. By Theorem (19.6), the general solution of the given equation is

$$y = C_1e^{x/2} + C_2e^{2x/3}.$$

The final case to consider is that in which the roots of the auxiliary equation $m^2 + bm + c = 0$ of $y'' + by' + cy = 0$ are conjugate complex numbers of the form

$$z_1 = s + ti \quad \text{and} \quad z_2 = s - ti,$$

where s and t are real numbers and $i^2 = -1$. We may anticipate, from Theorem (19.6), that the general solution of the differential equation is

$$y = C_1e^{z_1x} + C_2e^{z_2x} = C_1e^{(s+ti)x} + C_2e^{(s-ti)x}.$$

To define complex exponents, we must extend some of the concepts of calculus to include functions whose domains include complex numbers. Since a complete development requires advanced methods, we shall merely outline the main ideas.

In Section 11.8 we discussed how certain functions can be represented by power series. We can easily extend the definitions and theorems of Chapter 11 to infinite series that involve complex numbers. Since this is true, we shall use the power series representations (11.48) to *define* e^z, $\sin z$, and $\cos z$ for every complex number z as follows.

Functions of $z = a + bi$ **(19.8)**

(i) $e^z = 1 + z + \dfrac{z^2}{2!} + \cdots + \dfrac{z^n}{n!} + \cdots$

(ii) $\sin z = z - \dfrac{z^3}{3!} + \dfrac{z^5}{5!} - \cdots + (-1)^n \dfrac{z^{2n+1}}{(2n+1)!} + \cdots$

(iii) $\cos z = 1 - \dfrac{z^2}{2!} + \dfrac{z^4}{4!} - \cdots + (-1)^n \dfrac{z^{2n}}{(2n)!} + \cdots$

Using (19.8)(i) gives us

$$e^{iz} = 1 + (iz) + \frac{(iz)^2}{2!} + \frac{(iz)^3}{3!} + \frac{(iz)^4}{4!} + \frac{(iz)^5}{5!} + \cdots$$

$$= 1 + iz + i^2\frac{z^2}{2!} + i^3\frac{z^3}{3!} + i^4\frac{z^4}{4!} + i^5\frac{z^5}{5!} + \cdots.$$

Since $i^2 = -1$, $i^3 = -i$, $i^4 = 1$, $i^5 = i$, and so on, we see that

$$e^{iz} = 1 + iz - \frac{z^2}{2!} - i\frac{z^3}{3!} + \frac{z^4}{4!} + i\frac{z^5}{5!} - \cdots,$$

which may also be written in the form

$$e^{iz} = \left(1 - \frac{z^2}{2!} + \frac{z^4}{4!} - \cdots\right) + i\left(z - \frac{z^3}{3!} + \frac{z^5}{5!} - \cdots\right).$$

If we now use formulas (iii) and (ii) of (19.8), we obtain the following result, named after the Swiss mathematician Leonhard Euler (1707–1783).

Euler's formula **(19.9)**

> If z is a complex number,
> $$e^{iz} = \cos z + i \sin z.$$

The laws of exponents are true for complex numbers. In addition, formulas for derivatives may be extended to functions of a *complex* variable z. For example, $D_z\, e^{kz} = ke^{kz}$, where k is a complex number. If the auxiliary equation of $y'' + by' + cy = 0$ has complex roots $s \pm ti$, then the general solution of this differential equation may be written in the following equivalent forms:

$$y = C_1e^{(s+ti)x} + C_2e^{(s-ti)x}$$

$$y = C_1e^{sx+txi} + C_2e^{sx-txi}$$

$$y = C_1e^{sx}e^{txi} + C_2e^{sx}e^{-txi}$$

$$y = e^{sx}(C_1e^{itx} + C_2e^{-itx})$$

This can be further simplified by using Euler's formula. Specifically, from (19.9),

$$e^{itx} = \cos tx + i \sin tx$$

and

$$e^{-itx} = \cos tx - i \sin tx,$$

from which it follows that

$$\cos tx = \frac{e^{itx} + e^{-itx}}{2} \quad \text{and} \quad \sin tx = \frac{e^{itx} - e^{-itx}}{2i}.$$

If we let $C_1 = C_2 = \frac{1}{2}$ in the preceding discussion and then use the formula for $\cos tx$, we obtain

$$y = \tfrac{1}{2}e^{sx}(e^{itx} + e^{-itx})$$

$$= \tfrac{1}{2}e^{sx}(2 \cos tx) = e^{sx} \cos tx.$$

Thus, $y = e^{sx} \cos tx$ is a particular solution of $y'' + by' + cy = 0$. Letting $C_1 = -C_2 = i/2$ leads to the particular solution $y = e^{sx} \sin tx$. This is a partial proof of the next theorem.

Theorem (19.10)

> If the auxiliary equation $m^2 + bm + c = 0$ has distinct complex roots $s \pm ti$, then the general solution of $y'' + by' + cy = 0$ is
> $$y = e^{sx}(C_1 \cos tx + C_2 \sin tx).$$

EXAMPLE Solve the differential equation $y'' - 10y' + 41y = 0$.

SOLUTION The roots of the auxiliary equation $m^2 - 10m + 41 = 0$ are

$$m = \frac{10 \pm \sqrt{100 - 164}}{2} = \frac{10 \pm 8i}{2} = 5 \pm 4i.$$

Applying Theorem (19.10) yields the general solution of the differential equation:

$$y = e^{5x}(C_1 \cos 4x + C_2 \sin 4x)$$

EXERCISES 19.3

Exer. 1–22: Solve the differential equation.

1 $y'' - 5y' + 6y = 0$

2 $y'' - y' - 2y = 0$

3 $y'' - 3y' = 0$

4 $y'' + 6y' + 8y = 0$

5 $y'' + 4y' + 4y = 0$

6 $y'' - 4y' + 4y = 0$

7 $y'' - 4y' + y = 0$

8 $6y'' - 7y' - 3y = 0$

9 $y'' + 2\sqrt{2}y' + 2y = 0$

10 $4y'' + 20y' + 25y = 0$

11 $8y'' + 2y' - 15y = 0$

12 $y'' + 4y' + y = 0$

13 $9y'' - 24y' + 16y = 0$

14 $4y'' - 8y' + 7y = 0$

15 $2y'' - 4y' + y = 0$

16 $2y'' + 7y' = 0$

17 $y'' - 2y' + 2y = 0$

18 $y'' - 2y' + 5y = 0$

19 $y'' - 4y' + 13y = 0$

20 $y'' + 4 = 0$

21 $\dfrac{d^2y}{dx^2} + 6\dfrac{dy}{dx} + 2y = 0$

22 $\dfrac{d^2y}{dx^2} + 2\dfrac{dy}{dx} + 6y = 0$

Exer. 23–30: Find the particular solution of the differential equation that satisfies the stated conditions.

23 $y'' - 3y' + 2y = 0$; $y = 0$ and $y' = 2$ when $x = 0$

24 $y'' - 2y' + y = 0$; $y = 1$ and $y' = 2$ when $x = 0$

25 $y'' + y = 0$; $y = 1$ and $y' = 2$ when $x = 0$

26 $y'' - y' - 6y = 0$; $y = 0$ and $y' = 1$ when $x = 0$

27 $y'' + 8y' + 16y = 0$; $y = 2$ and $y' = 1$ when $x = 0$

28 $y'' + 5y = 0$; $y = 4$ and $y' = 2$ when $x = 0$

29 $\dfrac{d^2y}{dx^2} - 2\dfrac{dy}{dx} + 5y = 0$; $y = 0$ and $\dfrac{dy}{dx} = 1$ when $x = 0$

30 $\dfrac{d^2y}{dx^2} - 6\dfrac{dy}{dx} + 13y = 0$; $y = 2$ and $\dfrac{dy}{dx} = 3$ when $x = 0$

19.4 NONHOMOGENEOUS LINEAR DIFFERENTIAL EQUATIONS

In this section we shall consider second-order nonhomogeneous linear differential equations with constant coefficients—that is, equations of the form

$$y'' + by' + cy = k(x),$$

where b and c are constants and k is a continuous function.

It is convenient to use the differential operator symbols D and D^2 such that if $y = f(x)$, then

$$Dy = y' = f'(x) \quad \text{and} \quad D^2 y = y'' = f''(x).$$

Definition (19.11)

If $y = f(x)$ and if f'' exists, then the **linear differential operator** $L = D^2 + bD + c$ is defined by

$$L(y) = (D^2 + bD + c)y = D^2 y + b\,Dy + cy = y'' + by' + cy.$$

Using L, we may write the differential equation $y'' + by' + cy = k(x)$ in the compact form $L(y) = k(x)$. In Exercises 19 and 20 you are asked to verify that for every real number C,

$$L(Cy) = CL(y),$$

and that if $y_1 = f_1(x)$ and $y_2 = f_2(x)$, then

$$L(y_1 \pm y_2) = L(y_1) \pm L(y_2).$$

For the differential equation $y'' + by' + cy = k(x)$ (that is, $L(y) = k(x)$), the corresponding homogeneous equation $L(y) = 0$ is the **complementary equation**.

Theorem (19.12)

Let $y'' + by' + cy = k(x)$ be a second-order nonhomogeneous linear differential equation. If y_p is a particular solution of $L(y) = k(x)$ and if y_c is the general solution of the complementary equation $L(y) = 0$, then the general solution of $L(y) = k(x)$ is $y = y_p + y_c$.

PROOF Since $L(y_p) = k(x)$ and $L(y_c) = 0$,

$$L(y_p + y_c) = L(y_p) + L(y_c) = k(x) + 0 = k(x),$$

which means that $y_p + y_c$ is a solution of $L(y) = k(x)$. Moreover, if $y = f(x)$ is any other solution, then

$$L(y - y_p) = L(y) - L(y_p) = k(x) - k(x) = 0.$$

Consequently $y - y_p$ is a solution of the complementary equation. Thus, $y - y_p = y_c$ for some y_c, which is what we wished to prove. ■

If we use the results of Section 19.3 to find the general solution y_c of $L(y) = 0$, then according to Theorem (19.12) all that is needed to determine the general solution of $L(y) = k(x)$ is *one* particular solution y_p.

EXAMPLE 1 Solve the differential equation $y'' - 4y = 6x - 4x^3$ using the particular solution $y_p = x^3$.

SOLUTION The complementary equation is $y'' - 4y = 0$, which, by Theorem (19.6), has the general solution

$$y_c = C_1 e^{2x} + C_2 e^{-2x}.$$

Applying Theorem (19.12), we find that the general solution of the given nonhomogeneous equation is

$$y = C_1 e^{2x} + C_2 e^{-2x} + x^3.$$

If a particular solution of $y'' + by' + cy = k(x)$ cannot be found by inspection, then the following technique, called **variation of parameters**, may be employed. Given the differential equation $L(y) = k(x)$, let y_1 and y_2 be the expressions that appear in the general solution $y = C_1y_1 + C_2y_2$ of the complementary equation $L(y) = 0$. For example, as in (19.6), we might have $y_1 = e^{m_1x}$ and $y_2 = e^{m_2x}$. Let us now attempt to find a particular solution of $L(y) = k(y)$ that has the form

$$y_p = uy_1 + vy_2,$$

where $u = g(x)$ and $v = h(x)$ for some functions g and h. The first and second derivatives of y_p are

$$y'_p = (uy'_1 + vy'_2) + (u'y_1 + v'y_2)$$
$$y''_p = (uy''_1 + vy''_2) + (u'y'_1 + v'y'_2) + (u'y_1 + v'y_2)'.$$

Substituting these into $L(y_p) = y''_p + by'_p + cy_p$ and rearranging terms, we obtain

$$\begin{aligned} L(y_p) = u(y''_1 + by'_1 + cy_1) + v(y''_2 + by'_2 + cy_2) \\ + b(u'y_1 + v'y_2) + (u'y_1 + v'y_2)' + (u'y'_1 + v'y'_2). \end{aligned}$$

Since y_1 and y_2 are solutions of $y'' + by' + cy = 0$, the first two terms on the right side are 0. Hence, to obtain $L(y_p) = k(x)$, it is sufficient to choose u and v such that

$$\begin{cases} u'y_1 + v'y_2 = 0 \\ u'y'_1 + v'y'_2 = k(x) \end{cases}$$

We can show that this system of equations always has a unique solution u' and v'. We may then determine u and v by integration and use the fact that $y_p = uy_1 + vy_2$ to find a particular solution of the differential equation. Our discussion may be summarized as follows.

Variation of parameters **(19.13)**

If $y = C_1y_1 + C_2y_2$ is the general solution of the complementary equation $L(y) = 0$ of $y'' + by' + cy = k(x)$, then a particular solution of $L(y) = k(x)$ is

$$y_p = uy_1 + vy_2,$$

where $u = g(x)$ and $v = h(x)$ satisfy the following system of equations:

$$\begin{cases} u'y_1 + v'y_2 = 0 \\ u'y'_1 + v'y'_2 = k(x) \end{cases}$$

EXAMPLE 2 Solve the differential equation $y'' + y = \cot x$.

SOLUTION The complementary equation is $y'' + y = 0$. Since the auxiliary equation $m^2 + 1 = 0$ has roots $\pm i$, we see from Theorem (19.10) that the general solution of the homogeneous differential equation $y'' + y = 0$ is

$$y = C_1 \cos x + C_2 \sin x.$$

Thus, we seek a particular solution of the differential equation that has the form $y_p = uy_1 + vy_2$, where $y_1 = \cos x$ and $y_2 = \sin x$. The system of equations in Theorem (19.13) is, therefore,

$$\begin{cases} u' \cos x + v' \sin x = 0 \\ -u' \sin x + v' \cos x = \cot x \end{cases}$$

Solving for u' and v' gives us

$$u' = -\cos x, \quad v' = \csc x - \sin x.$$

If we integrate each of these expressions (and discard the constants of integration), we obtain

$$u = -\sin x, \quad v = \ln|\csc x - \cot x| + \cos x.$$

Applying Theorem (19.13), we find that a particular solution of the given equation is

$$y_p = -\sin x \cos x + \sin x \ln|\csc x - \cot x| + \sin x \cos x,$$

or, equivalently,

$$y_p = \sin x \ln|\csc x - \cot x|.$$

Finally, by Theorem (19.12), the general solution of $y'' + y = \cot x$ is

$$y = C_1 \cos x + C_2 \sin x + \sin x \ln|\csc x - \cot x|.$$

Given the differential equation

$$L(y) = y'' + by' + cy = e^{nx},$$

where e^{nx} is not a solution of $L(y) = 0$, we can reasonably expect that there exists a particular solution of the form $y_p = Ae^{nx}$, since e^{nx} is the result of finding $L(Ae^{nx})$. This suggests that we use Ae^{nx} as a trial solution in the given equation and attempt to find the value of the coefficient A. This is called the **method of undetermined coefficients** and is illustrated in the next example.

EXAMPLE 3 Solve the differential equation $y'' + 2y' - 8y = e^{3x}$.

SOLUTION The auxiliary equation $m^2 + 2m - 8 = 0$ of the differential equation $y'' + 2y' - 8y = 0$ has roots 2 and -4. By Theorem (19.6), the general solution of the complementary equation is

$$y_c = C_1 e^{2x} + C_2 e^{-4x}.$$

From the preceding remarks, we seek a particular solution of the form $y_p = Ae^{3x}$. Since $y_p' = 3Ae^{3x}$ and $y_p'' = 9Ae^{3x}$, substitution in the given equation leads to

$$9Ae^{3x} + 6Ae^{3x} - 8Ae^{3x} = e^{3x}.$$

Dividing both sides by e^{3x}, we obtain

$$9A + 6A - 8A = 1, \quad \text{or} \quad A = \tfrac{1}{7}.$$

Thus, $y_p = \frac{1}{7}e^{3x}$, and, by Theorem (19.12), the general solution is

$$y = C_1e^{2x} + C_2e^{-4x} + \tfrac{1}{7}e^{3x}.$$

Three rules for arriving at trial solutions of second-order nonhomogeneous differential equations with constant coefficients are stated without proof in the next theorem. (See texts on differential equations for a more extensive treatment of this topic.)

Theorem on particular solutions **(19.14)**

(i) If $y'' + by' + cy = e^{nx}$ and n is not a root of the auxiliary equation $m^2 + bm + c = 0$, then there is a particular solution of the form $y_p = Ae^{nx}$.

(ii) If $y'' + by' + cy = xe^{nx}$ and n is not a solution of the auxiliary equation $m^2 + bm + c = 0$, then there is a particular solution of the form $y_p = (A + Bx)e^{nx}$.

(iii) If either $\quad y'' + by' + cy = e^{sx}\sin tx$

or $\quad y'' + by' + cy = e^{sx}\cos tx$

and the complex number $s + ti$ is not a solution of the auxiliary equation $m^2 + bm + c = 0$, then there is a particular solution of the form

$$y_p = Ae^{sx}\cos tx + Be^{sx}\sin tx.$$

Part (i) of Theorem (19.14) was used in the solution of Example 3. Illustrations of parts (ii) and (iii) are given in the following examples.

EXAMPLE 4 Solve $y'' - 3y' - 18y = xe^{4x}$.

SOLUTION Since the auxiliary equation $m^2 - 3m - 18 = 0$ has roots 6 and -3, it follows from the preceding section that the general solution of $y'' - 3y' - 18y = 0$ is

$$y = C_1e^{6x} + C_2e^{-3x}.$$

Since 4 is not a root of the auxiliary equation, we see from Theorem (19.14)(ii) that there is a particular solution of the form

$$y_p = (A + Bx)e^{4x}.$$

Differentiating twice, we have

$$y_p' = (4A + 4Bx + B)e^{4x}$$
$$y_p'' = (16A + 16Bx + 8B)e^{4x}.$$

Substitution in the given differential equation produces

$$(16A + 16Bx + 8B)e^{4x} - 3(4A + 4Bx + B)e^{4x} - 18(A + Bx)e^{4x} = xe^{4x},$$

which reduces to

$$-14A + 5B - 14Bx = x.$$

Thus, y_p is a solution, provided

$$-14A + 5B = 0 \quad \text{and} \quad -14B = 1.$$

This gives us $B = -\frac{1}{14}$ and $A = -\frac{5}{196}$. Consequently

$$y_p = (-\tfrac{5}{196} - \tfrac{1}{14}x)e^{4x} = -\tfrac{1}{196}(5 + 14x)e^{4x}.$$

Applying Theorem (19.12), we find that the general solution is

$$y = C_1e^{6x} + C_2e^{-3x} - \tfrac{1}{196}(5 + 14x)e^{4x}.$$

EXAMPLE 5 Solve $y'' - 10y' + 41y = \sin x$.

SOLUTION In Example 4 of the preceding section we found the general solution

$$y_c = e^{5x}(C_1 \cos 4x + C_2 \sin 4x)$$

of the complementary equation. Referring to Theorem (19.14)(iii) with $s = 0$ and $t = 1$, we seek a particular solution of the form

$$y_p = A \cos x + B \sin x.$$

Since

$$y_p' = -A \sin x + B \cos x \quad \text{and} \quad y_p'' = -A \cos x - B \sin x,$$

substitution in the given equation produces

$$-A \cos x - B \sin x + 10A \sin x - 10B \cos x + 41A \cos x + 41B \sin x = \sin x,$$

which can be written

$$(40A - 10B) \cos x + (10A + 40B) \sin x = \sin x.$$

Consequently y_p is a solution, provided

$$40A - 10B = 0 \quad \text{and} \quad 10A + 40B = 1.$$

The solution of this system of equations is $A = \frac{1}{170}$ and $B = \frac{4}{170}$, and hence

$$y_p = \tfrac{1}{170} \cos x + \tfrac{4}{170} \sin x = \tfrac{1}{170}(\cos x + 4 \sin x).$$

Thus, the general solution is

$$y = e^{5x}(C_1 \cos 4x + C_2 \sin 4x) + \tfrac{1}{170}(\cos x + 4 \sin x).$$

EXERCISES 19.4

Exer. 1–10: Solve the differential equation by using the method of variation of parameters.

1 $y'' + y = \tan x$

2 $y'' + y = \sec x$

3 $y'' - 6y' + 9y = x^2e^{3x}$

4 $y'' + 3y' = e^{-3x}$

5 $y'' - y = e^x \cos x$

6 $y'' - 4y' + 4y = x^{-2}e^{2x}$

7 $y'' - 9y = e^{3x}$

8 $y'' + y = \sin x$

9 $\dfrac{d^2y}{dx^2} - 3\dfrac{dy}{dx} - 4y = 2$

10 $\dfrac{d^2y}{dx^2} - \dfrac{dy}{dx} = x + 1$

Exer. 11–18: Solve the differential equation by using undetermined coefficients.

11 $y'' - 3y' + 2y = 4e^{-x}$

12 $y'' + 6y' + 9y = e^{2x}$

13 $y'' + 2y' = \cos 2x$

14 $y'' + y = \sin 5x$

15 $y'' - y = xe^{2x}$

16 $y'' + 3y' - 4y = xe^{-x}$

17 $\dfrac{d^2y}{dx^2} - 6\dfrac{dy}{dx} + 13y = e^x \cos x$

18 $\dfrac{d^2y}{dx^2} - 2\dfrac{dy}{dx} + 2y = e^{-x} \sin 2x$

Exer. 19–20: Prove the identity if L is the linear differential operator (19.11) and y, y_1, and y_2 are functions of x.

19 $L(Cy) = CL(y)$

20 $L(y_1 \pm y_2) = L(y_1) \pm L(y_2)$

c **Exer. 21–22: A solution of a differential equation of the form $y'' = f(x, y)$ with initial conditions $y = y_0$ at $x = a$ and $y = y_{-1}$ at $x = a - h$ is sometimes approximated using the formula $y_{k+1} = 2y_k - y_{k-1} + h^2 f(x_k, y_k)$ for $k = 0, 1, 2, \ldots, n - 1$, where $h = (b - a)/n$ and $x_k = a + kh$. If $h \approx 0$, then y_k is an approximation to y at $x = x_k$. Use this formula, with $n = 4$ and $a = 0$, to approximate y at $x = \frac{1}{2}$ for the given differential equation and initial conditions.**

21 $y'' = (4x^2 - 2)y; \quad y_0 = 1, \quad y_{-1} = 0.984496$

22 $y'' = y - x; \quad y_0 = 1, \quad y_{-1} = 0.882823$

19.5 VIBRATIONS

FIGURE 19.7

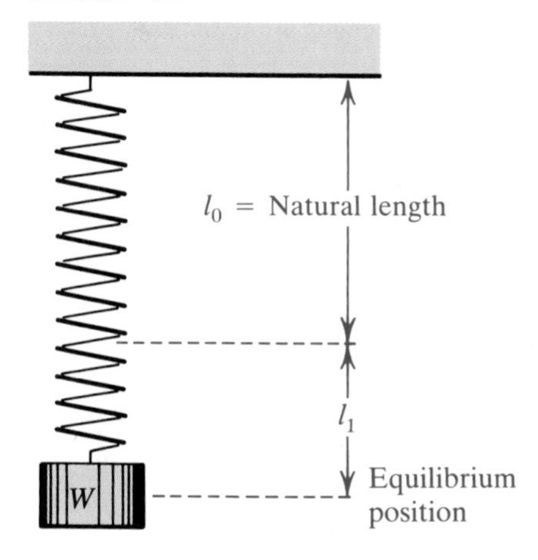

Vibrations in mechanical systems are caused by external forces. A violin string vibrates if bowed, a steel beam vibrates if struck by a hammer, and a bridge vibrates if a marching band crosses it in cadence. In this section we shall use differential equations to analyze vibrations that may occur in a spring.

According to Hooke's law, the force required to stretch a spring y units beyond its natural length is ky, where k is a positive real number called the *spring constant*. The **restoring force** of the spring is $-ky$. Suppose that a weight W is attached to the spring and, at the equilibrium position, the spring is stretched a distance l_1 beyond its natural length l_0, as illustrated in Figure 19.7. If g is a gravitational constant and m is the mass of the weight, then $W = mg$, and at the equilibrium position

$$mg = kl_1, \quad \text{or} \quad mg - kl_1 = 0,$$

provided that the mass of the spring is negligible compared to m.

FIGURE 19.8

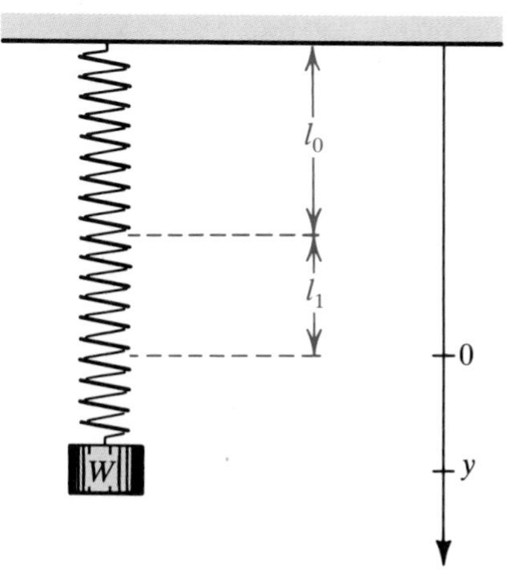

Suppose the weight is pulled down and released. Consider the coordinate line with origin at the equilibrium position and positive direction downward, as shown in Figure 19.8, where y is the coordinate of the center of mass of the weight after t seconds. The force F acting on the weight when its acceleration is a is given by Newton's second law of motion $F = ma$. If we assume that the motion is **undamped**—that is, there is no external retarding force—and the weight moves in a frictionless medium, we see that

$$F = mg - k(l_1 + y) = mg - kl_1 - ky = -ky.$$

Since $F = ma$ and $a = d^2y/dt^2$, this implies that

$$m\frac{d^2y}{dt^2} = -ky.$$

Dividing both sides of this equation by m leads to the following differential equation.

Free, undamped vibration **(19.15)**

$$\frac{d^2y}{dt^2} + \frac{k}{m}y = 0$$

EXAMPLE 1 If a weight of mass m is in free, undamped vibration, show that the motion is simple harmonic. Find the displacement if the weight in Figure 19.8 is pulled down a distance l_2 and released with zero velocity.

SOLUTION If we denote k/m by ω^2, then (19.15) may be written

$$\frac{d^2y}{dt^2} + \omega^2 y = 0.$$

The solutions of the auxiliary equation $m^2 + \omega^2 = 0$ are $\pm\omega i$. Hence, from Theorem (19.10), the general solution is

$$y = C_1 \cos \omega t + C_2 \sin \omega t.$$

It follows that the weight is in simple harmonic motion (see Exercise 26 of Section 4.7).

If the weight is pulled down a distance l_2 and is released with zero velocity, then at $t = 0$

$$l_2 = C_1(1) + C_2(0), \quad \text{or} \quad C_1 = l_2.$$

Since
$$\frac{dy}{dt} = -\omega C_1 \sin \omega t + \omega C_2 \cos \omega t,$$

we also have (at $t = 0$)

$$0 = -\omega C_1(0) + \omega C_2(1), \quad \text{or} \quad C_2 = 0.$$

Hence the displacement y of the weight at time t is

$$y = l_2 \cos \omega t.$$

FIGURE 19.9
Simple harmonic motion

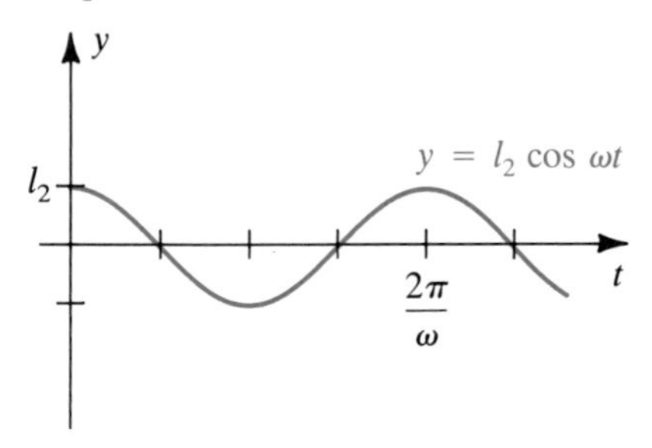

This type of motion was discussed earlier in the text (see Example 3 of Section 4.7). The *amplitude* (the maximum displacement) is l_2, and the *period* (the time for one complete vibration) is $2\pi/\omega = 2\pi\sqrt{m/k}$. A typical graph indicating this type of motion is illustrated in Figure 19.9.

EXAMPLE 2 An 8-pound weight stretches a spring 2 feet beyond its natural length. The weight is then pulled down another $\frac{1}{2}$ foot and released with an initial upward velocity of 6 ft/sec. Find a formula for the displacement of the weight at any time t.

SOLUTION From Hooke's law, $8 = k(2)$, or $k = 4$. If y is the displacement of the weight from the equilibrium position at time t, then, by (19.15),

$$\frac{d^2y}{dt^2} + \frac{4}{m} y = 0.$$

Since $W = mg$, it follows that $m = W/g = \frac{8}{32} = \frac{1}{4}$. Hence

$$\frac{d^2y}{dt^2} + 16y = 0.$$

As in the solution to Example 1, the general solution is

$$y = C_1 \cos 4t + C_2 \sin 4t.$$

At $t = 0$ we have $y = \frac{1}{2}$, and therefore

$$\tfrac{1}{2} = C_1(1) + C_2(0), \quad \text{or} \quad C_1 = \tfrac{1}{2}.$$

Since

$$\frac{dy}{dt} = -4C_1 \sin 4t + 4C_2 \cos 4t$$

and $dy/dt = -6$ at $t = 0$, we obtain

$$-6 = -4C_1(0) + 4C_2(1), \quad \text{or} \quad C_2 = -\tfrac{3}{2}.$$

Hence the displacement at time t is given by

$$y = \tfrac{1}{2} \cos 4t - \tfrac{3}{2} \sin 4t.$$

FIGURE 19.10
Damping force

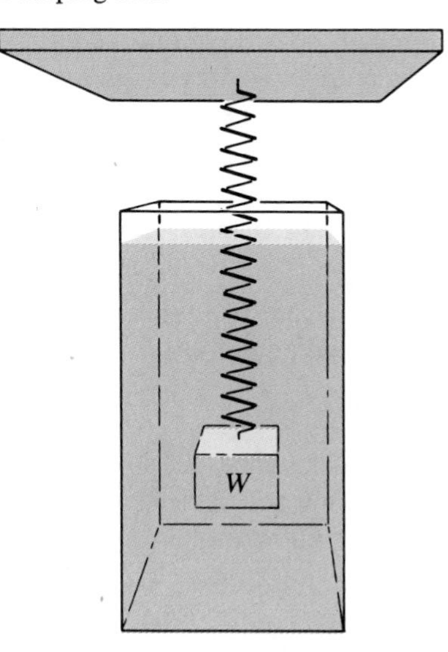

Let us next consider the motion of the spring when a *damping* (or frictional) force is present, as is the case when the weight moves through a fluid (see Figure 19.10). The situation of a shock absorber in an automobile is a good illustration. We shall assume that the direction of the damping force is opposite that of the motion and that the force is directly proportional to the velocity of the weight. Thus, the damping force is given by $-c(dy/dt)$ for some positive constant c. According to Newton's second law, the differential equation that describes the motion is

$$m\frac{d^2y}{dt^2} = -ky - c\frac{dy}{dt}.$$

Dividing by m and rearranging terms gives us the following differential equation.

Free, damped vibration **(19.16)**

$$\frac{d^2y}{dt^2} + \frac{c}{m}\frac{dy}{dt} + \frac{k}{m}y = 0$$

EXAMPLE 3 Discuss the motion of a weight of mass m that is in free, damped vibration.

SOLUTION As in Example 1, let $k/m = \omega^2$. To simplify the roots of the auxiliary equation, we also let $c/m = 2p$. Using this notation, we may write (19.16) as

$$\frac{d^2y}{dt^2} + 2p\frac{dy}{dt} + \omega^2 y = 0.$$

The roots of the auxiliary equation $m^2 + 2pm + \omega^2 = 0$ are

$$\frac{-2p \pm \sqrt{4p^2 - 4\omega^2}}{2} = -p \pm \sqrt{p^2 - \omega^2}.$$

The following three possibilities for roots correspond to three possible types of motion of the weight:

$$p^2 - \omega^2 > 0, \quad p^2 - \omega^2 = 0, \quad \text{or} \quad p^2 - \omega^2 < 0.$$

Let us classify the three types of motion as follows.

FIGURE 19.11
Overdamped vibration

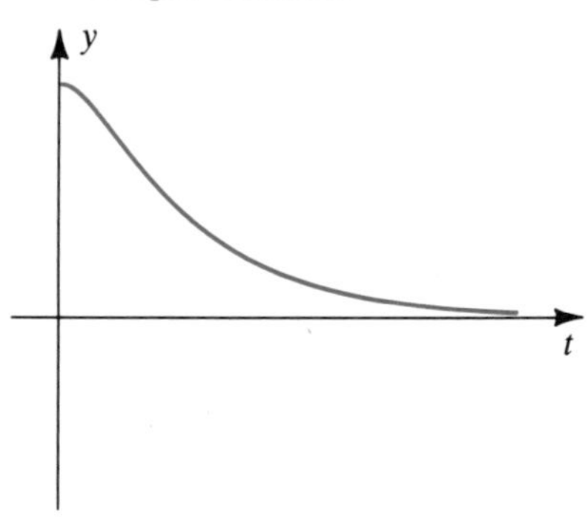

Case 1 $p^2 - \omega^2 > 0$ **(overdamped vibration)**
The roots of the auxiliary equation are real and unequal, and, by Theorem (19.6), the general solution of the differential equation is

$$y = e^{-pt}(C_1 e^{\sqrt{p^2-\omega^2}t} + C_2 e^{-\sqrt{p^2-\omega^2}t}).$$

In the case under consideration,

$$p^2 - \omega^2 = \frac{c^2}{4m^2} - \frac{k}{m} = \frac{c^2 - 4mk}{4m^2} > 0,$$

so $c^2 > 4mk$. This shows that c is usually large compared to k; that is, the damping force dominates the restoring force of the spring, and the weight returns to the equilibrium position relatively quickly. In particular, this happens if the fluid has a high viscosity, as is true for heavy oil or grease.

The manner in which y approaches 0 depends on the constants C_1 and C_2 in the general solution. The graph of a particular solution is sketched in Figure 19.11. (See also Figure 19.14.)

FIGURE 19.12
Critically damped vibration

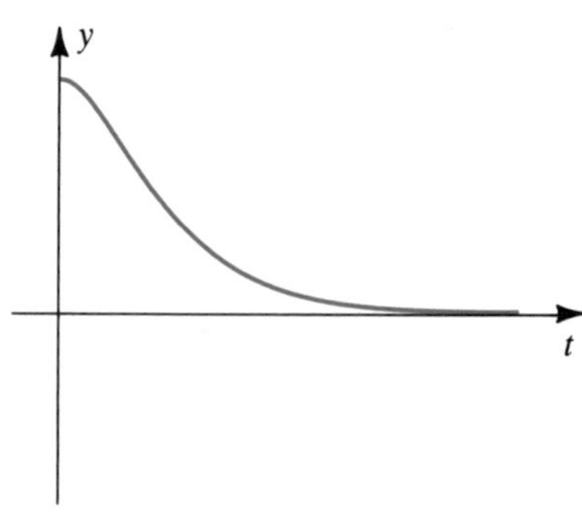

Case 2 $p^2 - \omega^2 = 0$ **(critically damped vibration)**
In this case the auxiliary equation has a double root $-p$, and, by Theorem (19.7), the general solution of the differential equation is

$$y = e^{-pt}(C_1 + C_2 t).$$

The graph of a special case of this equation is sketched in Figure 19.12. The graph is similar to that in Figure 19.11; however, it decreases more rapidly for values of x near 0. Any decrease in the damping force leads to the oscillatory motion discussed in the following case.

FIGURE 19.13
Underdamped vibration

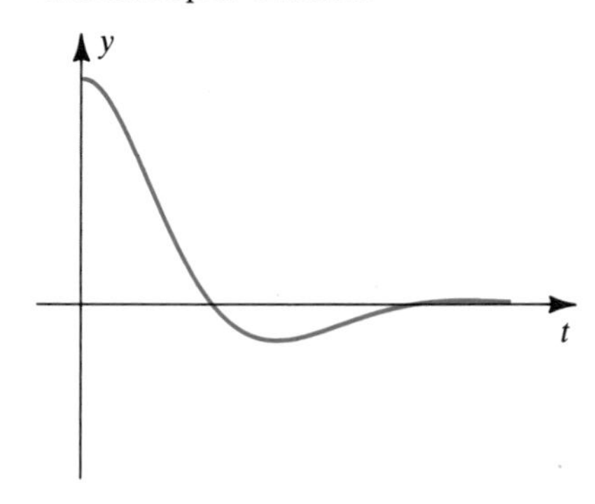

Case 3 $p^2 - \omega^2 < 0$ **(underdamped vibration)**
The roots of the auxiliary equation are complex conjugates $a \pm bi$, and, by Theorem (19.10), the general solution of the differential equation is

$$y = e^{-at}(C_1 \cos bt + C_2 \sin bt).$$

In this case c is usually smaller than k, and the spring oscillates while returning to the equilibrium position, as illustrated in Figure 19.13. Underdamped vibration could occur in a worn shock absorber in an automobile.

EXAMPLE 4 A 24-pound weight stretches a spring 1 foot beyond its natural length. If the weight is pushed downward from its equilibrium position with an initial velocity of 2 ft/sec and if the damping force is $-9(dy/dt)$, find a formula for the displacement y of the weight at any time t.

SOLUTION By Hooke's law, $24 = k(1)$, or $k = 24$. The mass of the weight is $m = W/g = \frac{24}{32} = \frac{3}{4}$. Using the notation introduced in Example 3, we have

$$2p = \frac{c}{m} = \frac{9}{\frac{3}{4}} = 12 \quad \text{and} \quad \omega^2 = \frac{k}{m} = \frac{24}{\frac{3}{4}} = 32.$$

Equation (19.16) is $$\frac{d^2y}{dt^2} + 12\frac{dy}{dt} + 32y = 0.$$

We can verify that the roots of the auxiliary equation are -8 and -4. Since the roots are real and unequal, the motion is overdamped, as in Case 1 of Example 3, and the general solution of the differential equation is

$$y = C_1e^{-4t} + C_2e^{-8t}.$$

Letting $t = 0$ gives us

$$0 = C_1 + C_2, \quad \text{or} \quad C_2 = -C_1.$$

Since

$$\frac{dy}{dt} = -4C_1e^{-4t} - 8C_2e^{-8t},$$

we have, at $t = 0$,

$$2 = -4C_1 - 8C_2 = -4C_1 - 8(-C_1) = 4C_1.$$

Consequently $C_1 = \frac{1}{2}$ and $C_2 = -C_1 = -\frac{1}{2}$. Thus, the displacement y of the weight at time t is

$$y = \tfrac{1}{2}e^{-4t} - \tfrac{1}{2}e^{-8t} = \tfrac{1}{2}e^{-4t}(1 - e^{-4t}).$$

FIGURE 19.14

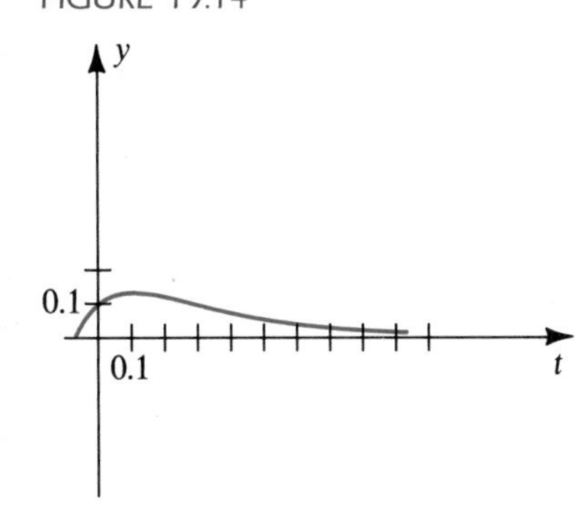

The graph is sketched in Figure 19.14. In Exercise 9 you are asked to find the damping force that will produce critical damping.

EXERCISES 19.5

1 A 5-pound weight stretches a spring 6 inches beyond its natural length. If the weight is raised 4 inches and then released with zero velocity, find a formula for the displacement of the weight at any time t.

2 A 10-pound weight stretches a spring 8 inches beyond its natural length. If the weight is pulled down another 3 inches and then released with a downward velocity of 6 in./sec, find a formula for the displacement at any time t.

3 A 10-pound weight stretches a spring 1 foot beyond its natural length. If the weight is released from its equilibrium position with a downward velocity of 2 ft/sec and if the frictional force is $-5(dy/dt)$, find a formula for the displacement of the weight.

4 A 6-pound weight is suspended from a spring whose spring constant is 48 lb/ft. Initially the weight has a downward velocity of 4 ft/sec from a position 5 inches below the equilibrium point. Find a formula for the displacement.

5 A 4-pound weight stretches a spring 3 inches beyond its natural length. If the weight is pulled down another 4 inches and released with zero velocity in a medium where the damping force is $-2(dy/dt)$, find the displacement y at any time t.

6 A spring stretches $\frac{1}{2}$ foot when an 8-pound weight is attached. If the weight is released from the equilibrium position with an upward velocity of 1 ft/sec and if the damping force is $-4(dy/dt)$, find a formula for the displacement y at any time t.

7 Describe a spring-weight system that leads to the following differential equation and initial values:

$$\frac{1}{4}\frac{d^2y}{dt^2} + \frac{dy}{dt} + 6y = 0$$

$$\text{at } t = 0, \quad y = -2 \quad \text{and} \quad \frac{dy}{dt} = -1$$

8 A 4-pound weight stretches a spring 1 foot. If the weight moves in a medium whose damping force is $-c(dy/dt)$, where $c > 0$, determine the values of c for which the motion is

(a) overdamped (b) critically damped

(c) underdamped

9 In Example 4 of this section, find the damping force that will produce critical damping.

10 Suppose an undamped weight of mass m is attached to a spring having spring constant k, and an external periodic force given by $F \sin \alpha t$ is applied to the weight.

(a) Show that the differential equation that describes the motion of the weight is

$$\frac{d^2y}{dt^2} + \omega^2 y = \frac{F}{m} \sin \alpha t,$$

where $\omega^2 = k/m$.

(b) Prove that the displacement y of the weight is given by

$$y = C_1 \cos \omega t + C_2 \sin \omega t + C \sin \alpha t,$$

where $C = F/[m(\omega^2 - \alpha^2)]$. This type of motion is called *forced vibration*.

19.6 SERIES SOLUTIONS

As shown in Chapter 11, a power series $\sum a_n x^n$ determines a function f such that

$$y = f(x) = a_0 + a_1 x + a_2 x^2 + a_3 x^3 + a_4 x^4 + \cdots$$

for every x in the interval of convergence. Moreover, series representations for the derivatives of f may be obtained by differentiating each term. Thus,

$$y' = a_1 + 2a_2 x + 3a_3 x^2 + 4a_4 x^3 + \cdots = \sum_{n=1}^{\infty} n a_n x^{n-1},$$

$$y'' = 2a_2 + 3 \cdot 2a_3 x + 4 \cdot 3a_4 x^2 + \cdots = \sum_{n=2}^{\infty} n(n-1) a_n x^{n-2},$$

and so on. Power series may be used to solve certain differential equations. In this case the solution is often expressed as an infinite series and is called a **series solution** of the differential equation.

EXAMPLE 1 Find a series solution of the differential equation $y' = 2xy$.

SOLUTION If the solution is given by $y = \sum a_n x^n$, then $y' = \sum n a_n x^{n-1}$, and substitution in the differential equation gives us

$$\sum_{n=1}^{\infty} n a_n x^{n-1} = 2x \sum_{n=0}^{\infty} a_n x^n = \sum_{n=0}^{\infty} 2a_n x^{n+1}.$$

It is convenient to change the summation on the left so that the same power of x appears as in the summation on the right. This may be accomplished by replacing n by $n + 2$ and beginning the summation at $n = -1$. Thus,

$$\sum_{n=-1}^{\infty} (n+2) a_{n+2} x^{n+1} = \sum_{n=0}^{\infty} 2a_n x^{n+1},$$

or

$$a_1 + 2a_2 x + \cdots + (n+2) a_{n+2} x^{n+1} + \cdots = 2a_0 x + \cdots + 2a_n x^{n+1} + \cdots.$$

Comparing coefficients, we see that $a_1 = 0$ and $(n+2)a_{n+2} = 2a_n$ if $n \geq 0$. Consequently the coefficients are given by

$$a_1 = 0 \quad \text{and} \quad a_{n+2} = \frac{2}{n+2} a_n \quad \text{if} \quad n \geq 0.$$

In particular,

$$a_1 = 0, \quad a_2 = a_0, \quad a_3 = \tfrac{2}{3} a_1 = 0, \quad a_4 = \tfrac{1}{2} a_2 = \tfrac{1}{2} a_0, \quad a_5 = \tfrac{2}{5} a_3 = 0,$$

$$a_6 = \tfrac{1}{3} a_4 = \frac{1}{2 \cdot 3} a_0, \quad a_7 = \tfrac{2}{7} a_5 = 0, \quad a_8 = \tfrac{1}{4} a_6 = \frac{1}{2 \cdot 3 \cdot 4} a_0,$$

etc. It can be shown that if n is odd, then $a_n = 0$, whereas $a_{2n} = (1/n!)a_0$ for every positive integer n. The series solution is therefore

$$y = \sum_{n=0}^{\infty} a_n x^n = a_0\left(1 + x^2 + \frac{1}{2!}x^4 + \cdots + \frac{1}{n!}x^{2n} + \cdots\right).$$

It follows from (19.8)(i) that the series solution in Example 1 can be written as $y = a_0 e^{(x^2)}$. This form may also be found directly from $y' = 2xy$ by using the separation of variables technique discussed in Section 19.1. The objective in Example 1, however, was to illustrate series solutions, not to find the solution in the simplest manner. In many instances it is impossible to find the sum of $\sum a_n x^n$ and the solution must be left in series form.

EXAMPLE 2 Solve the differential equation $y'' - xy' - 2y = 0$.

SOLUTION Substituting

$$y = \sum_{n=0}^{\infty} a_n x^n, \quad y' = \sum_{n=1}^{\infty} n a_n x^{n-1}, \quad \text{and} \quad y'' = \sum_{n=2}^{\infty} n(n-1)a_n x^{n-2},$$

we obtain

$$\sum_{n=2}^{\infty} n(n-1)a_n x^{n-2} - x\sum_{n=1}^{\infty} n a_n x^{n-1} - 2\sum_{n=0}^{\infty} a_n x^n = 0,$$

or

$$\sum_{n=2}^{\infty} n(n-1)a_n x^{n-2} = \sum_{n=0}^{\infty} n a_n x^n + \sum_{n=0}^{\infty} 2a_n x^n.$$

We next adjust the summation on the left so that the power x^n appears instead of x^{n-2}. This can be accomplished by replacing n by $n + 2$ and starting the summation at $n = 0$. This gives us

$$\sum_{n=0}^{\infty} (n+2)(n+1)a_{n+2}x^n = \sum_{n=0}^{\infty} (n+2)a_n x^n.$$

Comparing coefficients, we see that $(n+2)(n+1)a_{n+2} = (n+2)a_n$; that is,

$$a_{n+2} = \frac{1}{n+1}a_n.$$

Letting $n = 0, 1, 2, \ldots, 7$ leads to the following form for the coefficients a_k:

$$a_2 = a_0 \qquad a_3 = \frac{1}{2}a_1$$

$$a_4 = \frac{1}{3}a_2 = \frac{1}{3}a_0 \qquad a_5 = \frac{1}{4}a_3 = \frac{1}{2 \cdot 4}a_1$$

$$a_6 = \frac{1}{5}a_4 = \frac{1}{3 \cdot 5}a_0 \qquad a_7 = \frac{1}{6}a_5 = \frac{1}{2 \cdot 4 \cdot 6}a_1$$

$$a_8 = \frac{1}{7}a_6 = \frac{1}{3 \cdot 5 \cdot 7}a_0 \qquad a_9 = \frac{1}{8}a_7 = \frac{1}{2 \cdot 4 \cdot 6 \cdot 8}a_1$$

In general,

$$a_{2n} = \frac{1}{1 \cdot 3 \cdots (2n-1)} a_0, \qquad a_{2n+1} = \frac{1}{2 \cdot 4 \cdots (2n)} a_1 = \frac{1}{2^n n!} a_1.$$

The solution $y = \sum a_n x^n$ may, therefore, be expressed as a sum of two infinite series:

$$y = a_0 \left[1 + \sum_{n=1}^{\infty} \frac{1}{1 \cdot 3 \cdots (2n-1)} x^{2n} \right] + a_1 \sum_{n=0}^{\infty} \frac{1}{2^n n!} x^{2n+1}$$

EXERCISES 19.6

Exer. 1–12: Find a series solution for the differential equation.

1 $y'' + y = 0$

2 $y'' - 4y = 0$

3 $y'' - 2xy = 0$

4 $y'' + 2xy' + y = 0$

5 $\frac{d^2y}{dx^2} - x\frac{dy}{dx} + 2y = 0$

6 $\frac{d^2y}{dx^2} + x^2 y = 0$

7 $(x + 1)y' = 3y$

8 $y' = 4x^3 y$

9 $y'' - y = 5x$

10 $y'' - xy = x^4$

11 $(x^2 - 1)y'' + 6xy' + 4y = -4$

12 $y'' + y = e^x$

19.7 REVIEW EXERCISES

Exer. 1–34: Solve the differential equation.

1 $xe^y\,dx - \csc x\,dy = 0$

2 $x\,dy - (x + 1)y\,dx = 0$

3 $x(1 + y^2)\,dx + \sqrt{1 - x^2}\,dy = 0$

4 $y' + 4y = e^{-x}$

5 $y' + (\sec x)y = 2\cos x$

6 $(x^2 y + x^2)\,dy + y\,dx = 0$

7 $y \tan x + y' = 2\sec x$

8 $y' + (\tan x)y = 3$

9 $y\sqrt{1 - x^2}\,y' = \sqrt{1 - y^2}$

10 $(2y + x^3)\,dx - x\,dy = 0$

11 $e^x \cos y\,dy - \sin^2 y\,dx = 0$

12 $(\cos x)y' + (\sin x)y = 3\cos^2 x$

13 $y' + (2\cos x)y = \cos x$

14 $y'' + y' - 6y = 0$

15 $y'' - 8y' + 16y = 0$

16 $y'' - 6y' + 25y = 0$

17 $\frac{d^2y}{dx^2} - 2\frac{dy}{dx} = 0$

18 $\frac{d^2y}{dx^2} = y + \sin x$

19 $y'' - y = e^x \sin x$

20 $y'' - y' - 6y = e^{2x}$

21 $y' + y = e^{4x}$

22 $y'' + 2y' = 0$

23 $y'' - 3y' + 2y = e^{5x}$

24 $xe^y\,dx - (x + 1)y\,dy = 0$

25 $xy' + y = (x - 2)^2$

26 $\sec^2 y\,dx = \sqrt{1 - x^2}\,dy - x\sec^2 y\,dx$

27 $\frac{d^2y}{dx^2} + 5\frac{dy}{dx} + 7y = 0$

28 $\frac{d^2y}{dx^2} + y = \csc x$

29 $e^{x+y}\,dx - \csc x\,dy = 0$

30 $y'' + 10y' + 25y = 0$

31 $\cot x\,dy = (y - \cos x)\,dx$

32 $y'' + y' + y = e^x \cos x$

33 $y' + y\csc x = \tan x$

34 $y'' - y' - 20y = xe^{-x}$

35 The bird population of an island experiences seasonal growth described by $dy/dt = (3\sin 2\pi t)y$, where t is the time in years and $t = 0$ corresponds to the beginning of spring. Migration to and from the island is also seasonal. The rate of migration is given by $M(t) = 2000\sin 2\pi t$ birds per year. Hence the complete differential equation for the population is

$$\frac{dy}{dt} = (3\sin 2\pi t)y + 2000\sin 2\pi t.$$

Solve for y if the population is 500 at $t = 0$. Determine the maximum population.

36 As a jar containing 10 moles of gas A is heated, the velocity of the gas molecules increases and a second gas B is formed. When two molecules of gas A collide, two molecules of gas B are formed. The rate dy/dt at which gas B is formed is proportional to $(10 - y)^2$, the number of pairs of molecules of gas A. Find a formula for y if $y = 2$ moles after 30 seconds.

37 In chemistry, the notation $A + B \rightarrow Y$ is used to denote the production of a substance Y due to the interaction of two substances A and B. Let a and b be the initial amounts of A and B, respectively. If, at time t, the concentration of Y is $y = f(t)$, then the concentrations of A and B are $a - f(t)$ and $b - f(t)$, respectively. If the rate at which the production of Y takes place is given by $dy/dt = k(a - y)(b - y)$ for some positive constant k and if $f(0) = 0$, find $f(t)$.

38 Let $y = f(t)$ be the population, at time t, of some collection, such as insects or bacteria. If the rate of growth dy/dt is proportional to y—that is, if $dy/dt = cy$ for some positive constant c—then $f(t) = f(0)e^{ct}$ (see Theorem (7.33)). In most cases the rate of growth is dependent on available resources such as food supplies, and as t becomes large, $f'(t)$ begins to decrease. An equation that is often used to describe this type of variation in population is the *law of logistic growth* $dy/dt = y(c - by)$ for some positive constants c and b.

(a) If $f(0) = k$, find $f(t)$.

(b) Find $\lim_{t \to \infty} f(t)$.

(c) Show that $f'(t)$ is increasing if $f(t) < c/(2b)$ and is decreasing if $f(t) > c/(2b)$.

(d) Sketch a typical graph of f. (A graph of this type is a *logistic curve*.)

39 Use a differential equation to describe all functions such that the tangent line at any point $P(x, y)$ on the graph is perpendicular to the line segment joining P to the origin. What is the graph of a typical solution of the differential equation?

40 The following differential equation occurs in the study of electrostatic potentials in spherical regions:

$$\frac{d}{d\theta}\left(\sin\theta \frac{dV}{d\theta}\right) = 0$$

Find a solution $V(\theta)$ that satisfies the conditions $V(\pi/2) = 0$ and $V(\pi/4) = V_0$.

APPENDICES

I MATHEMATICAL INDUCTION

The method of proof known as **mathematical induction** may be used to show that certain statements or formulas are true for all positive integers. For example, if n is a positive integer, let P_n denote the statement

$$(xy)^n = x^n y^n,$$

where x and y are real numbers. Thus, P_1 represents the statement $(xy)^1 = x^1y^1$, P_2 denotes $(xy)^2 = x^2y^2$, P_3 is $(xy)^3 = x^3y^3$, and so on. It is easy to show that P_1, P_2, and P_3 are *true* statements. However, since the set of positive integers is infinite, it is impossible to check the validity of P_n for *every* positive integer n. To show that P_n is true requires the following principle.

Principle of mathematical induction

> If with each positive integer n there is associated a statement P_n, then all the statements P_n are true, provided the following two conditions are satisfied:
>
> **(i)** P_1 is true.
>
> **(ii)** Whenever k is a positive integer such that P_k is true, then P_{k+1} is also true.

To better understand this principle, consider a collection of statements

$$P_1, P_2, P_3, \ldots, P_n, \ldots$$

that satisfy conditions (i) and (ii). By (i), statement P_1 is true. By (ii), whenever a statement P_k is true, the *next* statement P_{k+1} is also true. Since P_1 is true, P_2 is also true, by (ii). However, if P_2 is true, then, by (ii), we see that the next statement P_3 is true. Once again, if P_3 is true, then, by (ii), P_4 is also true. If we continue in this manner, we can argue that if n is any *particular* integer, then P_n is true, since we can use condition (ii) one step at a time, eventually reaching P_n. Although this type of reasoning does not actually *prove* the principle of mathematical induction, it certainly makes it plausible. The principle is proved in advanced algebra using postulates for the positive integers.

When applying the principle of mathematical induction, we always follow these two steps:

Step 1 Show that P_1 is true.

Step 2 *Assume* that P_k is true and then prove that P_{k+1} is true.

Step 2 often causes confusion. Note that we do not *prove* that P_k is true (except for $k = 1$). Instead, we show that *if* P_k happens to be true, then the statement P_{k+1} is also true. We refer to the assumption that P_k is true as the **induction hypothesis**.

EXAMPLE 1 Prove that for every positive integer n, the sum of the first n positive integers is

$$\frac{n(n+1)}{2}.$$

SOLUTION If n is any positive integer, let P_n denote the statement

$$1 + 2 + 3 + \cdots + n = \frac{n(n+1)}{2}.$$

The following are some special cases of P_n:

If $n = 1$, then P_1 is

$$1 = \frac{1(1+1)}{2}; \quad \text{that is,} \quad 1 = 1.$$

If $n = 2$, then P_2 is

$$1 + 2 = \frac{2(2+1)}{2}; \quad \text{that is,} \quad 3 = 3.$$

If $n = 3$, then P_3 is

$$1 + 2 + 3 = \frac{3(3+1)}{2}; \quad \text{that is,} \quad 6 = 6.$$

If $n = 4$, then P_4 is

$$1 + 2 + 3 + 4 = \frac{4(4+1)}{2}; \quad \text{that is,} \quad 10 = 10.$$

Although it is instructive to check the validity of P_n for several values of n as we have done, it is unnecessary. We need only apply the two-step process outlined prior to this example. Thus, we proceed as follows:

Step 1 If we substitute $n = 1$ in P_n, then the left-hand side contains only the number 1 and the right-hand side is $\frac{1(1+1)}{2}$, which also equals 1. Hence P_1 is true.

Step 2 Assume that P_k is true. Thus, the induction hypothesis is

$$1 + 2 + 3 + \cdots + k = \frac{k(k+1)}{2}.$$

Our goal is to show that P_{k+1} is true—that is,

$$1 + 2 + 3 + \cdots + (k + 1) = \frac{(k + 1)[(k + 1) + 1]}{2}.$$

By the induction hypothesis, we already have a formula for the sum of the first k positive integers. Hence, to find a formula for the sum of the first $k + 1$ positive integers, we may simply add $(k + 1)$ to both sides of the induction hypothesis P_k. Doing so, we obtain

$$\begin{aligned} 1 + 2 + 3 + \cdots + k + (k + 1) &= \frac{k(k + 1)}{2} + (k + 1) \\ &= \frac{k(k + 1) + 2(k + 1)}{2}. \end{aligned}$$

We factor the numerator by grouping terms:

$$\begin{aligned} 1 + 2 + 3 + \cdots + k + (k + 1) &= \frac{(k + 1)(k + 2)}{2} \\ &= \frac{(k + 1)[(k + 1) + 1]}{2} \end{aligned}$$

We have shown that P_{k+1} is true, and therefore the proof by mathematical induction is complete.

Let j be a positive integer, and suppose that with each integer $n \geq j$ there is associated a statement P_n. For example, if $j = 6$, then the statements are numbered $P_6, P_7, P_8, \ldots$. The principle of mathematical induction may be extended to cover this situation. To prove that the statements P_n are true for $n \geq j$, we use the following two steps, in the same manner as we did for $n \geq 1$.

Extended principle of mathematical induction for P_k, $k \geq j$

(I) Show that P_j is true.

(II) Assume that P_k is true with $k \geq j$ and prove that P_{k+1} is true.

EXAMPLE 2 Let a be a nonzero real number such that $a > -1$. Prove that

$$(1 + a)^n > 1 + na$$

for every integer $n \geq 2$.

SOLUTION For each positive integer n, let P_n denote the inequality $(1 + a)^n > 1 + na$. Note that P_1 is *false*, since $(1 + a)^1 = 1 + (1)(a)$. However, we can show that P_n is true for $n \geq 2$ by using the extended principle with $j = 2$.

Step 1 We first note that $(1 + a)^2 = 1 + 2a + a^2$. Since $a \neq 0$, we have $a^2 > 0$ and therefore $1 + 2a + a^2 > 1 + 2a$, or $(1 + a)^2 > 1 + 2a$. Hence P_2 is true.

Step 2 Assume that P_k is true. Thus, the induction hypothesis is

$$(1 + a)^k > 1 + ka.$$

We wish to show that P_{k+1} is true—that is,

$$(1 + a)^{k+1} > 1 + (k + 1)a.$$

Since $a > -1$, we see that $1 + a > 0$. Hence multiplying both sides of the induction hypothesis by $1 + a$ does not change the inequality sign. This multiplication leads to the following equivalent inequalities:

$$(1 + a)^k(1 + a) > (1 + ka)(1 + a)$$
$$(1 + a)^{k+1} > 1 + ka + a + ka^2$$
$$(1 + a)^{k+1} > 1 + (k + 1)a + ka^2$$

Since $ka^2 > 0$,

$$1 + (k + 1)a + ka^2 > 1 + (k + 1)a$$

and, therefore, $$(1 + a)^{k+1} > 1 + (k + 1)a.$$

Thus, P_{k+1} is true and the proof is complete.

EXERCISES

Exer. 1–22: Prove that the statement is true for every positive integer n.

1 $2 + 4 + 6 + \cdots + 2n = n(n + 1)$

2 $1 + 4 + 7 + \cdots + (3n - 2) = \dfrac{n(3n - 1)}{2}$

3 $1 + 3 + 5 + \cdots + (2n - 1) = n^2$

4 $3 + 9 + 15 + \cdots + (6n - 3) = 3n^2$

5 $2 + 7 + 12 + \cdots + (5n - 3) = \frac{1}{2}n(5n - 1)$

6 $2 + 6 + 18 + \cdots + 2 \cdot 3^{n-1} = 3^n - 1$

7 $1 + 2 \cdot 2 + 3 \cdot 2^2 + \cdots + n \cdot 2^{n-1} = 1 + (n - 1) \cdot 2^n$

8 $(-1)^1 + (-1)^2 + (-1)^3 + \cdots + (-1)^n = \dfrac{(-1)^n - 1}{2}$

9 $1^2 + 2^2 + 3^2 + \cdots + n^2 = \dfrac{n(n + 1)(2n + 1)}{6}$

10 $1^3 + 2^3 + 3^3 + \cdots + n^3 = \left[\dfrac{n(n + 1)}{2}\right]^2$

11 $\dfrac{1}{1 \cdot 2} + \dfrac{1}{2 \cdot 3} + \dfrac{1}{3 \cdot 4} + \cdots + \dfrac{1}{n(n + 1)} = \dfrac{n}{n + 1}$

12 $\dfrac{1}{1 \cdot 2 \cdot 3} + \dfrac{1}{2 \cdot 3 \cdot 4} + \dfrac{1}{3 \cdot 4 \cdot 5} + \cdots + \dfrac{1}{n(n + 1)(n + 2)} = \dfrac{n(n + 3)}{4(n + 1)(n + 2)}$

13 $3 + 3^2 + 3^3 + \cdots + 3^n = \frac{3}{2}(3^n - 1)$

14 $1^3 + 3^3 + 5^3 + \cdots + (2n - 1)^3 = n^2(2n^2 - 1)$

15 $n < 2^n$

16 $1 + 2n \le 3^n$

17 $1 + 2 + 3 + \cdots + n < \frac{1}{8}(2n + 1)^2$

18 If $0 < a < b$, then $\left(\dfrac{a}{b}\right)^{n+1} < \left(\dfrac{a}{b}\right)^n$

19 3 is a factor of $n^3 - n + 3$.

20 2 is a factor of $n^2 + n$.

21 4 is a factor of $5^n - 1$.

22 9 is a factor of $10^{n+1} + 3 \cdot 10^n + 5$.

Exer. 23–30: Find the smallest positive integer j for which the statement is true. Use the extended principle of mathematical induction to prove that the formula is true for every integer greater than j.

23 $n + 12 \le n^2$

24 $n^2 + 18 \le n^3$

25 $5 + \log_2 n \le n$

26 $n^2 \le 2^n$

27 $2^n \le n!$

28 $10^n \le n^n$

29 $2n + 2 \le 2^n$

30 $n \log_2 n + 20 \le n^2$

31 Prove that if a is any real number greater than 1, then $a^n > 1$ for every positive integer n.

32 Prove that

$$a + ar + ar^2 + \cdots + ar^{n-1} = \frac{a(1 - r^n)}{1 - r}$$

for every positive integer n and real numbers a and r with $r \neq 1$.

33 Prove that $a - b$ is a factor of $a^n - b^n$ for every positive integer n.
(*Hint:* $a^{k+1} - b^{k+1} = a^k(a - b) + (a^k - b^k)b$.)

34 Prove that $a + b$ is a factor of $a^{2n-1} + b^{2n-1}$ for every positive integer n.

II THEOREMS ON LIMITS, DERIVATIVES, AND INTEGRALS

This appendix contains proofs for some theorems stated in the text. The numbering system corresponds to that given in previous chapters.

Uniqueness theorem for limits

If $f(x)$ has a limit as x approaches a, then the limit is unique.

PROOF Suppose $\lim_{x \to a} f(x) = L_1$ and $\lim_{x \to a} f(x) = L_2$ with $L_1 \neq L_2$. We may assume that $L_1 < L_2$. Choose ϵ such that $\epsilon < \frac{1}{2}(L_2 - L_1)$ and consider the open intervals $(L_1 - \epsilon, L_1 + \epsilon)$ and $(L_2 - \epsilon, L_2 + \epsilon)$ on the coordinate line l' (see Figure 1). Since $\epsilon < \frac{1}{2}(L_2 - L_1)$, these two intervals do not intersect. By Definition (2.5), there is a $\delta_1 > 0$ such that whenever x is in $(a - \delta_1, a + \delta_1)$ and $x \neq a$, then $f(x)$ is in $(L_1 - \epsilon, L_1 + \epsilon)$. Similarly, there is a $\delta_2 > 0$ such that whenever x is in $(a - \delta_2, a + \delta_2)$ and $x \neq a$, then $f(x)$ is in $(L_2 - \epsilon, L_2 + \epsilon)$. This is illustrated in Figure 1, with $\delta_1 < \delta_2$. If an x is selected that is in *both* $(a - \delta_1, a + \delta_1)$ and $(a - \delta_2, a + \delta_2)$, then $f(x)$ is in $(L_1 - \epsilon, L_1 + \epsilon)$ and also in $(L_2 - \epsilon, L_2 + \epsilon)$, contrary to the fact that these two intervals do not intersect. Hence our original supposition is false, and consequently $L_1 = L_2$.

FIGURE 1

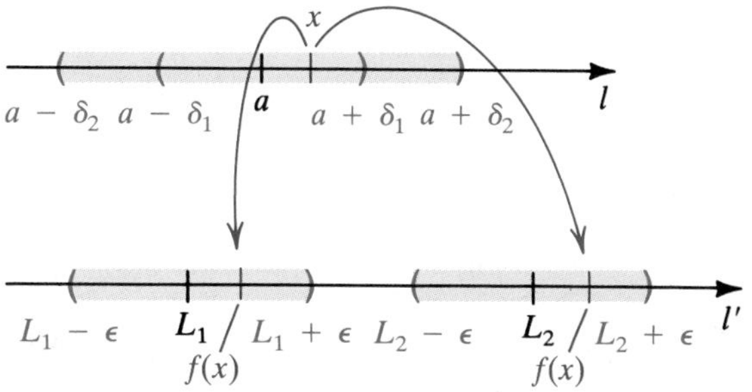

■

Theorem (2.8)

If $\lim_{x \to a} f(x)$ and $\lim_{x \to a} g(x)$ both exist, then

(i) $\lim_{x \to a} [f(x) + g(x)] = \lim_{x \to a} f(x) + \lim_{x \to a} g(x)$

(ii) $\lim_{x \to a} [f(x) \cdot g(x)] = \lim_{x \to a} f(x) \cdot \lim_{x \to a} g(x)$

(iii) $\lim_{x \to a} \left[\frac{f(x)}{g(x)}\right] = \frac{\lim_{x \to a} f(x)}{\lim_{x \to a} g(x)}$, provided $\lim_{x \to a} g(x) \neq 0$

PROOF Suppose that $\lim_{x \to a} f(x) = L$ and $\lim_{x \to a} g(x) = M$.

(I) According to Definition (2.4), we must show that for every $\epsilon > 0$ there is a $\delta > 0$ such that

$$\text{(1)} \qquad \text{if} \quad 0 < |x - a| < \delta, \quad \text{then} \quad |f(x) + g(x) - (L + M)| < \epsilon.$$

We begin by writing

$$\text{(2)} \qquad |f(x) + g(x) - (L + M)| = |(f(x) - L) + (g(x) - M)|.$$

Employing the *triangle inequality*

$$|b + c| \le |b| + |c|$$

for any real numbers b and c, we obtain

$$|(f(x) - L) + (g(x) - M)| \le |f(x) - L| + |g(x) - M|.$$

Combining the last inequality with (2) gives us

$$\text{(3)} \qquad |f(x) + g(x) - (L + M)| \le |f(x) - L| + |g(x) - M|.$$

Since $\lim_{x \to a} f(x) = L$ and $\lim_{x \to a} g(x) = M$, the numbers $|f(x) - L|$ and $|g(x) - M|$ can be made arbitrarily small by choosing x sufficiently close to a. In particular, they can be made less than $\epsilon/2$. Thus, there exist $\delta_1 > 0$ and $\delta_2 > 0$ such that

$$\text{(4)} \qquad \begin{aligned} &\text{if} \quad 0 < |x - a| < \delta_1, \quad \text{then} \quad |f(x) - L| < \epsilon/2, \quad \text{and} \\ &\text{if} \quad 0 < |x - a| < \delta_2, \quad \text{then} \quad |g(x) - M| < \epsilon/2. \end{aligned}$$

If δ denotes the *smaller* of δ_1 and δ_2, then whenever $0 < |x - a| < \delta$, the inequalities in (4) involving $f(x)$ and $g(x)$ are both true. Consequently, if $0 < |x - a| < \delta$, then, from (4) and (3),

$$|f(x) + g(x) - (L + M)| < \epsilon/2 + \epsilon/2 = \epsilon,$$

which is the desired statement (1).

(II) We first show that if k is a function and

$$\text{(5)} \qquad \text{if} \quad \lim_{x \to a} k(x) = 0, \quad \text{then} \quad \lim_{x \to a} f(x)k(x) = 0.$$

Since $\lim_{x \to a} f(x) = L$, it follows from Definition (2.4) (with $\epsilon = 1$) that there is a $\delta_1 > 0$ such that if $0 < |x - a| < \delta_1$, then $|f(x) - L| < 1$ and hence also

$$|f(x)| = |f(x) - L + L| \le |f(x) - L| + |L| < 1 + |L|.$$

Consequently,

$$\text{(6)} \qquad \text{if} \quad 0 < |x - a| < \delta_1, \quad \text{then} \quad |f(x)k(x)| < (1 + |L|)|k(x)|.$$

Since $\lim_{x \to a} k(x) = 0$, for every $\epsilon > 0$ there is a $\delta_2 > 0$ such that

$$\text{(7)} \qquad \text{if} \quad 0 < |x - a| < \delta_2, \quad \text{then} \quad |k(x) - 0| < \frac{\epsilon}{1 + |L|}.$$

If δ denotes the smaller of δ_1 and δ_2, then whenever $0 < |x - a| < \delta$, both inequalities (6) and (7) are true and consequently

$$|f(x)k(x)| < (1 + |L|) \cdot \frac{\epsilon}{1 + |L|}.$$

Therefore,

$$\text{if} \quad 0 < |x - a| < \delta, \quad \text{then} \quad |f(x)k(x) - 0| < \epsilon,$$

which proves (5).

Next consider the identity

$$f(x)g(x) - LM = f(x)[g(x) - M] + M[f(x) - L]. \tag{8}$$

Since $\lim_{x\to a} [g(x) - M] = 0$, it follows from (5), with $k(x) = g(x) - M$, that $\lim_{x\to a} f(x)[g(x) - M] = 0$. In addition, $\lim_{x\to a} M[f(x) - L] = 0$ and hence, from (8), $\lim_{x\to a} [f(x)g(x) - LM] = 0$. The last statement is equivalent to $\lim_{x\to a} f(x)g(x) = LM$.

(iii) It is sufficient to show that $\lim_{x\to a} 1/g(x) = 1/M$, for once this is done, the desired result may be obtained by applying (ii) to the product $f(x) \cdot 1/g(x)$. Consider

$$\left|\frac{1}{g(x)} - \frac{1}{M}\right| = \left|\frac{M - g(x)}{g(x)M}\right| = \frac{1}{|M|\,|g(x)|}\,|g(x) - M|. \tag{9}$$

Since $\lim_{x\to a} g(x) = M$, there exists a $\delta_1 > 0$ such that if $0 < |x - a| < \delta_1$, then $|g(x) - M| < |M|/2$. Consequently, for every such x,

$$\begin{aligned} |M| &= |g(x) + (M - g(x))| \\ &\le |g(x)| + |M - g(x)| \\ &< |g(x)| + |M|/2 \end{aligned}$$

and therefore

$$\frac{|M|}{2} < |g(x)|, \quad \text{or} \quad \frac{1}{|g(x)|} < \frac{2}{|M|}.$$

Substitution in equation (9) leads to

$$\left|\frac{1}{g(x)} - \frac{1}{M}\right| < \frac{2}{|M|^2}\,|g(x) - M|, \quad \text{provided} \quad 0 < |x - a| < \delta_1. \tag{10}$$

Again, since $\lim_{x\to a} g(x) = M$, it follows that for every $\epsilon > 0$ there is a $\delta_2 > 0$ such that

$$\text{if} \quad 0 < |x - a| < \delta_2, \quad \text{then} \quad |g(x) - M| < \frac{|M|^2}{2}\epsilon. \tag{11}$$

If δ denotes the smaller of δ_1 and δ_2, then both inequalities (10) and (11) are true. Thus,

$$\text{if} \quad 0 < |x - a| < \delta, \quad \text{then} \quad \left|\frac{1}{g(x)} - \frac{1}{M}\right| < \epsilon,$$

which means that $\lim_{x\to a} 1/g(x) = 1/M$. ■

Theorem **(2.13)**

If $a > 0$ and n is a positive integer, or if $a \leq 0$ and n is an odd positive integer, then

$$\lim_{x \to a} \sqrt[n]{x} = \sqrt[n]{a}.$$

PROOF Suppose $a > 0$ and n is any positive integer. We must show that for every $\epsilon > 0$ there is a $\delta > 0$ such that

$$\text{if} \quad 0 < |x - a| < \delta, \quad \text{then} \quad |\sqrt[n]{x} - \sqrt[n]{a}| < \epsilon,$$

or, equivalently,

(1) if $\quad -\delta < x - a < \delta \quad$ and $\quad x \neq a, \quad$ then $\quad -\epsilon < \sqrt[n]{x} - \sqrt[n]{a} < \epsilon.$

It is sufficient to prove (1) if $\epsilon < \sqrt[n]{a}$, for if a δ exists under this condition, then the same δ can be used for any *larger* value of ϵ. Thus, in the remainder of the proof $\sqrt[n]{a} - \epsilon$ is considered to be a positive number less than ϵ. The inequalities in the following list are all equivalent:

$$-\epsilon < \sqrt[n]{x} - \sqrt[n]{a} < \epsilon$$
$$\sqrt[n]{a} - \epsilon < \sqrt[n]{x} < \sqrt[n]{a} + \epsilon$$
$$(\sqrt[n]{a} - \epsilon)^n < x < (\sqrt[n]{a} + \epsilon)^n$$
$$(\sqrt[n]{a} - \epsilon)^n - a < x - a < (\sqrt[n]{a} + \epsilon)^n - a$$
$$-[a - (\sqrt[n]{a} - \epsilon)^n] < x - a < (\sqrt[n]{a} + \epsilon)^n - a$$

If δ denotes the smaller of the two positive numbers $a - (\sqrt[n]{a} - \epsilon)^n$ and $(\sqrt[n]{a} + \epsilon)^n - a$, then whenever $-\delta < x - a < \delta$, the last inequality in the list is true and hence so is the first. This gives us (1).

Next suppose $a < 0$ and n is an odd positive integer. In this case $-a$ and $\sqrt[n]{-a}$ are positive and, by the first part of the proof, we may write

$$\lim_{-x \to -a} \sqrt[n]{-x} = \sqrt[n]{-a}.$$

Thus, for every $\epsilon > 0$ there is a $\delta > 0$ such that

$$\text{if} \quad 0 < |-x - (-a)| < \delta, \quad \text{then} \quad |\sqrt[n]{-x} - \sqrt[n]{-a}| < \epsilon,$$

or, equivalently,

$$\text{if} \quad 0 < |x - a| < \delta, \quad \text{then} \quad |\sqrt[n]{x} - \sqrt[n]{a}| < \epsilon.$$

The last inequalities imply that $\lim_{x \to a} \sqrt[n]{x} = \sqrt[n]{a}$. ■

Sandwich theorem **(2.15)**

Suppose $f(x) \leq h(x) \leq g(x)$ for every x in an open interval containing a, except possibly at a.

$$\text{If} \quad \lim_{x \to a} f(x) = L = \lim_{x \to a} g(x), \quad \text{then} \quad \lim_{x \to a} h(x) = L.$$

PROOF For every $\epsilon > 0$ there is a $\delta_1 > 0$ and a $\delta_2 > 0$ such that

(1)
$$\text{if} \quad 0 < |x - a| < \delta_1, \quad \text{then} \quad |f(x) - L| < \epsilon, \quad \text{and}$$
$$\text{if} \quad 0 < |x - a| < \delta_2, \quad \text{then} \quad |g(x) - L| < \epsilon.$$

If δ denotes the smaller of δ_1 and δ_2, then whenever $0 < |x - a| < \delta$, both inequalities in (1) that involve ϵ are true; that is,

$$-\epsilon < f(x) - L < \epsilon \quad \text{and} \quad -\epsilon < g(x) - L < \epsilon.$$

Consequently, if $0 < |x - a| < \delta$, then $L - \epsilon < f(x)$ and $g(x) < L + \epsilon$. Since $f(x) \le h(x) \le g(x)$, if $0 < |x - a| < \delta$, then $L - \epsilon < h(x) < L + \epsilon$, or, equivalently, $|h(x) - L| < \epsilon$, which is what we wished to prove. ∎

Theorem **(2.18)**

> If k is a positive rational number and c is any real number, then
>
> $$\lim_{x \to \infty} \frac{c}{x^k} = 0 \quad \text{and} \quad \lim_{x \to -\infty} \frac{c}{x^k} = 0,$$
>
> provided x^k is always defined.

PROOF To use (2.16) to prove that $\lim_{x \to \infty} (c/x^k) = 0$, we must show that for every $\epsilon > 0$ there is a positive number N such that

$$\left| \frac{c}{x^k} - 0 \right| < \epsilon \quad \text{whenever} \quad x > N.$$

If $c = 0$, any $N > 0$ will suffice. If $c \ne 0$, the following four inequalities are equivalent for $x > 0$:

$$\left| \frac{c}{x^k} - 0 \right| < \epsilon, \qquad \frac{|x|^k}{|c|} > \frac{1}{\epsilon}, \qquad |x|^k > \frac{|c|}{\epsilon}, \qquad x > \left(\frac{|c|}{\epsilon} \right)^{1/k}$$

The last inequality gives us a clue to a choice for N. Letting $N = (|c|/\epsilon)^{1/k}$, we see that whenever $x > N$, the fourth, and hence the first, inequality is true, which is what we wished to show. The second part of the theorem may be proved in similar fashion. ∎

Theorem **(2.24)**

> If $\lim_{x \to c} g(x) = b$ and if f is continuous at b, then
>
> $$\lim_{x \to c} f(g(x)) = f(b) = f\left(\lim_{x \to c} g(x) \right)$$

FIGURE 2

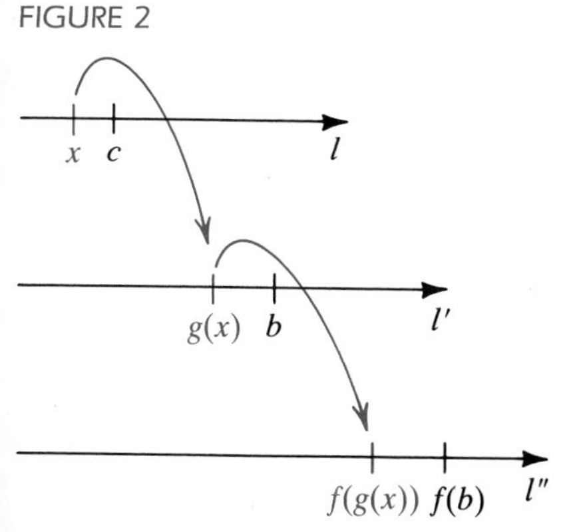

PROOF The composite function $f(g(x))$ may be represented geometrically by means of three real lines l, l', and l'', as shown in Figure 2. To each coordinate x on l there corresponds the coordinate $g(x)$ on l' and then, in turn, $f(g(x))$ on l''. We wish to prove that $f(g(x))$ has the limit $f(b)$ as x approaches c. In terms of Definition (2.4), we must show that for every $\epsilon > 0$ there exists a $\delta > 0$ such that

(1) $\qquad$ if $0 < |x - c| < \delta$, then $|f(g(x)) - f(b)| < \epsilon$.

Let us begin by considering the interval $(f(b) - \epsilon, f(b) + \epsilon)$ on l'', shown in Figure 3. Since f is continuous at b, $\lim_{z \to b} f(z) = f(b)$ and hence, as illustrated in the figure, there exists a number $\delta_1 > 0$ such that

(2) $\qquad$ if $|z - b| < \delta_1$, then $|f(z) - f(b)| < \epsilon$.

FIGURE 3

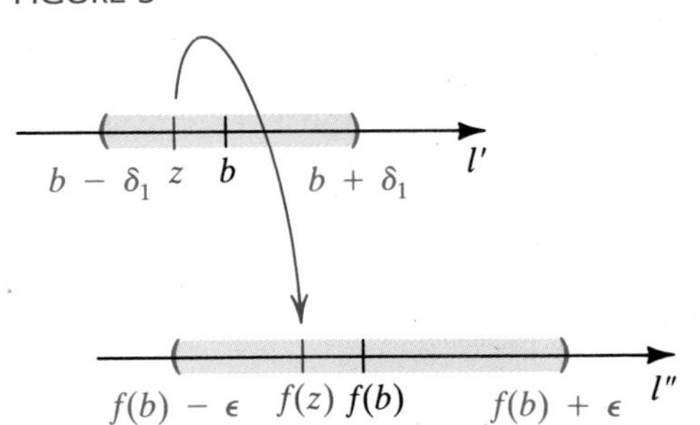

FIGURE 4

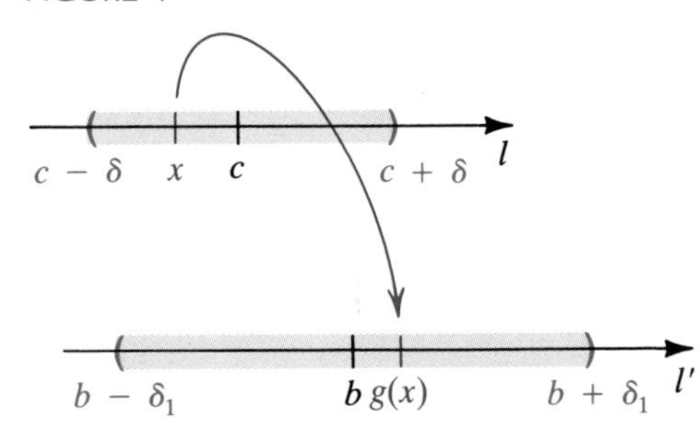

In particular, if we let $z = g(x)$ in (2), it follows that

(3) $$\text{if } \quad |g(x) - b| < \delta_1, \quad \text{then} \quad |f(g(x)) - f(b)| < \epsilon.$$

Next, turning our attention to the interval $(b - \delta_1, b + \delta_1)$ on l' and using the definition of $\lim_{x \to c} g(x) = b$, we obtain the fact illustrated in Figure 4—that there exists a $\delta > 0$ such that

(4) $$\text{if } \quad 0 < |x - c| < \delta, \quad \text{then} \quad |g(x) - b| < \delta_1.$$

Finally, combining (4) and (3), we see that

$$\text{if } \quad 0 < |x - c| < \delta, \quad \text{then} \quad |f(g(x)) - f(b)| < \epsilon,$$

which is the desired conclusion (1). ■

Theorem **(3.14a)**

If n is a positive integer and $f(x) = x^{1/n}$, then

$$f'(x) = \frac{1}{n} x^{(1/n) - 1}.$$

PROOF By Definition (3.5),

$$f'(x) = \lim_{h \to 0} \frac{(x + h)^{1/n} - x^{1/n}}{h}.$$

Consider the identity

$$u^n - v^n = (u - v)(u^{n-1} + u^{n-2}v + \cdots + uv^{n-2} + v^{n-1}).$$

If $u \neq v$, then

$$\frac{u - v}{u^n - v^n} = \frac{1}{u^{n-1} + u^{n-2}v + \cdots + uv^{n-2} + v^{n-1}}.$$

Substituting $u = (x + h)^{1/n}$ and $v = x^{1/n}$, we obtain

$$\frac{(x + h)^{1/n} - x^{1/n}}{(x + h) - x} =$$

$$\frac{1}{(x + h)^{(n-1)/n} + (x + h)^{(n-2)/n}x^{1/n} + \cdots + (x + h)^{1/n}x^{(n-2)/n} + x^{(n-1)/n}}.$$

Letting $h \to 0$, we have

$$f'(x) = \frac{1}{x^{(n-1)/n} + x^{(n-1)/n} + \cdots + x^{(n-1)/n} + x^{(n-1)/n}}$$

$$= \frac{1}{nx^{1-(1/n)}} = \frac{1}{n} x^{(1/n)-1}. \blacksquare$$

Chain rule **(3.33)**

> If $y = f(u)$, $u = g(x)$, and the derivatives $\dfrac{dy}{du}$ and $\dfrac{du}{dx}$ both exist, then the composite function defined by $y = f(g(x))$ has a derivative given by
>
> $$\frac{dy}{dx} = \frac{dy}{du}\frac{du}{dx} = f'(u)g'(x) = f'(g(x))g'(x).$$

PROOF If $y = f(x)$ and $\Delta x \approx 0$, then the difference between the derivative $f'(x)$ and the ratio $\Delta y/\Delta x$ is numerically small. Since this difference depends on the size of Δx, we shall represent it by means of the function notation $\eta(\Delta x)$. Thus, for each $\Delta x \neq 0$,

(1) $$\eta(\Delta x) = \frac{\Delta y}{\Delta x} - f'(x).$$

It should be noted that $\eta(\Delta x)$ does *not* represent the product of η and Δx, but rather states that η *is a function of* Δx, whose values are given by (1). Moreover, applying (3.27), we see that

(2) $$\lim_{\Delta x \to 0} \eta(\Delta x) = \lim_{\Delta x \to 0} \left(\frac{\Delta y}{\Delta x} - f'(x)\right) = 0.$$

The function η has been defined only for nonzero values of Δx. It is convenient to extend the definition of η to include $\Delta x = 0$ by letting $\eta(0) = 0$. It then follows from (2) that η *is continuous at* 0.

Multiplying both sides of (1) by Δx and rearranging terms gives us

(3) $$\Delta y = f'(x)\,\Delta x + \eta(\Delta x) \cdot \Delta x,$$

which is true if either $\Delta x \neq 0$ or $\Delta x = 0$. Since $f'(x)\,\Delta x = dy$, it follows from (3) that

(4) $$\Delta y - dy = \eta(\Delta x) \cdot \Delta x.$$

Let us now consider the situation stated in the hypothesis of the theorem,

$$y = f(u) \quad \text{and} \quad u = g(x).$$

If $g(x)$ is the domain of f, then we may write

$$y = f(u) = f(g(x));$$

that is, y is a function of x. If we give x an increment Δx, there corresponds an increment Δu of u and, in turn, an increment Δy of $y = f(u)$.

Thus,

$$\Delta u = g(x + \Delta x) - g(x)$$

$$\Delta y = f(u + \Delta u) - f(u).$$

Since dy/du exists, we may use (3) with u as the independent variable to write

(5) $$\Delta y = f'(u)\,\Delta u + \eta(\Delta u) \cdot \Delta u$$

for a function η of Δu such that, by (2),

(6) $$\lim_{\Delta u \to 0} \eta(\Delta u) = 0.$$

Moreover, η is continuous at $\Delta u = 0$ and (5) is true if $\Delta u = 0$. Dividing both sides of (5) by Δx gives us

$$\frac{\Delta y}{\Delta x} = f'(u)\frac{\Delta u}{\Delta x} + \eta(\Delta u) \cdot \frac{\Delta u}{\Delta x}.$$

If we now take the limit as Δx approaches zero and use the facts

$$\lim_{\Delta x \to 0} \frac{\Delta y}{\Delta x} = \frac{dy}{dx} \quad \text{and} \quad \lim_{\Delta x \to 0} \frac{\Delta u}{\Delta x} = \frac{du}{dx},$$

we see that

$$\frac{dy}{dx} = f'(u)\frac{du}{dx} + \lim_{\Delta x \to 0} \eta(\Delta u) \cdot \frac{du}{dx}.$$

Since $f'(u) = dy/du$, we may complete the proof by showing that the limit indicated in the last equation is 0. To accomplish this, we first observe that since g is differentiable, it is continuous and hence

$$\lim_{\Delta x \to 0} [g(x + \Delta x) - g(x)] = 0,$$

or, equivalently,
$$\lim_{\Delta x \to 0} \Delta u = 0.$$

In other words, Δu approaches 0 as Δx approaches 0. Using this fact, together with (6), gives us

(7) $$\lim_{\Delta x \to 0} \eta(\Delta u) = \lim_{\Delta u \to 0} \eta(\Delta u) = 0$$

and the theorem is proved. The fact that $\lim_{\Delta x \to 0} \eta(\Delta u) = 0$ can also be established by means of an ϵ-δ argument using (2.4). ■

Theorem **(5.22)**

> If f is integrable on $[a, b]$ and c is any number, then cf is integrable on $[a, b]$ and
>
> $$\int_a^b cf(x)\,dx = c\int_a^b f(x)\,dx.$$

PROOF If $c = 0$, the result follows from Theorem (5.21). Assume, therefore, that $c \neq 0$. Since f is integrable, $\int_a^b f(x)\,dx = I$ for some number I. If P is a partition of $[a, b]$, then each Riemann sum R_P for the function

cf has the form $\sum_k cf(w_k)\,\Delta x_k$ such that for every k, w_k is in the kth subinterval $[x_{k-1}, x_k]$ of P. We wish to show that for every $\epsilon > 0$ there is a $\delta > 0$ such that whenever $\|P\| < \delta$,

(1)
$$\left|\sum_k cf(w_k)\,\Delta x_k - cI\right| < \epsilon$$

for every w_k in $[x_{k-1}, x_k]$. If we let $\epsilon' = \epsilon/|c|$, then, since f is integrable, there exists a $\delta > 0$ such that whenever $\|P\| < \delta$,

$$\left|\sum_k f(w_k)\,\Delta x_k - I\right| < \epsilon' = \frac{\epsilon}{|c|}.$$

Multiplying both sides of this inequality by $|c|$ leads to (1). Hence

$$\lim_{\|P\|\to 0} \sum_k cf(w_k)\,\Delta x_k = cI = c\int_a^b f(x)\,dx. \quad ■$$

Theorem **(5.23)**

> If f and g are integrable on $[a, b]$, then $f + g$ is integrable on $[a, b]$ and
> $$\int_a^b [f(x) + g(x)]\,dx = \int_a^b f(x)\,dx + \int_a^b g(x)\,dx.$$

PROOF By hypothesis, there exist real numbers I_1 and I_2 such that

$$\int_a^b f(x)\,dx = I_1 \quad \text{and} \quad \int_a^b g(x)\,dx = I_2.$$

Let P denote a partition of $[a, b]$ and let R_P denote an arbitrary Riemann sum for $f + g$ associated with P; that is,

(1)
$$R_P = \sum_k [f(w_k) + g(w_k)]\,\Delta x_k$$

such that w_k is in $[x_{k-1}, x_k]$ for every k. We wish to show that for every $\epsilon > 0$ there is a $\delta > 0$ such that whenever $\|P\| < \delta$, $|R_P - (I_1 + I_2)| < \epsilon$. Using Theorem (5.11)(i), we may write (1) in the form

$$R_P = \sum_k f(w_k)\,\Delta x_k + \sum_k g(w_k)\,\Delta x_k.$$

Rearranging terms and using the triangle inequality, we obtain

(2)
$$\begin{aligned} |R_P - (I_1 + I_2)| &= \left|\left(\sum_k f(w_k)\,\Delta x_k - I_1\right) + \left(\sum_k g(w_k)\,\Delta x_k - I_2\right)\right| \\ &\le \left|\sum_k f(w_k)\,\Delta x_k - I_1\right| + \left|\sum_k g(w_k)\,\Delta x_k - I_2\right|. \end{aligned}$$

By the integrability of f and g, if $\epsilon' = \epsilon/2$, then there exist $\delta_1 > 0$ and $\delta_2 > 0$ such that whenever $\|P\| < \delta_1$ and $\|P\| < \delta_2$,

(3)
$$\left|\sum_k f(w_k)\,\Delta x_k - I_1\right| < \epsilon' = \epsilon/2$$

and

$$\left|\sum_k g(w_k)\,\Delta x_k - I_2\right| < \epsilon' = \epsilon/2$$

for every w_k in $[x_{k-1}, x_k]$. If δ denotes the smaller of δ_1 and δ_2, then whenever $\|P\| < \delta$, both inequalities in (3) are true and hence, from (2),

$$|R_P - (I_1 + I_2)| < (\epsilon/2) + (\epsilon/2) = \epsilon,$$

which is what we wished to prove. ■

Theorem **(5.24)**

> If $a < c < b$ and if f is integrable on both $[a, c]$ and $[c, b]$, then f is integrable on $[a, b]$ and
>
> $$\int_a^b f(x)\,dx = \int_a^c f(x)\,dx + \int_c^b f(x)\,dx.$$

PROOF By hypothesis, there exist real numbers I_1 and I_2 such that

$$\int_a^c f(x)\,dx = I_1 \quad \text{and} \quad \int_c^b f(x)\,dx = I_2. \tag{1}$$

Let us denote a partition of $[a, c]$ by P_1, of $[c, b]$ by P_2, and of $[a, b]$ by P. Arbitrary Riemann sums associated with P_1, P_2, and P will be denoted by R_{P_1}, R_{P_2}, and R_P, respectively. We must show that for every $\epsilon > 0$ there is a $\delta > 0$ such that if $\|P\| < \delta$, then $|R_P - (I_1 + I_2)| < \epsilon$.

If we let $\epsilon' = \epsilon/4$, then, by (1), there exist positive numbers δ_1 and δ_2 such that if $\|P_1\| < \delta_1$ and $\|P_2\| < \delta_2$, then

$$|R_{P_1} - I_1| < \epsilon' = \epsilon/4 \quad \text{and} \quad |R_{P_2} - I_2| < \epsilon' = \epsilon/4. \tag{2}$$

If δ denotes the smaller of δ_1 and δ_2, then both inequalities in (2) are true whenever $\|P\| < \delta$. Moreover, since f is integrable on $[a, c]$ and $[c, b]$, it is bounded on both intervals and hence there exists a number M such that $|f(x)| \le M$ for every x in $[a, b]$. We shall now assume that δ has been chosen so that, in addition to the previous requirement, we also have $\delta < \epsilon/(4M)$.

Let P be a partition of $[a, b]$ such that $\|P\| < \delta$. If the numbers that determine P are

$$a = x_0, x_1, x_2, \ldots, x_n = b,$$

then there is a unique half-open interval of the form $(x_{d-1}, x_d]$ that contains c. If $R_P = \sum_{k=1}^{n} f(w_k)\,\Delta x_k$, we may write

$$R_P = \sum_{k=1}^{d-1} f(w_k)\,\Delta x_k + f(w_d)\,\Delta x_d + \sum_{k=d+1}^{n} f(w_k)\,\Delta x_k. \tag{3}$$

Let P_1 denote the partition of $[a, c]$ determined by $\{a, x_1, \ldots, x_{d-1}, c\}$, let P_2 denote the partition of $[c, b]$ determined by $\{c, x_d, \ldots, x_{n-1}, b\}$, and consider the Riemann sums

$$\begin{aligned} R_{P_1} &= \sum_{k=1}^{d-1} f(w_k)\,\Delta x_k + f(c)(c - x_{d-1}) \\ R_{P_2} &= f(c)(x_d - c) + \sum_{k=d+1}^{n} f(w_k)\,\Delta x_k. \end{aligned} \tag{4}$$

Using the triangle inequality and (2), we obtain

$$\begin{aligned} |(R_{P_1} + R_{P_2}) - (I_1 + I_2)| &= |(R_{P_1} - I_1) + (R_{P_2} - I_2)| \\ &\le |R_{P_1} - I_1| + |R_{P_2} - I_2| \\ &< \frac{\epsilon}{4} + \frac{\epsilon}{4} = \frac{\epsilon}{2}. \end{aligned} \tag{5}$$

It follows from (3) and (4) that

$$|R_P - (R_{P_1} + R_{P_2})| = |f(w_d) - f(c)|\,\Delta x_d.$$

Employing the triangle inequality and the choice of δ gives us

$$\begin{aligned} |R_P - (R_{P_1} + R_{P_2})| &\le (|f(w_d)| + |f(c)|)\,\Delta x_d \\ &\le (M + M)[\epsilon/(4M)] = \epsilon/2, \end{aligned} \tag{6}$$

provided $\|P\| < \delta$. If we write

$$\begin{aligned} |R_P - (I_1 + I_2)| &= |R_P - (R_{P_1} + R_{P_2}) + (R_{P_1} + R_{P_2}) - (I_1 + I_2)| \\ &\le |R_P - (R_{P_1} + R_{P_2})| + |(R_{P_1} + R_{P_2}) - (I_1 + I_2)|, \end{aligned}$$

then it follows from (6) and (5) that whenever $\|P\| < \delta$,

$$|R_P - (I_1 + I_2)| < (\epsilon/2) + (\epsilon/2) = \epsilon$$

for every Riemann sum R_P. This completes the proof. ■

Theorem **(5.26)**

> If f is integrable on $[a, b]$ and $f(x) \ge 0$ for every x in $[a, b]$, then
> $$\int_a^b f(x)\,dx \ge 0.$$

PROOF We shall give an indirect proof. Let $\int_a^b f(x)\,dx = I$, and *suppose* that $I < 0$. Consider any partition P of $[a, b]$, and let $R_P = \sum_k f(w_k)\,\Delta x_k$ be an arbitrary Riemann sum associated with P. Since $f(w_k) \ge 0$ for every w_k in $[x_{k-1}, x_k]$, it follows that $R_P \ge 0$. If we let $\epsilon = -(I/2)$, then, according to Definition (5.15), whenever $\|P\|$ is sufficiently small,

$$|R_P - I| < \epsilon = -\frac{I}{2}.$$

It follows that $R_P < I - (I/2) = I/2 < 0$, a contradiction. Therefore the supposition $I < 0$ is false and hence $I \ge 0$. ■

Theorem **(7.6)**

> If f is continuous and increasing on $[a, b]$, then f has an inverse function f^{-1} that is continuous and increasing on $[f(a), f(b)]$.

PROOF If f is increasing, then f is one-to-one and hence f^{-1} exists. To prove that f^{-1} is increasing, we must show that if $w_1 < w_2$ in $[f(a), f(b)]$, then $f^{-1}(w_1) < f^{-1}(w_2)$ in $[a, b]$. Let us give an indirect proof of this fact. *Suppose* $f^{-1}(w_2) \le f^{-1}(w_1)$. Since f is increasing, it follows that $f(f^{-1}(w_2)) \le f(f^{-1}(w_1))$ and hence $w_2 \le w_1$, which is a contradiction. Consequently $f^{-1}(w_1) < f^{-1}(w_2)$.

We next prove that f^{-1} is continuous on $[f(a), f(b)]$. Recall that $y = f(x)$ if and only if $x = f^{-1}(y)$. In particular, if y_0 is in an open interval $(f(a), f(b))$, let x_0 denote the number in the interval (a, b) such that $y_0 = f(x_0)$, or, equivalently, $x_0 = f^{-1}(y_0)$. We wish to show that

$$\lim_{y \to y_0} f^{-1}(y) = f^{-1}(y_0) = x_0. \tag{1}$$

A geometric representation of f and its inverse f^{-1} is shown in Figure 5. The domain $[a, b]$ of f is represented by points on an x-axis and the domain $[f(a), f(b)]$ of f^{-1} by points on a y-axis. Arrows are drawn

FIGURE 5

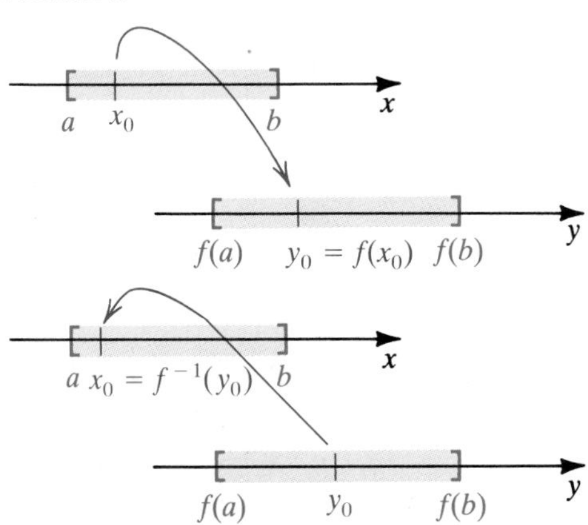

from one axis to the other to represent function values. To prove (1), consider any interval $(x_0 - \epsilon, x_0 + \epsilon)$ for $\epsilon > 0$. It is sufficient to find an interval $(y_0 - \delta, y_0 + \delta)$, of the type sketched in Figure 6, such that whenever y is in $(y_0 - \delta, y_0 + \delta)$, $f^{-1}(y)$ is in $(x_0 - \epsilon, x_0 + \epsilon)$. We may assume that $x_0 - \epsilon$ and $x_0 + \epsilon$ are in $[a, b]$. As in Figure 7, let $\delta_1 = y_0 - f(x_0 - \epsilon)$ and $\delta_2 = f(x_0 + \epsilon) - y_0$. Since f determines a one-to-one correspondence between the numbers in the intervals $(x_0 - \epsilon, x_0 + \epsilon)$ and $(y_0 - \delta_1, y_0 + \delta_2)$, the function values of f^{-1} that correspond to numbers in $(y_0 - \delta_1, y_0 + \delta_2)$ must lie in $(x_0 - \epsilon, x_0 + \epsilon)$. Let δ denote the smaller of δ_1 and δ_2. It follows that if y is in $(y_0 - \delta, y_0 + \delta)$, then $f^{-1}(y)$ is in $(x_0 - \epsilon, x_0 + \epsilon)$, which is what we wished to prove.

FIGURE 6

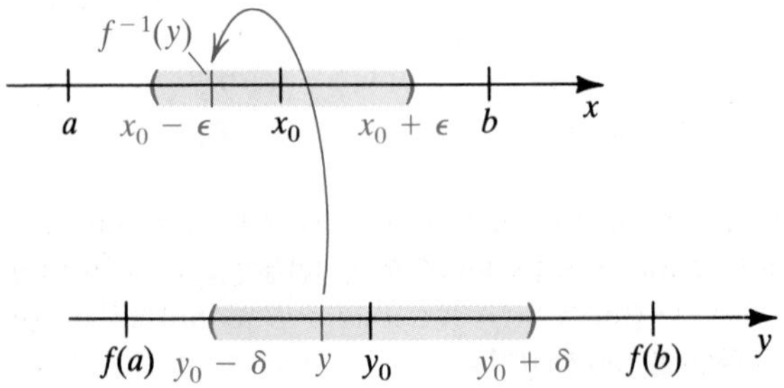

FIGURE 7

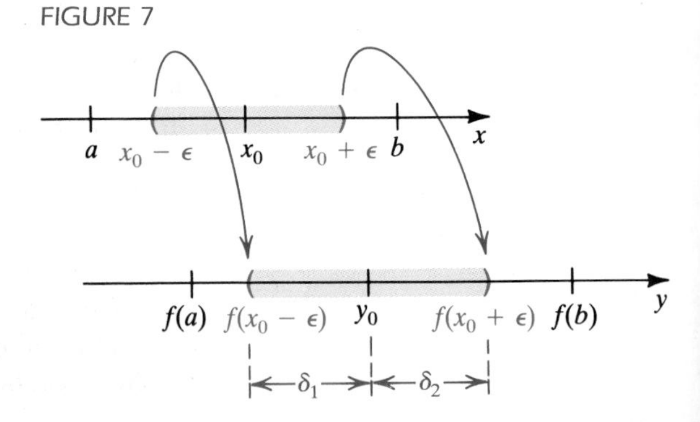

The continuity at the endpoints $f(a)$ and $f(b)$ of the domain of f^{-1} may be proved in a similar manner using one-sided limits. ■

Theorem **(17.35)**

If $x = f(u, v)$, $y = g(u, v)$ is a transformation of coordinates, then

$$\iint_R F(x, y)\,dx\,dy = \pm \iint_S F(f(u, v), g(u, v))\frac{\partial(x, y)}{\partial(u, v)}\,du\,dv.$$

As (u, v) traces the boundary K of S once in the positive direction, the corresponding point (x, y) traces the boundary C of R once in either the positive direction, in which case the plus sign is chosen, or the negative direction, in which case the minus sign is chosen.

PROOF Let us begin by choosing $G(x, y)$ such that $\partial G/\partial x = F$. Applying Green's theorem (18.19) with $G = N$ gives us

(1) $$\iint_R F(x, y)\,dx\,dy = \iint_R \frac{\partial}{\partial x}[G(x, y)]\,dx\,dy = \oint_C G(x, y)\,dy.$$

Suppose the curve K in the uv-plane has a parametrization

$$u = \phi(t),\ v = \psi(t); \quad a \le t \le b.$$

From our assumptions on the transformation, parametric equations for the curve C in the xy-plane are

(2) $$x = f(u, v) = f(\phi(t), \psi(t))$$
$$y = g(u, v) = g(\phi(t), \psi(t))$$

for $a \le t \le b$. We may therefore evaluate the line integral $\oint_C G(x, y)\,dy$ in (1) through formal substitutions for x and y. To simplify the notation, let

$$H(t) = G[f(\phi(t), \psi(t)), g(\phi(t), \psi(t))].$$

Applying a chain rule to y in (2) gives us

$$\frac{dy}{dt} = \frac{\partial y}{\partial u}\frac{du}{dt} + \frac{\partial y}{\partial v}\frac{dv}{dt} = \frac{\partial y}{\partial u}\phi'(t) + \frac{\partial y}{\partial v}\psi'(t).$$

Consequently

$$\oint_C G(x, y)\,dy = \oint_C H(t)\frac{dy}{dt}\,dt$$
$$= \int_a^b H(t)\left[\frac{\partial y}{\partial u}\phi'(t) + \frac{\partial y}{\partial v}\psi'(t)\right]dt.$$

Since $du = \phi'(t)\,dt$ and $dv = \psi'(t)\,dt$, we may regard the last line integral as a line integral around the curve K in the uv-plane. Thus,

(3) $$\oint_C G(x, y)\,dy = \pm\oint_K G\frac{\partial y}{\partial u}\,du + G\frac{\partial y}{\partial v}\,dv.$$

For simplicity we have used G as an abbreviation for $G(f(u, v), g(u, v))$. The choice of the $+$ sign or the $-$ sign is made by letting t vary from a to b and noting whether (x, y) traces C in the same direction or the opposite direction, respectively, as (u, v) traces K.

The line integral on the right in (3) has the form

$$\oint_K M\,du + N\,dv$$

with

$$M = G\frac{\partial y}{\partial u} \quad \text{and} \quad N = G\frac{\partial y}{\partial v}.$$

Applying Green's theorem, we obtain

$$\begin{aligned}
\oint_K M\,du + N\,dv
&= \iint_S \left(\frac{\partial N}{\partial u} - \frac{\partial M}{\partial v}\right) du\,dv \\
&= \iint_S \left(G\frac{\partial^2 y}{\partial u\,\partial v} + \frac{\partial G}{\partial u}\frac{\partial y}{\partial v} - G\frac{\partial^2 y}{\partial v\,\partial u} - \frac{\partial G}{\partial v}\frac{\partial y}{\partial u}\right) du\,dv \\
&= \iint_S \left[\left(\frac{\partial G}{\partial x}\frac{\partial x}{\partial u} + \frac{\partial G}{\partial y}\frac{\partial y}{\partial u}\right)\frac{\partial y}{\partial v} - \left(\frac{\partial G}{\partial x}\frac{\partial x}{\partial v} + \frac{\partial G}{\partial y}\frac{\partial y}{\partial v}\right)\frac{\partial y}{\partial u}\right] du\,dv \\
&= \iint_S \frac{\partial G}{\partial x}\left(\frac{\partial x}{\partial u}\frac{\partial y}{\partial v} - \frac{\partial y}{\partial u}\frac{\partial x}{\partial v}\right) du\,dv.
\end{aligned}$$

Using the fact that $\partial G/\partial x = F(x, y)$, together with the definition of Jacobian (17.34), gives us

$$\oint_K M\,du + N\,dv = \iint_S F(f(u, v), g(u, v))\frac{\partial(x, y)}{\partial(u, v)}\,du\,dv.$$

Combining this formula with (1) and (3) leads to the desired result. ■

III TABLES

Table A Trigonometric functions

Degrees	Radians	sin	tan	cot	cos		
0	0.000	0.000	0.000		1.000	1.571	90
1	0.017	0.017	0.017	57.29	1.000	1.553	89
2	0.035	0.035	0.035	28.64	0.999	1.536	88
3	0.052	0.052	0.052	19.081	0.999	1.518	87
4	0.070	0.070	0.070	14.301	0.998	1.501	86
5	0.087	0.087	0.087	11.430	0.996	1.484	85
6	0.105	0.105	0.105	9.514	0.995	1.466	84
7	0.122	0.122	0.123	8.144	0.993	1.449	83
8	0.140	0.139	0.141	7.115	0.990	1.431	82
9	0.157	0.156	0.158	6.314	0.988	1.414	81
10	0.175	0.174	0.176	5.671	0.985	1.396	80
11	0.192	0.191	0.194	5.145	0.982	1.379	79
12	0.209	0.208	0.213	4.705	0.978	1.361	78
13	0.227	0.225	0.231	4.331	0.974	1.344	77
14	0.244	0.242	0.249	4.011	0.970	1.326	76
15	0.262	0.259	0.268	3.732	0.966	1.309	75
16	0.279	0.276	0.287	3.487	0.961	1.292	74
17	0.297	0.292	0.306	3.271	0.956	1.274	73
18	0.314	0.309	0.325	3.078	0.951	1.257	72
19	0.332	0.326	0.344	2.904	0.946	1.239	71
20	0.349	0.342	0.364	2.747	0.940	1.222	70
21	0.367	0.358	0.384	2.605	0.934	1.204	69
22	0.384	0.375	0.404	2.475	0.927	1.187	68
23	0.401	0.391	0.424	2.356	0.921	1.169	67
24	0.419	0.407	0.445	2.246	0.914	1.152	66
25	0.436	0.423	0.466	2.144	0.906	1.134	65
26	0.454	0.438	0.488	2.050	0.899	1.117	64
27	0.471	0.454	0.510	1.963	0.891	1.100	63
28	0.489	0.469	0.532	1.881	0.883	1.082	62
29	0.506	0.485	0.554	1.804	0.875	1.065	61
30	0.524	0.500	0.577	1.732	0.866	1.047	60
31	0.541	0.515	0.601	1.664	0.857	1.030	59
32	0.559	0.530	0.625	1.600	0.848	1.012	58
33	0.576	0.545	0.649	1.540	0.839	0.995	57
34	0.593	0.559	0.675	1.483	0.829	0.977	56
35	0.611	0.574	0.700	1.428	0.819	0.960	55
36	0.628	0.588	0.727	1.376	0.809	0.942	54
37	0.646	0.602	0.754	1.327	0.799	0.925	53
38	0.663	0.616	0.781	1.280	0.788	0.908	52
39	0.681	0.629	0.810	1.235	0.777	0.890	51
40	0.698	0.643	0.839	1.192	0.766	0.873	50
41	0.716	0.656	0.869	1.150	0.755	0.855	49
42	0.733	0.669	0.900	1.111	0.743	0.838	48
43	0.750	0.682	0.933	1.072	0.731	0.820	47
44	0.768	0.695	0.966	1.036	0.719	0.803	46
45	0.785	0.707	1.000	1.000	0.707	0.785	45
		cos	cot	tan	sin	Radians	Degrees

Table B Exponential functions

x	e^x	e^{-x}	x	e^x	e^{-x}
0.00	1.0000	1.0000	2.50	12.182	0.0821
0.05	1.0513	0.9512	2.60	13.464	0.0743
0.10	1.1052	0.9048	2.70	14.880	0.0672
0.15	1.1618	0.8607	2.80	16.445	0.0608
0.20	1.2214	0.8187	2.90	18.174	0.0550
0.25	1.2840	0.7788	3.00	20.086	0.0498
0.30	1.3499	0.7408	3.10	22.198	0.0450
0.35	1.4191	0.7047	3.20	24.533	0.0408
0.40	1.4918	0.6703	3.30	27.113	0.0369
0.45	1.5683	0.6376	3.40	29.964	0.0334
0.50	1.6487	0.6065	3.50	33.115	0.0302
0.55	1.7333	0.5769	3.60	36.598	0.0273
0.60	1.8221	0.5488	3.70	40.447	0.0247
0.65	1.9155	0.5220	3.80	44.701	0.0224
0.70	2.0138	0.4966	3.90	49.402	0.0202
0.75	2.1170	0.4724	4.00	54.598	0.0183
0.80	2.2255	0.4493	4.10	60.340	0.0166
0.85	2.3396	0.4274	4.20	66.686	0.0150
0.90	2.4596	0.4066	4.30	73.700	0.0136
0.95	2.5857	0.3867	4.40	81.451	0.0123
1.00	2.7183	0.3679	4.50	90.017	0.0111
1.10	3.0042	0.3329	4.60	99.484	0.0101
1.20	3.3201	0.3012	4.70	109.95	0.0091
1.30	3.6693	0.2725	4.80	121.51	0.0082
1.40	4.0552	0.2466	4.90	134.29	0.0074
1.50	4.4817	0.2231	5.00	148.41	0.0067
1.60	4.9530	0.2019	6.00	403.43	0.0025
1.70	5.4739	0.1827	7.00	1096.6	0.0009
1.80	6.0496	0.1653	8.00	2981.0	0.0003
1.90	6.6859	0.1496	9.00	8103.1	0.0001
2.00	7.3891	0.1353	10.00	22026.0	0.00005
2.10	8.1662	0.1225			
2.20	9.0250	0.1108			
2.30	9.9742	0.1003			
2.40	11.0232	0.0907			

Table C Natural logarithms

n	0.0	0.1	0.2	0.3	0.4	0.5	0.6	0.7	0.8	0.9
0*		7.697	8.391	8.796	9.084	9.307	9.489	9.643	9.777	9.895
1	0.000	0.095	0.182	0.262	0.336	0.405	0.470	0.531	0.588	0.642
2	0.693	0.742	0.788	0.833	0.875	0.916	0.956	0.993	1.030	1.065
3	1.099	1.131	1.163	1.194	1.224	1.253	1.281	1.308	1.335	1.361
4	1.386	1.411	1.435	1.459	1.482	1.504	1.526	1.548	1.569	1.589
5	1.609	1.629	1.649	1.668	1.686	1.705	1.723	1.740	1.758	1.775
6	1.792	1.808	1.825	1.841	1.856	1.872	1.887	1.902	1.917	1.932
7	1.946	1.960	1.974	1.988	2.001	2.015	2.028	2.041	2.054	2.067
8	2.079	2.092	2.104	2.116	2.128	2.140	2.152	2.163	2.175	2.186
9	2.197	2.208	2.219	2.230	2.241	2.251	2.262	2.272	2.282	2.293
10	2.303	2.313	2.322	2.332	2.342	2.351	2.361	2.370	2.380	2.389

* Subtract 10 if $n < 1$; for example, $\ln 0.3 \approx 8.796 - 10 = -1.204$.

IV TABLE OF INTEGRALS

Basic forms

1 $\int u\,dv = uv - \int v\,du$

2 $\int u^n\,du = \frac{1}{n+1}u^{n+1} + C, \quad n \neq -1$

3 $\int \frac{du}{u} = \ln|u| + C$

4 $\int e^u\,du = e^u + C$

5 $\int a^u\,du = \frac{1}{\ln a}a^u + C$

6 $\int \sin u\,du = -\cos u + C$

7 $\int \cos u\,du = \sin u + C$

8 $\int \sec^2 u\,du = \tan u + C$

9 $\int \csc^2 u\,du = -\cot u + C$

10 $\int \sec u \tan u\,du = \sec u + C$

11 $\int \csc u \cot u\,du = -\csc u + C$

12 $\int \tan u\,du = -\ln|\cos u| + C$

13 $\int \cot u\,du = \ln|\sin u| + C$

14 $\int \sec u\,du = \ln|\sec u + \tan u| + C$

15 $\int \csc u\,du = \ln|\csc u - \cot u| + C$

16 $\int \frac{du}{\sqrt{a^2 - u^2}} = \sin^{-1}\frac{u}{a} + C$

17 $\int \frac{du}{a^2 + u^2} = \frac{1}{a}\tan^{-1}\frac{u}{a} + C$

18 $\int \frac{du}{u\sqrt{u^2 - a^2}} = \frac{1}{a}\sec^{-1}\frac{u}{a} + C$

19 $\int \frac{du}{a^2 - u^2} = \frac{1}{2a}\ln\left|\frac{u+a}{u-a}\right| + C$

20 $\int \frac{du}{\sqrt{u^2 - a^2}} = \ln|u + \sqrt{u^2 - a^2}| + C$

Forms involving $\sqrt{a^2 + u^2}$

21 $\int \sqrt{a^2 + u^2}\,du = \frac{u}{2}\sqrt{a^2 + u^2} + \frac{a^2}{2}\ln|u + \sqrt{a^2 + u^2}| + C$

22 $\int u^2\sqrt{a^2 + u^2}\,du = \frac{u}{8}(a^2 + 2u^2)\sqrt{a^2 + u^2} - \frac{a^4}{8}\ln|u + \sqrt{a^2 + u^2}| + C$

23 $\int \frac{\sqrt{a^2 + u^2}}{u}\,du = \sqrt{a^2 + u^2} - a\ln\left|\frac{a + \sqrt{a^2 + u^2}}{u}\right| + C$

24 $\int \frac{\sqrt{a^2 + u^2}}{u^2}\,du = -\frac{\sqrt{a^2 + u^2}}{u} + \ln|u + \sqrt{a^2 + u^2}| + C$

25 $\int \frac{du}{\sqrt{a^2 + u^2}} = \ln|u + \sqrt{a^2 + u^2}| + C$

26 $\int \frac{u^2\,du}{\sqrt{a^2 + u^2}} = \frac{u}{2}\sqrt{a^2 + u^2} - \frac{a^2}{2}\ln|u + \sqrt{a^2 + u^2}| + C$

27 $\int \frac{du}{u\sqrt{a^2 + u^2}} = -\frac{1}{a}\ln\left|\frac{\sqrt{a^2 + u^2} + a}{u}\right| + C$

28 $\int \frac{du}{u^2\sqrt{a^2 + u^2}} = -\frac{\sqrt{a^2 + u^2}}{a^2 u} + C$

29 $\int \frac{du}{(a^2 + u^2)^{3/2}} = \frac{u}{a^2\sqrt{a^2 + u^2}} + C$

Forms involving $\sqrt{a^2 - u^2}$

30 $\displaystyle\int \sqrt{a^2 - u^2}\, du = \frac{u}{2}\sqrt{a^2 - u^2} + \frac{a^2}{2}\sin^{-1}\frac{u}{a} + C$

31 $\displaystyle\int u^2\sqrt{a^2 - u^2}\, du = \frac{u}{8}(2u^2 - a^2)\sqrt{a^2 - u^2} + \frac{a^4}{8}\sin^{-1}\frac{u}{a} + C$

32 $\displaystyle\int \frac{\sqrt{a^2 - u^2}}{u}\, du = \sqrt{a^2 - u^2} - a\ln\left|\frac{a + \sqrt{a^2 - u^2}}{u}\right| + C$

33 $\displaystyle\int \frac{\sqrt{a^2 - u^2}}{u^2}\, du = -\frac{1}{u}\sqrt{a^2 - u^2} - \sin^{-1}\frac{u}{a} + C$

34 $\displaystyle\int \frac{u^2\, du}{\sqrt{a^2 - u^2}} = -\frac{u}{2}\sqrt{a^2 - u^2} + \frac{a^2}{2}\sin^{-1}\frac{u}{a} + C$

35 $\displaystyle\int \frac{du}{u\sqrt{a^2 - u^2}} = -\frac{1}{a}\ln\left|\frac{a + \sqrt{a^2 - u^2}}{u}\right| + C$

36 $\displaystyle\int \frac{du}{u^2\sqrt{a^2 - u^2}} = -\frac{1}{a^2 u}\sqrt{a^2 - u^2} + C$

37 $\displaystyle\int (a^2 - u^2)^{3/2}\, du = -\frac{u}{8}(2u^2 - 5a^2)\sqrt{a^2 - u^2} + \frac{3a^4}{8}\sin^{-1}\frac{u}{a} + C$

38 $\displaystyle\int \frac{du}{(a^2 - u^2)^{3/2}} = \frac{u}{a^2\sqrt{a^2 - u^2}} + C$

Forms involving $\sqrt{u^2 - a^2}$

39 $\displaystyle\int \sqrt{u^2 - a^2}\, du = \frac{u}{2}\sqrt{u^2 - a^2} - \frac{a^2}{2}\ln\left|u + \sqrt{u^2 - a^2}\right| + C$

40 $\displaystyle\int u^2\sqrt{u^2 - a^2}\, du = \frac{u}{8}(2u^2 - a^2)\sqrt{u^2 - a^2} - \frac{a^4}{8}\ln\left|u + \sqrt{u^2 - a^2}\right| + C$

41 $\displaystyle\int \frac{\sqrt{u^2 - a^2}}{u}\, du = \sqrt{u^2 - a^2} - a\cos^{-1}\frac{a}{u} + C$

42 $\displaystyle\int \frac{\sqrt{u^2 - a^2}}{u^2}\, du = -\frac{\sqrt{u^2 - a^2}}{u} + \ln\left|u + \sqrt{u^2 - a^2}\right| + C$

43 $\displaystyle\int \frac{u^2\, du}{\sqrt{u^2 - a^2}} = \frac{u}{2}\sqrt{u^2 - a^2} + \frac{a^2}{2}\ln\left|u + \sqrt{u^2 - a^2}\right| + C$

44 $\displaystyle\int \frac{du}{u^2\sqrt{u^2 - a^2}} = \frac{\sqrt{u^2 - a^2}}{a^2 u} + C$

45 $\displaystyle\int \frac{du}{(u^2 - a^2)^{3/2}} = -\frac{u}{a^2\sqrt{u^2 - a^2}} + C$

46 $\displaystyle\int \frac{u^2\, du}{(u^2 - a^2)^{3/2}} = \frac{-u}{\sqrt{u^2 - a^2}} + \ln\left|u + \sqrt{u^2 - a^2}\right| + C$

Forms involving $a + bu$

47 $$\int \frac{u\,du}{a+bu} = \frac{1}{b^2}(a + bu - a\ln|a+bu|) + C$$

48 $$\int \frac{u^2\,du}{a+bu} = \frac{1}{2b^3}[(a+bu)^2 - 4a(a+bu) + 2a^2\ln|a+bu|] + C$$

49 $$\int \frac{du}{u(a+bu)} = \frac{1}{a}\ln\left|\frac{u}{a+bu}\right| + C$$

50 $$\int \frac{du}{u^2(a+bu)} = -\frac{1}{au} + \frac{b}{a^2}\ln\left|\frac{a+bu}{u}\right| + C$$

51 $$\int \frac{u\,du}{(a+bu)^2} = \frac{a}{b^2(a+bu)} + \frac{1}{b^2}\ln|a+bu| + C$$

52 $$\int \frac{du}{u(a+bu)^2} = \frac{1}{a(a+bu)} - \frac{1}{a^2}\ln\left|\frac{a+bu}{u}\right| + C$$

53 $$\int \frac{u^2\,du}{(a+bu)^2} = \frac{1}{b^3}\left(a + bu - \frac{a^2}{a+bu} - 2a\ln|a+bu|\right) + C$$

54 $$\int u\sqrt{a+bu}\,du = \frac{2}{15b^2}(3bu - 2a)(a+bu)^{3/2} + C$$

55 $$\int \frac{u\,du}{\sqrt{a+bu}} = \frac{2}{3b^2}(bu - 2a)\sqrt{a+bu} + C$$

56 $$\int \frac{u^2\,du}{\sqrt{a+bu}} = \frac{2}{15b^3}(8a^2 + 3b^2u^2 - 4abu)\sqrt{a+bu} + C$$

57 $$\int \frac{du}{u\sqrt{a+bu}} = \frac{1}{\sqrt{a}}\ln\left|\frac{\sqrt{a+bu}-\sqrt{a}}{\sqrt{a+bu}+\sqrt{a}}\right| + C, \quad \text{if } a > 0$$
$$= \frac{2}{\sqrt{-a}}\tan^{-1}\sqrt{\frac{a+bu}{-a}} + C, \quad \text{if } a < 0$$

58 $$\int \frac{\sqrt{a+bu}}{u}\,du = 2\sqrt{a+bu} + a\int \frac{du}{u\sqrt{a+bu}}$$

59 $$\int \frac{\sqrt{a+bu}}{u^2}\,du = -\frac{\sqrt{a+bu}}{u} + \frac{b}{2}\int \frac{du}{u\sqrt{a+bu}}$$

60 $$\int u^n\sqrt{a+bu}\,du = \frac{2}{b(2n+3)}\left[u^n(a+bu)^{3/2} - na\int u^{n-1}\sqrt{a+bu}\,du\right]$$

61 $$\int \frac{u^n\,du}{\sqrt{a+bu}} = \frac{2u^n\sqrt{a+bu}}{b(2n+1)} - \frac{2na}{b(2n+1)}\int \frac{u^{n-1}\,du}{\sqrt{a+bu}}$$

62 $$\int \frac{du}{u^n\sqrt{a+bu}} = -\frac{\sqrt{a+bu}}{a(n-1)u^{n-1}} - \frac{b(2n-3)}{2a(n-1)}\int \frac{du}{u^{n-1}\sqrt{a+bu}}$$

Trigonometric forms

63 $\int \sin^2 u\, du = \frac{1}{2}u - \frac{1}{4}\sin 2u + C$

64 $\int \cos^2 u\, du = \frac{1}{2}u + \frac{1}{4}\sin 2u + C$

65 $\int \tan^2 u\, du = \tan u - u + C$

66 $\int \cot^2 u\, du = -\cot u - u + C$

67 $\int \sin^3 u\, du = -\frac{1}{3}(2 + \sin^2 u)\cos u + C$

68 $\int \cos^3 u\, du = \frac{1}{3}(2 + \cos^2 u)\sin u + C$

69 $\int \tan^3 u\, du = \frac{1}{2}\tan^2 u + \ln|\cos u| + C$

70 $\int \cot^3 u\, du = -\frac{1}{2}\cot^2 u - \ln|\sin u| + C$

71 $\int \sec^3 u\, du = \frac{1}{2}\sec u \tan u + \frac{1}{2}\ln|\sec u + \tan u| + C$

72 $\int \csc^3 u\, du = -\frac{1}{2}\csc u \cot u + \frac{1}{2}\ln|\csc u - \cot u| + C$

73 $\int \sin^n u\, du = -\frac{1}{n}\sin^{n-1} u \cos u + \frac{n-1}{n}\int \sin^{n-2} u\, du$

74 $\int \cos^n u\, du = \frac{1}{n}\cos^{n-1} u \sin u + \frac{n-1}{n}\int \cos^{n-2} u\, du$

75 $\int \tan^n u\, du = \frac{1}{n-1}\tan^{n-1} u - \int \tan^{n-2} u\, du$

76 $\int \cot^n u\, du = \frac{-1}{n-1}\cot^{n-1} u - \int \cot^{n-2} u\, du$

77 $\int \sec^n u\, du = \frac{1}{n-1}\tan u \sec^{n-2} u + \frac{n-2}{n-1}\int \sec^{n-2} u\, du$

78 $\int \csc^n u\, du = \frac{-1}{n-1}\cot u \csc^{n-2} u + \frac{n-2}{n-1}\int \csc^{n-2} u\, du$

79 $\int \sin au \sin bu\, du = \frac{\sin (a-b)u}{2(a-b)} - \frac{\sin (a+b)u}{2(a+b)} + C$

80 $\int \cos au \cos bu\, du = \frac{\sin (a-b)u}{2(a-b)} + \frac{\sin (a+b)u}{2(a+b)} + C$

81 $\int \sin au \cos bu\, du = -\frac{\cos (a-b)u}{2(a-b)} - \frac{\cos (a+b)u}{2(a+b)} + C$

82 $\int u \sin u\, du = \sin u - u \cos u + C$

83 $\int u \cos u\, du = \cos u + u \sin u + C$

84 $\int u^n \sin u\, du = -u^n \cos u + n\int u^{n-1} \cos u\, du$

85 $\int u^n \cos u \, du = u^n \sin u - n \int u^{n-1} \sin u \, du$

86 $$\int \sin^n u \cos^m u \, du = -\frac{\sin^{n-1} u \cos^{m+1} u}{n+m} + \frac{n-1}{n+m} \int \sin^{n-2} u \cos^m u \, du$$
$$= \frac{\sin^{n+1} u \cos^{m-1} u}{n+m} + \frac{m-1}{n+m} \int \sin^n u \cos^{m-2} u \, du$$

Inverse trigonometric forms

87 $\int \sin^{-1} u \, du = u \sin^{-1} u + \sqrt{1-u^2} + C$

88 $\int \cos^{-1} u \, du = u \cos^{-1} u - \sqrt{1-u^2} + C$

89 $\int \tan^{-1} u \, du = u \tan^{-1} u - \frac{1}{2} \ln (1+u^2) + C$

90 $\int u \sin^{-1} u \, du = \frac{2u^2-1}{4} \sin^{-1} u + \frac{u\sqrt{1-u^2}}{4} + C$

91 $\int u \cos^{-1} u \, du = \frac{2u^2-1}{4} \cos^{-1} u - \frac{u\sqrt{1-u^2}}{4} + C$

92 $\int u \tan^{-1} u \, du = \frac{u^2+1}{2} \tan^{-1} u - \frac{u}{2} + C$

93 $\int u^n \sin^{-1} u \, du = \frac{1}{n+1}\left[u^{n+1} \sin^{-1} u - \int \frac{u^{n+1}\,du}{\sqrt{1-u^2}}\right], \quad n \neq -1$

94 $\int u^n \cos^{-1} u \, du = \frac{1}{n+1}\left[u^{n+1} \cos^{-1} u + \int \frac{u^{n+1}\,du}{\sqrt{1-u^2}}\right], \quad n \neq -1$

95 $\int u^n \tan^{-1} u \, du = \frac{1}{n+1}\left[u^{n+1} \tan^{-1} u - \int \frac{u^{n+1}\,du}{1+u^2}\right], \quad n \neq -1$

Exponential and logarithmic forms

96 $\int ue^{au} \, du = \frac{1}{a^2}(au-1)e^{au} + C$

97 $\int u^n e^{au} \, du = \frac{1}{a} u^n e^{au} - \frac{n}{a} \int u^{n-1} e^{au} \, du$

98 $\int e^{au} \sin bu \, du = \frac{e^{au}}{a^2+b^2}(a \sin bu - b \cos bu) + C$

99 $\int e^{au} \cos bu \, du = \frac{e^{au}}{a^2+b^2}(a \cos bu + b \sin bu) + C$

100 $\int \ln u \, du = u \ln u - u + C$

101 $\int u^n \ln u \, du = \frac{u^{n+1}}{(n+1)^2}[(n+1)\ln u - 1] + C$

102 $\int \frac{1}{u \ln u} \, du = \ln |\ln u| + C$

Hyperbolic forms

103 $\int \sinh u \, du = \cosh u + C$

104 $\int \cosh u \, du = \sinh u + C$

105 $\int \tanh u \, du = \ln \cosh u + C$

106 $\int \coth u \, du = \ln |\sinh u| + C$

107 $\int \operatorname{sech} u \, du = \tan^{-1} \sinh u + C$

108 $\int \operatorname{csch} u \, du = \ln |\tanh \tfrac{1}{2}u| + C$

109 $\int \operatorname{sech}^2 u \, du = \tanh u + C$

110 $\int \operatorname{csch}^2 u \, du = -\coth u + C$

111 $\int \operatorname{sech} u \tanh u \, du = -\operatorname{sech} u + C$

112 $\int \operatorname{csch} u \coth u \, du = -\operatorname{csch} u + C$

Forms involving $\sqrt{2au - u^2}$

113 $\int \sqrt{2au - u^2} \, du = \dfrac{u - a}{2} \sqrt{2au - u^2} + \dfrac{a^2}{2} \cos^{-1}\left(\dfrac{a - u}{a}\right) + C$

114 $\int u\sqrt{2au - u^2} \, du = \dfrac{2u^2 - au - 3a^2}{6} \sqrt{2au - u^2} + \dfrac{a^3}{2} \cos^{-1}\left(\dfrac{a - u}{a}\right) + C$

115 $\int \dfrac{\sqrt{2au - u^2}}{u} \, du = \sqrt{2au - u^2} + a \cos^{-1}\left(\dfrac{a - u}{a}\right) + C$

116 $\int \dfrac{\sqrt{2au - u^2}}{u^2} \, du = -\dfrac{2\sqrt{2au - u^2}}{u} - \cos^{-1}\left(\dfrac{a - u}{a}\right) + C$

117 $\int \dfrac{du}{\sqrt{2au - u^2}} = \cos^{-1}\left(\dfrac{a - u}{a}\right) + C$

118 $\int \dfrac{u \, du}{\sqrt{2au - u^2}} = -\sqrt{2au - u^2} + a \cos^{-1}\left(\dfrac{a - u}{a}\right) + C$

119 $\int \dfrac{u^2 \, du}{\sqrt{2au - u^2}} = -\dfrac{(u + 3a)}{2} \sqrt{2au - u^2} + \dfrac{3a^2}{2} \cos^{-1}\left(\dfrac{a - u}{a}\right) + C$

120 $\int \dfrac{du}{u\sqrt{2au - u^2}} = -\dfrac{\sqrt{2au - u^2}}{au} + C$

A **Student's Solutions Manual** to accompany this text is available from your college bookstore. The guide, by Jeffery A. Cole and Gary K. Rockswold, contains detailed solutions to approximately one-third of the exercises as well as strategies for solving other exercises in the text.

ANSWERS TO ODD-NUMBERED EXERCISES

Answers are usually not provided for exercises that require lengthy proofs.

CHAPTER 1

EXERCISES 1.1

1 **(a)** -15 **(b)** -3 **(c)** 11

3 **(a)** $4-\pi$ **(b)** $4-\pi$ **(c)** $1.5-\sqrt{2}$ **5** $-x-3$

7 $2-x$ **9** $-\frac{6}{5}, \frac{2}{3}$ **11** $-\frac{9}{2}, \frac{3}{4}$ **13** $-2 \pm \sqrt{2}$

15 $\frac{3}{4} \pm \frac{1}{4}\sqrt{41}$ **17** $(12, \infty)$ **19** $[9, 19)$ **21** $(-2, 3)$

23 $(-\infty, -2) \cup (4, \infty)$ **25** $\left(-\infty, -\frac{5}{2}\right] \cup [1, \infty)$

27 $\left(\frac{3}{2}, \frac{7}{3}\right)$ **29** $(-\infty, -1) \cup \left(2, \frac{7}{2}\right]$ **31** $(-3.01, -2.99)$

33 $(-\infty, -2.001] \cup [-1.999, \infty)$ **35** $\left(-\frac{9}{2}, -\frac{1}{2}\right)$ **37** $\left[\frac{3}{5}, \frac{9}{5}\right]$

39 **(a)** The line parallel to the y-axis that intersects the x-axis at $(-2, 0)$
(b) The line parallel to the x-axis that intersects the y-axis at $(0, 3)$
(c) All points to the right of and on the y-axis
(d) All points in quadrants I and III
(e) All points below the x-axis
(f) All points within the rectangle such that $-2 \le x \le 2$ and $-1 \le y \le 1$

41 **(a)** $\sqrt{29}$ **(b)** $\left(5, -\frac{1}{2}\right)$

43 $d(A, C)^2 = d(A, B)^2 + d(B, C)^2$; area $= 28$

45

47

49

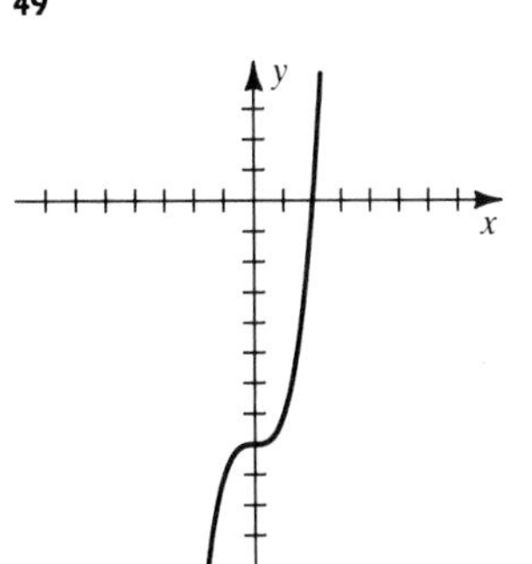

51

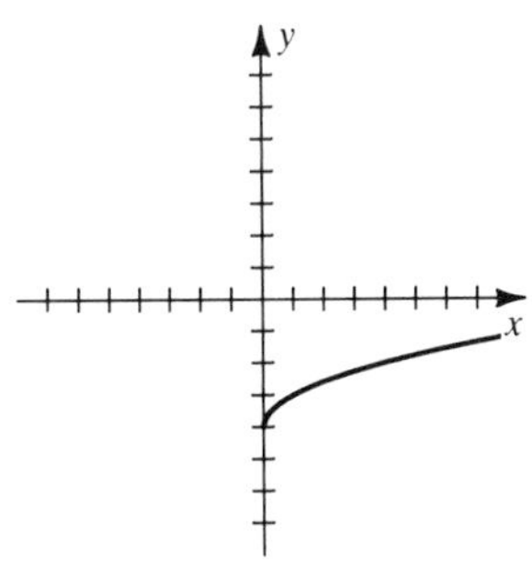

53

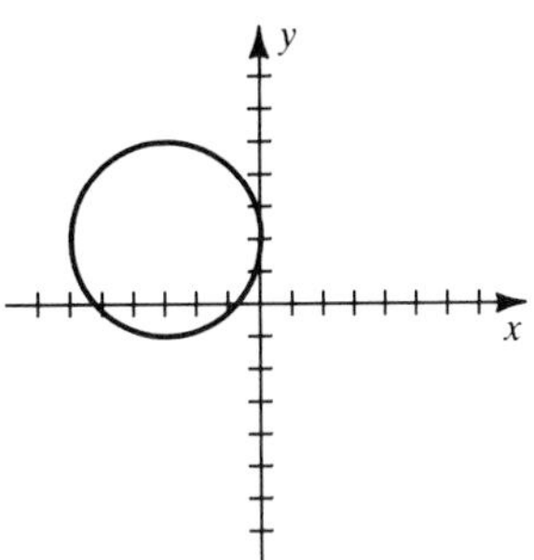

55

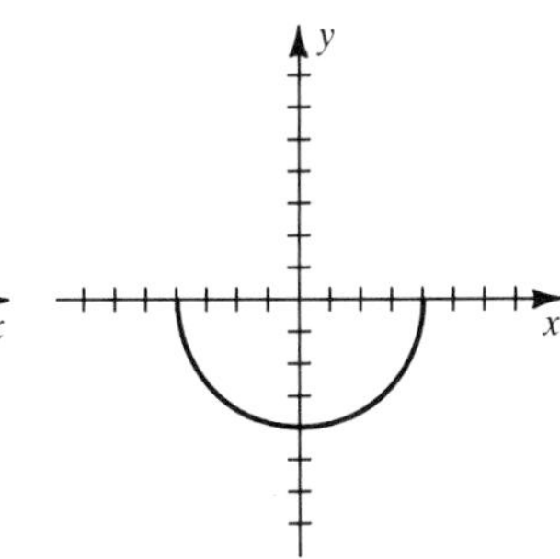

57 $(x-2)^2 + (y+3)^2 = 25$ **59** $(x+4)^2 + (y-4)^2 = 16$

61 $4x + y = 17$ **63** $3x - 4y = 12$ **65** $5x - 2y = 18$

67 $5x - 7y = -15$

69

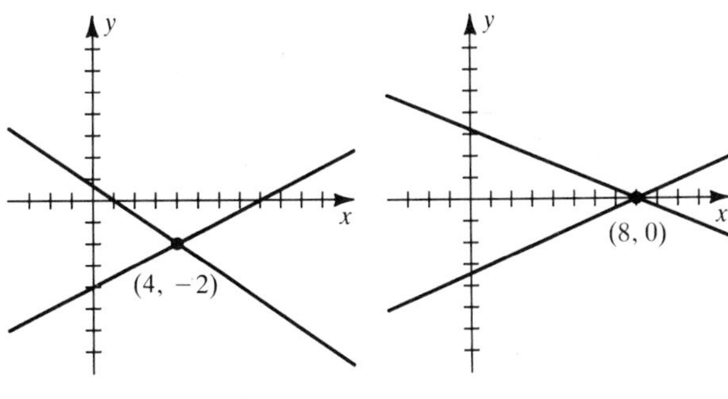

73 $x \approx 0.41$; $y \approx 0.15$

75 **(a)** 1 cm **(b)** Capsule: $\frac{11\pi}{96}$ cm^3; tablet: $\frac{\pi}{8}$ cm^3

77 $4 \le p < 6$ **79** $0 \le v < 30$

81 **(a)** 0 °C **(b)** $\frac{1}{273}$ **(c)** 163.8 °C

EXERCISES 1.2

1 -12; -22; -36

3 **(a)** $5a - 2$ **(b)** $-5a - 2$ **(c)** $-5a + 2$
(d) $5a + 5h - 2$ **(e)** $5a + 5h - 4$ **(f)** 5

5 **(a)** $a^2 - a + 3$ **(b)** $a^2 + a + 3$
(c) $-a^2 + a - 3$ **(d)** $a^2 + 2ah + h^2 - a - h + 3$
(e) $a^2 + h^2 - a - h + 6$ **(f)** $2a + h - 1$

7 All real numbers except -2, 0, and 2

9 $\left[\frac{3}{2}, 4\right) \cup (4, \infty)$ **11** **(a)** Odd **(b)** Even **(c)** Neither

13

15

17

19

21

23 **(a)**

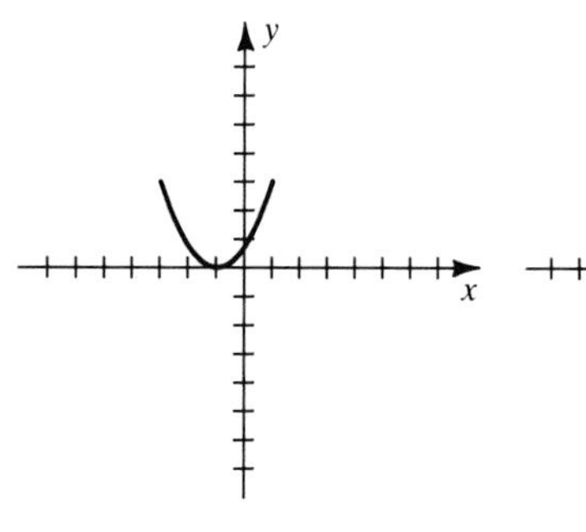

(b)

(c)

(d)

(e)

(f)

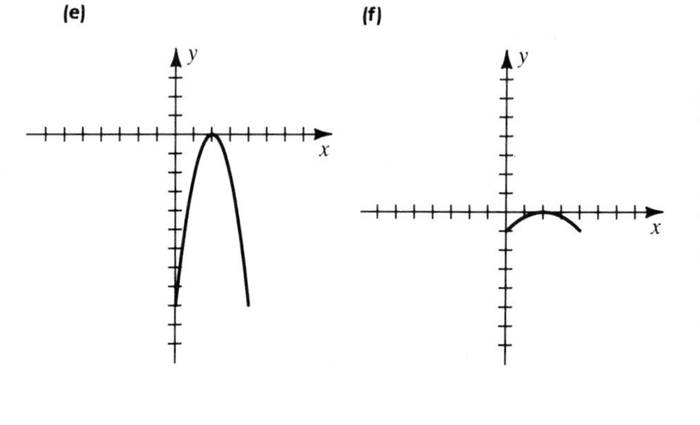

(g)

(h)

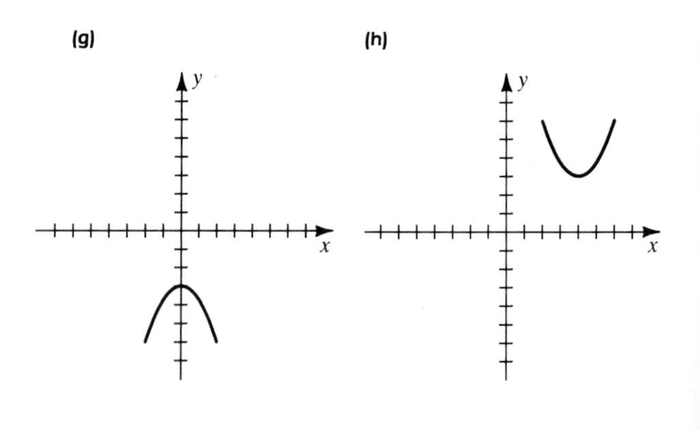

25

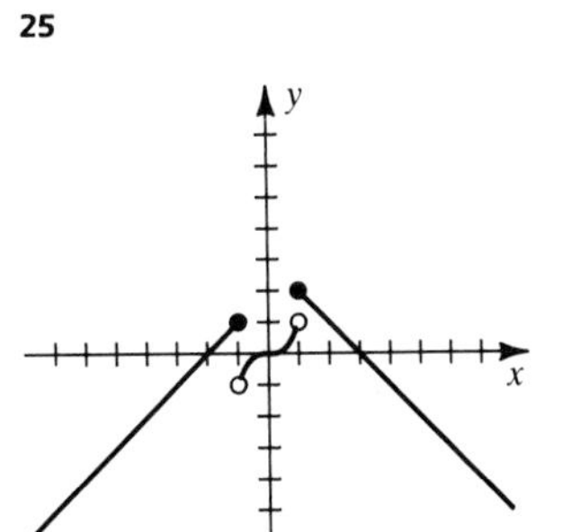

27

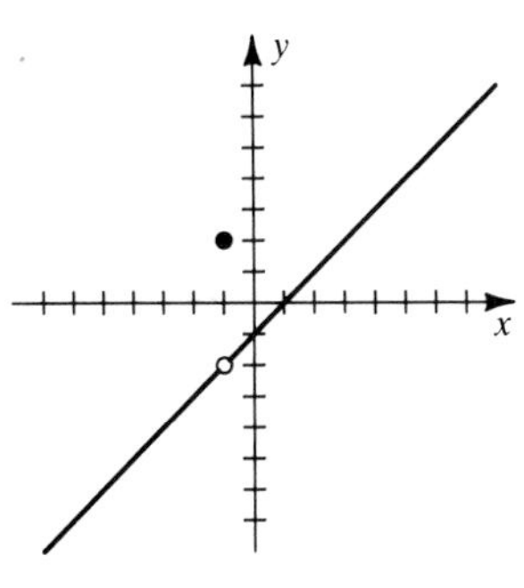

29 **(a)**

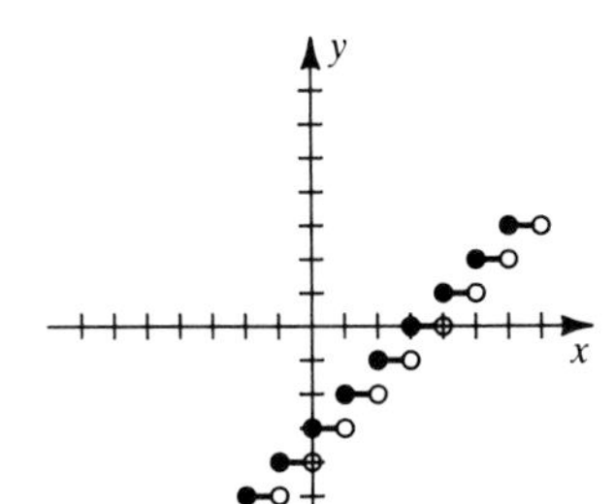

(b)

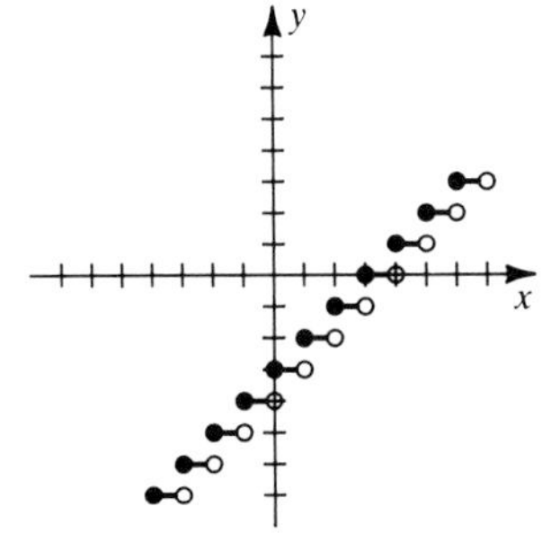

(c)

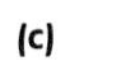

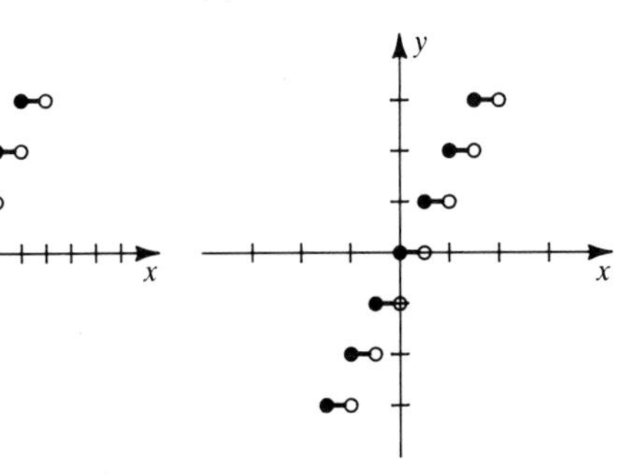

(d)

31 **(a)** $2\sqrt{x+5}$; 0; $x+5$; 1 **(b)** $[-5, \infty)$; $(-5, \infty)$

33 **(a)** $\dfrac{3x^2+6x}{(x-4)(x+5)}$; $\dfrac{x^2+14x}{(x-4)(x+5)}$; $\dfrac{2x^2}{(x-4)(x+5)}$; $\dfrac{2x+10}{x-4}$

(b) All real numbers except -5 and 4; all real numbers except -5, 0, and 4

35 **(a)** $x+2-3\sqrt{x+2}$; $[-2, \infty)$

(b) $\sqrt{x^2-3x+2}$; $(-\infty, 1] \cup [2, \infty)$

37 **(a)** $\sqrt{\sqrt{x+5}-2}$; $[-1, \infty)$ **(b)** $\sqrt{\sqrt{x-2}+5}$; $[2, \infty)$

39 **(a)** $\sqrt{28-x}$; $[3, 28]$ **(b)** $\sqrt{\sqrt{25-x^2}-3}$; $[-4, 4]$

41 **(a)** $\dfrac{1}{x+3}$; all real numbers except -3 and 0

(b) $\dfrac{6x+4}{x}$; all real numbers except $-\frac{2}{3}$ and 0

Exer. 43–50: Answers are not unique.

43 $u = x^2 + 3x$, $y = u^{1/3}$ **45** $u = x - 3$, $y = \dfrac{1}{u^4}$

47 $u = x^4 - 2x^2 + 5$, $y = u^5$ **49** $u = \sqrt{x+4}$, $y = \dfrac{u-2}{u+2}$

51 7.91; 5.05 **53** $V = 4x^3 - 100x^2 + 600x$

55 $d = 2\sqrt{t^2 + 2500}$ **57** **(a)** $y = \sqrt{h^2 + 2hr}$ **(b)** 1280.6 mi

59 $d = \sqrt{90{,}400 + x^2}$

61 **(a)** $y = \dfrac{bh}{a-b}$ **(b)** $V = \frac{\pi}{3}h(a^2 + ab + b^2)$ **(c)** $\dfrac{200}{7\pi}$ ft

EXERCISES 1.3

1 **(a)** $\dfrac{5\pi}{6}$ **(b)** $\dfrac{2\pi}{3}$ **(c)** $\dfrac{5\pi}{2}$ **(d)** $-\dfrac{\pi}{3}$

3 **(a)** $120°$ **(b)** $150°$ **(c)** $135°$ **(d)** $-630°$

5 $\dfrac{20\pi}{9}$ **7** $x = 8$, $y = 4\sqrt{3}$

Exer. 9–16: Answers are in the order sin, cos, tan, cot, sec, csc.

9 $\frac{3}{5}, \frac{4}{5}, \frac{3}{4}, \frac{4}{3}, \frac{5}{4}, \frac{5}{3}$ **11** $\frac{5}{13}, \frac{12}{13}, \frac{5}{12}, \frac{12}{5}, \frac{13}{12}, \frac{13}{5}$

13 $-\frac{3}{5}, \frac{4}{5}, -\frac{3}{4}, -\frac{4}{3}, \frac{5}{4}, -\frac{5}{3}$

15 $-\dfrac{7}{\sqrt{53}}, -\dfrac{2}{\sqrt{53}}, \dfrac{7}{2}, \dfrac{2}{7}, -\dfrac{\sqrt{53}}{2}, -\dfrac{\sqrt{53}}{7}$

17 **(a)** $\cot\theta = \dfrac{\sqrt{1-\sin^2\theta}}{\sin\theta}$ **(b)** $\sec\theta = \dfrac{1}{\sqrt{1-\sin^2\theta}}$

19 **(a)** $\tan\theta = \sqrt{\sec^2\theta - 1}$ **(b)** $\sin\theta = \dfrac{\sqrt{\sec^2\theta - 1}}{\sec\theta}$

Exer. 21–26: Answers are not unique.

21 $4\cos\theta$ **23** $\sin\theta$ **25** $\sin\theta$

27 **(a)** $\dfrac{\sqrt{3}}{2}$ **(b)** $\dfrac{\sqrt{2}}{2}$ **29** **(a)** $-\dfrac{\sqrt{3}}{3}$ **(b)** $-\sqrt{3}$

31 **(a)** -2 **(b)** $\dfrac{2}{\sqrt{3}}$

33 **(a)**

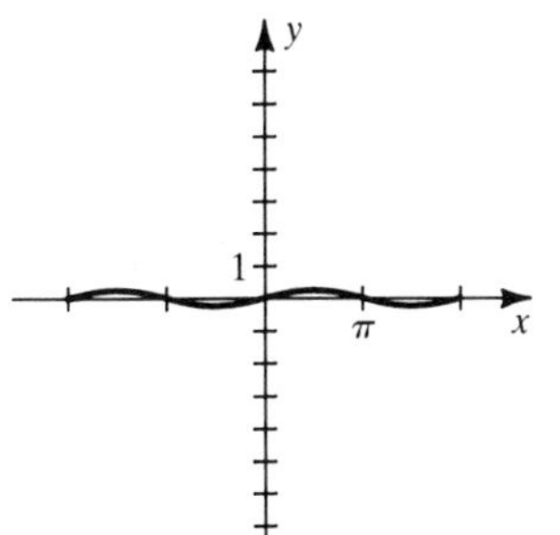

(b)

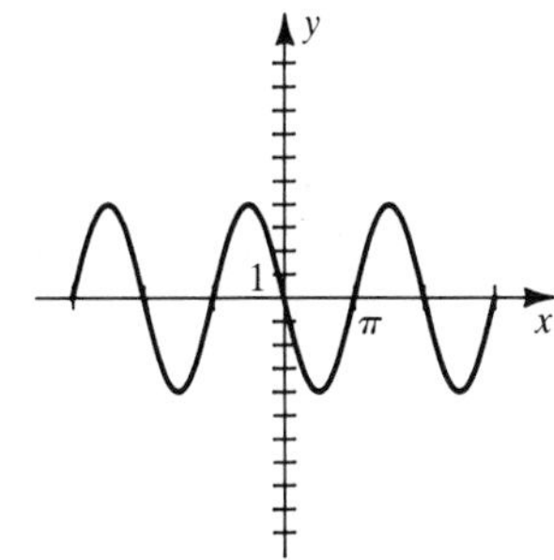

35 **(a)** **(b)**

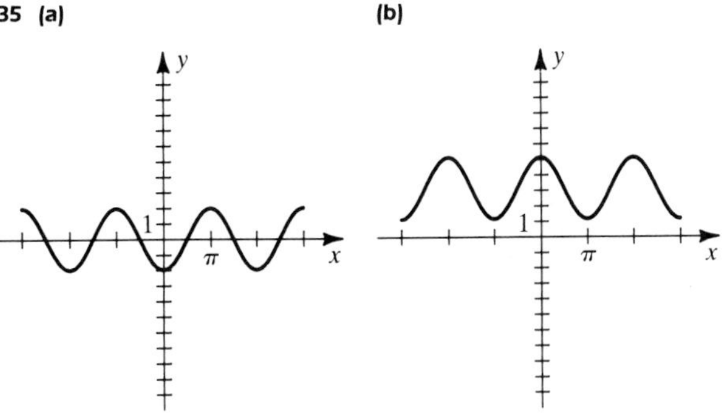

37 **(a)** **(b)**

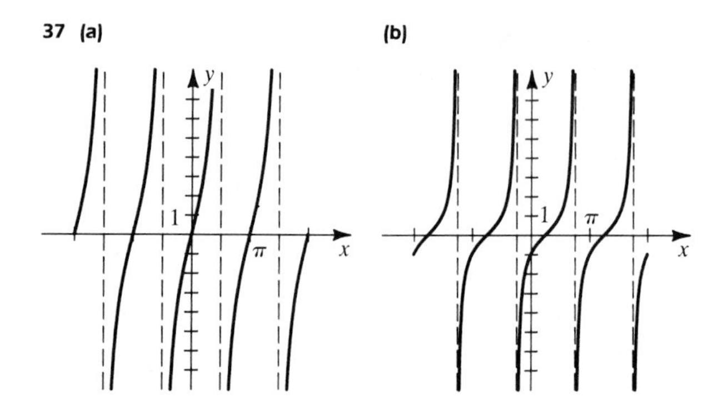

Exer. 39–42: Answers are not unique.

39 $u = \tan^2 x + 4,\ y = \sqrt{u}$ **41** $u = x + \frac{\pi}{4},\ y = \sec u$

43
$$\frac{f(x+h) - f(x)}{h} = \frac{\cos(x+h) - \cos x}{h}$$
$$= \frac{\cos x \cos h - \sin x \sin h - \cos x}{h}$$
$$= \frac{\cos x \cos h - \cos x}{h} - \frac{\sin x \sin h}{h}$$
$$= \cos x\left(\frac{\cos h - 1}{h}\right) - \sin x\left(\frac{\sin h}{h}\right)$$

Exer. 45–54: Typical verifications are given.

45 $(1 - \sin^2 t)(1 + \tan^2 t) = (\cos^2 t)(\sec^2 t)$
$= (\cos^2 t)(1/\cos^2 t) = 1$

47 $\dfrac{\csc^2\theta}{1 + \tan^2\theta} = \dfrac{\csc^2\theta}{\sec^2\theta} = \dfrac{1/\sin^2\theta}{1/\cos^2\theta} = \dfrac{\cos^2\theta}{\sin^2\theta} = \left(\dfrac{\cos\theta}{\sin\theta}\right)^2 = \cot^2\theta$

49 $\dfrac{1 + \csc\beta}{\sec\beta} - \cot\beta = \dfrac{1}{\sec\beta} + \dfrac{\csc\beta}{\sec\beta} - \cot\beta$
$= \cos\beta + \dfrac{\cos\beta}{\sin\beta} - \cot\beta = \cos\beta$

51 $\sin 3u = \sin(2u + u) = \sin 2u \cos u + \cos 2u \sin u$
$= (2\sin u \cos u)\cos u + (1 - 2\sin^2 u)\sin u$
$= 2\sin u \cos^2 u + \sin u - 2\sin^3 u$
$= 2\sin u(1 - \sin^2 u) + \sin u - 2\sin^3 u$
$= 2\sin u - 2\sin^3 u + \sin u - 2\sin^3 u$
$= 3\sin u - 4\sin^3 u = \sin u(3 - 4\sin^2 u)$

53 $\cos^4\dfrac{\theta}{2} = \left(\cos^2\dfrac{\theta}{2}\right)^2 = \left(\dfrac{1 + \cos\theta}{2}\right)^2$
$= \dfrac{1 + 2\cos\theta + \cos^2\theta}{4}$
$= \dfrac{1}{4} + \dfrac{1}{2}\cos\theta + \dfrac{1}{4}\left(\dfrac{1 + \cos 2\theta}{2}\right)$
$= \dfrac{1}{4} + \dfrac{1}{2}\cos\theta + \dfrac{1}{8} + \dfrac{1}{8}\cos 2\theta$
$= \dfrac{3}{8} + \dfrac{1}{2}\cos\theta + \dfrac{1}{8}\cos 2\theta$

55 $\dfrac{\pi}{12} + \pi n,\ \dfrac{11\pi}{12} + \pi n$, where n denotes any integer

57 $\dfrac{\pi}{6}, \dfrac{5\pi}{6}, \dfrac{3\pi}{2}$ **59** $\dfrac{\pi}{4}, \dfrac{5\pi}{4}$ **61** $0, \pi, \dfrac{2\pi}{3}, \dfrac{4\pi}{3}$ **63** $0, \pi$

65 $214°20', 325°40'$ **67** $70°20', 250°20'$ **69** $153°40', 206°20'$

71 $-0.7, 0.4$

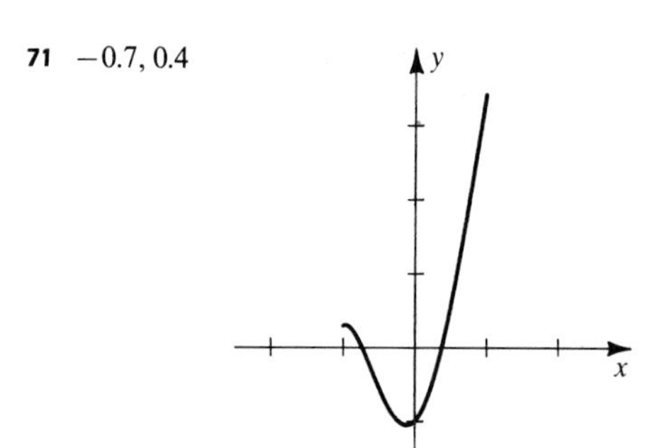

CHAPTER 2

EXERCISES 2.1

DNE denotes Does Not Exist.

1 -7 **3** 4 **5** 7 **7** π **9** -3 **11** $\frac{7}{2}$

13 4 **15** $\frac{1}{9}$ **17** 32 **19** $2x$ **21** 12

23 DNE **25** **(a)** -1 **(b)** 1 **(c)** DNE

27 **(a)** DNE **(b)** -6 **(c)** DNE

29 **(a)** DNE **(b)** DNE **(c)** DNE

31 **(a)** 3 **(b)** 1 **(c)** DNE **(d)** 2 **(e)** 2 **(f)** 2

33 **(a)** 1 **(b)** 1 **(c)** 1 **(d)** 3 **(e)** 3 **(f)** 3

35 **(a)** 1 **(b)** 0 **(c)** DNE **(d)** 1 **(e)** 0 **(f)** DNE

37 **(a)** DNE **(b)** DNE **(c)** DNE **(d)** DNE **(e)** 0 **(f)** DNE

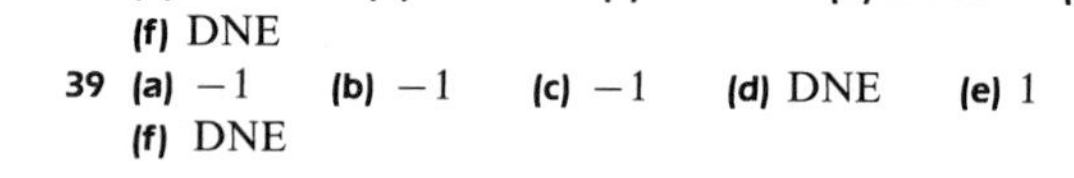

39 **(a)** -1 **(b)** -1 **(c)** -1 **(d)** DNE **(e)** 1 **(f)** DNE

41

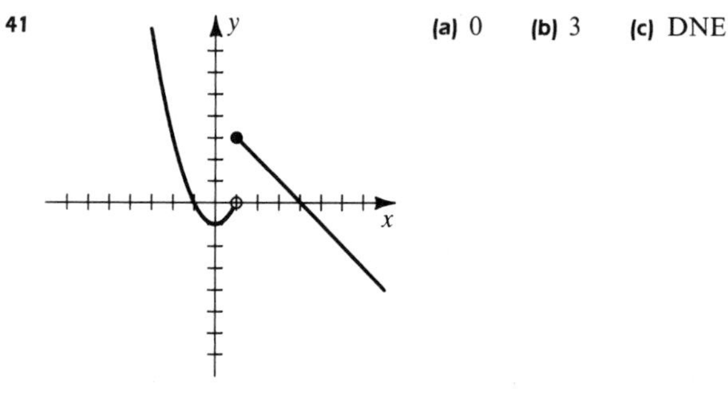

(a) 0 **(b)** 3 **(c)** DNE

43

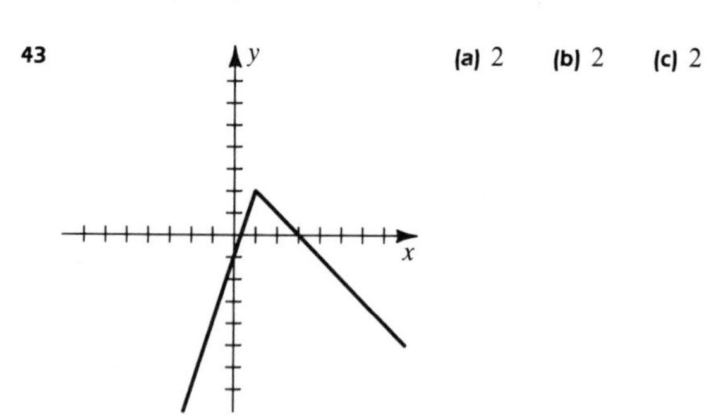

(a) 2 **(b)** 2 **(c)** 2

45 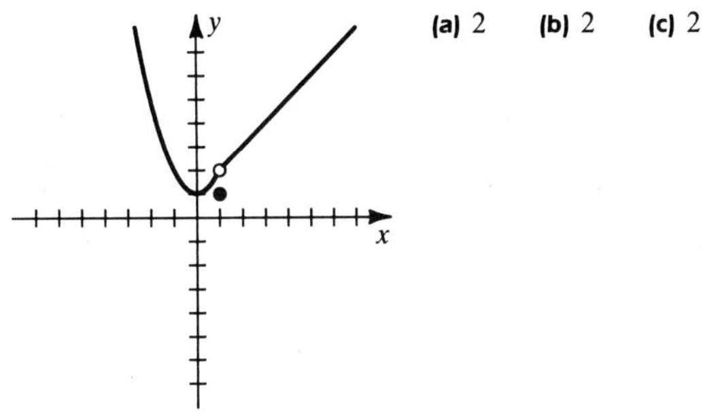

(a) 2 **(b)** 2 **(c)** 2

47 **(a)** $T(x) = \begin{cases} 0.15x & \text{if } x \le 20{,}000 \\ 0.20x - 1000 & \text{if } x > 20{,}000 \end{cases}$ **(b)** \$3000; \$3000

49 **(a)** $2g$'s, the g-force at liftoff
(b) Left-hand limit of 8—the g-force just before the second booster is released; right-hand limit of 1—the g-force just after the second booster is released
(c) Left-hand limit of 3—the g-force just before the engines are shut down; right-hand limit of 0—the g-force just after the engines are shut down

Exer. 51–56: A calculator cannot *prove* results on limits. It can only suggest that certain limits exist.

57 **(a)** Approximate values: 1.0000, 1.0000, 1.0000; -1.2802, 0.6290, -0.8913
(b) The limit does not exist.

EXERCISES 2.2

1 **(a)** $\lim_{t \to c} v(t) = K$ means that for every $\epsilon > 0$, there is a $\delta > 0$ such that if $0 < |t - c| < \delta$, then $|v(t) - K| < \epsilon$.
(b) $\lim_{t \to c} v(t) = K$ means that for every $\epsilon > 0$, there is a $\delta > 0$ such that if t is in the open interval $(c - \delta, c + \delta)$ and $t \ne c$, then $v(t)$ is in the open interval $(K - \epsilon, K + \epsilon)$.

3 **(a)** $\lim_{x \to p^-} g(x) = C$ means that for every $\epsilon > 0$, there is a $\delta > 0$ such that if $p - \delta < x < p$, then $|g(x) - C| < \epsilon$.
(b) $\lim_{x \to p^-} g(x) = C$ means that for every $\epsilon > 0$, there is a $\delta > 0$ such that if x is in the open interval $(p - \delta, p)$, then $g(x)$ is in the open interval $(C - \epsilon, C + \epsilon)$.

5 **(a)** $\lim_{z \to t^+} f(z) = N$ means that for every $\epsilon > 0$, there is a $\delta > 0$ such that if $t < z < t + \delta$, then $|f(z) - N| < \epsilon$.
(b) $\lim_{z \to t^+} f(z) = N$ means that for every $\epsilon > 0$, there is a $\delta > 0$ such that if z is in the open interval $(t, t + \delta)$, then $f(z)$ is in the open interval $(N - \epsilon, N + \epsilon)$.

7 0.005 **9** $\sqrt{16.1} - 4$ **11** $|(3.9)^2 - 16| = 0.79$

13 Given any ϵ, choose $\delta \le \epsilon/5$.

15 Given any ϵ, choose $\delta \le \epsilon/2$.

17 Given any ϵ, choose $\delta \le \epsilon/9$.

19 Given any ϵ, choose $\delta \le 2\epsilon$.

21 Given any ϵ, let δ be any positive number.

23 Given any ϵ, let δ be any positive number.

31 Every interval $(3 - \delta, 3 + \delta)$ contains numbers for which the quotient equals 1 and other numbers for which the quotient equals -1.

33 Every interval $(-1 - \delta, -1 + \delta)$ contains numbers for which the quotient equals 3 and other numbers for which the quotient equals -3.

35 $1/x^2$ can be made as large as desired by choosing x sufficiently close to 0.

37 $1/(x + 5)$ can be made as large (positively or negatively) as desired by choosing x sufficiently close to -5.

39 There are many examples; one is $f(x) = (x^2 - 1)/(x - 1)$ if $x \ne 1$ and $f(1) = 3$.

41 Every interval $(a - \delta, a + \delta)$ contains numbers such that $f(x) = 0$ and other numbers such that $f(x) = 1$.

EXERCISES 2.3

1 15 **3** -2 **5** 8 **7** $\frac{7}{5}$ **9** 81 **11** 0

13 -13 **15** $5\sqrt{2} - 20$ **17** $\pi - 3.1416$ **19** -23

21 -7 **23** DNE **25** $-\frac{3}{8}$ **27** $-\frac{1}{4}$ **29** 2

31 $\frac{72}{7}$ **33** -2 **35** -2 **37** $-\frac{1}{8}$ **39** $\frac{3}{5}$

41 -810 **43** 3 **45** 1 **47** $\frac{1}{8}$

49 **(a)** 0 **(b)** DNE **(c)** DNE

51 **(a)** 0 **(b)** 0 **(c)** 0

53 $(-1)^{n-1}; (-1)^n$

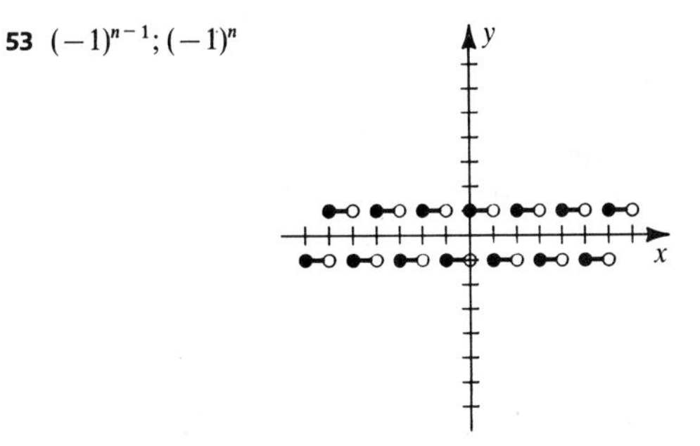

55 0; 0

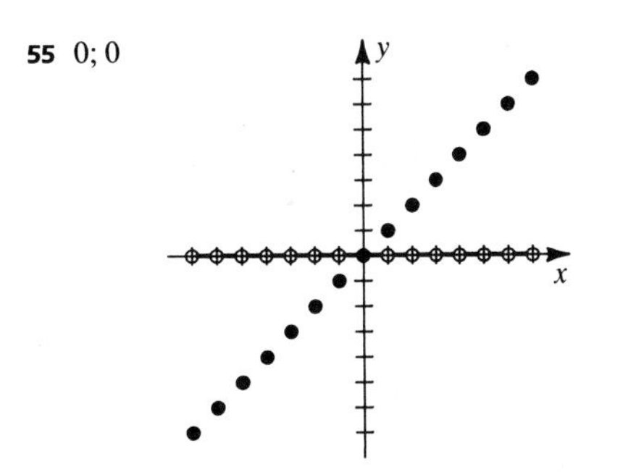

57 **(a)** $n - 1$ **(b)** n **59** **(a)** n **(b)** $n + 1$

65 *Hint:* Let $g(x) = cx^2$.

67 Because Theorem (2.8) is applicable only when the individual limits exist, and $\lim_{x \to 0} \sin \frac{1}{x}$ does not exist

69 **(a)** 0
(b) If $T < -273\,^\circ\text{C}$, the volume V is negative, an absurdity.

71 **(a)** DNE **(b)** The image is moving farther to the right.

EXERCISES 2.4

1 (a) $-\infty$ (b) ∞ (c) DNE
3 (a) $-\infty$ (b) ∞ (c) DNE
5 (a) $-\infty$ (b) $-\infty$ (c) $-\infty$
7 (a) ∞ (b) $-\infty$ (c) DNE
9 (a) ∞ (b) ∞ (c) ∞ **11** $\frac{5}{2}$ **13** $-\frac{7}{3}$
15 0 **17** $-\infty$ **19** ∞ **21** 1 **23** DNE
25 0.996664442, 0.999966666, 0.999999666, 0.999999996; the limit appears to be 1.
27 $x = -2$, $x = 2$; $y = 0$ **29** None; $y = 2$
31 $x = -3$, $x = 0$, $x = 2$; $y = 0$
33 $x = -3$, $x = 1$; $y = 1$ **35** $x = 4$; $y = 0$

Exer. 37–40: Answers are not unique.

37

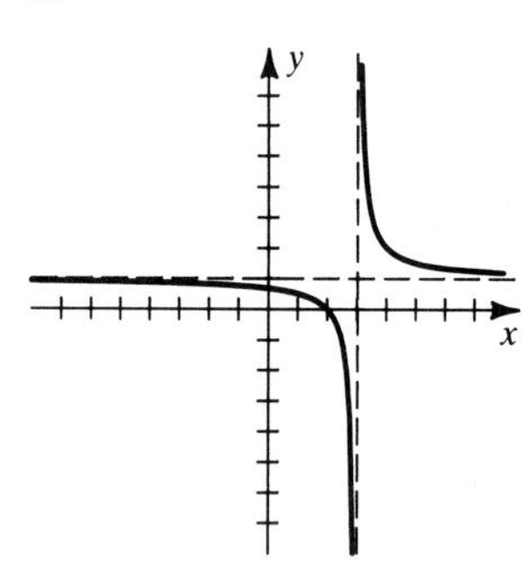

39

41 (a) $V(t) = 50 + 5t$; $A(t) = 0.5t$ (b) $c(t) = t/(10t + 100)$
(c) $c(t)$ approaches 0.1.

EXERCISES 2.5

1 Jump **3** Removable **5** Jump **7** Infinite
9 Removable **11** Jump **13** Removable
15 Removable **17** Removable
19 $\lim_{x\to 4} f(x) = 12 + \sqrt{3} = f(4)$
21 $\lim_{x\to -2} f(x) = 19 - \frac{1}{\sqrt{2}} = f(-2)$
23 f is not defined at -2. **25** $\lim_{x\to 3} f(x) = 6 \neq 4 = f(3)$
27 $\lim_{x\to 3} f(x) = 1 \neq 0 = f(3)$ **29** $\lim_{x\to 0} f(x) = 1 \neq 0 = f(0)$
31 $-3, 2$ **33** $-2, 1$
35 If $4 < c < 8$, $\lim_{x\to c} f(x) = \sqrt{c-4} = f(c)$.
$\lim_{x\to 4^+} f(x) = 0 = f(4)$ and $\lim_{x\to 8^-} f(x) = 2 = f(8)$
37 If $c > 0$, $\lim_{x\to c} f(x) = \frac{1}{c^2} = f(c)$. **39** $\left\{x : x \neq -1, \frac{3}{2}\right\}$
41 $\left[\frac{3}{2}, \infty\right)$ **43** $(-\infty, -1) \cup (1, \infty)$ **45** $\{x : x \neq -9\}$
47 $\{x : x \neq 0, 1\}$ **49** $[-5, -3] \cup [3, 4) \cup (4, 5]$
51 $\left\{x : x \neq \frac{\pi}{4} + \frac{\pi}{2}n\right\}$ **53** $\{x : x \neq 2\pi n\}$ **55** $c = \sqrt[3]{w-1}$
57 $c = \frac{1}{2} + \frac{1}{2}\sqrt{4w+1}$
59 $f(0) = -9 < 100$ and $f(10) = 561 > 100$. Since f is continuous on $[0, 10]$, there is at least one number a in $[0, 10]$ such that $f(a) = 100$.
61 $g(35°) \approx 9.79745 < 9.8$ and $g(40°) \approx 9.80180 > 9.8$. Since g is continuous on $[35°, 40°]$, there is at least one latitude θ between 35° and 40° such that $g(\theta) = 9.8$.

2.6 REVIEW EXERCISES

1 13 **3** $-4 - \sqrt{14}$ **5** $\frac{7}{8}$ **7** $\frac{32}{3}$ **9** ∞
11 3 **13** -1 **15** $4a^3$ **17** $\frac{1}{3}$ **19** $\frac{3}{2}$
21 0 **23** $-\infty$ **25** $-\infty$

27 (a) 6 (b) 4 (c) DNE

29 (a) $\frac{1}{11}$ (b) -1 (c) DNE

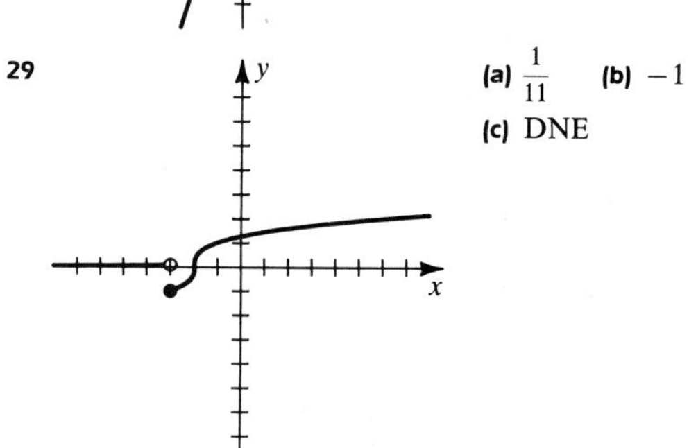

31 (a) 1 (b) 3 (c) DNE

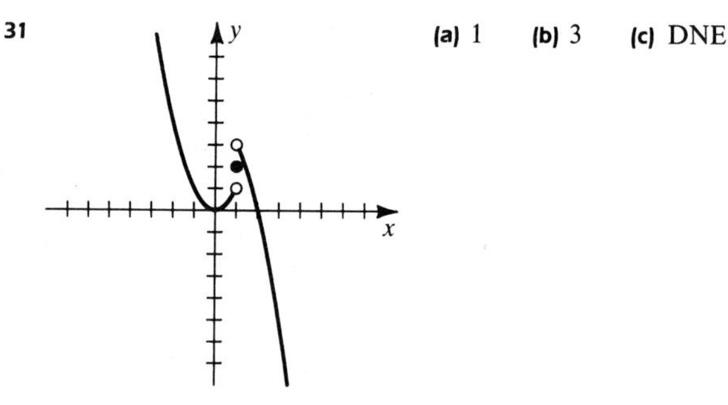

33 Given any ϵ, choose $\delta \leq \epsilon/5$. **35** ± 4 **37** 0, 2
39 $\mathbb{R}$ **41** $[-3, -2) \cup (-2, 2) \cup (2, 3]$
43 $\lim_{x\to 8} f(x) = 7 = f(8)$

CHAPTER 3

EXERCISES 3.1

1 (a) $10a - 4$ (b) $y = 16x - 20$
3 (a) $3a^2$ (b) $y = 12x - 16$
5 (a) 3 (b) $y = 3x + 2$

7 **(a)** $\dfrac{1}{2\sqrt{a}}$ **(c)**

(b) $y = \frac{1}{4}x + 1$

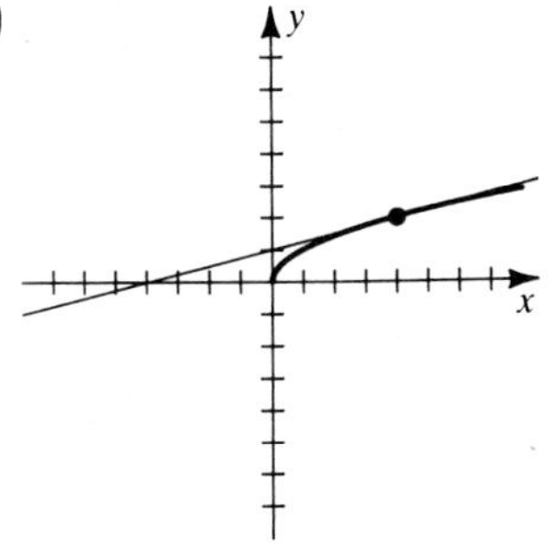

9 **(a)** $-\dfrac{1}{a^2}$ **(c)**

(b) $y = -\frac{1}{4}x + 1$

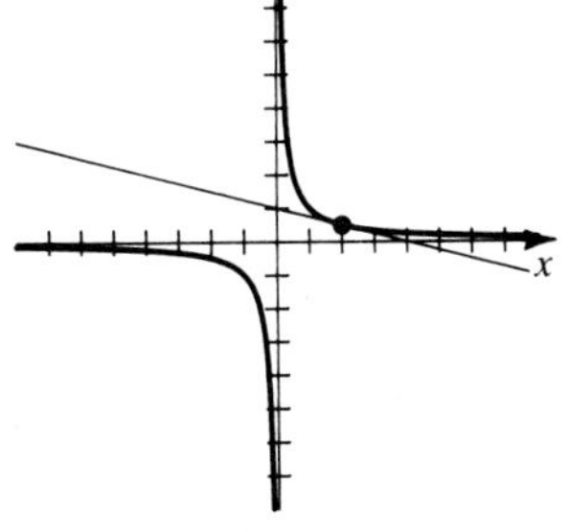

11 **(a)**

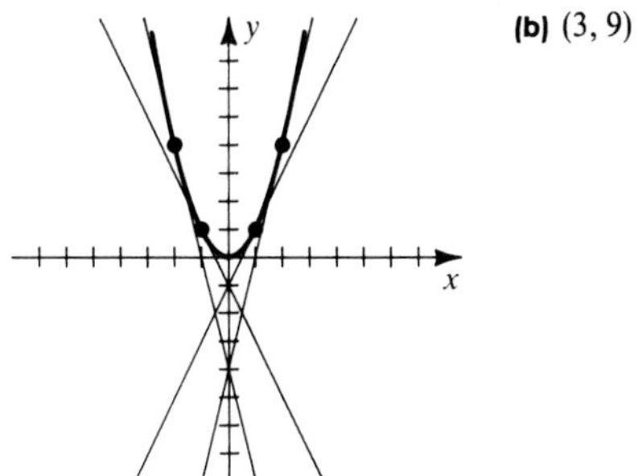

(b) $(3, 9)$

13 In cm/sec: **(a)** 11.8; 11.4; 11.04 **(b)** 11

15 In ft/sec: **(a)** -32 **(b)** $-32\sqrt{10}$

17 **(a)** Creature at $x = 3$ **(b)** No hit

19 **(a)** 6.5 **(b)** 6 **21** **(a)** $p_v = -\dfrac{200}{v^2}$ **(b)** -2

23

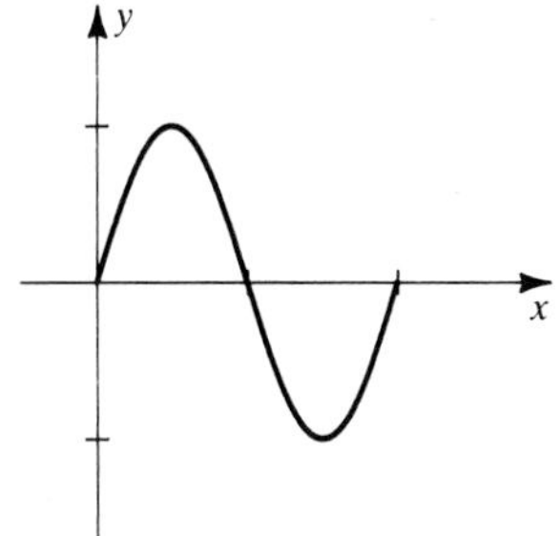

(a) -1
(b) -0.9703; -0.9713

25 In ft/sec: -0.06864; -0.06426; -0.6382

EXERCISES 3.2

1 **(a)** $-10x + 8$ **(b)** $\mathbb{R}$ **(c)** $y = 18x + 7$
(d) $\left(\frac{4}{5}, \frac{26}{5}\right)$

3 **(a)** $3x^2 + 1$ **(b)** $\mathbb{R}$ **(c)** $y = 4x - 2$ **(d)** None

5 **(a)** 9 **(b)** $\mathbb{R}$ **(c)** $y = 9x - 2$ **(d)** None

7 **(a)** 0 **(b)** $\mathbb{R}$ **(c)** $y = 37$ **(d)** All

9 **(a)** $\dfrac{-3}{x^4}$ **(b)** $(-\infty, 0) \cup (0, \infty)$ **(c)** $y = -\frac{3}{16}x + \frac{1}{2}$
(d) None

11 **(a)** $\dfrac{1}{x^{3/4}}$ **(b)** $(0, \infty)$ **(c)** $y = \frac{1}{27}x + 9$
(d) None

13 $18x^5$; $90x^4$; $360x^3$ **15** $6x^{-1/3}$; $-2x^{-4/3}$; $\frac{8}{3}x^{-7/3}$

17 $36t^{-1/5}$ **19** 0

21 **(a)** No, because f is not differentiable at $x = 0$
(b) Yes, because f' exists for every number in $[1, 3)$

23 **(a)** No **(b)** Yes **25** **(a)** Yes **(b)** No

27 **(a)** Yes **(b)** Yes **29** **(a)** No **(b)** No

31 $f'(-1) = 1$, $f'(1) = 0$, $f'(2)$ is undefined, $f'(3) = -1$

33 The right-hand and left-hand derivatives are unequal at $a = 5$.

35 The right-hand and left-hand derivatives are unequal at $a = 2$.

37 $\{x: x \neq 0\}$

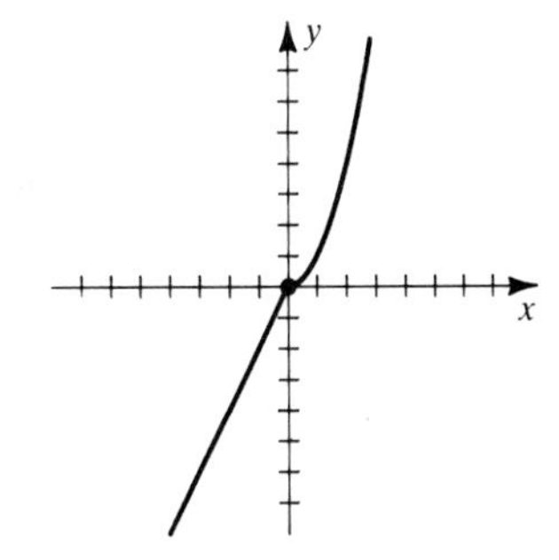

39 $\{x: x \neq -1\}$

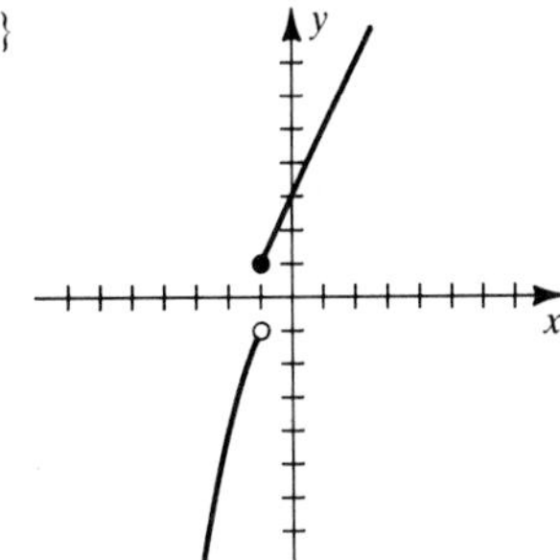

41 f is not differentiable at ± 1, ± 2.

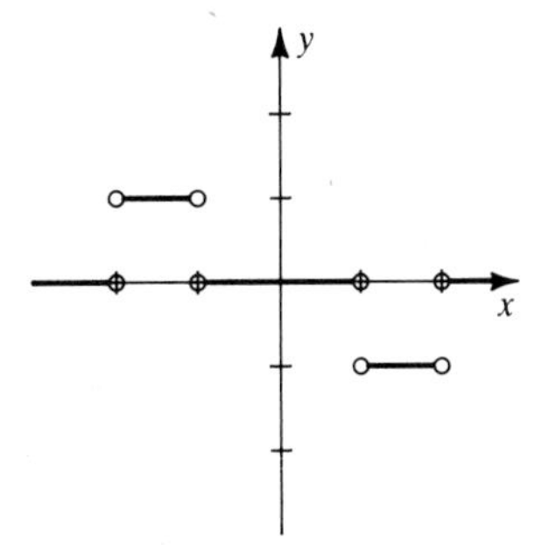

43 $\frac{1}{8}$ **45** $F_C = \frac{9}{5}$ **47** $V_r = 4\pi r^2$

49 **(a)** $A_r = 2\pi r$ **(b)** 1000π ft^2/ft

51 **(a)** The formula gives an approximation of the slope of the tangent line at $(a, f(a))$ by using the slope of the secant line through $P(a - h, f(a - h))$ and $Q(a + h, f(a + h))$.
(b) Subtract and add $f(a)$ in the numerator and consider two limits.
(c) -2.0406; -2.0004; -2.0000 **(d)** -2

53 **(a)** 53.2 ft/sec **(b)** 88.3 ft/sec

55 $x \approx -0.7$

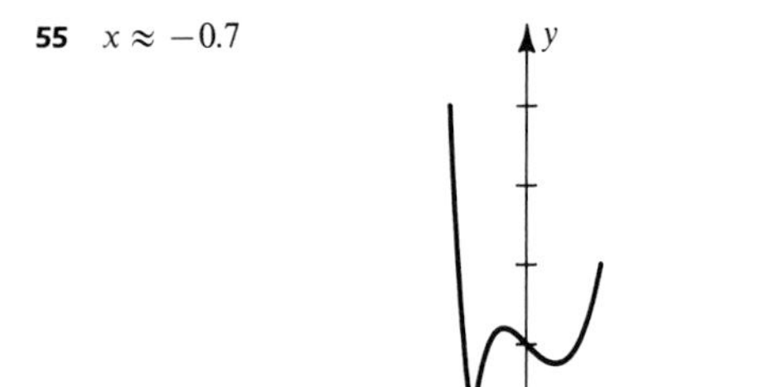

EXERCISES 3.3

1 $10t^{2/3}$ **3** $-20s^3 + 8s - 1$ **5** $6x + \frac{4}{3}x^{1/3}$

7 $10x^4 + 9x^2 - 28x$ **9** $\frac{5}{2}x^{3/2} + \frac{3}{2}x^{1/2} - 2x^{-1/2}$

11 $18r^5 - 21r^2 + 4r$ **13** $416x^3 - 195x^2 + 64x - 20$

15 $\dfrac{23}{(3x + 2)^2}$ **17** $\dfrac{-27z^2 + 12z + 70}{(2 - 9z)^2}$ **19** $\dfrac{6v^2}{(v^3 + 1)^2}$

21 $-\dfrac{3t + 10}{3\sqrt[3]{t}(3t - 5)^2}$ **23** $-\dfrac{1 + 2x + 3x^2}{(1 + x + x^2 + x^3)^2}$ **25** $\dfrac{-14x}{(x^2 + 5)^2}$

27 $2t - \dfrac{2}{t^3}$ **29** $-\dfrac{4}{81}s^{-5}$ **31** $10(5x - 4)$ **33** $\dfrac{-10}{(5r - 4)^3}$

35 $\dfrac{-21t^2 - 20t + 6}{5(2 + 7t^2)^2}$ **37** $2 - \dfrac{4}{x^2} - \dfrac{6}{x^3}$ **39** $\frac{1}{7}(6x - 5)$

41 $-2, 3$ **43** $0, 4$ **45** $-5, 3$ **47** $\dfrac{-3x + 2}{x^3}$

49 $\dfrac{4x - 3}{3\sqrt[3]{x^2}}$ **51** $\dfrac{2}{(x + 1)^3}$ **53** $y = \frac{4}{5}x + \frac{13}{5}$

55 **(a)** $-2, \frac{2}{3}$ **(b)** $-\frac{4}{3}, 0$ **57** $\left(\frac{1}{9}, -\frac{8}{27}\right), (1, 0)$

59 $x = 0$

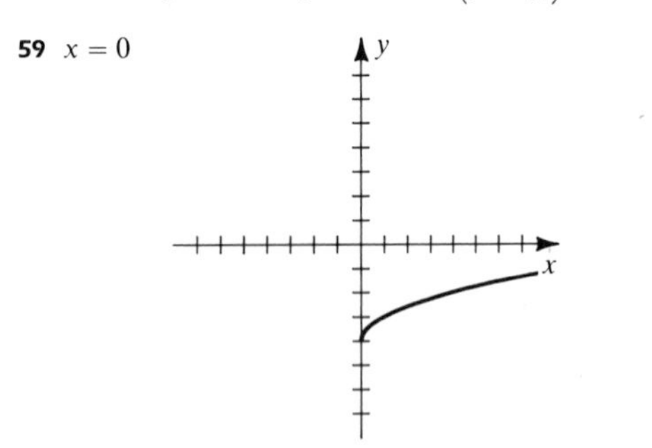

61 In ft/sec: **(a)** 4, 10, 18 **(b)** $6\sqrt{5}$ **63** $-\frac{1}{2}$

65 $y = 2x - 1$, $y = 18x - 81$

67 **(a)** 1 **(b)** -3 **(c)** -4 **(d)** 11 **(e)** $-\frac{1}{25}$ **(f)** $\frac{1}{9}$

69 **(a)** -4 **(b)** 1 **(c)** -20 **(d)** $-\frac{1}{4}$

73 $(8x - 1)(x^2 + 4x + 7)(3x^2) + (8x - 1)(2x + 4)(x^3 - 5) + 8(x^2 + 4x + 7)(x^3 - 5)$

75 $x(2x^3 - 5x - 1)(12x) + x(6x^2 - 5)(6x^2 + 7) + (2x^3 - 5x - 1)(6x^2 + 7)$

77 **(a)** $\frac{1}{4}$ cm/min **(b)** 36π cm^3/min **(c)** 12π cm^2/min

79 In cm^2/sec: **(a)** 3200π **(b)** 6400π **(c)** 9600π

81 **(a)** 1.01 **(b)**

(c) 1; l_2 is nearly parallel to the tangent line, but l_1 is not.

EXERCISES 3.4

1 $-4 \sin x$ **3** $5 \csc v(1 - v \cot v)$

5 $t^2 \sin t - 2t \cos t + 1$ **7** $\dfrac{\theta \cos \theta - \sin \theta}{\theta^2}$

9 $t^2(t \cos t + 3 \sin t)$

11 $-2x \csc^2 x + 2 \cot x + x^2 \sec^2 x + 2x \tan x$

13 $\dfrac{2 \sin z}{(1 + \cos z)^2}$ **15** $-\csc x(1 + 2 \cot^2 x)$

17 $-x \csc^2 x - \csc^3 x + \cot x - \csc x \cot^2 x$

19 $-\sin x$ **21** $\dfrac{\sec^2 x + x^2 \sec^2 x - 2x \tan x}{(1 + x^2)^2}$ **23** $-\csc^2 v$

25 $-\cos x - \sin x$ **27** $\sin \phi + \sec \phi \tan \phi$

29 $y - \sqrt{2} = \sqrt{2}\left(x - \frac{\pi}{4}\right)$; $y - \sqrt{2} = -\frac{1}{\sqrt{2}}\left(x - \frac{\pi}{4}\right)$

31 $\left(\frac{\pi}{4}, \sqrt{2}\right), \left(\frac{5\pi}{4}, -\sqrt{2}\right)$ **33** $\left(\frac{\pi}{4}, 2\sqrt{2}\right)$

35 **(a)** $\frac{\pi}{6} + 2\pi n, \frac{5\pi}{6} + 2\pi n$ **(b)** $y = x + 2$

37 **(a)** $\frac{\pi}{4} + 2\pi n, \frac{7\pi}{4} + 2\pi n$ **(b)** $y - 4 = \sqrt{3}\left(x - \frac{\pi}{6}\right)$

39 $x \approx 0.9, 2.4, 3.7$

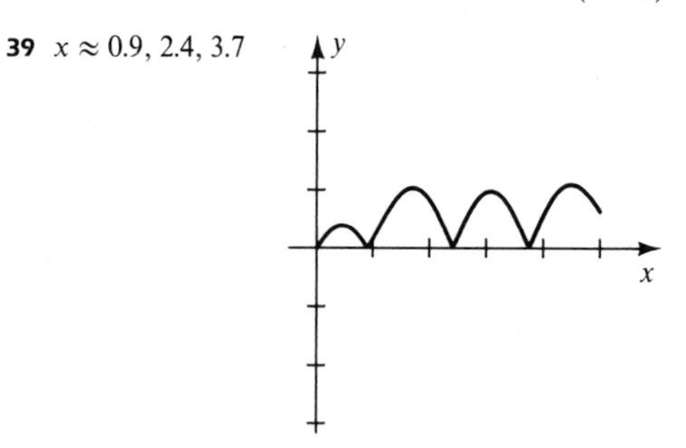

41 $\frac{\pi}{6} + 2\pi n, \frac{5\pi}{6} + 2\pi n$ **43** $(16, 96)$

45 **(a)** $-\sin x$; $-\cos x$; $\sin x$; $\cos x$ **(b)** $\sin x$

47 $2\sec^2 x\,(3\tan^2 x + 1)$

49 $D_x \cot x = D_x\left(\dfrac{\cos x}{\sin x}\right) = \dfrac{(\sin x)(-\sin x) - (\cos x)(\cos x)}{\sin^2 x}$

$= \dfrac{-1(\sin^2 x + \cos^2 x)}{\sin^2 x} = -\dfrac{1}{\sin^2 x} = -\csc^2 x$

51 $D_x \sin 2x = D_x\,(2\sin x\cos x)$

$= 2[\sin x\,(-\sin x) + \cos x\cos x]$

$= 2(\cos^2 x - \sin^2 x) = 2\cos 2x$

EXERCISES 3.5

1 **(a)** $(4x-4)\,\Delta x + 2(\Delta x)^2$; $(4x-4)\,dx$ **(b)** -0.72; -0.8

3 **(a)** $\dfrac{-(2x+\Delta x)\,\Delta x}{x^2(x+\Delta x)^2}$; $-\dfrac{2}{x^3}\,dx$

(b) $-\dfrac{7}{363} \approx -0.01928$; $-\dfrac{1}{45} = -0.0\overline{2}$

5 **(a)** $-9\,\Delta x$ **(b)** $-9\,dx$ **(c)** 0

7 **(a)** $(6x+5)\,\Delta x + 3(\Delta x)^2$ **(b)** $(6x+5)\,dx$ **(c)** $-3(\Delta x)^2$

9 **(a)** $\dfrac{-\Delta x}{x(x+\Delta x)}$ **(b)** $-\dfrac{1}{x^2}\,dx$ **(c)** $\dfrac{-(\Delta x)^2}{x^2(x+\Delta x)}$ **11** -3.94

13 0.92 **15** 1.80 **17** 2.12

19 **(a)** With $h = 0.001$, $y \approx -0.98451 - 0.27315(x - 2.5)$.

(b) -1.011825 **(c)** -1.011825

(d) They are equal because the tangent line approximation is equivalent to using (3.31).

21 ± 0.02; $\pm 2\%$ **23** ± 0.04; $\pm 4\%$ **25** 1.1

27 $\pm 45\%$ **29** ± 0.06

31 $\pm 1.92\pi$ in.^{2} $\approx \pm 6.03$ in.^{2}; ± 0.0075; $\pm 0.75\%$

33 30 in.^{3}; 30.301 in.^{3}

35 3301.661 ft^2; ± 11.464 ft^2; ± 0.00347; $\pm 0.347\%$

37 $\dfrac{1}{50\pi}$ cm ≈ 0.00637 cm **39** -1 cm **41** 40% increase

43 $\dfrac{5\sqrt{2}\pi}{81}$ lb ≈ 0.274 lb **45** $\pm\dfrac{\pi}{9}$ ft $\approx \pm 0.35$ ft **47** $\pm 0.19°$

49 *Hint:* Show that $v\,dp = -\dfrac{c}{v}\,dv$.

51 dA is the shaded region.

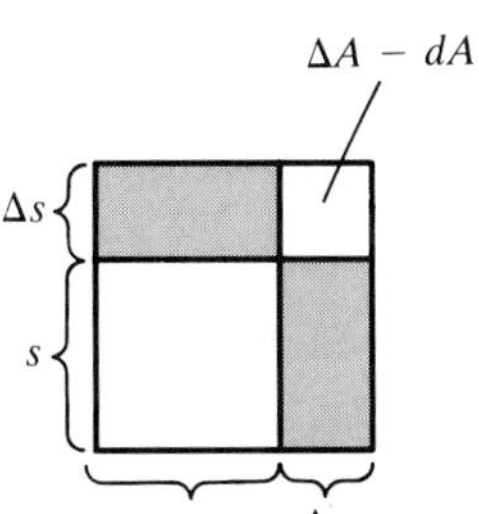

EXERCISES 3.6

1 $6x^2(x^3-4)$ **3** $\dfrac{-3}{2(3x-2)^{3/2}}$ **5** $6x\sec^2(3x^2)$

7 $3(x^2-3x+8)^2(2x-3)$ **9** $-40(8x-7)^{-6}$

11 $-\dfrac{7x^2+1}{(x^2-1)^5}$ **13** $5(8x^3-2x^2+x-7)^4(24x^2-4x+1)$

15 $17{,}000(17v-5)^{999}$

17 $2(6x-7)^2(8x^2+9)(168x^2-112x+81)$

19 $12\left(z^2-\dfrac{1}{z^2}\right)^5\left(z+\dfrac{1}{z^3}\right)$ **21** $8r^2(8r^3+27)^{-2/3}$

23 $-5v^4(v^5-32)^{-6/5}$ **25** $\dfrac{w^2+4w-9}{2w^{5/2}}$ **27** $\dfrac{6(3-2x)}{(4x^2+9)^{3/2}}$

29 $2x\cos(x^2+2)$ **31** $-15\cos^4 3\theta\sin 3\theta$

33 $4(2z+1)\sec(2z+1)^2\tan(2z+1)^2$

35 $(2-3s^2)\csc^2(s^3-2s)$ **37** $-6x\sin(3x^2) - 6\cos 3x\sin 3x$

39 $-4\csc^2 2\phi\cot 2\phi$ **41** $2z\cot 5z - 5z^2\csc^2 5z$

43 $2\tan\theta\sec^5\theta + 3\tan^3\theta\sec^3\theta$

45 $25(\sin 5x - \cos 5x)^4(\cos 5x + \sin 5x)$

47 $-9\cot^2(3w+1)\csc^2(3w+1)$ **49** $\dfrac{4}{1-\sin 4w}$

51 $6\tan 2x\sec^2 2x\,(\tan 2x - \sec 2x)$ **53** $\dfrac{\cos\sqrt{x}}{2\sqrt{x}} + \dfrac{\cos x}{2\sqrt{\sin x}}$

55 $\dfrac{8\cos\sqrt{3-8\theta}\,\sin\sqrt{3-8\theta}}{\sqrt{3-8\theta}}$

57 $x\sec^2\sqrt{x^2+1} + \dfrac{x\tan\sqrt{x^2+1}}{\sqrt{x^2+1}}$

59 $\dfrac{2\sec\sqrt{4x+1}\,\tan\sqrt{4x+1}}{\sqrt{4x+1}}$ **61** $-\dfrac{3\csc^2 3x\cot 3x}{\sqrt{4+\csc^2 3x}}$

63 **(a)** $y - 81 = 864(x-2)$; $y - 81 = -\dfrac{1}{864}(x-2)$ **(b)** $\dfrac{1}{2}, 1, \dfrac{3}{2}$

65 **(a)** $y = 32$; $x = 1$ **(b)** ± 1

67 **(a)** $y = 6x$; $y = -\dfrac{1}{6}x$ **(b)** $\dfrac{\pi}{3} + \dfrac{2\pi}{3}n$

69 $\dfrac{3}{2(3z+1)^{1/2}}$; $-\dfrac{9}{4(3z+1)^{3/2}}$ **71** $20(4r+7)^4$; $320(4r+7)^3$

73 $3\sin^2 x\cos x$; $6\sin x\cos^2 x - 3\sin^3 x$ **75** $4\dfrac{1}{48}$

77 $\dfrac{dK}{dt} = mv\dfrac{dv}{dt}$ **79** -0.1819 lb/sec **81** -4; 15

83 $-\dfrac{2}{5}$ **85** 4.91

87 **(a)** *Hint:* Differentiate both sides of $f(-x) = f(x)$ using the chain rule.

(b) *Hint:* Differentiate both sides of $f(-x) = -f(x)$ using the chain rule.

89 **(a)** $\dfrac{dW}{dt} = (1.644 \times 10^{-4})L^{1.74}\dfrac{dL}{dt}$ **(b)** 7.876 cm/month

91 **(a)** 60π cm^2; ± 1.508 cm^2 **(b)** $\pm 0.8\%$

EXERCISES 3.7

1 $-\dfrac{8x}{y}$ **3** $-\dfrac{6x^2+2xy}{x^2+3y^2}$ **5** $\dfrac{10x-y}{x+8y}$ **7** $-\sqrt{\dfrac{y}{x}}$

9 $\dfrac{-4x\sqrt{xy}-y}{x}$ **11** $\dfrac{1}{6\sin 3y\cos 3y - 1} = \dfrac{1}{3\sin 6y - 1}$

13 $\dfrac{-y\cot(xy)\csc(xy)}{1 + x\cot(xy)\csc(xy)}$ **15** $\dfrac{\cos y}{x\sin y + 2y}$

17 $\dfrac{4x\sqrt{\sin y}}{4y\sqrt{\sin y} - \cos y}$ **19** $-\dfrac{\sqrt{2}}{5}$ **21** -1 **23** 4

25 $-\dfrac{36}{23}$ **27** -2π **29** $-\dfrac{3}{4y^3}$ **31** $-\dfrac{2x}{y^5}$

33 $\dfrac{\sin y}{(1+\cos y)^3}$ **35** An infinite number **37** None

39 Let $f_c(x) = \begin{cases} \sqrt{x} & \text{if } 0 \le x \le c \\ -\sqrt{x} & \text{if } x > c \end{cases}$ for any $c > 0$

41 0.09 **43** **(a)** 1.28 **(b)** $c \approx 1.25$

EXERCISES 3.8

1 60 **3** $\frac{4}{15}$ **5** 3 **7** $-\frac{24}{17}$

9 $0.15\pi \approx 0.471 \text{ cm}^2/\text{min}$ **11** $\frac{20}{9\pi} \approx 0.707$ ft/min

13 $-\frac{3}{8}\sqrt{336} \approx -6.9$ ft/sec **15** $\frac{64}{11}$ ft/sec; $\frac{20}{11}$ ft/sec

17 $-7442\pi \approx -23{,}380 \text{ in.}^3/\text{hr}$ **19** $\frac{10}{3}$ ft/sec

21 $5 \text{ in.}^3/\text{min}$ (increasing) **23** $\frac{15}{32}\sqrt{3} \approx 0.81$ ft/min

25 $-\frac{\sqrt{2}}{5\sqrt[4]{3}} \approx -0.2149$ cm/min **27** π m/sec

29 $\frac{11}{1600} = 0.006875$ ohm/sec **31** $\frac{13.37}{112\pi} \approx 0.038$ ft/min

33 64 ft/sec **35** $\frac{180(6 + \sqrt{2})}{\sqrt{10 + 3\sqrt{2}}} \approx 353.6$ mi/hr

37 $-\frac{27}{25\pi} \approx -0.3438$ in./hr **39** $\frac{10{,}000\pi}{135} \approx 232.7$ ft/sec

41 $\frac{\pi}{10}\sqrt{3} \approx 0.54 \text{ in.}^2/\text{min}$

43 $\frac{2{,}640{,}000}{\sqrt{180{,}400}} \approx 6215.6$ ft/min ≈ 70.63 mi/hr

45 $\frac{1000\pi}{3}$ ft/sec ≈ 714.0 mi/hr

47 Ground speed is $\frac{175}{88}\frac{d\theta}{dt}$ mi/hr.

49 **(a)** $2v\frac{dv}{dt} = gr(1 + \sec^2\theta)\frac{d\theta}{dt}$

(b) $2v\frac{dv}{dt} = g\tan\theta\,(1 + \sec^2\theta)\frac{dr}{dt}$ **51** 19.25 mi/hr

3.9 REVIEW EXERCISES

1 $\frac{-24x}{(3x^2 + 2)^2}$ **3** $6x^2 - 7$ **5** $\frac{3}{\sqrt{6t + 5}}$

7 $\frac{2(7z - 2)}{3(7z^2 - 4z + 3)^{2/3}}$ **9** $-\frac{144x}{(3x^2 - 1)^5}$ **11** $-\frac{4(r + r^{-3})}{(r^2 - r^{-2})^3}$

13 $\frac{12}{5(3x + 2)^{1/5}}$ **15** $\frac{1024s(2s^2 - 1)^3(18s^3 - 27s + 4)}{(1 - 9s^3)^5}$

17 $3(x^6 + 1)^4(3x + 2)^2(33x^6 + 20x^5 + 3)$

19 $(9s - 1)^3(108s^2 - 139s + 39)$ **21** $12x + \frac{5}{x^2} - \frac{4}{3x^{5/3}}$

23 $\frac{-53}{2\sqrt{(2w + 5)(7w - 9)^3}}$ **25** $-\frac{\sin 2r}{\sqrt{1 + \cos 2r}}$

27 $12x^2 \sin 8x^3$ **29** $5\sec x\,(\sec x + \tan x)^5$

31 $2x(\cot 2x - x\csc^2 2x)$ **33** $\frac{2}{1 + \cos 2\theta}$

35 $\frac{(\cos\sqrt[3]{x} - \sin\sqrt[3]{x})^2(\cos\sqrt[3]{x} + \sin\sqrt[3]{x})}{\sqrt[3]{x^2}}$

37 $\frac{\csc u\,(1 - \cot u + \csc u)}{(\cot u + 1)^2}$ **39** $10\tan 5x \sec^2 5x$

41 $\frac{\tan^3(\sqrt[4]{\theta})\sec^2(\sqrt[4]{\theta})}{\sqrt[4]{\theta^3}}$ **43** $\frac{4xy^2 - 15x^2}{12y^2 - 4x^2y}$ **45** $\frac{1}{\sqrt{x}(3\sqrt{y} + 2)}$

47 $\frac{\cos(x + 2y) - y^2}{2xy - 2\cos(x + 2y)}$ **49** $y = \frac{9}{4}x - 3$; $y = -\frac{4}{9}x + \frac{70}{9}$

51 $\frac{7\pi}{12} + \pi n, \frac{11\pi}{12} + \pi n$

53 $15x^2 + \frac{2}{\sqrt{x}}$; $30x - \frac{1}{\sqrt{x^3}}$; $30 + \frac{3}{2\sqrt{x^5}}$

55 $\frac{5(y^2 - 4xy - x^2)}{(y - 2x)^3} = -\frac{40}{(y - 2x)^3}$

57 **(a)** $6x\,\Delta x + 3(\Delta x)^2$ **(b)** $6x\,dx$ **(c)** $-3(\Delta x)^2$

59 $\pm 0.06\sqrt{3} \approx \pm 0.104 \text{ in.}^2$; $\pm 1.5\%$ **61** -0.57

63 **(a)** 2 **(b)** -7 **(c)** -14 **(d)** 21 **(e)** $-\frac{10}{9}$

(f) $-\frac{19}{27}$

65 **(a)** Vertical tangent line at $(-1, -4)$
(b) Cusp at $(8, -1)$

67 2% **69** $\frac{68\pi}{5} \text{ ft}^2/\text{ft}$ **71** $\frac{5}{6} \text{ ft}^3/\text{min}$ **73** $\frac{dp}{dv} = -\frac{p}{v}$

75 **(a)** $h(t) = 60 - 50\cos\frac{\pi}{15}t$ **(b)** 10.4 ft/sec

CHAPTER 4

EXERCISES 4.1

1 Maximum of 4 at 2; minimum of 0 at 4; local maximum at $x = 2$, $6 \le x \le 8$; local minimum at $x = 4$, $6 < x < 8$, $x = 10$

3 **(a)** Min: $f(-3) = -6$; max: none
(b) Min: none; max: $f(-1) = \frac{2}{3}$
(c) Min: none; max: $f(-1) = \frac{2}{3}$
(d) Min: $f(1) = -\frac{2}{3}$; max: $f(3) = 6$

5

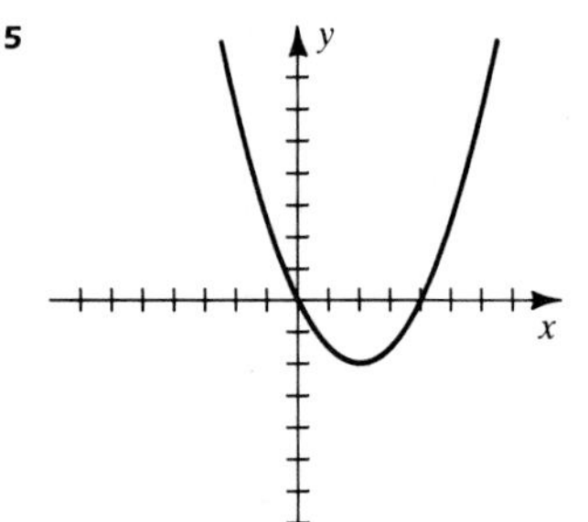

(a) Min: $f(2) = -2$; max: none
(b) Min: none; max: none
(c) Min: $f(2) = -2$; max: none
(d) Min: $f(2) = -2$; max: $f(5) = \frac{5}{2}$

7 Min: $f(-2) = f(1) = -3$; max: $f(-3) = f(0) = 5$

9 Min: $f(8) = -3$; max: $f(0) = 1$ **11** $\frac{3}{8}$ **13** $-2, \frac{5}{3}$

15 2 **17** ± 4 **19** $\frac{[illegible] + \sqrt{153}}{8}, \pm 2$ **21** $0, \frac{15}{7}, \frac{5}{2}$

23 None **25** $\pi n, \frac{2\pi}{3} + 2\pi n, \frac{4\pi}{3} + 2\pi n$

27 $\frac{\pi}{6} + 2\pi n, \frac{5\pi}{6} + 2\pi n, \frac{3\pi}{2} + 2\pi n$

29 $\frac{3\pi}{2} + 2\pi n$

31 πn **33** None

35 $0, \pm\sqrt{k\pi - 1}$ for $k = 1, 2, 3, \ldots$

37 **(a)** Since $f'(x) = \frac{1}{3}x^{-2/3}$, $f'(0)$ does not exist. If $a \neq 0$, then $f'(a) \neq 0$. Hence 0 is the only critical number of f. The number $f(0) = 0$ is not a local extremum, since $f(x) < 0$ if $x < 0$ and $f(x) > 0$ if $x > 0$.

(b) The only critical number is 0, for the same reasons given in part (a). The number $f(0) = 0$ is a local minimum, since $f(x) > 0$ if $x \neq 0$.

39 **(a)** There is a critical number, 0, but $f(0)$ is not a local extremum, since $f(x) < f(0)$ if $x < 0$ and $f(x) > f(0)$ if $x > 0$.

(b)

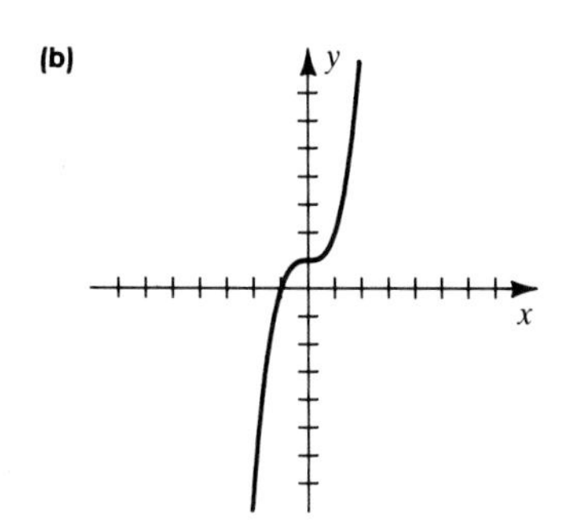

(c) The function is continuous at every number a, since $\lim_{x\to a} f(x) = f(a)$. If $0 < x_1 < x_2 < 1$, then $f(x_1) < f(x_2)$ and hence there is neither a maximum nor a minimum on (0, 1).

(d) This does not contradict Theorem (4.3) because the interval (0, 1) is open.

41 **(a)** If $f(x) = cx + d$ and $c \neq 0$, then $f'(x) = c \neq 0$. Hence there are no critical numbers.

(b) On $[a, b]$ the function has absolute extrema at a and b.

43 If $x = n$ is an integer, then $f'(n)$ does not exist. Otherwise, $f'(x) = 0$ for every $x \neq n$.

45 If $f(x) = ax^2 + bx + c$ and $a \neq 0$, then $f'(x) = 2ax + b$. Hence $-b/(2a)$ is the only critical number of f.

47 Since $f'(x) = nx^{n-1}$, the only possible critical number is $x = 0$, and $f(0) = 0$. If n is even, then $f(x) > 0$ if $x \neq 0$ and hence 0 is a local minimum. If n is odd, then 0 is not an extremum, since $f(x) < 0$ if $x < 0$ and $f(x) > 0$ if $x > 0$.

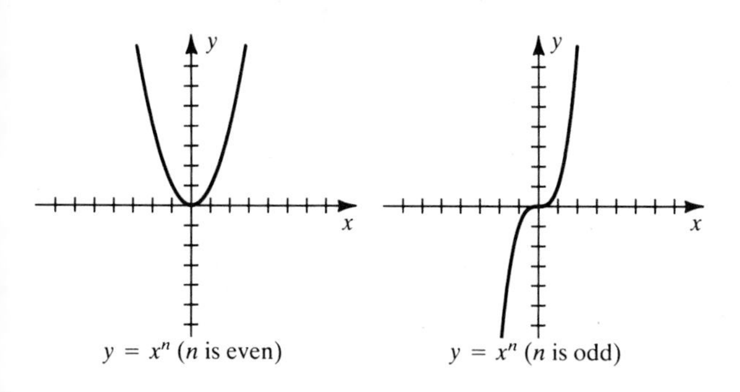

$y = x^n$ (n is even) $y = x^n$ (n is odd)

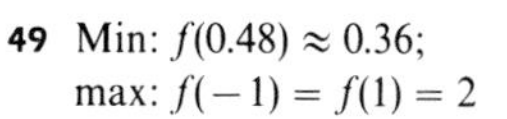

49 Min: $f(0.48) \approx 0.36$; max: $f(-1) = f(1) = 2$

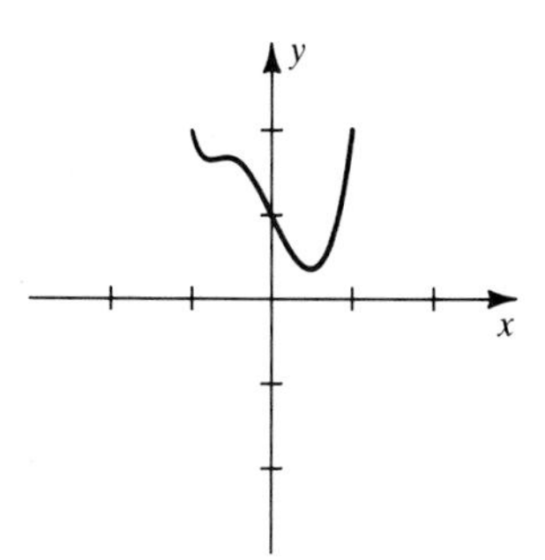

51 $-2.41, -0.92, 0.41, 1.41$

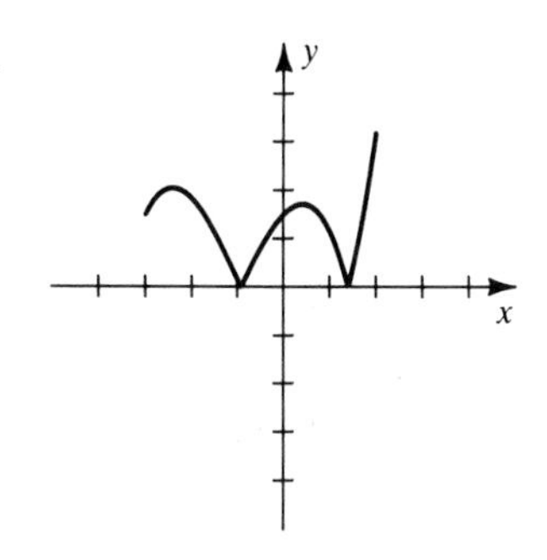

EXERCISES 4.2

1 3, 7 **3** 2 **5** 0 **7** $\frac{\pi}{4}, \frac{3\pi}{4}$ **9** 2

11 f is not continuous on $[0, 2]$.

13 f is not differentiable on $(-8, 8)$. **15** 2

17 $\frac{1}{3}(2 - \sqrt{7}) \approx -0.22$ **19** 2 **21** 2

23 The number c such that $\cos c = 2/\pi$ ($c \approx 0.88$)

25 $f(-1) = f(1) = 1$. $f'(x) = 1$ if $x > 0$, $f'(x) = -1$ if $x < 0$, and $f'(0)$ does not exist. This does not contradict Rolle's theorem, because f is not differentiable throughout the open interval $(-1, 1)$.

27 *Hint:* Show that $c^2 = -4$.

29 *Hint:* Let $f(x) = px + q$.

31 *Hint:* If f has degree 3, then $f'(x)$ is a polynomial of degree 2.

33 Let x be any number in $(a, b]$. Applying the mean value theorem to the interval $[a, x]$ yields $f(x) - f(a) = f'(c)(x - a) = 0(x - a) = 0$. Thus, $f(x) = f(a)$, and hence f is a constant function.

35 *Hint:* Use the method of Example 4.

37 *Hint:* Show that $dW/dt < -44$ lb/mo.

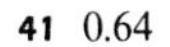

41 0.64

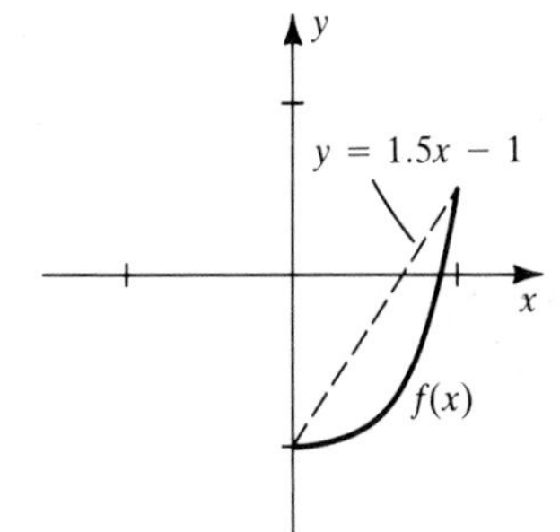

EXERCISES 4.3

1 Max: $f\left(-\frac{7}{8}\right) = \frac{129}{16}$; increasing on $\left(-\infty, -\frac{7}{8}\right]$; decreasing on $\left[-\frac{7}{8}, \infty\right)$

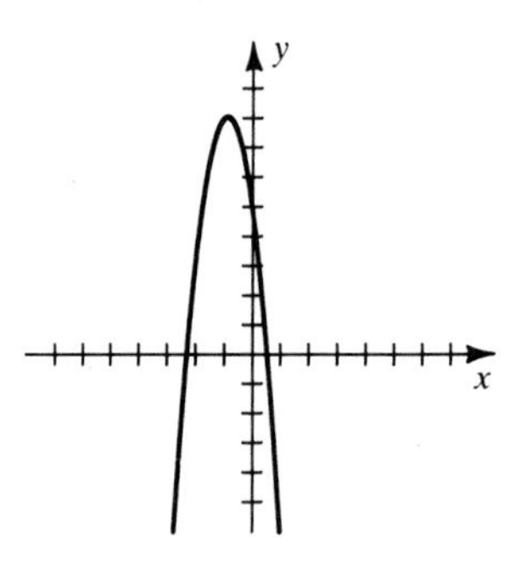

3 Max: $f(-2) = 29$; min: $f\left(\frac{5}{3}\right) = -\frac{548}{27}$; increasing on $(-\infty, -2]$ and $\left[\frac{5}{3}, \infty\right)$; decreasing on $\left[-2, \frac{5}{3}\right]$

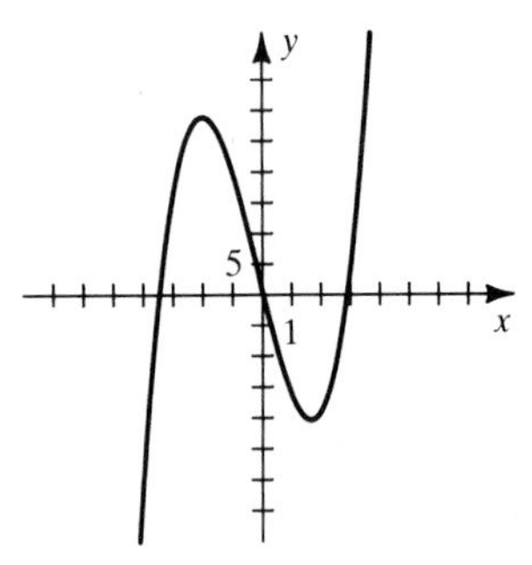

5 Max: $f(0) = 1$; min: $f(-2) = f(2) = -15$; increasing on $[-2, 0]$ and $[2, \infty)$; decreasing on $(-\infty, -2]$ and $[0, 2]$

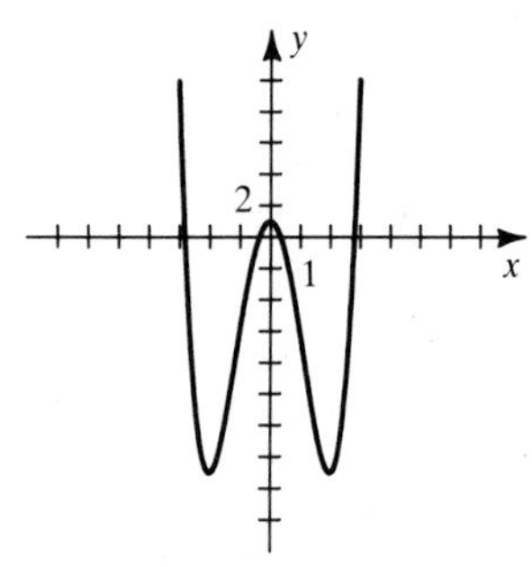

7 Max: $f\left(\frac{3}{5}\right) = \frac{216}{625} \approx 0.35$; min: $f(1) = 0$; increasing on $\left(-\infty, \frac{3}{5}\right]$ and $[1, \infty)$; decreasing on $\left[\frac{3}{5}, 1\right]$

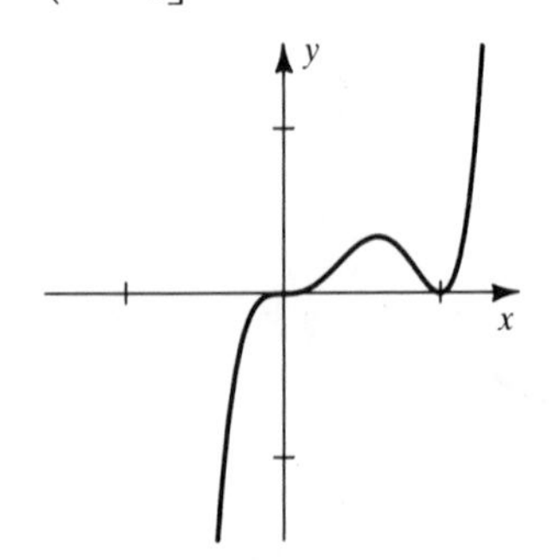

9 Min: $f(-1) = -3$; increasing on $[-1, \infty)$; decreasing on $(-\infty, -1]$

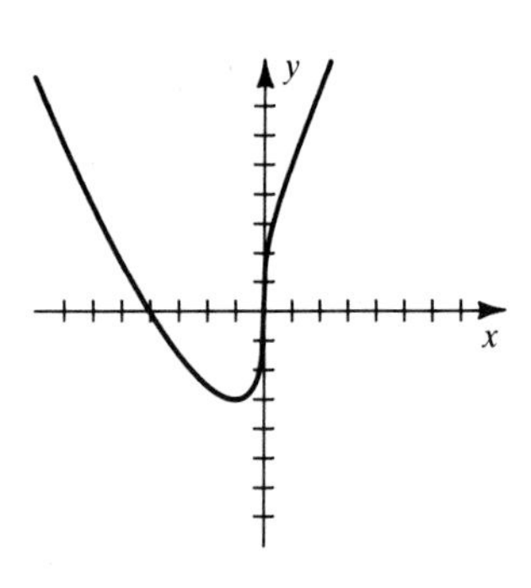

11 Max: $f\left(\frac{7}{4}\right) = \frac{441}{16}\sqrt[3]{\frac{49}{16}} + 2 \approx 42.03$; min: $f(0) = f(7) = 2$; increasing on $\left[0, \frac{7}{4}\right]$ and $[7, \infty)$; decreasing on $(-\infty, 0]$ and $\left[\frac{7}{4}, 7\right]$

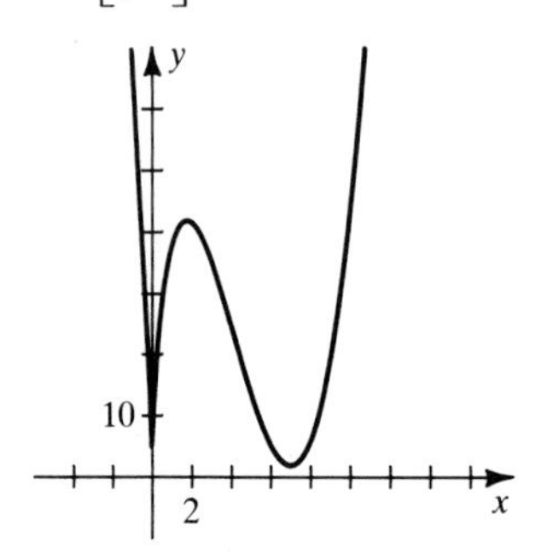

13 Max: $f(0) = 0$; min: $f(-\sqrt{3}) = f(\sqrt{3}) = -3$; increasing on $[-\sqrt{3}, 0]$ and $[\sqrt{3}, \infty)$; decreasing on $(-\infty, -\sqrt{3}]$ and $[0, \sqrt{3}]$

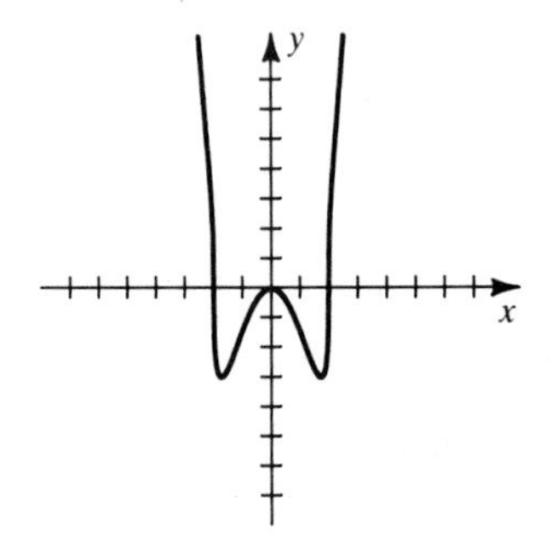

15 No extrema; increasing on $(-\infty, -3]$ and $[3, \infty)$

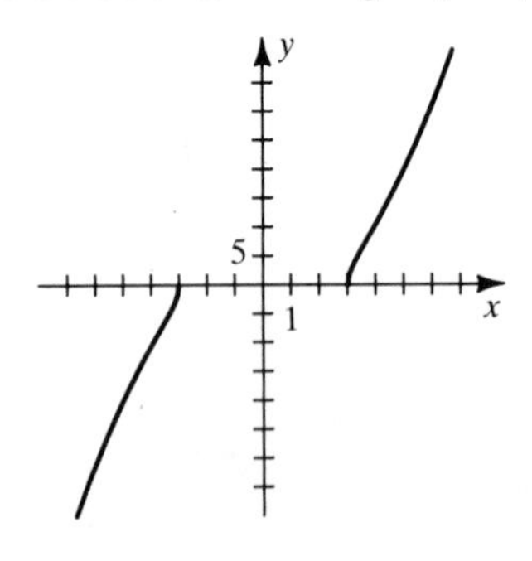

17 Max: $f\left(\frac{\pi}{4}\right) = \sqrt{2}$; min: $f\left(\frac{5\pi}{4}\right) = -\sqrt{2}$; increasing on $\left[0, \frac{\pi}{4}\right]$ and $\left[\frac{5\pi}{4}, 2\pi\right]$; decreasing on $\left[\frac{\pi}{4}, \frac{5\pi}{4}\right]$

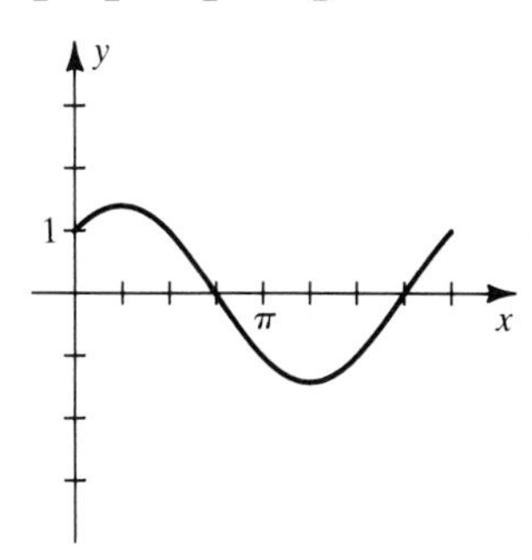

19 Max: $f\left(\frac{5\pi}{3}\right) = \frac{5\pi}{6} + \frac{\sqrt{3}}{2}$; min: $f\left(\frac{\pi}{3}\right) = \frac{\pi}{6} - \frac{\sqrt{3}}{2}$; increasing on $\left[\frac{\pi}{3}, \frac{5\pi}{3}\right]$; decreasing on $\left[0, \frac{\pi}{3}\right]$ and $\left[\frac{5\pi}{3}, 2\pi\right]$

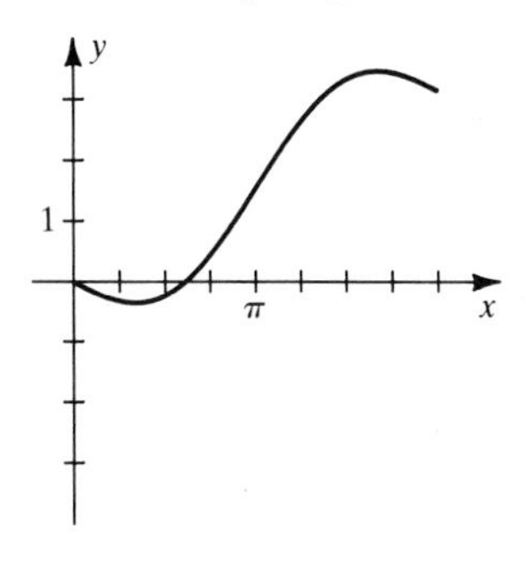

21 Max: $f\left(\frac{\pi}{6}\right) = \frac{3\sqrt{3}}{2}$; min: $f\left(\frac{5\pi}{6}\right) = -\frac{3\sqrt{3}}{2}$; increasing on $\left[0, \frac{\pi}{6}\right]$ and $\left[\frac{5\pi}{6}, 2\pi\right]$; decreasing on $\left[\frac{\pi}{6}, \frac{5\pi}{6}\right]$

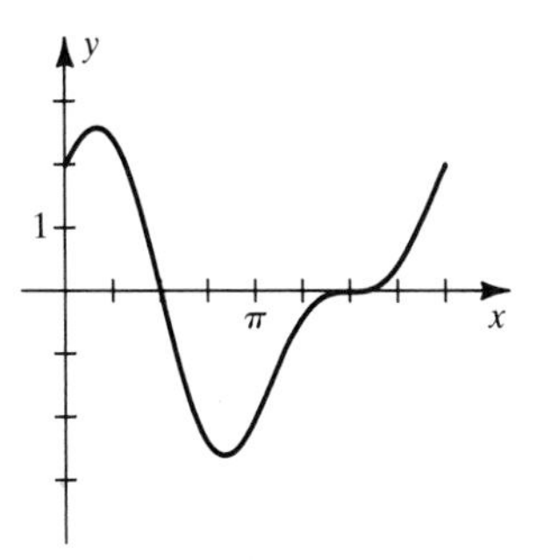

23 Max: $f(-\sqrt{3}) = \sqrt[3]{6\sqrt{3}} \approx 2.18$; min: $f(\sqrt{3}) = -\sqrt[3]{6\sqrt{3}}$

25 Max: $f(-1) = 0$; min: $f\left(\frac{5}{7}\right) = -\frac{9^3 12^4}{7^7} \approx -18.36$

27 Max: $f(4) = \frac{1}{16}$ **29** Min: $f(0) = 1$

31 Max: $f\left(\frac{\pi}{4}\right) = 1$ **33** $-\frac{11\pi}{6}, -\frac{7\pi}{6}, \frac{\pi}{6}, \frac{5\pi}{6}$

35 **37**

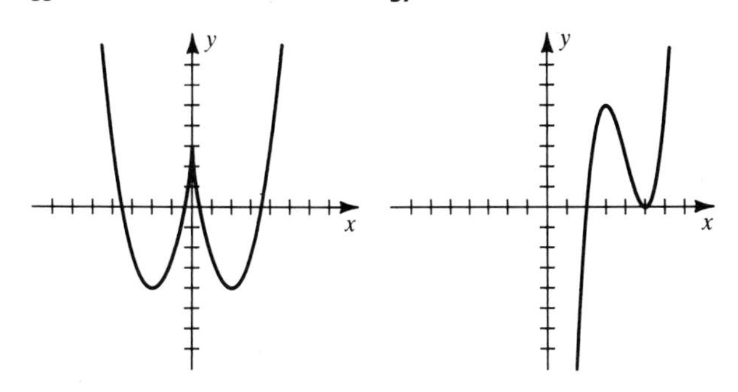

39

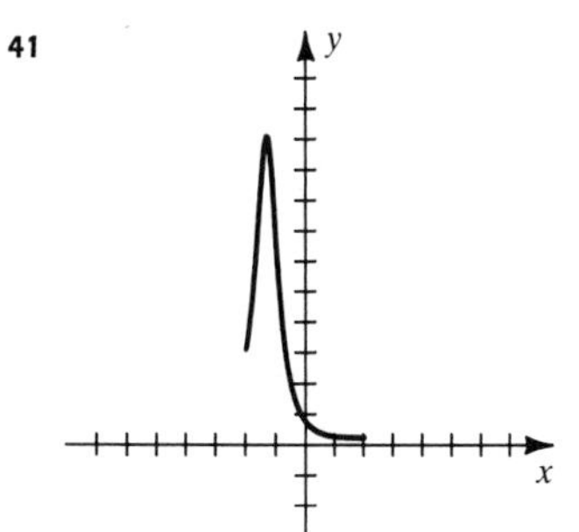

41

(a) Max: $f(-1.31) \approx 10.13$
(b) increasing on $[-2, -1.31]$; decreasing on $[-1.31, 2]$

43 Max at $x \approx -0.51$; min at $x \approx 0.49$

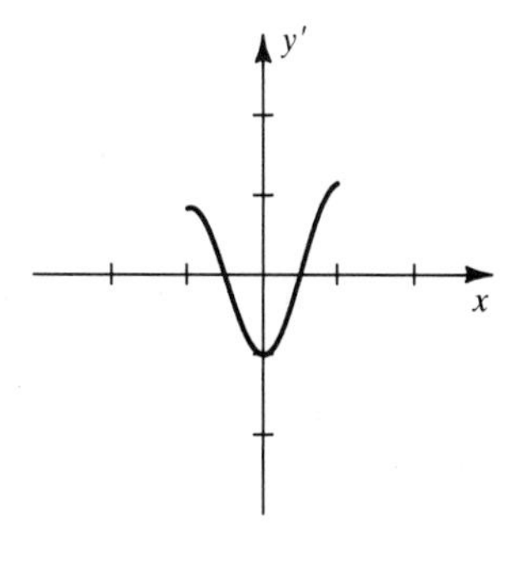

EXERCISES 4.4

1 Since $f''\left(\frac{1}{3}\right) = -2 < 0$, $f\left(\frac{1}{3}\right) = \frac{31}{27}$ is a maximum; since $f''(1) = 2 > 0$, $f(1) = 1$ is a minimum; CU on $\left(\frac{2}{3}, \infty\right)$; CD on $\left(-\infty, \frac{2}{3}\right)$; x-coordinate of PI is $\frac{2}{3}$.

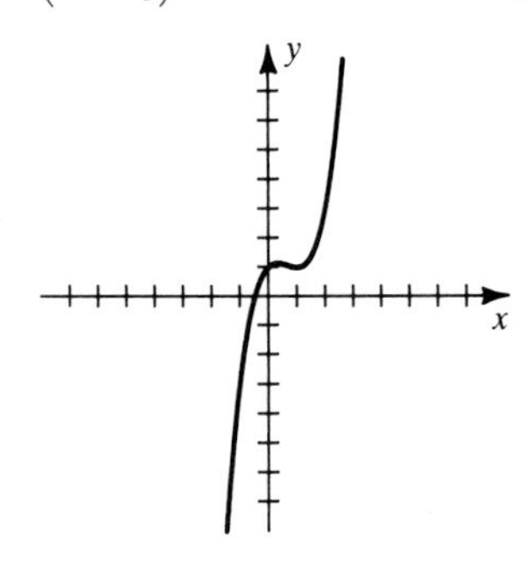

3 Since $f''(1) = 12 > 0$, $f(1) = 5$ is a minimum; CU on $(-\infty, 0)$ and $\left(\frac{2}{3}, \infty\right)$; CD on $\left(0, \frac{2}{3}\right)$; x-coordinates of PI are 0 and $\frac{2}{3}$.

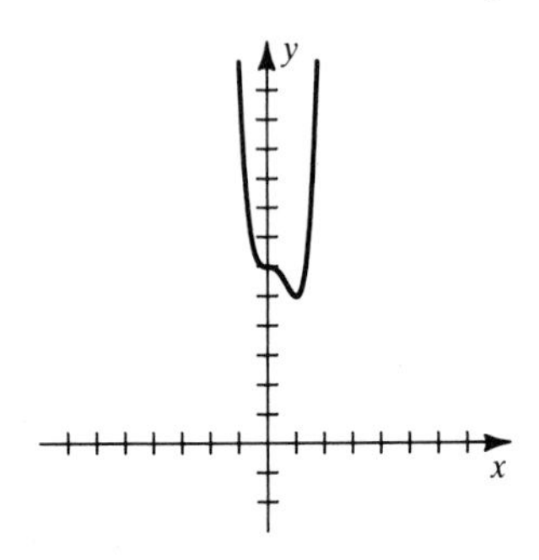

5 Since $f''(0) = 0$, use the first derivative test to show that $f(0) = 0$ is a maximum; since $f''(\pm\sqrt{2}) = 96 > 0$, $f(\pm\sqrt{2}) = -8$ are minima; CU on $\left(-\infty, -\sqrt{\frac{6}{5}}\right)$ and $\left(\sqrt{\frac{6}{5}}, \infty\right)$; CD on $\left(-\sqrt{\frac{6}{5}}, \sqrt{\frac{6}{5}}\right)$; x-coordinates of PI are $\pm\sqrt{\frac{6}{5}}$.

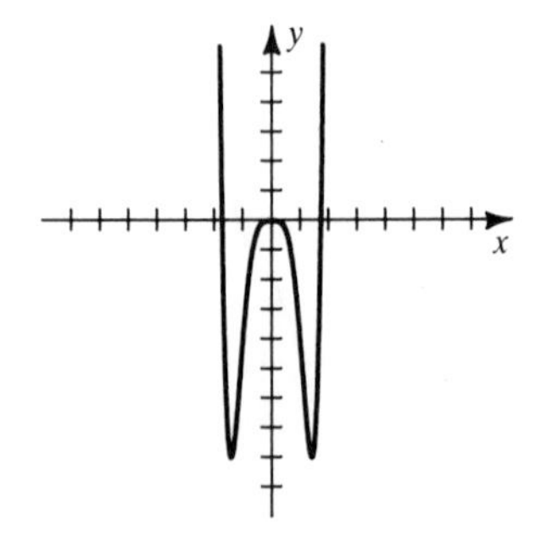

7 Since $f''(0) = -4 < 0$, $f(0) = 1$ is a maximum; since $f''(\pm 1) = 8 > 0$, $f(\pm 1) = 0$ are minima; CU on $\left(-\infty, -\sqrt{\frac{1}{3}}\right)$ and $\left(\sqrt{\frac{1}{3}}, \infty\right)$; CD on $\left(-\sqrt{\frac{1}{3}}, \sqrt{\frac{1}{3}}\right)$; x-coordinates of PI are $\pm\sqrt{\frac{1}{3}}$.

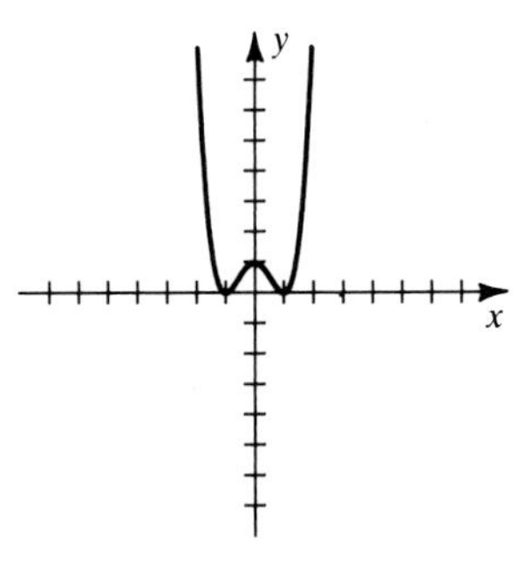

9 No local extrema; CU on $(-\infty, 0)$; CD on $(0, \infty)$; x-coordinate of PI is 0.

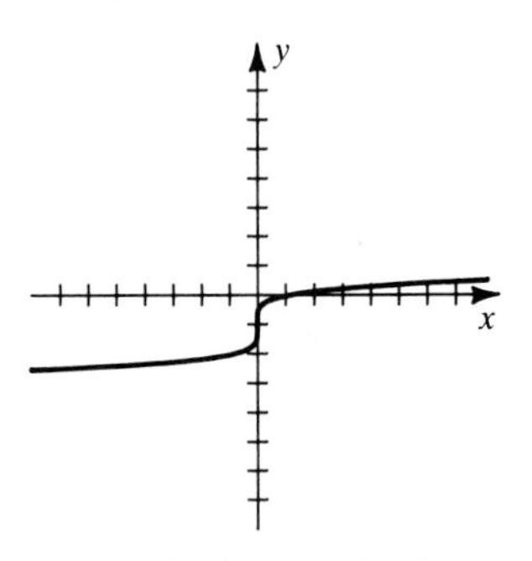

11 Since $f''\left(-\frac{4}{3}\right) < 0$, $f\left(-\frac{4}{3}\right) \approx 7.27$ is a maximum; since $f''(0) = 0$, use the first derivative test to show that $f(0) = 0$ is a minimum; CU on $\left(\frac{2}{3}, \infty\right)$; CD on $(-\infty, 0)$ and $\left(0, \frac{2}{3}\right)$; x-coordinate of PI is $\frac{2}{3}$.

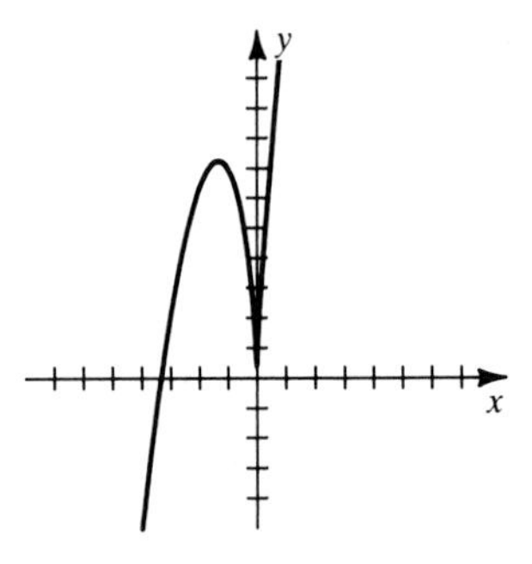

13 Since $f''(0) < 0$, $f(0) = 0$ is a maximum; since $f''\left(\frac{10}{7}\right) > 0$, $f\left(\frac{10}{7}\right) \approx -1.82$ is a minimum. Let $a = \frac{20 - 5\sqrt{2}}{14} \approx 0.92$ and $b = \frac{20 + 5\sqrt{2}}{14} \approx 1.93$.

CU on $\left(a, \frac{5}{3}\right)$ and (b, ∞); CD on $(-\infty, a)$ and $\left(\frac{5}{3}, b\right)$;
x-coordinates of PI are $a, \frac{5}{3}$, and b.

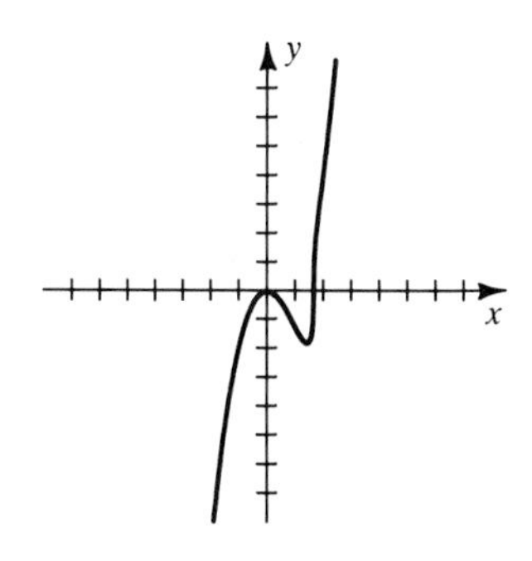

15 Since $f''(-2) > 0$, $f(-2) \approx -7.55$ is a minimum;
CU on $(-\infty, 0)$ and $(4, \infty)$; CD on $(0, 4)$;
x-coordinates of PI are 0 and 4.

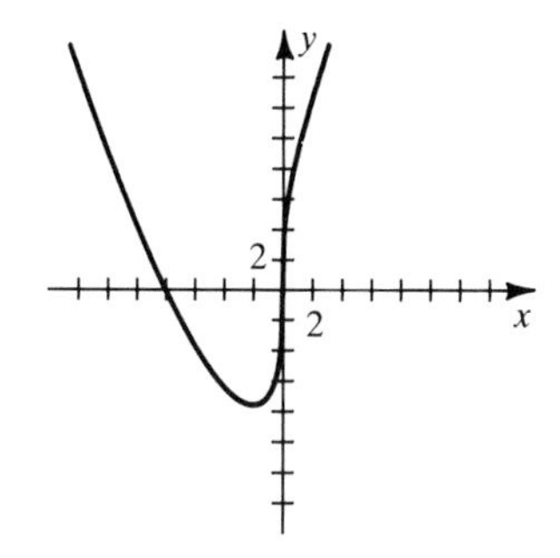

17 Since $f''(\pm\sqrt{6}) < 0$, $f(\pm\sqrt{6}) \approx 10.4$ are maxima;
since $f''(0) > 0$, $f(0) = 0$ is a minimum. Let
$a = -\frac{1}{2}\sqrt{27 - 3\sqrt{33}} \approx -1.56$ and $b = -a$.
CU on (a, b); CD on $(-3, a)$ and $(b, 3)$; x-coordinates of PI are a and b.

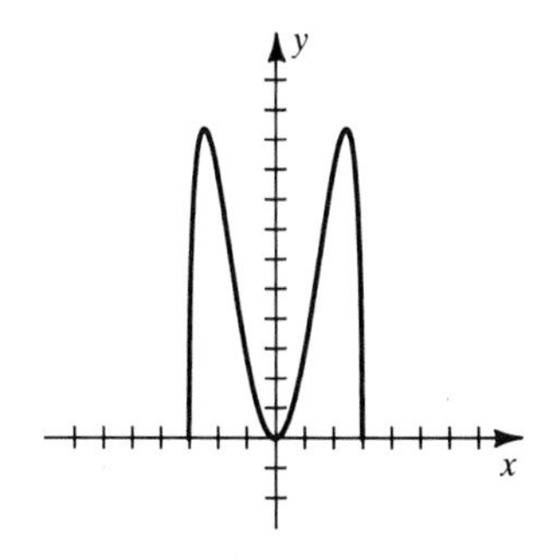

19 Since $f''\left(\frac{\pi}{4}\right) = -\sqrt{2} < 0$, $f\left(\frac{\pi}{4}\right) = \sqrt{2}$ is a maximum;
since $f''\left(\frac{5\pi}{4}\right) = \sqrt{2} > 0$, $f\left(\frac{5\pi}{4}\right) = -\sqrt{2}$ is a minimum.

21 Since $f''\left(\frac{5\pi}{3}\right) = -\frac{\sqrt{3}}{2} < 0$, $f\left(\frac{5\pi}{3}\right) = \frac{5\pi}{6} + \frac{\sqrt{3}}{2}$ is a maximum;
since $f''\left(\frac{\pi}{3}\right) = \frac{\sqrt{3}}{2} > 0$, $f\left(\frac{\pi}{3}\right) = \frac{\pi}{6} - \frac{\sqrt{3}}{2}$ is a minimum.

23 Since $f''\left(\frac{\pi}{6}\right) = -3\sqrt{3} < 0$, $f\left(\frac{\pi}{6}\right) = \frac{3\sqrt{3}}{2}$ is a maximum;
since $f''\left(\frac{5\pi}{6}\right) = 3\sqrt{3} > 0$, $f\left(\frac{5\pi}{6}\right) = -\frac{3\sqrt{3}}{2}$ is a minimum.

25 Since $f''(0) = \frac{1}{4} > 0$, $f(0) = 1$ is a minimum.

27 Since $f''\left(\frac{\pi}{4}\right) = -8 < 0$, $f\left(\frac{\pi}{4}\right) = 1$ is a maximum.

29 Since $f''\left(-\frac{11\pi}{6}\right) = f''\left(\frac{\pi}{6}\right) = -\frac{\sqrt{3}}{2} < 0$,
$f\left(-\frac{11\pi}{6}\right) \approx -2.01$ and $f\left(\frac{\pi}{6}\right) \approx 1.13$ are local maxima.
Since $f''\left(-\frac{7\pi}{6}\right) = f''\left(\frac{5\pi}{6}\right) = \frac{\sqrt{3}}{2} > 0$,
$f\left(-\frac{7\pi}{6}\right) \approx -2.70$ and $f\left(\frac{5\pi}{6}\right) \approx 0.44$ are local minima.

31 **33**

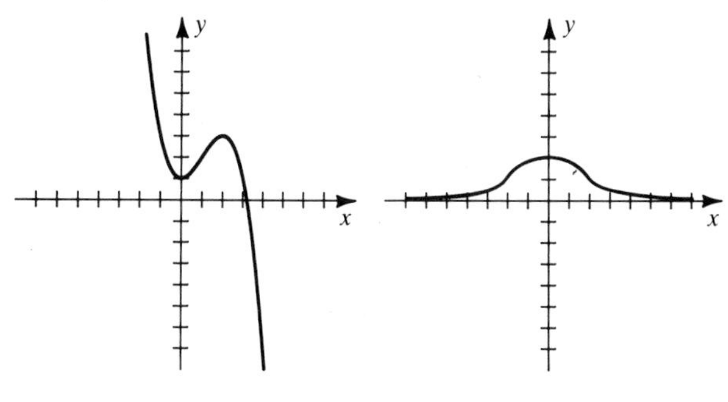

35 **37**

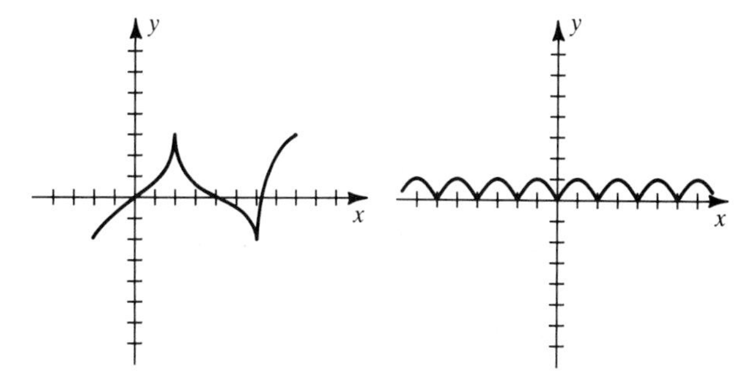

39 If $f(x) = ax^2 + bx + c$, then $f''(x) = 2a$, which does not change sign. Thus, there is no point of inflection.
(a) CU if $a > 0$. **(b)** CD if $a < 0$.

41

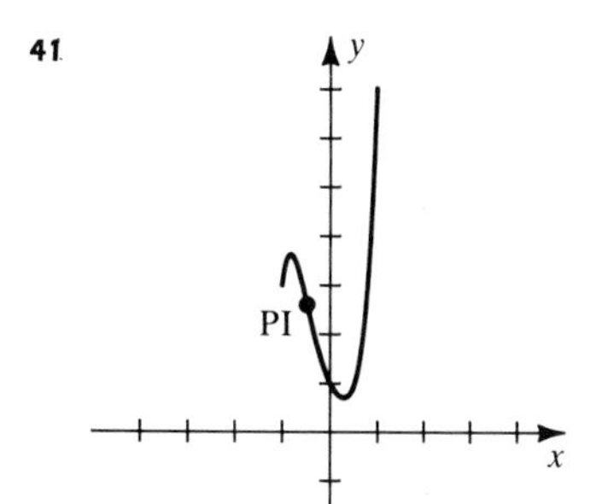

(a) CU on $(-0.48, 1)$;
CD on $(-1, -0.48)$
(b) -0.48

43 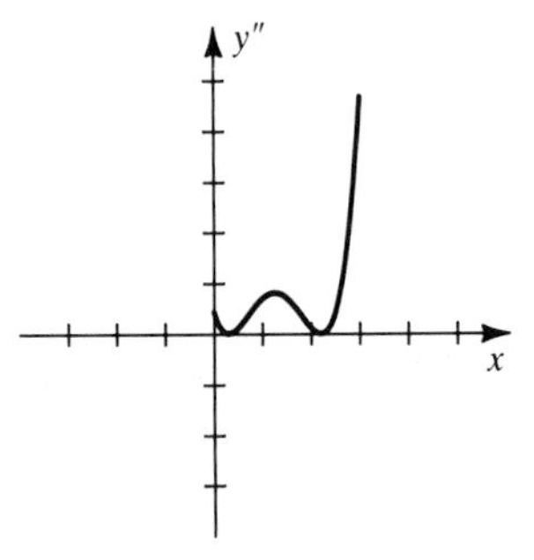

(a) CU on (0, 3)
(b) No PI on (0, 3)

9 Min: $f(4) = 4$

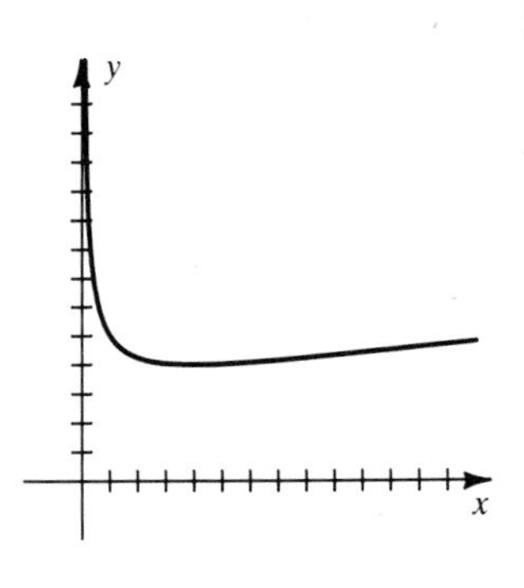

EXERCISES 4.5

1 No extrema

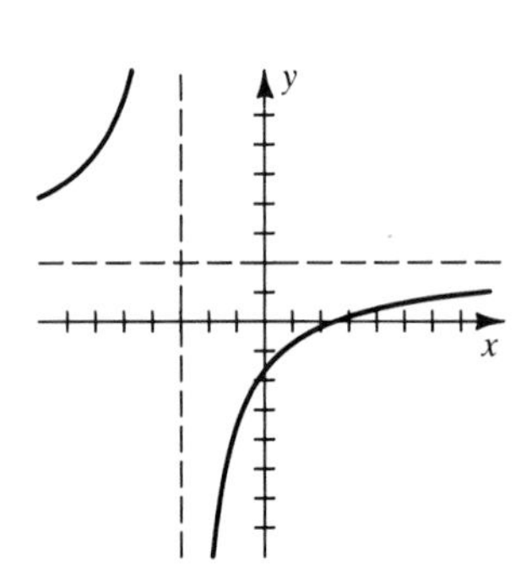

3 Max: $f(5 + 2\sqrt{6}) \approx 1.05$;
min: $f(5 - 2\sqrt{6}) \approx 5.95$

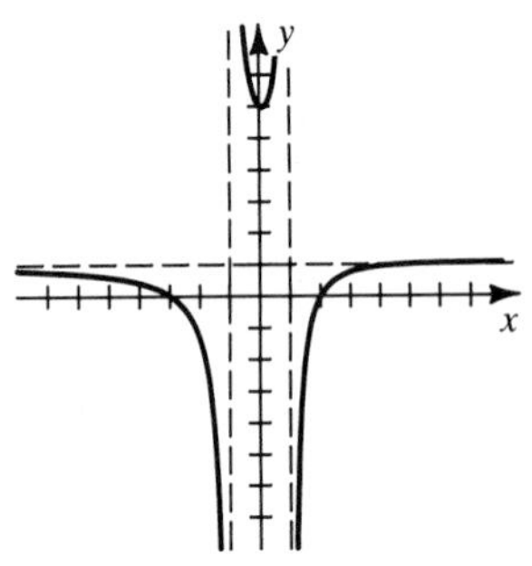

5 Max: $f(12 - 2\sqrt{30}) \approx 0.25$;
min: $f(12 + 2\sqrt{30}) \approx 2.93$

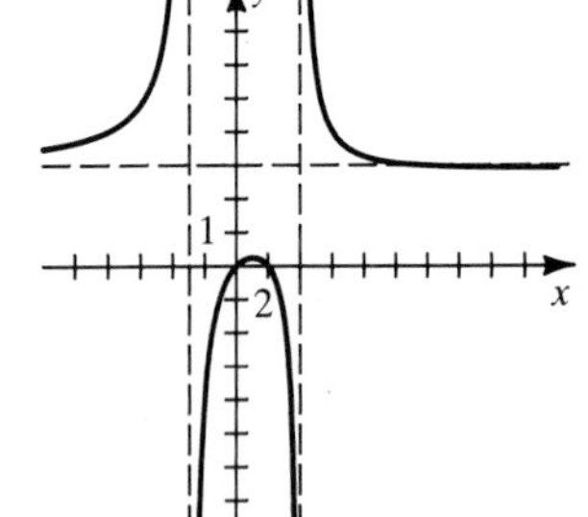

7 Min: $f(0) = 0$

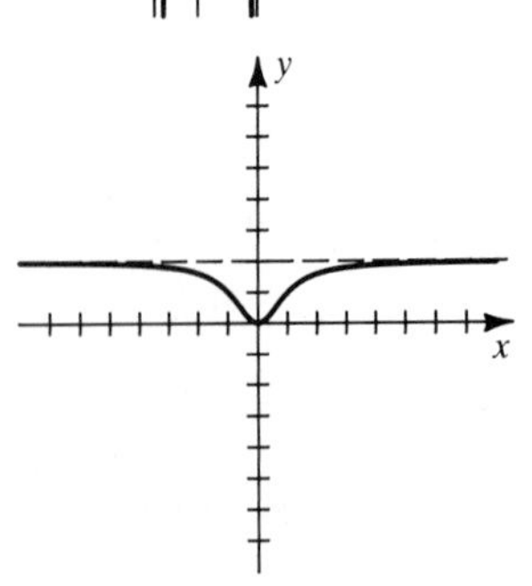

11 No extrema

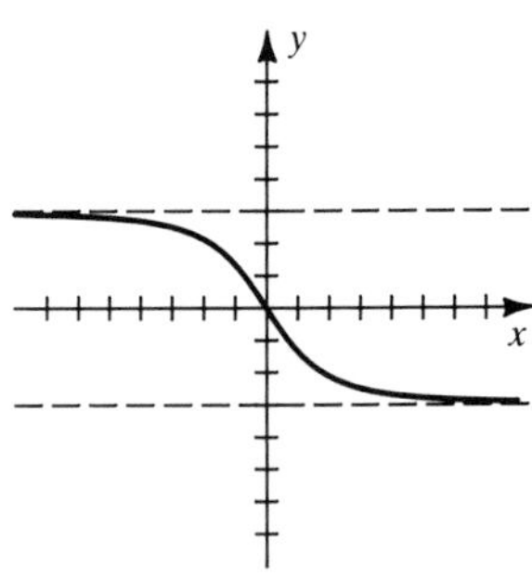

13 No extrema

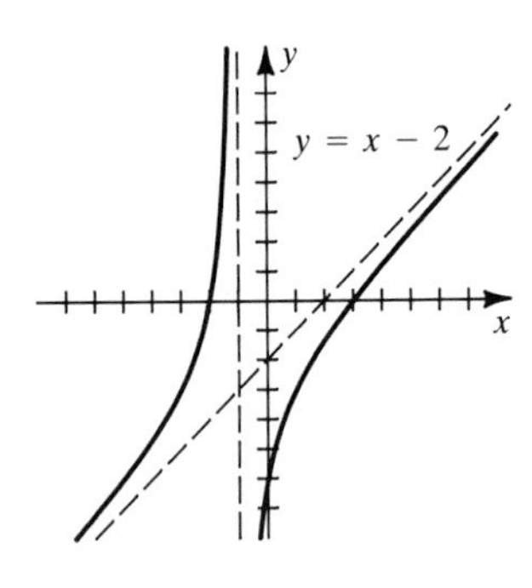

15 Max: $f(-2) = -4$;
min: $f(0) = 0$

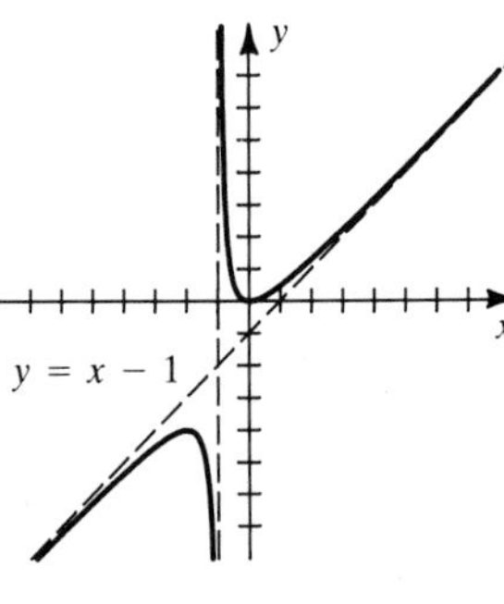

17 Max: $f(-3 + \sqrt{5}) \approx 1.53$;
min: $f(-3 - \sqrt{5}) \approx 10.47$

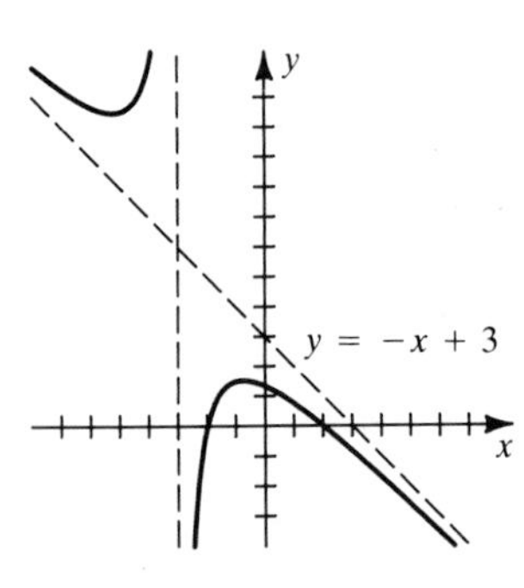

19 Max: $f(8) = \frac{3}{32}$; PI: $\left(16, \frac{1}{12}\right)$

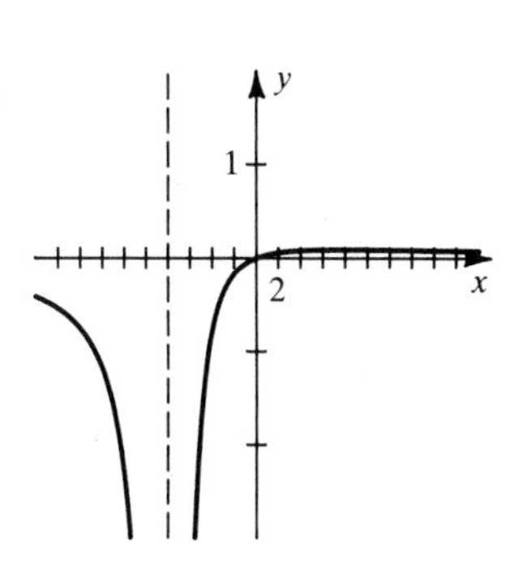

21 Max: $f(1) = \frac{3}{2}$;

min: $f(-1) = -\frac{3}{2}$;

PI: $\left(\pm\sqrt{3}, \pm\frac{3}{4}\sqrt{3}\right)$, (0, 0)

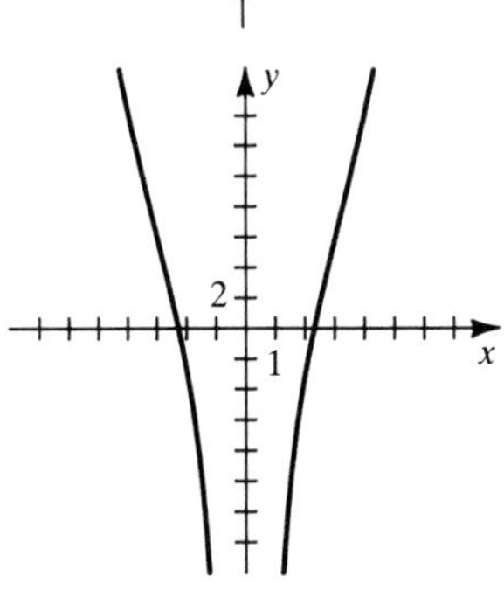

23 No extrema;
PI: $(\pm 3, 6)$

25

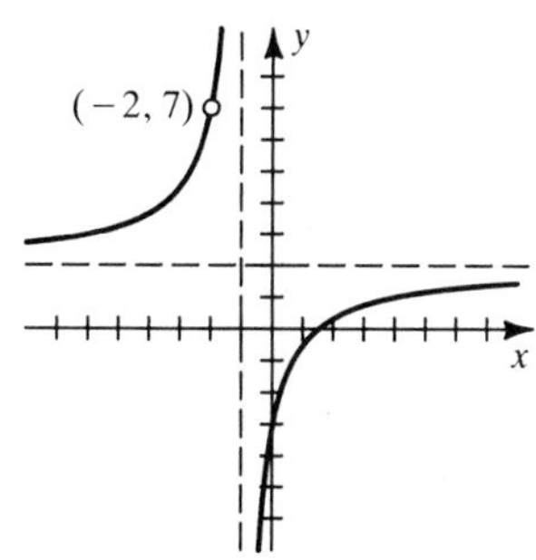

27

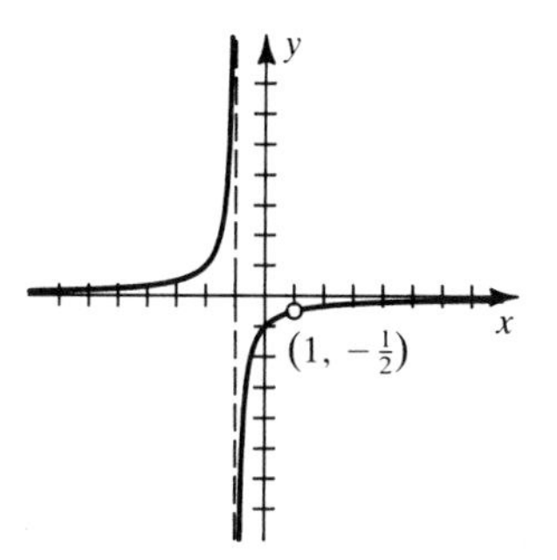

29

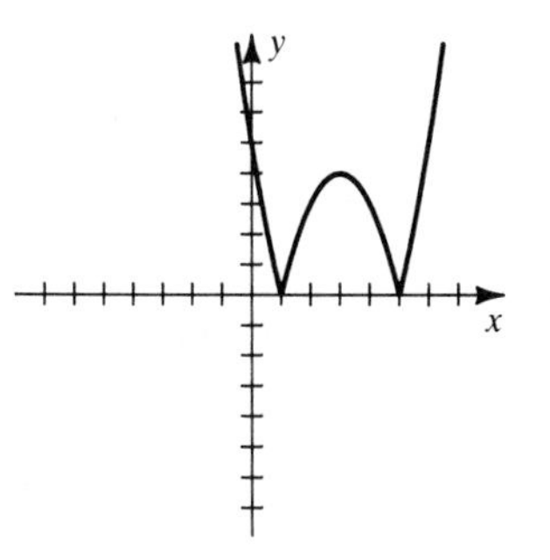

31

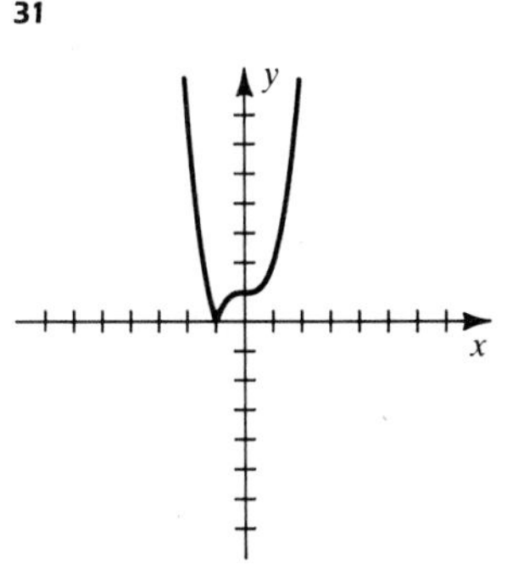

33

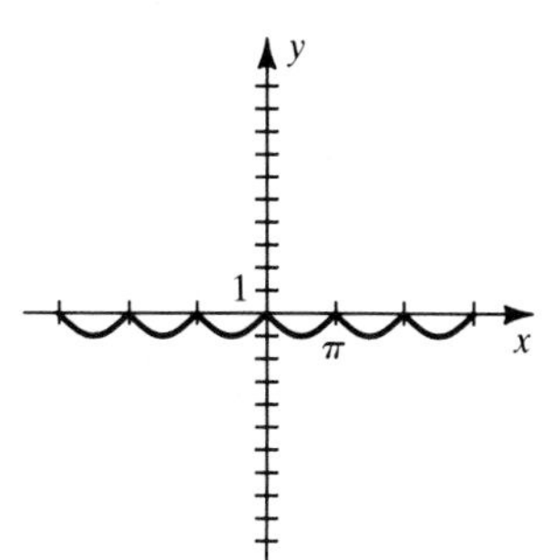

35

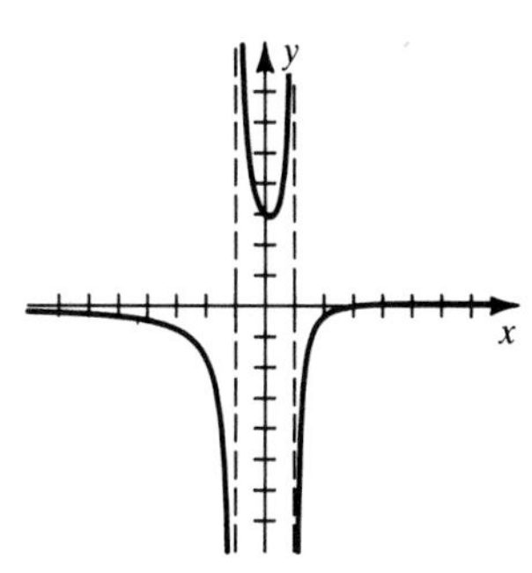

37

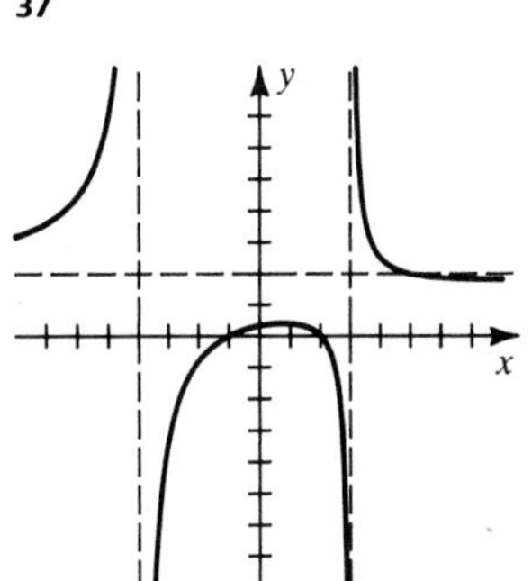

39 **(b)**

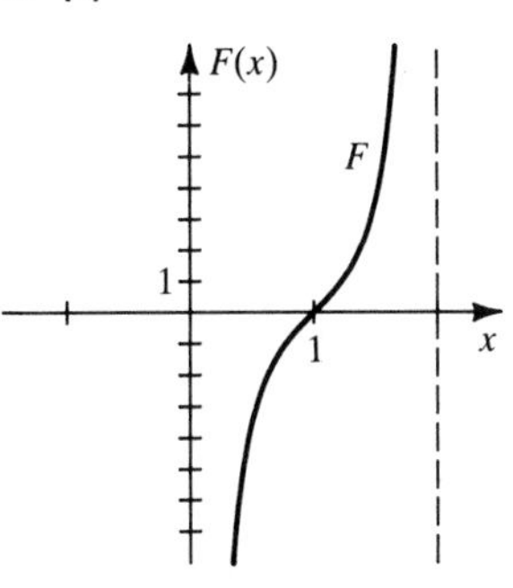

41

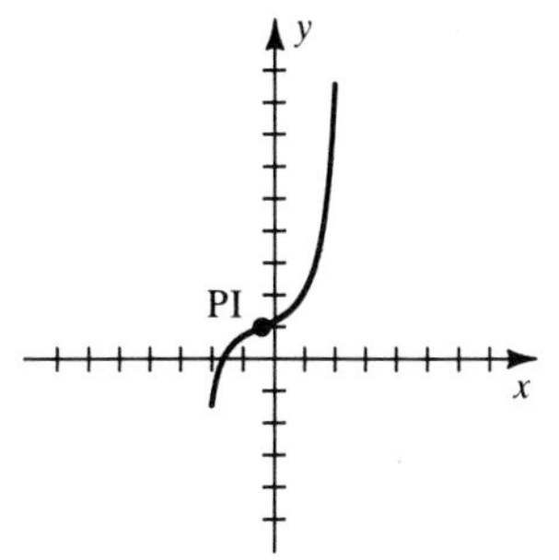

(a) CU on $(-0.43, 2)$;
CD on $(-2, -0.43)$
(b) -0.43

EXERCISES 4.6

1 Side of base = 2 ft; height = 1 ft

3 Radius of base = height = $\frac{1}{\sqrt[3]{\pi}}$

5 $x = 166\frac{2}{3}$ ft; $y = 125$ ft

7 Approximately 2:23:05 P.M. **9** $5\sqrt{5} \approx 11.18$ ft

11 Length $= 2\sqrt[3]{300} \approx 13.38$ ft; width $= \frac{3}{2}\sqrt[3]{300} \approx 10.04$ ft; height $= \sqrt[3]{300} \approx 6.69$ ft

15 55 **17** Radius $= \frac{1}{2}\sqrt[3]{15}$; length of cylinder $= 2\sqrt[3]{15}$

19 Length of base $= \sqrt{2}\,a$; height $= \frac{1}{2}\sqrt{2}\,a$ **21** $\frac{32}{81}\pi a^3$

23 (1, 2) **25** Width $= \frac{2}{\sqrt{3}}a$; depth $= \frac{2\sqrt{2}}{\sqrt{3}}a$ **27** 500

29 **(a)** Use $\frac{36\sqrt{3}}{2+\sqrt{3}} \approx 16.71$ cm for the rectangle.
(b) Use all the wire for the rectangle.

31 Width $= \frac{12}{6-\sqrt{3}} \approx 2.81$ ft; height $= \frac{18-6\sqrt{3}}{6-\sqrt{3}} \approx 1.78$ ft

35 37 **37** 18 in., 18 in., 36 in.

39 $\frac{4}{1+\sqrt[4]{\frac{1}{2}}} \approx 2.17$ mi from A

43 **(c)** $4\sqrt{30} \approx 21.9$ mi/hr

45 60° **47** $2\pi\left(1-\frac{1}{3}\sqrt{6}\right)$ radians $\approx 66.06°$

49 $\tan\theta = \frac{\sqrt{2}}{2}$; $\theta \approx 35.3°$

53 $\tan\theta = \sqrt[3]{\frac{4}{3}}$; $\theta \approx 47.74°$; $L = \frac{4}{\sin\theta} + \frac{3}{\cos\theta} \approx 9.87$ ft

55 **(b)** $\cos\theta = \frac{2}{3}$; $\theta \approx 48.2°$

EXERCISES 4.7

1 $v(t) = 6(t-2)$; $a(t) = 6$;
left in $[0, 2)$;
right in $(2, 5]$

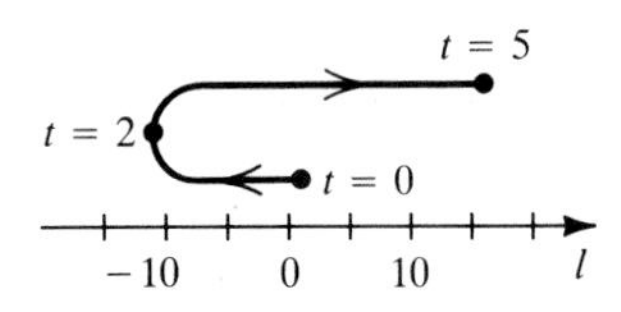

3 $v(t) = 3(t^2-3)$; $a(t) = 6t$;
right in $[-3, -\sqrt{3})$;
left in $(-\sqrt{3}, \sqrt{3})$;
right in $(\sqrt{3}, 3]$

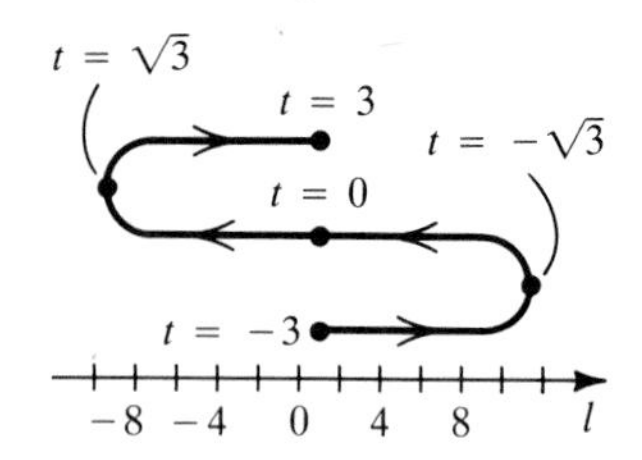

5 $v(t) = -6(t-1)(t-4)$; $a(t) = -6(2t-5)$;
left in $[0, 1)$; right in $(1, 4)$;
left in $(4, 5]$

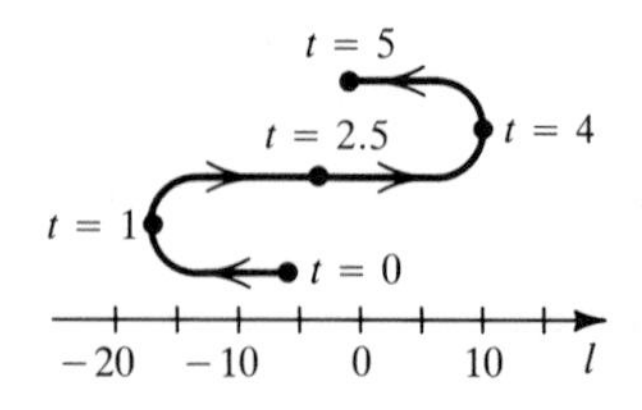

7 $v(t) = 4t(2t^2-3)$; $a(t) = 12(2t^2-1)$;
left in $\left[-2, -\sqrt{\frac{3}{2}}\right)$; right in $\left(-\sqrt{\frac{3}{2}}, 0\right)$;
left in $\left(0, \sqrt{\frac{3}{2}}\right)$; right in $\left(\sqrt{\frac{3}{2}}, 2\right]$

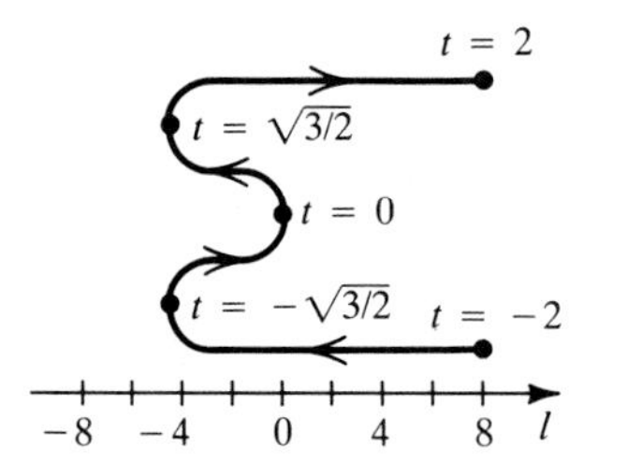

9 **(a)** 30 ft/sec **(b)** 2.8 sec

11 **(a)** $v(t) = 16(9-2t)$; $a(t) = -32$ **(b)** 324 ft **(c)** 9 sec

13 5; 8; $\frac{1}{8}$ **15** 6; 3; $\frac{1}{3}$ **17** $79{,}200\pi$; $3600\sqrt{2\pi}$

19 **(a)** $y = 4.5\sin\left[\frac{\pi}{6}(t-10)\right] + 7.5 = 4.5\sin\left(\frac{\pi}{6}t - \frac{5\pi}{3}\right) + 7.5$
(b) 1.178 ft/hr

21 **(a)** In in./sec: 0, $-\pi$, 0, π, 0
(b) $(n, n+1)$, where n is an odd positive integer

27

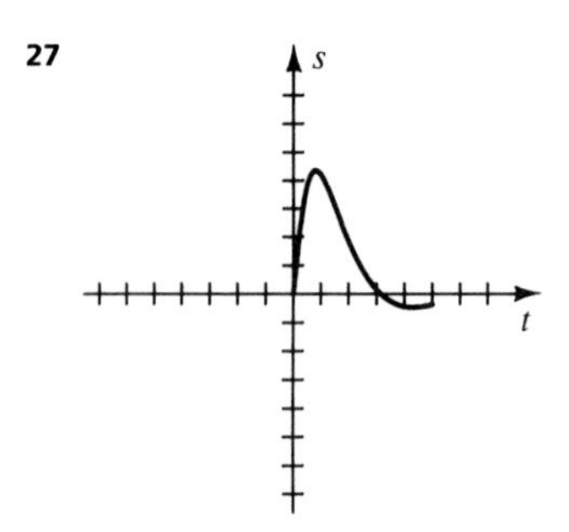

t	0	1	2	3	4	5
s	0	4.21	1.82	0.14	−0.45	−0.37
v	10	−1.51	−2.29	−1.07	−0.18	0.25
a	0	−5.40	1.11	1.12	0.66	0.20

29 **(a)** 806
(b) $c(x) = \frac{800}{x} + 0.04 + 0.0002x$; $C'(x) = 0.04 + 0.0004x$;
$c(100) = 8.06$; $C'(100) = 0.08$

31 **(a)** 11,250
(b) $c(x) = \frac{250}{x} + 100 + 0.001x^2$; $C'(x) = 100 + 0.003x^2$;
$c(100) = 112.50$; $C'(100) = 130$

33 $C'(5) = \$46$; $C(6) - C(5) \approx \$46.67$

35 **(a)** −0.1 **(b)** $50x - 0.1x^2$ **(c)** $48x - 0.1x^2 - 10$
(d) $48 - 0.2x$ **(e)** 5750 **(f)** 2

37 **(a)** $1800x - 2x^2$ **(b)** $1799x - 2.01x^2 - 1000$
(c) 100 **(d)** \$158,800

39 **(a)** 3990 mills **(b)** \$15,420.10

EXERCISES 4.8

1 1.2599 **3** 1.3315
5 -1.7321 **7** 4.6458
9 0.56 **11** 1.50
13 ± 3.34 **15** $-1, 1.35$
17 $-1.88, 0.35, 1.53$
19 2.71 **21** $-1.16, 1.45$
23 ± 2.99
25 **(a)** 3, 3.1425465, 3.1415927, 3.1415926, 3.1415926
(b) They approach 2π.
27 $f'\left(\frac{1}{2}\right) = 0$ and hence the expression for x_2 would be undefined.
29 **(a)** f: $x_1 = 1.1, x_2 = 1.066485, x_3 = 1.044237, x_4 = 1.029451$
g: $x_1 = 1.1, x_2 = 0.9983437, x_3 = 0.9999995, x_4 = 1.000000$
(b) Because $f'(1) = 0$
31 $x_5 = 0.525$
33 **(a)** 1.5 **(b)** 1.34

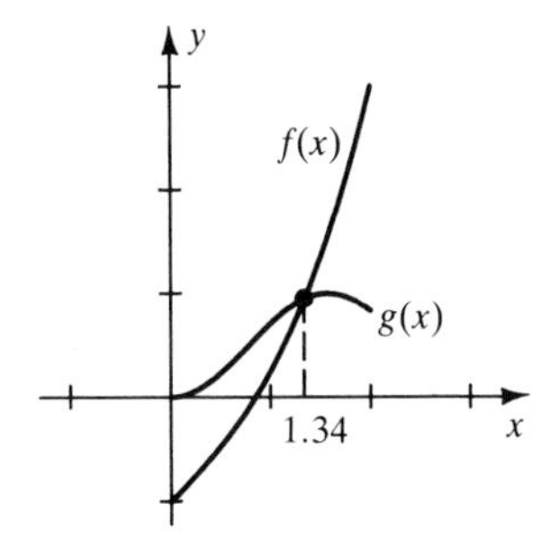

4.9 REVIEW EXERCISES

1 Max: $f(3) = 1$; min: $f(6) = -8$
3 $-2, -1, \frac{1}{3}$
5 Max: $f(2) = 28$; min: $f\left(-\frac{1}{2}\right) = -\frac{13}{4}$;
increasing on $\left[-\frac{1}{2}, 2\right]$;
decreasing on $\left(-\infty, -\frac{1}{2}\right]$ and $[2, \infty)$

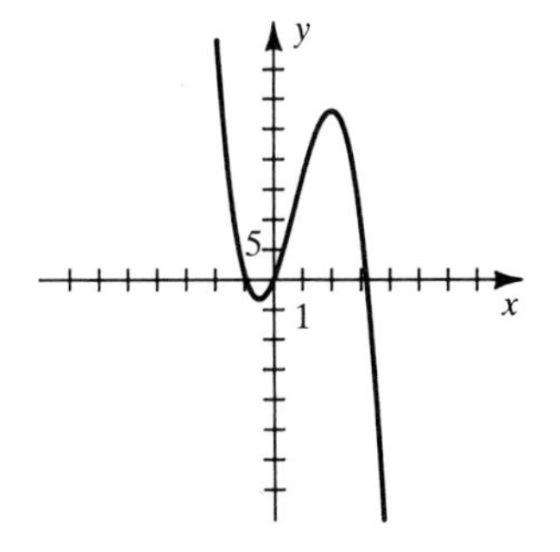

7 Max: $f(1) = 3$;
increasing on $(-\infty, 1]$;
decreasing on $[1, \infty)$

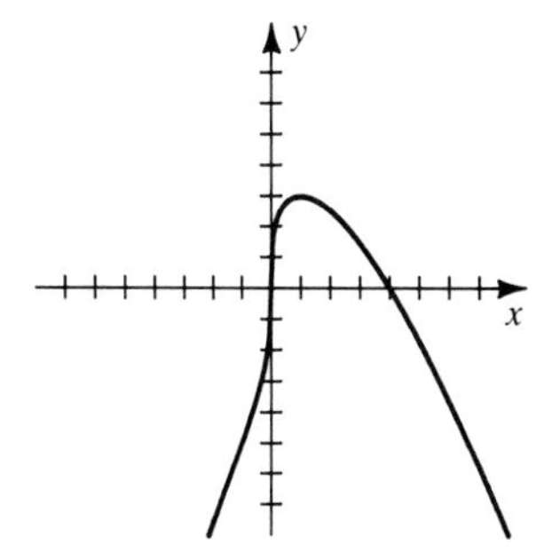

9 Since $f''(0) = 0$ and $f''(2)$ is undefined, use the first derivative test to show that there are no extrema; CU on $(-\infty, 0)$ and $(2, \infty)$; CD on $(0, 2)$; x-coordinates of PI are 0 and 2.

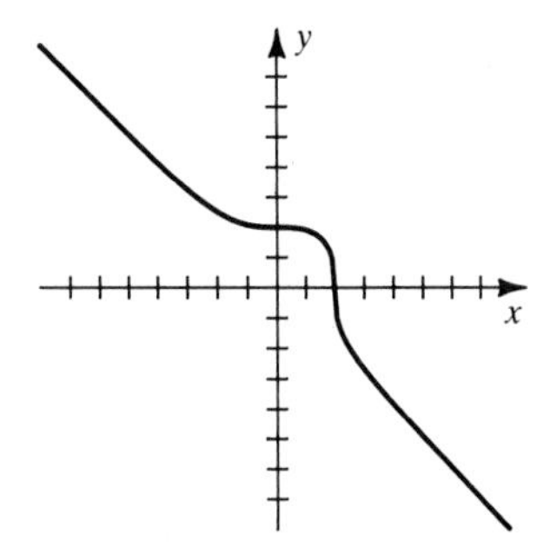

11 Since $f''(0) = -2 < 0$, $f(0) = 1$ is a maximum;
CU on $\left(-\infty, -\frac{1}{3}\sqrt{3}\right)$ and $\left(\frac{1}{3}\sqrt{3}, \infty\right)$;
CD on $\left(-\frac{1}{3}\sqrt{3}, \frac{1}{3}\sqrt{3}\right)$; x-coordinates of PI are $\pm\frac{1}{3}\sqrt{3}$.

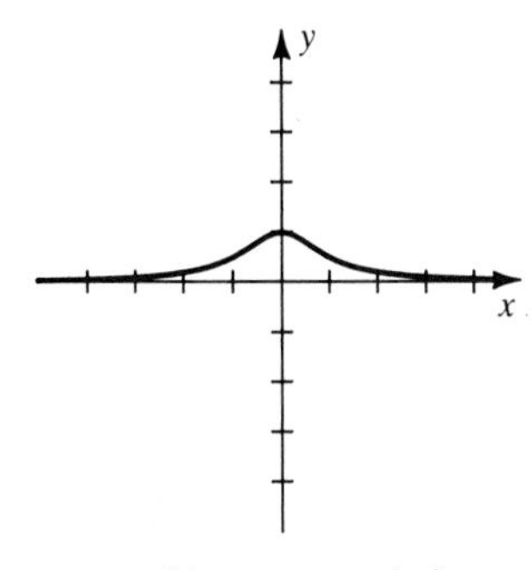

13 Max: $f\left(\frac{\pi}{2}\right) = 3$ and $f\left(\frac{3\pi}{2}\right) = -1$;
min: $f\left(\frac{7\pi}{6}\right) = f\left(\frac{11\pi}{6}\right) = -\frac{3}{2}$

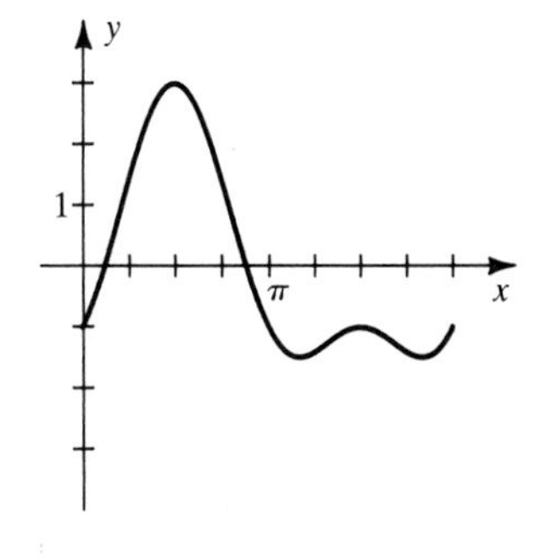

15

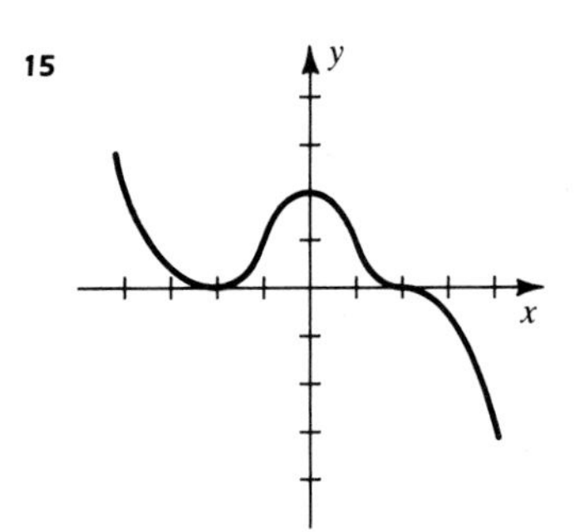

17 Max: $f(0) = 0$

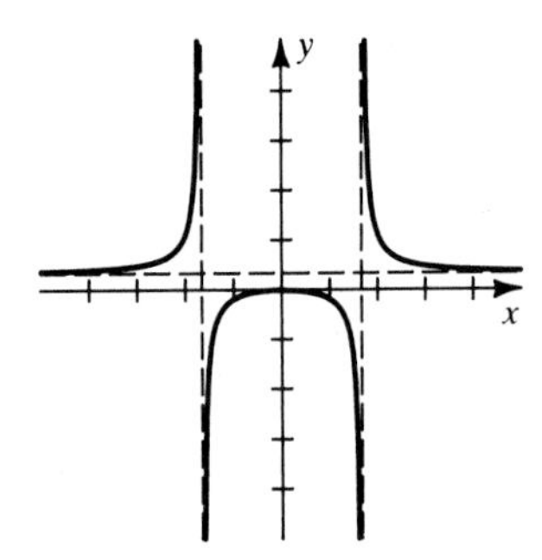

19 No extrema

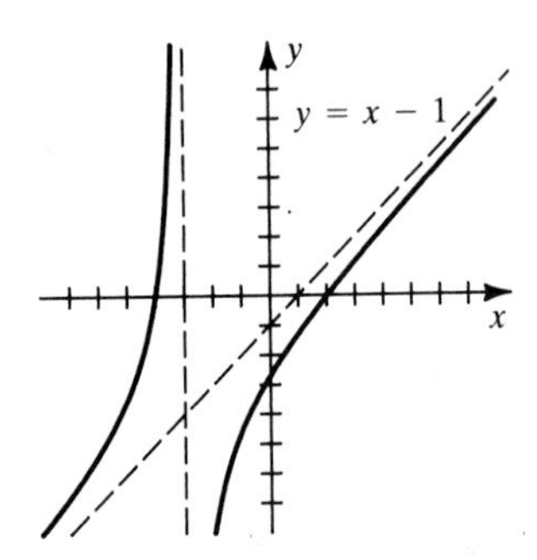

21 Max: $f(3 + \sqrt{7}) \approx 0.08$; min: $f(3 - \sqrt{7}) \approx 0.37$

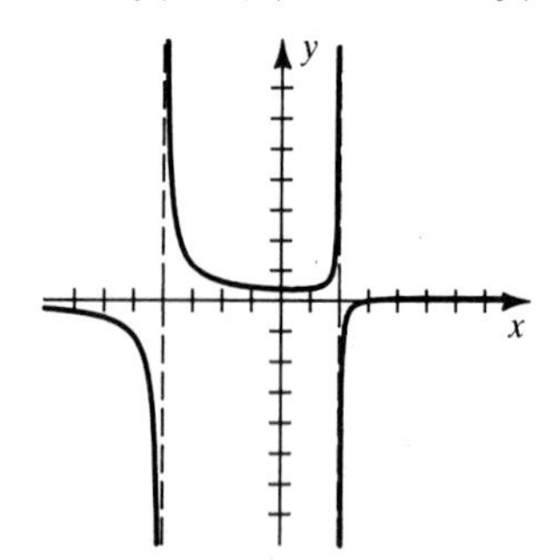

23 $\dfrac{\sqrt{61} - 1}{3}$ **25** 125 yd by 250 yd **27** $\dfrac{\pi}{2}$

29 Radius of semicircle is $\dfrac{1}{8\pi}$ mi, length of rectangle is $\dfrac{1}{8}$ mi.

31 **(a)** Use all the wire for the circle.

(b) Use length $\dfrac{5\pi}{4 + \pi} \approx 2.2$ ft for the circle and the remainder for the square.

33 $v(t) = \dfrac{3(1 - t^2)}{(t^2 + 1)^2}$; $a(t) = \dfrac{6t(t^2 - 3)}{(t^2 + 1)^3}$; left in $[-2, -1)$; right in $(-1, 1)$; left in $(1, 2]$

35 $C'(100) = 116$; $C(101) - C(100) = 116.11$

37 **(a)** $18x$ **(b)** $-0.02x^2 + 12x - 500$ **(c)** 300 **(d)** \$1300

39 4.493

CHAPTER 5

EXERCISES 5.1

1 $2x^2 + 3x + C$ **3** $3t^3 - 2t^2 + 3t + C$

5 $-\dfrac{1}{2z^2} + \dfrac{3}{z} + C$ **7** $2u^{3/2} + 2u^{1/2} + C$

9 $\dfrac{8}{9}v^{9/4} + \dfrac{24}{5}v^{5/4} - v^{-3} + C$ **11** $3x^3 - 3x^2 + x + C$

13 $\dfrac{2}{3}x^3 + \dfrac{3}{2}x^2 + C$ **15** $\dfrac{24}{5}x^{5/3} - \dfrac{15}{2}x^{2/3} + C$

17 $\dfrac{1}{3}x^3 + \dfrac{1}{2}x^2 + x + C$ **19** $-t^{-1} - 2t^{-3} - \dfrac{9}{5}t^{-5} + C$

21 $\dfrac{3}{4}\sin u + C$ **23** $-7\cos x + C$

25 $\dfrac{2}{3}t^{3/2} + \sin t + C$ **27** $\tan t + C$ **29** $-\cot v + C$

31 $\sec w + C$ **33** $-\csc z + C$ **35** $\sqrt{x^2 + 4} + C$

37 $\sin \sqrt[3]{x} + C$ **39** $x^3\sqrt{x - 4}$ **41** $\cot x^3$

43 $a^2x + C$ **45** $\dfrac{1}{2}at^2 + bt + C$ **47** $(a + b)u + C$

49 $f(x) = 4x^3 - 3x^2 + x + 3$ **51** $y = \dfrac{8}{3}x^{3/2} - \dfrac{1}{3}$

53 $f(x) = \dfrac{2}{3}x^3 - \dfrac{1}{2}x^2 - 8x + \dfrac{65}{6}$

55 $y = -3\sin x + 4\cos x + 5x + 3$ **57** $t^2 - t^3 - 5t + 4$

59 **(a)** $s(t) = -16t^2 + 1600t$ **(b)** $s(50) = 40{,}000$ ft

61 **(a)** $s(t) = -16t^2 - 16t + 96$ **(b)** $t = 2$ sec **(c)** -80 ft/sec

63 Solve the differential equation $s''(t) = -g$ for $s(t)$.

65 10 ft/sec^2 **67** 19.62

69 $C(x) = 20x - 0.0075x^2 + 5.0075$; $C(50) \approx \$986.26$

EXERCISES 5.2

1 $\dfrac{1}{44}(2x^2 + 3)^{11} + C$ **3** $\dfrac{1}{12}(3x^3 + 7)^{4/3} + C$

5 $\dfrac{1}{2}(1 + \sqrt{x})^4 + C$ **7** $\dfrac{2}{3}\sin\sqrt{x^3} + C$

9 $\dfrac{2}{9}(3x - 2)^{3/2} + C$ **11** $\dfrac{3}{32}(8t + 5)^{4/3} + C$

13 $\dfrac{1}{15}(3z + 1)^5 + C$ **15** $\dfrac{2}{9}(v^3 - 1)^{3/2} + C$

17 $-\dfrac{3}{8}(1 - 2x^2)^{2/3} + C$ **19** $\dfrac{1}{5}s^5 + \dfrac{2}{3}s^3 + s + C$

21 $\dfrac{2}{5}(\sqrt{x} + 3)^5 + C$ **23** $-\dfrac{1}{4(t^2 - 4t + 3)^2} + C$

25 $-\dfrac{3}{4}\cos 4x + C$ **27** $\dfrac{1}{4}\sin(4x - 3) + C$

29 $-\dfrac{1}{2}\cos(v^2) + C$ **31** $\dfrac{1}{4}(\sin 3x)^{4/3} + C$

33 $x - \dfrac{1}{2}\cos 2x + C$ **35** $-\cos x - \cos^2 x - \dfrac{1}{3}\cos^3 x + C$

37 $\dfrac{1}{3\cos^3 x} + C$ **39** $\dfrac{1}{1 - \sin t} + C$ **41** $\dfrac{1}{3}\tan(3x - 4) + C$

43 $\dfrac{1}{6}\sec^2 3x + C$ **45** $-\dfrac{1}{5}\cot 5x + C$

47 $-\dfrac{1}{2}\csc(x^2) + C$ **49** $f(x) = \dfrac{1}{4}(3x + 2)^{4/3} + 5$

51 $f(x) = 3\sin x - 4\cos 2x + x + 2$

53 **(a)** $\dfrac{1}{3}(x + 4)^3 + C_1$

(b) $\dfrac{1}{3}x^3 + 4x^2 + 16x + C_2$; $C_2 = C_1 + \dfrac{64}{3}$

55 **(a)** $\dfrac{2}{3}(\sqrt{x} + 3)^3 + C_1$

(b) $\dfrac{2}{3}x^{3/2} + 6x + 18x^{1/2} + C_2$; $C_2 = C_1 + 18$

59 474,592 ft^3 **61** **(a)** $\dfrac{dV}{dt} = 0.6\sin\left(\dfrac{2\pi}{5}t\right)$ **(b)** $\dfrac{3}{\pi} \approx 0.95$ L

63 *Hint:* (i) Let $u = \sin x$. (ii) Let $u = \cos x$.
(iii) Use the double angle formula for the sine. The three answers differ by constants.

EXERCISES 5.3

1 34 **3** 40 **5** 10 **7** 500

9 $\dfrac{1}{3}n(n^2 + 6n + 20)$ **11** $\dfrac{1}{12}n(3n^3 + 14n^2 + 9n + 46)$

Exer. 13–18: Answers are not unique.

13 $\sum_{k=1}^{5}(4k - 3)$ **15** $\sum_{k=1}^{4}\dfrac{k}{3k - 1}$

17 $1 + \sum_{k=1}^{n}(-1)^k\dfrac{x^{2k}}{2k}$ **19** **(a)** 10 **(b)** 14

21 **(a)** $\dfrac{35}{4}$ **(b)** $\dfrac{51}{4}$ **23** **(a)** 1.04 **(b)** 1.19

Exer. 25–30: Answers for (a) and (b) are the same.

25 28 **27** 18 **29** 6 **31** **(a)** 20 **(b)** $\dfrac{1}{4}(b^4 - a^4)$

EXERCISES 5.4

1 **(a)** 1.1, 1.5, 1.1, 0.4, 0.9 **(b)** 1.5
3 **(a)** 0.3, 1.7, 1.4, 0.5, 0.1 **(b)** 1.7

5 **(a)** 42 **(b)** 30 **(c)** 36 **7** $\dfrac{49}{4}$ **9** 79 **11** 0.28

13 $\int_{-1}^{2}(3x^2 - 2x + 5)\,dx$ **15** $\int_{0}^{4}2\pi x(1 + x^3)\,dx$

17 $-\dfrac{14}{3}$ **19** $\dfrac{14}{3}$ **21** $-\dfrac{14}{3}$

23 $\int_{0}^{4}\left(-\dfrac{5}{4}x + 5\right)dx$ **25** $\int_{-1}^{5}\sqrt{9 - (x - 2)^2}\,dx$

27 36 **29** 25 **31** 2.5 **33** $\dfrac{9\pi}{4}$ **35** $12 + 2\pi$

EXERCISES 5.5

1 30 **3** -12 **5** 2 **7** 78 **9** $-\dfrac{291}{2}$

11 Use Corollary (5.27). **13** Use Theorem (5.26).

15 Use Theorem (5.26). **17** $\int_{-3}^{1} f(x)\,dx$

19 $\int_{e}^{d} f(x)\,dx$ **21** $\int_{h}^{c+h} f(x)\,dx$ **23** **(a)** $\sqrt{3}$ **(b)** 9

25 **(a)** -1 **(b)** 2 **27** **(a)** 3 **(b)** 6

29 **(a)** $\sqrt[3]{\dfrac{15}{4}}$ **(b)** 14

31 1.426 **33** Use (5.22) and (5.23)(i).

EXERCISES 5.6

1 -18 **3** $\dfrac{265}{2}$ **5** 5 **7** $\dfrac{31}{32}$ **9** $\dfrac{20}{3}$ **11** $\dfrac{352}{5}$

13 $\dfrac{13}{3}$ **15** $-\dfrac{7}{2}$ **17** 0 **19** $\dfrac{10}{3}$ **21** $\dfrac{53}{2}$ **23** $\dfrac{14}{3}$

25 0 **27** $\dfrac{1}{3}$ **29** $\dfrac{5}{36}$ **31** $\dfrac{3}{2}(\sqrt{3} - 1)$ **33** $1 - \sqrt{2}$

35 0 **37** **(a)** $\sqrt{3}$ **(b)** $\dfrac{1}{2}$ **39** **(a)** $\dfrac{544}{225}$ **(b)** $\dfrac{38}{15}$

41 0 **43** $\dfrac{1}{x + 1}$ **47** **(a)** $\dfrac{6}{7}cd^{1/6}$

51 *Hint:* Use Part I of the fundamental theorem of calculus (5.30) and the chain rule.

53 $\dfrac{4x^7}{\sqrt{x^{12} + 2}}$ **55** $3x^2(x^9 + 1)^{10} - 3(27x^3 + 1)^{10}$

EXERCISES 5.7

1 **(a)** 10.75 **(b)** $10\frac{2}{3} \approx 10.67$ **3** **(a)** 0.96 **(b)** 0.96

5 **(a)** 1.41 **(b)** 1.39 **7** **(a)** 0.88 **(b)** 0.88
9 **(a)** 0.39 **(b)** 0.39 **11** **(a)** 2.24 **(b)** 2.34

13 **(a)** $\dfrac{2625}{256} \approx 10.25$ **(b)** $\dfrac{3125}{3072} \approx 1.02$

15 **(a)** $\dfrac{1}{2} = 0.5$ **(b)** $\dfrac{1}{6} \approx 0.17$ **17** **(a)** 6416 **(b)** 54

19 **(a)** 41 **(b)** 8 **21** **(a)** 6.5 **(b)** 6
23 **(a)** 8.65 **(b)** 8.59 **29** **(a)** 127.5 **(b)** 131.7
31 0.174 m/sec **33** 0.28 **35** 1.48

5.8 REVIEW EXERCISES

1 $-\dfrac{8}{x} + \dfrac{2}{x^2} - \dfrac{5}{3x^3} + C$ **3** $100x + C$

5 $\dfrac{1}{16}(2x + 1)^8 + C$ **7** $-\dfrac{1}{16}(1 - 2x^2)^4 + C$

9 $-\dfrac{2}{1 + \sqrt{x}} + C$ **11** $3x - x^2 - \dfrac{5}{4}x^4 + C$

13 $\dfrac{1}{6}(4x^2 + 2x - 7)^3 + C$ **15** $-\dfrac{1}{x^2} - x^3 + C$

17 $\dfrac{3}{5}$ **19** $\dfrac{1}{6}$ **21** $\sqrt{8} - \sqrt{3} \approx 1.10$ **23** $\dfrac{52}{9}$ **25** $-\dfrac{37}{6}$

27 $8\sqrt{3} + 16 \approx 29.86$ **29** $\dfrac{1}{5}\cos(3 - 5x) + C$

31 $\dfrac{1}{15}\sin^5 3x + C$ **33** $-\dfrac{1}{6\sin^2 3x} + C$

35 $\dfrac{2}{15}(16\sqrt{2} - 3\sqrt{3})$ **37** $\dfrac{1}{6}$ **39** $\sqrt[5]{x^4 + 2x^2 + 1} + C$

41 0 **43** $y = x^3 - 2x^2 + x + 2$ **45** $\dfrac{135}{4}$

47 Use Corollary (5.27). **49** $\int_a^e f(x)\,dx$

51 **(a)** $-16t^2 - 30t + 900$ **(b)** -190 ft/sec

(c) $\frac{15}{16}(-1 + \sqrt{65}) \approx 6.6$ sec

53 **(a)** 341.36 **(b)** 334.42

CHAPTER 6

EXERCISES 6.1

Exer. 1–4: Answers are not unique.

1 $\int_{-2}^{2} [(x^2 + 1) - (x - 2)]\,dx$ **3** $\int_{-2}^{1} [(-3y^2 + 4) - y^3]\,dy$

5 $\int_0^4 (4x - x^2)\,dx = \frac{32}{3}$ **7** $2\int_0^2 [5 - (x^2 + 1)]\,dx = \frac{32}{3}$

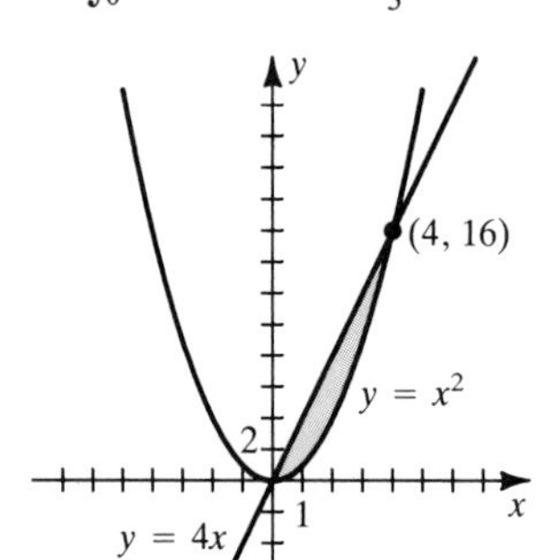

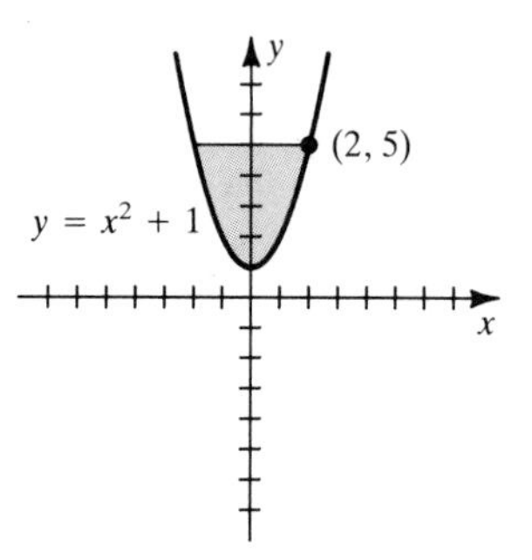

9 $\int_1^2 \left[\frac{1}{x^2} - (-x^2)\right] dx = \frac{17}{6}$

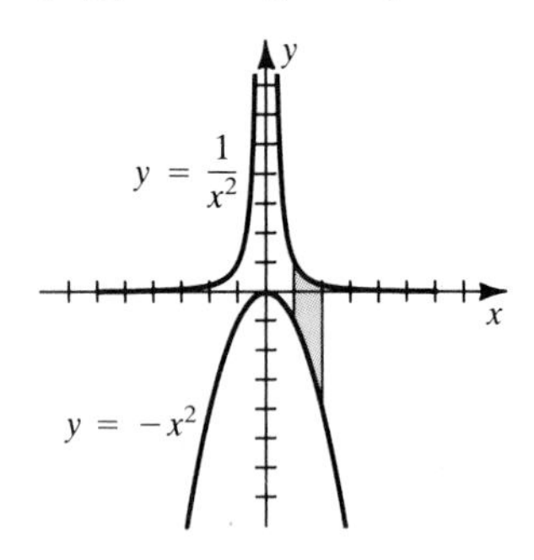

11 $\int_{-1}^{2} [(4 + y) - (-y^2)]\,dy = \frac{33}{2}$

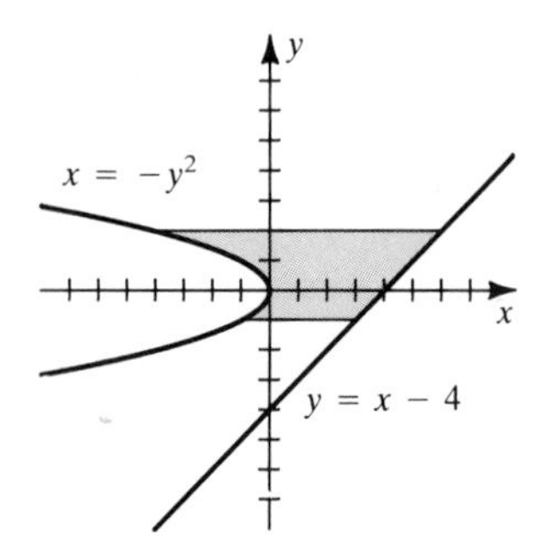

13 $2\int_0^{\sqrt{3}} [(2 - y^2) - (y^2 - 4)]\,dy = 8\sqrt{3}$

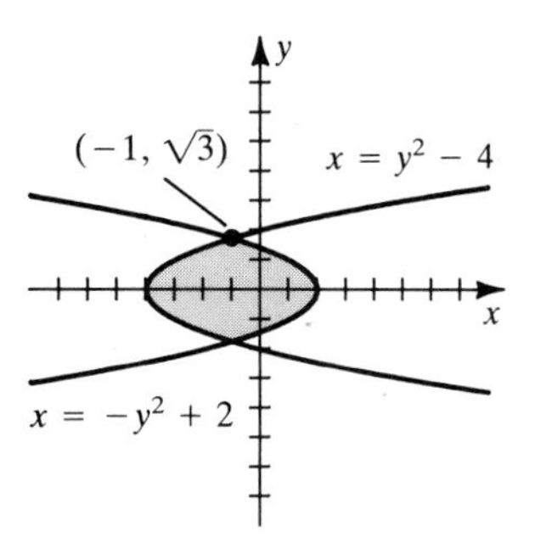

15 $2\int_0^2 [(4y - y^3) - 0]\,dy = 8$

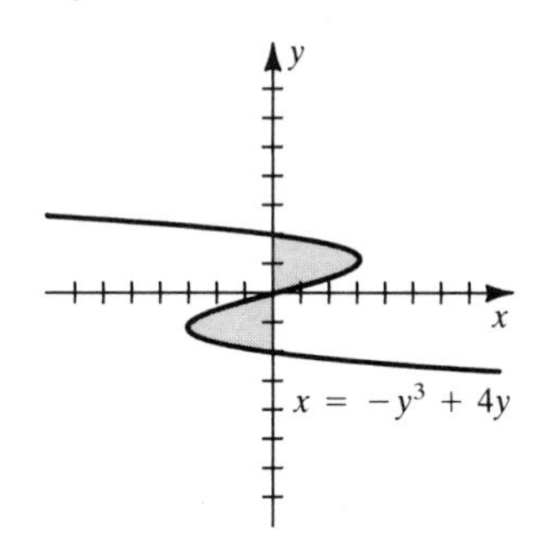

17 $2\int_0^1 [0 - (x^3 - x)]\,dx = \frac{1}{2}$

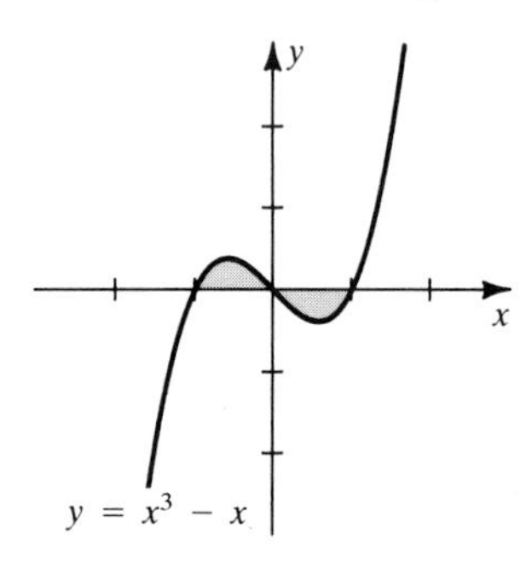

19 $\int_{-3}^{0} [(y^3 + 2y^2 - 3y) - 0]\,dy + \int_0^1 [0 - (y^3 + 2y^2 - 3y)]\,dy = \frac{71}{6}$

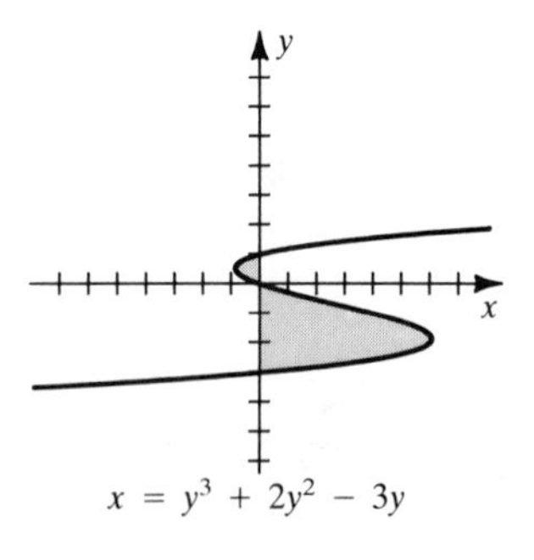

21 $2\int_0^2 x\sqrt{4-x^2}\,dx = \frac{16}{3}$

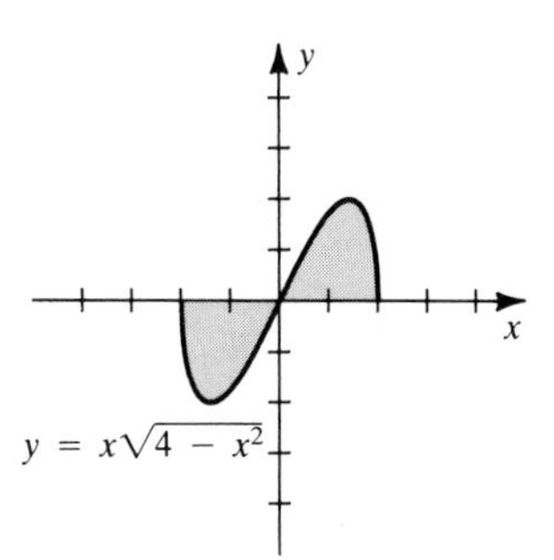

23 $\frac{1}{2}(2\pi + 3\sqrt{3}) \approx 5.74$

25 **(a)** $\int_0^1 (3x - x)\,dx + \int_1^2 [(4-x) - x]\,dx$

(b) $\int_0^2 \left(y - \frac{1}{3}y\right) dy + \int_2^3 \left[(4-y) - \frac{1}{3}y\right] dy$

27 **(a)** $\int_1^4 [\sqrt{x} - (-x)]\,dx$

(b) $\int_{-4}^{-1} [4 - (-y)]\,dy + \int_{-1}^{1} (4-1)\,dy + \int_1^2 (4 - y^2)\,dy$

29 **(a)** $\int_{-6}^{-1} [(x+3) - (-\sqrt{3-x})]\,dx + 2\int_{-1}^{3} \sqrt{3-x}\,dx$

(b) $\int_{-3}^{2} [(3-y^2) - (y-3)]\,dy$

31 9 **33** 12 **35** $4\sqrt{2}$

37 $\int_0^1 (x^2 - 6x + 5)\,dx + \int_1^5 -(x^2 - 6x + 5)\,dx + \int_5^7 (x^2 - 6x + 5)\,dx$

39 **(a)** 4.25 **(b)** 4.50

41 $\int_{-1.5}^{-1.1} -(x^3 - 0.7x^2 - 0.8x + 1.3)\,dx + \int_{-1.1}^{1.5} (x^3 - 0.7x^2 - 0.8x + 1.3)\,dx$

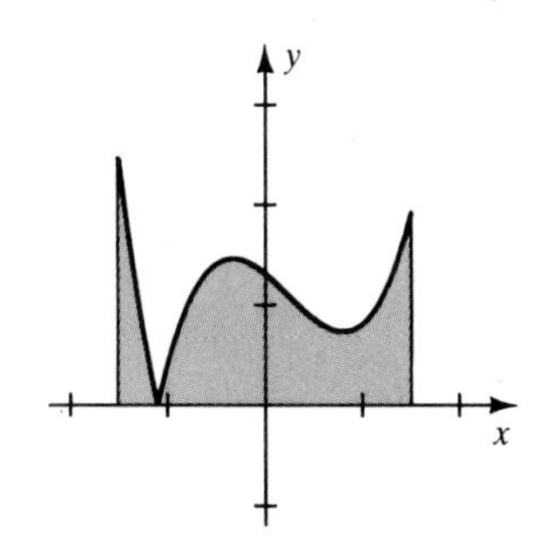

EXERCISES 6.2

1 $\pi \int_{-1}^{2} \left(\frac{1}{2}x^2 + 2\right)^2 dx$

3 $2 \cdot \pi \int_0^4 [(\sqrt{25 - y^2})^2 - 3^2]\,dy$

5 $\pi \int_1^3 \left(\frac{1}{x}\right)^2 dx = \frac{2\pi}{3}$

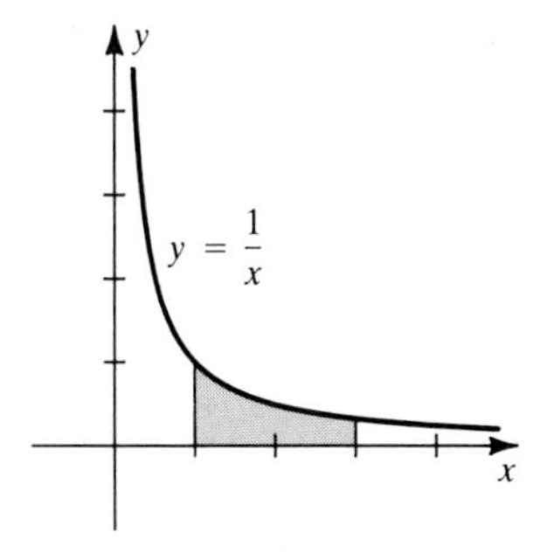

7 $\pi \int_0^4 (x^2 - 4x)^2\,dx = \frac{512\pi}{15}$

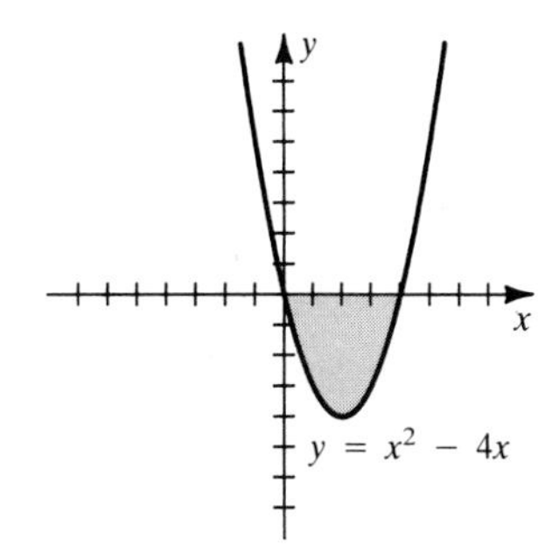

9 $\pi \int_0^2 (\sqrt{y})^2\,dy = 2\pi$

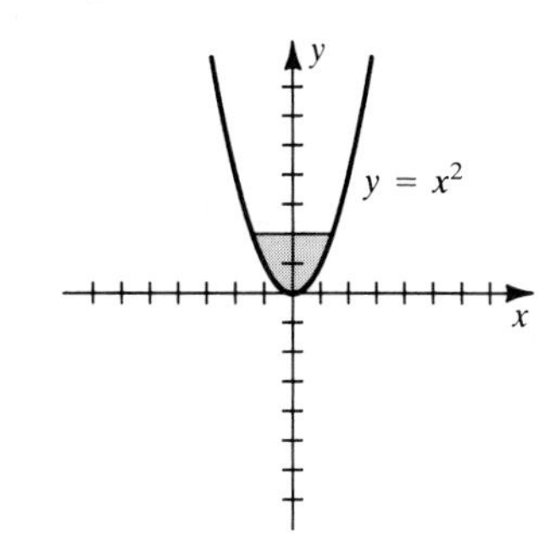

11 $\pi \int_0^4 (4y - y^2)^2\,dy = \frac{512\pi}{15}$

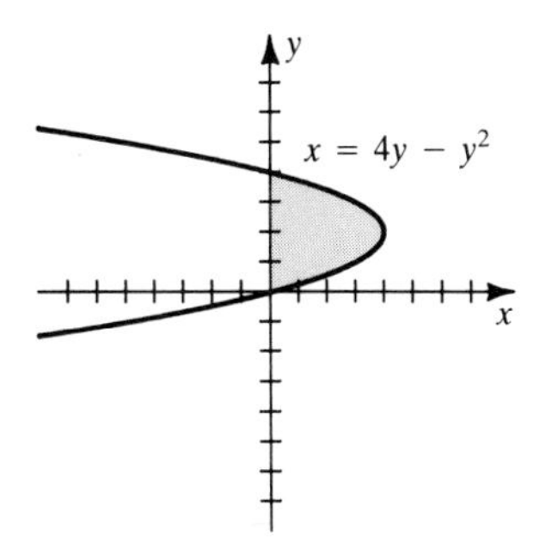

13 $2 \cdot \pi \int_0^{\sqrt{2}} [(4 - x^2)^2 - (x^2)^2]\,dx = \frac{64\pi\sqrt{2}}{3}$

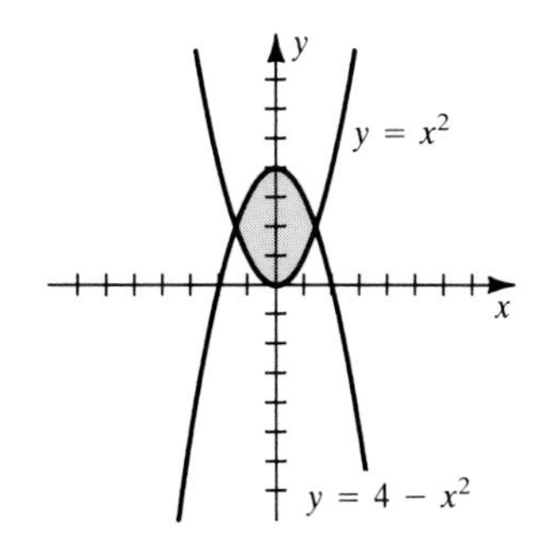

15 $\pi \int_0^2 [(4-x)^2 - (x)^2]\, dx = 16\pi$

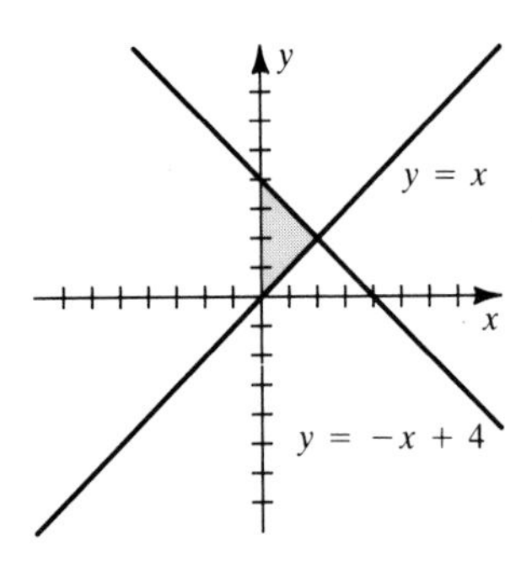

17 $\pi \int_0^2 [(2y)^2 - (y^2)^2]\, dy = \frac{64\pi}{15}$

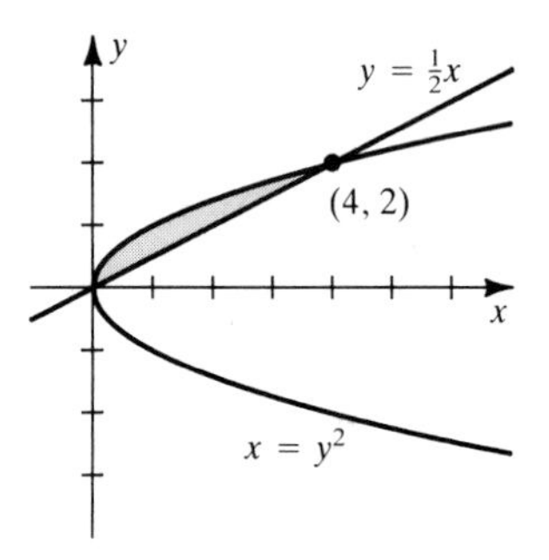

19 $\pi \int_{-1}^2 [(y+2)^2 - (y^2)^2]\, dy = \frac{72\pi}{5}$

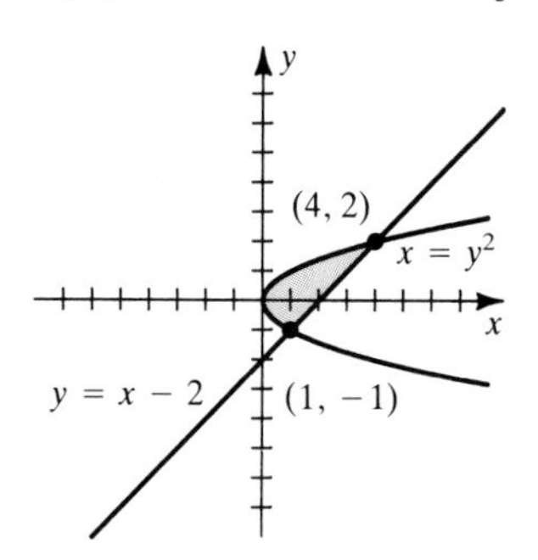

21 $\pi \int_0^\pi (\sin 2x)^2\, dx = \frac{1}{2}\pi^2$

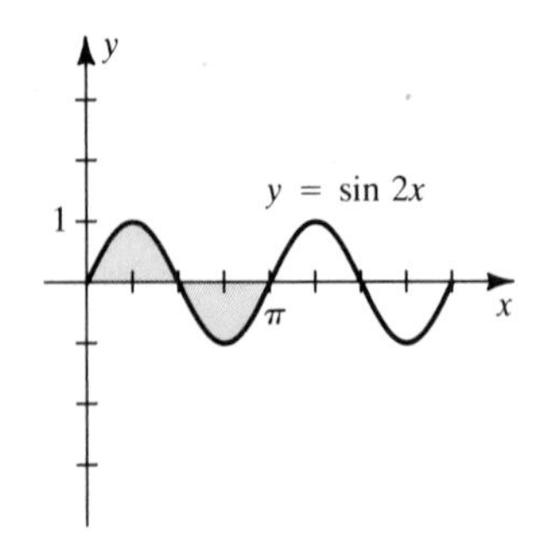

23 $\pi \int_0^{\pi/4} [(\cos x)^2 - (\sin x)^2]\, dx = \frac{\pi}{2}$

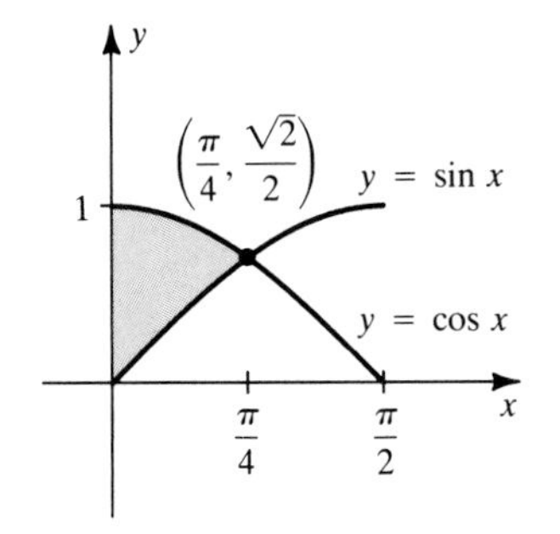

25

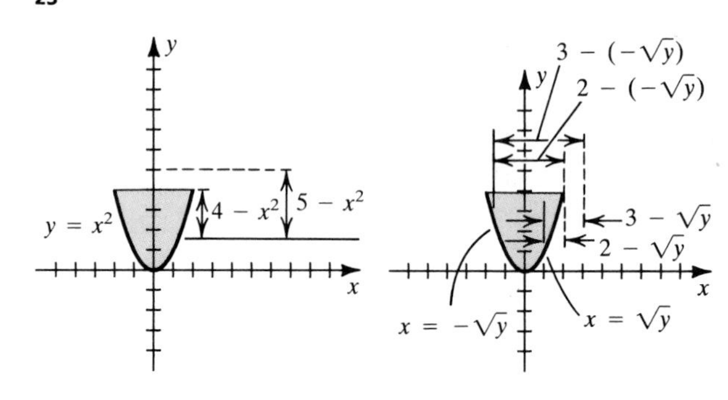

(a) $2 \cdot \pi \int_0^2 (4 - x^2)^2\, dx = \frac{512\pi}{15}$

(b) $2 \cdot \pi \int_0^2 [(5 - x^2)^2 - (5 - 4)^2]\, dx = \frac{832\pi}{15}$

(c) $\pi \int_0^4 \{[2 - (-\sqrt{y})]^2 - [2 - \sqrt{y}]^2\}\, dy = \frac{128\pi}{3}$

(d) $\pi \int_0^4 \{[3 - (-\sqrt{y})]^2 - [3 - \sqrt{y}]^2\}\, dy = 64\pi$

27 **(a)** $\pi \int_0^4 \left\{\left[\left(-\frac{1}{2}x + 2\right) - (-2)\right]^2 - [0 - (-2)]^2\right\} dx$

(b) $\pi \int_0^4 \left\{(5 - 0)^2 - \left[5 - \left(-\frac{1}{2}x + 2\right)\right]^2\right\} dx$

(c) $\pi \int_0^2 \{(7 - 0)^2 - [7 - (-2y + 4)]^2\}\, dy$

(d) $\pi \int_0^2 \{[(-2y + 4) - (-4)]^2 - [0 - (-4)]^2\}\, dy$

29 $\pi \int_{-2}^0 [(8 - 4x)^2 - (8 - x^3)^2]\, dx + \pi \int_0^2 [(8 - x^3)^2 - (8 - 4x)^2]\, dx$

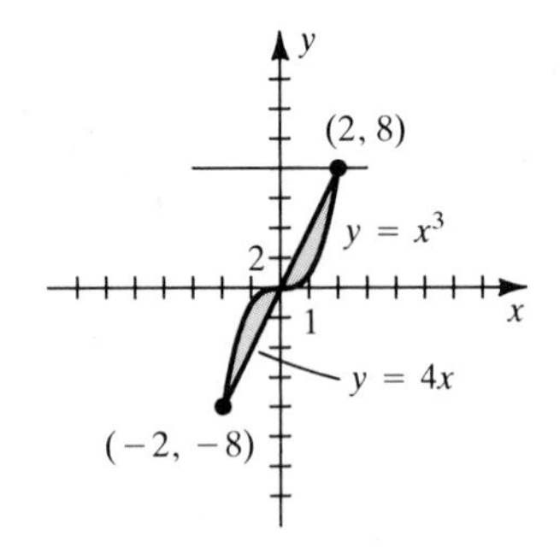

31 $\pi \int_2^3 \{[2-(3-y)]^2 - [2-\sqrt{3-y}]^2\}\, dy$

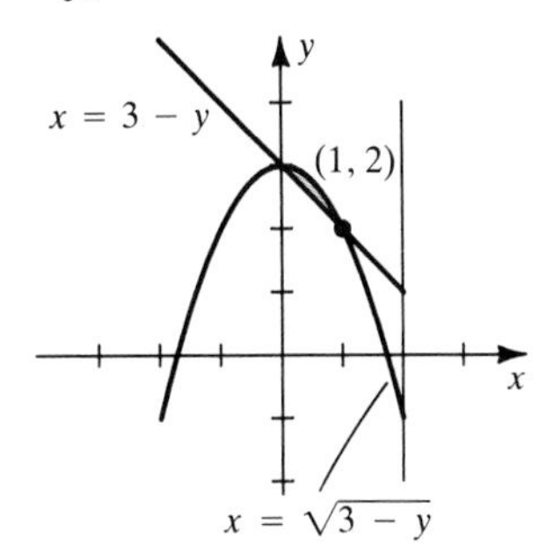

33 $2 \cdot \pi \int_0^1 \{[5-(-\sqrt{1-y^2})]^2 - [5-\sqrt{1-y^2}]^2\}\, dy$

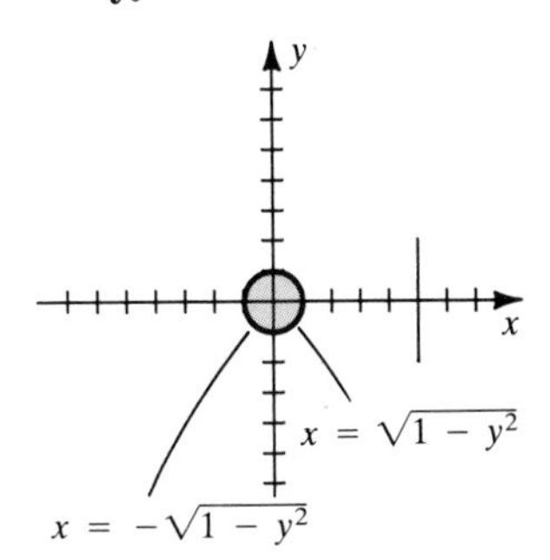

35 $\pi \int_0^h r^2\, dy = \pi r^2 h$ **37** $\pi \int_0^h \left(\frac{r}{h}x\right)^2 dx = \frac{1}{3}\pi r^2 h$

39 $\pi \int_0^h \left(\frac{R-r}{h}x + r\right)^2 dx = \frac{1}{3}\pi h(R^2 + Rr + r^2)$ **41** $\frac{63\pi}{2}$

43 **(a)** 0.45, 2.01 **(b)** 0.28

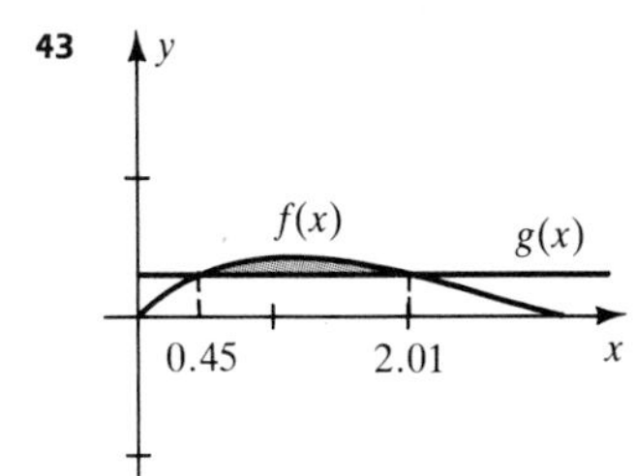

EXERCISES 6.3

1 $2\pi \int_2^{11} x\sqrt{x-2}\, dx$ **3** $2\pi \int_0^6 y\left(-\frac{1}{2}y + 3\right) dy$

5 $2\pi \int_0^4 x\sqrt{x}\, dx = \frac{128\pi}{5}$

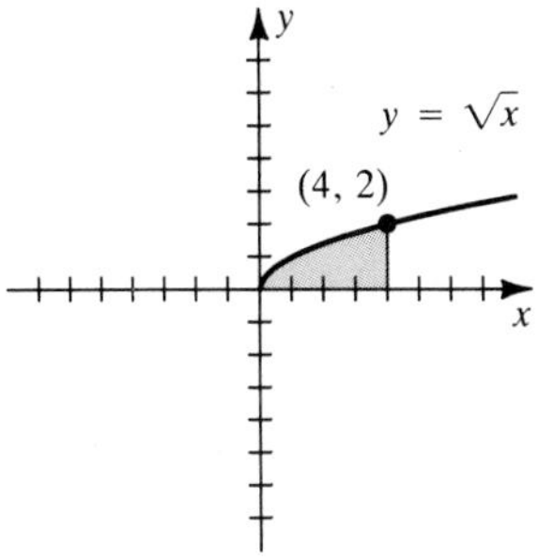

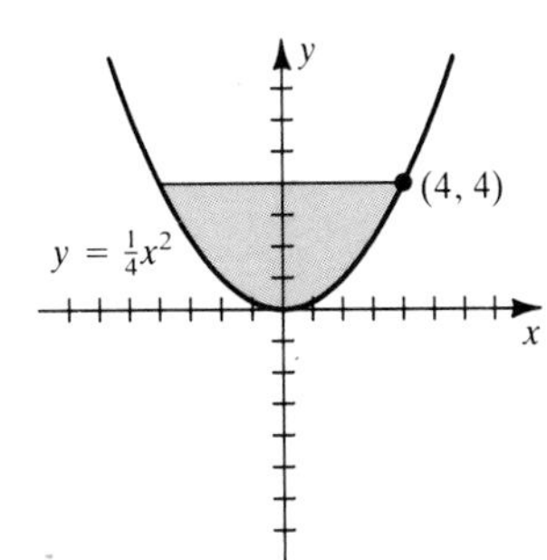

7 $2\pi \int_0^2 x(\sqrt{8x} - x^2)\, dx = \frac{24\pi}{5}$

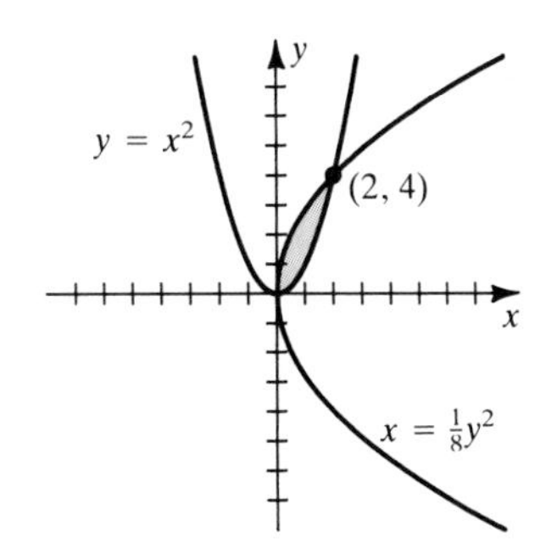

9 $2\pi \int_4^7 x\left[\left(\frac{1}{2}x - \frac{3}{2}\right) - (2x - 12)\right] dx = \frac{135\pi}{2}$

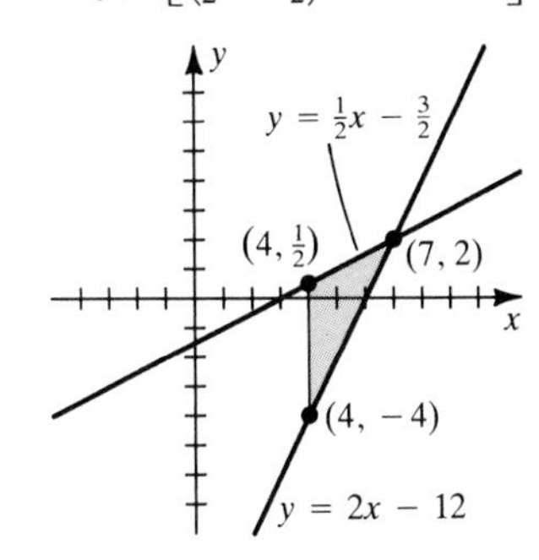

11 $2\pi \int_0^2 x[0 - (2x - 4)]\, dx = \frac{16\pi}{3}$

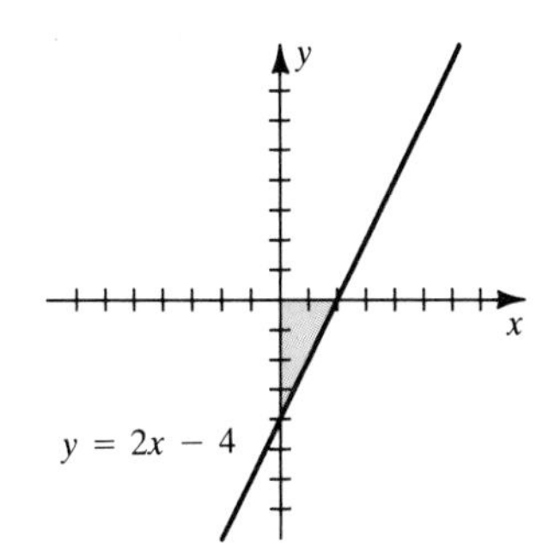

13 $2 \cdot 2\pi \int_0^4 y\sqrt{4y}\, dy = \frac{512\pi}{5}$

15 $2\pi \int_0^6 y\left(\frac{1}{2}y\right) dy = 72\pi$

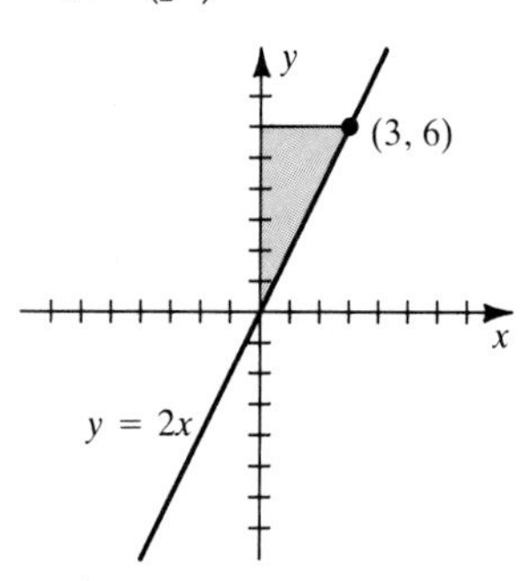

17 $2\pi \int_0^2 y[0 - (y^2 - 4)]\, dy = 8\pi$

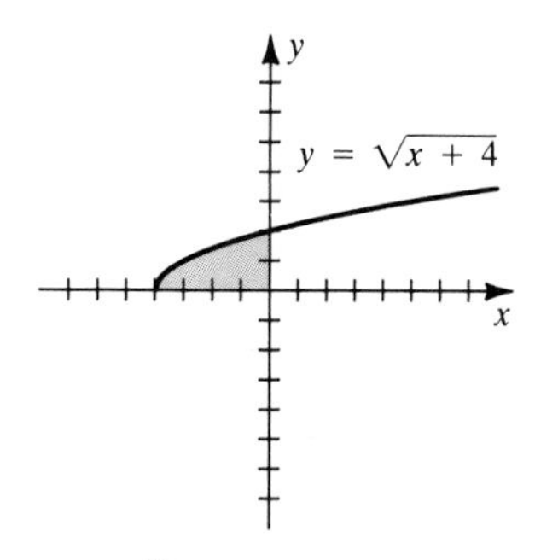

19 **(a)** $2\pi \int_0^2 (3 - x)(x^2 + 1)\, dx$

(b) $2\pi \int_0^2 [x - (-1)](x^2 + 1)\, dx$

21 **(a)** $2 \cdot 2\pi \int_0^4 (4 - y)\sqrt{y}\, dy$ **(b)** $2 \cdot 2\pi \int_0^4 (5 - y)\sqrt{y}\, dy$

(c) $2\pi \int_{-2}^{2} (2 - x)(4 - x^2)\, dx$

(d) $2\pi \int_{-2}^{2} [x - (-3)](4 - x^2)\, dx$

23 $2\pi \int_0^1 (2 - x)[(3 - x^2) - (3 - x)]\, dx$

25 $2 \cdot 2\pi \int_{-1}^{1} (5 - x)\sqrt{1 - x^2}\, dx$

27 **(a)** $2\pi \int_0^{1/2} y(4 - 1)\, dy + 2\pi \int_{1/2}^{1} y[(1/y^2) - 1]\, dy$

(b) $\pi \int_1^4 \left(\frac{1}{\sqrt{x}}\right)^2 dx$

29 **(a)** $2\pi \int_0^1 x(x^2 + 2)\, dx$

(b) $\pi \int_0^2 (1)^2\, dy + \pi \int_2^3 [(1)^2 - (\sqrt{y - 2})^2]\, dy$

31 76π

33

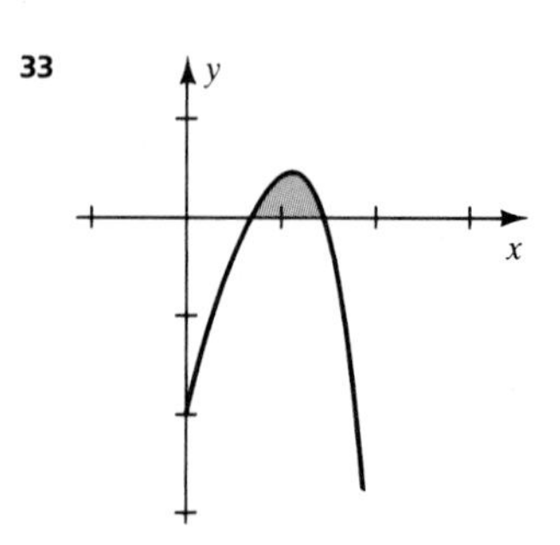

(a) 0.68, 1.44

(b) $2\pi \int_{0.68}^{1.44} x(-x^4 + 2.21x^3 - 3.21x^2 + 4.42x - 2)\, dx$

EXERCISES 6.4

Exer. 1–26: The first integral represents a general formula for the volume. In Exercises 1–8, the vertical distance between the graphs of $y = \sqrt{x}$ and $y = -\sqrt{x}$ is $[\sqrt{x} - (-\sqrt{x})]$, denoted by $2\sqrt{x}$.

1 $\int_c^d s^2\, dx = \int_0^9 (2\sqrt{x})^2\, dx = 162$

3 $\int_c^d \frac{1}{2}\pi r^2\, dx = \int_0^9 \frac{1}{2}\pi(\sqrt{x})^2\, dx = \frac{81\pi}{4}$

5 $\int_c^d \frac{\sqrt{3}}{4} s^2\, dx = \int_0^9 \frac{\sqrt{3}}{4}(2\sqrt{x})^2\, dx = \frac{81\sqrt{3}}{2}$

7 $\int_c^d \frac{1}{2}(B + b)h\, dx = \int_0^9 \frac{1}{2}\left[2\sqrt{x} + \frac{1}{2}(2\sqrt{x})\right]\left[\frac{1}{4}(2\sqrt{x})\right] dx = \frac{243}{8}$

9 $\int_c^d s^2\, dx = 2\int_0^a [\sqrt{a^2 - x^2} - (-\sqrt{a^2 - x^2})]^2\, dx = \frac{16}{3}a^3$

11 $\int_c^d \frac{1}{2}bh\, dx = 2\int_0^2 \frac{1}{2}\left[\frac{1}{\sqrt{2}}(4 - x^2)\right]\left[\frac{1}{\sqrt{2}}(4 - x^2)\right] dx = \frac{128}{15}$

13 $\int_c^d lw\, dx = \int_0^h \left(\frac{2ax}{h}\right)\left(\frac{ax}{h}\right) dx = \frac{2}{3}a^2h$

15 $\int_c^d \frac{1}{2}\pi r^2\, dy = 2\int_0^4 \frac{1}{2}\pi\left[\frac{1}{2}\left(4 - \frac{1}{4}y^2\right)\right]^2 dy = \frac{128\pi}{15}$

17 $\int_c^d lw\, dy = \int_0^a [\sqrt{a^2 - y^2} - (-\sqrt{a^2 - y^2})]y\, dy = \frac{2}{3}a^3$

19 $\int_c^d \frac{1}{2}bh\, dx = \int_{-a}^{a} \frac{1}{2}[\sqrt{a^2 - x^2} - (-\sqrt{a^2 - x^2})]h\, dx = \frac{1}{2}\pi a^2 h$

21 $\int_c^d \frac{1}{2}bh\, dx = \int_0^4 \frac{1}{2}\left(\frac{2}{4}x\right)\left(\frac{3}{4}x\right) dx = 4\ \text{cm}^3$

23 $\int_c^d \frac{1}{2}\pi r^2\, dy = \int_0^a \frac{1}{2}\pi\left[\frac{1}{2}(a - y)\right]^2 dy = \frac{\pi}{24}a^3$

25 The areas of cross sections of typical disks and washers are $\pi[f(x)]^2$ and $\pi\{[f(x)]^2 - [g(x)]^2\}$, respectively. In each case, the integrand represents $A(x)$ in (6.13).

EXERCISES 6.5

1 **(a)** $\int_1^3 \sqrt{1 + (3x^2)^2}\, dx$ **(b)** $\int_2^{28} \sqrt{1 + \left[\frac{1}{3}(y - 1)^{-2/3}\right]^2}\, dy$

3 **(a)** $\int_{-3}^{-1} \sqrt{1 + (-2x)^2}\, dx$ **(b)** $\int_{-5}^{3} \sqrt{1 + \left[\frac{1}{2}(4 - y)^{-1/2}\right]^2}\, dy$

5 $\int_1^8 \sqrt{1 + \left(\frac{4}{9}x^{-1/3}\right)^2}\, dx = \left(4 + \frac{16}{81}\right)^{3/2} - \left(1 + \frac{16}{81}\right)^{3/2} \approx 7.29$

7 $\int_1^4 \sqrt{1 + \left(-\frac{3}{2}x^{1/2}\right)^2}\, dx = \frac{8}{27}\left[10^{3/2} - \left(\frac{13}{4}\right)^{3/2}\right] \approx 7.63$

9 $\int_1^2 \sqrt{1 + \left(\frac{1}{4}x^2 - \frac{1}{x^2}\right)^2}\, dx = \frac{13}{12}$

11 $\int_1^2 \sqrt{1 + \left(-\frac{3}{2}y^{-4} + \frac{1}{6}y^4\right)^2}\, dx = \frac{353}{240}$

13 $\int_0^2 \sqrt{1 + \left(\frac{7}{2} - 3y^2\right)^2}\, dy$

15 $8\int_a^1 \sqrt{1 + [(-x^{-1/3})(1 - x^{2/3})^{1/2}]^2}\, dx = 6$, where $a = \left(\frac{1}{2}\right)^{3/2}$

17 **(a)** $s(x) = \frac{1}{27}[(9x^{2/3} + 4)^{3/2} - 13^{3/2}]$

(b) $\frac{1}{27}\{[9(1.1)^{2/3} + 4]^{3/2} - 13^{3/2}\} \approx 0.1196;\ \frac{\sqrt{13}}{30} \approx 0.1202$

19 $\sqrt{17}(0.1) \approx 0.4123$ **21** $\frac{\pi\sqrt{5}}{360} \approx 0.0195$ **23** 9.80

25 1.91 **27 (a)** 3.79 **(b)** It is smaller.

29 $2\pi \int_0^1 \sqrt{4x}\sqrt{1+(x^{-1/2})^2}\,dx = \frac{8\pi}{3}(2^{3/2}-1) \approx 15.32$

31 $2\pi \int_1^2 \left(\frac{1}{4}x^4+\frac{1}{8}x^{-2}\right)\sqrt{1+\left(x^3-\frac{1}{4}x^{-3}\right)^2}\,dx = \frac{16{,}911\pi}{1024} \approx 51.88$

33 $2\pi \int_2^4 \frac{1}{8}y^3\sqrt{1+\left(\frac{3}{8}y^2\right)^2}\,dy = \frac{\pi}{27}[8(37)^{3/2}-13^{3/2}] \approx 204.04$

35 $2\pi \int_4^5 \sqrt{25-y^2}\sqrt{1+[(-y)(25-y^2)^{-1/2}]^2}\,dy = 10\pi$

37 $2\pi \int_0^h \left(\frac{r}{h}x\right)\sqrt{1+\left(\frac{r}{h}\right)^2}\,dx = \pi r\sqrt{h^2+r^2}$

39 $2\cdot 2\pi \int_0^r \sqrt{r^2-x^2}\sqrt{1+[(-x)(r^2-x^2)^{-1/2}]^2}\,dx = 4\pi r^2$

41 *Hint:* Regard ds as the slant height of the frustum of a cone that has average radius x.

43 6.07

EXERCISES 6.6

1 (a) and **(b)** 6000 ft-lb **3 (a)** $\frac{128}{3}$ in.-lb **(b)** $\frac{64}{3}$ in.-lb

5 $W_2 = 3W_1$ **7** 36,945 ft-lb **9** 276 ft-lb **11** 2250 ft-lb

13 (a) $\frac{81\pi}{2}(62.5) \approx 7952$ ft-lb **(b)** $\frac{189\pi}{2}(62.5) \approx 18{,}555$ ft-lb

15 500 ft-lb **17** $575\left(\frac{1}{2}-40^{-1/5}\right) \approx 12.55$ in.-lb

19 $W = \frac{Gm_1m_2h}{(4000)(4000+h)}$ **21** 36.85 ft-lb

23 (a) $\frac{3}{10}k$ J (k a constant) **(b)** $\frac{9}{40}k$ J

EXERCISES 6.7

1 250; 140; 0.56 **3** 14; -27; -46; $\left(-\frac{23}{7}, -\frac{27}{14}\right)$

5 $\frac{1}{4}$; $\frac{1}{14}$; $\frac{1}{5}$; $\left(\frac{4}{5}, \frac{2}{7}\right)$ **7** $\frac{32}{3}$; $\frac{256}{15}$; 0; $\left(0, \frac{8}{5}\right)$

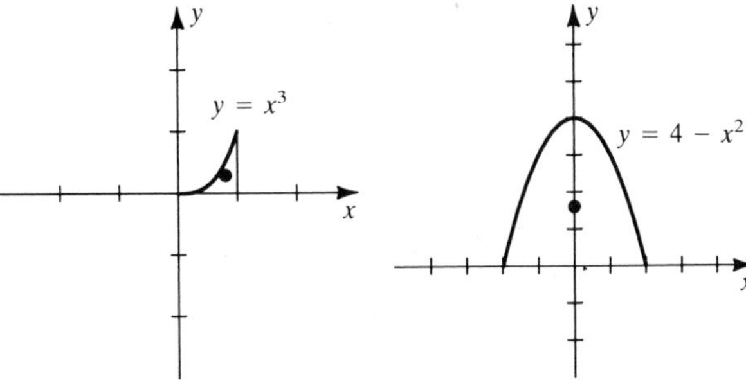

9 $\frac{4}{3}$; $\frac{4}{3}$; $\frac{32}{15}$; $\left(\frac{8}{5}, 1\right)$

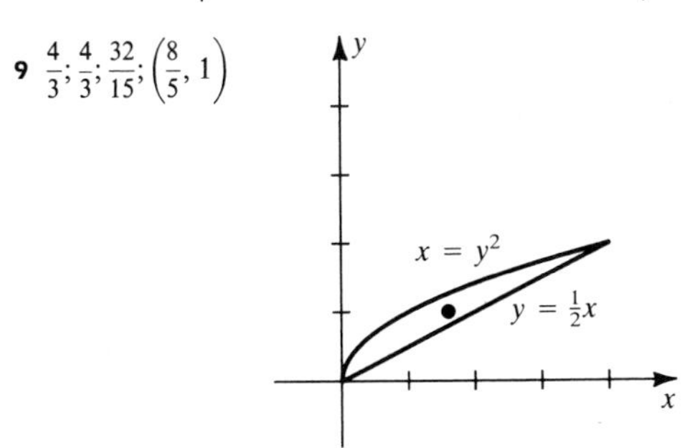

11 $\frac{9}{2}$; $-\frac{27}{10}$; $-\frac{9}{4}$; $\left(-\frac{1}{2}, -\frac{3}{5}\right)$ **13** $\frac{9}{2}$; $\frac{9}{4}$; $\frac{36}{5}$; $\left(\frac{8}{5}, \frac{1}{2}\right)$

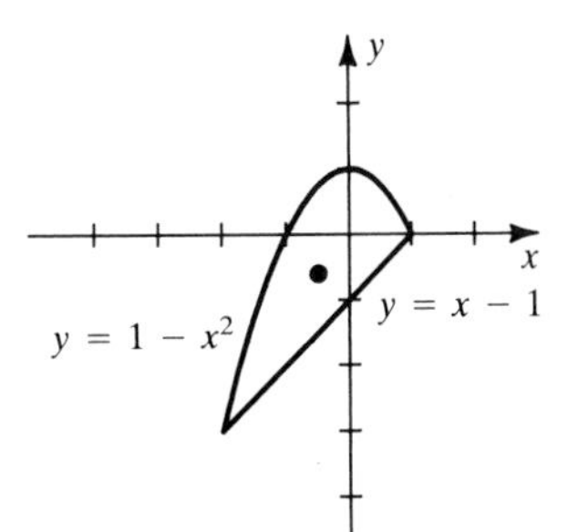

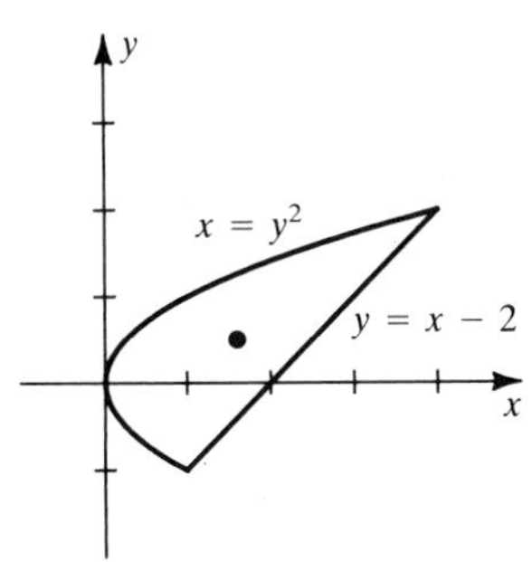

15 $\left(\frac{4a}{3\pi}, \frac{4a}{3\pi}\right)$

17 With the center of the circle at the origin, the centroid is $\left(0, -\frac{20a}{3(8+\pi)}\right)$.

19 Show that the centroid is $\left(\frac{1}{3}a, \frac{1}{3}(b+c)\right)$.

21 $(2\pi\cdot 3)(\sqrt{2}\sqrt{18}) = 36\pi$ **23** $\left(\frac{4a}{3\pi}, \frac{4a}{3\pi}\right)$

25

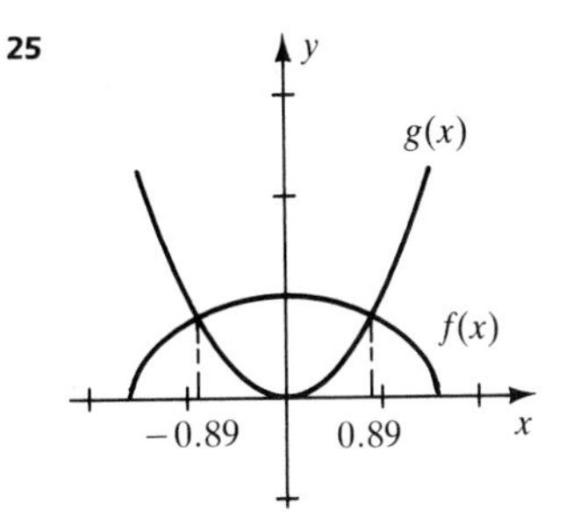

(a) $\rho \int_{-0.89}^{0.89} (\sqrt{|\cos x|} - x^2)\,dx$ **(b)** 1.19ρ

EXERCISES 6.8

1 (a) $\frac{1}{2}(62.5)$ lb **(b)** $\frac{3}{2}(62.5)$ lb

3 (a) $\frac{\sqrt{3}}{3}(62.5)$ lb **(b)** $\frac{\sqrt{3}}{24}(62.5)$ lb

5 $\frac{16}{3}(60)$ lb **7** $\frac{592}{3}(62.5)$ lb

9 (a) 90(50) lb **(b)** 54(50) lb; 36(50) lb **11** 1.56 L/min

13 In min: **(a)** 20 **(b)** 66 **(c)** 115 **(d)** 197

15 (a) $9[(601)^{2/3}-1] \approx 632$ min
(b) $2\cdot 9[(301)^{2/3}-1] \approx 790$ min

17 666 **19** 11 **21 (a)** and **(b)** 150 J

23 $9 - \frac{5\sqrt{5}}{3} \approx 5.27$ gal **25** 1.45 coulombs

27 (a) $\int_0^{1/30} 12{,}450\pi \sin(30\pi t)\,dt = 830\text{ cm}^3$
(b) It is not safe, since approximately 0.026 joule is inhaled.

29 32

6.9 REVIEW EXERCISES

1 **(a)** $2\int_0^2 [(-x^2) - (x^2 - 8)]\, dx = \frac{64}{3}$

(b) $4\int_{-4}^0 \sqrt{-y}\, dy = \frac{64}{3}$

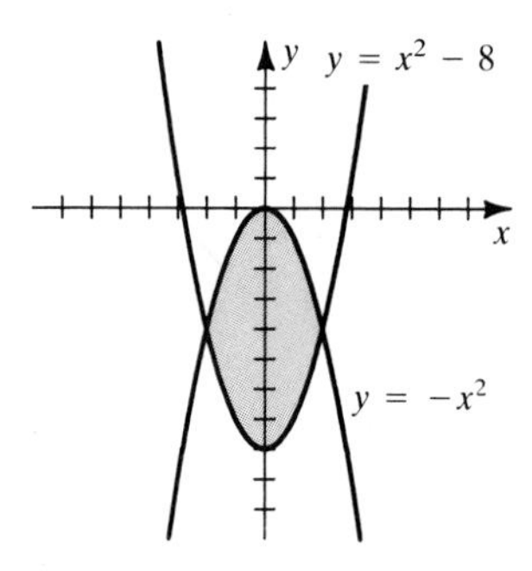

3 $\int_a^b [(1 - y) - y^2]\, dy = \frac{5\sqrt{5}}{6}$, where $a = \frac{1}{2}(-1 - \sqrt{5})$ and $b = \frac{1}{2}(-1 + \sqrt{5})$

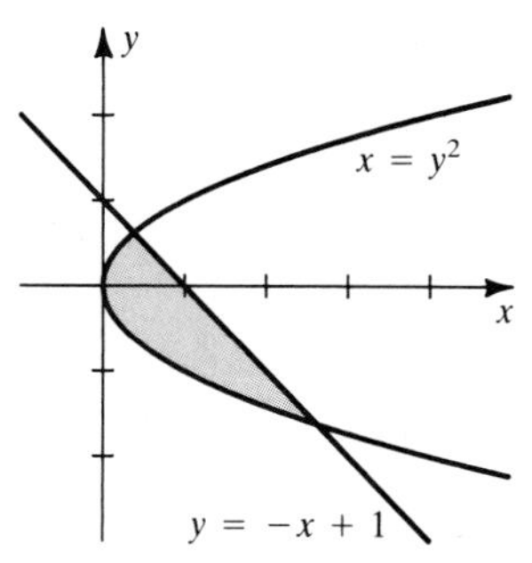

5 $\int_{\pi/3}^{\pi} \left(\sin x - \cos \frac{1}{2}x\right) dx = \frac{1}{2}$

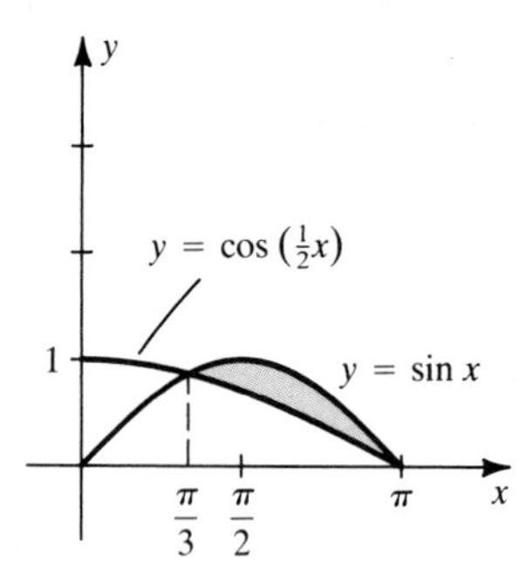

7 $\pi \int_0^2 (\sqrt{4x + 1})^2\, dx = 10\pi$

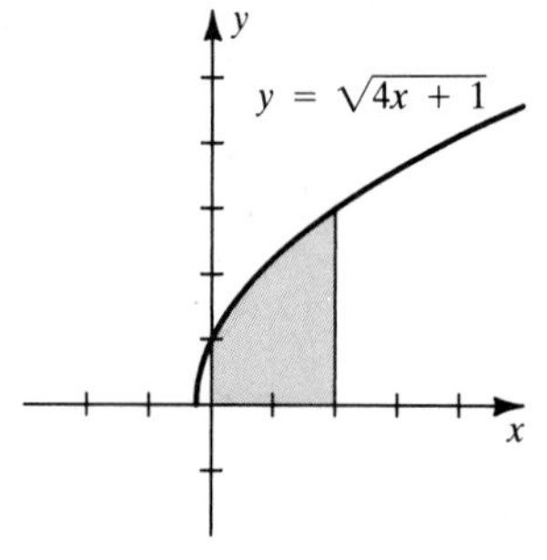

9 $2\pi \int_0^1 x[2 - (x^3 + 1)]\, dx = \frac{3\pi}{5}$

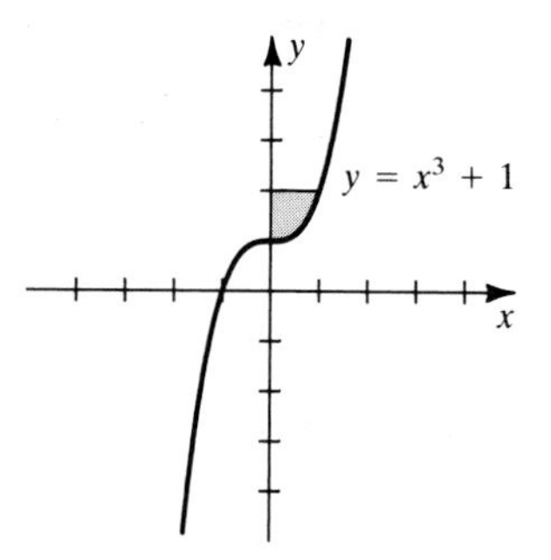

11 $2\pi \int_0^{\sqrt{\pi/2}} x \cos x^2\, dx = \pi$

13 **(a)** $\pi \int_{-2}^1 [(-4x + 8)^2 - (4x^2)^2]\, dx = \frac{1152\pi}{5}$

(b) $2\pi \int_{-2}^1 (1 - x)[(-4x + 8) - 4x^2]\, dx = 54\pi$

(c) $\pi \int_{-2}^1 \{(16 - 4x^2)^2 - [16 - (-4x + 8)]^2\}\, dx = \frac{1728\pi}{5}$

15 $\int_{-2}^5 \sqrt{1 + \left[\frac{1}{3}(x + 3)^{-1/3}\right]^2}\, dx = \frac{1}{27}(37^{3/2} - 10^{3/2}) \approx 7.16$

17 $\int_0^4 (5 - y)(62.5)\pi(6)^2\, dy = 432\pi(62.5)$ ft-lb

19 $\rho \int_0^{\sqrt{8}} (6 - y)2(\sqrt{8} - y)\, dy + \rho \int_{-\sqrt{8}}^0 (6 - y)2(y + \sqrt{8})\, dy = 96(62.5)$ lb

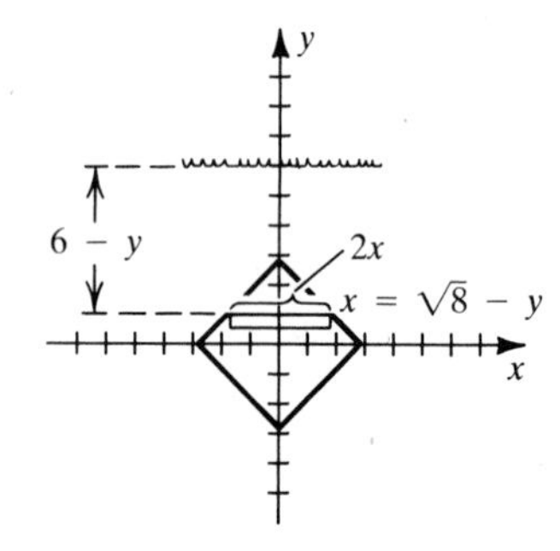

21 $4; -\frac{4}{21}; \frac{16}{15}; \left(\frac{4}{15}, -\frac{1}{21}\right)$

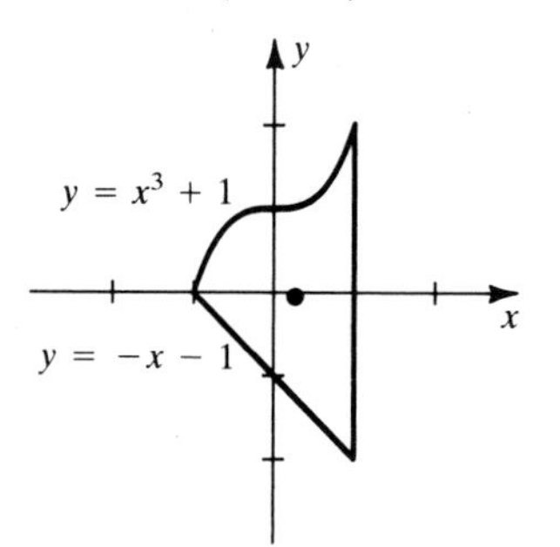

23 $2\pi \int_1^2 \left(\frac{1}{3}x^3 + \frac{1}{4}x^{-1}\right)\sqrt{1 + \left(x^2 - \frac{1}{4}x^{-2}\right)^2}\, dx = \frac{515\pi}{64} \approx 25.3$

25 900 ft

27 **(a)** The area under the graph of $y = 2\pi x^4$

(b) **(i)** The volume obtained by revolving $y = \sqrt{2}x^2$ about the x-axis
(ii) The volume obtained by revolving $y = x^3$ about the y-axis
(c) The work done by a force of magnitude $y = 2\pi x^4$ as it moves from $x = 0$ to $x = 1$.

CHAPTER 7

EXERCISES 7.1

1 $\frac{x-5}{3}$ **3** $\frac{2x+1}{3x}$ **5** $\frac{5x+2}{2x-3}$ **7** $-\frac{1}{3}\sqrt{6-3x}$

9 $3 - x^2, x \geq 0$ **11** $(x-1)^3$

13 **(a)** The graph of f is a line of slope $a \neq 0$ and hence is one-to-one. $f^{-1}(x) = \frac{x-b}{a}$
(b) No (not one-to-one)

15 **(a)**
(b) $[-1, 2]; \left[\frac{1}{2}, 4\right]$
(c) $\left[\frac{1}{2}, 4\right]; [-1, 2]$

17 **(a)**
(b) $[-3, 3]; [-2, 2]$
(c) $[-2, 2]; [-3, 3]$

19

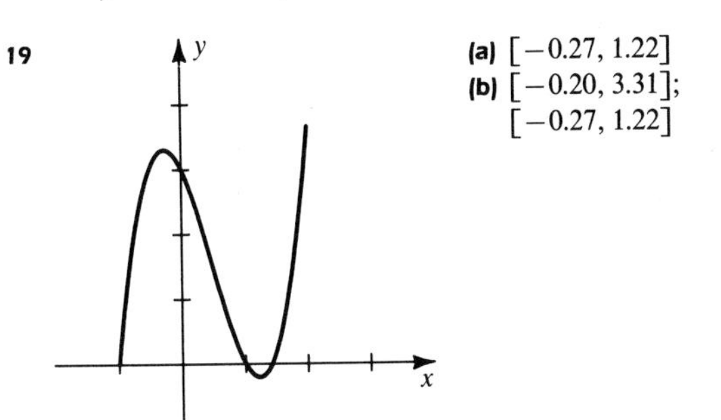

(a) $[-0.27, 1.22]$
(b) $[-0.20, 3.31]$; $[-0.27, 1.22]$

21 **(a)** f is increasing on $\left[-\frac{3}{2}, \infty\right)$ and hence is one-to-one.
(b) $[0, \infty)$ **(c)** x

23 **(a)** f is decreasing on $[0, \infty)$ and hence is one-to-one.
(b) $(-\infty, 4]$ **(c)** $-\frac{1}{2\sqrt{4-x}}$

25 **(a)** f is decreasing on $(-\infty, 0)$ and $(0, \infty)$ and hence is one-to-one.
(b) All real numbers except zero **(c)** $-\frac{1}{x^2}$

27 **(a)** f is increasing, since $f'(x) > 0$ for every x **(b)** $\frac{1}{16}$

29 **(a)** f is decreasing, since $f'(x) < 0$ for $x > 0$ **(b)** $-\frac{2}{7}$

31 **(a)** f is increasing, since $f'(x) > 0$ for every x **(b)** $\frac{1}{16}$

EXERCISES 7.2

1 $\frac{9}{9x+4}$ **3** $\frac{2(3x-1)}{3x^2-2x+1}$ **5** $\frac{2}{2x-3}$ **7** $\frac{15}{3x-2}$

9 $\frac{3x^2}{2x^3-7}$ **11** $1 + \ln x$ **13** $\frac{1}{2x}\left(1 + \frac{1}{\sqrt{\ln x}}\right)$

15 $-\frac{1}{x}\left[\frac{1}{(\ln x)^2} + 1\right]$ **17** $\frac{20}{5x-7} + \frac{6}{2x+3}$

19 $\frac{x}{x^2+1} - \frac{18}{9x-4}$ **21** $\frac{x}{x^2-1} - \frac{x}{x^2+1}$ **23** $\frac{1}{\sqrt{x^2-1}}$

25 $-2\tan 2x$ **27** $9\csc 3x \sec 3x$ **29** $\frac{2\tan 2x}{\ln \sec 2x}$

31 $\tan x$ **33** $\sec x$ **35** $\frac{y(2x^2-1)}{x(3y+1)}$ **37** $\frac{y^2 - xy\ln y}{x^2 - xy\ln x}$

39 $(5x+2)^2(6x+1)(150x+39)$ **41** $\frac{(14x+11)(x-5)^2}{\sqrt{4x+7}}$

43 $\frac{(19x^2+20x-3)(x^2+3)^4}{2(x+1)^{3/2}}$ **45** $y = 8x - 15$

47 $(10, 5\ln 10 - 5) \approx (10, 6.51)$; $y'' = -(5/x^2) < 0$ implies that the graph is CD for $x > 0$. **49** ± 0.73 yr

51 **(a)** $s'(0) = 0$ m/sec; $s''(0) = \frac{bc}{m_1 + m_2}$ m/sec^2
(b) $s'\left(\frac{m_2}{b}\right) = c\ln\left(\frac{m_1+m_2}{m_1}\right)$; $s''\left(\frac{m_2}{b}\right) = \frac{bc}{m_1}$

53 The graphs coincide if $x > 0$; however, the graph of $y = \ln(x^2)$ contains points with negative x-coordinates.

55 0.567 **57** **(a)** No **(b)** Yes **(c)** No

EXERCISES 7.3

1 $-5e^{-5x}$ **3** $6xe^{3x^2}$ **5** $\frac{e^{2x}}{\sqrt{1+e^{2x}}}$ **7** $\frac{e^{\sqrt{x+1}}}{2\sqrt{x+1}}$

9 $(-2x^2+2x)e^{-2x}$ **11** $\frac{e^x(x-1)^2}{(x^2+1)^2}$ **13** $12(e^{4x}-5)^2e^{4x}$

15 $-\frac{e^{1/x}}{x^2} - e^{-x}$ **17** $\frac{4}{(e^x+e^{-x})^2}$

19 $e^{-2x}\left(\frac{1}{x} - 2\ln x\right)$ **21** $5e^{5x}\cos e^{5x}$

23 $e^{-x}\tan e^{-x}$ **25** $e^{3x}\left(\frac{\sec^2\sqrt{x}}{2\sqrt{x}} + 3\tan\sqrt{x}\right)$

27 $-8e^{-4x}\sec^2(e^{-4x})\tan(e^{-4x})$ **29** $e^{\cot x}(1 - x\csc^2 x)$

31 $\frac{3x^2 - ye^{xy}}{xe^{xy} + 6y}$ **33** $\frac{e^x\cot y - e^{2y}}{2xe^{2y} + e^x\csc^2 y}$

35 $y = (e+3)x - (e+1)$

37 Min: $f(-1) = -e^{-1} \approx -0.368$; increasing on $[-1, \infty)$; decreasing on $(-\infty, -1]$; CU on $(-2, \infty)$; CD on $(-\infty, -2)$; PI: $(-2, -2e^{-2}) \approx (-2, -0.271)$

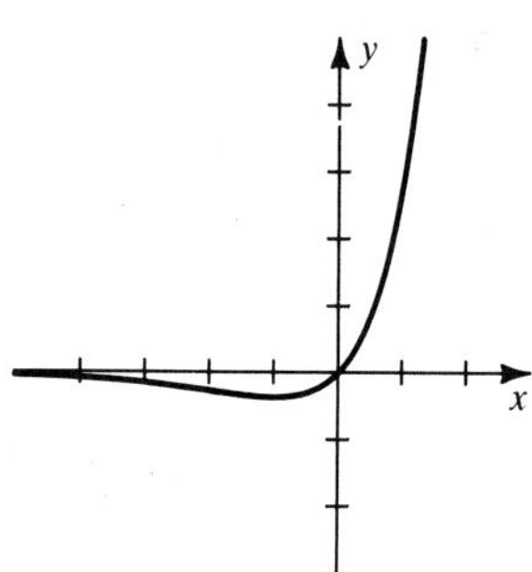

39 Decreasing on $(-\infty, 0)$ and $(0, \infty)$; CU on $\left(-\frac{1}{2}, 0\right)$ and $(0, \infty)$; CD on $\left(-\infty, -\frac{1}{2}\right)$; PI: $\left(-\frac{1}{2}, e^{-2}\right)$

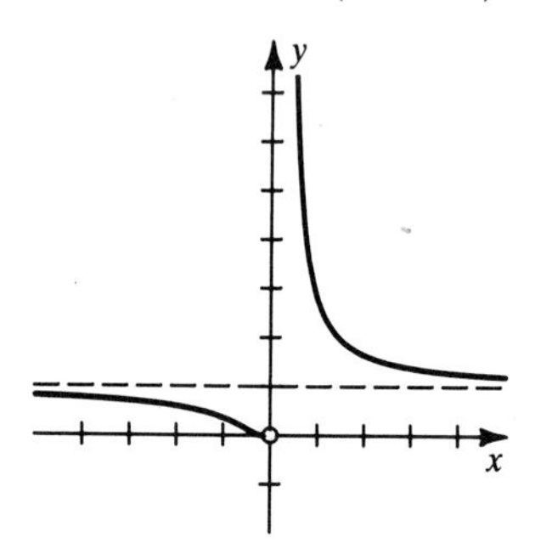

41 Min: $f(e^{-1}) = -e^{-1}$; increasing on $[e^{-1}, \infty)$; decreasing on $(0, e^{-1}]$; CU on $(0, \infty)$; no PI

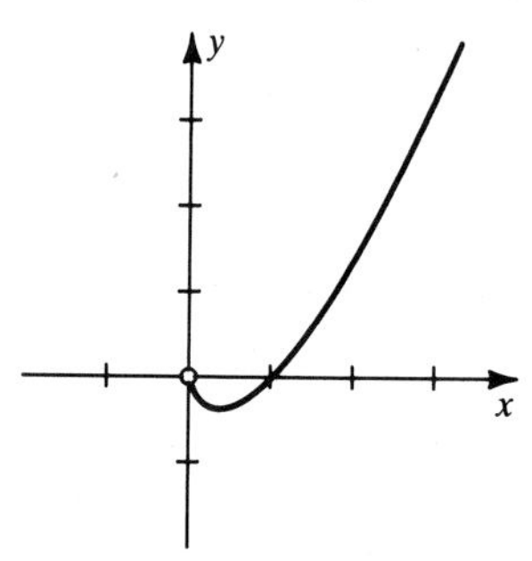

43 $q'(t) = -cq(t)$ **45 (a)** $\dfrac{\ln(a/b)}{a-b}$ **(b)** $\lim_{t\to\infty} C(t) = 0$

47 (a) 75.8 cm; 15.98 cm/yr **(b)** 3 mo; 6 yr

49 (a) $f\left(\dfrac{n}{a}\right)$ **(b)** At $x = \dfrac{2}{a}$

51

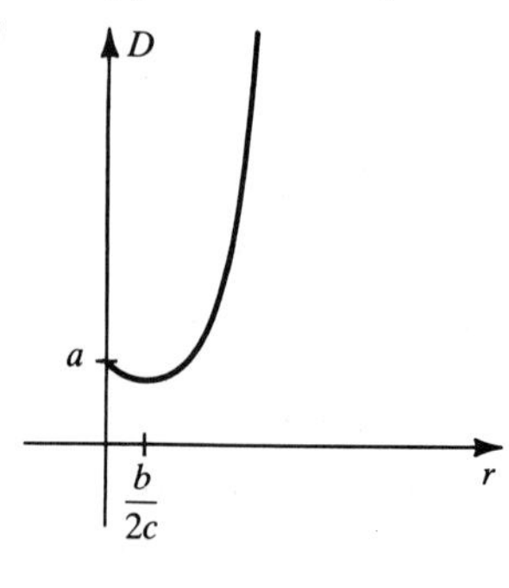

55 Max: $f(\mu) = \dfrac{1}{\sigma\sqrt{2\pi}}$; increasing on $(-\infty, \mu]$; decreasing on $[\mu, \infty)$; CU on $(-\infty, \mu - \sigma)$ and $(\mu + \sigma, \infty)$; CD on $(\mu - \sigma, \mu + \sigma)$; PI: $\left(\mu \pm \sigma, \dfrac{1}{\sigma\sqrt{2\pi e}}\right)$; both limits equal 0

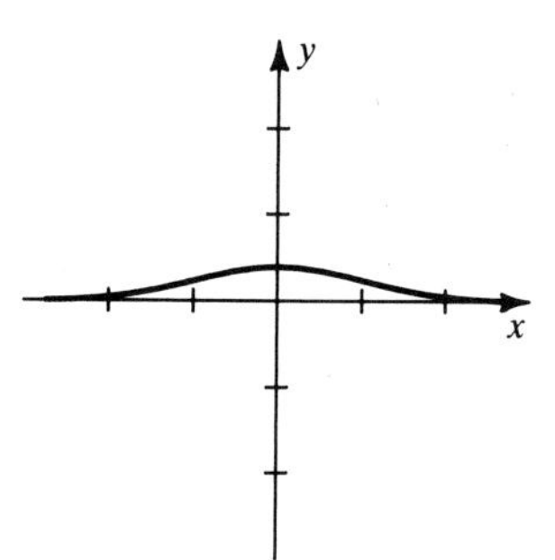

57 $e^{-1/2} \approx 0.607$ **59** 0.57

EXERCISES 7.4

1 (a) $\frac{1}{2}\ln|2x+7| + C$ **(b)** $\ln\sqrt{3}$

3 (a) $2\ln|x^2 - 9| + C$ **(b)** $\ln\frac{25}{64}$

5 (a) $-\frac{1}{4}e^{-4x} + C$ **(b)** $-\frac{1}{4}(e^{-12} - e^{-4})$

7 (a) $-\frac{1}{2}\ln|\cos 2x| + C$ **(b)** $\frac{1}{4}\ln 2$

9 (a) $2\ln\left|\csc\frac{1}{2}x - \cot\frac{1}{2}x\right| + C$ **(b)** $2\ln(2+\sqrt{3})$

11 $\frac{1}{2}\ln|x^2 - 4x + 9| + C$ **13** $\frac{1}{2}x^2 + 4x + 4\ln|x| + C$

15 $\frac{1}{2}(\ln x)^2 + C$ **17** $\frac{1}{2}x^2 + \frac{1}{5}e^{5x} + C$

19 $-\frac{3}{2}\ln|1 + 2\cos x| + C$ **21** $e^x + 2x - e^{-x} + C$

23 $\ln(e^x + e^{-x}) + C$ **25** $3\ln|\sin\sqrt[3]{x}| + C$

27 $\frac{1}{2}\ln|\sec 2x + \tan 2x| + C$ **29** $-\frac{1}{3}\ln|\sec e^{-3x}| + C$

31 $\ln|\csc x - \cot x| + \cos x + C$ **33** $\ln|\csc x| + C$

35 $x + 2\ln|\sec x + \tan x| + \tan x + C$ **37** 4

39 $\pi(1 - e^{-1})$ **41** $y = 2e^{2x} - \frac{3}{2}e^{-2x} + \frac{7}{2}$

43 $y = 3e^{-x} + 4x - 4$ **45** $\dfrac{2}{\ln(13/4)} \approx 1.697$

47 (a) 25 **(b)** 205 **(c)** 12 **49** $\Delta S = c\ln\dfrac{T_2}{T_1}$

51 (a) $\frac{5}{2}(1 - e^{-4t})$ **(b)** $\lim_{t\to\infty} Q(t) = \frac{5}{2}$ coulombs

53 (a) $s(t) = kv_0(1 - e^{-t/k})$ **(b)** $\lim_{t\to\infty} s(t) = kv_0$

55 0.731370; 0.742984; 0.746211; 0.746855; R

EXERCISES 7.5

1 $7^x \ln 7$ **3** $8^{x^2+1}(2x\ln 8)$ **5** $\dfrac{4x^3 + 6x}{(x^4 + 3x^2 + 1)\ln 10}$

7 $5^{3x-4}(3\ln 5)$ **9** $\dfrac{-(x^2+1)10^{1/x}(\ln 10)}{x^2} + (2x)10^{1/x}$

11 $\dfrac{30x}{(3x^2+2)\ln 10}$ **13** $\left(\dfrac{6}{6x+4} - \dfrac{2}{2x-3}\right)\dfrac{1}{\ln 5}$

15 $\dfrac{1}{x \ln x \ln 10}$ **17** $ex^{e-1} + e^x$

19 $(x+1)^x\left[\dfrac{x}{x+1} + \ln(x+1)\right]$ **21** $2^{\sin^2 x}(\sin 2x)\ln 2$

23 $x^{\tan x}\left(\sec^2 x \ln x + \dfrac{\tan x}{x}\right)$

25 **(a)** 0 **(b)** $5x^4$ **(c)** $\sqrt{5}x^{\sqrt{5}-1}$
(d) $(\sqrt{5})^x \ln\sqrt{5}$ **(e)** $(1 + 2\ln x)x^{1+x^2}$

27 **(a)** $\dfrac{7^x}{\ln 7} + C$ **(b)** $\dfrac{342}{49\ln 7} \approx 3.59$

29 **(a)** $\dfrac{-5^{-2x}}{2\ln 5} + C$ **(b)** $\dfrac{12}{625\ln 5} \approx 0.012$ **31** $\dfrac{10^{3x}}{3\ln 10} + C$

33 $\dfrac{-3^{-x^2}}{2\ln 3} + C$ **35** $\dfrac{\ln(2^x+1)}{\ln 2} + C$

37 $(\ln 10)\ln|\log x| + C$ **39** $-\dfrac{3^{\cos x}}{\ln 3} + C$

41 **(a)** $\pi^\pi x + C$ **(b)** $\dfrac{1}{5}x^5 + C$ **(c)** $\dfrac{x^{\pi+1}}{\pi+1} + C$
(d) $\dfrac{\pi^x}{\ln \pi} + C$ **43** $\dfrac{1}{\ln 2} - \dfrac{1}{2} \approx 0.94$

45 **(a)** \$0.05/yr **(b)** \$0.95

47 **(a)** In trout/yr: 95; 62; 53 **(b)** 9.36

49 pH ≈ 2.201; $\pm 0.1\%$

51 **(b)** $S = \dfrac{k}{x}$, where $k = \dfrac{a}{\ln 10}$;
$S(x) = 2\,S(2x)$ (twice as sensitive)

53 **(a)** With $n = r/h$,
$$\ln A = \ln\left[P(1+h)^{rt/h}\right] = \ln P + rt\ln(1+h)^{1/h}.$$
(b) Since $h = r/n$, $n \to \infty$ if and only if $h \to 0^+$. Thus,
$$\ln A = \lim_{h\to 0^+}\left[\ln P + rt\ln(1+h)^{1/h}\right] = \ln P + rt\ln e = \ln(Pe^{rt})$$
and $A = Pe^{rt}$.

55 Let $h = x/n$. Then
$$\lim_{n\to\infty}\left(1 + \frac{x}{n}\right)^n = \lim_{h\to 0^+}(1+h)^{x/h} = \lim_{h\to 0^+}\left[(1+h)^{1/h}\right]^x = e^x.$$

57

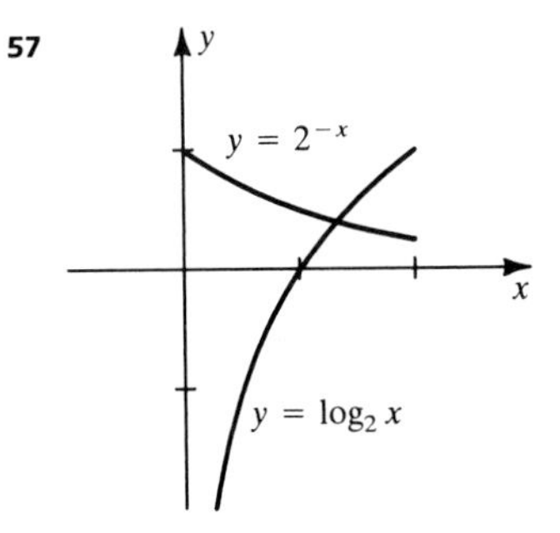

(a) 1.32 **(b)** $\pi\displaystyle\int_1^{1.32}\left[(2^{-x})^2 - (\log_2 x)^2\right]dx$ **(c)** 0.145

EXERCISES 7.6

1 $q(t) = 5000(3)^{t/10}$; 45,000; $\dfrac{10\ln 10}{\ln 3} \approx 20.96$ hr

3 $30\left(\dfrac{29}{30}\right)^5 \approx 25.32$ in.

5 $\dfrac{\ln(40/4.5)}{0.02} \approx 109.24$ yr after Jan. 1, 1980 (March 29, 2089)

7 $\dfrac{5\ln(1/6)}{\ln(1/3)} \approx 8.15$ min

9 Proceed as in the solution to Example 1.

11 $P(z) = \left(\dfrac{288 - 0.01z}{288}\right)^{3.42}$ **13** $\dfrac{29\ln(2/5)}{\ln(1/2)} \approx 38.34$ yr

15 $600\left(\dfrac{1}{2}\right)^{-3/16} \approx 683.27$ mg **17** $v(y) = \sqrt{2k\left(\dfrac{1}{y} - \dfrac{1}{y_0}\right) + v_0^2}$

19 $\dfrac{5700\ln(0.2)}{\ln(1/2)} \approx 13{,}235$ yr **21** Use Theorem (5.35).

23 $V(t) = \dfrac{1}{27}(kt + C)^3$

25 **(c)**

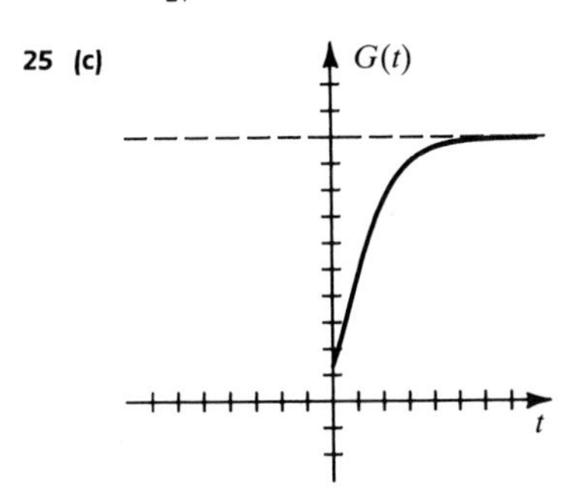

7.7 REVIEW EXERCISES

1 $\dfrac{10 - x}{15}$

3 f is decreasing, since $f'(x) < 0$ for $-1 \le x \le 1$; $-\dfrac{1}{8}$

5 $\dfrac{75x^2}{5x^3 - 4}$ **7** $-2(1 + \ln|1 - 2x|)$

9 $\dfrac{12}{3x+2} + \dfrac{3}{6x-5} - \dfrac{8}{8x-7}$ **11** $\dfrac{-4x}{(2x^2+3)[\ln(2x^2+3)]^2}$

13 $\dfrac{\ln x - 1}{(\ln x)^2}$ **15** $2x$ **17** $\dfrac{4e^{4x}}{e^{4x}+9}$

19 $\dfrac{10^x}{x\ln 10} + 10^x(\ln 10)\log x$ **21** $\dfrac{1}{4x\sqrt{\ln\sqrt{x}}}$

23 $2xe^{-x^2}(1 - x^2)$ **25** $\dfrac{3(e^{3x} - e^{-3x})}{2\sqrt{e^{3x} + e^{-3x}}}$

27 $\dfrac{10^{\ln x}\ln 10}{x}$ **29** $\dfrac{2\ln x\,(x^{\ln x})}{x}$ **31** $-\sec x$

33 $2e^{-2x}\csc e^{-2x}(\csc^2 e^{-2x} + \cot^2 e^{-2x})$

35 $-16\tan 4x$ **37** $(\sin x)^{\cos x}(\cos x\cot x - \sin x\ln\sin x)$

39 $-\dfrac{y}{x}$ **41** $\left[\dfrac{4}{3(x+2)} + \dfrac{3}{2(x-3)}\right](x+2)^{4/3}(x-3)^{3/2}$

43 **(a)** $-2e^{-\sqrt{x}} + C$ **(b)** $2(e^{-1} - e^{-2}) \approx 0.465$

45 **(a)** $-\dfrac{4^{-x^2}}{2\ln 4} + C$ **(b)** $\dfrac{3}{8\ln 4} \approx 0.271$

47 $-\dfrac{1}{2}\ln|\cos x^2| + C$ **49** $\dfrac{x^{e+1}}{e+1} + C$

51 $-\ln|1 - \ln x| + C$ **53** $-\dfrac{1}{2}e^{-2x} - 2e^{-x} + x + C$

55 $\dfrac{1}{2}x^2 - 2x + 4\ln|x+2| + C$ **57** $-\dfrac{1}{8}e^{4/x^2} + C$

59 $-\dfrac{1}{2(x^2+1)} + C$ **61** $\ln(1 + e^x) + C$

63 $\dfrac{(5e)^x}{\ln(5e)} + C$ **65** $2\ln 10\sqrt{\log x} + C$

67 $\cos e^{-x} + C$ **69** $-\ln|1 + \cot x| + C$

71 $-\dfrac{1}{4}\ln|1 - 2\sin 2x| + C$ **73** $-\ln|\cos e^x| + C$

75 $-\frac{1}{3}\cot 3x + \frac{2}{3}\ln|\csc 3x - \cot 3x| + x + C$

77 $\frac{1}{9}\ln|\sin 9x| + \frac{1}{9}\ln|\csc 9x - \cot 9x| + C$

79 $y = -\frac{1}{9}e^{-3x} + \frac{5}{3}x - \frac{8}{9}$ **81** $4e^2 + 12 \approx 41.56$ cm

83 $y - e = -2(1+e)(x-1)$

85 $\frac{\pi}{8}(e^{-16} - e^{-24}) \approx 4.42 \times 10^{-8}$ **87** $\frac{5\ln(1/100)}{\ln(1/2)} \approx 33.2$ days

89 **(a)** $\frac{3\ln(3/10)}{\ln(1/2)} \approx 5.2$ hr or 2.2 additional hr

(b) $10\left[1 - \left(\frac{1}{2}\right)^{7/3}\right] \approx 8.016$ lb

91 $100{,}000(2)^6 = 6{,}400{,}000$

CHAPTER 8

EXERCISES 8.1

1 **(a)** $-\frac{\pi}{4}$ **(b)** $\frac{2\pi}{3}$ **(c)** $-\frac{\pi}{3}$ **3** **(a)** $\frac{\pi}{3}$ **(b)** $\frac{\pi}{4}$ **(c)** $\frac{\pi}{6}$

5 **(a)** Not defined **(b)** Not defined **(c)** $\frac{\pi}{4}$

7 **(a)** $-\frac{3}{10}$ **(b)** $\frac{1}{2}$ **(c)** 14 **9** **(a)** $\frac{\pi}{3}$ **(b)** $\frac{5\pi}{6}$ **(c)** $-\frac{\pi}{6}$

11 **(a)** $-\frac{\pi}{4}$ **(b)** $\frac{3\pi}{4}$ **(c)** $-\frac{\pi}{4}$

13 **(a)** $\frac{\sqrt{3}}{2}$ **(b)** $\frac{\sqrt{2}}{2}$ **(c)** Not defined

15 **(a)** $\frac{\sqrt{5}}{2}$ **(b)** $\frac{\sqrt{34}}{5}$ **(c)** $\frac{4}{\sqrt{15}}$

17 **(a)** $\frac{\sqrt{3}}{2}$ **(b)** 0 **(c)** $-\frac{77}{36}$ **19** $\frac{x}{\sqrt{x^2+1}}$ **21** $\frac{3}{\sqrt{9-x^2}}$

23

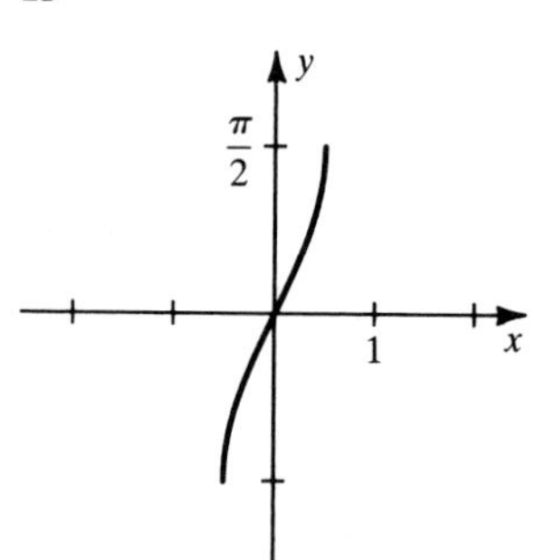

25

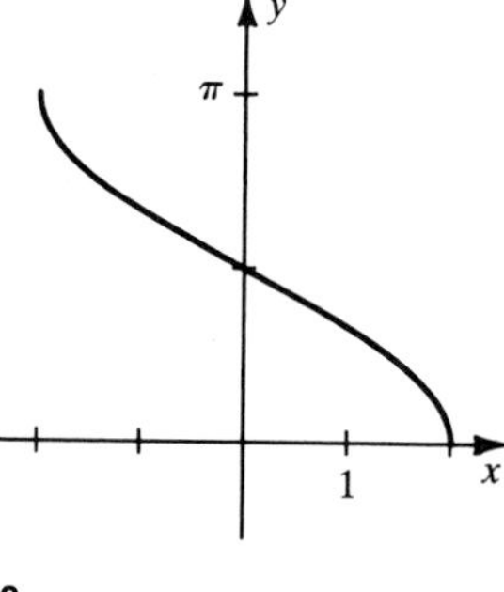

27

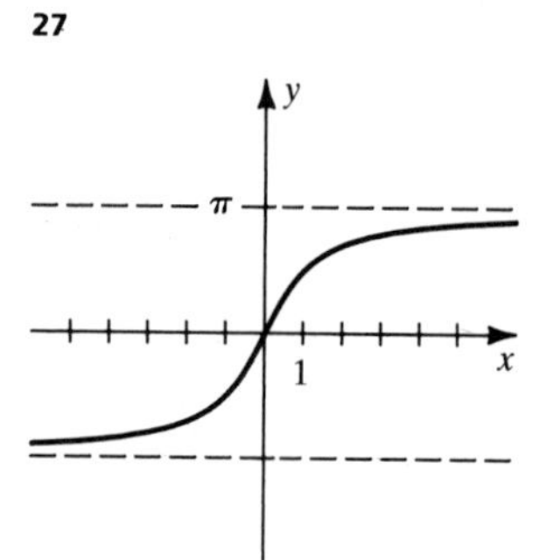

29

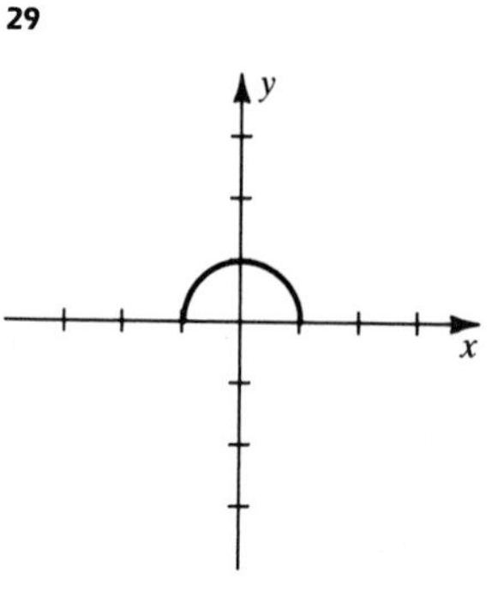

31 **(a)** $y = \cot^{-1} x$ if and only if $x = \cot y$ for $0 < y < \pi$.

(b)

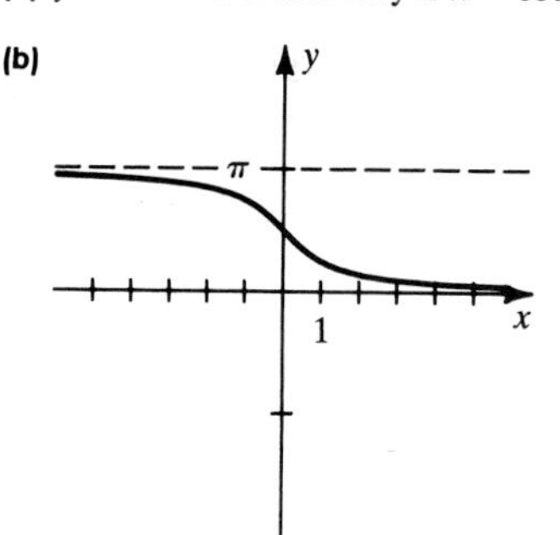

33 **(a)** $\arctan \frac{1}{4}(-9 \pm \sqrt{57})$ **(b)** $-1.3337, -0.3478$

35 **(a)** $\arccos\left(\pm\frac{1}{5}\sqrt{15}\right)$, $\arccos\left(\pm\frac{1}{3}\sqrt{3}\right)$

(b) 0.6847, 2.4569, 0.9553, 2.1863

37 **(a)** $\alpha = \theta - \sin^{-1}\frac{d}{k}$ **(b)** 40°

39 **(a)** $\sin^{-1} x = \tan^{-1}\frac{x}{\sqrt{1-x^2}}$ if $|x| < 1$

(b) $\cos^{-1} x = \tan^{-1}\frac{\sqrt{1-x^2}}{x}$ if $0 < x < 1$ and

$\cos^{-1} x = \tan^{-1}\frac{\sqrt{1-x^2}}{x} + \pi$ if $-1 < x < 0$

EXERCISES 8.2

1 $\frac{1}{2\sqrt{x}\sqrt{1-x}}$ **3** $\frac{3}{9x^2 - 30x + 26}$

5 $\frac{-e^{-x}}{\sqrt{e^{-2x}-1}} - e^{-x}\,\text{arcsec}\, e^{-x}$ **7** $\frac{2x^3}{1+x^4} + 2x\arctan(x^2)$

9 $\frac{x}{(x^2-1)\sqrt{x^2-2}}$ **11** $\frac{1}{\sqrt{1-x^2}(\sin^{-1} x)^2}$

13 $\frac{9(1+\cos^{-1} 3x)^2}{\sqrt{1-9x^2}}$ **15** $\frac{2x}{(1+x^4)\arctan(x^2)}$

17 $\left(\frac{1}{x^2}\right)\sin\left(\frac{1}{x}\right) + \sec x \tan x - \frac{1}{\sqrt{1-x^2}}$ **19** $3^{\arcsin(x^3)}\frac{(3\ln 3)x^2}{\sqrt{1-x^6}}$

21 $\frac{1 - 2x\arctan x}{(x^2+1)^2}$ **23** $\frac{1}{2\sqrt{x}}\left(\frac{1}{\sqrt{x-1}} + \sec^{-1}\sqrt{x}\right)$

25 $(\tan x)^{\arctan x}\left(\arctan x \cot x \sec^2 x + \frac{\ln \tan x}{1+x^2}\right)$

27 $\frac{ye^x - 2x - \sin^{-1} y}{\frac{x}{\sqrt{1-y^2}} - e^x}$ **29** **(a)** $\frac{1}{4}\tan^{-1}\left(\frac{x}{4}\right) + C$ **(b)** $\frac{\pi}{16}$

31 **(a)** $\frac{1}{2}\sin^{-1}(x^2) + C$ **(b)** $\frac{\pi}{12}$

33 $-\tan^{-1}(\cos x) + C$ **35** $2\tan^{-1}\sqrt{x} + C$

37 $\sin^{-1}\left(\frac{e^x}{4}\right) + C$ **39** $\frac{1}{6}\sec^{-1}\left(\frac{x^3}{2}\right) + C$

41 $\frac{1}{2}\ln(x^2+9) + C$ **43** $\frac{1}{5}\sec^{-1}\left(\frac{e^x}{5}\right) + C$

45 $\pm\frac{7}{3576}$ rad **47** $-\frac{25}{1044}$ rad/sec **49** $\sqrt{4800} \approx 69.3$ ft

51 $\frac{2\pi}{27} \approx 0.233$ mi/sec **53** 0.72

EXERCISES 8.3

1 **(a)** 27.2899 **(b)** 2.1250 **(c)** -0.9951 **(d)** 1.0000 **(e)** 0.2658 **(f)** -0.8509

3 $5\cosh 5x$ **5** $3x^2\sinh(x^3)$

7 $\dfrac{1}{2\sqrt{x}}(\sqrt{x}\,\text{sech}^2\sqrt{x}+\tanh\sqrt{x})$

9 $\left(\dfrac{1}{x^2}\right)\text{csch}^2\left(\dfrac{1}{x}\right)$

11 $\dfrac{-2x\,\text{sech}(x^2)[(x^2+1)\tanh(x^2)+1]}{(x^2+1)^2}$

13 $-12\,\text{csch}^2 6x\coth 6x$ **15** $2\coth 2x$

17 $\dfrac{4x\sinh\sqrt{4x^2+3}}{\sqrt{4x^2+3}}$ **19** $\dfrac{-\text{sech}^2 x}{(\tanh x+1)^2}$ **21** $-\dfrac{4x}{(1-x^2)^2}$

23 $e^{3x}\,\text{sech}\,x\,(3-\tanh x)$ **25** $-\text{sech}\,x$

27 $\dfrac{1}{3}\sinh(x^3)+C$ **29** $2\cosh\sqrt{x}+C$

31 $\dfrac{1}{3}\tanh 3x+C$ **33** $-2\coth\left(\dfrac{1}{2}x\right)+C$

35 $-\dfrac{1}{3}\text{sech}\,3x+C$ **37** $-\text{csch}\,x+C$ **39** $\ln|\sinh x|+C$

41 $\dfrac{1}{2}\sinh^2 x+C$, or $\dfrac{1}{2}\cosh^2 x+C$ **43** $\dfrac{1}{3}(\cosh 3-1)\approx 3.023$

45 $(\ln(2\pm\sqrt{3}),\ \pm\sqrt{3})$

47 Show that $A=\dfrac{1}{2}(\cosh t)(\sinh t)-\displaystyle\int_1^{\cosh t}\sqrt{x^2-1}\,dx$ and that $\dfrac{dA}{dt}=\dfrac{1}{2}$.

49 **(a)** $286{,}574\ \text{ft}^2$ **(b)** 1494 ft

51 **(a)** $\lim\limits_{h\to\infty} v^2=\dfrac{gL}{2\pi}$ **(b)** *Hint:* Let $f(h)=v^2$.

53 **(a)** 0.7 **(b)** 0.722

$y=\tanh x$, $y=\text{sech}^2 x$

55 $\cosh x+\sinh x=\dfrac{e^x+e^{-x}}{2}+\dfrac{e^x-e^{-x}}{2}=e^x$

57 $\sinh(-x)=\dfrac{e^{-x}-e^{-(-x)}}{2}=\dfrac{e^{-x}-e^{x}}{2}=\dfrac{-(e^x-e^{-x})}{2}=-\sinh x$

59 $\sinh x\cosh y+\cosh x\sinh y$

$$=\frac{(e^x-e^{-x})(e^y+e^{-y})}{4}+\frac{(e^x+e^{-x})(e^y-e^{-y})}{4}$$
$$=\frac{(e^{x+y}+e^{x-y}-e^{-x+y}-e^{-x-y})+(e^{x+y}-e^{x-y}+e^{-x+y}-e^{-x-y})}{4}$$
$$=\frac{2e^{x+y}-2e^{-x-y}}{4}=\frac{e^{x+y}-e^{-(x+y)}}{2}=\sinh(x+y)$$

61 $\sinh(x-y)=\sinh(x+(-y))$
$=\sinh x\cosh(-y)+\cosh x\sinh(-y)$ (Exer. 59)
$=\sinh x\cosh y-\cosh x\sinh y$ (Exer. 57 and 58)

63 $\tanh(x+y)=\dfrac{\sinh(x+y)}{\cosh(x+y)}=\dfrac{\sinh x\cosh y+\cosh x\sinh y}{\cosh x\cosh y+\sinh x\sinh y}$; divide the numerator and denominator by the product $\cosh x\cosh y$ to obtain $\dfrac{\tanh x+\tanh y}{1+\tanh x\tanh y}$.

65 Let $y=x$ in Exercise 59.

67 From Exercise 66,

$$\cosh 2y=\cosh^2 y+\sinh^2 y=(1+\sinh^2 y)+\sinh^2 y=1+2\sinh^2 y,$$

and hence

$$\sinh^2 y=\frac{\cosh 2y-1}{2}.$$

Let $y=\dfrac{x}{2}$ to obtain the identity.

69 Let $y=x$ in Exercise 63.

71 $\cosh nx+\sinh nx=\dfrac{e^{nx}+e^{-nx}}{2}+\dfrac{e^{nx}-e^{-nx}}{2}=e^{nx}=(e^x)^n=(\cosh x+\sinh x)^n$

EXERCISES 8.4

1 **(a)** 0.8814 **(b)** 1.3170 **(c)** -0.5493 **(d)** 1.3170

3 $\dfrac{5}{\sqrt{25x^2+1}}$ **5** $\dfrac{1}{2\sqrt{x}\sqrt{x-1}}$ **7** $\dfrac{4}{16x^2-1}$

9 $-\dfrac{2}{x\sqrt{1-x^4}}$ **11** $-\dfrac{|x|}{x\sqrt{1+x^2}}+\sinh^{-1}\left(\dfrac{1}{x}\right)$

13 $\dfrac{4}{\sqrt{16x^2-1}\cosh^{-1}4x}$ **15** $-\dfrac{1}{x^2+2x}$ **17** $-\dfrac{1}{2x\sqrt{1-x}}$

19 $\dfrac{1}{4}\sinh^{-1}\left(\dfrac{4}{9}x\right)+C$ **21** $\dfrac{1}{14}\tanh^{-1}\left(\dfrac{2}{7}x\right)+C$

23 $\cosh^{-1}\left(\dfrac{e^x}{4}\right)+C$ **25** $-\dfrac{1}{6}\text{sech}^{-1}\left(\dfrac{x^2}{3}\right)+C$

27 $y=\sinh 3t$

29 **31**

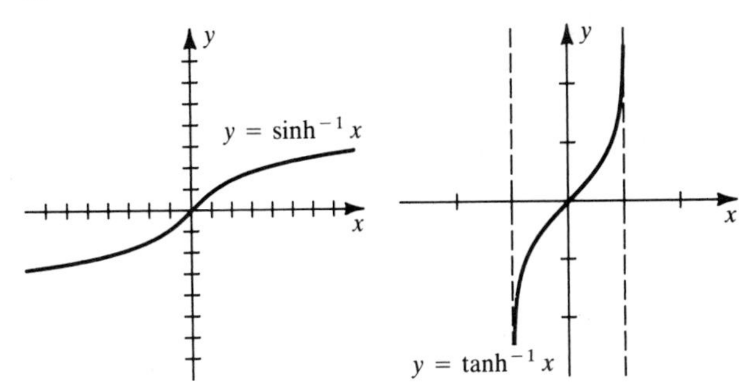

33 Let $u=x$ in Theorem (8.16) and differentiate $\cosh^{-1}u$.

35 Let $u=x$ in Theorem (8.16) and differentiate $\text{sech}^{-1}u$.

37 Differentiate the right-hand side.

39 Use a procedure similar to that given in the text for $\sinh^{-1}x$.

41 Use a procedure similar to that given in the text for $\sinh^{-1}x$.

8.5 REVIEW EXERCISES

1 $\dfrac{1}{2x\sqrt{x-1}}$ **3** $\dfrac{2x}{\sqrt{x^4-1}}+2x\,\text{arcsec}(x^2)$

5 $2^{\arctan 2x}\left(\dfrac{2\ln 2}{1+4x^2}\right)$ **7** $\dfrac{2x}{(1+x^4)\tan^{-1}(x^2)}$ **9** $\dfrac{-x}{\sqrt{x^2(1-x^2)}}$

11 $4(\tan x+\tan^{-1}x)^3\left(\sec^2 x+\dfrac{1}{1+x^2}\right)$

13 $\dfrac{1}{(1+x^2)[1+(\tan^{-1}x)^2]}$ **15** $-5e^{-5x}\sinh e^{-5x}$

17 $-e^{-x}(e^{-x}\cosh e^{-x}+\sinh e^{-x})$

19 $(\cosh x-\sinh x)^{-2}$, or e^{2x} **21** $\dfrac{2x}{\sqrt{x^4+1}}$ **23** $\dfrac{1}{3x^{2/3}}$

25 $\dfrac{1}{6}\tan^{-1}\left(\dfrac{3}{2}x\right)+C$ **27** $-\sqrt{1-e^{2x}}+C$

29 $\dfrac{1}{2}\sinh(x^2)+C$ **31** $\dfrac{\pi}{3}$ **33** $\cosh(\ln x)+C$

35 $\dfrac{1}{2}\sin^{-1}\left(\dfrac{2}{3}x\right)+C$ **37** $-\dfrac{1}{3}\operatorname{sech}^{-1}\left(\dfrac{2}{3}|x|\right)+C$

39 $\dfrac{1}{25}\sqrt{25x^2+36}+C$ **41** $\left(\pm\dfrac{4}{15},\ \sin^{-1}\left(\pm\dfrac{4}{5}\right)\right)$

43 Let $c=\tan^{-1}\dfrac{1}{2}$. Min: $f(c)=5\sqrt{5}$; increasing on $\left[c,\dfrac{\pi}{2}\right)$; decreasing on $(0,c]$.

45 **(a)** $\dfrac{1}{2}\tan^{-1}4+\dfrac{\pi}{2}n$ for $n=0,1,2,3$ **(b)** 0.66, 2.23, 3.80, 5.38

47 $\dfrac{1}{260}$ rad/sec $\approx 0.22°$/sec **49** $-\dfrac{800}{2581}\approx -0.31$ rad/sec

CHAPTER 9

EXERCISES 9.1

1 $-(x+1)e^{-x}+C$ **3** $\dfrac{1}{27}e^{3x}(9x^2-6x+2)+C$

5 $\dfrac{1}{5}x\sin 5x+\dfrac{1}{25}\cos 5x+C$

7 $x\sec x-\ln|\sec x+\tan x|+C$

9 $x^2\sin x+2x\cos x-2\sin x+C$

11 $x\tan^{-1}x-\dfrac{1}{2}\ln(1+x^2)+C$

13 $\dfrac{2}{9}x^{3/2}(3\ln x-2)+C$ **15** $-x\cot x+\ln|\sin x|+C$

17 $-\dfrac{1}{2}e^{-x}(\sin x+\cos x)+C$ **19** $\cos x\,(1-\ln\cos x)+C$

21 $-\dfrac{1}{2}\csc x\cot x+\dfrac{1}{2}\ln|\csc x-\cot x|+C$

23 $\dfrac{1}{3}(2-\sqrt{2})\approx 0.20$ **25** $\dfrac{\pi}{4}$

27 $\dfrac{1}{40{,}400}(2x+3)^{100}(200x-3)+C$

29 $\dfrac{1}{41}e^{4x}(4\sin 5x-5\cos 5x)+C$

31 $x(\ln x)^2-2x\ln x+2x+C$

33 $x^3\cosh x-3x^2\sinh x+6x\cosh x-6\sinh x+C$

35 $2\sqrt{x}\sin\sqrt{x}+2\cos\sqrt{x}+C$

37 $x\cos^{-1}x-\sqrt{1-x^2}+C$ **39** Let $u=x^m$.

41 Let $u=(\ln x)^m$.

43 $e^x(x^5-5x^4+20x^3-60x^2+120x-120)+C$

45 2π **47** $\dfrac{\pi}{2}(e^2+1)\approx 13.18$ **49** $\left(\dfrac{3\ln 3-2}{2},1\right)$

EXERCISES 9.2

1 $\sin x-\dfrac{1}{3}\sin^3 x+C$ **3** $\dfrac{1}{8}x-\dfrac{1}{32}\sin 4x+C$

5 $-\dfrac{1}{3}\cos^3 x+\dfrac{1}{5}\cos^5 x+C$

7 $\dfrac{1}{8}\left(\dfrac{5}{2}x-2\sin 2x+\dfrac{3}{8}\sin 4x+\dfrac{1}{6}\sin^3 2x\right)+C$

9 $\dfrac{1}{4}\tan^4 x+\dfrac{1}{6}\tan^6 x+C$ **11** $\dfrac{1}{5}\sec^5 x-\dfrac{1}{3}\sec^3 x+C$

13 $\dfrac{1}{5}\tan^5 x-\dfrac{1}{3}\tan^3 x+\tan x-x+C$

15 $\dfrac{2}{3}\sin^{3/2}x-\dfrac{2}{7}\sin^{7/2}x+C$ **17** $\tan x-\cot x+C$

19 $\dfrac{2}{3}-\dfrac{5}{6\sqrt{2}}\approx 0.08$ **21** $\dfrac{1}{2}\left(\dfrac{1}{2}\sin 2x-\dfrac{1}{8}\sin 8x\right)+C$

23 $\dfrac{3}{5}$ **25** $-\dfrac{1}{5}\cot^5 x-\dfrac{1}{7}\cot^7 x+C$

27 $-\ln(2-\sin x)+C$ **29** $-\dfrac{1}{1+\tan x}+C$

31 $\dfrac{3}{4}\pi^2\approx 7.40$ **33** $\dfrac{5}{2}$

35 **(a)** Use the trigonometric product-to-sum formulas.

(b) $\int\sin mx\cos nx\,dx$

$$=\begin{cases}-\dfrac{\cos(m+n)x}{2(m+n)}-\dfrac{\cos(m-n)x}{2(m-n)}+C & \text{if } m\neq n\\[2mm] -\dfrac{\cos 2mx}{4m}+C & \text{if } m=n\end{cases}$$

$\int\cos mx\cos nx\,dx$

$$=\begin{cases}\dfrac{\sin(m+n)x}{2(m+n)}+\dfrac{\sin(m-n)x}{2(m-n)}+C & \text{if } m\neq n\\[2mm] \dfrac{x}{2}+\dfrac{\sin 2mx}{4m}+C & \text{if } m=n\end{cases}$$

EXERCISES 9.3

1 $\dfrac{1}{2}\ln\left|\dfrac{2}{x}-\dfrac{\sqrt{4-x^2}}{x}\right|+C$ **3** $\dfrac{1}{3}\ln\left|\dfrac{\sqrt{x^2+9}}{x}-\dfrac{3}{x}\right|+C$

5 $\dfrac{\sqrt{x^2-25}}{25x}+C$ **7** $-\sqrt{4-x^2}+C$

9 $-\dfrac{x}{\sqrt{x^2-1}}+C$ **11** $\dfrac{1}{432}\left[\tan^{-1}\left(\dfrac{x}{6}\right)+\dfrac{6x}{x^2+36}\right]+C$

13 $\sin^{-1}\left(\dfrac{x}{3}\right)+C$ **15** $\dfrac{1}{2(16-x^2)}+C$

17 $\dfrac{1}{243}(9x^2+49)^{3/2}-\dfrac{49}{81}\sqrt{9x^2+49}+C$

19 $\dfrac{(3+2x^2)\sqrt{x^2-3}}{27x^3}+C$ **21** $-\dfrac{8}{x^2}+8\ln|x|+\dfrac{1}{2}x^2+C$

23 $25\pi[\sqrt{2}-\ln(\sqrt{2}+1)]\approx 41.85$

25 $y=\sqrt{x^2-16}-4\sec^{-1}\dfrac{x}{4}$

27 Let $u=a\tan\theta$. **29** Let $u=a\sin\theta$.

31 Let $u=a\sec\theta$.

EXERCISES 9.4

Answers are expressed as sums that correspond to partial fraction decompositions. Logarithms can be combined. Thus, an equivalent answer for Exercise 1 is $\ln|x|^3(x-4)^2+C$.

1 $3\ln|x|+2\ln|x-4|+C$

3 $4\ln|x+1|-5\ln|x-2|+\ln|x-3|+C$

5 $6\ln|x-1|+\dfrac{5}{x-1}+C$

7 $3 \ln|x-2| - 2\ln|x+4| + C$

9 $2\ln|x| - \ln|x-2| + 4\ln|x+2| + C$

11 $5\ln|x+1| - \dfrac{1}{x+1} - 3\ln|x-5| + C$

13 $5\ln|x| - \dfrac{2}{x} + \dfrac{3}{2x^2} - \dfrac{1}{3x^3} + 4\ln|x+3| + C$

15 $x + 4\ln|x| + \ln(x^2+4) - \dfrac{1}{2}\tan^{-1}\left(\dfrac{x}{2}\right) + C$

17 $\ln(x^2+4) + \dfrac{1}{2}\tan^{-1}\left(\dfrac{x}{2}\right) + 3\ln|x+5| + C$

19 $-\dfrac{1}{2}\ln(x^2+4) + \dfrac{1}{2}\tan^{-1}\left(\dfrac{x}{2}\right) + \dfrac{1}{2}\ln(x^2+1) + C$

21 $\ln(x^2+1) - \dfrac{4}{x^2+1} + C$

23 $\dfrac{1}{2}x^2 + x + 2\ln|x| + 2\ln|x-1| + C$

25 $\dfrac{1}{3}x^3 - 9x - \dfrac{1}{9x} - \dfrac{1}{2}\ln(x^2+9) + \dfrac{728}{27}\tan^{-1}\left(\dfrac{x}{3}\right) + C$

27 $2\ln|x+4| + \dfrac{6}{x+4} - \dfrac{5}{x-3} + C$

29 $2\ln|x-4| + 2\ln|x+1| - \dfrac{3}{2(x+1)^2} + C$

31 $\dfrac{1}{6}\ln|x-3| - \dfrac{7}{2(x-3)} + \dfrac{5}{6}\ln|x+3| - \dfrac{3}{2(x+3)} + C$

33 $\dfrac{1}{2a}(\ln|a+u| - \ln|a-u|) + C = \dfrac{1}{2a}\ln\left|\dfrac{a+u}{a-u}\right| + C$

35 $-\dfrac{b}{a^2}\ln|u| - \dfrac{1}{au} + \dfrac{b}{a^2}\ln|a+bu| + C = -\dfrac{1}{au} + \dfrac{b}{a^2}\ln\left|\dfrac{a+bu}{u}\right| + C$

37 $\dfrac{1}{2}\ln 3 \approx 0.55$ **39** $\dfrac{\pi}{27}(4\ln 2 + 3) \approx 0.67$

41 $-\dfrac{1}{b}\ln\left|a + \dfrac{b}{x}\right| + C$

EXERCISES 9.5

1 $\dfrac{1}{2}\tan^{-1}\dfrac{x+1}{2} + C$ **3** $\dfrac{1}{2}\tan^{-1}\dfrac{x-2}{2} + C$

5 $\sin^{-1}\dfrac{x-2}{2} + C$

7 $-2\sqrt{9-8x-x^2} - 5\sin^{-1}\dfrac{x+4}{5} + C$

9 $\dfrac{1}{2}\left[\tan^{-1}(x+2) + \dfrac{x+2}{x^2+4x+5}\right] + C$

11 $\dfrac{x+3}{4\sqrt{x^2+6x+13}} + C$ **13** $\dfrac{2}{3\sqrt{7}}\tan^{-1}\dfrac{4x-3}{3\sqrt{7}} + C$

15 $\ln\left(\dfrac{e^x+1}{e^x+2}\right) + C$ **17** $1 + \dfrac{\pi}{4} \approx 1.79$ **19** $\dfrac{\pi}{20} \approx 0.16$

EXERCISES 9.6

1 $\dfrac{3}{7}(x+9)^{7/3} - \dfrac{27}{4}(x+9)^{4/3} + C$

3 $\dfrac{5}{81}(3x+2)^{9/5} - \dfrac{5}{18}(3x+2)^{4/5} + C$ **5** $2 + 8\ln\dfrac{6}{7} \approx 0.767$

7 $\dfrac{6}{7}x^{7/6} - \dfrac{6}{5}x^{5/6} + 2x^{1/2} - 6x^{1/6} + 6\tan^{-1}(x^{1/6}) + C$

9 $\dfrac{2}{\sqrt{3}}\tan^{-1}\sqrt{\dfrac{x-2}{3}} + C$ **11** $\dfrac{3}{5}(x+4)^{5/3} - \dfrac{9}{2}(x+4)^{2/3} + C$

13 $\dfrac{2}{7}(1+e^x)^{7/2} - \dfrac{4}{5}(1+e^x)^{5/2} + \dfrac{2}{3}(1+e^x)^{3/2} + C$

15 $e^x - 4\ln(e^x+4) + C$

17 $2\sin\sqrt{x+4} - 2\sqrt{x+4}\cos\sqrt{x+4} + C$ **19** $\dfrac{137}{320}$

21 $\ln|\cos x| - \ln(1-\cos x) + C$

23 $\dfrac{1}{2}\ln|e^x - 1| - \dfrac{1}{2}\ln(e^x+1) + C$

25 $\dfrac{4}{3}\ln(4-\sin x) + \dfrac{2}{3}\ln(\sin x + 2) + C$

27 $\dfrac{2}{\sqrt{3}}\tan^{-1}\dfrac{2\tan(x/2)+1}{\sqrt{3}} + C$ **29** $\ln\left|\tan\dfrac{x}{2} + 1\right| + C$

31 $-\dfrac{1}{5}\ln\left|2\tan\dfrac{x}{2} - 1\right| + \dfrac{1}{5}\ln\left|\tan\dfrac{x}{2} + 2\right| + C$

EXERCISES 9.7

1 $\sqrt{4+9x^2} - 2\ln\left|\dfrac{2+\sqrt{4+9x^2}}{3x}\right| + C$

3 $-\dfrac{x}{8}(2x^2-80)\sqrt{16-x^2} + 96\sin^{-1}\dfrac{x}{4} + C$

5 $-\dfrac{2}{135}(9x+4)(2-3x)^{3/2} + C$

7 $-\dfrac{1}{18}\sin^5 3x\cos 3x - \dfrac{5}{72}\sin^3 3x\cos 3x - \dfrac{5}{48}\sin 3x\cos 3x + \dfrac{5}{16}x + C$

9 $-\dfrac{1}{3}\cot x\csc^2 x - \dfrac{2}{3}\cot x + C$

11 $\dfrac{2x^2-1}{4}\sin^{-1}x + \dfrac{x\sqrt{1-x^2}}{4} + C$

13 $\dfrac{1}{13}e^{-3x}(-3\sin 2x - 2\cos 2x) + C$

15 $\sqrt{5x-9x^2} + \dfrac{5}{6}\cos^{-1}\dfrac{5-18x}{5} + C$

17 $\dfrac{1}{4\sqrt{15}}\ln\left|\dfrac{\sqrt{5}x^2-\sqrt{3}}{\sqrt{5}x^2+\sqrt{3}}\right| + C$

19 $\dfrac{1}{4}(2e^{2x}-1)\cos^{-1}e^x - \dfrac{1}{4}e^x\sqrt{1-e^{2x}} + C$

21 $\dfrac{2}{315}(35x^3 - 60x^2 + 96x - 128)(2+x)^{3/2} + C$

23 $\dfrac{2}{81}(4 + 9\sin x - 4\ln|4 + 9\sin x|) + C$

25 $2\sqrt{9+2x} + 3\ln\left|\dfrac{\sqrt{9+2x}-3}{\sqrt{9+2x}+3}\right| + C$ **27** $\dfrac{3}{4}\ln\left|\dfrac{\sqrt[3]{x}}{4+\sqrt[3]{x}}\right| + C$

29 $\sqrt{16-\sec^2 x} - 4\ln\left|\dfrac{4+\sqrt{16-\sec^2 x}}{\sec x}\right| + C$

9.8 REVIEW EXERCISES

1 $\dfrac{1}{2}x^2\sin^{-1}x - \dfrac{1}{4}\sin^{-1}x + \dfrac{1}{4}x\sqrt{1-x^2} + C$

3 $2\ln 2 - 1 \approx 0.39$ **5** $\dfrac{1}{6}\sin^3 2x - \dfrac{1}{10}\sin^5 2x + C$

7 $\dfrac{1}{5}\sec^5 x + C$ **9** $\dfrac{x}{25\sqrt{x^2+25}} + C$

11 $2\ln\left|\dfrac{2-\sqrt{4-x^2}}{x}\right| + \sqrt{4-x^2} + C$

13 $2\ln|x-1|-\ln|x|-\dfrac{x}{(x-1)^2}+C$

15 $-5\ln|x-3|+2\ln|x+3|+2\ln(x^2+9)+\dfrac{1}{3}\tan^{-1}\dfrac{x}{3}+C$

17 $-\sqrt{4+4x-x^2}+2\sin^{-1}\dfrac{x-2}{\sqrt{8}}+C$

19 $3(x+8)^{1/3}+\ln[(x+8)^{1/3}-2]^2-$
$\ln|(x+8)^{2/3}+2(x+8)^{1/3}+4|-\dfrac{6}{\sqrt{3}}\tan^{-1}\dfrac{(x+8)^{1/3}+1}{\sqrt{3}}+C$

21 $\dfrac{1}{13}e^{2x}(2\sin 3x-3\cos 3x)+C$

23 $\dfrac{1}{4}\sin^4 x-\dfrac{1}{6}\sin^6 x+C$ **25** $-\sqrt{4-x^2}+C$

27 $\dfrac{1}{3}x^3-x^2+3x-\dfrac{1}{4}\ln|x|-\dfrac{1}{2x}-\dfrac{23}{4}\ln|x+2|+C$

29 $2\tan^{-1}\sqrt{x}+C$ **31** $\ln|\sec e^x+\tan e^x|+C$

33 $\dfrac{1}{125}[10x\sin 5x-(25x^2-2)\cos 5x]+C$

35 $\dfrac{2}{7}\cos^{7/2}x-\dfrac{2}{3}\cos^{3/2}x+C$

37 $\dfrac{2}{3}(1+e^x)^{3/2}+C$

39 $\dfrac{1}{16}[2x\sqrt{4x^2+25}-25\ln(\sqrt{4x^2+25}+2x)]+C$

41 $\dfrac{1}{3}\tan^3 x+C$ **43** $-x\csc x+\ln|\csc x-\cot x|+C$

45 $-\dfrac{1}{4}(8-x^3)^{4/3}+C$

47 $-2x\cos\sqrt{x}+4\sqrt{x}\sin\sqrt{x}+4\cos\sqrt{x}+C$

49 $\dfrac{1}{2}e^{2x}-e^x+\ln(1+e^x)+C$

51 $\dfrac{2}{5}x^{5/2}-\dfrac{8}{3}x^{3/2}+6x^{1/2}+C$

53 $\dfrac{1}{3}(16-x^2)^{3/2}-16(16-x^2)^{1/2}+C$

55 $\dfrac{11}{2}\ln|x+5|-\dfrac{15}{2}\ln|x+7|+C$

57 $x\tan^{-1}5x-\dfrac{1}{10}\ln(1+25x^2)+C$ **59** $e^{\tan x}+C$

61 $\dfrac{1}{\sqrt{5}}\ln|\sqrt{7+5x^2}+\sqrt{5}x|+C$

63 $-\dfrac{1}{5}\cot^5 x+\dfrac{1}{3}\cot^3 x-\cot x-x+C$

65 $\dfrac{1}{5}(x^2-25)^{5/2}+\dfrac{25}{3}(x^2-25)^{3/2}+C$

67 $\dfrac{1}{3}x^3-\dfrac{1}{4}\tanh 4x+C$

69 $-\dfrac{1}{4}x^2e^{-4x}-\dfrac{1}{8}xe^{-4x}-\dfrac{1}{32}e^{-4x}+C$

71 $3\sin^{-1}\dfrac{x+5}{6}+C$ **73** $-\dfrac{1}{7}\cos 7x+C$

75 $-9\ln|x-1|+18\ln|x-2|-5\ln|x-3|+C$

77 $x^3\sin x+3x^2\cos x-6x\sin x-6\cos x+\sin x+C$

79 $-\dfrac{\sqrt{9-4x^2}}{x}-2\sin^{-1}\left(\dfrac{2}{3}x\right)+C$

81 $24x-\dfrac{10}{3}\ln|\sin 3x|-\dfrac{1}{3}\cot 3x+C$

83 $-\ln x-\dfrac{4}{\sqrt[4]{x}}+4\ln(\sqrt[4]{x}+1)+C$ **85** $-2\sqrt{1+\cos x}+C$

87 $-\dfrac{x}{2(25+x^2)}+\dfrac{1}{10}\tan^{-1}\dfrac{x}{5}+C$ **89** $\dfrac{1}{3}\sec^3 x-\sec x+C$

91 $\dfrac{7}{\sqrt{5}}\tan^{-1}\left(\dfrac{x}{\sqrt{5}}\right)-\dfrac{3}{2}\tan^{-1}\left(\dfrac{x}{2}\right)+\ln(x^2+4)+C$

93 $\dfrac{1}{4}x^4-2x^2+4\ln|x|+C$ **95** $\dfrac{2}{5}x^{5/2}\ln x-\dfrac{4}{25}x^{5/2}+C$

97 $\dfrac{3}{64}(2x+3)^{8/3}-\dfrac{9}{20}(2x+3)^{5/3}+\dfrac{27}{16}(2x+3)^{2/3}+C$

99 $\dfrac{1}{2}e^{(x^2)}(x^2-1)+C$

CHAPTER 10

EXERCISES 10.1

1 $\frac{1}{2}$ **3** $\frac{1}{40}$ **5** $\frac{3}{13}$ **7** 0 **9** $-\frac{1}{2}$ **11** $-\frac{1}{2}$

13 $\frac{1}{6}$ **15** ∞ **17** $\frac{1}{3}$ **19** ∞ **21** 1 **23** 0

25 ∞ **27** $\frac{2}{5}$ **29** ∞ **31** 0 **33** ∞ **35** 2

37 DNE **39** $\frac{3}{5}$ **41** -3 **43** 0 **45** ∞ **47** ∞

49 2 **51** 1

53 0.9129, 0.9901, 0.9990, 0.9999; predict limit of 1

55 gt **57** $\frac{1}{2}A\omega_0 t\sin\omega_0 t$ **59** **(a)** 1 **(b)** $-\frac{1}{18}$

61 **(a)** 0.2499, 0.4969, 0.7266, 0.9045

(b)

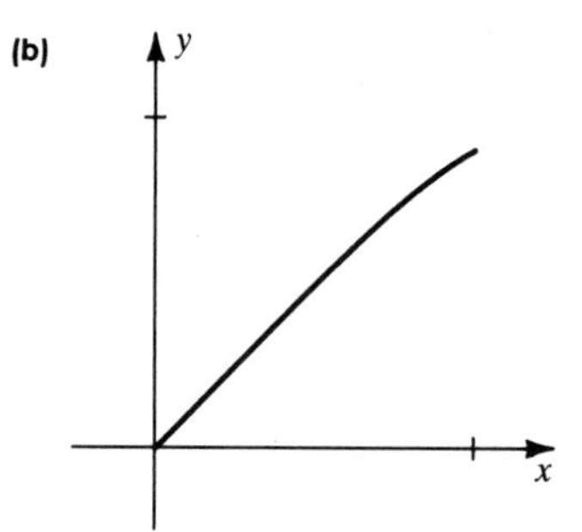

EXERCISES 10.2

1 0 **3** 0 **5** 0 **7** 0 **9** 1 **11** 0 **13** e^5

15 1 **17** 1 **19** ∞ **21** e^2 **23** 2 **25** 0

27 1 **29** $-\infty$ **31** $\frac{1}{2}$ **33** ∞ **35** e **37** ∞

39 $e^{1/3}$ **41** ∞

43 1

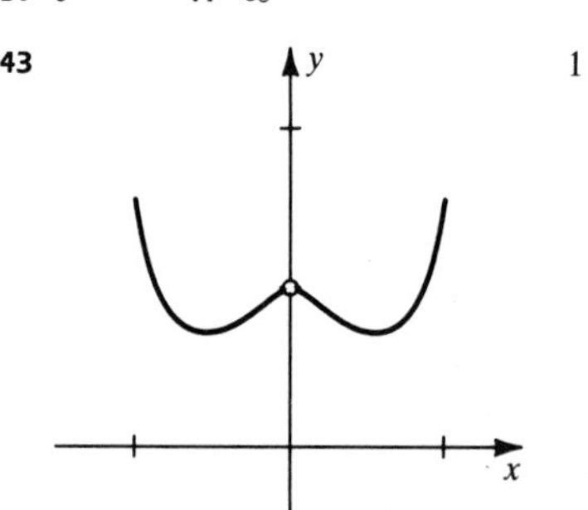

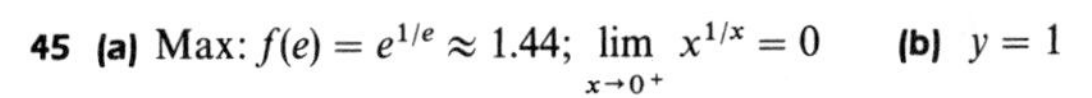

45 **(a)** Max: $f(e) = e^{1/e} \approx 1.44$; $\lim_{x \to 0^+} x^{1/x} = 0$ **(b)** $y = 1$

(c)

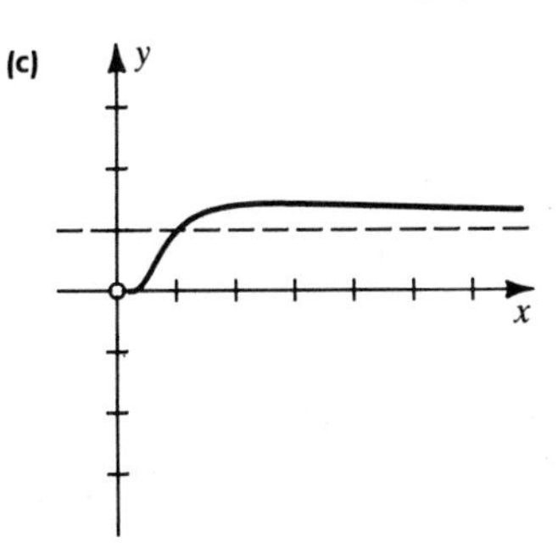

49 gt

EXERCISES 10.3

C denotes that the integral converges; D denotes that it diverges.

1 C; 3 **3** D **5** D **7** C; $\frac{1}{2}$ **9** C; $-\frac{1}{2}$
11 D **13** D **15** C; 0 **17** D **19** D
21 C; π **23** C; ln 2 **25** C **27** D
29 **(a)** Not possible **(b)** π **31** **(a)** 1 **(b)** $\frac{\pi}{32}$ **33** π
35 **(b)** No **37** If $F(x) = \frac{k}{x^2}$, then $W = k$.
39 **(a)** $\frac{1}{k}$ **(b)** No, the improper integral diverges.
41 **(b)** $c = \frac{4}{\sqrt{\pi}}\left(\frac{m}{2kT}\right)^{3/2}$ **43** $\frac{1}{s}$, $s > 0$ **45** $\frac{s}{s^2+1}$, $s > 0$
47 $\frac{1}{s-a}$, $s > a$
49 **(a)** 1; 1; 2 **(b)** *Hint:* Let $u = x^n$ and integrate by parts.
51 0.49

EXERCISES 10.4

1 C; 6 **3** D **5** D **7** D **9** C; $3\sqrt[3]{4}$ **11** D
13 C; $\frac{\pi}{2}$ **15** D **17** C; $-\frac{1}{4}$ **19** D **21** D
23 D **25** C; 0 **27** D **29** D **31** C **33** D
35 $n > -1$ **37** **(a)** 2 **(b)** Not possible
39 **(a)** Not possible **(b)** Not possible **41** 1.79
43 **(b)** $T = 2\pi\sqrt{\frac{m}{k}}$ **45** **(a)** t is undefined at $y = 0$.

10.5 REVIEW EXERCISES

1 $\frac{1}{2}\ln 2$ **3** ∞ **5** $\frac{8}{3}$ **7** 0 **9** $-\infty$ **11** e^8
13 e **15** 1 **17** D **19** D **21** C; $-\frac{9}{2}$ **23** D
25 C; $\frac{\pi}{2}$ **27** D **29** 0.14 **31** 0

CHAPTER 11

EXERCISES 11.1

1 $\frac{1}{5}, \frac{1}{4}, \frac{3}{11}, \frac{2}{7}; \frac{1}{3}$ **3** $\frac{3}{5}, -\frac{9}{11}, -\frac{29}{21}, -\frac{57}{35}; -2$
5 $-5, -5, -5, -5; -5$ **7** $2, \frac{7}{3}, \frac{25}{14}, \frac{7}{5}; 0$
9 $\frac{2}{\sqrt{10}}, \frac{2}{\sqrt{13}}, \frac{2}{\sqrt{18}}, \frac{2}{5}; 0$ **11** $\frac{3}{10}, -\frac{6}{17}, \frac{9}{26}, -\frac{12}{37}; 0$
13 1.1, 1.01, 1.001, 1.0001; 1 **15** 2, 0, 2, 0; DNE
17 C; 0 **19** C; $\frac{\pi}{2}$ **21** D **23** C; 0 **25** D **27** D
29 C; e **31** C; 0 **33** C; $\frac{1}{2}$ **35** D **37** C; 1
39 C; 0 **41** C; 0
43 **(b)** 10,000 on A; 5000 on B; 20,000 on C
45 **(a)** The sequence appears to converge to 1.
(b) Use mathematical induction; 1
47 **(a)** The sequence appears to converge to approximately 0.739.
49 **(a)** $x_2 = 3.5$, $x_3 = 3.178571429$, $x_4 = 3.162319422$,
$x_5 = 3.162277660$, $x_6 = 3.162277660$
51 **(a)** $B = \frac{1}{4}$ **(b)** 1.10

EXERCISES 11.2

1 **(a)** $-\frac{2}{35}, -\frac{4}{45}, -\frac{6}{55}$ **(b)** $-\frac{2n}{5(2n+5)}$ **(c)** C; $-\frac{1}{5}$
3 **(a)** $\frac{1}{3}, \frac{2}{5}, \frac{3}{7}$ **(b)** $\frac{n}{2n+1}$ **(c)** C; $\frac{1}{2}$
5 **(a)** $-\ln 2, -\ln 3, -\ln 4$ **(b)** $-\ln(n+1)$ **(c)** D
7 C; 4 **9** C; $\frac{\sqrt{5}}{\sqrt{5}+1}$ **11** C; $\frac{37}{99}$ **13** D **15** D
17 $-1 < x < 1$; $\frac{1}{1+x}$ **19** $1 < x < 5$; $\frac{1}{5-x}$ **21** $\frac{23}{99}$
23 $\frac{16,181}{4995}$ **25** C **27** C **29** D **31** D **33** D
35 Needs further investigation **37** D **39** D
41 C; $\frac{41}{24}$ **43** C; $\frac{6}{7}$ **45** C; $\frac{8}{7}$ **47** C; $\frac{5}{3}$
49 **(a)** 0.21037; 0.26720; 0.26940 **(b)** 0.265 **51** $S_{32} \approx 4.06$
53 Disprove; let $a_n = 1$ and $b_n = -1$ **55** 30 m
57 **(b)** $\frac{Q}{1 - e^{-cT}}$ **(c)** $-\frac{1}{c}\ln\frac{M-Q}{M}$ **59** **(b)** 2000
61 **(a)** $a_{k+1} = \frac{1}{4}\sqrt{10}a_k$
(b) $a_n = \left(\frac{1}{4}\sqrt{10}\right)^{n-1} a_1$; $A_n = \left(\frac{5}{8}\right)^{n-1} A_1$; $P_n = \left(\frac{1}{4}\sqrt{10}\right)^{n-1} P_1$
(c) $\frac{16}{4 - \sqrt{10}} a_1$; $\frac{8}{3}a_1^2$

EXERCISES 11.3

Exer. 1–12: (a) Each function f is positive-valued and continuous on the interval of integration. Since $f'(x)$ is negative, f is decreasing. (b) The value of the improper integral is given, if it exists.

1 **(a)** $f'(x) = \frac{-4}{(2x+3)^3} < 0$ if $x \geq 1$ **(b)** $\int_1^\infty f(x)\,dx = \frac{1}{10}$; C

3 **(a)** $f'(x) = \dfrac{-4}{(4x+7)^2} < 0$ if $x \ge 1$ **(b)** $\int_1^\infty f(x)\,dx = \infty$; D

5 **(a)** $f'(x) = x(2 - 3x^3)e^{-x^3} < 0$ if $x \ge 1$

(b) $\int_1^\infty f(x)\,dx = \dfrac{1}{3e}$; C

7 **(a)** $f'(x) = \dfrac{1 - \ln x}{x^2} < 0$ if $x \ge 3$ **(b)** $\int_3^\infty f(x)\,dx = \infty$; D

9 **(a)** $f'(x) = \dfrac{1 - 2x^2}{x^2(x^2-1)^{3/2}} < 0$ if $x \ge 2$ **(b)** $\int_2^\infty f(x)\,dx = \dfrac{\pi}{6}$; C

11 **(a)** $f'(x) = \dfrac{1 - 2x \arctan x}{(1+x^2)^2} < 0$ if $x \ge 1$

(b) $\int_1^\infty f(x)\,dx = \dfrac{3\pi^2}{32}$; C

Exer. 13–28: A typical b_n is listed; however, there are many other possible choices.

13 $b_n = \dfrac{1}{n^4}$; C **15** $b_n = \dfrac{1}{3^n}$; C **17** $b_n = \dfrac{\pi/4}{n}$; D

19 $b_n = \dfrac{1}{n^2}$; C **21** $b_n = \dfrac{1}{\sqrt{n}}$; D **23** $b_n = \dfrac{1}{n^{3/2}}$; C

25 $b_n = \dfrac{1}{e^n}$; C **27** $b_n = \dfrac{1}{\sqrt{n}}$; D **29** D **31** C

33 D **35** D **37** C **39** C **41** C **43** C
45 C **47** $k > 1$ **49** **(b)** $n > e^{100} - 1 \approx 2.688 \times 10^{43}$

51 Since $\lim_{n\to\infty} \dfrac{a_n}{b_n} = 0$, there is an M such that if $K > M$, then $\dfrac{a_k}{b_k} < 1$, or $a_k < b_k$. Since $\sum b_n$ converges and $a_n < b_n$ for all but at most a finite number of terms, $\sum a_n$ must also converge.

53 $\sum_{k=1}^{\infty} a_k = \sum_{k=1}^{n} a_k + \sum_{k=n+1}^{\infty} a_k$, where the error $E = \sum_{k=n+1}^{\infty} a_k < \int_n^\infty f(x)\,dx$. (See Figure 11.8.)

55 8

57 Since $\sum a_n$ converges, $\lim_{n\to\infty} a_n = 0$ and $\lim_{n\to\infty} \dfrac{1}{a_n} = \infty$. By (11.17), $\sum \dfrac{1}{a_n}$ diverges.

59 0.405

61 The series diverges for $k = 1, 2,$ and 3.

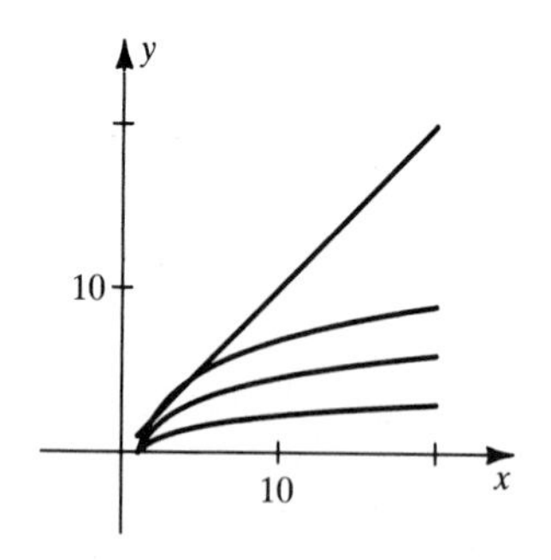

EXERCISES 11.4

1 $\frac{1}{2}$; C **3** $\frac{5}{3}$; D **5** 0; C **7** 1; inconclusive

9 ∞; D **11** 0; C **13** 2; D **15** $\frac{1}{3}$; C **17** $\frac{1}{2}$; C

19 C **21** C **23** C **25** C **27** D **29** C
31 D **33** C **35** D **37** D **39** D

EXERCISES 11.5

1 **(a)** Conditions (i) and (ii) are satisfied.
(b) Converges, by (11.30)

3 **(a)** Condition (i) is satisfied, but (ii) is not.
(b) Diverges, by (11.17)

5 CC **7** CC **9** D **11** AC **13** AC
15 D **17** CC **19** AC **21** D **23** CC **25** D
27 D **29** D **31** AC **33** 0.368 **35** 0.901
37 0.306 **39** 141 **41** 5

45 No. If $a_n = b_n = \dfrac{(-1)^n}{\sqrt{n}}$, then both $\sum a_n$ and $\sum b_n$ converge by the alternating series test. However, $\sum a_n b_n = \sum \dfrac{1}{n}$, which diverges.

EXERCISES 11.6

1 $[-1, 1)$ **3** $(-2, 2)$ **5** $(-1, 1]$ **7** $[-1, 1)$
9 $[-1, 1]$ **11** $(-6, 14)$ **13** Converges only for $x = 0$

15 $(-2, 2)$ **17** $(-\infty, \infty)$ **19** $\left[\dfrac{17}{9}, \dfrac{19}{9}\right)$ **21** $(-12, 4)$

23 Converges only for $x = 3$ **25** $(0, 2e)$ **27** $\left(-\dfrac{5}{2}, \dfrac{7}{2}\right]$

29 $(-\infty, \infty)$ **31** $\dfrac{3}{2}$ **33** $\dfrac{1}{e}$ **35** ∞ **37** Use (11.35).

39 $J_0(x) \approx 1 - \dfrac{x^2}{4} + \dfrac{x^4}{64} - \dfrac{x^6}{2304}$

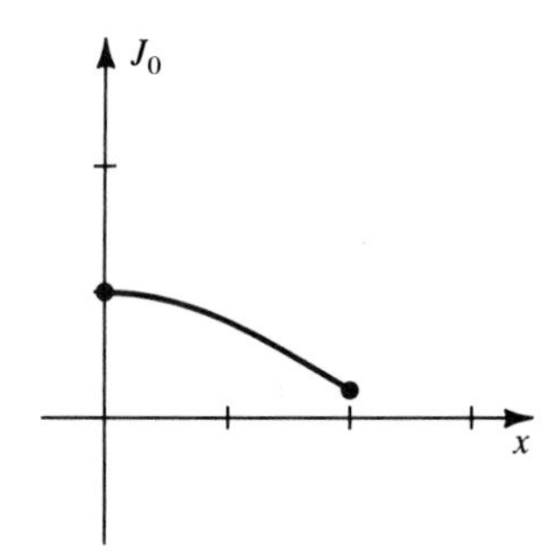

41 Use (11.35). **43** Use (11.37).

45 Assume that $\sum a_n x^n$ is absolutely convergent at $x = r$. Let $x = -r$. Then $\sum |a_n(-r)^n| = \sum |a_n r^n|$ is absolutely convergent, which implies that $\sum a_n(-r)^n$ is convergent. This is a contradiction.

EXERCISES 11.7

1 **(a)** $\sum_{n=0}^{\infty} 3^n x^n$ **(b)** $\sum_{n=1}^{\infty} n3^n x^{n-1}$; $\sum_{n=0}^{\infty} \dfrac{3^n}{n+1} x^{n+1}$

3 **(a)** $\dfrac{1}{2} \sum_{n=0}^{\infty} (-1)^n \left(\dfrac{7}{2}\right)^n x^n$

(b) $\dfrac{1}{2} \sum_{n=1}^{\infty} (-1)^n \dfrac{n7^n}{2^n} x^{n-1}$; $\dfrac{1}{2} \sum_{n=0}^{\infty} (-1)^n \dfrac{7^n}{(n+1)2^n} x^{n+1}$

5 $\sum_{n=0}^{\infty} x^{2n+2}$; $r = 1$ **7** $\sum_{n=0}^{\infty} \dfrac{3^n}{2^{n+1}} x^{n+1}$; $r = \dfrac{2}{3}$

9 $-1 - x - 2 \sum_{n=2}^{\infty} x^n$; $r = 1$ **11** **(b)** 0.183; 0.182321557

15 $\sum_{n=0}^{\infty} \frac{3^n}{n!} x^{n+1}$ **17** $\sum_{n=0}^{\infty} (-1)^n \frac{1}{n!} x^{n+3}$

19 $\sum_{n=0}^{\infty} (-1)^n \frac{1}{n+1} x^{2n+4}$ **21** $\sum_{n=0}^{\infty} (-1)^n \frac{1}{2n+1} x^{(2n+1)/2}$

23 $\sum_{n=0}^{\infty} \frac{-5^{2n+1}}{(2n+1)!} x^{2n+1}$ **25** $\sum_{n=0}^{\infty} \frac{1}{(2n)!} x^{6n+2}$ **27** 0.3333

29 0.0992 **31** 0.9677 **33** $\sum_{n=1}^{\infty} (2n)x^{2n-1}$

37 $-\sum_{n=1}^{\infty} \frac{1}{n^2} x^n$

EXERCISES 11.8

1 $\frac{3^n}{n!}$ **3** $a_n = 0$ if $n = 2k$, and $a_n = (-1)^k \frac{2^{2k+1}}{(2k+1)!}$ if $n = 2k+1$

5 $(-1)^n 3^n$ **7 (b)** $\sum_{n=0}^{\infty} (-1)^n \frac{1}{(2n)!} x^{2n}$

9 $\sum_{n=0}^{\infty} (-1)^n \frac{3^{2n+1}}{(2n+1)!} x^{2n+2}$ **11** $\sum_{n=0}^{\infty} (-1)^n \frac{2^{2n}}{(2n)!} x^{2n}$

13 $1 + \sum_{n=1}^{\infty} (-1)^n \frac{2^{2n-1}}{(2n)!} x^{2n}$ **15** $\sum_{n=0}^{\infty} \frac{(\ln 10)^n}{n!} x^n$

17 $\sum_{n=0}^{\infty} (-1)^n \frac{1}{\sqrt{2}(2n+1)!}\left(x - \frac{\pi}{4}\right)^{2n+1} + \sum_{n=0}^{\infty} (-1)^n \frac{1}{\sqrt{2}(2n)!}\left(x - \frac{\pi}{4}\right)^{2n}$

19 $\sum_{n=0}^{\infty} (-1)^n \frac{1}{2^{n+1}} (x-2)^n$ **21** $\sum_{n=0}^{\infty} \frac{2^n}{e^2 n!} (x+1)^n$

23 $2 + 2\sqrt{3}\left(x - \frac{\pi}{3}\right) + 7\left(x - \frac{\pi}{3}\right)^2$

25 $\frac{\pi}{6} + \frac{2}{\sqrt{3}}\left(x - \frac{1}{2}\right) + \frac{2}{3\sqrt{3}}\left(x - \frac{1}{2}\right)^2$

27 $-\frac{1}{e} + \frac{1}{2e}(x+1)^2 + \frac{1}{3e}(x+1)^3$ **29** 0.5; 0.125

31 0.9986; 3.13×10^{-7} **33** 0.0997; 2×10^{-6}

35 0.6667; 0.1 **37** 0.4969; 9.04×10^{-6} **39** 0.4864

41 0.4484

43 (a) 0.309524, −0.690476

(b)

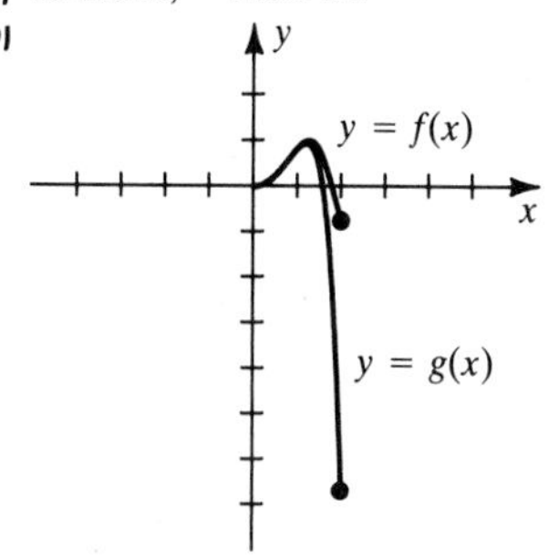

The first approximation is more accurate.

45 $2\sum_{n=0}^{\infty} \frac{1}{2n+1} x^{2n+1}$

47 (a) $\pi = 4\left[1 - \frac{1}{3} + \frac{1}{5} - \frac{1}{7} + \cdots + (-1)^n \frac{1}{2n+1} + \cdots\right]$

(b) 3.34 with an error of less than $\frac{4}{11}$ **(c)** 40,000

49 (c) 16.7 ft

EXERCISES 11.9

1 (a) x; x; $x - \frac{1}{6}x^3$

(b)

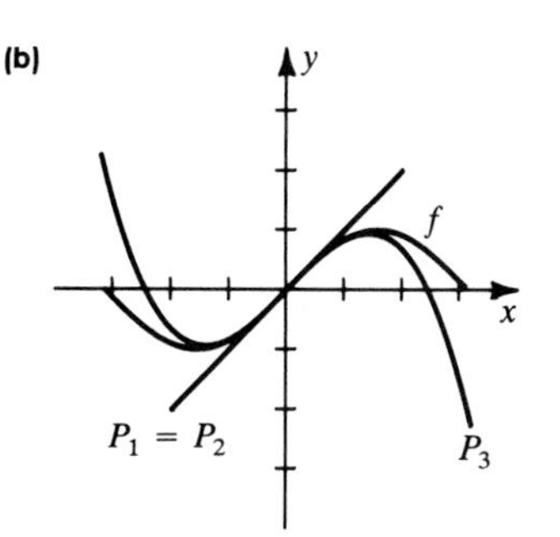

(c) 0.0500; 2.6×10^{-7}

3 (a) x; $x - \frac{1}{2}x^2$; $x - \frac{1}{2}x^2 + \frac{1}{3}x^3$

(b)

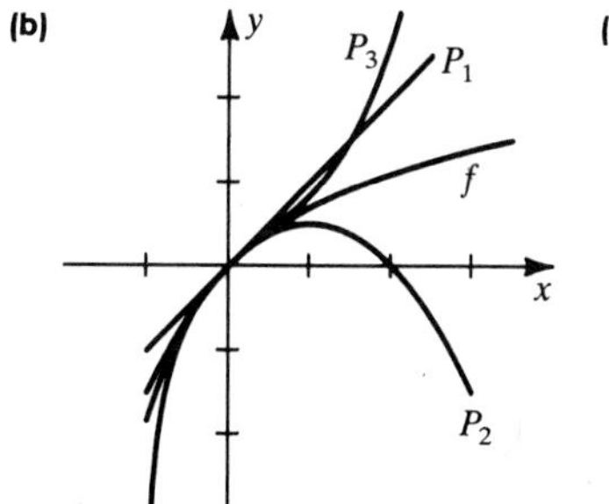

(c) 0.7380; 0.164

5

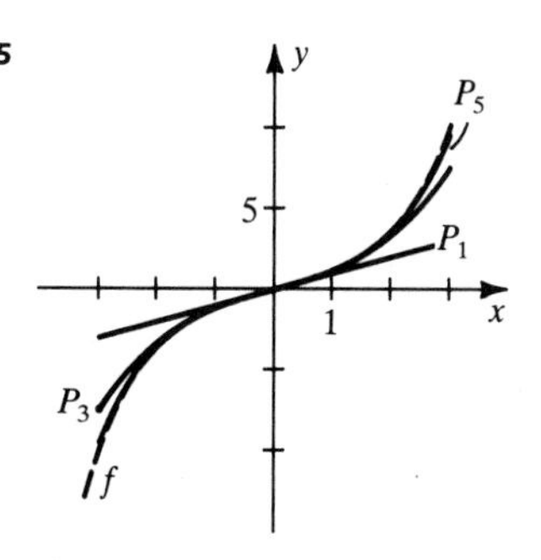

$P_1(x) = x$;

$P_3(x) = x + \frac{1}{6}x^3$;

$P_5(x) = x + \frac{1}{6}x^3 + \frac{1}{120}x^5$

7 $\sin x = 1 - \frac{1}{2}\left(x - \frac{\pi}{2}\right)^2 + \frac{1}{24}\sin z\left(x - \frac{\pi}{2}\right)^4$, z is between x and $\frac{\pi}{2}$.

9 $\sqrt{x} = 2 + \frac{1}{4}(x-4) - \frac{1}{64}(x-4)^2 + \frac{1}{512}(x-4)^3 - \frac{5}{128}z^{-7/2}(x-4)^4$, z is between x and 4.

11 $\tan x = 1 + 2\left(x - \frac{\pi}{4}\right) + 2\left(x - \frac{\pi}{4}\right)^2 + \frac{1}{3}(3\tan^4 z + 4\tan^2 z + 1)\left(x - \frac{\pi}{4}\right)^3$, z is between x and $\frac{\pi}{4}$.

13 $\frac{1}{x} = -\frac{1}{2} - \frac{1}{4}(x+2) - \frac{1}{8}(x+2)^2 - \frac{1}{16}(x+2)^3 - \frac{1}{32}(x+2)^4 - \frac{1}{64}(x+2)^5 + z^{-7}(x+2)^6$, z is between x and −2.

15 $\tan^{-1} x = \frac{\pi}{4} + \frac{1}{2}(x-1) - \frac{1}{4}(x-1)^2 + \frac{3z^2-1}{3(1+z^2)^3}(x-1)^3$, z is between x and 1.

17 $xe^x = -\frac{1}{e} + \frac{1}{2e}(x+1)^2 + \frac{1}{3e}(x+1)^3 + \frac{1}{8e}(x+1)^4 + \frac{ze^z + 5e^z}{120}(x+1)^5$, z is between x and −1.

Exer. 19–30: Since $c = 0$, z is between x and 0.

19 $\ln(x+1) = x - \frac{1}{2}x^2 + \frac{1}{3}x^3 - \frac{1}{4}x^4 + \frac{x^5}{5(z+1)^5}$

21 $\cos x = 1 - \frac{x^2}{2!} + \frac{x^4}{4!} - \frac{x^6}{6!} + \frac{x^8}{8!} - \frac{\sin z}{9!}x^9$

23 $e^{2x} = 1 + 2x + 2x^2 + \frac{4}{3}x^3 + \frac{2}{3}x^4 + \frac{4}{15}x^5 + \frac{4}{45}e^{2z}x^6$

25 $\frac{1}{(x-1)^2} = 1 + 2x + 3x^2 + 4x^3 + 5x^4 + 6x^5 + 7x^6(z-1)^{-8}$

27 $\arcsin x = x + \frac{1+2z^2}{6(1-z^2)^{5/2}}x^3$ **29** $f(x) = -5x^3 + 2x^4$

31 0.9998; $|R_3(x)| < 4 \times 10^{-9}$

33 2.0075; $|R_3(x)| < 3 \times 10^{-10}$

35 -0.454545; $|R_5(x)| \le 5 \times 10^{-7}$

37 0.223; $|R_4(x)| < 2 \times 10^{-4}$

39 0.8660254; $|R_8(x)| < 8.2 \times 10^{-9}$

41 Five decimal places, since $|R_3(x)| \le 4.2 \times 10^{-6} < 0.5 \times 10^{-5}$

43 Three decimal places, since $|R_2(x)| \le 1.85 \times 10^{-4} < 0.5 \times 10^{-3}$

45 Four decimal places, since $|R_3(x)| \le 3.82 \times 10^{-5} < 0.5 \times 10^{-4}$

47 If f is a polynomial of degree n, then the Taylor remainder $R_n(x) = 0$, since $f^{(n+1)}(x) = 0$. By (11.45), we have $f(x) = P_n(x)$.

EXERCISES 11.10

1 (a) $1 + \frac{1}{2}x - \frac{1}{8}x^2 + \sum_{n=3}^{\infty} (-1)^{n-1}\frac{1 \cdot 3 \cdot 5 \cdot \cdots \cdot (2n-3)}{2^n n!}x^n$; 1

(b) $1 - \frac{1}{2}x^3 - \frac{1}{8}x^6 - \sum_{n=3}^{\infty} \frac{1 \cdot 3 \cdot 5 \cdot \cdots \cdot (2n-3)}{2^n n!}x^{3n}$; 1

3 $1 - \frac{2}{3}x + \frac{5}{9}x^2 + \sum_{n=3}^{\infty} \frac{(-2)(-5)(-8)\cdots(1-3n)}{3^n n!}x^n$; 1

5 $1 - \frac{3}{5}x - \frac{3}{25}x^2 + \sum_{n=3}^{\infty} \frac{(3)(-2)(-7)\cdots(8-5n)}{5^n n!}(-x)^n$; 1

7 $1 - 2x + 3x^2 + \sum_{n=3}^{\infty} (-1)^n(n+1)x^n$; 1

9 $1 - 3x + 6x^2 + \sum_{n=3}^{\infty} (-1)^n \frac{1}{2}(n+1)(n+2)x^n$; 1

11 $2 + \frac{1}{12}x - \frac{1}{288}x^2 + 2\sum_{n=3}^{\infty} (-1)^{n-1}\frac{2 \cdot 5 \cdot 8 \cdot \cdots \cdot (3n-4)}{24^n n!}x^n$; 8

13 (a) $x + \sum_{n=1}^{\infty} \frac{1 \cdot 3 \cdot 5 \cdot \cdots \cdot (2n-1)}{2^n(2n+1)n!}x^{2n+1}$ **(b)** 1

Exer. 15–20: The error in the approximation is less than 0.5×10^{-3}.

15 0.508 **17** 0.198 **19** 0.296

21 $[-0.7, 1]$

23 (a) Use $(1-x)^{-1/2} \approx 1 + \frac{1}{2}x$ and a half-angle formula.

(b) $6.39\sqrt{\frac{L}{g}}$

11.11 REVIEW EXERCISES

1 C; 0 **3** D **5** C; 5 **7** D **9** AC **11** D
13 D **15** AC **17** D **19** D **21** AC **23** CC
25 C **27** C **29** C **31** CC **33** C **35** C

37 D **39** 0.158 **41** $(-3, 3)$ **43** $[-12, -8)$ **45** $\frac{1}{4}$

47 $\sum_{n=1}^{\infty} (-1)^{n+1}\frac{1}{(2n)!}x^{2n-1}$; ∞ **49** $\sum_{n=0}^{\infty} (-1)^n \frac{2^{2n}}{(2n+1)!}x^{2n+1}$; ∞

51 $1 + \frac{2}{3}x + 2\sum_{n=2}^{\infty} (-1)^{n-1}\frac{1 \cdot 4 \cdot 7 \cdot \cdots \cdot (3n-5)}{3^n n!}x^n$; 1

53 $e^{-x} = e^2 \sum_{n=0}^{\infty} (-1)^n \frac{1}{n!}(x+2)^n$

55 $\sqrt{x} = 2 + \frac{1}{4}(x-4) + \sum_{n=2}^{\infty} (-1)^{n-1}\frac{1 \cdot 3 \cdot 5 \cdot \cdots \cdot (2n-3)}{2^{3n-1}n!}(x-4)^n$

57 0.189 **59** 1.002

61 $\ln \cos x = \ln\left(\frac{1}{2}\sqrt{3}\right) - \frac{1}{3}\sqrt{3}\left(x - \frac{\pi}{6}\right) - \frac{2}{3}\left(x - \frac{\pi}{6}\right)^2 - \frac{4}{27}\sqrt{3}\left(x - \frac{\pi}{6}\right)^3 - \frac{1}{12}(\sec^4 z + 2\sec^2 z \tan^2 z)\left(x - \frac{\pi}{6}\right)^4$, z is between x and $\frac{\pi}{6}$.

63 $e^{-x^2} = 1 - x^2 + \frac{1}{6}(4z^4 - 12z^2 + 3)e^{-z^2}x^4$, z is between x and 0.

65 0.7314

CHAPTER 12

EXERCISES 12.1

1 $V(0, 0)$; $F(0, -3)$; $y = 3$ **3** $V(0, 0)$; $F\left(-\frac{3}{8}, 0\right)$; $x = \frac{3}{8}$

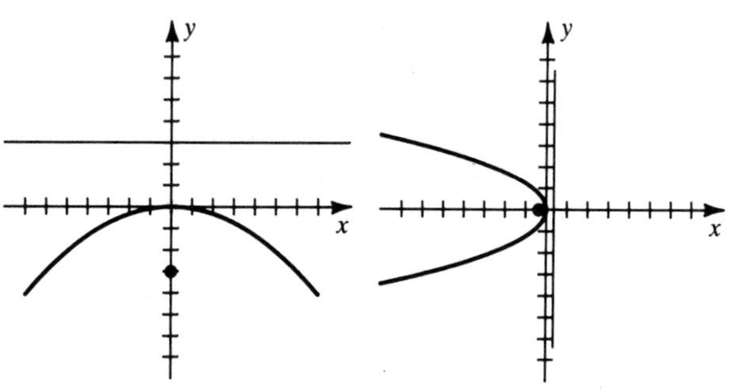

5 $V(0, 0)$; $F\left(0, \frac{1}{32}\right)$; $y = -\frac{1}{32}$ **7** $V(2, -2)$; $F\left(2, -\frac{7}{4}\right)$

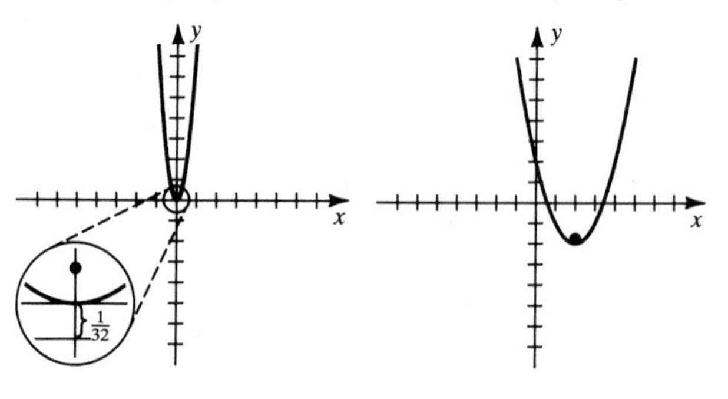

9 $V(-1, 0);\ F(2, 0)$ **11** $V(-4, 2);\ F\left(-\frac{7}{2}, 2\right)$

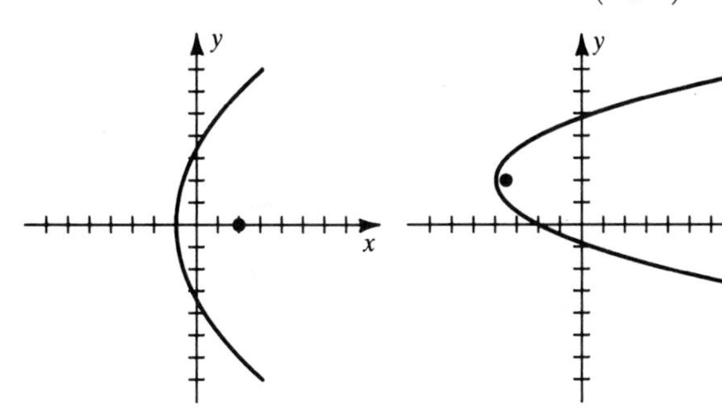

13 $V(-5, -6);\ F\left(-5, -\frac{97}{16}\right)$

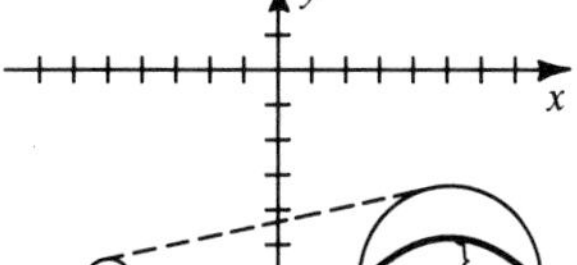

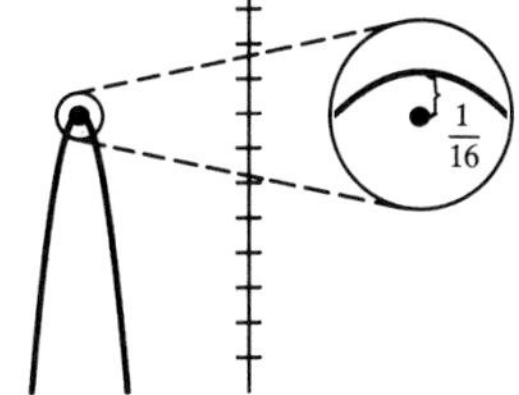

15 $V\left(0, \frac{1}{2}\right);\ F\left(0, -\frac{9}{2}\right)$

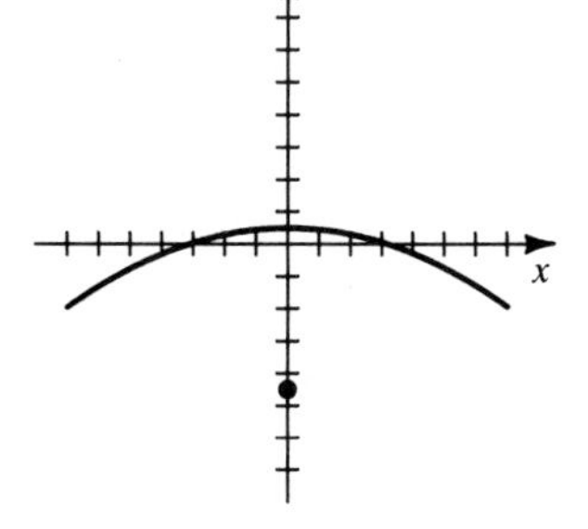

17 $y^2 = 8x$ **19** $(y + 5)^2 = 4(x - 3)$ **21** $y^2 = -12(x + 1)$

23 $3x^2 = -4y$ **25** $\frac{9}{16}$ ft from the vertex

27 **(a)** $\frac{8}{3}$ **(b)** 2π **(c)** $\frac{16\pi}{5}$ **29** **(a)** $p = \frac{r^2}{4h}$ **(b)** $\frac{1}{2}\pi r^2 h$

31 **(a)** $x^2 = 500(y - 10)$ **(b)** $\int_{-200}^{200} \sqrt{1 + \left(\frac{1}{250}x\right)^2}\, dx$

(c) 282 ft

33 *Hint:* Let the upper right corner of the rectangle be $(x, ax^2 + k)$.

35 **(a)** $y = -\frac{7}{160}x^2 + x$ **(b)** 17.5 ft

39 $(\pm 2.280, -0.3)$

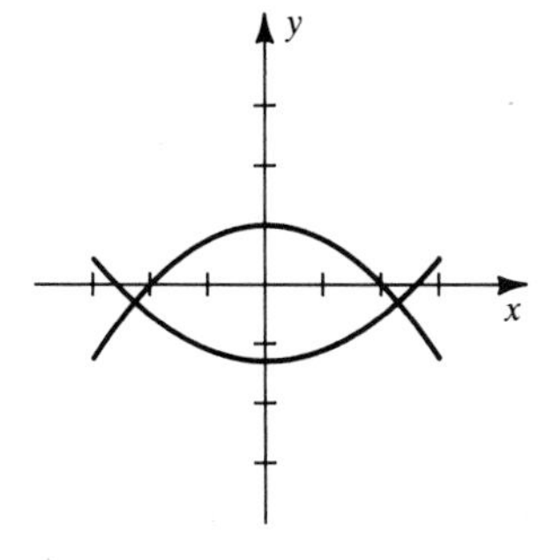

EXERCISES 12.2

1 $V(\pm 3, 0);\ F(\pm\sqrt{5}, 0)$ **3** $V(0, \pm 4);\ F(0, \pm 2\sqrt{3})$

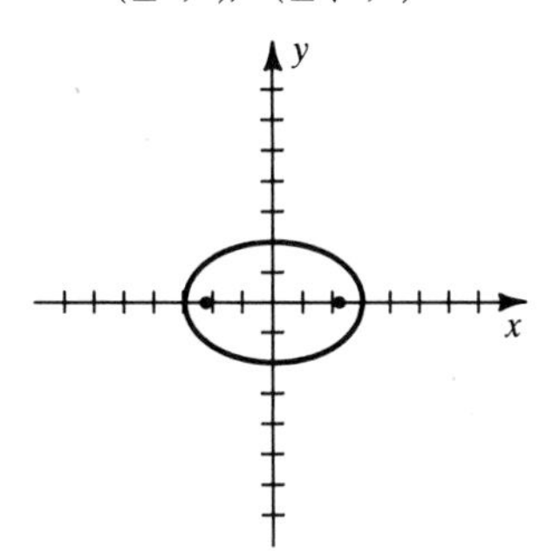

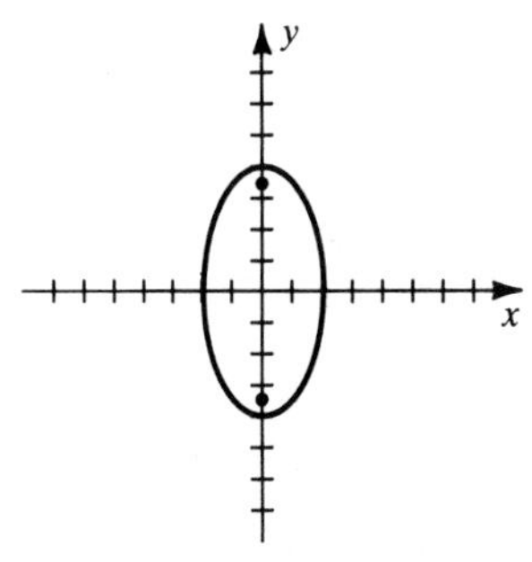

5 $V(0, \pm\sqrt{5});\ F(0, \pm\sqrt{3})$ **7** $V\left(\pm\frac{1}{2}, 0\right);\ F\left(\pm\frac{1}{10}\sqrt{21}, 0\right)$

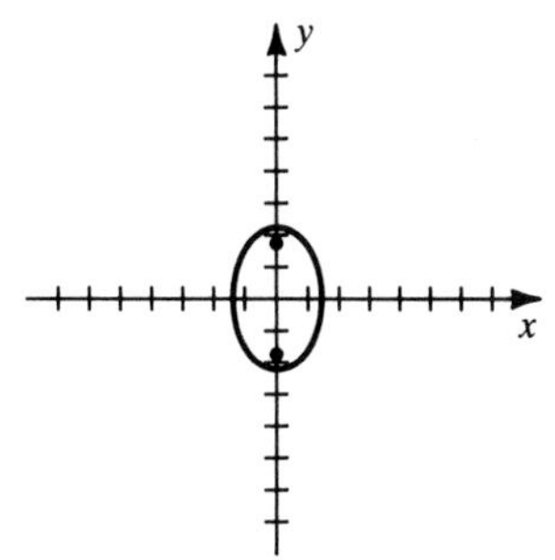

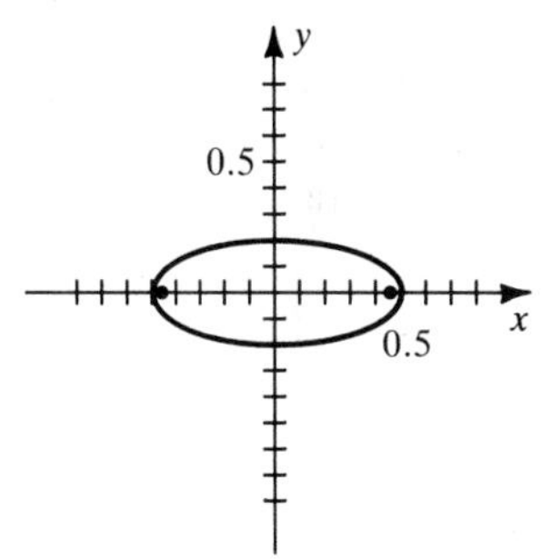

9 $V(4 \pm 3, 2);\ F(4 \pm \sqrt{5}, 2)$ **11** $V(-3 \pm 4, 1);\ F(-3 \pm \sqrt{7}, 1)$

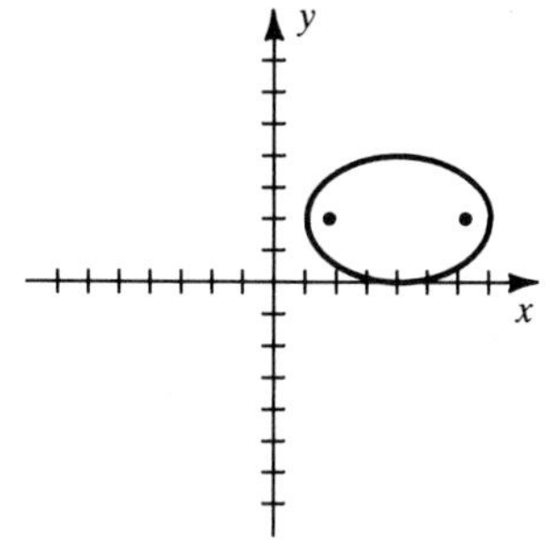

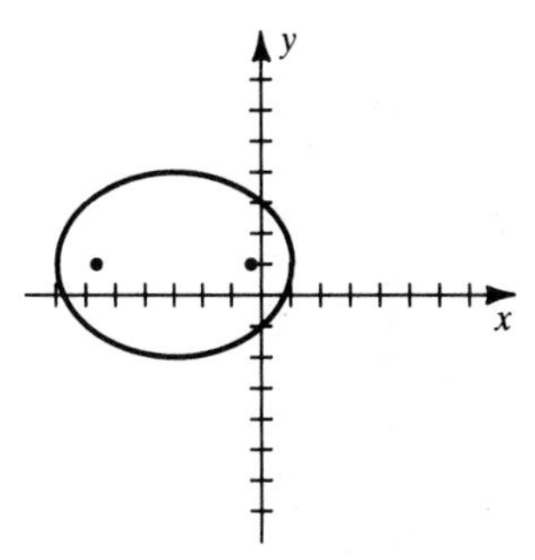

13 $V(5, 2 \pm 5);\ F(5, 2 \pm \sqrt{21})$

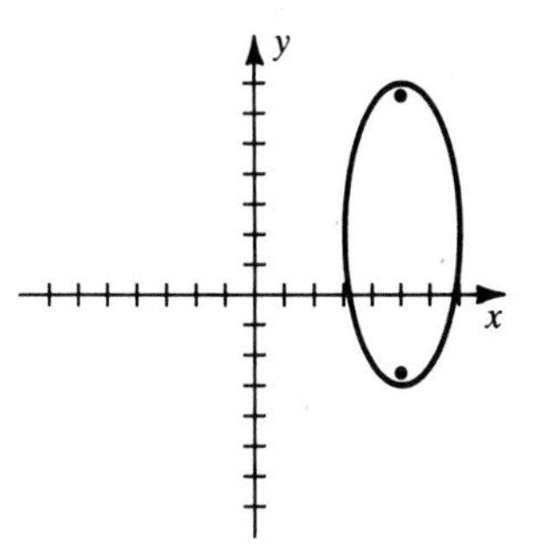

15 $\frac{x^2}{64} + \frac{y^2}{39} = 1$ **17** $\frac{4x^2}{9} + \frac{y^2}{25} = 1$ **19** $\frac{8x^2}{81} + \frac{y^2}{36} = 1$

21 $\frac{x^2}{7} + \frac{y^2}{16} = 1$ **23** $\frac{x^2}{4} + 9y^2 = 1$ **25** $\sqrt{84} \approx 9.165$ ft

29 **(a)** $\sqrt{\frac{2E}{m}}; \sqrt{\frac{2E}{k}}$ **(b)** $A\sqrt{mk} = 2\pi E$ **31** $\left(\pm\sqrt{3}, \frac{3}{2}\right)$

33 $5x - 6y = -28$; $6x + 5y = 3$ **35** $\frac{4}{3}\pi ab^2$

37 864 **41** 509×10^6 km^2

45 **(a)** $(\pm 1.540, 0.618)$

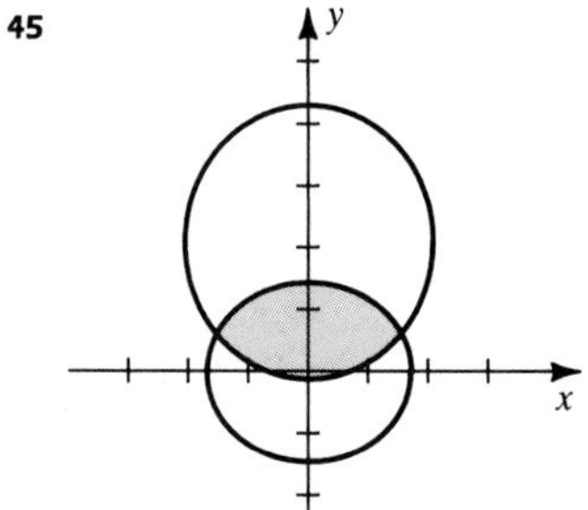

(b) $2\int_0^{1.54}\left[\sqrt{\frac{1}{2.9}(6.09 - 2.1x^2)} - \left(2.1 - \sqrt{\frac{1}{4.3}(21.07 - 4.9x^2)}\right)\right] dx$

EXERCISES 12.3

1 $V(\pm 3, 0)$; $F(\pm\sqrt{13}, 0)$ **3** $V(0, \pm 3)$; $F(0, \pm\sqrt{13})$

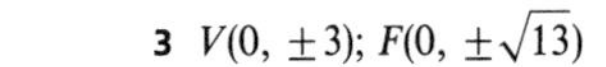

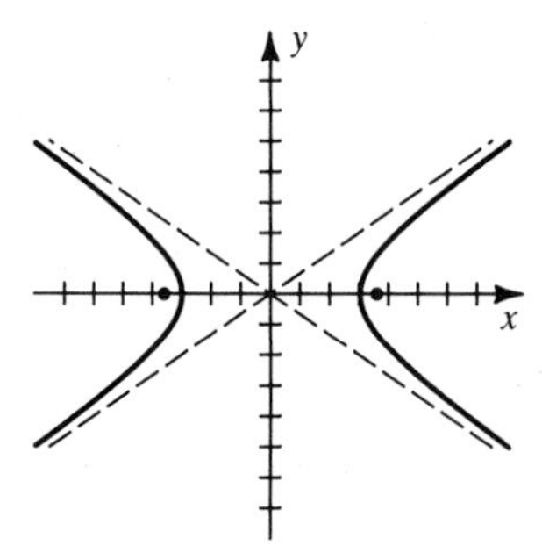

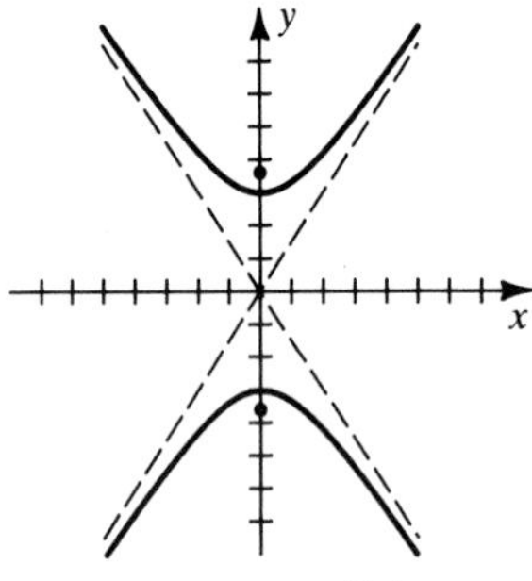

5 $V(0, \pm 4)$; $F(0, \pm 2\sqrt{5})$ **7** $V(\pm 1, 0)$; $F(\pm\sqrt{2}, 0)$

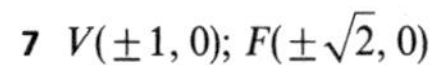

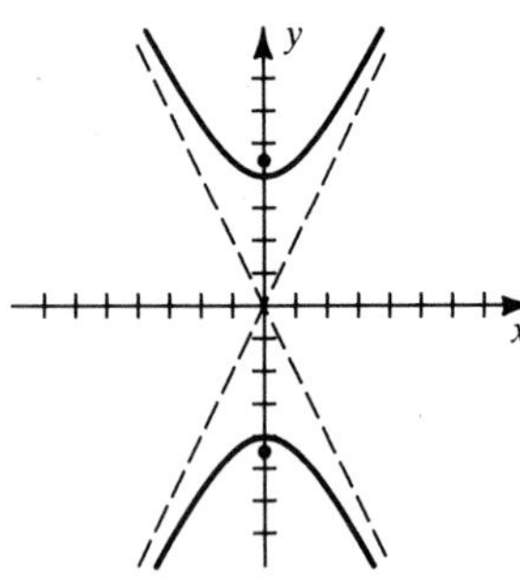

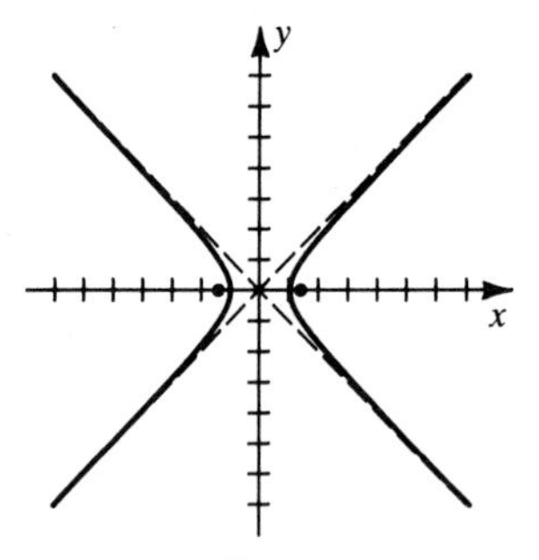

9 $V(\pm 5, 0)$; $F(\pm\sqrt{30}, 0)$ **11** $V(0, \pm\sqrt{3})$; $F(0, \pm 2)$

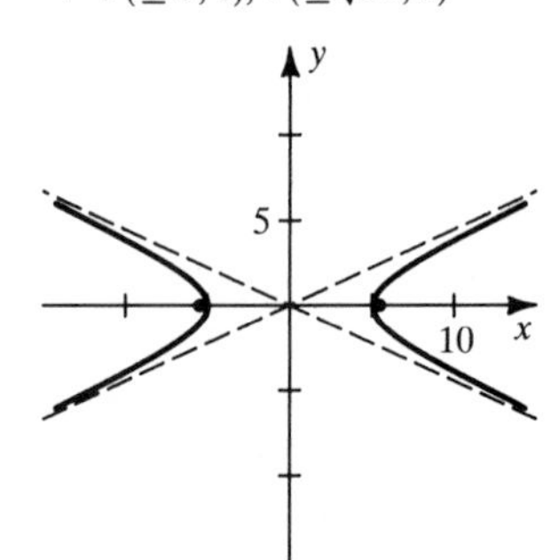

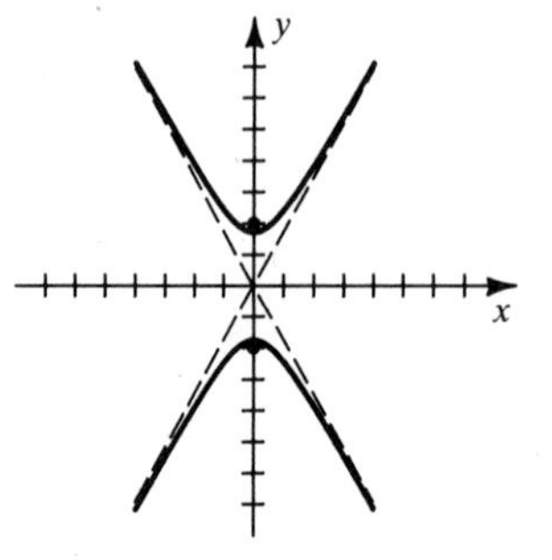

13 $V(-5 \pm 2\sqrt{5}, 1)$; $F\left(-5 \pm \frac{1}{2}\sqrt{205}, 1\right)$ **15** $V(-2, -5 \pm 3)$; $F(-2, -5 \pm 3\sqrt{5})$

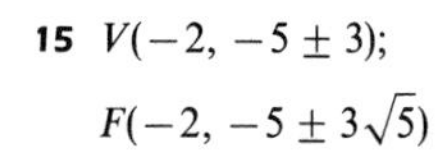

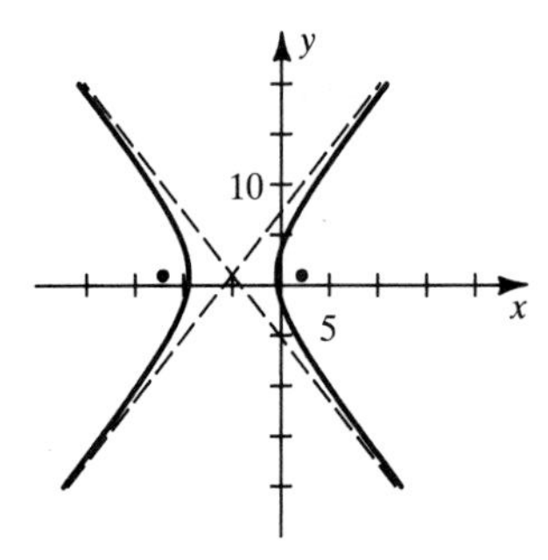

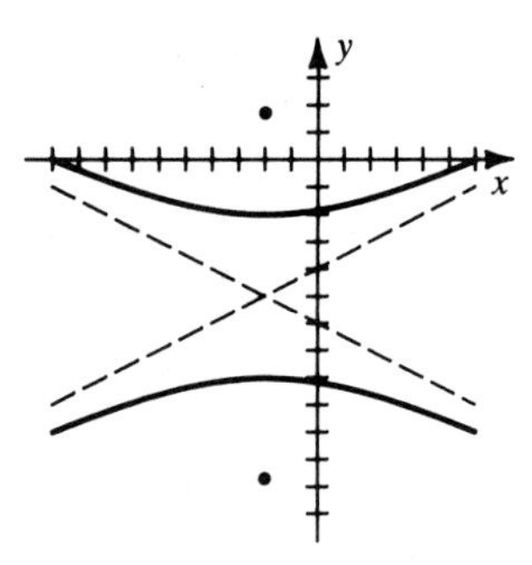

17 $V(6, 2 \pm 2)$; $F(6, 2 \pm 2\sqrt{10})$

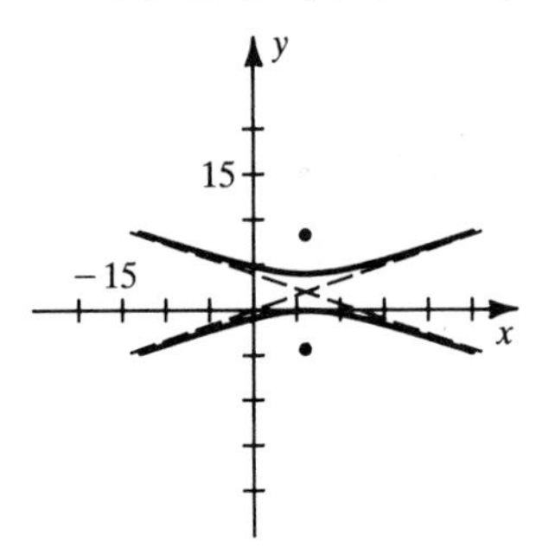

19 $y^2 - \frac{x^2}{15} = 1$ **21** $\frac{x^2}{9} - \frac{y^2}{16} = 1$ **23** $\frac{y^2}{21} - \frac{x^2}{4} = 1$

25 $\frac{x^2}{9} - \frac{y^2}{36} = 1$ **27** $\frac{x^2}{25} - \frac{y^2}{100} = 1$

29 The graphs have the same asymptotes.

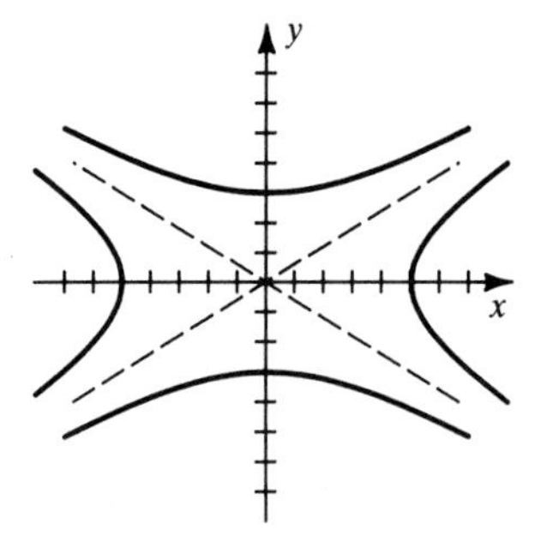

31 If a coordinate system like the one in Example 5 is used, then the ship's coordinates are $\left(\frac{80}{3}\sqrt{34}, 100\right) \approx (155.5, 100)$.

33 $(\pm 2\sqrt{2}, -6)$ **35** $4x + 5y = -3$; $5x - 4y = -14$

37 $\frac{b}{a}[bc - a^2 \ln(b + c) + a^2 \ln a]$

39 $4\pi\sqrt{2}\left(2\sqrt{3} - \sqrt{2} + \ln\frac{1 + \sqrt{2}}{2 + \sqrt{3}}\right) \approx 28.69$

41 $\frac{1}{2}$ AU

43 The rays will collect at the exterior focus of the hyperbolic mirror, located below the parabolic mirror in the figure.

45

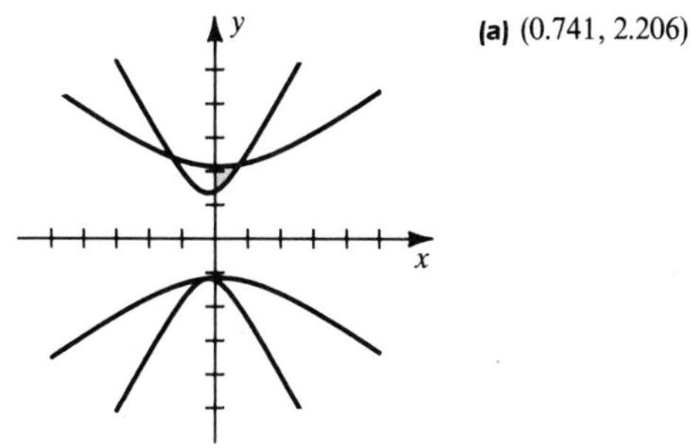

(a) (0.741, 2.206)

(b) $\int_0^{0.74} \left\{0.5 + \sqrt{\frac{1}{5.3}[14.31 + 2.7(x - 0.1)^2]} - [0.1 + \sqrt{1.6 + 3.2(x + 0.2)^2}]\right\} dx$

EXERCISES 12.4

Exer. 1–13: The answer in part (a) gives the value of $B^2 - 4AC$ in the identification theorem.

1 **(a)** 0, parabola **(b)** $(y')^2 = 2(x')$

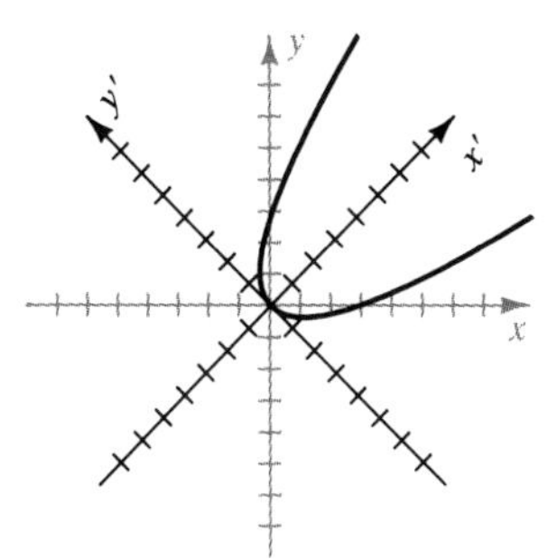

3 **(a)** -36, ellipse **(b)** $\frac{(x')^2}{9} + (y')^2 = 1$

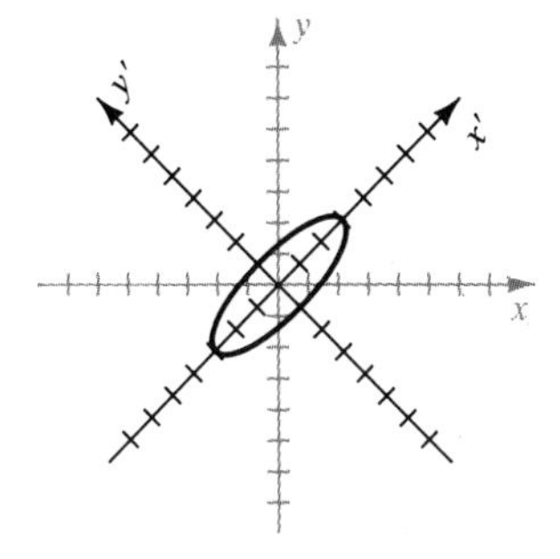

5 **(a)** 256, hyperbola **(b)** $\frac{(x')^2}{1/4} - (y')^2 = 1$

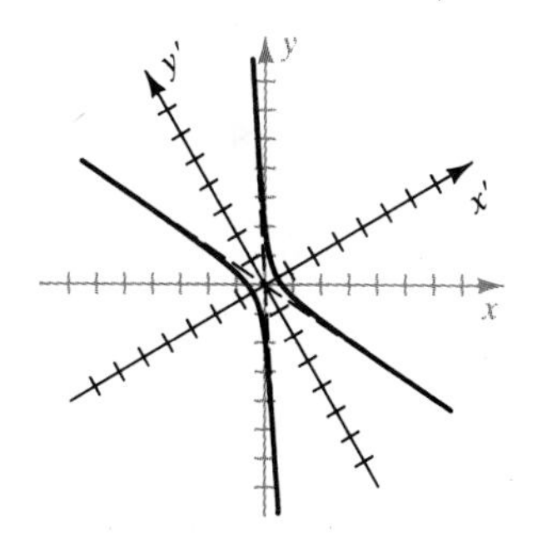

7 **(a)** 0, parabola **(b)** $(y')^2 = 4(x' - 1)$

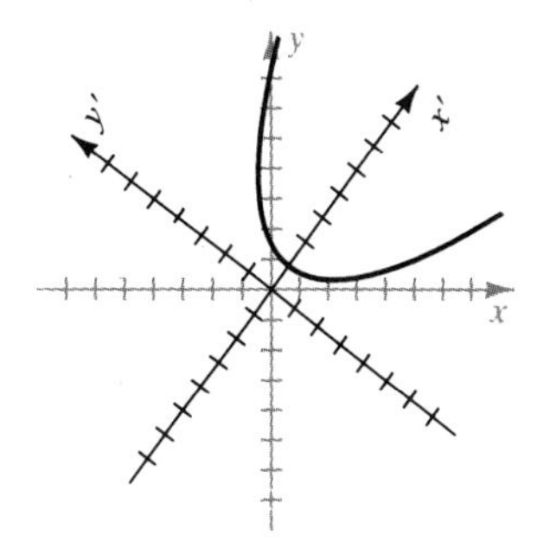

9 **(a)** -2704, ellipse **(b)** $\frac{(x' - 2)^2}{4} + (y')^2 = 1$

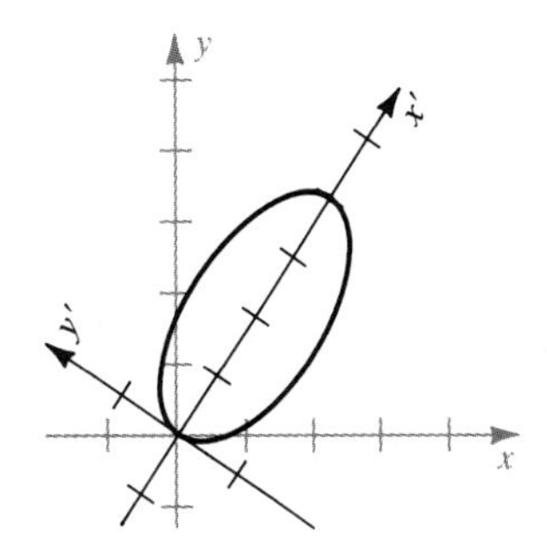

11 **(a)** 128, hyperbola **(b)** $(y' + 2)^2 - \frac{(x')^2}{1/2} = 1$

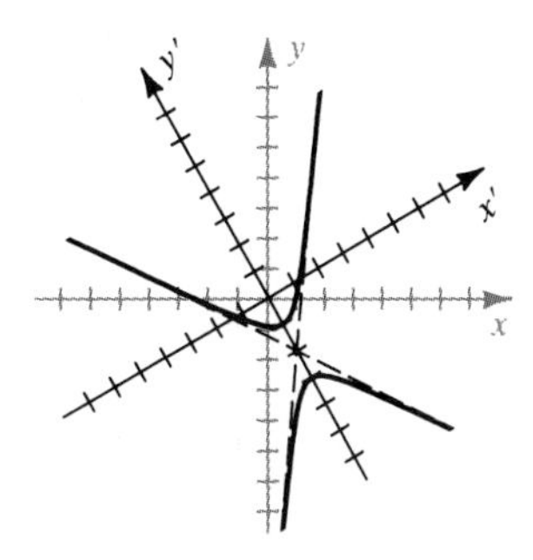

13 **(a)** -1600, ellipse **(b)** $\frac{(x')^2}{16} + (y')^2 = 1$

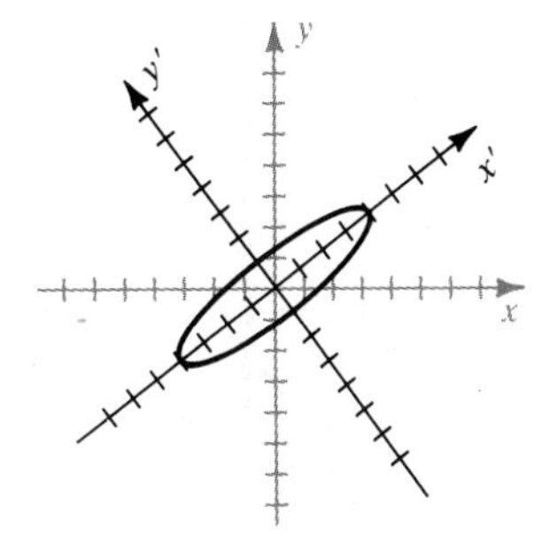

15

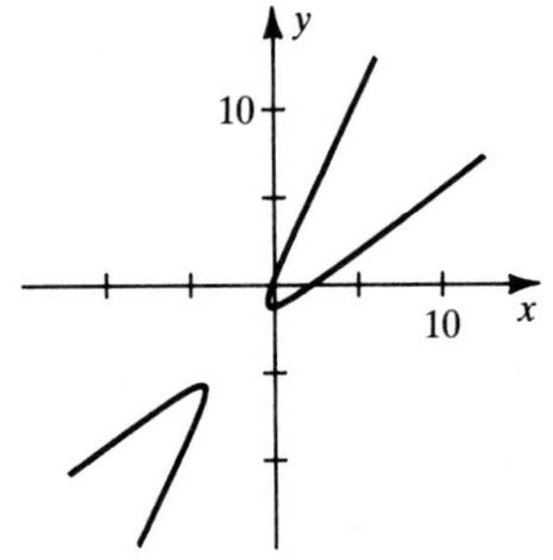

13 $V(2, -4)$; $F(4, -4)$

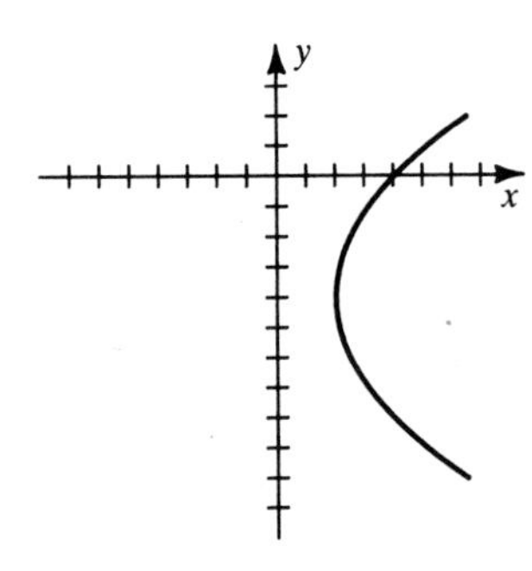

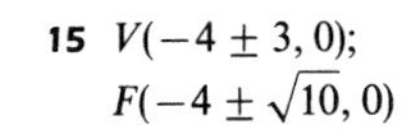
15 $V(-4 \pm 3, 0)$; $F(-4 \pm \sqrt{10}, 0)$

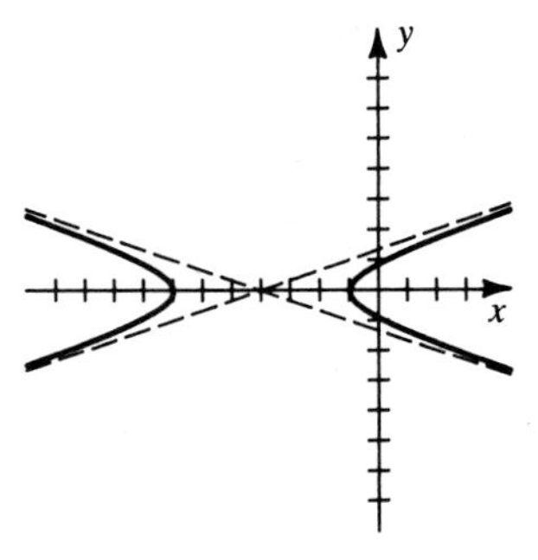

12.5 REVIEW EXERCISES

1 $V(0, 0)$; $F(16, 0)$

3 $V(0, \pm 4)$; $F(0, \pm\sqrt{7})$

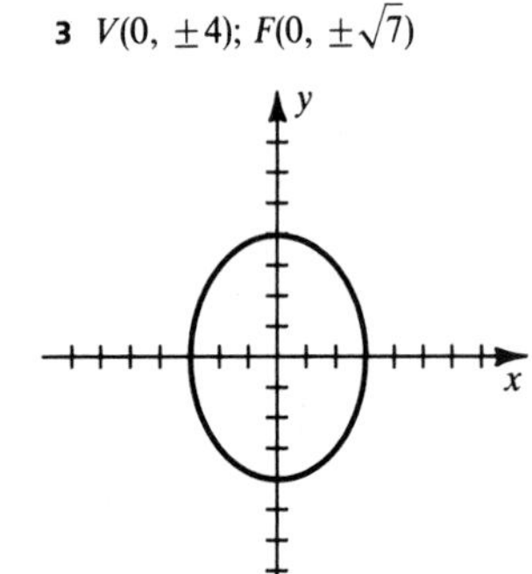

5 $V(\pm 2, 0)$; $F(\pm 2\sqrt{2}, 0)$

7 $V(0, 4)$; $F\left(0, -\frac{9}{4}\right)$

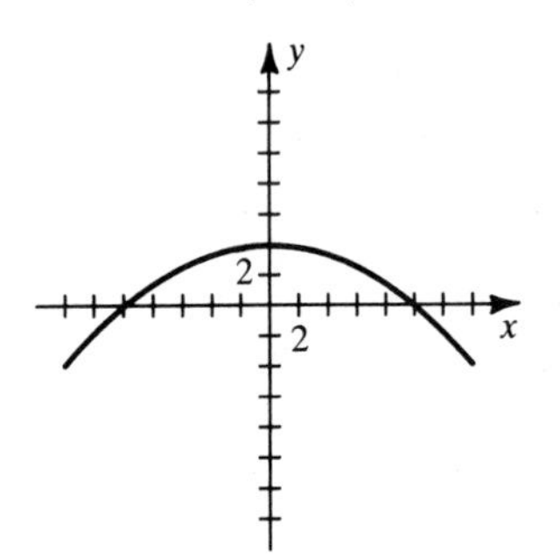

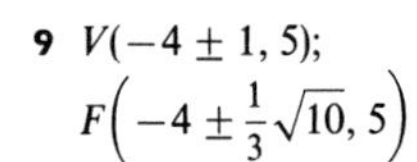
9 $V(-4 \pm 1, 5)$; $F\left(-4 \pm \frac{1}{3}\sqrt{10}, 5\right)$

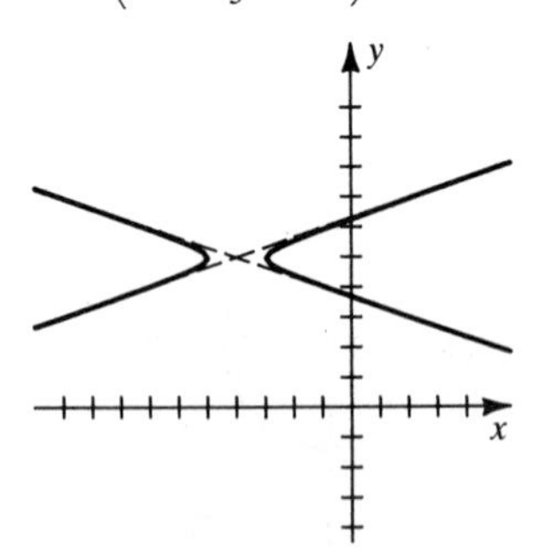

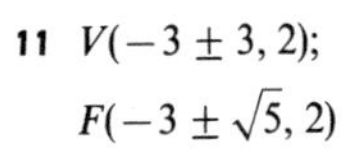
11 $V(-3 \pm 3, 2)$; $F(-3 \pm \sqrt{5}, 2)$

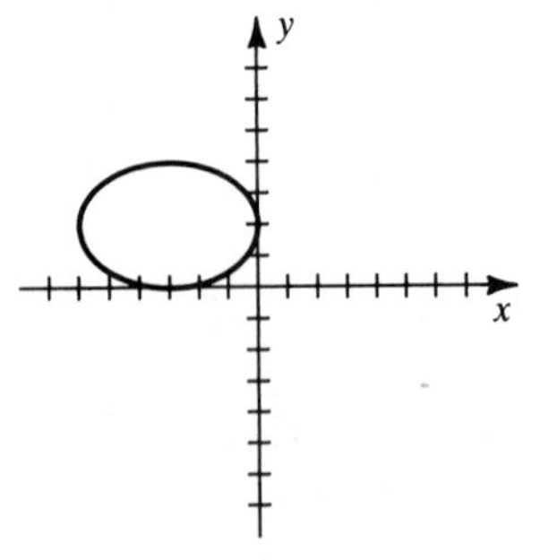

17 $\frac{y^2}{49} - \frac{x^2}{9} = 1$ **19** $x^2 = -40y$ **21** $\frac{x^2}{75} + \frac{y^2}{100} = 1$

23 $\frac{y^2}{36} - \frac{x^2}{4/9} = 1$ **25** $\frac{x^2}{25} + \frac{y^2}{45} = 1$

27 **(a)** $y = \sqrt{960\left(1 - \frac{x^2}{10{,}000}\right)}$ **(b)** $\sqrt{960} \approx 31$ ft

29 $y - 2 = -\frac{16}{15}(x + 3)$; $y - 2 = \frac{15}{16}(x + 3)$

33 $\frac{8}{3}p^2$ **35** $\frac{64\pi}{3}$

37 $\bar{x} = 0$; $\bar{y} = \frac{(2/3)b^3}{K}$, where $K = bc - a^2[\ln(c + b) - \ln a]$

39 $(y')^2 = 3x'$

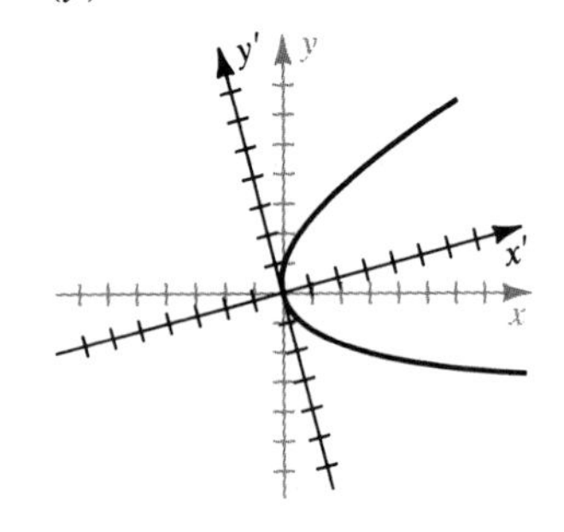

CHAPTER 13

EXERCISES 13.1

1 $y = 2x + 7$

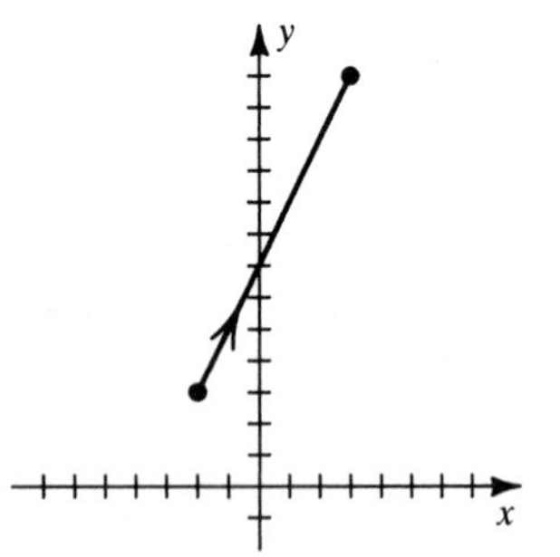

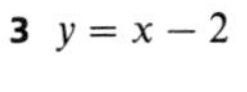
3 $y = x - 2$

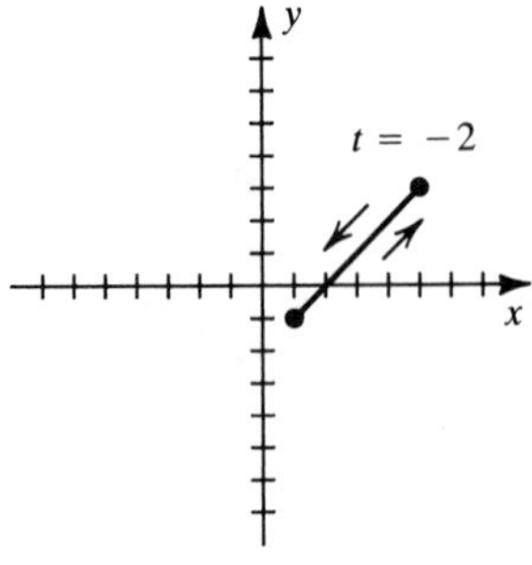

5 $(y-3)^2 = x+5$

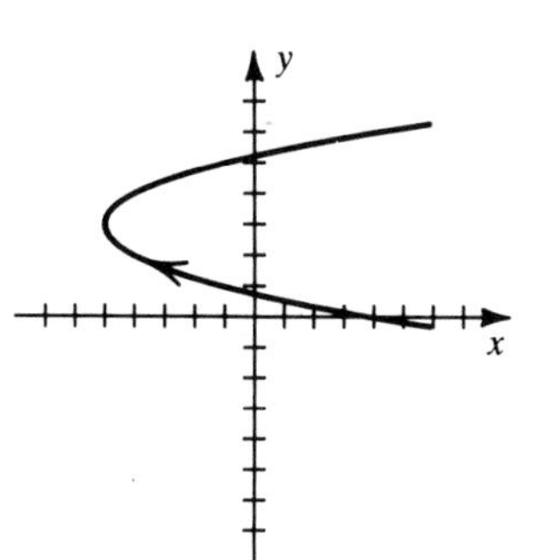

7 $y = \frac{1}{x^2}$

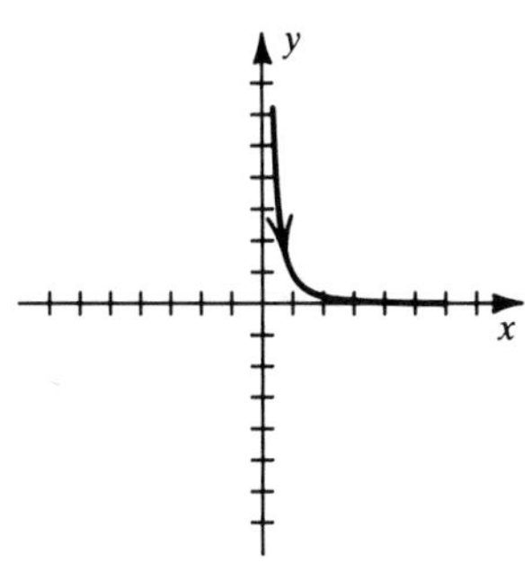

9 $\frac{x^2}{4} + \frac{y^2}{9} = 1$

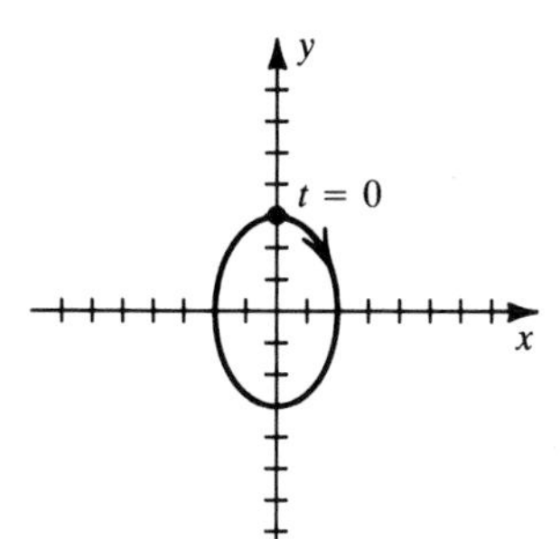

11 $x^2 - y^2 = 1$

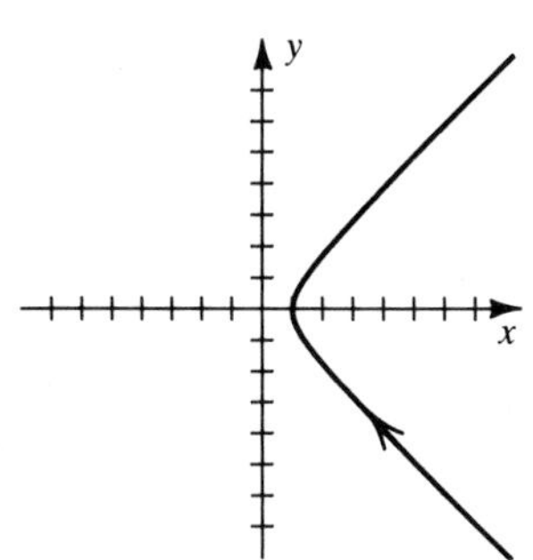

13 $y = \ln x$

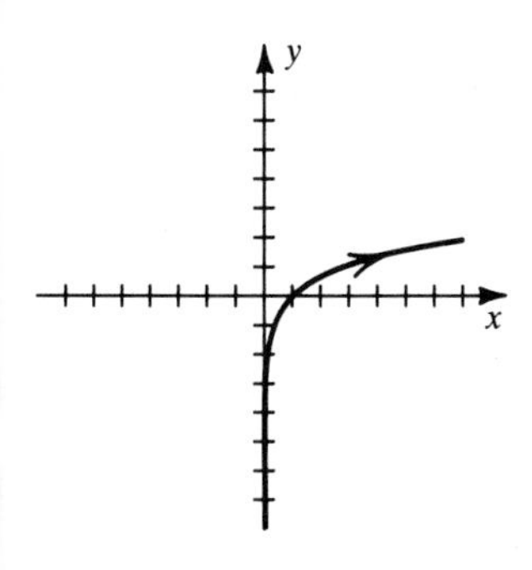

15 $y = \frac{1}{x}$

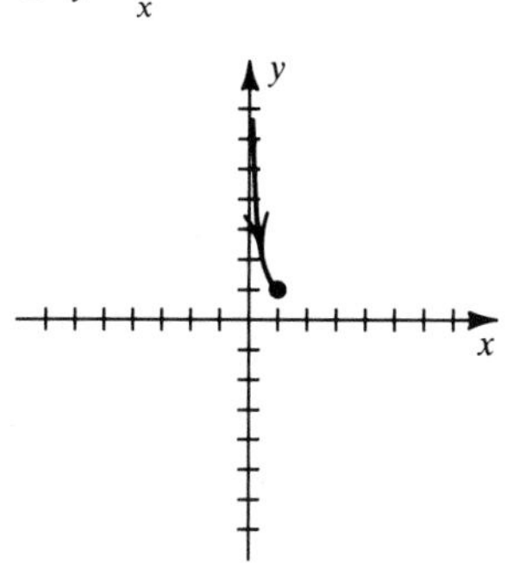

17 $x^2 - y^2 = 1$

19 $y = \sqrt{x^2 - 1}$

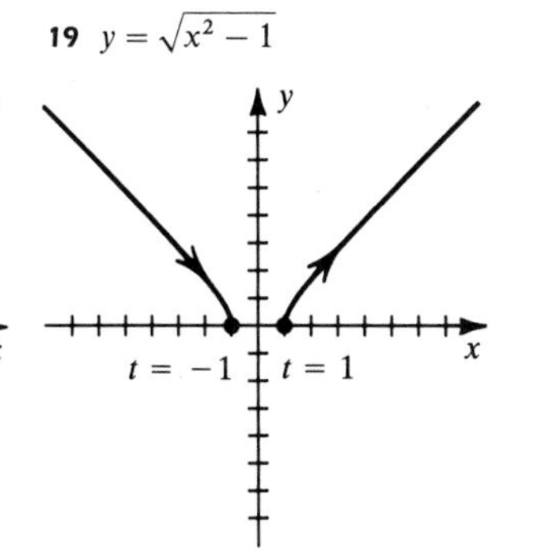

21 $y = |x-1|$

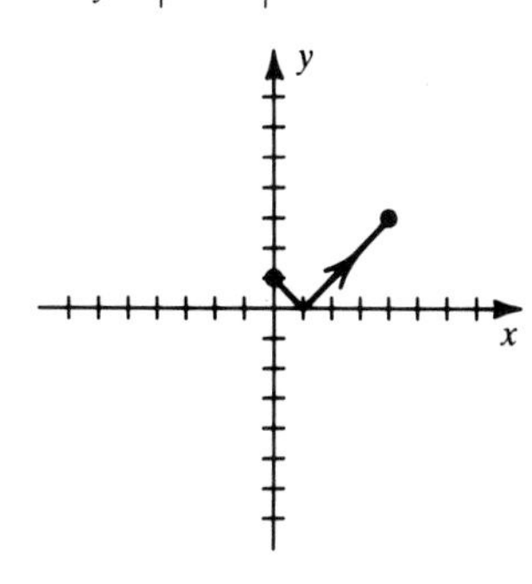

23 $y = (x^{1/3} + 1)^2$

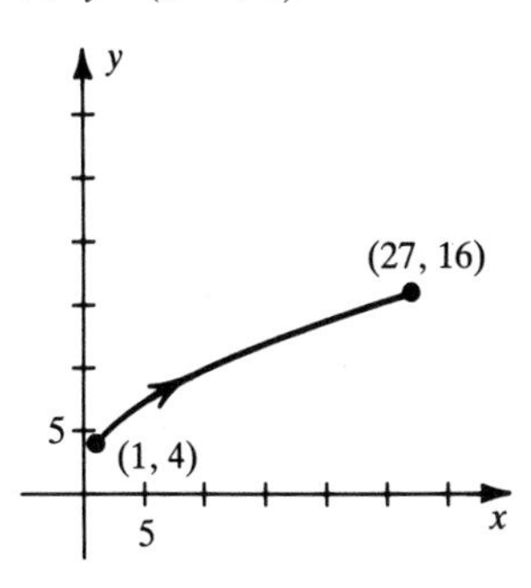

25 C_1

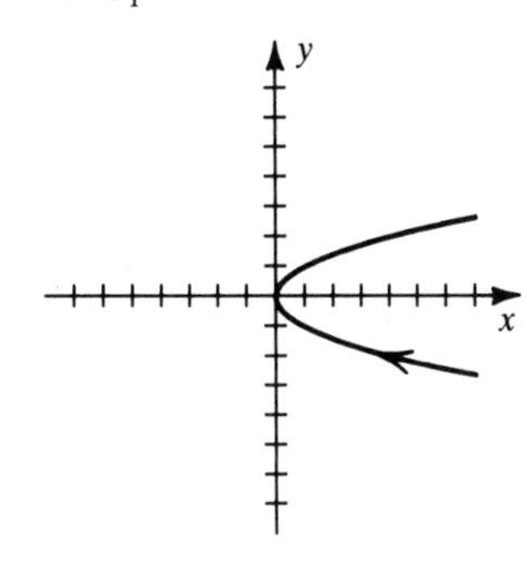

C_2

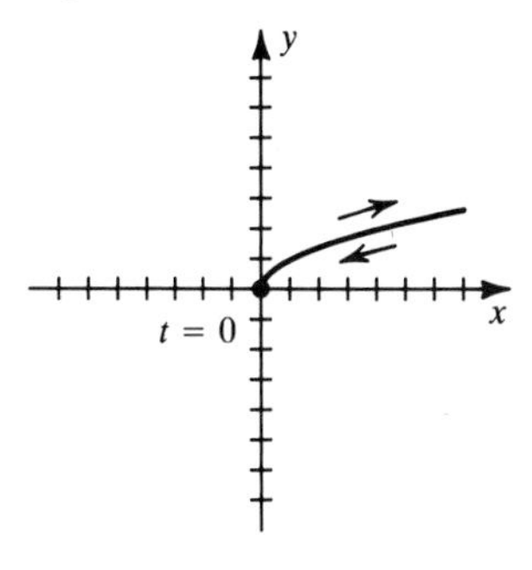

C_3

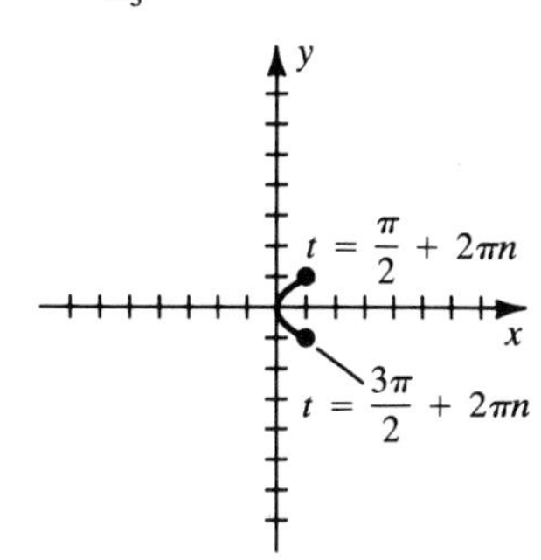

C_4

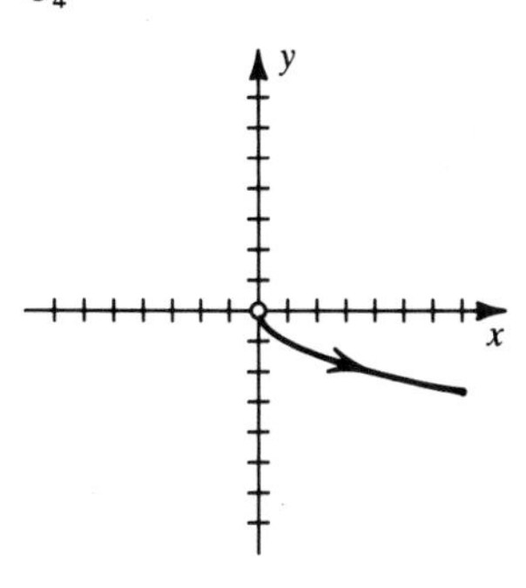

27 (a)

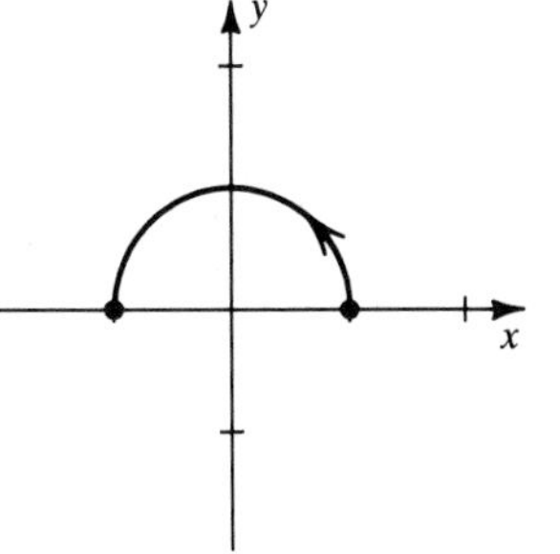

(b)

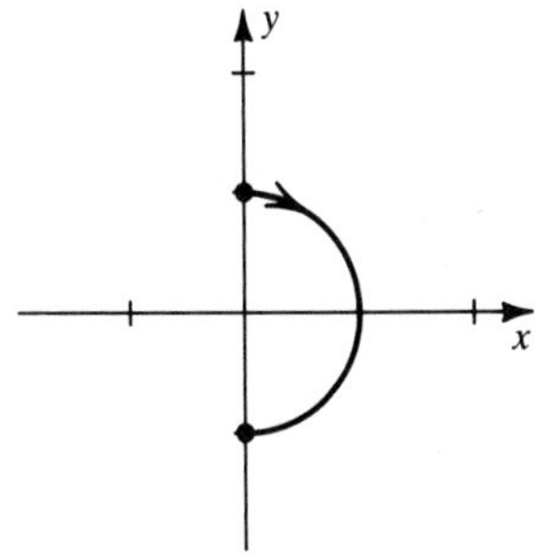

(c)

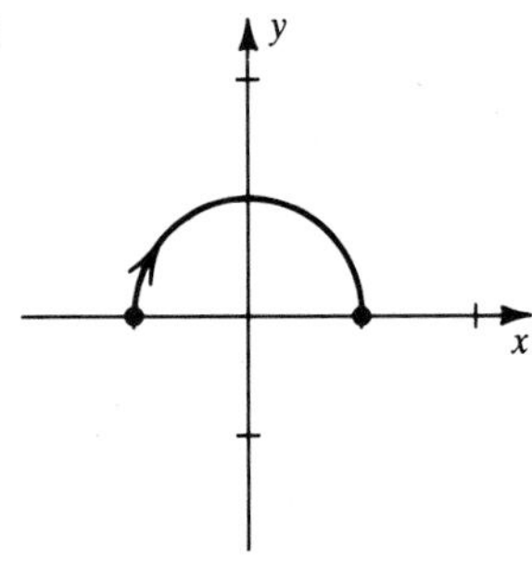

35 $x = 4b\cos t - b\cos 4t,\ y = 4b\sin t - b\sin 4t$

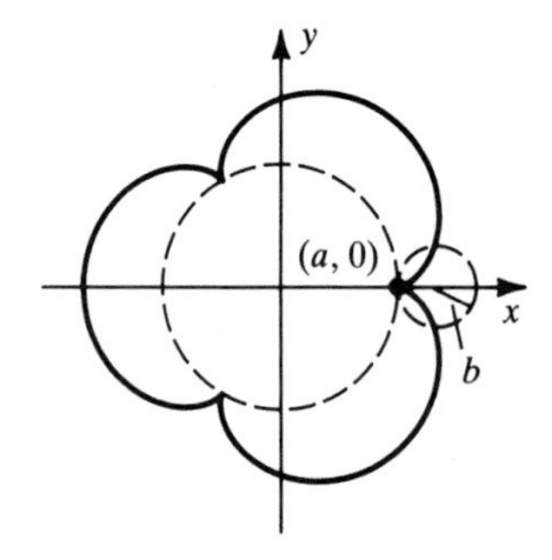

39 **(a)** The figure is an ellipse with center (0, 0) and axes of lengths $2a$ and $2b$.

41 **43**

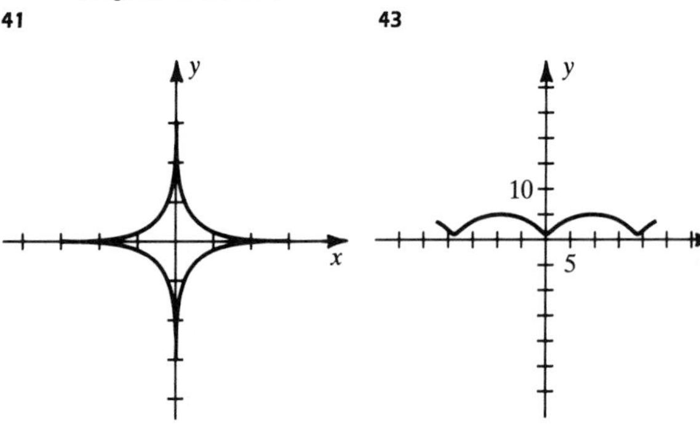

45 A mask with a mouth, nose, and eyes

47 The letter A

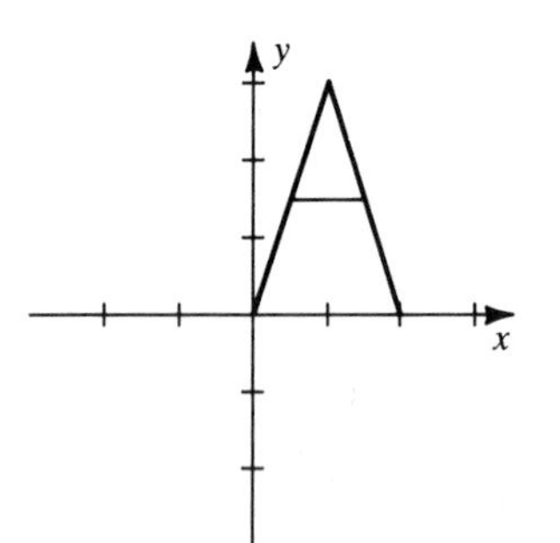

EXERCISES 13.2

1 1; -1 **3** $\frac{1}{4}$; -4 **5** $-\frac{2}{e^3}$; $\frac{1}{2}e^3$

7 $-\frac{3}{2}\tan 1 \approx -2.34$; $\frac{2}{3}\cot 1 \approx 0.43$ **9** $(-27, -108)$, $(1, 12)$

11 **(a)** Horizontal: $(16, \pm 16)$; vertical: $(0, 0)$ **(b)** $\frac{3t^2 + 12}{64t^3}$

(c)

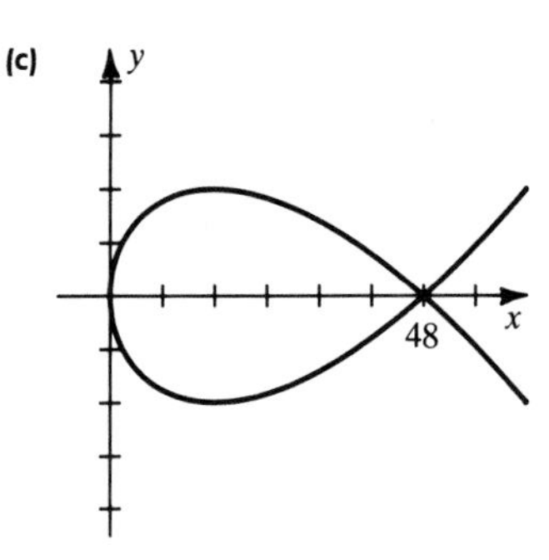

13 **(a)** Horizontal: $(2, -1)$; vertical: $(1, 0)$ **(b)** $\frac{-2t + 4}{9t^5}$

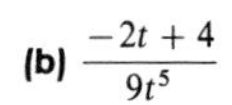

(c)

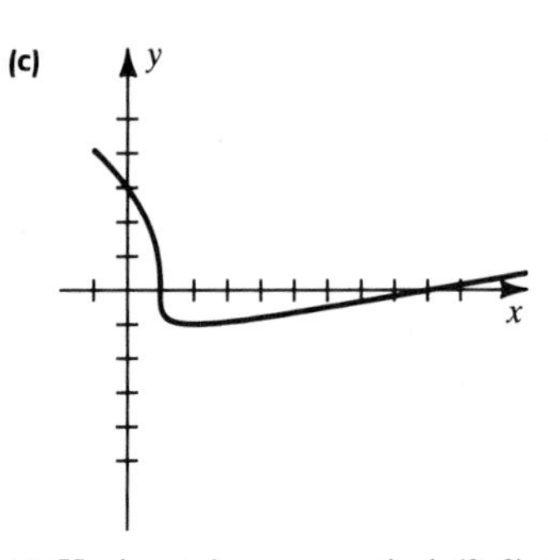

15 **(a)** Horizontal: none; vertical: $(0, 0)$, $(-3, 1)$

(b) $\frac{1 - 3t}{144t^{3/2}(t - 1)^3}$

(c)

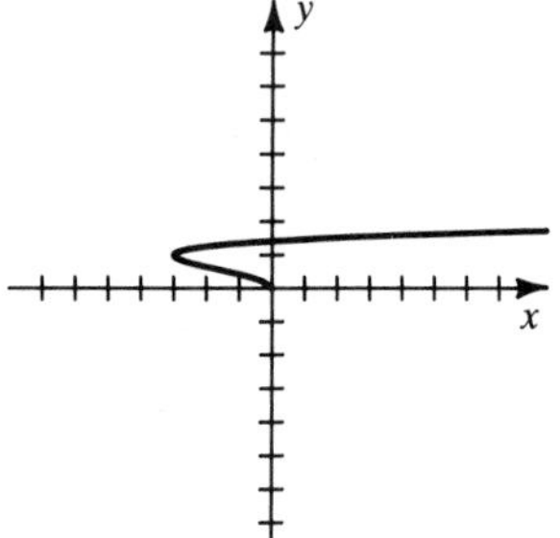

17 **(a)** Horizontal: $(\pm 1, 0)$; vertical: $(0, \pm 1)$ **(b)** $\frac{1}{3}\sec^4 t \csc t$

(c)

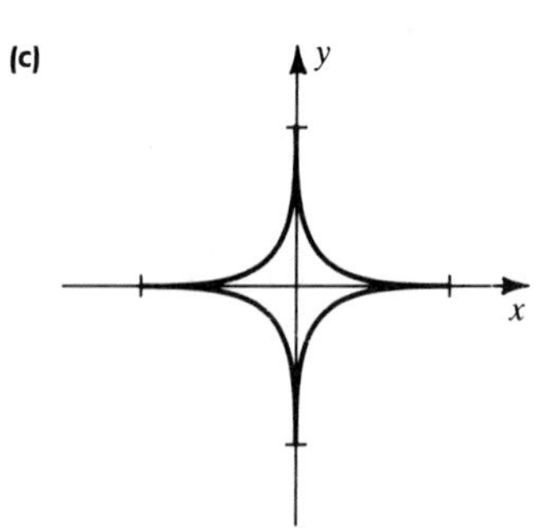

19 Horizontal: $(0, \pm 2)$, $(2\sqrt{3}, \pm 2)$, $(-2\sqrt{3}, \pm 2)$; vertical: $(4, \pm\sqrt{2})$, $(-4, \pm\sqrt{2})$

21 $\frac{2}{27}(34^{3/2} - 125) \approx 5.43$ **23** $\sqrt{2}(e^{\pi/2} - 1) \approx 5.39$

25 $\frac{1}{8}\pi^2 \approx 1.23$ **27** 15.9 **29** $\frac{8\pi}{3}(17^{3/2} - 1) \approx 578.83$

31 $\frac{11\pi}{9} \approx 3.84$ **33** $\frac{64\pi}{3} \approx 67.02$ **35** $\frac{536\pi}{5} \approx 336.78$

37 $\frac{2}{5}\sqrt{2}\pi(2e^{\pi} + 1) \approx 84.03$ **39** 2.2

EXERCISES 13.3

1  **3** 

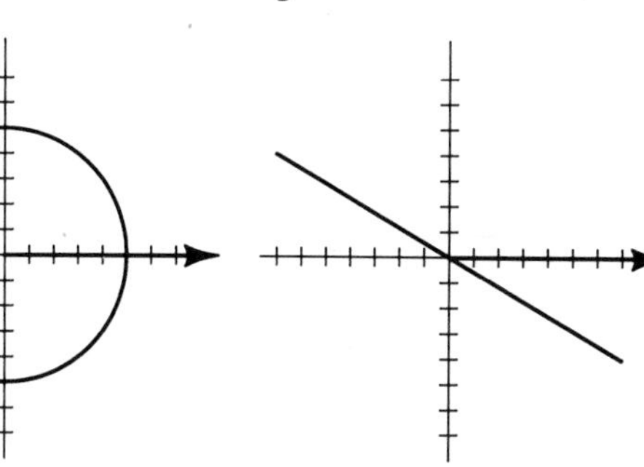

5 **7**

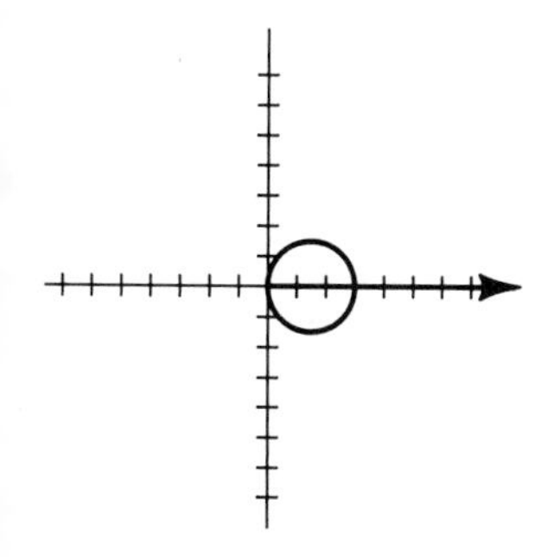

9 **11**

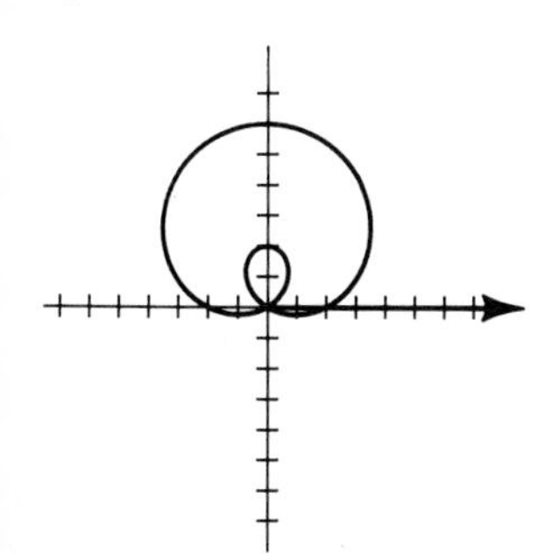

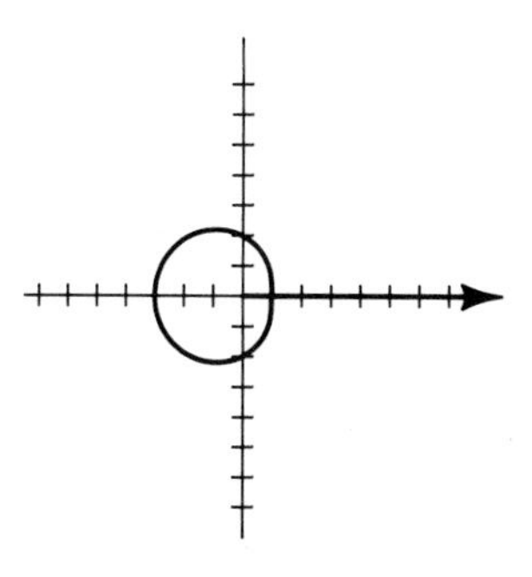

13 **15**

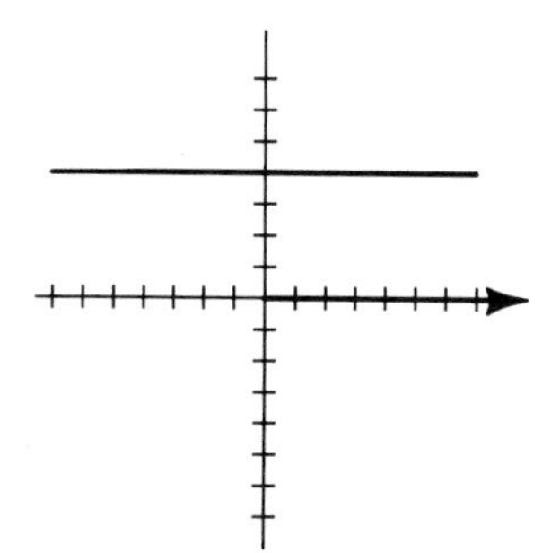

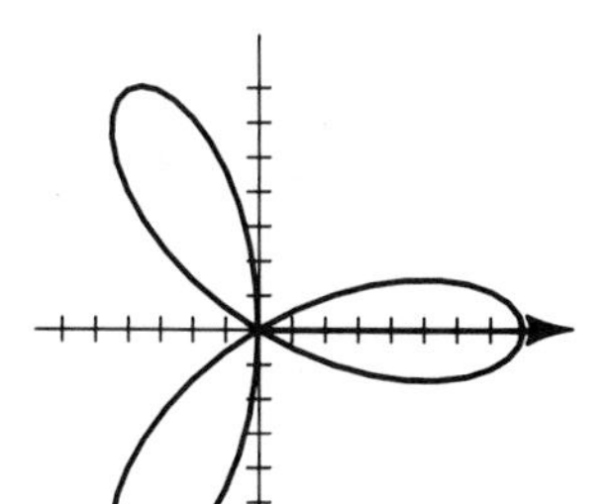

17 **19** 

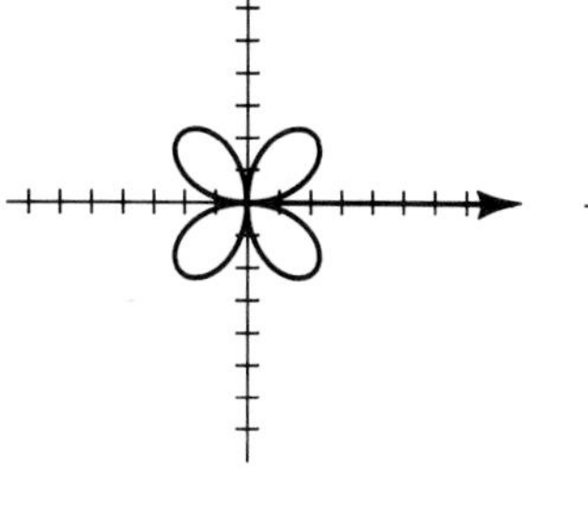

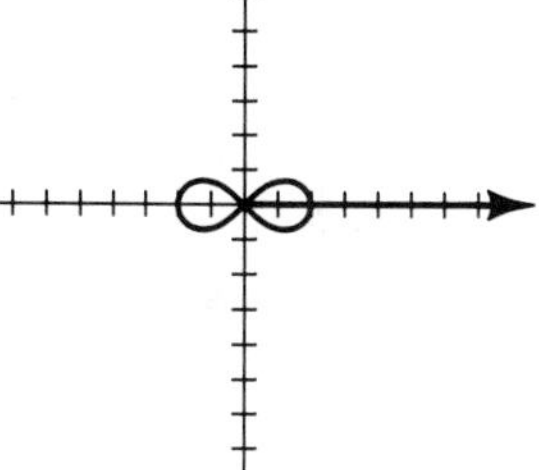

21  **23** 

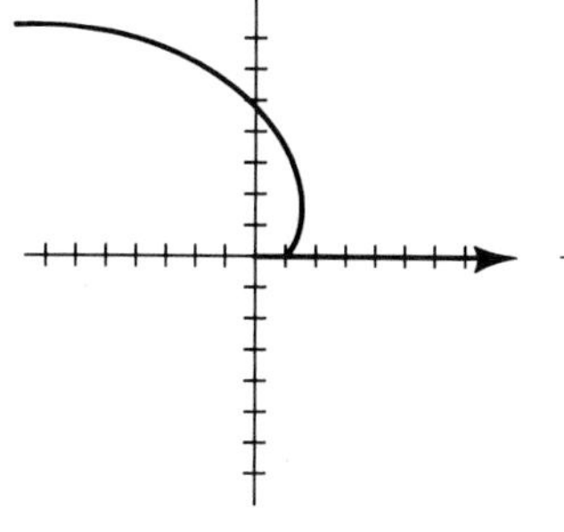

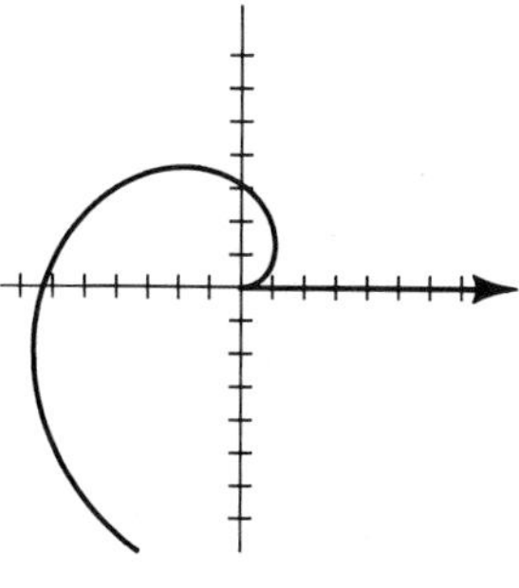

25

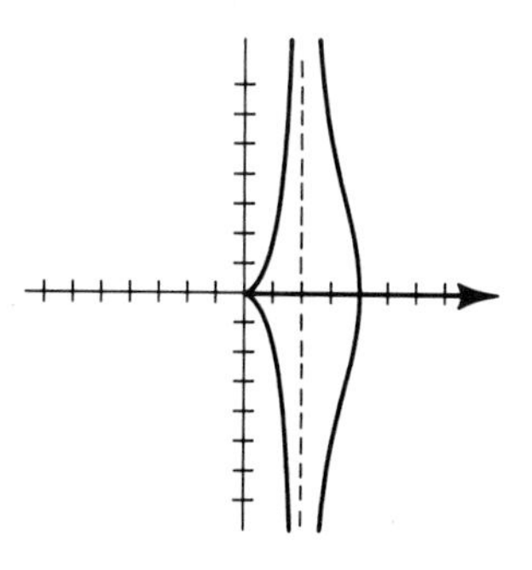

27 $r = -3 \sec\theta$ **29** $r = 4$ **31** $\theta = \tan^{-1}\left(-\frac{1}{2}\right)$

33 $r^2 = -4 \sec 2\theta$ **35** $r\theta = a \sin\theta$

37 $x = 5$

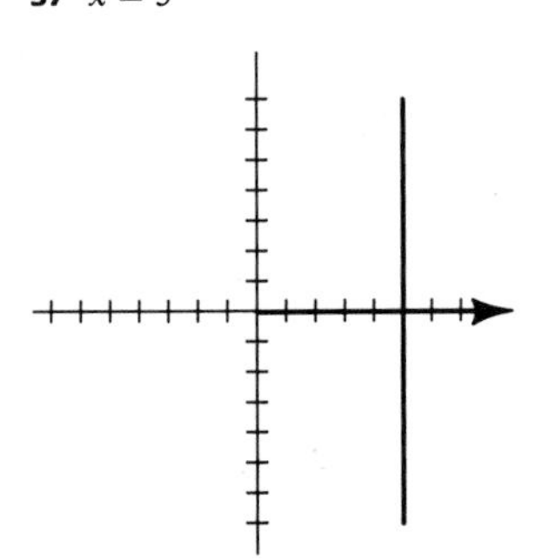

39 $y = -3$

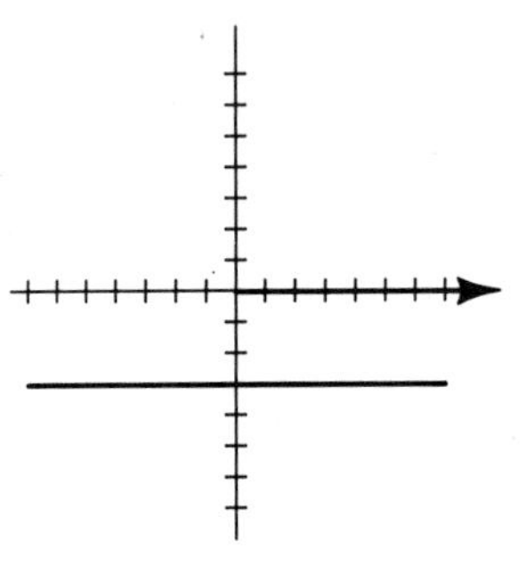

41 $x^2 - y^2 = 1$

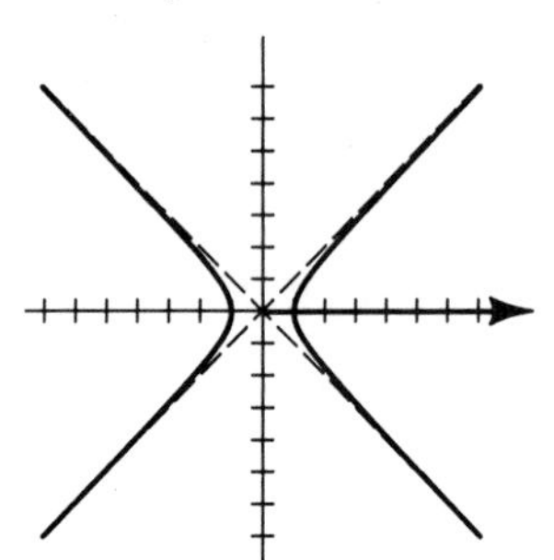

43 $y - 2x = 6$

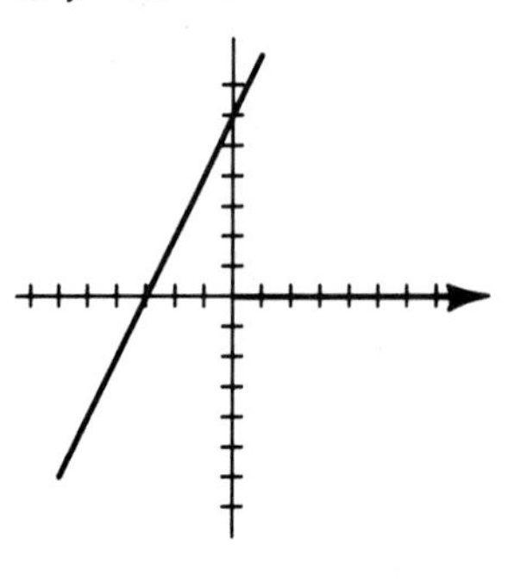

45 $y = -x^2 + 1$

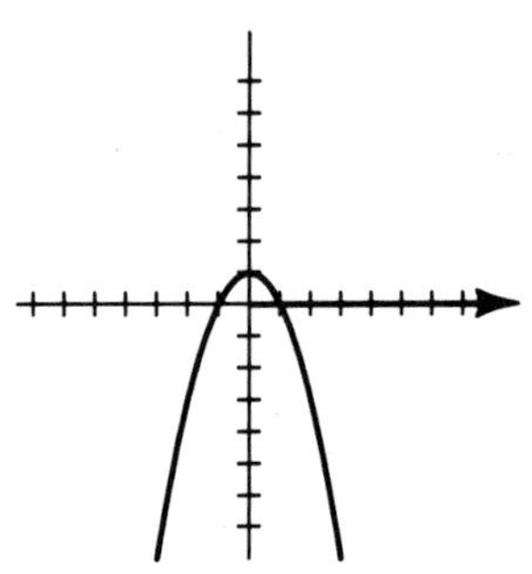

47 $(x + 1)^2 + (y - 4)^2 = 17$

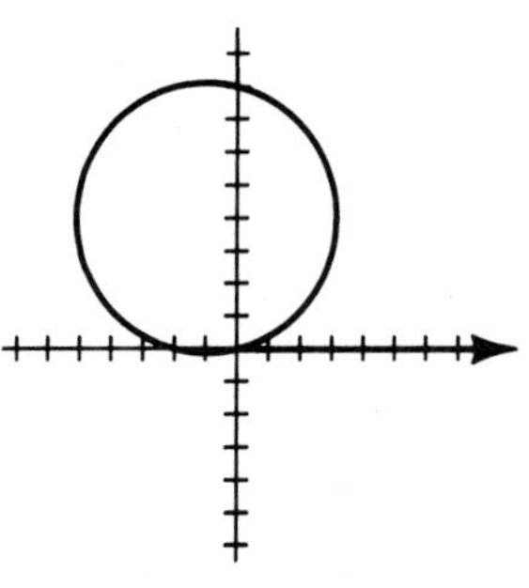

49 $y^2 = \dfrac{x^4}{1 - x^2}$

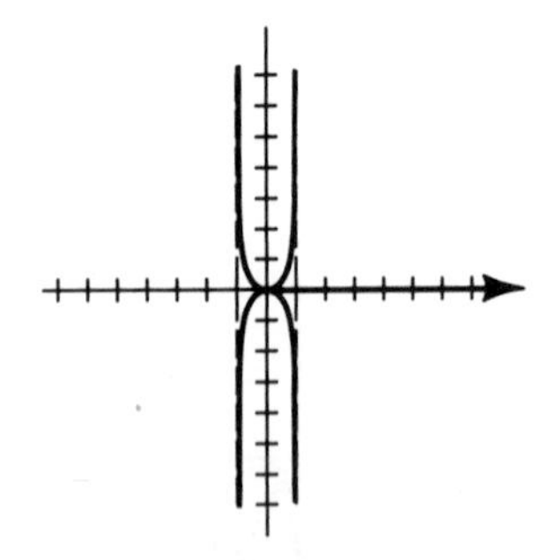

51 $\sqrt{3}/3$ **53** -1 **55** 2 **57** 0 **59** $\dfrac{1}{\ln 2}$

61 Let $P_1(r_1, \theta_1)$ and $P_2(r_2, \theta_2)$ be points in an $r\theta$-plane. Let $a = r_1$, $b = r_2$, $c = d(P_1, P_2)$, and $\gamma = \theta_2 - \theta_1$. Substituting into the law of cosines, $c^2 = a^2 + b^2 - 2ab\cos\gamma$, gives us the formula.

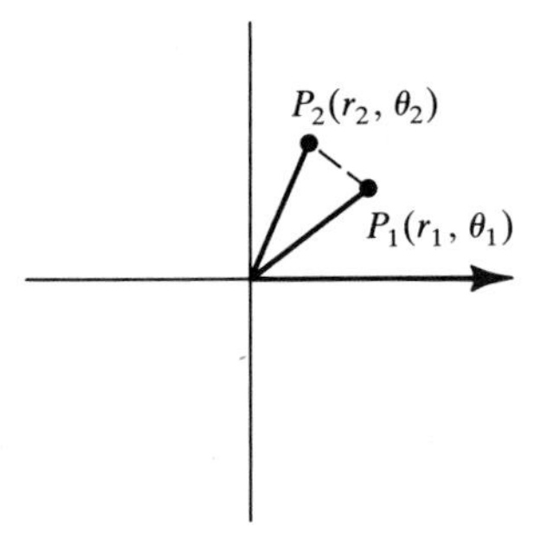

65 Use (13.10).

67 Symmetric with respect to the polar axis

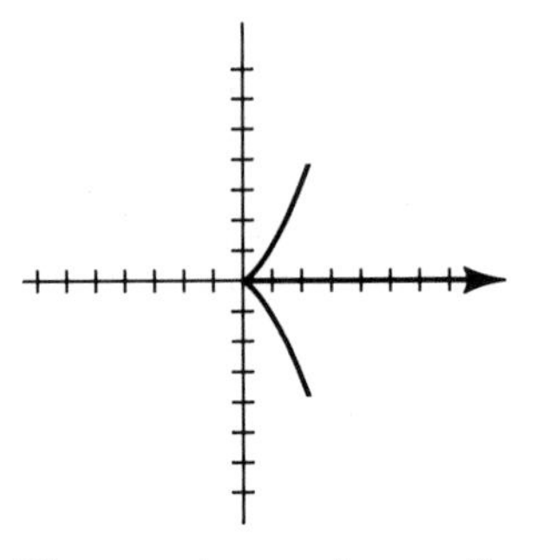

69 The approximate polar coordinates are $(1.75, \pm 0.45)$, $(4.49, \pm 1.77)$, and $(5.76, \pm 2.35)$.

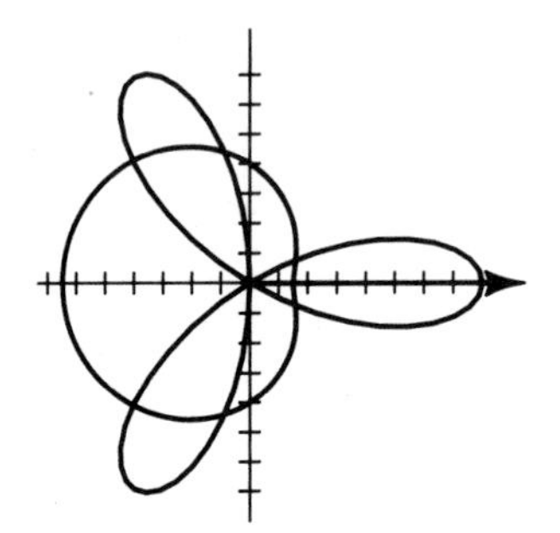

EXERCISES 13.4

1 π **3** $\dfrac{3\pi}{2}$ **5** $\dfrac{\pi}{2}$ **7** $\dfrac{1}{4}(e^{\pi} - 1) \approx 5.54$ **9** 2

11 $\dfrac{9\pi}{20}$ **13** $\displaystyle\int_0^{\arctan(3/4)} \frac{1}{2}(4\sec\theta)^2\,d\theta + \int_{\arctan(3/4)}^{\pi/2} \frac{1}{2}(5)^2\,d\theta$

15 $\displaystyle\int_{\pi/4}^{\arctan 3} \frac{1}{2}[(4\csc\theta)^2 - (2)^2]\,d\theta$

17 **(a)** $\displaystyle 8\int_0^{\pi/6} \frac{1}{2}[(4\cos 2\theta)^2 - (2)^2]\,d\theta$

(b) $\displaystyle 8\left[\int_0^{\pi/6} \frac{1}{2}(2)^2\,d\theta + \int_{\pi/6}^{\pi/4} \frac{1}{2}(4\cos 2\theta)^2\,d\theta\right]$

19 $2\pi + \dfrac{9}{2}\sqrt{3} \approx 14.08$ **21** $4\sqrt{3} - \dfrac{4\pi}{3} \approx 2.74$

23 $\frac{5\pi}{24} - \frac{1}{4}\sqrt{3} \approx 0.22$ **25** $\frac{3\pi}{4} + 11 \arcsin \frac{1}{4} - \frac{1}{4}\sqrt{15} \approx 4.17$

27 $\sqrt{2}(1 - e^{-2\pi}) \approx 1.41$ **29** 2 **31** $\frac{3\pi}{2}$ **33** 2.4

35 $\frac{128\pi}{5} \approx 80.42$ **37** $4\pi^2 a^2$ **39** 4.2 **41** $4\pi^2 ab$

43 $\frac{2}{5}\pi\sqrt{2}(2 + e^{-\pi}) \approx 3.63$

EXERCISES 13.5

1 $\frac{1}{3}$, ellipse **3** 3, hyperbola

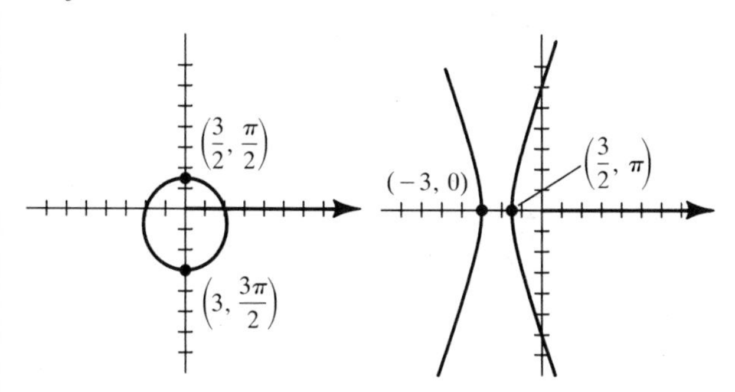

5 1, parabola **7** $\frac{1}{2}$, ellipse

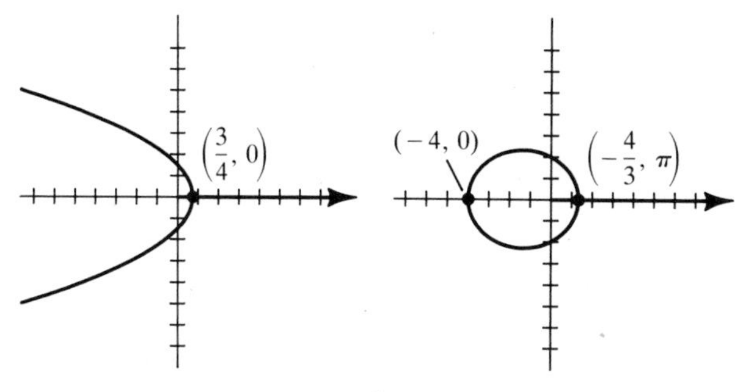

9 $\frac{3}{2}$, hyperbola

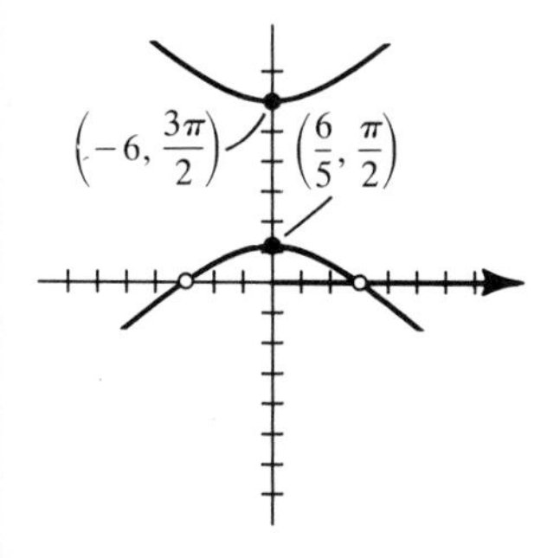

11 $9x^2 + 8y^2 + 12y - 36 = 0$ **13** $8x^2 - y^2 + 36x + 36 = 0$

15 $4y^2 + 12x - 9 = 0$ **17** $3x^2 + 4y^2 + 8x - 16 = 0$

19 $4x^2 - 5y^2 + 36y - 36 = 0$, with $(\pm 3, 0)$ excluded

21 $r = \frac{2}{3 + \cos \theta}$ **23** $r = \frac{12}{1 - 4 \sin \theta}$ **25** $r = \frac{5}{1 + \cos \theta}$

27 $r = \frac{8}{1 + \sin \theta}$ **29** $-\frac{3}{5}\sqrt{3}$ **31** 3

33 $\int_{\pi/6}^{\pi/3} \frac{1}{2}(2 \sec \theta)^2 \, d\theta$ **35** $\int_{\pi/4}^{5\pi/4} \frac{1}{2}\left(\frac{4}{1 - \cos \theta}\right)^2 d\theta$

39 **(b)** $a\sqrt{1 - e^2}$

13.6 REVIEW EXERCISES

1 **(a)** $y = \frac{2x^2 - 4x + 1}{x - 1}$ **(b)**

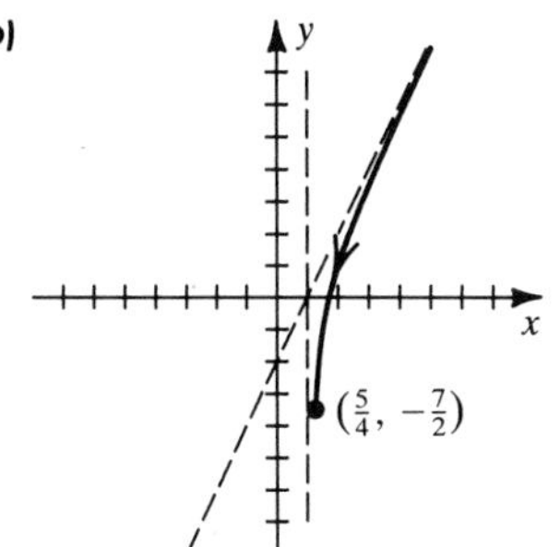

3 **(a)** $y = 2^{-x^2}$ **(b)**

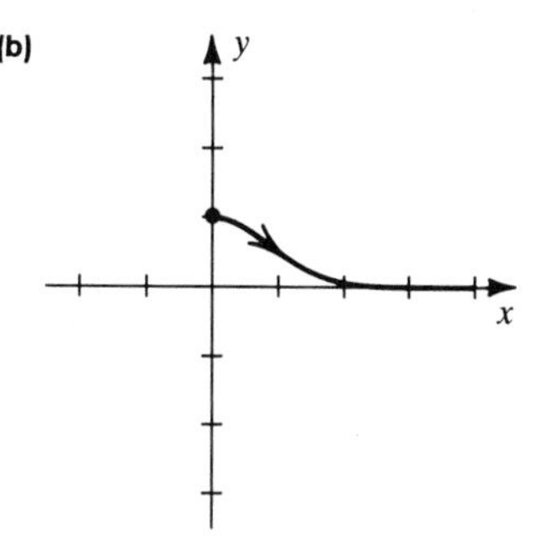

5 C_1 C_2

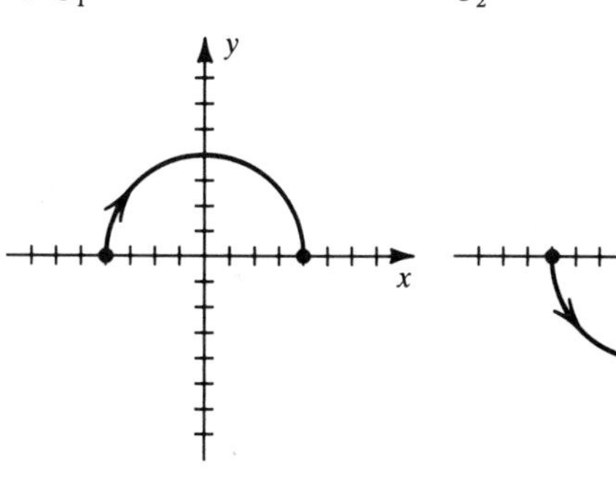

C_3 C_4

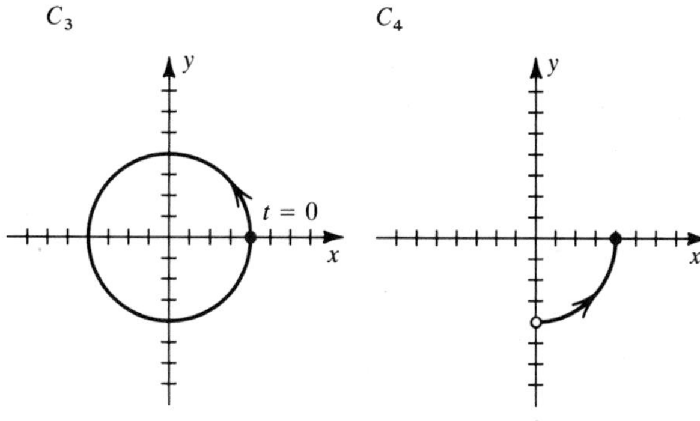

7 **(a)** $\frac{3t^2 + 2}{t}$ **(b)** Horizontal: none; vertical: 0 **(c)** $\frac{3t^2 - 2}{2t^3}$

9

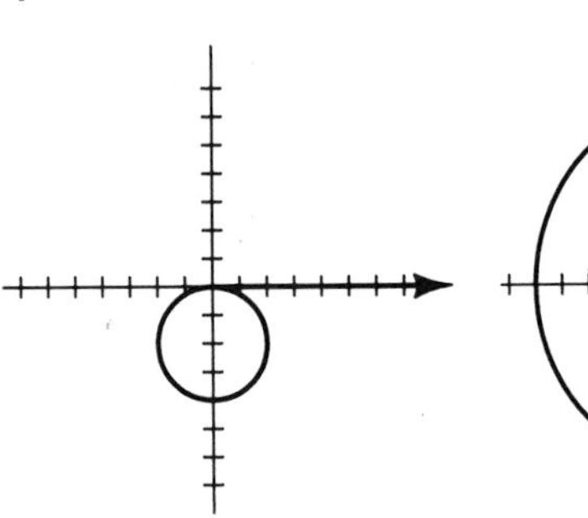

11

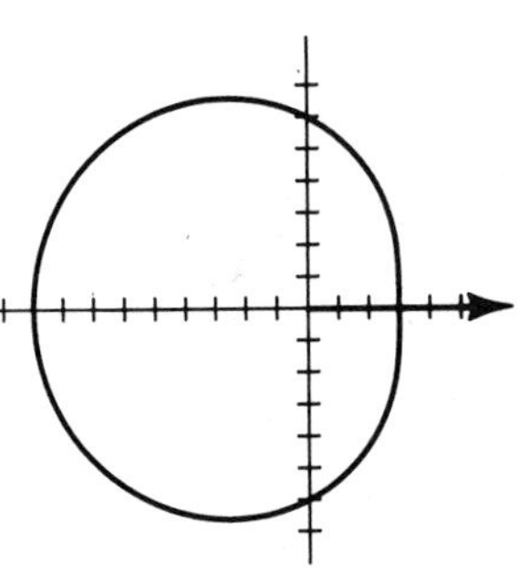

13

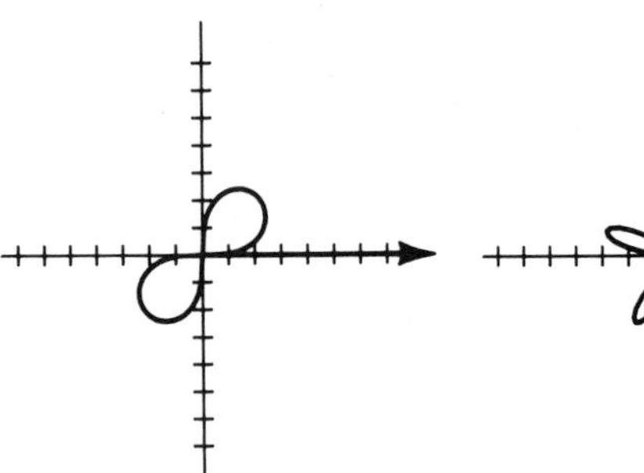

15

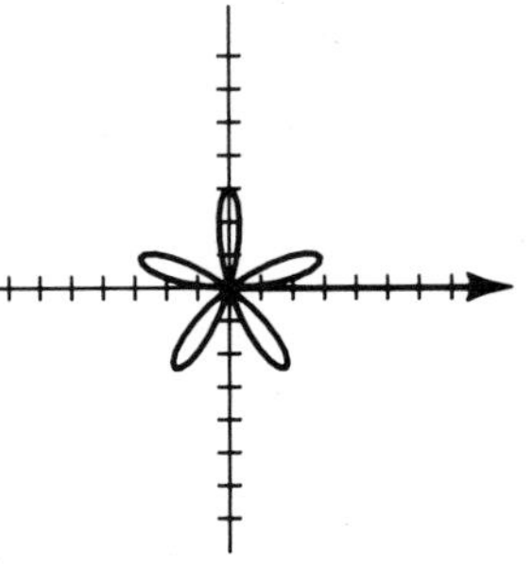

17 **19**

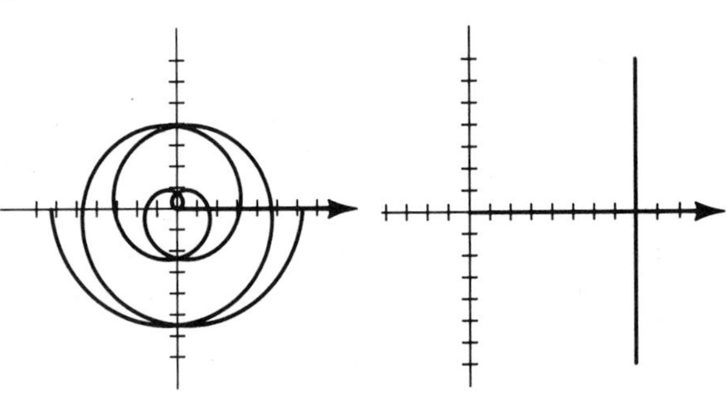

21 **23**

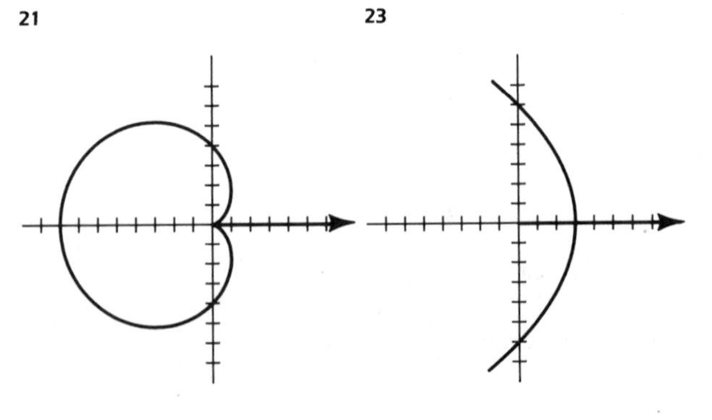

25

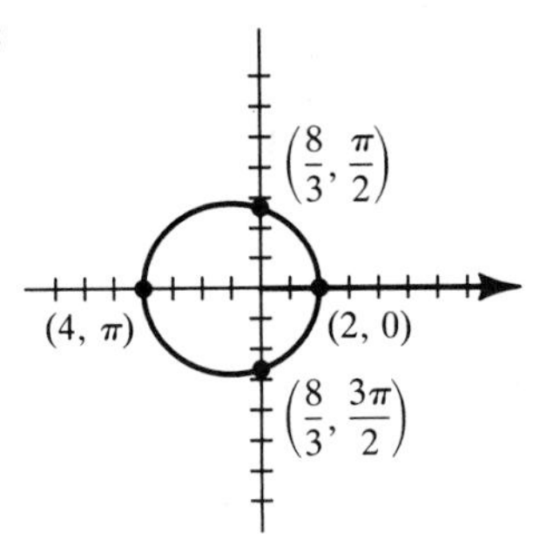

27 $r = 4 \cot \theta \csc \theta$ **29** $r(2 \cos \theta - 3 \sin \theta) = 8$
31 $r = 2 \cos \theta \sec 2\theta$ **33** $x^3 + xy^2 = y$
35 $(x^2 + y^2)^2 = 8xy$ **37** $y = (\tan \sqrt{3})x$ **39** -1 **41** 2
43 $\sqrt{2} + \ln(1 + \sqrt{2}) \approx 2.30$
45 $2\pi[5\sqrt{2} + \ln(1 + \sqrt{2})] \approx 49.97$ **47** $2\pi a^2(2 - \sqrt{2}) \approx 3.68a^2$

CHAPTER 14

EXERCISES 14.1

1 $\sqrt{29}$

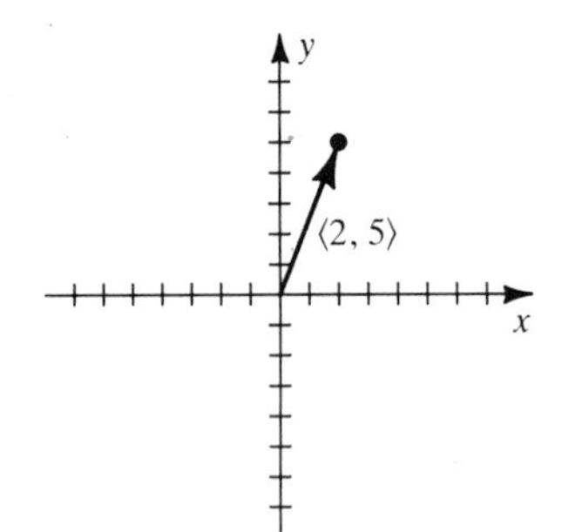

3 5

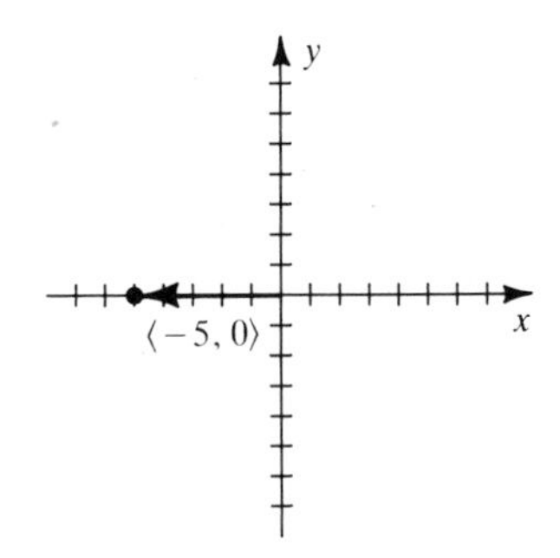

5 $\sqrt{41}$

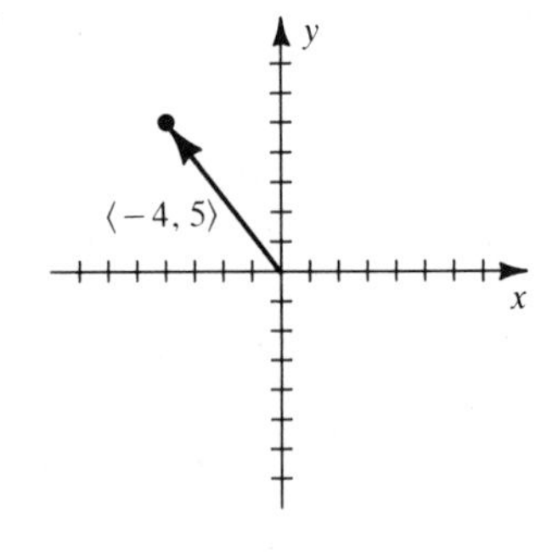

7 $\langle 3, 1\rangle, \langle 1, -7\rangle, \langle 4, -6\rangle, \langle -3, -12\rangle, \langle 3, -32\rangle$

9 $\langle -15, 6\rangle, \langle 1, -2\rangle, \langle -14, 4\rangle, \langle 24, -12\rangle, \langle 12, -12\rangle$

11 $2\mathbf{i} + 7\mathbf{j}, 4\mathbf{i} - 3\mathbf{j}, 6\mathbf{i} + 4\mathbf{j}, 3\mathbf{i} - 15\mathbf{j}, 17\mathbf{i} - 17\mathbf{j}$

13 $-\mathbf{b}$ **15** $\mathbf{f}$ **17** $-\frac{1}{2}\mathbf{e}$

19 $\langle 4, 7\rangle$

(4, 7)
Q(5, 3)
P(1, −4)

21 $\langle -6, 0\rangle$ **23** $\langle 9, -3\rangle$

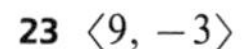

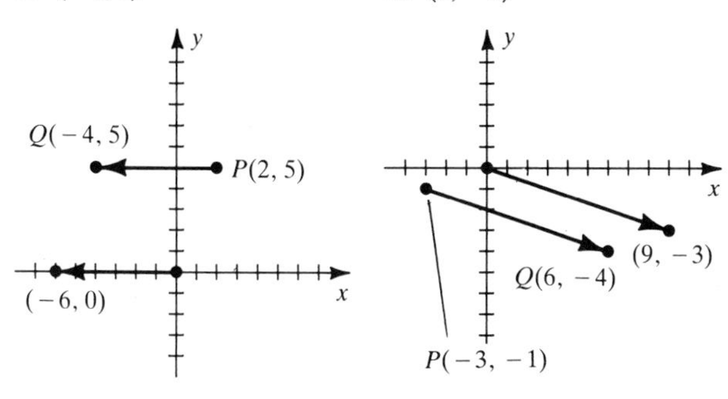

25 **(a)** $-\frac{8}{17}\mathbf{i} + \frac{15}{17}\mathbf{j}$ **(b)** $\frac{8}{17}\mathbf{i} - \frac{15}{17}\mathbf{j}$

27 **(a)** $\left\langle \frac{2}{\sqrt{29}}, -\frac{5}{\sqrt{29}} \right\rangle$ **(b)** $\left\langle -\frac{2}{\sqrt{29}}, \frac{5}{\sqrt{29}} \right\rangle$

29 **(a)** $\langle -12, 6\rangle$ **(b)** $\left\langle -3, \frac{3}{2} \right\rangle$ **31** $\frac{24}{\sqrt{65}}\mathbf{i} - \frac{42}{\sqrt{65}}\mathbf{j}$

33 **(a)** $\pm\frac{3}{5}$ **(b)** None **(c)** 0 **35** 40.96; 28.68

37 −6.180; 19.021 **39** 301.5 mi/hr; 144.3°

41 $\sin^{-1}(0.4) \approx 23.6°$ **43** $p = 2$; $q = -3$

45 If **a** and **b** have the same direction

47 The circle with center (x_0, y_0) and radius c

EXERCISES 14.2

1 **(a)** $\sqrt{104}$ **(b)** $(3, 1, -1)$ **(c)** $\langle 2, -6, 8\rangle$

3 **(a)** $\sqrt{53}$ **(b)** $\left(-\frac{1}{2}, -1, 1\right)$ **(c)** $\langle 7, -2, 0\rangle$

5 **(a)** $\sqrt{3}$ **(b)** $\left(\frac{1}{2}, \frac{1}{2}, \frac{1}{2}\right)$ **(c)** $\langle -1, 1, 1\rangle$

7 **(a)** $\langle 1, 3, 0\rangle$ **(b)** $\langle -5, 9, 2\rangle$ **(c)** $\langle -22, 42, 9\rangle$
(d) $\sqrt{41}$ **(e)** $3\sqrt{41}$

9 **(a)** $4\mathbf{i} - 2\mathbf{j} - 3\mathbf{k}$ **(b)** $2\mathbf{i} - 6\mathbf{j} - 7\mathbf{k}$ **(c)** $11\mathbf{i} - 28\mathbf{j} - 30\mathbf{k}$
(d) $\sqrt{29}$ **(e)** $3\sqrt{29}$

11 **(a)** $\mathbf{i} + \mathbf{k}$ **(b)** $\mathbf{i} + 2\mathbf{j} - \mathbf{k}$ **(c)** $5\mathbf{i} + 9\mathbf{j} - 4\mathbf{k}$ **(d)** $\sqrt{2}$
(e) $3\sqrt{2}$

13

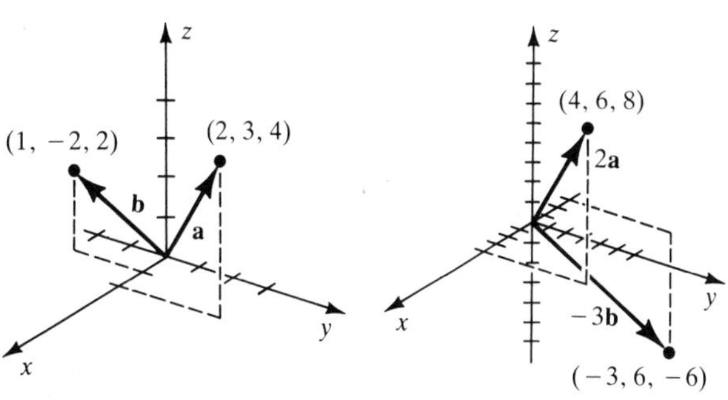

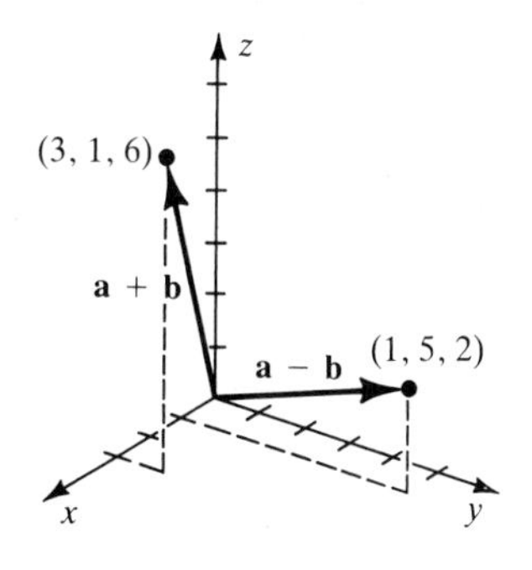

15 $\frac{1}{\sqrt{30}}\langle -2, 5, -1\rangle$

17 **(a)** $28\mathbf{i} - 30\mathbf{j} + 12\mathbf{k}$ **(b)** $-\frac{14}{3}\mathbf{i} + 5\mathbf{j} - 2\mathbf{k}$

(c) $\frac{2}{\sqrt{457}}(14\mathbf{i} - 15\mathbf{j} + 6\mathbf{k})$

19 $(x - 3)^2 + (y + 1)^2 + (z - 2)^2 = 9$

21 $(x + 5)^2 + y^2 + (z - 1)^2 = \frac{1}{4}$

23 **(a)** $(x + 2)^2 + (y - 4)^2 + (z + 6)^2 = 36$
(b) $(x + 2)^2 + (y - 4)^2 + (z + 6)^2 = 16$
(c) $(x + 2)^2 + (y - 4)^2 + (z + 6)^2 = 4$

25 $(x + 3)^2 + \left(y - \frac{5}{2}\right)^2 + z^2 = \frac{89}{4}$

27 $(x - 1)^2 + (y - 1)^2 + (z - 1)^2 = 1$ **29** $(-2, 1, -1)$; 2

31 $(4, 0, -4)$; 4 **33** $(0, -2, 0)$; 2

35 All points inside or on the sphere of radius 1 with center at the origin

37 All points inside or on a rectangular box with center at the origin and having edges of lengths 2, 4, and 6 in the x, y, and z directions, respectively

39 All points inside or on a cylindrical region of radius 5 and altitude 6 with center at the origin and axis along the z-axis

41 All points not on a coordinate plane

43 *Hint:* P is $\left(\frac{x_1 + x_2 + x_3}{4}, \frac{y_1 + y_2 + y_3}{4}, \frac{z_4}{4}\right)$.

EXERCISES 14.3

1 3 **3** **(a)** −12 **(b)** −12 **5** −99 **7** $-\frac{3}{\sqrt{30}}$

9 0 **11** $\arccos\frac{-3}{\sqrt{534}} \approx 97.5°$ **13** $\arccos\frac{6}{13} \approx 62.5°$

15 *Hint:* Use Theorem (14.21). **17** $-3, 5$ **19** 74

21 $\arccos \dfrac{37}{\sqrt{3081}} \approx 48.2°$ **23** $\dfrac{-82}{\sqrt{126}}$ **25** 1

27 $-4\sqrt{3} \approx 6.93$ ft-lb **29** $1000\sqrt{3} \approx 1732$ ft-lb

35 (a) *Hint:* $a_1 = \mathbf{a} \cdot \mathbf{i} = \|\mathbf{a}\| \|\mathbf{i}\| \cos \alpha$.

47 When **a** and **b** have the same or opposite direction

EXERCISES 14.4

1 $\langle 5, 10, 5 \rangle$ **3** $\langle -4, 2, -1 \rangle$ **5** $-6\mathbf{i} - 8\mathbf{j} + 18\mathbf{k}$

7 $0\mathbf{i} + 0\mathbf{j} + 0\mathbf{k} = \mathbf{0}$ **9** $0\mathbf{i} + 0\mathbf{j} + 0\mathbf{k} = \mathbf{0}$

11 *Hint:* Use Corollary (14.31).

13 $\langle 12, -14, 24 \rangle$; $\langle 16, -2, -5 \rangle$

Exer. 15–18: c is a nonzero scalar.

15 (a) $c\langle 13, 7, 5 \rangle$ **(b)** $\dfrac{9}{2}\sqrt{3}$

17 (a) $c\langle -10, -8, -20 \rangle$ **(b)** $\sqrt{141}$ **19** $\sqrt{\dfrac{282}{11}} \approx 5.06$

23 4

EXERCISES 14.5

In answers it is assumed that the domain of each parameter is $\mathbb{R}$.

1 $x = 4 + \frac{1}{3}t,\ y = 2 + 2t,\ z = -3 + \frac{1}{2}t$

3 $x = 0,\ y = t,\ z = 0$

5 $x = 5 - 3t,\ y = -2 + 8t,\ z = 4 - 3t$; $\left(1, \frac{26}{3}, 0\right), \left(\frac{17}{4}, 0, \frac{13}{4}\right), \left(0, \frac{34}{3}, -1\right)$

7 $x = 2 - 8t,\ y = 0,\ z = 5 - 2t$; $(-18, 0, 0)$, lies in xz-plane, $\left(0, 0, \frac{9}{2}\right)$

9 $x = -6 - 3s,\ y = 4 + s,\ z = -3 + 9s$ **11** $(5, -7, 3)$

13 Do not intersect

15 $\theta = \cos^{-1}\left(\dfrac{15}{\sqrt{38}\sqrt{18}}\right) \approx 55°$ and $180° - \theta$

17 $\theta = \cos^{-1}\left(\dfrac{71}{\sqrt{82}\sqrt{249}}\right) \approx 60°$ and $180° - \theta$

19 (a) $z = 4$ **(b)** $x = 6$ **(c)** $y = -7$

21 $6x - 5y - z = -84$ **23** $3x - y + 2z = -11$

25 $x + 42y - 5z = 8$ **27** $2x + y - 2z = -3$

29 (a)

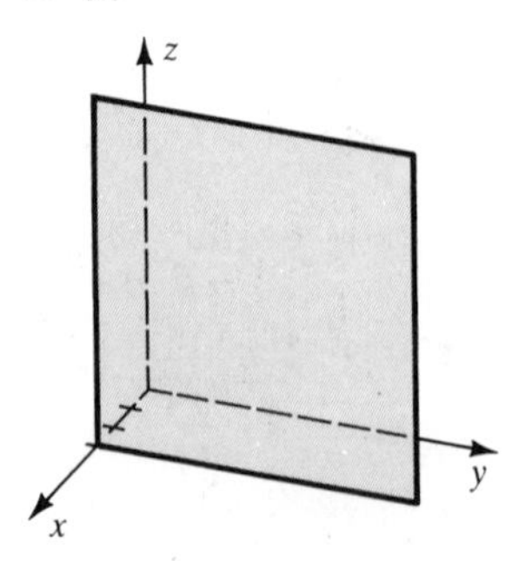

(b)

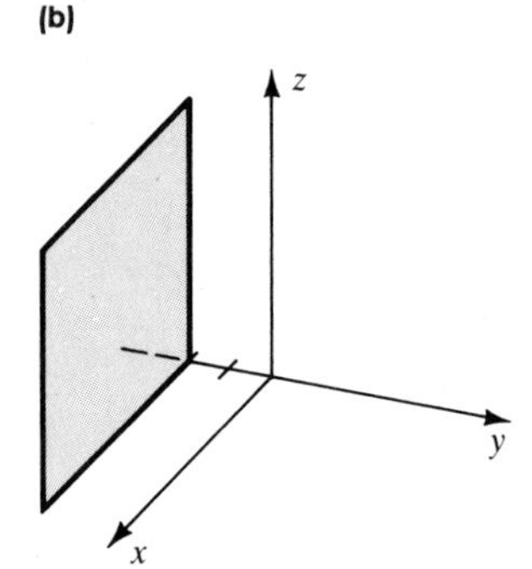

(c)

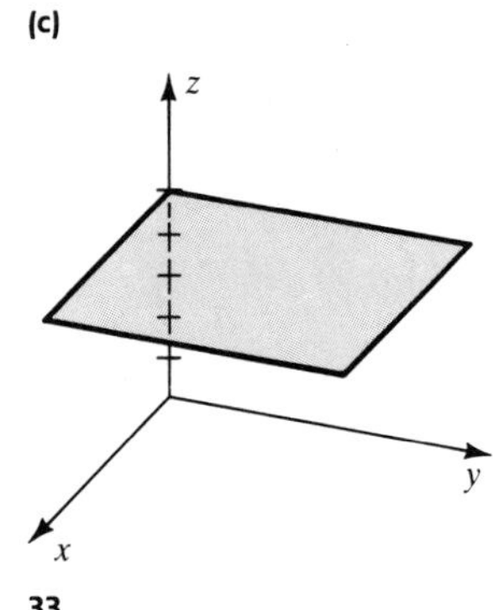

31

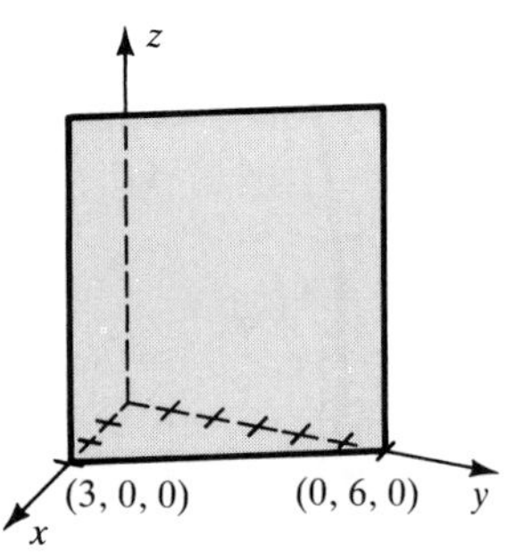

33

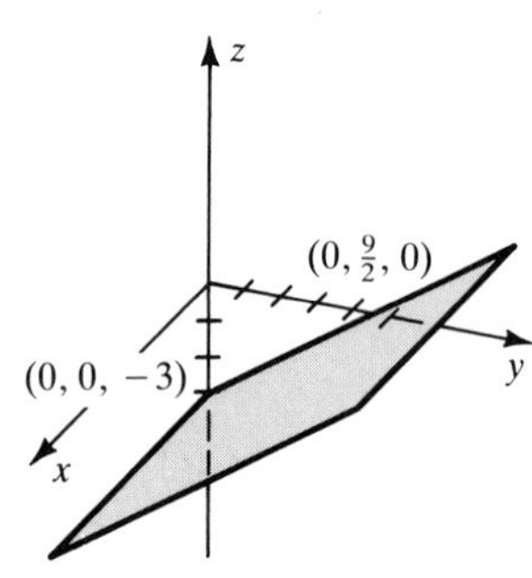

35

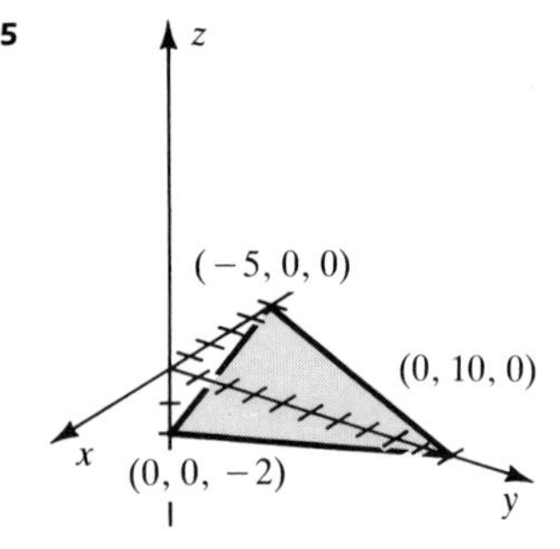

37 $x + z = 5$ **39** $3x + 2y = 6$ **41** $4x - y + 3z + 7 = 0$

43 $\dfrac{x-5}{-3} = \dfrac{y+2}{8} = \dfrac{z-4}{-3}$ **45** $\dfrac{x-4}{-7} = \dfrac{z+3}{8}$; $y = 2$

47 $x = 3 - t,\ y = 2 + 5t,\ z = t$ **49** $x = 3t,\ y = 4 - t,\ z = t$

51 $\dfrac{7}{\sqrt{59}} \approx 0.91$ **53** $\dfrac{17}{6\sqrt{14}} \approx 0.76$ **55** $\dfrac{89}{\sqrt{521}} \approx 3.90$

57 $6x + 11y + 4z = 38$ **59** $\sqrt{\dfrac{5411}{90}} \approx 7.75$

61 $\sqrt{\dfrac{474}{17}} \approx 5.28$ **63** $\dfrac{x}{3} + \dfrac{y}{-2} + \dfrac{z}{5} = 1$

65 $6x + 4y + 3z = 12$

67

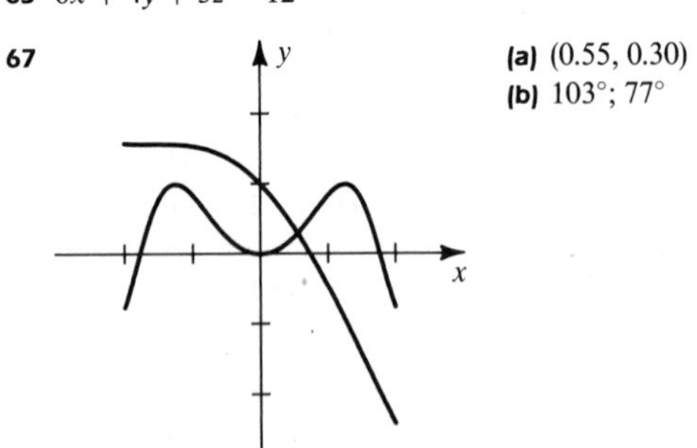

(a) $(0.55, 0.30)$
(b) $103°$; $77°$

EXERCISES 14.6

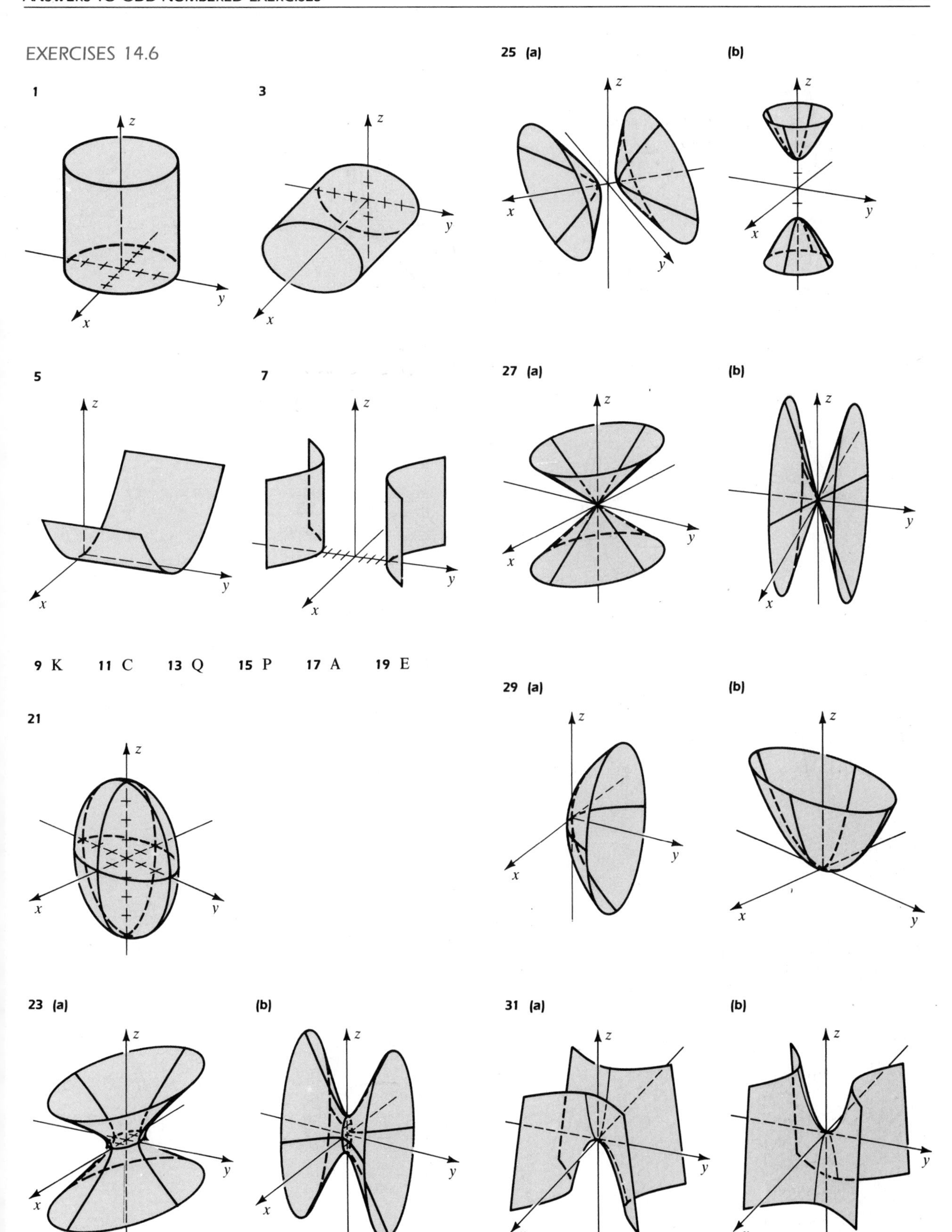

33 Hyperboloid of one sheet

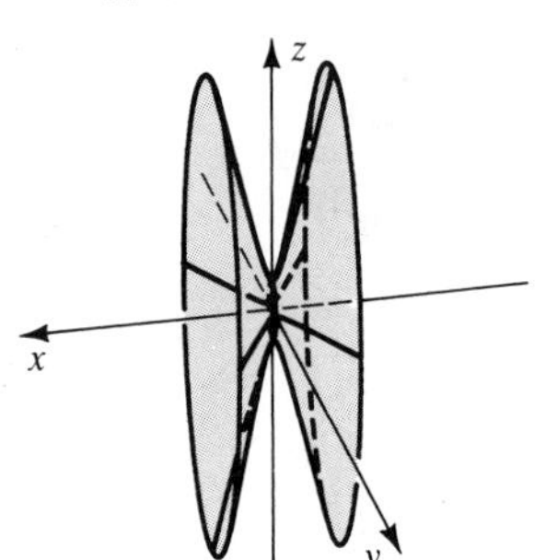

35 Paraboloid

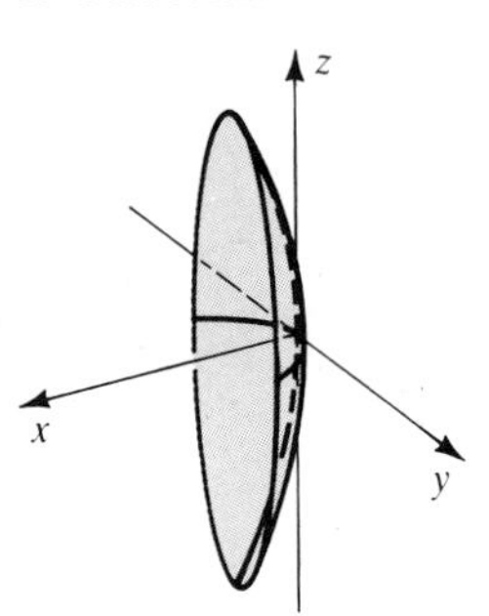

37 Cone

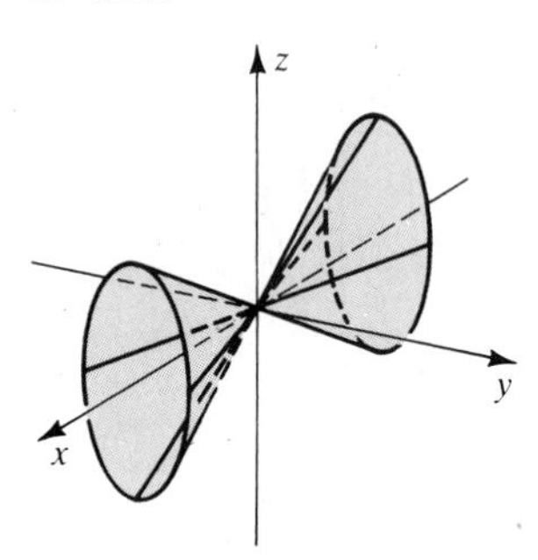

39 Ellipsoid

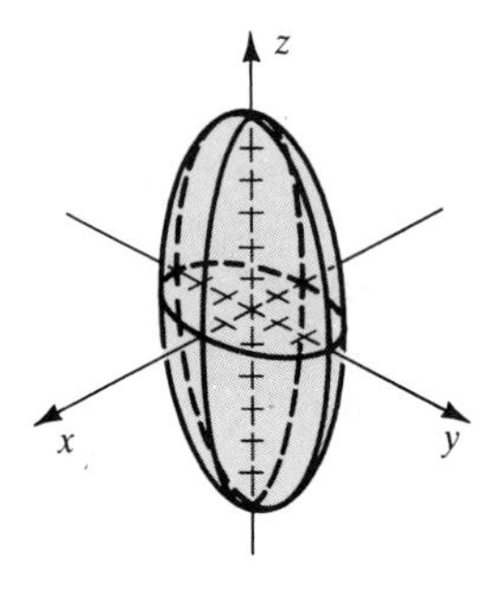

41 Exponential cylinder

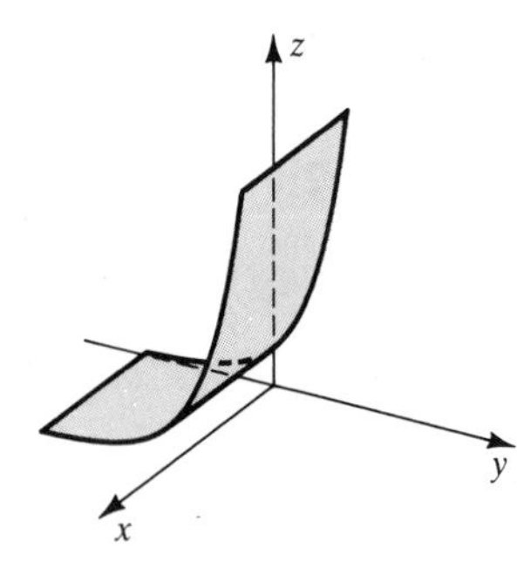

43 Plane

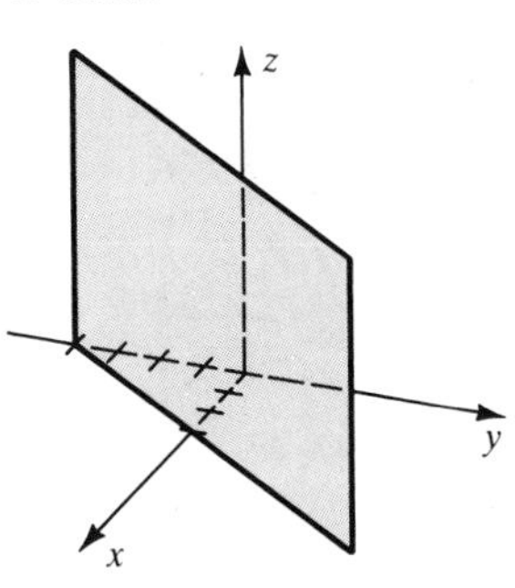

45 Hyperboloid of two sheets

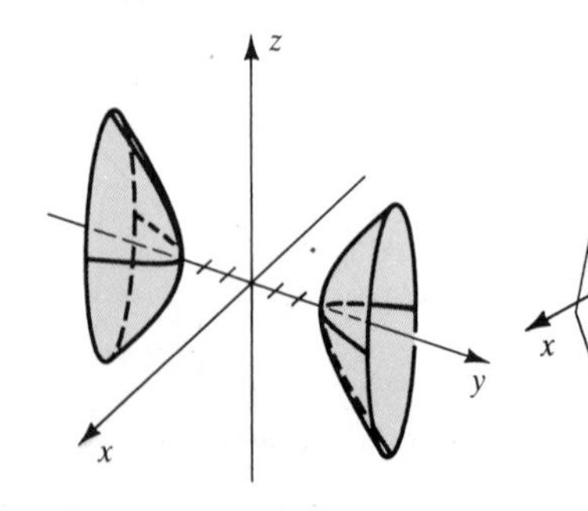

47

49

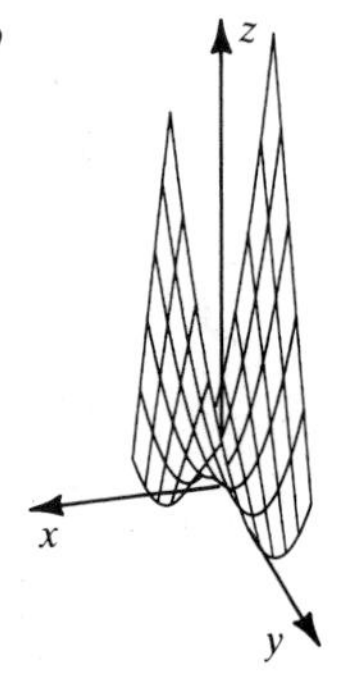

51 $x^2 + z^2 + 4y^2 = 16$ **53** $z = 4 - x^2 - y^2$

55 $y^2 + z^2 - x^2 = 1$

57 **(a)** The Clarke ellipsoid is flatter at the north and south poles. **(b)** Ellipses **(c)** Ellipses

14.7 REVIEW EXERCISES

1 $5\mathbf{i} - 13\mathbf{j} - 8\mathbf{k}$ **3** $3\sqrt{33}$ **5** 26

7 $\arccos \dfrac{-27}{\sqrt{962}} \approx 150.52°$ **9** $\dfrac{1}{\sqrt{26}}(3\mathbf{i} - \mathbf{j} - 4\mathbf{k})$

11 $22\mathbf{i} - 2\mathbf{j} + 17\mathbf{k}$ **13** $\dfrac{9}{\sqrt{33}} \approx 1.57$ **15** 156 **17** **0**

19 80 **21** *Hint:* Use Theorem (14.21).

23 **(a)** $\sqrt{38}$ **(b)** $\left(2, -\frac{7}{2}, \frac{5}{2}\right)$

(c) $(x + 1)^2 + (y + 4)^2 + (z - 3)^2 = 16$ **(d)** $y = -4$

(e) $x = 5 + 6t,\ y = -3 + t,\ z = 2 - t$ **(f)** $6x + y - z = 25$

25 $6x - 15y + 5z = 30$ **27** $x = -13t + 5,\ y = 6t - 2,\ z = 5t$

29 $4x + 3y - 4z = 11$ **31** $\dfrac{x^2}{64} + \dfrac{y^2}{9} + z^2 = 1$

33 **(a)** $\dfrac{1}{\sqrt{66}}\langle 1, 4, 7\rangle$ **(b)** $x + 4y + 7z = 5$

(c) $x = 2 + 7t,\ y = -1 - 7t,\ z = 1 + 3t$ **(d)** 59

(e) $\arccos \dfrac{59}{\sqrt{3745}} \approx 15.40°$

(f) $\sqrt{66} \approx 8.12$ **(g)** $\sqrt{\dfrac{264}{35}} \approx 2.75$

35 $x = 3 + 2t,\ y = -1 - 4t,\ z = 5 + 8t;\ x = -1 + 7t,$
$y = 6 - 2t,\ z = -\dfrac{7}{2} - 2t$

37 $\theta = \arccos \dfrac{-25}{\sqrt{2295}} \approx 121.46°$ and $180° - \theta$

39 **41**

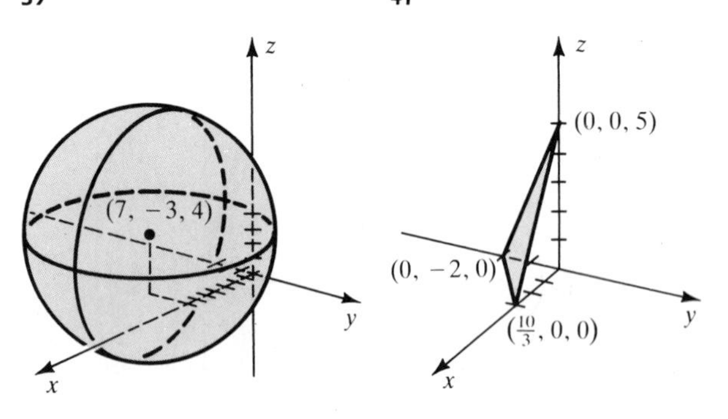

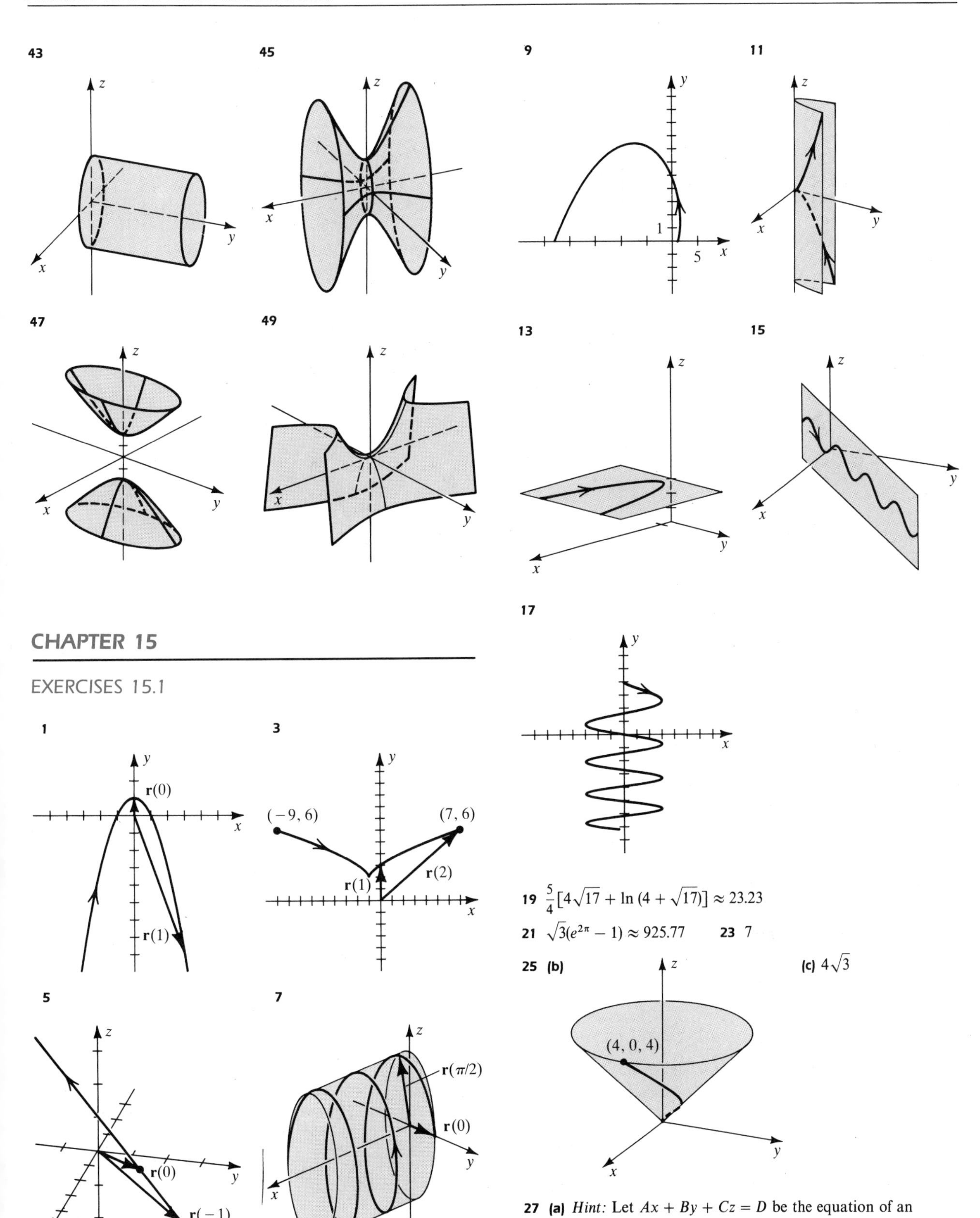

CHAPTER 15

EXERCISES 15.1

19 $\frac{5}{4}[4\sqrt{17} + \ln(4 + \sqrt{17})] \approx 23.23$

21 $\sqrt{3}(e^{2\pi} - 1) \approx 925.77$ **23** 7

25 (b) (see graph) **(c)** $4\sqrt{3}$

27 (a) *Hint:* Let $Ax + By + Cz = D$ be the equation of an arbitrary plane. **(b)** 7

EXERCISES 15.2

1 **(a)** $[1, 2]$
(b) $\frac{1}{2}(t-1)^{-1/2}\mathbf{i} - \frac{1}{2}(2-t)^{-1/2}\mathbf{j}$; $-\frac{1}{4}(t-1)^{-3/2}\mathbf{i} - \frac{1}{4}(2-t)^{-3/2}\mathbf{j}$

3 **(a)** $\left\{t : t \neq \frac{\pi}{2} + \pi n\right\}$ **(b)** $\sec^2 t\mathbf{i} + (2t+8)\mathbf{j}$; $2\sec^2 t \tan t\mathbf{i} + 2\mathbf{j}$

5 **(a)** $\left\{t : t \neq \frac{\pi}{2} + \pi n\right\}$ **(b)** $2t\mathbf{i} + \sec^2 t\mathbf{j}$; $2\mathbf{i} + 2\sec^2 t \tan t\mathbf{j}$

7 **(a)** $\{t : t \geq 0\}$ **(b)** $\frac{1}{2\sqrt{t}}\mathbf{i} + 2e^{2t}\mathbf{j} + \mathbf{k}$; $-\frac{1}{4t\sqrt{t}}\mathbf{i} + 4e^{2t}\mathbf{j}$

9 $-t^3\mathbf{i} + 2t\mathbf{j}$ **11** $-4\sin t\mathbf{i} + 2\cos t\mathbf{j}$

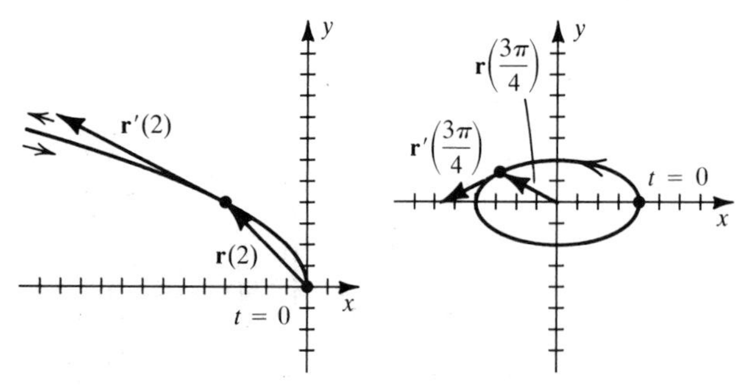

13 $3t^2\mathbf{i} - 3t^{-4}\mathbf{j}$

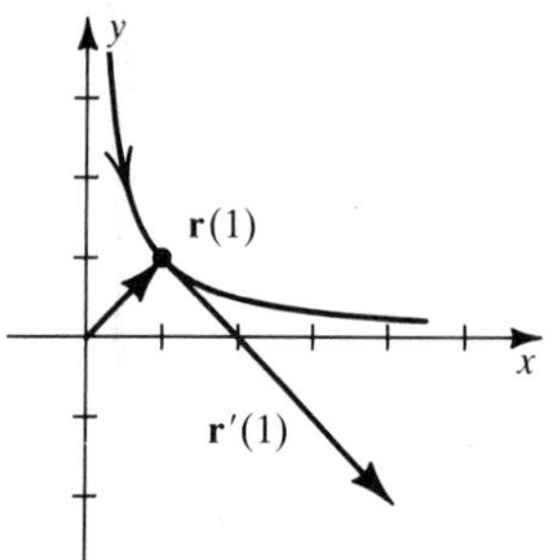

15 $2\mathbf{i} - \mathbf{j}$

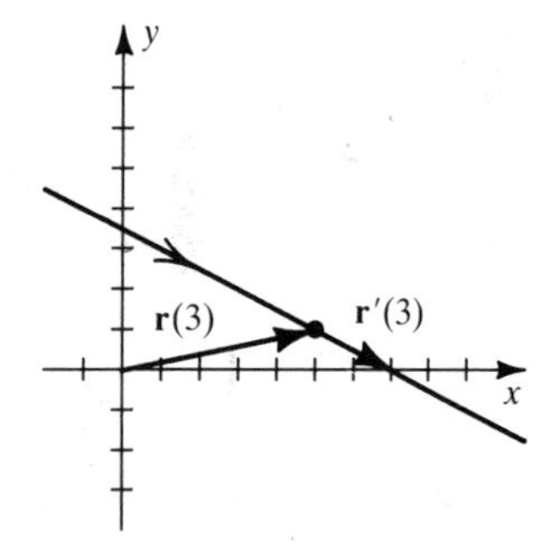

17 $x = 1 + 6t$, $y = -2 - 10t$, $z = 10 + 8t$

19 $x = 1 + t$, $y = t$, $z = 4$ **21** $\pm\frac{1}{\sqrt{5}}(2\mathbf{i} - \mathbf{j})$

25 $x = ae^t$, $y = be^t$, and $z = ce^t$ for constants a, b, and c; the graph is a half-line with endpoint O deleted.

27 $16\mathbf{i} - 8\mathbf{j} + 6\mathbf{k}$ **29** $\left(1 - \frac{1}{\sqrt{2}}\right)\mathbf{i} - \frac{1}{\sqrt{2}}\mathbf{j} + (\ln\sqrt{2})\mathbf{k}$

31 $\left(\frac{1}{3}t^3 + 2\right)\mathbf{i} + (3t^2 + t - 3)\mathbf{j} + (2t^4 + 1)\mathbf{k}$

33 $(t^3 + t + 7)\mathbf{i} + (2t - t^4)\mathbf{j} + \left(\frac{1}{2}t^2 - 3t + 1\right)\mathbf{k}$ **35** $x + y = 1$

37 $(1 + 5t^2)\sin t + (2t^3 + 3t)\cos t$; $[(t^3 + 4t)\sin t - t^2\cos t]\mathbf{i} + [(3t^2 - 2)\sin t + (t^3 - 2t)\cos t]\mathbf{j} + [-3t\sin t + (1 - t^2)\cos t]\mathbf{k}$

EXERCISES 15.3

1

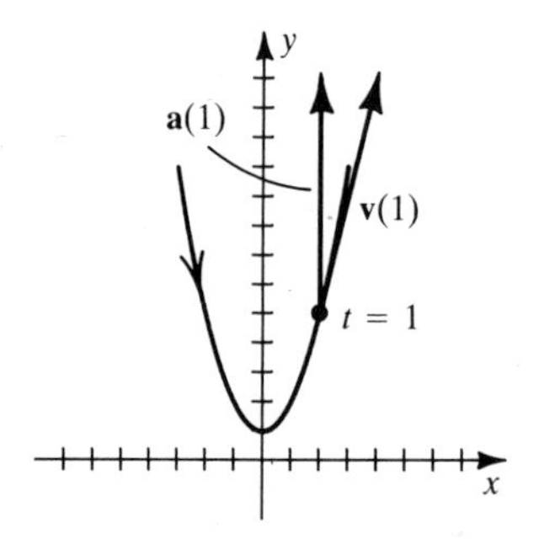

3

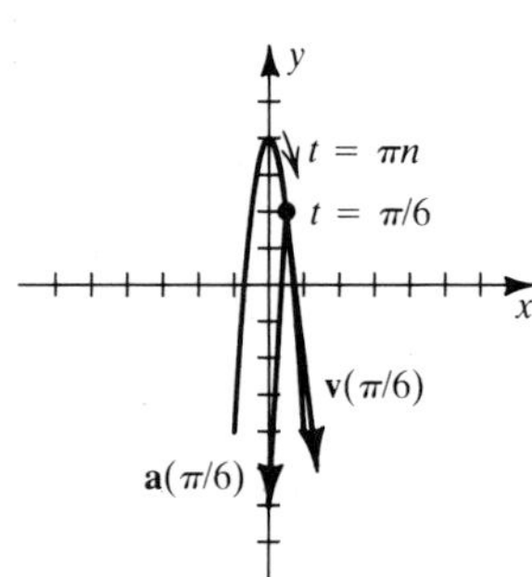

5

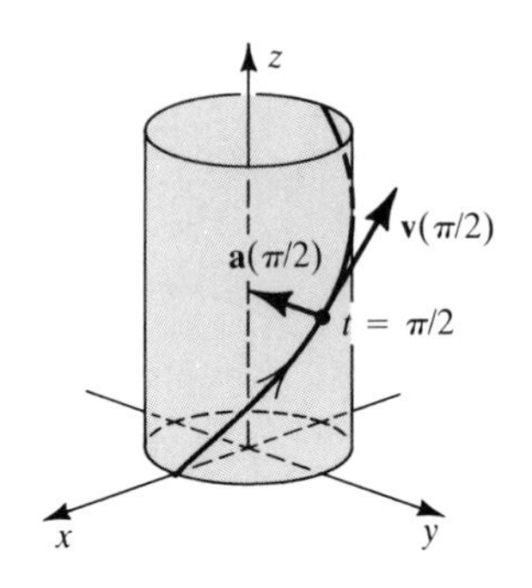

7

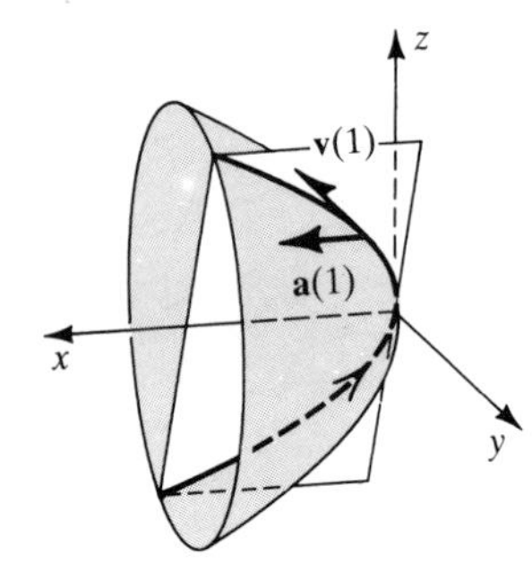

9 $-\frac{1}{2}\mathbf{i} - \frac{1}{3}\mathbf{j}$; $\frac{1}{2}\mathbf{i} + \frac{2}{9}\mathbf{j}$; $\frac{1}{6}\sqrt{13}$ **11** $2\mathbf{i} - \mathbf{j}$; $4\mathbf{i} + \mathbf{j}$; $\sqrt{5}$

13 $e^{\pi/2}(-\mathbf{i} + \mathbf{j} + \mathbf{k})$; $e^{\pi/2}(-2\mathbf{i} + \mathbf{k})$; $\sqrt{3}e^{\pi/2}$

15 $\mathbf{i} + 2\mathbf{j} + 3\mathbf{k}$; $\mathbf{0}$; $\sqrt{14}$ **19** **(a)** 18,054 mi/hr **(b)** 86.7 min

21 **(a)** $750\sqrt{3}\mathbf{i} + (-gt + 750)\mathbf{j}$ **(b)** $\frac{(1500)^2}{8g} \approx 8789$ ft

(c) $\frac{(1500)^2\sqrt{3}}{2g} \approx 60{,}892$ ft **(d)** 1500 ft/sec

23 $\sqrt{250g} \approx 89.4$ ft/sec **25** 0.46 rev/sec

29 2.51 ft **31** 0.14 (mi/hr)/ft

EXERCISES 15.4

1 **(a)** $\frac{1}{(1+t^2)^{1/2}}\mathbf{i} - \frac{t}{(1+t^2)^{1/2}}\mathbf{j}$; $-\frac{t}{(1+t^2)^{1/2}}\mathbf{i} - \frac{1}{(1+t^2)^{1/2}}\mathbf{j}$

(b)

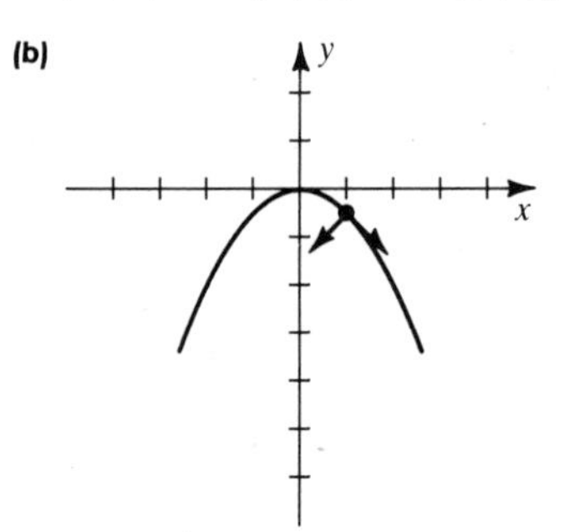

3 **(a)** $\dfrac{t^2}{(t^4+1)^{1/2}}\mathbf{i}+\dfrac{1}{(t^4+1)^{1/2}}\mathbf{j};\ \dfrac{1}{(t^4+1)^{1/2}}\mathbf{i}-\dfrac{t^2}{(t^4+1)^{1/2}}\mathbf{j}$

(b)

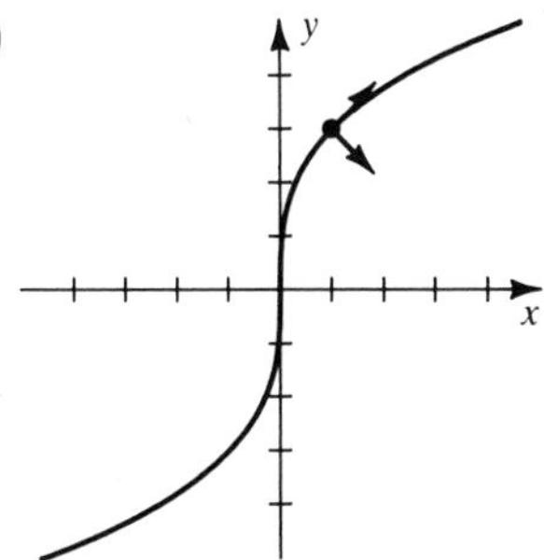

5 **(a)** $\cos t\mathbf{i}-\sin t\mathbf{k};\ -\sin t\mathbf{i}-\cos t\mathbf{k}$

(b)

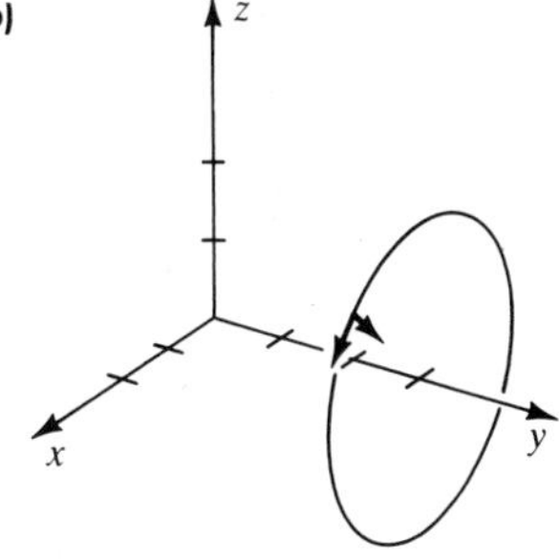

7 $\dfrac{6}{10^{3/2}}\approx 0.19$ **9** 2 **11** 4 **13** $\dfrac{2}{17^{3/2}}\approx 0.03$

15 0 **17** $\dfrac{48}{21^{3/2}}\approx 0.50$

19 **(a)** 1 **(b)** $\left(\dfrac{\pi}{2}, 0\right)$

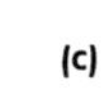
(c)

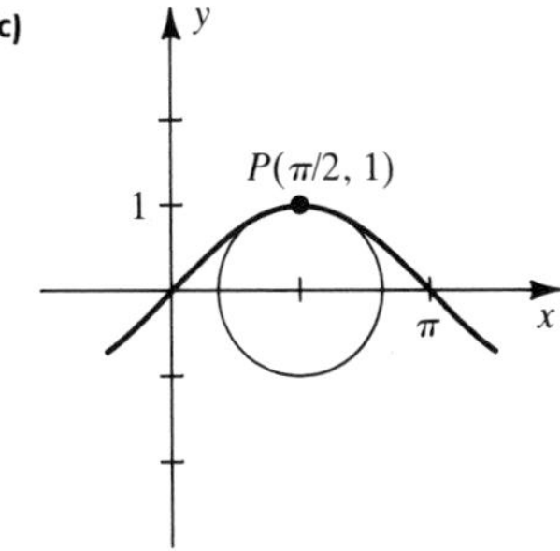

21 **(a)** $2\sqrt{2}$ **(b)** $(-2, 3)$

(c)

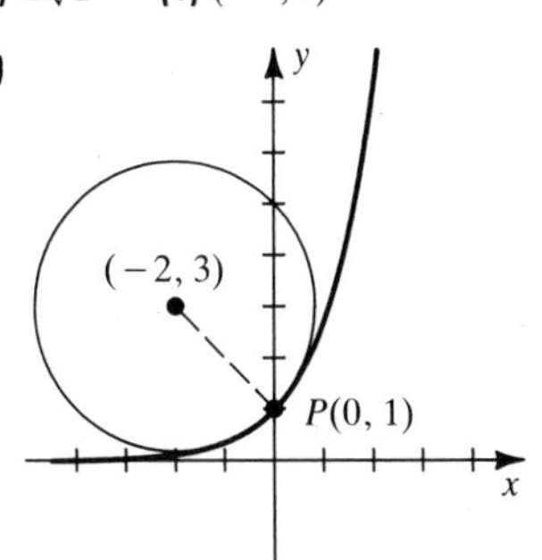

23 $\left(\ln\sqrt{2}, \dfrac{1}{\sqrt{2}}\right)$ **25** $(0, \pm 3)$ **27** $\left(\dfrac{1}{\sqrt{2}}, -\dfrac{1}{2}\ln 2\right)$

29 $(\pm\sqrt{2}, -20)$ **31** $(0, 0)$ **35** $\dfrac{8-3\sin^2 2\theta}{(1+3\cos^2 2\theta)^{3/2}}$

39 $\left(-\dfrac{14}{3}, \dfrac{29}{12}\right)$ **41** $(4, -2)$ **43** $(0, -4)$

49 $x=\dfrac{4}{5}s-3,\ y=\dfrac{3}{5}s+5;\ s\geq 0$

51 $x=4\cos\dfrac{1}{4}s,\ y=4\sin\dfrac{1}{4}s;\ 0\leq s\leq 8\pi$

EXERCISES 15.5

1 $\dfrac{4t}{(4t^2+9)^{1/2}};\ \dfrac{6}{(4t^2+9)^{1/2}};\ \dfrac{6}{(4t^2+9)^{3/2}}$

3 $\dfrac{6t(t^2+2)}{(t^4+4t^2+1)^{1/2}};\ \dfrac{6(t^4+t^2+1)^{1/2}}{(t^4+4t^2+1)^{1/2}};\ \dfrac{2(t^4+t^2+1)^{1/2}}{3(t^4+4t^2+1)^{3/2}}$

5 $\dfrac{t}{(1+t^2)^{1/2}};\ \dfrac{2+t^2}{(1+t^2)^{1/2}};\ \dfrac{2+t^2}{(1+t^2)^{3/2}}$

7 $\dfrac{-65\sin t\cos t}{(16\sin^2 t+81\cos^2 t+1)^{1/2}};\ \dfrac{(81\sin^2 t+16\cos^2 t+1296)^{1/2}}{(16\sin^2 t+81\cos^2 t+1)^{1/2}};$
$\dfrac{(81\sin^2 t+16\cos^2 t+1296)^{1/2}}{(16\sin^2 t+81\cos^2 t+1)^{3/2}}$

9 $\dfrac{36}{\sqrt{5}}\approx 16.10;\ \dfrac{18}{\sqrt{5}}\approx 8.05$

15.7 REVIEW EXERCISES

1

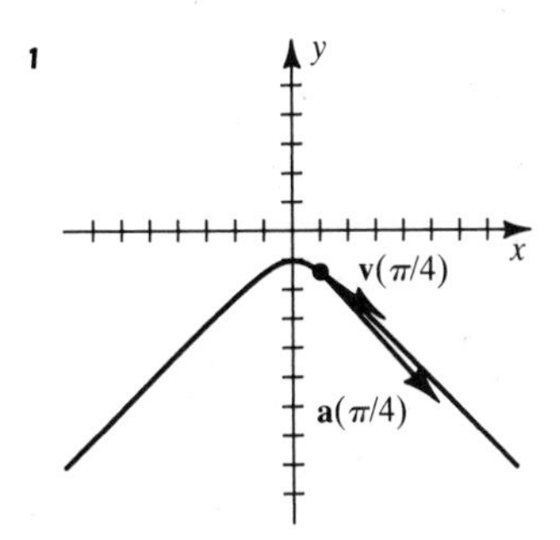

3 $2t\mathbf{i}+(8t-4t^3)\mathbf{j};\ 2\mathbf{i}+(8-12t^2)\mathbf{j};\ 2|t|\sqrt{17-16t^2+4t^4}$

5 **(a)** $\dfrac{1}{\sqrt{3}}(\mathbf{i}+\mathbf{j}+\mathbf{k})$ **(b)** $x=t,\ y=1+t,\ z=1+t$

7 $0, \dfrac{10}{3}$

9 $3\mathbf{i}+3t^2\mathbf{j}+4t^3\mathbf{k},\ 6t\mathbf{j}+12t^2\mathbf{k},\ \sqrt{34};\ 3\mathbf{i}+3\mathbf{j}+4\mathbf{k},\ 6\mathbf{j}+12\mathbf{k},\ \sqrt{34}$

11 $2\mathbf{i}+\dfrac{1}{4}\mathbf{j}-\mathbf{k}$

13 $\mathbf{r}(t)=(3t-4)\mathbf{i}+(2t^3-8t+9)\mathbf{j}+\left(\dfrac{5}{2}t^2-t\right)\mathbf{k}$

17 $\dfrac{108}{82^{3/2}}\approx 0.15$

19 $\pm\sqrt{\dfrac{36+\sqrt{3096}}{90}}\approx\pm 1.009$ **21** $\dfrac{5}{6}\sqrt{5}\approx 1.86$

23 $\dfrac{-8\cos 2t\sin 2t+\sin t\cos t}{(4\cos^2 2t+\sin^2 t)^{1/2}};\ \dfrac{2|\cos 2t\cos t+2\sin 2t\sin t|}{(4\cos^2 2t+\sin^2 t)^{1/2}}$

CHAPTER 16

EXERCISES 16.1

1 $\mathbb{R}^2$; $-29, 6, -4$ **3** $\{(u, v)\colon u \neq 2v\}$; $-\frac{3}{2}, \frac{4}{9}, 0$

5 $\{(x, y, z)\colon x^2 + y^2 + z^2 \leq 25\}$; $4, 2\sqrt{3}$

7

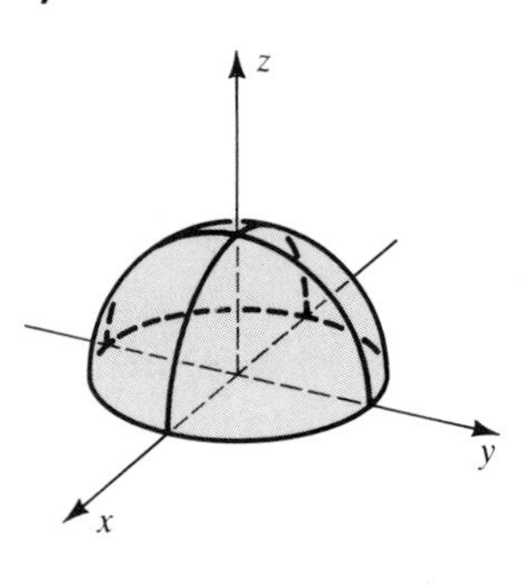

9

11

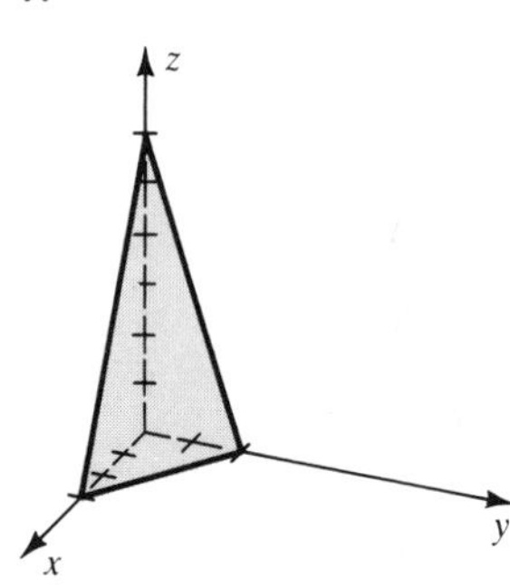

13

15

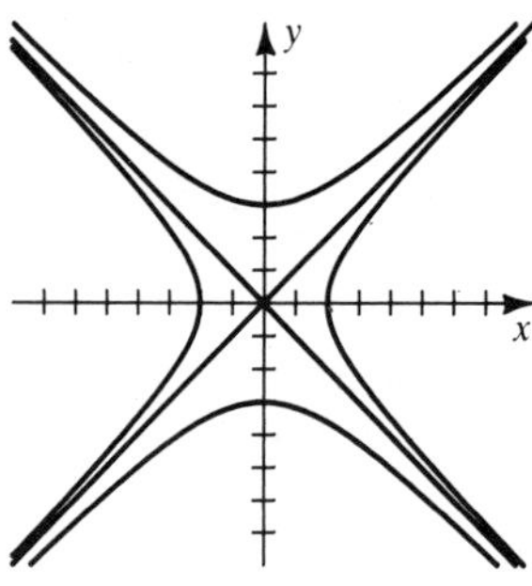

17

19

21 $y \arctan x = \pi$ **23** $x^2 + 4y^2 - z^2 = -1$

25 **(a)** Ellipses **(b)** None **(c)** None

27 **(a)** $k > 8$: none; $k = 8$: the point $(0, 0, 8)$; $0 < k < 8$: circles
(b) None **(c)** None

29 (d) **31** (a) **33** (f)

35

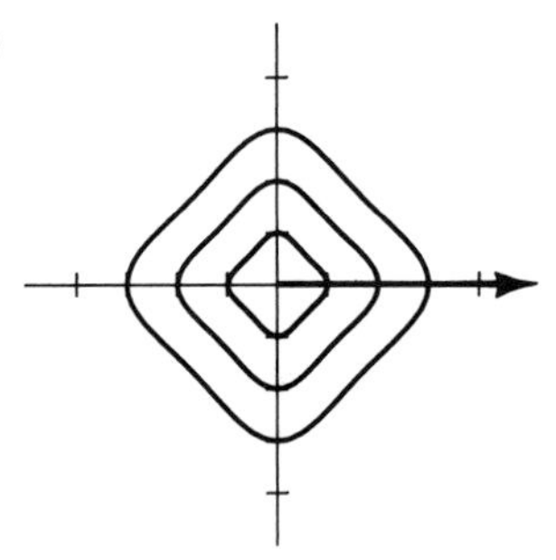

37 None; the origin; the sphere with center $(0, 0, 0)$ and radius 2

39 Planes with x-intercept k, y-intercept $\frac{k}{2}$, and z-intercept $\frac{k}{3}$

41 None; the z-axis; the right circular cylinder with the z-axis as its axis and radius 2

43 **(a)** Circles with center at the origin **(b)** $x^2 + y^2 = 100$

45 Five; spheres with centers at the origin; the force F is constant if (x, y, z) moves along a level surface.

47 **(a)** $P = kAv^3$ for $k > 0$
(b) A typical level curve (see figure) shows the combinations of areas and wind velocities that result in a fixed power $P = c$.

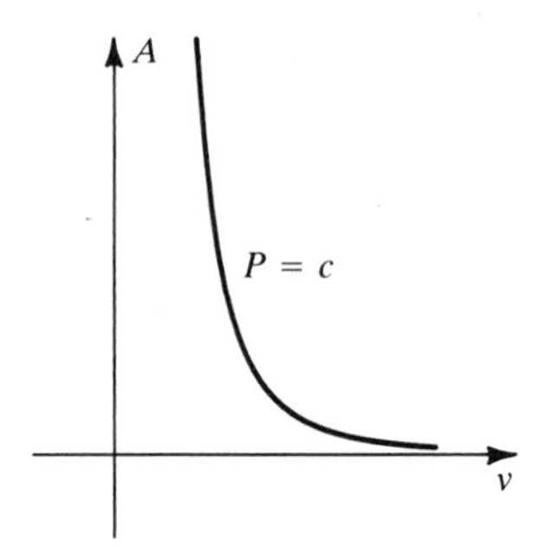

(c) $Av^3 = \frac{8}{3}\pi \times 10^5$

49 Example: 5′11″ and 175 lb are approximately 180 cm and 80 kg. From the graph, we have a surface area of approximately 2.0 m^2. Using the formula, we obtain $S \approx 1.996$ m^2.

EXERCISES 16.2

1 $-\frac{2}{3}$ **3** 1 **5** 0 **7** 0 **9** $\frac{1}{2}$

Exer. 11–20: The answer gives equations of possible paths, and their resulting values, to use in (16.4).

11 $x = 0, -\frac{1}{2}$; $y = 0, 2$ **13** $y - 2 = m(x - 1), \frac{m}{1 + m^2}$

15 $y = 2x, \frac{12}{31}$; $y = 0, 0$ **17** $x = y = 0, 0$; $x = y = z, 1$

19 $x = at,\ y = bt,\ z = ct,\ \dfrac{a^3 + b^3 + c^3}{abc}$ **21** 0 **23** 1

25 $\{(x, y): x + y > 1\}$ **27** $\{(x, y): x \geq 0 \text{ and } |y| \leq 1\}$

29 $\{(x, y, z): z^2 \neq x^2 + y^2\}$ **31** $\{(x, y, z): x \geq 2,\ yz > 0\}$

33 0 **35** $\dfrac{x^4 - 2x^2y^2 + y^4 - 4}{x^2 - y^2}$; $\{(x, y): x^2 \neq y^2\}$

37 $x^2 + 2x \tan y + \tan^2 y + 1$; $\left\{(x, y): y \neq \dfrac{\pi}{2} + \pi n\right\}$

39 e^{x^2+2y}; $(x^2 + 2y)(x^2 + 2y - 3)$; $e^{2t} + 2(t^2 - 3t)$

41 $2x^2 - 3xy - 2y^2 - x + 7y$

43 The statement $\lim_{(x,y,z,w)\to(a,b,c,d)} f(x, y, z, w) = L$ means that for every $\epsilon > 0$ there is a $\delta > 0$ such that if $0 < \sqrt{(x - a)^2 + (y - b)^2 + (z - c)^2 + (w - d)^2} < \delta$, then $|f(x, y, z, w) - L| < \epsilon$.

EXERCISES 16.3

1 $f_x(x, y) = 8x^3y^3 - y^2$; $f_y(x, y) = 6x^4y^2 - 2xy + 3$

3 $f_r(r, s) = \dfrac{r}{(r^2 + s^2)^{1/2}}$; $f_s(r, s) = \dfrac{s}{(r^2 + s^2)^{1/2}}$

5 $f_x(x, y) = e^y + y \cos x$; $f_y(x, y) = xe^y + \sin x$

7 $f_t(t, v) = -\dfrac{v}{t^2 - v^2}$; $f_v(t, v) = \dfrac{t}{t^2 - v^2}$

9 $f_x(x, y) = \cos \dfrac{x}{y} - \dfrac{x}{y} \sin \dfrac{x}{y}$; $f_y(x, y) = \left(\dfrac{x}{y}\right)^2 \sin \dfrac{x}{y}$

11 $f_x(x, y, z) = 6xz + y^2$; $f_y(x, y, z) = 2xy$; $f_z(x, y, z) = 3x^2$

13 $f_r(r, s, t) = 2re^{2s} \cos t$; $f_s(r, s, t) = 2r^2e^{2s} \cos t$; $f_t(r, s, t) = -r^2e^{2s} \sin t$

15 $f_x(x, y, z) = e^z - ye^x$; $f_y(x, y, z) = -e^x - ze^{-y}$; $f_z(x, y, z) = xe^z + e^{-y}$

17 $f_q(q, v, w) = \dfrac{v}{2\sqrt{qv}\sqrt{1 - qv}}$;

$f_v(q, v, w) = \dfrac{q}{2\sqrt{qv}\sqrt{1 - qv}} + w \cos vw$; $f_w(q, v, w) = v \cos vw$

19 $w_{xy} = w_{yx} = 4y^3 - 12xy^2$

21 $w_{xy} = w_{yx} = -6x^2e^{-2y} + 2y^{-3} \sin x$

23 $w_{xy} = w_{yx} = -\dfrac{2xz}{y^2} \sinh \dfrac{z}{y}$ **25** $18xy^2 + 16y^3z$

27 $t^2(\sec rt)(\sec^2 rt + \tan^2 rt)$

29 $(1 - x^2y^2z^2) \cos xyz - 3xyz \sin xyz$

31 $w_{rrs} = w_{rsr} = w_{srr} = 36r^2s^2t - 6st^2e^{rt}$

33 Show that $\dfrac{\partial^2 f}{\partial x^2} = \dfrac{y^2 - x^2}{(x^2 + y^2)^2} = -\dfrac{\partial^2 f}{\partial y^2}$.

35 Show that $\dfrac{\partial^2 f}{\partial x^2} = -\cos x \sinh y - \sin x \cosh y = -\dfrac{\partial^2 f}{\partial y^2}$.

37 Show that $\dfrac{\partial^2 w}{\partial x^2} = -\cos (x - y) - \dfrac{1}{(x + y)^2} = \dfrac{\partial^2 w}{\partial y^2}$.

39 Show that $w_{xx} = -c^2e^{-c^2t} \sin cx = w_t$.

41 Show that $\dfrac{\partial^2 v}{\partial t^2} = a^2[-k^2(\sin akt)(\sin kx)] = a^2 \dfrac{\partial^2 v}{\partial x^2}$.

43 Show that $u_x = 2x = v_y$ and $u_y = -2y = -v_x$.

45 Show that $u_x = e^x \cos y = v_y$ and $u_y = -e^x \sin y = -v_x$.

47 $w_{xx}, w_{xy}, w_{xz}, w_{yx}, w_{yy}, w_{yz}, w_{zx}, w_{zy}, w_{zz}$

49 In deg/cm: **(a)** 200 **(b)** 400

51 In volts/in.: **(a)** $-\dfrac{100}{9}$ **(b)** $\dfrac{50}{9}$ **(c)** $-\dfrac{50}{9}$

53 $\dfrac{\partial C}{\partial x} \approx -36.58\ (\mu g/m^3)/m$ is the rate at which the concentration changes in the horizontal direction at (2, 5).

$\dfrac{\partial C}{\partial y} \approx -0.229\ (\mu g/m^3)/m$ is the rate at which the concentration changes in the vertical direction at (2, 5).

55 **(a)** 3.57; 4.81; 4.98; $\lim_{p\to\infty} \dfrac{p}{0.8 + 0.2p} = 5$ **(b)** $\dfrac{p(p - 1)}{[q + p(1 - q)]^2}$

(c) $p(p - 1)$; as the number of processors increases, the rate of change of the speedup increases.

57 **(a)** $\dfrac{\partial T}{\partial t} = T_0\omega e^{-\lambda x} \cos (\omega t - \lambda x)$ is the rate of change of temperature with respect to time at depth x.

$\dfrac{\partial T}{\partial x} = -T_0\lambda e^{-\lambda x}[\cos (\omega t - \lambda x) + \sin (\omega t - \lambda x)]$ is the rate of change of temperature with respect to the depth at time t. **(b)** Show that $k = \dfrac{2\lambda^2}{\omega}$.

59 **(a)** $\dfrac{\partial V}{\partial x} = -0.112y$ ml/yr is the rate at which lung capacity decreases with age for an adult male.

(b) $\dfrac{\partial V}{\partial y} = 27.63 - 0.112x$ ml/cm is difficult to interpret because we usually think of adult height y as fixed instead of a function of age x.

61 $e_k = -a_{kk}$ for every k

63 $x = 1,\ y = t,\ z = -4t + 12$

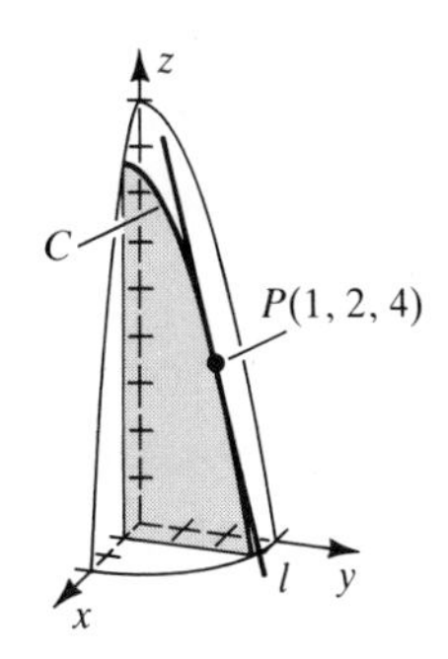

65 0.0079600438, 0.0597349919; 0.0079600333, 0.0598334499

67 1.8369; 4.1743

EXERCISES 16.4

1 **(a)** $10y\,\Delta y - x\,\Delta y - y\,\Delta x + 5(\Delta y)^2 - \Delta x\,\Delta y$

(b) $-y\,dx + (10y - x)\,dy$ **(c)** $\Delta x\,\Delta y - 5(\Delta y)^2$

Exer. 3–6: The expressions for ϵ_1 and ϵ_2 are not unique.

3 $\epsilon_1 = -3\,\Delta y$; $\epsilon_2 = 4\,\Delta y$

5 $\epsilon_1 = 3x\,\Delta x + (\Delta x)^2$; $\epsilon_2 = 3y\,\Delta y + (\Delta y)^2$

7 $(3x^2 - 2xy)\,dx + (-x^2 + 6y)\,dy$

9 $(2x \sin y)\,dx + (x^2 \cos y + 3y^{1/2})\,dy$

11 $xe^{xy}(xy + 2)\,dx + (x^3e^{xy} - 2y^{-3})\,dy$

13 $[2x \ln (y^2 + z^2)]\,dx + \left(\dfrac{2x^2y}{y^2 + z^2}\right) dy + \left(\dfrac{2x^2z}{y^2 + z^2}\right) dz$

15 $\left[\dfrac{yz(y + z)}{(x + y + z)^2}\right] dx + \left[\dfrac{xz(x + z)}{(x + y + z)^2}\right] dy + \left[\dfrac{xy(x + y)}{(x + y + z)^2}\right] dz$

17 $(2xz - z^2t)\,dx + (4t^3)\,dy + (x^2 - 2xzt)\,dz + (12yt^2 - xz^2)\,dt$

19 7.38 **21** 1.87 **23** **(a)** $\pm\frac{1}{4}$ ft² **(b)** $\pm\frac{47}{192}$ ft³

25 **(a)** 380 lb **(b)** $\pm 11.5\%$ **27** ± 0.0185

29 $\pm\frac{6W}{A-W}\%$ **31** $\pm 7\%$ **33** 2.96

35 Maximum error in z must not exceed ± 2.9 feet.

37 $\pm 1.7\pi$ in.^{2} **39** Use Corollary (16.19).

43 $(x_4, y_4) \approx (1.8460, 1.1546)$

45 $\begin{bmatrix} f_x & f_y & f_z \\ g_x & g_y & g_z \\ h_x & h_y & h_z \end{bmatrix}\begin{bmatrix} \Delta x \\ \Delta y \\ \Delta z \end{bmatrix} = \begin{bmatrix} -f \\ -g \\ -h \end{bmatrix}$

EXERCISES 16.5

1 $2x \sin(xy) + y(x^2 + y^2)\cos(xy)$; $2y \sin(xy) + x(x^2 + y^2)\cos(xy)$

3 $2r(\ln s)^2 + 8r \ln s + 2s \ln s$; $\frac{2r^2 \ln s}{s} + \frac{4r^2}{s} + 2r + 2r \ln s$

5 $3x^2e^{3y} + ye^x + 4x^3y^2$; $3x^3e^{3y} + e^x + 2x^4y$

7 $3 \ln(uvt) + 3 + \frac{vt}{u}$; $t \ln(uvt) + \frac{3u}{v} + t$; $v \ln(uvt) + \frac{3u}{t} + v$

9 $-34y + 6r - 24s$ **11** $\frac{-3(1+t^2)}{(t+1)^4}$

13 $4 \sin^3 t \cos t + \tan 4t \sin t - 4 \cos t \sec^2 4t$

15 $-\frac{6x^2 + 2xy}{x^2 + 3y^2}$ **17** $\frac{12\sqrt{xy} + y}{6\sqrt{xy} - x}$

19 $-\frac{2z^3 + 2xy^2}{6xz^2 - 6yz + 4}$; $-\frac{2x^2y - 3z^2}{6xz^2 - 6yz + 4}$

21 $-\frac{e^{yz} - 2yze^{xz} + 3yze^{xy}}{xye^{yz} - 2xye^{xz} + 3e^{xy}}$; $-\frac{xze^{yz} - 2e^{xz} + 3xze^{xy}}{xye^{yz} - 2xye^{xz} + 3e^{xy}}$

23 **(a)** $0.88\pi \approx 2.76$ in.^{3}/min **(b)** $0.3\pi \approx 0.94$ in.^{2}/min

25 $\frac{dT}{dt} = \frac{V}{k}\frac{dP}{dt} + \frac{P}{k}\frac{dV}{dt}$ **27** -6.4 in.^{3}/min

29 762.6 cm^2/yr **33** 3 **35** 0

EXERCISES 16.6

1 $-\frac{4}{5}\mathbf{i} + \frac{3}{5}\mathbf{j}$ **3** $3\mathbf{i} + 2\mathbf{j}$ **5** $-8\mathbf{i} + \mathbf{j} - 9\mathbf{k}$

7 $-\frac{10}{\sqrt{2}} \approx -7.07$ **9** $-\frac{1}{8\sqrt{13}} \approx -0.03$ **11** $\frac{67}{8\sqrt{26}} \approx 1.64$

13 $\frac{1}{2\sqrt{26}} \approx 0.098$ **15** $16\sqrt{14} \approx 59.87$ **17** $\frac{15e^{-2}}{\sqrt{35}} \approx 0.34$

19 $-\frac{12}{\sqrt{10}} \approx -3.79$

21 **(a)** $-\frac{28}{\sqrt{26}}$ **(b)** $\frac{1}{\sqrt{5}}\mathbf{i} - \frac{2}{\sqrt{5}}\mathbf{j}$; $\sqrt{80}$ **(c)** $-\frac{1}{\sqrt{5}}\mathbf{i} + \frac{2}{\sqrt{5}}\mathbf{j}$; $-\sqrt{80}$

23 **(a)** $-\frac{25}{7\sqrt{22}}$ **(b)** $-\frac{2}{\sqrt{14}}\mathbf{i} + \frac{3}{\sqrt{14}}\mathbf{j} + \frac{1}{\sqrt{14}}\mathbf{k}$; 1

(c) $\frac{2}{\sqrt{14}}\mathbf{i} - \frac{3}{\sqrt{14}}\mathbf{j} - \frac{1}{\sqrt{14}}\mathbf{k}$; -1

25 **(a)** $-\frac{28}{\sqrt{2}}$ **(b)** The direction of $-12\mathbf{i} - 16\mathbf{j}$

(c) The direction of $12\mathbf{i} + 16\mathbf{j}$ **(d)** The direction of $4\mathbf{i} - 3\mathbf{j}$

27 **(a)** $-\frac{178}{\sqrt{14}}$ **(b)** The direction of $4\mathbf{i} - 8\mathbf{j} + 54\mathbf{k}$

(c) $\sqrt{2996} \approx 54.8$

29 **(b)** $\frac{\partial T}{\partial r}$ is the rate of change of temperature in the direction normal to the circular boundary.

31 **(b)** $\nabla f(1, 2) \approx 1.00003333\mathbf{i} - 0.111112345\mathbf{j}$; $\nabla f(1, 2) = \mathbf{i} - \frac{1}{9}\mathbf{j}$

33 0.1294 **41** **(b)** $5 + \sqrt{3}$

EXERCISES 16.7

1 $16(x - 2) + 6(y + 3) + 6(z - 1) = 0$; $x = 2 + 16t$, $y = -3 + 6t$, $z = 1 + 6t$

3 $16(x + 2) + 18(y + 1) + (z - 25) = 0$; $x = -2 + 16t$, $y = -1 + 18t$, $z = 25 + t$

5 $4(x + 5) - 3(y - 5) + 20(z - 1) = 0$; $x = -5 + 4t$, $y = 5 - 3t$, $z = 1 + 20t$

7 $x + \sqrt{3}\left(y - \frac{\pi}{3}\right) + (z - 1) = 0$; $x = t$, $y = \frac{\pi}{3} + \sqrt{3}t$, $z = 1 + t$

9 $-x + \frac{1}{2}(y - 2) - (z - 1) = 0$; $x = -t$, $y = 2 + \frac{1}{2}t$, $z = 1 - t$

11 **13**

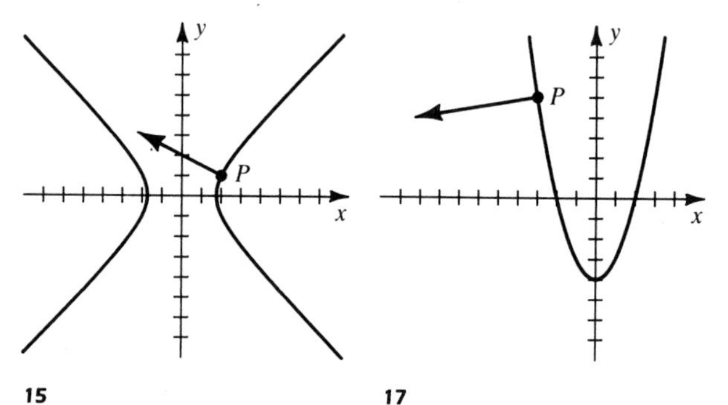

15 **17**

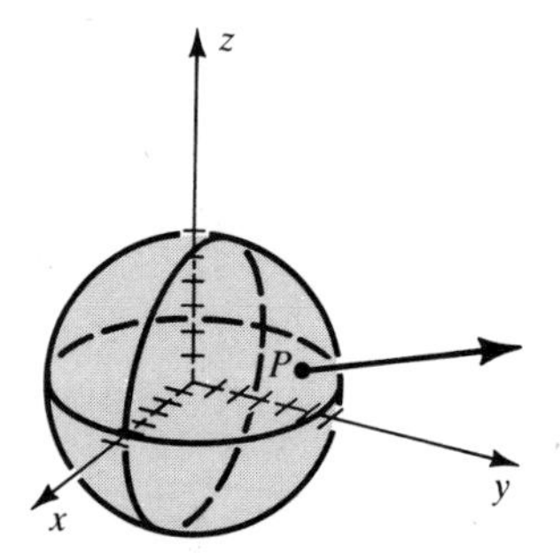

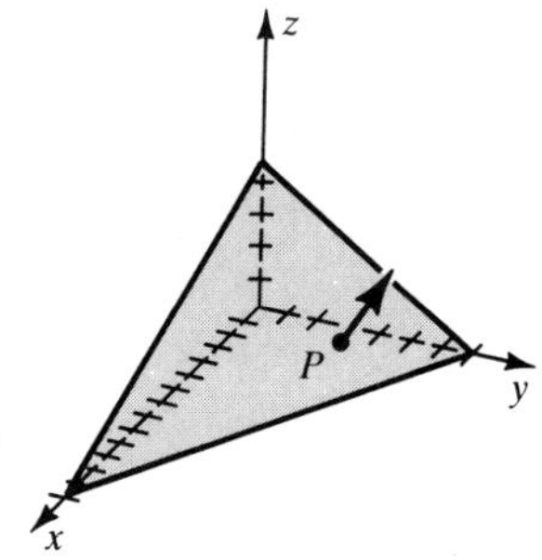

19

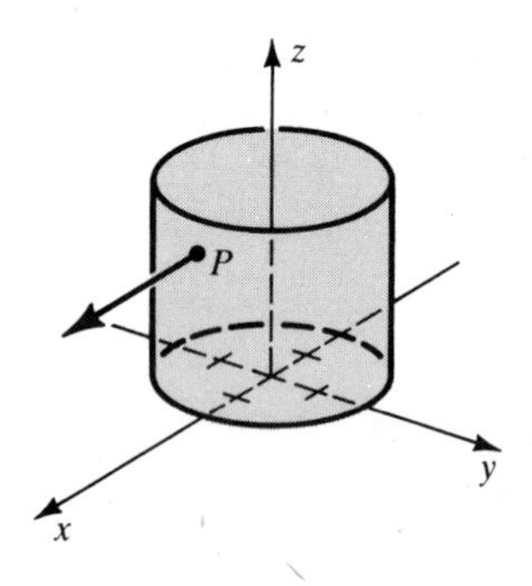

25 $\left(\frac{8\sqrt{2}}{\sqrt{5}}, \frac{2\sqrt{2}}{\sqrt{5}}, -\frac{2\sqrt{2}}{\sqrt{5}}\right), \left(-\frac{8\sqrt{2}}{\sqrt{5}}, -\frac{2\sqrt{2}}{\sqrt{5}}, \frac{2\sqrt{2}}{\sqrt{5}}\right)$
31 (1.2718, 1.1787)

EXERCISES 16.8

1 Max: $f(-2, 1) = 4$ **3** Min: $f\left(\frac{1}{2}, -\frac{1}{4}\right) = -\frac{1}{2}$
5 Min: $f(0, 0) = 0$ **7** SP: $(0, 0, f(0, 0))$; min: $f(1, -1) = -1$
9 SP: $(3, -2, f(3, -2))$
11 SP: $(2, 4, f(2, 4)), (-3, -4, f(-3, -4))$;
min: $f(2, -4) = -\frac{266}{3}$; max: $f(-3, 4) = \frac{617}{6}$
13 SP: $(0, 0, f(0, 0))$; min: $f(4, -8) = -64$, $f(-1, 2) = -\frac{3}{2}$
15 SP: $(-2, -\sqrt{3}, f(-2, -\sqrt{3}))$; min: $f(-2, \sqrt{3}) = -48 - 6\sqrt{3}$
17 No extrema or saddle points **19** Min: $f(\sqrt[3]{2}, 2\sqrt[3]{2}) = \frac{12}{\sqrt[3]{2}}$
21 $(0, 0), (\pm 1, 0), (0, \pm 1)$; min: $f(0, 0) = 0$; max: $f(0, \pm 1) = \frac{3}{e}$
23 Min: $f\left(\frac{1}{2}, -\frac{1}{4}\right) = -\frac{1}{2}$; max: $f\left(-\frac{1}{\sqrt{2}}, \frac{1}{2\sqrt{2}}\right) = 1 + \sqrt{2}$
25 Min: $f(0, 0) = 0$; max: $f(4, 3) = 67$
27 Min: $f(1, 2) = f(1, -1) = -1$; max: $f(-1, -2) = 13$
29 $\frac{1}{\sqrt{26}}$ **31** $\left(\frac{2}{\sqrt[4]{12}}, \sqrt[4]{12}, \pm\frac{2\sqrt{2}}{\sqrt[4]{12}}\right), \left(-\frac{2}{\sqrt[4]{12}}, -\sqrt[4]{12}, \pm\frac{2\sqrt{2}}{\sqrt[4]{12}}\right)$
33 Square base, altitude $\frac{1}{2}$ the length of the side of the base
35 $\frac{8}{\sqrt{3}}, \frac{6}{\sqrt{3}}, \frac{12}{\sqrt{3}}$ **37** $1, \frac{4}{3}, 4$
39 Square base of side $\sqrt[3]{4}$ ft, height $2\sqrt[3]{4}$ ft
41 (18 in.) × (18 in.) × (36 in.) **43** $\left(2 - \frac{2}{3}\sqrt{3}, 2 - \frac{2}{3}\sqrt{3}\right)$
47 $y = \frac{1}{2}x + \frac{8}{3}$
49 $y = mx + b$ with $m \approx 1.23$, $b \approx -18.09$; grade of 68
51 $\left(\frac{14}{3}, \frac{11}{3}\right)$
53 **(a)** $a + b + c = 2$, $a + 2b + 2c = 2$, $a + 2b + 4c = 3$
(b) $4x - y - 2z + 1 = 0$
55 $(-0.35, -0.17)$

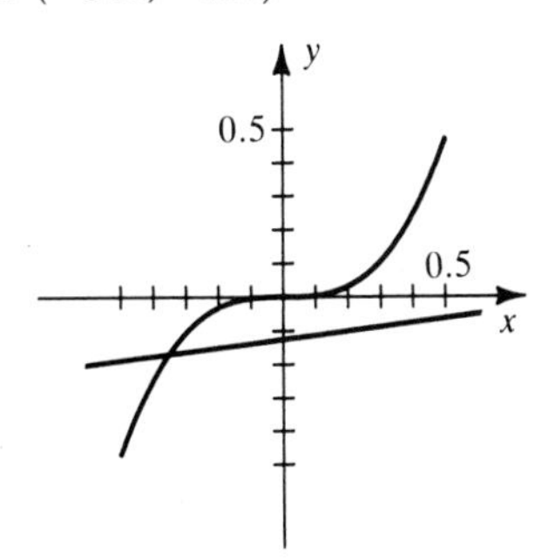

EXERCISES 16.9

1 Min: $f\left(\frac{1}{\sqrt{5}}, \frac{2}{\sqrt{5}}\right) = f\left(-\frac{1}{\sqrt{5}}, -\frac{2}{\sqrt{5}}\right) = 0$;
max: $f\left(\frac{2}{\sqrt{5}}, -\frac{1}{\sqrt{5}}\right) = f\left(-\frac{2}{\sqrt{5}}, \frac{1}{\sqrt{5}}\right) = 5$
3 Min: $f\left(-\frac{5}{\sqrt{3}}, -\frac{5}{\sqrt{3}}, -\frac{5}{\sqrt{3}}\right) = -5\sqrt{3}$;
max: $f\left(\frac{5}{\sqrt{3}}, \frac{5}{\sqrt{3}}, \frac{5}{\sqrt{3}}\right) = 5\sqrt{3}$
5 Min: $f\left(\frac{1}{3}, -\frac{1}{3}, \frac{1}{3}\right) = \frac{1}{3}$
7 Min: $f(0, -1, 0) = 1$ and $f(2, 1, 0) = 5$
9 Min: $f\left(1, \frac{2}{\sqrt{3}}, -1, \frac{8}{3}\right) = -\frac{16}{3\sqrt{3}}$;
max: $f\left(1, -\frac{2}{\sqrt{3}}, -1, \frac{8}{3}\right) = \frac{16}{3\sqrt{3}}$
11 $\left(\frac{6}{\sqrt{29}}, \frac{9}{\sqrt{29}}, \frac{12}{\sqrt{29}}\right)$ **13** Square base of side $\frac{2}{\sqrt[3]{7}}$, height $\frac{7}{2\sqrt[3]{7}}$
15 $\frac{8}{3}$ **17** Height is twice the radius.
21 Width $= 8\sqrt{3}$ in., depth $= \frac{16}{3}\sqrt{6}$ in.
23 Max: $f(0.97, 0.17) \approx 1.55$; min: $f(-0.87, -0.35) \approx -2.73$

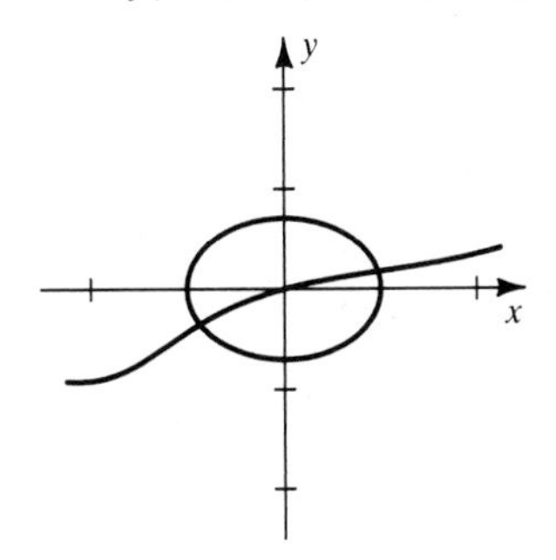

16.10 REVIEW EXERCISES

1 $\{(x, y): 4x^2 - 9y^2 \le 36\}$; the hyperbola $9y^2 - 4x^2 = 108$
3 $\{(x, y, z): z^2 > x^2 + y^2\}$;
the hyperboloid of two sheets $z^2 - x^2 - y^2 = 1$
5 $\frac{5}{4}$ **7** DNE **9** Lemniscates; DNE
11 $f_x(x, y) = 3x^2 \cos y + 4$; $f_y(x, y) = -x^3 \sin y - 2y$
13 $f_x(x, y, z) = \frac{2x}{y^2 + z^2}$; $f_y(x, y, z) = \frac{2y(z^2 - x^2)}{(y^2 + z^2)^2}$;
$f_z(x, y, z) = -\frac{2z(x^2 + y^2)}{(y^2 + z^2)^2}$
15 $f_x(x, y, z, t) = 2xz\sqrt{2y + t}$; $f_y(x, y, z, t) = \frac{x^2 z}{\sqrt{2y + t}}$;
$f_z(x, y, z, t) = x^2\sqrt{2y + t}$; $f_t(x, y, z, t) = \frac{x^2 z}{2\sqrt{2y + t}}$
17 $f_{xx}(x, y) = 6xy^2 + 12x^2$; $f_{xy}(x, y) = f_{yx}(x, y) = 6x^2 y - 9y^2$;
$f_{yy}(x, y) = 2x^3 - 18xy$
21 **(a)** $(2x + 3y)\,\Delta x + (3x - 2y)\,\Delta y + (\Delta x)^2 + 3\,\Delta x\,\Delta y - (\Delta y)^2$;
$(2x + 3y)\,dx + (3x - 2y)\,dy$ **(b)** -1.13; -1.1

25 $12x + 18y;\ 18x - 22y$ **27** $3e^{-t}(9t^2 \cos 3t^3 - \sin 3t^3)$

29 $\dfrac{z^3 - 2xy}{\cos y - 3xz^2};\ \dfrac{z \sin y - x^2}{\cos y - 3xz^2}$ **31** **(a)** $-\dfrac{14}{\sqrt{41}}$ **(b)** 14

33 $-16(x + 2) + 4(y + 1) - 7(z - 2) = 0;\ x = -2 - 16t,$ $y = -1 + 4t,\ z = 2 - 7t$

37

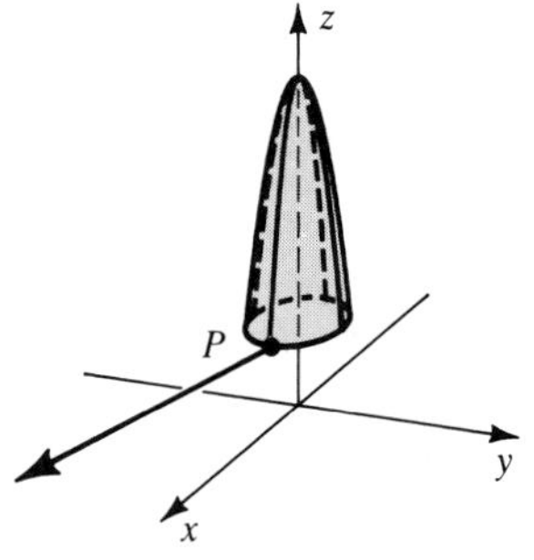

39 Min: $f(0, -1) = -2$

41 Min: $f\left(-\sqrt{\frac{8}{3}}, -\sqrt{\frac{2}{3}}, -\sqrt{\frac{4}{3}}\right) = f\left(-\sqrt{\frac{8}{3}}, \sqrt{\frac{2}{3}}, \sqrt{\frac{4}{3}}\right)$
$= f\left(\sqrt{\frac{8}{3}}, -\sqrt{\frac{2}{3}}, \sqrt{\frac{4}{3}}\right)$
$= f\left(\sqrt{\frac{8}{3}}, \sqrt{\frac{2}{3}}, -\sqrt{\frac{4}{3}}\right) = -\frac{8}{9}\sqrt{3};$

max: $f\left(\sqrt{\frac{8}{3}}, \sqrt{\frac{2}{3}}, \sqrt{\frac{4}{3}}\right) = f\left(-\sqrt{\frac{8}{3}}, -\sqrt{\frac{2}{3}}, \sqrt{\frac{4}{3}}\right)$
$= f\left(-\sqrt{\frac{8}{3}}, \sqrt{\frac{2}{3}}, -\sqrt{\frac{4}{3}}\right)$
$= f\left(\sqrt{\frac{8}{3}}, -\sqrt{\frac{2}{3}}, -\sqrt{\frac{4}{3}}\right) = \frac{8}{9}\sqrt{3}$

43 $(0, 4, 4)$ **45** $\sqrt{\dfrac{TC}{k_0 k_1}}$

CHAPTER 17

EXERCISES 17.1

1 R_x **3** R_y **5** Neither **7** R_x and R_y **9** Neither

11 **(a)** 39 **(b)** 81 **(c)** 60 **13** -36 **15** $\frac{163}{120}$ **17** $\frac{36}{5}$

19 $\frac{1}{2}(4e - e^4) \approx -21.86$

21

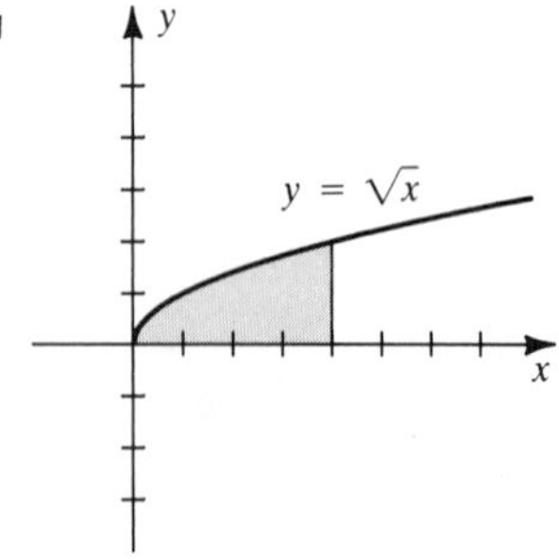

(a) $\int_0^4 \int_0^{\sqrt{x}} f(x, y)\, dy\, dx$
(b) $\int_0^2 \int_{y^2}^4 f(x, y)\, dx\, dy$

23

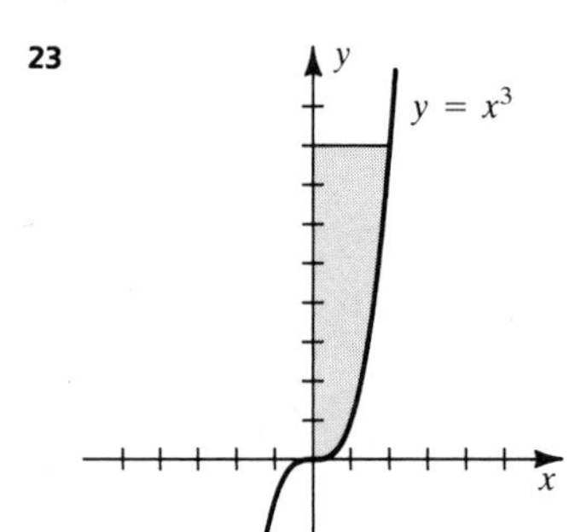

(a) $\int_0^2 \int_{x^3}^8 f(x, y)\, dy\, dx$
(b) $\int_0^8 \int_0^{y^{1/3}} f(x, y)\, dx\, dy$

25

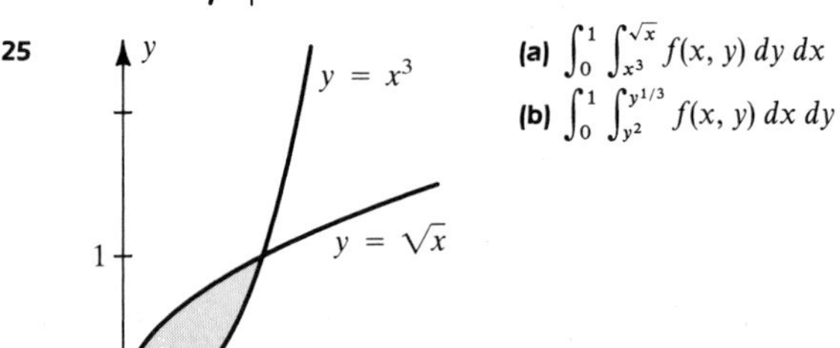

(a) $\int_0^1 \int_{x^3}^{\sqrt{x}} f(x, y)\, dy\, dx$
(b) $\int_0^1 \int_{y^2}^{y^{1/3}} f(x, y)\, dx\, dy$

Exer. 27–32: Answers are not unique.

27 $\int_{-1}^4 \int_{-1}^2 (y + 2x)\, dx\, dy = \frac{75}{2}$ **29** $\int_0^1 \int_{-2y}^{3y} xy^2\, dx\, dy = \frac{1}{2}$

31 $\int_0^2 \int_0^{x^2} x^3 \cos xy\, dy\, dx = \frac{1}{3}(1 - \cos 8) \approx 0.38$

33

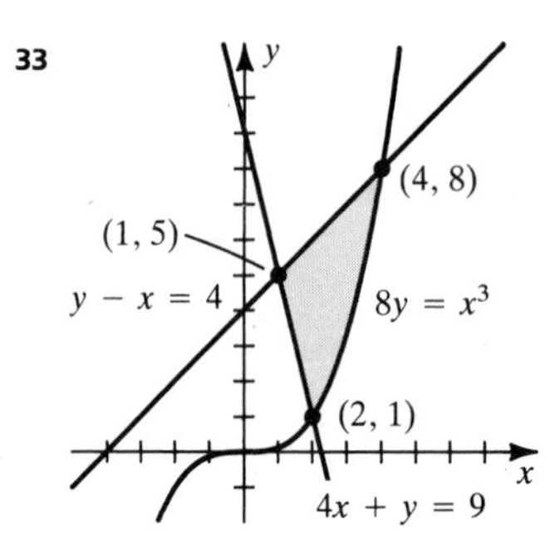

(a) $\int_1^2 \int_{9-4x}^{x+4} f(x, y)\, dy\, dx + \int_2^4 \int_{x^3/8}^{x+4} f(x, y)\, dy\, dx$
(b) $\int_1^5 \int_{(9-y)/4}^{2y^{1/3}} f(x, y)\, dx\, dy + \int_5^8 \int_{y-4}^{2y^{1/3}} f(x, y)\, dx\, dy$

35

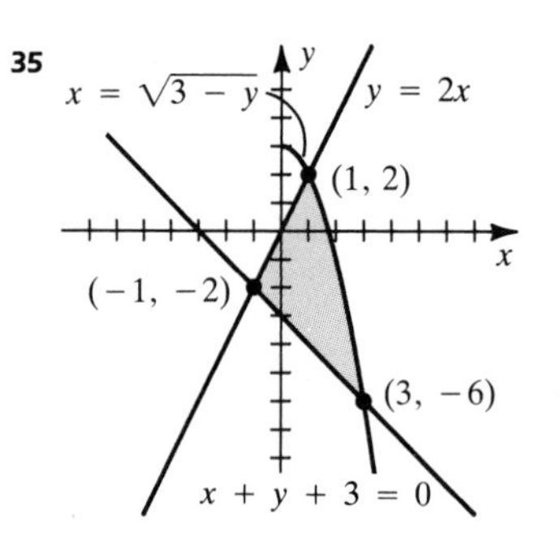

(a) $\int_{-1}^1 \int_{-x-3}^{2x} f(x, y)\, dy\, dx + \int_1^3 \int_{-x-3}^{3-x^2} f(x, y)\, dy\, dx$
(b) $\int_{-6}^{-2} \int_{-y-3}^{\sqrt{3-y}} f(x, y)\, dx\, dy + \int_{-2}^2 \int_{y/2}^{\sqrt{3-y}} f(x, y)\, dx\, dy$

37

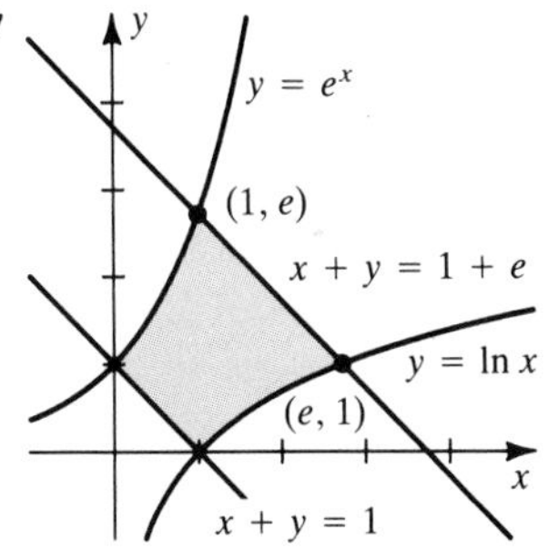

(a) $\int_0^1 \int_{1-x}^{e^x} f(x, y)\,dy\,dx + \int_1^e \int_{\ln x}^{1+e-x} f(x, y)\,dy\,dx$

(b) $\int_0^1 \int_{1-y}^{e^y} f(x, y)\,dx\,dy + \int_1^e \int_{\ln y}^{1+e-y} f(x, y)\,dx\,dy$

39

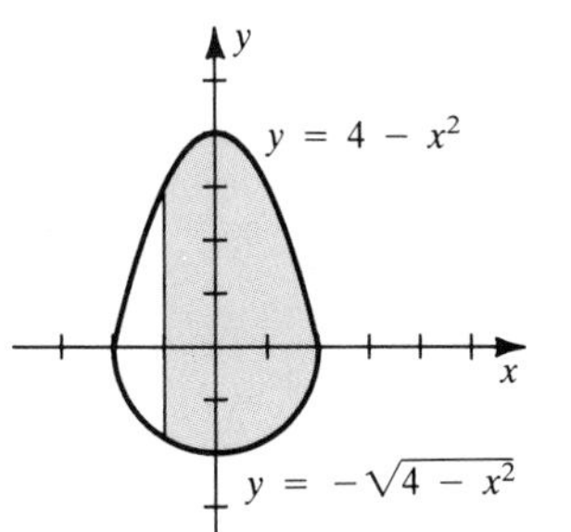

41

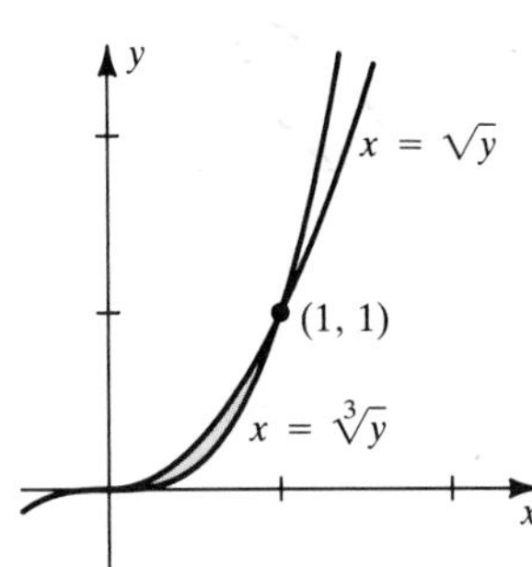

43

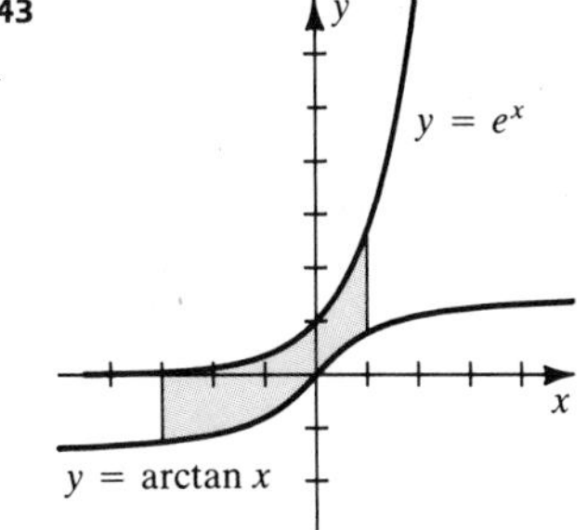

45 $\int_0^2 \int_0^{y/2} e^{(y^2)}\,dx\,dy = \frac{1}{4}(e^4 - 1) \approx 13.40$

47 $\int_0^4 \int_0^{\sqrt{x}} y \cos x^2\,dy\,dx = \frac{1}{4}\sin 16 \approx -0.07$

49 $\int_0^2 \int_0^{x^3} \frac{y}{\sqrt{16 + x^7}}\,dy\,dx = \frac{8}{7}$

51 1.16 **53** 0.75

EXERCISES 17.2

1 $\int_0^5 \int_{-x}^{4x - x^2} dy\,dx$

3 $\int_{-3}^3 \int_{-\sqrt{9-y^2}}^{(9-y)/2} dx\,dy$

5 $\int_1^2 \int_{-x^2}^{1/x^2} dy\,dx = \frac{17}{6}$ **7** $\int_{-1}^2 \int_{-y^2}^{y+4} dx\,dy = \frac{33}{2}$

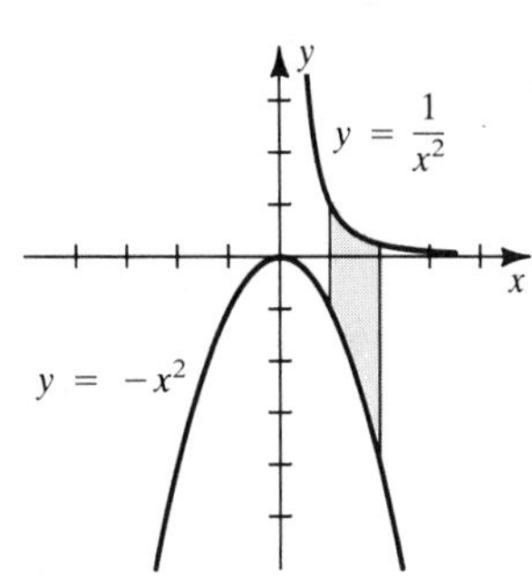

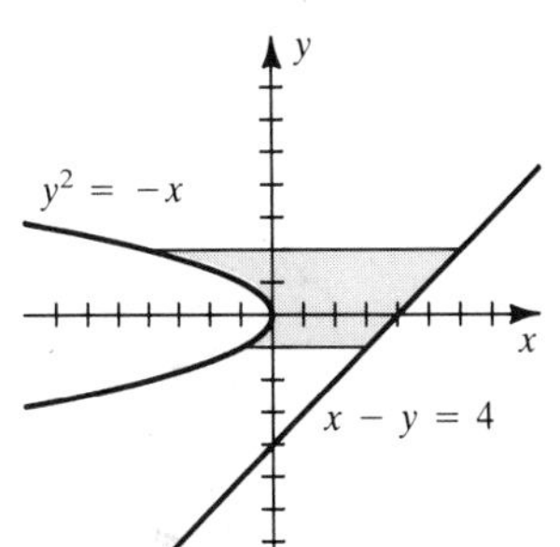

9 $\int_0^1 \int_x^{3x} dy\,dx + \int_1^2 \int_x^{4-x} dy\,dx = 2$

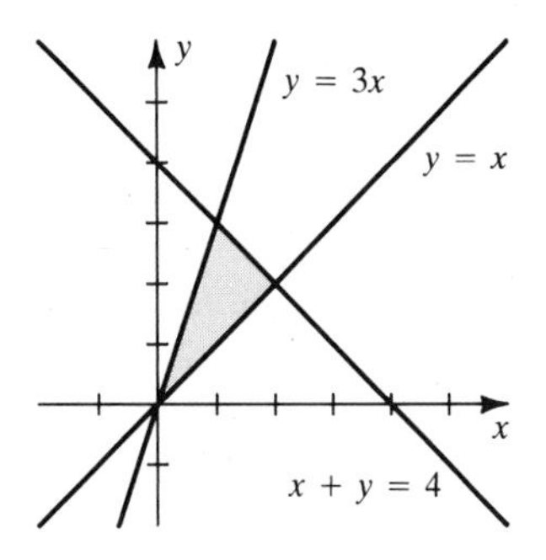

11 $\int_{-\pi}^{\pi} \int_{\sin x}^{e^x} dy\,dx = e^{\pi} - e^{-\pi} \approx 23.10$

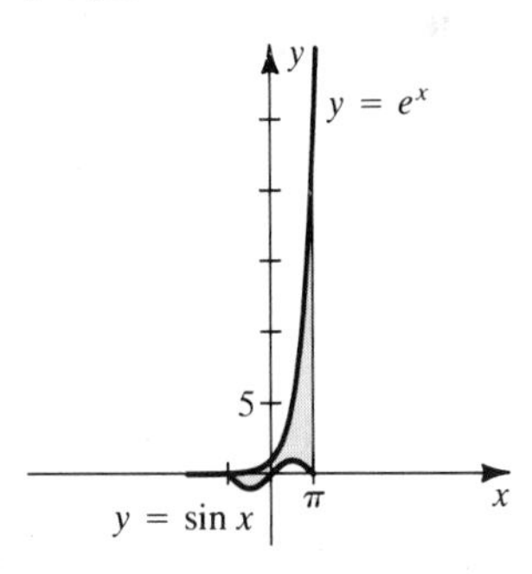

13 $\int_0^3 \int_0^{(12-4x)/3} \frac{1}{12}(60 - 20x - 15y)\,dy\,dx$

15 $\int_0^2 \int_0^{4-y^2} (6 - x)\,dx\,dy$

17 The plane $z = 3$

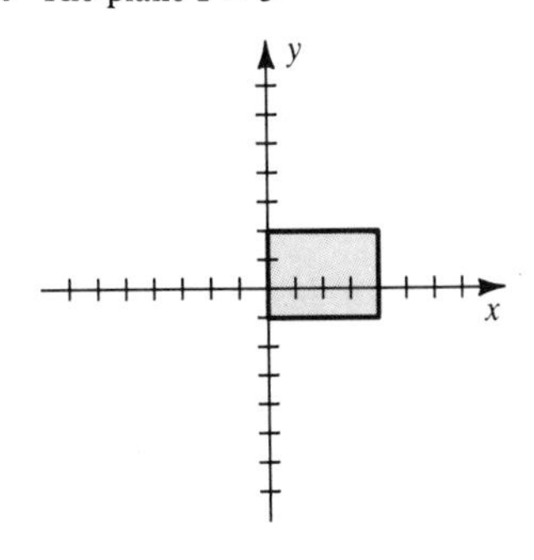

19 The paraboloid $z = x^2 + y^2$

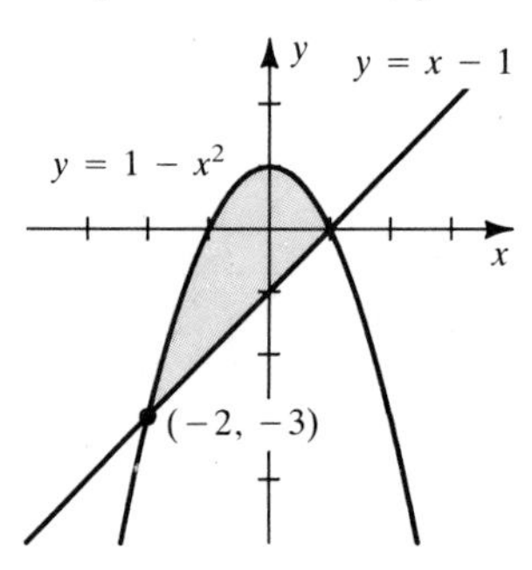

21 $\frac{34}{3}$

23 $\int_0^3 \int_0^{2x} (9 - x^2)^{1/2}\, dy\, dx = 18$

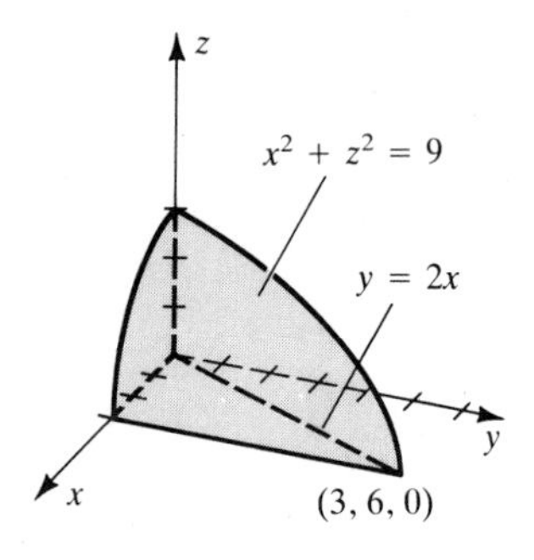

25 $\int_0^2 \int_0^{4-2x} (4 - 2x - y)\, dy\, dx = \frac{16}{3}$

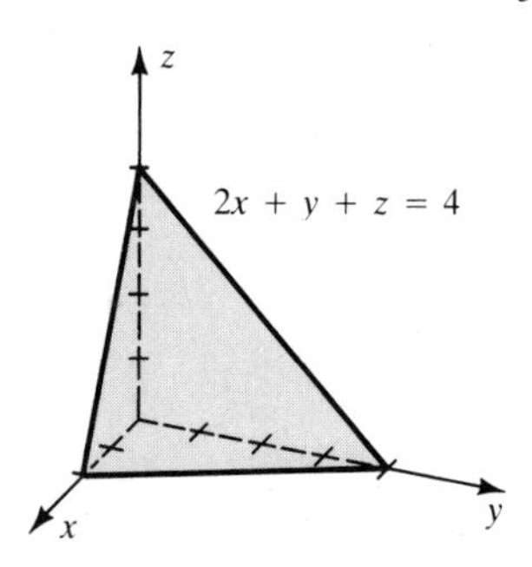

27 $\int_0^2 \int_0^{4-x^2} (x^2 + y^2)\, dy\, dx = \frac{832}{35}$

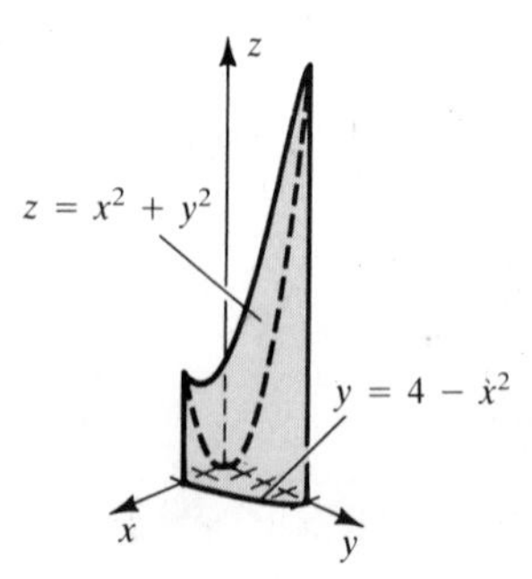

29 $\int_0^4 \int_{x^2/16}^{\sqrt{x}/2} x^3\, dy\, dx = \frac{128}{9}$

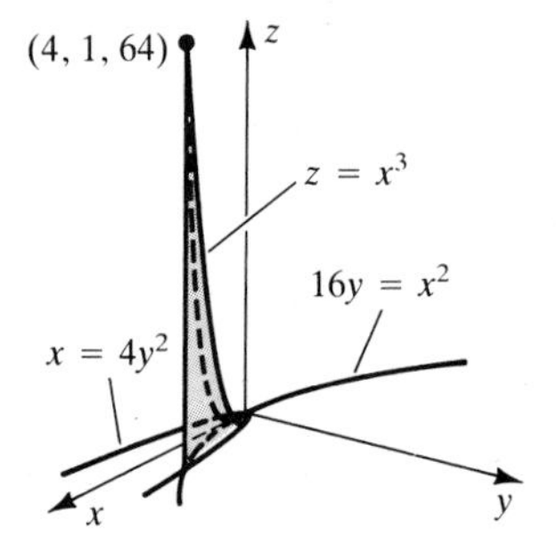

31 $\int_{-1}^{2} \int_{2-x}^{4-x^2} (x^2 + 4)\, dy\, dx = \frac{423}{20}$ **33** 0.42

EXERCISES 17.3

1 $2\int_0^{\pi/2} \int_0^{4\sin\theta} r\, dr\, d\theta$ **3** $2\int_0^{2\pi/3} \int_0^{1+2\cos\theta} r\, dr\, d\theta$

5 $4\int_0^{\pi/4} \int_0^{2\sqrt{\cos 2\theta}} r\, dr\, d\theta$ **7** $\int_0^{\pi/3} \int_0^{4\sin 3\theta} r\, dr\, d\theta = \frac{4\pi}{3}$

9 $2\int_{2\pi/3}^{\pi} \int_3^{2-2\cos\theta} r\, dr\, d\theta = \frac{9}{2}\sqrt{3} - \pi \approx 4.65$

11 $2\int_0^{\pi/4} \int_0^{3\sqrt{\cos 2\theta}} r\, dr\, d\theta = \frac{9}{2}$ **13** $\int_0^{2\pi} \int_0^2 (r^3)\, r\, dr\, d\theta = \frac{64\pi}{5}$

15 $\int_0^{2\pi} \int_a^b (\cos^2\theta)\, r\, dr\, d\theta = \frac{\pi}{2}(b^2 - a^2)$

17 $\int_0^{\pi/4} \int_0^{3\sec\theta} (r)\, r\, dr\, d\theta = \frac{9}{2}[\sqrt{2} + \ln(\sqrt{2} + 1)] \approx 10.33$

19 $\int_0^{\pi} \int_0^a e^{-r^2}\, r\, dr\, d\theta = \frac{\pi}{2}(1 - e^{-a^2})$

21 $\int_0^{\pi/4} \int_{\sec\theta}^{2\sec\theta} \left(\frac{1}{r}\right) r\, dr\, d\theta = \ln(\sqrt{2} + 1) \approx 0.88$

23 $\int_0^{\pi/2} \int_0^2 \cos(r^2)\, r\, dr\, d\theta = \frac{\pi}{4}\sin 4 \approx -0.59$

25 $8\int_0^{\pi/2} \int_3^5 (25 - r^2)^{1/2}\, r\, dr\, d\theta = \frac{256\pi}{3}$

27 $2\int_{-\pi/2}^{\pi/2} \int_0^{2\cos\theta} (r)\, r\, dr\, d\theta = \frac{64}{9}$

29 $4\int_0^{\pi/2} \int_0^{4\sin\theta} (16 - r^2)^{1/2}\, r\, dr\, d\theta = \frac{128}{9}(3\pi - 4) \approx 77.15$

31 π **33** 7.307

EXERCISES 17.4

1 $4\int_0^1 \int_0^1 \sqrt{\left(\frac{-x}{\sqrt{4 - x^2 - y^2}}\right)^2 + \left(\frac{-y}{\sqrt{4 - x^2 - y^2}}\right)^2 + 1}\, dy\, dx$

3

$4\int_0^3 \int_0^{\sqrt{9-x^2}} \sqrt{\left(\frac{8x}{3\sqrt{16x^2 + 9y^2 + 144}}\right)^2 + \left(\frac{3y}{2\sqrt{16x^2 + 9y^2 + 144}}\right)^2 + 1}\, dy\, d$

5 $\int_0^1 \int_0^1 \sqrt{(x)^2 + (1)^2 + 1}\, dy\, dx =$

$\frac{1}{2}[\sqrt{3} + 2\ln(1 + \sqrt{3}) - \ln 2] \approx 1.52$

7 $\pi c k^2 \sqrt{\left(\frac{1}{a}\right)^2 + \left(\frac{1}{b}\right)^2 + \left(\frac{1}{c}\right)^2}$ **9** $\frac{\pi}{6}(5^{3/2} - 1) \approx 5.33$

11 $2a^2(\pi - 2)$ **13** 247.4 ft^2 **15** 1.24

EXERCISES 17.5

1 $\frac{39}{2}$ **3** $-\frac{1}{12}$ **5** $\frac{513}{8}$

7 $\int_0^6 \int_0^{(6-x)/2} \int_0^{(6-x-2y)/3} f(x, y, z)\, dz\, dy\, dx$;

$\int_0^3 \int_0^{6-2y} \int_0^{(6-x-2y)/3} f(x, y, z)\, dz\, dx\, dy$;

$\int_0^3 \int_0^{(6-2y)/3} \int_0^{6-2y-3z} f(x, y, z)\, dx\, dz\, dy$;

$\int_0^2 \int_0^{(6-3z)/2} \int_0^{6-2y-3z} f(x, y, z)\, dx\, dy\, dz$;

$\int_0^6 \int_0^{(6-x)/3} \int_0^{(6-x-3z)/2} f(x, y, z)\, dy\, dz\, dx$;

$\int_0^2 \int_0^{6-3z} \int_0^{(6-x-3z)/2} f(x, y, z)\, dy\, dx\, dz$

9 $\int_{-3/2}^{3/2} \int_{-\sqrt{9-4x^2}}^{\sqrt{9-4x^2}} \int_0^{9-4x^2-y^2} f(x, y, z)\, dz\, dy\, dx$;

$\int_{-3}^{3} \int_{-\sqrt{9-y^2}/2}^{\sqrt{9-y^2}/2} \int_0^{9-4x^2-y^2} f(x, y, z)\, dz\, dx\, dy$;

$\int_{-3}^{3} \int_0^{9-y^2} \int_{-\sqrt{9-z-y^2}/2}^{\sqrt{9-z-y^2}/2} f(x, y, z)\, dx\, dz\, dy$;

$\int_0^9 \int_{-\sqrt{9-z}}^{\sqrt{9-z}} \int_{-\sqrt{9-z-y^2}/2}^{\sqrt{9-z-y^2}/2} f(x, y, z)\, dx\, dy\, dz$;

$\int_{-3/2}^{3/2} \int_0^{9-4x^2} \int_{-\sqrt{9-4x^2-z}}^{\sqrt{9-4x^2-z}} f(x, y, z)\, dy\, dz\, dx$;

$\int_0^9 \int_{-\sqrt{9-z}/2}^{\sqrt{9-z}/2} \int_{-\sqrt{9-4x^2-z}}^{\sqrt{9-4x^2-z}} f(x, y, z)\, dy\, dx\, dz$

11 $2\int_0^4 \int_0^{4-y} \int_0^{\sqrt{4-z}} dx\, dz\, dy = \frac{128}{5}$

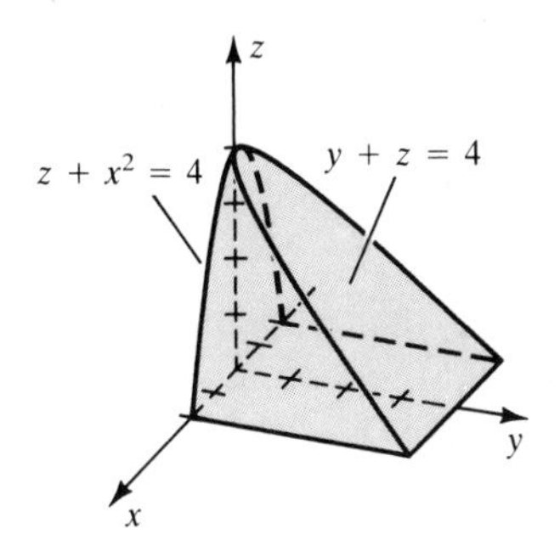

13 $\int_{-1}^{1} \int_{z^2}^{2-z^2} \int_0^{4-z} dx\, dy\, dz = \frac{32}{3}$

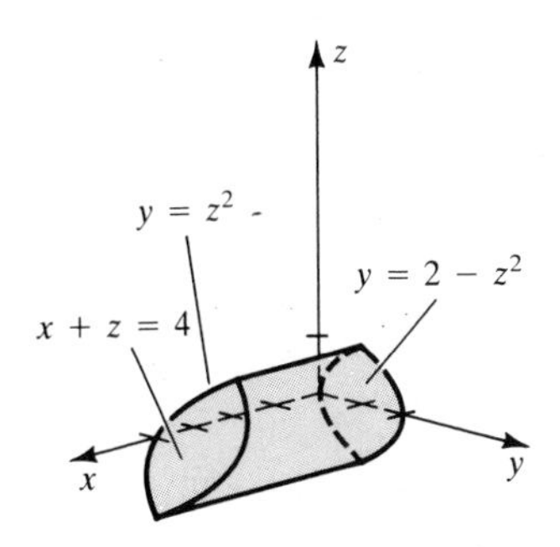

15 $\int_{-1}^{1} \int_{-\sqrt{1-y^2}}^{\sqrt{1-y^2}} \int_0^{2-y-z} dx\, dz\, dy = 2\pi$

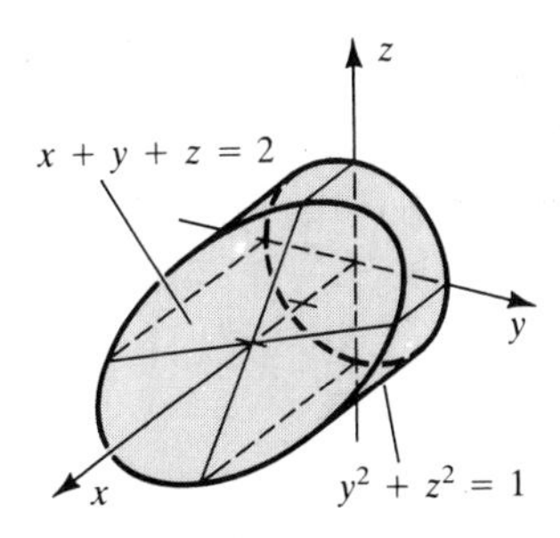

17 $\int_{-3}^{3} \int_{-1}^{2} \int_0^{9-x^2} dz\, dy\, dx = 108$

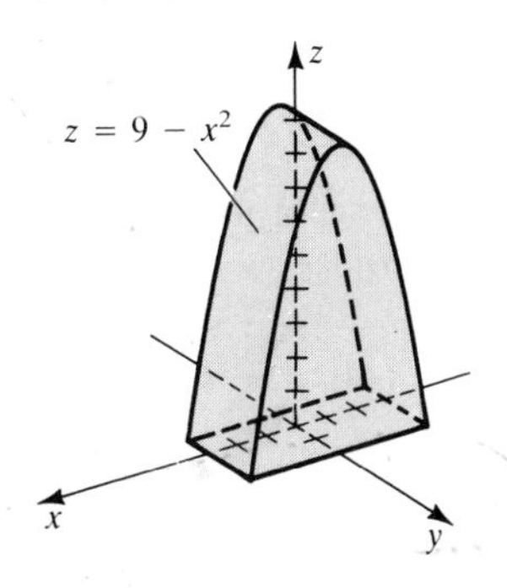

19 $\int_0^1 \int_{x^3}^{x^2} \int_0^{z^2} dy\, dz\, dx = \frac{1}{70}$

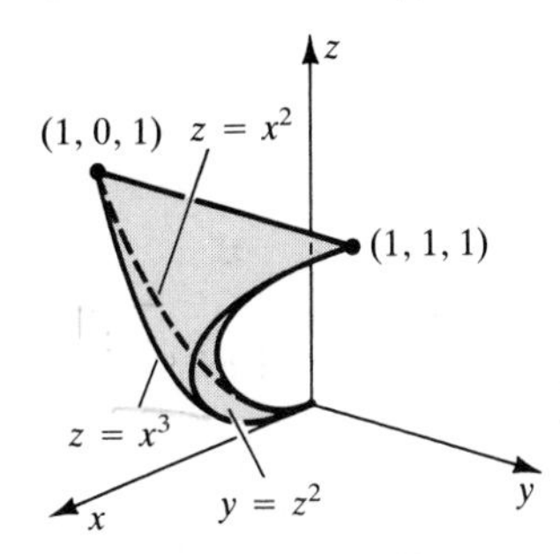

21 $\frac{1}{6}abc$

23 The region bounded by the planes $z = 0$, $z = 1$, $x = 2$, $x = 3$ and the cylinders $y = \sqrt{1 - z}$ and $y = \sqrt{4 - z}$

25 The region under the plane $z = x + y$ and over the region in the xy-plane bounded by the parabola $y = x^2$ and the line $y = 2x$

27 The region bounded by the paraboloid $z = x^2 + y^2$ and the planes $z = 1$ and $z = 2$

29 $\int_0^1 \int_0^{e^{-x}} y^2\, dy\, dx$ **31** $\int_0^2 \int_0^{4-2y} \int_0^{4-x-2y} (x^2 + y^2)\, dz\, dx\, dy$

33 1.1685×10^9 kg **35** 0.77

EXERCISES 17.6

1 $m = \frac{2349}{20}$; $\bar{x} = \frac{1290}{203}$, $\bar{y} = \frac{38}{29}$

3 $m = 8k$ (k a proportionality constant); $\bar{x} = 0$, $\bar{y} = \frac{8}{3}$

5 $m = \frac{1}{4}(1 - e^{-2}) \approx 0.22$; $\bar{x} = 0$, $\bar{y} = \frac{4(1 - e^{-3})}{9(1 - e^{-2})} \approx 0.49$

7 $m = 4 \ln(\sqrt{2} + 1) - 4 \ln(\sqrt{2} - 1) - \pi \approx 3.91$; $\bar{x} = 0$, $\bar{y} = \frac{16 - \pi}{4m} \approx 0.82$

9 $3^5\left(\frac{31}{28}\right)$; $3^7\left(\frac{19}{8}\right)$; $3^5\left(\frac{1259}{56}\right)$ **11** $64k$; $\frac{32}{3}k$; $\frac{224}{3}k$

13 (δ = density) **(a)** $\frac{1}{3}a^4\delta$ **(b)** $\frac{1}{12}a^4\delta$ **(c)** $\frac{1}{6}a^4\delta$ **15** $\frac{a}{\sqrt{3}}$

17 $\bar{x} = \bar{y} = \bar{z} = \frac{7}{12}a$ (with fixed corner at O)

19 $m = \int_0^4 \int_{-\sqrt{x}/2}^{\sqrt{x}/2} \int_{-\sqrt{x-4z^2}}^{\sqrt{x-4z^2}} (x^2 + z^2)\, dy\, dz\, dx$; the integrals for M_{yz}, M_{xz}, and M_{xy} have the same limits, but the integrands are $x(x^2 + z^2)$, $y(x^2 + z^2)$, and $z(x^2 + z^2)$, respectively.

21 $m = \int_{-a}^{a} \int_{-\sqrt{a^2-x^2}}^{\sqrt{a^2-x^2}} \int_0^{\sqrt{a^2-x^2-y^2}} dz\, dy\, dx$; the integral for M_{xy} has the same limits, but the integrand is z. By symmetry, $\bar{x} = \bar{y} = 0$.

23 **(a)** $m = \int_0^3 \int_0^{9-x^2} \int_0^{6-2x} dy\, dz\, dx$; the integrals for M_{yz}, M_{xz}, and M_{xy} have the same limits, but the integrands are x, y, and z, respectively.

(b) $m = \frac{135}{2}$; $M_{yz} = \frac{567}{10}$; $M_{xz} = \frac{729}{5}$; $M_{xy} = \frac{2673}{10}$; $\bar{x} = \frac{21}{25}$, $\bar{y} = \frac{54}{25}$, $\bar{z} = \frac{99}{25}$

25 $I_z = \int_{-a}^{a} \int_{-\sqrt{a^2-x^2}}^{\sqrt{a^2-x^2}} \int_{-\sqrt{a^2-x^2-y^2}}^{\sqrt{a^2-x^2-y^2}} (x^2 + y^2)(x^2 + y^2 + z^2)\, dz\, dy\, dx$

27 $I_z = \int_0^a \int_0^{b(1-x/a)} \int_0^{c(1-x/a-y/b)} (x^2 + y^2)\, \delta\, dz\, dy\, dx$

Exer. 29–32: Answers are not unique.

29 1.2 **31** 1.3

EXERCISES 17.7

1 **(a)** The right circular cylinder of radius 4 with axis along the z-axis **(b)** The yz-plane **(c)** The plane parallel to the xy-plane with z-intercept 1

3 The plane parallel to the yz-plane with x-intercept -3

5 A paraboloid with vertex $(0, 0, 0)$ and opening upward

7 The right circular cylinder with trace $x^2 + (y - 3)^2 = 9$ in the xy-plane

9 The cone $z^2 = 4x^2 + 4y^2$

11 The sphere with center at the origin and radius 3

13 The cylinder with trace $y^2 = 2x$ in the xy-plane and rulings parallel to the z-axis

15 $r^2 + z^2 = 4$ **17** $3r\cos\theta + r\sin\theta - 4z = 12$ **19** $r = 2$

21 $r = 2z$ **23** $r^2 \sin^2\theta + z^2 = 9$

25 $\int_0^{2\pi} \int_0^4 \int_0^{16-r^2} f(r, \theta, z)\, r\, dz\, dr\, d\theta$

27 $\int_0^{2\pi} \int_0^3 \int_0^4 f(r, \theta, z)\, r\, dz\, dr\, d\theta + \int_0^{2\pi} \int_3^4 \int_0^{\sqrt{25-r^2}} f(r, \theta, z)\, r\, dz\, dr\, d\theta$

29 **(a)** 8π **(b)** $\left(0, 0, \frac{4}{3}\right)$

31 **(a)** $\frac{1}{2}\pi h a^4 \delta$ **(b)** $\pi h a^2\left(\frac{1}{4}a^2 + \frac{1}{3}h^2\right)\delta$ **33** $\frac{1}{4}k\pi^2 a^4$

35 $\frac{1}{8}k\pi^2 a^6$ **37** 215,360 kg **39** $\frac{7\pi}{16}$

EXERCISES 17.8

1 **(a)** $(0, 2, 2\sqrt{3})$ **(b)** $\left(2, \frac{\pi}{2}, 2\sqrt{3}\right)$

3 **(a)** $\left(\sqrt{10}, \cos^{-1}\left(\frac{-2}{\sqrt{5}}\right), \frac{\pi}{4}\right)$ **(b)** $\left(\sqrt{2}, \frac{\pi}{4}, -2\sqrt{2}\right)$

5 **(a)** The sphere of radius 3 and center O **(b)** A half-cone with vertex O and vertex angle $\frac{\pi}{3}$ **(c)** A half-plane with edge on the z-axis and making an angle of $\frac{\pi}{3}$ with the xz-plane

7 The sphere of radius 2 and center $(0, 0, 2)$

9 The plane $z = 3$

11 The sphere of radius 3 and center $(3, 0, 0)$

13 The plane $y = 5$

15 The right circular cylinder of radius 5 with axis along the z-axis

17 The cone $x^2 + y^2 = 4z^2$ **19** The paraboloid $6z = x^2 + y^2$

21 $\rho = 2$ **23** $\rho(3\sin\phi\cos\theta + \sin\phi\sin\theta - 4\cos\phi) = 12$

25 $\rho = 2\csc\phi$ **27** $\tan^2\phi = 4$

29 $\rho^2(\sin^2\phi\sin^2\theta + \cos^2\phi) = 9$

31 $\frac{1}{2}k\pi a^4$ (k a proportionality constant); center of mass is $\frac{2}{5}a$ from base along the axis of symmetry.

33 $\frac{2}{9}k\pi a^6$ **35** $\frac{16\pi}{3}$ **37** $\frac{124}{5}k\pi$

39 $\frac{256\pi}{5}(\sqrt{2} - 1) \approx 66.63$

41 **(a)** $(-3\sqrt{2}, 3\sqrt{6}, -6\sqrt{2})$ **(b)** θ should be increased by 105°, ϕ decreased by $\left(45 + \arctan\frac{8}{\sqrt{128}}\right) \approx 80.26°$, and L increased by $\sqrt{192} - 12 \approx 1.86$ in.

EXERCISES 17.9

1 **(a)** Vertical lines; horizontal lines **(b)** $x = \frac{1}{3}u$, $y = \frac{1}{5}v$

3 **(a)** Lines with slopes 1 and $-\frac{2}{3}$ **(b)** $x = \frac{3}{5}u + \frac{1}{5}v$, $y = -\frac{2}{5}u + \frac{1}{5}v$

5 **(a)** Vertical lines; lines with slope -1 **(b)** $x = u^{1/3}$, $y = v - u^{1/3}$

7 **(a)** Vertical lines; horizontal lines **(b)** $x = \ln u$, $y = \ln v$

9 **(a)** The rectangle with vertices $(0, 0)$, $(6, 0)$, $(6, 5)$, $(0, 5)$ **(b)** The ellipse $\frac{u^2}{9} + \frac{v^2}{25} = 1$

11 **(a)** The triangle with vertices $(0, 0)$, $(-1, 3)$, $(2, 4)$ **(b)** The line $-u + 3v = 5$

13 $4u^2 + 4v^2$ **15** $2u(uv - 1)e^{-(2u+v)}$ **17** -6

19 $\int_0^2 \int_0^{2-u} (v - u)\, 2\, dv\, du$

21 $\int_{-1}^{1} \int_{-\sqrt{1-v^2}}^{\sqrt{1-v^2}} (u^2 + v^2)\, 6\, du\, dv$

23 $\int_1^3 \int_{-1}^{1} (u^2\cos^2 v)\frac{1}{2}\, du\, dv = \frac{1}{3} + \frac{1}{12}\sin 6 - \frac{1}{12}\sin 2 \approx 0.23$

25 $\int_1^2 \int_{\sqrt{u}}^{\sqrt{2u}} \left(\frac{u^2}{v^2} + 2v^2\right)\frac{1}{v}\, dv\, du = \frac{15}{8}$

27 $-\int_2^4 \int_{-u/2}^{2u} \left(\frac{v}{u}\right)\left(-\frac{1}{5}\right) dv\, du = \frac{9}{4}$ **29** 1.08×10^{12} km³

31 **(a)** $(1, 1, 1)$ **(b)** $\left(\frac{1}{\sqrt{2}}, \frac{1}{\sqrt{2}}, -\sqrt{2}\right)$

(c) $\left(\frac{1}{\sqrt{2}} - 1, \frac{1}{\sqrt{2}} + 1, 0\right)$

37

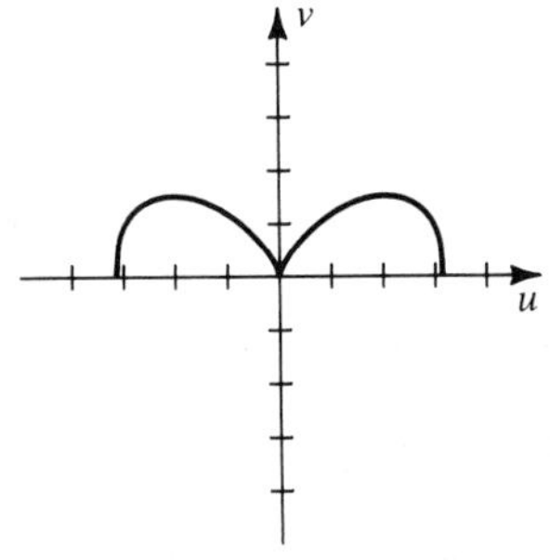

49 **(a)** $\int_0^4 \int_0^{\sqrt{25-x^2}} dy\,dx$ **(b)** $\int_0^3 \int_0^4 dx\,dy + \int_3^5 \int_0^{\sqrt{25-y^2}} dx\,dy$

(c) $\int_0^{\arctan(3/4)} \int_0^{4\sec\theta} r\,dr\,d\theta + \int_{\arctan(3/4)}^{\pi/2} \int_0^5 r\,dr\,d\theta$

51 **(a)** $\int_0^5 \int_0^{\sqrt{25-x^2}} \int_0^{\sqrt{25-x^2-y^2}} dz\,dy\,dx - \int_0^4 \int_0^{\sqrt{16-x^2}} \int_3^{\sqrt{25-x^2-y^2}} dz\,dy\,dx$

(b) $\int_0^{\pi/2} \int_0^4 \int_0^3 r\,dz\,dr\,d\theta + \int_0^{\pi/2} \int_4^5 \int_0^{\sqrt{25-r^2}} r\,dz\,dr\,d\theta$

(c) $\int_0^{\pi/2} \int_0^{\arctan(4/3)} \int_0^{3\sec\phi} \rho^2 \sin\phi\,d\rho\,d\phi\,d\theta + \int_0^{\pi/2} \int_{\arctan(4/3)}^{\pi/2} \int_0^5 \rho^2 \sin\phi\,d\rho\,d\phi\,d\theta$

53 $e - 1 \approx 1.72$

17.10 REVIEW EXERCISES

1 $-\frac{5}{84}$ **3** $\frac{63}{4}$ **5** $-\frac{107}{210}$

7 $\int_2^4 \int_{-\sqrt{x^2-4}}^{\sqrt{x^2-4}} f(x, y)\,dy\,dx$ **9** $\int_{-2}^2 \int_{y^2-4}^{4-y^2} f(x, y)\,dx\,dy$

11

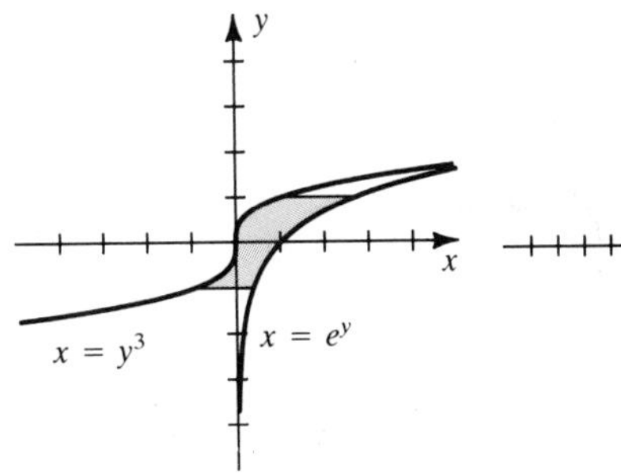

13

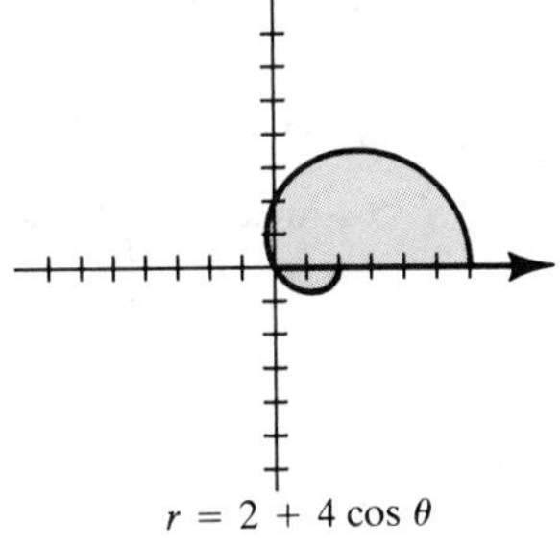

$r = 2 + 4\cos\theta$

15 $\frac{1}{4}(1 - e^{-81}) \approx 0.25$ **17** 13 **19** $4\sqrt{2\pi} \approx 17.77$

21 $2\ln 2 - \frac{3}{4} \approx 0.64$

23 $\int_0^{\pi/2} \int_0^{\pi/4} \int_0^{4\sqrt{2}} (\rho \cdot \rho^2 \sin\phi)\,d\rho\,d\phi\,d\theta = 64(2 - \sqrt{2})\pi \approx 117.78$

25 $9k$ (k a proportionality constant); $\left(\frac{9}{4}, \frac{27}{8}\right)$

27 $2\pi k$ **29** $\frac{1}{20}ab^4k$ **31** $\frac{256}{15}$; $\left(0, \frac{8}{7}, \frac{12}{7}\right)$

33 $I_y = \int_0^4 \int_{-\sqrt{1+y^2}}^{\sqrt{1+y^2}} \int_{-\sqrt{1-x^2+y^2}}^{\sqrt{1-x^2+y^2}} k(x^2 + z^2)\sqrt{x^2 + z^2}\,dz\,dx\,dy$

35 $\pi a^4 k$

37 Rectangular: $(-3\sqrt{2}, 3\sqrt{2}, 6\sqrt{3})$; cylindrical: $\left(6, \frac{3\pi}{4}, 6\sqrt{3}\right)$

39 $z = 9 - 3x^2 - 3y^2$; a paraboloid with vertex $(0, 0, 9)$ and opening downward

41 $x^2 + y^2 = 16$; a right circular cylinder of radius 4 with axis along the z-axis

43 $\sqrt{x^2 + y^2 + z^2}(\sqrt{x^2 + y^2 + z^2} - 3) = 0$; the sphere of radius 3 with center at the origin, together with its center

45 **(a)** $z = r^2 \cos 2\theta$ **(b)** $\cos\phi = \rho \sin^2\phi \cos 2\theta$

47 **(a)** $2r\cos\theta + r\sin\theta - 3z = 4$

(b) $2\rho\sin\phi\cos\theta + \rho\sin\phi\sin\theta - 3\rho\cos\phi = 4$

CHAPTER 18

EXERCISES 18.1

1

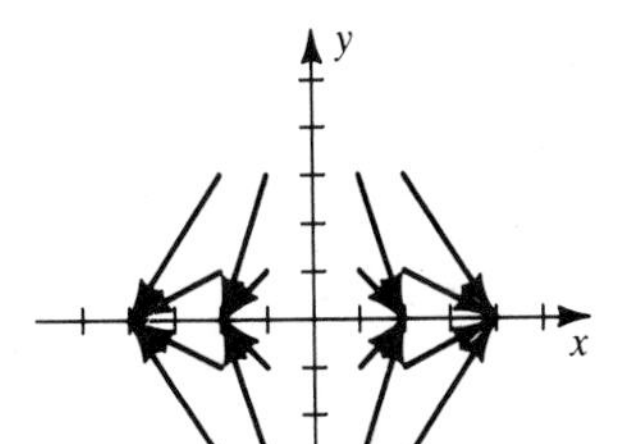

3

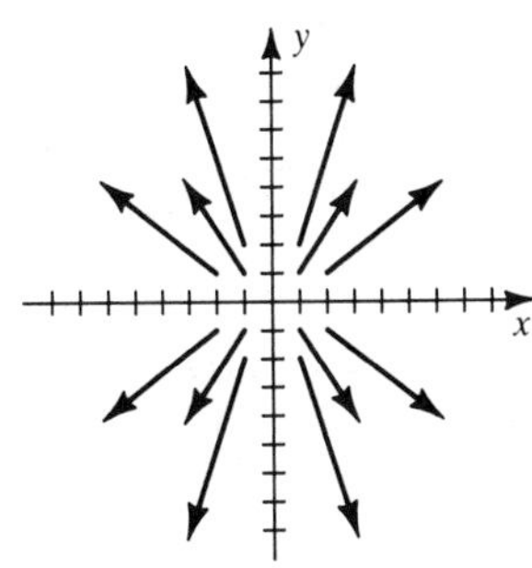

5

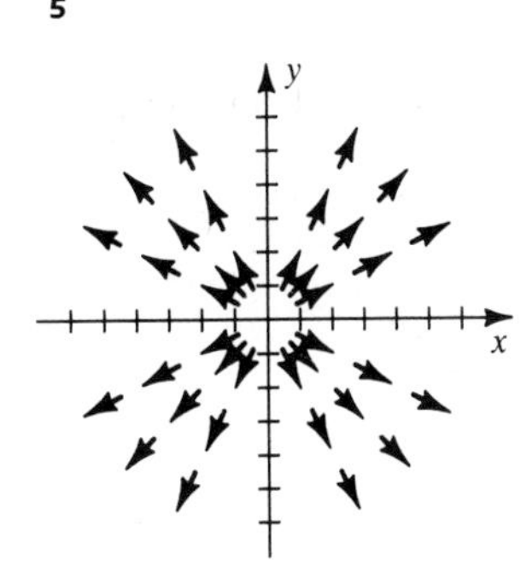

7

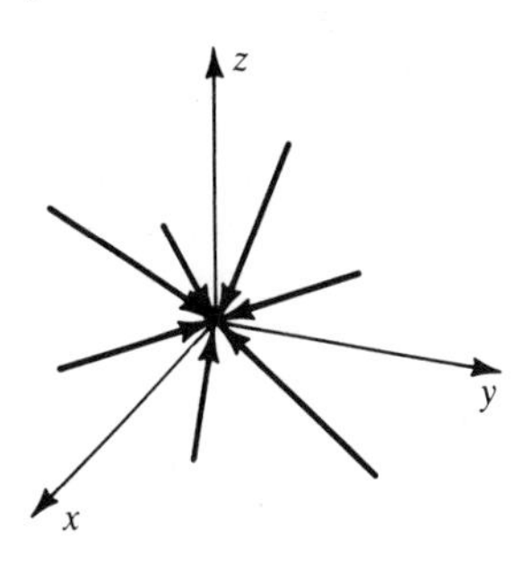

9

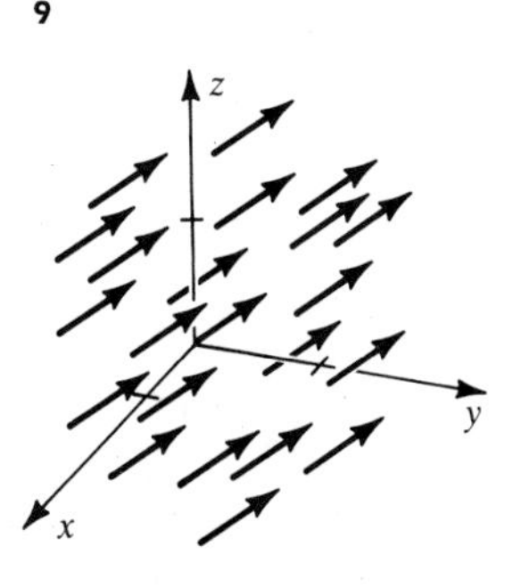

11 $\mathbf{F}(x, y, z) = 2x\mathbf{i} - 6y\mathbf{j} + 8z\mathbf{k}$ **13** $\mathbf{F}(x, y) = \dfrac{1}{1 + x^2y^2}(y\mathbf{i} + x\mathbf{j})$

15 $\mathbf{i} + x^2\mathbf{j} + y^2\mathbf{k}$; $2xz + 2xy + 2$

17 $-y^2 \cos z\mathbf{i} + (6xyz - e^{2z})\mathbf{j} - 3xz^2\mathbf{k}$; $3yz^2 + 2y \sin z + 2xe^{2z}$

37 0.145

EXERCISES 18.2

1 $14(2^{3/2} - 1) \approx 25.60$; 21; 14 **3** $\dfrac{34}{7}$ **5** $-\dfrac{16}{3}$

7 **(a)** $\dfrac{15}{2}$ **(b)** 6 **(c)** 7 **(d)** $\dfrac{29}{4}$

9 $\dfrac{1}{12}(3e^4 + 6e^{-2} - 12e + 8e^3 - 5) \approx 23.97$

11 **(a)** 19 **(b)** 35 **(c)** 27 **13** $\dfrac{3}{2}\sqrt{14}$

15 $\dfrac{9}{2}$ (for all paths) **17** 0 **19** $\dfrac{412}{15}$ **23** $\bar{x} = 0, \bar{y} = \dfrac{1}{4}\pi a$

27 $I_x = \dfrac{4}{3}ka^3$; $I_y = \dfrac{2}{3}ka^3$

29 If the density at (x, y, z) is $\delta(x, y, z)$, then $I_x = \int_C (y^2 + z^2)\, \delta(x, y, z)\, ds$, $I_y = \int_C (x^2 + z^2)\, \delta(x, y, z)\, ds$, and $I_z = \int_C (x^2 + y^2)\, \delta(x, y, z)\, ds$.

31 0.1554

EXERCISES 18.3

1 $f(x, y) = x^3y + 2x + y^4 + c$

3 $f(x, y) = x^2 \sin y + 4e^x + c$

5 $f(x, y) = 2y^3 \cos x + 5y + c$

7 $f(x, y, z) = 4x^2z + y - 3y^2z^3 + c$

9 $f(x, y, z) = y \tan x - ze^x + c$

Exer. 11–14: A potential function f is given along with the value of the integral.

11 $f(x, y) = xy^2 + x^2y$; 14

13 $f(x, y, z) = 3x^2y^3 + 2xz^2 + z$; -31

15 $\dfrac{\partial}{\partial y}(4xy^3) \neq \dfrac{\partial}{\partial x}(2xy^3)$ **17** $\dfrac{\partial}{\partial y}(e^x) \neq \dfrac{\partial}{\partial x}(3 - e^x \sin y)$

21 $\dfrac{\partial N}{\partial z} \neq \dfrac{\partial P}{\partial y}$

23 $f(x, y, z) = -\dfrac{1}{2}c \ln (x^2 + y^2 + z^2) + d$, where $c > 0$ and d is a constant

27 This does not violate (18.16) since D is not simply connected. M and N are not continuous at (0, 0).

29 $W = c\dfrac{d_2 - d_1}{d_1 d_2}$

EXERCISES 18.4

1 $-\dfrac{7}{60}$ **3** $\dfrac{2}{3}$ **5** π **7** -3 **9** 0 **11** 0

13 -3π **15** $\dfrac{128}{3}$ **17** $6 - \dfrac{5}{2}\ln 5 \approx 1.98$ **19** $\dfrac{3}{8}\pi a^2$

23 Green's theorem does not apply since M and N are undefined at (0, 0) and hence are not continuous everywhere inside the unit circle.

27 $\bar{x} = 0, \bar{y} = \dfrac{4a}{3\pi}$

EXERCISES 18.5

1 $\dfrac{2}{3}\pi a^4$ **3** $5\sqrt{14}$

5 **(a)** $\int_0^4 \int_0^{(12-3y)/4} \dfrac{1}{2}(12 - 3y - 4z)y^2z^3\left(\dfrac{1}{2}\sqrt{29}\right) dz\, dy$

(b) $\int_0^3 \int_0^{6-2z} x\left[\dfrac{1}{3}(12 - 2x - 4z)\right]^2 z^3\left(\dfrac{1}{3}\sqrt{29}\right) dx\, dz$

7 **(a)** $\int_0^8 \int_0^6 \left(4 - 3y + \dfrac{1}{16}y^2 + z\right)\left(\dfrac{1}{4}\sqrt{17}\right) dz\, dy$

(b) $\int_0^2 \int_0^6 [x^2 - 2(8 - 4x) + z]\sqrt{17}\, dz\, dx$

9 Since $\iint_S g(x, y, z)\, dS = c \iint_R dA$, the value of the integral equals the volume of a cylinder of altitude c, with rulings parallel to the z-axis, whose base is the projection of S on the xy-plane.

11 $2\pi a^3$ **13** 3π **15** 18 **17** 8

21 **(a)** $\dfrac{255}{2}\pi\sqrt{2}$; $\bar{x} = \bar{y} = 0$, $\bar{z} = \dfrac{1364}{425}$ **(b)** $1365\pi\sqrt{2}$

EXERCISES 18.6

1 24 **3** 20π **5** 0 **7** $\dfrac{136\pi}{3}$ **9** 24

11 Both integrals equal $4\pi a^3$. **13** Both integrals equal 4π.

27 625π lb upward

EXERCISES 18.7

1 Both integrals equal -1. **3** Both integrals equal πa^2.

5 0 **7** -8π

9 The curl meter rotates counterclockwise for $0 < y < 1$ and clockwise for $1 < y < 2$. There is no rotation if $y = 1$. curl $\mathbf{F} = 2(1 - y)\mathbf{k}$; $|(\text{curl } \mathbf{F}) \cdot \mathbf{k}| = |2(1 - y)|$ has a maximum value 2 at $y = 0$ and $y = 2$ and a minimum value 0 at $y = 1$.

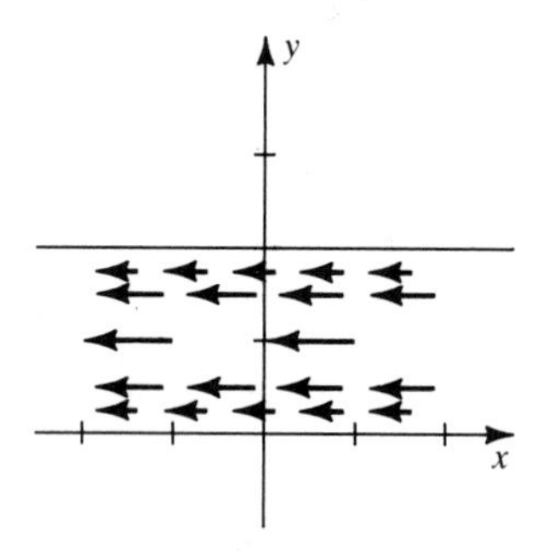

11 Typical field vectors are shown in Figure 18.5. A curl meter rotates counterclockwise for every $(x, y) \neq (0, 0)$. curl $\mathbf{F} = 2\mathbf{k}$; $|(\text{curl } \mathbf{F}) \cdot \mathbf{k}| = 2$ for every (x, y).

18.8 REVIEW EXERCISES

1

3

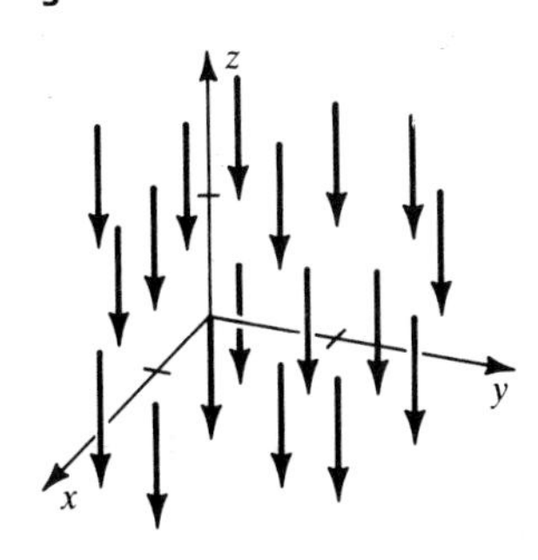

5 $\mathbf{F}(x, y) = (y^2 \sec^2 x)\mathbf{i} + (2y \tan x)\mathbf{j}$ **7** -8 **9** $-\frac{56}{5}$

11 $\frac{1}{144}(1025^{3/2} - 17^{3/2}) \approx 227.40$ **13** 70 **15** 0 **17** $\frac{25}{2}$

19 $f(x, y, z) = x^2e^{2y} + y^2 \cot z + c$ **21** $\frac{1}{6}$

23 $3x^2z^4 + xz^2$; $(2x^2y - 2xyz)\mathbf{i} + (4x^3z^3 - 2xy^2)\mathbf{j} + (yz^2)\mathbf{k}$

25 $\frac{1}{5}\sqrt{3}$ **27** $\frac{5\pi}{2}$ **29** Both integrals equal 8π.

CHAPTER 19

EXERCISES 19.1

1 **(a)** $y = x^3 + C$ **(b)** $y = x^3 + 2$

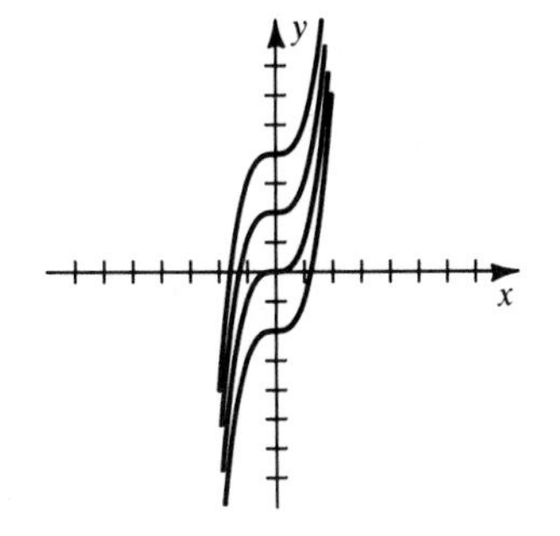

3 **(a)** $y = \sqrt{4 - x^2} + C$ **(b)** $y = \sqrt{4 - x^2}$

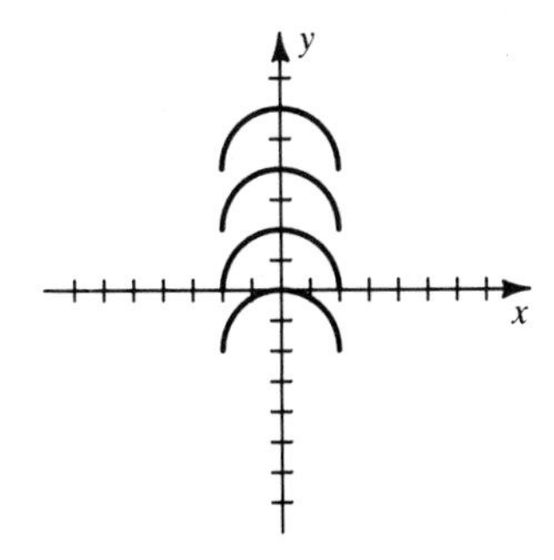

11 $y = Ce^{2 \sin x}$ **13** $y = Cx$ **15** $x^3e^yy^5 = C$

17 $y = -1 + Ce^{(x^2/2) - x}$ **19** $y = -\frac{1}{3}\ln(3C + 3e^{-x})$

21 $y^2 = C(1 + x^3)^{-2/3} - 1$ **23** $x \sin x + \cos x - \ln|\sin y| = C$

25 $\sec x + e^{-y} = C$ **27** $y^2 + \ln y = 3x - 8$

29 $y = \ln(2x + \ln x + e^2 - 2)$ **31** $y = 2e^{2 - \sqrt{4 + x^2}} - 1$

33 $\tan^{-1} y - \ln|\sec x| = \frac{\pi}{4}$ **35** $xy = k$; hyperbolas

37 $2x^2 + y^2 = k$; ellipses **39** $2x^2 + 3y^2 = k$; ellipses

41 **(a)** 1.524 **(b)** $y = e^{x^2/2}$; $e^{1/2} \approx 1.648721$ **43** 1.647355

EXERCISES 19.2

1 $y = \frac{1}{4}e^{2x} + Ce^{-2x}$ **3** $y = \frac{1}{2}x^5 + Cx^3$

5 $y = \frac{e^x}{x} - \frac{1}{2}x + \frac{C}{x}$ **7** $y = \frac{e^x + C}{x^2}$

9 $y = \frac{4}{3}x^3 \csc x + C \csc x$ **11** $y = 2 \sin x + C \cos x$

13 $y = x \sin x + Cx$ **15** $y = \left(\frac{1}{3}x + \frac{C}{x^2}\right)e^{-3x}$

17 $y = \frac{3}{2} + Ce^{-x^2}$ **19** $y = \frac{1}{2}\sin x + \frac{C}{\sin x}$

21 $y = \frac{1}{3} + (x + C)e^{-x^3}$ **23** $y = x(x + \ln x + 1)$

25 $y = e^{-x}(1 - x^{-1})$ **27** $Q = CV(1 - e^{-t/RC})$

29 $f(t) = \frac{80}{3}(1 - e^{-0.075t}) + Ke^{-0.075t}$

31 **(a)** $f(t) = M + (A - M)e^{k(1-t)}$ (k a constant) **(b)** 28 items

33 $y = y_L(1 - ce^{-kt})$, $k > 0$

35 **(b)** $y = \frac{I}{k}(1 - e^{-kt})$; $\frac{I}{k}$ **(c)** 0.58 mg/min **37** -3.23103

39 0.14817

EXERCISES 19.3

1 $y = C_1e^{2x} + C_2e^{3x}$ **3** $y = C_1 + C_2e^{3x}$

5 $y = C_1e^{-2x} + C_2xe^{-2x}$ **7** $y = C_1e^{(2+\sqrt{3})x} + C_2e^{(2-\sqrt{3})x}$

9 $y = C_1e^{-\sqrt{2}x} + C_2xe^{-\sqrt{2}x}$ **11** $y = C_1e^{-3x/2} + C_2e^{5x/4}$

13 $y = C_1e^{4x/3} + C_2xe^{4x/3}$

15 $y = C_1e^{(2+\sqrt{2})x/2} + C_2e^{(2-\sqrt{2})x/2}$

17 $y = e^x(C_1 \cos x + C_2 \sin x)$

19 $y = e^{2x}(C_1 \cos 3x + C_2 \sin 3x)$

21 $y = C_1e^{(-3+\sqrt{7})x} + C_2e^{(-3-\sqrt{7})x}$ **23** $y = -2e^x + 2e^{2x}$

25 $y = \cos x + 2 \sin x$ **27** $y = e^{-4x}(2 + 9x)$

29 $y = \frac{1}{2}e^x \sin 2x$

EXERCISES 19.4

1 $y = C_1 \cos x + C_2 \sin x - \cos x \ln|\sec x + \tan x|$

3 $y = \left(C_1 + C_2x + \frac{1}{12}x^4\right)e^{3x}$

5 $y = C_1e^x + C_2e^{-x} + \frac{2}{5}e^x \sin x - \frac{1}{5}e^x \cos x$

7 $y = \left(C_3 + \frac{1}{6}x\right)e^{3x} + C_2e^{-3x}$, where $C_3 = C_1 - \frac{1}{36}$

9 $y = C_1e^{-x} + C_2e^{4x} - \frac{1}{2}$ **11** $y = C_1e^x + C_2e^{2x} + \frac{2}{3}e^{-x}$

13 $y = C_1 + C_2e^{-2x} + \frac{1}{8}\sin 2x - \frac{1}{8}\cos 2x$

15 $y = C_1e^x + C_2e^{-x} + \frac{1}{9}(-4 + 3x)e^{2x}$

17 $y = e^{3x}(C_1 \cos 2x + C_2 \sin 2x) + \frac{1}{65}e^x(7 \cos x - 4 \sin x)$

21 0.776805

EXERCISES 19.5

1 $y = -\frac{1}{3} \cos 8t$ **3** $y = \frac{1}{8}\sqrt{2}e^{-8t}(e^{4\sqrt{2}t} - e^{-4\sqrt{2}t})$

5 $y = \frac{1}{3}e^{-8t}(\sin 8t + \cos 8t)$

7 If m is the mass of the weight, then the spring constant is $24m$ and the damping force is $-4m\frac{dy}{dt}$. The motion is begun by releasing the weight from 2 feet above the equilibrium position with an initial velocity of 1 ft/sec in the upward direction.

9 $-6\sqrt{2}\frac{dy}{dt}$

EXERCISES 19.6

1 $y = a_0 \sum_{n=0}^{\infty} \frac{(-1)^n}{(2n)!} x^{2n} + a_1 \sum_{n=0}^{\infty} \frac{(-1)^n}{(2n+1)!} x^{2n+1}$
$= a_0 \cos x + a_1 \sin x$

3 $y = a_0\left[1 + \sum_{n=1}^{\infty} \frac{2^n(3n-2)(3n-5)\cdots 7\cdot 4\cdot 1}{(3n)!} x^{3n}\right] + a_1\left[x + \sum_{n=1}^{\infty} \frac{2^n(3n-1)(3n-4)\cdots 8\cdot 5\cdot 2}{(3n+1)!} x^{3n+1}\right]$

5 $y = a_0(1 - x^2) + a_1\left[x - \sum_{n=1}^{\infty} \frac{(2n-3)(2n-5)\cdots 5\cdot 3\cdot 1}{(2n+1)!} x^{2n+1}\right]$

7 $y = a_0(x+1)^3$

9 $y = -5x + a_0 \sum_{n=0}^{\infty} \frac{x^n}{n!} + a_1 \sum_{n=0}^{\infty} \frac{(-x)^n}{n!} = -5x + a_0 e^x + a_1 e^{-x}$

11 $y = a_0 \sum_{n=0}^{\infty} (n+1)x^{2n} + a_1 \sum_{n=0}^{\infty} \left(\frac{2n+3}{3}\right)x^{2n+1} + \sum_{n=1}^{\infty} (n+1)x^{2n}$

19.7 REVIEW EXERCISES

1 $\sin x - x \cos x + e^{-y} = C$ **3** $y = \tan(\sqrt{1-x^2} + C)$

5 $y = \frac{2x - 2\cos x + C}{\sec x + \tan x}$ **7** $y = 2 \sin x + C \cos x$

9 $\sqrt{1-y^2} + \sin^{-1} x = C$ **11** $\csc y = e^{-x} + C$

13 $y = \frac{1}{2} + Ce^{-2 \sin x}$ **15** $y = (C_1 + C_2 x)e^{4x}$

17 $y = C_1 + C_2 e^{2x}$

19 $y = C_1 e^{-x} + C_2 e^x - \frac{1}{5}e^x(2 \cos x + \sin x)$

21 $y = \frac{1}{5}e^{4x} + Ce^{-x}$ **23** $y = C_1 e^x + C_2 e^{2x} + \frac{1}{12}e^{5x}$

25 $y = \frac{(x-2)^3}{3x} + \frac{C}{x}$

27 $y = e^{-5x/2}\left[C_1 \cos\left(\frac{1}{2}\sqrt{3}x\right) + C_2 \sin\left(\frac{1}{2}\sqrt{3}x\right)\right]$

29 $\frac{1}{2}e^x(\sin x - \cos x) + e^{-y} = C$ **31** $y = \frac{1}{2}\cos x + C \sec x$

33 $y = \frac{\ln|\sec x + \tan x| - x + C}{\csc x - \cot x}$

35 $y = \frac{3500}{3}e^{(3/(2\pi))(1 - \cos 2\pi t)} - \frac{2000}{3}$; 2366

37 $f(t) = \frac{ab[e^{k(b-a)t} - 1]}{be^{k(b-a)t} - a}$

39 $y\,dy + x\,dx = 0$; a circle with center at the origin

INDEX

D

E

T

U

V

ALGEBRA

EXPONENTS AND RADICALS

$a^m a^n = a^{m+n}$ $\qquad a^{m/n} = \sqrt[n]{a^m} = (\sqrt[n]{a})^m$

$(a^m)^n = a^{mn}$ $\qquad \sqrt[n]{ab} = \sqrt[n]{a}\sqrt[n]{b}$

$(ab)^n = a^n b^n$ $\qquad \sqrt[n]{\dfrac{a}{b}} = \dfrac{\sqrt[n]{a}}{\sqrt[n]{b}}$

$\left(\dfrac{a}{b}\right)^n = \dfrac{a^n}{b^n}$ $\qquad \sqrt[m]{\sqrt[n]{a}} = \sqrt[mn]{a}$

$\dfrac{a^m}{a^n} = a^{m-n}$ $\qquad a^{-n} = \dfrac{1}{a^n}$

ABSOLUTE VALUE ($d > 0$)

$|x| < d$ if and only if $-d < x < d$

$|x| > d$ if and only if either $x > d$ or $x < -d$

$|a + b| \le |a| + |b|$ (Triangle inequality)

$-|a| \le a \le |a|$

INEQUALITIES

If $a > b$ and $b > c$, then $a > c$

If $a > b$, then $a + c > b + c$

If $a > b$ and $c > 0$, then $ac > bc$

If $a > b$ and $c < 0$, then $ac < bc$

QUADRATIC FORMULA

If $a \ne 0$, the roots of $ax^2 + bx + c = 0$ are

$$x = \frac{-b \pm \sqrt{b^2 - 4ac}}{2a}$$

LOGARITHMS

$y = \log_a x$ means $a^y = x$ $\qquad \log_a 1 = 0$

$\log_a xy = \log_a x + \log_a y$ $\qquad \log_a a = 1$

$\log_a \dfrac{x}{y} = \log_a x - \log_a y$ $\qquad \log x = \log_{10} x$

$\log_a x^r = r \log_a x$ $\qquad \ln x = \log_e x$

BINOMIAL THEOREM

$$(x + y)^n = x^n + \binom{n}{1} x^{n-1} y + \binom{n}{2} x^{n-2} y^2 + \cdots + \binom{n}{k} x^{n-k} y^k + \cdots + y^n,$$

where $\dbinom{n}{k} = \dfrac{n!}{k!(n-k)!}$

ANALYTIC GEOMETRY

DISTANCE FORMULA

$$d(P_1, P_2) = \sqrt{(x_2 - x_1)^2 + (y_2 - y_1)^2}$$

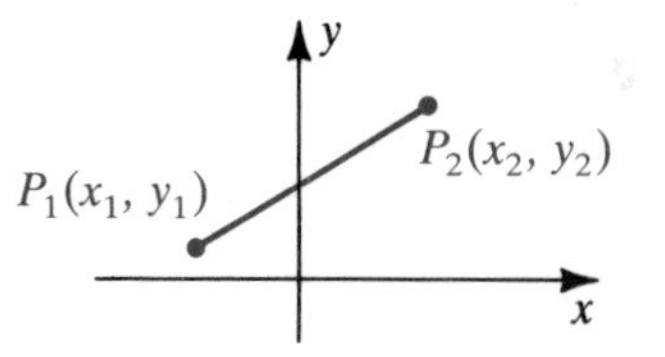

EQUATION OF A CIRCLE

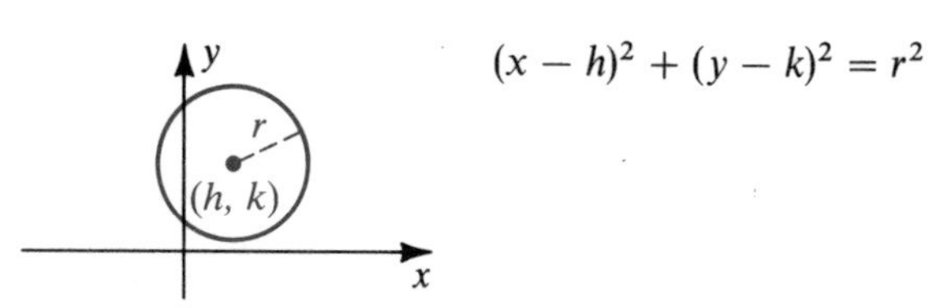

$$(x - h)^2 + (y - k)^2 = r^2$$

SLOPE m OF A LINE

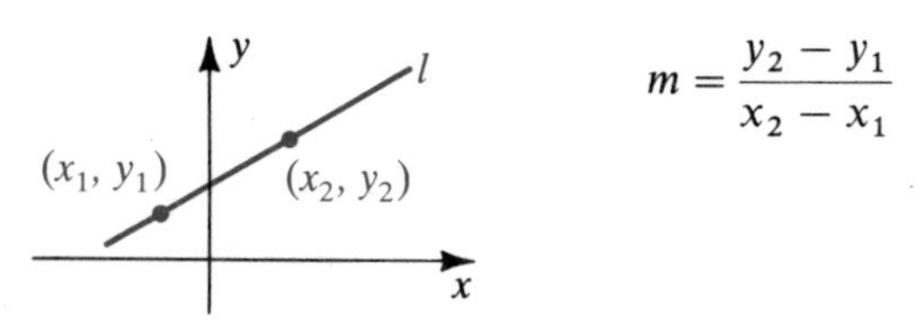

$$m = \frac{y_2 - y_1}{x_2 - x_1}$$

POINT-SLOPE FORM

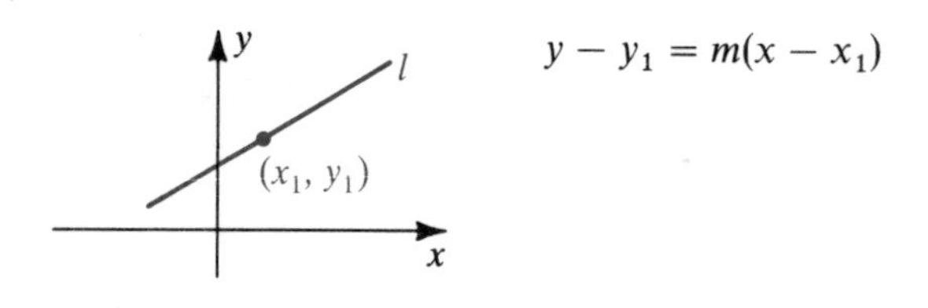

$$y - y_1 = m(x - x_1)$$

SLOPE-INTERCEPT FORM

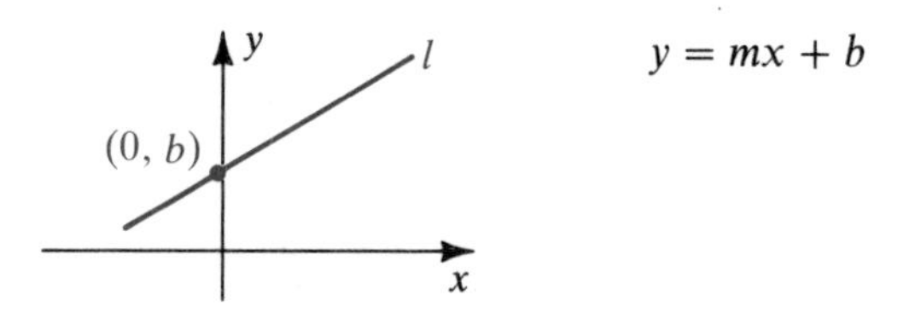

$$y = mx + b$$

GRAPH OF A QUADRATIC FUNCTION

$y = ax^2, a > 0$ $\qquad y = ax^2 + bx + c, a > 0$

y x $\qquad$ y x $-\dfrac{b}{2a}$